SOLAR ENERGY INTERNATIONAL PROGRESS

Pergamon Titles of Related Interest

Boer Sun II
Eggers-Lura Solar Energy for Domestic Heating and Cooling
Fazzolare and Smith Changing Energy Use Futures
Machovec Solar Energy Index
Messel Energy for Survival
deMontbrial Energy: The Countdown
Veziroglu Solar Energy and Conservation

Related Journals*

Energy
Energy Conversion
Solar Energy
Space Solar Power Review
Sun at Work in Britain
Sun World

*Specimen copies available upon request.

SOLAR ENERGY INTER-NATIONAL PROGRESS

Proceedings of the International Symposium-Workshop on Solar Energy, 16-22 June 1978, Cairo, Egypt

EDITED BY

T. NEJAT VEZIROGLU

Editorial Board:
H.W. Hiser
I.A. Sakr
J.W. Sheffield

VOLUME 3

PERGAMON PRESS

New York • Oxford • Toronto • Paris • Frankfurt • Sydney

Pergamon Press Offices:

U.S.A.	Pergamon Press Inc., Maxwell House, Fairview Park, Elmsford, New York 10523, U.S.A.
U.K.	Pergamon Press Ltd., Headington Hill Hall, Oxford OX3 0BW, England
CANADA	Pergamon of Canada, Ltd., Suite 104, 150 Consumers Road, Willowdale, Ontario M2J 1P9, Canada
AUSTRALIA	Pergamon Press (Aust.) Pty. Ltd., P.O. Box 544, Potts Point, NSW 2011, Australia
FRANCE	Pergamon Press SARL, 24 rue des Ecoles, 75240 Paris, Cedex 05, France
FEDERAL REPUBLIC OF GERMANY	Pergamon Press GmbH, Hammerweg 6, Postfach 1305, 6242 Kronberg/Taunus, Federal Republic of Germany

Library of Congress Cataloging in Publication Data

International Symposium-Workshop on Solar Energy,
Cairo, 1978.
Solar energy.

Presented by the Clean Energy Research Institute
of the University of Miami, Coral Gables, Fla.
Bibliography: p.
Includes index.
1. Solar energy—Congresses. I. Veziroglu,
T. Nejat. II. Miami, University of, Coral Gables,
Fla. Clean Energy Research Institute. III. Title.
TJ810.I63 1978 621.47 79-13656
ISBN 0-08-025077-7

Printed in the United States of America

SOLAR ENERGY: INTERNATIONAL PROGRESS

TABLE OF CONTENTS

VOLUME I

Page No.

CHAPTER 1 SOLAR RADIATION

Plane Incident Solar Radiation, Effects of Slope and Orientations — 1
J.I. Jimenez, C. Romeu-Nedwed, J.E. Llebot and J. Casas-Vezquez

Relationship Between Global and Net Radiation — 14
A. Guerrini, A. Lavagnini and F. Vivona

Atmospheric Pollution and Solar Radiation — 20
M.S. Abdel-Salam and N.A. Higazy

Solar Energy at Ground Surface — 41
N.H. Helwa

Contribution to the Study of Nocturnal Radiation Very Near the Ground: Effect of a Circular Screen 56
J. Brodin, A. Flourens, V.V. Pham and C. Sinnasse

New Formulation for the Factor Ψ for the Middle East Region 81
E.M. Abd El-Salam

Computation of the Diffuse Solar Radiation for Egypt 90
E.M. Abd El-Salam

Solar Data Application to Egypt 114
M.S. Abdel Salam and N.A. Higazy

Global Radiation in Italy 127
A. Guerrini, A. Lavagnini and F. Vivona

On the Distribution of Solar Energy in Greece 140
G.J. Macris

CHAPTER 2 FLAT PLATE COLLECTORS

The Flat Plate Solar Collector: Its Steady-State and Transient State Behaviour 159
E. Hahne, N. Fisch and A. Arafa

Flat-Plate Solar Collectors - Influence of Design Material and Configuration 187
S-E. Ransmark

My New Types of Flat Plate Solar Collectors 215
R. Jankovic

Measurements on a Flat Thermal Solar Energy Collector with a Cellular Structure After Francia 239
A. Bahr, H. Piwecki and L. Rigolini

A Simulated Comparison of the Useful Energy Gain in a Fixed and a Fully Thermal Tracking Flat Plate Collector 258
P. Drago

Flat Plate Boosters for Elevating Temperatures 274
I.A. Sakr and A.I. Hegazi

A Black Liquid Flat Plate Collector for Economical Applications 298
R.S. Daryan and P.K.C. Pillai

Passive Solar Wall Collectors 308
A. Faist and J.B. Gay

Energy Analysis Method for a Solar Panel 324
B. Ciborro

CHAPTER 3 CONCENTRATING COLLECTORS

A Two-Dimensional Finite-Difference Solution for the Transient Thermal Behavior of a Tubular Solar Collector 328
F.L. Lansing

Optical and Thermal Performance of a Half-Circular, Cylindrical Concentrator 351
K. Bordoloi, T. Murray, F. Bynum and E. Chaves

Modeling a V-Trough Collector in Terms of Zero, Single or Multiple Thermal Capacitance Nodes 378
K. Bordoloi, C. Hoang and F. Bynum

High-Efficiency Solar Concentrator 392
F.L. Lansing and J. Dorman

The Scheme of Automatic Adjustment of Accuracy of the Inflatable Solar Collector 404
A. Gokhman and N. Ozboya

Thermo-Fluid Analysis of a Cylindrical Solar Heat Absorber 412
M.F. Khalil

An Experimental Investigation of Comparative Performance of a Header-Type Compound Parabolic Concentrator and a Flat-Plate Collector 431
K. Krishna Rao, S. Basheer Ahmed, R. Natarajan and M.C. Gupta

CHAPTER 4 COLLECTOR TECHNOLOGY AND RESEARCH

Interferometric Investigations of Convection in Slat, Flat Plate and Vee-Corrugated Solar Collectors 446
B.A. Meyer, K.R. Randall, M.M. El-Wakil and J.W. Mitchell

Study of a Multi-Pass Radiation Fluid-Heater Exposed to Non-Uniform Radiation 463
N.M. Rafat, M.M. Elkotb and M.F. El-Refaie

Simulations of Solar Captation Systems: Methodology and Applications to Distillation, Gas and Water Heating 490
R. Rocaries, G. Olalde, P. Granier, B. N'Doye and M. Daguenet

Low Cost Spiral Shaped Solar Collector 512
P.K.C. Pillai and R.C. Agarwal

Energy Requirements of an Aluminum Solar Collector 522
J.E. Robinson and W. Ashton

Optimum Inclination of Linear Reflectors 529
B.A. Khalefa and A.S. Hegazy

Selective Surface Coating with Black Chrome 536
T. Vurgun

Cadmium Stannate Selective Coatings Prepared by Spray Pyrolysis 540
O.P. Agnihotri and B.K. Gupta

Phase Change Collector Wall Versus Water Collector Wall 549
L. Bourdeau and A. Jaffrin

On The Evaluation of the Feasibility of Solar Collectors for Electrical Power Generation Applications 565
A. Gokhman and N. Ozboya

Effect of Solar Radiation on Outdoor Bus Bar Design 577
M.M. Abdel-Aziz

VOLUME II

CHAPTER 5 THERMAL STORAGE

Performance of a Storage Unit Coupled to a Flat Plate Solar Collector 593
S.B. Ahmed and V. Sriramulu

Transient Characteristics of Vertical Porous Heat Storage 604
N. Seki, S. Fukusako and H. Inaba

Seasonal Regeneration Through Underground Strata 623
T. Yokoyama, H. Umemiya, T. Teraoka, H. Watanabe, K. Katsuragi and K. Kasahara

Storage Tank Efficiency as Simulated in a Markovian Model of Meteorology 654
R. Lestienne

Some Aspects of Energy Storage Systems Using Glauber Salt Mixtures 666
D.Y.S. Lou and A.J. Harrison

A Direct Contact Thermal Energy Storage System 680
A.S. Elasfouri

CHAPTER 6 HEATING

Solar Heating Systems with Air Collectors and Rock Bed Storage 697
W.L. Dutre, L. Cypers, J. Berghmans and A. Debosscher

Influence of Heat Exchanger Effectiveness on the Performance of Thermosyphon Double Loop Water Heating 731
A.F. Orlando, L. Goldstein, Jr. and D. Magnoli

Hot Water Production by Means of a Solar Thermosyphon Loop 745
W.L. Dutre, L. Cypers, J. Berghmans and A. Debosscher

Effectiveness of Packed-Bed Solar Air Heaters 767
S.C. Gupta

Performance Comparison of Different Types of Solar Water Heaters Under Egyptian Conditions 778
S.W. Youssef and I.A. Sakr

Solar Water Heaters 793
D. Buxey

Latent Heat Diode Wall 802
L. Bordeau and A. Jaffrin

CHAPTER 7 COOLING

Comparison Between Two Absorption Cooling Systems of the "Open" Type Under Different Climate Conditions 827
M. Bolzan and R. Lazzarin

Optimisation of Cooling Systems in Tropical Zones 855
J. Flechon

Analysis of Ammonia-Water Intermittent Solar Refrigerator Operating with a Flat Plate Collector 870
A. Venkatesh and M.C. Gupta

A Thermodynamic Analysis of a Solar-Powered Jet Refrigeration System 886
F.L. Lansing and V.W. Chai

Development of a Solar Desiccant Dehumidifier 898
Z. Lavan, D. Gidaspow, M. Onischak, S. Chaturvedi, B. Mathipraksam and S. Perkari

Liquid Desiccant Solar Heat Pump for Developing Nations 932
H. Robison

CHAPTER 8 APPLICATIONS IN BUILDINGS

A Review of the Historical Developments of the Use of Passive Solar Systems 939
C.L. Robbins

Natural Lighting in Buildings: A Passive Solar System 960
C.L. Robbins

Radiative Cooling of Structures 983
V.K. Gupta and H.R. Tyagi

Solar Collector Related Research and Development in the United States for Heating and Cooling of Buildings 991
R.K. Collier

California State Energy Efficient State Office Building 1029
B.S. Setty

CHAPTER 9 BIOLOGICAL AND CHEMICAL CONVERSION

A Quinone-Assisted Photoformation of Energy-Rich Chemical Bonds 1056
S.W. Fox, T. Adachi and W. Stillwell

Ethanol, A Clean Burning Bio-Solar Fuel - Its Production and Utilization 1075
H.B. Mathur

Electrolytic Hydrogen Production and Its Potential Utilization for Ship Propulsion 1094
C.A. Schmidt

Hybrid Chemical Concepts for Storage of Solar Thermal Energy 1109
A.C. Vialaron

Use of Renewable Fuels in the Rural Development of India 1119
G.C. Pandey

VOLUME III

CHAPTER 10 DIRECT ELECTRICAL CONVERSION

Physics and Economics of Photovoltaic Solar Energy Conversion 1132
Y. Marfaing

Realization of P-N Junction Solar Cells by an Ion Implantation Doping Procedure 1154
J.C. Muller, M. Hage-Ali and P. Siffert

Solar Energy Conversion by Fluorescent Materials for Utilization in Space Communication Systems 1167
A.K. Sen and J.S. Sehra

New Possibility of Solar Energy Conversion by Means of Refractory Thermoelements 1184
S. Abdalla, J.M. Dusseau, F. Roche and J.M. Darolles

On Direct Conversion of Sunlight to Electricity 1192
M.A. Melehy

Performance of the Noval Solar Cell Substitute for the Future 1210
S.C. Rastogi

MOS Solar Cells 1216
H.M. Kizilyalli

Some Optical and Electrical Properties of Selenium-Cadmium Sulphide Photovoltaic Cells 1230
K. Abdel-Hady, F. El-Akkad and J. Al-Dabbagh

Optimization of Graded Doping CdTe Solar Cells 1242
A.S. Bouazzi

Saur-Vidyut Kosh - The Solar Cells 1257
M. Sharon, S.G. Sharan, B.M. Prasad, A.P. Bholagir and R.R. Pradhananga

CHAPTER 11 THERMAL POWER

Solar Thermal Power Plant: Thermodynamic Analysis 1279
A.A. Samuel, U.S.P. Shet and M.C. Gupta

Solar Power Units with Screw Expanders 1293
J.M. Merigoux and P. Pocard

New Developments for Future Solar Power Plants 1318
J. Lorenz, J. Feustel and M. Kraft

Power from the Sun by Thermodynamics - French Tower Type Projects 1329
M.J. Bignon

Steam and Gas Turbines for Small Outputs Using Solar Energy 1338
K. Bammert and H. Posentrup

Turbine Selection for Small Capacity Solar-Power Generation 1351
A. Mobarek, N. Rafat and M. Saad

A Helioelectric Farm 1368
O.J.M. Smith and P.S. Smith

A Central Receiver Solar Power System for Remote Locations 1408
R.J. Holl

Investigation of Pool Boiling Heat Transfer in the Energy Carrying Liquids of Solar Energy Power Plants 1476
M.S. Abdelhafiez and M.A. Abdelsalam

The Use of Solar Furnaces in Materials Research 1443
D. Suresh and P.K. Rohatgi

CHAPTER 12 DISTILLATION AND DESALINATION

Efficiency of a Solar Still as a Function of Its Main Parameters: Production Estimation for Valencia, Spain 1474
J. Riera, J.A. Martinez-Lozano, F. Tejerina and F. Tena-Sanguesa

Solar Water Supply and Purification System for Remote Areas and Villages 1497
U.R. Nejib and H.C. Hallez

Solar Water Recovery from the Air 1526
W.G. Bulang

Double Exposure Solar Still 1546
M.M. Mousa, F.M. Abdelfattah and I.A. Sakr

Direct Solar Distillation 1561
I.D. Abdelrazek and F.H. Hammad

Solar Energy for Desalination 1570
A.M. El-Nashar

Development Plans and Conceptual Designs for Multi-Purpose Solar Desalination Units (MSDU) 1587
H.K. Abdel-Aal

CHAPTER 13 AGRICULTURAL AND RURAL APPLICATIONS

Review on Application of Solar Energy for Drying Food Stuffs 1600
K.V. Mallan, T.V. Balasubramanian and R.K. Bhardwaj

Solar Energy Utilization for Grain Drying 1615
VR. Muthuveerappan

Studies on Utilisation of Solar Energy in Hay Making of Barseem 1634
R.D. Verma and M.K. Trehan

Solar Tobacco Drying Plant 1645
L. Andreotti, A. Ciminale, G. Marchetti, D. Ruggi and A. Serravezza

Crop and Timber Drying in Malaysia Proposed Plans for Studies 1654
K.S. Ong

Energy Consumptions and Possibilities for the Applications of Solar Energy in Italian Agriculture 1666
L. Andreotti, G. Marchetti, F. Passaro, D. Ruggi, A. Serravezza, A. Di Vecchia and A. Fanchiotti

Solar Energy Utilization for Agricultural Purposes in Egypt 1677
M. Hassib

Solar Powered Water Pump for the Rural Third World 1681
J. Boldt

A Solar Battery System and Its Use for Irrigation in Egypt 1692
M.G. Osman

Utilization of Solar Energy in Dairies 1703
R.D. Verma, S.C. Sarma, V.K. Nayar and R.K. Hohli

Global Radiation Within a Crop: Pear Trees 1713
A. Guerrini, A. Lavagnini and F. Vivona

VOLUME IV

CHAPTER 14 COMMERCIAL AND DOMESTIC UTILIZATION

Street Lighting by Sunlight 1723
J.G. Sakas

The First Solar Energy System in West Germany for Industrial Utilization 1742
I. Hoffman

A Simple Water Pump, Powered by Solar Energy 1748
K. Speidel

Solar Cookers 1762
D. Buxey

Solar Energy for Cooking in Egypt 1768
M.G. Osman

CHAPTER 15 WIND ENERGY

Sow Wind Turbines and Reap the Wind 1796
O.J.M. Smith

Alternative Energy for Domestic Hot Water: Wind of Solar? 1834
A.L. Weisbrich

Comfortable Heating and Cooling by Solar Reflectors and Wind 1850
S.A. Husain

Combination of Wind and Solar Energies at the Northeast Coast of Egypt 1862
I.A. Sakr, M.S. Abdel Hafiez and A.S. Hegazy

A Low Cost Tilting Blade Vertical Axis Windmill for Rural Irrigation 1891
V. Mukhopadhyay

CHAPTER 16 NATIONAL AND INTERNATIONAL PROGRAMS

Solar Energy: UNIDO Programme of Action for Developing Countries 1901
S. Rao

Contemporary Research and Development for the Use of Solar Energy in Thailand 1911
V. Bhanthumnavin

End Use Energy in Brazil and the Solar Technologies 1928
I.C. Macedo

French Solar Energy Program and Its Application for International Cooperation 1937
M.J. Pheline

Solar Collector Development in Rwanda 1950
F. Kalos, C. Ntakirutinka, P. Mpawenayo and G. Saunier

The Impact of Solar Energy Integration on the Power System of the Kingdom of Saudi Arabia 1962
F. Sultan and F. Taher

Solar Energy for the Common Man in the Arab World 1971
D.N. Singh

An Assessment of Need for Developing and Implementing Technical and Skilled Worker Training for the Solar Energy Industry 1976
C.G. Orsak. R. Barnstone, Y. Jani, H. Gibson and J. Morehouse

The Design of the Two-Year Associate Degree Curriculum for Training Solar Engineering Technicians 2003
M.Z. Lowenstein

CHAPTER 17 ECONOMICS AND POLICIES

Economic Considerations of Solar Energy 2021
J.D. Parker and J.A. Tipton

The Comparative Economics of Passive and Active Systems 2044
F. Roach, S. Noll and S. Ben-David

An Economic Analysis of Solar Energy Alternatives 2084
D.A. Huettner, B.J. Taylor, C.A. Ingene and J.F. Horrell

Impact of Solar Heating and Cooling of Buildings on Electric Utilities 2106
A.S. Debs

Economic Analyses of Solar Water Heater and Solar Stills in Thailand 2126
P. Wibulswas

Evaluation of Solar Thermal Power Plants Using Economic and Performance Simulations 2138
N. El-Gabawali

The American Experience with Solar Energy in Centralized Electricity Generation: Technology and Economics 2162
D.L. Hurwood

Commercialization of Solar Energy Devices in Developing Countries - Problems and Prospects 2189
R.O. Onyeonwu

The Economics of Solar Space Heating in Zambia 2202
G. Lewis

Solar Energy - Will It Conserve Our Non-Renewable Resources? 2210
S. Baron

The Role of Information Media in Public Understanding of Solar Energy 2230
S. Galal

Community Characteristics Predicting Public Interest in Household Solar 2239
S. Warkov

Accelerated Commercialization of Solar Energy - A New Priority for International Action 2251
L. Rishnell and N.D. Jorstad

Why Not Solar Energy? 2270
T. Yarman

CHAPTER 18 WORKSHOP REPORTS

Report 1: Architectural Applications 2286
Group Leader: A. Bowen

Report 2: Agricultural and Industrial Applications 2288
Group Leader: E. Bilgen

Report 3: Energy Conversion 2290
Group Leader: Y. Marfaing

Report 4: Solar Planning 2292
Group Leader: A.J. Parker, Jr.

Report 5: Solar Energy Commercialization 2304
Group Leader: N.W. Lutkefedder

SOLAR ENERGY INTERNATIONAL PROGRESS

VOLUME 3

PHYSICS AND ECONOMICS OF PHOTOVOLTAIC SOLAR ENERGY CONVERSION

Y. Marfaing
Laboratoire de Physique des Solides C.N.R.S.
Bellevue, France

ABSTRACT

This paper presents principles of the photovoltaic effect. From this description are derived the basic properties required for the semiconducting materials to be used in photovoltaic solar converters. Then, the structure and the performances of various photovoltaic devices produced by industry or studied in the laboratory are presented. Finally, some considerations are given about the cost of electric energy supplied by photovoltaic generators for the present time and to be expected in the future.

PHYSICAL PRINCIPLES OF THE PHOTOVOLTAIC (PV) EFFECT

Two basic effects are involved as indicated in Fig. 1, which schematizes the structure of a photovoltaic cell (homojunction type):

- Absorption of photons in parts A of the cell
  This absorption occurs by electronic transitions from a band of filled electronic states (or valence band limited by energy E_V) to a band of empty electronic states (or conduction band limited by energy E_C). The difference $E_C - E_V = E_g$ corresponds to the forbidden band gap of the semiconductor considered. It also defines the energy threshold of the optical absorption process. Only photons of energy $h\nu \geqslant E_g$ have a chance to be absorbed.

The first result of this absorption process is the excitation of a valance electron to the conduction band, giving what is called an electron-hole pair. The hole is the particle with positive charge equivalent to the vacancy left in the valence band by the excitation of one electron.

Just after the excitation process, the internal energy of the electron-hole pair is equal to that of the absorbed photon, by way of energy conservation. But, very fast relaxation processes act to finally reduce the internal energy of the pairs to the threshold energy E_g.

- Collection and dissociation of the electron-hole pairs by part C of the cell
  A built-in electric field F exists in part C, represented by the gradient of electronic energy levels E_C and E_V, which is usually due to a doping gradient (e.g. transition from n-type to p-type conductivity).

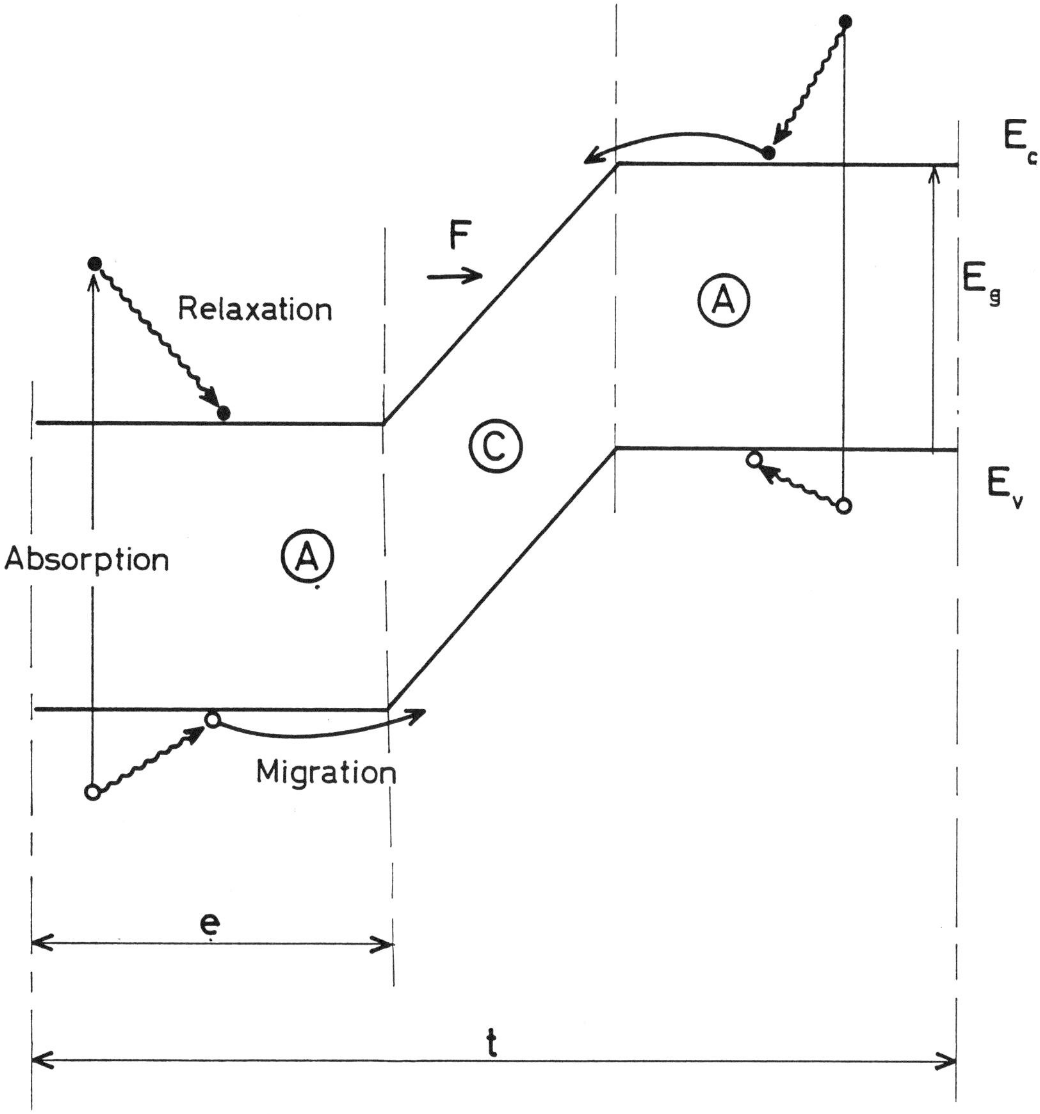

Fig. 1 Absorption and Collection in a Photovoltaic Cell (Homojunction Type) Represented by its Electronic Energy Diagram

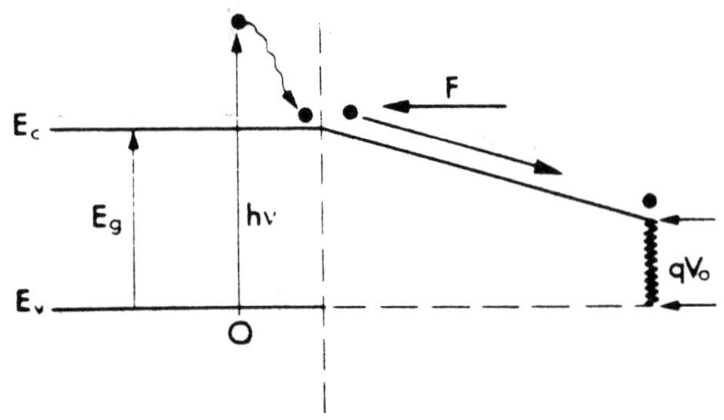

Fig. 2 Loss of Potential Energy in a Photovoltaic Cell

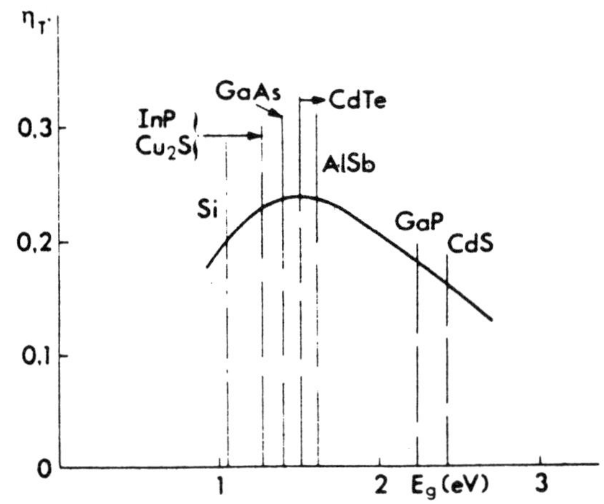

Fig. 3 Theoretical Energy Conversion Efficiency of a Photovoltaic Solar Cell As A Function Of The Energy Band Gap (Room Temperature Operation).

If the pairs are able to migrate up to part C before internal recombination occurs, they are dissociated by the built-in electric field to produce charged particles of opposite charges. In this way, an electric current (or photocurrent) is generated in the cell. Besides, an electric polarization builds which appears as a voltage across the terminals of the cell.

Presently, some considerations can be given about the energy efficiency of the PV effect, as a solar energy converter.

First, there are fundamental losses associated with those photons which are not absorbed in the material ($h\nu < E_g$) and with the relaxation process described above.

For instance, considering Si as an example (E_g = 1.1 eV) 23% of the total incident energy is not absorbed and 33% is lost during the relaxation processes. The ultimate conversion efficiency is thus 44%.

However, this value can never be reached for a single cell operating at room temperature or above. Other losses come into the picture due to the fact that electrons (or holes) accelerated by the built-in electric field lose a part of their potential energy (Fig. 2). In this way, the output potential energy is reduced to $q\ V_o$, a fraction of the internal energy of the electron hole pairs. In the case of Si, this fraction is

$$\frac{q\ V_o}{E_g} \simeq 0.5$$

The theoretical efficiency is thus only 22%. Finally, due to various secondary losses, the attainable efficiency for Si is limited to the range of 12-18%. (These values refer to the Aml solar spectrum available on earth).

Generalizing this type of analysis leads to the well-known curve giving the theoretical efficiency (at room temperature) as a function of the band gap E (Fig. 3). This curve is a guide for choosing semiconductor materials adapted to photovoltaic solar energy conversion.

MATERIALS REQUIREMENTS

Three main properties are worth considering for determining the efficiency of the photovoltaic effect.

Optical Absorption

It is measured by the absorption coefficient α which represents the attenuation factor of the photon flux J propagating into the solid

$$J\ (x) = Jo\ (1 - r)\ \exp\ (-\alpha\ x) \qquad (1)$$

r is the surface reflection coefficient.

The condition for an efficient optical absorption is

$$t > 1/\alpha \tag{2}$$

where t is the thickness of the cell along the light direction. α is wavelength dependent; however, average or mean values α_m can be defined inside the absorption range of the material. These values are given in Table I for several materials of interest. Si has a low mean absorption coefficient, the consequence of which is that a minimum thickness $t_m = 50\mu$ is required. On the other side, materials like GaAs, CdTe, and others have a high mean absorption coefficient. The required thickness is only 3μ which opens the possibility of a realization in thin films.

Electron-Hole Pair Collection

The ability of electron-hole pairs to migrate up to the built-in field region is estimated from the "diffusion length", which is a measure of the mean free path before recombination. The diffusion length L_D expresses as a function of the mobility μ and the lifetime τ of the minority carriers in the region considered (the less numerous type of carriers in that region)

$$L_D = \left[\frac{kT}{q} \mu \tau\right]^{\frac{1}{2}} \tag{3}$$

k is the Boltzman constant and q is the magnitude of the electronic charge.

This quantity is not an intrinsic one and depends strongly upon the purity and the cristalline perfection of the material as shown in Fig. 4 for Si [1]. Representative values are given in Table I.

After Fig. 1, the condition for an efficient collection is

$$e < L_D \tag{4}$$

But as e is related to t, the total thickness, one can write a generalized condition, combining (2) and (4)

$$1/\alpha_m < t < L_D \tag{5}$$

Thus, the smaller the mean absorption coefficient, the greater the required diffusion length. This put severe conditions on the use of materials like Si.

Built-In Field Or Potential Barrier Region

The region where the electric field is established (part C of Fig. 1) corresponds to a potential energy jump or potential barrier for charged particles. It can be realized in different ways:

- a junction between two pieces of the same semiconductor with opposite types of conductivity (n-type and p-type). This structure corresponds to the energy diagram of Fig. 1.

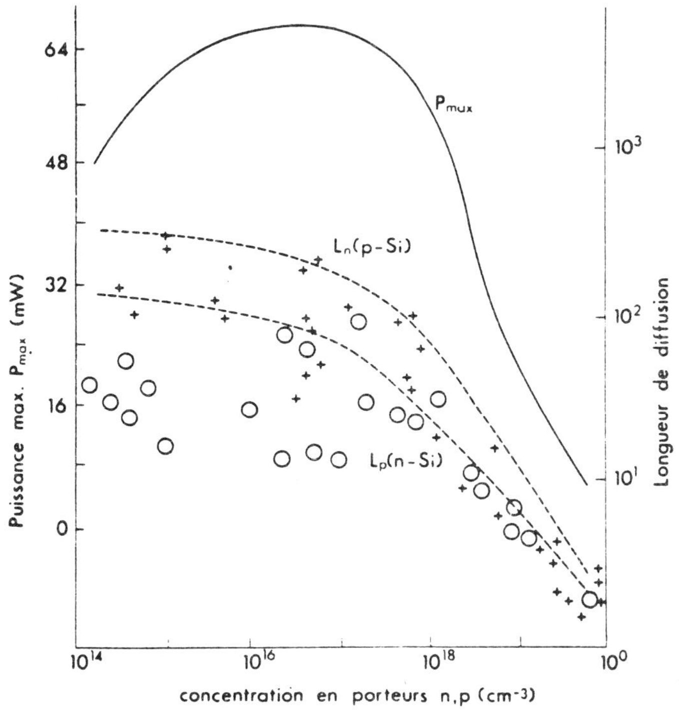

Fig. 4 Diffusion Lengths in Si (n-type and p-type) As A Function of the Majority Carrier Concentration (Or Impurity Doping) (From [1])

TABLE I

REPRESENTATIVE VALUES

Material	Energy band gap 300'K (eV)	Mean Absorption Coefficient (cm^{-1})	Electron Diffusion Length (μm)	Hole Diffusion Length (μm)
Si	1.1	200	50 - 200	20 - 100
GaAs	1.45	3.10^3	2 - 10	1 - 5
InP	1.35	3.10^3	5	3
CdTe	1.5	3.10^3	0,5 - 2	0,3 - 1
Cu_2S	1.21	10^4	5	
$CuInSe_2$	1.04	3.10^3		

- a heterojunction combining two different materials with opposite types of conductivity

- a contact between a metal and a semiconductor (Shottky barrier)

Depending on the material considered, its possible availability under the two types of conductivity, one of these different structures can be preferred. The physical principles of operation are basically the same in all cases. Detailed analysis can be found, for example, in [2].

PRACTICAL REALIZATIONS

We first describe the most advanced device, the single crystal Si solar cell, and then we present the other types of cells which are presently under development or under study in the laboratory.

Single Crystal Si Solar Cell
The structure of a p-n junction cell is shown in Fig. 5.

The base of the cell is a circular wafer of single crystal Si, between 100 μ and 300 μ thick, doped with boron impurities to give p-type conductivity. The other side of the p-n junction is made by diffusing phosphorus impurities to form a thin layer - 0.4 μ thick - with n-type conductivity. The cell is covered with a soldered metallic grid which constitutes one of the electrodes, the other one being realized by metallizing the back side of the cell. Finally, an anti-reflecting coating is formed on the front surface by putting a transparent oxide layer (TiO_2, SiO) with a refractive index around 2. The current voltage output of the cell under solar illumination is shown in Fig. 6. The current flowing under short-circuit conditions is directly proportional to the flux of incident photons

$$I_{cc} = K \int_{E_g}^{\infty} Jo\ (h\nu)\ d(h\nu) \tag{6}$$

K is the photoelectric efficiency or collection efficiency which accounts for transmission losses, reflection losses and recombination losses.

In good cells, one has:

$$K = .75$$
$$I_{cc} = 35\ \text{mA/cm}^2$$

The open circuit voltage V_{oc} is related to I_{cc} by the expression

$$V_{oc} = \frac{kT}{q} \ln \left[\frac{I_{cc}}{I_o(E_g, L_d)} \right] \tag{7}$$

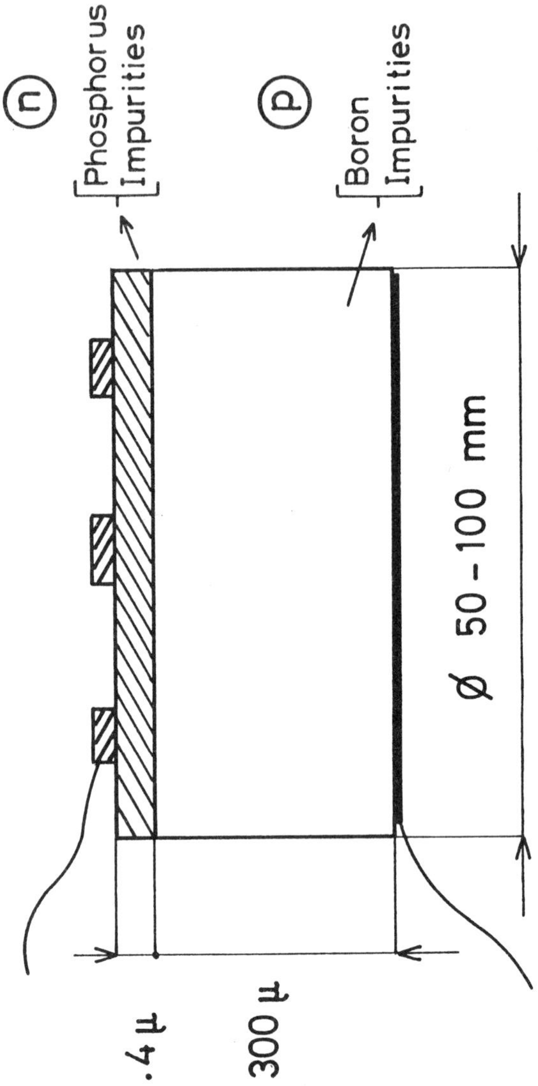

Fig. 5 Structure of a p-n Junction Single Crystal Si Cell

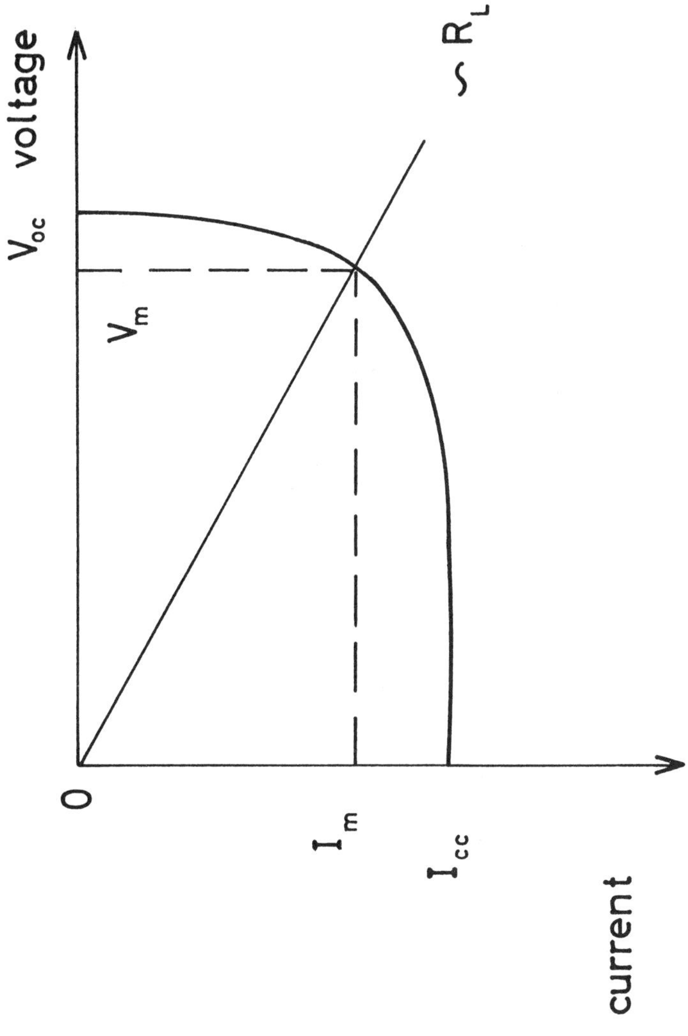

Fig. 6 Current Voltage Output of a Si Solar Cell

V_{OC} increases with the photon flux in a logarithmic way. I_O is a reference current, actually the dark current flowing under a reverse bias. Its expression is of the form

$$I_o = \frac{kT\ \mu}{L_D\ N_I}\ N_c\ N_v\ \exp\left(\frac{-E_g}{kT}\right) \tag{8}$$

N_I is the impurity concentration in doped regions. N_C and N_V are effective densities of available states in conduction band and valence band respectively. From (6) and (7) one obtains:

$$V_{oc} = \frac{E_g}{q} - \frac{kT}{q}\ \ln\left[\frac{kT\ N_c\ N_v}{I_{cc} L_D\ N_I}\right] \tag{9}$$

It is verified that V_{OC} is always smaller than E_g/q for $T > 0$. In good cells $V_{OC} \simeq .55$ volts. To supply electric power, the cell is connected to a load resistor R_1. The current flowing in the load is related to the voltage across it by:

$$A\ I = \frac{V}{R_1} = A\ I_o\ (\exp \frac{qV}{kT} - 1) \tag{10}$$

where A is the total cell area. The value of R_1 has to be adjusted to give a maximum power P. It is usual to write

$$P_m = V_m\ I_m = f\ V_{oc}\ I_{cc} \tag{11}$$

The energy efficiency is then

$$\eta = \frac{P_m}{E_o} = f\ \frac{V_{oc}\ I_{cc}}{E_o} \tag{12}$$

where E_O is the solar constant. f is called filling factor and measures the relative area of the rectangle $I_m\ V_m$ inscribed inside the current voltage curve. f depends essentially on I_O and on an eventual series resistance (not taken into account here). The measured energy efficiency is around 14% for industrially produced cells. In the laboratory, efficiencies up to 18% (always under AM1 illumination) have been obtained [3].

Given a material, such as Si, other types of photovoltaic structures are possible, as indicated above. One of the most extensively studied is the MIS cell, the energy diagram of which is depicted in Fig. 7. This is basically a metal-semiconductor contact, or Schottky barrier, with a thin insulating interfacial layer (SiO_2). Illumination strikes the metal layer which is also very thin in order to reduce the optical absorption in it. Efficiencies up to 12% have been measured on MIS silicon structures [4]. The main advantage of this cell lies in that the elaboration process is a low temperature one, avoiding diffusion of impurities at high temperatures.

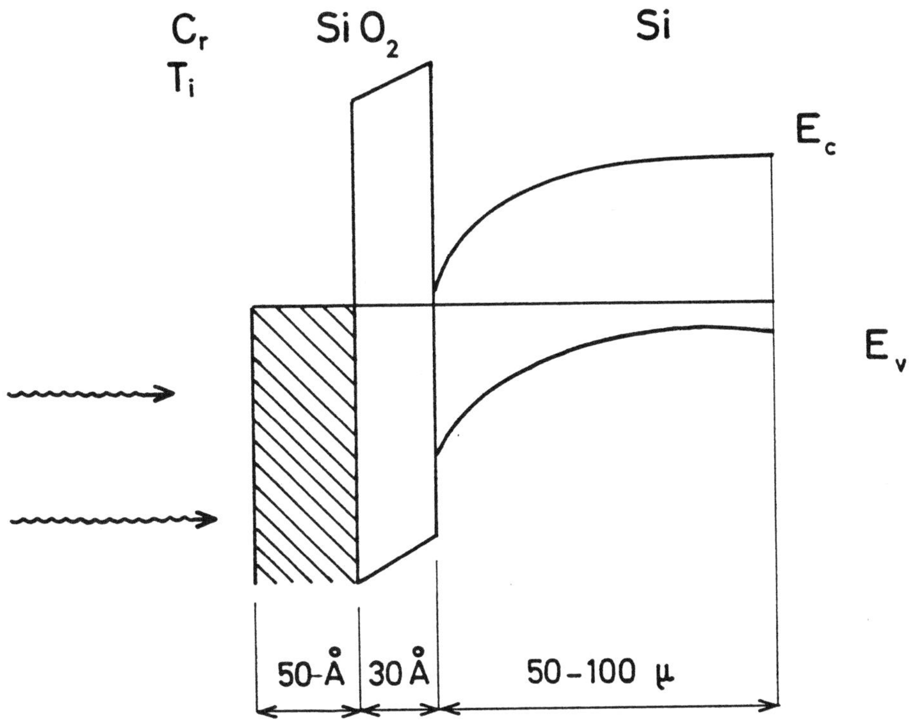

Fig. 7 Energy Diagram Of A MIS Structure

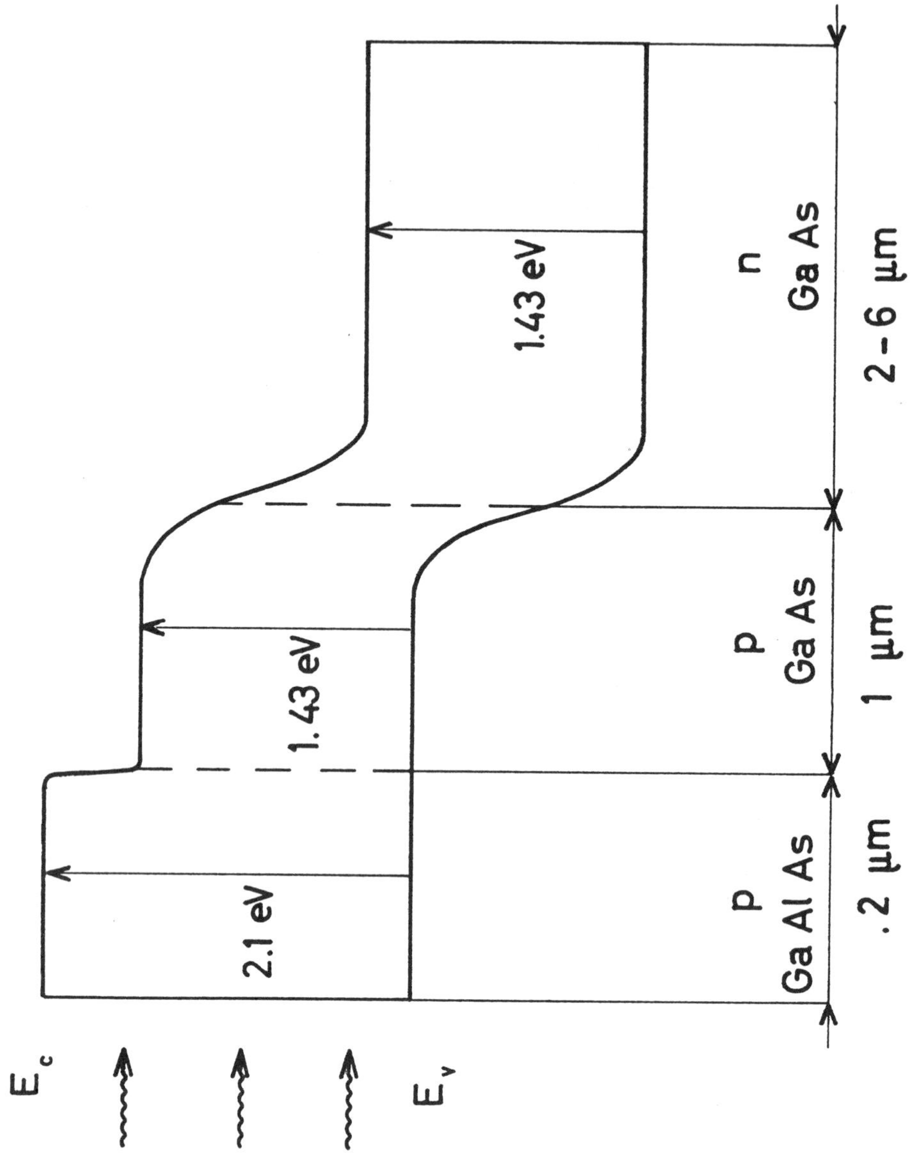

Fig. 8 Energy Diagram of A Ga As - Ga Al As Heterojunction

Ga As - Ga Al As Heterojunction [2]

The limit efficiency of Ga As is higher than that for Si (Fig. 3). An efficient structure using Ga As is the heterojunction between Ga As n-type and the derived alloy Ga Al As p-type. The corresponding energy diagram is shown in Fig. 8. Ga Al As constitutes a high bandgap window (E_g = 2.1 eV) which transmits photons to Ga As (E_g = 1.43 eV) and protects the surface of this material. The lattice match between the two materials is excellent. The entire structure is fabricated by vapor or liquid epitaxy on a single crystal Ga As substrate. This structure has delivered the highest efficiency measured up to now on a PV device: 23% [5]. The cost of this cell is much higher than the cost of Si cells. Therefore, it is considered as being utilized under concentrated sunlight. In these conditions, the efficiency is a function of both the concentration ratio C and the temperature reached by the cell under illumination (Fig. 9) [6]. It is seen that the efficiency can be kept high even with a cell at high temperature. In addition, the heat generated in the cell can be removed by a cooling system and utilized for several purposes.

Thin Film Cells

The search for low cost solar cells invites one to consider structures made in an economic way with thin film materials. The essential requirements are to obtain polycristalline films with a grain diameter several times the diffusion length in the material, or, after (4), of the order of thickness of the films. The minimum thickness of the films is basically fixed from the mean absorption coefficient (Table I).

The prototype of the thin film cells is the Cu_2S/CdS heterojunction (Fig. 10). The absorbing part of the cell is a thin layer of Cu_2S p-type ($E_g \simeq$ 1.2 eV) formed by a chemical reaction on a thicker CdS polycristalline layer of opposite conductivity type. The latter is usually prepared by vacuum evaporation of CdS on a metallized glass substrate. Simpler methods of preparation are under study, such as the pyrolysis of chemical compounds ($CdCl_2$ + thiourea)(400°C). As shown in Fig. 10, the Cu_2S/CdS interface extends deep into the grain boundaries of the CdS polycristalline film. The grain diameter is of several microns. Other properties of this system are presented in Table II. Two quantities are worth considering in designing heterojunctions: the first one is electron affinity which has to be the same in the two materials to give a good electronic match between the conduction bands; the second one is lattice parameter which has also to be the same for a good lattice match at the interface. With regard to this point, the Cu_2S/CdS heterojunction is not perfect because of a relative mismatch of 4%. Nevertheless, a maximum efficiency of 8.5 % has been measured in the laboratory [7]. Table II also presents the properties of other photovoltaic systems studied in the perspective of a thin film realization. Three of them are heterojunctions based on the use of CdS as the n-type side of the junction: CdS/InP, $CdS/CuInSe_2$ [9], CdS/CdTe [10],[11], the last column refers to homojunctions between CdTe n and CdTe p [12]. In the heterojunctions, CdS is considered to act as a window, transmitting photons to the underlying photosensitive material (InP, $CuInSe_2$, or CdTe) and protecting the surface of the latter.

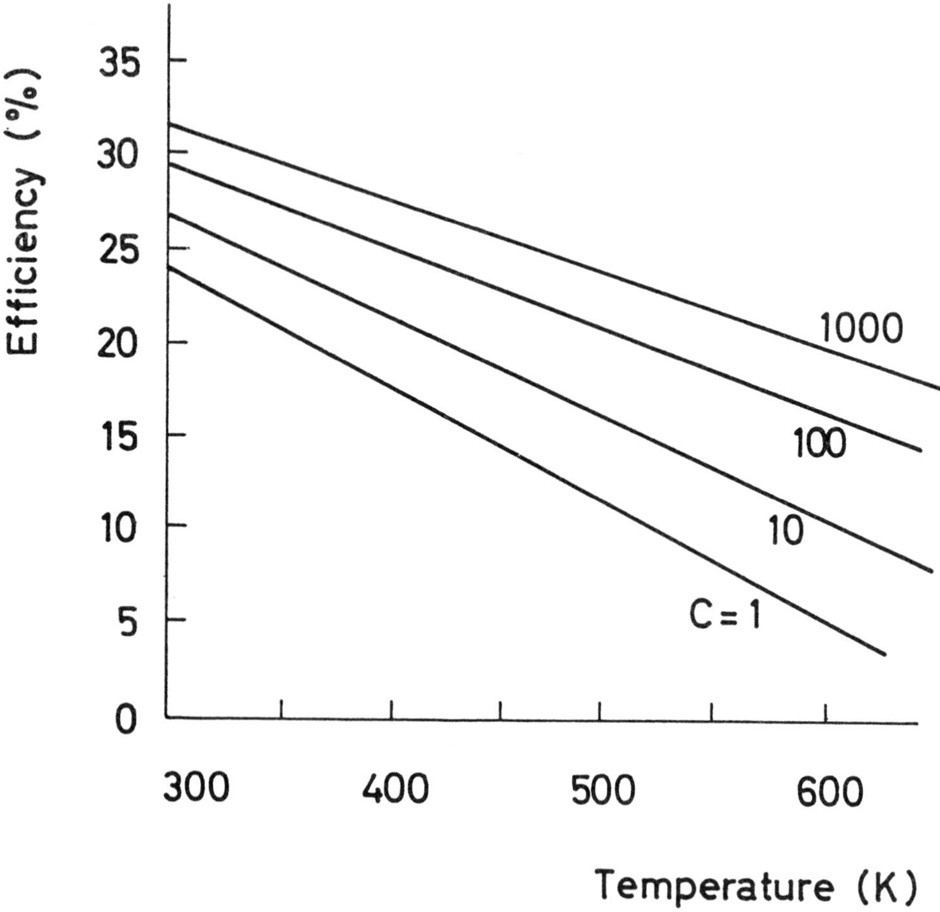

Fig. 9 Theoretical Efficiency Of A Ga Al As - Ga As Cell As A Function Of Concentration Ratio And Cell Temperature (From [6])

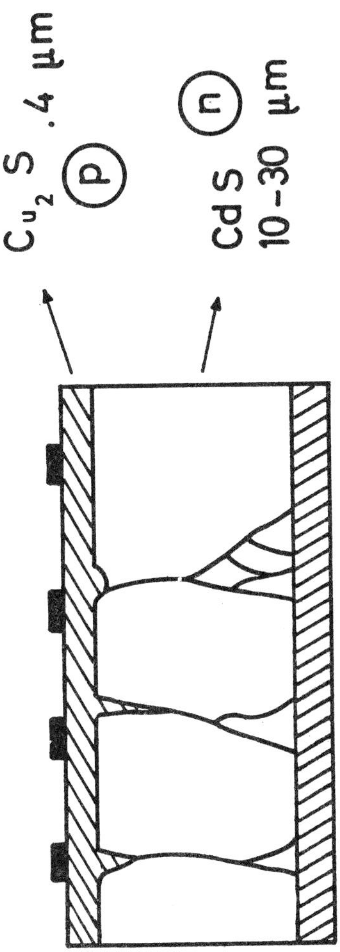

Fig. 10 Cross Section Of A Cu_2S/CdS Heterojunction

TABLE II

PROPERTIES OF PHOTOVOLTAIC SYSTEMS IN THE PERSPECTIVE OF THIN FILM REALIZATION

	CdSn	Cu_2Sp	InP p	$CuInSe_2$ p	CdTe p	CdTe n
Band gap (eV)	2.42	1.2	1.35	1.04	1.5	1.5
Electron Affinity (eV)	4.5	4.2	4.38		4.4	4.4
Lattice Constant Å	4.14 (Hex)	3.96 (Hex)	5.87 (cub)	5.78 (cub)	6.48 (cub)	6.48 (cub)
$\frac{\Delta a}{a}$ %		4 %	1 %	1 %	9 %	
η (%) Single Crystal			14	12	10	
η (%) Thin film		8.5	5.5		8	8

Up to now, these cells have mostly been studied in a single crystal form and promising efficiencies have been measured. The present objective is to reproduce such results using thin film materials.

ECONOMICS OF SOLAR CELLS

The possibilities of utilizing solar cells to generate electricity in real applications depend upon the cost of the produced energy.

The cost is a function of several factors: cost of the cells, expressed for example, in dollar per peak watt (the power generated under standard AM1 isolation) or in dollar per unit panel area, concentration ratio used, mode of operation (either direct utilization of the generated electricity or with temporary storage) and expected lifetime of the panels before degradation imposes replacement.

The most important factor is the cost of the cells. Silicon solar cells are the only commercially available cells. The present price is around 10-15 \$/watt, which corresponds to 1000-1500 \$/panel square meter. It is expected (ERDA) that by 1982 the price will go down to 2 \$/watt and to .5 \$/watt in 1986. This price reduction would be obtained by extrapolation of the present silicon technology and automation of the processing.

The cost of electrical energy can now be readily computed, taking as a conservative figure a panel lifetime of 15 years. Fig. 11 displays the relation between energy cost, energy efficiency and panel price, calculated for the simplest operation conditions: no concentration, no storage. To reach a given level of electricity cost, to which corresponds a possible application, the efficiency has to be kept above a minimum value because there exist uncompressive charges per m^2 beside the panel cost (land, frame, connecting network, etc.). This minimum value is thought to be between 8 and 10%.

The price of presently available Si cells set them just outside the scale. However, by using concentration of the sunlight, it is possible to get a significant reduction in energy cost. This is shown in Fig. 12 for two Si technologies. Roughly, the cost can be cut down by a factor of 10 with only a medium concentration ratio. This means that several applications of solar cells are already economically viable, including water pumping and electricity supply in isolated areas.

Finally, Fig. 13 tentatively shows the expected evolution of the solar electricity cost for different solar conversion modes. Right now, solar conversion using concentration is the more economic way to generate electricity, either in the thermodynamic mode (distributed concentrators or central tower) or in the photovoltaic mode.

Photovoltaic generation without concentration is the most evolutive technology and can lead to the highest cost reduction. In the long term, this technology could be competitive with centralized thermodynamic conversion

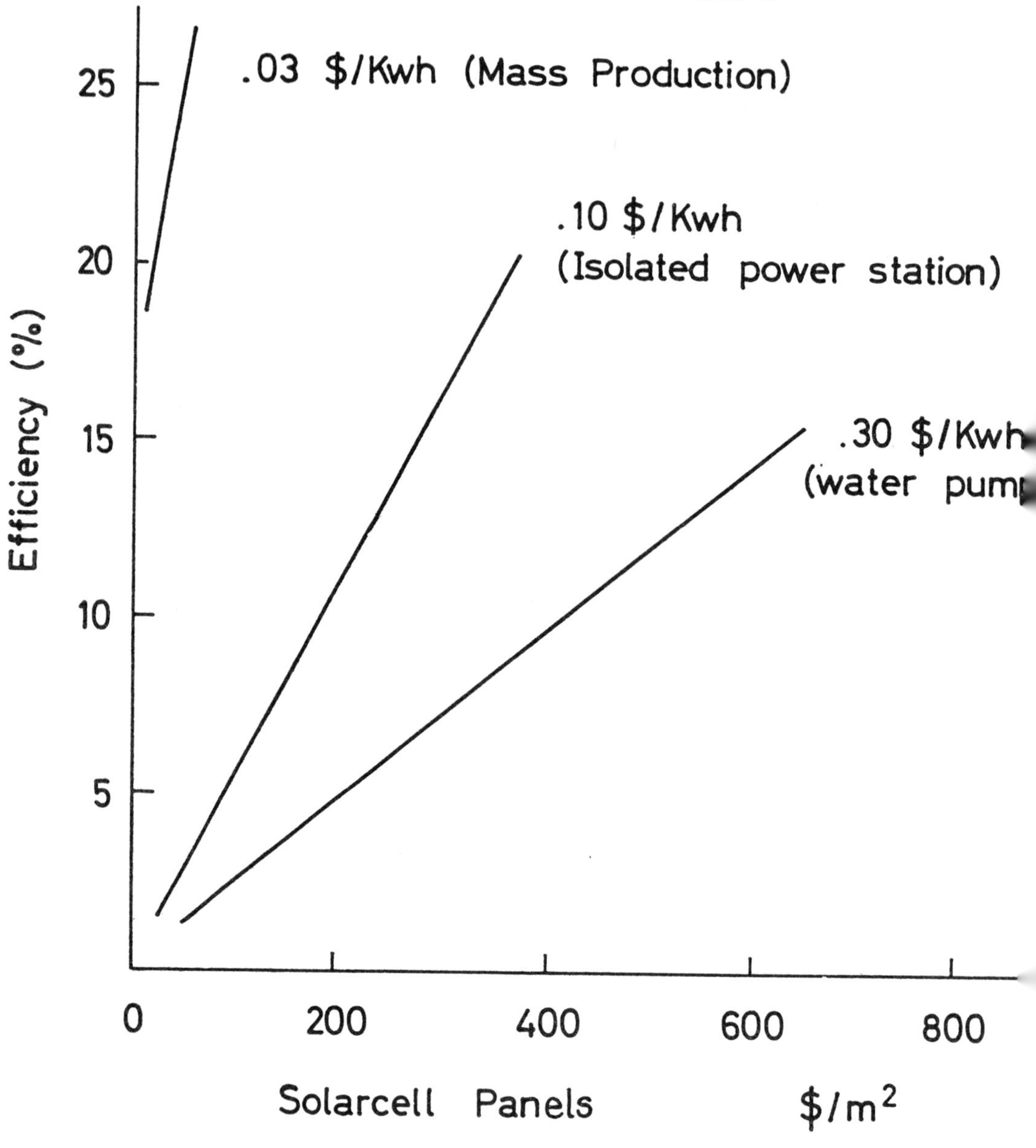

Fig. 11 Relation Between Electricity Cost, Efficiency And Cost Of The Solar Cell Panels (No Concentration, No Storage) (From [13])

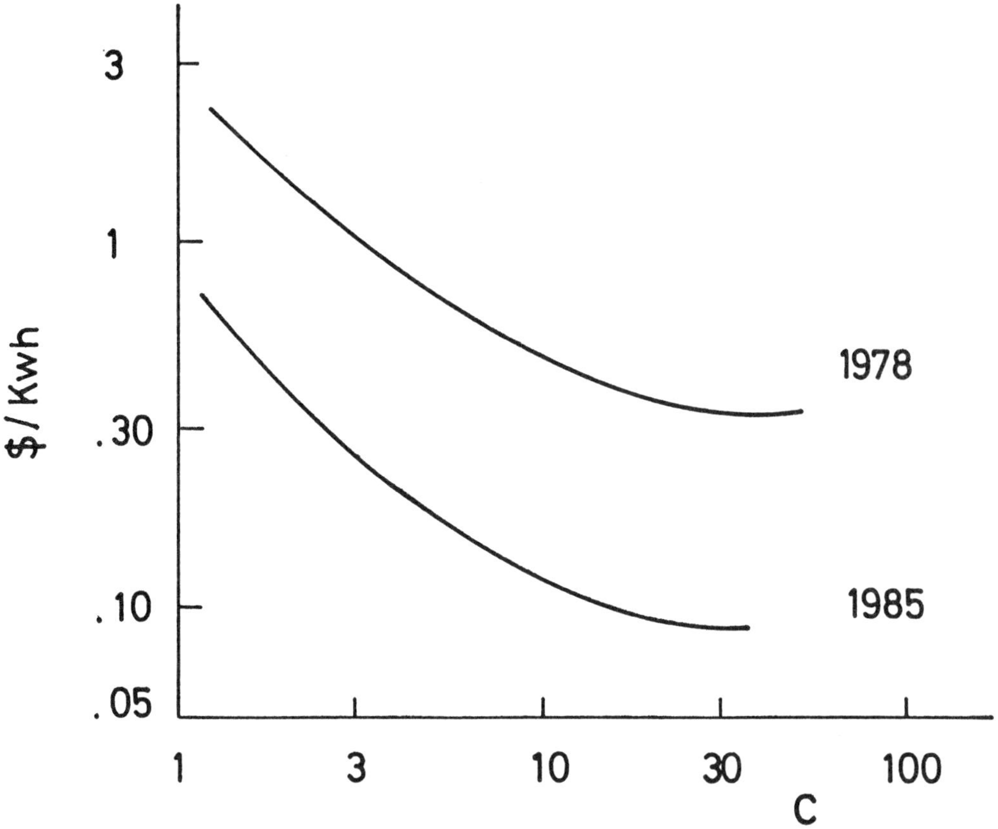

Fig. 12 Cost Of Electrical Energy Supplied By Si Solar Cells As A Function Of Concentration

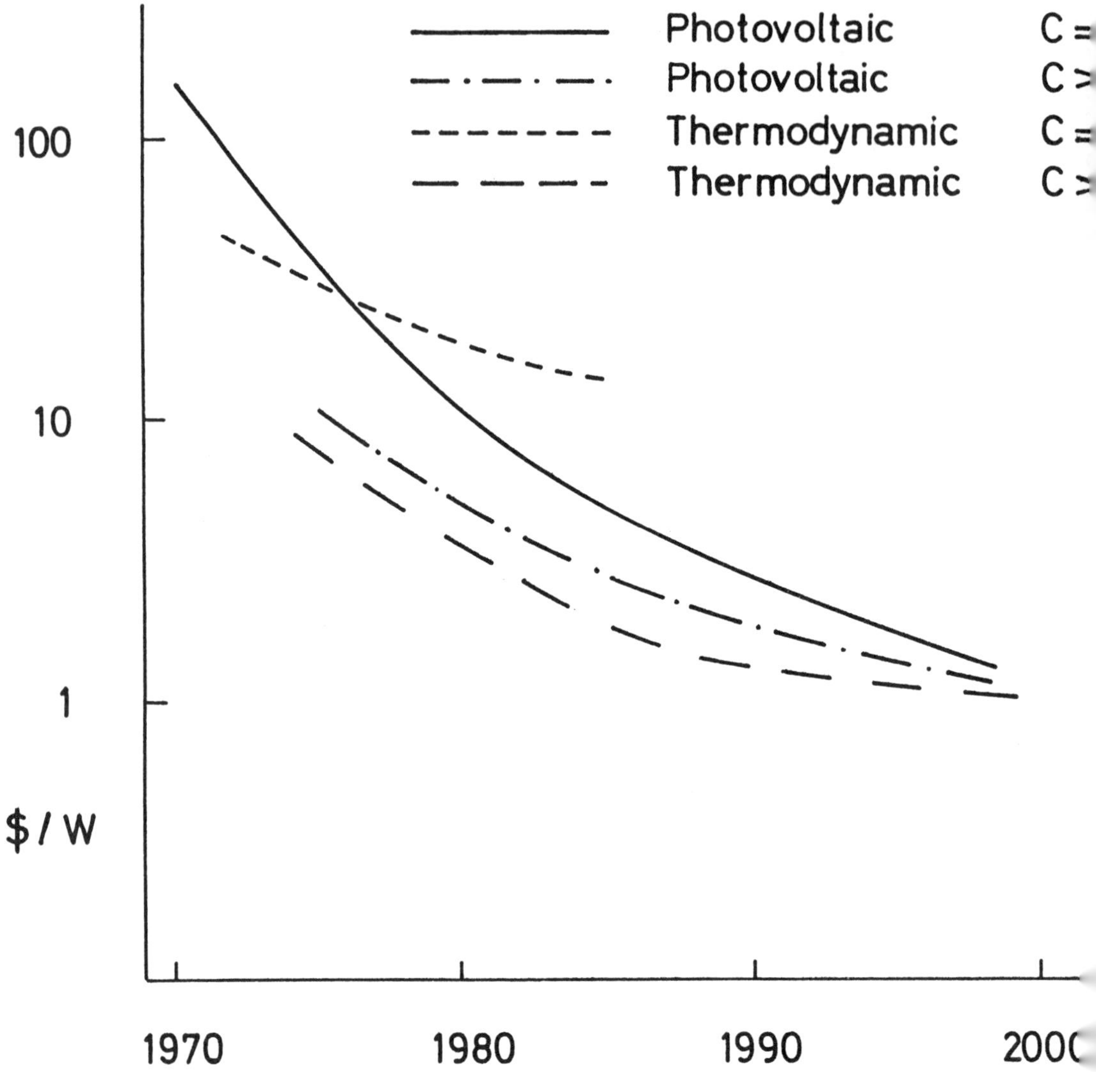

Fig. 13 Prospective Variations Of Nominal Power Cost For Several Solar Conversion Technologies

for mass production of electricity.

CONCLUSIONS

Photovoltaic solar energy conversion for terrestrial applications is now a reality. Silicon solar cells are industrial products with large possibilities of technical development and price reduction. There is also a promising area of research for new materials and new technologies. Some applications are already within the reach of existing solar cells, and the time has come for various tests and demonstrations in real fields. In this way, the exact possibilities of photovoltaic converters will appear, and both market and production will rapidly expand.

REFERENCES

1. Iles, P.A. and Soclof, S.I., Proceedings 11th I.E.E.E. Photovoltaic Spec. Conf., Phoenix, Arizona, U.S.A. (1975).

2. Hovel, H.J., "Solar Cells" in Semiconductors and Semimetals, Vol. II Ed. R.K. Willardson, A.C. Beer, Academic Press, 1975).

3. Lindmayer, J. and Allison, J.F., Proceedings 9th I.E.E.E. Photovoltaic Spec. Conf., Silver Springs, Maryland, U.S.A. (1972).

4. Fabre, E., Michel, J. and Baudet, Y., Proceedings 12th I.E.E.E. Photovoltaic Spec. Conf., Baton Rouge, Louisiana, U.S.A. (1976).

5. Woodall, J.M. and Hovel, H.J., Appl. Phys. Lett. 30, 492 (1977).

6. Hovel, H.J., I.B.M. J. Res. Develop. 22, 112 (1978).

7. Barnett, A.M., Meakin, J.D. and Rothward, A., Proceedings Photovoltaic Solar Energy Conference, Luxembourg (1977), p. 535.

8. Wagner, S. et al, Appl. Phys. Lett. 26, 229 (1975).

9. Shay, J.L. et al, Appl. Phys. Lett. 27, 89 (1975).

10. Yamaguchi, K., Matsumoto, H., Nakayama, N. and Ikegami, S., Japan J. Appl. Phys. 15, 1575 (1976).

11. Yamaguchi, K. et al, Japan J. Appl. Phys. 15, 2281 (1976).

12. Mimila-Arroyo, J., to be published in Solar Energy Materials.

13. Rodot, M., Barbe, M. and Dixmier, J., Rev. Phys. Appl. 12, 1223 (1977).

REALIZATION OF P-N JUNCTION SOLAR CELLS BY AN ION IMPLANTATION DOPING PROCEDURE

J.C. Muller, M. Hage-Ali and P. Siffert
Cnetre de Recherches Nucleaires
Groupe de Physique et Applications des Semiconducteurs (PHASE)
67037 Strasbourg-Cedex, France

ABSTRACT

The possibility of using a low cost ion implantation procedure for the preparation of junction solar cells has been investigated.

The method employs a D.C. glow discharge ion source and a short post acceleration structure, without any mass separation. Preparation of the cells in a continuous way is possible at competitive speeds since the ion beam current density reaches 1 mA/cm^2.

The properties of silicon cells, obtained by discharge bombardment in BF_3 or PF_5 atmosphere followed by recristillisation of the damaged layer either by thermal annealing or fast surface laser pulses, have been investigated. Rutherford backscattering, SIMS, electrical measurements have been used. Finally, characteristics and performance of the devices are presented.

INTRODUCTION

Ion implantation is now widely used as a doping method of semiconductors, since it has many advantages, which are now well known. However, the high cost and rather low current density of the conventional machines, the necessity to anneal the radiation induced defects, explain the small interest for preparing solar cells.

We propose to realize the junction of these cells by using a low cost high current ion implantation procedure, followed by a quick annealing by a short laser pulse. The implant machine presented in the first part of this paper, is based on a glow discharge in a gas atmosphere containing the dopant [1,2]. Since no mass separation is used, the cost of the apparatus is strongly reduced (absence of magnet, as well as of stabilized power supplies, no lenses...) and the molecular ions increase the concentration of dopant atoms in the surface vicinity (zone between the projected maximum ion range R_p and the surface). However, this procedure creates high defect densities, in excess to that existing after classical implantation. These defects can be annealed by thermal heating at temperatures up to 1000°C. However, according to recent work [3-5] complete recristallisation does not occur. Therefore, we investigated the possibility to use laser pulse annealing, which was found in the literature [6-10] to be very efficient for low dose implantation, even for solar cells [11].

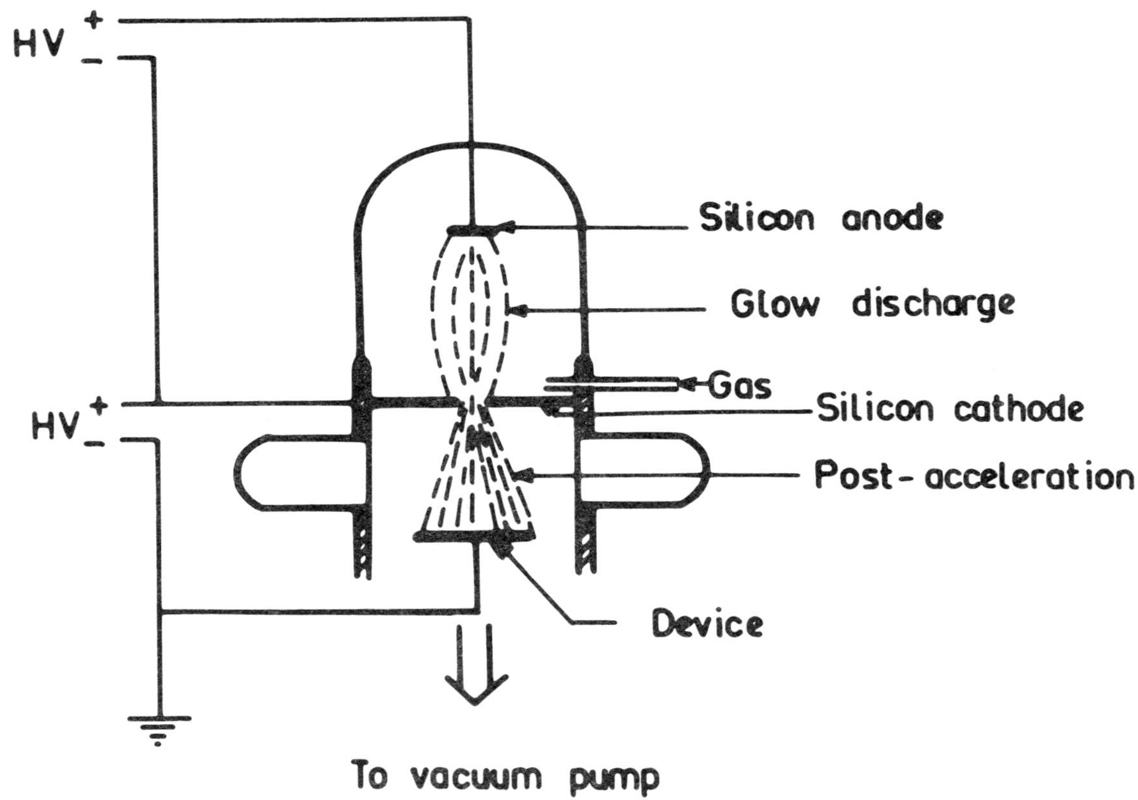

Fig. 1 Schematic of the implantation machine using a D.C. glow discharge source with a short post acceleration structure (no magnet for mass selection).

In a second part, the properties of the implanted layers will be investigated by various techniques, including Rutherford backscattering (RBS), secondary ions mass spectroscopy (SIMS), surface conductivity measurements and, finally, investigation of the cell's performance.

EXPERIMENTAL PROCEDURE

Ion Implantation

Implantation has been performed with the apparatus shown in Figure 1. The ion source consists of a glow discharge plasma, created under 6 kV D.C. voltage, in a glass chamber in which a gas containing the dopant (BF_3 for B, PF_5 for P) is introduced at a pressure of 10^{-1} - 10^{-2} torr. The ion beam is extracted and accelerated at energies ranging from 20 to 80 kV, directly towards the sample. The current densities lie between 0.1 - 1 mA/cm^2. In this procedure, the junction location is determined by the energy of the lightest ion (B^+ in the case of BF_3 gas), whereas a high concentration of dopants is introduced closest to the surface by the heavier molecular ions (BF_3^+, BF_2^+...).

Annealing

a) Thermal: Thermal annealing is performed under reduced pressure (10^{-6} torr) at temperatures up to 1000°C for 30 minutes, in general.

b) Laser: Laser annealing has been performed with a glass neodym laser, emitting at 1.06μ, having the following characteristics: pulse width 35 ns, spot diameter 16 mm, repetition rate 1 pulse/min., maximum overage power incident on the sample 1.5 joule per pulse/cm^2. The sample is mounted in a vacuum chamber ($p \simeq 10^{-2}$ torr) with a glass window for the irradiation by the laser pulse.

PROPERTIES OF THE IMPLANTED LAYER

The implanted surface layers obtained after BF_3 or PF_5 glow discharge have been studied by various techniques in order to determine their main characteristics. Here, we restrict ourselves to the results obtained after BF_3 discharge.

Crystal Damage

To determine the distribution of the damaged zone, we made use of the RBS technique under channeling conditions. $^4He^+$ ions of energy 1-2 MeV hit the sample mounted on a 3-axis goniometer. The backscattered particles are detected at an angle of 80° with respect ot the beam direction ("glancing angle geometry"). This geometrical arrangement leads to a depth resolution of 70 Å.

In order to compare the direct glow discharge induced damage to convention-

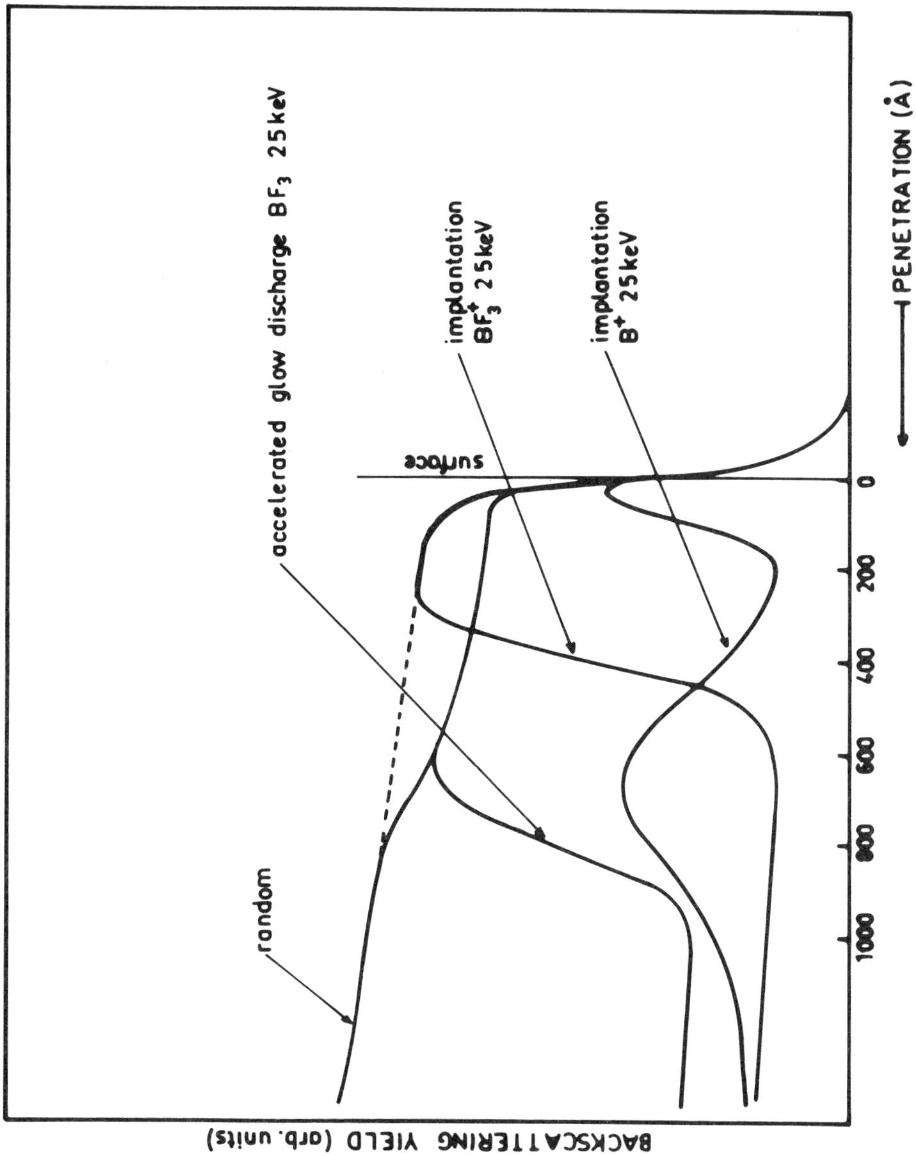

Fig. 2 Backscattering spectra of 1MeV[4] ions under channeling conditions for a classical ion implantation of boron and BF_3^+ and for an accelerated glow discharge of BF_3 at the same energy.

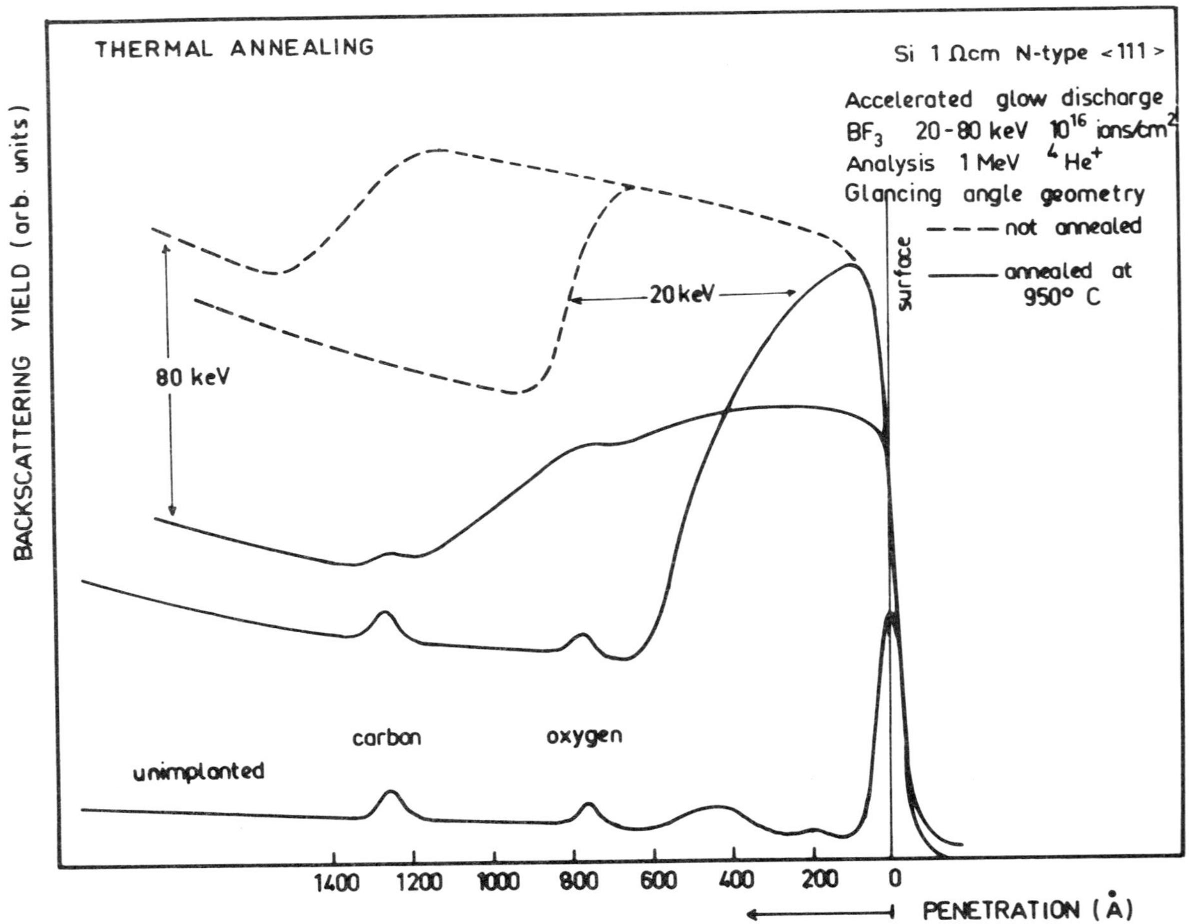

Fig. 3 Backscattering spectra in the same conditions before and after BF_3 accelerated glow discharge at energies 20 and 80 keV. The effect of thermal annealing under vacuum at 950°C during 30 minutes is

ally implanted sample damage, we investigated in a conventional implant machine both B^+ and BF_3^+ mass separated ions. The results are reported in Figure 2.

It appears that the defects extend to a depth corresponding to the range of the lightest element; the amorphous level is reached for the two heavier mass doping. The shoulder observed by accelerate glow discharge is probably due to recoil or to sputtering silicon atoms. Figure 3 shows the effect of thermal annealing on BF_3 implanted layers at two different enregies (20 and 80 KeV). It becomes clear that this procedure leads only to partial recovery of the cristallinity, even at temperatures of annealing as high as 950° C. Epitaxial regrowth from the undamaged bulk material towards the surface occurs, but it is not sufficient to restore the initial crystal perfection.

The result obtained after laser annealing is reported in Figure 4. It appears that the recristallization is considerably better when compared to thermal annealing, even after a pulse as short as 35 ns. Only a surface damage peak of 100 Å extension in depth remains.

Doping Profiles

SIMS has been employed to analyze both the positive and negative ions emitted under argon ion bombardment before and after annealing. The ions of interest here, have been profiled in Figure 5. It appears that:

- the boron distribution after laser annealing is very close to the surface when compared to the distribution from thermal annealing. A similar result has been reported for low dose rate implantation in Ref. [12]. The high power density of the laser used here is sufficient to melt the silicon at the surface due to the high absorption coefficient of light in the amorphous material. This molten material regrows epitaxially on the undamaged silicon substrate.

- the high flourine concentration after implantation, resulting from our particular conditions, is strongly reduced after annealing, probably as a result of desorption from the hot zone.

Sheet Resistance

The evolution of the sheet resistance, as measured by the Van der Pauw technique, for the two annealing methods, are reported in Figure 6. It appears that after a laser pulse annealing at energies higher than 0.8 Joule/pulse.cm^2, the sheet resistance is identical to that of a thermal treatment at 1000°C for 30 minutes.

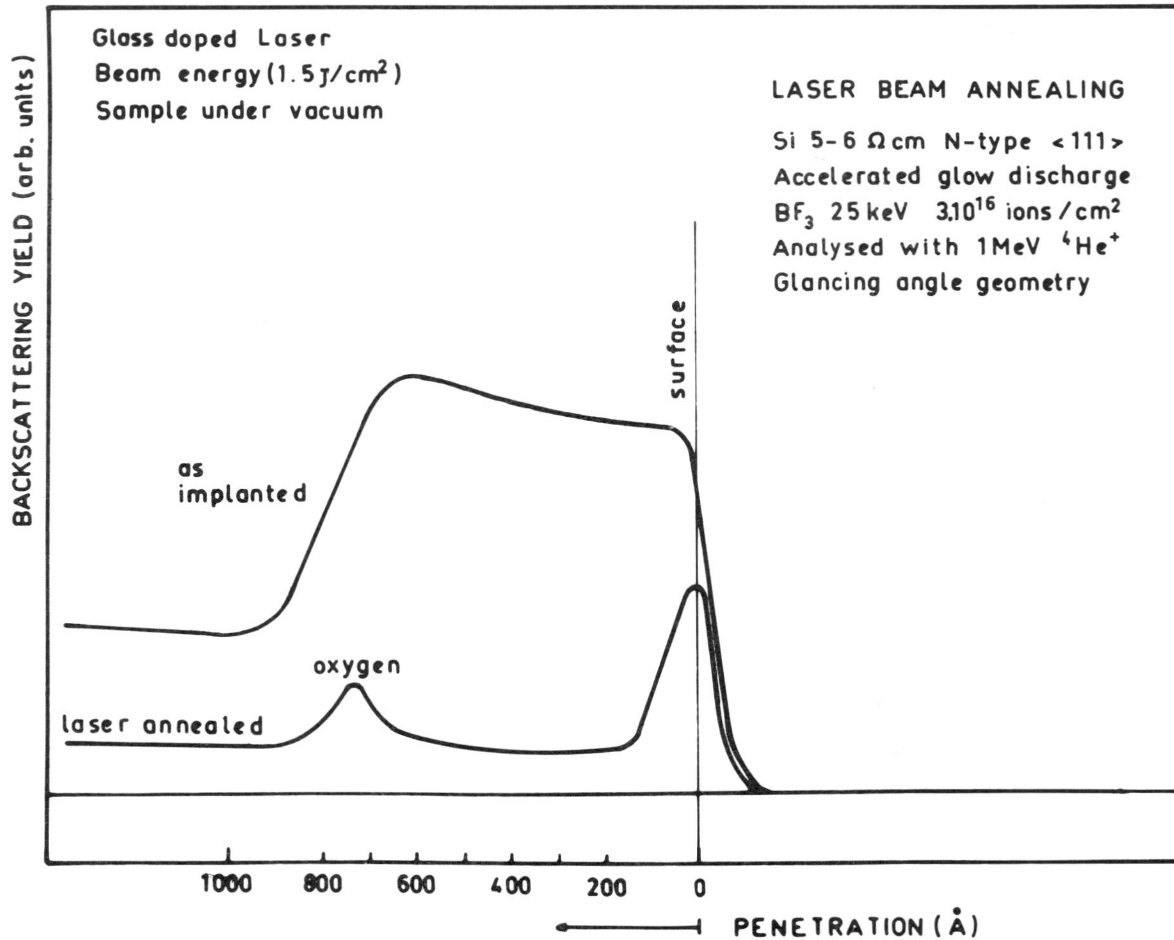

Fig. 4 Backscattering spectra of the damaged layer annealed with a laser

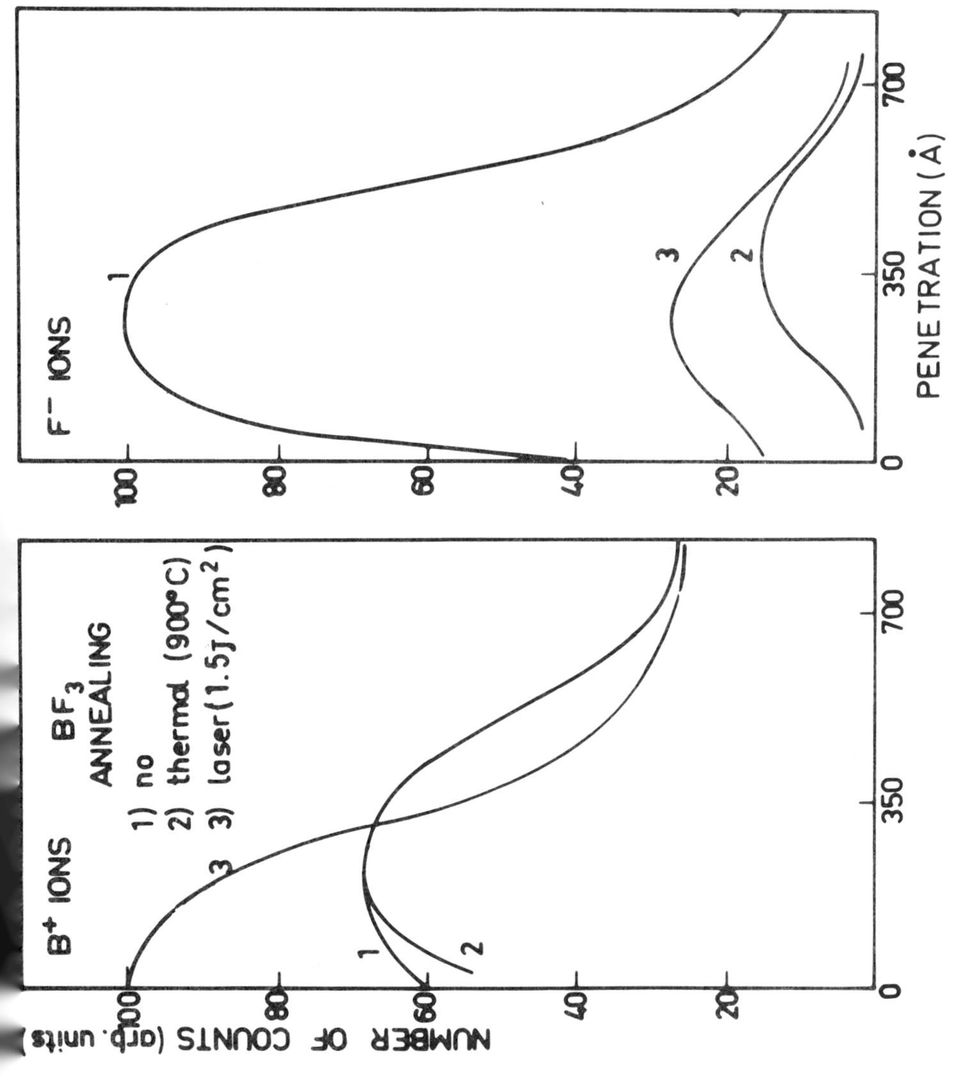

Fig. 5 Distributions of boron and fluorine ions as measured by SIMS after 1) implantation 2) thermal 3) laser annealing.

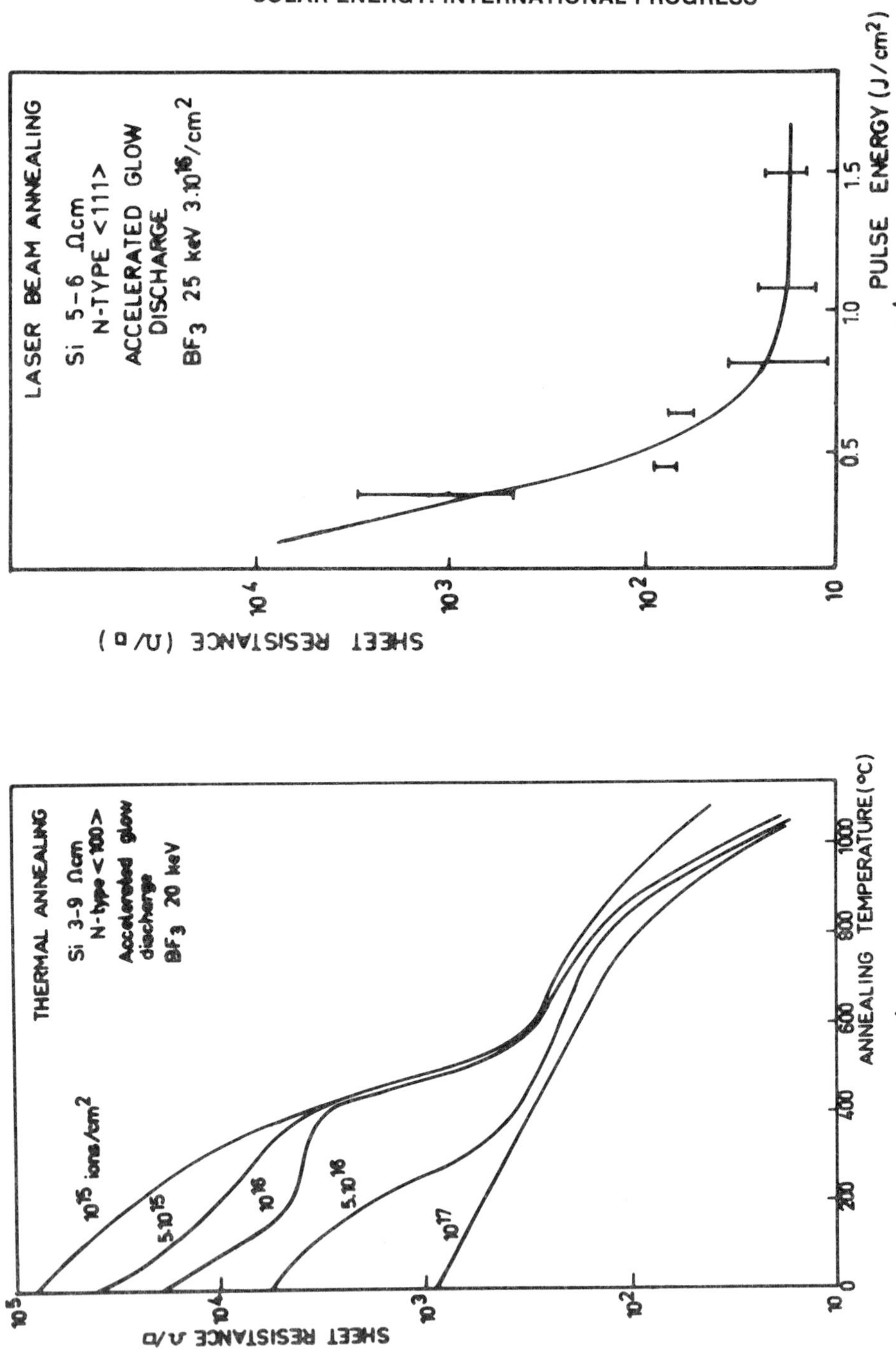

Fig. 6 Evolution of the sheet resistance versus (a) the annealing temperature, (b) the laser pulse energy.

CELL CHARACTERISTICS

Device Preparation

Monocristalline silicon samples of N and P-Type having either <111> or <100> orientations and resistivities ranging between 1 and 10Ω cm. have been employed. The implantations were generally performed at energies around 25 keV for BF_3 and 45 deV for PF_5 at doses in excess of 10^{16} cm^{-2} over an area of 1-8 cm^2. The layers are annealed either by heating or by laser irradiation. Due to some inhomogeneous photon distribution in the pulse, some scattered surface damage can be created.

The back contact consists of vacuum evaporated aluminum, without any sintering. The grids on the rectifying contact consist of gold, which has been deposited also under vacuum by evaporation. Anti-reflective coating is obtained by evaporation of a 700 Å thick SiO film.

Special Response

Figure 7 shows the spectral response of a commercial diffused P-N cell (fabricated by RTC-France) as well as BF_3 and PF_5 discharge cells. After normalization on the maximum current value, the response of all cells appear to be quite identical, the implanted cell shows a little advantage in the IR spectrum side, whereas the laser beam annealed cell exhibits a strong absorption in the blue part of the spectrum, probably, as a result of some surface contamination during the experiment.

I-V Characteristics

The I-V characteristics have been measured both under Strasbourg sun and tungsten lamp illumination up to 100 mW/cm^2 with a water filter, to simulate roughly AM1 conditions.

a) Thermal annealing: A typical I-V characteristic of a PF_5 cell is shown in Figure 8. The poor back contact explains the poor fill factor (FF $\simeq$ 0.65) observed for 100 mW energy illumination density. However, for lower illumination conditions under Strasbourg sun, efficiencies up to 13-14% have been reached for BF_3 doping [13].

Laser beam annealed samples are under investigation, no results are available now.

CONCLUSION

Comparable performance with standard diffusion cells can be obtained with our simple implant technique followed by a thermal annealing. If the laser annealing will give all its possibilities an effective automatic continuous processing can be developed, operating close to room temperature at speeds which will be compatible with the ribbon EFG growth procedure.

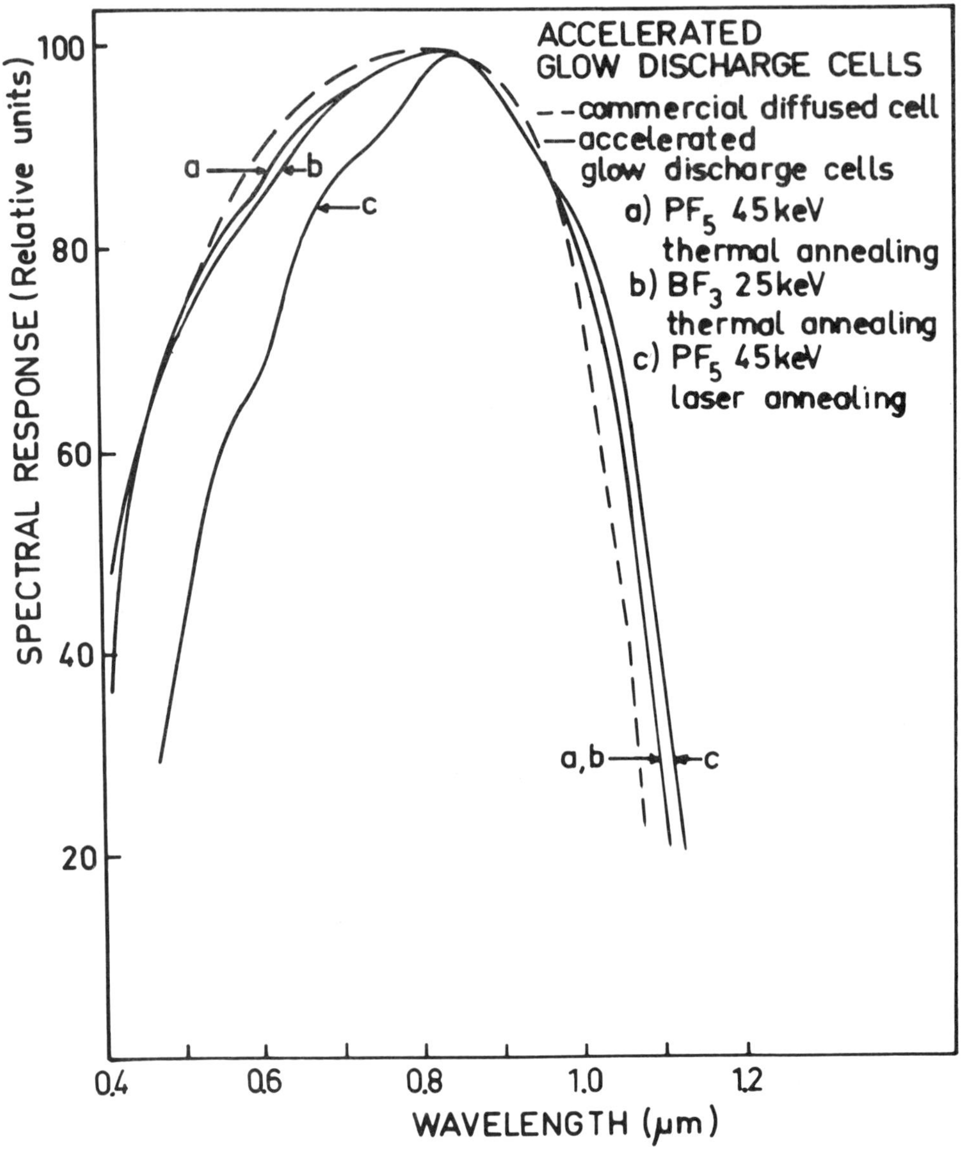

Fig. 7 Spectral response of a commercial N-P diffused cell and accelerated glow discharge PF_5 or BF_3 cells with (a) and (b) classicll thermal annealing and (c) laser pulse annealing.

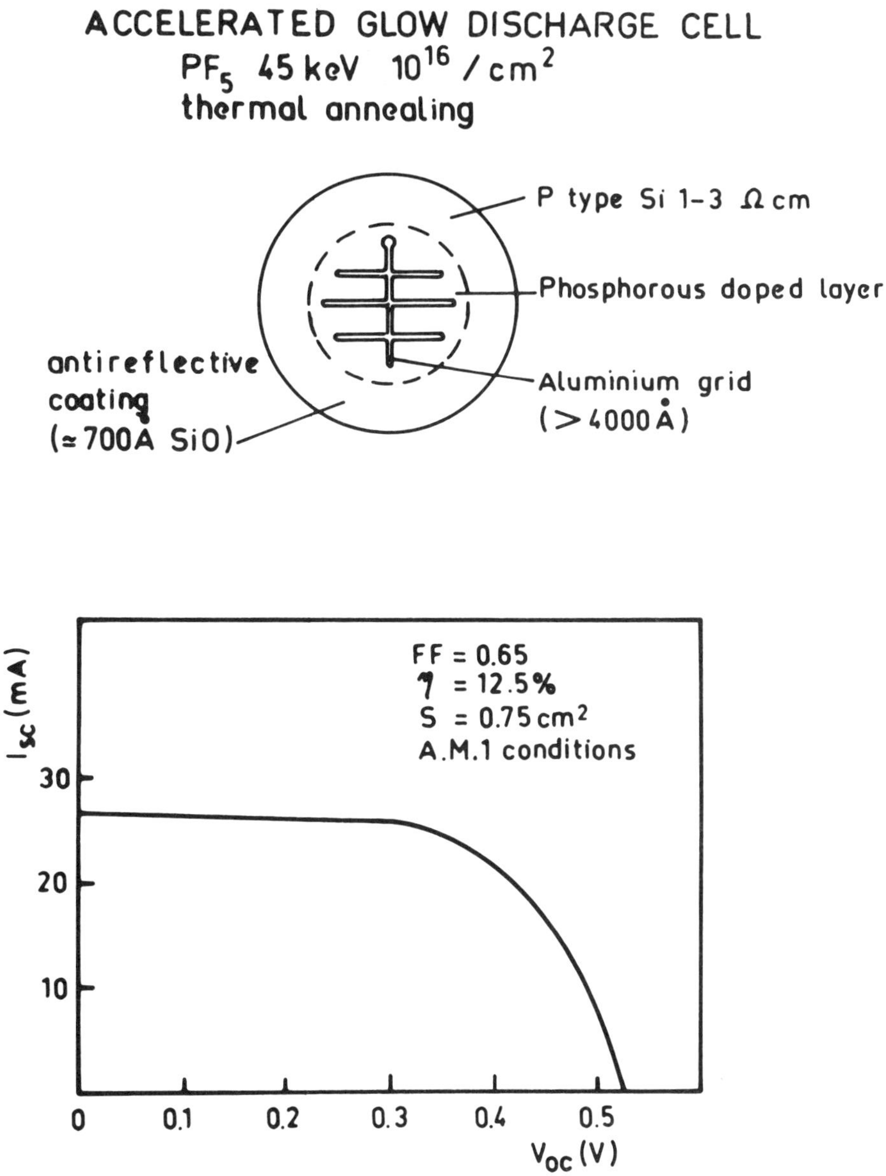

Fig. 8 Typical I-V characteristics of a PF_5 cell test under a tungsten lamp at 100 mW/cm^2 with a water filter.

REFERENCES

[1] J.P. Ponpon, P. Siffert, 11th Photovoltaic Specialists Conference, Phoenix, Arizona USA, 1975.

[2] R. Wichner, 11th Photovoltaic Specialists Conference, Phoenix, Arizona USA, p. 243, 1975.

[3] Keiichiroh Uda, Mototaka Kamoshida, J. Appl. Phys. 48, 18, 1977.

[4] L. Csepregi, J.W. Mayer, T.W. Sigmon, Appl. Phys. Lett. 29, 92, 1975.

[5] L. Csepregi, E.F. Kennedy, S.S. Lau, J.W. Mayer, Appl. Phys. Lett. 29, 645, 1976.

[6] J. Krynicki, J. Suski, S. Ugniewsi, R. Grotzschel, R. Klabes, U. Kreissig, J. Rudiger, Phys. Lett. 61A, 181, 1977.

[7] G.A. Kachurin, V.A. Bogatyrev, S.I. Romanov, L.S. Smirnov, Ion Implantation in Semiconductors, Ed. F. Chernow, J.A. Borders, D.K. Brice, p. 445, 1976.

[8] E.I. Shtyrokov, I.B. Khaibullin, M.M. Zaripov, M.F. Galyatudinov, R. Bayazitov, Sov. Phys. Semicond. 9, 1309, 1976.

[9] R.T. Young, C.W. White, G.J. Clark, J. Narayan, W.H. Christie, M. Murakami, P.W. King, S.D. Kramer, Appl. Phys. Lett. 32, 139, 1978.

[10] A. Gat, J.F. Gibbons, Appl. Phys. Lett., 32, 142, 1978.

[11] R.T. Young, C.W. White, G.J. Clark, J. Narayan, R.D. Westbrook, Photovoltaic Solar Energy Conference, Ed. D. Reidel, p. 861, Luxembourg, 1977.

[12] O.G. Kutokova, L.N. Strel'Tsov, Sov. Phys. Semicond., Vol. 10, 265, 1976.

[13] J.C. Muller, J.P. Ponpon, J.J. and A. Grob, R. Stuck, P. Siffert, Photovoltaic Solar Energy Conference, Ed. D. Reidel, p. 897, Luxembourg, 1977.

SOLAR ENERGY CONVERSION BY FLUORESCENT MATERIALS FOR UTILIZATION IN SPACE COMMUNICATION SYSTEMS

A.K. Sen and J.S. Sehra
Centre of Advanced Study in Radiophysics and Electronics
University of Calcutta
Calcutta, India

ABSTRACT

In space communication, solar energy is widely employed to obtain the electrical power required to operate various electrical systems in a spacecraft. However, the efficiency of conversion of solar energy to electrical energy is typically under 10%. A major part of the electrical power obtained is utilized for the operation of the onboard communication systems. Direct utlization of solar energy for communication may increase the overall efficiency of the energy conversion involved. There are two ways in which such direct utilization can be affected. Firstly, solar energy can be utilized directly as the carrier of intelligence and secondly, solar energy may be utilized directly as a pump source for an onboard laser. The use of fluorescent materials may further extend the possibilities of using lasers at longer wavelengths for which the pump is derived from the fluorescent emission. There is also a possibility of utilizing the fluorescent emission directly as a carrier of intelligence. All these possibilities have been studied in this paper with special reference to space communication systems.

INTRODUCTION

In space applications, silicon solar cells are widely employed as a primary source of supply [1]. The weight, size and cost leads to a limitation of the power output of a spacebourne transmitter. This in turn tends to limit the radiated power which still remains one of the outstanding problems in space communication. While attempts to reduce the weight, size and cost of solar cells by increasing the efficiency of energy conversion is still going on, radically different methods of energy conversion are also being investigated [2,8]. Table 1 shows a comparison of the capabilities of typical sources of supply for space applications. From the table it appears that the specific weight of the various energy sources are of the same order of magnitude. The life is of the order of years for the usual sources of energy such as solar cells. For lives of the order of 100 years, nuclear energy sources possess potentialities. So far as the cost is concerned, silicon solar cells are cheap, while the efficiency of energy conversion is typically under 10%. In space applications, solar energy is mostly used to operate various electrical systems in a spacecraft. Direct utilization of solar energy for communication may increase the overall efficiency of the energy conversion involved. It is the purpose of this paper to outline the various possibilities of direct utilization of solar energy with special

TABLE 1

TYPICAL PERFORMANCE CHARACTERISTICS OF THE SOURCES OF SUPPLY IN SPACE [1-8]

Device	Energy source	Effect	Unit capacity	Maximum efficiency (Theoretical efficiency) Percent	Specific power Watts/ Kg.	Life	Use
Thermionic converter	Caesium vapour type (heat)	Edison effect (Thermionic emission)	300 W	10-20 (40)	5-10	1-2 yrs.	Space probe
MHD (Magneto-hydro-dynamic generator)	High velocity gas flow (heat)	Faraday law	20 MW	60 (85-90) overall generating plant			Space platform and extra-terrestrial stations
Fuel cell	Hydrogen oxygen	Electrolysis	15 KW	90 (92)	5-50	1-10 days	Pioneer & Mercury satellites
	Hydrogen carbon air	Electrolysis		35-75 (90)			Gemini flights & Lunar excursion module

* SNAP - Systems for Nuclear Auxiliary Power started in 1955.

TABLE 1 (continued)

TYPICAL PERFORMANCE CHARACTERISTICS OF THE SOURCES OF SUPPLY IN SPACE [1-8]

Device	Energy source	Effect	Unit capacity	Maximum efficiency (Theoretical efficiency) Percent	Specific power Watts/ Kg.	Life	Use
Nuclear radio-isotope	Depends upon isotope	Nuclear fission		6-10	5-50	100 yrs.	SNAP*
Thermo-electric generator	Any heat source or fuel	Seebeck effect	5000 W	6-12 (35-50)	5-15	1-2 yrs.	Space probe
Solar cells (silicon)	Unlimited energy source (sunlight)	Photo-voltaic effect	250 W	10-14 (25)	5-15	1-2 yrs.	Usual space probe

reference to space communication systems.

TYPICAL SOLAR CHARACTERISTICS

The sun emits both electromagnetic and corpuscular radiations [9,11]. The later constitutes what is called the solar wind which is the steady output of the low energy plasma from the sun. There are no more than a few 10 particles cm^{-2} sec^{-1} in the exospheric regions. These are Van Allen radiation belts which are trapping zones for electrons and protons in the earth's magnetic field [9,11]. There are galactic and solar cosmic rays also. The solar constant, of course, relates to the electromagnetic part of the radiation, the spectrum of which is shown in Figure 1. The sun's surface is at a temperature of about 6000°K due to the fusion process by which hydrogen isotopes are changed into helium. Above the earth's atmosphere, spectral radiancy, i.e., the amount of energy received per second per unit wavelength interval, corresponds to the spectral radiancy of a black body radiating at a temperature of 5900°K. Electromagnetic energy emitted by the sun is in a broadband of wavelength (Figure 1). More than 95% of this energy is concentrated in the narrow span of 0.24μm to 2.5μm peaking at green. The mean solar radiation bombarding our outer atmosphere at normal incidence is 1.35 KW/m^2 [11,13]. The sun is one of the brightest radiators, giving off 7000 Watts per square centimeter of its surface spread over all wavelengths. In the region of peak-energy output, namely 4800°A (green-light), the sun produces only 0.00001 Watts per square centimetre. So, sunlight is incoherent and has inherent low intensity at any given wavelength [14]. This solar energy can be directly utilized for communication systems. Typical solar characteristics are given in Table 2 [15], while coherence and radiance capabilities of the sun are compared with other sources in Table 3 [16,19].

METHODS OF DIRECT UTILIZATION OF SOLAR ENERGY IN SPACE COMMUNICATION SYSTEMS

There are two ways in which direct utilization of solar energy for space communication may be affected. Firstly, the solar energy can be utilized directly as the carrier of intelligence and secondly, the solar energy may be utilized as a pump source for an onboard laser [4,16,20,21].

Carrier Of Intelligence

Modulation is the process of imposing a message signal on a carrier. In order to have a particular carrier frequency, some sort of mirror lenses or filters are needed [4,16,20,21]. The frequency bandwidth of a message remains the same no matter what the carrier frequency is. The frequency of the visible spectrum of light from 4000 to 7000°A (hundreds of millions of MHz) is so high that each message or band of frequencies is a very small percentage of the carrier frequency. Thus, many more messages can be sent. Hence, it can accommodate as many as 80 million TV channel [22]. Modulation of light can be accomplished by internal control of light source or by external means. Many modulation methods have been investigated including

TABLE 2

TYPICAL SOLAR CHARACTERISTICS [12,15]

Mass (M) = 1.99×10^{30} Kg

Radius (R_o) = 6.96×10^{8} m

Mean distance from the earth = 1.496×10^{11} m = 215 R_o

Mean angular diameter as seen from the earth = 31.54"

Solar constant beyond the terrestrial atmosphere (AMO) = 1.95 cal cm^{-2} min
= 0.136 W cm^{-2}.

Solar constant at sea level (AMI) = 0.12 W cm^{-2}

Total luminosity = 3.9×10^{24} Joules/s.

Solar radiation reaching the outer atmosphere = 1.5×10^{18} KWh yr^{-1}.

Surface temperature = 5750°K

Velocity of Escape = 617 Km/s

Mean sidereal period of axial rotation = 25.36 days

TABLE 3

COHERENCE AND RADIANCE CAPABILITIES OF VARIOUS SOURCES [16,19]

Source	$\Phi_{\lambda COH}$ Coherent spectral flux (W/μ)	Φ_{COH} Coherent flux (W)	L_λ Spectral Radiance ($W/mm^2.Sr$)	L Radiance ($W/mm^2.Sr$)	M_λ Spectral Emittance ($W/mm^2.\mu$)	M Emittance (W/mm^2)
Tungsten @ 2800°K	$1.4x10^{-7}$		0.25	0.38	0.8	1.2
Sun	$2.8x10^{-6}$		2.2	2.0	68	61
Continuous laser (He-Ne, 0.633μ) at 77°K	$1.3x10^{7}$	$1.6x10^{-3}$	$8.7x10^{12}$	$1.1x10^{3}$	$1.1x10^{7}$	$1.4x10^{-3}$
Pulsed Laser	$1.5x10^{13}$	$1.5x10^{8}$	$7.5x10^{16}$	$7.5x10^{11}$	$7.5x10^{10}$	$7.5x10^{5}$

electro-optic, acoustic and magneto-optic techniques. The most successful method is the electro-optic method [20,23,24]

All electro-optic modulators can be described as essentially functioning because they cause a change in phase shift δ with length L such that [20]

$$\frac{d\delta}{dL} = \beta E^m$$

where β is a constant, E is the electric field, and m is a constant which, for different electro-optic methods, may take values of one or two. At low modulating frequencies where transit time effects are not important, we find [20]

$$\delta = \beta E^m L$$

At microwave frequencies of modulation, E is a function of L because of transit time effect.

The linear electro-optic effect can be said to be the linear dependence of the index of refraction on an external electric field. If the electro-optic material is suitable, the index of refraction will change at the modulation rate of the electric field. This effectively modulates the incident light. Electro-optic crystals are Cucl, Zns, $N_4(CH_2)$(HMTA), ZnSe, KTH. Kerr electro-optic modulator material carbon disulphide has recently been shown to possess better properties for HF modulation [25]. Special electro-optic modulators, viz., interference filter, Fabry-Perot modulator [26], multi-crystal, travelling-wave modulator (phase) [27], Kerr Travelling Wave Effect microwave modulator, microwave envelope phase shifter, internal electro-optic modulator [28], external piezoelectric modulator [29], external modulation of light using shift of optical absorption edge [30], etc., are developed for amplitude/frequency modulation. In addition, techniques such as modulation using Raman scattering [31], ultrasonics [32], magneto-optic Kerr Effect [33] in thin magnetic films, Zeeman [34], stark and magnetostrictive [35] effect and others [36,42], are proposed and attempted.

Since light is directly utilized as a carrier of intelligence, the efficiency of direct utilization of solar energy is excessively high.

Pump Sources

Laser communication is getting more attention than radio communication because of an overcrowding of channels [1]. Solar energy can be utilized as a pump source to on onboard laser. Since the time in 1960 when Dr. Maiman and Schawlow and Townes (1958) demonstrated the first ruby laser [20,22], laser action has been observed in a variety of solids (crystalline and non-crystalline insulators and semiconductors), organic liquids, gases and vapors, and thousands of discrete wavelengths varying from the vacuum ultraviolet to the submillimeter portion of the infrared have been attained. The quest for new methods of generating coherent radiation, at even higher frequencies is still in effect, with emphasis at present on the generation of coherent radiation in the far ultraviolet and soft x-ray regions [20,22]. As the radiation falls on the material and excites some of the electrons by absorption of radiations, it pushes them to higher orbits. As these electrons fall back to their normal levels, the absorbed light is re-emitted, giving rise to well-known fluorescence phenomena. The absorbed radiation, which is proportional to the population of the lower state, will always exceed stimulated emission, which is proportional to the population of the higher state. A method must be provided for exciting the participating atoms (or molecules) to invert the population. Such excitation is frequently called pumping. Once an inverted population has been produced and the medium iş irradiated with the corresponding wavelength, there will be the net stimulated emission in addition to spontaneous emission. Hence, with lasers, we can obtain coherent, well collimated and directional light beams which can be utilized to accommodate numerous messages [14,16,20,22].

Fluorescent Materials

Utilization of fluorescent materials may further extend the possibility of using the lasers at longer wavelengths for which the pump is derived from the fluorescent emission. There is also a possibility of utilizing the fluorescent emission directly as a carrier of intelligence. By fluorescence, we mean the emission following immediately on excitation [16]. Luminescence is defined radiation, non-thermal in origin [16]. Excitation refers to the raising of an electron, associated with an atom, or group of atoms, to a higher energy state. The falling back to a lower state, when accompanied by the generation of electro-magnetic radiation, is called emission. Luminescence is classified according to the method of excitation, but we are only concerned here with photoluminescence, which is due to visible (or ultraviolet) radiation. Typical photoluminescence efficiencies of fluorescent materials [16,43,44] are given in Table 4.

All the phosphors emit peak wavelength between 0.3μ to 0.7μ. In Table 4, η_q is quantum efficiency and quantum yield [44] and is defined as the ratio of the number of emitted quanta to the absorbed quanta and it is assumed that all the emitted quanta will escape, that is, none are absorbed in the phosphor. The quantum yield is defined as the ratio of the incident to emitted quanta. η_e is energetic efficiency and refers to the ratio of the output to input energy. It is frequently referred to as radiant efficiency or intrinsic efficiency when only an absorbed part of the incident energy is considered. Thus in photoluminescence, the efficiency given is (emited)/(absorbed) radiation. When the phosphor obeys Stokes Rule, the η_e will of course be lower than the η_q.

$$\frac{\eta_e}{\eta_q} = \frac{\lambda_a}{\lambda_e}$$

where η_e and η_q are energetic and quantum efficiency and λ_a, λ_e are the wavelengths of the absorbed and emitted radiations. η_e = luminous efficiency (lumens-emitted/watts incident or absorbed). In principle, any

TABLE 4

TYPICAL PHOTOLUMINESCENCE EFFICIENCIES OF FLUORESCENT MATERIALS [16,43,44]

Number	Sample	Reflectance	Efficiencies		
			η_e	η_q	η_1
1021	$Zn_2\ SiO_4$: Mn	0.36	0.35	0.70	175
1026	$Ca\ Wo_4$: Pb	0.05	0.42	0.75	65
1027	$Mg\ Wo_4$	0.07	0.44	0.84	115
1028	$Zn_2\ SiO_4$: Mn	0.08	0.33	0.68	180
1029	$Ca\ SiO_3$: Pb, Mn	0.17	0.29	0.68	80
1030	$(Mgo)_x\ (AS_2O_5)$ y : Mn	0.05	0.29	0.73	23
1031	$3Ca_3\ (PO_4)_2$ $Ca\ (F,\ Cl)_2$: Sb, Mn	0.23	0.34	0.71	125
1032	$Ba\ SiO_2\ O_5$: Pb	0.35	0.55	0.75	
1033	$Ca_3\ (PO_4)_2$: Tl	0.15	0.49	0.56	

aggregats of suitably excitable molecules could be placed inside a cavity and used as a laser. Table 5 shows the most typical laser lines [16,18,19, 45].

Let us select a laser material glass or $CaWO_4$ which can be operated with the photoluminescent emission as the pump source. From Table 4 quantum efficiency of $CaWO_4$ is 75% and for $MgWO_4$ is 84%. Table 6 shows the materials in which stimulated emission was obtained by Electron Beam Pumping (EBP) or optical pumping [21].

The maximum theoretical power conversion efficiency [46] in CdS is a 27% external (35% internal [47]). Taking typical efficiency of laser materials to be operated with fluorescent emission to be 30% and that of fluorescent material to be 80%. The overall efficiency of energy conversion for direct utilization of solar energy is ~25% and can be further increased.

Dye Lasers

Laser action has also been reported from a solution containing a fluorescent dye which requires an optical pumping [21,48]. According to Witt [49], all coloured organic compounds (called chromogens) containe certain unsaturated chromophoric groups (e.g., $-NO_2$, -N, = N-, CO) which are responsible for the color; should these compounds contain certain auxochromic groups (e.g. $-NG_2$ and -OH), then they possess dyeing properties, i.e., all the organic compounds having an intense color (a strong, broad visible absorption) which would be imparted more or less permanently to other materials. Materials which are called dyes today can absorb and emit in the ultraviolet and near infrared as well as in the visible. Although thousands of dyes are known, only a fraction fluoresce in solution and it is among these that lasing dyes are found. The most strongly fluorescent dyes belong to the oxazole ($C_{24}H_{16}N_2O_2$ or P- Bis [2-(5-phenyloxazolyl)] benzene), Xanthene ($C_{28}H_{31}ClN_2O_3$ - Rhodamine B), anthracene ($C_{26}H_{18}$ -9, 10-Diphenylanthracene), coumarin ($C_{14}H_{17}NO_2$-7-Diethylamino-4-methylcoumarin), acridine ($C_{13}H_9N$-Acridine), azine, phthalocyanine and polymethine ($C_{23}H_{29}IN_2S_2$-3, 3-Diethylthiatricarbocyanine iodide) families [21]. In general, the Xanthene dyes absorb and emit light in the visible part of the spectrum. Two dyes in which laser action is most easily obtained, Rhodamine 6G and Rodamine 2B, are Xanthene dyes which emit between ~5500 and 5900°A and 5800 and 6300° A. The anthracene, oxazole, and coumarin dyes absorb in the near UV and emit in the violet or blue-violet regions of the spectrum [21].

After absorbing light, the molecules are energy-rich and they may re-emit the absorbed photon by emitting a photon of light. This radiative relaxation is called fluorescence. The energy of the emitted light is less than that of the absorbed light; the fluorescence is stokes shifted to a longer wavelength from the absorption. Some variation of the radiative life time for a particular dye (typically $\sim 5 \times 10^{-9}$ sec is observed by changing the solvent, the concentration or the temperature of the solution. The ratio of the number of emitted photons to the number of absorbed pump photons is called the fluorescent quantum efficiency and can be as high as unity for some dye solutions and as low as 0.01 for others. However, in this case

TABLE 5

TYPICAL LASER LINES [16,18,19,45]

Ion	Host	$\lambda(\mu)$
Cr	AlO_3	0.6943 0.6934 at 77°K
Nd	Glass or $CaWO_4$	1.06
Ne		0.2358[illegible]
Ne	He	0.6328 1.1523 3.39
Ar		0.4579 0.4765 0.4880 0.4965 0.5145
CO_2	He, N	~ 10.6

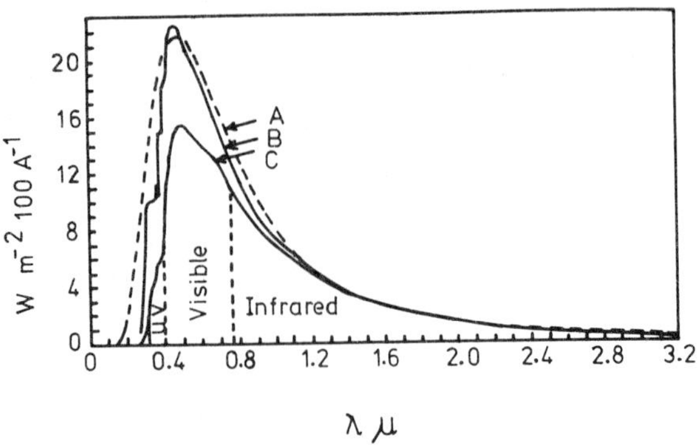

Fig. 1 Showing typical spectral distribution of solar radiation. A: Energy Curve for black body at 6000°K; B: Solar Energy Curve outside atmosphere; C: Solar Energy Curve at sea level.

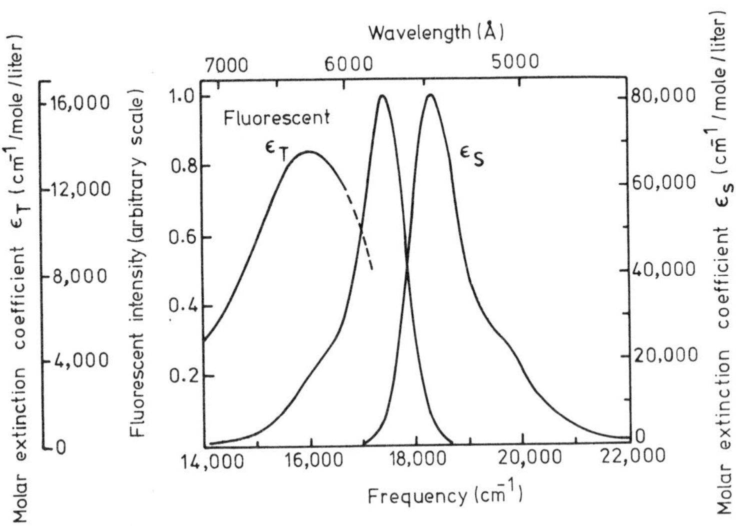

Fig. 2 Typical absorption and fluorescence spectra. The subscripts S and T denotes singlet and triplet plates, respectively. (Levine, A.K. and DeMaria, A.J. Lasers. Marcel Dekker, Inc., New York (1971) p. 273 [21,52]

inversion required for lasing is small [21].

A significant advantage of dye lasers is the possibility through proper dye selection, of obtaining intense, coherent light at almost any wavelength from 3400 to 11750°A. In addition, the lasing wavelength of a single dye laser is continuously tunable over a few hundred angstroms up to 400° A, either by varying the solution or cavity parameters, or by using an intra-cavity wavelength selector such as diffraction grating [21]. With transverse pumping by a pulsed N_2 laser, a tuning range of 1760°A was obtained (from 3910 to 5670°A [50,51]. A typical dye has a fluorescent life time of ~5 x 10^{-9} sec, the optical pumping may be utilized. Power efficiency is 10% and 25% with laser pumped and 0.4% with flashlight pumped.

Most dye solutions have the following spectral properties which are similar in many respects to those of Rhodamine B given in Figure 2. (i) The widths of the principle absorption and emission bands are generally on the order of 1000 of cm^{-1}. One or more absorption bands may be found toward the UV from the principle absorption. (ii) The fluoresence peak occurs at longer wavelengths than the principle absorption peak. The displacement is called the stokes shift of fluorescence from absorption. The extent of this stokes shift and width of the fluorescence substantially overlaps the long wavelength of the absorption. (iii) The fluorescence spectrum is generally a mirror image of that of the principle absorption band. (iv) The fluorescent life time is typically ~5 x 10^{-9} sec. (v) Excited-state absorption bands may be observed on a transient basis by monitoring the solution transmission following exposure to light in the principle absorption band. A triplet-triplet absorption band may overlap the fluorescence band and may persist for as long as 10^{-3} sec. depending on the solution treatment. (vi) After optical irradiation, new long-lived absorption bands due to new chemical species formed by photochemical processes may sometimes be observed. These may be transient if the new species reform into the original dye molecules or permanent if the decomposition is irreversible [21].

From Figure 2 when the light falls, it can be observed by means of filters in the different absorption bands so that fluorescence takes place at the peak. There is a strong correlation between quantum efficiency and lasing efficiency for polymetine dyes in differenct solvents. On the other hand, laser action is easily obtained with solution of Xanthene dyes, even though the life time and quantum efficiencies vary widely. As an example, Rhodamine B in a water solution has a quantum efficiency of 25% and a life time of 0.94 x 10^{-9} sec; in a methanol solution the quantum efficiency are 62% and 2.0 x 10^{-9} sec., respectively. Laser action is easily obtained with both solutions; although their lasting efficiency may differ. Thus, if we use light, i.e., sunlight directly as a pump source, to dye laser typical efficiencies of the order of 60% may be attained and η_f (DTTC in DMSO) = 1.0. If we use the fluorescent emission as the pump source, then overall quantum efficiency of energy conversion is 80% x 60% = 50%. Thus with dye lasers, we may obtain an energy conversion efficiency of up to 50% typical or more [21].

The quantum efficiency of a dye solution, (the wavelength displacement are often correlated with a change solvent dielectric constant and the index of refraction) and the life time of fluorescent are all functions of the solvent used and pH of the solution. When the concentration of dye molecules is too high, there can be large absorptive losses due to the overlap of the long wavelength tail of the absorption with the fluorescence [21].

The efficiency of a dye laser is a function of the level of population, which changes with time during optical pumping, the frequency as determined by the emission and absorption spectra and other parameters such as temperature, concentration and length of the active medium. As the temperature is decreased, the absorption and fluorescence spectra narrow and the amount of overlap decreases, thereby causing the lasing frequency to move toward the peak of the fluorescence curve [21].

CONCLUSION

Different methods of direct utilization of solar energy for communication are discussed. Direct utilization of the solar energy for communication may increase the overall efficiency of the energy conversion to typically 25%~50% or more with suitable combinations of fluorescent materials, fluorescent emission of which may be used as a pump source for the onboard laser.

ACKNOWLEDGEMENTS

The authors are thankful to Professor M.K. Das Gupta, Head of the Department and Director Centre of Advanced Study in Radiophysics and Electronics, University of Calcutta for helpful discussions and encouragements.

REFERENCES

1. Sen, A.K.; Sehra, J.S.; Trehan, S.K.; Saha, B.; Sekhar De, S. and Saha, S.K. Communication Systems for Deep Space Missions. Proceedings of the All India Interdisciplinary Symposium on "Recent Trends of Research and Development in Digital Technique and Pattern Recognition" February 15-17, 1977, held at Indian Statistical Institute, Calcutta, India (in press).

2. Seth, S.P. Direct Energy Conversion - A Materials Problem. J. Institute of Electronics & Telecommunication Engineers, 20 (1974) 19-24.

3. Soo, S.L. Direct Energy Conversion. Prentice-Hall (1968).

4. Kettani, M.L. Direct Energy Conversion. Addison-Wesley (1970).

5. Wysocki, J.J. and Rappapart, P.J. Applied Physics 31 (1960) 571.

6. Krassner, G.N. and Michaels, J.V. Introduction to Space Communication Systems. McGraw-Hill Book Company (1964) 207-374.

7. Snyder, N.W. Energy Conversion for Space Power. Academic Press, Inc., New York (1961).

8. Snyder, N.W. Space Power Systems. Academic Press, Inc. New York (1961)

9. Filipowsky, R.F. and Muehldorf, C.I. Space Communication Systems. Prentice-Hall, Inc., Englewood Cliffs, New Jersey (1965).

10. Langton, N.H. Space Research and Technology Vol. 1 The Space Environment. University of London Press Ltd. (1969) 94-220.

11. Obayashi, T. and Hakura, Y. Enhanced Ionization in the Polar Ionosphere and Solar Corpuscular Radiation in Space Research. Proceedings of Internal Space Symposium, Nice, France 1960. Interscience Publishers, New York (1960) 665.

12. Deb, S. Solar Energy - Its Nature and Possibilities as a Source of Power for Terrestrial Use. J. Institution Electronics & Telecommunication Engineers 23 (1977) 659-675.

13. Jain, G.C. Trends in Silicon Solar - Photovoltaic Cells. An invited talk at International Solar Energy Congress 1977, January 16-22, 1978, Vigyan Bhawan, New Delhi, India.

14. Pike, C.A. Lasers and Masers. W. Foulsham & Co. Ltd. (1967) 154.

15. Kondratyev, K. Ya. Radiation in the Atmosphere. Academic Press, New York (1969).

16. Levi, L. Applied Optics: A Guide to Optical System Design/Vol. 1.

John Wiley & Sons, Inc., New York (1968) 578.

17. Way Nant, R.W., Cullom, J.H., Basil, I.T. and Baldwin, G.D. "Beam Divergence Measurements for Switched Ruby Lasers" Applied Optics 4 (1965) 1648-1651.

18. Snitzer, E. Glass Lasers. Applied Optics 5 (1966) 1487-1499.

19. Cheo, P.K. and Cooper, H.G. Ultraviolet Ion Laser Transitions Between 2300-4300°A. J. Applied Physics 36 (1965) 1862-1865.

20. Ross, M. Laser Receivers: Devices, Techniques, Systems. John Wiley & Sons, Inc. (1967).

21. Levine, A.K. and DeMaria, A.J. Lasers. Marcel Dekker, Inc., New York (1971).

22. Jacobwtiz, H. Electronics Made Simple. Vakils, Feffer and Simons Pvt. Ltd. (1969) 189-191.

23. Billings, B.H. J. Optical Society Am. 39 (1949) 797.

24. Billings, B.H. ibid 39 (1949) 802.

25. Holshouser, D, Gaddy, O.and Stanfield, R. Third International Quantum Electronics Conference, February 1963, Columbia University Press, New York (1964).

26. Gordon, E.I. and Rigdon, J.D. Bell Systems Technology Journal 42 (1963) 155.

27. Peters, C.J. Proceedings IEEE, 51 (1963) 147.

28. Gurs, K. and Muller, P. in Proceedings of Symposium on Optical Masers, Polytechnic Press, New York (1963).

29. Seraphin, B.O., McCauley, P.G. and Marco, L.G. in Proceedings of Symposium on Optical Masers, Polytechnic Press, New York (1963).

30. Eden, R.C. and Coleman, P.D. Proceedings IEEE 51 (1963) 1776.

31. Brewer, R.G. J. Applied Physics 33 (1962) 1606.

32. Cummins, H., Knable, N. Gampel, L. and Yeh, Y. Applied Physics Letter 2 (1963) 62.

33. Chen, D. in Proceedings of Symposium on Optical Masers, Polytechnic Press, New York (1963).

34. Kiss, Z. Applied Physics Letter 2 (1963) 61.

35. Bennett, W.R. Jr, Kindlman, P. J. Rev. Science Instr. 33 (1962) 601.

36. Bloembergen, N., Pershan, P.S. and Wilcox, L.R. Physics Review 120 (1960) 2014.

37. Dow, H.A. J. Optical Society Am. 53 (1963) 915.

38. Kaya, P. Proceedings IRE 50 (1962) 323.

39. Pankove, J.I. and Berkeyheiser, J.E. Proceedings IRE 50 (1962) 1976.

40. Rigrod, W.W. and Kamnow, I. Proceedings IEEE 51 (1963) 137.

41. Johnson, K. Microwave Journal (August 1964).

42. Myers, R.A. Technical Report 433, Cruft Lab, Harvard University Contr. Hour - 1866(16), NR-372-012.

43. Kallman, H.P. and Spruch, G.M. Luminescence of Organic and Inorganic Materials. Wiley, New York (1962). A Bril. pp. 479-493.

44. Tregellas-Williams, J. A Review of the Measurement of the Quantum Efficiency of Inorganic Phosphors. J. Electrochemical Society 105 (1958) 173-178.

45. Bevolo, A.J. and Barker, W.A. Laser Emission Lines and Materials. Applied Optics 4 (1965) 531-543.

46. Klein, C.A. IEEE Journal Quantum Electronics 5 (1968) 186.

47. Hurwitz, C.E. Applied Physics Letters 9 (1966) 420.

48. Sorokih, P.P. and Lankard, J.R. IBM Journal of Research Development 10 (1966) 162.

49. Witt, O.N. Journal of Chemical Society (London) (1876).

50. Shank, C.V. Dienes, A. Trozzolo, A.M. and Myer, J.A. Applied Physics Letter 16 (1970) 405.

51. Dienes, A., Shank, C.Y. and Trozzolo, A.M. Applied Physics Letter 17 (1970) 189.

52. Weber, M.J. and Bass, M. IEEE Journal of Quantum Electonics 5 (1969) 175.

NEW POSSIBILITIES OF SOLAR ENERGY CONVERSION BY MEANS OF REFRACTORY THERMOELEMENTS

S. Abdalla, J.M. Dusseau, F. Roche and J.M. Darolles
Centre d'Etudes d'Electronique des Solides
(associé au C.N.R.S.)
Montpellier Cédex, France

ABSTRACT

We show that high efficiency thermoelectric conversion can be obtained by using refractory compounds with low thermal conductivity and high thermoelectric power. We deal with boron silicon compounds with a high boron concentration such as $B_{14}Si$. Samples are obtained by chemical vapor deposition by pyrolisis of mixture of BBr_3 and $SiBr_4$ under reduced pressure.

Experimental results show that the thermoelectric figure of merit is $3.75.10^{-3}$ degrees^{-1} and zT_H = 7,5 at 1700°K; moreover, the total conversion efficiency would exceed 20%. This is particularly interesting for solar energy conversion.

LIST OF SYMBOLS

T_H = Temperature of hot source

T_C = Temperature of cold source

η_c = Theoretical Carnot cycle Efficiency

η = Conversion efficiency

z = Figure of merit

S = Thermoelectric power

σ = Electrical conductivity

$\varkappa$ = Thermal conductivity

r = Ratio of the load resistance to the element resistance

INTRODUCTION

The poor values obtained for thermoelectric conversion using two different metals (Seebeck Effect in 1882), and the wide use of the semiconductors in the past years did not promote the development of this technique, comparing with the developments from that known as the photovoltaic effect.

The conversion efficiency of a thermoelectric element converter could be written in the form

$$\eta = \eta_c f(z\ T_H)$$

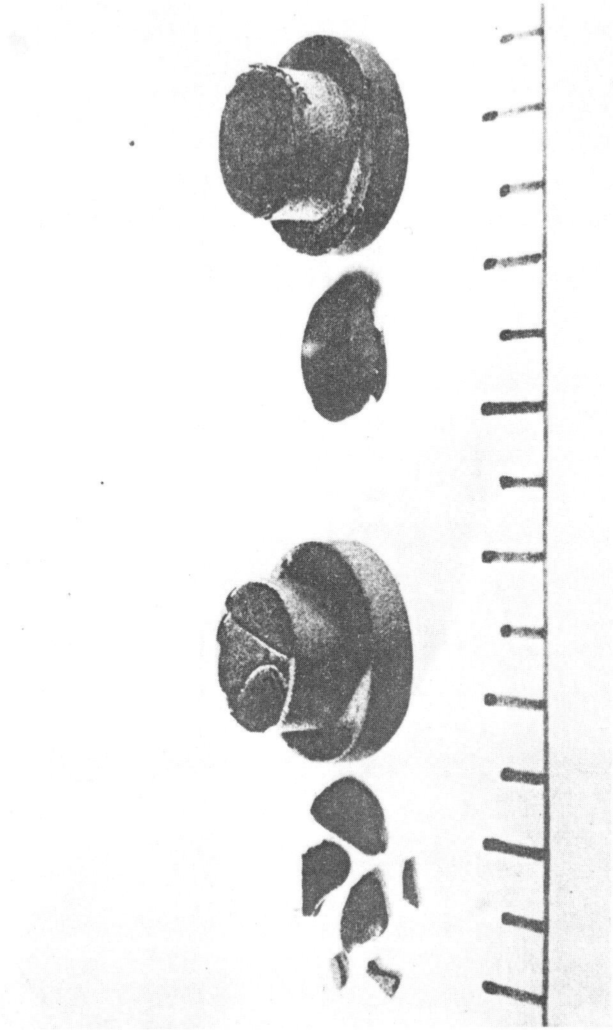

Figure 1 Samples and Graphite Substrates

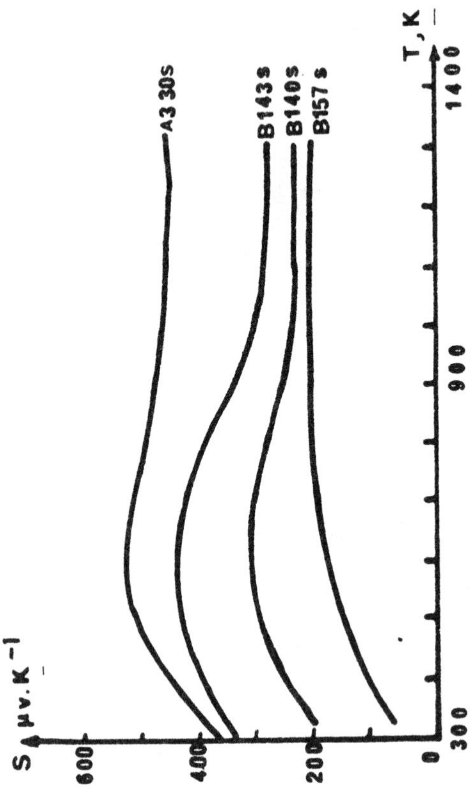

Figure 2 Thermoelectric Power Vs. Temperature

where $\eta_C = \frac{T_H - T_C}{T_H}$, is the Carnot Cycle Efficiency.

One notices that this efficiency exceeds 80% if one uses the refractory thermoelements operating between 1800°K and 300°K.

In addition,

$$F(zT_H) = \frac{r/(1 + r)}{1 + \frac{1 + r}{zT_C} - 1/2\ \eta_C \frac{1}{1+r}}$$

where r is the load to element resistance ratio and z is figure of merit coefficient which is characteristic of the studied material. One has

$$z = \frac{\sigma S^2}{\chi} .$$

So, it is evident that a high value of z will be obtained for small values of χ. Moreover, χ is the thermal conductivity summation of both lattice and electronic thermal conductivities. After the Wiedemann-Franz Law, the latter on is directly proportional to the electronic conductivity. Taking into account the thermoelectric power expression deduced from the Boltzmann equation, one can show that for a semiconductor, the maximal value of zT_H product is approximately unity; that explains the poor efficiencies observed in the classical semiconductors. On the other hand, it seems to be possible to the borides to exceed this limit. Indeed, Golikova et al [1] have shown that the thermal conductivity of boron and α-AlB_{12} changes slowly and even decreases with increasing temperature.

In addition, as the electric conductivity of those borides is increasing, the thermoelectric power maintains at a high value [1]. Consequently, these results lead us to study the compound $B_{14}Si$, synthesized at Laboratoire C.N.R.S. d'Odeillo - Font Romeu, France by B. Arnas and C. Combescure [2]. Briefly, $B_{14}Si$ is obtained by pyrolysis of a BBr_3 and a $SiBr_4$ mixture under reduced pressure. X-ray diffraction patterns correspond to the interstitial compound deduced by Giese et al [3]. The deposits obtained on a graphic substrate (Figure 1) are homogeneous and have a density near the theoretical value (2.51 gm Cm^{-3}). Samples, disk-formed (10 to 15 mm diameter), are some hundreds of microns to several millimeters thick. The obtained disks permit the easy measure of physical properties as well as realizing thermoelectric cells.

ELECTRICAL PROPERTIES

For several samples, thermoelectric power is given in Figure 2. First, one may notice that after an initial increase in the low temperature range, the thermoelectric power stays approximately independent of T until about 1500°K; then the thermoelectric power raises with the purity of the samples.

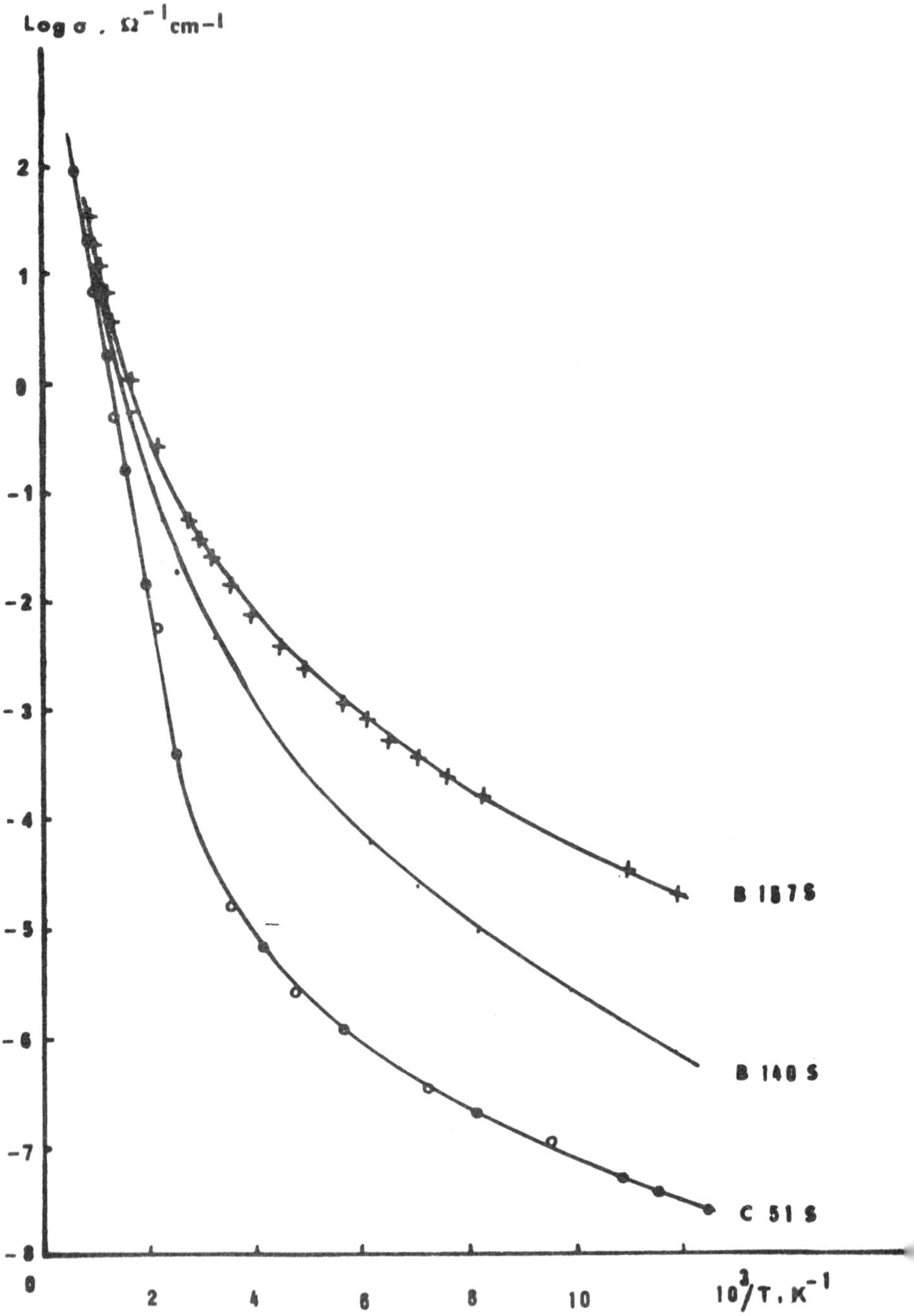

Figure 3 Electrical Conductivity Temperature Dependence

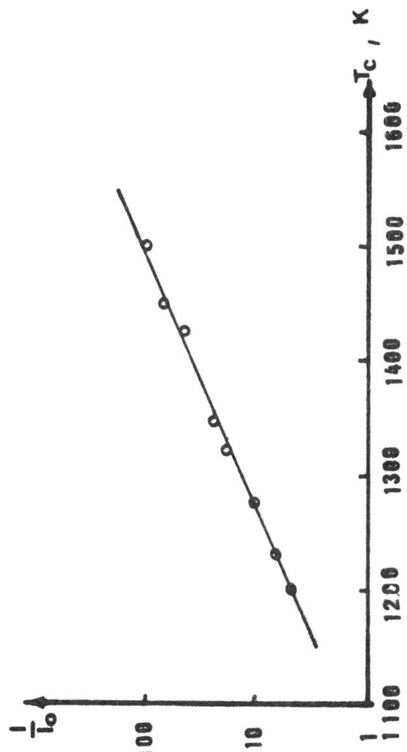

Figure 4 Current Delivered as Function of T_H

Moreover, the electric conductivity increases rapidly with temperature (Figure 3). Since thermoelectric power remains constant, it is likely that this variation is not principally due to an increase in the free carriers number. This could be explained in that the electronic part of the thermal conductivity remains negligible. Due to the very complicated structure of the crystal lattice, it is likely that lattice thermal conductivity is nearly independent on the atoms arrangement in the elementary cell. So, amorphous conditions are approached, which explains why the thermal conductivity of the borides is small (for $B_{14}Si$ χ = 0.02 W/cm K).

FIGURE OF MERTI OF $B_{14}Si$ COMPOUND

In the figure of merit expression z, and beyond T_H = 500°K, only the electrical conductivity increases rapidly with temperature. Previous experimental results lead to:

$S = 500\ \mu V/degree$	$\chi = 0.02$ W/cm degree
$\sigma = 300\ S\ cm^{-1}$	at 1700°K
$\sigma = 80\ S\ cm^{-1}$	at 1250°K

One will deduce:

$z = 3.75\ 10^{-3}/degree$	and $zT_E = 7.5$	at 1700°K
$z = 10^{-3}/degree$	and $zT_H = 1.25$	at 1250°K

Consequently, the figures of merit and zT_H products are particularly strong at high temperatures and highly exceed the obtained values with classical semiconductors. Nevertheless, this advantage disappears for a hot source temperature lower than nearly 1200°K (~900°C). Studies are presently underway to obtain a good zT_H product for the lower temperatures. For instance, a possible solution consists in disposing an element of $B_{14}Si$ and another one of GeSi in series on the thermal circuit.

CURRENT DELIVERED BY A $B_{14}Si$ SHEET FUNCTIONING AS CONVERTER

The experimental measures for the net conversion efficiency has not been realized until now because we did not dispose, for the moment, an n-type material. On the other hand, its theoretical evaluation is complicated because z varies with the conductivity from the hot source to the cold one.

Yet one can estimate the total efficiency to about 20% which will constitute a considerable progress referring to the actual situation.

Figure 4 represents the variation of the current generated by a sheet of $B_{14}Si$ under a temperature difference of 500 degrees as a function of hot temperature source T_H. It is observed that the current increase by a factor of 10 for a hot source temperature increase of 250 degrees, from where, the interest is to operate these elements at a very high temperature. The short circuit current density (experimental corrected values taking into account the series resistances of the device) would be of 100 A/cm^2 at T_H = 1500°K with open circuit voltage of 0.16 volts. Measures at higher temperatures are actually being attempted.

CONCLUSION

Solar energy conversion by means of a thermoelectric converter may be carried out with satisfactory efficiency if refractory compounds with a high number of atoms in the elementary cell are used providing that both electrical conductivity and thermoelectric power are sufficiently high. Our results for $B_{14}Si$ are encouraging and this study opens the way for looking after other refractory materials with higher electrical conductivity, other parameters being equal.

REFERENCES

1. Golikova, O.A., Zaitzev, V.K., Petrov, A.V. Stileans, L.S., and Tkalenko, E.N., Soviet Physics Semiconductors, Vol. 6, No. 9 (1973).

2. Arnas, B., Combescure, C., Dusseau, J.M., Lepetre, T., Robert, J.L., and Pistoulet, B., Journal of Non-Common Metals, 47, 135 (1976).

3. Giese, R.F., J. Economy, V-I Matkovich Z-Fur Krist 122, 1 and 2, 144 (1965).

ON DIRECT CONVERSION OF SUNLIGHT TO ELECTRICITY

M. A. Melehy
Department of Electrical Engineering and Computer Science
and Institute of Materials Science
University of Connecticut
Storrs, Connecticut 06268, U. S. A.

ABSTRACT

A new physical theory for homojunction and heterojunction solar cells is presented. The theory relates the voltage-current characteristics of such cells to the device physical parameters and the spectral distribution of the incident radiation. The derived formula is shown to be accurately in agreement with experiment for Si solar cells.

In the treatment, we consider the photon-generated electron-hole pairs and determine their transport through the device and its battery (or load). The analysis is based on this author's thermodynamic theory of generalyzed fields (TTGF). The TTGF is a theory particularly developed for interfacial phenomena and more generally for all systems which have gradients of any physical parameters, when the particle fluxes are vanishing in the system.

INTRODUCTION

If the hope in increasing the standard of living in the world is to be realized, and if the global environment is to be simultaneously preserved, abundant clean energy sources must be developed. Direct conversion of sunlight to electricity may provide the ultimate means for production of unlimited amounts of pollution-free energy.

In many parts of the world, solar energy is available in staggering amounts. In Phoenix, Arizona, U.S.A., for example, accounting for weather and climatic conditions, the amount of solar energy falling annually on an area of about 808 square kilometers is equivalent to the entire electric energy consumed in the U.S.A. in 1976. This energy totalled [1] about 1.85×10^{12} kilowatt-hours. In the Sahara Desert, and the Arabian Peninsula, for example, and in many parts of Australia and South America, the same amount of solar energy is received annually by even less than 808 square kilometers.

Direct conversion of sunlight to electricity may be achieved by an ever-growing family of solid-state devices, which are generally known as solar cells. These devices may be fabricated by any one, or a pair, of a large variety of semiconductor materials. Regardless of their materials, however, the underlying principles of operation of all these devices are essentially the same.

In this paper, we shall explain these principles. Furthermore, we shall present a new physical theory which expresses mathematically these principles and relates the solar-cell voltage-current characteristics to the physical parameters of the device and its temperature and to the spectral properties of the incident solar radiation.

The method of analysis to be presented in this paper is based on this author's [2,3,4] thermodynamic theory of generalized fields (TTGF). How does the TTGF differ from the existing theories?

Classical thermodynamics and statistical mechanics consider an equilibrium state to be uniform. In treating nonuniform cases at equilibrium, ordinarily the system is approximated by dividing it into small parts, each of which is considered exactly uniform. Nonequilibrium is but a departure from equilibrium. Thus classical nonequilibrium thermodynamics, and statistical-mechanics are exact for systems exactly uniform in equilibrium, and are otherwise approximate formulations. How good the approximations are in the classical theories will depend on how weak the nonuniformity is in a system when it is in equilibrium. For interfacial systems, the approximations may become excessive.

By contrast, the new thermodynamic theory of generalized fields (TTGF), in its phenomenological and statistical-mechanical versions, treat nonuniformity in an exact way, and hence account for the gradients of all parameters, even at equilibrium. Therefore, the TTGF reduces to the classical theories for systems which are uniform in equilibrium. However, generally and for interfacial phenomena particularly, new fundamental consequences follow such as: (1) the necessary use of extreme generalizations of Kirchhoff's laws of circuit theory at least for all stationary processes; and (2) the introduction of new significant quantities of transport energy.

Applying the TTGF to p-n junctions (under dark conditions), for example, has led [5,6] to three voltage terms, rather than one as obtainable in the conventional theories. These voltage terms account for the transport energy associated with the three processes: diffusion and drift of carriers in the junction, and the diffusion of minority carriers in the bulk. The resulting voltage-current equation has accurately predicted the voltage characteristics [5,6,7] reported by numerous authors on p-n junctions and heterojunctions in the period 1951-75. In some experiments, the current variation extended over a range whose maximum-to-minimum ratio is greater than 10^{10} to 1. The samples were studied at temperatures in the range 4.2-800°K. The diodes were fabricated from semiconductor crystals having energy gaps in the range 0.17-2.8eV.

For an appreciation of our treatment, the reader should be familiar with the fundamental TTGF concepts and relationships, which are thoroughly covered in the literature, particularly in Ref. 3.

GENERAL FEATURES IN SOLAR CELLS

Fig. 1 schematically represents a solar cell, which is either a homojunction, or a heterojunction. Regardless of the cell type, a number of conditions should be therein fulfilled.

First, the front region, which receives the incident radiation, should be almost entirely transparent to the radiation so that nearly all the energy is allowed to penetrate through the junction and reach the base region. Thus, if the cell is a p-n homojunction, the front region (commonly called the surface layer) must be sufficiently thin. In case of Si solar cells, for example, the thickness of the surface layer is about 0.2 microns. According to the optical attenuation measurements reported by Dash and Newman [8], it is estimated that this particular thickness would absorb only about 4% of the incident radiation.

In heterojunction devices, however, the transparency of the front region (commonly called window region) to the incident radiation is achieved by selecting the material of that region to have an energy gap sufficiently wider than that of the base region. Thus in heterojunction cells, the front region can be considerably wider than its counterpart in homojunctions. This is an interesting advantage in heterojunctions, particularly for high intensity applications.

The second feature in a solar cell is that the base region should be sufficiently thick so that nearly all the useful photons would be totally absorbed therein. Photons of energy less than that of the gap (pertaining to the base region) are somewhat useless. In fact, these photons are even undesirable, for they can only produce heat in the device. A rise in the solar-cell temperature automatically reduces the cell efficiency.

THE SOLAR CELL UNDER OPEN-CIRCUIT CONDITIONS

Consider a solar cell (Fig. 1) under open-circuit conditions. As the incident photons penetrate through the junction, they get absorbed in the base region. Consequently, electron-hole pairs are generated in that region. Under such conditions, the n-type base region will resemble in a number of ways a uniformly doped crystal of the same doping level and type, which is subjected to the same radiant energy. Specifically, in both cases, the net current density vanishes everywhere. Consequently, there is no reason why radiant energy would not decay in an identical manner in both cases. Thus, we may expect to find the profile of excess minority carriers in a uniform crystal identical with that of the base region.

But from the experimental measurements of Dash and Newman on Si and Ge crystals [8], it is known that monochromatic light is attenuated in a

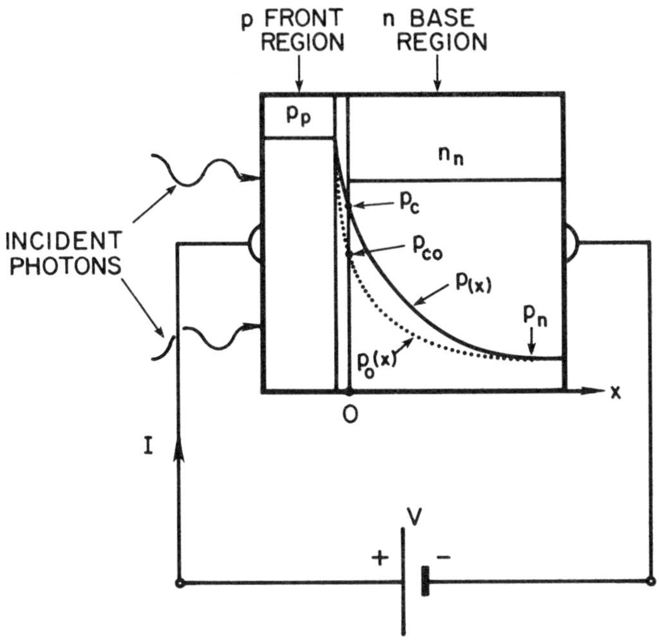

Fig. 1. Distribution of relevant concentrations of majority and minority carriers in a homojunction, or heterojunction solar cell. The concentration of minority holes p_o and p in the base region correspond, respectively, to open-circuit conditions and I > 0. The figure is not to-scale.

simple exponential manner, and the attenuation constant depends on the wavelength, that is the photon energy. A similar result, therefore, occurs in the base region of the cell.

Accordingly, if the incident radiation is monochromatic, by balancing through any differential volume in the base region the time rate of absorption of optical energy with the net rate of recombination of electron-hole pairs, it can be shown that

$$p_o(x) = (p_{co} - p_n)e^{-x/L_o} + p_n \tag{1}$$

Here p_o is the concentration of minority holes in the base region at any point x away from the junction (Fig. 1); p_n is the value of p_o at thermal equilibrium (dark conditions): p_{co} is the value of p_o at $x = 0$, that is at the junction-base boundary, and L_o is what may be called the optical attenuation length.

The validity of Eq. (1) requires that the concentration of majority electrons n_n in the base region satisfies the inequality

$$n_n >> p_{co} \tag{2}$$

DISTRIBUTION OF MINORITY CARRIERS IN THE PRESENCE OF ELECTRICAL CONDUCTION

It will become apparent that in Si solar cells when subjected to sunlight, no matter how strong, or weak the intensity may be, the following is always satisfied:

$$p_o(0) = p_{co} >> p_n \tag{3}$$

Let the open-circuit voltage when the device is subjected to radiation be V_{oc}.

Imagine now that we connected a battery of voltage V (Fig. 1) to the cell. Our immediate objective is to find how, under such a condition, the open-circuit distribution $p_o(x)$ will be modified to $p(x)$, if $V \neq V_{oc}$.

At this point, one may be able to guess that if $V > V_{oc}$ then $I > 0$, and vice versa; I being the cell current (Fig. 1). This result will indeed be confirmed after we describe the problem quantitatively.

If we now assume that the device is isothermal and minority holes constitute a classical gas, then we may express one-dimensionally the active hole diffusion field f_{dao} [3] corresponding to open-circuit conditions by

$$f_{dao} = -(1/p_o)d(kp_oT)/dx = -kT\ d(\ln p_o)/dx \tag{4}$$

Here T is the device absolute temperature and k is Boltzmann's constant.

If $I \neq 0$, then p_o changes to p. Correspondingly, the hole diffusion field in the base region will be

$$f_{da} = -kT\ d(\ln p)/dx \tag{5}$$

If $V > V_{oc}$, then near the junction (Fig. 1), $p > p_o$ and $f_{da} > f_{dao}$. The difference between f_{da} and f_{dao} will support the hole passive diffusion field f_{ds}. Specifically,

$$f_{ds} = f_{da} - f_{dao} = -kT\ d[\ln(p/p_o)]/dx \tag{6}$$

It is a fundamental consequence of the TTGF that the hole diffusion current density J_p will be given by

$$J_p = \mu_p p\ f_{ds} \tag{7}$$

where μ_p is the mobility of minority holes in the base region.

Since minority holes are assumed to be a classical gas, the Einstein diffusion-mobility relationship holds, and we have

$$D_p/\mu_p = kT/e \tag{8}$$

Here D_p is the diffusion constant for minority holes in the base region and e is the magnitude of the electron charge.

From Eqs. (6), (7), and (8), it follows that

$$J_p = -eD_p p\ d[\ln(p/p_o)]/dx \tag{9}$$

It is particularly interesting to observe that if the cell is in total darkness, then $p_o = p_n$, and Eq. (9) reduces to

$$J_p = -eD_p(dp/dx) \tag{10}$$

which is the familar form of Fick's law. Yet when $p = p_o$, $J_p = 0$ everywhere, and this result remains valid for any distribution of $p_o(x)$.

If we now consider the process of generation and recombination of electron-hole pairs, from basic principles, it can be shown that the continuity of minority holes may be expressed by:

$$\partial p/\partial t = -(1/e)(\partial J_p/\partial x) - (p-p_o)/\tau_p \tag{11}$$

Here τ_p is the lifetime of minority holes in the base region.

Under steady-state conditions, $(\partial p/\partial t) = 0$. Therefore, from Eqs. (9) and (11), we get

$$p\, d^2[\ln(p/p_o)]/dx^2 + (dp/dx)\, d[\ln(p/p_o)]/dx - (p-p_o)/L_p{}^2 = 0 \quad (12)$$

$$L_p{}^2 = D_p \tau_p \quad (13)$$

The quantity L_p is the diffusion length for minority holes in the base region.

Equation (12) may be solved through the substitution:

$$p(x) = p_o(x)\ F(x) \quad (14)$$

where F(x) is some function of x to be determined.

In general, Eq. (12) is nonlinear. However, for values of x satisfying the inequality:

$$\exp(x/L_o) \ll (p_{co} - p_n)/p_n \quad (15)$$

and when Eq. (14) is substituted into Eq. (12), we get the linear ordinary differential equation:

$$F'' - (1/L_o)F' - (1/L_p{}^2)F + (1/L_p{}^2) = 0 \quad (16)$$

For intensities of incident radiation several orders of magnitude higher, or lower than that of ordinary sunlight, condition (15) will be satisfied at least in the range: $0 \leq x \leq 10\ L_o$. This range of x is more than needed to determine with accuracy the cell voltage current relationship from p(x) and $J_p(x)$.

For values of x = 0, and x = ∞, we respectively have

$$p(0) = p_c,\ p(\infty) = p_n \quad (17)$$

Therefore, from Eq. (14) and the solution of Eq. (16) subject to these two boundary conditions, we have

$$p(x) = p_o(x)[1 - (p_{co}-p_c)\ e^{-\alpha x}/p_{co}] \quad (18)$$

$$\alpha = \tfrac{1}{2}\ [(1/L_o{}^2 + 4/L_p{}^2)^{\frac{1}{2}} - (1/L_o{}^2)] \quad (19)$$

In single-crystal solar cells, τ_p is relatively large and so will be D_p and L_p (Eq. 13). In this case, usually

$$L_p \gg L_o \quad (20)$$

Under such a condition α reduces simply to

$$\alpha = L_o/L_p^2 \tag{21}$$

For values of x satisfying Condition (15), $J_p(x)$ may be found from Eqs. (1), (9) and (18) to be

$$J_p(x) = eD_p\alpha\,[(p_c-p_{co})/p_{co}]\;p_o(x)\;e^{-\alpha x} \tag{22}$$

As we discussed earlier, the absorption of photons in the front layer in a typical cell should be about 4% of the incident ones, or less. Thus the cell terminal current density $J = I/A \simeq J_p(0)$; I being the cell terminal current and A the cross-sectional area of the device. From Eq. (22), it follows that

$$J = I/A = eD_p\alpha(p_c-p_{co}) \tag{23}$$

Defining

$$J_s = I_s/A = eD_p p_{co}\alpha \tag{24}$$

we then get from Eq. (23)

$$p_c = p_{co}(I + J/J_s) = p_{co}(1 + I/I_s) \tag{25}$$

As we shall see I_s represents the magnitude of the cell reverse-saturation current, which is commonly called the short-circuit current.

WORK DONE ON A COMPOSITE CARRIER BY THE DIFFUSION AND ELECTROSTATIC FORCE FIELDS

In the last section, the concentration p_c of minority holes collected at the base-junction boundary has been related to the device terminal current. In order to determine the dependency of p_c on the device terminal voltage V, we must invoke the law of conservation of energy around a loop [3,9] comprising the cell and whatever is connected to it. In order to accomplish this fundamental energy balancing step, which is to be completed in the next section, we must calculate the work done on carriers in each region of the cell by various force fields.

Under stationary conditions, any cross-section in the loop comprising the cell and the battery (or resistive load) carries the same current I. However, depending on the location of the cross-section, I may be carried by electrons only, holes only, or by both electrons and holes in varying proportions. We may then represent I by a ficticious carrier whose charge is e. This charge is transported around the loop linking the cell and the battery (Fig. 1). Further discussion of the concept of a composite carrier may be found in Ref. 3.

Work Done on a Composite Carrier by the Active Diffusion Field in the Base Region

For simplicity let $V > V_{oc}$, so that $I > 0$. This mode of operation is not proper for a solar cell; but it is easier for writing the equations.

Imagine now a composite (positive) carrier leaving the junction and entering the base region. The carrier then will be subjected to two force fields. One is the active diffusion field f_{dao}, which exists when $I = 0$. The other is the passive hole diffusion field f_{ds}, which is associated with $J_p(x)$.

Let W_{dao} be the work done in electron volts on a composite carrier by f_{dao} through the base region. We then have

$$W_{dao} = (1/e) \int_0^{\infty} f_{dao} \, [J_p(x)/J_p(0)] \, dx \tag{26}$$

Substituting Eqs. (4) and (22) into (26), we obtain

$$W_{dao} = -(V_T/p_{co}) \int_0^{\infty} e^{-\alpha x} \, (dp_o/dx) \, dx \tag{27}$$

where

$$V_T = kT/e \tag{28}$$

The integral in Eq. (27) can be evaluated by considering Eq. (1). The result will be

$$W_{dao} = V_T(p_{co} - p_n)/[p_{co}(1 + \alpha L_o)] \tag{29}$$

For intensities of incident radiation even orders of magnitude less than ordinary sunlight, still

$$p_{co} >>> p_n \tag{30}$$

Thus

$$W_{dao} \simeq V_T/(1 + \alpha L_o) \tag{31}$$

Since, ordinarily, Condition (20) and Eq. (21) are satisfied, Eq. (31) then reduces to

$$W_{dao} \simeq V_T \tag{32}$$

Now let W_{ds} be the work done in electron volts by f_{ds} on a composite carrier through the base region. We then have

$$W_{ds} = \int_0^\infty f_{ds} \, [J_p(x)/J_p(0)] \, dx \qquad (33)$$

In view of Eqs. (1), (6), (18), and (22), and Inequality (30), Eq. (33) may approximately be expressed as

$$W_{ds} \simeq V_T(a\alpha) \int_0^\infty e^{-(\alpha + 1/L_o)x}[1 + ae^{-\alpha x}]^{-1} \, dx \qquad (34)$$

$$a = (p_c - p_{co})/P_{co} \qquad (35)$$

As we shall see, when the cell current is delivering power, i.e. whenever $-I_s < I < 0$, the factor a satisfies the condition

$$-1 < a < 0 \qquad (36)$$

Eq. (34) may be rewritten in the form

$$W_{ds} = \rho V_T \qquad (37)$$

where the factor ρ is given by

$$\rho = a\gamma \int_0^\infty e^{-(1+\gamma)x} \, [1 + a e^{-\gamma x}]^{-1} \, dx \qquad (38)$$

$$\gamma = \alpha L_o \qquad (39)$$

If $L_p >> L_o$, and $I \leq 0$, that is Condition (36) is satisfied, then it can be shown that

$$\rho = \gamma \sum_{n=1}^{\infty} (-)^{n-1} \, [a^n/(1+n\gamma)] \qquad (40)$$

For cells in which $L_p >> L_o$, Eqs. (21), and (39) lead to

$$\gamma = (L_o/L_p)^2 \qquad (41)$$

Work Done on a Composite Carrier by the Active Electrostatic and Active Diffusion Fields in the Junction

Since the junction width is much smaller than a diffusion length, the electron-hole generation-recombination process is negligible in the junction. Accordingly, the work done by the active diffusion field W_{daj} and the active electrostatic field W_{eaj} in this space-charge region will be in the same form as in p-n junctions and heterojunctions which are subjected to total darkness. By comparison with such diodes [2,3,5], it follows that in electron volts

$$W_{daj} = \phi_o - V_T \ln(p_c/p_n) \tag{42}$$

$$W_{eaj} = \psi_o + \delta\psi \tag{43}$$

$$\delta\psi = \delta\psi^*\{[\ln^{-2}(p_p/p_c) - \ln^{-2}(p_p/p_n)] - [\ln^{-2}(p_p/p_c) + \ln^{-1}(p_p/p_c)](p_c/p_p)\} \tag{44}$$

Here ϕ_o and ψ_o are, respectively, the values of W_{daj}, and W_{eaj} at thermal equilibrium. Thus $\delta\psi$ represents the departure of W_{eaj} from its value at thermal equilibrium.

The quantity $\delta\psi^*$ is a constant at any one temperature and is, therefore, independent of current. In principle, it should be possible to calculate theoretically that parameter. For simplicity, $\delta\psi^*$ has been determined at one particular point on the voltage-current experimental curve.

All the terms which can possibly appear in the expression of $\delta\psi$ have been retained in Eq. (45), except two terms multiplied by the extremely-small ratio: (p_n/p_p).

THE SOLAR CELL VOLTAGE-CURRENT RELATIONSHIP

Our primary objective next is to relate the solar cell voltage to its current. To do so, we shall invoke a fundamental TTGF rule which may be stated as follows: around any closed loop in any system containing stationary processes, the work done on a composite carrier by all the active force fields less than that done by all the passive force fields must vanish.

The preceding statement is but an expression of the law of conservation of energy in the framework of the TTGF. It is obvious that our rule bears some resemblance to Kirchhoff's law of electric circuit theory. The validity of our rule can be shown to hold for all stationary processes and for any system containing any finite number of constituents.

Now consider Fig. 1. For a composite carrier making a complete loop in the clockwise direction, our statement of the law of conservation of energy, by virtue of Eqs. (32), (37), and (43) may be written as

$$V + \phi_o - V_T \ln(p_c/p_n) - (\psi_o + \delta\psi) + V_T(1 - \rho) + \phi_c + \psi_c + V_b = 0 \tag{45}$$

Here ϕ_c and ψ_c, respectively, represent the work done on a composite carrier by the active diffusion and electrostatic force fields in the Ohmic contacts, in the absence of incident radiation. The quantity V_b

represents the back-ε.m.f. contributed by the surface layer as a result of the photons absorbed therein, and the junction between the electrode and that surface layer.

Since minority carriers are generated in relatively small amounts in the front layer, V_b remains substantially current-independent for constant intensities of incident radiation.

Under thermal equilibrium conditions (total darkness), $V = 0$, $p_c = p_n$, $\delta\psi = 0$, $V_b = 0$, and the term $V_T(1-\rho)$ vanishes as may be seen in the corresponding exact Eqs. (26) and (33), when substituting $J_p(n) = 0$, and $f_{ds} = 0$.

As is well known, in the absence of radiation, the Ohmic contacts can have no injection of minority carriers, regardless of the intensity of current therein. Thus ϕ_c, and ψ_c will remain unchanged regardless of I. Of course, we may only have a passive component of work, i.e. an Ohmic voltage, which is here neglected. Accordingly, at thermal equilibrium, Eq. (45) reduces to

$$\phi_o + \psi_o + \phi_c + \psi_c = 0 \tag{46}$$

Subtracting Eq. (46) from Eq. (45) and solving for V, we get

$$V = V_T \ln(p_c/p_n) + \delta\psi + V_T(\rho-1) + V_b \tag{47}$$

In summary, the equations derived earlier expressing the dependency of V on I and the cell ambient conditions are:

$$V = V_T\{\rho-1) + \ln[(p_{co}/p_n)(1 + I/I_s)]\} + \delta\psi + V_b \tag{48}$$

$$V_T = kT/e, \; p_{co} = J_s/(eD_p\alpha) \tag{49}$$

$$\delta\psi = \delta\psi^*\{[\ln^{-2}(p_p/p_c) - \ln^{-2}(p_p/p_n)]$$

$$- [\ln^{-2}(p_p/p_c) + \ln^{-1}(p_p/p_c)](p_c/p_p)\} \tag{50}$$

$$p_c = p_{co}(1 + I/I_s), \tag{51}$$

If $L_p > 10L_o$, or more preferably $L_p > 50L_o$, as it is the case in single-crystal Si solar cells, then Eq. (41) holds and

$$p_{co} \simeq J_s L_p^2/(eD_p L_o) = J_s \tau_p/eL_o \tag{52}$$

Furthermore, it can be shown that

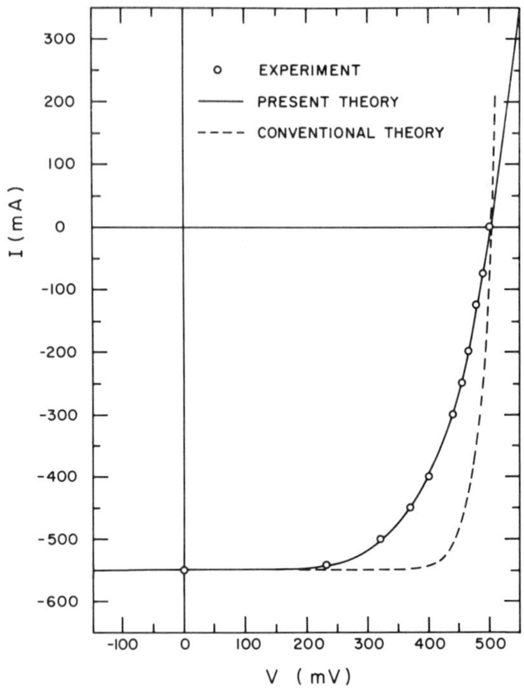

Fig. 2. Present and conventional theories are compared with experiment for a silicon homojunction solar cell. The device surface area A = 21.8 cm^2, and the temperature of the device during the experiment was 46.4°C. The measurements were conducted on a clear day at about 10:06 a.m., E.S.T., in Storrs, Connecticut, U.S.A., on 22 August, 1974.

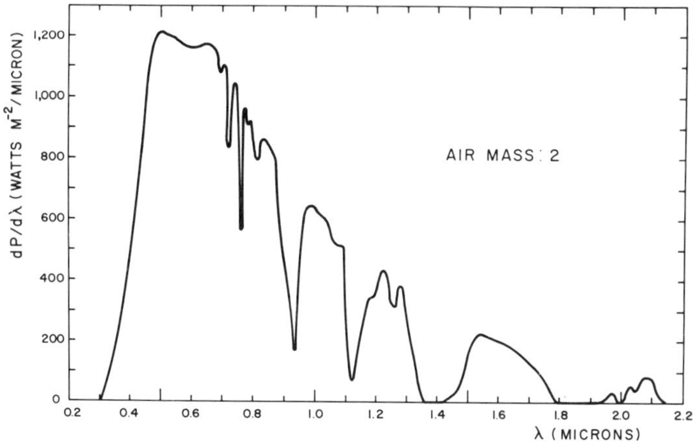

Fig. 3. Spectral distribution of the solar power density corresponding to air mass 2.

$$0 < \rho << 1, \text{ whenever } L_p >> 10\, L_o,\ I > -0.95\, I_s \qquad (53)$$

The factor ρ diminishes rapidly for (L_p/L_o) exceeding 50 and whenever $I > 0.7I_s$, and ρ almost vanishes for all values of $I > 0$.

COMPARISON OF THEORY AND EXPERIMENT

Theory and experiment are compared in Fig. 2 for a Si solar cell which is made by Solar Power Corp. The theoretical curves are either based on our theory, or the conventional equations

$$I = I_s'(e^{V/V_T}-1) - I_L,\ I_s' = eD_p p_n A/L_p \qquad (54)$$

where I_L is a parameter which is supposed to depend on the incident radiation, and all other symbols are as defined earlier.

It should be noticed here that I_s' is completely different from I_s (Eq. (24)), which pertains to our theory.

The plot of the dependency of V on I according to our theory is computed from Eqs. (48), (50), and (51). Values of the various constant physical parameters used in the computations are: $\rho = 0$, $p_p = 4x10^{20}/cm^3$, $n_n = 1.6x10^{15}/cm^3$, $p_n = 3.88x10^6/cm^3$, $p_{co} = 1x10^{15}/cm^3$, $I_s = 550mA$, $\delta\psi^* = 64.8V$, and $V_b = -0.33V$. Here n_n is the concentration of majority carriers in the base region, and all symbols are as defined earlier. Other information relevant to the ambient conditions, the device, and the experiment may be found under Fig. 2. Values of the doping levels were obtained from the manufacturer.

It should be possible to calculate V_b theoretically. However, we have not done so in this paper. Rather, we simply calculated that constant parameter at one experimental point.

INFLUENCE OF SPECTRAL PROPERTIES ON THE DEVICE ELECTRICAL PERFORMANCE

In outer space, the solar spectral properties should substantially remain constant throughout the year. But terrestially, these properties change according to the season, time of day, and geographic location, not to speak of the sky conditions, which introduce an additional random factor.

Under clear skies, the solar spectral distribution changes because the thickness of the atmospheric air mass penetrated by sunlight can vary widely depending on the angle of incidence. The thickness of this layer becomes least, of course, when sunlight falls perpendicular to the horizontal plane. Such a condition corresponds to what is called air mass 1. Above the atmosphere, the air mass is zero.

Fig. 3 shows the spectral power density distribution of sunlight at air

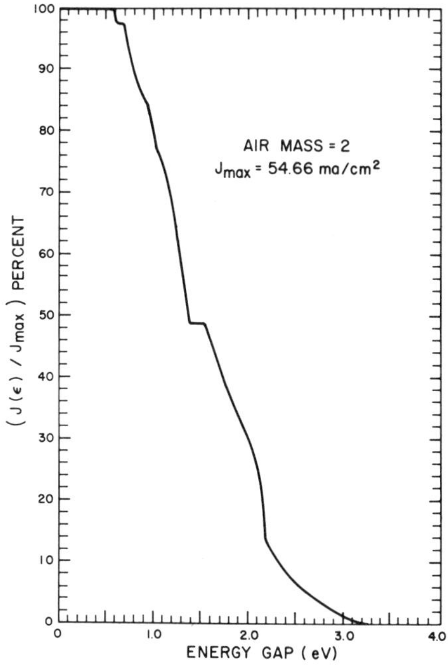

Fig. 4. Dependency of the upper limit of the solar-cell reverse saturation current density J_s on the energy gap of the base region at air mass 2.

mass 2, which corresponds to an angle of incidence (measured from the vertical) of 60°. These data were calculated from those published by Moon [10].

According to our theory, J_S can be calculated from the spectral distribution of the power density of those photons which have energy higher than the energy gap pertaining to the base region, and which penetrate through the junction to the base.

Fig. 4 represents the ideal dependency of J_S on the base energy gap if two conditions are satisfied: (1) the entire sunlight penetrates through the junction to the base region where it is entirely absorbed; and (2) if each photon of energy higher than that of the base energy gap produces one electron-hole pair. The data of Fig. 4 are calculated from those of Fig. 3.

We can now estimate J_S pertaining to our experiment from Fig. 4, and the measurements by Dash and Newman [8] on the optical attenuation in Si.

This estimation should take into account: (1) the device surface reflectivity; (2) the portion of the solar flux absorbed in the surface layer, which generates a current opposed to the terminal current; and (3) the low-energy photons which cannot be absorbed in the base region.

The steps just described give a value approximately equal to the measured value of J_S which is about 25.2 mA/m^2. This suggests that for every photon absorbed there is an electron-hole pair generated.

It is interesting, therefore, to note that J_S depends entirely on the spectral distribution of the incoming solar radiation, the device geometry, semiconductor material, and surface reflectivity.

From Eqs. (48), (49), (50), (51) and (52), it follows that for fixed values of I and J_S, a higher value for V and for the overall cell efficiency will result, if p_{co} can be increased. This increase can be achieved by increasing τ_p and decreasing L_o. These two conditions can be fulfilled by choosing the proper materials.

REFERENCES

1. Edison Electric Institute Statistical Yearbook of the Electric Utility, EEI Publication No. 7737, New York, 1977.

2. M. A. Melehy, "On the Theory of Generalized Fields in Nonequilibrium Thermodynamics: I. Some General Aspects of the Steady State; II. Unified Theory of Electrical Conduction in p-n Junctions, Heterojunctions, and Schottky Diodes," Proceedings of the Pittsburgh 1969 International Symposium on Thermodynamics, pp. 377-405, edited by E. B. Stuart, B. Gal-Or, and A. J. Brainard, MONO Book Corp.,

Baltimore, Md., 1970.

3. M. A. Melehy, "Foundations of the Thermodynamic Theory of Generalized Fields," MONO Book Corp., Baltimore, Md., 1973.

4. P. van Heerden, "Foundations of the Thermodynamic Theory of Generalized Fields," American Journal of Physics, Vo. 44, pp. 895-896, 1976.

5. M. A. Melehy, "An Introduction to the Generalized Field Theory: I. Forward Conduction in p-n Junctions and Heterojunctions," International Journal of Electronics, Vol. 24, pp. 41-68, 1968.

6. Ref. [3], Chapter 7, pp. 151-185.

7. F. C. Jain and M. A. Melehy, "High-Forward-Voltage Phenomenon in Injection BaAs/Ge Heterojunctions," Applied Physics Letters, Vol. 27, pp. 36-38, 1975.

8. W. C. Dash and R. Newman, "Intrinsic Optical Absorption in Single-Crystal Germanium and Silicon at 77°K and 300°K," Physical Review, Vol. 99, pp. 1151-1155, 1955.

9. M. A. Melehy, "On the Theory of Homojunction and Heterojunction Solar Cells," International Journal of Electronics, Vol. 44, pp, 211-217, 1978.

10. P. Moon, "Proposed Standard Solar-Radiation Curves for Engineering Use," Journal of Franklin Institute, Vol. 230, pp. 583-617, November, 1940.

PERFORMANCE OF THE NOVAL SOLAR CELL SUBSTITUTE FOR THE FUTURE

S.C. Rastogi
M.M.M. Engineering College
Gorakhpur, (U.P.), India

ABSTRACT

This paper deals with a future solar cell, in which solar radiations obtained from the sun have been utilized for the phase transformation of the electrolytic crystal by developing a potential difference between the electrodes placed in different phases (viz., solid, liquid).

The paper also reviews the earlier investigations made by the author regarding the working principle, experimental observations and design considerations. Further investigation has been carried out on long-life performance of the cell in respect to power. An effort has also been made for improvement in the design of the proper electrode. The investigations show a great potential of such a cell in the future.

INTRODUCTION

At present, throughout the world, there is an energy crisis. It has also become an established fact that the present global conventional reserves of energy are not sufficient to meet our demand in the 21st century unless an alternative source of energy is discovered. If fusion reactors become possible, there may be a small amount of relief. However, hydrogen and solar energy seem to be an alternative as a future source of energy. These are non-pollutant, inexhaustible and dependable sources of energy. The difficulty remains in harnessing this potential energy properly and efficiently. Many attempts have already been made towards this but there remains a large scope in exploring the hidden energy. One of the promising methods of extracting this energy is discussed here with its working principle, operation and characteristics. Although the power output of the cell is poor, improvements in the design of the electrode, the electrolyte, and the series-parallel combinations of such a cell may yield more power.

WORKING PRINCIPLES

The principle of operation is such that phase transformations are accompanied by effects of the production of static electrification. A potential difference of measurable magnitude (hundreds of millivolts) is developed, when:

1. solvent freezes from a very dilute solution

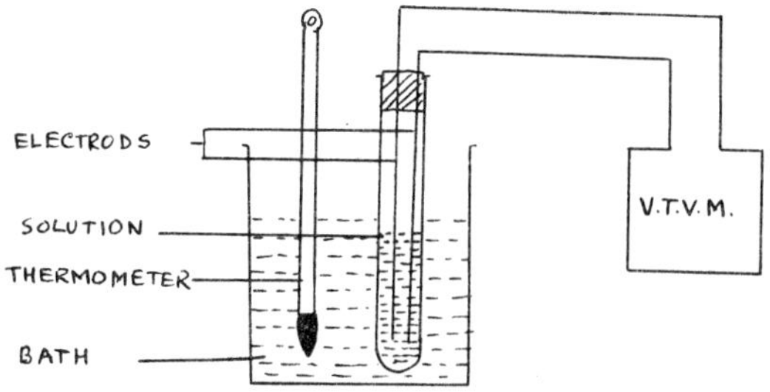

FIG. 1. EXPERIMENTAL DETERMINATION OF PRECIPITATION AND DISSOLUTION POTENTIAL

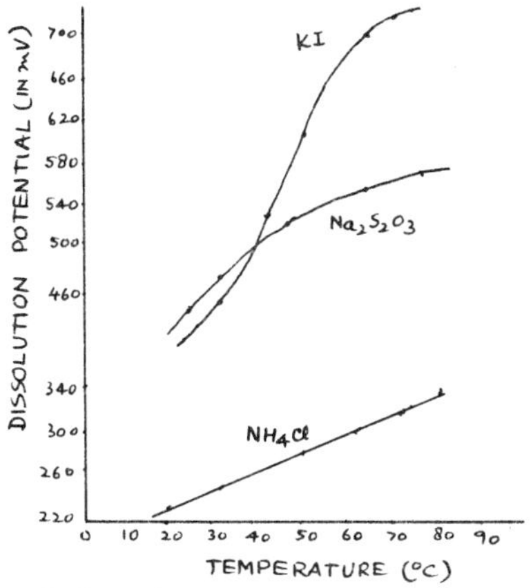

FIG. 2. TEMPERATURE DEPENDANCE OF DISSOLUTION POTENTIAL FOR NH_4Cl, $Na_2S_2O_3$, KI

2. precipitation occurs from the molten electrolyte

3. dissolution of the electrolyte in water takes place

4. precipitation of the electrolyte occurs from the saturated solution.

DISSOLUTION AND PRECIPITATION POTENTIALS

The dissolution and precipitation of NH_4Cl, $Na_2S_2O_3$ and KI has been studied. The experimental set-up is shown in Figure 1. Figure 2 shows the temperature dependence of the dissolution potential. $Na_2S_2O_3$ has been chosen for the study because of (1) low melting point, (2) fairly high dissolution potential, (3) low change of potential difference with the change of temperature, and (4) temperature dependence of precipitation potential is the same as the dissolution potential.

The observations of precipitation and dissolutions potential were as follows:

(a) Potential difference is developed as soon as a tiny crystal is formed on the electrode. In the same manner, the dissolution potential is developed as soon as the first crystal goes into the solution.

(b) Precipitation and dissolution potential vary with the temperature.

(c) The electrode from which the crystallization or dissolution takes place is always negative.

(d) The potential difference is more if the electrodes are clean and there is no nucleation present.

PERFORMANCE OF THE CELL

The above stated some of the features which led to the utilization of solar radiations for the purpose of dissolution and precipitation of $Na_2S_2O_3$ and consequently the generation of voltage. A solar cell using sodium thiosulphate has been studied. It has been observed that the output of the cell is very small which is due to many factors. Typical open circuit voltage, short circuit current and internal resistance was of the order of 400 mV, 25 μA and 5 $K\Omega$ respectively. However, a series-parallel combination helped in increasing the voltage and current, and in decreasing the resistance (Figure 3). Still, some combination may be developed to reduce the internal resistance by adding some impurity in the $Na_2S_2O_3$.

The plot shown in Figure 4 shows the time performance of the solar cell. Two probes from the electrodes of the cell were fed to the recorder for recording the E.M.F. The developed sensitivity and speed of the recorder was

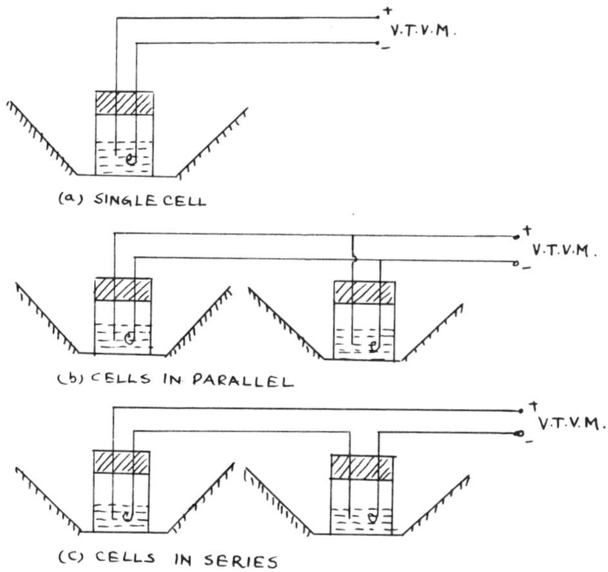

FIG. 3. ARRANGEMENT OF CELLS

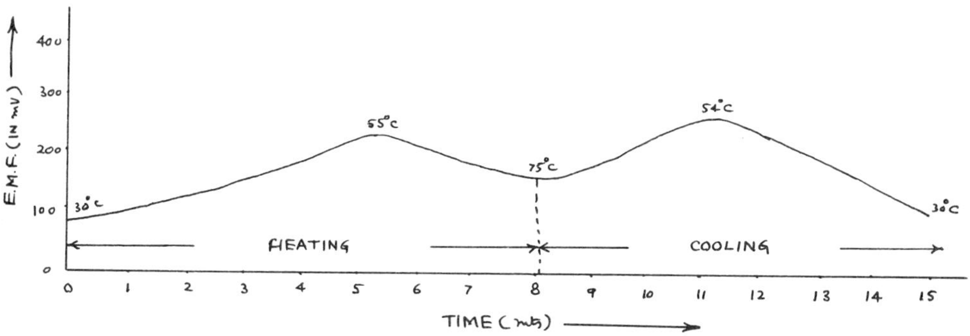

FIG. 4. E.M.F. RECORDER PLOT

set to 0.2 volt/cm and 0.125 mm/sec. The experiment was performed at 30°C. The plot shows the assymmetry potential at the start and the finish of the experiment. The variation of dissolution and precipitation is found to be about 140 mV and this has occurred at 55°C.

ELECTRODES

Electrodes of Cu, Al, and Pt were made for the study and it was found that the dissolution and precipitation potential measured were approximately the same. However, it is suggested that the platinum electrode should be used for the purpose. As Al and Cu are oxidized, forming an oxidized coating over the electrodes, a lesser efficiency and erratic behavior of the cell is caused.

It is also suggested that only the small portion of the electrode should be dipped in the electrolyte in order to avoid the large area where undesired nucleation may occur in crevices.

It has also been observed that if the electrodes are simultaneously heated and cooled by means of a concentrator and passing cooled water into the tubular electrode, the results would be better.

CONCLUSIONS

1. The investigations support a great potential for such a cell.

2. Sodium thiosulphate is deliquescent in moist air, therefore, an air-tight vessel should be used for the cell or some arrangement should be made to remove the moisture.

3. The melting point of thiosulphate is 48°C and if heater to greater temperatures, it can decompose. Therefore, the temperature of an electrolyte cell should not exceed more than 55°C.

4. The electrode should be made of platinum. However, other materials may be explored in order to subsitute it for the platinum, due to its high cost.

5. Only a small portion of the electrodes should be dipped.

6. Efforts can be made in determining a suitable electrolyte which will yield a better performance by adding some impurities to sodium thiosulphate, thereby reducing internal resistance.

7. P.B. was obtained practically in small as compared to the observed value, because of (a) simultaneous crystallization on both electrodes, and (b) owing to the large electrode area and roughness, there is the distinct possibility of nucleation occurring in the crevices.

ACKNOWLEDGEMENTS

Thanks are due to the authorities of the Chemistry Department, Gorakhpur University and Dr. B.N. Sharma, Principal, M.M.M. Engineering College, Gorakhpur, who provided facilities and encouragement for this work.

REFERENCES

1. Rastogi, S.C., Goel, R.C., and Sharma, B.N., "A Noval Solar Cell", paper presented at the conference on Utilization of Solar Energy, 1974, Allahabad, paper published in Electrical India, October 1975.

2. Rastogi, S.C., Goel, R.C., and Sharma, B.N., "Performance of the Noval Solar Cell", paper presented in the 8th Meeting of the AllaIndia Solar Energy Working Group, and the Conference of Utilization of Solar Energy, Jadavpur University, Calcutta, India, 1976.

3. Glasstone, S., "An Introduction to Electrochemistry".

4. Shukla, R.D., "Precipitation and Dissolution Potentials", Ph.D. Thesis, Gorakhpur University, 1970.

MOS SOLAR CELLS

H.M. Kızılyallı
Physics Department
Middle East Technical University
Ankara, Turkey

ABSTRACT

The solid state physical principles of MOS structure and their electrical characteristics is reviewed for its application to solar cells. The various parameters like insulator thickness, resistivity of the substrate, insulator charge, surface-state density, minority carrier lifetime, and crystal orientation that affects the performance of MOS solar cells will be discussed. The paper is concluded with a summary of recent developments in the field.

INTRODUCTION

Although the direct photoconversion of solar energy to electricity is a very enticing prospect for clean production of electrical power, the main limitations presently are high costs of solar cells and low efficiency of conversion. MOS solar cells are of current interest because they represent a potentially low-cost method for fabricating large scale solar energy conversi devices. It is conceivable that this structure can be adapted to polycrystalli and amorphous film materials where unwanted effects as grain-boundary diffusio of the dopant in p-n configuration may be eliminated. Conventional Schottky barrier devices always have an oxide layer present, to some degree, between the semiconductor and the metal. The insertion of an interfacial layer between the metal and the semiconductor, if properly chosen, can have one or more of the following advantageous effects: it can (1) control transport, (2) shape the electric field in the device, and (3) introduce surface state charges and furthermore (4) influence the nucleation and growth of the metal electrode film directly influencing the amount of light transmitted to the semiconductor [1]. An alternate point of view is advanced by Green et al. [2], and more recently by Card [3], and Shewchun et al. [4] who suggest that the mechanism for the current conduction is tunneling through the oxide layer of a MOS device that is formed when these oxide layers are created. Basically, a field-induced junction at the surface of the semiconductor under the oxide layers and coupling to this junction is achieved by tunneling, allowing the metal to be contacted to this field-induced junction.

Two oxide thickness regions are distinguishable: (1) If the thickness of the insulating layer in a MOS structure, $d > d_{cr}$ the semiconductor is essentially in "thermal equilibrium", tunnel currents are of small magnitude. (2) If the oxide thickness $d < d_{cr}$, then the tunnel currents are large enough to disturb the semiconductor from equilibrium, hence nonequilibrium MOS configurations are formed. This occurs at around 28 Å for Al-SiO_2- pSi [4] and 50 Å for

Au - nGaAs [5]. Nonequilibrium devices can be further classified as, majority carrier, surface-state, or minority carrier diodes depending on whether the dominant tunnel current flow near zero bias is between the metal and the majority carrier band, the surface states, or the minority carrier band. Which device is formed depends on whether the insulator-semiconductor interface at zero bias is accumulated, depleted or in strong inversion [6]. Minority carrier devices can show I-V characteristics which obey the ideal Shockley diode equation and are suitable for solar photovoltaic energy conversion. If the thickness of the oxide is below about 10 Å for Si (and 22 Å for GaAs), the concept of a tunnel MOS diode becomes invalid, fundamentally based on experimental evidence. Below about mentioned minimum oxide thickness d_{min} it appears to be more proper to describe such structures as Schottky barriers.

Only the minority-carrier nonequilibrium MOS structure will be discussed here because it is of specific interest for the solar cell application. The minority carrier quasi-Fermi level of semiconductor is pinned to the metal Fermi level, and the majority carrier flow through the oxide is small. If the majority-carrier component cannot be kept small, V_{oc} will always be smaller than expected for a comparable diffused p-n junction solar cells.

Since it has been found that the interface behavior of MOS contacts forward-biased by optical illumination is qualitatively different from that of the same contacts biased in the dark by an applied forward voltage [7], therefore the theory of Card and Rhoderick [8] can not be used to explain the increases observed in the V_{oc} of the cell due to interfacial layer. In the optical bias case, the potential drop in the oxide layer will be negligibly small due to the interface states taking on the opposite charge as a result of communication mainly with the minority carriers. In the latter case, the ideality factor n will be substantially increased by the presence of oxide layer, since the potential drop V_{ox} across this layer is large as a consequence of the charge induced in the interface states by the applied forward bias. Since the interfacial layers formed on semiconductor are usually oxides, the terms oxide, insulator and interfacial layer are used synonymously.

THEORY

The exact role of interfacial layer is not clearly understood yet. First a summary of the electrical characteristics of the semiconductor-oxide system will be presented, and then theories of Fonash and Card will be reviewed. Finally, the parameters that affect the performance of MOS Solar Cells will be discussed.

Semiconductor-Oxide System

The various types of charges and states associated with this system are: fast surface states, mobile impurity ions, traps ionized by irradiation and finally, surface-state charge [9].

i. Fast surface states are located at the oxide-semiconductor interface and are due to disruption of the periodicity of the lattice at the surface, a a high density of states N_{st} will be introduced into the forbidden gap near a semiconductor surface and their number should be approximately equal to the number of surface atoms.

ii. Mobile impurity ions, such as sodium ionic contamination, may cause the formation of the space charges within the oxide layer that may induce its image charges in the semiconductor and in the metal.

iii. A positive space-charge in SiO_2 layer may result from exposure to ionizing radiation such as gamma ray, x-ray irradiation. Trapping of the holes near the SiO_2-Si interface will result in a build up of a space charge, if there is an electric field in the oxide to seperate the electron-hole pairs generated in the oxide by irradiation.

iv. Fixed surface-state charge located in the oxide near the interface between oxide and semiconductor. This charge is identified by a parallel translation of the capacitance-voltage characteristics along the voltage axis and its density per unit area is designated by Q_{ss}. It cannot be discharged and charged and Q_{ss} is not significantly affected by the oxide thickness or impurities in Si [10] and GaAs [11] but has a strong dependence on the oxidation and annealing conditions and the orientation of these crystals [4].

Theory of Fonash

The advantageous effects of the insulating layer, as stated in the introduction, will be discussed in detail.

i. Control of transport: In general transport through such a thin-film layer could involve variable range hopping [12], [13], small-polaron hopping [14], impurity conduction [15], etc. The simplest control for the purpose of reducing its value could be achieved if the conduction mechanism across the interface layer was direct tunneling. If we consider on n-type semiconductor device, reduction can be obtained by using an oxide layer such that the conduction band of oxide would be substantially above the conduction band of the semiconductor,while the valence band of the oxide would be comparable in energy to the valence band of the semiconductor.

ii. Field shaping: Beneficial field shaping can be achieved by introducing localized states near the semiconductor-oxide boundary whose occupancy tends to follow the imref of the majority carriers of the semiconductor. For example, for n-type semiconductor, one would want to insert localized states which have a relatively high capture cross section for electrons σ_n and low capture cross section for holes σ_p. These states will fill as the device develops a voltage due to solar irradiation and become more negative thus enhancing the total voltage developed. The insertion of these localized states, which are negative when occupied, will decrease the dark current density J_D that consists of the thermionic and diffusion components towards the diffusion limit. It is beneficial to use field shaping to keep the electric field in the semiconductor as large as possible for effective separation of light-generated carriers. Employing Au-Oxide-GaAs structures, Stirn and Yeh have used MOS configuration to obtain a 15 % efficiency and V_{oc} as high as 700 mV. in which case the oxide layer is believed to be introducing field shaping [1].

iii. Insertion of surface state charge: It is obvious that using the oxide layer to insert surface state charge is equivalent to modifying the ϕ_M according to

$$\phi_M^{eff} = \phi_M - \frac{Q_{ss} d_{eff}}{\epsilon_i \epsilon_o} \quad (1)$$

where Q_{ss} is surface state charge density per unit area (positive or negative) in the oxide. It may be considered to be located at an effective distance d_{eff} from the metal. This influence of Q_{ss} on effective metal work function ϕ_M^{eff} provides the possibility of enhancing the barrier height ϕ_B. If Q_{ss} is assumed to be negative, which is advantageous case for an n-type semiconductor, and since $Q_{ss} = q\ N_{ss}$ where N_{ss} is fixed charge number density, then the barrier height changes as a function of N_{ss}. Thermally-oxidized p-type silicon has a sheet of positive charge located in the oxide near the oxide-silicon interface [16]. The inherent interface charge will induce an n-type surface inversion layer [17], [18], or a field-induced n^+-p junction. For either Au or Al on n-type silicon, ϕ_B usually decreases with d [19], since contamination introduces positive oxide charge, and will have a negative effect on V_{oc} and should be avoided. The nature of the positive and negative charges which occur in SiO_2 grown thermally on Si, which are responsible for Anderson localization of electrons in the inversion layer and for the slow states, are lately examined by Mott [20].

Theory of Card

The transport processes in the semiconductor were formulated in a very general way by using the definitions and the set of differential equations which form the basis of semiconductor device physics: The quasi-Fermi level for either carrier is defined as the quantity which, when substituted into the place of fermi level, gives the concentration of that carrier under nonequilibrium conditions. Thus the concentration of electrons under nonequilibrium conditions is given by

$$n(x) = N_c \exp\left(\frac{E_{Fn}(x) - E_c(x)}{kT}\right) \quad (2)$$

where $E_{Fn}^{(x)}$ is the quasi-Fermi level for electrons, and similarly, the concentration of holes under nonequilibrium conditions

$$p(x) = N_V \exp\left(\frac{E_V(x) - E_{Fp}(x)}{kT}\right) \quad (3)$$

where E_{Fp} is the quasi-Fermi level for holes. In the semiconductor, the electron and hole current densities may be written as

$$J_n = q\ D_n \frac{dn(x)}{dx} + qn\mu_n\varepsilon \quad (4)$$

and

$$J_p = -q\, D_p \frac{dp(x)}{dx} + qp\mu_p\varepsilon \tag{5}$$

substituting eqns.(2) and (3) into (4) and (5) respectively and using [21]

$$\varepsilon = \frac{1}{q}\frac{dE_v(x)}{dx} = \frac{1}{q}\frac{dE_c(x)}{dx} \tag{6}$$

one obtains

$$J_n = n\mu_n \frac{dE_{Fn}}{dx} \tag{7}$$

and

$$J_p = p\mu_p \frac{dE_{Fp}}{dx} \tag{8}$$

where all symbols have their usual meanings. In the steady state, the currents given by eqn.(7) and (8) are equal to the tunnel currents through the insulator layer.

Fig.1 shows the energy band diagram for an n-type MOS structure in the dark. The majority carrier electron tunnel current through the insulator is given by [8]

$$J_n = \frac{4\pi m^*_{te} q(kT)^2}{h^3} \exp\,(-\chi_e^{1/2} d)\exp\left(\frac{E_{Fn}(0)-E_c(0)}{kT}\right) \tag{9}$$

but

$$\exp\left(\frac{E_{Fn}(0)-E_c(0)}{kT}\right) = \frac{n(0)}{N_c} \tag{10}$$

hence

$$J_n = \frac{4\pi m^*_{te} q(kT)^2}{h^3 N_c} \cdot n(0)\exp(-\chi_e^{1/2} d) \tag{11}$$

where m^*_{te} is the effective mass for electrons with momentum transverse to the barrier, χ_e is the average barrier height of oxide to electrons

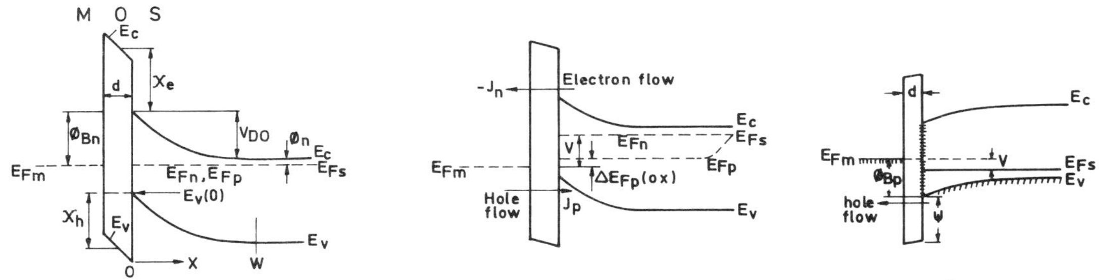

Fig. 1 Energy band diagrams for n-type MOS structure in the dark (a) V = 0, (b) applied voltage V > 0, (c) p-type MOS contacts, under applied forward bias V.

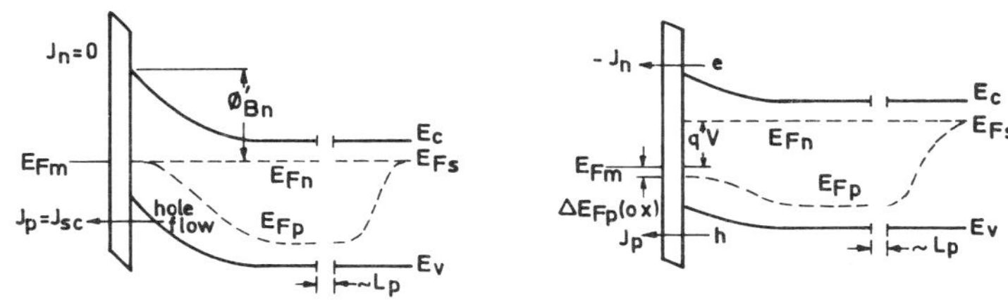

Fig. 2 Energy band diagrams for n-type MOS solar cell under solar illumination (a) V = 0, short-circuit conditions, external load R = 0, (b) V > 0

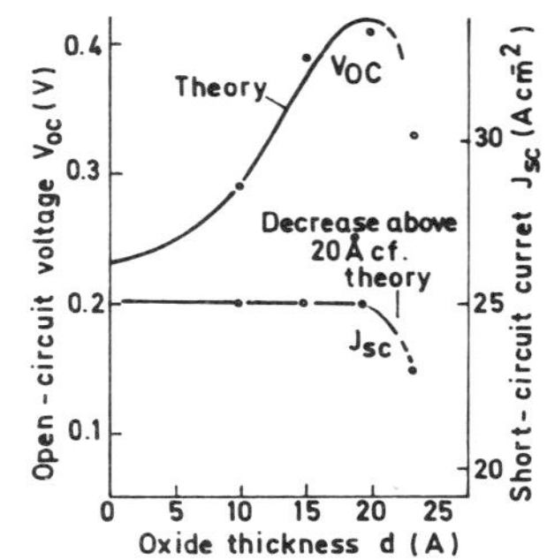

Fig. 3 Open circuit voltage V_{OC} and short-circuit current density J_{SC}, for silicon MOS solar cells; open circles, experiment; solid line, theory [After Card and Yang]

tunneling into the metal from the semiconductor conduction band, and $\exp(-\chi_e^{\frac{1}{2}}d)$ is the transmission coefficient of the oxide. The applied bias V will cause a change ΔV_D in zero-bias value of the diffusion potential V_{DO} and defining

$$n = -V/\Delta V_D \tag{12}$$

then

$$V_D = V_{DO} + \Delta V_D = V_{DO} - \frac{V}{n} \tag{13}$$

using the zero-bias condition

$$V_{DO} + \phi_n = \phi_{Bn} \tag{14}$$

Combining eqns. (13) and (14) we get

$$V_D + \phi_n = \phi_{Bn} - \frac{V}{n} \tag{15}$$

on the other hand,

$$E_c(0) - E_{Fn}(0) = q(V_D + \phi_n) \tag{16}$$

hence,

$$E_{Fn}(0) - E_c(0) = -q(\phi_{Bn} - \frac{V}{n}) \tag{17}$$

finally by letting,

$$A = \frac{4\pi m_{te}^* q k^2}{h^3}$$

we obtain

$$J_n = A\,T^2 \exp(-\chi_e^{\frac{1}{2}}d)\exp(-\frac{q\phi_{Bn}}{kT})\exp(\frac{qV}{nkT}) \tag{18}$$

where the ideality factor

$$n = 1 + \frac{d/\epsilon_i[(\epsilon_s/W) + qD_{sb}]}{1+(d/\epsilon_i)qD_{sa}} \tag{19}$$

and ϵ_s is the permittivity of the semiconductor, ϵ_i is the permittivity of the insulator, W is the depletion width in the semiconductor, D_{sa} is the density of interface states in equilibrium with the metal, D_{sb} is the density of interface states in equilibrium with the semiconductor, and A is the modified Richardson constant.

For the n-type MOS structure, J_n is the dominant component of the dark current. There are also small contributions from the injection of minority carriers into the neutral region of semiconductor J_{pi}, and from recombination in the depletion region J_{rec}.

Let us consider the energy-band diagram of MOS structure given in Fig.2. and the behavior of the quasi-Fermi level for holes E_{Fp} under illuminated conditions. The holes are transported by diffusion to the space charge region, and are swept into the metal. As the thickness of the oxide layer, d, increases it provides more of a barrier to tunneling, and in order to pass the same hole current, as the thinner layer, the change in E_{Fp} across the oxide, $\Delta E_{Fp}(ox)$ must increase. The maximum hole current is obtained under the assumption that $p(W) = 0$. Proviso that this remains a valid assumption, as d increases the hole current from the neutral region stays at the maximum value obtained in the absence of the oxide layer. The evidence of this is suggested by the constancy of the short-circuit (minority carrier) hole current J_{sc} which are photogenerated in the semiconductor. This can be better appreciated from a theoretical view point by making use of the eqns.(3) and (8) with $J_p = J_{sc}$. This current is given by [22]

$$J_p = \frac{4\pi m_{th}^* q (kT)^2}{h^3 N_v} p(0) \exp(-\chi_h^{\frac{1}{2}} d)[\exp(-\Delta E_{Fp}(ox)/kT) - 1] \qquad (20)$$

where m_{th}^* is the effective mass for holes with momentum transverse to the barrier. Since the parameters m_e^*, m_h^*, χ_e and χ_h are not known accurately and could provide a source of asymmetry between the tunneling probabilities of electrons and holes, hence it has been assumed that $m_e^* = m_h^* = m_o$ and $\chi_e = \chi_h = \chi$, so that the tunneling transmission probabilities of the oxide is the same for both electrons and holes, $\exp(-\chi^{\frac{1}{2}} d)$.

The total current in the MOS solar cell under illumination is

$$J = J_{sc} - J_{no} \exp(-\chi^{\frac{1}{2}} d) \exp(qV/n'kT) - J_{rec} \qquad (21)$$

where

$$J_{no} = A\, T^2 \exp(-q\phi'_{Bn}/kT) \qquad (22)$$

Space charge recombination current under illumination,

$$J_{rec} \cong J'_r \exp(qV/2kT) \qquad (23)$$

where ϕ'_{Bn} and n' are the barrier height and ideality factor value under illumination, which are in general different from their dark values ϕ_{Bn} and n. Under solar irradiation J_n and J_{rec} oppose the photogenerated current since for a given voltage, they flow in the opposite direction through the device, the minority carrier injection current J_{pi} is neglected.

It has been shown that the values of J_n and J_{rec} for a given voltage may be appreciably changed by illumination [23]. The current densities in forward bias are

$$J_{dark} = J_n + J_{rec}(dark) \cong J_o \exp(eV/kT) + J_r(dark)\exp(eV/2kT)$$

and

$$J_{illum.} = J_n + J_{rec}(illum) - J_{sc} \cong J_o \exp(eV/kT)+J_r(illum)\exp(eV/2kT)-J_{s}$$

In the dark, J_{rec} is negligible but under illumination, it provides contribution.

The dark recombination current density is given by [24]

$$J_{rec} = \frac{1}{2} q \frac{n_i}{o} W \exp(qV_F/2kT)$$

where τ_o is the effective lifetime within the depletion region and n_i is the intrinsic carrier concentration of the semiconductor. Letting

$$J_r = \frac{qn_iW}{2\tau_o}$$

one obtains

$$J_{rec} = J_r \exp(qV/2kT)$$

Under illumination, we must replace W by an enlarged W' as a result of the increased hole concentration in the depletion region as indicated by E_{Fp} in Fig.2.

Under illumination the charge in the interface states will be changed by photoionization of these states. In addition, the surface concentrations of holes p_s and electrons n_s, and particularly minority carriers in the device, will be increased by illumination. As the oxide thickness d increases, the surface concentration of holes $p_s = n(0)$ must increase to maintain the same photocurrent of holes J_p tunneling into the metal as can be seen from eqn.(20). This means that the charge in interface states will also be changed under illumination by the change in the relative rates of capture of holes and electrons, which is a linear function of p(0) and n(0) [25].

We set the total current, given by eqn.(21), to zero, to be able to obtain the open-circuit voltage V_{oc}. The recombination current J_{rec} will be negligible due to large voltages such as V_{oc}, hence substituting J = 0 and $V = V_{oc}$, we get

$$J_{sc} = J_{no} \exp(-\chi^{\frac{1}{2}}d)\exp(qV_{oc}/n'kT)$$

or

$$V_{oc} = (n'kT/q)(\ln(J_{sc}/J_{no}) + \chi^{\frac{1}{2}}d) \qquad (24)$$

where

$$J_{no} = A\, T^2 \exp(-q\phi'_{Bn}/kT)$$

Fig.3 shows the dependence of the V_{oc} and J_{sc} for Au-nSi MOS solar cells, on the insulating layer thickness d as measured by Lilington and Townsend [26] and the theoretical curve [7]. The argreement with the experimental data is observed.

The short-circuit current J_{sc} for the oxide thickness of interest is the same current as that without the oxide and is given by [27]

$$J_p = J_m + J_i + J_d + J_b$$

where J_m is the current density due to absorption in the metal film and is a very small fraction of the total photogenerated current. The components J_i and J_d arise from photon absorption within the inversion and depletion layers respectively and are given by

$$J_i(\lambda) = qN(\lambda)T(\lambda)[1-\exp(-\alpha(\lambda)x_i)]$$

and

$$J_d(\lambda) = -qN(\lambda)T(\lambda)[\exp(-\alpha(\lambda)x_i)-\exp(-\alpha(\lambda)W)]$$

where N(λ) is the incident photon flux, T(λ) is the transmittance, x_i is the inversion layer thickness, W is the depth of the depletion region and α is the semiconductor absorption coefficient. The bulk component of the photocurrent J_b is

$$J_b(\lambda) = \frac{-q\alpha(\lambda)N(\lambda)T(\lambda)L_p}{\alpha^2 L_p{}^2-1}\left\{(\alpha L_p-1)\exp(-\alpha W) + \frac{2[\exp(-\alpha L)-\exp(-\alpha W)\exp(\frac{W-L}{L_p})]}{\exp(\frac{L-W}{L_p})-\exp(\frac{W-L}{L_p})}\right\}$$

where L_p is the diffusion length of the holes and L is the semiconductor thickness.

The oxide layer has an associated tunneling probability $\exp(-\chi^{\frac{1}{2}}d)$ and this reduces the majority current in the forward biased MOS structure. Due to the reduction in majority current, V_{oc} is increased, hence the conversion efficiency of the solar cell is improved by an oxide layer of about 18-20 Å thick.

PARAMETERS

The various parameters that affect the performance of MOS solar cells are oxide charge, conductivity of the substrate, minority carrier lifetime, oxide thickness, fast surface state density N_{st} and fixed surface state density Q_{ss} and crystal orientation. We will elaborate on these furthermore.

The linear dependence of V_{oc} on oxide thickness d can be explained by the following expression which relates the effective metal work function ϕ_M^{eff} to the actual value ϕ_M:

$$\phi_M^{eff} = \phi_M - \frac{qN_{st} + Q_{ss}}{\epsilon_i \epsilon_o} d$$

The open-circuit voltage V_{oc} depends on ϕ_M^{eff} by [29]

$$V_{oc} = A - \phi_M^{eff} - Bn$$

where A and B are constants.

The substrate conductivity is linearly related to the doping density. The conversion efficiency increases with doping density which can be explained by thinking of the device as an p^+-n junction diode. As the donor impurity concentration is increased, the barrier height ϕ_{Bn} increases and V_{oc} increases accordingly [30]. A competing effect is the decrease in short-circuit current J_{sc} due to degradation of the diffusion length and minority carrier lifetime. Eventually this decrease in J_{sc} becomes stronger than the enhancement in V_{oc} and the efficiency falls down. The conversion efficiency of an Al-SiO_2-pSi device reaches a peak around $N_A = 3x10^{23}$ m^{-3}(or ρ=0.014 Ωcm) and then descends [4].

The fast surface states act as recombination-generation centers and furnish tunneling paths between the semiconductor and metal. A net current flows to the metal from surface states considered localized at the semiconductor-oxide interface. The effective current flows due to the interchange of the charge between the valence and conduction bands of the semiconductor and these states by recombination and generation processes. Hence, J_{sc} will also be influenced by fast surface states in addition to surface state charge.

The density of the fast surface states N_{st} depends on crystollographic orientation: N_{st}<111>> N_{st}<110>> N_{st}<100>in Si [4]. Stirn and Yeh have shown that V_{oc}, n, ϕ_B and J_{no} of GaAs MOS solar cells have strong crystallographic orientation effect [11]. The density of fast surface states N_{st} and surface state charge Q_{ss} can both be increased by annealing in O_2, N_2 and minimized by a high-temperature hydrogen heat treatment [10]. It has been shown that these parameters can play a significant role in optimizing the conversion efficiency of MOS configuration and about 21 % efficiency is possible under AM2 illumination with high substrate doping and low defect density for the Al-SiO_2-pSi device [4].

The conventional approach to fabricating diffused p-n junctions for the potential barrier in GaAs has not been succesful for obtaining high-efficiency solar cells due to the fact that nearly all the minority carriers are generated within several microns from the surface and excessive recombination at the surface and within the diffused region seriously limit the collection efficiency of the minority carriers. In addition, the structure of the MOS solar cells is most effective when used with a direct bandgap material such as GaAs, specifically in thin film form. GaAs has a bandgap which is better suited than that of Si for matching solar spectrum. But the research has been performed on the GaAs-oxide system is relatively new and considerably complex compared to the Si-oxide system.

RECENT DEVELOPMENTS

The low temperatures and small number of processing steps involved in a MOS device fabrication make this type of solar cell extremely competitive when future large scale manufacture is considered. Furthermore, this structure can be adapted to polycrystalline and amorphous materials in thin film form. Townsend and Lillington [31] fabricated MOS solar cells giving an overall efficiency of 8.6 % using cast polycrystalline silicon substrates. Green and Godfrey [32] has also carried out experimental work on polycrystalline substrates and obtained efficiencies around 10 %. Their replacement of back ohmic contact by insulator-metal contact produced an accumulated surface that has the properties of a p^+-p contact . This blocks minority carrier flow and improves the collection efficiency.

Delahoy et al [33] fabricated MOS solar cells on ribbon silicon which is pulled from a silicon melt via a capillary die (the edge-defined, film-fed growth method). The as-grown ribbon is sufficiently smooth and thin as to render the expensive processes of wafer sawing and polishing unnecessary.

Anderson et al [34] proposes the exploitation of the work-function modification by oxidation or chemisorption of transition-metal elements, and Fabre et al [35] reports the use of such reactions in the chromium and titanium MOS solar cells.

In general, the insulating layers have been oxide layers which were grown by thermal, plasma and anodic oxidation, in addition to the unintentional oxide layers that exist both on Si and GaAs practical surfaces. Lately, Bonnet [36] has used ZnSe film as an insulating layer in the fabrication of thin film CdSe MIS solar cell. Although the efficiency achieved is low (J_{sc} = 5 mA/cm^2 and V_{oc} = 600 mW with a direct solar radiation intensity of 80 mW/cm^2), the work is encouraging since CdSe can be manufactured as thin film with good electronic transport properties and has a direct energy gap of 1.7 eV.

The physics of solar photovoltaic energy conversion by MOS structure discussed here can serve as a basis for understanding of their operation, regardless of the composition of the solar cells. The outlook for future development of MOS solar cells for terrestrial use in energy conversion lies in the fields of processing, materials and concentrators.

REFERENCES

1. S.J.Fonash, (1976) J.Appl.Phys., 47, 3597
2. M.A.Green, F.D.King and J.Shewchun, (1974) Solid State Electron., 17,551
3. H.C.Card, (1977) Solid State Electron., 20, 971
4. J.Shewchun, R.Singh and M.A.Green, (1977) J.Appl.Phys., 48, 765
5. R.J.Stirn and Y.C.M.Yeh, (1975) Appl.Phys.Lettl, 27,95
6. A.S.Grove, (1967) Physics and Technology of Semiconductor Devices, p.266, Wiley, New York
7. H.C.Card and E.S.Yang, (1976) Appl.Phys.Lett., 29,51
8. H.C.Card and E.H.Rhoderick, (1971) J.Phys., D4, 1589
9. A.S.Grove, ibid., p.334
10. T.I.Kamins and B.E.Deal, (1975) J.Electrochem.Soc., 122,557
11. R.J.Stirn and Y.C.M.Yeh, (1977) IEEE Trans.Electron Devices, ED-24,476
12. N.F.Mott, (1969) Phil.Mag., 19,835
13. N.Apsley and H.P.Hughes, (1974) Phil.Mag., 30,963
14. A.J.Bosman and H.J.van Daal, (1970) Adv.Phys., 19,1
15. A.Miller and E.Abrahams, (1960) Phys.Rev.120,745
16. A.S.Grove, ibid., p.343
17. G.C.Salter and R.E.Thomas, (1977) Solid State Electron., 20,95
18. M.A.Green and R.B.Godfrey, (1975) Appl.Phys.Lett. 29,610
19. H.C.Card, (1974) Solid State Commun., 14, 1011
20. N.F.Mott., (1977) Adv.Phys., 26,363
21. A.S.Grove, ibid., p.152
22. H.C.Card and E.H.Rhoderick, (1971) J.Phys., D4,1602
23. P.Panayatos, H.C.Card and E.S.Yang, (1977) Proc.IEEE, 65,1213
24. A.S.Grove, ibid., p.187
25. A.S.Grove, ibid., p.131
26. D.R.Lilington and W.G.Townsend, (1976) Appl.Phys.Lett. 28,97
27. R.F.McQuat and D.L.Pulfrey, (1975) Proc.11th IEEE Photovoltaic Spec. Conf., 371
28. D.L.Pulfrey, (1976) IEEE Trans.Electron Devices, 23,587
29. H.M.Kızılyallı, (1978) Proc.İzmir International Symposium on Solar Energy
30. M.Wolf, (1971) Energy Convers., 11,63
31. W.G.Townsend and D.R.Lillington, (1978) Proc.Photovoltaic Solar Energy Conf., 207
32. M.A.Green and R.B.Godfrey, ibid., p.299

33. A.E.Delahoy, W.A.Anderson and J.K.Kim, ibid., p.308
34. W.A.Anderson, J.K.Kim, and A.E.Delahoy, (1977) IEEE Trans.Electron Devices, ED-24, 453
35. E.Fabre, J.Michel and Y.Baudet, (1976) Proc. 12 th IEEE Photovoltaic Specialists Conf., 904
36. D.Bonnet, (1978) Proc.Photovoltaic Solar Energy Conf., 630

SOME OPTICAL AND ELECTRICAL PROPERTIES OF SELENIUM-CADMIUM SULPHIDE PHOTOVOLTAIC CELLS

K. Abdel-Hady, F. El Akkad and J. Al-Dabbagh
National Research Centre
Cairo, Egypt

ABSTRACT

Selenium-Cadmium sulphide photovoltaic cells were prepared using the vacuum deposition technique. The current-voltage characteristics in the dark were investigated at different temperatures. The illumination characteristics were also studied. The logarithm of the forward current was found to increase linearly with voltage, while the reverse current shows a power law dependence on voltage. The results can be interpreted in terms of a self-consistent model in which the current is carried by injection over a potential barrier in the presence of a conduction band spike. The diffusion potential was determined to be 1.6 - 2.1 V. The spectral sensitivity curve was determined and found to be similar to that of the human eye.

INTRODUCTION

The importance of the photovoltaic properties of Se stems from their use in photocontrolling apparatus and measuring instruments. Due to the difficulty of obtaining a p-n junction on Se, other photovoltaic structures are usually used [1,2] (e.g., shottky barrier or heterojunction with n-type material). In this respect, CdS appears to be a suitable material for heterojunction action with Se since it has the same crystal structure (hexagonal) with reasonable lattice mismatch (4%) [3,4].

To the author's knowledge, the only published work on Se-CdS cells was reported by Kunioka and Sakai [5]. They have given a short account on the preparation procedure and some properties of the cell.

In the present work, we provide further study of the optical and electrical characteristics of this type of cells. We analyzed our results on the basis of current transport theories for heterojunctions in an attempt to elucidate the transport mechanism.

EXPERIMENT

The cells were prepared by depositing CdS films on hexagonal Se layers using vacuum evaporation technique. The hexagonal Se was prepared by annealing an amorphous layer (120 min. at 70°C) deposited on a quartz substrate at room temperature. Au and Al evaporated films were used as contacts for Se and CdS, respectively. The cell configuration is shown

TABLE I

VALUES OF η AT DIFFERENT TEMPERATURES CALCULATED FROM FIGURE 3

Temperature °K	η
303	11.5 ± 0.4
323	12.0 ± 0.1
333	13.0 ± 0.3
343	14.8 ± 0.8

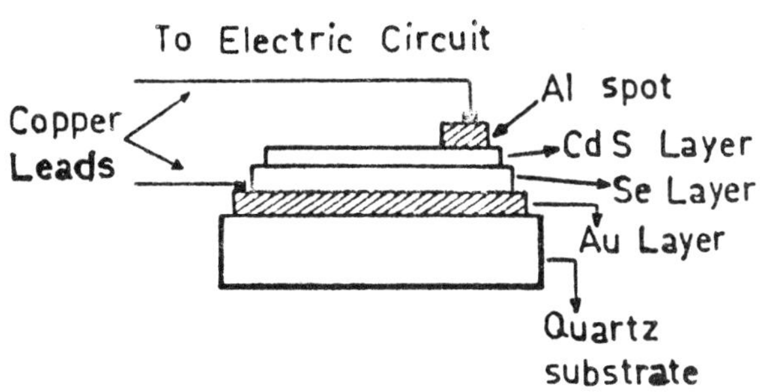

Figure 1 Configuration of the Prepared Se-CdS Heterojunction (not to scale)

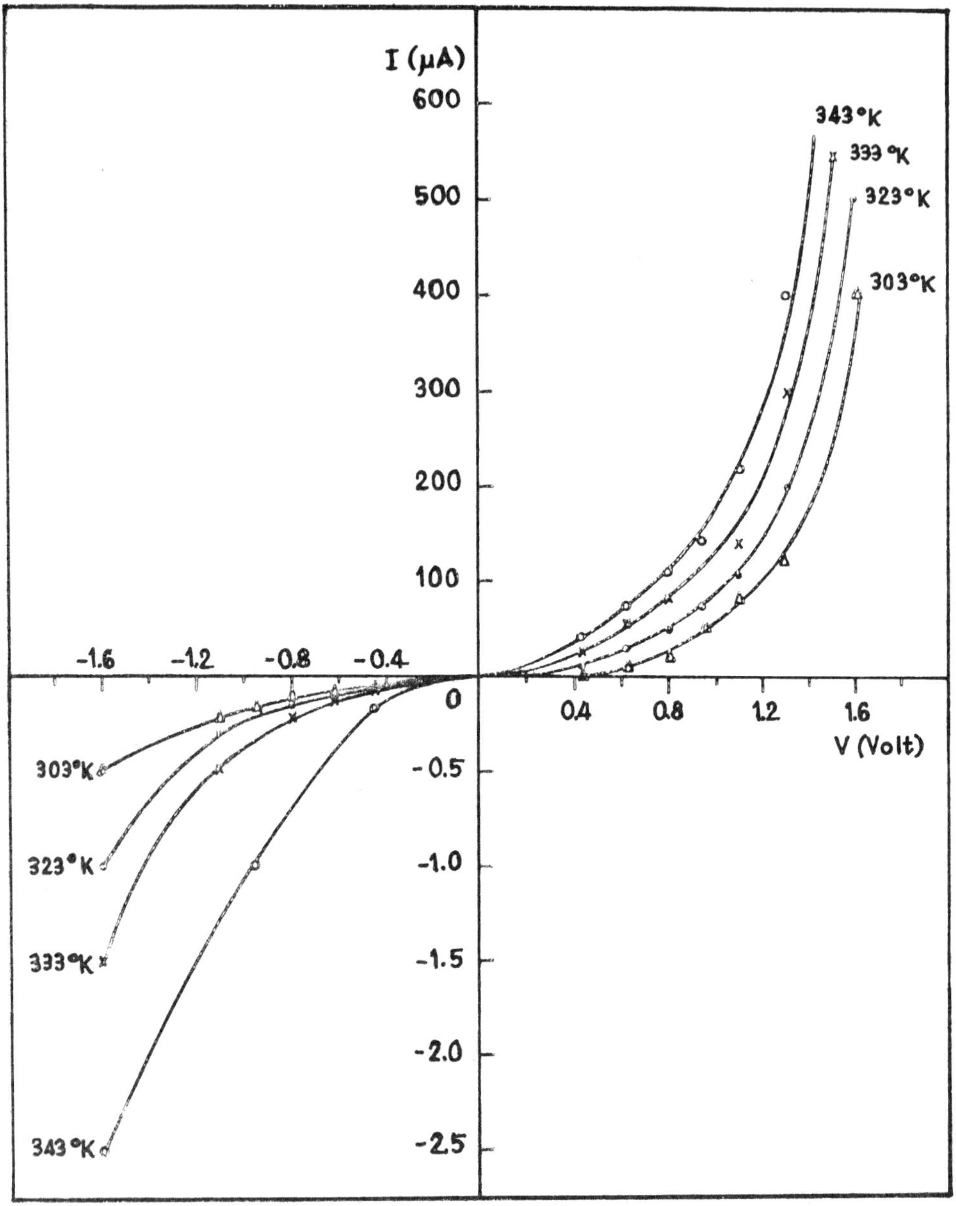

Figure 2 Current-Voltage Characteristics at Different Temperatures for Se-CdS Heterojunction

schematically in Fig. 1. For the illumination characteristics, the cell was illuminated from the CdS side using a tungsten lamp. The maximum power received by the sample surface was 0.45 mW/cm^2.

GENERAL CHARACTERISTICS

The current-voltage characteristics in the dark at different temperatures are shown in Fig. 2. The law of variation of I vs. V was reproducible on different cells, although for a given cell some instabilities of characteristics were observed. Such instabilities have been also reported by previous authors [5].

The semi-logarithmic plot of the forward characteristics shows a linear behaviour as shown in Fig. 3. Hence, the dependence of I on V can be expressed as:

$$I = I_o \exp \frac{eV}{\eta kT} \quad (1)$$

The values of η, determined from the high voltage region of the curves, are given in Table I.

In the lower voltage region, the curves are characterized by lower values of η. The break in the I-V characteristics, shown in Fig. 3, has also been observed on other heterojunctions (e.g., Ge - GaAs (6)) and attributed to a change in the transport mechanism.

The reverse - current voltage characteristics at different temperatures are shown in Fig. 4. These characteristics are usually represented by a power law $I - V^{\eta}$ for heterojunctions. In our case, η increases gradually and approaches the value 3 at high voltages. Similar results have been obtained by Knauss and Breslow [7] on tunneling junctions and interpreted by Simmons [8] and Holm [9] using the following theoretical expression:

$$I = \beta \ (V + \gamma V^3)$$

where β and γ are constants. Therefore, the theory predicts that η varies from 1 to 3 with increasing voltage as observed experimentally (Fig. 2).

The semilogarithmic dependence of the short circuit current I_{sc}, on the open circuit voltage V_{oc}, yields a straight line, as shown in Fig. 5, where $\eta = 1.85$. The transport mechanism in this range of V_{oc} values is expected to be different from that operative at high voltages, as indicated from the dark characteristics. Also, the photo-induced carriers are expected to influence η due to, for example, the photoconductivity of Se. This explaines the lower values of η determined from I_{sc} vs. V_{oc} curve, as compared with that of the dark characteristics (Table I).

The short circuit current, I_{sc}, is proportional to the light intensity I_L. Also, the open circuit voltage depends in a logarithmic way on I_L (Fig. 5).

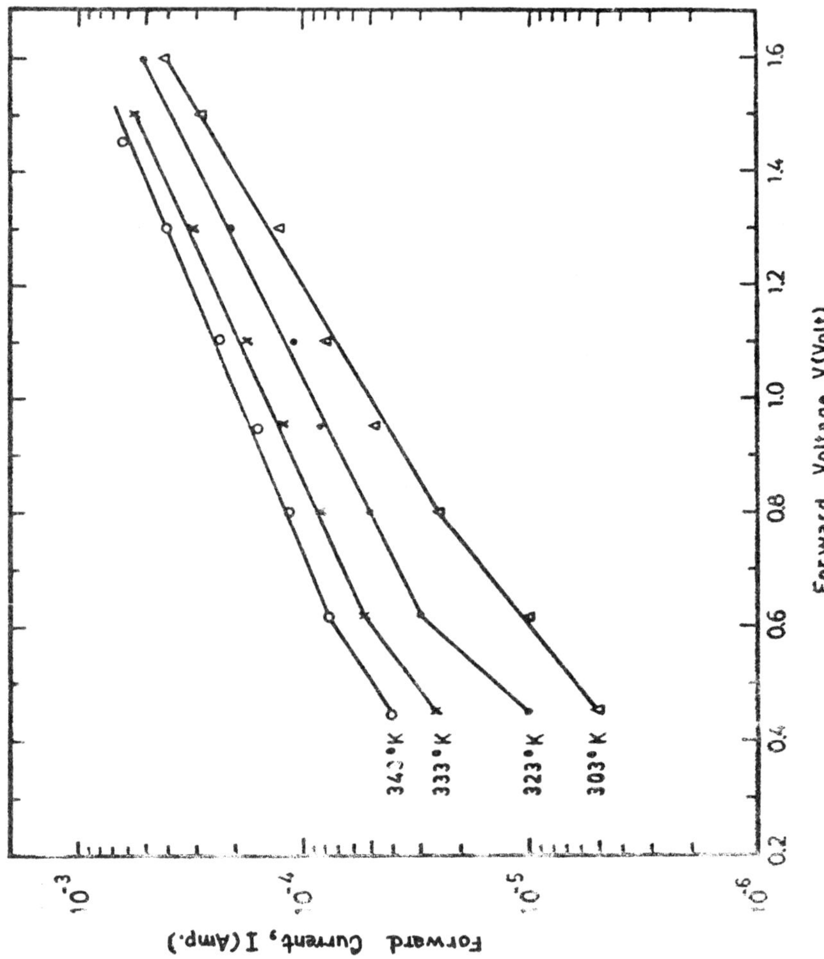

Figure 3 Forward Current-Voltage Characteristics at Different Temperatures for Se-CdS Cell

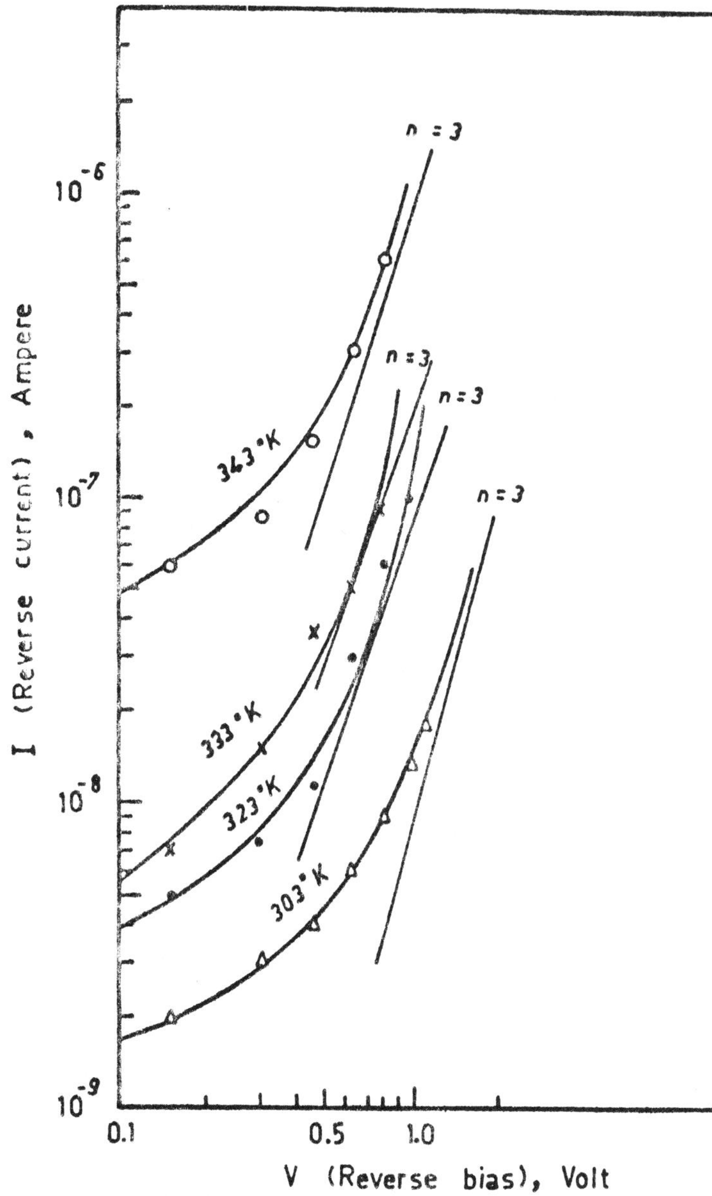

Figure 4 Reverse Current-Voltage Characteristics at Different Temperatures for Se-CdS Heterojunction

These dependencies are typical for abrupt p-n junctions. Over a wide range of light intensities

$I_f >> I_s$, V_{oc} is given by

$$V_{oc} = \frac{\eta kT}{e} \; i_n \left(\frac{I_f}{I_s} + 1\right)$$

where I_f is the photo-current and I_s is the dark saturation current. Taking into account that $I_f = B\ I_L$, where B is the proportionality factor, the expression (2) gives

$$V_{oc} = \frac{kT}{e} \; i_n \frac{BI_L}{I_s} \qquad (3)$$

The dependence of V_{oc} on the logarithm of light intensity (Fig. 6) yields a value of $\eta = 1.92$ in good agreement with that calculated from the I_{sc} vs. V_{oc} curve.

The short circuit current measured, while illuminating the sample with monochromatic light, is shown in Fig. 6 as a function of wavelenght λ. The spectral sensitivity is limited by photons with energies between the forbidden gaps of Se and CdS. The maximum response is obtained at $\lambda = 575$ mμ, very near to the maximum sensitivity of the human eye which is significant for practical purposes.

THE TRANSPORT MECHANISM IN Se-CdS HETEROJUNCTIONS

The current-voltage characteristics presented in Figs. 2-4 can shed some light on the transport mechanism in Se-CdS heterojunctions. The fact that the slope d (log I)/dV is temperature-dependent indicates a thermal transport mechanism; while the high values of η suggests the presence of positive reverse barrier. For this mechanism, the current-voltage reaction is given by

$$I = A \exp \frac{eK_2 (V - V_D)}{kT} \qquad (4)$$

where $$\frac{1}{K_2} = \eta = 1 + \frac{\varepsilon_2 N_{D_2}}{\varepsilon_1 N_{A1}}$$

where K_2V is the fraction of voltage V appearing on the high conductivity side of the junction (in our case CdS); N_{A1} and N_{D2} are the acceptor and donor concentrations in Se and CdS, respectively; ε_1 and ε_2 are the dielectric constants of Se and CdS, respectively, and V_D is the diffusion potential.

Following this interpretation, the diffusion potential V_D can be determined

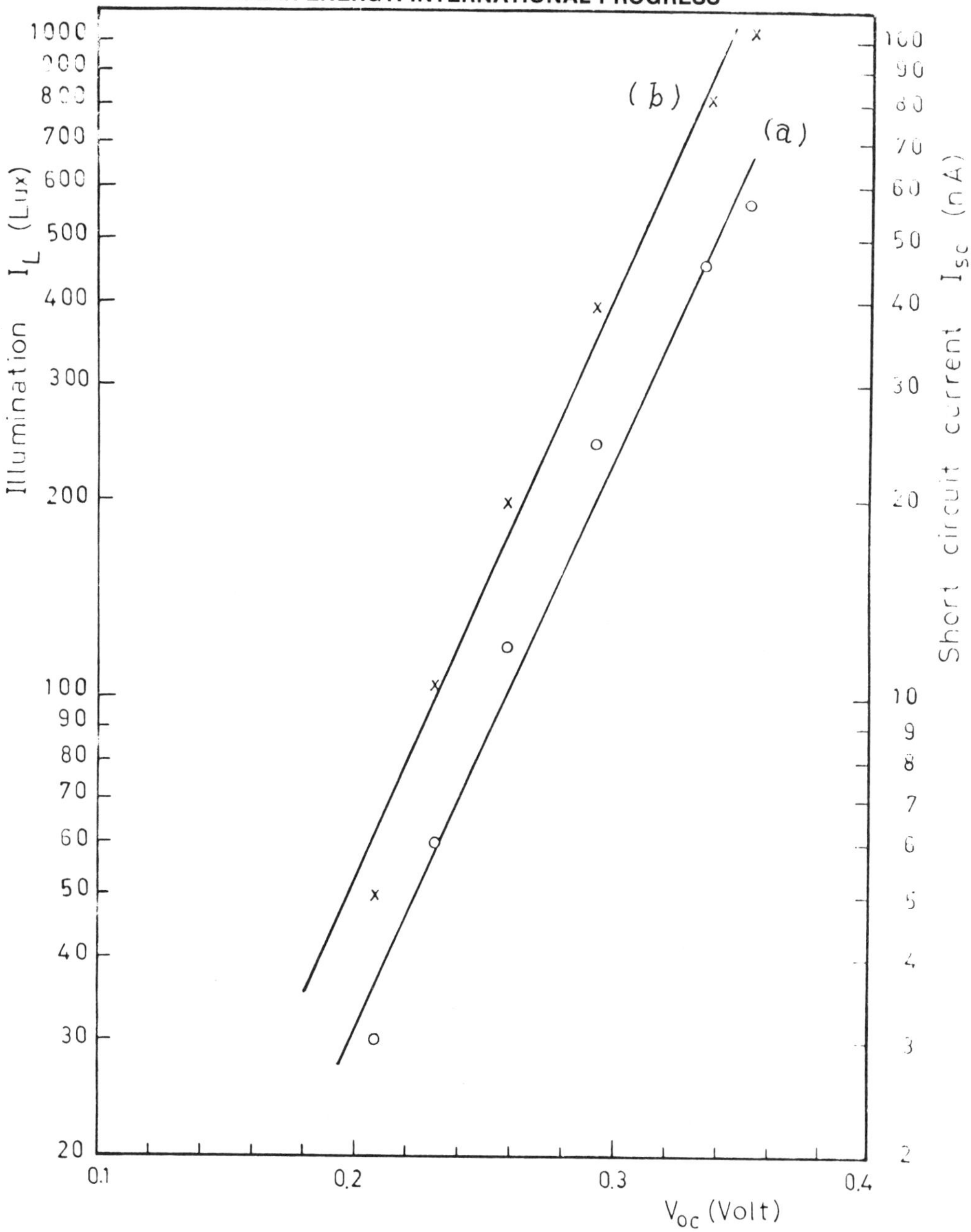

Figure 5(a) Variation of the Open Circuit Voltage V_{oc} With Illumination I_L

(b) Variation of the Short Circuit Current I_{sc} with Open Circuit Voltage V_{oc}

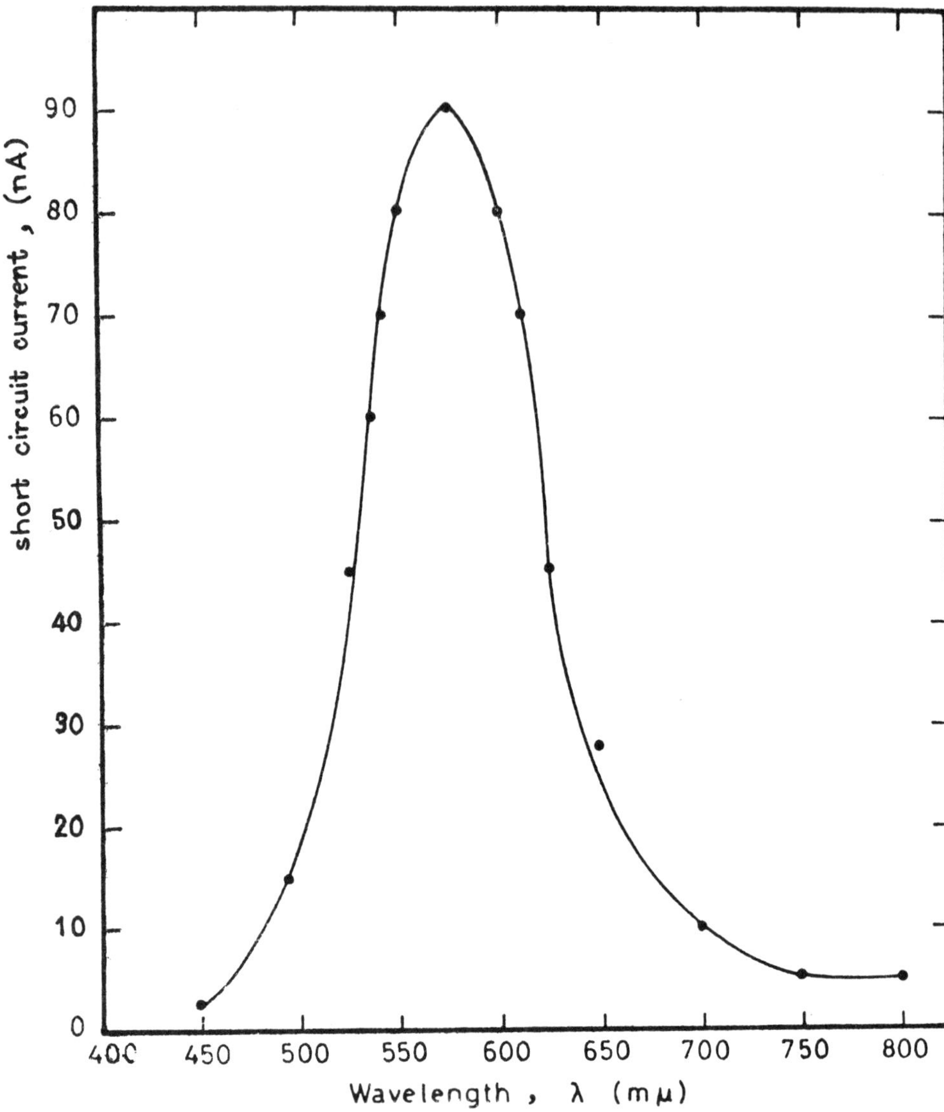

Figure 6 Spectral Response Curve For Se-CdS Heterojunction

Following this interpretation, the diffusion potential V_D can be determined experimentally by plotting log I against $1/\eta$ T. According to equation (4), this relation yields a straight line with slope $\frac{e}{K}$ $(V - V_D)$. The experimental lines in Fig. 7 give the values of V_D between 1.6 and 2.1 V.

A similar interpretation has been given by Anderson [6] for Ge - GaAs heterojunctions. In this case, η = 2.1 and 16 in the low and high voltage regions respectively. He attributed the value η = 2.1 to the recombination of carriers in the transition region in the presence of the negative reverse barrier. The value η = 16 is due to the injection of the carriers over a potential barrier in the presence of the positive reverse barrier (equation 4). Evidently, this interpretation implies the presence of a conduction bank spike. Unfortunately, we cannot provide a direct verification of this possiblity, since the electron affinity χSe of the hexagonal is not known. However, the criteria for the conduction band spike is χSe > χCdS, i.e.

$$eV_D = (\chi Se - \chi CdS) - (\delta_n + \delta_v) + E_{gSE} \quad (5)$$

so that the proposed model would be self-consistent if

$$eV_D > E_{gSE} - (\delta_n - \delta_v) \quad (6)$$

where δ_n and δ_v are the Fermi level depths in CdS and in Se, respectively. Inequality (6) can be tested as follows:

For non-degenerate semiconductors, as in our case, we have

$$\delta_i = kT \text{ Lu } \frac{n_i}{N_i}$$

where c and v denote the conduction and valence bands respectively, n_i is the carrier concentration and N_i the density of states given by

$$N_i = \frac{2\ (2\P\ m^*\ kT)^{\frac{3}{2}}}{h^3}$$

where m* is the effective mass.

The carrier concentration is related to the conductivity σ by

$$n_i = \frac{\sigma_i}{e\ \mu_i}$$

Therefore, using the experimental values of σ with equations (7) and (8), δ_i can be determined, knowing m*, and using reasonable values of μ.

For Se, using $m^* = 0.8\ m_o$, $\mu = 10\ cm^2$/volts/sec.,

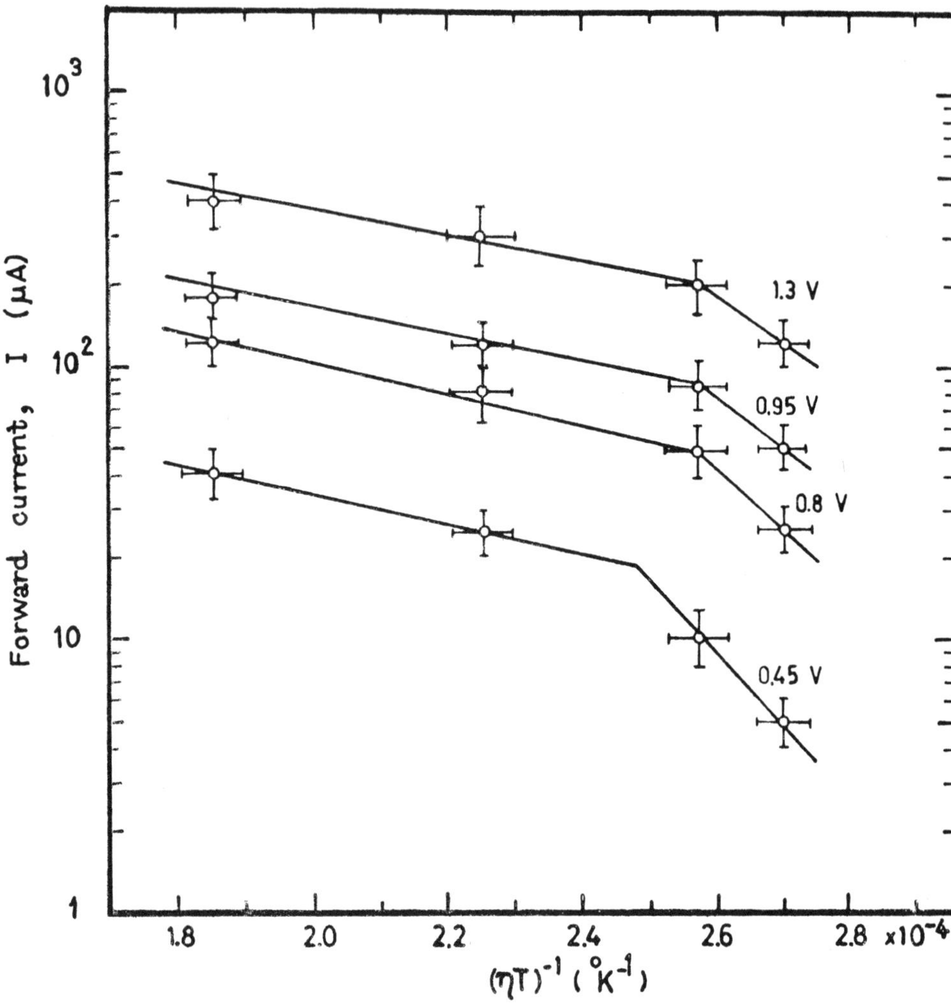

Figure 7 Dependence of Forward Current on $(\eta T)^{-1}$ for Different Values of Applied Voltage

$\sigma = 5 \times 10^{-6}\ \Omega - 1\ cm^{-1}$ and T = 300°K, we find $\delta_V = 0.40$ eV.

For CdS, using m* = 0.204 mo, $\mu = 340\ cm^2$/volts/sec., $\sigma = 10^{-3}\ \Omega^{-1}\ cm^{-1}$ and T = 300°K, we find $\delta_n = 0.30$.

Introducing these values of δ_V and δ_n into inequality (6), and using E_{gSE} = 1.9 eV yields $E_{gSE} - (\delta_V + \delta_n) = 1.4$ eV.

Since V_D = 1.63 to 2.15 eV as obtained experimentally, inequality (6) is verified and the proposed model is self-consistent.

To summarize, we have shown that assuming the existence of a conduction bank spike and taking into account the fact that the conductivity of CdS is higher than that of Se, the temperature dependence of the slope d(Log.I)/dV and the high value of n (11-15) could be explained. Using the same model, a value of $V_D > E_{gSE}\ (\delta_V + \delta_n)$ was determined which is consistent with the assumption of a conduction bank spike.

CONCLUSION

Se-CdS photovoltaic cells have been prepared using the vacuum evaporation technique and the current voltage characteristics in the dark have been investigated. Also, the short circuit current and the open circuit voltage have been studied as a function of illumination. The spectral sensitivity curve has also been recorded.

The results could be explained in terms of a self-consistent model in which the current is carried out by injection over a potential barrier (1.6 to 2.1 eV) in the presence of a conduction bank spike. The validity of this model may be confirmed by further investigation covering wider temperature range of measurements and including other characteristics.

REFERENCES

1. Preston, J.S., Proc. Roy. Soc. London 202A, 449, 1950.

2. Sakai, Y. and Kunioka, A., J. Inst. Elect. Engrs., Japan 82, 743, 1962.

3. Milnes, A.G., Feucht, D.L., Heterojunction and Metal Semiconductor Junctions, New York, 1972.

4. Aboullayev, G.B., Asadov, Y.G. and Mamedov. "The Physics of Selenium and Tullurium".

5. Kunioka, A. and Sakai, Y., Solid State Elect. 8, 961, 1965.

6. Anderson, R.L., Solid State Elect. 5, 341, 1962.

7. Knauss, H.P., Breslow, R.A., Proc. IRE, 50, 1843, 1962.

OPTIMIZATION OF GRADED DOPING CdTe SOLAR CELLS[+I]

A.S. Bouazzi
Laboratoire d'Energie Solaire
Belvedere, Tunis

ABSTRACT

From the CdTe band gap value (1.5 eV), one predicts a high efficiency for CdTe solar cells. The high surface recombination velocity and the small minority carrier lifetime make the present efficiency below 7% [1]. The existence of an electric field in the frontal region of CdTe solar cells can counterbalance the recombination effects by accelerating the minority carriers towards the junction. We study here a p-n CdTe junction with a graded frontal doping; its efficiency is computed to fix the optimum depth of the junction and the optimum frontal electric field resulting from the graded doping; two recombination parameters are taken into account: the surface recombination velocity and the minority carrier lifetime. The quantum responses of the optimized structure are simulated by the computer and the results are compared to those of the CdTe homojunction and to the graded band gap CdHgTe heterojunction.

INTRODUCTION

One of the directions in research in the photovoltaic domain is the realization of a structure with high efficiency as a goal to compensate for the high price. The efficiency of II-VI [2] and III-V [3] heterostructures - is studied. Internal electric fields or quasi-fields are used in these photovolataic structures to counterbalance the high recombination rate at the surface and in the bulk; and thus, the efficiency is higher. A gradient of doping in the frontal part of a photovoltaic structure gives rise to such an electric field; we study here a model of p-n CdTe photocell where the frontal doping is varying from the surface to the junction.

THE MODEL

The model diagram is represented in Figure 1; in the frontal region, the acceptor concentration is varying linearly from $N_A(O)$ at the surface to $N_A(d)$ at the junction which is at the depth d, while the donor concentration is constant at the back region. Some assumptions are taken to study this structure

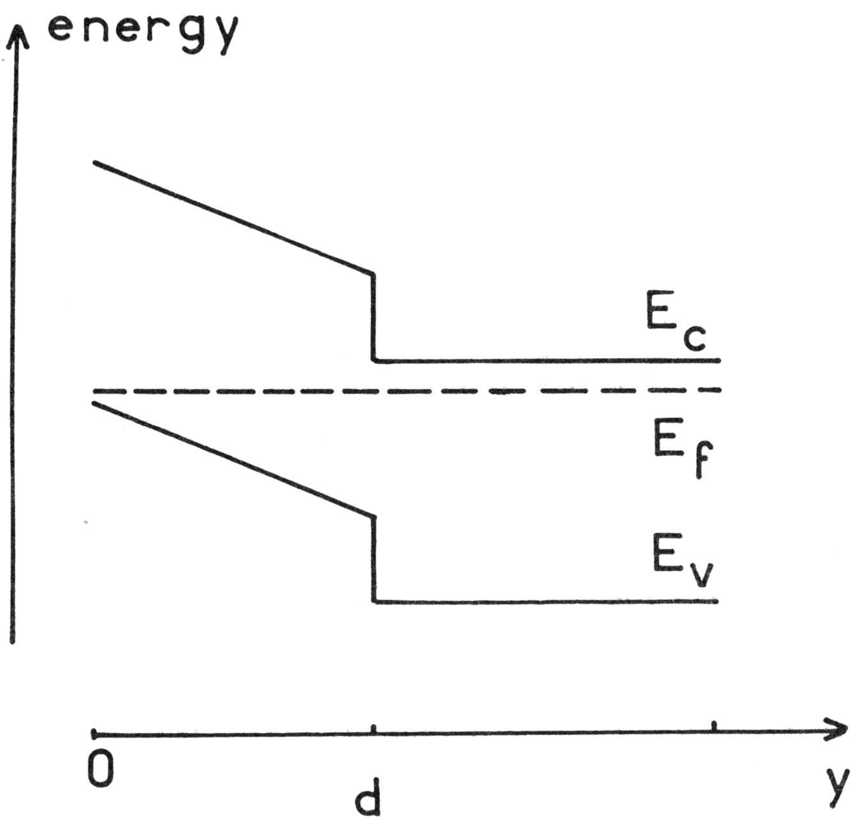

Figure 1 Band Diagram Of Graded Doped P-N CdTe Junction

model; we consider that the minority carrier lifetime does not vary with the doping and neither does the mobility, we neglect the space charge region phenomena.

The variation of the doping gives rise to an electric field E as:

$$E = \frac{1}{q}\frac{dE_c}{dy} = \frac{1}{q}\frac{dE_v}{dy}$$

if the electrical neutrality is conserved, we have:

$$-(E_v - E_{Fo}) = \frac{kT}{q} \text{Log} \frac{N_v}{N_A}$$

so we have:

$$E = \frac{E_v(d) - E_v(0)}{d}$$

and then:

$$E = \frac{kT}{q}\frac{1}{d} \text{Log} \frac{N_A(d)}{N_A(0)}$$

This electric field is not very strong and does not give rise to a space-charge concentration, but accelerates the minority carriers towards the junction.

To increase the electric field, we enhance the photocurrent. We also have to diminish the acceptor concentration $N_A(d)$ at the junction, but the less doped side imposes the saturation current. So, a low value of $N_A(d)$ will give us a low value of the open circuit voltage.

The photocurrent and the photovoltage vary in opposite ways with the electric field, and we need an optimization to fix the optimum electric field to have the maximum efficiency.

PHOTOCURRENT CALCULATION IN THE P-REGION

The minority carrier concentration is determined from the resolution of the diffusion equation. We have in the stationary case:

$$D_n \frac{\partial^2 n}{\partial y^2} + \mu_n \frac{\partial n}{\partial y} E - \frac{n}{\tau_n} + G_n(y) = 0$$

where n is the minority carrier concentration in excess, Dn, μ_n and τ_n are respectively, the diffusion constant, the mobility and the lifetime of the minority carriers (the electrons), and $G_n(y)$ is the generation function.

The general solution of this equation is:

$$n = C_1 \exp \Lambda\, y + C_2 \exp \Lambda_2 y + \frac{1}{\Lambda_1 - \Lambda_2} [F_1(y) \exp \Lambda_1 y + F_2(y) \exp \Lambda_2 y]$$

where

$$\Lambda_i = -\frac{\mu_n E}{2D_n} - (-1)^i [(\frac{\mu_n E}{2D_n})^2 + \frac{1}{L_n{}^2}]^{\frac{1}{2}}$$

$$\Lambda_i = \frac{E}{2kT} - (-i)^i \Delta$$

$$i = 1 \text{ or } 2 \text{ and } L_n = \sqrt{D_n \tau_n}$$

the functions $F_1(y)$ and $F_2(y)$ are given by:

$$F_i = (-1)^i \frac{1}{D_n} \int_o^y G(y) \exp (-\Lambda_i) \, dy$$

C_1 and C_2 are integrating constants determined by the boundary conditions:

$$\text{at } y = 0 \qquad q(\mu_n nE + D_n \frac{\partial n}{\partial y}) = qnS$$

$$\text{at } y = d \qquad n = n_{Po}[\exp \frac{qV}{RT} - 1]$$

where S is the superficial recombination velocity, $n_{Po} = \frac{n_i{}^2}{N_a}$ ni is the intrinsic carrier concentration and N_A is the acceptor carrier concentration in the p region; V is the electrostatic difference through the junction.

The current density under illumination is:

$$J_n = J_{1n} - J_{sn} (\exp \frac{qV}{kT} - 1)$$

J_{1n} is the photocurrent and J_{an} is the saturation current; J_n is determined by the equation:

$$J_n = q(D_n \frac{\partial n}{\partial y} + n\mu_n E) \quad \text{for } y = d$$

After simplification, we obtain the expression of the photocurrent and the saturation content in the P region as:

$$J_{1n} = \frac{qD_n}{2} \exp\left(-\frac{Ed}{kT}\right)\{F_1(d)\ [\frac{E}{1kT} + \Delta - \frac{S}{D_n}]$$

$$+ F_1(d)\ [\frac{E}{2kT} - \Delta - \frac{S}{D_n}]\}/\{(\frac{S}{D_n} - \frac{E}{1kT})$$

$$x\ sh\Delta d + \Delta ch\Delta d\}$$

and:

$$J_{sn} = qD_n \frac{n_i^2}{N_A} [\Delta \frac{S}{D_n} ch\Delta d + (\Delta^2 + \frac{ES}{2kTD_n}$$

$$-(\frac{E}{1kT})^2\ sh\Delta d]/\{(\frac{S}{D_n} - \frac{E}{2kT})\ sh\Delta d + \Delta ch\Delta d\}$$

We notice that if E = 0 in these equations, one obtains the classical equations of the p-n homojunction.

To determine $F_i(d)$, we have to determine $G_n(y)$ from the equation:

$$G_n(y) = -\frac{dI}{dy} = \alpha I$$

so we can write the photon flux versus y

$$I = I_o \exp\left(-\int_o^y x\ (y)\ dy\right)$$

$$I = I_o \exp(-\alpha y)$$

I_o is the incident photon flux and the reflection losses are neglected.

The absorption coefficient $\alpha(h\nu)$ is taken from the experimental curve of Kireev et al [4] and the curve of Lavagna et al [5]. The curve has been divided into three regions and an exponential approximation has been taken in each part like:

$$\alpha = \alpha_o \exp\left(\frac{h\nu - Eg}{\varepsilon}\right)$$

where α_o (cm^{-1}) and ε (eV) are the coefficients of the table I.

$$G_n(y) = \alpha I_o \exp\left(- \alpha y\right)$$

and

$$F_i(d,h\nu) = (-1)^i \frac{I_o(h\nu)}{D_n} \int_o^d \alpha e^{-\Lambda iy} e^{-\alpha y} dy$$

$$F_i(d,h\nu) = (-1)^{i+1} \frac{I_o(h\nu)}{D_n} \frac{\alpha}{\Lambda_i + \alpha} (\exp(-(\Lambda_1 + \alpha)d)-1)$$

PHOTOCURRENT DENSITY IN THE N-REGION

The expression of the hole current density is:

$$J_\rho = q\mu_\rho PE - qD_p \frac{\partial\rho}{\partial y}$$

With E = 0, the continuity equation gives:

$$\frac{\partial\rho}{\partial t} = G_\rho(y) - \frac{p}{\tau_\rho} - \frac{1}{q}\frac{\partial J_\rho}{\partial y}$$

in the stationary case:

$$D_\rho \frac{\partial^2\rho}{\partial y^2} - \frac{P}{\tau_\rho} + G_\rho(y) = 0$$

G(y) is the generation function.

One takes the boundary conditions as:

$$y = d, \quad \rho(d) = \frac{n_i^2}{N_D} (\exp \frac{qV}{kT} - 1)$$

and y = 1 ρ (1) = 0 (the superficial recombination velocity is taken infinite at the back side of the cell), and taking

$$J_\rho = -qD_\rho \frac{\partial\rho}{\partial y} \quad \text{for } y = d,$$

one obtains the expression of the hole photocurrent and saturation current

$$J_{1_\rho} = \frac{\alpha 1_p I_o(h\nu)q}{\Lambda - \alpha^2 1_\rho^2} e^{-\alpha(h\nu)d}$$

$$x\ (\coth\ (\frac{\Lambda-d}{l_\rho}) - \alpha(h\nu) l_\rho - \frac{\exp\ (-\alpha(h\nu)(l-d))}{\sin h\ (\frac{l-d}{l_p})}$$

and

$$J_{s_\rho} = \frac{qD_\rho}{l_\rho}\frac{n_i 2}{N_D} \coth \frac{l-d}{l_p}$$

OPTIMIZATION AND DISCUSSION

We calculate J_l as:

$$J_l(h\nu) = J_{l_p}(h\nu) + J_{ln}(h\nu)$$

and

$$J_l = \sum_{h\nu} J_l\ (h\nu)$$

with I_O taken for air mass one illumination.

The expression of the saturation current is:

$$J_s = J_{sp} + J_{sn}$$

The current voltage characteristic is given by:

$$J = -J_s\ (\exp \frac{qv}{kT}) - 1) + J_l$$

the product JxV has a maximum for V_M as:

$$(\Lambda + \frac{qV_M}{kT}) \exp \frac{qV_M}{kT} = \frac{J_l}{J_s} + \Lambda$$

and $J_M = J(V_M)$.

The maximum power delivered by the photocell is J_M x V_M and the efficiency is:

$$\eta = \frac{J_M \times V_M}{\text{incident energy}}$$

The numerical values used are the same used in reference [2]; μ_p = 40 $cm^2V^{-1}sec^{-1}$ and μ_n = 400 $cm^2V^{-1}sec^{-1}$ for high doped

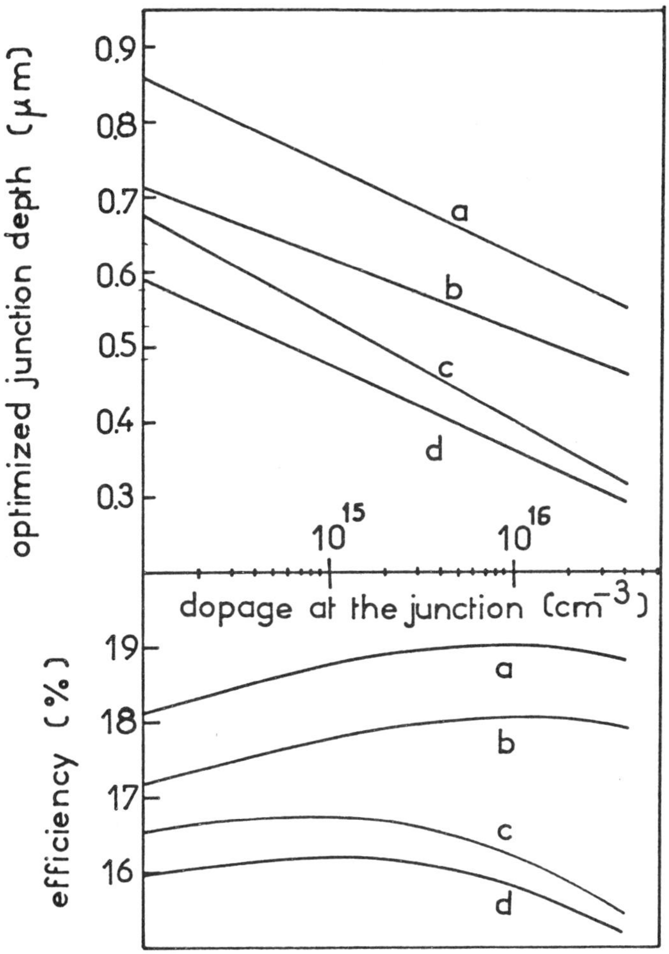

Figure 2 Optimum Doping At The Junction Vs. The Junction Depth And Corresponding Efficiency (Lower Part)

a) $S = 10^5$ cm sec^{-1}, $\tau_n = 10^{-9}$ sec

b) $S = 10^5$, $\tau_n = 5.10^{-10}$ sec

c) $S = 10^6$, $\tau_n = 10^{-9}$ sec

d) $S = 10^6$, $\tau_n = 5.10^{-10}$ sec

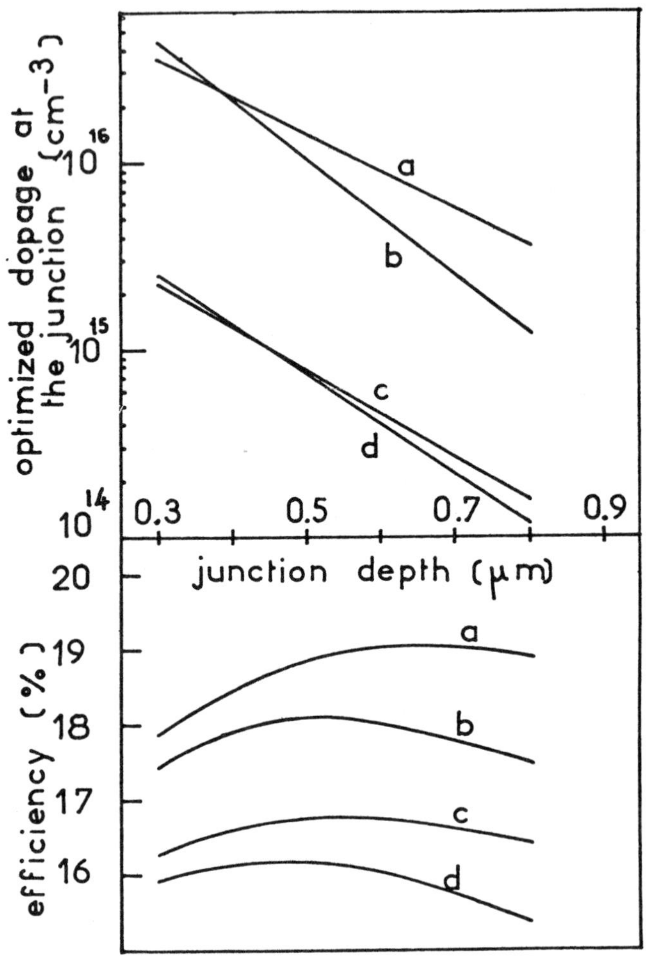

Figure 3 Optimum Junction Depth Vs. Doping At The Junction And Corresponding Efficiency

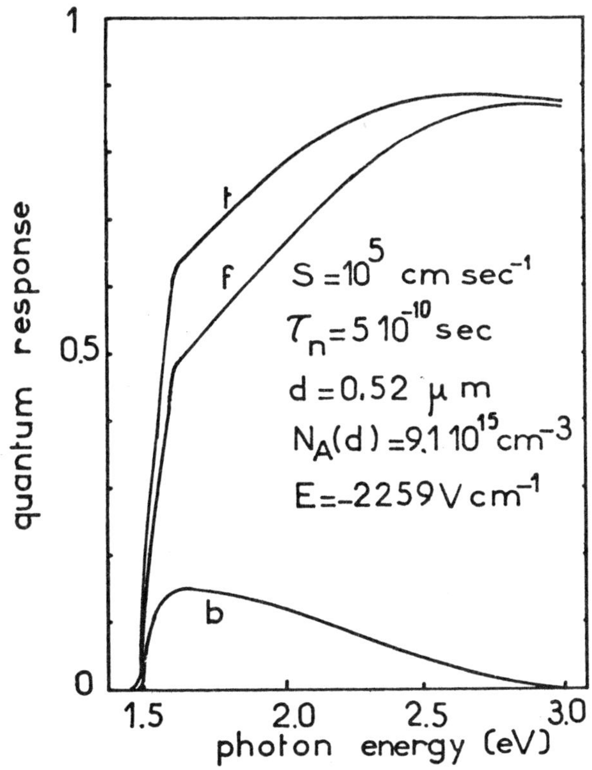

Figure 4 Quantum Response Of Graded Doped Junction

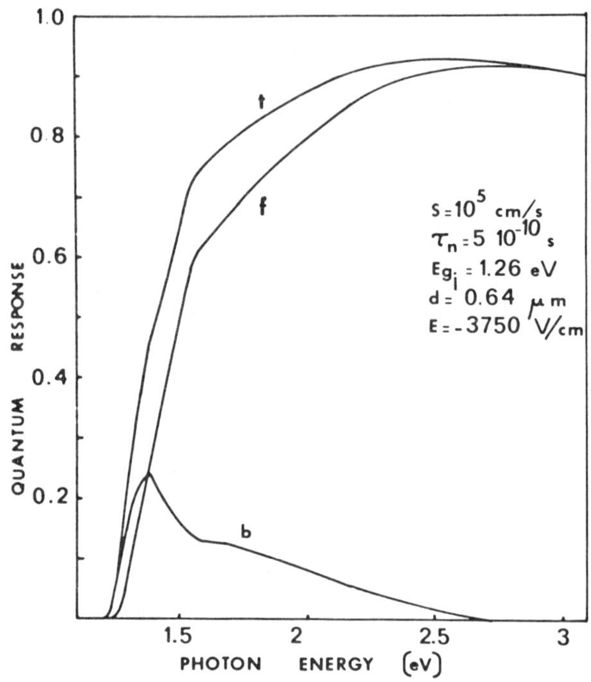

Figure 5 Quantum Response Of Graded Band Gap CdHgTe Structure

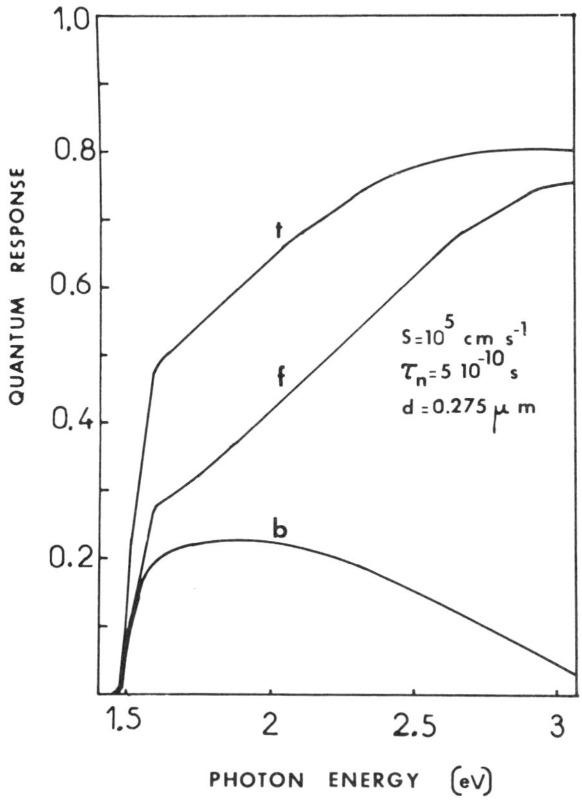

Figure 6 Quantum Response of CdTe Homojunctio

CdTe, the thickness of the back region is ℓ = 100 μm where the donor concentration is 5.5 $10^{17} cm^{-3}$.

The acceptor concentration at the front surface is $10^{18} cm^{-3}$ which is the limit of solubility of impurities.

The efficiency expression is computed with different values of S ($\Lambda 0^5$ - $\Lambda 0^6$ cm sec^{-1}) and τ_n (10^{-9} - 510^{-10} sec) with the goal being to optimize the thickness of the frontal region and the acceptor concentration at the junction; these two values fix the electric field in the frontal region.

The optimization is conducted in two steps. In the first step, we calculate the optimum doping $N_A(d)$, versus depth d (see Figure 2), and in the second step, we calculate the optimum depth d, versus doping $N_A(d)$ as shown in Figure 3.

For different recombination cases, the optimum couple ($N_A(d)$, d), which gives the maximum efficiency, is given by the intersection of the upper curves of Figures 2 and 3; the results are reported in Table II, where we report the results of the homojunction optimization.

The quantum response of the case b is computed and we obtain the Figure 4, which is to be compared to Figures 5 and 6, which are the quantum responses of graded band gap structure and homojunction, respectively.

We can conclude from Table II that the graded doped cell is more efficient than homojunction, and, when the recombination conditions get worse, the efficiency decreases in the second case more rapidly.

In the quantum responses, we remark that the graded doped cell is less sensitive to recombination for energetic photons, but in comparison with the graded band gap cell, we have lower quantum efficiency around the band gap value in the graded doped case; and, the quantum response for photon energy lower than 1.5 eV in the graded band gap case gives rise to an additional efficiency.

However, the graded doped structure can be realized using close-space-vapor-deposition (C.S.V.A.) method. To make the thin p-CdTe film we add an appropriate treatment to realize the doping gradient.

REFERENCES

1. Mimila-Arroyo, J, Bouazzi, A and Cohen Solal, G, J. Phys. Appl. 12, 2, 423 (1977).

2. Bouazzi, A., Mimila-Arroyo, J. and Marfaing, Y., Rev. Phys. Appl. 13, 3, 53 (1978).

3. Hutchby, J.A. and Fudurich, R.L., J. Phys. Appl. 47, 7, 3140 (1976).

4. Kireev, P.S., Volkova, L.V., Volkov, V.V. and Platonov, Y.V., Sov. Phys. Semic. 6, 1, 109 (1972).

5. Lavagna, M. and Picque, J.P., Rapport de stage de DEA, Lab. de Phys. des Solides CNRS - Meudon, Meudon, France (1975).

This work was partially carried out in Laboratoire de Physique des Solides CNRS - Meudon, Meudon, France.

TABLE I

$h\nu$ (eV)	α_0 (cm^{-1})	ε (eV)
$h\nu - E_g \ll 0.004$	2630	0.00986
$0.004 \ll h\nu \ll E_g$ 0.1	3741	0.0752
$0.1 \ll h\nu - E_g$	12300	0.719

TABLE II

EFFICIENCIES AND PARAMETERS OF THREE PHOTOCELLS OPTIMIZED ON COMPUTER

	Graded band gap CdHgTe structure						Graded doped CdTe structure				Homo-junction CdTe	
CASE	S (cm/s)	τ_n (sec)	d_{op} (μm)	$(E_{gj})_{op}$ (eV)	E (eV/cm)	η (%)	d_{op} (μm)	$N_A(\delta d)_{op}$ (cm^{-3})	E (V/cm)	η (%)	d_{op} (μm)	η (%)
a	10^5	10^{-9}	0.78	1.26	-3077	19.75	0.64	7.610^{15}	-1906	19.05	0.325	15.00
b	10^5	5.10^{-10}	0.64	1.26	-3750	18.75	0.52	9.110^{15}	-2259	18.05	0.275	14.65
c	10^6	10^{-9}	0.65	1.196	-4677	18.50	0.53	6.410^{14}	-3469	16.75	0.100	11.43
d	10^6	5.10^{-10}	0.54	1.208	-5407	17.90	0.45	1.010^{15}	-3838	16.20	0.100	11.43

SAUR-VIDYUT KOSH - THE SOLAR CELL

M. Sharon, S.G. Sharan, B.M. Prasad,
A.P. Bholagir and R.R. Pradhananga
University of Poona
Pune, India

ABSTRACT

For the first time, a fully solar galvanic cell has been developed which is reversible with solar radiation and has been named "Saur-Vidyut Kosh", meaning, by solar battery. The different types of photochemical reactions have been studied and concentration cells have been made. These cells can be charged by the sun within one hour and can be used for drawing the current. The following types of cells have been made successfully:

(1)	$\overset{+}{C}$	‖ KI, H^+, Grafite powder exposed	‖ KI, H^+, Grafite powder dark	‖ C	0.03 volt 4-6 mA
(2)	$\overset{+}{C}$	‖ KI, H^+, Grafite powder exposed	‖ Zn^{++}	‖ Zn	1.25 volt 60-100 mAmP
(3)	$\overset{+}{C}$ or Pt	‖ KI, Fe^{2+} H^+, Fe^{3+} exposed	‖ KI, Fe^{2+} H^+, Fe^{3+}	‖ C or Pt	40 mv 10-20 AmP
(4)	Pt	‖ Ce^{4+}, Ce^{3+} H^+ exposed	‖ Ce^{3+}, H^+ Ce^{4+}	‖ Pt	20 mv 30-35 AmP

A matrix of 7 cells (parallel) x 7 cells (in series) of type No. 2 gives 9.0 volts and 300 AmP current which is enough to run transistorized equipment.

INTRODUCTION

Rabinowitch [1,3] was the first person to demonstrate the possibility of a solar galvanic cell. When a solution of ferrous sulfate (10^{-2} M), H_2SO_4 (0.1 M) and thionine (10^{-4} M), exposed to 590 nm is combined, a colourless solution is obtained due to the following reaction:

$$Fe^{2+} + Th + H^2 \rightleftarrows Fe^{3+} + \text{semi-thionine}$$

By solvent extraction (ether/water phase), Rabinowitch separated semi-thionine while it was being formed in light from the thionine. The ether layer contained the colourless species of semi-thionine, whereas the water layer contained all inorganic ions and unconverted thionine. Then they joined the two solutions through a salt bridge and a potential was obtained through platinum electrode dipped in each solution. This gave a potential of 0.3585 volts. After drawing the current, two solutions were mixed and re-exposed to regenerate colourless semi-thionine. In this way, he could continue to repeat the cycles. After this historic work, there have been a series of publications on the thionine system. But unfortunately, all these publications concentrated their work by using an artificial light source giving a wavelength (590 nm) of range equivalent to absorbance wavelength of thionine. It would be quite interesting to discuss their results [4,9], but a doubt comes as to whether thionine would behave in the same manner if it is exposed to solar rays; because, solar rays have a band spectrum of which 590 nm is about 15% of the total spectrum. Nevertheless, Clark and Eckert [9] have postulated the following reaction with thionine:

$$TH^+ \xrightarrow{hr} TH^+ \text{ (excited)} \tag{1}$$

$$TH^+ \text{ (excited)} + H^+ + Fe^{2+} \xrightarrow{hr} {}^\circ TH_2^+ + Fe^{3+} \tag{2}$$

$$^\circ TH_2^+ + {}^\circ TH_2^+ \rightarrow TH^+ + TH_3^+ \quad \text{dark} \tag{3}$$

$$Fe^{3+} + TH_3^+ \rightarrow {}^\circ TH_2^+ + H^+ + Fe^{2+} \tag{4}$$

where TH^+ is purple colour thionine, $^\circ TH^2$ is colourless semithionine and TH_3^+ is a colourless lencothionine. In order to keep reaction (2) under control, reaction (3) and (4) should be avoided. In the paper, the authors have used a monochromatic light of 578 nm, with an intensity of 18.5 mW/cm^2 for the experiment on this type of system. The efficiency of the cell was reported to be 10^{-3}%. They have not, however, mentioned the reversibility of their results over continuous long hours of exposure (i.e., in terms of hours or days of irradiation). Peter et al [10] have used SnO_2/Pt electrode to improve the efficiency of the thionine system. They also used a monochromatic light from 150 w Xenon Camp. Tadayoshi et al [11] have also shown the current characteristics of a thionine system using a Xenon lamp. They have shown a sharp decrease in current upon shutting off the electric lamp.

Finally, Terence and Yim [12] have used a ferrous/ferric system using Rhodamine-B dye. They also used the tungsten lamp for their work. They have reported that one can constitute a solar galvanic cell with ferric (0.1 M) ferrous (1.3×10^{-4} M) and Rhodamine-B (1.17×10^{-3} M) in HCl solution (104 Molar). They also observed a potential rise in 25.0 minutes of irradiation.

Since all these works, including the work done by Rabinowitch, are performed by an artificial light of a particular wavelength, a doubt develops whether all these observations are valid if experiments are to be carried out in solar rays. Because, when the cell is manufactured, it will have to be used in solar rays and not in a tungsten lamp or a Xenon lamp.

Furthermore, the thionine system does not give any storage property to the cell, which means that a solar cell developed on ferrous-thionine cannot be completed with a photovoltaic cell, because the latter has a 2-300 mv and 500 mA current with an efficiency of 15-18%. Thionine reaction, being a very fast one, will have to design a method to separate the leuco (or semi) thionine from thionine [1,3]. If this is not possible, then one has to find a different type of photochemical reaction which is stable and has a storage property. The present paper is aimed to clarify these two aspects.

MATERIALS

A.R. grade chemicals were used without further purification, except for thionine, which was purified by the method discussed by Clark [9]. Rhodamine-B was used without any purification.

EXPERIMENT AND RESULTS

Our first object was to reproduce the iron/thionine work of Clark [9], Peter et al [10], Tadayoshi et al [11], Rabinowitch [1], Hardwick [5] and of Ainsworth [8] using solar rays directly. We regret to report that none of these experiments, except of Rabinowitch's, could be reproduced in solar rays. After half an hour of prolonged exposure, the entire colourless solution became purple-blue, similar to the original colour. This solution could not then be converted to colourless semi-thionine form. The concentration of ferrous, ferric and thionine, were kept the same as reported by these authors. However, results could be reproduced if the solutions were exposed to sunlight for 1.0 minute and then covered in the dark. Therefore, intermittent exposure to sun rays could show a current; but , thionine became blue after 3-4 days of continuous intermittent exposure. We think that the rate of back reaction is faster than the rate at which one could draw the current from this type of reaction. Hence, the system becomes ineffective after a prolonged exposure. The back reaction of this system is very fast. Therefore, this system cannot have a strong capacity unless by some way, the back reaction 2 could be made very slow, and reaction 3, 4 and 5 could be either completely minimized or completely prevented. We are working on these lines. It is also obvious that in the future, we should concentrate our work by using solar rays only and not by using any artificial source of light.

It was further observed that when ferrous (10^{-2} M) thionine (10^{-5} M) at pH 3, was exposed to Hg - UV light, the purple colour became colourless; but after a long exposure, the blue colour does not reappear after keeping the solution in the dark. In fact, few precipitations are observed after one week's exposure, suggesting that thionine may be getting completely degraded by UV light. This suggests that thionine should not be exposed to UV light. Therefore, a cell designed to work on thionine should be able to filter out UV light.

In order to investigate the effect of solar photon/or/UV photon radiation

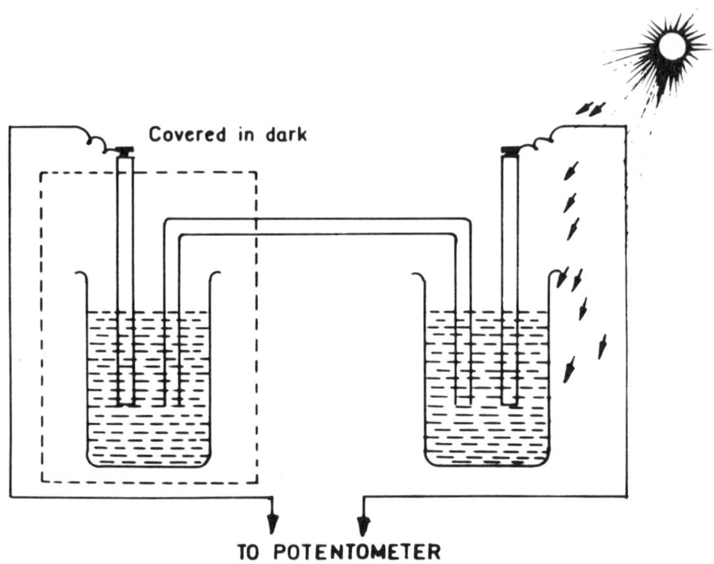

Fig. No. 1 A solar cell to study the photo chemical reactions and to measure the potential developed due to such reactions.

on a few more photochemical reactions, the following cell was designed, as shown in Figure 1. Two glass beakers were taken and 50 ml of the redox solution of the same concentration was added. A salt bridge was used to connect these two beakers. C or Pt electrodes were used as electrodes. One beaker was covered with black paper and the other beaker was exposed to either sun rays or to a UV (60 watt) bulb. The potential across the two platinum electrodes was measured. At the beginning of the experiment, the potential was almost zero, which increased on exposing one beaker to sunlight or to the UV light. This increase is due to the production of other oxidised/or reduced species of the chemical used in the beaker.

Ceric/Cerous Solution

Ce^{4+} ammonium sulfate saturated solution and 1.0 ml of $HClO_4$ acid were used for the purpose. The potential developed due to UV exposure is higher than the sunlight exposure, see Figure 2. But, the electrode potential fell down after 2-3 hours of exposure. This suggests that in the beginning, Ce^{3+} is produced. But, its conversion to Ce^{4+} is faster than the production of Ce^{3+} ions. This hinders the attainment of an equilibrium condition. Rudolph and Henry [13] have suggested the following reactions:

$$2\ Ce^{4+} + H_2O \xrightarrow[hv]{4000\ \mathring{A}} 2\ Ce^{3+} + 2\ H^+ + \tfrac{1}{2}\ O_2 \qquad (5)$$

$$Ce^{3+} + H_2O \xrightarrow[HClO_4]{2537\ \mathring{A}} Ce^{4+} + OH^- + \tfrac{1}{2}\ H_2 \qquad (6)$$

In the present work, no visible evolution of O_2 or H_2 was observed. Therefore, unless the rate of production of Ce^{3+} is made equal to its oxidation to Ce^{4+}, this redox reaction cannot be used for photogalvanic cells.

Ferrous/Ferric System

Ferrous ammonium sulfate solution (10^{-2}) and 1.0 ml $HClO_4$ acid were used for this system. The rate of increase of electrode potential with time is shown in Figure 2. The magnitude of electrode potential is very low, suggesting that oxidation of Fe^{2+} is not very efficient by UV or solar rays. However, the saturation point suggests that there is an equilibrium condition between Fe^{3+}/Fe^{2+}. This redox chemical can therefore be used for making a photogalvanic cell. The initial potential is not zero, because carbon electrodes were not of a very identical nature.

D-Tartaric Acid/L-Tartaric Acid

D-tartaric acid saturated solution in water was taken in both beakers and the electrode potentials were measured with time, see Figure 2. There is a large fluctuation in this electrode potential; nevertheless, this system does show some saturation value. The mechanism of such equilibrium is difficult to explain.

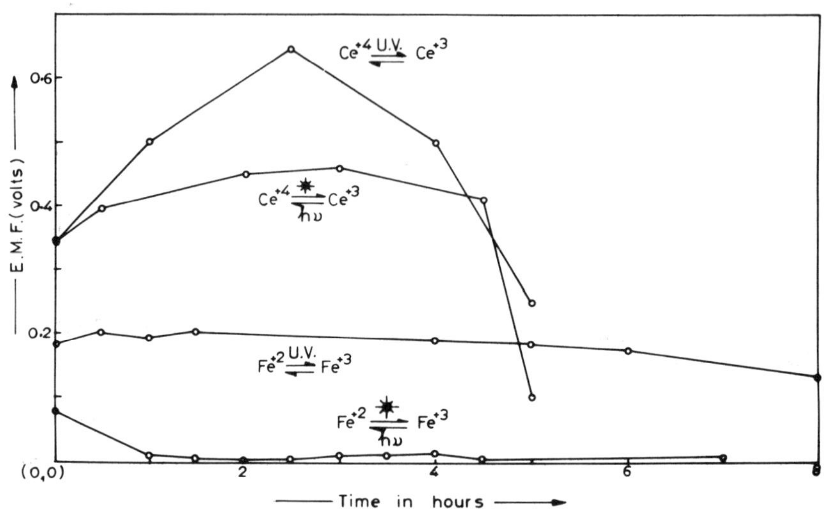

Fig. No. 2 Photochemical reactions

$Ce^{4+} \xrightarrow{hv} Ce^{3+}$

$Fe^{2+} \xrightarrow{hv} Fe^{3+}$

Iodide/Iodine System

A saturated solution of KI, along with 10.0 ml in $Na_2S_2O_3$ solution, was taken in one beaker (dark). In the other beaker, only a saturated solution of KI was added. The cell was assembled and the potential was measured with the time of photon irradiation, see Figure 4. The graph shows a saturation value after about 3-4 hours or irradiation. The high electrode potential suggests that oxidation of iodide is favoured by UV light. However, the oxidation with a solar photon is not bad. If photogalvanic cells are made of an iodine/iodide system, one can expect to get an equilibrium condition within 3 hours of irradition, which is quite satisfactory.

The possible mechanism of the reaction can be given as:

$$KI \rightarrow K^+ + I^- \tag{8}$$

$$I^- \xrightarrow[2400-4000\ A]{h\nu} I^- \tag{9}$$

$$I^{*-} \longrightarrow \tfrac{1}{2} I_2 + e^- \tag{10}$$

Ferrous Iodine/Ferric Iodide System

A solution of 10^{-2} M ferrous salt, iodide (10^{-2} M), ferric (10^{-2} M), were kept in two beakers, and one beaker was exposed to sunlight. The increase potential was measured. The yellow colour of solution changed to a colourless one within five minutes of exposure. The increased potential is shown in Figure 5. This experiment was done with a carbon electrode as well as with a platinum one. The platinum electrode gave more reproducible results; hence, Figure 5 is used with a platinum electrode.

Design Of A Photogalvanic Cell

From these redox reactions, it can be concluded that Fe^{2+}/Fe^{3+}, iodide/iodine and Fe^{2+}, $I_2Fe^{3+}I^-$ system would be the best for making a photogalvanic cell. Various designs were tried to make an iodine/iodide cell, but the following design gave the best results. A slight addition of thallium ion (in microgram) accelerated the production of iodine. Addition of electrolytic grade grafite helped to decrease the internal resistance of the cell, as well as helping in distributing electrons homogenously throughout the cell. Two types of photogalvanic cells were prepared, an electrochemical cell (using $Zn/AnCl_2$ as one of its half cell), and a second one as a concentration cell.

Electrochemical Cell

The design of the cell is shown in Figure 6. The half cell is made of a 100 ml beaker containing 50 ml of saturated solution of KI, grafite powder and 1.0 ml conc. HCl. In this beaker, a clay cylinder is dipped, containing a zinc rod and zinc chloride solution. In this beaker, a carbon electrode is used. This cell gives a potential of 1.29 volts and current of 60-100

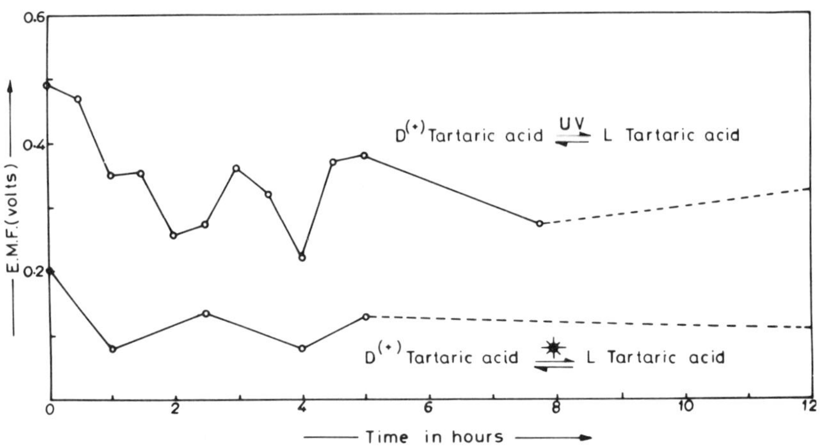

Fig. No. 3 **Photo chemical reactions**

(D) tartaric acid $\xrightarrow{hv}$ (L) Tartaric acid

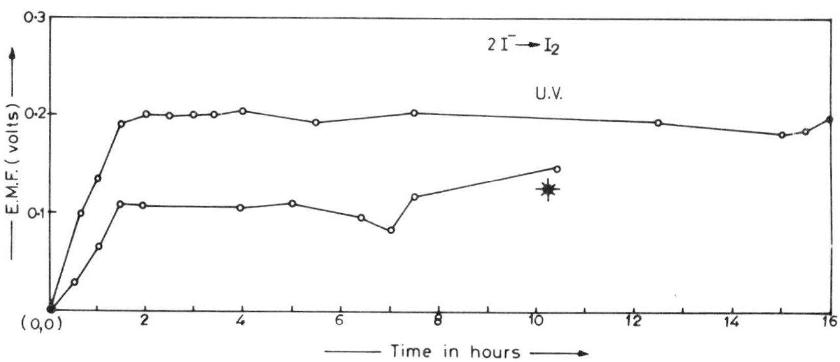

Fig. No. 4 Photo chemical reactions

$$I^- \xrightarrow[H^+]{hv} \tfrac{1}{2} I_2$$

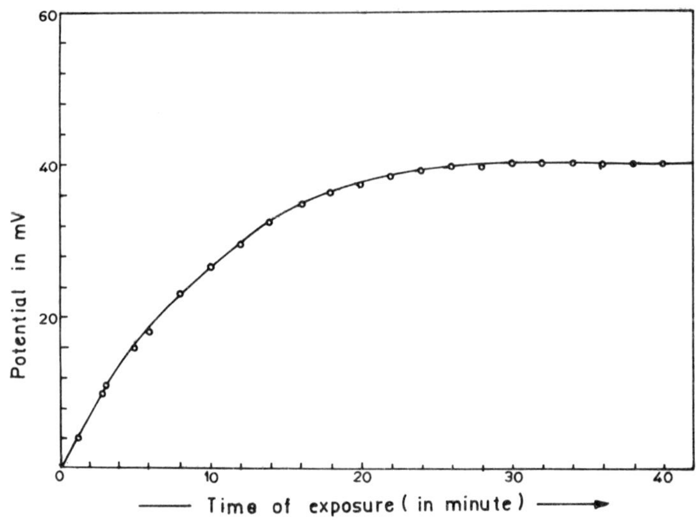

Fig. No. 5 Photo chemical reaction

$$Fe^{2+} + \frac{1}{2}I_2 \xrightarrow{h\nu} Fe^{3+} + I^-$$

mAmp. Theoretically, one should expect a potential of 1.30 volts (0.563 due I_2/I^- + 0.763 due Zn/Zn^{2+}). The characterisitc of current and potential of this cell is given in Table 1. When a matrix of 7 cell (parallel) and 7 cell (in series) is made, it gives a potential of 9.0 volts and the current of about 300 mAmP, see Figures 7 and 8. This was enough to drive the transistor radio or transistorised tape recorder. The mechanism of reaction can be written as:

$$I^- \xrightarrow{hv} I^{-}* \qquad (11)$$

$$*I^- \xrightarrow[H^+]{} \tfrac{1}{2} I_2 \qquad (12)$$

or

$$Tl \xrightarrow{hv} *Tl^+ \qquad (13)$$

$$*Tl + I^+ \longrightarrow *I^+ + Tl^+ \qquad (14)$$

$$*I^- \xrightarrow[H^+]{} \tfrac{1}{2} I_2 \qquad (15)$$

Electrochemical reaction can be written as

$$I_2 + 2e^+ \longrightarrow 2I^- \quad E^\circ + 0.56 \qquad (16)$$

$$Zn \longrightarrow Zn^{2+} + 2e^- \quad E^\circ + 0.76 \qquad (17)$$

Reaction (16) is reversible with a solar photon, whereas reaction (17) can be made reversible by the d.c. electroplating of Zn^{2+} into a Zn plate.

The potassium iodide compartment is fully reversible and can be charged by solar photons to recovert the converted iodide ions to iodine atom. When the zinc rod has been completely consumed, the zinc chloride can be taken out of the half cell and zinc can then be deposited back onto an electrode. These two reversible processes make this solar cell entirely chargeable.

If the potassium iodide compartment is sealed to stop the evaporation of iodine into the atmosphere, and if the semi-permeable membrane can stop the diffusion of zinc chloride into the iodide compartment, then the life of the cell can be prolonged for many years.

Concentration Cell

The design of the concentration cell was similar to what has been described earlier, except that the zinc/zinc chloride solution was replaced by saturated thiosulfate solution and a carbon electrode. Thiosulfate solution is used to reduce the iodine to iodide ion so that the concentration of iodine is maintained at almost zero in a clay cylinder and a high concentration in potassium iodide solution. Such a cell gave a potential of 0.15 volts and a current of about 4-8 mAmp, depending upon the internal resistance of the cell. The internal resistance of the cell can be further decreased, but the main drawback of this cell is that one has to continuously add sodium thiosulfate in the clay cylinder. Moreover, the porous

TABLE NO. 1

Material of the cell	Potential volts	Current mAmP	Remarks
Zn | $ZnCl_2$ | Water HCl | C^+	0.80	0.4	Current Falls down rapidly
$\overline{Zn}$ | $ZnCl_2$ | Water HCl Grafite | C^+	1.0	2.0	Current Falls down rapidly
$\overline{Zn}$ | $ZnCl_2$ | Water HCl Grafite KI | C^+	1.25	3.0	But on exposure to strong sun light the current increases to a maximum value of 60 M mAmP, depending upon the concentration of iodine produced.

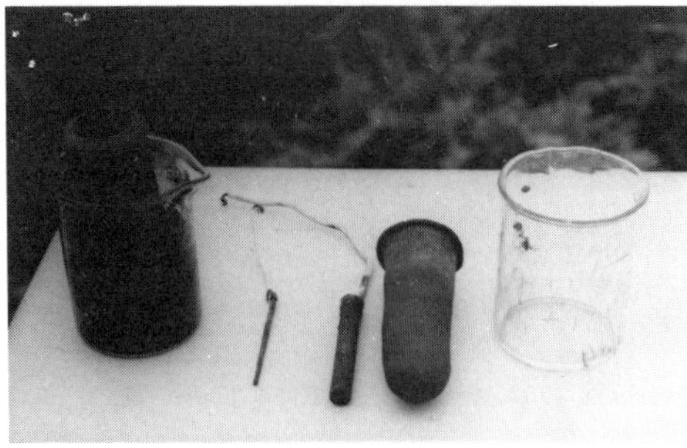

Fig. No. 6 Photograph of a $Zn / ZnCl_2 // KI$, Grafite power, H^+ / C cell using a clay cylinder as a membrane to separate the two half cells.

Fig. No. 7 A 7 x 7 matrix cell for C / KI, Grafite powder, $H^+ /$ Clay $/ ZnCl_2 / Zn$ using a individual cell

Fig. No. 8 A matrix of 7 x 7 cell for

C / KI, Grafite powder, H^+ / Clay / $ZnCl_2$ / Zn

using 7 pots in each pot there are 7 cells dipped in on KI solid vatt.

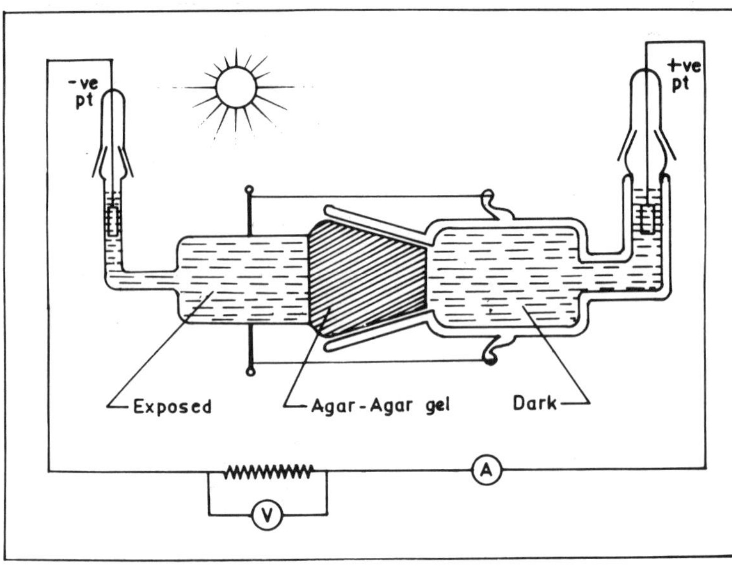

Fig. No. 9 A concentration SAUR VIDYUT KOSH of Fe^{2+} + I_2 system.

cylinder should be such that the thiosulfate ion does not diffuse out into the potassium iodide compartment.

Ferrous/Iodine System

With the $C/KI/Zn^{2+}$ Zn cell, we concluded that the iodine/iodide system can be used as a reversible system,and,if coupled with a suitable redox system, one can hope to get a full redox type cell. The only progress required in this direction is to find a photochemical redox system, which could be coupled with an iodide/iodine cell. We have had some success in this direction, the details of which are under preparation for publication.

The other redox system which has been tried is the ferrous/iodine system. Ridcal [14] and Kistiakowsky [15] studied the photochemical reactions

$$2\ Fe^{2+} + I_2 \underset{\text{dark}}{\overset{h\nu}{\rightleftarrows}} 2\ Fe^{3+} + 2I^- \qquad (18)$$

where they used ferric ammonium sulfate, H_2SO_4, ammonium sulfate, KI solution, and exposed it to a light of wavelength, 366, 436 and 546 nm. On exposure with these lights, ferric and iodide ions were formed. It was therefore thought to carry out this experiment in sun rays in order to observe the feasibility of this reaction and its durability. The experiments were carried out to find out the best concentration range of each species of these reactions, and, to find out any suitable photoactivator dyes to trigger the reaction in a short period of solar exposure. Two dyes were tried, thionine and Rodamine-B. However, thionine was found to be ineffective because it does not work in the presence of iodine.

For this experiment, a cell, as shown in Figure 1, was constituted. The solutions of ferrous, iodine, H_2SO_4 and dyes were equally divided and kept in the two beakers. One beaker was covered with black paper and the other exposed to sunlight. The potential across the two electrodes was measured. A current was also measured through a load of 3 K ohm. The increase of current with time of exposure, as well as the decrease of current with time (in the dark), was measured.

A series of combinations of concentrations of Fe^{3+}, Fe^{2+}, I^- and H_2SO_4 were tried and the maximum current obtained with each set was measured. These experiments were repeated in the presence of Rhodamine-B dye for several days. The results are shown in Table 2. From this data, it is clear that Fe^{3+} (10^{-2} M), Fe^{2+} (10^{-2} M), KI (10^{-2} M), Rhodamine (10^{-6} M) and H_2SO_4 (10^{-1} N) are the best combination for this work. All other combinations either give low current, or after long exposure, the solution becomes yellow (thus showing the generation of excess iodine or I_3^- species). This yellow colour decreases the current of the cell. The increase in current after exposure and its decrease in the dark, see Figure 10, shows the storage capacity of this cell, which is much greater than what one can get with thionine systems. Rhodamine-B helps in the formation of Fe^{3+} and I^- with a faster rate and it also helps in stabilising in the Fe^{3+}/I^- ions in the dark. The role of Rhodamine-B can be understood by taking the

TABLE NO. 2

No. Obs.	Conc. of Fe^{+2} in Molarity	Conc. of Fe^{+3} in Molarity	Conc. of I^- in Molarity	Conc. of Rh(B) in Molarity	Max. current in Micro Ampere	Max. Boten-tial in Milli volts	Time of attaining max. currenţ in minutes	Time of decay in minutes
1	2×10^{-2}	1×10^{-3}	1×10^{-1}	2.5×10^{-6}	7.5	20	12	12
2	2×10^{-2}	1×10^{-3}	1×10^{-1}	-	7.5	18	12	-
3	1×10^{-2}	1×10^{-1}	1×10^{-1}	2.5×10^{-6}				
4	1×10^{-2}	1×10^{-2}	1×10^{-2}	5.0×10^{-6}	10-13	35-45	10-25	170
5	1×10^{-2}	1×10^{-3}	1×10^{-3}	2.5×10^{-6}	0.25	-	60	-
6	2×10^{-2}	1×10^{-3}	1×10^{-1}	5.0×10^{-6}	8	25	15	
7	1×10^{-2}	1×10^{-2}	1×10^{-1}	5×10^{-6}	2	-	7	-
8	1×10^{-2}	1×10^{-2}	1×10^{-2}	2.5×10^{-6}	11-13	40	11.30	
9	1×10^{-2}	1×10^{-2}	1×10^{-2}	1×10^{-4}	12		12.25	-
10	1×10^{-2}	1×10^{-2}	1×10^{-2}	1×10^{-5}				
11	1×10^{-2}	1×10^{-2}	1×10^{-2}	1×10^{-6}	12-13	37.5	14	-
12	1×10^{-2}	1×10^{-2}	1×10^{-2}	-	11.5	37.5	15	-
13	1×10^{-1}	1×10^{-1}	1×10^{-1}	5×10^{-6}	8	2	30	-
14	1×10^{-1}	1×10^{-1}	1×10^{-2}	3×10^{-6}	13	37.5	45	-
15	1×10^{-2}	1×10^{-2}	1×10^{-3}	1×10^{-6}	2-3	-	15	60

MS

TABLE NO. 3

Substance Rhodamine-B	λmax nm	Remarks
Rhodamine-B	550	It has other absorption peak of small magnitude in UV also.
Thionine	590	
Iodine	352 and 283	
Iodide	-	
Fe^{3+}	-	
Ferrous	225 and 303	

spectra of Fe^{3+}, Fe^{2+}, I^-, I_2 and Rhodamine-B. The absorption band spectra of these species are shown in Table 3.

We know from the solar spectrum [9] that the maximum intensity of solar rays includes 450-700 nm and below 450 nm, the photons are of low intensity.

Considering this factor, Rhodamine-B would be the main species out of all present in this system, which would absorb the maximum intensity of solar rays. This energy, after absorption, excites the Rhodamine-B which then carried the following reactions:

$$Rh - B \xrightarrow{hv} Rh - B^* \quad (19)$$

$$I_3^- + Rh - B^* \longrightarrow I_3^{-*} + Rh - B \quad (20)$$

$$Fe^{2+} + Rh - B_1^* \longrightarrow Fe^{3+} + Rh - B \quad (21)$$

$$I_3^{-*} + Fe^{2+} \longrightarrow Fe^{3+} + \tfrac{1}{2} I_2 + 2I^- \quad (22)$$

$$Rh - B^* + \tfrac{1}{2} I_2 \longrightarrow Rh - B + \tfrac{1}{2} I_2^* \quad (23)$$

$$\tfrac{1}{2} I_2^* + Fe^{2+} \longrightarrow Fe^{3+} + I^- \quad (24)$$

$$Rh - B^* + Rh - B \longrightarrow Rh - B + heat \quad (25)$$

Quickenden [12] has shown the possibility of reaction by using a tungsten lamp, but we have not been able to see this experiment in solar rays. The reaction (22) suggests that if I_3^-* and Fe^{2+} concentration is the same and all Rh - B* has been used up by I_3^- (reaction 20), then one can hope to generate ½ I_2 species in the solution. Therefore, in order to prevent the formation of ½ I_2 species, Rh - B* should be high; or, I_3^- species should be low in the system.

If a higher concentration of Rh-B is used, then excited Rh-B* may interact by equation (25), which will make it ineffective, and therefore, Rh-B should be low enough to prevent reaction (25). The alternative is to lower the I_3^- concentration (i.e., to lower iodide ion concentration). This is clear from Table 2, that whenever iodide ion concentration is greater than the ferrous ions, the solution gives lower current and after a prolonged exposure becomes yellow in colour, whereas when all concentrations are in the range of 10^{-2} M Rhodamine (10^{-5} - 10^{-6} M). After a continuous exposure, the solution becomes pink coloured (Rhodamine-B colour) and the dark solution shows an intense yellow colour.

It is therefore concluded that Fe^{2+}/I_2 system is a fully reversible photochemical reaction. A cell of this type, shown in Figure 9, is designed, and, we have been reproducing the same cell for the last six months. This is made out of two β-34 cone/socket cylinders. The cone is filled with agar-agar black, containing saturated KCl. One compartment is blackened by black paint. Two platinum electrodes are dipped in two half cells.

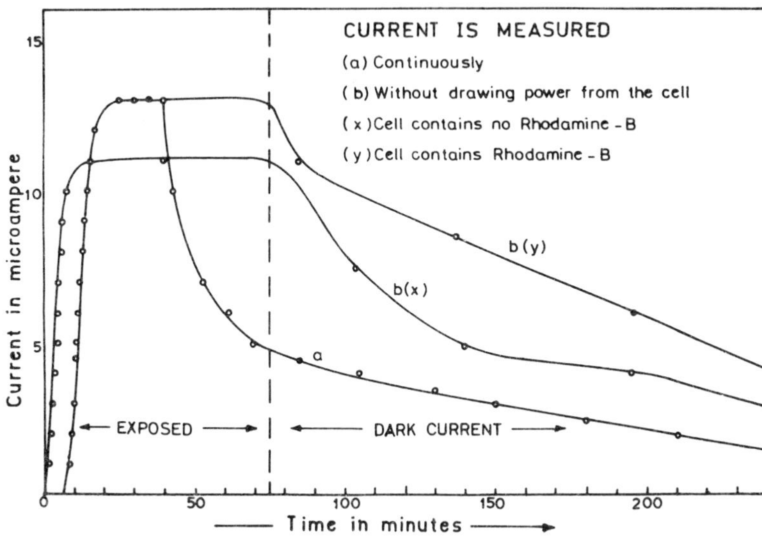

Fig. No. 10 Current characteristics of the Saur Vidyut Kosh using Fe^{2+}/I_2 system.

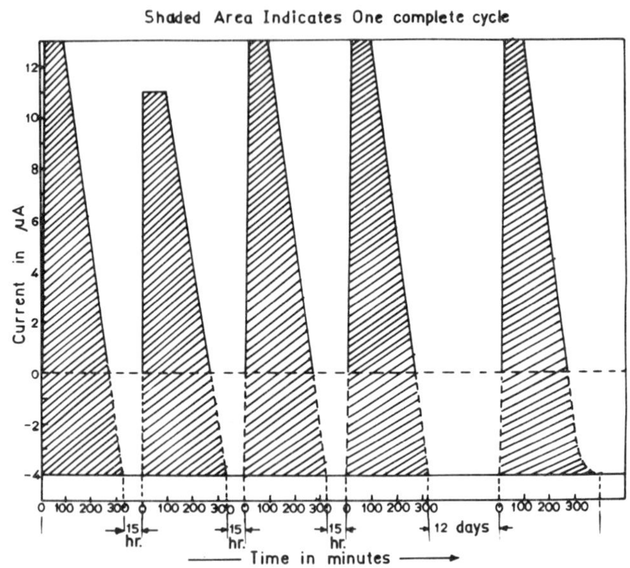

Fig. No. 11 Current characteristics for continuous several exposures - Dark cycles.

A solution mixture of effective concentration Fe^{2+} (10^{-2} M), Fe^{3+} (10^{-2} M), KI (10^{-2}) H_2SO_4 (10^{-1} N) and Rhodamine-B (10^{-6} M) is added to both glass half cells. This cell is kept directly in the sun for exposure. The current characteristics of this cell are shown in Figure 10. In Figure 11, the current time graph is shown for several exposure - dark cycles. This cell has an open voltage of 40-45 mv and it gives a current of 10-15 μAmp. This cell is independent of an exposed area, as it depends upon the concentration of a chemical photochemically produced in the light compartment. Therefore, one cannot calculate the efficiency as people do for a photovoltaic cell. The power of this cell is 30 μwatt. The authors have named this cell as SAUR (the solar) VIDYUT (electricity) KOSH (battery).

ACKNOWLEDGEMENTS

The authors are thankful to Tata Energy Research Institute, Bombay, India, who have given a grant to proceed with this work. One of us, S.G. Sharon, is grateful to TERI for providing a research assistantship to carry out this work. We would like to also thank Professor V.K. Phansalkar for providing the research facilities to carry out the work in this department.

REFERENCES

1. Rabinowitch, E., J. Phys. Chem. 8, 551-59, 1940.

2. Rabinowitch, E. , Ibid. 8, 560-566, 1940.

3. Mathai, K.G. and Rabinowitch, E., J. Phys. Chem. 66, 663-664, 1962.

4. Michaalis, L. and Schubert, M.P., An. Chem. Soc. 62, 204-210, 1940.

5. Hardwick, R., J. Phys. Chem. 66, 349-350, 1962.

6. Weber, K., Z. Physik Chem. 18, 15, 1931.

7. Weiss, J., Nature 136, 794, 1935.

8. Ainsworth, S., J. Phys. Chem. 64, 715-722, 1960.

9. Clark, W.D.K. and Eckert, J.A., Solar Energy 17, 147-150, 1975.

10. Wildes, P.D., Brown, K.T., Hoffman, M.Z., Lichtin, N.N. and Hall, D.E., Solar Energy 19, 579-582, 1977.

11. Sakata, T., Suda, Y., Tanaka, J and Tsubomura, H., J. Phys. Chem. 81, 537-542, 1977.

12. Quickenden, T.I. and Yim, G.K. (nee Tan), Solar Energy 19, 283-289, 1977.

13. Rudolph, J.M. and Henry, C.W., J. Solar Energy Science and Engineering, 4, 1961.

14. Rideal, E.K. and Williams, E.C., J. Chem. Soc. 127, 258, 1925.

15. Kistiakowsky, G.B., J. Am. Chem. Soc. 49, 976, 1927.

SOLAR THERMAL POWER PLANT : THERMODYNAMIC ANALYSIS

Anand A. Samuel, U.S.P.Shet and M.C. Gupta
Solar Energy Division
Indian Institute of Technology
Madras - 600036, India

ABSTRACT

There have been several proposals for solar thermal power plants using a Rankine cycle with a low boiling point fluid as the working medium. Organic Rankine-cycle engines are particularly suited for a number of reasons such as, high thermal efficiency and reliability, relatively uncomplicated mechanical components and adaptability for use over a wide power range.

In this study three working fluids: butane (C_4H_{10}), dichloro tetrafluoroethane (F114) and trichloro trifluoroethane (F113) are examined. These three fluids were chosen in view of their desirable properties such as high molecular weight, low boiling point and the slope of their vapour saturation curve.

INTRODUCTION

The solar thermal power system discussed in this paper consists of four sub-systems: collector system, storage system, prime mover and generator. The key components are the collectors and the prime mover. The thermal performance of the flat plate collector system and Rankine cycle are discussed in detail.

For the range of temperatures encountered in such solar thermal systems (60 to 100°C), a working fluid other than steam is preferred. But water is found to be the best sensible heat collection media. Hence, dual-fluid cycle is used for the analysis, with water as the collector fluid and one of the refrigerants as the working fluid. It is therefore necessary to consider the effect of working fluid on the thermal efficiency of the cycle.

FLAT-PLATE COLLECTORS

Flat plate collectors, which collect both direct and diffuse radiation, are economically best suited for power plants. The collector performance is a function of emissivities, number of glazings used and mass flow rate of the collector fluid. A computer programme has been developed and performance characteristics are predicted. The energy balance equations for the collector can be written as,

$$A_c . H . \tau\alpha = Q_u + Q_l + Q_s$$

$$Q_u = A_c . F_R . [S - U_L (T_{Fi} - T_a]$$

The heat transfer coefficients for conduction and convection in the gap between optical elements and outside and the over all heat losses are evaluated by standard thermodynamical method for different collector configurations. The analysis shows that at low emissivities evacuated collectors are found to be more efficient than the collector with 2 glazings. For normal emissivities, collector with two glazings may be used effectively. The variation of collector efficiency with the outlet temperature, for collectors with single and double glazings are shown in Fig.8a.

THERMODYNAMIC CYCLE ANALYSIS

In solar energy applications, the rate of heat supply is dependent on the variable nature of the insolation and collector configuration. Hence, the problem essentially centres around the choice of suitable thermodynamic heat cycle, for the low temperature, low pressure applications. For optimum results, the performance of the thermodynamic cycle should be such as to make the best use of the available heat source.

The two possible cycles considered here are Rankine cycle and Stirling cycle. From Fig.2, it is clear that in Rankine cycle system the higher performance of cycle is achieved through the better match with heat source. The work ratio which is defined as the ratio of net work output of the cycle to the expansion work of the cycle, is nearly unity for Rankine cycle. This makes it clear that Rankine,cycle is the one best suited for solar thermal power plants.

Fig.1 shows the schematic diagram of the solar thermal power plant incorporating a Rankine cycle. The representation of Rankine cycle on T-s and p-v diagrams are shown in Fig.2. In order to achieve high thermal efficiency the design criteria should consider factors, such as, the enthalpy drop in the turbine, mass flow rates and boiler heat supply. For small power plants the condenser temperature is dictated by ambient temperature and it cannot be varied. Applying the trilateral cycle concept [2] in the T-s diagram, it can be inferred that higher the operating temperature, better is the cycle efficiency. So the maximum temperature that could be obtained with flat plate collectors is determined and taken as the operating temperature for the cycle. A computer programme is developed to predict the performance of the cycle for variations in the parameters, such as turbine inlet and outlet temperature, etc.

The effects of turbine inlet (T_1) and outlet (T_o) temperatures on the thermal efficiency (η_{TH}) of the cycle are shown in Fig.3. An increase of two degree in the inlet temperature causes approximately 1 per cent rise in efficiency. Fig.4 explains the variation of another important parameter mass flow rate $\dot{m}^*$ of the working fluid with the turbine inlet temperature.

As the flow rate reduces the pumping power is also reduced and the cycle efficiency increases. Heat rate is calculated and used in the design of turbines and heat exchangers. The variation of heat rate with turbine inlet temperature is shown in Fig.5.

As the condenser pressure increases the pump work decreases for the same boiler pressure and this is shown in Fig.6. But, higher condenser pressure reduces the cycle efficiency. Hence a compromise is to be made keeping in view the size of the plant. Small power plants cannot afford to have large pumps.

The steep increase in cycle efficiency with increase of super heat is shown in Fig.7. This shows that a high degree of super heat is desirable.

SYSTEM ANALYSIS

The combined system of collector and Rankine cycle is analysed here. Fig.8a shows the increase of cycle efficiency with peak operating temperature and decrease of collector efficiency η_c with the outlet temperature. It should be noted that the collector efficiency decides the number of collectors needed to get the required energy for the heat engine.

In this analysis the over all system efficiency is considered as the product of cycle and collector efficiencies. In Fig.8a, the collector efficiency is plotted against maximum operating temperature for the optimum value of mass flow rate and an insolation of 750 kcal/hr m^2. For cycle efficiency the ideal Rankine cycle with F114 as working fluid is chosen. Fig.8b shows that when flat plate collector with single glazing is used in the system, the optimum operating temperature is found to be 92°C, while for two glazings, an improvement of 4°C is seen. The corresponding efficiencies for the cycle

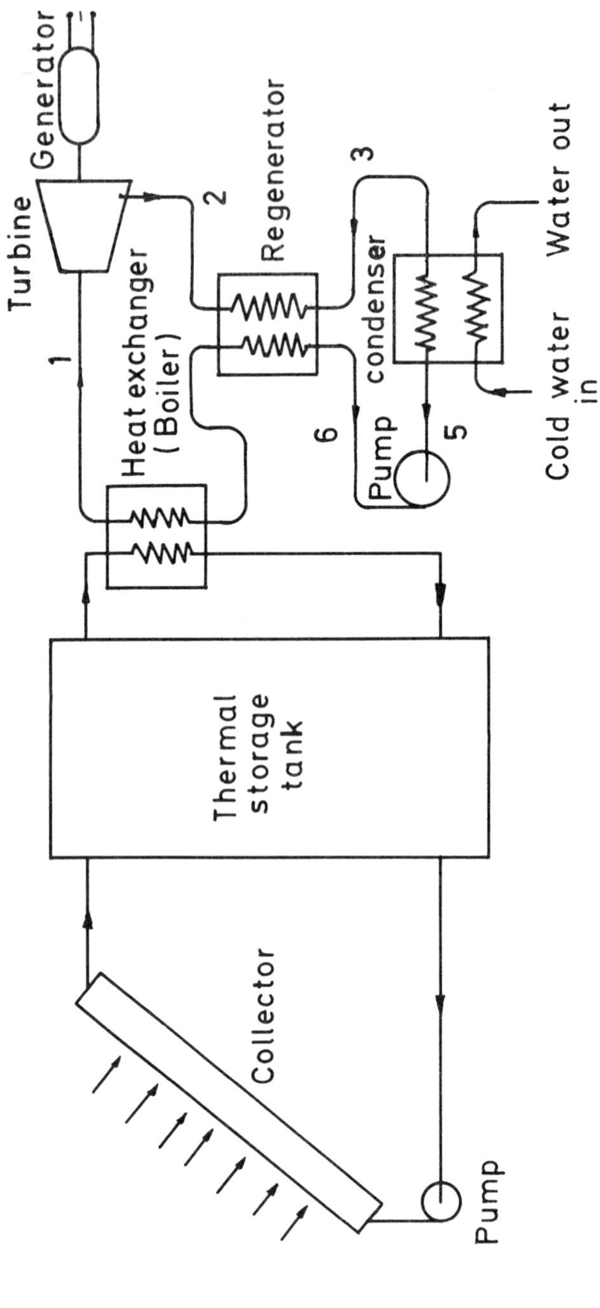

FIG.1. SCHEMATIC DIAGRAM OF THE SOLAR THERMAL POWER PLANT

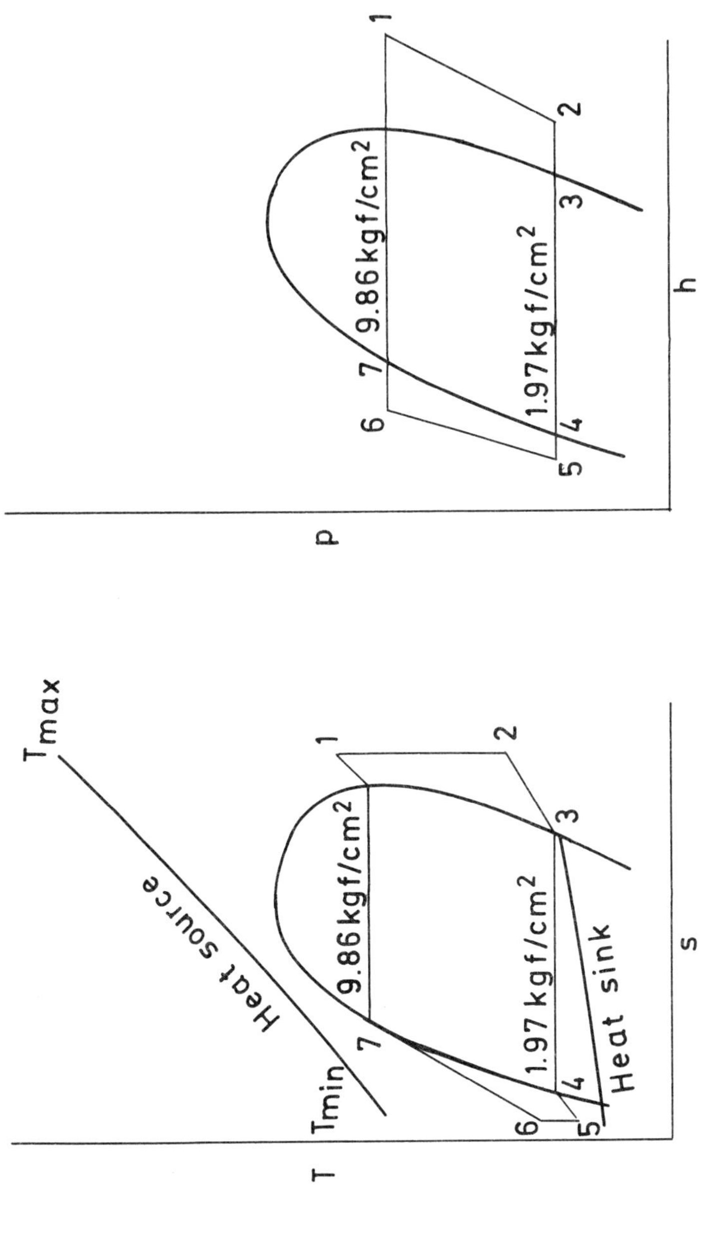

FIG.2. RANKINE CYCLE (WORKING FLUID: F114)

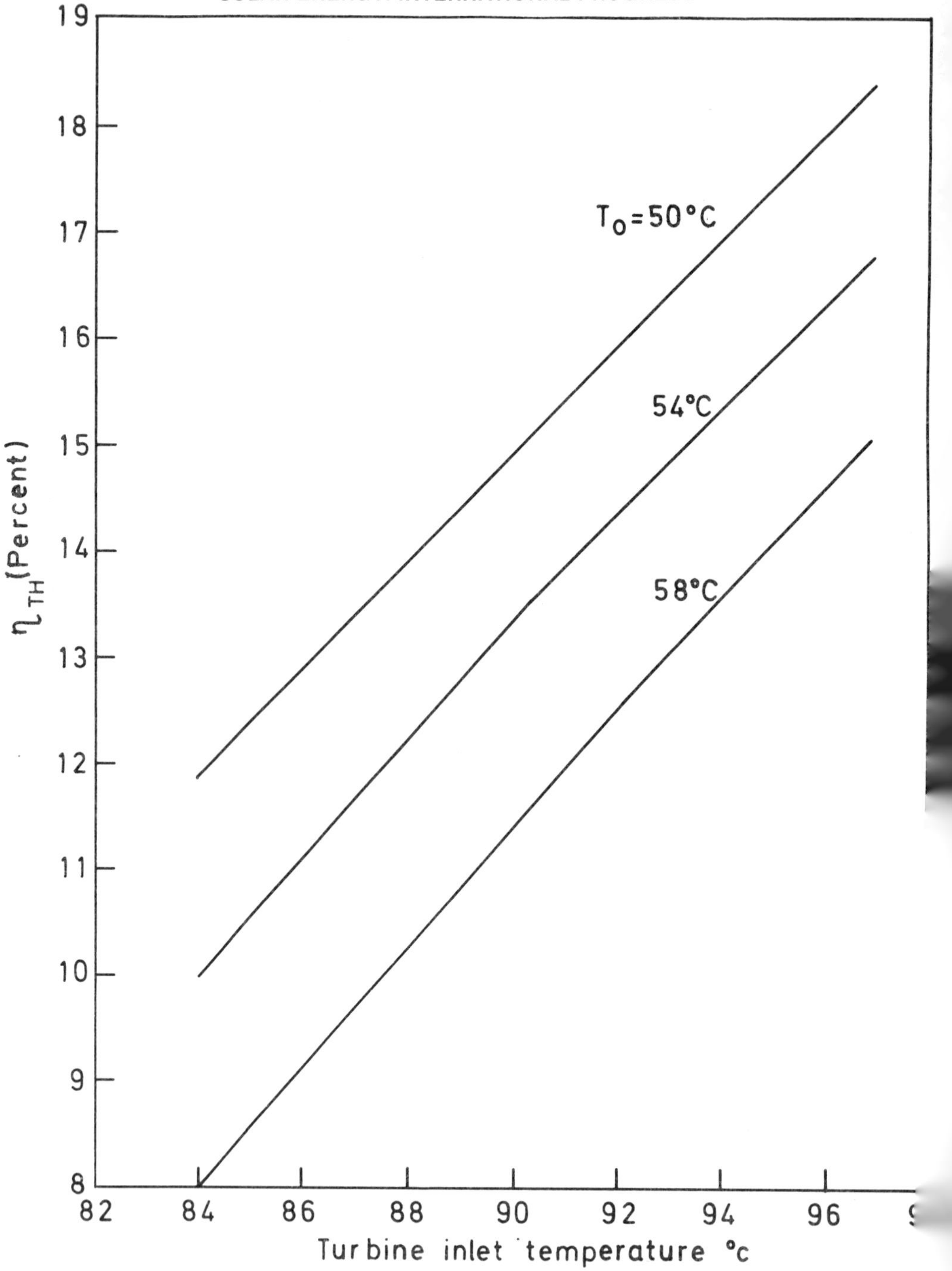

FIG.3. EFFECT OF T_o ON η_{TH}

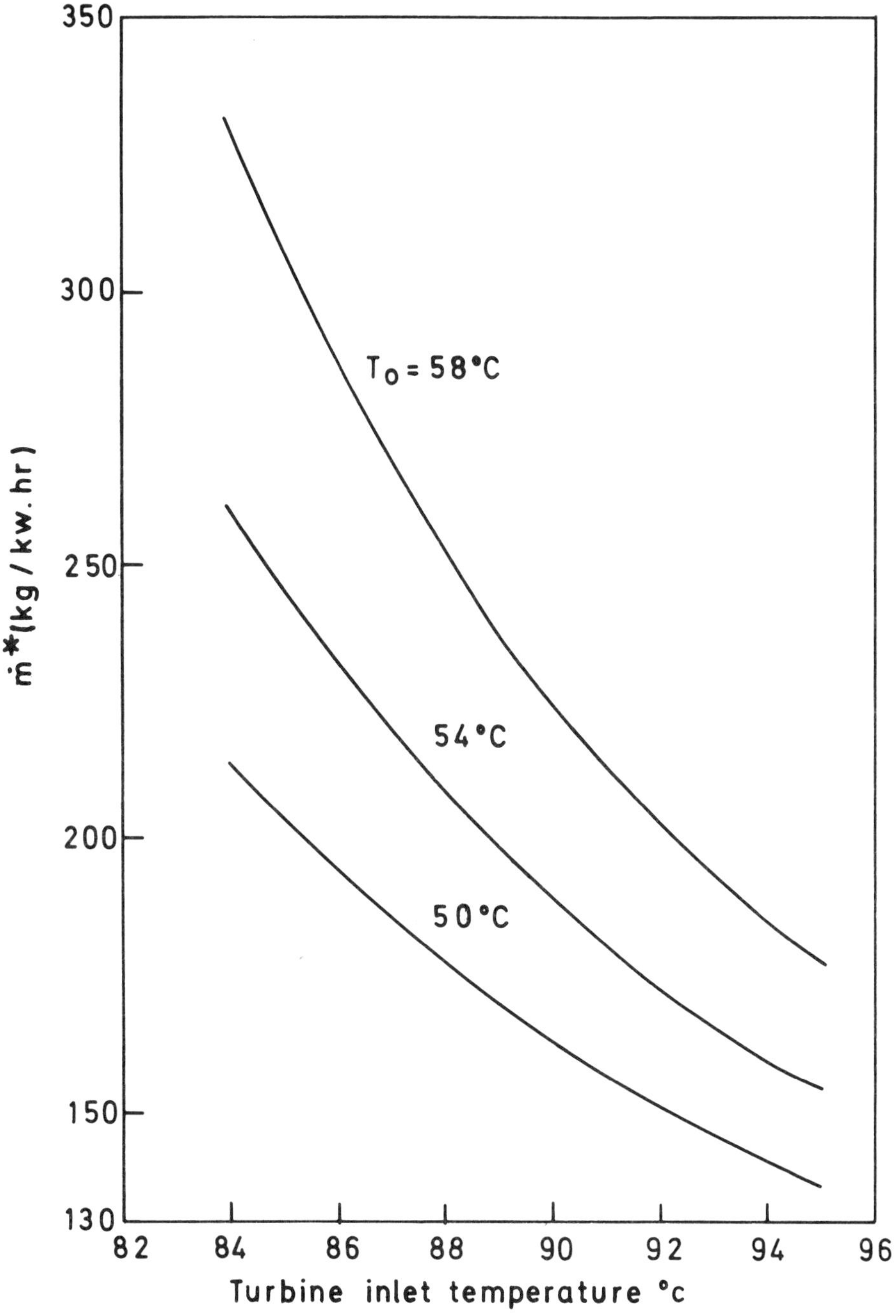

FIG.4. EFFECT OF T_0 ON $\dot{m}^*$

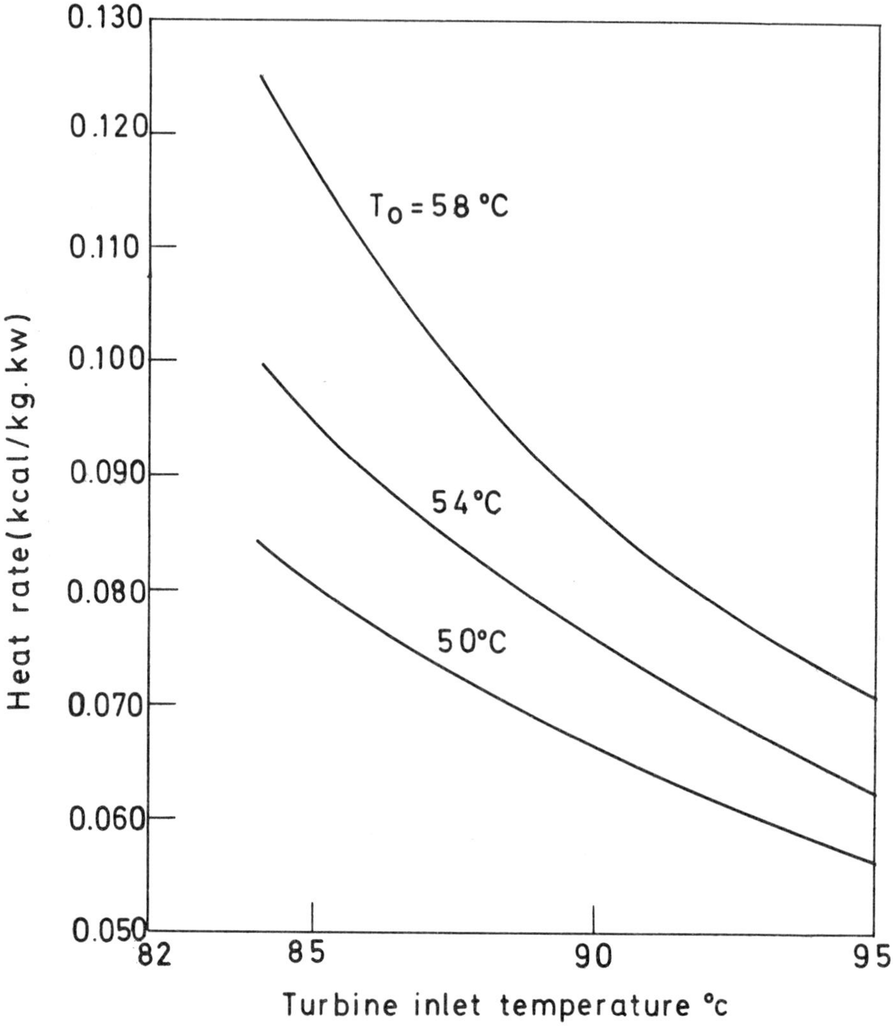

FIG.5. EFFECT OF T_0 ON HEAT RATE

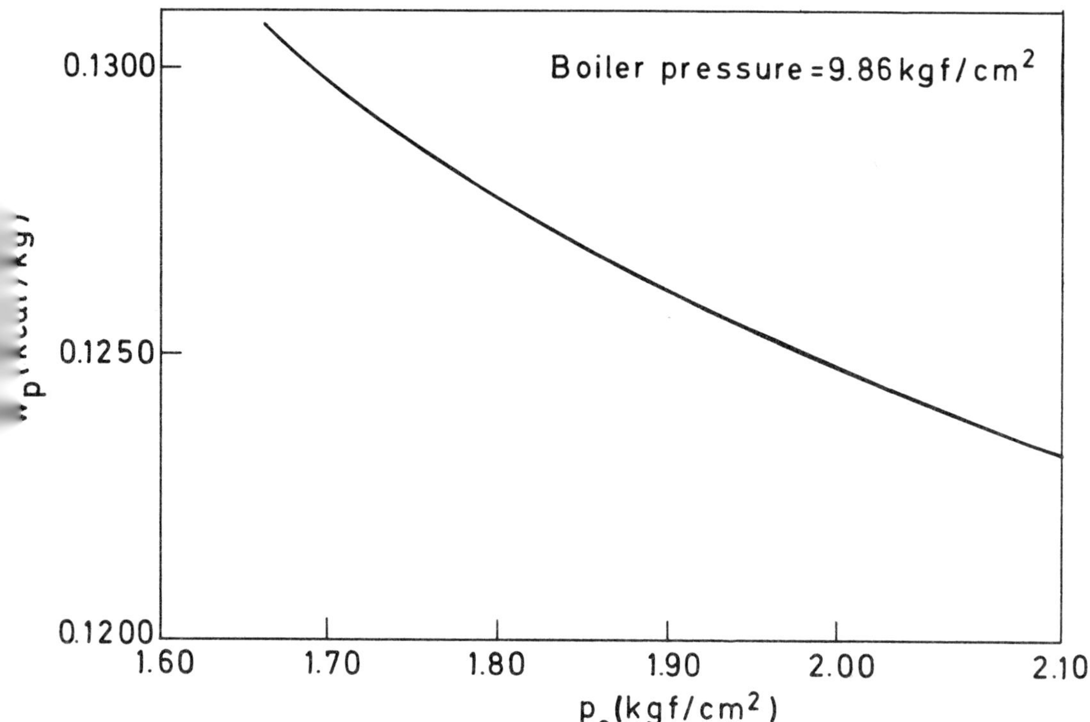

FIG.6. EFFECT OF p_c ON w_p

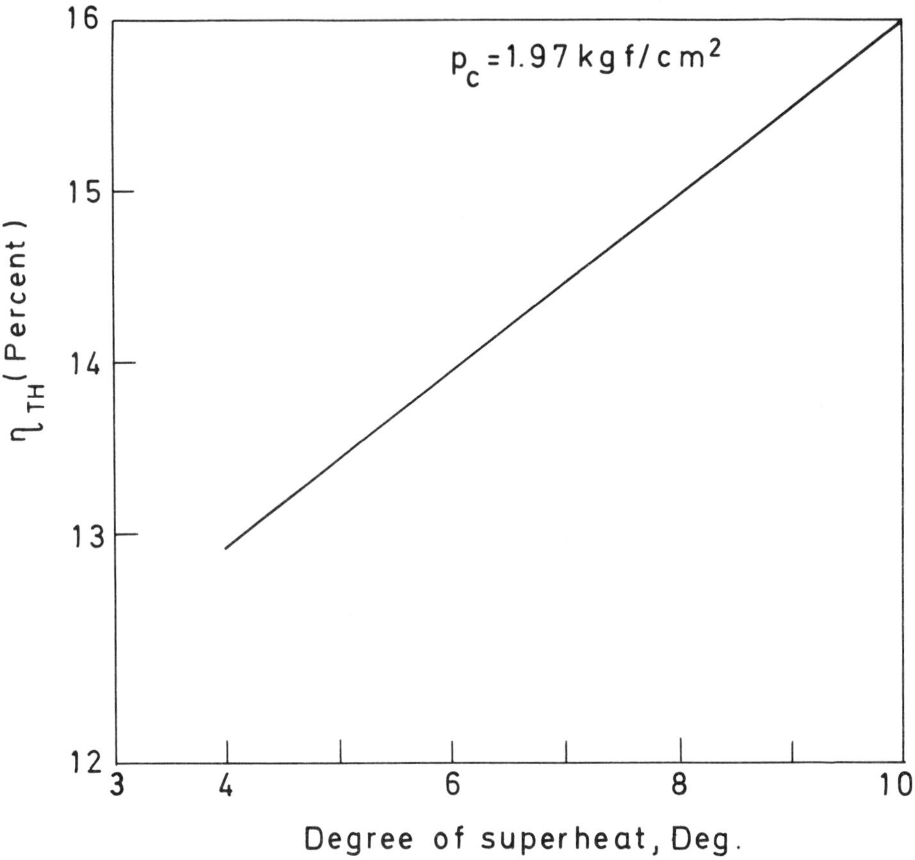

FIG.7. EFFECT OF DEG. OF SUPER HEAT ON η_{TH}

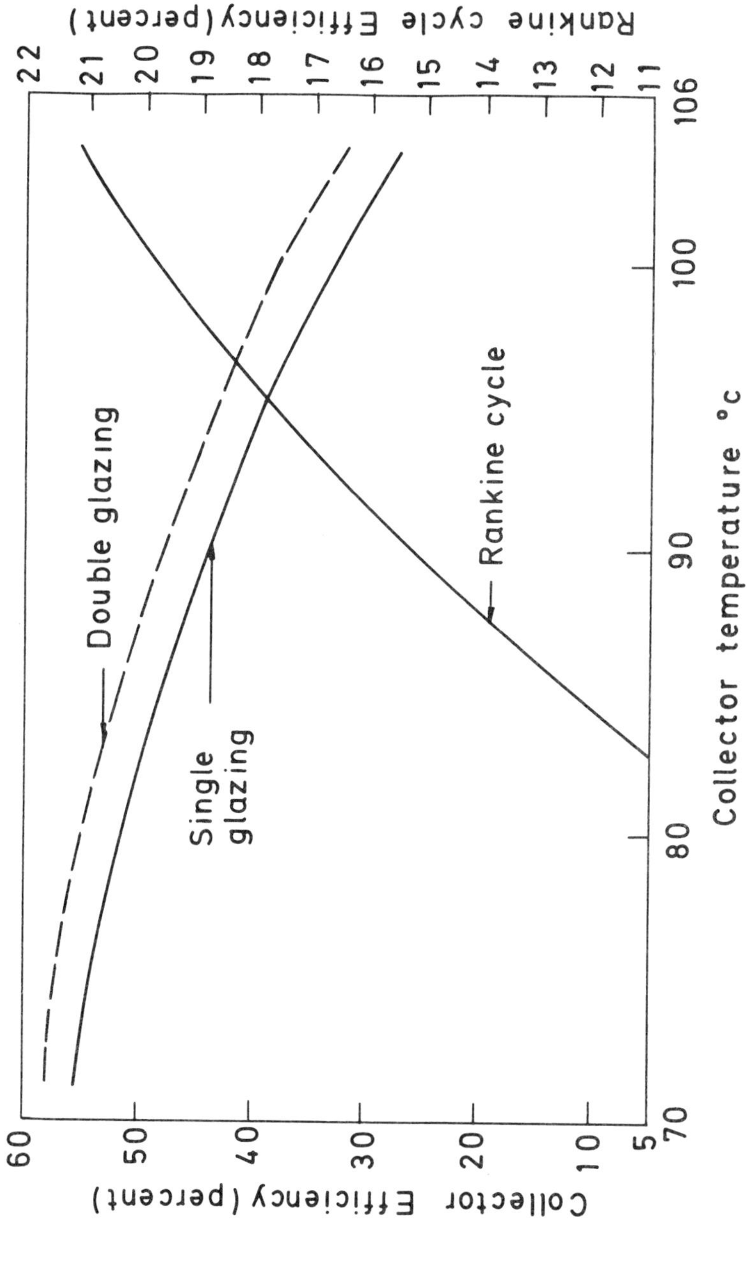

FIG. 8a. COLLECTOR AND CYCLE EFFICIENCY Vs COLLECTOR TEMPERATURE

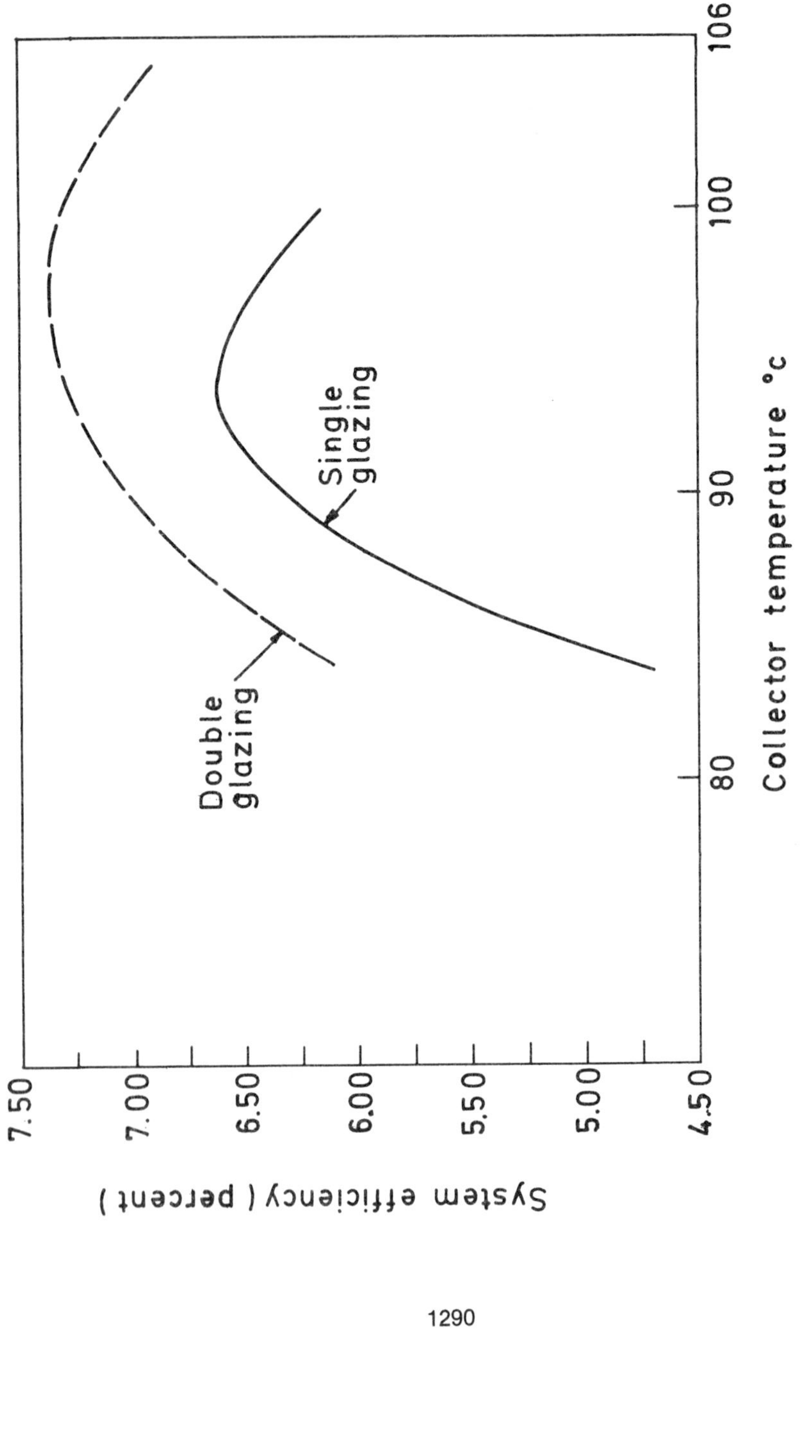

FIG.8b. SYSTEM EFFICIENCY Vs COLLECTOR TEMPERATURE

and collector clearly indicate that it is better to choose collectors with 2 glazings, and the marginal gain in collector efficiency by additional glazings does not justify the increased cost involved.

CONCLUSION

The present analysis makes it clear that the solar thermal power plant has a tremendous potential and is thermodynamically viable even with flat plate collectors. With F114 as working fluid, the power cycle can operate efficiently at turbine inlet temperatures ranging from 90°C to 95°C.

ACKNOWLEDGEMENT

One of the authors (A.A.S.) gratefully acknowledges the financial support received from the Council of Scientific and Industrial Research during the course of this research work.

NOMENCLATURE

A_c	-	Collector area [m^2]
F_R	-	Collector heat removal factor
H	-	Incident solar radiation [kcal/hr m^2]
$\dot{m}^*$	-	Mass flow rate of working fluid [kcal/kw.hr]
p_b	-	Boiler pressure [kgf/cm^2]
p_c	-	Condenser pressure [kgf/cm^2]
Q_L	-	Heat loss through the cover and the sides [kcal/hr]
Q_s	-	Heat stored in the collector [kcal/hr]
Q_u	-	Heat gained by the fluid [kcal/hr]
S	-	$H.\tau\alpha$
T_a	-	Ambient temperature (°C)
T_{Fi}	-	Fluid inlet temperature (°C)
T_i	-	Turbine inlet temperature (°C)
T_o	-	Turbine outlet temperature (°C)
U_L	-	Over all loss coefficient [kcal/hr m^2]
w_p	-	Pump work [kcal/kg]
τ	-	Transmitivity of glass plate
α	-	Absorbtivity of collector plate
η_{TH}	-	Thermal efficiency of the cycle (%)

-AAS

η_c - Collector efficiency (%)

REFERENCES

1. Deane H.Kihara and Paul S.Fukunaga, Working fluid selection and preliminary heat exchanger design for Rankine cycle Geo-Thermal Power Plant, University of Hawaii, Hawaii 96822, U.S.A.

2. Willson,S.S., and Radwan,M.S., Appropriate Thermodynamics for Heat Engines Analysis and Design, IJMEE, Vol.5, No.1, p.68-80 (1977).

3. Selcuk,M.K., Preliminary assessment of flat plate collector solar thermal power plants, Jet propulsion Laboratory, California, U.S.A.

4. Beckmann,W. and Duffie,J.A., Solar thermal processes, Willey and Sons, N.Y. 1976.

SOLAR POWER UNITS WITH SCREW EXPANDERS

J. M. MERIGOUX and P. POCARD
ALSTHOM-ATLANTIQUE - DME
91680 BRUYERES-LE-CHATEL FRANCE

ABSTRACT

After a general survey of the "ALSTHOM-ATLANTIQUE" activity in solar energy, the authors describe a particular application : a screw expander low power solar unit for electricity production. For that use, flat plate collectors and low concentration cylindrical collectors, air condenser and water condenser are compared. Electricity production is estimated for mediterranean and subtropical regions.

Some screw expander experimental data are given. An usual application of that engine in a solar pumping unit is shown.

PRESENTATION

The activity of ALSTHOM-ATLANTIQUE in the field of solar energy begun several years ago. So, a 30 kW solar engine built by RATEAU (subsidiary of ALSTHOM-ATLANTIQUE) with control devices produced by ALSTHOM is in operation since September 1975, at SAN LUIZ DE LA PAZ (MEXICO), another french firm (SOFRETES) assuming the engineering and the manufacturing of the flat plate collectors.

The electric turbogenerator shown Fig. 1 is an hermetic unit, revolving on gas bearings ; the mechanical power is produced by an high efficiency radial inflow turbine ; the working fluid is FREON 11.

The turbine is now developped in the 50 - 500 kW power range, in a more usual device (see Fig. 2) with oil bearings, gearbox and electric generator apart.

These expanders are used in an organic Rankine cycle with boiler and condenser.

For that cycle, in the 5 - 50 kW power range ALSTHOM-ATLANTIQUE has choosen the screw expander, due to the large experience obtained with CIT-ALCATEL, during ten years in the screw compressors.

An other way for ALSTHOM-ATLANTIQUE in the field of solar energy is to manufacture gas turbines for open Brayton cycle.

We can see on Fig. 3 a 130 kW electric generator with subsidiary combustion chamber and recuperator ; the cycle used is shown Fig. 4. That unit is the preview of the 1 MW units developped with DOLLER gas turbine, an industrial engine built by ALSTHOM-ATLANTIQUE and S.A.C.M. MULHOUSE (see Fig. 5).

These applications concern tower type solar power plants with high temperature cycles.

INTRODUCTION

For the low temperature cycles, with organic working fluid, ALSTHOM-ATLANTIQUE has developped 5 - 50 kW solar power plants with screw expander and low concentration cylindrical collectors. This choice comes from an economical and technical optimisation due to the wide range of operation in various climatic conditions and high reliability for that engine which is a low speed rotating motor, continiously lubricated by an important oil flow.

There are many uses for this expander but one of the most difficult applications is the elelctricity production for isolated telecommunication centers where no sufficient cooling water is available to assume the working fluid condensation.

CYCLE CHOICE

The cycle is shown on the sketch of Fig. 6.

The broken line circuit is the transfer fluid circuit with pressurized water or low temperature GILOTHERM boosted by a pump. The transfer fluid passes through the solar collectors and takes ther thermal energy ; it leaves that energy in the boiler which transfers it to the working fluid.

The continuous line circuit is the working fluid circuit. The boiler is fed with liquid freon by the freon pump. Heated in the boiler, the freon phasis changes from liquid to vapor. Its pressure rises and, then, expands in the screw motor ; freon becomes again liquid at the exit of the air condenser.

The expander drives an A.C. generator equipped with a rectifier.

The current so obtained is stored in 48 V batteries.

The studies have been made with two different types of collectors :
- fixed flat plate collectors
- concentration collectors with tracking of the sun.

The results are given in table 1.

MOTOR CHOICE

For low power, the high reliability solution is the oil injected screw expander.

The screw expander is a volumetric, rotative engine with only two rotating parts running at low peripherical speed (25 m/s). The term "screw expander" comes, as shown on Fig. 7 form the two helical rotors that engage into one another.

Large oil injection assumes good thermal exchanges, uniform temperature and rotor lubrication.

The sketches of Fig. 8 show the different phases of the expansion.

From the right to the left, we can see :
- At first (sketch n° 1), fréon fills a little volume, streaming through the inlet port.
- Then, due to the vapor pressure, it thrusts the twin rotors, each one into the directions shown by the arrows ; the volume inproves (sketches 2 and 3) ; freon expands.
- At last, fréon goes out of the motor through the outlet port.

To lubricate the rotors and to assume the minimum leakage between the working volumes, oil is injected into the expander inlet owing to obtain high expansion ratios and good efficiencies. The oil is recovered, at the motor exit, in an unsophisticated effective separator.

Before going into the expander, oil is preheated through a tubing by the transfer fluid coming from the collectors. That device allows the engine to take quikly its working temperature during starting an to avoid condensation of working fluid at the level of the rotors during running. Oil volume is kept between the working fluid and the outside, at the level of the shaft outlet to avoid freon leakage through the magnetic seal. This assemby is shown on Fig. 9. On Fig. 10, we can see the components of the expander, with, from the right to the left :
- the high pressure side with inlet parts
- the cylinder containing the rotors
- the outlet side ball bearings.

For the application described in this paper, the engine is a 102 mm diameter screw expander.

Its characteristics, at the nominal point (P = 750 W/m2), are given on Table 2.

We can see the handicap due to the use of the air condenser.

This handicap lies on two main points :
- Shaft output power.
  Heat exchanges are not as good in an air condenser as in a water condenser ; so the cold source temperature is higher for the air condenser and the temperature drop between hot and cold sources is smaller. The power output proportional to this difference of temperature is consequently smaller.

- Losses.
  The power required to drive the fan is relatively important especially at part load conditions.

HOT SOURCE EFFECT

To minimize the effect due to the air condenser , it is necessary to increase the temperature of the hot source using concentration collectors instead of flat collectors (i.e.as shown on Fig. 11).

This type of collector, cylindrical parabolic design, is equipped with a tracking device of the sun, from East to West. This collector, maximum temperature protected, assumes the regulation of the maximum power.

This product is manufactured by POLYSOLAR.

Low concentration ratio (6 to 8) collectors allow to optimize the maximum temperature of the cycle at values slightly above 100°C for a 750 W/m2 solar radiation with a 40°C temperature at the cold source and at higher values for better conditions of solar radiation ; that can be obtained without any special treatment of the absorber.

The benefit of the concentration collector appears on table 2 ; from an economic point of view 25 to 45 % of the price of kW/h can be saved compared to the use of the flat type collector. This benefit will still increase when its industrialisation reaches the level of the flat collectors one.

EFFECT OF THE SITE

Fig. 12, established from the data of the "METEOROLOGIE FRANCAISE", shows the great monthly variation of the insolation going from subtropical to mediteranean zone.

Fig. 13 gives for the plant described in this paper, an estimation of the average daily production of electrical energy for the different months of the year on a mediteranean climate.

Fig. 14 gives the same results for a plant located in a subtropical zone.

Theses calculations including the effect of cloudiness show that the annual power produced by a concentration collector plan can be 2,5 time higher than power produced by a flat collector plant.

Fig. 15 shows a picture of the electrical generating set with screw expander and air condenser.

ENGINE EXPERIMENTAL DATA

On the basis of experimental results given by the collector manufacturer, we have built a complete solar engine including screw expander, boiler, condenser and A.C. generator.

The solar collectors were simulated by an electric heater, the power of the heater being adjustable.

Fig. 16 is a view of this testrig. To define the plant design with accuracy, the following data must be known :
- screw expander adiabatic efficiency
- screw expander filling factor which is the ratio between the gas volume flow and the geometrical volume of the engine
- the built-in ratio as a function of the expansion ratio and the thermodynamic characteristics of the working fluid.

With a 6 expansion ration screw expander, 102 mm rotor diameter, we have obtained the performances plotted on Fig. 17.

Now the "Délégation Française à la Recherche Scientifique et Technique" (D.G.R.S.T.) participates to the development of this engine to higher expansion ratios (up to 10) and various built-in ratios.

This research finds a large application in the larger temperature drop cycle for pumping where pumped water is used to condense the working fluid.

Fig. 18 shows a view of a pumping solar unit for deep well with :

$Q\ h = 6000$
Where Q is the water flow pumped in m3 per day
h is the total manometric height in meters of water.

CONCLUSIONS

The screw expander is well adapted to the various applications of the solar energy in the low power range.

This is due to :
- the small amount of variation in efficiency for a large range of expansion ratios and rotating speed
- its ability to be used in connection to low concentration collectors due to its high expansion ratio possibility.
- its aptitude to operate with usual organic working fluids and the cheapest transfer fluids.

TABLE 1
RESULTS

	COLLECTOR TYPE	
	FLAT PLATE	CONCENTRATION
WORKING FLUID..........	R 113	R 113
MOTOR INLET PRESSURE...	2.2 ATA	5 ATA
MOTOR OUTLET PRESSURE..	1 ATA	1 ATA
MOTOR INLET TEMPERATURE	74°C	107°C
CONDENSER OUTLET TEMPERATURE...........	47°C	47°C
COLLECTOR OUTLET TEMPERATURE...........	88°C	125°C
AIR CONDENSER INLET TEMPERATURE...........	25°C	25°C
AIR CONDENSEUR OUTLET TEMPERATURE...........	40°C	40°C
EVAPORATOR POWER.......	66000 KCAL/H	70600 KCAL/H
CONDENSER POWER........	62700 KCAL/H	64200 KCAL/H
COLLECTOR ACTIVE AREA..	200 M2	200 M2

TABLE 2
CHARACTERISTICS

	COLLECTOR TYPE	
	FLAT PLATE	CONCENTRATION
NOMINAL SPEED.............	4200 R.P.M	4200 R.P.M
SHAFT OUTPUT POWER........	3800 W	7380 W
MECHANICAL LOSSES		
- FREON PUMP..............	130 W	450 W
- OIL PUMP................	100 W	340 W
NET OUTPUT POWER..........	3570 W	6590 W
ELECTRICAL POWER..........	2860 W	5270 W
ELECTRICAL LOSSES		
- FAN.....................	880 W	900 W
- TRANSFER FLUID PUMP.....	280 W	290 W
NET ELECTRIED POWER.......	1700 W	4080 W

FIG. 1

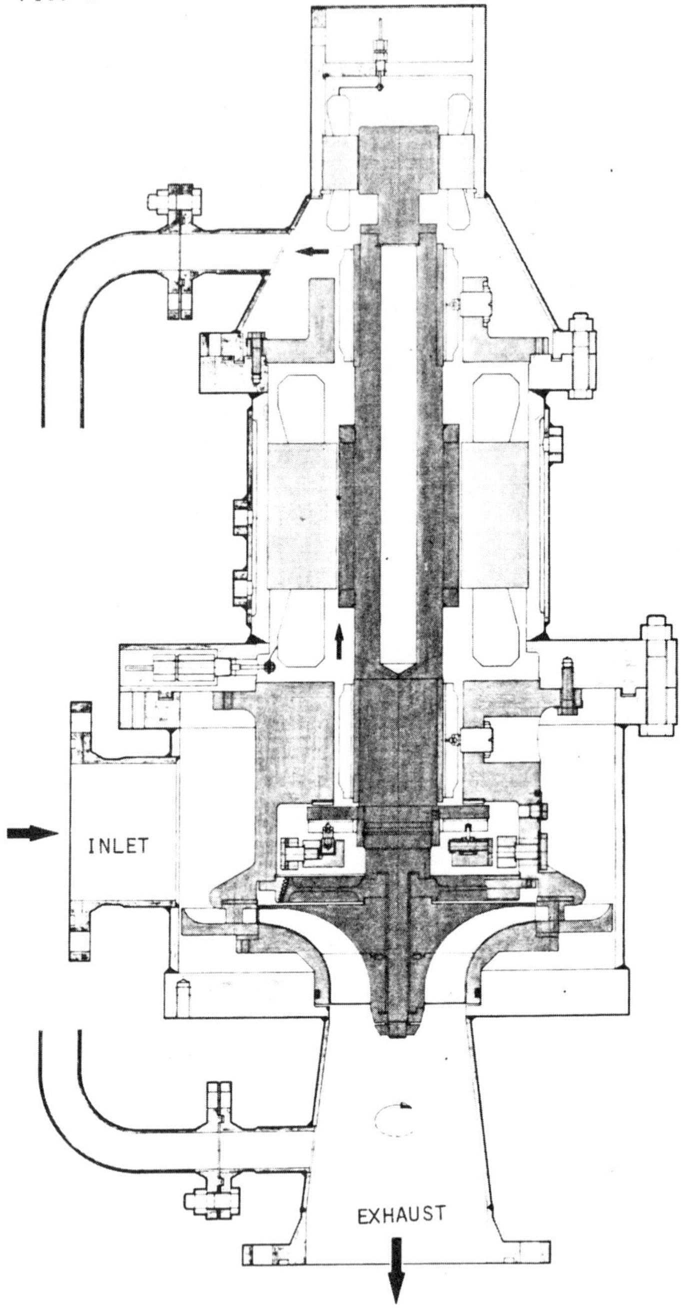

30 KW TURBO GENERATOR

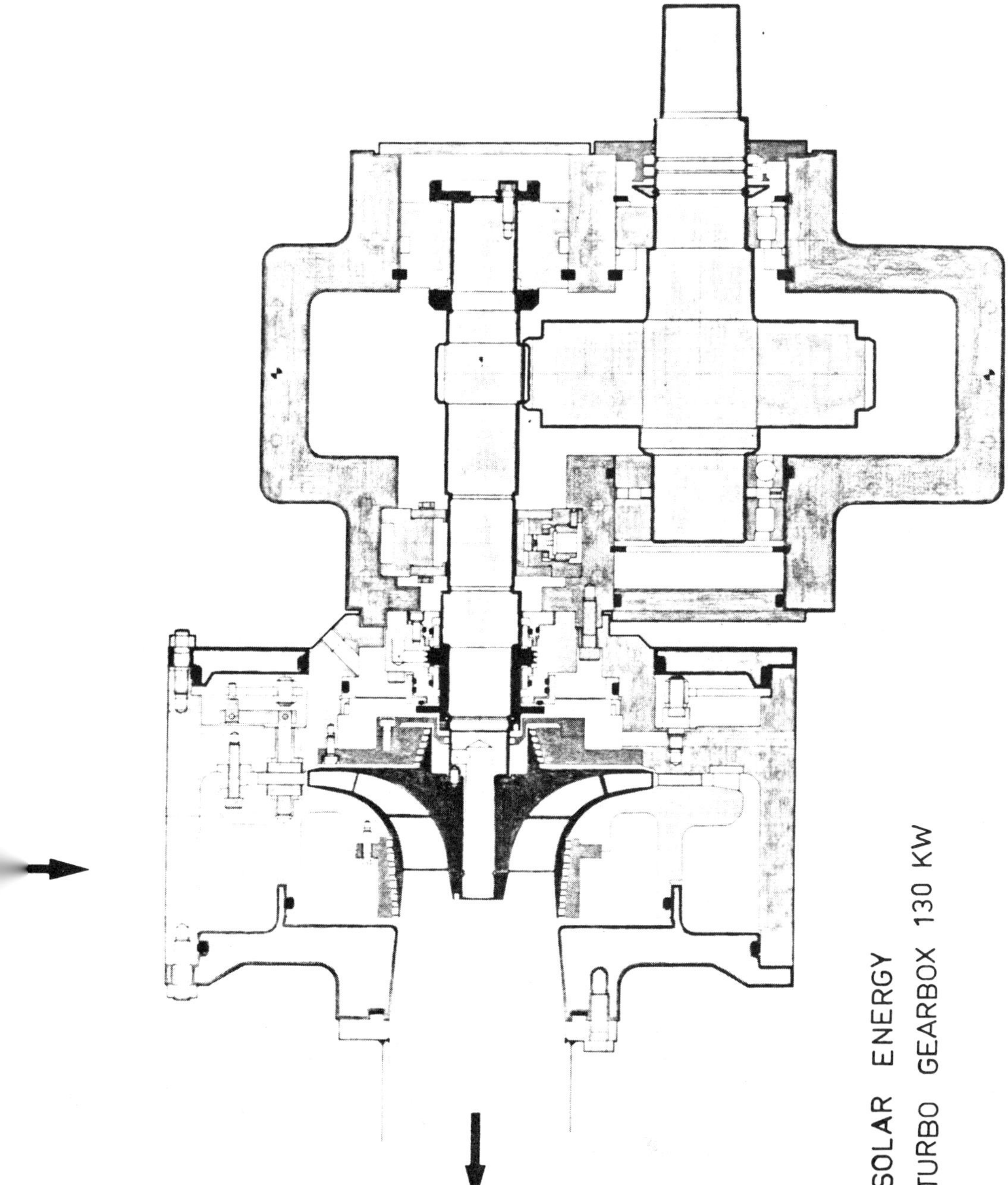

SOLAR ENERGY
TURBO GEARBOX 130 KW

FIG. 3

ELECTRICITY GENERATOR SET

BRAYTON CYCLE

AND SUBSIDIARY COMBUSTION CHAMBER

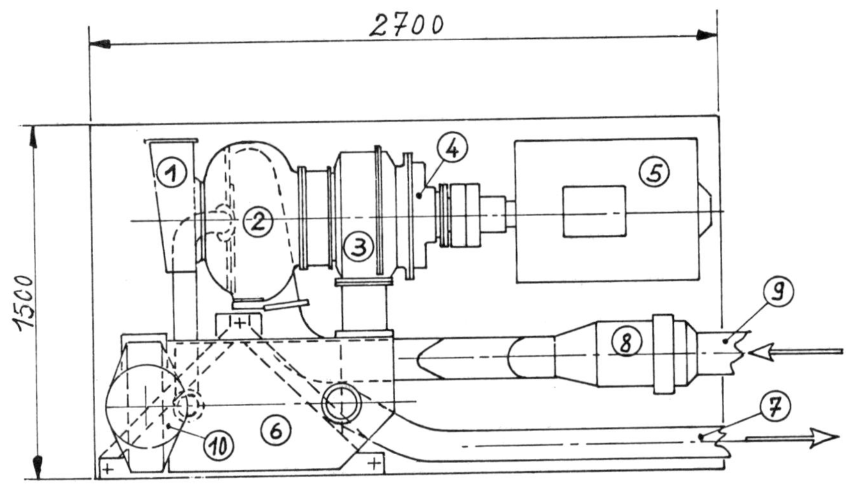

(1) Compressor inlet

(2) Gas generator housing

(3) Power turbine housing

(4) Gear box

(5) a.c. (asynchronous) generator

(6) Heat exchanger - Recuperator

(7) Air duct to solar furnace

(8) Combustion chamber

(9) Air duct from solar furnace

(10) Exit heat exchanger duct (chimney)

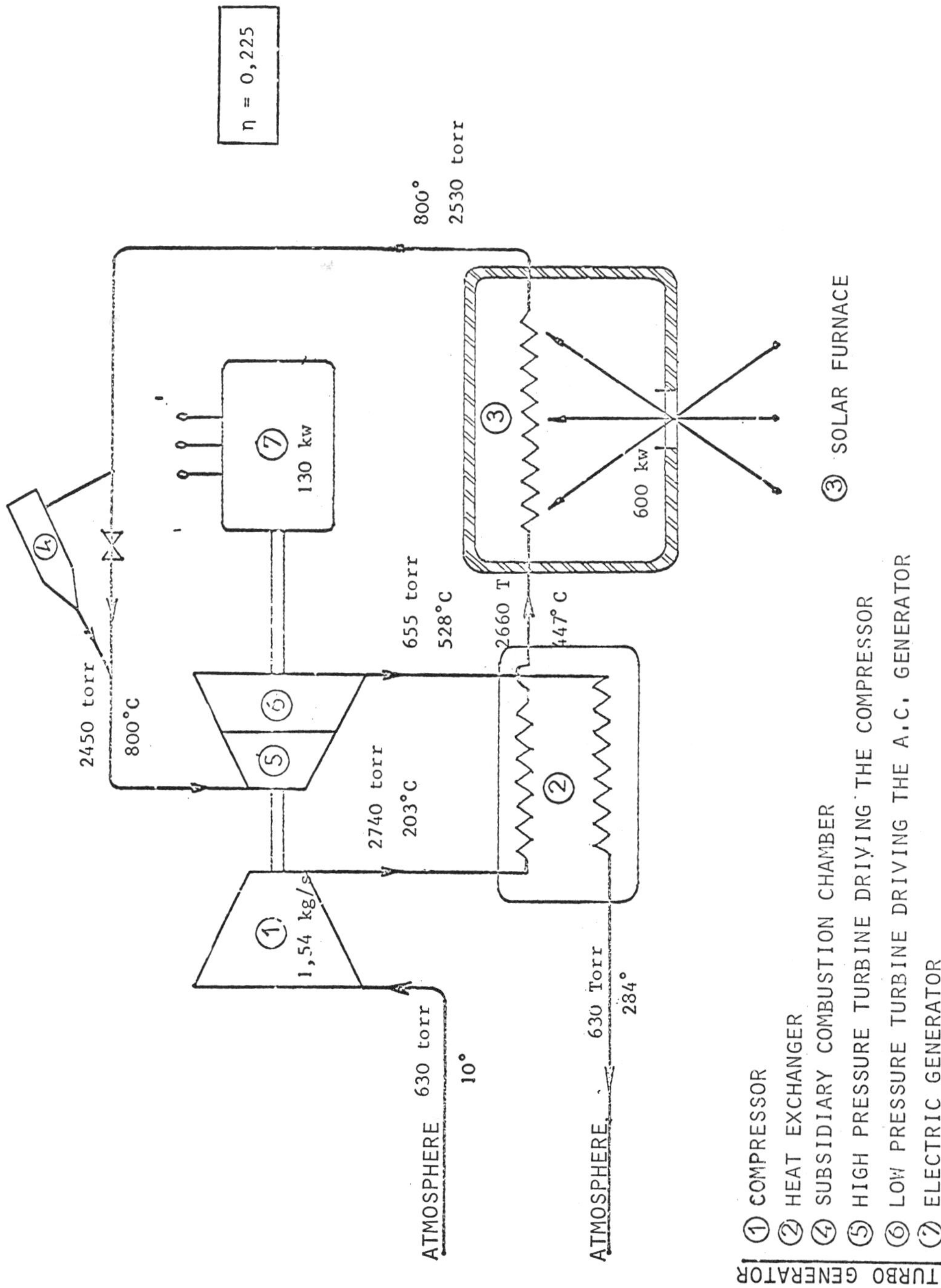
η = 0,225
800°
2530 torr
2450 torr
800°C
655 torr
528°C
2660 T
447°C
2740 torr
203°C
1,54 kg/s
130 kw
600 kw
ATMOSPHERE 630 torr
10°
ATMOSPHERE 630 Torr
284°
③ SOLAR FURNACE
TURBO GENERATOR
① COMPRESSOR
② HEAT EXCHANGER
④ SUBSIDIARY COMBUSTION CHAMBER
⑤ HIGH PRESSURE TURBINE DRIVING THE COMPRESSOR
⑥ LOW PRESSURE TURBINE DRIVING THE A.C. GENERATOR
⑦ ELECTRIC GENERATOR

FIG. 4

FIG. 5

Industrial engine built by ALSTHOM-ATLANTIQUE and S.A.C.M. MULHOUSE.

ELECTRIC GENERATOR SET WITH SCREW EXPANDER AND RANKINE CYCLE

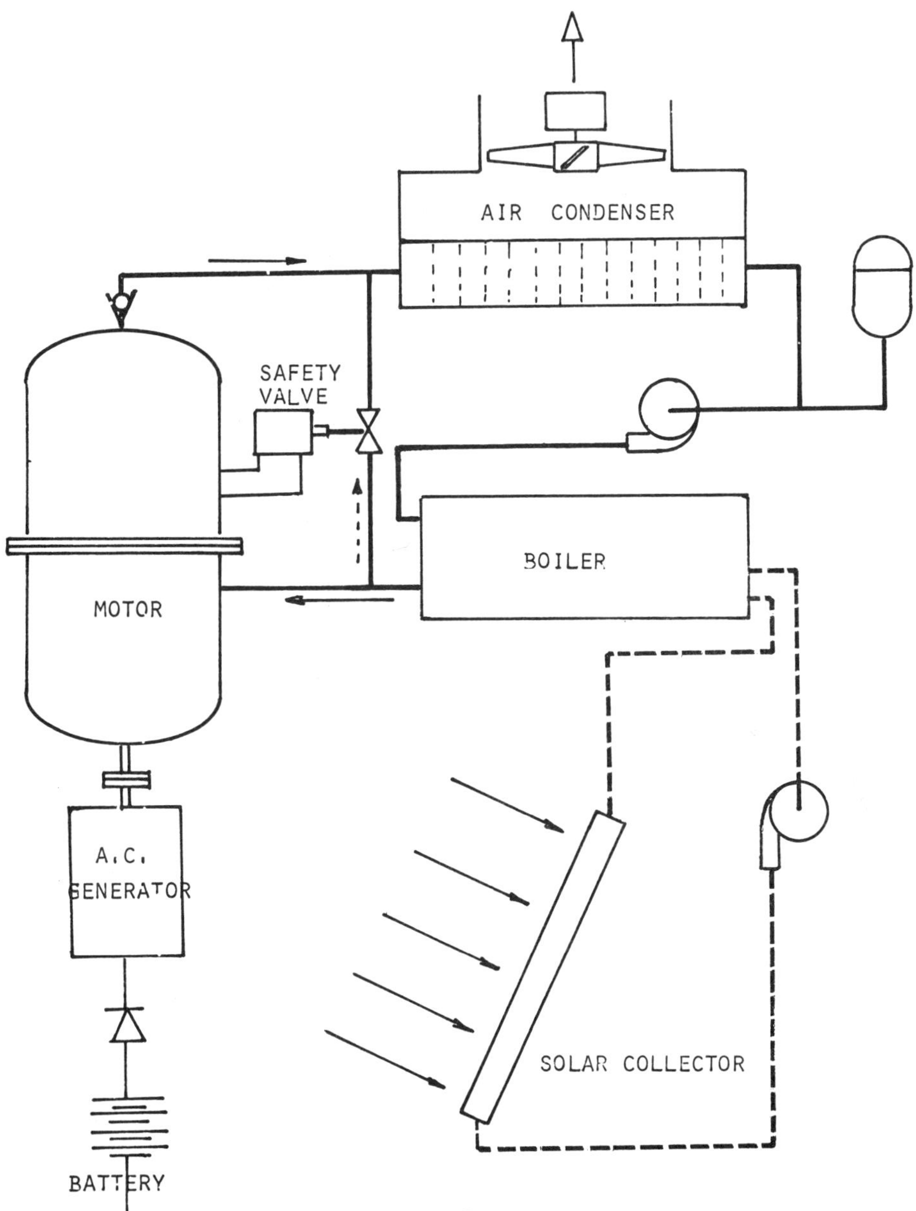

FIG. 6

FIG. 7

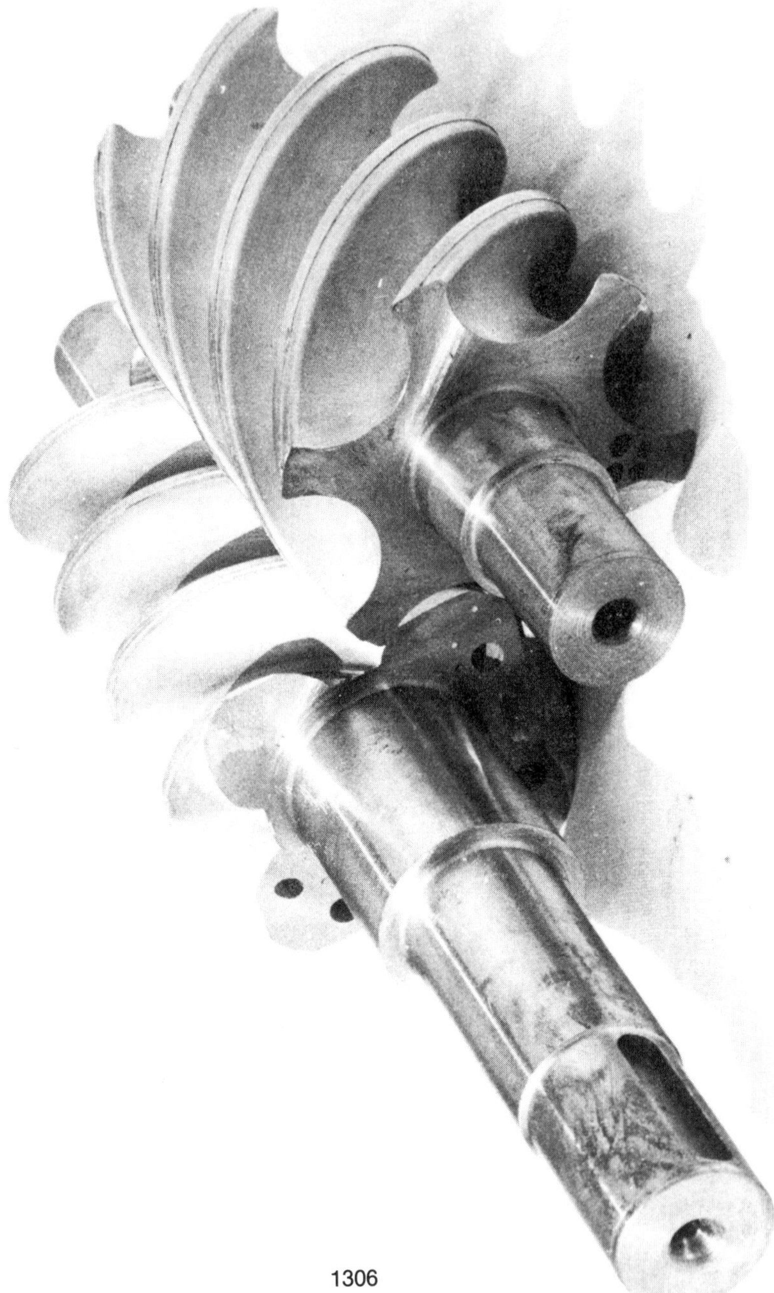

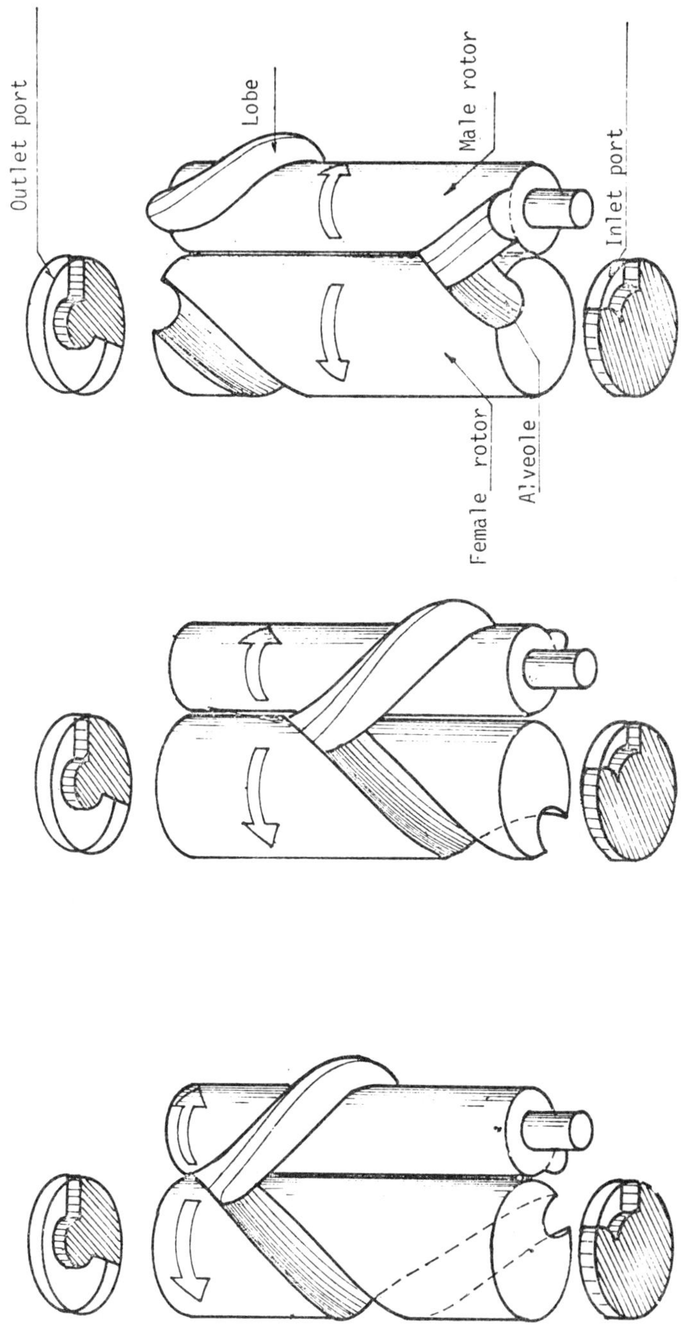

SCREW MOTOR
Different phases of expansion

FIG. 9

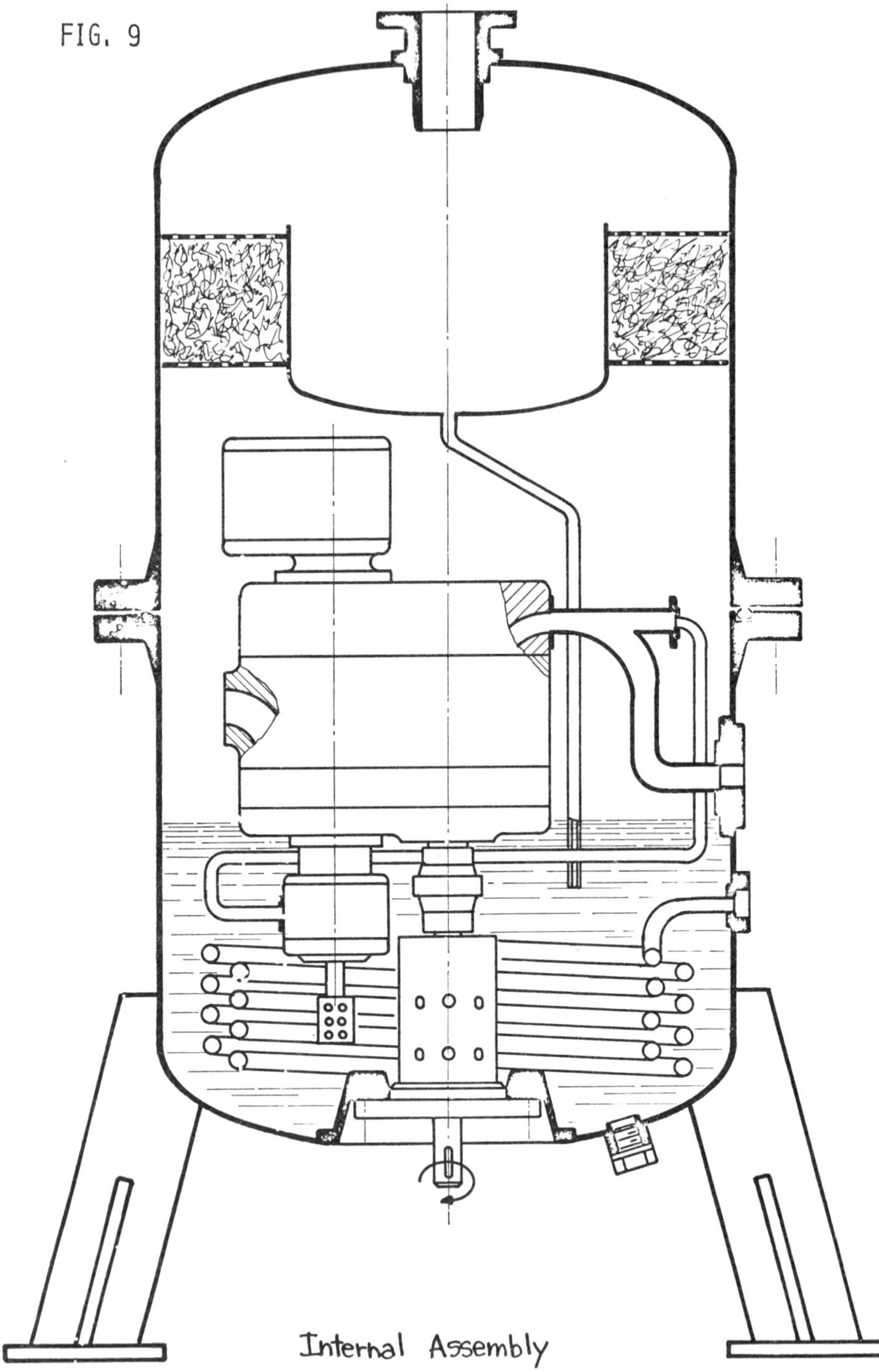

Internal Assembly

FIG. 10

FIG. 10 COMPONENTS OF THE EXPANDER

FIG. 11

CONCENTRATION COLLECTORS

FIG. 12

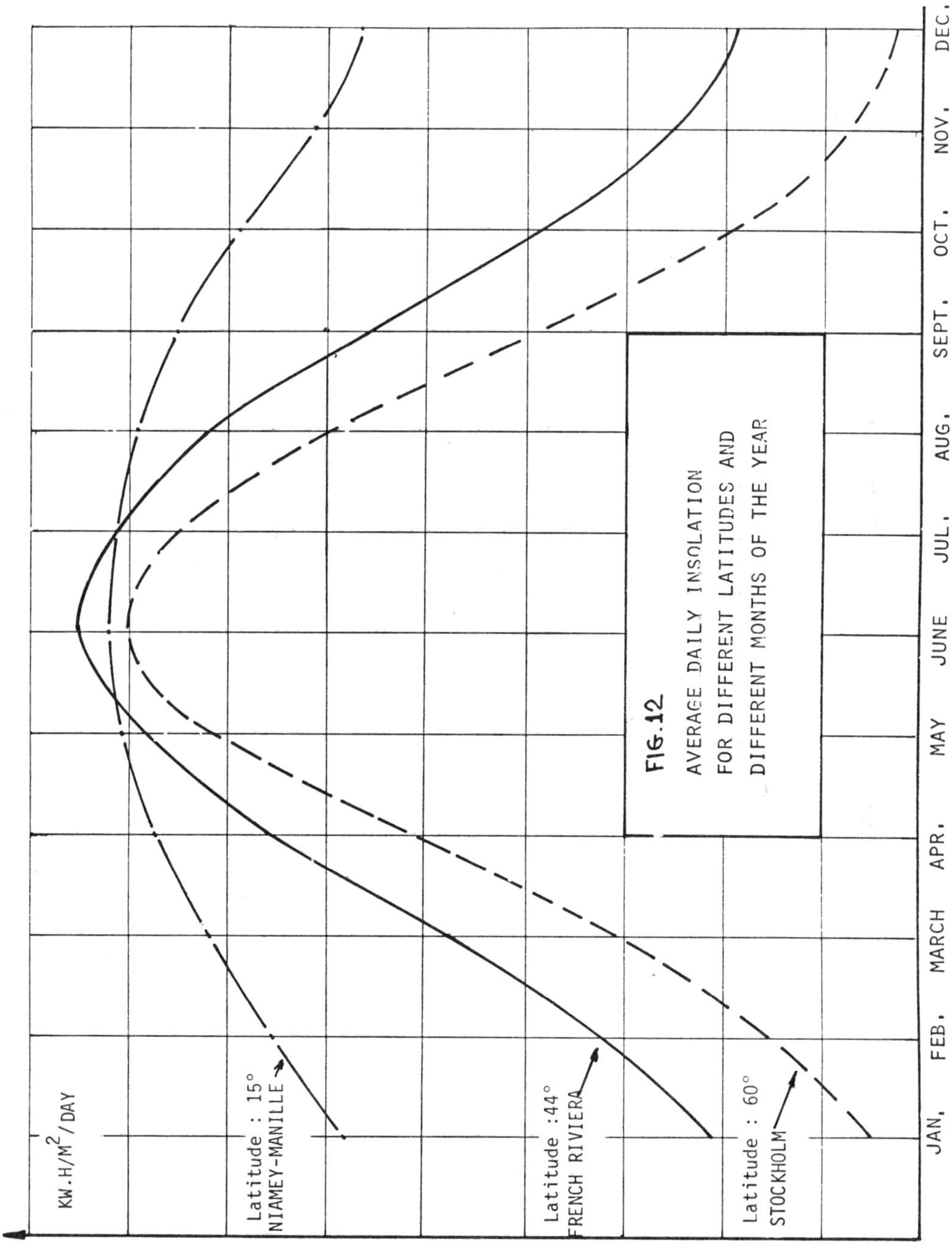
KW.H/M2/DAY
0
1
2
3
4
5
6
7
8
JAN.
FEB.
MARCH
APR.
MAY
JUNE
JUL.
AUG.
SEPT.
OCT.
NOV.
DEC.
Latitude : 15°
NIAMEY-MANILLE
Latitude :44°
FRENCH RIVIERA
Latitude : 60°
STOCKHOLM
FIG.12
AVERAGE DAILY INSOLATION
FOR DIFFERENT LATITUDES AND
DIFFERENT MONTHS OF THE YEAR

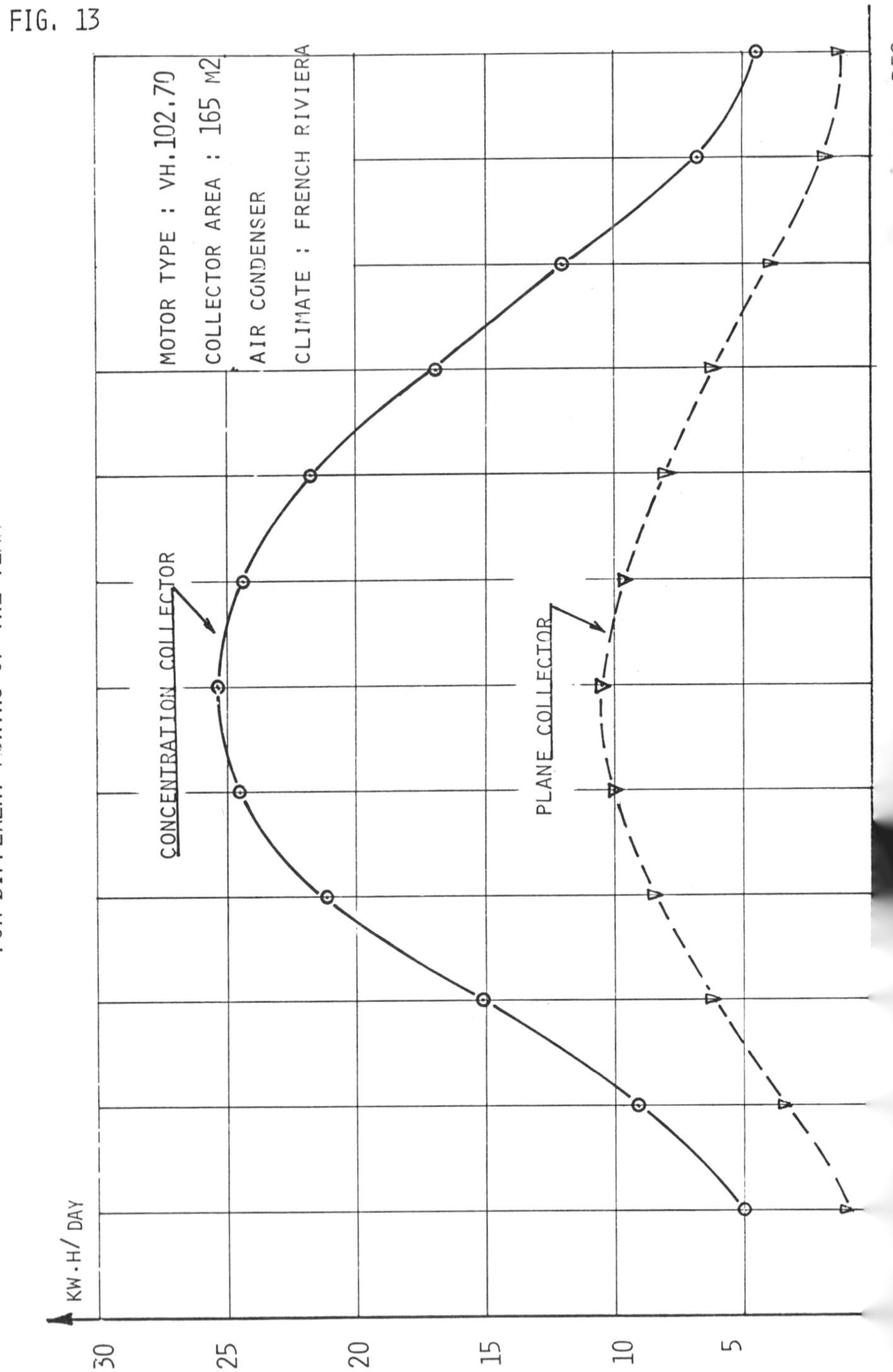

FIG.13 AVERAGE DAILY PRODUCTION OF ELECTRICAL ENERGY FOR DIFFERENT MONTHS OF THE YEAR

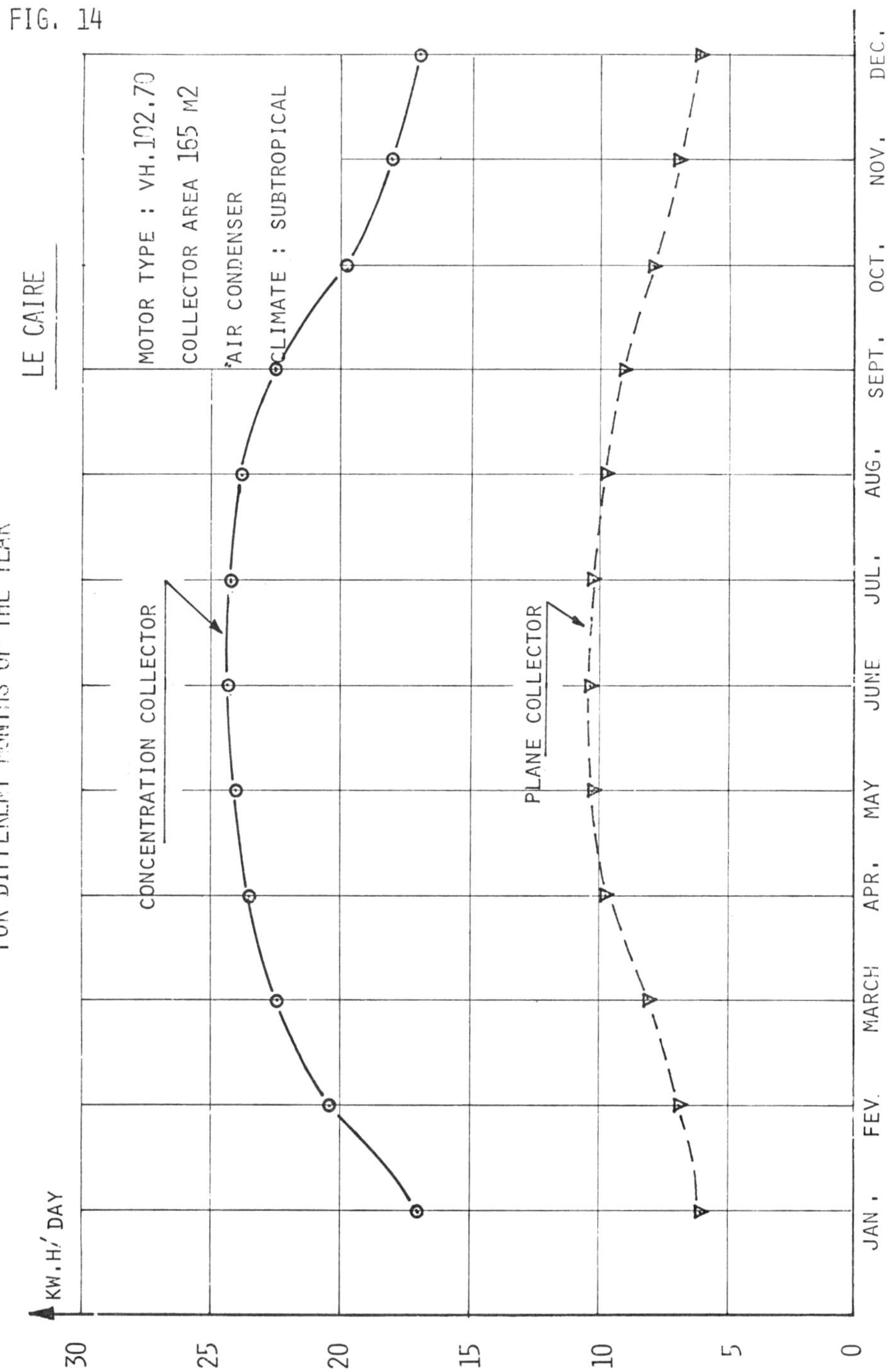

FIG. 14 AVERAGE DAILY PRODUCTION OF ELECTRICAL ENERGY FOR DIFFERENT MONTHS OF THE YEAR

FIG. 15

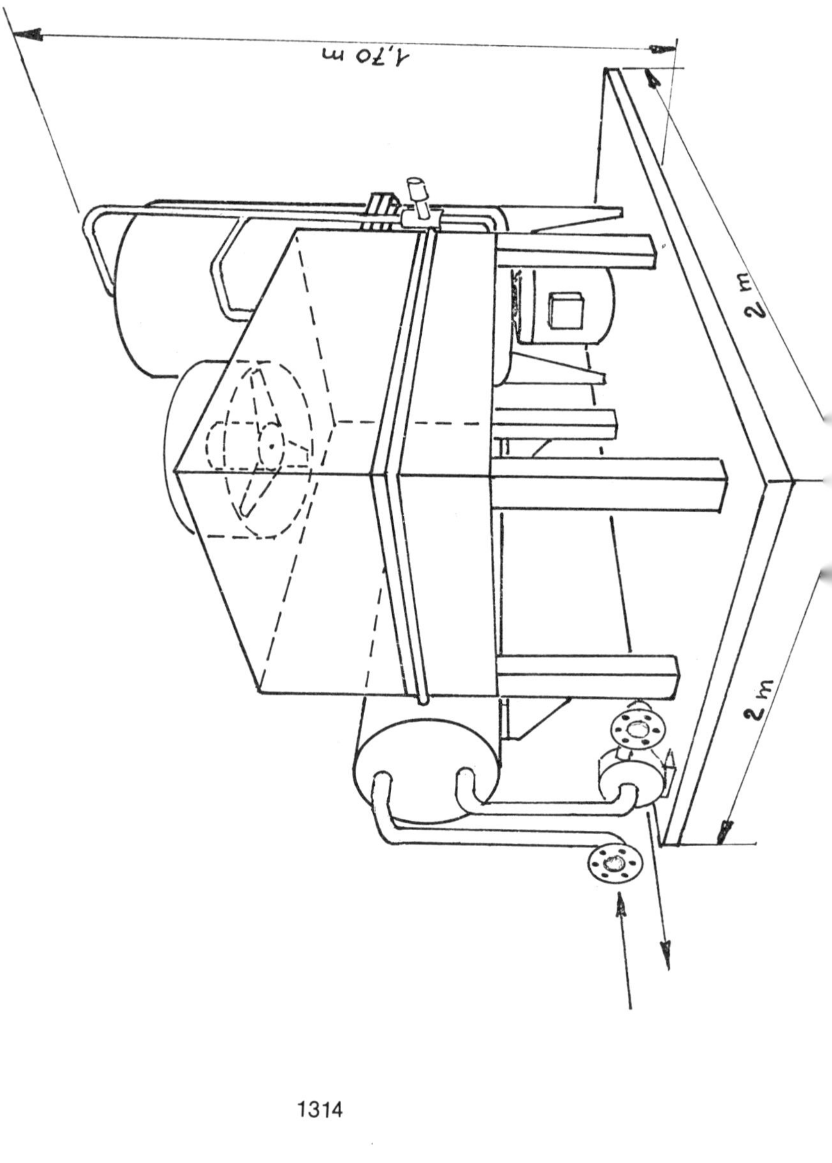

FIG. 15 ELECTRICAL GENERATING SET WITH SCREW EXPANDER AND AIR CONDENSER

FIG. 16

VIEW OF THE TEST RIG

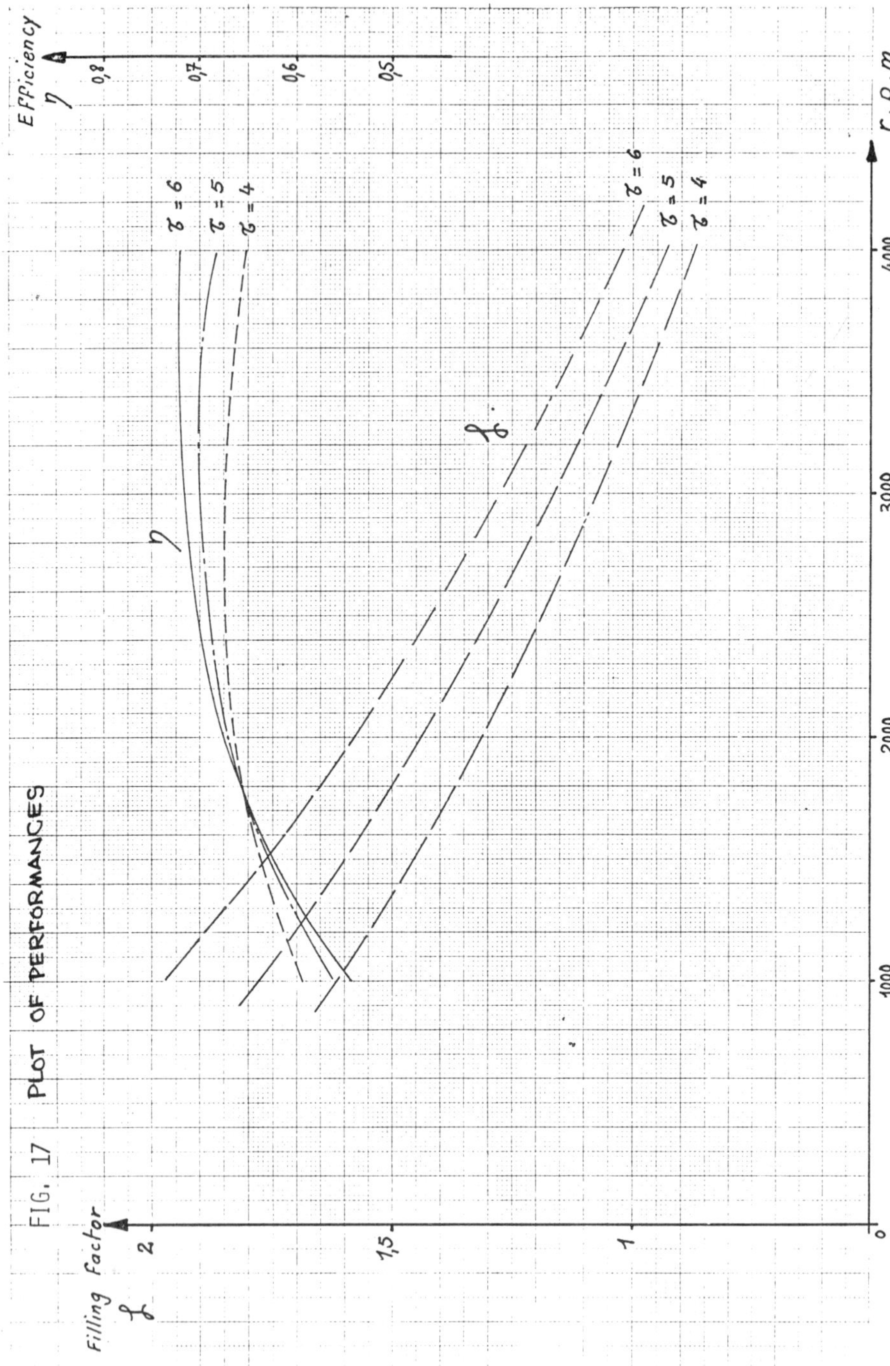

FIG. 17 PLOT OF PERFORMANCES

FIG. 18

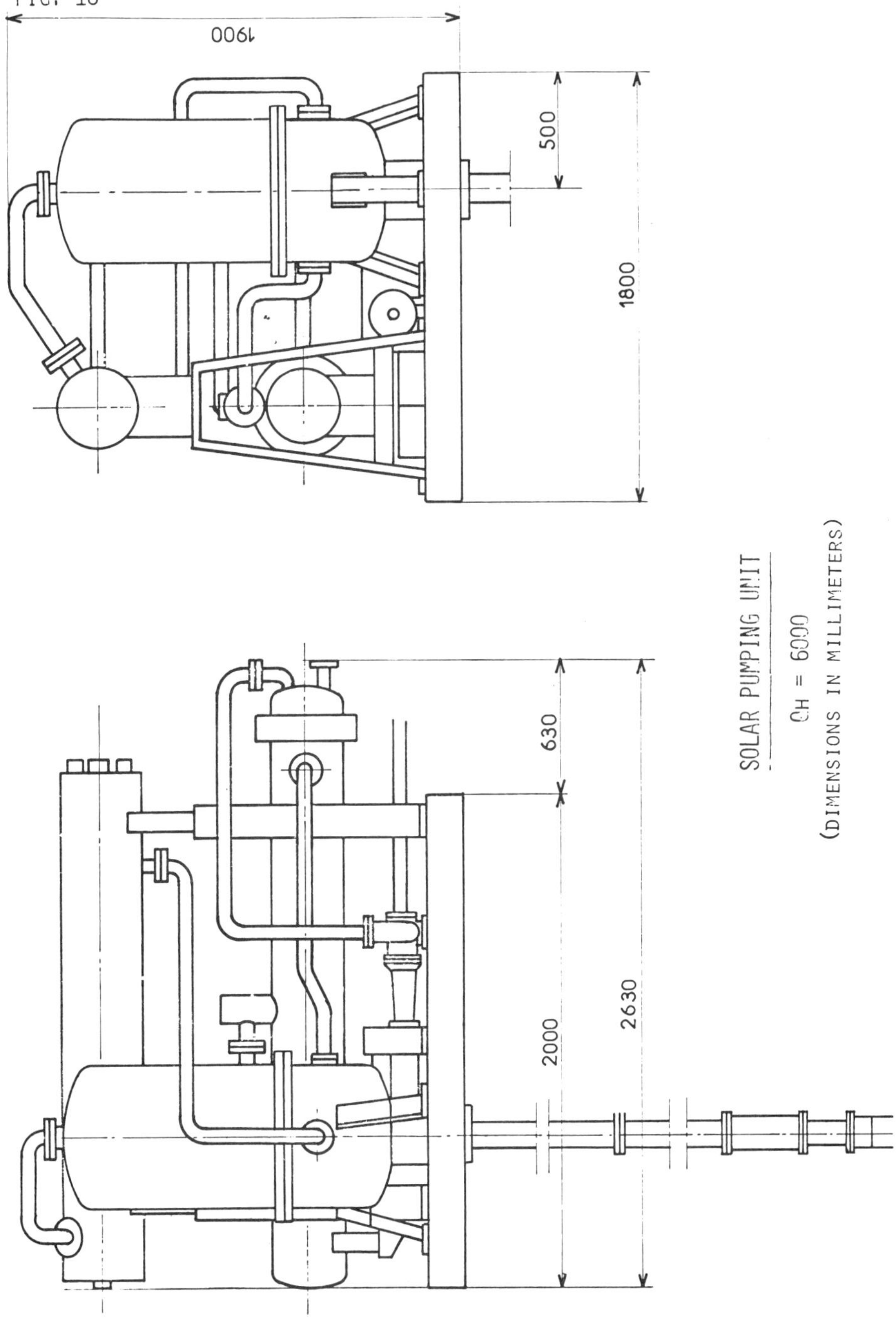

SOLAR PUMPING UNIT

Q_H = 6000

(DIMENSIONS IN MILLIMETERS)

NEW DEVELOPMENTS FOR FUTURE SOLAR POWER PLANTS

J. Lorenz, J. Feustel, M. Kraft
M.A.N. Corporation, Advanced Technology,
Munich, West Germany

ABSTRACT

It is the objective of this project described here to develop solar farm plants for efficient generating of mechanical and electrical energy in the range of 15 to 500 kW for decentralized energy supply in countries with high rates of direct insolation and with the possibility to use the waste heat if required for air conditioning or desalination.

DESCRIPTION OF BASIC PRINCIPLE

Heat of 200 °C to 300 °C generated in suntracking parabolic cylinder collectors having a concentration ratio of 16 to 40. The heat is led off into a storage system by means of thermo-oil. A steam generator then produces high-pressure steam which can be used to drive multi-stage turbines, screw-type engines or piston engines. The waste steam is condensed in the cooling circuit and the condensation heat led off in a cooling tower. The low-temperature heat can be diverted at various temperatures, depending on type of machine, and utilized by an absorption refrigeration plant for air conditioning or refrigeration. It can also be used to drive a desalinization plant, for example. The plant consists of four basic modules: the collector, the heat store, the energy converter and, if required, the waste-heat utilizer, Fig. 1.

OPTIMALIZATION OF OVERALL SYSTEM

Two temperature-related opposing tendencies can be seen as important factors determining rating and efficiency of solar farm plants.

Collector efficiency sinks with increasing temperature whereas the efficiency of the machine circuit increases. This necessitates optimalization to obtain the most favourable process temperatures. Besides the major elements, collector and energy converter, the other components, such as heat storage module, generator and auxiliary drive determine the system efficiency.

Fig. 2 shows the results of this optimalization for a plant of a capacity of about 50 kW where system efficiency ratings of 9 to 12 %, depending on the type of machine, can be expected with collector output temperatures of 300 °C.

Capital investment and operating costs are largely determined, especially in the case of high-rating plants, by the collector area.

A plant with 30 kW nominal rating at a collector temperature of 300 °C requires a collector area of 330 m^2. With a collector final temperature of 100 °C - the maximum achievable with flat collectors - a system efficiency of 3 to 4 % is attained, and thus an area of about 1150 m^2 is required. A comparison of areas required by plants with high and those with low process temperatures speaks clearly in favour of the more efficient parabolic cylinder collectors, in spite of their higher cost. This tendency is reinforced by the much lower land requirements and by the possibilities of considerably increasing efficiency by selective use of waste heat. The energy flux diagram shown in figure 3 for the plant discussed here makes this apparent.

PROTOTYP AND COMPONENT DEVELOPMENT

The first prototyp plant shown in Fig. 4 served to obtain basic data and had the following specifications:

platform size 60 m^2, heliostatic tracking, east-west orientation of collectors with a concentration ratio of C = 16. The collector, dimensioned 1,2 x 2,4 m is of frame and stringer type design with a reflecting surface of acrylic material or aluminium foil. The absorber tube is coated with black chromium ($\alpha/\varepsilon = 1$) and protected against convection losses by cladding with a glass tube. The collector dish is swivel-mounted around the absorber tube on a structure of props and struts. The entire structure is swivel-mounted on a ring with a 4 m diameter which can be pivoted around the vertical axis. With this plant collector output temperatures of 230 °C at efficiency rating of about 50 % had been measured.

In order to fulfill all requirements in regard to collector temperature, maximum energy exploitation and low manufacturing costs an improved collector platform had been conceived and constructed on the basis of systems optimalization and the results obtained from the first prototyp plant.

Based on this concept the 2,5 m wide collectors with a concentration factor of about 40 are joined together to form collector modules ranging from 15 to 150 m^2. Fig. 5.

The collector modules are rigidly arranged in north-south direction pivots around the vertical axis. The reflector consists of 4 - 5 mm hot formed glass of commercial type with a reflection of 85 %. With collectors of this size precision of manufacture and assembly is important since efficiency losses which are not related to temperature are largely determined by this factor. For the absorber steel tubes with selective coating, clad in glass tubes are used.

The galvanized black chromium gives a α/ε value of about 4 at the envisaged operating temperature of 300 °C and efficiency ratings of ~60 %. A ceramic filler is introduced between steel tube and glass cladding tube to reduce radiation losses from that part of the absorber tube which is not exposed to insolation. Operating safety and cost considerations spoke against evacuating the absorber cladding tube.

Fig. 6 shows the efficiency curve plotted for the selected collector in comparison with a collector used in the prototype plant.

ADVANCED DEVELOPMENTS

Collector Field

When developing the new concept, particular emphasis was placed on efficient manufacture through standardization of single components wherever possible, long service life by selection of suitable materials and manufacturing processes as well as easy exchangeability of damaged or defective components. The following concise description of concept will demonstrate this:

Fig. 7
Each single reflector with a reflecting area of 1 m^2 is equipped with 4 metal plates attached to the back. Rubber elements riveted to the metal plates serves as a connection to the main structure ensuring that varying thermal expansion and deformations of reflector and structure are absorbed. The supporting structure of the reflector consists of a transverse profile and 4 longitudinal profiles which have ends designed to receive the rubber elements and which are connected to the central tube.

Fig. 8 shows the assembly scheme for the 16 m^2 module.

These so-called beam-structures may be matched to form platforms of varying sizes which can be connected to the standardized tracking unit to fit all platform sizes.

Fig. 9 shows five examples for varions platform sizes of 16, 24, 32, 48 and 72 m^2.

The tracking unit fitted with two swivel mounts for azimuth and elevation with the azimuthal tracking unit being equipped with two motors one for solar tracking and one for fast return.

The fast return mechanism primarily serves to turn the platform away from the sun to prevent overheating in case of failure of the circuit pump or in emergency situations, and to return the platform to its nocturnal position. The platform may also be placed in a horizontal position, i.e. the reflecting area will be facing downwards to protect the reflectors from being soiled by dew or from being damaged when wind velocity is 130 km/h or above.

The electrical/electronic part of the tracking unit is so designed as to allow platforms of small plants with a low number of platforms to be controlled individually by sun sensors, while platforms of large plants are controlled by a central control and monitoring unit by means of microprocessors.

Studies of oscillation carried out on single components showed that resonance frequencies are negligible which means that oscillation excited by the wind is not to be expected. Inherent damping of the structure is sufficient to quickly absorb oscillation that might occur.

Static test confirmed our calculations to the extent that structure and reflectors will not be destroyed when placed in a vertical position in wind velocity of 130 km/h and when placed in a horizontal position if velocity is above that value.

Fig. 10 shows a complete collector module of about 32 m^2 collector area of the described design.

Energy-conversion Circuit

The energy-conversion circuit comprises the steam generator, expansion machine, generator, condensor, cooling tower and a number of control, monitoring and auxiliary devices.
Fig. 11 shows the temperature/entropy diagram for the circuit chosen for the solar farm. The water is heated in the steam generator from 60 °C to about 250 °C and converted into steam under a pressure of 40 bars and superheated to about 280 °C. The expansion from 40 bars to 0,2 bar with about 15 % moisture content takes place either in a fast running multi-

stage steam-turbine or in a 2 or 3-stage screw-type expander, dry rotors being used for the upper temperature range and wet rotors for lower temperatures. Whereas multi-stage turbines of relatively high efficiency ($\eta \geq 60$ %) in the upper performance range are already available on the market, screw-type machines of simple design are presently being qualified for high temperature operation under relevant development programmes, and would appear to offer advantages, especially in the lower and medium performance ranges of solar farm plants. M.A.N. has a wide programme of screw-type compressors and is presently developing special screw engines having ratings of 15 to 200 kW and operating at a temperature of about 300 °C, i.e. suitable for solar farm plants. This development locks promising since with an efficiency of 60 to 70 %, much lower speeds - about 25 % of that of a corresponding turbine - very simple construction and safe operation can be expected.

Fig. 12 gives an impression of the simple construction of the machine envisaged. The relatively low speed of the screw makes possible the coupling of a three-phase alternator either direct or via lowreduction gearing.

The choice of static frequency changer depends on the solar farm's mode of operation. If the generated power is to be fed into the grid the static frequency changer is grid controlled, a frequency setting not being necessary. In the case of island operation the static frequency changer must fulfill the conditions normally applying to emergency power plants. Power plants having to meet specific consumer demands can be allowed greater frequency and voltage fluctuations, depending on operating conditions.

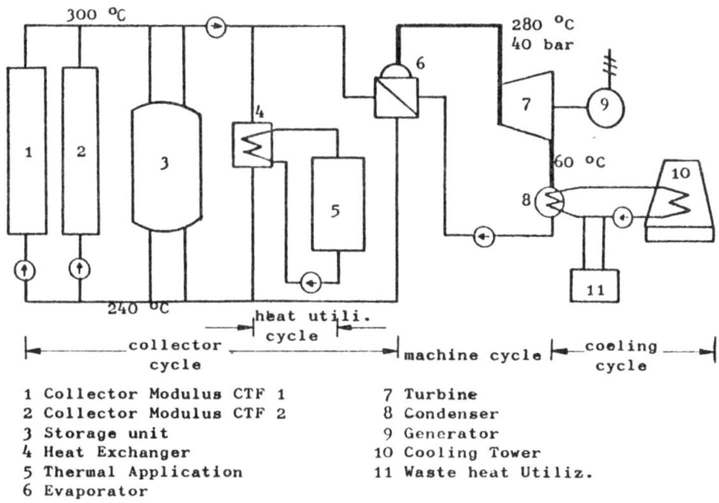

Fig. 1 Thermodynamic Circuit

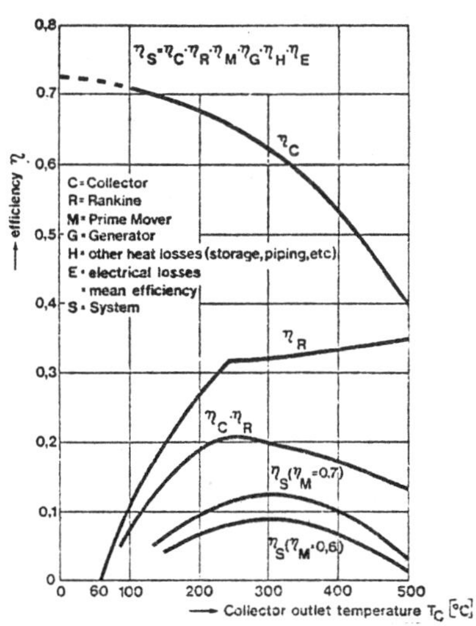

Fig. 2 Efficiency of M.A.N.-Solar Farm

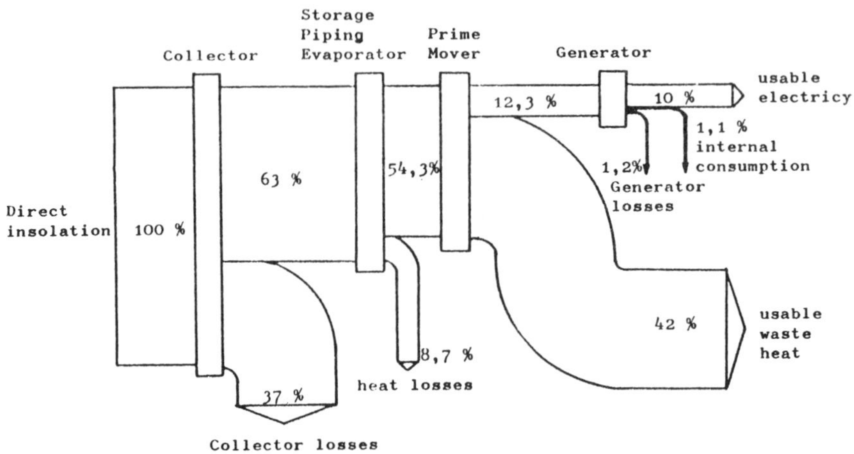

Fig. 3 Energy flow Diagram

Fig. 4 Prototyp Plant 60 m^2

Fig. 5 Improved Collector Platform 90 m^2

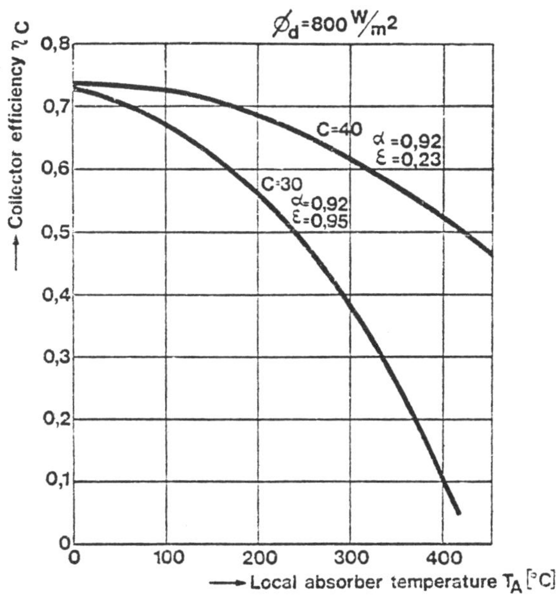

Fig. 6 Collector Efficiency of M.A.N.-Collector

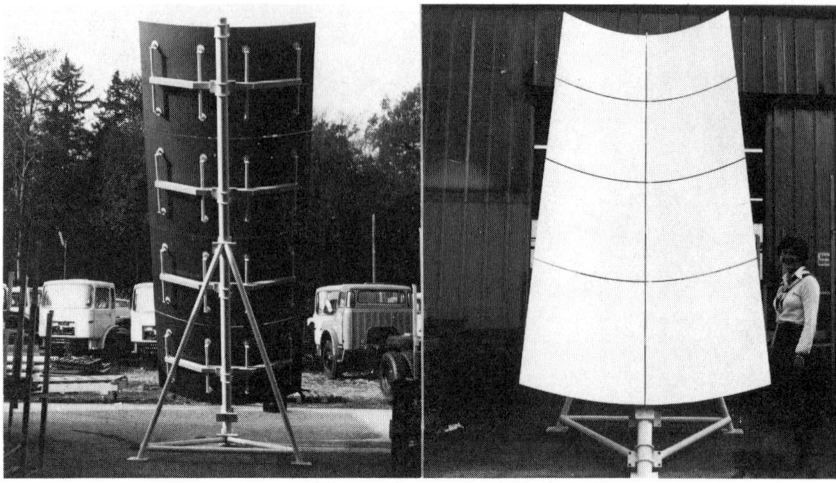

Fig. 7 Back and front View of Collector Module

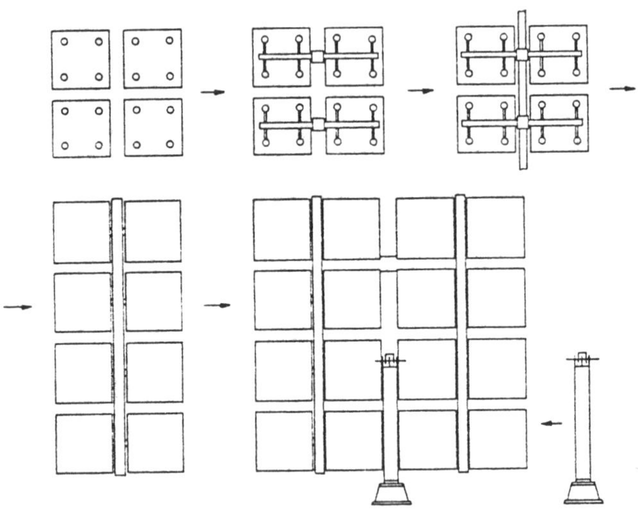

Fig. 8 Assembly Scheme of Solar Farm Modules

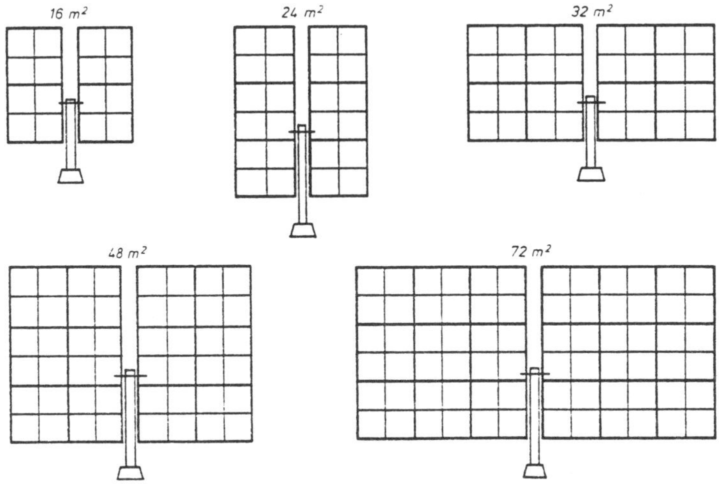

Fig. 9 Module Sizes

Fig. 10 Collector Module 32 m^2

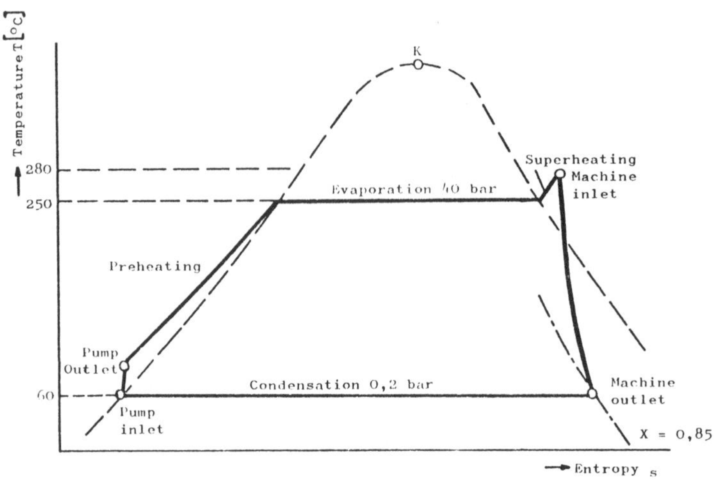

Fig. 11 Temperature/Entropy Diagram

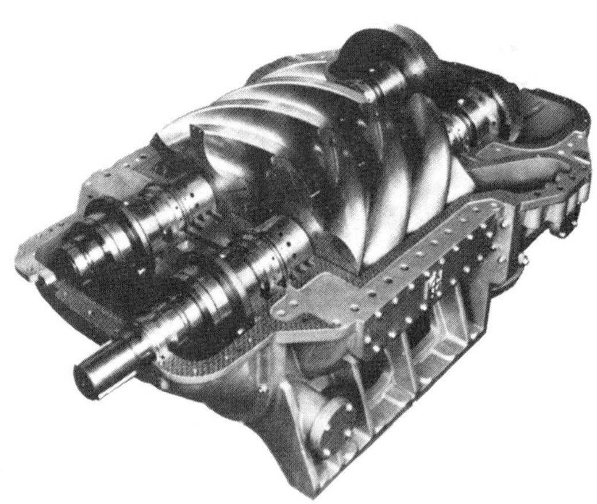

Fig. 12 Screw Expander

POWER FROM THE SUN BY THERMODYNAMICS FRENCH TOWER TYPE PROJECTS

M.J. Bignon
CETHEL
Paris, France

ABSTRACT

After successfully commissioning a 60 kWe unit at Odeillo, France is pursuing three tower-type projects.

Multi-Megawatt Project

Themis is jointly developed by C.N.R.S., E.d.F. and the private group CETHEL. The prototype 2 MWe unit is due for start-up in Targasonne near Odeillo in 1980. Component development is well under way. Main features are cavity receiver with molten salt heat transfer and storage.

Intermediate Projects

500 to 1000 kWe

Inti 800[R] is a 800 kW sophisticated unit for low industrial background areas, using Themis-type mirrors and Odeillo thermodynamics. Systems engineering is proceeding under a government research contract.

EEC research pilot plant, 100 kWe, is in the design stage for 1980 completion in Sicily. The supplier is a three partner consortium from French, West German and Italian private industry. Funding is 50% national and 50% European. Mixed technology for mirrors, anti-radiative direct superheated steam receiver and short duration storage are the main features.

INTRODUCTION

The CNRS solar furnace at Odeillo was built about 1967 under the direction of Professor Felix Trombe and it has set a long-lasting world record with 1 MW (thermal) capacity and temperatures above 4000°K.

In 1975, the CNRS decided to use it as a proving ground for the concept of a single receiver, high concentration solar thermodynamic power plant (flow sheet no. 1).

The high concentration factor and the one-sided field made a cavity type receiver very attractive. An organic fluid was selected as the transfer and storage medium in order to avoid the engineering problems associated with:

- the low transfer factors of the superheated system
- the design and operating problems of a solidifiable medium

The maximum temperature was thus limited to about 330°C (bulk), but this is perfectly suitable for the generation of medium pressure, slightly superheated steam to feed the kind of plant we had in mind. The main constraint on the unit was indeed prompt delivery rather than high efficiency and for that reason, the turbine had to be obtained "off the shelf".

A thermal storage with a 2-hour full-power capacity was incorporated to investigate the operating conditions of a stratification vessel and allow for an operation of the turbine when there is no sunshine.

The unit was ordered in March 1976, and started up in October of the same year.

The receiver was designed and supplied by Fives-Cail Babcock, the thermal conversion system by Caliqua, a member of the Heurtey group.

The operation of the plant proved that the design is sound and easy to control. It has also revelaed some of the practical problems associated with the closed loop control of the mirrors and the heat conservation in the storage.

During the year 1976, following a suggestion by the late M. Gregory, head of the C.N.R.S., the industrial partners who had taken part in the Odeillo plant entered a joint venture called CENTHEL with the purpose of further developing solar thermodynamic tower-type projects.

The partners are:

Air-Industrie, a member of the Saint-Gobain-Pont-A-Mousson group, glass specialists since the 17th century.

Caliqua, a member of the Heurtley engineering group.

Fives-Cail Babcock, thermal equipment and heavy machinery suppliers.

Seri, the engineering branch of the Renault national car manufacturer.

There was indeed at about that time a flurry of projects, three of which have materialized and will be described hereafter.

Project THEM - THEMIS

Project THEM is part of the C.N.R.S.-Pirdes program which has been described by another paper in this Symposium. Its purpose is to develop high power systems in the multi-megawatt range with emphasis on efficiency as a condition for achieving competitive prices in the long run. For such a project, an association with the nationalized utility authority E.d.F. was an obvious choice and this association selected CETHEL, first as a preferred partner for the development of components, then as an associated industrial architect for the construction of the first prototype unit THEMIS.

Mirrors are the first major componentsof a tower-type project. Even before CETHEL was formed, two members had performed basic design work on mirror construction and control which culminated in the delivery of two first generation heliostat prototypes to be tested at the E.d.F. facility, in Martiques near Marseilles, in the summer of 1977. At the same time another prototype, developed by C.N.R.S. themselves, was also tested.

The CETHEL heliostats have reflective areas of about 50 m2 each and they use the principle of focussing by single direction bending long rectangular mirrors that are slightly offset to tangent the theoretical profile of the optimum reflective surface. They use 6 mm back-silvered float glass. The mounting is altazimuthal and the tracking by open loop control from a central mini-computer.

The C.N.R.S. prototype has about 44 m2 made of flat facets, laser adjusted to tangent the theoretical sphere. The control is a closed loop with individual sensors and low definition override by open loop for non-sunshine periods and other modes of operation.

All three prototypes have demonstrated their ability to feed a cavity-type receiver at a distance of 300 m and over and they have supplied very useful information for the design of second generation prototypes which are presently in the development stage for testing this summer.

The receiver will be a cavity-type, single phase system. It was felt that the advantage in low thermal losses was to be preferred to those of light weight and adaptability to surrounding mirror field for the latitude of southern France and the power level of 40 MW (th) maximum, contemplated for a module.

Originally there were two design alternates: one with high pressure saturated steam, one with molten salt. Direct solar superheating of the steam was deliberately rejected in view of design and start-up problems.

In order to conserve the high potential of the solar heat, the temperature must be as high as possible, which means a high steam pressure, tentatively 180 bar or 357°C. The receiver had to be heavy, with consequences on the supporting tower, and expensive. It was nevertheless developed to have an alternate choice in case the molten salt concept would not prove to be acceptable.

In the meantime, the technical references of using sodium-potassium nitrite-nitrate eutectic for heat transfer were investigated and a test loop for components installed. Successful applications in the chemical industry (caustic soda, phtalic anhydride) were found at a power level commensurate with the project, so the molten salt design has become the preferred choice for the first prototype unit.

The salt is stable up to about 500°C, which allows for a definite improvement in the available steam conditions. However, it is not clear how large amounts of salt, as those required for the heat storage, will behave. It is expected that the portion of the salt subjected to the solar flux in the receiver will suffer most and therefore that part of the circuit will be made of higher alloy so as not to be a limiting factor when progressively raising the operating conditions of the unit.

The THEMIS project will embody the results of the fundamental and applied research and provide an opportunity for practically testing the components and the operating modes of a solar unit when coupled to a utility grid. At the same time, it will be a reference for the solar technology at the multi-megawatt level.

The decision to proceed with the construction of a prototype unit was announced by the public powers in September 1977 and funds were earmarked for that purpose in the 1978 budget.

A site was selected in December 1977 at Targasonne very close to the Spanish border, a few miles from Odeillo.

The design capacity of the plant is 2 MW (e) with a peak at 3 MW. The mirror field will be comprised of 350 heliostats and enjoy 2,600 hours of direct solar illumination per year. The high altitude (1,700 m) provides a high solar constant, with peaks well above 1 kW/m2 normal area. Weather conditions are mild for such a mountain site and the wind specification is 50 km/h (14 m/s) for normal operation. Precautions are specified however for snow and freezing.

The molten salt receiver is one typical feature of the project. It is designed for 335°C inlet, 450°C outlet and a salt flow rate of 280 tons/hour.

The selection of focussing mirrors provides a high concentration of energy at the focal plane, where the opening of the receiver is located. This keeps the thermal losses low. The concentration is somewhat less at the receiver exchange surface: the heat transfer problems and associated risk of local overheating are thus alleviated.

The other problems of the salt system are:

the control of the hydrostatic pressure in the circuits at ground level and, more specifically, in the heat storage. This is done by dividing the system into two loops:

One primary loop including the receiver, the downcomer, the exchanger and the riser, with a circulating pump and controls

One secondary loop, including the shell-side of the exchanger, the storage tanks and the steam generating exchangers

The high flux transfer at high temperature is thus confined to the primary loop.

The storage, containing 600 T of salt, has a 50 MW (th) capacity.

- The prevention of the solidification which occurs at about 150° when the salt is new. This temperature goes up with salt aging and is the usual reason for salt regeneration. Electric tracing of the pipes for start-up and the presence of an auxiliary fired heater on the salt circuit should be an answer to any foreseeable situation.

The steam generator, heated by the salt, will supply 50 bar, 410°C superheated steam to the turbine. The design of the machine is not completely frozen at this time. Consideration is given to radial flow vs. axial flow.

The condenser will be direct air-cooled as a forerunner of plants designed for desertic areas. By the same token, freezing problems on cold winter days are alleviated.

The program of plant operations includes

- Run-of-the-sun production of power, without making use of the storage
- Peak hour production, supplementing the sun with power draw from storage.
- Round-the-clock production, with optimized management of the storage to fit the demand curve.

It is expected that the plant will yield about 3,500 MWh (e) per year.

Project INTI 800[R]

In April 1976, the Delegation for New Energies sent out a "call for ideas" for an intermediate power solar unit to be used in isolated areas.

Drawing upon the design work made for Odeillo and for THEM, CETHEL quoted a 800 kW tower-type plant using THEM type mirrors with Odeillo type thermodynamics. Its registered trade name is derived from the name of the ancient Peruvian sun-god.

This project was submitted to the technical commission who retained it for further design work with a subsidy from the Research Fund of the D.G.R.S.T.

A report is due next month and it will develop the emphasis given to the following features:

- non-sophisticated design, wherever possible. If not, like the mirror control, readily exchangeable components.
- conventional equipment, for which operators and maintenance shops can be found anywhere in the world

- stand-alone basis, allowing installation in remote areas where no other electric power is available. Sun energy, depending on availability, may be the major source of just a fuel-saver.
- Upper and lower limits for technical feasibility and economic competitiveness will be investigated.

Such a unit could already compete with Diesel units in high fuel cost areas. Its development depends on the existence of a prototype plant to demonstrate the economics of the operation. The new public agency called Solar Energy Committee is considering the possible funding of the project in the near future.

The EEC Project

The Commission of the European Communities has investigated the feasibility of a 1 MW(e) pilot plant since 1976 with the help of a British consulting company who based their report on French, German and Italian technologies.

The report was considered satisfactory and the three countries were asked to nominate an industrial partner who would have to do the work and be supplied with the funding to cover 50% of that work. The other 50% is to be covered by the Commission.

France selected CETHEL upon recommendation by the Delegation for New Energies and the DGRST promised to cover the national share of the funds,

Germany selected Messerschmitt-Bolkov-Blohm on a similar basis,

Italy selected Ansaldo in an association with the national utility Authority ENEL, who were to act as a host partner.

Conversations between the partners led to an acceptable sharing of the work and confirmed the budget. Thus, on November 15, 1977, a contract was signed to cover phase B - design, and phase C - construction, of the project. This made it, at that time, the largest running solar project.

The mirror field will be supplied partly by CETHEL, with a technology derived from THEM, partly by M.B.B. This will give an opportunity to compare heliostats of very different designs. The control will be all open loop for the French part of the field and it may include a closed loop system for the M.B.B. mirrors. The field will be comprised of two separate subfields according to the mirror design, one East, one West. For obvious operating and safety reasons, there will be a common control.

The receiver is derived from the experimental work performed by Professor Francia at the Sant Ilario Solar Station. Making use of the anti-radiative structure that was developed and proved at a lower power level, this receiver is able to produce superheated steam from cold water in one sequence. The superheater is not subjected to the high direct solar flux, but to a re-radiated low flux which avoids the heat transfer problems usually associated with the poor characteristics of the superheated steam.

The purpose of the plant is deliberately experimental, even though the power produced will be fed into the Italian national grid. To keep the investment low, the thermal storage design is reduced to the absolute miniumum required to protect the machinery against thermal shocks. Half an hour coasting power is available from a superheated water vessel which produces steam by flashing at a sliding pressure. This steam is superheated by exchange with a small amount of molten salt which is kept at high temperature by exchange with a slip-stream of steam during normal operation.

The site has been selected on the southern slopes of the Aetna Mountain in Sicily. It enjoys a good amount of sunshine, a favorable soil and the vicinity of plentiful cooling water.

The design phase of the project is proceeding on schedule and it will supply by the end of this year, a firm base, both technically and financially, to decide the execution of the construction phase. According to the present plan, the unit should be ready within two years after this decision. It will be operated by ENEL and available to all three partners for experimental purposes during a few years, after which it will become the property of ENEL for a nominal fee.

CONCLUSION

The main features of the three tower-type projects presently designed in France are collected in the following table.

THEMIS is a reduction of what could be a power station of tomorrow, taking its share in the baseload or intermediate base of a large power grid. High efficiency conversion is a necessity, even at the cost of sophistication.

INTI 800^R is a prototype of a stand-alone unit designed for remote areas with scarce industrial background. Ease of operation and maintenance is paramount.

EEC is a proving ground where varied technologies will be compared. Flexibility is important. The low storage capacity will be tested for compatibility with the requirements of a grid connection.

Each of them will bring specific information for the needs of future developments.

TABLE I

FRENCH TOWER-TYPE PROJECTS - MAIN FEATURES

	THEMIS	INTI 800[R]	EEC
Power (MWe)	2	0.8	1.0
Mirror Field	North	North	North
Unit Area (m2)	50	50	25 and 50
Tracking	not frozen	open loop	closed loop & open loop
Tower Height (m)	80	50	50
Receiver Type	cavity	cavity	anti-radiative
Fluid	molten salt	H.T. organic	steam-water
T max (bulk) °C	430	330	510
Storage Type	molten salt	H.T. organic	pressurized water & molten salt
Capacity (MWh)	50	9	~0.5
Steam Cycle			
Inlet Pressure	50 bar	26 bar	90 bar
Temperature	410°	320°	510°
Condenser Pressure	0,1 bar	0,15 bar	0,06 bar
Purpose	Power Generation	Demonstration	Experimental

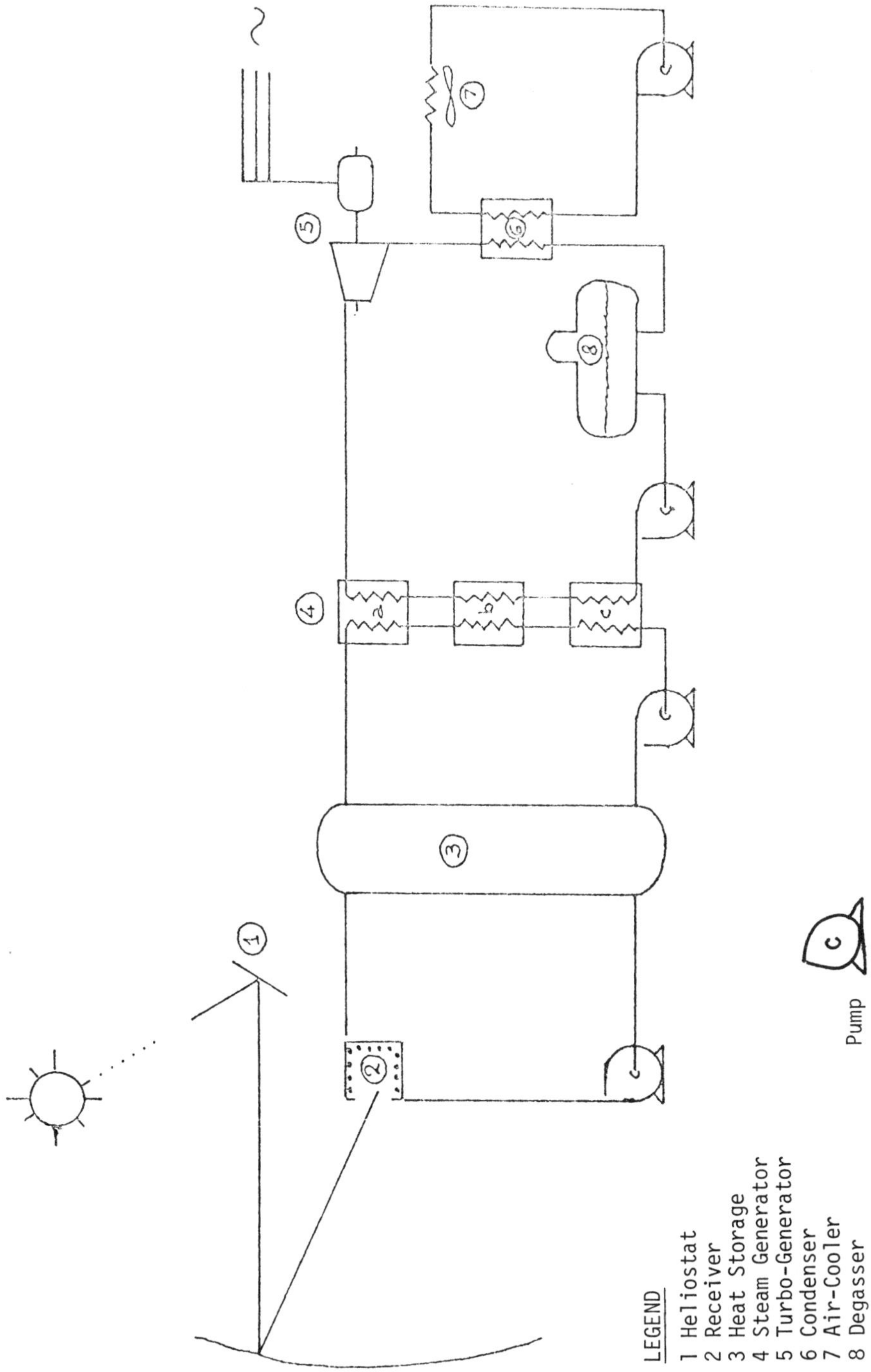

Figure 1 Odeillo Solar Energy Experiment - Flow Diagram - Main Circuits - CETHEL

STEAM AND GAS TURBINES FOR SMALL OUTPUTS USING SOLAR ENERGY

K. Bammert
Institute for Turbomachinery and Gas Dynamics
Hanover University

H. Pösentrup
Maschinenfabrik Augsburg-Nürnberg Aktiengesellschaft
Advanced Technology Division, Munich

Federal Republic of Germany

ABSTRACT

The conversion of the solar energy into electrical or mechanical energy in turbine power plants requires new design criteria. The choice of the working fluid and the configuration of the cycle are of great importance. Not every variant is suitable for the solar farm and the solar tower concept in the same way. The design of the turbomachinery and the other components of the cycle do strongly influence each other. An important point of view in designing a small solar power station is the simplicity of the construction so that no special service is required in remote locations. The efficiency should be as high as possible. Some important criteria for the choice and design of small power turbine plants using solar energy are discussed. The advantages of open cycle air turbines are compared with steam turbines. Indications to the choice of suitable cycles and turbomachinery are given.

INTRODUCTION

Electric current can be generated by solar energy. Feeding this current into grids supplied by fossil or nuclear operated power plants is not yet economical. The investment costs per kilowatt are several times those of modern thermal power plants. This situation can change in the foreseeable future if manufacturing costs of solar power plants will fall due to greater demand and the prices of the present primary energy carriers rise still further.

Far better prospects of utilizing the solar radiation energy exist at the present juncture in places where it does not have to compete directly with other energy sources and where one succeeds in matching the energy supply with demand. Such conditions obtain in remote locations, with no developed infrastructure, where mostly only small amounts of power are required. Solar energy can in these places be utilized, for example, for irrigation pumps, mills for grain grinding and tools for craft industries, or for the supply of simple lighting, television, appliances or medical instruments. Reliable and simple power packages without great demands on maintenance have excellent possibilities of application in such places.

RECEIVER AND CYCLE CONCEPTS

For conversion of energy of the solar irradiation into electric current with thermal power plants only concentrating systems come into question. The working medium or the heat carrier can be heated up either in a central receiver (solar tower) or in distributed collectors (solar farm). The concept of the heater plays a decisive role in the choice of the cycle and working medium

With the solar farm concept the maximum process temperature is approx. 600 K ($\approx$ 325 ^{o}C) as, on the one hand, the degree of concentration of the collectors is not very high and, on the other, the losses due to radiation and convection increase substantially with rising temperatures. As the ambient temperature in most cases will be of the order of 300 K ($\approx$ 30 ^{o}C) or more, an acceptable efficiency can be expected only in case of steam cycles with compression in the fluid phase. Water, or other suitable agents, such as refrigerants or organic fluids, could in this case act as working medium.

The solar tower concept, on the contrary, makes higher degrees of concentration possible for the solar energy irradiated in the receiver. In this case steam temperatures comparable to those in conventional steam power plants can be obtained. If the heat is transferred to the working medium in a suitably designed cavity receiver, the radiation losses of the receiver can be reduced to such an extent that working medium temperatures of up to 1100 K ($\approx$ 825 ^{o}C) necessary for gas turbine cycles can be attained, without reducing the receiver efficiency appreciably.

For thermal solar power plants of small output not all possible types will come into question. Should the cycle data be applied only in relation to an optimum cycle efficiency, and without taking into account the demands of the turbomachinery, this could easily lead to unrealistic results.

STEAM TURBINE PLANTS

An important parameter while designing turbines is the volume flow rate at inlet. Steam cycles, especially with water as working medium, are characterized by a large, useful enthalpy drop. A small output therefore requires a small mass flow. The steam data of modern, conventional power plants would, with small outputs as for example 50 kW, result in such small flow rates that no proper turbine could be built for them. Though solar farm plants involve comparatively low working medium temperatures and more unfavourable cycle efficiencies they offer more suitable boundary conditions for the turbines and probably require, for small outputs, less investment costs than solar tower plants.

As mentioned already, while stipulating cycle data the requirements of the turbomachines must be duly taken into account in order to arrive at realisable designs. Fig. 1 shows for various steam turbine types the minimum volume flow $\dot{V}_{min}$ at turbine inlet as a function of the difference of total enthalpy Δh_{tot} of the first stage. The limiting values of parameters applied to curves a, b and c such as rotor diameter, circumferential speed, ratio of blade height to rotor diameter, ratio of circumferential speed to absolute velocity, admission ratio, angle of in-flow, etc. have been chosen in accordance with the modern literature [1, 2]. Some of the boundaries mentioned there can no doubt be displaced slightly, but thereby leading soon to unfavourable machines with greatly reduced efficiency. The curves shown should therefore represent a usable indication for design boundaries of steam turbines. If care is taken not to fall short of $\dot{V}_{min}$ as per Fig. 1, then, with the help of the power output and the enthalpy drop of the first stage, one can determine the mass flow, the density, and thus the steam data at turbine inlet. Inversely, one can also proceed with the steam data attainable in the solar farm plants in order to select an appropriate type of turbine, if one considers again the adequate values of $\dot{V}_{min}$ in Fig. 1.

The partial-admission impulse turbine requires the least steam through-puts, utilizing at the same time a large enthalpy drop. With increasing total enthalpy drop of the stage the minimum flow increases only very gradually. Impulse wheels are added mostly as a regulating stage for multi-stage steam turbines. As even with a small through-put a large enthalpy drop can be utilized in them, more favourable conditions for the subsequent turbine portion result because of the reduced pressure and the larger volume flow. The curve a of Fig. 1 is based on an admission ratio of 0.1, the value representing a lower limit.

Curve b for a centripetal rotor is valid for a reaction stage which accordingly must have full admission. The minimum volume through-puts are therefore larger than with the partial-admission impulse stage. In addition, they increase faster for greater enthalpy drop. The minimum volume flow of an axial or radial reaction stage increases even more steeply with increasing enthalpy drop. As all curves shown are based on a wheel diameter of 0.1 m, a speed of 60,000 rpm corresponds to a circumferential speed of 314 m/s.

The intimated results will now be applied to a solar farm plant with an effective output of 50 kW. Fig. 2 depicts the schematic of the circuit of a simple thermal solar power plant. Steam is generated in the collector system a, expanded in turbine b and condensed in condenser c; the condensate will again be brought to the collector pressure by pump d. At first it must be established which through-put volumes are to be expected at turbine inlet with different maximum operating pressures and what inferences are to be drawn thereby. Fig. 3 shows schematically in an h,s-diagram for steam different expansion curves for a turbine inlet temperature of 573 K ($\approx$ 300 °C) and a condensation temperature of 333 K ($\approx$ 60 °C). The turbine efficiency for the variation with an inlet pressure of 20 bar was assumed at 65 %. It represents the quotient out of the enthalpy drop actually utilized in the turbine and the isentropic enthalpy drop with same initial and final pressures and with same initial temperature of the expansion. The cycle efficiency then amounts to 18.7 % without taking the pumping work into consideration. The cycle efficiency is defined as the ratio of the available net turbine output to the amount of heat absorbed by the working medium in the receiver, and not to the solar energy irradiated in the mirrored area. The efficiencies for the other turbine inlet pressures were determined such that the total efficiency of the cycle remained constant. Fig. 4 shows which turbine efficiencies and which volume flow rates are required at the turbine inlet in order to realize a constant cycle efficiency for different maximum process pressures.

At a pressure of 20 bar the volume flow rate at turbine inlet falls short of the corresponding value of $\dot{V}_{min}$ of the limit curve a (Fig. 1). There are thus present unfavourable conditions and it appears questionable if a turbine efficiency of 65 % can be achieved. At a pressure of 10 bar, the volume flow rate at turbine inlet is twice as much. The turbine efficiency of 74.8 % necessary for a similar cycle efficiency appears to be more realistic than the assumed 65 % at 20 bar. Besides the larger volume flow rate at inlet the lower steam moisture at outlet and the smaller expansion ratio also favour the lower process pressure. The latter is to be preferred also because of the movable joints necessary between the machinery and the collectors moving with the sun. The heat transfer to the water and the steam and the efficiency of the collectors need not be all too adversely affected because of the lower pressure level. At 10 bar the ratio of superheat to warming-up heat and heat of evaporation amounts to 0.106 as against 0.086 at 20 bar. The deliberations set forth here for a plant of a capacity of 50 kW can be applied in a similar manner for higher capacities also. Because of the greater mass through-put in the latter case the optimum process pressure will be shifted to higher values.

It has been suggested to use other working media for steam cycle instead of water, such as the organic substances benzene or toluene. However, for the very small outputs and turbines acting as prime movers as in the present case, no advantages are to be expected from these as the volumes at turbine inlet would tend to be even lower than with steam.

GAS TURBINE PLANTS

The high turbine inlet temperatures necessary for adequate efficiency of a gas turbine plant can be achieved in case of a solar power plant only in a central receiver (solar tower). For small outputs this may possible lead to greater construction expenditure as compared to a solar farm plant. On the other hand the gas turbine offers some substantial advantages as an energy conversion system, such as, above all, the better cycle efficiency, the simple set-up, and the possibility of steady transition to higher capacites in closed-cycle plants without additional development work. Especial advantages of open-cycle gas turbines are the doing away with all cooling as well as the possibility of running the plant without solar energy by installing combustion chambers fired by fossil fuel [3].

Fig. 5 shows a circuit diagram for a simple, open-cycle gas turbine. Ambient air is sucked in by the compressor a, compressed and conveyed to the turbine d via the recuperator b and the receiver c. After expansion the air is exhausted to the atmosphere through the recuperator. The cycle efficiency of the plant depends on the ratio of the pressures at the high and low pressure sides of the cycle. In Fig. 6 the cycle efficiencies for plants with and without recuperator have been shown as a function of pressure ratio. Suction temperature of compressor has been assumed as 313 K ($\approx$ 40 ^{o}C) and inlet temperature at the turbine as 1073 K ($\approx$ 800 ^{o}C). Relative pressure losses for the entire cycle have been taken as 11 % with recuperator and 6 % without recuperator. The compressor efficiency amounts to 80 %, that of turbine to 82 %.

As to be expected, the cycle efficiency for the plant with recuperator is substantially higher. As the cost of the turbine plant would probably amount to less than half of the total cost of a solar power plant, the additional expenditure for the recuperator would be worthwhile. A temperature difference of 40 K between the hot and cold sides was taken for the above calculations.

For a new plant one would select a pressure ratio corresponding to the optimum efficiency or slightly higher, as with smaller pressure ratios the efficiency falls steeply. Corresponding to the curves shown in Fig. 6 the inlet volumes of the machines amount to approx. 0.5 to 1.0 m^3/s at optimum pressure ratio. According to the results shown in Fig. 1 these values lie above the lower limits applicable to turbines. Also the compressor through-put has an acceptable range. For small volume flows radial compressors are specially suitable.

As long as the boundary conditions can be retained, there is a linear relationship between the volume flow rates in the cycle and the output of the plant. If during transition to substantially higher outputs the size of the machines and equipment of the open gas turbine cycle becomes too large, one could turn to the closed-cycle process by inserting a cooler. By a proper selection of the pressure level in the closed system one achieves a higher output directly corresponding to the increase of pressure, with machines of same size as in open-cycle plant, so that additional development costs need not be incurred [4].

SUMMARY

For solar heated thermal power plants of small output steam turbine plants with distributed collectors (solar farm) and gas turbine plants with central receiver (solar tower) can be used. The solar farm concept implies moderate maximum process temperatures. For small outputs the inlet volume of the steam turbine lies at the lower limit of the possible range. Therefore, careful deliberations for selecting the maximum process pressure are necessary. Only moderate cycle efficiencies can be attained because of the low turbine inlet temperature and the high condensation temperature. At the same time the technological demands are low. For small outputs the collector system of the solar farm concept should be less expensive than that of the solar tower concept.

The gas turbine requires a central receiver because of the high inlet temperature necessary. The construction expenses for the open-cycle type suitable for small and medium outputs are relatively low. Special advantages of the open-cycle gas turbine lie in the fact that no cooling of the working medium is necessary and that, by installing a combustion chamber for fossil-fuel heating, operation without solar irradiation is also possible. With a plant output of 50 kW sufficient volume through-puts for turbomachines are achieved. An extrapolation for higher outputs is possible without difficulty.

BILBIOGRAPHY

1 Traupel, W.: Thermische Turbomaschinen. Vol. I, 3rd edition, Berlin Heidelberg New York, Springer-Verlag 1977.

2 Bammert, K.: Lecture on Steam Turbines. Published by G. Wasmund, Cologne 1963, pp. 1/84.

3 Bammert, K.: The Gas Turbines in Solar Power Plants. The Middle East Solar Technology Conference "SOLTECH '78", Bahrain, 24-27 April 1978.

4 Bammert, K.: A General Review of Closed-Cycle Gas Turbines Using Fossil, Nuclear and Solar Energy. Pocket edition no. 57. Published by Karl Thiemig, Munich 1975.

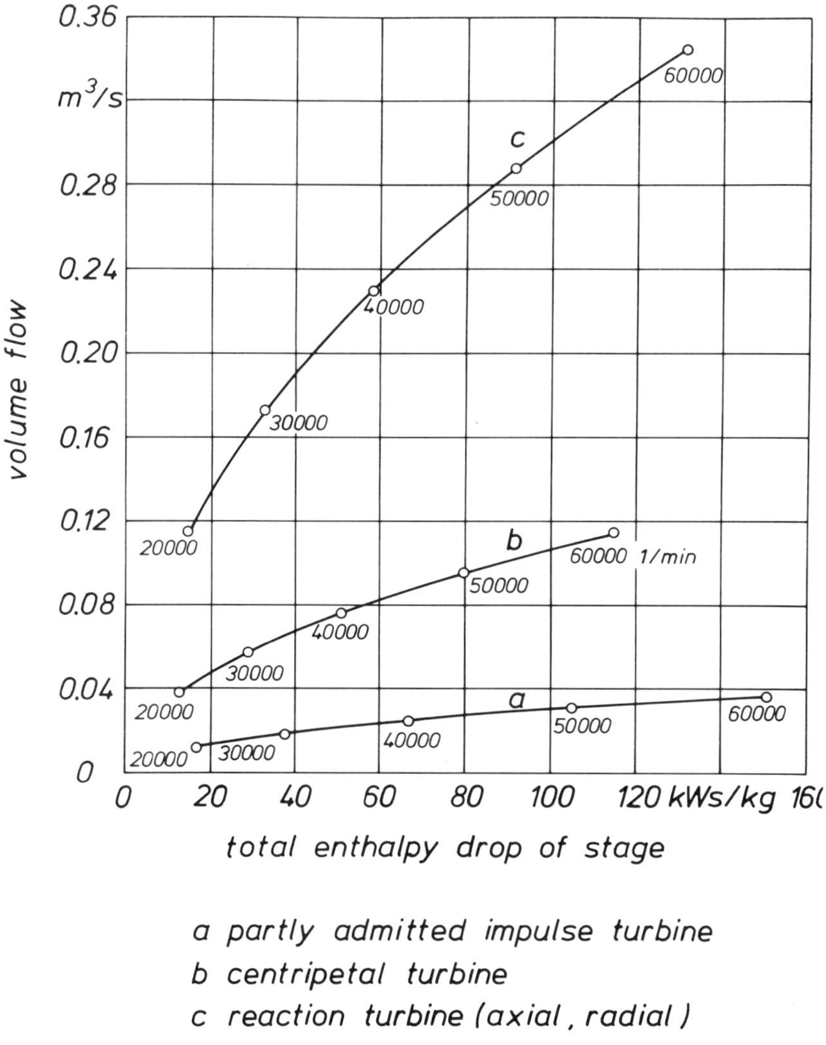

Fig. 1 Minimum volume flow at the inlet of different types of turbines

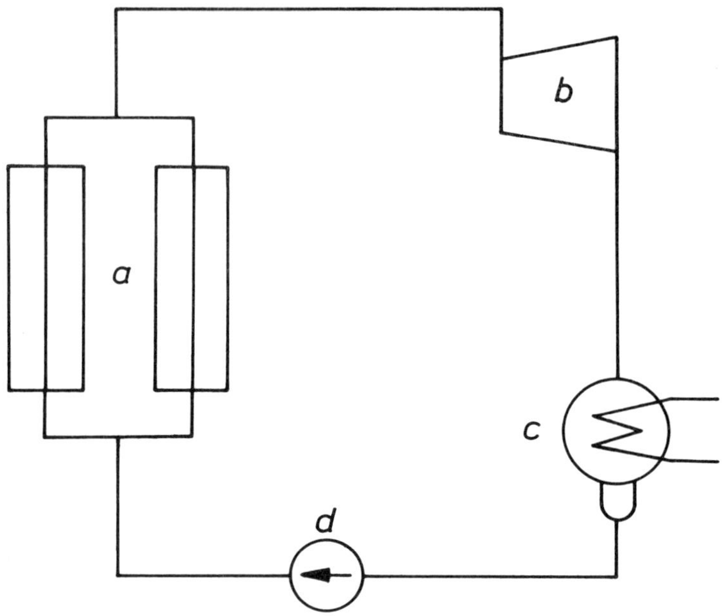

a collector system

b turbine

c condenser

d pump

Fig. 2 Diagram of a solar power plant with steam turbine

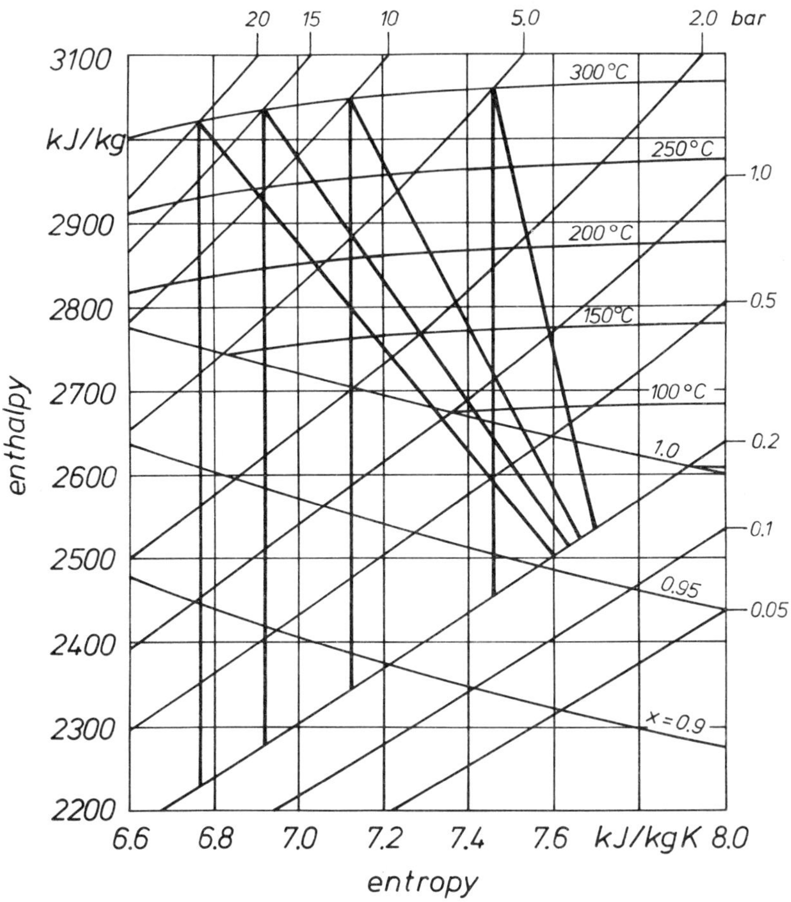

Fig. 3 Enthalpy-entropy diagram for expansion in turbine (steam)

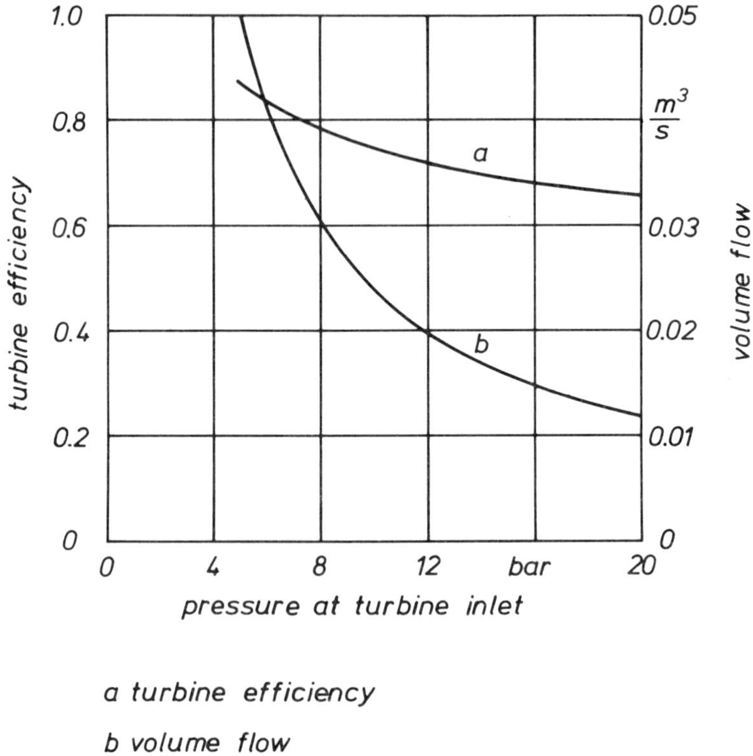

Fig. 4 Turbine efficiency and volume flow at turbine inlet (output 50 kW)

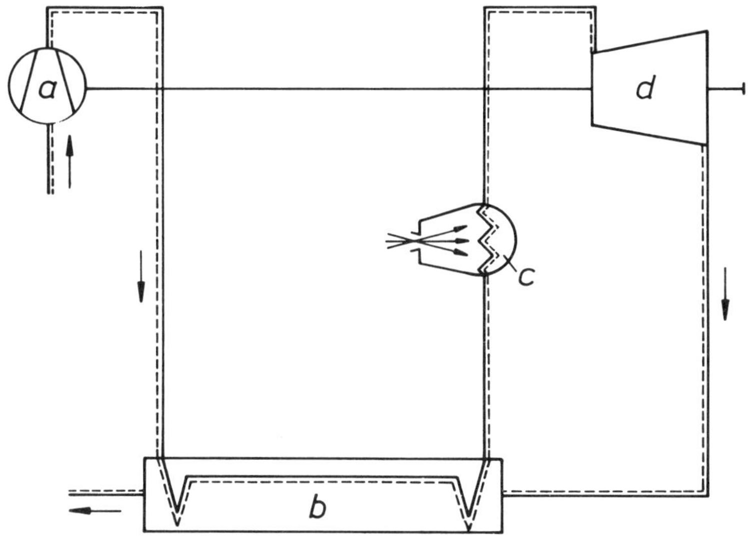

a compressor

b recuperator

c receiver

d turbine

Fig . 5 Diagram of a solar power plant with gas turbine

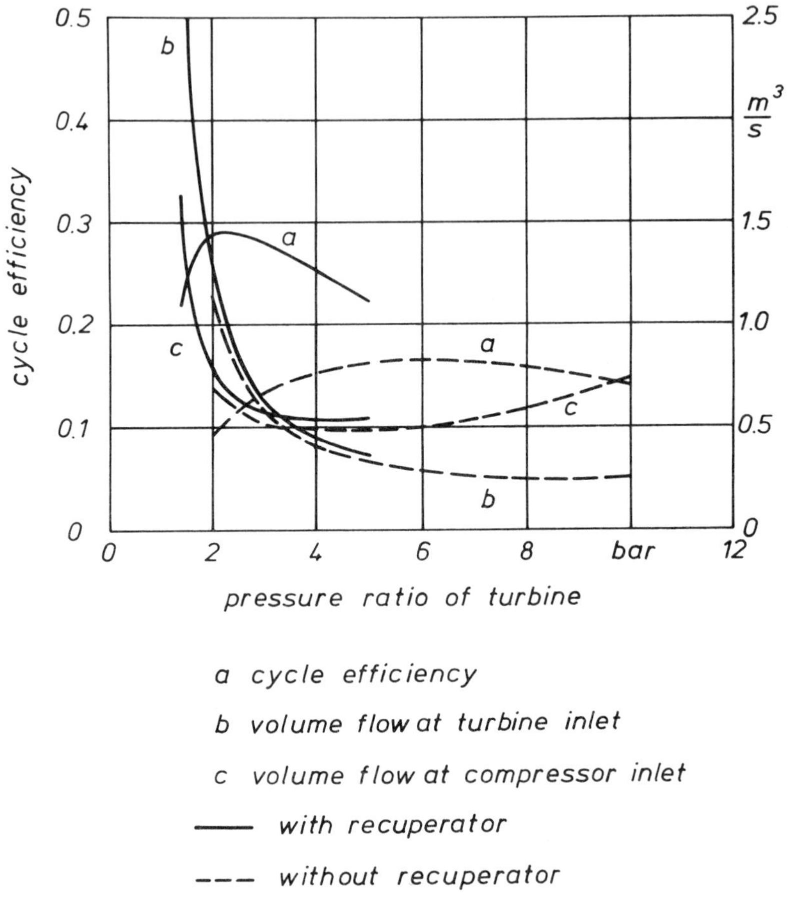

Fig. 6 Cycle efficiency and volume flow of an open-cycle gas turbine (output 50 kW)

TURBINE SELECTION FOR SMALL CAPACITY SOLAR-POWER GENERATION

A. Mobarak, N. Rafat
Cairo University
and
M. Saad
Unesco Regional Office
Cairo, Egypt

ABSTRACT

In developing countries lying within the solar belt, there is an urgent need for power generation systems of sufficient capacity to supply power to small communities in remote areas. The turbine for small power units in the range from 1 to 100 KW. is not of the conventional type and requires special design. To attain high cycle efficiency (Rankine cycle) for power systems of small capacity, the enthalpy drop should be as high as possible whereas the mass flow rate is small. This leads to problems such as partial admission, multistaging, high rotative speeds, etc. The present study indicates that the most suitable choice of turbine is a multistage radial Ljungström-type turbine.

INTRODUCTION

Nowadays, it is universally realised that in the near future and very probably during the first part of the next century, man will have to depend for his energy requirement on solar radiation as a complementary source of energy. This is necessitated by both the impending depletion of fossil fuels and by the recent awareness of the hazards of pollution. The impending depletion of fuels is accelerated by the ever increasing energy demands, and also by the imposed limitation of the efficiency of the presently existing means of energy conversion. Pollution of the atmosphere is a result of the unconverted portion of the energy input into useful form. In the case of presently used power systems, this portion represents approximately two thirds of input energy. Solar energy offers the answer to both problems, since it is unlimited, free and produces no pollution. Another advantage in the application of solar radiation for power generation, is the matching of distributed nature of solar energy with the distributed pattern of energy consumption. That is, the power generating system and the load can be located at the same place, thereby, reducing the need for energy transmission and distribution with the associated losses and cost. Fortunately, this happens to follow the thesis that for small capacity solar power generating systems, the cost per unit power is comparable with that produced by a large solar power system. The substituion, however, is not simple because solar energy has its inherent disadvantages, its availability is intermittent, it is seasonal and its intensity is low. With these factors in mind, the present conversion of solar energy into electric power, which represents the ultimate goal in solar energy utilisation, cannot compete favorably with conventional systems. Forecasts however indicate that a good percentage of energy needs will come from solar and other renewable sources by the year 2000.

In developing countries, especially in remote areas, there is a strong need for small power units. Until recently, this has not been actively shared by developed countries because of the high capital cost per unit power produced with the low intensity of sun's energy. This attitude is gradually changing as a result of the general awareness of the obvious advantages of solar energy. Chatel [1], gives a brief review of international activities in this field and recommends developed topics and possible research most suitable for international cooperation in solar energy. Solar power units may be used for pumping water from underground wells and/or for producing electricity.

CHOICE OF THE THERMODYNAMIC CYCLE

The choice of a suitable thermodynamic cycle is an important factor in the design of a small capacity solar power system which is simple, reliable and of low cost. A number of thermodynamic cycles using different working fluids such as steam [2], gas [3] and freon [4] were proposed for solar power generation. These cycles are either of the single or double loop type, that is, in which either a single or dual-fluid system is used. They may be of the condensing or the non-condensing types.

In general, the main consideration in the choice of both the cycle and the working fluid is to attain a reasonably high thermal efficiency. This also satisfies the low cost criterion since the collector is one of the most expensive components of the plant. The solar focussing collector area for a solar constant of 1 kW/m^2 is related to the thermal efficiency as:

$$\text{Collector area} = \frac{\text{power generated in KW}}{\eta_{thermal}} \qquad (1)$$

For the sake of simplicity, reliability and ease of maintenance, it is preferable to use steam or air as the working fluid in a vapour cycle or in a gas turbine cycle respectively.

Contrary to conventional practice in steam power generation, which has its upper temperature limit governed by metallurgical consideration, the common practice in the design of solar concentrators is to limit the target temperature to 250° C [5]. Above this temperature, the collector losses become excessive and difficult to control, resulting in a lower overall performance of the collector.

For the low power range under consideration (1-100 KW), and for maximum temperature range below 250° C, the simple open or closed cycle gas turbine will never compete favourably with the small steam power plant as regards thermal efficiency. Therefore, unless sophisticated types of concentrators with high concentration ratio are developed, the most suitable system for solar power gneration in this small power range and in the prevailing conditions under consideration would be that using steam as the working fluid.

Figure 1 is a schematic diagram for a solar power generating plant. Solar radiation is received by cylindrical parabolic collectors which concentrate the energy on targets in the form of piping placed at the focal plane. Water flows under pressure through these pipes where the concentrated energy acts to generate steam. The steam flows into a receiver vessel and then to a turbo-generator before going to the condenser. The condensate is pumped back to the collectors to complete the cycle.

Based on the above limit of temperature, the choice of pressure is of no appreciable consequence as could be seen from the steam property-diagrams. The choice of turbine inlet pressure is based on the following argument. The use of solar energy for power generation is naturally applicable mostly in warm climates. Under these conditions, the condenser side of the cycle will be subjected to a rather high value of condenser water inlet temperature, which in turn will impose a rather high limit to the condenser pressure. Taking a maximum value of 40° C as the temperature of the available condensing water, and assuming a temperature rise of 15° C through the condenser, the temperature of the steam should be in the neighbourhood of 65° C. This corresponds to a pressure of 0.25 atm. Assuming an allowable steam quality of 90%, the state of the steam at the exit of the turbine can be determined. Taking a turbine efficiency of 70%, determines the state of the steam at the turbine inlet. Considering simplicity, and choosing saturation conditions at the turbine inlet, gives the inlet state of the steam as 5 atm. (and 150° C).

Based on the above, the designer of the solar power plant should put sufficientcare on the condenser part of the cycle, since a substantial part of the enthalpy drop would take place by the incorporation of the condenser. An additional side advantage of using a condenser, other than for thermodynamic considerations, is the fact that in locations where solar energy applications are most desirable, a supply of fresh water for steam raising is naturally scarce and the use of a condenser would be necessary to have a closed cycle. Figure 2 is an enthalpy-entropy diagram of the proposed cycle. It is noted that the isentropic enthalpy drop in the turbine is 482 kj/kg ($\simeq$ 115 kcal/kg) and the actual enthalpy drop is 363 kj/kg ($\simeq$ 87 kcal/kg). This value is considered a limiting value in order to achieve a reasonable thermodynamic efficiency of the cycle. Higher enthalpy drops are to be considered if higher thermal efficiency is to be achieved. When no condenser is used, a superheater is then justifiable in order to obtain a comparable value of enthalpy drop.

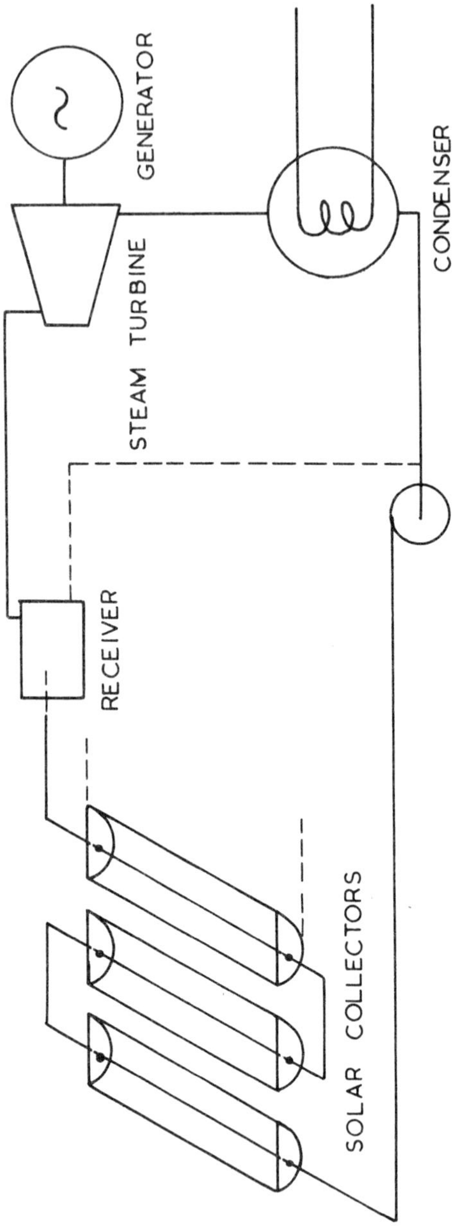

Fig. 1 SCHEMATIC DIAGRAM OF SOLAR POWER GENERATING PLANT

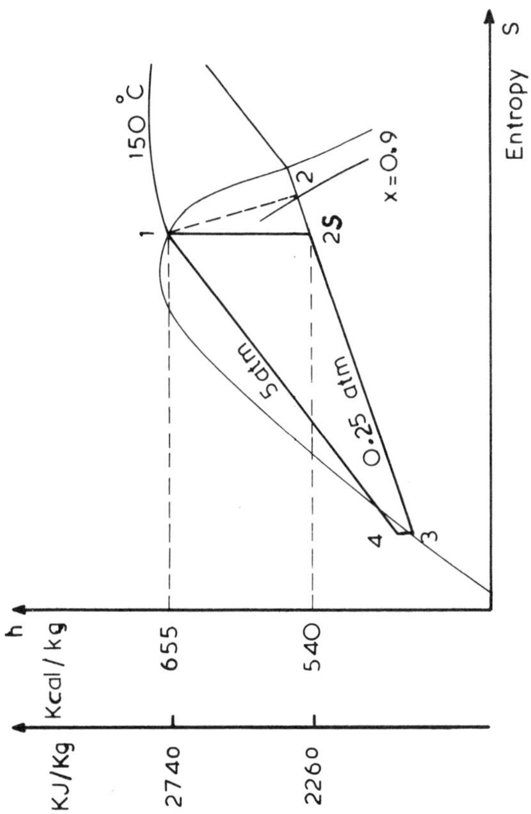

Fig. 2 h_s DIAGRAM FOR CYCLE

Based on the above enthalpy drop, the mass rate of flow per KW output can be calculated:

$$\dot{m} = \frac{\text{power output}}{\Delta h_{actual}} \simeq 10 \text{ kg/kWh},$$

and the Rankine efficiency is 20%

TURBINE SELECTION

The process of design of a turbine to work in conjunction with the above proposed cycle is one of matching and optimization. Referring to the simple Rankine cycle shown in Fig. 2, its thermal efficiency is:

$$\eta_{th} = \frac{W}{Q_{input}} = \eta_t \cdot \Delta h_s / Q_{input} \qquad (2)$$

where Q_{input} is of the order of (2000-2400 kj/kg),($\simeq$ 480-575 kcal/kg). The turbine efficiency η_t is considered to be of the order of 0.6-0.75 for such small size turbine of high enthalpy drop [6,7]. Therefore in order to attain an overall thermal efficiency of 15-20%, the enthalpy drop Δh_s should be of the order of (420-500 kj/kg),(100-120 kcal/kg). This corresponds to an ideal exit nozzle velocity of 900-1000 m/s in a single stage impulse turbine. The mass flow rate $\dot{m}$ would then be calculated according to the equation:

$$\text{Power} = \dot{m} \cdot \eta_t \cdot \Delta h_s \qquad (3)$$

For the power range under consideration, $\dot{m}$ is of the order of 12-1200 kg/hr or 0.0033-0.33 kg/s. Such mass flow rates are very small as compared with those of usual practice, and partial admission becomes unavoidable in these cases. Moreover, since the enthalpy drop is relatively high, the turbine should either be of the impulse or of the multistage type. Accordingly the most suitable type of turbine for the present application seems to fall within the following group of turbines which are investigated hereafter for proper selection.

1. The simple impulse turbine, Fig. 3a:For such a turbine the velocity ratio $\nu = u/c_1$, where u is the peripheral speed and c_1 is the velocity at nozzle exit, is of the order of 0.45 or less. This implies that the peripheral speed should be of the order of 500 m/s, which lies outside the applicable range of most modern practice.

2. Velocity compounded turbine:This may be either of the Curtis type

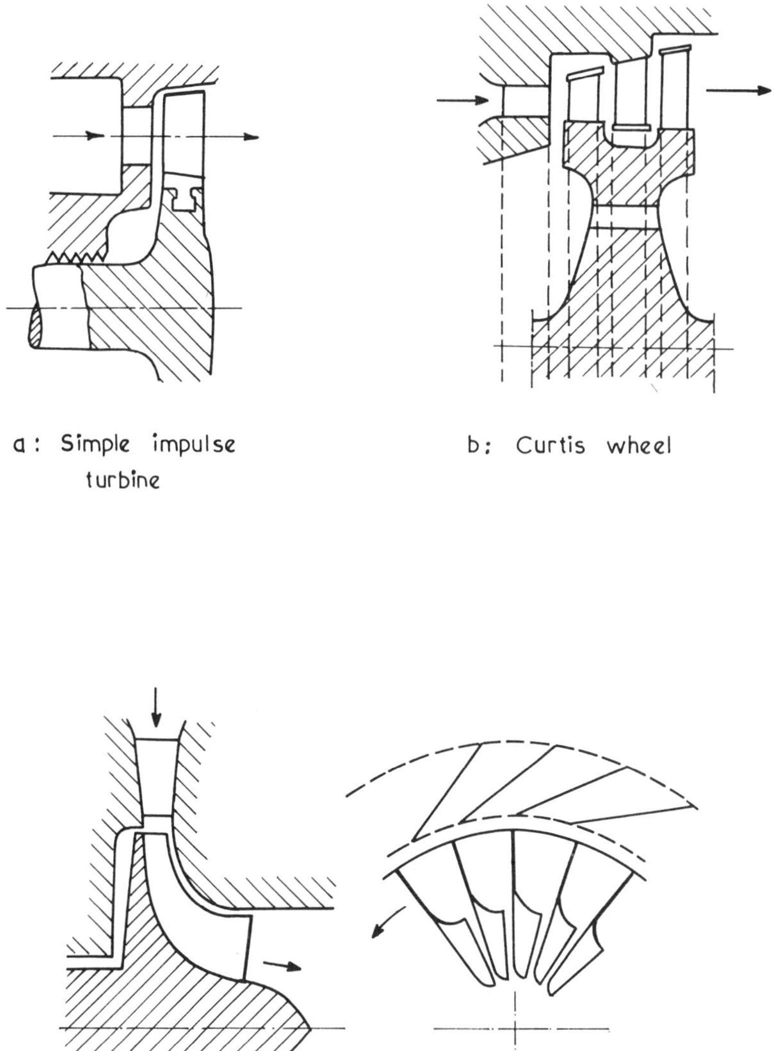

a : Simple impulse turbine

b: Curtis wheel

C : Centripetal turbine

Fig. 3 Different types of turbines

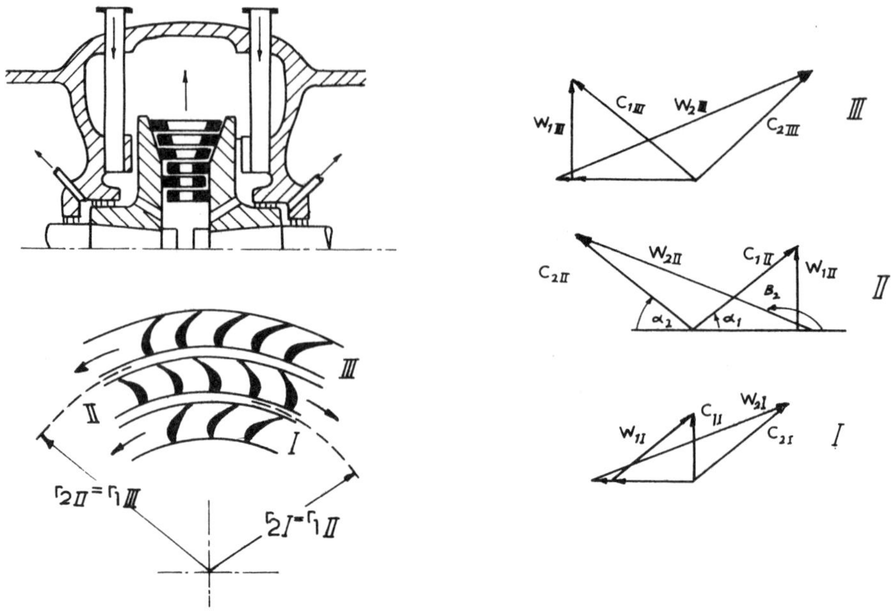

d : Ljungström turbine

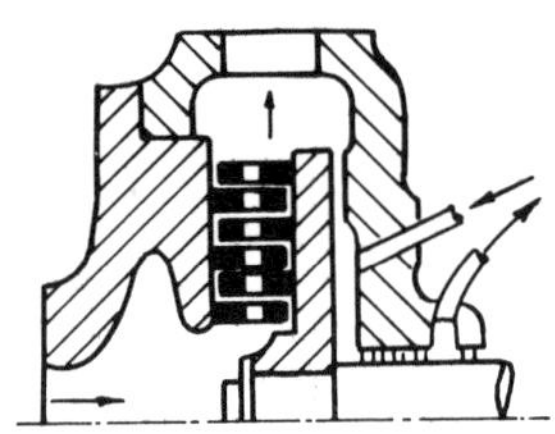

e : Multistage radial turbine

Fig. 3 Different types of turbines

Fig. 3b or of the Terry type turbines. This type is an impulse turbine with several re-enteries of the fluid. The peripheral speed of such turbines, where ν = 0.22 could then be brought down to acceptable values in the order of 250 m/s. However with such a value, the relative Mach number $M_{w1} = w_1/a_1$, where w_1 is the relative velocity at inlet and a_1 is the sonic velocity before the rotor, will be exceedingly higher than 1.0. The supersonic relative flow in this case combined with the necessity of using partial admission and the moisture content associated with these conditions will result in a very low efficiency, particularly when the turbine would have to run at off-design conditions. Supersonic relative flow in turbine cascades even under the most favourable conditions is very difficult to cope with.

3. Centripetal turbine: Although this type of turbine could be designed with full admission, it has to run at high peripheral speeds (and hence high R.P.M.). This follows from the relation between the work developed by such a turbine and its peripheral speed $w = u_1^2$, where u_1 is the peripheral speed at the outer diameter d_1. This equation is valid for the conventional radial bladed turbine rotor (Fig. 3c). Hence this type of turbine would not be applicable for the present situation.

4. Multistage radial turbine: This type of turbine seems to offer the proper solution for our problem of turbine selection. A turbine of this type may have two counter rotating rotors with no stator wheels, the Ljungström turbine (Fig. 3d), or may be a conventional multistage radial, outward flow turbine as shown in Fig. 3e.
Factors for the above mentioned types of turbines are presented in Table 1 based on a 100 kW output and an enthalpy drop of 418 kj/kg (100 kcal/kg). The maximum diameter of the turbine wheel is limited to 0.6m and the minimum diameter, whenever relevant, is limited to 0.2m. For the axial machines the blade height is conveniently chosen as 1 cm, whereas for the radial machines it is chosen to be 0.5 cm. The table gives a basic comparison between the turbines and provides means for selecting the most appropriate turbine for the present application. On the basis of the lowest speed of revolutions the Curtis turbine seems attractive. The inlet Mach number however is 1.855 and the wheel has to be designed for supersonic turning of the flow. This, together with the necessity of having partial admission to the wheel, raises difficulties in the design. If the Mach number is limited to lower values, the total enthalpy drop would have to be reduced and this leads to a still lower thermal efficiency. The table also indicates that the highest rotative speed belongs to the centripetal turbine followed by the simple impulse turbine. This high speed of revolution would still be augmented if a lower tip diameter is resorted to in order to reduce the disc friction and the losses accruing from using partial admission. An estimate of these losses may be obtained from the following relation:

$$\text{Power loss} = f\,(\varepsilon,\ 1/d,\ Re)\ N^3 d^5 \qquad (3)$$

TABLE 1 - COMPARISON OF DIFFERENT TYPES OF TURBINES (100 kW)

Factor \ Type	Impulse	Curtis	Centripetal	Ljungström	Multistage Radial
$\lambda = 2W/u_2^2$	5	15	10	2.9	5.5
$\phi_2 = cm_2/u_2$	0.45	0.45	0.45	0.45	0.45
$\nu = u_1/c_1$	0.45	0.225	0.9	1.1	0.5
No. of Stages	1	1	1	10	5
No. of Cascades	2	4	2	10	10
Admission	partial	partial	full	full	full
$U_{Max.}$, m/s	411	205	647	260	267
r.p.m.	13320	6660	20600	8300	8500
Mw_1	1.335	1.855	0.647	0.325	0.85
η_0	0.66	0.62	0.68	0.71	0.69

λ = stage loading coefficient

ν = velocity ratio

η_0 = overall turbine efficiency

ϕ = flow coefficent

Mw_1 = relative Mach No. at inlet

where ε is partial admission ratio, l/d is blade height to diameter ratio, Re is the Reynolds number and N is the r.p.m. From the above equation it appears that for a certain peripheral speed ($\pi d N/60$), it is more expedient to work with higher N and lower d. Such procedure would be the more called for since for the small power range (1-100 KW) considered here, the disc friction and partial admission loss may consume a large portion if not all of the generated power.

Figure 4 shows the characteristics of an impulse turbine depicting its performance for different speeds of revolution, and different partial admission ratios. These curves were constructed for a 0.2m wheel diameter and a blade height of 1.0 cm, steam inlet conditions are as designated in Fig. 2. It could be seen from Fig. 4 that the greater part of the low speed of revolution range (6000-15000 R.P.M.) is associated with a very low efficiency (less than 3%). Hence it is clear that partial admission impulse turbines are not the answer as discussed previously.

THE RADIAL MULTISTAGE TURBINE

On the basis of the above analysis and discussion, it seems that the multistage radial outflow turbine offers the optimum choice for small solar power generation systems. Turbines of this type may either have counter rotating rotors as in the Ljungström type or have a pure radial outflow blade wheel with fixed nozzles as shown in Fig.3d,e. The Ljungström turbine with its two shafts may, if required, be used to drive two machines simultaneously, otherwise the simple radial turbine with a single shaft may be used. The dimensions, design and number of cascades of the two types for the same rotational speed and power output would be very nearly the same. Compared with the Ljungström turbine which is a 100% reaction turbine, the radial turbine is designed as an impulse turbine with almost double the stage loading coefficient.

Summing up, the multistage radial turbine could be seen to offer the following advantages:

1. Partial admission is avoided since admission takes place at the smallest raduis.

2. Steam flow is outwards from one stage to the next into successively increasing areas. This goes in unison with the expansion of steam.

3. Multistaging allows the proper choice of the rotational speed in a convenient working range.

4. In special cases as in rural areas the turbine with two counter rotating rotors could be made to drive an electric generator and a pump simultaneously.

5. When the double machine simultaneous-drive does not present itself, the

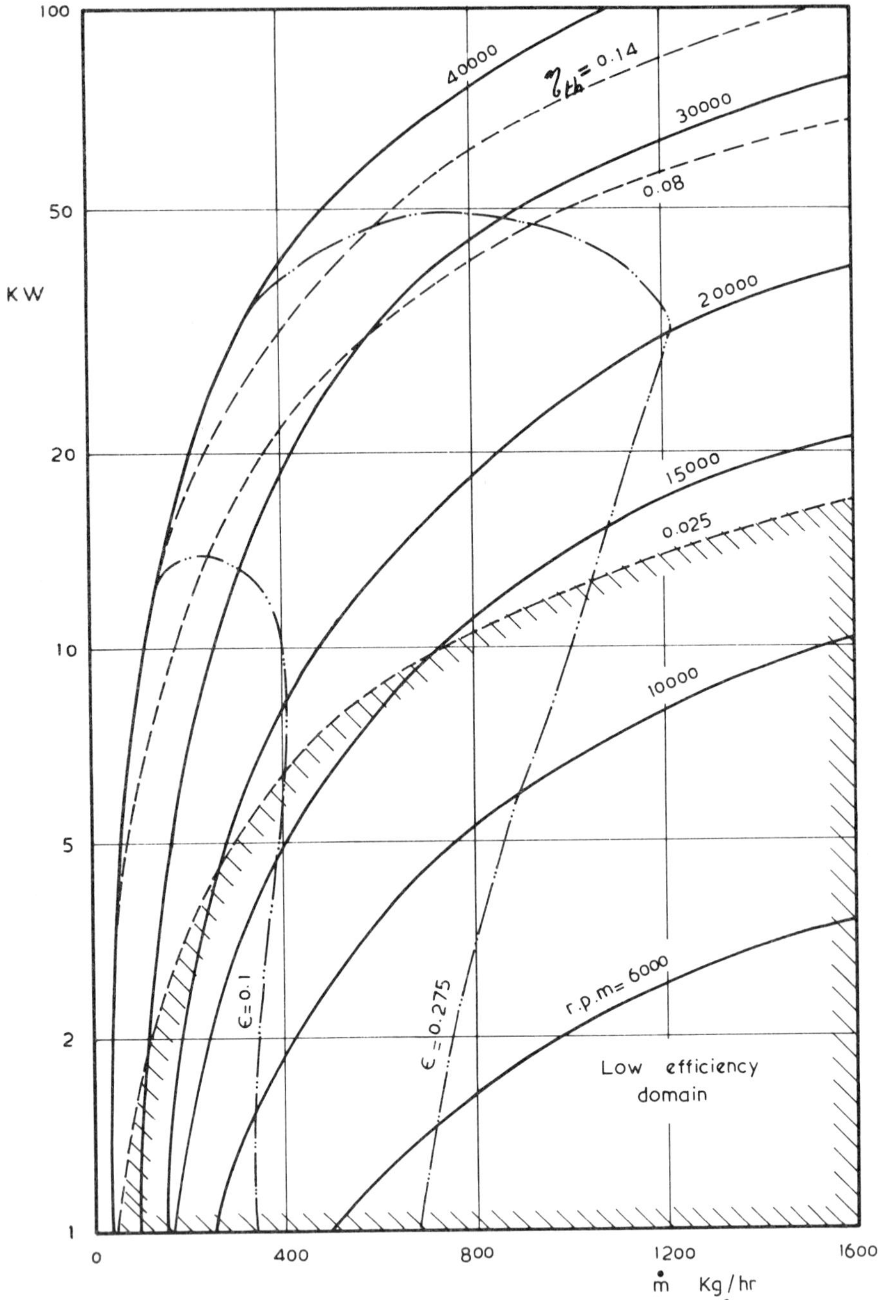

Fig.4 Turbine power (impulse) versus mass flow rate ṁ for different r.p.m. intial condition: saturated steam at 150 °C, wheel diameter 20 Cm.

pure radial turbine with fixed blades could be preferably used, thus avoiding the problems of the complicated construction of the counter-rotating turbine parts.

Now referring to the velocity diagrams of the Ljungström turbine Fig. 3d, the work in any stage is given by:

$$w = u_1 c_{u1} - u_2 c_{u2} \qquad (4)$$

where c_u is the peripheral component of the absolute velocity, subscripts 1 and 2 refer to the cascade inlet and outlet conditions respectively. This could be reduced to the form which gives the stage loading coefficient as:

$$\lambda = \frac{w}{u_2^2/2} = -2\,(D^2 + 1)(1 + \Phi_2 \cot \beta_2) \qquad (5)$$

where β_2 is the exit blade angle, $\Phi_2 = c_{r2}/u_2$ is the flow coefficient, c_{r2} is the radial velocity and $D = d_1/d_2$ is the diameter ratio. Equation (5) is used for any repeated stage. For the first stage,which could be considered as a special case, the loading coefficient is given by:

$$\lambda = -2\,(1 + \Phi_2 \cot \beta_2) \qquad (6)$$

The total work developed by the turbine per unit mass flow rate is:

$$W = \sum_1^z w = \lambda\, u_{2I}^2 \left[\frac{1}{D^2+1} + \left(\frac{1}{D}\right)^2 + \left(\frac{1}{D}\right)^4 + \ldots + \left(\frac{1}{D}\right)^{2(z-1)} \right] \qquad (7)$$

where u_{2I} is the peripheral velocity at the exit of the first rotor. For $D = 0.9$, $\Phi_2 = 0.45$ and $\beta_2 = 166^\circ$ [8,9] ,

$$\lambda = -2\,(D^2 + 1)(1 + \Phi_2 \cot \beta_2) = 2.913 \qquad (8)$$

The net power of the turbine would then be given by:

$$\text{Net Power} = \dot{m}W - \text{Power Loss} \qquad (9)$$

where the disc friction power loss is calculated from equation [10]:

$$\text{Power Loss} = 1074 \times 10^{-9} u_2^3 d_2^2\, \rho_2 \qquad (10)$$

where ρ_2 is the density of the fluid at turbine exit. The thermal efficiency is then given by:

$$\eta_{th} = \frac{\text{net power}}{\dot{m}\,(h_o - h_{f_2})} \tag{11}$$

where h_o is the total enthalpy at inlet to the turbine (for the proposed case at 150° C and 5 bar), and h_{f2} is the enthalpy of condensate at the condenser exit.

Figure 5 shows the variation of turbine specific work with the number of stages for different speeds of rotation. Lines of constant thermal efficiency are also shown. The thermal efficiency appears to depend only on the total enthalpy drop. Hence the thermal efficiency reaches its maximum value when the maximum specific work is obtained.

Figure 6 shows the variation of the turbine power with the number of stages z (or d_2 the outer diameter). The mass flow rate which is proportional to the rotational speed is estimated for an effective blade height of 4 mm. Although such a low value has not been so far a usual practice in turbine design, it should not raise up any uncertainties since turbines with low aspect ratio have been already investigated lately with satisfactory results for the overall efficiency (0.6-0.75), [6,7]. Higher efficiencies may again be expected as a result of the inherent reheating in the turbine. Moreover, it is worth mentioning, that while working with a higher value of enthalpy drop which naturally requires a proportionally higher number of stages, the favourable effect on the value of disc friction due to the decrease in density is more than compensates the unfavourable effect of diameter increase. This has been observed when working with high vacuum.

However, in order to introduce into practice radial turbines of the configuration discussed here, they preferably will have to be mass produced with the application of precision die-casting. By this very simple technique of casting the combination of disc and blades in one operation, the blade rows will form short cantilevers protruding from the disc. Otherwise, if conventional blade fastenings are used, the expense of manufacturing the turbine would be prohibitive.

CONCLUSION

The aim of the present study is to investigate the engineering creteria for the selection of the prime-mover of a solar power plant of limited output (up to 100 kW) to be used in developing countries. The study has shown that such a prime mover would not be of the conventional type. Since the overall efficiency of solar power plants is low, all efforts expended in raising the efficiency in the design of the plant should be given the highest consideration. Hence the enthalpy drop across the turbine should

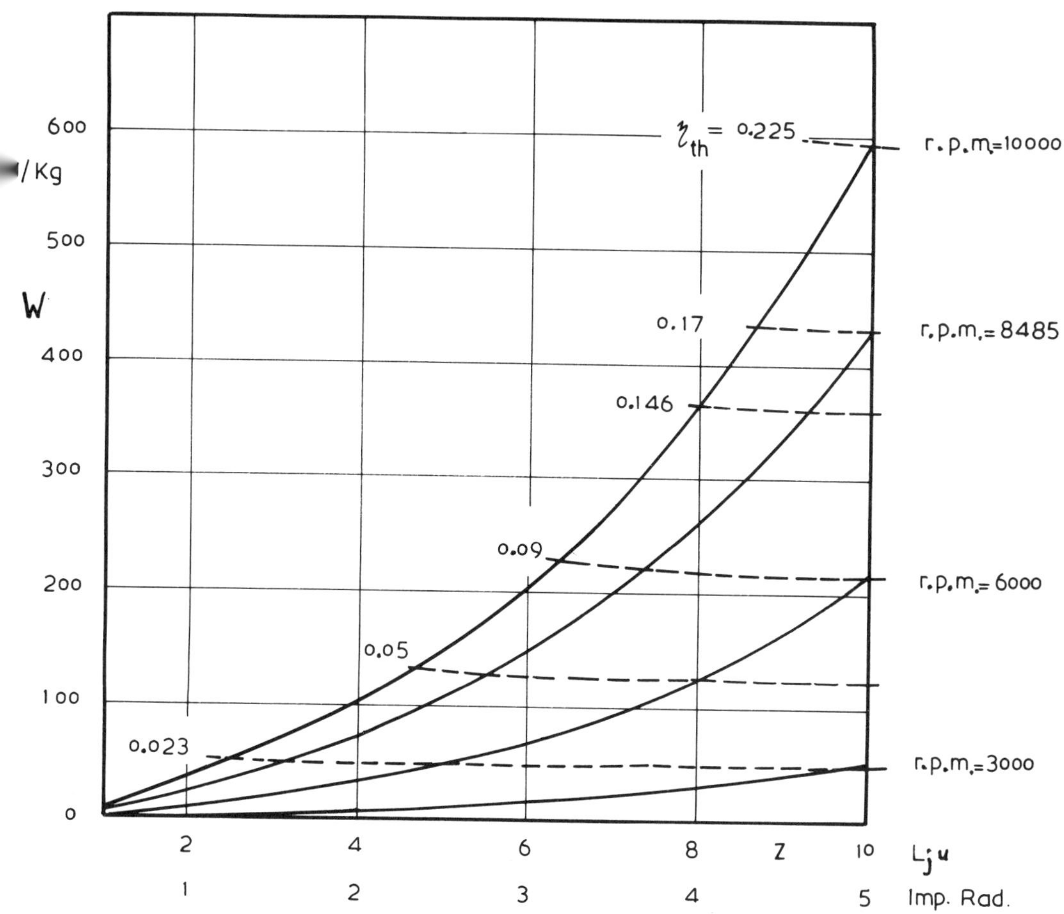

Fig.5 Turbine work against number of stages for different r.p.m. thermal efficiency lines are indicated.

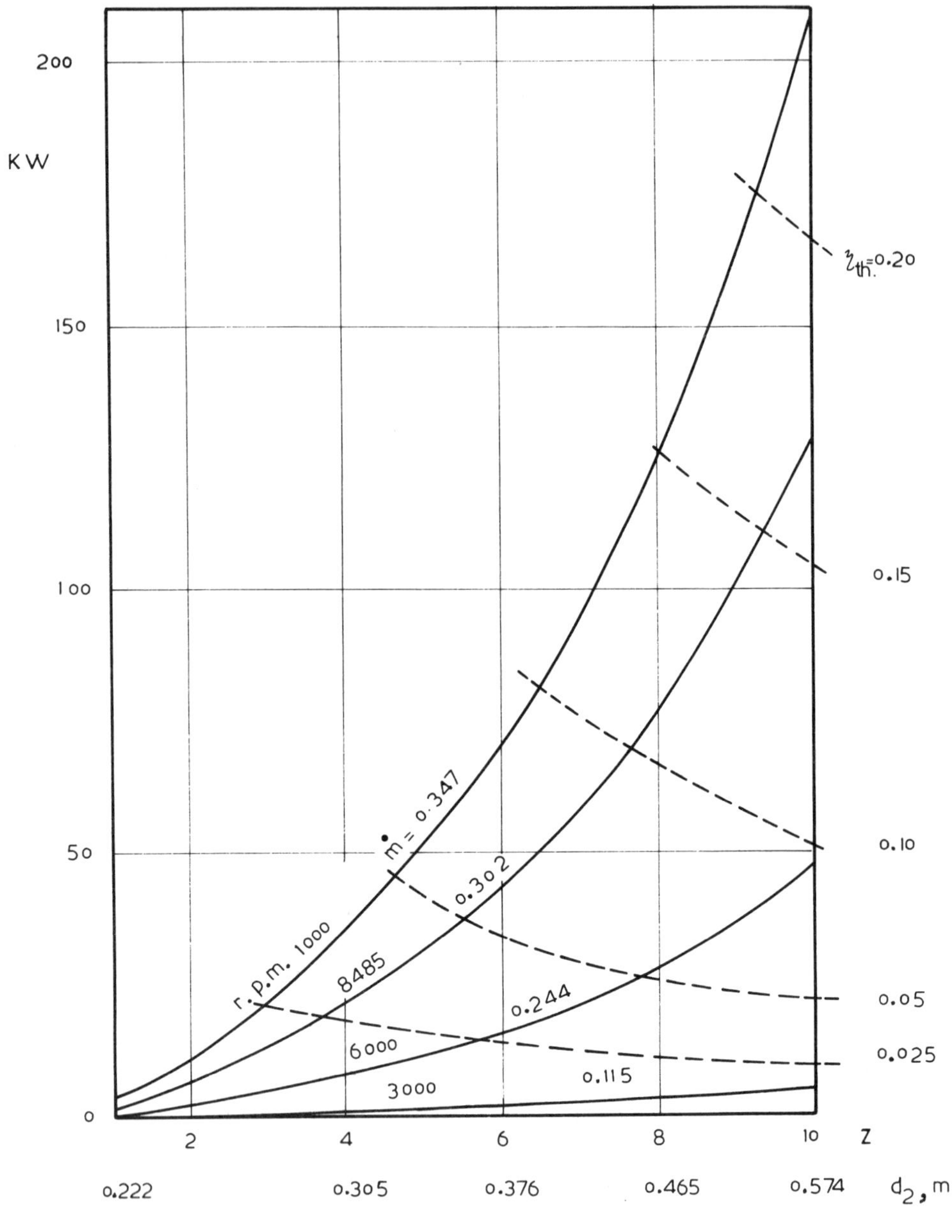

Fig. 6 Turbine power KW versus number of stages for different r.p.m., constant thermal efficiency lines are also shown.

be as high as possible. The mass flow rate being already small as a result of the small power range, will still be reduced because of the necessity of using a high value of enthalpy drop. Under these conditions it was shown that certain types of turbines would require partial admission and have to run at high rotative speeds. It was also shown that the radial outflow turbine and the Ljungström turbine would allow full admission at moderate speed and would represent the best selection. The Ljungström turbine, however, offers more flexibility in using the power output. It is recommended that more development work should be carried out in the design and the manufacturing of such a small capacity turbine.

REFERENCES

1. B.H. Chatel, "New Potentialities for International Co-operation in the Field of Solar Energy and its Application", Solar Energy Vol. 18, pp. 69-71, Pergamon Press 1976.

2. A. Sobin, W. Wagner and C. Easton, "Central Collector Solar Energy Receivers", Solar Energy Vol. 18, pp. 18-30, Pergamon Press 1976.

3. K. Bammert, "Solar Power Plants with Closed Cycle Gas Turbines" Tenth World Energy Conference, September 1977, Istanbul, Turkey.

4. W. Dickinson, A. Clark and L. Wouters, "The Shallow Solar Pond Energy Conversion System", Solar Energy Vol. 18, pp. 3-10, Pergamon Press 1976.

5. G. Löf and S. Karaki, "A Rational Method for Evaluating Solar Power Generation Concepts", Internation. Congress - The Sun in the Service of Mankind, Paris 1973.

6. G.O. Ohlsson, "Low Aspect Ratio Turbines", Journal of Engineering for Power, Trans. ASME, pp. 13-17, January 1964.

7. H. Linnardt and D. Silvern, "Analysis of Partial Admission Axial Impulse Turbines" pp. 297-308, ARS Jour., March 1961.

8. W. Traupel "Thermische Turbomaschinen, Springer Verlag". Berlin/Gottingen/Heidelberg, 1962.

9. W. Kearton;"Steam Turbine Theory and Practice" Seventh Edition, Pitman & Son Ltd., London 1961.

10. D. Trojanovskij, "Untersuchung und Berechnung axialer Turbinen Stufen," VEB Verlag Technik Berlin 1973.

A HELIOELECTRIC FARM

Otto J.M. Smith and Phyllis S. Smith
612 Euclid Ave.,
Berkeley, California, 94708, U.S.A.

ABSTRACT

A multimodule solar thermal electric plant has been designed to use conventional materials and engineering techniques. Each module occupies only 0.8 hectare and contains 500 mirrors of 2.4 square meters each. 600 modules, each with a heat receptor on a short tower, collect solar energy for a 100 MW generator operating 12 hours a day.

Sophisticated computer programs are used for both design and plant operation. The design program minimizes bus bar cost by finding the most cost-effective mirror dimensions, spacing, module dimensions, tower height, and receptor dimensions. The operation program completely surveys the solar field, calibrates all shaft encoders and control devices, and aims the reflected solar light toward the tower all by computer software. No engineering adjustments or manual check-out is needed in the solar farm.

A land area of 550 hectares is sufficient to produce 500,000 megawatt-hours of electric energy each year. This can be a 100 MW, 50-percent duty factor plant, or 70-MW, 70-percent duty factor, or 50 MW base load plant, with no significant change in megawatt-hour cost. 480 hectares is used for heliostat fields, and 70 hectares is used for radiative heat-rejection panels. The latter eliminate the need for a dry cooling tower, and partially compensate for heat pollution from fossil-fuel plants.

The 300,000 small mirrors and 600 short towers permit mass production techniques to significantly reduce the manufacturing cost. The mirror costs are a predominant part of the total system cost, and minimizing the mirror cost produces an economically competitive power plant. Estimated manufacturing and installation cost including interest during construction is 175 million dollars for each 100 MW power plant. Estimated operations, maintenance, and repair costs are 4 million dollars per year.

Half of the construction cost is direct labor wages, which would make a desirable impact on employment and economy. A building program of 1000 solar megawatts each year would support 40,000 maintenance laborers with wages of 400 million dollars per year (in 1978 dollars). These workers would be employed on a solar farm of less than 30 km square.

The Smith Multimodule Solar Thermal Electric System uses small hexagonal interlocking modules that minimize piping and land costs. Smith's system used a molten salt for a heat transfer fluid (HTF) which is pumped from the solar receptors on 27-meter tall towers into the power house where it boils water for a steam turbine. The hot HTF is stored in tanks with granite balls to provide heat for the night-time operation of the boiler and turbine.

These solar plants would be built in 100 MW units, and consequently the capital required for each unit is much less than for a 1000-MW plant. Experience obtained with the first few 100 MW plants can be used to improve subsequent units.

INTRODUCTION

Solar power plants can be built immediately to reduce our dependence on imported energy, to provide needed employment, and to save depletable resources. They can be built using conventional engineering techniques and conventional materials. The design can be versatile with alternatives available for almost every component. The plant can be build so that it will be easy to retrofit: to convert to improved or lower cost mirrors, to convert to improved heat exchange fluids, or to install additional heat storage and the associated additional mirror fields.

In such a design, which we have developed, the solar field is divided into a multitude of identical modules, so that one or more modules can be removed from service for maintenance without disturbing the operation of the remainder. This results in high reliability and high "on-line-availability". This design has been simulated on a digital computer to minimize cost.

Cost minimization is the key to making solar power competitive with other energy sources. The costs must include societal costs for pollution, for depletion of scarce resources, for safety, and for decomissioning. Considering all of these, solar power is easily the most economic of all possible sources of new electricity.

The initial cost of a solar power plant includes the life-time fuel cost, which is not subject to inflation, whereas labor and fuel costs will continue to rise. Consider ten 100 MW solar plants financed by bonds of 1.8 billion dollars fully amortized in 40 years at 7.5% interest. Annual bond payments will be 142 million dollars. Repairs, maintenance, and overhead will cost 40 million dollars per year, initially. Because most of the life-time cost of the plants is purchased at the time of construction, these plants rapidly become more cost effective as inflation increases the value of the electricity, more rapidly than it is increasing labor costs. The initial bus-bar cost will be 36 mils per KWH in 1978 dollars.

MODULAR SYSTEM

In this 100-megawatt modular system, the electrical generator is driven by a conventional steam turbine whose heat is obtained from heat exchange fluids from a solar farm. The solar farm area, 5.5 square kilometers, is divided into 600 almost identical modules. Each module area is 9170 square meters containing 500 small mirrors of 2.4 square-meters area each which reflect the sunlight and concentrate it through the window of a receptor on a short tower (27 meters high) to the south of these mirrors. Each receptor receives 750 KW of thermal power from its associated module of mirrors. The receptor is an insulated cavity containing pipes through which heat-exchange fluids are pumped. These fluids leave the receptors at high temperature, are collected at the powerhouse where they pass through heat exchangers and are cooled off, and then are pumped back into the solar field.

Each module has a hexagonal ground plan for the optimum utilization of the land area. The hexagons are interlocking like a honeycomb, and the tower for the mirrors in one hexagon has its base in the middle of a different hexagon to the south. This field pattern minimizes the cost of piping; it also minimizes the losses due to the angle of the mirrors with respect to the incident sun rays and losses due to reflection at the cavity receptor window.

The modular design permits the removal of one or more modules from service without disturbing the remainder of the solar field. All scheduled maintenance will be done at night, but unscheduled repairs could be carried out during the day.

COST SENSITIVITY

The mirrors and their controls cost 39% of the turn-key power plant cost. The solar energy receiver, tower, and piping cost only 23%. The powerhouse with heat exchanger, turbine, generator and switchyard cost 15%. The efficiency of the mirror system is 84%; that of the receiver and heat transport is 89%; and that of the powerhouse is 37%. It can be seen that a change of design that would save 1% of the mirror cost (0.39% of system cost) would be economical even if it increased the receiver heat system cost by 1.7% (0.39% of system cost), or would be economical even if it increased the receiver cavity cost by 4.5%. Consequently, the minimum cost design is one which minimizes the mirror system cost per square meter of mirror surface, and maximizes efficiency in the receiver and turbine.

The usefulness of a mirror is proportional to its area, the square of a linear dimension, whereas the cost of a mirror is proportional to its mass, the cube of a linear dimension, plus the control system cost. The cost per unit of usefulness has a minimum near two square meters of area. The tower height and mass cost is proportional to the cube of the focal length, whereas the useful receptor area is proportional to the square of the focal length. Maintenance costs vary as a power of height. Short focal lengths imply more towers and higher piping costs. The composite cost per unit usefulness has a minimum which is near a focal length of 70 meters and a tower height of 27 meters.

Mirror flatness and the surface imperfections increase the cone of the angle of reflected light to more than the 0.5 degrees of the sun diameter. Less expensive and less precise mirrors require a larger receptor aperture and lower concentration ratio. High efficiency can still be achieved at the low concentration ratios by using either a deep cavity or an aperture window.

MIRROR CONSTRUCTION

The mirror platen is fiberglass polyester two meters wide by 1.2

meters high cast over a paraboloid mold. The mirror surface is metallized polyester which is held down on the platen. With this construction, the mirror surface can be replaced when damaged by sandstroms. Suitable surfaces are chromed polyester or aluminized polyester.

The mirror platen is mounted on a one-meter high pedestal with steping-motor controls to adjust both the angle of elevation and the azimuth angle to keep the reflected sunlight aimed at the receptor window. The low profile of the mirror field minimizes wind forces.

RECEPTOR AND STEAM SYSTEM

Each receptor has a double-paned window which is cooled by triethylene glycol between the panes. The cool glass minimizes heat loss by convection and infrared reradiation. The triethylene glycol is circulated to the powerhouse and provides preheat of the boiler feedwater.

In the center of the receptor cavity are black pipes at 400°C with a heat-exchange fluid, and at the back of the cavity are additional black pipes at 600°C. Partherm 290 or Hitec is an acceptable heat exchange fluid to bring 550°C heat to the central powerhouse for the superheater and reheater heat exchangers and for heat storage.

HEAT REJECTION

All steam turbines need a condenser which cools the exhaust steam and condenses it back to water so that it can be repressurized for boiler feedwater again. For every megawatt of electrical generation, there are approximately two megawatts of rejected heat which the condenser must dispose of. In our design, the purified spray water used in the condenser is cooled by secondary water in a closed reservoir system. This secondary water is circulated underneath 70 hectares of metal panels which are painted with titanium dioxide. This paint is black for infrared and white for visible light, so that it radiates an average of 150 MW thermal continuously away from the earth into space. A depth of one meter underneath the 70 hectares will provide a thermal time constant of approximately five days.

MIRROR FIELD PREPARATION AND INSTALLATION

A machine drives through the solar field and digs the foundation holes and inserts a rebar form at regular internals. A second machine pours the concrete. After a month, the mirror pedestals are bolted down and the computer and controls plugged in. Only visual alignment of the pedestal to the vertical axis is made by unskilled workers.

At night, with a full field of mirrors, a laser beam below the closed window protective cover, under the control of a computer, locks on to mirror number one, and this mirror centers its reflected light on the receptor under the control of photo-receptors on the protective cover. the computer reads the horizontal and vertical shaft encoders of the laser and mirror, and stores this information as the coordinates and length of a base line in the field. This procedure is automatically repeated for every mirror in the field. During one day, one mirror tracking the sun with the photocells on the protective cover provides the true north information, so that the field of mirrors has been completely surveyed and calibrated without any manual operation or adjustment. Thereafter, all mirrors can be computer controlled to track the sun.

This procedure minimizes the installation and engineering adjustment costs. Separately controlling each mirror is less expensive than mounting many mirrors rigidly together on a large frame and attempting to align all of them to the same focal point by an expensive rigid structure that will not bend under thermal and wind forces, and by an expensive engineering optical adjustment of each mirror panel with respect to the other mirror panels.

COMPARISONS

The Smith Multimodule Solar-Electric System is unique and superior to other proposed systems due to the inclusion of the following cost-effective engineering advancements:

1. All field surveying, alignment, and check-out calibrations can be performed automatically by computer, instead of expensive manual field surveying and adjustment.

2. The mirrors are connected together by computers instead of a large physical structure.

3. The hexagonal field boundaries for the heliostat fields minimizes piping cost.

4. The cooled-window heat receptor has high efficiency with large receptor aperture and even with low concentration ratio.

5. There is no change of phase (boiling) in the heat receptor.

6. Power plant performance can be optimized by computer on-line by controlling the flow rates of the fluids in the receptor. The geometry does not physically constrain the flow rates to narrow ranges.

7. Sensible heat is stored in separate chambers for separate functions: e.g., preheat, boiling, and superheat.

8. The spray condenser yields superior performance.

9. Radiative heat rejection into the night sky minimizes heat pollution.

10. A cool pond is used to decouple the condenser from the heat rejection. The heat rejection pumping can be wind-augmented.

11. The mirror surface is easily replacable.

12. Module costs are minimized by a mirror assembly production line for large quantities of small items in which nearly all labor is within the factory, and not in the field.

13. Short tower heights reduce the long-term maintenance and repair costs.

14. High thermodynamic efficiency is achieved by using both high pressure and low pressure turbines with a steam reheat before the low pressure turbine.

15. A minimum cost-per-performance design has been determined with an optimization computer program.

ENGINEERING DESIGN

The engineering design is detailed in Tables I through V, and Figures 1 through 14.

Figure 1 is the elevation view of a heliostat (concentrator).

Figure 2 is the plan view of a heliostat.

Figure 3 is an axonometric view of the heat receptor on the tower, showing the reflecting shields and the array of photocells for calibration purposes.

Figure 4 is an elevation view of the heat receptor showing the laser gun and the minimum and maximum angles of the reflected laser beam, (14° and 34° respectively).

Figure 5 is a plan view of the receptor showing the hinged doors that insulate it at night, and which carry the calibrating photocells.

Figure 6 is a front elevation view of the photocell array.

Figure 7 is a plan view of a module of 500 mirrors.

Figure 8 is a perspective array of 11 hexagonal modules and their relationship to their associated towers.

Figure 9 is an elevation view of four towers.

Figure 10 is the piping detail of four branches with 2 towers per branch.

Figure 11 is the HTS piping system for the North-East quarter of a 600-module solar thermal electric plant.

Figure 12 is a simplified steam flow diagram.

Figure 13 is the zone melting and zone freezing controls for the eutectic alloy fusion system to hold turbine inlet conditions constant.

Figure 14 is the computer system for the 100-MW Smith Multi-Module Solar Thermal Electric Power Plant.

Figure 15 is the computer control system interconnections for one heliostat for surveying, calibrating, check out, and control.

PERFORMANCE CALCULATIONS

Figure 15 is the flow chart for a digital computer program which calculates the performance of every mirror in a module as a function of time for all selected days during the year, and calculates the capital cost and maintenance cost of every component as a function of design dimensions, and modifies all design dimensions to minimize the total cost per megawatt hour. The design dimensions given in Figures 1 through 11 and Tables 1 through V were obtained by the modifications from this computer program.

Table I has the dimensions of the collector subsystem.

Table II has the dimensions of the receiver subsystem.

Table III has the dimensions of the thermal storage subsystem.

Table IV is the cost estimate per heliostat.

HELIOSTATS

At the bottom of Table IV is given $135.72 as my conservative estimate of the heliostat cost leaving the assembly production line. This is carried forward into Table XII, one module costs. Shaft encoders in large quantity can be obtained for six dollars each. Stepping motors sufficient for stowing the mirrors in one minute are nine dollars each. Figure 14 shows the microprocessor system for computer control. This will cost approximately six dollars per mirror. Instead of an incremental shaft encoder, the actual pulses to the stepping motors could be used when combined with an index mark.

TOWERS

The towers in this system are 27 meters high, because the computer optimization chose this as the optimum minimum-cost design. The short tower results in only a modest cost. This is included in Table XII as 6,000 dollars. Also included is $25,000 for the windowed receiver.

Diminishing the height of the tower will reduce the tower cost, at a corresponding increase in the blocking of the reflected light from the most distant mirrors at noon, reducing the electrical output, or requiring a greater distance in the north-south direction between the most northern rows of heliostats, increasing the land and piping costs.

HELIOSTAT INSTALLATION

The next items in Table XII are the installation costs for mirrors and mirror controls.

The field installation above includes the adjustment and checkout. The installation consists of removing the pedestal from a truck, setting the pedestal on the four studs in its base, and tightening the four nuts shown in Figures 1 and 2. This should be five minutes per mirror. An electrician walks through the completed module and plugs in all of the pre-

CONCENTRATOR MODULE ELEVATION

FIGURE 1

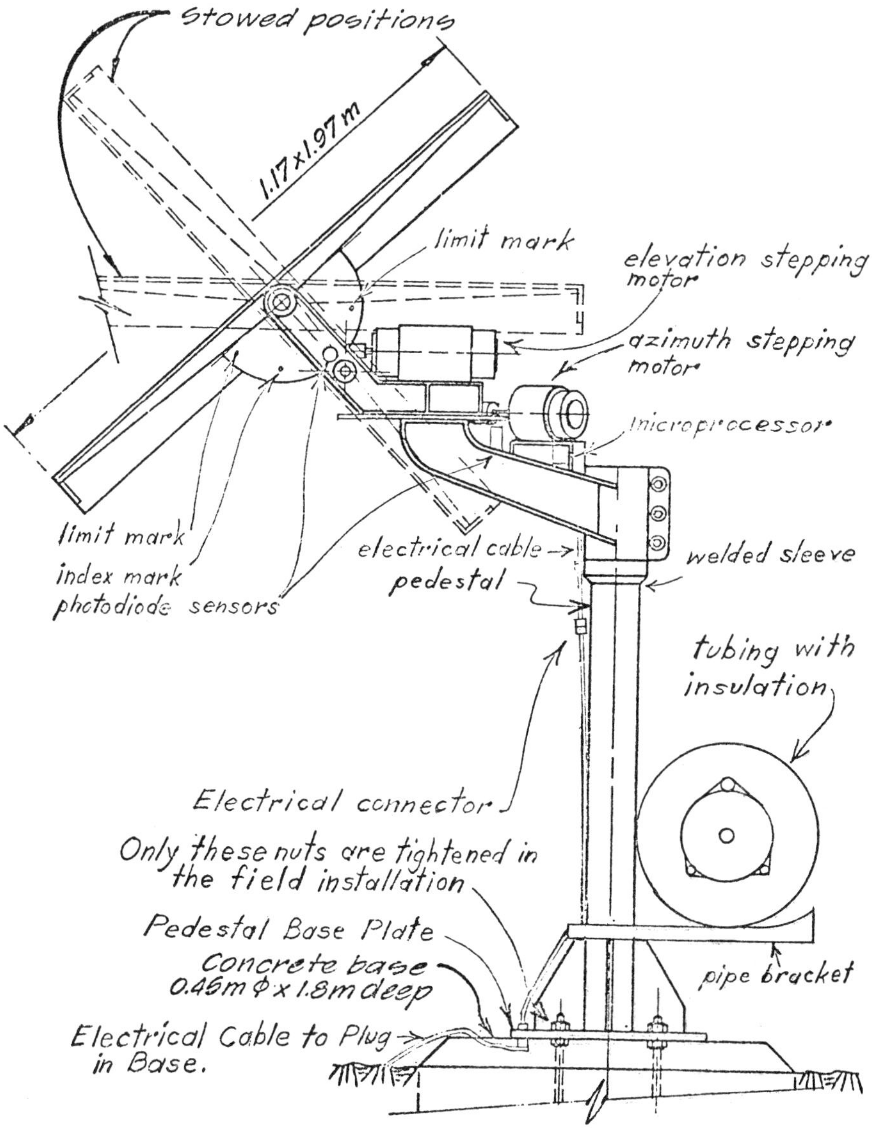

FIG. 1
OTTO J.M. SMITH
11 FEB, 1978

CONCENTRATOR MODULE PLAN

FIGURE 2

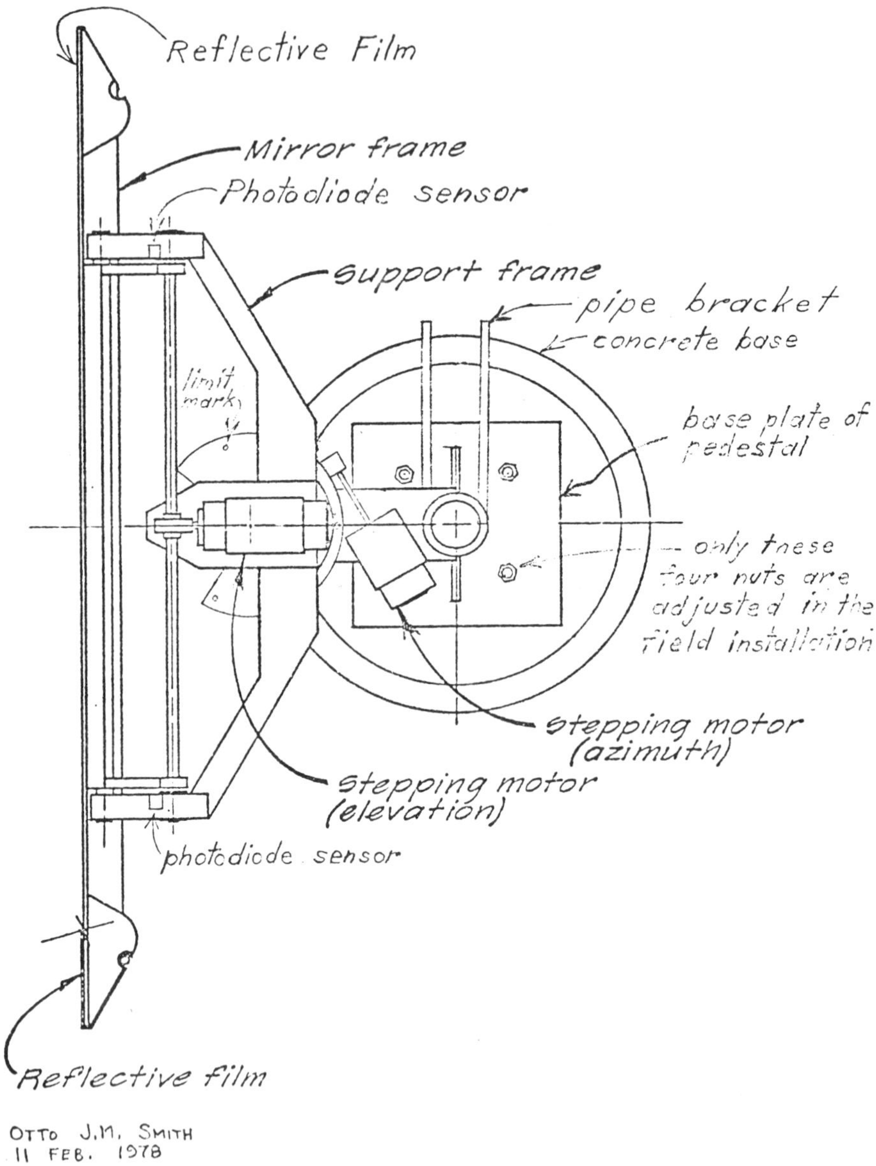

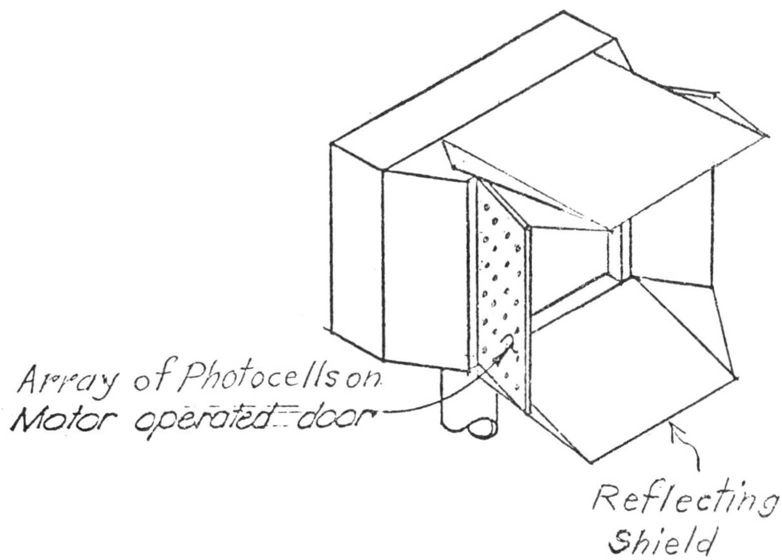

NORTHEAST VIEW OF HEAT RECEPTOR

AXONOMETRIC VIEW

FIGURE 3

RECEIVER MODULE VERTICAL SECTION

FIGURE 4

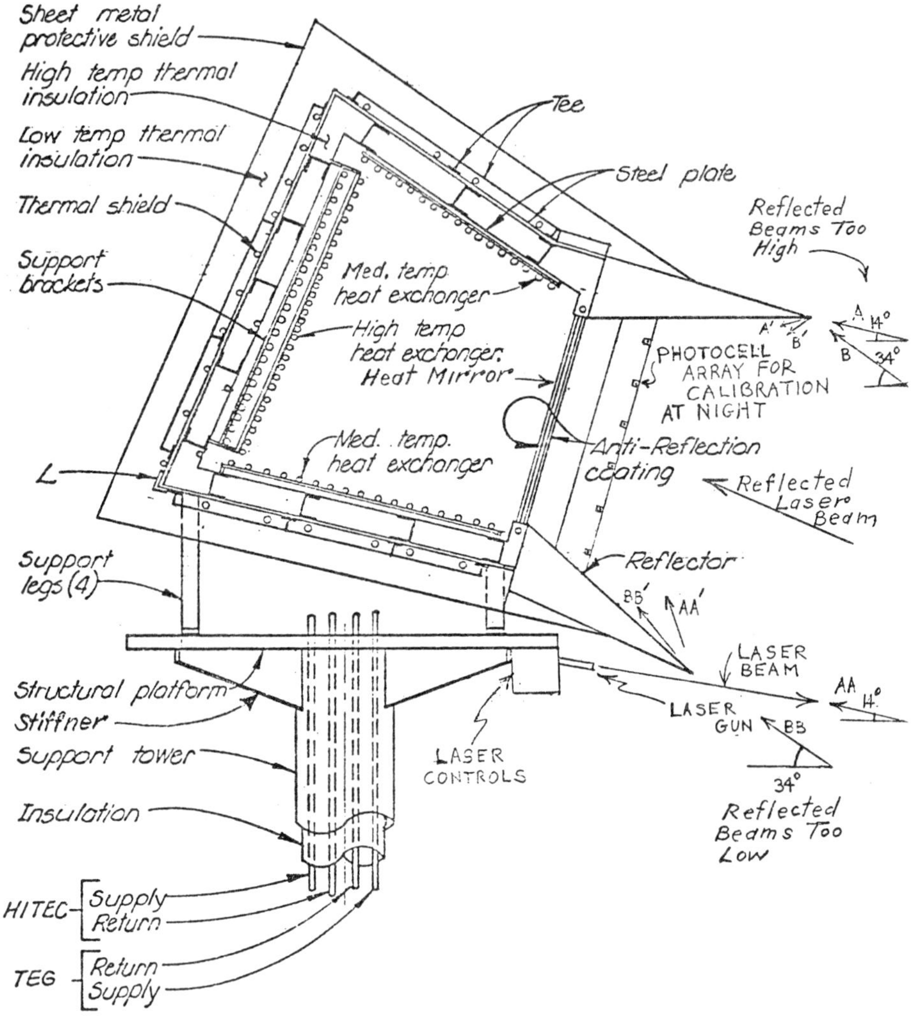

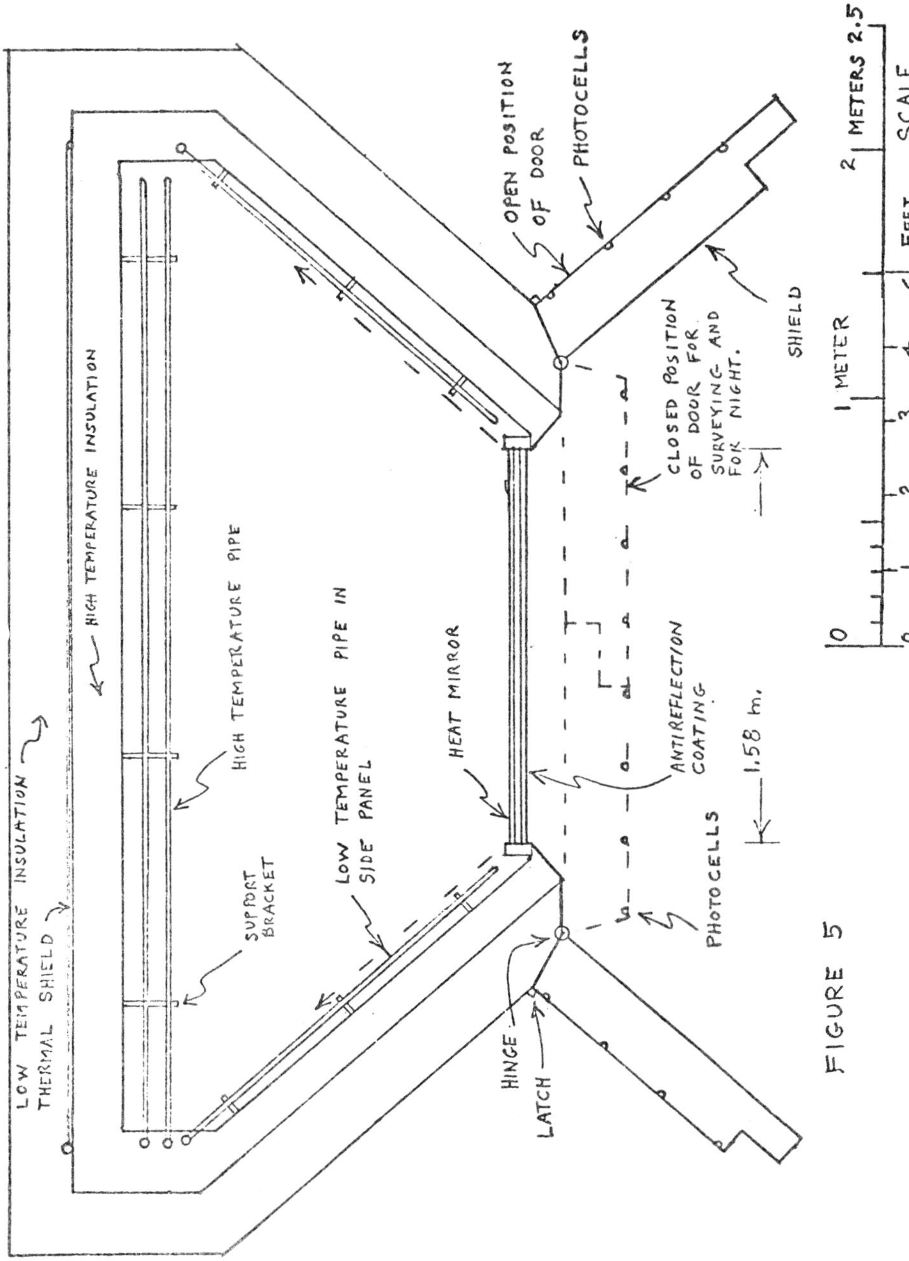
LOW TEMPERATURE INSULATION
THERMAL SHIELD
HIGH TEMPERATURE INSULATION
HIGH TEMPERATURE PIPE
SUPPORT BRACKET
LOW TEMPERATURE PIPE IN SIDE PANEL
HEAT MIRROR
ANTIREFLECTION COATING
1.58 m.
PHOTOCELLS
HINGE
LATCH
CLOSED POSITION OF DOOR FOR SURVEYING AND FOR NIGHT.
OPEN POSITION OF DOOR
PHOTOCELLS
SHIELD
0
1 METER
2
METERS 2.5
0
1
2
3
4
5
FEET
SCALE

FIGURE 5

FIGURE 6
NORTH ELEVATION OF WINDOWED RECEPTOR WITH REFLECTING SHIELDS
SMITH DESIGN NO. 116 AND 117.

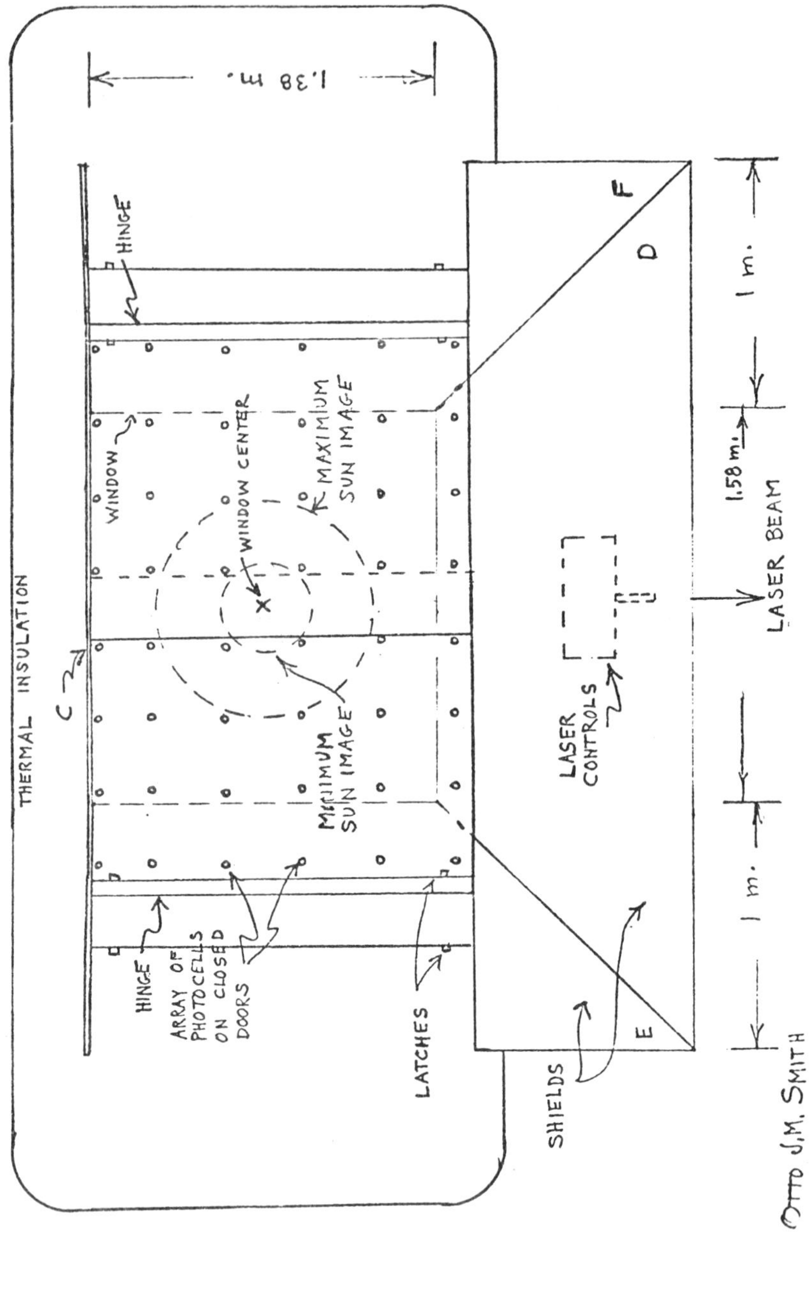

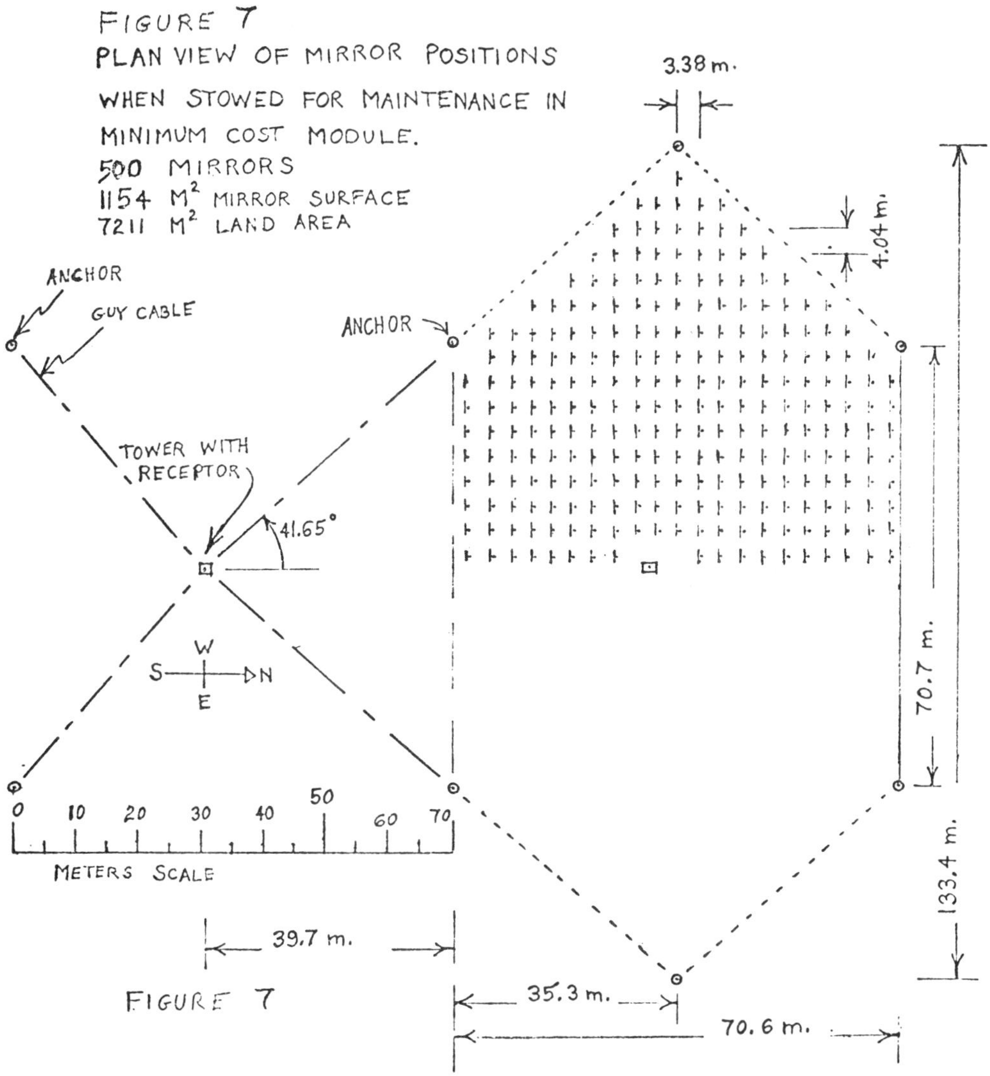
FIGURE 7
PLAN VIEW OF MIRROR POSITIONS
WHEN STOWED FOR MAINTENANCE IN
MINIMUM COST MODULE.
500 MIRRORS
1154 M² MIRROR SURFACE
7211 M² LAND AREA
ANCHOR
GUY CABLE
ANCHOR
TOWER WITH
RECEPTOR
41.65°
W
S
N
E
0
10
20
30
40
50
60
70
METERS SCALE
39.7 m.
3.38 m.
4.04 m.
70.7 m.
133.4 m.
35.3 m.
70.6 m.
FIGURE 7

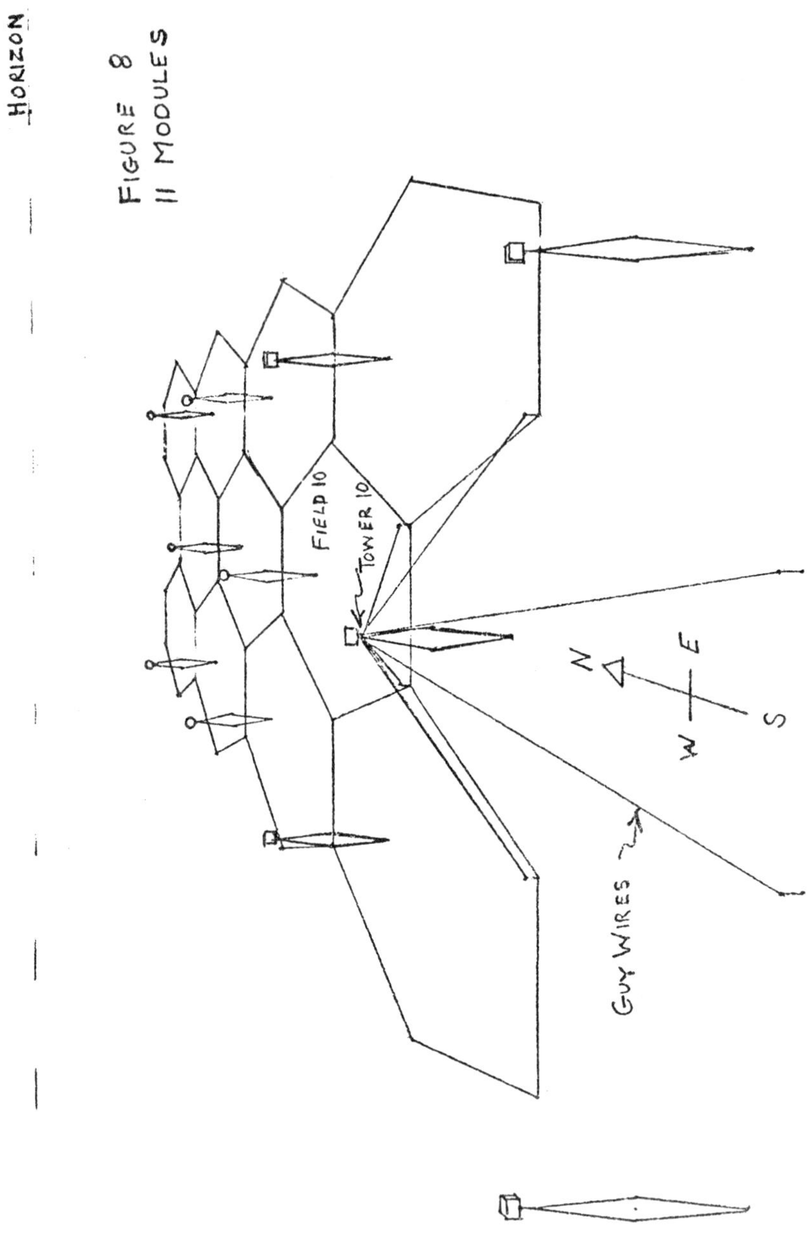
HORIZON
FIGURE 8
11 MODULES
FIELD 10
TOWER 10
N
W
E
S
GUY WIRES

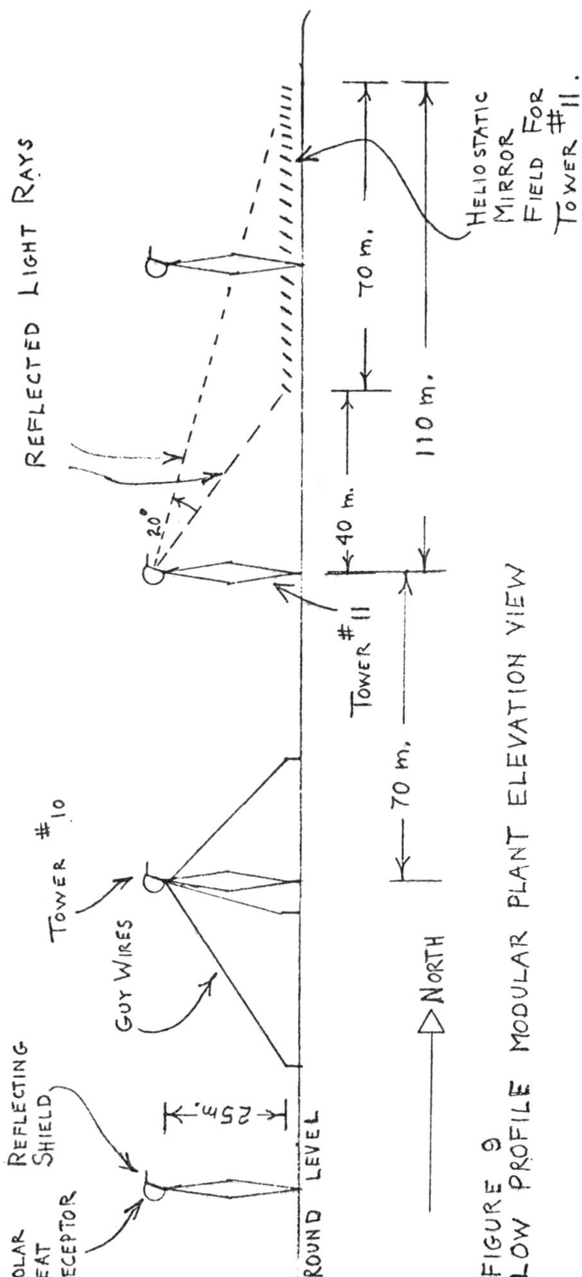

FIGURE 9
LOW PROFILE MODULAR PLANT ELEVATION VIEW

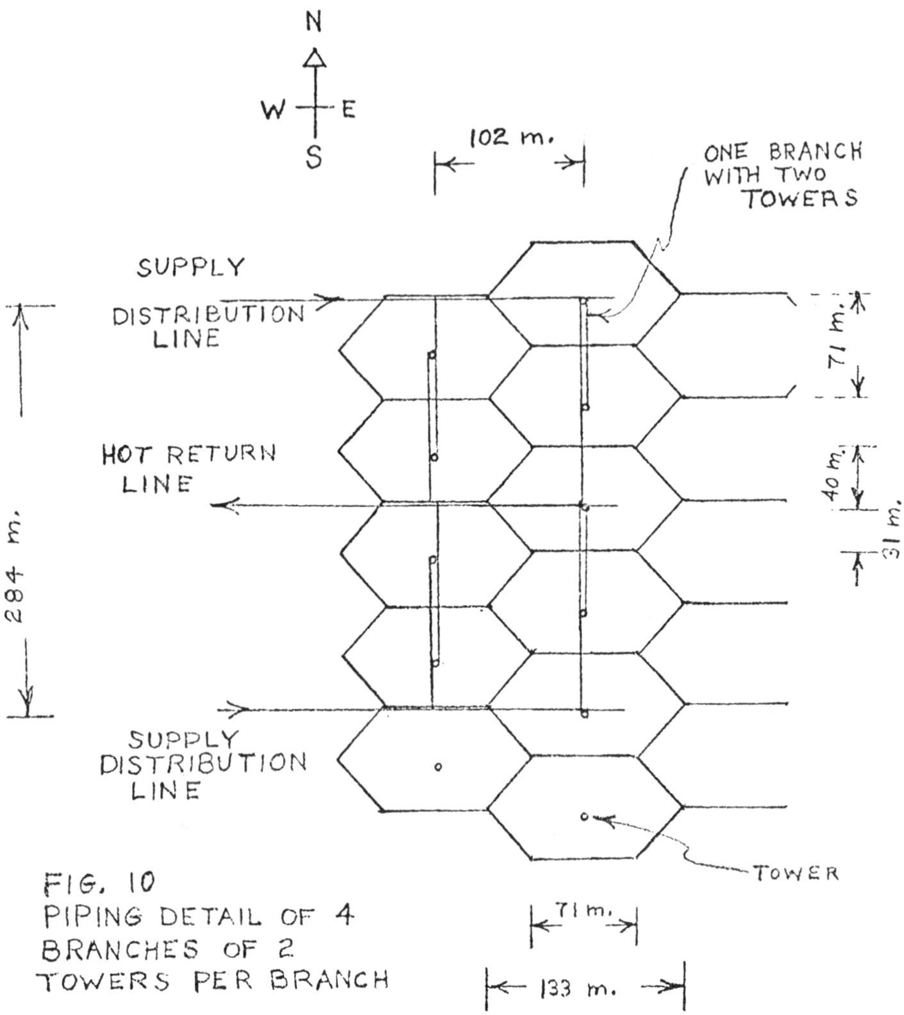

FIG. 10
PIPING DETAIL OF 4 BRANCHES OF 2 TOWERS PER BRANCH

ONE PIPE SYSTEM FOR HTS.

SECOND PIPE SYSTEM FOR TEG.

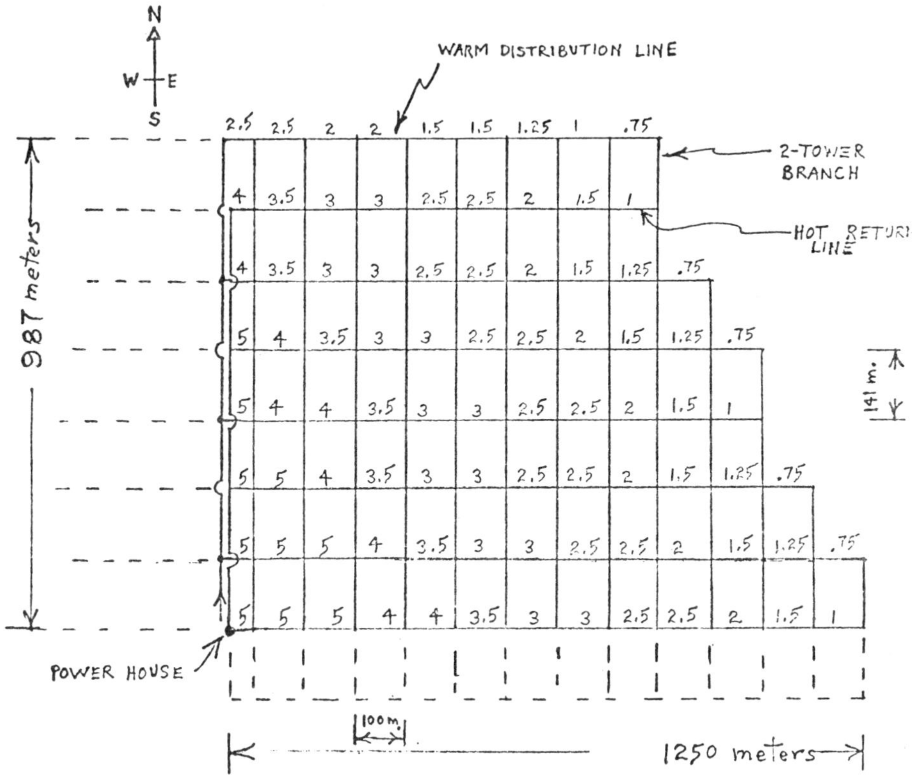

FIGURE 11

HTS PIPING SYSTEM FOR NORTH-EAST QUARTER

OF A 600-MODULE SOLAR THERMAL ELECTRIC PLANT

ALL BRANCHES HAVE 0.75" PIPE

PIPE SIZES FOR DISTRIBUTION LINES ARE SHOWN IN INCHES

FIGURE 12
THERMODYNAMIC SYSTEM

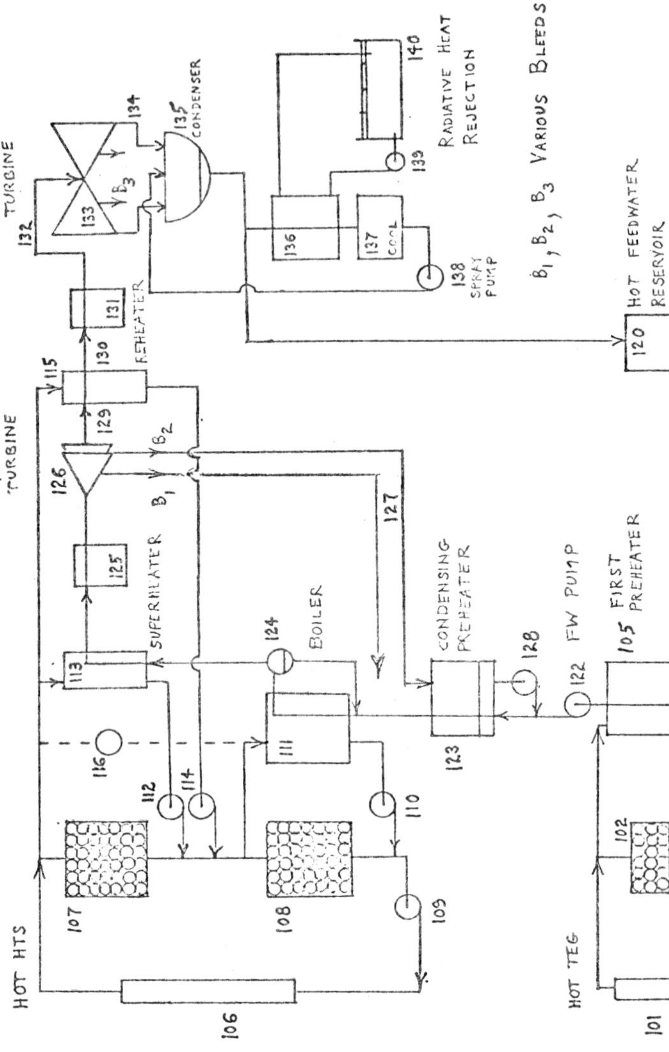

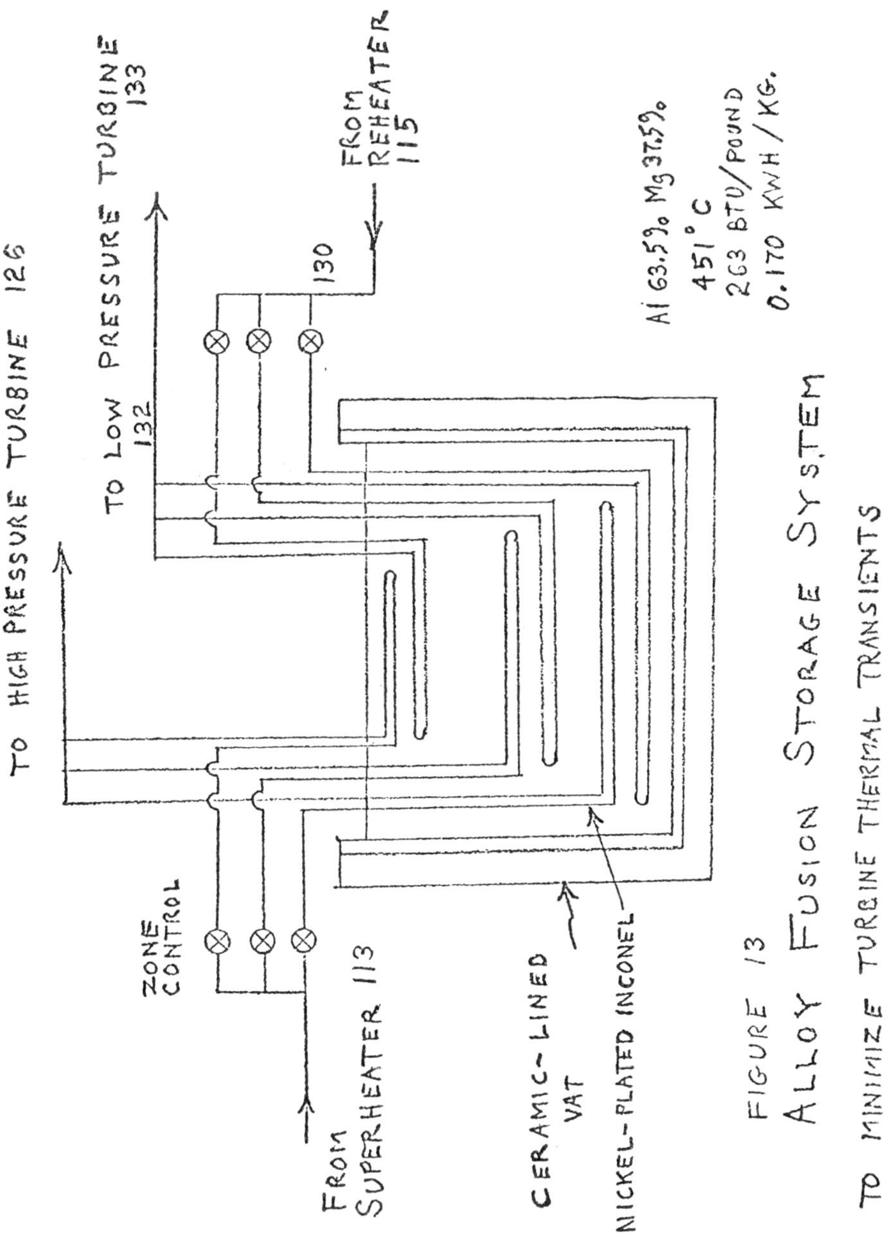

FIGURE 13
ALLOY FUSION STORAGE SYSTEM
TO MINIMIZE TURBINE THERMAL TRANSIENTS

FIGURE 14

COMPUTER SYSTEM FOR 100-MW SMITH MULTI-MODULE SOLAR THERMAL ELECTRIC POWER PLANT

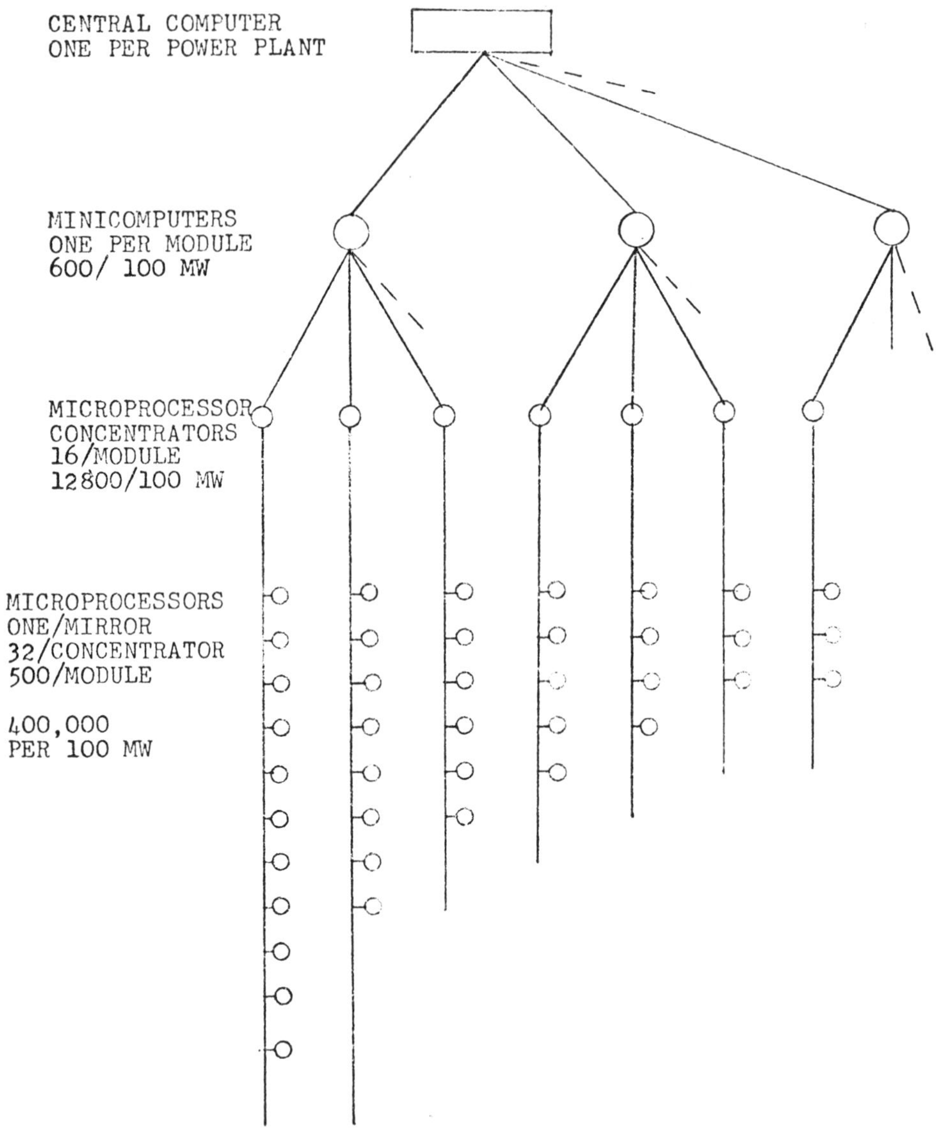

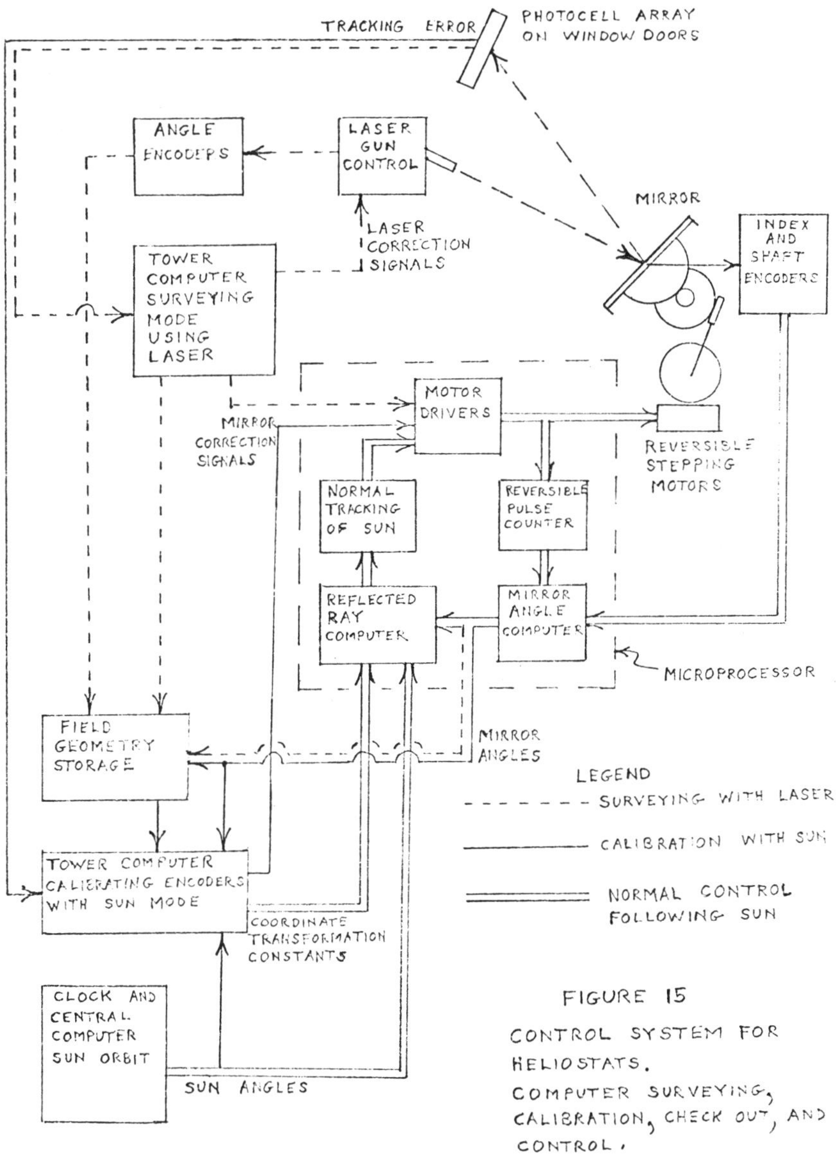

FIGURE 15

CONTROL SYSTEM FOR HELIOSTATS.
COMPUTER SURVEYING, CALIBRATION, CHECK OUT, AND CONTROL.

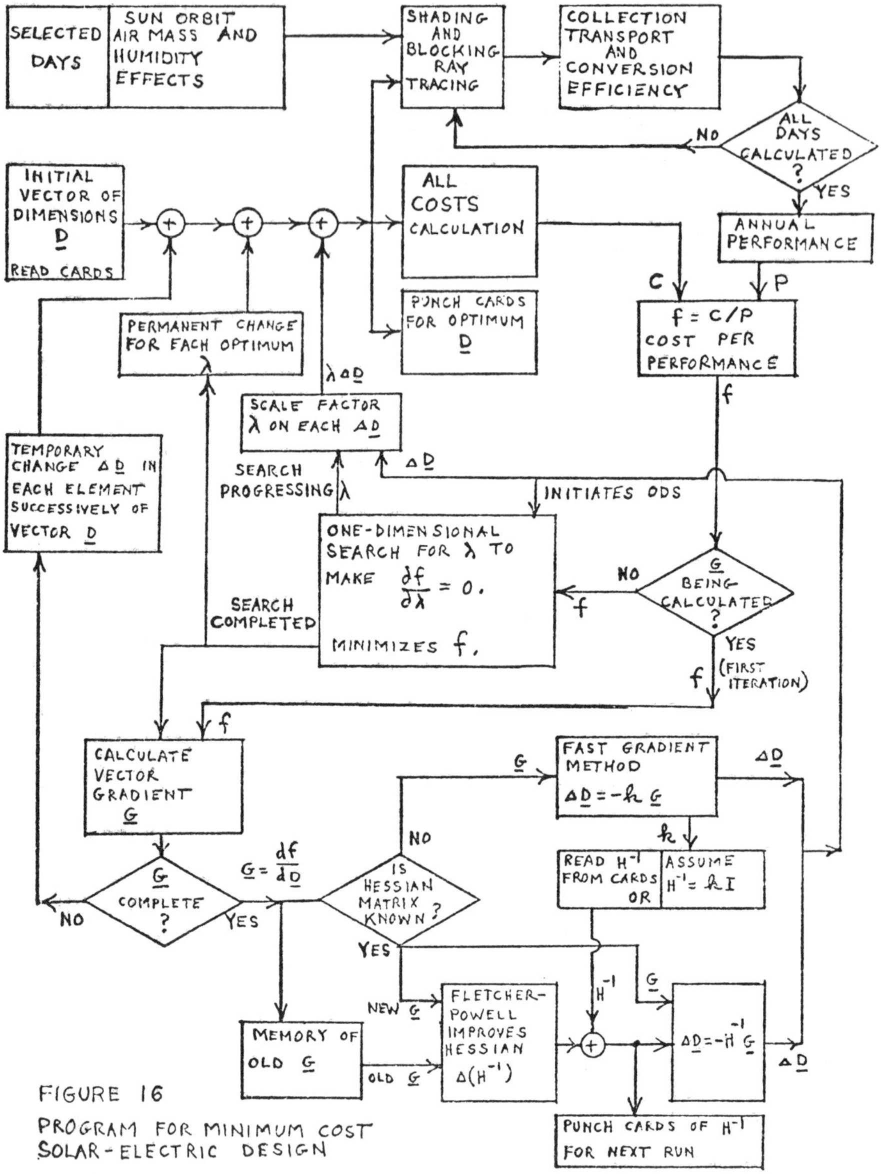

FIGURE 16

PROGRAM FOR MINIMUM COST SOLAR-ELECTRIC DESIGN

wired harnesses on the pedestals to the harnesses laid out in the field. There is no further labor required in the field. All surveying, checking, calibration, adjustment, and control is performed from the computer console in the central power house. The flow chart for this is shown in Figure 15. A 3-man crew costing $500 per day can install 36 heliostats per day, at a cost of $14.00 per heliostat. An electrician costing $12.00 per hour can connect 30 controls per hour, at a cost of 0.40 per control. The adjustment and checkout computer operator time required for individually checking every operation mode of each heliostat is one dollar per heliostat (12 per hour). This is done by the computer system in Figure 14 and the program in Figure 15.

The receiver checkout requires 5 days of a 3-man crew, and will cost $2500.

These costs are shown in Table XII. These items are $10,230 for 500 heliostats.

PIPE SYSTEM

Schedule 10 stainless pipes as small as 19 min inside diameter may be used for the towers. These pipes would have insulation about 7" thick. The cost of the insulation is not insignificant. However, the insulation, if it is sufficiently thick, not only provides higher efficiency for the performance of the piping system during the daytime, but has a bearing upon selection of the method of operation at night. At night, the receptor shutters should be closed across the windows in the towers and very low pumping rates should be maintained to hold the HTS slightly above its melting point. No attempt should be made to purge the system at night for the purpose of minimizing heat losses from the cavities and windows on the towers. If this is the policy of operation that is adopted, then there is an advantage in having a surplus of insulation around the HTS (Hitec) pipes. This would be of the order of magnitude of 7" thick insulation, on both the smallest pipes, and on the larger pipes.

The pipes will be assembled in bundles of two or four tubes per bundle, as shown in Figure 1, so that the insulation is effective for the entire bundle. These bundles can be prefabricated and brought out to the field in long pipe carriers.

The piping bundle supports are at the bases of the few heliostats in the particular lines desired, so that no special foundations have to be provided.

The piping layout is shown in Figure 11. Each branch passes through the base of two towers in the north-south direction. This layout is for modules of 500 mirrors each. A similar system of smaller pipes would be used for the tri-ethylene-glycol.

The table of appropriate quantity quotations for HTS piping is shown in Table V. The total is 1.237 million dollars. The TEG piping system is listed in Table VI. The total cost is $844,000.

Table VII lists the number of brackets required. They will cost $153,000.

Tables VIII and IX list the insulation requirement. It is 13,000 cubic meters and will cost 2.8 million dollars. The pipe installation cost is shown in Table X. This is 2.731 million dollars.

The installation costs are from the Lee Saylor handbook, which are possibly 20 percent too high. The total of all of these components is $7,831,000, which is $13,052 per module.

PIPE HEATING

We propose to heat the piping by passing electrical current throughout the entire system when it is cold and prior to charging it with HTS. The thermal insulation is electrically insulating also. Brackets in solid thermal contact with the pipes are avoided because these represent thermal losses. A good thermally insulated system will automatically have high electrical resistance between pipes and ground and between one pipe system and another pipe system, compared to the electrical resistance down the pipe. Consequently, the heating power will be concentrated in the pipes, and where leakage currents flow through brackets, the heating power in the brackets will be desirable to compensate for the thermal losses through the same brackets. Consequently, it is expected that conventional practice will result in a system that is ideally suited to "electrical heat tracing."

The electrical current is used only to warm up the cold empty pipe system before first charging it with molten HTS. The electrical current is not used during normal operation.

With regard to backup power, in order to start up this solar power plant, the station power should be sufficient for the computers, for the heliostat tracking controls, and for the pumps, so that the power plant can start without reliance on an exterior power source. A small gas turbine generator can provide station electrical power for auxiliaries and for the "electrical heat tracing." The turbine exhaust through a heat exchanger can provide process steam. Biomass fuel or fossil fuel can be stored for this emergency and start-up power.

HTS DRAINING

It is not intended that the HTS be drained each day. The high level of thermal insulation on the pipes has been chosen to provide sufficient insulation so that a very small flow rate of HTS at night will be sufficient to keep the entire system above 400°F. There are insulated doors provided for each receiver window, and these doors would be closed whenever there was no sunshine. The flow controls that keep the HTS returning from the solar field at a constant 930°F during the day would continue to function at night with the set point lowered to 450°F. The economic way is to con-

TABLE I

COLLECTOR SUBSYSTEM

Number of Modules for 100 MW(e)	600	
Number of Heliostats (Mirror Assemblies) Per Module	500	
Reflective Area Per Heliostat	2.3	Square Meters
Reflective Area Per Module	1,150	Square Meters
Total Number of Heliostats for 100 MW(e)	300,000	
Total Reflective Surface	690,000	Square Meters
Field Area Per Module	7,211	Square Meters
Total Collector Field Area (600 Modules)	4,327,000	Square Meters
Field Density (Reflective Area ÷ Field Area)	0.16	
Construction	Fiberglass polyester cast over parabaloid mold. Metal-ized polyester surface held down on platen.	
Control	Azimuth and Elevation	
Field Array	Hexagonal Modules	
Control Mechanism	Stepping Motors	
Reflectivity	91%	
Mirror System Efficiency	84%	

TABLE II

RECEIVER SUBSYSTEM

Receiver Type	Horizontal Cavity, with Triethylene Glycol Cooled Glass Cover
Tower Type	Guyed Steel Truss
Working Fluid	HTS Molten Salt[1]
HTS Inlet Temperature to Solar Field	320°C (608°F)
HTS Outlet Temperatures from Solar Field	600°C (1,112°F)
Total HTS Mass Flow Rate in Solar Field	1,016 Kilograms Per Second (9,049,800 Pounds Per Hour)
HTS Mass Flow Rate Per Receptor	0.924 Kilograms Per Second (7,318 Pounds Per Hour)
Triethylene Glycol Inlet Temperature to Solar Field	143°C (290°F)
Triethylene Glycol Outlet Temperature from Solar Field	200°C (392°F)
Total Triethylene Glycol Mass Flow Rate in Solar Field	323 Kilograms Per Second (2,549,800 Pounds Per Hour)
Triethylene Glycol Mass Flow Rate Per Receptor	0.293 Kilograms Per Second (2,318 Pounds Per Hour)
Volume of HTS in Collector Field Piping	746 Cubic Meters (197,000 Gallons)
Volume of Triethylene Glycol in Collector Field Piping	204 Cubic Meters (54,000 Gallons)
Receiver Aperture Size	1.38 Meters by 1.58 Meters
Approximate Length of Piping	
Triethylene Glycol System	138 Km (86 Miles)
HTS System	138 Km (86 Miles)
Total	322 Km (200 Miles)

[1]Partherm 290 is a tradename for HTS sold by Park Chemical Co.; HITEC is a tradename for HTS (heat transfer salt) developed by E.I. DuPont.

TABLE III

THERMAL STORAGE SUBSYSTEM

Storage Concept	3-Stage, Sensible Heat
Storage Tank Operating Pressure	HITEC - 60 PSIG TEG - 35 PSIG Due to Static Head in 20m Towers
Net Electrical Output and Duration	100 MW(e) for 4 Hours
Thermal Storage Units	
1st Preheat	
Media	TEG
Volume	4,110 m^3 (1,086,000 Gallons)
2nd Preheat	
Media	HTS
Volume	1,836 m^3 (485,000 Gallons)
Steam Generator/Boiler	
Media	HTS
Volume	2,048 m^3 (541,000 Gallons)
Superheater/Reheater	
Media	HTS
Volume	2,392 m^3 (632,000 Gallons)
Fusion Storage System for Turbine Inlet Temperature Control	
Eutectic Alloy Media	63.5% Aluminum 37.5% Magnesium
Fusion Storage Concept	Ceramic lined vat filled with eutectic alloy. Nickel-plated inconel pipes buried in the eutectic solution extracts latent (phase change) heat from the solution as it changes phases.

TABLE IV

COST PER HELIOSTAT, DOLLARS

Item	Cost
Structural Frame	$ 25.35
Reflective Surface and Support	12.36
Mechanical Drive (Stepping Motors)	19.52
Shaft Encoders	12.00
Microprocessors	6.50
Total Cost Per Heliostat Using Mass Production	$ 75.72
Margin for Unexpected Cost Overruns	60.00
Conservative Total	$135.72

TABLE V

HTS PIPING SYSTEM

STAINLESS STEEL 304, WELDED, SCHEDULE 10, CARLOAD LOTS

Millimeters Nominal Diameter	KG/M	$/ Meter	Meters Length	Metric Tons	Dollars
19	1.3	4.72	98,000	127	462,349
25	2.1	7.08	1,400	3	9,907
32	2.7	8.69	2,000	5	17,376
38	3.1	9.76	3,400	11	33,197
51	4.0	12.07	3,400	13	41,028
64	5.3	15.79	5,800	31	91,568
76	6.5	20.40	5,400	35	110,164
89	7.5	23.52	2,600	19	61,160
102	8.5	26.56	3,360	28	89,239
127	11.7	36.52	2,500	29	91,296
152	14.0	41.61	560	8	23,301
203	20.2	59.36	1,120	23	66,480
254	28.1	77.54	560	16	43,418
305	36.4	101.24	840	31	85,042
Total			130,940	379	1,225,525
Plus 5% for Expansion Omegas:			137,537	398	1,286,801

Average Cost	3230 dollars per m. ton
Smallest Pipe	3640 dollars per m. ton
Largest Pipe	2780 dollars per m. ton
Length of Pipe	137 km.
Weight of Pipe	400 m. tons
Cost of Pipe	1.287 million dollars

TABLE VI

TEG PIPING SYSTEM

Pipe Size, MM	Length Meters	Metric Tons	Dollars
13	98,000	99	373,435
19	1,400	2	6,605
25	5,400	11	38,216
32	3,400	9	29,538
38	4,200	13	41,009
51	8,000	32	96,537
64	7,960	40	117,774
76	560	4	11,422
89	560	4	13,171
127	1,120	13	40,906
152	840	12	34,957
Total	130,940	239	803,570
Plus 5% for Expansion Omegas	137,537	251	843,749

TABLE VII

PIPE SUPPORT BRACKETS ON PEDESTALS

In Branches:	
42 Brackets Per Branch 300 Branches	12,600
In Distribution Runs:	
26 Brackets Per 100 Meters 311 Distribution Runs of 100m.	8,086
Total Number of Brackets	20,686
Cost at $5.00 Each	$153,430

TABLE VIII

INSULATION

Center Pipe Size	High Temp. Insulation Cu. Mt.	High Temp. Insulation Cu. Ft.	Low Temp. Insulation Cu. Mt.	Low Temp. Insulation Cu. Ft.
0.75	1,525	53,872	5,593	197,529
1.0	36	1,278	128	4,509
1.25	55	1,933	186	6,585
1.5	98	3,468	324	11,438
2.0	110	3,869	339	11,974
2.5	208	7,348	607	21,422
3.0	210	7,421	591	20,872
3.5	111	3,908	295	10,421
4.0	153	5,411	395	13,949
5.0	129	4,563	314	11,094
6.0	32	1,142	75	2,645
8.0	82	2,886	173	6,093
10.0	48	1,683	95	3,366
12.0	82	2,886	157	5,532
	2,879	101,668	9,272	327,430
Plus 5% for Joints and 5% for Expansion	3,023	111,835	9,735	360,173

Length of 4-Pipe Bundles	35.28 kms.	115,748 feet
Length of Smallest 2-Pipe Bundles	27.44 kms.	90,026 feet
Length of All Other 2-Pipe Bundles	32.94 kms.	108,071 feet
Total Length of All Bundles	95.66 kms.	313,845 feet

TABLE IX

INSULATION COST

Pipe Size	Mineral Wool Insulation I.D.	Thickness	O.D.	Dollars Per Meter
19mm	25mm	76mm	180mm	$ 5.9
	200mm	50mm	300mm	7.1
	300mm	50mm	400mm	9.7
Total				$22.7

Composite Cost Dollars Per Cubic Meter	200	
Total Volume Cubic Meters	13,336	
Insulation Cost		$2,659,320
Length of Bundles, Meters	95,660	
Aluminum Sheet Covering, Dollars Per Meter	1.64	
Aluminum Sheet Covering		156,923
Total		$2,816,243

TABLE X

INSTALLATION COST

Central Pipe Size MM ID	Length Meters	Installation Cost $/Meter	Installation Cost Dollars
19	98,000	15.12	1,482,216
25	1,400	25.16	35,228
32	2,000	25.59	51,184
38	3,400	26.25	89,240
51	3,400	27.85	94,706
64	5,800	29.46	170,880
76	5,400	30.97	167,248
89	2,600	32.81	85,300
102	3.360	36.02	121,044
127	2,500	40.16	100,392
152	560	42.59	23,844
203	1,020	53.05	54,105
254	560	75.89	42,490
305	840	98.43	82,680
Total			2,600,558
Plus 5% for Expansion Omegas			2,730,586
Total Installation Cost of Piping			2.731 Million Dollars

TABLE XI

PIPING COST PER MODULE

	Dollars
HTS Pipe, From Table V	$ 2,145
TEG Pipe, From Table VI	1,406
Brackets, From Table VII	256
Insulation, From Table IX	4,694
Installation, From Table X	4,552
Total, Dollars Per Module	$13,052

TABLE XII

100 MW SMITH MULTI-MODULE SOLAR POWER PLANT

DOLLARS PER MODULE

Number of Mirrors	(500)
COMPONENT COST PER MODULE	
Mirror System, From Table IV	67,860
Towers - 27 Meters High	6,000
Windowed Receiver	25,000
Subtotal	98,860
INSTALLATION COST PER MODULE	
Roads	9,600
Clearing and Grading	32,000
Concrete Footings of Heliostats	34,100
Towers and Foundation	20,000
Heliostat Wiring	3,833
Piping, From Table XI	13,052
Mirror Installation	7,000
Mirror Controls	230
Receiver Adjustment and Checkout	3,000
Subtotal	122,815
TOTAL COST PER MODULE	221,675

TABLE XIII

COST SUMMARY, DOLLARS

100 MW POWER PLANT COST

600 Modules	133,005,000
Heat Storage	4,000,000
Heat Exchangers, Turbine, Generator, Condenser, Cooling, Transformer, Switch Yard	20,000,000
Central Computer System	1,000,000
Subtotal	158,005,000
Interest During Construction	17,227,000
Total	175,232,000
Cost Per Nameplate KW	1,752

TABLE XIV

COST BREAKDOWN BY PERCENTAGES

	Dollars	Percent
Land Preparation	24,960,000	14.2
Heliostats and Installation	67,813,800	38.7
Towers and Installation	32,400,000	18.5
Piping	7,831,200	4.5
Power House	25,000,000	14.3
Interest	17,227,000	9.8
Total	175,232,000	100.0

TABLE XV

LABOR IMPACT OF 100 MW PLANT

AMOUNTS IN MILLION DOLLARS

	Direct Labor Wages	Indirect Labor Expenses	Materials and Components Purchased	Total
Heliostats				
Frame	2.85	0.95	3.80	7.60
Mirror	2.09	0.70	0.93	3.71
Drives			5.86	5.86
Encoders			3.60	3.60
Microprocessors			1.95	1.95
Margin	9.00		9.00	18.00
Towers	1.35	0.45	1.80	3.60
Receivers	5.63	1.88	7.50	15.00
Roads	4.32	1.44		5.76
Grading	14.40	4.80		19.20
Heliostat Footings	7.67	2.56	10.23	20.46
Tower Installation	6.75	2.25	3.00	12.00
Wiring	.86	.29	1.15	2.30
Piping	2.05	.68	5.10	7.83
Mirror Installation	3.15	1.05		4.20
Controls Installation	0.10	0.04		0.14
Adjustment	1.35	.45		1.80
Power House	4.50	1.50	19.00	25.00
Total	66.06	19.04	72.91	158.01

Table XVI

ENVIRONMENTAL IMPACT

Collector Field Area (100 MW)	4,327,000 Square Meters
Heat Rejection Area	610,000 Square Meters
Total Area	4,937,000 Square Meters
Reflective Surface	690,000 Square Meters

tinue to pump HTS.

STEAM SYSTEM

The TEG will provide only a small portion of the feedwater preheat. It is expected that five turbine bleeds would be used for condensing preheaters in this 100 MW plant.

TOTAL COSTS

The total computed costs from Tables IV through XI have been divided by 600 modules to obtain the costs per module entered in Table XII. The total cost per module is $221,675.

Table XIII gives the 100 MW power plant cost. The total capital and interest cost is 175 million dollars.

Table XIV shows the cost breakdown by percentages. The heliostats are 39 percent of the total cost. The towers and receivers are 18.5 percent of the total cost. The heat transport piping systems are 4.5% of the total cost. Land preparation is 14 percent; the power house is 14 percent; and interest is 10 percent.

Annual payments to retire a 7.5%, 40 year bond of 175 million dollars are 13.84 million dollars. This is a capital cost of $27.70 per mega-watt hour, or 2.77 cents per kilowatt-hour.

MINIMUM COST DESIGN

Figure 16 is the flow chart for an optimization computer program which we have written and which was used to obtain the optimum dimensions for this design.

Given accurate cost estimates, this program will change the dimensions of the design to minimize the cost per generated mega-watt-hour.

For a feasibility study, we suggest that the entire plant design be subdivided into three overlapping phases: (a) the solar field including the tower and heat receptor; (b) the piping system including the heat receptor, number of modules, heat storage, heat exchangers, and size of turbine; and (c) the power house with heat exchangers, turbine, bleeds, condenser and heat rejection system.

Individual sequential optimization runs for each of these subsystems will probably be less expensive than combining all of the systems into one computer program. The result of this study will be the minimum cost design, and will yield the number of modules, number of mirrors, mirror size, tower height, piping dimensions, volume of heat storage, number and position of turbine bleeds, and area of heat rejection panels. Maintenance costs can be included in this optimization.

This will give a solid engineering base for the construction of a pilot plant with cost effective components.

OPTIMIZATION

1. The capital cost is computed as a function of a vector of adjustables, the dimensions of the mirrors, receptor, tower, field boundaries, and spacings in the field.

2. The annual productivity is computed as a function of the same vector of adjustables, and considers optical reflectivities, shading, blocking, and cosine factors for every mirror for every half hour of the year.

3. The life cycle cost in mills per kilowatt hour, hereinafter called "cost", is computed.

4. The cost is iteratively minimized by changing the vector of adjustables to obtain the optimum minimum cost design. In the present program this is accomplished by a fast gradient step with hard limiting, a cubic one-dimensional-search, updating the Hessian curvature matrix by a modified Fletcher-Powell algorithm, a second-order minimization step, another cubic one-dimensional-search, and a second updating of the Hessian matrix, with this sequence iterated until the improvement in cost is negligible.

The program includes daily atmospheric conditions in the insolation subroutine. Shading is calculated only for low sun angles and only for the mirrors expected to have shading losses.

LABOR IMPACT

Table XV shows that 38% of the total plant cost is direct labor wages, and 43% of the total cost is direct and indirect labor expenses. These labor wages will have a desirable impact in the region where the components are manufactured. The high labor costs are concurrent with a low materials demand and a low drain on scarce resources.

ENVIRONMENTAL IMPACT

In this desert area, the heat that remains is 81% of the undisturbed desert heating. 6.5% of the undisturbed desert heat is exported into the power system as electricity. 15% of the undisturbed desert heat is radiated into the sky. The net effect is that the collector field will be slightly cooler than the undisturbed desert.

REFERENCES

Benning, Alber C., "Black Chrome - A Solar Selective Coating," American Electro-platers' Society, Inc., Coatings for Solar Collectors Symposium, Atlanta, Georgia, 9 November 1976.

Benning, A.C., "Black Chromium - A Solar Selective Coating," The Harshaw Chemical Company, Division of Kewanee Industries, Inc., 1945 East 97th Street, Cleveland, Ohio 44106.

Birchenall, C.E., "Thermal Storage in Metals," Sharing the Sun 1976 Conference ISES and SESC; Winnipeg, Volume 8, pp. 123-137, 15 August, 1976.

Borzoni, John T., "Comparison of Three Solar Selective Absorber Coatings," Energy Resources Center, Honeywell, Inc., Minneapolis, Minnesota.

Clarkson, Clarence W. and James S. Herbert, "Transparent Glazing Media for Solar Energy Collectors," International Solar Energy Conference, Fort Collins, Colorado, 21 August 1974. (ASG Industries, Inc., P.O. Box 929, Kingsport, Tennessee, 37662.

"Glycols" Booklet F-41515A, 7/71-12M. Union Carbide Corporation, Chemicals and Plastics Division, 270 Park Avenue, New York City, 10017.

Milleron, Norman, "Large-Scale Energy Sources," Chemical and Engineering News, Letters Section, 2 August, 1976, page 3.

Riaz, M., P.L. Blackshear Jr., and H.O. Pfannkuch, "High-Temperature Energy Storage in Native Rocks," 1976 Conf., ISES and SESC., Winnipeg, Volume 8, pp. 123-137, 15 Aug. 1976.

Smith, Otto J.M., Calin Mihaileanu, and Mihai Pop, "Centrala Solaroelectrica Functionind pe baza Ciclului Rankine," Energetica, Bucuresti, Romania, Anul 23, Nr. 1, January, 1975.

Smith, Otto J.M., "Low Cost, Low Technology Solar Electric Plant," Report on the Symposium and Workshop on the 5 MWT Solar Thermal Test Facility, ERDA, Houston, Texas, Contract E(29-2) 3701, 17 May 1976. pp. 264-265.

Smith, Otto J.M., "Smith Multimodule Solar-Electric Plant," ASCE Specialty Conference on Solar Energy. Colorado State University, Ft. Collins, Colo., 80523, 20-21 July, 1976.

Smith, Otto J.M., "Smith Multimodule Solar-Electric Plant," Sharing the Sun 1976 Joint Conference of the American Section of the International Solar Energy Society and the Solar Energy Society of Canada, Inc., Winnipeg; Volume 5, pp. 99-116, 15 August, 1976.

Smith, Otto J.M., "Multimodule Practical Solar-Thermal-Electric Power Plants," Proceedings of Frontiers of Power Technology Conference, Oklahoma State University, Stillwater; pp. 7-1 to 7-18, 27 October 1976.

Smith, Otto J.M., "Solar is Practical," Testimony to Consumer Affairs/

Special Impact Advisory Committee to the Federal Energy Administration, 3 and 4 March, 1977, 1200 Pennsylvania Avenue, N.W., Washington, D.C. 20461.

Smith, Otto J.M., "Near Term Solar Electricity," monograph, Electrical Engineering and Computer Sciences Department, University of California, Berkeley, California, 94720, 9 April, 1977. Text only published in Solarcon Times 77, pages 7-9, August 19, 1977.

Smith, Otto J.M., with Marty Baron, R.G. Albrecht, Ronald B. Robie, et. al., "Proposal to Jet Propulsion Laboratory in Response to RFP-EL-2-8978-209 by Wismer and Becker Contracting Engineers and California Department of Water Resources," November 11, 1977, Volume I, The Technical Volume.

Smith, Otto J.M. and Phyllis S. Smith, "The Smith Helioelectric Farm," Second International Helioscience Institute Conference on Alternate Energy, Palm Springs, California, 10 April 1978.

A CENTRAL RECEIVER SOLAR POWER SYSTEM FOR REMOTE LOCATIONS

R. J. Holl
McDonnell Douglas Astronautics Company
Huntington Beach, California U.S.A.

ABSTRACT

A solar power plant has been designed to provide electrical and/or thermal power for locations isolated from a utility grid. Up to 1,000 KW electric or 4,000 KW thermal or various combinations can be provided. Applications can include domestic, agricultural, and light industrial use with different amounts of electricity and thermal power available upon demand.

A small central receiver is employed for solar energy collection and steam is used as the working fluid. Modular thermal storage matches various specific applications without changing the balance of the system. All equipment is designed for easy transportability and minimum on-site assembly.

Attractive economics are attained through: (1) use of a common, prefabricated receiver/tower assembly for all installations, (2) common usage of heliostats under development for the United States Large Central Receiver Thermal Electric Program, (3) simplicity in system design and maximum use of commercial equipment, and (4) skill requirements and costs for operations and maintenance approximating those of a similar size oil-fired plant.

INTRODUCTION

As a nondepletable resource, solar energy is acknowledged to have an important role in helping solve our future energy needs. The question is, when? The answer must be based, to some extent, on economics. Although the production of power using solar energy is attractive for environmental and other reasons, it must be affordable. This is particularly true if it is to receive widespread application. Its earliest economic justification will occur for selected applications in the more favorable solar energy regions of the world. Factors favoring solar energy economics are:

- Regions of high annual solar radiation
- Land available at reasonable cost
- High cost alternative energy source
- Efficient collection of solar energy
- Low life cycle solar plant costs
- Maximum utilization of collected energy

Rural areas, many in developing nations, share the first three characteristics. Small power plants, isolated from a utility grid, will generally burn high-priced oil at relatively low efficiency. Remote sites will also impose high oil transportation costs. Additionally, many of these regions have little existing energy infrastructure. This allows the power infrastructure to be built initially on a nondepletable energy source assuring its relevance in the future.

The central receiver has long been recognized as an efficient means of collecting and concentrating the sun's energy [1, 2, 3, 4]. The principle of this type plant is illustrated on Figure 1. Heliostats, two-axis tracking mirrors, reflect and concentrate the sun's rays on a tower-mounted receiver. A fluid is circulated through the receiver to extract the collected energy and power the plant. The central receiver also offers good prospects for low life-cycle costs. The high concentration of solar energy provides the high temperatures needed for modern, efficient conversion to mechanical and electrical energy, the central receiver is compatible with plant sizes large enough (2-4 MW thermal) to amortize certain fixed plant costs and operations and maintenance costs over a significant amount of energy. For these size plants, transmission of energy as reflected sunlight is more economical than collecting fluid heated throughout the field. Additionally, since large scale deployment will be essential in achieving low solar plant costs, plant designs should be capable of meeting different needs with common equipment.

The multipurpose plant was selected to maximize the use of the collected solar energy.

M74

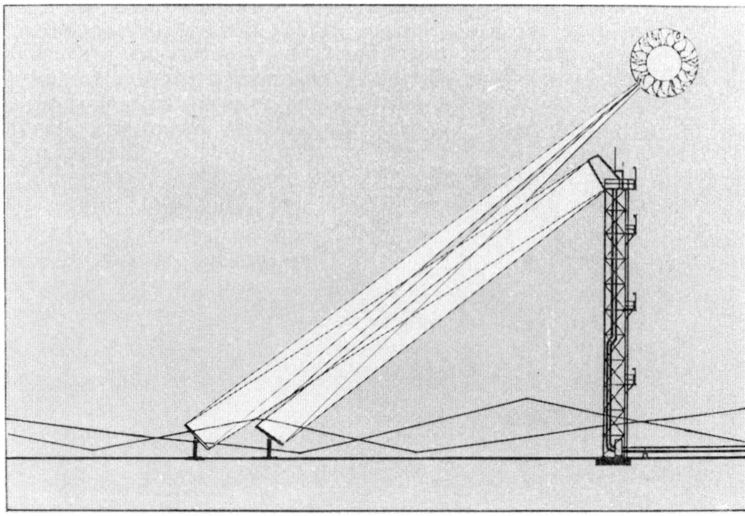

Figure 1. Central Receiver Principle

M74

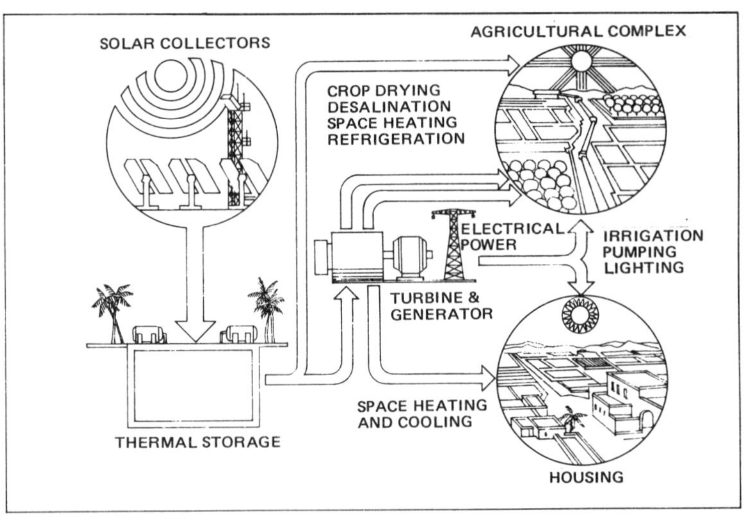

Figure 2. Multipurpose Solar Power Plant

THE MULTIPURPOSE PLANT CONCEPT

The multipurpose plant is based on use of the "total energy" collected from the sun. An old concept for fossil-fueled plants, the use of total energy received limited application in the era of cheap fuel. Today it is receiving much more attention in a variety of ways [5, 6].

The classical "total energy" system makes use of the thermal energy that would otherwise be rejected to the environment in the thermodynamic production of electricity by reducing the temperature difference used for the generation of electricity. Although it reduces the electrical output somewhat, it yields the balance of thermal energy (typically 2/3 to 3/4 of the total) at a useful, elevated temperature rather than being wasted.

Figure 2 illustrates this approach with a central tower plant. The heliostat field reflects and concentrates the solar energy onto the tower-mounted receiver. A thermal transport fluid is heated and stored in a thermal storage unit until needed. This stored energy can be used for the production of electricity or used directly as thermal energy. Provisions can be made to extract variable amounts of the turbine working fluid at an intermediate or set base temperature.

Potential applications of the electrical and thermal energy are illustrated in the figure. The plant produces electricity to meet all community needs and utilizes the thermal energy that would otherwise be rejected to the environment. The thermal energy is available at temperatures suitable for such applications as cooling or refrigeration, desalination, food processing, crop drying, heating of buildings, and production of domestic hot water. This full utilization of the collected energy produces economies when compared with the use of oil in meeting all needs separately.

The most important factors in achieving total energy utilization are (1) selecting a plant concept and configuration that best meets the electrical and thermal power demands of the user, and (2) designing the plant with sufficient flexibility to meet various demands throughout the year as well as the demands from different sites.

SYSTEM DESCRIPTION

The plant employs existing state-of-the-art equipment and technology combined with that developed under the United States solar tower program.

System Configuration

The system schematic is shown on Figure 3. A molten heat transfer salt (HTS)* is used as both the receiver coolant and the thermal storage fluid. HTS is employed for these functions because of (1) safety, (2) good heat

*53% KNO_3, 7% $NaNO_3$, 40% $NaNO_2$

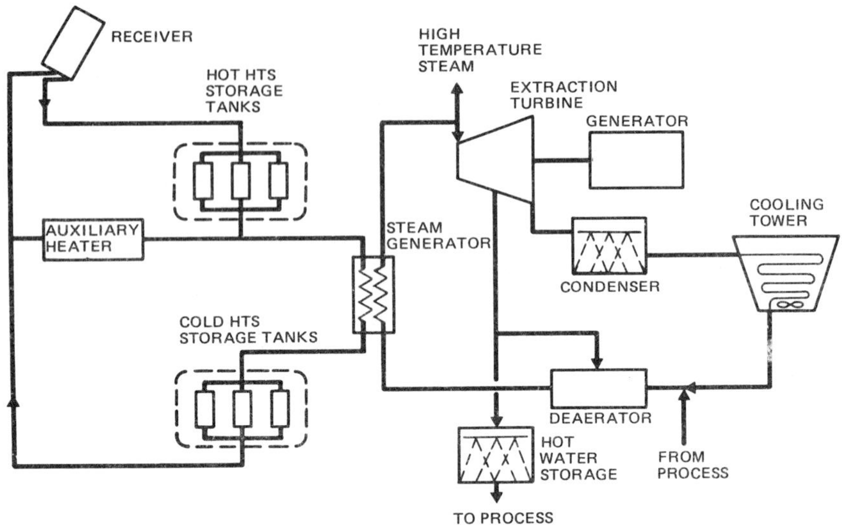

Figure 3. System Schematic

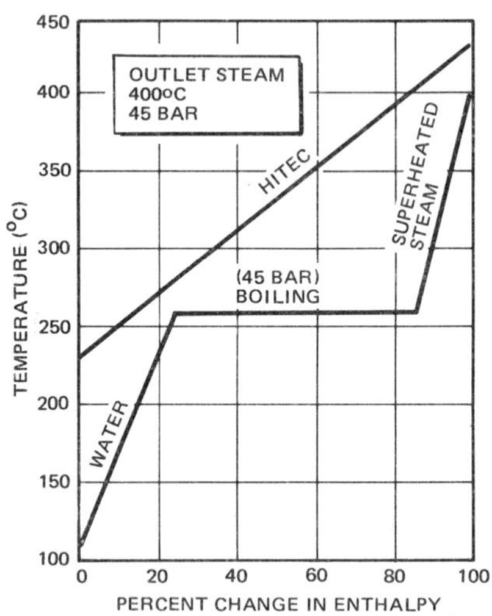

Figure 4. Steam Generation Process

transfer properties, (3) moderate cost, (4) low vapor pressure, and (5) more than 25 years of industrial experience. Its high freezing temperature (142°C for fresh salt) imposes the requirement to trace-heat all lines and provide immersion heaters in the storage tanks.

HTS enters the receiver at 230°C and exits at a temperature of 430°C. The hot salt is routed to the high-temperature thermal storage tanks where it is accumulated or passed on to the steam generator. It then serves as a steady state source of thermal energy for the steam generation process. The energy collection and high-temperature storage capacities are defined so that sufficient thermal energy is available to buffer out solar-induced transients. By the end of an average day, the high-temperature tanks are sufficiently full to provide several hours of system operation after sunset. The actual sizing of the equipment will depend on the specific application and the demand cycle. An oil-fired HTS heater is provided for periods of low insolation or high power demand.

A steam Rankine power conversion cycle is employed with either a multistage extraction turbine (as shown) or a back pressure turbine to supply low temperature thermal energy.

To achieve full performance potential, the multistage turbine is designed to provide steam at the maximum acceptable temperature and pressure, 400°C and 46 bar, and can be used with a variety of turbine back-pressure conditions. Turbine speed will be in the range of 4500-5200 rpm, geared down to the 1500 rpm speed of a 4-pole synchronous generator. Turbines in this size range are available with expansion efficiencies between 60 and 70 percent. The corresponding conversion efficiency to electricity is about 25 percent. The value may vary somewhat depending on the number of stages, tolerances, and interstage sealing techniques selected. In general, the refinements improving performance have a direct impact on hardware cost.

The relationship between the steam generation process and the HTS temperature is shown in the pinch point diagram depicted in Figure 4. A minimum salt temperature of 230°C is required to ensure that a positive temperature gradient exists between the salt and the water/steam at all points in the steam generator. The 430°C hot HTS temperature represents a reasonable and well established level for a molten salt system.

The plant provides electricity, high pressure steam, and hot water. Modular HTS and hot water storage tanks are provided to match different user requirements.

Plant Layout

An overall view of the layout for a 4 MWth plant is shown on Figure 5. The heliostat field configuration, tower height, and receiver aperture size and tilt were selected together on the basis of a combined optimization. The procedure considered heliostat cost, tower cost vs height, sun angle cosines, heliostat blocking and shadowing, the fraction of reflected energy intercepted by the receiver, and receiver thermal losses.

M74

Figure 5. Plant Layout

M74

Figure 6. Heliostat Under Test at China Lake

Annual energy collection was determined by integrating performance factors over the year using one day per month at 10 minute intervals throughout the day. Minimum cost for the annual energy collected was the selection criterion.

The resulting optimum field layout is a north field of 173 heliostats facing a receiver mounted on top of a 46 m tower tilted downward 30 degrees toward the field. The field is contained in a 90 degree arc truncated at the sides toward the back. The maximum field depth is 185 m from the base of the tower, and the maximum field width is 178 m. The heliostats are in a radial-stagger array in which they are located on circular arcs about the tower with heliostats in successive rows staggered to "see through" the row in front. This arrangement, minimizing blocking of the reflected beam, was found to be most cost effective.

The power plant and balance of system equipment are located at the base of the tower to minimize piping, pumping power, and thermal losses.

Heliostats

Heliostats are under various stages of design or development in several countries [7, 8, 9, 10]. The one selected for the United States 10 MWe Pilot Plant at Barstow [11] was evaluated and found suitable for this plant. Figure 6 shows the heliostat under test in the Mohave Desert at China Lake, California.

The requirements used as the basis for the heliostat design are:

A. Survive winds up to 40 m/s at a 10 degree angle of attack, in any preferred orientation, and up to 22.4 m/s in the most adverse orientation

B. Operation in winds gusting to 16 m/s

C. Survive temperatures from -30°C to 49°C

D. Survive earthquakes with 0.25 g peak acceleration

E. Survive hail 19 mm in diameter at velocities of 20 m/s, in any orientation

F. Heliostat field availability > 0.99

Since the requirements were developed on the basis of conditions that exist in the U.S. southwest desert, it is felt they are representative of conditions for many areas of application.

From an accuracy standpoint, the heliostat is designed for a total beam error (pointing accuracy plus slope error) of about 3.5 mr in reflected beam coordinates under the most adverse combination of winds, temperatures, and gravity loading. The value for average wind conditions is 2.5 mr.

M74

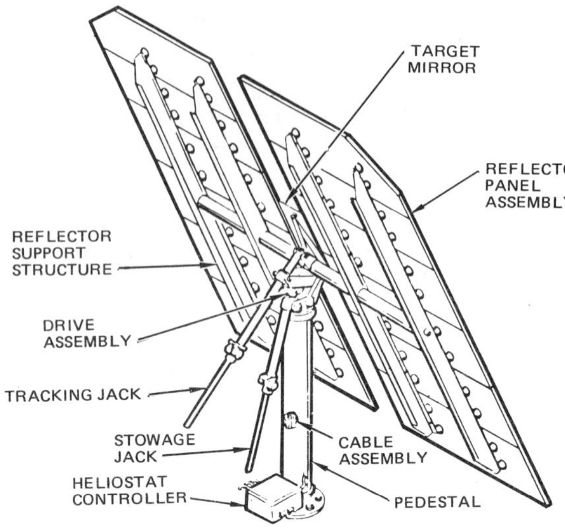

Figure 7. Heliostat Assembly

M74

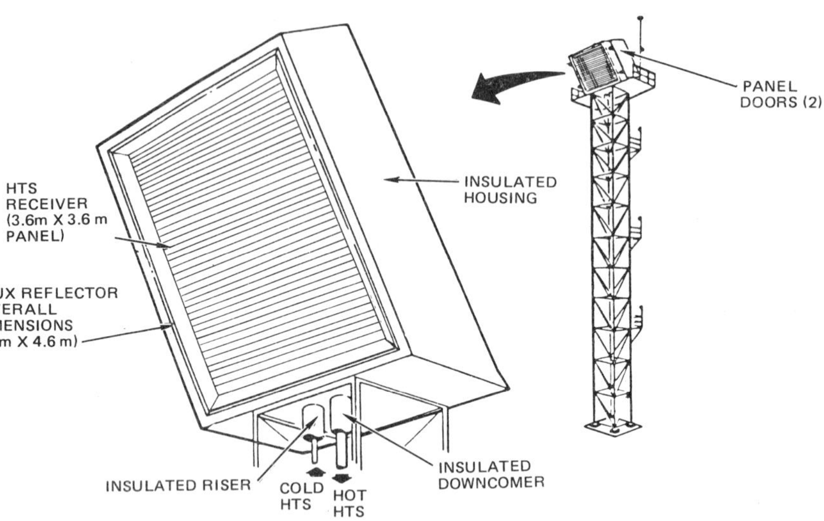

Figure 8. Receiver

Figure 7 shows the heliostat assembly. The reflector consists of 12 panels, approximately 1.1 by 2.9 m, plus a central mirror on the drive unit to provide a reflective surface area of 38 m^2. Each panel subassembly is made up of a second surface silvered mirror which is bonded to a foam core. The foam core in turn is bonded to a thin galvanized steel back sheet. The resulting reflector sandwich is then bonded to the support structure.

The drive unit employs azimuth and elevation drives. The azimuth drive uses a 240 V three-phase motor with a series worm gear/harmonic drive output that achieves a speed reduction of 45,000:1. Two linear actuators are employed for the elevation drive, one for tracking and one for inverted stowage.

The heliostat is supported on a standard 50 cm welded steel pipe. A flange interface bolted to the drive unit and the foundation allows removal of the drive unit for repair. The foundation is a precast reinforced concrete cone preloaded in a drilled hole.

An open loop control system is employed. Gimbal axes absolute encoders are used to accurately locate reference points on the gimbal axes. Motor shaft incremental encoders are used to count motor revolutions and interpolate between bit changes on the absolute encoders.

Receiver

A conceptual definition of the receiver is shown in Figure 8. It consists of an absorbing panel, mounted in an insulated box-like housing. The panel is tipped 30 degrees toward the heliostat field. The reflected sunlight strikes the absorbing panel, providing the source of energy for the plant. Because all beams converge at the center of the panel, a high heat-flux condition exists at that point. The ratio of maximum to average heat flux on the panel is 2.5.

In order to improve the capture of the low-flux perimeter energy and reduce convective losses, actively cooled reflector elements are installed along the panel edges. These reflectors are positioned so that incident power is reflected onto the central, high temperature panel.

A schematic of the receiver fluid loop is shown in Figure 9. The incoming HTS first passes through the reflector panels which surround the receiver. The flow is then split into two paths by a three-way diverting valve. The two flow branches are routed to the sides of the absorbing panel where they enter the panel as four parallel 25 mm tubes (not shown). The initial flow passes are through the central high-heat-flux portion of the receiver. As the fluid temperature increases, it continues to pass toward the top or bottom portions of the panel. As indicated in the schematic, the diverting operation is controlled by an outlet temperature sensor in each leg and total flow is controlled by a mix outlet temperature sensor.

The benefits gained from the flow strategy are to minimize both the local film temperatures in the HTS and heat losses from the surface. In addition,

M74

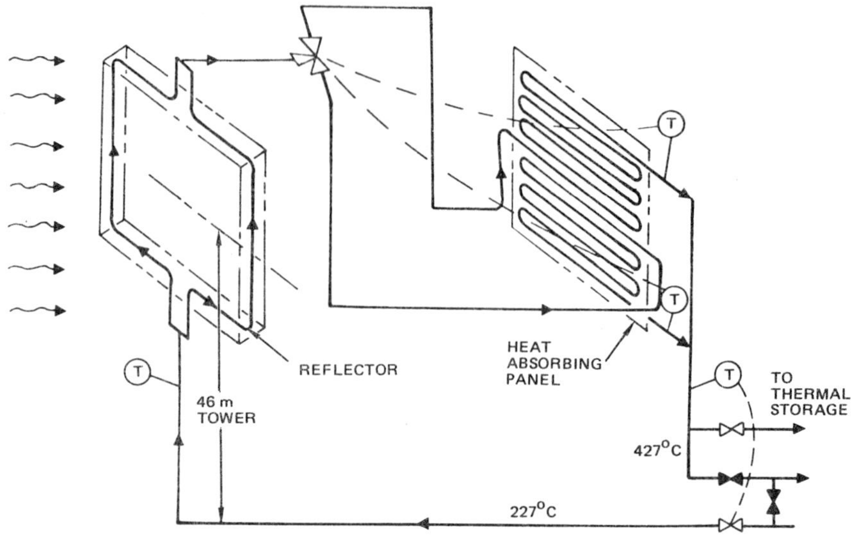

Figure 9. Receiver Fluid Loop

M74

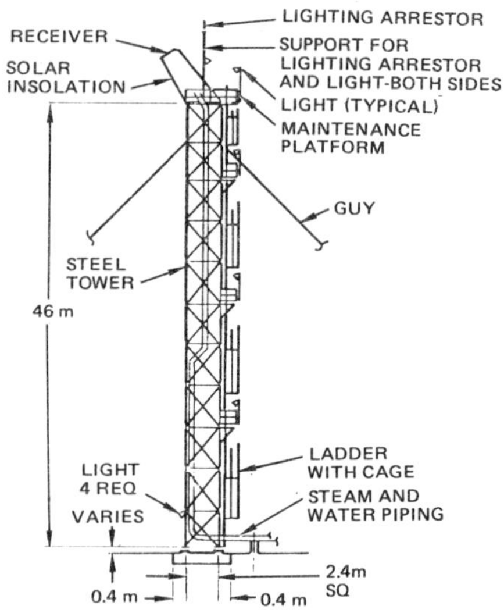

Figure 10. Guyed Receiver Tower

the horizontal routing permits draining of the receiver. Because of the high freeze point of HTS, doors are included in the design to close over the absorbing panel during nonoperating periods. In addition, provisions are included for trace heating of all HTS lines and equipment.

In order to provide the most suitable receiver with minimum maintenance requirements, efforts were made to define a simple design based on existing HTS technology. The simplicity occurs not only in the straightforward panel design, but also in the sensors and flow-control equipment. The other overriding factor in the design effort was to minimize receiver weight which impacts not only the receiver cost, but the cost of the supporting tower.

Tower

The tower is built to provide reliable support for the receiver within specified sway limits during system operation. It is capable of surviving maximum wind and seismic conditions anticipated for the selected site. The tower also serves as the mounting structure for the riser and downcomer, housing the circulating receiver coolant. It must also permit access to the receiver for maintenance.

Because tower structures of the height and load-bearing capacity for this application are well within the state-of-the-art, extensive data and experience are available. With this data base, it is possible to readily select the preferred configuration.

A guyed-steel tower was selected based on a conceptual design analysis. Candidates considered in the study included concrete, guyed, and free-standing steel structures. Concrete was eliminated because of its larger mass and high costs. Steel towers are much lighter, minimizing foundation requirements and allowing prefabrication. The guyed-steel tower was the most economic candidate because of less tower structure, the use of prefabricated sections, and minimum field erection.

The baseline tower design shown in Figure 10 consists of four prefabricated tower sections. These welded tower sections would be shipped to the site and require only final placement, connection and painting in the field.

The design includes the necessary access ladders and catwalk landings to provide for receiver access and maintenance. Lightning protection and warning lights are also included.

The tower riser and downcomer pipes are mounted on the inside of the tower to reduce overturning moments. The lines are permitted to expand downward from the receiver with an intermediate bend to aid in the accommodation of thermal expansion.

M74

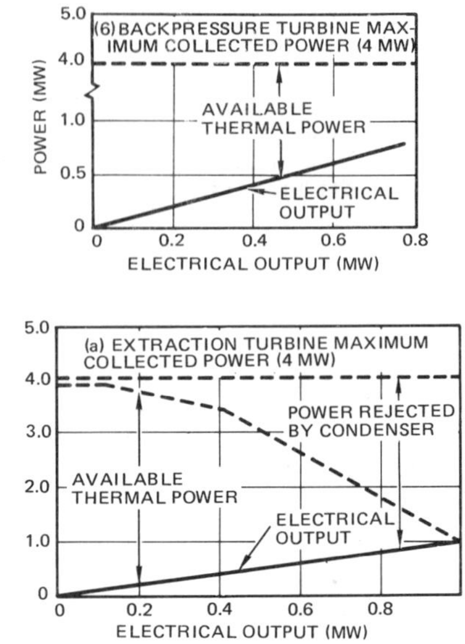

Figure 11. Performance Characteristics

M74

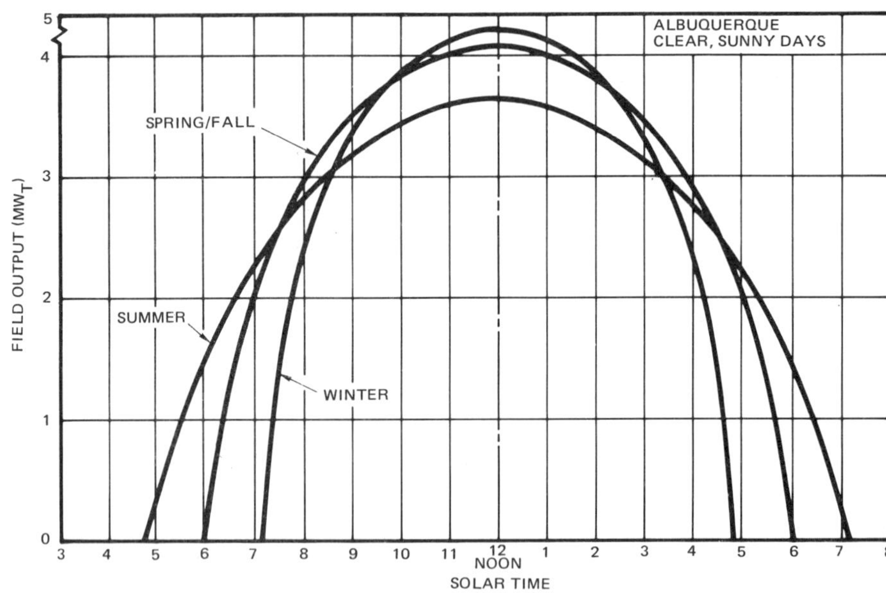

Figure 12. Daily Energy Profiles, Four Seasons

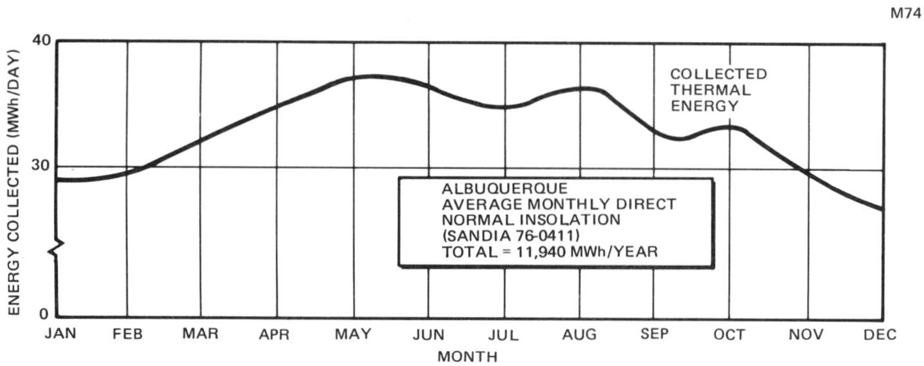

Figure 13. Annual Energy Collected

TABLE 1

COLLECTION EFFICIENCIES

Mirror Reflectivity	0.88
Loss of Effective Mirror Area Due to Shadowing and Blocking	0.95
Effective Area Loss Due to Off-Normal Incidence	0.88
Tower Shadow and Atmospheric Attenuation	0.98
Fraction of Redirected Beam Intercepted by Receiver	0.98
Receiver Thermal Efficiency	0.86
Overall	0.61

PERFORMANCE

The collection efficiency factors for the heliostat field and the receiver are shown in Table 1. It can be seen that the overall efficiency from the normal component of solar radiation to heat absorbed in the receiver coolant is 0.61.

The mirror reflectivity represents an average operating value allowing for a three percent degradation due to dust buildup.

Shadowing is defined as the loss of effective mirror area due to shadowing of the mirrors by other heliostats. Blocking is defined as the loss of effective mirror area due to interception of the reflected beam by another heliostat. The degree of shading and blocking will vary with sun angle. The factor given in the table is for the annual average.

Since the heliostat mirrors must bisect the angle between the sun and the receiver, there is a loss in effective area due to off-normal incidence. The value shown is the annual average for sun elevations above 15 degrees.

Tower shadowing and atmospheric attenuation are small effects on an annual average basis. The rural location is expected to have relatively clear air. This factor could vary, however, depending upon the specific location.

The fraction of the beam intercepted by the receiver, including the terminal concentrator, is relatively high in the optimized design. Losses from the concentrator are included under receiver thermal efficiency.

Standard heat transfer techniques were used to compute the radiation losses from the terminal concentrator and the high temperature panel. Convection losses were computed for an average wind velocity of 16 km/hr directed parallel to the heated surface. This is a conservative upper limit neglecting the benefits of the terminal concentrator. A lower limit, corresponding to natural convection, would be about one third of the value assumed. Conduction losses through the insulation and supports were calculated with standard techniques and were small compared to radiative and convective losses.

The performance characteristics for conversion of the collected thermal energy into electricity are shown in Figure 11. It can be seen that the extraction turbine, view (a), has a higher potential electrical output (1 MWe) compared with 0.8 MWe for the back pressure turbine, view (b). However, at the higher electrical output, much less thermal energy is available for use in the complex. Either configuration could be employed with the decision dependent upon the electrical and thermal power requirements of the user.

The performance of the energy collection portion of the system, which will be available in some combination of electrical and thermal energy, is summarized in Figures 12 and 13 for data taken at Albuquerque, New Mexico.

These data for 35-degree north latitude are probably representative of many sunny regions for application. Figure 12 shows the daily energy collection profiles for each of the four seasons. Applying this performance data to the available direct insolation, the resulting annual energy collection characteristics are shown in Figure 13.

CONCLUSIONS

Existing technology on heat transfer salt systems and steam power conversion can be combined with solar equipment under development to produce a multi-purpose solar power plant with attractive performance characteristics. Full utilization of the collected energy is important to system economics and is enhanced with the provision of modular thermal storage at both high and low temperature.

Earliest economic viability may occur for rural sites isolated from an electric grid. The support infrastructure must be developed as an integral part of the system.

REFERENCES

1. F. Trombe, "High Temperature Furnaces," Proceedings of World Symposium on Applied Solar Energy. Phoenix, Arizona. Nov. 1955. pp. 63-72.

2. V. A. Baum, "Prospects for the Application of Solar Energy and Some Research Results in the U.S.S.R.," Proceedings of World Symposium on Applied Solar Energy. Phoenix, Arizona. Nov. 1955. pp. 289-298.

3. L. L. Vant-Hull and A. F. Hildebrandt, "Tower Top Collection of Redirected Solar Radiation," Proceedings of ISES Annual Meeting of U.S. Section. Cleveland, Ohio. Oct. 1973. pp. 64-66.

4. C. R. Easton, R. W. Hallet, S. Gronich, and R. L. Gervais, "Evolution of a Central Receiver Solar Tower Power Plant," Proceedings of 9th Intersociety Energy Conversion Engineering Conference (IECEC). San Francisco, California. Aug. 1974. pp. 271-276.

5. G. S. Leighton, "The Modular Integrated Utility System - A New Approach to the Supply of Utility Services," Proceedings of 8th IECEC. Philadelphia, Pennsylvania. Aug. 1973. pp. 648-659.

6. R. B. Pope, W. P. Schimmel, Jr., D. O. Lee, W. H. McCulloch, and B. E. Bader, "A Combination of Solar Energy and the Total Energy Concept - The Solar Community," Proceedings of 8th IECEC. Philadelphia, Pennsylvania. Aug. 1973. pp. 304-311.

7. R. W. Hallet, Jr., and R. L. Gervais, "A Central Receiver Solar Thermal Power System," Proceedings of DFVLR International Symposium on Solar Thermal Power Stations. Cologne, West Germany. April 11-13, 1978. Paper No. 21.

8. J. Temboury, "The Project CESA-1, A 1 MW Solar Power Plant in Almeria," Proceedings of DFVLR International Symposium on Solar Thermal Power Stations. Cologne, West Germany. April 11-13, 1978. Paper No. 17.

9. C. Etievant, F. Pharabod and M. Bignon, "Themis 2 MW Central Receiver Power Plant," Proceedings of DFVLR International Symposium on Solar Thermal Power Stations. Cologne, West Germany. April 11-13, 1978. Paper No. 18.

10. J. Hofmann and C. Kindermann, "Heliostat - Fields for Central Receiver Solar Power Plants in the 1 MW Range," Proceedings of DFVLR International Symposium on Solar Thermal Power Stations. Cologne, West Germany. April 11-13, 1978. Paper No. 7.

11. A. C. Skinrood, "Issues and Methodology for the Selection of a Conceptual Design for a Solar Central Receiver Pilot Plant," Proceedings of DFVLR International Symposium on Solar Thermal Power Stations. Cologne, West Germany. April 11-13, 1978. Paper No. 20.

INVESTIGATION OF POOL BOILING HEAT TRANSFER IN THE ENERGY CARRYING LIQUIDS OF SOLAR ENERGY POWER PLANTS

N.S. Abdelhafiez
University of Alexandria
Alexandria, Egypt

M.A. Abdelsalam
University of Helwan
Cairo, Egypt

ABSTRACT

Refrigerants are now being widely used as energy carrying mediums in solar energy devices for power production on a large scale. Thermal properties of major refrigerants are well-known, but some problems of heat transfer during their phase-change are not thoroughly studied. One of the most important problems, to which the solar devices are usually subjected under certain operation conditions, is the pool boiling. A lot of research has been published in this field, giving different formulae, which are valid only in specified conditions.

The present paper suggests a generalized equation which gives the heat flux of the pool boiling of refrigerants from horizontal surfaces at the normal earth gravity. The suggested relationship was obtained by using the dimension analysis theory and a developed method of regression analysis. A large volume of experimental data, up to 600 points collected from 12 different researches in the range of $0.003 < P/Per < 0.78$, was used in the analysis with the aid of a digital computer. Verification of the obtained equation shows that the absolute average deviation of the Nusselt number calculated by it, from that obtained by experimentation, doe not exceed 10%.

INTRODUCTION

Presently, the low thermal potential devices, such as flat plate collectors or solar farms, are successfully used for power production are commercially available now, and producted by well-known firms such as Dornier, M.A.N., Sulzar and Fiat. In the Solar Energy Laboratory of the Academy of Science of Egypt, a unit of such a category is effeciently operating. The majority of these installations are working on the basis of well-known vapour cycles or their modifications using specified liquids, an energy carrying medium. These liquids must go with the low heat flux input to the absorbers of the above described solar devices. It was found that the refrigerants achieve this goal.

The efficiency of low thermal potential solar energy power plants is mainly influenced by the processes of heat transfer during the refrigerants phase-change. Any improvements in this way are urgently required.

One of the most important heat transfer problems during the phase-change of liquid is the problem of pool boiling.

The pool boiling of the majority of known refrigerants are experimentally studied [1-7]. Much research has been published, but the absence of a generalized heat transfer equation describing the mechanism of heat transfer is the common feature of such research. However, some formulae contain constants, which need experimentation for their determination [1-4], and others are applicable only for specific conditions [3-7].

This may be due to the fact that the factors affecting the pool boiling are much greater than those affecting the convective heat transfer. The last depends mainly upon the physical properties of fluids such as viscosity μ, density ρ, thermal conductivity λ coefficient of thermal expansion of the fluid, and also on the geometry of the heating surface. But, the pool boiling is affected, and moreover the mentioned factors, by the thermodynamical properties, such as saturation temperature, the latent heat of evaporation, density at saturation temperature, surface tension of fluid and its vapour, and also by thermal properties of the heating surface such as thermal conductivity, specific heat capacity, and surface thickness and roughness, besides the boundary conditions such as gravitational acceleration and existence of a magnetic field.

Hence, analysis of the large amount of accumulated experimental data in literature, taking into consideration the factors influencing the pool boiling, and obtaining a generalized formula describing the mechanism of heat transfer, is the main goal of the present paper.

REVIEW OF PREVIOUS WORK

The boiling phenomena is not clearly described in the literature, but there are many trials to explain it by the aid of dimension analysis. One of the most important trials is done by Stefan [2], who has used the parameters: $\dot{q}$, λ', ρ', C_p, σ, T_s, ρ'', r, ΔT_s, η, g, d, f, R_b to describe the heat transfer in pool boiling. Using the theory of dimension analysis, he has constructed the following dimensionless groups:

$$N_u = f(K_1, K_2, \dots \dots, K_7, K_9) \tag{1}$$

where,

$$K_1 = \frac{qd}{T_s \lambda'} ; \quad K_2 = \frac{(fd)^2 \rho' d}{\sigma} \quad \text{(Weber number)}$$

$$K_3 = \frac{C_p T_s}{(fd)^2} ; \quad K_4 = \frac{r \rho''}{(fd)^2 \rho'}$$

$$K_5 = \frac{\nu'}{a'} \text{ (Prandtl-Number)}; \quad K_6 = \frac{(fd)\ d}{\nu'} \text{ (Reynolds Number)}$$

$$K_7 = \frac{(fd)^2}{dg} \text{ (Froude Number)}; \quad K_8 = \frac{\Delta T}{T_s}$$

$$K_9 = \frac{R_p}{d}.$$

Equation (1) describes the heat transfer by pool boiling. It contains 8 dimensionless groups; which is too much by comparison with that used in solving problems of heat transfer by convection.

The derivation of a formula for heat transfer in pool boiling, using equation (1), needs 5 experimental points [1,8] for the evaluation of each dimensionless number, making the total required experimental points $5^8 = 390625$. That reveals clearly how the volume of experimentation will be, and the difficulty of deriving a correct and accurate formula.

Many authors have tried to overcome this problem by reducing the number of dimensionless groups by the selection of exponents of some dimensionless groups from published experimental data such as boundary conditions [2, 9-14]. For example, the author [2] suggested the relation between q and α as $\alpha \sim q^{0.7}$ for a horizontal pipe and $\alpha \sim q^{0.8}$ for a flat horizontal plate. Moullier [9] assumed that the relation is $\alpha \sim q^{0.7}$. In [10], the relationship has the form $\alpha \sim q^{0.62}$.

Another relationship between $\alpha = f(Pr)$, $\alpha = f(\sigma)$, was proposed by other authors [11-13].

In some cases, the splitting of some dimensionless groups, the neglecting of or adding of some parameters to them was done to simplify the analysis [9,10].

An examination of equation (1) also shows that the thermal properties of the heating surface are not considered, although in references [11-13], have been proven experimentally that the thermal properties of the heating surface greatly affect the heat transfer coefficient.

Finally, it is easy to conclude that all reviewed approaches to the established goal in this paper tends to a formulae of unallowable descrepancies.

THE PRESENT ANALYSIS

In the present research, another approach is used. The essence is the addition of new dimensionless numbers describing the thermal properties of the surface to the dimensionless groups of equation (1) after exchanging f by a/d^2 and inserting the dimensionless number $\rho' = \rho''/\rho$. Then, the general dimensionless equation expressing the heat transfer of pool boiling will be:

$$N_u = f(X_1, X_2, \ldots, X_{13}) \qquad (2)$$

where

$$X_1 = \frac{\dot{q}\, d}{\lambda' T_s} \qquad X_2 = \frac{a'^2\, \rho'}{\sigma\, d}$$

$$X_3 = \frac{C'_p\, T_s\, d^2}{a'^2} \qquad X_4 = \frac{r\, d^2}{a'^2}$$

$$X_5 = \frac{\rho''}{\rho'} \qquad X_6 = \frac{\rho^1 - \rho''}{\rho'_2}$$

$$X_7 = \frac{\nu'}{a'} \qquad X_8 = \frac{d'^2}{d^3 g}$$

$$X_9 = \frac{R_p}{d} \qquad X_{10} = \frac{(\rho\, C_p\, \lambda)c}{(\rho' C_p\, \lambda')}$$

$$X_{11} = \frac{(\rho\, C_p\, \lambda)S}{(\rho' C'_p\, \lambda')} \qquad X_{12} = \frac{a_c}{a'}$$

$$X_{13} = \frac{ds}{a'}$$

For obtaining an optimal relationship, containing the most important dimensionless numbers of equation (2) by the aid of the accumulated experimental data in the literature, a method of regression analysis was developed. It has been used to solve similar physical problems in thermodynamics for the first time by Wagner [15,16]. The method is thoroughly explained by the authors in [17].

Based on statistical mathematics, the method selects the valid experimental points from the large amount of available experimental data, and determines the most significant variables from the possible parameters describing the physical phenomena under study.

To apply the above mentioned method, a large amount of experimental data was collected from the literature concerning the conditions of pool boiling from a horizontal surface at normal earth gravity. The collected data represents the commonly used refrigerants R-11, R-12, R-21, R-22, R-113, R-114, RC-318, propan, n-butan and CO_2.

The collected experimental data were taken from the tables, diagrams, curves and appendices included in the published research. Some of these diagrams were magnified to increase the accuracy of readings. The selected experimental points are conditioned by the line of minimum square error which was plotted as if it was in absence. From this line, we have selected nearly four points, which will be called hereafter characteristic points. They cover the total region of nucleate boiling.

Accordingly, 389 characteristic points were chosen from more than 600 experimental points representing all collected data. The validity of characteristic points are tested by the method regression analysis.

It is not correct to cancel some of the characteristic points at the beginning of the study, owing to their wide deviation from the majority of the selected characteristic points. That may be due to the absence of exact criterion, e.g., an equation, to calculate heat transfer by pool boiling, giving us a fair decision of canceling. So, all the selected 389 points have been used, and the regression analysis flow has been applied several times.

In the first flow, we have used all collected characteristic points. After obtaining a preliminary equation, some characteristic points were canceled, which are not supported by the bulk of characteristic points. Then, the regression analysis flow is repeated until the final optimum equation is obtained. The number of the characteristic points used in the second regression analysis flow are 231, which represents 611 experimental points from 12 researches in the range of $0.003 < P/P < 0.78$.

Table 1 gives enough information about the 231 points used in the second regression analysis. It provides the name of the author, research reference, the pressure range, the art of heating, dimension, material, geometry, roughness of heating surface, and the symbols of the used points in Figure 1. The figure between brackets in the same table refers to the total number of collected experimental points.

Table 2 summarizes the results of applying the successive regression analysis steps and shows the obtained equations in the end of each step.

RESULTS AND CONCLUSION

The last equation in Table 2 represents the required generalized equation of heat transfer by pool boiling for the refrigerants mentioned hereabove. The equation may be written as follows:

$$\frac{\alpha d}{\lambda'} = 20.77 \left(\frac{\dot{q}\ d}{T_s}\right)^{0.7451} \left(\frac{\rho''}{\rho'}\right)^{0.581} \left(\frac{\nu'}{d'}\right)^{0.533} \qquad (3)$$

The above obtained equation covers the pressure region defined by P_S/Pcr ranging from 0.003 to 0.78. It is also valid for the following ranges of shown dimensionless groups:

From 0.138×10^{-1} to 9.66 For $\frac{\dot{q}\ d}{\lambda' T_s}$

From 0.518×10^{-3} to 0.264 (ρ''/ρ')

From 1.84 to 7.7 For (ν'/d')

The absolute derivation percentage of the Nusselt number calculated by equation (3) from that obtained by experimentation, was plotted on Figure 1. It shows clearly that more than 76% of the used characteristic points has an average absolute derivation not more than 10%. This deviation is not very much from the point of view of practical applications. It will be acceptable if we take into consideration that the measurements in heat transfer experiments have an uncertainty of ±15%, and that the values of physical thermodynamical properties of refrigerants differ slightly from one reference to another in the literature. So, we may conclude that the obtained equation achieves the established goal in the present paper.

For practical use, equation (3) is reduced to the following simple form:

$$\alpha = C \cdot \dot{q}^{\,0.7451} \tag{4}$$

where α - coefficient of heat transfer in $W/(m^2 \cdot K)$

$\dot{q}$ - heat flux in W/m^2

C - constant, represents the values product of physical and thermal properties given by equation (3) for a certain refrigerant at certain pressure conditions.

The values of constant C are drawn on Figure 2 and Figure 3 for the refrigerants R-21, R-133, R-114, CO_2, RC-318, R-11, R-12, R-22, C_3H_8, n-c_4, H_{10} against a wide range of pressures.

We have also calculated the constant C for other refrigerants like NH_3, R-40, R-13, R-13-B1 which have no experimental points, or that their published results are limited for certain pressure regions.

The simplified equation (4) in parallel with the values of C from Figure 3 and Figure 4 are very helpful for determination of the coefficient of heat transfer corresponding to any heat flux.

For comparison, equation (4) is also plotted on Figure 4, together with the equations of other authors which are tabulated in [18] for RC-318. The curves from 1-6, shown in Figure 4, represent equations for general liquids and curves from 7-9 and are especially for refrigerants. Curve 10 represents the plot of the authors equation (4).

The comparison shows clearly that curve 10 may be the average curve, and gives a satisfactory result for coefficient of heat transfer.

ACKNOWLEDGEMENTS

The authors are deeply indebed to Professor K. Stefan of the Institute for Heat Transfer and Thermodynamics at Rohr University, for his fruitful discussions and interesting comments about the present paper. To him, we extend our sincere appreciation.

REFERENCES

1. Stefan, Hess, G., Grundlagen des Warmeuberganges bei der Verdampfung, VDI- Bildungswerk (1970) 1-9.

2. Stefan, K., Beitrag Zur Thermodyamik des Warmeuberganges beim Seiden, Abh. Dt. Kaltetechn, Ver. Nr. 18, C.F. Mueller, Karlsruhe (1964).

3. Fedders, H., Messung des Warmeuberganges beim Blasensieden Von Wasser an metallisch Rohren Diss. Tu Berlin (1970).

4. Cryder, D.S., Gilliland, E.R., Heat Transmission from Metal Surfaces to Boiling Liquids, Industrial and Engineering Chemistry, 24, 12 (1932) 1382-1387.

5. Huber, D.A., Hoehne, T.C., Pool Boiling of Benzene, Diphenyl, and Benzene-Diphenyl Mixture Under Pressure, Journal of Heat Transfer 85.3 (1963) 215-220.

6. Haffner, H., Warmeubergang an Kalemittel bei Blasenverdampfung, Filmverdampfund und uberkritischen Zustnad des Fluids. Bundesministerium fun Bildung und Wissen-Schaft, Forchungsbericht K 70-24, Kerforschung (1970).

7. Yamagate, K. Nishikawa, K., Studies on Heat Transfer in Nucleate Boiling, Memoirs F. Fac. Engng. Kyushu Univ. 16 (1965) 1/28.

8. Adler, Y.P., Markova, E.V., Granovsky, Y.V., The Design of Experiments to Find Optimal Conditons, Mir Publishers, Moscow, 1975.

9. Mueller, F., Warmeubergang bei der Verdampfund Unter Hohen Drucken, VDI-Forschungsheft 522, VDI-Dusseldorf (1967).

10. Vaininger, D., Beitrag Zur Bestimmung des Druckeinflusses auf den Warmeubergang bei ausgebildeter Blasenver Dampfung. Diss. TH Wurch (1972).

11. Grigorev, V.A., Porlov, U.M., Ametistov, E.V., Correlation Experimental Data on Heat Transfer With Pool Boiling of Several Cryogenic Liquids, Thermal Engng. 20, 9 (1973) 81-89.

12. Marto, P.J., Moulson, J.A., Maynard, M.D., Nucleate Pool Boiling of Nitrogen With Different Surface Condition, Journal of Heat Transfer 90, 4 (1968) 437-444.

13. Jergel, M., Stevenson, R., Heat Transfer to Boiling Helium from Aluminum Surfaces, Cryogenics, August (1972) 312-313.

14. Magrini, U., Nannei, E., On the Influence of The Thickness And Thermal Properties of Nucleate Pool Boiling, Journal of Heat Transfer, May (1975) 173-178.

15. Wagner, W., Eine Mathematisch Statistische Methods Zum Aufstellen Thermodynamischer Geichungen - Gexeight am Beispiel der Dampf Druckkurven Reiner Fluider Stoffe, Fortschr-Ber VDI-Z, Reihe 3, Nr. 39 (1974).

16. Wagner, W., A New Correlation Method for Thermodynamic Data Applied to the Vapour Pressure Curve of Argon, Nitrogen and Water, IUPAC Thermodynamic Tables Project Centre, Department of Chemical Engineering and Chemical Technology, Imperial College of Science and Technology, London SW7 2BY England (1977).

17. Abdelsalam, M.A., Berchung des Warmeuberganges bei der Blasenverdamfung Reiner Stoffe Mit Hilfe Einer Regressions Analyse. Diss. Bochum Ruhr Universitate 1977.

18. Danilova, G., Kupriyanova, A., Boiling Heat Transfer to Freons C-318 and -21. Heat Transfer Soviet Research 2, 2 (1970) 79-83.

19. Labuntsov, D.A., Teploernergetika, No. 12.

20. Mikheyer, M.A., Heat Transfer Fundamentals, 2nd Edition, Mir Publishers, Moscow (1968) 126.

21. Tolubinskiy, V.I., Glushchenko, L.F., Boiling Heat Transfer to a Subcooled Fluid in an Annulus. In the collection: Thermal Physics and Heat Engineering, Kiev, Nau Kova Dumka Press (1964).

22. Aladyev, I.V., Teploenergetika, No. 4 (1963).

23. Kutateladze, S.S., Fundamentals of Heat Transfer, R.D. Cess, Ed., Academic Press (1963).

24. Rohsenow, W.M., A Method of Correlating Heat Transfer Data for Surface Boiling of Liquids, Trans. ASME, 74 (1952) 969-967.

25. Borishanskiy, V.M., Kozyrev, A.P., IFZh 5, 12 (1962).

26. Ratiani, G.V., Avaliani, D.I., Warmeubergang und Kritische Warmestromdichte bei Siedenden Frigenen, Kholodilnaya Teknika 3 (1965) 23.

27. Danilova, G.N., Kholadilnaya Teknika 2 (1965).

28. Hesse, G., Warmeubergang bei Blasenverd Ampfung bei Maximaler Warmestromdichte und im Ubergangsbereich Zur Filmverdamp fung. Diss. Tu Berlin (1972).

29. Wickenhauser, G., Einflu der Warmedichte und des Siededruckes auf dem Warmeubergang beim Blasen sieden Von Kaltemitteln. Diss. Uni. Munchen (1967).

30. Gorenflo, D., Zur Druckabh Angingkeit des Warmeuberganges an Siedende Kaltmettel Chemi-Ing. Techn. 40, 15 (1968) 757-762.

31. Henrici, H., Hesse, G., Untersuchngen uber den den Warmeubergang beim Verdampfung Von R-114 und R-114-01-Gemischen an einen Horizontalen Glatlrohr. Kaltechnik-Klimatisierung 23, 2 (1971) 54-58.

32. Schroth, H., Ein Beitrag Zur Verdampfung an uberfluteten Glatt-Und Rippen-Rohren. Luft-Kaltechnik, 5 (1968) 212-218.

33. Stefan, K., Einflub des Ols auf den Warmeubergang von Siedendem Frigen 12 und Frigen 22. Kaltechnik 16, 6 (1964) 162-166.

34. Happel, O., Warmeubergang bei der Verdampfung binarer Gemische im Gebiet des Blasen - und Ubergangssiedens. Diss. Ruhr Universitate Bochum (1975).

35. Danilova, C., Influence of Pressure and Temperature of Heat Exchange to Boiling of Freons, Journal of Refrigeration 8, 12 (1965) 395-398.

36. Sciance, C., Colver, C., Sliepcevich, C., Pool Boiling of Methane Between Atmospheric Pressure and the Critical Pressure. International Advances in Cryogenic Engineering, 12 (1967) 395-408.

37. Abadzic, E., Warmeubergang beim Sieden in der Nahe des Kritischen Punktes. Diss. Uni. Munchen (1967).

NOMENCLATURE

a	thermal diffusivity in m^2/S
A	surface area in cm^2
b	$\sqrt{\dfrac{2\sigma}{g(\rho' - \rho'')}}$
c	constant defined in equation (1)
cp	specified heat at constant pressure KJ/Kg·K°
d	diameter of bubble in m
f	bubble frequency 1/S
g	gravitational acceleration m/S^2
k	constant
$K_1 \ldots K_9$	dimensionless groups given in equation (1)
L	length of heating surface in cm
P_s	saturation pressure in bar
P_{cr}	critical pressure in bar
P_r	P_s/P_{cr} pressures ratio

$\dot{q}$	heat flux in W/m^2
r	latent heat in KJ/Kg
t	surface thickness in cm
T_s	saturation temperature in °K
T_w	wall temperature in °K
$T_s T_w - T_s$	temperature difference in °K
α	coefficient of heat transfer in $W/(m^2 \cdot K)$
α_s	coefficinet of heat transfer at saturation temperature in $W/(m^2 \cdot K)$
α_{Po}	coefficient of heat transfer at pressure P_o (bar) in $W/(m^2 \cdot K)$
β	angle of contact is grad
η	dynamic viscosity in $(NS)/m^2$
λ	thermal conductivity in W/(mk)
ν	kinematic viscosity in m^2/S
ρ	density in Kg/m^3
σ	surface tension N/m

Indices

/	denotes properties of saturated liquids
//	denotes properties of saturated vapour
o	denotes initial conditions
c	denotes covering of surface of heat transfer
s	denotes surface of heat transfer

TABLE 1

SUMMARY OF THE USED EXPERIMENTAL DATA IN THE PRESENT PAPER

Author	Name of Refrigerants	Number of Points	Pressure in bars	Art of heating	Heating surface characteristics: Geometry	Dimensions	Materials	Roughness	% of characteristic points	Symbol
Hesse [1972]	R 12 R 114 R 113	6[¤] 28 11 × (204)	7,0 - 14,0-30,0 3,0 - 6,0-9,0 -12,0-15,0-20,0 0,5 - 1,0	Electric.	Tube	D_o = 1,4 (cm) t =0,075 (cm) L =35,0 (cm)	Nickel	R_p = 0,61	63,0 %	◪ ● ▶
Wickenhäuser [1972]	R 113 RC 318	8 10 (36)	1,02 - 1,55-3,21 3,65 - 7,03-13,82	Electric.	Tube	D_o = 0,08 (cm) L = 27,0 (cm)	Copper	R_p = 0,4 R_p = 0,9	82,0 %	□ △
Stephan [1964]	R 11	16 (28)	1,31	Electric.	Tube Plate	D = 3 (cm) L =50 (cm)	Copper	R_p = 7,9 - 4,4 - 1,4 0,51 - 0,15	65,0 %	▲
Gorenflo [1968]	R 11 R 113	9 6 (31)	1,3 - 2,0 - 3,0 0,1 - 0,4	"	" "	D = 1,2 (cm)	" " "	R_p = 0,4, glatt gezogen	79,0 %	▼ +
Hesse [1971]	R 114	13 (33)	0,37 - 1,28 - 2,52	"	Tube	D = 3 (cm) t = 0,04 (cm) L = 47,0 (cm)	Copper	R_p = 0,2 (poliert)	100,0 %	*
Schroth [1968]	R11 R12	9 8 (56)	0,537-1,22 - 2,24 2,0 -2,67 - 5,02	"	"	D = 2,5 (cm) L = 40,0 (cm)	Steel 35hb	R_p = 9,0	71,0 %	⊙ ◓
Stephan [1964]	R 12	13 (20)	1,63 - 2,85 - 3,51 - 5,02	"	Plate	D = 13,0 (cm) t = 2,5 (cm)	Copper	R_p = 1,0	100,0 %	▷
Happel [1975]	R 113	5 (10)	1,0	"	Tube	D = 1,4 (cm) t = 0,075 (cm) L = 40,0 (cm)	" "	R_p = 0,43	83,0 %	⌖

TABLE 1 (continuation)

SUMMARY OF THE USED EXPERIMENTAL DATA IN THE PRESENT PAPER

Author	Name of Refrigerants	Number of Points	Pressure in bar	Art of heating	Heating surface characteristics			Roughness	% of characteristic points	Symbol
					Geometry	Dimensions	Materials			
Danilova and Kupriyonova [1970]	R 21 RC 318	10 8 (33)	0,71 - 1,04 - 2,15 3,65 - 13,82	Elec.	Tube	D = 0,51 (cm) t = 0,02 (cm) L = 9,1 (cm)	Stainless steel	Rz = 3,84 μ, 9,1 μ	90.0 %	◇ ♠
Danilova [1965]	R 12 R 22 R 113	8 23 3 (76)	5,68 - 9,59 3,55 - 4,99- 6,81 - 9,10- 15,32-24,24 1,09	Electric.	Tube	D = 1,25 (cm) t = 0,02 (cm) L = 25,3 (cm)	" " " "	outer surface not treated	80,0 ¤	■ ○ ▽
Sciance et. al [1967]	Propan n-Butan	17 12 (63)	2,13 - 4,26- 8,52 -12,78- 21,3 -25,56 1,9 - 5,7 - 11,4	Electric.	Cylinder	D = 2,06 (cm) L = 10,16 (cm)	ARAMCO- iron covered with gold.		81,0 ×	◁ ◀
Abadzic [1967]	Kohlendioxid CO_2	6 (21)	55,7-57,4	" "	Wire	D = 0,01 (cm) L = 10,0 (cm)	chrome-plated		67,0 ×	⌽

¤ Total number of characteristic points

× Total number of Experimental points

TABLE 2

THE RESULTS OF EACH STEP IN THE REGRESSION ANALYSIS

The t - value by aprobability S(99.00 %) and n (231) = 2.597

+ The calcultated $\hat{t}$ - value after equation of t test [17]

Reg. Step	Regression equation	S_F
1	$Nu = 43.13 \quad X_1^{0.8005}$ (76.15)+ (25.39)+	
2	$Nu = 392.1 \quad X_1^{0.7324} \quad X_5^{0.5411}$ (120.37) (74.59) (46.7)	100.00 %
3	$Nu = 206.77 \quad X_1^{0.7451} \quad X_5^{0.581} \quad X_1^{0.553}$ (125.9) (128.47) (82.25) (20.82)	100.00 %
3*	$\frac{\alpha d}{\lambda^l} = 206.77 \quad (\frac{\dot{q} \; d}{T_s})^{0.7451} \quad (\frac{\rho''}{\rho})^{0.581} \quad (\frac{\nu'}{a'})^{0.553}$ (125.9) (128.47) (82.25) (20.82)	

* The optimal heat transfer equation

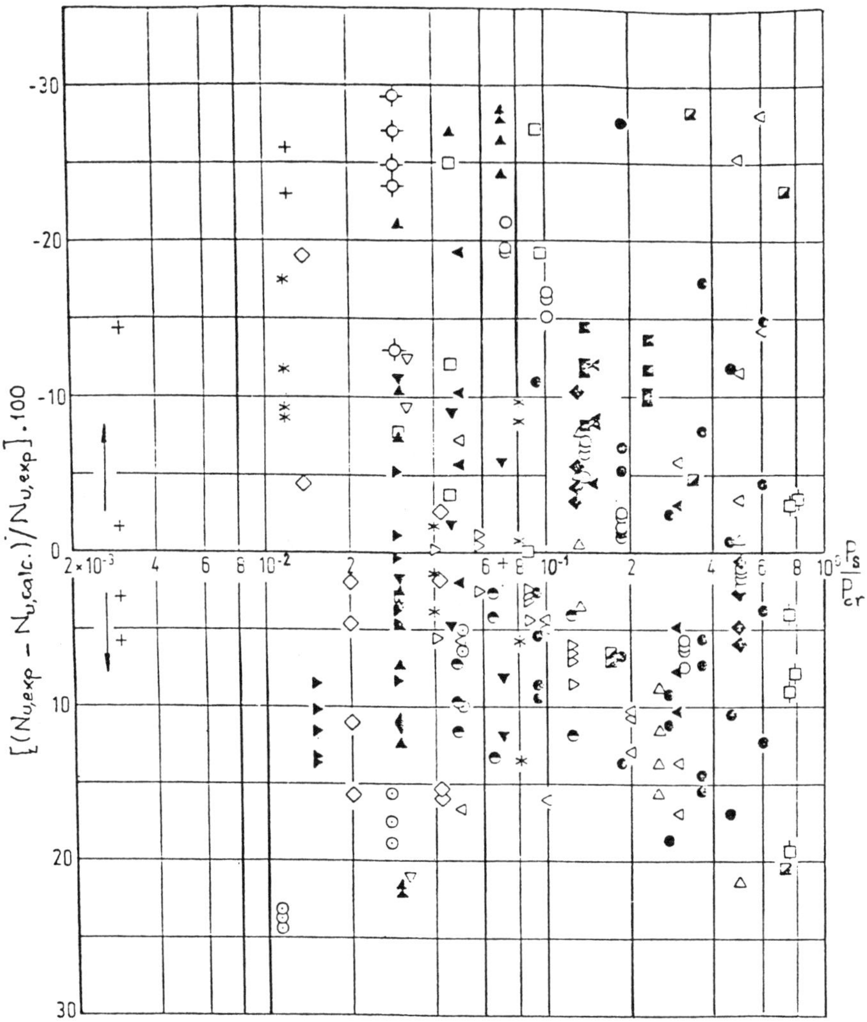

Figure 1 Percentage Deviation of the Calculated N_u Number By Equation (3) From the Measured N_u Number for The Toal P_s/P_{cr}

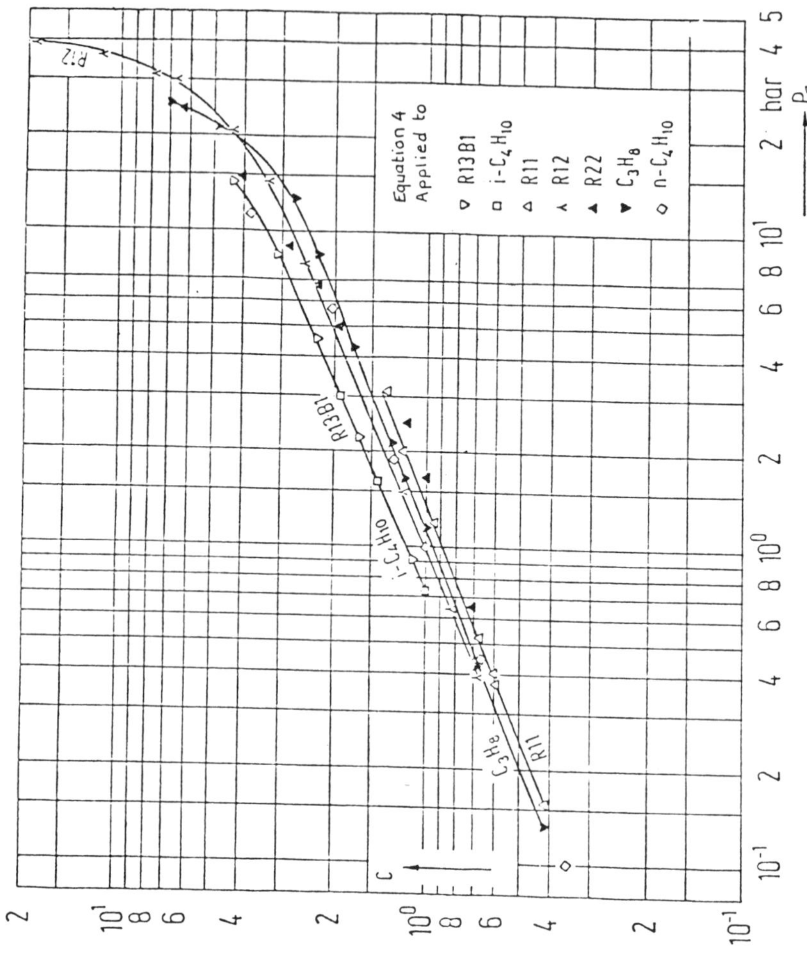

Figure 2 Values of Constant C of Equation (4) Against Saturation Pressure Ratio For Shown Refrigerants

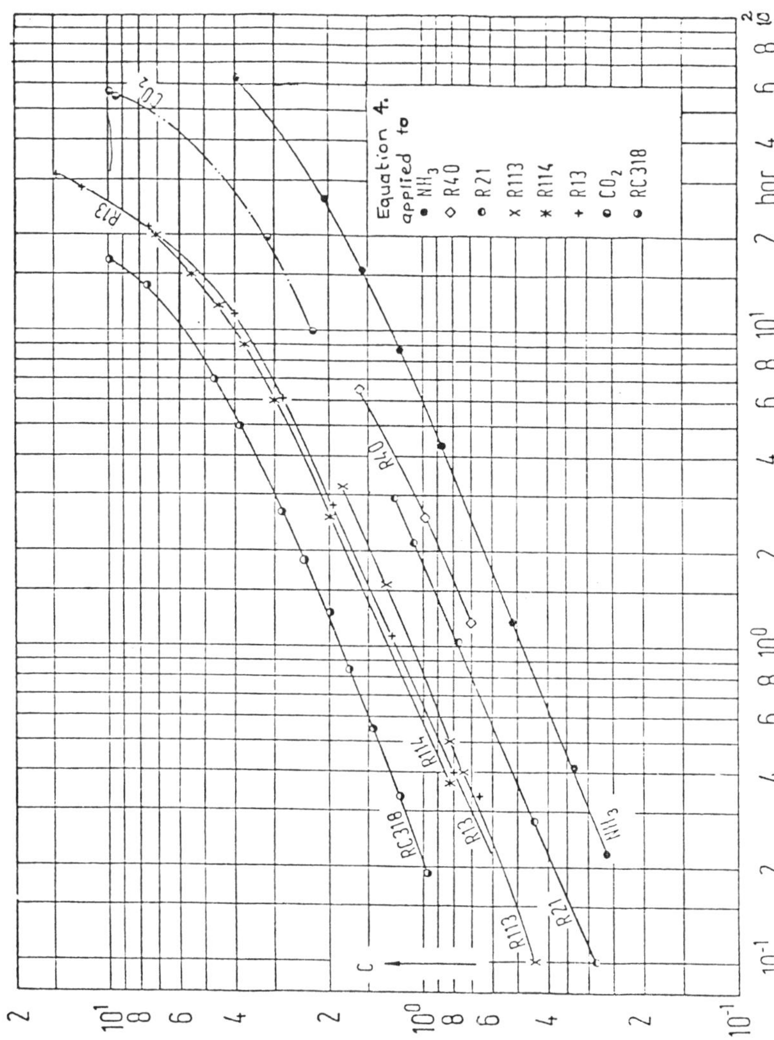

Figure 3 Values of Constant C of Equation (4) Against Saturation Pressure Ratio For Shown Refrigerants

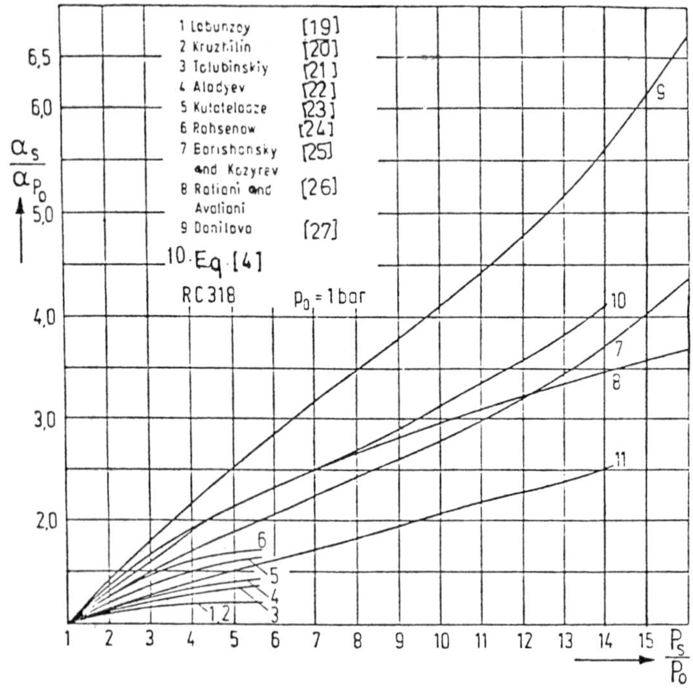

Figure 4 Comparison Between Equation (4) Applied to RC-318 and Equations of Other Authors Applied to the Same Refrigerant

THE USE OF SOLAR FURNACES IN MATERIALS RESEARCH

D.Suresh
Department of Mech. Engg.
Indian Institute of Science
Bangalore-560 012, India.

and

P.K. Rohatgi
Director
C.S.I.R. Complex
Trivandrum, India.

ABSTRACT

Solar furnaces are promising tools for atudying ultra high temperature properties of materials. In this paper, we have reviewed the applications of solar furnaces in the field of materials research in detail and have briefly projected some of their industrial applications. Research on crystal growth, high temperature phase diagrams of oxide systems, ignition of materials, purification and stabilization of refractories is being conducted utilizing the very high temperatures attainable in solar furnaces, without any contamination. Specific future opportunities for research in these areas using solar furnaces have been identified. Some of the future prospects of solar furnaces could be the study on freezing of superheated alloys, zone-refining of large quantities of materials, growing new single crystals for semi-conductor and laser applications, vacuum welding and materials research in outer space.

INTRODUCTION

Extraction and processing of materials consumes a very large fraction of world's nonrenewable energy sources like coal, gas and oil. In fact, in future, the shortage of these nonrenewable energy sources will restrain the production of metals and materials. This restraint can be overcome to a great extent by the utilization of Solar Energy for extraction and processing of materials. In recent years increasing efforts are being made to utilize solar furnaces for research as well as for industrial applications. The French CNRS 1000 kW solar furnace is a landmark of achievement in this direction. Solar furnaces offer the advantage of attaining very high temperatures more than 3500°C, without any contamination, magnetic or electric field interference and with the possibility for measuring the temperature very accurately. Spot melting in a cavity by using the sample itself as crucible, and workability in any desired atmosphere or in vacuum are the major attractions of these furnaces, in addition to, of course the advantage of free fuel and negligible maintenance costs. Solar furnaces are very useful in the study of high temperature properties of materials, such as phase studies, crystal growth, purification and stabilization of refractories and in testing of materials at very high temperatures. Industrial applications like processing of materials, pre-heating scrap, upgrading the raw materials, extractive metallurgy, melting and casting of metals in rural workshops could also be possible. Solar welding and brazing in vacuum could become a major technology of the future. Space factories, in future with a solar furnace on-board could perform materials research and produce new materials for use in electronics and other applications. At present, France and the U.S.S.R are engaged in exploring large scale industrial utilisation of solar furnaces, and Chile and several countries are about to join them in the near future. Japan is involved extensively in basic research with small solar furnaces and the U.S.A is operating military furnaces for material testing and research. There is also considerable amount of properietory work going on in several countries which is not reported in open literature.

CRYSTAL GROWTH

Demand for solar cells and several other requirements for ultra pure single crysta[l] can be met with, in the near future, by growing crystals in solar furnaces. In conventional methods such as metal flux method and rf-heated floating zone method, it is sometimes difficult to grow high quality single crystals, because of the hig[h] melting point and reactivity of the melt. Solar furnaces are more suitable for these applications because of the ability to operate in clean and controlled atmospheres(Fig.1). Under these conditions, it will be possible to grow many sing[le] crystals which would otherwise decompose or vary in stoichiometry under normal heating conditions. The major results obtained by previous workers in crystal growth using solar furnaces are summarized in Table I.

In 1956 Laszlo (1) reported the growth of single crystals of ZrO_2 in a solar furna[ce]. Later, Farber (2) grew perfect crystals of Al_2O_3, CaO, MgO, HfO_2, TiC and TiB_2 by vapour deposition and puddle melt methods, in a solar furnace. The growth of alum[ina] and ruby crystals may be further investigated for use in laser applications. A technique to grow crystals in a desired crystallographic axis could be developed, with suitable temperature control, perhaps using a laser as a stand-by source. Laszlo et al (3) have grown thoria single crystals from the vapour phase in a 1.5 solar furnace. Due to vapour fractionation, magnesium, the major impurity in the solid thoria, was completely eliminated, while silicon and iron contents were red[uced]. Crystals grown by condensation from vapours clearly exhibited the characteristic crystal habit. Crystals as large as 3 mm long and 1 mm in cross section were obt[ained]. The growth rate was 0.5 mm/min at the highest flux available, ie., 500 cal/sq.cm/[...]. Even at the lowest flux level, crystal growth occurred due to the fact that vapou[r] pressure of the solid was sufficient to cause nucleation. Laszlo et al have also suggested the possible mechanisms for crystal growth out of a continuously recedi[ng] molten base. The mechanisms of growth of crystals in solar furnaces must be ascertained by systematic studies of varying different parameters in a controlled manner.

Sakurai et al (4) have studied the growth of UO_2 crystals by a solar furnace, in helium atmosphere. They have suggested a mechanism of this growth, which is not satisfactory and needs further explanation. The crystals grown had good stoichio[-] metry with a pyramidal form and optically flat surfaces of size about 0.15 mm. Sakurai and Ishigame (5) have reported the growth of nickel oxide crystals in a 10 m solar furnace, for use in their optical and electrical investigations. Optically flat crystals with surface area of about 40 sq.mm were obtained. The resistivity of the specimen obtained from the melt was found to be reduced dras[ti-] cally, possibly due to the boiling off of certain impurities and specific lattic[e] re-arrangements. The stoichiometry of these crystals also showed deviation. The study on the variation of stoichiometry after treatment in a solar furnace will be an interesting area for future research. Crystals may be prepared with desir[ed] resistivities from the same starting material by this method, which cannot be easily done using other methods.

Ishigame et al (6) grew Ni^{2+} doped CaO single crystals which were a few mm in gr[ain] size, in a solar furnace. Single crystals of CaO were also grown as a reference for optical measurements. Pure CaO crystals were colourless while Ni^{2+} doped Ca[O] crystals were pink in colour and the transmittance was measured in IR, visible

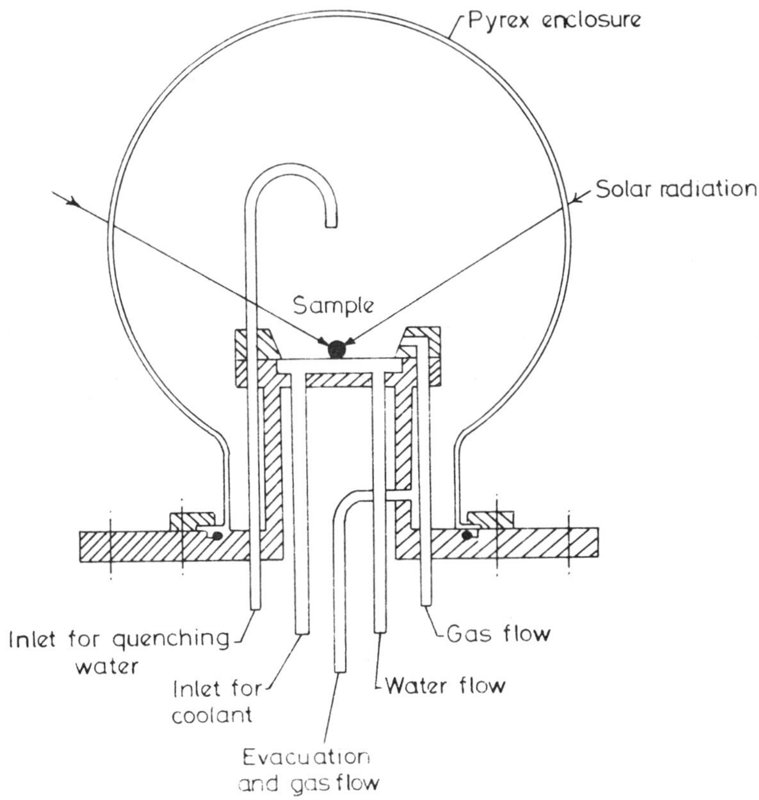

FIG. 1 CONTROLLED ATMOSPHERE CHAMBER
USED IN A SOLAR FURNACE

TABLE I

SUMMARY OF WORK DONE IN CRYSTAL GROWTH USING SOLAR FURNACES

Sl. No.	Investigators	Ref.	Year	Crystal growth	Nature of crystals
1.	T.S. Laszlo	(1)	1956	ZrO_2	Single crystal
2.	E.A. Farber	(2)	1964	Al_2O_3, CaO, MgO HfO_2, TiC, TiB_2	Single crystal
3.	T.S. Laszlo et al	(3)	1967	ThO_2	Single crystals 3 mm X 1 mm X 1 mm
4.	T. Sakurai et al	(4)	1968	UO_2	Crystals with pyramidal form of size 0.15 mm flat faces.
5.	T. Sakurai et al	(5)	1968	NiO	Polycrystalline, optically flat crystals of 40 mm^2 area
6.	M. Ishigame et al	(6)	1968	Ni^{2+} doped CaO	Single crystals-colourless-few mm in grain size
7.	Y. Nigara et al	(7)	1971	Y_2O_3	Single crystal-size few mm

TABLE II

SUMMARY OF THE WORK DONE IN PHASE STUDIES OF DIFFERENT SYSTEMS USING SOLAR FURNACES

Sl.No.	Investigators	Ref.	Year	System studied	Results obtained
1.	W.M. Conn	(17)	1954	Al_2O_3-SiO_2	Liquidus curve presented-upto 48% of alumina clear glass resulted and above 56% of alumina devitrified glass.
2.	T. Noguchi et al	(18)	1967	Al_2O_3-Y_2O_3	Tentative phase diagram presented-Peak between 40 and 60 moel% Y_2O_3
3.	M. Mizuno et al	(19)	1975	Al_2O_3-La_2O_3	Tentative phase diagram presented-New orthorhombic phase at 80 mole% La_2O_3.
4.	M. Mizuno et al	(19)	1975	Al_2O_3-CeO_2	Tentative phase diagram given-New orthorhombic phase at 90 mole% CeO_2
5.	M. Mizuno et al	(20)	1975	Al_2O_3-Ga_2O_3	Tentative phase diagram given-Two phase region between 15 to 25 mole% Ga_2O_3 - αAl_2O_3, β-Ga_2O_3 phases found
6.	J.P. Coutures et al	(22)	1975	$(1-x)Al_2O_3$-$x\ Nd_2O_3$	Amorphous phase recrystallised at 900°C.

TABLE II (CONTD.)

Sl.No.	Investigators	Ref.	Year	Systems studied	Results obtained
7.	J.P. Coutures et al	-	1976	Al_2O_3-Nd_2O_3	Tentative phase diagram given-A new phase Nd_2O_3.11 Al_2O_3 observed.
	M. Mizuno et al	(23)	1977		Reversible phase transition of $NdAlO_3$ at 1100°C-Two eutectic points identified.
8.	M. Mizuno et al	(21)	1977	Al_2O_3-Pr_2O_3	Tentative phase diagram presented- $PrAlO_3$ identified in two forms-Cooling curve of Pr_2O_3 in air showed 'rochage' phenomenon
9.	M. Mizuno et al	(24)	1977	Al_2O_3-Sm_2O_3	Tentative diagram given-Single monoclinic phase of $2Sm_2O_3.Al_2O_3$ observed-Three eutectics identified
10.	M. Mizuno et al	(25)	1977	Al_2O_3-Eu_2O_3	Tentative diagram given-Single monoclinic phase $2Eu_2O_3.Al_2O_3$ observed-Three eutectics identified.

TABLE II (CONTD.)

Sl.No.	Investigators	Ref.	Year	System Studied	Results obtained
11.	M. Mizuno et al	(25)	1977	Al_2O_3-Gd_2O_3	Tentative diagram given-Single stable monoclinic phase $2Gd_2O_3.Al_2O_3$ observed-Three eutectics identified.
12.	M. Mizuno et al	(26)	1978	Al_2O_3-Dy_2O_3	Three stable and single phases identified.
13.	Pol Duwez et al	(10)	1957	ZrO_2-ThO_2	Tentative phase diagram given-Thermal expansion measured-Complete phase diagram given.
	T. Sakurai et al	(11)	1975		
14.	P.E. Evans	(12)	1960	ZrO_2-UO_2	Tentative phase diagram given.
15.	T. Noguchi et al	(13)	1967	ZrO_2-CaO	Tentative phase diagram given-Anomalies found at 3.5,7.0,10.0 mole% CaO.
16.	T. Noguchi et al	(14)	1968	ZrO_2-MgO	Phase diagram incomplete-A new cubic like phase at 12.5 mole% MgO-Phase modification at 1300°C.
17.	A. Rouanet et al	(--)	1968	ZrO_2-La_2O_3	Phase diagram given-A new phase- Two eutectics identified.

TABLE II (CONTD.)

Sl.No.	Investigators	Ref.	Year	System Studied	Results obtained
18.	A. Rouanet et al	---	1968	ZrO_2-Sm_2O_3	Phase diagram given-Eutectic point at 75 mole% Sm_2O_3- New phase at high temperatures in Sm_2O_3 side.
19.	A. Rouanet et al	---	1968	ZrO_2-Gd_2O_3	Phase diagram given-A eutectic point and a New phase at high temperatures on Gd_2O_3 side.
20.	T. Noguchi et al	(13)	1968	ZrO_2-TiO_2	Tentative diagram given- Anomaly at 17.5 mole% TiO_2.
21.	T. Noguchi et al	(13)	1969	ZrO_2-SrO	Tentative phase diagram given- Freezing point of 60 mole% SrO, sensitive to stoichiometry. Two complex compounds observed.
22.	T. Noguchi	(15)	1970	ZrO_2-Y_2O_3	Liquidus curve given-Anomalies at 7.5, 15.0 and 20.0 mole% Y_2O_3-Biaxial negative crystals at 15 mole% Y_2O_3.
23.	A. Rouanet	(16)	1971	ZrO_2-Ln_2O_3	Phase diagrams given

TABLE II (CONTD.)

Sl.No.	Investigators	Ref.	Year	System Studied	Results obtained
24.	G. Benezech et al	(30)	1971	Yb_2O_3-Cr_2O_3	Phase diagram given.
25.	J. Coutures et al	(31)	1974	Nd_2O_3-Y_2O_3	Phase diagram determined by Cofusion method.
26.	J. Coutures et al	(27)	1974	La_2O_3-Y_2O_3	Phase diagram given-Several extended solid solutions at high temperatures.
	M. Mizuno et al	(28)	1976		Liquidus curve determined-Tentative phase diagram given-Two forms of $LaYO_3$ established.
27.	F. Sibieudu et al	(32)	1975	ThO_2-Ln_2O_3	Tentative phase diagrams given.
28.	M. Yoshimura et al	(33)	1976	Ce_2O_3-WO_3	Six stable phase of cerium tungstates identified.
29.	J.P. Coutures	(29)	1977	i) Ln_2O_3-A_2O_3 ii) Ln_2O_3-Ln_2O_3	Phase diagrams presented
	F. Sibieudu et al		1975	iii) Ln_2O_3-AO_2 iv) Ln_2O_3-$LnAO_3$	
30.	G. Brauer et al	(9)	1960	TiO_2-Ti_2O_3	Phase diagram presented.
31.	M. Yoshimura et al	(34)	1977	La_2O_3-WO_3	Rapid quenching using hammer anvil device-Phase diagram presented-Glass formation near the eutectic.

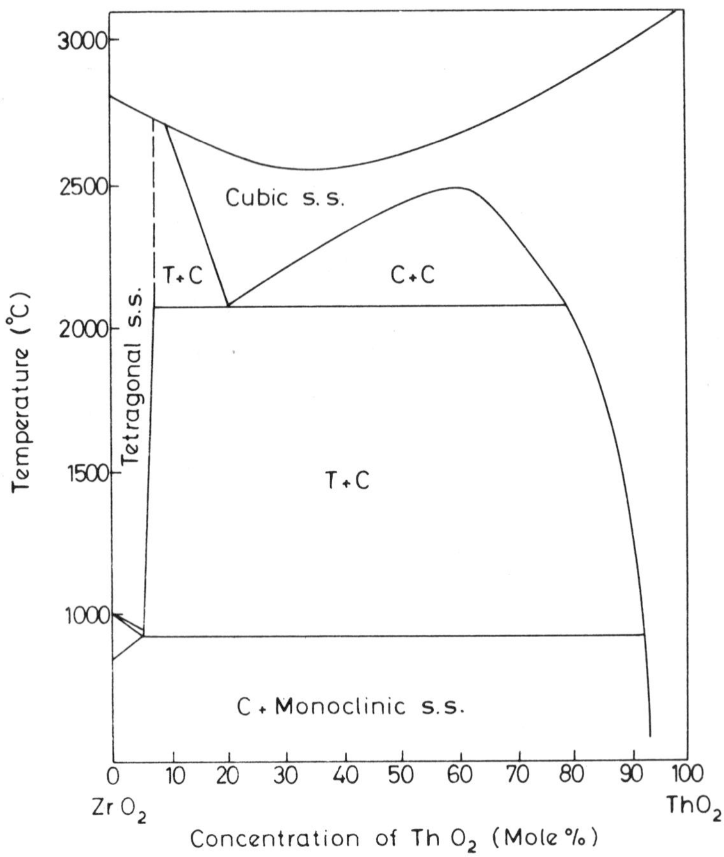

FIG. 2. PHASE DIAGRAM OF ZrO_2 - ThO_2

and UV regions. One may try to grow silicon crystals doped with suitable elements for use in electronics industry by a technique utilizing solar zone-refining.

Nigara et al (7) have grown yttrium oxide crystals in a solar furnace and the reflectivity was measured in the IR region. The crystals were only a few mm in size. The mechanism of growth and the presence of hexagonal structure may be taken up for further research. Payne (8) has remarked that solar furnaces are not suitable for drawing silica fibres because temperature control in solar furnaces is not easy, which is of paramount importance to ensure drawing fibres with good diameter control. The major question before solar scientists is the development of suitable instrumentation to go with a solar furnace which gives controlled growth.

PHASE DIAGRAM DETERMINATION FOR OXIDE SYSTEMS

Study of high temperature phase diagrams with solar furnaces offers the possibility of quenching the fused specimens in air at cooling rates of about 1800-2300°C/sec. and the possibility for splat cooling. The phase diagram of polycomponent systems can be estimated in a solar furnace, using it as a static as well as a dynamic system. In 1960, Brauer (9) published a phase diagram for TiO_2-Ti_2O_3 system. Other than this, we can broadly classify the systems investigated so far using solar furnaces into, (a) zirconia, (b) alumina and (c) lanthanide based systems. Table II shows the summary of work done by previous investigators.

a) Zirconia Systems

Pol Duwez et al (10) investigated the phase relationships in the system ZrO_2-ThO_2. The work of Duwez with a solar furnace established the existence of a cubic CaF_2 type solid solution, for thoria concentrations above 17.5 mole%. Only after prolonged heating at 2000°C and at high temperatures, this solid solution was stable. At 1350°C, it decomposed into tetragonal zirconia and thoria-rich phase containing atleast 99 mole% thoria. During subsequent cooling, tetragonal zirconia transformed into monoclinic form, through an atomic shear mechanism which could not be suppressed even if the rate of cooling was as high as 1500°C/sec. Since the polymorphic inversion of zirconia will always be present, these bodies will be very sensitive to thermal shock. Because of the inadequate data only a tentative diagram was presented. Later Sakurai et al (11) have given a complete phase diagram of this system studied in a solar furnace (Fig. 2).

Evans (12) gave a tentative phase diagram for ZrO_2-UO_2 system. Melting in a solar furnace avoided tungsten contamination as occurred in previous work. From 1350°C, the specimens were cooled relatively slowly and the position of the eutectoid (circa 940°C) could not be established because of the transformation from tetragonal to monoclinic form in the furnace-cooled specimens. Establishing the eutectic point by a modified method may be undertaken to develop a complete phase diagram.

Noguchi and his coworkers (13) have studied the ZrO_2-TiO_2, ZrO_2-SrO, and ZrO_2-CaO systems using a solar furnace. In the ZrO_2-TiO_2 system studied by them, the quenched specimen showed $ZrTiO_4$, monoclinic ZrO_2 and tetragonal TiO_2 as incongruently melting compounds. An anomaly in the liquidus curve of the system had been found at 17.5 mole% TiO_2, in the ZrO_2-rich region. Zirconia had shown two exothermic peaks in the cooling curve due to structure modification from cubic to tetragonal and it was found that quenching could not freeze the high temperature ZrO_2 modification. Stabilization studies may further be extended which could result in a different class of refractories.

In the ZrO_2-CaO system, anomalies in the liquidus curve were noticed for 3.5, 7.0 and 10.0 mole% CaO at 2603°, 2622° and 2578°C respectively. The melt was analysed to contain only rhombic form. But the relation between cubic ZrO_2 and cubic zirconia solid solution, in and above the temperature range of 2200° to 2800°C need further investigation. The crystal structure of zirconia above 2500°C, at which a new stable zirconia solid solution may exist, is yet to be studied. In the case of ZrO_2-SrO system, the freezing points of various compositions were examined, and it was found that the freezing point of the composition 60 mole% SrO was very sensitiv to variation in stoichiometry. A tentative phase diagram was given. The existence of complex compounds $Sr_nZr_{n-1}O_{3n-2}$ (n = 2,3,4) and $SrZrO_3$ was confirmed.

In the ZrO_2-MgO system studied by Noguchi et al (14) the composition of 12.5 mole% MgO quenched from the melt showed a new cubic like phase, which easily decomposed into monoclinic and tetragonal forms by reheating at 1300°C. To present a complete phase diagram, the crystal modification above 2300°C has to be investigated. The presence of compounds like $Mg_2Zr_3O_8$ and $MgZr_3O_7$ was not confirmed satisfactorily in this study.

Noguchi and his coworkers (15) have also studied the ZrO_2-Y_2O_3 system. The composition of 15.0 mole% Y_2O_3 quenched from the melt showed a new cubic phase. Yttria showed an exothermic peak in the cooling curve at about 2283°C in the solid state, due to the cubic hexagonal transition. Rouanet (16) gave the phase diagrams for ZrO_2-La_2O_3, ZrO_2-Pr_2O_3, ZrO_2-Nd_2O_3, ZrO_2-Tb_2O_3, ZrO_2-Dy_2O_3, ZrO_2-Ho_2O_3, ZrO_2-Er_2O_3, ZrO_2-Y_2O_3 and ZrO_2-Yb_2O_3 systems.

b) Alumina Systems

Conn (17) studied the system Al_2O_3-SiO_2 and gave the liquidus curve, which agreed with the results obtained by previous workers who used a different method.

Noguchi and Mizuno (18) presented the phase diagram for Al_2O_3-Y_2O_3 systems. Mizuno et al (19) determined the phase diagram for the Al_2O_3-La_2O_3 system at high temperatures in argon atmosphere and the presence of a new orthorhombic phase was established at a composition of 80.0 mole% La_2O_3 (Fig. 3). A high temperature phase diagram was presented for the system Al_2O_3-CeO_2, establishing a new orthorhombic phase at a composition of 90.0 mole% CeO_2. They (20) have given a tentative phase diagram for Al_2O_3-Ga_2O_3 system (Fig. 4), after estimating the liquidus curve. The solid solution between α- Al_2O_3 and β-Ga_2O_3 contained a two phase region at the composition range of 15 and 25 mole% Ga_2O_3. Quenched β-Ga_2O_3 gave the needle like crystals. Same authors (21) gave a tentative diagra at high temperatures, for the Al_2O_3-Pr_2O_3 system (Fig. 5). They have established rhombohedral and cubic forms for the compound $PrAlO_3$ at 950°C. Cooling curve of Pr_2O_3 in air showed the rochage phenomenon, for which the reason has to be established. The presence of $PrO_{1.83}$ as suggested by them has yet to be verified.

Coutures (22) published the results on the system (1-x) Al_2O_3-xNd_2O_3. An amorphou phase was obtained by rapid quenching with a hammer-anvil device (Fig. 6) in a solar furnace. Their study showed that the amorphous phase was indeed a glass,

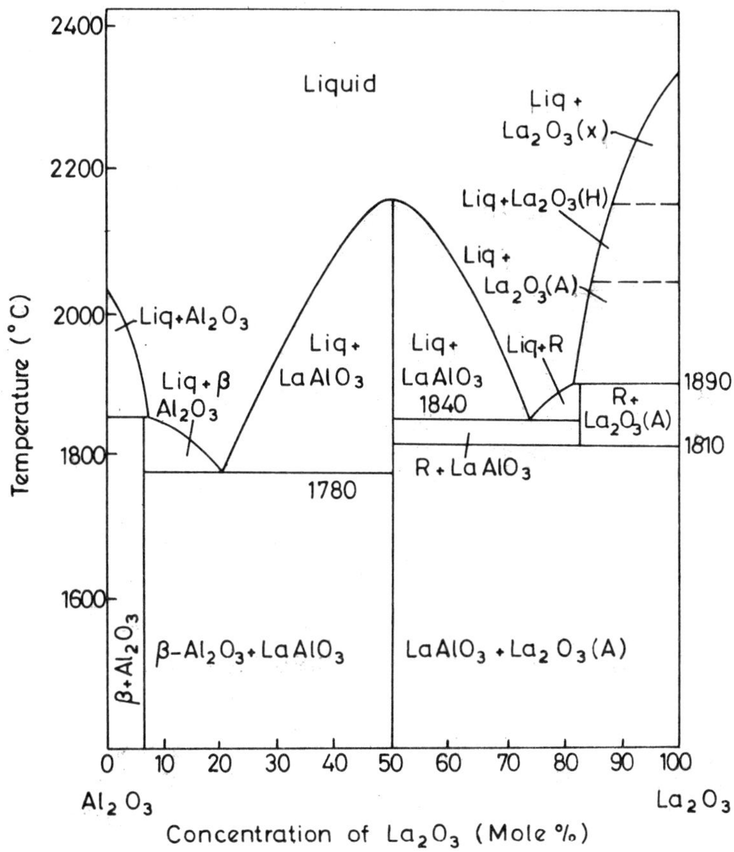

FIG. 3. PHASE DIAGRAM OF Al_2O_3-La_2O_3 IN ARGON.

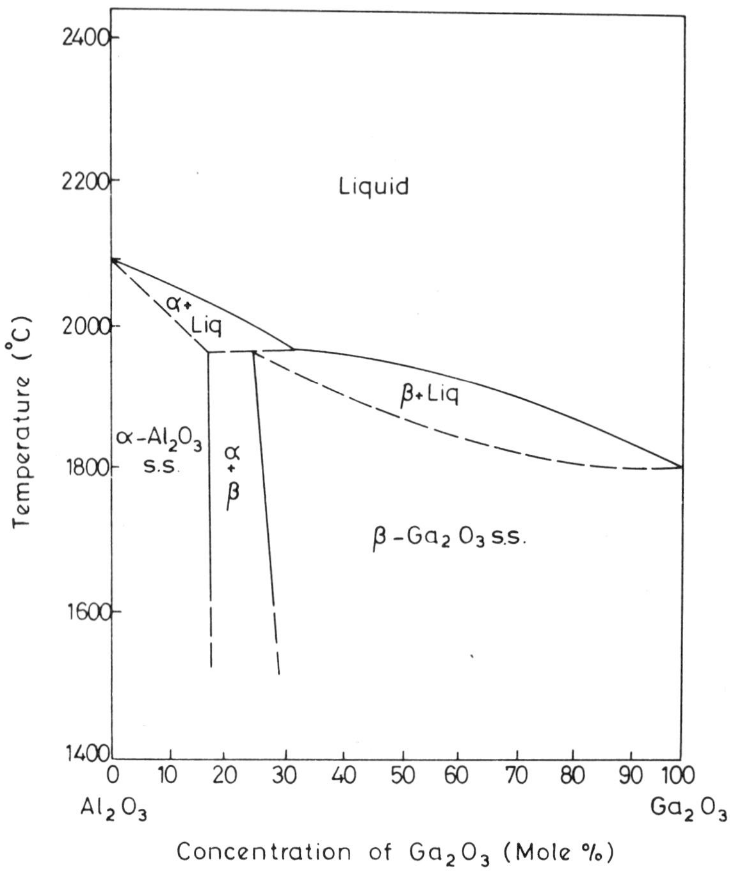

FIG.4. PHASE DIAGRAM OF Al_2O_3 - Ga_2O_3 IN AIR.

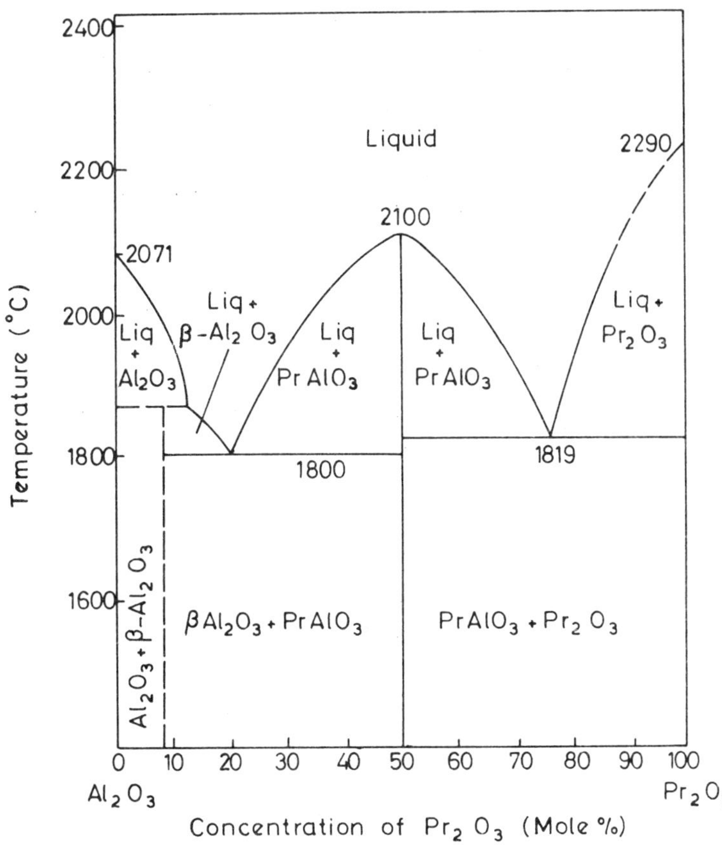

FIG. 5. PHASE DIAGRAM OF Al_2O_3-Pr_2O_3 IN AIR

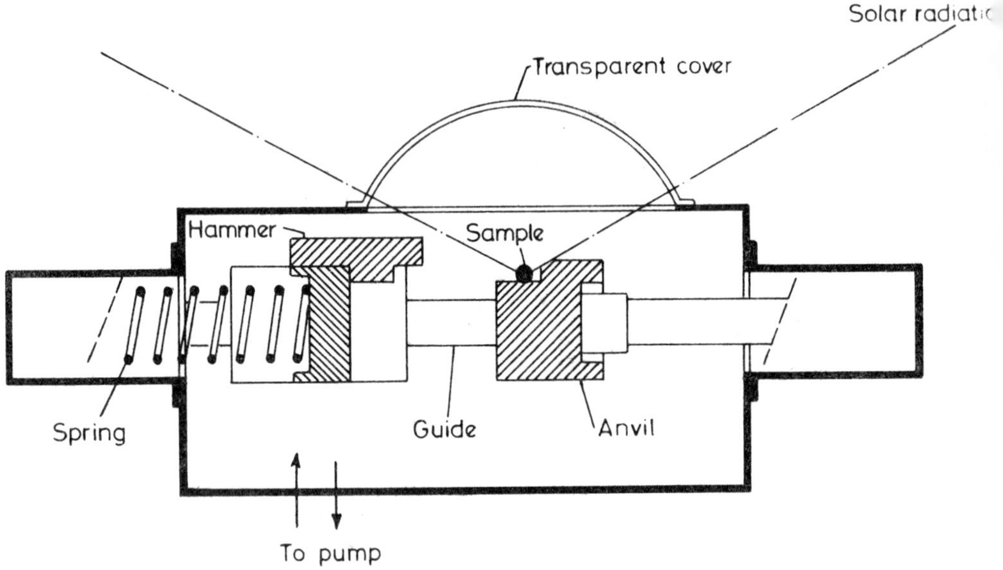

FIG.6. HAMMER-ANVIL DEVICE USED IN A SOLAR FURNACE.

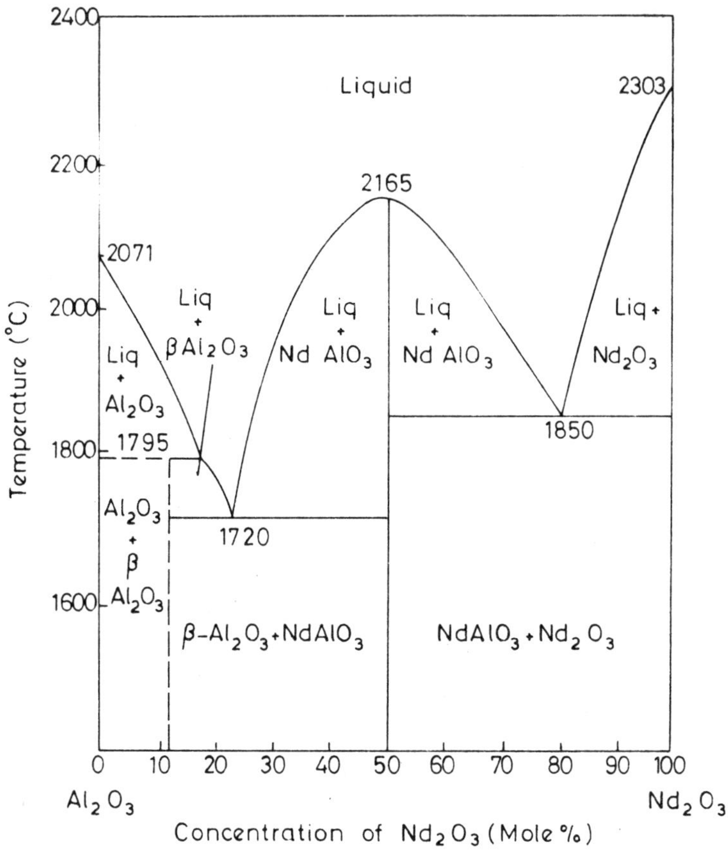

FIG.7. PHASE DIAGRAM OF Al_2O_3-Nd_2O_3 IN AIR.

which recrystallized at about 900°C. In another study using splat cooling device a new phase was obtained in the neodymium rich part of Al_2O_3-Nd_2O_3 system. In 1977 Mizuno et al (23) proposed a tentative phase diagram of the same system (Fig.7) from the cooling curve measurements. Two eutectic points were observed with 23 and 80 mole% Nd_2O_3 at 1720°C and 1850°C respectively. They observed the presence of $Nd_2O_3.11Al_2O_3$, in a fused specimen, coexisting with $NdAlO_3$. At 1100°C, $NdAlO_3$ was seen to undergo reversible phase transition from rhombohedral to cubic. Later on these authors (24) published a tentative phase diagram for Al_2O_3-Sm_2O_3 system at high temperatures. A single, monoclinic phase of $2Sm_2O_3.Al_2O_3$ with freezing point at 1977°C was observed. Three eutectic points at 1775°, 1910°, 1880°C respectively at 24,61 and 80 mole% Sm_2O_3 were observed. At about 750°C, there was a reversible phase transition between orthorhombic and rhombohedral forms. They (25) recently gave tentative phase diagrams for the systems Al_2O_3-Eu_2O_3 and Al_2O_3-Gd_2O_3. Each of the quenched specimen from the melt showed single monoclinic phases of $2Eu_2O_3$. Al_2O_3 and $2Gd_2O_3$. Al_2O_3 respectively. These compounds showed no phase transition in repeated heating and cooling cycles. Al_2O_3-Eu_2O_3 system showed three eutectics at 1710°, 1900°, and 1860°C with 24, 62 and 77 mole% Eu_2O_3. In the Al_2O_3-Gd_2O_3 system, three eutectics at 1720°, 1920° and 1930°C with 25, 65 and 73 mole% Gd_2O_3 were observed. Recently Mizuno et al (26) gave the results of their study on Al_2O_3-Dy_2O_3 system. Quenched specimens from the melt showed three stable and single phases, namely $3Dy_2O_3$. $5Al_2O_3$, $DyAlO_3$ and $2Dy_2O_3$. Al_2O_3.

c) Lanthanide Systems

Coutures et al (27) have shown that the La_2O_3-Y_2O_3 system formed several extended solid solutions at high temperatures, the phases of which were very different from those observed with products cooled down to room temperature. A phase diagram was presented with no experimental data on liquidus curves. Mizuno et al (28) have given the liquidus curve for this system with the investigation on crystal modification of quenched specimens from the melt with a solar furnace. A tentative phase diagram at high temperature was presented (Fig. 8). The presence of orthorhombic and monoclinic forms of $LaYO_3$ was established.

Coutures (29) has obtained the phase diagrams for the rare-earth sesquioxide systems at high temperatures. For a better knowledge of these diagrams, splat cooling process has been adopted. The systems were analysed using X-ray diffractometry and thermogravimetric differential scanning calorimetry. The systems studied were Ln_2O_3-A_2O_3 (Ln = La, Nd, Sm, Gd, Dy, Er, Yb ; A = Al, Cr), Ln_2O_3-Ln_2O_3, Ln_2O_3-AO_2 (A = Zr, Th) and Ln_2O_3-$LnAO_3$ (A = Cr, W). Benezech et al have given a high temperature phase diagram on Yb_2O_3-Cr_2O_3 system melted in air in a solar furnace (Fig. 9).

In 1974, Coutures et al (31) have given the phase diagrams for Nd_2O_3-Y_2O_3 and Nd_2O_3-Yb_2O_3 systems studied by cofusion method in air in a 2 kW solar furnace.

Sibieude and Foex (32) have studied the various phases and phase transitions in the systems ThO_2-Ln_2O_3 (where Ln = La, Nd, Sm, Gd, Ho, Dy, Er, Y) from ambient temperature upto the melting point. Tentative phase diagrams have also been presented. As an example, the high temperature phase diagram of the system ThO_2-Y_2O_3 is cited in Fig. 10.

In 1976 Yoshimura et al (33) studied the system Ce_2O_3-WO_3 in a 2 kW solar furnace using cofusion method. Molten balls were cooled at a rate of 10^5 K/sec by splat cooling method using a mechanical hammer (Fig. 11). High temperature X-ray methods were employed for phase studies. Six stable phases of cerium tungstates were identified. They could not obtain any phase other than the monoclinic, at room temperature even by a splat cooling method for $Ce_2O_3.3WO_3$. Further studies for single crystals at high temperatures could be done. They have also studied the sub-liquidus phase relation in the ternary system CeO_2-Ce_2O_3-WO_2. A recent work of Yoshimura et al (34) has been on the system La_2O_3-WO_3 melted in a solar furnace and rapidly quenched. Using a hammer-anvil device, a cooling rate of 10^5 °C/sec was achieved. A phase diagram has been presented. Near the eutectic composition, glass formation was observed. The colour of glass phase ranged from colourless to pale green. Further study on the viscosity of the melts could be done. The low temperature structures of $La_2O_3.2WO_3$ and $La_2O_3.WO_3$ could be determined. A study on various factors affecting the glass formation such as the mechanical stress by contact with metal walls could be conducted in future.

IGNITION STUDIES

For ignition of any material, say a propellant, unless energy exceeding the threshold value is supplied, ignition will not occur. Correlation study between time and energy will help to predict for example the likely flame spread by irradiation due to nuclear explosions. Military furnaces like the French LCA solar furnace and the U.S. Army WSSF are used to test materials in a simulated nuclear explosion environment. Historically, solar ignition studies started in the Archimedian age. But a real scientific basis for using the solar furnace in ignition tests, was laid by Newton. Lomonosov (35) constructed a catadioptric igniting instrument.

In 1954, Gardon (36) reported the result of ignition studies on materials in the MIT solar furnace. Relatively large area had to be irradiated to treat the heat transfer problem as one dimensional. When wood was subjected to ignition tests, the smoke and decomposition products formed a protective layer and eventually the effective radiation was reduced. Wells (37) has also experienced a similar difficulty during his studies on a different system. When the sample enclosed in glass melted, the vapour clouded the enclosure and turned off the heat. Probably this phenomenon could be used in vacuum coating of materials over a substrate for example, refractories on plastic.

MATERIAL TESTING AND STUDIES ON THERMOPHYSICAL PROPERTIES AT HIGH TEMPERATURES

Solar furnaces are important in studying the behaviour of materials subject to very severe thermic impulses or cycles in any ambient condition. The development of a suitable material which will withstand very high temperatures and thermal shocks has great military and scientific importance, and finds immediate application in several fields. U.S. National Bureau of Standards (38) set up a 5 ft (1.5 m) solar furnace in the 1950's for studies on temperature-resistant materials to withstand the extreme conditions found in atomic reactors, aircraft engines and guided missiles. The laboratoire Central de l'Armement has built on the Pyrenees Orientals, a complex comprising a main solar furnace of

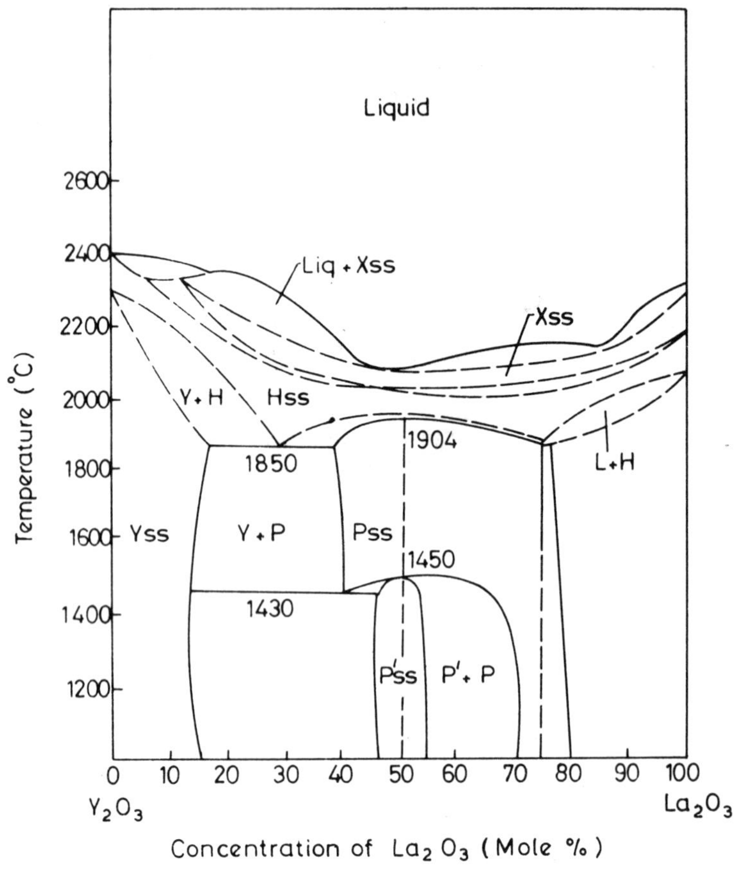

FIG. 8. PHASE DIAGRAM OF Y_2O_3-La_2O_3 IN AIR

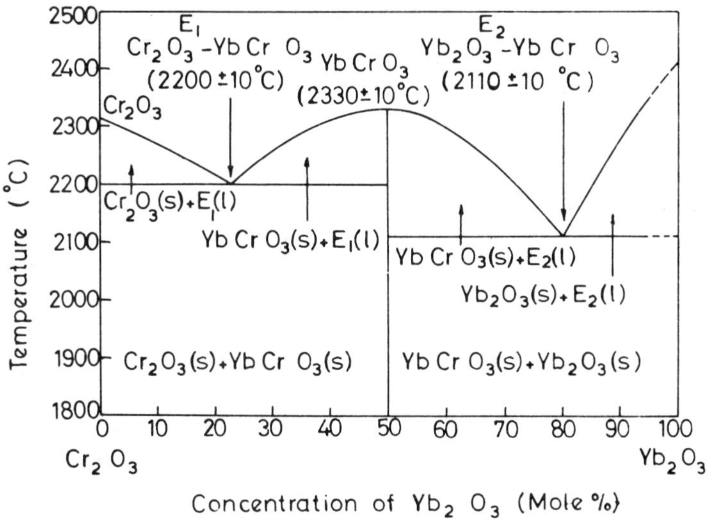

FIG. 9. PHASE DIAGRAM OF Yb_2O_3-Cr_2O_3 IN AIR.

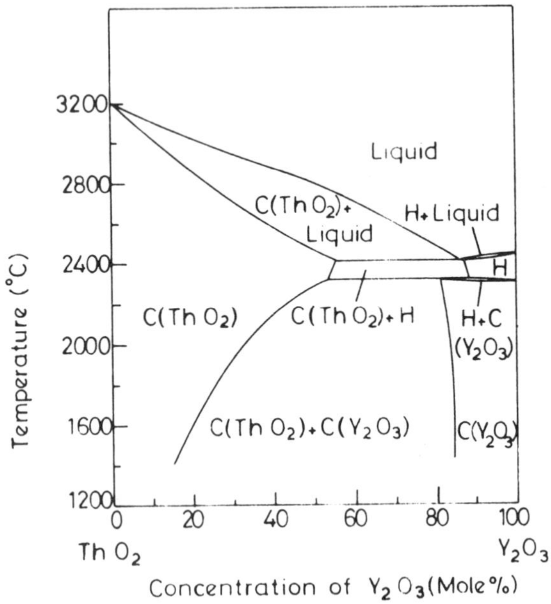

FIG.10. PHASE DIAGRAM OF ThO_2-Y_2O_3.

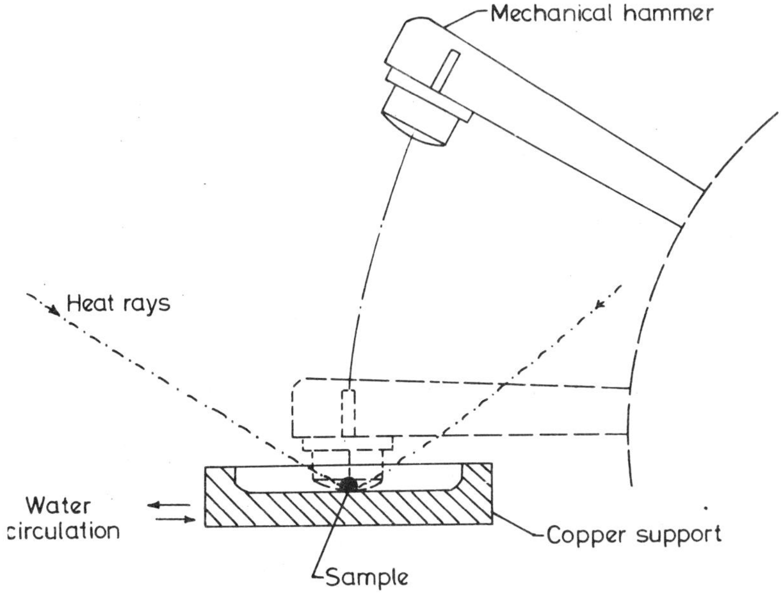

FIG.11. MECHANICAL HAMMER FOR SPLAT COOLING IN A SOLAR FURNACE.

42 kW power and a secondary furnace of 400 W. These LCA furnaces have been designed to deliver brutal thermic impulses onto materials (39). The speed of application and amount of energy provided to the target are the important parameters in such a pyrogenic study. U.S. Army solar furnace at the White Sands Missile Range is also being used to test materials in a simulated radiant thermal energy environment.

Otts et al (40) have reported some results of material testing in the 5 mW Solar Thermal Test Facility. It has been found that concrete will withstand 500 to 1000 suns and white, exposed-agregate concrete over 1000 suns for upto 2 minutes without severe melting. Commercial alumina fire bricks can withstand upto 2500 suns for upto 2 minutes while alumina-silica based ceramic insulating materials can withstand 1500 suns for over one minute. Zirconia based ceramic insulating materials can withstand 2500 suns for over 4 minutes. A similar test-result was announced earlier by Sakurai et al (41). A refractory brick (SK35) melted even in feeble sunlight within a minute, in their 70 kW solar furnace.

Coutures (29) has reported the results of a study in a 2 kW solar furnace on thermophysical properties of refractory materials involving rare-earth sesquioxides, which could be used as electrodes for the MHD process. The MHD materia must have good mechanical and electrical properties such as pure electronic conduction with good emission. These MHD electrode materials must also have goo refractairity and must resist vaporization and corrosion at high temperatures. The solid solutions of zirconia-yttrium sesquioxide series are a promising class of materials for MHD applications, because of their high refractory nature. Addition of rare earth oxides produces varying degrees of electronic conductivit for these solid solutions. Noguchi et al (13) have studied the stoichiometric compound $SrZrO_3$ for use as MHD insulator and conductor material. A study on thermophysical properties of ternary Y_2O_3-ZrO_2-Ln_2O_3 cubic solid solutions on th zirconia rich side is being conducted at Odeillo in collaboration with the USSR Institute of High Temperature.

OTHER APPLICATIONS OF SOLAR FURNACES

Bigg et al (42) studied the ice-nucleating properties of meteoritic material in solar furnace. Their attempt was to duplicate the small particles formed by evaporation and recondensation during the flight of meteors in the atmosphere, heating meteors at a low pressure in a 12 ft (3.6 m) solar furnace.

Trombe et al produced sintered refractories such as zirconium oxide, calcium zirconate and alumina in solar furnaces. Sintering of ferrite compacts in a solar furnace has been done at Keio University. High temperature mechanical, thermal, optical and electrical properties of various materials have been studi with solar furnaces (43, 44, 45, 46).

The preparation of the hot-junction bead of thermocouple was suggested by Laszl Solar welding is an important application of solar furnaces, as given by Uglov He has given a calculation of the power required for welding a device. A gene thermal balance equation is derived on the assumption of a linear and slow-mov source, full penetration of the weld, circular weld pool with diameter equal t weld width which is identical on both faces, and a mean pool temperature of 1.

times the melting temperature. Frantsevich et al (48) have very recently reported the results of their studies on technological processes like welding in the SG1-5 solar furnace of 2 m aperture under cosmic simulated conditions. The radiant flow ranged from 150-1900 W/sqcm. Various alloys in the form of 0.25 - 2 mm thick flat plates (or cylinders) with weld joints were used. The effects of radiant flow efficiency, process speed, and joint type on the size and shape of the weld pool were studied. Solar vacuum welding and brazing could be a very powerful technique in the near future. Solar furnaces operating in space factories can be used for obtaining new materials with special characteristics (49).

Very recently some work has been undertaken at the General Electric Company, by Cline and Anthony (50). They have utilized a 5.5 m solar furnace to process silicon wafers by thermomigration, which requires an intense, uniform source of heat with a high colour-temperature. But they feel that the energy savings afforded by the use of free sunlight is incidental relative to the value added to the silicon wafer by the thermomigration process.

Solar furnaces have also been used in photo-chemistry. Guillemonat (51) used a 18.4 m diameter solar furnace for studying nitrogen fixation process. Marcus et al (52) reported an exploratory effort at Stanford Research Institute to study the photochemical reactions in liquids in open or closed vessels, irradiated by keeping them at the focus of a solar concentrator. But, for other chemical processes the intermittant operation of solar devices might be a handicap. There may however be some intermittant chemical processes which could yet be studied with solar furnaces.

Solar Chemical Engineering is a novel field which has been initiated by the French Scientists. Flamant et al (53) have reported a preliminary design and investigation of solar chemical reactors. In a fluidised bed reactor of 2 kW solar furnace, silica sand fluidized particles were heated and some applications such as decomposition of $CaCO_3$, and thermochemical cycles were realized as feasible. Other industrial applications of solar furnaces, for example in extractive metallurgy, mineralogy, refractory industries, and others will be presented elsewhere (54).

SUMMARY AND CONCLUSIONS

In this paper we have summarized the possible applications of solar furnaces in various fields of materials research and industry, and tried to identify the opportunities for further research. Crystal growth in a solar furnace is still a developing field and there are several important research problems yet to be solved. Ishigame's work on Ni^{2+} doped CaO crystals has revealed that there is a possibility of growing doped crystals with desired resistivity. The change in resistivity, the deviation from stoichiometry of the resulting crystals and the mechanism of such a crystal growth can be studied in future, by varying the parameters such as surrounding atmosphere, irradiation, quenching rates and the initial thermal history of the specimen. There is ample scope for detailed study on single crystal growth of semiconductor materials like silicon doped with suitable materials which will find immediate applications in electronics. Coupling solar zone-refining techniques, this method could be useful for obtaining crystals of pure alumina and zirconia which may be used in electrical and optical studies. New uses of the crystals of titania grown in a solar furnace may be thought of. With innovations in temperature control and instrumentation solar furnaces can become an important tool for crystal growth. Extensive work on high temperature phase diagram determination has been done using solar furnaces. However most of the diagrams are tentative and need to be finalized through further research. These diagrams are incomplete either because of insufficient data or due to unavoidable polymorphic inversions taking place and lack of refinement in experimental techniques. In the case of zirconia based systems, especially ZrO_2-ThO_2, ZrO_2-UO_2 and ZrO_2-TiO_2 the phase transformations can not be suppressed and these systems are sensitive to therma shock. Stabilization of such systems over a wide range of temperature will be an interesting research on a new class of refractory materials. Noguchi's work on ZrO_2-SrO system has shown that the freezing points of the composition 60 mole% SrO is very sensitive to variation in stoichiometry. This may be a basic test for estimating the various chemical compositions. Coutures has coupled splat cooling with a hammer-anvil device in a solar furnace to study the phase diagrams of a series of lanthanide systems. An amorphous glass which recrystallizes at 900°C has been obtained in the case of $(1-x)Al_2O_3 - x\ Nd_2O_3$ system. The uses of such new amorphous glasses may be an area for future research. Stable, single and monoclinic phases have been identified in Al_2O_3-Gd_2O_3 and Al_2O_3-Eu_2O_3 systems. Rare-earth sesquioxides are found to be very useful as high temperature materials usable inside atomic reactors as duct walls and as conductor materials in MHD applications and the may be made in solar furnaces.

Purification of materials in a solar furnace is one of the most promising fields for future research. Walter's work on the relative volatility of seven oxides from a melt has shown that silicon volatility is very different from that of aluminium which is the least volatile component. The dependence of alkali volatility on oxyge pressure has also been found. This study can be extended for instance, in purifying Al-Si alloys. These alloys could be superheated either in vacuum or in any desired atmosphere and splat cooled in a solar furnace. This superheating would result in the boiling off the impurities from the liquid metal. Because of a rapid cooling it may be possible to produce ultrafine grain materials, amorphous materials and eve materials with different crystal structures. The vacuum evaporation can also effectively be used for vacuum deposition of a variety of refractories on metals an plastics. Graphitizing of coconut shells and fibres and pineapple and jute fibres using solar furnaces could be other areas of possible research.

Brauer found that the melting of TiO_2 resulted in a non-stoichiometric composition, depending upon temperature and oxygen pressure. Study on titania which is abundant available in India may be undertaken. Solar melting of metal-metal, metal-ceramic and ceramic-ceramic eutectic systems and directional solidification of reinforced eutectics formed in a solar furnace are the other promising fields.

There are several opportunities for industrial uses of solar furnaces. Flamant's recent work in extractive metallurgy, such as thermal decomposition of phosphate ores in a solar furnace and beneficiation of the by-products, can lead to large-scal industrial utilization of solar furnaces for mineral beneficiation. Heating clay, ilmenite, monazite and other minerals in a desired atmosphere to ultra high temperatures in a solar furnace, direct extraction of metals may be achieved. Sola smelting units (as installed in the USSR) could become widely used in future. Sola vacuum welding is another good prospect. This could be a powerful technique in electronics industry, and for welding reactive metals like titanium for applicatio in vacuum sealing. Foundry operations, sintering of refractory materials, metals and compounds, pre-heating scrap and charge materials, heat-treatment, melting and casting, and dehydration of ores can be done by establishing small scale solar furnaces throughout the country. As Sakurai has envisaged, it will be possible in near future to operate a solar furnace in outer space. Metals would melt in high vacuum, without any necessity for a container or holder, and be crystallised free from any strain which might be introduced by the differences of specific gravity between liquid and solid states. Thus solar furnaces stationed at the moon or space factories could produce new materials hitherto unknown.

ACKNOWLEDGEMENTS

We wish to express our sincere thanks to Mr. M. Jose for the preparation of illustrations and to Tata Energy Research Institute, India, for financial support. We are grateful to Dr. C.S. Menon, Dr. P.N. Mohandas and Mr. G. Natrajan for their comments on the manuscript.

REFERENCES

1. Laszlo, T.S., 'Solar furnace in high-temperature research', Sciences, Vol. 124, 1956, pp. 797-800.

2. Farber, E.A., 'Crystals of high-temperature materials produced in the solar furnace', Solar Energy, Vol. 8, No. 1, 1964, pp. 38-42 (Ceram. Abs. Aug. 1964, 230b).

3. Laszlo, T.S. et al, 'Thoria single crystals grown by vapour deposition in a solar furnace', J. Phy. Chem. Solids, Vol. 28, 1967, pp. 313-316.

4. Sakurai, T. et al, 'Uranium dioxide crystals grown by a solar furnace', J. Crys. Gr., Vol. 2, 1968, pp. 326-327.

5. Sakurai, T. et al, 'Growth of nickel oxide crystals by solar furnace fusion J. Crys. Gr., Vol.2, 1968, pp. 284-286.

6. Ishigame, M. et al, 'Electronic absorption spectrum of Ni^{2+} in CaO ' J. Phy. Soc. Japan, Vol. 25, No.6, 1968, pp. 1629-1632.

7. Nigara, Y. et al, 'Infrared properties of Yttrium oxide', J. Phy. Soc.Japan Vol. 30, No. 2, 1971, pp. 453-458.

8. Payne, D.N., University of Southampton, England, Personal Communication, Nov. 1977.

9. Brauer, G. et al., 'Uber den schmelzpunkt und die thermische dissoziation von titandioxyd', J. Inorg. Nucl. Chem., Vol. 16, 1960, pp. 67-76.

10. Duwez, P. et al., 'Phase relationships in the system zirconia-thoria', J.Am. Ceram. Soc., Vol. 40, No. 9, 1957, pp. 321-324.

11. Sakurai, T. et al, 'Phase relationships in the system ZrO_2-ThO_2', Rev. int. Htes. Temp. et Refract., Vol. 12, No. 1, 1975, pp. 74-77.

12. Evans, P.E., 'The system UO_2-ZrO_2', J. Am. Ceram. Soc., Vol. 43, No. 9, 196 pp. 443-447.

13. Noguchi, T., 'High temperature phase studies with a solar furnace', Advance in high temperature chemistry, 2, 1969, pp. 235-262.

14. Noguchi, T. et al., 'Liquidus curve measurements in the ZrO_2-MgO system wi the solar furnace', Bull. Chem. Soc. Japan, Vol. 41, 1968, pp. 1583-1587.

15. Noguchi, T., 'The liquidus curve of the ZrO_2-Y_2O_3 system as measured by a solar furnace', Bull. Chem. Soc. Japan, Vol. 43, 1970, pp. 2614-2616.

16. Rouanet, A., 'Contribution a l'etude des systemes zirconeoxydes des lanthanides an voisinage de la fusion', Rev. Int. Hautes Temper. et Refract., Vol. 8, 1971, pp. 161-180.

17. Conn, W.M., 'Use of solar furnace for studying the system alumina-silica' Ceram. Bull., Vol. 33, No. 3, 1954, pp. 69-72.

18. Noguchi, T. et al, 'Phase diagram of the system $Al_2O_3-Y_2O_3$', Kogyo Kagaku Zasshi, Vol. 70, 1967, pp. 834-837.

19. Mizuno, M. et al, 'High temperature phase studies on the system $Al_2O_3-Ln_2O_3$ with a solar furnace', Paper presented at the ISEC, Los Angeles, 1975, pp.1-6.

20. Mizuno, M. et al, 'The liquidus curve in the system $Al_2O_3-Ga_2O_3$ as measured with a solar furnace', Reports of the GIRIN, Vol. XXIV, No. 6, 1975, pp.184-187.

21. Mizuno, M. et al, 'Phase diagram of the system $Al_2O_3-Pr_2O_3$ at high temperature,' Yogyo-Kyokai-shi, Vol. 85, No. 1, 1977, pp. 24-29.

22. Coutures, J.P. et al, 'Obtention et études des verres refractaires a base d'alumine et de sesquioxyde de neodyme', Mat. Res. Bull., Vol. 10, No. 6, 1975, pp. 539-546.

23. Mizuno, M. et al, 'Phase diagram of the system $Al_2O_3-Nd_2O_3$ at high temperature', Yogyo-Kyokai-Shi, Vol. 85, No. 2, 1977, pp. 90-95.

24. Mizuno, M. et al, 'Phase diagram of the system $Al_2O_3-Sm_2O_3$ at high temperatures', Yogyo-Kyokai-Shi, Vol. 85, No. 8, 1977, pp. 374-379.

25. Mizuno, M. et al, 'Phase diagrams of the system $Al_2O_3-Eu_2O_3$ and $Al_2O_3-Gd_2O_3$ at high temperatures,' Yogyo-Kyokai-Shi, Vol. 85, No. 11, 1977, p.543-548.

26. Mizuno, M. et al, 'High temperature studies on the system $Al_2O_3-Ln_2O_3$ with a solar furnace,' Paper presented the ISEC'77, New Delhi, Jan. 1978, pp.1682-1683.

27. Coutures, J. et al, 'Etude a haute temperature du diagramme d'equilibre du systeme forme par le sesquioxyde de lanthane avec le susequoxyde d'yttrium', J. Sol. State. Chem., Vol. 11, No. 4, 1974, pp. 294-300.

28. Mizuno, M. et al, 'Phase diagram of the system $La_2O_3-Y_2O_3$ at high temperatures', Yogyo-Kyokai-Shi, Vol. 84, No. 7, 1976, pp. 342-347.

29. Coutures, J.P., 'Solar furnaces experiments for thermophysical properties studies of rare-earth oxides MHD materials,' Paper presented at the 13th Rare-Earth Research Conference, Oct. 1977.

30. Benezech, G. et al, 'Dispotifs d' analyse thermique pour l'etude des transformations cristallines et des changements de phase a haute temperature', Colloques Internationaux (CNRS) No. 205, Odeillo, Sep. 1971, pp. 57-69.

31. Coutures, J. et al, 'Etude a haute temperature des systems formes par le sesquioxyde de neodyme avec les sesquioxydes d'yttrium et d'ytterbium', Mat. Res. Bull., Vol. 9, No. 12, 1974, pp. 1603-1612.

32. Sibieude, F. et al, 'Phases et transitions de phases a haute temperature observes dan les systems $ThO_2-Ln_2O_3$ (Ln = Lanthanide et yttrium)', J. Nuc. Mat., Vol. 56, 1975, pp. 229-238.

33. Yoshimura, M. et al, 'Identification of binary compounds in the system $Ce_2O_3-WO_3$', J. Sol. St. Chem., Vol. 16, No. 34, 1976, pp. 219-232.

34. Yohimura, M. et al, 'Rapid quenching of melts in the system La_2O_3-WO_3', J. Mat. Sci., Vol. 12, 1977, pp. 415-417.

35. Veinburg, V.B., 'Optics in equipment for the utilization of solar energy', Book, State Publishing House of Defence Industry, Moscow, 1959.

36. Gardon, R., 'Segmented-mirror solar furnace for high intensity thermal radiation studies,' Rev. Sci. Instr., Vol. 25, No. 5, 1954, pp. 459-463.

37. Wells, R.R., Albany Metallurgy Research Centre, U.S.A., Personal communication, Dec. 1977.

38. 'Notes from the National Bureau of standards : Solar furnace', J. Franklin Inst., Vol. 264, 1957, p. 216.

39. Jaymes, M. 'Solar furnaces of Laboratoire Central de l'Armement', Rev. Int. Hautes Temp. Refract., Vol. 12, No. 4, 1975, pp. 301-306.

40. Otts, J.V. et al, 'The USA 5MW solar thermal test facility', Paper presented at the ISEC'77, New Delhi, Jan. 1978.

41. Sakurai, T. et al, 'Construction of a large solar furnace', Solar Energy, Vol.8 No. 4, 1964, pp. 117-127.

42. Bigg, E.K. et al, 'Ice nucleating properties of meteoritic material, J. Atmos. Sci., Vol. 24, No. 1, 1967, pp. 46-49.

43. Leon, G.S. et al, ' Analysis of solar furnace performance in mechanical testing at extremely high temperatures', J. Engg. Power, Vol. 82, No. 4, 1960, pp. 325-332.

44. Walton, J.D., Jr. 'High temperature evaluation of radomes using the 1000 KW French solar furnace', Am. Ceram. Soc. Bull, Vol. 52, No. 4, 1973, pp. 424.

45. Tanemura, S. et al, 'Spectroscopic properties of SiC at high temperatures', Paper presented at the International Solar Energy Congress and Exposition, Rockville, 1975, pp. 159-160.

46. Afzal, F.A. et al, Measurement of electrical conductivity at high temperatures using a solar furnace,' Paper presented at the ISES Conference, Melbourne, 197

47. Uglov, A.A., 'Energy Conditions of welding with solar radiation', Welding Production, Vol. 23, 1976, pp. 1-3 (IAA, Vol. 17, No. 20, Oct. 15, 1977).

48. Frantsevich, I.N. et al, 'Studies of technological processes by solar energy under Cosmic simulated conditions,' Paper presented at the International Astronautical Congress 28th, Prague, Oct. 1977. (IAA Vol. 17, No. 24, Dec. 15 1977).

49. Hoez, M. et al, 'Remarks concerning solar furnaces in space', Paper presented at the ISES Conference, Melbourne, 1970.

50. Cline, H.E. et al, 'Thermomigration of silicon wafers in a solar furnace,' Solar Energy, Vol. 19, 1977, pp. 715-719.

51. Guillemonat, A., '50 kW solar furnace at Bauzareah', Industries et travaux d'outre-Mer, Vol. 3, No. 23, 1955, pp. 614-618.

52. Marcus, R.J. et al, 'Photochemistry in the solar furnace', Ind. Eng. Chem., Vol.51, No. 11, 1959, pp. 1335-1338.

53. Flamant, G. et al, 'Applications of concentrated solar energy in the field of mineralurgy and extractive metallurgy,' Paper presented at the ISEC'77 New Delhi, Jan. 1978.

54. Suresh, D. et al, 'Potential applications of solar furnaces in materials industry', Paper to be presented at the 2nd International Solar Forum, Hamburg, July 1978.

EFFICIENCY OF A SOLAR STILL AS A FUNCTION OF ITS MAIN PARAMETERS. PRODUCTION ESTIMATION FOR VALENCIA, SPAIN

J.Riera, J.A.Martínez-Lozano, F.Tejerina and F.Tena-Sanguesa
Faculty of Physics
Valencia (Spain)

ABSTRACT

The basic equations of the thermal performance of a solar still as found in previous papers has been considered.
An experimental study of the solar destilation process has been carried out in Valencia (Spain). With this purpose two different stills have been constructed, one of them provided with a floating absorber and prepared to minimize the thermal inertia.
The studied parameters have been distributed in two groups:
a) outer parameters: solar radiation, ambient temperature and wind speed
b) inner parameters: water temperature, cover temperature and theirs difference
A relationship between production and each one of these groups of parameters has been developed. With the aid of available outer parameters data for the last thre years in Valencia, we estimate the annual production of both stills for this location.

INTRODUCTION

Solar destillation is one of the solar energy applications that have been known for a longer time. Towards the end of the last century a desalinization plant was built in Las Salinas (Chile) with a production of $20m^3$/day during the summer.
Nevertheless it is still an important area nowadays and investigations are going one. The constant population increasing involves a constant increasing of drinking water needs. In an -- other hand the actual energy crisis draws attention on the necessity of new energy sources.
In order to improve the performance of a solar still we may consider two different aspects: increasing of productivity and cost decreasing. From all different kinds of solar stills developed up to now, the greenhouse still has been the one with the best solution in both directions.
We may put the several papers on the subject in two different groups:

a) Theorical works with the only purpose of establishing mathematical expressions that can show in an aproximated way the physical processes having place in the inside of the solar still. We may deduce from them a digital simulation model of the still performance.

b) Experimental works showing the influence of the various parameters with an influence on the still performance. In any case, the final aim is to find an expression to evaluate the solar still output in a fixed place.

THEORETICAL BASIS

The greenhouse solar still combines the energy collecting by the basin water and its subsequent distribution to other parts of the still or to the surroundings. The distilled water output is convected with evaporation, one of the three mechanism of heat transfer from brine to cover.

If we call $\alpha_w tH_s$ to the absorbed energy by the salt water, $\alpha_g H_s$ the one absorbed by the cover, q_{ga} energy lost through the cover q_b the one lost through the bottom and edges and c_{wg} to the thermal capacity of water, the energy balance of the whole sistem may be expressed as follows:

$$\alpha_g H_s + \alpha_w tH_s = q_{ga} + q_b + c_{wg}(dT_w/dt) \qquad (1)$$

The heat transfer between the cover and the brine is $q_r+q_c+q_e$, while the heat flow to the surroundings is the sum of these heat fluxes, plus the solar energy absorbed by the glass:

$$q_{ga} = q_r + q_c + q_e + \alpha_g H_s \qquad (2)$$

where we assume that the heat transfer by diffusion and conduction is negligible having in mind that the distance between cover and basin water is over 6 cm. [1] .

Dunkle [2] gets the estimated values:

$$q_r = 0,9\ (T_w^4 - T_g^4) \qquad (3)$$

$$q_c = 8,84.10^{-4} \left[T_w - T_g + \frac{(P_w - P_g)T_w}{268,9.10^3 - P_w} \right]^{1/3} (T_w - T_g) \qquad (4)$$

$$q_e = 0,047(P_w - P_g)h_w \qquad (5)$$

$$q_{ga} = q_{ra} + q_{ca} = 0,9\ T_g^4 - (T_a - 12)^4 + h_{ga}(T_g - T_a) \qquad (6)$$

The establishment of these mathematical relations led to the elaboration of different digital and graphic simulations of transient solar still operation [3], [4] .

EXPERIMENTAL RESEARCH

An experimental study of the solar distillation processes has been carried out in Valencia (Spain). The main purposes of this study have been:

- To compare the performance of a low thermal inertia still provided with a floating absorber [5], with the performance of a conventional still.
- To evaluate the influence of some especific parameters on the still efficiency.
- To obtain the thermal efficiency of the still.
- To get a general expression that should describe the still output as a function of the most important variables. Through this expression it is posible to make an evaluation of the annual production of such still in Valencia.

In order to get to these purposes two different stills have been constructed. One of them provided with a floating absorber and ready to minimize the thermal inertia.

Both stills have been put to work for a period of two months. The water level in the basin was changed from day to day using the values of 2,3 and 4cm. Daily amont of distilled water was measured every day.

The variables whose influence on the still performance were higher have been classified in two groups:

a) External variables: daily solar radiation, H_s; ambient temperatura, T_a; wind speed, v_w; water level, $\bar{h}$; and sealing.
b) Internal variables: mean water temperature, T_w; mean cover temperature, T_g; and the difference between the temperatures of the water and cover, T_w-T_g.

The values of H_s and v_w have been obtained from the "Centro Meteorologico de Levante", in Valencia.

Through a measurement system designed in our laboratory we have been able to get a hourly record of T_a, T_w and T_g values.

EXPERIMENTAL STILLS AND INSTRUMENTATION

As it has been previously seaid, two different stills have been built and tested. The main common characteristics are:

- A simple slope roof, oriented E-W, with only one transparent glass surface, and 20°of inclination.
- Framework and isolation with expanded polystyrene , 5cm. thick.
- The distillated chanels forming an integral part of the canopy, covered with butyl rubber.
- The back wall of the canopy lined with reflecting tape.
- Silicona seal.

The main characteristics of the low thermal inertia still are:

- A black absorber floating on the water surface.
- Reflectant cover of the basin instead of the black lining of butyl rubber.

In order to be able to get periodic and simultaneus temperature measurements in the different parts of the still we have used a

TABLE I

N°	Day	Level (cm)	v_w (Km/h)	T_a (°C)	H_s (MJm^{-2})	CONVENTIONAL STILL T_w (°C)	T_g (°C)	Output (Kgm^{-2})	η (%)	STILL WITH FLOATING ABS. T_w (°C)	T_g (°C)	Output (Kgm^{-2})	η (%)	Observ.
1	16 VIII	2	4,42	25	24,520	41	30	5,139	39	39	28	4,552	34	
2	17	4	10,63	28	26,320	40		5,270	37	38	28	3,959	28	
3	18	4	5,96	25	23,310	40	32	4,730	38	39	29	4,055	32	
4	22	2	6,63	22	13,810	31	24	2,143	29	29	24	1,289	17	
5	24	4	6,79		25,730	43	29	4,755	34	40	27	3,984	29	
6	25	2	7,63	22	24,140			5,061	39			3,998	31	
7	26	4	6,25	22	21,260	35	26	3,689	32	33	25	3,154	27	
8	29	2	7,08		24,140	35	25	4,664	36	33	25	3,388	26	(*)
9	30	4	5,38	21	24,810	39	27	5,086	38	38	27	4,257	32	(**)
10	31	2	6,46		18,160	36	26	3,316	34	36	27	3,271	33	
11	1 IX	4	4,25	22	20,840	36	27	3,889	35	36	26	3,879	34	
12	2	4	4,50	22	20,630	37	27	4,002	36	37	27	3,907	35	
13	5	4	4,04		22,510	39	27	4,461	37	38	27	4,300	35	
14	6	2	3,96	23	22,130	38	28	4,857	41	37	28	4,777	40	

(*) Sealing improvement of the conventional still

(**) Sealing improvement of the still with floating absorber

TABLE I (Cont.)

N°	Day	Level (cm)	v_w (Km/h)	T_a (°C)	H_s (MJm^{-2})	CONVENTIONAL STILL T_w (°C)	T_g (°C)	Output (Kgm^{-2})	η (%)	STILL WITH FLOATING ABS. T_w (°C)	T_g (°C)	Output (Kgm^{-2})	η (%)	Observ.
15	7 IX	2	4,46	23	21,130	38	28	4,823	42	37	27	4,741	42	
16	8	4	5,13	23	23,510	39	28	4,630	36	39	29	4,499	35	
17	9	4	4,25	22	21,670	39	28	4,528	39	38	28		36	
18	10	4	4,75	23	22,510	38	28	4,625	38	38	28		26	
19	11	4	5,17	23	22,180	39	28	4,725	39	39	28		26	
20	12	2	4,54	23	21,420	37	27	4,815	42	36	27	4,313	37	
21	15	2	5	22	15,360	35	26	3,268	39	33	25	2,989	36	(*)
22	19	2	5,25	17	6,900	23	18	0,796	21	23	18	0,872	23	
23	20	4	4,25	20	18,490	35	25	3,397	34	35	25	3,702	37	
24	21	3	6,17	21	12,890	31	24	2,036	29	31	24	2,080	30	
25	22	3	4,71	19	13,770	30	22	2,452	33	30	22	2,594	35	
26	23	3	4,96	21	17,820	35	25	3,631	38	35	25	3,664	38	
27	24	3	5,08	24	12,430	32	26	1,982	30	32	26	2,125	32	
28	25	3	7,63	22	16,530	35	26	3,296	37	34	25	3,332	37	
29	26	3	4	20	17,660	35	25	3,617	38	34	25	3,605	38	

(*) Sealing improvement of the still with floating absorber

TABLE I (Cont.)

N°	Day	Level (cm)	v_w (Km/h)	T_a (°C)	H_s (1 Jm^{-2})	CONVENTIONAL STILL T_w (°C)	T_g (°C)	Output (Kgm^{-2})	η (%)	STILL WITH FLOATING ABS. T_w (°C)	T_g (°C)	Output (Kgm^{-2})	η (%)	Observ.
30	27 IX	2	5,08	22	18,830	35	26	4,116	41	34	26	4,175	41	
31	28	4	4,17	21	15,150	33	25	2,729	33	32	25	2,826	35	
32	29	3	3,58	21	10,790	31	23	1,972	34	31	23	1,952	34	
33	30	4	6	22	18,070	36	26	3,651	37	35	26	3,770	39	
34	3 X	2	4,17	21	15,820	33	25	3,466	41	33	25	3,458	41	
35	4	3	3,67		11,550	27	23	1,702	27	27	23	1,661	27	
36	5	4	5	22	15,060	35	26	3,300	41	34	26	3,252	40	
37	6	3	10,25	19	8,540			1,264	27					
38	7	3	13,29	17	14,060	31	21	3,103	41	30	21			
39	8	4	14,96		11,380	28	22	1,229	20	28	22	1,229	20	
40	10	3	5,79	18	17,490	28	21	3,489	37	28	21	3,571	38	
41	11	2	4,33	18	16,740	31	23	3,664	41	31	22	3,507	39	
42	12	3	2,92	19	15,560	32	23	2,839	34	32	23	2,661	37	
43	13	3	3,04	19	14,600	32	23	2,993	38	32	23	2,946	37	
44	14	3	3,96	19	10,420	27	22	1,904	34	28	22	2,048	36	

system of termistors connected to a plotter that registred the calibrated values hourly.

RESULTS

The daily values of the different variables and of the output obtained through the test period (17 August to 14 October,1977) are shown in TABLE 1. In the observations column are expressed the days in that the sealing of any of the stills was modified. It is to be observed that the 2nd cange made on the sealing of the low thermal inerce still, on the 15th of September, was completed with a change of the floating absorber. As it will be explaned lately, only the data obtained from this day on have been used for the analysis of the corresponding still. The blank spaces in the data of the TABLE 1 correspond to values that was not possible to obtain owing mainly to failures in the data recording sistem or in the water collecting sistem.

Influence of solar radiation on the still output. Comparison between both stills performance.

The performance of the conventional still may be divided in two different periods. The first one includes from the beginning to the 29th of August. In this day a deficiency in the sealing was observed and mended. The data obtained can be seen in figure 1. The linear regression analysis gives the values

$$Y = 0.2342\ X - 0.6801 \qquad (7)$$

(Y: Output, $Kg.m^{-2}$; X: Solar Radiation, $MJ.m^{-2}$)
with a correlation coefficient of $r = 0.9671$.
The performance of the low thermal inertia still may be divided in three periods. A first one including the first days until the 30th of August that the sealing had to be changed. A second one in which the floater was altered owing to the heat and had to be changed on the 15th of September. Athird one with continuous and correct performance. Only the data obtained in this third period have been used to get a relation between output and solar radiation. These data can be seen in figure 2. The linear relation is now:

$$Y = 0.2772\ X - 1.2329 \qquad (8)$$

(Same units as in (7)), being the linear correlation coefficient $r = .9637$ in good agreement with the one corresponding to the conventional still. Using the whole amount of data we get $r = 0.8653$ ehich is much lower.
The correlation obtained is very high. Higher even than the one showed by some other papers. It could be explaned by the isola-

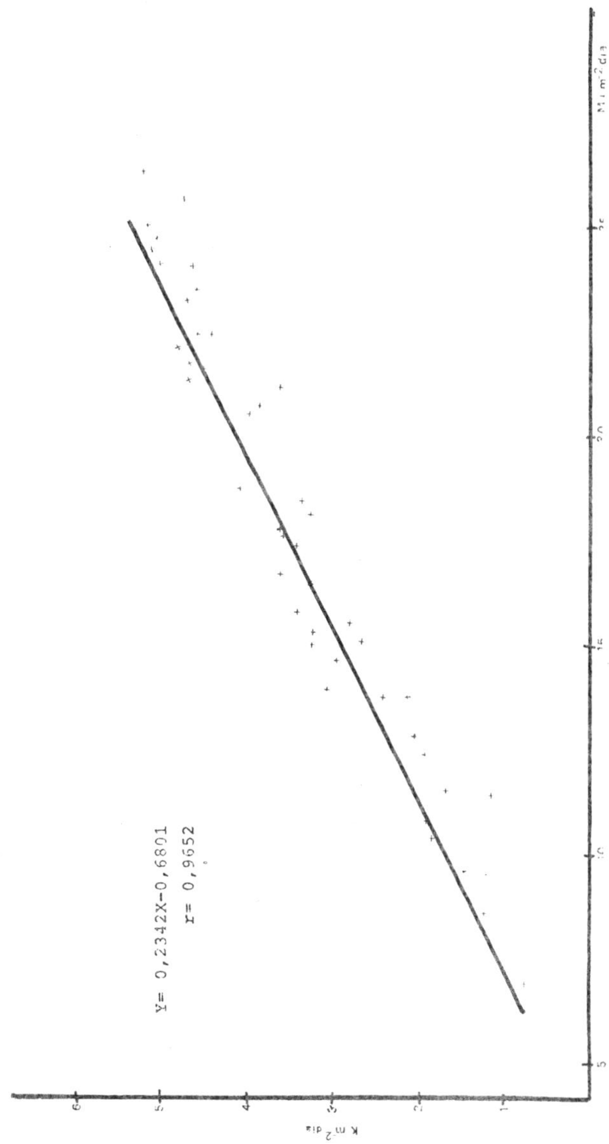

FIG. 1.-RELATIONSHIP BETWEEN SOLAR RADIATION AND STILL OUTPUT (CONVENTIONAL STILL)

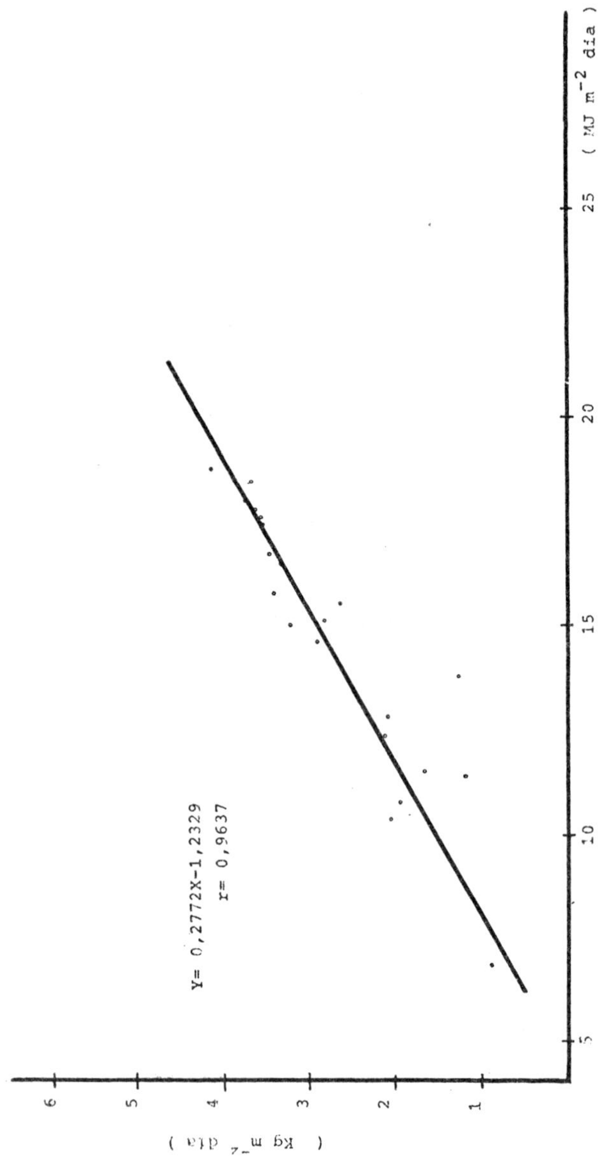

FIG.2.-RELATIONSHIP BETWEEN SOLAR RADIATION AND STILL OUTPUT (STILL WITH FLOATING ABSORBER)

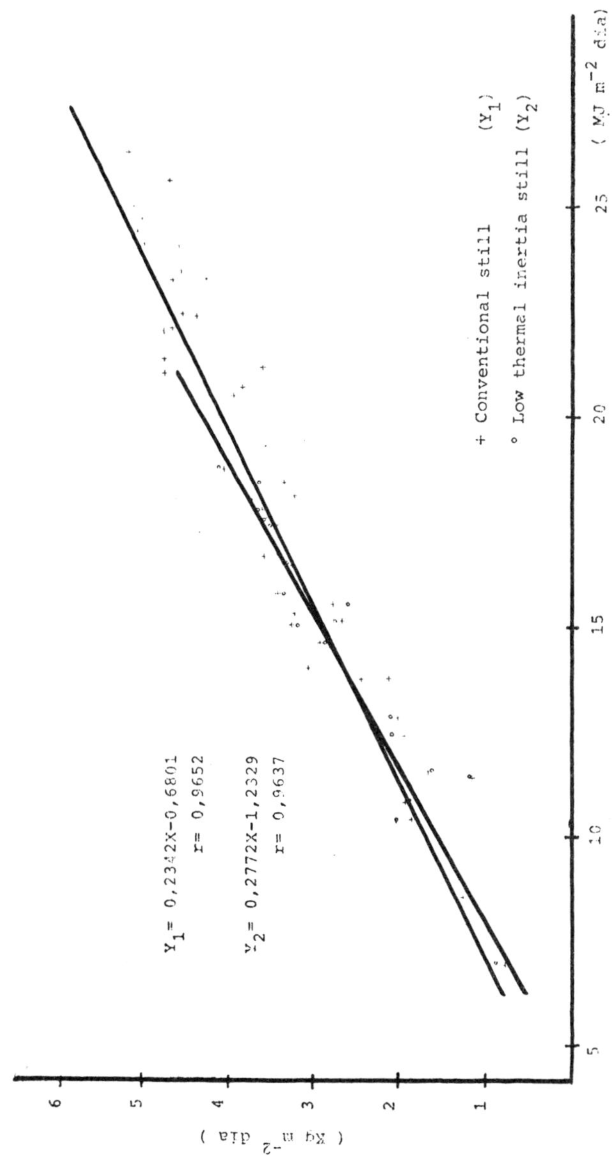

FIG.3.-RELATIONSHIP BETWEEN SOLAR RADIATION AND STILL OUTPUT

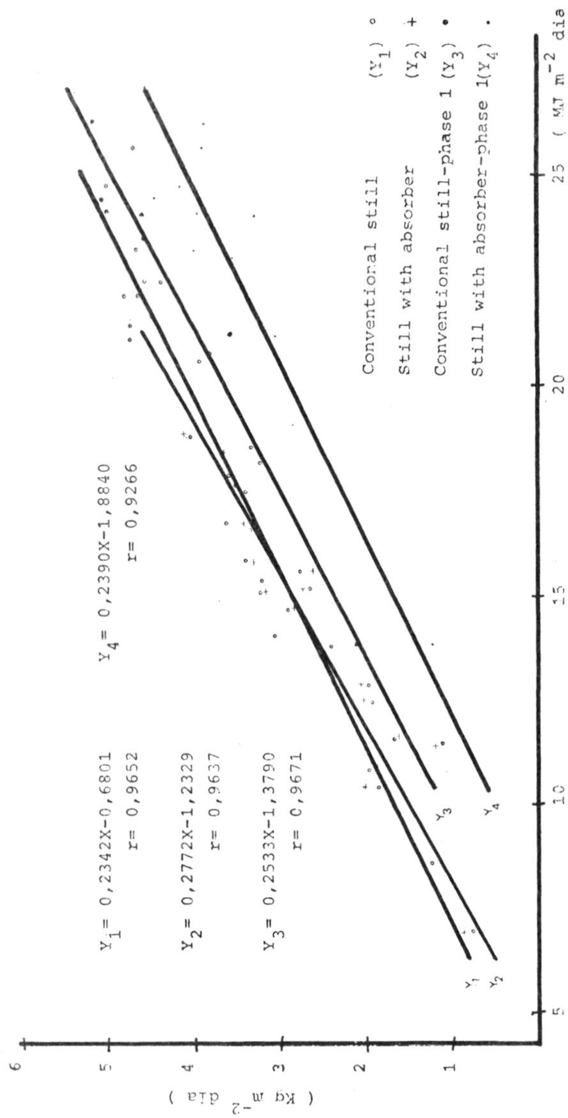

FIG.4.-SEALING INFLUENCE ON THE DAILY OUTPUT

tion in which the stills have been working so that the influence of some other parameters such as T_a, has been lower.
Through the expresions (7) and (8) we may compare the performance of both stills.
The mean value of the output obtained from the still provided with a floating absorber has been 4% higher than the obtained from the conventional still. Nevertheles before deducing a definitive conclusion we must observe the following details:

a) The computed radiation values for the low thermal inertia still have not a high value in any case.

b) To evaluate the expresion (7) corresponding to the conventional still, data from the first period have been used.

c) The floating absorber, even in the regular first period has shown dry areas.

Anyway there is no doubt about the good prospects involved in the use of the floating absorber.

Influence of different factors on the daily water production. External parameters.

a) In order to evaluate the influence of the correct sealing of the stills on the water production, we evaluate this production as a function of the solar radiation for the first period of both stills.
For the conventional still we get:

$$Y = 0.2533\ X - 0.6801 \qquad r = 0.9652 \qquad (9)$$

For the low thermal inertia still:

$$Y = 0.2390\ X - 1.8840 \qquad r = 0.9266 \qquad (10)$$

If we compare these expresions with the (7) and (8),(see fig.4) we get an increasing of 15% for the conventional still and an increasing of 66% for the other still, as a result of improving of sealing. The big difference in the second case has to do also with the anomalous performance of the floating absorber in the first period and justifies the decission of not using those data for the global calculations.
Anyway, the influence of a good sealing is made patent in the previous considerations.

b) To the evaluation of the influence of the water level on water production we have considered three groups according to such levels.For everyone of these groups a linear correlation has been obtained a follows:

$$h=2cm. \qquad Y = 0.2461\ X - 0.7484 \qquad r = 0.9663 \qquad (11)$$

$$h=3cm. \qquad Y = 0.2581\ X - 0.9978 \qquad r = 0.9235 \qquad (12)$$

$$h = 4cm. \quad Y = 0.2366\ X - 0.8794 \quad r = 0.9520 \qquad (13)$$

which can be seen in figure 5.
We may observe that production increases with the decreasing of water level. This increasing is more important when the solar radiation is high. As a consecuence of diminishing the water level from 4 cm. to 2 cm., the mean value of production increases in 10%.
The anomalous appearance of the regression line corresponding to the 3 cm. level might be explained if we have in mind that none of the values in this particular case correspond to the first period. Moreover among the days selected for this water level there are seldom anyone with a high solar radiation. We could confirm this by comparing the performance of the three different water levels for days with similar solar radiation characteristics.

c) In fig. 6 may be observed the influence of the surroundingstemperature on water production. The correspondind regression line is

$$Y = 0.7495\ X - 3.1571 \qquad r = 0.6151 \qquad (14)$$

where X is T_a (°C) and Y output ($Kg.m^{-2}$).
The correlation coefficient is very low if compared with the case of solar radiation. The great isolation of the sistem may explain this low correlation.
Anyway, if we compare water production for the days with the same surroundings temperature we get to the conclusion that the increment of one degree in the temperature implies aproximately an increment of 0.30 $Kg.m^{-2}$ in the daily water production.

d) Wind speed seems to have a negative effect on production. mevertheless the low correlation obtained means that no valid hypothesis may be deduced from the set of data used for this purpose.

Influence of different factors on the daily water production. Internal parameters.

a) In fig. 7 is shown the influence of the mean water temperature. The regresion line may be expressed as:

$$Y = 0.2390\ X - 4.7083 \qquad r = 0.8971 \qquad (15)$$

As it was to be expected, there is an important correlation between both variables. An increasing of 1°C in the water temperature means an increasing of about 0.24 $Kg.m^{-2}$ in water production.

b) If we refer to the mean temperature of the cover the equation obteined (see fig.8) is:

$$Y = 0.3323\ X - 4.9354 \qquad r = 0.7963 \qquad (16)$$

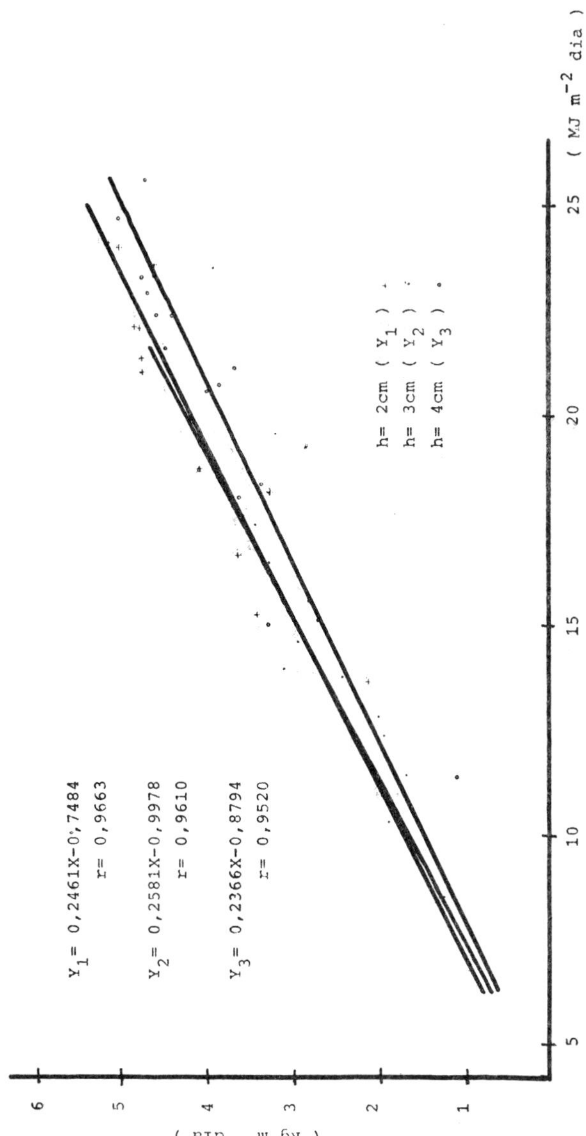

FIG.5.-INFLUENCE OF THE WATER LEVEL ON THE DAILY OUTPUT

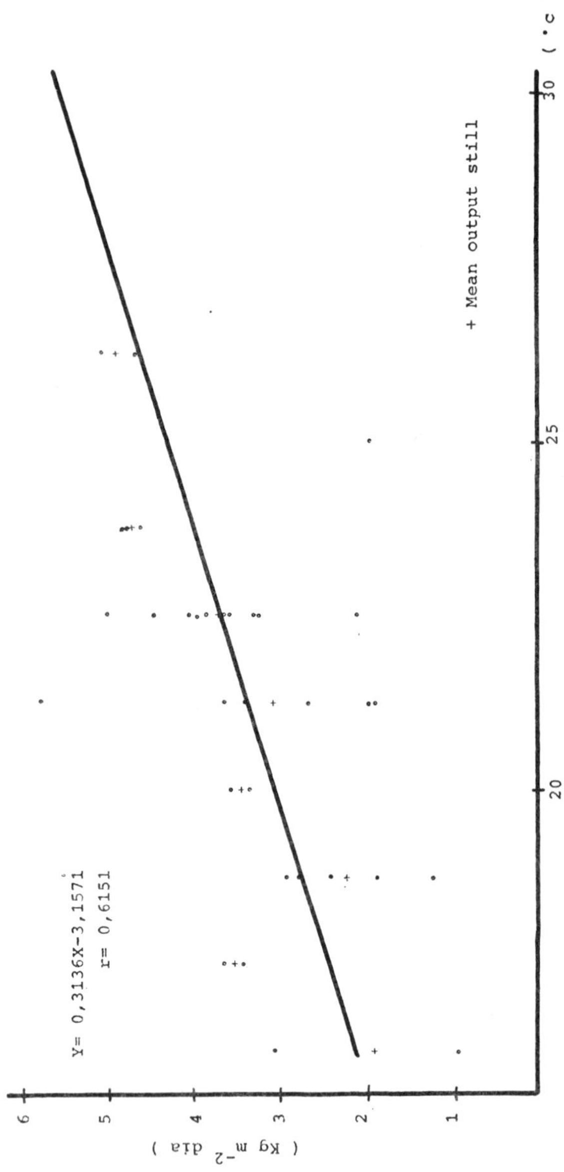

FIG.6.-RELATIONSHIP BETWEEN AMBIENT TEMPERATURE AND STILL OUTPUT

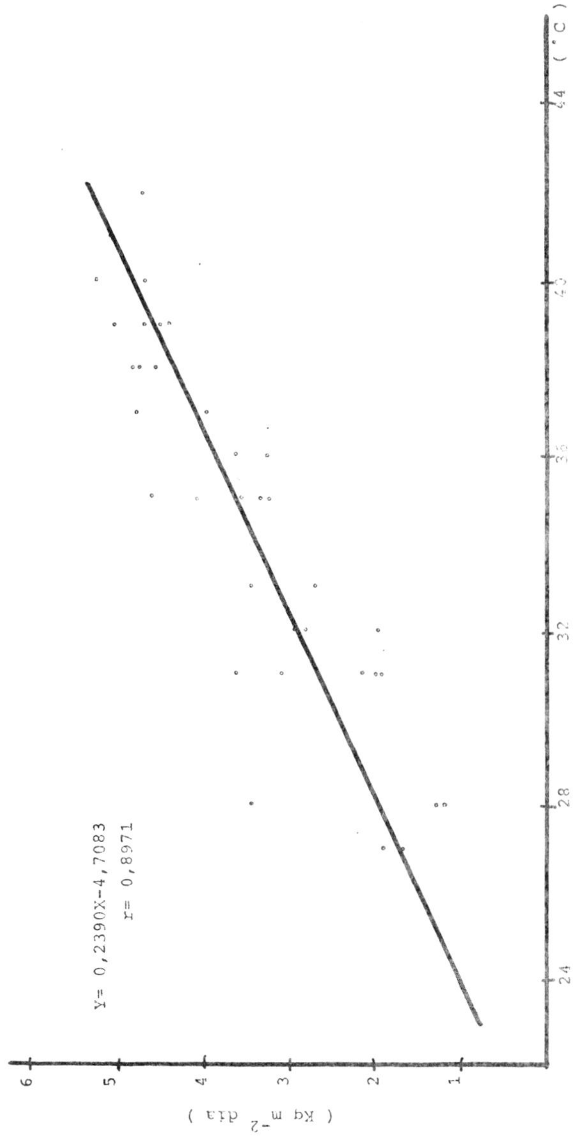

FIG.7.-RELATIONSHIP BETWEEN WATER TEMPERATURE AND STILL OUTPUT

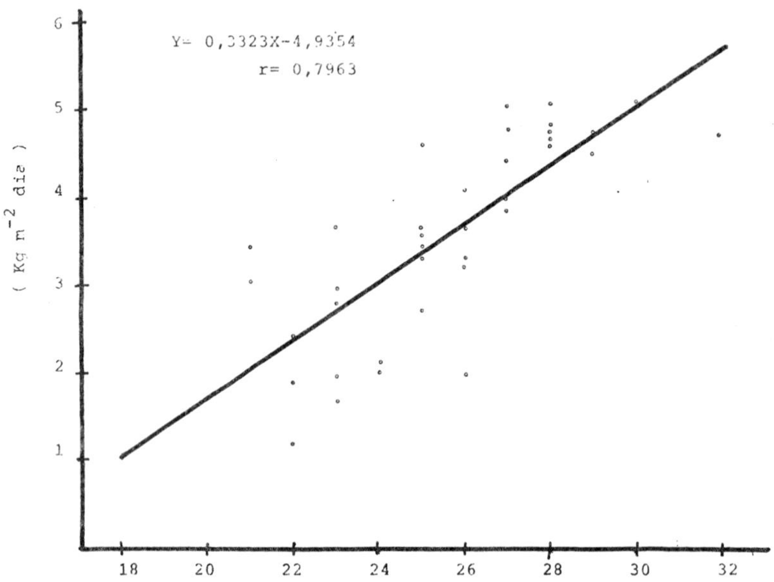

FIG.8.-RELATIONSHIP BETWEEN COVER TEMPERATURE AND STILL OUTPUT

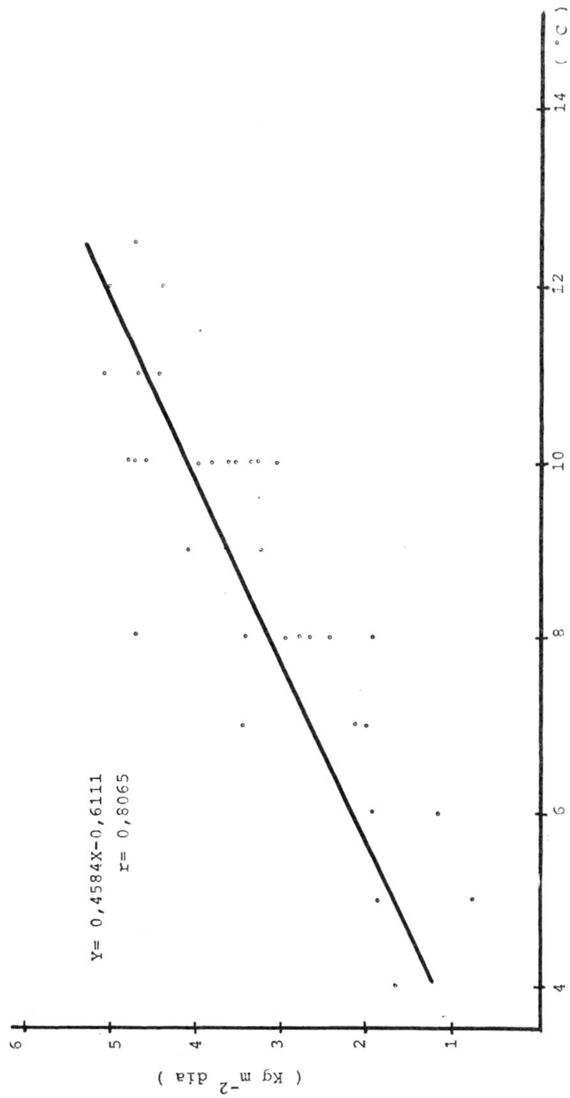

FIG.9.-RELATIONSHIP BETWEEN ΔT (T_w-T_g) AND STILL OUTPUT

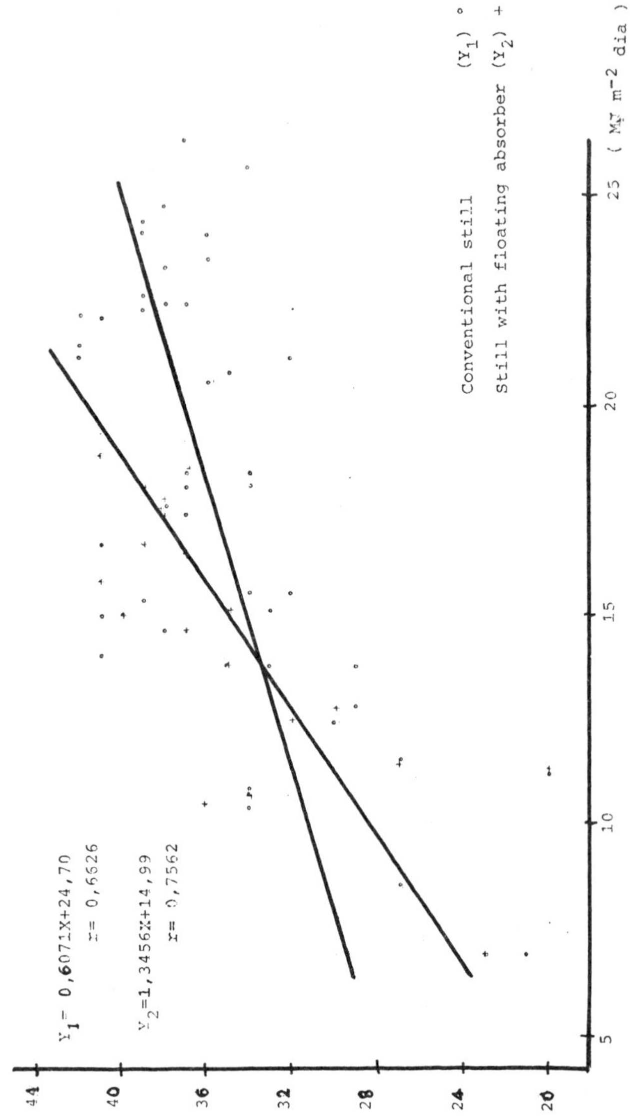

FIG.10.-RELATIONSHIP BETWEEN SOLAR RADIATION AND STILL EFFICIENCY

What means also an increasing of water production with cover temperature. It is not because the high cover temperature involves a high water production but because of the evident correlation between cover temperature and the parameters T_w, T_a and H_s.

c) The validity of this affirmation is shown by representing (fig. 9) the water prodxction as a function of the temperature differences ($T_w - T_g$), getting:

$$Y = 0.4584\ X - 0.6111 \qquad r = 0.8065 \qquad (17)$$

We may, then, decide that the water production can be increased more by means of increasing the temperature differences between cover and basin water than by increasing in the same amount the cover temperature.

Evaluation of the still efficiency

The still efficiency is defined as:

$$\eta = (Dh_w/H_s).100 \qquad (18)$$

being: D : the daily water production ($Kg.m^{-2}$)
H_w: the latent heat of vaporization ($MJ.Kg^{-1}$)
H_s: the daily solar radiation ($MJ.m^{-2}$)

Following the general use among the different papers, we refer the still efficiency to the glass surface, what in our case implies to multiplie the expression (18) by the factor (cover surface/water surface) = 0.7937.
Using as a value for h_w the one corresponding to the 73°C [6] we get the values that can be seen in TABLE 1.
In figure 10 is shown the relation between still efficiency and solar radiation, being the corresponding equation of the regression lines:

Conv. still: $Y = 0.6070\ X + 24.70 \qquad r = 0.6626 \qquad (19)$

L.T.I.still: $Y = 0.1346\ X + 14.99 \qquad r = 0.7562 \qquad (20)$

What means that the stills efficiency encreases with solar radiation. An increasing of 10 $J.m^{-2}$ in the daily solar radiation may produce an increasing of 6 to 13 units in the still efficiency.
If we compare the mean value of the efficiency for both stills we get to the conclusion that it is about 6% lower in the case of the conventional one.

General expression giving the water production as a function of the external parameters. Yearly production estimation in Valencia(Spain).

In order to get a better description of the still performance as a function of the external parameters, a multiple linear correlation has been deduced between these parameters and water production. The expresion obtained is:

$$Y = 0.2439\ X_1 - 0.0294\ X_2 - 0.0073\ X_3 - 0.1222 \quad (21)$$

where: Y : water production ($Kg.m^{-2}$)
X_1: daily solar radiation ($MJ.m^{-2}$)
X_2: mean surroundings temperature (°C)
X_3: mean wind speed ($Km.h^{-1}$)

The multiple correlation coefficient is r = 9792.
Through the equation(21) we are able to try an evaluation of the yearly water production in Valencia. With this purpose, we have evaluated the mean values of the parameters H_s, T_a and v_w as measures for the last three years. They are:

$H_s = 15.10\ MJ.m^{-2}$
$T_a = 17°C$
$v_w = 6.73\ Km.h^{-1}$

They give as mean daily value a production of 3.01 $Kg.m^{-2}$ and a yearly production of 1.1 $Tm.m^{-2}$.
The higher and lower values of water production are held in July and December, being 4.75 and 0.87 $Kg.m^{-2}$ respectively.

CONCLUSIONS

As a result from the experimental procedures we may deduce the following conclusions:
The low thermal inertia still has a good prospect having in mind that its production has been over 4% higher than the conventional one.
The effect of external parameters on production has been as follows:

a) Solar radiation is, without any doubt, the ono with the higher influence. The correlation in this case is over 96%

b) The sealing has proved to be very important, since with an improvement of sealing the production has also improved in as much as 15%.

c) Termal inertia has also a considerable effect. In fact, a reduction of the water level from 4 to 2 cm. meant an improvement of the water production of about 10%, which can evenbe higher in days with a high solar radiation.

d) Production also increases with surroundings temperature although the sealinglessenes its influence and has proved to be more important.

e) Wind speed has not a clear influence on production and the low correlation obtained in this case means that more experimentation is needed before making any conclusions.

The effect of internal parameters may be resumed:

a) A very important relation exists between production and water temperature.

b) Production also increasing with increasing cover temperature and with increasing difference between cover and water temperature. Knowing that the effectof the water temperature is much more important, we may conclude that for the same value of this magnitude, an increasing of the cover temperature (usually lower than the water one) means a disminution of the production

The relation that best describe production (Y) as a function of solar radiation (X_1), surroundings temperature (X_2), and wind speed (X_3) for the experimental period is:

$$Y = 0.2440\ X_1 - 0.0294\ X_2 - 0.0073\ X_3 - 0.1222$$

with a multiple correlation coefficient of 0.9792.

Through this relation and considering the climatic conditions of Valencia, as characterized by the anual mean values of H_s, T_a, and V_w, we get for this particular place the yearly production, being 1.1Tm m^{-2}, equivalent to a daily production of 3.01Kg m^{-2}.

REFERENCES

2.-R.V.Dunkle:"Solar water distillation:the roof-type still and in multiple-effect diffusion still".International developements in heat transfer,p895-912.The American Society of Mechanical.Engineers (1961).

1.-M.Bahodori,F.E.Edlin:"Improvement of solar stills by the surface treatment of the glass".Solar Energy 14,339 (1973).

3.-P.I.Cooper:"Digital simulation of transient solar still processes".Solar Energy 12,313 (1969).

4.-R.N.Morse,W.R.W.Read:"A rational basis for the engineering development of a solar still".Solar Energy 12,5 (1968).

5.-W.Szulmayer:"Solar stills with low thermal inertia".Solar Energy 14,415 (1973)

6.-P.I.Cooper:"The maximun efficiency of single-effect solar stills".Solar Energy 15,205 (1973).

NOMENCLATURE

c_{wg}	especific heat of basin water
h_{ga}	convective heat transfer coefficient from glass
h_w	latent heat of evaporation
P_g	partial pressure of water vapour at T_g
P_w	partial pressure of water vapour at T_w
q_c	convective heat transfer from water
q_{ca}	convective heat transfer from cover
q_e	evaporative heat transfer from water
q_r	radiative heat transfer from water
q_{ra}	radiative heat transfer from cover
t	transmittance of glass for solar radiation
α_w	absorptance of water for solar radiation
α_g	absorptance of cover for solar radiation

SOLAR WATER SUPPLY AND PURIFICATION SYSTEM FOR REMOTE AREAS AND VILLAGES

U. R. Nejib and H. C. Hallez
Wilkes College
Wilkes-Barre, Pennsylvania U.S.A.
STR Corporation
P.O. Box 1442, Kingston, Pennsylvania U.S.A.

ABSTRACT

Substantial betterment in the quality of life can be realized in remote villages by the availability of a potable water supply system of modest capability. The main obstacle is the lack of adequate power because of remoteness or other feasibility reasons. Solar cells are available power sources for such areas, in spite of their relatively high expense.

The Solar Water Supply and Purification System (SWPS) proposed is simple in design, of very low maintenance, and is economically feasible. Cost will vary depending on water quality and location. The system utilizes solar power packs with a short duty cycle. To further reduce the electric load, gravity flow techniques and hydraulic valves are incorporated.

A moderate SWPS of 10,000 gallons weekly capacity will ensure a supply of 7 gallons/person/day to a village of 200, or any other combination. This quantity is more than sufficient for drinking and cooking. For a TDH of 50 feet, the capital cost is about $30,000 at moderate insolation. Automatic operation and modular construction minimize labor and maintenance needs. These are vital factors which effect successful operation of any project in remote areas.

INTRODUCTION

The problem of adequate potable water supply for domestic use is perhaps as ancient as the dawn of civilization. In an arid climate it can determine to a great extent the general well-being of a small community. Modern technology has in the past offered some partial solutions in the form of hydromechanical systems which can transport and purify sufficient quantities of water to help alleviate the potable water supply problem. However, these systems normally need fuel and lubricant to operate an engine driven pump and are consequently characterized by high maintenance. As a resuly, their use has been limited to sizeable communities where these inconveniences can be justified.

FIGURE 2. A RECOMMENDED MULTIPLE USE FOR THE S W P S. NOTE REMOTE LOCATION ON RIVER.

Steady progress over the past few years with photovoltaic converters has opened another approach to the supply of potable water in reasonable quantities for small communities. It is both possible technically and economically to use the SWPS as a viable source to deliver necessary water for villages and areas where electric power is not available. The system may be dedicated to a single village or to serve many small communities from a single source (see Figure 1 and 2).

Briefly, this approach involves the use of a moderate number of photovoltaic converters to charge a battery bank continuously, which in turn supplies energy to periodically drive a pump. The pump output is stored in an elevated tank of sufficient size to allow a daily ration of water to be drawn for domestic purposes. When the tank is depleted beyond a preset level and the batteries are sufficiently charged the pump will be allowed to cycle again (see Figure 3).

A well designed system will of course be able to purify the water before consumption. In the worst case, this will involve the removal of solids and disinfection of pathogenic bacteria. This case involves the greatest expense of the various possibilities.

More favorable circumstances can materially reduce the cost of a system. For example, the system may be installed in an area with solids-free water, but where the need for disinfection still exists. In this case, fewer solar panels will be required to supply the needed energy, with a consequent savings in cost. A carefully designed system incorporating these ideas can supply potable water at reasonable cost and with excellent reliability to a small community. For example, a 14,000 gallon/week system will cost about $55,000 at moderate insolation and a TDH of 100 feet, using a poor quality water source. Reflected over the life of the system (10 - 12 years) the cost per gallon is about.85¢.

The analysis shows that the cost of treated water decreases as the solar insolation increases. However, the surprising result is that higher horsepower systems supplying a fixed quantity of water offer a more economical solution. A second result of this study is the confirmation that intermittent operation of a large pump is far more economical than continuous operation of a small pump when large capacities are considered. The data indicates a decrease in unit costs as system capacity, water quality and insolation increase and as total dynamic head decreases.

All of these factors make the SWPS a possible solution to the adequate potable water supply shortages in remote areas and villages.

BASIC REQUIREMENTS

In general, water contaminants can be classified into three distinct groups; non-soluble solids, pathogenic organisms, and chemical impurities. The production of potable water involves the elimination of one or more of these contaminants. By far the most difficult to treat are the chemical impurities. Fortunately the absence of industry in rural settings virtually precludes chemical contaminants from local water sources. The task then becomes the elimination of solids and pathogenic material to yield potable water. The process can obviously be divided into two steps; first, the elimination of solids and second, the removal of pathogens.

Filtration is one accepted method of achieving the first goal. Needless to say, the type of filtration used will depend on the type of solids to be removed. Such items as particle size and concentration, density, compacting or not, biodegradable or not, all need to be considered to determine filter bed media. Thus anything from pebble/sand combinations, diatematious earth, or porous membranes can be a candidate for various applications; the final selection being dictated by the source. The mechanical construction of the device can remain relatively uniform. Size of the separation unit is likewise a function of pollutant concentration, capacity or flow, and desired maintenance interval. A judicious choice, however, can be made to handle a wide variety of possible cases and afford substantial standardization.

The presence of pathogenic material presents a different problem. For very low flow rates these undesirable bacteria can be mechanically removed by a microfilter. High flow rates, however, are most commenly treated by chlorination of the supply. This technique is virtually universal and is employed in individual residences with less than perfect water supplies to large municipal water works supplying millions of gallons of water daily.

Essentially, chlorine in the form of sodium hypochloride or similar compound is added to the water and allowed to remain in solution for a specified minimum amount of time during which all pathogenic bacteria are destroyed. This is followed by a secondary treatment to remove the chlorine and any other odorous organic material. The resultant treated water is entirely safe to drink and in a properly adjusted system quite palatable and fresh. Other methods of purification are available including ozone treatment and UV irradiation. However, their technology is not as widely known or accepted as the more traditional chlorination.

LIMITATIONS AND CONSTRAINTS

Remote areas are characterized by the absence of readily available electric power, industry, maintenance supplies and facilities, and often poor or nonexistent surface transportation. Yet these areas support small communities and villages at various levels of prosperity. Human biological needs demand the location of small communities near water sources of adequate capacity. The quality of the source, however, can range from crystal clear mountain spring water to the highly contaminated solids-laden variety, barely fit for human consumption even after boiling. Presumably most supplies fall somewhere between these extremes. A proposed potable water system for these areas should be able to cope with considerable dirt pollution, as well as, the presence of pathogenic bacteria; still producing an effluent safe for human consumption.

In the past, technology has offered some partial solutions in the form of hydro-mechanical systems which can transport and purify adequate quantities of water to help appease the potable water supply problem. Normally, engine driven pumps with appropriate filtering devices have been the basis of these solutions. But these systems need fuel and lubricant for operation and are consequently characterized by high maintenance requirements. As a result, their use has been limited to sizeable communities where these demands can be met. The absence of maintenance facilities and supplies mandates the design of a reliable system of the greatest allowable simplicity. The system should be automatic and simple so as to minimize human interferance and misinterpretation of operating status.

The total cost of a given system will in large measure be determined by the quality of water the system is expected to supply. A reasonable assessment of the needs of the beneficiaries must be made along with a determination of the ultimate benefit a potable water system might have on a remote community. Certainly the per capita water consumption of remote villages is far less than that of urban areas. The absence of water-consuming convenience appliances alone will account for a large portion of the disparity, in addition the fact that most domestic water must be transported manually dictates frugality with respect to that resource.

From a practical viewpoint, modular construction to facilitate erection and minimize complications of installation is virtually a must. The sub-assemblies so designed, however, should be small enough for easy transportation and handling. To minimize cost and contribute to reliability, all major components should be commercially available items.

SWPS DESIGN CRITERION

One of the peculiarities of municipal water systems, large or small, is their intermittent use of pumps to fill elevated water tanks. The high location of these tanks in turn provides a relatively uniform hydraulic pressure at the many discharge points they serve, without relying on the continuous operation of pumps. Conceptually these systems are prime candidates for operation by an intermittent power source.

The commercially available photovoltaic converters can be incorporated as the power source for a small municipal water works. Briefly, a feasable system could be assembled utilizing a nominal number of solar panels to charge a large battery bank over an extended period. The energy so stored will then be released to drive a large capacity pump for limited time, the output of which will be stored in a large elevated reservoir. Consumers can then draw water from this tank over a period of time coincident with the time needed to recharge the batteries.

The conceptual deliberations of the previous section establishes hardware guidelines for the real system. These guidelines are quite loose and admit a very broad range of designs. However, the economic optimization of a technically feasable system requires the meticulous balancing of all operating variables and their effect on cost. These variables include pump size, pump power consumption, pump operating head and output, battery size, insolation, solar panel specifications, filter element characteristics, water source quality and location, treated water discharge point, capacity etc.

Capacity is normally dictated by community population. To provide this supply at the most economical cost involves a careful selection of components. The large number of interacting variables necessitates the use of computer methods to help maximize the economic effectiveness of the possible systems. Results of this process are included in the data presented.

FIGURE 3. POWER SYSTEM OUTLINE

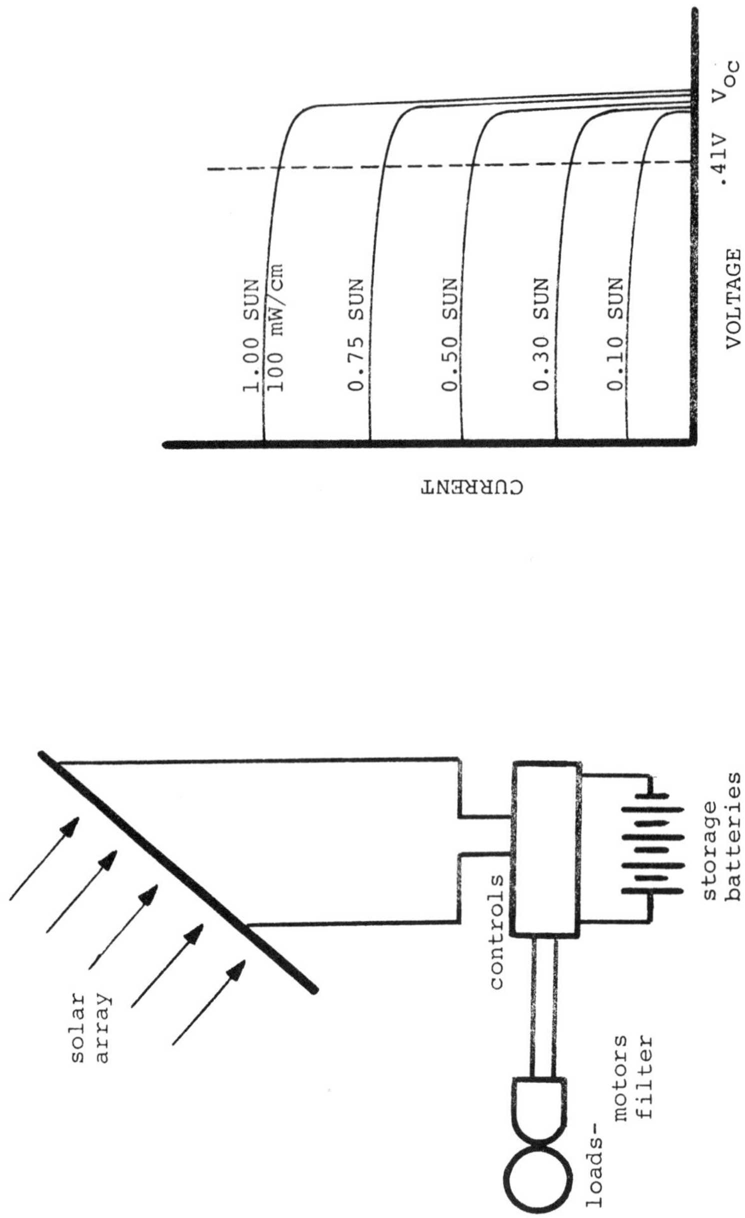

SYSTEM OUTLINE

SOLAR CELL CHARACTERISTICS

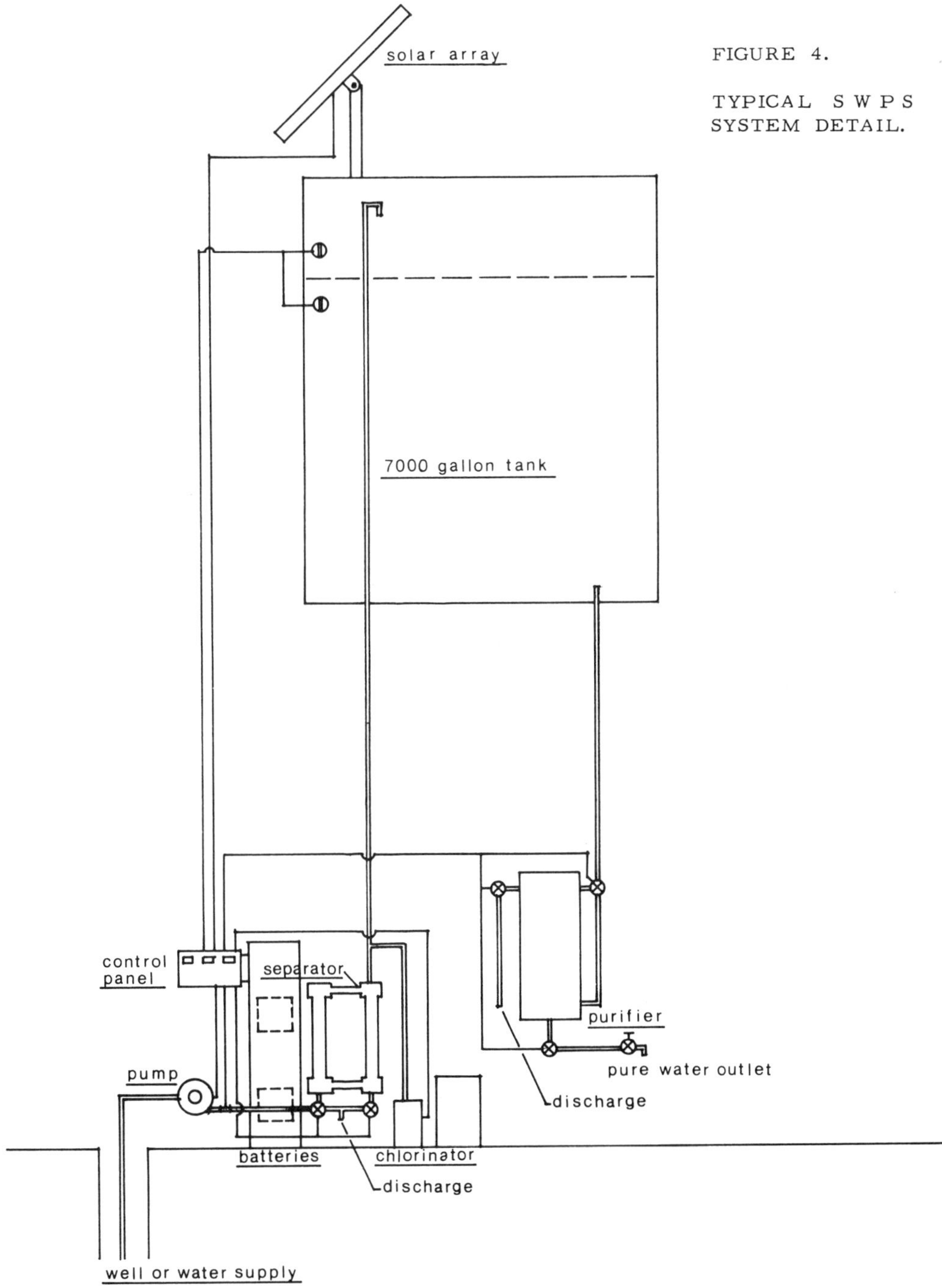

FIGURE 4.

TYPICAL SWPS SYSTEM DETAIL.

FIGURE 5. PUMP CHARACTERISTICS

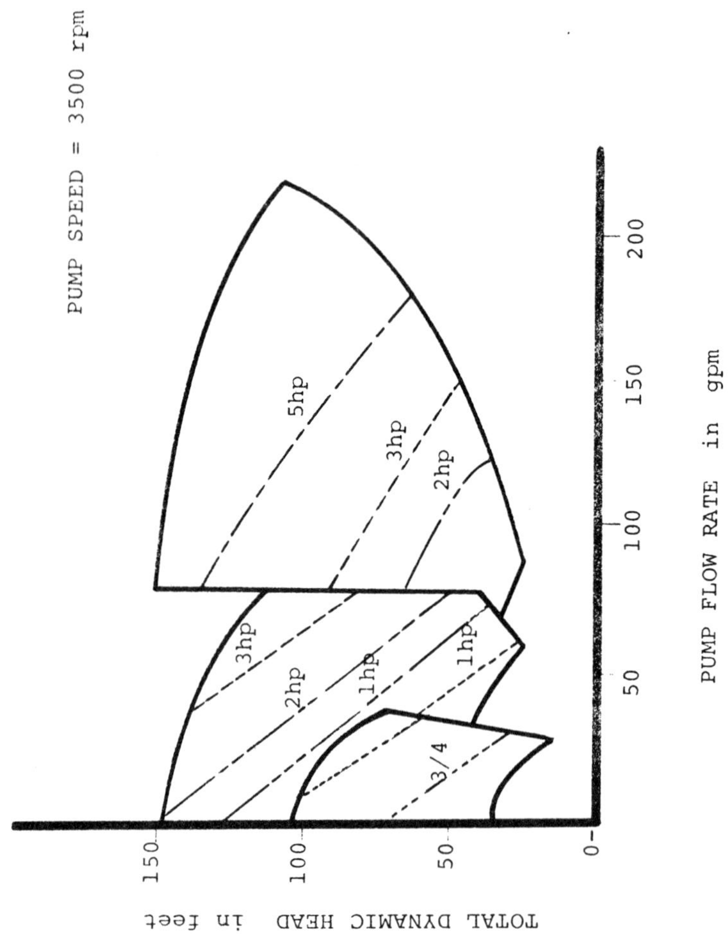

FIGURE 6. FILTERATION SYSTEM CHARACTERISTICS

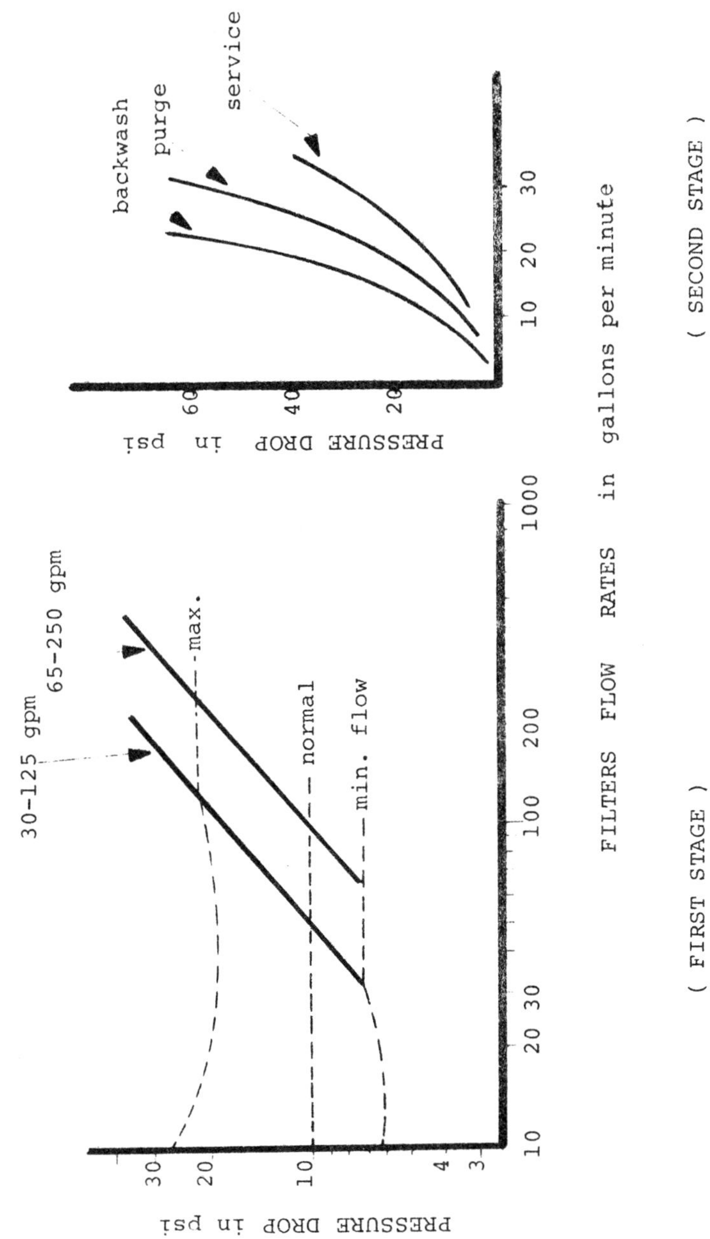

SWPS SYSTEM DESCRIPTION

The system developed incorporates large storage capacity in the form of lead-acid batteries and a minimum number of solar panels to continuously charge the batteries. A water tank large enough to hold a seven day supply is filled by a large pump operating intermittently (see Figures 3 & 4). Its operating cycle is determined by the water level in the tank and the charge level of the batteries. Normally the pump operates within 7 days, however, the controls are such that more frequent operation is automatically provided if the solar insolation has been sufficient, conversely in periods of prolonged foul weather capacity will be reduced. In this sense the system is self-limiting. Fortunately this is an inherently favorable characteristic in that it will provide more water in the summer when the demand is greater and less in the winter when demand is reduced.

A 3500 rpm centrifugal pump, driven by a dc motor is used. Pump sizes and flow rates are given in pump characteristics shown in Figure 5 (see Figure 5). Note design changes for a given horsepower pump. The system uses solar panels capable of delivering approximately 20 watts peak to the storage batteries. A redundant pump configuration is adopted to ensure utmost reliability and service. The control system will ensure alternating operation among the two pumps.

Filtration requisites of the system are entirely stimulated by water source contamination levels. If total treatment is necessary, then a three stage process is mandated. The first stage involves mechanical filtration to remove suspended solids from source water before this water is stored in the reservoir. Typically the filtration device removes particles of sand, clay, earth, gravel, lava, shale etc. to leave the water relatively clear (see Figure 6). The pathogenic bacteria admit a reasonable simple solution through chlorination. A mild solution of sodium hypochloride is injected into the holding tank filler pipe whenever the tank is being filled. The amount is metered by a small pump and adjusted for local conditions. Temperature and water source contamination all have a bearing on this variable. The adjustment procedure is not difficult and is accomplished with the aid of commercially available testing kits which are easy to use.

The size of the storage tank ensures an adequate retention time for the chlorine to destroy all pathogenic matter. Before the water is discharged for consumption it passes through a final stage purifier which eliminates residual chlorine taste, odor and chlorinated hydrocarbons. Dissolved minerals, however, are unaffected by the purifier and will remain dissolved leaving the water pleasant tasting and fresh.

The purifier is sized so that a single charge of the active material will treat 3.7 million gallons of chlorinated water. At anticipated consumption rates this becomes over a ten year supply.

The efficient and proper operation of the system is governed by fully automatic control systems. Since reliability and longevity are of prime importance, the system is designed to achieve all necessary monitoring and control functions with utmost simplicity using solid state circuitry. The basic inputs are the monitoring of battery charge level, water tank level, pump output, filter status, and chlorinator performance. This information is processed and distributed to the various components. For example, if the water level is low enough, and the battery charge is high enough, the pump will be allowed to cycle. Note that the water tank need not be completely empty for the pump to cycle. In addition, the occurance of certain malfunctions and system conditions will be signaled on an appropriate panel to help prevent damage to the system, facilitate proper maintenance and minimize down time.

The system employs modular construction which implies that erection entails a minimum of technical skill as far as electronics or hydraulics are concerned. Still a solid foundation for the system will be necessary to support the weight of a full water tank, not a particularly difficult problem, but something that cannot be ignored. Other major components are all preassembled as a unit, requiring only simple connection of pipes and wires. Thus installation should be quick and simple.

SWPS ANALYSIS AND EVALUATION

The system analysis is considered over a rather wide range. Data accumulated deals with practically all aspects of system operation and costing, including comparisons. The cost analysis conducted does not include system salvage value which may well be quite ample, or transportation and labor costs. Although the latter is minimized due to the design of the system. It must be noted that the analysis reflects early 1977 component prices.

The operating range presented here covers capacities from 7,000 - 20,000 gallons/week, dynamic heads from 20 - 100 feet, and pump sizes from 0.75 - 3.00 HP.

The detailed analysis shows that for the given capacities, the most effective pump sizes are the 2 HP and the 3 HP pumps. This can be concluded by examining Figures 7 and 8, respectively (see Figures 7 & 8).

A typical SWPS operation chart using a 2 HP pump is outlined in Figure 9 (see Figure 9). These charts give complete costing and sizing data for a given system. For example, an 8,000 gallon tank system (approximately 30,000 liter/week) will require a capital cost of $43,000. Projected over a 12 years service (twice the battery life expectancy), the cost per gallon will be about.83¢. The necessary current load to be supplied by the solar array will be then about 230 Amp. hours/week @36-volts. This data is for a dynamic head of 100 feet and a sunshine of 25 hours/week.

Unit cost comparisons for the 2 and 3 HP systems, operating at 90 and 100 feet heads are given in Figures 10 - 13 (see Figures 10 - 13). The running variable in these graphs is insolation which varies 20 - 35 hours/week of sunshine. The data discloses that the systems are very cost sensitive to operating head and solar insolation. The data also indicates that the unit cost decreases as the system capacity increases. This might imply that this decrease will continue for larger capacities. However, this is not the case. If capital is not a limitation, then the cost-effective system will be in the region where the rate of change of unit cost decrease per capital exhibits maximum drop. This means that for extra expenditures of capital minimum cost reduction per unit is realized. Figures 14 - 17 illustrate these trends(see Figures 14 - 17). It is interesting to notice that the 2 HP and 3 HP systems at 90 foot head and insolation of 20 are very close in cost effectiveness.

The impact of dynamic head variation on system performance and economics is illustrated in Figures 18 and 19 (see Figures 18 & 19). Note the pump change effect of Figure 18. For the same amount of energy, a different pump design was used for high dynamic heads. This design change ensured a more efficient 3 HP system operation. On the other hand, no such change was made in the 2 HP system. Figure 19 exhibits the extremely low cost for the 2 HP system operating at low heads (see Figure 19). It will cost about·17¢ per gallon for a 2 HP system operating at a 20,000 gallon capacity with 20 hours/week insolation.

This cost reduction could be attained not only through dynamic head reduction, but also, if the source water is of good quality. This is shown in Figure 20 (see Figure 20). Approximately a unit cost reduction of nearly 100% is achieved if the source was relatively clean.

It is important to point out the advantages that this system has in terms of salvage value. It is very much possible that if the area or the village is ultimately electrified then the SWPS is well suited to form the basis of a municipal water works. With the substitution of motors and connection to power lines, the capacity of the system will increase by fourfold and is limited only by the filtration apparatus.

CONCLUSIONS

This paper attempted to present in detail the possibility of utilizing solar energy for a water purification system. The study demonstrated that the system is cost effective and simple to install and maintain. What is important about the SWPS is its impact on remote areas and villages all over the world.

The incorporation of such a system by a community would undoubtedly increase the economic efficiency by diverting time and efforts previously spent combating various problems due to the absence of a potable water source. The general state of health of the community alone is sufficient reason to warrant serious consideration of small potable water plants for remote community use.

To quantify the economic effects of a pure water supply system is difficult and certainly goes well beyond the primary benefit of increasing productive time of the inhabitants. It extends to improvement in the education of the inhabitants, their mental and physical welfare and the development of new and effective attitudes to solve many problems which existed or may arise in the future.

ACKNOWLEDGMENT

The authors would like to acknowledge the cooperation and support extended by the STR Corporation of Kingston in the development of this work.

FIGURE 7. SWPS LOAD OPERATION

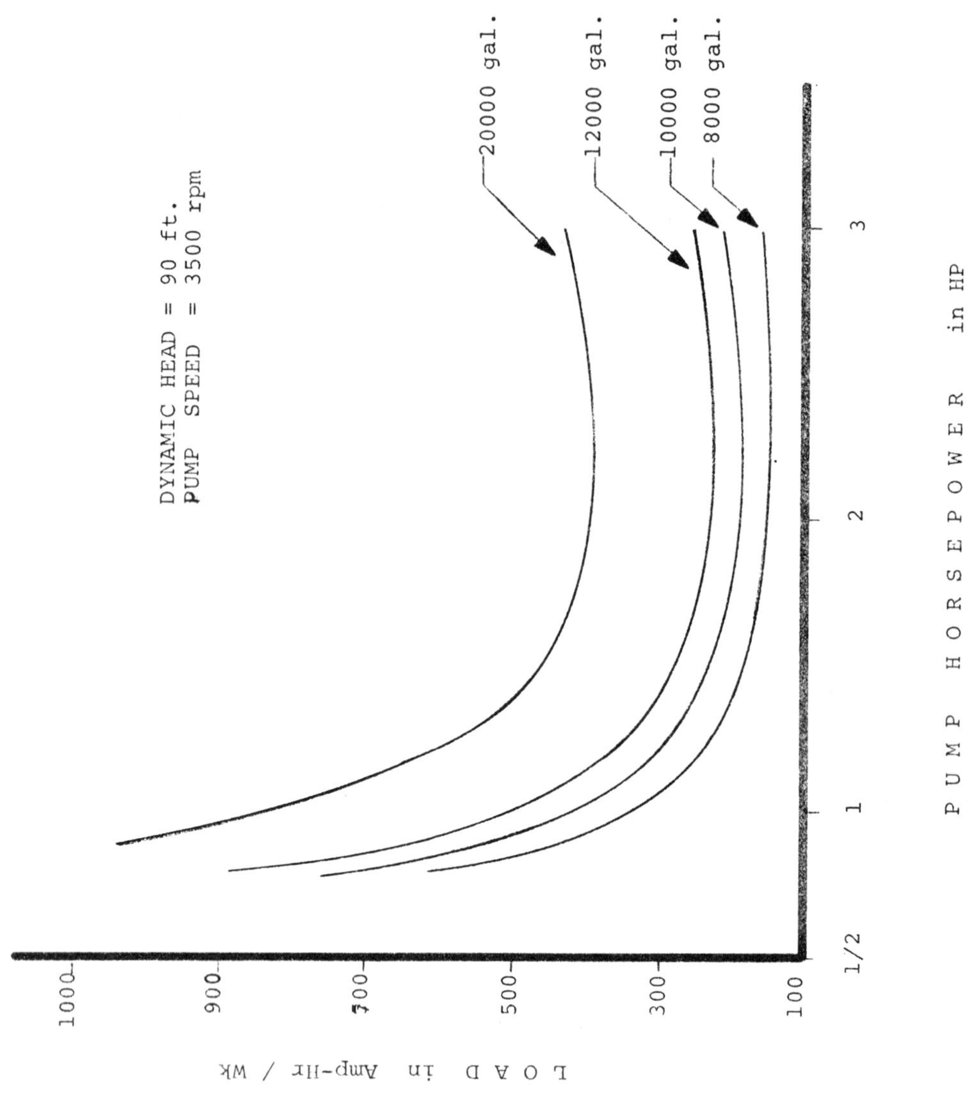

FIGURE 8. SWPS RUNNING-TIME OPERATION

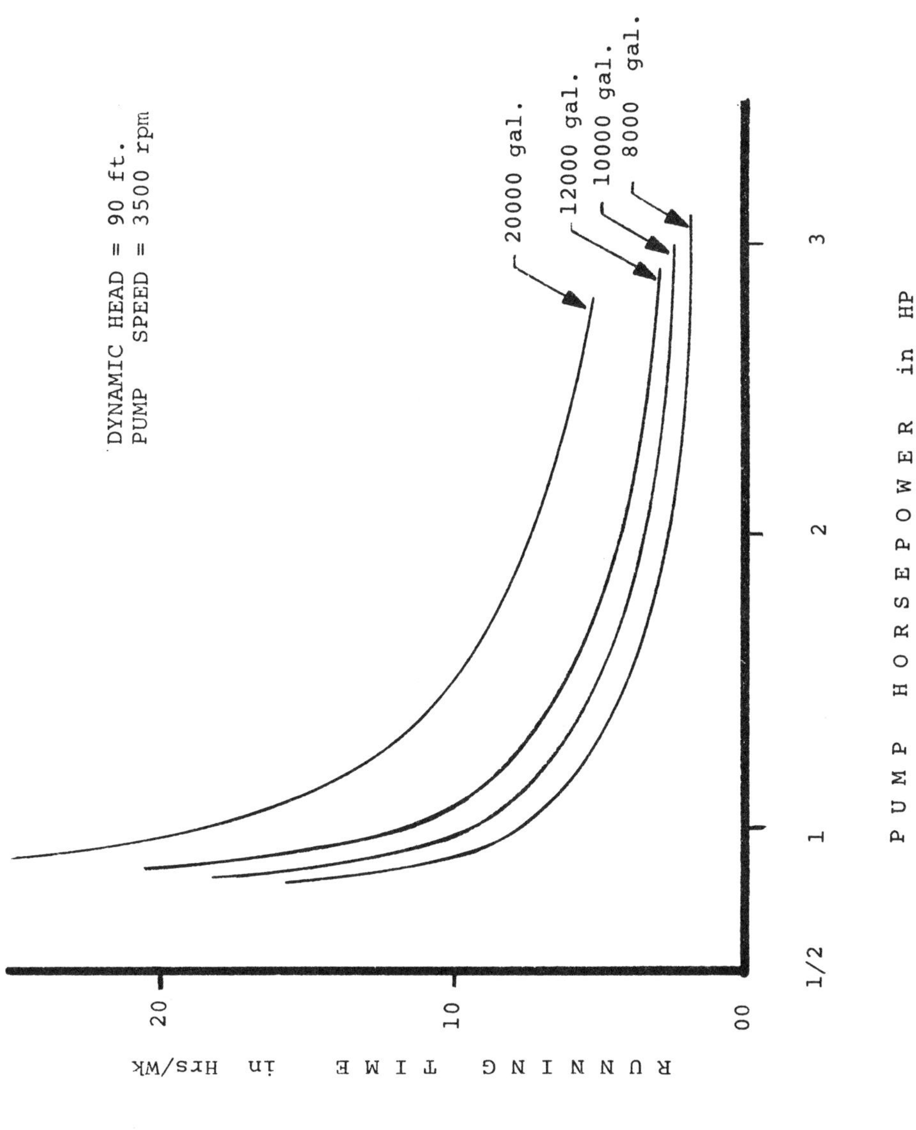

FIGURE 9. 2-HP SWPS OPERATION CHART

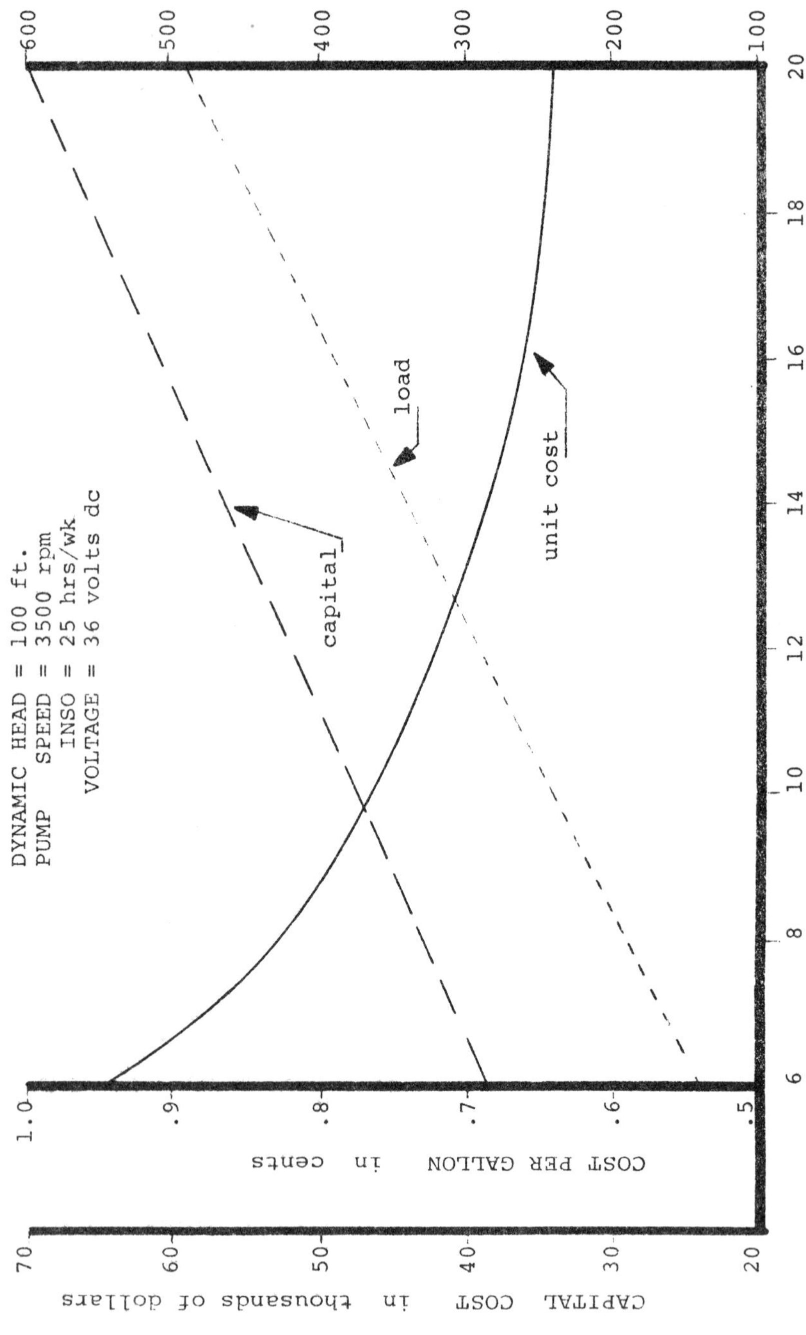

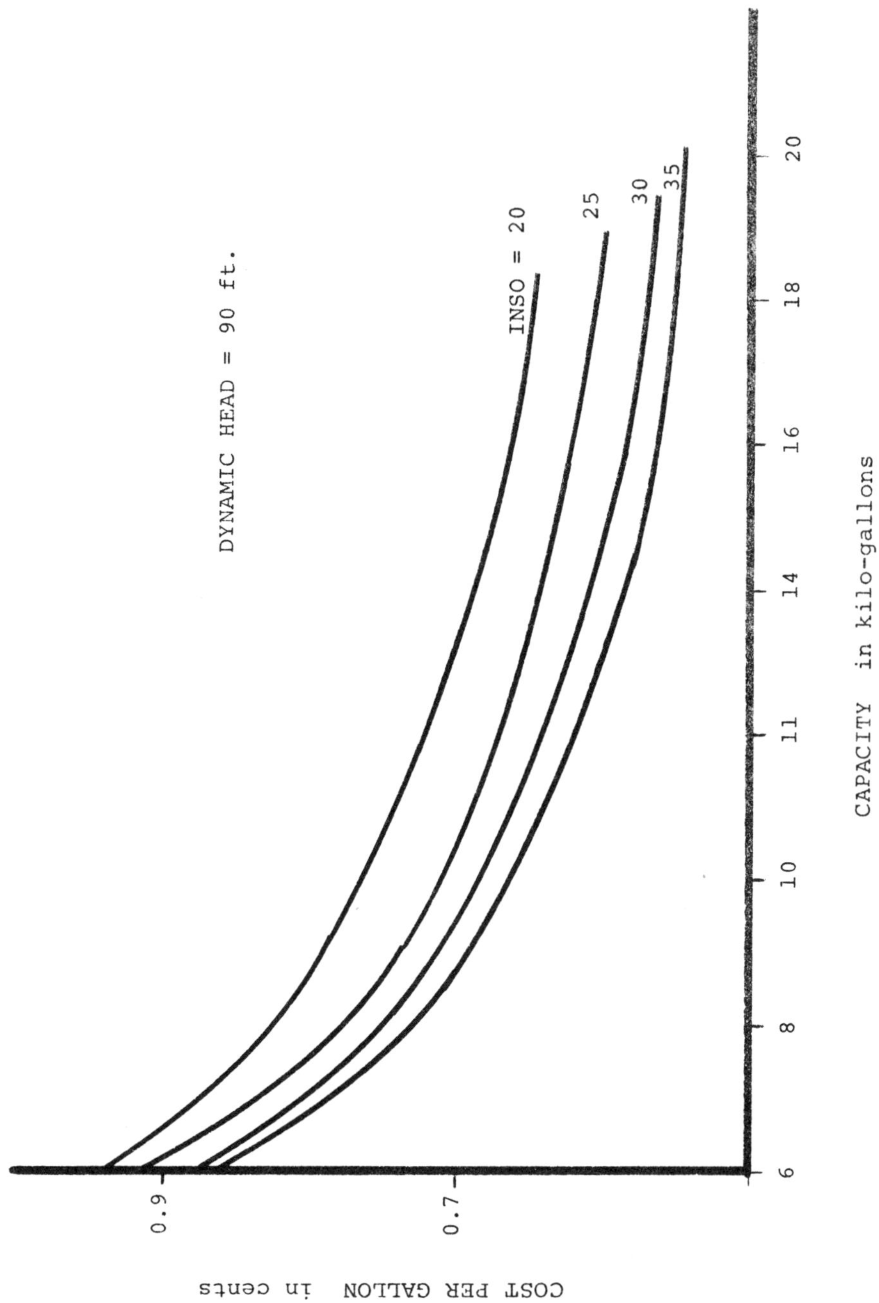

FIGURE 10. COST PER GALLON OF A 2-HP S W P S

FIGURE 11. COST PER GALLON OF A 3-HP S W P S

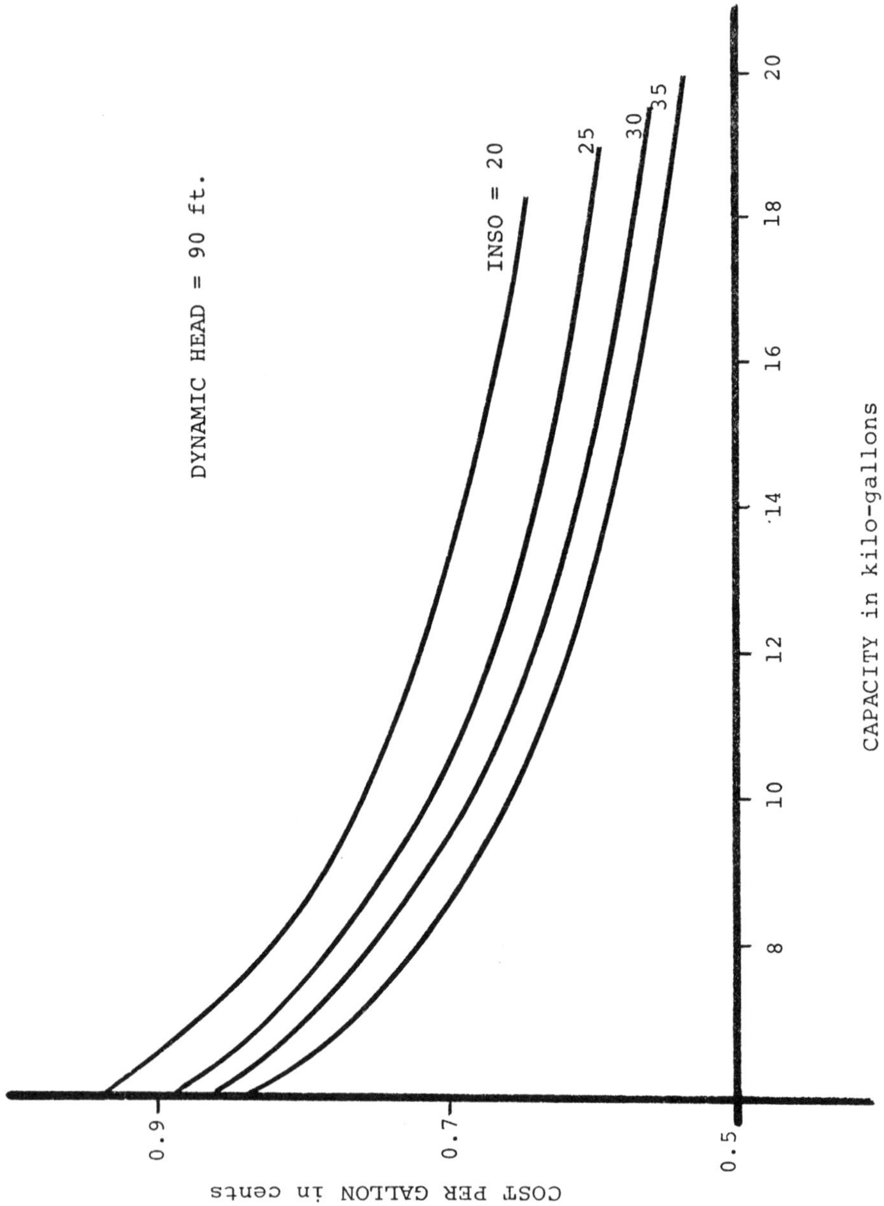

FIGURE 12. COST PER GALLON OF A 2-HP SWPS

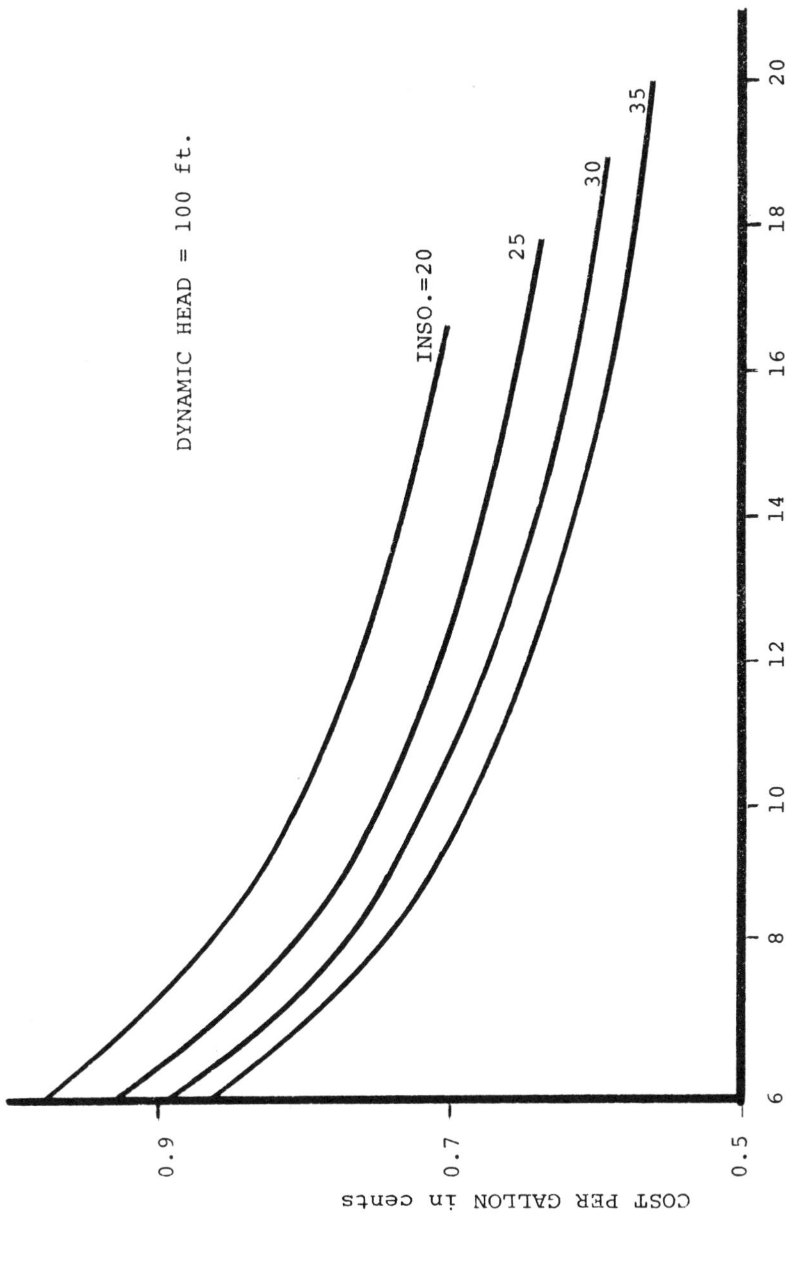

FIGURE 13. COST PER GALLON OF A 3-HP SWPS

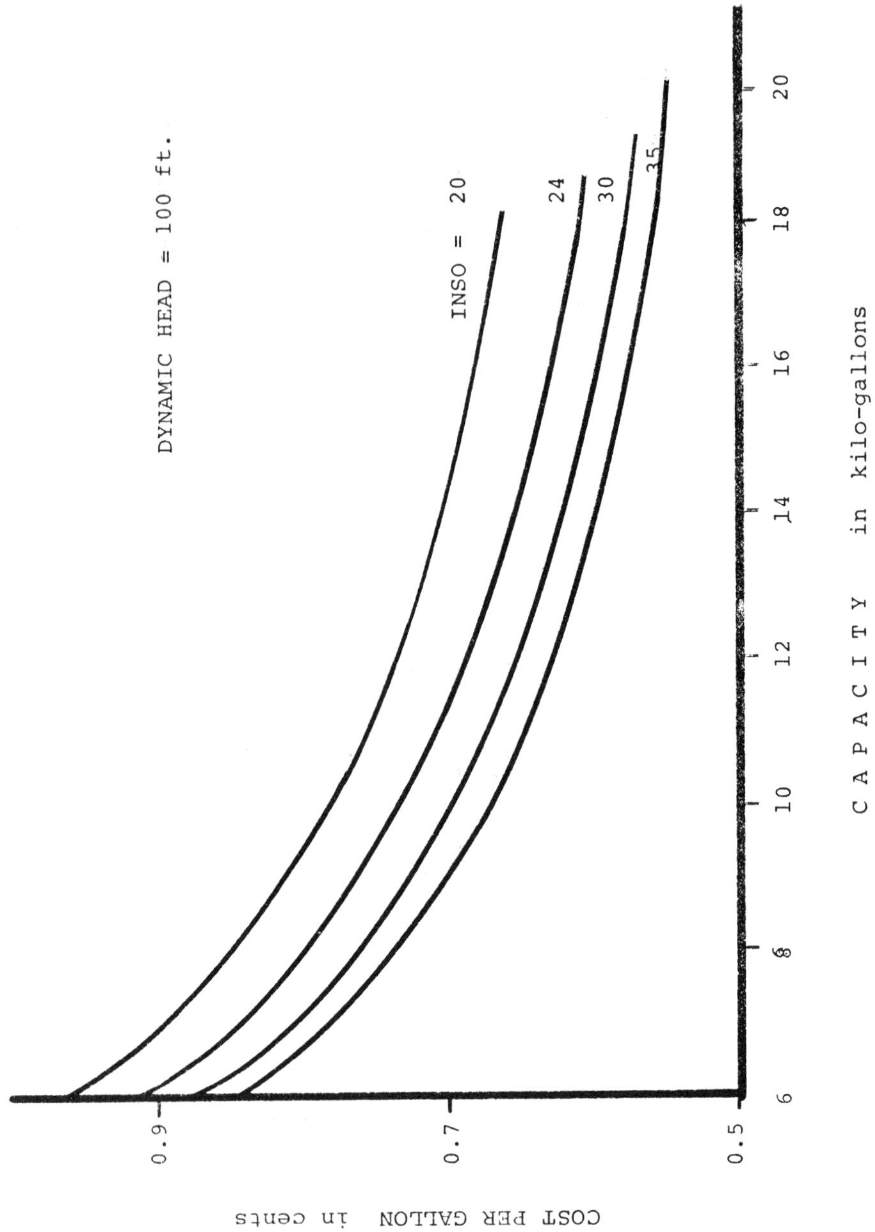

FIGURE 14. 2-HP S W P S COSTING OPTIMIZATION CHART

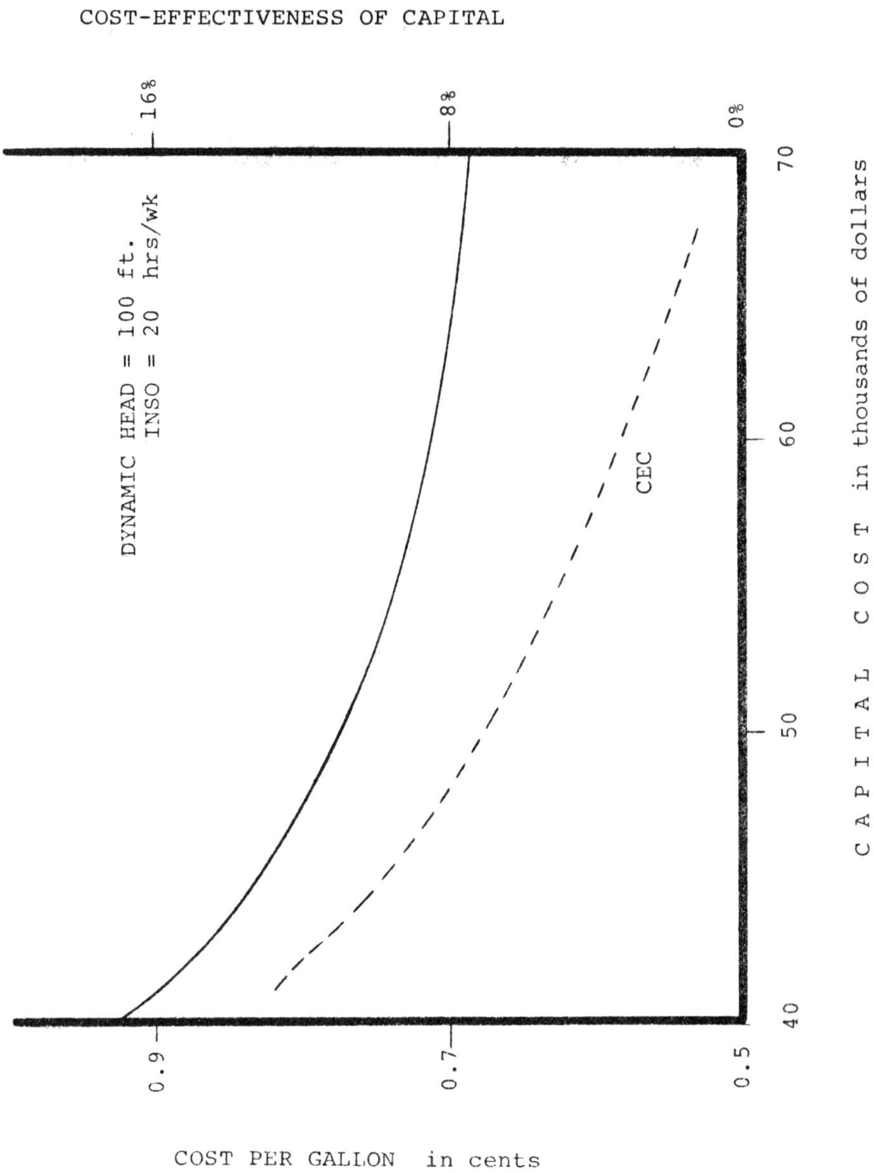

FIGURE 15. 2 & 3-HP S W P S COSTING OPTIMIZATION CHART

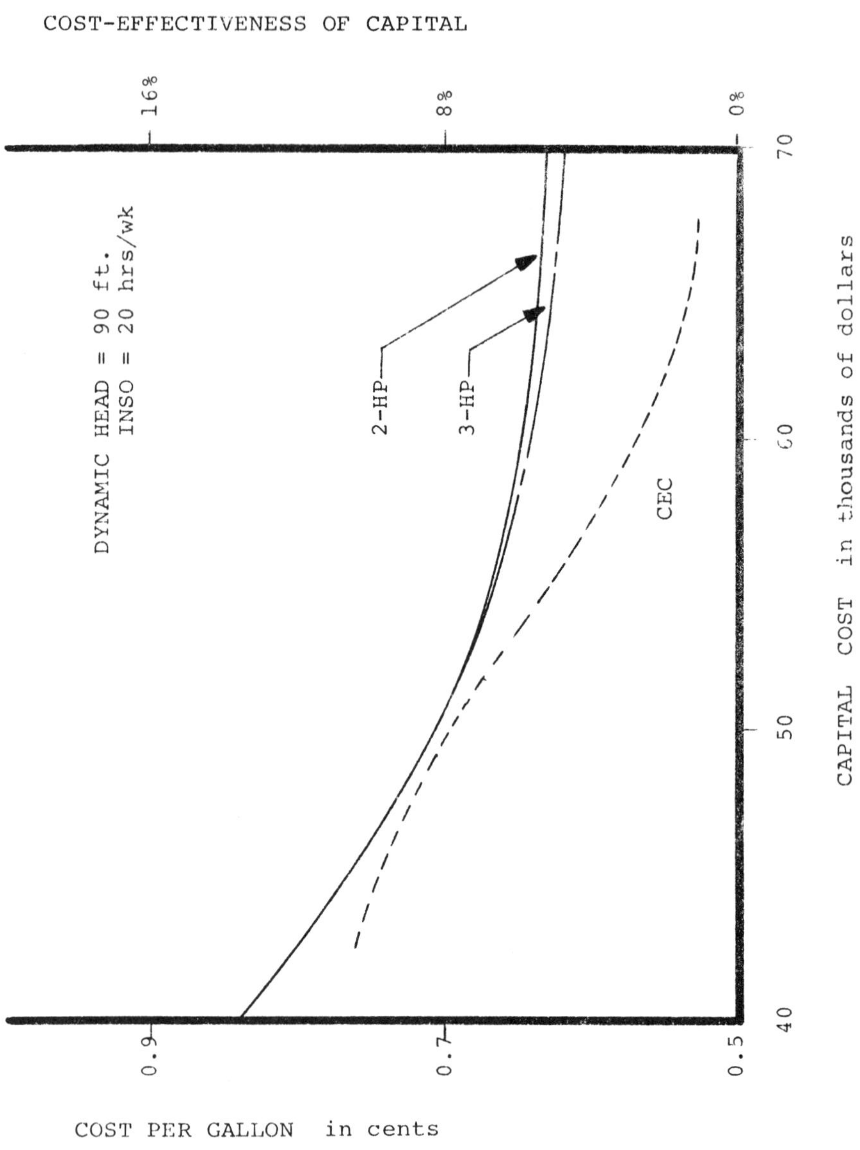

FIGURE 16. 2 & 3-HP SWPS COST COMPARISON

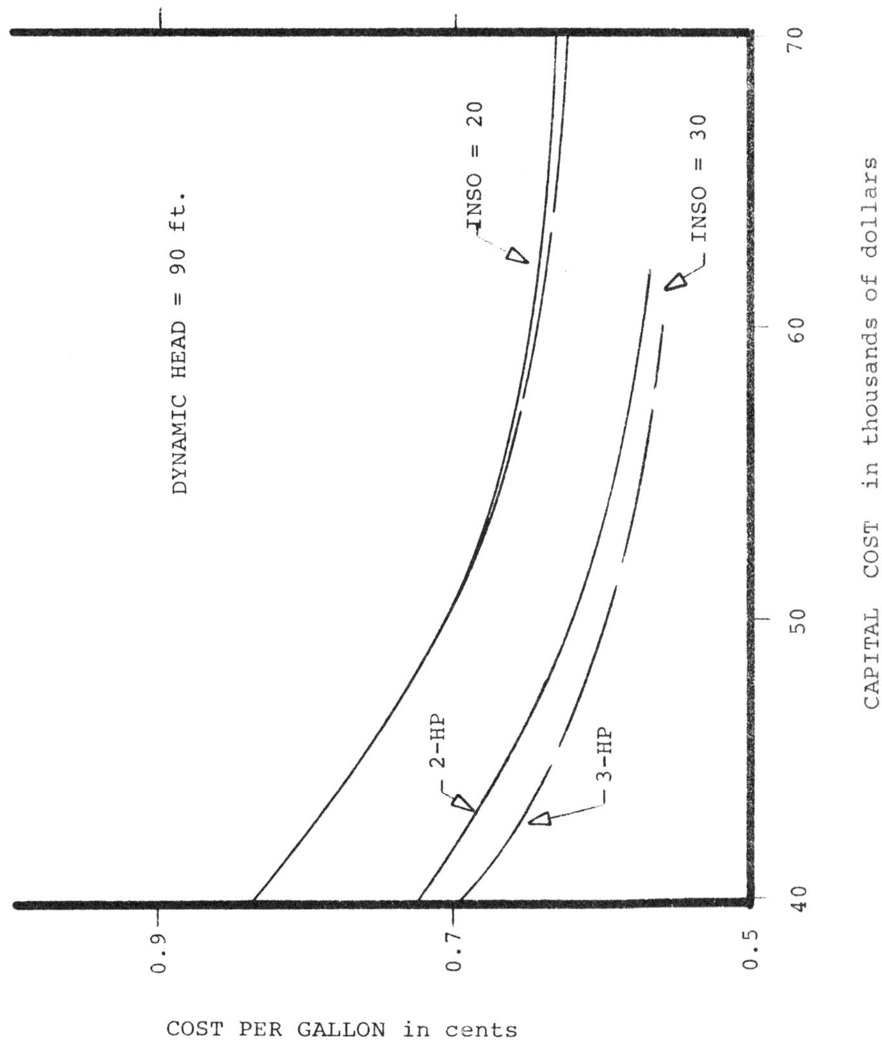

FIGURE 17. 2 & 3-HP SWPS COST COMPARISON

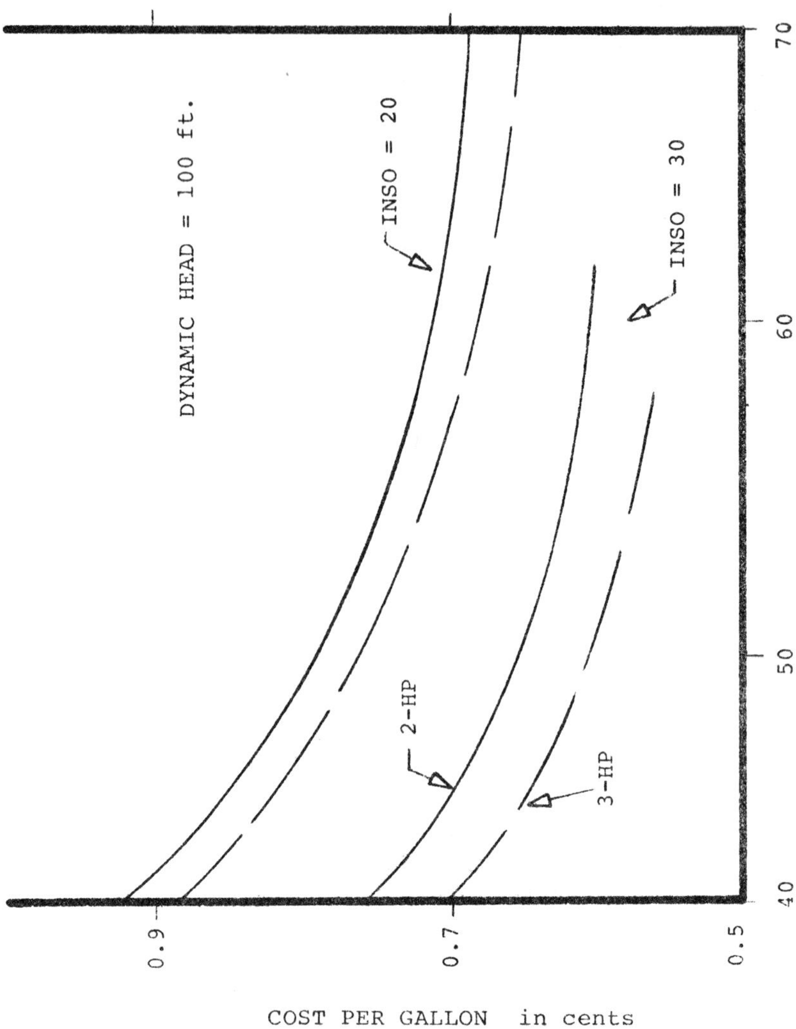

FIGURE 18. 2 & 3-HP SWPS OPERATING HEAD AND PUMP VARIATION EFFECTS

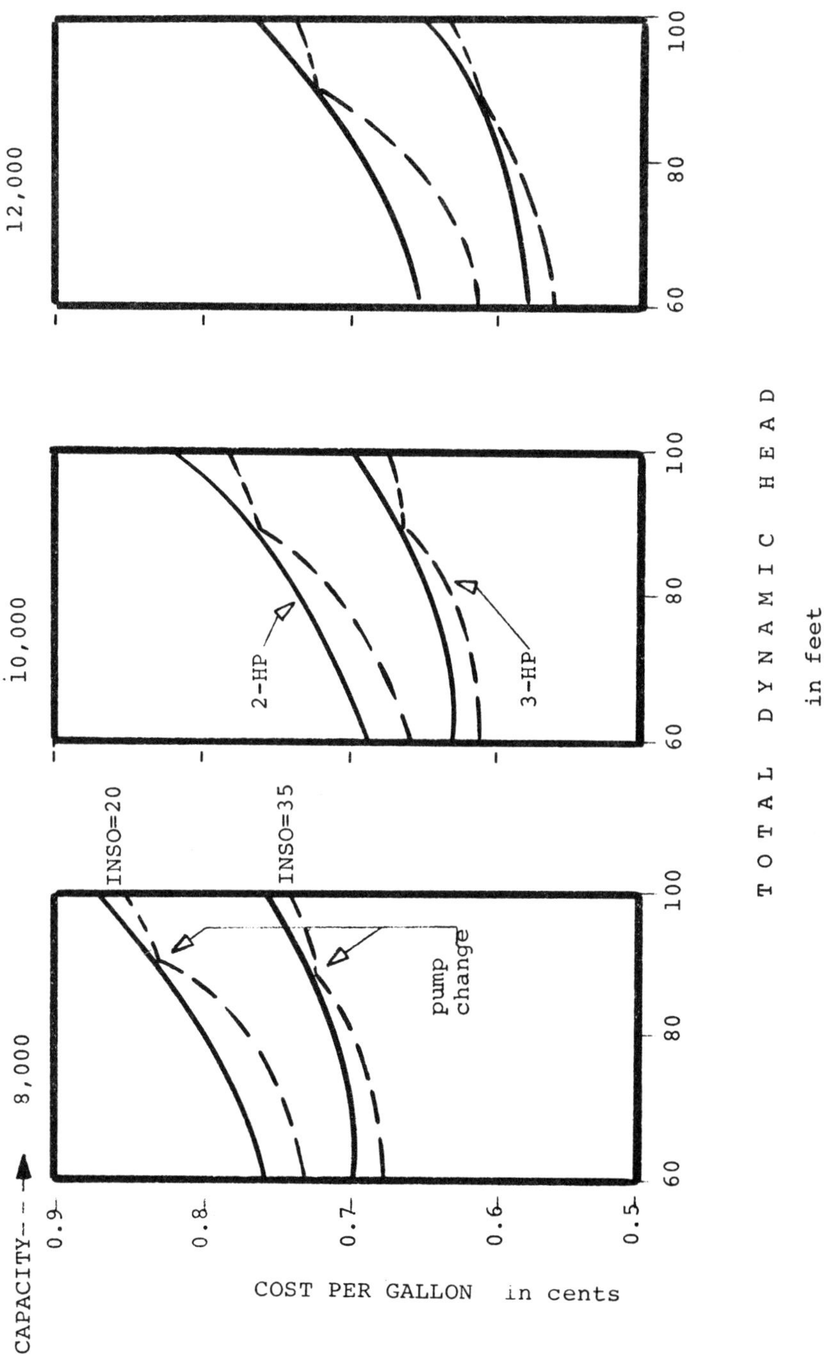

FIGURE 19. 2-HP HIGH FLOW SWPS

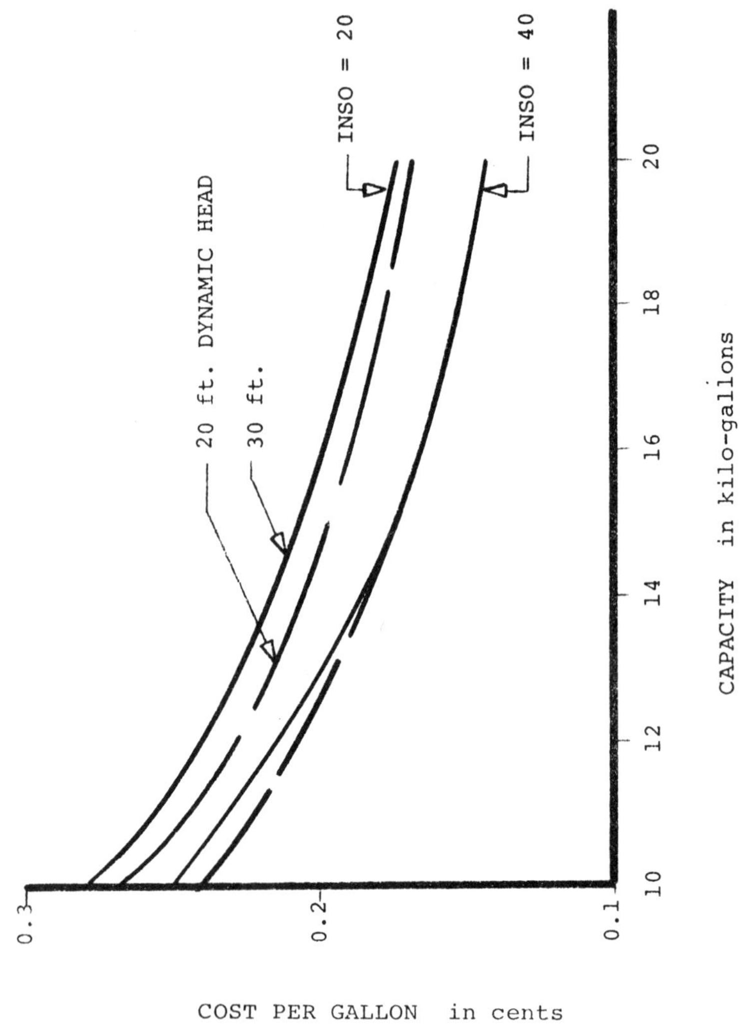

FIGURE 20. EFFECTS OF SOURCE WATER QUALITY ON 2-HP S W P S

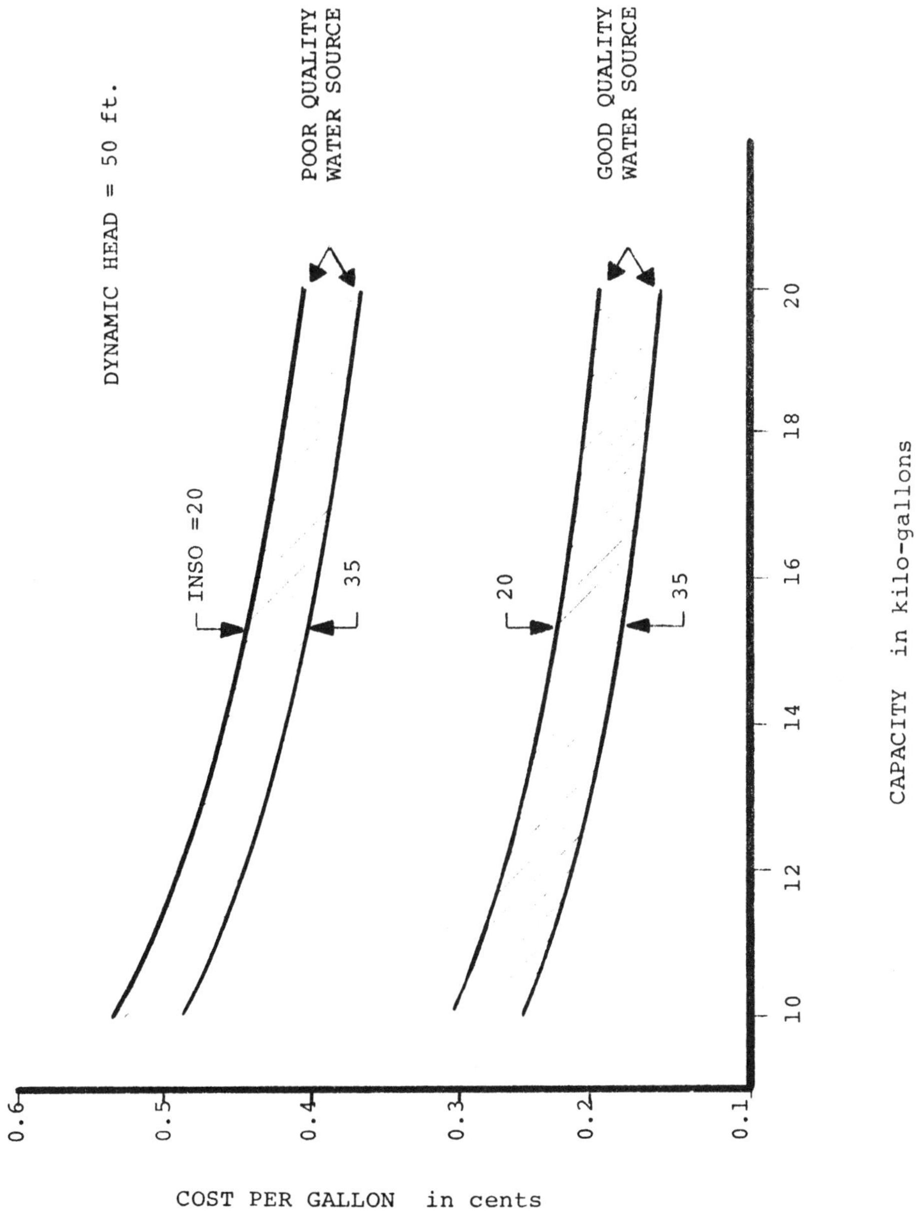

SOLAR WATER RECOVERY FROM THE AIR

W. G. Bulang
M.A.N. - Neue Technologie
Munich Germany Federal Republic

ABSTRACT

From the known gas drying processes which in principle might be used for water recovery the adsorption technique seems to be the most appropriate. Because of the low water content of the air which in arid zones is of the order of 10 grammes per cubic meter large volumes of air have to be treated. The installations required for the recovery of usable amounts of water are large but on the other hand cheap low grade solar heat can be used. Some additional mechanical energy is required for the air circulation. The meteorological premises of the water recovery process which takes advantage of the natural heat exchange between earth surface and adjoining atmospheric layers influenced by solar radiation and heat emission and the thermodynamic principle have been analysed.

The process may be of importance to supplement existing water resources.

INTRODUCTION

Water precisely fresh water is essential for live. Fresh water is accomplishing a multiple and sensitive cycle between ocean, atmosphere and continents. Only fourty per cent of continental surface is humid enough to allow agriculture. Measurements of radiation by meteorological satelites as well as recent insight in energy and water economy of the earth show that we are facing world-wide an accelerating hydrological cycle above land due partly to human activity against the remaining forests and moist areas accompanied by an increasing demand of fresh water. It is not only necessary to convert arid areas to fertile arable soil by irrigation but to prevent cultivated areas to become arid.

Increasing evaporation and pollution entails diminution of usable near surface water therefore fossil underground water reserves are utilized for drinking water supply or irrigation but these are mostly unrenewable resources.

Underground water and in some cases water from Rivers and artificial irrigation systems due to mineral salt content in connection with evaporation has resultet in salt enrichment and hence sterilisation of arable soil.

Besides the ecological doubtful redistribution of surface water by storage ponds and canal systems remain two locations to tap natural water cycle to cover fresh water demand, the desalination of seawater und the recovery of the evaporated water from the atmosphere.

In countries adjacent to coasts seawater desalting plants are installcd with increasing overall capacities especially in the states possessing raw materials as oil.

The artificial fresh water production by desalting processes implies the utilization of energy. Energy consumption is the most important expenditure not only in relation to costs but also in regard to the limited resources of fossil primary energy. Groundwater pumping and seawater desalination will enter into competition with the increasing demand for fossil fuel and fossil raw material for chemical products and protein production. This underlines the necessity to conserve energy or at least to emphasize efforts to consume primary energy efficiently.

THE NATURAL WATER CYCLE AND SOLAR RADIATION

The connection between fresh water and energy does not only exist in the desalination technology but is inherent in all natural fresh water which has been distilled by solar radiation energy. Solar radiation is the motor of the global evaporation process in nature which is responsible for the multiple atmospheric water cycle between oceans and continents. The almost constant flux of solar radiation does unfortunately not result in a homogeneous distribution of precipitation on the continents. The uneven distribution is furthermore subject to periodical as well as irregular fluctuations regarding areas as well as seasons.

In subtropical countries fresh water sources have always strongly influenced and limited expansion of population and way of live. In these areas far higher precipitations are required for plant live and agriculture than in temperate zones. Another important point to consider is the difference in the water cycle itself. In temperate climate precipitation seeps into a porous stratum and is collected by rivulets and rivers and flows back to the sea whereas in subtropical

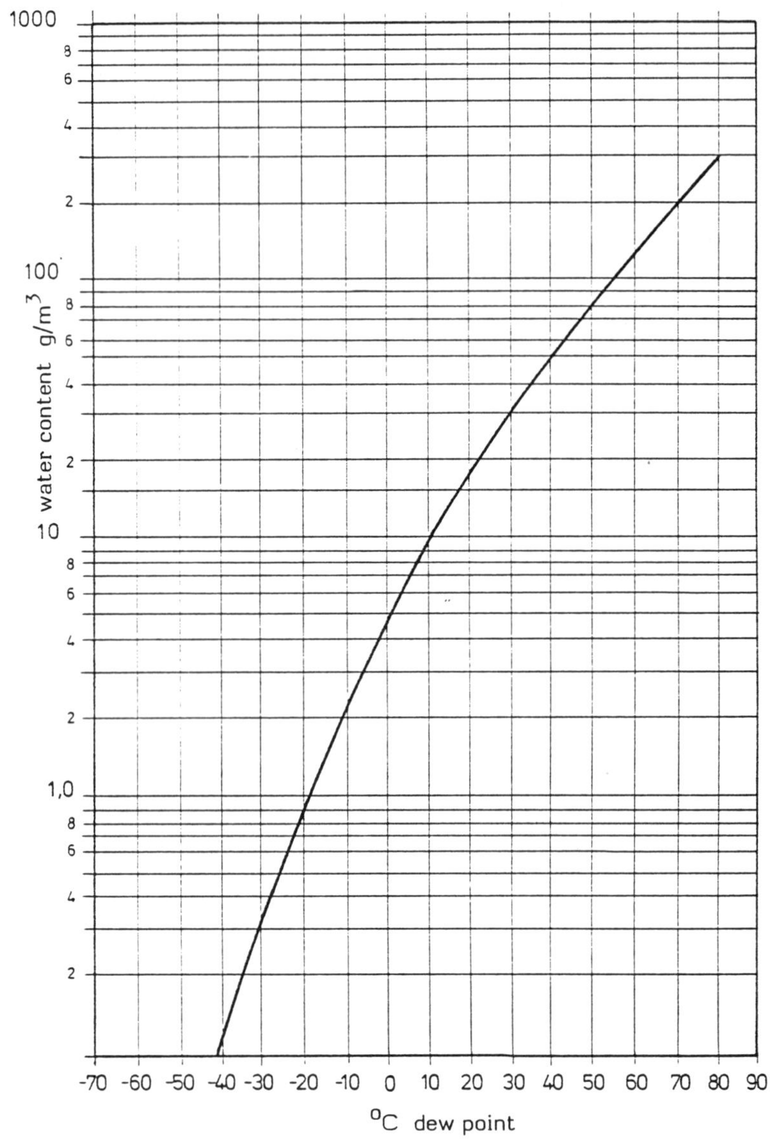

FIG. 1 WATER CONTENT IN MOIST AIR

climate irrigation and precipitation is mostly evaporated sometimes it sinks into deep aquifers and joints the fossil water tables.

There are several ways by which solar energy may help to produce fresh water. Solar pumps or solar stills are considered or in utilization where groundwater, brackish water or sea water are available.

Due to the adverse effect of solar radiation evaporation takes place where surface water occures or where plant live exists. By the evaporation of surface water there is water vapor present in the atmospheric air. Mostly more water per cubic meter in hot zones than in temperate or cold zones. Nevertheless, hot zones are often considered as dry and cold zones as moist. This is due to the ability of air to contain water depending of air temperature, Fig. 1.

If, for instance, we compare absolute humidity, which is the amount of water contained per unity of volume of air between In Salah and Paris we find nearly the same values. But air is extremely dry and evaporation as high as 5200 mm per year in In Salah.

Precipitation only occurs if saturation of air with water vapor is achieved and condensation is strong enough to produce large droplets. Saturation of moist air in nature is connected with rising air streams due to instability or forced updrafts by mountains. Stabilisation of the atmosphere by downward motion of global air masses often prevents destabilization and updrafts generated by surface insolation for instance in the subtropical belts on both sides of the tropics.

THERMODYNAMICS OF WATER SEPARATION

In terms of thermodynamics condensation of water as well as separation of air and water vapor requires the removal of entropy. This in nature occurs by adiabatic cooling of air masses in updrafts or by condensation on cooled surfaces. Updrafts represent mechanical work produced by solar energy, cooled surfaces are generated by emission of heat radiation into space by which entropy of the system is rejected. A third relevant process which occurs in nature is adsorption or absorption or chemical reaction. In the later cases the liberation of water as such requires again entropy removal. There exist

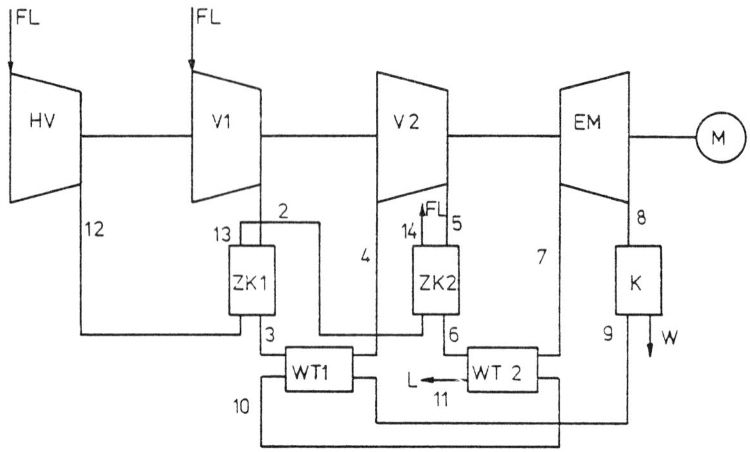

FIG. 2 WATER RECOVERY BY COMPRESSION AND ADIABATIC EXPANSION

V	Compressor
HV	Compressor
EM	Expansion turbine
M	Motor
FL	Moist air
L	Dry air
W	Water
1 - 14	Thermodynamic states
ZK	Intercooler
WT	Heat exchanger
K	Condenser

different techniques to separate water vapor and gases which could be considered to recover water from the air.

Compression and adiabatic cooling precipitation

Water condensation often occurs if atmospheric air is compressed for technical purposes. To run such a process effectively and continuously moist air has to be compressed cooled to ambient temperatures and then expanded adiabaticly. Here mechanical work is used to remove the mixing entropy of the air water vapor system as well as entropy generated by irreversibilities in the process steps.

If water recovery is anticipated the amount of mechanical work required has to be kept small by carefully designing the process which has to be run as reversible as possible. A process flow sheet of the process which is investigated at M.A.N. is shown in Fig. 2. Coupling arrangement of compressors and expansion turbine has been provided for as well as counterflow heat exchangers.

In this case solar radiation as primary energy source seems to be too costly if it is transformed to heat and subsequently to motion but the utilization of wind if available could prove to be viable.

Separation of air and water by refrigeration

Separation of water vapor and air can readily be achieved. Energy mechanical work or high temperature heat is required to provide the entropy transfer.

Absorption refrigeration by solar energy is advancing. Water vapor condensation could to a certain amount be utilized beside room cooling especially in hot and damp climates. However, water recovery in a large scale will require a costly equipment and vast areas of solar collectors.

For refrigeration by mechanical compression primary energy consumption will be 600 kWh/m^3 fresh water favourable conditions assumed. Compressors powered by diesel engines will require roughly 60 kg of fuel oil to recover one cubic meter of fresh water, if night operation is considered and high relativ humidity is prevailing. Wind might be considered as energy source.

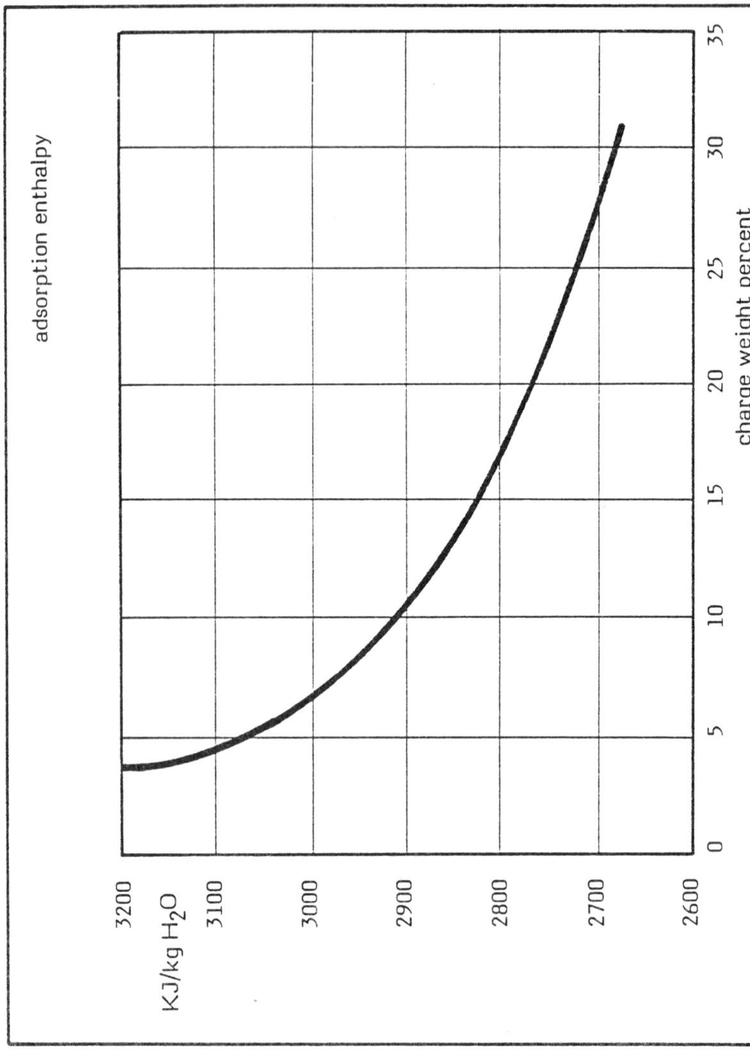

FIG. 3 ADSORPTION ENTHALPY

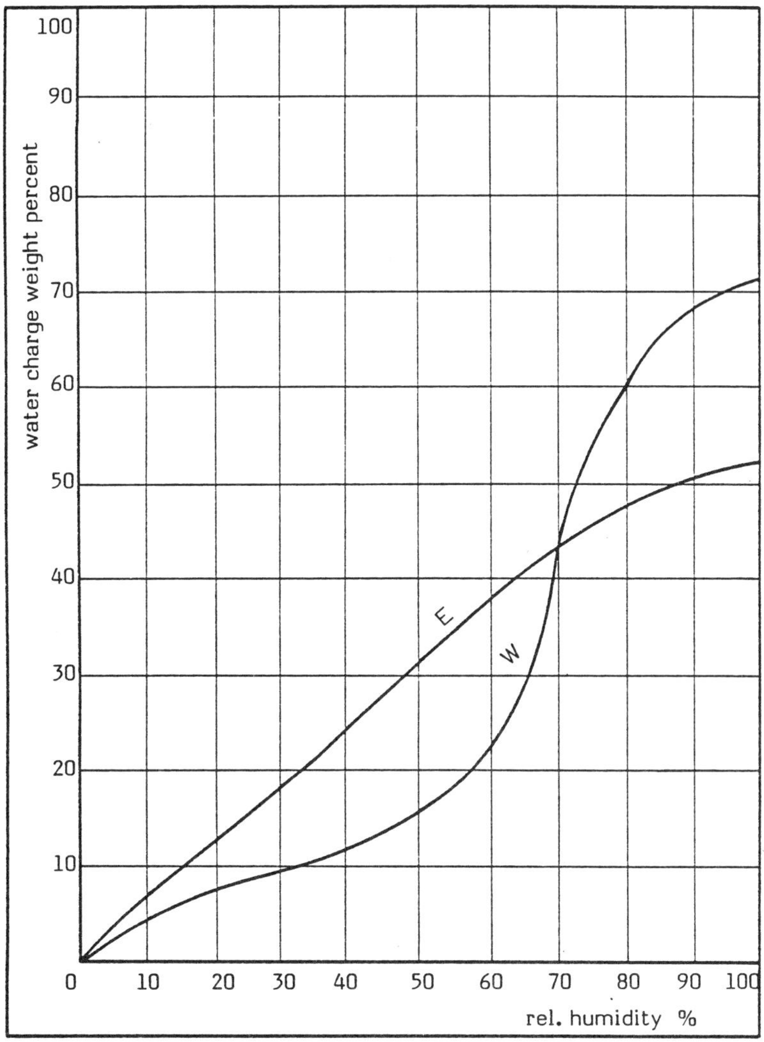

FIG. 4 ADSORPTION ISOTHERMES WATER VAPOR ON SILICAGEL AT 25°C

Adsorption technique for air and water separation

Separation of water and air and the recovery of condensed water requires a reversible reaction between the water vapor and some adsorbing material. For the forward reaction to take place an enthalpy change ΔH larger than absolute temperature times entropy change $T \cdot \Delta S$ at ambient temperatures is required. The reverse reaction takes place if the enthalpy change ΔH is smaller than $T \cdot \Delta S$.

If temperatures near 20°C are considered for the forward adsorption reaction for instance night temperatures in subtropical areas and if heat has to be used at temperatures which are efficiently achievable by flat solar collectors for the reverse desorption reaction, silicagel has been found to be the most appropriate adsorbent existing. Fig. 3 shows the adsorption enthalpy for silicagel and water vapor which comprises evaporation enthalpy and binding enthalpy as a function of the charge ratio (grammes of water adsorbed per 100 gramms of gel).

Typical adsorption isotherms for silicagel and water are shown in Fig. 4. Two different mechanism of adsorption can be distinguished from the shapes of the isotherms. The first is physical or Langmuir adsorption for which the very large internal surface is responsible. A first molecular layer of water molecules is strongly bonded partly by hydroxyl groups. Additional layers are condensed subsequently. These are less strongly bonded as is seen by the decreasing adsorption enthalpy (Fig. 3). About 30 grammes of water can be adsorbed by 100 grammes of silicagel in equilibrium with the water vapor partial pressure, if relative humidity amounts to 60% or more. The second mechanism is capillary condensation which only occurs at lower temperatures and large relative humidities as is to be seen by the curve labelled w.

As it is easily to be seen relative humidities of at least 40% are required to get sufficient charge percentage. Desorption to low remaining water content requires low relative humidity - 10% or less and temperatures of 80oC or above corresponding to the increasing adsorption enthalpy.

In gas drying techniques a small pore gel type is used which requires regeneration temperatures of 150oC in order to achieve very low dew points of the dried gases.

In the case of water recovery lower regeneration temperatures are sufficient as 5 - 10% of residual water charge must be tolerated because of the higher enthalpy values and hence higher temperatures required for the desorption of these last quantities of water.

Fig. 5 presents a microphotograph by SEM of a split face through a silicagel pearl. Magnification is 54 000. The gel consists of small spherical particels about 400 Å in diameter.

Regeneration of the adsorber is achieved by passing a stream of heated air through the silicagel bed. If the rate of air is properly controlled air can be saturated with water upon leaving the gel bed. During the desorption of water by the air stream sensible heat is compensating the adsorption enthalpy.

SOLAR WATER RECOVERY PROCESS

Besides the adsorber bed described in the previous section the recovery process installation possesses solar collectors for air heating condensers and air fans. As condensation has to take place during the desorption period a heat storage material has to be used to absorb the condensation heat. Cooling of the storage material has to be achieved by heat exchange with cold air during the night. Stones, pebbles and caned water as well as phase change materials as salt hydrates can be used.

As high values of specific vapor concentrations and high air temperatures reduce the amount of air circulating during the desorption process and the amount of heat storage material required for the water condensation because of the larger temperature change it is advantageous to add heat to the system silicagel-air-vapor during the desorption process. This can be done by using an integrated adsorber-collector arrangement or by circulating the airstream between adsorber and solar collector until its temperature and water content is high enough. In both cases only a small air stream charged with water vapor is circulated through the condenser.

Air ventilation is provided by fans. Large amounts of air have to be transported both for cooling down the heat storage material in the condenser and to charge the adsorber with water. At least the air passed through the silicagel bed for its charging with water has to be filtered to achieve long live times.

FIG. 5 SILICAGEL STRUCTURE

Care has to be taken to design the circulating system for pressure differentials as low as possible. Fig. 6 shows a schematic cross section through a recovery installation the flow schema of the charging and cooling air streams are to be seen. Fig. 7 shows the same installation during the desorption and water condensation period. The air circulation is closed to prevent loss of water vapor.

Our laboratory arrangement by which the feasibility of the recovery process has been demonstrated is shown in Fig. 8 - 10.

Work is done now to construct a small plant to be installed at a site where the process can be studied at genuine meteorological and solar radiation conditions.

Development so far does not provide enough information for a final judgement of the economic viability. Major cost items are:

1. Silicagel: 5 - 10 t per production unit (t/d)
2. Solar collectors: 150 m^2 " "
3. Heat storage material:
   a) Pebbles or stones 200 t " "
   b) Water contained 30 - 40 t " "
4. Blower air ducts
   Dust filters
5. Site areas

Beside the solar energy transformed to heat mechanical work is required to move the large volumes of air through the adsorber bed and the heat storage condenser. Provided the pressure differential can be kept as low as 100 pascales by construction means 2 - 3 kg of oil per ton of fresh water are needed. Wind energy may be used in some places.

UTILISATION OF THE RECOVERY PROCESS

The water recovery processes described not only depend on the thermodynamics and the efficiency of the plant equipment but they also depend strongly on climate in context with the demand for fresh water and its utilisation as well.

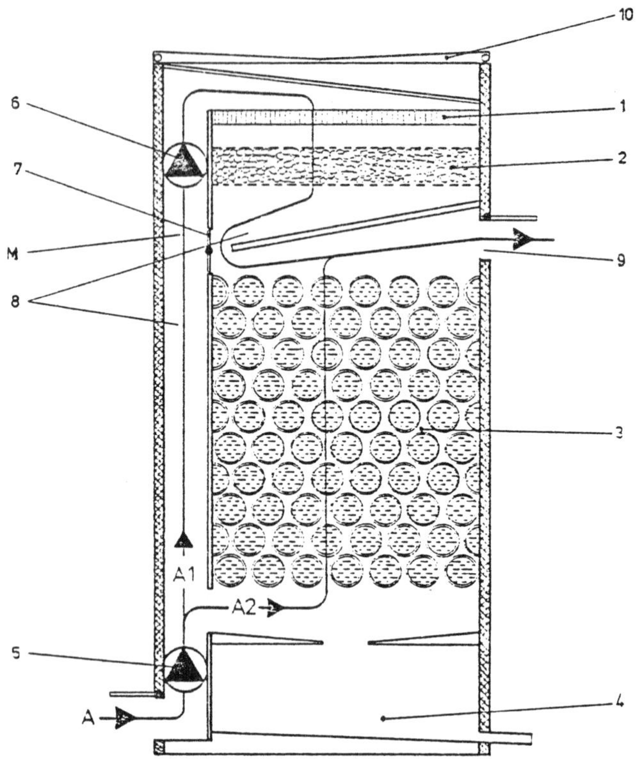

FIG. 6 SOLAR WATER RECOVERY PLANT CROSS SECTION ADSORPTION AND COOLING PHASE

1.	Solar collector	7.	Valve
2.	Adsorber	8.	Air ducts
3.	Condenser	9.	Air exit
4.	Water chamber	10.	Mirror
5. + 6.	Blowers	A, A1, A2	Air streams

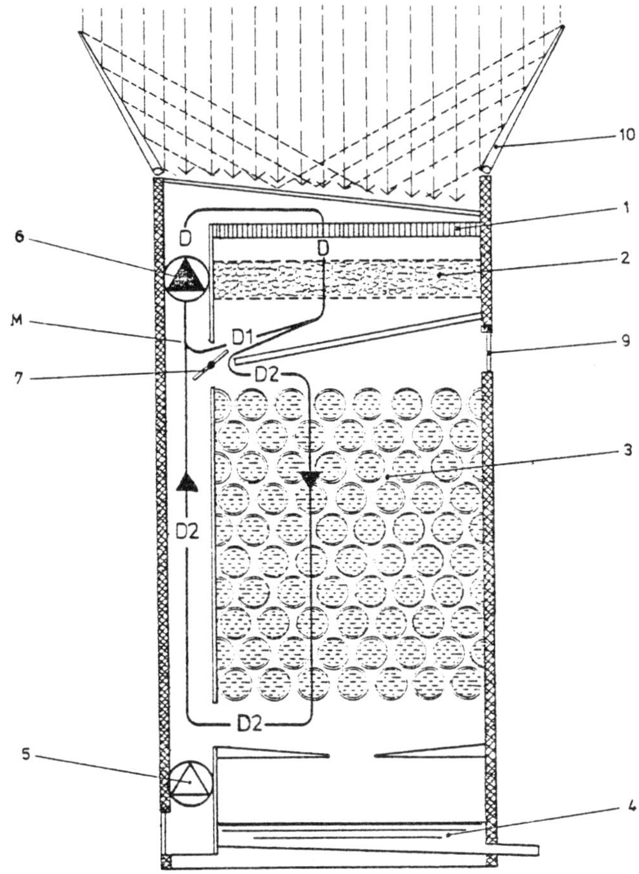

FIG. 7 SOLAR WATER RECOVERY PLANT CROSS SECTION CONVECTION HEATING AND BY-PASS CONDENSATION

1. Solar collector
2. Adsorber
3. Condenser
4. Water chamber

5. + 6. Valve
9. Air exit
10. Mirrors
D, D1, D2 Air streams

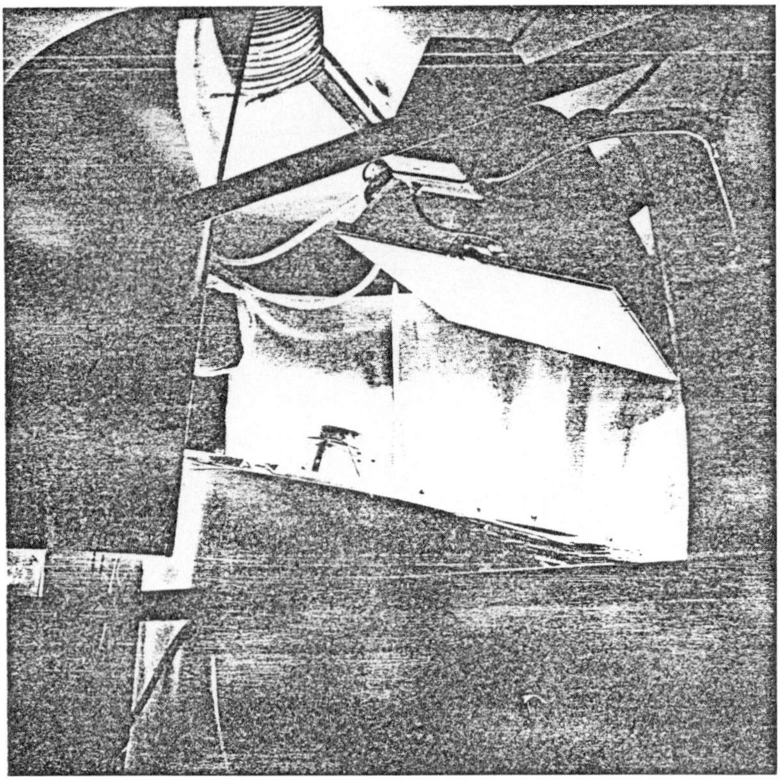

FIG. 8 SIMULATION OF SOLAR RADIATION

FIG. 9 SOLAR COLLECTOR OF EXPERIMENTAL PLANT

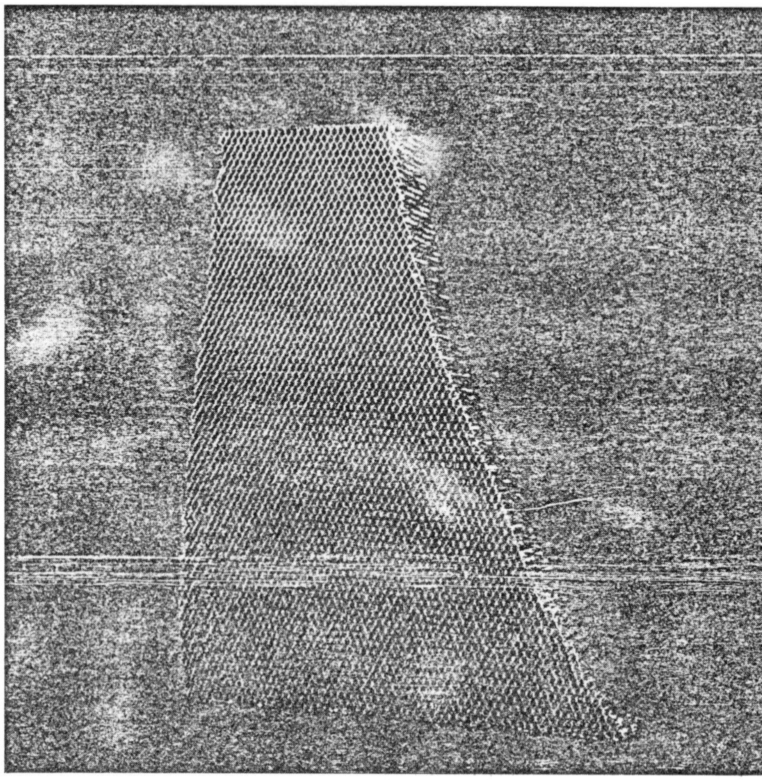

FIG. 10 HEXEL STRUCTURE ABSORBER HEAT EXCHANGER

As long as experiences with a test facility at an interesting and suitable site do not exist a final market analysis cannot be made. A first study about climate, natural water resources, geology and economic situations in some countries of the northern subtropics has shown that favourable conditions for the process do exist.

The central part of the Sahara desert on Fig. 11 shown in white is not suitable for water recovery of the air at least during the hot period of the year due to the hyper-arid climate. Humidity is to low to charge silicagel sufficient. On the other hand extremely high evaporation rates exclude agriculture by irrigation.

More interesting are the arid, semi-arid and semi-humid areas situated in the southern part of the Sahara. There the water vapor partial pressure is high enough for the charging of silicagel. Fig. 12 shows a humidity profile through the Sahara in terms of partial pressures for summer and winter time.

From the desert and semi-desert regions of the earth especially the zones with Monsun influences are favourable for the water recovery process as long as sunshine is not reduced by cloud formation. Certain parts of Saudi Arabia have proved to have favourable conditions as well. In the central Iranian highlands the conditions for the water recovery process are very unfavourable.

CONCLUSION

Water recovery of the atmospheric air deserves further investigation at least in view to future fresh water requirements in certain areas of the world despite of the fact that water recovery from the air is thermodynamicly far more difficult than desalination of sea or brackish water. Solar energy can be used to regenerate adsorbent material releasing water vapor in more concentrated form so that condensation can take place at cooled heat storage material. Water recovery should be investigated in context with special meteorological considerations and fresh water demand structure analysis.

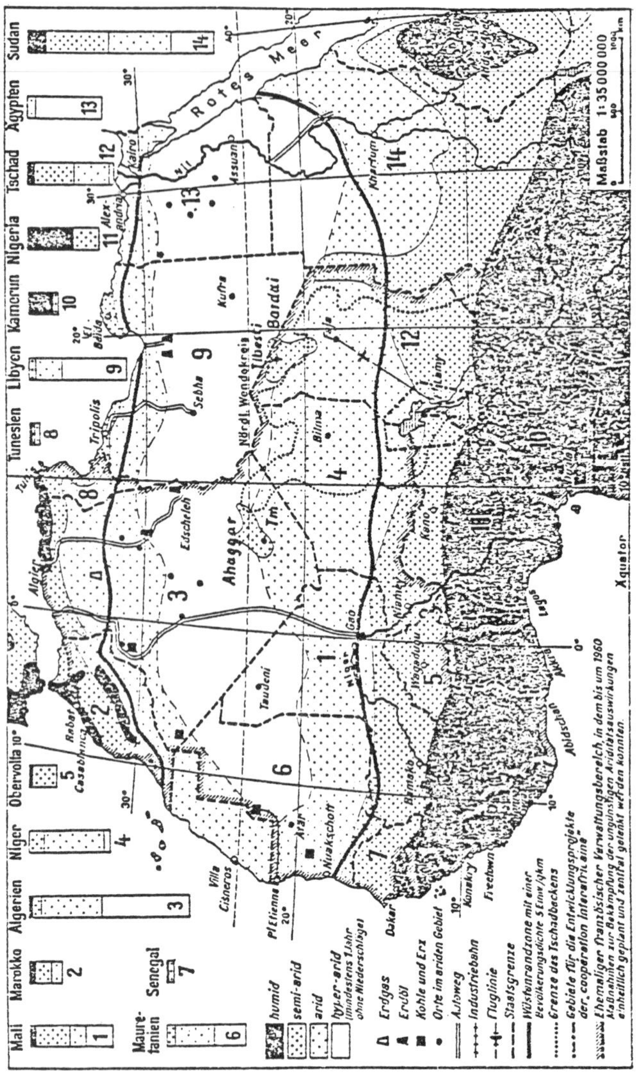

FIG. 11 CLIMATIC ZONES OF SAHARA

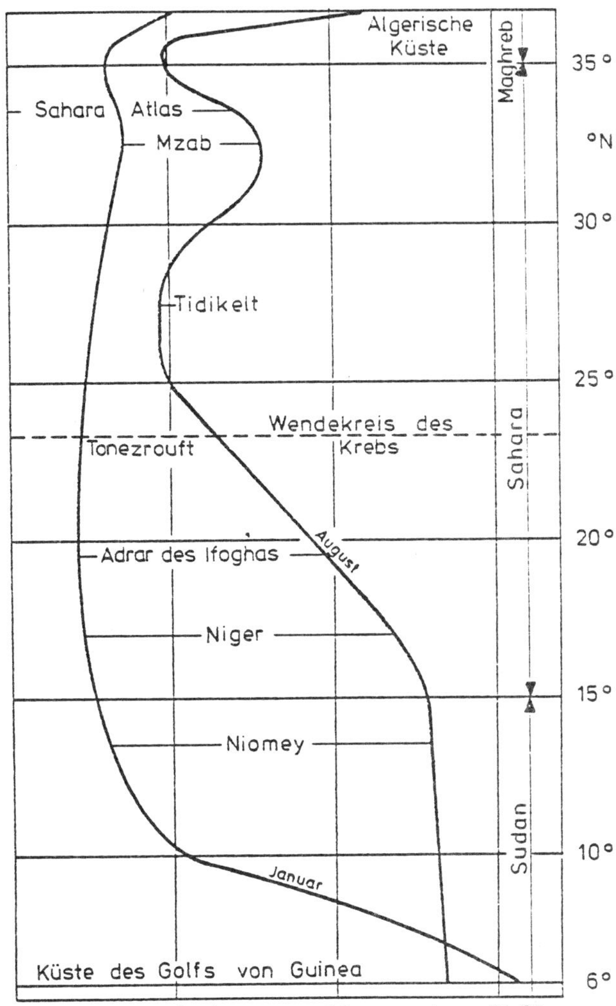

FIG. 12 PROFILE OF WATER VAPOR PRESSURE FOR THE 2° E MERIDIAN

DOUBLE EXPOSURE SOLAR STILL

M.M. Mousa
Zagazig University
Shoubra, Cairo, Egypt

F.M. Abdelfattah
Port-Said Canal University
Port-Said, Egypt

I.A. Sakr
Solar Energy Laboratory
National Research Centre
Dokki, Cairo, Egypt

ABSTRACT

Desalination of sea water through solar still utilization is considered as a long-term solution for the fresh water shortage problem, especially in arid areas. Improvement of solar still productivity, besides their design simplicity, are the most important aims of the new studies in the solar distillation field.

This work deals with a new method of increasing the productivity of a simple solar still. It is based on raising the evaporator temperature inside the still as a result of increasing the quantity of solar radiation incident on the evaporator. This effect is carried out by exposing the water basin evaporator to the solar radiation at both its upper and lower surfaces. Therefore, the unit is called a Double Exposure Solar Still. The new distillation unit has been constructed and is being run at Cairo High Polytechnical Institute, or at the Shoubra Faculty of Engineering. Outdoor experiments were carried out on both basic and modified units to study the effect of natural conditions on the performance.

INTRODUCTION

As population increases, agricultural and industrial extension in desert and arid areas is needed. This leads to an increase in the per capita consumption of fresh water. It has been proven that the desalination of sea water by solar radiation is a successful and economical process due to the following advantages:

(1) Solar distillation depends upon an abundant and permanent

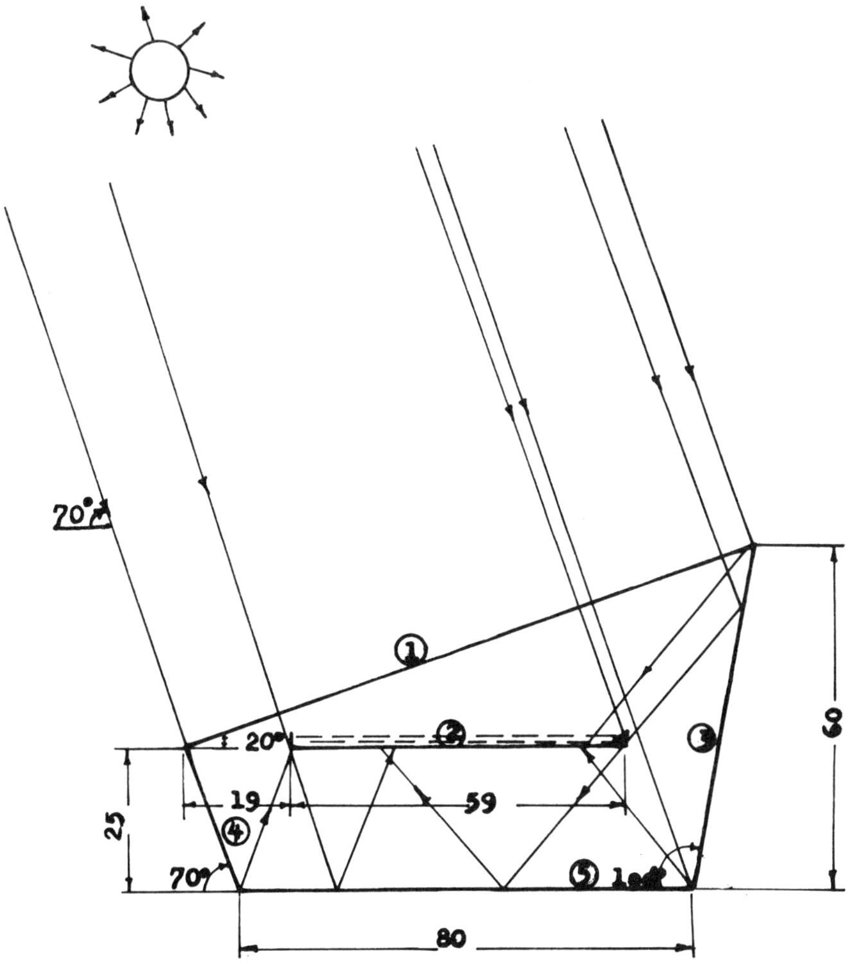

1 - Glass - cover
2 - Evaporator
3 - R.H.S. mirror
4 - L.H.S. mirror
5 - Bottom mirror

Figure 1 Optical Analysis of the D.E.S.S.

-7-

(2) Solar distillation takes place without air pollution.

(3) No need for movable parts, generation or distribution of energy cost.

(4) Solar stills may be designed to produce different scales of fresh water.

There are many factors influencing solar still productivity, such as solar radiation intensity, site ambient conditions, design factors and operational factors. The most important factor is the saline or the basin water temperature.

This is an attempt to improve the solar still performance by suggesting, constructing and testing a modified still unit called a Double Exposure Solar Still. Its idea is based on increasing productivity by an increase of the basin water temperature as a result of exposing the still evaporator to solar radiation not only at its upper surface, but at its lower one as well.

Design Of The Still Unit

An optical analysis is carried out in order to get the shape which will be able to capture the largest possible value of solar fluxes as follows:

(1) The still cover is taken to be inclined by an angle of 20° on a horizontal. It was found by Sakr [1] that this angle is appreciable in order to obtain a compacted still volume without falling back and without a reflux of condensate.

(2) The right-hand side wall is obtained by trials to get a 100% exposure ratio, i.e., complete double exposure at the instance of normal inclination between sun rays and still cover as shown in Figure 1.

where:

$$\text{Exposure Ratio} = \frac{\text{Area of Lower Exposed Surface}}{\text{Area of Upper Exposed Surface}}$$

$$= \frac{59 \times \text{Length of the Evaporator}}{59 \times \text{Length of the Evaporator}}$$

$$= 100\%$$

The complete double exposure is obtained only under the above shown position. But, it is partially obtained according to time and place of the still site. At Cairo, the place of our experiments (30°N, 31°E), the complete double exposure may be obtained instantaneously every day during the period from 1st

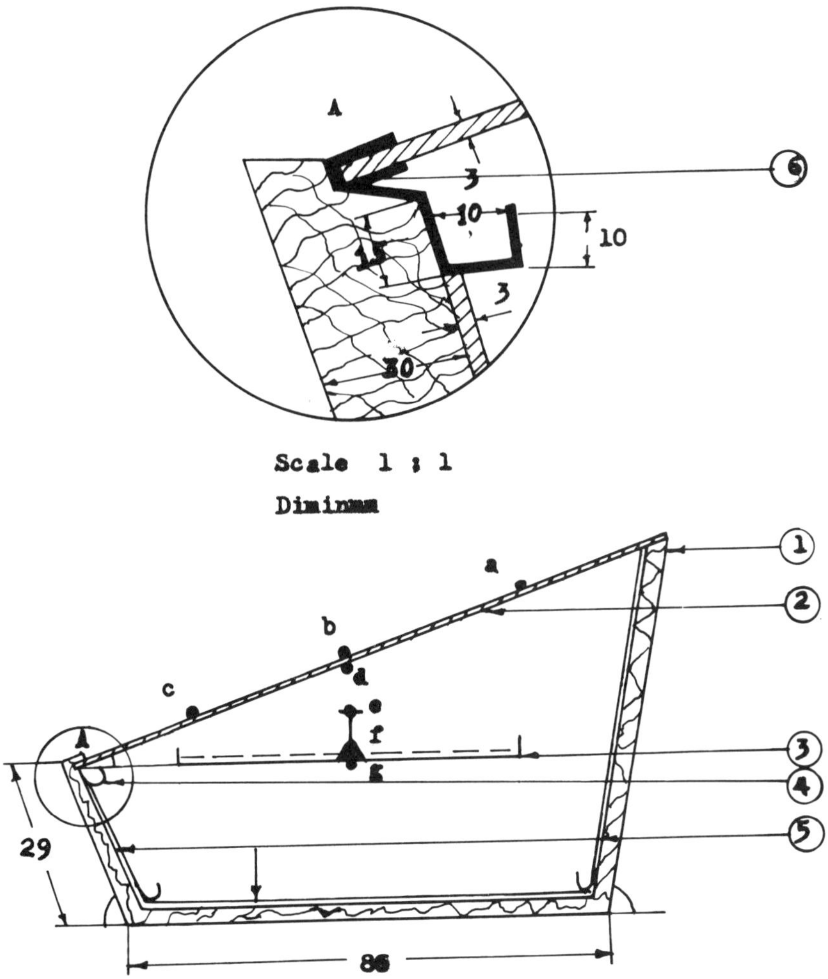

1 - Wooden box
2 - Glass cover
3 - Evaporater
4 - Traughs
5 - Three flat mirrors
6 - Rubber frame
7 - a,b,c,d,e,f and g are thermo-couples possitions.

Figure 2 Still Unit and Thermo-Couples Location

April to 15th September, due to the fact that the sun altitude exceeds 70° during that period.

The cost or base insulation is saved by using a wooden frame. In addition to the fixed mirrors on the base, two side walls lower the heat loss.

EXPERIMENTAL WORK

Outdoor experiments were carried out at Cairo High Polytechnical Institute as follows:

(1) Using the modified solar still, shown in Figure 2, with an evaporator of 59 x 95.5 surface area and a 1 cm. water depth.

(2) Using the basic unit, i.e., by cancelling the effect of mirrors, and with an evaporator of 77 x 95.5 cm^2 and a 1 cm. water depth.

(3) The following data were measured: relative humidity (Φ), ambient air temperature (t_a), feeding water (t_f), outside glass surface (t_{go}), inside glass surface (t_{gi}), basin water (t_{bw}), basin dry surface (t_{bd}), and the temperature of the upper still volume (t_v), in addition to solar radiation intensity (I), and the still yield (Y).

DISCUSSION OF THE RESULTS

Effect Of The Still Design

The following values were obtained:

(1) Average yield of the modified unit is 5.35 lit/m^2 per day under an average solar radiation intensity of 6400 $Kcal/m^2$ per day; while that of the basic unit is 4.25 lit/m^2 per day under an average solar intensity of 6000 $Kcal/m^2$ per day, i.e., yield is improved by 26% with solar intensity difference of only 7%.

(2) Improvement of C.O.P. from 40.44% in the case of the basic unit to 47.26% in the case of the modified one, i.e., C.O.P. is improved by a value of 17%.

Variation Of Parameters With Time

(1) It is found that all parameters, except relative humidity of air, increase with time approaching maximum values and start to decrease towards sunset time, as shown in Figures 3 and 4.

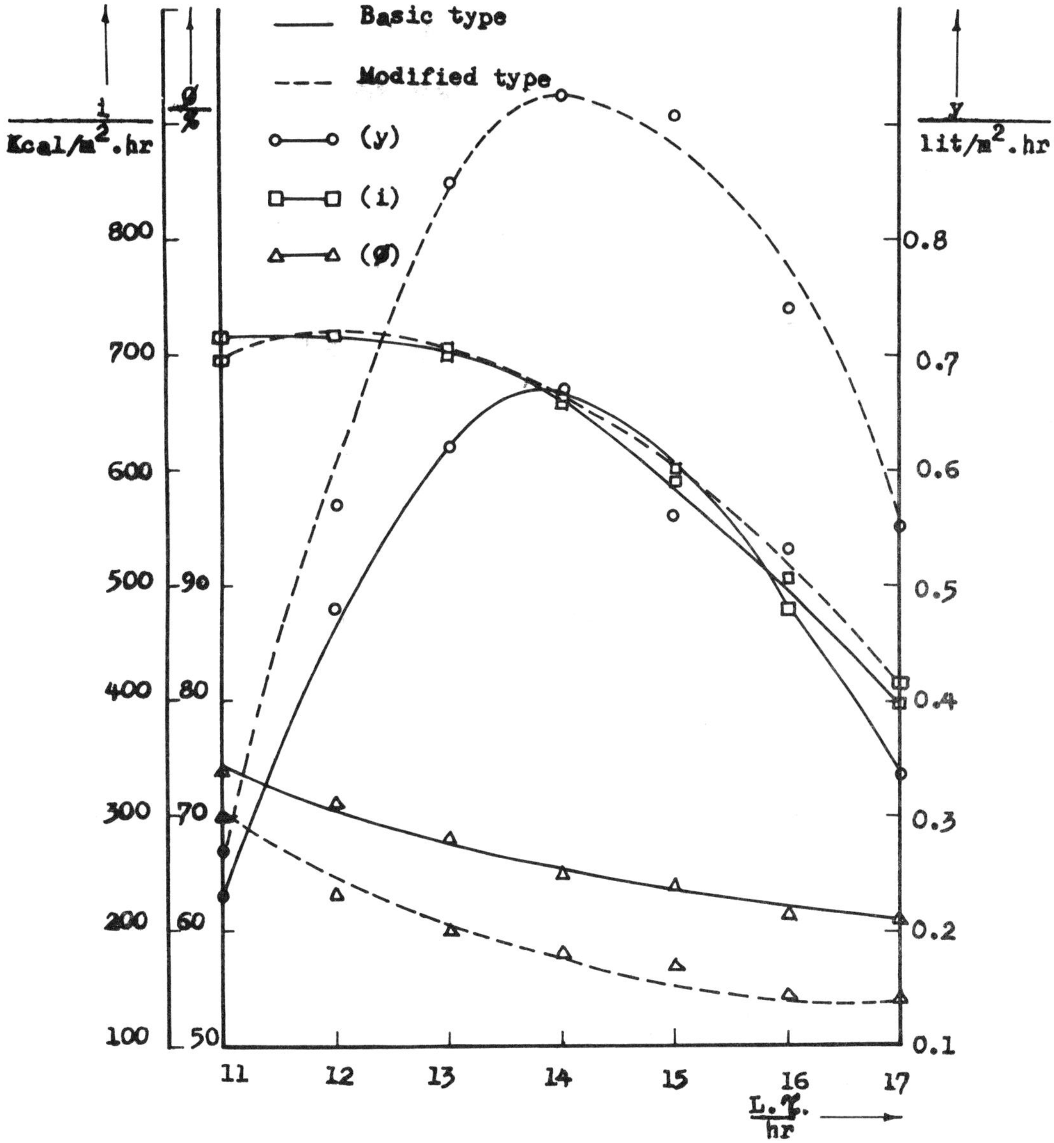

Figure 3 Variation of Specific Yield, Solar Intensity and Humidity With Time For The Two Types

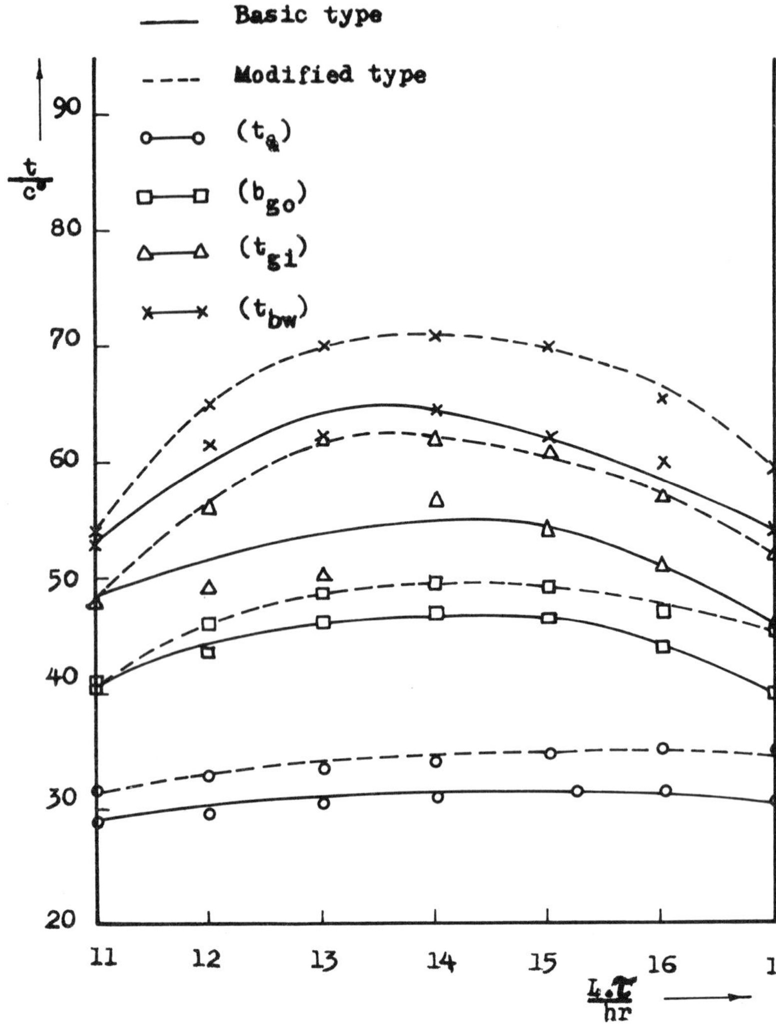

Figure 4 Variation of Temperatures t_a, t_{go}, t_{gi} and t_{bw} With Local Time

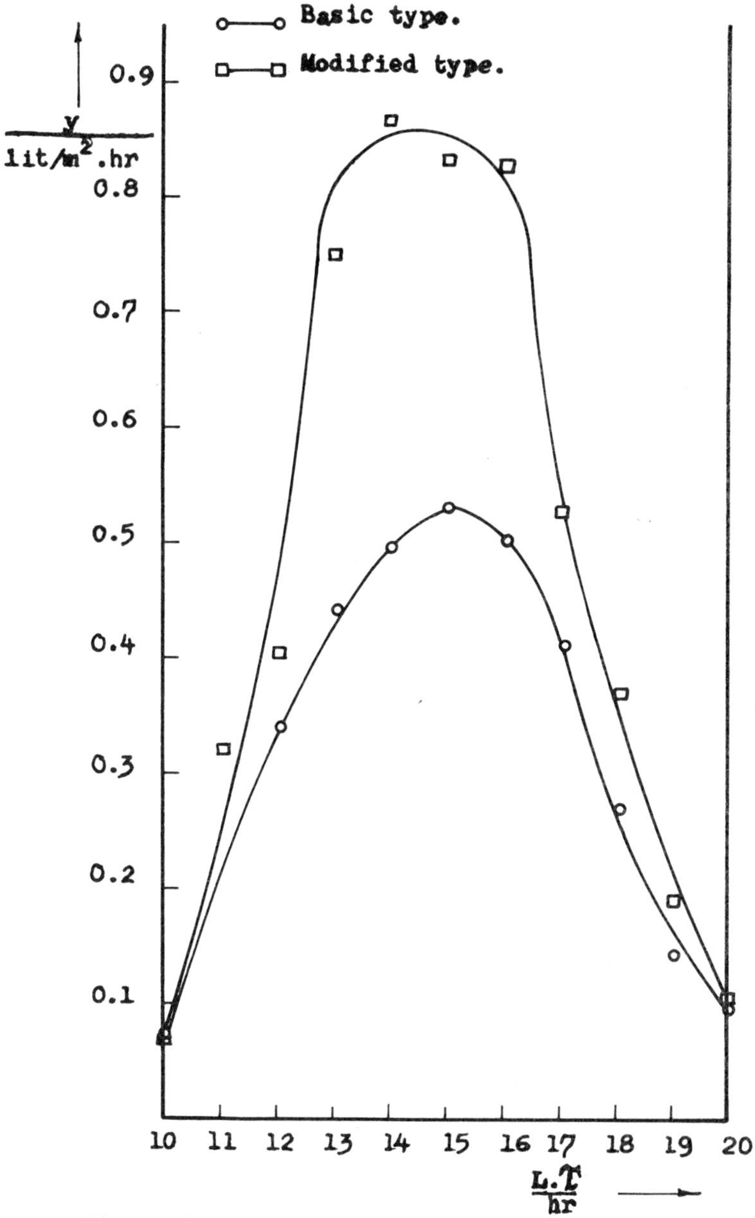

Figure 5 Variation of Specific Yield (y) With Local Time

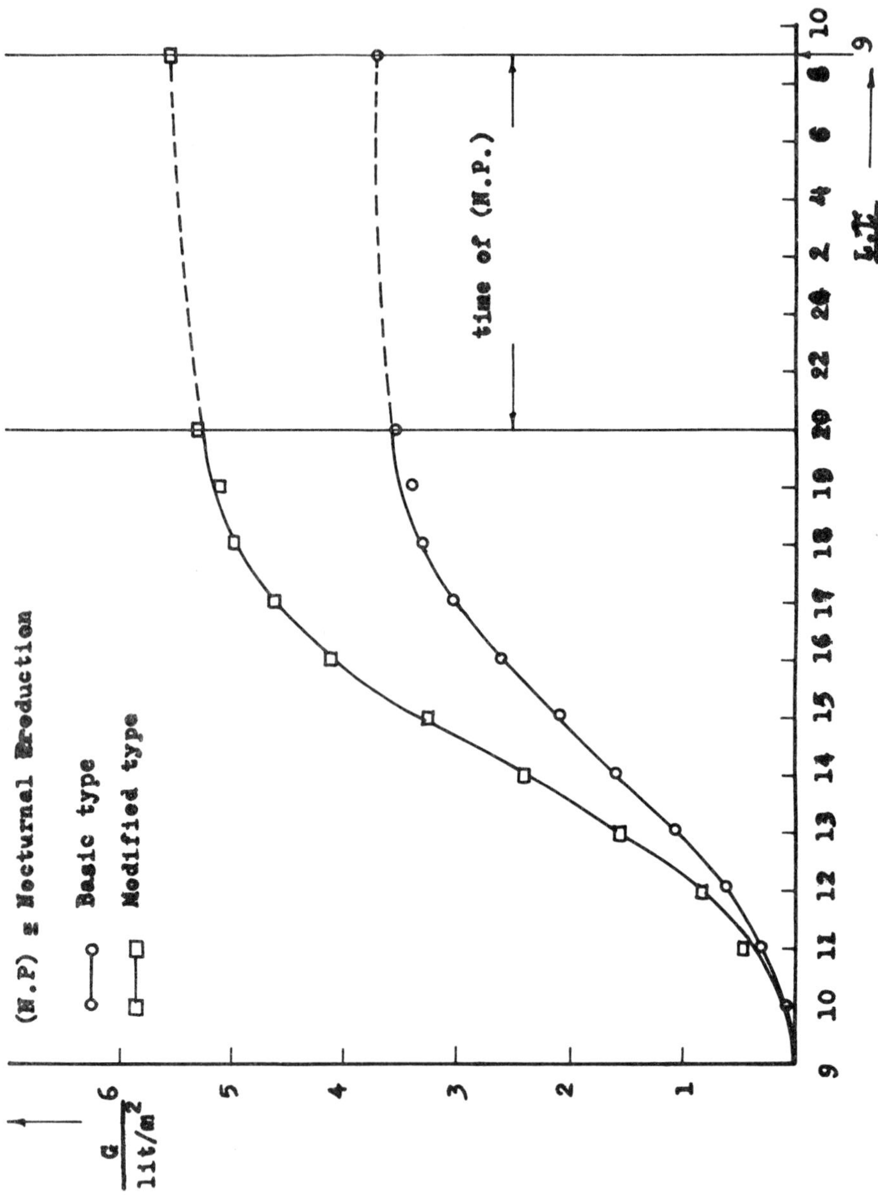

Figure 6 Variation of Accumulated Production With Time

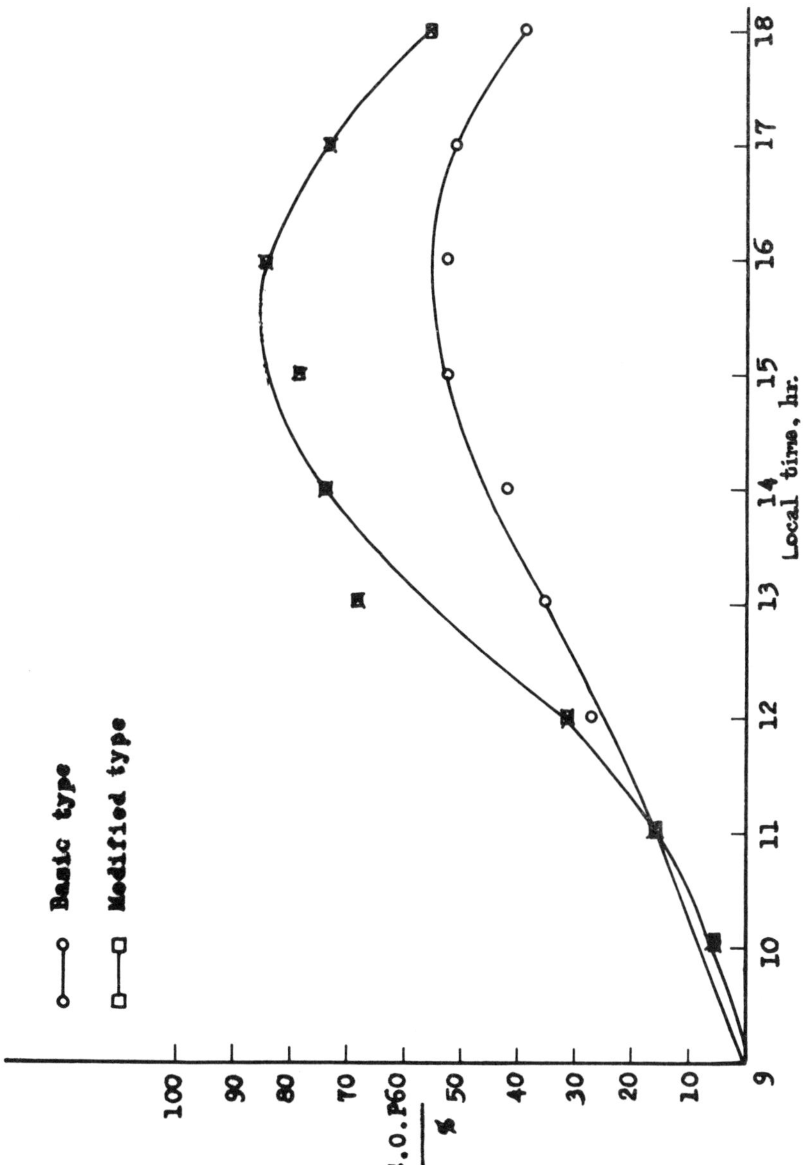

Figure 7 Variation of C.O.P. With Time

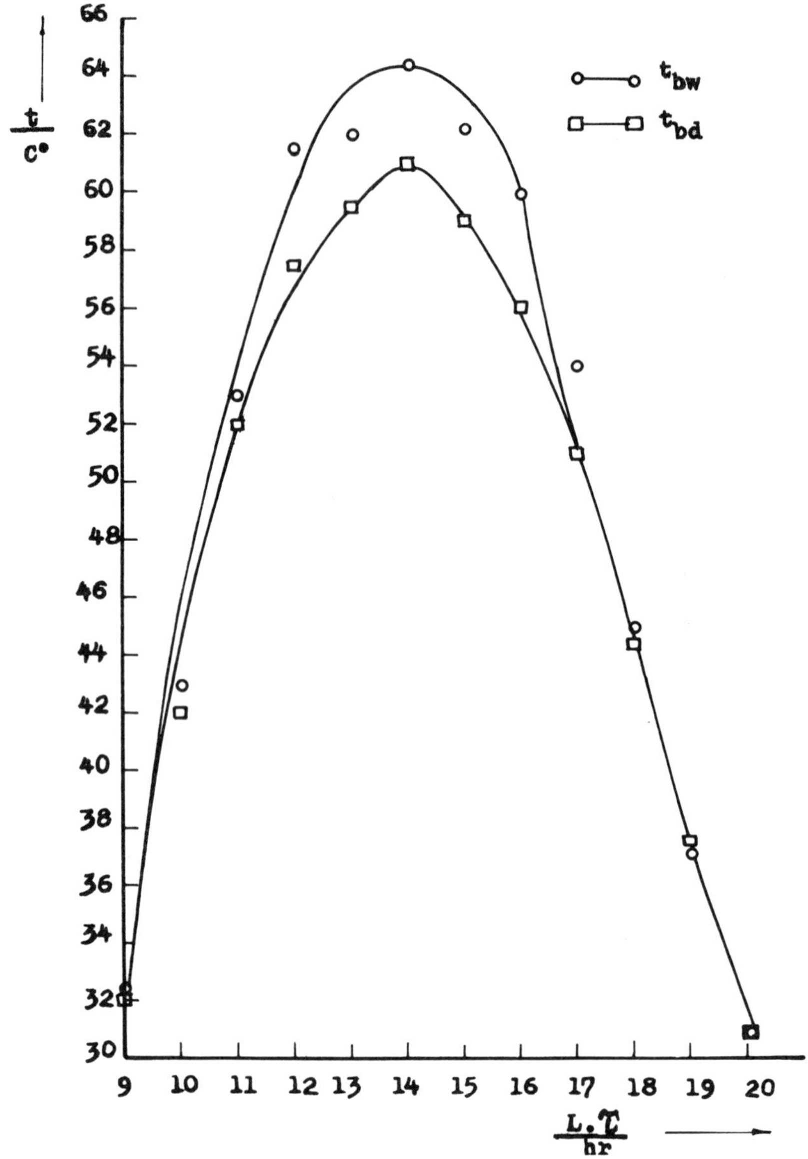

Figure 8 Variation of Basin Temperature With Time For The Basic Type

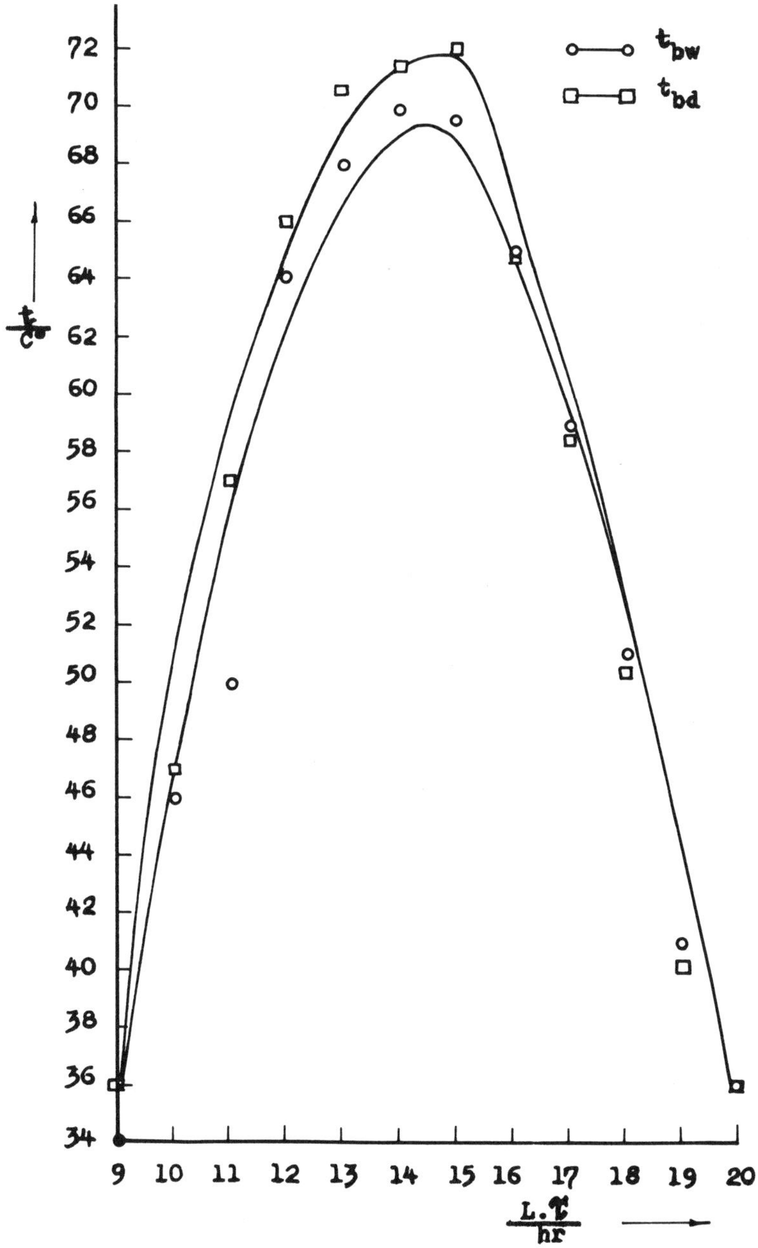

Figure 9 Variation of Basin Temperature With Time For The Modified Type

(2) Figure 3 shows that the maximum values of (Y) are at about 14 o'clock, while maximum value of (I) is earlier than that by about two hours. This lagging is due to thermal inertia of the still. The rate of increase for (Y) of the modified type is higher than that of the basic one until the peak point, after which the opposite occurs.

Effect of Double Exposure On The Still Productivity

It is shown from Figures 5 and 6 that:

(1) Specific yields of both basic and modified types recorded at 9, 10 and 11 o'clock A.M. are approximately equal. That is because this period is the time of temperature build-up in the two units.

(2) After this period, the yield of the modified unit starts to be higher than the basic one until 14 o'clock, where the maximum difference is reached, after which the difference begins to decrease.

(3) The nocturnal production per unit area of the still evaporator is higher in the case of the Double Exposure Solar Still than that of a single one due to higher energy stored after sunset.

Effect of Double Exposure On The Still Performance

Figure 7 shows that:

(1) C.O.P. of the two stills increases gradually until they reach the maximum value at about 15 o'clock.

(2) Maximum C.O.P. of the modified unit reaches 84.5%, while that of the basic one reaches to 53.5%.

Effect of Double Exposure On The Basin Temperature

Figure 8 shows that:

(1) In the case of the basic type, (t_{bw}) is always higher than (t_{bd}), but in the case of the modified one, the case is not so.

(2) For the modified type, (t_{bd}) is higher than (t_{bw}) by a value which increases with time until the peak point; after which it decreases until the two will be equalized, and then begins to be similar to the basic type.

(3) This can be explained by the fact that in the case of the Double Exposure Solar Still, the morning period is the build-up period and the evaporator is subjected to energy reflected from the mirrors in the underneath, causing a rapid

temperature rise as shown in Figure 9. Consequently, the energy reflected decreases at the same time; the rate of cooling of the evaporator is higher than the rate of receiving energy causing the continuous decrease of (t_{bd}) then (t_{bw}) the point of intersection of the two curves.

CONCLUSIONS

(1) The idea of double exposure is a successful technique to improve both productivity and performance of solar stills.

(2) The effect of double exposure is more effective at the first half of the day.

(3) The glass yield obtained is very high compared to the other surfaces.

(4) A disadvantage is the rapid spoiling of the used mirrors.

REFERENCES

1. Sakr, I.A., "Performance Characteristics of a Plastic Conical Solar Still", 1973.

NOMENCLATURE

English Alphabetical:

A	area in m^2
G	accumulated production in lit/m^2
I	intensity of solar radiation in kcal/m^2/hr
L	latent heat in Kcal/kg
q	rate of heat transfer in Kcal/hr
R	reflectivity
T	absolute temperature, trasmissivity in °K
t	temperature in °C
Y	still yield in lit/m^2/day
y	specific yield of the still in lit/m^2/hr

Greek Letters:

α	absorbtivity
ε	emmisibility
Φ	relative humidity in %
τ	time in hours

Indices:

a	ambient air
av	average value
b	brine, saline water, basin
c	convective, condensation
d	dry
e	evaporative
f	feed water
g	glass cover
i	inner, intermediate
l	lower
m	mirror
o	outer
r	radient
v	volume

DIRECT SOLAR DISTILLATION

I.D. Abdelrazek and F.H. Hammad
Department of Metallurgy, Nuclear Research Centre
Atom Energy P.O. Box, Cairo, Egypt.

ABSTRACT

The process of desalination by direct solar distillation of sea water to provide remote coastal area in Egypt with fresh water is discussed. A review to some types of solar stills is presented with emphasis on performance, efficiency and operational problems. A simple design of a proven solar still is adopted as the unit cell in a proposed 20 m^3 per day distillation plant. A cost estimate of distilled water from the proposed plant is given.

INTRODUCTION

The Egyptian coasts stretch for more than 3500 km along the Mediterranean and the Red seas. Practically uninhibited due to lack of fresh water for plantation and human needs. Desalination of sea water may be the key to develop new urban areas and tourist resorts as well. Solar distillation of sea water represents one of the uses of the low density, but cheap and clean, solar energy. The solar distillation process may be carried out in either horizontal basin-type or wicked - type stills. In basin-type stills, saline water is supplied continuously or intermittently to a pool having depths ranging from about 1 inch to 1 ft. The bottom of the pool has a black surface to absorb solar energy. A drain is provided for continuous or intermittent discard of brine. Supported above the basin is a transparent inclined cover of glass sheets or a plastic film. This transparent cover transmitts solar energy to the brine and acts as a condensing surface. The pool itself is adequately insulated to minimize heat losses to the surroundings. In all designs of basin-type stills, the horizontal surface of both water and bottom intercepts less solar radiation than those tilted

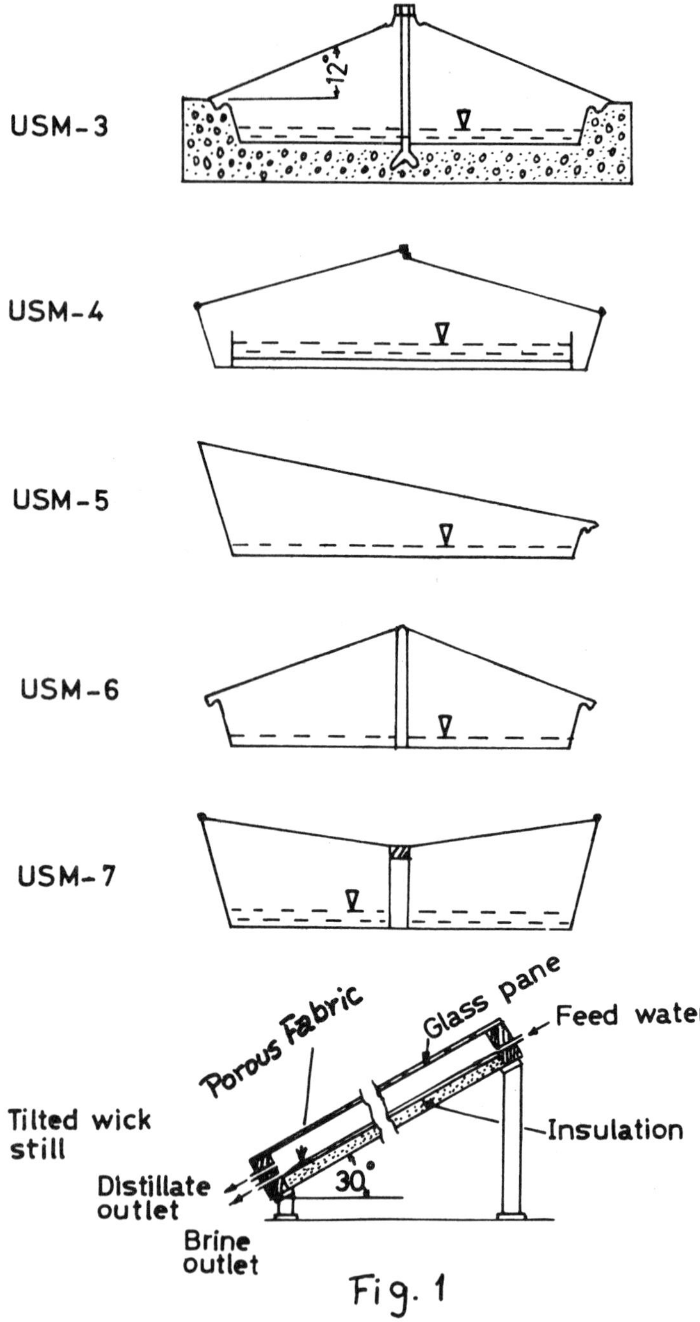

Fig. 1 Schematic Diagram For Some of the Known Solar Stills

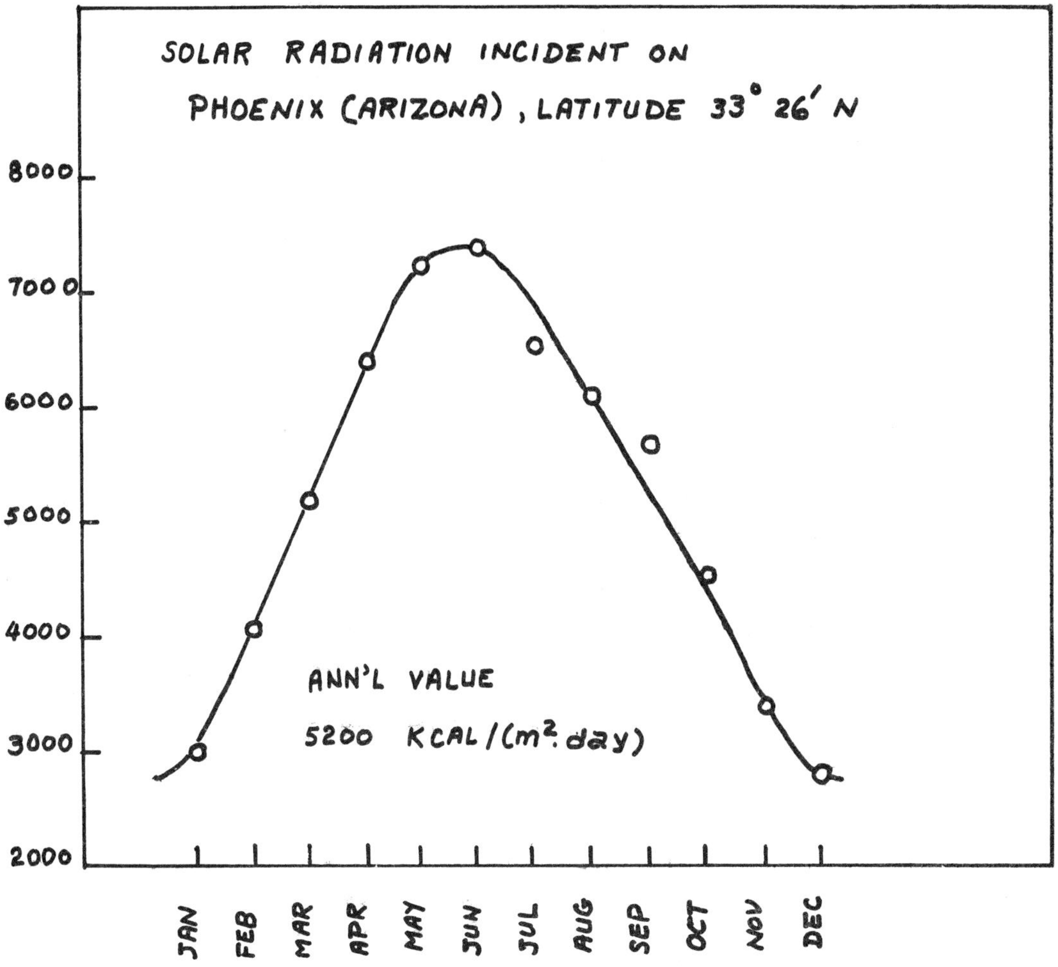

Fig. 2.

Fig. 2 Variation of the Amount of Solar Insolation Along the Year For the Case of Phoenix, Arizona, U.S.A.

towards the sun. Tilted solar stills utilise a porous wick[1] or a series of cascade troughs[2] to contain the water on the inclined surface. Although they have greater output per unit area than horizontal solar stills, yet due to greater constructional costs and frequent repair , the final water cost is not appreciably reduced (3). Fig. 1 shows a schematic diagram for some of the known solar stills. The discussion , here , will be restricted to basin-type stills with imprevious liner.

PERFORMANCE AND EFFICIENCY

The amount of solar irradiation incident on horizontal surfaces depends on the time of the year, the time of day and the location expressed by the latitude. Even on the same day, solar insolation varies along the hours after sunrise. Both variations assume a sinusoidal course, although different in amplitude and frequency (3 , 4). Fig. 2 shows the variation of the amount of solar insolation along the year for the case of phoenix (Arizona, U.S.A.) as an example (5). Solar radiation measurements around the world have been compiled by Löf et al. (6) as well as others in the form of monthly maps of the normal total solar and sky radiation on horizontal surface. Hence for design purposes, these maps, together with the data collected from pilot plant units may be used to predict the productivity and performance of new solar distillation plants.

Not all of the incident radiation on a solar still is utilized, due to reflection at basin water surface and basin liner. The effective absorbance reported for latitudes from 0 to 45° and cover slopes from θ to 60° is about 0.85 (3). By improving the operating conditions such as frequent cleaning of the glass cover and prevention of formation of layers of salt on the basin water and surface, close to maximum absorbance could be attained.

Even the energy absorbed in the basin water and liner of the still is dissipated either by unproductive conductive losses to the surroundings or by combined convection, evaporation and radiation to the inside glass surface. An estimated increase of 15 per cent in productivity was achieved for 1 inch of polystyrene insulation beneath the basin liner. The evaporative fraction of the energy transfered could be appreciably increased by running the still at high mean temperature and would result in increasing the still efficiency. A situation which could be achieved by lowering the water depth in the basin (3). This "theoritical" increase in efficiency is offset by higher thermal losses from the sides and bottom of the basin. Besides, operating at high temperature renders the glass more hydrophobic. This when coupled with high distillation rates, promotes partial drop-wise condensation, thus increasing the reflection losses from the glass. Higher distillation rate will also result in faster deposition of salts on the water surface and basin liner. To secure a smooth running and reasonably economical process, (a) the basin should be thermally insulated and (b) the saline water flow through the still should be kept at a rate almost twice the maximum distillation rate . When these precautionary measures were taken, the reparted measured efficiency barely exceeded 50 per cent (3).

The production rates for all types of solar stills (both basin - and tilted - types) were reported to vary linearly with the incident solar radiation (1 , 3). The relating equation is

$$D = \frac{Q}{a} - b \qquad dm^3 / (m^2.day)$$

Where

D = Production rate
Q = Incident solar energy, kcal /(m^2.day)
a , b = Parameters of the solar still and vary from one design to another.

This linear relation was reported to hold for Q = 3000 to 6000 K cal / (m^2.day) . Which practically covers the range of experiments performed on all types of solar stills(1).

PROPOSED PLANT

Increasing the productivity of a solar distillation plant means building additional stills. This would result in

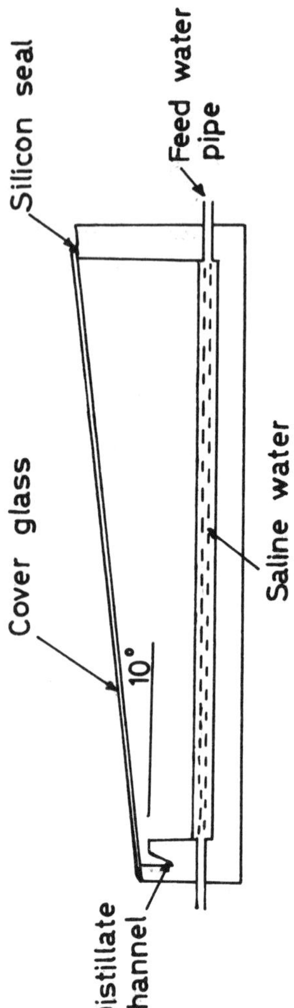

Fig. 3

Fig. 3 Cross-Section Design of a Solar Still

a slight decrease in the cost of the produced dislilled water due to purchasing of larger quantities of materials. The cost of the water produced by solar distillation is reported to be competitive with the cost of water produced by other methods of desalination , only in the size range below about 5000 - 10000 gal / day (7).

A 20 m^3 / day (5200 gal / day) size solar distillation plant should satisfy the needs of a reasonable sized community (500 persons) in a remote costal area (7). The basin still chosen as the basic unit for the proposed 20 m^3/day plant is similar to the CRISO MK VI model designed by Cooper (3) , as with proper insulation an efficiency of about 50 per cent could be attained. Design simplicity , in addition, would allow hiring of unskilled labor for maintenance and operation. Fig 3 is a schematic diagram for the solar still under consideratinn. The still itself is a 0.15 m thick concrete container having an available area 6.25 m^2 for evaporation (6 m X 1.04m). The long sides are 0.48 and 0.30 high. The provision allow to mount a glass cover, that slopes 10°, at the top . Two pipes are fitted for feed water and brine removal. The distillate is collected in a side trough.

Design criterion 20 m^3/day (5260 gal/day)
Local total and sky radiation on horizontal surface,
in June = 6500 - 7000 Kcal /(m^2 day)
in December= 2500 - 3000 " " ")
Yearly averaged value = 5000 " " .
Productivity (for 45% efficiency) = 4.05 kg / (m^2.day)
Area for 20 m^3 / day = 4938 m^2
= 5000 m^2
(50 batteries each comprises 16 standard units)
Total area = 50 X 16 X 6.25 m^2 = 5000 m^2
Glass cover 12 sheets of 4 mm thick glass, each measures 1.15 X 0.50 m
concrete thickness = 0.10 m
" volume = 0.180 m^3/m^2.

COST OF THE DISTILLED WATER

For the proposed 5000 m^2 plant, the cost of the

distilled water may be estimated according to the current local prices of materials and labor rates as :

	LE/m^2
Excavation work	1.00
Concrete	16.21
Metallic frame for glass	2.00
Glass	1.50
Pipes, fittings, tanks	2.00
Seams, junctures, others	1.00
Insulation and black,....	1.00
	24.56

Considering 8% interest per year
5% amortization
3% maintenance and others
16% capital factor

The annual cost is 35.60 X 0.20 = 3.93 LE / (m^2.yr)
= 5.61 $ / (m^2.yr)

The average daily production of dislilled water
= 4.05 dm^3 / (m^2.day)

The average yearly " " " "
= 1.215 m^3 / (m^2. yr)

Cost of water $= \frac{3.93}{1.478} = 3.23$ LE/m^3.
$= 4.62$ $ / m^3.

This estimation is based on prices quoted by local contractors. The concrete prices form 66 per cent of the total cost and hence reduction in concrete cost would lower the cost of the produced distilled water. A 20 per cent reduction in the prices of concrete would result in 13 per cent lowering of the produced water whereas a 30 per cent reduction would cut the water prices by almost 20 per cent. The use of prefabricated slabs of concrete and assembling at the plant site is a possible approach for achieving lower price distilled water.

CONCLUSIONS

Solar distillation process still the cheapest among all desalination methods in the size range from 20 to 40 m^3/day, although the estimated price of water produced by solar distillation is almost three times of the price reported before 1973. Proper installing, insulating and operating of the proposed plant may prolong the plant life time and thus a further decrease in water prices.

AKNOWLEDGMENT

This study was supported by the Nuclear Research Centre in Cairo, to whom the authors are deeply indebted.

REFERENCES

(1) J.R.Hirschmann, Solar Desalination in Chile, Desalination, 17, 17 - 30 (1975) .

(2) S.Satcunanathan and H.P. Hansen, An investigation of some of the parameters involved in solar desalination, solar Energy, 14, 353-363 (1973).

(3) P.I. Cooper, The maximum efficiency of single-effect solar stills, Solar Energy, 15, 205-217 (1973).

(4) J.F. Kreider and F.Kreith, Solar Heating and Cooling, McGraw-Hill, New York, First Edition, 1975, PP. 47.

(5) Andrew Porteous, Saline water Distillation Processes, Longman, 1975 , PP. 99.

(6) G.O.G.Löf et al, Univ. Wisconsin Eng. Exptl. Stat. Bull, 21 , 1965.

(7) E.D. Howe and B.W. Tleimat, Twenty years of work on solar distillation at the University of California, Solar Energy, 16, 97-105 (1974).

(8) Ibid, Solar distillers for use on Coral Islands, Desalination, 2, 109 - 115 , 1967.

SOLAR ENERGY FOR DESALINATION

Ali M. El-Nashar
University of Mansoura, El-Mansoura, Egypt

ABSTRACT

With the existing shortages in water and energy in many parts of the world, the use of solar energy for desalination and water reuse is examined. The desalination processes of current interest are reviewed and their energy demand is compared with the purpose of using the most energy efficient ones in combination with solar energy systems.

INTRODUCTION

Shortages of water and energy are now experienced or anticipated by many countries. Several countries in the Middle East and North Africa are currently suffering from inadequate fresh water supplies. Recent severe drought conditions in Western Africa, Western United States, and England has necessitated a reexamination of our water supply - demand balance on a global basis.

A survey of the earth's energy resources indicates that there is a need for new energy sources. Solar energy could meet this need if a program of solar energy system design and development is initiated. The direct solar energy irradiating a square kilometer of ground at noon on a clear day in the desert south of 35 N latitude produces over 400 MW th (megawatt thermal) of heat even in midwinter [1].

Though two third of our planet is covered with water, 97% of this water lies in the oceans, 2% is frozen in the glaciers and ice caps, and only 0.0001% as fresh water flowing in rivers; see Table 1. The average instantaneous fresh water in all rivers and fresh water lakes are 250,000 cubic kilometers. This has been the principal sources of water for domestic, agricultural, and industrial activities. On a world wide per capita basis, this is equivalent to about 60,000 cubic meters. However, world population, which is now approaching the four billion mark, is estimated to grow at an annual rate of 2.0%. At this rate, the population may be eight billion by 2000. This means that in the early part of the 21st century, only 30,000 cubic meters of fresh water will be available to meet all the human needs. This quantity would have been adequate if the fresh water sources were evenly distributed in the world. Unfortunately this is not true. The fresh water sources of the world's continents are given in Table 2. The disparity is even greater within the different

TABLE 1

ESTIMATED WORLD WATER SUPPLY

WATER IN LAND AREAS	VOLUME (1,000) cubic kilometers	PERCENT OF TOTAL WATER
Fresh water lakes	125	0.0009
Saline lakes and inland seas	104	0.008
Rivers	125	0.0001
Ground water	8350	0.61
Soil moisture, icecaps, and glaciers	29267	2.145
Sub-total:	37,800	2.8
Oceans	1,320,000	97.3
Total:	1,360,000	100

TABLE 2

PER CAPITA WATER RESOURCES

CONTINENT	POPULATION MILLIONS 1969	STABLE STREAM FLOW CU. METER/YR
EUROPE	642	2,100
ASIA	2,040	1,960
AFRICA	345	5,500
N. AMERICA	312	7,640
S. AMERICA	185	21,100
AUSTRALIA	18	27,500

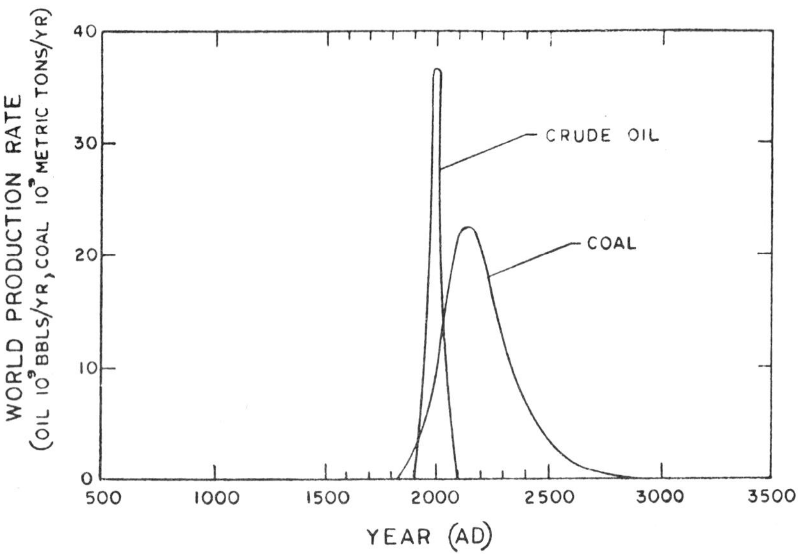

FIGURE 1 World fossil fuel utilization rate

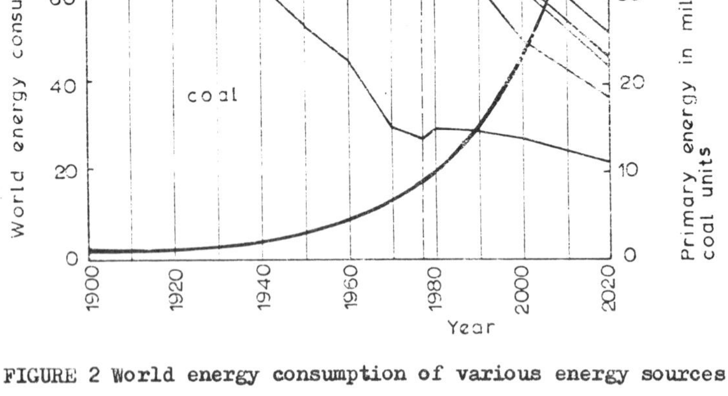

FIGURE 2 World energy consumption of various energy sources

regions of the world, as for example in the Arab countries. Egypt as one of the Arab countries has a per capita fresh water available from natural resources of about 2000 cubic meter per year while Kuwait, another Arab country, has a per capita water supply close to zero.

Several alternative methods has been considered for achieving the water supply - demand balance. These include : 1- desalination of brackish and sea water; 2- reuse of industrial waste water; 3- reuse of municipal waste water; and 4- reuse of agricultural irrigation waste water.

Our natural fossil fuels, we all know, are limited and will run out at sometime if unrestricted consumption continues. The alarming picture of Figure 1 indicates that for coal, the earth's supplies are sufficient to serve as the major source of energy for two or three centuries. The corresponding period for crude oil is only about 100 years.

Uranium 235 has been badly depleted and serious shortages are expected within the next two decades [1]. Serious problems are facing the breeder reactor program.

With the anticipated shortages in fossil and fissile fuels, solar energy is expected to play a more active role in supplying the required energy for mankind. The forecast in the world use of solar energy was made up to the year 2020, see Figure 2. It can be seen that about 25% of the world energy consumption is expected to be provided by solar energy by 2020.

SIMPLE SOLAR STILLS

A simple solar still is shown in Figure 3. This is the familiar type of solar still. Sunlight enters the still through a transparent cover made of either glass or plastic. Most of the solar radiation is absorbed on the black bottom of the still. The absorbed energy is released as heat which warms the saline water contained in the basin. Water vapor saturates the air above the water and condenses on the cooler transparent surface and runs down into the condensate collection trough. Solar stills have been built and operated in the Chilean Andes, Greece, Egypt, and other countries.

The simple solar still suffers from the limitation that it has a low radiation interception efficiency, ranging from 30% to 60%, and, therefore, its daily water production is only about 3 liters/day m^2 for an irradiation intensity of 4200 k cal /day m^2 [2]. Since the latent heat of evaporation of water is 540 k cal /kg, then 3 liters would require 1620 k cal for evaporation, hence the efficiency is 39%. The fact that the interception efficiency is low requires a relatively large area for the solar collector and, therefore, a relatively high capital investment.

DESALINATION PROCESSES AND THEIR ENERGY CONSUMPTION

The separation of water or salt from a salt solution requires energy. For sea water, the absolute minimum separation energy amounts to

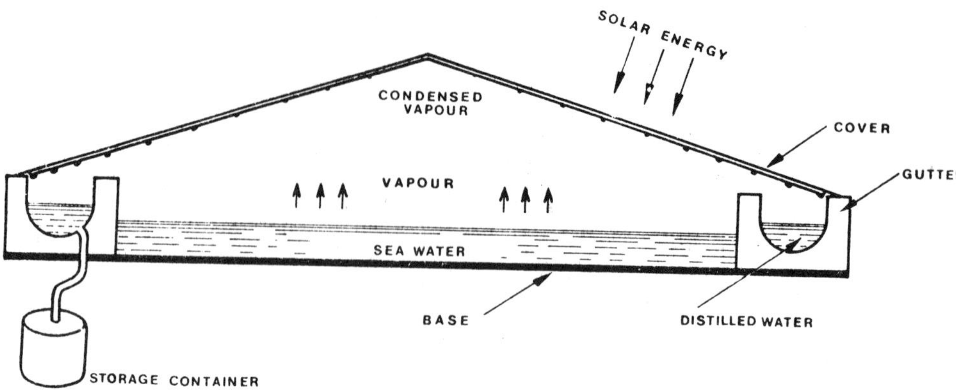

FIGURE 3 Simple solar still

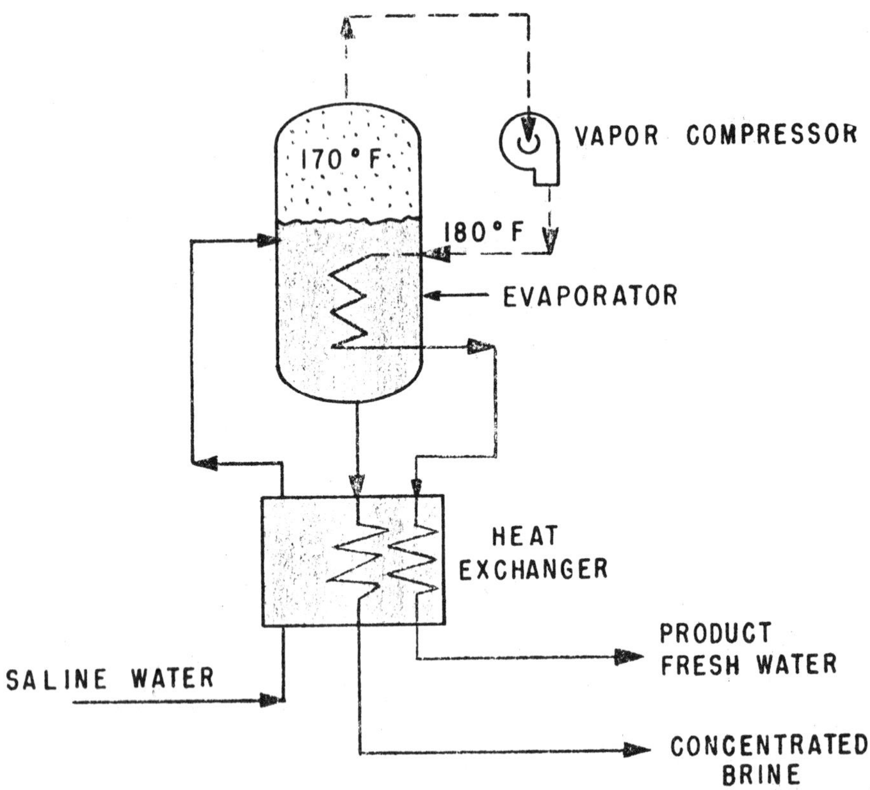

FIGURE 4 Schematic of a vapor compression desalination plant

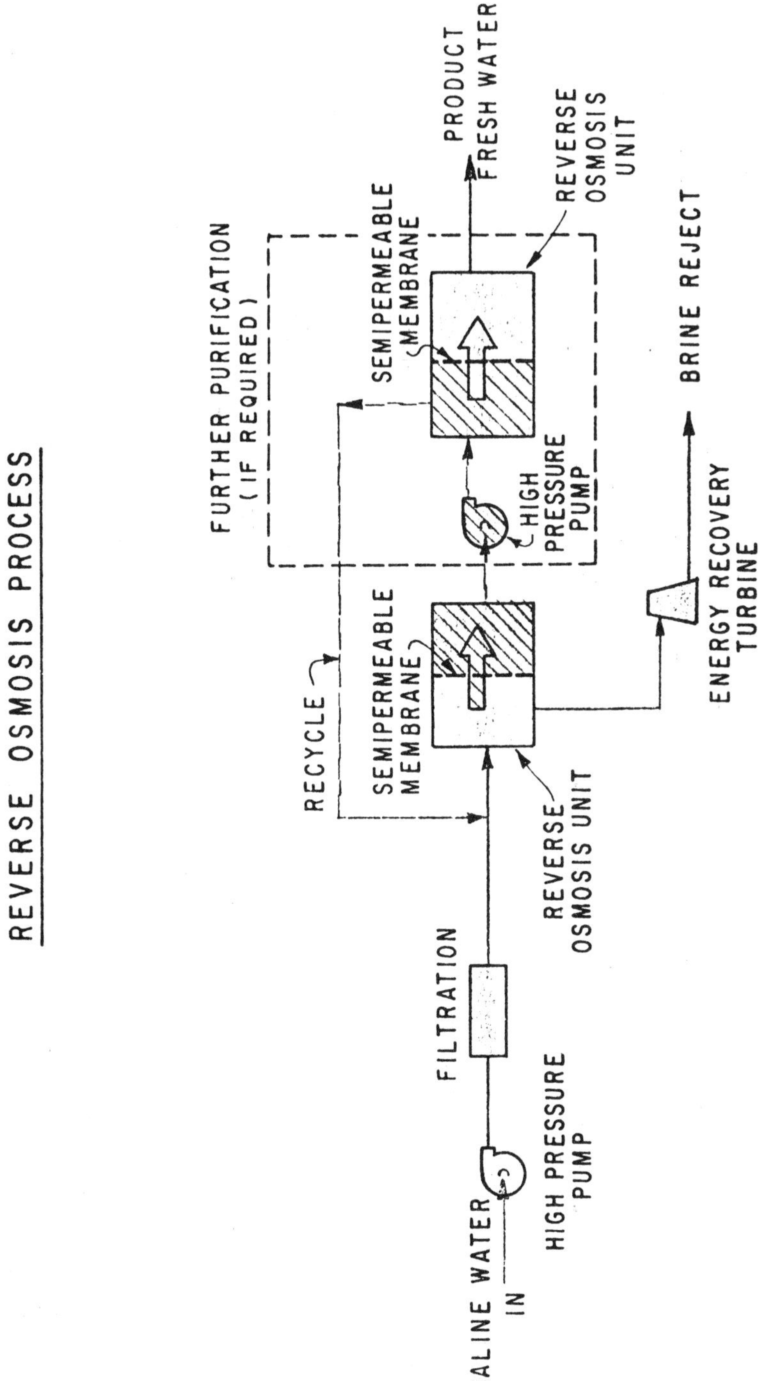

FIGURE 5 Reverse osmosis plant with power recovery

TABLE 3

CLASSIFICATION OF MAJOR DESALINATION PROCESSES

A. PROCESSES INVOLVING DISTILLATION

- Simple Solar Stills
- Multistage Flash Process
- Multieffect Multistage Flash Process
- Long Tube Vertical (LTV) Multieffect Process
- Vapor Compression Process

B. PROCESSES INVOLVING CRYSTALLIZATION

- Indirect Freezing Process
- Direct Vacuum Freezing Process

C. PROCESSES INVOLVING MEMBRANE SEPARATION

- Reverse Osmosis
- Electrodialysis

TABLE 4

CURRENT ENERGY REQUIREMENT FOR SEAWATER DESALINATION BY THE MAJOR PROCESSES

PROCESS	POWER CONSUMPTION	
	kWh/k gal.	k cal./lit.
PROCESSES USING THERMAL ENERGY		
Simple Solar Still	2400	540
Multistage Flash	220	50
Multieffect Multistage Flash	110	25
Long Tube Vertical Multieffect	240	54
PROCESSES USING ELECTRICAL ENERGY*		
Vapor Compression	55	12
Direct Freezing	45	10
Reverse Osmosis (with power recovery)	75	17
Reverse Osmosis (without power recovery)	110	25
Electrodialysis	120	27

* The energy values required by the processes using electrical energy are the thermal energy equivalent using a thermal efficiency of 33%.

2.8 kWh/1000 gal (0.64 k cal/lit) of product water. Less energy would be required for separation of less saline waters. In any practical process the energy required is related to the potential causing transport of the salt or water, and will be much larger than the theoretical minimum.

The desalination processes of current interest may be classified into three categories as shown in Table 3. Some of these processes require thermal energy (e.g. Multistage Flash and Long Tube Vertical Multieffect distillation processes) and some require electrical energy (e.g. vapor compression and reverse osmosis).

The current energy requirement for the major desalination processes for sea water are given in Table 4. Among these processes, vapor compression, direct freezing, and reverse osmosis (with power recovery) are the most energy efficient. Vapor compression and reverse osmosis processes seem to be in a more advanced stage of development than the direct vacuum freezing process. Schematics of the vapor compression and reverse osmosis processes are shown in Figures 4 and 5.

An energy efficient desalination process would mean a smaller and, therefore, less expensive solar collector.

SOLAR ENERGY CONVERSION SYSTEMS

Solar energy may be used to generate heat or electric power which can then be used for operating any desalination process. The energy input section and the desalination section may be combined in one piece of equipment as in the case of the simple solar still. They can also be separate as in the other desalination processes.

Processes employing heat or thermal energy can be supplied with steam from solar boilers heated by concentrated solar energy collectors. Processes using electrical power could be operated by electricity generated in solar heat engines or photovoltaic systems. The pumping power requirements in thermal desalination processes could be provided by a steam turbine. A schematic diagram of a solar energy conversion system which may provide both steam and pumping power for a thermal desalination process is shown in Figure 6. This system uses parabolic solar collectors and heat pipes. Heat exchangers are used to transfer the heat from the heat pipe fluid to the working fluid. As the working fluid passes through the successive heat exchangers, it is initially heated to saturation temperature and subsequently boils.

A recuperated Brayton cycle heat engine was suggested[3] for the conversion of solar energy into electrical power required by such desalination processes as reverse osmosis and vapor compression, see Figure 7. Thermal efficiencies greater than 40% can be practically achieved with this engine. Helium is a suitable working fluid for this engine.

Direct conversion of solar energy into electricity can be achieved by the

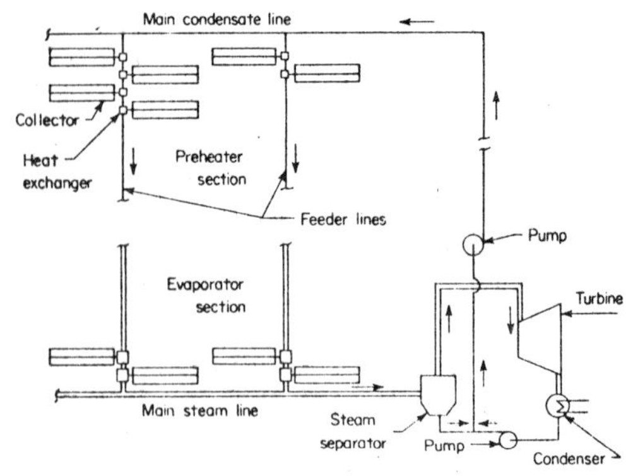

FIGURE 6 Solar energy conversion system

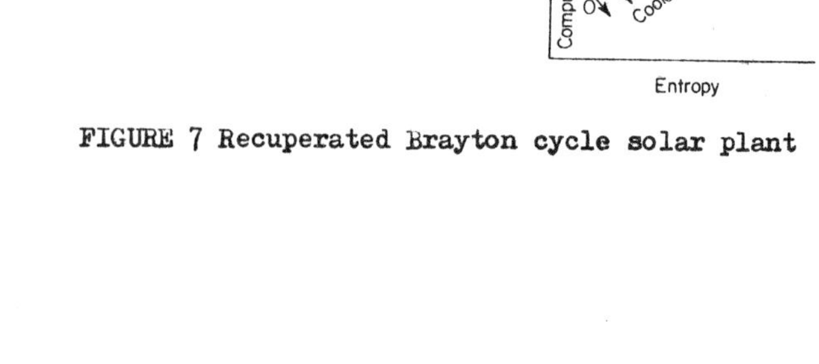

FIGURE 7 Recuperated Brayton cycle solar plant

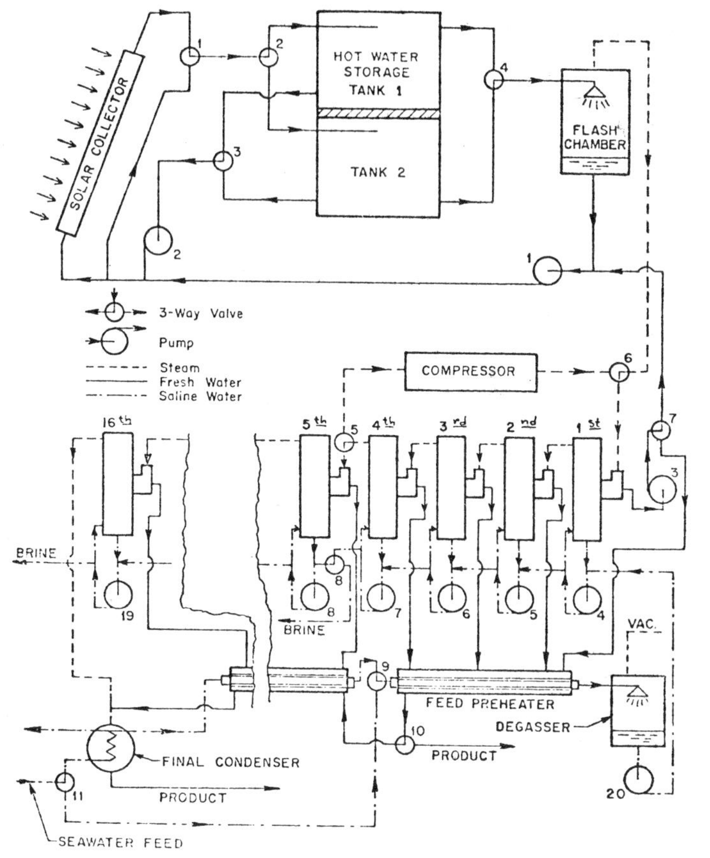

FIGURE 8 Solar energy assisted multieffect distillation plant with vapor compression option

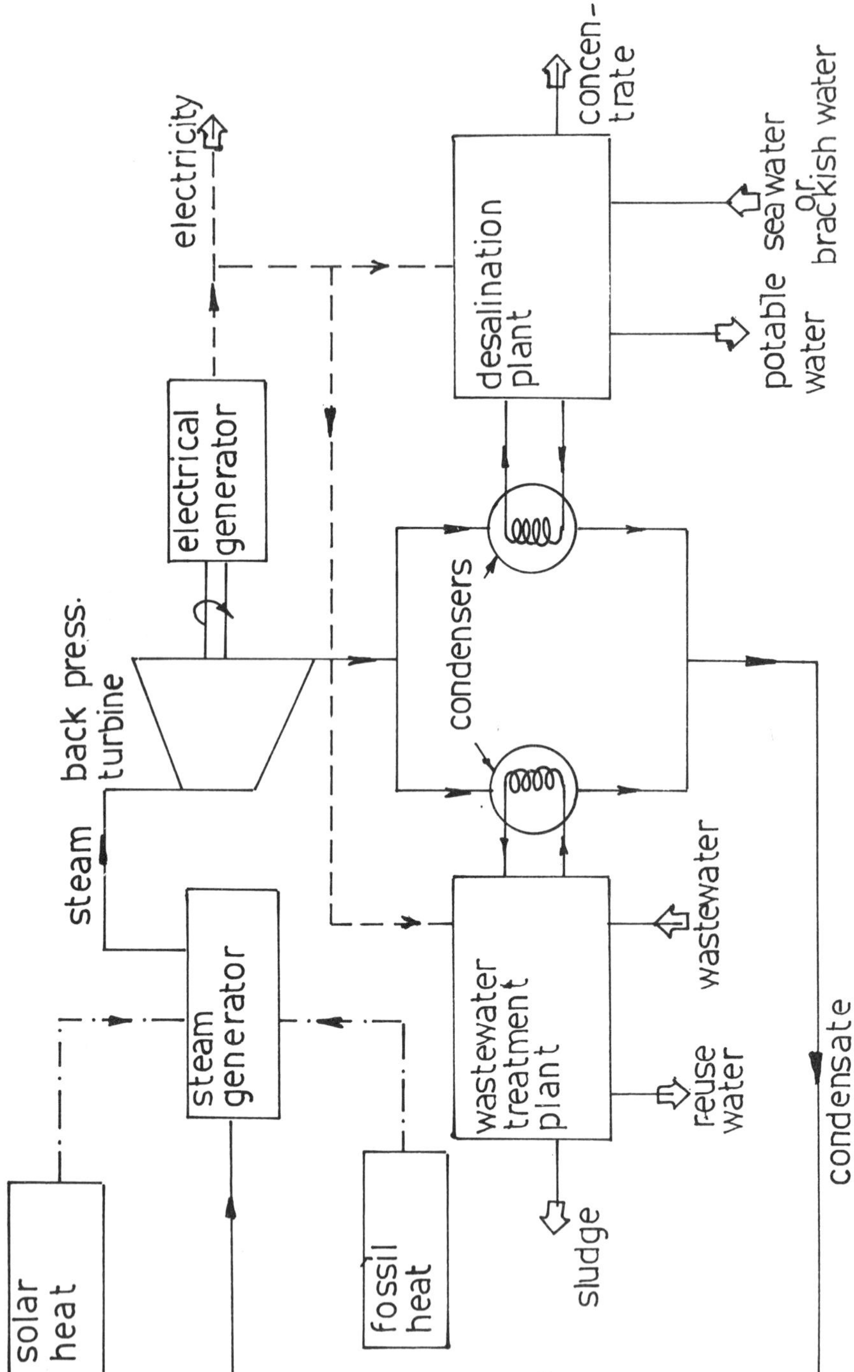

FIGURE 9 Proposed electrical power plant, desalination plant, and wastewater treatment plant complex

use of silicon solar cells. The efficiency of these cells can be as high as 20% and their cost is expected to be $ 1.00 / watt by 1980 [4].

SOLAR ENERGY ASSISTED DESALINATION PLANTS

In order to enable a desalination plant to produce water during extended sunless periods, an extra source of power (either thermal or electrical) should be used in these periods. A solar energy - assisted multieffect distillation plant with vapor compression option was suggested in [2] and its flow schematic is shown in Figure 8. In this system it is proposed to produce steam from a solar boiler using flat plate collectors. The vapor compressor is manifolded between the first and the fourth effects of the 16 effects employed.

During sunny periods, solar heated water from tank 2; see Figure 8; is passed into a flash chamber to produce steam. This steam passes into the first effect where it condenses and is returned into the steam producing section. Vapor produced in the first effect is introduced into the second effect which is maintained at a slightly lower pressure than the first effect. The vapor produced in the second effect is passed to the third effect and so on, until the last effect from which the vapor produced is passed into the final condenser.

During periods of sunless or cloudy weather, the temperature of the solar heated water drops and this may automatically switches on the vapor compressor.

SOLAR ENERGY FOR WASTE WATER RENOVATION

Any total water resource management program in a water shortage region should consider the treatment for reuse of industrial, municipal, and agricultural wastewaters. This treatment may be simple conventional primary and secondary treatment or a more sophisticated treatment which include tertiary treatment processes such as activated carbon, ion exchange, and reverse osmosis. In an energy shortage region , solar energy may provide part or all of the energy required by these processes.

One possible use of solar energy in waste water treatment is in increasing the pollutant removal efficiency of the treatment process by optimizing the waste water temperature. In a ferric chloride, powdered activated carbon (PAC) and an aluminum sulphate, PAC waste water treatment pilot plant [5] , it was found that optimal removals of COD, suspended solids, turbidity, and phosphorus in sewage waste water occured between 50 - 60°C.

A solar energy assisted power plant, desalination and waste water treatment complex may be proposed for medium to large size desert communities which is located either on sea coast or in an inland location where brackish water with high salinity exists. Figure 9 is an outline of such a system. In this system, a back pressure steam turbine may be used. The thermal energy required by the desalination and waste water treatment sections are provided by two steam condensers. The electrical energy required

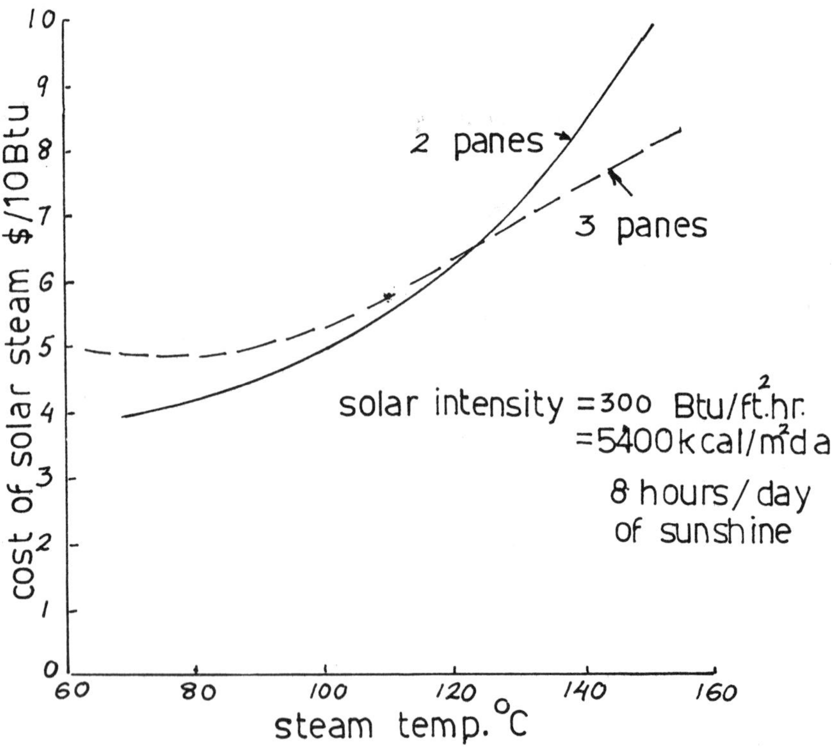

FIGURE 10 Effect of steam temperature on the cost of steam produced by solar boilers

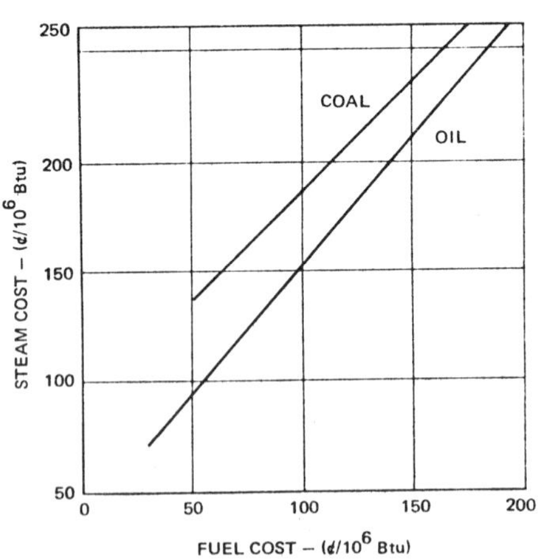

Figure 11 EFFECT OF FUEL COST ON STEAM COSTS

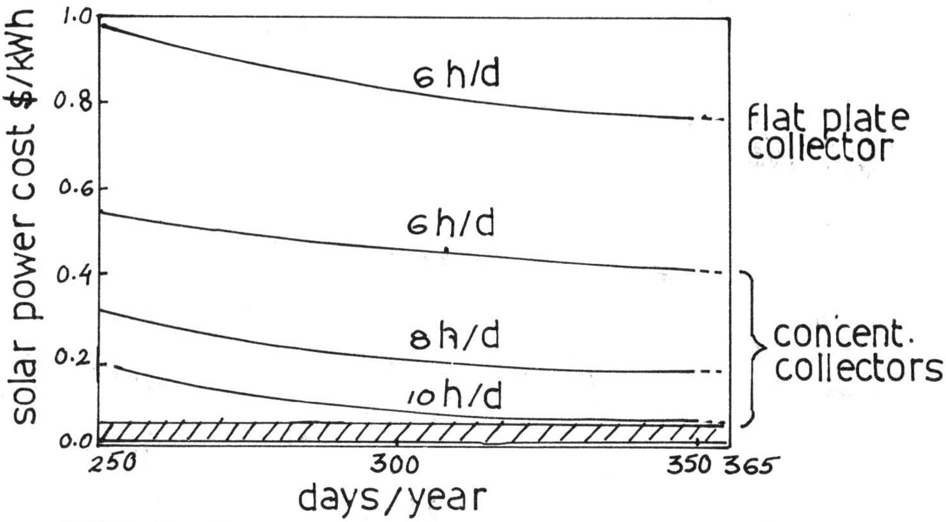

FIGURE 12 Estimated cost of electrical energy from solar plants

by both sections are provided from the main electrical output.

ECONOMIC CONSIDERATIONS

The economics of using solar energy for desalination and waste water treatment depend largely on the type of solar system used, the climatic conditions, and the cost of conventional energy. Solar energy cost either in the form thermal or electrical energy can be used as an index for comparison with the cost of conventional energy. This cost is usually given in US \$/10^6Btu for thermal energy, and ¢/kWh for electrical energy. This cost is usually defined as the annual capital charges of the solar equipment (first cost times amortization rate) divided by thesolar heat collected and utilized.

The cost of steam from a solar boiler is dependent on its temperature. Flat plate collectors could be used to produce steam for the distillation processes at temperatures of 100°C or more. However, these collectors suffer from the disadvantage that their efficiency drops as the temperature of the collector increases. Multistage distillation processes have higher energy efficiency when using heating steam at higher temperature. The steam temperature used depends mainly on economic considerations.

Using a selective surface absorber (one that has a high solar absorptance but low infrared emittance) similar to that used by Hottel et al. [6], it is possible to make an approximate estimation of the cost of steam from solar boilers using flat plate collectors. The following cost parameters are used to calculate the steam cost :

Capital cost of flat solar collectors are ,

\$ 55.00 /m^2 (\$ 6.00 /ft^2) 2-panes
\$ 70.00 /m^2 (\$ 8.00 /ft^2) 3-panes

Cost of solar boiler = \$ 10,000 /1000 lb steam per hour
= \$ 22,000 /1000 kg steam per hour

Amortization = 20 years at 10 %.

The collector efficiency values of reference [6] are used in the calculations and the cost figures are given in Figure 10. It is clear from the graph that the cost of steam produced by solar boilers increases as its temperature increases. These costs do not include any operating costs.

The cost of steam produced in conventional steam generators depends heavily on the cost of fossil fuel used. Figure 11 shows the effect of the cost of oil and coal on the cost of steal produced. In many parts of the world the cost of oil is now over \$ 2.00 /10^6Btu and the resulting cost of steam is over \$ 2.50 /10^6 Btu. With the expectation that conventional energy cost will, on an average, increases at a faster rate than the solar equipment costs, the economic attractiveness of solar energy use is expected to improve

The cost of electricity generated by solar power plants as given in [7] is shown in Figure 12. In this Figure, the cost of electricity in \$/kWh is given for flat plate collectors and concentrators as a function of the annual clear days per year, and for 6,8,and 10 hours of operation per day. The dashed area represents the range of cost of electricity produced by conventional power plants. With a favorable climatic conditions , solar

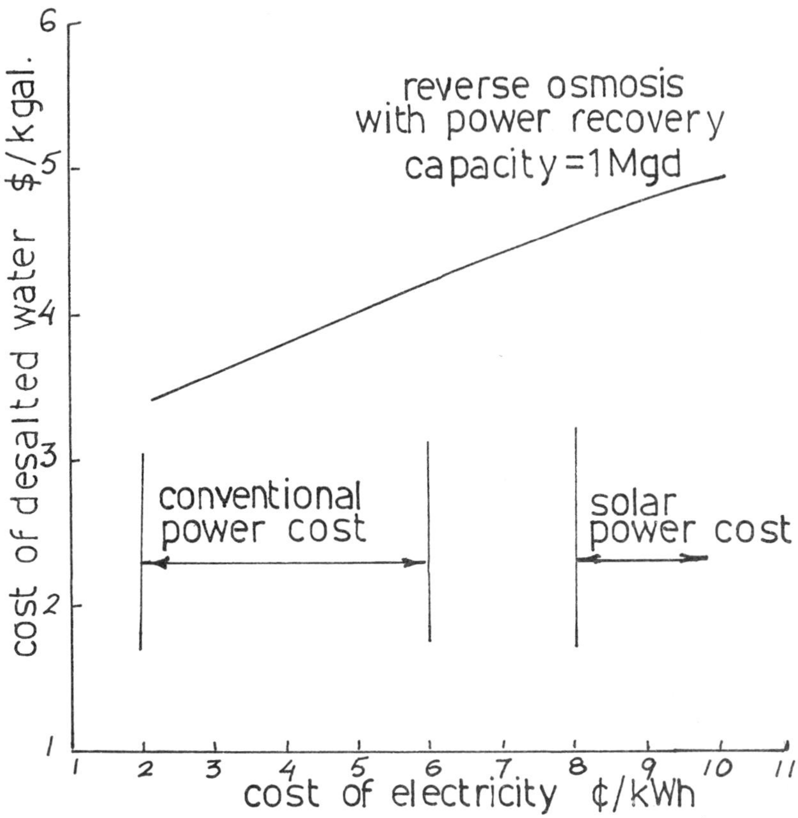

FIGURE 13 The effect of the cost of energy on the cost of desalted water by reverse osmosis

power plants using concentrating collectors are expected to produce electrical energyfor less thal 0.1 $/kWh. However, this cost is substantially higher than the cost of electrical power produced now by conventional plants.

The effect of the cost of electrical power in ¢/kWh on the cost of water produced by a reverse osmosis sea water plant having a capacity of 1 Mgd (million gallon per day) is shown in Figure 13. It is clear that the use of electrical power generated by solar plants will cause a moderate increase in the cost of desalted water produced. Similar trend occurs with other desalination processes. Generally, the effect of the cost of energy on the cost of water produced is not substantial.

CONCLUSION

It is believed that the combination of solar plants with certain energy efficient desalination processes like reverse osmosis and vapor compression, can be a promising development for medium and large size communities in arid areas. The economic feasibility of this combination is expected to be improved in the next few years as the conventional fuels - both fossil and fissile - become scarce and more expensive.

REFERENCES

1 Ralph, E.L.,"A plan to utilize solar energy as an electric power source", Eighth IEEE Photovoltaic Specialists Conference, Aug. 1970

2 Tleimat, B.W.;Howe, E.D.; and Laird, A.D.K.," Solar assisted still with vapor compression option", Proceedings of the International Congress on Desalination and Water Reuse, Tokyo, Nov. 1977.

3 Woodcock, G.R., and Gregory,D.L.,"Heat engines for solar power", Energy Technology Handbook, D.M. Considine Editor, McGraw Hill Inc. 1977.

4 Oman, H., and Gelzer, J.W.,"Solar cells and arrays", Energy Technology Handbook, D.M.Considine Editor, McGraw Hill Inc. 1977.

5 University of Arizona,"Feasibility study of a nuclear power-sewage treatment system for the conservation and reclamation of water resources", Water Reuse Highlights, AWWA Research Foundation, Jan. 1978.

6 Hottel and Unger, Solar Energy, Oct. 1959

7 Delyannis, E., and Delyannis, A.,"Solar applications in desalting", Proceedings of the International Congress on Desalination and Water Reuse, Tokyo, Nov. 1977.

8 Savage, W.F., and Reed, S.A.,"Desalting cost update- 1977", Fifth Annual NWSIA Conference, San Diego, July 1977.

****

DEVELOPMENT PLANS AND CONCEPTUAL DESIGNS FOR MULTI-PURPOSE SOLAR DESALINATION UNITS (MSDU)

H.K. Abdel-Aal
University of Petroleum & Minerals
Dhahran, Saudi Arabia

ABSTRACT

Solar distillation occupies a favored position amoung other conventional desalination process in capacity range up to 50,000 or perhaps 100,000 GPD. For higher range output (in the neighborhood of millions of GPD), other functions have to be sought for the solar evaporators in order to make them competitive with ones run by fuels.

This paper outlines some of the design features of a Multi-Purpose Solar Channel-Evaporator that would serve to produce high-purity sodium chloride and magnesium chloride salts in addition to fresh water. The progress of water evaporation accompanied by the preferential precipitation of the salts is simulated in this work, utilizing "the inclined-flat-surface-model". The velocity distribution, the residence time and the depth of water along different channels are determined by the use of the "equations of change" for a flow system.

Process analysis is included as well, for adopting the MSDU to existing conventional desalination plants whose rejected brines would serve as feed stock for the MSDU.

INTRODUCTION

The purpose of this research is to introduce and develop the Multi-Purpose Solar Desalination Units (MSDU), whereby effluents from conventional desalination plants, sea water, and brackish water could be converted to fresh water along with some valuable mineral products. Water thus produced will be provided for its customary beneficial uses especially in arid and semi-arid regions of the world. The vision of "making the desert bloom" in a country like Saudi Arabia should have its role in saline water conversion development as one of those hyperboles by which man is inspired to reach ample fresh water supply.

For the saline water conversion to be a practical reality, its cost should not be greatly out of line with the highest cost for fresh water obtained from conventional sources. The cost of energy has been always a major component of the cost of distilled water when using fossil fuels. The question, therefore, arises as to whether we can lower such operating costs if we use the sun as a source of energy?

When talking about the cost of fresh water produced by desalination methods in general, one has to be aware of the fact that the total product cost is the sum of:

- Fixed costs which include interest and depreciation on the capital investment.
- Operating costs which consist of cost of energy (fuel) and labor.

By using solar energy instead of conventional fuels, the unit cost (capital investment) for solar stills are acceptable for units of very small capacities and are less than that of very small fuel-operated distillers of comparable sizes [1]. Increasing the size of the solar stills does not, however, decrease the capital cost appreciably. Thus it is found that the cost of large-size ones are relatively more costly than competitive fuel-operated distillation plants.

In order for solar distillers to become economically attractive for any but the very small installations, now commercially available, there has to be a significant decrease in the cost of water produced by these large units. It is along these lines that this work is undertaken for the recovery of some of the valuable mineral salts along with producing fresh water.

What Is A Multi-Purpose Solar Distillation Unit (MSDU)?

In this part we shall endeavour to explain a very basic concept, namely that concerned with the terminology of MSDU. Such proposed units are recommended to be tailored to conventional desalination plants run by fossil fuels. This, however, does not exclude using sea water and brackish water as a feed. If this is the case, such MSDU are envisaged to function along the following lines:

1. Disposal of saline water conversion wastes [2]
2. Production of fresh water
3. Recovery and extraction of mineral salts [3]

As far as function number 1, it is worth mentioning that the disposal of saline-water conversion wastes presents a crucial problem in this regard. Brines rejected from distillation plants come in the form of simple, inorganic compounds. They can not be reduced any further to a simpler form. Under the present known conditions it is not economically feasible to separate the brines from the water either to extract them as valuable salts or for the reclamation of the water itself. The disposal costs incurred in any of the well-known schemes make the production of potable water both from saline and brackish water less attractive.

Using a MSDU would thus eliminate such a disposal problem while producing fresh water and valuable mineral salts.

Description Of MSDU

The plant will consist of a number of shallow inclined basins (10 m wide and 50 m long) arranged in a stepped series as shown in Figure 1. Each basin is covered with a transparent plastic cover. Pre-concentrated saline water is fed to the upper basin, from which it over-flows to the next and so on.

Solar energy absorbed by the water and by black polyethylene films covering the bottom, supplies heat for evaporation of water. Water condenses on the inside of the cover and to be collected in a trough leaving each basin. Mineral salts will preferentially precipitate at the bottom of different basins in accordance with their solubility as shown in Table 1.

Flow conditions and level of saline water along the different basins (channels) are controlled to assume optimum performance of operation as shown in the next part.

Application Of Equations Of Change For Design Calculation

Using the inclined-flat-surface model described by Bird & Associates [5], it has been possible to calculate the velocity distribution (θ) and the depth of water attained in these basins (H) for the distillation process of saline water. The evaporation along the MSDU takes place under the influence of solar flux at the rate of 3.5 m/year [3].

The basic equations used are:

1- The average velocity

$$\langle v \rangle = \frac{\rho g \, (H)^2 \, \text{Cos } \beta}{3\mu} , \text{ (cm/sec)}$$

2- Volumetric flow rate

$$Q = \frac{\rho g \, H^3 \, W \cos \beta}{3\mu} , \text{ (cm}^3\text{/sec)}$$

3- Residence time

$$\theta = \frac{L}{\langle v \rangle} , \text{ (sec)}$$

4- Hydraulic Diameter

$$D_H = \frac{4 \text{ x cross sectional area}}{\text{wetted perimeter}} , \text{ (cm)}$$

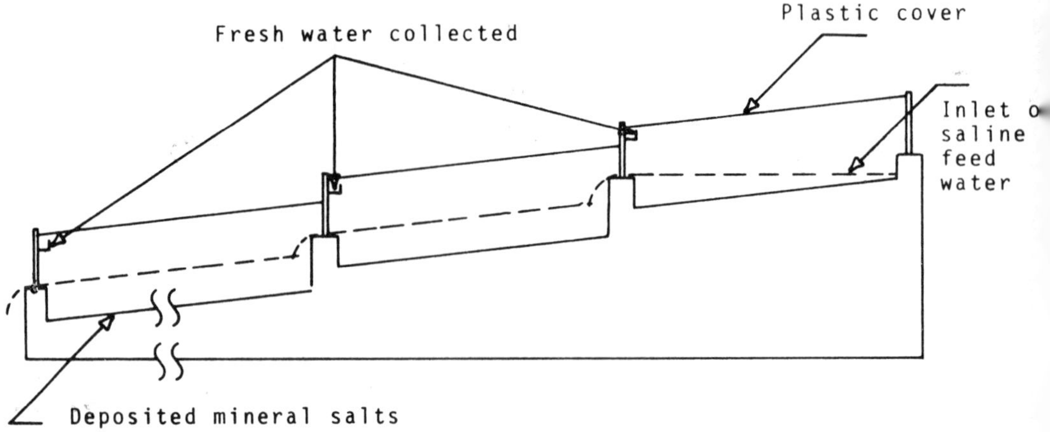

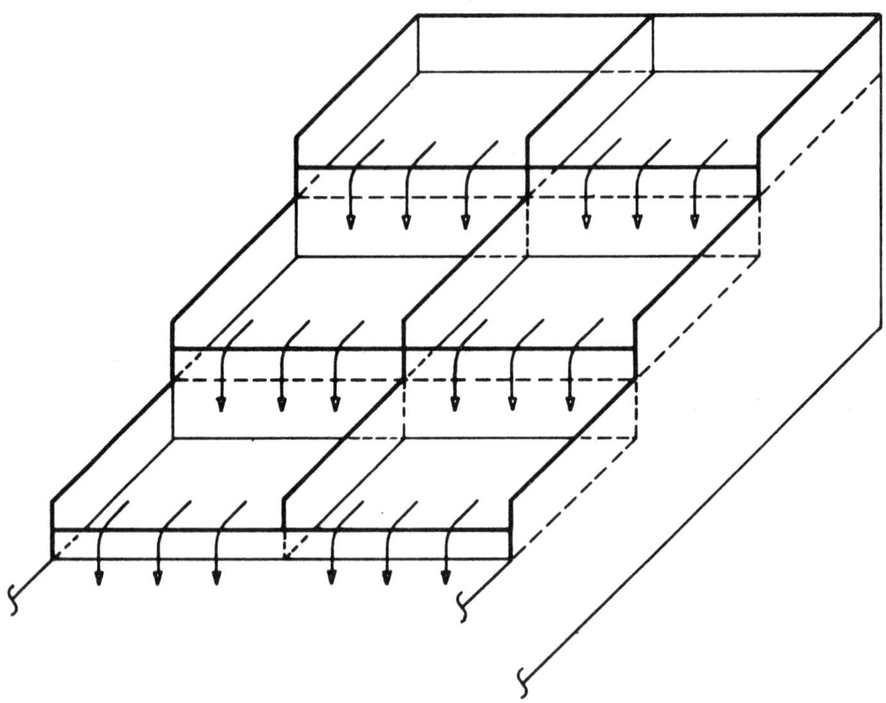

Figure (1): Stepped multipurpose solar Distillation Unit (MSDU).

TABLE 1

PROGRESS OF THE SEPARATION OF SALTS ALONG THE CHANNEL EVAPORATOR [4]

(Preferential Separation Method Using Solar Energy)

Element	Salt	Sp. Gravity	Channel No.	Range of PPM
Ca^{++}	$CaCO_3$ + Ca SO_4	1.130	6-8	2100 590 1000
Na^{+}	Nacl	1.200	17-19	85,000 15,300 55,000
Mg^{++}	(i) Mg SO_4	1.29 - 1.33	8-11	14,400 1,850 10,000 29,000
	(ii) Carnallite (Kcl Mg Cl_2 6 H_2))	-	12-17	10,000 14,000
	(iii) Mg Cl_2	1.33 - 1.37	17-19	42,000 (Viscus pdt.)

Note: The final product leaving the evaporator contains Mg Cl_2 is a highly concentrated form (23 times its initial intake in the sea water)

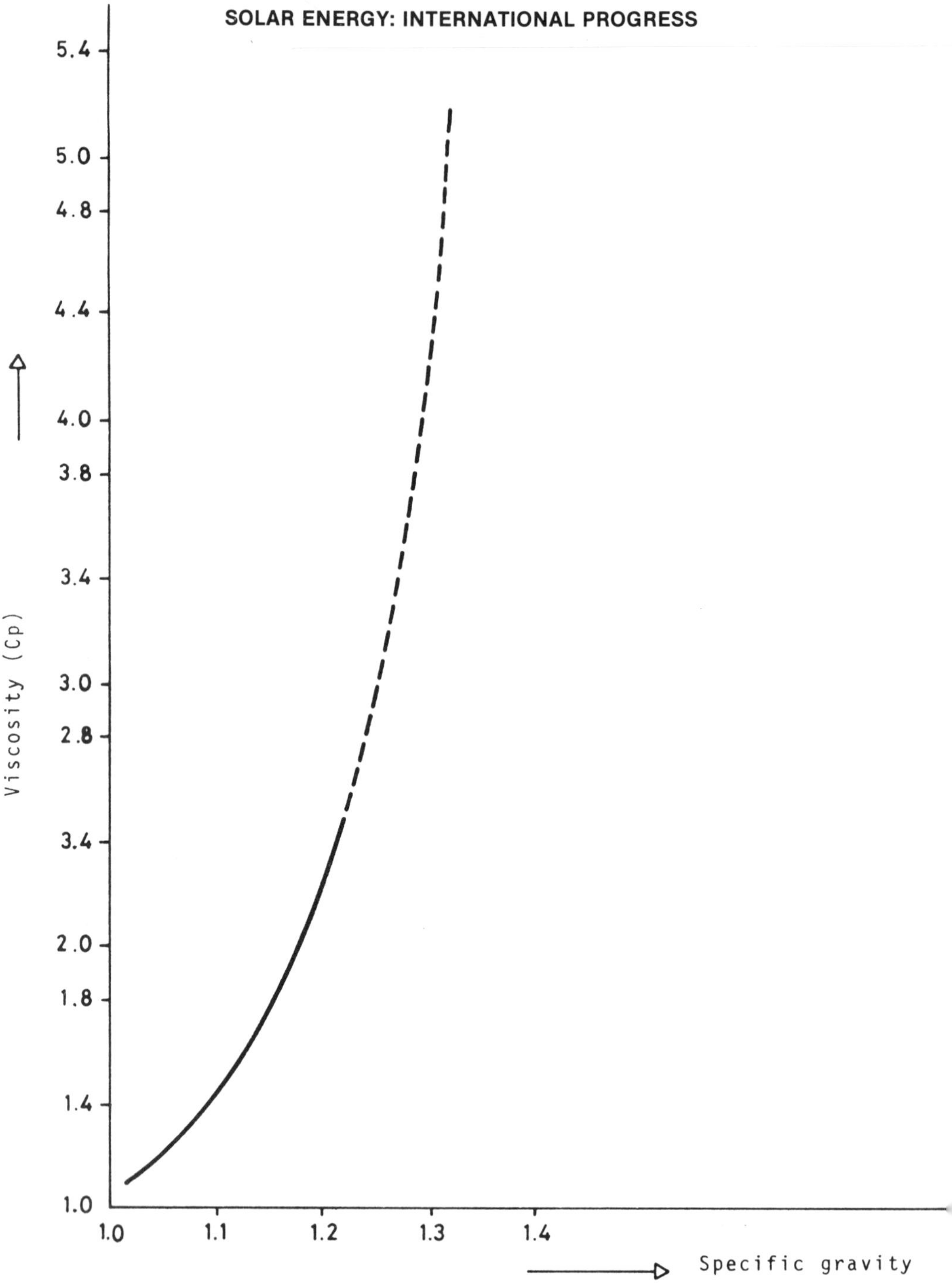

Figure (2) Change of Viscosity with Specific gravity for sea water (enchanced evaporation)

where:

ρ, μ = density and viscosity of the fluid

= 1.3 gm/cm^3 and 4×10^{-2} gm/cm-sec respectively for this calculation [6]. See Figure 2.

g = gravitational force 980 cm/sec^2

β = the inclination of the basin with the vertical direction

= 89°, $\cos \beta = 0.0175$

W, L = width and length of the basin

= 10 m and 50 m respectively

H = height of the fluid in the basin

= 2 cm (initially)

Calculations

The caluculational procedure used in this work is schematically outlined in Figure 3.

(i) Initial Average velocity

$$\langle v \rangle = \frac{1.3 \times 980 \times (2)^2 \ (.0175)}{3 \times 4 \ 10^{-2}} = 7.4317 \times 10^2 \text{ cm/sec} = 7.43 \text{ m/sec}$$

ii) Initial volumetric flow rate

$$Q = \frac{1.3 \times 980 \times (2)^3 \ (1000) \ (.0175)}{3 \times 4 \times 10^2} = 1.48633 \times 10^6 \text{ cm}^3\text{/sec}$$

$$= 1.48633 \times 10^3 \text{ lit/sec}$$

$$Re = \frac{1.3 \ (7.43 \times 10^2) \ (1.99)}{4 \times 10^{-2}} = 4.81 \times 10^4$$

iii) Calculation of the residence time in the channels:

Ist channel

$$\theta \ \frac{L}{\langle v \rangle} = \frac{50\text{m}}{7.43 \text{ m/sec}} = 6.728 \text{ sec}$$

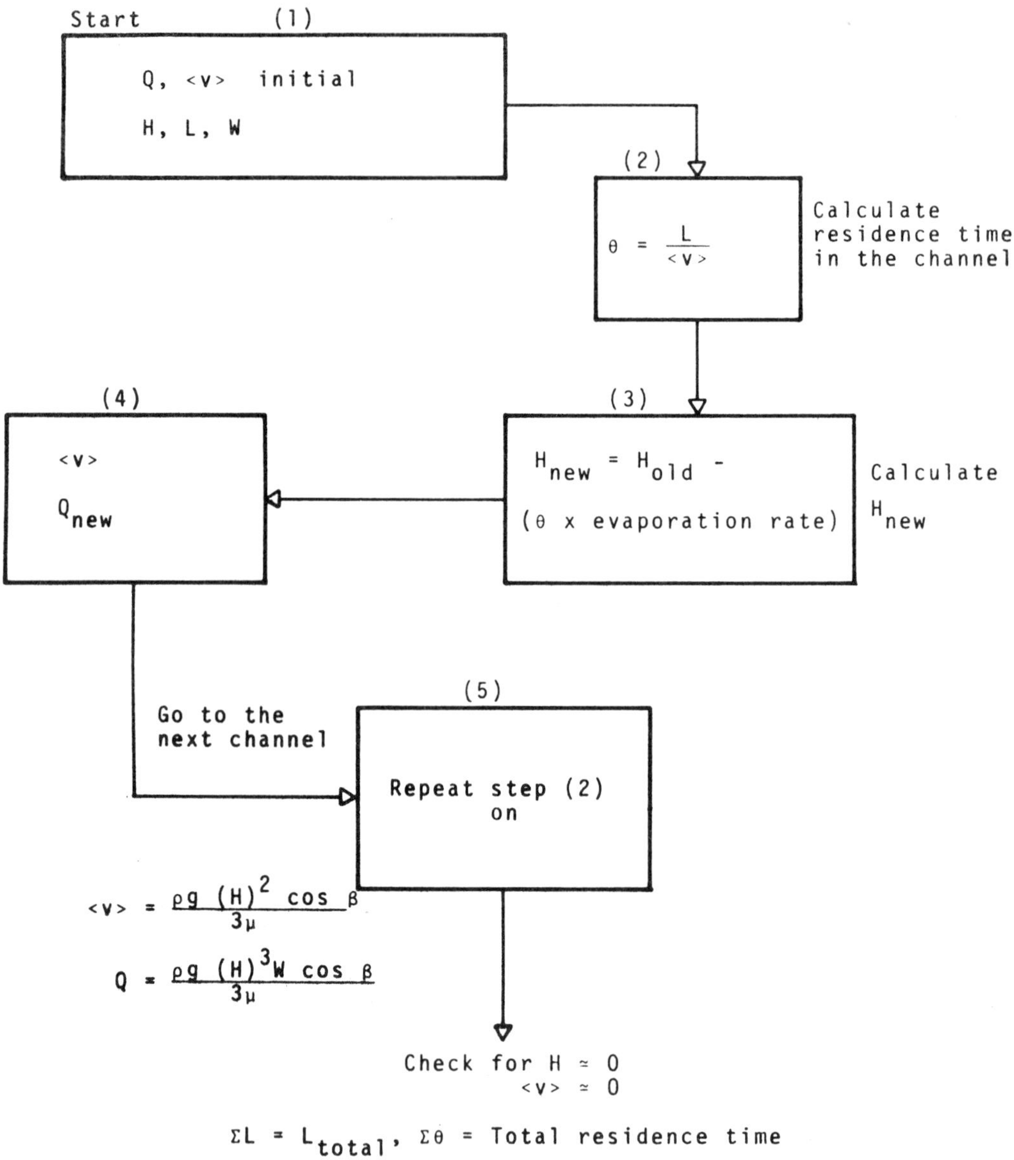

Figure: (3) Schematic Procedure for channel-to-channel calculation using flat surface model.

Now assuming the rate of evaporation by solar energy:

1mm/hr (2.778×10^{-4} mm/sec), based on 3.5 m/year average solar evaporation rate in Saudi Arabia [3], and assuming 300 working day and an average 10 hours of sunshine, we get the following results:

The height of fluid entering the second channel is:

$$H_2 = (20 \text{ mm} - 6.728 \text{ sec} \times 2.778 \times 10^{-4} \frac{\text{mm}}{\text{sec}}) = 19.998 \text{ mm}$$

The velocity profile, flow rate, residence time for channel 2 are:

$$\langle v \rangle = \frac{\rho g \, (H)^2 \, \text{Cos } \beta}{3\mu} = \frac{1.3 \times 980 \, (1.9998)^2}{3 \times 4 \times 10^{-2}} (.0175)$$

$$= 7.4302 \times 10^2 \text{ cm/sec}$$

$$Q = \frac{\rho g \, (H^3) \, W \cos \beta}{3\mu} = 1.4858 \times 10^6 \text{ cm}^3\text{/sec}$$

$$Q = \frac{L}{\langle v \rangle} = \frac{50\text{m}}{7.4302 \text{ m/sec}} = 6.7293$$

Similarly H_3 could be calculated and another cycle could be made for the third channel.

Calculation for the first five basins are shown in Table 2.

Channel-to-channel calculations shows that average values could be conservatively predicted as follows:

Average rate of evaporation ΔH, 0.002 mm per channel, average residence time θ = 10 sec for a viscosity in the range of 4 - 4.5 cp. This is estimated for the preliminary stage in distillation. With further evaporation, the viscosity of the saline water increases to about 5.5. cp which will change the residence time profile and consequently change both $\langle v \rangle$, and ΔH (θ = 14 sec, v =360 cm/sec, ΔH = 0,004 mm). These conditions are typical for the first stage in actual evaporation. Finally, extrapolating the value of viscosity from Figure 2 gives μ = 8 cp. corresponding to a specific gravity of 1.4 at which (θ = 50, v - 100 cm/sec, ΔH = 0.02 mm). Such a stage characterizes the presence of a very viscous flow. This stage could very well continue until separation of solid magnesium chloride occurs as an end product in the last few basins. Based on the data given for the three above identified regions: preliminary stage in distillation, first stage in actual evaporation, and viscous flow stage, overall average values for θ and H were calculated and found to be:

θ = 31 seconds per basin

TABLE 2

SAMPLE CALCULATION FOR THE FIRST FIVE BASINS

1	2	3	4	5	6
Channel#	H(mm)	θ(sec)	<v>($^{cm}/sec$)	Q(cm^3/sec)	DQ cm^3
1	20	6.728	7.4317×10^2	1.48633×10^6	
					530
2	19.998	6.7293	7.4302	1.4858	
					358
3	19.996	6.7306	7.4287	1.4854	
					417
4	19.994	6.7319	7.4273	1.4852	
					660
5	19.992	6.7333	7.4258	1.484579	

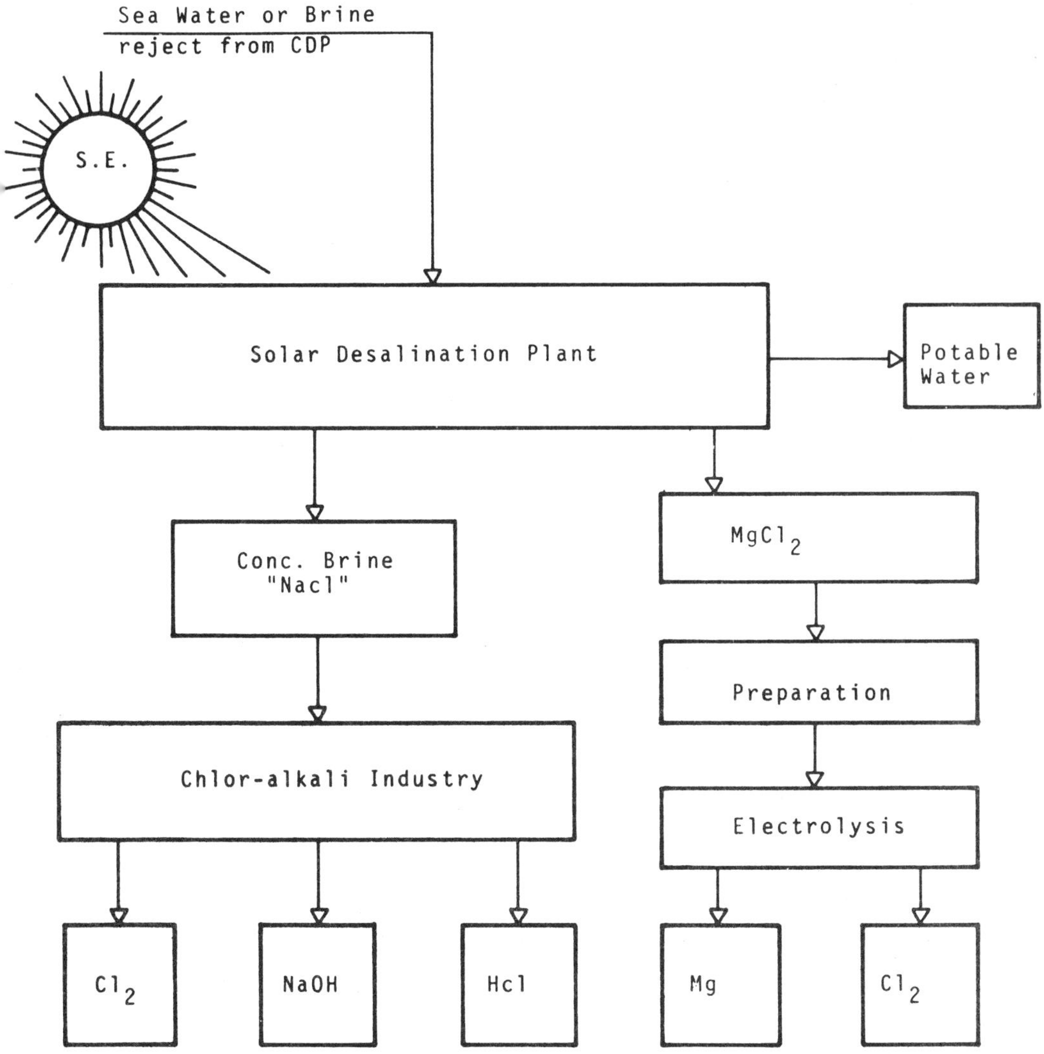

Figure: (4) Solar Desalination Plant (SDP) provides fresh water, conc. Brine, Nacl & $MgCl_2$

H = 0.0115 mm per basin

Accordingly the number of basins required for complete evaporation of the 2 cm initial intake water (Q = 1.5 x 10^3 lit/sec) will be 1750 and the surface area occupied by them will be 875,000 m^2, which is little less than 1 Km^2. The total residence time spent in evaporating this much saline water is about 15 hours.

The amount of sodium and magnesium salts separated along these evaporators are readily calculated and found to be 1350 tons/hr and 160 tons/hr respectively.

CONCLUSION

Because of the emphasis on understanding the basic physical principles underlying the separation process of both fresh water and mineral salts by solar distillation of saline water, we feel there is a very definite need for experimental work in this area. Currently, experimental equipment is being designed at our chemical engineering laboratories. In the proposed research, the change of density and viscosity of the evaporating fluid along the channels will be taken into consideration. Different forms of the equations of change have to be used [6]. The use of boundary layer thickness (δ) in correlating the height of the evaporating liquid (H) is anticipated in this work.

The proposed scheme of MSDU could very well serve the purpose of supplying pure sodium chloride solution for the chlor-alkalai industry. Magnesium chloride on the other hand, is the raw material for magnesium metal. Chemical products manufactured from these mineral salts are shown in Figure 4.

As far as the operating scheme of the MSDU, it is to be noted that the saline water goes through two cycles:

1. is to preconcentrate the saline water by going once through the unit. This concentrates the saline water up to a specific gravity of 1.3. In this cycle separation of salts other than magnesium takes place. This is a concentration of about ten times the original one.

2. is to allow the concentrated brine (specific gravity 1.3) to go through the unit once more. Here we are having separation of magnesium chloride as the end product.

REFERENCES

1. Howe, Everette D. "Fundamentals of Water Desalination", Marcel Dekker Inc., New York, 1974.

2. Abdel-Aal, H.K. "The Disposal of Saline Water Brines Is Not A Burden Anymore", Pure Water (IDEA), Vol. III, Nos. 2 & 3, May 1975.

3. Kettani, M. Ali and H.K. Abdel-Aal "Production of Magnesium Chloride From The Brines Of Desalination Plants Using Solar Energy", Proceedings of the 4th International Symposium On Fresh Water From The Sea, Vol. 2, 509-516, 1973.

4. Abdel-Aal, H.K. and F. Al-Somait, "Solar Energy Prospects In Saudi Arabia", Energy Communications, Vol. 4, No. 2, 1978.

5. Bird, R., Byron & Associates, "Transport Phenomena", John Wiley & Sons, Inc. 1960.

6. Al-Nemer, M. Thesis Research in Progress. Chemical Engineering Department, UPM, Dhahran, Saudi Arabia.

REVIEW ON APPLICATION OF SOLAR ENERGY FOR DRYING FOOD STUFFS

K.V. Mallan, T.V. Balasubramanian
and R.K. Bhardwaj
Motilal Nehru Regional Engineering College,
Allahabad - 211 004
India

ABSTRACT

A review is given on the application of solar energy for drying food stuffs. Sun-drying was the first application of solar energy and was practised since the early days of civilization. Slow drying rates, poor quality product and contamination are inherent drawbacks of this method. Solar drying is recognised as a superior method to sun-drying. But sun-drying is still the largest method of preserving food stuffs. In order to bring out the potentialities of solar drying and the limitations of sun-drying, both these methods are discussed The progress in the solar dryer research is hindered due to the lack of a mathematical model to compute and compare the performance of different solar dryers. The necessity to initiate studies on the influence of various combinations of air velocity, temperature, and humidity on drying rate is emphasized.

INTRODUCTION

Drying of food stuffs was the prime application of solar energy since the early days of civilization. It is evident from the existence of storage pits in the three ancient valley civilizations of Indus, Nile and Tigris-Euphrates, that sun-dried grains were stored as early as 3,000 B.C. [1,2]. The process of drying food stuffs by spreading them on open ground, directly exposed to the Sun and wind constitutes sun-drying. Sun-drying is practiced in many parts of the world, where solar radiation and climatic conditions are favourable. But many limitations of the sun-drying process and a better understanding of the methods of utilizing solar energy to advantage, have given rise to a scientific method called solar drying. Whereas there is no control over the drying rate and quality of product in sun-drying, better quality of product and higher drying rates are possible in solar drying. Solar dryer research and development has been expanding rapidly in the past few years, spurred on mainly by the potential use of these dryers for large-scale applications.

This paper is a review of the status of solar dryers at present. However, sun-drying also is discussed, as it is reported to be more appropriate in some cases. This paper divides the field of application of solar energy for drying of food stuffs into four main areas: Sun-drying, cabinet solar dryers, batch solar dryers and continuous flow solar dryers.

For more details on principles of drying, the reader is referred to Refs:[3-6] and an earlier review paper [7].

SUN-DRYING

The efforts in studies on sun-drying were diverted in two directions : to improve drying rate by understanding various parameters controlling it and to determine the relative advantages of the two processes - sun-drying and solar drying. Wilson [8] and Szulmayer [9] studied the effect of direct radiation and other parameters on sun-drying of grapes. In both cases, grapes were spread on open-tired-racks, of almost similar dimensions (45 m long and 1.2 m wide), consisting of 8-10 tiers. They were so arranged that direct sun-exposure occured only during early mornings and late afternoons,while during the hottest period of the day the products were shaded by the tiers just above and by an over- hanging roof. Wilson [8] had observed that the temperature of the untreated grapes exposed directly to solar radiation was 6-8°C above the ambient temperature, while that of treated grapes was only 4°C. This temperature increase was attributed to absorption of solar radiation in excess of that required to provide latent heat of vaporization of water and convection heat losses to air. Simultaneous studies on two different designs of solar dryers have helped in establishing the superiority of sun-drying of grapes over solar drying. This was because the drying rate in the solar dryers were poor and the grapes remained green even after drying was completed, eventually getting infected by moulds. It was concluded that drying by hot air is equally good as the drying rate was almost equal, irrespective of whether the grapes in the racks were in sun or shade. Szulmayer [9] had studied the influence of air temperature and humidity on the drying rate of sun-dried grapes. The sorption isotherms of grapes were superimposed on a psychrometric chart in order to simplify the analysis of drying process. If the moisture content of grapes and air temperature and humidity are known, it is possible to assess the feasibility of drying with these charts. However, the usefulness of this method is limited because the data on sorption isotherms of other crops are not readily available.

It was concluded that heat alone is not the only factor governing drying rate and that ambient humidity has good control over it. It was observed that resorption had occured when the atmospheric humidity was high (about 70 per cent).

Singhal et al [10] studied sun-drying of paddy on black painted and unpainted surfaces. The drying rate from unpainted surface was observed to be much higher than that on black painted surface.

A general conclusion of the above studies can be given as follows:

1) Sun-drying is an appropriate method under favourable atmospheric conditions.

2) Drying by hot air is equally good.

3) The temperature of products exposed to direct radiation will be higher than that of ambient air.

4) Resorption may occur during night time if exposed, and can be avoided if secluded from surroundings by simple means such as covering with a plastic cover.

5) Proper color development requires direct exposure to solar radiation

and 6) the absorptivity of the product and not that of the surface, controls drying.

Many investigators [10-16] had studied sun-drying to assess the relative merits of solar drying, invariably on cabinet solar dryers. In all such investigations different products including grapes were dried simultaneously by both the methods. It was concluded that solar drying is a better process, requiring less time and giving a product of better quality.

CABINET SOLAR DRYERS

They can be considered as the first step towards a scientific method for utilization of solar energy for drying. From the review paper by Löf [7] it appears that cabinet solar dryers were studied as early as 1957 [11]. The potentialities of solar drying were recognised, in 1958, by a committee of International Solarists (Solar scientists) and since then research activity in this direction has increased. Many studies [11-16] were initiated to assess the various advantages over sun-drying and to evolve a scientific method for better utilization of solar energy.

The cabinet solar dryers constituted of a small wooden box (around 1 m in length and 0.3 m in width) with length about 3 times the width to minimise the shading effect of the side pannels. The base and side planks served as insulation. The top covers were inclined at certain angle to horizontal. In later studies inclination was equal to the Latitude of the location [14-16]. Top covers were provided with one or more glass panes. The dryers were placed facing the equator, so that maximum radiation is transmitted through the glass panes. The products to be dried were placed in containers having perforations/wire mesh at the bottom. The trays in turn were placed a few cm above the bottom in the cabinet dryers, so that radiation falls on the products. The glass allows only the short wave solar radiation to pass through it, causing a temperature rise in the cabinet. In order to facilitate moisture removal, holes were provided at the bottom and vertical sides, except the short front side. The holes on the vertical sides were at the top, a few cm below the cover. As the ambient air temperatures were far below that inside the cabinets, air entered through the holes at bottom and escaped out through the holes on the vertical sides setting in convection currents carrying moisture along with them.

Slight variation from this design are noticed in two cases[15, 17]. Ghosh [15] had employed the cabinet-dryer principle for drying cocoa on a large scale in a glass-roof drier. The dryer was a small shed with sloping roofs, oriented lengthwise along the north-south axis. The opening on the eastern and western walls, above and below the level of platforms containing the cocoa beans, were regulated to control the air flow through it. Moist air was vented through an opening at the centre of the roof. Headly et al [17] had employed an inclined collector, a separate drying cabinet and a dehumidifier in the dryer design, all connected by air passages. Hot air from collector is diverted on to the dryer cabinet by vanes. The air in thedryer cools down and convection currents are set in due to the difference in density of air. The dehumidifier absorbs moisture in the air and the same air flows to the collector.

In all cases, except [17], the rate of air flow had to be controlled by regulating the openings on the sides to suit the climatic conditions. It is reported that low air circulation had resulted in fogging [13] of the lower side of the inside glass cover. A product of better quality, saving in drying time and energy, and less contamination were the advantages reported in favour of cabinet solar dryers.

Cost analysis [13,14,15] show that they can economically be used if the life is more than 5 years, and where drying is possible for about two thirds of an year. The effect of dehumidifier [17] on drying rate is inconclusive, as comparative data are not available.

Eventhough cabinet solar dryers are advantageous in many respects, they are unsuitable for large scale applications, because of their low drying rates governed by convection currents and due to large area requirement per unit mass of product dried, compared to batch solar dryers.

BATCH SOLAR DRIERS

These dryers are similar in construction and operation to batch dryers employed in process industries [3,4] except that source of heat energy is solar radiation. All batch solar driers investigated [8,10,18-28] consisted of three basic units: an energy collection unit (collector), an air handling unit (Blower or fan), and a dryer unit (Bin or shelf for holding the food stuffs). Atmospheric air heated in the collector was forced through the products by the air handling unit. The air, while passing through the drier unit, transfers heat for the vaporization of water and carries with it the moisture released.

In early studies [18,19,20] the roof formed of metallic sheets served as collector plate and the space between this sheet and ceiling below served as a passage for the air flow. Plastic or glass cover above the sheet metal absorber were used to reduce thermal losses [18,19] from the absorber plate. In some tests, transparent cover [18,19] or black painted cloth[8] put a few cm above ground served as the heat collector. Finned or flat plate collectors were generally employed [8,24,27,28]. But with the aim of improving the collector efficiency and hence the performance of the dryer, either two pass air heaters [29] or permiable media collector construction [21,23] were used. In order to make the construction of collector simple and cheap, Wrubleski et al [25] used a black painted plywood instead of a metallic collector plate. Either to minimise the heat losses from the bottom of collector [22] or to save space [8,20,24,26,27,28] dryer units were housed below the collector unit.

The dryer units were either bins [18-20,24-28] with perforations for air passage at the bottom or wooden shelves [21-23] covered from all sides and provided with openings for air to enter and exit from them. In the shelf type units trays were used to hold the products. Generally, the side covers were opaque, serving as insulation too. In two unconventional designs, the dryer units were provided with translucent covers on eastern and western walls [21,23], so that direct exposure of goods to undesirably high radiation intensities could be prevented. This partial exposure to radiation was to enhance proper colour development, due to the decomposition of residual chlorophyll in the fruit tissues.

Three studies [18,19,25] were aimed at economically augmenting the drying rate in forced circulation natural air drying by the use of cheap solar collectors. Determination of optimum bed thickness of products [28], heat and mass transfer coefficients during constant rate of drying [23] and optimum dryer design [24,26,27] were the specific aims of other investigators.

Depending on the collector area, and air flow rates different products, mostly grains [18,19,23,25-28], weighing from 50 to 50,000 kg were dried. The inclusion of the air handling unit had resulted in enhanced drying rates. Hence, the total drying time to process the product between known limits of moisture content is generally given as an index of dryer performance. As the drying time depends on the design and meteorological parameters, a comparison of performance of different dryers is not possible, as discussed in a later section.

CONTINUOUS FLOW SOLAR DRYERS

The continuous solar dryers also have the three basic units constituting a batch solar dryer. The only difference is that a continuous flow of wet products into the dryer unit and dried products out of it is maintained in this design. In the first study reported [30] the dryer unit consisted of two chambers, one surrounding the other. The chamber walls are perforated. Grains flow into the outer chamber and out of it, was manually controlled. The inner chamber was fed with hot air from a collector by a blower. The air had to pass through the perforations on the walls of the inner and outer chambers and the product in the outer chamber, before it could find its way out. High drying rates claimed are attributed to through mixing and slow movement of grains in the dryer.

Generally all solar drying methods have several advantages over sun-drying. Considerable saving in drying time and energy, product of better quality [10-28] longer storage life [13,30] large saving in electricity cost compared to systems e using electricity [16,20,25] and feasibility of drying even under adverse atmospheric conditions [13,21,25] are the plus points reported. Many of these are indicative of better utilization of solar energy. From the available data on solar dryers, commonly reported parameters pertaining to their performance are summarised in table 1. It reveals that most of the investigators had recorded the drying time and either the initial and final mass of products or the initial and final moisture content of products to indicate the performance of dryers. This is probably because of the lack of a mathematical model to compute performance as well as the difficulty in expressing it in any other fashion. A measure of drying time would have been an appropriate index of performance if the magnitude of other parameters controlling it were maintained within a generalised range in all the studies. As the corresponding data on the magnitude of these parameters are usually not given, comparison of performance of different dryers is not possible as discussed below.

This discussion is limited to batch solar dryers only because cabinet dryers are not suitable for large scale applications and data on continuous flow dryers are very few. A good dryer suitable for large scale application has to be the one which can dry a large mass of product in the least possible time. Hence dryer performance may be treated as a function of drying rate.

In process industries, the dryer performance is assessed from a plot of moisture content of product vs time elapsed. The graph is a straight line till critical moisture content has reached. Beyond this point, diffusion and capilliarity control drying rate and hence is an exponential curve. From this plot the drying rates can be computed. Even if drying time is taken as an index of performance, it's computation [3,5] emphasizes that steady state inlet condition for the air supply be maintained. It is reported [23] that total drying time can be computed from the following equation, even under unsteady inlet condition of air supply to dryer.

$$\frac{M - M_E}{M_o - M_E} = e^{-Kt} \qquad \ldots(1)$$

where, M is Moisture content of the material at time t. (dry basis) in percentage,

M_E is equilibrium Moisture content (dry basis) of the material in reference to the drying air, percentage,

M_o is moisture content (dry basis) at t equals zero, percentage,

t is total drying time, hr,

and K is constant dependent on the material, 1/hr.

In the above equation, the value of K depends on the product dried, and varies as a function of temperature, roughly proportional to the saturation water vapor pressure at the given temperature. Unfortunately, this equation is applicable to a thin layer analysis and is not applicable to a deep layer analysis [5]. As steady-state-conditions can rarely be expected in solar dryers either total drying time or rate of drying will not give a true index of their performance. Hence a comparison based on drying time cannot be justified.

It may be noted that the requirements for high drying rate are opposed to those for a high collector efficiency. At low temperature rise and at large flow rates the collector will have high efficiency [24,31-35]. On the other hand, higher drying rates require a high velocity, a safe high temperature and a low humidity air [3,4]. In other words, the performance of a dryer will depend on the quality and rate of energy supply. This dependence of dryer performance on collector design is the main deterrent on the path of technical advancement and commercialization of solar dryers.

Commercialization of batch solar dryers involve the analysis of economic feasibility of the systems and processes. Even though, solar energy with all its limitations is freely available, the first cost of solar energy collection equipment are exhorbitant-ranging from Rs* 435 to Rs. 695/- per square meter of collector area depending on the application [36,37,38]. At the present rate, the cost of collectors suitable for dryers is about Rs. 470/- per square meter. The total cost of collector (of area 50.75 square meter), and supporting structure for a paddy drying unit, in Tamil Nadu, India [38] of capacity one tonne per day is reported to be Rs. 23,500/-. The processing cost in this unit are not known. But the processing cost per tonne of paddy in a batch solar dryer, provided with a collector of 100 square meter area is reported to be Rs. 14.35 [10]. An optimistic cost analysis on a large scale unit [39] for processing 10 tonnes of grains per day shows that the processing cost can be as little as Rs. 6/- per tonne of paddy, to reduce moisture content from 22 to 13%. This analysis assumes that the system will be operational for 300 to 320 days in a year, and total investment for the whole plant will be

* 1 U.S $ = 8.70 Indian Rupees

around Rs. 2 lakhs. This compares favourably with the present processing cost of Rs. 12.44 per tonne of sun-dried paddy, dried over an area of 200 square meter [10]. A further decrease in processing cost can be expected if cheap collectors [31,40,41] having higher collection efficiencies are employed.

The collector efficiency, in turn, depends on meteorological, working and collector parameters [32,33,34,42]. These have necessitated the inclusion of auxiliary heating equipment in the solar dryer designs [15,20,21,23,24], in order to make the drying process continuous. Generally electric heating [15,20,24] or hot flue gas from a butane gas burner system were used as auxiliary sources of energy. Energy storage systems is another alternative suggested [35,41,43] to make the drying process continuous. Storage of heat in the form of sensible heat or latent heat [44] are two of the many other methods proposed. A solar water heater coupled to a liquid to gas heat exchanger cum storage tank is suggested by Khanna [43]. Rock absorption and storage collector system was studied by Schlag et al [41] so that total system cost can be minimized, making them suitable for agricultural applications. As more heat can be stored per kilogram of substance used in the latent heat storage methods [44], it is considered to be most suitable method for energy storage. In both the above cases of auxiliary heating or storage systems, further invesment will be uneconomical. These economic factors together with the lack of a suitable and reliable method to assess the performance of solar dryers have restrained their commercialization.

Recognizing the influence of collector design on the performance of dryers many investigators aiming at optimum designs suitable for large scale applications have diverted their efforts toward the analysis of different aspects of collectors. Determination of heat transfer coefficients [28,45,46] and performance of collectors [35,41] are the different aspects being studied at present.

A better understanding of the heat transfer coefficients and collector performance will definitely help to improve the performance of dryers. But it is felt that better performance can be achieved if more attention is paid to the parameters determining the drying rate, which depends on the heat transfer in the dryer. Improvement in dryer performance can be expected if conditions conducive to higher drying rates are maintained at the dryer inlet. Even though air humidity is the major factor determining the drying rate, usually no control is exercised on this because it will be uneconomical to do so. The other parameters which govern drying rate are temperature and the velocity of air. These two can be controlled within certain limits, depending on the collector design and meteorological parameters. As the humidity at the inlet to dryer will be equal to that in the ambient air, to improve the drying rate both temperature and velocity have to be increased [3,4] . But any increase in the magnitude of one will correspondingly decrease the magnitude of the other. Thus, at low temperature the kinetic and heat energy associated with the air at the exit of the collector will be more and at high temperature the case will be just the reverse. In both these cases nothing can be said at present about the heat transfer, as in the former case a lower temperature and in the later case a lower velocity may reduce the heat transfer, and hence the drying rate. But to what extend each combination affects heat transfer has not yet been investigated. If investigations are carried out in this direction, it will not only help in selecting optimum conditions for the operation of the dryer depending on the ambient air humidity, product to be dried, and solar insolation, but also in optimizing dryer design.

CONCLUSION

In view of the discussions in the previous sections it is concluded that solar drying is definitely a better method with regard to energy utilization and product quality. At present, large scale applications of solar dryers are limited in number because of the lack of a design procedure and high initial cost on solar collection equipment. Selection of an optimum dryer design is not possible due to the lack of a mathematical model to compute and compare the performance of different models. Under these circumstances, it would seem highly desirable that an attempt be made to obtain accurate experimental data on the effect of various combinations of air velocity, temperature and humidity on drying rate. The task of accumulating experimental data with a high degree of accuracy will be huge one. The justification for attempting the same lies mainly in an improvement in the dryer performance.

REFERENCES

1 . George Borgstorm, Principles of Food Science, The Macmillan Company, New York, Vol 1, 1968.

2 . Radhakrishnan,S., Cultural Heritage of India, 110,1958 .

3 . Robert H. Perry and Cecil H. Chilton, Chemical Engineers' Handbook, McGraw-Hill Kogakusha,ltd, New Delhi, Fifth edition, 1973.

4 . Robert E. Treybal, Mass - Transfer Operations, McGraw-Hill Kogakusha, Ltd., New Delhi,1968.

5 . ASHRAE Handbook of Fundamentals, ASHRAE, New York, 1972.

6 . ASHRAE Handbook and Product Directory 1974, Applications, ASHRAE, New York, 1974.

7 . G.O.G.Lof, Recent Investigations in the Use of Solar Energy for the Drying of Solids, Solar Energy,Vol 6, 122, 1962.

8 . B.W.Wilson, The Role of Solar Energy in the Drying of Vine Fruit, Aust. J. Agric: Res:,Vol: 13, 662, 1962.

9.. W. Szulmayer, Thermodynamics of sun-drying , Paper No. V 24, ISEC , Paris, 1973.

10 . O.P.Singhal and C.P.Gupta, Use of Solar Energy for Para-boiling and Drying of Paddy, ISEC, New Delhi, Paper No. 0362, 1978.

11 . Ismailova, A.A., Possibility of Applying Solar Energy to drying fruits and vegetables, Teploenergetica, Vol.1, 1957.

12 . Khan, E.U., Practical Devices for the Utilization of Solar Energy, Solar Energy, Vol. 8, 17, 1964.

13 . Lawand, T.A., A Solar Cabinet Drier, J. Solar Energy Vol. 10, 158, 1966.

14 . S.G.Kapoor and H.C.Agrawal, Solar Dryers for Indian Conditions, ISEC, Paris, Paper No.V29, 1973.

15 . Biswa Nath Ghosh, A New Glass-roof Dryer for Cocoa Beans and other crops, ISEC, Paris, Paper No. V30,1973.

16 . V.R.Muthuveerappan, G.Ambalavanan, M. Kathiresan and P.Gandhidasan, Solar Drying, Proceedings of the Fifth

Meeting of AISEWG and Conference, I.I.T., Madras, 51,1973

17 . Oliver St. C. Headley and Basil G.F. Springer, A Natural Convection Solar Crop Dryer, ISEC, Paris, Paper No. V26, 1973.

18 . F.Buelow, Drying Crops with Solar Heated Air, U.N. Conference on New Sources of Energy, Rome, Paper No. 35/S/17, 1961.

19 . Davis, C and Lipper, R, Solar Energy Utilization for Crop Drying, U.N.Conference on New Sources of Energy, Rome, Paper No: 35/S/53, 1961.

20 . A.L. Phillips, Drying Coffee with Solar Heated Air, Solar Energy, Vol. 9, 213, 1965.

21 . Akyurt, M and Selcuk, M.K, A Solar Dryer Supplemented with Auxiliary Heating Systems for Continuous Operation, Solar Energy, Vol. 14, 313, 1973.

22 . Suppramaniam Satcunanathan, A Crop Drier Utilizing a Two-pass Solar Air Heater, ISEC, Paris, Paper No.V27, 1973.

23 . Selcuk. M.K., O Ersay and M.Akyurt, Development, Theoretical Analysis and Performance Evaluation of Shelf Type Solar Dryers', Solar Energy, Vol:16, 81, 1974.

24 . V.R.Muthuveerappan, G.Ambalavanan and M.Kathiresan, Utilization of Solar Energy for Paddy Drying, Proceedings of the Sixth Meeting of AISEWG, Motilal Nehru Regional Engineering College, Allahabad, Section III, 21, 1974.

25 . M.Wrubleski and P.J.Catania, Experience with a Solar Collector for grain Drying in Saskatchewan, Paper presented at the CASE Annual Meeting, Paper No. 77-106, 1977.

26 . B.Kempe Gowda, T.M. Basava Kumar and M.Ramachandra, Solar Heated Air for Drying Paddy, ISEC, New Delhi, Paper No. 0110, 1978.

27. D.K.Biswas and S.K.Tandon , Performance Characteristics of Solar Grain Drying System, ISEC, New Delhi, Paper No. 0244, 1978.

28 . V.R.Muthuveerappan, G.Ambalavanan and G. Kamaraj, Batch Drying Studies on Solar Paddy Dryer-one ton per day capacity and experimental investigation and heat transfer

Studies on one ton per day Solar Paddy Dryer, ISEC, New Delhi, Paper No. 0261, 1978.

29 . S.Satcunanathan and S.Deonarine, A Two-pass Solar Air Heater, Solar Energy, Vol. 15, 41, 1973.

30 . S. Pattanayak, P.Sengupta and B.C. Raychaudhuri, Continuous Solar Grain Dryer, ISEC, New Delhi, Paper No. 0020, 1978.

31 . V.R.Muthuveerappan and K.Iynkaran, Cost Effective Optimum Design of Solar Air Heater, ISEC, New Delhi, Paper No. 0255, 1978.

32 . Garg, H.P., Optimum Performance Prediction of Solar Air Heaters, Proceedings of the Fifth Meeting of AISEWG I.I.T., Madras, 34, 1973.

33 . James E.Hill, John P.Jenkins and Dennis E.Jones, Testing of Solar Collectors According to ASHRAE Standard 93-77, ISEC, New Delhi, Paper No. 1121, 1978.

34 . Whiller, A, Performance of Black Painted Solar Air Heaters of Conventional Design, Solar Energy, Vol.8, 31, 1964.

35 . P.W.Niles, E.J. Carnegie, J.G. Pohl and J.M. Cherne, Design and Performance of an Air Collector for Industrial Crop Dehydration, Solar Energy, Vol. 20, 19,1978.

36 . T.Tani, S.Sawata, T.Tanaka, K.Sakuta, T. Horigome and S.Karaki, Estimation of Collector Area and Electrical Cost for STEPS in Japan, ISEC, New Delhi, Paper No. 0021, 1978.

37 . J.J.Kowalczewski and J.C.Sheridan, Cost of Solar Energy,ISEC, New Delhi, Paper No. 0368, 1978.

38 . Annamalai University, Solar Paddy Dryer, Private Communication, 1978.

39 . National Industrial Development Corporation Limited, Utilization of Solar Energy, NIDC Brochure, 1978.

40 . S.P.Sabberwal and S.S.Mathur, Some Aspects Towards the Performance Evaluation and Ensuing Design Improvements of Solar Collector Systems, ISEC, New Delhi, Paper No. 0269, 1978.

41 . J.H.Schlag, D.C.Ray, A.P. Sheppard and J.M.Wood, Inexpensive Solar Collectors for Agricultural Requirements, Solar Energy, Vol. 20, 89, 1978.

42 . A.Arafa, N.Fisch and E Hahne, A Parametric Investigation on Flat Plate Solar Collectors, ISEC, New Delhi, Paper No. 0191, 1978.

43 . Khanna, M.L., and Mohansingh.N, Industrial Solar Drying, Solar Energy, Vol. 11,87, 1967.

44 . T. Tanaka, T.Tani, S.Sawata, K.Sakuta and T. Horigome, Fundamental Studies on Heat Storage of Solar Energy, Solar Energy, Vol. 19, 415,1977.

45 . M.A.S.Malik and F.H.Buelow, Heat Transfer Characterestics of Solar Dryers, ISEC,Paris,Paper No. V 25, 1973.

46 . Malik, M.A.S, and Buelow, F.H., Analytical Model for Determining Heat Transfer Coefficents in no glass Flat Plate Collectors used for Drying Agricultural Crops, ISEC, New Delhi, Paper No. 0200, 1978.

TABLE - 1 PERFORMANCE DATA OF DRYERS

First Author	Dryer Type*	Dryer Area[1] m^2	Flow Rate[2]	Weight of Product, kg		Drying time hr	Moisture content %		Moisture Removed %	Crops dried
				Initial	Final		INITIAL	FINAL		
Buelow	B	n.r[3]	3.05	n.r.	n.r.	-	20	10	10	Grain
Davis	B	n.r	4.6	n.r.	n.r	-	20	10	10	Grain
Wilson	B	102	0.33	n.r.	n.r.	-	-	-	65	Grapes
Khan	A	1.11	-	2.83	0.25	11	-	-	91	Vegetables
Phillips	B	55.6	1.22	n.r.	n.r.	8-24	55	12	33	Coffee
Lawand	A	1.10	-	5.7	1.2	n.r.	n.r.	n.r.	79	Fruits
Akyurt	B	1.00	n.r.	n.r.	n.r.	n.r.	n.r.	n.r.	n.r.	-
Szulmayer	-	-	-	n.r.	n.r.	72	n.r.	13	n.r.	Grapes
Satunanathan	B	1.534	0.19	6.76	4.52	n.r.	-	-	33.1	Cocoa
Kapoor	A	1.30	-	0.28	0.09	2.5	-	-	-	Vegetables
Ghosh	A	n.r.	-	-	-	56	47	8	39	Cocoa
Muthuveerappan	B	7.00	0.19	-	-	n.r.	30	14	16	Paddy
Selçuk	B	-	-	4.9	1.91	4	-	-	-	Grapes
Muthuveerappan	B	7.00	0.805	50	n.r.	1	30	14	16	Paddy
Wrubleski	B	43.7	1.95	5×10^4	n.r.	47	16.5	14.5	2	Wheat
Kempegowda	B	5.26	1.09	825	n.r.	n.r	24	11	13	Paddy
Pattanayak	C	1.86	n.r.	1000	n.r.	n.r	n.r.	n.r	-	Paddy
Singhal	B	100	n.r.	1000	n.r.	8	n.r.	n.r.	-	Paddy

* A is cabinet solar dryer, B: Batch Solar Dryer and C: Continuous Flow Solar Dryer.

1 : Dryer Area for A or Collector Area for B and C

2 : Flow Rate in m^3/min. per m^2 of Collector Area.

3 : n.r. : not reported.

SOLAR ENERGY UTILIZATION FOR GRAIN DRYING

VR. Muthuveerappan
Annamalai University
Annamalainagar-608101, INDIA.

ABSTRACT

A roof-cum-solar collector was successfully designed, fabricated and tested for one year under actual working conditions for grain drying applications at the Annamalai University. This paper presents the description of one ton per day solar paddy dryer and analyses the year round experimental results obtained on it. The process of parboiling and drying of the paddy and the advantages and economics of using solar energy for drying are discussed. The design details which are required to fix the dimensions of the dryer and the construction and instrumentation details are briefly specified. The difficulties encountered during the fabrication and year round operation and maintenance are dealt with. The improvements and the possible designs for the future applications are specified at the end of this paper.

INTRODUCTION

Grain drying has become a major agricultural field using solar energy, and modern rice mills and other conventional mills are using furnace oil to dry parboiled paddy. Par - boiling is a process of pre soaking the paddy in water and then steaming. The above process gelatinises the starch and hardens the rice grains. The parboiled paddy contains higher nutrional content and more yield than raw paddy. Also it has better keeping quality as it is more resistant to insect attack than raw rice. After parboiling the moisture content of the paddy is about 25 to 30 percent. This is to be reduced by drying to about 13 to 14 percent before milling. For this, solar dryers are found to be most suitable. Hot air at low relative humidily is used for drying.

It is evident from psychrometric chart that increasing the

temperature of air by 5 C, the relative humidity of air reduces by 20 percent. A flat plate collector can economically produce hot air at a temperature of 15 C - 20 C higher than the ambient temperature and there by the relative humidity can be reduced considerably. This hot air at low relative humidity is sufficient for grain drying. It can be recommended for rural applications as well as industrial uses wherever hot air at a temperature around 55 C is required for process heating as listed below

1. Drying parboiled paddy in rice mills.

2. Drying of harvested paddy and hay.

3. Drying of other cereals after harvest.

4. Drying of Tea leaves, coffee beans, fruits etc.

5. Drying of fish, Yarn, cloth, ceramic articles, paper in paper industries, cashew nut, Ground nut etc.

6. Pre heated air for combustion in Thermal power stations and similar plants.

DESIGN OF THE PLANT

Selection of blower

First, the amount of air required has to be fixed to dry one ton of paddy per day capacity from an initial moisture content of 30 percent to the final amount of 14 percent (wet basis). The following are the assumptions.

i. The total number of sun shine hours is eight per day

ii. Annual average atmospheric temperature is 27.7^{o}C

iii. Average relative humidity of atmospheric air is 74 percent.

The amount of moisture in one ton of paddy (initial) is 300 Kg. and the moisture after drying (final) is 114 Kg. Moisture to be removed in 8 hours in 186 Kg. and for one hour it is 23.25 Kg. Assuming entering air temperature as 27.7 C, and temperature rise of air as 10 C, the temperature of air entering the dryer is 37.7 C. Relative humidity of air after passing through the batch dryer is assumed as 90 percent. Using psychrometric chart moisture taken by 1 Kg.

of dry air is calculated as 0.004 Kg/minute. Air required to remove 23.25 Kg. of water is 97 Kg. per minute. and volume of air required is 85 m^3/min. Hence a blower of 85 m^3/min. capacity should be used.

From the experience on the pilot plant, the following results were obtained. Quantity of paddy dried is 50 Kg. in 2 hrs. Quantity of air utilised is 10.53 m^3 /m. and area of collector is 4.27 sq.m.

Fixing of collector area

Based on the pilot plant studies the area of collector required to dry one ton of paddy is 21.37 sq.m. Annual average daily solar intensity is 470 Kcal/hr.m^2. The average daily solar intensity is least in the month of December and its value is 313 K.cal/Hr. m^2 The ratio of annual average intensity to the least average intensity of radiation is 1.5. Hence giving an allowance for winter season, the actual area of collector is 32.05 sq.m.

Fixing the total number of bays

The flow is fully developed at a duct length of about seventy times the equivalent diameter of the duct. Based upon the above result the length of the collector is fixed of 7.3 m and width of the collector is 4.75 m. Assuming width of each bay as 61 cm, total number of bays is 8. To be on the safer side 9 bays were used which will give 39.48 sq.m. including the projections over the inlet and outlet ducts. Leaving projections the area of collector which receives insolation will be 395 sq.ft. which will be sufficient. The design is to have a roof-cum-solar collector, so that the collector can be fixed on the top of the building.

Slope of the collector

The collector is oriented at a slope of 10° south facing according to the rule that the collector should have a slope of 0.9 times the latitude for year round operation. The latitude of the site is 11° North

Design of rafters

The size of the teak wood rafter chosen is 15 cm x 5 cm The height of the rafter 15 cm is chosen for the following reasons. The insulation thickness at the bottom is selected as 5 cm of fibre glass. A duct of 62 mm size i.e. 31 mm in between absorber plate and bottom glass and 31 mm in between rear plate and absorber plate is selected. There is a stagnant air gap in between the two glass covers. A two-glass cover collector is selected. Above the top glass a clearance is given to fix the top glass in position. The height of the rafter is taken as 15 cm.

Selection of material for absorber, rear and bottom plates

Aluminium sheets have a good heat transfer characteristics but the cost is higher than that of G.I. sheets. Hence G.I. sheets are selected.

Selection of insulation

Fibre glass insulation is selected as the best material for our solar collector taking into consideration the cost and thermal conductivity values. Since the temperature empolyed in the collector is low, it is decided to have 5 cm thickness of insulation which would be sufficient.

Bottom plate and rear plate construction

The width of each bay is designed to suit the availability of the materials.

The bottom plate is bent to 2.5 cm on both sides(lengthwise) and is fixed in between the rafters such that the bottom of the rafter and bottom plates are in a plane. At a height of 5 cm from the bottom, a groove is cut in the rafters. The rear plate is bent on both sides lengthwise and this plate is kept in position in the groove cut and a reaper is fixed in the groove to keep the rear plate in position by means of case nails. This arrangement is to avoid any leakage of water in to the insulation. Both bottom plate and rear plate are kept as a box like arrangement.

Fibre glass is cut to the required size and is kept over the bottom plate.

Painting of absorber plate and rear plate:

The top surface of the rear plate is painted with glazing paint in order to improve the performance. Since the air duct passage is in between the rear and absorber plates, this will aid in multi reflections inside the air duct passage. The top surface of the absorber plate is painted with blackboard paint to increase the absorptivity of the surface. To increase the adhesive property of the paint one coat of red oxide is given to the plate prior to painting with the white paint or black board paint.

Fixing of glass plates:

The size of glass plate used is 91 cm x 61 cm Reapers are fixed to the rafter to support the glass on lenthwise direction and cross pieces are provided to support the glass plate in the crosswise direction. Allowance is given for expansion. The cross pieces are of teakwood when fixing the top glass cover, cycle rubber tubes are used for the purpose of air tight sealing.

The cross piece is provided with small holes for the purpose of draining water if the top glass breaks. Over the cross piece, there is a reaper whose top surface is made semi-circular for the easy flow of water, above the top glass.

The entry and exit of the collector is made as bell mouths. This has the main advantage of less pressure loss at entry and exit.

All the bays have two glass covers except the 7th bay. At the entry section only one glass cover is provided for all the bays, as the projection at the northern side is ineffective. At the southern end, the last panel consists of two glass covers. The top glass projects beyond the collector to drain rain water.

The description of each bay is as follows:

Bay No. 1 : It is designed such that air is drawn on both sides of absorber plate. The air enters through the bell mouth and divides its passage at the absorber plate. The air travels for a length of 5.8 m and then flows into the horizontal duct.

Turbulance creators :

In all the 9 bays, to create turbulence aluminium expanded wire meshes are fixed normal to the flow of air, at 3 intervals. In each collector they are fixed below the cross pieces and there are eight wire mesh obstructions.

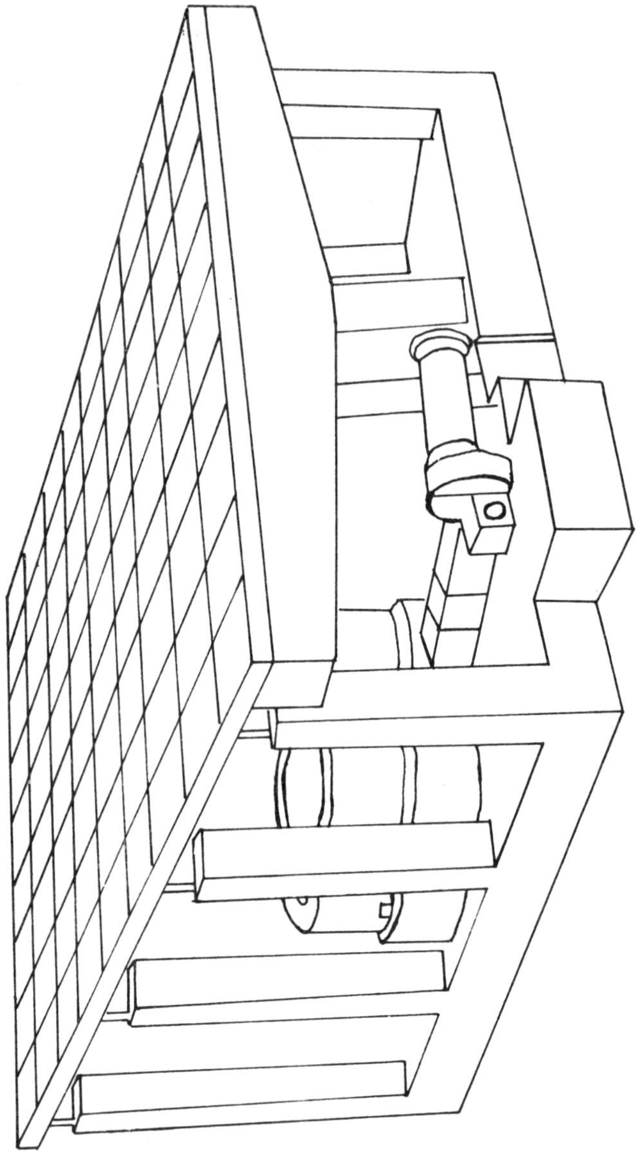

Figure 1 Isometric View of the Solar Paddy Dryer

The first bay consists of two glass covers with a stagnant air gap in between them. This stagnant air gap reduces convection and radiation losses from the collector. The flow is pulsating in nature when it flows above the absorber plate because of protruding cross pieces in the air gap between the top glass and absorber plate.

Bay No. 9 : The design is same as that of the first bay except that the absorber plate is punched with 3mm and 6mm holes. The burrs are left as it is, to promote turbulence.

Details of other bays :

Bay No. 2 : The design is same as that of Bay No.1.

Bay No. 8 : Here the air duct depth is changed and air is drawn only below the absorber plate.

Bay No. 3 : This is similar to Bay No. 8

Bay No. 7 : The construction is same as that of Bay No. 3 except that the bottom glasses are absent.

Bay No. 4 : The design is same as that of Bay No. 3

Bay No. 6 : The design is same as that of Bay No. 3 except that the duct depth is decreased.

Bay No. 5 : The design is same as that of Bay No. I except that the absorber plate is of aluminium.

Air after flowing through all the nine bays, enters the horizontal duct which collects the air and directs it to the vertical duct. Surrounding the horizontal duct, another duct is fitted with fibreglass insulation in between them. The ir which is collected in the horizontal duct flows into a vertical square duct which is again made of G.I. sheet. The air then takes a 90 o bend and flows through a circular duct connected to the blower. The air from the blower is taken to a rectangular duct. The isometric view of the plant with duct arrangements are shown in Fig.1.

Batch Dryer : Since it is a low temperature collector, it is decided to have batch dryer. The capacity of the dryer is to hold 2.25 tons of paddy at a time. Inside of the dryer, along the circumference a projection is given to keep the ring over which G.I. wire mesh is to be fixed to support paddy.

TABLE I
MONTHLY PERFORMANCE CHART

Area of the collector : 36.6 m^2

S.No.	Months	Average Collector efficiency in percent
1.	Oct. 1976	62.875
2.	Nov. 1976	64.1
3.	Dec. 1976	67.74
4.	Jan. 1977	74.00
5.	Feb. 1977	65.6
6.	Mar. 1977	69.0
7.	Apr. 1977	63.06
8.	May 1977	63.6
9.	Jun. 1977	65.33
10.	Jul. 1977	54.12
11.	Aug. 1977	55.5
12.	Sep. 1977	....

INSTRUMENTATION

The temperature measurements are being mady by copper constatan thermocouples and altogether there are 220 thermo couple points. The velocity and pressure are measured by hot wire anemometer and Inclined thetube Manometer, respectively.

TEST DATA AND RESULTS

Efficiency of the collector varies from 54 % to 74 % (Table I) and the performance coefficient varies from 9.5 to 15.2. The following are the drying results. Weight of the paddy dried is 500 Kg./batch in the time duration of 160 min-210 min. Air flow rate is 50.4 m^3 /min which is constant. Hot air temperature is 42.5 C to 47.2 C and temperature rise above ambient is 10.5 C to 15.2 C. The wind velocity is 1.5 m/s to 1.7 m/s. Initially the moisture content of the paddy is 30 percent (wet basis) and finally it is brought to 14.2 percent (wet basis). The dryer performance curve and the collector performance curve are given in Fig.2, and Fig.3. respectively. The energy comparison graph is shown in Fig. 4.

TECHNICAL DETAILS OF THE SOLAR PADDY DRIER

The collector is of the roof-cum collector type the gross area and nett area of the collector are 50.75 m^2 and 36.6 m^2 respectively. The number of bays is 9. Width is 0.61 m and length is 7.8 m. The glass covers are of window glass type and of 3 mm thickness. In 8 bays, they are 2 in numbers for each bay and 1 in number for the remaining one bay. The insulation is provided by means of fibre glass of 51 mm thickness. The absorber plates are plain G.I. sheets (24 gauge) in 8 bays and plain Aluminium sheet (24 gauge) in 1 bay. Black board paint is used as absorber coating. The drier is of batch type, the dia. of which is 2.12 m and height is 700 m.m. Blower capacity is 79.2 m^3/min the head is 51 m.m. (static) and the motor power is 2 H.P. Excluding blower motor and instrumentation, the cost of the collector is Rs. 560 per square metre. The solar paddy drier is located at Modern Rice Mill, Manalur, South Arcot District, Tamilnadu, India.

DRYING CHARACTERISTICS

It is noted that the moisture content of the paddy decreases initially at a rapid rate, remains almost the same and then falls at a lower rate. During the initial period the removal

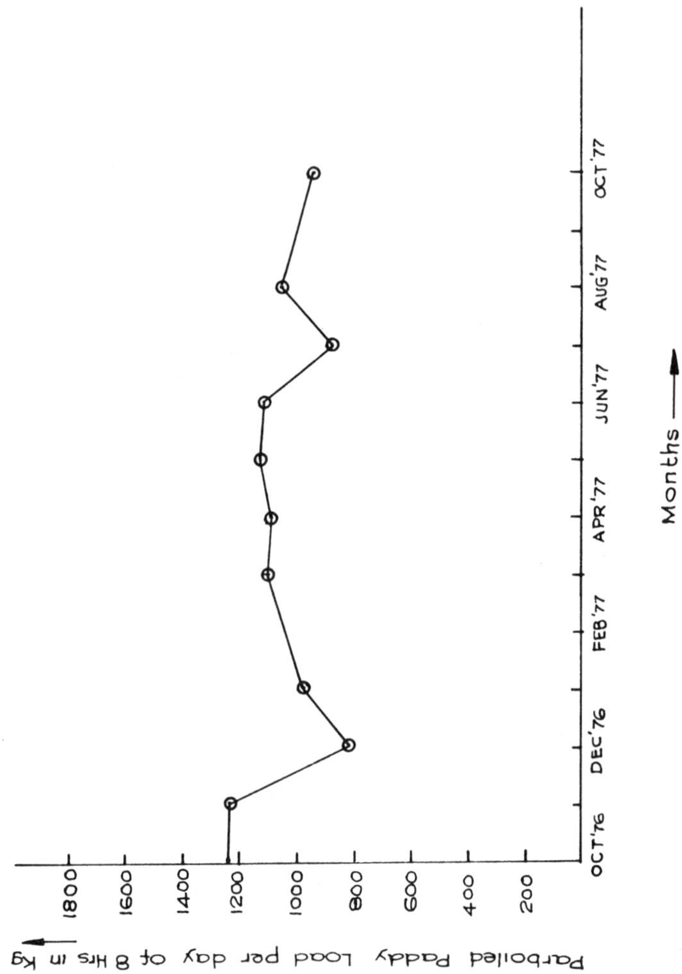

Figure 2 Dryer Performance Curve

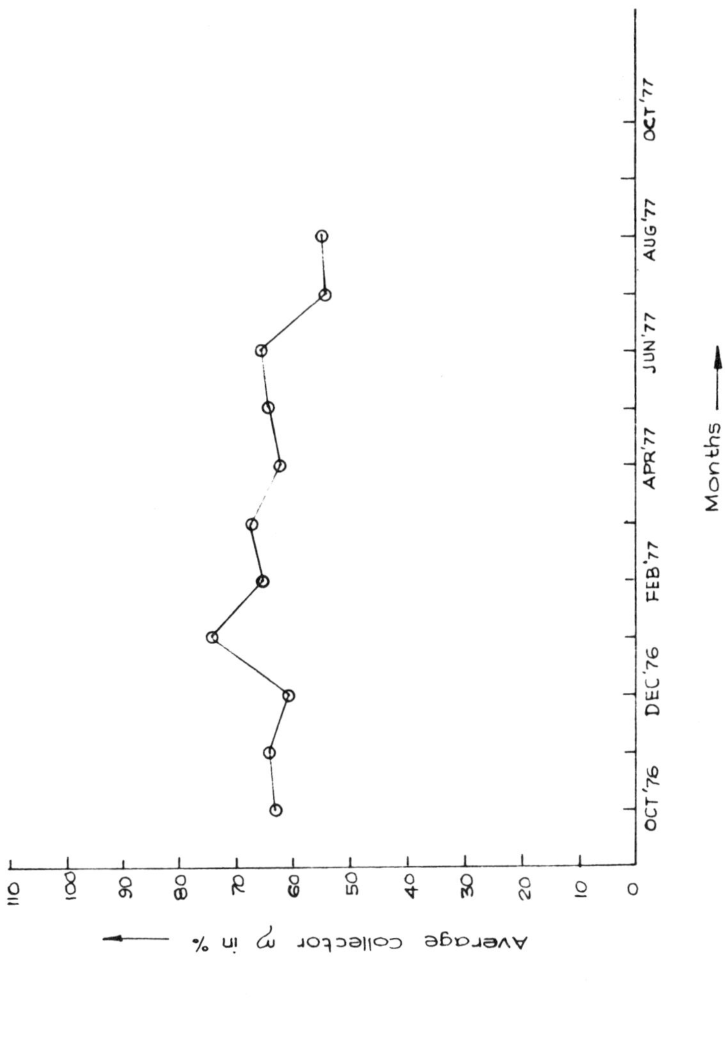

Figure 3 Collector Performance Curve

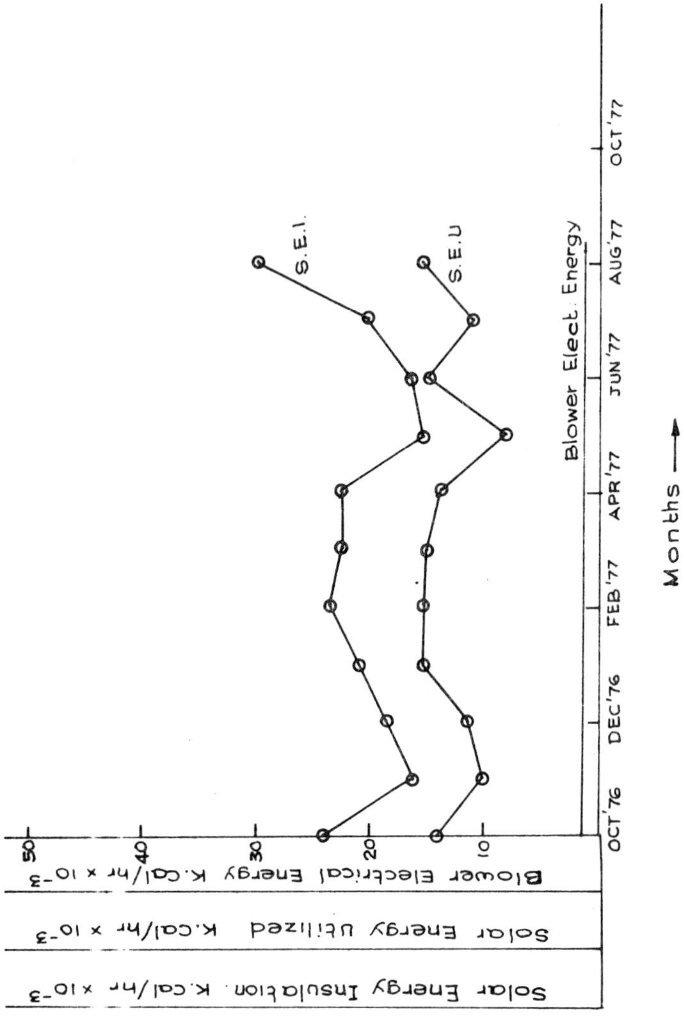

Figure 4 Energy Comparison Graph

of the surface moisture of the paddy takes place and the latent heat of vaporisation is supplied by the hot air. Once the surface has been dried, the migration of moisture inside the Kerhel requires time (tempering time). During this migration of the moisture to the surface layer of the Kernel, there is no appreciable drying. Once the movement of the moisture is complete, thereafter the moisture evaporation takes place at a slower rate when the paddy is in contact with the hot air. During the tempering period, supply of air is not required and the solar heated air can be profitably used by charging the bin with fresh parboiled paddy or tempered paddy for further drying. Table (II) gives the time taken to reduce the moisture content from 30 percent to 13 percent for various depths of paddy. In all experiments the same quantity of air is used. It can be concluded by comparing the time of drying and quantity of paddy dried in each test, that a 150 mm depth of paddy seems to be an optimum one, since the maximum quantity of paddy was dried for the given time. It is also interesting to observe that the time of drying of parboiled paddy reduces considerably if the drying operations are carried out without lapse of time after parboiling.

FUEL OIL SAVED

From the energy balance chart (Table III) the average fuel oil saved is 1.6353 Kg./Hr. Fuel oil saved for a day of 8 hours is 13.0824 Kg. and for 300 days of sun shine in a year it is 3924.72 Kg. Total cost is Rs. 4700.

Hence Rs. 4700.00 can be saved by using the solar collector in one year.

ECONOMICS OF THE SOLAR COLLECTOR

Cost of the solar collector is Rs. 23,600. Furnace oil saving per year is Rs. 4,700. Maintenance of the collector is 3 percent and interest rate is 12 percent.

$$\text{Capital recovery factor} = \frac{4700}{23600} - 0.03$$

$$= 0.17$$

From interest tables, number of years required for recovery is 11 years.

TABLE II
TRIALS CONDUCTED ON THE SOLAR PADDY DRYER

S.No.	Depth of Paddy	No.of batches dried in a day	Quantity of paddy dried per batch Kg.	Total quantity dried Kg.	Time taken per batch min.	Total time taken min.
1.		First	380		180	510
	0.015 m	Second	380	1140	135	
		Third	380		195	
2.		First	500		255	515
	0.020 m	Second	500	1000	260	
3.	0.025 m	One batch	625	625	370	370
4.	0.030	One	750	750	495	495

TABLE III
Net Area of the Collector = 36.6 m^2
MONTHLY ENERGY BALANCE CHART

MONTHS	Oct. 1976	Nov. 1976	Dec. 1976	Jan. 1978	Feb. 1977	Mar. 1977
Average Solar Energy Insolation in K.cal/hr.	23.41 x 10^3	16.02 x 10^3	18.42 x 10^3	21.45 x 10^3	24.2 x 10^3	22.5 x 10^3
Average Solar Energy Utilised in Kcal./Hr.	14.42 x 10^3	10.19 x 10^3	10.96 x 10^3	15.33 x 10^3	15.7 x 10^3	15.05 x 10^3
Blower Elect Energy in Kcal/Hr.	1.28 x 10^3	1.28 x 10^3	1.28 x 10^3	1.28 x 10^3	1.28 x 10^3	1.28 x 10^3
Fuel oil saved* in Kg./Hr.	1.80	1.27	1.37	1.91	1.96	1.88

TABLE III (Contd...)

MONTHS	Apr. 1977	May 1977	Jun. 1977	Jul. 1977	Aug. 1977	For a Year
Average Solar Energy Insolation in K.cal/Hr.	22.5 x 10^3	15.7 x 10^3	18.52 x 10^3	20.76 x 10^3	30.03 x 10^3	21.23 x 10^3
Average Solar Energy Utilised in Kcal./Hr.	14.01 x 10^3	9.47 x 10^3	12.34 x 10^3	11.23 x 10^3	15.2 x 10^3	13.08 x 10^3
Blower Elect. Energy in K.cal/Hr.	1.28 x 10^3	1.28 x 10^3	1.28 x 10^3	1.28 x 10^3	1.28 x 10^3	1.28 x 10^3
Fuel oil saved* in Kg./Hr.	1.75	1.18	1.54	1.40	1.90	1.6353

* Assuming Combustion 80 percent and Calorific Value of the fuel as 10,000 Kcal/Kg.

TABLE IV
PARBOILED PADDY DRYING PERFORMANCE

MONTHS		Total paddy load in Kg.	Total No. of working hours	Paddy load per hour in Kg.	Paddy load per day of 8 hour in Kg.
Oct.	1976	1900	12.25	155.1	1240.81
Nov.	1976	3450	22.5	153.33	1226.66
Dec.	1976	5600	53.25	103.286	826.29
Jan.	1977	7200	58.75	122.5	980.42
Feb.	1977	9700	75.5	130.2	1041.61
Mar.	1977	7868	57.	138.03	1104.28
Apl.	1977	12630	93.5	135.08	1080.64
May	1977	21790	154	141.49	1131.9
Jun.	1977	9500	68	139.7	1117.64
July	1977	1325	12	110.41	883.33
Aug.	1977	7600	57.5	132.173	1057.39
Sep.	1977	3000	25.75	116.5	932.03

Total parboiled paddy load : 91,463 Kg. (From Oct.1976 to Aug. 1977)
Total Hours worked : 690 Hours

Load/Hr. : $\frac{91463}{690}$ = 132.55 Kg.

Load/day of 8 hours : 132.55 x 8 = 1060.44 Kg.
i.e. Load/day : 1.06 Ton.

COEFFICIENT OF PERFORMANCE

The Coefficient of performance is the ratio of solar energy utilised to blower energy used and this works out to 9.5 to 15.2 during the year of operation.

FUTURE DESIGNS

Apart from theoretical and experimental studies the following practical points are worth mentioning.

1. The two glass covers were found to be good to increase the temperature of air without increasing blower running cost. But in industrial atmosphere, particularly in the rice mill area where this collector is located, lot of dust is inevitable and it has been noted, dust is settling in between the glass covers which cannot be removed. Moreover, it has been found that the inner layers of glass covers gave way for a number of times due to higher temperature conditions. Hence it is preferable to have single glass cover in the air heater designs.

2. It has been found to be advantageous to have air flow both above and below the absorber plate than flow below the absorber plate alone.

3. Uniform air distribution in all bays is another essential requirement for better utilisation of the collector.

4. More breakage of glasses are due to over heating during idle conditions. Suitable ventilation should be provided to prevent over heating during stagnation periods during day time.

CONCLUSION

Two drying bins for each solar collector are recommended so as to pass the hot air alternatively allowing tempering time in one of the bins while drying is carried in another bin. This will increase the drying capacity for the same collector. Since optimum thickness of paddy bed is found to be 0.15 m as against 0.3 m in the original design, the resistance of flow of air through the bed is correspondingly reduced to half. This will allow the use of black coated iron filings in the air ducts of the solar collector in such a way that the pressure loss due to these filings is equal to 0.0125 m of water column, which is saved by the reduction in the bed thickness. These filings increase the turbulence and thereby better heat transfer rate for the same blower power is achieved. Matching of the solar collector with the bin dryer incorporating supplementary heating is an important study to ensure constant output of paddy dried through out the year to meet the industrial requirements. These investigations helped us in arriving at a suitable design of solar paddy dryer of ten ton per day capacity useful for small rice mills in rural areas.

STUDIES ON UTILISATION OF SOLAR ENERGY IN HAY MAKING OF BARSEEM

R.D. Verma and M.K. Trehan
National Dairy Research Institute
Karnal, Haryana, India

ABSTRACT

The dairy animals are fed with green fodder and concentrates to keep their food balanced. The cross-breeding of animals is being carried out in a big way in the country to increase the milk production. The green fodder is scarce in summer months. Therefore, to keep the dairy animals in good shape for production of milk, they have to be fed with hay prepared from green fodder when available so that it can be used in the period of scarcity. The present paper deals with the studies carried out to collect basic information on drying characteristics of barseem when dried under different working conditions. The comparison of characteristics is made when dried, by making use of solar energy and without making use of direct radiation. This information can be used for design and development of a hay dryer by making use of solar energy.

INTRODUCTION

Barseem is one of the major green fodder which is fed to the dairy animals. In India, barseem crop is ready normally in the month of October and is available up to the end of April. This is not available during summer months but it is essential to feed the same to the milking animals to keep their feed balanced. Therefore, the barseem is to be preserved for the period of scarcity. One way of preserving barseem is to make hay and the other to make silage. The present study is carried out to determine the drying characteristics of barseem for hay making. This data is essential for the design and development of a hay dryer. At present, there is no equipment available for the purpose, but has a great scope in a country like ours where the tremendous efforts are put to increase the milk production. This type of equipment will certainly find a definite place at the dairy farm. Keeping these points in view, the present study was undertaken.

NEED FOR DRYING

At the time of harvesting, most of the agricultural crops have

Table 1.

Summary of Recommendations for Drying*
Grain with Natural Air and Heated Air

	Ear Corn	Shelled Corn	Wheat	Oats	Barley	Grain Sorghum	Soya beans	Rice	Pea-nuts
1. Maximum moisture contents of crop at harvesting for satisfactory drying:									
With Natural air, percent	30	25	20	20	20	20	20	25	45-50
With heated air, per cent	35	35	25	25	25	25	25	25	45-50
2. Maximum moisture contents of crop for safe storage in a tight structure per cent	11	11	11	11	11	10	9	10	11
3. Maximum relative humidity of air entering crop which will dry crop down to safe storage level when natural air is used for drying, per cent.	60	60	60	60	60	60	65	60	75
4. Maximum safe temperature of heated air entering crop for drying when crop is to be used for seed, °F.	110	110	110	110	105	110	110	110	90
Sold for commercial use, °F a.	130	130	140	140	105	140	120	110	90
Animal feed, °F b.	180	180	180	180	180	180			

* Based on Agricultural Engineer's Hand Book 1961 Table 49-6.

higher moisture contents. Crops with higher moisture content after harvesting deteriorate from several causes such as growth of micro-flora, particularly the aerobic, species of molds. The respiration of the micro-flora as well as increased respiration of agricultural produce due to higher moisture content, produces heat and more moisture to cause further damage to the crop. Such self heating of forage crops stored at too high a moisture content can result in uncontrolled temperature rises to the point of ignition and spontaneous combustion. For safe long-term storage of agricultural produce, maximum moisture contents have been determined, known as safe storage moisture content, below which, produce can be stored for definite duration without chances of spoilage at ambient temperatures. Table 1 gives the summary of recommendations for drying grains with natural air and heated air. Good quality hay has been shown to be a well-balanced feed. To produce good quality hay, the crop must be cut a the proper stage of maturity, cured as fast as possible to avoid loss of leaves and nutrients due to weather, dried to a proper moisture content and stored under proper conditioning. Barseem contains 70 to 85% moisture when harvested. This moisture content must be reduced to 20 to 25% for safe storage. Excessive moisture in stored hay will cause mold development and spoilage of hay.

DRYING CHARACTERISTICS

Drying is a heat and mass transfer process involving the vaporization of water from the liquid state, mixing the vapour with the drying air and removal of the vapours by carrying away the mixture. Considerable heat is needed for vapourization of the moisture. Heat must be supplied by reducing the sensible heat of the drying air or applying heat directly to the product. The moist material loses moisture whenever its water vapour pressure exceeds the partial pressure of water vapours in the drying air. In the initial stages, the moisture removal rate is quite rapid since moisture on the surface of the product presents a wet surface to the drying air. Drying thus can be accelerated by increasing the flow rates of drying air. This state is known as surface drying, when the surface moisture has been removed, further drying depends upon the rate at which moisture within the product moves to the outer surface by diffusion process. This may be a slow or a rapid process depending upon the type of material being dried. The rising of temperature of the product rather than the large flow rates of drying air is likely to improve the moisture removal process at this stage.

Under direct radiation, part of the radiation may penetrate the material and be absorbed, thus generating heat within the material as well as the surface. This process raises the temperature of the product above that of the surrounding air, and thus, raises its vapour pressure. The rate of drying of barseem is normally controlled on the basis of the quality of dried hay, even though it may be desirable to dry the products at the maximum drying rates. In the case of barseem, rapid drying may result in the formation of hard drying surfaces, considerably reducing the flow of moisture from the inside of the stem to the surface. Hence it is necessary to provide controls for varying the rate of drying, either by controlling the heat supply or by controlling the humidities of the surrounding air or both.

The drying of a product by aeration only and without any heat supply is known as adiabatic drying. The latent heat of vapourization of moisture is extracted from the air surrounding the product and thus lowering the temperature of the surrounding air. The drying air progressively becomes cooler and more humid as it progresses through the material being dried. Since latent heat of vapourization is quite high, compared to the heat capacity of air, relatively large volumes of air, at low relative humidity must be used in this type of drying.

DRYING BY SOLAR ENERGY

Solar energy has been used for drying of various agricultural products like grains, fruits, vegetables, tea leaves, coffee and fodder. The produce to be dried is normally spread on a suitable ground or on plate-form and exposed to direct sunshine. This method of drying is quite simple and cheap where labour costs are not high. It is frequently required to turn the produce for uniform drying. This method can be successfully used for drying fodder crops like barseem. The drawbacks of this method are contamination by dirt, insect and bacteria and also loss due to wetting by rain. The product is in general not homogenious and of high quality because of no control over the drying conditions. However, moisture content of the produce is reduced to a safe storage moisture content.

Two types of solar drying systems can be used.

1. Directly lined glass covered solar dryers. In this system, the solar energy absorber acts as a dryer for barseem. This fodder receives direct heat from the sun. Since the absorbtivity of barseem is quite good, product temperatures are higher than the surrounding air. This helps in further drying.

2. Indirectly heated solar dryers in which air, heated in a solar air heater, is supplied to a separate drying chamber.

This system has been studied for drying of grains, coffee, timber, etc. This can be used for hay making of barseem. A supplementary source of heat can be provided in this system to increase the efficiency.

Merits of Solar Drying

The advantages of solar drying are summarized as follows:

1. In comparison with the drying system, utilizing fuel, solar heated units can effect saving in fuel.

2. In comparison with the natural drying, i.e., material being exposed directly in the sun and air, a substantial saving in time is realised.

3. Limited space available in houses can be effectively used by solar drying.

4. Hot air produced by solar air heaters can be used for any dehydration process.

5. There is no risk of fire with solar units.

6. Solar energy is clean, there is no danger of contamination of the product, whether for animal or human consumption.

7. Fabrication of solar air heaters is very simple and material required for the fabrication are locally available.

8. As the solar unit is locally assembled and built, it can be repaired and serviced locally.

RESULTS AND DISCUSSION

The trials were conducted at Dairy Engineering Division, National Dairy Institute, Karnal, to observe the drying pattern of the barseem by putting the samples in the sun to receive direct radiation during the daytime. These samples were put inside a covered shed during the night. Simultaneously, the samples were put in the covered shed where direct solar radiation was not available. The samples of fresh-cut barseem from the National Dairy Institute farms, of 5 kg each were taken. The barseem was baled into rectagular bales of a size of 18"x12"x12". The loss of weight of the sample every two hours was recorded along with the dry bulb temperature and the wet bulb temperature.

The percentage loss of moisture was calculated and tabulated as shown in the Tables 2 and 3 respectively for the samples

Table 2.

Observations for the sample exposed to DIRECT RADIATION

Sr. No.		Time (Days)	Weight of Sample(kg)	Loss of Weight(kg)	% Loss of Moisture %
1.		0	5.000	-	-
2.	Sample-1.	1	2.000	3.000	68.2
3.	Atmospheric	2	1.200	3.800	86.4
4.	Conditions:-	3	0.850	4.150	94.3
5	w.b.t=23.5°C	5	0.600	4.400	100
6	D.b.t. = 31.5°C	7	0.600	4.400	100
1.		0	5.000	-	-
2.		1	4.330	0.670	15.4
3.		2	3.930	1.070	24.6
4.		3	3.280	1.720	39.6
5.	Sample-II	5	2.730	2.270	52.2
6.	Atmospheric	7	1.780	3.220	74.1
7.	Conditions:-	8	1.480	3.520	81.0
8.	W.b.t. = 12.5°C	9	1.180	3.820	87.9
9.	D.b.t. = 19.5°C.	10	0.930	3.070	93.7
10.		11	0.930	4.070	93.7
11.		14	0.655	4.345	100
12.		15	0.655	4.345	100

Table 3.

Observations for the sample put in the covered shed

Sr. No.		Time (Days)	Weight of sample(Kg)	Loss of weight(Kg.)	% loss of moisture %
1.		0	5.000	-	-
2.	Sample-I.	1	3.400	1.600	36.4
3.	Atmospheric	2	2.700	2.300	51.8
4.	Conditions:- W.b.t. = 23.5°C	3	1.850	3.150	71.6
5.	D.b.t. = 30.5°C.	5	1.100	3.900	88.6
6.		7	1.100	3.900	88.6
1.		0	5.000	-	-
2.		1.	4.680	0.320	7.3
3.	Sample - II	2	4.380	0.620	14.3
4.	Atmospheric	3	3.930	1.070	24.6
5	Conditions:-	5	3.380	1.620	37.3
6.	W.b.t. = 12.5°C	7	2.730	2.270	52.2
7.	D.b.t. = 22°C.	8	2.480	2.520	58.0
8.		9	1.930	3.070	70.6
9.		10	1.880	3.120	71.8
10.		11	1.780	3.220	74.1
11.		14	1.080	3.920	90.2
12.		15	1.080	3.920	90.2

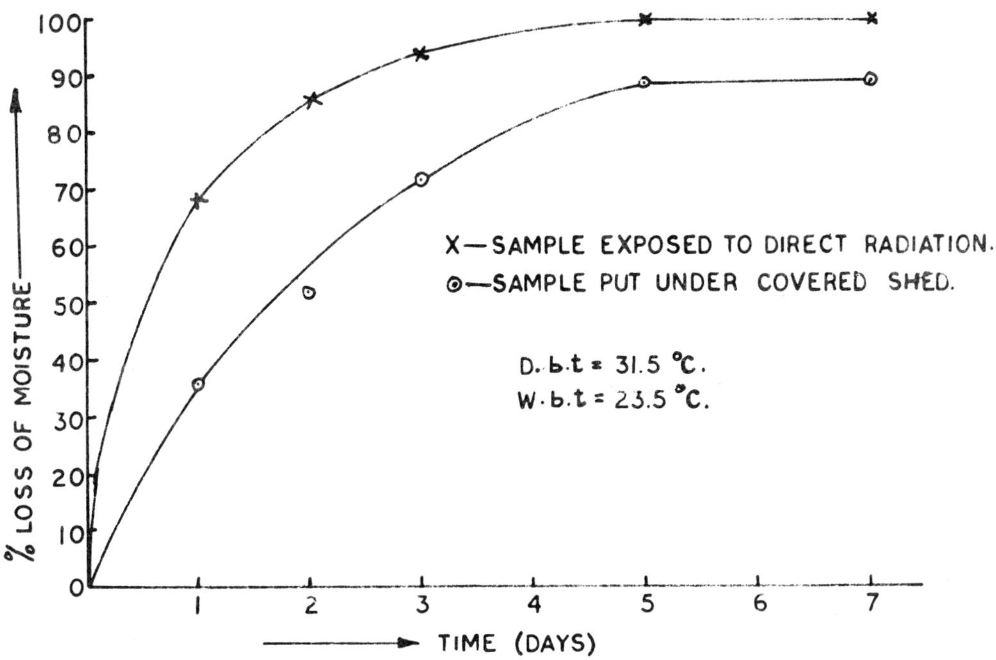

FIG. 1. DRYING CURVE.

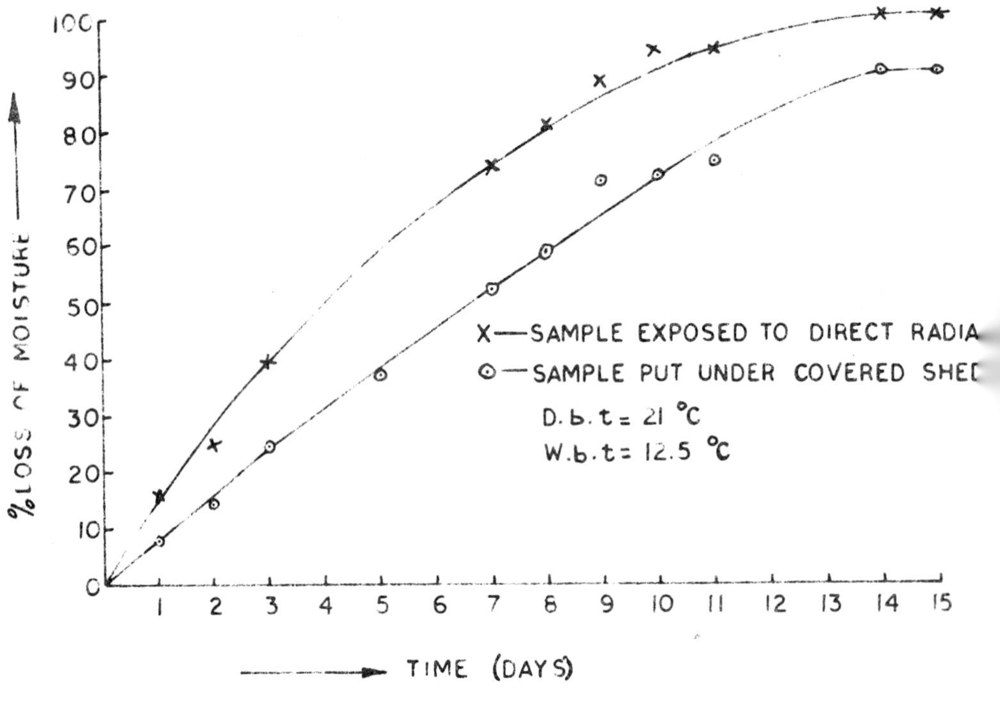

FIG. 2. DRYING CURVE.

which were exposed to direct radiation and put under the shed. The results obtained were summarized in the form of graphs. Figure 1 and 2 shows the relation between time and percentage loss of moisture for the different samples studied at 31.5°C wbt., 23.5°C wbt., and dbt 21°C, Wbt = 12.5°C. The loss of moisture from both the samples exposed to direct radiation and under a covered shed is very fast in the beginning, as indicated in Figure 1. The percentage moisture loss is as high as 68% on the first day for the sample to direct radiation which has reduced to only 6 per cent between the third to the fifth day and the moisture loss was negligible between the fifth to the seventh day. The similar trend was observed for the samples put under a covered shed. However, the percentage moisture loss was only 35% on the first day and 18% between the third and fifth day. The moisture removal was very slow between the fifth to seventh day. At the end of seven days, this sample was still left with 11% moisture where as the sample exposed to direct radiation was completely dry after the fifth day. Similar behaviour was observed for different samples which were studied at conditions 21°C dbt., and 12°C wbt., as shown in Figure 2. From the results obtained, it is evident that the moisture removal rate is quite rapid in the initial stages because the barseem presents a wet surface to the drying air. When the moisture is removed from the surface, further drying depends upon the rate at which the moisture is moved to the surface by diffusion which is very slow.

CONCLUSIONS

80% moisture can be removed from barseem by exposing the bales to direct radiation in about two days when the environmental conditions are 31.5°C dbt. and 23.5°C wbt. where as it takes about a week to remove the same percentage of moisture from barseem at 21°C wbt. Therefore, the hay making process of barseem is very quick somewhere in the end of the month of March at Karnal. Hence, it is recommended that hay from barseem should be made towards the end of March or early April to store the same for the period of scarcity. These drying characteristics can be used for designing a suitable hay dryer utilising solar energy.

REFERENCES

1. Bruhn, H.D., Flow of Air Through Chopped Hay in How Dries, Agr. Eng. 28:202-204, 207, 1947.

2. Srivastva, R.K., Analysis of the Air Flow Through Baled Hay in a Batch Drier, Part I. Journal of Agri. Eng. Vol. IV, No. 2, September 1969.

3. Bwlow, F.H., Drying Crops With Solar Heated Air, Presented at the United Nations Conference on New Sources of Energy, April 1961.

4. Roy, E.W. and Biswas, D.K., Utilisation of Solar Energy For Drying of Farm Production, Presented at the Symposium on Post Harvest Technology, held in Delhi, India, 1972.

5. Shedd, C.K., Resistance of Hay to Air Flow and Its Relation to Design of Barn-Hay Curing Equipment, Agr. Eng. 27(4): 169-170.

6. Malik, M.A.S. and Buelow, F.H., Heat Transfer Characteristics of a Solar Dryer, International Congress - The Sun in The Service of Making, Paris, July 1973, V-25.

7. Mennan, K.D., Solar Driers, Presented at the Seminar on Industrial Uses of Solar Energy, June 1975.

8. Davis, C. and Lipper, R., Solar Energy Utilisation for Crop Drying. E/Conf. 35/S/53, United Nations Conference on New Sources of Energy, 1961.

9. Verma, R.D. and Trehan, M.K., Drying Characteristics of Barseem for Hay Making, Inst. of Engineers (India), September 1977, UDC 633. 3:631-562.

10. Muthuveerappan, V.R., Solar Drying, Presented at the Seminar on Industrial Use of Solar Energy, June 1975.

SOLAR TOBACCO DRYING PLANT

L. Andreotti, A. Ciminale, G. Marchetti,
D. Ruggi, and A. Serravezza
CTIP Solar, S.p.A.
Rome, Italy

ABSTRACT

The paper describes the solar section of a tobacco curing facility. Bulk curing of tobacco lasts seven days, during which the temperature of the hot air ranges between 35° C and 70° C.

The solar section is made up of 74 m^2 of concentrator collectors, 2 tanks for heat storage with a total capacity of 6000 litres and a water-air heat exchanger with an area of 60 m^2.

The solar energy system has been designed to produce 40% of the total heating load with 15% of this thru heat recovery and 25% by the direct conversion of solar energy.

INTRODUCTION

CTIP Solar S.p.A. has designed and constructed the solar energy retrofit of a curing facility for Virginia Bright Tobacco under the program "Agricultural Mechanization" founded by the National Research Council, the Italian governmental R & D agency.

The curing facility is located in Santa Fista, near Pistrino (Perugia).

The construction of the solar section was finished only in August 1977 so the system has been tested only in the late part of the 1977 tobacco season.

In late 1978, the system will also be used to heat a greenhouse in order to improve the rate of utilization, since in Italy Virginia Bright tobacco is cured from late July to mid October.

Only the presence of various agricultural users at different times of the year, indeed, can justify our choice of an integrated system with a central thermal plant; if tobacco curing were the only form of energy consumption, the solar components would rather be integrated in the dryer structure.

Our system has been designed to dry corn in the same chamber that is used to dure tobacco.

Tobacco Cure

The cure for Virginia Bright Tobacco lasts an average of seven days and is performed in a closed chamber with heated air.

The first phase, lasting 2 to 3 days, is the yellowing of the leaves which is accomplished by circulating in a closed loop air at 35° C.

During the second phase the temperature rises gradually to 75°C, and this temperature is maintained for one day or two.

The curing process, which reduces tobacco to 15% of its initial weight requires without the solar retrofit between 1.1 and 1.2 kg of diesel oil to produce each kg of dry tobacco.

The chamber characteristics for which the solar energy system has been designed are:

Tobacco before cure	6000 kg/cure
Tobacco after cure	800 kg/cure
Diesel oil consumption	880 kg/cure
Heating load	5,270,000 kcal/cure
Air flow	3,000-15,000 m^3/h
Temperature range	35-75° C

A normal season usually consists of ten cures.

Description of the Solar System

The collection area is divided into two banks, each one made up of twelve Swiss-made parabolic through concentrator collectors connected in series while the two banks are connected in parallel.

	T Min	T Max	Cloudiness Factor	SOLAR Radiation*
	^{o}C	^{o}C	-	$\frac{Kwh}{m^2\ g}$
JANUARY	0.2	8.0	.26	4.098
FEBRUARY	-0.5	9.3	.38	5.634
MARCH	0.5	11.7	.37	7.330
APRIL	3.5	16.4	.39	8.786
MAY	7.3	22.1	.47	9.573
JUNE	10.3	24.8	.52	9.906
JULY	12.2	28.3	.62	9.762
AUGUST	12.3	28.6	.61	9.192
SEPTEMBER	9.8	23.3	.49	8.001
OCTOBER	4.8	17.6	.49	6.276
NOVEMBER	2.1	11.7	.30	4.524
DECEMBER	0.1	7.4	.25	3.672

TABLE 1 WEATHER DATA

(*) For a 45° tilted surface, tracking the sun E-W in a clear day.

COLLECTOR TEMP. = 70°C

	CLEAR DAY Kwh/m^2	AVERAGE	η %
JANUARY	2.476	0.644	60.4
FEBRUARY	3.472	1.319	61.6
MARCH	4.482	1.658	61.1
APRIL	5.032	1.962	57.3
MAY	5.505	2.587	57.5
JUNE	5.866	3.050	59.2
JULY	5.739	3.550	58.8
AUGUST	5.394	3.290	58.7
SEPTEMBER	4.901	2.401	61.2
OCTOBER	3.956	1.938	63.0
NOVEMBER	2.792	0.838	61.7
DECEMBER	2.187	0.547	59.6

TABLE 2 THEORETICAL COLLECTOR PERFORMANCES

The primary closed loop is formed by the collectors, the pumps and two heat exchangers which are located in two identical steel tanks. The water rate for the primary loop ranges between 2,000 and 4,000 m^3/hour.

The two tanks have a capacity of 3 m^3 each and are connected to an water-to-air exchanger in order to heat the incoming air used for the cure.

A three way valve controls the flow of water to the heat exchanger to avoid over heating the air.

An air-to-air heat exchanger is located upstream to recover heat from the outgoing air.

The presence of the two tanks gives users the freedom to experiment with two different heat storage capacities, depending on whether one or both tanks are used. The characteristics of the system are:

collector area	74 m^2
storage capacity	3 - 6 m^3
primary loop heat exchanger	10 m^2
primary loop water flow rate	2 - 4 m^3/hour
secondary loop water flow rate	3 m^3/hour
primary loop temperature maximum	150°C
secondary loop maximum temperature	95° C
area of air to air heat exchanger	45 m^2
air flow	3,000 - 15,000 m^3/h

System Performance

The theoretical performance of the system has been computed given the climatic data(table 1) and the collector performances (table 2).

The system could give a percentage of the heating load varying between 17% and 30% with an overall seasonal average of 25%.

These values take into consideration a cloudiness factor.
With the cloudiness factor at zero the system could give a maximum

of 41% of the heating load in August and 36% in October. The theoretical efficiency of the collector at 70°C ranges between 58.7% and 63%.

The temperature in the two tanks could rise from 20°C to 95°C in the one and one half days available between two cures in the presence of a clear sky. The heat thus stored could provide one half day of heat load (438 kwh at a temperature of 35° C).

Heat recovery will represent 12 - 15% of the total heating load.

Experimental results and plans

During the initial phase in September 1977, data were recorded only during clear days, measuring the temperature in the tanks.

The energy collected varied between 290 kwh/day and 372 kwh/day depending on the working temperature and the weather.

A full scale recording of data will be done during the tobacco season of 1978.

REFERENCES

Proposta per la realizzazione di un impianto a collettori concentratori solari per l'essiccazione del tabacco. CTIP Solar Spa, Roma,Italy, April 1977.

	CURED TOBACCO	ENERGY INPUT (AVERAGE) (1)		ENERGY SUPPLIED BY THE SOLAR PLANT (2)		(2)/(1)
		Kwh/day	Kwh-Total	Kwh/day	Kwh/Total	%
August	3.200	814	24.400	245	7.570	30
September	3.200	814	24.400	178	5.328	22
October	1.600	814	12.200	141	2.109	17

- 8 -

TABLE 3 ENERGY REQUIRED FOR THE CURE

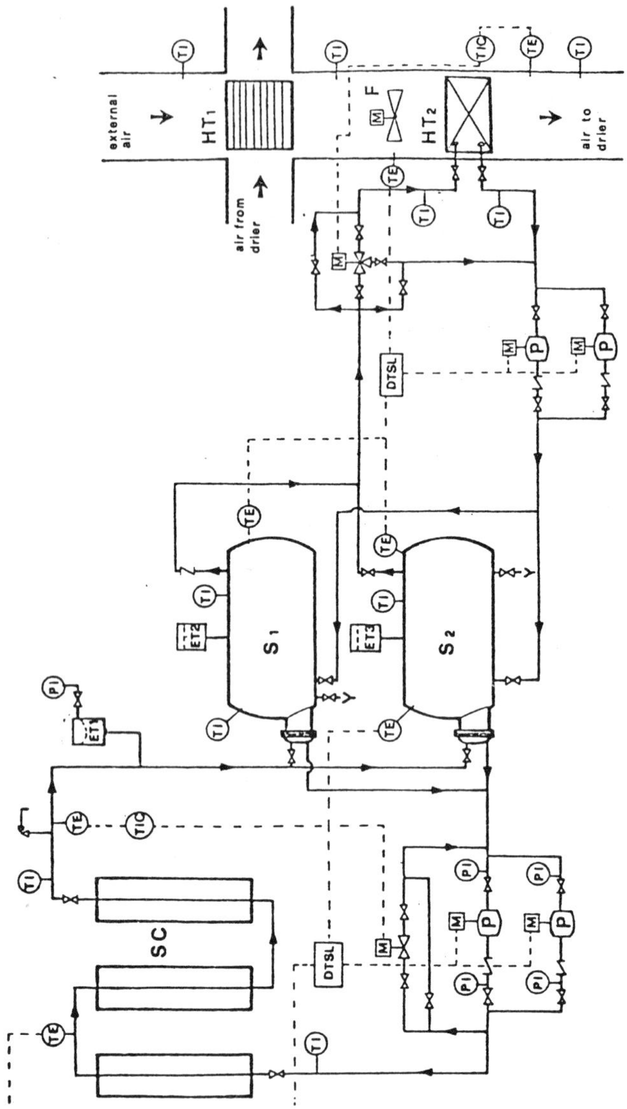

FIGURE 1 SCHEME OF THE SOLAR ENERGY SYSTEM

FIGURE 2 PHOTO OF THE COLLECTORS IN FRONT OF THE CURING FACILITIES

CROP AND TIMBER DRYING IN MALAYSIA PROPOSED PLANS FOR STUDIES

Dr. K.S. Ong
University of Malaya
Faculty of Engineering
Pantai Valley, Kuala Lumpur 22 - 11,
MALAYSIA.

ABSTRACT

Solar energy provides a source of non-polluting and virtually inexhaustible energy supply which could be harnessed relatively cheaply and easily for domestic, commercial, industrial and agricultural applications. The practical utilisation of solar energy in rural farming areas in Malaysia is mainly for crop and timber drying. Some feasible solar crop drying systems are discussed. A programme for research and development on solar drying technology for rural development is outlined.

INTRODUCTION

Solar energy provides a source of non-polluting and virtually inexhaustible energy supply which could be harnessed relatively cheaply and easily for direct heating of water and air for domestic, commercial, industrial and agricultural applications. The field of study of solar energy utilisation is very wide ranging from fuel production by photosynthesis (energy plantations and biomass energy) to power generation by photovoltaics (solar-electric) or thermal heating (solar-thermal conversion). Although technical feasibility has been domonstrated for a number of solar energy systems their eventual successful implementation would be decided mainly by economics and level of development of the country involved.

In the more developed and industrialised nations, a considerable amount of research is being conducted on converting solar energy to electrical and mechanical power. Developing nations with limited financial resources could ill-afford to spend large sums of money on these projects. At best they can only hope to import and adapt the technology developed by the rich nations. However, there are a number of low-technology solar energy applications which are best to be studied by the intending users themselves, especially for rural agricultural use. In order to be able to adapt imported solar energy technology for local use an effort should be made towards setting up a research body to consider the technical and economic feasibilities of the various possible ways in which solar energy could be utilised to aid in the development of the country. The research body could identify and give priority ratings to the possible projects and provide financial funds for their studies. The purpose of the paper is geared towards the call of setting up such an organisation. Towards this end, some proposed plans for studies are outlined. The present paper focuses on the potential of solar crop and timber drying in Malaysia. Other potential applications have been described in Reference [1].

EXAMPLES OF SOLAR ENERGY UTILISATION

Solar energy could be used for :

(i) space heating and crop drying using solar-heated air
(ii) hot water heating
(iii) solar distillation for producing potable water for human and animal consumption from brackish water
(iv) solar pumps for rural irrigation
(v) energy conservation in greenhouse design
(vi) biogas production and energy plantations for fuel
and (vii) rural electrification.

Of the above, only crop drying and water heating are practical and economical at the present time. In Malaysia, where there is a vast amount of solar energy available throughout the year, the largest potential use to which it could be put to for rural development is for crop and timber drying.

AVAILABILITY OF SOLAR ENERGY

Malaysia is situated a few degrees north of the Equator. The climate is uniform throughout the country with hot sunny days and occasional rain at different periods of the year. Daily temperatures range from 20° to 35°C (65°-95°F) and the nights are usually warm. The average mean daily total solar radiation level is about 4.5 kW h m^{-2} (1420 Btu ft^{-2}), with minor monthly variations throughout the year. The actual instantaneous radiation intensity pattern fluctuates quite rapdily over a wide range throughout the day due to cloud cover. Rainfall which usually occurs in the late afternoons or evenings does not constitute a major obstacle to the utilisation of solar energy locally. Seldom does rain fall continuously for long periods. At most, day-long overcast skies or heavy rainfall occurs for maybe something like one or two weeks in a year. The rest of the year is bright and sunny. Apart from the intermittent nature of radiation intensity and high humidity, local climatic conditions favour the use of solar energy both as a primary or as a supplementary heat-source in conjunction with conventional fuel.

SOLAR DRYING

Drying by direct exposure to the sun in an open space is one of the oldest form of solar energy utilisation. This traditional method of drying has many disadvantages, mainly contamination by:

(i) dirt, dust and wind-blown debris
(ii) insect infestation
and (iii) human and animal interference .

Other factors which tend to reduce the speed of drying and quality of dried products are:

(iv) degradation by overheating
(v) intermittent sunshine
and (vi) interruption and wetting by rain.

Although the spreading of the crop on the ground or on a platform and drying it directly by the sun is cheaply and successfully employed for many products throughout the world, the above mentioned factors could lead to a loss in production. Also, it is labour intensive. The drawbacks mentioned above could be overcome by providing a sheltered drying area or chamber in which the crops to be dried are stored. A stream of air heated by solar energy to reduce its relative humidity could then be passed over the crops. This form of solar drying could:

(i) improve the quality of the crop to be dried
(ii) reduce spoilage by contamination and local overheating
(iii) reduce spillage losses
and (iv) speed up the drying process.

The overall drying cost may be lowered because of the above.

SOLAR CROP DRYERS

The drying of a wide variety of crops using solar heated air has been demonstrated in the United States [2,3] and elsewhere [4]. Coffee, hay, tobacco, rice, maize, wheat, cottonseed, timber, and a host of other agricultural products including meat have been dried and/or are stored using solar heated air drying or ventilated sheds.

There are basically two types of solar dryers; viz

(i) a passive or natural system which operates only on solar energy alone without supplementary power source
and (ii) an active or forced system which requires a supplementary power source to circulate the air within the dryer unit.

Solar dryers could be further classified as direct or indirect types. The former involves directly exposing the material to the sun while in the latter the material is dried by circulating hot air over it without directly exposing the material to the sun. The merits of any solar dryer would depend upon the type and quantity of product to be dried. Sometimes a passive system incorporates a wind-driven fan for the air circulation.

A survey of solar agricultural dryers could be obtained in reference [4]. The rest of this paper is devoted to a general and basic description of the types of solar dryers which could be used in our rural areas for crop and timber drying.

HOT BOX DRYER

The hot box solar dryer shown in Fig.1 is a simple type of direct and passive solar drying unit which can be made very easily and cheaply. It is sometimes called the cabinet or tray-type dryer and is essentially a solar hot box. It consists of a transparent glass sheet or plastic film covering an insulated box made from cheap and easily available materials like plywood, brick, concrete, galvanised iron or aluminium sheets. The material to be dried is placed on wire mesh or steel netting located a few centimeters

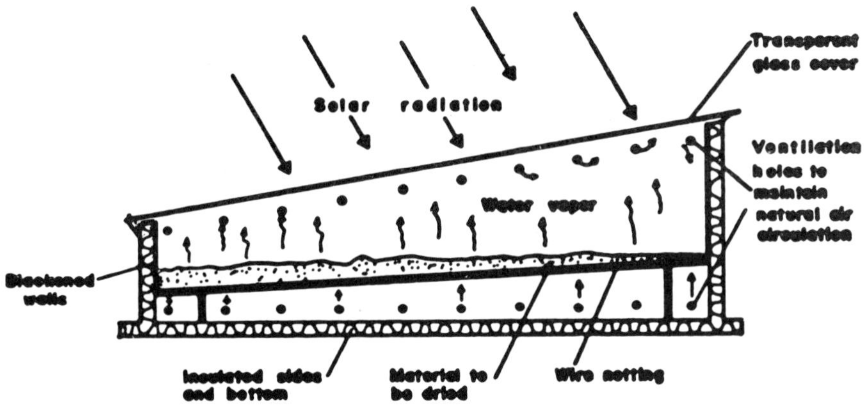

FIG. I HOT BOX DRYER

above the floor of the box. Ventilation holes are provided at the top and bottom of the box. Solar radiation is transmitted through the roof and absorbed on the blackened interior surface. As the internal temperature rises, warm air passes out of the ventilation holes by natural convection. In doing so, fresh air is drawn up through the base. As a result, there is a constant flow of air over the material to be dried. Temperature differentials between internal and external ambient air of about $40^{o}C$ ($104^{o}F$) could be obtained with this type of solar dryer. Drying times could be considerably reduced with resulting superior product quality when compared to drying under the open sun. This type of dryer has been successfully employed in many locations throughout the world for batch drying of small quantities of fruits, vegetables and grains.

A PASSIVE (NATURAL CONVECTION) SOLAR DRYER

A simple low-cost solar crop dryer is shown in Fig. 2. This consists of a conventional type of solar air heater collector connected to a drying chamber. The former could be fabricated from galvanised iron or aluminium sheets rivetted together to form a sandwich with air passages in between. The bottom and sides of the collector are insulated for better performance and could be supplied with plastic film or glass sheets as top cover. Solar radiation falling on the collector plate heats up the air inside it. The warm air rises and discharges into the drying chamber. Ambient air enters through the open bottom end of the collector. The air is thus circulated via natural convection.

The drying chamber consists of a vertical stack of trays on which the material to be dried is placed. To increase air circulation rates, a wind operated ventilator could be placed on top of the chamber. If electric power supply is available a small electric fan could be substituted. This type of dryer is suitable for small-scale use and could be made mobile. It could be rolled out for use when required. The solar collector panels could be uncoupled quite simply from the drying chamber and stored separately.

AN ACTIVE (FORCED CONVECTION) SOLAR DRYER-CUM-STORAGE SHED

A more permanent and larger scale use of the solar drying and storage shed system is shown in Fig.3. Here, the solar heat absorbing panel utilises the whole roof area and constitutes the roof of the drying chamber. Air circulation is by means of an electrically driven fan. Ambient air is drawn in from one end of the solar absorber panel roof. It is heated up by solar energy as it passes along the length of the roof. The hot air is then discharged into the drying chamber. The product to be dried could be placed in bins with perforated bottoms or in sacks in a pre-arranged order inside the shed. In the latter, the shed would have to incorporate a perforated floor. A supplementary heater could be installed to boost the air temperature, or, if a 24-hour drying process is required.

If the crops could be safely stored overnight by forced circulation of unheated air alone, supplementary heating would be unnecessary. Intermittant drying is thus employed. The supplementary heat source could be

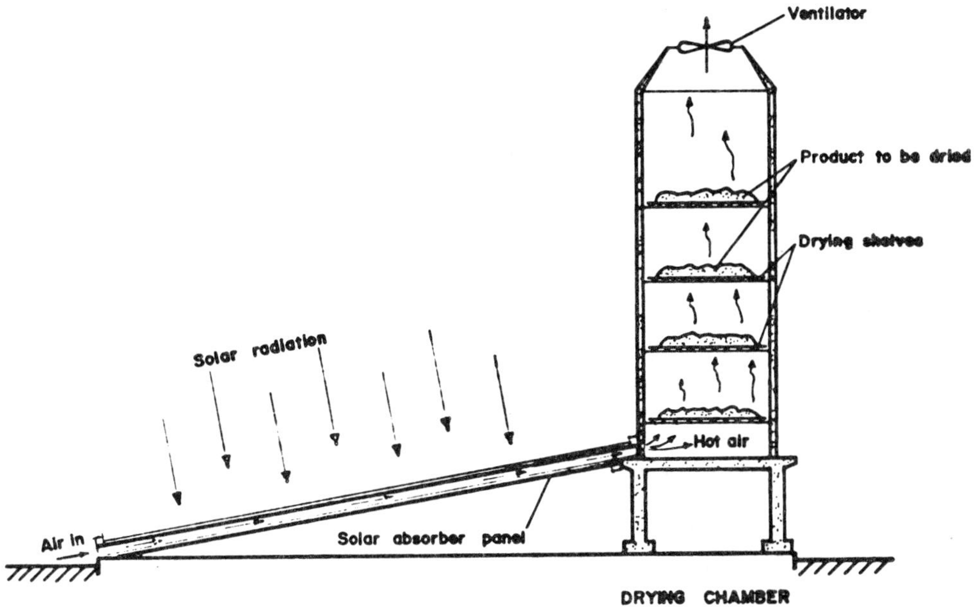

FIG. 2 A SIMPLE SOLAR CROP DRYER

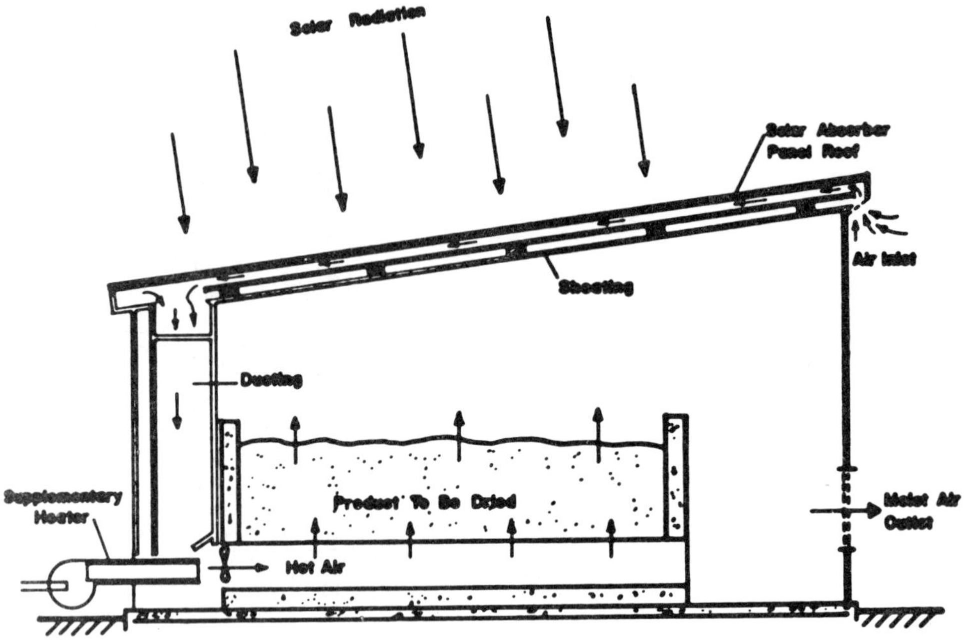

FIG. 3 SOLAR CROP DRYER AND STORAGE BUILDING

obtained from conventional oil burners or from burning the waste material obtained from left-over stalks or shells of the crop concerned. The system just described is ideal for grain drying. A slightly different design would have to be used for timber drying.

A SOLAR TIMBER DRYER

Solar timber dryers have been successfully used in Australia, India and also in other parts of the world [5,6,7]. A schematic arrangement of the Australian solar timber drying kiln is shown in Fig.4. It is very similar to that of the previous forced circulation solar dryer-cum-shed arrangement. The main differences lie in the air circulation mode and the arrangement of the timber stack.

Hot air supplied from the solar heater is circulated among sawn timber planks arranged neatly in a stack in the drying chamber. A reversible recirculating fan mounted on the top of a false ceiling is installed to assist in the air circulation within the kiln. It could be reversed to ensure uniform drying of the stack. The side walls are provided with air vents which can be opened or closed depending upon the direction of air circulation and whether air is recirculated within the stack and/or the collector.

Various other designs have been tried. In Fig. 4, the solar air heater is shown as a separate unit. It could be incorporated as part of the roof of the drying chamber as in Fig.3. Alternatively, instead of having a seperate flat plate collector unit, the roof and walls of the drying chamber could be replaced with glass or plastic sheets [6]. The interior of the chamber which is painted black then acts as the collector surface, as in the dryer of Fig.1.

POTENTIAL OF SOLAR DRYING RICE AND TIMBER IN MALAYSIA

Rice is the second largest crop in Malaysia with over 90,000 areas of land under cultivation. Sun drying has been the traditional practice for many years in the country. The requirements for artificial dryers did not arise until a double cropping scheme was introduced in recent years. The padi, been harvested during the rainy season is wet with moisture contents well over 25%. Studies have shown that padi at this moisture content will begin to germinate within 48 hours after harvest unless dried very soon. For safe storage, the moisture content must not exceed 14%. Padi can be dried by passing heated air or dehumidified (unheated) air through it. Past records show that a substantial proportion of sun-dried padi is cracked due to stresses set up in the grain when the temperature rises above 45°C(113°F). Hot air between 38° to 55°C (100° - 130°F) at a relative humidity between 30 - 40% should be quite ideal for artificial drying of padi. This can be achieved very simply using a flat plate solar air heater of the type discussed in Reference [8].

Yeo [9] demonstrated the effectiveness of a solar-electric padi dryer operated in Malaysia and showed that solar energy can supply most of the

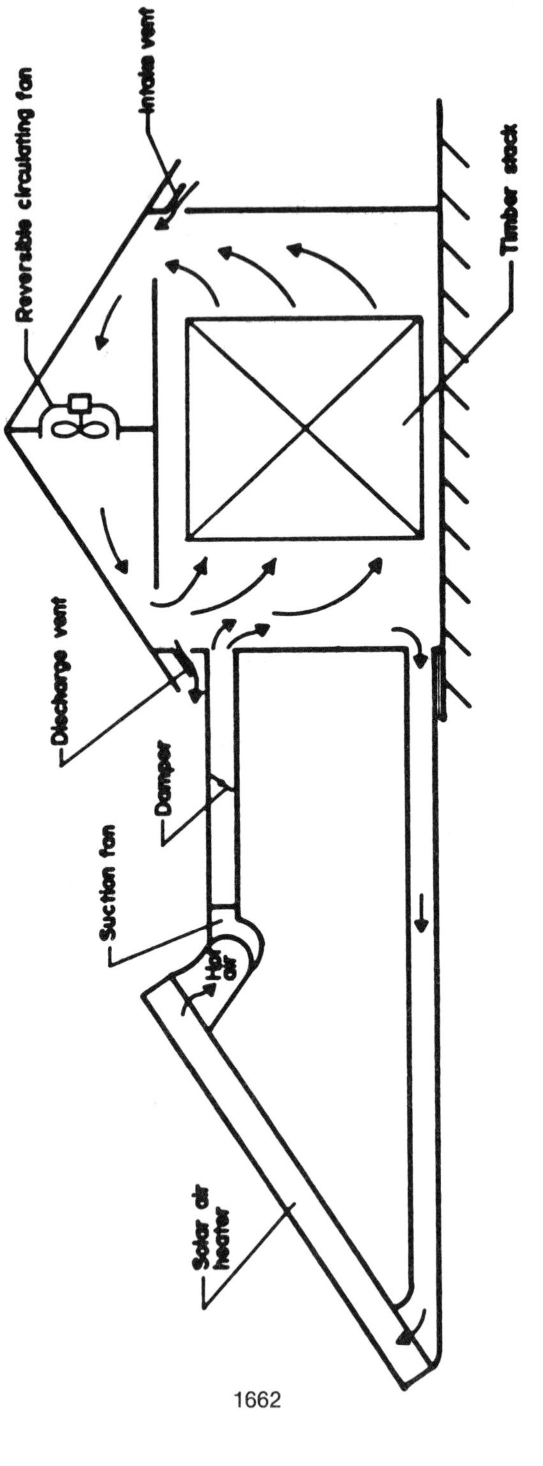

FIG. 4 SCHEMATIC DIAGRAM OF SOLAR TIMBER DRYING KILN

heat required at 40°C (104°F). At this temperature, however, drying rate was slow. With supplementary heating up to 60°C (140°F), the drying rate was very much faster. Other solar grain drying experiments were performed by Williams et.al. /10/ on drying tapioca.

The drying of timber using electrically-heated hot air at 50°C (122°F) for 8 hours a day was investigated at the Faculty of Engineering /11/. The intermittent batch drying process was employed to simulate solar drying conditions. It was shown that meranti pipit could be dried from 44.1% moisture content (oven-dry basis) to 14.8% in 8 days and Keledang from 101.4% to 7.5% in 10 days. The energy input to the circulating fan was minimal when compared to the solar energy which could be collected and utilised in place of the electrical heating elements.

INTERMITTENT BATCH DRYING

Most agricultural products can be dried more effectively by intermittent batch drying operation. This would depend upon the drying characteristics of the crop. A discussion of grain drying theory is beyond the scope of this paper. Suffice it is to outline the stages involved. At the start of the drying period moisture is first driven off from the surface of the grain. This initial rate of drying could be accomplished very quickly and at a low temperature. Diffusion of water from the center of the grain to the surface then takes place. This is a slow process and depends upon the nature of the material being dried. Usually a higher temperature or a lower ambient relative humidity is required. Intermittent or multi-stage drying could prove to be more effective than continuous drying from the point of effective heat utilisation especially for less porous products like rice and timber. Here, solar drying comes into its own. Drying could be carried out for a few hours in the day when sunshine is available. The dryer is non-operative at night. During this time, moisture diffusion will take place. Drying is then recommenced the next morning.

PROPOSED PLANS FOR STUDIES

An outline for the above would include the following steps to be taken:

(i) The agricultural products to be dried have to be identified first. At the present moment, for Malaysia, crops like rice, cocoa and pepper and sawn timber seem to be the most potential large-scale "users" of solar drying. Preliminary studies on solar padi and timber dryers /9,11,12/ have produced promising results.

(ii) The drying characteristics of these materials have to be determined in order to optimise the drying processes. Intermittent batch drying with multi-stage passes seems to be more advantageous. For the latter, the overall drying period may be shorter but at the expense of higher capital and operating fuel cost. The main criteria for comparison should be on an energy saving basis, i.e. weight of material dried/fuel consumed, all other factors being made equal.

(iii) At the same time, the performance of solar air heaters have to be investigated in order to determine their operating characteristics. Of these, the flat-plate type of solar heat collector incorporating two metal sheets rivetted together with rear insulation and top plastic or glass cover seems to be the most economical form of construction. The heat-collection ability in terms of air mass flow rate and temperature rise would indicate its effectiveness.

(iv) Solar radiation data would have to be collected over a period of time in order to determine the availability of incident solar energy. The data would have to be taken as close as possible to the areas where the dryers are to be used as seasonal variations are possible in different parts of the country.

(v) Experimental solar dryers would have to be constructed and evaluated over a period of time in order to evaluate their performance. Two types of dryers which hold promise are the smaller, natural convection dryer and the larger, forced convection dryer of Figs. 2 and 3 respectively. The natural convection dryer could be built very cheaply by the padi farmer and sited alongside his crops. After the harvest, the padi could be pre-dried to about 20% moisture content before being sold and sent to the larger milling and drying complexes. The natural convection dryer would have a capacity of about 1 ton/day. With pre-drying, the farmer is able to store his harvest longer and also to fetch a higher price. At the large complexes, 25 ton/day solar dryer-cum-storage sheds could be employed for intermediate drying down to about 16-18% moisture content for temporary storage with aeration for some weeks until drying down to 12% moisture content could be completed by conventional continuous flow kiln dryers. This intermediate drying and storage stage could extend the operating time of existing kiln dryers.

(vi) Prototype large-scale dryers could then be developed and tested at the intended sites in order to determine their technical and economic feasibilities.

CONCLUSION

A considerable amount of solar drying research has to be undertaken locally in order to optimise systems design and to make them economically viable. The materials to be dried have to be identified first. Then the drying characteristics of these materials have to be determined. Finally, prototype systems have to be designed, built and tested in order to evaluate their performances both economically and technically. Compared to the other fields of solar energy utilisation, solar drying is relatively new. Although some work has been carried out in this field, most of them have been on an ad-hoc basis. What is required is a long term plan with proper identification of objective, procedure, and execution of the planned projects.

REFERENCES

1. Ong K.S., Utilisation of solar energy in Malaysia - some prospective applications. 3rd IES/IEM Engineering Convention on energy resources, utilisation and conservation, Singapore, April 1977.

2. Foster G.H. and Peart R.M., Solar grain drying - Progress and Potential, Agriculture Information Bulletin No. 401, Nov. 1976.

3. Solar Grain Drying Conference, Illinois, Jan 1977.

4. A survey of solar agricultural dryers. Technical Report T99, Brace Research Inst., Canada, Dec. 1975.

5. Read E.R.W., et al., A solar timber kiln. Int. Congress "The Sun in the Service of Mankind", Paris, July 1973.

6. Sharma S.N., et. al., A solar timber seasoning kiln. Jour. of Timber Development Association of India, Vol. 17, No.2, April 1972.

7. Tschernitz J.L. and Simpson W.T., Solar Kilns - Feasibility of utilising Solar Energy for drying lumber in developing countries, US Dept. of Agriculture, Forest Products Lab., Madison, Jan. 1977.

8. Ong K.S. and Than C.F., Report on the performance characteristics of a flat plate solar air heater. Jour. Faculty of Engineering, University of Malaya, Vol. 15, June 1976.

9. Yeo S.H., A preliminary investigation in use of heated air from flat plate for padi drying. Thesis, Faculty of Agriculture, University of Malaya, 1974.

10. Williams S,N., et. al., Solar heat dryer for crops and other products. Tropical Agriculture, Vol. 46, 1969.

11. Cheah W.B. and Chuah C.J., Preliminary investigation on kiln drying of timber - Parts I and II. Thesis, Faculty of E n gineering, University of Malaya, 1977/78.

12. Ooi S.H., Solar kiln for timber seasoning, Timber Trade Review, Vol.7, No.1, 1978.

ENERGY CONSUMPTIONS AND POSSIBILITIES FOR THE APPLICATIONS OF SOLAR ENERGY IN ITALIAN AGRICULTURE

L. Andreotti, G.Marchetti, F. Passaro,
D. Ruggi, and A. Serravezza
CTIP Solar, S.p.A.
Rome, Italy

A. Di Vecchia
CTIP International Inc.
New York, New York U.S.A.

A. Fanchiotti
University of Venice
Venice, Italy

ABSTRACT

This paper summarizes data and conclusions of a study completed by CTIP Solar S.p.A. on the possibilities of solar energy applications in Italian Agriculture. In order to identify those sectors where solar energy could be applied, it was necessary to have a detailed picture of energy inputs in Italian Agriculture; the analysis of energy consumptions was extended to those primary transformation processes which are traditionally performed on the farm. A survey was then made of the systems that are currently in use in agriculture throughout the world. Conclusions were finally drawn about the potential of penetration of the different techniques in agriculture.

INTRODUCTION

It has always been thought that for the complete definition of an agricultural system, the use of three main parameters - land, labour and capital - was sufficient. It was a different relationship among these three elements which made the alimentary system of the Eskimos so different from that of the Chinese and that of the Dutch polders.

Only recently has energy become another important element in agriculture.

SOLAR ENERGY: INTERNATIONAL PROGRESS

The massive introduction of synthetic fertilizers and the intensive use of agricultural machines started in the U.S. when energy of fossil origin, then extraordinarily cheap, made it advantageous to replace human work with technologically sophisticated means (as an example, ten years ago, when crude oil cost only $1.50/barrel, one dollar provided an amount of energy equivalent to 4,000 hours of human work).

This led to the possibility of getting larger yields from cultivated lands - the same amount of food produced by a Masai on 100 hectars of land is produced in Europe on a tenth of a hectare - and also of utilizing some of the people once employed in the primary sector (direct production of raw materials) in other areas (such as food processing).

Nowadays, however, agriculture cannot keep on developing by following this pattern, because the energy needs of society grow continually in every sector, while, on the contrary, the shortage of traditional energy sources is worsening dramatically.

Since it cannot be thought possible to return to the primitive agricultural systems because of the rapid increase in population relative to the available land, the possibility of utilizing alternative energy sources (wind, waves, sun, etc.) in various agricultural activities must be considered more seriously. These sources were in the past disregarded because the cost of their direct competitors was extraordinarily low.

In this paper, one of these alternative energy sources, the sun, and its possible applications in Italian agriculture are analyzed. The general advantages offered by solar energy in comparison with other energy sources are well known:

- it is available in large quantities;
- it is available in every location;
- it is free;
- its exploitation has no negative impact on the environment.

Moreover other specific elements make agriculture one of the most promising sectors for the use of solar energy:

- the peculiar feature of solar radiation, that it is not concentrated, while negative in other fields, is adequate to the energy needs of agriculture, where small productive units scattered on large areas are needed more than highly powerful units;

- most of the solar energy systems employed in agriculture call for very simple technologies and can be operated by unskilled staff.

For these reasons, in Autumn 1976, CTIP Solar elaborated a proposal for a multiclient study on the state of the art of the solar technologies which may apply to Italian agriculture. The study was completed in January 1978 and the final report, consisting of three volumes for a total of about 1500 pages, was submitted to the various public bodies and private groups who acted as sponsor 1 .

The information given in this paper is drawn from the conclusions of that study.

ENERGY CONSUMPTIONS IN ITALIAN AGRICULTURE

In order to identify the sectors where solar energy could be profitably used, it was essential to have a detailed picture of the present energy consumptions in Italian agriculture. Since these data did not exist in useful form, it was necessary to start with a detailed analysis of the various components of agriculture energy balance.

As a beginning, the official sources (ISTAT, UMA, ENEL) were consulted. However, since the data they provided were not complete, it was necessary to integrate them with data obtained from various public bodies and private groups, as well as data collected by field surveys in farms located in several Italian regions. Data thus collected were then verified by random spot checks in six Italian regions.

Final results relative to 1974 are shown in table 1 : the total consumption of 2.3×10^{8} GJ (=6.4×10^{10} kwh or 5.5×10^{13}kcal) is equivalent to 5% of the final energy consumption in Italy in that year.

Comparing these ata with those relative to other productive sectors, it can be seen that every worker employed in agriculture makes use of a direct energy input of 35 GJ (not including chemicals and domestic consumptions) while every worker employed in industry makes use of 189 GJ and every worker employed in services of 110 GJ. This shows the low technological level of agriculture as well as the need to increase its energy input.

	GJ/Year	Percentage
Rural house heating	3.5×10^7	15%
Crop Processing	2.0×10^7	9%
Livestock shelter heating	1.4×10^7	6%
Crop drying	5.2×10^6	2%
Greenhouse heating	4.7×10^6	2%
Irrigation	3.9×10^6	2%
Various electrical applications	3.1×10^6	2%
Fuel for machines	5.8×10^7	25%
Chemical fertilizer	8.4×10^7	37%
Total	2.3×10^8	100%

Table 1 - Energy consumption in Italian Agriculture (1974).

The increasing cost of conventional energy sources, however, makes increase in its consumption impossible. The utilization of "alternative" energies is therefore the only existing possibility of increasing the yields of this sector.

SYSTEMS UTILIZING SOLAR ENERGY IN AGRICULTURE

The second part of the study aimed to identify which solar technologies could be profitably utilized in Italian agriculture. CTIP Solar made contact with over a hundred institutions in sixty countries all over the world, which supplied it with information regarding theirexperiences in the applications of solar energy to agricultural uses (whether experimental or at an advanced project stage). Before estimating their economic feasibility, the systems were analyzed from a technological point of view in order to verify their adaptability to Mediterranean agricultural conditions.

Information on the most feasible systems (more than 100) were given in the study in standardized forms, developed for this purpose, that allow a rapid comparison among the various types of solar energy techniques.

Systems used in agriculture may be divided into four main groups; conservative, home-made flat plate collector, industrial flat plate collector, concentrating collector. Applications of the four types to various agricultural operations were represented in matrix form (table 2).

A definition of the most appropriate systems in absolute was deliberately omitted because of several political and economic reasons, such as the coexistence of modern and backward techniques in Italian agriculture; the programmatic presence of Government; the deep changes to which agricultural policies are presently being subjected not only in Italy, but in the European Common Market; and finally the increasing interdependence among the Mediterranean agricultural systems. All these elements cause a continuous evolution in the Italian agricultural system and do not allow the choice of single optimal solutions.

There are also more strictly "solar" reasons for this fact, such as the variability of climate and insolation conditions in the different Italian regions, that makes a definition of systems suitable for the whole country not feasible.

Systems / Application	T (C) (*)	Conservative		Home-made flat-plate collectors		Industrial flat-plate collectors		Concentrating collectors	
		New	Retrofit	New	Retrofit	New	Retrofit	New	Retrofit
Greenhouses	5 + 25	x	x	x	x	x	x		x
Crop drying									
Corn	+2 +20			x	x		x		x
Rice	+ 6				x				
Wheat	+ 20			x					
Peanuts	30			x	x				
Forage	+10			x			x		
Burley tobacco	35			x					
Bright tobacco	30-80			x					x
Fruits	40-75			x		x			
Other	40-70			x	x	x	x		
Livestock shelter									
Poultry	30	x		x	x				
Swine	20-30	x		x	x				
Cattle	40-90				x		x		x
House heating	20-60	x				x	x		
Residence + greenhouse htg	20-60	x	x			x	x		

Table 2 - Thermal application for which solar energy systems are reported.

(*) When there is a "+" mark this means a difference regard to the environment.

Agricultural Applications

The most common solar energy applications in agriculture are those which use heat producing systems. Power generation is necessary only for a few applications (for example for irrigation).

The sector with the largest potential in Italy is the climatization of greenhouses. The area covered with greenhouses, more than 12,000 hectares, is one of the largest in the world. Solar climatization of greenhouses would not only allow a reduction of energy consumptions, but would also make it possible to accelerate the ripening of fruits and vegetables. From a technical point of view, one has to consider that most of the greenhouses in Italy are in the central and southern regions, where technologies are much less advanced than in the North; therefore, greenhouses should be designed to optimize energy balance at a low level of sophistication.

The most energy-consuming sector in agriculture is at present the climatization of rural houses, where, however, the variety of typologies, construction techniques and population distribution patterns make it rather difficult to define typical characteristics. Solar systems for this purpose are already becoming popular in the U.S., but then are less crucial in Italy where energy consumption per person is lower.

Another important sector in Italian agriculture is crop-drying, especially for some particular applications. On the one hand, there are a large number of specific crops, for instance tobacco, which require sophisticated processes (and consequently complex installations), while on the other hand, other crops, such as fruit and vegetables, can use very simple systems. A large number of solar installations have been built in the livestock sector, especially for dairy cows (milking parlors, stable washing and winter-time heating). There are also some experimental installations in the hog raising sector; they are used to prepare mashes for animal feeding and to heat pig nurseries. Crop processing, such as oil, wine, and cheese production is finally suited to the introduction of solar energy. In Italy, in fact, such processes are often carried out at a cooperative level near production farms.

Economic Evaluations

The costs of every analyzed installation were indicated in the information sheet describing it; general and long-term economic evaluations, however, were not possible, as these costs generally referred to prototypes.

In the next few years a great development of the industry of solar components may be expected; according to U.S. evaluations, that could create a market of about one billion dollars per year, allowing a considerable reduction of costs.

If economic evaluations are made with today's costs, they usually give negative results: in other words, there are few commercial applications where solar energy is at present competitive in the agricultural sector. The main reason for this fact is that it is not possible in a short time to pay back the expensive investments for solar systems, which usually work (and save conventional fuels) only part of the year. Moreover, while negative economic aspects are felt at a microeconomic level, the positive elements, especially the social ones, involve the macroeconomic level.

Integrated Systems

An alternative which consistently yields positive economic results is to diversify the utilization of each solar system by using it for several applications and, thus, for longer periods throughout the year.

A detailed analysis of a farm is needed in order to design an integrated system suited to it: the thermal load curves of the different applications must be charted, and the difficult problems relative to the interconnection of the applications must be solved. The most sophisticated integrated system make use of solar thermal plants that serve several final energy-consuming activities.

Social Elements

Beside the economic aspect, there are a large number of social elements in favour of an increase of the utilization of solar technologies:

Trade balance for fuels

The use of solar systems in agriculture would allow an increase of energy consumptions per person without affecting the balance of trade.

Employment

The use of solar systems may lead to an increase of the number of people employed, especially in farms located in marginal areas that would otherwise return to a very low technological level of agriculture or would be completely abandoned. Moreover the production of solar components may require an increase of work in the industrial sector. The final result would be that the utilization of solar energy would substitute energy which is at present imported with both manpower and energy locally available.

Ecology

Solar energy systems cause much less pollution in the surrounding environment than the conventional energy ones. Furthermore, by the reclamation of marginal areas, solar energy may contribute to avoid the decay of bio-systems.

Reorganization of agriculture

The use of solar energy can allow an approach to agriculture different from the one used in the past, that consisted chiefly in an attempt of applying a model created for American conditions to the completely different situation of Italian agriculture.

In any case the quantification of all the above social elements was not an object of the study.

International Analogies

The results of this study can be applied to most of the Mediterranean countries, whose conditions are similar to those of central and southern Italy. Furthermore the survey on solar installations was carried out on a world-wide basis: information about systems constructed more than 20 countries was presented.

CONCLUSIONS

An increase of energy input is necessary for Italian agriculture, in order to develop it to the technological level suited to an advanced society. Solar technologies, which could contribute considerably to this increase, involve high installation costs, which make them unprofitable at a micro-economic level.

Nevertheless, since the social benefits which solar energy could provide counter-balance the high investment costs at a macroeconomic level, we may expect a large increase in Italy (and in other countries with similar conditions) of the demand for solar systems for agriculture, especially if incentives are granted by governmental agencies. As a consequence, a considerable reduction of the costs of components may be expected, which could finally take solar energy beyond the threshold of economic competitivity.

REFERENCES

Studio "Multiclient" sulle possibilita' di utilizzazione dell'energia solare nell'agricultura italiana, Final Report, CTIP Solar Spa , Rome, Italy, January 1978.

ESSICCATOIO PER TABACCO BRIGHT			N° 21. d
PROGETTO : CTIP Solar S. p. A. Roma, Italia			ANNO 1977
RIF. BIBLIOGRAFICI			
LOCALITÀ Pietrino (Perugia), Italia.			
LATITUDINE 43°30'N	ALTITUDINE	INSOLAZIONE 3,58 kWh/m²g	

DESCRIZIONE SISTEMA

L'impianto, (finanziato dal C.N.R.), attualmente comprende una cella convenzionale per l'essiccazione del tabacco Bright e una centrale termica solare.

La celladi essiccazione è stata opportunamente modificata con l'aggiunta di canali esterni in cui sono sistemati uno scambiatore acqua-aria per il riscaldamento dell'aria esterna e uno scambiatore di recupero di calore aria-aria.

La centrale termica è composta di due batterie di collettori solari concentratori del tipo "parabolico-cilindrico", dotata di dispositivi automatici di inseguimento del sole; di due serbatoio.di accumulo di calore (6.000 l) di un circuito primario ed un circuito secondario.

Nel circuito primario (collettori-accumulatori) circola una soluzione di acqua e glicole, sotto pressione, che trasferisce l'energia raccolta dai collettori all'acqua contenuta nei due accumulatori.

Nel circuito secondario (accumulatori termici, scambiatore acqua-aria) circola acqua a pressione atmosferica che trasferisce il calore accumulato all'aria esterna e/o di ricircolo utilizzata per la cura del tabacco "Bright".

Dispositivi di regolazione assicurano il funzionamento completamente automatico dell'impianto.

L'impianto verrà completato con un sistema riscaldamento di una serra (200 m²) per la produzione di pomodori. Area totale collettori : 74 m².

FABBISOGNO ENERGETICO

Dati relativi ad una cella di essiccazione convenzionale :

tabacco verde trattato	6.000 kg/cura
tabacco secco prodotto	800 kg/cura
consumo di combustibile	1,1 kg/kg secco
quantità di calore necessaria	5.270.000 kcal/cura
temperatura aria di essiccazione	32° - 75°C
numero di cure a stagione	10

La stagione di cura va da Agosto alla metà di Ottobre.

PRESTAZIONI N° 21. d/2

Secondo i dati progetto la centrale termica solare dovrebbe fornire il 30% del carico termico necessario.

Il sistema di recupero del calore dell'aria di scarico dovrebbe consentire un ulteriore risparmio energetico del 15%.

CONTRIBUTO SOLARE

Energia fornita da 1 m² di collettore esposto a sud, inclinato di 45° nell'orizzontale e con temperatura di lavoro di 70°C.

mese	kWh/m²g	mese	kWh/m²g
gennaio	2,3	luglio	5,4
febbraio	3,2	agosto	5,1
marzo	4,2	settembre	4,8
aprile	4,8	ottobre	3,7
maggio	5,2	novembre	2,5
giugno	5,5	dicembre	1,9

VALUTAZIONI ECONOMICHE

I costi dell'impianto non sono significativi perchè lo stesso è stato realizzato con tutta una serie di dispositivi sperimentali atti a verificare diverse condizioni di funzionamento. Tali dispositivi nelle successive installazioni potranno essere eliminate riducendo di molto i costi d'investimento.

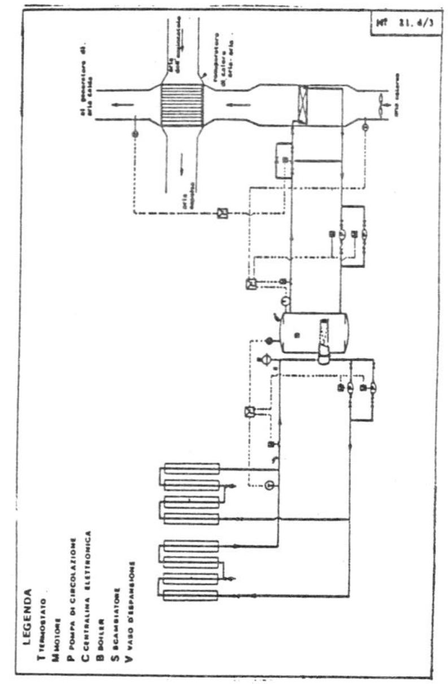

FIGURE 1 TYPICAL FORM USED FOR THE SURVEY

SOLAR ENERGY UTILIZATION FOR AGRICULTURAL PURPOSES IN EGYPT

M. Hassib
National Research Centre
Dokki, Cairo, Egypt

ABSTRACT

With the increasing foodstuff consumption in Egypt emerges the application of solar energy for agricultural purposes as a highly recommended resolution. Corresponding water sources for desalination are abundant, and installations can be, to great extent, locally supplied. Nutrient-film technique in polyethylene coated greenhouses is very appreciable. Growing vegetables, expensive crops, raising fishes and poultry on own plant wastes can be relatively competitive and even vital.

INTRODUCTION

Egypt is facing in the near future a challenge which may be the worst in its whole history. The tremendously increasing population is evoking the serious danger of famine, especially for a country which is not capable of industrial competition on the consuming foreign markets in order to integrate its alimentation demands through exchange of its own industrial products against food requirements. The population has exceeded now 38 million and is expected to reach 70 million in the year 2000. The local wheat production is already, and remains, very far behind the consumption, unless triticales are going to be grown. The imports of other cereals, fats and newly sugars are growing continuously. The only rescue out of this affliction can be through the use of the vast deserts, 97% of the total area in Egypt, at locations where soft water can be gained and using it very carefully raising relatively expensive crops, such as vegetables and fruits and small animals. International experiences and practices are now too far to be reliable exhibiting that the use of non-conventional water sources, with differing grades of salinity, in non-rainy regions can only be possible through desalination depending upon the use of the free source of energy, namely solar energy. The needed initial installations may not be that cheap such as conventional farming, but following a proper plan in crop selection and management and an automation system to avoid the recent high salaries of the agrarian labour, should lead to competitive prices of the second products.

PROPOSED LOCATIONS

Certain suppositions must be taken in consideration for the choice of the suitable locations. Water as a matter of fact is the main element for agricultural establishments, yet almost soft water, up to 2000 p.p.m., is too precious to be elaborated for such purposes, then it can be used even for growing several salt-tolerant crops and serve some other objects, without the required expenses of any of the desalination processes. Hence, it is clear for Egypt that raising of desalination establishments can take place anywhere at the long coastal front, over 3000 kilometers, at both Mediterranean and Red Seas or at sites where saline water can be gained from near of deep wells. Such sites are abundantly existing in the northern regions of the Sahara, Sinai, and in the eastern desert. The regularly flowing amount/annum differs between 5000-300,000 tons from such wells, with a salinity ranging between 2000-10,000 p.p.m. Yet, there are some other aspects which must be taken in consideration, namely the kind and size of community that may inhabit the regions where this relatively expensive, till now, crop growing can be executed. Real probabilities are expected to be found at isolated sites, with intensive mining operations, tourism assemblies, considerable seaports, and army units. It is then preferable if soft water is gained secondarily as a waste in a procedure during which also energy is generated and can be possibly stored in order to offer, simultaneously, several services levelling the balance between expenses and income. Speaking about energy, it is sometimes necessary to have additional devices for energy, for moveable parts, and in this case it is advisable to spare the own gained energy and substitute it through other means, such as wind whose velocity is ranging between 5-45 K/h. on the Red Sea Coast. Additively to crop growing, raises the necessity for heat production or cooling systems for processing and storage of excessive agricultural products, respectively. Such locations exist already in Egypt at sites with mass production which is not all thought to be immediately consumed or with non-convenient transportation.

APPROPRIATE SYSTEMS

Specially, in the case of Egypt, an appropriate system must exhibit certain attributes in order to be principally acceptable. Egypt is spoiled through the free excessive supply of Nile water, mild weather, and fertile soil. What may be good enough for parched, and though a rich country on the Persian Gulf does not, compularily,have the same advantages for countries like Egypt. For these countries an appropriate system whether for crop growing, processing of agricultural products or raising animals must be cheap in acquisition, serving numerous purposes, having a simple automation to deal with, and with low running costs. Luckily, it is already so far that the developments of such equipments has produced several models which are approaching the figured objective. After Hodges and Hodge [1] greenhouse construction should be simple, fairly inexpensive, and rely on using the locally available material as much as possible. Plastic-covered greenhouses proved to be successful in intensive agricultural operations in several arid areas including Abu Dhabi, Kuwait, Iran, Mexico and Arizona. The substructure frame can be constructed of wood, iron, or aluminum. Inflated greenhouses using polyethylene film (12 mil Monsanto film made of light-stabilized plastic sheets with a yield strength of about 1400 psi and with approximately 12-14 months life span in warm arid climate, or any similar films). Newly produced coated membranes can last for more than 15 years. A greenhouse with a closed environment and air-locks seems to be very proper. In this case CO_2 enrichment leads to better yields. Hodges and Hodge estimated the costs of producing 2 million pounds of greenhouse harvested, 20 folds yield as conventional farming [2], vegetables per year of a 10 acre unit in Abu Dhabi to be less than 20 cents per pound (compared to $1.50/pound or more for imported vegetables). Under the different examined models of greenhouses, the soil-less construction seems to be the most suitable for Egyptian conditions, by which the available amount of irrigation water is not always high enough. Crop growing by the intensive methods of soil-less culture is becoming, everywhere, quite prevalent in recent agricultural and horticultural operations. These methods provide an immediate source of income, almost like instant returns, while the longer-term projects are underway. Hydroponics, i.e., growing plants in water-cultures placed in containers where the nutrietient solution is recirculated, were first used by American research workers in the early part of this century. In the last episode, modified hydroponics were constructed, with the so-called nutrietient-film technique (NFT) and some are called the hydroponic layflats or the hydrocanal system. Some recent designs have been developed to accommodate the need for speedy construction, light weight components, easy portability, and immediate crop production. By all these new models, the nutritient film is flowing past the roots and ensures enough moisture for all roots, grown in plastic tubes. This method has the advantage of providing simplicity of operation, can tolerate fluctuations in the concentration of the nutrient solution and thus one formula of the nutritient solution can be used for all stages of growth of all crops and under all climates in various countries. Thereby, such a model can be highly recommended for large-scale application in Egypt.

PROPOSED CROPS

As for vegetables and small trees, it must be taken in consideration that the choice of the crops has to care for the commercial point of view, and vital necessity at some sites in Egypt. The recommended crops must be those who can be sold for high prices, such as vegetables in the off-season, strawberries, and some exotic spices. Speaking about vital necessity in isolated sites with bad or impossible transportation, vegetables, salad plants, and small trees such as vines can be adopted. In addition to these conventional objects, algae mass-culture for protein production, fish basins supported with own or additive feeds, poultry and sheep fed on wastes of higher plants, can be also raised. Productive high-protein field crops such as soya been can be grown as well to integrate the nutritional supply of the inhabitants in some locations.

REFERENCES

1. Hodges, C.N. and C.O. Hodge, Hortiscience 6:33, 1971.
2. Modest Technologies, Mosaic 8:44, 1977.

SOLAR POWERED WATER PUMP FOR THE RURAL THIRD WORLD

J. Boldt
Department of Mechanical Engineering
Laboratory for Energetics
Technical University of Denmark
Lyngby, Denmark

ABSTRACT

A vapor actuated pump containing no other moving parts than two check valves has been developed. The vapor is to be produced directly in a flat-plate solar collector (about 25 square metres) with plane reflectors along its sides.

The yield is in the order of 5-20 cubic metres per day. The investment cost is relatively low - less than one U.S. dollar per capita - and the cost of the water is in the order of 30-100 dollars per 1000 cubic metres, assuming a 10 year life-time. The pump can only be used for shallow wells (about 0-10 metres depth).

The pump has been developed with special reference to the technological resources of the developing countries, thus being based on the concept of appropriate technology. It is robust, simple, and suitable for manufacture and maintenance by local workshops from readily available materials, requiring no hign standards of accuracy. Its operation is reliable under laboratory conditions and is expected to be so in the field. The production process is labor intensive, thus providing local opportunities for employment.

INTRODUCTION

A major part of the World's population is still without safe water supply. Many have little or no ready access to any water at all. The lack of water causes victims of debilitating diseases or death. Lack of water also results in economic and social stagnation of entire communities and regions. Often the population - especially women and children - spend a great deal of their time on manual handling of water at the expense of other productive activities or education.

Developing water supply systems is therefore one of the major objectives in raising the Third World's living standard.

It is in the light of this situation that the Laboratory for Energetics at the Technical University of Denmark has developed a prototype water pump. The pump is the so-called thermopump, as patented by Kleen |1|. It was later followed up by Coleman |2| and MacCracken |3|, but since then no further development seems to have taken place. The thermopump appears interesting, however, and in 1975 a project was started in our laboratory with the aim of developing a pump that could become of use to the rural population in the developing countries.

The pump has been developed with special reference to the technological resources of the rural areas. It is thus based on the concept of "intermediate" or "appropriate technology", which is a set of technologies lying between the capital intensive technologies usually exported to the developing countries by the industrialized nations, and the indigenious technologies that the developing countries already possess. The aim of the appropriate technology is to raise the levels of employment, productivity and technological skills of the inhabitants.

THE BASIC PRINCIPLE OF THE THERMOPUMP

From Figure 1 it can be seen that the thermopump basically consists of seven parts: A heat source, cylinder, vapor tube, condenser, delivery-refill tube, a pair of check valves, and a foot valve.

From the beginning the pump is filled with water. As vapor is formed in the cylinder, or supplied from a solar collector, water equal in amount to the volume of vapor is expelled through the outlet check valve. When the vapor

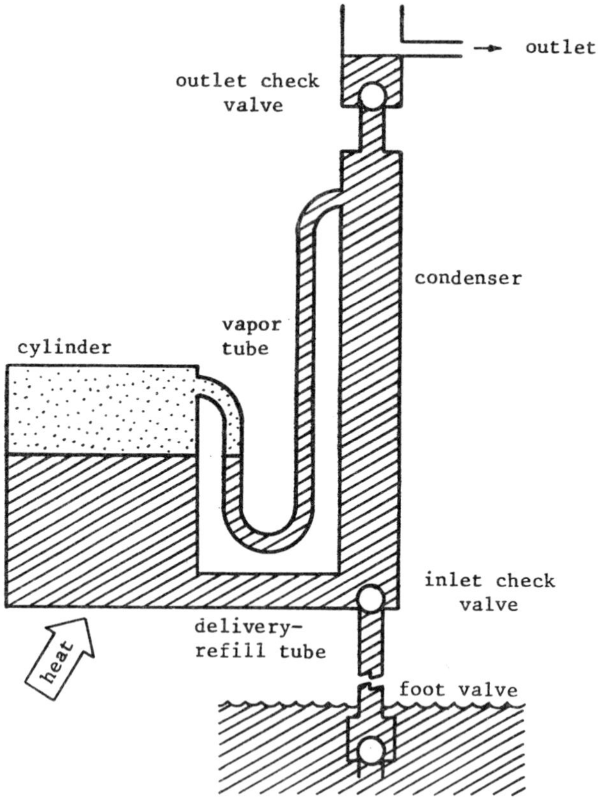

Figure 1: Basic thermopump diagram

volume enlarges so that the water level comes below the base of the "U" of the vapor tube, a triggering action takes place which siphons all the vapor from the cylinder into the condenser. As the vapor hits the mass of cool water in the condenser, it condenses. The water that has been expelled through the outlet check valve cannot return and fresh water is sucked in through the inlet check valve refilling the pump. This terminates the pumping cycle, and the pump starts over again.

The maximum obtainable suction lift is 7-8 metres (depending on the temperature of the water and the amount of dissolved air), but if mounted in a well the total lift can be larger.

UTILIZING SOLAR ENERGY

The pump can be operated by any heat source. As the developing countries often are characterized by a dry and sunny climate, it appears natural to investigate the possibilities of harnessing solar energy. It has a principal advantage of being well suited to the dispersed communities of the sunny rural areas. Furthermore the need for water is greatest during periods of sunny weather.

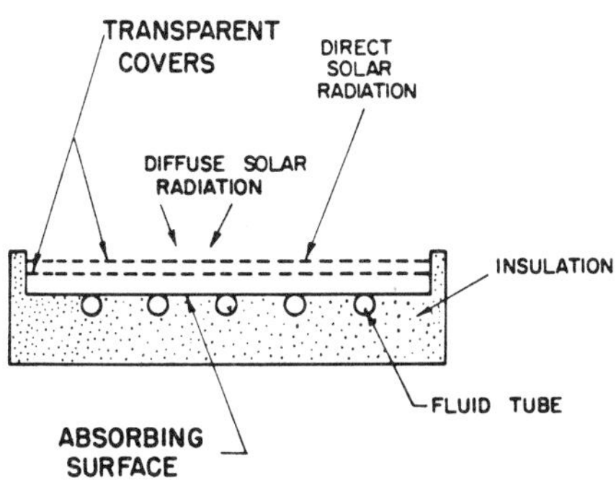

Figure 2: Typical flat-plate collector

The collection of the solar radiation is done by a flat-plate collector (see Figure 2). The flat-plate collectors have the advantages of being able to utilize diffuse solar radiation, not requiring orientation towards the sun, and requiring little maintenance.

For several reasons it appears advantageous to mount the solar collector as an integral part of a roof structure. Firstly, the construction becomes cheaper. Secondly, the risk that the surroundings will throw shadows on the collector will be less, thus increasing the power output. Thirdly, the collector will be cooling the rooms below. And finally, the collector will be less accessible to damage caused by children's pranks. Figure 3 displays the general set-up of the pump system.

As vaporis formed in the absorber, the water level in this is lowered. To prevent the absorber from running dry, a storage tank is mounted in parallel with it. Following a day's operation the absorber is cooled. The vapor accumulated in the absorber and the storage will then condensate, thus creating a partial vacuum that causes a refilling of water from the well.

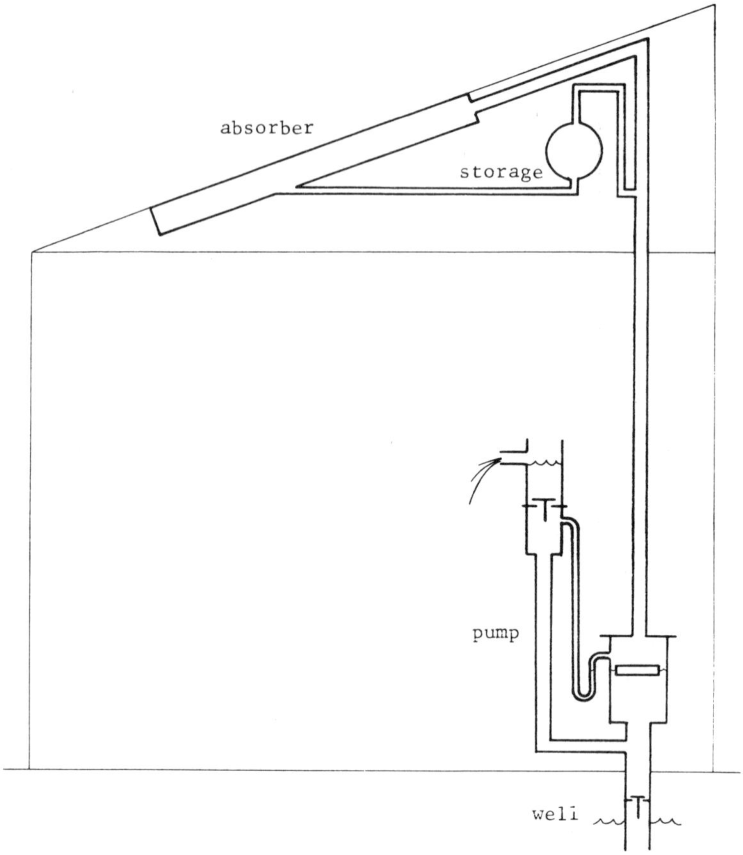

Figure 3: Principle diagram of the solar powered thermopump

If the distance between the absorber and the water level in the well is too large, the refilling can be carried out by means of a tube and check valve from the storage down to the reservoir catching the pumped water.

The configuration shown in Figure 3 differs from the basic configuration (Figure 1) in three ways: A) the water inlet has been moved from the condenser to the cylinder, B) the inlet check valve has been removed, and C) an insulating float is used in the cylinder to reduce heat losses.

Figure 4: The prototype pump. The solar collector is electrically simulated.

PERFORMANCE

Boiling water in flat-place collectors is only possible a few hours per day. To increase the working period and the capacity one could use another working medium with a lower boiling point. For example using ethyl alcohol (boiling at 78°C) would approximately double the output. The working medium and the water can be separated by a rubber membrane (such as a rubber-tube from a vehicle) suspended between the inlet and outlet check valves in Figure 1.

However ethyl alcohol has one serious disadvantage: It is drinkable, and hence the pump might be vulnerable to sabotage. In addition, other fluids with suitable boiling points are poisonous, unstable, corrosive, too expensive or the like.

For the present another improvement is therefore chosen: To place reflecting plane surfaces (for example white-painted metal plates) along the sides of the absorber (see Figure 5).

Figure 5.

A computer programme that determines the energy balance of this system and compares it with the experimental results from the pump has been developed. Taking the climate of Dar Es Salaam, Tanzania, as the reference case for our calculations, the performance of the pump has thus been determined.

Figure 6 shows some of the calculated results. The pumped output is shown when 4 different solar collectors are being used: A flat-plate collector, a flat-plate collector with stationary reflectors along its sides, and two cases with quasi-stationary reflectors (tilts are adjusted 4 times annually). For the three cases, where a reflector system is being used, the reflector areas are: 265% (quasi-stationary, large), 139% (quasi-stationary, medium), and 178% (stationary) of the absorber area. The dashed part of the upper curve is for net usable power greater than 14 kilowatts - this is the maximum allowable power input to the protype pump.

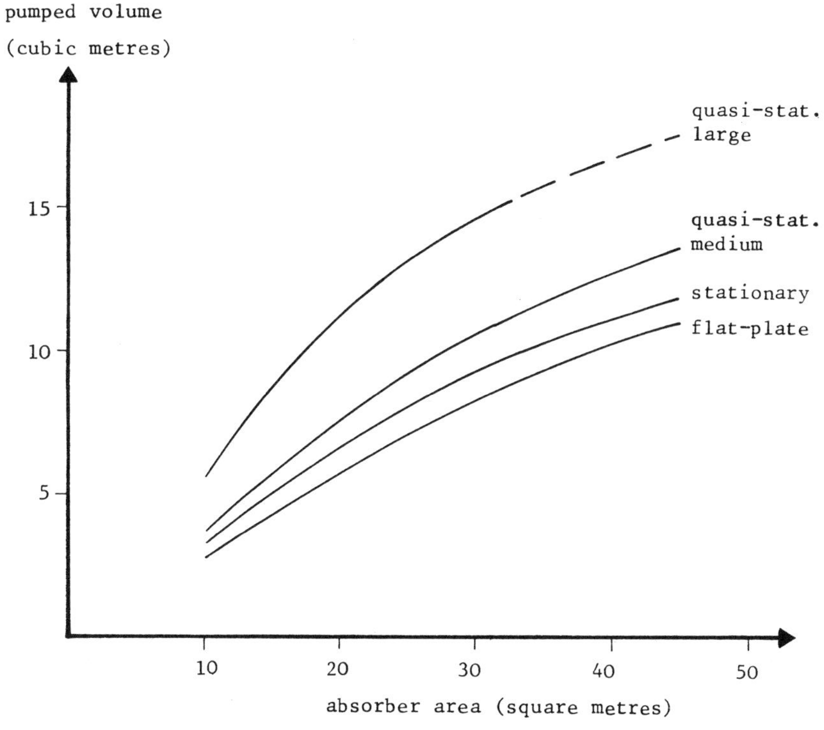

Figure 6: Daily pumped volume of water on a clear December day (15th) at Dar Es Salaam, Tanzania. See text for further explanation.

Figure 7 shows the influence of the number of glass covers on the water price (assuming constant weather conditions similar to those of Figure 6).

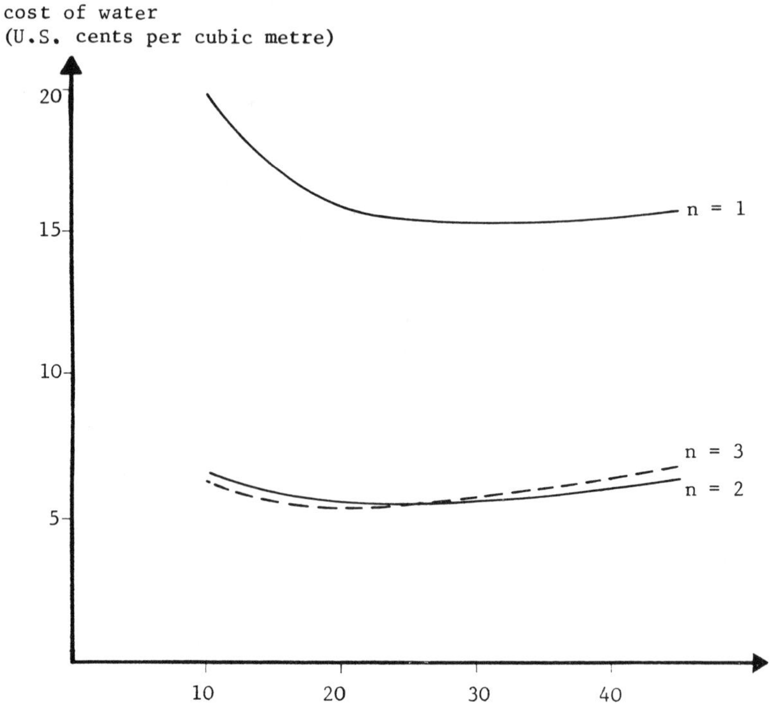

Figure 7: The price of water using a medium-size quasi-stationary reflector system with different numbers (n) of glass covers

The difference between 2 and 3 covers is almost nil, while 1 glass is uneconomical.

The absorber tilt has been optimized to gain the maximum yield during the dry season (June to October). With a stationary reflector system and an absorber area of 25 square meters, the daily output on an annual basis is in the order of 5-17 cubic metres - depending on the season and the climate (to simulate non-clear skies, intermittent clouds have been superposed the insolation functions for clear skies).

For domestic purposes it is common to design new community supplies for delivery of an average of 10-25 litres daily for people using public standpipes |4|. As the average number of inhabitants in a tanzanian village is 2000 |5|, the average designed daily water requirement is 2-5 cubic metres per village. The thermopump thus seems well applicable with respect to yield.

ECONOMY

Tanzanian material prices and salaries have been used for the economic calculations. For other communities the prices might deviate considerably. Furthermore the costs have been calculated in fixed prices (1977), using an effective interest rate of 5 per cent and a lifetime of 10 years.

On this basis the cost of the water is in the order of 4-11 U.S. cents per cubic metre, using a 25 square metre absorber with stationary reflectors. An analysis prepared for the United Nations Water Conference, 1977 |6|, places the basic cost of groundwater supply for domestic use and stock watering between 1 and 10 U.S. cents per cubic metre in the majority of cases, although there are many instances outside this range.

The per capita investment is 70-80 U.S. cents. In a WHO survey, the per capita cost at 1970 price levels for new community water supplies under government auspices was found to range on the average over large areas from 6 to 24 U.S. dollars for public standpipes in rural areas |4|. In view of this the thermopump appears to be economically very favorable.

CONSTRUCTION AND MAINTENANCE

Constructing the pump requires welding and soldering facilities. Beyond this no advanced equipment is required. The pump should therefore be suitable for manufacture by local workshops. The production technique is rather labor intensive, thus providing a large number of work places relative to the invested capital.

Without moving parts except the two valves the pump should be reliable in operation, although it is not maintenance-free.The absorber glass should be washed once or twice annually, and the absorbing surface plus the reflectors should be repainted after about 5 years. Furthermore the scale deposits in the absorber pipes should be removed regularly - about every second year, depending on the water quality.

If a quasi-stationary reflector system is being used, the tilt of the reflectors should be adjusted about four times a year.

The required maintenance is relatively simple, but as one normally cannot expect very much with respect to maintenance standard, thermopumps should not be built without the existance of a viable maintenance system. The success of rural water supplies of all kinds hinges not only on the establishment of water supply equipments in the rural areas, but also on the fact that they are fully operational and properly maintained. This calls not only for sufficient funds for operation and maintenance to be made available but also for district-level mobile teams of trained operators and mechanics in the rural areas to carry out this important work.

EVALUATION

Compared to traditional water lifting devices of similar yield such as the rope-and-bucket lift and modern pumps, the thermopump appears promising for the following reasons:

1) While many cases of illness and death from water-born diseases are caused by open-well systems, the thermopump is easily made with a closed well.
2) It is not dependent on fuel supplies or electrical supply lines and therefore it is well applicable to the hundreds of thousands of isolated villages without commercial energy supplies.
3) It is robust and reliable in operation.
4) It can be produced locally from readily available materials, requiring no high standards of accuracy.
5) It can be maintained by any village craftsman.
6) It is cheap.
7) The production process is labor intensive, thus providing local opportunities for employment.
8) The import share is low - if not nil - and the energy source must not be imported. This advantage is further enlarged by the rising oil prices.

On the other hand the range of applications of the thermopump is limited. Because of its relatively low output, it is best suited for use in rural districts (where the majority of the population lives), and its use is limited to shallow wells of 15 metres or less.

The educational effects of the solar powered thermopump lies in three fields:

- increasing acquaintance with automatic devices, thus making the concept of machines more familiar, especially the use of vapor.
- introducing the concept of utilizing solar energy.
- making local craftsmen more familiar with soldering and welding techniques.

As a long term solution the capacity of the thermopump is probably too low, but as with other intermediate technologies the main purpose has been to develop a technology that brings the optimum benefit with respect to technological and social development on the basis of the present premises.

Much of the post-war effort to help the developing countries has failed to achieve its objectives. Part of the reason is the assumption in which many aid policies are based that the technologies used by industrialised countries can also - with little modification - benefit societies which are still partly in a pre-industrial state.

From the experience of such failures the concept of the intermediate technology has evolved: To satisfy each type of environment and to help communities to progress from one device to a more sophisticated type, a range of devices is desirable, varying in complexity and standard of workmanship required.

In this view the thermopump seems very promising.

References

1. Kleen, N.E.: "Thermally Actuated Pump", U.S. patent No. 2,553,817, 1951, pp.8
2. Coleman, R.E.: "Thermopump", U.S. patent No. 2,744,470, 1952, pp.5
3. MacCracken, C.D.: "Vapor Actuated Pump System", U.S. patent No. 2,927,434, 1960, pp. 6, Followed up by patents No. 2,969,747 (1961) and No. 3,073,257 (1963). Only minor changes
4. "Resources and Needs: Assessment of the World Water Situation". Paper 76-96746. United Nations Water Conference, Argentina 1977.
5. Müller, J.: "Preliminary review of the farm implement supply sector in Tanzania". CDR project paper D.77.4. Center for Development Research, Copenhagen, 1977, pp. 32.
6. "Water for Agriculture". Paper 76-26517. United Nations Water Conference, Argentina 1977.

A SOLAR BATTERY SYSTEM AND ITS USE FOR IRRIGATION IN EGYPT

M. G. OSMAN
EL-MANSOURA UNIVERSITY
EL-MANSOURA, EGYPT

SUMMARY

A solar battery for irrigation in Egypt is discussed keeping in view the motor rating, fluctuations of insolation and matching and the use of a concentrator stressed upon. A concentrator using an array of plane mirrors is examined and the trends of variation of its performance with the solar altitude are stated. The fall in the cell efficiency at high insolation is examined and its use for countering the effect of varying AM-value suggested. Expressions for the internal battery resistance and other design parameters are derived and discussed.

INTRODUCTION

Solar energy constitutes an ideal source for generation of electrical power for terrestrial use (1,2). Various schemes have already been considered for realisation of this goal. An interesting field of power, so generated, is the driving of agro-irrigational pumps in remote regions of sun-rich countries (3.4). This is essentially the case of regions which are not yet served by conventional transmission grids. As such, the generator required should be of small scale, feeding a particular type of load. This latter would be a motor driving, in turn, a pump. The specifications for the usual problems of storage and smoothing, to take care of the fluctuating nature of sun light, in this case, are also somewhat different from those of generators intended for industrial and household uses. It is, therefore, necessary to spell out the various peculiar features of such a generator in concrete terms and to examine the ways these determine the ultimate design of the generator/motor systems. Russian workers, in particular, have contributed some fruitful results, in these regards, for specific system (5). In the present paper an attempt has been made to look into the system in general terms and to bring out the broad considerations which should dictate its design. The system is first presented in block diagram. The peculiar demands of the load, as against the basically fluctuating nature of the generator output,

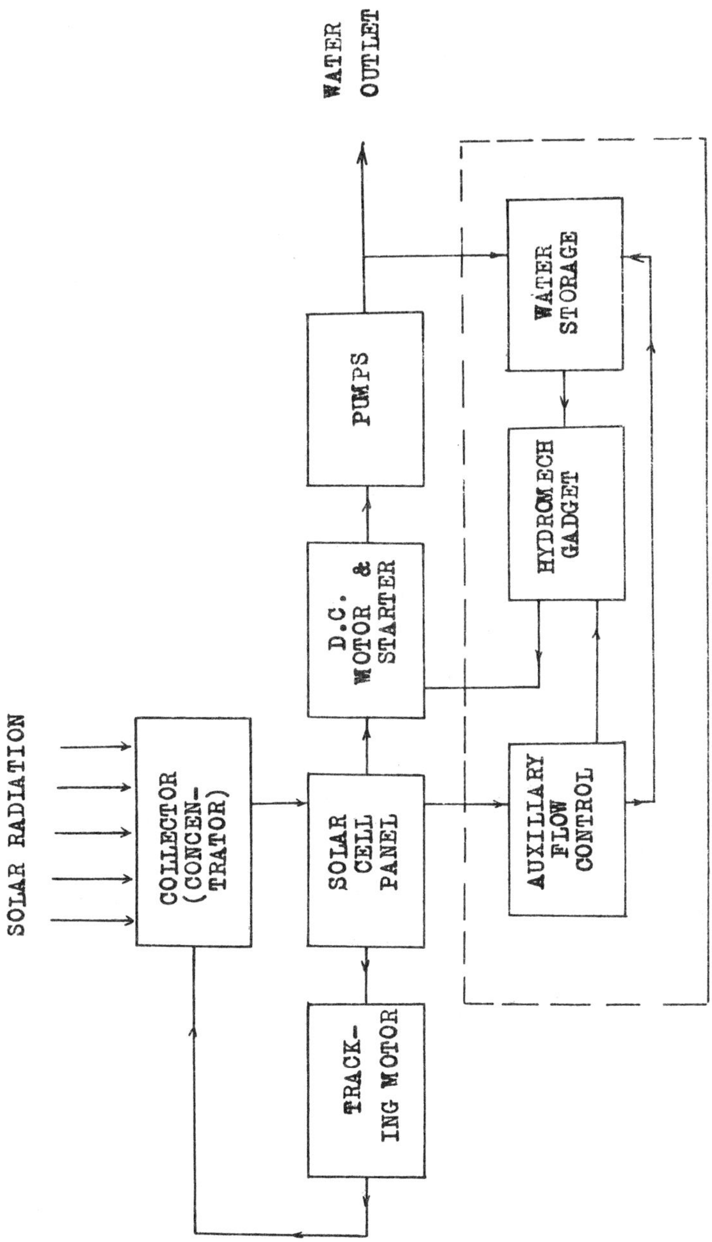

Fig. 1. Showing the basic arrangement in the proposed system.

are examined, The need for the use of a concentrator for the system is discussed and the working of one which meets these, to a large extent, outlined. The adverse effect of concentration on the performance of solar cells is studied and the possible use of this for smoothing of output suggested. Some basic relations governing the input resistance and the design of the battery are also given and the possible compatibility of the requirements for matching and smoothing discussed.

THE BASIC SYSTEM

The basic system involved in the solar-battery-fed irrigation pump may be represented by a block diagram as shown in Fig. 1. In general, the collector may be a suitable type of concentrator that focusses the solar beam onto the solar cell panel which feeds the motor and the starter. The motor drives the pump forcing up sub-soil water which may, in the general case be also stored in a suitable overhead reservoir before being drained off for use. In addition, feedback link might be provided from the water reservoir to the motor for ensuring the safety and evenness of operation of the same. A part of the battery output has also to be diverted to run an auxiliary motor for tracking the sun.

THE POWER RATING

The power rating of the battery is apparently determined by the depth of the subsoil water and the required discharge rate of the pump. However, additional considerations arise, in this regards, because of the (i) peculiar demand of the motor and (ii) the fluctuations in sun shine as are normally encountered. For running a pump with the help of the d.c. output of the solar battery, the obvious choice is a shunt-wound d.c. motor. As is well-known, these have to be used in conjuction with a starter and at the initial stage the total power consumption of the system tends to be about 150% the normal motor rating.

A more vexing problem is that arising out of occasional cloudness and the diurnal variation of the AM-value. While the latter is predictable. the former is hardly so. During cloudy hours the collector would be served, largely, by diffuse radiation.

Evidently, in cloudy days, especially during the winter time, the proportion of diffuse radiation increases. Apparently, this should mean a lesser limit of fluctuations in intensity. However, in the present context, it is not the mean value, but the maximum change in insolation following a clouding, that really matters and this latter may be more than 50% in the presence of patches of dense clouds. Even if, for considerations arising out of starting, the generator rating is made 150% the motor rating the output might go down considerably below this during occasional cloudings. If the variation in intensity is so large that the internal impedance of the generator is changed, then the power delivered to the motor will be further affected through mismatching. The total of these effects might reduce the power below the specified operating limit of the motor, and provision for switching the motor off, when this occurs, must be made unless the generator rating is so high as to preclude such an eventuality. But a high enough rating, taking care of the extreme swing in intensity, would be too expensive a proposition with the present high cost of solar cells. As such, we limit considerations to one having a capacity 150% of the motor ratings, i.e., the same limit as is required from the consideration of motor starting. Even this moderately excess capacity should increase the battery cost considerably if it were to be achieved with a flat collector. The use of a concentrator should help to check this rise in cost. An auxiliary battery, charged by the solar battery when the pump is not operating, has been used by Russian workers to meet the starting problem (5). But in the present case, the normal motor rating being in the 5-10 KW range, an auxiliary battery with 50% this rating would be a much more costly proposition compared to concentrator system.

THE PROBLEM OF MATCHING

A possible problem associated with the use of an earth-based solar battery, that requires close scrutiny, is the problems of matching. For best utilisation of a solar cell, this has to be operated at the maximum output point. This is a point at which the load resistance (R_L) becomes equal to the internal resistance (r_{in}) which is determined by the junction resistance (R_j); the resistance contributed by the layers (R_s) and the leakage resistance (R_{sh}). For good cells, the effect of R_{sh} is small even at normal insolation and the condition of matching reduces to

$$r_{in} = R_j + R_s = -V_m/I_m, \qquad (1)$$

I_m being the current and V_m the voltage at the maximum power point. R_j decreases with increases in intensity. Hence, even with an invariant R_s it will be difficult to maintain matching with a fixed external load R_L, if insolation varies due to occasional clouding etc. This means, that apart from the low available power, clouding will cause a mismatching resulting in a sharp drop in the power delivered to the load. If, however, R_s does not vary appreciably with intensity, and R_j is made $\ll R_s$ by increasing insolation, then matching may be maintained at least over a certain limit of fluctuations in intensity. This, of course requires the use of a concentrator. The advantage of using concentration under the aforesaid condition, is further illustrated in Fig. 2 showing the performance of cell with $R_s = 0.1\,\Omega$ for three different values of concentration, viz., 1, 40, 60 and a load mismatched by 20%. Here the power delivered (P) and the load voltage (V), measured in terms of the corresponding parameters (P_{AMI} and V_{AMI}) under the AMI conditions, have been presented as functions of the intensity incident on the collector, The tendency towards a better degree of constancy of both with higher concentration is quite evident.

A SIMPLE CONCENTRATOR AND ITS PERFORMANCE

The advantages of using a concentrator to take care of the problems of starting the motor, occasional clouding and matching have been pointed out. However, a concentrator for use in a small scale battery unit operating in remote regions should be inexpensive and simple. One such system (7) might consist of a collection of plane mirrors arranged in linear rows in the East-West and in linear columns in the North-South direction (Fig. 3) simulating a parabolic reflector. Mirrors on a column are connected to a common shaft rotating at half the angular speed of the sun about a N-S axis. The mirrors are initially so positioned that the reflected beams cover a solar cell module, held horizontally above the ground and proportional in size to a mirror. Each column has its own module - the modules being inter-connected in a suitable manner. The mirrors in a row are instantaneously parallel to each other but those in a column have varying orientations depending upon the rows to which these belong.

The system has an ideal concentration C_i given by the total mirror area in a column to that of the module. The actual concentration will be less than this by a factor ρ the so-called 'mirror utilisation coefficient - taking into account the inclination of the ray path to the mirror and the module. ρ will be a function of both the solar altitude α and the position of the mirror. If the instantaneous average value for a column is ρ_α then the effective concentration C' may be taken as $C_i\ \rho_\alpha\ S_\alpha/S_1$, were S_α is the actual insolation at the time and S_1 the AMI insolation. This definition of C' enables us to treat the incident insolation always as the AMI value.

Again, to avoid shadowing of a mirror by the adjacent ones,

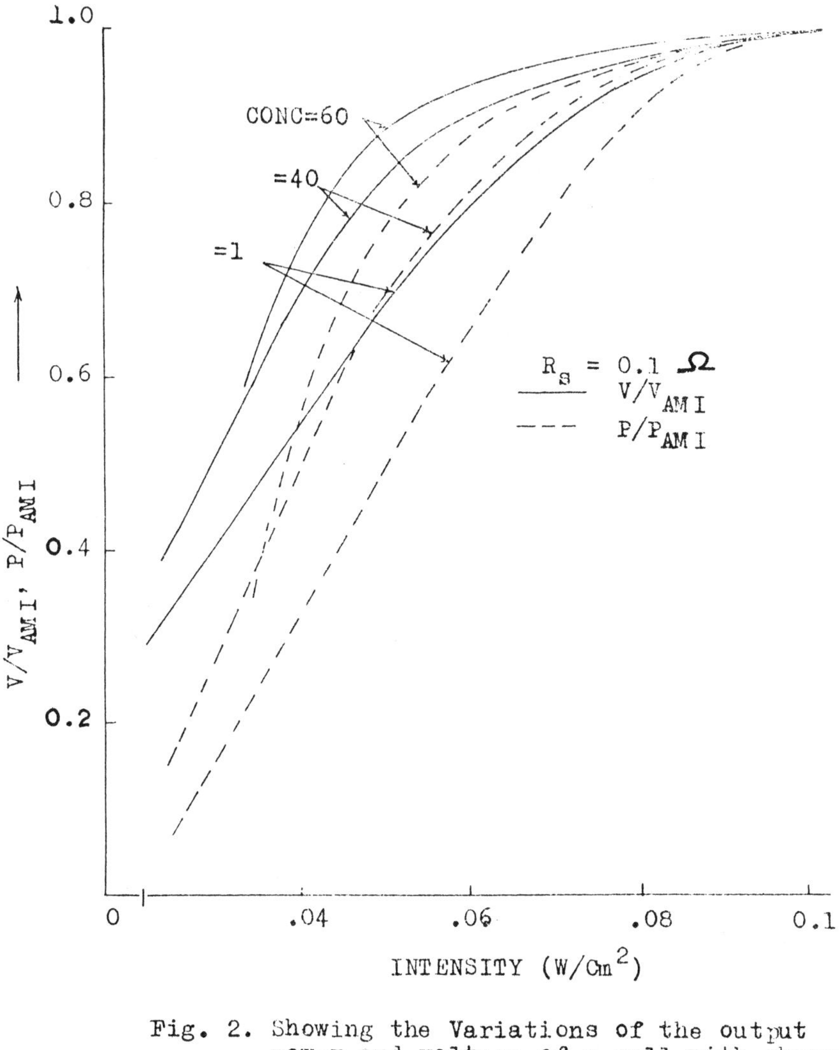

Fig. 2. Showing the Variations of the output power and voltage of a cell with changes in insolation at different values of concentrations. $R_s = 0.1\,\Omega$.

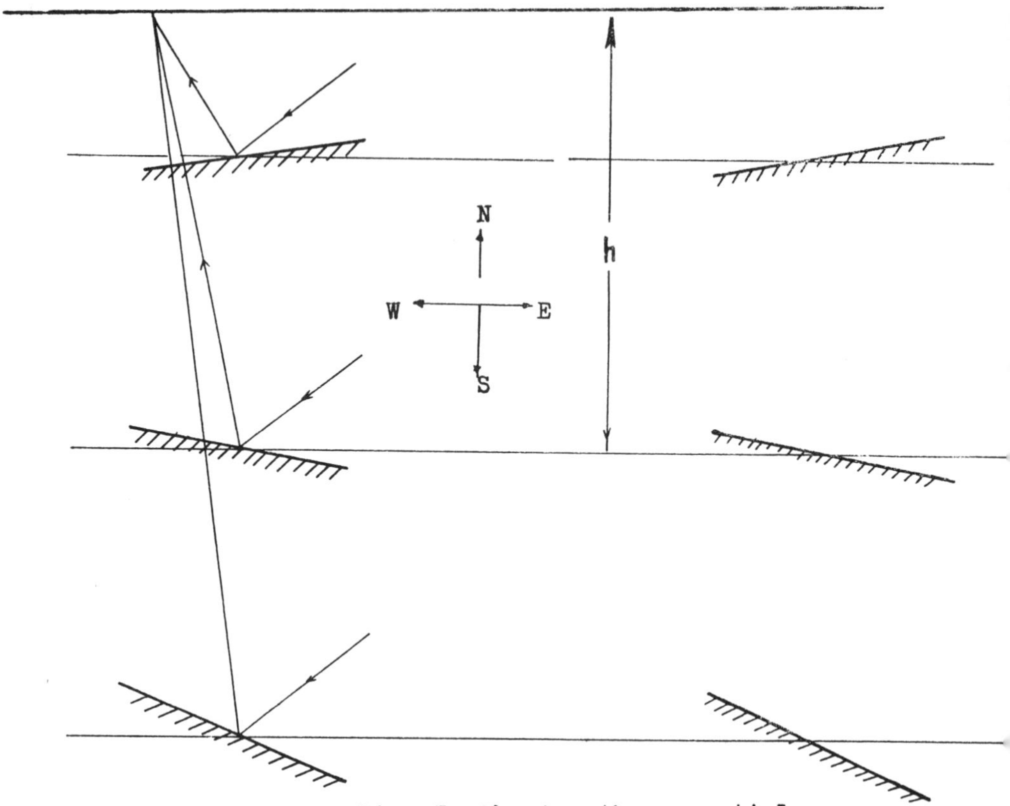

Fig. 3. Showing the essential details of the concentrator under discussion.

idle land will have to be provided between them. The area coverage factor δ is the ratio of the total mirror area to the land area required to just avoid shadowing. δ should be given by $\delta 1$, $\delta 2$, δ 1 and δ 2 arising out of considerations relating to E-N and the N-S direction and depend on $\propto$ and mirror position. Their instantaneous average values for a column may be written as $\delta_1\alpha$ and $\delta_2\alpha$. Obviously, the least values of these have to be considered in designing the concentrator.

Applying simple geometrical optical considerations and assuming the mirror inclination to vary in continuous fashion one can obtain expressions for P_α, $\delta_{1\alpha}$ and $\delta_{2\alpha}$. This has been done by Deb and Sha (8). The results are given in term of α and ϕ m, the minimum value of the angle between a reflected ray and the horizontal line in the plane of incidence. For ϕ min the range 30° - 60°, $P_\alpha \geqslant 0.7$, increasing with increasing α and ϕ_m. $\delta_{1\alpha}$ depends very little on ϕ m, especially for large value of α, but varies sharply with α - being extermely small for low values of the same. $\delta_{2\alpha}$ varies in the same fashion as P_α, but for its opposite trends of variation with ϕ m for low and high values of α. The low $\delta_{1\alpha}$ values demands cheap availability of land and/or utilisation of the idle space for other purposes for the concentrator to be economically feasible. Fortunately, both these conditions may be met in the remots regions of developing countries.

PROS AND CONS FOF A CONCENTRATOR

The use of intense concentration generally leads immediately to (i) a large voltage drop across R_s and (ii) an increased operating temperature due to a high dissipation density. The combined effect of these is an working efficiency η that is lower than η_1, the value at the AMI insolation. The ratio $\eta/\eta_1 = k$ is the so-called 'efficiency factor'. The temperature rise may be minimised through the use of a good heat sink. The effect arising out of R_s may be estimated by the method given by Wolf and Raushenback (9) assuming R_s to be independent of intensity

SMOOTHING OF OUTPUT

The possible effects of the fluctuating of the input solar energy on the performance of a shunt-wound D.C. motor have been pointed out already. As for the outflow of water for irrigation, quite a substantial degree of functuations may be tolerated and the use of an inexpensive overhead tank surely constitutes a satisfactory answer even if water is required during the dark hours.

The variation of insolation during the sun-lit hours is countered appreciably through the use of the aforesaid steering arrangement. But then, there remains the effect of the hour-to-hour variation of the AM-value, which decreases with the increase in the solar altitude in a manner depending upon the geographical location and the time of the year. As stated already, the effective concentration at any insolation S_α, may be written as $C' = C_i\, P_\alpha \frac{S_\alpha}{S_1}$. The resultant fluctuations in the output may, however, be checked to a certain extent by the deterioration of the cell performance at higher intensity. Thus, if the C_1 and the ϕ m-values are such that the working point on the P_1 vs. C' curve, in the early morning hour, is to the left and that at noon to the right of the maximum, the p.c. variation of the output in a day may be substantially reduced.

INTERNAL RESISTANCE OF BATTERY AND DESIGN EQUATIONS

Let the battery utilise a collector of effective concentration C' and a solar panel consisting of n parallel branches of m cells in series and that C' is such that the internal cell resistance $\simeq R_s$. Hence, the battery resistance

$$R_{in} \simeq \frac{m}{n} R_s , \qquad (2)$$

and the power delivered

$$P_o = m\, n\, C'\, S_1\, k\, \eta_1\, A , \qquad (3)$$

A being the area of a cell. m should be such that the required operating voltage V_o of the motor is obtained under condition of matching, i.e., if V'_m is the voltage at the maximum power point of a cell at the given concentration C', $m = V_o/V'_m$. As the efficiency drop, due, essentially to the lowering of the output voltage we may write

$$V'_m = k\, V_m . \qquad (4)$$

$$\text{Hence, } m = \frac{V_o}{kV_m} \qquad (5)$$

From (2) - (5) we get,

$$n = \frac{P_o\, V_m}{V_o\, C' S_1 \eta_1\, A} = \frac{I_o V_m}{C' S_1\, \eta_1\, A} \qquad (6)$$

$$\text{and } R_{in} = \frac{C' S_1 n_1 A}{k V_m^2} R_L R_s \qquad (7)$$

k being a decreasing function of C', we find that for a given output R_{in} should increase with increasing concentration. For matching $R_{in} \simeq R_L$

$$\text{and } \frac{k}{C'} = \frac{S_1 n_1 A R_s}{V_m^2} \qquad (8)$$

Eqns. (5) - (8), enable one to design the battery, utilising the basic cell parameters. However, the question remains as to if one could secure smoothing of output, as outlined in the preceding section, as well, using the value of C' thus obtained. Perhaps a few calculations would be in order in this connection. Typically, A = 2 cm x 1 cm, S_1 = 0.1 W/cm For η_1 = 10% and V_m = 0.5 volt, we then have

$$R_{in} = 8 \times 10^{-2} \frac{C'}{k} R_L R_s \qquad (9)$$

and for n_1 = 5%, V_m = 0.4 volt,

$$R_{in} = 6 \times 10^{-2} \frac{C'}{k} R_L R_s \qquad (10)$$

Evidently, Eqn. (9) should be approximately valid for typical Si cells and Eqn. (10) for the best of Cu_2S-Cds cells. The values of C'/k for Si cells with different values of R_s' as given by Eqn. (8) for matching, are given in column 2 in Table I, and given in column 3 of the same Table. As for smoothing, operation has to be maintained around C_m, we conclude that the condition for matching and that for smoothing are almost compatible.

TABLE I
Values of C'/k and C'_{m}/k
for S_i cells

R_s(Ω)	C'/k for matching	C_m/k
0.1	125.0	90.0
0.2	62.5	50.0
0.35	35.0	40.0

CONCLUSIONS

It is found that for a solar battery driving a shunt-wound d.c. motor as is usually employed for pumping of water, considerable constraint on design is exercised by the motor characteristics, fluctuations of insolation and matching, eventhough for irrigation in remote sun-rich regions, storage and smoothing of output do not pose much of a problem. There appears to be a good case for the use of a concentrator. A version of this, consisting of an array of plane mirrors, is discussed and is found to be worth-considering for remote regions whereland is cheap. Concentration, of course, leads to a deterioration in the efficiency - the value being reduced to 60% the normal when the output is maximum. But a possibility exists of the use of this effect for smoothing of fluctuations arising out of the diurnal variations of the AM-value. Subject to certain simplifying assumtions, simple equations can be derived for the internal resistance R_{in} and the design of the battery. R_{in} is found to increase with increasing concentration. For S_1 cells, at least, the conditions of matching and the aforesaid smoothing of the fluctuations are shown to be mutually compatible. Mention has been made of the use of a storage battery to take care of the problems of starting and clouding.

REFERENCES

1. R.L. Ralph, Solar Energy, 14, 11 (1972)

2. V.A.Baum, R.R.Aparasc and B.A.Graf, ibid, 1, 6 (1957)

3. S.Deb and M.K.Mukherjee, Jour. Sci and Ind. Res., 28, 81 (1969)

4. S.Deb, The Photovoltaic Power and its Applications in Space and on Earth, International Congress on "The Sun in the Service of Mankind", Paris, July 2-3, 1973, p. 633, C.N.E.S., France (1973).

5. V.A.Baum, Editor Semiconducting Solar Energy Converter, Consul ants Bureau, New York, 1969.

6. A. Mani and O. Chacko, Ind. J. Met. and Geophysics, 14, 270 (1963).

7. A.F.Hildebrandt, G.M.Hass, W.R.Jenkins and J.P.Colaco, Transaction Amer. Geophys. Union, 53, 684 (1972).

8. S.Deb and H.Saha, Solar Energy,

9. M.Wolf and H.Raushenback, Advanced Energy Conversion, 3, 455 (1963).

UTILIZATION OF SOLAR ENERGY IN DAIRIES

R.D. Verma, S.C. Sarma, V.K. Nayar and R.K. Hohli
National Dairy Research Institute
Karnal-132001, (Haryana), India

ABSTRACT

With the present crisis of different fuels producing thermal energy, use of solar energy is being widely emphasized. Water and air heating by solar heat from low to medium temperature range is an economical proposition provided some basic points we kept in mind at the time of design of heat collectors. In the dairies, on the milk production side, the solar energy is utilized by nature for growth of green fodder. The solar energy is also used for hay making of green fodder so that the same may be utilized in the period of scarcity of green fodder. The hot water is needed for washing the milk containers, testes and udder of the animals before milking. The water needed for these operations are at moderate temperature which can be easily obtained by solar water heaters. The milk at the processing plant is received through a network of chilling centers located in the villages as close to the milk producers; the milk at these centers is to be chilled immediately after reception. This can be done by using vapour absorption refrigeration systems by making use of solar energy in the generator. On the milk processing side, hot water is needed for washing and sterilization of cans, bottles and other containers. Hot water is also used for processing of milk, e.g., pasteurization and sterilization, etc. This can be supplied by suitably designed solar water heaters. The solar air heaters can work as preheaters for heating air, required for production of milk powder. This paper deals with the studies carried out to utilize solar energy for some of the operations carried out in dairies.

INTRODUCTION

The milk is to be cooled immediately after milking. This cooling of milk is normally carried out in bulk milk coolers, plate chillers and can coolers, etc., by making use of vapour compression systems for producing the refrigeration required for the purpose. In villages where most of the milk is produced by the farmers, it is very difficult and impractical for them to run vapour compression machines for chilling milk and then maintain these machines in proper working condition. Moreover, due to the present crisis of different fuels producing thermal energy, we have to think of some alternative source of energy to meet the demand and for this purpose the use of solar energy is increasing day by day. The milk chilling unit working on the vapour absorption system of refrigeration using solar energy for heating in the generator can easily replace the existing systems and reduce the cost of chilling to a great extent. On the milk processing side, hot water and hot air is needed for operations such as can washing, bottle washing, sterilization, pasteurization of milk, cheese making and in a spray dryer for making milk powder. The hot water and hot air required for these processes is at medium temperatures which are specified for each of these processes and thus can be achieved by making use of suitably designed solar water and solar air heaters. There may be flat plate collectors or parabolic conical concentrators.

FLAT PLATE COLLECTOR

These are used for heating air or water to lower temperature ranges of 50° to 80° C. This type of collector utilizes both direct and diffused radiation incident on them. These are cheaper and simple in construction. They need no tracking device and are highly efficient in utilizing solar energy for heating air or water or any other fluid. A flat plate collector comprises of a radiation absorbing plate beneath a transparent cover, with the underside of the absorbing plate being insulated to reduce heat losses. In order to remove the collector heat, there is a provision for circulating a heat removing fluid through the system. One such collector fabricated for heating water is shown in Figure 3.1.

CONCENTRATORS

These become of interest when temperatures exceeding 80° C are required. A concentrator consists of a reflecting surface which focusses the sun's rays on to a small area called an absorber. All concentrators concentrate direct radiation only and the diffuse radiation is lost. The concentrating system has to be moved from time to time to maintain focus of the solar image on the absorbing surface. The construction and operation of concentrator is difficult as compared to flat plate collectors and the cost is also high. The conical and parabolic concentrators were fabricated for heating water.

CONICAL CONCENTRATORS

Two conical concentrators of aperture diameter of 75 centimeters and 150 centimeters weredesigned and fabricated. Cyclindrical absorbers of size 8.7 cms. diameter and 27 cms. length and 11.2 cms. diameter 73 cms. length were provided respectively. Aluminimized mylar paper was pasted on G.I. sheet of 24 gauge to provide the reflecting surface for fabrication of conical concentrators. The absorbers are of mild steel which was painted with black paint to absorb maximum energy. The absorber was fitted with an inlet and outlet tube and the flow rate was controlled by means of valves. Sensitive thermometers were provided to measure the inlet and outlet temperatures and thermo-couples were soldered on the surface of the absorber to measure the surface temperature.

The absorber was fitted at the center of the cone where maximum radiation can be collected. The conical concentrators were fitted on metallic stands

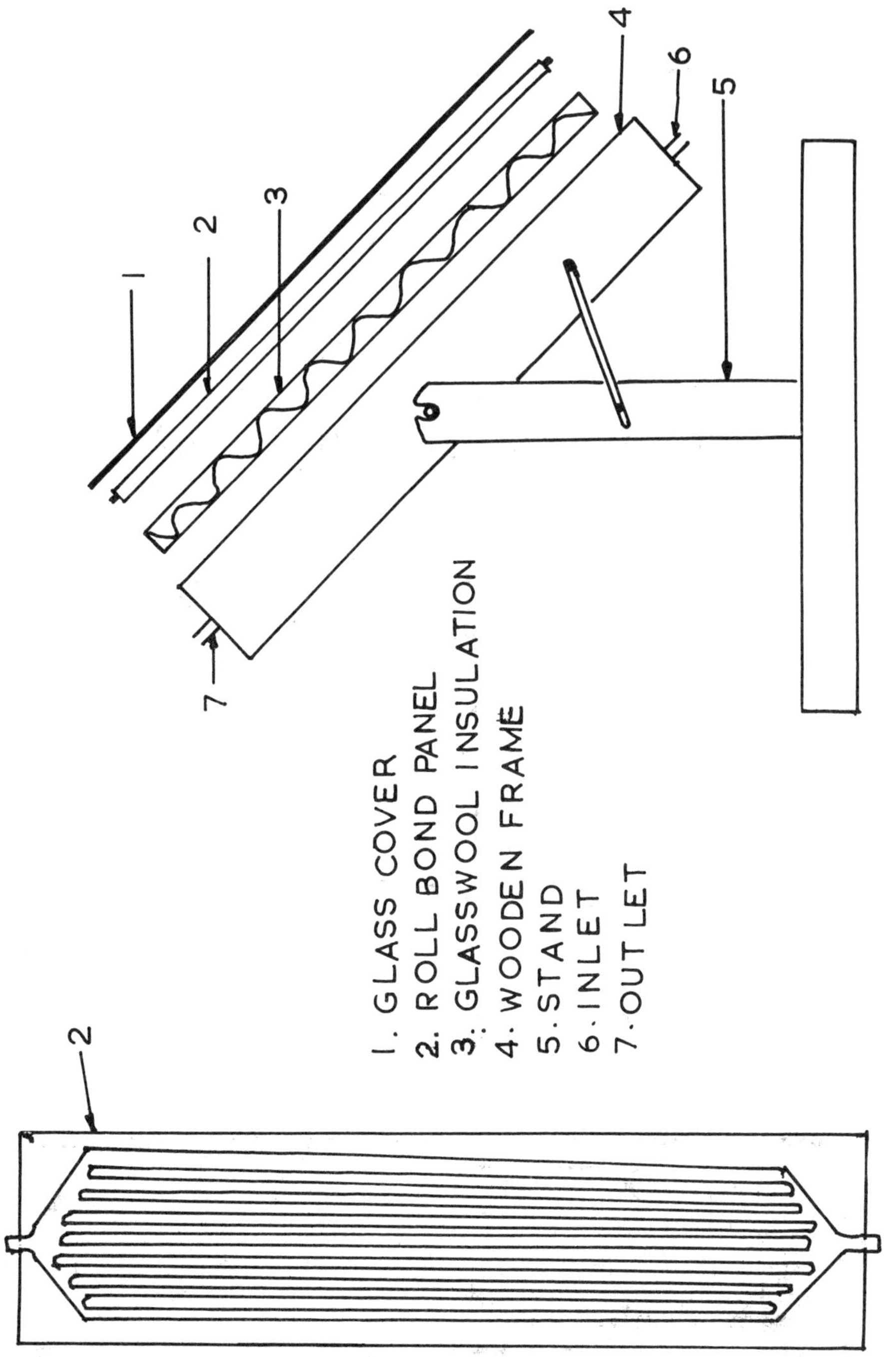

FIG.3.1. FLAT PLATE COLLECTOR.

which have provision to track horizontally as well as vertically. The conical concentrators fabricated are shown in Figure 3.2.

PARABOLIC CONCENTRATOR

A parabolic concentrator of aperture area 167 centimeters x 120 centimeters was designed and fabricated with the help of G.I. sheet of 24 gauge. The reflecting surface was provided with the help of aluminum mylar paper pasted on the surface. An absorber of 167 centimeters length and 11.2 centimeters diameter was fitted at the focal point of the concentrator. A provision was made to adjust the absorber to fix at a correct focal point. The absorber was provided with an inlet and outlet tubes and valves were fitted to adjust the flow rate. Sensitive thermometers were used to measure the inlet and outlet temperatures. Thermocouples were soldered on the surface of the absorber to measure the surface temperature. The concentrator was fitted on a metallic stand which has the provision to track the concentrator vertically. The cylindrical parabolic concentrator is shown in Figure 3.3.

To sterilize milk, a stainless steel capsule of 2.5 centimeters diameter and 52.5 centimeters length was fabricated. The capsule was provided with a cap and the whole capsule was fitted into the absorber as shown in Figure 3.4 during sterilization.

A tubular heat exchanger of 2.5 centimeters diameter and 1.5 centimeters length was designed and fabricated of stainless steel to pasteurize the milk with the help of the water to be supplied by the conical concentrator. The exchanger was of counter-flow type. A 2.5 centimeter diameter stainless steel pipe was fitted into a 3.75 centimeter steel pipe. The inlet and outlet were provided to both the pipes and was fitted with valves to regulate the flow rate. Sensitive thermometers were provided to measure the water as well as milk temperatures at different places. The heat exchanger is shown in Figure 3.5.

PASTEURIZATION OF MILK BY SOLAR ENERGY

The tubular heat exchanger designed for pasteurization of milk was filled with 450 millimeters of raw milk. The milk was held in the heat exchanger in which the hot water from the conical concentrator was supplied. The milk was held in the heat exchanger until the temperature reached 72°C and kept for 15-20 seconds. Then this pasteurized milk was taken out and samples were drawn from this for chemical and bacteriological tests and fresh raw milk was filled in the heat exchanger to repeat similar trials.

STERILIZATION OF MILK BY SOLAR ENERGY

200 millimeters of milk was teken in the stainless steel capsule and tightened the lid. The capsule was fitted into the absorber. The absorber was

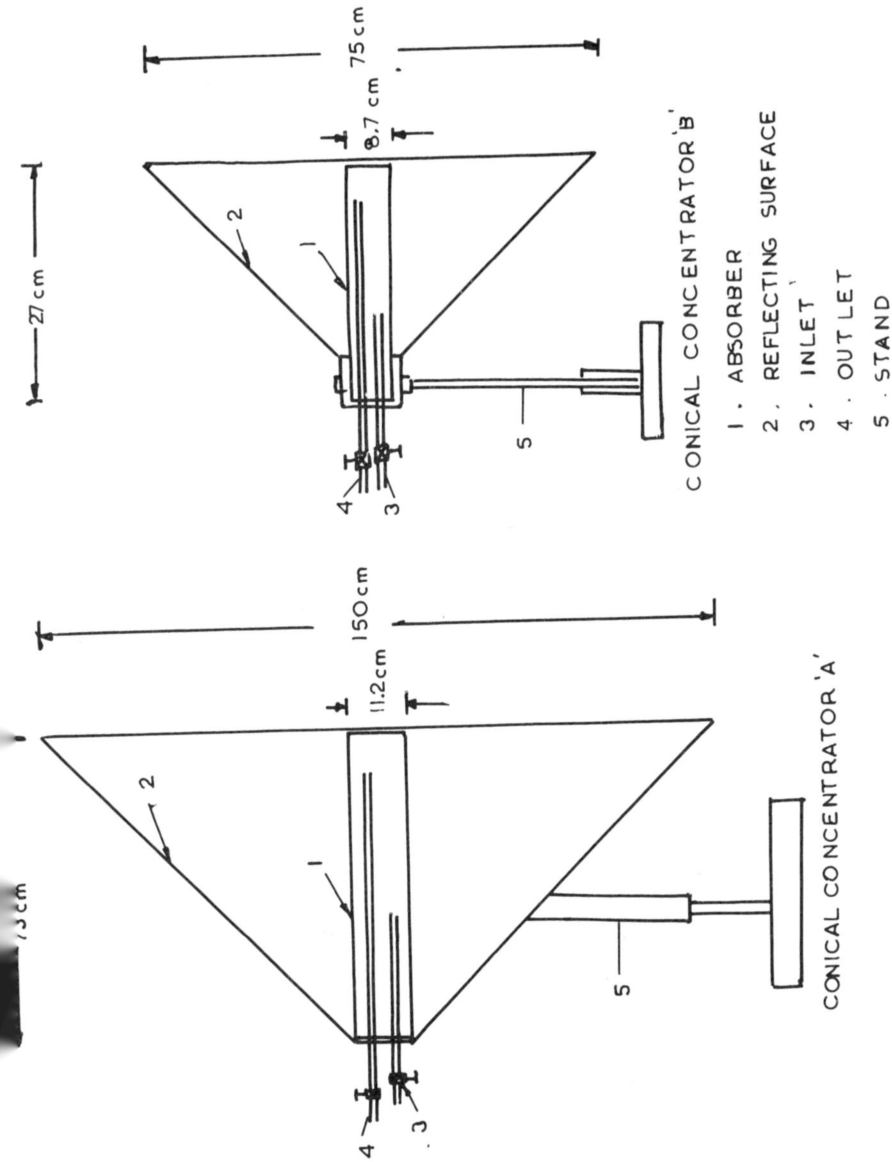

FIG. 3.2. CONICAL CONCENTRATORS

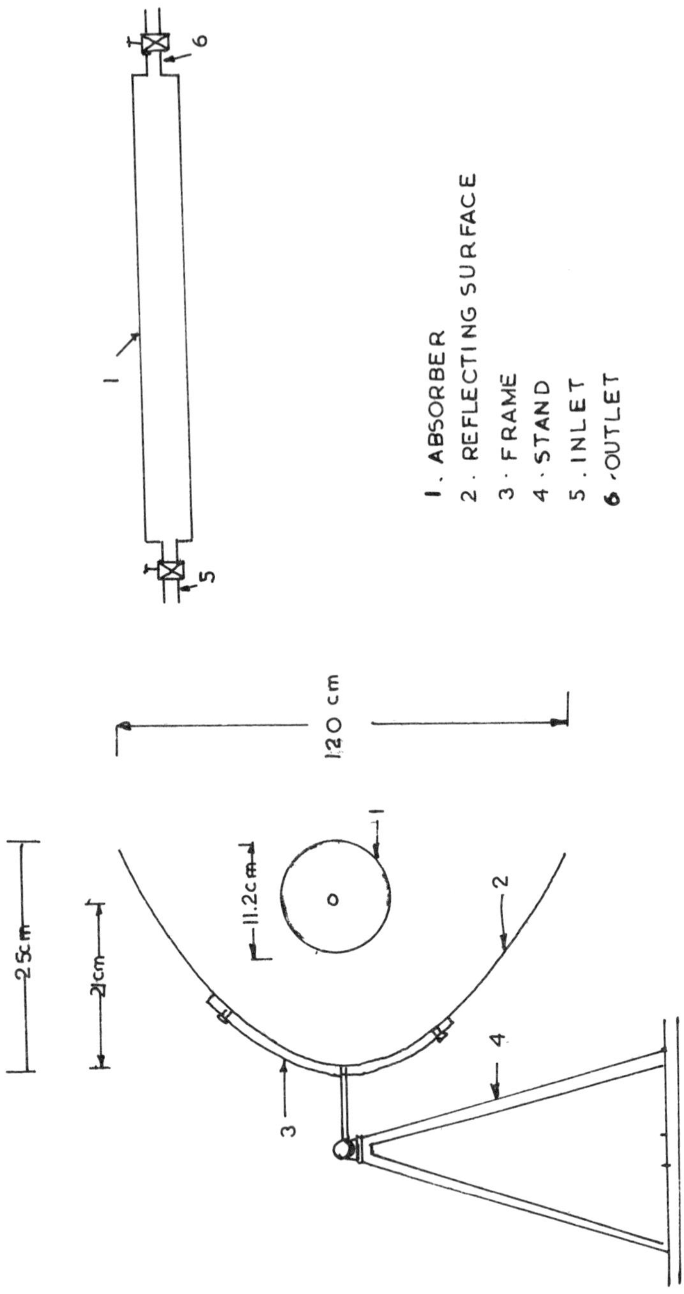

FIG. 3.3. PARABOLIC CONCENTRATOR

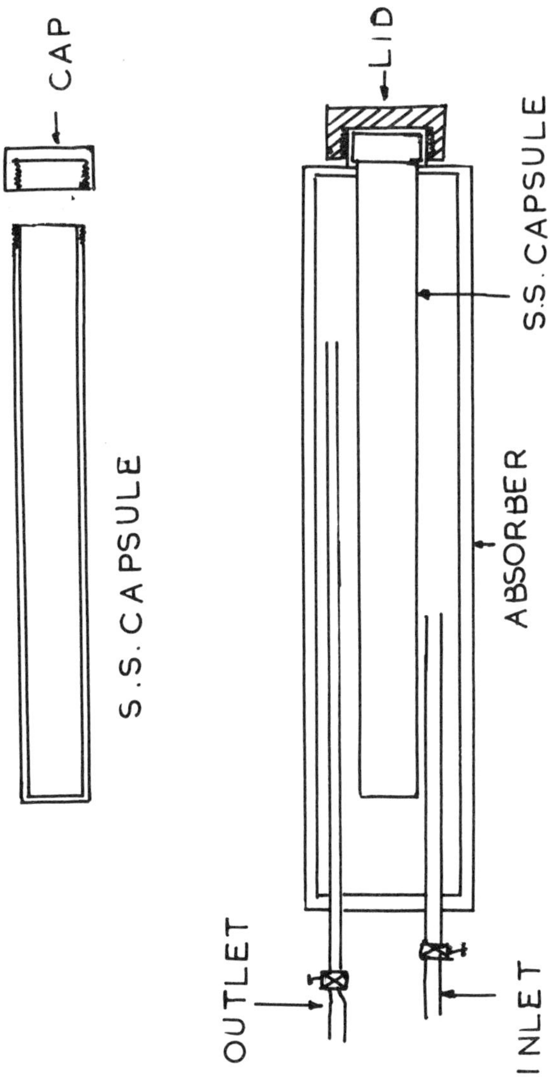

FIG.3.4. ASSEMBLY OF CAPSULE IN THE ABSORBER.

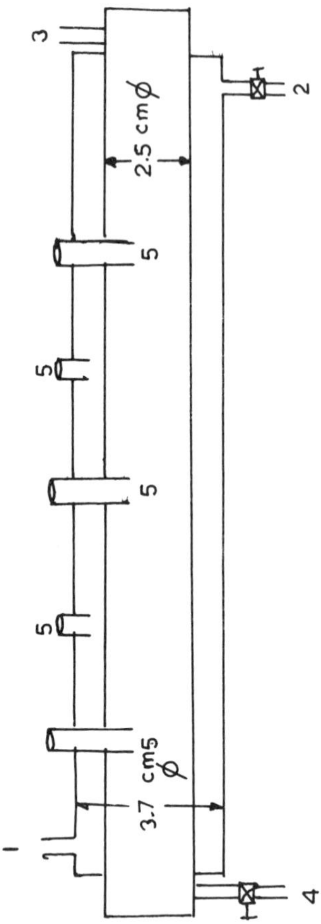

1 INLET OF HOT WATER
2 OUTLET OF HOT WATER
3 INLET OF RAW MILK
4 OUTLET OF PASTEURISED MILK
5 SENSITIVE THERMOMETER POCKETS

FIG.3.5. TUBULAR HEAT EXCHANGER

TABLE .1.

CHEMICAL AND BACTERIOLOGICAL TESTS OF PASTEURIZED MILK

SAMPLES	STANDARD PLATE COUNT		COLIFORM BACTERIA		PHOSPHATASE TEST	
	RAW MILK	PASTEURIZED MILK	RAW MILK	PASTEURIZED MILK	RAW MILK	PASTEURIZED MILK
SAMPLE 1	Above 10^6	19,000	20,000	Nil	+ ve	- ve
SAMPLE 2	Above 10^6	22,600	22,600	Nil	+ ve	- ve
SAMPLE 3	Above 10^6	18,500	19,500	Nil	+ ve	- ve
SAMPLE 4	Above 10^6	25,650	24,500	Nil	+ ve	- ve
SAMPLE 5	Above 10^6	17,500	18,500	Nil	+ ve	- ve

filled with water and tracked the concentrator. This was held for about two hours until the temperature had risen to 110°C. After this, the capsule was held for 30 minutes in the absorber. Chemical and bacteriological tests were carried out for this sterilized milk. The samples were stored at room temperature and the chemical and bacteriological tests were carried out at regular intervals.

PASTEURIZATION OF MILK

The results of bacteriological and chemical tests of the milk pasteurized by solar energy is shown in Table 1. The standard plate count of the samples were 30,000; tests for coliform bacteria showed nil and the phosphates tests were negative. The results obtained have indicated that the tests of standard plate count, coliform bacteria and phosphates were within the standards prescribed by ISI.

STERILIZATION OF MILK

The results of bacteriological and chemical tests of the milk sterilized by solar energy have shown initially nil spore count and tubidity tests have also shown negative results. The acidity of milk was 0.15. The samples were tested for acidity on the third, seventh and tenth day after storing the sample in an incubator at a temperature of 55 ± 1°C. The acidity of the samples tested after storage up to the tenth day was found to be the same. The results obtained were in agreement with ISI standards.

REFERENCES

1. Panday, M.M. and Gupta, C.P., "Pasteurization of Milk By Solar Energy", presented in the International Solar Energy Congress held at New Delhi, India in January, 1978.

2. Reddy, J. Manohar and Verma, R.D., "Sterilization of Milk By Solar Energy", presented at the Mechanical Engineering Conference, held in New Delhi, India in December, 1977.

GLOBAL RADIATION WITHIN A CROP: PEAR TREES

A. Guerrini, A. Lavagnini
and F. Vivona
Istituto di Fisica della
Atmosfera, del CNR
Roma, Italia

ABSTRACT

Measurements of global radiation carried out inside the vegetation in locality of S. Pietro Capofiume (Po Valley) are presented and discussed.
After a brief description of the type of vegetation under consideration, mention is made of the equipment employed during the measurement survey.
Finally the results achieved are presented and commented on.

INTRODUCTION

The global radiation behaviour inside the vegetation is strongly influenced by the type of vegetation and by its composition.
In the present work we want to examine how the radiation behaves inside an orchard.
The measurements were carried out in locality of S. Pietro Capofiume province of Ferrara (Po Valley). The climatology of the area is reported in (5).
The orchard of 18 years old is formed by pear - trees and it extends over an area of 2 Ha facing North/East.
The trees are palmed like cultivated in the direction South/East with a format of 2.25 m. on the files and 4.5 among the files.
The variety planted on two different stocks are four and precisely: Williams Guyot (on seedgling), Decana d'Inverno (on seedling), Passa Crassana (on quince), Abate Fétel (on quince).
The Decana d'Inverno and Abate varieties acts as pollinators. Along the seasonal time period the harvest of the fruit starts in June for the Williams Guyot variety, followed by the Abate Fétel in September and by Passa Crassana in October and Decana d'Inverno in the late October or in the early November.
The average productions per "tornatura" (2080 mq) are 200 q, 150 q, 200 q, 30 q, respectively.

EQUIPMENT DESCRIPTION

The measurement equipment (fig.1) consists of a dexion pylon which supports orthogonally with respect to the orchard line, direction disposed on six different levels at 0.5; 1.5; 2.0; 2.5; 4.0 m from the ground (direction South/West and a series of pyranometers (direction North/East).
The anemometers employed of the Casella type, record on mechanical counter the pulses emitted by a photocell with revolving of the cups. By the number of the turns the velocity in m/s is obtained.

The psychrometers, working on the principle of the temperature difference between a dry bulb and a wet bulb of a thermometer were designed and carried out in our Institute to satisfy the specific requirements. The temperature sensors were thermoresistors of the PT100. We don't want to dwell on the description of this type of equipment because data provided by this instrumentation will be not taken into account in the present paper.
The global radiation is measured by 6 pyranometers of the Lintronic type whose characteristics are, on the average, as follows:

Sensitivity	0.25
Temperature coefficient	-0.3% °C
Time response	20 sec. to 66%; 3 min to 99%
Cosine response	± 2%, 0.65° incidence; ± 4%, 65°-80° incidence

The sensor is composed of a 80 junction thermophile protected by a frosted glass dome generating a flat response in the 0.3 ÷ 3.5 mμ spectral range.
The thermopile output is sent to an integrator which provides 100 runs/hour for each millivolt in input.

RESULTS

The measurement period goes from mid-June to the end of August 1975 for a total of 44 days of measurements. These were carried out at 8 a.m., 12 (noon), 5 p.m. for a duration of 10 minutes each. In such a manner, even if in an approximate way, we could take into account the radiation behaviour inside the vegetation during the most significant hours.
As a reference for the vegetation canopy, the total foliage area of each layer was assumed. Each layer is 50 cm thick. The two trees near that were the sensors were located, have been examined.
To determine the leaf area of each layer we counted the number of leaves present in the layer, took a certain number of leaves and measured the area by planimeter and making the average. Such a value multiplied by the number of leaves present in the layer gives the total area of the layer itself. The area of the standard layer is obtained by averaging the values relative to the two trees.
In table 1 are shown the values in cm^2 of the mean area of leaves and of the total area of the layer.
To obtain the radiation mean behaviour inside the vegetation we need to consider those case in which the global radiation incident on the top of the vegetation at the same hour, has the same intensity (within a fixed interval). For all these cases we calculated the mean values of I (global radiation) as a function of Z (sensor level).
As a value of radiation intensity incident on the top of the vegetation was assumed the one relative to the highest level (in fact the pyranometer was located just above the vegetation). The following radiation intervals were found (in cal. cm^{-2} min^{-1}): 2.00 ÷ 1.51; 1.50 ÷ 1.01; 1.00 ÷ 0.51; 0.50 ÷ 0.01.
In table 2 are reported, for each interval and hour, the mean values of I relative to each level.

where

ΔI_6 = intensity interval corresponding to I_6
h = hour during which the measurement was carried out
$\underline{n}$ = number of cases
$\bar{I}$ = global radiation mean value
σ = standard deviation

In fig.2 are reported the I profiles as a function of Z with the relative standard deviations.

COMMENTS

The pyranometers are located just out of the vegetation: data are therefore relative to 8 a.m., to the radiation incident on the orchard side exposed to North/East, while at 5 p.m. the data refer to the radiation incident an the side towards South/West and which goes lengthwise through the vegetation (direction West/East). The data of 12.00 (noon) represent the radiation with the sun at the Zenith.
We must remember that the Sun height on the horizon varies from 78° (21 June) to 45° (21 September).
As a first approximation we can represent the behaviour of I with two straight lines, one relative to levels I -V and the other to levels V-VI.

$$I = a + b\ Z$$

where I is expressed in $cal./cm^2/min.$ and Z in meters.
In table 3 are shown the coefficients $\underline{a}$ and $\underline{b}$ and the corresponding straight lines are drawn in fig.2.
The straight lines were determined by the least square method. They seem to represent fairly well the I behaviours at 8 a.m. and 5 p.m., where as it is evident at noon a certain deviation from the linear trend.

CONCLUSIONS

The mean profiles of the global radiation incident on an orchard consisting of palmed like cultivated pear-trees were determined experimentally for the hours 8 a.m., 12.00 (noon) and 5 p.m.
As a first approximation such behaviours may be represented by two straight lines.
For high intensity incident radiation we observe at noon a certain deviation from the linear behaviour.

Acknowledgements- We thank M.Mastrangelo and M.Roselli for translation;P.Mollo for diagrams;BelardinelliF.,Bimbi R.,Di Menno I. and Sciattella C. for setting up of measurement equipment;Bimbi R. and Roselli M. for measurements ; F. Antonelli for data elaboration.

References

1) Idso, S.B. and de Wit, C.T. 1970 "Light Relations in Plant Canopies" Appl. Opt. 9, 177 - 184.

2) Monteith, J.L. 1959 "The Reflection of Shortwave Radiation by Vegetation". Quart. J. Roy. Met. Soc. 85, 386-392

3) Kyle W.J., Davies J.A. and Nunez M. 1977 "Global Radiation within Corn" Boundary-Layer Meteorology 12, 25-35

4) Sinclair, T.R., and Lemon, E.R. 1974"Penetration of Photosynthetically Active Radiation in Corn Canopies" Agron. J. 66, 201-205

5) Severini M. 1977 "La base micrometeorologica di S. Pietro Capofiume" IFA - Nota Interna n.12.

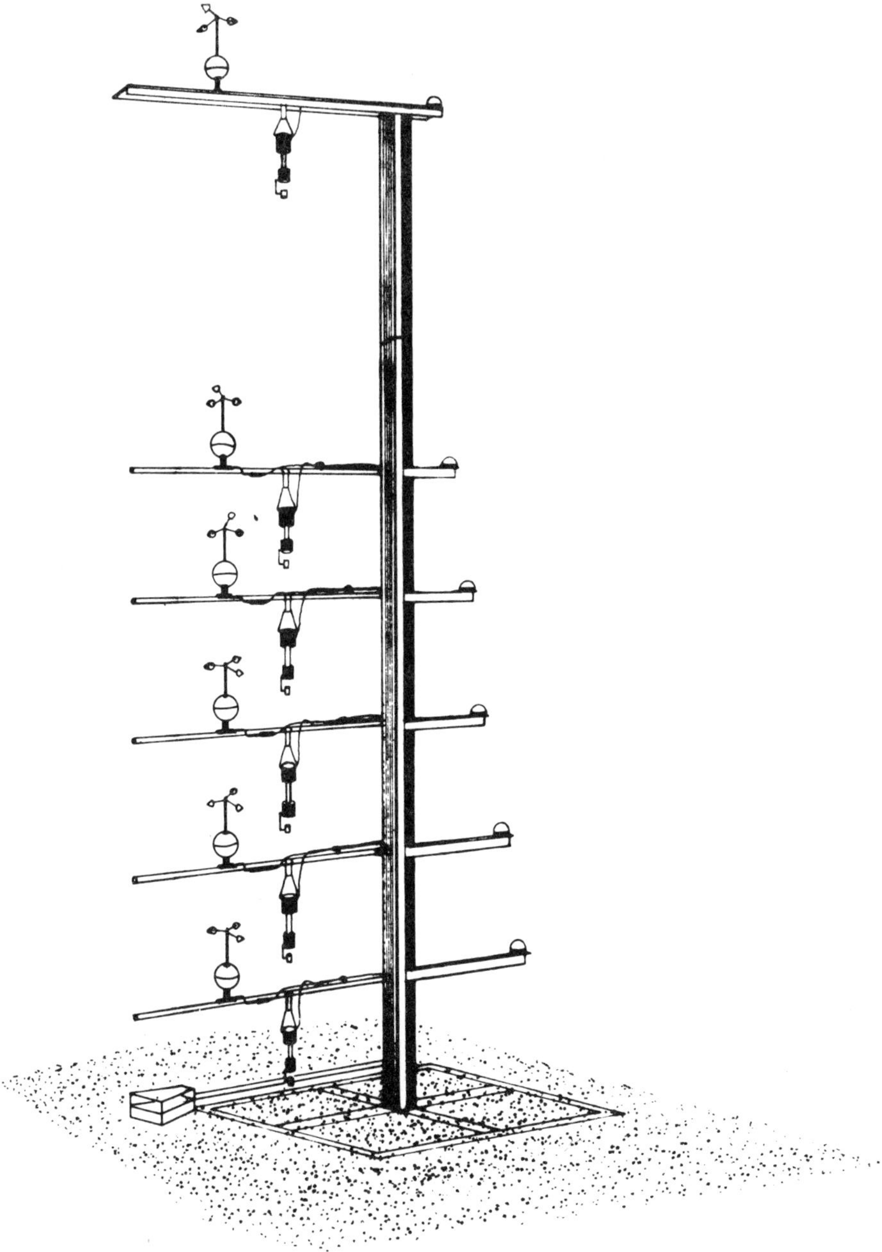

fig. 1

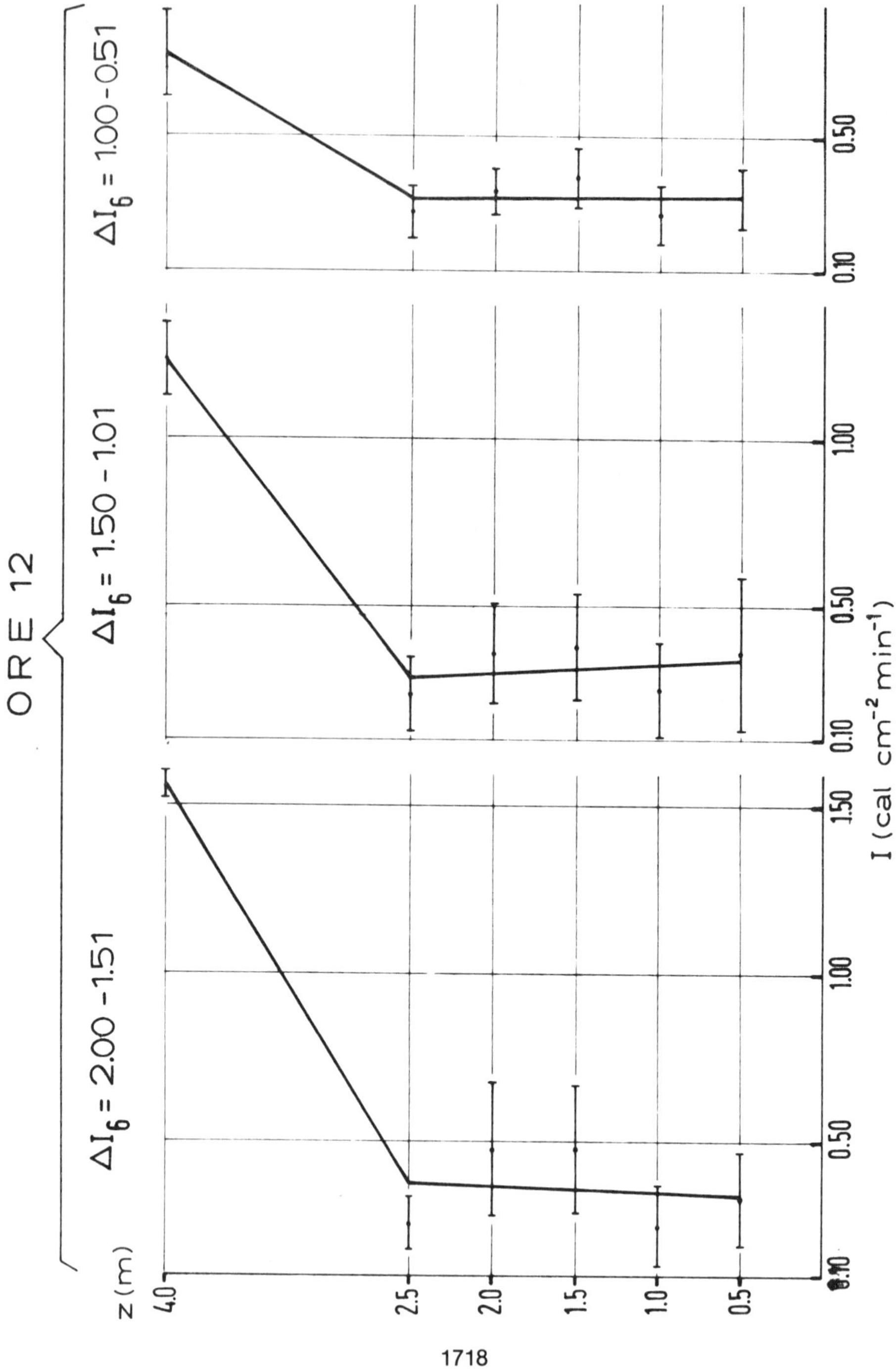
ORE 12
ΔI6 = 2.00 -1.51
ΔI6 = 1.50 - 1.01
ΔI6 = 1.00-0.51
z (m)
4.0
2.5
2.0
1.5
1.0
0.5
0.10
0.50
1.00
1.50
I (cal cm-2 min-1)

FIG. 2a

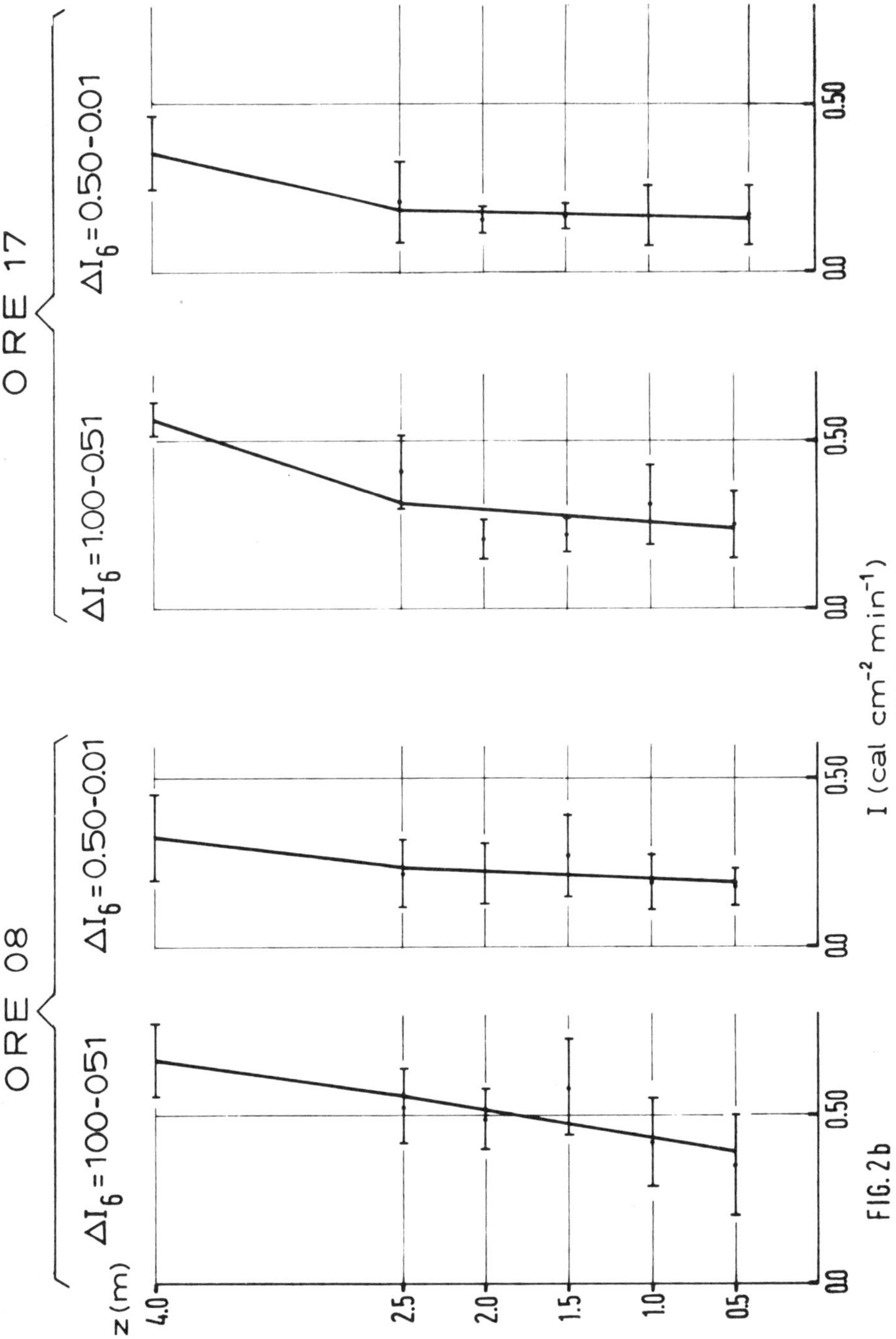
ORE 17
ORE 08
ΔI6=0.50-0.01
ΔI6=1.00-0.51
ΔI6=0.50-0.01
ΔI6=100-051
z (m)
4.0
2.5
2.0
1.5
1.0
0.5
0.0
0.50
I (cal cm-2 min-1)

FIG. 2b

TABLE 1

LEAVES MEAN AREA AND LAYER TOTAL AREA (cm^2)

Level	I case		II case		Standard case	
	$\overline{A}$	A(tot)	$\overline{A}$	A(tot)	$\overline{A}$	A(tot)
II	26	818	17	2735	22	1776
III	20	2782	21	4101	21	3441
IV	27	4588	15	3008	21	3798
V	27	3846	16	3567	22	3706
VI	30	2050	14	1853	22	1951

TABLE 2

MEAN VALUES OF I WITH STANDARD DEVIATIONS ($w\ m^{-2}$)

			0.5		1.0		1.5		2.0		2.5		4.0	
ΔI_6	h	n	$\bar{I}$	σ	$\bar{I}$	σ	$\bar{I}$	σ	$\bar{I}$	σ	$\bar{I}$	σ	$\bar{I}$	σ
2.00-1.51	8	0	0.00	0.00	0.00	0.00	0.00	0.00	0.00	0.00	0.00	0.00	0.00	0.00
2.00-1.51	12	5	0.33	0.14	0.25	0.12	0.48	0.19	0.48	0.20	0.26	0.08	1.57	0.04
2.00-1.51	17	0	0.00	0.00	0.00	0.00	0.00	0.00	0.00	0.00	0.00	0.00	0.00	0.00
1.50-1.01	8	0	0.00	0.00	0.00	0.00	0.00	0.00	0.00	0.00	0.00	0.00	0.00	0.00
1.50-1.01	12	23	0.36	0.23	0.25	0.14	0.38	0.16	0.36	0.15	0.24	0.11	1.24	0.11
1.50-1.01	17	1	0.50	0.00	0.83	0.00	0.57	0.00	0.80	0.00	1.12	0.00	1.41	0.00
1.00-0.51	8	35	0.35	0.15	0.42	0.13	0.57	0.16	0.49	0.09	0.53	0.11	0.67	0.11
1.00-0.51	12	9	0.32	0.09	0.27	0.09	0.38	0.09	0.34	0.07	0.28	0.08	0.75	0.13
1.00-0.51	17	14	0.25	0.10	0.31	0.12	0.22	0.05	0.21	0.06	0.41	0.11	0.57	0.05
0.50-0.01	8	9	0.17	0.06	0.19	0.08	0.27	0.12	0.22	0.09	0.22	0.10	0.33	0.13
0.50-0.01	12	4	0.19	0.13	0.21	0.17	0.21	0.14	0.18	0.12	0.15	0.11	0.33	0.15
0.50-0.01	17	26	0.17	0.09	0.17	0.09	0.17	0.04	0.16	0.04	0.21	0.12	0.36	0.11

TABLE 3

COEFFICIENT CORRELATION STRAIGHT LINES $I = a + b Z$

		Level			
		I - V		V - VI	
ΔI_6	h	a	b	a	b
2.00÷1.51	12	.33	.02	-1.60	.79
2.00÷1.51	12	.36	-.03	-1.29	.63
2.00÷1.51	12	.32	0	-.40	.29
1.50÷1.01	8	.34	.09	.38	.07
1.50÷1.01	17	.21	.04	-.10	.17
0 .50÷0.01	8	.18	.03	.09	.06
0 .50÷0.01	17	.16	.01	-.09	.11

SOLAR ENERGY INTERNATIONAL PROGRESS

Pergamon Titles of Related Interest

Boer Sun II
Eggers-Lura Solar Energy for Domestic Heating and Cooling
Fazzolare and Smith Changing Energy Use Futures
Machovec Solar Energy Index
Messel Energy for Survival
deMontbrial Energy: The Countdown
Veziroglu Solar Energy and Conservation

Related Journals*

Energy
Energy Conversion
Solar Energy
Space Solar Power Review
Sun at Work in Britain
Sun World

*Specimen copies available upon request.

SOLAR ENERGY INTERNATIONAL PROGRESS

Proceedings of the International Symposium-Workshop on Solar Energy, 16-22 June 1978, Cairo, Egypt

EDITED BY

T. NEJAT VEZIROGLU

Editorial Board:

H.W. Hiser

I.A. Sakr

J.W. Sheffield

VOLUME 2

PERGAMON PRESS

New York • Oxford • Toronto • Paris • Frankfurt • Sydney

Pergamon Press Offices:

U.S.A.	Pergamon Press Inc., Maxwell House, Fairview Park, Elmsford, New York 10523, U.S.A.
U.K.	Pergamon Press Ltd., Headington Hill Hall, Oxford OX3 0BW, England
CANADA	Pergamon of Canada, Ltd., Suite 104, 150 Consumers Road, Willowdale, Ontario M2J 1P9, Canada
AUSTRALIA	Pergamon Press (Aust.) Pty. Ltd., P.O. Box 544, Potts Point, NSW 2011, Australia
FRANCE	Pergamon Press SARL, 24 rue des Ecoles, 75240 Paris, Cedex 05, France
FEDERAL REPUBLIC OF GERMANY	Pergamon Press GmbH, Hammerweg 6, Postfach 1305, 6242 Kronberg/Taunus, Federal Republic of Germany

Library of Congress Cataloging in Publication Data

International Symposium-Workshop on Solar Energy,
Cairo, 1978.
Solar energy.

Presented by the Clean Energy Research Institute
of the University of Miami, Coral Gables, Fla.
Bibliography: p.
Includes index.
1. Solar energy—Congresses. I. Veziroglu,
T. Nejat. II. Miami, University of, Coral Gables,
Fla. Clean Energy Research Institute. III. Title.
TJ810.I63 1978 621.47 79-13656
ISBN 0-08-025077-7

Printed in the United States of America

SOLAR ENERGY: INTERNATIONAL PROGRESS

TABLE OF CONTENTS

VOLUME I

Page No.

CHAPTER 1 SOLAR RADIATION

Plane Incident Solar Radiation, Effects of Slope and Orientations 1
J.I. Jimenez, C. Romeu-Nedwed, J.E. Llebot and J. Casas-Vezquez

Relationship Between Global and Net Radiation 14
A. Guerrini, A. Lavagnini and F. Vivona

Atmospheric Pollution and Solar Radiation 20
M.S. Abdel-Salam and N.A. Higazy

Solar Energy at Ground Surface 41
N.H. Helwa

Contribution to the Study of Nocturnal Radiation Very Near the Ground: Effect of a Circular Screen 56
J. Brodin, A. Flourens, V.V. Pham and C. Sinnasse

New Formulation for the Factor Ψ for the Middle East Region 81
E.M. Abd El-Salam

Computation of the Diffuse Solar Radiation for Egypt 90
E.M. Abd El-Salam

Solar Data Application to Egypt 114
M.S. Abdel Salam and N.A. Higazy

Global Radiation in Italy 127
A. Guerrini, A. Lavagnini and F. Vivona

On the Distribution of Solar Energy in Greece 140
G.J. Macris

CHAPTER 2 FLAT PLATE COLLECTORS

The Flat Plate Solar Collector: Its Steady-State and Transient State Behaviour 159
E. Hahne, N. Fisch and A. Arafa

Flat-Plate Solar Collectors - Influence of Design Material and Configuration 187
S-E. Ransmark

My New Types of Flat Plate Solar Collectors 215
R. Jankovic

Measurements on a Flat Thermal Solar Energy Collector with a Cellular Structure After Francia 239
A. Bahr, H. Piwecki and L. Rigolini

A Simulated Comparison of the Useful Energy Gain in a Fixed and a Fully Thermal Tracking Flat Plate Collector 258
P. Drago

Flat Plate Boosters for Elevating Temperatures 274
I.A. Sakr and A.I. Hegazi

A Black Liquid Flat Plate Collector for Economical Applications 298
R.S. Daryan and P.K.C. Pillai

Passive Solar Wall Collectors 308
A. Faist and J.B. Gay

Energy Analysis Method for a Solar Panel 324
B. Ciborro

CHAPTER 3 CONCENTRATING COLLECTORS

A Two-Dimensional Finite-Difference Solution for the Transient Thermal Behavior of a Tubular Solar Collector 328
F.L. Lansing

Optical and Thermal Performance of a Half-Circular, Cylindrical Concentrator 351
K. Bordoloi, T. Murray, F. Bynum and E. Chaves

Modeling a V-Trough Collector in Terms of Zero, Single or Multiple Thermal Capacitance Nodes 378
K. Bordoloi, C. Hoang and F. Bynum

High-Efficiency Solar Concentrator 392
F.L. Lansing and J. Dorman

The Scheme of Automatic Adjustment of Accuracy of the Inflatable Solar Collector 404
A. Gokhman and N. Ozboya

Thermo-Fluid Analysis of a Cylindrical Solar Heat Absorber 412
M.F. Khalil

An Experimental Investigation of Comparative Performance of a Header-Type Compound Parabolic Concentrator and a Flat-Plate Collector 431
K. Krishna Rao, S. Basheer Ahmed, R. Natarajan and M.C. Gupta

CHAPTER 4 COLLECTOR TECHNOLOGY AND RESEARCH

Interferometric Investigations of Convection in Slat, Flat Plate and Vee-Corrugated Solar Collectors 446
B.A. Meyer, K.R. Randall, M.M. El-Wakil and J.W. Mitchell

Study of a Multi-Pass Radiation Fluid-Heater Exposed to Non-Uniform Radiation 463
N.M. Rafat, M.M. Elkotb and M.F. El-Refaie

Simulations of Solar Captation Systems: Methodology and Applications to Distillation, Gas and Water Heating 490
R. Rocaries, G. Olalde, P. Granier, B. N'Doye and M. Daguenet

Low Cost Spiral Shaped Solar Collector 512
P.K.C. Pillai and R.C. Agarwal

Energy Requirements of an Aluminum Solar Collector 522
J.E. Robinson and W. Ashton

Optimum Inclination of Linear Reflectors 529
B.A. Khalefa and A.S. Hegazy

Selective Surface Coating with Black Chrome 536
T. Vurgun

Cadmium Stannate Selective Coatings Prepared by Spray Pyrolysis 540
O.P. Agnihotri and B.K. Gupta

Phase Change Collector Wall Versus Water Collector Wall 549
L. Bourdeau and A. Jaffrin

On The Evaluation of the Feasibility of Solar Collectors for Electrical Power Generation Applications 565
A. Gokhman and N. Ozboya

Effect of Solar Radiation on Outdoor Bus Bar Design 577
M.M. Abdel-Aziz

VOLUME II

CHAPTER 5 THERMAL STORAGE

Performance of a Storage Unit Coupled to a Flat Plate Solar Collector 593
S.B. Ahmed and V. Sriramulu

Transient Characteristics of Vertical Porous Heat Storage 604
N. Seki, S. Fukusako and H. Inaba

Seasonal Regeneration Through Underground Strata 623
T. Yokoyama, H. Umemiya, T. Teraoka, H. Watanabe, K. Katsuragi and K. Kasahara

Storage Tank Efficiency as Simulated in a Markovian Model of Meteorology 654
R. Lestienne

Some Aspects of Energy Storage Systems Using Glauber Salt Mixtures 666
D.Y.S. Lou and A.J. Harrison

A Direct Contact Thermal Energy Storage System 680
A.S. Elasfouri

CHAPTER 6 HEATING

Solar Heating Systems with Air Collectors and Rock Bed Storage 697
W.L. Dutre, L. Cypers, J. Berghmans and A. Debosscher

Influence of Heat Exchanger Effectiveness on the Performance of Thermosyphon Double Loop Water Heating 731
A.F. Orlando, L. Goldstein, Jr. and D. Magnoli

Hot Water Production by Means of a Solar Thermosyphon Loop 745
W.L. Dutre, L. Cypers, J. Berghmans and A. Debosscher

Effectiveness of Packed-Bed Solar Air Heaters 767
S.C. Gupta

Performance Comparison of Different Types of Solar Water Heaters Under Egyptian Conditions 778
S.W. Youssef and I.A. Sakr

Solar Water Heaters 793
D. Buxey

Latent Heat Diode Wall 802
L. Bordeau and A. Jaffrin

CHAPTER 7 COOLING

Comparison Between Two Absorption Cooling Systems of the "Open" Type Under Different Climate Conditions 827
M. Bolzan and R. Lazzarin

Optimisation of Cooling Systems in Tropical Zones 855
J. Flechon

Analysis of Ammonia-Water Intermittent Solar Refrigerator Operating with a Flat Plate Collector 870
A. Venkatesh and M.C. Gupta

A Thermodynamic Analysis of a Solar-Powered Jet Refrigeration System 886
F.L. Lansing and V.W. Chai

Development of a Solar Desiccant Dehumidifier 898
Z. Lavan, D. Gidaspow, M. Onischak,
S. Chaturvedi, B. Mathipraksam and S. Perkari

Liquid Desiccant Solar Heat Pump for Developing Nations 932
H. Robison

CHAPTER 8 APPLICATIONS IN BUILDINGS

A Review of the Historical Developments of the Use of Passive Solar Systems 939
C.L. Robbins

Natural Lighting in Buildings: A Passive Solar System 960
C.L. Robbins

Radiative Cooling of Structures 983
V.K. Gupta and H.R. Tyagi

Solar Collector Related Research and Development in the United States for Heating and Cooling of Buildings 991
R.K. Collier

California State Energy Efficient State Office Building 1029
B.S. Setty

CHAPTER 9 BIOLOGICAL AND CHEMICAL CONVERSION

A Quinone-Assisted Photoformation of Energy-Rich Chemical Bonds 1056
S.W. Fox, T. Adachi and W. Stillwell

Ethanol, A Clean Burning Bio-Solar Fuel - Its Production and Utilization 1075
H.B. Mathur

Electrolytic Hydrogen Production and Its Potential Utilization for Ship Propulsion 1094
C.A. Schmidt

Hybrid Chemical Concepts for Storage of Solar Thermal Energy 1109
A.C. Vialaron

Use of Renewable Fuels in the Rural Development of India 1119
G.C. Pandey

VOLUME III

CHAPTER 10 DIRECT ELECTRICAL CONVERSION

Physics and Economics of Photovoltaic Solar Energy Conversion 1132
Y. Marfaing

Realization of P-N Junction Solar Cells by an Ion Implantation Doping Procedure 1154
J.C. Muller, M. Hage-Ali and P. Siffert

Solar Energy Conversion by Fluorescent Materials for Utilization in Space Communication Systems 1167
A.K. Sen and J.S. Sehra

New Possibility of Solar Energy Conversion by Means of Refractory Thermoelements 1184
S. Abdalla, J.M. Dusseau, F. Roche and J.M. Darolles

On Direct Conversion of Sunlight to Electricity 1192
M.A. Melehy

Performance of the Noval Solar Cell Substitute for the Future 1210
S.C. Rastogi

MOS Solar Cells 1216
H.M. Kizilyalli

Some Optical and Electrical Properties of Selenium-Cadmium Sulphide Photovoltaic Cells 1230
K. Abdel-Hady, F. El-Akkad and J. Al-Dabbagh

Optimization of Graded Doping CdTe Solar Cells 1242
A.S. Bouazzi

Saur-Vidyut Kosh - The Solar Cells 1257
M. Sharon, S.G. Sharan, B.M. Prasad, A.P. Bholagir and R.R. Pradhananga

CHAPTER 11 THERMAL POWER

Solar Thermal Power Plant: Thermodynamic Analysis 1279
A.A. Samuel, U.S.P. Shet and M.C. Gupta

Solar Power Units with Screw Expanders 1293
J.M. Merigoux and P. Pocard

New Developments for Future Solar Power Plants 1318
J. Lorenz, J. Feustel and M. Kraft

Power from the Sun by Thermodynamics - French Tower Type Projects 1329
M.J. Bignon

Steam and Gas Turbines for Small Outputs Using Solar Energy 1338
K. Bammert and H. Posentrup

Turbine Selection for Small Capacity Solar-Power Generation 1351
A. Mobarek, N. Rafat and M. Saad

A Helioelectric Farm 1368
O.J.M. Smith and P.S. Smith

A Central Receiver Solar Power System for Remote Locations 1408
R.J. Holl

Investigation of Pool Boiling Heat Transfer in the Energy Carrying Liquids of Solar Energy Power Plants 1476
M.S. Abdelhafiez and M.A. Abdelsalam

The Use of Solar Furnaces in Materials Research 1443
D. Suresh and P.K. Rohatgi

CHAPTER 12 DISTILLATION AND DESALINATION

Efficiency of a Solar Still as a Function of Its Main Parameters: Production Estimation for Valencia, Spain 1474
J. Riera, J.A. Martinez-Lozano, F. Tejerina and F. Tena-Sanguesa

Solar Water Supply and Purification System for Remote Areas and Villages 1497
U.R. Nejib and H.C. Hallez

Solar Water Recovery from the Air 1526
W.G. Bulang

Double Exposure Solar Still 1546
M.M. Mousa, F.M. Abdelfattah and I.A. Sakr

Direct Solar Distillation 1561
I.D. Abdelrazek and F.H. Hammad

Solar Energy for Desalination 1570
A.M. El-Nashar

Development Plans and Conceptual Designs for Multi-Purpose Solar Desalination Units (MSDU) 1587
H.K. Abdel-Aal

CHAPTER 13 AGRICULTURAL AND RURAL APPLICATIONS

Review on Application of Solar Energy for Drying Food Stuffs 1600
K.V. Mallan, T.V. Balasubramanian and R.K. Bhardwaj

Solar Energy Utilization for Grain Drying 1615
VR. Muthuveerappan

Studies on Utilisation of Solar Energy in Hay Making of Barseem 1634
R.D. Verma and M.K. Trehan

Solar Tobacco Drying Plant 1645
L. Andreotti, A. Ciminale, G. Marchetti, D. Ruggi and A. Serravezza

Crop and Timber Drying in Malaysia Proposed Plans for Studies 1654
K.S. Ong

Energy Consumptions and Possibilities for the Applications of Solar Energy in Italian Agriculture 1666
L. Andreotti, G. Marchetti, F. Passaro, D. Ruggi, A. Serravezza, A. Di Vecchia and A. Fanchiotti

Solar Energy Utilization for Agricultural Purposes in Egypt 1677
M. Hassib

Solar Powered Water Pump for the Rural Third World 1681
J. Boldt

A Solar Battery System and Its Use for Irrigation in Egypt 1692
M.G. Osman

Utilization of Solar Energy in Dairies 1703
R.D. Verma, S.C. Sarma, V.K. Nayar and R.K. Hohli

Global Radiation Within a Crop: Pear Trees 1713
A. Guerrini, A. Lavagnini and F. Vivona

VOLUME IV

CHAPTER 14 COMMERCIAL AND DOMESTIC UTILIZATION

Street Lighting by Sunlight 1723
J.G. Sakas

The First Solar Energy System in West Germany for Industrial Utilization 1742
I. Hoffman

A Simple Water Pump, Powered by Solar Energy 1748
K. Speidel

Solar Cookers 1762
D. Buxey

Solar Energy for Cooking in Egypt 1768
M.G. Osman

CHAPTER 15 WIND ENERGY

Sow Wind Turbines and Reap the Wind 1796
O.J.M. Smith

Alternative Energy for Domestic Hot Water: Wind of Solar? 1834
A.L. Weisbrich

Comfortable Heating and Cooling by Solar Reflectors and Wind 1850
S.A. Husain

Combination of Wind and Solar Energies at the Northeast Coast of Egypt 1862
I.A. Sakr, M.S. Abdel Hafiez and A.S. Hegazy

A Low Cost Tilting Blade Vertical Axis Windmill for Rural Irrigation 1891
V. Mukhopadhyay

CHAPTER 16 NATIONAL AND INTERNATIONAL PROGRAMS

Solar Energy: UNIDO Programme of Action for Developing Countries 1901
S. Rao

Contemporary Research and Development for the Use of Solar Energy in Thailand 1911
V. Bhanthumnavin

End Use Energy in Brazil and the Solar Technologies 1928
I.C. Macedo

French Solar Energy Program and Its Application for International Cooperation 1937
M.J. Pheline

Solar Collector Development in Rwanda 1950
F. Kalos, C. Ntakirutinka, P. Mpawenayo and G. Saunier

The Impact of Solar Energy Integration on the Power System of the Kingdom of Saudi Arabia 1962
F. Sultan and F. Taher

Solar Energy for the Common Man in the Arab World 1971
D.N. Singh

An Assessment of Need for Developing and Implementing Technical and Skilled Worker Training for the Solar Energy Industry 1976
C.G. Orsak. R. Barnstone, Y. Jani, H. Gibson and J. Morehouse

The Design of the Two-Year Associate Degree Curriculum for Training Solar Engineering Technicians 2003
M.Z. Lowenstein

CHAPTER 17 ECONOMICS AND POLICIES

Economic Considerations of Solar Energy 2021
J.D. Parker and J.A. Tipton

The Comparative Economics of Passive and Active Systems 2044
F. Roach, S. Noll and S. Ben-David

An Economic Analysis of Solar Energy Alternatives 2084
D.A. Huettner, B.J. Taylor, C.A. Ingene and J.F. Horrell

Impact of Solar Heating and Cooling of Buildings on Electric Utilities 2106
A.S. Debs

Economic Analyses of Solar Water Heater and Solar Stills in Thailand 2126
P. Wibulswas

Evaluation of Solar Thermal Power Plants Using Economic and Performance Simulations 2138
N. El-Gabawali

The American Experience with Solar Energy in Centralized Electricity Generation: Technology and Economics 2162
D.L. Hurwood

Commercialization of Solar Energy Devices in Developing Countries - Problems and Prospects 2189
R.O. Onyeonwu

The Economics of Solar Space Heating in Zambia 2202
G. Lewis

Solar Energy - Will It Conserve Our Non-Renewable Resources? 2210
S. Baron

The Role of Information Media in Public Understanding of Solar Energy 2230
S. Galal

Community Characteristics Predicting Public Interest in Household Solar 2239
S. Warkov

Accelerated Commercialization of Solar Energy - A New Priority for International Action 2251
L. Rishnell and N.D. Jorstad

Why Not Solar Energy? 2270
T. Yarman

CHAPTER 18 WORKSHOP REPORTS

Report 1: Architectural Applications 2286
Group Leader: A. Bowen

Report 2: Agricultural and Industrial Applications 2288
Group Leader: E. Bilgen

Report 3: Energy Conversion 2290
Group Leader: Y. Marfaing

Report 4: Solar Planning 2292
Group Leader: A.J. Parker, Jr.

Report 5: Solar Energy Commercialization 2304
Group Leader: N.W. Lutkefedder

PERFORMANCE OF A STORAGE UNIT COUPLED TO A FLAT PLATE SOLAR COLLECTOR

S.B.AHMED and V.SRIRAMULU
Department of Mechanical Engineering
Indian Institute of Technology
Madras, INDIA

ABSTRACT

A storage unit is an indispensable component of a solar power generating unit. A thermal storage unit coupled to a system of flat plate collectors is described. In a thermal storage unit stabilization effects are always present. The processes occurring in such a system have been simulated on a digital computer in order to determine the performance characteristics of the unit for a given load pattern. The analysis considers a selectively coated surface employing water as the working medium. The temperature distribution in the storage unit undergoes a change when the water flow rate is altered. The storage volume does not affect the performance significantly at least, in the range of volumes investigated.

INTRODUCTION

Utilization of solar energy for power generation is being considered an alternative to the existing conventional methods. Although solar power generation is not going to replace these conventional methods in the immediate future, every endeavour must be made to hasten its introduction in the commercial field. But solar power production is beset with a number of formidable problems, one of them being the temporal nature of sun's energy. Unlike a steam power plant which burns the fossil fuel at a steady rate, in the case of a solar power plant changes with time. Apart from having suitable controls incorporated into it, the plant must also have a storage unit so that energy can be stored during the time when the sunshine is plentiful. It can be used subsequently when the sky is overcast or when power is needed at nights.

The output temperature of a solar collector will depend very much on the type of collector used. The present work deals with a system of selectively coated flat plate collectors that are coupled to a thermal storage system.

Of particular significance in any storage system are those

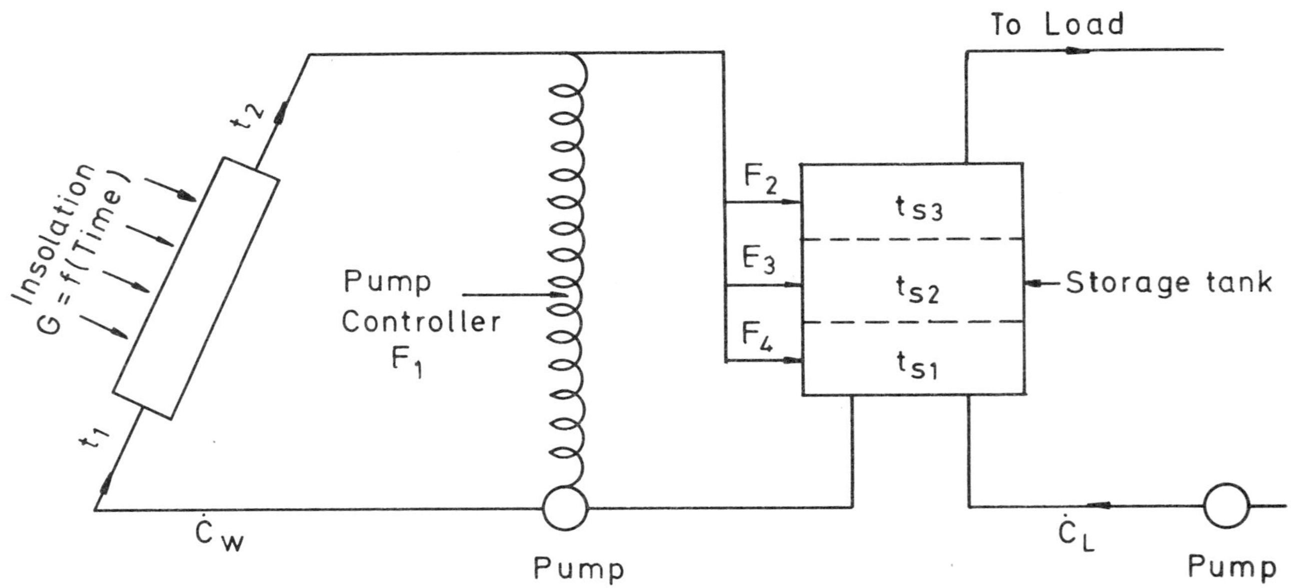

FIG.1. SOLAR COLLECTOR - STORAGE SYSTEM

factors that influence the operation of the collector. Since the useful gain from the collector decreases as the average plate temperature increases, the objective behind the design of a storage unit should be to maximise the gain.

Since solar energy is stochastic in nature, the study of solar devices has to be made with respect to time. Detailed studies of major constituents of solar systems have been reported in [1, 2] . The performance of a collector storage unit can be described by convenient mathematical models [2,4,5] . The necessary meteorological data for the study of its performance can be expressed as trignometric functions[6] . A full descriptions of the mathematical model is found in [5] .

STRATIFIED TANK

The schematic of the collector-storage unit is shown in Fig.1. Water heated in the flat plate collector is fed to the storage tank, if the collector outlet temperature is greater than the lowest temperature of the tank contents, so that energy can be added to the tank. This process is simulated in the analysis by control functions.

If the collector outlet temperature is greater than tha temperature of water in the tank, the circulating pump is turned on; otherwise, it is turned off. In the latter case, the stagnant water in the collector gets heated up until its temperature has risen to a value at which it can be pumped into the tank.

The storage is divided into three sections and water from the collector is assumed to be fed to the appropriate section. Each section is assumed to be at a uniform temperature. The outlet from the collector is connected to the three sections, maintained at different temperatures, through valves. Functions F_2, F_3 and F_4 control the operation of the valves that feed the water heated in the collector to the tank. They take the values 0 or 1, the value 1 indicating that the valves are open. The logical sequence that governs the operation of the valves is given in [6] . Function F_1 represents the pump controller.

The performance of a collector-storage unit depends upon the load pattern imposed on the plant. The load pattern considered in the analysis consists of three loads as shown in Fig.2. The peak load occurs around 12 noon. Operation of the plant in the night is rendered possible by the energy stored in the unit. An important consideration in

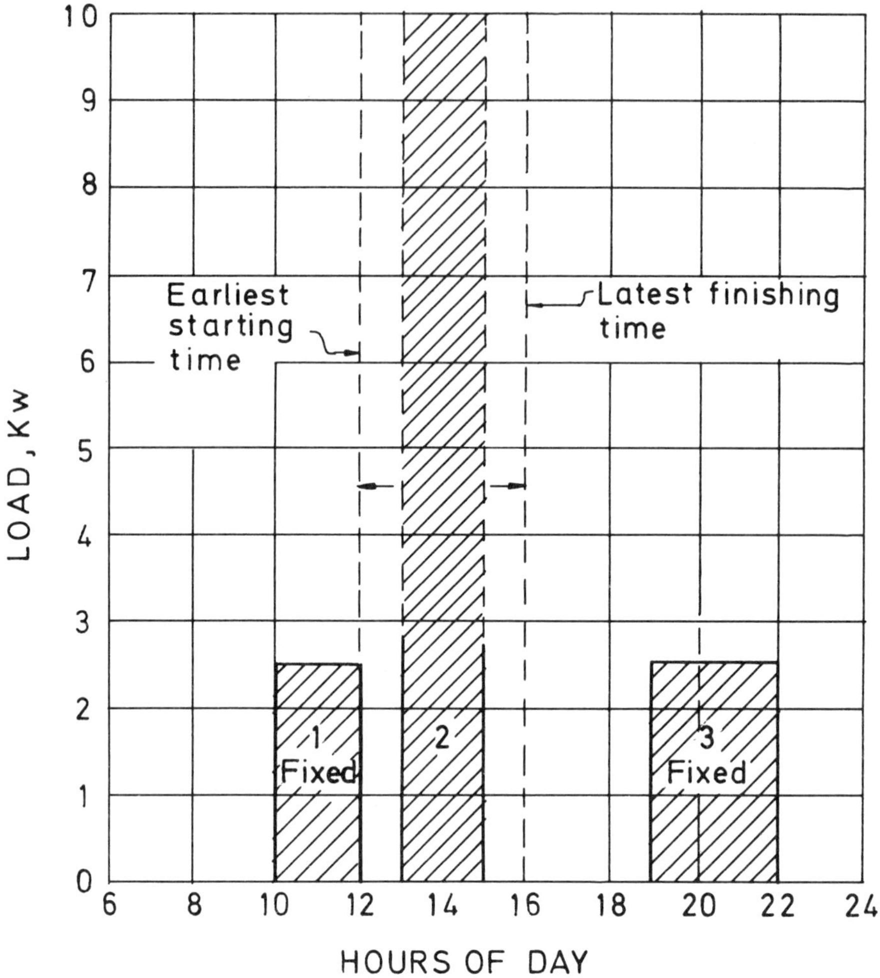

FIG. 2. LOAD PATTERN

design of thermal storage unit is the overnight temperature drop. If the fall in temperature were to be excessive, the plant would not be ready for operation the next day unless the unit is equipped with an auxiliary heating system.

The following equation, which includes the thermal capacitance of the collector, describes the collector performance

$$c_a \frac{dt_m}{d\theta} = \tau G\alpha - U_L(t_m-t_a) - F_1\ 2\ \dot{c}_W(t_m-t_{s1}) - F_1(1-F)\ (\tau\alpha G-U_L\ (t_m-t_a)$$

$$t_2 = t_m + F_1\ (t_m-t_{s1})$$

The transmittivity τ of the glazing depends upon the incidence angle of the sun rays. The collector considered is selectively coated so as to achieve an output temperature of 95°C at a flow rate of 50 kg hr^{-1} m^{-2}. The meteorological inputs have been simulated by trignometric functions. The differential equations which describe the transient nature of the three sections are given by,

$$c_{s3}\frac{dt_{s3}}{d\theta} = F_1F_2\ \dot{c}_W\ (t_2-t_{s3}) - F_5\ \dot{c}_L\ (t_{s3}-t_{s2}) - U_{s3}\ A_{s3}\ (t_{s3}-t_a)$$

$$c_{s2}\frac{dt_{s2}}{d\theta} = F_1F_3\ \dot{c}_W(t_2-t_{s3}) + F_1F_2\ \dot{c}_W\ (t_{s3}-t_{s2}) - F_5\ \dot{c}_L\ (t_{s2}-t_{s1}) - U_{s2}\ A_{s2}\ (t_{s2}-t_a)$$

$$c_{s1}\frac{dt_{s1}}{d\theta} = F_1F_4\ \dot{c}_W(t_2-t_{s1}) + F_1F_3\ \dot{c}_W\ (t_{s2}-t_{s1}) + F_1F_2\ \dot{c}_W\ (t_{s2}-t_{s1}) - F_5\ \dot{c}_L\ (t_{s1}-t_r) - U_{s1}\ A_{s1}\ (t_{s1}-t_a)$$

F_5 represents the control function for the load pump circulating water from the storage through the load. It serves the same purpose as F_1. The total energy delivered to the load during the day or over a longer period can be written as

$$\dot{Q}_r = \dot{c}_L \int_\theta (t_{s3}-t_r)\ d\theta$$

RESULTS AND DISCUSSION

A preliminary investigation has been carried out to predict the performance of the storage unit coupled to a flat plate collector(black coated) when meeting just the domestic needs of hot water. Assuming that the domestic load is comprised of three units one starting at 8 AM, the second at 12 Noon and the third at 6 PM, computations have been carried out using a step interval of 0.01 hr. The results obtained are shown in Fig.3, which, incidentally also shows the quantum of load drawn during the three periods. It is discernible that the variation in temperature with time is different for the three sections.

The analysis has been extended to include solar power generation, the load pattern for which is shown in Fig.2. Computations have been carried out for typical winter and summer conditions. Fig.4 shows the variation in temperature of the tank contents as well as of the collector output temperature for winter.

The maximum temperature attained by the water is about 85°C compared with 95°C obtainable in summer. Since the overall plant performance depends very much on the temperature, any reduction in the operating temperature would entail a loss in the overall efficiency. The overall efficiency is the products of the collector efficiency, heat transport efficiency and the Rankine cycle efficiency. Every endeavour, therefore, must be made to maintain the Rankine cycle efficiency and collector efficiency high. Heat transport cannot be increased without increasing the initial investment cost which arises from better insulation to be provided to the pipes carrying the hot fluids.

It has been found that the flow rate of water being circulated from the storage tank through the load has to be varied depending upon the time of operation in order to maintain the power output constant. On the other hand if the flow rate were maintained constant, the power output would vary continuously because of the change in temperature occurring in the storage unit.

The effect of changing the flow rate through the load has been investigated and the results obtained are shown in Fig.5. Whenever load is drawn from the unit, significant changes occur in the temperature distribution.

The effect of storage volume on the performance has also been investigated. The range of storage volumes studied

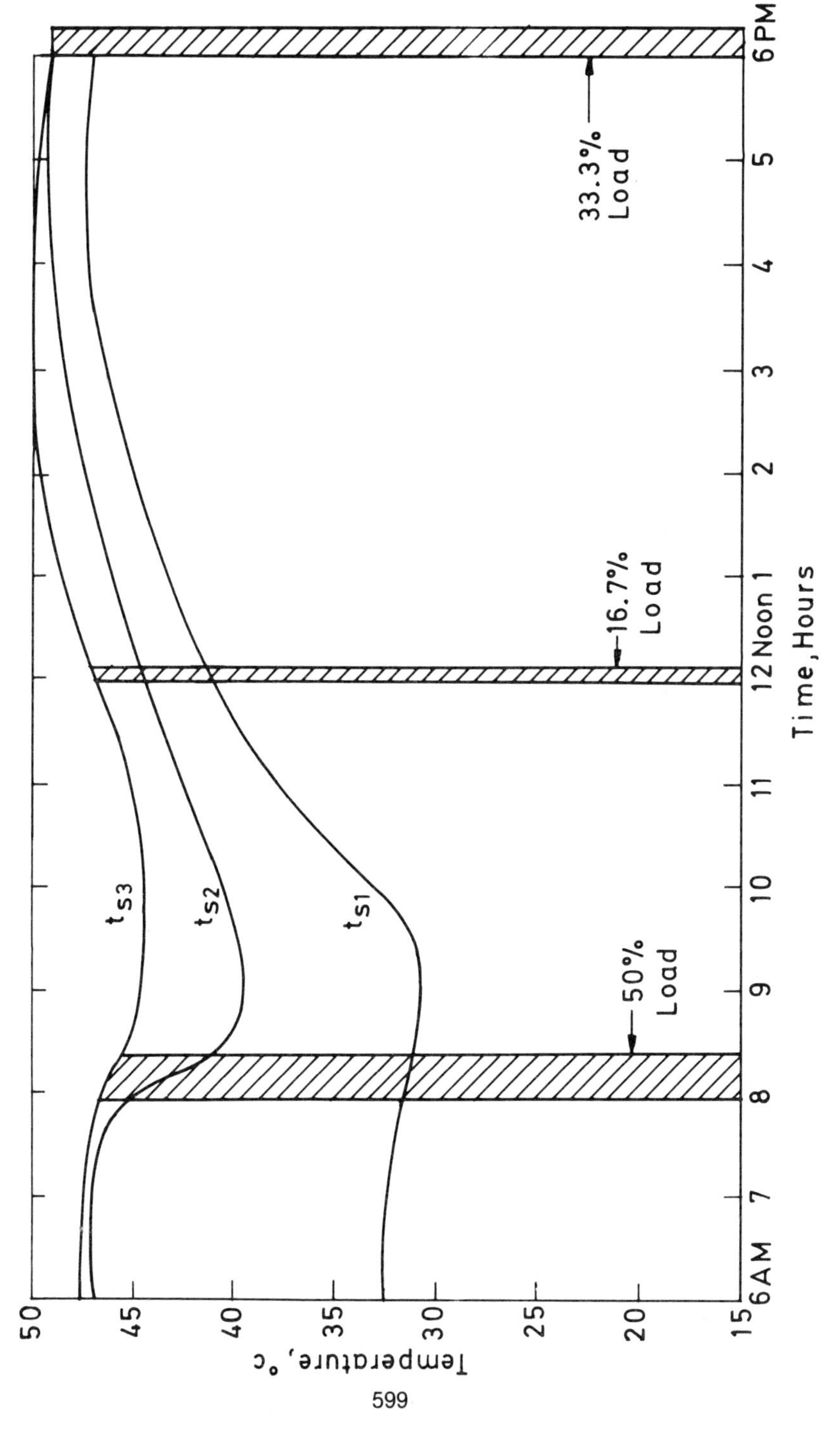

FIG.3. CHARACTISTIC OF STORAGE TANK

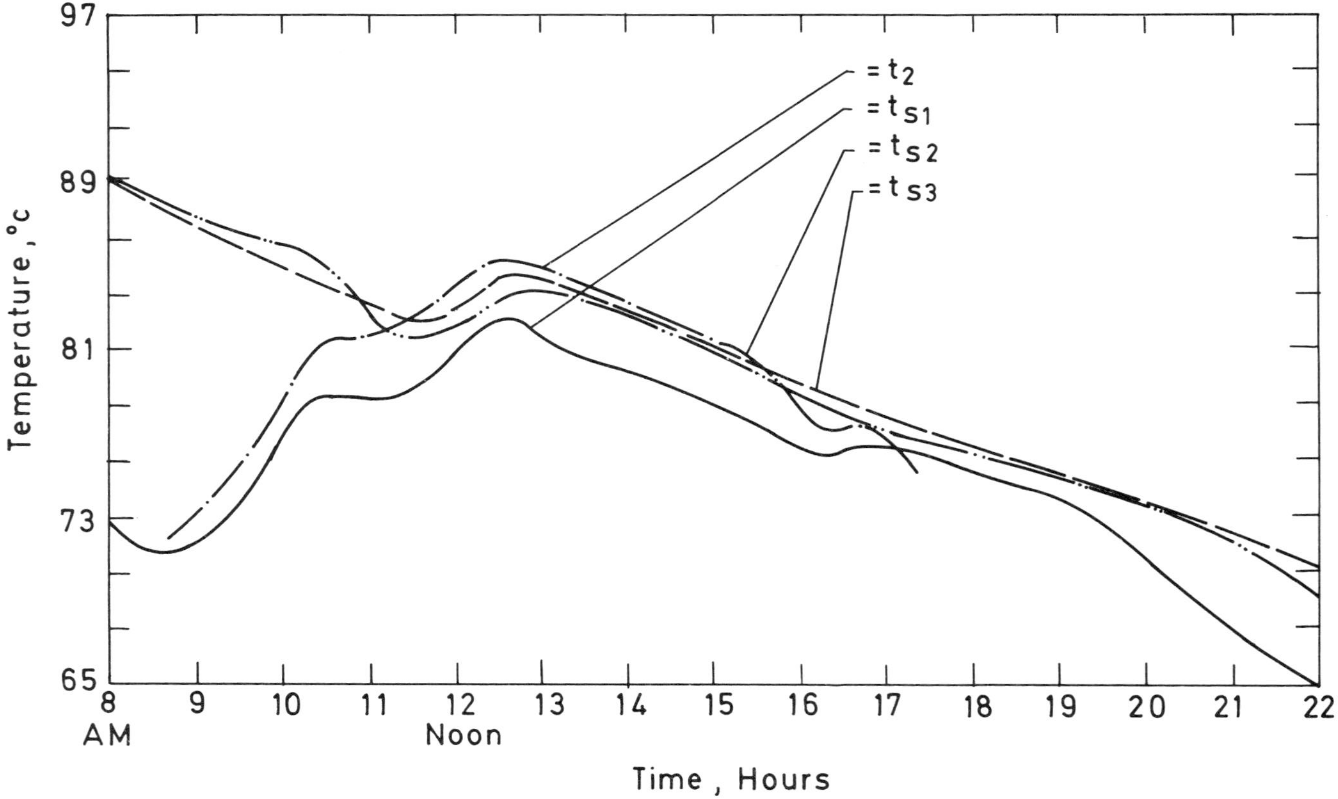

FIG.4. PREDICTED PERFORMANCE IN WINTER

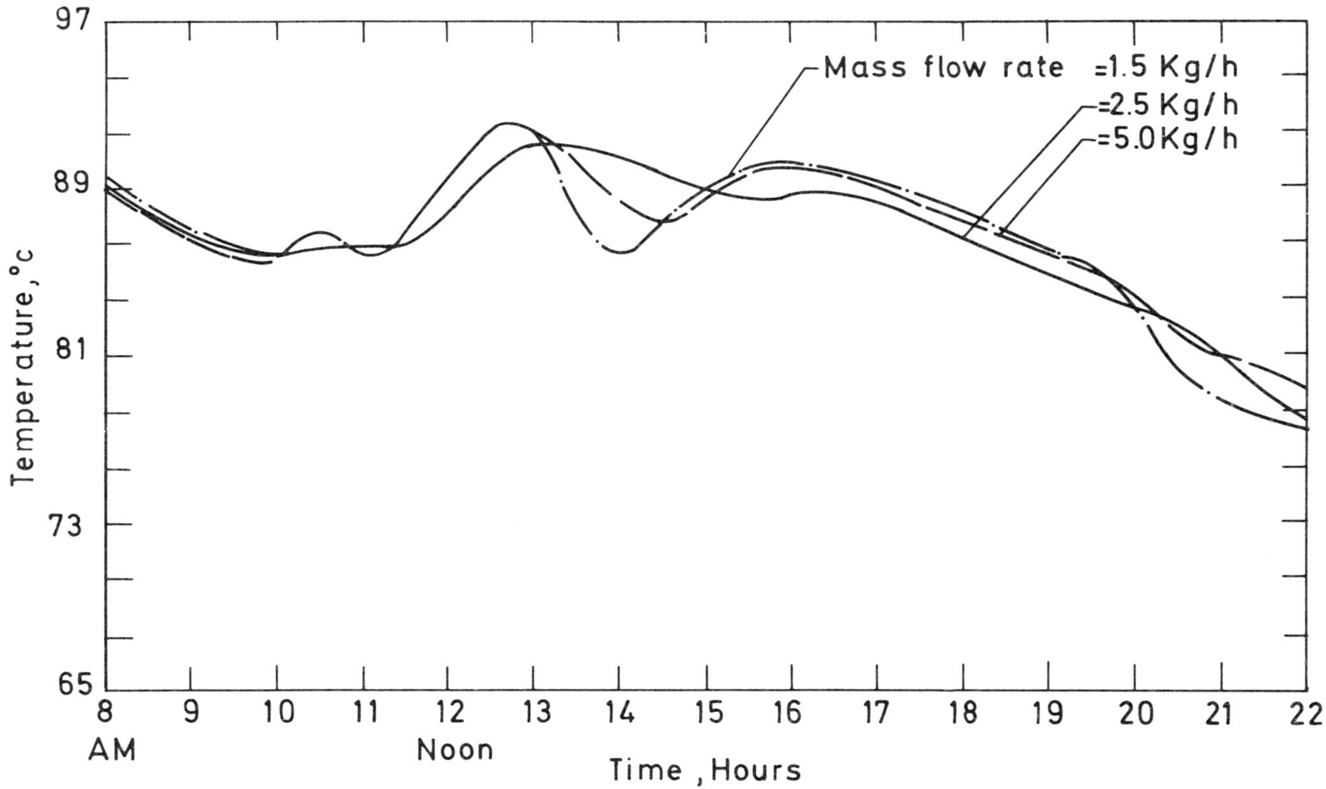

FIG.5. EFFECT OF FLOW RATE THROUGH LOAD ON COLLECTOR OUTPUT TEMPERATURE

is 2.5 to 4.5 m^3/m^2 of collector area. The variation in performance was found to be insignificant. But, the maximum temperature attained was the highest for 2.5 m^3/m^2 of collector area.

NOMENCLATURE

A_{s3}	Surface area of top tank section
A_{s2}	Surface area of middle tank section
A_{s1}	Surface area of bottom tank section
c_a	Thermal capacity of collector
c_{s3}	Thermal capacity of top tank section
c_{s2}	Thermal capacity of middle tank section
c_{s1}	Thermal capacity of bottom tank section
$\dot{c}_L$	Thermal capacity rate of fluid transferred to load
$\dot{c}_W$	Thermal capacity rate fluid passing through collector
F	Plate efficiency factor
G	Insolation
Q_u	Useful energy collection rate
$\dot{Q}_r$	Energy transferred to load
t_a	Ambient temperature
$t_{a\ max}$	Maximum daily ambient temperature
$t_{a\ min}$	Minimum daily ambient temperature
t_{s3}	Top tank temperature
t_{s2}	Middle tank temperature
t_{s1}	Bottom tank temperature
t_r	Load temperature
U_L	Collector loss coefficient
U_{s3}	Loss coefficient for top section
U_{s2}	Loss coefficient for middle section
U_{s1}	Loss coefficient for bottom section
α	Absorptivity of plate
F_1, F_2, F_3, F_4, F_5	Control functions
θ	Time
τ	Transmittivity of cover

REFERENCES

1. 'Low Temperature Engineering Applications of Solar Energy', ASHRAE, 1967
2. Duffie,J.A. and Beckman,W.A. "Solar Thermal processes" 1974, John Wiley & Sons.
3. Bliss, R.W. "The derivations of several plate efficiency factors useful in the design of flat plate solar heat collectors", Solar Energy 3, 1959,p55.
4. Sheridan, N.R. et, al "Study of solar processes by analog computer", Solar Energy, 11, 1967, p69.
5. Close,D.J. "A design approach for solar processes", Solar Energy, 11, 1967, p112.
6. Sriramulu,V. and Ahmed,S.B. "Analysis of thermal storage unit for solar energy", International Solar Energy Congress, New Delhi, India, Jan.16-21, 1978,

TRANSIENT CHARACTERISTICS OF VERTICAL POROUS HEAT STORAGE

N. Seki, S. Fukusako and H.Inaba
Dept. of Mechanical Engng., Faculty of Engng.
Hokkaido University, Sapporo 060, Japan

ABSTRACT

In the district where the amount of solar energy is limited by the time of insolation, it is important problem to store as much the energy as possible in the short period. This study deals with the transient characteristics of heat transfer in the vertical porous heat storage saturated with water and trasnsformer oil as a moving medium. The present experiments are made for various combinations of solid particles (glass beads and iron balls) and the above-mentioned media. Moreover, in the present analysis, the validity of numerical model adopted is investigated by especially introducing the heat transfer between particle and fluid into the governing equations. A comparison of the experimental data with the numerical predictions shows a good correlation as a whole.

INTRODUCTION

In recent years, it is well known that one's attention has focussed on the effective use of solar energy due to the shortage of energy resources on the earth. In this field, many researches has recently been reported on the problems of solar heat collector and the collector design techniques seem to be well developed. However, there is a only few reports dealing with the problems of heat storage for solar energy. In the district where the amount of solar energy is limited due to the insolating of time, the development of heat storage as well as solar collector is an important problem from the standpoint of effective use of solar energy. This study deals with the transient heat transfer characteristics of vertical rectangular porous heat storage, which composed of fluid and solid particles, for the purpose of improving the dynamic thermal characteristics of heat storage. Hitherto in many engineering fields the heat storage composed of a single phase medium has usually been used and its investigations have widely been performed as the problem related to heat conduction. However, the heat transmission in the single phase heat storage is restricted depending mainly on the thermal conductivity of the material adopted and also the amount of heat stored in the storage is limited according to the heat capacity of the material adopted. On the other hand, referring to the heat storage composed of fluid and solid particles, that is, porous media, it might be expected that the fluid transmits the incident heat to the solid particles in a shorter period of time due to the natural convection of fluid occurring in the storage.

Most of the previous investigations [1-7] relating to the heat transfer on the present problem have been carried out under a steady state condition, for example, the research of natural convection in a porous bed of volcanic debris in the geophysical field [8] or the effect of natural convection on porous insulation for nuclear power reactor core [9].
On the contrary, only few researches for unsteady state condition have so far been reported. P.H.Holst et al.[10-11] studied experimentally and analytically the effects of temperature-dependent fluid properties and the three dimensional fluid motion on heat transfer characteristics in a horizontal porous media. J.W.Elder [12] predicted analytically the behaviors of fluid motion in the horizontal porous media. Moreover, M.A.Combarnous et al. [13] reported the onset of natural convection in a horizontal porous media heated from below. However, those investigations were restricted to obtain the correlation between Nusselt number and Rayleigh number in a confined vessel whose two opposing walls are isothermal, and did not refer to the influences of geometrical aspect, solid diameter or porosity and the properties of porous media on transient characteristics of heat transfer in a porous layer.

This study is to discuss the influence of porosity related to the diameter of solid particles and imposed heat flux on transient heat transfer characteristics of vertically-positioned rectangular porous heat atorage under several combinations of water, transformer oil and glass beads, iron balls. Moreover, the numerical analysis adopted in the present study is preliminarily examined by comparing the predictions obtained under isothermal boundary condition of two opposing vertical walls with the results obtained experimentally. After confirming an agreement between them, the effects of aforementioned parameters on transient heat transfer characteristics of the present storage are numerically discussed under the condition of one constant-heat-flux vertical wall and the other insulated walls.

EXPERIMENTS

To examine preliminarily the validity of the present analytical model to be discussed in the following section, the experiments are performed by using a rectangular test apparatus of 38 mm (width W) x 571 mm (height H) in section area and 250 mm in depth, whose two opposing vertical walls are kept isothermally while the other walls insulated.
The schematic diagram of the experimental apparatus is depicted in Fig.1-(a). The main parts of the experimental apparatus consist of a test section and heating and cooling parts. The test section in which solid particles are packed and fluid is filled is indicated at the middle of the test apparatus shown in Fig.1-(a). For the isothermal opposing vertical walls to be maintained at different temperature, the vertical hot and cold walls are constructed by using copper plates of 5 mm in thickness.

Surface temperature of the hot wall is maintained at an uniform temperature by using five controllable main heaters. The guard heaters are mounted outside the main heaters across a bakelite plate in order to minimize the heat-loss from the main heaters to the environment.

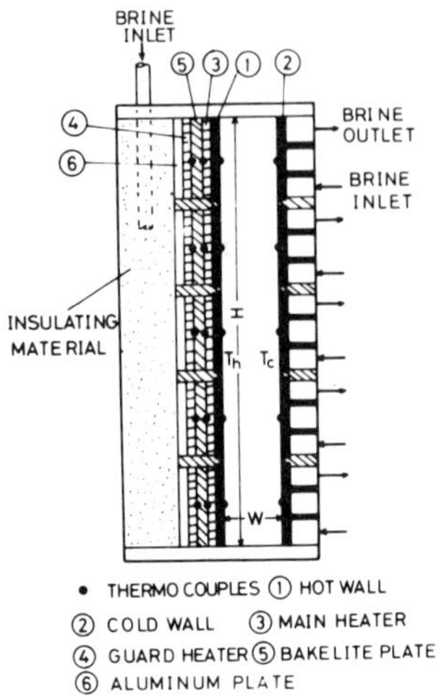

Fig.1-(a) Schematic view of experimental apparatus

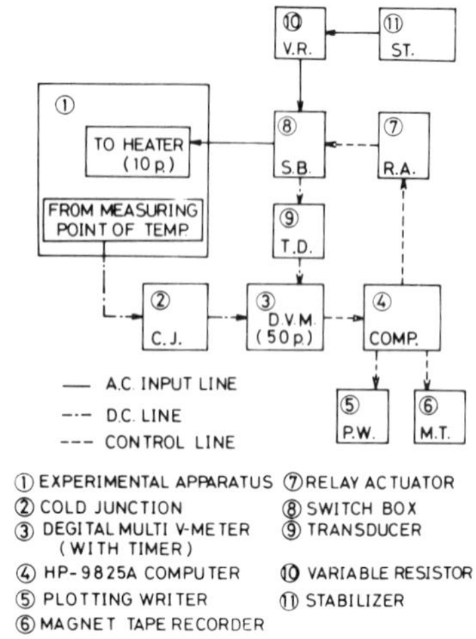

Fig.1-(b) Schematic diagram of the data treatment system

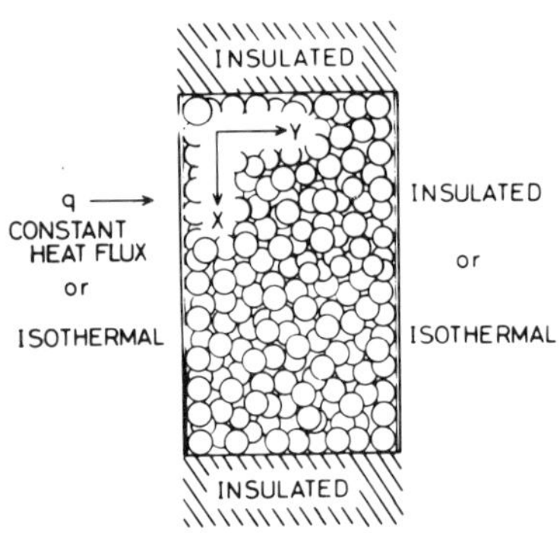

Fig.2-(a) The physical system as modelled mathematically

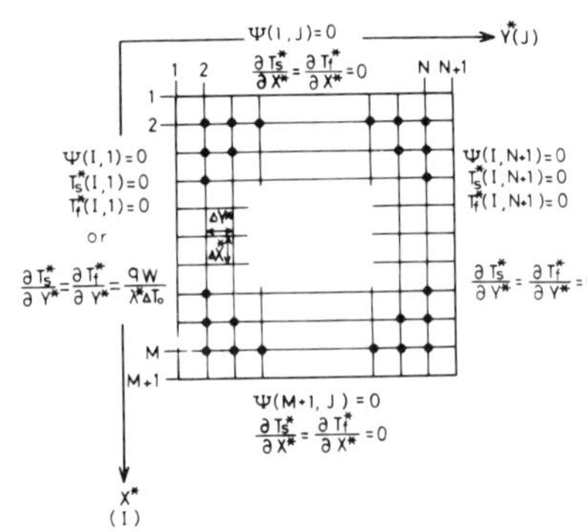

Fig.2-(b) Grid model

The power input into the main heaters and guard heaters is automatically controlled by the switch on-off system of the temperature controller as shown if Fig.1-(b). Surface temperature of the cold wall is kept at an uniform temperature by inducing temperature-controlled coolant (brine) into each of five separate chambers attached to the outside of the cold wall. Eighteen Cu-Co thermocouples (0.3 mm in diameter) are attached on the hot and cold walls, moreover, additional ten thermocouples are inserted between the main and guard heaters. Five small probes (stainless steel pipe of 0.8 mm in diameter) with Cu-Co thermocouples of 0.1 mm in diameter are in the Y direction at the middle height of the test section to measure the fluid temperature distribution. Moreover, in order to measure the temperature distributions of soild particles, Cu-Co thermocouples of 0.1 mm in diameter are embedded in the center of solid particles positioned in the Y direction at the middle height of the test section. Measurements are made for two combinations of water-glass beads and transformer oil-iron balls. The porosity ε obtained in the present study is in the range of $0.38 \leqq \varepsilon \leqq 0.42$. The fluid used is carefully injected from a nozzle at the bottom wall into the test-section in order to avoid the mixing of undesirable air bubbles. The experimental apparatus is covered with an insulating material (100 mm in thickness) and placed inside a temperature-controlled room to prevent a thermal disturbance from the environment.

The test runs are carried out after all the temperatures inside the test-section are confirmed to be kept uniformly at the temperature of cold wall. To evaluate the heat-loss form the apparatus during the runs, the careful experiments are preliminarily performed by comparing the power input to the test-section with the heat flow at the cold side. From the results of above-mentioned experiments, the heat-loss from the apparatus is estimated to be within ± 5 % of the power input.
The reading, recording and processing of the present experimental data are carried out with the aid of a Howlett-Packard data treatment system. The central element of the system is a H.P.9825 A digital computer with 6.8 K core capacity as shown in Fig.1-(b). The physical properties of porous media assumed in order to perform the present numerical calculations are evaluated by adopting the physical ones corresponding to the average value of the surface temperatures of hot and cold walls.

DESCRIPTION OF THE PRESENT NUMERICAL ANALYSIS

When a fluid phase moves through a porous medium, a distinction is made between the temperatures of fluid and solid phase. Especially, in setting up the mathematical model of transient heat transfer in a confined porous layer, it is important to take account of the heat transfer between solid and fluid phases. In the present numerical analysis, a mathematical model is obtained through energy balances taken for both the fluid phase and the solid phase by means of a heat transfer coefficient α between the solid and the fluid phases.

Moreover, the present momentum equations are expressed by taking account of the friction resistance proportional to a square of fluid velocity reported by S.Ergun [14]. The physical system and grid network as shown in Fig.2-(a) and 2-(b) can be modelled mathematically by writing the appropriate mass, momentum and energy balances on a differential volume of a porous medium. In the present study, the two-dimensional laminar equations are simplified by neglecting the temperature dependency of the fluid properties except density. The governing equations can be performed as follows:

Mass balance;

$$\frac{\partial U}{\partial X} + \frac{\partial V}{\partial Y} = 0 \tag{1}$$

Momentum balance;

$$\frac{\rho_f}{\varepsilon}\frac{\partial U}{\partial t} = -\frac{\partial p}{\partial X} + (\rho_f - \rho_{fc})g - \frac{\mu_f}{k}U - G\,U^2 \tag{2}$$

$$\frac{\rho_f}{\varepsilon}\frac{\partial V}{\partial t} = -\frac{\partial p}{\partial Y} - \frac{\mu_f}{k}V - G\,V^2 \tag{3}$$

where the value of G is a constant reported by S.Ergun [14].

Energy balance of fluid phase;

$$\varepsilon(\rho c_p)_f\frac{\partial T_f}{\partial t} + (\rho c_p)_f\left(U\frac{\partial T_f}{\partial X} + V\frac{\partial T_f}{\partial Y}\right) = \lambda_f\left(\frac{\partial^2 T_f}{\partial X^2} + \frac{\partial^2 T_f}{\partial Y^2}\right) - \alpha\,(T_f - T_s) \tag{4}$$

Energy balance of solid phase;

$$(1-\varepsilon)(\rho c_p)_s\frac{\partial T_s}{\partial t} = \lambda_s\left(\frac{\partial^2 T_s}{\partial X^2} + \frac{\partial^2 T_s}{\partial Y^2}\right) - \alpha\,(T_s - T_f) \tag{5}$$

Correlation for predicting the volumetric heat transfer coefficient, α, may be found in reference [15].

Introducing stream function ψ ($U = \frac{\partial\psi}{\partial Y}$, $V = -\frac{\partial\psi}{\partial X}$) and following non-dimensional variable into above-mentioned governing equations:

$$X^* = X/W, \quad Y^* = Y/W, \quad U^* = UW(\rho c_p)_f/\lambda_f, \quad V^* = VW(\rho c_p)_f/\lambda_f,$$

$$T^*_{s,f} = (T_{s,f} - T_2)/\Delta T_o \text{ or } (T_{s,f} - T_c)/(T_h - T_c), \; Pr = \nu_f/a_f,$$

$$\Psi = \psi/a_f, \quad \tau = \lambda_f\, t/\varepsilon(\rho c_p)_f\, W^2, \quad Ra^* = g\beta_f\, kqW^2/\nu_f\, a_f\lambda^* \text{ or}$$

$$g\beta_f\, kW(T_h - T_c)/\nu_f\, a_f, \quad E = (1-\varepsilon)/\varepsilon, \quad L = (\rho c_p)_s/(\rho c_p)_f,$$

$$\Lambda = \lambda_s/\lambda_f, \quad D = \alpha W^2/\lambda_f \tag{6}$$

One obtains the following nondimensional equations to be analyzed:

$$\frac{\partial^2\Psi}{\partial X^{*2}} + \frac{\partial^2\Psi}{\partial Y^{*2}} = 0 \tag{7}$$

$$\frac{\partial}{\partial\tau}\left(\frac{\partial^2\Psi}{\partial X^{*2}} + \frac{\partial^2\Psi}{\partial Y^{*2}}\right) - A\frac{\partial T_f^*}{\partial Y^*} + B\left(\frac{\partial^2\Psi}{\partial X^{*2}} + \frac{\partial^2\Psi}{\partial Y^{*2}}\right)$$
$$+ C\left(U^*\frac{\partial U^*}{\partial Y^*} - V^*\frac{\partial V^*}{\partial X^*}\right) = 0 \tag{8}$$

where $A = \varepsilon^2\, W^2\, Pr\, Ra^*/\, k$, $B = \varepsilon^2\, W^2 Pr\, /\, k$, $C = 0.0234d\, W\varepsilon^2\, /\, k(1-\varepsilon)$.

Energy balance of fluid phase;

$$\frac{\partial T_f^*}{\partial\tau} + \left(U^*\frac{\partial T_f^*}{\partial X^*} + V^*\frac{\partial T_f^*}{\partial Y^*}\right) - \left(\frac{\partial^2 T_f^*}{\partial X^{*2}} + \frac{\partial^2 T_f^*}{\partial Y^{*2}}\right) + D(T_f^* - T_s^*)$$
$$= 0 \tag{9}$$

Energy balance of solid phase;

$$E\,L\,\frac{\partial T_s^*}{\partial \tau} - \Lambda\left(\frac{\partial^2 T_s^*}{\partial X^{*2}} + \frac{\partial^2 T_s^*}{\partial Y^{*2}}\right) + D\left(T_s^* - T_f^*\right) = 0 \qquad (10)$$

Initial condition ($\tau = 0$):
The temperature distribution of porous layer is equal to the temperature of the cold wall. The mass flux distribution is equal to zero.

Boundary conditions are as follows ($\tau > 0$):
for two opposing vertical walls, isothermal,

$$Y^* = 0\ ;\quad \frac{\partial \Psi}{\partial X^*} = \frac{\partial \Psi}{\partial Y^*} = \Psi = 0\ ,\quad T_{s,f}^* = 1$$

$$Y^* = 0\ ;\quad \frac{\partial \Psi}{\partial X^*} = \frac{\partial \Psi}{\partial Y^*} = \Psi = 0\ ,\quad T_{s,f}^* = 0$$

$$X^* = 0\ ;\quad \frac{\partial \Psi}{\partial Y^*} = \frac{\partial \Psi}{\partial X^*} = \Psi = 0\ ,\quad \frac{\partial T_{s,f}^*}{\partial X^*} = 0$$

$$X^* = H/W\ ;\quad \frac{\partial \Psi}{\partial Y^*} = \frac{\partial \Psi}{\partial X^*} = \Psi = 0\ ,\quad \frac{\partial T_{s,f}^*}{\partial X^*} = 0$$

In practice, there are many cases that solar energy penetrates into the heat storage through the surface. However, in the present study constant-heat-flux is assumed to be impinged on the heating side to simplify the boundary condition.
For one vertical wall heated with constant-heat-flux and the opposing vertical wall insulated,

$$Y^* = 0\ ;\quad \frac{\partial \Psi}{\partial X^*} = \frac{\partial \Psi}{\partial Y^*} = \Psi = 0\ ,\quad \frac{\partial T_{s,f}^*}{\partial Y^*} = \text{constant}$$

$$Y^* = 1\ ;\quad \frac{\partial \Psi}{\partial X^*} = \frac{\partial \Psi}{\partial Y^*} = \Psi = 0\ ,\quad \frac{\partial T_{s,f}^*}{\partial Y^*} = 0$$

$$X^* = 0 \; ; \quad \frac{\partial \Psi}{\partial Y^*} = \frac{\partial \Psi}{\partial X^*} = \Psi = 0 \; , \quad \frac{\partial T^*_{s,f}}{\partial X^*} = 0$$

$$X^* = H/W; \quad \frac{\partial \Psi}{\partial Y^*} = \frac{\partial \Psi}{\partial X^*} = \Psi = 0 \; , \quad \frac{\partial T^*_{s,f}}{\partial X^*} = 0.$$

In view of the mathematically complexity to solve these equations, these differential equations are mainly solved by the Alternative Direction Implicit (ADI) procedure. The number of grids in the horizontal direction (Y^*) is 21, while that in the vertical direction (X^*) is decided corresponding to the aspect-ratio H/W. The line iterative method is performed until two successive iterative values of Ψ, T^*_s and T^*_f are attained within 0.1 per cent. In case of a rectangular confined cavity packed with spherical solid particles, the porosity ε ranges form 0.26 to 0.48 theoritically. However, the ε value in the present computation is extended to the range of $0.05 \leqq \varepsilon \leqq 0.6$. The temperature difference, ΔT_O, between two opposing vertical walls, which one vertical wall is heated with constant-heat-flux and the opposing vertical wall insulated, is set at ΔT_O = 50 °C in order to express the actual temperature difference as nondimensional form in this calculation. The computations are performed with a FACOM 230-75 Digital Computer Center of Hokkaido University.

RESULTS AND DISCUSSION

In order to clarify analytically the transient heat transfer characteristics of the present rectangular heat storage which composed of fluid and solid particles and is placed vertically, it is necessary to preliminarily examine the validity of the analytical model adopted by comparing the numerical and experimental results. Fig.3 for H/W = 15 demonstrates a comparison between the numerical and experimental results under the condition of two opposing isothermal vertical walls (surface temperature of hot wall T_h = 50 °C and surface temperature of cold wall T_c = 5 °C) for the combination of water (λ_f = 0.511 kcal/mh°C at 20 °C, ρc_p = 997 kcal/m^3°C) and glass beads (λ_s = 0.65 kcal/mh°C, ρc_p = 540 kcal/m^3°C). A comparison between the numerical results obtained and experimental ones under the above-mentioned boundary condition for the combination of transformer oil (λ_f = 0.107 kcal/mh°C, ρc_p = 391 kcal/m^3°C) and iron balls (λ_s = 42.0 kcal/mh°C, ρc_p = 861 kcal/m^3°C) is shown in Fig.4 for H/W = 15. These figures show the relationship between the mean Nusselt number Nu^*_h at the hot side and nondimensional time τ, and it might be understood that there is a good agreement between the numerical and experimental results except early time of τ after starting of the test run. These differences between numerical and experimental results for small time of τ could be considered to be caused by the reason that the heating wall has comparatively a large heat capacity while the input power to

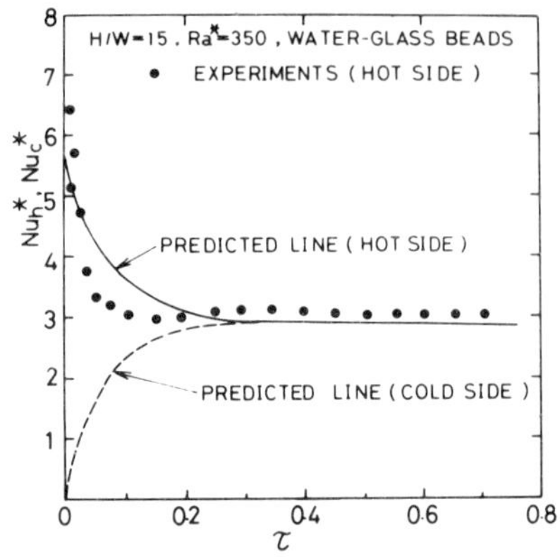

Fig.3 Comparison of analysis and experiments (two opposing isothermal walls,water-glass beads,H/W = 15, T_h = 50 °C and T_c = 5 °C)

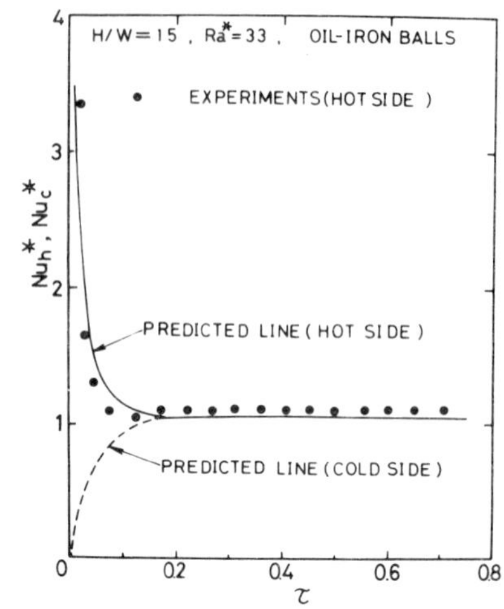

fig.4 Comparison of analysis and experiments (two opposing isothermal walls,trans.oil-iron balls,H/W = 15, T_h = 50 °C and T_c = 5 °C)

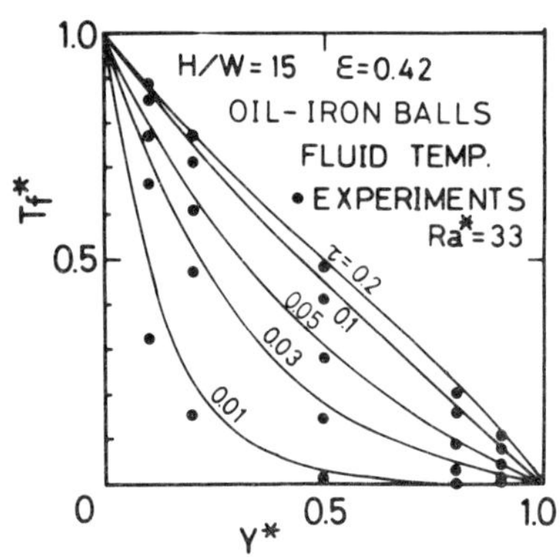

Fig.5 Comparison of analysis and experiments for fluid temperature T_f^* (two opposing isothermal walls, trans.oil-iron balls,H/W = 15, T_h = 50 °C and T_c = 5 °C)

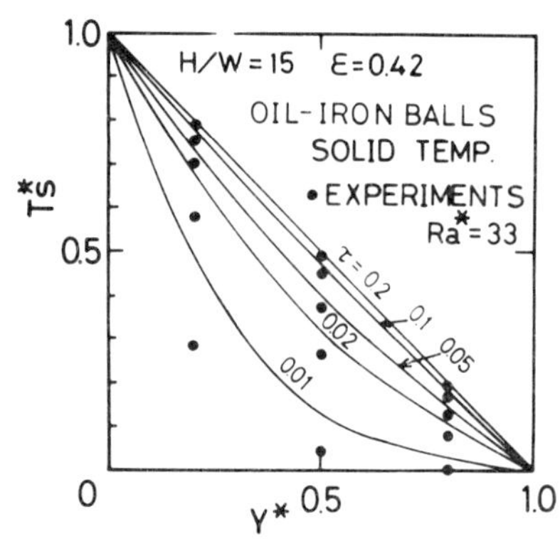

Fig.6 Comparison of analysis and experiments for solid temperature T_s^* (two opposing isothermal walls, trans.oil-iron balls,H/W = 15, T_h = 50 °C and T_c = 5 °C)

the heater is a little fluctuated by the switch on-off system of the temperature controller. However it is understood that such a difference in Fig.4 for iron balls with large heat capacity is comparatively small. Fig.5 and Fig.6 show the comparison between numerical and experimental results of nondimensional fluid and solid temperature distributions, T_f^* and T_s^*, in the horizontal direction (Y^*) at the middle height of porous layer ($X^* = 0.5\ H/W$) under the same condition as adopted in Fig.4. There are a little difference of T_s^* between both results due to the contact heat resistance between heating wall and solid particles adopted. However, it is clear that a good coincidence is obtained between the numerical and experimental results as the time τ elapses. From these results, it might be understood that the prediction obtained by using the analytical model assumed in the present study shows very well the experimental tendency of transient heat transfer in the media.

Fig.7 demonstrates the transient heat characteristics of vertical heat storage composed of water and glass beads, which are predicted by numerical computation based on the above-mentioned confirmation of the validity of the present analytical model. This figure shows the nondimensional temperature difference between the heating and insulating vertical walls, ΔT_s^* and ΔT_f^*, against the nondimensional time τ, which is obtained numerically for $H/W = 1$. In this case, one vertical wall is heated with constant-heat-flux q = 600 kcal/m^{2}h and the other insulated. These thermal assumptions may be similar to the conditions for solar heat storage. The porosity ε of confined cavity packed with spherical solid particles ranges form 0.26 to 0.48 theoritically, but in the present study the value of ε is extended to the range of $0.05 \leq \varepsilon \leq 0.6$. From this figure, it can be seen that as the value of ε increases, that is, the volumetric rate of fluid to the porous media increases, the value of ΔT_s^* approaches thermally to the quasi-steady state in a shorter period of time after starting of the run. Especially in case of $\varepsilon = 0.6$, this value approaches to the quasi-steady state at $\tau = 0.2$. Moreover, the value of ΔT_s^* increases gradually with elapsing of time τ. Though the tendency of ΔT_s^*, it is interesting that the values of ΔT_f^* are lower than those of ΔT_s^* except for $\varepsilon = 0.05$. The difference between ΔT_s^* and ΔT_f^* in case of $\varepsilon = 0.05$ may be mainly caused by the reason that the heat capacity of solid phase in the porous heat storage is extremely larger than that of fluid phase. Figs.8 and 9 for $\varepsilon = 0.2$ and 0.6 respectively show the relationship between nondimensional temperature distributions and τ. As the porosity ε increases, the value of T_f^* in the vicinity of insulated vertical wall becomes larger than that of T_s^* with elapsing of time τ. Moreover, from these transient behaviors of T_s^* and T_f^*, it should be noticed that the existence of fluid in the storage is remarkably effective on the uniformness of the temperature distributions in the heat storage.

Fig.10 illustrates the transient characteristics of ΔT_s^* and ΔT_f^* in the porous heat storage for the combination of water and iron balls. Though the predicted transient tendency of ΔT_s^* and ΔT_f^* is similar to that for the above-mentioned combination (water-glass beads), one should be noted that the uniformness of the temperature distributions in the heat storage for the combination of water and iron balls is obtained more earlier than that for the combination of water and glass beads, and also the values of ΔT_s^* and ΔT_f^* for the former combination are smaller than those for the

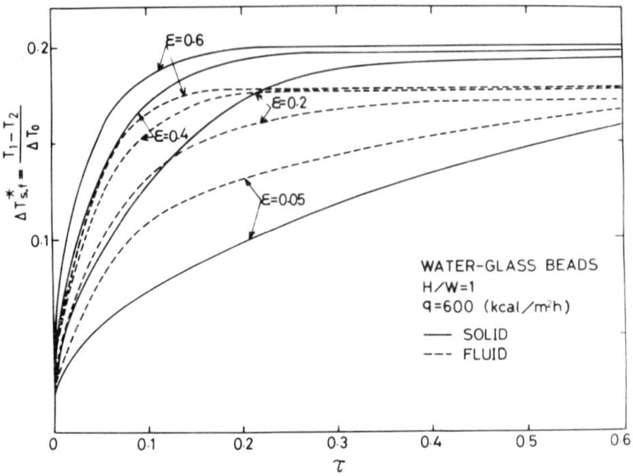

Fig.7 Relationship between ΔT_s^* or ΔT_f^* and τ (one wall heated with constant-heat-flux, the opposing wall insulated, water-glass beads, H/W = 1, q = 600 kcal/m²h)

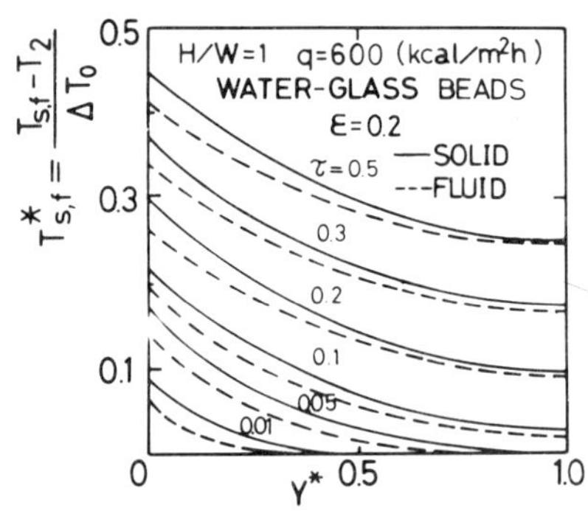

Fig.8 Relation between T_s^* or T_f^* and Y^* (one wall heated with constant-heat-flux, the opposing wall insulated, water-glass beads, ε = 0.2)

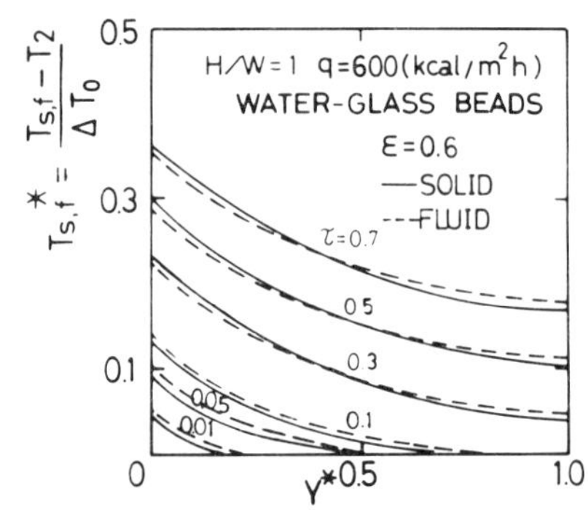

Fig.9 Relation between T_s^* or T_f^* and Y^* (one wall heated with constant-heat-flux, the opposing wall insulated, water-glass beads, ε = 0.6)

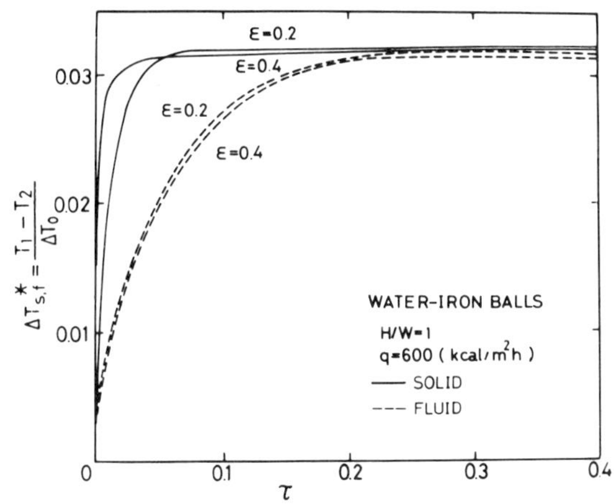

Fig.10 Relationship between ΔT_s^* or ΔT_f^* and τ (one wall heated with constant-heat-flux, the opposing wall insulated, water-iron balls, $H/W = 1$, $q = 600$ kcal/m^{2}h)

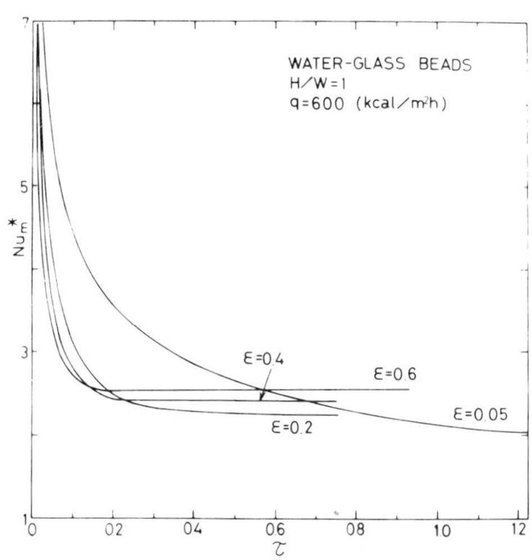

Fig.11 Relationship between Nu_m^* and τ (one wall heated with constant-heat-flux, the opposing wall insulated, water-glass beads, $H/W = 1$, $q = 600$ kcal/m^{2}h)

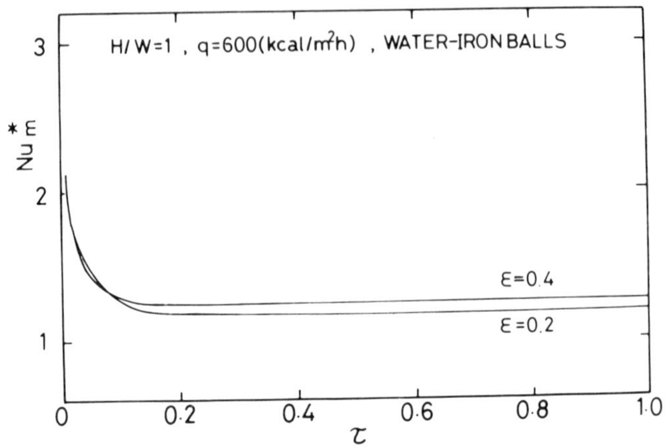

Fig.12 Relationship between Nu_m^* and τ (one wall heated with constant-heat-flux, the opposing wall insulated, water-iron balls, H/W = 1, q = 600 kcal/m^{2}h)

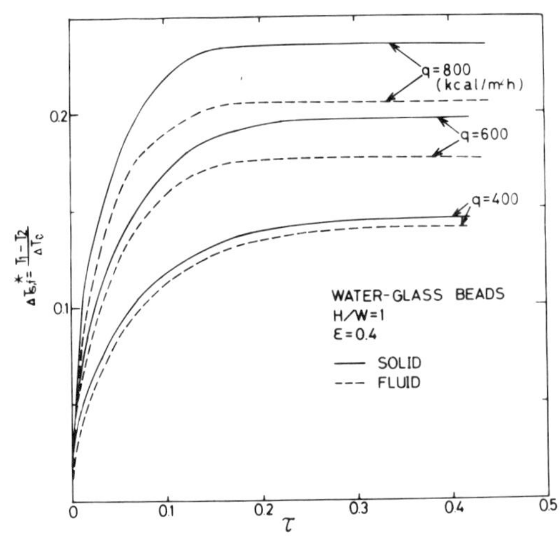

Fig.13 Relationship between ΔT_s^* or ΔT_f^* and τ (one wall heated with constant-heat-flux, the opposing wall insulated, water-glass beads, H/W = 1, ε = 0.4)

latter one. Figs.11 and 12 show the relationship between the mean Nusselt number Nu_m^* at the heating side and τ. From this figure, the behavior of heat transmission through the porous heat storage can be understood more clearly. Fig.13 presents the relationship between ΔT_s^* or ΔT_f^* and τ in the range of heat flux q = 400 ∿ 800 kcal/m^2h for the combination of water and glass beads, ε = 0.4 and H/W = 1, which clarifies the transient behavior of heat transmission into the heat storage. It is intriguing that as q increases, the uniformness of the temperature distributions in the storage becomes gradually difficult to obtain even by the aid of natural convection in the storage and also the value of ΔT_s^* or ΔT_f^* increases gradually. Moreover, it is interesting that the difference between the uniformness of the solid temperature (ΔT_s^*) and fluid temperature (ΔT_f^*) increases with elapsing of time τ.
Fig.14 shows the relationship between Nu_m^* and τ obtained under the same condition as assumed in Fig.13. Fig15 ∿ Fig.18 show several examples of transient flow patterns for q = 400 and 800 kcal/m^2h in the heat storage modelled in the present study. From these figures, it might be understood that the magnitude of natural convection in the storage increases with increasing heat flux q and also ΔT_s^* and ΔT_f^* increase with increasing heat flux q. However, the temperature difference between ΔT_s^* and ΔT_f^* also increases with increasing q. This means the uniformness of temperature distribution of solid phase is difficult to reach as compared with that of fluid phase. Moreover, it is interesting that the center of the flow pattern with a little strong strength of eddy shifts to the center region of the storage after increased time of τ.
In Fig.19 the relationship between τ and the mean solid temperature $T_{s,m}^*$ for q = 600 kcal/m^2h, which corresponds to the amount of heat stored in the solid phase, is demonstrated by using ε as a parameter. From this figure, it is clear that the value of $T_{s,m}^*$ increases with elapsing of time τ and ε and also its increasing rate for iron balls having large thermal conductivity is remarkably large compared with that for glass beads under the same condition. Fig.20 shows the relationship between $T_{s,m}^*$ and τ for the combination of transformer oil and glass beads, which is obtained under the same condition as adopted in the aforementioned prediction of transient thermal characteristics for the combination of water and glass beads. Fig.21 shows the predicted relationship between ΔT_s^* or ΔT_f^* and τ obtained under the same condition as adopted in Fig.20. From these figure, it should especially be noticed that the period of time to reach the quasi-steady state for this combination is distinguishably short and also the increasing rate of $T_{s,m}^*$ with time τ is remarkably large compared with those for the combination of water and glass beads. These results might be mainly caused by low heat capacity of transformer oil adopted as a moving fluid in this case.

In the present study the numerical analyses are carried out for H/W = 1 ∿ 15 in order to investigate the effect of H/W on transient heat transfer characteristics of the present modelled heat storage. However, the obtained numerical predictions show little effect of H/W on transient heat transfer characteristics of the storage. General speaking, it might be expected that the heat transmission in the present heat storage becomes more complicated due to the variation of natural convection mode as H/W and q increase.

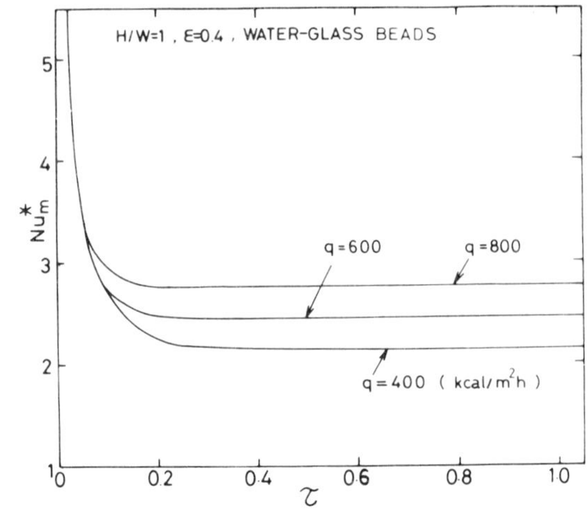

Fig.14 Relationship between Nu_m^* and τ (one wall heated with constant-heat-flux, the opposing wall insulated, water-glass beads, H/W = 1, ε = 0.4)

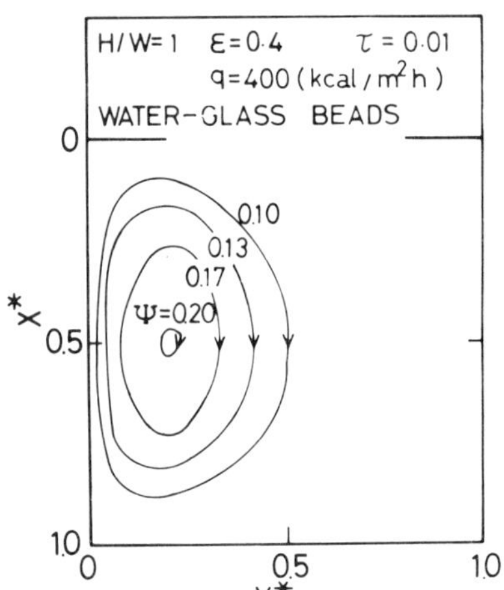

Fig.15 Flow pattern predicted for q = 400 kcal/m^{2}h (one wall heated with constant-heat-flux, the opposing wall insulated, water-glass beads , H/W = 1, ε = 0.4, τ = 0.01)

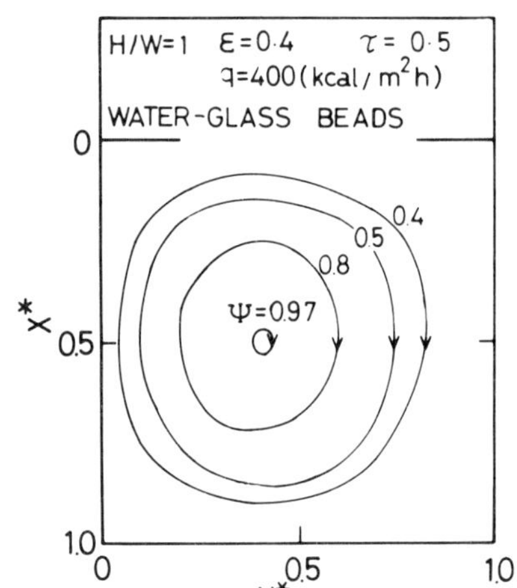

Fig.16 Flow pattern predicted for q = 400 kcal/m^{2}h (one wall heated with constant-heat-flux, the opposing wall insulated, water-glass beads , H/W = 1, ε = 0.4, τ = 0.5)

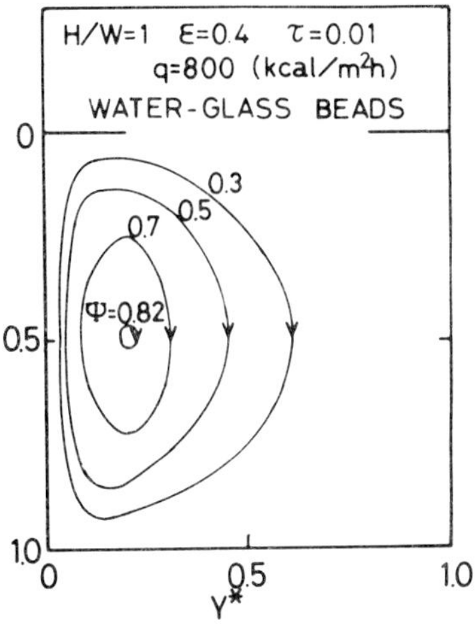

Fig.17 Flow pattern predicted for q = 800 kcal/m^{2}h (one wall heated with constant-heat-flux, the opposing wall insulated, water-glass beads , H/W = 1, ε = 0.4, τ = 0.01)

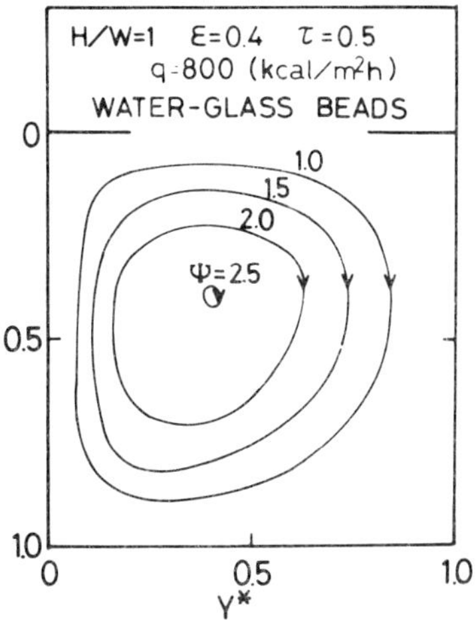

Fig.18 Flow pattern predicted for q = 800 kcal/m^{2}h (one wall heated with constant-heat-flux, the opposing wall insulated, water-glass beads , H/W = 1, ε = 0.4, τ = 0.5)

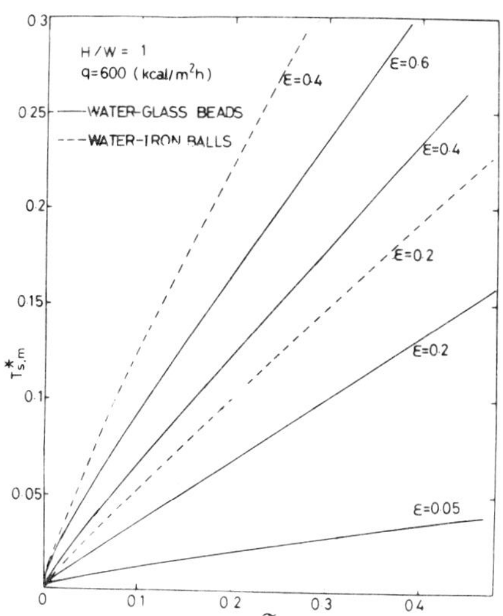

Fig.19 Relationship between $T^*_{s,m}$ and τ (one wall heated with constant-heat-flux, the opposing wall insulated, water-glass beads and water-iron balls, H/W = 1, q = 600 kcal/m^{2}h)

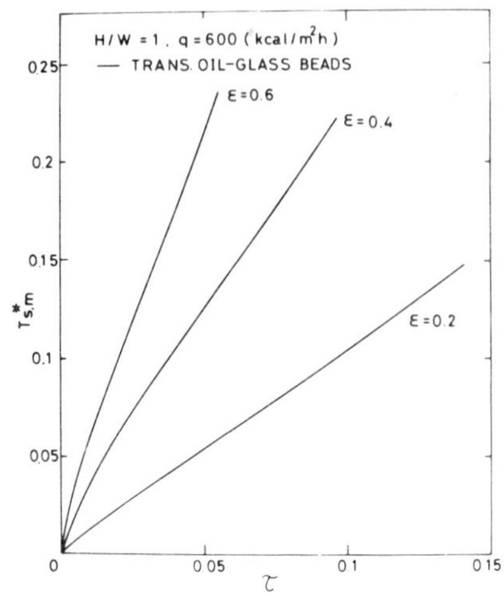

Fig.20 Relationship between T^*_s,m and τ (one wall heated with constant-heat-flux, the opposing wall insulated, trans.oil-glass beads, H/W = 1, q = 600 kcal/m^{2}h)

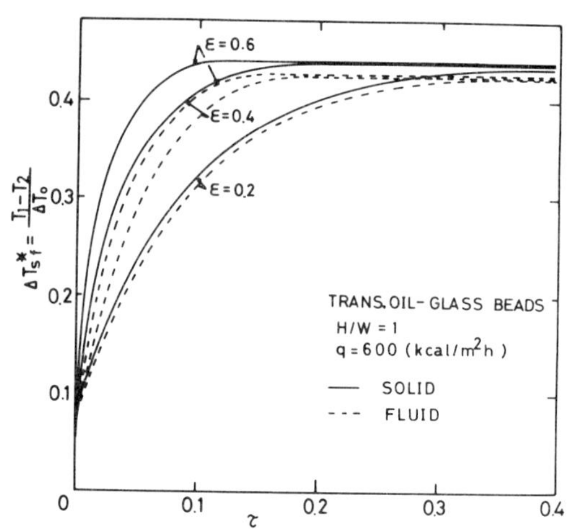

Fig.21 Relationship between ΔT^*_s or ΔT^*_f and τ (one wall heated with constant-heat-flux, the opposing wall insulated, trans.oil-glass beads, H/W = 1, q = 600 kcal/m^{2}h)

CONCLUSION

In this study the transient heat transfer characteristics of rectangular porous heat storage, which composed of fluid and solid particles and positioned vertically, are investigated for the combinations of water,transformer oil and glass beads, iron balls. From the standpoint of the effective use of incident heat energy such as solar energy, the validity of the present analytical model assumed is preliminarily confirmed by comparing the transient temperature distributions and Nusselt number predicted numerically under the boundary condition of two opposing isothermal vertical walls with those obtained experimentally.
Referring to porous heat storage having one vertical wall heated with constant-heat-flux and the other walls insulated, the effects of porosity related to the diameter of solid particles and imposed heat flux on transient heat transfer characteristics of the storage for the combination of water, transformer oil and glass beads, iron balls might be clarified numerically in this study. From the standpoint of engineering design, these factors should be properly selected in taking account of insolating period of heat energy as solar energy to the heat storage adopted.

NOMENCLATURE

a, thermal diffusivity;
c_p, specific heat at constant pressure;
d, diameter of solid particle;
g, gravitational acceleration;
H, height of rectangular cavity
k, permiability;
Nu^*, Nusselt number, defined by Eq.(6);
p, pressure;
Pr, Prandtl number;
Ra^*, Rayleigh number, defined by Eq.(6);
t, time;
T, temperature;
ΔT_o, temperature difference between constant-heat-flux hot wall and insulated cold wall;
U, V, velocity;
W, width of rectangular cavity;
X, Y, distance from upper wall, distance from hot wall.

GREEK SYMBOLS

α, heat transfer coefficient;
β, cubical expansion coefficient;
λ, thermal conductivity;
λ^*, thermal conductivity of porous medium (without convection);
ν, kinematic viscosity;
ρ, density;
ε, porosity;

ψ, stream function;
τ, nondimensional time, defined by Eq.(6)

SUBSCRIPTS

c,h, cold wall (isothermal), hot wall (isothermal);
f,s, fluid phase, solid phase;
m, mean value;
1,2, hot wall (constant-heat-flux), cold wall (insulated).

SUPERSCRIPT

$*$ nondimensional.

REFERENCES

1. K.J.Schneider, 11th International Congress of Refrigeration (Munich, West Germany), Paper No.LL-4, 1963, p.247.
2. T.Kaneko, M.F.Mohtadi and K.Aziz, Int.J.Heat Mass Transfer, Vol.17, 1974, p.485.
3. C.G.Bankvall, Wärme- und Stoffübertragung, Vol.7, 1974, p.22.
4. S.A.Bories and M.A,Combarnous, J.Fluid Mech., Vol.57, 1973, p.63.
5. M.P.Vlasuk, 4th All-Union Heat and Mass Transfer Conference (Minsk), Paper No.1, 1972, p.49.
6. J.E.Weber, Int.J.Heat Mass Transfer, Vol.18, 1975, p.567.
7. N.Seki, S.Fukusako and H.Inaba, Int.J.Heat Mass Transfer, to be published.
8. R.A.Wooding, J.Fluid Mech., Vol.2, 1957, p.273.
9. B.K.C.Chan, C.M.Ivey and J.M.Barry, J.Heat Transfer, Transaction of the ASME, Vol.92, 1970, p.21.
10. P.H.Holst and K.Aziz, The Canadian Jounal of Chemical Engineering, Vol.50. 1972, p.232.
11. P.H.Holst and K.Aziz, Int. J.Heat Mass Transfer, Vol.15, 1972, p.73.
12. J.W.Elder, J.Fluid Mech., Vol.27, 1967, p.609.
13. M.A.Combarnous and S.A.Bories, Advance in Hydroscience (book), 1976, p.231.
14. S.Ergun, Chem. Eng., Prog., Vol.48, 1952, p.89.
15. M.A.Combarnous and S.A.Bories, Int.J.Heat Mass Transfer, Vol.17, 1974, p.505.

SEASONAL REGENERATION THROUGH UNDERGROUND STRATA

T. Yokoyama, H. Umemiya, T. Teraoka, H. Watanabe
University of Yamagata, Yonezawa, Japan

K. Katsuragi
Nippon Chikasui-Kaihatsu Co., Ltd., Japan

and

K. Kasahara
Mayekawa Mfg. Co., Ltd., Tokyo, Japan

ABSTRACT

Different from present Solar energy systems, which use only heat enerty and regenerate daily at most, our system makes good use of not only heat but also cold, which are stored until the next available season, namely "Seasonal Regeraration".

Inhabitants of Temperate Zones, such as Japan, struggle with cold weather in winter, while they suffering from hot sultry weather in summer. From another point of view, in this area the Sun supplies an enormous amount of heat energy in summer, and in winter delivers an enormous amount of cold energy, the "Condensed Cold" of snow and ice. Therefore, both hot and cold sources are available in the same places.

It seems to us that underground strata stores each season's heat source until the next available season because of its insulation composed of impervious layers and its huge heat capacity. This enables us to obtain originated in summer and cold originated in winter, provided that the waste "Cool Heat" (in winter) and "Hot Heat" (in summer) are alternately stored, and these heat sources are regenerated permanently. In addition to air conditioning, we can also use this energy for agriculture and fish culturing.

In fact, our field experiments and theoretical analysis shows this to be possible.

1. INTRODUCTION

Large amounts of low concentration Solar energy reach the earth. Temperatures are neither high nor low and furthermore the energy varies by region and season so we cannot make good use of it, but have suffered from its influences.

Especially in the Temperate Zone, e.g. Japan such an enormous amount of heat energy is received in summer that cooling is needed for many months. In winter cold weather or what we might call "Condensed Cold" i.e. snow necessitates heating for many months. Presently most heating is done by petroleum. In Japan for example, 7 ~ 10% of all oil is used for heating. Oil heating is relatively inexpensive, but in future shortages of petrodeum will require us to heat without oil.

From another point of view, the weather of each season e.g. heat in summer and cold in winter represent possibile permanent heat sources and heat sinks not found anywhere else. Furthermore where Solar energy systems cannot be used in winter, we may collect heat energy by Solar system or refrigerator during summer satisfactorily for heating in winter.

Different from the usual Solar Energy System of only using a heat source and at most daily regeneration, if each season's weather is regenerated until available next season, we can make good use of each season. To be concrete, the regenerated hot heat originating in summer carries on heating in winter for fish cultivation, agriculture, etc. directly or by heat pump, then snow melting. In return, the regenerated cool heat originating in winter carries on cooling directly or by refrigerator.

Underground near the earth's surface is known to be of constant temperature throughout the year. Thsi characteristic is due to the fact that the earth consists of less theremally diffusivity material such as soil. In the absence of underground water flow, we can show this insulating effect is sufficient for seasonal regeneration by Carslow's vertical one dimensional solution [1]. From injection of a heat source with $\Theta j^{o}c$ into the layer of thickness 2b initially $\Theta\infty^{o}c$, then unsteady temperature distribution is given below,

$$\frac{\Theta a-\Theta\infty}{\Theta ij-\Theta\infty} = \frac{1}{2}\left(\text{erf}\,\frac{b-z}{2\sqrt{\kappa t}} + \text{erf}\,\frac{b+z}{2\sqrt{\kappa t}}\right) \quad \text{------} \quad (1)$$

where thermal diffusivity of the soil is generally very small, at most $10^{-3}m^2/h$ order, so the temperature drop of the injected heat source over a half year period is,

$$\tilde{\Theta}a \equiv \frac{1}{b}\int_0^b \frac{\Theta a(z)-\Theta\infty}{\Theta i\quad -\Theta\infty}\,dz \simeq 0.7 \sim 0.8 \quad \text{------} \quad (2)$$

also shown in Fig. 1, where 2b = 19m, and $\frac{kt}{b^2} \simeq 0.04$. So this insulating condition causes surface water penetrating with yearly cyclic temperature to be constant. Namely, a constant underground temperature is the result of natural seasonal regeneration.

Man has always used wells, springs, etc. for drinking water supply, cooling and other purposes because of their good quality and constant temperature. But as we have not paid attention to recharging, we have suffered from land subsidence and sea water intrusion, not to mention water level lowering here and there. For this reason we must in future discharge these water sources paying attention to natural as well as artificial recharging.

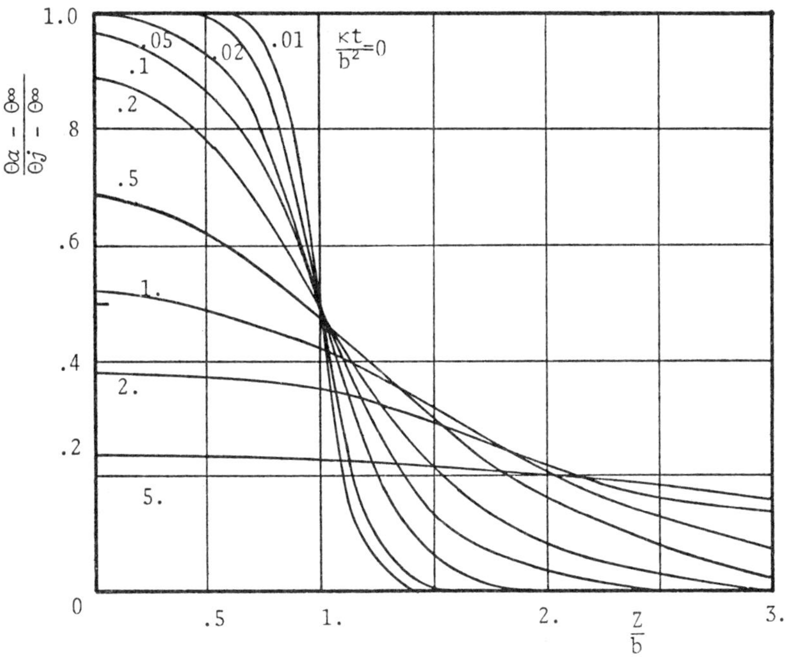

Fig. 1 Temperature Propagation with $\frac{\kappa t}{b^2}$ (Carslow, et al. [1])

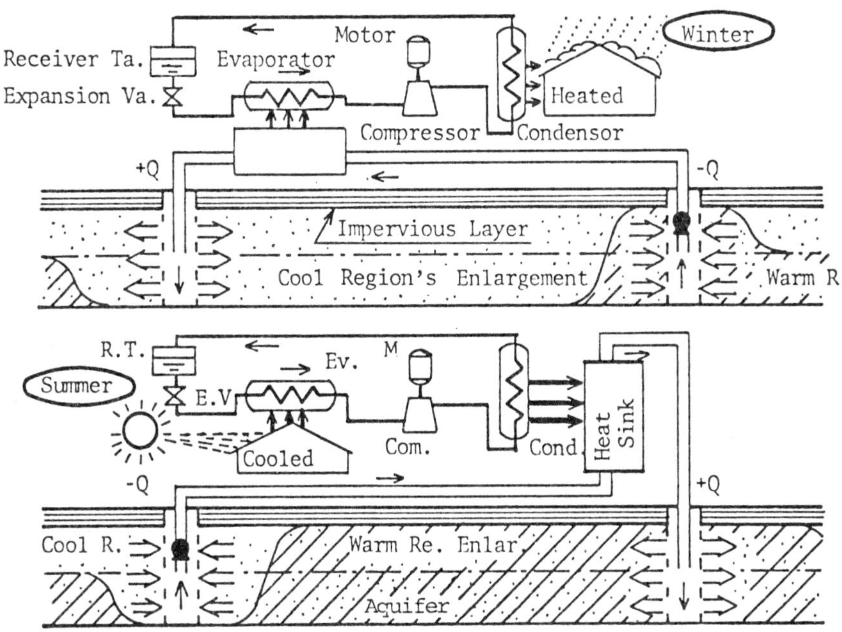

Fig.2 Scheme of Regenerative Heat Pump with Aquifer

From our idea, we discharge from one well and simultaneously recharge the waste water into the other well for storing of water. Adding unconcentrated each season's heat source spreading in field to recharge water, regenerating in the aquifer until available next season, finally we can use the recharged water and regenerated heat source or sink more effectively, for example as regenerative heat pump illustrated in Fig. 2.

But some problems resulting from artificial recharge arise. The first one is chemical pollution, and the second thermal pollution in the aquifers. The former can be prevented by purification, and with less effort in the case of waste water from closed system. But for the latter to pretemper recharge water equally to the native is no worth regeneration. Therefore the methods by which we prevent the thermal pollution and use the water for regeneration are that under the given natural condition such as natural flow, aquifer depth, properties of layers, etc., we establish suitable distance between the recharge well and the discharge well. For example, in the area of alluvial cones or basins where natural flow is not significant, thermal diffusion is not rapid and directed to pump up well. Therefore we need a greater distance, but can expect long-run effective regeneration during seasons. On the contrary, under the rapid natural flow cross to main well flow, thermal pollution can be easily cleared and we can use underground water as a constant heat source or sink.

In this paper, as the model field experiment on the above mentioned, we mainly testify that the seasonal regeneration is less affected by natural flow and thermal pollution by simultaneous recharge, on the supposition of our practical building heating by heat pump in winter, and direct cooling or if need by refrigeration in summer corresponding to our heavy snow fall area.

In addition to experiments, by the assumption of simple underground strata-model and its numerical analysis based on the divisions along with complex potential function, we try to simulate these seasonal regenerating phenomena.

2. PREVIOUS WORK

2-1 Artificial Recharge

In many seaside countries such as Netherlands, surface spread recharge is successful against sea water intrusion into coastal aquifers. In Europe, surface recharge is often used for purification of water quality management [2].

Other places where they are suffering from water shortage, underground is used for important water storage. Especially in Israel, water is resourced by well artificial recharge amounting to 10% of the annual fresh water production [3]. Most preventive matter is clogging around the well, and this clogging phenomena as to aquifers consisted of fractures and holes in rock such as limestone, sandstone, basalt, etc. was studied by Schwartz [4] both practically and with models.

In other countries such as Japan, rich in water resources but paying no attention to saving water source and conserve underground atrata, land subsidences and sea water intrusions in Quaternary deposit are occuring. Artificial recharge from wells is very useful as well as control of discharge amount, for it is direct water dilivering to the layer subject to subsidence. In fact, artificial recharge makes the piezometric surface up-lift as shown by Yokoyama [5]. Therefore at some places in Japan artificial recharge installations are working. In unconsolidated aquifer composed of alluvial deposits, artificial recharge has been widely studied by the American Geological Survey and also Japanese Geological Survey, then clogging phenomena of unconsolicated aquifer is cleared by Price [6], and a recovery method for clogged well proposed by Nagai, et.al. [7].

Consequently we reached to be able to recharge if the quality of water is not so different from the native one's and if the flow rate of recharge is not forced.

2-2 Thermal Pollution Followed by Artificial Recharge

If non-fresh water such as sewage is recharged, chemical pollution might appear. By the same taken, if the temperature is very different from the native, in time thermal pollution occurs especially in the case of well recharge. For different from geothermal we cannot expect any heat source in aquifer not mention to heat sink. From this reason we need to seat a suitable distance between the recharge well and the discharge well.

Under the parallel uniform flow caused by a group of injection wells of hot water into oil sand layer and the other group of production wells, Lauwerier [8] solved one dimensional unsteady state temperature distribution. Furthermore, Gringorton et.al. [9] derived approximate solution for any given flow pattern. Especially for pair wells of same strength of sink and source he grasped the non dimensional parameters dominating temperature propagation coinciding with that of Yokoyama, et.al. [10]. Furthermore, Yokoyama tested his analysis in practical experiments and in scaled down model experiments [11]. Therefore we can predict temperature propagation and estimate satisfactory distance of pair wells under simple conditions.

2-3 Seasonal Regeneration

Constant temperature of underground water, which means truly natural seasonal regeneration, has been used as the heat source of heat pumps or heat sinks of refrigerators. Due to wasting of water, however, we have lost the benefit of underground water.

For the purpose of replenishment of underground water, this wasted hot or cool water has been recharged, but has been given up because of thermal pollution. But from our opinion, recharge with yearly cyclic temperature, 1) from an exclusive recharge well, 2) from alternately exchanging wells, let us not only avoid thermal pollution but also make good use of these regenerated heat sources or sinks.

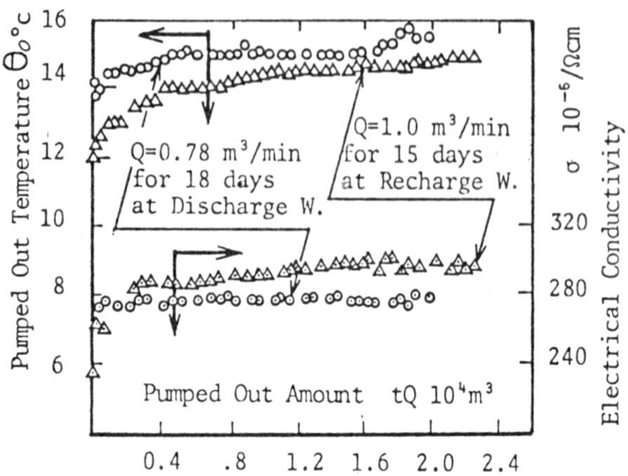

Fig. 3 Gained Cool Water after Regeneration during one Year

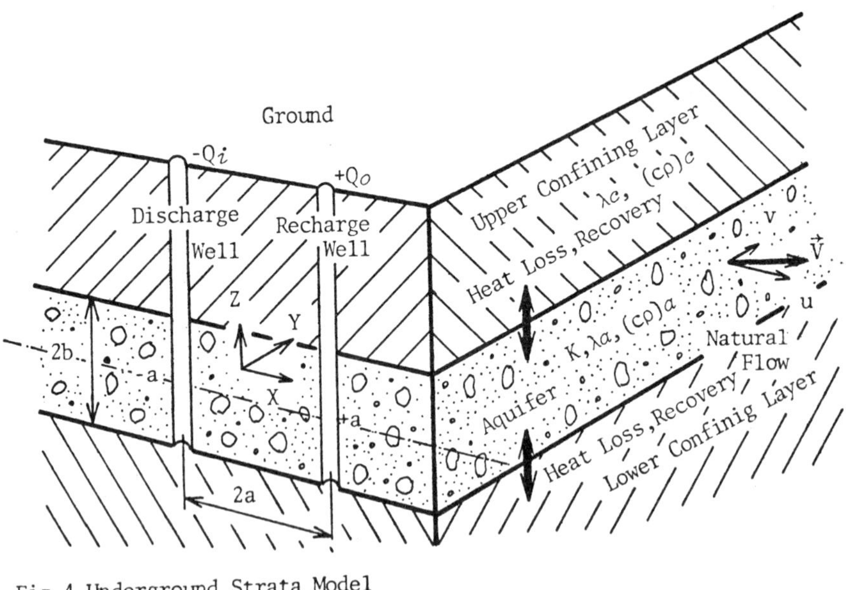

Fig.4 Underground Strata Model

In Israel 10% of all fresh water is obtained by artificial recharge mainly in winter, then the temperature of discharge water in summer is cooler than surface water. Yokoyama, et.al. repumped in summer $7^{o}c$ water recharged the previous winter, and after the regeneration during months got the discharge temperature shown in Fig. 3. But no one has experimented with effective seasonal regeneration with quantitative data and a reasonable theory.

Regrettably, Gringorton's appromimate solution cannot be applied to entire seasonal regeneration, for its initial condition is limited to uniform temperature and its discharge well boundary condition is constant, etc. Furthermore, a problem such as seasonal regeneration involves flow pattern change. Therefore it is difficult to predict seasonal regeneration through underground strata quantitatively.

3. ASSUMPTION AND MODELING OF UNDERGROUND STRATA, AND FUNDAMENTAL EQUATION

Practically, the underground strata of the alluvial epoch is spreading almost horizontally but at more or less non-uniform thickness. Due to the fact of that in confined aquifer, natural vertical flow, which is mainly caused by leakage of confining layers, is negligible compared with horizontal flow, in addition to the assumption of uniform thickness of a aquifer confined imperviously up and down as illustrated in Fig. 4, which is broadly adapted in hydrogeological analysis, we can use the following simple Laplace equation for potential field in confined aquifers.

$$\frac{\partial^2\phi}{\partial\chi^2} + \frac{\partial^2\phi}{\partial y^2} = 0 \quad \text{--------} \quad (3)$$

Under natural horizontal flow, when one sink of $Qi/2b$ strength corresponding to discharge well of Qi m^3/h is placed at (-a, 0), and one source of $QO/2b$ strength corresponding to recharge well of QO m^3/h is simultaneously placed at (a, 0) as illustrated in Fig. 4, water level distribution is given below,

$$h(\chi,y)\equiv K\phi = \frac{1}{K}(u\chi + vy) + \frac{Qi}{8\pi bk}\,ln[(\chi+a)^2+y^2] - \frac{Qo}{8\pi bk}\,ln[(\chi-a)^2+y^2] \quad \text{------} \quad (4)$$

Therefore, the complex potential function of this flow pattern is,

$$W = \phi+i\psi= sU + \frac{Qi}{4\pi b}\,ln(s+a) - \frac{Qo}{4\pi b}\,ln(s-a) \quad \text{-------} \quad (5)$$

Where $s=x+iy$ is complex viriable, and $U=u+iv$ corresponds to natural flow complex potential.

By the way, as the heat transfer phenomena in aquifers is dominated mainly by convectional term forced by well horizontal flow but negligibly by conduction and vertical natural convection term because of vertical small temperature difference distinguished from Sahrock [12], only horizontal two directions are considered satisfactory for thermal diffusion of aquifers. In additions, to estimate heat loss into confining layers or heat recovery from them, vertical heat conduction at the interface of aquifer and confining layers are considered also. Then we can assume uniform thermal properties adn the symmetry of vertical distribution, so the energy equation of aquifer is given below,

$$\lambda a\left(\frac{\partial^2\Theta a}{\partial\chi^2}+\frac{\partial^2\Theta a}{\partial y^2}\right)+c\rho\left(\frac{\partial\phi\partial\Theta a}{\partial\chi\partial\chi}+\frac{\partial\phi\partial\Theta a}{\partial y\partial y}\right)+2\lambda v\frac{\partial\Theta a}{a\partial z}\Bigg|_{z=b}=(c\rho)a\frac{\partial\Theta a}{\partial t} \quad \text{----} \quad (6)$$

In most cases, horizontal heat conduction term is negligible compared with the convectional term, as simulated by YOKOYAMA, et. al. [10]. Under the typical conditions such as same strength of sink and source $Qi=Qo$, stagnant natural flow U=0, uniform initial temperature of the aquifer, and constant recharging temperature, Gringorton, et. al. [9] gained approximate solution. Beside them, YOKOYAMA, et. al. [10] obtained independently non-dimensional relation, then ensured it by a series of numerical analysis. Consequently, these results combined by Gringorton and YOKOYAMA can be shown below and in Fig. 5,

$$\frac{\Theta a(\chi,y,t)-\Theta\infty}{\Theta o\quad-\Theta\infty}=G\left\{\frac{x}{a},\frac{y}{a},\frac{tQc\rho}{a^2 2b(c\rho)a},\frac{c\rho(c\rho)aQ2b}{\lambda c(c\rho)ca^2}\right\} \quad \text{----------} \quad (7)$$

Where each term on the right hand is a position relative to the well, non-dimensional time corresponding to apparent injected heat capacity ratio to overall heat capacity of aquifer, and coefficient of heat loss, respectively. In this equation, tQ indicates total recharged or discharged amount from 5=0, and tQ can be substituted by summation ΣtQ of each interval amount t_1Q_1, t_2Q_2, ---, as long as daily interval, namely,

$$\Sigma tQ = t_1Q_1 + t_2Q_2 + \text{---} + t_nQ_n \quad \text{--------} \quad (8)$$

But on the problem of un-equal strength of sink and source and with natural flow, furthermore, involving flow pattern changes such as our seasonal regeneration, we can hardly take in hand any approximate solution not to mention exact solution. From this reason, we try to solve fundamental equation (6) by numerical method.

While in confining layers where convectional heat transfer does not exist, three dim ensional heat conduction flow must be considered. Supposing that thermal properties are constant, the equation is given below,

$$\lambda c\left(\frac{\partial^2\Theta c}{\partial\chi^2}+\frac{\partial^2\Theta c}{\partial y^2}+\frac{\partial^2\Theta c}{\partial z^2}\right)=(c\rho)c\frac{\partial\Theta c}{\partial t} \quad \text{--------} \quad (9)$$

4. NUMERICAL ANALYSIS

4-1 Division Based on Complex Potential Function

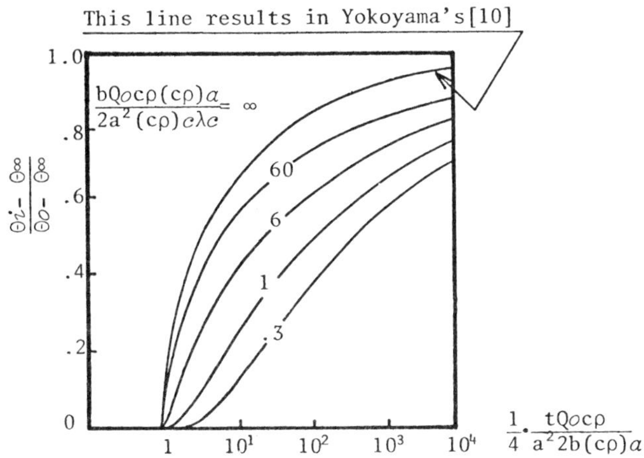

Fig. 5 Discharged Temperature by Gringorton [9]

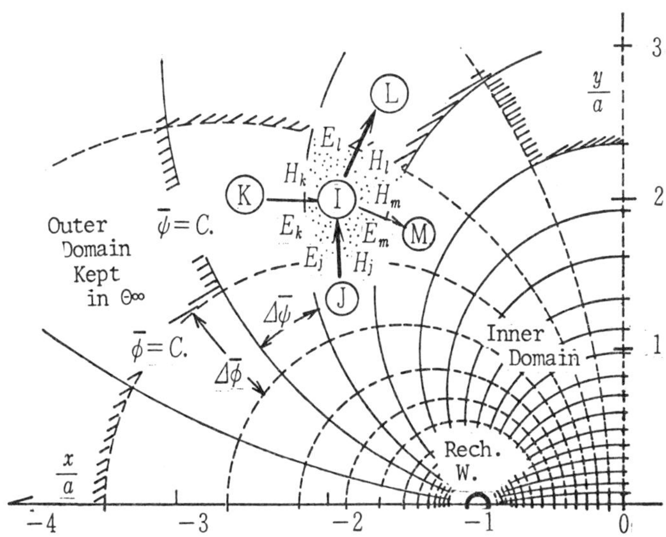

Fig.6 Mapping and Division Based on $\bar{W} = \bar{\phi} + i\bar{\psi}$

In the case of involving convectional term, Up Wind Difference Method is of considerable meaning in physics and stability as appointed out by ROACHE (13). Also in our equation (6), convectional heat transfer is dominating temperature diffusion. Therefore, when we simulate by Up Wind Difference Method, division is not simple equal division but based on complex potential function, $W=\phi+i\psi$.

KROEGER [14] used changeable conformal mapping based on the variable interface of solid and liquids and fixed surface, then he solved the two dimensional propagation problem of interface with high accuracy. Different from him, Gringorton, et. al. [9] obtained an approximate solution with coordinates based on complex potential functions.

Also in our numerical analysis, being based on the complex potential function of this flow, temperature propagation analysis is greatly simplyfied, and accuracy of numerical value is raised, furthermore we can shorten computing time, as appointed by YOKOYAMA, et. al. [11].

But when flow pattern changes are involved such as seasonal regeneration, we cannot avoid temperature replacing errors caused by sudden change of conformal mapping corresponding to flow pattern change. Therefore we had better fix one conformal mapping most suitable for every flow pattern. From this reason, we selected the next conformal mapping equation (10), as indicated in Fib. 6.

$$\bar{W} = \bar{\phi} + i\bar{\psi} = ln\frac{s-a}{s+a}, \qquad s=\chi+iy \quad \text{--------} \quad (10)$$

Where $\bar{W}$ is the complex potential function for same strength of sink and source at (-a, 0), (a, 0). In this case, surroundings of wells are divided into 44 cells, namely approximate equi-potential line $\bar{\phi}$ is divided into 44 parts, and the any space between wells, namely approximate flow line $\bar{\psi}$ are also divided into 44 parts.

By the way, as vertical heat transfer is simply only conduction, the number of vertical divisions is only 4 with equal span. Finally we divided inner domain of whole the strata into 44 x 44 x 4 ≃ 7700 cells.

4-2 Upwind Finite Difference Equation

We must prepare 2 types of equation,namely for the aquifer and for the confining layers. Among them equations of the aquifer have convectional term forced by potentional field eq. (4), then these are complicated. On the contrary, the other equations of confining layers have simple conducting term. Heat balance in un-equal division of the aquifer illustrated by Fig. 6 is,

$$A(i)a^2(c\rho)a\,\frac{\Theta a^{\Delta t}(i)-\Theta a(i)}{\Delta t}=Ej+Ek-El-Em+Hj+Hk-Hl-Hm-2Hc \quad \text{-------} \quad (11)$$

Where E indicates convectional term of which temperature is given upside of the potential flow, H indicates heat conduction flow, and subscript c indicates the component associated with confining layers.

E and H are estimated lineally by difference and distance between nodal points, but are modified to have true flux because of non-rectangular divisions. Then the equation of the aquifer by Up Wind Difference method is given by,

$$\Theta a^{\Delta t}(i)=\{1-Ma(i)\}\Theta a(i)+Ma(j)\Theta a(j)+Ma(k)\Theta a(k)+Ma(l)\Theta a(l)+Ma(m)\Theta a(m)+2Mc(i)\Theta c(i) \quad \text{--------} \quad (12)$$

Notice that the temperature coefficient M is varied by the situation and flow pattern, resultantly this equation corresponds to a heterogeneous and unisotropic heat conduction problem (YOKOYAMA, et. al. [11]).

In each, stability conditions are determined by the individual cell as below,

$$1 \geqq Ma(i) \quad \text{--------} \quad (13)$$

But the overall stability condition, namely step time Δt, is decided at the cells surrounding the discharge well.

4-3 Flow Pattern for Each Season

In this method, though any flow pattern can be considered, with our experiments, where one sink and one source are considered in addition to uniform natural flow, then the location of sink and source are exchanged alternatively and these strengths are variable according to each season.

4-4 Boundary and Initial Condition of Temperature

As illustrated in Fig. 7, vertical end planes are assumed to be kept in $\Theta\infty$. Here, vertical end planes are determined by one dimensional Carslow's solution eq. (1), sufficiently 2b away from aquifer.

Side planes are also assumed to be kept in constant $\Theta\infty$, and this assumption is practically correct. Then, these horizontal ends are determined along with $\bar{W}$, eq. (10), about 3a away from each well by test-run simulation.

Each recharge well for summer or winter is kept in constant Θs or $\Theta\omega$. Initial temperature over the entire domain, except recharge well, is to be $\Theta\infty$.

4-5 Computation

The flow chart of computation is illustrated in Fig. 8. At the first stage unequal division based on $\bar{W}$, eq. (10), the second, area A $[i]$, nodal point, etc. of each cell, the third, potential field Θ corresponding to each season, the fourth, the step time Δt, and the last, temperature coefficient $M[i]$ of each cell is given. Finally, an unsteady state temperature field is computed explicitly.

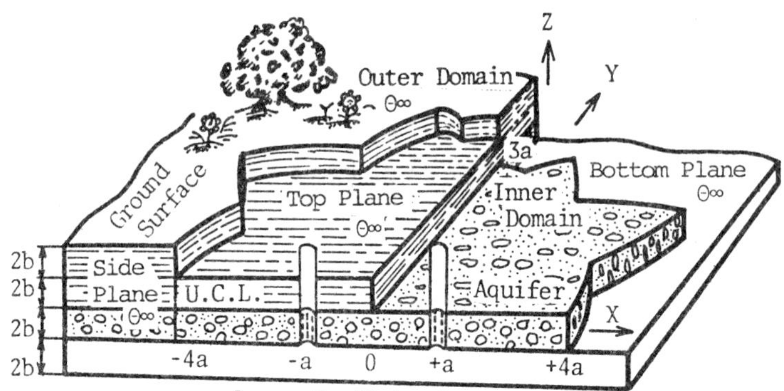

Fig.7 Simulation Domain and Boundary Condition of Strata

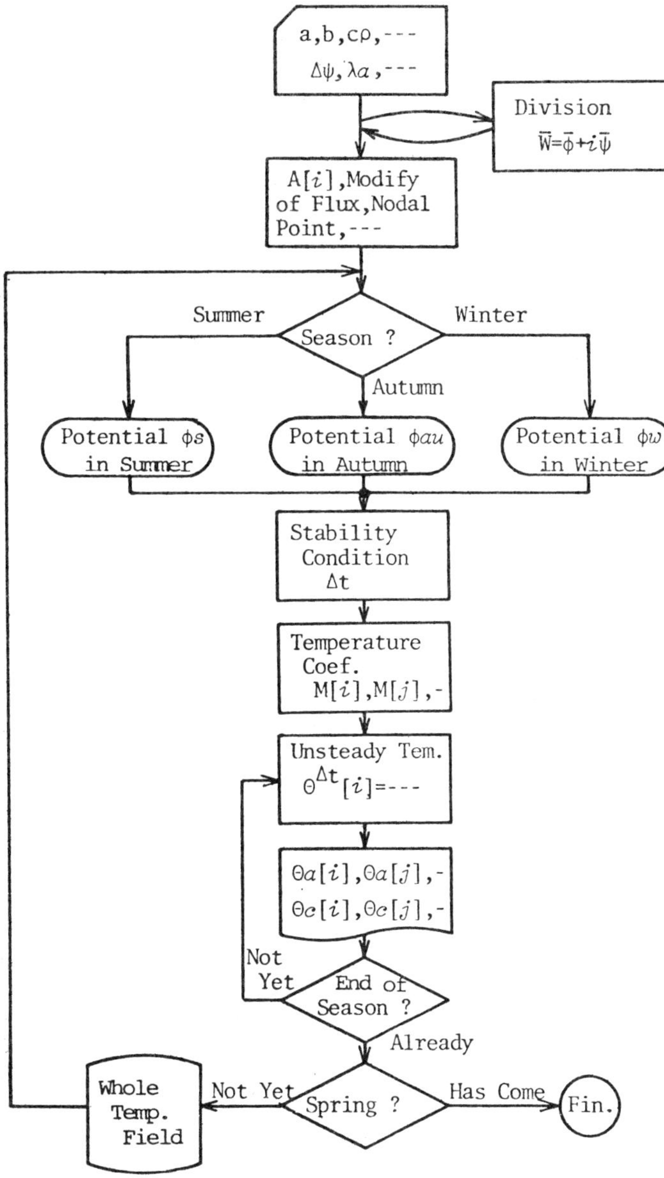

Fig.8 Flow Chart of Computation

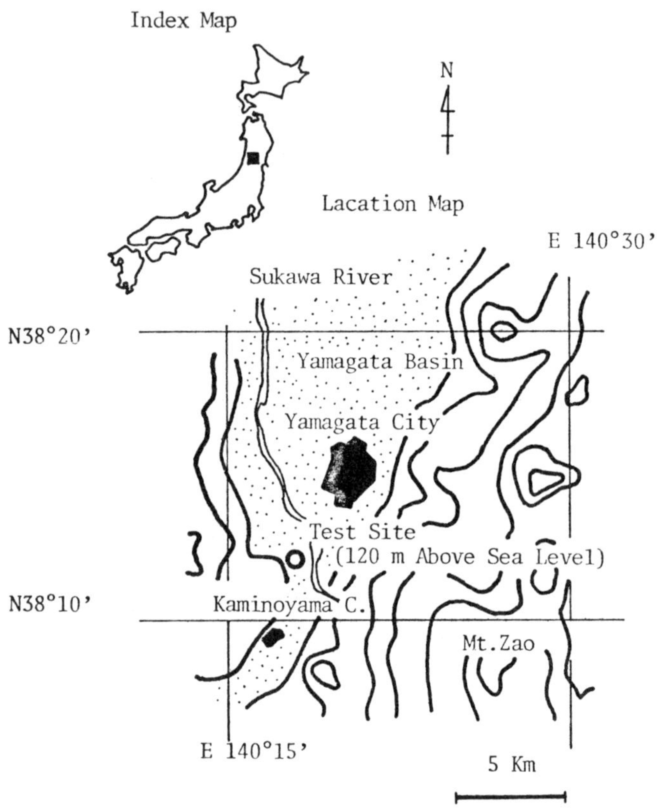

Fig.9 Test Site Location In Yamagata Basin

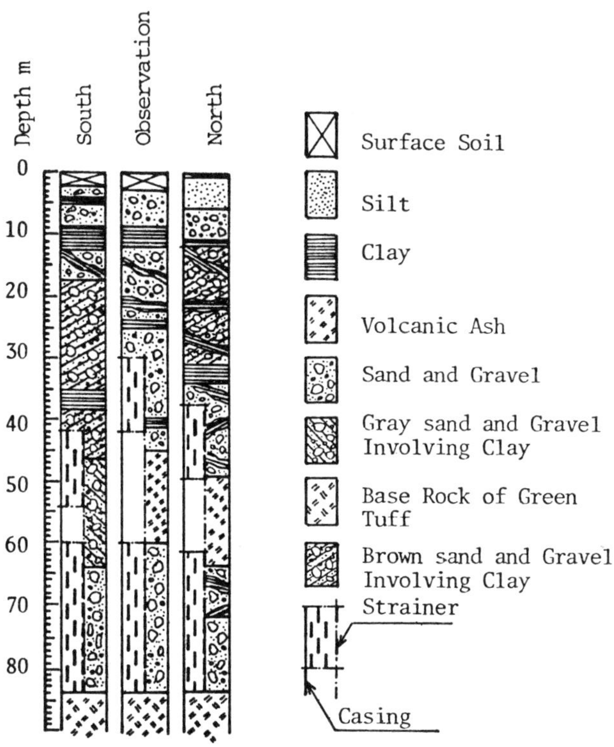

Fig.10 Geologic Columns of Wells at Nippon Chikasui Kaihatsu Co., Ltd. in Yamagata Basin.

5. EXPERIMENTAL FIELD APPARATUS AND ENVIRONMENTAL CONDITIONS

5-1 Natural Environment

Our test site is located in Yamagata Basin as mapped in Fig. 9 and surrounded by mountains. The underground strata is composed of alluvial deposits based on green tuff rocks of the tertiary period as shown in Fig. 10. Layers combining volcanic ash and mud flow originating from Mt. Ryusan. Layers are not uniform and as well there are other usual edge deposits of alluvial corn. The wells have two strainer parts but the main permeable aquifer is limited in sand and gravel aquifer near 80m. We therefore regard the mean thickness of aquifer to be 19m. Lower confining layer is surely a bed of green tuff, but the upper one is not a complete slit layer but somewhat leaky. Nevertheless, we can regard the permeability of such as, brown sand and gravel and volcanic ash, as less than 10^{-1} times that of the main aquifer. Furthermore for practical analysis, we should assume the aquifer is confined and no leakage exists.

The ground is close to the horizontal but slightly inclined toward the northwest. Besides the two dual purpose wells without extraordinary construction for recharge well, one observation well was drilled at the corner of the 22.4 m equilateral triangle formed by the three. Natural underground-water flow is estimated as almost constant as illustrated Fig. 11, by the three wells' static water levels in 1974 and 1977, and by permeability measuring of its aquifer tests. That is,

$$U = u + iv = -6.7\text{m/y} - i16\text{m/y} \quad \text{--------} \quad (14)$$

which is a suitable value for this area.

Our test building, NIPPON CHIKASUI KAIHATSU is not pile mounted on the land, but floats. It consits of three floors with a total area of 1000 m^2. The ground and first floors are offices for about 50 men and then the second floor is often used for meetings.

5-2 Climate of This District

The average yearly temperature is 11°c and precipitation is 1200 mm of which 300 mm is snow in winter. In August the average temperature is 24.5^{o}c, and a Summer day, defined as days when the highest daily temperature is more than 25°c is 100 days per year. In January teh average temperature is -1.2^{o}c, and a Winter day, defined as a day when the lowest temperature is less than 0°c total 110 days. Therefore we need heating for about 5 months and cooling for 2 months. For example, though our building requires 160 ℓ/day x 5 months/y heavy oil, namely 1.5 x 10^8 Kcal/y considering boiler efficiency 0.7 for heating, we require cooling also.

5-3 Requirements of Warm Water and Cool Water Regarding Heat Pump and Direct Cooling

In the case of temperature lowering of underground water from 20°c to 10°c

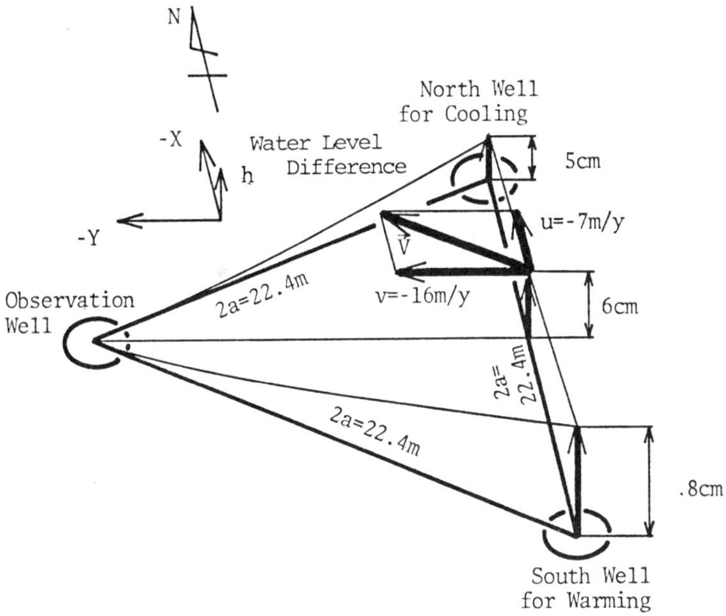

Fig.11 Well Situation and Estimated Natural Flow
where $K=1.6\times10^{-4}$ m/s

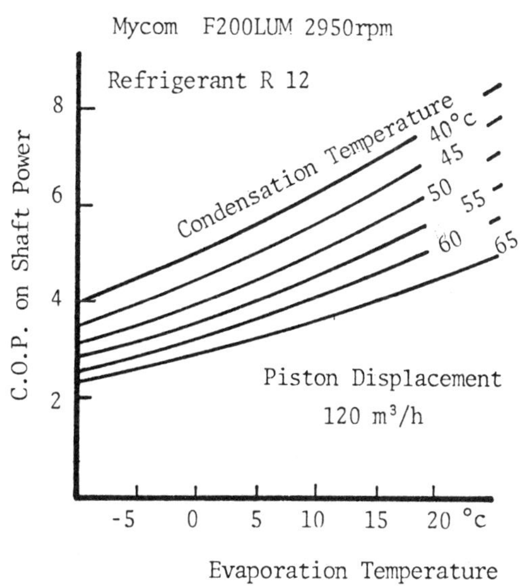

FIG. 12 COEFFICIENT OF PERFORMANCE OF HEAT PUMP

we can set heat pump evaporative temperature at 5°c and condensing temperature at 50°c. We then can gain a coefficient of performance based on shaft power of more than 4.0 as indicated in Fig. 12. Therefore running costs are cheaper than for an oil heater. The requirement for warm water totaled 9400 ~ 11000 m^3 for winter heating source, which supplied 3/4 of the required heat 1.5 x 10^8Kcal, and the remaining 1/4 is given by heat pump itself.

By the way, presumed heat loss is 20 ~ 30% by conduction plus further added natural flow out, etc., so perhaps a total of 50% of initial temperature difference (Θs-$\Theta \infty$) may be lost. Then we should recharge about 10000 m^3 with 24°c.

But the well distance is too close to avoid thermal pollution more than 4000 m^3 recharge amount, for 4000 m^3 recharge amount corresponds to $tQc\rho/2a^2b(c\rho)a$ = 3, as indicated in Fig. 5. Avoiding it, we must gain 36°c hot water and decrease total amount to 4000 m^3. But, beside our experimental limitation, to verify thermal pollution, we recharge about 10000 m^3 at 24°c in easy heating manner.

On the contrary, if 10°c cool water is gained in summer, we can cool directly with the temperature increasing from 10°c to 16°c. Therefore the requirement for cool heat is,

$$0.023 \text{ USRT/m}^2 \times 6 \times 10^2 \text{ m}^2 \times 10 \text{ h/d} \times 2 \text{ months} = 0.62\text{x}10^8\text{Kcal} \quad \text{----} \quad (15)$$

where 1 USRT = 3024 Kcal/h. Then the requirement for cool water is,

$$\frac{0.62 \times 10^8 \text{ Kcal}}{6^{o}\text{cx}10^8\text{Kcal/m}^{3o}\text{c}} = 10^4 \text{ m}^3 \quad \text{--------} \quad (16)$$

Then for the purpose of experiment on thermal pollution, we recharge about 10000 m^3 total amount with 6°c in winter on the assumption of an identical heat loss phenomena.

In this case amounts of recharge water are balancing, and regenerated heat is also almost balancing favourably.

5-4 Experimental Apparatus and Measurements

As shown in Fig. 13, cooling water is pumped from the north well and, after direct cooling, waste water is sprindled on the roof for heat collect by convection and radiation without any ingenious equipment. Flowing through the filtration tower packed with fine sand and active carbon, water is additionally heated by a heat exchanger, and then recharged from another south well not forced. (In winter, vise versa as shown later)

As the power of the submerged pump was too strong, about 8 Kw, discharged water was heated negligibly but, in the case of a discharging rate of less than 100 l/min., to the negligible extent of about 1°c. Flow rates are measured by the volume type flow meters, and water levels of wells are measured by probe-sensor type level meters. Water quality is inspected by elec-

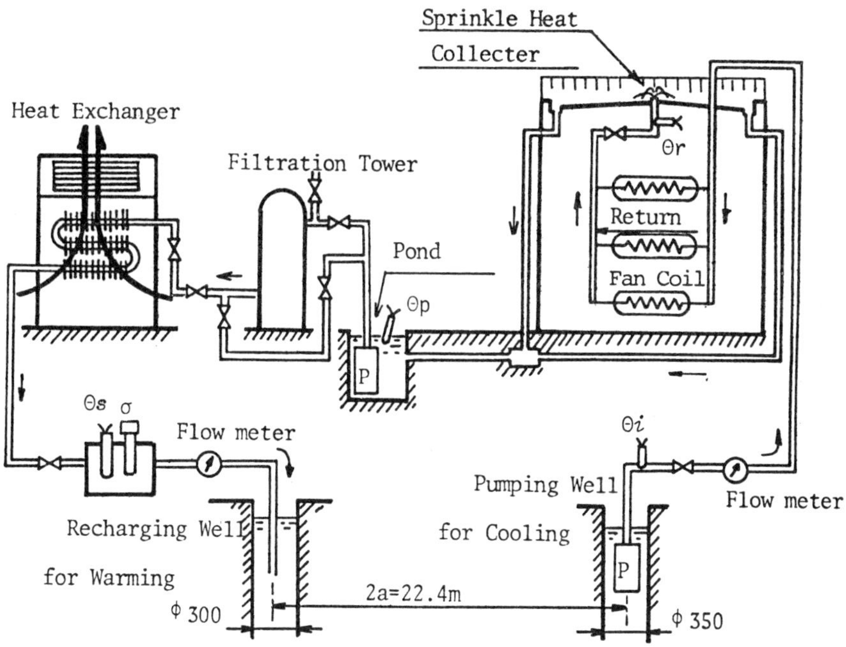

FIG. 13 MODEL APPARATUS IN SUMMER

Table 1 Material Properties and Conditions for Simulation

$(c\rho)_a$:	660 Kcal/m^3°c	,predicted by measuring properties of resembling aquifer samples
$(c\rho)_c$:	300 Kcal/m^3°c	,refer to that of JSME*
$Q_o/2b$: summer;	0.303 m^2/h	,mean value of total amount divided by the total hours of each season, including intervals between experiment
:autumn;	0	
:winter;	0.255 m^2/h	
$Q_i/2b$: summer;	0.340 m^2/h	,same above mentioned
:autumn;	0.0147m^2/h	
:winter;	0.304 m^2/h	
u :	-0.18×10^{-2}m/h	,observation
v :	-0.76×10^{-3}m/h	,observation
λ_a :	0.5 Kcal/mh°c	,refer to that of JSME*
λ_c :	1.0 Kcal/mh°c	,refer to that of JSME*

* JSME, Material of Heat Transfer, 1973, p.259

tric conductivity meter and sampling checks are carried out monthly. Temperature at any point is measured with 0.5 mm C-C or Ar-Cr thermo-couples.

Data is gathered into the laboratory and recorded. Climate data such as atmospheric temperature, dew point, overall isolation, etc. are obtained from the Yamagata Meteorological Observation Center. It is situated several Km. away from our test site, so we can assume the same weather conditions. Besides the main pump, another small pump of 1 Kw. for drinking and living matters was included in the north well until autumn with about 3 m^3 daily discharge. In winter, this small pump is moved to the south well.

6. WARM WATER RECHARGE AFTER COOLING IN SUMMER

6-1 Recharging and it's effect

Experiments were carried out from July 16 to September 18 usually in the daytime, but sometimes at night. Experimental conditions and material properties are listed in Table 1.

Recharging is very favorably going and total amount is 8843 m^3 but which is a little less than total amount of discharge 9930 m^3 and that of plan, as shown in Fig. 14. This is caused by that the amount of discharge contains daily 3 m^3 of living water and backwash water for the filtration tower.

In spite of daily discharging, the water level recovery after experiments is smooth and we have kept water level constant as indicated in Fig. 15 due to the effect of recharging.

Clogging of aquifer is not in appearance and apparent permeability by equation (4) is not decreasing but remains constant as shown in Fig. 16. Namely, filtration tower trapped most organics (algae) and detritus, so we needed back wash daily. Quality of the recharge water has been kept nearly natural through all seasons and suitable for drinking. Only zinc (Zn), but was increased from 0.1 ppm (natural) to 0.6 ppm (recharge) by flowing pipes within allowances for drinking.

6-2 Heating Pattern

As shown in Fig. 17, heat exchanger is useful at night when there is no isolation, but cannot increase temperature far over dew point much less to atmospheric temperature. On the contrary, the sprinkling, which is simplest, can heat water warmer over dew point under isolation. In addition to this, from another experiments of heating with only exchanger but except sprinkling, absorption of isolation is ensured to be more important for heating rather than the convectional heating. Nevertheless we have recharged 8843 m^3 with 23.7°c mean temperature, close to the expected 24°c value.

6-3 Stored Heat Energy and Thermal Pollution

Stored heat energy heated from 16°c are shown daily in Fig. 18, and total 0.672×10^8 Kcal (6720 l of oil) with temperature up from 16°c to 23.7°c

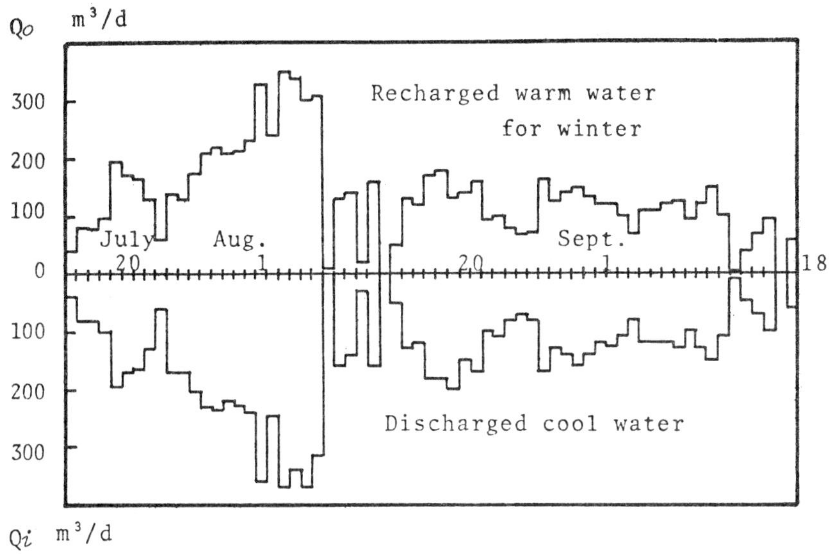

Fig.14 Recharge and Discharge In Summer,1977

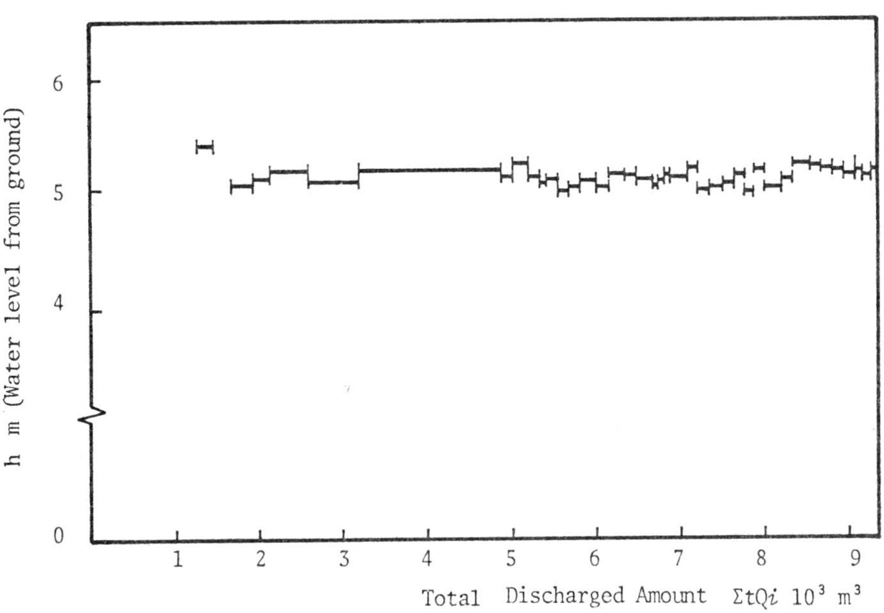

FIG. 15 RECOVERY EFFECT OF WATER LEVEL AFTER PUMPING,IN SUMMER (1977)

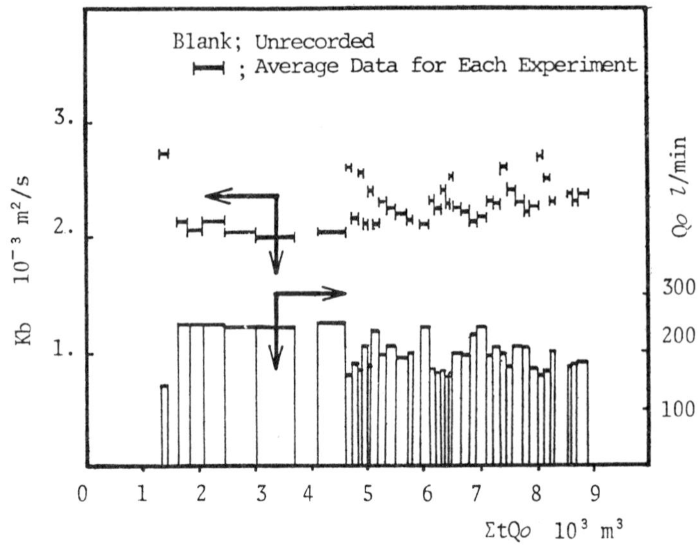

FIG.16 CONSTANT CORRELATION APPARENT PERMEABILITY & TOTAL RECHARGED AMOUNT, IN SUMMER (1977)

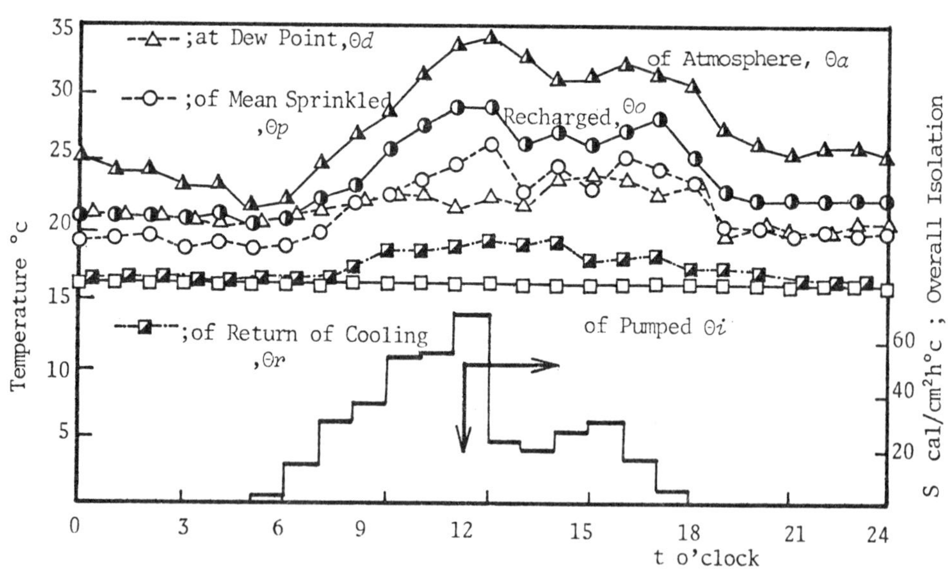

FIG. 17 TYPICAL TEMPERATURE CHANGE PATTERN THROUGH AUGUST 5 (1977)

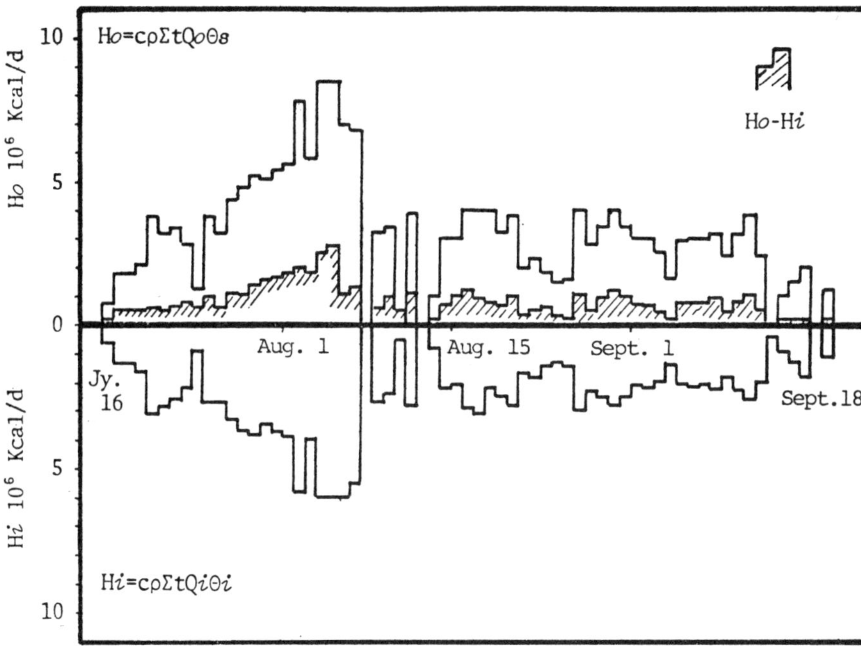

FIG.18 Regenerated Heat In Summer(1977)

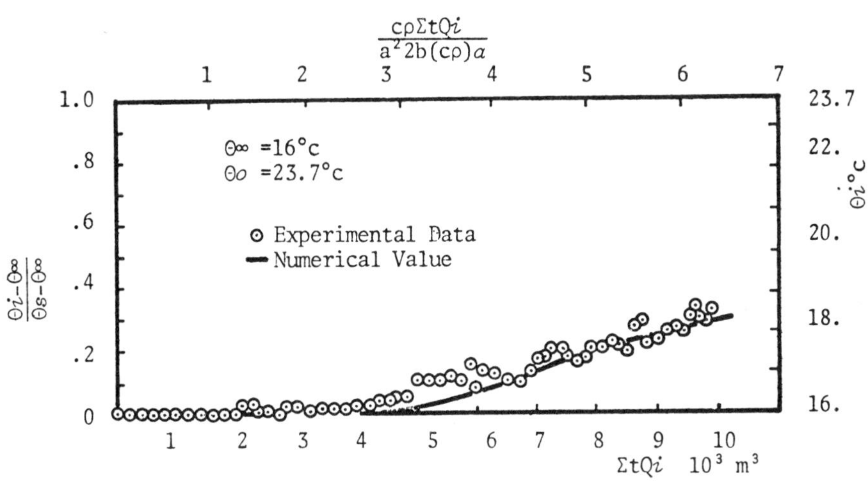

Fig.19 Thermal Pollution into Discharge Well, Summer, 1977

over the summer. If temperature drop until winter is within 50%, corresponding to 20°c from 23.7°c, it is expected to supply 0.88 x 10^8 Kcal heat source of heat pump using the heat of lowering from 20°c to 10°c, but a little less than 1.1 x 10^8 Kcal of the plan.

Thermal pollution followed by warm water recharge have appeared to discharge well of north cool well, as indicated in Fig. 19, since the time of 4000 m^3 discharged amount as predicted by previous equation (7) and Fig. 5.
In spite of fluctuation, experimental data coincided with the numerical value analyzed under the same conditions as experiments and with the rough assumptions for underground strata.

7. REGENERATION DURING AUTUMN

In this period from September 19 to December 25 recharge is not going but regeneration phenomena is experimented. While living water at least with daily 3 m^3, totaled 660 m^3, is discharged from north well.

From the numerical analysis, the temperature of the aquifer is lowering slowly day after day by heat conduction to confining layers and diffusion on the natural flow or by discharge for living water. By the way, since the arrival of natural cool water with Θ_∞=16°c, temperature is suddenly lowering in acceleratation, especially at upstream of natural flow, namely south-east area around south well. Therefore the temperature at south well is lowering in acceleratation since November 10, while the temperature at the north well is slowly decreasing yet as indicated in Fig. 20. But area with nearly 23.7°c still remains around the downstream area between wells.

Experimental data is not satisfactory to compare with numerical analysis, for temperature of discharge, which is measured in tank, is heated not only while the passing through pipes in the building but also while staying in tank under the sun-shine, owing to small flow rate. Nevertheless experimental data are at most 20% higher (about 1°c) than numerical value and have same tendency of lowering with time as numerical analysis.

8. RE-PUMPED WARM WATER IN WINTER AND RECHARGE OF WASTE COOL WATER

8-1 Model Apparatus in Winter

Inversely to that of summer, we re-pump seasonaly regenerated warm water from the south well, then recharge waste cool water into north the well as shown in Fig. 21.

Instead of heat pump, we use warm water for snow melting on the roof with a 300 m^2 area. The nozzles of spray type are replaced for the sprinkler type. Therefore sprayed water is satisfactorily cooled without heat exchanger. The other parts of apparatus and measurements are same as for summer.

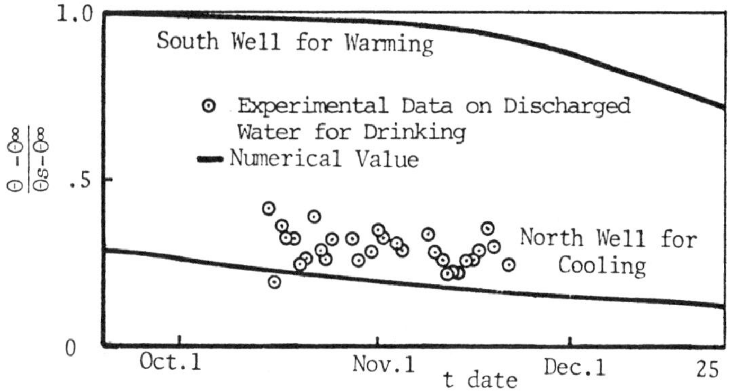

Fig.20 Regeneration During Autumn(1977)

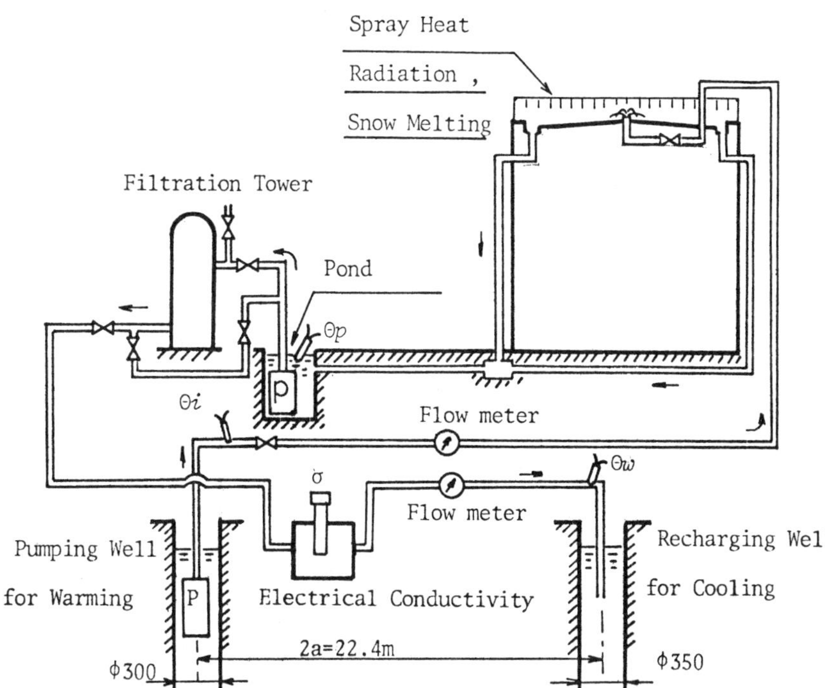

Fig.21 Model Apparatus in Winter

8-2 Re-pumped Warm Water and Recharge of Waste Cool Water

Warm water has been re-pumped from south well for snow melting since December 26, daily 150 m^3, totaled 11200 m^3 including living water. In return, waste cool water with 5.3°c mean temperature has been simultaneously recharged into the north well, daily 140 m^3, totaled 9430 m^3, favorably as indicated in Fig. 22 but with slightly additional viscous resistance due to low temperature.

As shown in Fig. 23, with 20°c at the first, we have gotten warm water with about 19°c till the end of January corresponding to 4000 m^3 discharged amount. Though re-pumped temperature in numerical analysis has lowered in acceleration affected by recharged cool water since the time corresponding to 5000 m^3 re-pumped amount, in the experiment the temperature has not lowered as acceler-ately as the numerical value, but only slightly. The disagreement at the end of season looks to be caused not only by non-uniformity of the aquifer but also by underestimating heat loss into confining layers or heat recovery from these layers. Nevertheless the numerical value coincided with experi-mental data within a 10% error overall the seasons. So this numerical analy-sis with division based on complex potential function and with rough assump-tions is suitable for long period and wide space heat conduction problem dominated by convectional term such as seasonal regeneration through under-ground strata.

8-3 Recovering Heat Energy and Suitable Distance of Well

Until the time when the discharging well is affected by recharged cool water, re-pumped water temperature and its heat energy have been kept at least 40% of that original temperature difference ($\Theta s-\Theta\infty$) or original injected heat energy in the previous summer, which can be called Recovery Coefficient, in spite of about half year regeneration. But since arrival of recharged cool water, the recovery coefficient has been decreasing as indicated in Fig. 23. Resultantly we have gotten warm water at 18.3°c mean temperature overall re-pumped amount, more than recharged amount in summer, then 0.26 x 10^8 Kcal among injected heat 0.672 x 10^8 Kcal in summer, namely 38%, is recovered. The warm water with 18.3°c, totalled 12000 m^3, corresponds to 0.91 x 10^8 Kcal for heat source of heat pump of the case of temperature lowering from 18.3°c to 10°c; nearly this heat energy is satisfactory for our objective amount of 1.1 x 10^8 Kcal.

So we propose the suitable distance of wells to be not affecting distance yet from recharge well, which is estimated by equation (7) and Fig. 5 as testified already in summer. That is,

$$\frac{\Sigma tQc\rho}{a^2 2b(c\rho)a} \leqq 3 \quad \text{--------} \quad (17)$$

For example, at our test site, let ΣtQ 11000 m^3, substituting material pro-perties of this test site listed in Table 1, then suitable well distance well distance without affecting is from eq. (17)

$$2a \geqq 40 \text{ m} \quad \text{--------} \quad (18)$$

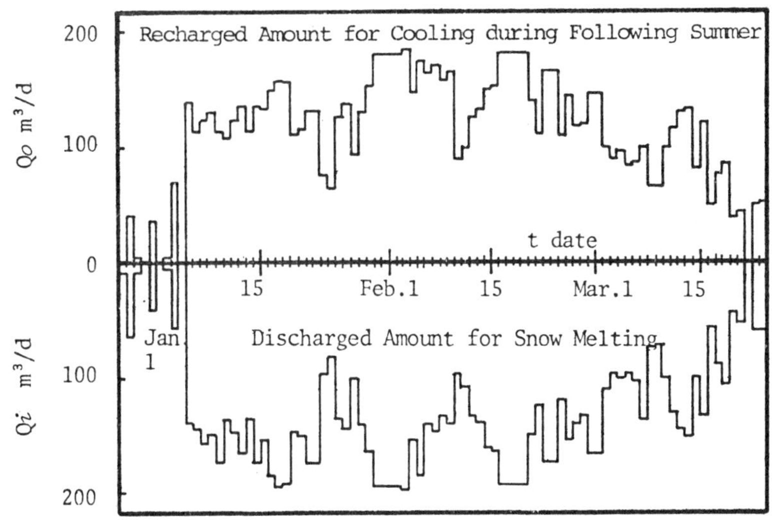

Fig. 22 Daily Discharged and Recharged Amounts, Winter, 1978

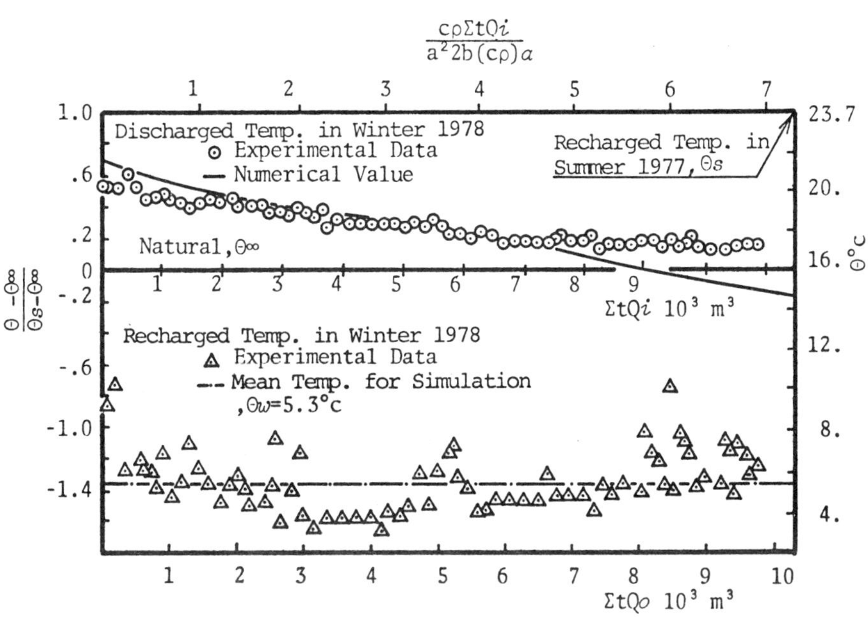

Fig.23 Discharged Warm Water and Recharging Cool Water after Snow Melting, Winter, 1978

In this case, if heated temperature in summer is higher than 23.7°c, the distance is less than 40 m.

9. DISCUSSION

Residual heat in the aquifer and confining layers is partly diffusing or directly mixing with the natural cool water and diminishing its temperature, then partly might be recovered in next winter.

If natural flow is more significant than the present, seasonal regeneration is not effective but we can more easily avoid thermal pollution. Vice versa being more stagnant natural flow we can more effectively regenerate seasonaly and so get recovery coefficient more than 40%, but we must pay more attention to thermal pollution.

By the way this recharged cool water in winter is expected to be re-pumped with 10°c within 4000 m^3 next summer, and used for direct cooling not to mention for heat sink of refrigerator.

Furthermore if high temperature water is originated by Solar Heating or refrigerator in summer, for example 50°c, we can get hot water of more than 30°c in winter even if the surface is covered with snow, or even if Solar Heating System is not used due to absence of climate. Therefore Seasonal Regeneration is very useful for saving energy and can be applied for many aspects especially in Temperate Zone.

10. CONCLUSIONS

Though the experiment is not sufficient for generalization and for quantitative analysis, recovery coefficient of nearly 40% coinciding with our numerical analysis is almost satisfactory to indicate that the seasonal regeneration through underground strata is possible and expected to many applications.

We have arrived at following conclusions.

(1) Underground water can be successfully used as a low concentrated heat source without bad influences due to shortage of underground water if its waste water is recharged. Supposing that well distance 2a is chosen as below under the given conditions such as thickness of aquifer 2b, etc.,

$$\frac{\Sigma tQc\rho}{a^2 2b(c\rho)a} \leqq 3 \quad \text{--------} \quad (17)$$

then its thermal pollution into pumping well can also be avoided. And over this distance, the most effective seasonal regeneration can be expected.

(2) In the field where slow natural flow, compared to main well flow, horizontally and vertically thermal diffusion are hardly appearing, therefore we can seasonably regenerate enormous amounts of heat sink or source wasted by heat pump, originated by Solar Heating, etc.

(3) In spite of rough assumptions for underground strata, numerical analysis based on division along with complex potential function carries on higher accuracy and less efforts for long period and wide space thermal problem dominated by convectional term.

NOMENCLATURE

English

a	:	2a = distance between discharge and recharge well	m
A	:	Aa^2 = Area of each cell	m^2
b	:	2b = Thickness of aquifer	m
$c\rho$	:	Heat capacity of water	$Kcal/m^3{}^\circ c$
E	:	Convectional heat transfer rate of each cell	Kcal/h
G	:	Temperature function	-
h	:	Water level	m
H	:	Conduction heat flow rate of each cell	Kcal/h
K	:	Hydraulic conductivity	m/s
M	:	Temperature coefficient of each cell	-
Q	:	Well flow rate	m^3/min
s	:	$= x + iy$, complex variable	m
S	:	Isolation	$cal/cm^2h^\circ c$
t	:	Time	h
u	:	x direction component of natural underground water flow	m/y
U	:	$= u + iy$, complex potential of natural flow	m/y
v	:	y direction component of natural underground water flow	m/y
W	:	$= \phi + i\psi$, complex potential function	m^2/h
(x, y, z)	:	Three dimensional Cartesian coordinates	m

Subscript

a	;	of aquifer
au	;	of autumn
c	;	of confining layer
i	;	of sink, namely discharge
j	;	of injection, namely recharge
i, j, k, l, m	;	of cells of I, J, K, L, M
o	;	of source, namely recharge
p	;	of sprinkling
r	;	of return of cool water
s	;	of summer
w	;	of winter
Δ	;	step value
$-$	;	of approximate complex potential function
∞	;	infinitive condition

Greek

Θ	:	Temperature	°c
κ	:	Thermal diffusivity	m^2/h
λ	:	Thermal conductivity	Kcal/mh°c
ϕ	:	Potential function	m^2/h
ψ	:	Flow line function	m^2/h
σ	:	Electrical conductivity	$\mu/\Omega cm$

REFERENCE

(1) Carslow and Jaeger, Conduction of Heat in Solid, Oxford University Press, p. 68, (1946)
(2) Klaus R. Inhoff, Water Quantity and Quality Management in the Ruhr Valley, Journal W.P.C.P., vol 46-7, July 1974
(3) J. Schwarz, Artificial Ground Water Recharge in Israel, Belgian-Israel Symposium on Groundwater Quality Control and Management, November 1974
(4) J. Schwarz, Clogging Processes Resulting from Artificial Recharge with Lake Kinneret Water, Belgian-Israel Symposium on Groundwater Quality Control and Management, November 1974
(5) Takao Yokoyama, Hiromichi Umemiya and Hiroto Abiko, Heat Exchange of Storage Type through Underground Strata by Artificial Recharge, Japanese Asociation of Groundwater Hydrology, vol 17-2, September 1975 (Japanese except Abstract)
(6) Price, Don, Hart, D.H., and Forworthy, B.L., Artificial Recharge in Oregon and Washington, U.S. Geol. Survey Water Supply Paper 1594-c, 1962
(7) S. Nagai and Murushita, Clogging and Renovation of Artificial Recharge Water Well, Water for Industry, vol 170, 1972 (Japanese)
(8) H.A. Lauwerier, The Transport of Heat in an Oil Layer Caused by the Injection of Hot Fluid, A.S.R., A-5, p.145, 1955
(9) A. C. Gringorton et J. P. Sauty, Exploritation de L'Energie Calorifique des Aquifers, Bereau de Recherchis Geologique et Minieres, sgn/ame-JPS/GP/356, 1975
(10) T. Yokoyama, H. Umemiya, K. Katsuragi, et. al., Thermal Pollution Followed with Artificial Recharge (Simultaneous Recharge and Discharge with Same Amount), Japanese Asociation of Groundwater Hydrology, vol 17-3, December 1975 (Japanese except Abstract)
(11) T. Yokoyama, T. Sasaki, et. al., Model Experiment and Numerical Analysis for the Purpose of the Cyclic Heat Generative through Underground Strata, Refrigeration, Vol. 52-602, December 1977 (Japanese except Abstract)
(12) V. E. Shrock and A.D.K. Laird, Physical Modeling of Combined Forced and Natural Convection in Wet Geothermal Formations, ASME, Journal of Heat Transfer, May 1976
(13) P. J. Roache, Numerical Solution of Laminar Separated Flow, AIAA. J., 8 (1970)
(14) P. G. Kroeger and S. Ostroh, the Solution of a Two Dimensional Freezing Problem Including Convection Effects in the Liquid Region, Int. J. Heat and Mass Transfer, vol 17, 1974

STORAGE TANK EFFICIENCY AS SIMULATED IN A MARKOVIAN MODEL OF METEOROLOGY

R. LESTIENNE
CNRS - PIRDES
282, bd Saint-Germain 75007 PARIS

ABSTRACT

Our aim is to identify and analyse, using a very simple mathematical model, the main characteristics of the behavior of a storage tank coupled to a solar power plant with fluctuating daily production. The positive correlations between the daily production at successive days is accounted for by using fundamental two states first order markovian process, ruled by the value greater or lower than 0,5 of the ratio R of the insolation duration to the astronomical day length, and by a production density probability which is exponentially decreasing for "bad" days and gaussian for "fine" days. The efficiency of the storage, as mesured by the ratio of the solar energy actually used to satisfy the demand to the total energy needs, is examined as a function of the storage capacity, in the hypothesis of a constant daily demand and in typical meteorological conditions corresponding to the southern part of France. Rules for the best compromise between storage cost and energy savings are proposed and illustrated by a specific example.

INTRODUCTION

In order to account for the existing correlations between the type of wheather at days d and d + k in a simple way, pratical (although certainly not perfect) for solar energy applications, we have proposed [1] to use a two states first order markovian model of meteorology. The two states are labelled f ("fine wheather") and b ("bad wheather"). The model is completely determined by the two transition probabilities

$$\alpha = \Pr\{f(d+1) \mid f(d)\}$$

and

$$\delta = \Pr\{b(d+1) \mid b(d)\}$$

which are the probabilities of fine (bad) wheather at day d + 1 knowing that the wheather is fine (bad) at day d. Of course we have also the transition probabilities $f \rightarrow b$ and $b \rightarrow f$, the value of which is $\beta = 1 - \alpha$ and $\gamma = 1 - \delta$ respectively.

An inspection of the insolation data at several sites in France have led us to adopt for criterium to distinguish between "fine" and "bad" days the ratio R of the insolation dura-
.../...

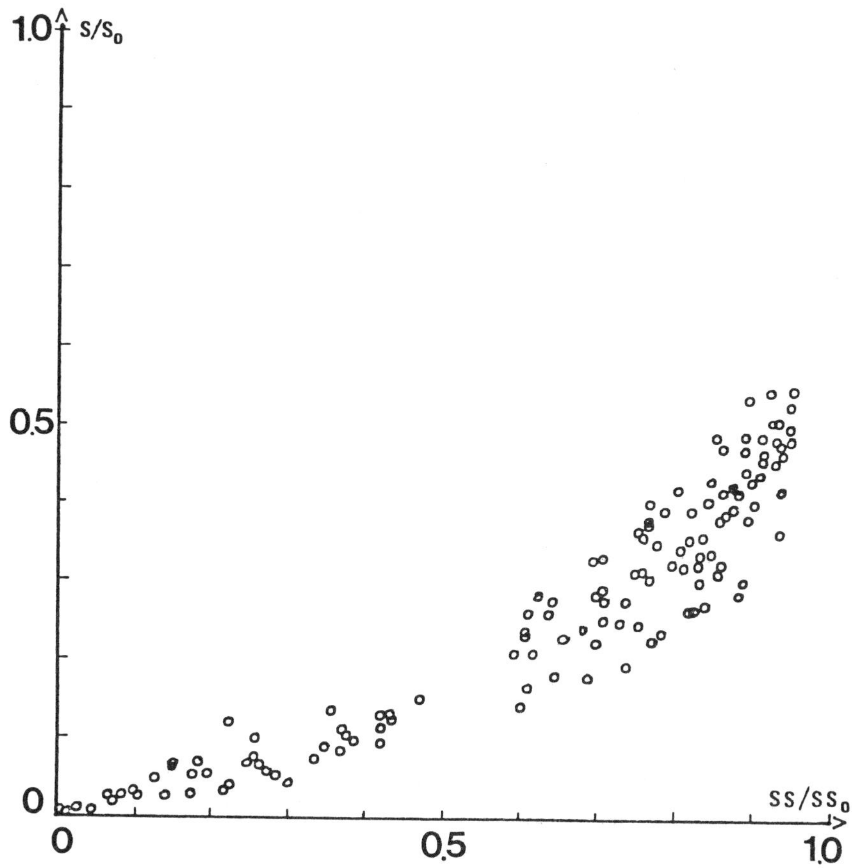

Figure 1. Relative insolation duration vs direct daily radiation. The data are taken in september at Carpentras [1]

tion to the astronomical day leugth. For R $\geqslant$ 0,5 the state of wheather is said "fine", and "bad" otherwise. Note that this cut does not imply that the meteorological variables assume only two values.

Our analysis has also shown that the α and δ parameters are varying according to seasons. For the purpose of the present study, we shall use the values α = 0,7 and δ = 0,5 as representative of typical mean yearly wheather conditions in the southern France. This corresponds to a moderate correlation between the type of wheather at consecutive days: C_1 = 0,2.

Using these data, we shall examine the behavior of the energy storage tank of a solar plant, assuming furthermore that the daily energy production of the plant is proportionnal to the direct radiation collected during that day.

CORRELATION BETWEEN DIRECT RADIATION AND INSOLATION

This correlation has been studied by M. VACHAUD and coworkers using a six years sample at Carpentras (Southern France) [2]. The figure 1 shows a typical two dimensional diagram of relative insolation vs relative daily direct radiation, both variables being relative to their maximum astronomical value. The correlation between these variables is very strong. The probability densities of daily direct radiation in the case of "bad" or "fine" days are therefore very different and display a very small overlap between them. Indeed (figure 2) in the first case the observed repartition has an exponential shape, continuously decreasing from its maximum value at S/So = 0, while in the case of fine days one observe a gaussian shaped distribution, whose average value $\langle S/So \rangle$ is about 4 to 6 times greater than the average value observed in the case of bad days. The monthly repartitions in the two cases were observed to be compatible with each other within a given season, and therefore added (starting december 1st) in the figure 2.

Adopting as unit of daily direct radiation the mean value $\langle p \rangle_b = b$ of direct radiation collected at a day of bad wheather, the mean value $\langle p \rangle_f = f$ and standard deviation σ_f of the distribution relative to the direct radiation collected at a day of fine wheather are :

.../...

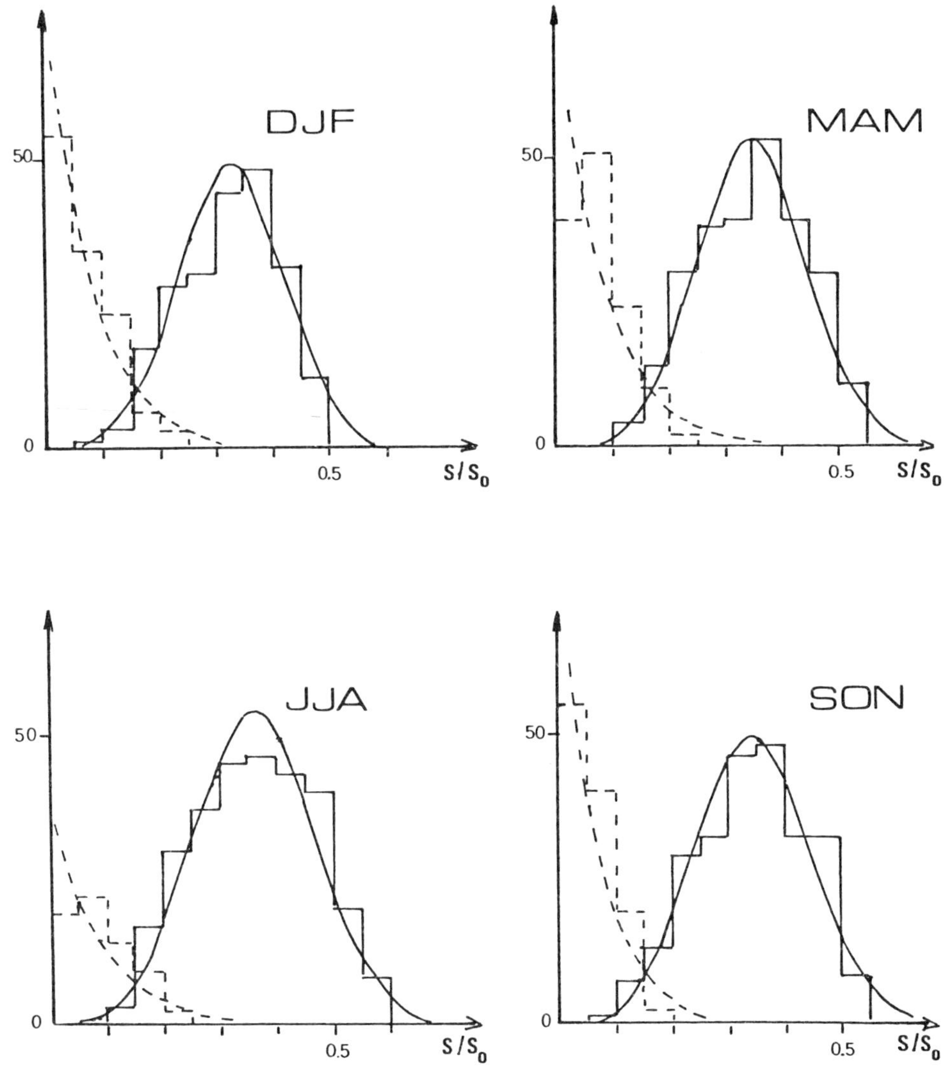

Figure 2. Repartition of the daily direct radiation as observed in days of fine wheather (full lines) and bad wheather (dashed lines) at Carpentras. The curves are the best gaussian (respectively exponential) fit to the data.

Season	f	σ_f
DJF	4,56	1,23
MAM	4,36	1,21
JJA	4,05	1,19
JON	5,57	1,62

SIMULATION OF THE DAILY PRODUCTION OF A SOLAR PLANT

As stated above, we assume here that the daily production of the solar plant is proportionnal to the daily direct radiation. This is not understood as a realistic assumption but as scholar assumption for the purpose of identification of the main characteristics of the storage behavior. Adopting the values $\alpha = 0.7$, $\beta = 0.5$, $f = 4.80$ b, $\sigma_f = 1.20$ b, we have generated random series up to 10000 days ruled by the Markov's process above defined. At bad days, the production p of the plant was sorted according to the law :

$$P_r(p)\,dp = \frac{1}{b}\exp(-p/b)$$

while at fine days, it was sorted according to :

$$P_r(p)\,dp = 1./(\sigma_f\sqrt{2\pi})\exp(-p^2/2\sigma^2).$$

We have checked that the repartition of daily production thus obtained was compatible with these laws and with the predicted percentages and clustering of fine and bad days. These repartitions are quite similar to those observed in figure 2.

THE STORAGE TANK BEHAVIOR

Let us now adopt as unit of energy the daily demand, which we assume to be constant. Let So be the maximum capacity of the storage tank (in this unit) and S (d) ($0 \leq S(d) \leq S_o$) the stored energy in the tank at the beginning of the day d. Let p (d) the production of the solar plant during that day. In general, the stored energy at the beginning of day d + 1 is :

$$S(d+1) = S(d) + p(d) - 1 \qquad (1)$$

except if the r.h.s of (1) is $<$ o or $>$ So

- if it is $<$ o, we add to the deficit record table its absolute value D (d) = $|S(d) + p(d) - 1|$, which is the extra supply energy that one has to afford in order to satisfy the demand. Thus one has S (d + 1) = o.
- if it is $> S_o$, the energy produced in excess is lost. Thus in this case S (d + 1) = So.

The efficiency η of the storage tank is defined as $\eta = 1 - x$, where x is the limit, for infinitely long series, of the ratio between the total deficit $D = \Sigma D(d)$ and the total energy produced $P = \Sigma p(d)$.

The figure [3] shows the variation of x with the storage capacity S_O, for several values of the mean daily production π = b $\beta/(\beta+\gamma) + f \times \gamma/(\beta+\gamma)$ [1] $\simeq 3.375$ b, taking into account the numeral values above mentioned for β, $\delta = 1-\gamma$ and f. A glance at this figure shows that the behavior of the

the storage tank efficiency is quite different, whether the mean daily production π is greater or smaller than the daily demand (=1) :

- for $\pi <$ 1, any stored energy is rapidly used to satisfy the demand, so that the capacity of the storage tank has little influence, if any, on x.
- for $\pi > 1$, the energy produced in excess with respect to the demand is cumulating in the storage tank, where it constitutes a reserve for the case of long series of days with insufficient production. The probability of such a series being roughly exponentially decreasing with its length, x (So) is roughly exponentially decreasing. Such a behavior is indeed suggested by the analogy with the problems of gambler's ruin, , when the gambler with an initial stake So plays against the bank.

A closer inspection of the figure [3] shows also that :

1 - The fluctuation of the daily production about the mean value b (in the case of bad weather) or f (in the case of fine wheather) lowers the efficiency η of the storage by a multiplicative, roughly constant factor, without altering the shape of the function x (So). The dotted lines show the variation x (So) obtained when one fixes the daily production to b or f according to the wheather.
According to the previous remark, we shall reason in the following in the frame of a non fluctuating daily production.

2 - x is maximum at S_o = o. In this case, one has to afford by fuel or otherwise, the energy needed for satisfying the demand, each time that the produced energy is insufficient. In case of bad wheather the shortage is 1 - b, and in case of fine weather, 1 - f if and if only $f < 1$. Thus :

$$x\,(So = o) = \frac{\text{mean daily shortage}}{\text{mean daily production}}$$

$$= \frac{\beta}{\beta+\gamma}\;\frac{1-b}{\pi} \quad \text{if } b<1<f$$

$$= \frac{\beta}{\beta+\gamma}\;\frac{1-b}{\pi} + \frac{\gamma}{\beta+\gamma}\;\frac{1-f}{\pi} \quad \text{if } b<f<1$$

3 - If $b < f < 1$ the storage tank is empty every evening whatever the wheather and whatever the storage capacity might be, apart from a possible transient initial stage. Thus x (So) is constant with S_o

4 - If $b < 1 < f$ but $\pi < 1$, the tendancy is to deplete the storage tank. In the stationnary state the energy stored during a sequence of fine days (mean length $1/\beta$ cf. [1]) is used for suppling the energy lacking in the sequence of bad days that follows (mean length 1/). Thus for So $\gg 1/\gamma \cdot (1-b)$

$$x \simeq \frac{\frac{1}{\gamma}(1-b) - \frac{1}{\beta}(1-f)}{\left(\frac{1}{\gamma}+\frac{1}{\beta}\right)\pi} \qquad (2)$$

So that x (So) is decreasing from the value x(o) above defined towards the assymptotic value (2).

5 - For $b < 1 < f$ but $\pi > 1$, the tendancy is to accumulate energy in the storage tank, and the behavior of x (So) approaches the above mentionned exponential law.

6 - For $1 < b < f$, x (So) = o $\forall$ So.

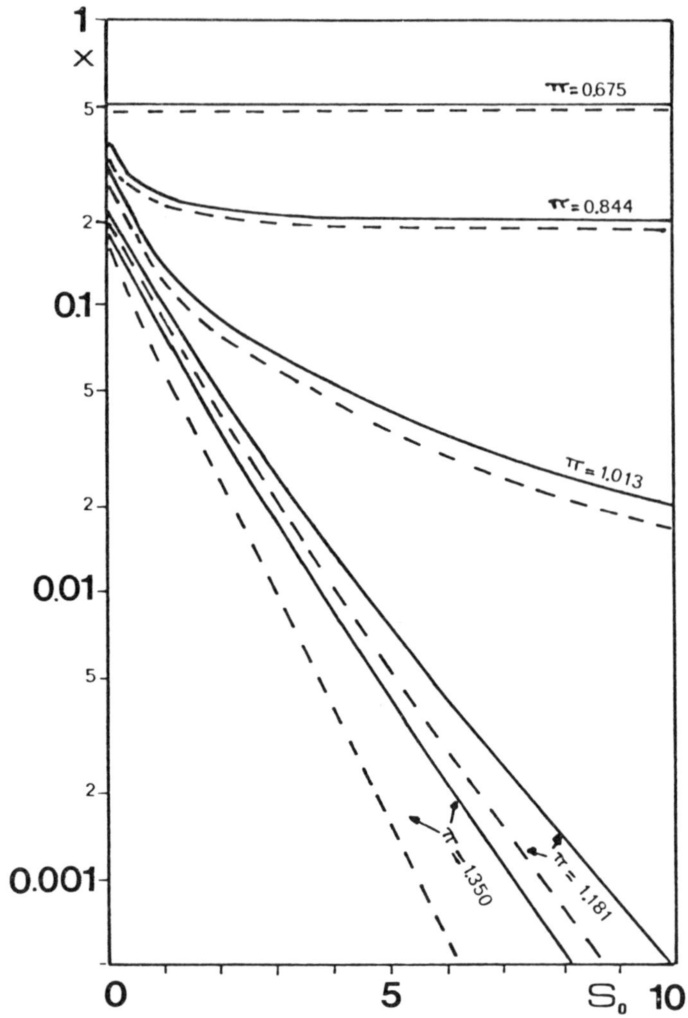

Figure 3. Relative deficit x as a function of the storage capacity S_0 in various cases of average daily production π (in units of daily demand). The full lines curves refer to fluctuating daily production, and dashed lines to fixed daily production in a given wheather state.

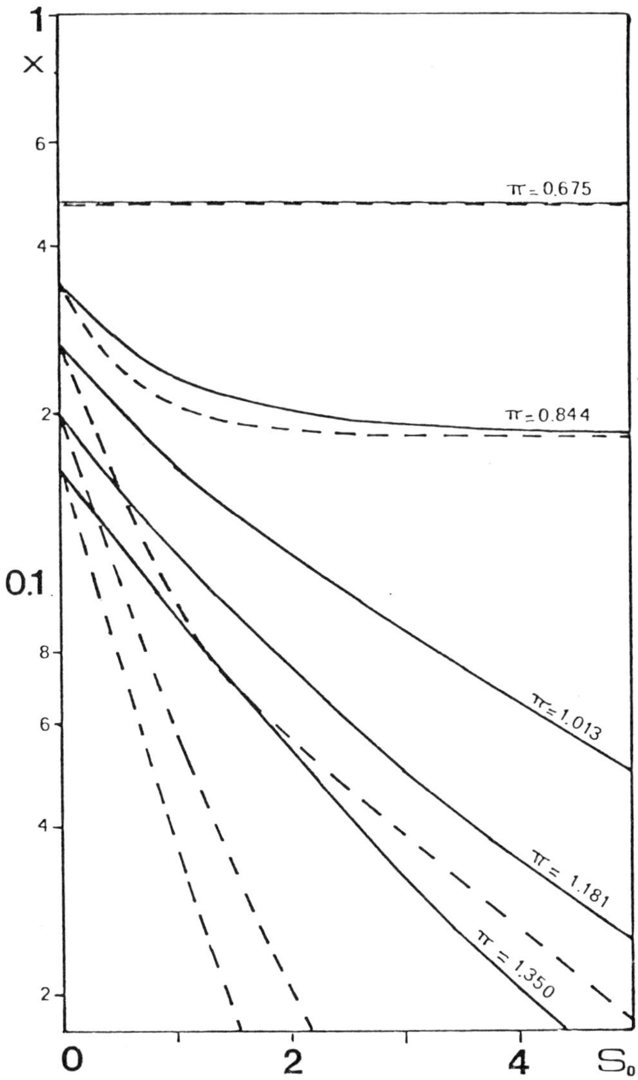

Figure 4. Relative deficit x as a function of the storage capacity S_o for two values of the wheather correlation between successive days : C_1 = 0.5 (full lines), C_1 = 0.0 (dashed lines).

EFFECT OF THE CORRELATIONS BETWEEN SUCCESSIVE DAYS

The correlation between successive days obviously does not influence the value of x when no storage is available (equation $x(S_o = o)$ above)
It does not influence either the assymptotic value (2), at fixed percentage of fine days, as a straight forward calculation shows it. In all other situations, the positive correlation coefficient between successive days introduced by the Markov process enhances the relative deficit and decreases the storage efficiency. To display this effect, we have drawn in figure 4 the relative deficit for $\alpha = .625$, $\delta = .375$ on one hand and $\alpha = .8125$, $\delta = .6875$ on the other hand, values which represent two climatic conditions with the same probabilities of fine or bad days and the same daily average production, but with much different correlations between successive days ($C_1 = o$ and $C_1 = 0,5$ respectively).

OPTIMUM SIZE OF THE STORAGE TANK

To determine the economic optimum of a storage tank in a given climatic condition is on extremely complex task, because one has to incorporate in the calculation many parameters that can influence the result : thermal losses in the storage, structure and fluctuations occuring in the production and in the demand within the day, etc...

For the sake of mathematical exercice, we have studied the following situation :

- constant daily demand equal to 1.4 during the first five days of the week, and 0 otherwise. In applications of industrial heat this will be a rather frequent situation, because the demand will be zero or low during the week end, when the factory does not work, allowing the system to refill the storage tank during these days. This "week end effect" on the efficiency of the storage tank may be seen in figure 5 , where the curves x (So) of figure 3 (no effect) are drawn as narrow continuous lines and those obtained with weakly demand modulation are drawn as dotted lines. The relative deficit x(So) is enhanced at low S_o values, due to the increase of the demand from 1 to 1.4, but this increase is rapidly balanced by the week end refilling for So $\sim$ 3 to 5.

- no storage tank is perfectly energy-tight. A reasonnable value of daily energy leakage for good storage tanks in future applications of industrial solar heat might be 10 % about. The effect of this energy leakage may be seen also in figure 5 , where the curves x(So) obtained under this condition are drawn as dashed lines. The effect of energy leakage is of course to limit the gain of efficiency at large S_o values.

The combination of the two preceeding effects still enhances the relative deficit and flatten the curve x(So) (thick continuous lines).

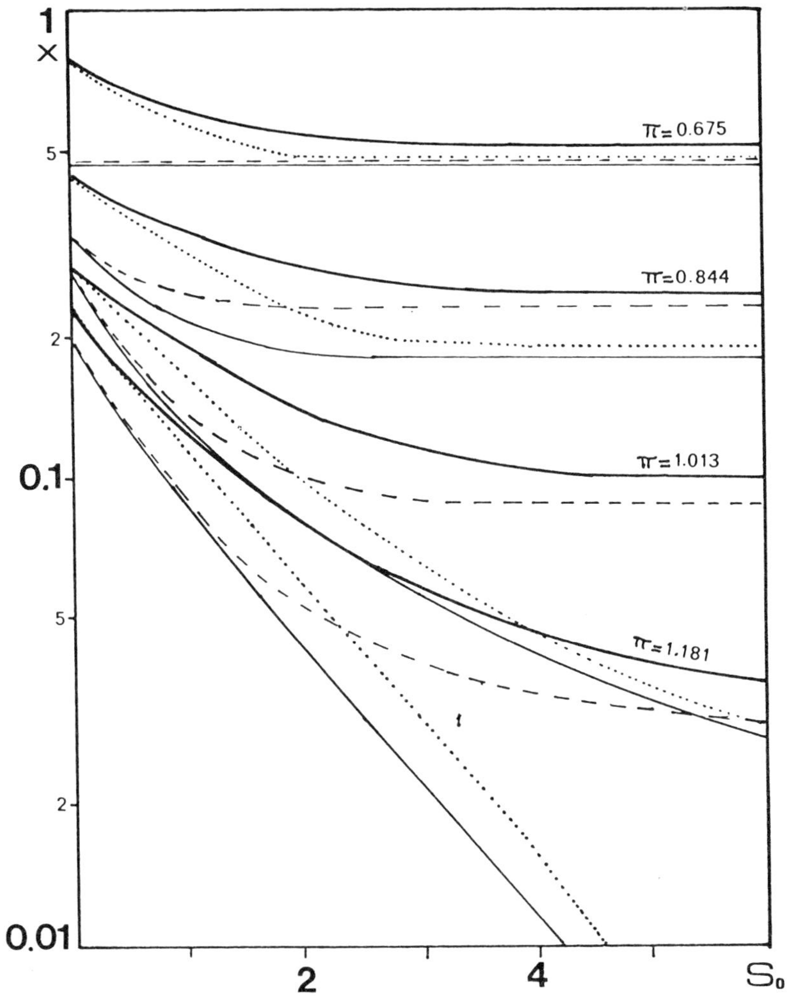

Figure 5. Relative deficit vs the storage capacity So affected by "week end effect" (dotted lines), "leakage effect" (dashed lines) and combined effect (thick lines) as compared to the values of figure 3 (narrow lines).

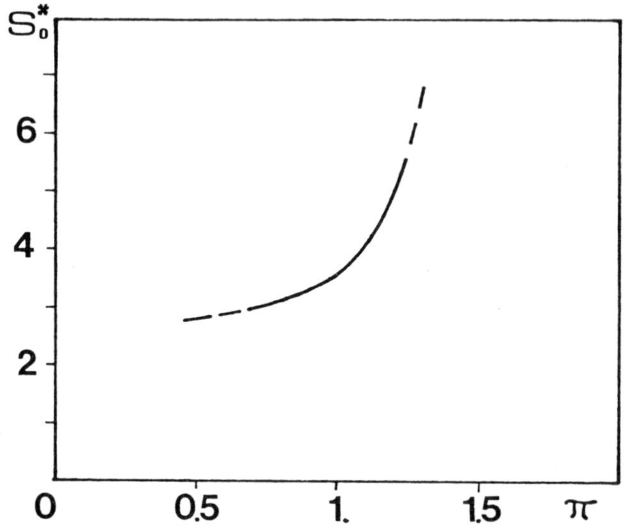

Figure 6. Example of the variation of the optimum size of the storage tank capacity S_0^* as a function of the mean daily production π (in units of mean daily demand) around $\pi = 1$ in the conditions stated in the text.

We have also incorporated in the exercise some economical considerations, namely :

- a life-time T of the storage tank equal to 15 years.
- a ratio R between the installation cost of one kWh of storage capacity to one kWh of relief fuel equal to 20 [3]
- a monetary actualisation rate of 10 %, and an increase in value of the relief fuel of 1 % a year [4], so that the actualised expenditure over 15 years of relief fuel is $\sum_{i=0}^{14} \frac{(1.01)^i}{(1.10)^i} \simeq$ 8.82 times the first year expenditure.

Note that, on the other hand, we are neglecting the maintenance cost of the storage tank.

Under all the above mentioned conditions, one obtains an optimum size of the storage tank S_0^* given by :

$$\left(\frac{dx}{dS_0}\right)_{S_0^*} \simeq - \frac{R}{3220} \simeq 0.0062$$

The solution of which is increasing from $S_0^* \sim 3$ to $S_0^* \sim 5$ when π is increasing from 0.675 to 1.181 (figure 6).

FOOTNOTES AND REFERENCES

1 - R. LESTIENNE, Modèle Markovien simplifié de météorologie à deux états ; l'exemple d'Odeillo. La Météorologie (to be published).

2 - P. BOIS, J. GOUSSEBAILE et M.J. MEJEAN, private communication.

3 - This will be the case, for instance, if the today installation cost of a storage tank is estimated at 40 $ par kWh capacity, whereas the today cost of the relief fuel at the site is 2 $ /kWh.

4 - For other works using the same values, see for instance the MITRE Corporation, METREK Division documents.

SOME ASPECTS OF ENERGY STORAGE SYSTEMS USING GLAUBER SALT MIXTURES

D. Y. S. Lou and A. J. Harrison
University of Delaware
Newark, Delaware 19711 U.S.A.

ABSTRACT

An investigation on the phase-transformation process of the Glauber salt mixture has been performed. Results on the visual observations of the crystallization process, measurement of the crystallization rate and its dependence upon the degree of undercooling are reported in this paper. Based on experimental investigation, it is concluded that the crystallization process is limited by the rate of thermal diffusions from the growing interface.

I. INTRODUCTION

This paper describes the work to investigate the thermal and physical properties of the Glauber salt heat storage mixture developed by Dr. M. Telkes[1,2]. The work includes visual observations ot the crystallization process, measurement of the crystallization rate and its dependence upon the degree of undercooling.

The mixture under investigation[3] consists of Glauber salt ($Na_2SO_4 \cdot 10H_2O$) (88% by weight) a thixotropic thickening agent (an attapulgite clay) (~9% by wt.) and borax (~3% by weight). The Glaubers salt is the phase change T.E.S. material. It partially melts in a peritectic transformation at 32 C (90 F) with a heat of transformation 250kJ/Kg (108 Btu/lb). The partial melting results in a saturated aqueous solution containing 36% by wt. of Na_2SO_4 and excess anhydrous sodium sulfate. The anhydrous sodium sulfate, being denser than the saturated solution, segregates out and is thus unable to recombine readily with the solution on cooling. This naturally reduces the heat storage capabilities of the salt. The thixotropic thickening material is added to produce a gel which inhibits settling of the anhydrous particles on melting, enabling their reconversion to Glauber salt and the recovery of the stored heat of fusion. The borax crystals are added as a suitable nucleating agent to minimize supercooling of the mixture. The borax is denser than the saturated solution and is also supported by the gel on a finely dispersed scale.

The peritectic reaction which occurs on solidification is not fully understood and the development of a basic understanding of this and of the physical aspects of the behaviour of the gel will probably be a vital step in optimizing the performance of the mixture. The crystallization and thermal properties mutually determine the rate at which heat can be input to or recovered from the mixture and are important in the design and sizing of heat exchange and container systems. Some data are available for the crystallization from saturated solutions of Glauber salt[4] and for the thermal conductivity of solutions of various compositions[5]. We are concerned with how these are modified in the presence of the thickening agent. It will be shown that the solidification process is distinctly different in the thickened material, and it is obvious that even if the thermal conductivities of thickened and unthickened materials do not differ drastically, convective heat transfer in the thickened material will be considerably reduced.

II. EXPERIMENTAL PROGRAM

Crystal Growth Observations on an Unthickened Glauber Salt Solution

Glauber salt crystallizes from a super-saturated solution as colorless, long, needle-shaped, monoclinic crystals which may be 2 cm or more in length. Using a cell of cylindrical cross-section and with radial heat flow, Glauber salt crystals were observed to nucleate at a borax film around the periphery of the cell and to grow generally in the radial direction. However, a significant proportion of the crystals had their axes deviate as much as 60° from the radial direction. This suggests that the direction of growth is crystallographically controlled provided that a component of heat flow exists along the growth axis.

Crystal Growth Observations on Thickened Mixtures

In contrast to the unthickened mixtures the large monoclinic crystals are no longer observed but are replaced by a more homogeneous solidification. The presence of the greyish colored thickener makes visual observations very difficult and a suitably thin cell had to be constructed to enable observations to be made using transmitted light.

The Crystallization Cell

A cell (Figure 1) was designed to enable nearly unidirectional heat flow so that its use could be extended to the measurement of the crystallization rate. The cell has internal dimensions

3" x 2" x 1/4" and has 1/2" thick plexiglas walls to minimize heat losses through the sides. The 1/4" thickness is near the maximum that can be used if satisfactory observations are to be made using transmitted light. Further insulation of the cell walls is provided by an air gap between the wall and a second wall which is press fitted into position and sealed with vacuum grease. Filling apertures are located at both ends of the cell. The heat flow is controlled by brass chill plates on the top and bottom which are sealed by O-rings. The plates are thermostated by a flow of water through copper tubes, soldered to the chill plates, from constant temperature recirculating water baths. The plate temperatures are monitored using iron-constantan thermocouples.

Procedure

The cell is warmed and filled with the liquid mixture (which has been stored in a thermostated bath at 34 C) using a jet suction pump to draw in the mixture. The filling tubes are sealed with rubber membranes to allow for expansion on solidification. The chill plates are maintained at a temperature just above the transition temperature (34 C) for several hours to allow thermal equilibrium to be achieved. One or both of the chill plates are then cooled to a desired temperature. A microscope lamp with an infra-red filter, supplying back illumination for the cell, is switched on only when observations are being made to minimize distortion of the crystallization by heat from the lamp.

Observations

After cooling the chill plate there is usually a delay of 5-10 minutes before any solidification is observed. Following this delay, the initial crystallization occurs very rapidly and is probably associated with a rapid return to the equilibrium temperature after an initial supercooling. Steady growth of the 'crystallization' front which is virtually linear (sometimes curved because of end effects) then proceeds. The crystallization front is defined by the increased light transmission through the solidified material and by the dark band of segregated thickener that is observed to develop just ahead of the interface. In most cases, the top plate was maintained at 34 C and the bottom plate at some lower temperature. Solidification proceeded until some equilibrium position was reached near the top plate. When cooling was effected from top and bottom, simultaneously the two solidification fronts coalesced near the center line of the cell leaving a dark line of segregated thickener.

Figure 2 shows the cell after equilibrium has been achieved in an experiment where cooling was effected from the bottom plate.

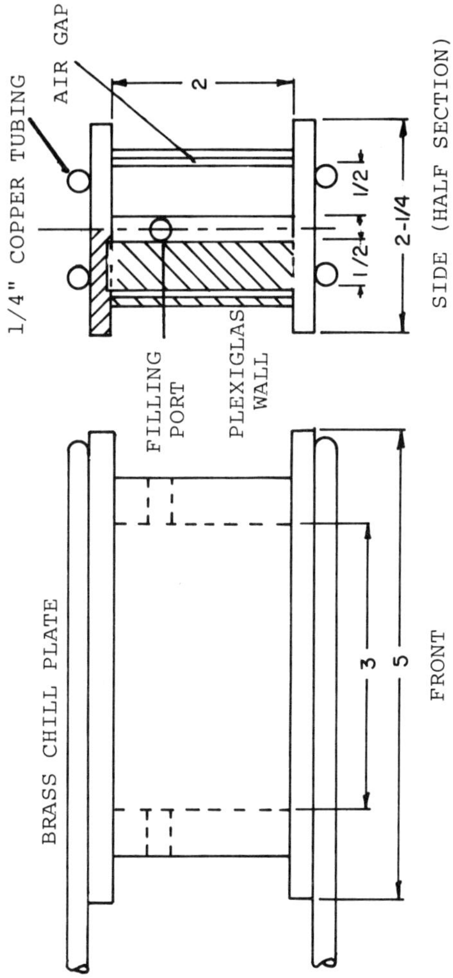

Figure 1. Schematic diagram of crystallization cell #1.

The long needle-shaped crystals are replaced by a cellular structured solid which is produced by microsegregation of the thickener into the crystal grain boundaries. The crystallite or cell size is in the early stages less than 0.5mm dia. but this increases as the solidification proceeds and as the rate slows down. The dark band at the top of the cell is the unsolidified mixture which now contains an increased proportion of thickener.

After solidification has been completed the cooling plate was increased to a temperature above 32 C to remelt the mixture. No apparent visual change occurred after several hours at 40 C since the mixture was found to retain its cellular structure in the melted state. The structure broke down very easily when probed through the filling port to yield a mixture which was homogeneous in appearance. It is probable that surface adhesion to the cell walls contributed to the stability of the structure, but on disassembly of the cell the structure appeared to be continuous throughout the cell. From the point of view of the stability of the mixture this result is very encouraging, but it showed that it would be impossible to make visual measurements of the melting rate.

On refreezing the mixture, very little change was observed in the appearance of the cell. The development of the solidification front could be observed in some cases because of the faint band of thickener that developed ahead of the interface. After the second cycle it was impossible to discern the position of the growing solid-liquid interface.

Measurement of the Crystallization Velocity

To determine whether the crystallization rate is controlled by the thermal diffusion rate or by the crystallization kinetics, it was studied as a function of the degree of undercooling. Information on the crystallization velocity is also important in the determination of the maximum thickness of salt mixture that can be used for a given practical set of heat retrieval conditions.

A series of preliminary measurements were made using cell #1 which was shown in Figure 1. In each case the cell was first equilibrated at 34 C and then the top plate maintained at 34 C while the bottom plate was reduced to the required cooling temperature. Three nominal cooling temperatures 23, 26 and 29 C were investigated. Measurement of the interface position was recorded approximately every 2 minutes using a vernier travelling microscope.

Figure 3 summarizes the results of these studies. The

UNSOLIDIFIED MIXTURE TOP PLATE 34 C

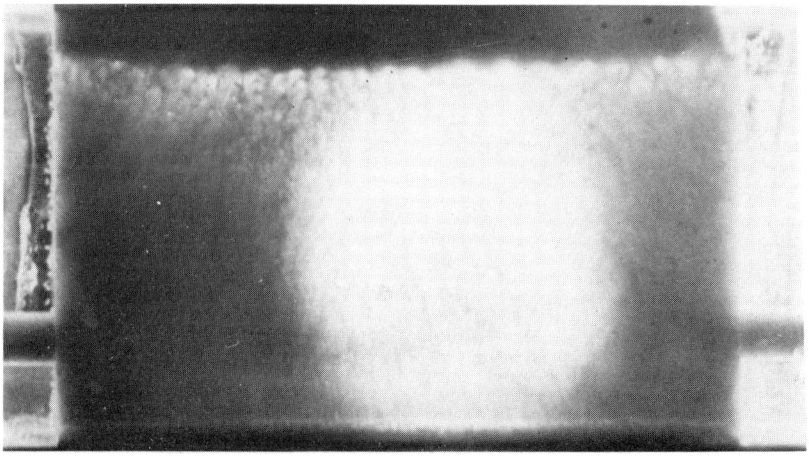

CELLULAR BOTTOM PLATE 26 C

A. MIXTURE SOLIDIFIED

>40 C

B. REMELTED

Figure 2. Photographs of the cell using back illumination

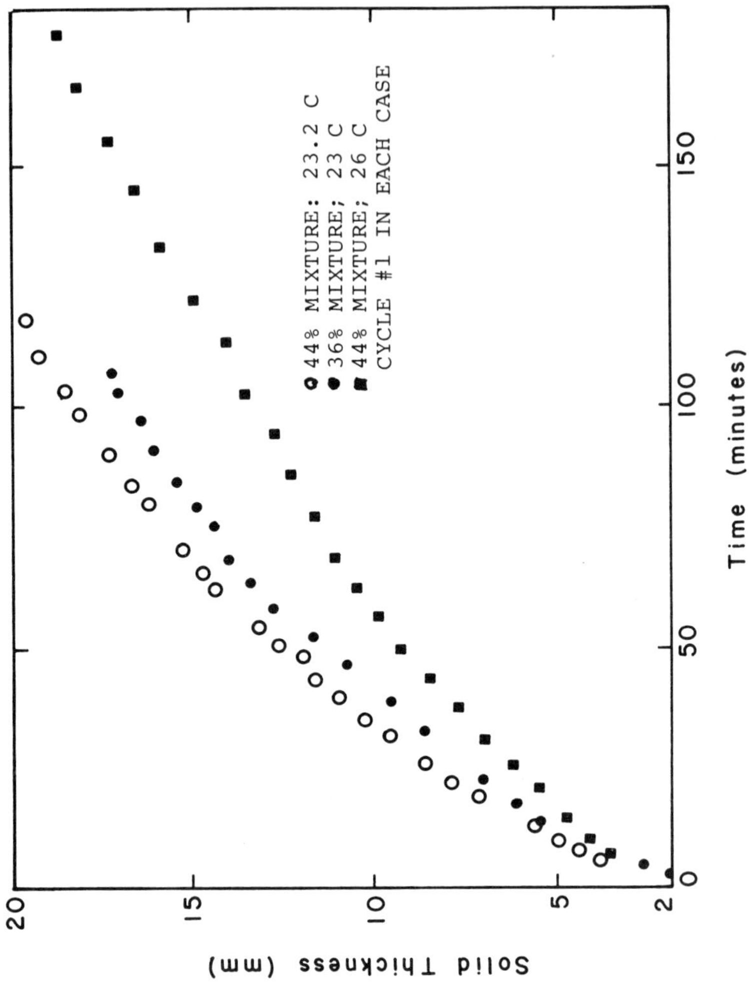

Figure 3. Effect of cooling temperature on crystallization rate.

crystallization rate at 23 C was faster than that at 26 C. Crystallization could not be initiated by maintaining the bottom plate at 29 C. The growth rate appears to be parabolic, the growth rate decreasing with increasing thickness of solid. These observations are consistent with a crystallization process that is controlled by the rate of thermal diffusion from the solidification front.

The observation that crystallization could not be initiated by a chill plate at 29 C (even after several hours) suggests that at least 3 C of supercooling is required even in the presence of the borax nucleating agent. For most practical applications this would not impose a serious limitation on the use of the salt mixture.

Figure 3 also shows the effect on the crystallization rate of using a lower proportion of Na_2SO_4 in the mixture to give the equivalent of a thickened saturated solution. Crystallization occurs more slowly from the 36% mixture than from the 44% mixture. The reduction in rate may be caused by a lower interface temperature which results from the progressive reduction in concentration of the liquid mixture as crystallization proceeds.

Multiple Heating and Cooling Cycles

The stability of the mixture under the effect of successive solidification and melting cycles is a prime factor in the determination of its suitability for practical applications. The cellular structure of the material that is maintained on melting makes it difficult to observe multiple cycles. The addition of a small amount of green food dye enabled 5 freezing cycles to be observed because the dye was pushed ahead of the interface on each freezing cycle. The increase in optical density caused by the addition of the dye made the first solidification cycle harder to observe.

The results showing the effect of added dye and of multiple cycles are shown in Figure 4. There was no significant change in the crystallization rate over the first 5 cycles when cooled at 26 C.

Estimation of the Ratio of the Solid and Liquid Thermal Conductivities

The equilibrium interface position was recorded for four cooling plate temperatures with the top plate maintained at ~40 C. The ratio of the thermal conductivities of the solid and the liquid were then computed using the assumptions that linear temperature profiles existed and that the interface was at the

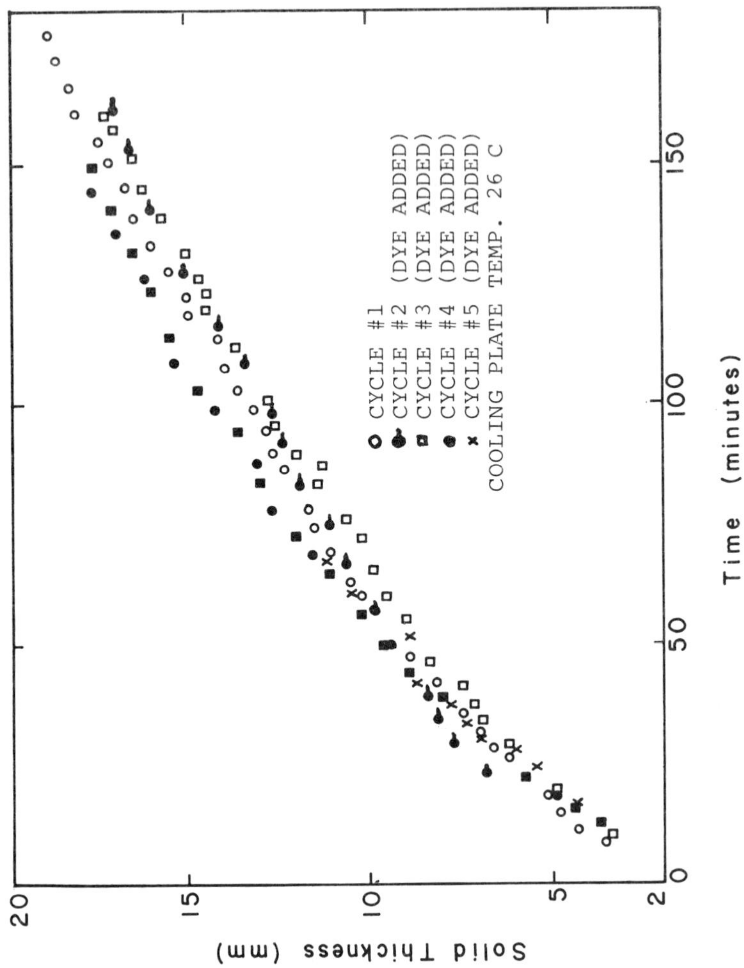

Figure 4. Effect of cycling on crystallization rate.

transformation temp (31.7 C). Values for k_s/k_l of 1.11, 0.97, 1.00 and 1.04 were obtained. These results were somewhat surprising (c.f k(ice)/k(water) ~4.3) and suggested that the temperature distribution within the mixture was being modified by conduction vertically through the thick plexiglas walls. To what extent this would affect the growth rate has not been determined.

Modified Cell and Vacuum Technique

To minimize conduction through the cell walls a new thin walled cell was constructed which it is hoped will yield more quantitatively reliable crystal growth rate data. Heat loss by convection from the side walls is reduced by carrying out the experiment inside a vacuum chamber which is evacuated to ~100 microns. A cell wall thickness of 1/8" and a reduced cell height to 1-1/8" were required to prevent distortion of the cell through internal pressure. The sample thickness was increased from 1/4" to 3/8" to increase the sample/wall thickness ratio.

The plexiglas to brass seal was achieved using an epoxy cement and the filling ports (which are threaded internally) are plugged with brass screws and sealed with silicone rubber. The cell is supported in the vacuum bell jar by the copper tubes which supply the thermostated water to the chill plates.

Only one set of data has been obtained with this system so far (Figure 5). This cannot be compared directly with the data from cell #1 because of the reduced cell height. However the new data lies between the data in Figure 3 as expected from the cooling temperature (24.3 C cf 23 and 26 C) and the parabolic shape is retained.

Simple Analysis of the Crystal Growth Rate Data

One can represent the interface growth in the cell at any given time (t) by a simple model as shown in Figure 6. In overly simplified model, we have assumed that linear temperature profiles are maintained and will consider the heat flow to be unidirectional. Let the interface velocity be v cm/sec. when the solid thickness is x. Then, for unit cross-sectional area:

$$\text{Rate of formation of solid} = v\rho_s \ \text{gm/cm}^2\text{/sec.}$$

$$\text{Rate of heat release at interface} = v\rho_s \Delta H_c \ \text{J/sec/cm}^2$$

(ρ_s = solid density and ΔH_c is the heat of fusion of the solid mixture in J/gm).

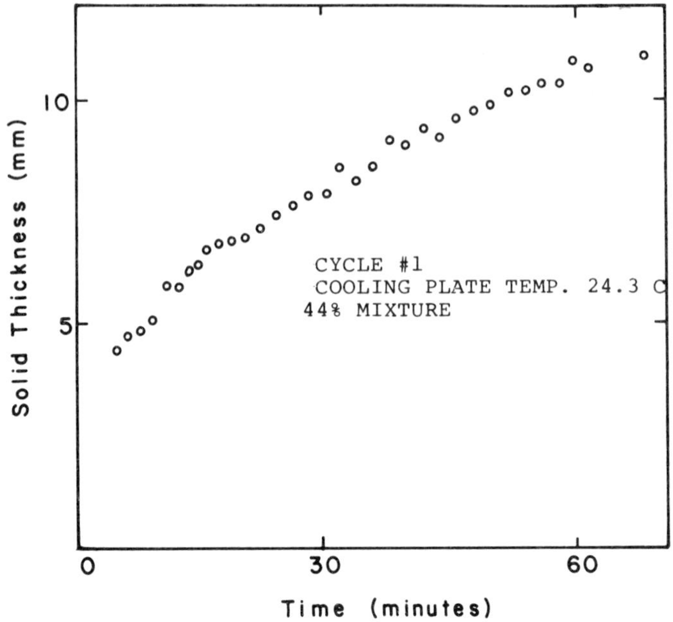

Figure 5. Data from thin walled cell

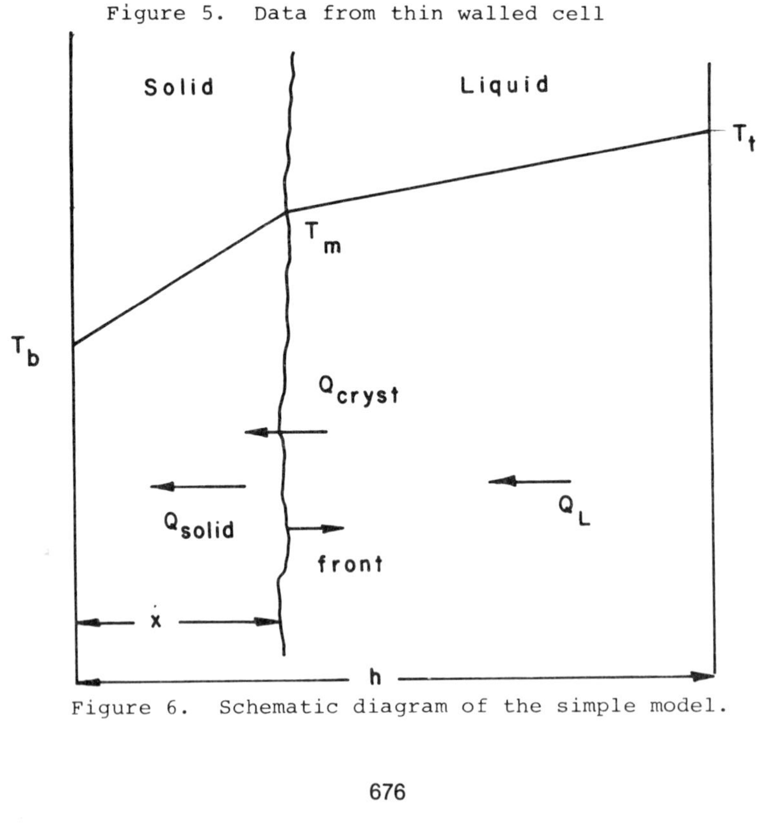

Figure 6. Schematic diagram of the simple model.

Heat flow through the solid $Q_s = \frac{k_s(T_m - T_b)}{x}$ J/sec/cm^2

Heat flow through the liquid $Q_l = \frac{k_l(T_t - T_m)}{(h-x)}$ J/sec/cm^2

If there is no heat flow through the side or end walls of the cell

$$Q_s = Q_{cryst} + Q_l$$

$$v\rho_s \Delta H_c = \frac{k_s(T_m - T_b)}{x} - \frac{k_l(T_t - T_m)}{(h-x)}$$

$$\Delta H_c = \frac{1}{v\rho_s}\left\{\frac{k_s(T_m - T_b)}{x} - \frac{k_l(T_t - T_m)}{(h-x)}\right\}$$

The data from 3 experiments using the 44% thickened Glaubers salt mixture were treated as shown above. The first two sets of data were those obtained from cell #1 (Figure 3) and the third set was from the modified thin-walled vacuum cell. Experimental heats of fusion (or crystallization) were obtained. The interface temperature was taken to be 31.7 C (89 F) and the density of the solid taken to be that of Glauber salt (1.46 g/cm^3).

The experimental data were fitted to 3rd order polynomial expansions (this gave a better fit than a 2nd order polynomial) and the solid thickness after 25, 50, 100 and 150 minutes was calculated from these. Data from the early part of the crystallization were not considered since the temperature profiles would not be fully developed in the early stages. Differentiation of the polynomial expansions permitted the corresponding velocities d_x/d_t to be calculated.

Results of Analysis

Case #1 Cell #1, h = 4.763cm, T_t = 34 C, T_b = 26 C x(mm) = 2.2827 + 0.15597t - 4.995 x $10^{-4}t^2$ = 7.8243 x $10^{-7}t^3$ (t in mins)

Case #2 Cell #1, h = 4.763cm, T_t = 34 C, T_b = 23.2 C x(mm) = 2.45518 + 0.25929t - 1.349 x $10^{-3}t^2$ = 3.2329 x $10^{-6}t^3$ (t in mins)

Case #3 Cell #2, h = 2.619cm, T_t = 34 C, T_b = 24.25 C x(mm) = 3.34357 + 0.22268t - 2.395 x $10^{-3}t^2$ + 1.2373 x $10^{-5}t^3$ (t in mins)

Time (mins)	ΔH (cal/grm)		
	Case #1	Case #2	Case #3
25	40.0	29.2	39.0
50	29.8	26.8	45.4
100	25.3	26.0	
150	25.3	23.6	

All the results show a lower experimental ΔH than the theoretical 55 cal/grm (reduced from the 60 cal/grm of pure G.S. by the addition of 'inert' thickener + borax). It is likely that the thermal conductivity of the solid will be higher than that of the liquid or saturated solution and this may be a partial explanation for the low measured value. If heat is lost through the cell walls this will increase the crystallization velocity and so reduce the calculated heat of fusion. The latter explanation is consistent with the higher ΔH values calculated from the data from the thin walled vacuum cell. The fact that the calculated ΔH values tend to decrease with increasing solid thickness for cases 1 and 2 has not been satisfactorily explained.

It must be emphasized that the calculated ΔH is not necessarily a measure of the total heat storage capability of the salt since we have not considered the unsolidified material nor the effect of any crystallization that occurs in the solid behind the interface; it is merely a device to analyze simply the process occurring in the cell. The accurate determination of the thermal conductivity of the solid and the acquisition of a complete and consistent set of data from the thin walled cell will enable a more precise and meaningful analysis of the solidification of the mixture.

III. CONCLUSIONS

(1) Solidification of the thickened mixture results in a cellular structure which is caused by microsegregation into the crystal grain boundaries.
(2) The cellular structure is retained on melting and refreezing.
(3) Some macrosegregation of the thickener occurs ahead of the growing interface. This appears to be significant only on the first cycle.
(4) The crystallization is limited by the rate of thermal diffusion from the growing interface.
(5) The crystallization rate is the same for the first 5 cycles when a cooling temperature of 26 C is used.
(6) A thickened saturated solution of Glaubers salt (36%) crystallizes more slowly than the 44% mixture when cooled at ~23 C.

REFERENCES

1. M. Telkes, Thermal Storage for Solar Heating and Cooling, Workshop NSF-RANN-75-04 ASHRAE, N. Y., p. 17 (1975).

2. M. Telkes, Applications of Solar Energy for Heating and Cooling of Buildings, ASHRAE, GRP 170, VII-4 (1977).

3. Final report ERDA Contract No. E(11-1)-4042 to be submitted by the Institute of Energy Conversion, University of Delaware. (1977).

4. G. H. Montillon and W. L. Badger, Ind. Engng. Chem. 19 809, (1927).

5. Gmelin, 8th edition Ergänzungsband Na-3, 1966.

ACKNOWLEDGEMENT

This work is done under a contract No. E (11-1)-4042 from the Department of Energy.

A DIRECT CONTACT THERMAL ENERGY STORAGE SYSTEM

A. S. Elasfouri
Cairo University
Guiza, Egypt

ABSTRACT

A direct contact thermal energy storage system is suggested. In this system, the heat storing material would be divided into small droplets. During the energy extraction process, the droplets while being in their liquid state, would be allowed to fall freely in the heat carrying fluid, and meanwhile to solidify.

Hereafter, the solidification of these droplets during the energy extraction process is studied under certain assumptions. Considering low Archimedes number combinations, solutions are given corresponding to different ranges of Reynolds number. The factors affecting the performance of the system during the energy extraction process are presented, and the possibilities of obtaining compact heat exchanger are investigated and discussed.

NOTATIONS

A : projected area in a plane normal to the direction of motion
C_D : drag coefficient
D : diameter
g : acceleration due to gravity
H : height of the heat exchange column
h : heat transfer coefficient
k : thermal conductivity
L : latent heat of solidification
m : mass
$\dot{Q}$: rate of heat transfer
r : radius
T : temperature
ΔT_f: temperature rise of the heat carrying fluid
t : time
u : velocity
v : volume
θ : difference between the temperature of the droplet and

that of the fluid
μ : dynamic viscosity
ν : kinematic viscosity
ρ : density
τ : period of the energy discharging process

Dimensionless Numbers

Ar : Archimedes number = $\frac{g D^3}{\nu_f^2} \frac{\rho_s - \rho_f}{\rho_f}$

Bi : Biot number = $\frac{h r_0}{k_s}$

Re : Reynolds number = $\frac{\rho_f u D}{\mu_f}$

Ste.Fo: Stefan number x Fourier number = $\frac{k_s \Delta T_f \tau}{\rho_s L r_o^2}$

Subscripts

f : fluid
i : final inner
o : outer
s : sphere

INTRODUCTION

Some materials characterized by having relatively high values of latent heat of solidification can be used to store thermal energy [1]. During the energy charging process the heat storing material gains heat and stores it on the form of latent heat of melting, and thus is transformed to the liquid state. When heat is extracted from it during the energy discharging process, the material; loosing its latent heat of solidification; returns to its solid state.

In this paper, it is proposed to carry out the energy discharging process by direct contact heat exchange between the heat storing material and the heat carrying fluid. This could be accomplished by dividing the heat storing material into small liquid droplets and letting them to solidify while falling freely downward through the heat carrying fluid which will flow in the opposite (upward) direction.

ANALYSIS

Here, the analysis of the energy discharging process is reduced to the study of the solidification of a spherical liquid droplet of the heat storing material. This analysis is made under the following assumptions :

1. No chemical reaction occurs between the heat storing material and the heat carrying fluid. Moreover, no surfactant effect is considered.
2. The phase change of the droplet from liquid to solid phase takes place at one fixed temperature denoted the temperature of phase change T_s.
3. The system; heat storing material-heat carrying fluid; has low Archimedes number.
4. Since the heat storing materials are characterized by having very small Stefan numbers [2], the undercooling of the solidified layers below the temperature of phase change is negligible [3].
5. The properties of the heat storing material at the temperature of phase change rest the same whether it is in the liquid or solid state.
6. The droplet is sufficiently small [4] that it maintains its spherical shape throughout the solidification process, and that heat is assumed to flow through it in the radial direction only.

With these assumptions, the governing equations are as follows :

I. Momentum Equation

The first equation to be written is the equation of motion of the droplet through the heat carrying fluid, in which the forces considered are :

1. the gravity force;
2. the buoyancy force; and
3. the drag force.

Thus, the balance of these forces may be expressed as;

(weight) - (buoyancy force) - (drag force)

= (mass x acceleration)

Or;

$$mg - V\rho_f g - \frac{1}{2}\rho_f C_D A u^2 = m\frac{du}{dt} \tag{1}$$

which may be rewritten on the form;

$$V\rho_s g - V\rho_f g - \frac{1}{2}\rho_f C_D \pi r_o^2 u^2 = V\rho_s \frac{du}{dt} \tag{2}$$

or;

$$\frac{4}{3}\pi r_o^3 g(\rho_s - \rho_f) - \frac{1}{2}\rho_f C_D \pi r_o^2 u^2 = \frac{4}{3}\pi r_o^3 \rho_s \frac{du}{dt} \tag{3}$$

and finally the equation reduces to;

$$(\rho_s - \rho_f)g - \frac{3}{8}\rho_f C_D \frac{u^2}{r_o} = \rho_s \frac{du}{dt} \tag{4}$$

Different cases may now be considered [5, 6] :

a. The fluid is moving upward with a constant velocity, but the maximum Reynolds number of the system; based on the relative velocity; is less than 0.2. In this case the coefficient of drag is given by :

$$C_D = \frac{24}{Re} = \frac{12\mu_f}{\rho_f u r_o}$$

Substituting in Eq.(4), and integrating to have;

$$-\frac{2}{9}\frac{\rho_s r_o^2}{\mu_f} \ln\left[\frac{(\rho_s - \rho_f)g}{\rho_s} - \frac{9}{2}\frac{\mu_f}{\rho_s r_o^2} u\right] = t + C \tag{5}$$

where C is a constant of integration.

With zero initial sphere velocity, the expression of this velocity will be;

$$u_s = \frac{2}{9}\frac{r_o^2 g}{\mu_f}(\rho_s - \rho_f)\left[1 - \exp.\left(-\frac{9}{2}\frac{\mu_f}{\rho_s r_o^2} t\right)\right] \tag{6}$$

and, with a sphere terminal velocity of;

$$u_s^* = \frac{2}{9}\frac{r_o^2 g}{\mu_f}(\rho_s - \rho_f) \tag{7}$$

or, in a dimensionless form;

$$Re_{,s}^* = \frac{1}{18} Ar \tag{8}$$

where $Re_{,s}^*$ is Reynolds number based on the terminal velocity of the sphere.

It should be noted that the coefficient of drag depends on the relative velocity, and not on the absolute velocity of the sphere.

b. The fluid is moving upward with a constant velocity sufficiently high; compared with that of the sphere; that the Reynolds number of the system is not affected by the velocity of the droplet. In this case the coefficient of drag will have a given value dependent on the velocity of the fluid only, and thus it will be constant in Eq.(4), as we are solving for the absolute velocity of the sphere.

Under this condition, the integration of Eq.(4) yields;

$$\sqrt{\frac{3}{8}\frac{C_D \rho_f}{g(\rho_s-\rho_f)r_0}}\tanh^{-1}\sqrt{\frac{3}{8}\frac{C_D \rho_f}{g(\rho_s-\rho_f)r_0}}\cdot u = \frac{3}{8}\frac{C_D \rho_f}{\rho_s r_0}t + C_1 \tag{9}$$

where C_1 is a constant of integration.

Again, with zero initial absolute sphere velocity, the expression for the sphere velocity will be;

$$u_s = \sqrt{\frac{8}{3}\frac{g(\rho_s-\rho_f)r_0}{C_D \rho_f}}\tanh\left[\frac{\sqrt{3C_D \rho_f g(\rho_s-\rho_f)/8r_0}}{\rho_s}\right]t \tag{10}$$

and the sphere terminal velocity will be;

$$u_s^* = \sqrt{\frac{8}{3}\frac{g(\rho_s-\rho_f)r_0}{C_D \rho_f}} \tag{11}$$

or, in a dimensionless form it is;

$$Re_{,s}^* = \left(\frac{4}{3C_D}Ar\right)^{1/2} \tag{12}$$

Figure 1 shows the variation of $Re_{,s}^*$ as a function of Archimedes number. The above considered two cases are represented on the same figure; i.e.; Eq.(8) corresponding to small values of Reynolds number of the system ($Re \lesssim 0.2$), and Eq.(12) corresponding to high values of Reynolds number of the system compared with $Re_{,s}^*$, drawn for some chosen

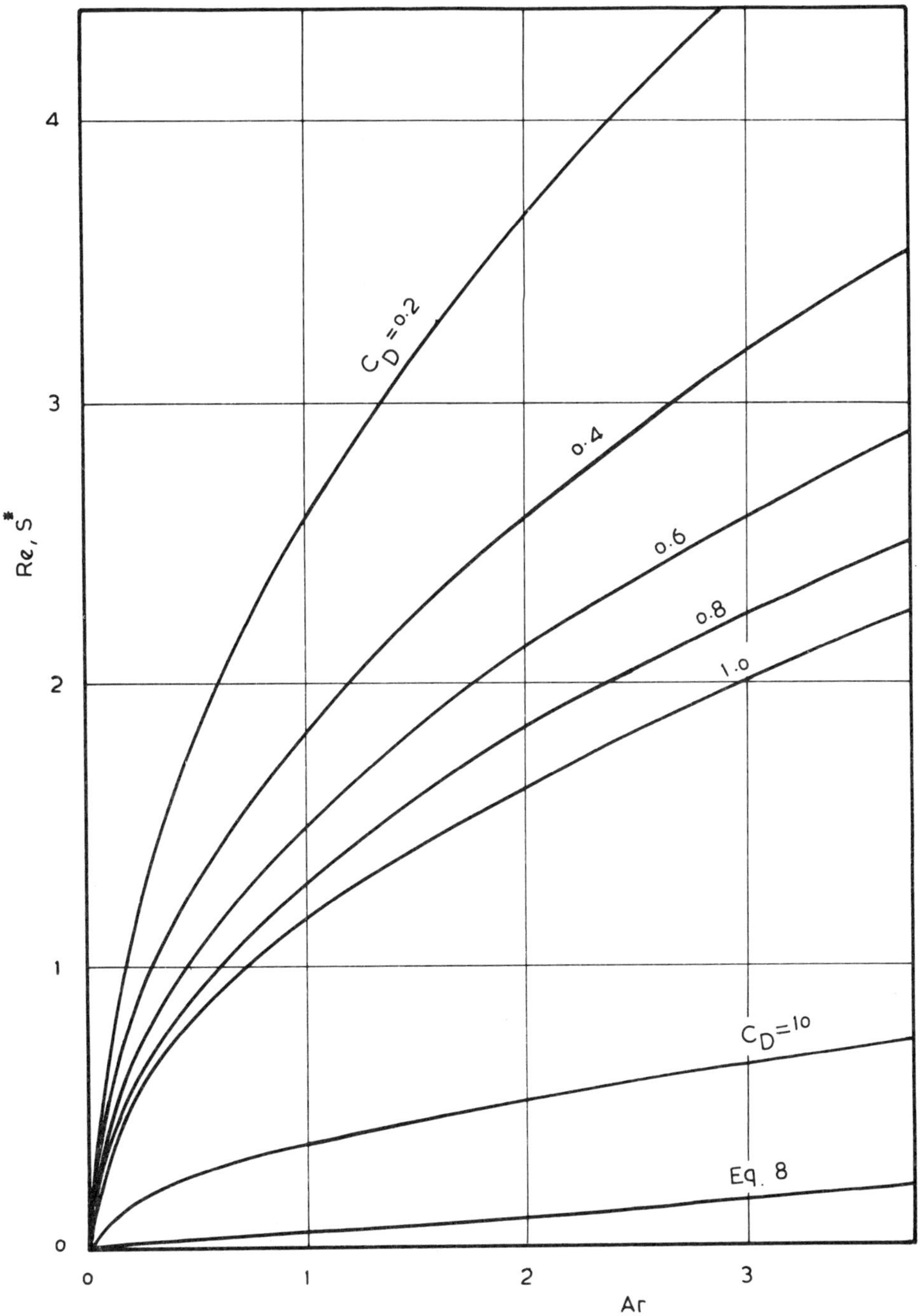

Fig 1

values of the coefficient of drag. It should be remarked that the coefficient of drag for the sphere [6] is a decreasing function with the increase of Reynolds number, reaching a lowest value of 0.1 for values of Reynolds number greater than 300,000. Thus; low values of the coefficient of drag correspond to large values of the Reynolds number of the system.

11. Heat Transfer Equation

Considering the sphere at any instant during solidification, the rate of heat transferred from the sphere to the surrounding fluid is related to the latent heat of solidification, and to the rate of solidification according to;

$$\dot{Q} = \rho_s L \frac{dv_s}{dt} \tag{13}$$

where v_s is the volume of the solidified spherical shell.

As solidification starts at the outside surface and moves inward radially, Eq.(13) becomes;

$$\dot{Q} = -\rho_s L \, 4\pi r^2 \frac{dr}{dt} \tag{14}$$

Also; this same rate of heat transfer may be expressed as;

$$\dot{Q} = 4\pi \frac{(T_s - T_f)}{\left(\frac{1}{r} - \frac{1}{r_o}\right)/(k_s) + 1/(hr_o^2)} \tag{15}$$

This last expression was written after taking into account the assumption that the subcooling of the solidified part below the temperature of phase change is neglected.

Now; introducing the quantity θ for the difference between the temperature of the sphere and that of the heat carrying fluid, Eq.(15) becomes;

$$\dot{Q} = \frac{4\pi\theta}{\left(\frac{1}{r} - \frac{1}{r_o}\right)/(k_s) + 1/(hr_o^2)} \tag{16}$$

Here; h will be considered constant; as the temperature of the heat carrying fluid will be allowed to vary in a very limited range; and also because we are considering systems of low Archimedes numbers; i.e.; the variations in the absolute velocity of the sphere will not change appreciably the Reynolds number of the system, and hence the coefficient of heat transfer.

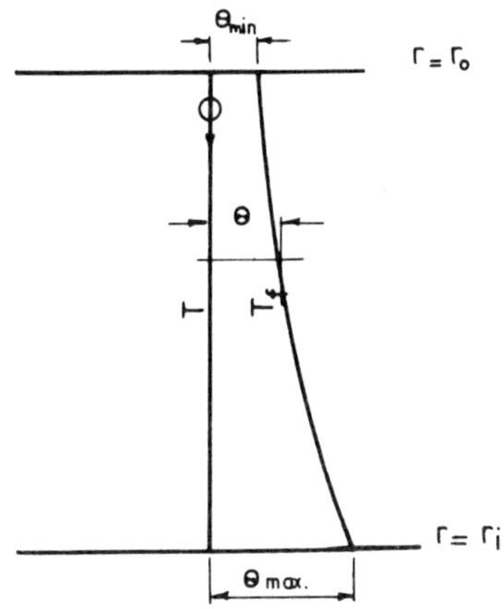

Fig. 2

Assuming that the temperature field within the heat carrying fluid is established, it is evident from Fig. 2 that θ can be written as;

$$\theta = \theta_{min} + \Delta T_f\,(r_o^3 - r^3)/(r_o^3 - r_i^3)$$

So, Eq.(16) becomes;

$$-\rho_s L r^2 \frac{dr}{dt} = \frac{\theta_{min} + \Delta T_f\,(r_o^3 - r^3)/(r_o^3 - r_i^3)}{(1/r - 1/r_o)/k_s + 1/(h r_o^2)} \qquad (17)$$

This differential equation can be integrated, noting that the initial condition of solidification is that the inner radius of the solidified spherical shell is equal to the outer radius of the droplet.

Thus, the solution of Eq.(17) will be;

$$\begin{aligned}\frac{k_s \Delta T_f \tau}{\rho_s L r_o^2} &= \frac{1}{3}\left(1 - \frac{1}{h r_o/k_s}\right) \ln\left(1 - \Delta T_f/\theta_{max}\right)\\ &+ \frac{1}{6}\ln\frac{(a^2 + a r_o + r_o^2)(a - r_i)^2}{(a^2 + a r_i + r_i^2)(a - r_o)^2}\\ &+ \frac{1}{\sqrt{3}}\left(\tan^{-1}\frac{2r_i + a}{a\sqrt{3}} - \tan^{-1}\frac{2r_o + a}{a\sqrt{3}}\right)\end{aligned} \qquad (18)$$

where;

$$a^3 = r_o^3\left\{1 + \frac{\theta_{min}}{\Delta T_f}\left[1 - \left(\frac{r_i}{r_o}\right)^3\right]\right\}$$

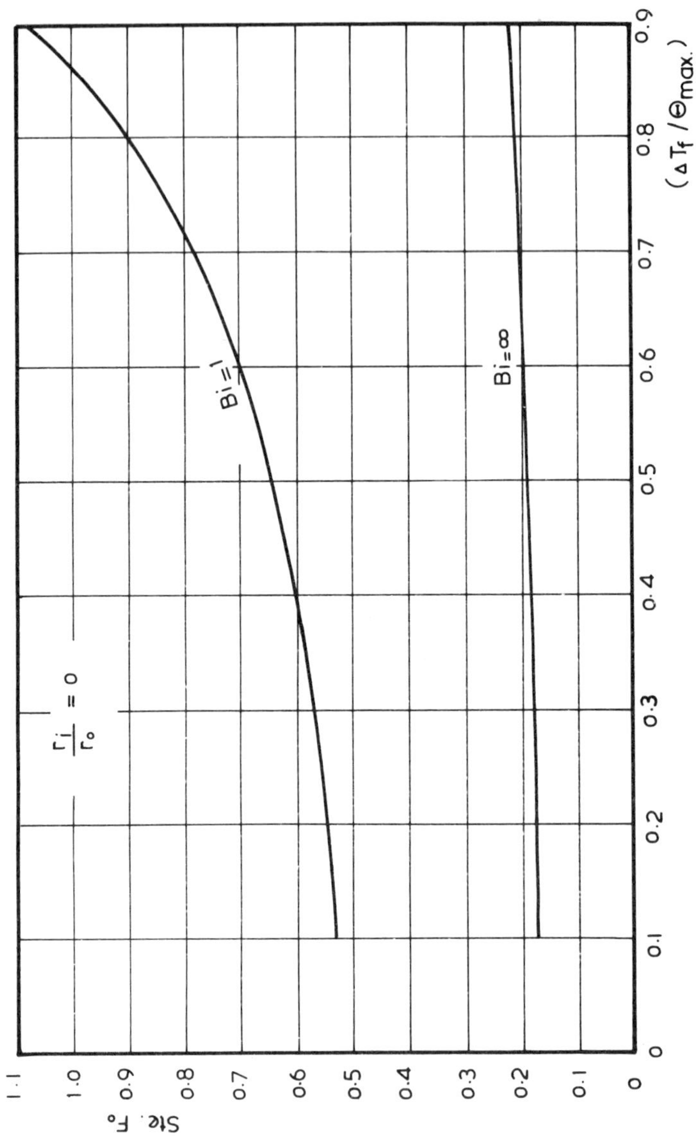

Fig 3

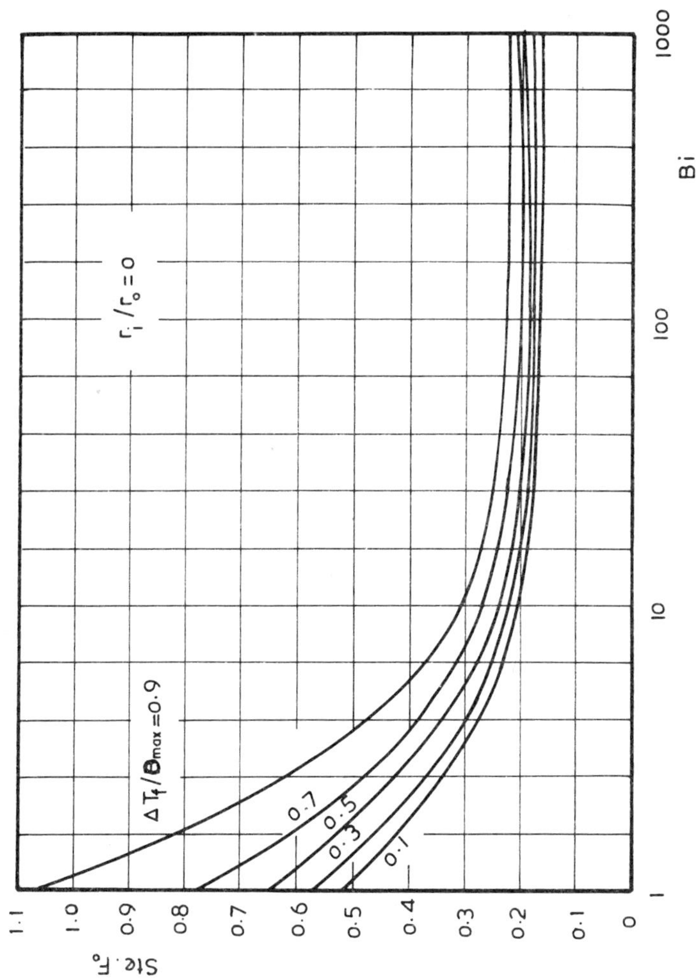

Ste.Fo
Bi
1
10
100
1000
0
0.1
0.2
0.3
0.4
0.5
0.6
0.7
0.8
0.9
1.0
1.1
ri/ro=0
ΔTf/θmax=0.9
0.7
0.5
0.3
0.1

Fig 4

In dimensionless form, Eq.(18) means that Ste.Fo is a function of the quantities : Bi, $\Delta T_f/\theta_{max}$, and r_i/r_o.

Figure 3 shows the variations of the quantity Ste.Fo as a function of $\Delta T_f/\theta_{max}$ for the limiting values of Biot number; i.e., for Bi = 1, and for Bi = ∞, in the case of allowing complete solidification of the droplet; i.e., for r_i/r_o = 0. It should be noted that Bi = 1 corresponds to the case of equal internal (conduction) and external (convection) thermal resistances, while Bi = ∞ corresponds to the case of predominant internal thermal resistance compared with the external one. The figure shows clearly that much can be gained as regards decreasing the period of the energy extraction process through the increase of Biot number, specially for high values of $\Delta T_f/\theta_{max}$ (the practical range).

Figure 4 shows another plot of Ste.Fo against Biot number for different values of $\Delta T_f/\theta_{max}$. The curves show that practically for Bi = 100 no decrease in Ste.Fo can be obtained by still increasing Biot number, which means that in this range (Bi > 100) the above equation of heat tranfer can be used even if the velocity of the droplet varies in a wide range, since the total period of the energy extraction process becomes independent of the Biot number.

On the other hand, the decrease of the total period of the energy extraction process may be obtained also if partial solidification of the droplet is allowed. The effect of allowing only partial solidification on the decrease of the total period of the energy extraction process is shown in Fig. 5, which is a plot of Ste.Fo against r_i/r_o for $\Delta T_f/\theta_{max}$ = 0.9. It should be noted that r_i/r_o = 0 means complete solidification, while r_i/r_o = 1 indicates no solidification. From the figure it can be seen that the gain obtained through partial solidification is much more pronounced in systems which have low Biot numbers.

This gain, however, will be on the expense of the total quantity of energy exchanged during the heat discharging process between the droplet and the heat carrying fluid. The quantities of heat corresponding to different values of r_i/r_o expressed as percentages of the total quantity of heat exchanged in the case of complete solidification are shown on the same figure by the dotted line. It can be seen that the decrease in the quantity of heat exchanged with the decrease of the solidified volume follows approximately the decrease in the dimensionless period of the heat extraction process (Ste.Fo).

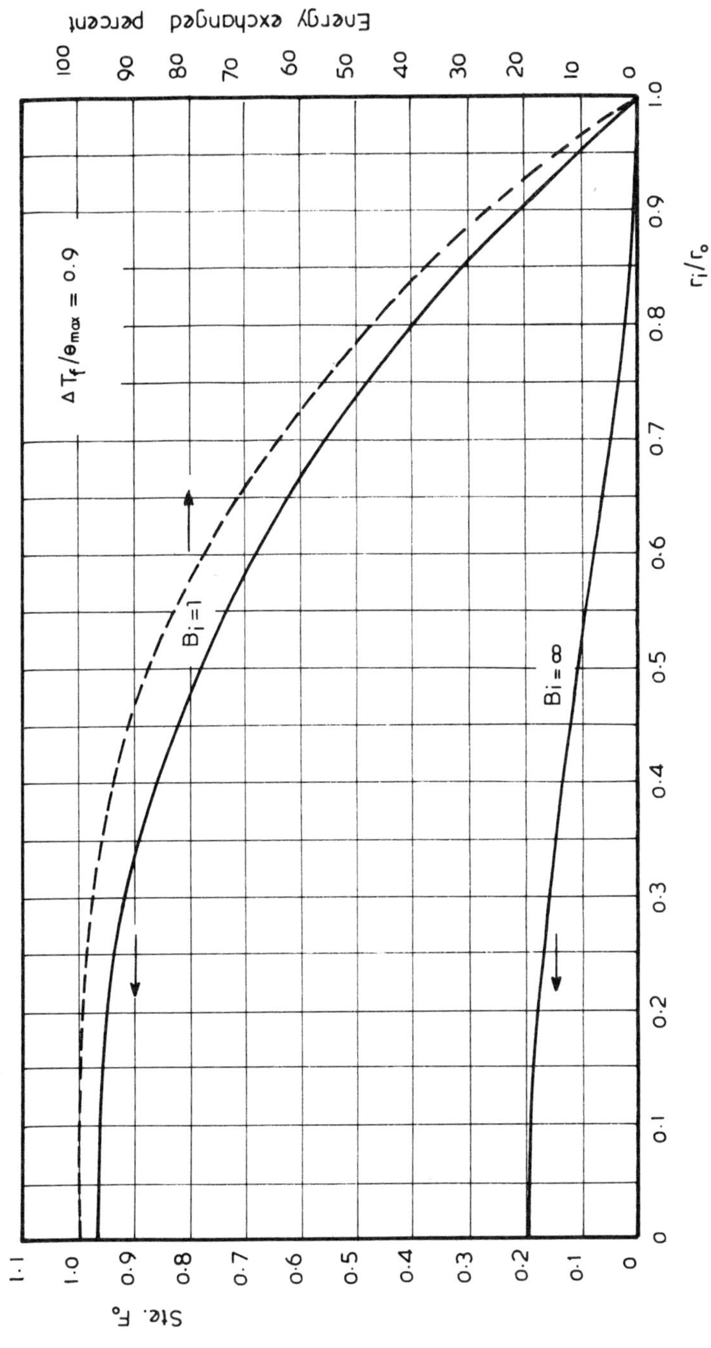

Fig 5

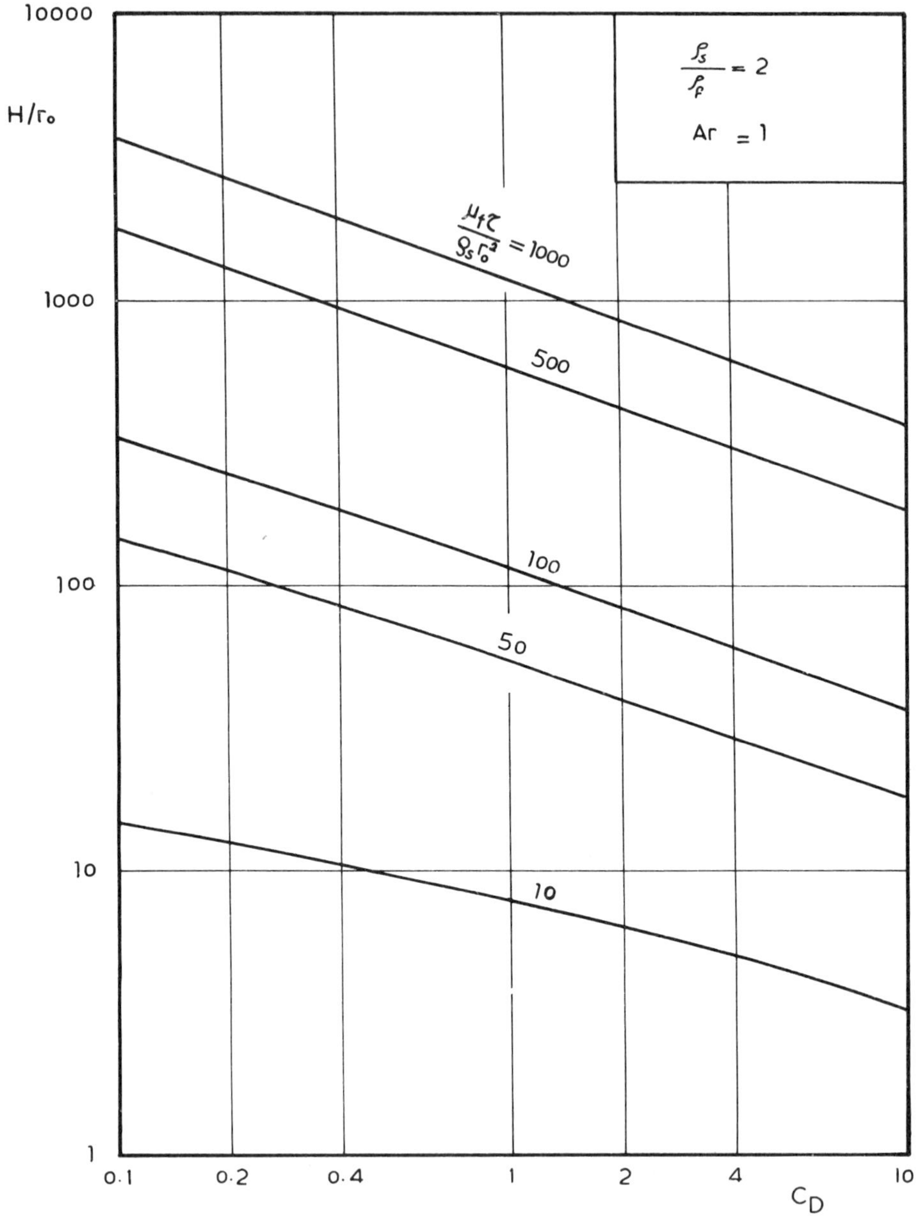

10000
H/r₀
1000
100
10
1
0.1
0.2
0.4
1
2
4
10
C_D
$\rho_s / \rho_f = 2$
Ar = 1
$\frac{\mu_f \tau}{\rho_s r_o^2} = 1000$
500
100
50
10

Fig 6

Height of Heat Exchange Column

From the above relations, the height of the heat exchange column can be calculated as;

$$H = \int_0^{\tau} u_s \, dt \tag{19}$$

where u_s is the absolute velocity of the sphere.

The value of the height of the heat exchange column will depend on the case to be considered. Two cases have been considered above. In case a above, where the maximum Reynolds number of the system would not be allowed to exceed 0.2, the absolute velocity of the sphere is given by Eq.(6). Substituting this expression of u_s in Eq.(19) and integrating, we will have;

$$H = \frac{2}{9} \frac{r_o^2 g}{\mu_f} (\rho_s - \rho_f) \left\{ \tau - \frac{2}{9} \frac{\rho_s r_o^2}{\mu_f} \left[1 - \exp\left(-\frac{9}{2} \frac{\mu_f \tau}{\rho_s r_o^2}\right)\right]\right\} \tag{20}$$

and after arrangement of terms;

$$\frac{H}{r_o} = \frac{1}{36} \frac{\rho_s}{\rho_f} Ar \left\{ \frac{\mu_f \tau}{\rho_s r_o^2} - \frac{2}{9} \left[1 - \exp\left(-\frac{9}{2} \frac{\mu_f \tau}{\rho_s r_o^2}\right)\right]\right\} \tag{21}$$

this means that in this case, the dimensionless height H/r_o is a function of the dimensionless quantities: ρ_s/ρ_f, Ar, and $\mu_f \tau/\rho_s r_o^2$.

It is found that for values of $\mu_f \tau/\rho_s r_o^2$ greater than 4, Eq.(21) can be put on the form;

$$\frac{H}{r_o} = \frac{1}{36} \frac{\rho_s}{\rho_f} Ar \left(\frac{\mu_f \tau}{\rho_s r_o^2} - \frac{2}{9} \right) \tag{22}$$

In the case b, the upward velocity of the heat carrying fluid is large compared with the downward velocity of the droplet, so that the Reynolds number of the system is hardly affected by the absolute velocity of the sphere. In this case, the absolute velocity of the sphere is given by Eq. (10). Using this relation in Eq.(19) and integrating, we finally have for the height of the heat exchange column;

$$H = \frac{8}{3} \frac{\rho_s r_o}{\rho_f C_D} \ln \left(\cosh \frac{1}{\rho_s} \sqrt{\frac{3 C_D \rho_f g (\rho_s - \rho_f)}{8 r_o}} \; \tau \right) \tag{23}$$

which may be put on the form;

$$\frac{H}{r_o} = \frac{8}{3}\frac{\rho_s}{\rho_f}\frac{1}{C_D} \ln\left[\cosh\left(\frac{1}{8}\frac{\mu_f\tau}{\rho_s r_o^2}\sqrt{3C_D Ar}\right)\right] \qquad (24)$$

Here; the dimensionless height H/r_o is a function of the dimensionless quantities : ρ_s/ρ_f, C_D, $\mu_f\tau/\rho_s r_o^2$, and Ar.

In the above two expressions for the height of the heat exchange column, it is evident that in order to obtain a short heat exchange column, systems should have small values of : ρ_s/ρ_f, $\mu_f\tau/\rho_s r_o^2$, and Ar.

In the second expression; Eq.(24); however, we have as well the coefficient of drag as one of the variables on which depends the dimensionless height H/r_o. In order to study the effect of the coefficient of drag on the height of the heat exchange column, Eq.(24) was drawn. Figure 6 is a graphical representation of Eq.(24) drawn as H/r_o against C_D on a log-log paper, for $\rho_s/\rho_f = 2$, and for $Ar = 1$. It can be seen that H/r_o decreases with the increase of C_D. The equation tends to a straight line relation on the log-log paper for high values of $\mu_f\tau/\rho_s r_o^2$, as in this case Eq.(24) tends to;

$$\frac{H}{r_o} = \frac{8}{3}\frac{\rho_s}{\rho_f}\frac{1}{C_D}\left(\frac{1}{8}\frac{\mu_f\tau}{r_o^2\rho_s}\sqrt{3C_D Ar} - \ln 2\right) \qquad (25)$$

which is approximately a straight line relation on a log-log paper.

CONCLUSION

A direct contact thermal energy storage system is suggested, and the factors affecting its performance is assessed. The treatment of the energy discharging process is reduced to the study of the solidification of a liquid droplet of the heat storing material which would be allowed to fall freely downward through the heat carrying fluid. This study is made after making certain simplifying assumptions. Under these assumptions, expressions have been obtained which may be used in the design of such a system.

The means of obtaining compact heat exchange column through the decrease of the absolute velocity of the droplet and/or the decrease of the total period of the heat extraction process are discussed, and their limitations are presented. Compact heat exchange column may be obtained through the decrease of all or one of the following quantities : Ar, ρ_s/ρ_f, and $\mu_f\tau/\rho_s r_o^2$; and/or through the increase of both or one of the quantities : Bi, and C_D. Shorter heat exchange column may be obtained also by allowing only partial solidification, however, in this case this will be on the expense of the total quantity of heat exchanged during the heat extraction process.

REFERENCES

1. Telkes, M., "Thermal Energy Storage", 10th Intersociety Energy Conversion Engineering Conference, Newark, Delaware, August 18-22, 1975, pp. 111-115.
2. Shamsundar, N., and Sparrow, E. M., "Storage of Thermal Energy by Solid-Liquid Phase Change - Temperature Drop and Heat Flux", Journal of Heat Transfer, Nov. 1974, pp. 541-543.
3. Carslaw, H. S., and Jaeger, J. C., "Conduction of Heat in Solids", Oxford University Press, Amen House, London, 1959.
4. Harper, J. F., "The Motion of Bubbles and Drops through Liquids", Adv. Appl. Mech., 12, 1972, pp. 59-129.
5. Pruppacher, H. R., Le Clair, B. P., and Hamielec, A. E., "Some Relations between Drag and Flow Pattern of Viscous Flow past a Sphere and a Cylinder at Low and Intermediate Reynolds Numbers", J. Fluid Mech. (1970), vol. 44, part 4, pp. 781-790.
6. Streeter, V. L., "Handbook of Fluid Dynamics", Mc Graw Hill Book Co., New York, 1961.

SOLAR HEATING SYSTEMS WITH AIR COLLECTORS AND ROCK BED STORAGE

W.L.Dutré, L.Cypers, J.Berghmans and A.Debosscher
Katholieke Universiteit te Leuven
Afdeling Toegepaste Mechanika en Energiekonversie
Laboratorium voor Warmteoverdracht en Reaktorkunde
Celestijnenlaan 300 A, B-3030 Heverlee, Belgium

ABSTRACT

A domestic air heating system based on air collectors and rock bed storage is studied in this paper. For the system being considered a dynamic model has been developed, including the regulation criteria and the dynamic behaviour of the building. Making use of the climatological data for temperature and solar radiation as half-hour averages for the Belgian climatological reference year, the year round performance of the system has been calculated over a wide range of the most important design parameters, such as collector area, thermal capacity of the storage system, the total heat demand and its distribution. Results of these calculations are reported in this paper and the relative influence of system component size as well as the importance of each of the possible operation modes under the climatic conditions being considered are discussed.

INTRODUCTION

Solar heating systems as presently being developed can be devided into two main categories according to the kind of fluid used to withdraw the heat from the collectors, which is either a liquid or a gas. In the latter case it is quite obvious to use air if heating systems for dwellings are considered.

Although air collectors are presently much less developed than liquid collectors, they offer some specific advantages when air is the only heat transporting fluid being used throughout the complete solar heating system. Obviously, in such systems there is no need to protect the system against freezing and thightness is not a very critical condition since small leakages from an air system can never have serious consequences. It can also be expected that air collectors are much less sensitive to corrosion such that their life-time expectancy can be rated higher than for water collectors. From a system point of view, an interesting feature of air systems consists of the possibility to heat the air withdrawn from the dwelling, directly in the collector and return it to the dwelling, whenever the operating conditions are in favour of such a direct loop. The considerably lower apparent heat capacity of air collectors as compared to water collectors is of interest to climatological situations in which partially clouded skies cause the solar radiation intensity to be very intermittent. In fact, this characteristic of air collectors results in a fast dynamic response to insolation variations. Air systems also offer the possibility

SOLAR AIR HEATING SYSTEM

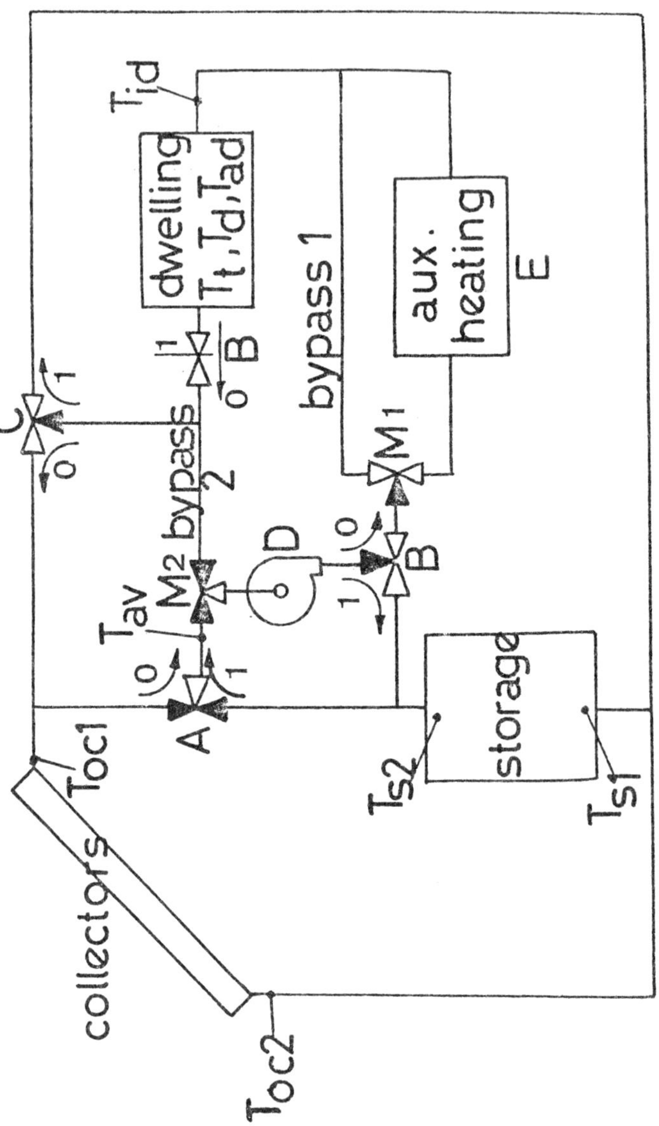

figure 1

to use a regenerative rock bed storage system and thus eliminate the need of an intermediate heat exchanger between the collector loop and the storage as well as between the storage and the dwelling air circuit. However, to obtain the same storage capacity, the use of a rock bed storage at least doubles the occupied volume as compared to a water tank storage. The poor heat transfer between the absorber and the air flow in the collectors causes air collectors on the average to have a lower efficiency than liquid collectors, unless special design features are introduced to increase the internal heat transfer area or to generate turbulent flow in the air collector duct. The latter naturally increases the head loss across the collector array. Unavoidable in air systems is certainly the larger power being required to circulate the fluid and accordingly a higher electrical energy consumption. Compared to water collector systems with an air heat distribution system, this difference becomes less significant.

For liquid as well as for air collector systems, a large number of different systems can be designed, depending mostly on the number of system components such as storage vessels, heat exchangers, ventilators, valves,..., and the number of possible operational modes being desired. The system considered in this paper consists of an air collector array, only one rock bed storage, one ventilator, an auxiliary heater and the necessary control devices.

Within these limitations, the number of possible operating modes has been choosen as high as possible, such that the system can operate in the most efficient way under all circumstances. Increasing the number of components, for instance by the addition of a second storage reservoir or a heat pump, may yield additional energy savings, however at the expense of higher complexity of the system and its regulation.

System Description.

The solar air heating system considered in this paper is represented schematically in figure 1. The three-way valves A, B and C are controlled by binary logical control functions and are either in position 0 or position 1 as indicated in the figure. The ventilator operation is ON-OFF controlled and is characterised by the control variable D which equals 0 in the OFF position and 1 if the ventilator has to be turned ON. The auxiliary heating is commanded by the logical variable E which equals 0 if no auxiliary heating is needed, 1 in the opposite case. The position of the valves M1 and M2 is continuously variable to regulate the air flow through the corresponding bypasses in order to avoid the air temperature T_{id} in the dwelling inlet duct to exceed the nominal value T_n for which the heating system has been designed. This is achieved by M1 on bypass 1 if overheating results from the auxiliary heating system delivering additional heat to the air having been preheated in the collectors or in the storage, while M2 on bypass 2 limits T_{id} whenever the collector or storage outlet temperature exceeds T_n. If the auxiliary heating consists of electrical heating elements with temperature commanded power regulations, M1 and bypass 1 can be omitted.

The operational modes to which this system can be switched by the control functions A, B, C and D are as follows :

1. Mode OFF (D = 0, E = 0) : the system is turned off if the house does not demand heat and no solar energy can be stored.

2. Mode C-S, (collector-storage loop, with A = 0, B = 1, C = 1, D = 1, E = 0) storage loading mode : heat is transferred from the collectors to the storage if no heat is required for the house and the collector outlet temperature T_{OC1} sufficiently exceeds the temperature T_{S1} in close to the appropriate storage outlet sufficiently exceeds the temperature T_{s1}

3. Mode C-D, (collector-dwelling loop, with A = 0, B = 0, C = 1, D = 1 , E = 0) : heat is supplied directly from the collectors to the dwelling whenever a heat demand occurs which can be supplied completely by the collectors or to which the collectors can contribute by increasing the air temperature to a level above the storage temperature T_{S2}.

4. Mode C-S-D (collector-storage-dwelling loop with A = 1, B = 0, C = 0, D = 1, E = 0) : heat is withdrawn from the collectors as well as from the storage when a heat demand occurs to which the collectors can contribute only partially and an additional temperature increase can be obtained from the storage.

5. Mode S-D, (storage-dwelling loop, with A = 1, B = 0, C = 1, D = 1, E = 0) : heat is withdrawn from the storage only when a heat demand occurs, the collectors cannot contribute to the required heat supply and the storage temperature T_{S2} sufficiently exceeds the dwelling air temperature.

6. Mode A-D, (auxiliary-dwelling loop, with A = 0, B = 0, C = 0, D = 1, E = 1) : if neither the collectors nor the storage can contribute to an occuring heat demand, the auxiliary heating system only will supply the required heat to the dwelling.

7. In each of the modes C-D, C-S-D or S-D, the auxiliary energy source will supply additional energy if the temperature T_{av}, available from the solar system, is not sufficient to meet the heat demand within a reasonable time, thus giving rise to three additional possible situations :
   Mode C-A-D, with A = 0, B = 0, C = 1, D = 1, E = 1
   Mode C-S-A-D, with A = 1, B = 0, C = 0, D = 1, E = 1
   Mode S-A-D, with A = 1, B = 0, C = 1, D = 1, E = 1
   The minimum required air temperature at the air distribution duct inlet below which auxiliary heat supply becomes necessary, depends on the heating system design as well as on the indoor and outdoor temperatures. In this study, this minimum required heating air temperature T_{ma} , has been considered to be a simple function of the indoor air temperature T_{ad} only.

MINIMUM REQUIRED VALUE FOR T_{id}

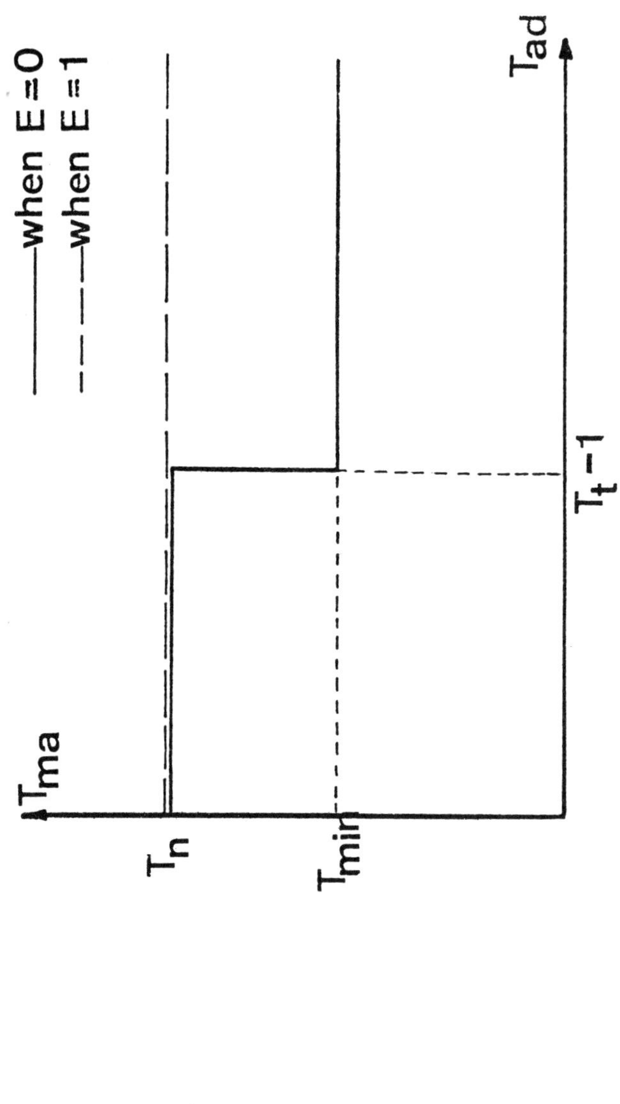

figure 2

LOGICAL STATE VARIABLES

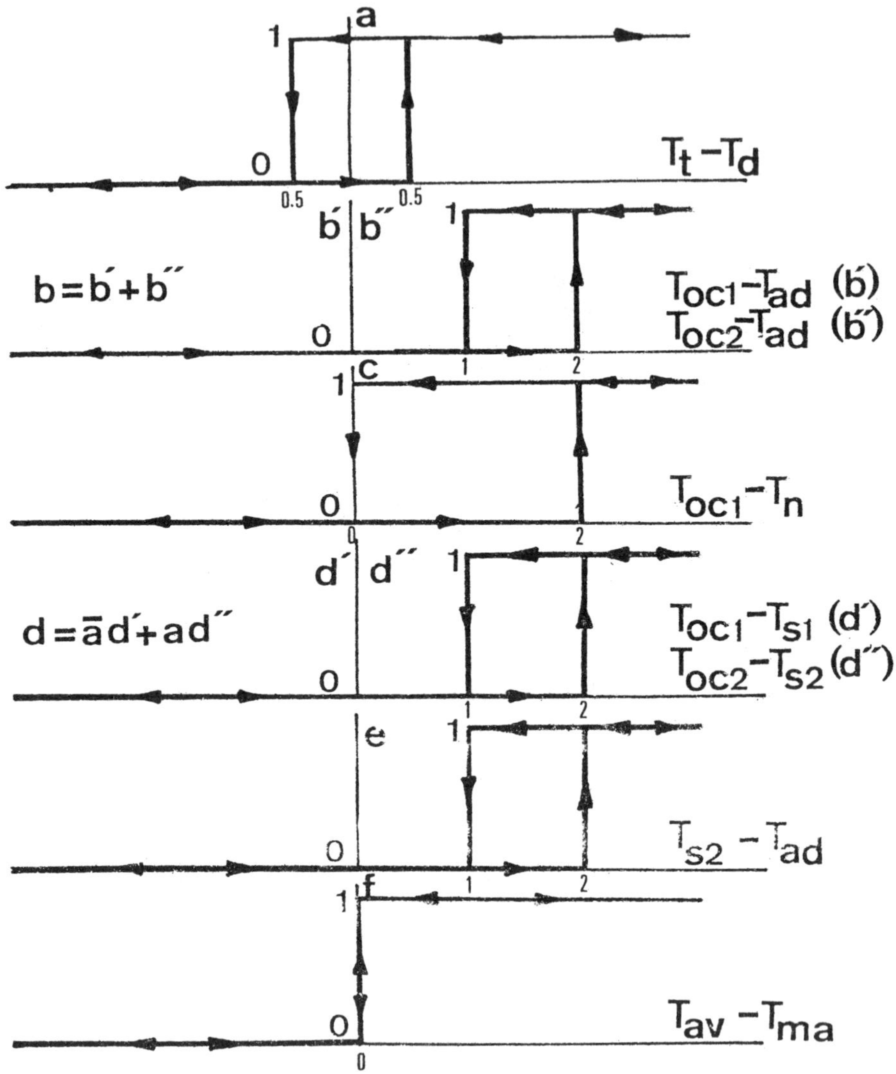

figure 3

However,, whenever auxiliary heat is used, the air is assumed to be heated to the design temperature T_n. Naturally, T_{ma} is subjected to a lower limit T_{min}, sufficiently above the thermostat setting point. Figure 2 represents T_{ma} as a function of T_{ad}.

The regulation system to select the desired operating mode is based on a number of logical state variables a, b, ... f, assigned to physical quantities, representing measurable temperature differences. The value of each of these logical state variables is represented as a function of its corresponding temperature difference in figure 3, where the necessary hysteresis of the regulation system is taken into account. The temperatures being used in this control system are :

T_d : the effective indoor dwelling temperature, which is considered to be a linear combination of the indoor air temperature and the inside wall temperature.

T_{ad} : indoor air temperature

T_t : thermostat setting point

T_{oc1} and T_{oc2} : collector outlet temperatures

T_{s1} and T_{s2} : storage temperatures

T_{av} : available solar system outlet temperature

T_n : nominal design temperature for the air flow to the dwelling, for which the auxiliary system and the distribution system has been designed.

T_{ma} : minimum required temperature for the air flow to the dwelling.

T_{id} : air temperature in the dwelling inlet duct.
$T_{id} = T_{av}$ if E = 0 (i.e. no auxiliary heat supply, occuring if $T_{ma} \leq T_{av} \leq T_n$);
$T_{id} = T_n$ otherwise.

If T_{av} would exceed T_n, the valve M2 is modulated such that $T_{id} = T_n$.
If energy is supplied from the auxiliary heating system, T_{id} is limited to the value T_n by the action of valve M1.

In figure 4, the left part of the table indicates the required combination of the logical valve control variables A, B and C, corresponding to each of the operating modes described above. The right hand side of this table gives the combinations of the logical state variables a, b, c, d and e as defined in figure 3, which are mutually exclusive and to which the indicated combinations of the variables A, B and C should be made to correspond in order to switch to the desired operating mode.
The Boolean expressions giving the correspondance between the state variables and the control variables can be written as follows :

$A = a\,b\,\bar{c}\,\bar{d} + a\,\bar{b}\,e$ or $\bar{A} = \bar{a} + bc + bd + \bar{b}\bar{e}$

$B = \bar{a}$ or $\bar{B} = a$

$C = \bar{a} + bc + bd + \bar{b}e$ or $\bar{C} = ab\bar{c}\bar{d} + a\bar{b}\bar{e}$

LOGICAL VALVE CONTROL VARIABLES

OPERATING MODE	VALVE CONTROL VARIABLES			STATE VARIABLES				
	A	B	C	a	b	c	d	e
C—S	0	1	1	0	—	—	1	—
C—(A)—D	0	0	1	1	1	1 0	— 0	—
C—S-(A)—D	1	0	0	1	1	0	1	—
S-(A)-D	1	0	1	1	0	—	—	1
A—D	0	0	0	1	0	—	—	0

figure 4

Equivalent analogue circuit for a 3-point house model

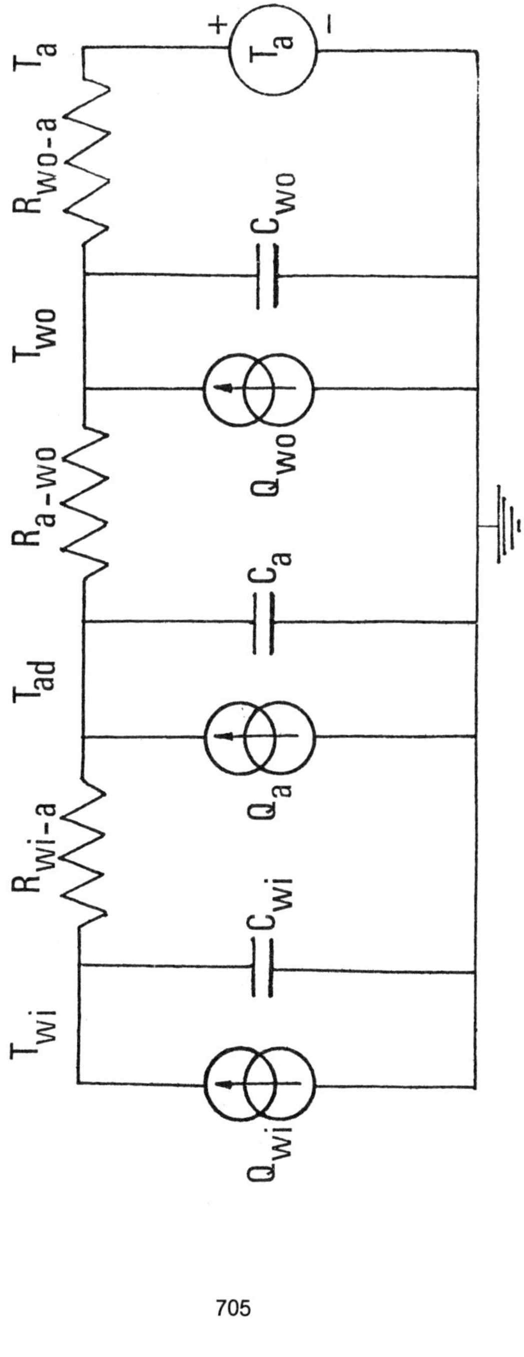

Figure 5

The ventilator ON-OFF control function D is given by :

$D = a + d$ or $\overline{D} = \overline{a}\,\overline{d}$

and the auxiliary heating ON-OFF control function E is given by :

$E = a\,\overline{f}$ or $\overline{E} = \overline{a} + f$

If some of the operational modes are excluded, the control system can be simplified. From mathematical simulations as well as from the practical experience obtained thus far on a small test facility with 15.6 m^2 of air collectors, it follows that the combined operating modes C-S-D and C-S-A-D occur only during a very limited number of hours a year under Belgian climatological conditions. Excluding these modes from the list of possible system operations does then not affect noticeably the system performance but results in a considerable simplification of the control system.

The regulation system described above can be built completely with conventional temperature sensors, regulators, relays, servomotors, etc., but the conversion of the analog values of the temperature differences to the logical state variables a, b, .., e and the subsequent conversion of a, b, ..., e to the logical control variables A,B,C,D and E can better be accomplished by a special designed electronic device. Logical integrated circuit elements or a simple microprocessor can be used for this purpose.

Mathematical Simulation Model

A mathematical model has been developed to simulate dynamically the behaviour of the system described above, including the regulation criteria. In this simulation programme, the system differential equations are solved, using half-hour averaged meteorological data. Since the operating mode to be selected at the beginning of each time step also depends on the thermal condition of the dwelling, the thermal behaviour of the building must also be simulated dynamically. This simulation incorporates the hysteresis around the thermostat setting point, the thermal capacities of the structure, the incident solar radiation on the building, internal heat dissipation and outdoor temperature as a function of time. For the purpose of parametric studies of the annual system performance a simplified dynamic thermal model of the building can be used.
A three point model as represented by the equivalent analogue circuit of figure 5, has been used in the calculations reported hereafter. This model distinguishes between the outer wall temperature T_{wo}, the inner wall temperature T_{wi} and the indoor air temperature T_{ad}. T_a is the ambient temperature, represented by a voltage generator.
R and C represent thermal resistances and thermal capacities respectively. The current generators have the following physical significance

Q_{wo} : the thermal power supplied to the outer walls, during periods of insolation.

Daily thermostat setting cycle

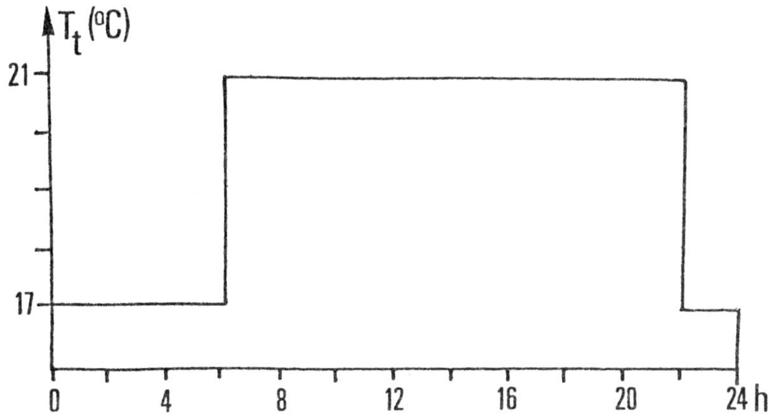

Figure 6

Daily internal heat dissipation cycle

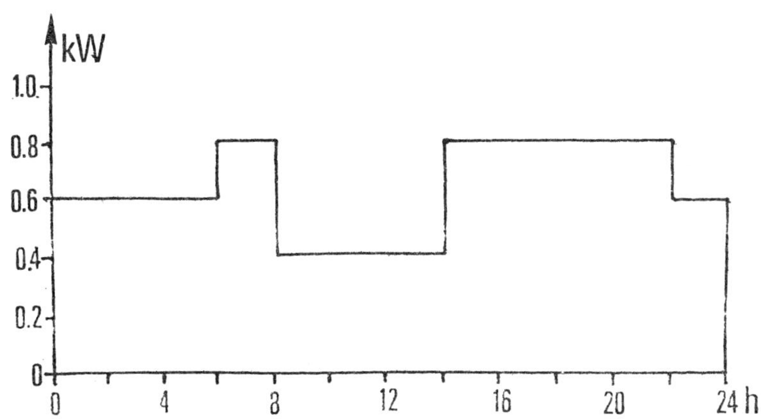

Figure 7

Overall storage characteristics

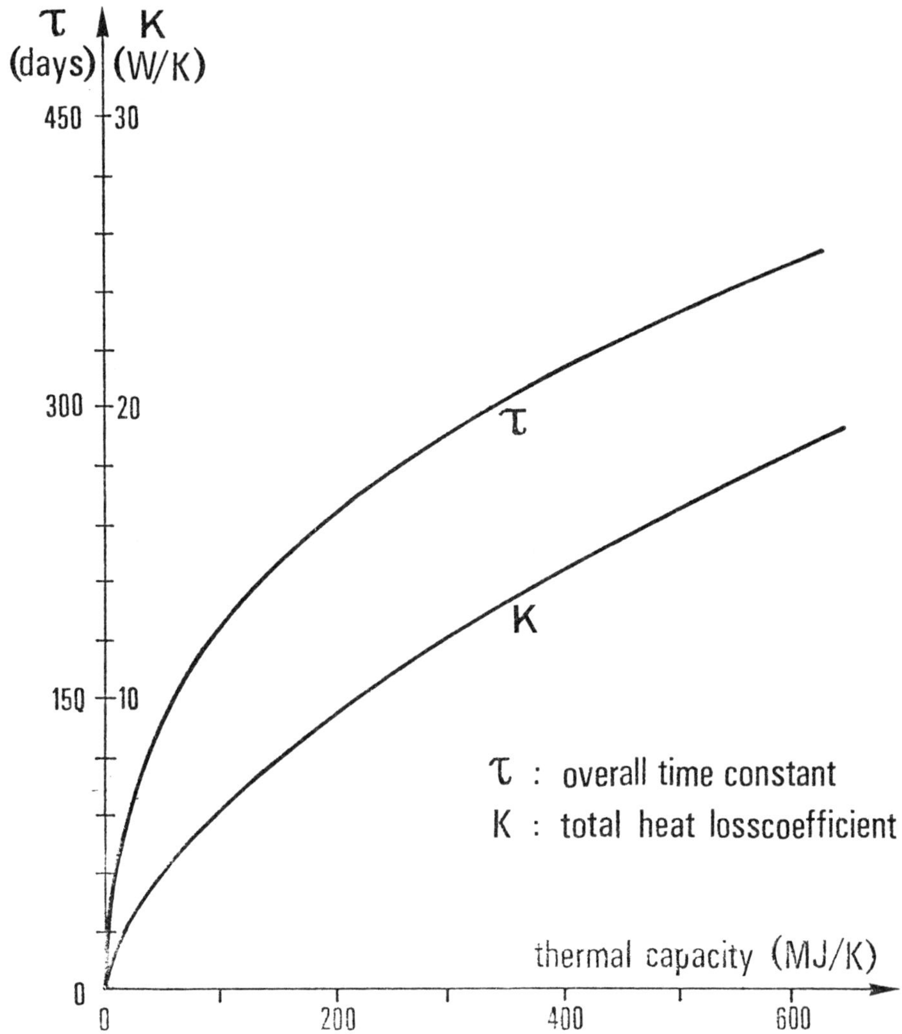

Figure 8

Double glass collector efficiency curve

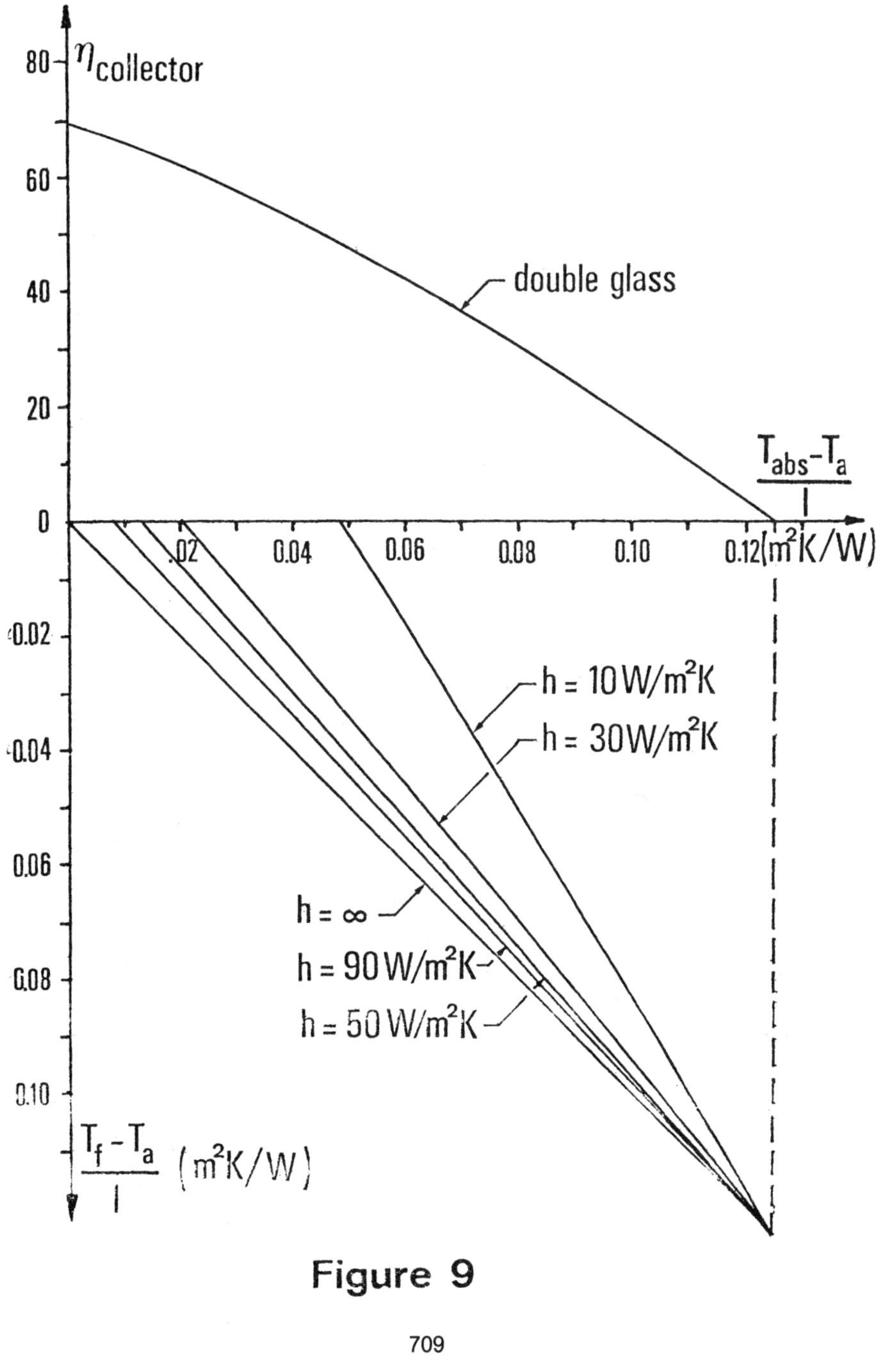

Figure 9

Q_{wi} :the thermal power supplied to the inner walls by incident solar radiation through the windows and heat losses from the heat distribution ducts to the floor and ceiling in which they are mounted.

Q_a :internal heat dissipation in the indoor air from lighting,occupants, ...,ventilation losses and the heat supply from the heating system.

To satisfy the periodicity requirement when complete year simulations are performed,several consecutive runs are needed in the case of seasonal storage systems.In practice,two runs appear to be sufficient if the simulation is initiated after the heating season has expired.

An experimental test facility has been build to validate the simulation model.Results of the comparison between theoretically calculated and measured values will be shortly available.

Results

Using the computer simulations programme,the overall performance of systems as described above,has been calculated for a wide range of the most important system design parameters,such as total collector area,thermal storage capacity,total heat demand of the dwelling as well as its temporal distribution.
In these calculations,the hysteresis loops of the regulators as represented in figure 3,a daily thermostat setting cycle as represented in fig.6 and an internal heat dissipation power cycle as represented in fig.7,have been used.The climatological data which has been used as input,represents the Belgian climatological reference year,consisting of half-hourly averaged real data for each month.Each of these months has been selected by the Belgian Royal Meteorological Institute out of its observations over a twenty year period and based on an appropriate Chi-square selection criterion.
The rock bed storage system is assumed to have a specific heat capacity of 2 MJ/m^3K,an average rock to air heat transfer coefficient of 1000 W/m^3K and an overall heat loss coefficient and time constant as represented in figure 8 as a function of its total thermal capacity.The storage system is assumed to loose heat to an environment at a constant temperature of 10°C. The nominal design temperature T_n of the heating system has been taken to be 35°C.
Unless otherwise specified,all the calculations have been performed using double glass-black paint air collectors with an average stationary efficiency curve as represented in figure 9 versus $(T_{abs}-T_a)/I$,where T_{abs} is the absorber temperatureT_a the ambient temperature and I the solar insolation on the collector per unit area.In the same figure,the approximate correspondance between the absorber temperatur T_{abs} and the average fluid temperatur T_f is represented for different values of the collector internal absorber plate to air heat transfer coefficient. If not specified,a value of 50 W/m^2K has been used for this heat transfer coefficient.The collector orientation,if not specified,has been taken

Heat demand distribution types

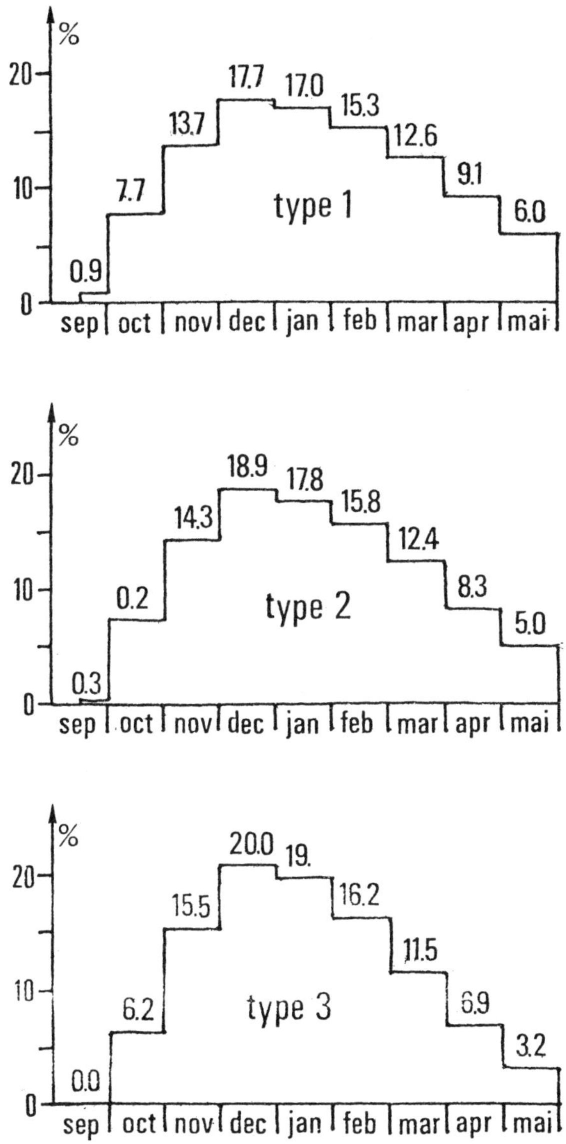

Figure 10

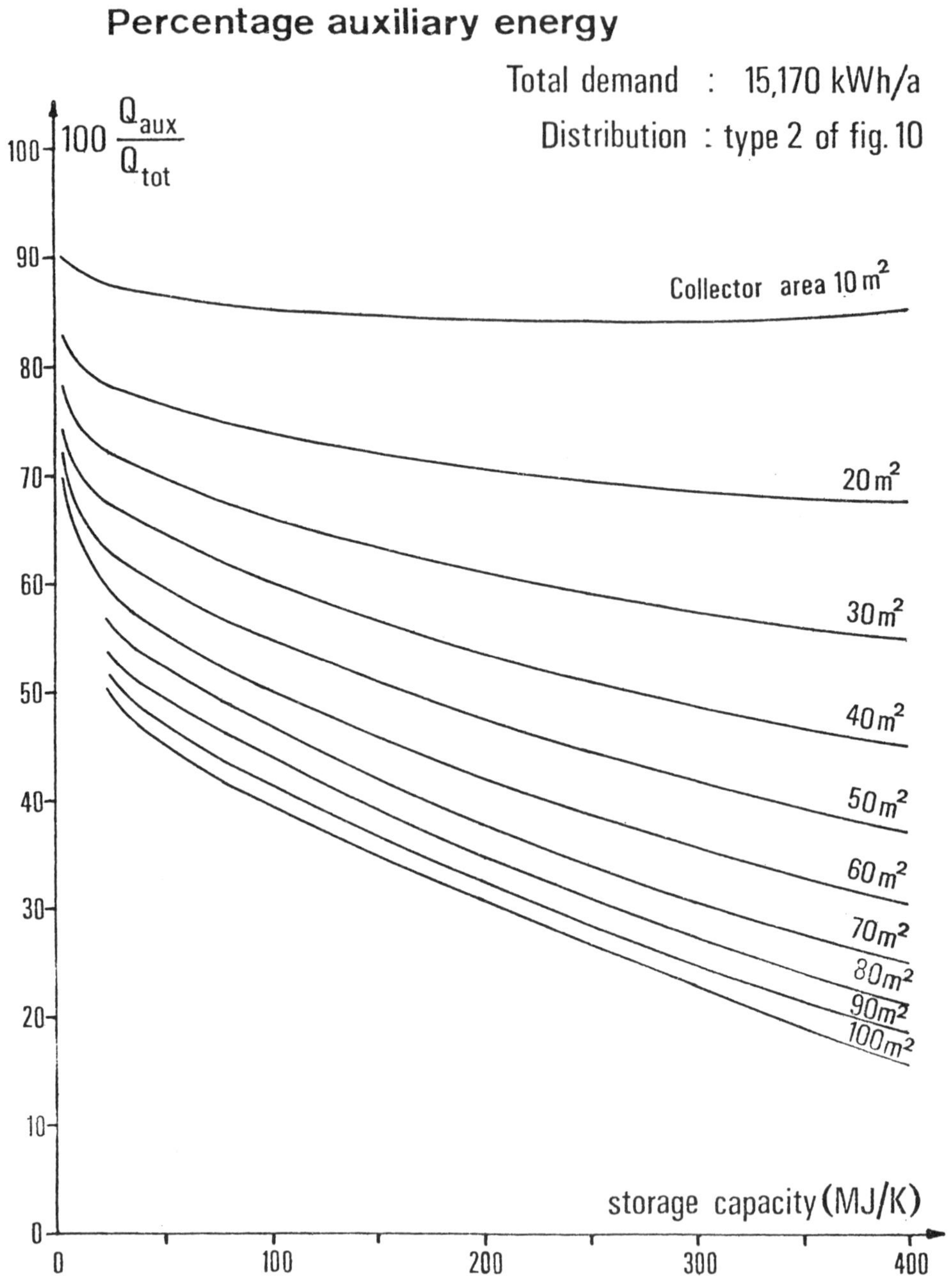
Percentage auxiliary energy
Total demand : 15,170 kWh/a
Distribution : type 2 of fig. 10
$100 \frac{Q_{aux}}{Q_{tot}}$
Collector area 10 m²
20 m²
30 m²
40 m²
50 m²
60 m²
70 m²
80 m²
90 m²
100 m²
storage capacity (MJ/K)
0
10
20
30
40
50
60
70
80
90
100
0
100
200
300
400

Figure 11

Curves of constant solar energy contribution

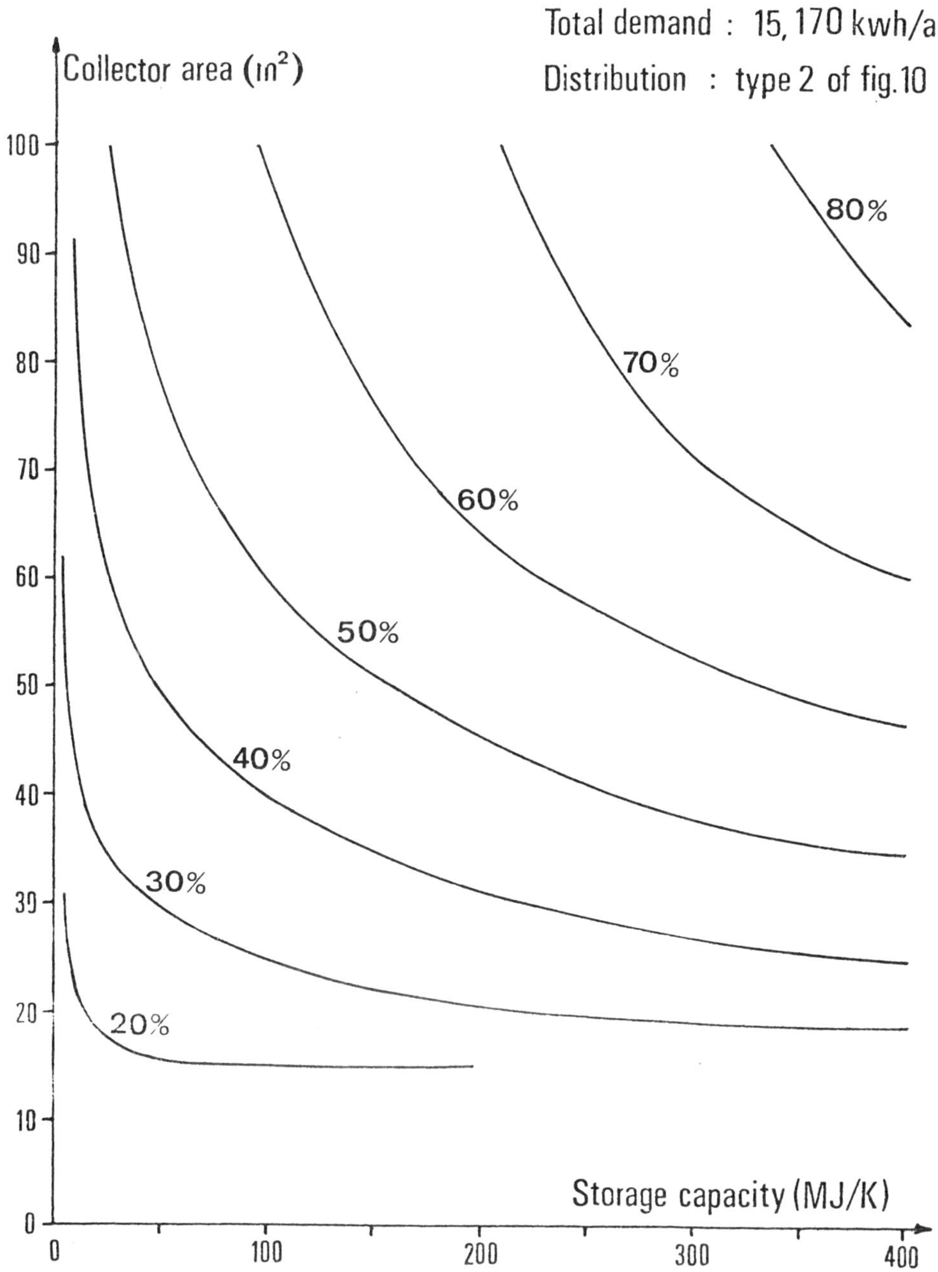

Figure 12

Influence of demand distribution on the system performance

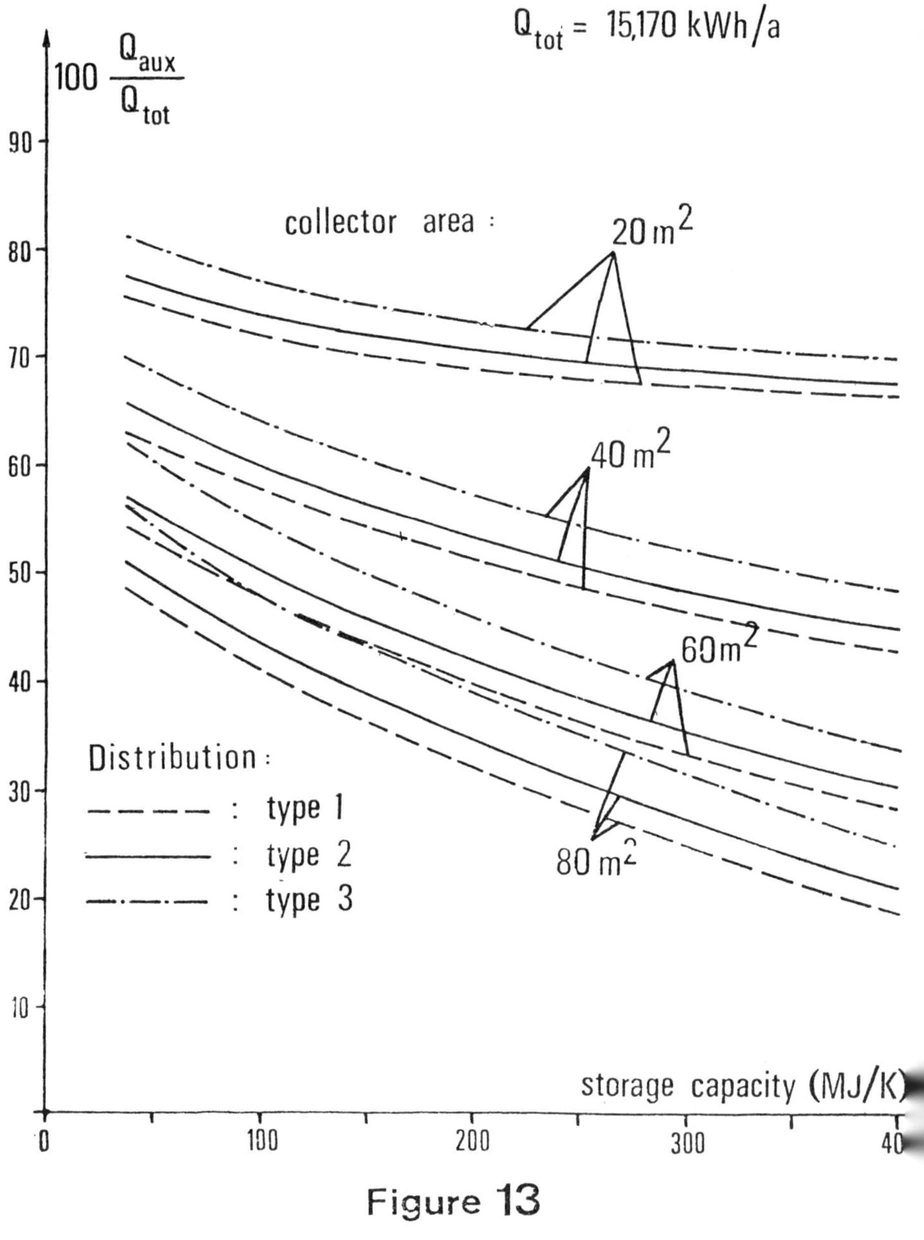

Figure 13

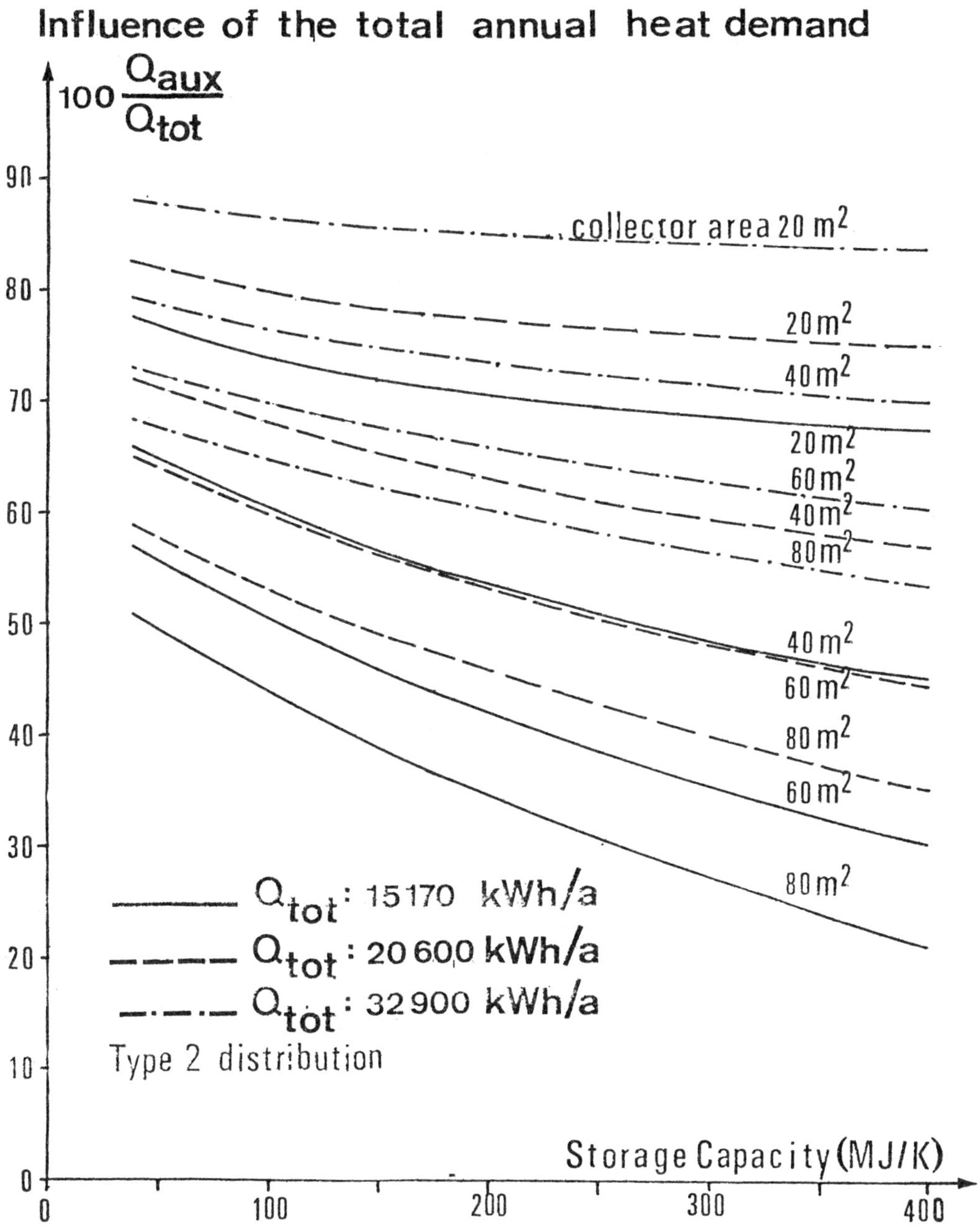
Influence of the total annual heat demand
100 Qaux/Qtot
collector area 20 m2
20 m2
40 m2
20 m2
60 m2
40 m2
80 m2
40 m2
60 m2
80 m2
60 m2
80 m2
Qtot: 15170 kWh/a
Qtot: 20600 kWh/a
Qtot: 32900 kWh/a
Type 2 distribution
Storage Capacity (MJ/K)
0
10
20
30
40
50
60
70
80
90
100
200
300
400

Figure 14

to be south,with a 50° inclination angle.

Three different types of houses have been considered,which can be characterised by the distribution of the heat demand.The monthly histogrammes of the heat demand of each type is represented in figure 10.The heating season extends from September 15 to May 31.The influence of other parameters,such as total annual demand,storage insulation,internal collector heat transfer coefficient,collector type and collector orientation,has been investigated,assuming the type 2 demand distribution. In all the results reported hereafter,the electrical energy consumption needed to drive the ventilator,is never included in the auxiliary energy percentage.

For a dwelling with a total annual heat demand Q_{tot}=15,170 kWh,the net auxiliary energy to be supplied,denoted by Q_{aux}, is given in the tables 1,2 and 3 for each of the three types of heat demand distributions respectively,and for different values of the collector area and the storage capacity.The figures in the tables give Q_{aux} as a percentage of the total demand,the efficiency of the auxiliary heating system not being included.For the mean distribution type(i.e. type 2 in figure 10),these results (listed in table 2) are also represented graphically in figur 11.Curves of constant percentage of solar energy are shown in figure 12,in a collector area versus storage capacity diagramme.For different values of the collector area,figure 13 shows the influence of the heat demand distribution on the system performance.It can be observed that peaking of the distribution increases the auxiliary energy required to meet the heat demand.Naturally,this influence slightly decreases with increasing storage capacity.The difference in system performance between the type 1 and type 3 distribution cases,is about 5 %for small collector areas and increases to about 7 % for larger systems.
The influence of the total annual heat demand on the system performance is illustrated in figure 14,where the percentage of auxiliary energy is plotted versus storage capacity for different collector areas and three different values of the total heat demand,15,170 kWh/a,20,600 kWh/a and 32,800 kWh/a,the distribution being of the type 2 in each case.It can be observed from these results that the absolute value of the solar energy contribution to be expected from a particular system,naturally increases with increasing total demand.

The thermal insulation of the storage system has a considerable influence on the system performance.To evaluate this influence,calculations have been performed for other storage insulation characteristics than those represented in figure 8.Figure 15 represents the effect on the system performance as a results of a doubling of the storage heat loss coefficient K as compared to the values given in figure 8.Figure 16 represents the effect of infinite storage insulation as compared to the reference cases with the storage characteristic as in figure 8.Figure 16 thus represents the physical limit of what can be gained from a better storage insulation,taking the results of table 2 or figure 11 as a

Annual Percentage Auxiliary Energy

Total energy demand : 15,170 kWh/year
Distribution : type 1 of figure 10
Collector orientation : South, 50° inclination
Location : Brussels, Belgium

Table 1

Storage Capacity (MJ/K)	Collector area (m^2)									
	10	20	30	40	50	60	70	80	90	100
4	88.4	80.7	75.1	71.5	68.6	66.0				
8	87.6	78.8	72.2	67.8	64.4	61.8				
16	87.0	77.3	70.4	65.8	61.7	58.2				
24	86.6	76.5	69.6	64.6	60.2	56.5				
32	86.3	75.8	68.9	63.6	59.0	55.2				
40	86.0	75.2	68.2	62.7	58.0	53.9	50.7	48.1	45.8	43.7
80	84.8	72.9	65.2	59.2	54.0	49.6	46.0	43.1	40.6	38.8
120	84.2	71.1	62.6	56.2	50.7	45.8	41.9	39.0	37.0	35.3
160	83.8	69.8	60.5	53.5	47.7	42.6	38.7	35.4	33.5	31.6
200	83.6	68.8	58.9	51.3	45.1	39.8	35.4	32.3	30.2	28.4
240	83.6	68.2	57.1	49.2	42.7	36.9	32.4	29.2	26.8	24.9
280	83.6	67.4	55.8	47.4	40.4	34.4	29.7	26.3	23.9	22.0
320	83.8	66.8	54.6	45.7	38.2	32.2	27.2	23.7	21.1	18.7
360	84.0	66.6	53.6	44.2	36.5	30.1	24.9	21.1	18.3	16.1
400	84.1	66.0	52.8	42.8	34.7	28.1	22.6	18.7	15.9	13.5

Annual Percentage Auxiliary Energy

Total energy demand : 15,170 kWh/year
Distribution : type 2 of figure 10
Collector orientation : South, 50° inclination
Location : Brussels, Belgium

Table 2

Storage Capacity (MJ/K)	Collector area (m^2)									
	10	20	30	40	50	60	70	80	90	100
4	90.0	82.9	78.1	74.3	72.0	69.7				
8	89.0	80.9	75.2	71.2	67.9	65.3				
16	88.2	79.4	73.4	69.0	64.9	61.8				
24	87.6	78.7	72.3	67.7	63.4	59.6				
32	87.3	77.8	71.5	66.5	62.2	58.0				
40	87.0	77.2	70.6	65.5	60.9	56.8	53.6	50.6	48.4	46.4
80	85.6	74.7	67.4	61.8	56.5	52.3	48.6	45.7	43.2	41.5
120	84.9	72.9	64.7	58.5	53.2	48.4	44.6	41.6	39.2	37.7
160	84.5	71.5	62.6	55.9	50.1	45.0	41.0	38.1	35.9	34.1
200	84.2	70.5	60.7	53.4	47.5	41.9	37.6	34.7	32.4	30.8
240	84.2	69.6	59.1	51.2	44.8	39.3	34.5	31.5	29.3	27.3
280	84.1	68.7	57.7	49.3	42.6	36.6	31.9	28.5	26.3	24.3
320	84.3	68.3	56.6	47.7	40.6	34.3	29.6	26.0	23.1	21.1
360	84.4	67.8	55.6	46.2	38.4	32.1	27.0	23.3	20.5	18.1
400	84.6	67.3	54.5	44.9	36.9	30.1	24.7	21.0	19.1	15.5

Influence of the storage insulation on the system performance

Storage loss coeff. : 2 x K of fig. 8

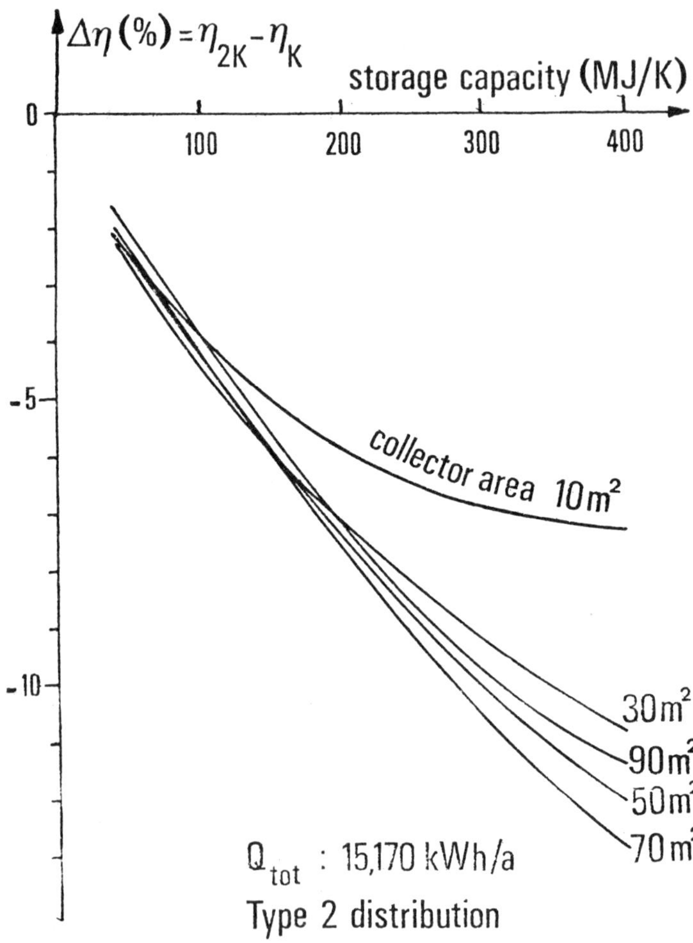

Figure 15

Influence of the storage insulation on the system performance

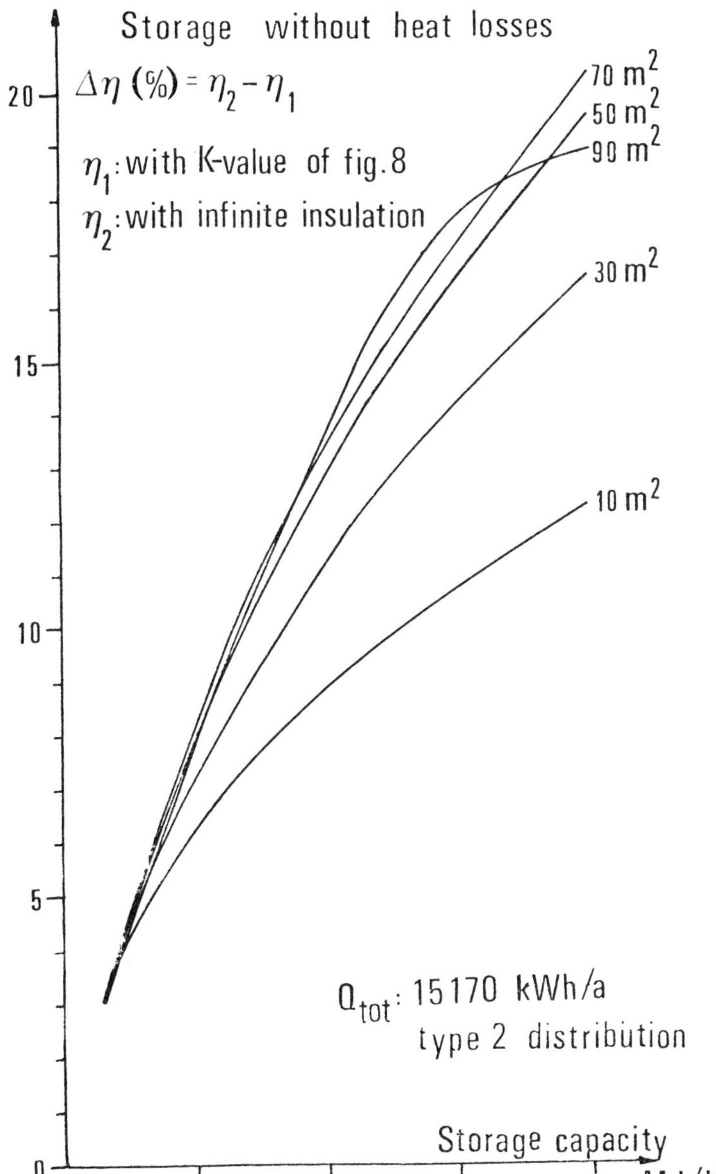

Figure 16

reference.It appears that the effect of storage insulation is very important and increases with increasing storage capacity.For sufficiently large collector areas,the effect is limited of course by the fact that with increasing storage capacity the system can finally meet the total demand if the storage is perfectly insulated.

As mentioned above,generally,the lower internal heat tranfer coefficient from the absorber to the cooling fluid in air collectors as compared to liquid collectors,lowers the collector efficiency at a specified fluid temperature.This heat transfer coefficient h is defined per unit area of the flat collector,such that an increase of the inner heat exchange surface per unit collector area,increases the effective value of h.In order to evaluate the importance of this parameter,which can be influenced by the design of the air collectors as well as by changing the mass flow rate,system calculations have been performed for different values of this heat tranfer coefficient.For the 15,170 kWh/a case with the type 2 distribution,the results are plotted in figure 17 versus storage capacity,for different values of the collector area and the heat transfer coefficient.This figure shows the relative difference in solar energy converted between the cases with the indicated h values and the cases given in table 2 and figure 11.It is clear from figure 17 that an increase of the effective value of the heat transfer coefficient in the air collector increases the overall system efficiency.However, above 30 W/m^2K, the resulting additional energy savings become rather negligible, such that air collectors should not be designed for still higher values of h.

If the system is considered to be equipped with single glass-black paint collectors having an average efficiency curve as represented in figure 18,instead of the double glass collectors of figure 9,it is of interest to evaluate the corresponding influence on the overall system efficiency. This effect is shown in figure 19 for some values of the collector area, as a function of the storage capacity.The influence increases with the collector area but within the range of systems being considered,it is limited to about 5 %.

All the results mentioned thus far, correspond to south oriented collectors at a 50° inclination angle.The dependence of the system performance on collector orientation and inclination has been studied up to ± 20° deviation from the south direction and from 30° to 80° inclination angles.For different collector areas and storage capacities,the results are summarised in the tables 4,5,6,7 and 8.It appears that the system performance depends very little on collector orientation.The effect due to a non optimal collector orientation is small as compared to most of the other influences which have been studied.For the climatological conditions for which the calculations have been performed,a slightly east oriented collector (about 10° east) gives a somewhat higher energy yield as compared to a south oriented collector.

Influence of the collector internal heat transfer coefficient on the system performance

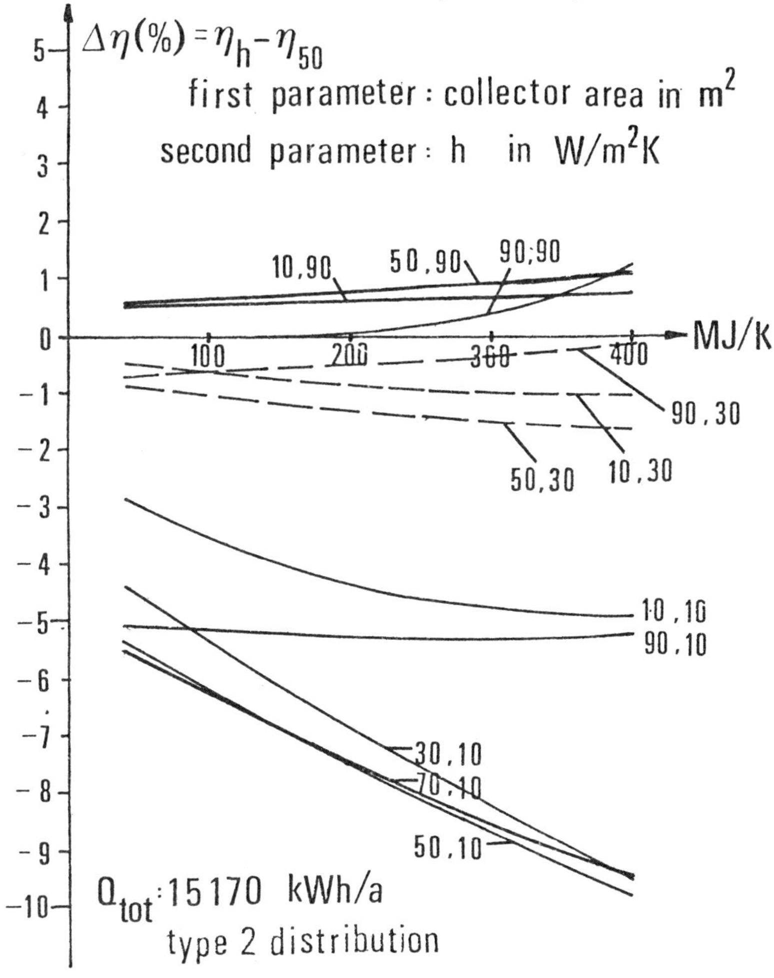

Figure 17

Single glass collector efficiency

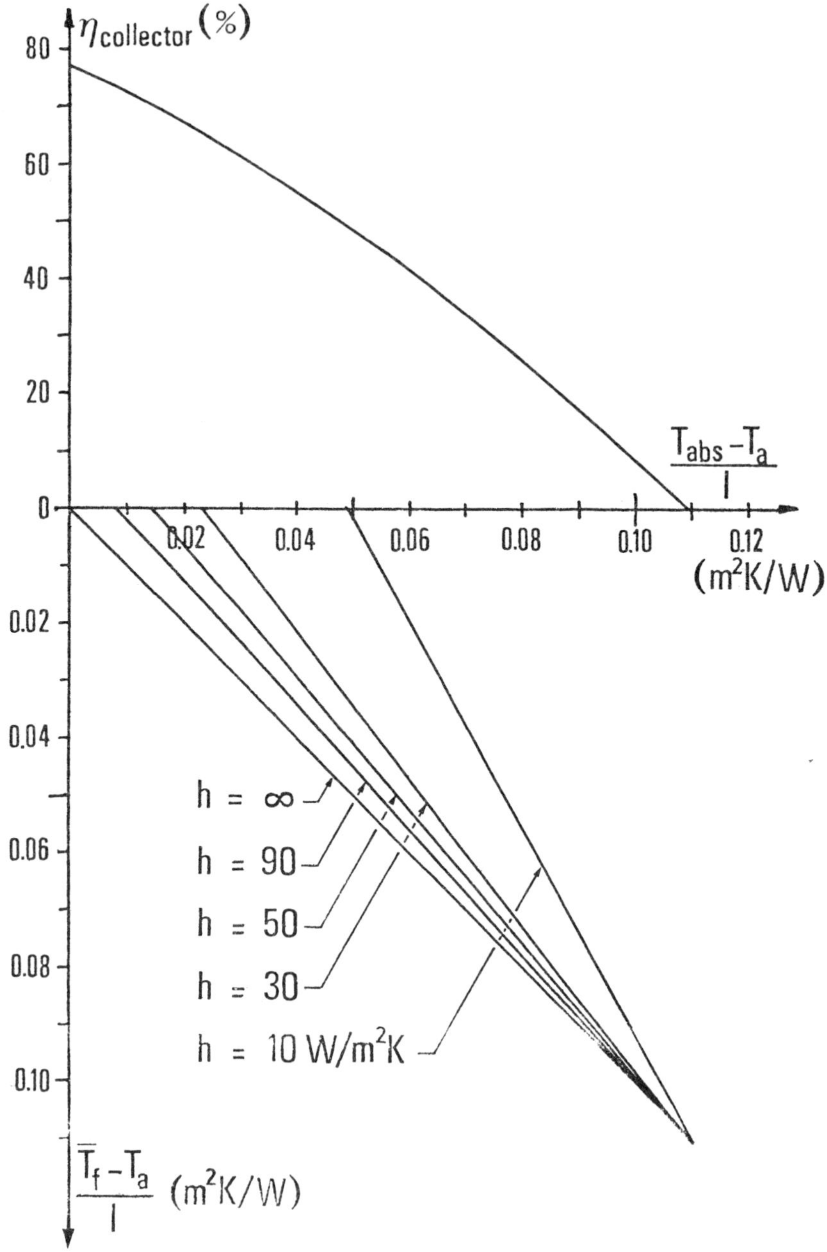

Figure 18

Influence of collector efficiency on the system performance

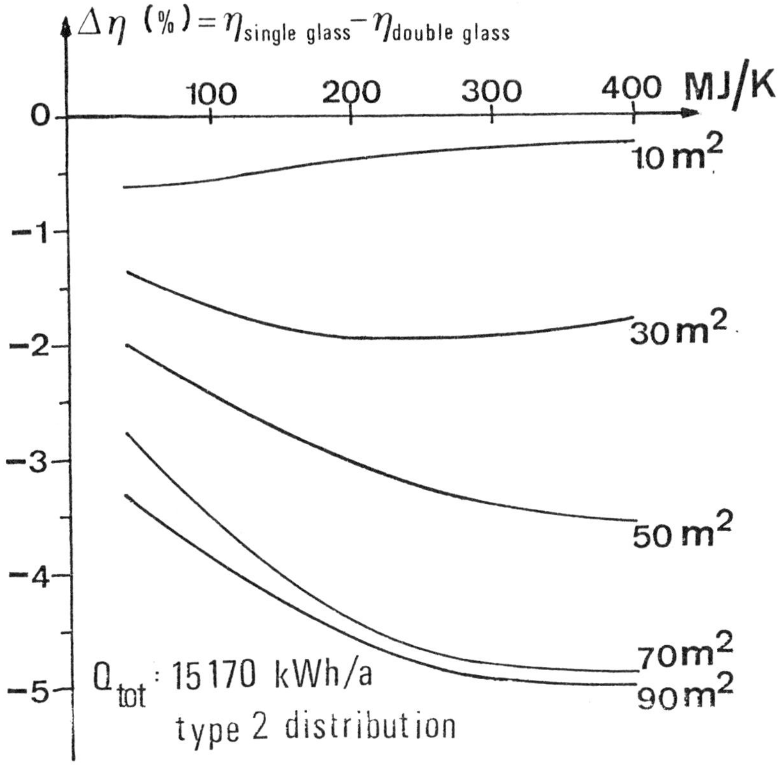

Figure 19

Annual Percentage Auxiliary Energy

Total energy demand : 15,170 kWh/year
Distribution : type 3 of figure 10
Collector orientation : South, 50° inclination
Location : Brussels, Belgium

Storage Capacity (MJ/K)	Collector area (m^2)									
	10	20	30	40	50	60	70	80	90	100
4	91.6	86.2	82.7	80.1	78.0	76.5				
8	90.5	84.4	80.3	77.0	74.4	72.3				
16	89.7	82.9	78.1	74.0	70.6	67.8				
24	89.0	82.0	76.8	72.0	68.2	65.0				
32	88.5	81.4	75.6	70.9	66.5	63.0				
40	88.1	80.7	74.8	69.7	65.2	61.6	58.7	55.9	53.6	51.7
80	86.7	78.0	71.4	65.6	60.7	56.6	53.0	50.2	47.9	46.3
120	85.9	76.0	68.6	62.5	57.1	52.4	48.6	46.2	44.0	42.5
160	85.5	74.4	66.3	59.7	53.8	49.1	45.1	42.6	40.6	38.9
200	85.3	73.2	64.4	57.2	51.0	45.8	41.7	39.2	37.1	35.2
240	85.1	72.2	62.6	55.0	48.5	42.9	38.6	36.1	33.7	31.9
280	85.1	71.4	61.2	53.0	46.1	40.5	35.9	32.9	30.5	28.5
320	85.0	70.8	59.8	51.3	44.2	38.0	33.1	30.1	27.6	25.6
360	85.1	70.0	58.7	49.8	42.1	35.7	30.8	27.4	24.9	22.7
400	85.1	69.6	57.7	48.2	40.4	33.7	28.6	25.0	22.3	19.8

Table 3

Annual Percentage Auxiliary Energy : Dependence on Collector Orientation

Collector area : 10 m^2

Storage Capacity (MJ/K)	Collector Inclination (degrees)	Collector Orientation				
		Degrees East		South	Degrees West	
		20	10		10	20
40	30	87.5	87.4	87.6	87.8	88.0
	40	87.0	87.0	87.1	87.3	87.7
	50	86.8	86.8	87.0	87.2	87.5
	60	86.9	86.8	87.0	87.3	87.7
	70			87.4		
	80			87.9		
80	30	86.2	86.2	86.3	86.5	86.7
	40	85.7	85.6	85.7	86.1	86.4
	50	85.4	85.5	85.6	85.9	86.4
	60	85.6	85.6	85.7	86.0	86.6
	70			86.1		
	80			86.9		
120	30	85.4	85.4	85.5	85.7	86.0
	40	84.9	84.8	85.0	85.3	85.6
	50	84.7	84.8	84.9	85.2	85.6
	60	84.9	84.8	85.0	85.4	85.9
	70			85.5		
	80			86.4		
200	30	84.6	84.7	84.8	85.0	85.3
	40	84.2	84.2	84.3	84.7	85.0
	50	84.1	84.0	84.2	84.6	85.1
	60	84.3	84.3	84.6	84.8	85.4
	70			85.1		
	80			86.0		
400	30	84.9	84.9	85.0	85.3	85.6
	40	84.4	84.4	84.6	84.9	85.3
	50	84.4	84.4	84.6	85.0	85.5
	60	84.7	84.7	84.9	85.4	85.9
	70			85.6		
	80			86.6		

Table 4

Annual Percentage Auxiliary Energy : Dependence on Collector Orientation

Collector area : 30 m^2

Storage Capacity (MJ/K)	Collector Inclination (degrees)	Collector Orientation				
		Degrees East		South	Degrees West	
		20	10		10	20
40	30	72.5	72.6	72.7	73.2	73.6
	40	71.3	71.3	71.5	72.0	72.6
	50	70.6	70.5	70.6	71.3	71.9
	60	70.2	70.2	70.4	71.0	71.9
	70			70.6		
	80			71.3		
80	30	69.5	69.5	69.6	70.1	70.6
	40	68.2	68.0	68.3	68.8	69.6
	50	67.2	67.2	67.4	68.0	69.0
	60	66.9	67.0	67.3	68.1	68.9
	70			67.6		
	80			68.5		
120	30	66.8	66.7	67.0	67.2	67.9
	40	65.4	65.5	65.5	66.1	66.8
	50	64.6	64.5	64.7	65.4	66.2
	60	64.4	64.3	64.7	65.5	66.4
	70			65.1		
	80			66.4		
200	30	62.6	62.5	62.8	63.3	64.0
	40	61.2	61.2	61.5	62.1	62.8
	50	60.4	60.5	60.7	61.4	62.4
	60	60.4	60.4	60.8	61.5	62.7
	70			61.5		
	80			63.1		
400	30	56.2	56.3	56.4	56.9	57.8
	40	54.7	54.7	55.1	55.8	56.8
	50	54.2	54.2	54.5	55.5	56.5
	60	54.5	54.6	54.8	55.9	57.3
	70			56.2		
	80			58.2		

Table 5

Annual Percentage Auxiliary Energy : Dependence on Collector Orientation

Collector Area : 50 m^2

Storage Capacity (MJ/K)	Collector Inclination (degrees)	Collector Orientation				
		Degrees East		South	Degrees West	
		20	10		10	20
40	30	63.6	63.4	63.8	64.2	65.1
	40	61.7	61.8	62.1	62.5	63.5
	50	60.5	60.5	60.9	61.6	62.6
	60	60.0	60.0	60.3	61.1	62.5
	70			60.4		
	80			61.3		
80	30	59.5	59.5	59.6	60.3	61.1
	40	57.6	57.5	57.8	58.5	59.5
	50	56.3	56.3	56.5	57.5	58.5
	60	55.9	55.7	56.2	57.1	58.4
	70			56.6		
	80			57.6		
120	30	56.0	55.9	56.2	56.7	57.5
	40	53.9	53.9	54.3	55.0	56.0
	50	52.7	52.7	53.2	53.9	55.2
	60	52.3	52.4	52.8	53.7	55.0
	70			53.4		
	80			54.6		
200	30	50.3	50.1	50.5	50.9	52.0
	40	48.2	48.0	48.4	49.3	50.2
	50	46.9	47.1	47.5	48.2	49.6
	60	46.7	46.7	47.2	48.3	49.7
	70			47.9		
	80			49.7		
400	30	39.5	39.5	39.6	40.4	41.6
	40	37.5	37.4	37.6	38.6	40.0
	50	36.4	36.2	36.9	37.8	39.5
	60	36.5	36.4	37.0	38.3	40.0
	70			38.3		
	80			41.1		

Table 6

Annual Percentage Auxiliary Energy : Dependence on Collector Orientation

Collector Area : 70 m^2

Storage Capacity (MJ/K)	Collector Inclination (degrees)	Collector Orientation				
		Degrees East		South	Degrees West	
		20	10		10	20
40	30	56.8	56.8	57.1	57.7	58.5
	40	54.6	54.6	55.2	55.7	56.6
	50	53.1	53.2	53.6	54.5	55.6
	60	52.4	52.5	52.8	53.8	55.2
	70			53.0		
	80			53.9		
80	30	52.3	52.2	52.6	53.0	53.9
	40	50.0	49.9	50.3	51.0	52.0
	50	48.3	48.3	48.6	49.5	50.9
	60	47.7	47.7	48.1	49.2	50.6
	70			48.3		
	80			49.3		
120	30	48.2	48.1	48.3	49.1	50.0
	40	45.8	45.7	46.2	47.0	48.1
	50	44.2	44.1	44.6	45.6	47.1
	60	43.6	43.4	44.1	45.2	46.8
	70			44.4		
	80			45.9		
200	30	41.3	41.2	41.3	42.4	43.3
	40	38.7	38.6	39.0	39.9	41.3
	50	37.4	37.3	37.6	39.0	40.2
	60	36.9	36.8	37.4	38.3	40.2
	70			38.1		
	80			40.0		
400	30	27.8	28.0	28.1	29.2	30.3
	40	25.4	25.4	25.7	26.6	28.4
	50	24.1	24.0	24.7	25.9	27.7
	60	24.1	24.2	24.9	26.1	28.1
	70			26.0		
	80			28.7		

Table 7

Annual Percentage Auxiliary Energy : Dependence on Collector Orientation

Collector Area : 90 m^2

Storage Capacity (MJ/K)	Collector Inclination (degrees)	Collector Orientation				
		Degrees East		South	Degrees West	
		20	10		10	20
40	30	52.1	52.1	52.2	53.0	53.8
	40	49.8	49.7	50.1	50.8	51.9
	50	48.2	48.0	48.4	49.4	50.7
	60	47.3	47.2	47.7	48.7	50.1
	70			47.7		
	80			48.5		
80	30	47.1	47.1	47.3	48.1	49.0
	40	44.5	44.4	44.7	45.4	46.8
	50	42.9	42.8	43.2	44.0	45.7
	60	42.2	42.0	42.5	43.5	45.1
	70			42.3		
	80			43.6		
120	30	43.0	42.9	43.0	43.7	44.7
	40	40.6	40.5	40.8	41.5	42.6
	50	39.0	39.0	39.2	40.4	41.6
	60	38.4	38.3	38.8	39.7	41.1
	70			38.8		
	80			39.8		
200	30	36.0	35.8	36.1	37.1	37.9
	40	33.4	33.3	33.9	34.8	36.0
	50	32.1	31.9	32.4	33.3	34.7
	60	31.7	31.6	32.2	33.1	34.6
	70			32.4		
	80			33.8		
400	30	21.2	22.1	21.2	22.3	23.3
	40	18.9	18.8	19.1	20.1	21.4
	50	17.7	17.5	19.1	19.0	20.7
	60	17.8	17.4	18.1	19.3	20.9
	70			19.4		
	80			21.4		

Table 8

CONCLUSIONS

A solar air heating system and its control system to allow for different operation modes has been described and a computer simulation model, describing the complete dynamic behaviour of the system has been developed.
The simulation programme has been used to investigate the dependence of the overall system efficiency on the basic design parameters and some secondary parameters.From the results it appears that starting from a basic design,the system performance can be improved considerably mainly by additional storage insulation and improved internal heat transfer in the air collectors.However,the system performance also depends on the heat demand distribution.
An increase of the system efficiency by a specified amount can be achieved by different measures,but the answer to the question which measure should be preferred,can be given only by economic considerations. The results reported in this paper give sufficient basic data to perform such a complete economic evaluation.

ACKNOWLEDGEMENT

This work was supported by the Belgian government as part of the Belgian R&D programme on solar energy applications.

INFLUENCE OF HEAT EXCHANGER EFFECTIVENESS ON THE PERFORMANCE OF THERMOSYPHON DOUBLE LOOP WATER HEATING SYSTEM

A.F.Orlando, L.Goldstein Jr., D.Magnoli
Centro de Tecnologia Promon - CTP
Rio de Janeiro, Brasil

ABSTRACT

A double loop thermosyphon solar heater has been developed and its performance measured under different conditions.

Absorbers in a collector are non selective and of the roll-bond aluminum type. For the required consumption hot water temperature range (40-60°C), single glazing collectors are experimentally shown to be more efficient than double glazing collectors.

Heat exchanger design is found to be critical to the technical and economic performance of the water heater. The influence of the heat exchanger has been analysed in terms of the coefficient of performance-COP. It is defined as the ratio of the usable heat when the heat exchanger is used, and when the consumption water is heated directly in the collectors. The COP is smaller for higher hot water temperatures.

Finally, the system is found to be more efficient for larger ratios between heat exchanger surface and collector surface.

INTRODUCTION

Water heating is one of the most commercially favorable solar energy applications. Its economic feasibility varies widely in Brasil, according to the availability of alternative energy sources, whose average costs (March 1978 US dollars) are:

	Cents/kWh (e) or (th)
. Residential electricity	6.1
. Industrial electricity	2.6
. Town gas	3.2
. LPG-Liquefied Petroleum Gas	2.7
. Fuel Oil	0.5

For residential consumption, electricity, town gas and LPG are normally used. For industrial purposes, fuel oil and electricity are mainly used.

The world's seven largest countries by surface area receive the following estimated extraterrestrial incident solar energy radiation on a horizontal surface in a one year period in 10^{15} kWh (1)

. USSR	46
. Brasil	29
. China	27
. USA	25
. Australia	25
. Canada	20
. India	11

The actual solar radiation at the ground level, however, could be 40-60% lower than the values shown above due to atmospheric absorption and reflection. Furthermore most of the Brasilian surface is located between the Equator and the Tropic of Capricorn. The solar radiation is thus less attenuated and more evenly distributed all over the year. Therefore, Brasil is certainly one of the richest countries in the world with respect to solar energy potential and could derive large benefits from its applications.

Under present energy price structure, the feasibility of utilizing solar water heater in industry depends heavily upon the economics of solar heating systems.

The use of cheap collectors can substantially reduce the system cost. Aluminum is potentially a good candidate as an absorber material, because it is relatively cheap and available in Brasil. However, aluminum collectors must be used with corrosion inhibited water in a closed loop system.

Double loop systems are in general less efficient than open loop systems, because collectors must operate at higher temperatures. For residential size water heating applications, double loop systems are presently not competitive with open loop systems (employing copper collectors) due to the additional cost of the heat exchanger and other components, even if cheap aluminum collectors are used. However, larger double loop systems can be cheaper per unit area of collectors, because the cost of the heat exchanger and other components do not increase appreciably with capacity.

Large thermosyphon systems (up to possibly 20 square meters) (2) may find application in regions of Brasil with limited energy supply. When aluminum collectors are used in a double loop system, the water heater system could perhaps be cost effective with respect to open loop systems using copper collectors.

This article investigates the use of one and two covers for the collectors, together with the loss of efficiency associated with the use of a heat exchanger, in a thermosyphon loop.

SYSTEM DESCRIPTION

The double loop thermosyphon system is described in detail elsewhere (3).

Figure 1 shows the system schematic.

Treated water plus corrosion inhibitor (fluid A) circulates naturally in a closed primary loop between the solar collectors, where it is heated, and a heat exchanger, where heat is transferred to cold consumption water (fluid B). Fluid B pressure is reduced to near atmospheric at the pressure reduction tank, which is located above the heat exchanger; then, by gravity, fluid B is fed to heat exchanger, where it is heated, and is stored in the hot water storage tank, below the collector level, ready for consumption. A thermostatic valve, at the exit of the heat exchanger, keeps the fluid B temperature constant (40-60°C) by varying the flow rate.

Collectors

Absorbers in a collector are two 0.40 x 1.80 m roll bond type aluminum plates, making up 1.44 m^2 of effective collector area. The system has 3 such collectors, inclined at 23°30' (local latitude of São Paulo), thus adding up 4.32 m^2 of collector surface. Each plate has six 46.5 mm^2 (cross section) ducts, properly enlarged to minimize the influence of the pressure drop on the thermosyphon flow rate. The calculated flow rate differences between ducts are less than 10%. Absorber surfaces are non selective, painted flat black, with a solar absorptance of 0.95. Collectors have a 4 mm thick glass plate as an upper cover and a 100 microns Mylar type plastic film (sold by RHODIA) as an alternative second cover, thus protected against UV radiation. The plastic cover has a measured solar transmittance of 0.85 and an infrared transmittance (100°C) of 0.18. Insulation is a 7.82 cm thick fiber glass blanket.

Working Fluid (fluid A)

A small percentage of MAXLUB (made by BARDAHL) in a 10% aqueous solution was added to the primary water in this development.

Heat Exchanger

It is an off-the-shelf 0.32 m^2 heat transfer surface shell-and-tube heat exchanger. Maintenance was minimized by circulating clean fluid A through the tubes.

Thermostatic Valve

Continously regulating valves have been used in this development, keeping the consumption water (fluid B) constant in the 40-60°C range.

Three different ratios of collector area to heat exchanger surface were used: 4.5 : 1, 9.0 : 1 and 13.5 : 1, respectively with one, two and three collectors connected to the thermosyphon primary loop. The heat exchanger surface has been kept the same for all experiments. In single glazing collector tests, only the upper glass cover has been used.

SYSTEM PERFORMANCE EVALUATION

Forced circulation system performance can be obtained by means of available standards (4). However, test procedures for thermosyphon systems are not well developed mainly due to the difficulties in measuring the flow rate.

The double loop thermosyphon water heating system, described in this work, can be easily evaluated by measuring directly the energy flow rate to fluid B. The measurements made during the tests do not necessarily follow standard accuracy values in this type of test procedure; however, they provide the required preliminary information for system design.Improvements are underway to reduce the uncertainty of the measurements. The following quantities have been directly measured (3): incident radiation, consumption water (fluid B) flow rate, inlet and outlet temperatures from collectors and heat exchanger secondary loop (fluid B). Fluid A flow rate through collectors was calculated assuming no heat loss to ambient from pipes and other components.

System Efficiency (η)

The system efficiency is defined as the ratio between the actual energy flow rate to fluid B and the total incident radiation on the collector plane. The energy flow rate ($\dot{Q}$) can be calculated by the following expression:

$$\dot{Q} = \dot{m}_A C_{p,A} (T_{h,A} - T_{c,A}) = \dot{m}_B C_{p,B} (T_{h,B} - T_{c,B}) \qquad (1)$$

where

$\dot{m}_A$, $\dot{m}_B$ = mass flow rate of fluid A and fluid B respectively, kg/seg

$C_{p,A}$, $C_{p,B}$ = specific heats of fluid A and fluid B respectively, j/kg°C

$T_{h,A}$, $T_{h,B}$ = temperatures of fluid A leaving collector and fluid B entering storage tank respectively, °C

$T_{c,A}$, $T_{c,B}$ = temperatures of fluid A entering collector and fluid B entering heat exchanger, respectively, °C

For temperature differences larger than 10°C, the efficiency uncertainty is estimated as in the 10-15% range.

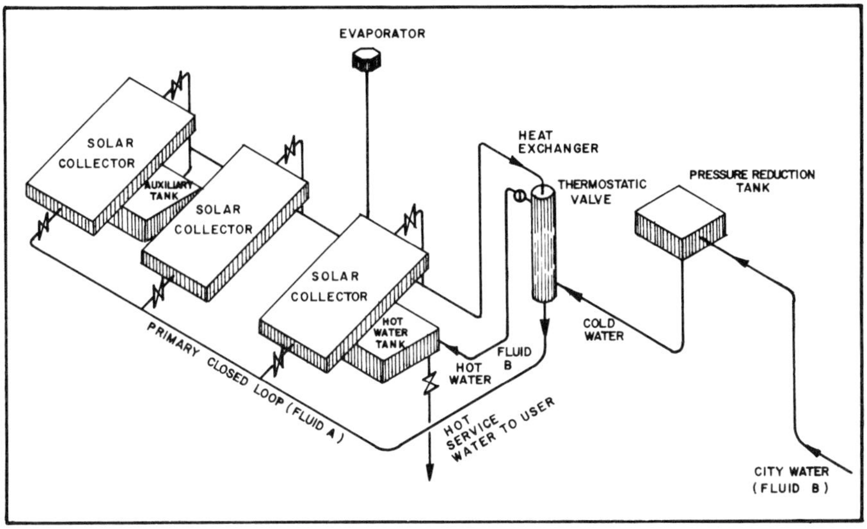

Fig. 1 - Schematic of thermosyphon solar water heater

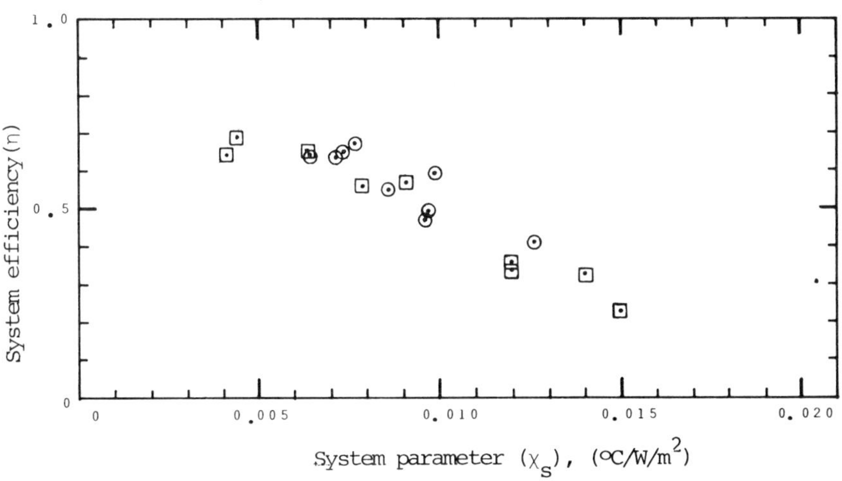

Fig. 2.- Efficiency data : 3 collectors (A_c/A_t= 13.5)

⊙ single glazing collectors: 13/4, 14/4, 18/4

⊡ double glazing collectors: 4/5, 5/5

System Parameter (X_s)

Can be defined as:

$$X_s = \left(\frac{T_{h,B} + T_{c,B}}{2} - T_{amb}\right) / I \quad (°C/W/m^2) \tag{2}$$

where

T_{amb} = ambient temperature, °C

I = incident radiation on collector plane, W/m^2

The uncertainty of X_s is estimated to be in the 10-20% range.

Collector Parameter (X_c)

Can be defined as:

$$X_c = \left(\frac{T_{h,A} + T_{c,A}}{2} - T_{amb}\right) / I \quad (°C/W/m^2) \tag{3}$$

The uncertainty of X_c is estimated to be in the 5-15% range.

INFLUENCE OF COLLECTOR COVER ON SYSTEM EFFICIENCY

Figure 2 shows the efficiency data points for single glazing and double glazing collectors. The collector surface to heat exchanger surface ratio is 13.5 : 1, that is, all three collectors are connected to the thermosyphon loop. The scatter in the data is due to the uncertainty of efficiency measurements and possibly to the system thermal inertia. Although more data should be required to establish a more precise functional relationship between the system efficiency and the system parameter (X_s), it is clear that single glazing collectors are more efficient than double glazing ones, for fluid B temperature range ($T_{h,B}$) of 40-60°C.

The efficiency curves for single glazing and double glazing collectors, are shown in Figure 3 and Figure 4 for two and one collectors, respectively.

Single glazing collectors are shown to be more efficient. Also, there is a general trend for the two curves, in each figure, to cross over, denoting that double glazing collectors can be more efficient at higher temperatures.

Standard collector calculation procedures (5) show that this is actually the case. This is confirmed by some collector evaluation curves (2).

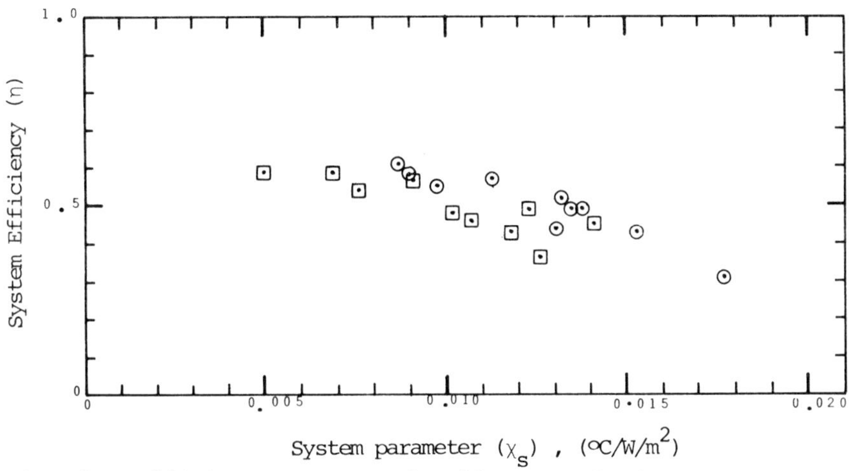

Fig. 3 - Efficiency data : 2 collectors (A_c/A_t= 9.0)

⊙ single glazing collectors: 9/5, 10/5
⊡ double glazing collectors: 26/4, 3/5

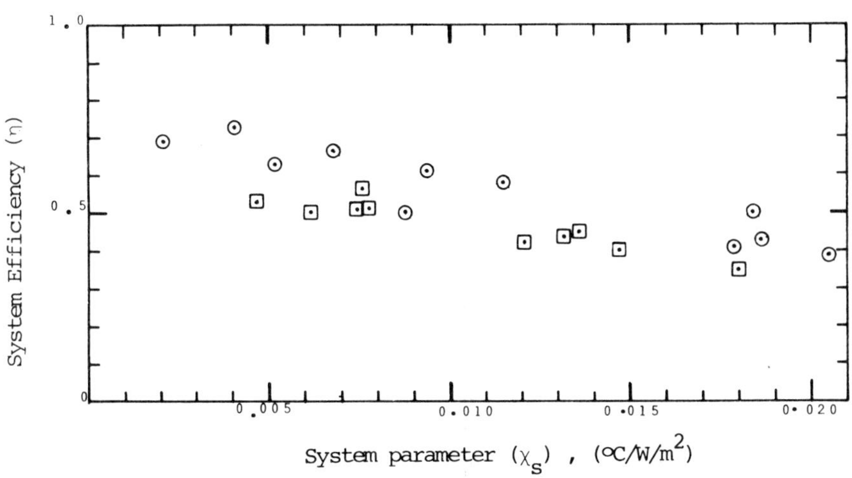

Fig. 4 - Efficiency data : 1 collector (A_c/A_t= 4.5)

⊙ single glazing collectors: 11/5, 12/5
⊡ double glazing collectors: 27/4, 28/4

It is thus concluded that single glazing collectors are more efficient for heating water in the 40-60°C temperature range.

Figure 5 shows the evaluation curve for single glazing collectors, that is, the collector efficiency (η) as a function of the collector parameter (X_c). The scatter is again due to uncertainty in efficiency measurement and possibly to the system thermal inertia. However, a linear least square fit to the data points results in the following equation.

$$\eta = 0.76 - 5.7\,X_c \qquad (4)$$

This approximates the following theoretical equation obtained by standard collector calculation procedures (5):

$$\eta = 0.73 - 5.5\,X_c \qquad (5)$$

It is thus assumed that equation (4) represents the performance of the studied single glazing collector.

INFLUENCE OF HEAT EXCHANGER SURFACE ON SYSTEM EFFICIENCY

The usable heat to the consumption water (fluid B) can be calculated by the following expressions:

$$\dot{Q} = IA_c\,\eta \qquad (6)$$

$$\dot{Q} = IA_c\,(0.76 - 5.7\,X_c) \qquad (7)$$

$$\dot{Q} = UA_t\,(LMTD) \qquad (8)$$

where

A_c = collector surface, m^2

A_t = heat exchanger surface, m^2

LMTD = logarithmic mean temperature difference, °C

For the majority of the system operating conditions, the LMTD can also be written, to within 2% as:

$$LMTD = \frac{T_{h,A} + T_{c,A}}{2} - \frac{T_{h,B} + T_{c,B}}{2} \qquad (9)$$

adding and subtracting T_{amb} to the right hand side of equation (9) the following equation results:

$$\eta = \frac{UA_t}{A_c}\,(X_c - X_s) \qquad (10)$$

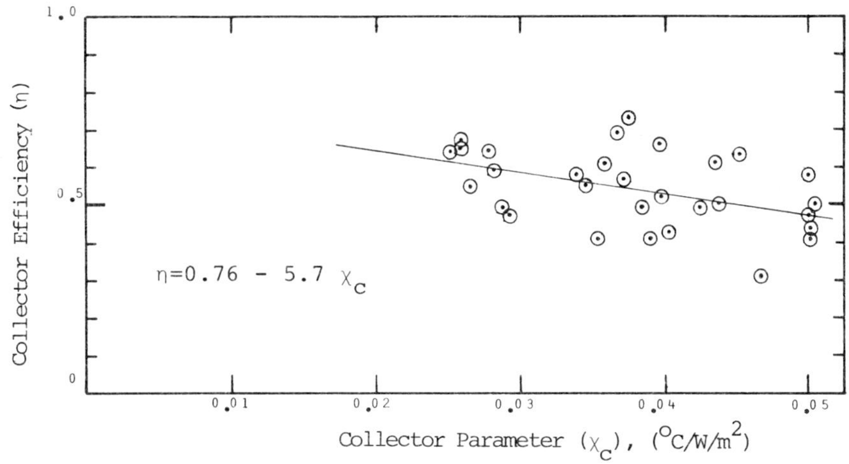

Fig. 5 - Collector Efficiency curve - single glazing

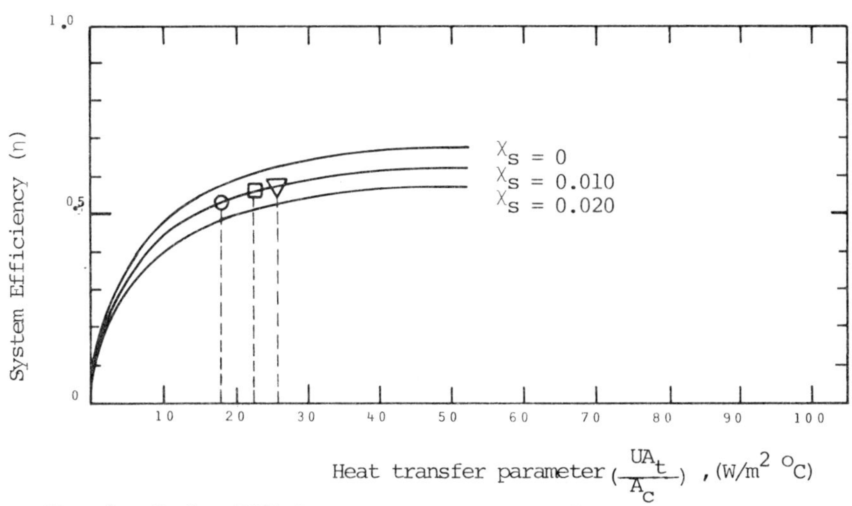

Fig. 6 - System Efficiency versus Heat Transfer Parameter for different values of χ_s

○ 3 collectors (A_c/A_t=13.5)

□ 2 collectors (A_c/A_t= 9.0)

▽ 1 collector (A_c/A_t= 4.5)

Equations (4) and (10) can further be combined to give

$$\eta = 0.76 - 5.7 \cdot \frac{0.76 + \frac{UA_t}{A_c} X_s}{5.7 + \frac{UA_t}{A_c}} \tag{11}$$

Figure 6 shows the calculated efficiency (Eq. (11)) as a function of the heat transfer parameter UA_t/A_c, for several values of X_s. Higher values of X_s require an efficient heat exchange (high UA_t/A_c).

Figure 6 also shows that when the collector area to heat exchanger surface ratio (A_c/A_t) decreases, the heat transfer parameter UA_t/A_c gets larger and the system efficiency increases. This fact can be explained by the following reasoning. When the collector surface decreases, the flow rate through the constant area heat exchanger also decreases. Heat transfer coefficients are known to vary with the flow rate by a power less than 1. The overall heat transfer coefficient U varies less than the area ratio (A_c/A_t) and the heat transfer parameter UA_t/A_c becomes smaller. The experimental data obtained by this investigation confirm this fact, as shown in Figure 6.

It is therefore concluded that for a given X_s the smaller the area ratio (A_c/A_t) the higher the system efficiency.

COEFFICIENT OF PERFORMANCE FOR THE HEAT EXCHANGER

Double loop systems are less efficient than open loop systems because the collectors must operate at higher temperatures with reduced efficiency. When the heat exchanger is used the following expressions can be written for the usable heat in the collectors:

$$\dot{Q} = IA_c \eta \tag{12}$$

$$\eta = a - b \left\{ \frac{T_{h,A} + T_{c,A}}{2} - T_{amb} \right\} / I \tag{13}$$

or

$$\dot{Q} = A_c \left\{ aI - b \left(\frac{T_{h,A} + T_{c,A}}{2} - T_{amb} \right) \right\} \tag{14}$$

An energy balance in fluid A and fluid B sides of the heat exchanger shows:

$$\dot{Q} = (C)_A (T_{h,A} - T_{c,A}) \tag{15}$$

$$\dot{Q} = \varepsilon (C)_{min} (T_{h,A} - T_{c,B}) = (C)_B (T_{h,B} - T_{c,B}) \quad (16)$$

where

$(C)_A = \dot{m}_A C_{p,A}$ (W/°C)

$(C)_B = \dot{m}_B C_{p,B}$ (W/°C)

$(C)_{min} = \min \{(C)_A, (C)_B\}$

ε = heat exchanger effectiveness

Equation (15) and (16) can be combined to give:

$$T_{c,A} = T_{c,B} + \left(\frac{Q}{\varepsilon (C)_{min}} - \frac{Q}{(C)_A} \right) \quad (17)$$

$$T_{h,A} = T_{h,B} + \left(\frac{\dot{Q}}{\varepsilon (C)_{min}} - \frac{\dot{Q}}{(C)_B} \right) \quad (18)$$

substituting (17) and (18) into (14) the following equation can be obtained:

$$\dot{Q} = (COP)\ A_c \{aI - b\ (\frac{T_{h,B} + T_{c,B}}{2} - T_{amb})\} \quad (19)$$

Where (COP) is the coefficient of performance and can be given by the following expression:

$$COP = \left\{ 1 + \frac{b\ A_c}{2} \{ (\frac{1}{\varepsilon\ (C)_{min}} - \frac{1}{(C)_A}) + (\frac{1}{\varepsilon\ (C)_{min}} - \frac{1}{(C)_B}) \} \right\}^{-1} \quad (20)$$

If an open loop system had been used for heating water from $T_{c,B}$ to $T_{h,B}$, the system usable heat ($\dot{Q}_{ideal}$) would be:

$$\dot{Q}_{ideal} = A_c \{aI - b\ (\frac{T_{h,B} + T_{c,B}}{2} - T_{amb})\} \quad (21)$$

The coefficient of performance can actually be interpreted as the fraction of the ideal usable heat to be obtained from systems using a heat exchanger. Figure 7 shows the calculated COP from Eq.(20) for the tested thermosyphon double loop system using single glazing collectors. Smaller area ratios (A_c/A_t) result is higher COP, which decreases when X_s increases. This fact again confirms earlier conclusions of this study.

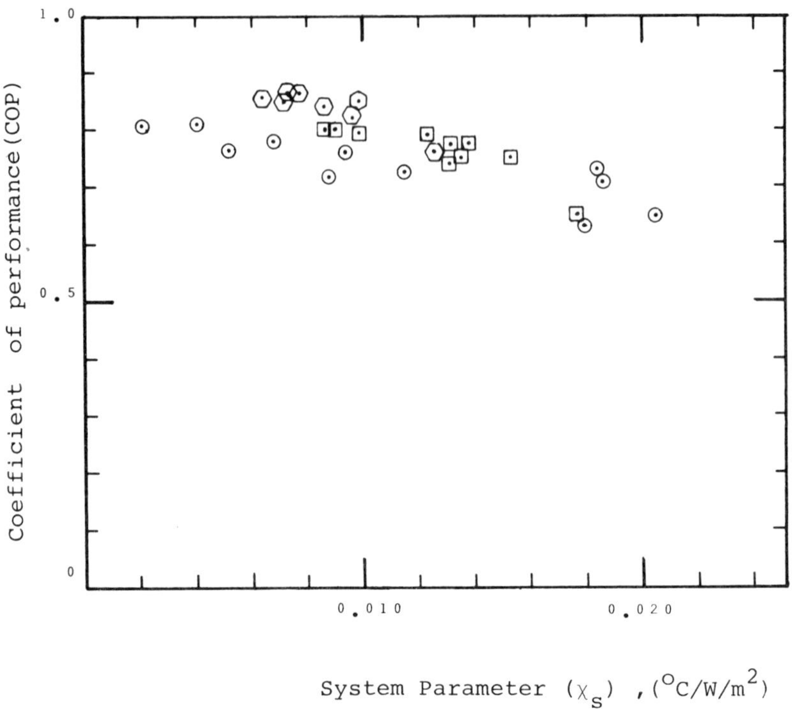

Fig. 7 - Coefficient of Performance versus system parameter

⊙ 3 collectors (A_c/A_t= 13.5)

⊡ 2 collectors (A_c/A_t= 9.0)

⎔ 1 collector (A_c/A_t= 4.5)

Equation (20) shows that higher flow rates (forced circulation) result in better coefficient of performance. Double loop system are therefore more adequate for forced circulation systems.

Parameter b is related to the collector heat loss. Double glazing collectors thus have higher coefficient of performance, as confirmed by this investigation. However, because of their lower efficiency for the temperature range of 40-60°C, it is suggested that single glazing collectors with selective surface absorbers can in fact provide higher efficiency of solar energy conversion, together with minimizing the influence of the heat exchanger in the system.

CONCLUSIONS

A double loop thermosyphon solar water heater has been analysed with respect to the collector covers and the system heat exchanger.

Single glazing collectors are concluded to be more efficient than double glazing ones for the temperature range of water heating utilization (40-60°C).

The influence of the heat exchanger on the system performance can be minimized by decreasing collector area to heat exchanger surface ratio and possibly by the use of a selective absorber. Higher flow rates result in higher coefficient of performance. Forced circulation systems are therefore more adequate to double loop configurations.

AKNOWLEDGEMENT

The authors wish to thank FINEP - Financiadora de Estudos e Projetos, of Brasil, for a soft loan granted to CTP - Centro de Tecnologia Promon for the development of this research program which was carried out under contract with Promon Engenharia S.A. of São Paulo, Brasil.

REFERENCES

1 - ORLANDO, A.F. et al. Geração de vapor a baixa pressão via energia solar/Low pressure solar steam generation/ In: SEMINÁRIO DE UTILIDADES, 2., IBP, São Paulo, 1977. Anais. Rio de Janeiro, IBP, 1977. v.2, p.2-9.

2 - MORSE, R.N. & CLOSE, D.J. Solar water heating. In: JORDAN, R.C. & LIU, B.Y.H. Applications of solar energy for heating and cooling of buildings. New York, ASHRAE, 1977. (ASHRAE GRP 170).

3 - ORLANDO, A.F.; MAGNOLI, D.; GOLDSTEIN, L. Thermosyphon solar water heating system under brasilian conditions. Rio de Janeiro, 1978. To be presented at the 13th. Intersociety Energy Conversion Engineering Conference. IECEC. Meeting. San Diego, California, Aug. 1978.

4 - HILL, J.E. et al. Development of proposed standards for testing solar collectors and thermal storage devices. Washington, National Bureau of Standards, 1976. 258p. (NBS. Technical note 899).

5 - DUFFIE, J.A. & BECKMAN, W.A. Solar energy thermal processes. New York, J.Wiley, 1974. 386p.

HOT WATER PRODUCTION BY MEANS OF A SOLAR THERMOSYPHON LOOP

W.L.Dutré, L.Cypers, J.Berghmans and A.Debosscher
Katholieke Universiteit te Leuven
Afdeling Toegepaste Mechanika en Energiekonversie
Laboratorium voor Warmteoverdracht en Reaktorkunde
Celestijnenlaan 300A, B-3030 Heverlee, Belgium

ABSTRACT

The results of a study of hot water production using a solar thermosyphon loop, are reported. A dynamic model has been developed describing the behaviour of the loop consisting of a number of solar collectors and a water tank placed above the collectors. The loop is segmented and for each segment the time dependent mass conservation, momentum and energy equations are satisfied. The solution of these equations is obtained numerically and is compared to temperature distributions determined experimentally. It is found that good agreement can be obtained between theoretical predictions and experiment.
Using the temperature and solar radiation data for a typical Belgian year, the dynamic model allows one to calculate the energy produced by the thermosyphon system as a function of system parameters : collector size, storage tank size, daily consumption rate, etc...
Energy savings with respect to an all-electric heater are reported as a function of the system parameters.

INTRODUCTION

Among the thermal solar energy applications of interest in the near future, solar water heaters are very likely the most important for North European countries. For this application, two different principles can be used, i.e. the forced flow system and the natural flow based on buoyancy forces only. Both principles exhibit specific advantages and should be developed. In fact the applicability of either one system or the other depends upon local conditions and requirements on the site and upon the building design aspects.
Thermosyphon loops require larger piping diameters in the collector circuit, the storage tank inlet to be positioned above the collector outlet and heat exchangers should be designed for natural circulation flow conditions. On the other hand, there is no circulation pump and no control divices since a thermosyphon loop is completely self-regulating. If it is designed such that the temperature can never exceed its maximum allowed level, the system can operate completely autonomously with a high degree of reliability.

The system considered in this paper consists of a small array of collectors being connected in parallel to a hot water tank with an internal heat exchanger. Since the sanitary hot water is separated from the collector circuit by the heat exchanger wall, the system can be simply protected against freezing by the use of a glycol-water mixture in the primary circuit.

This paper presents a mathematical model to simulate the time dependant behaviour of such systems, allowing for the temperature dependence of the fluid thermophysical properties and the real time dependant meteorological data to be used. The validity of this simulation model has been checked experimentally, using the measurements performed on a full scale experimental solar thermosyphon loop as described above.

The Mathematical Model

An outline of the thermosyphon loop is given in fig.1. For the mathematical simulation the loop is segmented.
Provided that the piping is well insulated, the following segmentation proved to be sufficiently accurate :

- the collector array : three segments of equal length in the flow direction (segments 1, 2 and 3 in figure 1);
- the tank and its internal heat exchanger : three segments in the primary loop (segments 6, 7 and 8 in figure 1) and three corresponding segments in the storage volume (segments 11, 12 and 13 in fig.1);
- the hot leg as well as the cold leg, two segments each (segments 4,5, 9 and 10 in fig.1).

Each segment is completely characterised by a single temperature and hence by a single density of the fluid, which in the case being considered is a mixture of glycol and water.
The simplified one dimensional momentum equation describing the fluid motion in the collector loop can be written as follows :

$$\frac{\partial(\rho v)}{\partial t} = -\frac{\partial p}{\partial s} - \Delta p - g\rho\frac{\partial z}{\partial s} \tag{1.1}$$

where :

ρ = fluid density
v = fluid velocity
p = static pressure
Δp = friction head losses per unit length
g = gravitational constant
s = the coordinate in the flow direction in the closed loop
z = the vertical coordinate, opposite to the gravitational force

The local instantaneous mass flow rate $\dot{m}(t,s)$ is given by :

$$\dot{m}(t,s) = v(t,s)\,\rho(t,s)\,S(s) \tag{1.2}$$

where :

$S(s)$ = the local cross section of the loop.

Since the relative density variations are small, the mass flow rate can be considered to be a function of time only :

$$\dot{m}(t) = v(t,s)\,\rho(t,s)\,S(s) \tag{1.3}$$

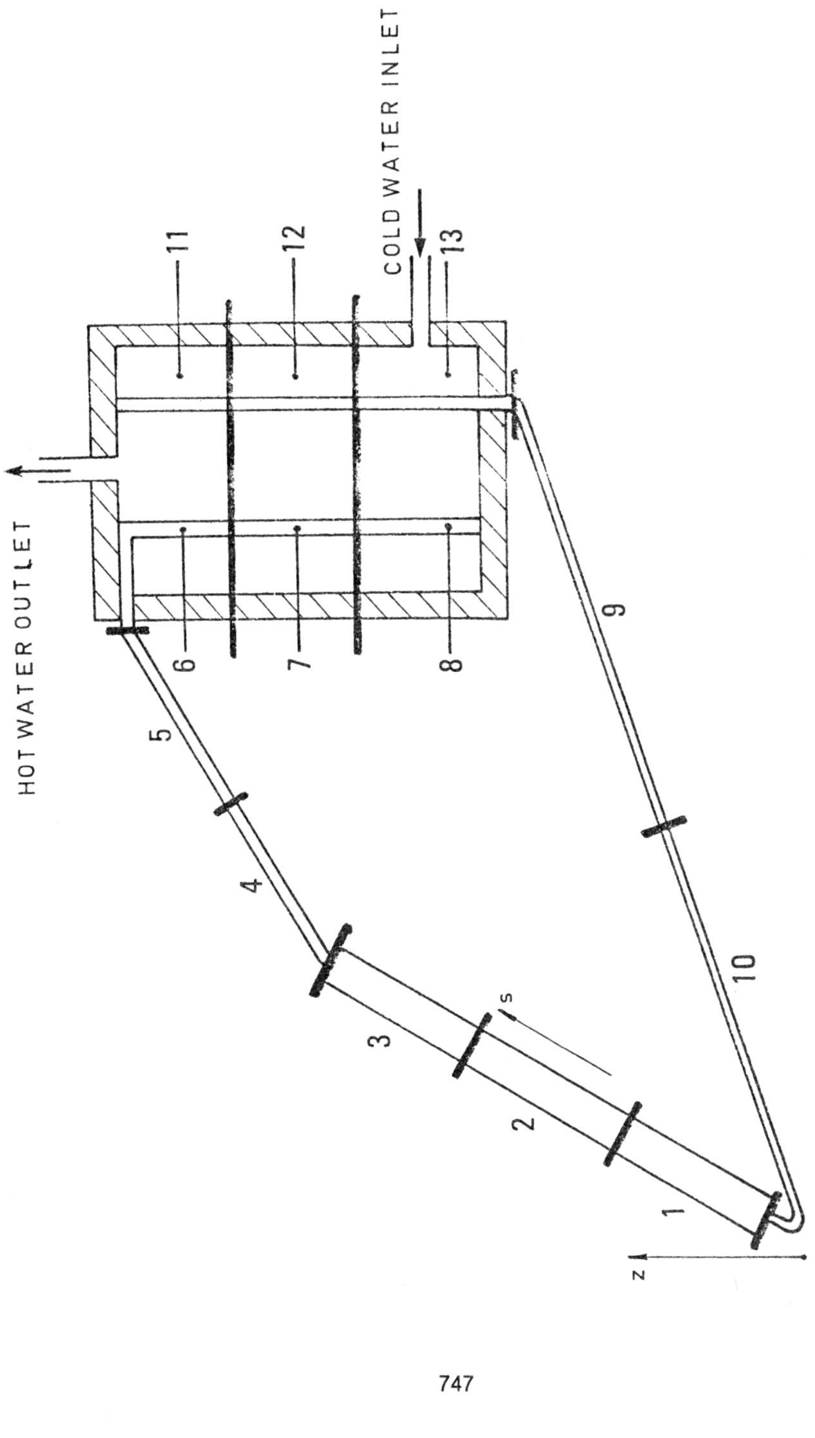

FIG.1: SOLAR THERMOSYPHON LOOP

Substitution of Eq.(1.3) into Eq.(1.1) results in :

$$\frac{\partial}{\partial t}\left[\frac{\dot{m}(t)}{S(s)}\right] = -\frac{\partial p}{\partial s} \quad -\Delta p \; - g\rho\frac{\partial z}{\partial s} \tag{1.4}$$

Integration of (1.4) along the closed loop yields :

$$\frac{d}{dt}\dot{m}(t)\oint\frac{ds}{S} = -\oint \Delta p\, ds \; - g\oint \rho\, dz \tag{1.5}$$

The integral $\oint \frac{ds}{S}$ is a purely geometrical characteristic of the loop.

$\oint \Delta p\, ds$ is the total head loss, consisting of two components :

- A laminar frictional term which can be easily evaluated as a function of the mass flow rate $\dot{m}(t)$, for each of the loop segments. The loop integral can then be replaced by the corresponding sum of terms.
- The losses, due to changes in the flow direction and changes in cross section. These head losses are assumed to be proportional to the square of the mass flow rate; the proportionality coefficient can be determined experimentally.

$g\oint \rho\, dz$ is the driving force, due to the density changes along the loop. After segmentation this integral is also to be replaced by the corresponding summation.

The energy balance equation for each of the segments can be written as follows :

$$c_j\,\rho_j\,\Delta s_j\,S_j\left(\frac{dT_j}{dt}\right) = a_j\Big[-|\dot{m}|c_j T_j + 0.5(\dot{m}+|\dot{m}|)c_{j-1}T_{j-1} - 0.5(\dot{m}-|\dot{m}|)c_{j+1}T_{j+1} + q_j\Big] \tag{1.6}$$

where :

c_j = the specific heat of the fluid in segment j
Δs_j = the length of segment j in the loop direction
T_j = the fluid temperature in segment j
q_j = the heat gain of segment j, due to insolation, heat losses to the ambient and/or heat transfer in the boiler
a_j = a factor smaller than 1. This factor allows for the thermal capacity of the structural material of segment j.

The energy balance is specified for a mass flow rate in either direction, thus also for reverse flow. This explains terms like $0.5(\dot{m}+|\dot{m}|)$

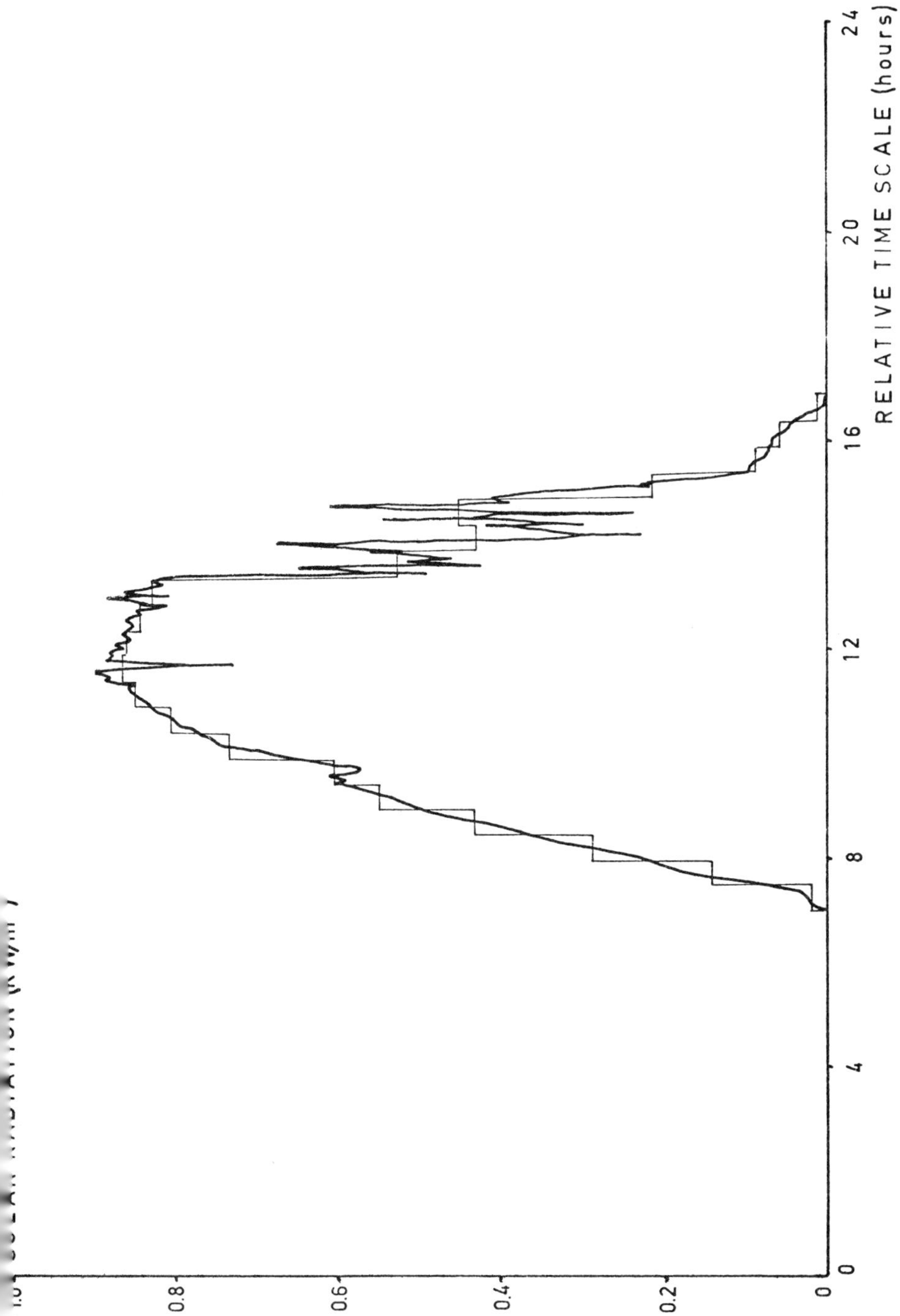

FIG.2a: SOLAR RADIATION

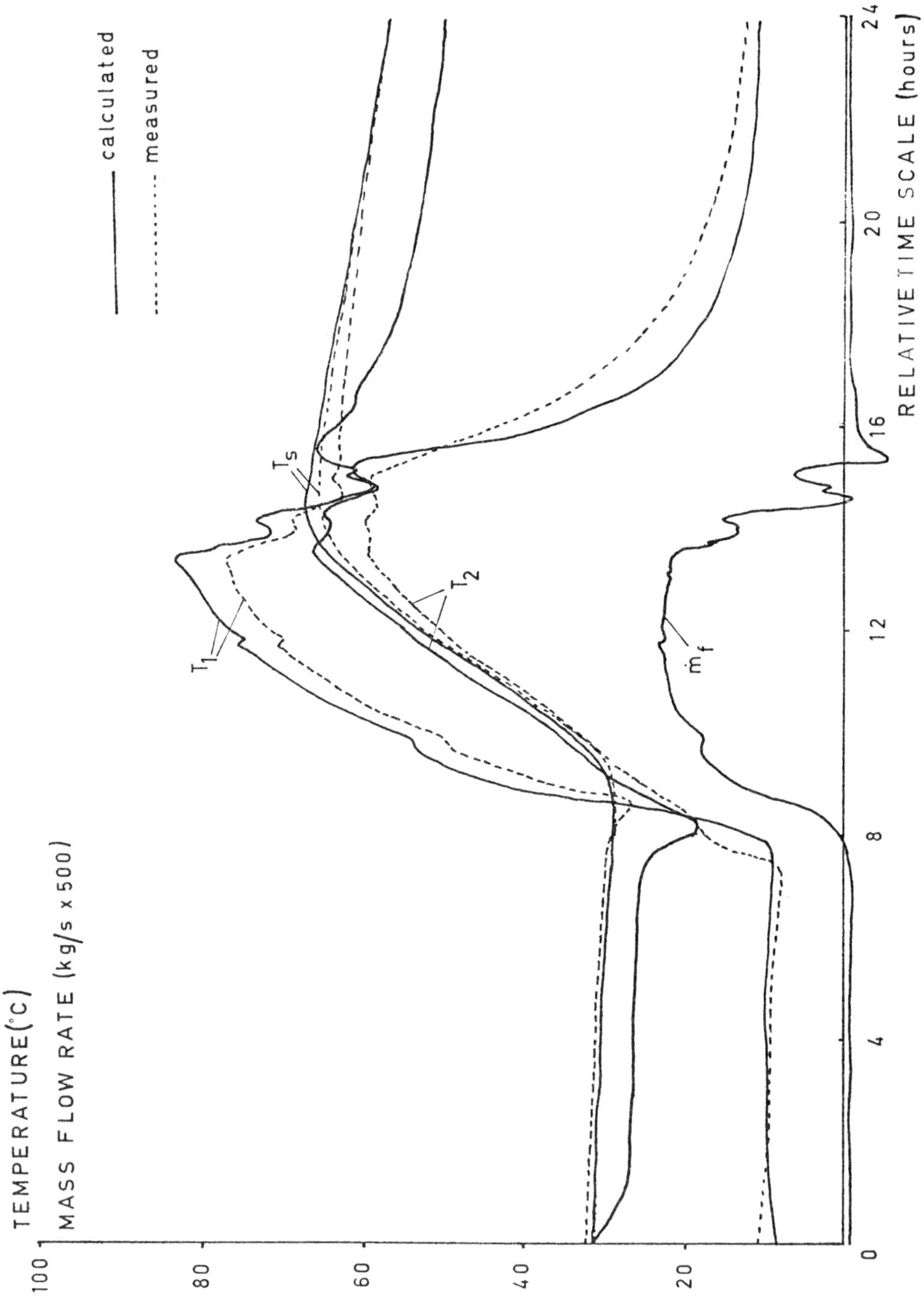

FIG. 9: COMPARISON THEORY - EXPERIMENT

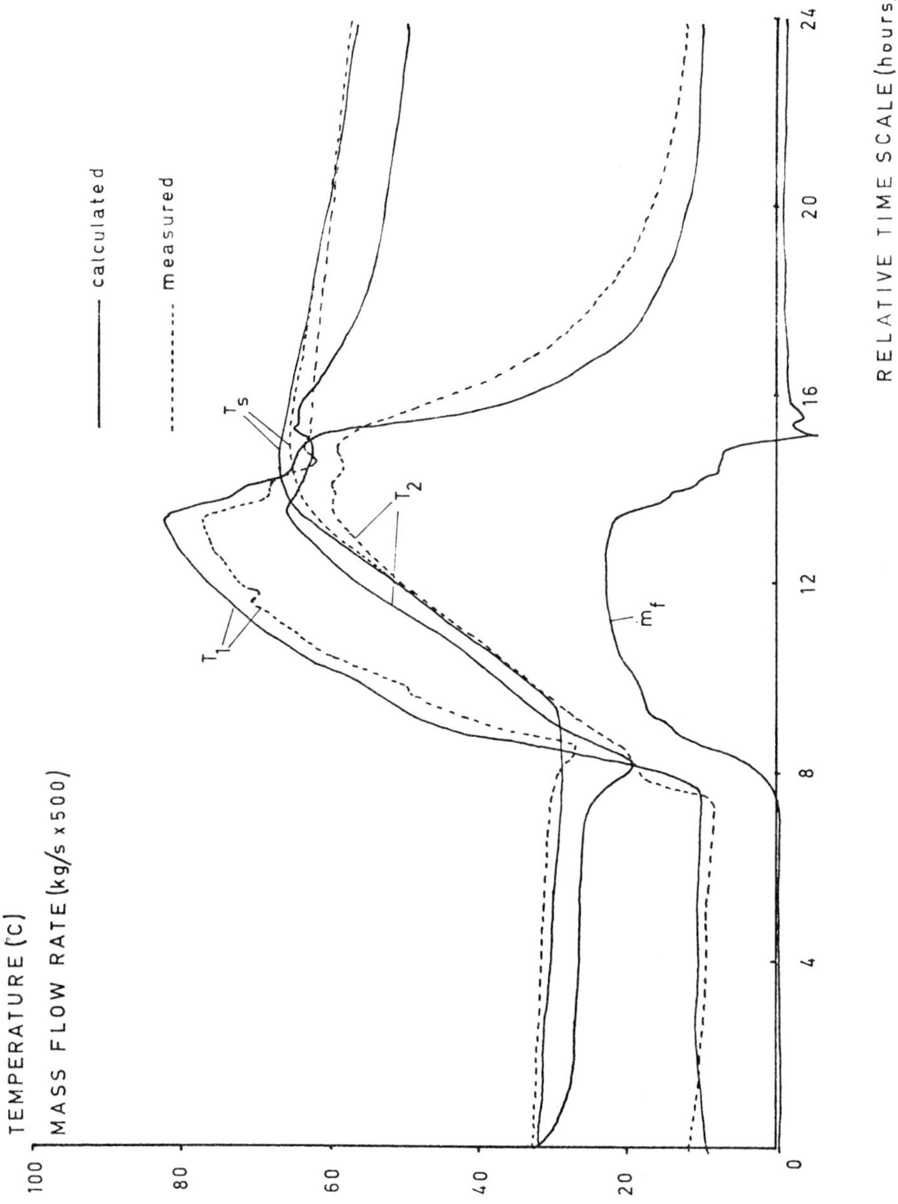

FIG.2c: COMPARISON THEORY-EXPERIMENT

The heat transfer from the primary loop to the tank is taken to be proportional to the temperature difference between the tank segment and the corresponding loop segment (e.g. 11 and 6). The heat transfer coefficient can be determined experimentally as a function of the loop mass flow rate.

Experimental Verification

An experimental set up with a gross collector area of 14.2 m^2 and a storage tank of 300 l has been built. Every two minutes the following data are measured and stored on minicomputer disc : three temperatures in the tank, in such positions that the average value of these measurements represents the mean storage temperature, entrance and outlet temperature of the heat exchanger and the temperature difference between in and outlet, ambient temperature and incident solar radiation intensity on the collectors. The measured incident solar radiation and the ambient temperature are used as input data for the mathematical simulation model. The validity of the model is then investigated by comparison of the measured and calculated temperatures as a function of time.
Some results of such a comparison are given in the figures 2 and 3 for two rather different daily radiation profiles.

The measured radiation profiles are represented in the figures 2a and 3a , the origin of the time scale being arbitrary. Also the half-hour by half-hour histogramme of the solar radiation curve is represented in these figures. Since the available climatological data in Belgium consists of half-hour average values, the influence of using half-hour averaged input data has been investigated by comparison of the temperature profiles resulting from calculations with such averaged data as well as from calculations based on more detailed data (values every two minutes) with the experimentally observed temperature profiles. The theoretical curves represented in the figures 2b and 3b result from the use of the detailed weather data, the theoretical curves in the figures 2c and 3c are obtained from the calculations based on half-hour average data.

In both cases, the heat exchanger inlet temperature T_1 and the outlet temperature T_2 show some discrepancy between the measured and calculated curves, the calculated curves giving higher temperatures. Most of these differences are due to insufficient thermocouple immersion in the fluid resulting in a systematic error by extended surface effect.
For the mean tank temperature T_s , resulting from measurements with better thermocouple mounting conditions, the agreement between theory and experiment is satisfactory.
In the same figures, also the theoretically calculated mass flow rates are plotted versus time, although without comparison to an experimental curve. It can be seen that a small reverse flow occurs from sunset to sunrise, resulting in additional heat losses of the system to the ambient.

The model represents very well the dynamic behaviour of the loop. The fast fluctuations of the solar radiation input result in rapid changes of the heat exchanger temperature. The theoretical curves in figures 2b and 3b

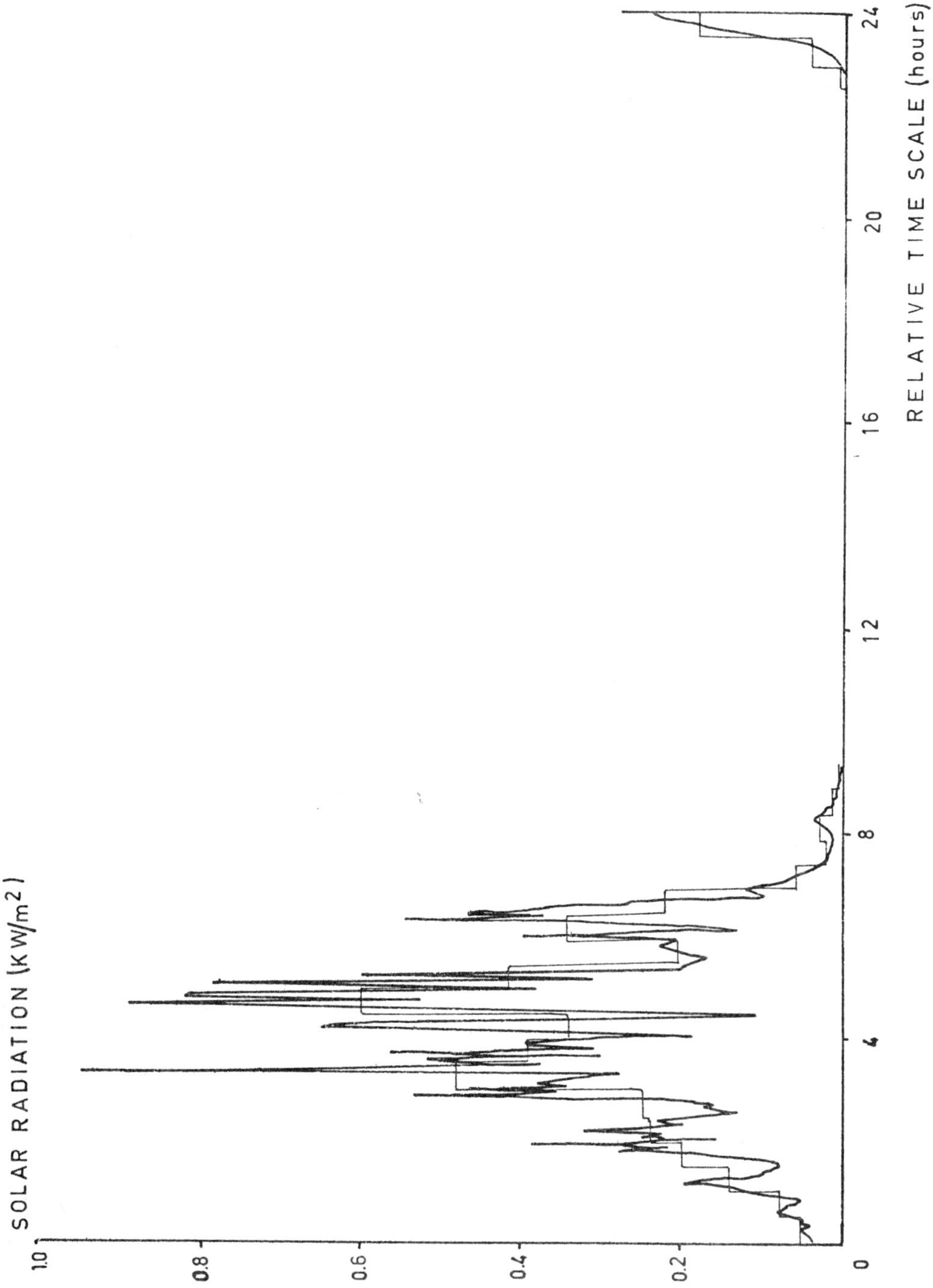

FIG.3a: SOLAR RADIATION

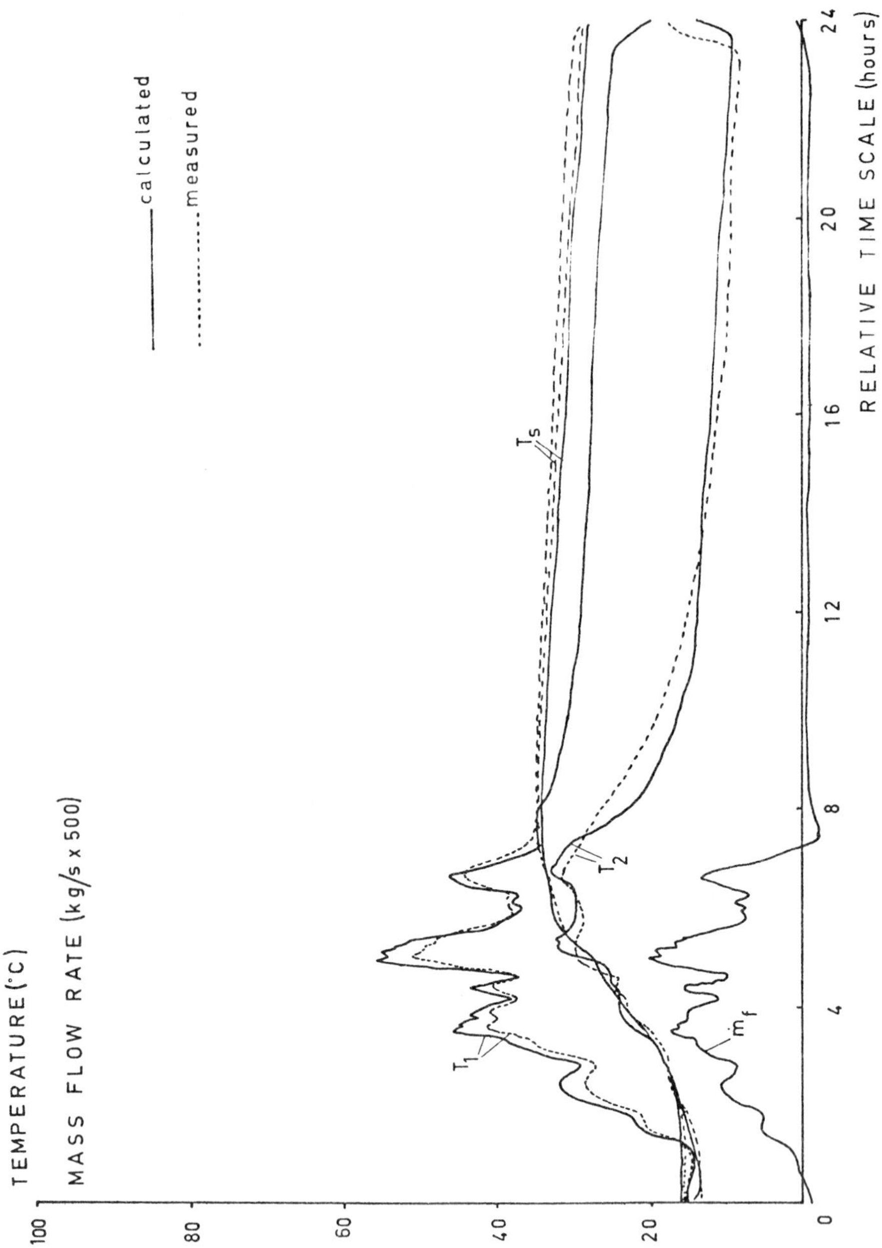

FIG.3b: COMPARISON THEORY-EXPERIMENT

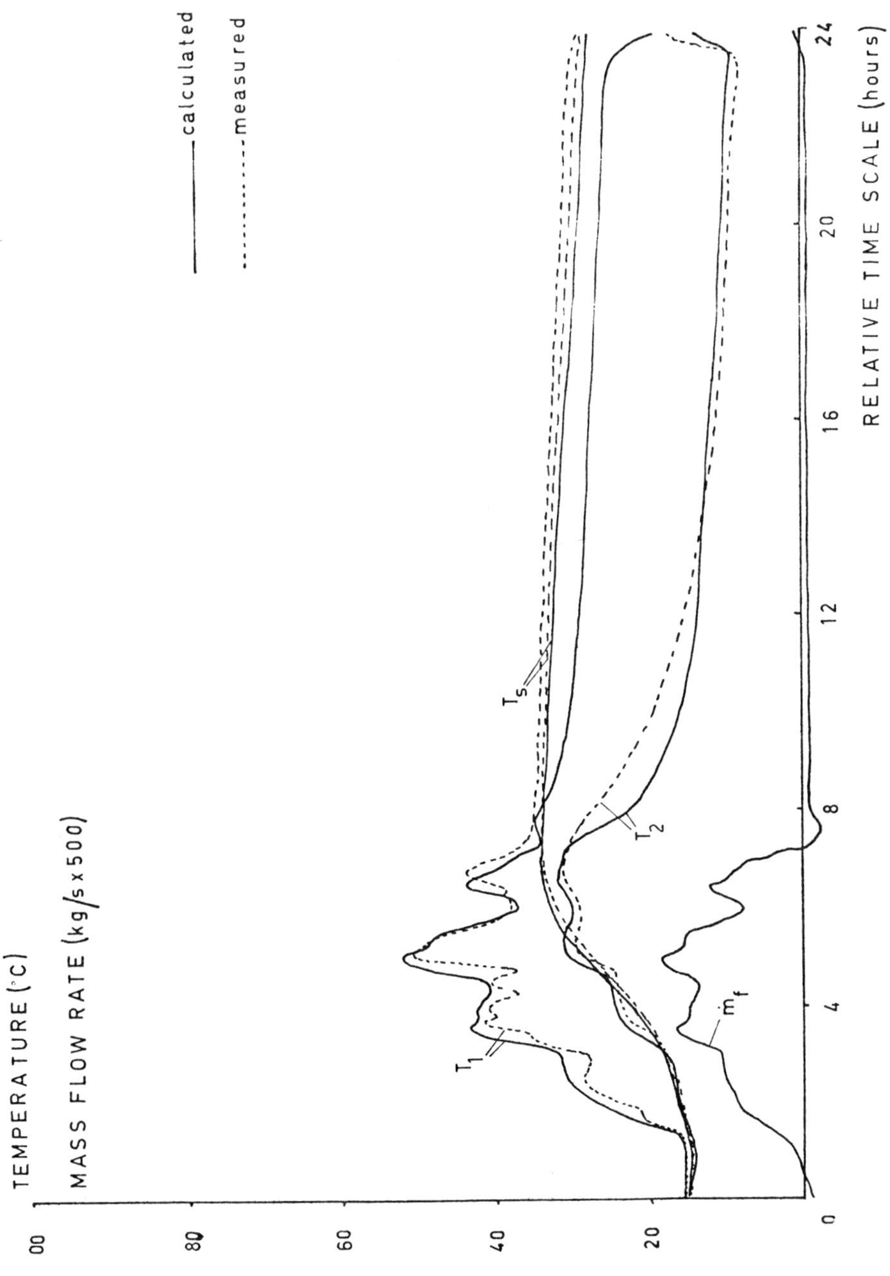

FIG.3c: COMPARISON THEORY-EXPERIMENT

calculated from two-minute by two-minute weather data, in fact represent very well the small irregularities corresponding to solar radiation fluctuations.
Comparison of the figures 2b and 3b with 2c and 3c respectively, shows that the average tank temperature T_s is not noticeably influenced by the use of half-hour averaged climatic data. The detailed behaviour of the loop is of course more influenced; the rapid response of the temperatures and mass flow rate to the sudden changes in solar insolation have disappeared since such fluctuations do not occur in the averaged input data. However, in overall system efficiency calculations, only the whole history of the tank temperature influenced the amount of energy that can be extracted from the system, such that it can be concluded that the use of half-hour averaged values of the climatological data is sufficiently accurate to calculate the overall energetic performance of the system.
Comparisons of the experimentally and theoretically calculated mean tank temperature over longer periods of time have also been made. Such comparisons are shown in the figures 4 and 5, each for a one week period. Figure 4 shows a one week period without hot water drawn off. The curves in figure 5 show a one week period with hot water drawn off three times a day, each time 100 liters.

From the good agreement between measured values and the storage tank temperature resulting from the simulation model, it can be concluded that the model can be used in a parametric study of thermosyphon solar water heaters and to predict the system yearly performance under given climatological conditions with sufficient accuracy.

Results

Half-hourly values of insolation and temperature in Belgium are available for a period of almost 20 years.
A reference year was constituted by selecting for each month, the one out of that 20-years observation period, that approximated best the mean month. The selection is based on the Khi-square criterion, considering mean insolation, mean temperature, variations of insolation and temperature around the mean [1].
Using this climatological reference year, a large number of complete year simulations of the thermosyphon system have been performed.
In fig.6 results of calculations are given for the annual energy savings as a function of collector area for different amounts of daily water consumption. The distribution of the hot water demand is as follows :
25 % from 7.00 a.m. to 7.30 a.m.
37,5 % from 12.00 a.m. to 0.30 p.m.
37,5 % from 5.30 p.m. to 6.00 p.m.
The system temperature is assumed to be the outside temperature, a conservative estimate. Cold water from the main supply is assumed to be at 10°C and the required temperature of the hot water consumption is at 50°C. When the tank temperature is higher than 50°C the drawn of water is mixed with the cold water at 10°C to obtain 50°C; the flow rate through the solar storage tank is then correspondingly decreased. The tank

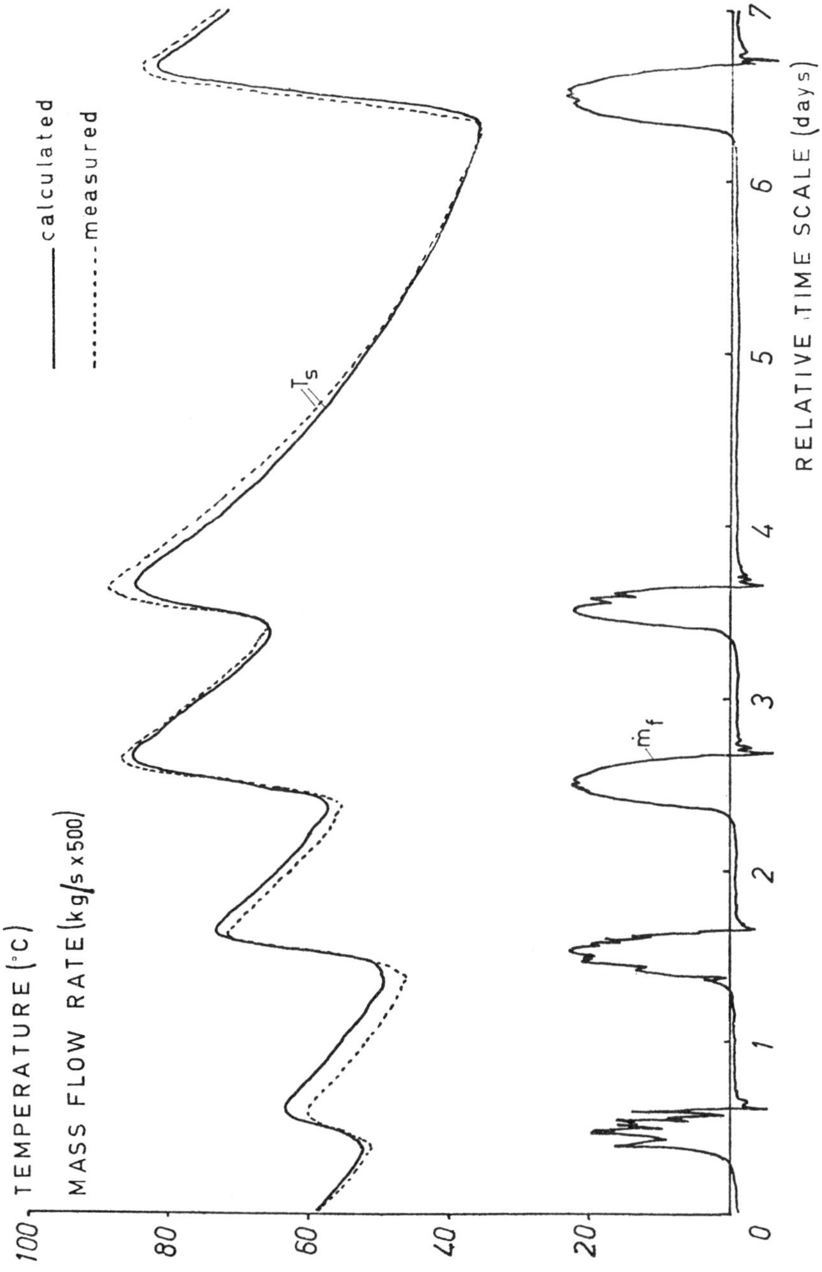

FIG. 4: COMPARISON THEORY-EXPERIMENT

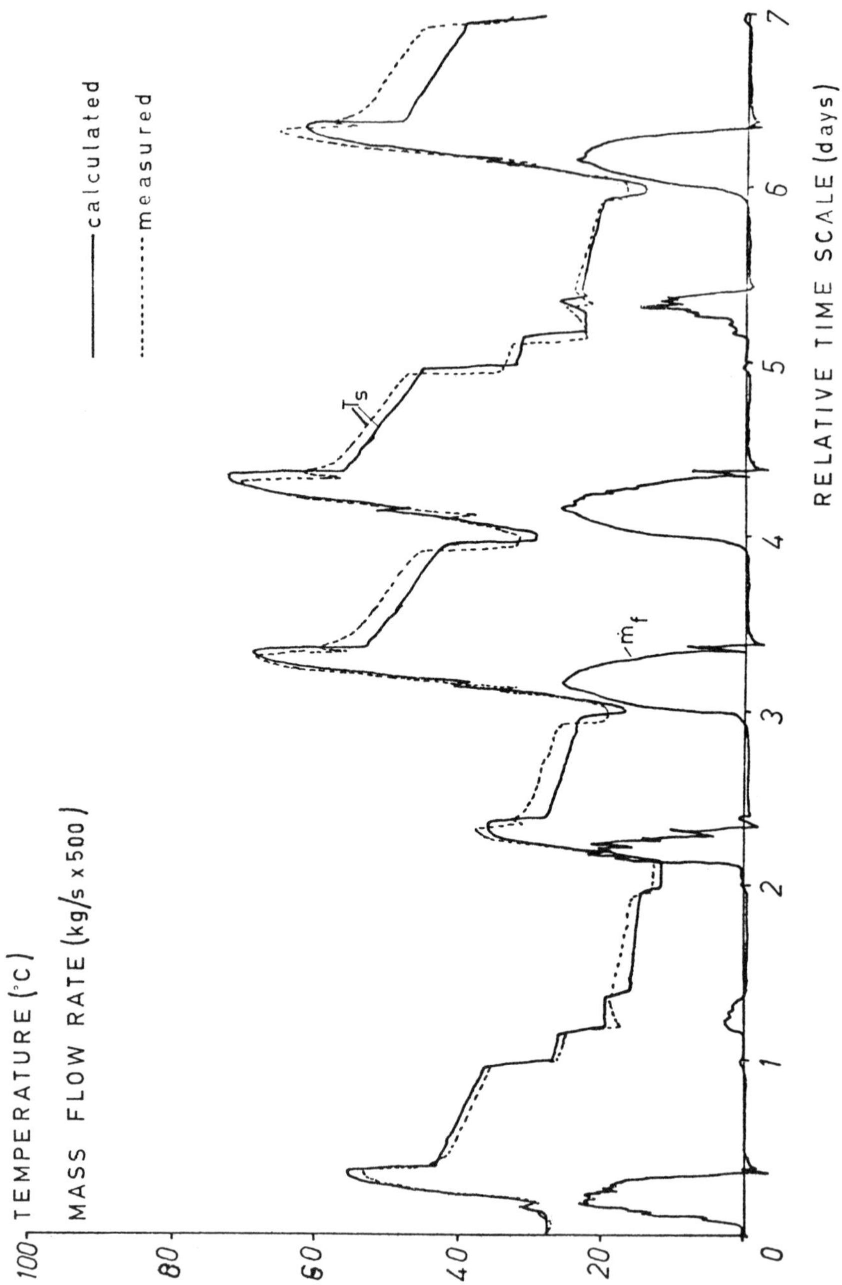

FIG.5:COMPARISON THEORY-EXPERIMENT

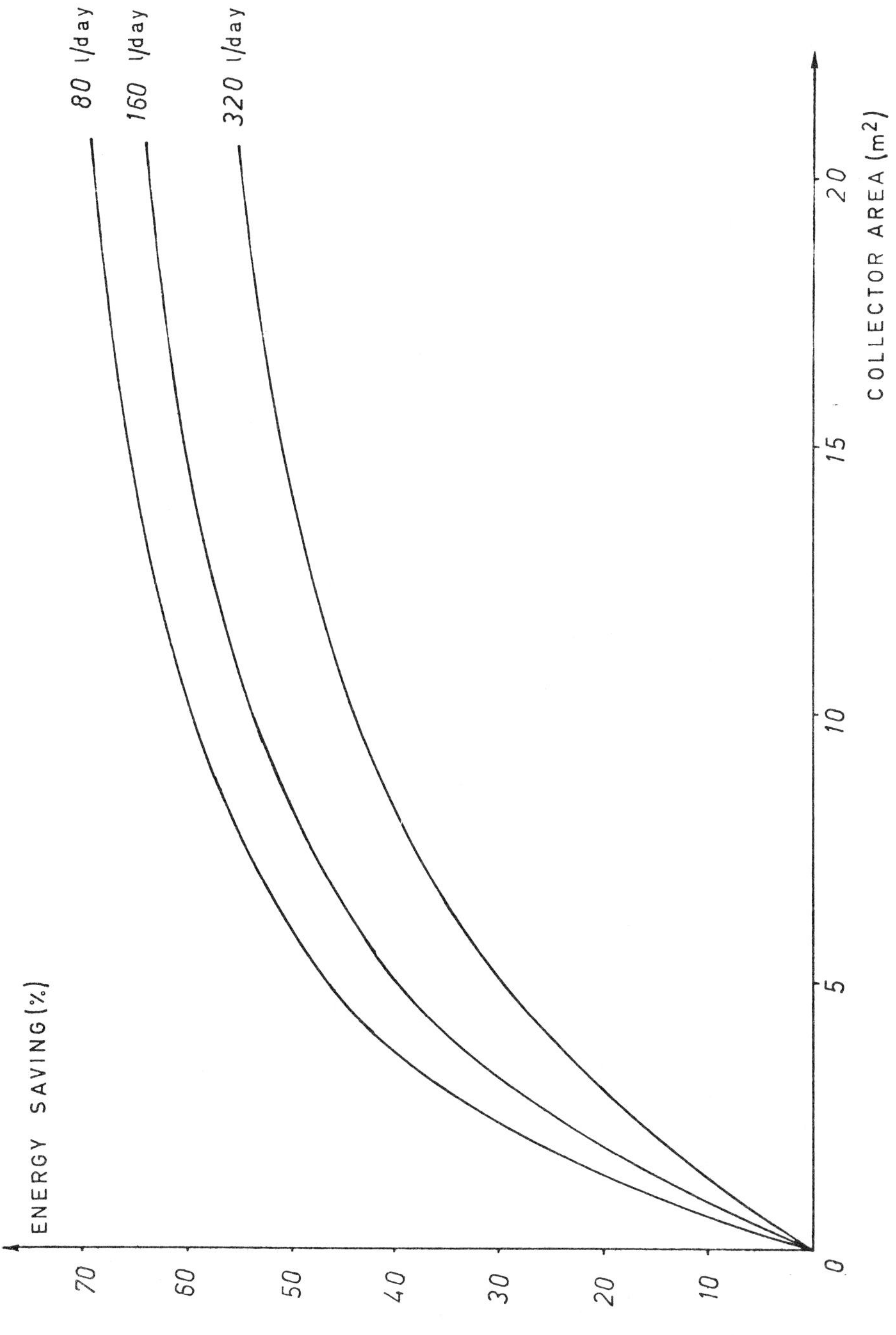

FIG.6: ANNUAL ENERGY SAVING

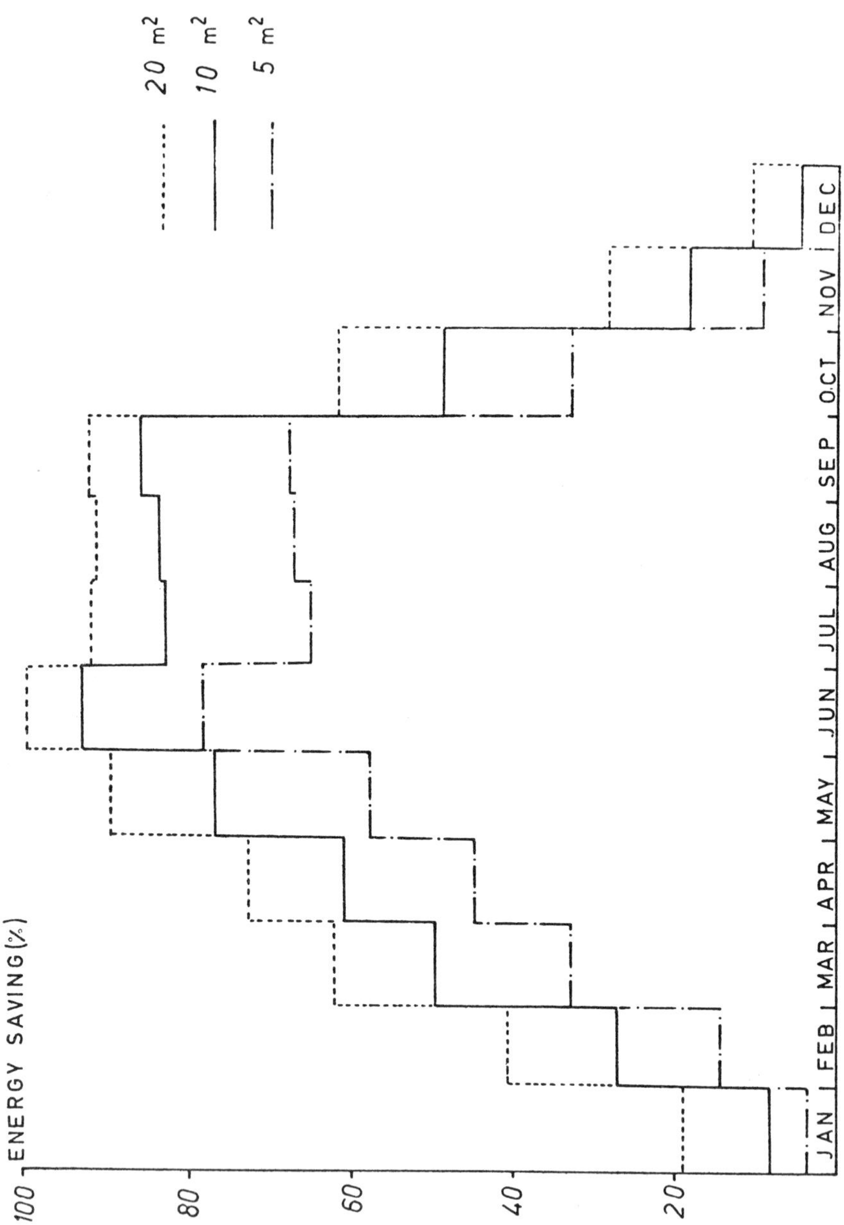

FIG. 7: MONTHLY ENERGY SAVING

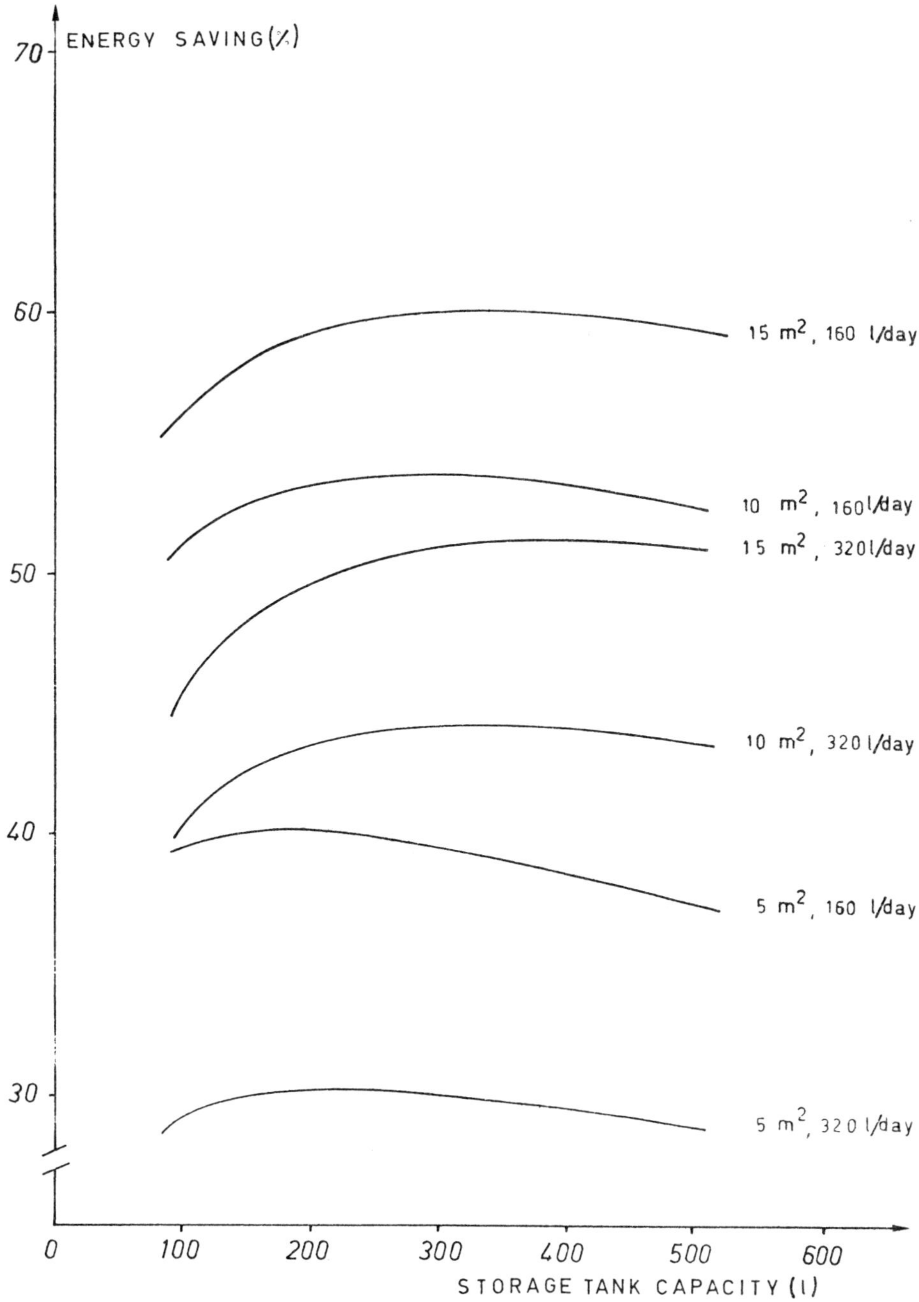

FIG. 8: ANNUAL ENERGY SAVING

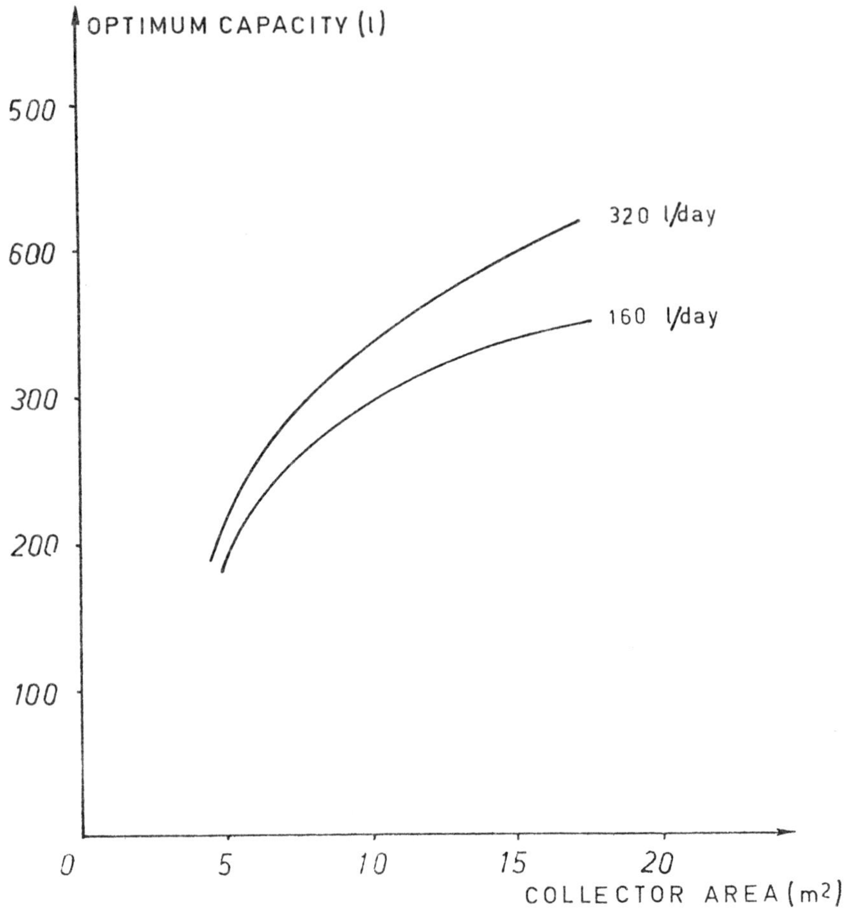

FIG.9: OPTIMUM STORAGE TANK CAPACITY

volume is 300 l and the collectors were oriented to the south at an inclination of 50°.
The efficiency curve of the solar collectors is approximated by :

$$\eta = 0.70 - 3.42\left(\frac{T_f - T_{amb}}{I}\right) - 16.8\left(\frac{T_f - T_{amb}}{I}\right)^2 \qquad (3.1)$$

for $I \geq 250\ W/m^2$

Where I is the solar insulation on the plane of the collectors, T_f is the fluid temperature, being different for each of the segments 1,2 and 3 and T_{amb} is the ambient temperature.

For $I < 250\ W/m^2$ the efficiency curve is :

$$\eta = 0.70 - 4\left(\frac{T_f - T_{amb}}{I}\right) - 20\left(\frac{T_f - T_{amb}}{I}\right)^2 \qquad (3.2)$$

The histogrammes in fig.7 show the monthly savings for three different collector areas. In fig.8 the yearly energy savings are given as function of the storage tank volume for different collector areas and daily hot water consumption quantities. It can be seen from these curves that in each case an optimum storage tank volume exists, which however is not very critical. The optimum value of storage tank volume is represented in fig.9 as a function of the collector area for two different daily hot water demand values. All the results shown in fig.7, 8 and 9 correspond to south oriented collectors with a 50° inclination angle.
For two typical cases table 1 gives the influence of collector orientation on the overall energy savings. Due to a predomination of the sunny hours before noon, an orientation of some 15 degrees to the east is slightly better than an exact south orientation. However, orientation can vary within large values without seriously affecting the annual savings.
Table 2 gives the influence of a non-return valve to avoid reverse flow and of a perfectly insulated storage tank.
In table 3, the annual savings are given for the same total daily hot water demand (320 l) but with different distributions. One can notice from these values that a predominant draw off in the morning is very disadvantageous to the system performance.

CONCLUSIONS

The mathematical model, simulating the time dependant behaviour of a thermosyphon solar water heater, has been shown to be sufficiently accurate to predict the long term system performance under given climatological conditions.

Using the temperature and solar radiation data of the Belgian reference year for solar system calculations, a parametric study of the thermosyphon solar water heater has been performed to investigate the influence of the

Table 1 : Influence of Collector Orientation

Case : daily use of 160 l
10 m^2 collector area
the values give the yearly savings in energy (%)

	Orientation				
	Degrees to the east		South	Degrees to the west	
Inclination	30°	15°	-	15°	30°
20°			51.5		
30°			52.8		
40°			53.5		
50°	53.8	54.1	53.7	53.1	51.8
60°			53.7		
70°			53.0		
80°			51.7		

Case : daily use of 320 l
10 m^2 collector area
the values give the yearly savings in energy (%)

	Orientation				
	Degrees to the east		South	Degrees to the west	
Inclination	30°	15°	-	15°	30°
20°			42.4		
30°			43.5		
40°			44.1		
50°	44.2	44.5	44.0	43.4	42.3
60°			43.8		
70°			42.9		
80°			41.3		

Table 2 : Influence of a non-return valve and of the insulation of the boiler.

Case : daily use of 160 l
10 m^2 collector area

Case	Yearly savings in energy (%)
standard	53.7
with non-return valve	57.4 (+ 3.7 %)
perfect insulation	59.1 (+ 5.4 %)
non-return valve + perfect insulation	64.6 (+ 10.9 %)

Table 3 : Influence of the daily distribution of hot water consumption.

Case : daily use of 320 l
10 m^2 collector area

Distribution			Annual savings (%)
from 7.00 to 7.30	from 12.00 to 12.30	from 17.30 to 18.00	
320	0	0	37.0
240	80	0	39.6
240	0	80	39.9
160	160	0	41.2
160	80	80	42.2
160	0	160	42.4
80	240	0	41.9
80	160	80	43.7
80	120	120	44.0
80	80	160	44.5
80	0	240	44.4
0	320	0	41.7
0	240	80	44.3
0	160	160	45.8
0	80	240	46.4
0	0	320	46.0

main design parameters on the system annual performance, such as collector area, storage tank size, daily hot water demand and its distribution, collector orientation, the influence of a one way valve in the collector loop and storage tank insulation. The results of these calculations are shown in the figures and tables and can be summarized qualitatively as follows :

- A fifty percent energy saving on hot water production in a one family house in Belgium is possible with a thermosyphon solar water heater with a collector area ranging from 5 to 10 m^2.
- The system performance is not very sensitive to collector orientation. An inclination of about 50° and slightly to the east is optimal.
- The distribution of the daily hot water demand has a considerable influence on the system performance.
- The optimal storage tank volume is a function of the collector area and demand parameters, but the optimum is not very pronounced.
- A non-return valve and better storage tank insulation have an important influence on the system performance.

ACKNOWLEDGEMENT

The work reported in this paper was supported by the Belgian governement as part of the Belgian National Research and Development Programme on Solar Energy Applications.

REFERENCES

1. R.Dogniaux, M.Lemoine, R.Sneyers
Analyse de données météorologiques relative au rayonnement solaire et à la température de l'air en vue de la sélection des mois-types constitutifs d'années-types pour le traitement de problèmes de captation d'énergie solaire.
Programme R.D.Energie (1977)

EFFECTIVENESS OF PACKED-BED SOLAR AIR HEATERS

S.C. Gupta
Agr. Engg. Deptt.
Allahabad Agricultural Institute
Allahabad, India.

ABSTRACT

Investigations on quite a few types of air heaters like flat plate absorbers as well as packed-bed absorbers have been reported. A single glass cover Solar air heaters with packed-bed absorber made out of indigenously available cheap materials like iron shavings from machining operations on lathe and another of iron wires, scrap of iron and aluminium have been fabricated and tested for their performance. The test procedure recommended by National Bureau of Standards, U.S.A. has been followed during experimentation.

Comparative efficiencies of both the air heaters have been evaluated with and without using metallic tray with absorbers. Effect of mass flow rate, aspect ratio, length of absorber, air temperatures and Solar insolation has also been studied. It has been found that efficiencies of collection of these packed-bed absorbers are as high as 70 to 75 percent at outlet air temperatures of 50 to 53°C.

INTRODUCTION

Heated air can economically be used for many applications e.g. crop drying, timber drying, space heating etc., where the temperature requirements are within the range which can be effectively achieved using flat plate collectors. Air heaters can be built significantly cheaper than water heaters and are relatively free from corrosion problems, and small amount of leakage is also not a problem. They operate at lower temperatures and flat plate collectors may be used to receive direct, diffuse and reflected radiation. Such air heaters are simple in construction, so easy to fabricate and operate with low initial and operative costs. Air as a heat extracting medium passes over an absorber beneath a glass cover of covers. All around insulation except the top of the absorber reduces heat losses and heat equal to the difference between the incoming radiation to the absorber and the outward losses from the absorber by radiation and convection is transferred to the air stream.

The performance of a Solar air heaters depends upon the orientation of the heater, insolation, insulation around the absorber, retative properties of transparent covers, stagnant air natural convection barriers between absorber and transparent covers, ambient temperature, mass flow rate of the air, aspect ratio of the duct through which air flows, heat ransfer coefficient between absorber and air stream and radiative properties of the absorber (1, 2). For an air heater higher performance results by increasing air flow rates through the heater. The orientation of the air heater for its best average performance depends upon the seasonal requirements and latitude angle. For year round operation, the maximum Solar energy is collected when the tilt is 0.9 times the latude angle (3). Radiative properties of transparent covers have minor effect on the performance of the air heater since retardation from the absorber is only a part of the total heat loss. Glass cover with low iron contact will increase the efficiency of the collector. Stagnant air gaps provides high resistance to heat transfer by convection between the absorber and transparent cover. The heat loss due to convection and radiation can be further reduced by the provision of multiple stagnant layer but at the cost of the energy falling on the absorber. Ambient temperature will affect the performance of air heater because the heat losses from the air heater are proportional to the temperature difference between the absorber and the ambient temperature. For given outlet temperature, low ambient temperature results in low efficiency and vice versa.

All above parameters, of course, affects the performance of air heaters but their influence is almost identical for all different/various types of air heaters and are less dependent on controllable operative parameters. In designing a air heater, the factors like aspect ratio of the duct and dust length through which air flows, mass flow rate of the air, relative properties of the absorber, and heat transfer coefficient between absorber and air stream are more dependent on the type of absorber and flow system. Outlet temperature of the air depends on duct length and also on mass flow rate and turbulance extent etc. An increase in air flow rates results in lower outlet air temperature, lower temperature in air duct, less heat loss flux and thus increases the heat collection efficiency of the air heater but with an increase in fan running cost(4). To have more heat transfer area of the absorber, turbulent flow with moderate mass flow rate of the air, packed-bed type air heater with some cheaper, loose, packing materials to act as absorber will prove efficient.

Investigations on quite a few types of air heaters like flat plate absorbers as well as packed bed absorbers such as flat or corrugated iron sheet, V-shaped troughts, slit and expanded aluminium foil matrix, overlapping glass strips and wire mesh etc. have been reported (5, 6, 7). Heat transfer coefficient between absorber and air stream considerable affects the performance of the air heaters. With increasing heat transfer coefficient the temperature difference between outlet air stream and absorber plate temperature decreases. In the present study Solar air heaters using indigenously available cheap materials as packed-bed absorbers have been experimented for their performance. The effect of inluencing parameters viz mass flow rate, aspect ratio, thickness of the absorber, compactness, etc. have been studied and feasibility of using two different materials to form packed-bed absorbers has been analysed.

THEORY

In a single exposure one pass Solar air heater, the temperature distribution along the length L of the air heater may be expressed by writing the energy balance for air heater.

If $dx = U_l (b\ dx)(t_{ab} - t_a) + h_c (b\ dx)(t_{ab} - t)$ and

$$\dot{m}\, C_f \frac{dt}{dx}\, dx = h_c (b\ dx)(t_{ab} - t).$$

Eliminating temperature tab from above two equations and solving the result putting boundary conditions that $t = t_{fi}$ when $x = 0$ and afternon dimensioning the solution, one will get

$$U = \left(1 - \frac{\alpha}{\beta}\right) e^{-\beta \bar{x}} + \frac{\alpha}{\beta}$$

where $U = \dfrac{t_f - t_a}{t_{fi} - t_a}$

$$\alpha = \frac{I.\ f.\ h}{c\ (t_{fi} - t_a)\ U_l},$$

$$\beta = \frac{h}{c},$$

$$C = \frac{m\ C_f}{bL}, \text{ and}$$

$$h = \frac{h_c - U_l}{h_c + U_l}$$

Further, the efficiency of the air heater may be expressed by

$$\eta = \frac{\int_0^{\tau} - mC_f\ (t_{fi} - t_{fe})\ d\tau}{\int_0^{\tau} I\ d\tau}$$

For the absorber when scrap material of good size is used, absorber-to-fluid film heat transfer coefficient may be replaced by volumetric heat transfer coefficient expressed as

$$hv = 650 \left(\frac{G}{D}\right)^{0.7}$$

where $G = \dfrac{\dot{m}}{bL}$

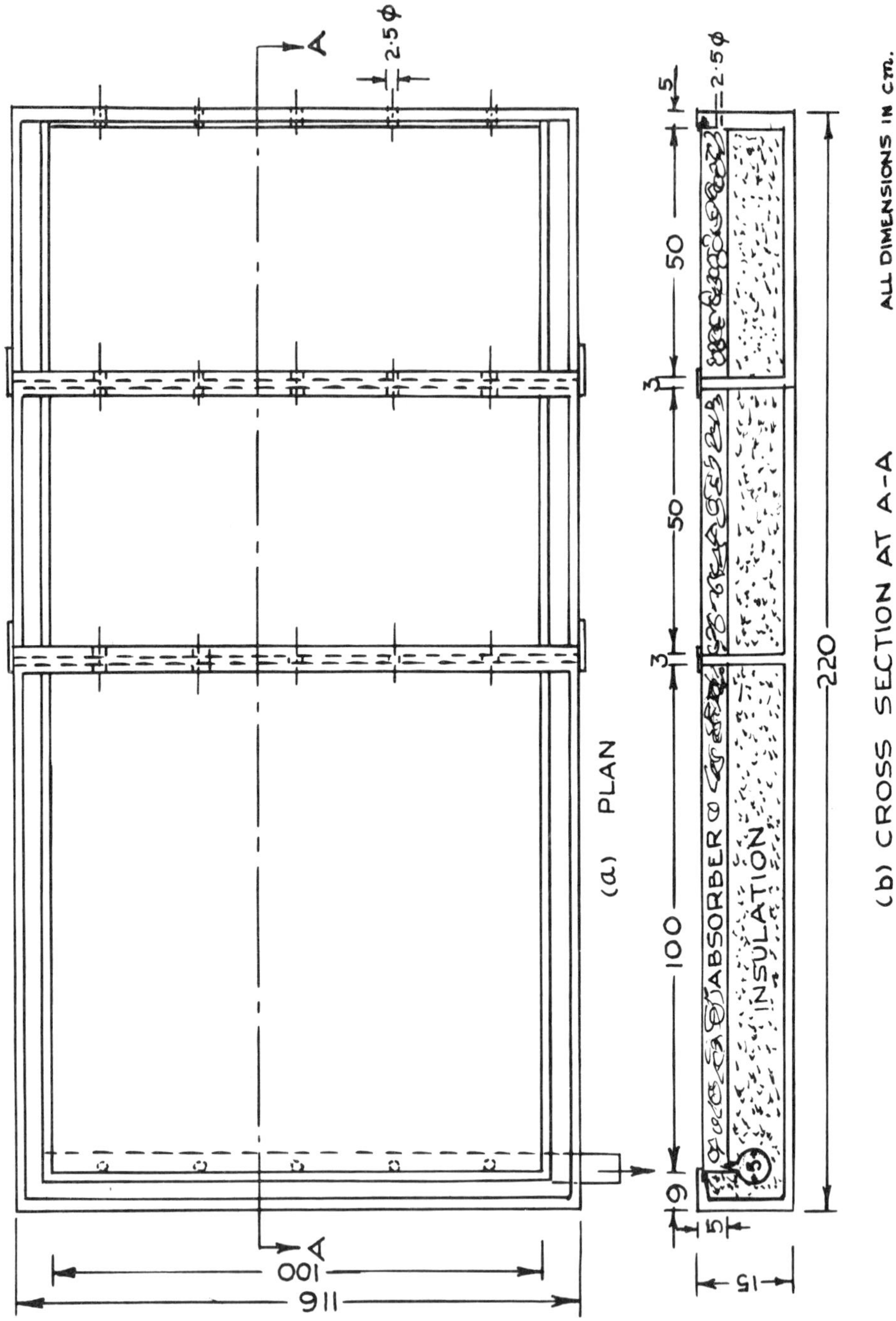

FIG.1- AIR HEATER

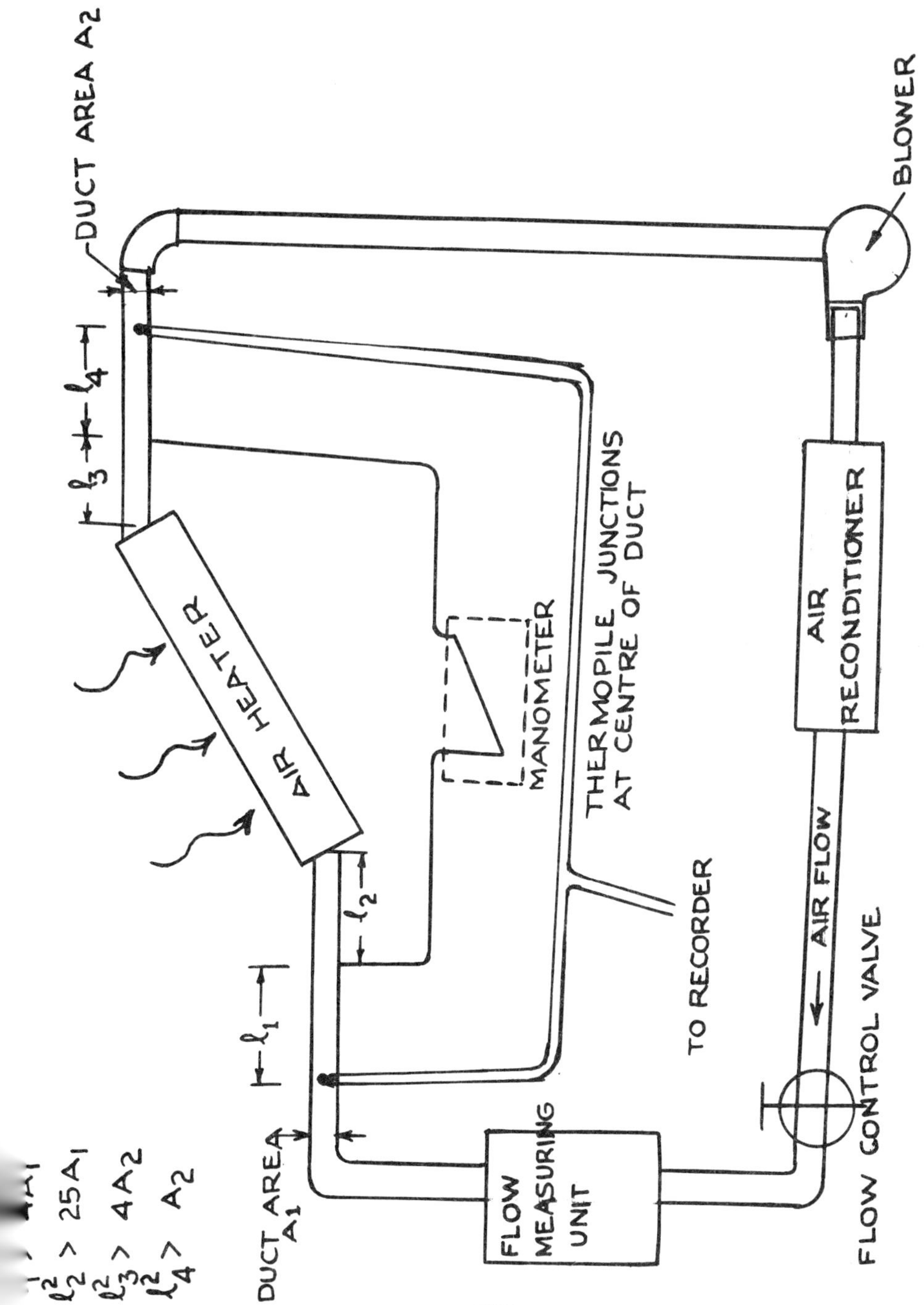

FIG.2- SOLAR AIR HEATER TESTING CONFIGURATION.

SYSTEM DESCRIPTION

Indigenously available cheap material, iron shavings from machining operation from lathe machine is used to construct the packed-bed absorber for Solar air heater. A layer of iron shavings was packed in a galvanised iron sheet tray. This tray is fixed in a wooden frame such that the back and sides of the tray are insulated with glass wool and top of the absorber is covered with a single ordinary window glass plate. The size of the absorber was 100 x 100 x 5 cm while the overall size of the air heater is 116 x 116 x 15 cm. At two opposite sides of the absorber tray, five holes of 2.5 cm each and at a center to centre distance of 20 cm, were made to create the uniform flow of air through whole of the absorber. Both air inlet as well as outlet headers made out of 5 cm diameter pipe and placed along the sides of the tray were provided corresponding five openings each. These header openings were connected to respective tray holes with the help of 2.5 cm diameter pipes. The absorber tray fixed with the wooden frame was made replaceable. In order to study the comparative performance of another packed-bed absorber material a duplicate air heater was fabricated. This air heater was neither provided inlet air header not it was insulated at inlet end. A simple wooden plank with five holes in alignment with the tray holes was placed at this end. Two more trays with half of this heater absorber area and one side open were fabricated and placed in the wooden frame to make identical cross-section. The provisions were made to attach either or both of these additional trays with the air heater tray to increase the length of the heater. The constructional details of this air heater with both the extended trays attached to it are shown in figure 1. In order to evaluate the performance of these packed-bed absorbers, testing configuration is shown in figure 2. The unit consists of a blower to maintain the flow, an air-reconditioning chamber so as to maintain the heater inlet air temperature, and a flow control value to maintain the flow. A differential manometer is used to measure the pressure difference across the air heater during testing. At inlet and outlet ends, thermocouple junctions are provided at the centre of the duct. Ordinary black borad paint was sprayed at the absorber after laying in the tray.

OBSERVATIONS AND RESULTS

In order to evaluate the performance of the packed-bed absorbers in the Solar air heaters, two types of the experiments were conducted. In the first experimentation, testing of 116 x 116 x 25 cm air heater with iron shavings absorber (absorber I) was done as per test procedure recommended by National Bureau of Standards, U.S.A. Inlet as well as outlet air headers were connected to the air-ducts to recirculate the air. An electric centrifugal pump was placed in line to maintain forced air circulation. Outlet air from air heater was reconditioned before recirculating it. A flow measuring device along with a control value was made to regu-

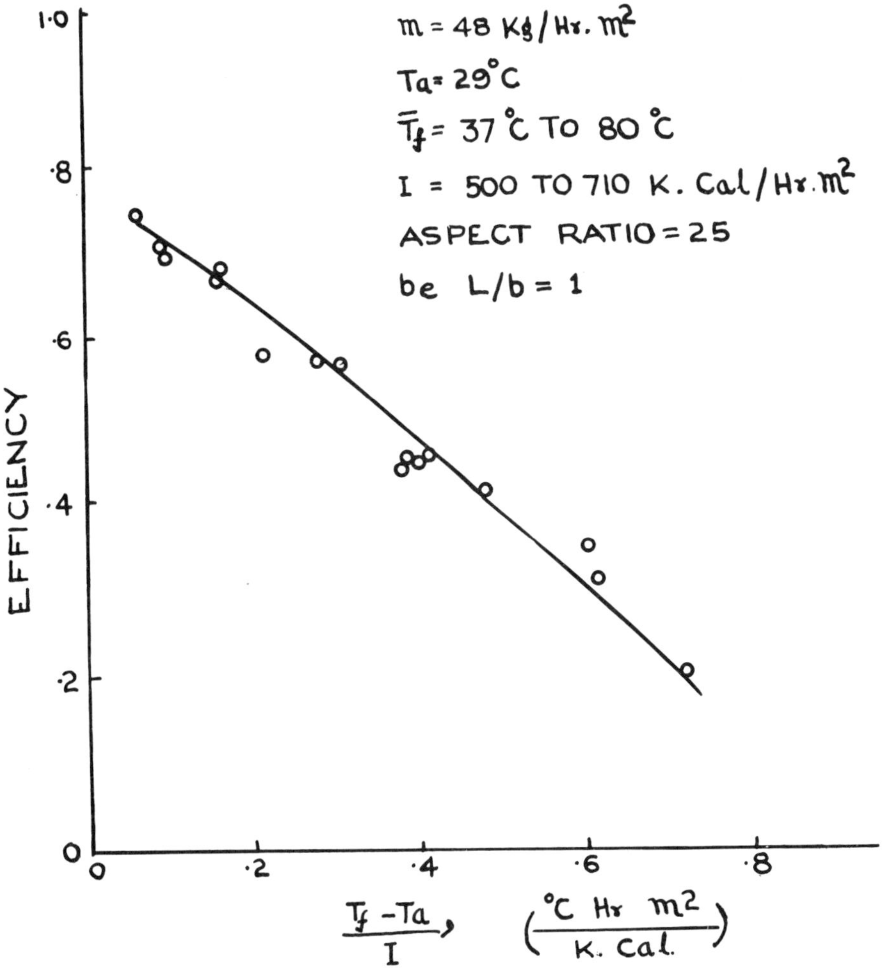

FIG.3- DATA FOR A IRON SHAVING PACKED BED SINGLE COVER COLLECTOR.

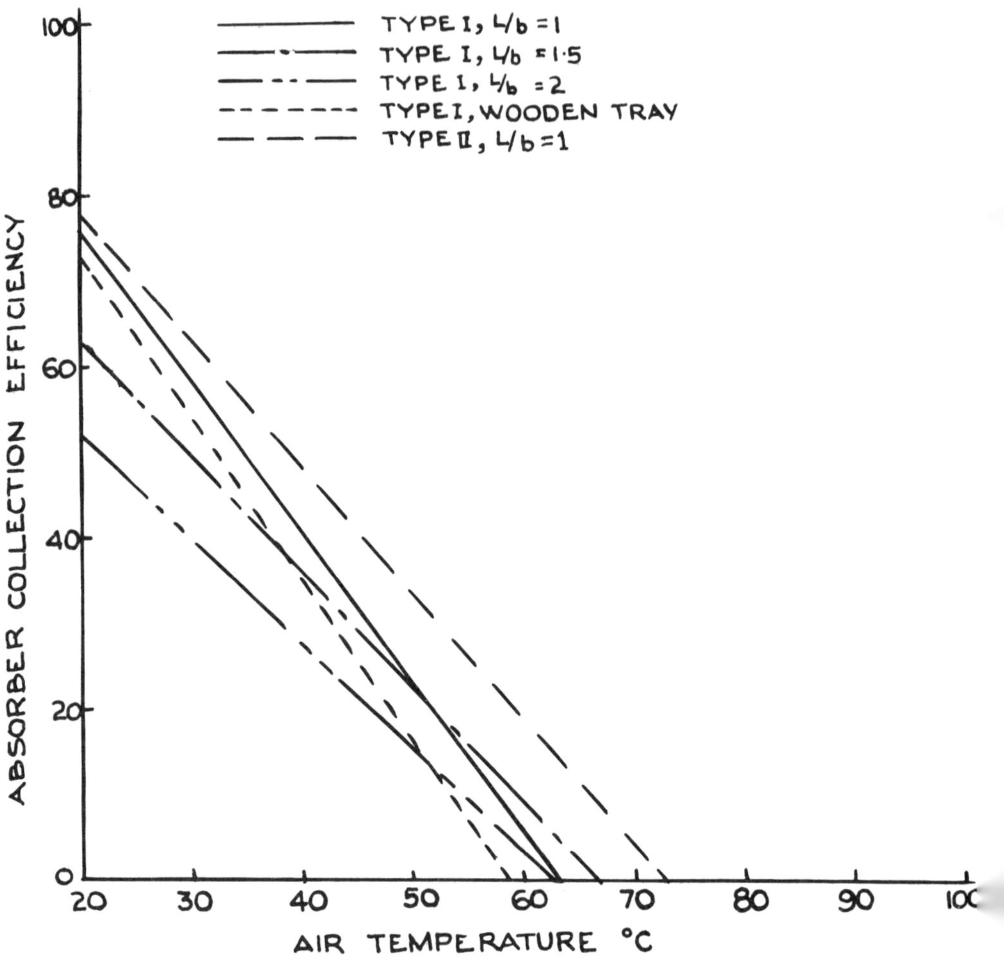

FIG.4- ABSORBER COLLECTION EFFICIENCY AT VARIOUS TEMPERATURERS.

late the air flow rate constant as per desire. Four thermocouples were embedded in the absorber at different locations in the flow of air. The unit was equipped with a differential manometer to measure the pressive difference across the air heater. Incident energy was determined by measurement with pyrometer placed alongwith the absorber top cover. The energy of the air entering and leaving the collector is determined by making appropriate measurements of mass flow rate, pressure differences, temperatures of ambient atmosphere as well as of the absorber at different locations, time and temperature difference of entering and outcoming air at absorber ends.

A series of similar tests was conducted to determine the average efficiency for 15 minutes periods each over a range of temperature differences between the average fluid (air) temperature and the ambient air. For each 'data point' during the test period of 15 minutes, the average insolation range selected was more than the limiting value of 540 k.cal/hr m^2 and ambient temperature range was always less than 30°C. Collector efficiency was calculated by camparing the energy carried by the transfering fluid, air, to the energy incident upon the collector. These results in the dimensionless form are plotted in figure 3.

In the second experimentation, the performance of iron shaving absorber is compared with the equal dimentioned absorber made out of iron wires placed with 30 percent (by weight) of alumnium scraps (absorber II). Both of these air heaters were experimented under identical conditions. Air was sucked through the air heaters with the electric centrifugal Blowers connected to the outlet headers. Same air flow rate was maintained in both the heaters and data was collected throughout the day regarding Solar radiation, rate of flow of air, pressure drop, inlet and outlet temperatures of the air, temperature of the absorber at different locations. In the second stage of comparative study both the air heaters were having similar packed-bed absorbers but the length of one of the absorber and thus of the air heater was increased by joining the half length tray unit/units. All the above experiments were repeated to compare the performance of air heater for length to width ratio of 1.5 and 2 as compared with that of ratio 1.

Later on to evaluate the comparative performance of these absorbers without the G.I. tray, packed-bed absorbers were put in wooden structure and similar experiments were repeated. The comparative performance of these packed-bed air heaters and the influence of various absorber parameter is shown in figure 4.

CONCLUSIONS

Solar air heaters with packed-bed absorbers work satisfactorily indicating a good heat transfer by fluid, air, as compared to other types of air heaters. Outlet air temperatures from iron shavings absorber are found quite closer to the average absorber temperatures at lower mass flow rates. When the mass flow rate was raised as high as 50 kg/hr m^2 there was considerable difference between the absorber temperature and outgoing air temperature at quasi-state conditions, though the overall efficiency of the air heater is improved at higher mass flow rates. Efficiencies of collection are as high as 70 - 75% at outlet temperature of 50 - 53°C. Any increase in length of the packed-bed absorbers in Solar air heater decreased the collection efficiency while outgoing air temperature was accor dingly increased at the same mass flow rate.

Comparing the performance of iron shaving absorber air heater with thealuminium waste blended iron wires absorber, latter is found better. Under the identical conditions this difference in collection efficiencies was almost negligible at lower range of inlet air temperature and lower mass flow rates but went on increasing with the increase of inlet air temperatures. The comparative cost of type II absorber under Indian conditions is 5 - 6 percent higher as compared to type I absorber made of iron shavings. When the G.I. tray was replaced by wooden tray collection efficiency decreased but this decrease was insignificant wherever the thickness of packed-bed absorber was more than 10 cm. The performance of packed-bed air heaters with G.I. tray remained un-affected with the minor changes in aspect ratio at its higher values while wooden tray absorber performance was decreased considerably. At lower values of aspect ratios the performance of air heaters with either type of trays was almost the same. Use of wooden tray is more advantageous as it decreased the constructional cost and with a great ease in manufacturing.

NOMENCLATURE

m	=	mass flow rate
I	=	Radiation rate incident on collector
L	=	Length of collector
b	=	Width of collector
C_f	=	Specific heat of air
U_l	=	Loss coefficient
f	=	Fraction of effective incident radiation
hc	=	Absorber-to-fluid film heat transfer coefficient
hv	=	Absorber to fluid Volumetric head transfer coefficient
Ta, ta	=	Ambient temperature
tab	=	Absorber temperature
T_f, t_f	=	Average air temperature inside collector
T_{fi}	=	Collector inlet air temperature
T_{fe}	=	Collector outlet air temperature
G	=	Superficial mass velocity
$\bar{x}$	=	$\frac{x}{L}$

x = Distance from air inlet end of the absorber.

REFERENCES

1. C.L. Gupta and H.P. Garg, Performance studies on Solar air heaters, Solar Energy 11, 1966, pp. 25-30.
2. V.D. Bevill and H. Brandt, "A Solar energy collector for heating air," Solar Energy 12, 1968, pp. 19-29.
3. H.P. Garg and C.L. Gupta, "Optimization of the tilt of flat plate Solar Collectors for India," J. Institution of Engineers (India) 48, 1967, pp. 21-28.
4. D.J. Close, "Solar air heaters for low and moderate temperature applications," Solar Energy 7, 1965, pp. 117-124.
5. A. Willier, "Performance of black painted Solar air heaters of conventional design," Solar Energy 8, No. 1, 1964, pp. 31-37.
6. P. Singh, "Cheap packed-bed Solar air heaters" First National Solar Energy Convention, Calcutta, India, 1976.
7. R.K. Swartman and O. Ogunlade, "An investigation of packed-bed collectors," Solar Energy 10, 1966, pp. 106-110.
8. M.J. Shoemaker, "Notes on Solar Collector with unique air permeable media." Solar Energy, 5, 1961, pp. 138-141.

PERFORMANCE COMPARISON OF DIFFERENT TYPES OF SOLAR WATER HEATERS UNDER EGYPTIAN CONDITIONS

S.W. Youssef and I.A. Sakr
National Research Centre
Solar Energy Laboratory
Cairo, Egypt

ABSTRACT

The solar collector is the essential item of equipment which transforms solar radiant energy to some other useful energy form. Solar collectors differ in several respects in shape. This paper reports the design details of four different types of solar water collectors, suitable to provide enough hot water for the domestic demand. The collectors are designed in such a way to be simply manufactured with local materials and technology, also to operate under the climatic conditions of Egypt. Tests are carried out on these different types, under identical conditions to investigate the performance of each. An economical study is carried out to determine the proper available materials used in solar water heaters, which makes the cost cheaper, so it can be easily used on a widespread basis. The results obtained are analyzed to select the suitable type to operate under local conditions instead of other conventional systems of heating depending on fossil fuels, which become costly and which are being depleted after one or two generations.

INTRODUCTION

In all solar water heaters of the flat plate type, problems are encountered in the design materials, insulation and inclinations, to obtain the best performance of such apparatus with minimum possible cost. To achieve this purpose, trails are made to use the available materials in the local market in manufacturing the different parts of the solar water heaters. Various materials are used in manufacturing the collectors, namely galvanized iron, brass, and copper. Window glass is provided to minimize the cost instead of any special type which will be costly. Improvement is made to the black coating, to raise its absorptivity. Tests are carried out on these heaters under the same climatic conditions with an inclination angle equal to 30°, which is found to be the best tilt angle throughout the year, to obtain the best performance of each according to the recorded results. Therefore, the best one can be selected to be manufactured on a large scale to provide hot water for domestic uses. Dust precipitation effect is studied and a recommendation time is given to clean the glass periodically.

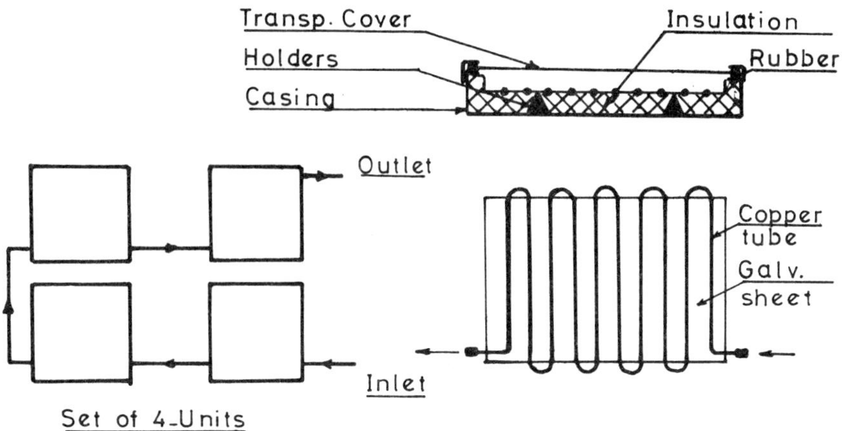

Fig. (1-a) Collector I Sinusoidal Type

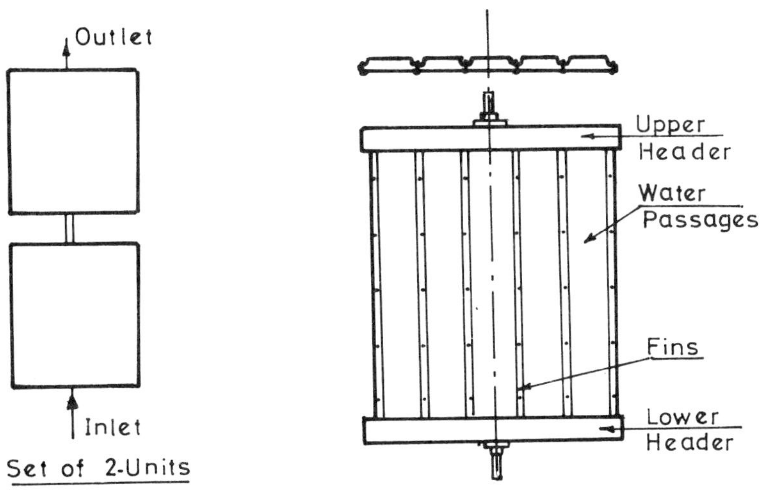

Fig. (1-b) Collector II Corrugated Sheet

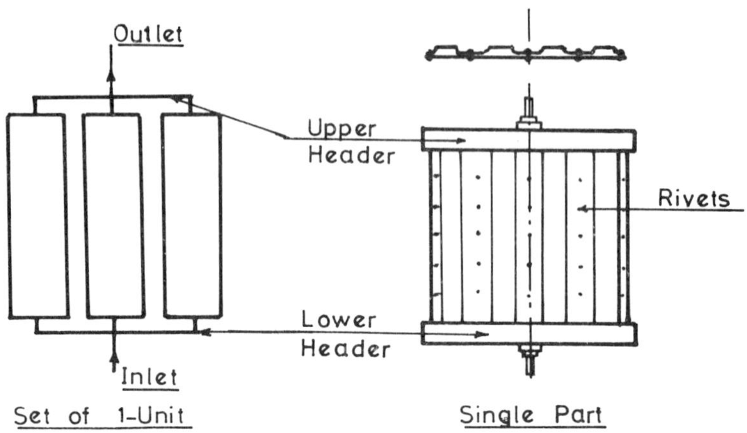

Fig.(1-C) Collector III Separate Corru. Parts

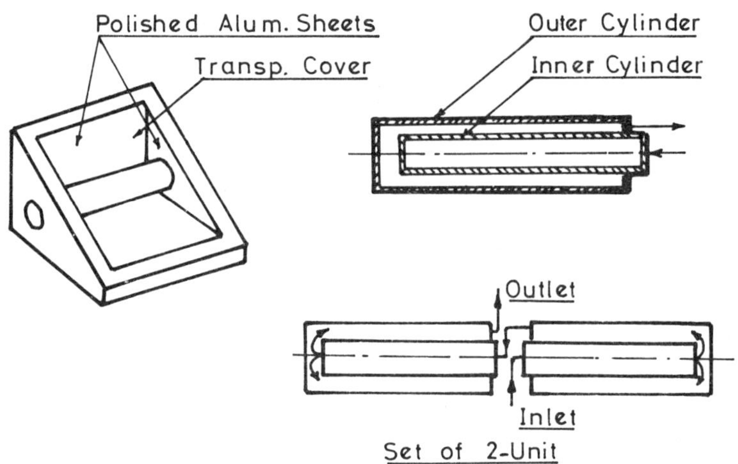

Fig.(1-d) Collector IV Cylindrical Type

DESCRIPTION OF THE COLLECTORS

Four different types are designed and manufactured to predict the best performance. The following is the description of each.

Collector I (Sinusoidal Type)

It is composed of a corrugated galvanized iron sheet (0.8 mm thickness) with dimensions of 500 x 600 mms forming an area equal to 0.3 m^2. A copper tube of 15 mms in diameter is bent sinusoidally, and is embedded in the corrugated galvanized iron sheet as shown in Fig. 1-a. The tube is attached to the plates by means of clamps, also soldered from its sides. The absorber is enclosed in a galvanized iron casing having the dimensions of 600 x 700 x 110 mms, insulated from bottom and sides, by 50 mms of glass wool, beneath a transparent cover of ordinary window glass (3 mms in thickness). Four identical units are set in series in our test forming a total area of 1.2 m^2 and are mounted on a 30° angle with horizontal frame. Moreover, they can be set parallel if it is desired.

Collector II (Corrugated Sheet)

It is formed from a galvanized iron sheet having an area of 0.3 m^2, with dimensions of 500 x 600 mms, which is attached to a corrugated sheet, forming together passages for water circulation and flat parts to behave as fins as shown in Fig. 1-b. Two headers are formed from galvanized iron, and are also welded to the plates from up and down, for hot water and cold water respectively. The absorber is enclosed in a galvanized iron casing having the dimensions of 600 x 700 x 110 mms. Two identical units are set in series forming a total area of 0.6 m^2.

Collector III (Separate Corrugated Parts)

The collector is composed of three separate identical parts forming a total area of 0.3 m^2. Each part is composed of a flat galvanized sheet (0.8 mm thickness), attached to another corrugated sheet by means of spot welding, forming in between passages for water circulation and fins, which are welded along its length as shown in Fig. 1-c. Two common headers formed by a copper tube are located down and up for inlet and outlet water. The absorber is enclosed in a galvanized iron casing with dimensions of 600 x 700 x 110 mms. During the testing period, only one unit at a time is tested.

Collector IV (Cylindrical Type)

It is composed of two concentric cylinders of 600 mms length, formed by a brass sheet of 0.4 mm thickness. The diameter of the outer cylinder is 120 mms, and the inner is 80 mms. The inner cylinder is closed from the entry and outlet by a flange welded to it, having a hole of 2 mms in diameter to provide water from the main portion to be heated in the space between the two cylinders as shown in Fig. 1-d. The surface of the outer cylinder is painted by a black coating to absorb incident and reflected heat. The cylinders are mounted in a galvanized iron casing of 0.8 mms thickness. The transparent side of the casing is inclined by an angle equal to 30° with the horizontal, and is covered by a window glass sheet. The inside surfaces of the casing are covered with polished aluminum sheets to form together with the transparent cover a system for reflecting and concentrating the rays upon the surface of the outer cylinder. Two repeated units are set in series to allow the hot water produced by the first unit to enter the second one to attain higher temperatures.

INSULATION

Different insulating materials are used in solar water heaters. Properties of most of these insulating materials are given in Table 1. It is seen that the best material which has the minimum thermal conductivity, and consequently, transfer the minimum amount of heat, is polyurethane. Instead of it, glass wool (which is not available on the market) is used with a thickness equal to 50 mms, to prevent heat from escaping outside. Moreover, it has a reasonable price on the market with huge quantities. The edges are also insulated as recommended by Tabor [2] to use edge insulation about the same thickness as the bottom insulation.

ORIENTATION

All the collectors are tilted facing south with an angle equal to 30° which equals the latitude of the place. This optimum angle is obtained after studying the variation of the daily total solar energy radiation for surfaces facing south, tilted at different angles. The study showed that the maximum amount of solar energy throughout the year for a fixed surface is for a surface tilted at an angle equal to 30° with the horizontal.

CIRCULATION

The circulation system is selected to be a natural themosiphon. The storage tank is positioned such that the level of the water is always above the top of the collector panel. The water which has been heated in the collector becomes lighter and rises, while the colder and heavier water will be drawn from its place in the bottom of the tank. Thermosiphon systems generally have low flow rates through the collector, but the fluid is undergoing a larger temperature rise.

COATING

The collector surfaces are painted with ordinary black paint after adding talc powder with equal weights to reduce its reflectivity. Adding talc powder raises its absorptivity ($\alpha = 0.96$), but its emissivity is about $\varepsilon = 0.76$. An advantage of this coating is that it does not peel off due to the repeated thermal expansion and contraction of the surface of the collector plate after a long run.

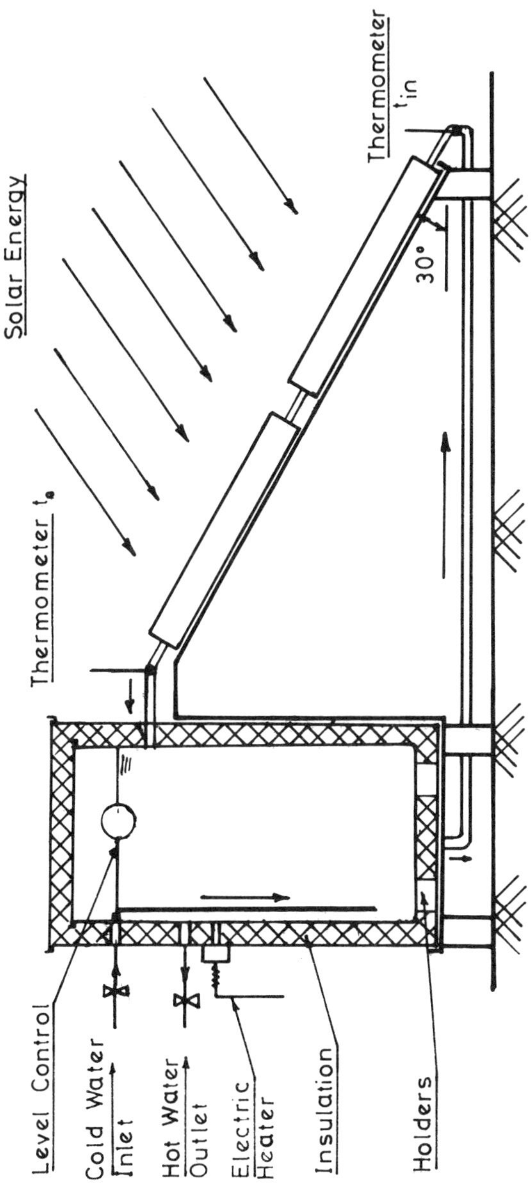

Fig.(2) Solar Water Heater with Hot Water Storage Tank

HOT WATER STORAGE TANK

Collectors I, II, III, are connected separately to a storage tank as shown in Fig. 2. The storage tank used is of a double wall galvanized iron sheet (0.8 mm thickness). It has an inner and outer rectangular cross-section of 33 x 66 cm and 43 x 76 cm, respectively, with a height of 80 cm. The tank is insulated by 5 cm of glass wool inside the double walls, and also from up and down. A level controller is provided inside to maintain a constant water level, by allowing the cold water to replace the hot. Two openings are also made, one for the hot water demand, and the other for an electric heater to heat the water on the overcast days, or during heavy demands. Collector IV has a tank for the outlet hot water while the inlet to the collector is from the main supply.

Test Procedure

The following quantities are measured and recorded for half an hour to get:

- incident solar radiation at the collector surface, (tilt 30°), H in kcal/m^2/hour
- inlet and outlet water temperatures, t_{in}, t_o in °C, respectively
- the mass flow rate in kg/hr, which is made constant over the day and is changed from day to day

A measure of the collector performance is the coefficient of effectiveness defined as the ratio of the useful gain over any time period to the incident solar energy over the same time period.

$$\text{C.O.E.} = \frac{\int \frac{Q_u}{A_c} \, d\tau}{\int (H)_T \, d\tau} \tag{1}$$

or

$$\text{C.O.E.} = \frac{q_u}{(H)_T} \tag{2}$$

where q_u = useful heat gained by the flowing water in the collector per unit area in cal/Cm^2/hr and is calculated from the following equation:

$$q_u = \frac{\dot{m} \; C_p \; \Delta t}{A_c} \quad \text{Kcal/m}^2\text{/hr} \tag{3}$$

where $\dot{m}$ = the flow rate of water in kg/hr

C_p = mean specific heat of water and is equal to 1 Kcal/kg/°C

Δt = difference between outlet and inlet water of the collector in °C and is equal to $(t_o - t_{in})$

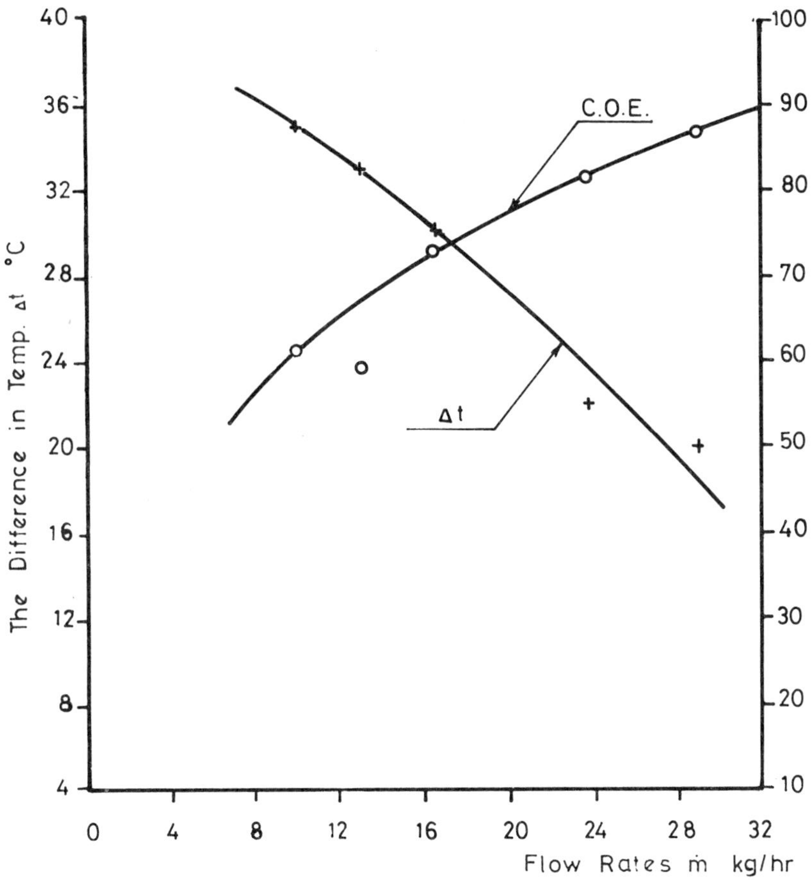

Fig. (3) Variation of % C.O.E. & Δt °C with ṁ kg/hr for Collector I (Area = 1.2 m^2)

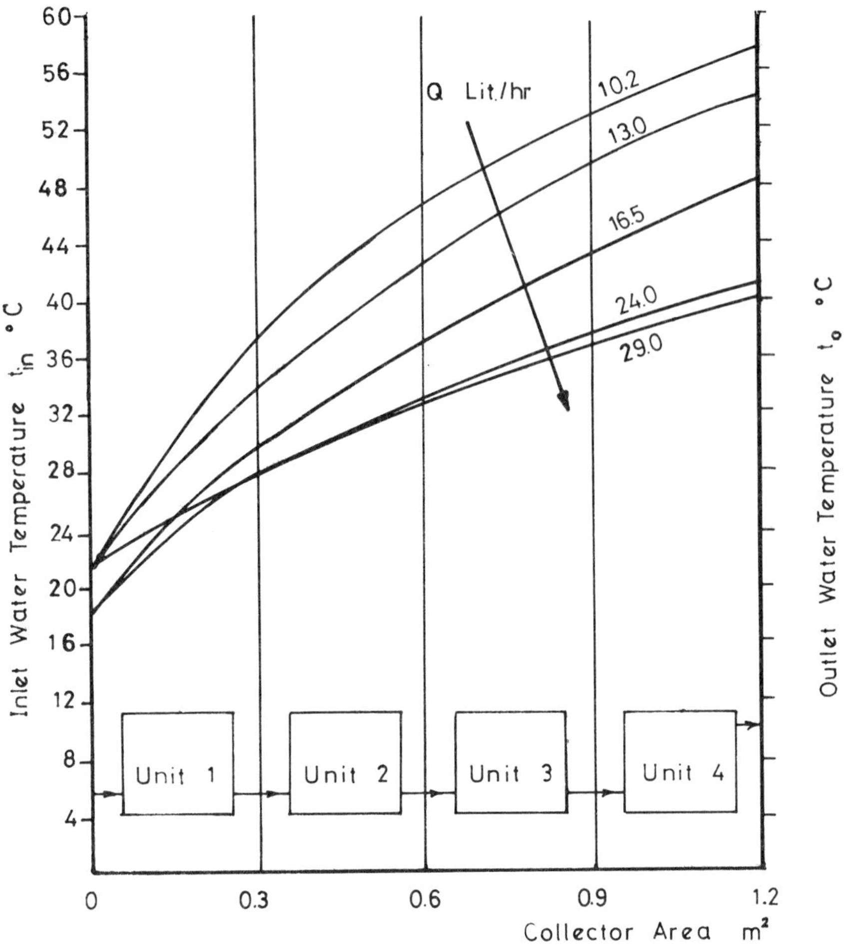

Fig.(4) Variation of t_0 in every Unit Collector Area with Different Flow Rates

$(H)_T$ = incident rate of solar energy on the collector in $Kcal/m^2/hr$

DISCUSSION OF THE RESULTS

The daily average C.O.E. for each heater is calculated from the measured data as given in the test procedure.

Collector I: The relation between the flow rate and both the average C.O.E. and the difference in temperature Δt between outlet and inlet as shown in Fig. 3. The relation between every unit collector area subjected to solar radiation and inlet and outlet water temperature is shown in Fig. 4. One can conclude the following:

- The outlet water temperature decreases as the flow rate is increased and vice versa.
- The coefficient of effectiveness of the heater increases by increasing the flow rate until it reaches a maximum value of 90% at flow rate 32 lit/hr per m^2 (Fig. 3), but the corresponding temperature difference has a value of 15°C, which is relatively low. On the conrary, the C.O.E. decreases at the low flow rates, and the difference in temperature Δt increases and may reach 35°C at flow rate value of 10 lit/hr.
- The outlet temperature of water from each unit increases as it passes in a set of series collectors as shown in Fig. 4. The rise in temperature of the first is higher than the second and so on. The effect of the flow rate on the temperature difference of each unit is comparatively low for flow rates higher than 24 lit/hr.

Collector II: The relation between the flow rate and both the average C.O.E., and the difference in temperature Δt is shown in Fig. 5. From this figure it is seen that:

- The C.O.E. of the heater increases by increasing the flow rate until it reaches a maximum value of 25 lit/hr, which is found to be the optimum value.
- Every square meter of the solar collector can provide an average daily amount equal to 150 lit., with a temperature rise of 20°C, or 90 lit/day of hot water with a temperature rise of 30°C.

Collector III: The relation between the flow rate and both average C.O.E., and the difference in temperature Δt is shown in Fig. 6. The following is concluded:

- The C.O.E. of the heater increases by increasing the flow rate until it reaches a value equal to 8 lit/hr (for a unit area of 0.3 m^2), then the C.O.E. decreases. Therefore, this value is the optimum to operate such a unit to attain a temperature rise equal to 17°C, and

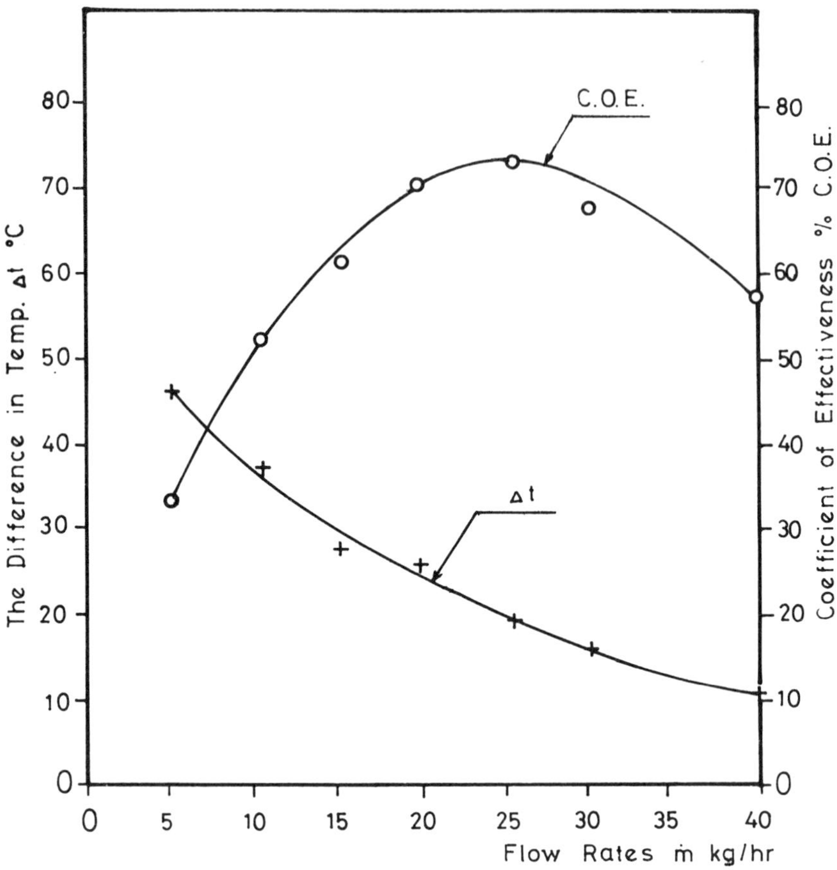

Fig.(5) Variation of % C.O.E. & Δt °C with ṁ kg/hr for Collector II (Area = 0.6 m^2)

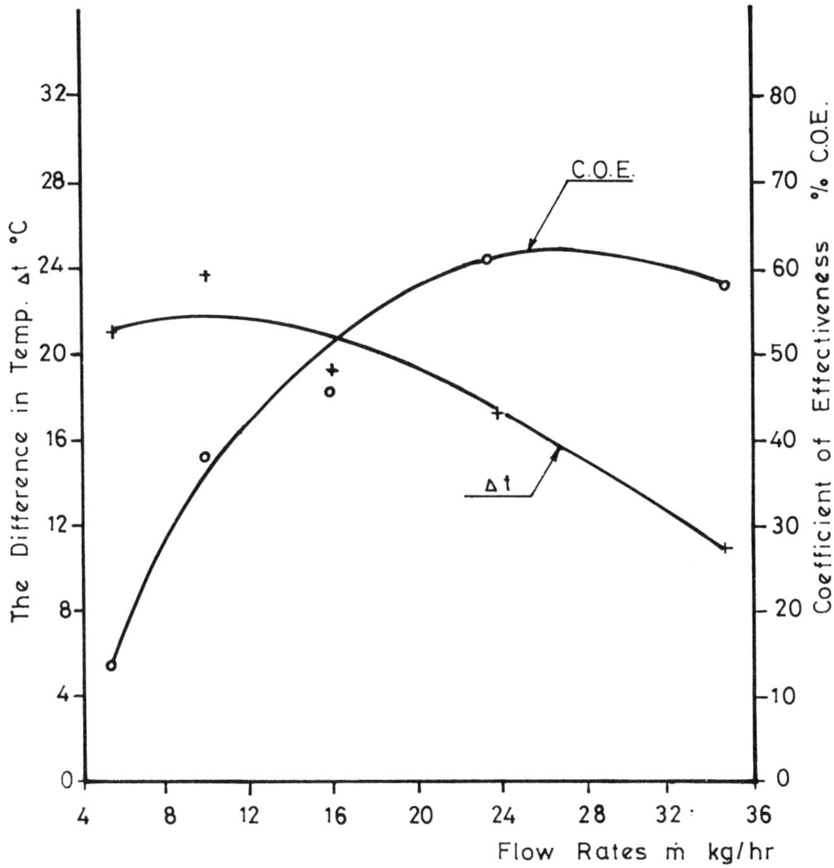

Fig.(6) Variation of % C.O.E. & Δt °C with ṁ kg/hr for Collector III (Area = 0.3 m^2)

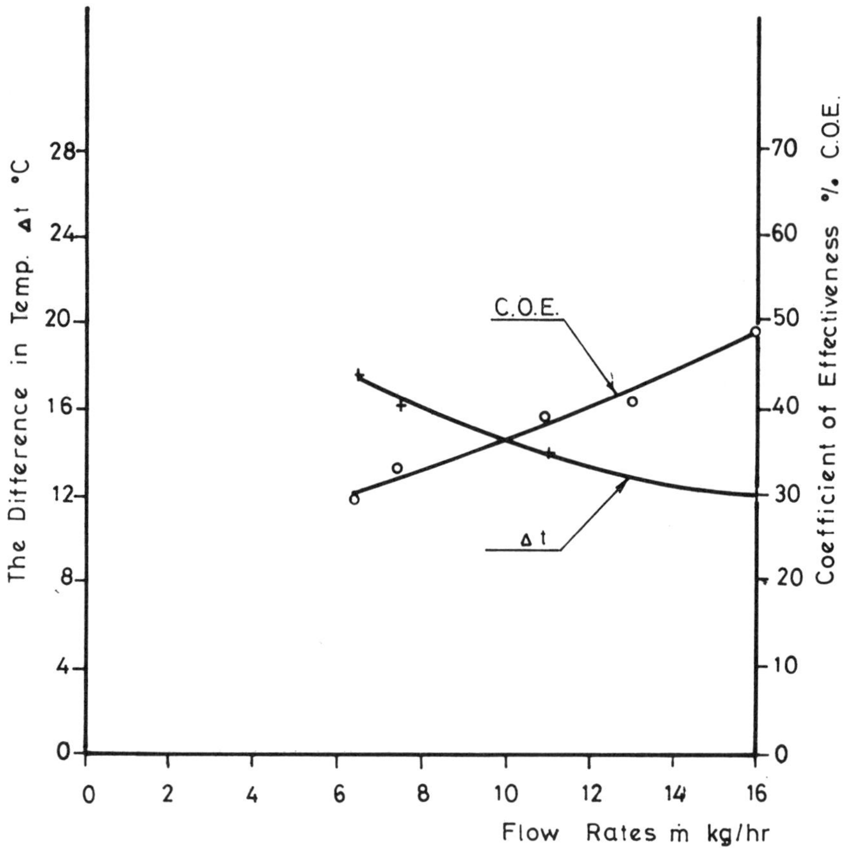

Fig.(7) Variation of % C.O.E. & Δt °C with ṁ kg/hr for Collector IV (Area = 0.72 m^2)

and C.O.E. equal to 62%.

Collector IV: The relation between the flow rate and the average C.O.E. and Δt is shown in Fig. 7. The conclusion obtained is that:

- The C.O.E. of the heater increases until it reaches a value equal to 50%, by increasing the flow rate to 16 lit/hr for an area of 0.72 m^2 (the area of the two units in series), also Δt is equal to 12°C.
- At a flow rate equal to 6.5 lit/hr, the temperature difference is 17°C, and the C.O.E. has a value of 30%.

DUST PRECIPITATION EFFECT

The effect of dust on glass transmissivity is difficult to be generalized, since it is mainly concerned with the environmental conditions which surrounds the solar water heaters, mainly the magnitude and direction of wind speed, rains, nature of the ground whether it is dust or sand, etc.

To study the dust effect on the energy transmitted through the glass covers, an experiment was carried on five similar pieces of the same glass used in solar water heaters, over a period of 30 days, having a dimension of 100 x 100 mms and an inclination angle of 30° with the horizontal. One of them was kept clean throughout the period of the test, while the rest were exposed to different weathering conditions. The reduction on energy transmitted was measured at noon time. The results showed gradual reduction in the first days, then this reduction became severe due to the accumulation of dust until the percentage reduction in the energy transmitted was 16% after one month. So, it is recommended to clean the glass weekly, at least, to avoid the reduction in the energy transmitted. Therefore, to obtain the best coefficient of effectiveness.

CONCLUSION

By making a comparison between results obtained from the curves drawn between the flow rates and both the average C.O.E. and the difference in temperature Δt, for the four different types of solar water heaters, it is seen that, the maximum C.O.E. obtained is that of collector I (sinusoidal type), while the minimum is the cylindrical type. Also, the maximum rise in temperature is obtained by collector II (corrugated sheet), while it is minimum in the cylindrical type.

Therefore, if the design is based on the average C.O.E., the sinusoidal type is the most suitable one, especially when higher flow rates are required.

If the choice is based on the maximum Δt, the corrugated sheet is more suitable. However, the manufacturing, installation, and the cost of the sinusoidal type make it preferable to be manufactured in mass production to be installed in hospitals, hotels, clubs and for domestic uses, instead of the conventional types of heating.

REFERENCES

1. Duffie, I.A., Beckman, W.A., Solar Energy Thermal Processes, 1974.

2. Tabor, H., "Radiation, Convection, and Conduction Coefficients in Solar Collectors", Bull. Res. Coun. Israel, 6C, 155, 1958.

3. Khatlab, N., M.Sc. Thesis, 1976, Faculty of Engineering, Cairo University, Cairo, Egypt.

4. Final Report on the Solar Water for Domestic Uses Project 1975/1 to the Academy of Science and Technology, Cairo, Egypt.

TABLE I

COMMONLY USED INSULATION MATERIALS IN FLAT PLATE COLLECTORS

Name	Thermal Conductivity K(W/m°C)
Mineral Wool (fibre-glass, rock-wool)	0.0332 - 0.0404
Hair Felt	0.0389
Granulated Cork	0.0476
Compressed Cork	0.0418 - 0.0462
Straw	0.0576
Sawdust	0.0649
Vermiculite (granulated)	0.0721
Polyurethane (foam rigid)	0.0245
Polystyrene (expanded)	0.0303

SOLAR WATER HEATERS

D. BUXEY
Dr. Beck & Co. (India) Ltd
61, Advent Building
General Bhonsle Marg, Bombay.

ABSTRACT

This report summarizes briefly the salient features of world developments and new developments in the field of solar heaters, considering the importance of saving hundreds of thousand tons of fuel per year.

In this report, the principles are studied so as to adapt the heater to varying needs. The needs vary not merely due to locale (North India requirements are quite different from the requirements of South India), but to habits also in general. All these factors are taken into account.

The first major industrial installation in a hotel in Matheran, near Bombay has been described. This has gone commercial.

The absorbtion and storing in same and seperate units, and the dual circulation system are discussed.

The various factors affecting design, such as heat removal, orientation and transmittance of glass covers etc are gone into various types of absorbers are gone into: the closed membrane type seems cheapest and best, but the natural circulation type seems better, if installation costs can be lowered. Where efficiency counts and not costs (in India, efficiency does not count, as one has all the time in the world), the pan type is better, but not so for this country.

The introduction of a 1 mm. thick polyethylene film increases the life, but also the cost.

Trends in the improvements of the designs are discussed. Auxiliary heating is also considered and the design of a solar switch given.

INTRODUCTION

This paper summarises briefly the salient features of world developments and new developments in the field of solar water heaters considering the importance of saving hundreds of thousand tons of fuel per year.

The second consideration for this study from the point of view of importance, is that it (solar water heaters) represent the most widespread direct use of solar energy at the present time.

In this report, the principles are studied (briefly - due to shortage of space) so as to adapt the heater to varying needs. The needs vary from country to country. In U.S.A., the prime need is for rapid heating of swimming pool water. In India, where the bulk of its population lives in villages, and where, at the time of writing, thousands of villages do not even have drinking water and are migrating elsewhere, the needs may be for hot water for washing and cooking. Again, the needs for Northern India vary from the needs of Southern India.

The emphasis in this paper - though principles are being studied only - will be more on rural needs rather than urban.

A preliminary study of Indian conditions has revealed 2 important needs :

1. for domestic heating, where the quantity of hot water needed per day varies somewhat from 75 - 200 litres per day, and

2. water is required for large scale purposes like in buildings, schools, colleges and dormitaries, in public bath houses, hotels, etc let alone laundries and hospitals.

It has been found by a study of solar energy availability, with reference to water heating, that this is amply available and the project is practical.

For instance, in Israel where so many water heaters are used, only 65 days (as against 60 in Cairo) are cloudy or semi-cloudy. All regions in the solar belt (latitude 45° N and 45° S) having more than 2000 hours of sunshine a year are suitable for use of such heaters.

In Delhi, it is 2844 hours

The problem lies in the hill stations of the Himalayas, such as Simla, where in winter months, due to sub-zero temperatures, the water freezes in the pipes and storage tanks, causing complications. These complications can be obviated by certain adjustments making solar heaters once again practical, but due to higher costs, such heaters find themselves as commercially non-acceptible. There is thus a clash between these two parameters.

In our research study, we concentrate on principles, and take technical considerations into our study, leaving aside the economic viability.

<u>Different hot water systems</u> :

There are basically three types :

1. Absorbing and storing type in the same unit
2. Absorbing and storing type in seperate units
3. Dual circulation system

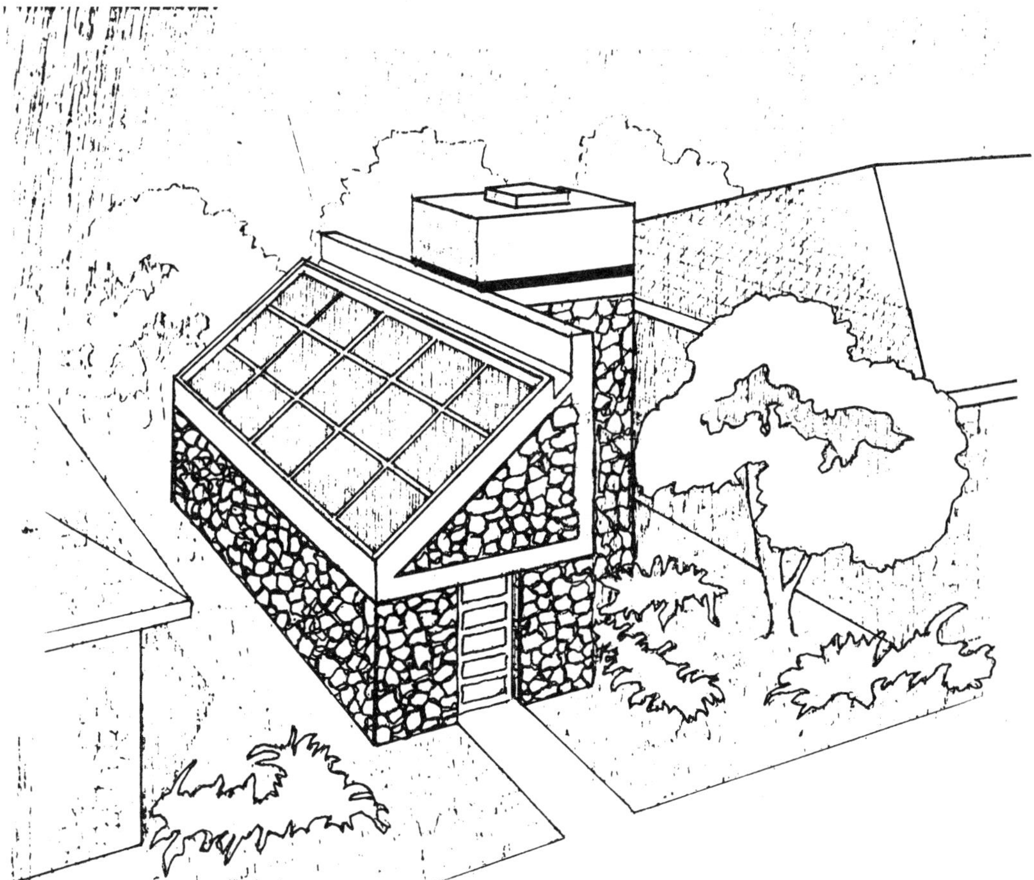

PROPOSED SOLAR HOT WATER INSTALLATION AT MATHERAN FOR " RUGBY HOTEL"

FIGURE 1

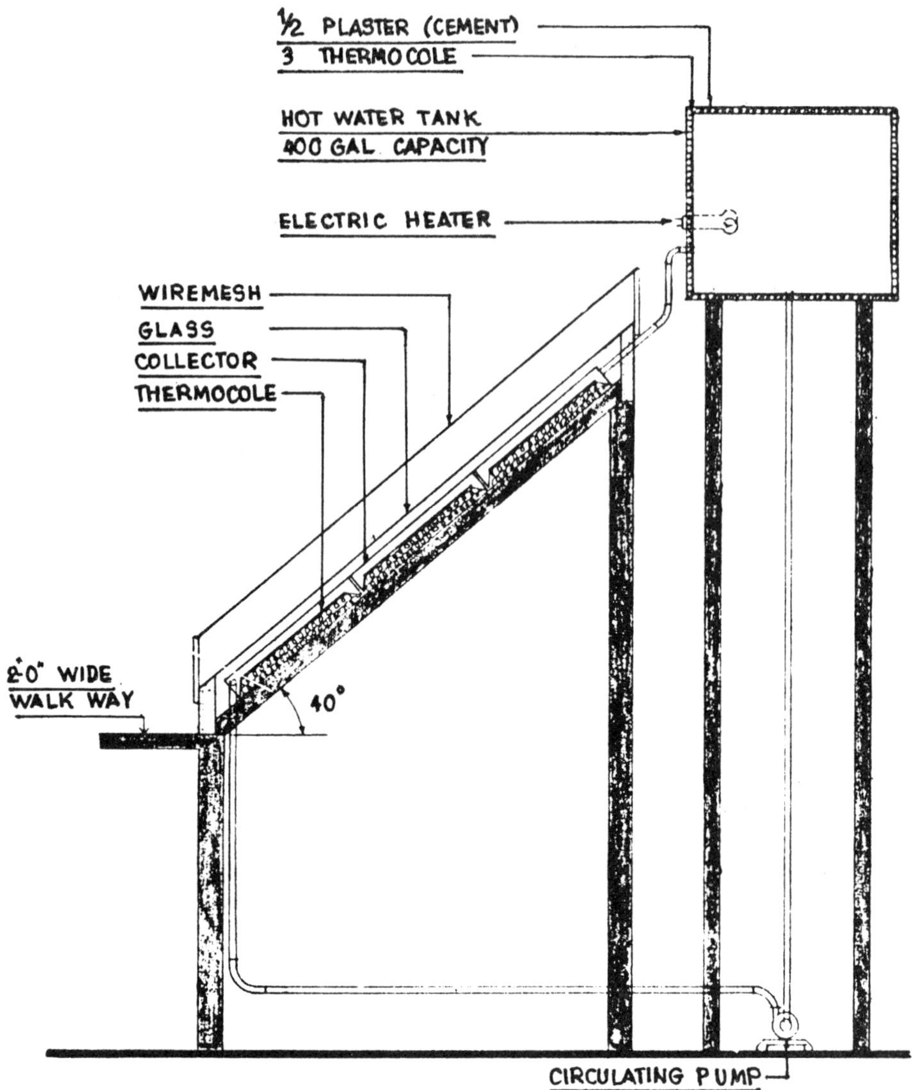

FIGURE 2.: SCHEMATIC DIAGRAM OF SOLAR INSTALLATION AT MATHERAN

Design criteria :

In design, the following five considerations are important :

1. orientation of absorber
2. effect of number of glass covers
3. absorber size and design
4. storage tank size, and
5. storage tank location

It is well known now that for the best all round performance, the absorber should point south (in the Northern Hemisphere) inclined at about 10° plus the geographical latitude.

Installation criteria :

The points to be taken into consideration on installation are :

1. aesthetic design
2. determination of area
3. location of storage tank
4. choice of thermo siphon or forced circulation, and
5. location of absorbers

The providing of simple mountings or logical placing of absorbers can have a profound effect on the wider acceptance of such heaters.

Efficiencies :

These have been tabulated in various senses.
For instance, they can be :

a) physical
b) economic

In the physical sense, it is the ratio of heat collected in hot water to integrated insulation on surface of absorber.

Efficiency at noon time is high and a figure of 70 % is easily reached.

Over an entire area. a typical value for a heater with storage tank lies between 30 - 40 %.

On the other hand, if efficiency drops from 40 % to 30 %, thereby halving the cost, the heater is also deemed efficient.

In the economic sense, it is measured as quantity of heat energy obtained for unit investment in a solar water heater.

It may be relevant to mention here that the first major industrial installation of such a heater was in Rugby Hotel, Matheran, delightful hill station near Bombay. The heater serves 6/7 months and has been deemed by the hotel authorities as eminently practical, both from the point of view of its working (where the above design and installation crieterial have been taken into account) and from the economic view point viz. returns on capital.

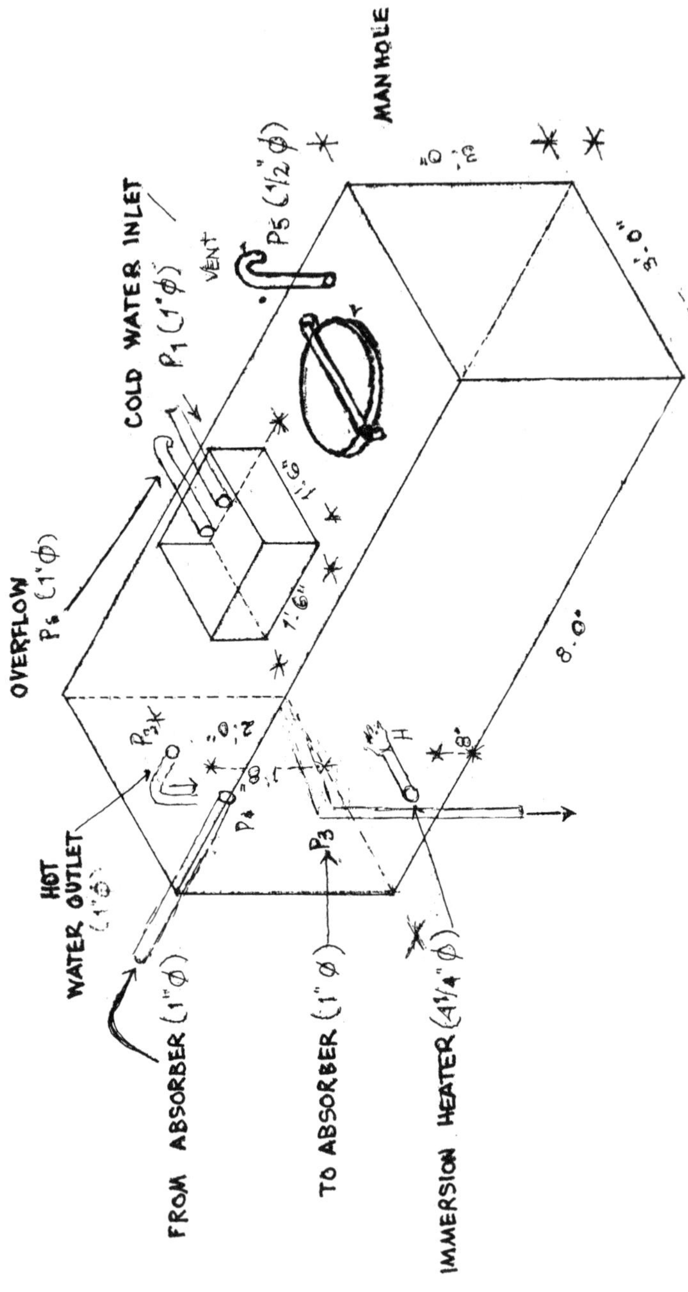

FIGURE 3: ISOMETRIC VIEW OF THE TANK SHOWING THE INLETS & OUTLETS & THEIR DIAMETERS, FOR PROP. SOLAR HOT WATER INSTALLATION AT MATHERAN FOR RUGBY HOTEL

The Company endeavoured to install a similar heater in Khandala Hotel, Khandala, another beautiful hill-station near-by, but completely failed to convince the authorities on the economic view point. Here, the heater had to serve all the 12 months, as the hill station of Khandala, unlike Matheran, is open during monsoon months. The electric cost was over Rs. 2000/- per month. But the economist considered the project a failure. Their chief point was that Rugby Hotel could also not be achieving savings, as it was situated in forest area, and twigs (not coal) would have to be burnt. Obviously, ecological considerations come into play. This point is important and worth mentioning. Even if by wood collection, the ordinary system is economic, continously increasing de-forestation (due to increasing popularity of hill stations) would lessen the charm of the hill stations and therefore, their attraction to city dwellers. With less tourists coming in, the economics suffer. Hence, in all hill stations, solar water heaters has to be a must, leaving aside merely technological considerations or their economy merely from the point of view of return on capital invested.

It is in this way, that the research scholars of this company view the importance of this first major industrial installation in Matheran, as it provided a model for all hill stations to follow.
The manufacturing company is in Ahmedabad, Gujarat and the development of India would be reflected in the development of this country, so far as such installations in strategic areas are concerned. (The impact on ecological development is considerable)

Cost Reduction :

Study has also been given to the various factors which would have to be taken into consideration to bring about reduction in costs.

These would be :

1. the introduction of new materials such as synthetics, plastics, selective surfaces etc.

2. by standardization viz. mass production

3. New design improvements, which also include new methods of installation etc. and

4. future plans for developments

The closed membrane type of heater seems cheaper and best, but the natural circulation type seems better if installation costs can be lowered.

Where efficiency counts and not costs (in India, efficiency does not seem to count, as one has all the time in the world), the pan type is better but not for this country.

The introduction of the 1 mm thick polyethylene film which is very strong and hardly ever breaks, increase the life span (but unfortunately, also, the cost).

In the case of the first commercial and industrial installation in India, the cost was Rs. 17,000/- but it cuts two ways :

a) panels (collectors) plus water tank : Rs. 7000

b) construction (optional) : Rs.10,000

It will thus be seen that the actual cost was only Rs. 7000/-. This comes down with bulk production (unless design is changed).

The installation could have been on the floor or the roof. It was not put on the floor, but on a height so as to get some advantage of lift. It could have been put on the roof, which would have serviced a number of rooms of the cottage where it was installed, but the hotel was made of a number of small cottages and it was desired that other rooms (8 rooms in all) in other cottages should be serviced, hence the construction was optional.

But, our study indicated, that the dual circulation system, comprising two circuits, the primary and secondary was more convenient for up-hill stations like Darjeeling and Simla, Mussourie etc, where water can freeze and damage the absorber.

The primary circuit consists of :

a) absorber
b) tank

both operating at atmospheric pressure

The secondary circuit consists of

a) an inside tank submerged in the primary tank

The inside tank and primary tank is covered with tank insulation and can be put on floor or roof, reducing costs.

But this modification is due to the hill stations of North India facing in winter months, sub-zero temperatures unlike hill stations of Central and South India, where even the winter months are relatively hot. Here, the absorber cannot be damaged. In the North, will modification design, and because of the ethylene glycol (anti-freeze) mixtures, such damage is prevented.

Talking of cost and designs, it must be noticed that basically there are only five types of designs :

a) open type
b) closed type
c) closed membrane type
d) natural circulation type and
e) once-through type

Other papers have treated of these designs in detail, suffice to state that item (c) is the cheapest and used in large numbers, followed next by item (a) and then (b).

The future lies with class (d), though this design will take time to enter the commercial market.

SOLAR SWITCH

In all these designs, auxiliary heating, it is found, cannot be dispensed with.

A solar switch can be used and the best design is of G. Neeman. American experience indicates that with the average service life of the heater taken as 8 years (for depreciation etc), solar heating becomes economical when cost of power is above 3 cents per KWH. As the cost of power drops, it becomes necessary to save on auxiliary heating.

If auxiliary heating is used for more than 15 % of total heating, solar heating cost will be higher than electric heating, even at cheaper night rates, so this mean auxiliary heating should be automatically controlled. This is American experience.

In India, the same pattern is noticed, as due to the hill-stations where we are concentrating on, the power if 5 times more than in other places. If the cost of switch is kept at 3 % of total cost of heater, the additional cost will be recovered within a year through economy in power.

There are 3 basic designs for such a switch, with a photo-electric cell, the thermostat inside flat plate collector and the thermostatic solar switch. We will not digress into these path-ways and bye-ways of such discussion, as this paper, though discussing principles and neglecting details of construction and material, has already been heavily loaded here.

Suffice to say, sufficient experiments have not been done with absorbers here as in other countries. Rural Japan concentrates on the pan type, with improved Russian design, while the sinusoidal type are more widely used in South America and Spain; a firm in Poona has come out with straight ducts with headers as used in Russia and Israel, and partly by Japan (on roof tops). The flat plate type, with upper plates ribbed are also used more in Japan, where the water passes through 2 paralled plates and this has all the advantages of the straight tube, closely spaced unit and is less expensive, and shows potential in rural India.

In conclusion, we may state that universal consensus of opinion calls for a world standard test. The South African Council for Scientific & Industrial Reasearch in Pretoria has set up such a test. Suffice to say, that the cost of the equipment (including fan, head tank, immersion heater, thermostat, pipes etc) comes to slightly less than Rs. 1000/- and is favourable.

LATENT HEAT DIODE WALL

L. Bourdeau and A. Jaffrin
Laboratoire CNRS d'Ecothermique Solaire
Observatoire de Nice
06 - FRANCE

ABSTRACT

A composite wall with separate collector side and storage volume operating as a thermal diode is proposed. The storage uses a latent heat material called chliarolithe. Laboratory experiments performed on the storage side show that it is possible to define satisfactory conditions both for the collector side and for the heat restitution : relatively low temperature in collector, little external losses, storage equivalent to 75 cm of concrete in the form of 5 cm of chliarolithe, excellent temperature control on the internal side.

INTRODUCTION

The present prototype of latent heat diode wall is the result of an attempt to combine the favorable thermal performances of a latent heat material, developped by the Laboratoire d'Ecothermique Solaire under the name "Chliarolithe" (I), and the basic ideas of passive solar heating of houses. On one hand, the solar heating of houses requires low temperature calories and a good control, on the other hand solar calories are easier to collect at low temperature but are not easy to store. This gap was bridged by the development of a latent heat material ("Chliarolithe", or warm stone) characterised by a low temperature of fusion (28° C), a relatively high thermal capacity (32 Kcal/Kg in latent heat) and a sufficiently high thermal conductivity (0.5 w/m° C).

Storing calories without raising too much the temperature is then possible and the benefits that can be extracted from the use of such a material are obvious. It is shown in a separate communication (2), that the collector temperature can be kept at a lower level than in usual sensible heat devices, so that the capture efficiency improves by a factor up to two; this allows then to take full advantage of the large specific thermal capacity of the material. Most of all, the fixed point of phase change offers a very efficient and simple control of ambiant temperature.

These considerations naturally lead to a concept of phase change "Trombe wall". If one now wants to reduce the external losses in the absence of sun, the phase-change Trombe-wall has to be splitted into two parts, so that the collector face is disconnected from the storage volume.

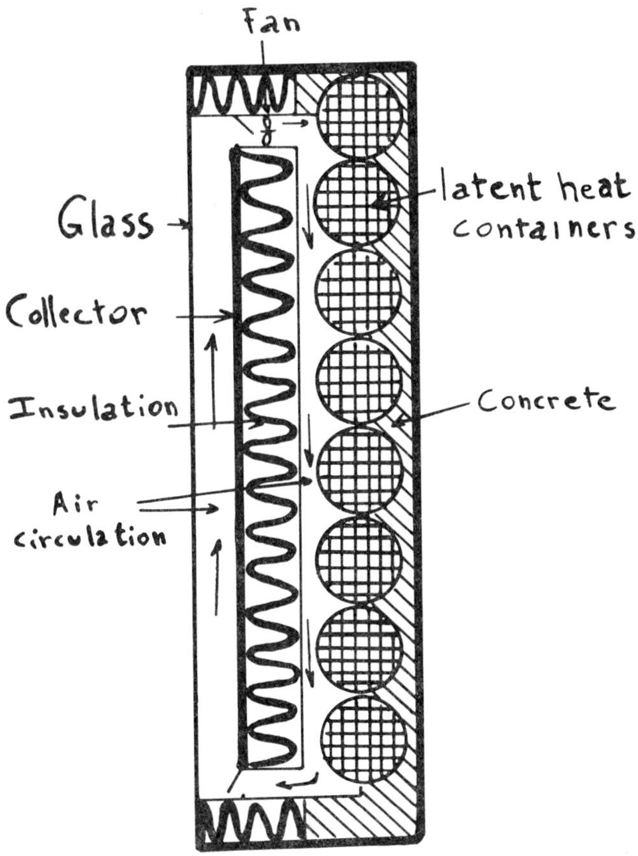

Figure I : Geometry of latent heat diode wall .

The transfer of calories is then achieved by the circulation of a fluid which can be interrupted at night. The diode wall is therefore a composite wall with internal circulation of fluid and a high thermal capacity on its internal side.

DESCRIPTION OF THE LATENT HEAT DIODE WALL

The proposed geometry of the latent heat diode wall prototype is sketched in fig. I. The circulating fluid is air. The collector side of the wall is a dark surface of relatively low thermal capa--city protected by a single glazing (it does not have to be effi--ciently protected at night because of the diode function). The heat transfer to the storage is achieved by the air flow heated along by the absorbing surface. In the figure, a simple system is considered where the air circulates between the surface and the glazing; in a more sophisticated solution,the front volume would be closed to minimize heat transfers to the glass and a circula--tion of air would be organized behind the collector surface like in most air solar collectors. Then the collector side is separated from the rear volume by a thick insulation which reduces the con--duction losses from the storage to the front side during the cooling phase. A forced circulation of air is created by a fan so that the air temperature stabilizes at the desired value (usually much lower than the one resulting from the natural thermo-circula--tion in a Trombe wall). Valves are sketched on fig. I to show how a reversed air circulation is prevented at night; but a similar valve effect could be obtained by creating a sufficient head-loss (IO to 20 mm of water) in the circuit at the base and top of the wall : this head-loss is in fact necessary to regularize the air flow activated by the fan.

On the storage side, the warm air losses its calories at the con--tact of the latent heat containers which are represented here by horizontal cylinders. Various geometries for the containers could be chosen; the only constraints on the choice are the questions of being waterproof (the chliarolithe does not accept water admixtures without degradations of its thermal properties) and of presenting the largest surface for heat transfer with the air. Calories are stored by the process of solid to liquid phase change of the chliarolithe and the fusion front progresses inside the cylinders. The heat flux conducted through the phase changing material reaches the rear side of the wall where it is transmitted to the house interior by convection and radiation. The temperature on this side of the wall cannot exceed, or even approach, 28° C before a comple--te fusion of the chliarolithe has occured.

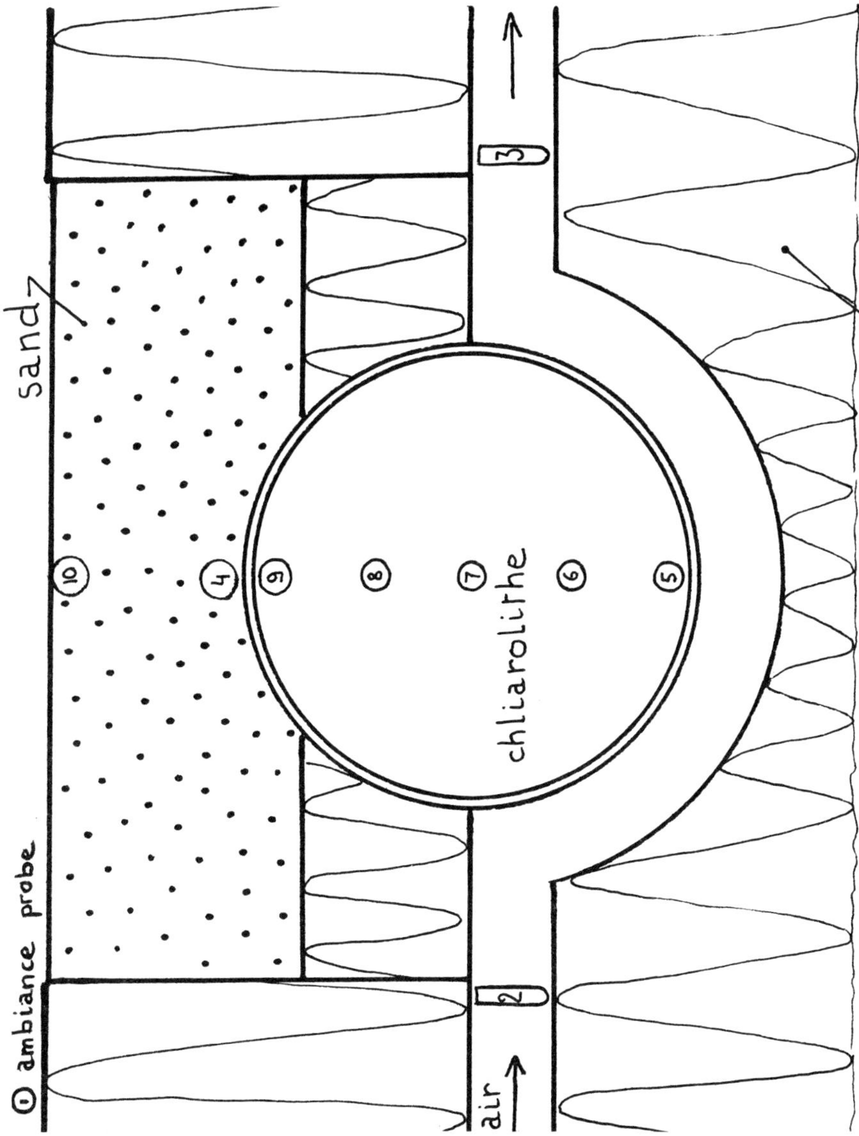

Figure 2 : Simplified prototype experimented in laboratory . Temperature probes are indicated by numbers I to IO .

DIODE WALL ELEMENT TESTED IN LABORATORY

The complete latent heat diode wall has not been yet experi--mented. The problem, at the present time, has been splitted in two : on one hand, conditions for collecting and storing solar calories in a simple latent heat Trombe wall have been studied in real weather and the results are presented in a separate com--munication (2). On the other hand, the storage side of the diode wall has been studied in laboratory conditions by simulating solar gains. The structure of the wall has been simplified and for commodity the vertical rear air circulation was replaced by an horizontal flow. Fig. 2 shows the simplified prototype.
A regular and horizontal stream of air with a mean velocity of 2 m/s is created by an centrifugal fan. An electric resistance placed in the stream heats the air. The warm air stream propaga--tes through an insulated channel of section 2 cm x 40 cm before reaching the outer surface of a storage element. This element is a 8 cm diameter cylinder of 40 cm length made of I,5 mm thick PVC plastic. It is filled with 2,8 Kg of chliarolithe. The heat exchange occurs along the lower half part of the cylinder. The upper part of the cylinder is in contact with a material which imitates the internal concrete or plaster side of a wall; for commodity reasons, solid plaster has been here replaced with sand to access to temperature probes. The heat transfer from the storage to the ambiance takes place on a horizontal surface (I5 cm x 40 cm) instead of a vertical surface. But the quantita--tive aspects of the transfer are not really affected by that situation. The whole prototype is placed in a room where the air and radiation temperatures are kept in a very narrow range around I7 ° C, because of a huge thermal inertial (3 meter thick walls without windows).

TEMPERATURE MEASUREMENTS

Temperatures are measured with IOO Ω platinum resistances. The incoming air temperature is measured, as shown in fig. 2, just before the heat exchange with the storage, and the outgoing air temperature just after. To minimize the effect of inhomogeneities in the air flow, a profile of temperatures along a section of the stream was recorded so that the indication of one probe can be analysed in terms of a mean temperature of the flow.
The process of phase change in the storage tank is studied by means of five probes placed along a vertical diameter in the cylinder. Probes are placed in the sand layer, one close to the storage tank, another one at the outer surface, a last probe measures the ambiant temperature. All platinum probes were care--fully calibrated and the uncertainty in the knowledge of the small temperature drop of the air was reduced to a minimum (O.I° C).

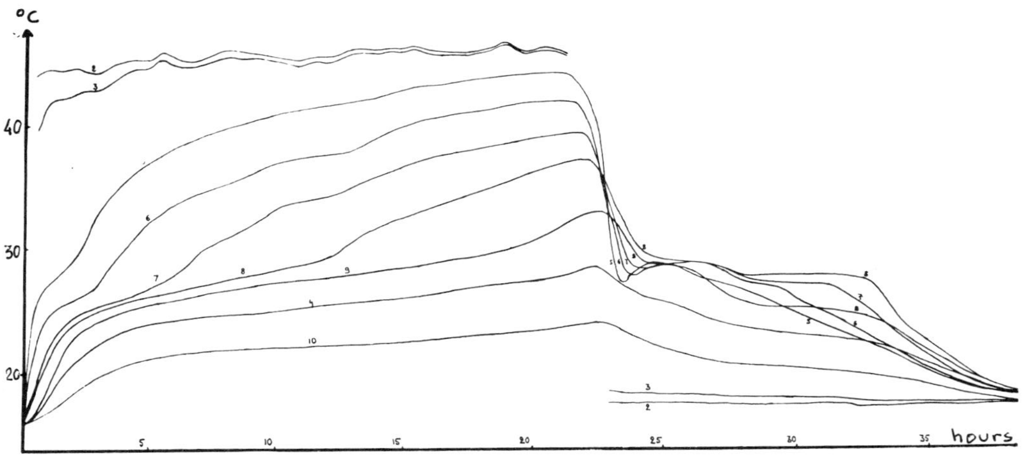

Figure 3 : Temperature variations during a complete heating and cooling cycle performed with 46° C hot air and permanent ventilation .

The temperatures were recorded on a MECI Speedomax at the rate of one set of points every IO minutes.

DESCRIPTION OF THE EXPERIMENTAL RUNS

Several runs were performed with various incoming air temperatures ranging from 34° C to 77° C. The heating phase was maintained during a sufficiently long period so that all probes in the latent heat material indicated temperatures above fusion. The cooling phase begins immediately after the electric resistances are switched off. Several configurations were experimented during the cooling phase : cooling phase with the same ventilation (2 m/s air velocity) as the heating phase; cooling phase with open air vents but no ventilation; cooling phase with closed air vents.
Fig. 3 shows the temperature variations during a complete heating and cooling cycle with permanent ventilation,a 46° C air temperatu--re for heating and a fairly constant I7,5° C ambiant air temperatu--re. Curves 2 and 3 indicate the activated air temperatures before and after thermal exchanges with the storage tubes. Curves 5 to 9 indicate temperatures inside the storage material at positions shown on fig. 2, curves 4 & 10 temperatures of the sand layer, on the storage side and on the ambiant air side respectively. During the cooling phase the restitution of heat to the ambiance takes place through the temperature gradient existing between curves 2 and IO. The main features are the following : although the heating air temperatures measured by probes 2 and 3 fluctuate, the distance between the two curves does not fluctuate and the amount of heat transfered to the storage can be extracted with a fair precision. This amount of course slowly decreases as the storage temperature approches the air temperature. The ambiant air temperature,which is not indicated in the figure during the heating phase,stays close to I7,5° C. Each curve denoted from 5 to 9 exhibits a pla--teau around 28° C corresponding to the fusion stage. This pla--teau appears to be longer for probes further away from the heat source, but this reflects the temperature stabilization of the previous fusion zone. The end of fusion is easier to locate : it corresponds to the next rise in the slope of the curves. It is clear for example that the last probe n° 9 indicates a complete fusion in the storage tank after I9 hours only.
The cooling process shown in fig. 3 takes place with constant ventilation so that the hierarchy of temperature curves is inversed; one notices a small surfusion of 2° C, and a long pla--teau of solidification for all probes except n° 5 placed in a zone which loses rapidly its calories. (probe n° 9 happens to be in the air gap created by the volume contraction of the mate--rial and this affects its response).

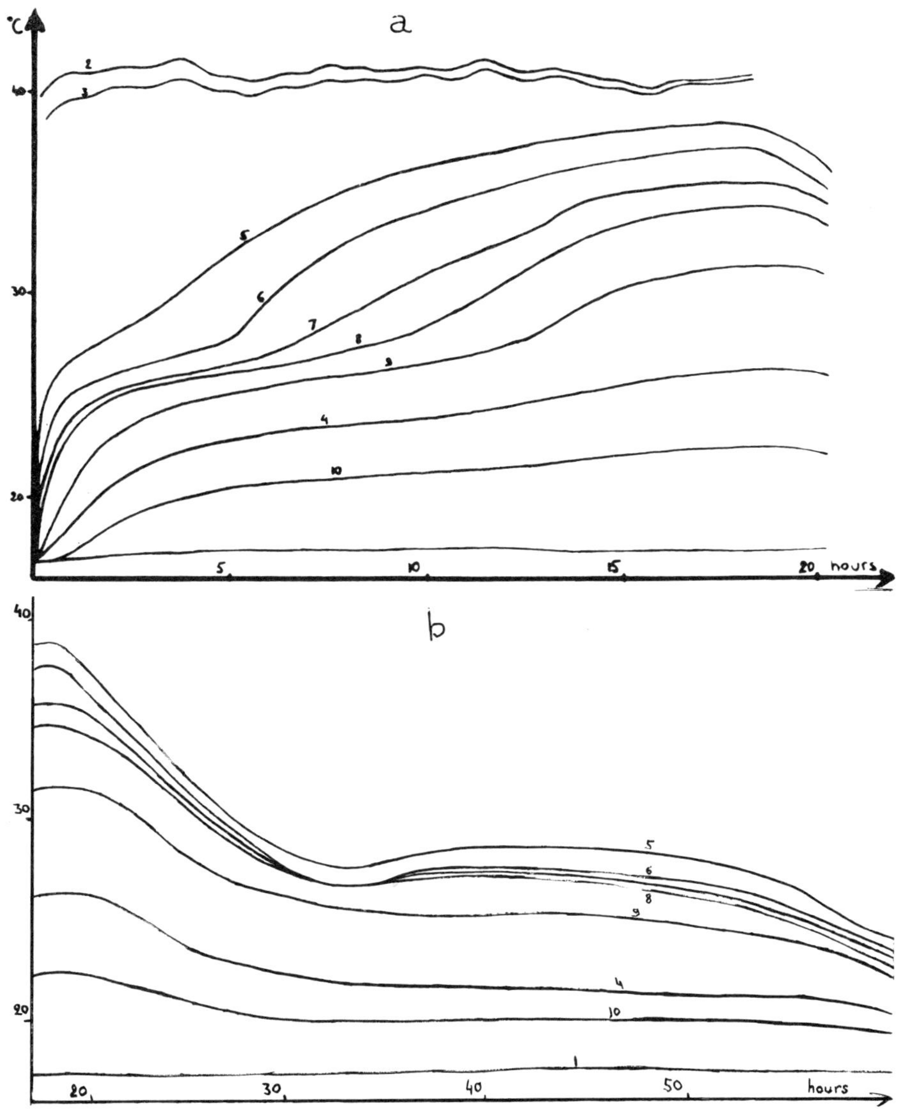

Figure 4 : Temperature variations during a complete heating cycle (a) with 4I° C air temperature and cooling cycle (b) with no ventilation and closed air vents .

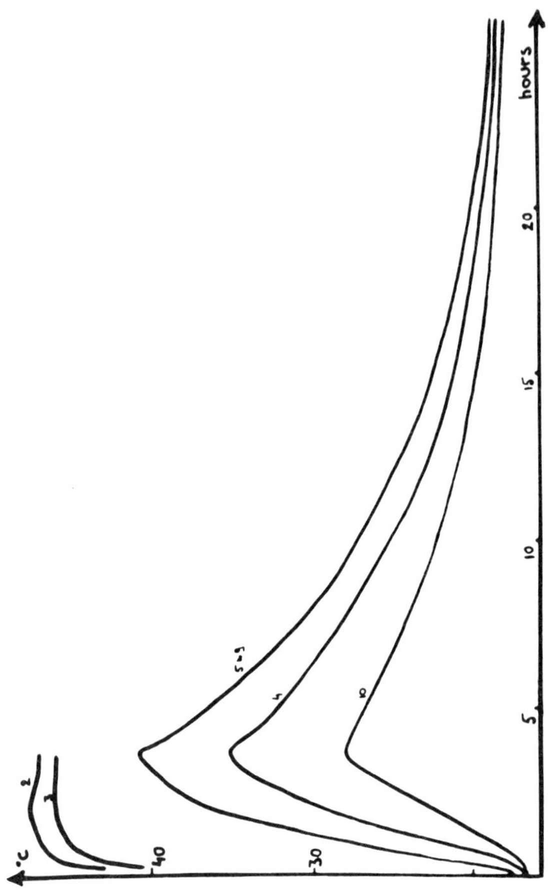

Figure 5 : Comparative run performed with a water storage tank .

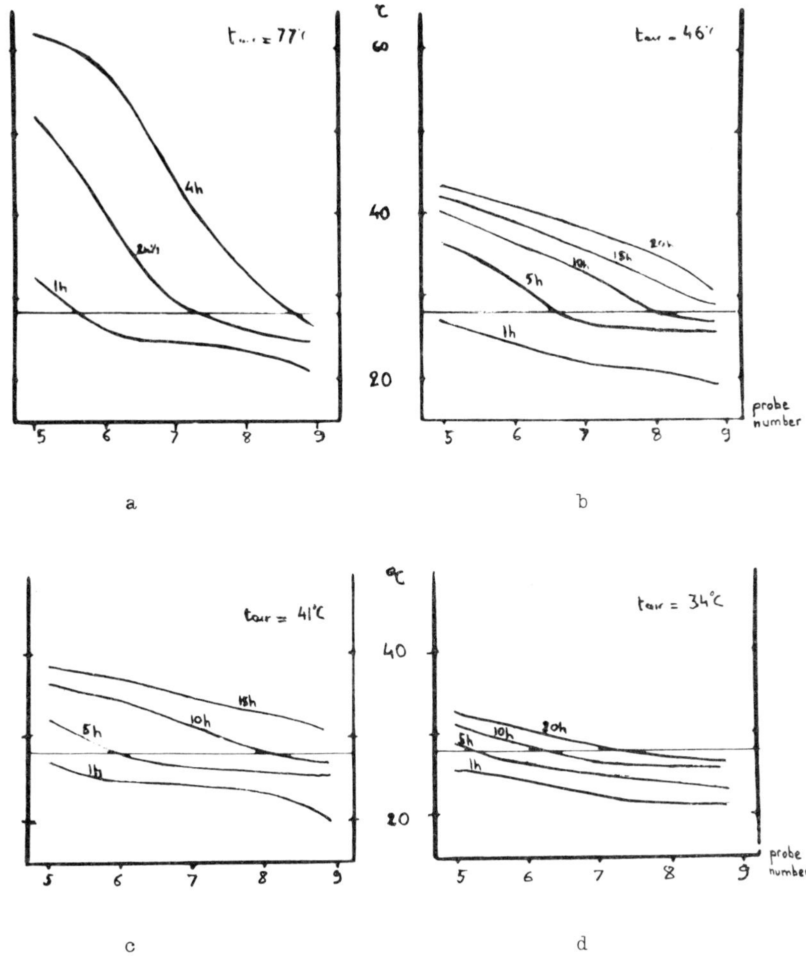

Figure 6 : Analysis of the progression of the fusion front inside the cylindrical storage tank (an horizontal line gives the melting point temperature).

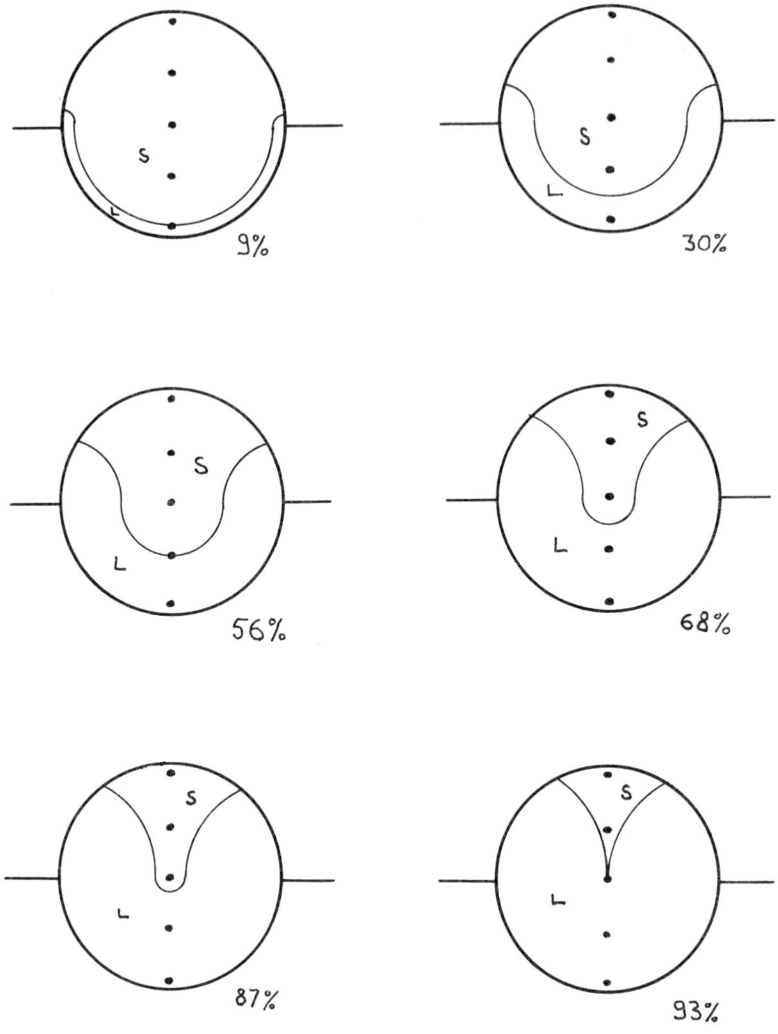

Figure 7 : Melting rates based on the hypothesis of a symmetrical fusion front initiated along the lower half cylinder .

The cooling plateau lasts only for IO hours because the ventilation takes away most of the stored heat.
Fig. 4a and 4b show the result of a run performed with 4I° C hea- -ted air, and a cooling process with no ventilation and closed air vents. The pattern is very similar but the cooling plateau is much longer (30 hours) and temperatures keep their initial hierarchy.
All the heat restitution takes place by conduction through the sand. A remarkable temperature stabilization around 20° C is achieved at the surface of sand. A comparitive run was performed with a water storage tank with 46° heated air. The result is shownin fig. 5. In that case water acts as a very good conductor because of free inter- -nal convection inside the tank. Therefore all probes 5 to 9 indica- -te the same temperature. All temperature curves can be understood as two arcs of hyperbola characteristic of high conductibility sen- -sible heat materials. No good temperature stabilization is obser- -ved on the internal side of the wall. Latent heat storage materials are thus excellent temperature filters. A similar conclusion was ob- -tained in a different kind of collector wall described in a separate communication (2).

DISCUSSION OF THE RESULTS

The results of other experiments at various temperatures have been summarized in a series of figures. Fig. 6a,b,c,d, analyse the progres- -sion of the fusion front inside the cylindrical storage tank. The fusion temperature at 28° C is indicated by probes 5 to 9 for various time intervals. In order to deduce from these curves the fraction of melted chliarolithe, it was necessary to make a hypothesis on the geometry of the fusion front (because of the complicated boundary conditions, no exact calculation was undertaken). We assume that the heated surface of the tube is at a constant temperature, and that the isotropic heat conduction pushes the fusion front so as to remain the envelope of all cylindrical waves emitted by the heated surface. This is sketched on Fig. 7. This is how the melted regions associa- -ted to each temperature probe response are guessed. We shall see later that the assumption of constant temperature on the exterior half surface of the cylinder is not really correct (there is a temperature drop along the air flow). Now, making use of curves of Fig. 6 and 7, we can deduce the percentage of melted chliarolithe as a function of time. The surprising result is that more than 90% of the chliaroli- -the has already melted when the middle probe n° 7 indicates the transition temperature of 28° C. Half of the latent heat has been used once the second probe n° 6 goes through the transition tempera- -ture.
The amount of sensible heat stored in the tank, at the end of a long heating period and a complete fusion, has been roughly estimated by considering that the homogeneous storage medium is in a thermal situa- -tion close to a stationary state. Then the isothermal lines are no longer represented by the curves of Fig. 7. These isothermal lines were obtained by electrical analogy using a conducting paper repro- -ducing the boundary conditions of the diode wall.

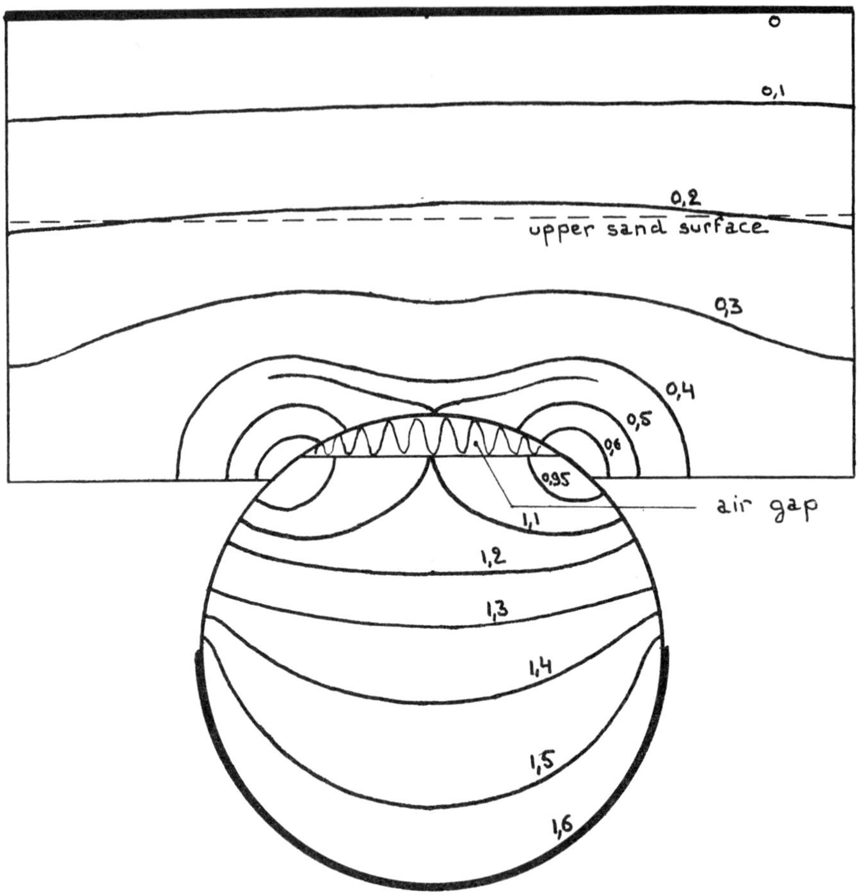

Figure 8 : Isothermal lines in the chliarolithe and the sand corresponding to a stationary thermal state. The heat exchange coefficient on the surface of sand is simulated by an equivalent sand layer .

The resulting curves are shown on Fig. 8.
There is of course an intermediate period during which the isothermal lines change their shape and any quantitative estimation becomes hazardeous. The amount of heat transfered by the air to the wall is easy to calculate, and good estimate of the various losses can be made : a horizontal wall to air exchange coefficient k= 9,5 w/m2° C was measured in a separate experiment done in similar conditions; this actually corresponds to a complete absence of wind and can be also accepted for a vertical wall. Heat loss through the sides of the prototype wall was calculated from the heat conductivity of the insulating material (polystyrene) and the temperature gradients. A summary of all contributing heat flux is listed in table I for the heating phase , and in table II for the cooling phase. We see that the uncertainty due to simplified assumptions does not seriously affect the accuracy of the balance.
The fact that the experimented prototype wall is reduced to a single storage element of 40 cm length instead of the complete array of long tubes shown in fig. I does make a difference in terms of transfered heat flux : the first difference is the existence in the small prototype of a large ratio of external surface to internal volume; this means that the amount of heat lost through the sides is overestimated compared to a complete wall. Instead of the present 25% of stored heat the actual losses should not exceed I0%, which is a tremendous improvement over a standart Trombe wall. The second difference is due to the fact that successive cylindrical storage tanks will receive an air flow of lower and lower temperature so that the global heat transfer to the storage will be a superposition of the different heat transfers affecting each storage tube.
Fig. 9 (a to d) shows a tentative estimate of the various heat contents and heat losses in the wall as a function of time. The latent heat content of the wall is specified by curve C. This content is expressed in terms of percentage in Fig. 9 e for various hot air temperatures. The full curves result from the assumption that the melted volumes of chliarolithe are symmetric (like in Fig. 7); the dotted curves correspond to the actual response of the probes. The difference between the full and dotted curves means that the fusion is faster on the side of the incident hot air, and this is not unexpected. The main feature is however common to both curves : there is a saturation for each temperature showing it is not worth seeking a complete fusion of the chliarolithe. After 6 hours of 46° hot air, 90% of the latent heat has been actually used and it would take another 5 hours to increase this percentage by 5%. For 41° C the saturation occurs after 8 hours and for 34° C after I3 hours.

As a sunny day in winter can only provide 5 hours of solar energy collection, let us study how much heat can be transfered and stored for various temperatures during that period . The result is presented in Fig. I0 a and I0 b in terms of percentage of melted chliarolithe, and the temperature drop of the incident hot air is shown in Fig. I0 c .

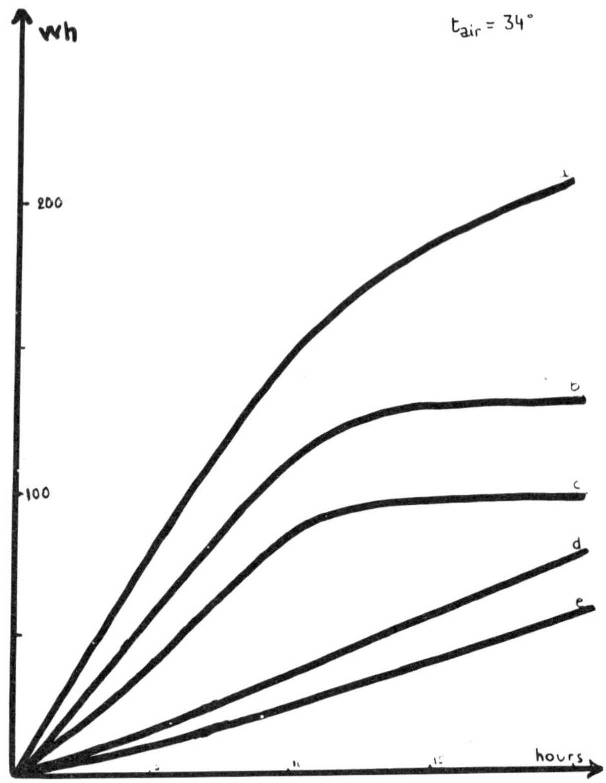

Figure 9 A : Tentative estimate of the various heat contents of the wall , and heat exchanges as a function of time during the heating phase. T_{air} = 34 °C .
Curve a : total accepted heat; curve b:total stored heat; curve c : latent heat ; curve d:total losses ; curve e : wall to ambiance transfer .

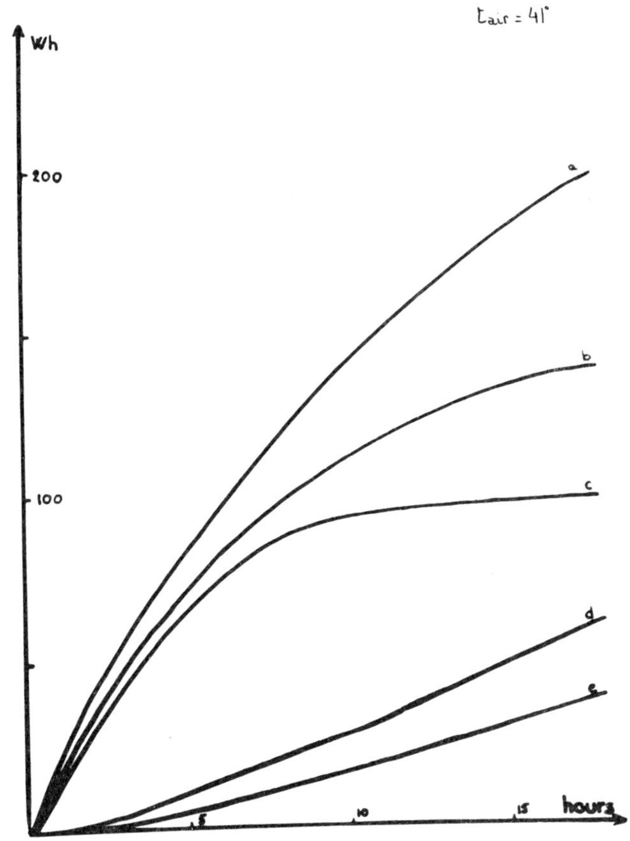

Figure 9 B : Same as figure 9 A , with T_{air} = 4I°C .

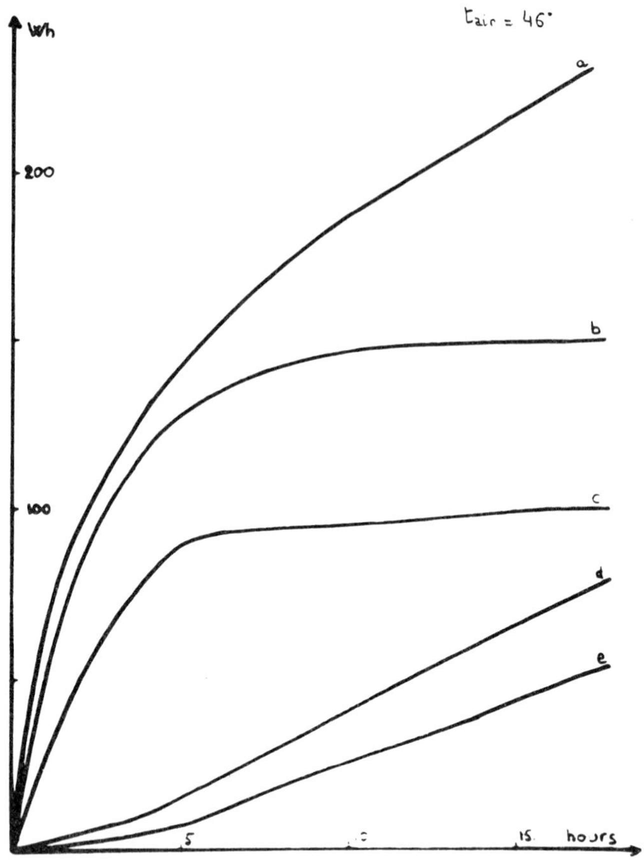

Figure 9 C : Same as figure 9 A , with T_{air} = 46°C.

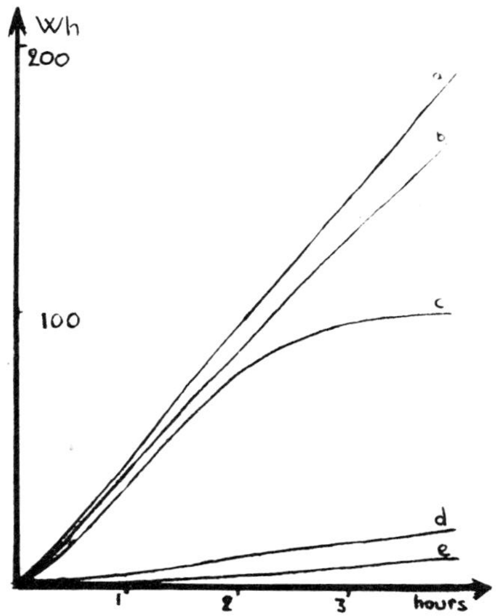

Figure 9 D : Same as figure 9 A with T_{air} = 77°C .

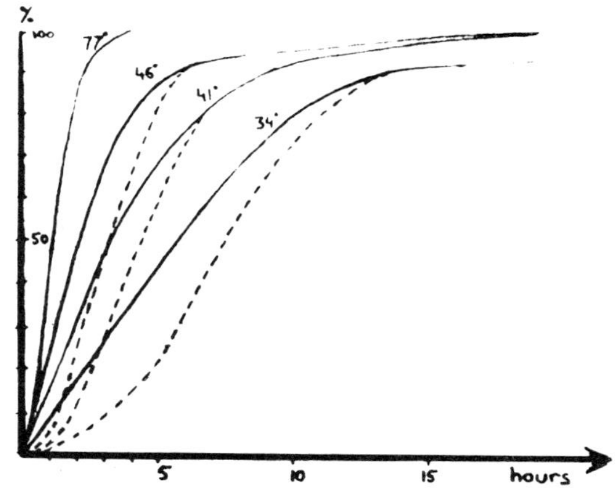

Figure 9 E : Melted chliarolithe fraction as a function of time for various hot air temperatures.

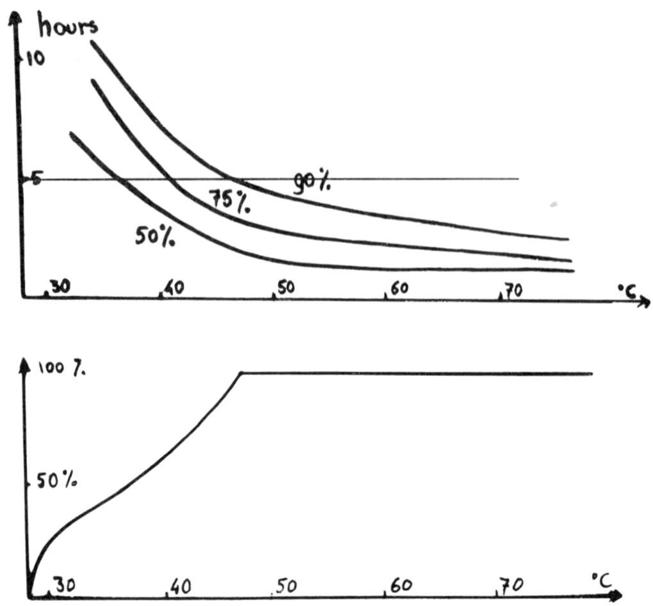

Figure IO a and b : Melted fraction of chliarolithe as a function of time and temperature (IO a) ;as a function of temperature after a 5 hour heating period (IO b) .

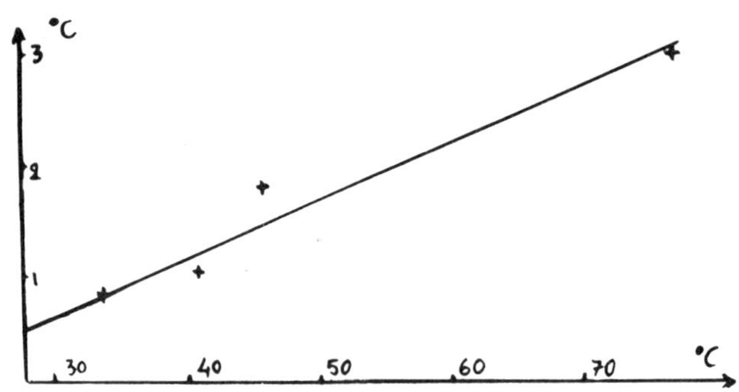

Figure IO c : Air tempearture drop after exchange with the storage as a function of initial temperature .

The preceeding discussion shows that it is reasonnable to use storage tank cylinders of diameter 8 cm for heating periods of five hours and incoming air temperatures ranging from 34° to 46° C, which correspond to solar heating contribution on the collector side of one to two days and a 50% to 90% use of the storage capacity.

Fig. II visualizes the temperature filtration obtained in a permanent regime of 5 hours of solar heat per day(46° C incoming air) for a few days : the interior wall surface temperature stays 3 to 4° C above the ambiant air temperature on a very smooth fashion. A completely passive temperature control is therefore achieved through the filtration ope--rated by the latent heat material.
A similar run was made where the heating period consisted of IO sequen--ces of I5 minutes 46° C hot air separated by 9 sequences of I5 minu--tes without heat input. The filtration process is such that every--thing behaves as if one had 5 hours of heat with a lower temperature air (around 34° C).

EXTRAPOLATION TO A COMPLETE DIODE WALL

We propose here to estimate the actual heat transfers which would take place in a 2.6 in high diode wall used as a collector and heat storage system for a house. As mentioned before, the experimental conditions are no longer those of the laboratory test : the ambiant air tempera--ture in the house is supposed to be 20° C instead of I7° C in the laboratory; successive horizontal storage tubes receive a heat flux at a temperature decreasing with height.
Assuming a collected solar heat flux of 400 w/m^2 during 5 hours and air velocity 2 m/sec in the storage air channel, the air temperatu--re rises by 26° C an the collector side (20° C for 300 w/m^2).Using Fig. IO , an estimation of the air temperature drop on the storage side as a function of height is made in Fig. I2 , thin curves (full line for 400 w/m^2, dotted line for 300 w/m2)assuming a linear de--crease along the surface of one tube; thick curves show the amount of melted chliarolithe in each tube. The maximum amount of stored latent heat is 2,3 Kwh/m^2 20% of sensible heat can be added to that value. We see that the bottom tubes receive much less calories that the top tubes; this leaves room for possible excess of collected heat in mid season without raising too much the temperature of the admitted air in the collector.
It would be anyway favorable to prevent over-heating the higher tubes by showing down the air flow along their surface (by enlarging the upper air channel).
A second aspect of the thermal behavior of the diode wall is the heat restitution to the house at 20° C. This implies a new estimate of the heat transfer between the chliarolithe at 28° C and the air at 20° through concrete or plaster which are better heat conductors than sand. Let us assume stationary heat exchange between the wall and the house (which is not far from truth, given the very slow variations). The equation of transfer reach.

$$R = k\,(T_f - T_s) = h\,(T_w - T_a)$$

TABLE I : Summary of all contributing heat flux and heat contents during the heating phase of the diode wall .

Storage material	Water	Chliarolithe				
air stream temperature (°C)	47°	77°	46°	46°	4I°	35°
ambiant temperature	I7°	I6°	I6°2	I6°5	I7°,2	I5°6
experiment duration	3 h I/2	4 h I/2	21 h	25 h	I8 h	21 h
latent heat (wh)	0	I00	I00	I00	I00	94
chliarolithe sensible heat (wh)	49	53,5	38,5	43	32	25
sand sensible heat (wh)	I5	9	I0	II	7	6,5
wall to ambiance transfer (wh)	I0,8	8,3	66	67	38	55
other losses (wh)	5	I2	32	38	22,5	20,5
total heat (wh)	84	I89	250	245	I83	I90

TABLE II : Summary of all contributing heat flux and heat contents during the cooling phase of the diode wall .

Storage material	Water	Chliarolithe		
wall configuration	closed air vents	ventila- -tion	closed air vents	open air vents
initial stored heat (wh)	64	148,5	I39	125,5
experiment duration	24 h	I7 h	42 h	29 h
final stored heat (wh)	4	I,5	I2	14
wall to ambiance transfer (wh)	48	36	97	68
other losses (wh)	I7	I22	30	44

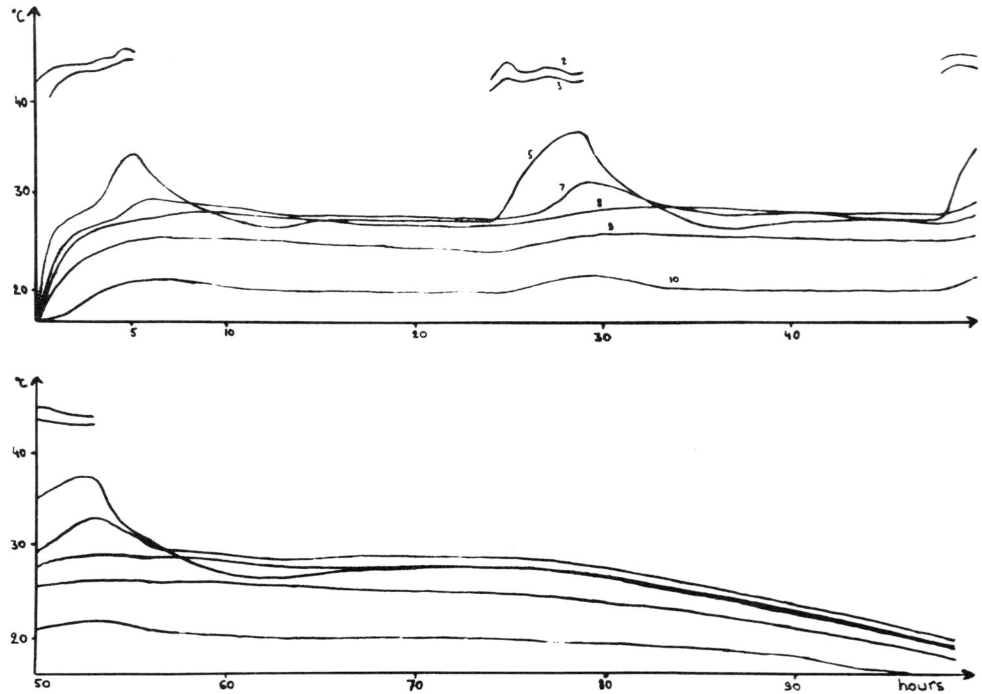

Figure II : Study of a quasi permanent regime simulating a realistic situation. Three successive 5 hour heating periods with 46°C hot air are separated by three successive I9 hour cooling periods with no ventilation and closed air vents .

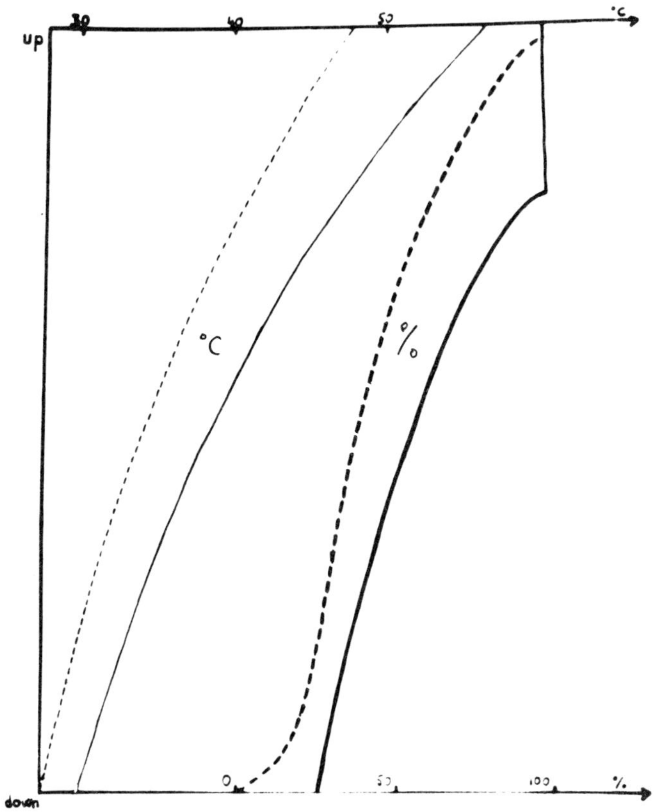

Figure I2 : Estimation of hot air temperature (thin curves) and percentage of melted chliarolithe (heavy curves) as a function of height in a 2.6 meter high diode wall submitted to 5 hours of solar radiation. Full lines refere to 400 W/m^2 and dotted lines to 300 W/m^2 collected solar energy.

R is the radiative and convective heat flux from the wall to the room, $h \simeq 10 w/m^{2\circ}$ C the radiative-convective coefficient of the wall, k the conductive coefficient of the wall (relative to 4 cm of sand or concrete, the PVC plastic tube, the plastic-sand contact and a mean thicknen of chliarolithe); T_f = 28° C the melting point of chliarolithe, T_w the internal surface temperature of the wall, T_i = the temperature in the room. With T_i = I7° C one measures R = $40w/m^2$ so that one finds the coefficient k = 5,3 $w/m^{2\circ}$ C. In realistic conditions the sand is replaced with concrete so that $k \simeq 10$ $w/m^{2\circ}$ C and T_i = 20° C; one finds then again a heat transfer $R \simeq 40$ w/m^2 and a surface temperature of the wall T_w = 23° C. This heat transfer to the house volume has to be compared with its heating requirements. Let us assume a heated space of 300 m^3, 45 m^2 of diode wall and a coefficient of insulation G = I,2 $w/m^{3\circ}$ C for the house, the indoor-outdoor difference of temperature being I0° C (which is a mean value observed in Nice in january). Then the total heat requirement amount to 86 Kwh, but the diode wall can only pro--vide 43 Kwh, i.e. 50% of the needs. One can say therefore that such a latent heat diode wall is not sufficient by itself to heat the whole house in january; additional heat has to be provided by other means during the day, and direct gain systems could do it; but the wall stores sufficient heat to maintain this heat flux for 36 hours. If now the heat restitution is increased by natural or forced con--vection between the storage volume and the house volume, it is easy to meet the total heat requirements but the storage capacity can only cover a I8 hour period following the 5 hour heating phase. The house will then be self-sufficient during consecutive sunny days. Earlier or later in winter, the heat requirements being lower, it will be favorable to operate with reduced heat transfer from the diode wall to prevent over-heating the house (no strong over heating can occur anyway since the wall temperature is limited below 28° C). The house can then be self sufficient for more than a day. Such thermal storage performance can only be approached by a 75 cm thick concrete Trombe wall with its lower collecting efficiency and too large restitution delay.

CONCLUSION

The present experiment and analysis show that a latent heat diode wall can bring decisive improvements in the concept of passive solar heating of houses : it improves collector efficiency, stores large amounts of heat in a reduced volume without risks of unconfortable high temperatures, can operate with incoming heat flux much weaker that its maximum capacity, and constitutes a relatively light com--posite structure that can be produced on an industrial level.

References

I. ANVAR - C.N.R.S. Report n° 77.36.216

2. "Phase change collector wall versus water wall", L. Bourdeau and A. Jaffrin, Section 6c of present Conference.

COMPARISON BETWEEN TWO ABSORPTION COOLING SYSTEMS OF THE "OPEN" TYPE UNDER DIFFERENT CLIMATE CONDITIONS

M. Bolzan and R. Lazzarin
Università di Padova
Istituto di Fisica Tecnica
via Marzolo 9 - Padova, Italy

ABSTRACT

Solar cooling is obtainable by traditional absorption devices: these devices are rather expensive and require two glazed or selective solar collectors.
The absorption systems of the open type seem more promising in order to reach a solar cooling economically reasonable.
Two typical possibilities are taken into account: Baum's system where the device is formed only of evaporator and absorber and the generator is in the open air, and the dehumidification-humidification system which is only a distant relation, or no relation at all, of the traditional absorption machine.
The two systems differ appreciably each other, but the regeneration is quite similar for both.
The performance of the regenerator is a function of the ambient conditions: insolation, outdoors temperature, relative humidity, wind velocity. It is also dependent on solution flow rate and its concentration. Three typical climate conditions during summer are taken into account: hot dry, temperate, hot humid.
The required dimensions of the regenerator are compared for every climate both in Baum's and for the most suitable dehumidification-humidification system.
The comparison is carried out also as regards water consumption for every situation.
Finally advantages and disadvantages of the various systems are examined with reference to the cooling temperature range, the working easiness and complexity.

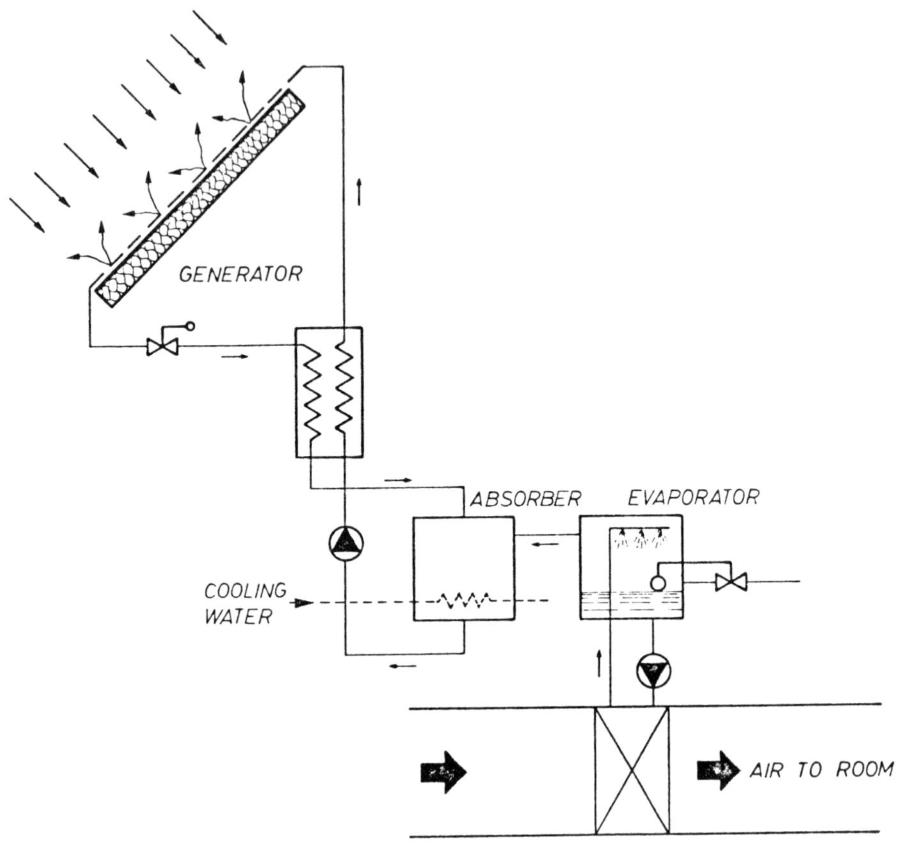

Fig. 1 - Block diagram of Baum's cycle

INTRODUCTION

Many solar air conditioning systems have been experimented: a great deal utilize solar heat to power a traditional absorption device; otherwise mechanical refrigeration devices have been used driven by solar heat engines. Both these possibilities are affected by two main disadvantages: (a) the required working temperature must be at least 80-90 °C; thus two glazed or selective solar collectors are needed to obtain a reasonable efficiency. (b) Heat is always supplied through a heat exchanger: system efficiency lowers for every further heat exchange /1/. Finally both absorption and compression devices are rather expensive; therefore they often require unacceptable initial costs.

These difficulties can be overcome to some extent by solar-oriented systems, better suitable to the particular characteristics of solar energy: "open" cycle devices belong to this family. These are much more influenced by climatic factors than traditional ones: moreover the performance varies appreciably from one "open" system to another according to the climatic conditions.

In this paper two very different "open" cycle systems have been selected as representative: Baum's and dehumidification-humidification.

BAUM'S SYSTEM

This system is a relative of traditional absorption system. The block diagram is the same with the only difference of being without condenser (Fig. 1). Absorber and evaporator are quite similar to traditional ones. The solution is regenerated flowing on a free surface heated by solar radiation: this surface acts at the same time as generator of the absorption device and solar collector; it may be of any material, thus eliminating every problem of corrosion /2,3/. Cooling water is required only at the absorber, whereas water must be supplied continously at the evaporator to restore the amount evaporated at the generator.

The refrigeration temperature is 5-10 °C. Thus the air will follow the usual treatment of traditional air conditioning:

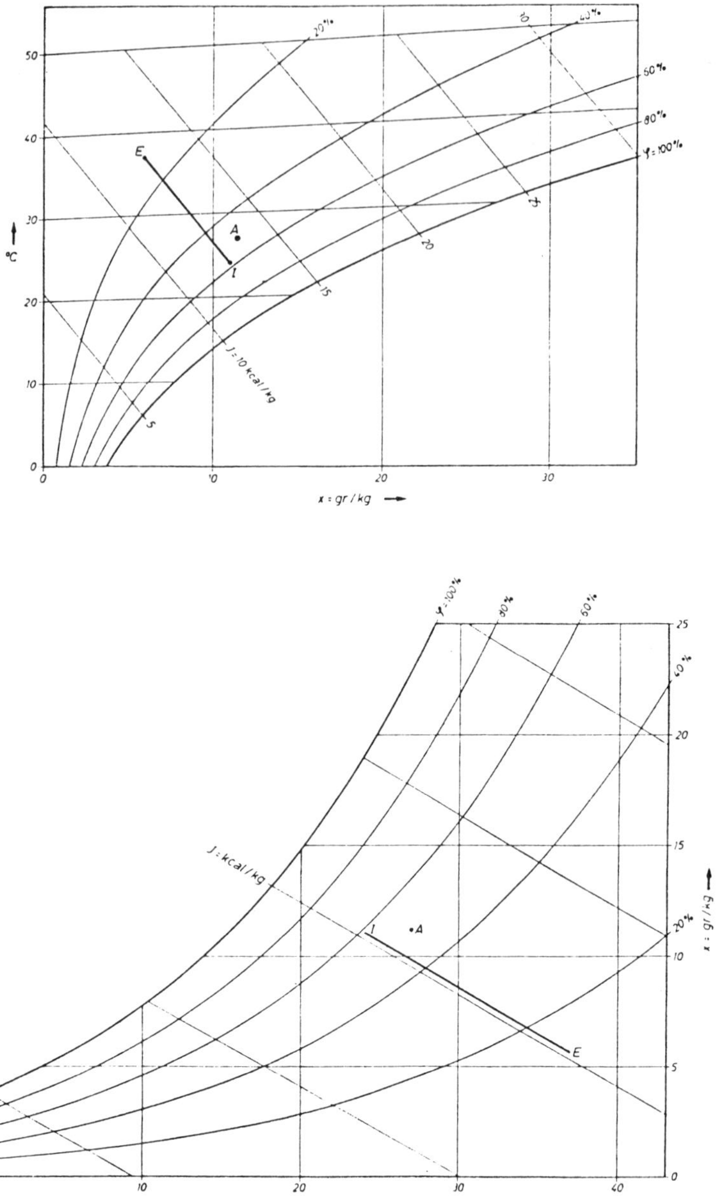

Fig. 2 - Evaporative cooling: (a) Mollier; (b) ASHRAE

for example dehumidification takes place cooling the air through a cooling coil. Many indirect contact heat exchanges are still present in this device and this lowers of course system efficiency.

THE DEHUMIDIFICATION-HUMIDIFICATION CYCLE

This rather unfavourable situation may be surmounted cooling the conditioned air by direct contact. The suggested method is characterized by very low initial costs and refrigeration temperatures closed to the ambient, increasing thus system efficiency.

The basic idea is bound with the coupling of two well-known refrigeration systems. The former is evaporative cooling /4/: for hot dry climates an evaporative cooling of the outdoors air, increasing its absolute humidity, is enough to obtain the desired refrigerating effect (Fig. 2a,b). The latter is dehumidification cooling /5/: for hot humid climates a great deal of the heat load is due to dehumidification of the outdoors fresh air. This is easily obtained through a dehumidifier where some liquid desiccants are circulated, such as mixtures of water and $CaCl_2$ or LiBr or LiCl. During this process the air increases its temperature, but it may be easily cooled e.g. by ambient air. Desiccants are regenerated by heating. These two systems are very simple, but they may be employed only in very particular climates. A suitable coupling of these methods allows to utilize the system in every climate /6,7/.

The simplest scheme is the following (Fig. 3a,b,c): the air to be treated - point 2 - (recirculated room air - point 1 - and outdoors fresh air - point 0 - suitably mixed) enters the dehumidifier: its absolute humidity lowers and its dry bulb temperature increases - point 3 -. The air is then cooled by indirect contact in a heat exchanger supplemented of water from cooling tower - point 4 -. The dehumidification process is here considered as adiabatic; this means to neglect the heat of solution which is usually a fraction of the heat of condensation. Anyhow the real process is one of simultaneous dehumidifying and cooling, that is dehumidifier and heat exchanger work in a single unit as a traditional absorber. The

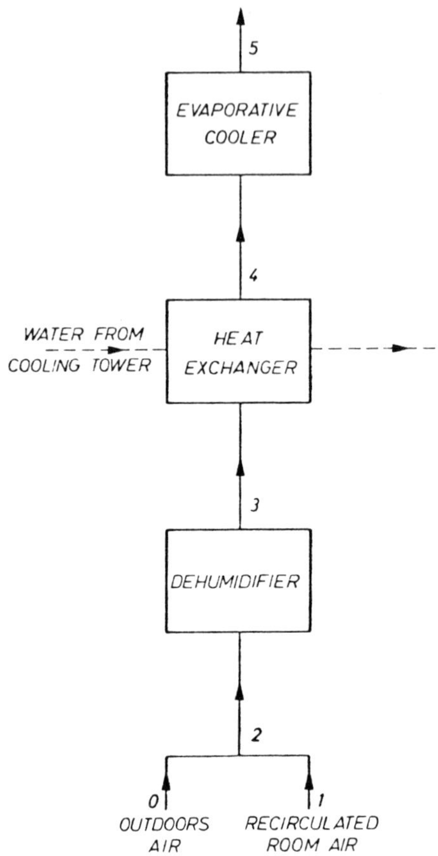

Fig. 3 - Dehumidification-humidification ideal cycle, scheme I: (a) block diagram

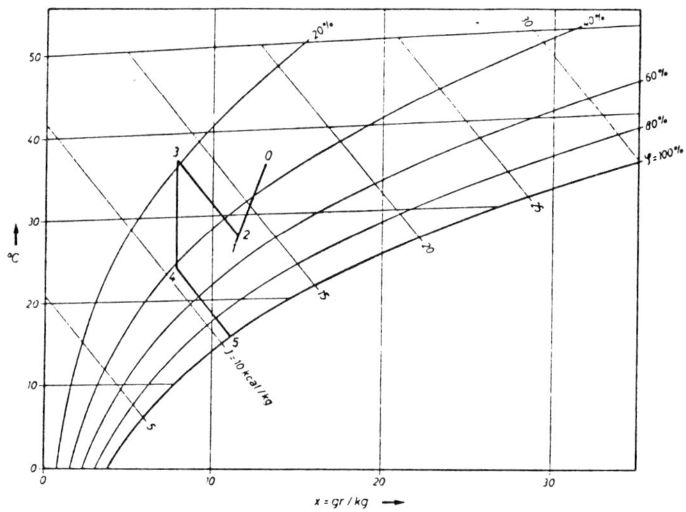

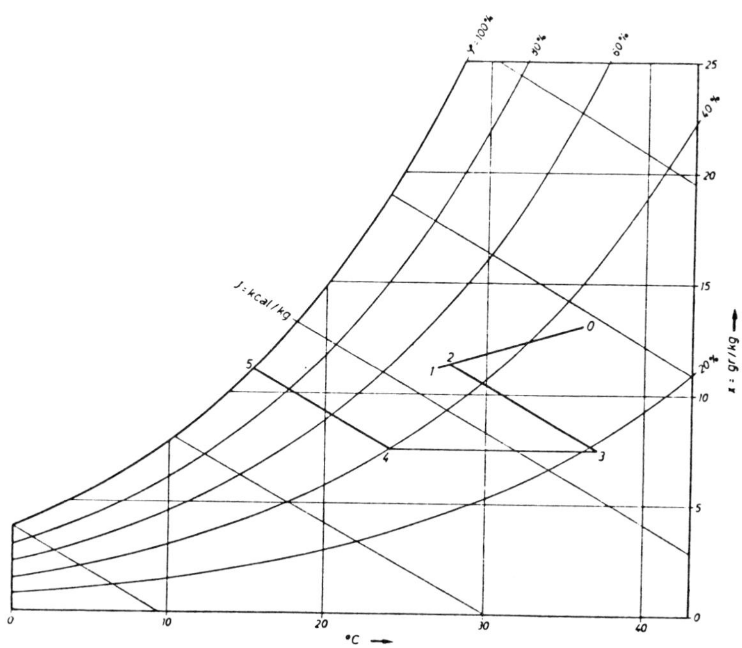

Fig. 3 - Dehumidification-humidification ideal cycle, scheme I: (b) Mollier; (c) ASHRAE

process line is thus 2-4 directly.

In Fig. 3b,c an ideal process is considered for a moderatly dry climate: the air leaves the heat exchanger at the outdoors wet bulb temperature. Finally the air enters a humidifier where its dry bulb temperature lowers and absolute humidity increases - point 5 -. When the air leaves the dehumidifier, it may be directly introduced into the room.

In Fig. 4 is represented the process for a hot humid climate. It is easily recognized that even in ideal conditions a stronger dehumidification is requested in order to get the same inlet point as before in Fig. 3; the dehumidification becomes prohibitive in real conditions because this implies a greater difficulty to regenerate the desiccants. The dehumidification may be limited cooling the dried air not only by water from cooling tower but also by air to be treated - point 2 -; the dotted line in Fig. 4 represents the new transformation. Otherwise a more suitable scheme may be used (Fig. 5a,b,c).

The air to be treated is first cooled by adiabatic saturation till point 3: an important fraction of sensible heat in 2 is turned into latent heat in 3; the latent heat is more easily handled by dehumidification. Dehumidification takes place after the air has left a heat exchanger where the dried air (5), first cooled by tower water (6), goes through a second cooling till point 7.

A more limited dehumidification is required by this scheme and weaker concentrations are needed at the dehumidifier; moreover the air comes out from the first saturation washed from dust and this allows a better working of the desiccants.

The regeneration of the desiccants is an important feature of all these schemes. It may be obtained like in Baum's system /8/: the solution flows on a free surface heated by solar radiation. The regenerated solution is cooled with benefit in counterflow by the solution from the dehumidifier and then it may be cooled further on by tower water (Fig. 6).

The great influence of the climate over all these systems is easily recognized: solar radiation, ambient temperature and humidity, wind velocity play a major role in the regeneration of desiccants. Cooling water temperature is strongly influenced by the outdoors wet bulb temperature. Moreover all the

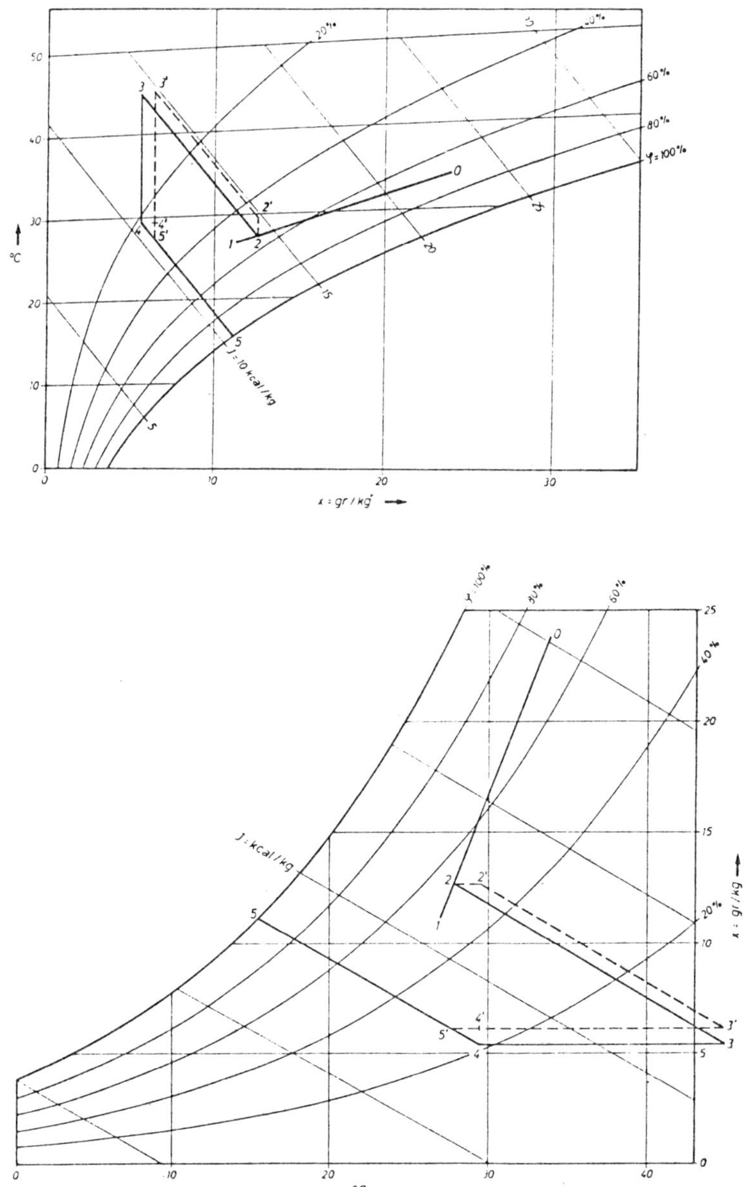

Fig. 4 - Dehumidification-humidification ideal cycle, scheme I for hot humid climate: (a) Mollier; (b) ASHRAE

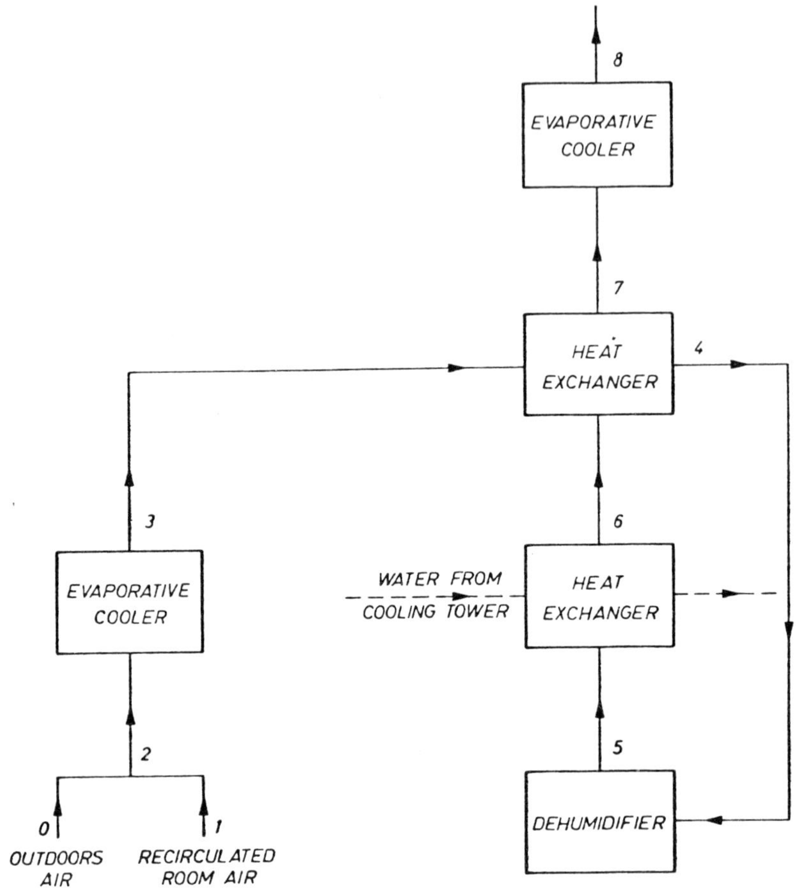

Fig. 5 - Dehumidification-humidification ideal cycle, scheme II: (a) block diagram

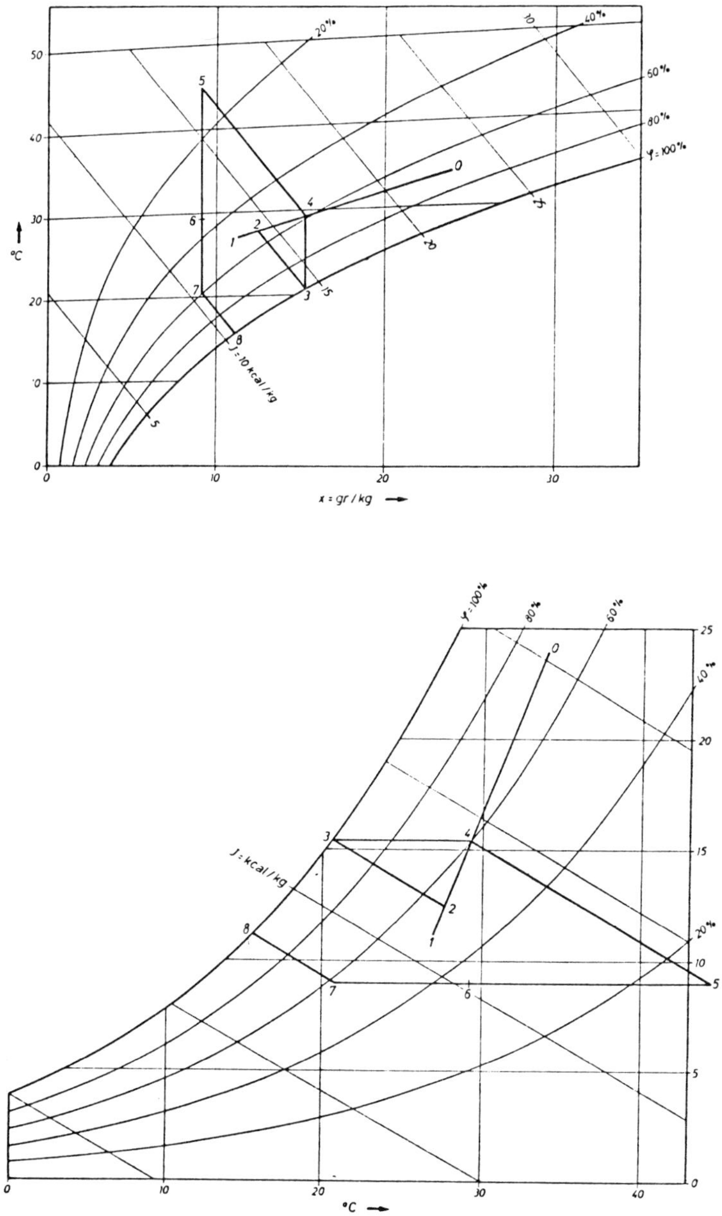

Fig. 5 - Dehumidification-humidification ideal cycle, scheme II: (b) Mollier; (c) ASHRAE

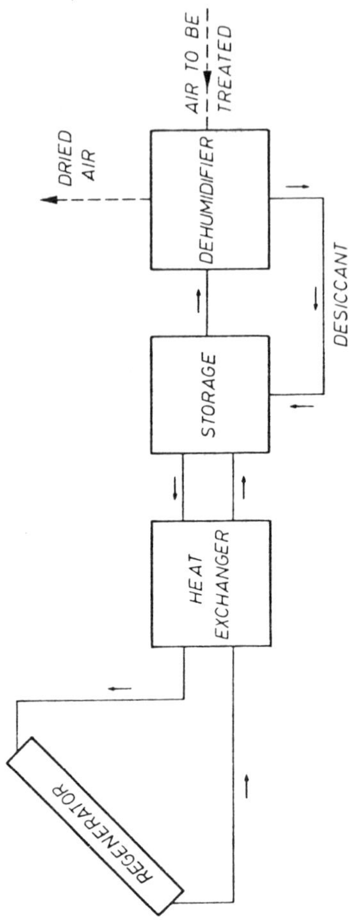

Fig. 6 - Block diagram for the regeneration of desiccant

dehumidification-humidification schemes assume different configurations according to outdoors conditions. Therefore it is worth comparing the different systems here reported in order to select the most suitable under the various climates.

Real processes must be of course considered: heat exchangers are here characterized by a 0.75 effectiveness, the evaporative coolers by a 0.9 approach to the full saturation along an adiabatic line. The tower water is disposable at a temperature of 3 °C above wet bulb ambient temperature.

The comparison is carried out as regards the dimensions of the regenerator and the water consumption under three typical climate conditions during summer: hot dry, temperate, hot humid. Both room parameters and fresh air flow rate are fixed in order to stress the comparison on climate influence. Instead of considering a constant temperature difference between room and inlet air, the conditioned air flow rate is kept constant to simulate better real working.

It is useless to consider instantaneous efficiency in these systems, since concentrated desiccants can work a long time before regeneration is required. Therefore it was necessary to perform the comparison during at least a whole clear day for every system, imposing a same concentration of the desiccant at dawn and at sunset, in order to be able to start in identical conditions on the following day.

THE CLIMATES

Three different typical climatic locations have been taken into account: Alger (37° N) for hot dry, Naples (41° N) for mediterranean temperate and Madras (13° N) for hot humid.

The hot dry climate is characterized by the following diurnal extreme conditions for dry bulb temperature and relative humidity in the sample day: 40 °C and 15%, 29 °C and 30%; the daily variation of dry and wet bulb temperatures is represented in Fig. 7a; in Fig. 7b is pictured the solar radiation on a horizontal surface. The temperate (32 °C and 40%, 23 °C and 70%) and the hot humid (40 °C and 40%, 32 °C and 64%) climates are reported by Fig. 8a,b and 9a,b. Wind velocity has been considered 3 m/s in every location for sake of uniformity.

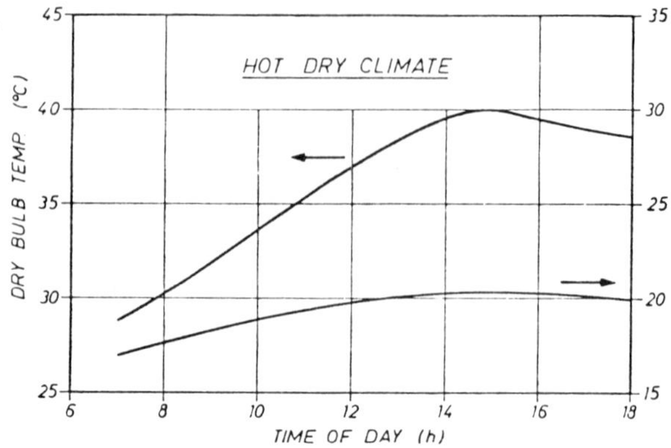

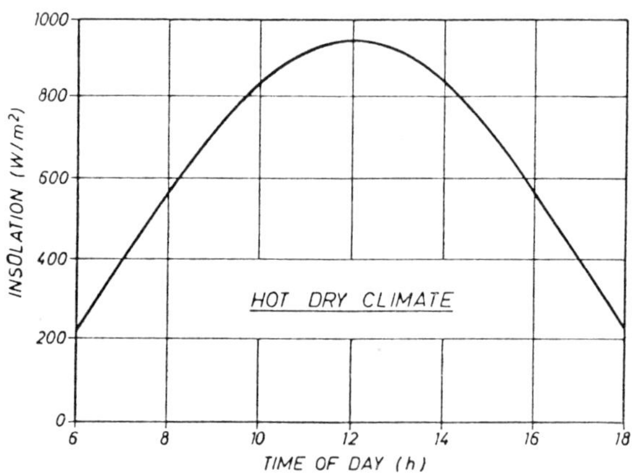

Fig. 7 - Daily variation for hot dry climate:
(a) dry and wet bulb temperatures;
(b) solar radiation

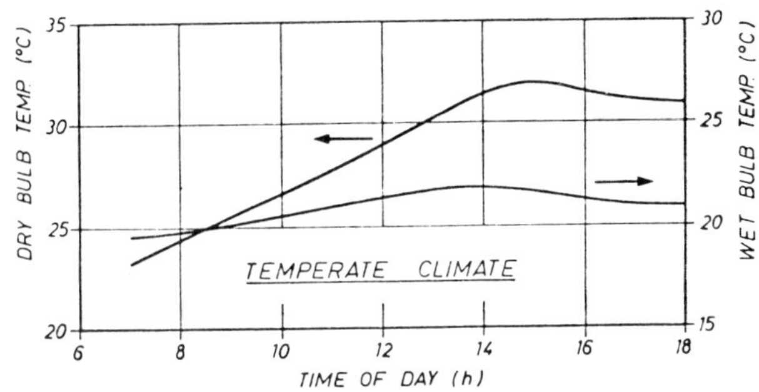

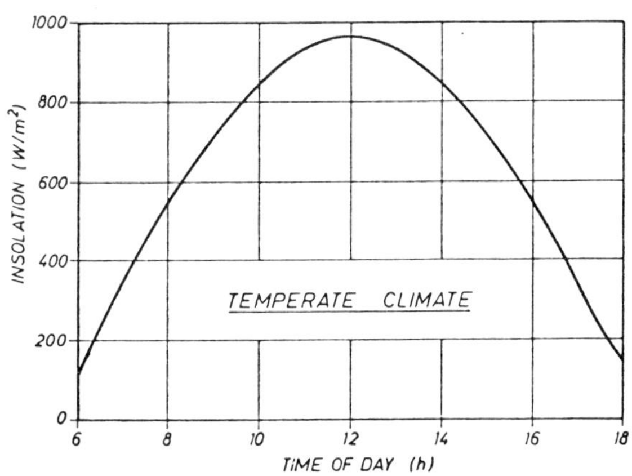

Fig. 8 - Daily variation for temperate climate:
(a) dry and wet bulb temperatures;
(b) solar radiation

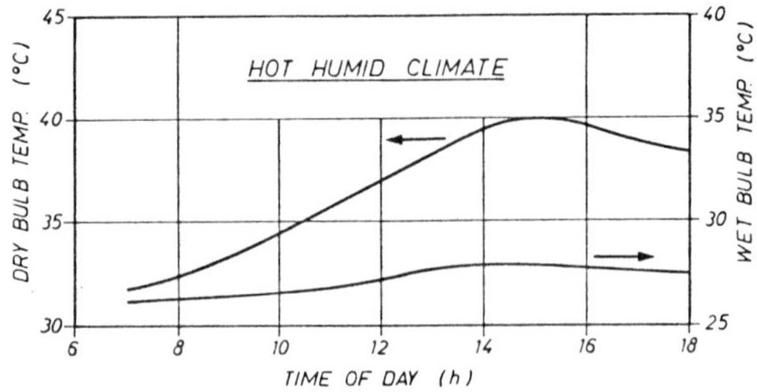

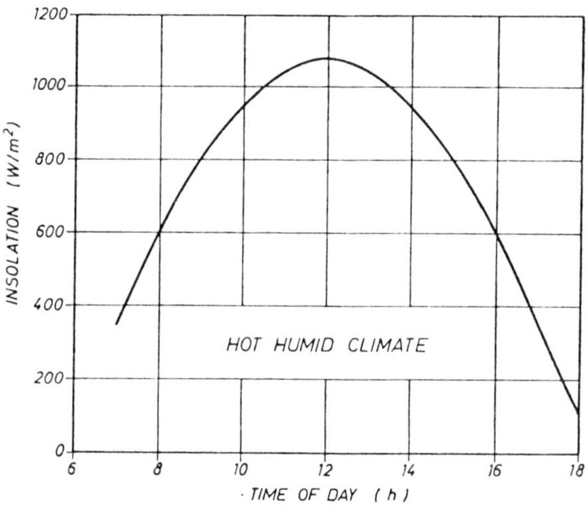

Fig. 9 - Daily variation for hot humid climate:
(a) dry and wet bulb temperatures;
(b) solar radiation

THERMAL LOAD

The same thermal load has been taken for every climate to achieve homogeneity in the comparison: the diurnal variation is represented in Fig. 10. It is the typical heat load of a single room office: the daily load is 24000 kcal per day (100000 kJ per day). The conditioned air flow rate has been established in 1000 kg/h (100 kg/h of outdoors fresh air). The internal latent load is supposed in 300 kcal/h.

STORAGE AND CONTROL PROBLEMS

Baum's system can be provided with a cold storage: control may be then obtained varying refrigerated water flow rate in the cooling coils. Otherwise a concentrated solution storage allows to increase the absorber concentration if a greater cooling capacity is required.

The dehumidification-humidification system cannot be provided of a cold storage. A dehydrated desiccant storage is perhaps the best way to obtain dehumidification during periods when little solar heat is available.

The control may be achieved either by varying the solution-to-air mass flow ratio in the dehumidifier (thus getting a different humidity at the end of the drying process) or controlling the evaporative cooling process.

A further method, which has been long utilized in evaporative cooling, is a two-step variation of the conditioned air flow rate.

RESULTS AND COMPARISONS

Since the two systems are quite similar as regards the dehydration of the desiccants, the comparison may be carried out through the required regenerator area.

The amount of the evaporated water at the regenerator lies not only on ambient conditions, but also on solution type, flow rate and concentration. A H_2O-LiBr mixture has been considered as reliable equilibrium diagrams are available for

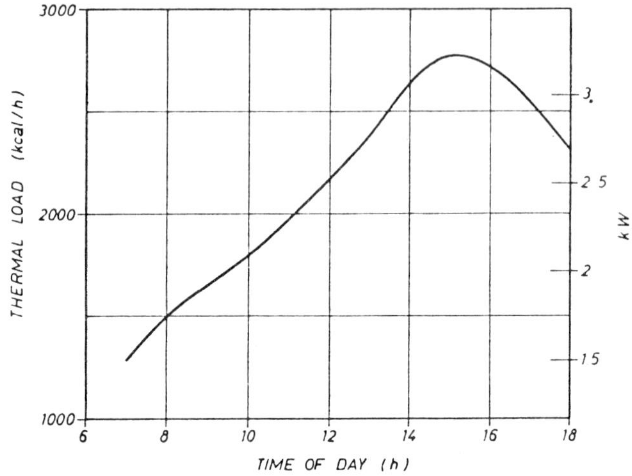

Fig. 10 - Daily variation of the assumed cooling load

TABLE I - COMPARISON BETWEEN THE TWO SYSTEMS FOR THE THREE DIFFERENT CLIMATES

	DEHUM.-HUMID.			BAUM		
	TEMPERATE	HOT DRY	HOT HUMID	TEMPERATE	HOT DRY	HOT HUMID
solution-to-air mass flow ratio (kg_{sol}/kg_{air})	0.27	0.23	0.45	-	-	-
specific solution flow rate per unit area (kg/m^2h)	10	20	11	8	15	9
working concentration (%)	49	51	51	51	50	53

TABLE II - COMPARISON BETWEEN THE TWO SYSTEMS FOR THE THREE DIFFERENT CLIMATES WITH REFERENCE TO REQUIRED AREA, OVERALL EFFICIENCY AND EVAPORATED WATER AT THE REGENERATOR

	DEHUM.-HUMID.			BAUM		
	TEMPERATE	HOT DRY	HOT HUMID	TEMPERATE	HOT DRY	HOT HUMID
required area (m^2)	27	11.5	41	26	11	40.5
overall efficiency	0.132	0.290	0.089	0.137	0.300	0.091
evaporated water at the regenerator (kg per day)	53	41	70	40	40	40

it. Two different effects must be taken into account in order to fix concentration and flow rate: higher concentrations mean easier dehumidification with more limited solution-to-air mass flow ratios, but higher concentrations imply also a more difficult regeneration. Moreover the concentration must exceed a minimum under which dehumidification does not take place.

A concentration 2-3% greater than this minimum is deemed a good compromise between solution-to-air mass flow ratio and easiness of regeneration.

In Table I are listed concentration, solution-to-air mass flow ratio and solution specific flow rate per unit area of regenerator (solution flow rate being equal in the dehumidifier and in the regenerator) in the three climates under conditions of maximum load. The scheme more suitable according to climate has been considered for the dehumidification-humidification system: this is scheme I for hot dry (Fig. 11a,b) and scheme II both for hot humid (Fig. 12a,b) and temperate (Fig. 13a,b).

Under these assumptions a comparison can be developed in Table II, where appear the required regenerator area, the overall daily efficiency of the system i.e. the ratio between refrigerating effect (removed heat load) and the daily solar radiation on the regenerator and the daily amount of the evaporated water.

The regenerator area is strongly dependent on climate, because water is evaporated at the regenerator not only by solar radiation, but also by the difference between water partial pressure in the solution and in the outside air. The good performance of both the systems in hot dry climates is then easily understood; of course performances get worse on passing to temperate and hot humid climates.

The hot humid climate requires a rather large regenerator area: this is likely to be reduced by regenerating the desiccants with spray chamber through air heated in air solar collectors /9,10/. The related higher cost may be justified even in a temperate climate where air solar collectors will be utilized for winter heating.

A rough comparison may lead to a wrong conclusion, i.e. that

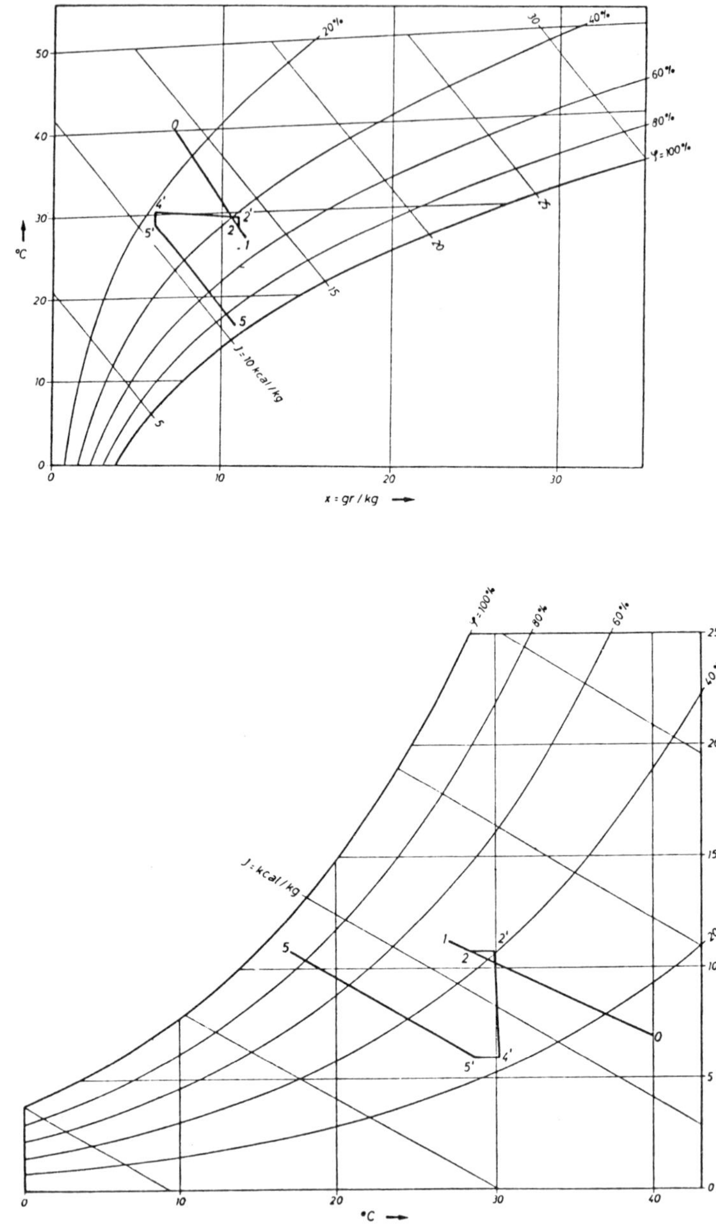

Fig. 11 - Dehumidification-humidification real cycle for hot dry: (a) Mollier; (b) ASHRAE

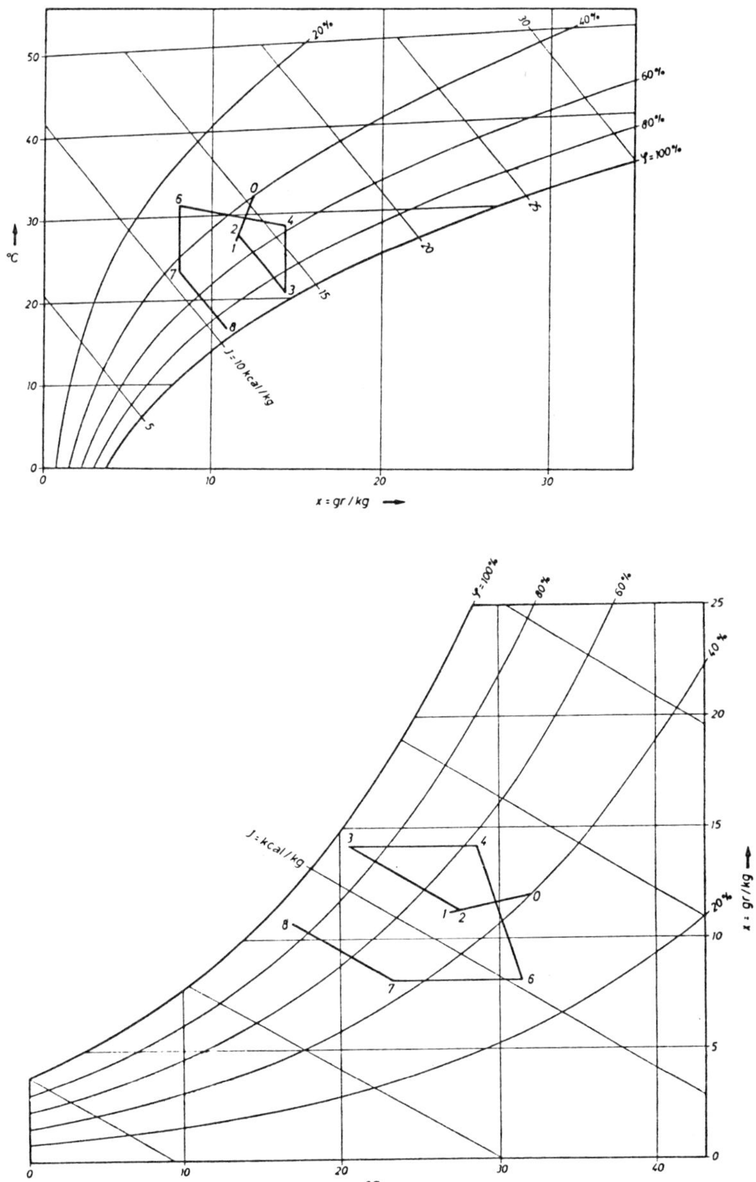

Fig. 12 - Dehumidification-humidification real cycle for temperate: (a) Mollier; (b) ASHRAE

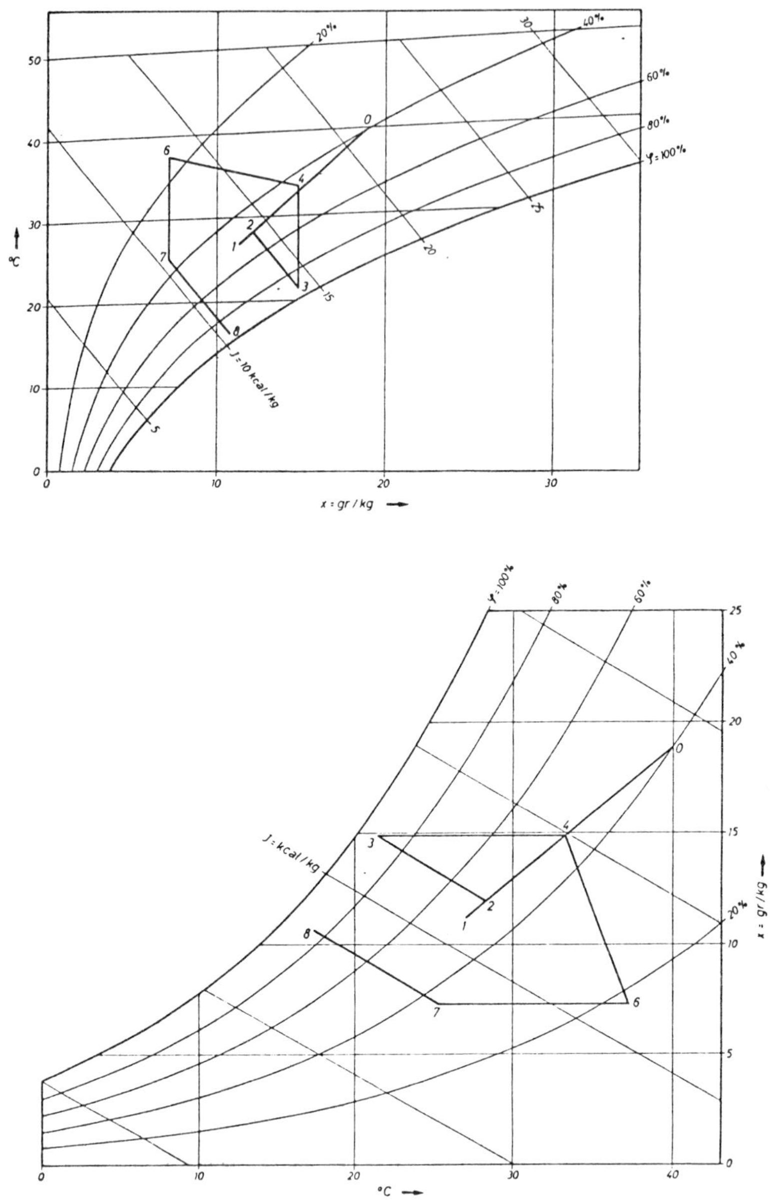

Fig. 13 - Dehumidification-humidification real cycle for hot humid: (a) Mollier; (b) ASHRAE

the two systems are equivalent. On the contrary one must keep in mind that the air after the dehumidification-humidification treatment can be introduced into the rooms directly, whereas in Baum's system only the cooling step is accomplished and a reheating should be necessary before introducing it into the rooms. Reheating does not mean only a direct expense of energy, but also an equal addition to thermal load, i.e. a greater regenerator area; the daily amount of the reheating may easily exceed some 30-40% of the thermal load.

Furthermore the greater complexity of Baum's system should be considered with respect to dehumidification-humidification: this is connected especially to the vacuum maintenance in the evaporator-absorber group.

WATER CONSUMPTION

Both the systems are characterized by a continuous water consumption: water is required at the evaporator in Baum's system, at the evaporative cooler in the dehumidification-humidification system, at the cooling tower in both.

Water consumptions are listed in Table III for every climate. The required amount of water is greater in temperate and hot humid climates, but it is of the same order as the amount needed by a traditional mechanical compression system with a water cooled condenser; in this respect the consumption may be considered a negligible problem.

On the contrary water consumption in hot dry climate is worthy of attention even if it is the lowest of the three, since scarcity of water is likely in a location with such a climate. Therefore at least a partial recover of water may be advisable under these conditions /11,12/.

Again one should not be deceived by too a hurried comparison: water consumption is often lower in Baum's cycle than in the other, but it is a qualitativly different consumption. As a matter of fact in Baum's system vacuum maintenance needs water at a satisfying level of purity at the evaporator.

TABLE III - COMPARISON BETWEEN THE TWO SYSTEMS FOR THE THREE DIFFERENT CLIMATES WITH REFERENCE TO WATER CONSUMPTION

WATER CONSUMPTION (kg/day)	DEHUM.-HUMID			BAUM		
	TEMPERATE	HOT DRY	HOT HUMID	TEMPERATE	HOT DRY	HOT HUMID
cooling tower	53	41	70	48	48	55
evaporative cooler	49	39	61	-	-	-
evaporator	-	-	-	40	40	40
TOTAL	102	80	131	88	88	95

CONCLUSIONS

"Open" type cooling devices seem very promising to achieve an economical solar cooling. The performances of these systems are strongly dependent on ambient conditions; therefore it is worth studying attentively the characteristics of every system in order to select the best for the various requirements. Dehumidification-humidification systems have proved better in all the three considered climates, since they allow a direct treatment of the conditioned air; however this is not enough to discard Baum's cycle. Baum's cycle shows at least the following advantages: (a) it cools till temperatures few degrees above 0 ^{o}C, whereas the other system can reach only a value which is a function of outdoors conditions and anyhow very close to ambient temperature; (b) it allows to select the most suitable system for the distribution of air conditioning, whereas the other obliges to choose an all-air system; (c) it requires air flow rate more limited than the other, i.e. smaller channels.

Baum's system is more flexible, but more expensive and a careful balance of advantages and disadvantages must be built up whenever a design is conceived in a particular condition. As a matter of fact, working with the sun, "the more flexible the more expensive" is almost a rule.

REFERENCES

1 - R. Lazzarin, E. Rizzon, M. Sovrano, B. Boldrin and G. Scalabrin, "Performance predictions of a LiBr absorption air conditioner utilizing solar energy", 41-5/0160, ISEC 77, New Delhi, 1978.

2 - A. Kakabaev and A. Khandurdyev, "Absorption solar refrigeration unit with open regeneration of solution", Geliotekhnika, vol.5,n.4,pp.28-32, 1969.

3 - V. Baum and A. Kakabayev, "Utilizing solar energy for creating comfortable room conditions in Turkmenia", 41-5 /1185, ISEC 77, New Delhi, 1978.

4 - M. Roubinet, "Economie d'énergie par l'utilization de l' humidification et du refroidissement par evaporation",

Revue pratique du froid and du conditionnement d'air, pp. 49-53, 85-94, 1975.

5 - D. M. Valentine, "Liquid absorption dehumidification systems", ASHRAE Symposium on Sorption Dehumidification, pp. 615-621, ASHRAE Trans., I , 1975.

6 - G. Grossman and I. Shwarts, "An open absorption system utilizing solar energy for air conditioning", Proc. Int. Sem. Heat Transfer in Buildings, Dubrovnik, I.C.H.M.T., 1977.

7 - H. Buchberg and N. Lassner, "Performance of an experimental regenerative evaporative cooler compared with predictions", Proc. Int. Sem. Heat Transfer in Buildings, Dubrovnik, I.C.H.M.T., 1977.

8 - M. C. Gupta and P. Gandhisan, "Open cycle 3-ton solar air-conditioner: concept, design and cycle analysis", 51-4/0325, ISEC 77, New Delhi, 1978.

9 - A. Kakabaev and A. Rakhmanov, "An absorption solar cooling system with spray chamber: description and test results", Geliotekhnika, vol.7,n.4, pp.38-43, 1971.

10 - A. Kakabaev and M. Golaev, "Properties of refrigerating installations with spray chamber and solar heaters", Geliotekhnika, vol.7,n.5, pp.49-56, 1971.

11 - A. Kakabaev and M. Golaev, "Glazed flat surface as a solution regenerator for use in an absorption solar cooling system", Geliotekhnika, vol.7,n.4, pp.44-49, 1971.

12 - B. Boldrin, G. Scalabrin, R. Lazzarin and M. Sovrano, "A new continuous solar still", 51-5/0161, ISEC 77, New Delhi, 1978.

OPTIMISATION OF COOLING SYSTEMS IN TROPICAL ZONES

J. Fléchon
Laboratory of Metallic Deposit Physics
Université de Nancy I
Case Officielle 140,
54037 Nancy-Cédex, France.

ABSTRACT

We examine the possibilities of solar refrigeration by the use of plane collectors associated with a boiler containing a binary solution NH_3, H_2O or NH_3, NaSCN. We show, starting from the diagrams 1/T, Log P, that in tropical zones, dry absorption ($CaCl_2$, $8NH_3$) seems the best solution.

INTRODUCTION

The use of solar energy allows us to hope for its application in domestic heating in the temperate zones, in refrigeration and in climatisation in the tropical zones.

In both cases we must create, basing ourselves on radiant energy, as great a temperature difference as possible between the ambient temperature and that of the closed environment that we wish to heat or cool. Strangely, the same problems are encountered in relation not to the quantity of calories, which can always be increased by multiplying the collecting surfaces, but to their quality which is characterised by the extent of their differences at the ambient.It is, in fact, the value of the temperature gradient between the external environment and the area studied which will define the interest of a heating or cooling system.

We wish to realise a refrigerator or climatisor functioning without exterior energy supply other than that of solar radiation. It must be reliable, that is able to function while remaining identical to itself for an indefinite time. The minimum time cycle will be of 24 hours. A support system could be envisaged for days of insufficient sunlight. In extreme cases, a uniformally-warm environment under a covered sun excludes all hope of using this method for refrigeration. On the other hand, intense diurnal radiation coupled with a low nocturnal temperature defines the most favourable conditions for the obtaining and conserving of cold. This is precisely the régime found in desert zones. In tropical regions the nocturnal temperature remains high. This imposes a careful choosing of the binary mixture used to collect the calories.

CONDITIONS IMPOSED BY SOLAR RADIATION

Only machines of intermittent absorption seem adapted to the intermittence of the energy supply. It is necessary therefore, to condense during the daytime a maximum quantity of a pure, volatile body whose latent vaporisation heat will provide during the nighttime the frigories used for refrigeration or climatisation. We must insist on the criterion of the refrigerating agent's purity. For example, the vaporisation of a water - ammonia mixture considerably reduces the machine's performances. A reversal of the expected effect may even be observed. The distension of ammonia in the presence of water heats the tubes instead of cooling them : the realisation of the mixture is exothermic.

The imposed time cycle leads us to make sufficient provision for the vessels so as to ensure a high-rate circulation of the cooling fluid. Equally, it demands particularly careful isolation: thermal losses threatening to compensate the effects of evaporation. Since the limiting temperature of the cold source depends on the pressure, it is not possible to use a capillary, an essential element in compressor machines. The evolution of the radiant energy being quasi-reversible, the fundamental parameter of solar refrigerators remains the vaporisation speed of the fluid, directly linked to the effect of cooling which takes place at nighttime whilst the binary mixture or the ammoniacate absorbing the gas, regenerates itself.

Independently of the number of frigories hoped for from a given mass of ammonia liquid, two opposite kinetics influence in inverse manner the temperature of the evaporator : the thermal losses which tend to raise it, the evaporation of the ammonia which reduces it. Thus, the mass of cooling fluid liberated in the course of insolation corresponds to a potential possibility of refrigeration, the effective lowering or the temperature depends on the liquid's disappearance rate. At the extreme, if one phenomenon compensates the other, the evaporator's temperature does not change.

We find with refrigeration the same problems as with insolation : a thermostatic study associating on the one hand the quantity of incident energy to the maximum limiting temperature possible from the collector (1) (2) (3), on the other hand the mass of condensed coolant to the quantity of potential coolant available; a thermokinetic study (4) associating,similarly,the incident energy flux to a relaxation time at the end of which a fraction of this limiting temperature may be attained in such a way that the evaporation speed of the condensed fluid determines the maximum temperature reduction permitted by thermal losses.

At the end of the insolation, the weaker the generating solution - or the more dissociated the ammoniacate - the greater the speed of reabsorption increases at nighttime and, thus, the more the effect of refrigeration is accentuated. But the exhaustion of the binary is only valid if the incident radiant intensity is sufficient. We might try to increase it by a rough concentration of simple implementation which will raise the limiting temperature.

AIM

The aim of the present text is to show that solar absorption refrigerators associated to a plane collector can only function validly within vigorously defined limits.

In addition, from the divergent appreciations (5), (6) being formulated on the comparitive performances in the binary mixtures ammonia - water, ammonia - sodium sulfocyanide it is well to make an assessment based on indisputable thermodynamic data.

PROBLEM

It is necessary to create a machine functioning without exterior intervention and without a mobile mechanism which might consume energy. We must use, without intermediary, the calories caught by the plane insolator on the vaporisation of a pure body which, condensed under pressure, will evaporate when the environment which has liberated it under the action of heat, is ready to reabsorb it once its temperature is lower.

These conditions oblige us to retain the intermittent absorption system which is of two types : wet absorption,
dry absorption.

The pure fluid cooling body is generally ammonia. Wet absorption corresponds to a physical phenomenon of the simple dissolution of ammonia in, for example, water, lithium nitrate $LiNO_3$ or sodium thiocyanate NaSCN. Dry absorption is based on a chemical phenomenon : the formation and the dissociation of a solid ammoniacate using calcium chloride $CaCl_2$ or strontium chloride $SrCl_2$.

Although the phenomenon is simple, since it uses the latent heat coming from vaporisation, dissolution and dissociation, it seems, according to Eggers-Lura et al (6) that in spite of very numerous laboratory studies, no industrially-practicable realisation usuable with solar energy has yet been developed. This means that beyond the principles, certain apparently minor parameters are decisive. One such, for example, in real working conditions, is ammonia's fixation speed at nighttime. If this is too slow the efficiency is derisory and the machine of no use whatever.

STUDY METHOD

Let us first examine wet absorption in the case of water and sodium sulfocyanide. We set the start of the cycle at the beginning of the day when the binary solution in the boiler attains its maximum concentration of ammonia, that is C_o . C_o depends on the ambient temperature taken to be equal to that of the boiler Ta_o and of T_e , the temperature of the evaporator. In fact, the diagram $\frac{1}{T}$ log p for the water - ammonia mixture of Bosnjakovic and Wucherer (7) and the results of George C. Blytas and Farrington Daniels (8) for the mixture NH_3, NaSCN which permits the associating of the pressure of pure NH_3 to T_e , is equal to that of the mixture at Ta_o of concentration C_o . We thus know C_o from the concentrated solution. Figure 1, classical, allows us by simple reading to determine C_o : we have drawn up figure 2 by using the numerical results of Blytas and by translating them following the rules retained on figure 1 so as to render the comparisons immediate.

Solar energy heats the solution by means of the insolator. Let T_m be the maximum temperature attained by the mixture. The vaporisation of the ammonia is followed by a liquefication at the supposed condensor at the temperature T_c . As before, the same diagrams allow us to associate with T_c the maximum pressure of pure NH_3 , a pressure equal to that found in the boiler which now contains the solution weak in NH_3 and of concentration C_1. For the machine to work $\Delta C \% = C_o - C_1$ must be positive within the conditions of the experiment.

Thus, the couples Ta_o , T_e and T_m , T_c impose rigorously on ΔC % a fixed value supposing the thermodynamic balance to have been attained. In fact, this value represents the most that can be hoped for from the system. As we have shown in the study of plane collectors, the static considerations are largely influenced by the kinetic phenomena of which we have said that the most important here is the speed of ammonia absorption by the solution at nighttime.

DATA OF THE PROBLEM IN TROPICAL ZONES

Taking as reference the statistics compiled over 20 years by the meteoroloical station at Bamako (Mali), we note that the hot season takes place between March and June with a maximum ambient average of 40° C, and with a minimum ambient average of 26° C.

During this period insolation is not always great, aerosols and sand - mist being abundant.

We are able , from field experiments conducted there, to retain for :

Ta_o	20° ,	25° ,	30° C	
T_e	-10° ,	-5° ,	0° C	
T_m	90° ,	100° ,	120° ,	140° C
T_c	30° ,	40° ,	50° C	

RESULTS FOR THE BINARY LIQUIDS

They are immediately obtained by simple consultation of the diagrams mantioned and can take two forms :

a) General Result

Figure 3 gives us the richness of the NH_3 , H_2O (I) and NH_3 , NaSCN (II) solutions in terms of the maximum temperature attained by the insolator - boiler when the condensor is brought successively up to 3 temperatures defined above (weak solutions).

On the same diagram are noted the concentrations in terms of Ta_o and T_e (rich solutions) corresponding to 3 values of Ta_o and to 3 values of T_e .

We note that at 90° the apparatus is unusable in the hot season. We have drawn a similar conclusion for heating in the East of France in winter (9).

On the one hand, it is necessary to have solar radiation intense enough to comfortably surpass 100° C (3) and a not excessive diurnal temperature to obtain a significant liquefaction of ammonia ; on the other hand, there must be a nocturnal temperature low enough to permit a rapid absorption in a sufficiently concentrated solution. In addition, the comparison of the slopes relative to the (I) NH_3 , H_2O mixture and those of (II) NH_3 , NaSCN, immediately show that in spite of the inconvenienceof the drawing - off of water by the ammonia during its evaporation and of the apparent advantage of the weak steam - tension of NaSCN, mixture I remains preferable to II. This conclusion conforms to that of A. Eggers-Lura et al (6) when he defined the performance coefficients of the 2 systems. The experimental work of Swartman et al (5) leads to different results. Let us note here that figures 1 and 2 alone,corresponding to commonly-admitted measurements,suffice to define qualitatively the mixture to be retained in preference to realise the proposed aim.

b) Extreme conditions and ΔC % :

Figure 4, obtained from the earlier figures, visualises the same remarks and gives them a quantitive meaning.

Comparitive study on an example : let us set the temperatures

Ta_o = 30° C T_e = -10° C T_m = 110° C T_c = 40° C

and let us suppose that we have initially 2 Kg 500 of water and 2 Kg 500 of NaSCN.

Figure 3 and an immediate calculation permit us to define in the 2 cases :

- m_o : total initial mass of the rich mixture,
- m_1 : initial mass of ammonia in the mixture,
- m_2 : mass of liquefied ammonia in the imposed conditions.

	NH_3 , H_2O	NH_3 , NaSCN
C_o %	45	44
C_1 %	36	39.5
m_o Kg	4.5	4.5
m_1 Kg	2	2
m_2 Kg	0.6	0.36

In fact, in the case (NH_3 , H_2O) let x be the mass in Kg of ammonia initially dissolved in the water.

$$0.45 = \frac{x}{x + 2.5} \qquad x = 2.04 \simeq 2 \text{ Kg}$$

let y be the mass in Kg of ammonia in the water solution

$$0.35 = \frac{y}{y + 2.5} \qquad y = 1.35 \simeq 1.4 \text{ Kg}$$

the mass of vaporised ammonia is, therefore, about 0.6 Kg. A similar calculation leads to 0.36 Kg if the binary NaSCN, NH_3 is used.

Thus, by imposing on the 2 binaries identical conditions, we can foresee, a priori, that one is better adapted than the other to refrigeration. Let us add, to give a complete picture, that the steam tension of the water necessitates the use of an ascending coolant which is no longer necessary with the sodium sulfocyanide, a salt whose steam tension is negligeable, but the thermodynamic data alone are favourable to the use of water.

If we wish to translate, in the example mentioned, the conclusions into frigories, we note that the enthalpy of NH_3 gas at 40° C (7) is 262.85 Kcal per Kg, that of liquid NH_3 145.52, and at -10° C 89.03, that is $\Delta H = 56.49$ Kcal which only leaves as a net frigorific power 200 Kcal / Kg NH_3 or in the example studied 120 Kcal in the case of the water, and 72 Kcal with the sodium sulfocyanide. This, without taking into account the thermal losses or the calorific capacity of the empty evaporator. NaSCN seems absolutely inadapted to such a usage in spite of apparently favourable characteristics (5).

SOLID PHASE CHEMICAL PROCESS : THE AMMONIACATES

It was especially in Germany during the inter-war that the absorbant properties of alkalino - terrous salts were examined. In particular, Plank (10), Linge (11), Altenkirsch (12), Mehl (13) contributed to the realisation of household cupboards with an intermittent cycle of absorption by dry means.
The practical use for many years now of machines based on this principle is reassuring, the essential question being to define how the distribution of solar energy and the performances necessary are compatible with a domestic exploitation of solar cold.

According to Eggers-Lura (6), the 2 salts best adapted to this purpose are calcium chloride $CaCl_2$ and strotium chloride $SrCl_2$. If we take into account the fact that $CaCl_2$ is a by-product currently unusable by the Solvay industries and of the chemical similarity of Ca and Sr, we can restrict ourselves to $CaCl_2$ alone.

The octoammoniacate $CaCl_2$, $8NH_3$ forms at the ambient. By heating we obtain a progressive dissociation following the schematic diagram (14).

$$CaCl_2 , 8\ NH_3 \xrightarrow{95°} CaCl_2 , 4\ NH_3 \xrightarrow{106°} CaCl_2 , 2\ NH_3$$

p(= 16 atm) (condensor at 40°) + 4 NH_3 p(= 16 atm) + 2 NH_3

which liberates 6 NH_3 per initial mole of $CaCl_2$ when the imposed conditions are those defined above. The separation of the remaining 2 NH_3 is only possible beyond 180° C (14), therefore there can be no question of envisaging its use for refrigeration.

In total 102 g of ammonia are usable for 111 g of calcium chloride. If we refer to the comparison seen in wet absorption to 2 Kg 500 of $CaCl_2$, there is a corresponding 2 Kg 300 of ammonia, thus, taking account of the earlier dissociation these are 460 Kilocalories gross which characterise the cooling power of the salt, or 4 times more than with the binary water - NH_3.

It should be noted that these results only have meaning in equilibrium, for the absorbing liquid as well as for the solid. In fact, as we have already pointed out, the essential parameter remains the absorption speed of the ammonia at nighttime.

In reality, the reactions are never complete, either because of a lack of time needed for the regeneration of the ammoniacate or because of the diurnal temperature at the collector. It would be interesting on this point to consult reference works. Degoin (15) considers that cold saturation is more or less terminated with 600 g of NH_3 per Kg of $CaCl_2$ which keeps about 200 g of NH_3 hot, thus, only 400 g of coolant are available per Kg of anhydrous salt, which represents 1 Kg of ammonia in the example studied, much superior to what can be hoped of the best of the binary liquids at equilibrium.

The same author evaluates at 1200 frigories the frigorific capacity per 24 hours of a domestic cooling-cupboard, which, according to the data, requires the use of a minimum of 15 Kg of $CaCl_2$ as it is advisable to take account of the thermal losses and of the calofific capacity characteristic of the equipment. From our own studies (2), about $4m^2$ of plane

Légendes des figures

Figure 1 : Diagramme 1/T ,Log P pour le mélange H_2O ,NH_3
Diagram 1/T ,Log P for the mixture H_2O ,NH_3

Figure 2 : ---------------------------------- NaSCN,NH_3

Figure 3 : Concentration des mélanges binaires dans les conditions d'emploi 1 H_2O,NH_3 2 NaSCN,NH_3
Concentration of binary mixtures in the employ conditions 1 H_2O,NH_3 2 NaSCN,NH_3

Figure 4 : Fraction d'ammoniac utilisable dans les conditions extrêmes
Fraction of ammonia available in the extreme conditions

Titre abrégé : Réfrigération solaire zone tropicale

Diagramme 13

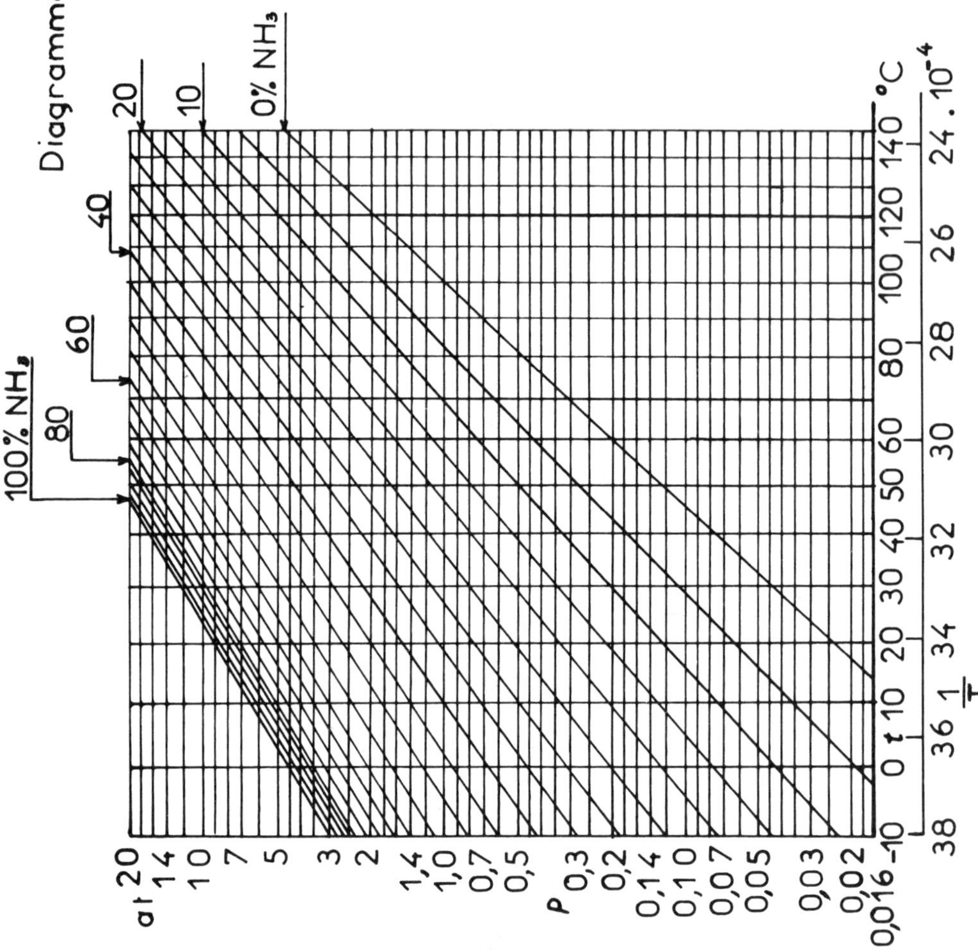

Diagramme $\frac{1}{T}$ log. p pour le mélange Eau Ammoniaque

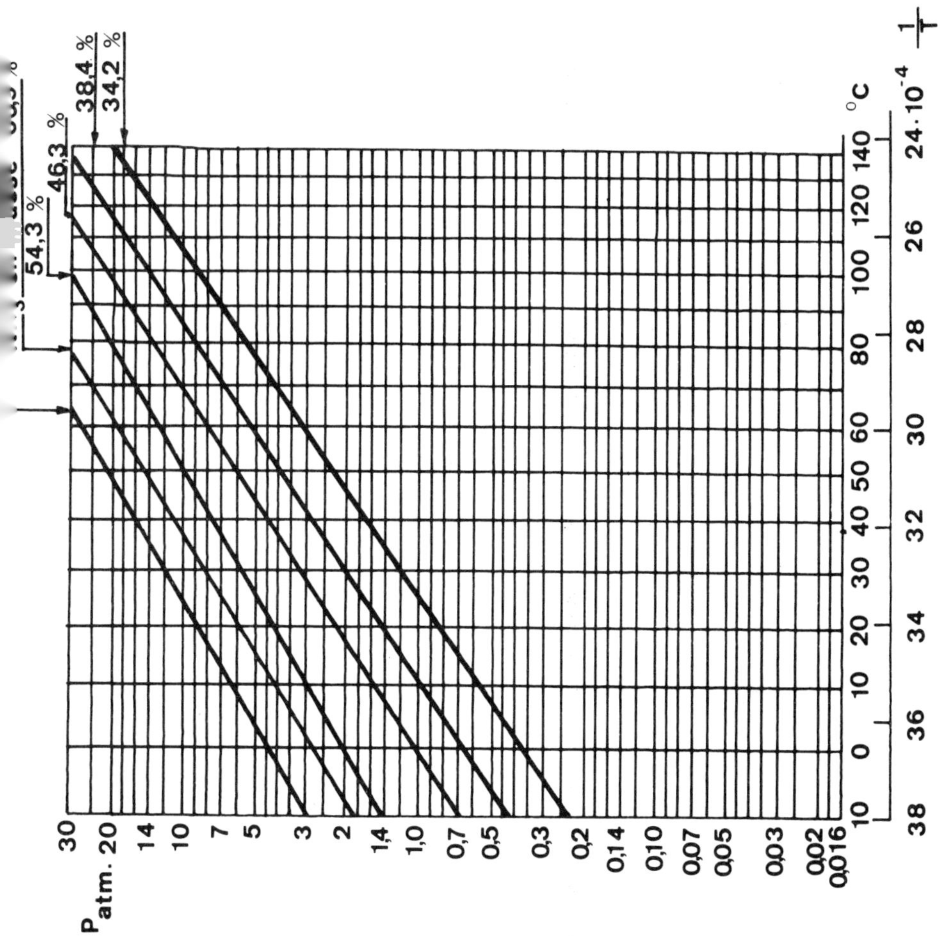

– DIAGRAMME $\frac{1}{T}$ Log P POUR LE MELANGE

Na SCN NH_3 (d'après BLYTAS).

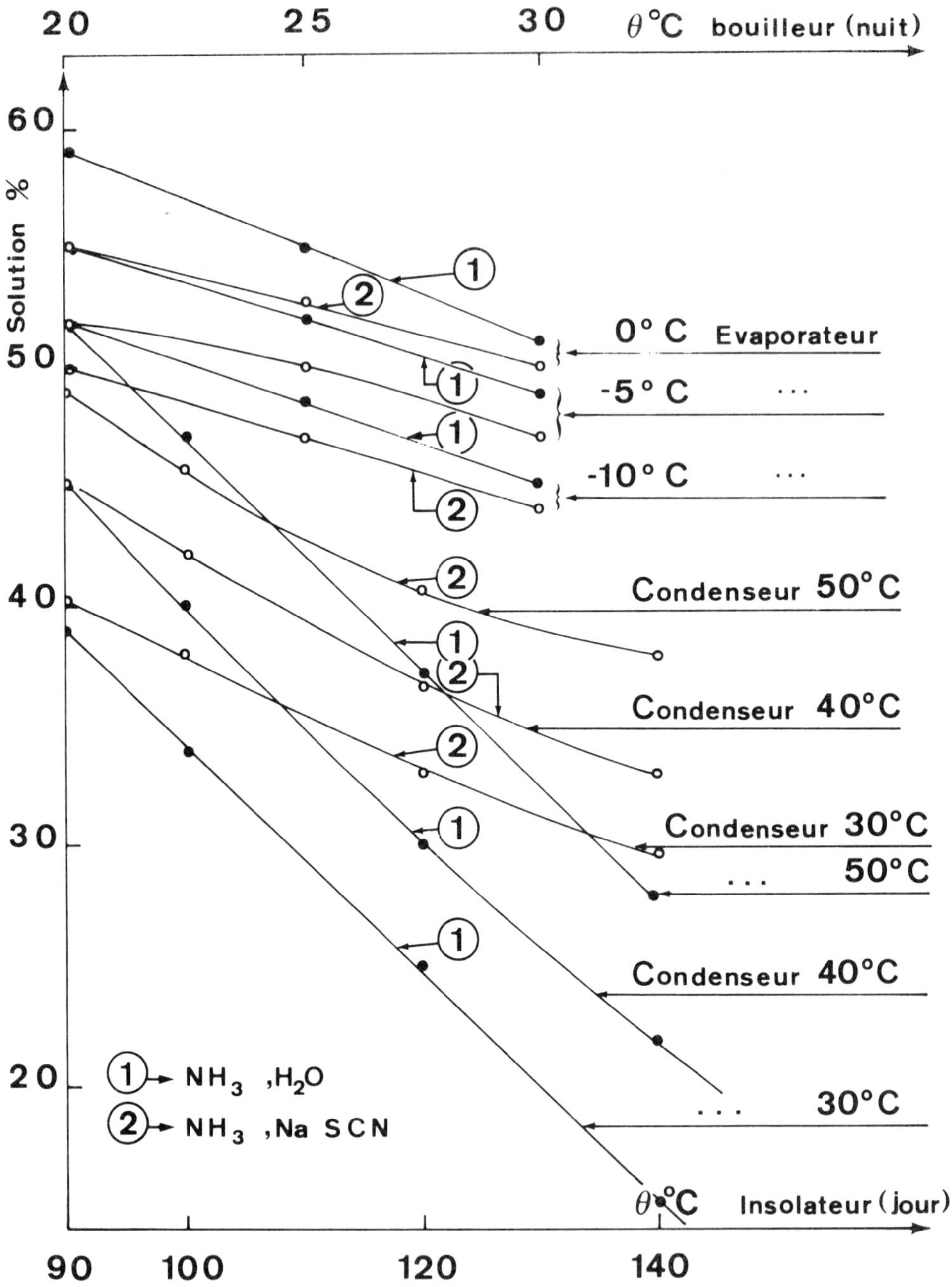
20
25
30
θ °C bouilleur (nuit)
60
50
40
30
20
Solution %
1
2
0° C Evaporateur
-5° C ...
-10° C ...
Condenseur 50°C
Condenseur 40°C
Condenseur 30°C
... 50°C
Condenseur 40°C
... 30°C
θ °C Insolateur (jour)
90
100
120
140
1 → NH_3 ,H_2O
2 → NH_3 ,Na SCN

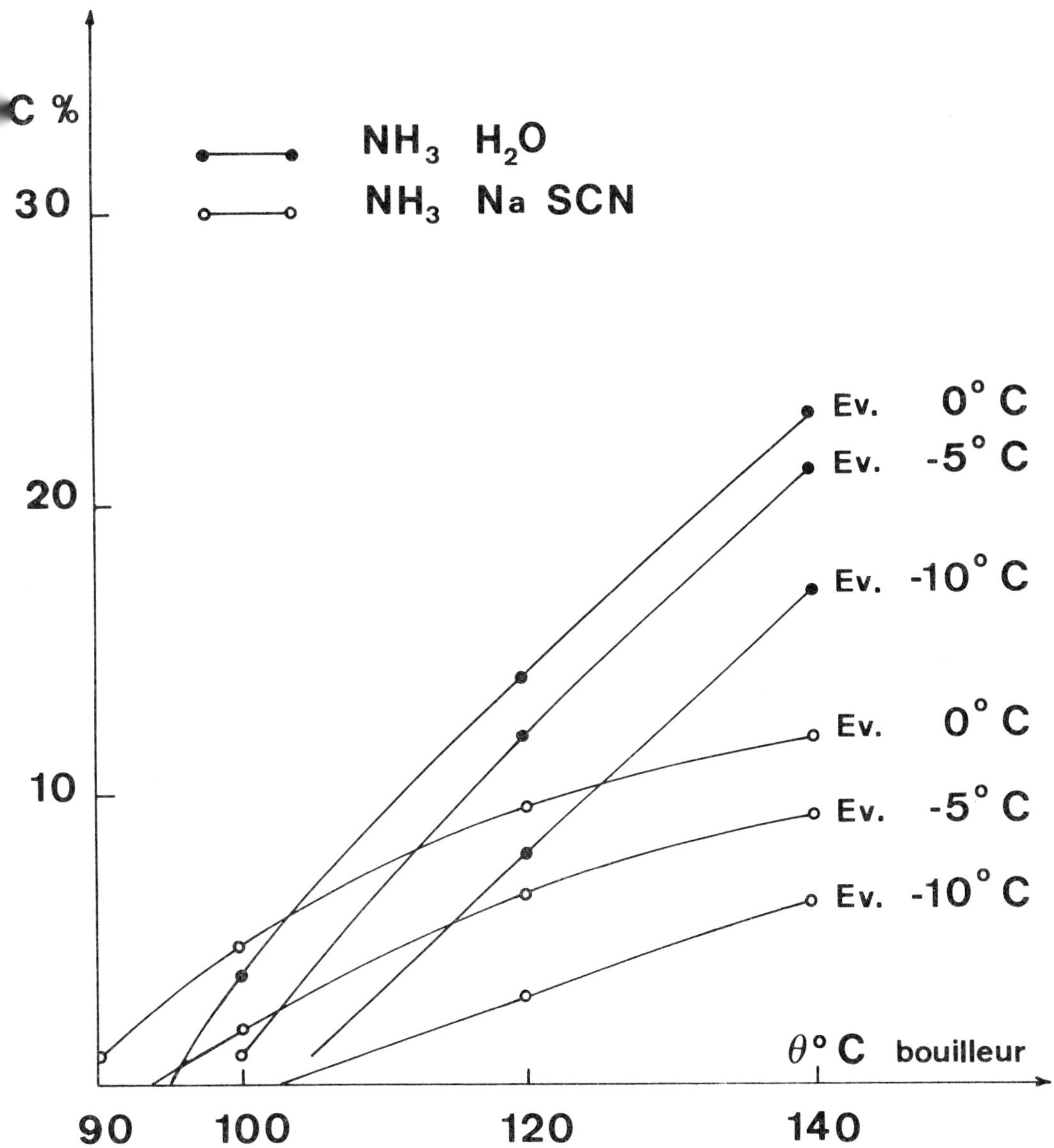
C %
NH3 H2O
NH3 Na SCN
30
20
10
Ev. 0° C
Ev. -5° C
Ev. -10° C
Ev. 0° C
Ev. -5° C
Ev. -10° C
θ° C bouilleur
90
100
120
140

insolators would be needed to heat the salt in times compatible with the number of efficient hours of sunlight.

CONCLUSION

We are of the opinion that, in the case of wet absorption, the conditions imposed which are not exceptional in tropical countries, lead to a usable energy efficiency comparable to that of the photovoltaic effect, and that it is important to raise the temperature of the boiler if one wishes to improve this efficiency.

This leads us firstly to consider, for refrigeration as well as for heating, solar radiation as a localised source of energy which would complement that of gas and oil in places without electricity.

Let us next note that with reference to the characteristics indicated by Eggers-Lura (6), the use of $CaCl_2$ would lead, in the conditions mentioned earlier, to the liberating for 2 Kg 500 of $CaCl_2$, 2 Kg 300 of liquid NH_3, or, 4 times more than the better of the 2 earlier mixtures. Let us recall that these performances are limits and that problems such as the considerable swelling of the salt at regeneration, the threat of clogging, the speed of octohydrate formation, remain unanswered.

However, the characteristics of the ammoniacate studied elsewhere (14), and without the direct preoccupation of their application to refrigeration,are promising enough for us to envisage, within the framework of the research work that we have set ourselves, prior to any realisation in situ, a systematic laboratorystudy of the $CaCl_2$, NH_3 group in conditions near to the climatic demands that characterise the tropical zone. Whilst awaiting in-depth verification, the chemical dissociation of the ammoniacates seem better adapted to extreme conditions than the simple physical phenomenon of the dissolution of a gas in a liquid phase. Its efficiency is better. There remains to be studied, however, under these conditions, the key parameter : the regeneration speed of the ammoniacate.

REFERENCES

(1) J. FLECHON, G. MARTIN, I. TOURE, Comptes rendus, 280, série B, 1975, p. 211.

(2) J. FLECHON, Comptes rendus, 283, série B, 1976, p. 127.

(3) J. FLECHON, Revue de Physique Appliquée, 11, 1976, p. 743-50.

(4) J. FLECHON, R. WERTWIJN et A. DIALLO, Comptes rendus, 281, série B, 1975, p. 9.

(5) R.K. SWARTMAN et al, Colloque UNESCO, Paris, Juillet 1973, E 5.

(6) A. EGGERS-LURA et al, Heliotechnique and Development, Ed. Kettani, Soussou 1976, p. 83-104.

(7) Règles pour Machines Frigorifiques (I.I.F.), 5e ed. p. 114-15.

(8) GEORGE C., BLYTAS, FARRINGTON, DANIELS, J. of Amer. Chem. Soc., 84-71, 1962, p. 1075-83.

(9) J. FLECHON, Journées Internationales sur le Chauffage Solaire, Liège, Sept. 1977.

(10) R. PLANK and J. KUPRIANOFF, Die Kleinkältemaschine, Spring Verlag Berlin, 1960 p. 351.

(11) K. LINGE, Uber Periodische Absorptionskältemaschinen, D. Sc. Thesis Karlsruhe, May 1929.

(12) E. ALTENKIRSCH, Absorptionskältemaschinen, Verlag Technik Berlin, 1954.

(13) W. MEHL, Die Thermischen Eigenschaften zeitschrift für die Gesamte Kälteindustrie, Heft 11, Vol. 41, p. 186, 1934.

(14) P. PASCAL, Traité de Chimie Minérale T. IV, p. 316, Masson 1956-70.

(15) BEGOIN, Réfrigération électrique automatique Girardot (out of print).

ANALYSIS OF AMMONIA-WATER INTERMITTENT SOLAR REFRIGERATOR OPERATING WITH A FLAT PLATE COLLECTOR

A.Venkatesh and M.C.Gupta
Solar Energy Division
Energy Research Centre
Indian Institute of Technology
Madras 600 036 INDIA

ABSTRACT

The intermittent solar refrigerator consists of three major components, generator-cum-absorber, rectifier and condenser-cum-evaporator. During the generation process vapours pass through the rectifier and are condensed in the condenser. During the refrigeration process vaporising liquid ammonia in the evaporator provides cooling and the vapours are absorbed by the weak solution in the absorber. The paper deals with the thermodynamic analysis of an ammonia-water intermittent solar refrigerator that uses a simple flat plate collector for generation of ammonia vapours from the ammonia-water solution. The processes that constitute the cycle have been simulated on a digital computer. For different initial solution concentrations and generator temperatures the following parameters have been evaluated and discussed. 1) Heat absorbed during generation. 2) Amount of vapours generated and the amount of ammonia condensed. 3) Minimum evaporator temperature. 4) Effective refrigeration and 5) Coefficient of performance.

INTRODUCTION

Escalating oil prices, fast depleting fossil fuels and combustion generated pollution on a large scale have forced mankind to look to new sources of energy such as, sun, wind, ocean thermal gradients, tidal power, etc. Amongst all these new sources, solar energy holdsout greater promise to the mankind as it is abundant, non-polluting and available free of cost. Of all possible direct uses of solar energy, the one that will be accepted with great enthusiasm is solar refrigeration. India is basically an agricultural country and the economy of the vast population living in the rural and costline areas is inextricably linked with agricultural produce and marine food products. In such an economy, preservation of these products before they are despatched to urban ares becomes a vital factor. In this context an intermittent ammonia-water solar refrigerator operating with a flat plate collector appears to have a good potential. It has no moving parts and hence it is easy to maintain. Also, it has no pumps, fans or any gadgets that requires electricity.

THEORETICAL CYCLE

Fig.1 shows the theoretical constant temperature refrigeration cycle. Solar radiation striking the generator raises the pressure and temperature of the ammonia-water solution as indicated by the path 1-2. At 2 the system pressure is equal to the saturation pressure of ammonia corresponding to the condensing temperature. Generation continues from 2 to 3 at constant pressure as solar heating is continued. The generated vapours pass to a water cooled condenser through a rectifier and ammonia is condensed. State 3, the end of the generation process depends upon the maximum attainable temperature in the flat plate collector. At this instant when the state of liquid ammonia in the condenser is at 2*3* and the state of solution in the generator is at 3,the generator and the condenser are isolated. The generator is cooled to ambient conditions so that the solution state reaches 4. At 4, the solution concentration, temperature and pressure are low and hence the generator is ready to operate as an absorber. The communication between the condenser and the absorber is now restored. A fraction of liquid ammonia in the condenser at 2*3* flashes adiabatically into the absorber to change the solution state to 5. As a consequence of flashing the temperature of the remaining quantity of ammonia in the condenser decreases to bring its state to 5*. The condenser is now ready to opearte as the evaporator. From 5* to 1* the evaporator absorbs energy from the surroundings to provide the effective refrigeration, and the rising vapours are reabsorbed in the absorber until the solution reaches the initial state 1.

THERMODYNAMIC ANALYSIS OF THE CYCLE

Heat added during generation

Heat added to the ammonia-water solution during the generation process 1-2-3 (Fig.1) is given by [1,2],

$$q_g = m_3 h_3' - m_1 h_1' + \int_3^1 dm\, h'' \tag{1}$$

where dm is the differential mass of solution distilling out of the generator. Eq.(1) can be approximated to

$$q_g = m_3 h_3' - m_1 h_1' + \int_3 dm\, h'' \tag{2}$$

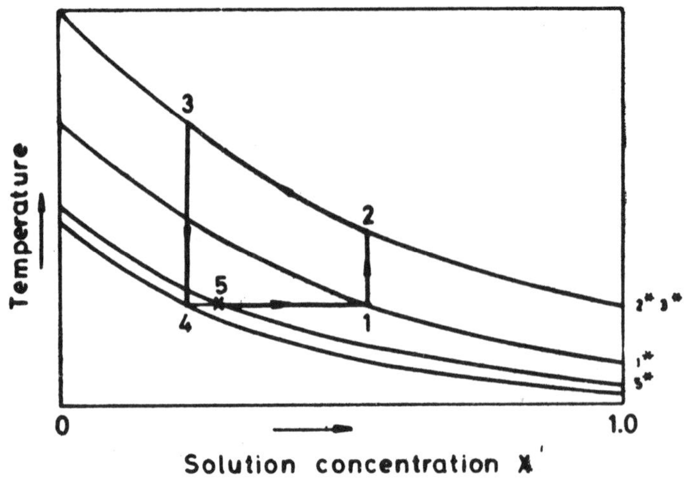

1, 2, 3, 4, 5	Solution states in the Generator - cum - absorber
1*, 2*, 3*, 5*	States of pure NH_3 in Condensor - cum - evaporater

FIG.1. THERETICAL CYCLE FOR SOLAR REFRIGERATOR

TABLE 1

$p = 11.9\ kgf/cm^2$
$t_c = 30^oC$

$t_g(^oC)$	x'	x"	x"-x'	1/(x"-x')	$\frac{x"-x'}{1-x'}$
130	0.2	0.782	0.582	1.72	0.728
108	0.3	0.910	0.610	1.64	0.870
88	0.4	0.968	0.568	1.77	0.947
68	0.5	0.99	0.49	2.04	0.980
55	0.6	0.996	0.396	2.52	0.990
45	0.7	0.998	0.298	3.36	0.993
39	0.8	0.9995	0.1995	5.0	0.998

as during the process 1-2 most of the heat added goes to rise the internal energy of the solution.

Amount of vapours generated

The constant pressure distillation process 2-3 begins at state 2 with m_2 kg of solution at a temperature of t_2 and solution concentration of x_2'. It may be pointed out here that $m_2 \approx m_1$ and $x_2' = x_1'$. At any instant during the process 2-3 the distilled vapours are in equilibrium with the solution. Also, the mass of distilled vapours equals the disappearing liquid mass. It can be shown by material and ammonia balance, that

$$m_3 = m_2 \exp \int_{x_2'}^{x_3'} \frac{dx'}{x'' - x'}. \tag{3}$$

As m_{vg}, the amount of vapours generated is the difference in solution masses between states 2 and 3,

$$m_{vg} = m_2 - m_3 \tag{4}$$

$$\text{i.e. } m_{vg} = m_2\left(1 - \exp \int_{x_2'}^{x_3'} \frac{dx'}{x'' - x'}\right) \tag{5}$$

The amount of vapours generated from a given mass of solution depend only on x_2' and x_3'. As pressures at states 2 & 3 are same (saturation pressure of ammonia corresponding to condensing temperature) x_2' and x_3' depend only on t_2 and t_3. It is reasonable to assume that the average temperature of cooling water in tropics is about 30°C. The saturation pressure of ammonia corresponding to this temperature is 11.9 kgf/cm^2. As the condenser is water cooled, the process 2-3 occurs at 11.9 kgf/cm^2.

To evaluate the integral in Eq.(5) values of x" corresponding to x' are obtained from ammonia-water property Tables [3] for various solution concentrations lying between 0.1 and 0.8 which covers the range of practical interest. These values are presented in TABLE 1. A plot of $1/(x''-x')$ vs x' is then made. The area enclosed between the curve, the ordinates at the initial and final solution concentrations during generation and the abscissa gives the required integral.

Amount of ammonia vapours condensed during generation

The vapours generated will pass through a rectifier before entering the condenser. Rectification causes the mass of

TABLE 2

h in kcal/kg

Datum

$(h_1)_{NH_3} = 0$ at $t = 0^oC$

$(h_1)_{H_2O} = 0$ at $t = 0^oC$

t_5^* (^oC)	h'	h"	h_{fg}	$\Delta h'$	$h_{fg_{av}}$	exp $h'/h_{fg_{av}}$
30.0	33.87	307.27	273.4	0	273.4	1.0
26.7	30.07	306.77	276.7	-3.8	275.0	0.986
21.16	23.67	305.97	282.3	-10.2	277.8	0.964
15.6	17.37	304.97	287.6	-16.5	280.5	0.943
10.05	11.07	303.77	292.7	-22.8	283.5	0.922
4.45	4.97	302.57	297.6	-28.9	285.5	0.904
- 1.05	- 1.23	301.17	302.4	-35.1	287.9	0.881
- 6.64	- 7.33	299.67	307.0	-41.2	290.2	0.867
-12.15	-13.33	298.07	311.4	-47.2	292.4	0.851
-17.7	-19.43	296.27	315.7	-53.3	294.5	0.835
-23.25	-25.43	294.47	319.9	-59.3	296.6	0.818
-28.8	-31.33	292.57	323.9	-65.2	297.6	0.804
-34.35	-37.33	290.57	327.9	-71.2	300.6	0.788
-40.0	-43.23	288.47	331.7	-77.1	302.5	0.775

$\Delta h' = (h')_{t_5^*} - (h')_{30^oC}$; $h_{fg_{av}} = \left[(h_{fg})_{t_5^*} + (h_{fg})_{30^oC}\right]/2$

vapours condensed, m_{vc}, to be less than the generated mass m_{vg}. Assuming 100% rectification, by applying mass balance to rectifier, it can be shown that

$$m_{vc} = m_{vg} \left(\frac{x'' - x'}{1 - x'} \right) \tag{6}$$

The quantity in brackets in Eq.(6) can be evaluated with the help of a plot of $(x''-x')/(1-x')$ vs generator tempeature.

Adiabatic flashing of ammonia during refrigeration process

As explained earlier at the beginning of the refrigeration process a fraction of liquid ammonia at state 2*3* in the condenser flashes into the absorber which contains the weak solution at state 4. Assuming the flashed vapours to be absorbed at constant temperature, the state of solution in the absorber changes from 4 to 5. Correspondingly the state of liquid ammonia remaining in the condenser undergoes achange from 2*3* to 5*. Finally the pressures are equalized ($p_5 = p^*_5$). During the adiabatic flashing it can be shown that

$$m^*_5 = m^*_2 \exp\left(\frac{h^*_5 - h^*_2}{h_{fg_{av}}} \right) \tag{7}$$

As the amount of ammonia flashed, m_{fl}, is equal to $m^*_2 - m^*_5$ and $m^*_2 = m_{vc}$

$$m_{fl} = m_{vc} \left[1 - \exp\left(\frac{h^*_5 - h^*_2}{h_{fg_{av}}} \right) \right] \tag{8}$$

TABLE 2 shows the evaluated values of $(h^*_5 - h^*_2)/h_{fg_{av}}$ for various assumed temperatures after adiabatic flashing.

Minimum evaporator temperature

The minimum evaporator temperature t^*_5 is the temperature of liquid ammonia in the condenser at the end of the adiabatic flashing process. The quantity of ammonia flashed m_{fl}, is responsible for the change of states from 2*3* to 5* (of liquid ammonia in the condenser) and from 4 to 5 (of solution in the absorber). It is discernible that the quantity of ammonia in the solution at state 5 differs from that state 4 by m_{fl}. Hence it can be shown that

$$m_{fl} = \frac{m_4 (x'_5 - x'_4)}{1 - x'_5} \tag{9}$$

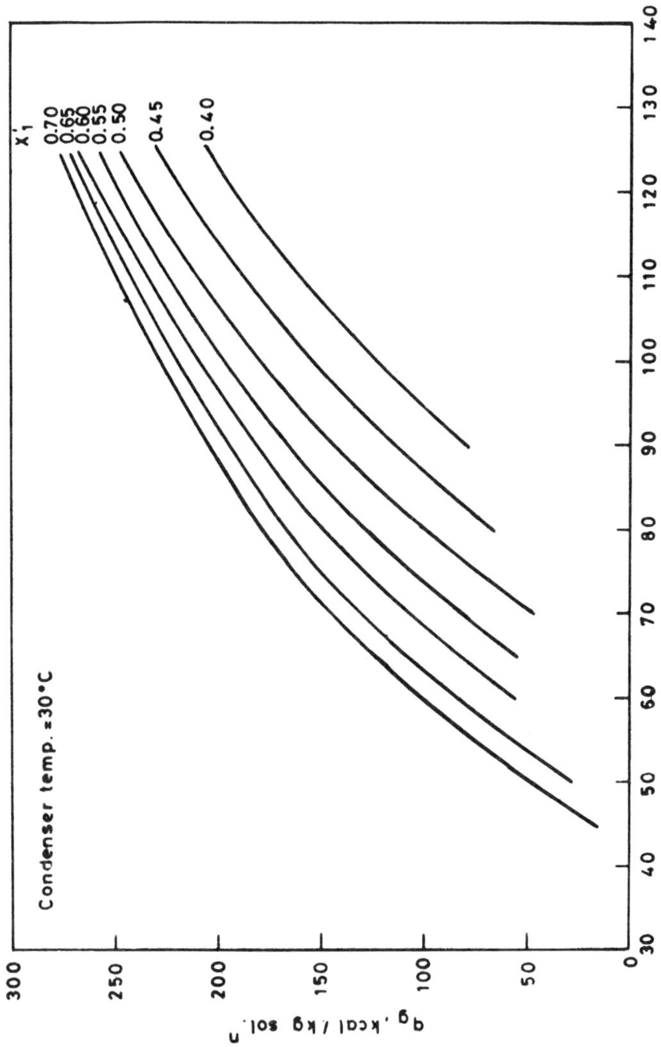

FIG. 2. EFFECT OF t_g AND X_1' ON HEAT ADDED

As states 5 and 5* have same pressure and that the change of state from 4 to 5 occurs at constant temperature it can be shown with the help of property tables that for any minimum evaporator temperature t_5^*

$$x_5' = 0.007804\ t_5^* + 0.53847 \tag{10}$$

and

$$\exp\left(\frac{h_5^* - h_2^*}{h_{fg_{av}}}\right) = 0.00331\ t_5^* + 0.893 \tag{11}$$

With the help of Eqs.(8),(9),(10) and (11) a quadratic equation in t_5^* can be obtained in terms of m_{vc}, m_4 and x_4' on solving which the value of t_5^* can be obtained for a given set of conditions.

Effective cooling

Effective cooling starts at the temperature t_5^* with a quantity of $(m_{vc} - m_{fl})$ kg of liquid ammonia in the evaporator and ends at state 1* when the last fraction of liquid ammonia vaporises. Hence the effective cooling q_c is given by

$$q_c = (m_{vc} - m_{fl})\left(\frac{h^*_{fg_5} + h^*_{fg_1}}{2}\right) \tag{12}$$

Coefficient of performance

The theoretical coefficient of performance is defined as

$$\text{c.o.p} = q_c/q_g \tag{13}$$

RESULTS AND DISCUSSIONS

The parameters discussed above have been computed with the aid of an IBM 370 computer. The step interval chosen for the analysis is 1°C. All the quantities required for the analysis have been expressed as functions generator temperature to facilitate computations.

Fig.2 shows the variation of q_g with t_g for various initial solution concentrations. The following observations can be made: 1) For a given initial solution concentration heat added in the generator increases with increase in generator temperature. 2) At any initial solution concentration , the

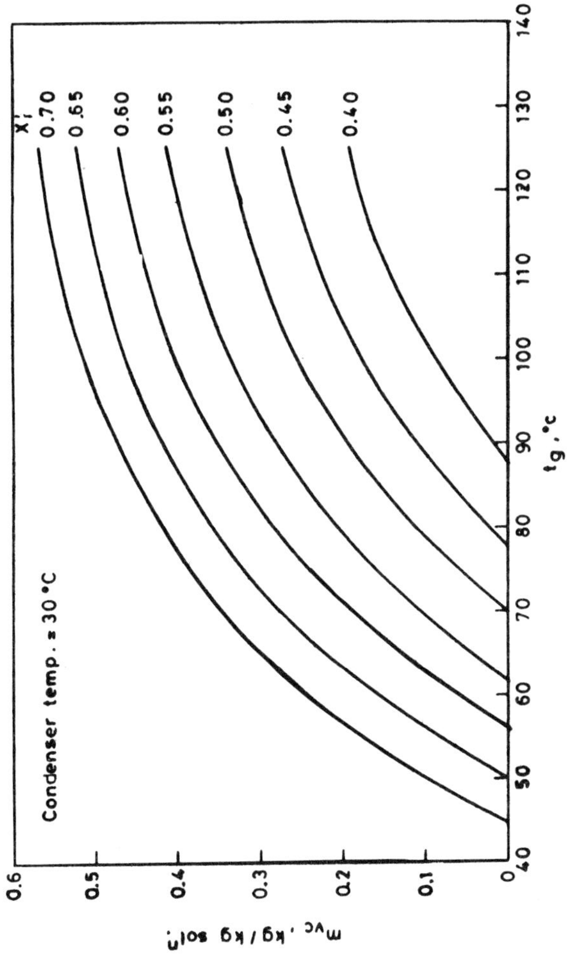

FIG.3. EFFECT OF t_g AND X'_1 ON m_{vc}

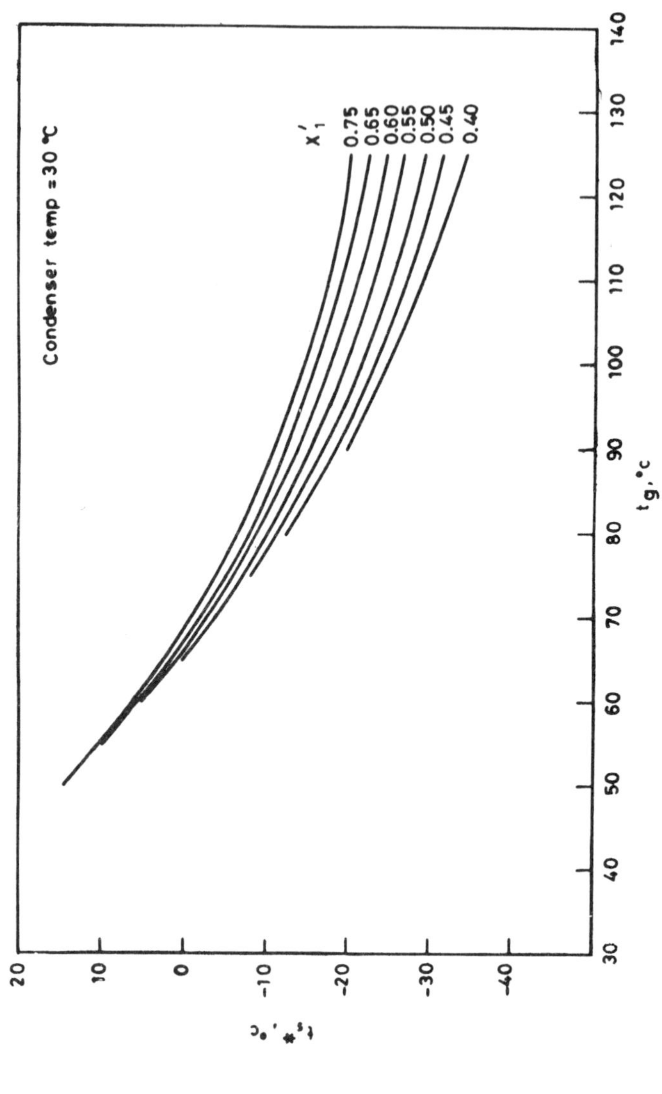

FIG. 4. EFFECT OF t_g AND X_1' ON MINIMUM EVAPORATOR TEMP t_5^*

heat transfer required to bring about a certain change in the generator temperature is smaller at higher generator temperatures, for at higher temperatures the solution mass and concentration will have reached smaller values which in turn reduces the vapours distilling off. 3) As one would expect, the heat transfer to the generator increases with increase in initial solution concentration for a given maximum generator temperature. As the initial concentration increases although state 2 shifts to lower temperatures as a consequence of which the sensible heating of the solution between the states 1 and 2 is reduced, the quantity of vapours distilling off between states 2 and 3 increases which requires higher heat transfer during 2-3, the net result being increased heat transfer.

Fig.3 shows the variation in the quantity of ammonia condensed during the generation process plotted against the generator temperature for different initial solution concentrations. The mass of condensate increases with temperature for any initial solution concentration. Also, the increase in the condensate for a given Δt_g is smaller at higher tempertures. This effect is more pronounced at higher initial solution concentrations, which arises on two counts; 1) the higher the temperature the lower is the solution concentration which reduces the generation of vapours, 2) vapours generated at higher temperatures carry away more water than at low temperatures, which on rectification produces a smaller mass of condensate.

Fig.4 shows the minimum evaporator temperature plotted against generator temperature for various initial solution concentrations. For a given initial solution concentration the minimum evaporator temperature decreases with increase in generator temperature. As the generator temperature is increased, state 3 shifts to lower concentrations as a consequence of which state 4 shifts to lower pressures and hence the pressure difference ($p_2^* - p_4$) available for flashing increases. The increased pressure difference causes a larger fraction of the condensed ammonia to flash thereby reducing the temperature of the remaining ammonia in the condenser to a lower value. Also, for a given genertor temperature the minimum evaporator temperature increases with increase in initial solution concentration. As the initial concentration is increased the mass of condensed ammonia increases. For a fixed maximum generator temperature state 4 is fixed. Hence for different initial solution concentrations, the pressure difference ($p_2^* - p_4$) responsible for flashing remains the same. As the initial solution concentration increases the condensate increases, and since the quantity of ammonia flashing is mainly influenced by

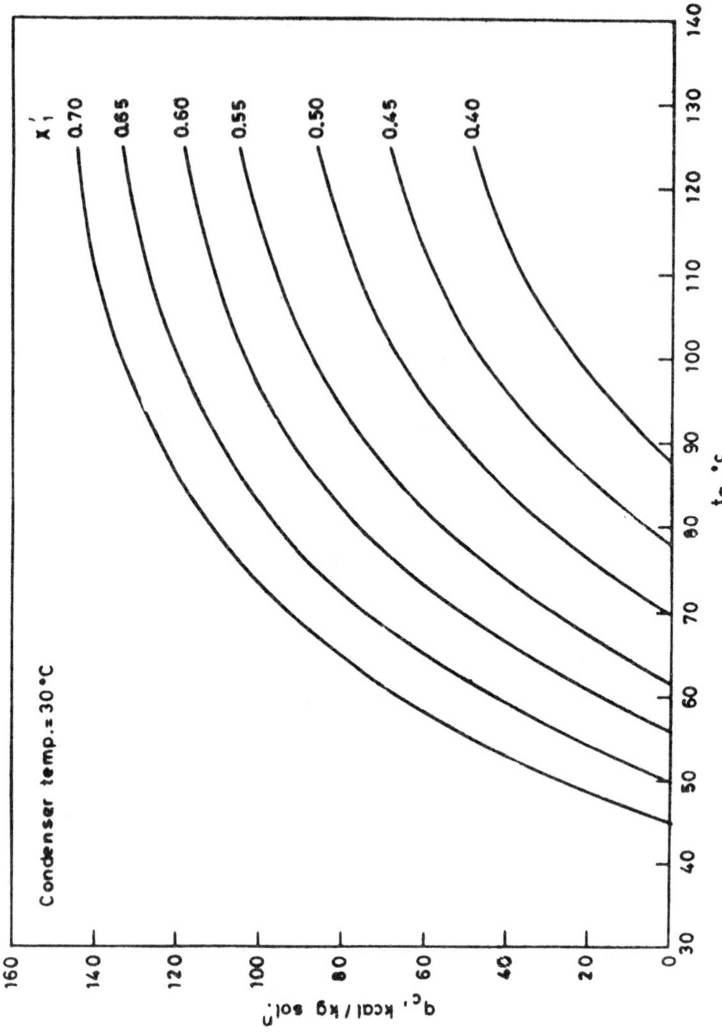

FIG.5. EFFECT OF t_g AND X_1' ON EFFECTIVE REFRIGERATION

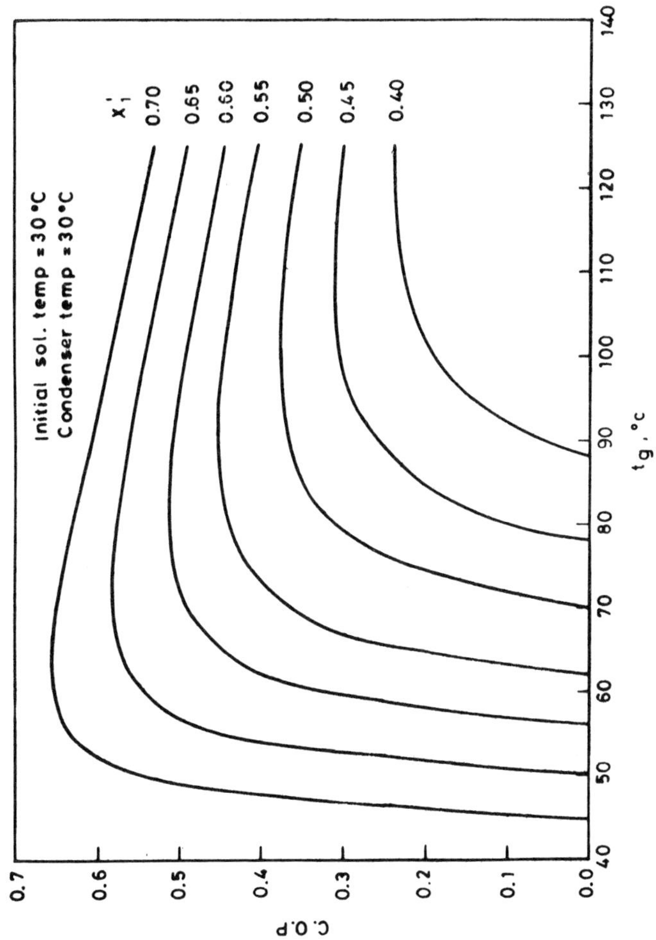

FIG.6. EFFECT OF t_g AND X'_1 ON C.O.P.

the pressure difference, the latent heat of vaporisation is now borne by a larger mass of condensate. Hence as the initial solution concentration increases the minimum evaporator temperature increases.

Fig.5 shows effective refrigeration which follows adiabatic flashing, as a function of generator temperature for different initial solution concentrations. It increases with increase in initial solution concentration for a given generator temperature and also with increase in generator temperature for a given initial solution concentration, the reason for which can be attribited to the increased mass of condensate available for effective refrigeration.

Fig.6 shows the variation of theoretical cop with the generator temperature for different initial solution concentrations. At a given initial solution concentration cop increases rapidly at first and then slowly until the maximum is reached beyond which it decreases gradually. The reason for this trend is the nature of variation of q_g and q_c shown in Figs.2 and 5. The results of the theoretical analysis as seen from Figs.2 to 6 clearly demonstrate that in order to obtain higher cop (higher effective refrigeration) higher generator temperatures and higher initial solution concentrations are imperative. However, attainment of low evaporator temperatures dictates the use of low initial solution concentrations. Hence a compromise has to be struck as regards the initial solution concentration for obtaining optimal performance. In practice the minimum evaporator required depends on the purpose for which cooling is required. For example, if the cooling is needed for freezing water or preservation of food, subzero temperatures are preferred. On the other hand if cooling is used for getting chilled water the evaporator temperature can be higher.

NOMENCLATURE

cop or c.o.p	coefficient of performance
h	enthalpy of liquid ammonia
h'	enthalpy of ammonia-water solution
h"	enthalpy of ammonia-water vapour
m	mass
m_{fl}	mass of ammonia flashed
m_{vc}	mass of ammonia condensed
m_{vg}	mass of ammonia-water vapours generated
p	pressure
q_c	effective refrigeration
q_g	heat added during generation

t	temperature
x'	solution concentration
x"	vapour concentration

<u>Subscripts</u>

1,2,3,4 & 5	properties corresponding to the states in Fig.1
1*,2*,3* & 5*	properties of pure ammonia corresponding to states in Fig.1.
g	properties of solution in the generator

REFERENCES

1. Chinnappa,J.C.V., " Experimental Study of the Intermittent Absorption Refrigeration Cycle Employing the Refrigerent-absorbent Systems of Ammonia-water and Ammonia-Lithium Nitrate", Solar Energy,Vol.5,1961.

2. Sargent,S.L., & Beckman,W.A.," Theoretical Performance of an Ammonia-Sodium Thiocyanate Intermittent Absorption Refrigeration Cycle", Solar Energy,Vol.12,1968.

3. George Scatchard, Epstein,L.F., James Warburton,Jr., and Cody,P.J.," Thermodynamic Properties of Saturated Ammonia-Water Mixtures", Journal of ASRE, 1947.

A THERMODYNAMIC ANALYSIS OF A SOLAR-POWERED JET REFRIGERATION SYSTEM

F. L. Lansing and V. W. Chai
Jet Propulsion Laboratory
Pasadena, California, U.S.A.

ABSTRACT

The article describes and analyzes a method of using solar energy to drive a jet refrigeration system. A new technique is presented in the form of a performance nomogram combining the energy and momentum equations to determine the performance characteristics. A numerical example, using water as the working fluid, is given to illustrate the nomogram procedure. The resulting coefficient of performance was found comparable with other refrigeration systems such as the solar-absorption system or the solar-Rankine turbocompressor system.

I. INTRODUCTION

As part of the Deep Space Network energy conservation project made for the Deep Space Communication Complex (DSCC) at Goldstone, California, many energy-conservation money-saving ideas are being studied to reduce the complex dependence on fossil fuel or electric-powered devices. The existing heating, ventilation, and air-conditioning equipment (HVAC) presents a sizable share of the total energy consumption of the complex, and the concept of utilizing the abundant solar energy falling on the site is under investigation.

The present state-of-the-art of solar-powered cooling devices is still limited to several options. Examples are: (1) a solar-powered Rankine engine driving a vapor compression refrigeration machine, (2) solar-powered photovoltaic array (concentrated or nonconcentrated) driving a vapor compression refrigeration machine, and (3) a solar-powered absorption refrigeration. The first option employs two rotating parts such as turbines (or engines), and compressors (excluding the pump) that have to be specially designed to adapt the low temperature nature of solar energy. The second option, though directly generates electricity with less mechanical parts, is neither economically competitive nor has the same conversion efficiency as the first. The third option possesses the highest coefficient of performance, but has some size, maintenance, and operation difficulties.

One of the least commonly used refrigeration techniques today is the jet refrigeration system as in Refs. [1], [2] and [3]. This system has not enjoyed wide application despite its advantages of simplicity, ruggedness of design, vibration-free, reliability, and its low operation and maintenance cost. Coupling this system with solar-energy as a driving source instead of conventional gas-fired boilers adds a new option to solar-powered cooling devices. This work intends to introduce the concept, describe the relevant parameters, and present the new analytic methodology used in performance determination.

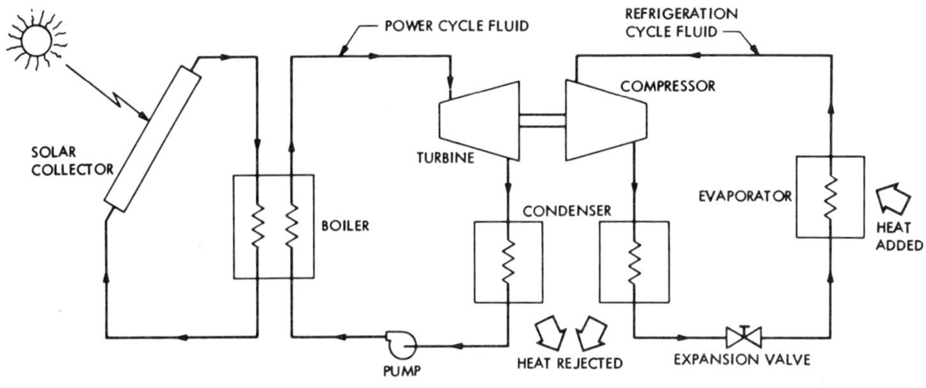

Fig. 1. Double fluid turbocompressor system

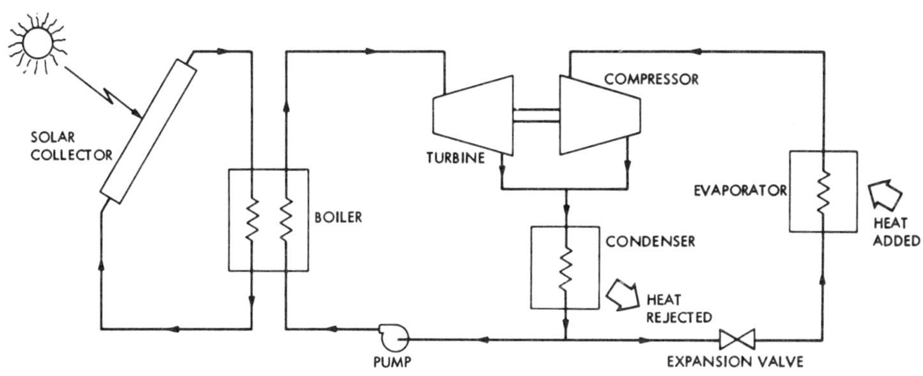

Fig. 2. Single fluid turbocompressor system

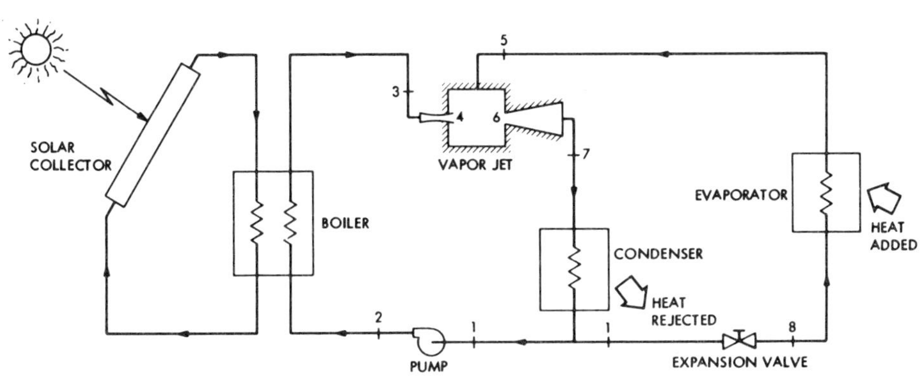

Fig. 3. Vapor jet refrigeration system

II. SYSTEM DESCRIPTION

The basic concept of a solar-powered vapor compression refrigeration system, as illustrated in Fig. 1, is to use a solar-Rankine power cycle to drive a vapor compression refrigeration cycle. Solar energy is used to vaporize the working fluid of the power cycle to saturated vapor at the boiler. Nonconcentrating flat-plate solar collectors or medium-concentration types can be used for this task to provide fluid temperatures that range from 90 to 160°C. For most working fluids, this range of temperatures is sufficient for vaporization. The vapor expands in the turbine and produces the mechanical work that is needed to drive the refrigeration compressor. The combination of the turbine and the compressor on a single shaft is sometimes called a turbo-compressor unit, Ref. [4] .

The working fluids and the temperatures used in both cycles so far can be independent of each other. In each cycle, the choice of the working fluid depends on performance, stability, economics, and safety factors.

Figure 2 shows a modified solar-powered vapor compression refrigeration system where both power and refrigeration cycles are employing the same working fluid and have a common condenser. The common condenser acts for both cycles as a heat sink at a constant temperature. This configuration with a single working fluid has been recommended, Ref. [5] , for its compactness and low maintenance cost benefits, though it does not have the freedom of choice of many working fluids due to chemical and thermal stability in the combined cycle.

The only role of the turbo-compressor unit, Figs. 1 and 2, is to expand in the turbine the vapor from the boiler pressure down to the condenser pressure and to compress in the compressor the vapor from the evaporator pressure up to the condenser pressure, thus exchanging two equal but opposite mechanical work. This energy exchange, from thermodynamics veiwpoint has a null effect on the overall cycle performance. The same function of the turbo-compressor can be replaced and made simpler by using the ejector scheme as shown in Figs. 3 and 4. The convergent-divergent nozzle replaces the turbine function in converting the enthalpy drop between the boiler pressure and the evaporation pressure to kinetic energy. In practice, the pressure at the nozzle exit (state 4) is made slightly lower than the evaporator pressure (state 5) to overcome the pressure drop in the line. In the mixing chamber, entrainment with momentum exchange occurs between the vapor with supersonic speed leaving the nozzle and the vapor with subsonic speed leaving the evaporator. The mixture (state 6), still with supersonic velocity, will emerge into the diffuser section where it converts its kinetic energy back into an enthalpy increase, thus raising its pressure to the condenser pressure (state 7). The diffuser replaces the compressor function. The condensate from the condenser (state 1) is divided to two branches; one branch goes to the power cycle loop through the pump (state 2) and the solar collector/boiler section to state (3) as vapor, and the other branch goes to the refrigeration cycle loop through an expansion valve (state 8) to the evaporator.

Figure 5 presents the sequence of events for the jet refrigeration scheme on the pressure-enthalpy diagram. Flow irreversibilities in the nozzle and the diffuser occur through the deviation of the isentropic states (state 4_{isen} and state 7_{isen}) from the actual ones (states 4 and 7).

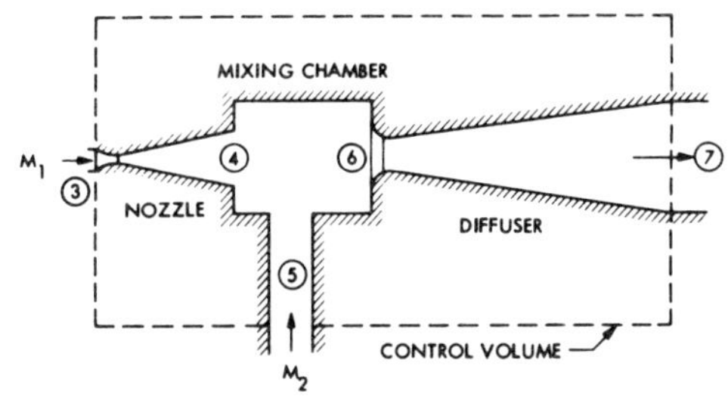

Fig. 4. Ejector control volume

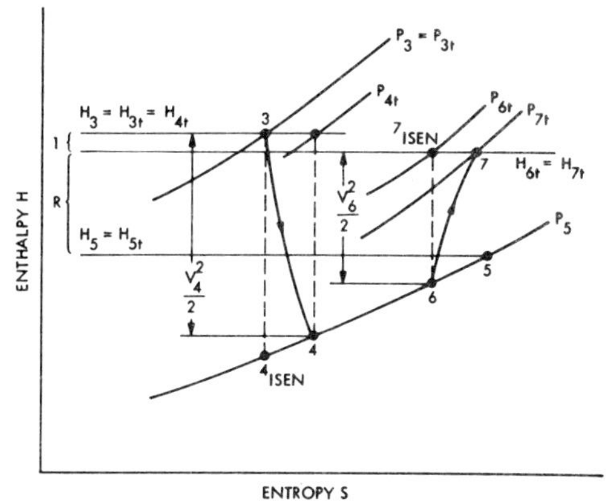

Fig. 6. Ejector presentation on the enthalpy-entropy (Mollier) diagram

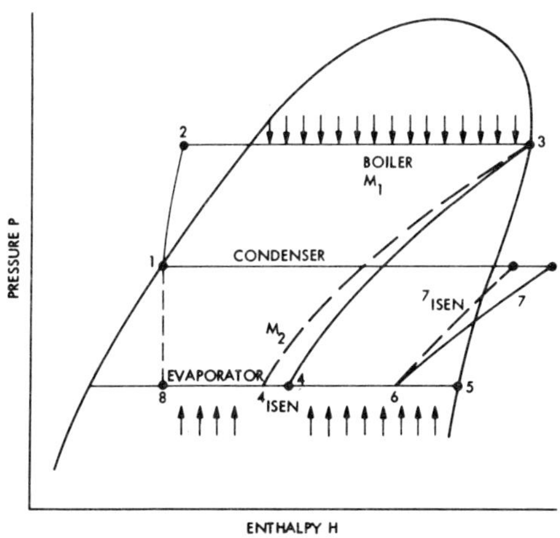

Fig. 5. Pressure-enthalpy diagram showing the complete jet refrigeration cycle

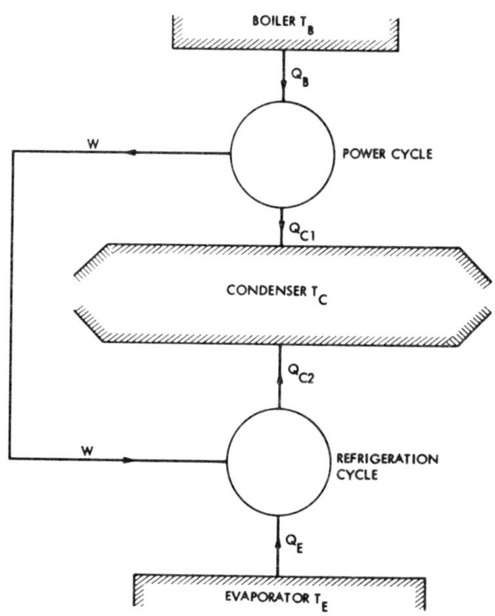

Fig. 7. Carnot cycle idealization for the jet refrigeration scheme

The jet-refrigeration system has many advantages besides operating without gross moving parts (except for the small circulating pump). It is vibration- and noise-free, relatively simple and inexpensive to manufacture with a well-developed technology of nozzles, diffusers, and heat exchangers. The system is low in operation and maintenance cost and its performance is highly competitive with solar-powered turbo-compressor units and solar powered absorption refrigeration units, as will be shown later.

III. ASSUMPTIONS AND GOVERNING EQUATIONS

The adiabatic control volume including the nozzle, the mixing chamber, and the diffuser is shown schematically in Fig. 4. Some assumptions are made in the analysis and are listed as follows:

1. The velocity of the fluid entering the nozzle (state 3) is negligible.
2. The velocity of the fluid leaving the evaporator (state 5) is negligible.
3. The velocity of the fluid leaving the diffuser (state 7) is negligible.
4. The pressure drop through the heat exchangers, and lines is negligible.
5. Saturated conditions for states 1, 3, and 5 are assumed.
6. Expansion and compression processes in the nozzle and the diffuser, respectively, are adiabatic.
7. Heat exchange processes in the evaporator and the condenser are ideal with no thermal losses by radiation.

According to assumptions 1, 2, and 3, the total and static enthalpy of states 3, 5, and 7, as depicted in Fig. 6, will coincide, i.e.,

$$\begin{aligned} H_3 &= H_{3t} \\ H_5 &= H_{5t} \\ H_7 &= H_{7t} \end{aligned} \tag{1}$$

Applying the first law of thermodynamics on the control volume of Fig. 4 gives:

$$M_1H_3+M_2H_5 = (M_1+M_2)H_7 \tag{2}$$

where M_1 is the mass flow rate for the power cycle and M_2 is the mass flow rate for the refrigeration cycle. Defining a mass ratio R as

$$R = \frac{M_1}{M_2} \tag{3}$$

Equation (2) can be rewritten in either one of the following forms:

$$R = \frac{H_7 - H_5}{H_3 - H_7}$$

$$\frac{R}{R + 1} = \frac{H_7 - H_5}{H_3 - H_5} \qquad (4)$$

Equation (4) indicates that the total enthalpy H_7 is located between the enthalpies H_3 and H_5 and divides their enthalpy difference $(H_3 - H_5)$ inversely by the ratio R, as shown in Fig. 6. Furthermore, applying the first law of thermodyanmics and the irreversibility relationships for the nozzle, gives

$$\frac{V_4^2}{2} = (H_3 - H_4) = (H_3 - H_{4,\, isen}) \cdot \eta_n \qquad (5)$$

and for the diffuser

$$\frac{V_6^2}{2} = (H_7 - H_6) = \frac{1}{\eta_d} (H_{7,isen} - H_6) \qquad (6)$$

where η_n and η_d are the isentropic efficiencies for the nozzle and the diffuser, respectively.

The momentum equation in the constant pressure mixing chamber alone would give:

$$M_1 \cdot V_4 = (M_1 + M_2) \cdot V_6$$

or $\qquad (7)$

$$\frac{V_6^2}{2} = \left(\frac{R}{R+1}\right)^2 \cdot \frac{V_4^2}{2}$$

The thermodynamic properties of states 1, 3 and 5 are completely determined once the three independent temperatures of the boiler, evaporator, and condenser are given. The solution of the set of Eqs. (4), (5), (6), and (7) requires the use of one of the property diagrams together with the design values of η_n and η_d. Instead of the usual trial and error procedure, a performance nomogram will be introduced to match the operating conditions.

A first manipulation of Eqs. (5), (6) and (7) give the following relationship:

$$(H_{7,isen} - H_6) = \left(\frac{R}{R+1}\right)^2 \cdot \eta_n \cdot \eta_d \, (H_3 - H_{4,\, isen}) \qquad (8)$$

The left-hand side of Eq. (8) represents the ideal compression work needed if the diffuser were replaced by an adiabatic compressor. On the other hand, the enthalpy difference $(H_3 - H_{4,\, isen})$ represents the ideal expansion work obtained if the nozzle were replaced by an adiabatic expander. The unique turbocompressor coupling effect made by the jet refrigeration scheme is clearly shown by the extra quantity $(R/R+1)^2$ in Eq. (8). The fact that the quantity $(R/R+1)^2$ appears is due to the entrainment phenomenon in the mixing chamber and the momentum exchange.

The overall coefficient of performance (COP) for the jet refrigeration system is defined as

$$COP = \frac{\text{refrigeration effect}}{\text{external energy input}} \tag{9}$$

The external energy added to the system (Fig. 3) is composed of both the heating part from state (2) to state (3) and the pump work from state (1) to state (2). Accordingly,

$$COP = \frac{(H_5 - H_1)}{R(H_3 - H_1)} \tag{10}$$

IV. CARNOT'S CYCLE IDEALIZATION

The maximum coefficient of performance of the jet refrigeration scheme can be determined using Carnot's principles as shown in Fig. 7. The Carnot's power cycle shown works between the boiler temperature (T_B) and the condenser temperature (T_c) as its limits. The Carnot's refrigeration cycle works between the condenser temperature (T_c) and the evaporator temperature (T_E) as its limits. Both cycles exchange exactly the same amount of mechanical work, which can be written either as

$$W = Q_B \left(1 - \frac{T_c}{T_B}\right)$$

or

$$W = Q_E \left(\frac{T_c}{T_E} - 1\right) \tag{11}$$

where Q refers to the amount of heat exchanged as shown in Fig. 7. The maximum coefficient of performance can then be written as

$$(COP)_{max} = \frac{Q_E}{Q_B} = \frac{(T_B - T_c)}{(T_B)} \cdot \frac{(T_E)}{(T_c - T_E)} \tag{12}$$

Combining Eqs. (10) and (12) together, will give the minimum mass ratio R for a given set of enthalpies (H_1, H_3, and H_5) as

$$R_{min} = \frac{(H_5 - H_1)}{(H_3 - H_1)(COP)_{max}} \tag{13}$$

V. PERFORMANCE NOMOGRAM

This is a graphical presentation of the system equations presented in Section III including energy, momentum, and continuity equations. The nomogram is illustrated in Fig. 8 and is made up of four curves, one for each quadrant. Knowing the three independent temperatures of the system (for the boiler, the condenser and the evaporator) together with the fifth assumption of saturated conditions, the thermodynamic properties of states 1, 3 and 5 are completely determined. Curve 1 can be constructed using the ejector energy equation, Eq. (4), to plot the mass ratio R as an ordinate versus H_7 as an abscissa. Equation (4) shows that R ranges from zero (at $H_7 = H_5$) to infinity (at $H_7 = H_3$). The Carnot's principle further limits the mass ratio R to only above or at least equal to R_{min} as given by Eq. (13). Accordingly, curve 1 will be divided to two regions: an upper region (above R_{min}), where the performance does not violate the second law of thermodynamics, and the lower region (below R_{min}), which is forbidden. Combining the energy and momentum Eqs. (4),(5), (6) and (7) to eliminate the mass ratio R, the resulting equation will be in the form

$$H_6 = H_7 - \left(\frac{H_7 - H_5}{H_3 - H_5}\right)^2 \cdot \eta_n \cdot \left(H_3 - H_{4,\ isen}\right) \tag{14}$$

The construction of the set of curves (2) can then follow using curve 1 and Eq. (14), with the nozzle efficiency (η_n) as a parameter. Curve (2) is a plot of the enthalpy H_6 versus the mass ratio R as shown in Fig. 8. Starting with a suitable value of H_7 at a given η_n, the value of R is determined graphically from curve 1, or numerically by Eq. (4), and together with Eq. (14) a point can be located on curve 2. The nozzle efficiency can then be varied and the set of curves (2) will be completed.

The above method in constructing curve (2) is found more straightforward in computations than other methods. For example, if the enthalpy H_7 was eliminated from Eqs. (4), (5), (6) and (7) instead of R, the follow - ing quadratic equation would be solved for a given value of H_6 instead of Eq. (14).

$$\begin{aligned} &R^2 \left(H_6 - H_3 + \eta_n (H_3 - H_{4,\ isen})\right) \\ &-R(H_3 + H_5 - 2H_6) - (H_5 - H_6) = 0 \end{aligned} \tag{15}$$

The set of curves (3) in Fig. 8, is plotted with H_6 and H_7 as coordinates, using Eq. (6) and one of the property diagrams with the diffuser efficiency (η_d) as a parameter. The nomogram is then completed by drawing a straight line in the fourth quadrant that passes through the origin with a 45-degree slope.

The performance characteristics are now ready to be determined for the given boiler, condenser, and evaporator temperatures and the nozzle and diffuser isentropic efficiencies. The path may start, for example, from point A Fig. 8 at an arbitrary value of H_7, down to curve (4), horizontally to meet curve (3) at point C at the selected diffuser efficiency, up to meet curve (2) at point D at the selected nozzle efficiency, and horizontally to meet curve (1) at point E. At point E the enthalpy H_7 should match the first value chosen at point A, otherwise the loop is repeated until the solution converges.

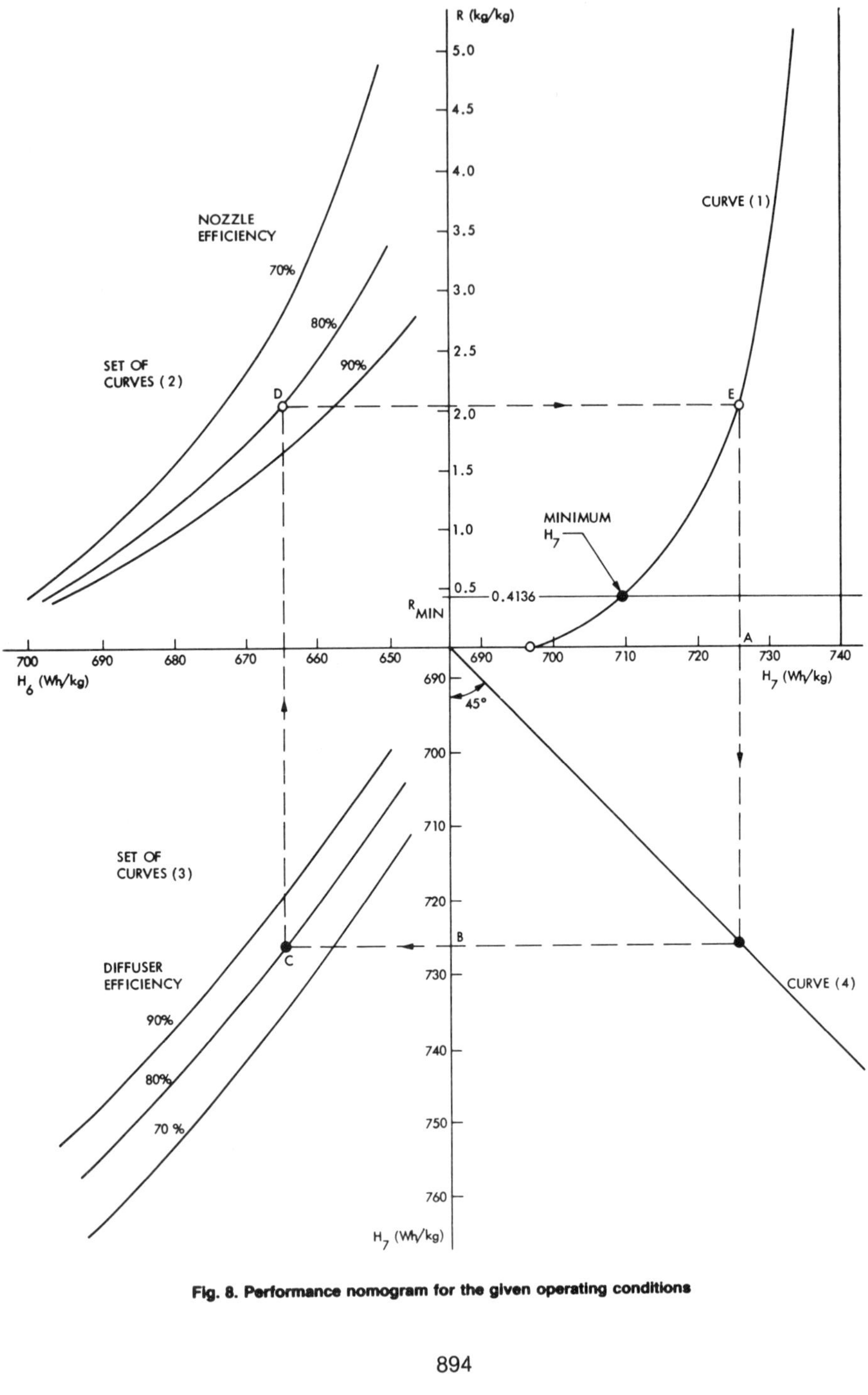

Fig. 8. Performance nomogram for the given operating conditions

VI. NOMOGRAM CASE EXAMPLE

The following is a numerical example selected for a jet refrigeration system using water as the working fluid. The following five independent input conditions are sufficient to describe the system. Numerically they are assigned as

1 Boiler temperature 93.33^{o}C(200°F)

2 Condenser temperature 26.67^{o}C(80°F)

3 Evaporator temperature 4.44^{o}C(40°F)

4 Nozzle efficiency 0.80

5 Diffuser efficiency 0.80

The maximum (COP) is calculated from Eq. (12) as 2.27 and the minimum mass ratio (R_{min}) as 0.414. Curves 1, 2, 3 and 4 are constructed as shown in Fig. 8. The coefficient of performance for the above example converges to 0.465 corresponding to a mass ratio of R of 2.02. Also, from the nomogram, the coefficient of performance drops to 0.289 corresponding to a mass ratio of 3.25 if the nozzle and diffuser efficiencies become 0.7 each. On the other hand, the coefficient of performance increases to 0.618 corresponding to a mass ratio of about 1.52 if the nozzle and diffuser efficiencies become 0.9 each. The above indicates how sensitive the performance is to the quality of manufacturing the nozzle and the diffuser. Also, it shows that the performance is comparable to other heat-powered refrigeration techniques.

VIII. SUMMARY AND REMARKS

The following points can be made from this first phase study of solar-powered jet refrigeration and the given example of performance:

1 A new technique is presented in the form of a performance nomogram combining energy and momentum equations to describe the performance characteristics.

2 The coefficient of performance obtained in the cited example with a nozzle efficiency of 0.80 was 0.465. This compares favorably with about 0.6 for the solar-powered absorption unit and with about 0.5 for solar-powered Rankine driving a vapor-compression refrigerator. The jet refrigeration performance has improved to about 0.62 when the nozzle and diffuser efficiencies increased to 0.90. This means that an improvement of the product efficiency from 0.64 to 0.81 (an increase of 27%) has produced a performance increase of about 50%.

3 The numerical example given can also be very helpful in sizing a 3-ton refrigeration unit (36,000 Btu/hr) to fit a conventional residence house. Based on the mass flow ratio R of 2.02, the mass flow rates entering the evaporator and the nozzle are 15.84 kg/hr

and 32.48 kg/hr, respectively, and the boiler heat rate would be about 23 Kwh_t (78,610 Btu/hr). If the local solar radiancy is taken as 0.8 Kw_t/sq.m (2.54 Btu/hr-ft^2), the collector area needed at 45% collector efficiency would be about 64m^2 ($\sim$ 690 ft^2); an area that is adequate for most south-facing roofs. The ejector unit itself for this type of flow conditions does not exceed 50 cm in length according to manufacturer's specifications.

4 Performance optimization studies are underway to analyze the effects of different types of working fluids, or the effects of different operating temperatures for the solar collector/boiler, condenser and evaporator. These will be reported later as a second phase of the program.

ACKNOWLEDGEMENT

This paper presents the results of one phase of research carried out at the Jet Propulsion Laboratory, California Institute of Technology, under Contract No. NAS 7-100, sponsored by the National Aeronautics and Space Administration.

DEFINITION OF TERMS

COP	Coefficient of performance
COP_{max}	Maximum coefficient of performance
H	Enthalpy
M_1	Mass flow rate for the power cycle
M_2	Mass flow rate for the refrigeration cycle
R	Mass flow ratio
R_{min}	Minimum flow ratio by Carnot principle
T	Temperature
V	Velocity
η	Efficiency

Subscripts

B	Boiler
c	Condenser
d	Diffuser
E	Evaporator
n	Nozzle
t	Total(stagnation)
isen	Isentropic

REFERENCES

1. Anderson, H., "Assessment of Solar Powered Vapor Jet Air-conditioning Systems," presented at 1975 International Solar Energy Congress and Exposition (ISES) held at UCLA, Los Angeles, California, p. 408.

2. Threlkeld, J.L., Thermal Environmental Engineering, Sect. 6, Prentice-Hall, Inc., New Jersey, 1962

3. ASHRAE Guide and Data Book, "Steam Jet Refrigeration Equipment," Chap. 12, pp. 133-138, 1967.

4. Barber, R. E., "Solar-air Conditioning Systems Using Rankine Power Cycles-Design and Test Results of Prototype Three-ton Unit," in Proceedings of the Institute of Environmental Science, Vol. 1, pp. 170-179, 1975.

5. Biancardi, F. R., et al., "Design and Operation of a Solar-powered Turbo-compressor Air-conditioning and Heating System," in IECEC '75 Record pp.186-194.

DEVELOPMENT OF A SOLAR DESICCANT DEHUMIDIFIER

Z. Lavan, D. Gidaspow, M. Onischak,
S. Chaturvedi, B. Mathiprakasam, and S. Perkari
Illinois Institute of Technology
IIT Center
Chicago, IL 60616

ABSTRACT

A preliminary design for a cross-cooled two ton silica gel dehumidifier is presented. The system consists of two identical fixed bed units. One dehumidifies while the other is being regenerated. The cross-cooling is achieved with room or ambient air flowing through rectangular channels. The process stream flows in perpendicular channels which are lined with paper-like sheets consisting of micron size silica gel particles held in a Teflon web. These sheets are made by a specially developed manufacturing procedure. The silica gel sheets were tested in an isothermal channel and were found to perform to our satisfaction. Detailed calculations have shown that the low regeneration temperatures of the silica gel result in sufficient dehumidification for air conditioning applications. The low regeneration temperature permits the use of flat plate solar collectors. A prototype dehumidifier was constructed and is presently being tested.

DESICCANT MATERIALS FOR SOLAR POWERED AIR CONDITIONING

The recent awareness of the energy crisis gave renewed impetus to solar powered desiccant dehumidification and air conditioning. Although many desiccant materials can be used for adsorption of water vapor from air, attention is being focused on molecular sieve and silica gel.

The equilibrium adsorption curves for molecular sieve and silica gel are given in figure 1. Molecular sieve has better adsorption characteristics at low relative humidities or at high temperatures. Hence dehumidifiers using molecular sieves can operate successfully in the adiabatic mode where both the dehumidified air and the desiccant bed are at high temperatures. Molecular sieves, however, require high regeneration temperatures (150°C) for desorbing. In the context of solar energy application it means using costly high performance collectors, and/or low collection efficiencies.

Silica gel, on the other hand, can be desorbed at relatively

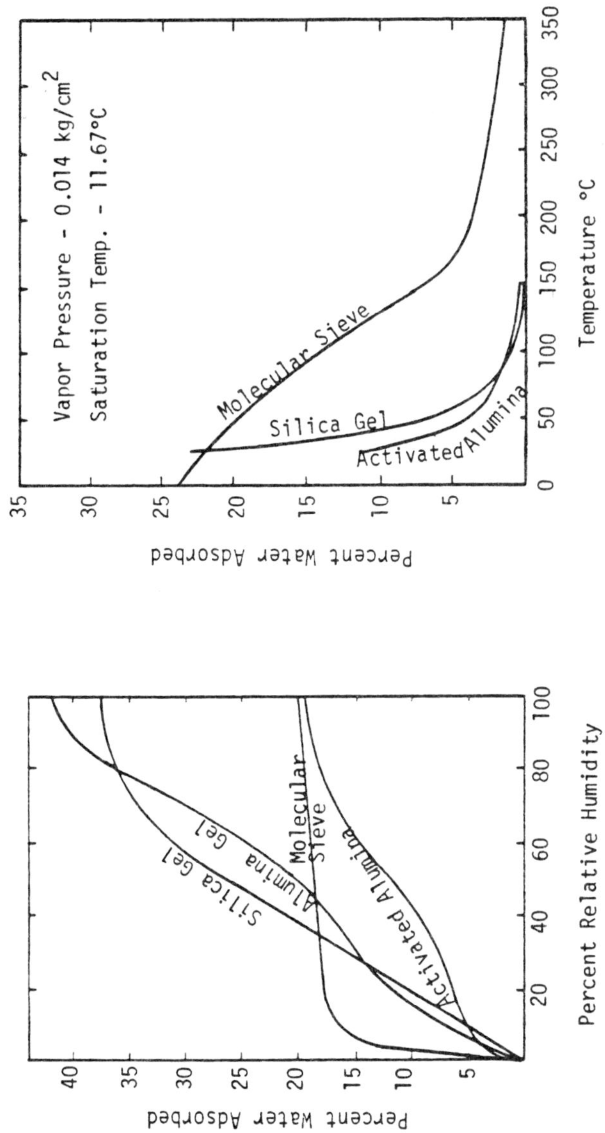

a) Adsorption vs Relative Humidity

b) Adsorption vs Temperature

Fig. 1. Water Vapor Equilibrium Curves for Desiccants

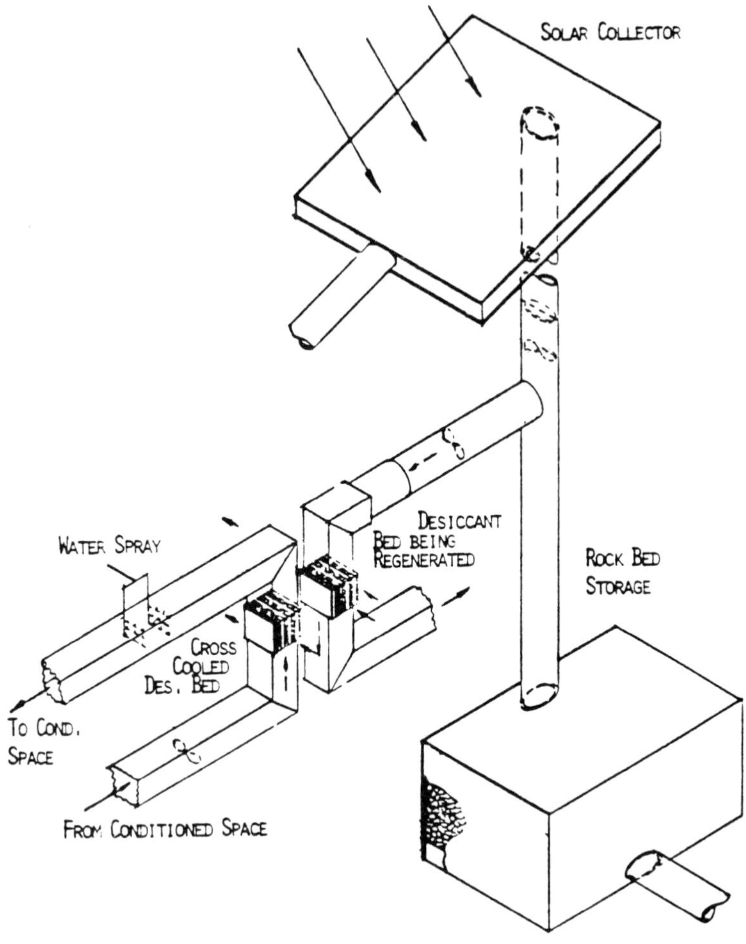

Fig. 2. Schematic of Solar Powered Desiccant Cooling System

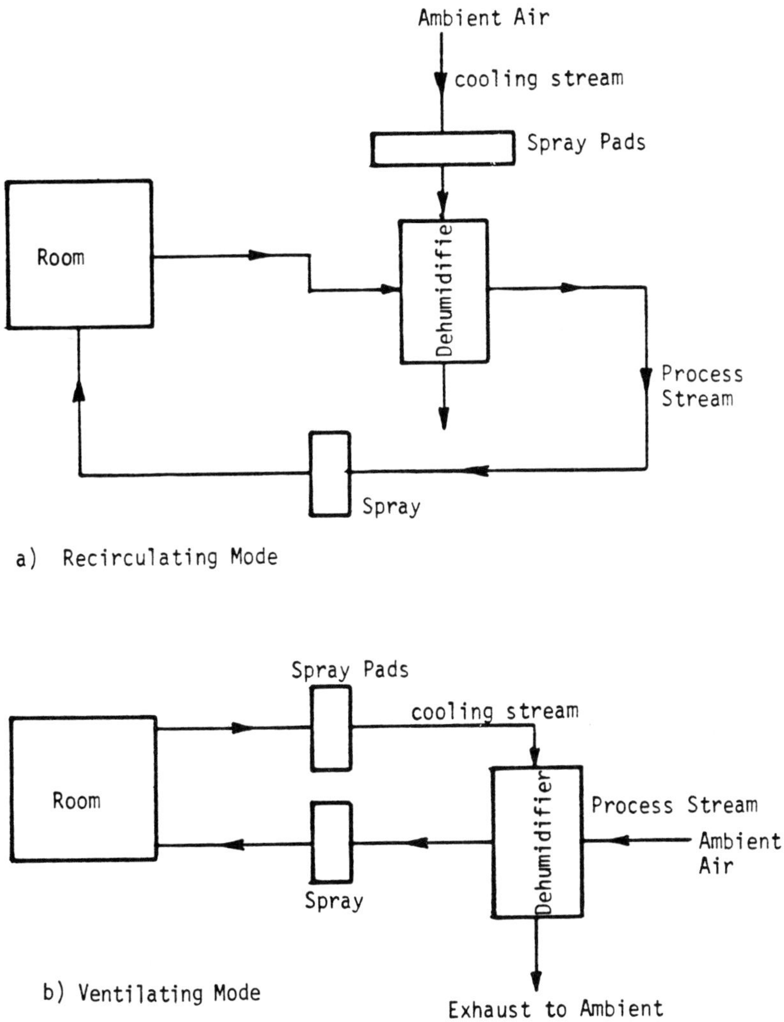

Fig. 3. Schematic of Operation In Circulating and Ventilating Modes

low temperatures. Figure 1 shows that the equilibrium water capacity of silica gel at say 82°C is very small. Hence silica gel dehumidifiers can utilize simple flat plate collectors for regeneration. However, silica gel has good adsorption capacity only at low temperatures. The adsorption capacity decreases rapidly as the bed temperature increases. It is therefore necessary to keep the adsorbent at a low temperature during dehumidification by removing heat.

DESCRIPTION OF THE SYSTEM

The present system consists of a cross-cooled dehumidifier constructed of paper-like sheets of silica gel. Detailed calculations have shown that regeneration temperatures below 82°C [1] (in conjunction with cross-cooling) results in sufficient dehumidification for air conditioning applications.

Figure 2 shows a schematic of our proposed solar energy cooling system. It consists of two dehumidifiers of equal size. One removes humidity from the air while the other is being regenerated with solar energy. The air leaving the cross-cooled dehumidifier working in the dehumidification mode is dry and only moderately warm. Cooling is achieved by evaporation of water, shown as a water spray in figure 2. In the adsorbing cycle the air from the conditioned space is passed through the process channels of the dehumidifier. The cross-cooling is accomplished either by ambient air or by room air. In the desorbing cycle the saturated bed is regenerated by hot air from flat plate solar collectors. A rock bed storage system is provided for storing the excess collected energy. There is also the possibility of storing excess cooling capacity in a rock bed storage.

The system can operate in a recirculating or in a ventilating mode as shown in figure 3. In the recirculating mode the air from the room is passed through the process channels, which are lined with silica gel sheets, to achieve the required dehumidification. The cross-cooling stream is humidified ambient air. The process stream leaving the dehumidifier is sprayed with water before being admitted to the conditioned space. In the ventilating mode, the ambient air is passed through the process channels and humidified room air is passed through the cross-cooling channels.

MANUFACTURING AND TESTING OF THE SILICA GEL SHEETS

Sheet Manufacturing Procedure

The silica gel sheet making process involves blending of micron sized silica gel with 200% of previously ground ammonium bicarbonate and about 5% Teflon in an adequate quantity of Shelsol. The resulting paste is filtered to form a "cake" which is physically worked by roller-reduction to form a web structure of Teflon. Sheets of silica gel of the required thickness are produced in the final reduction. The wet sheets are air dried overnight and the ammonium bicarbonate is decomposed by heating the sheet at 140°C for half an hour.

Successful silica gel sheets were made using Davison Syloid 63 nine micron size particles as the starting material, [2]. The dimensions of the first sheet tested were 39.68 cm by 7.3 cm by 1.52 mm thick. The weight of the sheet was 25.1 grams and it contained 4.7% Teflon. An electron microscopic examination of the sheet showed that the nine micron silica gel particles were held by means of thin Teflon strands forming a net like structure.

Figures 4(a) and 4(b) show the micron size silica gel powder and the manufactured sheet at a magnification of 1000. The appearance of the silica gel particles are identical in both cases. The size distribution is fairly wide. At this magnification the Teflon binder is almost indiscernable.

Figures 4(c) and 4(d) show the "cake" (wet mixture of silica gel and Teflon powder) before rolling and the final sheet at a magnification of 5000. The Teflon web can be clearly seen in 4(b). It is not present in the "cake" since the web is formed by the rolling process.

Desiccant Sheet Test Section

The performance of sheets of silica gel is studied isothermally in a parallel plate sorber shown in figure 5. The sheet is held in a shallow depth pan over which a mixture of air and water vapor is passed. The pan and the sheet form a part of a rectangular channel which has three functional sections: entrance and preheat, sorption, and exit sections. The channel is equipped with an inlet sampling system and an outlet mixing chamber so that inlet and outlet humidities could be continuously measured. A dew point hygrometer is used to measure humidities. The sorption section can be maintained at a constant temperature by electric heaters or cooled by air or water. Previously Onischak and Gidaspow

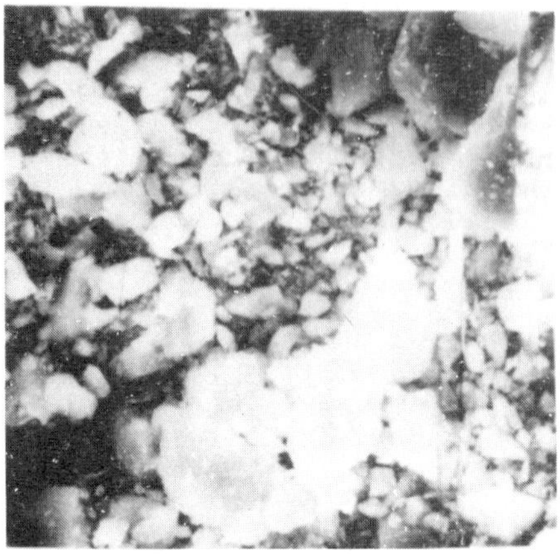

a) Silica Gel Powder 1000X

b) Silica Gel Sheet 1000X

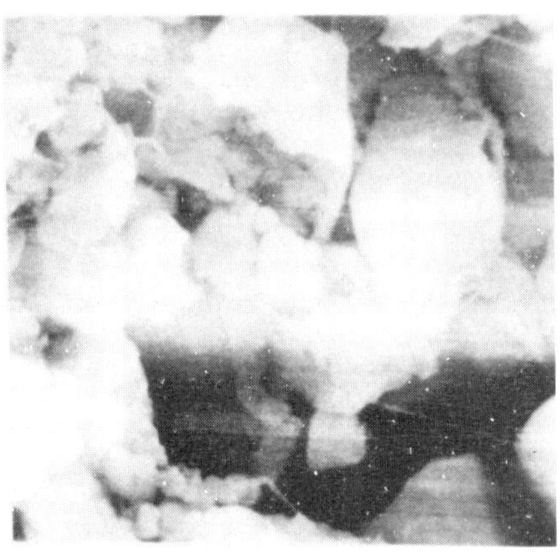

c) Silica Gel and Teflon Before Rolling 5000X

d) Silica Gel Sheet 5000X

Fig. 4. Electron Microscope Photographs of the Desiccant

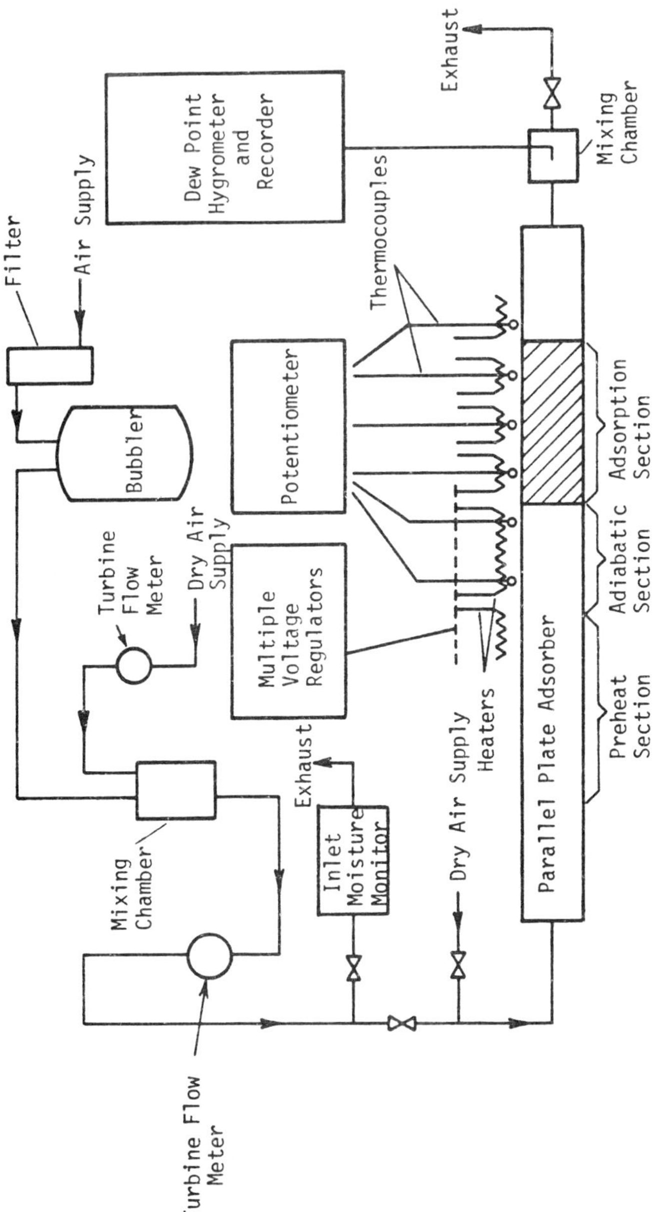

Fig. 5. Schematic Diagram of Adsorber Test Setup

used such a setup to study adsorption of water by anhydrous lithium chloride [3] and the kinetics of the removal of carbon dioxide [4]. The latter paper gives a more complete description of the apparatus and the method of data analysis. The cross section of the channel used here is 7.62 cm by 3.18 mm. The dimensions of the plate supporting the silica gel sheet are 39.68 cm by 7.62 cm.

Performance of Silica Gel Sheet

Two thicknesses of silica gel sheets were tested. Data from the thicker sheet (1.52 mm) is shown in figures 6-9 and data from the thinner sheet (0.71 mm) is given in figures 10-13.

Figure 6 shows a plot of the outlet humidity for an inlet dew point of air of 16.7°C for several air velocities with sorption occurring at room temperature. A thermocouple imbedded in the plate holding the sheet showed only a small rise in temperature due to the large heat capacity of the plate. The dimensionless outlet concentration indicated in figure 6 is the outlet humidity divided by the inlet humidity. Data of the type shown in figure 6 are needed for a systematic analysis of the sheet performance. These types of calculations for related systems were illustrated by Gidaspow [5] for analysis of reaction rate data in channels and in greater detail by Onischak and Gidaspow [4]. The present data will allow us to determine the controlling mechanisms needed for a rational dehumidifier design and scale-up. Channel widths and sheet thicknesses are important variables to be optimized. These variables determine pressure drops and sheet areas needed for a given cooling load.

Figure 6 shows that at low air velocities nearly dry air leaves the channel for a long time. There is only a slow rise in the humidity at an air velocity of 0.338 m/sec. At the intermediate flow rate a characteristic sygmoid type curve is obtained. Duplicate runs were made at each flow rate. Figure 7 shows the calculated water removed as a function of time. The final water removed was 0.325 g water/g silica gel. This value is only 21% smaller than the reported value for silica gel nine micron particles [2] for a relative humidity of 77%. It is felt that the loss in equilibrium capacity is small and that this sheet appears to be a practical candidate for use in the construction of a small cooling unit.

Figure 8 shows a calculation of water removed as a function of surface temperature. There is the expected drop in capacity with temperature. Figure 9 shows the dimensionless outlet humidity as a function of time for two different inlet humidities. As expected, at the low inlet dew point little

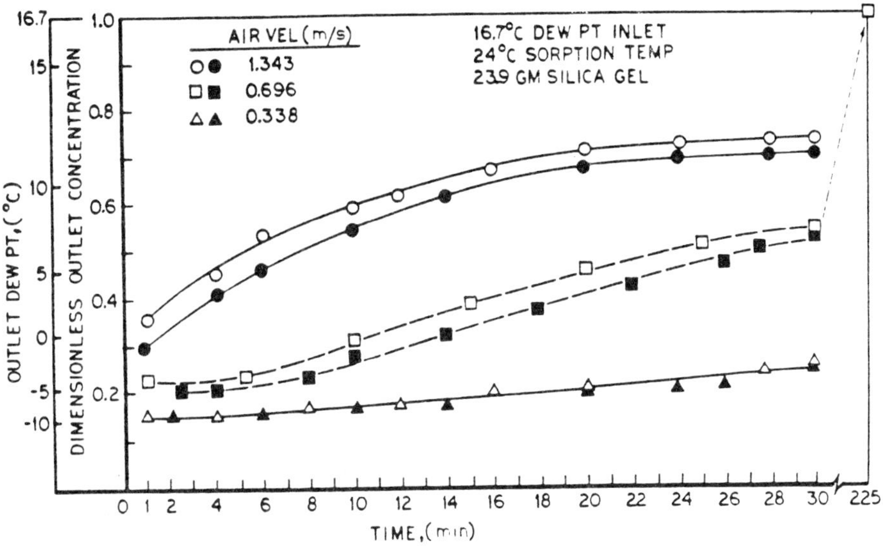

Fig. 6. Outlet Concentration and Dew Point vs Time (Silica Gel Sheet 1.52 mm thick, 39.68 x 7.3 cm)

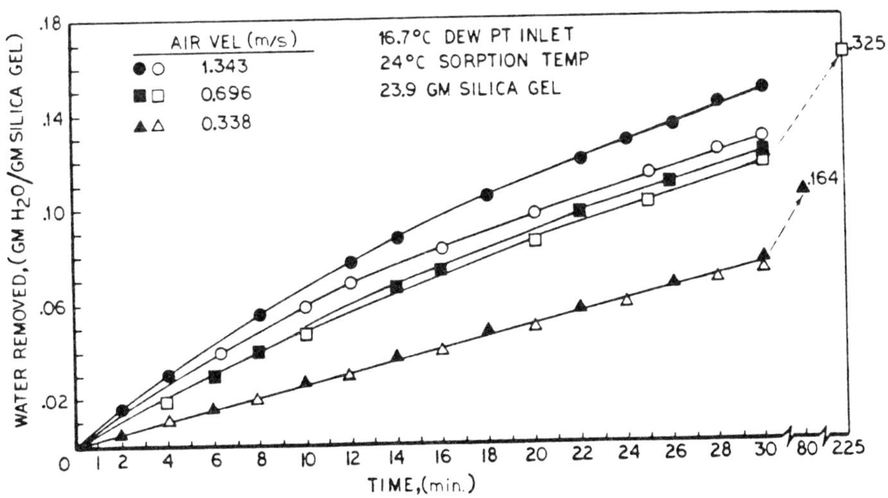

Fig. 7. Water Removed vs Time (Silica Gel Sheet 1.52 mm thick, 39.68 x 7.3 cm)

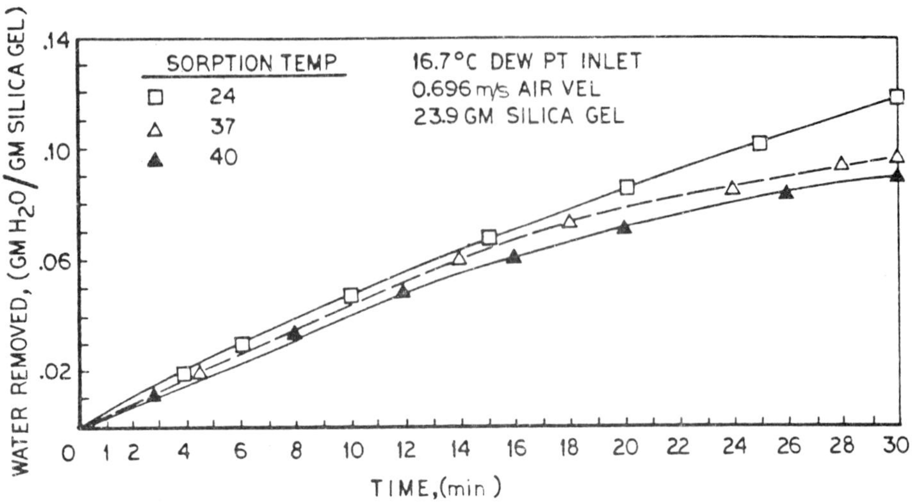

Fig. 8. Water Removed vs Time (Silica Gel Sheet 23.9 Gm, 1.52 mm Thick, 39.68 x 7.3 cm)

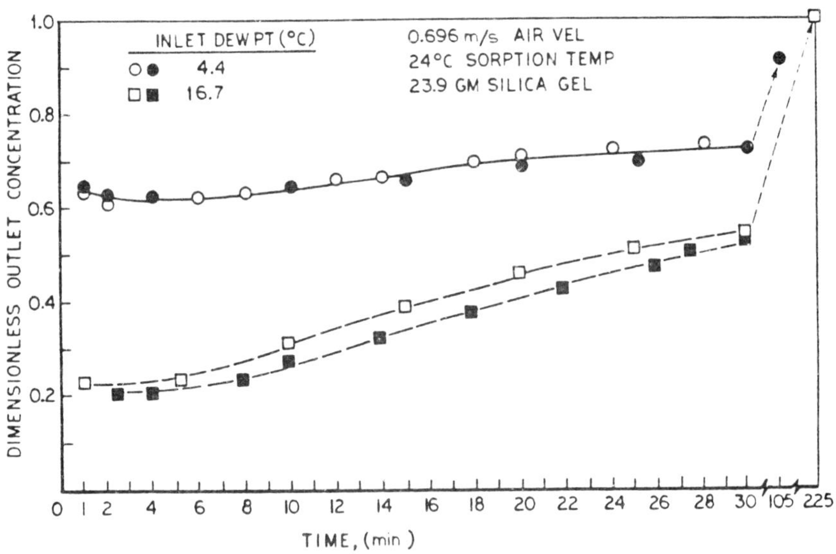

Fig. 9. Outlet Concentration and Dew Point vs Time (Silica Gel Sheet 23.9 Gm, 1.52 mm Thick, 39.68 x 7.3 cm)

water is removed.

Figures 10 and 11 show data for the thin silica gel sheet at several air velocities. Figure 10 gives the outlet concentration as a function of time and figure 11 shows the amount of water removed versus time.

The data shown in figures 10 and 11 were analyzed to give wall concentrations, rates of diffusion into the sheet and mass transfer Nusselt number as a function of contact time, x. Figure 12 shows the calculated wall and bulk concentrations as a function of contact time and figure 13 gives the calculated Nusselt numbers. These calculated Nusselt numbers are close to the theoretical asymptotic Nusselt number for a constant flux boundary condition on one wall and zero flux at the other. More data of this type are being taken and analyzed for various sheet thicknesses to determine the mass transfer role played by the silica gel sheet.

It appears that at the same inlet temperature, inlet dew point and sorbent temperature, the rate of adsorption depends in a complex manner on the sheet thickness. This suggest that the internal resistance to mass transfer must be considered along with the film resistance.

MATHEMATICAL MODEL

The cross-cooled dehumidifier is represented schematically in figure 14 which shows the coordinates and the dimensions for a typical unit of the dehumidifier. The sides of the walls of the process stream channel are made of silica gel. The performance of such a dehumidifier was studied analytically by Roy and Gidaspow [1].

The cooling stream, flowing in the channel perpendicular to the process stream, exchanges heat only with the solid wall separating them. The rate of exchange of heat or mass between the wall and the fluid streams can be calculated from the Nusselt and Sherwood numbers which are assumed to be constant in this analysis. Such assumptions are usually made in the literature. (See for example, Dunkle, [6], Chi and Wasan [7], Banks, et al, [8].)

Material Balance

The mass balance of the species in the process stream and in the solid, a fraction, f, of which is made up of the relative

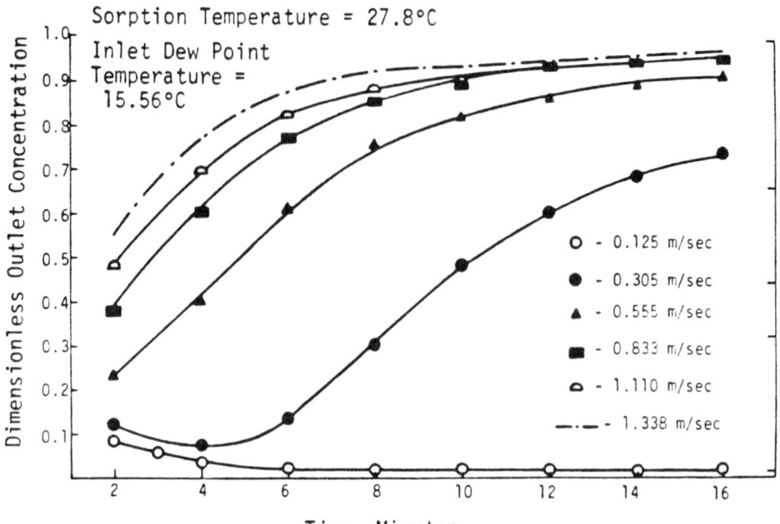

Fig. 10. Outlet Cencentration and Dew Point vs Time at Several Flow Rates, (Silica Gel Sheet 8.686 Gm, 0.71 mm Thick, 39.7 x 7.62 cm)

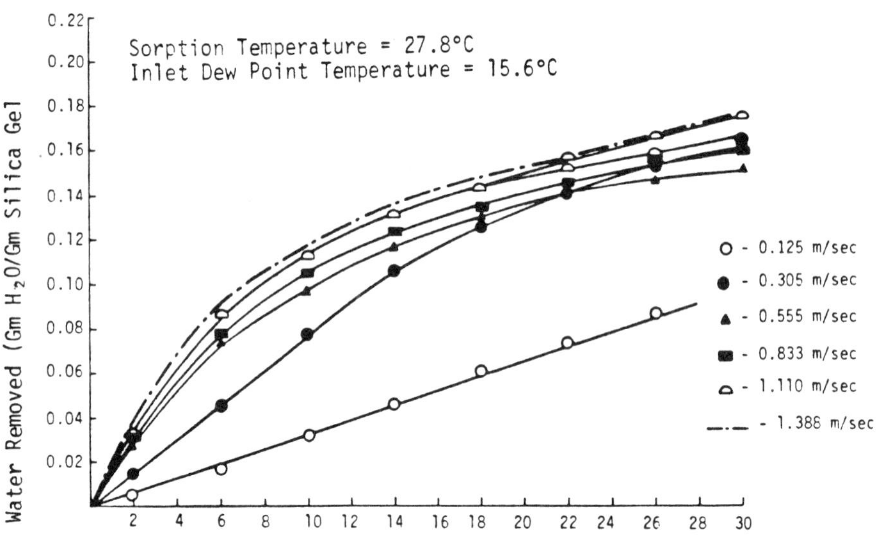

Fig. 11. Water Removed vs Time (Silica Gel Sheet 8.686 Gm, 0.71 mm Thick, 39.7 x 7.62 cm)

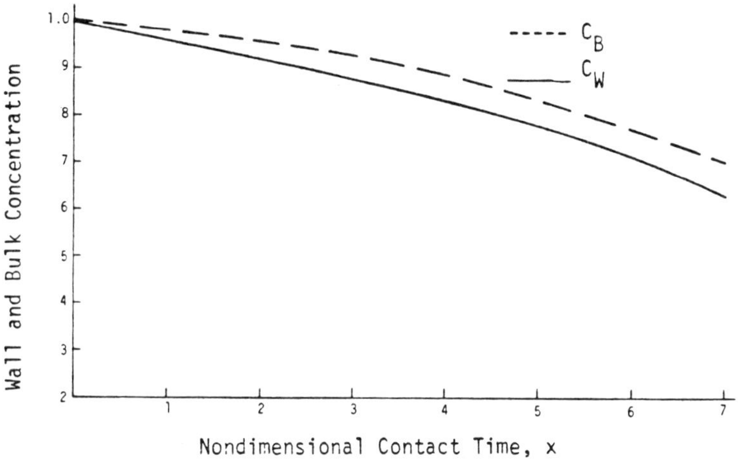

Fig. 12. Wall and Bulk Concentration vs Contact Time, (Silica Gel Sheet 8.686 Gm, 0.71 mm Thick, t = 28 min)

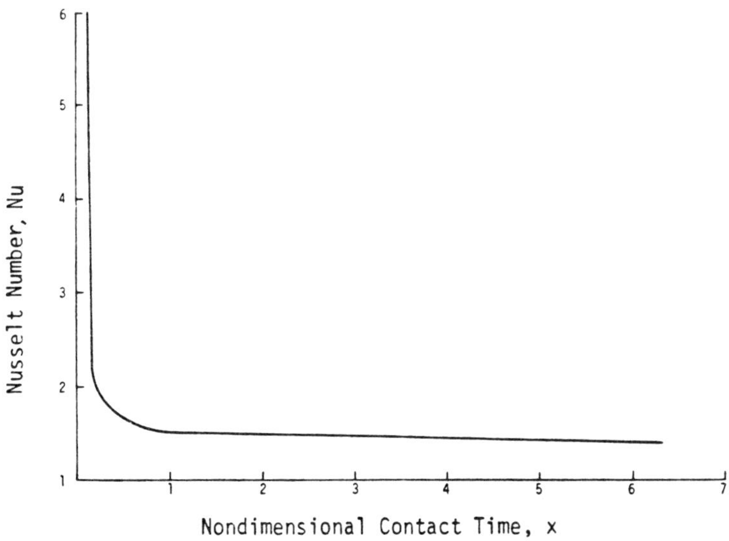

Fig. 13. Mass Transfer Nusselt Number vs Nondimensional Contact Time, (Silica Gel Sheet, 8.686 Gm, 0.71 mm Thick, t = 28 min)

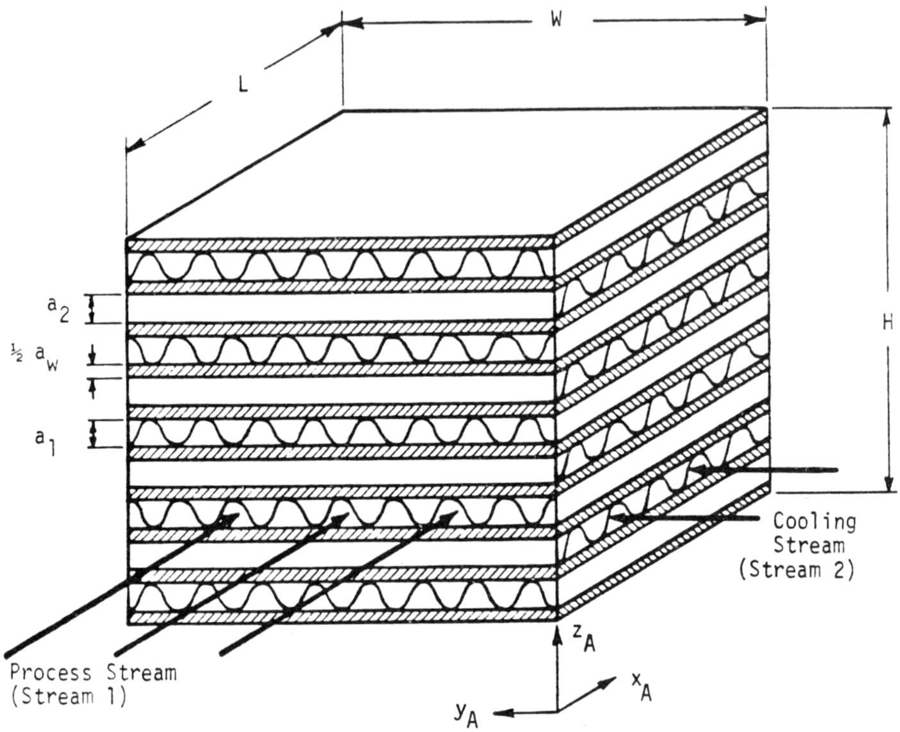

Fig. 14. Schematic of Cross-Cooled Dehumidifier

substance, can be formulated by the following equations.

Process stream

$$a_1\rho_1\left[\frac{\partial Y}{\partial t_A} + u_1\frac{\partial Y}{\partial x_A}\right] = K_Y(Y_w - Y) \qquad (1)$$

Wall

$$fa_w\rho_w\frac{\partial w}{\partial t_A} = K_Y(Y - Y_w) \qquad (2)$$

where K_Y is the mass transfer coefficient for the transfer of the species from the fluid to the solid and Y_w is the concentration in the fluid stream at the fluid-solid interface. Later we will assume thermodynamic equilibrium at the fluid-solid interface which will enable us to relate Y_w to the concentration of the species in the solid and the temperature of the solid at this point.

Heat Balance

The heat balance for the process and cooling streams, and for the wall can be formulated by the following system of equations.

Process stream

$$a_1\rho_1c_1\left[\frac{\partial T_1}{\partial t_A} + u_1\frac{\partial T_1}{\partial x_A}\right] + h_1(T_1 - T_w) = 0 \qquad (3)$$

Cooling stream

$$a_2\rho_2c_2\left[\frac{\partial T_2}{\partial t_A} + u_2\frac{\partial T_2}{\partial y_A}\right] + h_2(T_2 - T_w) = 0 \qquad (4)$$

Wall

$$a_w\rho_wc_w\frac{T_w}{\partial t_A} + h_1(T_w - T_1) + h_2(T_w - T_2) = QK_Y(Y - Y_w) \qquad (5)$$

In writing the heat balance for the wall we have assumed the heat effect associated with the sorption process to be denoted by Q, the conductive transport in the flow directions to be negligible, and the conductivity in the transverse directions, z_A, to be very high. These are the usual underlying assumptions in formulating the regenerator problem.

Equilibrium Relationship

The equilibrium relationship for the silica gel desiccant sheets gives the water concentration in air at the wall as a function of the wall temperature and the water concentration in the sorbent.

$$Y_w = Y_w(T_w, w) \tag{6}$$

DESIGN AND PERFORMANCE OF DEHUMIDIFIER

Computational Procedure

The system of equations (1) through (6) is solved using the method of Green's functions. The differential equations were converted into a system of nonlinear integral equations using the previously determined (Roy and Gidaspow, [9]) Green's function for a cross-flow regenerator. Simpson's rule was used to approximate integrals. A block by block integration technique was used. This method offers a considerable reduction in computer time and thus permits a systematic parametric study of the cross-cooled dehumidifier.

The input to the computer program is x, y, b, t, physical properties and initial and inlet temperatures and concentrations of the wall and the two streams. The output is Y, T_1, T_2 and T_w. The program also calculates the moisture gain by the desiccant and the moisture loss by the process stream. These two were found to be in excellent agreement.

The computer model described above permits a sufficient parametric study to assure optimum design. Figure 15 shows typical numerical results for various degrees of cross-cooling. It is clear that cross-cooling offers a considerable advantage over an adiabatic dehumidifier. The performance of the cross flow dehumidifier improves with increasing b. However large values of b imply large pressure losses as can be seen from figure 16 which will be discussed later in detail. One should therefore refrain from using very large values of b. Fortunately the difference in performance between say

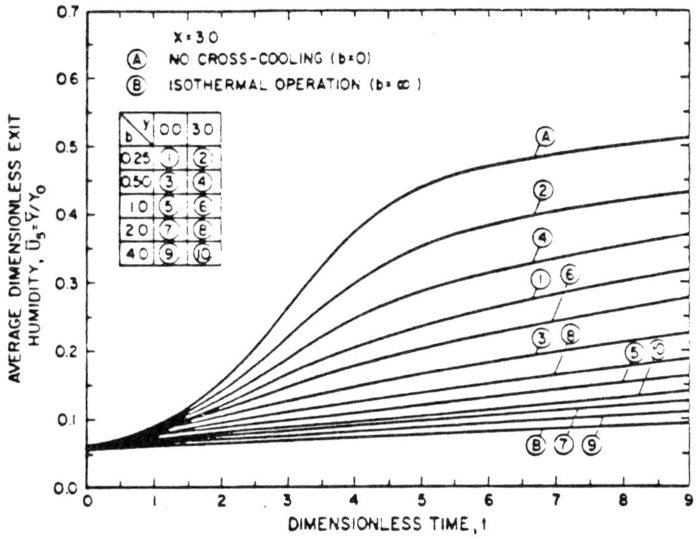

Fig. 15. Effect of Cross-Cooling on the Average Exit Humidity for the Cold Bed Operation(T_{wi} = 26.7°C) of the Silica Gel-Air-Water System

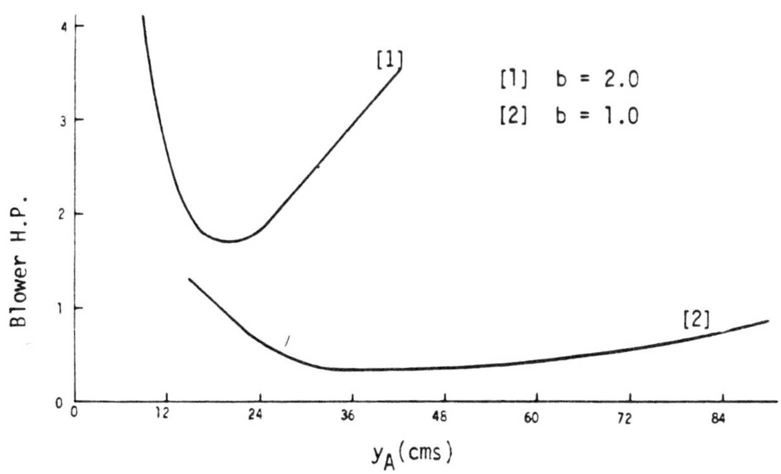

Fig 16. Blower Hp vs y_A (x = 3.0, $a_w = a_1$ = 0.132 mm, $\dot{V}_1$ = 27.5 m^3/min, V_s = 0.08 m^3)

$b = 1$ and $b \to \infty$ is not very large.

Short Time Performance

We chose $b = 1$ and 2 as reasonable values and calculated the performance for short time at various conditions and modes of operation. Room conditions were 23.9°C and 50% R.H and the ambient was taken as 32.2°C and 30%, 50% and 70% R.H.

These runs suggest that the cooling loads can be met with $x = y = 3$ and $b = 1$ or 2 for operation in the recirculating mode at all ambient conditions. For operation in the ventilating mode the cooling load is met when the ambient R.H is less than say 50%. Figures 17 and 18 serve as examples of the short time performance calculations.

Optimal Design of the Cross-Cooling Dehumidifier

For given dimensional variables x, y, b and t various combinations of dimensional parameters can be selected. Our approach is to first establish the physical constraints imposed by the problem and then to determine the remaining parameters which minimize the pressure drop.

An optimally designed dehumidifier should satisfy the following criteria: 1) The dehumidifier should meet the required cooling load over a period of time, t. This condition is ensured by proper choice of x, y, and b. 2) The overall dimensions of the dehumidifier should be reasonable. 3) The power requirements to overcome the pressure drop in both process and cross-cooling channels should not be excessive. In the optimization study we assume that a_w, a_1, $\dot{V}_1$, H, x, y and b are known. We seek to find x_A, y_A, u_1 and u_2 which minimize the pressure drop using the process stream volume flow rate and the silica gel mass as constraints. The latter lead to

$$\dot{V}_1 = \frac{H}{a_w+a_1+a_2} y_A a_1 u_1 \qquad (7)$$

$$V_s = n x_A y_A a_w \qquad (8)$$

We also use the definitions of the nondimensional parameters x, y and b

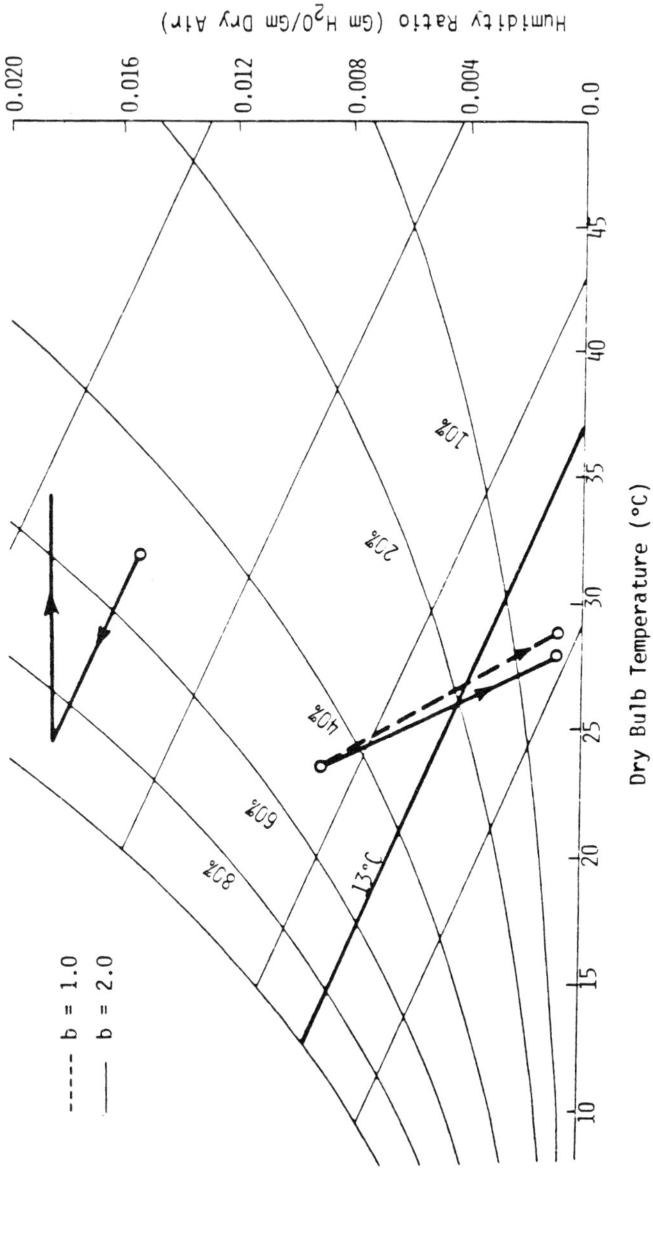

Fig. 17. Short Time Performance of Recirculating Mode ($x = 3.0$, $y = 3.0$, $t = 1.0$, $T_{wi} = 26.7°C$, $T_{DB\infty} = 32.2°C$, $R.H._{\infty} = 50\%$)

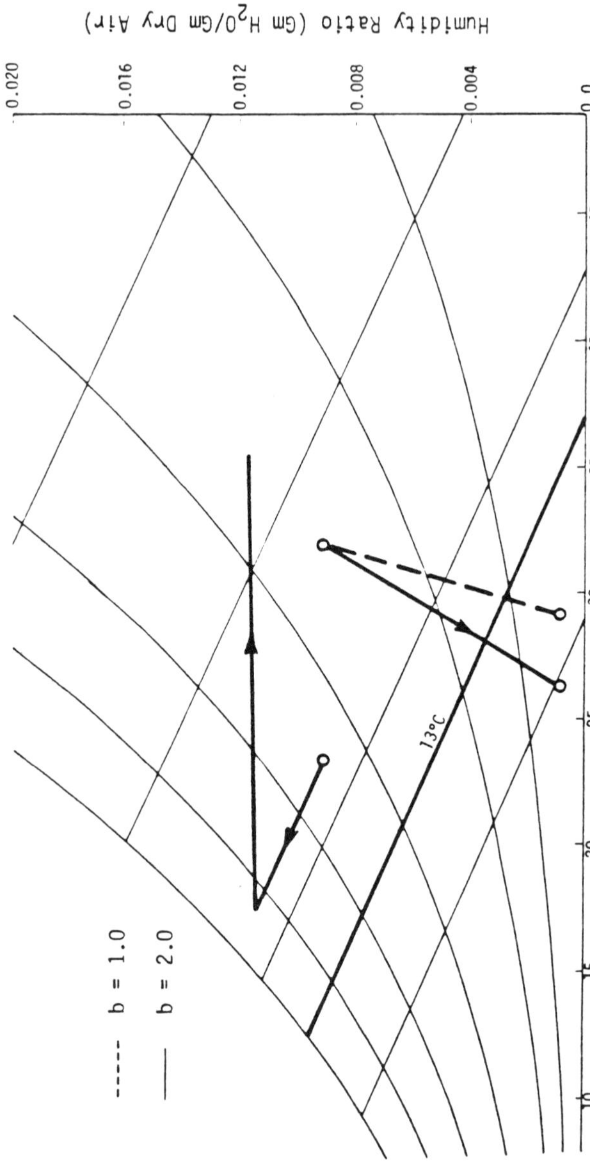

Fig. 18. Short Time Performance of Ventilating Mode (x = 3.0, y = 3.0, t = 1.0, T_{wi} = 26.7°C, T_{DB_∞} = 32.2°C, R.H.$_\infty$ = 30%)

$$x = \frac{h_1 x_A}{a_1 \rho_1 c_1 u_1} \tag{9}$$

$$y = \frac{h_2 y_A}{a_2 \rho_2 c_2 u_2} \tag{10}$$

$$b = \frac{h_2}{h_1} \tag{11}$$

Assuming the flow in both channels to be laminar and fully developed, the expression for the total pressure drop in rectangular channels can be written as

$$\Delta p = \frac{12}{a_1^{\,2}} \left[\mu_1 x_A u_1 + y_A u_2 \mu_2 b^2\right] \tag{12}$$

Using equations (9 - 11), we can express x_A, u_1 and u_2 as functions of y_A. Thus Δp is a function of y_A only. To minimize Δp set

$$\frac{d(\Delta p)}{dy_A} = 0 \tag{13}$$

to get

$$(y_A)_{opt} = \left[\frac{\mu_1 yx}{b\mu_2}\right]^{1/4} \frac{\dot{V}_1}{H} \frac{a_1}{r}\left[(a_w + a_1 + a_2)\right]^{1/2} \tag{14}$$

The blower electrical power requirement to overcome the pressure drop in the dehumidifier is

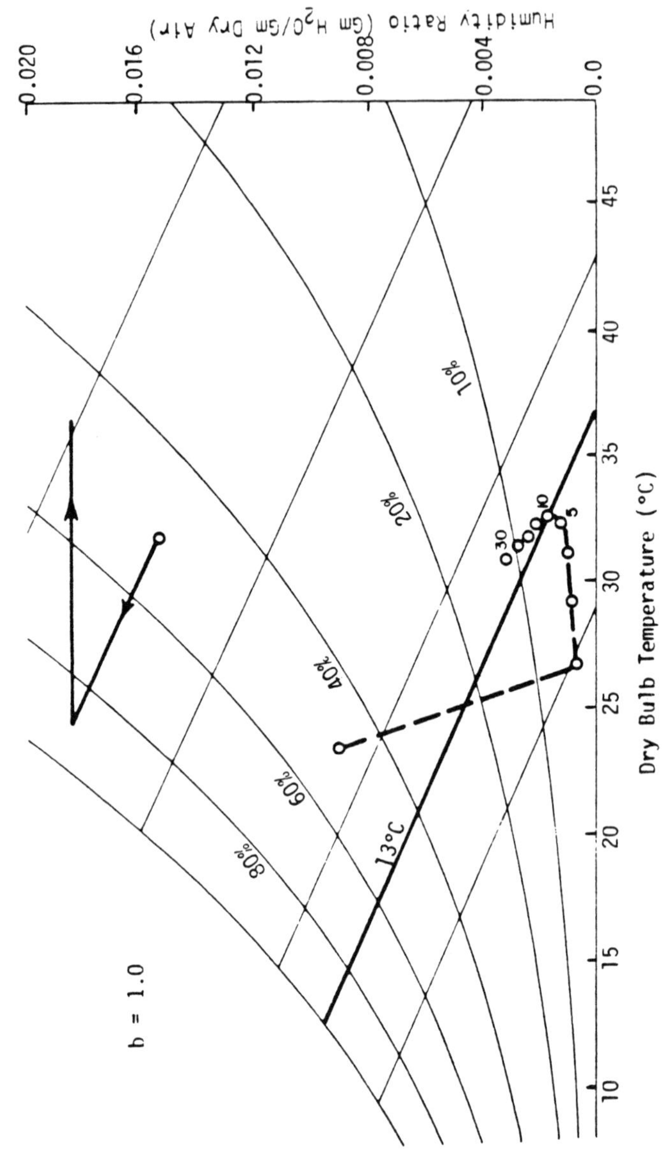

Fig. 19. Long Time Performance of Recirculating Mode (x = 3.0, y = 3.0, t = 1.0, T_{wi} = 26.7°C, $T_{DB\infty}$ = 32.2°C, R.H.$_{\infty}$ = 50%)

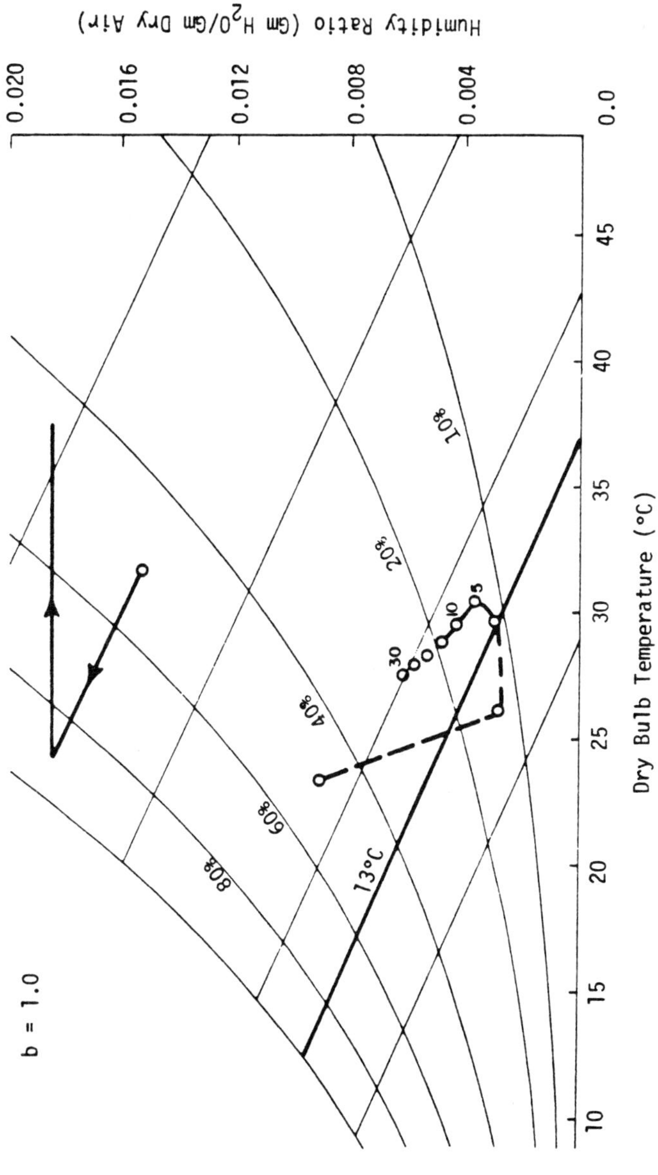

Fig. 20. Long Time Performance of Recirculating Mode ($x = 1.2$, $y = 1.2$, $t = 30.0$, $T_{wi} = 26.7°C$, $T_{DB\infty} = 32.2°C$, $R.H._{\infty} = 50\%$)

$$P_{electrical} = \frac{\dot{V}_1(\Delta p)_1 + \dot{V}_2(\Delta p)_2}{\eta_{Blower}} \qquad (15)$$

Figure 16 shows horsepower requirement as a function of y_A for b = 1 and 2. It is clear that b = 2 implies excessive power requirements hence b = 1 was chosen for subsequent long time calculations.

Long Time Calculations

The computer program was next run to predict the dehumidifier performance for b = 1 and t = 30.0. For the parameter chosen nondimensional time of 30 is approximately 30 minutes. Calculations were performed for operation in both the recirculating and ventilating modes. Room conditions for both cases are 23.9°C and 50% R.H.

Recirculating Mode

For the recirculating mode we chose the ambient as 32.2°C and 50% R.H. The results for these runs are shown in figures 19-21. Figure 19 shows the state of the air on a psychrometric chart after leaving the dehumidifier for x = y = 3. It is clear that on the average over 30 minutes the dehumidifier meets the cooling load as designated by the 13.3°C wet bulb temperature line. However for x = 1.2, y = 1.2 and b = 1.0, figure 20 shows that the dehumidifier does not meet the cooling load over a 30 minute period. Figure 21 shows the cooling capacity in tons as a function of nondimensional time, t, for different volume flow rates. 38.2 m^3/min meets the design cooling capacity of two tons. Figure 22 shows the moisture removed from process stream during dehumidification and the moisture removed from the desiccant during desorption. Air at 82°C (from solar collectors) is used to regenerate the silica gel.

Ventilating Mode

As an example of operation in the ventilating mode we chose ambient conditions of 32.2°C and 30% R.H as the starting state for the process stream. Figure 23 shows the state of the process stream leaving the dehumidifier during 30 minutes of operation. Figure 24 shows the cooling capacity versus time for this case. Curve (a) represents the conventional cooling load and curve (b) is the cooling load for a situation in which 100% fresh air is a requirement.

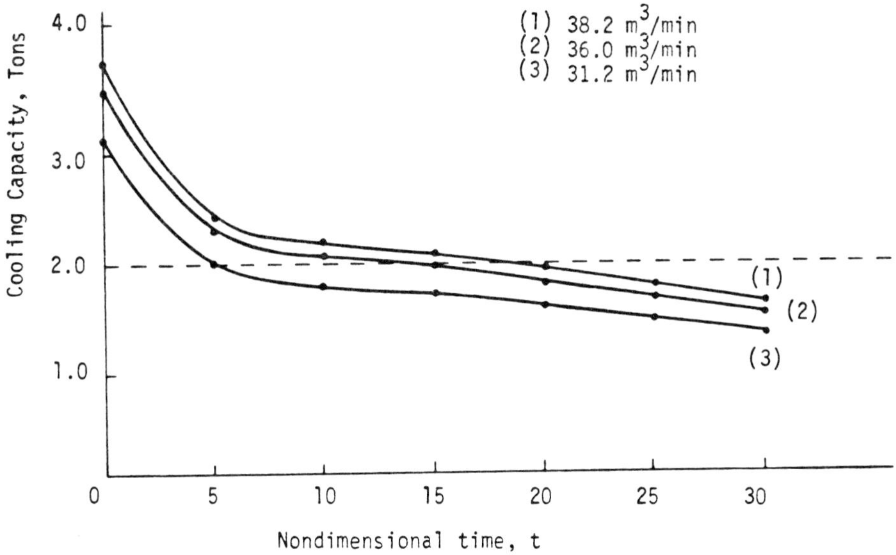

Figure 21. Recirculating mode performance vs time at Different Flow rates ($x = 3.0$, $y = 3.0$, $b = 1.0$, $T_{DB\infty} = 32.2°C$, R.H.$_{\infty}$ = 50%)

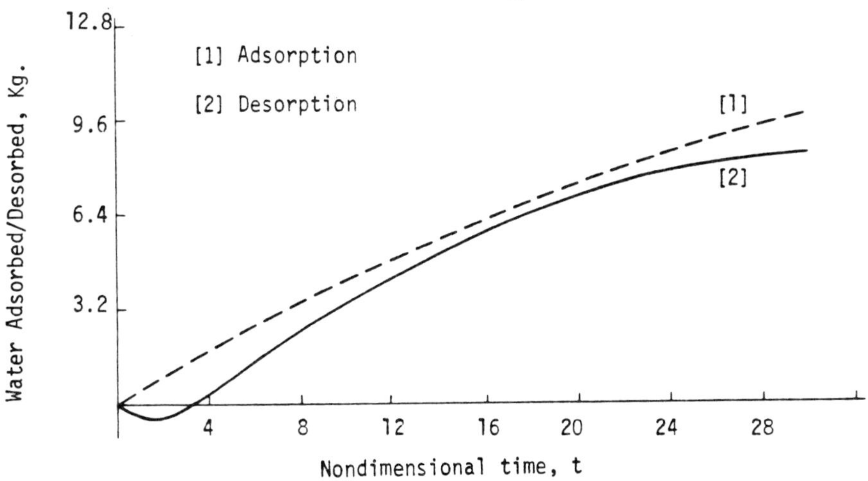

Fig. 22. Water Adsorbed/Desorbed vs Time ($x = 3.0$, $y = 3.0$) For Adsorption $b = 1.0$, $T_i = 23.9°C$, $Y_i = 0.009$, $T_{wi} = 26.7°C$, $w_i = 0.02$. For Desorption $b = 0.0$, $T_i = 82.2°C$, $Y_i = 0.0152$, $T_{wi} = 31.1°C$, $w_i = 0.08355$

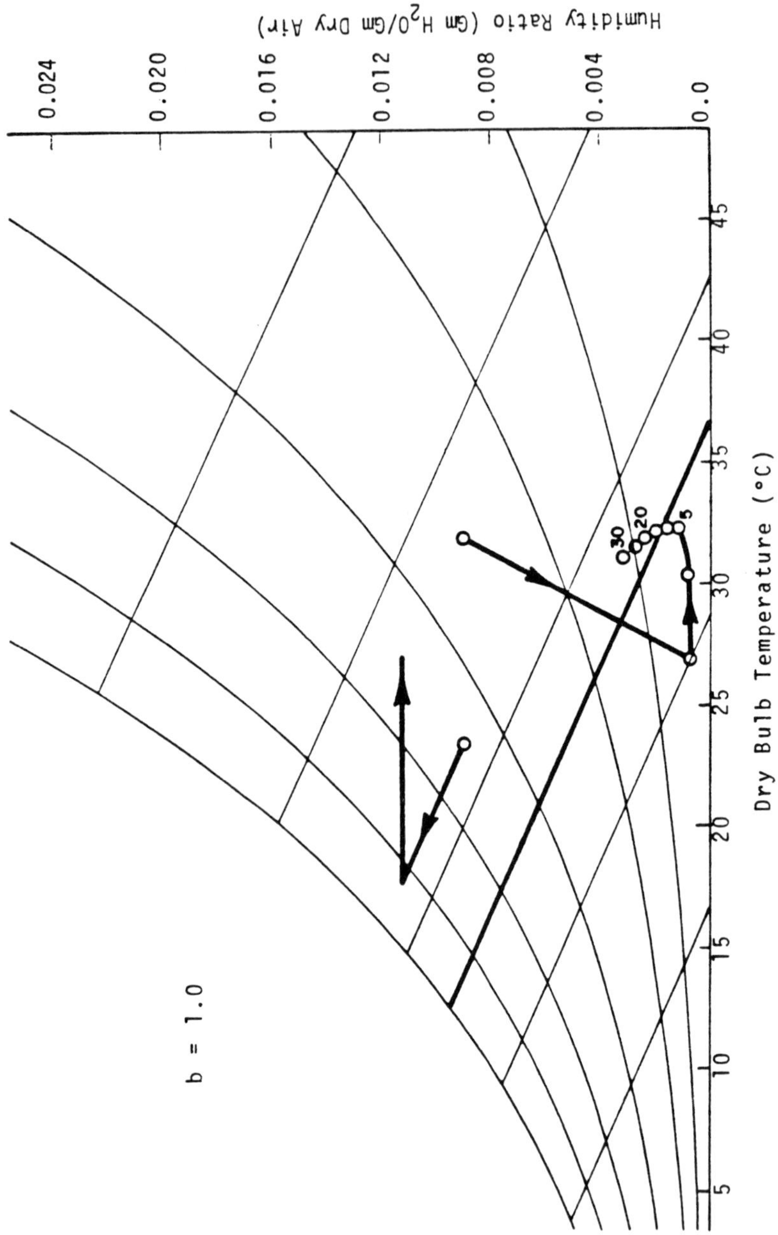

Fig. 23. Long Time Performance of Ventilating Mode (x = 3.0, y = 3.0, t = 30.0, T_{wi}, 26.7°C, T_{DB_∞} = 32.2°C, R.H$_\infty$ = 30%)

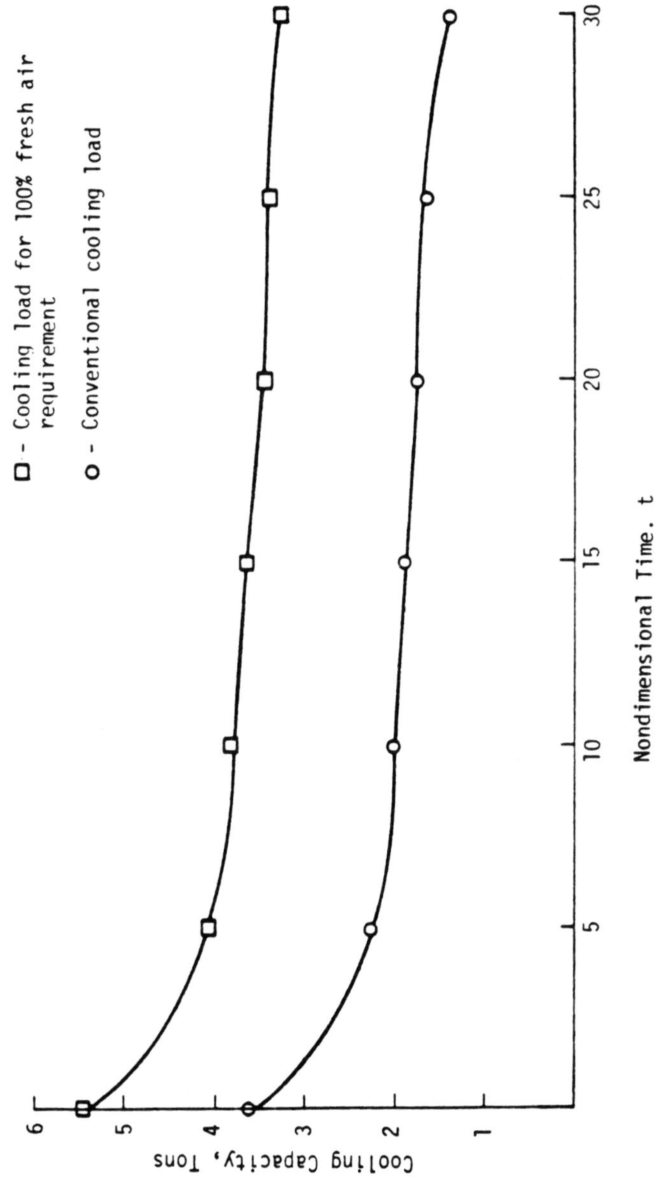

Fig. 24. Cooling Capacity in Ventilating Mode (x = 3.0, y = 3.0, b = 1, $T_{DB\infty}$ = 32.2°C, $R.H._{\infty}$ = 30%, Flow Rate = 38.2 m^3/min)

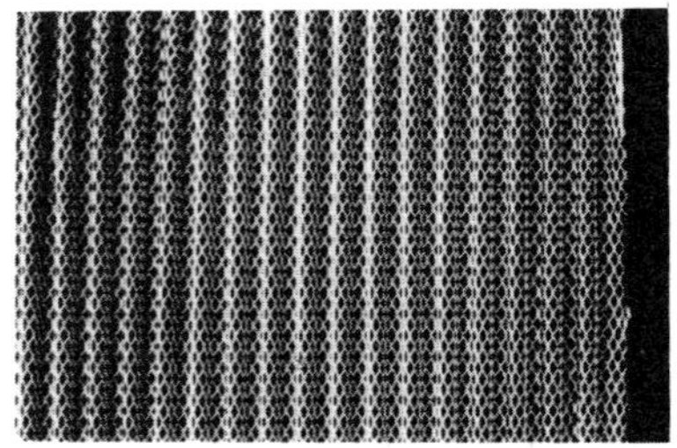

a) Corrugated Expanded Aluminum

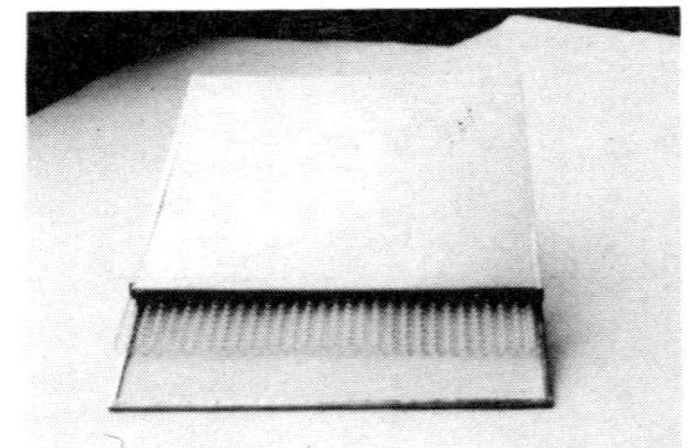

b) Process Stream Flow Channel

c) Enlarged Front View of Process Stream Channels

d) Dehumidifier Prototype

Fig. 25. Dehumidifier Prototype Showing Overall View and Components

THE CROSS-COOLED DEHUMIDIFIER

Preliminary calculations for a two ton unit suggest a 0.6 m x 0.6 m x 1.5 m dehumidifier with 1 mm thick silica gel sheets and 2 mm channel width. The air velocities are about 2 m/s and 0.4 H.P. are expended by the blowers to overcome the pressure drop in the process and cross-cooling channels. Cycling time is 30 minutes and regeneration is achieved with 82°C air. The electric coefficient of performance is over 24 and the thermal COP is between 0.35 and 0.60 depending on the extent of energy recovery. The thermal COP is nearly twice as large if 100% fresh air is a requirement.

Dehumidifier Prototype

A scaled down prototype of the cross-cooled dehumidifier was constructed and is presently being tested. The prototype, 15 cm x 15 cm x 15 cm is shown in figure 25(d). The process stream channels shown in figure 25(b) consist of two silica gel sheets 0.71 mm thick bonded to thin aluminum plates. The silica gel sheets face each other and are separated by a supporting corrugated structure. The latter, shown in figure 25(a), was formed by running thin expanded aluminum through a pinion wire assembly. The thin aluminum plates which are bonded to the silica gel sheets provide barriers between the process and the cooling streams while offering very little thermal resistance. The cooling stream channels are formed by placing corrugated expanded aluminum (running perpendicular to those in the process stream) between adjacent aluminum plates. The entire cross-cooling unit is placed in an aluminum sheet metal housing as shown in figure 25(d), and the edges of the housing openings are sealed with a caulking compound to prevent mixing of the process and cooling air streams. Figure 25(c) shows an enlarged front view section of the process stream channels.

NOMENCLATURE

a_1 - width of process channel

a_2 - width of cooling channel

a_w - thickness of walls in process stream

$b \equiv \frac{h_2}{h_1}$

C_W - concentration of water in process stream in equilibrium with wall

C_B - cup mixing concentration of water in process stream

c_1 - specific heat of process stream

c_2 - specific heat of cooling stream

f - fraction of silica gel in wall

H - dehumidifier height

h_1 - heat transfer coefficient of process stream

h_2 - heat transfer coefficient of cooling stream

K_y - mass transfer coefficient

L - length of process stream channels

Nu - ha/2k

n - number of process stream channels

P - power

p - pressure

Q - heat of sorption

$r = \frac{Nuk}{\rho c}$

T_1 - temperature of process stream

T_2 - temperature of cooling stream

T_w - wall temperature

T_{wi} - wall temperature at time t = 0

T_{DB_∞} - dry bulb temperature of ambient air

t - $h_1 t_A / a_w \rho_w c_w$

t_A - time

$\bar{U}_5$ - $\bar{Y}/Y_o$

u_1 - velocity of process stream

u_2 - velocity of cooling stream

$\dot{V}_1$ - volume flow rate of process stream

$\dot{V}_2$ - volume flow rate of cooling stream

V_s - volume of silica gel

w - concentration of water in sorbent

W - length of cooling stream channel

x $\equiv h_1 x_A / a_1 \rho_1 c_1 u_1$

x_A - distance in direction of process channel

Y - concentration of water in process stream

$\bar{Y} = \frac{1}{W} \int_o^W Y(x_A, y_A, t_A)\, dy_A$

Y_o = some reference concentration of water in process stream usually inlet concentration

Y_w - equilibrium concentration of water in air at wall

y $\equiv h_2 y_A / a_2 \rho_2 c_2 u_2$

y_A - distance in direction of cooling stream

z_A - distance at right angles to both cooling and process stream

ρ_1 - density of process stream

ρ_2 - density of cooling stream

ρ_w - density of the wall

μ_1 - dynamic coefficient of viscocity of process stream

μ_2 - dynamic coefficient of viscocity of cooling stream

REFERENCES

1. D. Roy and D. Gidaspow (1974): "Nonlinear Coupled Heat and Mass Exchange in a Cross-Flow Regenerator," Chem. Eng. Science, pp. 2101-2114, 1974.

2. Grace (1978): Davison Chemicals Division, Industrial Chemicals Department, Baltimore, MD 21203, Family of Syloid[R] Silicas at Work, 1978.

3. M. Onischak and D. Gidaspow (1969): "Reaction of Water with Anhydrous Lithium Chloride in a Parallel Plate Flow Reactor," Can. J. Ch.E., 47, pp. 98-99, 1969.

4. M. Onischak and D. Gidaspow (1972): "Regenerative Sorption of CO_2," Part II, "Separation of Gaseous Mixtures by Regenerative Sorption on Porous Solids," Recent Developments in Separation Science. Vol. II, Chemical Rubber Co., Cleveland Ohio, pp. 71-93, 1972.

5. D. Gidaspow (1971): "Green's Functions for the Graetz Problem and Calculation of Interfacial Concentrations," A.I.Ch.E. Journal 17, No. 1, pp. 19-24, 1971.

6. R.V. Dunkle (1965): "A Method of Solar Air Conditioning," Mech. & Chem. Eng. Trans., I.E. Aust., 1. pp. 73-78, 1965.

7. C.W. Chi and D.T. Wasan (1970): "Fixed Bed Adsorption Drying," A.I.Ch.E. Journal Vol. 16 pp. 23-31, 1970.

8. P.J. Banks, D.J. Close and I.L. Maclaine-Cross (1970): "Coupled Heat and Mass Transfer in Fluid Flow through Porous Media - An Analogy with Heat Transfer," Heat Transfer, Vol. VII, CT3. 1, (Elsevier Publishing Co. Amsterdam) 1970.

9. D. Roy and D. Gidaspow (1972): "A Cross Flow Regenerator - A Green's Matrix Representation," Chem. Eng. Science, pp. 779-793, 1972.

ACKNOWLEDGEMENT

This work is being supported by DOE Contract Number EG-77-C-03-1589

LIQUID DESICCANT SOLAR HEAT PUMP FOR DEVELOPING NATIONS

H. Robison
Coastal Carolina Energy Laboratory
Department of Physics
University of South Carolina
Conway, South Carolina 29626, U.S.A.

ABSTRACT

The U.S.C. liquid sorbent solar air conditioner has been redesigned to include a heating cycle, while at the same time to be more economical to construct and to operate. The triethylene glycol desiccant has been replaced with a calcium chloride solution. Concentration of the desiccant by direct solar radiation absorbed by the colored liquid allows collector operation at low temperature and high efficiency. The easily constructed open-channel trickle collector system is not used for transfer of sensible heat, but rather it is used for mass transfer of water absorbate from the absorbent solution to the natural ambient air stream. The hardware necessary to build this system is readily available worldwide.

In the cooling mode, the process air is cooled and dehumidified by contact with concentrated sorbent solution flowing across cooling coils. Sensible heat as well as the heat of sorption is absorbed by shallow well-water in rural regions, while in urban areas, municipal water is used for cooling and then returned to the water mains where the heat is dissipated into the earth. Final cooling to a temperature lower than the well-water is achieved by humidification of the cool, dry airstream.

In the heating mode, the process air is warmed and humidified by contact with the dilute sorbent solution as the air flows through the solar collectors. Final heating to a temperature higher than the storage temperature is achieved by the heat of sorption furnished by the dehumidfication of the warm, wet airstream.

INTRODUCTION

A HVAC system utilizing solar energy should be capable of year-round operation if it is to be cost effective. Furthermore, because of the dilute nature of sunshine, very large and usually very expensive collector arrays are necessary for operation of efficient air conditioning systems. Therefore, in order to provide more universal all-year use under various climatic conditions, the liquid desiccant solar air conditioner designed by the Coastal Carolina Energy Laboratory of the Physics Department of the University of South Carolina has been redesigned to operate as a heat pump.

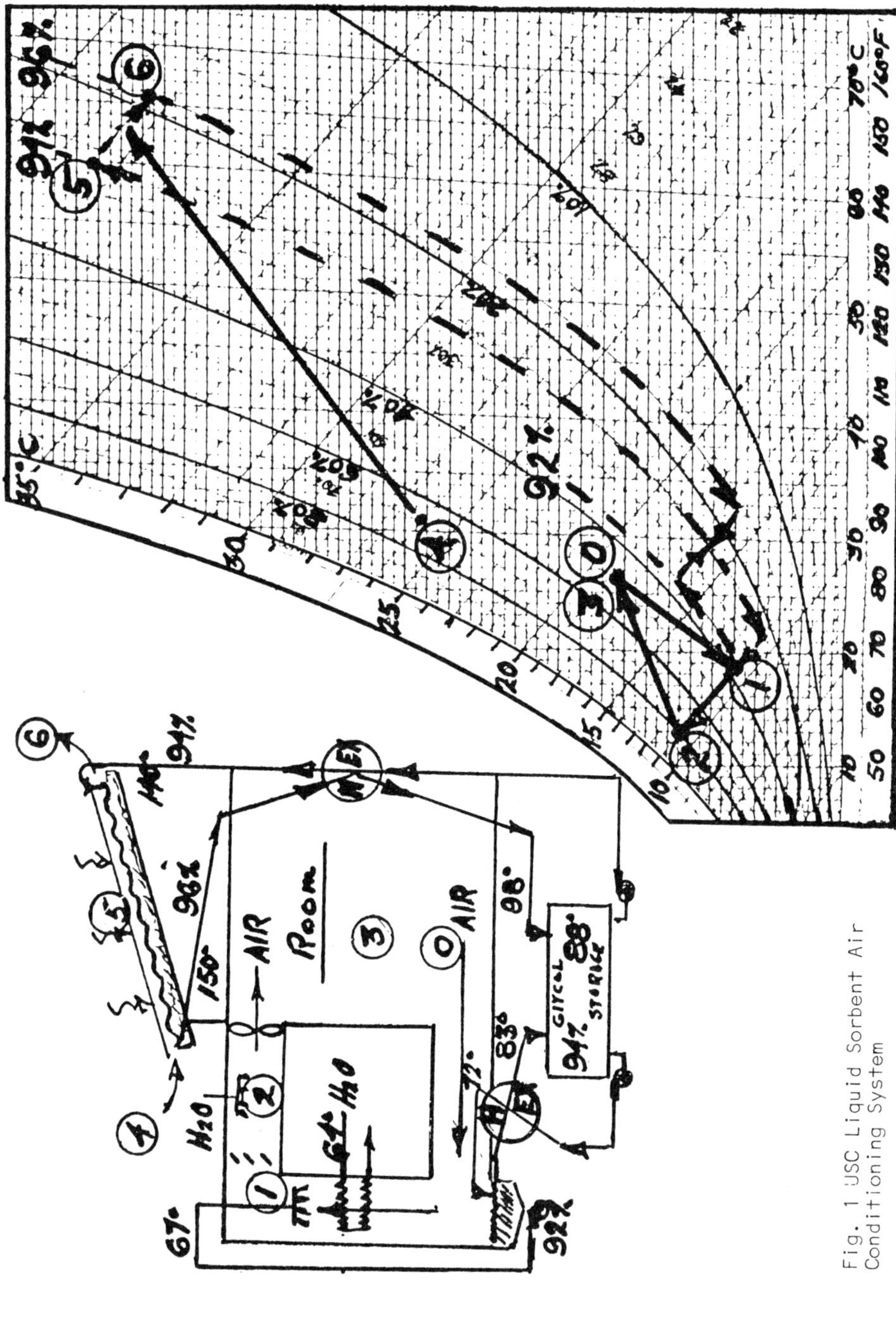

Fig. 1 USC Liquid Sorbent Air Conditioning System

Other changes have been incorporated in the redesign to assure that the materials needed for construction and operation of the system are available worldwide at reasonable prices. The original design has been described previously in some detail 1 . Figure (1) illustrates the system and shows its operating cycle.

SYSTEM REDESIGN

The triethylene glycol used as a desiccant in the original system has two serious disadvantages. It is expensive, and it does vaporize. Therefore, it has been necessary to choose another sorbent to eliminate loss during the heating cycle. Calcium chloride has been chosen as a suitable material. The research of Mullick and Gupta 2 indicates that this desiccant whould function well.

The Mullick-Gupta open channel trickle collectors, which utilize direct solar radiation, act as regenerators for the dilute desiccant solution in the original design. Although the collector components are very inexpensive and no skilled labor is required for construction, the new design has eliminated both the corrugated metal collector plates as well as the glass covers, to achieve even greater economy.

It is proposed to allow the dilute sorbent solution to flow down across standard asphalt roofing material with a light colored sand surface. The solution will contain a black coloring agent which will allow the direct absorption of the solar energy by the liquid, thus eliminating the heat transfer interface between the absorber plate and the liquid. Although the final choice of coloring agent has not been made at this time, adequate information to allow a proper choice is available from the publications of Trentleman 3 , Deminet 4 , Gillette 5 , Beverly 6 , and Minardi 7 . Direct absorption will allow collector-regenerator operation at a low temperature and at a high efficiency for transfer of the water absorbate from the dilute sorbent solution to the ambient airstream.

The glass collector cover is being replaced with a plactic bubble packing material from Baxter Corporation 8 , material which has been modified to be ultraviolet resistant. The material consists of a 3 mil polyethylene-nylon-polyethylene flat film laminated to a bubble material to provide $\frac{1}{2}$ in. (130 mm.) high by 1 in. (250 mm.) diameter bubbles. This material is avialable for less than 10¢/ft^2 (\$1.00/m^2).

HEATING CYCLE

A unique feature of a liquid sorbent air conditioner is the use of a small concentrated sorbent energy storage system. This storage may be accomplished at ambient temperatures or it is not a sensible heat system. Mullich and Gupta 2 , Duffie and Sheridan 9 , and Close and Dunkle 10 have described concentrated solution storage systems for

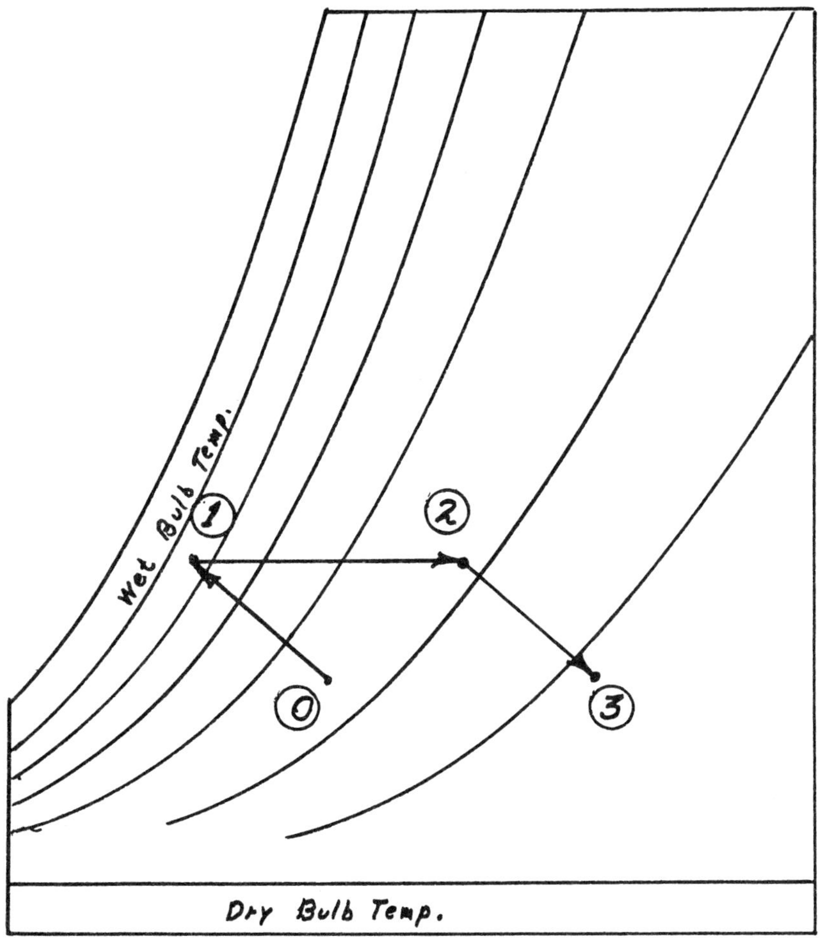

Fig. 2 Theoretical Heating Cycle

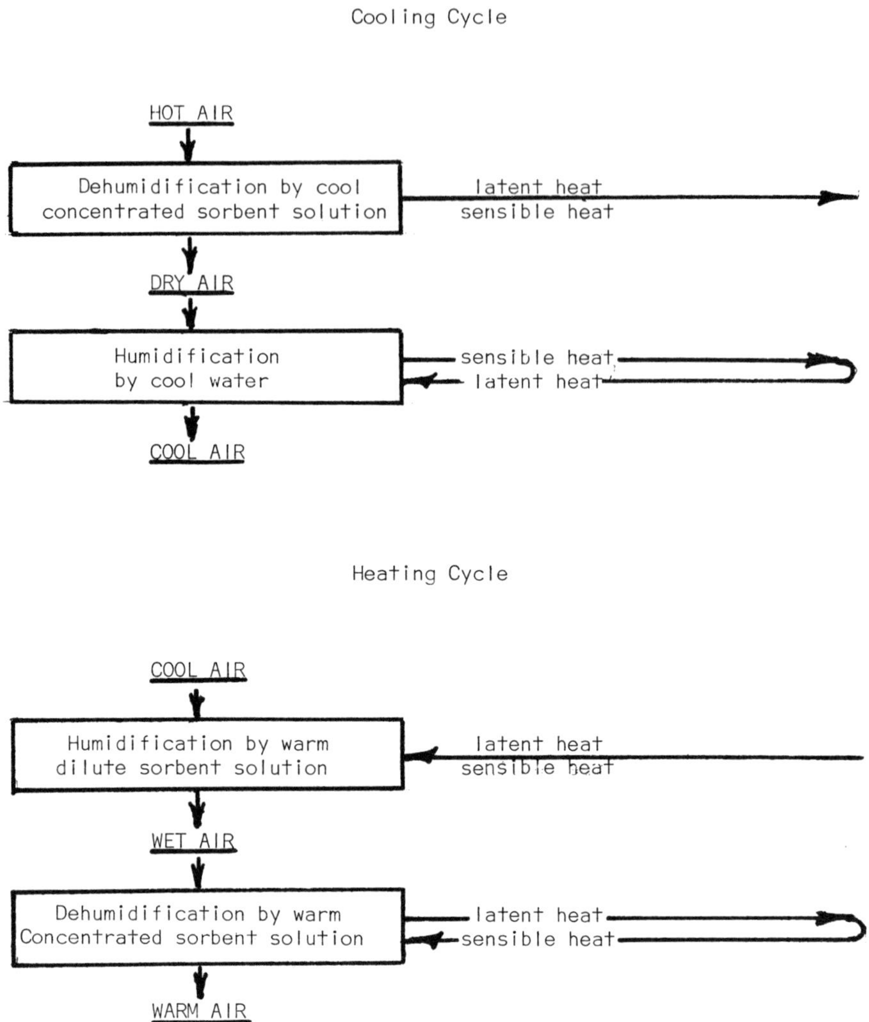

Fig. 3 Heat Pump Flow Diagram

air conditioning and Baugn and McDonald 11 and Robison 1 have suggested use of a similar system for a heating cycle as well.

It is theoretically possible to achieve a heating cycle as shown in figure (2). Air at state (0) is adiabatically cooled by humidification to state (1) and then heated sensibly to state (2). At state (2) the air is then adiabatically heated by dehumidification to achieve a final dry bulb temperature at state (3) that is higher than the temperature of the heater used for process (1-2).

In the redesigned system, the process air is both warmed and humidified as it passes through the collector-regenerator. This operation also concentrates the dilute desiccant solution. Final heating is achieved by dehumidification of the moist air.

Figure (3) shows the flow design for both the heating and cooling cycles of the heat pump.

CONCLUSION

Only thorough testing will determine if this inexpensive system can achieve true heat-pump performance.

REFERENCES

1. Robison, H. I., "Liquid Sorbent Solar Air Conditioner," Alternative Energy Sources, ed. T.N. Veziroglu, McGraw-Hill, New York, N. Y., (1978)
2. Mullick, S. C. and Gupta, M. C., "Solar Air Conditioning Using Absorbents," Use of Solar Engergy for the Cooling of Buildings, eds. de Winter, F. and De Winter, J. W., ERDA DOC, NTIS # SAN/1122-76/2, (1976)
3. Trentleman, J. and Wojciechowski, P. H., "Performance analysis of a Black Absorbing Collector (BLAC)," Proceedings of the 1977 Annual Meeting, American Section of the International Solar Energy Society, Vol. 1, eds. Beach, C. and Fordyce, E., (1977)
4. Deminet, C. and Beverly, W. D., All-Glass Collectors in Solar Utilization, " Ceramic Bulletin, pp 1058-1059 and 1067, Vol. 56, No 12, (1977)
5. Gillelte, R. B., Deminet, C. and Beverly, W. D. "Glass Solar Heat Collector Development," Radiative Transfer and Thermal Control, pp 291-314, Vol . 49, Progress in Astronautics and Aeronautics, ed. M Summerfield, American Institute of Aeronautics and Astronautics, New York, (1975)
6. Beverly, W. D., Gillette, R. B. and Libbey, B. D., "Glass Solar Collector Develeopement, "NSF report # NSF/RA-76-0035, (1976)
7. Minardi, J. E., and Chuang, H. N., Solar Engergy, Vol. 17, pp 179, Great Britain, (1958)
8. Baxter Corporation, Astro-Packaging Div., 100 Thomas Rd. North, Hawthorne, N. J. 07507
9. Duffie, J. A. and Sheridan, N. R., "Lithium Bromide Water Refrigerators for Solar Operation," Mechanical and Chemical Engineering Transactions of the Institution of Engineering, Australia MCL (1), (1965)
10. Close, D. J. and Dunkle, R. V., "Use of Absorbent Beds for Energy Storage in Drying of Heating Systems," Solar Energy, Vol. 19, pp 233-238, Great Britain, (1977)
11. Baugn, J. W. and McDonald, J. J., Theorectical "Modeling of an Ammonia/Water Absorption Cycle with Solar Energy Storage," Proceedings of the 1977 Annual Meeting, American Section of the International Solar Energy Society, Vol. 1, eds, Beach, C. and Fordyce, E., (1977)

A REVIEW OF THE HISTORICAL DEVELOPMENTS OF THE USE OF PASSIVE SOLAR SYSTEMS

C. L. Robbins
Florida Agricultural and Mechanical University
Tallahassee, Florida U.S.A.

ABSTRACT

The first well documented report of an effort to purposely design a convective self-flow solar system was by Pope [1] in 1903. However, there are countless earlier venacular and indigenous examples going back to 3000 BC.

A review of the historical developments of the use of passive solar systems can be divided into two major areas of study. These are:

1. Convective self-flow thermal systems.
2. Passive (natural) lighting systems.

Within the framework of convective self-flow thermal systems there are three distinct types of convective systems: mass, liquid (thermosiphon), mass plus air. Similarly, there are two distinct types of passive lighting systems: direct beam illumination, indirect diffuse illumination.

Although both passive thermal systems and passive lighting systems are concerned with the same basic electromagnetic phenomenon, each category has developed into a seperate field of study independent of the other.

This paper presents a basic review of each category with some examples of each that are considered in terms of system design, performance characteristics and energy conservation criteria.

BASIC CONVECTIVE SELF-FLOW THEORY: THERMAL

In most simple terms, a passive solar thermal system is one in which there is no need for mechanical energy. In all convective thermal systems, solar radiation is absorbed as heat energy which can then be convected throughout the system. There are three specific convective self-flow systems. These are:

1. Convective self-flow using mass as thermal storage.

2. Convective self-flow using water (i.e.: thermosiphons) as thermal storage.
3. Convective self-flow using mass, in combination with air, using mass as thermal storage.

Each has specific uses and operational characteristics.

Convective Self-Flow Using Mass As Thermal Storage

Mass systems are used primarily for space heating. Building materials which have a good thermal conductivity, and can be formed into a monolithic structure, may be used as a thermal storage mass in such a convective system. The actual convection does not occur through the material, rather, it occurs at the interior surface, between the mass and the interior space.

The conduction through the mass can be expressed: [2]

$$(1)\quad \frac{Q}{A} = \sum_{i=0}^{n} \{\kappa\}\{\Delta t\}$$

Where: Q = conductive heat flow
A = surface area
κ = conductivity of material
Δt = temperature differential, outside to inside

In many such systems, it is important to know the time it takes for the heat to flow through the mass, that is, the lag ( ) time. This can be roughly estimated as: [3]

$$(2)\quad \delta = \frac{L}{\kappa}\ (Q_{\kappa})$$

Where: δ = lag time
L = thickness of material
κ = conductivity of material
Q_{κ} = conductive heat flow

Convective Self-Flow Using Water As Thermal Storage

Water systems are used primarily for domestic hot water heating. The density differential created by temperature gradients is used to cause the fluid being heated by solar radiation to flow. The effect of this type of convective self-flow is usually termed the thermosiphon effect. The magnitude of the effect and the velocity of the flow can be expressed:

(3) $v_f = \frac{gdh^2}{h}$

Where: v_f = velocity of flow
g = gravitational acceleration constant
h = height from ground to bottom of water storage
dh = height of water in storage system

The location of the storage tank above the collector is a requirement. Further, cold water that is supplied from the storage tank to the collector must leave the storage tank from the bottom. [4]

It is also possible to develop a reverse thermosiphon. That is, to store cold water which has been cooled by (nighttime) radiant heat loss from the collector to the sky. [5]

Convective Self-Flow Using Mass in Combination With Air

There is not mathematical difference between a mass plus air combination and the system described in the section titled: Convective Self-Flow Using Mass As Thermal Storage.

The only distinction between the two is the manner in which air is used to convect heat through a space. In the previous section, the mass elements convected heat directly into the space. In the mass plus air combination, an effort is made to pass room air around the thermal storage mass. This air might be ducted to the room, or from the room, to the thermal mass. It is not necessary for the mass to be directly connected to the space being heated, although this is common.

The mass plus air combination is a very common passive solar system. Figure 1 shows a Trombe wall, a typical mass plus air passive thermal system. [4]

BASIC PASSIVE LIGHTING THEORY

A passive lighting system is defined as one that does not require any artificial lighting to provide reasonable levels of light to perform visual tasks. There are two specific types of passive lighting systems. These are: [6]

1. Daylight system: utilizing direct beam illumination in combination with diffuse illumination (i.e.: sunlight + skylight + other reflected light).
2. Skylight system: utilizing only diffuse illumination (i.e.: skylight + other reflected light).

RESEARCH, SYSTEM OR BUILDING	LOCATION	DATE	PROCESS DOCUMENTED
Unknown	Mesopotamia	3000 BC	solar engine [10]
Unknown	Egypt	1455 BC	singing statues [13]
Unknown	Egypt	1200 BC	passive cooling [11]
Archimedes	Greece	212 BC	solar engine [13]
Sinaqua and Anasazi Indians	USA	700 AD	passive heating using mass [14]
Pueblo Indians	USA	1540 AD	passive heating using mass
Pope	USA	1886, 1903	passive heating using mass; passive heating using mass in combination with air theory [1]
Hottinger	Germany	1935	solar house heating [4]
Hoteel	USA	1939	solar house heating [4]
Duncan-Keck House	USA	1940	passive heating using glazing and mass [15]
Hutchinson-Purdue University House	USA	1945	passive heating using glazing and mass [16]
MIT House #2	USA	1947	passive heating using glazing and mass [17]

Figure 1a: Historical Development of Passive Thermal Systems

RESEARCH, SYSTEM OR BUILDING	LOCATION	DATE	PROCESS DOCUMENTED
Sander's Weston House	USA	1960	passive heating using glazing and mass [18]
Coos Bay House	USA	1961	passive heating and cooling using water as storage; passive domestic hot water [21]
Zomeworks	USA	1971	passive heating and cooling, using water as storage; passive domestic hot water [22]
Skytherm	USA	1973	passive heating and cooling [23]
Crowther House	USA	1974	gravity convection heating [18]
Guadalupe #1	USA	1975	gravity convection heating [24]
Terry House	USA	1975	passive heating using skylights and mass [25]
Archisanti Solar Chimney	USA	1975	passive swimming pool heating [24]
Gaudenzi	Italy	1975	reverse thermosiphon
Cassalonga	France	1977	convective self-flow through mass and liquid theory

Figure 1b: Historical Development of Passive Thermal Systems

Each have specific operational characteristics.

Daylight Systems

The mathematics used to describe artificial lighting systems is not applicable to passive (natural) lighting systems as these systems do not follow the inverse-square law. [7] However, research, primarily in England [8], over the past 50 years has developed mathematics for 'large light' sources.

Following those mathematics, total clear sky illumination striking a given point can be expressed as:

(4) $\log_{10}$ DAYLIGHT = 4.1 - (0.1) (Cosec ϕ) for lat. less than 20°

(5) $\log_{10}$ DAYLIGHT = 3.95 - (0.1) (Cosec ϕ) for lat. = 20°+

Where: ϕ = solar altitude

The direct component of daylight (i.e.: sunlight) changes rapidly during the day. In addition, the sunlight includes the major proportion of the solar heat gain factor. [9] The indirect component (i.e.: skylight) changes depending upon the orientation in relation to four grogh quadrants.

The skylight can be determined by the expression:

(6) $$\text{Skylight} = (B_z) \left(\frac{1 + 2\,\text{Sin}\ \phi}{3}\right)$$

Where: B_z = liminance at zentih

ϕ = solar altitude

Skylight Systems

Skylight systems are very similar to Daylight Systems, with the exception that the direct component (i.e.: sunlight) is severely limited or totally blocked from the interior space.

The advantages of skylight systems over daylighting systems is twofold. First, they do not generate solar glare. That is, such systems tend to be more visually comfortable for people. Second, the skylight systems do not provide as much solar heat gain as the daylight systems.

Both the daylight and the skylight systems are affected by sky clearness and cloud cover. Sky clearness can increase or decrease the total available out-door illumination, depending on the prevailing atmospheric conditions. Cloud cover causes a decrease in the total daylight as compared with clear sky conditions. However, skylight systems are not as adversely affected as daylight systems. [9]

HISTORICAL DEVELOPMENT OF PASSIVE THERMAL SYSTEMS

The history of the use of solar energy is quite long although not always well documented. Early prehistoric cultures that developed some form of solar energy system were also cultures that worshiped the sun as a diety. Cultures that did not worship the sun did not develop their own solar systems, although they may have copied ideas from sun worshiping cultures.

3000 BC - 700 AD

One of the earliest recorded uses of solar energy was the development of a polished (gold) circular reflector in Mesopotamia. This solar engine was used to light alter fires by temple priestesses (cr: 3000 BC). Unfortunately, so few records remain about these reflectors that their description only comes to us from Greek writings from a much later period in time. [10]

During the New Empire Period of Egypt (1580-322 BC), there was a period of time in which the sun god 'Ra' was worshiped. During this period, while Amenkhotep III was Pharoh (cr: 1455 BC), a series of sounding statues were built. The sounding statues, which were called 'singing statues', were designed in such as way that when they were struck by sunlight it would cause the upper portions of the statues to heat up more rapidly than the lower portions. This would create a natural stack effect causing the air to rise rapidly upward through the statues. The rapid movement of the air escaping from the statues would cause a tone to be sounded much in the same manner as air passing through the tone pipes of a pipe organ. [10] [13]

In addition to the 'singing' statues, the Egyptians developed a passive cooling system. A large intake opening would draw air from above the roof line of a residence into a space between exterior and interior walls. In this space would be placed rows of porous pottery jars. The jars would evaporate water into the air stream as well as drip water onto a charcoal bed, placed on a grating. The air drawn into the intake would be drawn past the jars and through the charcoal. The cooled air would assist in drawing more air into the room, creating a reverse 'stack' effect. [11]

Probably the best known case of solar energy usage during this period of time is the case of Archimedes (287 - 212 BC). In the year 212 BC, he developed a group of solar reflectors, in the shape of Greek war shields, and used these devices to destroy a Roman fleet. After Archimedes there is little progress in the field of solar energy, although the Greeks 'borrowed' the concept of the sounding statue from the Egyptians during the Ptolemaic Period (332 - 30 BC) when Alexander the Great conquered Egypt from the

Persians. [13]

With the exception of the passive cooling technique, none of these early examples deal with passive solar systems for heating or cooling buildings.

700 - 1600 AD

Solar research seems to have declined during most of this period. During the latter part of this period, as the Renaissance began in Europe (cr: 1450 AD), some solar research was conducted in the areas of solar furnace and solar engine design. However, in the Americas, the native indians developed a number of indigenous examples of passive thermal systems. Three of these examples have been well documented.

The Sinaqua and Anasazi Indians developed a series of multistory cliff and cave dwellings near Camp Verde, Arizona. Two of these cliff dwellings, Mesa Verde and Montezuma's Castle, have thermal envelopes that suggest that a great deal of care was taken to respond to the local climate.

Using a combination of rock, mud and brush (held in place by timber crossbracing) the natives were able to develop a passive thermal system with the wall masses acting as thermal storage. Using trial and error methods over a period of almost 300 years, the indians were able to develop a wall construction that would provide a 12 hour thermal lag. Archiological excavations indicate that once the passive solar system was perfected, there is little evidence of continued use of auxiliary heating systems (i.e.: wood fires). With a 12 hour lag, the natives were able to heat most of the rooms from sundown to sunup, summer and winter, as well as keep the rooms cool during the day. [14]

Near Taos, New Mexico, is the San Geronimo de Taos Pueblo, a structure that has been occupied since it was first built in 1540. Freestanding, the complex is five stories high in some parts. Tree trunks were used as a structural foundation. Around these tree trunks adobe mud brick walls were built, sometimes as much as 36" thick. The upper stories were recessed, allowing each level to receive a maximum amount of solar radiation on both walls and roof areas.

The lower story had no doors or windows. This provided some means of protection as the only entrance to each dwelling of the complex was at the second or third level. However, the lower story also acted as a primary thermal storage. With no openings, heat that was conducted through the walls or roof would be convected through the air to the upper levels. With no cliffs or caves to offer some protection, the Pueblo was not as successful year around as the examples at Mesa Verte or Montezuma's Castle. However, a 10 hour lag was achieved, which would provide heat at night with only a minimal need for an auxiliary system.

Although successful as passive systems, the long slow trail and error process seems to have made this type of solution limited to a few examples from a limited, and interactive, number of cultures. It would not be until there was another resurgance in solar research, beginning at the end of the nineteenth century, that a theoretical approach would be applied to developing passive solar systems.

1900 - 1960

It is during this period of time that passive solar systems expand to include water heating. Although there may be some examples of passive water heating prior to this time, the documentation is poor. Overall, there are only a limited number of examples of passive buildings being constructed during this time frame. The majority of those were experimental structures, built by universities or research institutes. However, there was a great deal of work being conducted developing passive solar theory.

Pope as early as 1886, began writing about passive thermal systems. It was in 1903 that he applied a scientific approach to the design of a passive thermal system. [1] The results of his work not only helped to explain the Mesa and Pueblo examples, but also led him to develop an improvement upon these indigenous heating techniques. He theorized that if a wall mass could convect heat through a room then it would be much more efficient, if the air from the room were forced to pass around the mass. He further theorized that it would be possible to develop a duct system and heat rooms that were not adjacent to the thermal storage mass.

In 1935, at the Zurich Institute of Technology, and in 1939 at the Massachusetts Institute of Technology the first use of solar energy to heat space was attempted. These experiments were conducted by Hottinger and Hottel, respectively. [4] Although neither of these were passive, they provided the incentive for the development of a passive solar building. The first well documented passive building was constructed in 1940. This was the Duncan-Keck house of Flossmoor, Illinois (USA). [15]

The Duncan-Keck house did not use either the previously attempted mass system, or the mass in conbination with air system as suggested by Pope. Instead, a modified mass system was used. South facing glazing would allow radiation to pass through it and heat the walls and floor of the interior spaces. During the daytime, the solar heat gain on the glass would heat the house, while at night, the convected heat from the walls and floor provided heat. To avoid summer radiation, the entire south facade of the house had wide overhanging roof areas. Auxiliary heat was provided by a fireplace and (later) an oil burning furnace. [15]

The use of south facing glazing that could heat interior walls and floors, as well as provide daytime heat gain, was attempted in three other houses during this time period. The details of their construction varried some, but the passive solar system employed in each was based upon the same principal as the Duncan-Keck house. These other buildings were: MIT house #2 (1947) [16]; the Hutchinson-Purdue University house (1945) [17]; and the Sander's Weston house (1960) [18].

The best examples of a mass in combination with air system, commonly called a Trombe wall, occur in the Bliss Amado House (1954) [19] and the Odeillo Solar House (1956) [20].

The Bliss Armado House uses a passive solar collector detached from the structure. Heated air flows from the collector into a rock bed storage, and from there into a duct system that distributes the heated air throughout the house. Electric resistance heaters were provided for auxiliary heating, but were never used. [19]

Cooling was also provided during the summer. Radiant heat loss through the collectors at night would cause the rock storage to cool. The drawing of heat from storage back to the collector would cause an air movement through the distribution system that would help cool the building. [19]

The Odeillo Solar House utilized a basic Trombe wall. Air circulating from the building would be heated as it passed up the surface of the storage wall. The air was then recirculated throughout the building by natural convection. Auxiliary heating was provided by electric strip heaters. Cooling was not provided, however, the Trombe wall assisted the natural ventilation by the stack effect it created. [20]

Since 1960

Since 1960 there have been a variety of innovative passive solar systems developed, ranging from water storage systems for space heating to swimming pool heaters. However, there is no single passive theory or system that dominates the buildings being constructed during this time frame.

There are a number of examples of passive solar systems that make use of water as the storage medium. Storage tanks were large (10,000 gallons +) and domestic hot water heating often accompanied the space heating. Three key examples are: Coos Bay House (1961) [21]; Zomeworks (1971) [22] and the Skytherm House (1973) [23].

The Coos Bay House had a water type collector at the roof ridge and storage in an 8000 gallon tank in the basement. A small circulating pump was used to move the heated water from the collector to the tank. Air coming in contact with the tank would then be convected throughout the building. As the heated air would cool, it would fall back to the storage tank to be re-heated. In addition, the storage tank pre-heated the domestic hot water supply. [21]

Zomeworks (1971) is a very innovative building that makes use of a variety of passive techniques for heating, cooling and water heating. The heating system is composed of 90 56-gallon water barrels, located on the vertical face of the south wall. Heat is convected directly from the barrels into the rooms of the house during the heating season. A movable wall can close off the exterior portion of the barrels at night during the winter. During the summer, at night, the barrels would lose heat to the outside which would help cool the building. Hot water was provided by a seperate flat-plate thermosiphon. [22]

Skytherm (1973) uses four 8' by 32' water filled bags for both collector and storage. The bags were placed on the roof of the structure. The roof, being a steel deck, would allow the heat to flow through the deck to radiate and convect through the building. A reverse action would occur during the summer to help cool the building at night. [23]

Heating by simple gravity convection is used in two well known examples; the Crowther House (1974) [18] and the Guadalupe Project #1 (1975) [24]. Both of these buildings use flat-plate air heaters which are connected to a rock bed storage. Distribution is achieved by natural convection of the heated air from storage into the house and then back to the collector for re-heating. [18] [24]

The Terry Residence (1975) is another example of an innovative approach to passive system design. Instead of using south facing glazing, it uses skylights. The skylights allow radiation to flow into the building and heat the interior walls and floor, which have been designed as thermal storage masses. Heated air is then distributed throughout the building by natural radiation and convection. Solar heat gain on the glazing provides the daytime heating, while the heat stored in walls and floor provide the nighttime heating. [25]

The Archsanti Solar Chimney is a passive swimming pool heating system. The swimming pool is at the top of a hill. At the foot of the hill is a greenhouse which uses a plastic membrane to create the greenhouse effect. This membrane is extended from the greenhouse to the base of the swimming pool where it attaches to a rock storage system that acts as the foundation for the pool. There is a venting system to keep the pool from being heated during the summer. [24]

Passive Thermal Systems: Conclusion

Although passive solar systems seem to enjoy a popularity greater now than at any time during the past, there seems to be a limit to the type of applications. Historic examples included residences, public and religious building, whereas the modern examples seem to be mostly limited to residential systems.

Modern systems seem to be more efficient than past examples, but they are also more complex. This additional complexity means that they need more than 'passive' maintenance. However, the majority of the maintenance is limited to 'housekeeping' rather than equipment repair. With few working parts, passive solar systems should require only a limited amount of repair.

Passive thermal systems do not offer solutions that have been tried for commercial or industrial applications. In general, passive thermal systems offer the following advantages over active solar systems:

1. Low cleaning maintenance.
2. Minimal repair maintenance.
3. Lower initial cost.
4. Lower life-cycle cost.

THE HISTORICAL DEVELOPMENT OF PASSIVE LIGHTING SYSTEMS

The history of the use of passive lighting systems is very long as all cultures have used passive lighting systems, in one form or another, for daytime lighting. Certainly, prior to the invention of any efficient artificial lighting systems, daylight was the only means of lighting the interior of buildings.

However, even the early systems were sophisticated. That is, specific lighting levels were achieved at specific places in buildings. Different lighting levels were achieved, often not to facilitate visual comfort, but to create specific phsychological moods. This particular use of natural light has been evident throughout its history.

1600 BC - 600 AD

Documented examples of natural lighting systems prior to 1600 BC are rare. This is not because there were few examples, but because the majority of the permanent buildings were tombs, which had little need for any light.

There are many examples of buildings that made extensive use of planned passive lighting systems from about 1600 BC and 600 AD, because they represent clear examples of a variety of different natural lighting systems and techniques. These examples are: the Great Temple of Ammon at Karnak, Egypt (cr: 1530 BC); the Forum of Trajan, Rome (98 AD); the Pantheon, Rome (120 AD) and Hagia Sophia, Constantinople (532 AD). [26]

The Great Temple of Ammon at Karnak is considered one of the grandest of all of the Egyptian temples. The natural lighting systems were carefully planned to provide very low levels of light, thereby, making the interior to the Hypostyle Hall look enormous and appear to extend infinantly in all directions. [26] Other parts of the Temple are lit to much higher levels of illumination, especially the living quarters and work areas of the Priests.

Lighting techniques employed by the builders of the Great Temple include clerestory, roof monitors and a very unique light hole design in the Hypostyle Hall. Although it is unclear whether or not the builders had any comprehension of the physiological effects if natural light, it can be assumed that they understood the psychological impact. It is also clear from studying the Great Temple that it is sophisticated building in terms of its lighting systems. Obviously, a great deal of experimenting had been done long before the Great Hall was built.

RESEARCH, SYSTEM OR BUILDING	LOCATION	DATE	PROCESS DOCUMENTED
Great Temple of Ammon	Egypt	1530 BC	clerestory, monitor and special passive lighting [26]
Forum of Trajan	Rome	98 AD	clerestory, internal reflected light [26]
Pantheon	Rome	120 AD	'eye' skylight
Hagia Sophia	Constantinople	532 AD	'floating dome'
S. John's Chapel	England	1086	vaulted clerestory
Notre Dame (Paris)	France	1163	lancet and rose windows
Great Chalfield Manor House	England	1450	multi directional high wall side lighting
Palazza Farnese	Italy	1547	uni-directional low wall side lighting
S. Pauls Cathedral	England	1675	light wells and light lanterns
Crystal Palace	England	1850	glass building
University Museum, Oxford	England	1855	glass roof
Blondell	Germany	1912	indirect passive lighting theory [27]

Figure 2a: Historical Development of Passive Lighting Systems

RESEARCH SYSTEM OR BUILDING	LOCATION	DATE	PROCESS DOCUMENTED
Building Research Station (BRS)	England	1939–present	direct and indirect passive lighting systems
Hopkinson	England	1947, 1966	direct and indirect lighting theory [30]
Lynes	England	1969	design techniques [7]
Pleijel	Sweden	1972	design techniques [31]
Koenigsberger, et. al.	Africa	1973	design techniques
Robbins	USA	1975–present	daylighting theory; passive lighting; design criteria for USA [32] [33] [34]

Figure 2a: Historical Development of Passive Lighting Systems

The Forum of Trajan, (98 AD) is an excellent example of the use of Roman arches and vaults to create clerestory passages in order to allow light to be reflected throughout the building. Different levels for lighting were provided for different functions within the Forum. The Great Hall of the Markets, located near the center of the building, was lit almost entirely by internally reflected light, that is, there was no direct sunlight reaching this space. The many shops and businesses that surround the Great Hall each seem to be designed with specific lighting concepts. Some spaces that were believed to be used as 'offices' were only side lit. This provides the best light for writing. Other spaces that were used for storage or work areas were top lit by skylights. This is because skylights provide a even light throughout the room.

The Pantheon, in Rome, (120 AD) is one of the best known examples of the use of a single 'eye' skylight to provide illumination for the interior of a building. The eye is 27'-0" in diameter and is centered in the top of a 148'0" rotunda. Recesses were carefully placed in the walls of the rotunda so that at certain times of the year, beams of sunlight will light the recesses. There are seven such recesses placed at different locations around the rotunda.

Hagia Sophia is a monumental building that was initially built as a Christian church, but is currently a Muslem mosque. Throughout the ancient world there are writings that describe the famous 'floating dome' of the central narthex. An outstanding example of a natural lighting system in which glare was taken advantage of in designing the lighting system. The great dome in the center narthex is surrounded by 40 windows which, when struck by direct sunlight, cause specular reflections to beam off of the gold inlay of the dome. The specular reflection was so great that it made the support columns 'disappear'. Or, as Procopius, a philosopher of the time, wrote, the dome was 'as if suspended by a chain from heaven'. [26]

600 AD - 1800 AD

A definite distinction can be seen in the lighting of secular buildings as compared to non-secular buildings during this time period. Secular architecture was more dimly lit, with extensive use of stained and colored glazings. Non-secular buildings were, in general, bettin lit and had simpler lighting plans. The use of glass is a major innovation during this era. Examples can easily be found where the Master Builders experimented with different glazings in the same building.

In addition, candles and fireplaces would provide light at night and replace natural light in areas of a building where the sunlight could not reach.

Secular examples to be reviewed are: S. Johns Chapel, Tower of London, (1086); Notre Dame, Paris (1163); and S. Pauls Cathedral, London (1675). Non-secular examples are: Great Chalfield Manor House, Wiltshire England (1450) and Palazzo Farnese, Caprarala, Italy (1547).

S. Johns Chapel is of interest because it breaks the tradition of dimly lit specular architecutre. However, it should be understood that S. Johns Chapel was used as a meeting place and a place where sentences were executed. It uses a vaulted ceiling with clerestory lighting. There are little changes in the lighting techniques used to light this building over those of the past, however, it should be remembered that the 'Dark Ages' of Europe were not a time onnovation in the fields of architecture, art or science.

Notre Dame in Paris is one of the oldest French Gothic Cathedrals. The lighting consists of numerous 'long windows' and round 'rose' windows. The lancet windows were two to three stories tall, and rounded at the top. The circular 'rose' triforium windows were used as clerestore type lighting. Most all of the windows in the Cathedral were stained and leaded glass. The light level was not very high, but because there was no one predominant color in the stained glass, the combination of all of the colors produced a light that was very close to normal daylight.

S. Pauls Cathedral, London, has a variety of clerestory lights much like previous examples, however, it also uses a unique light well and light lantern system in the dome. Light entered through a series of light wells near the top of the dome, this light was joined by light entering through a high light lantern at the apex of the dome. The light passed through a variety of interior structures and partitions and eventually through a second interior dome. The use of the light wells and the lantern is unique to this Cathedral. [26]

In general, all of the secular examples had elaborate and ornate lighting systems, indicating that there was a firm understanding of the possible systems of natural lighting.

Both of the non-specular examples have very simple lighting solutions. The Great Chalfield Manor House typically uses large windows that are above eye level, thereby providing a great deal of light and very little glare. Windows appear on all sides of the room, whenever possible. Most rooms are never more than 30 to 35' wide in order to allow for complete penetration of light throughout the room.

The Palazzo Farnese is a pentagonal building with a circular cortile in the center. The unusual shape of the building has caused the designer to only allow natural light to enter each room from one direction. It was, therefore, necessary to use at least three different types of windows to achieve the desired lighting levels in various rooms. There is also a number of clerestory windows on the top (i.e.: third) floor of the building. These are used to light the attic.

1800 - 1900

The nineteenth century was most important in the field of passive lighting. It is during this period that there are clear examples of an attempt to develop new techniques for lighting buildings and break away from old traditional lighting techniques. In general, buildings begin to expand the amounts of glazed areas that they have. Although at that time there was no attempt to corelate the natural lighting with task lighting. That is, as more and more efficient artificial lighting systems were developed, there was less of a reliance upon natural lighting to provide task lighting. Windows were kept in buildings for aesthetic reasons rather than their operational characteristics.

Even as late as the nineteenth century there still was very little science in the development of natural lighting systems. That is, the same lighting systems were being used that had been used for the past 4000 years. By the nineteenth century the reasons why or how these systems were used had been lost or forgotten. The result was a series of buildings that had poor or unsuccessful lighting solutions.

Two examples of this are: the Crystal Palace, London (1850) and the University Museum, Oxford (1855).

The Crystal Palace was built to house the Great Exposition of 1851. This Exposition was the forerunner of the modern World's Fairs. The building was constructed along the same lines as a huge conservatory. But neither the lighting nor the heat gain problem could be totally resolved. Basically, the Crystal Palace provided too much light, without the proper contrast or freedom from glare.

The University Museum was constructed in a much more traditional manner, except that it had a glass roof. Again, the light was not correct for an exhibit space. The result of the lighting errors that occurred in these buildings, and similar lighting errors in other buildings, was the decision to formally study the phenomenon of daylight. This research would not begin until the early Twentieth century.

<u>1900 - Present</u>

It has only been since the beginning of the Twentieth century that any systematic attempt has been made to quantify and analyze the amount of daylight that exists in and around buildings.

Early research by Blondell (1912) better defined daylight photometry and suggested ways in which it was possible to mathematically determine daylight levels. [27] Kimball, in 1921, did a great deal of research in the city of Chicago. He mathematically determined how to better find solar location (i.e.: solar altitude and solar azimuth) and did research related to the reduction of available sunlight, and radiation, because of air pollution. [28]

With the establishment of the Building Research Station (BRS) in 1939, there began a unified and systematic study of daylight. By 1947, because of the work at the BRS, England was able to include specific natural lighting requirements into their building codes. [29] [30]

SOLAR ENERGY: INTERNATIONAL PROGRESS

During the 1960's and early 1970's a series of major breakthroughs were made in the study of the calculation of available daylight during clear and overcast sky conditions, and in the area of system design methodology. This research, conducted mainly in England [31], Africa [12] and Sweden [32] has been instrumental in the development of techniques and methodologies for studying natural light phenomenon in the United States. [33] [34] [35]

THE HISTORICAL DEVELOPMENTS OF PASSIVE LIGHTING SYSTEMS: CONCLUSION

By proper evaluation of the benefits and advantages derived from planned natural light utilization, it is possible to design an effective passive lighting system in terms of costs, energy conservation, and maintenance characteristics.

Unlike passive thermal systems, natural light cannot be stored. However, it provides better operational characteristics on overcast days than does passive (or active) solar thermal systems, therefore, storage is not a critical issue.

The light from passive lighting systems has been shown to be many times more effective than equal levels of artificial illumination. Further, when a unit of electric light is replaced by a unit of natural light, the savings in basic energy at the generating plant may be as much as three times the electric energy it took to produce the unit of artificial illumination.

In conclusion, a list of the advantages of passive lighting systems would include: [34]

1. Energy considerations:

   a. Located at energy load point.
   b. Possible reduction in electrical transmission and distribution system.
   c. Reduction in peak electrical loads.
   d. Useable in residential, commercial and industrial settings.

2. Cost considerations:

   a. Low initial costs.
   b. Low operating and maintenance costs.

3. Light Quality considerations:

   a. Better color rendering than artificial lighting systems.
   b. Better light color.
   c. Improved visual efficiency.

REFERENCES

1. Pope, C.H.; Annual Report, Smithsonian Institute; Washington DC: United States Government Printing Office, 1903

2. ASHRAE; Algorithms for Building Heat Transfer Subroutines; New York: ASHRAE, 1975

3. ASHRAE; Handbook of Fundamentals; New York: ASHRAE, 1977

4. Meinel, A.B. and Meinel, M.P.; Applied Solar Energy; Reading, Massachusetts: Addison-Wesley Publishing Co., 1977

5. Gaudenzi, P.; Lectures, International College of Applied Physics, Nerano, Italy, 1975

6. Robbins, C.L.; "Natural Lighting: New Research for an Old Environmental Concept"; EDRA-9 Conference, April 1978, University of Arizona, Tuscon

7. Lynes, J.; Principles of Natural Lighting; London: Elseven, 1969

8. Various Publications of the Building Research Station (BRS), Department of Environment, England

9. Robbins, C.L.; Natural Lighting in Buildings: A Passive Solar System; International Symposium-Workshop on Solar Energy; June, 1978, Cairo, Egypt

10. Veinberg, V.B.; Optics in Equipment for the Utilization of Solar Energy; Moscow: State Publishing House, 1959
Translation: AEC-tr-4471, USAEC, OTS, Dept. of Commerce, Washington, DC

11. Oakley, D.J.; Tropical Houses; London: Batsford, 1961

12. Koenigsberger, O.H., et. al.; Manual of Tropical Housing and Building; London: Longman, 1973

13. Fletcher, Sir B; A History of Architecture; New York, C. Scribner's Sons, 1963

14. National Park Service; "Montezuma Castle National Monument"; Washington, DC: United States Government Printing Office, National Park Service Historical Handbook #27, 1958

15. Keck, G.F.; "Design and Construction of Solar Homes"; MIT Symposium Proceedings, 1950

16. Hutchinson, F.W.; "The Solar House: A Research Report"; Heating and Ventilation, March, 1946

17. Dietz, A.H. and Czapek, E.L.; "Solar Heating of Houses by Vertical South-Wall Storage Panels"; Heating, Piping and Air Conditioning, March 1950

18. Shurcliff, W.A.; "Solar Heated Buildings-A Brief Survey", 6th Edition, Solar Energy Digest, October, 1974

19. Donovan, M.K. and Bliss, R.W.; "Solar House Heating--A Panel Discussion"; Proceedings, 1955 Phoenix Symposium

20. Staff writer; "The Trombe-Michel Solar Wall"; Architectural Design, Volumn XLII, October, 1973

21. Reynolds, J.S.; "Solar Energy for Pacific Northwest Buildings"; Center for Environmental Research, University of Oregon, 1974

22. Baer, S.; "Drumwall"; Albuquerque, NM: Zomeworks, Inc., 1974

23. Hay, H.R.; "California Solarchitecture House"; Paper EH-73, Paris International S.E.S., 1973

24. Aiello, D.; "Report on the Guadalupe Project"; Arizona State University, College of Architecture, 1975

25. Yellott, J.I.; Solar-Oriented Architecture; Washington, DC: AIA Research Corp., 1975

26. Fletcher, Sir B.; A History of Architecture on the Comparative Method; New York: Charles Scribner's Sons, 17th Ed., 1963

27. IES; IES Lighting Handbook; New York: Illuminating Engineering Society, 5th Ed., 1972

28. Kimball, H.H. and Hand, I.F.; "Sky Brightness and Daylight Illumination Measurement"; Monthly Weather Review, Vol. 48, 1921

29. Department of Environment, Building Research Station; Garston, Watford, WD2 7JR, England

30. Hopkinson, R.G.; Peterbridge, P. and Longman, J.; Daylighting; London: W. H. Heinemann, 1966

31. Pleijel, G.; "The Computation of Natural Radiation in Architecture"; Meddelande No. 25; Stockholm: Statens Namnd Fon Byggnadsforgking, 1972

32. Robbins, C.L.; The Architecture Use of Natural Light; Ames, IA: Iowa State Press, 1976

33. Robbins, C.L.; "Natural Lighting: New Research for an Old Environmental Concept"; Proceedings from the Environmental Design and Research Association Conference (EDRA-9); University of Arizona, Tuscon, April, 1978

34. Robbins, C.L.; The Assessment of Natural Light as a Resource in Florida; State of Florida Grant #28-2021-001

NATURAL LIGHTING IN BUILDINGS: A PASSIVE SOLAR SYSTEM

C. L. Robbins
Florida Agricultural and Mechanical University
Tallahassee, Florida U.S.A.

ABSTRACT

This paper reviews the issue of using natural lighting in buildings. In general, there are four basic reasons why natural lighting systems are required by building codes in many parts of the world. These reasons are:

1. To facilitate the performance of visual tasks and ensure visual comfort.
2. To provide visual communication channels between people and their outdoor environment.
3. To provide psychological impact to lighting schemes.
4. To conserve lighting energy, during daylight hours, and help reduce the total energy requirements in buildings.

The realization that there is a need to conserve energy in the United States has brought about a reviewed interest in the use of natural lighting systems to provide light within buildings during daylight hours. This interest has caused the development of new techniques to study daylight as well as new natural lighting systems called Planned Supplemental Lighting Systems (PSLSystems).

REASONS FOR USING NATURAL LIGHTING SYSTEMS

It has only been during the past seventy-five years that any systematic attempt has been made to try to quantify and analyze the amount of daylight available at any given time around and within a building.

Many people interested in building science and technology view the building envelope as a barrier, seperating the external (i.e.: uncontrolled) environment from the internal (i.e.: controlled) environment. It should be understood that this barrier is really a filter. That is, it excludes those things that are unwanted but admits those which are considered desirable. One of those desirable effects is natural light.

There are many people, especially in the United States, that argue that daylight is one of those unwanted and uncontrollable conditions that needs to be excluded, or certainly minimized, in buildings. In order to answer that argument, it is first necessary to briefly review the reasons for using a natural lighting system.

There are four general reasons for having a natural lighting system in buildings. These are: [1] [2]

1. To facilitate the performance of visual tasks and ensure visual comfort.
2. To provide visual communication channels between people and their outdoor environment.
3. To provide drama, mood, and emotional effects in lighting.
4. To conserve lighting energy and help reduce the total energy requirements in buildings.

To Facilitate the Performance of Visual Tasks

The use of natural light to facilitate the performance of visual tasks and to ensure visual comfort has been a long standing argument. Certainly, prior to the invention of any artificial lighting systems, natural light was the only source of light to aid in the performance of visual tasks. However, with the development of efficient artificial lighting systems, this function has been relegated to the artificial systems rather than the natural lighting systems. [3]

Recently this lighting design philosophy has been reconsidered. That is, as research in daylight photometry has progressed, it has better defined manners and techniques in which natural lighting systems may be used in lieu of, or to supplement, artificial lighting systems. [4] [5] This type of designed natural lighting system has been named a Planned Supplemental Lighting System (PSLSystem) in the U.S. [2] In England, because they supplement natural light by adding artificial light, such an approach has been named: Permanent Supplementary Artificial Lighting of Interiors (PSALI). [6]

A similar reliance on natural light to aid in the performance of visual tasks can be seen in different parts of the world. This is illustrated in Figure 1. [3] [7] [8]

Visual Communication Channels

It has been recognized that people like, and need, to 'see' what is happening outside, even if they do not want to experience it. Thus, many people like to look outside and see that it is raining without going outside and getting wet. [9] Sidelights (fenestration) and toplights (skylights) can provide that visual link to the outside.

However, it has been argued that it is possible to have such a link to the outside without using it as a light source. This argument has been used to counter the trend of designing 'window-walls'. That is, it was hoped that by arguing against large glazed areas in favor of small (out of the way) glazed areas, it would be possible to significantly improve the thermal efficiency of the building envelope. Certainly, all other

RECOMMENDED LIGHT LEVELS (FC)				
	LIGHT LEVELS		SUPPLEMENTAL LIGHT	
Country	Severe Tasks (1m/f2)	Very Severe Tasks (1m/f2)	Daylight Requirements (1m/sf)	Codified
USA	150	500-1000	0	no
UK	60	200-300	30-90	yes
Hungary	15-30	30-50	50+	yes
USSR	5-15	15-30	75+	yes

Figure 1: Lighting Requirements in Various Countries [3] [7] [8]

EXAMPLE BUILDING	PSALI SYSTEM (BTU/SF-YR)	ARTIFICIAL LIGHTING SYSTEM (BTU/SF-YR)
Lighting	22,700	44,500
Convenience	11,700	11,700
Fans, Pumps, Motors	10,400	15,500
Heating/Cooling	31,500	40,800
Average Annual Consumption	76,300 (BTU/SF-YR)	112,500 (BTU/SF-YR)

Figure 2: Annual Energy Usage for a Test Building [24]

things staying the same, by reducing the area of glazing struck by direct solar radiation it is possible to improve the thermal efficiency of the building envelope. However, a visual link with the outside should serve the double function of acting as a communications channel and providing light for performing tasks. [10] [11] The reason that this approach is now being recommended is because any visual link with the outside is de facto providing light for seeing. It is, therefore, important to design such openings as a natural lighting system in order for them to function properly as a light source as well as a visual link. [12]

Psychological Impact

Natural lighting in buildings has often been described in terms of its psychological impact on individuals. [13] [14] Therefore, architects have often used natural lighting because of the sense of mood it can help create. In addition, daylighting can help generate sharp visual contrasts. Emotional impact and visual contrasts are both part of each persons individual visual perception. [15]

There has been a great deal of research that links individual emotional impact and visual perception. Much of this research seems to indicate that impact is dependent upon the combination of an individual's perception of an area, or subject, being lit, and their individual memories of similar (or the same) subject or lighting conditions. [16] [17]

Much of the research seems to indicate that the greatest possible emotional stimulation can be accomplished by natural lighting rather than artificial lighting. [15] [16] [17] Further, individuals interviewed in both Australia and England comparing natural lighting and artificial lighting have indicated that natural lighting is more 'restful', and 'less bothersome'. [18] Studies in Germany and the United States have linked natural lighting to changes in a variety of personal work attitudes including absenteeism, sick leave and lighting provoked headaches. [16] [19]

Energy Conservation

Probably the most important reason for using natural lighting systems, in terms of current thinking, is its capacity to conserve energy. It is generally accepted that there are five categories of electricity demand in buildings. These are: [20]

1. Lighting
2. Convenience Electricity
3. Fans, Pumps and Motors
4. Cooling
5. Heating

Designed natural lighting systems (i.e.: PSLSystems or PSALI systems) have the capability to significantly reduce electrical energy demand in four of these categories: [21]

1. Lighting;
2. Fans, Pumps and Motors;
3. Cooling;
4. Heating;

in terms of the total electrical energy consumption (BTU/SF-YR) PSALI and simple artificial lighting systems compare as shown in Figure 2.

It is important to understand why the PSLSystems or the PSALI systems can reduce the total electrical energy consumption in a building. This occurs for two reasons. First, these systems reduce the thermal exchanges normally associated with daylighting techniques. A reduction in heat exchanges expected to occur through glass surfaces can reduce the size and capacity of both the heating and cooling systems. Second, there is a significant reduction in the daytime electricity needed for lighting. Because of the reduction in artificial lighting, there is also a reduction in the amount of heat generated by the luminaires that must be removed during the cooling season.

DETERMINING NATURAL LIGHTING CRITERIA FOR THE USA

During the 1960's and the early 1970's a series of major breakthroughs were made in the areas of the calculation of available daylight during clear and overcast sky conditions, and daylight systems design methodology and techniques. This research, conducted mainly in England [24], Africa [4] and Sweden [23] has been instrumental in the development of techniques and methodologies for studying natural light phenomenon in the United States. [1] [2] [25]

There are three major subject areas into which the study of natural lighting criteria can be grouped. These are:

1. Descriptive characteristics of solar flux (i.e.: scalar illumination)
2. Statistically established sky characteristics (i.e.: Design Sky)
3. Ergometric criteria (Sunlight and Skylight Factors)

This information, in turn, is then used to develop supplemental daylight lighting systems (i.e.: PSLSystems, PSALI Systems).

Scalar Illumination: Clear Sky

It should be understood that scalar illumination (i.e.: daylight) is often described in terms very similar to solar radiation. This is very logical as scalar illumination and solar radiation are the same electromagnetic phenomenon. However, the mathematics used to describe light flux are not applicable to solar phenomenon as such sources do not follow the inverse-square law. [24]

Luminance At Zenith (B_Z)

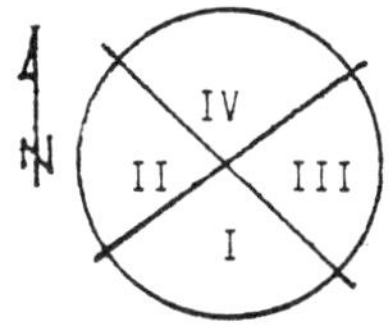

I = Facing Sun Quadrant
II = 90 Degrees To Sun Quadrant
III = 90 Degrees To Sun Quadrant
IV = Opposite Sun Quadrant

LUMINANCE AT ZENITH (B_z); FACING SUN QUADRANT (in fL.)

LATITUDE...	0°	4°	8°	12°	16°	20°	24°	28°	32°	36°	40°	44°	48°	52°	56°	60°
22 June	2500	2500	2500	2500	2500	2500	2500	2500	2500	2500	2500	2500	2500	2500	2500	2500
15 May /30 July	2467	2463	2458	2453	2449	2445	2441	2436	2432	2427	2423	2418	2413	2409	2404	2400
15 Apr./30 Aug.	2435	2427	2418	2408	2399	2390	2381	2372	2363	2354	2345	2336	2327	2318	2309	2300
21 Mar./23 Sept.	2400	2395	2382	2369	2355	2342	2329	2315	2302	2289	2275	2263	2250	2226	2213	2200
28 Feb /15 Oct.	2367	2321	2285	2250	2214	2179	2143	2107	2072	2036	2001	1975	1939	1904	1868	1833
28 Jan./15 Nov.	2334	2306	2246	2186	2126	2066	2006	1946	1886	1826	1766	1706	1646	1586	1526	1466
22 Dec.	2300	2220	2140	2060	1980	1900	1820	1740	1660	1580	1500	1420	1340	1260	1180	1100

LUMINANCE AT ZENITH (B_z); 90° TO SUN QUADRANT (in fL.)

LATITUDE...	0°	4°	8°	12°	16°	20°	24°	28°	32°	36°	40°	44°	48°	52°	56°	60°
22 June	1500	1461	1434	1408	1381	1355	1328	1302	1285	1259	1232	1206	1179	1153	1126	1100
15 May /30 July	1367	1342	1319	1296	1272	1249	1226	1202	1179	1136	1132	1109	1086	1026	1039	1016
15 Apr./30 Aug.	1234	1213	1193	1173	1153	1133	1113	1093	1073	1053	1033	1013	993	973	953	933
21 Mar./23 Sept.	1100	1082	1066	1049	1033	1016	1000	982	966	949	933	916	900	883	867	850
28 Feb./15 Oct.	1016	998	981	963	945	927	910	891	873	856	838	821	803	785	767	750
28 Jan./15 Nov.	933	911	892	873	855	836	818	798	779	761	742	724	705	687	668	650
22 Dec.	850	830	810	790	770	750	730	710	690	670	650	630	610	590	570	550

LUMINANCE AT ZENITH (B_z); OPPOSITE SUN QUADRANT (in. fL.)

LATITUDE...	0°	4°	8°	12°	16°	20°	24°	28°	32°	26°	40°	44°	48°	52°	56°	60°
22 June	1500	1442	1389	1336	1283	1230	1177	1124	1071	1018	965	912	859	806	753	700
15 May /30 July	1267	1218	1175	1132	1089	1047	1005	961	918	876	832	789	746	702	659	616
15 Apr./30 Aug.	1034	1001	967	934	901	867	834	800	767	734	701	667	634	600	567	534
21 Mar./23 Sept.	800	777	754	731	707	684	661	637	614	591	568	544	521	497	473	450
28 Feb./15 Oct.	682	665	648	632	615	599	582	566	549	533	516	500	483	467	450	434
28 Jan./15 Nov.	566	556	546	536	526	516	506	496	486	476	466	456	446	436	426	416
22 Dec.	450	447	443	440	437	433	430	427	423	420	417	413	410	407	403	400

Figure 3: Skylight at Zenith [1]

The total clear sky scalar illumination striking a given point can be expressed as:

(1) $\log_{10}$ DAYLIGHT = 4.1 - (0.1) (Cosec ϕ) Lat < 20°

(2) $\log_{10}$ DAYLIGHT = 3.95 - (0.1) (Cosec ϕ) Lat = 20+

Where: DAYLIGHT = sunlight + skylight, lm/f^2
ϕ = Solar altitude

Values for daylight can be quite high, with a maximum of 10,000 lm/f^2 (107,600 lux) occurring on the equator at solar noon.

The direct component of daylight (i.e.: sunlight) changes quite dramatically as the sun moves through the sky. Because of these large changes in sunlight, it is not considered as a stable source for interior light. In addition, the sunlight includes the major proportion of the thermal characteristics of solar flux.

Skylight values change depending upon the orientation of the measuring instrumentation in relationship to four ground quadrants and the location of the sun in the sky. Maximum skylight values occur when the sun is at zenith, thereby causing a zenith skylight (B_z). [5] Figure 3 shows the ground quadrants and the associated zenith skylight values. These values were first determined by Hopkinson [5] in 1966 and have been more exactly established in more recent work by Robbins. [2]

At times other than zenith, the skylight can be determined by the expression:

(3) Skylight = $(B_z) \left(\frac{1 + 2 \sin \phi}{3}\right)$

Where: B_z = Luminance at zenith, fLamberts

ϕ = Solar altitude

Figure 4 shows values for clear sky Daylight, Sunlight and Skylight for several locations.

Scalar Illumination: Overcast Sky

Scalar illumination is affected by both sky clearness and cloud cover. Sky clearness numbers cause a reduction, or addition, in the total daylight as a result of industrial/non-industrial atmospheric conditions. [26] Cloud cover causes a major decrease in the total daylight compared with clear sky conditions.

Although it is very difficult to accurately determine the illumination under overcast conditions, average overcast illumination levels have been emperically derived from 30 years worth of observations. [4] [5] Two general groups of overcast conditions are

LOCATION	LATITUDE	TOTAL DAYLIGHT (lm/f2)	SUNLIGHT (lm/f2)	SKYLIGHT QUADRANTS (fLambert) I	II/III	IV
COPENHAGEN	55.41N	6781	4534	2247	1015	680
LONDON	51.29N	6849	4558	2291	1059	740
NEW YORK	41.30N	7009	4579	2430	1194	936
TOKYO	35.41N	7036	4579	2457	1239	1003
CAIRO	29.52N	7067	4579	2485	1288	1086
BOMBAY	18.54N	9989	7498	2494	1364	1247
NAIROBI	1.16S	9999	7500	2499	1498	1493

Figure 4: Daylight, Sunlight and Skylight on 21 June, 12 Noon

LOCATION	LATITUDE	CLEAR SKY DAYLIGHT (lm/f^2)	AVG. 50% OVERCAST (fLambert)	AVG. 100% OVERCAST DAYLIGHT (fLambert)
COPENHAGEN	55.41N	6781	2900	1160
LONDON	51.29N	6849	3105	1242
NEW YORK	41.30N	7009	3605	1442
TOKYO	35.41N	7036	3840	1536
CAIRO	29.52N	7067	4145	1658
BOMBAY	18.54N	9989	4225	1690
NAIROBI	1.16S	9999	4495	1798

Figure 5: Clear Sky and Overcast Illumination on 21 June, 12 Noon

considered. These are:

1. 50% Overcast (i.e.: 4-7 cloud cover factor)
2. 100% Overcast (i.e.: 8-10 cloud cover factor)

All overcast conditions that have a cloud cover factor of less than 4 are considered clear sky conditions. However, for 50% and 100% illumination levels to be measured, it is necessary that the solar disc be completely obscured by cloud cover. Even if the cloud cover factor is greater than 4, if the solar disc is not hidden, it is necessary to consider the condition as clear sky. [2] [5]

Average 50% and 100% overcast sky illumination can be expressed emperically as:

(4) 50% OVERCAST ILLUMINATION = (50) (ϕ)
(5) 100% OVERCAST ILLUMINATION = (20) (ϕ)

Where: ϕ = solar altitude

A comparison of average values for clear sky daylight, 50% and 100% overcast illumination are displayed in Figure 5.

Daylight Factor

Because of the wide range of variation in outdoor light levels it is very difficult to determine corresponding interior natural lighting levels. However, it is possible to establish a ratio of the interior illumination to the simultaneous outdoor illumination. This ratio can be taken as a constant. Expressed as a percentage it is called the daylight factor (DF). [5] This can be expressed:

(6) $$DF = \frac{E^1}{E_2} \times 100$$

Where: E_1 = interior natural illumination, lm/f^2
E_2 = outdoor natural illumination, lm/f^2

The daylight factor concept has proven to be valid (i.e.: the ratio remains constant) only during conditions in which there is no direct sunlight. Therefore, a key to developing PSLSystems is to avoid using direct sunlight for interior natural illumination. Instead, three other sources of outdoor light will be used. These are:

1. Skylight
2. Externally reflected light
3. Internally reflected light

The magnitude of each of these components depends upon a variety of design variables that include such variables as: window size, glazing type, maintenance, size and location of external surfaces, and room configuration.

When the daylight factor for a given point has been established, like those in Figure 6, it can be converted into an illumination value, if the exterior (i.e.: outdoor illumination) is known.

Design Sky

Exterior illumination can be established by the statistical evaluation of long-term illumination data or, in lieu of existing long-term data, the following expression:

$$\text{(7)} \quad MDS = (\partial)\ (\rho_1) + (\psi)\ (\rho_2) + (\gamma)\ (\rho_3)$$

Where:
∂ = Average clear sky daily luminance
ψ = Average 50% overcast daily luminance
γ = Average 100% overcast daily luminance
ρ = Percentile of sky cover

Note: Luminance values calculated on an hourly basis from 8 a.m. to 4 p.m.

This value, called a Monthly Design Sky (MDS) represents the average outdoor diffuse illumination. A yearly Design Sky (DS) which is excepted as the minimum allowable outdoor design illumination, is a value that must be exceeded by 85% of the monthly Design Sky values throughout the year. [1] [5] [28]

Figure 7 shows typical Design Sky data as it is being collected for the U. S. A.

When both the Design Sky and the interior natural illumination levels are known, it is possible to ascertain the daylight factor for a given area. It is then possible to manipulate the design variables to achieve a given daylight factor.

This method of designing indoor natural illumination ensures that a minimum amount of outdoor light will be achieved during 85% of the daylight hours throughout the year.

PLANNED SUPPLEMENTAL LIGHTING SYSTEMS

The PSLSystem is based upon the concept that during daylight hours a designed natural lighting system can provide background lighting in lieu of using artificial lighting to provide background lighting. Specifically, the PSLSystem should be designed to provide light during 85% of all daytime conditions. It is assumed that the remaining 15% of the time that there is not enough skylight available to light background areas and the artificial lighting systems would be used.

SPACE OR VISUAL TASK	DESIRABLE MINIMUM LONDON, ENG. (5) DF = 500 lm/f^2	DAYLIGHT FACTOR (DF) JACKSONVILLE, FL (2) DF = 750 lm/f^2
Circulation	1.0	1.5
Classrooms	2.0 - 4.0	3.0 - 5.0
Laboratories	3.0 - 6.0	4.0 - 6.0
Offices (background)	1.0	2.5
Drafting (on work)	6.0	5.0
Drafting (background)	2.0	2.5
Typing (on work)	4.0	3.0
Typing (background)	2.0	1.5
Factories	5.0	4.0
Hospital (wards)	2.0	1.5
Food Prep Areas	2.0	1.5
Sleeping Areas	0.5	1.0
Gathering Areas	1.0	1.0

Figure 6: Typical Daylight Factors

Note: No DF greater than 2.5 should be provided from a uni-directional source. [25]

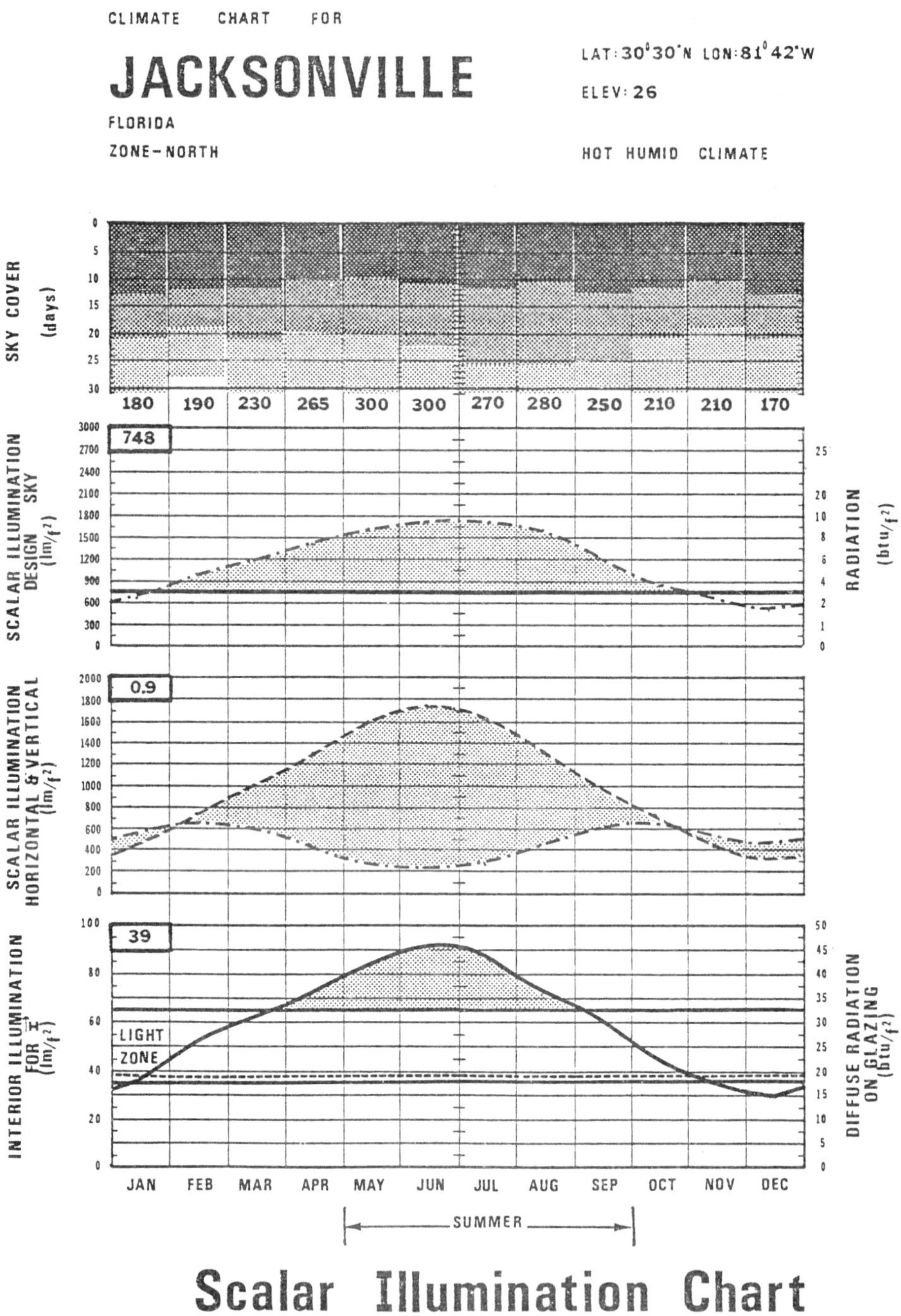

Figure 7a: Design Sky Data

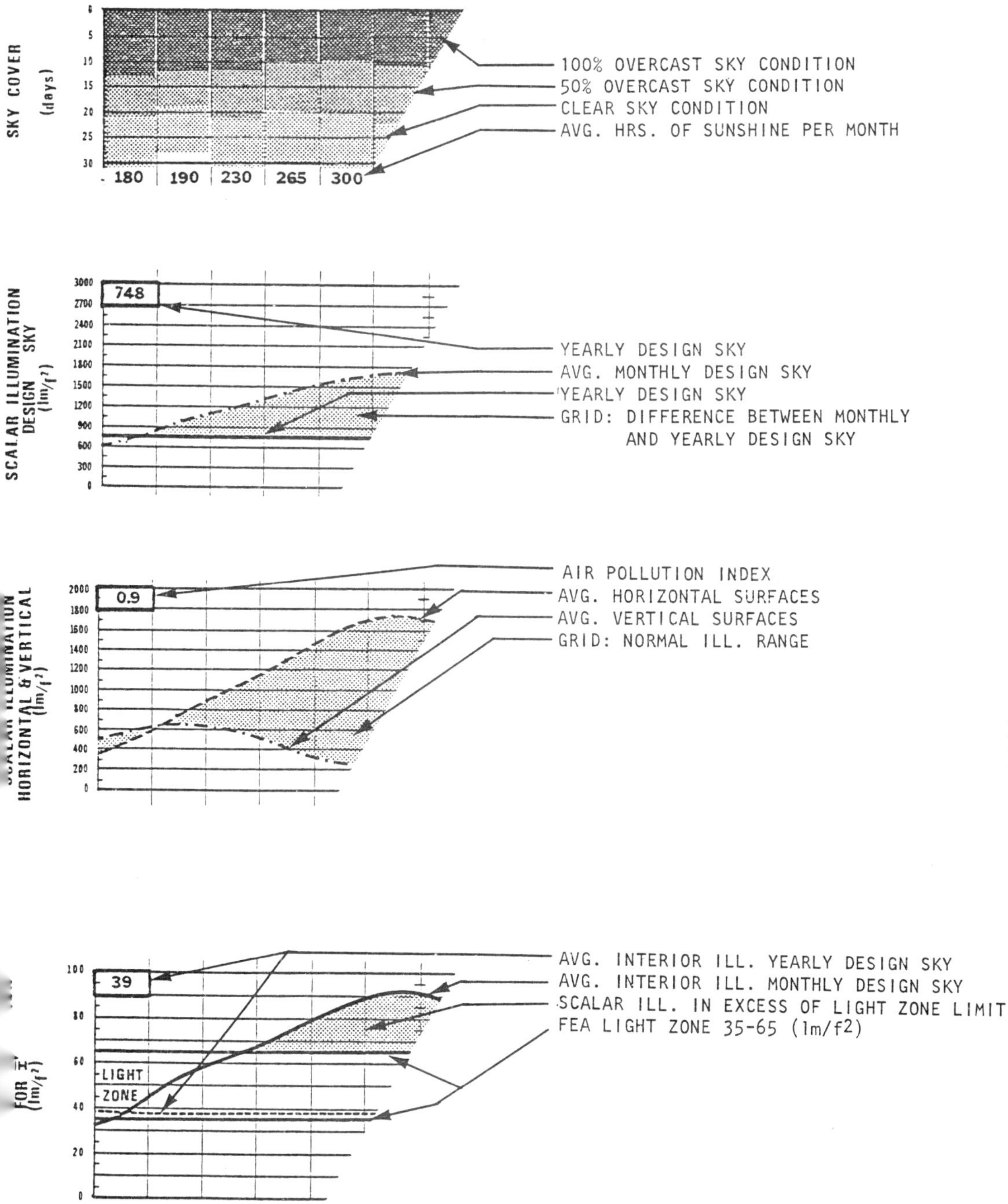

igure 7b: Key to Design Sky Data Chart

Room heights and depths may require this guideline to be altered. Further, specific micro-climatic conditions may require variations in the guideline. In general, natural lighting from a uni-directional source has significant light penetration to a point about 25' from the source. Spaces more than 50' from a uni-directional source are not appreciably affected by the light. [2] [5] It should be remembered that no distinction is being made, at this point, between side and top lighting. Either may be considered a uni-directional source of light.

PSLSystems are a combination of natural and artificial lighting. The natural light provides the background illumination while the artificial light provides supplemental on task light. Both the IES [3] and ASHRAE [29] suggest that the ratio between task and background illumination should be about 3 to 1. For example, a task that requires 150 footcandles of light should be surrounded by an area that is lit to 50 footcandles.

The IES and the Federal Energy Administration (FEA) have made recommendations as to proper task lighting levels. Figure 8 and 9 show how these recommendations can be supplemented by a natural lighting system.

Advantages of the PSLSystems

There are a variety of advantages in making use of planned supplemental natural lighting systems. These advantages, which relate to energy, cost, and light quality are:

1. Energy Advantages
   a. Located at load point
   b. Possible reduction in transmission and distribution system
   c. Reduce peak electrical loads
   d. Useable in residential, commercial and industrial settings
2. Cost Advantages
   a. Low initial cost
   b. Low operating and maintenance costs
3. Light Quality Advantages
   a. Color rendering
   b. Light color
   c. Visual efficiency

CATEGORY OF VISUAL TASK	EXAMPLE OF TASK	RECOMMENDED TASK ILLUMINATION (lm/f^2)	RECOMMENDED TASK BACKGROUND RATIO minimum back-ground illumination = 20	AVERAGE BACKGROUND ILLUMINATION PROVIDED BY NATURAL LIGHT (lm/f^2)
Casual	Hallway	20	1:1	20
Ordinary	Waiting Room	70	3:1	25
Moderately Difficult	Drafting	150	3:1	50
Difficult	Dentist Chair	500	10:1	50
Severe	Watch Repair	800+	10:1	not recommended

Figure 8: IES Task Lighting Recommendations [25]

CATEGORY OF VISUAL TASK	EXAMPLE OF TASK	RECOMMENDED TASK ILLUMINATION (lm/f^2)	RECOMMENDED TASK BACKGROUND RATIO minimum back-ground illumination = 20	AVERAGE BACKGROUND ILLUMINATION PROVIDED BY NATURAL LIGHT (lm/f^2)
Casual	Hallway	10	1:1	10
Ordinary	Waiting Room	35	3:1	20
Moderately Difficult	Reading, Writing	50	3:1	20
Difficult	Sewing	90	3:1	30
Severe	Repair Work	125	3:1	40

Figure 9: FEA Task Lighting Recommendations [25]

Energy Advantages

Like all solar energy systems, natural light is a problem solution located at the source of the problem. In terms of electrical energy, it is located at the load point. Unlike electricity, it does not have to be transmitted and distributed through the use of complex, and expensive, equipment. Because peak electrical demand occurs during the daytime hours between 8:00 a.m. and 4:00 p.m., a PSLSystem helps reduce the daily peak electrical usage.

Cost Advantages

The cost advantages are displayed in Figure 10. The PSLSystems include, as initial costs, both the cost of the artificial lighting system and the cost of side or top lights. None of the artificial systems included in Figure 10 include any costs for side or top lights, even though it could be assumed that such glazed areas would probably be included in a building using these lighting systems.

Initial costs, in Figure 10, include materials, labor, profit, and the cost of money. Operating and maintenance costs include such variables as the cost to remove heat, lighting energy, lamp replacement, maintenance labor, repair and repair parts, as well as other variables that are included in life cycle costing. [31] [32]

Light Quality Advantages

In lighting design, the designer should ensure that a lighting source be used that is both adequate and suitable for the visual task. Suitability, in this context, includes the concepts of color rendering, the color of the light. It has long been recognized that sunlight or skylight are the best source of 'color correct' light for tasks. Further, the human eye is 'designed' to view objects under daylight conditions. [33] [34]

NATURAL LIGHTING AS A PASSIVE SOLAR SYSTEM: CONCLUSION

By proper evaluation of the benefits and advantages from planned natural light utilization, a well designed PSLSystem can be shown to be an effective solar system in terms of costs, energy conservation and maintenance characteristics.

Unlike solar thermal systems, (i.e.: solar heating, cooling and domestic hot water) natural light cannot be stored. However, it provides better operational characteristics on overcast days than does solar thermal systems, therefore, storage is not a critical issue.

The light from natural lighting systems may be many times more effective than equal levels of illumination from conventional (artificial) lighting systems. When a unit of illumination of electric light is replaced by a unit of illumination of natural light,

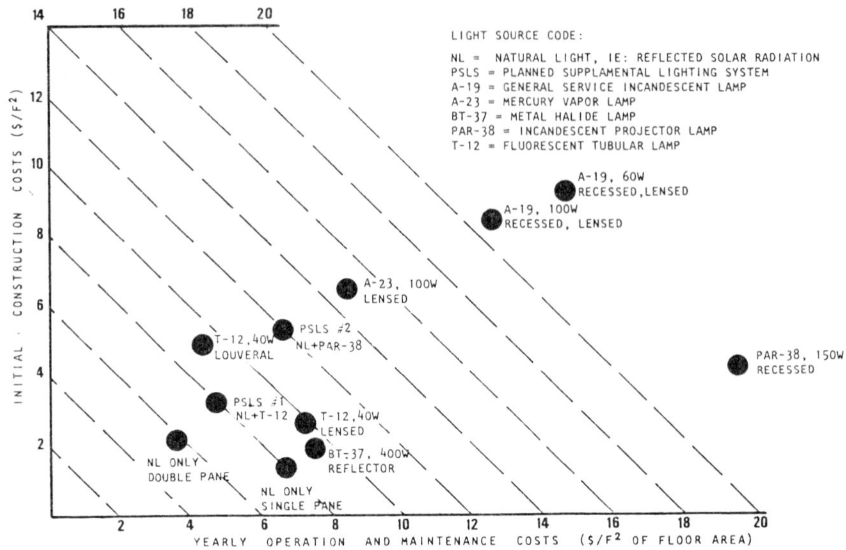

Figure 10: Initial Operating and Maintenance Costs for Various Lighting Systems

the savings in basic energy at the generating plant may be as much as three times the electric energy it took to produce the unit of artificial illumination. [35]

However, it must be stressed that natural lighting systems must be planned with as much care as is used to design and plan solar thermal systems. Just as collector slopes can be optimized, it is necessary to optimize the location of natural lighting openings in buildings and to analyze the entire natural lighting systems on some cost-benefit basis.

In addition, natural lighting systems are form defining, in architectural terms. That is, the size, shape and location of side and top lights has a definite effect upon the overall architectural solution. Although solar thermal systems may also be form defining, it is also possible to remove the solar system from the building completely. This is not possible with the natural lighting system.

Although natural lighting systems may require more time to design and analyze than conventaional lighting systems, the resulting benefits in terms of energy savings and human comfort are substantial and well worth the effort.

REFERENCES

(1) Robbins, C.L.; Architectural Use Of Natural Light; Ames, IA: Iowa State Press, 1976

(2) Robbins, C.L.; 'Assessment of Natural Light as a Resource in Florida'; current research, FAMU, 1977. Funded by the State of Florida

(3) Kaufman, J.E. and Christensen, J.F., eds.; IES Lighting Handbook; New York: IES, 5th edition, 1972

(4) Koenigsberger, et.al.; Manual of Tropical Housing and Building, Part 1: Climatic Design; London: Longman, 1973

(5) Hopkinson, R.G., Petergridge, P. and Longman, J.; Daylighting; London: W.H. Heinemann, 1966

(6) Department of Environment; Sunlight and Daylight: Planning Criteria and Design of Buildings; London, HMSO, 1971

(7) IES (UK); Recommendations For Lighting Building Interiors; London: IES (UK), 1968

(8) Farago, G. and Maroti, G.; Vilaritastechnika; Budapest: Muszaki Konyvkiado, 1962

(9) Sampson, F.K.; Contrast Rendition in School Lighting; New York: Educational Facilities Laboratory, 2nd ed., 1972

(10) Lighting Industry Federation; Interior Lighting Design; London: HMSO, 1970

(11) Hopkinson, R.G.; Architectural Physics: Lighting; London HMSO, 1969

(12) Paix, D.; The Design Of Buildings For Daylight; Commonwealth Experimental Building Station, Sidney (AUS), et seq.; Bulletin No. 7, 1962

(13) Hamlin, T.; Architecture Through The Ages; New York: John Putnam's Sons, Inc., 1953

(14) Jordan, R.F.; A Concise History Of Western Architecture; New York: Harcourt, Brace and World, 1970

(15) Gregory, R.L.; The Eye and the Brain; London: World University Library, 1966

(16) Lam, W.M.C.; Perception and Lighting as Formgivers for Architects; New York: McGraw-Hill Book Company, 1977

(17) Held, E. and Richards, W., ed.; Perception: Mechanisms and Models; San Francisco: W.H. Freeman and Co., 1971

(18) Smith, P.R.; The Use of Lighting Appraisal Sessions in Architectural Education; Physical Environment Report PR 8, 1975, University of Sidney, Australia

(19) Bartenbach, C. and Daniels, K.; Integrierte Licht-Klima-Deckensysteme; Berlin: CIE (GERMANY), 1977

(20) Stein, R.; Energy in Building Construction; Report of study under ERDA contract E (11-1)-2791; University of Illinois: Center for Advanced Computations, 1974

(21) Courtney, R.G.; Energy Conservation in the Built Environment; London: CIB Press, 1977

(22) Hopkinson, R.G. and Collins, J.B.; The Ergonomics of Light; London: MacDonald Technical and Scientific, 1970

(23) Pleigel, G.; The Computation of Natural Radiation in Architecture; Meddelande No. 25; Stockholm, Sweden: Statens Namnd Fon Byggnadsforgking, 1962

(24) Lyles, J.; Principles of Natural Lighting; London: Elseven, 1969

(25) Robbins, C.L.; "Natural Lighting: New Research for an Old Environmental Concept", EDRA-9 Conference; University of Arizona, April 1978

(26) ASHRAE, Algorithms for Building Heat Transfer Subroutines; New York: ASHRAE, 1975

(27) Department of the Environment; Sunlight and Daylight; London: HMSO, 197

(28) Robbins, C.L. and Mukerjea, R.N.; "The Architectural Use of Natural Light as a Means to Energy Conservation"; Iowa Energy Policy Council, 1977

(29) ASHRAE; ASHRAE Standard 90-75, Energy Conservation in New Building Design; New York: ASHRAE, 1975

(30) Office of Conservation and Environment, Federal Energy Administration; "Lighting and Thermal Operations"; Washington: USGPO, 1976

(31) Marshall, H.E. and Ruegg, R.T.; 'Energy Conservation Through Life-Cycle Costing"; JAE, Vol. XXX, No. 3, Spring 1977

(32) Abrahamson, D.E. and Emming, S.; "Energy Conservation: Implications for Building Design and Operation"; University of Minnesota, Conference Proceedings, 1974

(33) Eÿzaguirre, C.; Physiology of the Nervous System; Chicago: Year Book Medical Publishers, 1969

(34) Porter, T. and Mikellides, B.; Color for Architecture; New York: Van Nostrand Reinhold Co., 1976

(35) Griffith, J.W.; "Daylight Utilization in Passive Solar Systems Using Life-Cycle Cost-Benefit Analysis"; Proceedings of the Passive Solar Heating and Cooling Conference, May 18-19, 1976. University of New Mexico, Albuquerque, New Mexico

RADIATIVE COOLING OF STRUCTURES

V. K. Gupta and H. R. Tyagi
Department of Physics
J.N. Agricultural University
Jabalpur, 482004, India.

ABSTRACT

The sky under cloud-free, low humidity conditions is about 15°C cooler than the ambient air temperature. A black-body radiator, facing the sky and shielded from direct radiation, will cool down by emitting thermal radiation as per Stefan-Boltzman law. The radiative cooling potential is available for cooling structures during summer and becomes more effective during nights. The cloud-less and dust-free atmosphere makes the radiative cooling quite effective and may even cause freezing of water during the nights. However, even under worst atmospheric conditions of clouds, dust and humidity, a heat transfer rate of 40 kcal/hr m^2 is obtainable.

A radiative cooler of 2.2 m^2 radiative surface, facing north sky 60° above horizon, was used to evaluate the rate of nocturnal radiative cooling in April at Jabalpur (India) which is at 23° Lat N. Air was blown through the cooler at 2.5 m^3 per minute and air temperatures were measured at the inlet and outlet. An average temperature drop of the order of 7°C was obtained throughout the night. The radiative heat transfer rate was found to be 138 kcal/m^2 hr under clear sky conditions and initial relative humidity of 40 percent.

INTRODUCTION

The devices at present available for cooling of buildings and houses are costly and their operation is also quite expensive because of the high cost of electricity. Therefore, particular importance is attached now-a-days to the study of passive cooling of houses. There are several techniques of passive cooling: i.e. wind towers; utilization of drop in temperature during nights; evaporation of liquid; and radiative cooling by exposing a black surface to the cool atmosphere (1,2).

Radiative cooling results if the black-body radiator exposed to sky does not interact appreciably with the ambient air through conduction and convection, but exchanges infra-red electromagnetic radiations with the atmosphere as per Stefan-

Boltzman law. Since the atmosphere has radiative temperature lower than the ambient air, the radiative heat loss from the radiator is more than the heat radiation it receives from the atmosphere which also acts as a black-body. As such the net effect is the cooling of the black radiator. This radiative cooling is considerably reduced if the air is humid. However, even with complete cloud cover, heat transfer rate of 40kcal/ hr m^2 has been observed (3,4).

The temperature of a black-body radiator(and the consequent radiative cooling) is very much modified by the transfer of heat by convection and conduction from the warm ambient air to the cooled radiative surface. The phenomenon of black-body radiation can be enhanced considerably if the air adjacent to the black-body radiator is absolutely still. Once the wind is still, the layer of air adjacent to radiative surface cools down by conducting its heat to the ground. This thin but heavy air layer forms an insulation and prevents the convective heat transfer and also minimizes the conductive heat transfer from the warm atmosphere; but it still allows the infra-red radiation emitted by the black-body. However, a slight breeze will disturb the air insulation and transfer heat from the warm environment to the black-body radiator, thus reducing its efficacy (5).

To reduce convective losses in areas where considerable breeze blows, the radiator surface is covered with a sheet of material having high transparency to heat energy, leaving about 4 cm thick layer of air in-between.Generally this material is not perfectly transparent to infrared radiation; consequently a part of the heat energy is radiated back, thus decreasing cooling efficiency. Polypropylene and polythene have excellent infra-red transmission, but both of these materials deteriorate quickly. Teflon which has high resistance to weathering and has 70 percent transparency for heat energy through thickness of 0.012 mm, is considered to be best wind screen for radiative cooling (6).

The performance of a black-body radiator can be improved further by using a selective coating whose optical properties match with those of the atmosphere. If the selective radiator is specular for almost all wavelengths except for those of 8-13 microns for which it behaves as a perfect black-body under operating temperatures in passive cooling systems, then out of the radiations emitted by the atmosphere, only a part falling in the 8-13 microns region will be absorbed by it. But the atmosphere is partially transparent to infra-red radiations, specially in the 8-13 microns region, it does not emit these wavelengths. Also the space behind the atmosphere has an equivalent radiative temperature of only a few degrees kelvin. As such the space in fact becomes a pure sink of infrared radiations from the black-body radiator. This will present two major advantages. First the equilibrium temperature of the radiator will be considerably lower; and secondly this temperature will not be modified by diffused solar radiation. The cooling effect can thus be achieved during the day also (7)

The temperature of the black-body radiator is also modified by the direct solar radiation falling on it. This can be minimized by shading the device from the direct solar radiation without preventing it from facing the sky for diurnal performance; and for nocturnal operation, the problem of direct or diffused radiation does not arise.

The day-time cooling of black-body radiator is thus possible only when (i) either it has a selective black surface as discussed above, or (ii) it is shielded from direct solar radiation. In central India during summer, the sun reaches an altitude of almost 90° above the horizon. So the shielding of black-body radiator becomes extremely difficult in the case of large cooling devices. As such in these areas in the absence of selective coating, the advantage of radiative cooling can be taken only during nights.

The nocturnal radiative cooling of buildings and structures has great potential in dry and hot climates of central India. Jabalpur city at 23° Lat. N. typically represents the climate of central India and is quite hot during summer. From April to middle of June, the day and night temperature varies between 20-45 degrees celsius. The ralative humidity is comparatively low and generally its maximum value is 50 % during the morning hours. However, during the nights the ralative humidity is about 45%. Wind is almost still during the nights.

Clear sky, low relative humidity and almost still wind are quite favourable conditions available during nights in central India for applications of nocturnal radiative cooling. With this view point an experiment was designed to evaluate the potential of cooling warm air by radiative cooling during nights.

SET-UP

Now-a-days structures that may allow year-round operations are recommended for solar energy utilization. The devices that utilize solar energy for heating during winter may, with little modification, be used for radiative cooling during summer (8).

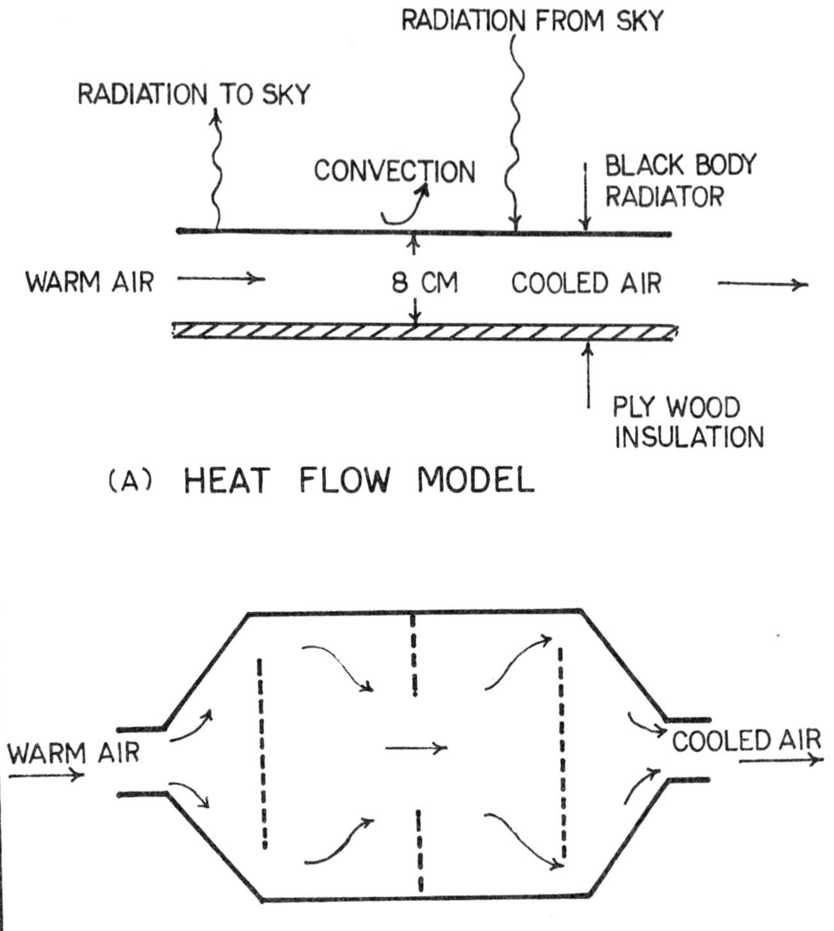

FIGURE 1. OPERATIONAL DETAILS OF RADIATIVE COOLER.

An experimental black-body radiator was built to evaluate its heat transfer rate and to find its capacity to cool the warm air. The radiator was a 2.2 m^2, dull- black mild steel sheet of 24 gauge. It covered a 8 cm deep box made of 2.5 cm thick wooden sides and 4 mm thick ply-wood base. Two holes of 8 cm diameter were drilled at the centres of the two opposite wooden sides; one for entry of warm air and second for the exit of the cooled air. The warm air was cooled by blowing it over the interior side of the radiator. To ensure that the air passes over the maximum surface area of the radiator, air distributors were installed inside.

The complete arrangement of the set-up is shown in Fig. 1(A) and Fig.1 (B). The temperatures of ambient air, warm air at outlet of blower, cooled air and radiator surface were measured with chromel-alumel thermocouples and a pyrometer. The portable blower had a capacity 2.5 m^3/min. The radiator was kept in open facing the sky and avoiding heat radiations from buildings and trees.

Wind screen was not used to cover the black-body radiator because of the following two reasons :

(i) In central India, the wind is generally still during summer nights. On the three nights when the experiment was being conducted the wind was almost calm and its velocity throughout the night was less than 1.5 km/hr. Therefore, it was considered that heat transfer rate due to convection from environment to the radiator would be smaller than the loss in its efficacy due to reflection of infra-red radiations by the wind screen.

(ii) Due to low humidity, clear sky and almost still wind, radiative cooling is quite massive during the nights. So the air in-between the radiator and the wind screen gets saturated quite quickly and condensation of water vapour takes place on the lower surface of the wind screen. This immediately reduces the radiative cooling rate because of its opacity to heat radiations.

RESULTS

The average values for three nights of temperatures of ambient air, air at the inlet of radiator, cooled air and radiator surface are shown in Table 1. All the heat losses undergone by the air blowing over the interior side of the radiator could be considered equal to the heat radiated to the night sky by the black-body radiator. Its calculated values are also shown

TABLE 1.

Heat transfer rate and temperatures of ambient air, air at outlet of blower, cooled air and radiator surface.

Time	Temp. of ambient air in °C.	Temp. of air at outlet of blower in °C.	Temp. of cooled air in °C.	Temp. of radiator surface in °C.	Heat transfer rate per hour in kcal.
1900 Hrs.	30.6	34.4	27.8	26.1	304.1
2000 Hrs.	30.0	33.9	26.7	25.6	331.8
2100 Hrs.	28.9	33.3	26.1	24.5	331.8
2200 Hrs.	27.8	32.2	25.6	25.0	304.1
2300 Hrs.	26.1	31.1	24.4	23.9	308.7
2400 Hrs.	25.6	30.6	23.9	22.8	308.7
0100 Hrs.	25.0	29.4	22.8	22.2	304.1
0200 Hrs.	25.6	29.4	22.8	22.2	304.1
0300 Hrs.	25.6	29.4	23.3	22.2	281.1
0400 Hrs.	25.0	30.0	23.3	22.2	308.1
0500 Hrs.	26.1	31.7	25.6	23.9	281.1
0600 Hrs.	27.8	32.2	26.1	24.4	281.1

in Table 1. The ambient air on leaving the blower was slightly warmed up.

Under the clear sky conditions and a relative humidity of 40 percent, the temperature of the warm air dropped by over 6.6° C. The average temperature of the surface of the radiator was 3.3°C lower than that of the ambient air and 7.7°C lower than that of warm air entering the radiator. The radiative heat transfer rate is found to be 138 kcal/m^2 hr.

The cooling performance of conventional cooling appliances are generally evaluated by calculating its coefficient of performance (COP) i.e. the ratio of heat transfer rate to power input. The cooling performance of black-body radiators can also be tested by determining their COP. In the present experimental study the blower that forced the warm air inside the radiator consumed the power at the rate of.17 kw-hr and the radiative heat transfer rate was .35 kw-hr. Thus the average value of COP comes out equal to 2.1 which is quite close to the peak COPs of 2 - 2.5 for conventional air cooling machines (9).

DISCUSSION

Although the radiative cooling phenomenon of black surface exposed to a clear sky can be used as an energy source, but still sufficient attention has not been devoted to this possibility because of the following two diffculties; the incomplete transparency of the atmosphere and considerably low power density involved in this phenomenon. However, this method is quite advantageous for moderate coolings because it does not require the use of an engine for lowering the temperature and its design is also quite simple. Another advantage of this system is that it can operate effectively during day and night if either selective radiator is used or the black radiator is shielded from direct solar radiation for day operations. For these reasons black-body radiator can be employed to cool the buildings and structures where low to moderate cooling is required. However, before putting it for mass-use its efficiency is to be increased. An extensive research in the areas of their design, wind screen, selective coating etc. is to be undertaken for this purpose.

ACKNOWLEDGEMENT

Kind encouragement and facilities offered by Prof. S.V. Arya, Dean, Faculty of Agricultural Engineering and help by Mr. N.K. Tiwari, Asstt. Prof. in Physics, is acknowledged with thanks.

REFERENCES

1. F. Daniels, "Direct Use of the Sun's Energy" Yale University Press, New Haven (1964).

2. M. N. Bahadori, "Thermal Performance of Old Buildings in Central Iran Employing Air Towers", Presented at the Passive Solar Heating and Cooling Conference, Los Alamos Scientific Laboratory, Albuquerque, New Mexico, May 1976, pp 157-159 of the Proceed. LA-6637-C.

3. R. W. Bliss, "Atmospheric Radiation Near the Surface of the Earth", Solar Energy 5, 103-120 (1961).

4. K. Ya Kondratyev, "Radiative Heat Exchange in the Atmosphere", Pergamon Press, Oxford (1965).

5. H. R. Hay and J. I. Yellot, "Natural Airconditioning with Roof Ponds and Movable Insulation". Trans ASHRAE 75, Part 1, 165-177 (1969).

6. T. E. Johnson, "Radiation Cooling of Structures with Infrared Transparent Wind Screens", Solar Energy 17, 173-178 (1975).

7. B. Bartoli, S. Catalanotti, B. Coiuzzi, V. Cuomo, V. Silvestrini and G. Troise, "Nocturnal and Diurnal Performances of selective Radiators", Applied Energy (3) (1977).

8. N. Laing, "The Use of Solar and Sky Radiation for Air-Conditioning of Pneumatic Structures", Proc. First International Colloquium on Pneumatic Structures, International Assoc. for SHELL structures, p. 103 (1967).

9. Handbook of Air Conditioning Fundamentals. ASHRAE, New-York (1972).

...

SOLAR COLLECTOR RELATED RESEARCH AND DEVELOPMENT
IN THE UNITED STATES
FOR HEATING AND COOLING OF BUILDINGS*

R. K. Collier
Los Alamos Scientific Laboratory
Los Alamos, NM 87545

ABSTRACT

This paper describes some of the research funded by the Research and Development Branch of the Heating and Cooling Division of Solar Energy of the United States Energy Research and Development Administration.† Specifically, this paper deals with collector and collector materials research reported on during FY-1977. The R&D Branch has funded research in open and closed cycle liquid heating flat plate collectors, air heating flat plate collectors, heat pipe collectors, concentrating collectors, collector heat transfer studies, honeycomb glazings, evacuated tube collectors, ponds both salt gradient and viscosity stabilized, materials exposure testing, collector testing standards, absorber surface coatings, and corrosion studies. A short description of the nature of the research is provided as well as a presentation of the significant results.

INTRODUCTION

The following is a description of research funded by the Research and Development Branch of the Heating and Cooling Division of Solar Energy of the United States Energy Research and Development Administration. Specifically, that research dealing with collectors and collector materials which was reported on during FY-77. A short description of the nature of the research is provided as well as a presentation of the significant results.

Los Alamos Scientific Laboratory[1]

This collector was developed as a roof integrated structure for the National Security and Resources Study Center (NSRSC) in Los Alamos. It consists of two steel sheets seam-welded together and then separated by high pressure air to form a quilted pillow-like device. An electroplated black chrome selective surface is applied to that portion utilized as the absorber with the remainder formed such that it also serves as

*Work performed under the auspices of the U. S. Department of Energy.

†Presently Research and Development for Heating and Cooling, Office of the Assistant Secretary for Conservation and Solar Applications, U. S. Department of Energy.

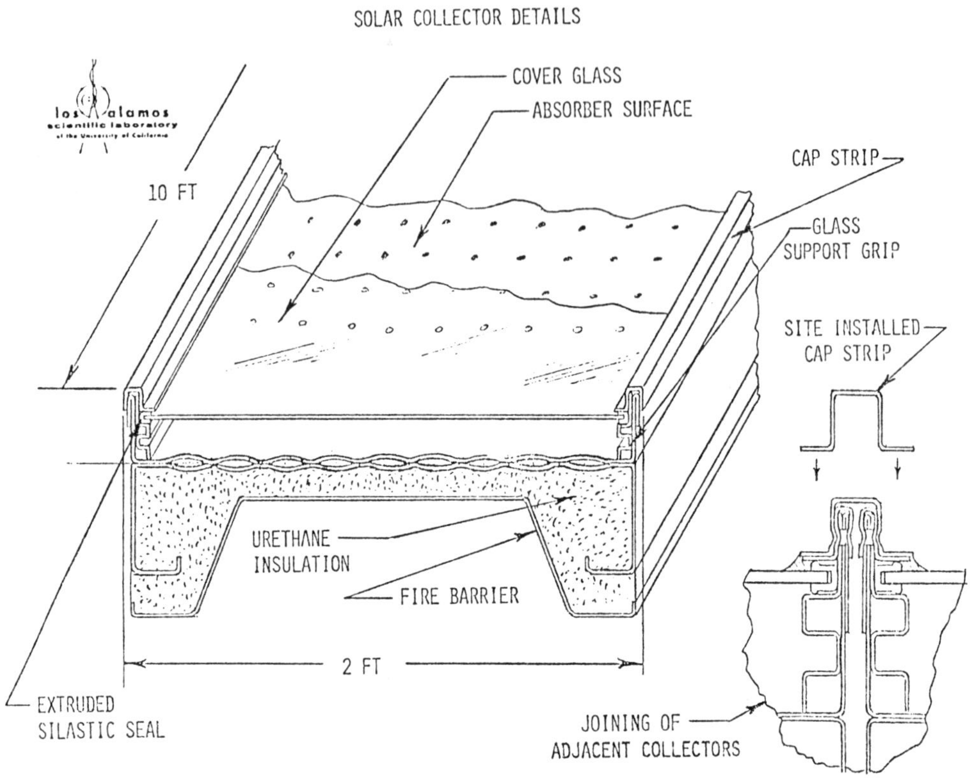

Fig. 1. Cross-sectional view of the LASL roof integrated solar collector.

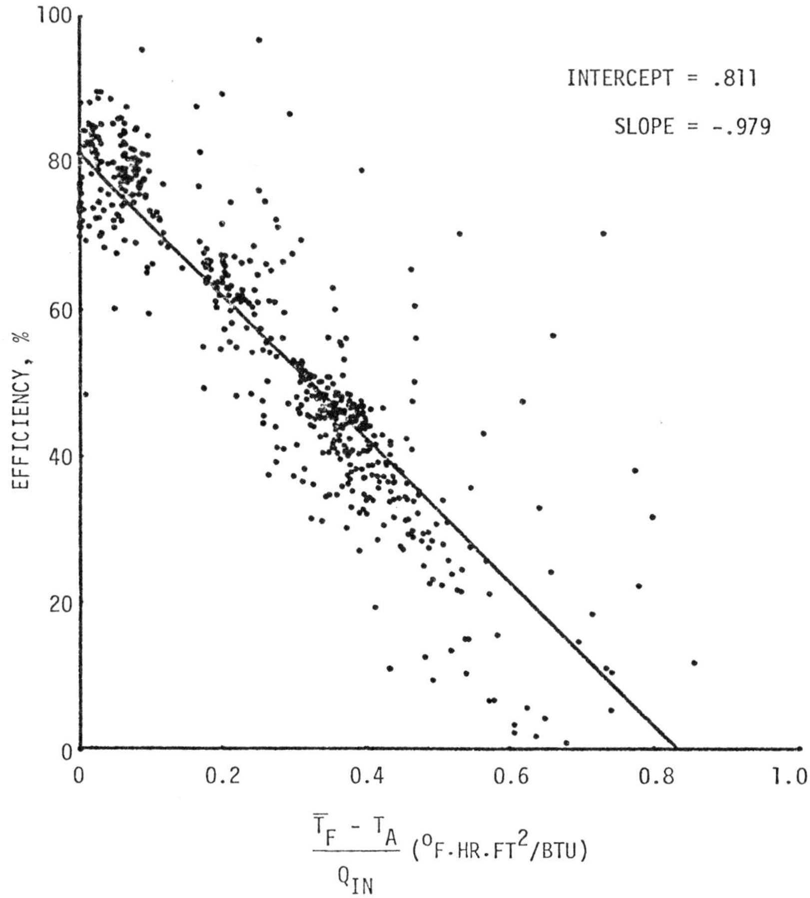

Fig. 2. Instantaneous thermal performance of the LASL roof integrated solar collector.

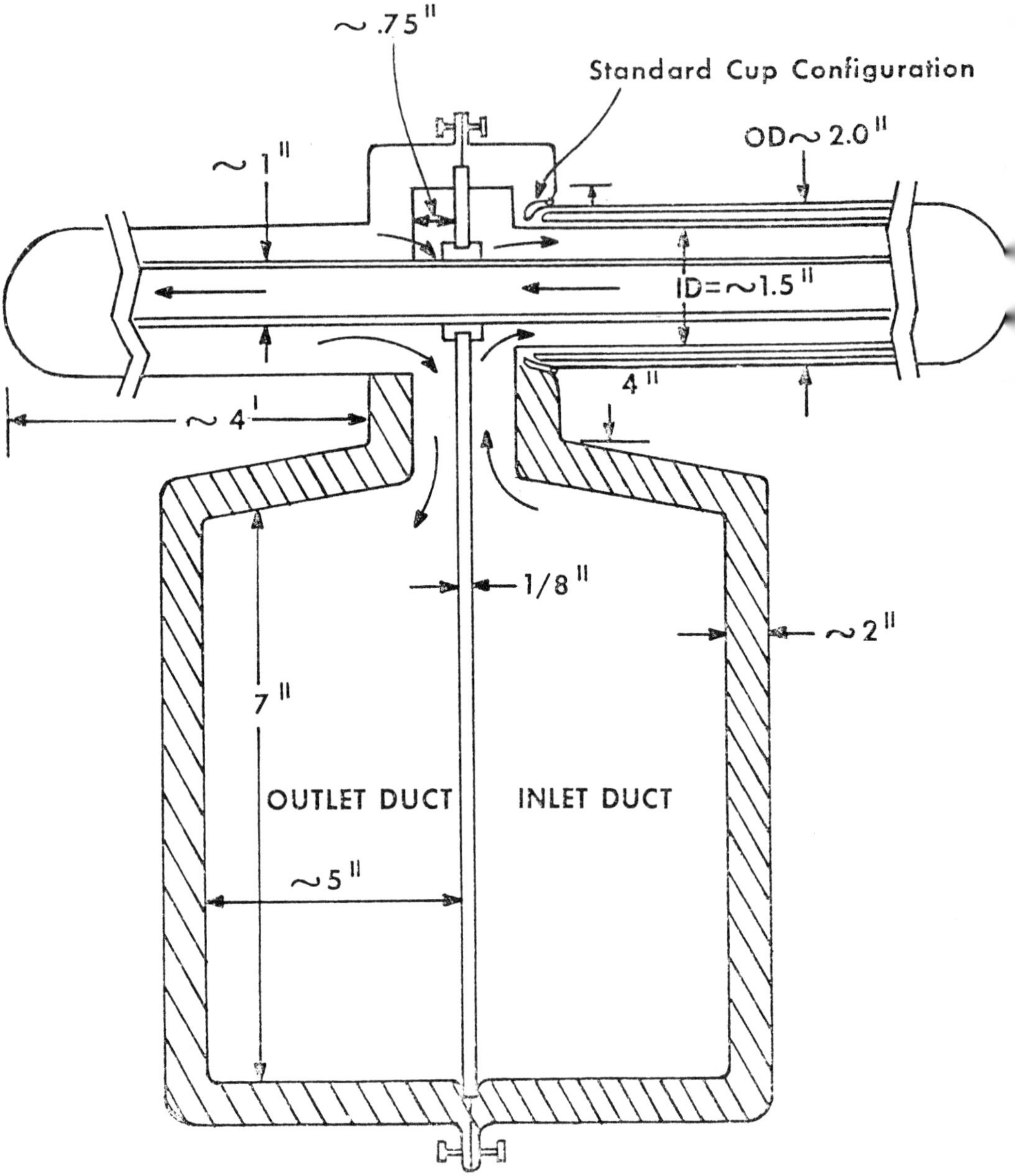

Fig. 3. Cross-sectional view of the Owens-Illinois air heating collector tube pair and manifold.

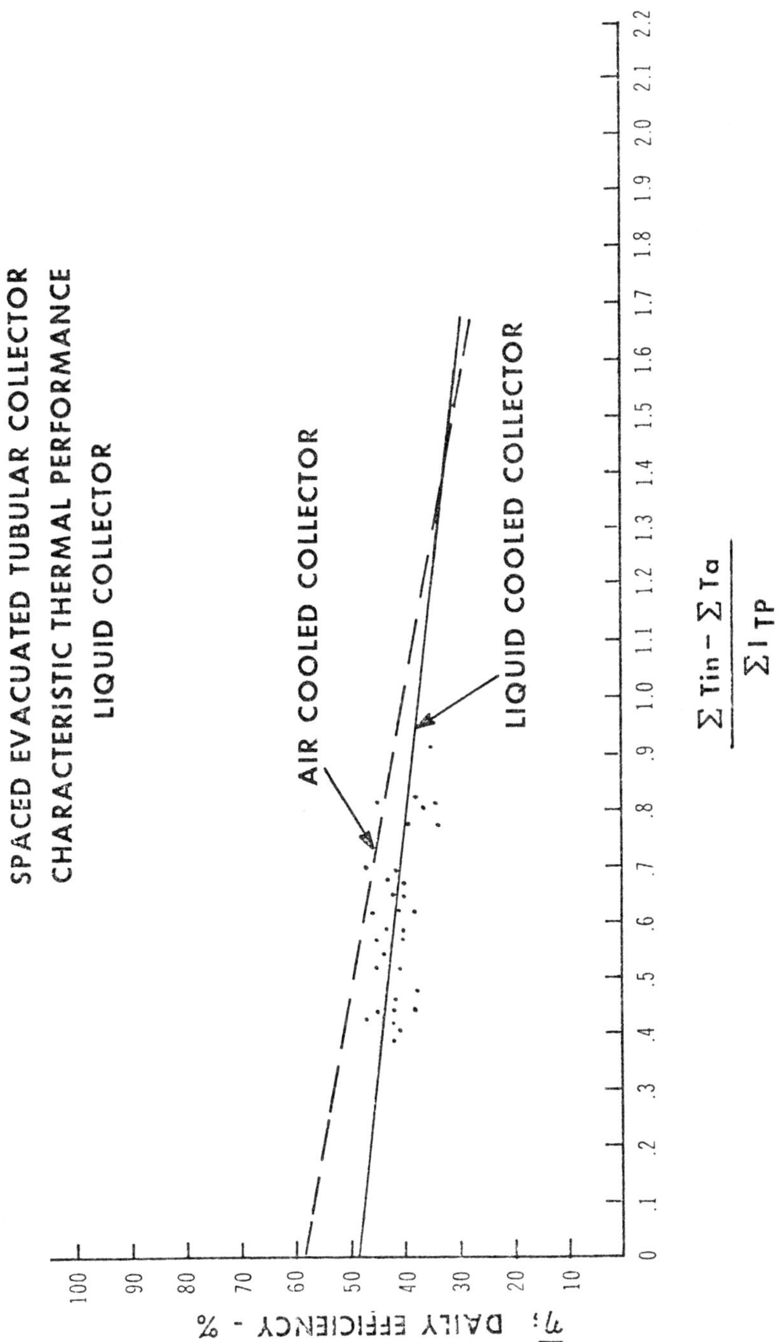

Fig. 4. Comparison of the daily thermal performance between the Owens-Illinois air and water heating collectors.

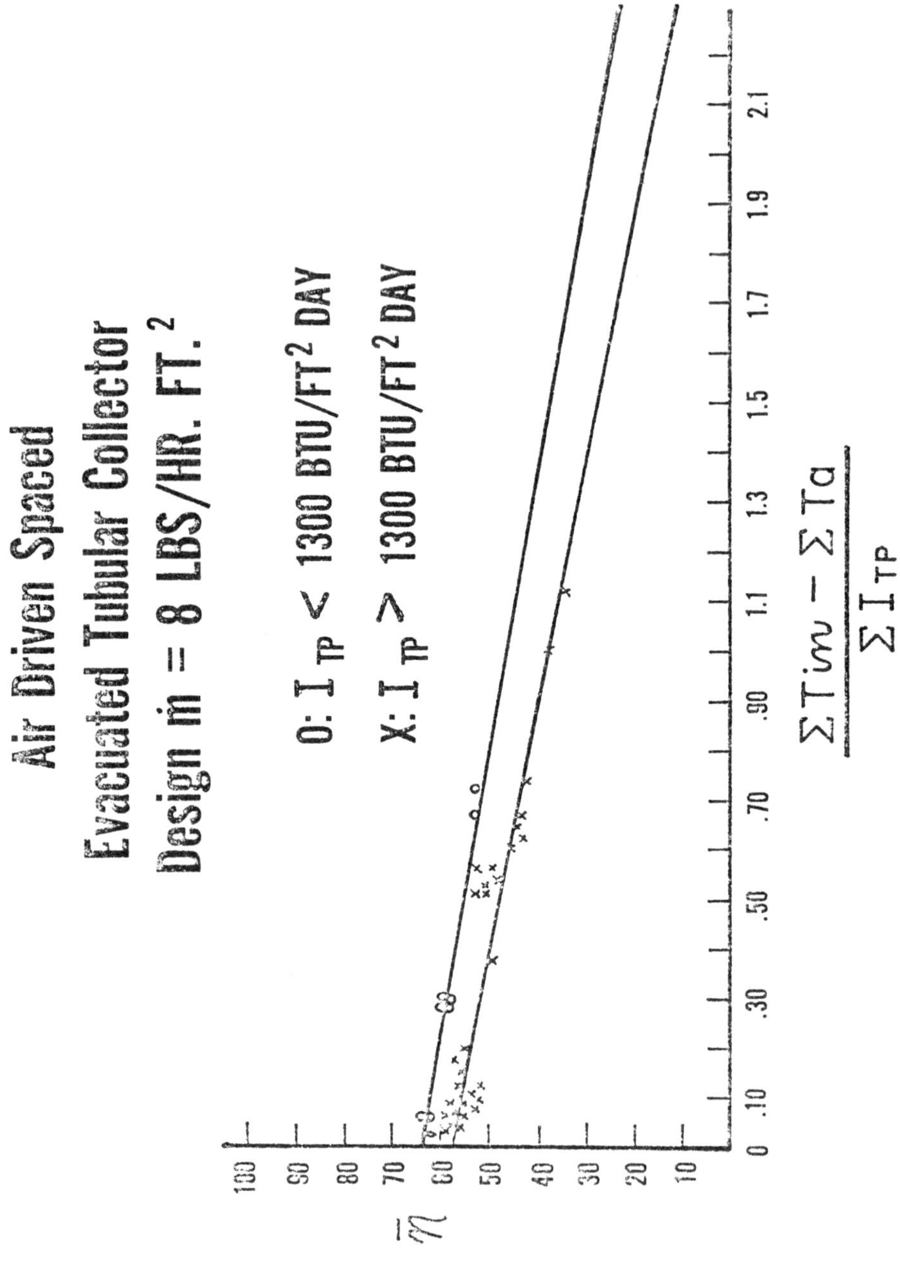

Fig. 5. Effect of insolation on the daily thermal performance of the Owens-Illinois air heating collector

the building roof when properly assembled together. A cross-sectional view of the collector as well as its instantaneous thermal performance is shown in Figs. 1 and 2, respectively.

Owens-Illinois[2]

Owens-Illinois has completed testing of their evacuated tube air heating collector array. The design is essentially identical to their Sunpak^R liquid heating collector except that the delivery tube has been increased to 1" o.d. The material for the delivery tube is still borosilicate glass. A cross-sectional view of a tube pair and manifold arrangement is shown in Fig. 3.

The thermal performance of the air cooled collector compared to the liquid cooled collector is shown in Fig. 4. Due to the fact that these arrays utilized a diffuse background, the collector array will perform better in the morning and afternoon than at noon and will perform better in diffuse radiation than in direct. This is shown in Fig. 5. Notice that the efficiencies shown are daily efficiencies.

Corning Glass Works[3]

Corning has completed the design, manufacture and testing of an evacuated glass tube collector utilizing a copper heat pipe absorber and an internal cusp type reflector. A cross-sectional view of the concept is shown in Fig. 6. The instantaneous thermal performance is shown in Fig. 7 compared to that of the conventional Corning evacuated flat plate design.

Lockheed[4,5]

Lockheed has developed plastic honeycombs for use as convection suppression in flat plate solar collectors. They examined two processes and three materials. The processes examined were the Hexcell expanded cell process and the Norfield thermal forming process. The materials investigated were mylar, lexan and teflon. Of the above combinations, the expanded cell lexan was found to be the most cost effective. A comparison of both instantaneous and daily efficiencies for different collector designs are shown in Figs. 8 and 9. The superior performance of the flat black absorber with lexan honeycomb even during off-axis illumination should be noted. An unresolved problem is the lifetime of the plastic honeycomb under actual operating conditions.

UCLA[6]

The UCLA effort is parallel with that of Lockheed except that glass honeycombs are being developed rather than plastic honeycombs. Initial studies were performed using glass tubes standing on end and sintered together to form a honeycomb structure. Later studies were performed using very thin sheet formed into a wavey pattern and then stacked on end to form a honeycomb structure. A comparison of the instantaneous thermal performance of various collector designs is shown in Fig. 10. Again, the honeycomb, flat

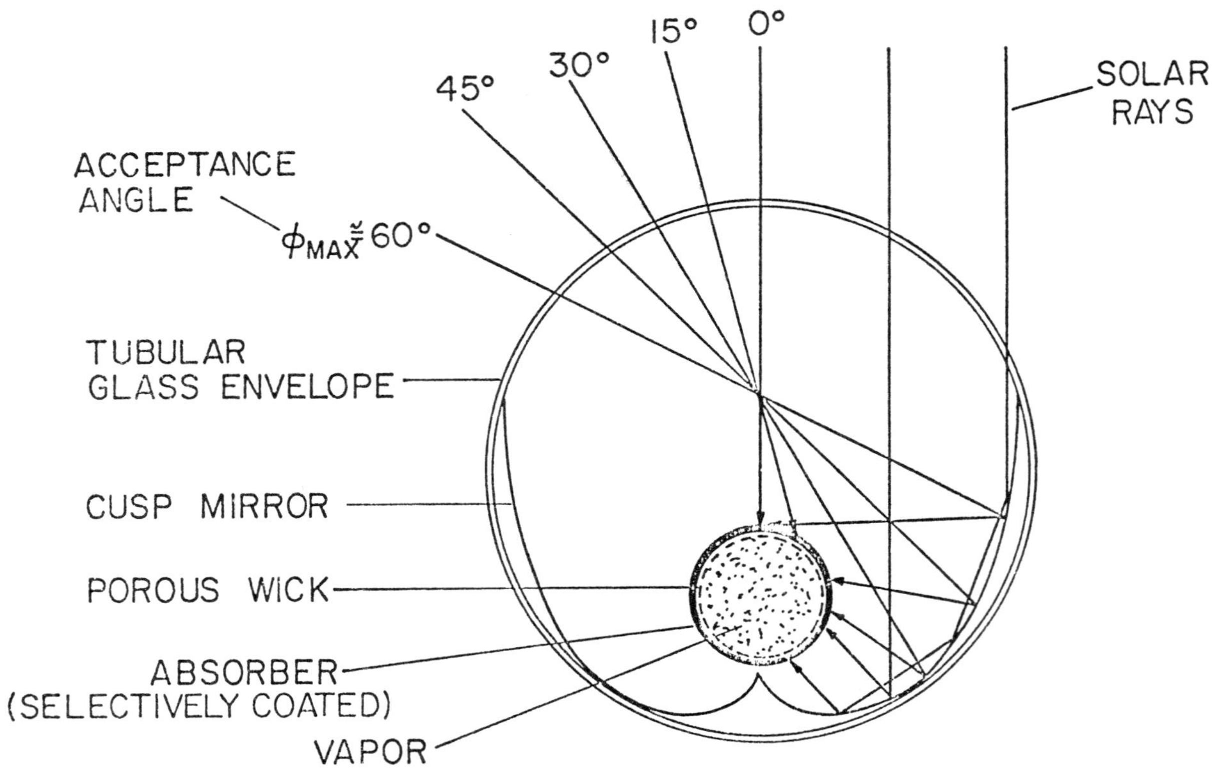

Fig. 6. Cross-sectional view of the Corning tubular cusp collector with heat pipe absorber.

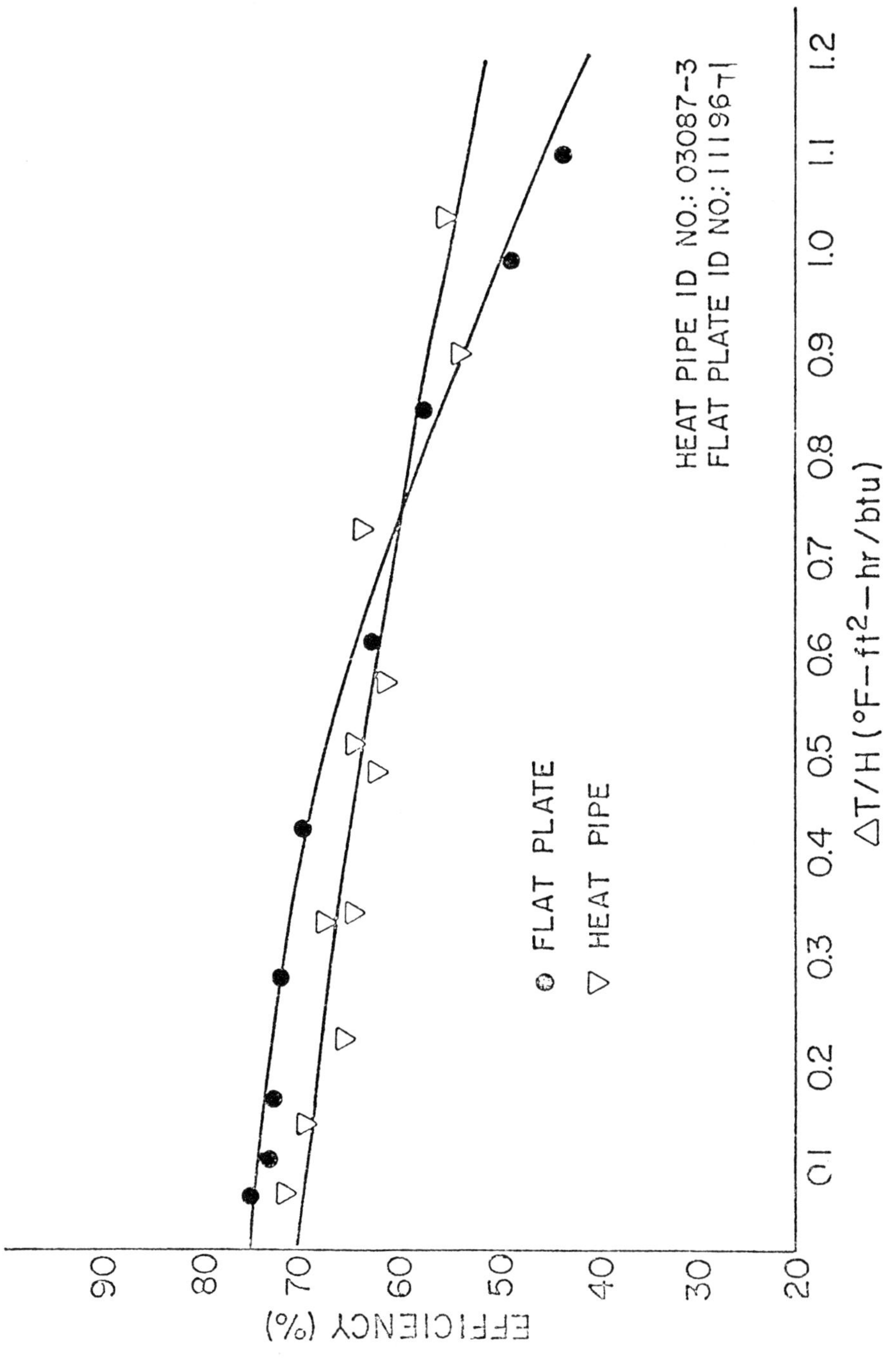

Fig. 7. Comparison of the instantaneous thermal performance between the Corning evacuated flat plate collector and the cusp-heat pipe collector.

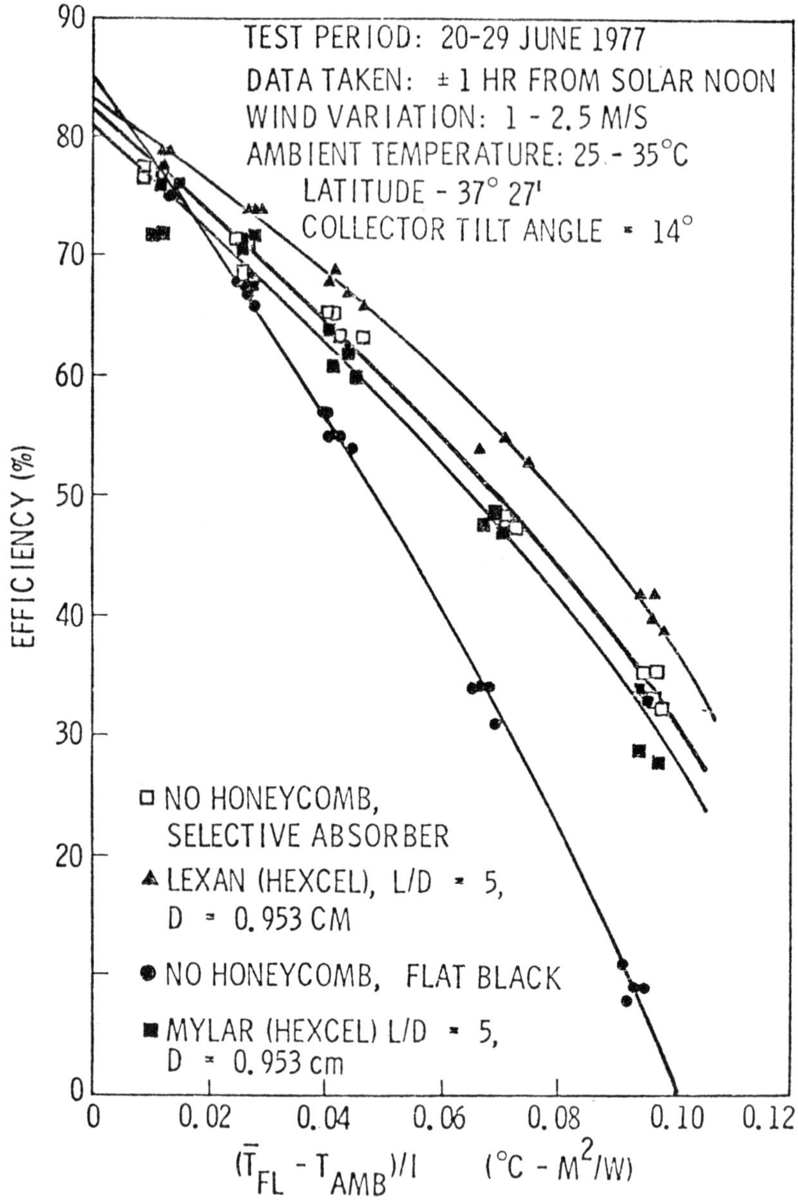

Fig. 8. Comparison of efficiency between a glass covered flat-black collector with and without Hecel lexan honeycomb and a selective black no honeycomb collector.

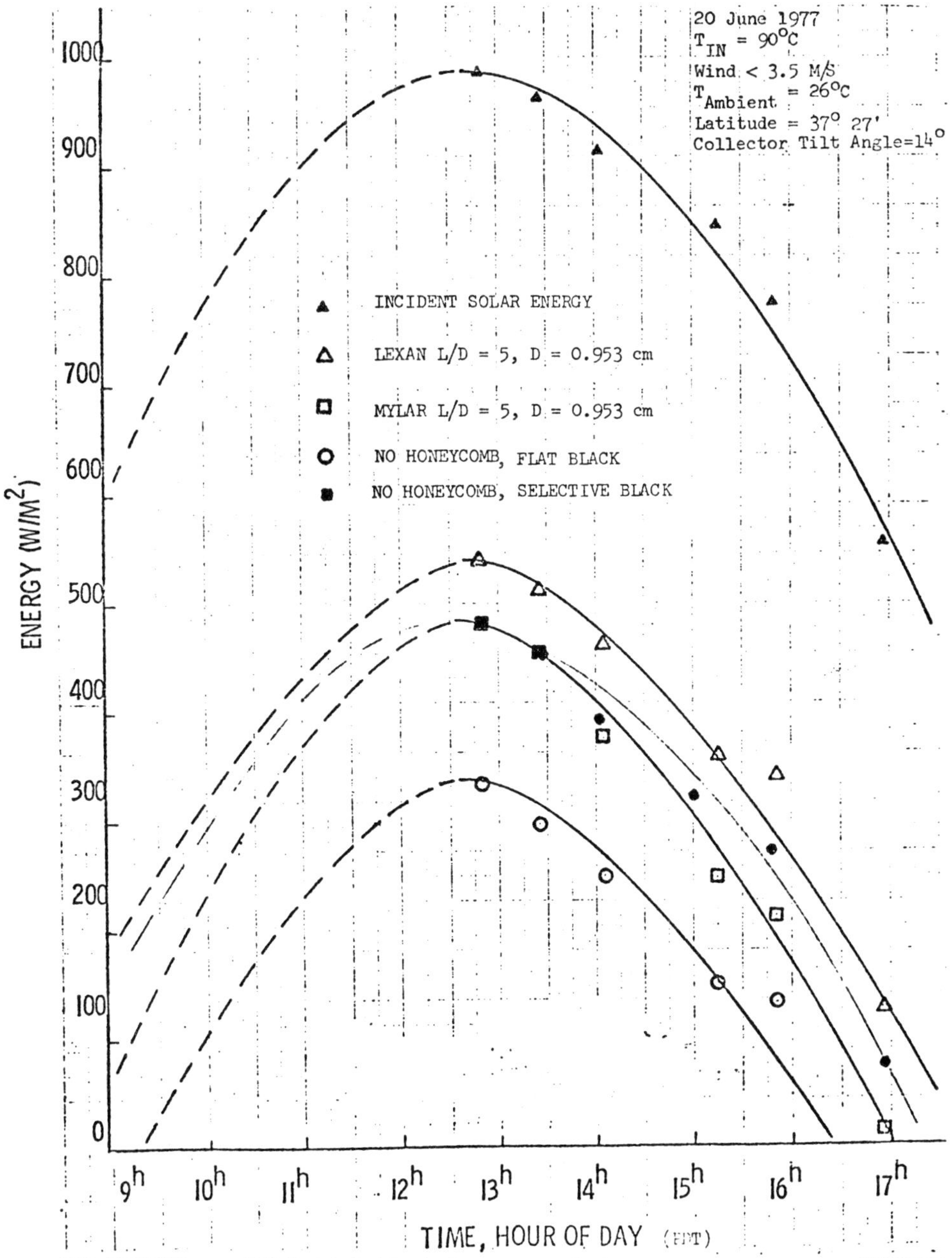

Fig. 9. Comparison of diurnal performance between glass covered flat black honeycomb collector, and a selective black no honeycomb collector.

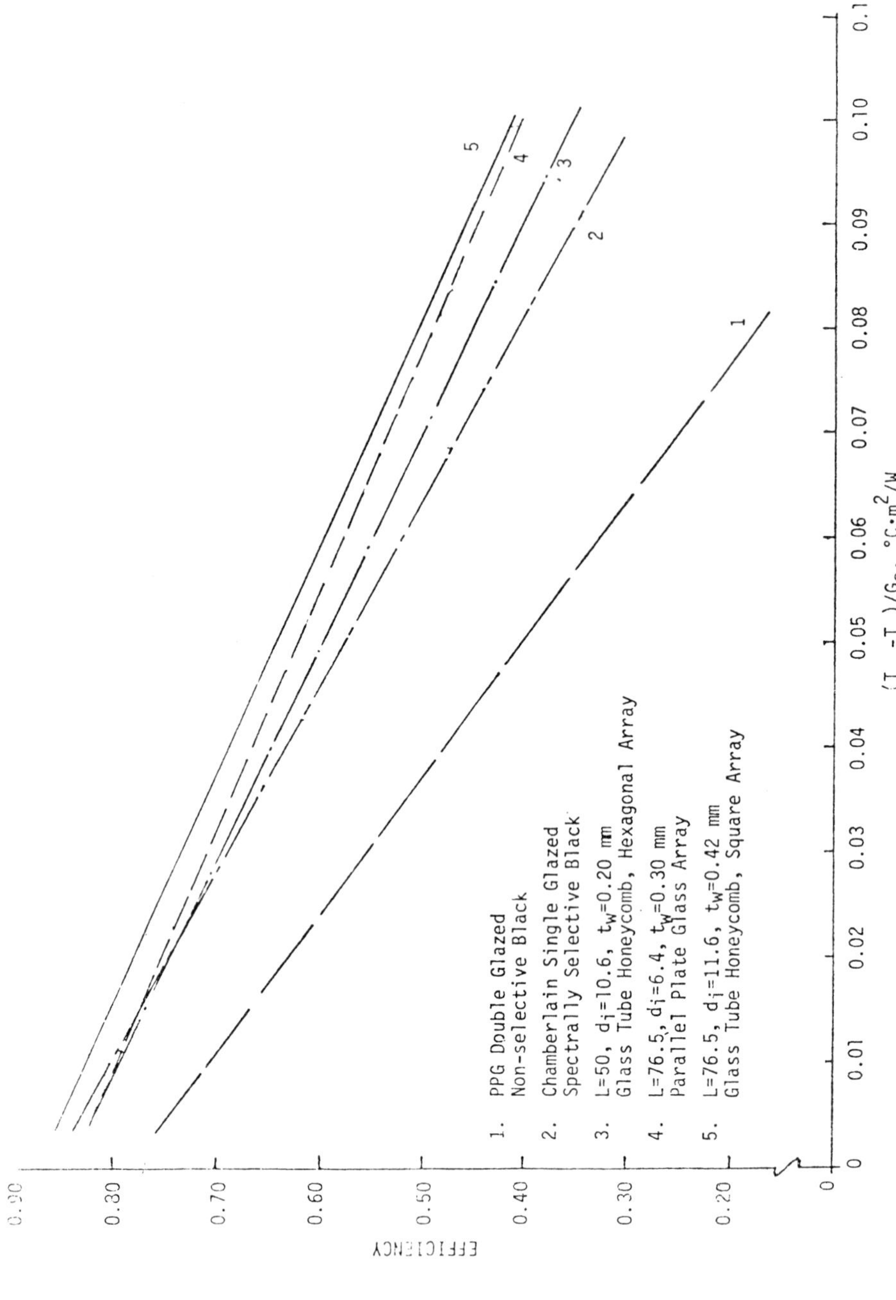

Fig. 10. Comparison of the instantaneous thermal performance of various collectors with and without glass honeycombs.

TABLE I

OVERALL CORROSION RATES OF ALUMINUM

AT 210°F UNDER LAMINAR FLOW

Solution	Alloy	R_{corr} (mpy)
Pure E.G./H_2O (35 Vol.%)	1100	0.19
	99.9%	0.16
	3003	0.13
	3004	0.10
E.G./H_2O + Cl^- (200 ppm)	1100	8.5
	99.9%	7.0
	3003	9.9
	3004	3.5
E.G./H_2O + Cu^{+2} (200 ppm)	1100	5.5
	99.9%	17
	3003	4.5
	3004	14
E.G./H_2O + Fe^{+3} (200 ppm)	1100	30
	99.9%	18
	3003	14
E.G./H_2O + Fe^{+3} + Cl^- + Cu^{+2} (200 ppm each)	99.9%*	71
	3003	74
	3004	120

* Under turbulent flow.

TABLE 2

PITTING CORROSION OF ALUMINUM IN AQUEOUS ETHYLENE GLYCOL SOLUTIONS AT 210°F UNDER LAMINAR FLOW

Solution	Alloy	$K(\mu/day^{1/3})$	Estimated Penetration Depth for Actual Plant in 20 Yrs. (mils)*
Pure E.G./H_2O	1100	2	15
(35 vol. %)	99.9%	3	20
	3003	1	15
	3004	0	0
E.G./H_2O + Cl^-	1100	37	300
(200 ppm)	99.9%	2	15
	3003	24	190
	3004	17	140
E.G./H_2O + Cu^{+2}	1100	47	380
(200 ppm)	99.9%	36	290
	3003	19	150
	3004	171	1400
E.G./H_2O + Fe^{+3}	1100	5	42
(200 ppm)	99.9%	5	42
	3003	3	25
E.G./H_2O + Cl^- + Cu^{+2} + Fe^{+3}	99.9%	132	1100
(200 ppm each)	3003	187	1500

*Based on a plant with 1.5×10^4 cm^2 of wetted surface area.

black design is shown to have superior performance. By using glass honeycombs, the problem with long-term durability is solved. Studies are continuing to develop cost effective glass honeycomb fabrication techniques.

Argonne National Laboratory[7]

Argonne has pioneered concentrating collector development of the Winston "ideal" designs. Their latest effort has been to design, build and test two collector module designs. One is a 3X CPC design and the other is a 1.5X cusp design. The 3X module utilizes aluminized plastic reflectors and Corning evacuated flat plate receivers while the 1.5X module utilizes Kinglux[R] aluminum reflectors and G.E. TC-100 evacuated tube receivers. The instantaneous thermal performance for each design is shown in Figs. 11 and 12, respectively. The possible performance of the 1.5X design with improved receiver tubes is shown in Fig. 13.

Jet Propulsion Laboratory[8]

JPL has designed, built and tested a concentrating collector of nominally 2X concentration which utilizes flat trough type reflectors which can be reversed twice a year to give year round solar acceptance. A schematic of the design is shown in Fig. 14. The reflectors in their reversed position are shown by the dotted lines.

Corning flat plate evacuated collectors are used as receivers in this collector. Performance curves for various reflector materials are given along with the performance of a "bare" tube in Fig. 15. It should be noted that the comparison is made on aperture area. Therefore it would take twice as many evacuated tubes to form the same area array as that for the concentrating collectors. The thermodynamic breakeven point may be at 300°F temperature difference but the cost effective breakeven point may be much lower. Data collection on the present test array is continuing.

Giner, Inc.[9]

Giner has determined the corrosion resistance of aluminum alloys in water/ethylene glycol mixtures under experimental conditions directly relevant to the operation of existing solar thermal energy collector systems. An aqueous ethylene glycol solution with 35% ethylene glycol by volume was chosen as the typical heat transfer medium. The results of overall and pitting corrosion rates on various aluminum alloys are shown in Tables 1 and 2. Notice that the pure solution yields very low corrosion rates while the presence of ions (Cl, Cu^{+2} and Fe^{+3}) increases corrosion, especially pitting, manyfold. Glycol decomposition products do not seem to accelerate the corrosion rate. However, the corrosion rate is pH sensitive and should be checked periodically. Organic heat transfer fluids, such as Therminol 55 are inert with respect to aluminum. Even with the addition of the deleterious contaminants, such as Cu^{+2}, Cl' and Fe^{+3} ions at 200 ppm each, minimal corrosion was encountered. The next phase of this project is to develop corrosion inhibitor schemes to make

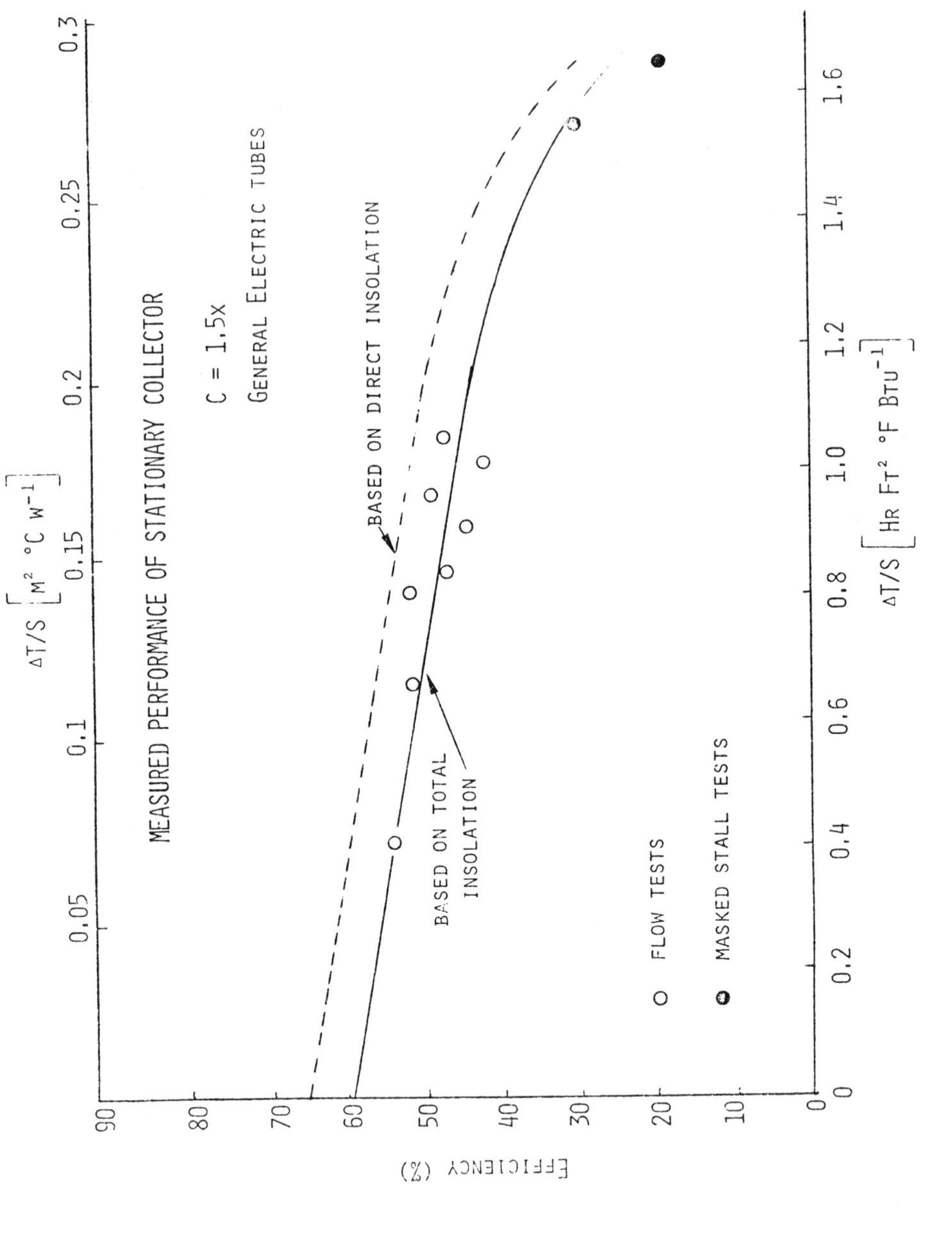

Fig. 11. Performance curve for 1.5 X cusp collector.

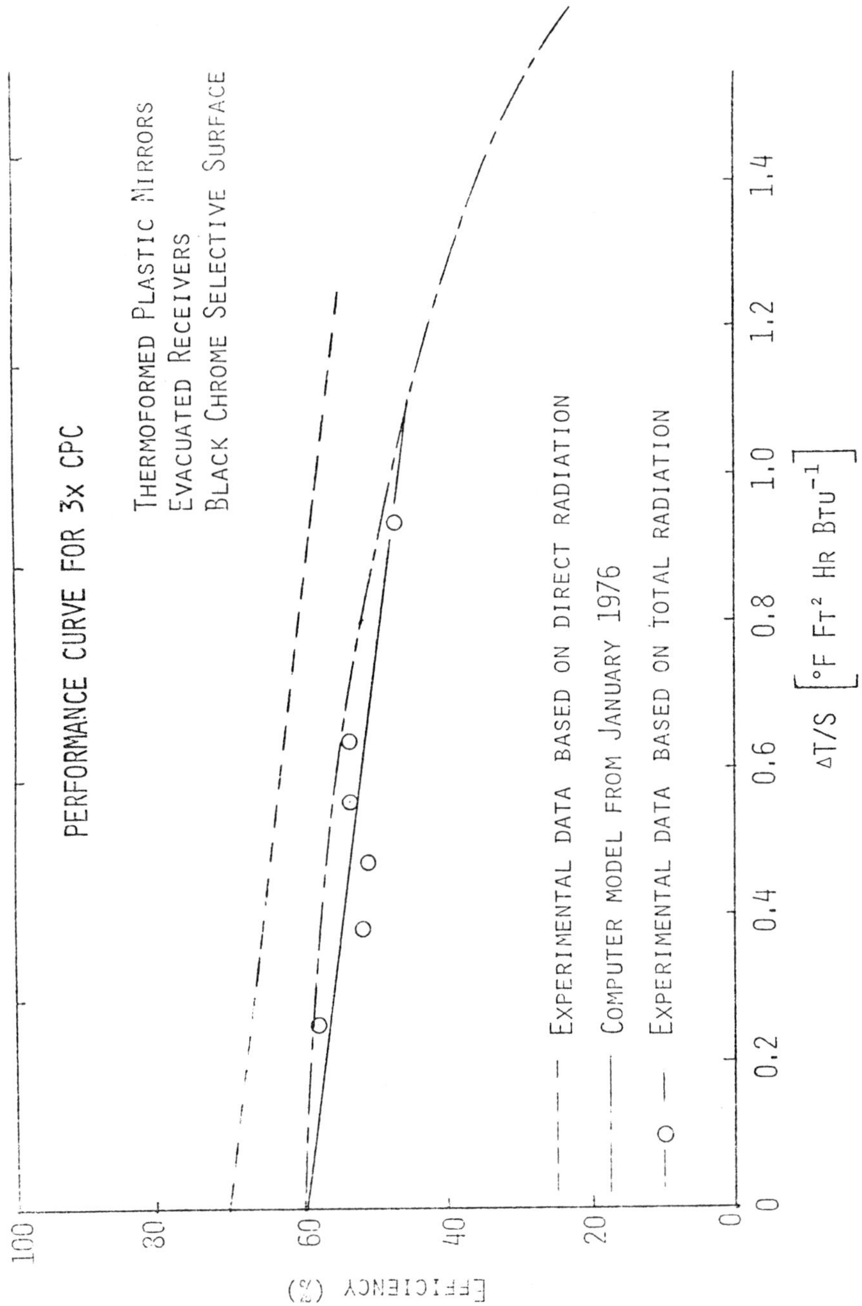

Fig. 12. Performance curve for 3X CPC Collector.

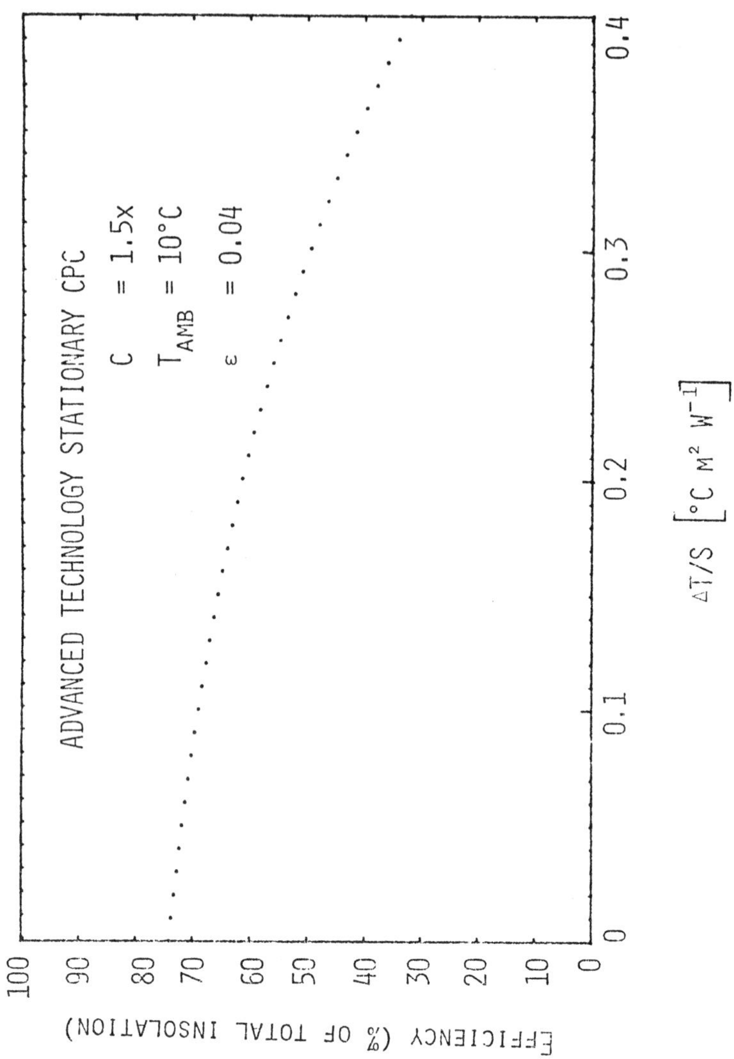

Fig. 13. Predicted performance of an "advanced design" 1.5X cusp collector.

Dimensions of vacuum tube receiver courtesy of Corning Glass Works, Corning, N.Y.

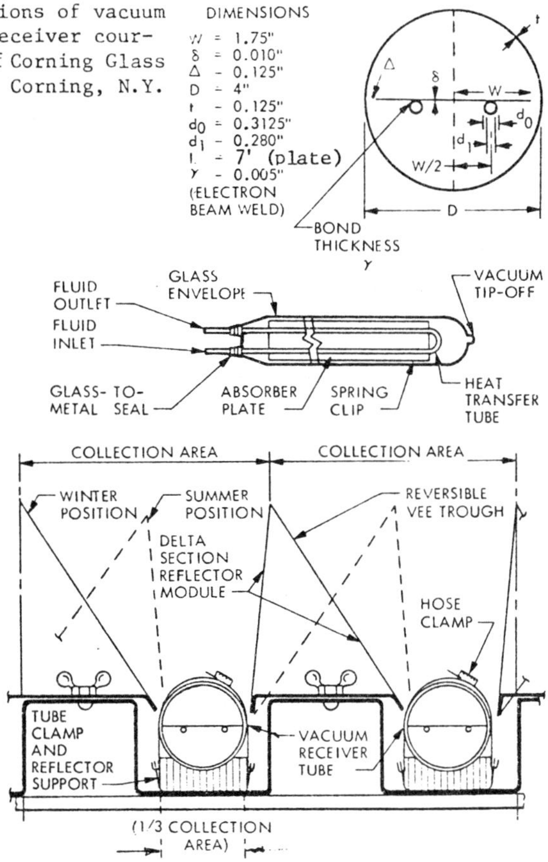

Fig. 14. Reversible asymmetric vee-trough collector with vacuum tube.

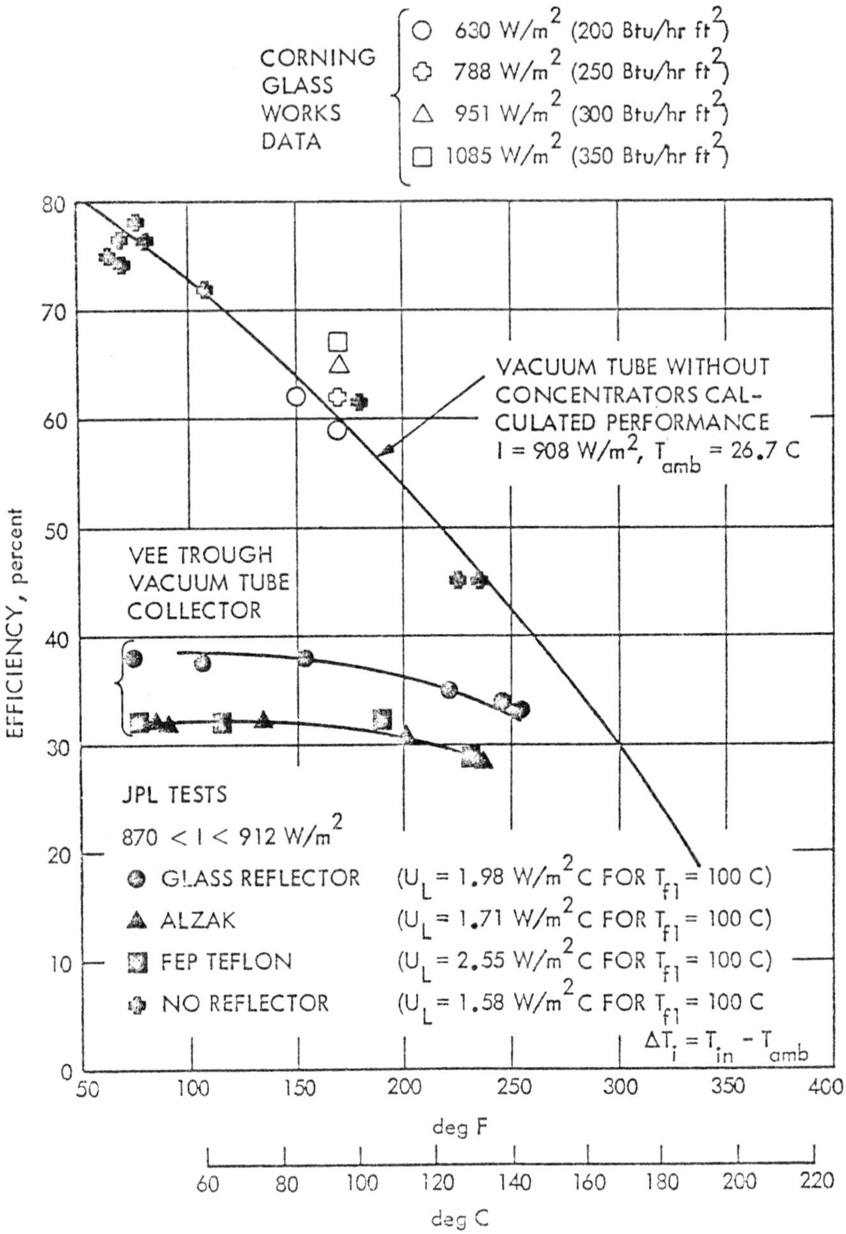

Fig. 15. Test data for collector efficiency versus temperature of the reversible vee-trough collector.

the aluminum/water/ethylene glycol system work satisfactorily as components of a practical and economical solar collector system.

University of Wisconsin[10]

Local and average heat transfer coefficients for natural convection between parallel plates have been experimentally studied using interferometric techniques. The experimental conditions were similar to those existing in flat plate solar collectors. Figure 16 shows a schematic of the test facility that provides a two-dimensional enclosure with isothermal surfaces. Figure 17 shows the results for the conduction regime. That is, the Grashoff number is sufficiently low such that no mass transfer occurs across the spacing. Notice that considerable convection takes place at the leading edge. Figure 18 shows typical results for the free convection regime as a function tilt angle. The effect of aspect ratio was investigated. The local Nusselt number was found to change but the integrated average Nusselt number was insensitive to aspect ratio. For the range of tilt angles between 45 and 90 degrees, the empirical correlation

$$\overline{N_{UL}} = 0.118\, [R_{aL} \cos^2 (\phi - 45°)]^{0.29}$$

adequately represents the data.

H = plate length	N_{UL} = average Nusselt number
A = aspect ratio, H/L	R_{aL} = Rayleigh number
L = plate spacing	ϕ = tilt angle

Lehigh University[11]

Two solar collectors using a pebble bed design were constructed and tested using air as the heat exchange medium. One collector had pebbles of metallurgical grade coke, and the other had a frothy volcanic material called scoria. The pebble size for both materials ranged from 6 to 19 mm in diameter. The pressure drop and mass transfer data for the two collectors are shown in Table 3. Typical collector efficiencies are shown in Figs. 19 and 20.

University of Waterloo Research Institute[12]

Two studies were performed both of which are related to methods for improving the efficiency of flat plate solar collectors.

The first study dealt with the free convective heat loss from a V-corrugated absorber plate to a plane transparent (glass) cover. Nusselt numbers were determined as a function of Rayleigh number for angles of inclination of the plate from the horizontal of 0, 30, 45 and 60 degree. The values of the depth ratio (the ratio of the average depth of the air layer to the depth of the V's) investigated were 1, 2.5 and 4; the angle

TABLE 3

Pressure Drop and Mass Transfer for Collectors

	Low Air Flow		High Air Flow	
	Scoria	Coke	Scoria	Coke
Pressure Drop				
in H_2O	0.06	0.24	0.42	0.59
Mass Transfer				
$(m^3/s.m^2).10^{-2}$	1.90	2.35	4.20	3.65

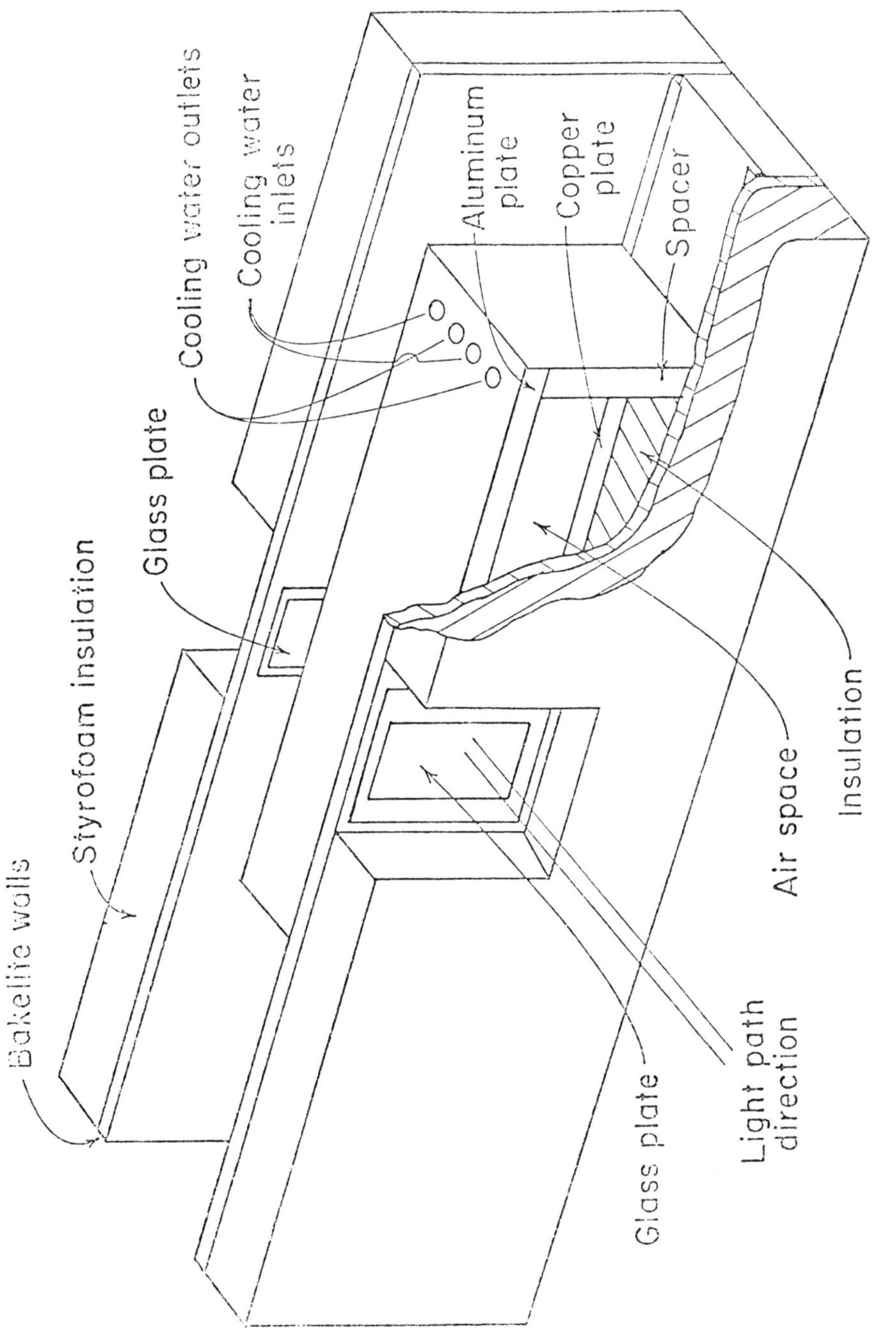

Fig. 16. Schematic of the University of Wisconsin test apparatus.

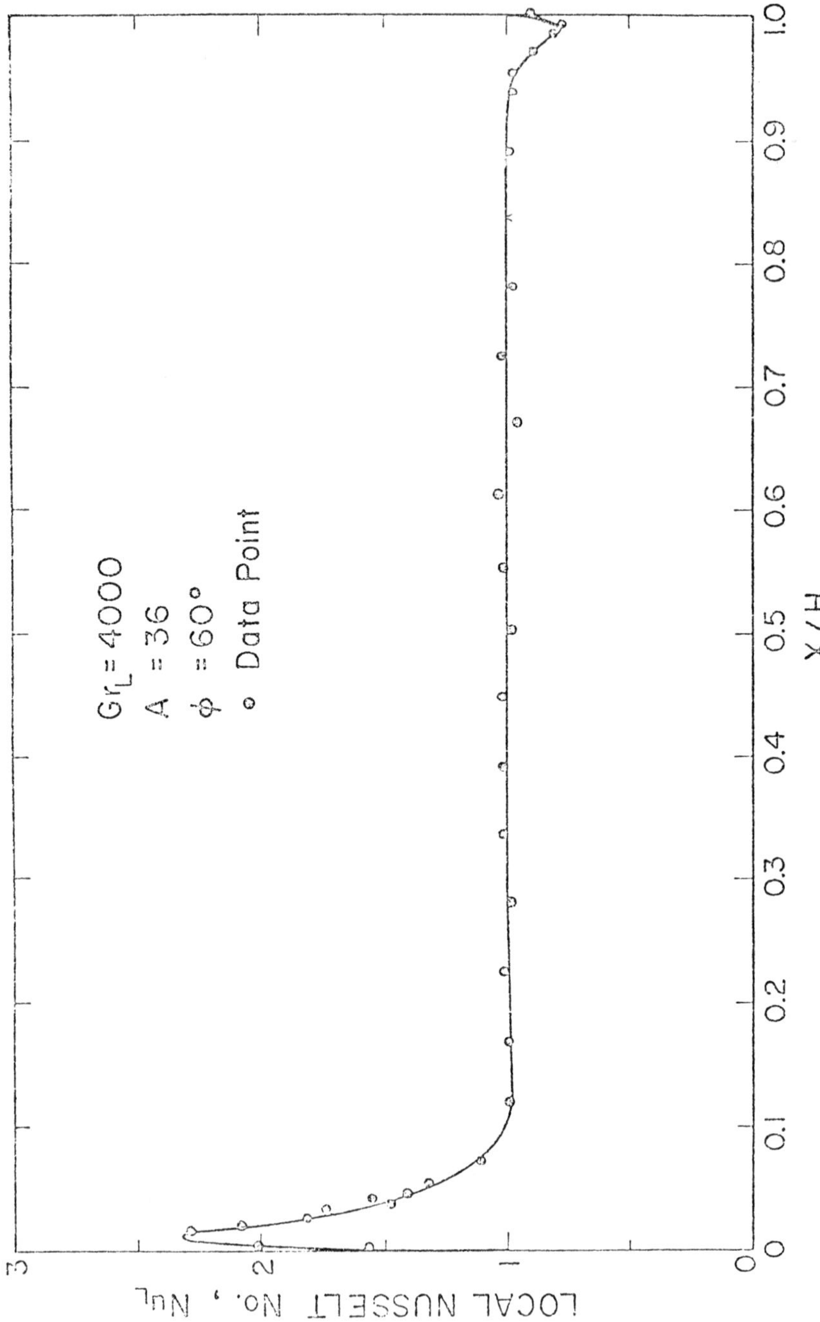

Fig. 17. Local Nusselt number as a function of non-dimensional plate length for the conduction

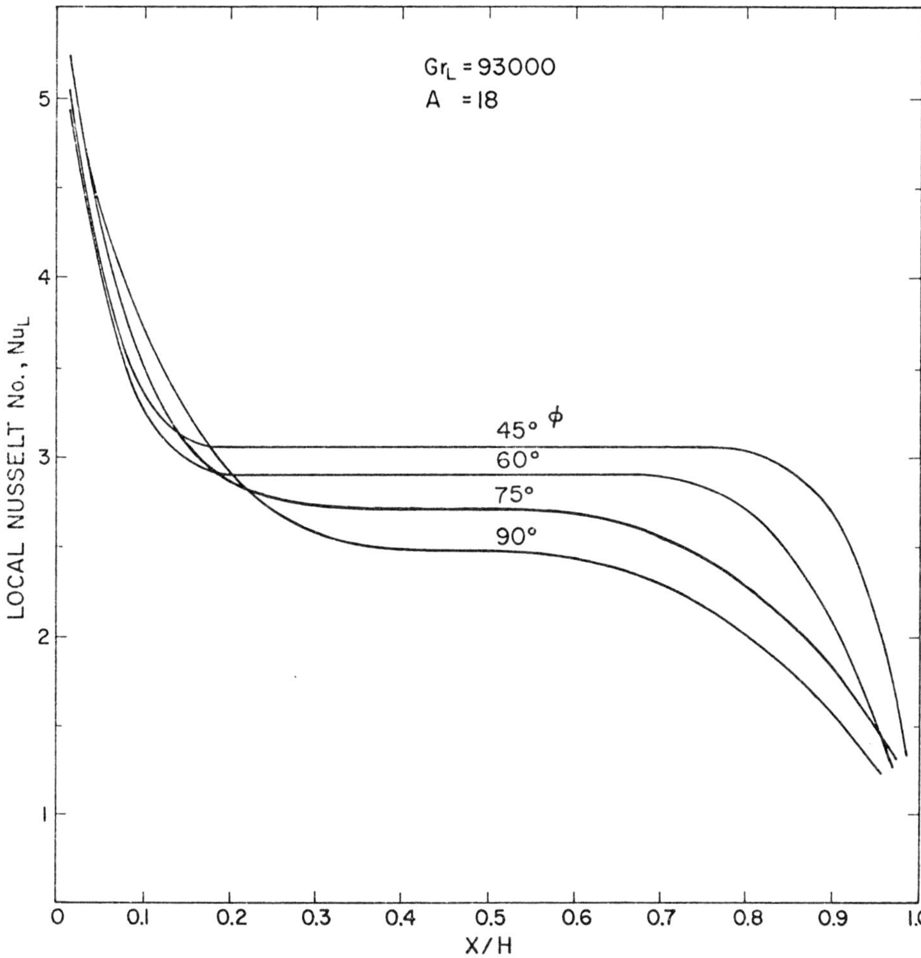

Fig. 18. Local Nusselt number as a function of non-dimensional plate length for various values of tilt angle.

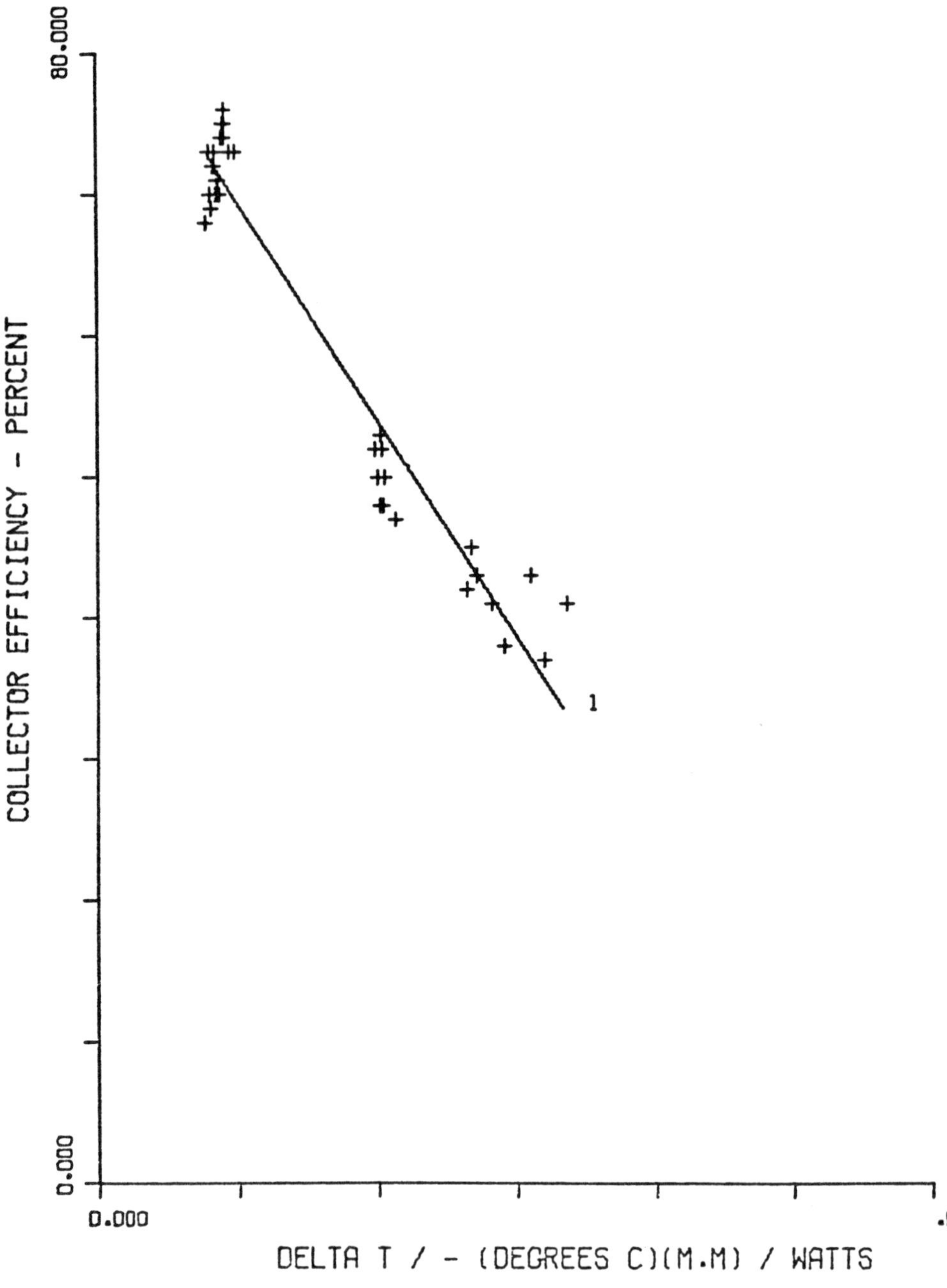

Fig. 19. Instantaneous thermal efficiency for scoria at high mass flow rate.

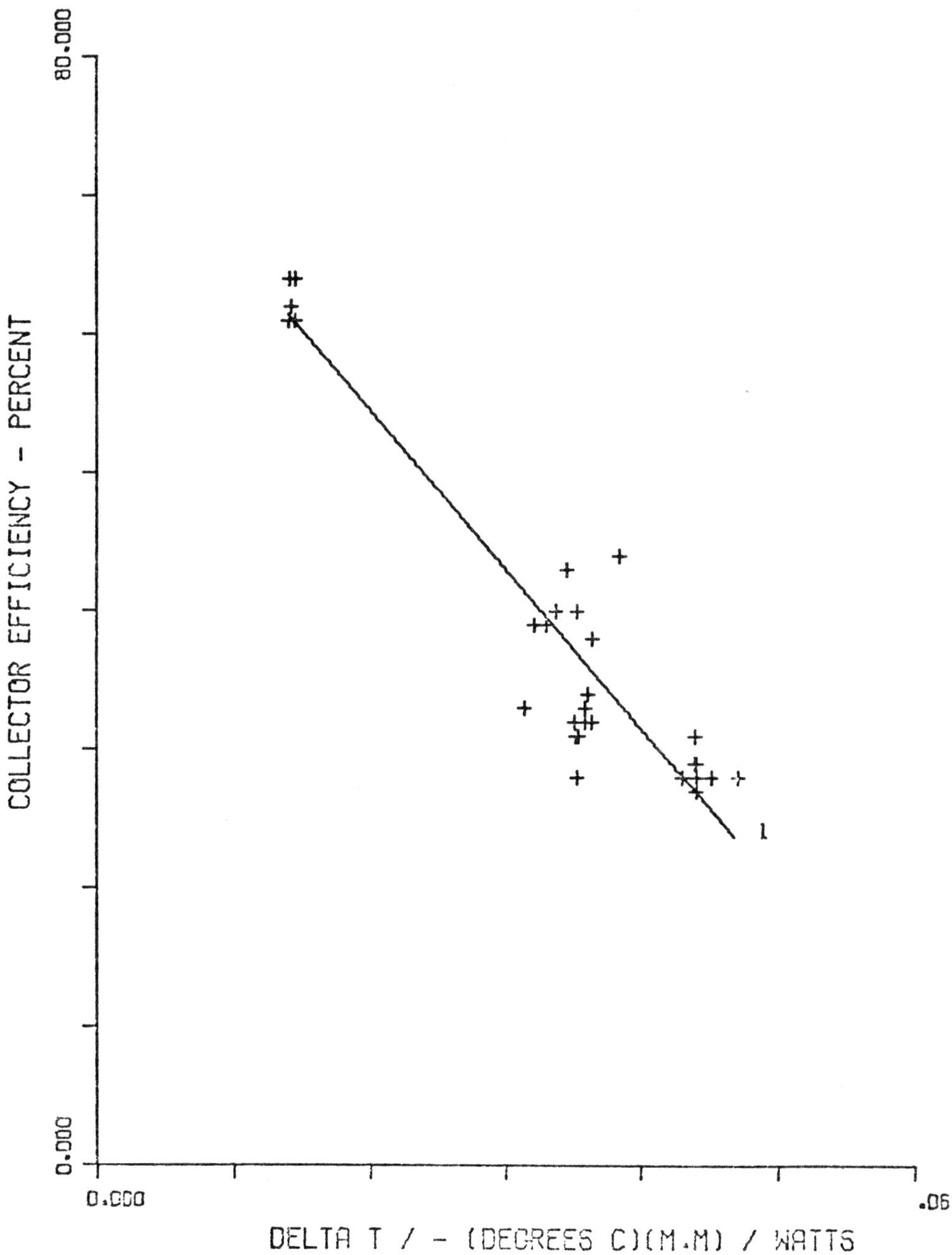

Fig. 20. Instantaneous thermal efficiency for coke at the low mass flow rate.

of opening of the V's was 60°. It was concluded that the heat transfer is greater than that for two parallel plane plates having the same average spacing, by up to 50% for a depth ratio of unity. However, for the larger values of the depth ratio, the increase in heat transfer is slight.

The second study dealt with the free convective heat transfer occurring in a honeycomb solar collector in which the honeycomb consists simply of a set of horizontal partitions (or slits) oriented normal to the plate. Aspect ratios (plate spacing to distance between partitions) of 5 and 10 were investigated, as were slit materials both opaque and transparent to thermal radiation. A significant conclusion is that for angles of inclination of the plate from the horizontal ranging from 70 to 90 degree, the horizontal slit honeycomb gives superior convection suppression to a square-celled honeycomb, for the same amount of honeycomb wall material.

University of Virginia[13]

The "solaris" solar collector is a patented design utilizing open channel flow of the heat transfer fluid over a blackened absorber with a single glazing. The University of Virginia has completed both a theoretical and an experimental evaluation of this collector design. A cross-section of the collector tested and analyzed is shown in Fig. 21. A comparison of the thermodynamic collector efficiency between the theoretical model and the experimental results are shown in Fig. 22. The two variables which influenced collector performance most were ambient temperature and wind. These two variables determine the amount of water droplet condensation on the inner glazing. This formation of this condensate was shown to reduce the transmittance of the inner glazing by 10 to 20 percent. Various techniques such as changing glazing materials and varying the number of glazings were experimented with in order to reduce the "drop-wise" condensation.

Honeywell, Inc.[14]

The development of a jet-impingement solar air heater was the objective of this work. The intention was to increase the efficiency of the air heater, over that of a "conventional" parallel-plate type, by increasing the absorber plate-to-air stream heat transfer coefficient. A comparison was made, both theoretically and experimentally, between the two designs. A schematic of the two designs compared is shown in Fig. 23. A comparison of the thermodynamic collection efficiency is shown in Fig. 24. For fixed pumping power, it is doubtful that the nominal increase in collector efficiency is enough to compensate for added complexity of the air jet system.

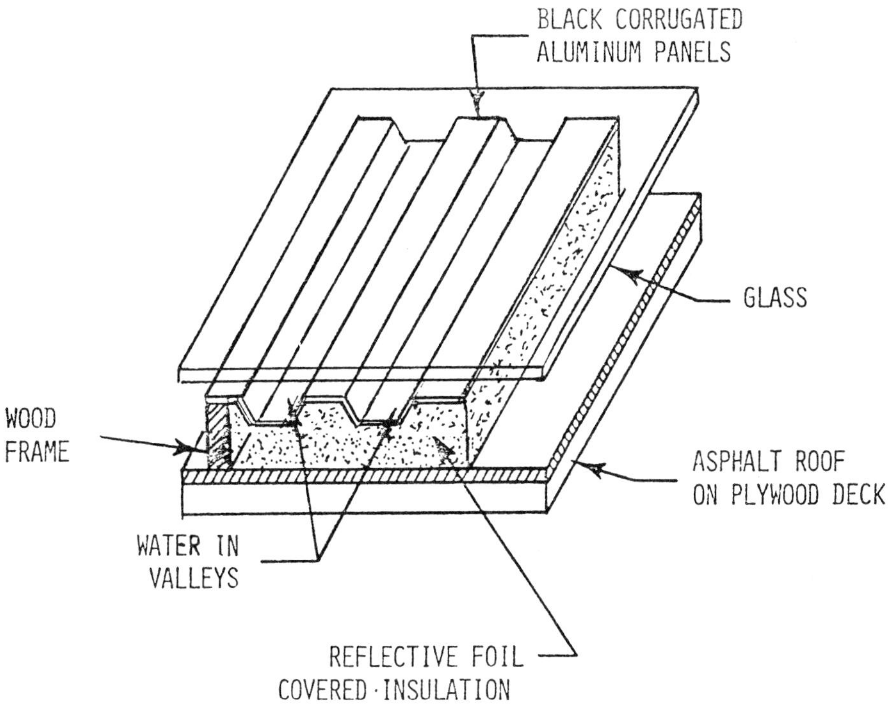

Fig. 21. Cross-sectional view of the "Solaris" collector.

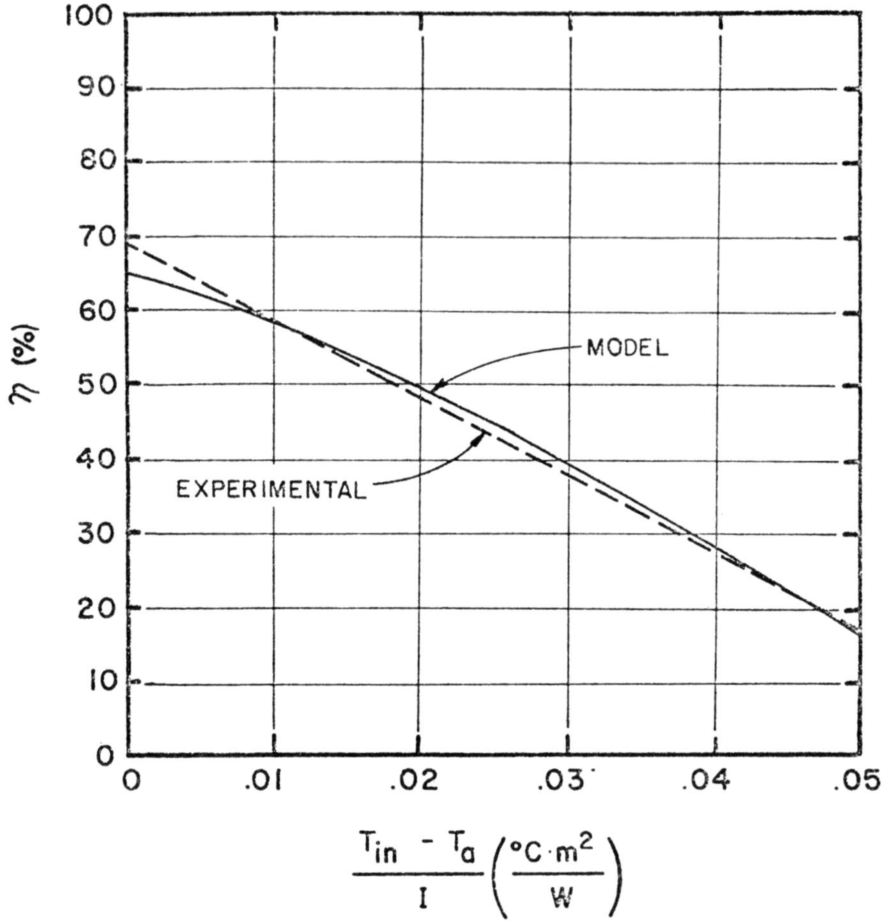

Fig. 22. Experimental and model predicted performance curves for the "Solaris" collector.

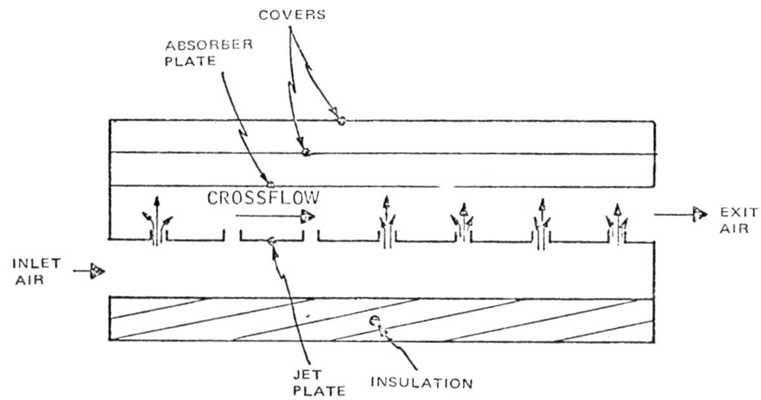

Schematic of Jet-Impingement Concept Collector
Fig. 23(a)

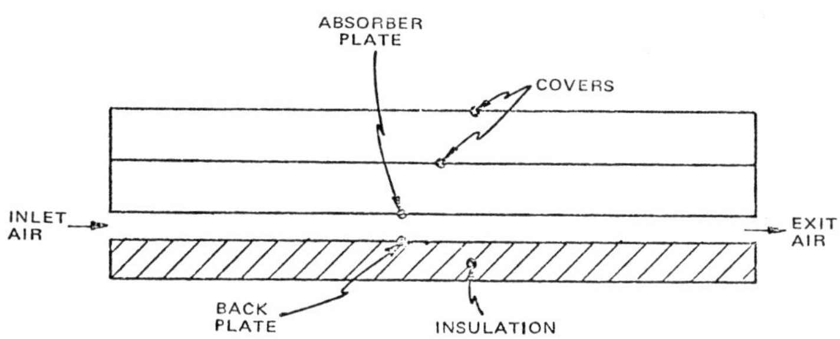

Schematic of Baseline Parallel-Plate Collector
Fig. 23(b)

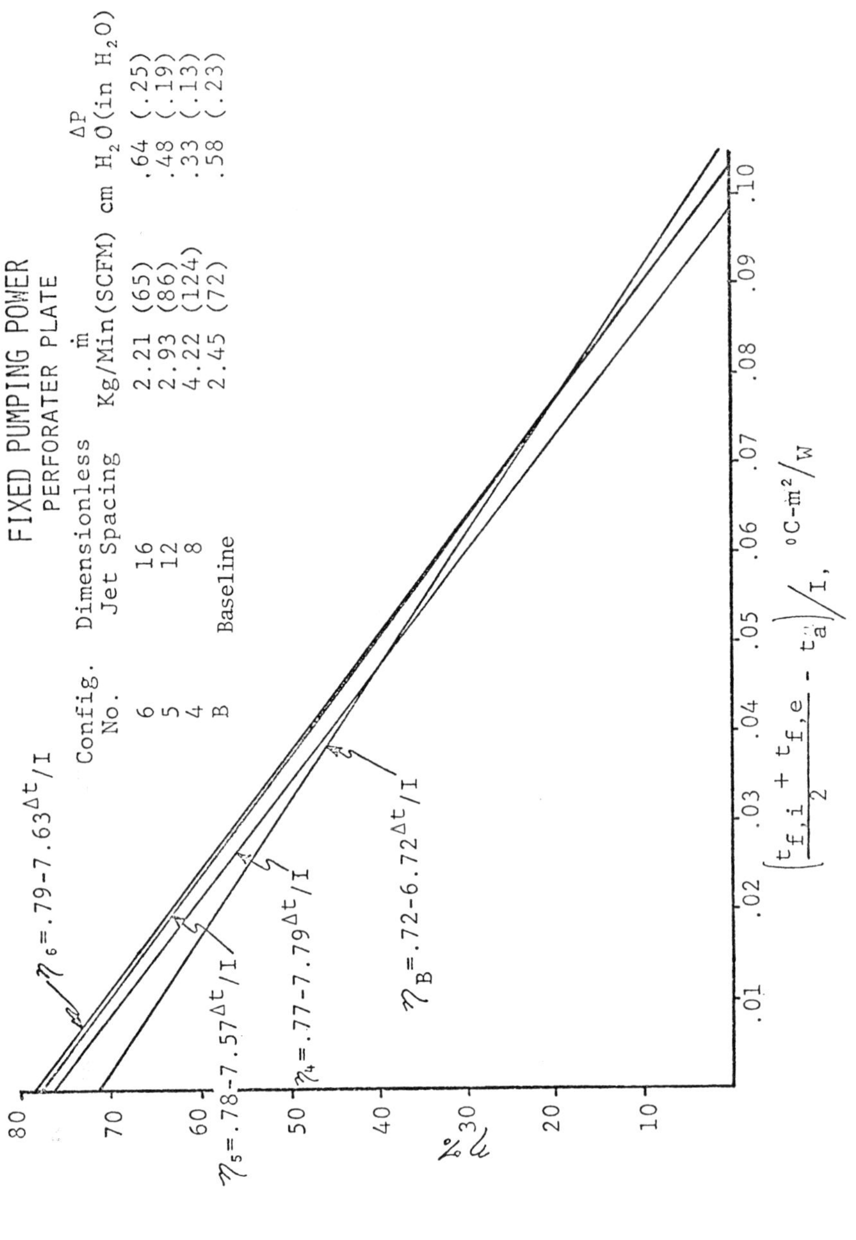

Fig. 24. Performance comparison between various jet designs and the baseline configuration.

Honeywell[15*]

A selective paint coating has been developed with solar absorptance (α) of 0.92 and room temperature emittance (ε) of 0.10. It was made through the optimization of coating thickness, pigment volume concentration, and pigment particle size. The selective paint has also been successfully scaled up to a sample size of 1 ft x 1 ft. The coating cost (materials and fabrication) was estimated at about 5¢/sq ft, a factor of 10 lower than the selective plated coatings investigated.

Accelerated humidity and outdoor exposure tests have been conducted for black chrome, black nickel and black iron coatings. The black chrome coating is most durable and its life was estimated to be greater than 30 years. The use of surface roughening to improve the optical properties of black chrome resulted in a sample with $\alpha = 0.95$ and $\varepsilon(100°C) = 0.09$. A method has been developed which reduced the cost of black chrome by about 30 percent.

Dipped AR coatings on glass using low refractive index Teflon FEP resulted in a transmittance increase of about 4.9 percent over the 0.35 μm to 0.75 μm region. The feasibility of AR coating Tedlar by glow-discharge polymerization was investigated. Strength and durability of etched AR glass were tested. Studies on the chemistry of glass etching for better process control have also been initiated.

National Bureau of Standards[16,17]

Test standards have been developed for liquid heating flat plate collectors. Presently, standards are being developed for air heating collectors and thermal storage devices.

IIT Research Institute[18]

A large number of actual and potential materials for solar energy utilization were exposed to natural weathering at three sites: Chicago, IL; Phoenix, AZ; and Miami, FL. The materials were exposed to conventional weathering simultaneously at three sites for periods of 3, 6, and 12 months. Accelerated testing (exposed to multiple suns) was conducted at the Phoenix, AZ site. Tests performed included:

1) Visual inspection.
2) Normal spectral transmittance of transparent materials.
3) Hemispherical spectral reflectance of opaque materials.
4) Hemispherical spectral reflectance of selective absorber coatings.
5) Dynamic tensile strength and elongation measurements of plastics.

*Material taken directly from Ref. 15.

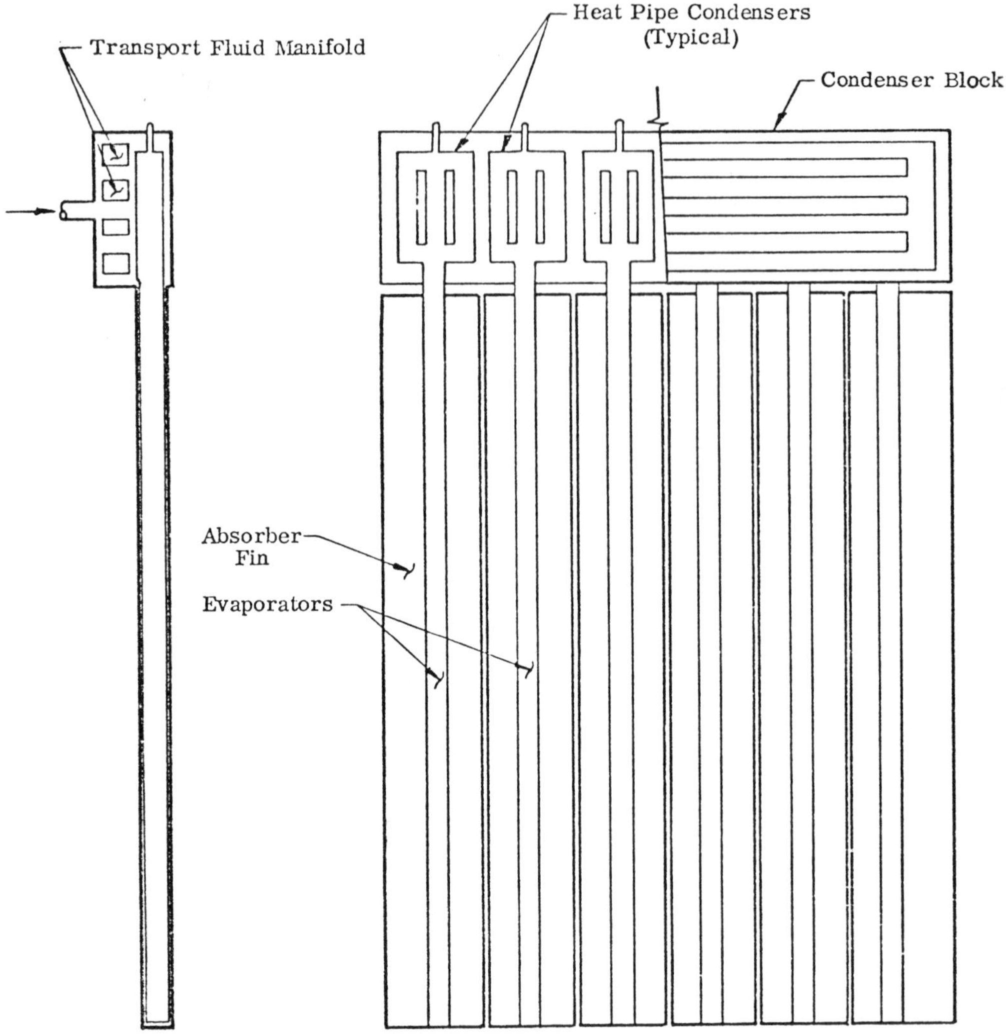

Fig. 25. Heat pipe roll bond collector.

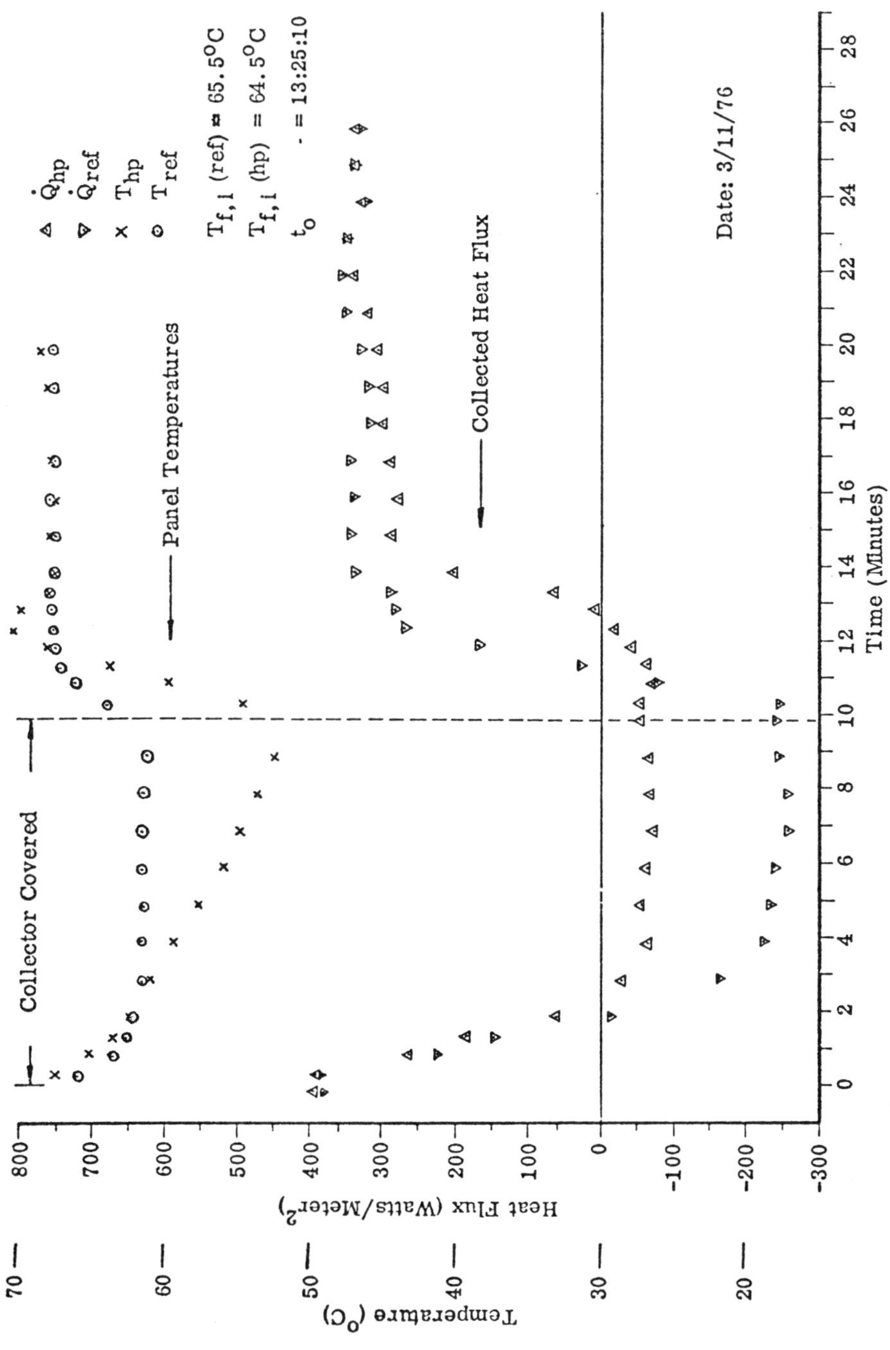

Fig. 26. Performance of reference and second heat pipe solar collectors during intermittent insolation.

6) Izod impact tests on all transparent cover materials.
7) Abrasion resistance tests on rigid plastic cover materials.
8) Bidirectional reflectance measurements on selective solar absorbers and mirrors.
9) Static dimensional stability tests on all plastic cover materials.

The materials chosen are too numerous to itemize here. They include a wide variety of glazing, absorber and reflector materials.

Dynatherm[19]

The objective of this program was to investigate the use of heat pipes as heat transport elements in flat-plate solar collectors. During the early phases of the program, analytical models for heat pipe collectors were developed and configuration and integration studies were performed. The feasibility of wickless heat pipes, fabricated from commercial Roll-Bond material, was investigated in the laboratory. These heat pipes performed satisfactorily with Freon 11 as the working fluid. Two complete heat pipe solar collectors, which utilized converted Roll-Bond panels as heat pipes, were tested along with a conventional reference collector of the same basic design. The second roll bond heat pipe absorber is shown in Fig. 25.

With steady solar input, the heat pipe and the reference collector yielded approximately the same instantaneous efficiency ($\sim$60%). But with intermittent insolation, such as on a partly cloudy day, the heat pipe collector collected up to twice as much energy as the reference collector. This effect is shown in Fig. 26. This improvement is attributable to the thermal diode capability of the heat pipes. Heat is transported only when the absorber temperature exceeds the fluid temperature in the manifold. During periods of low insolation, the heat pipe collector is automatically inactivated.

References

1. J. D. Balcomb, J. C. Hedstrom, S. W. Moore, and K. C. Herr, "Research on Integrated Solar Collector Roof Structures," presented at the 1975 International Solar Energy Congress and Exposition, UCLA Campus, July 28-August 1, 1975.

2. K. L. Moan, Evaluation of an All-Glass, Evacuated, Tubular, Nonfocusing, Nontracking Solar Collector Array, Final Report for DOE Contract No. EY-76-C-02-2919, Report No. COO-2609-3, July and September, 1977.

4. K. N. Marshall, R. K. Wedel and K.G.T. Hollands, "Economic Studies of Plastic Honeycomb Solar Collectors," Proceedings of the Flat-Plate Solar Collector Conference, ERDA Sponsored, February 28-March 2, 1977, Orlando, FL.

5. K. N. Marshall, R. K. Wedel and R. E. Dammann, Development of Plastic Honeycomb Flat-Plate Solar Collectors, Final Report for ERDA Contract E(04-3)1081, ERDA Report No. SAN/1081-76/1, April 1976. The 1977 Final Report is still in the draft stages.

6. H. Buckberg, D. K. Edwards, and J. D. MacKenzie, Transparent Glass Honeycomb Structures for Energy Loss Control, DOE Contract No. EY-76-G-03-1084, Report No. SAN/1084-77/1, August 1977.

7. W. W. Schertz, R. Winston, and R. G. Matlock, Status Report for Nonimaging Collectors, ERDA Contract No. W-31-109-ENG-38, Argonne Report No. ANL-SOL-77-01, January 1977.

8. M. Kudret Selcuk, A Fixed Tilt Solar Collector Employing Reversible Vee-Trough Reflectors and Vacuum Tube Receivers for Solar Heating and Cooling Systems, Final Report, Publication No. DOE/JPL-1024-77/1, December 1977.

9. J. Giner, F. H. Cocks and D. Wong, Study of Corrosion and its Control in Aluminum Solar Collectors, Annual Progress Report, June 1, 1976-May 31, 1977, ERDA Contract No. EY-76-C-02-2934-000.

10. K. R. Randall, M. M. El-Wakil and J. W. Mitchell, "Natural Convection Characteristics of Flat Plate Collectors," Proceedings of the Annual Meeting, American Section of the International Solar Energy Society, Orlando, FL, 1977.

11. D. R. Simpson, Low Cost Solar Collector of a Packed Bed Design, Final Report for ERDA Contract No. EY-76-S-02-2972-000, Report No. COO-2972-2, July 1977.

12. K.G.T. Hollands, G. D. Raithby and T. E. Unny, Studies on Methods of Reducing Heat Losses from Flat Plate Solar Collectors, Phase 2, Annual Progress Report for ERDA Contract No. EY-76-C-02-2597-000, Report No. COO-2597-3, January 1977.

13. J. T. Beard, et al., "Design and Operational Influences on Thermal Performance of 'Solaris' Solar Collectors," ASME Paper No. 77-WA/Sol-2.

14. D. R. Rask, et al., Low-Cost Solar Air Heater, Final Report for ERDA Contract No. EY-76-C-02-2929-000, Report No. COO-2929-13, August 22, 1977.

15. R.J.H. Lin and P. B. Zimmer, Optimization of Coatings for Flat Plate Solar Collectors, Final Report for ERDA Contract No. EY-76-C-02-2930-000, Report No. COO-2930-12, July 1977

16. ASHRAE Standard 93-77, Method of Testing Solar Collectors Based on Thermal Performance, Standards Committee, ASHRAE, 345 E. 47th St., New York, NY, June 1976.

17. D. E. Jones and J. E. Hill, "Testing of Flat-Plate Air Heaters According to ASHRAE Standard 93-77," Proceedings of the Annual Meeting, American Section of the International Solar Energy Society, Orlando, FL, 1977.

18. J. E. Gilligan, et al., "Handbook of Materials for Solar Energy Utilization--Parts I and II," ERDA Contract No. EY-76-C-04-0578-034-A002, Report No. COO/0578/034-8, February 1978.

19. W. B. Bienert and D. A. Wolf, Heat Pipes Applied to Flat-Plate Solar Collectors, Final Report for ERDA Contract No. E(11-1)-2604, Report No. COO/2604-76/1, May 1976.

CALIFORNIA STATE ENERGY EFFICIENT STATE OFFICE BUILDING

B. S. Setty
Benham-Blair & Affiliates, Inc.
Falls Church, Virginia 22151, U.S.A.

ABSTRACT

Advanced energy technology, utilizing concentrated solar collectors and ice storage system was employed to achieve the lowest energy consumption and lowest life cycle cost. The paper will discuss the functions of the concentrating collectors and its integration with ice storage systems as below:

- Utilization of concentrating solar collectors passive systems in order to save the non-renewable sources of energy.
- Modes and tracking mechanism of the concentrated solar collectors.
- Concentrating collector system interaction with high temperature storage system.
- Conversion of solar energy into ice, its implications and its interaction with the ice storage system.
- System's flexibility to provide solar heat at maximum efficiency.

A. INTRODUCTION

Depletion of energy reserves and pollution of the environment are the two major concerns which demanded the State of California's attention for developing an energy efficient State Office building. The State chose to have an energy competition enabling forty-one firms to enter the competition. Benham-Blair's California Division was among the entrants with extensive experience and was declared as the winner of this competition for developing the most energy efficient state building and for its "innovative architectural design and advanced state of the art solar technology."

The State Office building is approximately 250,000 gross square feet with an office area of 185,000 square feet, the remaining being parking and circulation area. The building is located in the Sacramento downtown area, with varying climatic conditions ranging from 104° F. in the summer to about 32° F. in the winter. The sun insolation values in summer reach 715 langleys/day with an average summer sunshine of 95%, which is equivalent to that of Albuquerque, New Mexico. Even though many facets of energy conservation techniques are employed in this particular building, only the nondepletable solar energy systems are the subject of this particular

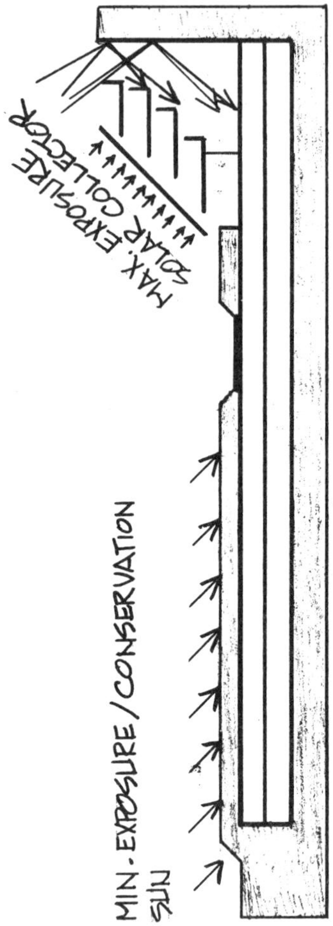

FIG.-1 MAXIMUM EXPOSURE TO SOLAR GENERATION

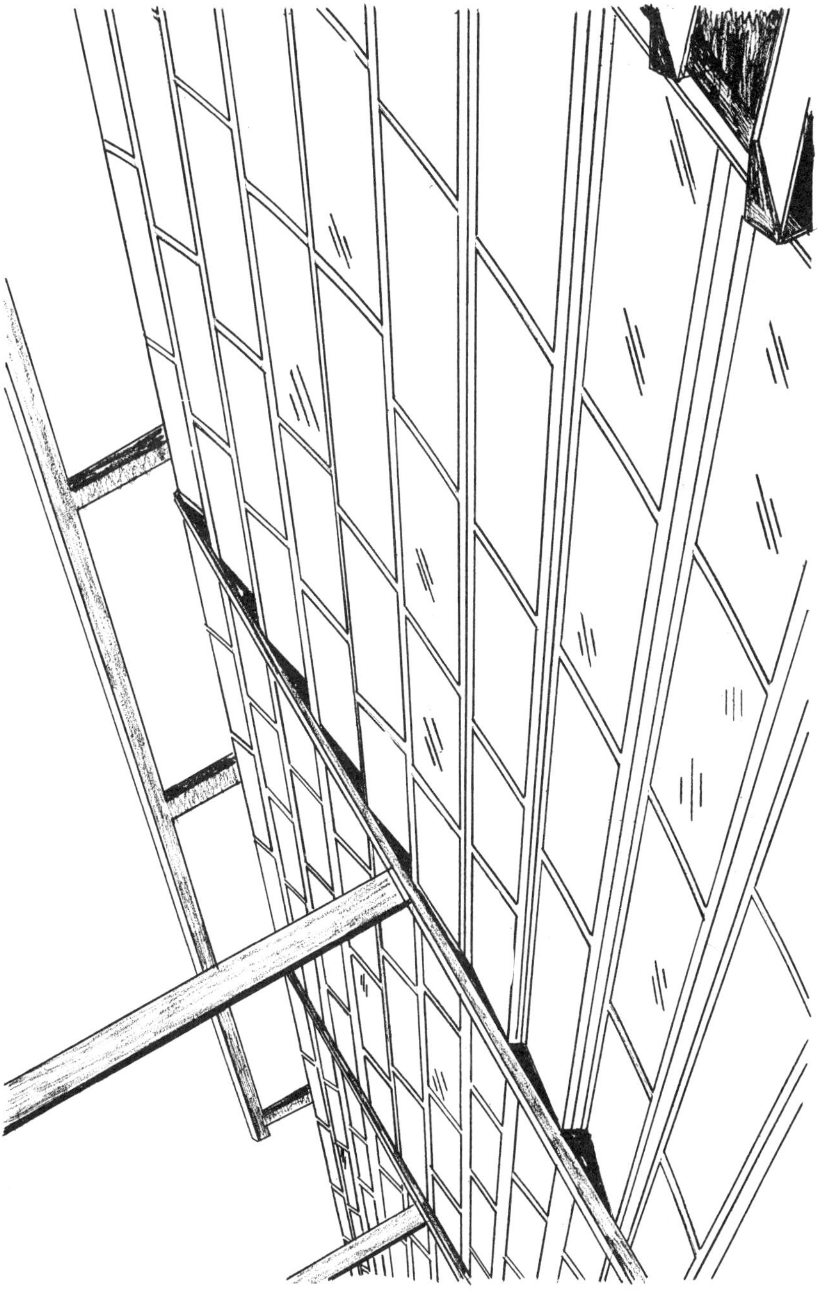

FIG-2 CONCENTRATED SOLAR COLLECTORS

paper because they have been instrumental in achieving the most energy efficient State Office building.

A multidisciplined approach was employed for developing the fully integrated "energy conscious design" right from the conceptual stage of building design until it became structure. After a sensitive site analysis, maximum energy conservation with minimum exposure to solar generation emerged as the optimum solution. The building was conceived primarily with heavy thermal mass and berms on top of the highest floor for achieving maximum energy conservation. The site, with maximum open space and within existing environmental attributes, lent itself to the development of a negative architectural design which was manifested in an underground structure with light wells spaced at different intervals. The maximum exposure to solar generation (Fig.-1), the climatic conditions, the sun insolation values in Sacramento, California, and the negative architecture dictated the use of solar collectors. By placing the largest vertical mass on the north side, i.e., exposing the south side, it provided a sun oriented vertical mass at an ideal position on the site for harnessing the solar energy. In order to meet the complete heating and cooling demands of the conceived building and to store the solar energy at different energy levels and to save the depletable forms of energy as much as possible, tracking concentrated collectors were selected for collection of solar energy (Fig.-2). This unique concept permitted the storing of the solar energy at different temperature levels in different storage tanks for performing multiple functions economically at different periods of the year.

B. SOLAR-PASSIVE SYSTEMS

The building was conceived with several passive systems which played a dominant role in saving the energy, even though it was not accounted for in the demonstrated energy consumption.

First, the solar slab facing south has half of the area for direct sunlight penetration and natural lighting and the other half for collection of solar energy through concentrating solar collectors. Double glazing with night operated venetian blinds allows the minimizing of night losses (Fig.-3) in winter. The direct sunlight passing through the double glazing heats up the floor thermal mass in the exposed structure and releases the stored heat for night heating in winter. The heated slab of the modular office space will act as a "LIVE IN SOLAR COLLECTOR."

Second, an indirect solar-passive system, being incorporated in this particular design, is the spray pond above the exposed solar slab (Fig.-4). The spray pond system consumes the least amount of energy at a very high coefficient of performance of 35 to 40, and the energy consumed is the energy at the spray nozzles required for pumping the water on an intermittent basis. The spray nozzles are controlled thermostatically by the thermostat, embedded in the concrete roof slab of the exposed structure.

Third, creation of a park (Fig.-5) above the subterranean structure during the planning of the building provided not only a good human scale and open

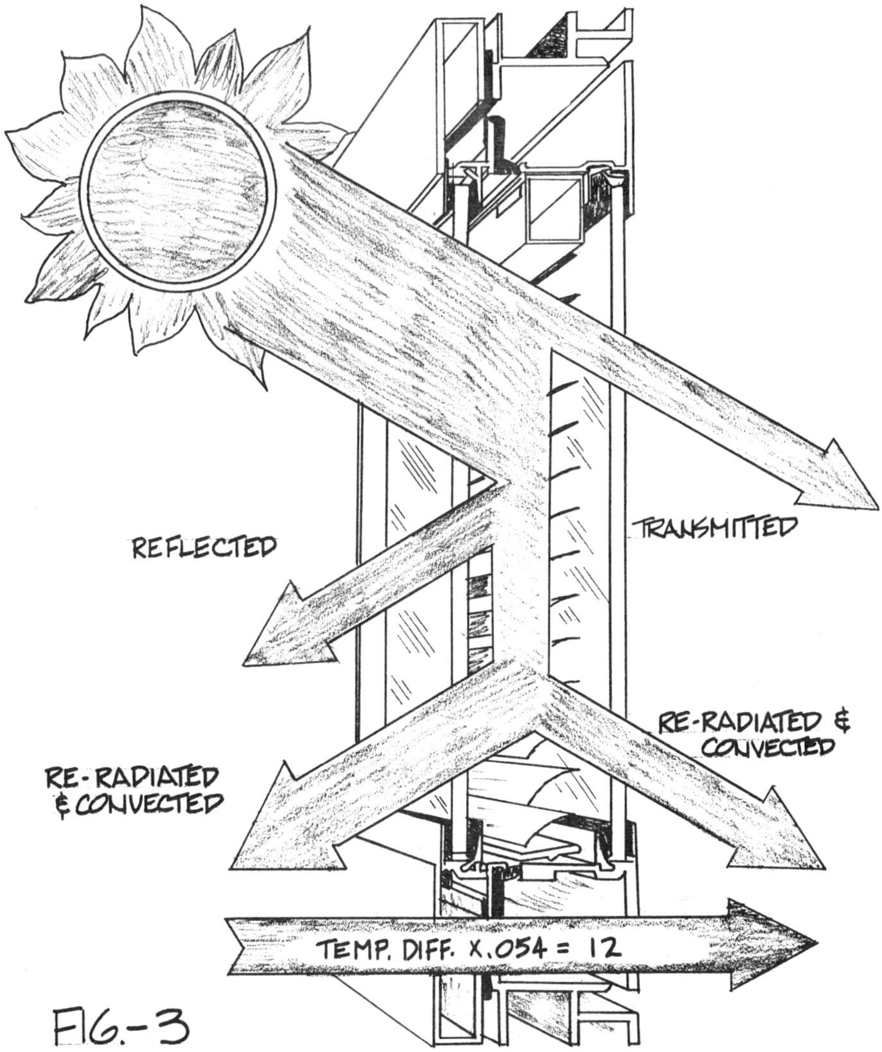

FIG.-3

DOUBLE GLAZING WINDOW W/LOUVERS

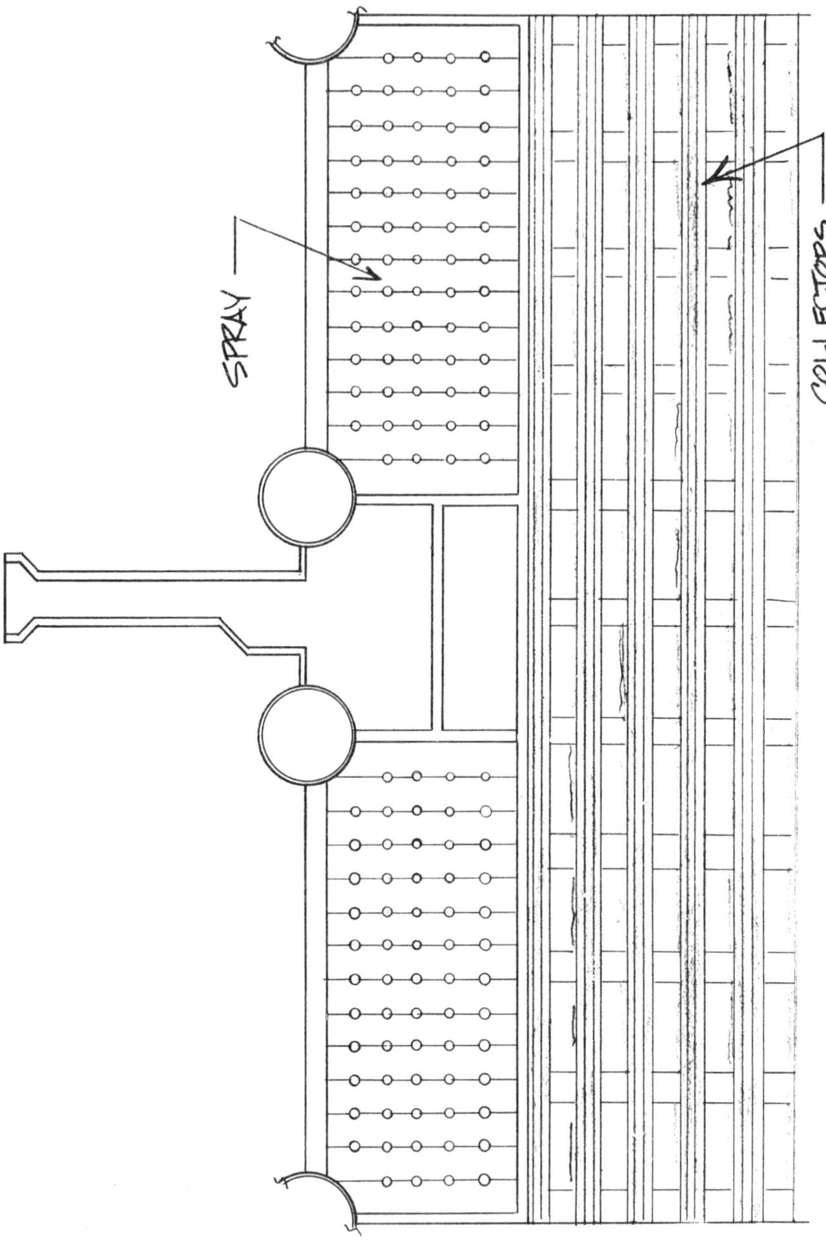

FIG.-4 SPRAY POND ON ROOF

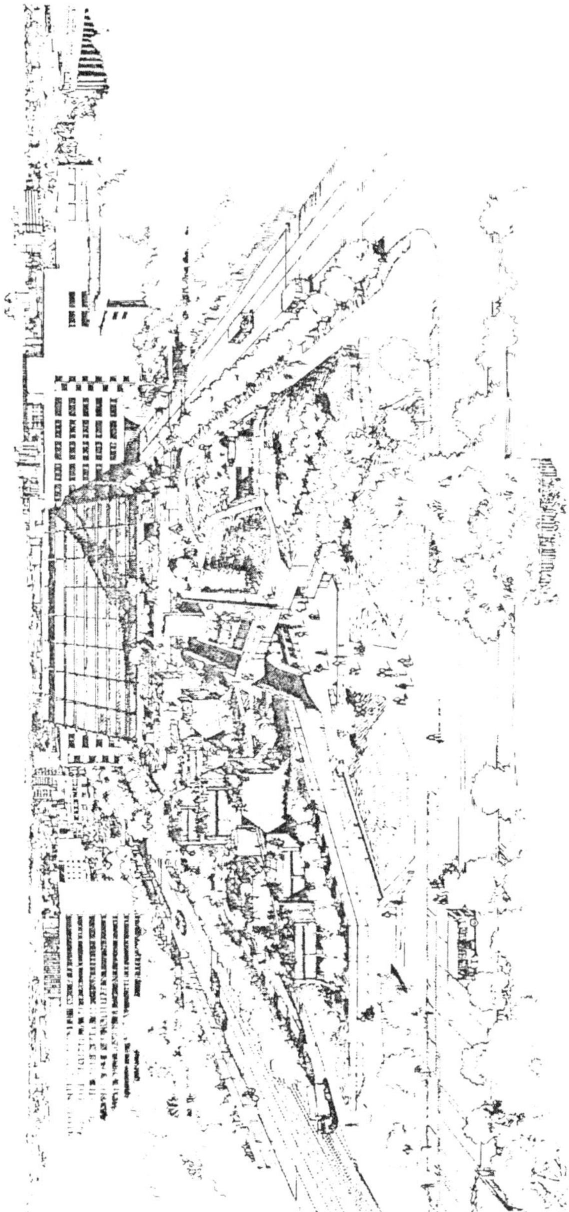

FIG-5 VIEW OF PARK

space but also a good thermal barrier for the solar radiation and convection effects. The surrounding air, in the vicinity of the open space is cooled by the evaporation process and enables the cutting down of both the convective and conductive losses that normally occur in the building and maintains comfortable surroundings. Large quantities of landscaping through trees, shrubs and overhangs provided a shading to the exposed walls in the light wells for reducing heat gains in summer.

Fourth, the solar facade, acting as an instrument for solar generation, creates a chimney effect along the subterranean mall, thus drawing the southerly air from the south side of the building, creating a temperature of approximately 75° F. to 80° F. in the open subterranean mall (Fig.-6), and 130° F. to 140° F. near the exposed solar slab. The vapor pressure difference between the subterranean mall and the exposed solar slab creates the chimney effect mentioned earlier and has a substantial impact on energy savings.

C. SOLAR HEATING AND COOLING SYSTEMS

High temperature solar utilization was the main objective and it was on a twofold basis: one, to store as much energy as possible at the highest temperature in the order of 350° F. without penalizing the cost of the solar system and two to store the maximum amount of energy within a minimum amount of space. Harnessing solar energy at a very high temperature led to the decision of using concentrated solar collectors (Fig.-7). Because of the nature of the building already conceived and the quest for achieving maximum energy conservation, the use of slats with a fixed absorber-receiver collector system was the most feasible and economical solution. The south facing solar slab which runs approximately 300' long and 60' high in an east-west direction, makes for a feasible proper integration of concentrated collectors with the exposed structure for solar generation. The computer and economic analyses indicated that 12,000 square feet of collectors will be sufficient to meet the space heating and cooling demand year round. The collectors are mounted on an east-west axis, and the sun is tracked in a north-south direction. The collector array consisting of 6 rows of collectors and 180' long are mounted continuously on to the solar slab for reducing the end losses that normally occur in a system without a continuous fixed absorber system. The shading of one row of the reflector system with another row is eliminated by a tracking mechanism and the angle of rotation of the receivers. Parabolic concentrators were considered and discarded in the earlier stages and had the disadvantage of shading one collector with another in addition to involving heavy structural design requirements. The energy transport system connected with the collector system is designed in a camouflaged way for transporting the collected energy in a very efficient way and for reducing the transportation losses. The solar computer analysis indicated that a north-south horizontal axis of collector intercepts more energy over a year than does a collector about a horizontal east-west axis. In the preliminary analysis, the availability of both direct and total solar radiation for both east-west tracking and north-south tracking collector systems were considered, and in this particular instance because of the complications involved in installing the

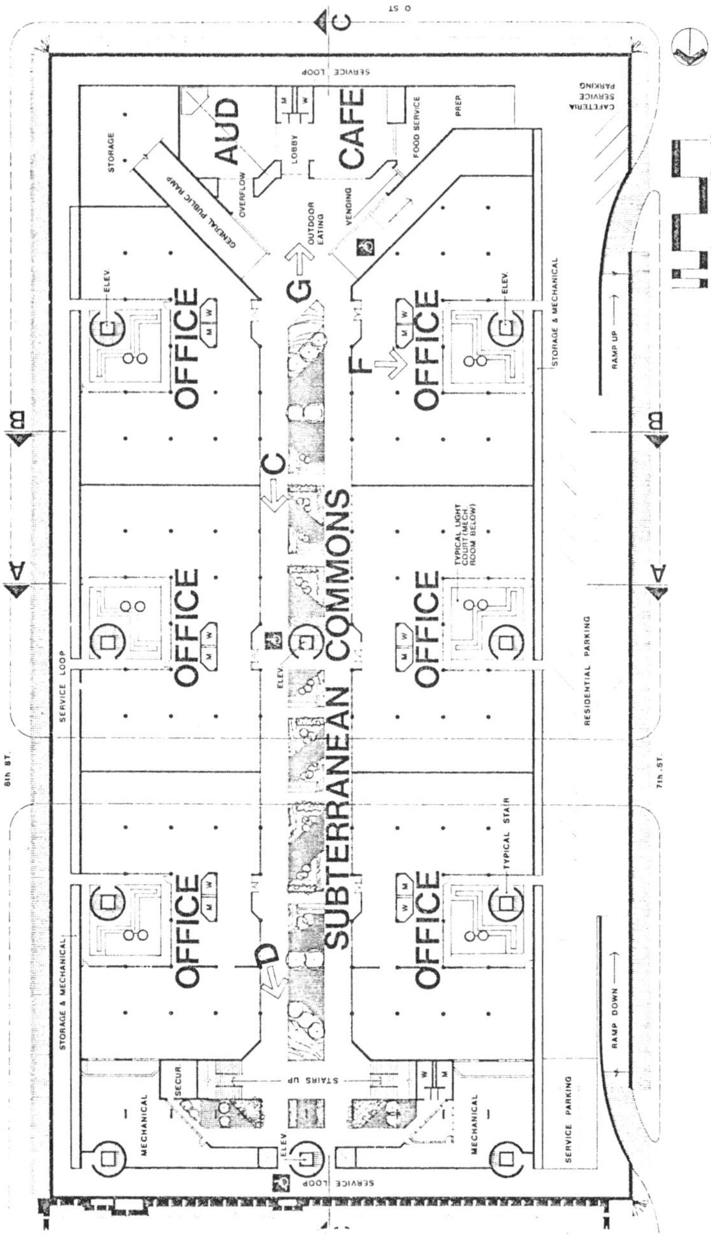

FIG-6 SUBTERRANEAN MALL

FIG-7 CONCENTRATED SOLAR COLLECTORS

field, an east-west axis system with a north-south tracking proved to be more beneficial and economical.

a) Tracking System

The collector tracking mechanism consists of a sun sensor and sun tracking and driving mechanism hardware. Tracking is accomplished in groups to get modularity. The sun sensor mounted on a representative receiver within each group tracks the sun. Sun availability is sensed by two sun sensors contained with the system and are activated by the following demand conditions:

- Storage temperature is below maximum or absorption machine or heating machine calling for cooling or heating will enable the concentrated solar collectors to position for operation and collect solar energy (Fig.-8).

- Through a network of gear mechanism, the segment drive motors cause the segments to rotate until the reflected light is centered on the receiver and tracking assembly. Loss of sun, lack of demand malfunction occurrence will activate gear mechanisms and the motor to the face down position (Fig.-9).

b) Solar Collector Supporting Structures

The supporting structure for the field requires a thorough analysis because of the many variables involved, such as the OSHA safety requirements for maintaining a fixed absorber system which is hanging in the air 12' away from the fixed solar slab, and cleaning of the collectors periodically for maintaining the collection efficiency at different insolation values during any time of the year. Various techniques, both manually operated systems and mechanized systems were considered as to their technical and economic feasibility. The custom-made mechanized system has a movable motor system on a track of rails at the top of the solar slab with intermittent tracks between the rows, and it has a platform moving longitudinally and transversely. It was finally decided to be very expensive and cumbersome. A simple manual cleaning system (Fig.-10), which is self-integrated with the solar slab, provided the operational flexibility and easy maintenance. The system was feasible and also met all of the OSHA safety standards.

D. SOLAR-STORAGE SYSTEMS

To store solar energy at a high temperature level for cooling in summer and to optimize the storage tank size and to store excess energy in an ice storage tank, and in a condenser water storage tank, the following temperature levels in different tanks were considered:

- Tank No. 1 - The storage temperature ranging from 120° F. to 350° F. in the high temperature storage tank.

- Tank No. 2 - The storage temperature ranging from 120° F. to 50° F. in

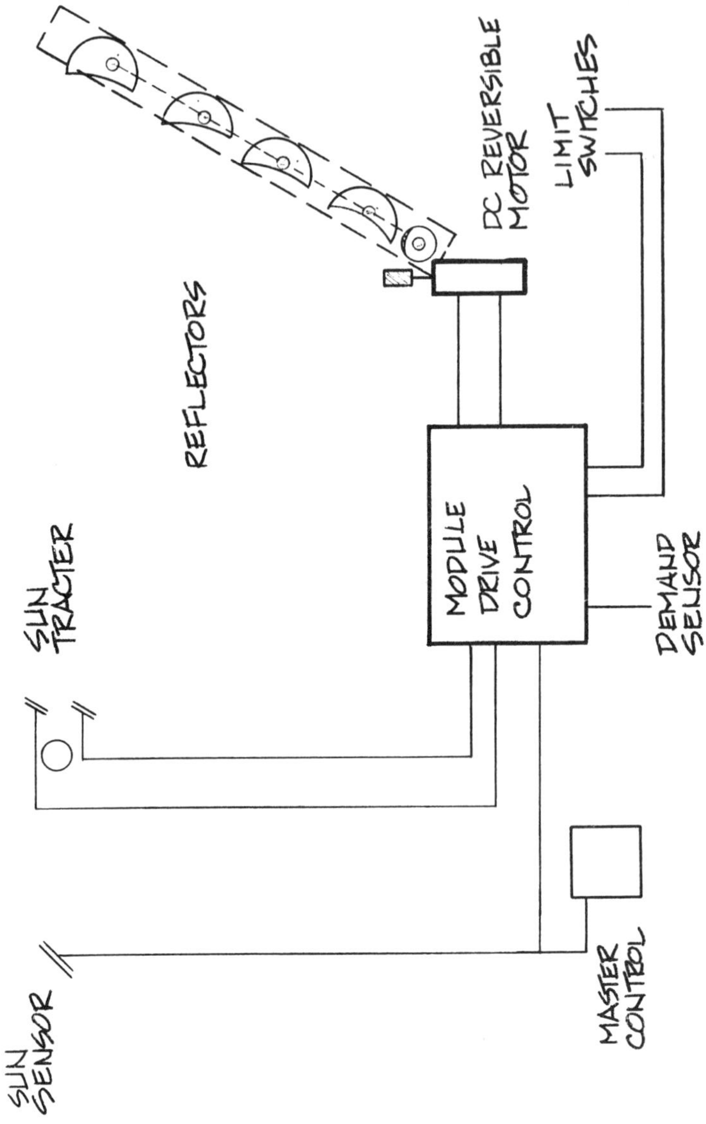

FIG-8 DRIVE MECHANISM

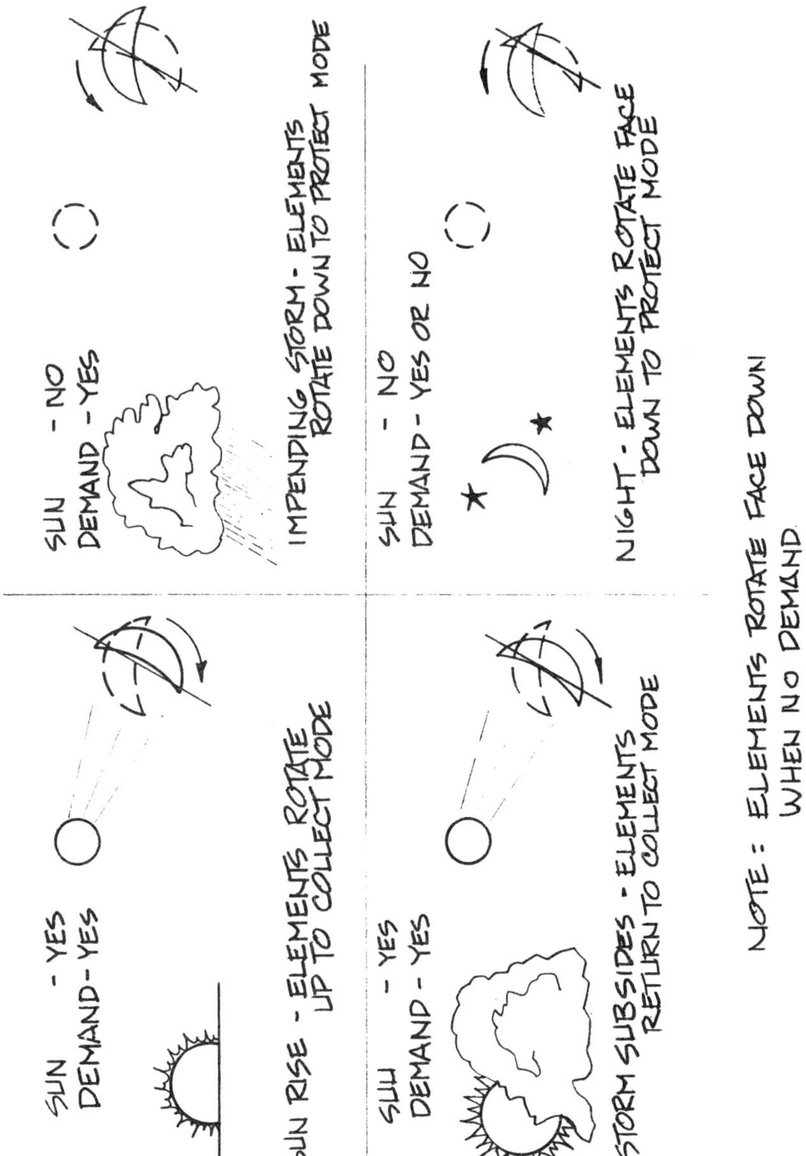

FIG.-9 COLLECTOR OPERATIONAL MODES

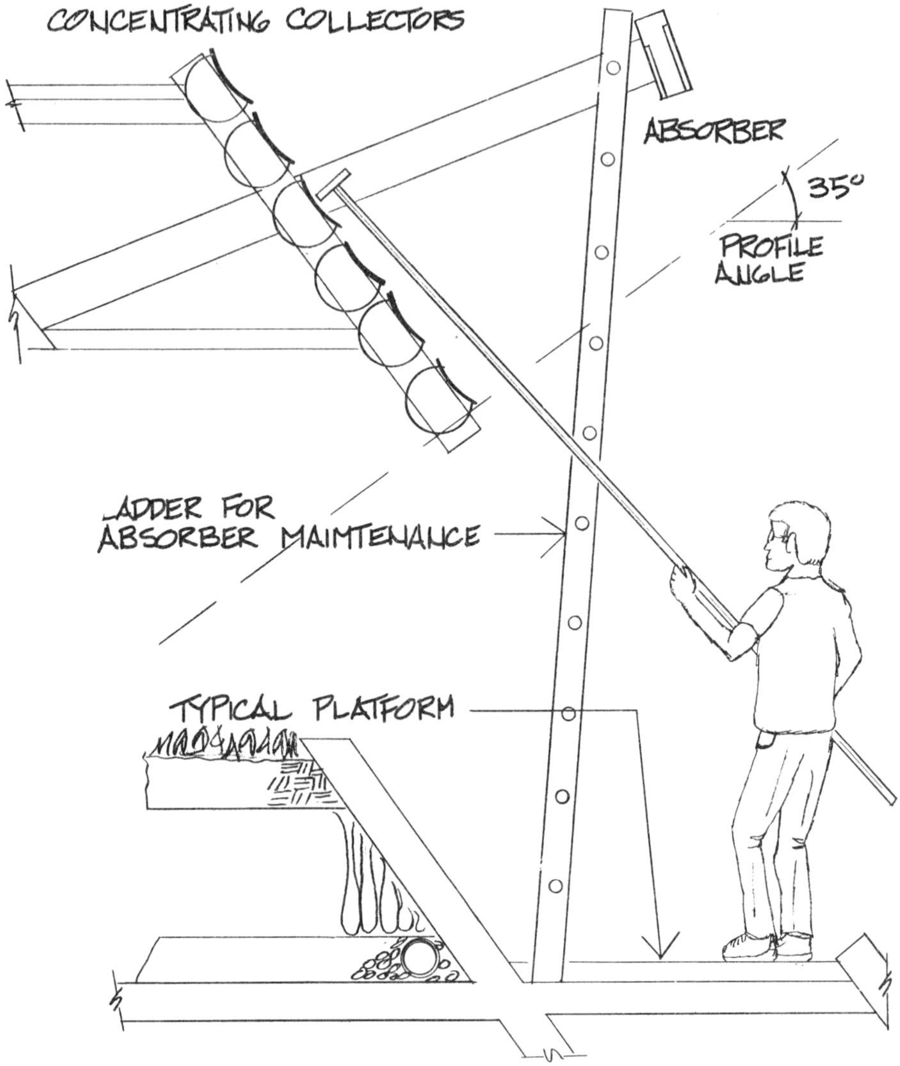

FIG.-10 MANUAL CLEANING SYSTEM

the medium temperature storage tank.

- Tank No. 3 - The storage temperature ranging from 50° F. to 0° F. in the ice storage tank.

The wide range of flexibility of storage tanks temperature in the tanks allows the solar systems to be optimized and to occupy less space (Fig.-11). The unstored or unutilized solar energy is transferred to the central utility system either as chilled water or as steam, if required, which in turn increases the overall annual efficiency of the system. In this case the annual efficiency of the solar utilization system is defined as the ratio of total annual usage over the total annual collection as noted below:

$$\text{Annual Efficiency of Solar Utilization} = \frac{\text{Annual Solar Usage}}{\text{Annual Solar Collection}}$$

The annual efficiency of solar utilization can be increased because:

- Stored energy is utilized from the storage tank as long as it is available from cooling and heating from temperatures ranging from 350° F. to 0° F.

- Not more energy is extracted from storage tanks than is required.

- If the maximum temperature level in one is reached, the other tanks operate in sequence for storing solar energy at different temperature levels of utilization.

- If no more solar storage is required, under those circumstances solar energy is transferred to central utility network

By employing these salient features a very high annual efficiency of utilization is achieved.

a) High Temperature Solar Storage System

The high temperature solar system designed (Fig.-12) is to provide high temperature water within the shortest time possible by incorporating a small storage tank in series with the large storage tank. This allows the flexibility of using the high temperature solar energy immediately without heating up the large storage tank without encountering any losses in the transit. The efficiency of solar utilization is increased significantly by quick collection and dispersion of the collected energy. A fifteen thousand gallon storage tank proved to be of adequate capacity for storing high temperature energy, and burying the tank below grade was considered for minimizing the losses. For reason of maintenance, possible corrosion and buoyancy problems, an exposed tank with insulation was determined to be more economical requiring less maintenance. After careful evaluation the stratification in the tank was eliminated by introducing turbulence into the storage tank through a pumping system. A spray tube with

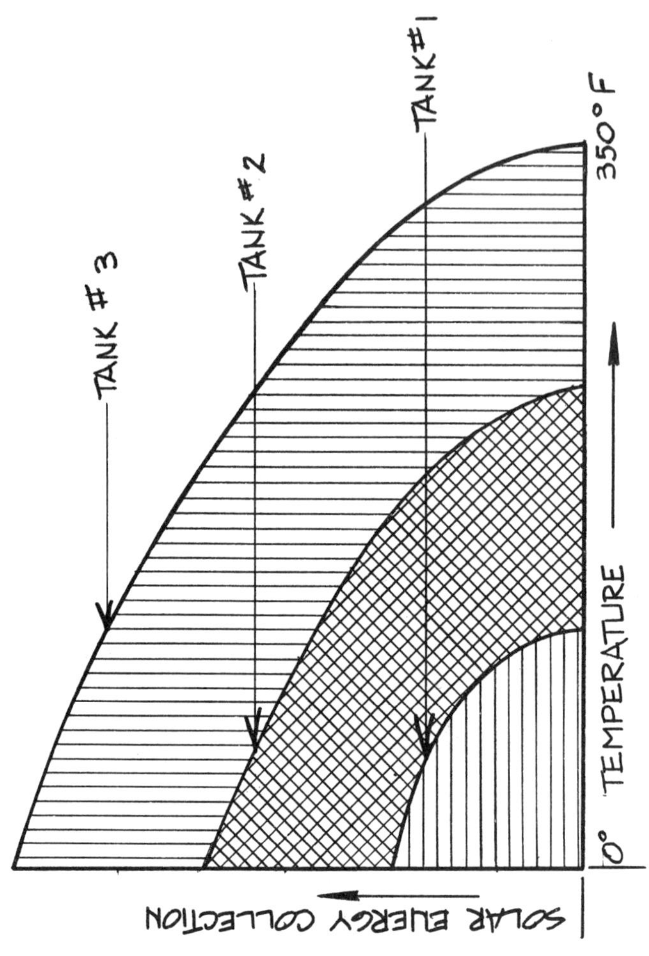

FIG.-11 TANKS TEMPERATURE DISTRIBUTION

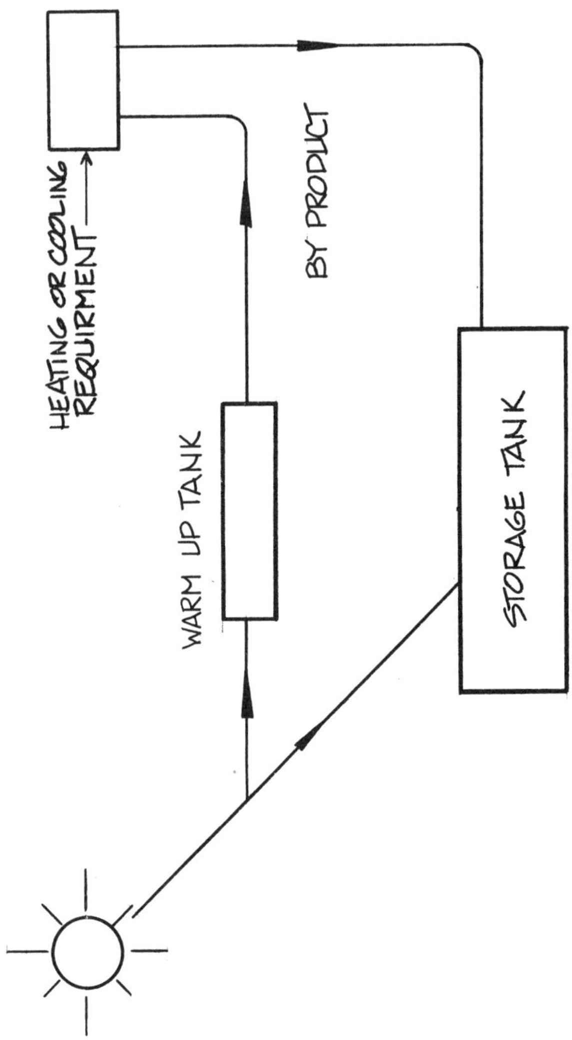

FIG.-12 HIGH TEMPERATURE DISTRIBUTION

perforated holes located at the bottom of the tank provided the required convection currents in the system for eliminating the stratification (Fig.-13). The tank is able to store solar energy from 120° F. to 350° F., which will provide space cooling through an absorption machine and space heating through a set of heat exchangers and control valves. The tank has the flexibility to drop in temperature to 60° F. At that time it can be used directly for heat pumping. The system is designed not to dump any auxiliary heat into the storage tank when solar energy is not available (Fig.-14). The tank is designed to hold a total solar collection of up to 25 million B.t.u.

b) Solar-assisted Ice Storage System

An ice storage system, even though it may be perceived as new in the industry, is not a new system. It is basically an old system which was used quite ordinarily in the refrigeration industry. The old ice storage technology and the new solar technology are integrated for developing an energy efficiency "SOLAR-ASSISTED ICE STORAGE SYSTEM." The ice storage system provides the saving in cost for storing the cold energy and at the same time the system provides the opportunity for harnessing the solar energy and storing it in the form of ice (Fig.-15), at low demand rates. Technically the system is feasible. Operationalwise the system is slightly more expensive than the most energy efficient system. However, it is cheaper than the conventional refrigeration system. The ice storage system concept was conceived for reducing the enormous size of the tank, otherwise required for storing chilled water energy. The harnessed energy of the sun can be stored directly in the high temperature storage tank till the tank temperature reaches 350° F. and if storage is still required the solar energy is transformed into ice by operating both the absorption machine and the screw machine in sequential order. This transformation is economically feasible and consumes the same amount of electrical energy as any other conventional refrigeration system. The electrical energy consumed to store the solar energy in the form of ice is approximately 1.1 horsepower per ton of refrigeration. Normally, the ice storage system is connected with the screw compressor system (Fig.-15), which operates at night when the outside wet bulb temperature and demand rates for energy consumption are very low. The chilled energy is stored in the form of ice by circulating the brine solution in the ice storage tank. The ice storage system with the ice bank coils is sized so as to take care of all the inefficiencies that are created due to the loss in thermal conductivity across the ice bank coils. The coils are spaced every 1/2" to 3/4" so as to limit the thickness of the ice being formed between the coils and for enhancing the thermal transfer efficiency across the coil bank. At the same time, this type of arrangement increases the heat transfer efficiency of the charging and discharging of the storage tank. Another advantage of having a storage tank is that the storage tank temperature can be shifted to a temperature level of 40° F. when the harnessed solar energy can be used directly as a chilled water system without undergoing a phase change. The initial cost of the ice storage system proved to be less than the conventional chilled water storage system for storing the same amount of energy. In this particular

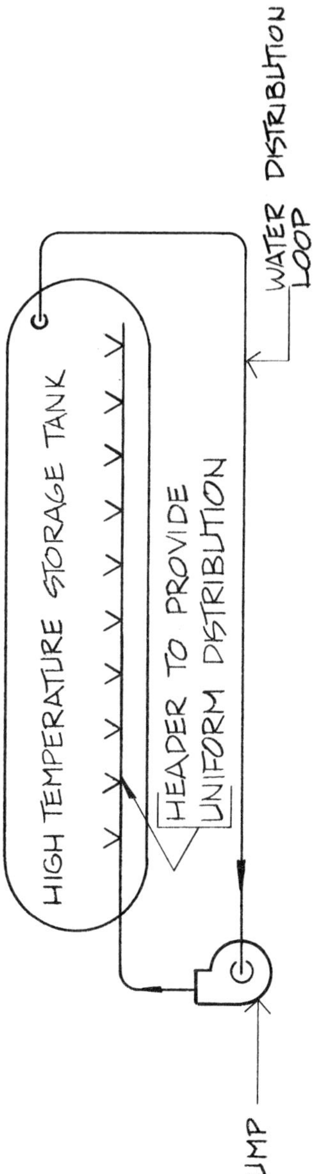

FIG-13 WATER SPRAY DISTRIBUTION

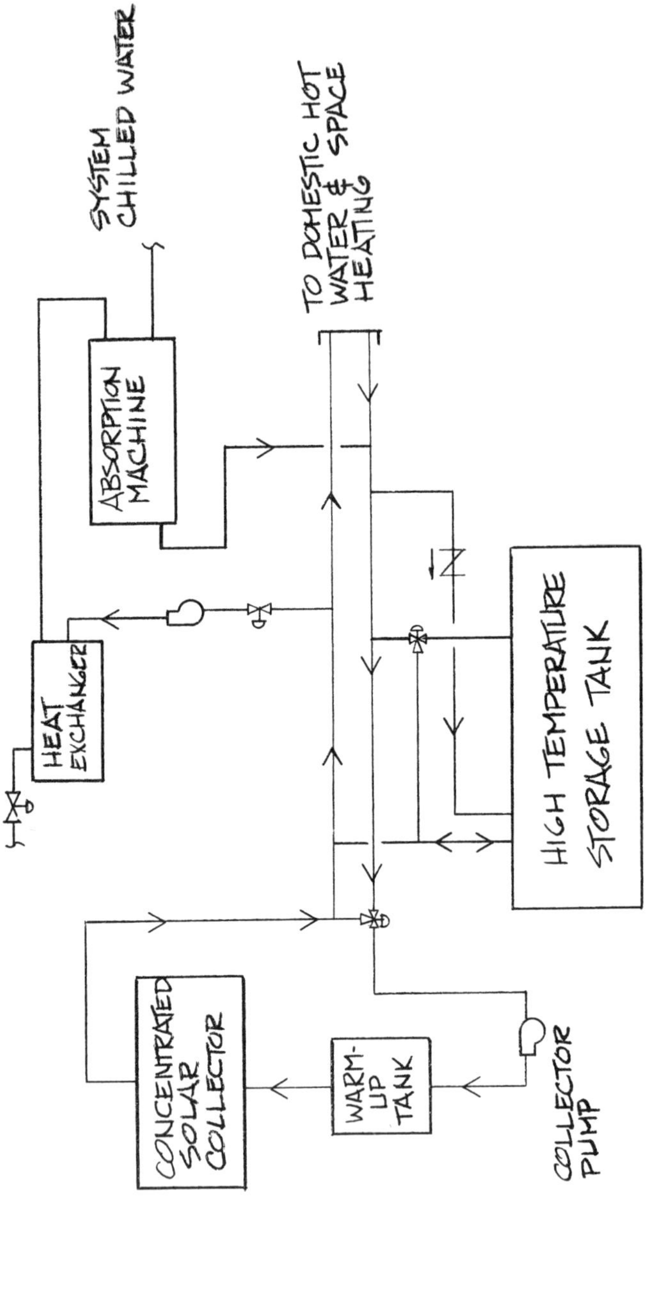

FIG-14 HIGH TEMPERATURE STORAGE TANK SYSTEM

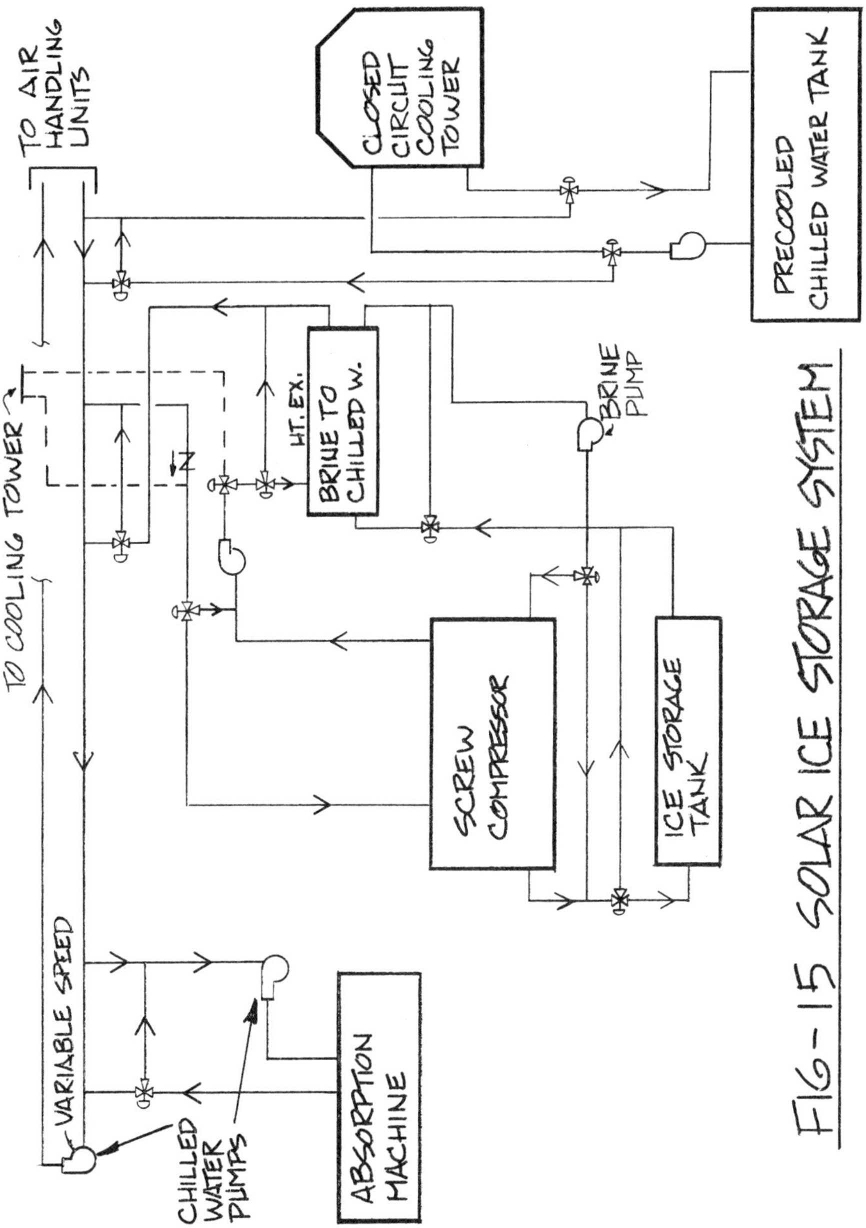

FIG-15 SOLAR ICE STORAGE SYSTEM

instance an 8,000 gallon ice storage tank was equivalent to a 90,000 gallon chilled water storage tank.

E. SOLAR DIRECT-HEATING SYSTEM

The Fig.-16 indicates the usage of the solar energy through the system of heat-exchangers and condenser piping loop to conventional air handling systems for providing heating if the solar energy is not available. The deficit is made up by the auxiliary steam energy from central utility systems. The solar condenser water piping is used at a low temperature level for providing direct heating to the air handling units. A condenser storage tank is introduced into this circuit for storing sun energy at a temperature of 120° F. The main function of the condensor water storage subsystem is to recycle the waste heat and also to act as a transportation system for solar water from a high tempeature storage tank which is supplied at a low energy level for meeting the heating demands in the building. The advantage of using a CONDENSOR-SOLAR INTEGRATED SYSTEM is that the solar energy, after being stored in the high temperature storage tank, can be stored further in the condenser storage tank in winter at a different energy level. The storage tank is performing a dual function of storing recycled waste heat and also the solar energy at a temperature level of 120° F. for space heating. It supplements the high temperature storage system and ice storage system for storing solar energy in sequential order.

F. SOLAR-LIGHTING

The negative architectural design was conceived with six light wells to allow natural light and solar energy to the work areas for reducing the use of artificial lighting. Since daylighting is to be the predominant energy conservation feature, a building model was constructed, and the daylight contours (Fig.-17) were simulated for determining the savings in artificial energy. Exterior surface treatments are given very high importance for achieving high levels of daylighting. The materials used are white stucco exterior wall surfaces with cylindrical towers and light concrete walkways extending around the entire perimeter of the courtyards. All soffits were closed with horizontal white or polished aluminum or stainless steel panels. The interior surface at fixed partition walls abutting the glass perimeter walls at right angles should be finished with a white matt surface. Perimeter floor areas near the glass are covered with a very light colored vinyl asbestos tile throughout the entire partition areas. Fixed partition walls parallel to the exterior glass should be full length clear plate glass to allow some light penetration. This enhances visual communications and also the savings in artificial energy. Sacramento is located at a north latitude of 38°34' and during a clear day the sky luminance is about 1300 FL and the presence of direct sunlight is another 3000 FL. A foam model was constructed and an artificial sky was simulated with the design sky luminance of 1300 FL and the estimated grand total displaced energy savings from daylighting was found to be in excess of 100,000 KWH on an annual basis.

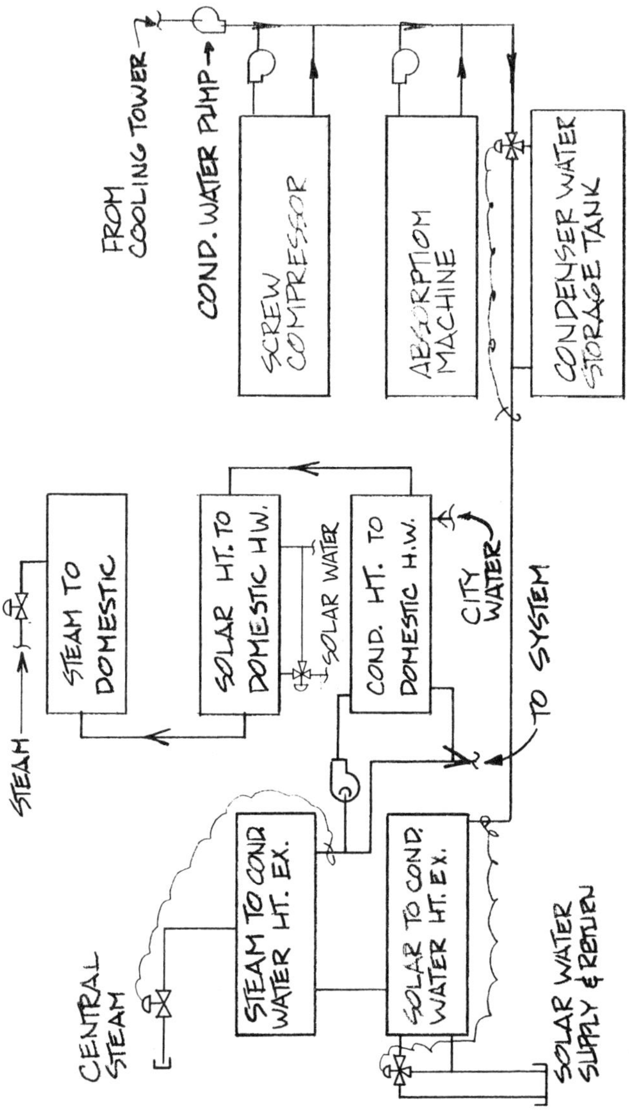

FIG-16 SOLAR DIRECT HEATING SYSTEM

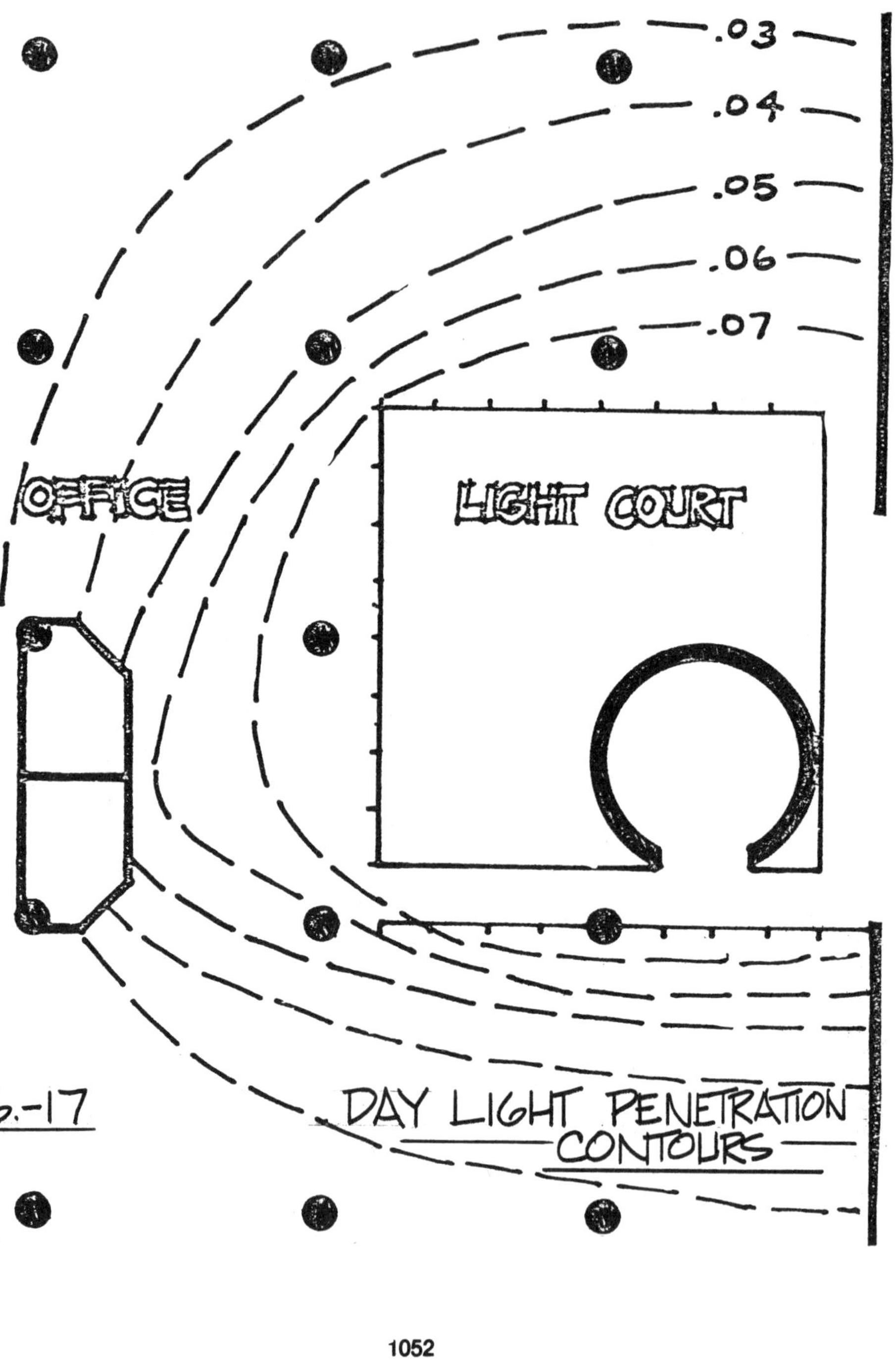

FIG.-17 DAY LIGHT PENETRATION CONTOURS

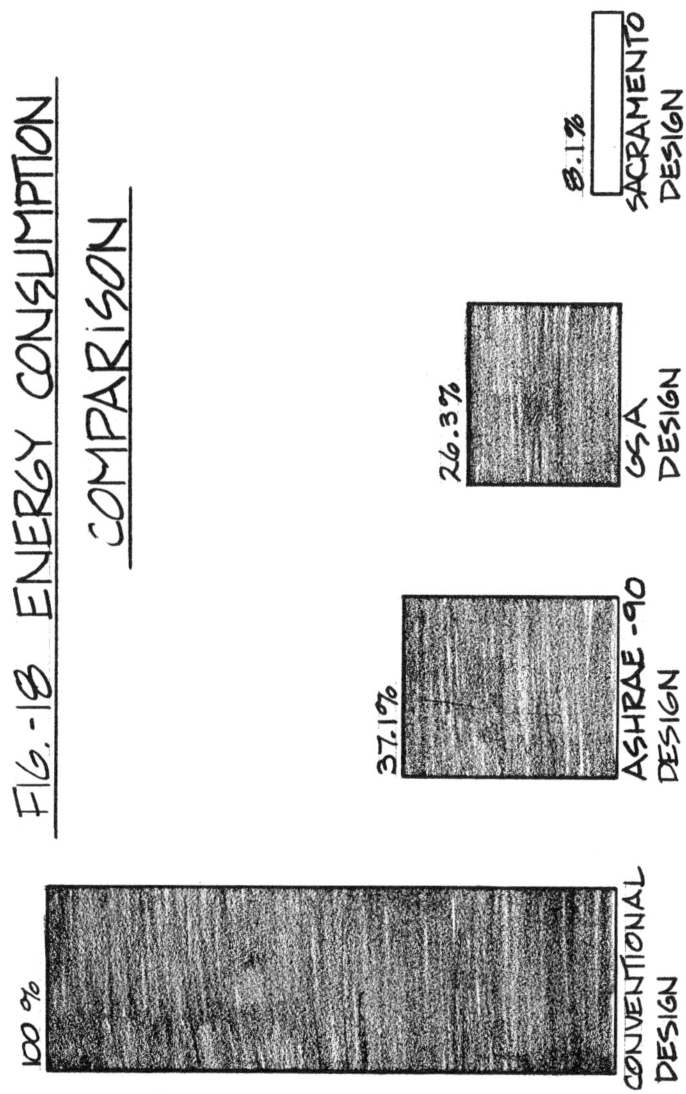
FIG.-18 ENERGY CONSUMPTION COMPARISON
100 %
CONVENTIONAL DESIGN
37.1%
ASHRAE-90 DESIGN
26.3%
GSA DESIGN
8.1%
SACRAMENTO DESIGN

FIG.-19 ENERGY CONSUMPTION COMPARISON BETWEEN CONVENTIONAL AND ENERGY EFFICIENT DESIGN

	CONVENTIONAL DESIGN	GSA DESIGN	ASHRAE DESIGN	SACRAMENTO DESIGN
ANNUAL BTU/SF AT BUILDING BOUNDARY	175200	55000	67072	19880
ANNUAL BTU/SF AT POWER PLANT	350000	100000	134000	662666
ANNUAL OIL CONSUMPTION/SF AT POWER PLANT IN GALS.	2.4-2.77	0.65	0.9321	0.46
ANNUAL GAS CONSUMPTION IN CU.FT./SF.	350-400	100	134-144	67
ANNUAL KW CONSUMPTION/SF AT POWER PLANT	30.8-35.2	8.79	118-12.7	5.82

CONCLUSION

The decision to create a negative architectural design and to generate solar power through concentrated solar collectors was the key in capturing approximately 4200 B.t.u.'s per square foot per year of sun's energy and the building utilized total energy consumption of 19,800 B.t.u.'s per square foot per year at the building boundary as compared to 175,000 B.t.u.'s per square foot per year for a 1975 conventional office building which is 8% of the energy consumed in conventional buildings (Figs.-18,-19). The energy savings achieved by this design approach is in the order of 10,000 barrels of oil per year for this building. This means if this design approach is utilized every year it would be possible to consume 1 million barrels of oil less per 100 buildings in the U.S.!

A QUINONE-ASSISTED PHOTOFORMATION OF ENERGY-RICH CHEMICAL BONDS

S. W. Fox, T. Adachi, and W. Stillwell
Laboratory of the National Foundation for Cancer Research
at the Institute for Molecular and Cellular Evolution
University of Miami
Coral Gables, Florida, U.S.A.

ABSTRACT

ADP and inorganic phosphate are converted to ATP by white light in the nonaqueous solvent, dimethylformamide, in the presence of tetrachloro-p-quinone or ubiquinone. The identity of the ATP has been verified by a number of tests. The conversion is somewhat more efficient at 10° than at 20°, 30°, or 40°C.

Conversion of ADP to ATP has been accomplished in aqueous suspension by use of cell-like structures aggregated from poly(aspartic acid, glutamic acid, tyrosine). This is believed to occur through formation of dopaquinone in the peptide structure during illumination. The way in which the quantitative yield of ATP has been influenced by pH and by added substances, such as $FeCl_2$, has been studied.

The yield in conversion of ADP → ATP in suspensions of microspheres is much less than in dimethylformamide suspension, probably due to attenuation of light by the microspheres, and by the production of water as a byproduct.

INTRODUCTION

During the opening ceremony, the statement was made that this civilization has moved from muscle power to energy. Such a statement is historically appropriate in the shadow of the pyramids of Giza; one cannot imagine the erection of such structures nowadays without the help of bulldozers, trucks, and cranes. To a biochemist, however, muscle power was already "energy". All biochemists would consider muscle action to be an expression of chemical energy; many would consider it to be electrical energy also [1].

Muscle power was shown many years ago, by Englehardt, Szent-Györgyi and others [2] to depend specifically upon a universal energy-rich compound of all living cells, adenosine triphosphate (ATP). Broda [3] has calculated that the metabolism of

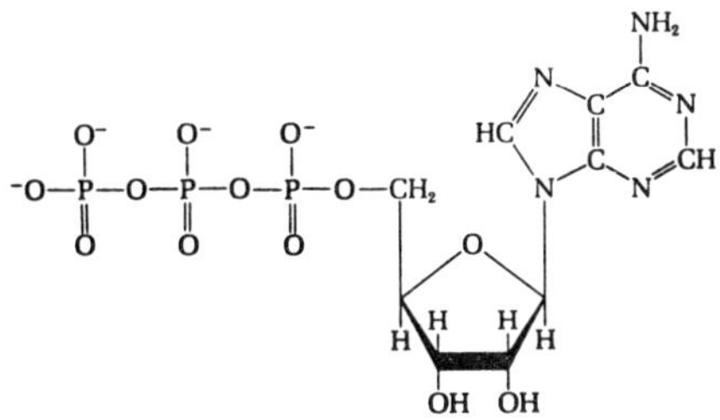

Fig. 1. Structure of ATP, in the ionized form at pH 7.0.

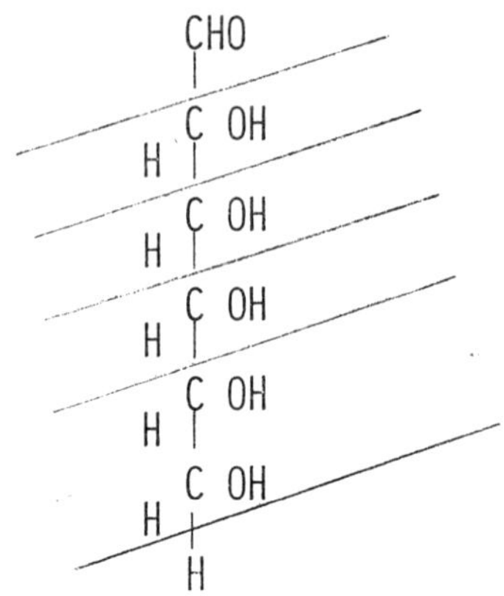

HCHO(HCOH)

Fig. 2. Formaldehyde, and glucose depicted as a polymer of formaldehyde.

a typical human being generates each day more ATP than the weight of that human being! ATP is a compound nicely balanced for the transfer of energy. It is composed of adenine (a nitrogen base) with the sugar, D-ribose, to which are attached three phosphoric acid residues in a triphosphate structure (Fig. 1). Under the control of specific enzymes in specific cases of utilization, the terminal phosphate is transferred with the release of approximately 7.3 K cal. per mole (standard value). This energy is used for the synthesis of protein from amino acids in the cell, for the synthesis of DNA and RNA, for various steps in carbohydrate metabolism, etc. The most efficient green-plant convertors of solar energy: sugarcane, corn, sorghum, etc. reduce carbon dioxide to glucose with the aid of ATP [4]. Schematically, the reaction is a reduction of CO_2 to formaldehyde, HCHO, of which glucose is a polymer (Fig. 2).

In several conferences, biochemists, chemical engineers, and others have discussed how to industrialize photosynthesis and how to industrialize the ATP cycle that operates in all cells. These discussions recognized that much of the biochemical formation of ATP is extremely complex, and that the natural sequence of conversions from sunlight, CO_2, H_2O, etc. to muscle power involves a long and inefficient (by industrial standards) food chain.

While our biochemical interest in solar energy is in the cellular synthesis and cellular utilization of ATP [1,3], the focus of our attention is more specifically protobiochemical [5]. If we can infer by experiments any of the processes by which the first cells on Earth made and used ATP, we believe those processes would have been simple and the compounds involved would have been relatively stable. Processes that would occur on the primitive Earth without the aid of a chemist would necessarily be simple; similarly the first compounds used would need to serve the function of charge transfer. Simplicity of reactions and stability of compounds are of course qualities appreciated in industrial processes.

While ATP is a cellularly universal energy transfer substance, quinones are also widely found as transporters of electrons and ions. One quinone is sufficiently ubiquitous to have earned the name ubiquinone (Fig. 3). We have used this quinone successfully in our experiments. It is, however, sufficiently expensive that we employ mostly other quinones. Of several tested, the tetrachloro-p-benzoquinone (chloranil; Fig. 3) functions the best. Chloranil is a common industrial product and inexpensive.

While we have looked especially at quinones and ATP, the

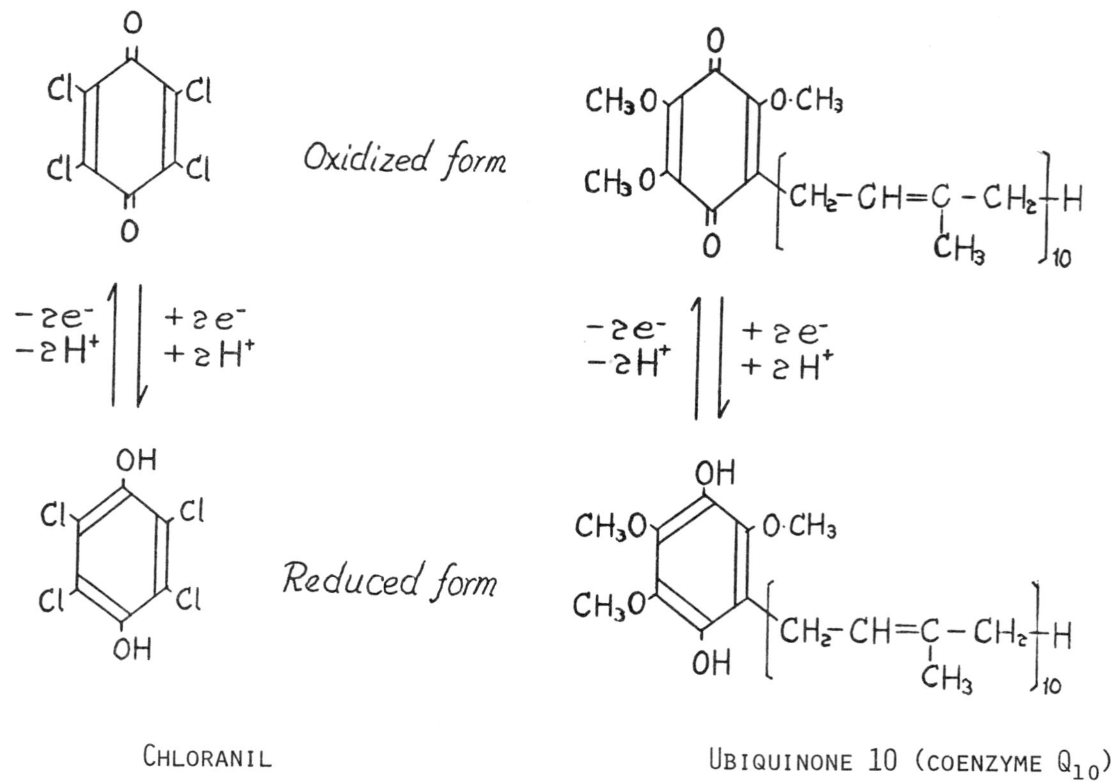

Fig. 3. Oxidized and reduced forms of chloranil (tetrachloro-p-benzoquinone) and ubiquinone 10.

TABLE I
IDENTIFICATION OF ATP
SYNTHESIZED IN DIMETHYLFORMAMIDE
BY WHITE LIGHT

Methods	Results
Chromatography in 5 systems	Agreement of R_f with that of standard ATP
Coincidence chromatography in 5 systems	Unchanged R_f when standard ATP added
Enzymic breakdown	ADP,AMP by potato apyrase ADP,AMP by pork ATPase ADP,AMP by luciferase
Luminescence in flash test with luciferin-luciferase	2×10^6 counts *vs.* 7×10^3 counts in control

potential for photosensitive processes among the substances that compose organisms is manifold [6]. In constructing experiments to simulate protophotophosphorylation in the earliest cells on Earth [7], we have sought mechanisms akin to those used by all cells now. This emphasis is motivated by the belief that the products most likely to have survived early stages of molecular evolution are those most likely to be ubiquitous now [cf. 4, p. 1038]. Among the charge transfer substances in this category, and also capable of absorption of light, the quinones especially invited attention [8].

In the experiments, radioactive ADP has typically been illuminated by white light in dimethylformamide (DMF) saturated with inorganic phosphate [9] with a form of chloranil present.

In the initial studies, specially prepared chloranil anion radical [10] or the quinol phosphate of chloranil [8] was used. The more recent experiments demonstrated that the simple quinone could be used [9]; it was evidently converted to quinol phosphate [8] by light. In some of the tests reported here, it has been found that the added chloranil is superfluous in experiments performed in the presence of tyrosine residues in peptides. The later design based on inclusion of tyrosine residues was tested because tyrosine is convertible to dopaquinone [11,12], as an <u>ortho</u>quinone moiety in peptides composing membranes.

ATP SYNTHESIS IN DIMETHYLFORMAMIDE SOLUTION [9]

The time-course of the synthesis in the nonaqueous solvent is shown in Fig. 4. The Na salt of ADP first dissolves in DMF before phosphorylation occurs. In other words, the formation of ATP occurs in two steps: (a) solution of ADP-Na salt in DMF and (b) phosphorylation of the ADP. The synthesis of ATP does not occur unless the inorganic phosphate is present. No ATP is found without illumination.

The ATP was identified as such in five solvent systems, by breakdown by three kinds of ATPase, and by flash test (Table I [9]).

Fig. 5 shows the effect of varied temperature on the conversion of ADP to ATP and AMP. The extent of conversion of ADP increases by about 35% from 10° to 40°. The yield of ATP decreases by almost 50% as the temperature is raised in the same range from 10°C to 40°C. This relative independence of temperature is characteristic of photochemical reactions [13]. While this inference is complicated by the fact that solution of ATP-Na salt in DMF must precede phosphorylation, the pro-

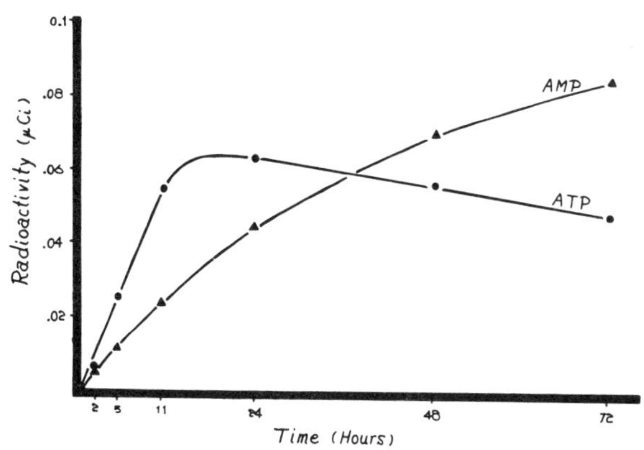

Fig. 4. Formation of ATP and AMP from ADP in dimethylformamide as a function of time of illumination.

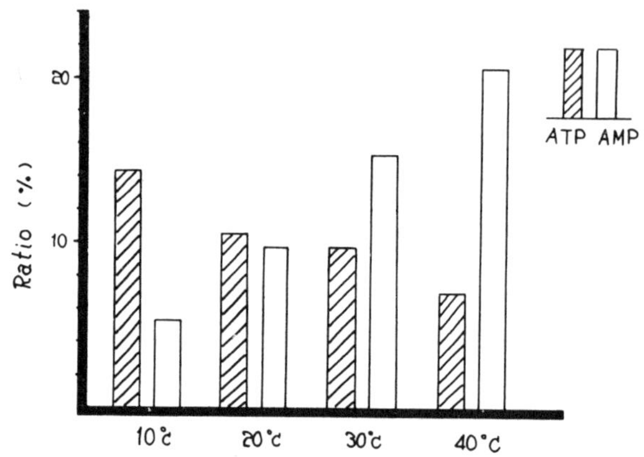

Fig. 5. Effect of temperature (±2°C) on formation of ATP and AMP by illumination of ADP in dimethylformamide. After 20 hr. of illumination.

cesses of solution and phosphorylation are both speeded by increase in temperature. The process that is increased with elevated temperature is that of hydrolysis of ATP to AMP; this is known to be a ready chemical process. Some ATP can be hydrolyzed by the water formed as byproduct. Fig. 5 suggests that ATP is formed first; hydrolysis follows at a lower rate at the beginning.

All of the results indicate that ADP and Pi can be converted to ATP by white light in an essentially nonaqueous solvent, DMF. For either a theory of protobiogenesis or a utilitarian process, a cyclical set of reactions would offer advantages. The outline of an essential cycle is depicted in Fig. 6.

ATP SYNTHESIS IN AN AQUEOUS SUSPENSION OF ARTIFICIAL CELLS

Theoretical

Contemporaneously, the biosynthesis of ATP is carried out in biological cells, in the presence of water as the sole liquid. The pathways are extremely complex [4]. The reactions occur in the hydrophobic regions of the membranes of cells; the membranes protect from water in the cell.

Simpler systems that are capable of direct synthesis of ATP have been reconstituted from components of cells. Some of the experiments and conditions have been described by Yoshida *et al.* [14] and by Baskin [15].

In our attempts to incorporate a quinone structure into polyamino acids that can be aggregated into cell-like structures, tyrosine residues appeared especially attractive. One reason for this is that tyrosine is susceptible, as stated, to conversion to the corresponding quinone through several stages of oxidation:

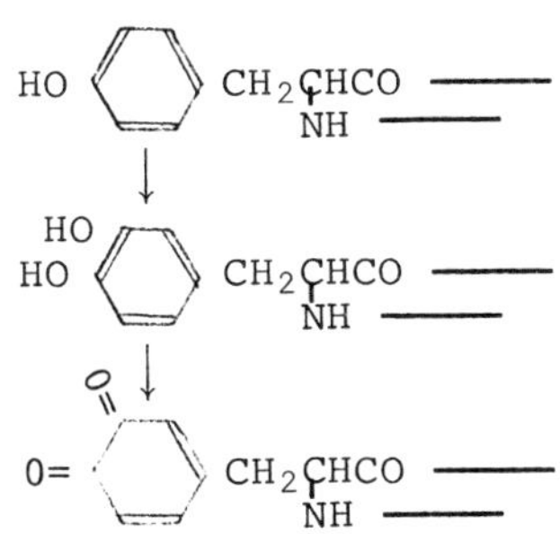

tyrosine residue

dopa residue

dopaquinone residue

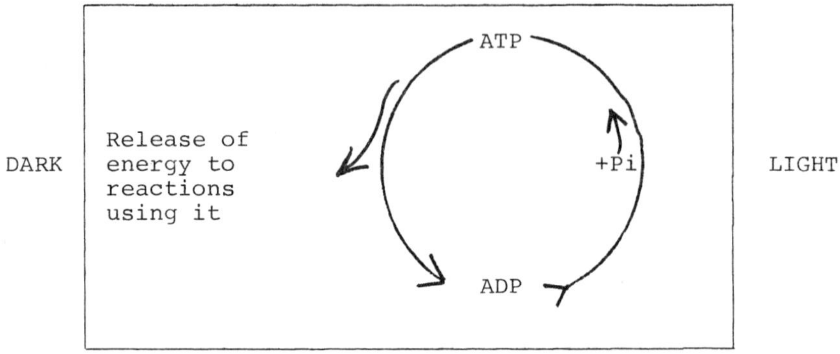

Fig. 6. Formation of energy-rich ATP by light, followed by release of energy in the dark.

In each of the intermediates depicted, and others not shown, the aryl moiety is a substructure in a polymer that contains tyrosine or a derivative. The convertibility of tyrosine to an orthoquinone has been inferred from the evidence accumulated by Raper, Mason, and others [11,12]. The probability of the occurrence of tyrosine on the primitive Earth follows from two laboratory studies [16,17], while the availability of aspartic acid and glutamic acid is known from studies of lunar samples and of meteorites [7].

Experimental

A description of one of the experimental sequences (no. 123) follows. An equimolar mixture of L forms of aspartic acid, cystine, glutamic acid, leucine and tyrosine was heated in a bath at 195-200° for 6½ hr. The crude product was stirred into water and the suspension was dialyzed in the coldroom for three days, and then filtered. Analysis revealed molar ratios of approximately 2 asp:1 glu:1 cys:2 leu:1 tyr in the undissolved solid.

Microspheres were made by heating 30 mg of solid in 1.0 ml of water and allowing to cool to room temperature (Fig. 7). Most of the polymer stuck to the walls of the tube. After cooling, 0.5 ml of the suspended microspheres was transferred and treated with 20 mg of $FeCl_2 \cdot 4H_2O$, 40 mg of KH_2PO_4, and 100 mg of nonlabelled ADP sodium salt. A quantity of 2.0 μCi of labelled ADP (8-^{14}C) in a cuvette of size 1x10x40 mm was desiccated to dryness. To this was added 0.40 ml of the first solution. The cuvette was sealed with paraffin.

The cuvette was illuminated for 14 hr. by a 120 v, 160 W Westron bulb at 10 cm distance. A sample was chromatographed on Whatman 3 MM paper with a solvent of 100 parts isobutyric acid: 60 parts 1 M NH_4OH:6 parts 0.1 M EDTA.

The chromatograms revealed a peak in the ATP region (Fig. 8). Yields of ATP from ADP in experiments of this type were 0.05-0.12%. The quantum yield must also be very small. However, the yield corrected for the small volume of microparticles would be considerably larger. Moreover, the illumination of microspheres en masse must result in attenuation of the illumination as it passes through succeeding layer of microspheres. Also, how much ATP is within the particles is not known.

In order to identify the ATP by enzymes, as was done with the product obtained in DMF, it was evidently necessary to carry out the synthesis in the absence of $FeCl_2$. The reason for this is that both $FeCl_2$ and proteinoids were found to inhibit the activity of the ATPases used for identification of the product

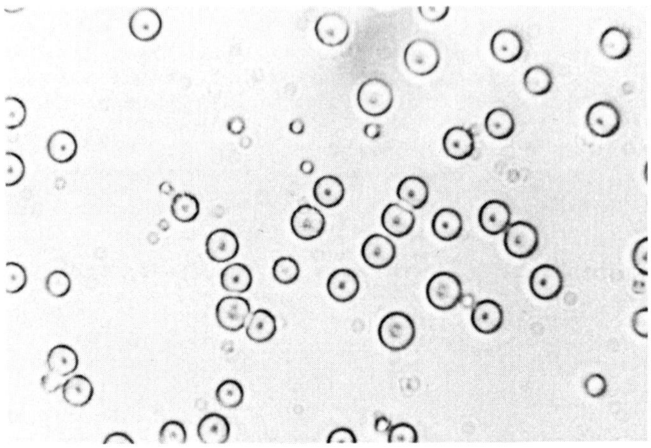

Fig. 7. Photomicrograph of microspheres made from poly(asp,glu,cys,leu,tyr). Approx. 1 μm in diameter.

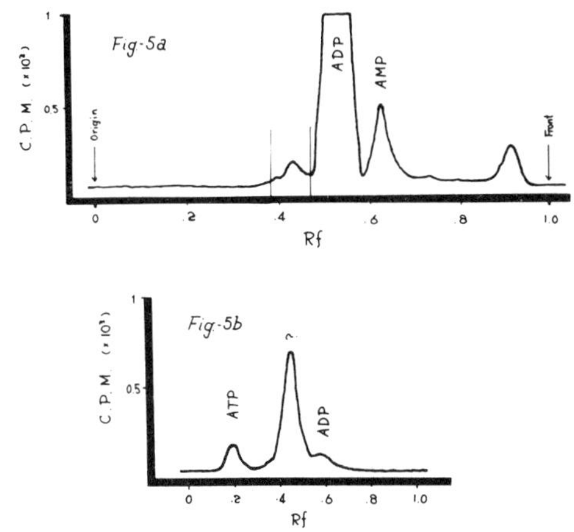

Fig. 8. Chromatograms showing ATP produced in a suspension of poly(asp,glu,cys,leu,tyr) microspheres in an aqueous solution of ADP and inorganic phosphate. 8a) Chromatogram on paper with isobutyric acid: NH_4OH:EDTA solvent. 8b) Rechromatogram of ATP fraction on PEI sheets (thin layer) with $LiCl_2$: acetic solvent.

produced in the presence of water (Fig. 9). The iron salt is added in some experiments because it was found to promote the production of the ATP peak.

In one of these experiments (no. 143), 100 mg of copoly(asp, cys,glu,leu,tyr) was heated with 1.0 ml of water at 80° for 10 min., and allowed to cool. The suspension was reheated at 100°, filtered rapidly and treated with 25 mg of lys-rich proteinoid [18] in 1.0 ml of water, and 20 mg of KH_2PO_4 and 20 mg of finely powdered chloranil (subl. pt. 286° uncorr.) were added to 0.5 ml of the suspension.

This suspension was added to the tube in which was already present 4.0 μCi of labelled ADP ($8-^{14}C$) trisodium salt in the dry state. The tube was illuminated for 67 hr. at 21°C.

The supernatant portion of the reaction solution was sampled for thin-layer chromatography on PEI-cellulose sheets with 2 M LiCl:2 N acetic acid (1:1). Another sample was treated with luciferase(Sigma). Since ATP could not be eluted from PEI-cellulose sheets, the effect of enzyme was studied directly by comparing results on sheets. They showed that the product was ATP.

Variations in the experiments yielding ATP included nine kinds of proteinoid or combinations thereof; presence or absence of $FeCl_2$; presence of large proportions of nonradioactive ADP; presence or absence of chloranil or of ubiquinone. Each of four polymers lacking either tyrosine or dopa failed to yield ATP.

EFFECT OF pH

When the process is carried out under controlled conditions, it becomes possible to study variables in the photochemical synthesis of ATP. This has been done with pH as a variable. The optimal pH was found to be between 2 and 3. The pH values of 5,6, and 7 were also tested. No yield was observed at these latter values. This result deserves further study with microspheres made from varying ratios of acidic polymer : basic polymer.

SUMMARY AND OUTLOOK

ADP and inorganic phosphate have been converted into ATP in a nonaqueous solution in dimethylformamide, in the presence of quinone and white light. The yield of dissolved ADP ---> ATP attains values of 20 percent.

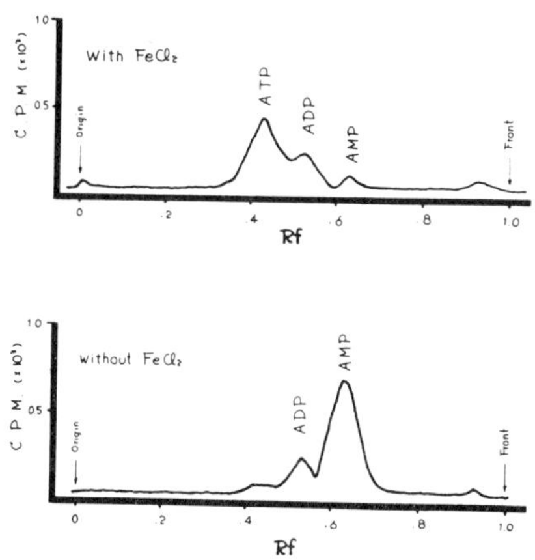

Fig. 9. Inhibition by $FeCl_2$ of breakdown of ATP by luciferase. Purchased ATP submitted to enzyme action during 1 hr. Proportions are 2.52 μg ATP tetrasodium salt to 2.0 mg $FeCl_2 \cdot 4H_2O$. $FeCl_2$ inhibits action of enzyme breaking ATP down to ADP and AMP.

The conversion occurs also in an aqueous suspension containing artificial cells composed of thermal copolyamino acids containing either tyrosine or dopa. Presumably dopaquinone is formed within the peptides during illumination. The conversion is aided by ferrous chloride, and appears to be increased by added chloranil, and by increased concentration of ADP. The yield is much smaller than in the DMF solution; this can be understood as due to the attenuating action of light of layers of extrasolution particles, and to accumulation of water. Amounts of ATP within microspheres are unknown.

Tests on a number of other polymers are now in order. Various ways of increasing hydrophobicity of the particle are being studied. Whether the synthesis is occurring in or on the particles has yet to be examined. For understanding the origin of cyclical bioenergetic processes through physical models, it is necessary to perform processes in the presence of an aqueous medium. The study of effects in stirred suspensions or single layers of microspheres is also on the agenda.

ACKNOWLEDGMENTS. We thank Mr. David Rose and the National Aeronautics and Space Administration (Grant No. NGR-10-007-008) for financial support. Contribution no. 324 of the Institute for Molecular and Cellular Evolution.

References

1. Lehninger, A. L., Bioenergetics, 2nd edition (W. A. Benjamin, Inc., Menlo Park, 1971).

2. Szent-Györgyi, A., The Living State (Academic Press, New York, 1972) pp. 19-36.

3. Broda, E., The Evolution of the Bioenergetic Process (Pergamon Press, Oxford, 1975) pp. 40-41.

4. Lehninger, A. L., Biochemistry, 2nd edition (Worth and Co., New York, 1975) pp. 636-639.

5. Fox, S. W., The origin and nature of protolife, in The Nature of Life, W. H. Heidcamp, ed. (University Park Press, Baltimore, 1978) pp. 23-92.

6. Erlanger, B. F., Photoregulation of biologically active macromolecules, Ann. Rev. Biochem. 45 267-283 (1976).

7. Fox, S. W. and Dose, K., Molecular Evolution and the Origin of Life, revised edition (Dekker, New York, 1977).

8. Stillwell, W., On the origin of photophosphorylation, J. Theor. Biol. 65 479-497 (1977).

9. Fox, S. W., Adachi, T., Stillwell, W., Ishima, Y., and Baumann, G., Photochemical synthesis of ATP: protomembranes and protometabolism, in Light Transducing Membranes: Structure, Function, Evolution, D. W. Deamer, ed. (Academic Press, New York, 1978).

10. Lichti, F. and Lucy, J. A., Reactions of vitamin A with acceptors of electrons. Formation of radical anions from 7,7,8,8-tetracyanoquinodimethane and tetrachloro-1,4-benzoquinone, Biochem. J. 112 221-229 (1969).

11. Happold, F. C. and Raper, H. S., XIV. The tyrosinase-tyrosine reaction. III. The supposed deaminizing action of tyrosinase on amino acids, Biochem. J. 19 92-100 (1925).

12. Mason, H. S., The chemistry of melanin. III. Mechanism of the oxidation of dihydroxyphenylalanine by tyrosinase, J. Biol. Chem. 172 83-89 (1948).

13. Kistiakowsky, G. B., Photochemical Processes (Chemical Catalog Co., New York, 1928).

14. Yoshida, M., Sone, N., Hirata, H., Kagawa, Y., Takeuchi, Y., and Ohno, K., ATP synthesis catalyzed by purified DCCD-sensitive ATPase incorporated into reconstituted purple membrane vesicles, Biochem. Biophys. Res. Commun. 67 1295-1300 (1975).

15. Baskin, R. J., The Ca^{2+}-Mg^{2+} ATPase protein of sarcoplasmic reticulum, in Membrane Proteins and their Interactions with Lipids, R. A. Capaldi, ed. (Dekker, New York, 1977) pp. 151-188.

16. Harada, K. and Fox, S. W., The thermal synthesis of amino acids from a hypothetically primitive terrestrial atmosphere, in The Origins of Prebiological Systems, S. W. Fox, ed. (Academic Press, New York, 1965) pp. 187-194.

17. Hayatsu, R., Studier, M. H., and Anders, E., Origin of organic matter in early solar system-IV. Amino acids: confirmation of catalytic synthesis by mass spectrometry, Geochim. Cosmochim. Acta 35 939-951 (1971).

18. Fox, S. W., Harada, K., and Rohlfing, D. L., The thermal copolymerization of α-amino acids, in Polyamino Acids, Polypeptides, and Proteins, M. Stahmann, ed. (University of Wisconsin Press, Madison, 1962) pp. 47-54.

ETHANOL, A CLEAN BURNING BIO-SOLAR FUEL - ITS PRODUCTION AND UTILIZATION

Prof. (Dr) H.B. Mathur
Head of the Department of
Mechanical Engineering
Indian Institute of Technology
Hauz Khas, New Delhi-110029
India

ABSTRACT

Ethanol's use as a fuel can be regarded as a direct method of obtaining energy from sun without the intermediary of storage in earth for long period of time. As long as the sun shines plants will perform their synthesis of starch from the abundant carbon dioxide and water that bathe our planet. From this annually renewed store of raw materials ethanol can be readily produced in quantities sufficient to meet the world demands.

Ethanol can be obtained from an astonishingly wide range of raw materials some one or the other of which is found in every habitable part of the globe. Bio-conversion of agricultural waste, farm crops, tropical grasses, utilisation of molasses - a by -product of sugar industry, sulphite liqour - a waste effluent of paper and pulp industry, to name a few, are some of the innumerable means of production of ethanol - an unstored, clean burning synthetic fuel.

Extensive experimental tests and theoretical investigations carried out by the author in the last ten years in the mechanical engineering department of I.I.T. Delhi have shown that ethanol can be utilised with great advantage in the existing automobiles/scooter/motorcycle engines, in the form of ethanol gasoline blends, while future 'Ethanol Engines' can be developed for 'total ethanol operation' with consequent high specific output and ten to fifteen percent higher thermal efficiency as compared to the present day engines. Its use to meet the transport energy needs of a developing country like India provides an immediate though partial solution of the twin problems of fuel oil scarcity and growing vehicular exhaust pollution.

INTRODUCTION

The fuel resources of this world are vast but not unlimited. These fuels can be classified into two categories namely the 'Stored' fuels and the 'Unstored' fuels. 'Stored' fuels are the fuels, which as the name suggests, have been in the storage of earth for very long periods of time. Major fuels of this type are coal and petroleum, the reserves of which are vast but not unlimited. With our present known resources and the rate of consumption, it is feared that these 'stored' fuels will be exhausted soon.

No doubt in future alternate sources of energy will be exploited; tidal, the world internal heat, and the most potent of all atomic energy. Nevertheless our present economy demands liquid fuels and there are various reasons to believe, that for a long time to come small independent power units fed by liquid fuels will continue to be used. Solid fuels are inconvenient in use and must be replaced by liquid fuels, even at a greater cost.

It is therefore desirable to search out and investigate the possibility of the use of 'Unstored' liquid fuels for the future. Unstored fuels are the ones that can be produced from raw materials available in abundance in nature and which do not depend on storage under the surface of earth for long duration for their production.

Ethanol is one such fuel (1). It can be produced in sufficient quantities from an astonishingly wide range of raw materials, some of which are to be found in every habitable country on the globe. They include vegetable matter, growing crops, waste organic matter, such as straw and saw dust, tropical grasses, farm waste, water gas, coal and petroleum. Its use can be regarded as a direct method of obtaining energy from sun without the intermediary of storage in earth for long period of time. As long as the sun shines plants will perform their synthesis of starch from the abundant carbon dioxide and water that bathe our planet. From this annually renewed store of raw materials ethanol can be readily produced. It thus possesses attributes of perennial renewal

PRODUCTION OF ETHANOL

Ethanol can be commercially produced from a variety of methods which can be broadly classified into two categories namely, (i) The fermentation methods and the (ii) synthetic methods. The former utilises naturally occuring raw materials which are renewable annually whereas the later makes use of stored materials such as coal and petroleum, hence is not of great utility in view of the non-renewable nature of these materials, hence not considered here.

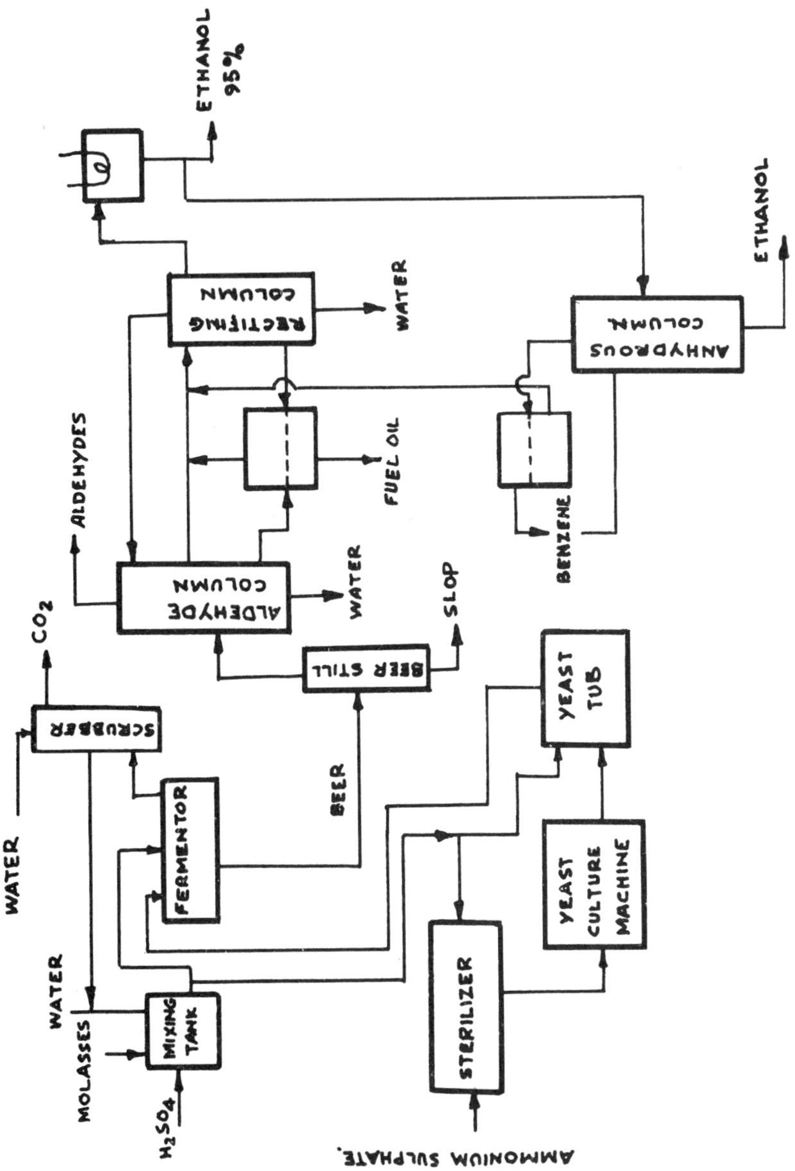

FIG.1. PRODUCTION OF ETHANOL FROM MOLASSES

Fermentation Process

Any material which is capable of being fermented by enzymes can serve as a source for ethanol production. These raw material can be categorised into three groups, namely (i) sugar containing materials, such as sugear-beet and molasses, (ii) starch containing materials such as potatoes, corn, grain, and (iii) cellulosic materials and by products such as wood and wood wastes, saw dust and saw mill waste, waste sulphite pulp liqours from paper and pulp industry.

Production is easiest from uniform materials such as sugar beet and potatoes, but they provide a much higher return when sold as human and animal food stuffs. For that reason special attention should be paid to the use of by products from other industries such as (i) molasses - a by product from sugar industry (ii) sulphite liqour - a waste product from paper and pulp industry (iii) saw dust and wood waste from saw mills (iv) various tropical grasses which have no other commercial value.

Ethanol from Molasses

Molasses is a by product from sugar refineries. It is a viscous fluid which can be stored indefinitely and is available in large amounts from a given site namely sugar refinery. As it has no alternative large scale outlet its initial cost is extremely low. Being uniform in composition it is cheap to handle and process.

Fig. I shows a flow diagram of the process of making ethanol from molasses by fermentation. In the fermentation process, molasses is first diluted to a 14-18% sugar concentration, and then acidified with sulphuric acid to form a mash. Yeast culture is added to the mash until a 3-5 percent yeast concentration is obtained. The most commonly used yeast in the distilling industry is Saccharomyces cerevisiac since it can reproduce very well under normal conditions, found in the distilleries, has a high fermentation rate and efficiency, and can tolerate a relatively high ethanol concentration. Many strains exist, but suitable ones are selected according to congeners which are produced, and to meet the end use of ethanol.

Yeast contains the enzyme invertase and zymase. One of the keys to the fermentation process is that invertase catalyzes the hydrolysis of sucrose in mash to invert sugar. Yeast itself is not consumed or changed during fermentation, and merely functions as a producing agent for the enzymes that catalyse the conversion.

$$C_{12}H_{22}O_{11} + H_2O \xrightarrow{\text{Invertose}} \underset{\text{glucose}}{C_6H_{12}O_6} + \underset{\text{Fructose}}{C_6H_{12}O_6}$$

Invert sugars are subsequently converted into about equal parts of ethanol and carbon dioxide by the action of zymase enzymes.

$$\underset{\text{Invert sugar}}{C_6H_{12}O_6} \xrightarrow{\text{Zymase}} \underset{\text{ethanol}}{2\,CH_3CH_2OH} + 2\,CO_2$$

A fermentation efficiency of about 95% can be obtained from sugar. During the fermentation the carbon dioxide is removed, liquified under pressure and marketed in cylinders, or else solidified, forming a white solid with a temperature of minus 70oC and used in many ways to provide a clean simple form of refrigeration (dry ice). The residue left after distilling the fermented volatile material, forms after drying and treatment an excellent cattle food known as "Distillers Grain".

The ethanol so produced has strength not exceeding 95%. It can be dehydrated in a dehydration plant so designed that ethanol containing less than 0.1 percent water can be obtained at a cost little greater than the erstwhile 95% product. This dehydration is achieved by using solid or liquid drying agents and or azeotropic methods. Trichlorethylene CH Cl: CCl_2(T.C.E) benzene etc. can be used as dehydrating agents with azeotropic distillation to produce 100 percent ethanol from "mash" containing 6 - 7 percent in one continuous operation.

Ethanol from Starchy Material

There is an abundance of farm crops capable of being used as raw materials in the fermentation process leading to ethanol production. Variety of raw materials drawn from the vegetable world, widely grown crops such as those of grain and barley are capable of fermentation into ethanol.

Fig. 2 shows a flow diagram of the process of making ethanol from farm crop such as potato or gram. They are steamed for one or two hours at two to three atmospheres pressure to gelatinise the starch present.

In case of potato the resultant pulp is cooled to about 50°C and an equal volume of water and ten percent malt is added. This mixture is allowed to stand for a short while, when the starch is converted to dextrin. The temperature is gradually raised to about 55°C while dextrin is converted first to maltose and then to dextrose.

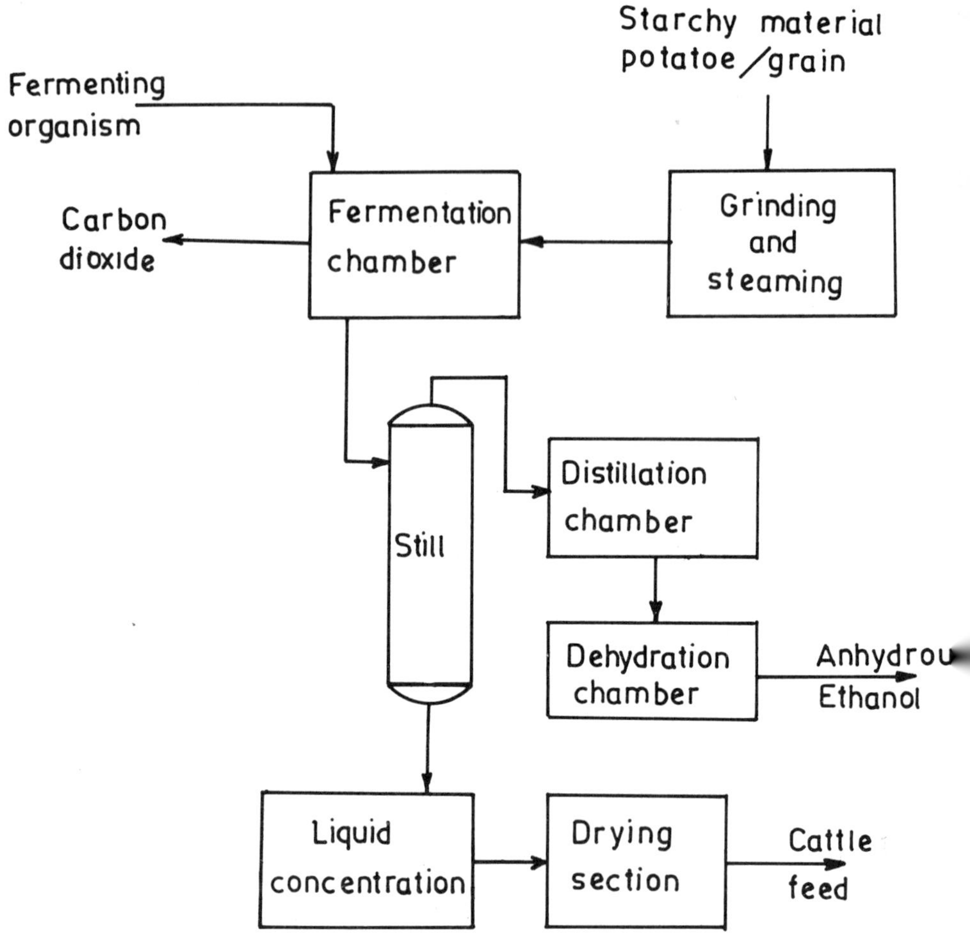

Fig. 2. BLOCK FLOW DIAGRAM FOR ETHANOL PRODUCTION FROM STARCHY MATERIAL

In case of grain as raw material, an enzyme known as amylase is used (amylase is produced from the growth of mold or a small quantity of grain). Amylase converts the starch in the grain pulp into sugar.

Yeast is then added to the liquid mash, and fermentation allowed to proceed for a couple of days. The yeast converts sugar to ethanol. The carbon dioxide formed can be recovered and marketed as "dry ice". When ethanol has reached its optimum percentage, corresponding to 50 percent conversion of the starch, the "mash" is distilled so that ethanol-water mixture is separated from the mash residue which is dried and marketed as cattle feed, equivalent in value weight for weight as oil cake. The ethanol water mixture is sent through additional distillation columns where ethanol concentration is increased to 190 proof (95 percent ethanol). This in turn is dried through an extra active distillation process to yield 200 proof ethanol.

On an average a ton of potato yields 22 gallons of ethanol, over 100 lbs of dry ice and some 200 lb of dried cattle feed, 10 bushels of corn yields 26 gallons of ethanol, over 163 lb of dry ice and some 186 lb of dried cattle feed.

Ethanol from Cellulosic Material

Cellulosic material such as vegetable waste matter, tropical grasses, straw etc can be fermented to yield ethanol. However the conversion of cellulosic material to ethanol is rather more complex than starchy or sugar containing material discussed earlier.

There are, in general, three types of cellulosic compounds present in varying amounts, in natural products, namely cellulose, lignocellulose and hemi-cellulose. Ligno cellulose is the main constituent of wood. Hemi-cellulose are the main constituents of grasses. They are readily hydrolised to pentoses, which in turn may be converted to ethanol by fermentation.

Fig. 3 shows a flow diagram of the process of making ethanol from cellulosic materials. Vegetable waste matter, such as tropical grass, straw etc. is chapped to a convenient size by a chopping machine. The choppings are soaked in dilute sulphuric acid kept at a certain temperature. After the optimum time of soaking and agitation the steeped material is passed through rollers and to remove the excess acid. The residual mass is steamed for a couple of hours to convert the insoluble hemi-cellulose into fermentable sugar. The steamed mass is nuetralised with lime suspension and fermenting organism is introduced to carry out fermentation over a couple of days during which about one-third of the available carbohydrates is converted into ethanol. The ethanol containing mash is distilled and ethanol concentrated in the usual manner The yield of ethanol is between 10 to 25 gallons per ton of raw material. This process with some modification, can be used to obtain ethanol from sulphite liqour - an unpleasant by product of paper and pulp industry. The production of one ton of paper pulp by, sulphite process, produces nearly ten tons of sulphite liqour, containing 2-4 percent total sugar and 0.3 percent sulphur. This is normally run to waste, polluting rivers. This can be treated to nuetralise the sulphite and remove sulphur dioxide. It can then be fermented to yield about thirty gallons of ethanol per ton of sulphite liqour.

ETHANOL FUEL - ITS SUITABILITY FOR AUTOMOBILE ENGINES

Ethanol if used as an automobile engine fuel offers certain advantages, the chief amongst them is the possibility of high compression operation without knock. It is an antiknock fuel. Its octane rating is around 106 and it can operate safely at compression ratios above 12.

It has a high latent heat of vaporisation. This property can be utilised to achieve lower charge temperature during induction, higher charge density, hence higher volumetric efficiency and cooler engine operation.

Absence of pre-ignition and knock resistence makes it suitable for high output engines and opens the option of supercharging of the engine which is seriously restricted with gasoline fuel.

Ethanol is safer than gasoline due to its higher flash point. Its vapours are not quite half as heavy as those of gasoline, so that it does not flow and accumulate in dangerous quantities at low levels, and a higher propor-tion is needed to form an explosive mixture with air. Hence the fuel tank is less likely to catch fire in case of accident.

Its uniformity of composition and much cleaner combustion are some other advantages of this fuel. It tends to produce less carbon deposits than normal gasoline and that the deposits are safter and easier to remove.

The aforesaid advantages apparently make ethanol a very suitable spark ignition engine fuel. However, it has certain limitations which have restricted the use of ethanol as a complete fuel for automobile engines. Its calorific value is about two-thirds that of gasoline. As a practical consequence of this it will give fewer miles per gallon (2). Its higher surface tension renders it less easy to atomise and its high solvent

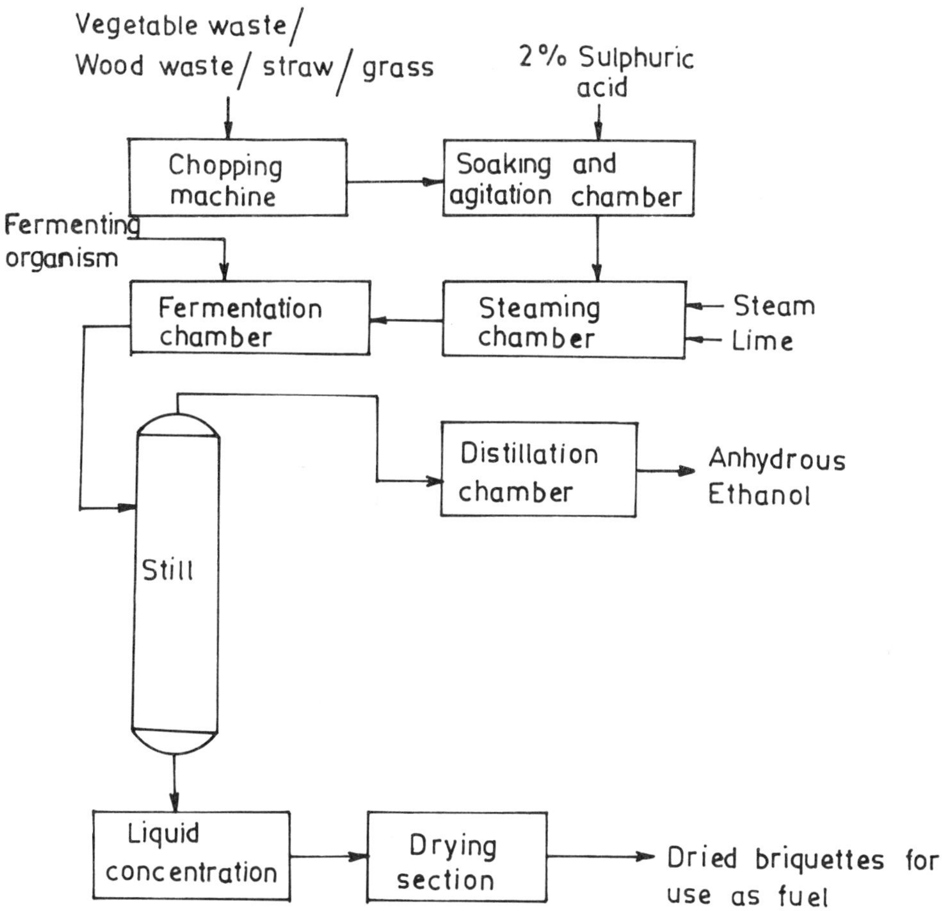

Fig. 3. BLOCK FLOW DIAGRAM FOR ETHANOL PRODUCTION FROM CELLULOSIC MATERIAL

power could create practical difficulties of storage etc. Its presence in large quantities could be condusive to corrosion.

Keeping these limitations in mind it may not be advisable to use ethanol neat (unblended) as an automobile fuel. It can be best utilised as a blend constituent. When used in ethanol gasoline blends it could, while retaining all advantages of gasoline in the blend, impart some of its useful properties to the resulting mixture. It might permit higher compression operation without knock, greater fuel economy, lower exhaust temperature and improved over all performance.

Fig. 4 shows the effect of adding ethanol to a typical regular gasoline with a research octane member of 91.0. Addition of ethanol upto ten percent raises the octane rating of gasoline by 3 while blending twenty five percent ethanol raises the octane number of gasoline from 91 to 99. The octane number appears to increase linearly with the increase of ethanol content in the blend upto twenty five percent ethanol addition. It has also been observed that this useful effect of ethanol blending is more marked if the starting stock is low grade gasoline while the improvement in octane rating is relatively lesser in case of high octane gasoline (3)

The power output of the engine also increases with the ethanol content of the blend. Fig. 5 shows this effect as observed in the laboratory with the Ricardo variable compression research engine. At any given compression ratio the engine brake mean effective pressure (indicative of engine power output) increases as the ethanol content increases from zero to thirty percent. The increase in power output that is obtained as the compression ratio of the engine is increased is also apparent. When the compression ratio reaches 9:1 the octane number of the regular grade of fuel, is insufficient to prevent knocking. Hence the power falls off rapidly in this area of knock with ethanol concentrations of less than about twenty percent.

Fig. 6 shows the effect of ethanol concentration on the specific fuel consumption of the engine. There is a clear improvement in fuel consumption associated with increasing the compression ratio of the engine made possible by ethanol blending. It may be observed from the figure that the fuel consumption is initially reduced as ethanol concentration is raised to a level of 10 to 20 percent. At compression ratio 9:1 the detrimental effect of knock is again apparent as indicated by the curve at ethanol concentrations of less than about 20 percent.

This trend of ethanol blending on the output and specific

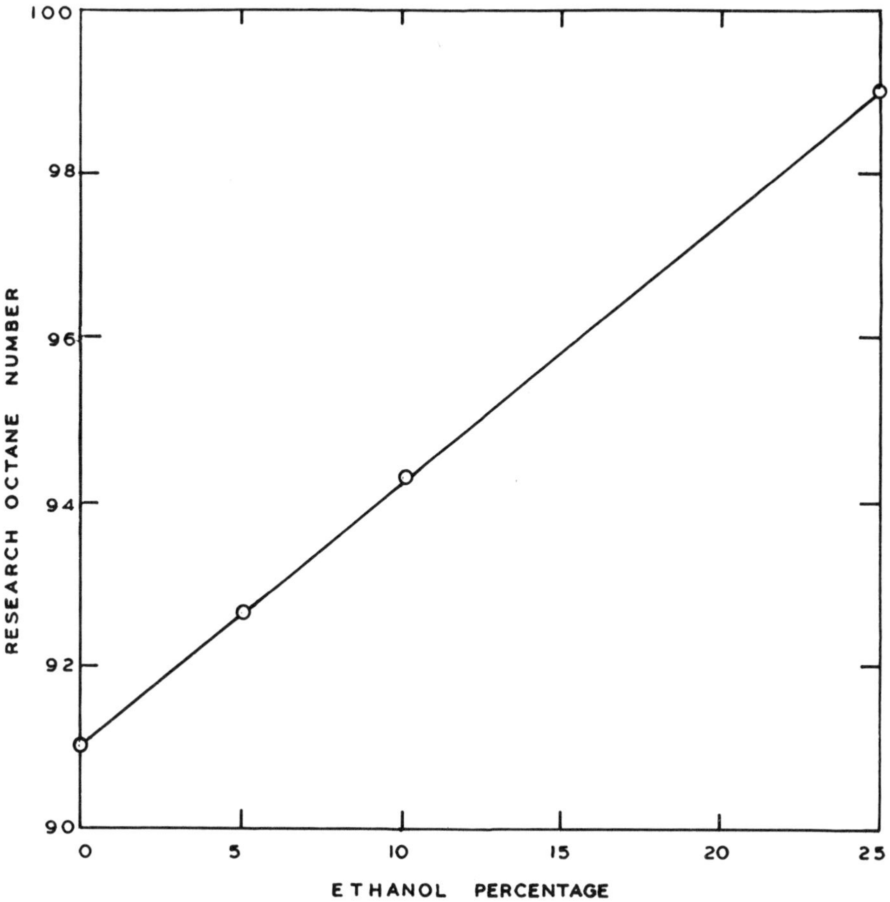

FIG. 4 VARIATION OF OCTANE RATING WITH THE ETHANOL PERCENTAGE IN THE BLEND.

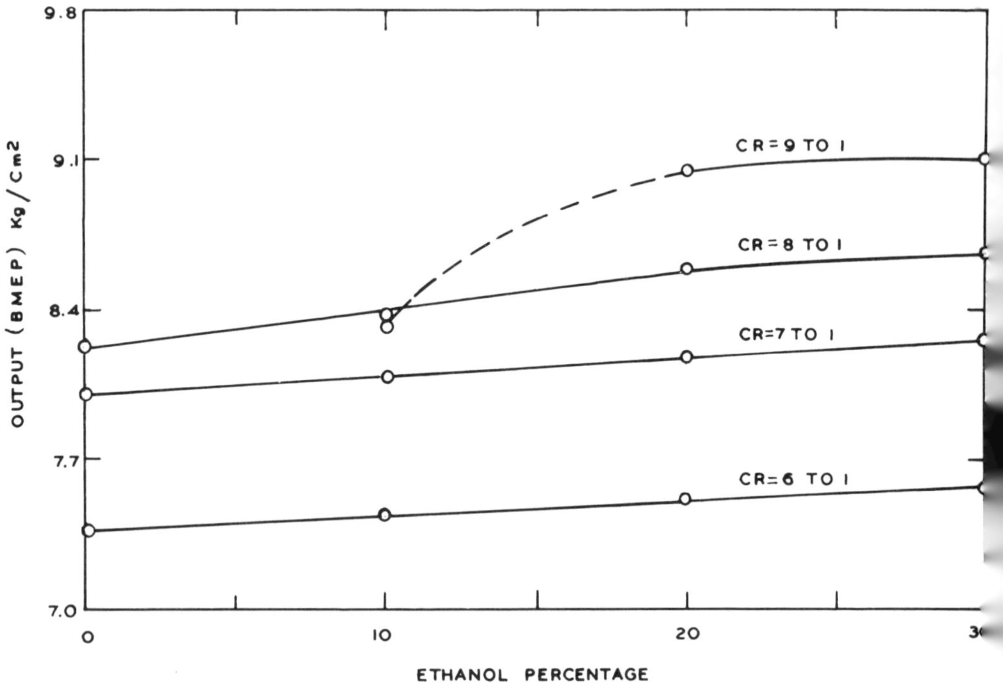

FIG. 5 VARIATION OF ENGINE POWER OUTPUT WITH ETHANOL PERCENTAGE IN THE BLEND AT VARIOUS COMPRESSION RATIOS.

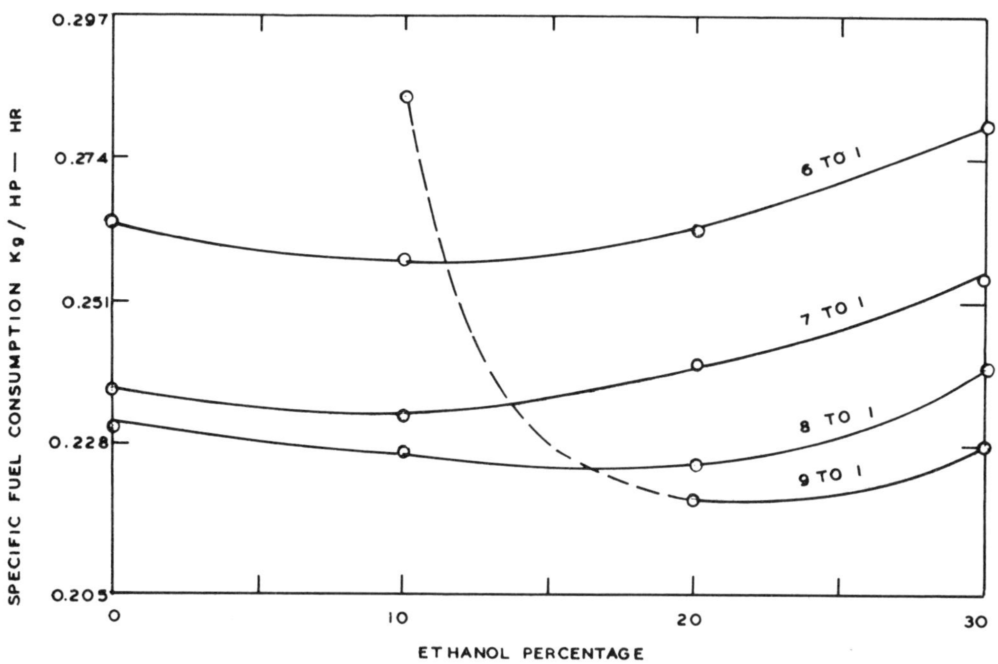

FIG. 6 VARIATION OF SPECIFIC FUEL CONSUMPTION WITH ETHANOL PERCENTAGE IN THE BLEND AT VARIOUS COMPRESSION RATIOS.

fuel consumption of Ricardo research engine has been confirmed in dynamometer tests on four cylinder B.M.C engine and road tests with typical automobiles.

Extensive experimental tests and theoretical investigations carried out by the author in the mechanical engineering department of the Indian Institute of Technology, Delhi, in the last ten years, have shown that ethanol can be utilised with great advantage in automobile, scooter, motor cycle engines in any of the following three ways.

1. In the form of ethanol gasoline blends in the existing automobile engines without any engine modifications. Use of ethanol gasoline blends containing upto twenty five percent ethanol by volume give no greater fuel consumption per mile than ordinary gasoline when used in privately owned cars. The "Optimum blend" increases engine power output of the gasoline engine by 6 to 8 percent. Use of ethanol gasoline blends also reduces the concentration of major pollutants present in the engine exhaust effluent. They produce 10 to 35 percent less exhaust hydrocarbons and upto 60 percent less exhaust carbon-monoxide. Their use can thus reduce the extent of air pollution caused by automobile engine exhaust effluent and at the same time add more and energy to the automobile fuel supply and stretch the fast dwindling gasoline supplies.

2. As a complete fuel in future "Ethanol Engines" designed to operate at higher compression ratios of the order of 12 or so. Such engines would have specially designed carburettor and fuel induction system to suit the characteristics of ethanol and achieve its vaporisation and combustion. Computerised studies show that such automobile engines would have about 18 percent more output, higher thermal efficiency and smoother operation as compared to gasoline engine.

3. As a duo-fuel by separate injection when required. The ordinary automobile engine makes full use of high octane rating of the fuel only during a small proportion of its service. If it were operated on a cheaper fuel of relatively low octane rating, but with automatic arrangement for ethanol injection under high loads and speeds, the combination might well make for great economy. This would permit greater use of low octane gasoline which is cheaper and would also reduce exhaust emission of lead salts.

ETHANOL FUEL - ITS UTILISATION IN HEAVY VEHICLES ENGINES (C.I. ENGINES)

Experimental studies have shown that use of ethanol in heavy

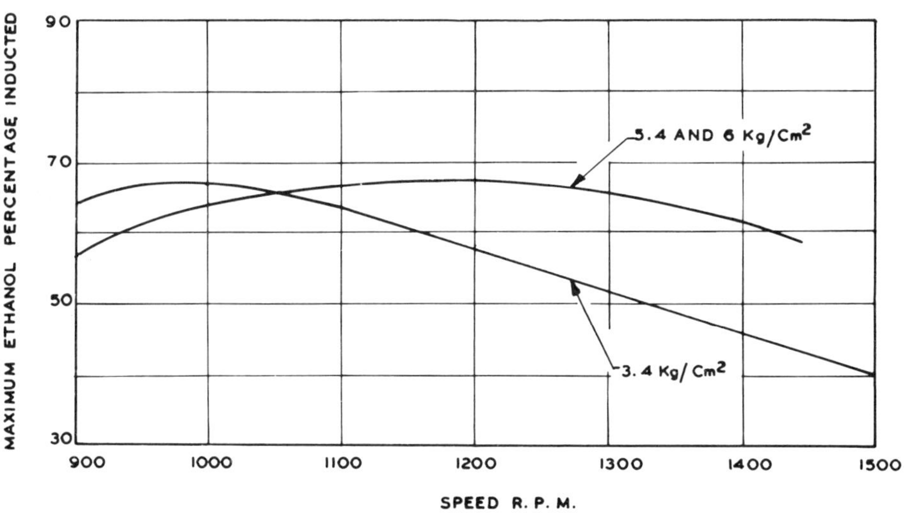

FIG. 7 VARIATION OF MAXIMUM ETHANOL PERCENTAGE INDUCTED AT VARIOUS SPEEDS.

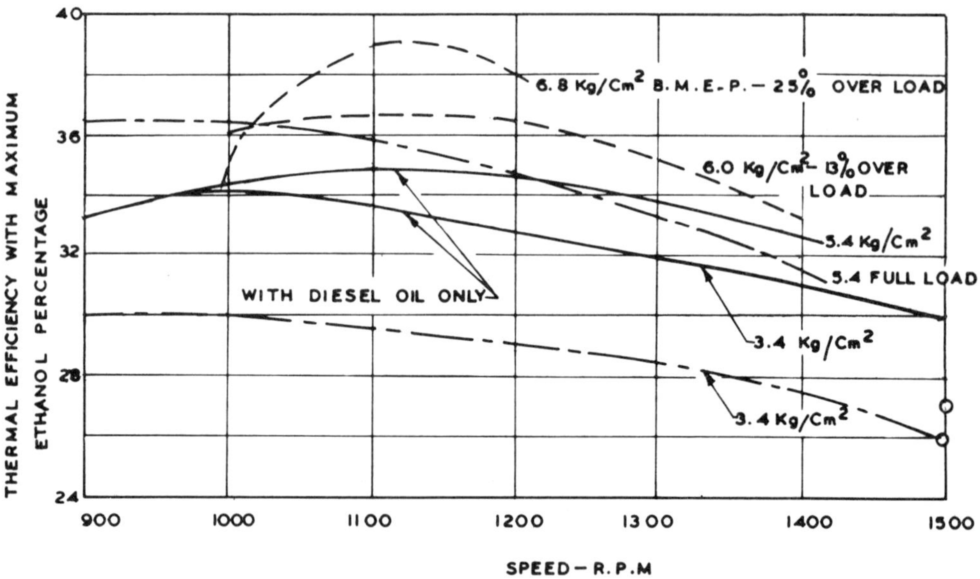

FIG. 8 PERFORMENCE OF A FOUR CYLINDER DIRECT INJECTION ENGINE WITH ETHANOL INDUCTION

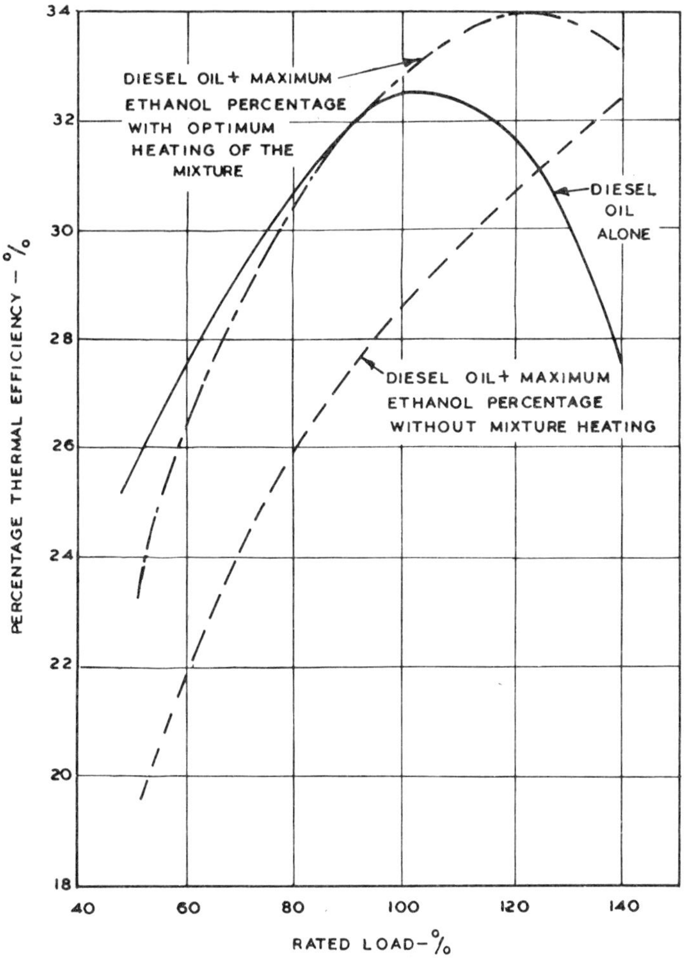

FIG. 9 EFFECT OF MIXTURE HEATING (WITH INDUCTION OF MAXIMUM QUANTITY OF ETHANOL) ON ENGINE PERFORMENCE.

vehicles powered by compression ignition engines is beneficial since it reduces the vehicular exhaust smoke density, increases the overload capacity of the engine and improves the thermal efficiency as well as the air utilisation factor. Best results are obtained with direct injection engines. By inducting ethanol so as to have bifuel operation of the engine, engine deposits are reduced andthere is no increase of wear of the engine components. By judicious application of mixture heating, normal thermal efficiency can be maintained at part loads and it can be surpassed at full load and overload conditions.

Fig. 7 and Fig. 8 show one of the typical test results obtained from a four cylinder direct injection automotive diesel engine having ethanol induction. Ethanol was carburatted and tests were conducted to cover the entire load speed range. It can be seen from fig. 7 that the percentage of ethanol inducted at full load is sensibly the same over most of the speed range. Figure 8 shows that above half load operation, the thermal efficiency is not adversely affected by induction of ethanol over the working speed range. It was observed during the tests that the overload capacity of the engine as well as its thermal efficiency increased appreciably with ethanol induction. In one case thermal efficiency as high as 43 percent was obtained under overload conditions. However, it was noted that the part load performance of the engine deteriorated with ethanol induction.

Presence of ethanol in the inducted charge increases the delay period considerably at part load impairing the combustion process and hence engine performance. This malady can be overcome by addition of heat to the air ethanol mixture before admission to the engine. Heating of the inducted mixture to the optimum value restores the delay period almost to the normal value. By judicious application of mixture heating on lines similar to that in S.I. engines the normal thermal efficiency can be maintained at part loads and surpassed at full and overload conditions; as shown in figure 9 which is based on results obtained in test trials carried out with induction heating.

CONCLUSIONS

Detailed studies and extensive tests on ethanol based fuels, carried out by the author in the last ten years have convincingly shown that in ethanol we have a fuel, which despite its drawbacks can provide the low knock fuel, with power to stand high compression ratios.

Capable of being produced in any country, that can support growing crops, within access to by products of tropical zones, it enables countries with no indigenous oil deposits to provide a satisfactory motor spirit at minimum cost in imported low knock fuel.

Its use in present day automobiles in the form of ethanol gasoline blends can be an immediate short term measure which can stretch the available fuel supplies and help meet the energy crisis. As a long term solution to the present day fuel oil scarcity, vehicular engines can be designed to accomodate ethanol neat as fuel.

Carburation of ethanol in heavy vehicular engines powered by diesel engines, can increase their power ratings by 20% to 50% depending upon the type of combustion chamber. This has an added attraction for road transport engines in that it reduces noise level and exhaust smoke density.

Ethanol can thus go a long way to provide a solution to the twin problems of fuel oil scarcity and growing air pollution due to vehicular engine exhaust emissions.

REFERENCES

1. Mathur, H.B., "Fuels of Future for Spark Ignition Engines" Proc. National Seminar on Fuel Efficiency, Oct. 1970, N.P.C., P. 41-49.

2. Shulman, W., "Physical Properties of Ethanol in Rapid Round up" - Chem. Engr., Vol. 68 May 15, 1961.

3. American Petroleum Institute, "Use of Alcohol in Motor Gasoline - A Review", Publication No.4082, Aug 1971.

4. Mathur, H.B., "Effects of Ethanol Blending on the Performance and Exhaust Emission of Spark Ignition Engines" Ph.D. Thesis, Deptt. of Mechanical Engineering, Indian Institute of Technology, New Delhi, June 1971.

5. Mathur, H.B., Mahajan, S.D., "Bi-Fuel Operation of Diesel Engine", Paper based on M.Tech Project, Mechanical Engineering Deptt., Indian Inst. of Tech., New Delhi, 1976 (under publication)

ELECTROLYTIC HYDROGEN PRODUCTION AND ITS POTENTIAL UTILIZATION FOR SHIP PROPULSION

C.A. SCHMIDT
Atomic Energy Commission
Santiago, Chile

ABSTRACT

The electrolytic Hydrogen Technology available today, and the one expected to be developed within the next 20 years, is examined from a technical and the economical point of view.

Light Water Reactors (LWR) for the near term and liquid Metal Fast Breeder Reactors (LMFBR) as primary energy sources are used to produce electrolytic hydrogen.

Potential utilization of hydrogen as a fuel for ships, is examined with some practical examples in WSEV.

1.- INTRODUCTION

The growing shortages of conventional fossil fuels have prompted increased efforts to develop synthetic fuel substitutes. Among them, by their nature, Hydrogen as a fuel, fulfill the requirements of storability and transportability, and is easily utilized in conventional and non-conventional engines. Also Hydrogen is potentially a more flexible energy carrier and energy storage medium than anything now known (3) and can be derived electrically or thermally from water a unexhautable energy source Hydrogen provides the greatest energy release per unit mass (51,600 BTU/lb) of any chemical fuel.

Hydrogen is a particulary attractive fuel for the following reasons:

a) It is assentially clean burning, the main combustion product is water.

b) It may be sustituted for nearly all fuel uses.

c) It is available from a renewable and universal raw material water.

d) Nearly all primary sources, nuclear, solar etc. may be used in its production.

The main drawbacks are:

a) High cost relative to even current fossil fuel prices.

b) A low density or a cryogenic fluid must be handled.

Hydrogen offers an expected option for fueling ships although there still ramains no clear solution as to its mode of on board storage and handling. Reviewing the technology available to deal with hydrogen, it appears that the best way for future hydrogen production is to use lectrolysis plants coupled to nuclear power reactors. In particular, it appears very attractive to employ hydrogen production as a means of levelizing the load fo nuclear reactors - a problems which must be faced once nuclear units account for more than about 50% of installed capacity. Under these conditions cost is controlled by fuel cost, which are relatively low for nuclear units. In particular when Breeders dominate the fuel costs will be exceptionally low (2.). Based on this reasoning two methods for hydrogen production will be addressed in this work: one based on LWR reactors for the near and the other based on LMFBR for the far term.

Although it is extremely early to assess the real value of hydrogen production at sea, some general ideas about this time form of production will be outline in one of the sections. Finally, potencial capabilities of hydrogen fueled ships will be briefly mentioned in the last section.

2.- ELECTROLYTIC TECHNOLOGY

With the exhaustion of fossil fuels the only available source for hydrogen production will be water. Electrolysis is the most likely process for practical large scale production of hydrogen. It offers the greatest potencial for meeting necessary capital and operating cost objetives without requiring a major technological breakthrough (11).

Electrolytic descomposition of water is accomplished by passing a direct current between two electrodes immersed in an electrolyte (usually potassium hydroxide solution.). The rate of hydrogen production is directly proportional to the current passing between the electrodes and is given by Faraday's Law as 1 lb. of H_2 per 12,060 Amp-hr. The theoretical energy input is 18 kw-hr per 1 lb. of H_2 at a voltage of 1.482 volts. Under usual operating conditions, commercial electrolysis plants require much higher power levels, due to greater than 20% power losses in electrolyte or in the electrodes themselves. Efficiency of water electrolysis should be defined as the energy stored as chemical energy in the hydrogen divided by the electrical energy required to produce that hydrogen. Commercially available electrolysis plants operate at electrical efficiencies varying between 57% and 72%. The best demostrated efficiency for advanced electrolysis cells is aproximately 80% (7). The principal operating parameters for electrolyzer modules are temperature, pressure, and most importantly, current density, which determines the production rate and power requirements at a given temperature and pressure (4). Today's capital cost of large-scale electrolysis plants is aproximately 170 \$/kw (127 \$/lb.-H_2 - day or 24.8 \$/SCF - hr), current densities are limited to 100-200 amp/ft^2 and power consumption is 120-140 kw-hr/MSCF for cell efficiencies of 56-65% (5.,7.). The most advanced electrolytic cells in the US. are those developed by Allis Chalmers and the General Electric Company. Allis Chalmers has developed a Cell for current densities of 800-1600 Amp/ft^2 and high pressure operation at 300 psi. While high current density is desirable for improved plant efficiency, high pressure becomes necessary to overcome the increased internal resistance resulting from gas voids produced at such densities. Operating at 250 °F this cell achieves an efficiency of 68%, requiring 115 kw-hr/MSCF H_2. Teledyne Isotopes has claimed that with about one year of moderate research the efficiency of the just mentioned Allis Chalmers Cell could be increased to 75%, concurrently reducing the capital cost to about 74 \$/Kw (55 \$/lb-H_2-day or 10.9 \$/SCF-hr) (5.). Conservatively it can be expected that this cell will be ready in 1985. General Electric using Solid Polymer Electrolyte (SPE) technology has proposed a cell that, working with an efficiency of 87%, will cost 35 \$/kw (26 \$/lb-H_2-day or 5.2 \$/SCF-hr) (5.,6.) It appears that this General Electric Cell requires 10 to 15 years to achieve commercial operation (5.); again to be conservative we will postulate that this cell can be in full commercial operation in the year 2000. A summary of present and projected electrolysis technology is presented in Table 1.

Table 1.-

Summary of Present and Projected

Electrolysis Technology

Cell Property	1978	Year 1985	2000
Current Density (Amp/ft^2)	100	800	2000
Power Consumption (Kw-hr/10^{3}SCF)	140	105	90
Efficiency (%)	56	75	87
Pressure (psia)	15	300	-
Capital Cost ($/Kw)	170	74	35

Note: Cost in 1978 dollars.

3.- HYDROGEN PRODUCTION AND DISTRIBUTION

3.1.- Shore Production and Distribution

With todays technology, the only method that appears feasible for hydrogen production is the use of land based electrolysis Plants coupled to nuclear power reactors. A preferred strategy is to use off-peak power from nuclear reactors to produce hydrogen when the normal electric load decreases. This strategy appear to be attractive due to the following facts:

a) Electric load varies with the time of day, time of week and seasonally.

b) When nuclear reactors are forced to load follow it puts an added strain on their fuel, which increases the incentive to lovelized load.

c) If a transcontinental H_2 pipe line is used, the peak load duration can be shifted from region to region; the pipe line also provides substantial storage capacity.

d) The projected use to load-level is larger than the shipboard market, hence one can be parasitic on broader based, larger scale technology.

e) If this comes into more widespread use then fuel cell development for reconversion to electricity will be accelerated benefits would then spread to their marine use.

Levelizing the load of nuclear reactors producing hydrogen, it is possible to obtain low fuel costs, reducing the production cost of the hydrogen. In order to put in practice this strategy, it is possible to consider LWR reactors and conventional technology for H_2 production during the near term (1978 to about the year 1990), and then switch to more advanced technology using LMFBR's coupled to advanced electrolysis plants. LWR and LMFBR reactors have been chosen as the main source of energy because the first is a very well proven current technology that offers relatively low fuel costs, and the second will start comercial operation in the 1990's, offering exceptionally low fuel cost (about 1/3 that of a LWR) (2.).

To distribute the hydrogen produced in each electrolysis plant, it is necessary to consider a distribution network running from the production plant to harbors or ship terminals. Two different networks can be considered: gas pipelines or liquid pipelines.

If hydrides are used to store H_2 on board, then gas pipeline networks are more attractive since they have much lower transmission and storage costs than liquid systems.

In order to assess the economic potential for the near and far term of the postulated methods for H_2 production using either gaseous or liquid systems for storage and distribution, a brief economic study will be performed to identify the costs, that added together, give the final H_2 production cost.

The following presents the calculation procedures to obtain the production cost of electrolytic hydrogen using LWR (Light Water Reactor) reactors for the near term and LMFBR (Liquid Metal Fast Breeder Reactor) rectors for the far term. The energy from nuclear reactors is only used during off-peak periods to increase the reactor's load factor and take advantage of the low fuel cost of nuclear power plants, thereby reducing Hydrogen production costs.

The technology described in section two is used to perform the following calculations:

Near Term Assumption (Conventional Technology):

Operations begin: year 1978

Energy source: Nuclear reactor type LWR

Electrolysis Plant: Conventional electrolytic Cell

Nuclear fuel cost (e): 4.24 mills/Kwh (e) (2.).

Cost of H_2 Plant (I): 170 $/Kw

Electrolysis Plant Load Factor (L): 0.4
(note that this must be low because off-peak power is used)

Electrolysis Plant Power Requirement (P): 140 Kwhr/10^3 SCF.

Capital Charge (C.CH.): 15%

Gaseous Storage Tank Cost (S): 1,305 $/MBTU (9.)

Storage time (t): 1 day

Transmission distance: 100 miles trhrough 36 inch pipe in gaseous phase.

Capital Cost (Cp):

$$Cp = 0.0334 \times \frac{C.CH.}{L} \times I \ (\$/MBTU)$$

$$Cp = 0.0334 \times \frac{0.15}{0.4} \times 170 = 2.12 \ (\$/MBTU)$$

Operation & Maintenance Cost ($C_{0\&m}$):

$$C_{o\&m} = 0.07 \ (\$/MBTU) \ (12.)$$

Fuel Cost (C_F):

$$C_F = 3.077 \times 10^{-3} \times e \times P \ (\$/MBTU)$$

$$C_F = 3.077 \times 10^{-3} \times 4.24 \times 140 = 1.82 \ (\$/MBTU)$$

Storage Cost (C_{st}):

$$C_{st} = 2.74 \times 10^{-3} \times \frac{C.CH.}{L} \times S \times t \ (\$/MBTU)$$

$$C_{st} = 2.74 \times 10^{-3} \times \frac{0.15}{0.4} \times 1305 \times 1 \ (\$/MBTU)$$

$$C_{st} = 1.34 \ (\$/MBTU)$$

Cost of Transmission (C_{tr}):

Considering a delivery distance of 100 miles using 36 in. pipe

$$C_{tr} = 0.09 \ (\$/MBTU) \ (9.)$$

Total Production Cost (C_t):

$$C_t = C_p + C_{o\&m} + C_F + C_{st} + C_{tr}$$

$$C_t = 2.12 + 0.07 + 1.82 + 1.34 + 0.09 \ (\$/MBTU)$$

$$C_t = 5.44 \ (\$/MBTU)$$

Note: Cost in 1978 dollars.

Far Term Assumptions (Advanced Technology)

Operation begins: year 1990

Energy source: Nuclear reactor, type: LMFBR

Electrolysis Plant: Advanced electrolytic Cells

Nuclear fuel Cost (e): 2.3 mills/Kwh (2.)

Cost of H_2 Plant (I): 74 \$/Kw

Electrolysis Plant Power Req. (P): 105 Kw-hr/10^3 SCF

Capital Charge (C.CH.): 15%

Gaseous Storage Tank Cost (S): 1,305 \$/MBTU (9.)

Storage time (t): 1 day

Transmission distance: 100 miles through 36 in pipe in gaseous phase.

With these assumptions the following costs are obtained:

C_p = 0.93 \$/MBTU
$C_{o\&m}$ = 0.07 \$/MBTU
C_F = 0.74 \$/MBTU
C_{st} = 1.34 \$/MBTU
C_{tr} = 0.09 \$/MBTU
C_t = 3.17 \$/MBTU

Considering Cryogenic Storage and Transmission:

The preceding costs are increased due to the additional costs for compression and liquefaction. Also the cost of transmission is more expensive. The costs subject to variation are presented in the folowing calculation.

Cost of compression = $C_{compr.}$ = 0.57 (\$/MBTU) (9.)
Cost of liquefaction= $C_{liq.}$ = 1.71 (\$/MBTU) (9.)

$$C_{st} = 2.74 \times 10^{-3} \times \frac{C.CH.}{L} \times S \times t \quad (\$/MBTU)$$

C.CH. = 0.15 L = 0.4 S = 544 \$/MBTU t = 1 day

$$C_{st} = 2.74 \times 10^{-3} \times \frac{0.15}{0.4} \times 544 \times 1 = 0.56 \ (\$/MBTU)$$

C_{tr} = 0.11 (\$/MBTU) (Considering 100 miles by 11 x 10^5 Gal. Barge).

$$C_t = C_p + C_{o\&m} + C_F + C_{compr.} + C_{liq.} + C_{st} + C_{tr}$$

Near term: $C_t = 6.96$ (\$/MBTU)

Far term: $C_t = 4.69$ (\$/MBTU)

Considering Mg. Hydride Storage:

The reduction in the production cost of the H_2 using hydrides is mainly due to the relatively low cost of this storage system and the consideration of no cost for transportation.

$$C_{st} = 2.74 \times 10^{-3} \times \frac{0.15}{0.4} \times S \times 1$$

$$S = 290 \text{ \$/MBTU} \quad (9.)$$

$$C_{st} = 2.74 \times 10^{-3} \times \frac{0.15}{0.4} \times 290 = 0.30 \text{ \$/MBTU}$$

$$C_t = C_p + C_{o\&m} + C_F + C_{st}$$

Near term: $C_t = 4.31$ \$/MBTU

Far term: $C_t = 2.04$ \$/MBTU

Note: Cost of energy removal/addition for load and release of the H_2 from Hydride not included.

Table 2. summarizes the results from calculations already performed. From these results it is possible to conclude that the most attractive method for the production and distribution of H_2, is for the near and far term, the one that uses gaseous storage and transmission networks. The production cost of H_2 stored and distributed in the gaseous phase (5.44 \$/MBTU), is 22% cheaper than that produced and distributed in the liquid phase. This is mainly due to the fact that a considerable amount of energy has to be used to compress and liquify the hydrogen; also liquid pipe networks are more capital intensive than gas pipeline networks and more expensive to operate. The future looks promising for H_2 production and transmission because capital costs and fuel costs appear amenable to substantial reduction (~ 41%), decreasing the total production cost to 3.17 \$/MBTU for the gaseous case and 4.69 \$/MBTU for the liquid case.

A conclusion which can be drawn from the cost figures in table

Table 2.-

Electrolytic Hydrogen Production Costs

	CONVENTIONAL TECHNOLOGY			ADVANCED TECHNOLOGY		
	Gaseous st.	Cryogenic st.	Hydride st.	Gaseous st.	Cryogenic st.	Hydrides st.
(I) Cost of plant ($/Kw)	170	170	170	74	74	74
(P) Power required (Kw hr/10^3 SCF-H_2)	140	140	140	105	105	105
(e) Fuel Cost (Mills/Kw-hr)	4.24	4.24	4.24	2.3	2.3	2.3
(s) Storage tank cost ($/MBTU)	1,305	544	290	1,305	544	290
(C_p) Capital cost ($/MBTU)	2.12	2.12	2.12	0.93	0.93	0.93
($C_{o\&m}$) O & M cost ($/MBTU)	0.07	0.07	0.07	0.07	0.07	0.07
(C_F) Fuel cost ($/MBTU)	1.82	1.82	1.82	0.74	0.74	0.74
(C_{st}) Storage cost ($/MBTU)	1.34	2.84*	0.30	1.34	2.84*	0.30
(C_{tr}) Transmission cost	0.09	0.11	-	0.09	0.11	-
(C_t) Total cost ($/MBTU)	5.44	6.96	4.31	3.17	4.69	2.04

* Includes cost of compression plus liquefaction.

Note: Costs given in 1978 dollars.

2. is that storage costs are still too high (~40 to 60% of production cost for advanced technology); hence it will be necessary to reduce this component substantially-an objective which suggest a substantial increase in the R & D in this area. Storage costs are mainly responsible for the fact that H_2 still cannot compete with fossil fuels. At 5.44 $/MBTU the current electrolysis concept is uneconomical: this price corresponds to fuel oil at about 32 $/bbl-far above the present petroleum price of ~ 16 $/bbl. The advanced concept price of 3.17 $/MBTU (18 $/bbl of equivalent oil) appears to be more attractive providing H_2 storage and transmission costs are reduced. From the cost figures tabulated in Table 2. it is important to recognize two important facts. The first, is that it is too early to start producing and transporting H_2 for ship propulsion because H_2 production costs are not competitive with today fossil fuel costs, and second, to decrease these costs it is necessary to concentrate R & D efforts on more efficient storage and distribution systems.

3.2.- On Board Production and Distribution

As has been seen in the preceding section it is very expensive and not presently competitive to produce and distribute to ships hydrogen produced in land based plants. Production and distribution on board ships appears to be even less favorable, because in using on board reactors (relatively small compared with land based reactors used to generate electricity) the fuel cost increases, and storage and handling systems are not developed to the extent for which it is possible to postulate that H_2 is a tractable on board fuel fulfilling minimum on board safety requirements.

Although the production of H_2 at sea in not presently attractive for most applications it may be worthwhile to evaluate a system to produce H_2 for a fleet of WSEV (Water Surface Effect Vessels) in which a "mother ship" is configured to carry a nuclear reactor and associated storage tanks to serve as a "tanker". Thus the requisite Liquid Hydrogen could be manufactured on site, with the tanker possessing the same range and mobility characteristics as the other elements of the fleet. To reduce vulnerability, the tanker could be provided with separatable storage and transfer vehicles.

4.- HYDROGEN SHIPS

Although the state-of-the-art in the field of hydrogen-fueled ships is in the first stages of development, hydrogen appears to be very attractive for ships which are primarily weight critical (8.). Mainly these vehicles are the high-speed, high performance marine vehicles such as water surface effect vessels (WSEV), hovercraft and hydrofoils. Also hydrogen looks attractive to fuel small deep-submergence submarines equiped with hydrogen-oxygen fuel cells because the high efficiency of the fuel cell minimizes energy storage requirements (1.).

Hydrofoil ships are those in which the hull rises above the surface of the water during operation and is supported on struts attached to submerged wings which produce the required lift by virtue of their speed through the water. Hovercraft vessels are those in which the vehicle rises above the surface of the water due to the lifting force obtained by blowing air through the botton. Different propulsion systems are under intesive study and development to propel WSEV; among the most developed are water-jets, marine propellers (supercavitating and superventillated) and air propulsion systems. WSEV displacement ranges from a few tons to 2,000 tons and maximun speeds range typically between 80 to 100 knots (8.). The US. Navy is developing a transoceanic WSEV of 2000 tons with a range of 4000 nautical miles and capable of achieving its optimum cruise speed in sea state 3.

Because the weight is critical, the most suitable engines for WSEV are gas turbines. General Electric and Pratt and Whitney have available a number of different models for different power, weights and specific fuel consumption (8.). Some of these models can be adapted to burn hydrogen. Hydrogen having almost 3 times as much energy per unit of mass as Jet Fuel can increase the payload available, or nearly double the range of action (10.). A Hovercraft of 4000 tons with a range of 4000 nautical miles at 80 knots, equiped with 3 gas turbines, model FT 9 D, that give 105,000 SHP with an SFC of 0.40 lbs/SHP-hr consumes, when these turbines are connected to a "water jet" propulsion system, 1903 tons of fuel to travel its full range. Adding this fuel weight to the 313 ton-weight of the propulsion machinery, a total of "not-useful" weight of 2216 tons, corresponding to 55% of the total weight of the craft, is obtained (3.). If instead of burning Jet fuel, hydrogen is used, the fuel required would decrease to 678 tons, increasing the useful weight to 3009 tons, correspondant to 75% of the total displacement of the vessel. Some problems envisioned for the operation of future H_2 fueled ships come directly from hydrogen characteristics. Because H_2 has such a low ignition energy, wide flamability limits, and high volatility and diffusivity, a completely new set of detection methods and safety procedures will have to be developed for the marine environment.

5.- CONCLUSIONS

Analyzing the considerations presented in this work, it is possible to conclude the following:

(a) At the present the most technologically developed method on which to rely for future H_2 production is the use of electrolysis plants coupled to LWR reactors for the near term and LMFBR's for the far term. In particular, it appears very attractive to employ H_2 production as a means of levelizing the load of nuclear reactors - a problem which must be face once nuclear units account for more than about 50% of installed capacity.

(b) The just mentioned strategy appears to be very attractive due to the following facts:

1) Electric load varies with time of day, time of week and seasonally.

2) When nuclear reactors are forced to load follow it puts and added strain on their fuel, which increases the incentive to levelize load.

3) If a transcontinental H_2 pipe line is used, the peak load duration can be shifted from region to region.

4) The projected use to load-level is larger than the shipboard market, hence one can be parasitic on broader based, larger scale technology.

(c) There is considerable incentive to develope Hydrides to the point where they become the preferred system to store H_2 on board ships; the recommend system for distribution of H_2 from land based production plants to ship terminals is the gas pipeline network.

(d) Future capital and fuel costs for H_2 production may decrease as much as ~ 41% as soon as Breeders reactors are forced into a load-following mode of operation, and when more efficient alectrolytic cells become available. This may not occur, however, until well into the next century.

(e) H_2 storage costs are still too high (~40 to 60% of H_2 production cost using advance technology) being mainly responsible for tha fact that H_2 still cannot compete with fossil fuels.

(f) H_2 storage is a key problem that requires intensive R & D efforts to improve present systems substantially, reducing costs and making possible the use of H_2 on board ships.

(g) The present H_2 prices of 5.44 $/MBTU (32 $/bbl of equivalent oil) are not competitive with the present prices of petroleum (~ 16 $/bbl). The advanced concept price of 3.17 $/MBTU (18 $/bbl of equivalent oil) appears to be more attractive if H_2 storage and transmission costs are reduced.

(h) It is to early to start producing (either land based or at sea) and transporting H_2 for ship propulsion unless better storage and distribution systems are developed, allowing cheaper costs and increased safety.

(i) Hydrogen fueled ships have not been developed yet, but the more suitable vessel for H_2 application appear to be WSEV marine vehicles. These ships are weight-critical, hence using gas turbines (low weight per unit power) fueled with Hydrogen (high energy per unit mass) it should be possible to double their range of action.

(j) It would appear that just as submarines led the way with respect to nuclear propulsion plants on board ships because of the large benefits achievable, WSEV may become the means for introduction of H_2 as a fuel for ships, again because of the large potential improvement in performance.

REFERENCES

1.- Buechler L. and Klotz C. "Fuel cells and their Application to Undersea Power", Symposium on "Marine Power Plants", Spring Meeting of the SNAME. Philadelphia, P.A., May 11-13, 1966.

2.- Benedict, M., Course Notes for M.I.T. course 22.32, "Economics of Nuclear Power", Nuclear Engineering Dept., M.I.T., Fall 1975.

3.- Barr R.A. and Etter R.J., "Selection of Propulsion Systems for High-Speed Advanced Marine Vehicles", Marine Technology vol. 12 # 1, Jan. 1975.

4.- Costa R. y Grimes P., "Electrolysis as a Source of Hydrogen and Oxygen", Chemical Engineering Progress, Vol. 63, # 14, April 1967.

5.- Fein Elihu, "A Hydrogen Based Energy Economy", Report 69-08-10 prepared by Futures Group for Northeast Utilities Service Company, October 1972.

6.- Feher Ronald D., "Energy Storage Alternatives for a Nuclear Total Utility System", Special problem for M.I.T. course 22.90, July 3, 1974.

7.- LM 2500 Marine Turbine Installation Design Handbook, MID-IDM-5000-2, General Electric, Marine and Industrial Dept., Cincinnati, Ohio, April 1970.

8.- Jane's Surface Skimmers, Hovercraft and Hydrofoils 1975-1976, compiled and adited by Roy Mc Leavy, London-England.

9.- Michel John, "Hydrogen and Synthetic Fuels for the Future", 166th National Meeting of the American Chemical Society, Chicago, Illinois, August 26-31, 1973

10.- Quandt E. "Investigation of Hydrogen Fuel for Naval Vehicles" Hydrogen Energy, part "B", T. Nejat Veziroglu, Plenum Press, 1975

11.- Russell, Nuttall and Fickelt "Hydrogen Generation by Solid Polymer Electrolyte Water Electrolysis", 166th National Meeting of the American Chemical Society, Div. of Fuel Chemistry, Chicago, Illinois 26-31, August, 1973.

12.- Winsche, W.E. Hoffman and Salzano, "Hydrogen: Its Future Role in the Nation's Energy Economy" A special Science compendium, Philip H. Abelson, AAAS Miscellaneous Publication 74-15.

HYBRID CHEMICAL CONCEPTS FOR STORAGE OF SOLAR THERMAL ENERGY

A. C. Vialaron
C.N.R.S./P.I.R.D.E.S.
Toulouse, 31 - FRANCE

ABSTRACT

We intend to give a short summary of proposed chemical cycles for storage of solar thermal energy (heat in $\rightarrow$ heat out) with constraints and expected practical performances. In competition with these concepts, hybrid devices (heat in $\rightarrow \Delta G°$ out) are proposed. As a typical exemple, the practical storage performance of one of these hybrid cycles, based on sulfur oxides and magnesium oxide, is discussed, in connection with electrolytical splitting of water into hydrogen. This system shows a practical energy density in the range of 300-330 KWh/m^3 released as hydrogen gas, at a very low cost.

INTRODUCTION

It is now obvious that solar energy must be stored, to be economically usable : technical arguments are numerous to confirm this necessity. Rather than making this inventory one time more, we will content ourselves with adding an argument of social order : solar energy is unequally distributed among humans, and has to be collected, condensed and redistributed.

The search of a physical or chemical devices, capable of storing this energy for recovery at wanted time, is certainly the most difficult step to get over, towards solar energy development.

Nature fills up our needs, at present time, with one of the most condensed storages, i.e. petroleum, but the numerous searches already done by chemists and physicists show that one will be satisfied with much less attractive solutions : only, favourable economics may offset this handicap.

In the list of criteria considered by W. E. WENTWORTH and E. CHEN [1] for the selection of suitable reactions for storing solar thermal energy, the availability and low cost of compounds and engineering are not forgotten.

We have got some help from G. ERVIN [2] and R. BARKER [3] papers, for taking stock of practical results excepted of some selected chemical cycles.

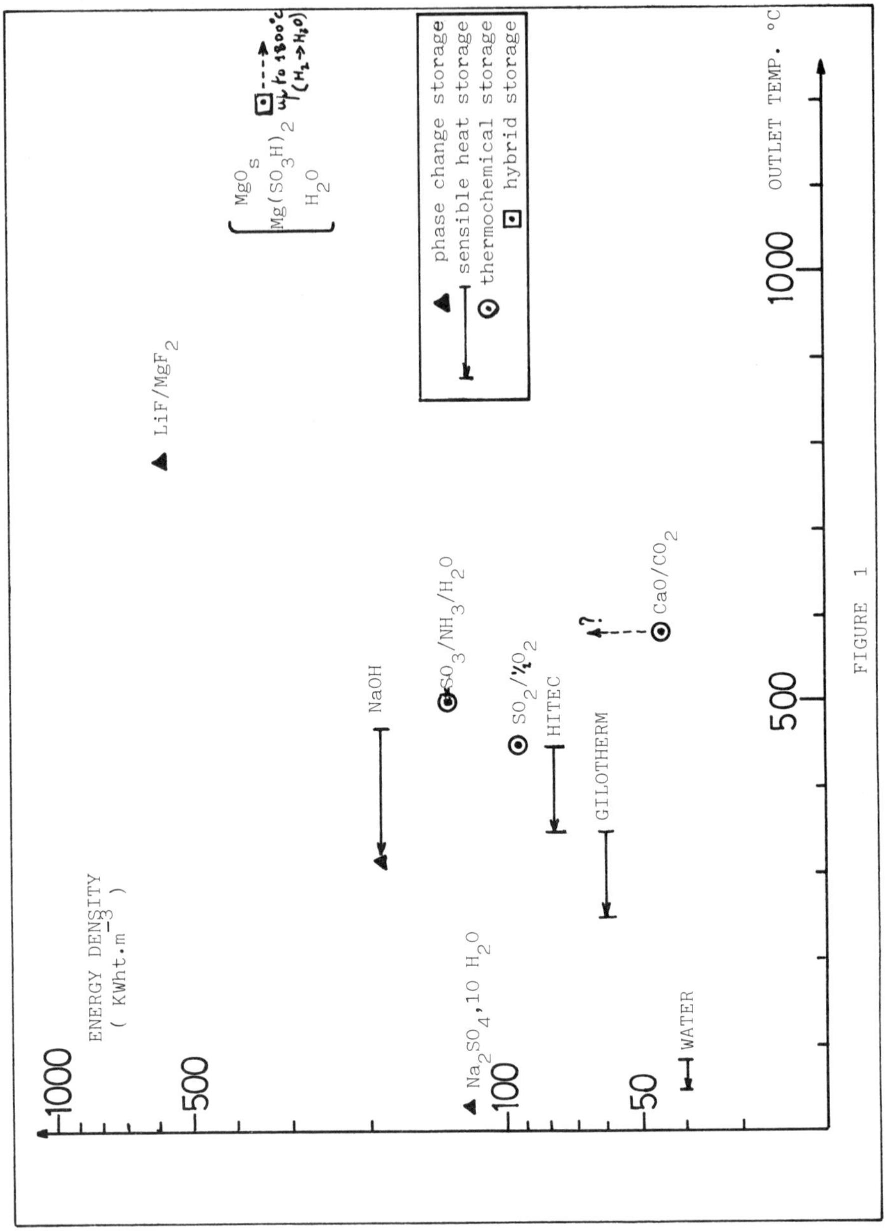
ENERGY DENSITY (KWht.m^{-3})
1000
500
100
50
OUTLET TEMP. °C
1000
500
Na_2SO_4,10 H_2O
WATER
NaOH
$SO_3/NH_3/H_2O$
$SO_2/½O_2$
HITEC
GILOTHERM
CaO/CO_2
?
LiF/MgF_2
MgO_s
$Mg(SO_3H)_2$
H_2O
up to 1800°C
(H_2 → H_2O)
phase change storage
sensible heat storage
thermochemical storage
hybrid storage

FIGURE 1

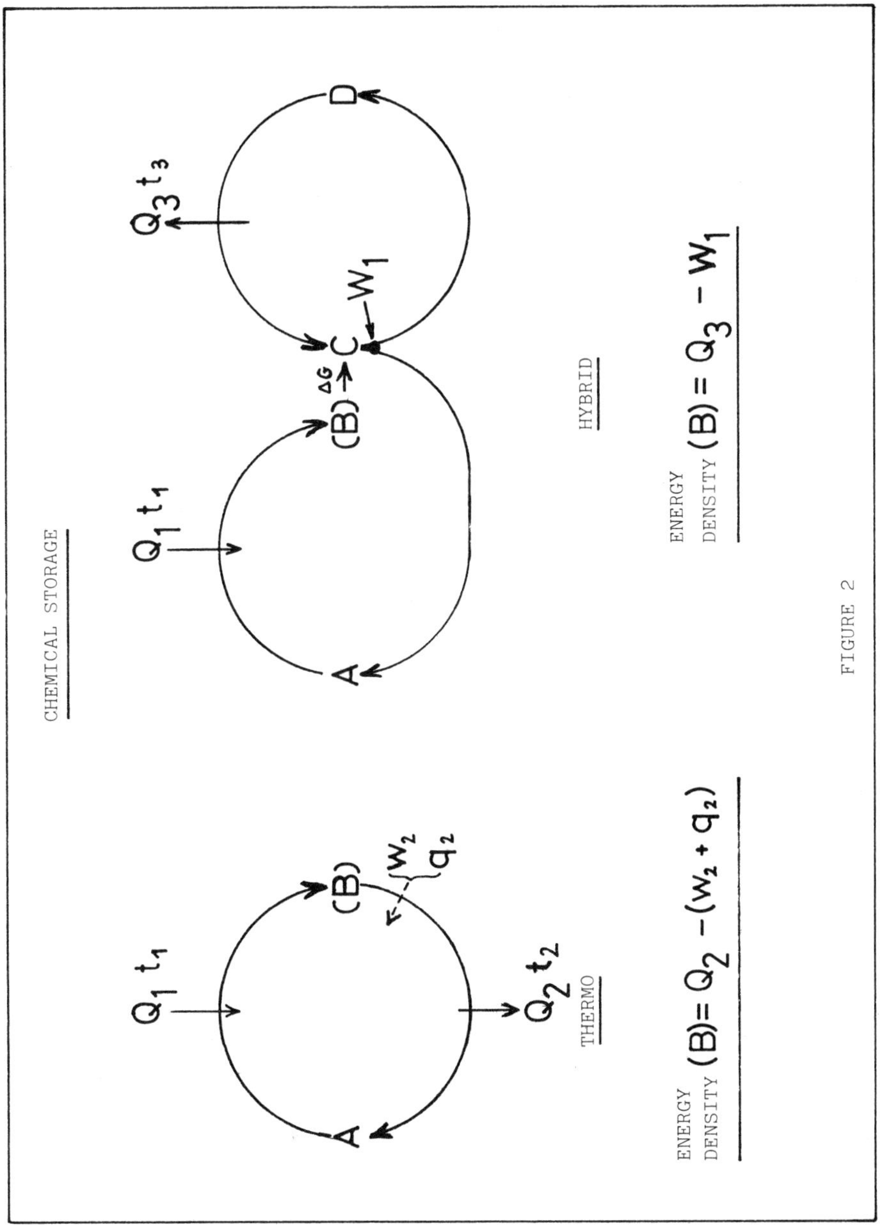
CHEMICAL STORAGE
THERMO
Q_1 t_1
Q_2 t_2
A
(B)
w_2
q_2
ENERGY DENSITY (B) = $Q_2 - (w_2 + q_2)$
HYBRID
Q_1 t_1
Q_3 t_3
A
(B)
ΔG
C
D
W_1
ENERGY DENSITY (B) = $Q_3 - W_1$

FIGURE 2

Figure 1 summarize practical performances of a few thermochemical systems, compared to physical processes, using heat capacity, or better, latent heat of melting of products.

In comparison, we have plotted the energy density of an hybrid storage, described in more details further.

PRINCIPLE OF AN HYBRID STORAGE (Figure 2)

In a thermochemical cycle for energy storage, one may consider that stored heat, say Q_1 at t_1, will be practically released, in equal quantity at t_2, but undervalued of possible energy consumption for melting and/or vaporizing condensed phases.

If liquid CO_2 or NH_3, for example, may be vaporized with ambiant heat (air or water), it is not the same with liquid SO_2, and, of course, water : the energy density of the storage will be equal to $Q_2 - (w_2 + q_2)$. Moreover, a subsequently thermodynamical use of heat, may take into account a lowering of the Carnot yield of the cycle between inlet thermal level t_1, outlet level t_2 and a cold temperature t_0.

Finally, the total efficiency of the system will be :

$$\eta_t = \frac{Q_2 - (w_2 + q_2)}{Q_1} \cdot \frac{\eta_{t_2} \quad t_0}{\eta_{t_1} \quad t_0}$$

We will define an hybrid chemical storage, as a combined cycle, in which input energy is thermal (say Q_1, t_1), like before, but <u>recovery of stored energy needs a further input of energy</u> (may be, work W_1 or thermal).

This complement is no more an consumption for modifying physical form of products, but really, a contribution to a second chemical reaction.

In fact, hybrid chemical storages are found in cycles of coupled reactions, which are numerous in the field of chemical splitting of water into hydrogen.

Output energy, Q_3 at t_3, released by the reversed second reaction may be approximatively the sum of input energies of the system, i.e Q_1 t_1 + W_1 undervalued of internal consumptions.

Energy density of the hybrid storage will be equal to $Q_3 - W_1$. The following example, will make more clear this concept of hybrid storage.

EXAMPLE OF AN HYBRID STORAGE, BASED ON MAGNESIUM OXIDE AND SULFUR

The hybrid chemical cycle, based schematically on the following reactions :

$$\text{I/}\quad SO_2 + 2\,H_2O \longrightarrow H_2SO_4 + H_2 \qquad \text{(electrochemical)}$$

$$\text{II/}\quad H_2SO_4 \longrightarrow SO_2 + H_2O + \tfrac{1}{2}\,O_2 \qquad \text{(thermochemical)}$$

and studied by Westinghouse [4] for nuclear applications, may be modified to comply with specifications of solar energy, by using metallic sulfate instead of sulfuric acid.

Global reactions are, with divalent metal :

$$\text{I/}\quad SO_2 + MO + H_2O \longrightarrow MSO_4 + H_2 \qquad \text{(electrochemical)}$$

$$\text{II/}\quad MSO_4 \longrightarrow MO + SO_2 + \tfrac{1}{2}\,O_2 \qquad \text{(thermochemical)}$$

By choosing Magnesium for M, many criteria are positive as :

- atomic weight : 24,3
- availability : very common (8th element of the lithosphere)
- sulfate dissociation t° : 1200 °C
- solubility of SO_2 in aqueous bisulfite solution : up to 8 moles of SO_2/liter solution.

This chemical cycle presents attractive features, especially concerning chemical storage of energy.

Figure 3 summarizes the steps of this process, which are in investigation for present time (the results of this research will be published soon [5]).

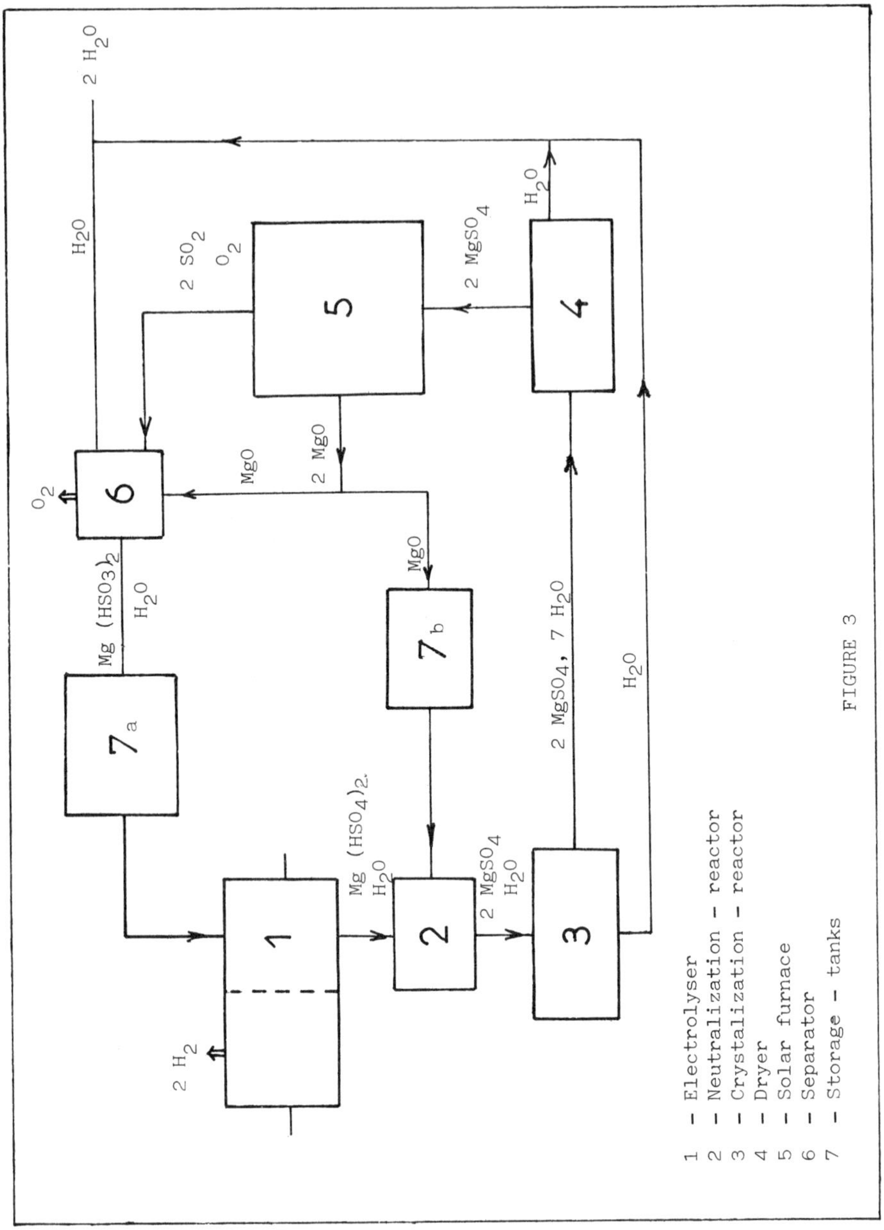
2 H_2O
H_2O
2 SO_2
O_2
2 $MgSO_4$
H_2O
O_2
MgO
2 MgO
MgO
Mg $(HSO_3)_2$
H_2O
2 $MgSO_4$, 7 H_2O
H_2O
Mg $(HSO_4)_2$
H_2O
2 $MgSO_4$
H_2O
2 H_2
1 - Electrolyser
2 - Neutralization - reactor
3 - Crystalization - reactor
4 - Dryer
5 - Solar furnace
6 - Separator
7 - Storage - tanks

FIGURE 3

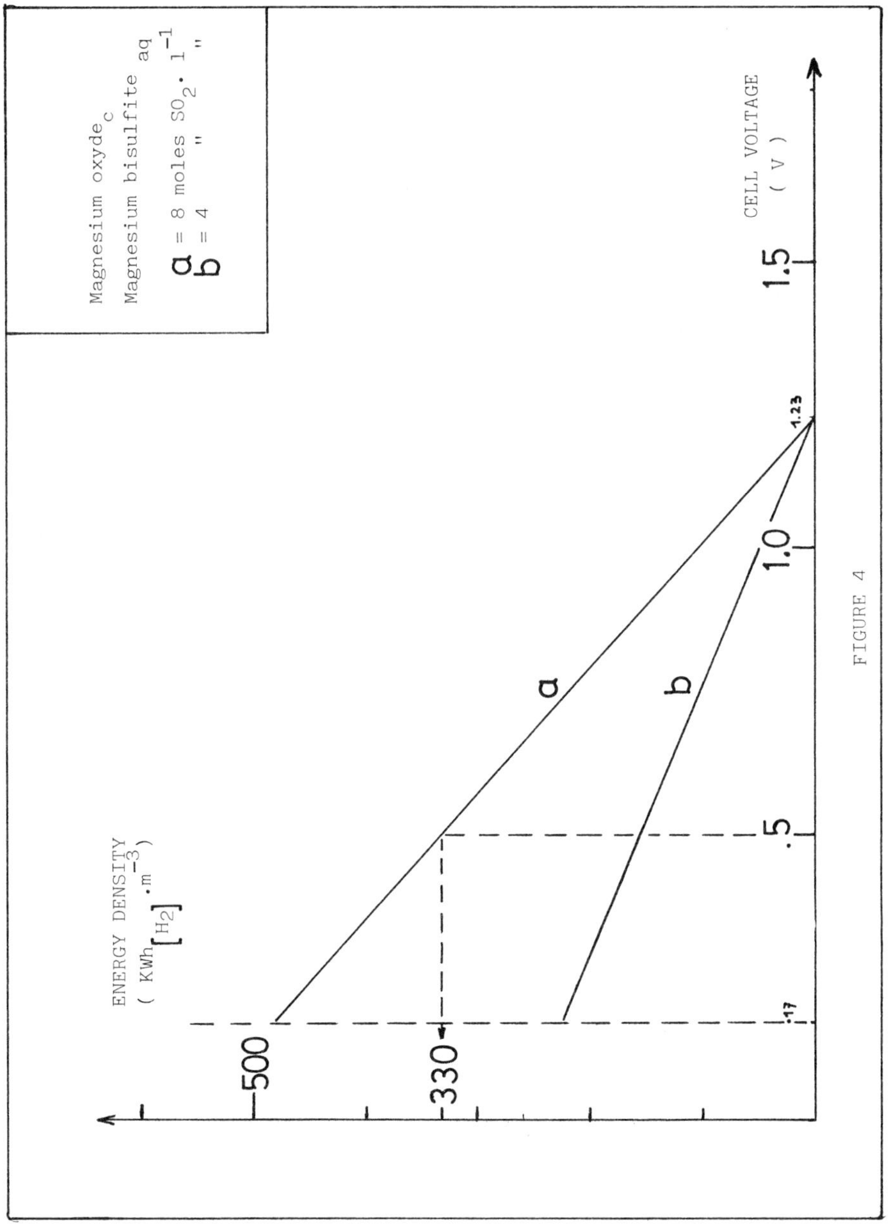
Magnesium oxyde c
Magnesium bisulfite aq
a = 8 moles SO2 · l-1
b = 4 " "
CELL VOLTAGE
(V)
ENERGY DENSITY
(KWh [H2]-1 · m-3)
500
330
.17
.5
1.0
1.23
1.5
a
b

FIGURE 4

In figure 4, we have plotted the practical energy density, expressed in KWh ($\rightarrow \Delta H^\circ$ of H_2 combustion) per m^3 of a two-tank storage (see figure 3 also) :

7a - Magnesium bisulfite aqueous solution

- curve a = 8 moles of SO_2 l^{-1} (d = 1,6)
- curve b = 4 moles of SO_2 l^{-1} (d = 1,25)

7b - Magnesium oxide (bulk density : 1,6)

With an almost quantitative faraday yield, and a voltage of 0.5 V on the electrochemical cell, the energy of the hybrid storage is about 330 KWh m^{-3}, at the maximum of SO_2 solubility.

This value is to be compared to other solutions (figure 1), with two remarks :

1 - Hydrogen combustion may give output heat up to 1800 °C (but we think that better uses are possible, for example synthetic fuels production).

2 - Scenarios of recovery of energy stored are to be found, with the problem of electrical energy (W_1) availability.

CONCLUSION

In a manner of conclusion, we will suggest two scenarios, which may have some interest : (figure 5).

Nota : these scenarios are based on a thermal efficiency of the chemical process of about $\eta = 0,5$.

- first scenario, all solar, with a field of photovoltaic cells, and a tower plant for thermochemistry (THEMIS french project size) : in the proposal, the hybrid storage will give adjustment between global solar input on solar cells (direct and diffused radiation) and direct solar input on heliostats and tower.

In south of France, for example, with global annual solar energy in the range of 1500 KWh m^{-2} (usable), and only 800 KWh m^{-2} direct energy collection, the hybrid storage will help to save on cells investment : without chemical assistance, the same hydrogen production (300 T H_2 for year) would need, at 1.8 V voltage, about 16 GWhe and 100 000 m^2 (instead of 30 000 m^2) of cells.

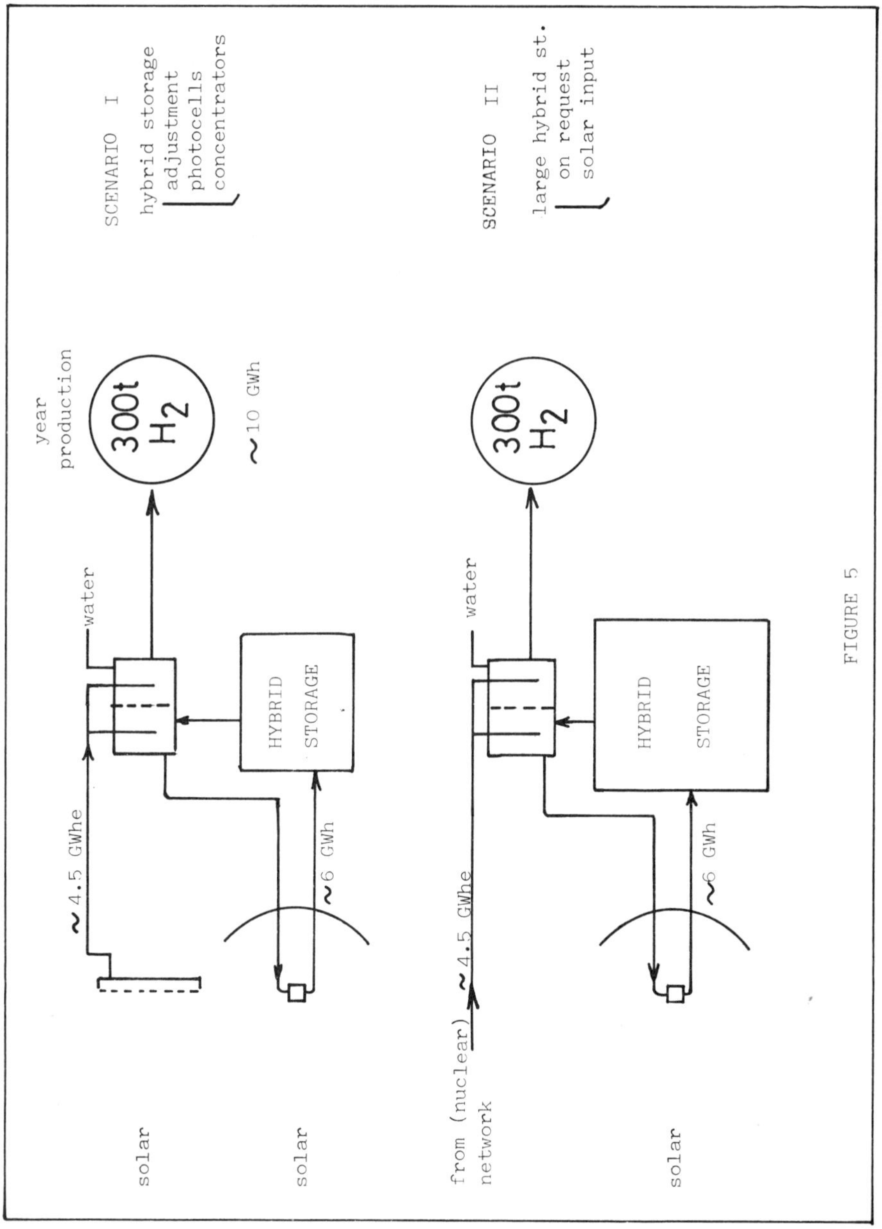
SCENARIO I
hybrid storage
adjustment
photocells
concentrators
year production
300t H_2
~10 GWh
water
HYBRID STORAGE
~4.5 GWhe
~6 GWh
solar
solar
SCENARIO II
large hybrid st.
on request
solar input
300t H_2
water
HYBRID STORAGE
~4.5 GWhe
~6 GWh
from (nuclear) network
solar

FIGURE 5

- the second scenario, is proposed as a solar complement to a conversion of nuclear electrical energy into hydrogen. A large tank, for example 1600 m^3, is capable of storing about 30 days of solar energy input to a THEMIS tower plant, and consequently to deliver this energy to hydrogen production at a constant or on request rate.

REFERENCES

1 - W. E. WENTWORTH and E. CHEN
Simple thermal decomposition reactions for storage of solar thermal energy - Solar Energy, vol. 18, 1976, pp. 205-214

2 - G. ERVIN
Solar heat storage using chemical reactions
Journal of solid state chemistry, vol 22, 1977, pp. 51-61

3 - R. BARKER
The reactivity of calcium oxide towards carbon dioxide and its use for energy storage
Journal of appl. chem. Biotechnology, vol 24, 1974, pp. 221-227

4 - L. E. BRECHER, S. SPEWOCK and C. J. WARDE
The Westinghouse sulfur cycle for the thermochemical decomposition of water
Inter. J. of Hydrogen Energy, vol 2, n° 1, 1977

5 - D. STEINMETZ and A. C. VIALARON
under publication.

USE OF RENEWABLE FUELS IN THE RURAL DEVELOPMENT OF INDIA

G.C. Pandey
Jawaharlal Nehru University
New Delhi 110067, India

ABSTRACT

Due to the recent energy crisis and an increasing awareness towards environmental conservation, most of the nations have begun looking for non-fossil forms fo energy which are relatively environmentally sound and easily available. The technology however is not developed to the extent that it can be used in the tropical and semitropic countries. In this communication, an attempt has been made to analyse the application of these renewable sources of energy in the domestic sector, especially in rural areas in tropical and semitropical countries using available technology.

A qualitative analysis with respect to economics of the community biogas and alternative ways to supplement the increased capital cost have been suggested. The cost foactor in the availability of solar cookers for use by villagers has also been discussed.

INTRODUCTION

Preferential utilization of renewable energy resources is necessary in order to reduce the burden on non-renewable energy resources. In India, as in most other developing countries, people have been using solar energy indirectly to the extent of about 50% of their total energy consumption. However, the efficient use of solar had been and is still a technological problem. An analysis of the consumption pattern of various fuels in the domestic sector indicates that although the percentage contribution of renewable fuels to the total energy consumed has gone down marginally, their actual amount has increased as indicated in Table I. It may also be noted that the percent decrease in contribution of non-commercial fuels is relatively very low as shown in Table II. The data indicate that during the last 18 years, contribution of non-commercial fuels has gone down by only 9%. This indicated that the major users of non-commercial fuels belong to the rural sector. The corresponding increase in commercial fuel is attributed to the change in the fuel use pattern and migration to urban sectors from rural areas. Thus, this increased use of renewable fuels is a matter of concern and needs attention with respect to better utilization of these fuels. This would require the application of appropriate technology. In my view, one of the main drawbacks with most developing countries, including India, is that most of the technologies that have been developed or are being developed have the urban orientation, i.e., only sophisticated technologies requiring high energy systems. However, the same technology even at the developmental stage may prove to be of great value if applied to the low

TABLE I
ACTUAL AND PERCENTAGE SHARE OF COMMERCIAL AND NON-COMMERCIAL FUELS

Fuel/year	1953-54		1960-61		1970-71	
	Actual in mtcr	Per-cent	Actual in mtcr	Per-cent	Actual in mtcr	Per-cent
Commercial fuels	62.67	31.1	101.16	40.6	197.19	52.4
Non commercial fuels	138.84	68.9	147.67	59.4	179.41	47.6

Source: Fuel Policy Committee Report 1974.

TABLE II
PERCENTAGE SHARE OF COMMERCIAL AND NON-COMMERCIAL FUELS CONSUMED IN DOMESTIC SECTOR

Fuel/year	1953-54	1960-61	1970-71
Commercial	7.4	12.4	16.5
Non commercial	92.6	87.6	83.5

Source: Fuel Policy Committee Report , 1974

TABLE III
ESTIMATED DISTRIBUTION OF FUEL DEMAND IN DIFFERENT ASPECTS OF ENERGY UTILIZATION IN DOMESTIC SECTORS (BASED ON 1970-1971 DATA)

		Rural Sector		Urban Sector	
Purpose	Item	Percentage Share of the Total Used in Domestic Sectors	Actual Amount in mtcr	Percentage Share of the Total Used in Domestic Sectors	Actual Amount in mtcr
Lighting					
	Kerosene	55%	15.17	5%	1.38
	Electricity	0.5%	0.02	65%	2.49
Subtotal for Lighting			15.19		3.87
Cooking					
	Non-Commercial Fuels	75%	134.56	8%	14.35
	Coal	5%	0.20	95%	3.87
	Electricity	-	-	9.5%	0.36
	Kerosene	-	-	30%	8.27
			134.76		26.85
Subtotal for Cooking			(83.4%)		(16.61%)
Percentage of fuel used for cooking w.r.t. total domestic fuel use			82.6%		17.4%
Cooling					
	Electricity	-	-	22.5%	0.86
Heating					
	Electricity	-	-	2.5%	0.10

energy systems such as rural areas. In the case of the developed world, this is not applicable because energy-wise, there is no or very little difference between the rural and urban systems.

In the present paper, an attempt has been made to analyse some of the possible applications of the existing technologies towards a better utilization of fuels in the domestic sector, especially in rural areas. This would also help in improving the quality of life of an average villager. The analysis is based on the data for India which may also be extended to other developing countries.

In India, a large part of the population (78-80%) lives in the rural sector. The living conditions of an average villager are not satisfactory. It has been estimated that the daily per capita income of a large part of the population (i.e., about 40-42%) is "only one which is not enough even to fulfill their basic needs of food and clothing". [1]

Table III summarizes the contribution of different fuels used in the domestic sector in rural areas. This table indicates that about 75% of the total non-commercial fuels are used in the rural areas mainly for cooking purposes, especially because of their availability at zero or very low cost. One of the ways of obtaining these is by gathering or by cutting trees from the nearby areas or forests. The composition of different non-commercial fuels (Table IV) indicates that wood is a major contributor (65%) to the non-commercial fuels. This large percentage of wood comes from the indiscriminate cutting of trees. This increased use of wood thus obviously may result in an ecological disturbance, especially when we have a greater demand for wood in other more useful areas, such as the paper industry, etc. The data of 1970-1971 indicates that the amount of wood used in the industrial sector (9.1 m cum) is much less than that used as fuel (13.7 m cum) [2]. Table V gives the percentage use of wood for different industrial purposes.

Similarly, dung and vegetable wastes can provide fertilizers in addition to fuel gas. Some aspects of how the existing fuel use pattern in the domestic sector can be improved using present day technology has been discussed with a special focus on the efficient utilization of renewable fuels in the domestic sector in rural India. Two main alternatives are: (1) based on a complete change-over to biogas or (2) partial use of solar cookers to supplement fuel requirements in the domestic cooking situation have been suggested.

Distribution of population based on income group indicates that a major part of the population has a very low income (about 250 m), i.e., "day and any programmed even with incentives will not affect their life style". [1] However, there is another group which includes lower middle and middle class people, which is about 240 m in India, who do have to depend upon the purchase of fuels for their domestic cooking. It is this class which can afford to use new technology provided it can bring down their fuel bills. Therefore, the alternatives suggested in this article can only be of help to this category and not the poor people. Different aspects of existing

TABLE IV
ESTIMATE OS DOMESTIC CONSUMPTION OF DIFFERENT NON-COMMERCIAL FUELS
(in mtcr)

Year/Fuel	Total non-commercial fuel	Fire wood	Cow dung	Vegetable waste
1960-61	147.67	95.99 (65%)	22.15 (15%)	29.53 (19.99%)
1965-66	163.43	106.23 (65%)	24.51 (14.99%)	32.69 (20%)
1970-71	179.41	116.62 (64.89%)	26.91 (15%)	35.88 (20%)

Source: Fuel Policy Committee Report 1974

TABLE V

SECTOR-WISE USE OF INDUSTRIAL WOOD 1970-1971 (in m cum)

Railways sleepers etc	0.20
Pulp and paper board industry	1.02
Furniture (50%)	4.55
Construction material (30%)	2.73
Miscellaneous	0.60
Total	9.10

Source: India Reference - Annual (1976)

fuel use pattern and the savings through introduction of new technology has been analysed below:

During the year, effective solar energy is available for about 300 days.

On an average per capita fuel consumption/day = 1 kg

The total amount of non-commercial fuels required annually
240 x 300 x 1 = 72 m kg of fuel

Out of total non-commercial fuels, wood accounts for about 65%.

Therefore, the actual amount of wood required annually

$$= \frac{72000 \times 65}{100} = 46800 \text{ m kg of wood}$$

Thus a complete change-over to biogas will be able to save 72000 m kg of non-commercial fuel or 46800 m kg of wood exclusively. On the other hand, savings due to partial use of solar cookers in domestic cooking will come to about 25-30% of this amount.

BIOGAS TECHNOLOGY

Mr. Makhijani in one of his articles on "Energy Policy For Rural India", writes "Technically and financially, it is not difficult to meet the basic needs of the poor for food, domestic water and fuel, clothing, sanitation and health. A need-based approach to rural development can help generate the resources for the relatively large investments required for the wide sprad development of irrigation and fuel supply". [3]

As already mentioned, a large part of the population is living in villages and is using non-commercial fuels to the extent of 75% to meet their domestic fuel requirements and wood contributes to about 55% of it, which is obtained from indiscriminate tree cutting with the result that our forest reserves are reducing which has obvious repercussions. To overcome this problem, some alternatives need to be thought of. Mr. Mukhijani has also mentioned in his paper about the community biogas plants to feed the requirements of the village population. This idea in principle is nice, but the basic problems still remained to be solved, i.e., capital cost for plant erection, pipeline for gas distribution, maintenance, etc. In addition, there is one very important factor, i.e., cultural factors, which does not allow any new technology to enter into the life-style of an average villager. This problem can be solved only through repeated trials, i.e., stepwise introduction of the technology may be able to solve the problem - although it is not so easy, at least not in India. This seems to be one reason why biogas became so popular in China in such a short time while India has not been able to popularize even gobar gas plants.

As far as the cost is concerned, only a few villagers can afford this on their own, even on a community basis. Therefore, general government should subsidize the cost and also provide loans for the rest of the cost through its agencies or banks.

In India, K.V.I.C. has been one of the active agencies for providing partial financial assistance for setting up gobar gas plants in rural areas. The scheme in principle is very attractive, but due to the restrictions of minimum cattle-head requirements, and also relatively large amounts of money required, could not become popular in spite of many incentives. Table VI gives an idea of the relatively low cost one has to pay in cash at the time of purchase for a gobar gas plant [4].

Once the basic problem of finance is solved, it may then be possible to go for a community biogas plant based on the following suggestions.

(1) Supply line to homes: There are various types of well-known supply pipes, i.e., metallic, PVC or polythene, etc. All these have one or the other merits, but cost-wise, they are expensive. One suggestion could be the use of processed bamboo pipes for the purpose especially because of its property - that if kept underground, it would not be affected by weathering. Its availability (production) in rural areas and minor processing by local craftsmen, makes this as a supply pipe very cheap and attractive.

(2) Size of plants: The plants of suitable size may be chosen based on the population to be served. It is suggested to have one plant at a suitable place to serve a part of the nearby group of houses instead of having a very large plant with longer service points. It may be mentioned that in most of the cases, it is expected that the locally available dung and vegetable wastes are not enough and therefore, it is advisable to go for a community biogas plant instead of gobar gas alone. The excreta collection may be done manually which will be the cheapest.

Maintenance

Since the technology of biogas plants is not very complex the local people may be given a short-term training by experts during the time of plant erection, and they should be made responsible for the maintenance. A regular check-up by experts will make the plant run continuously. The management may be taken over by the local village bodies such as village Panchayat.

Economic Aspect

The cost aspect of this planning includes expenditures on:

(1) Capital Cost - for setting up of the plant and supply line, etc.

(2) Maintenance Cost
   (i) Cost of collection of raw material, i.e., payment of labour

TABLE VI
SUBSIDIES AND LOANS PROVIDED FOR GOBAR GAS PLANT

Size of the plant in (cubic meters)	Estimated cost as on Feb, 1975	Approximate number of animals required	Subsidy provided by Govt of India		Loans provided to the extent of 75%
			General Section (25%)	Economically weaker section (75%)	
			in Rs	in Rs	in Rs
2	2332.00	2-3	583.00	1749.00	1749.00
4	3359.00	4-6	839.75	2519.25	2519.25

charges and sometimes dung and vegetable waste charges.

(ii) Cost towards wear and tear of the plant, etc.

(iii) Cost towards the labour charges and maintenance personnel,

and the recovery of capital through recovery:

(i) Through charges of gas supply which may be fixed on the basis of the number of people in a house or on any other basis.

(ii) and the sale of manure obtained as a result. This will give a good return because of its ability to replace the synthetic fertilizers.

In this analysis, I would like to emphasize two points; one is that the government should subsidize the capital cost because ultimately this process will replace large amounts of synthetic fertilizers which is a petroleum-based product, thereby reducing the import of oil (an important commodity). Another point is that although it may be difficult initially to popularize, but once it became popular, it will save large amounts of wood which is of great value.

Uses Other Than Cooking

The biogas may also be used for home and street lighting in the area at almost no extra cost.

The appropriate and cheap ovens for using the gas can be developed without much difficulty.

In summary, this approach will provide a good substitute for firewood, an important local resource and have large amounts of oil products and oil needed for fertilizers and kerosene for the lighting purposes. In addition, it will help in the improvement of the living conditions of an average villager.

PARTIAL USE OF SOLAR COOKERS

With our country being a tropical one, the solar energy is available for at least 300 days effectively during the year with varying intensities. Though a complete change-over to solar cooking may take a long time, the use of solar cookers partially in the domestic cooking area will certainly result in large amounts of fuel savings. Under the present conditions, the cost of a solar cooker (suitable for a family of four persons) comes to about 80-85%. Large scale production and local labour (craftsmen) and raw material may bring down this cost to about 50-60%. In these cookers, glass reflectors which are used, account for about 40-50% of the total cost. Thus, a substitute for a glass reflector may be able to bring down the cost, e.g., use of aluminum foil as a reflector will certainly bring down the cost considerably. The life of an aluminum reflector will be relatively much less, however, that will then only require a recurring expenditure of Rs. 10/ - say after every 3 months. This brings to a recurring expenditure of Rs. 3.33 per month which can easily be afforded by a family of four persons. Thus, if aluminum foil becomes effective, then the corresponding costs will come down to Rs. 40-50, in the case of the individual and to about Rs. 30/ - in large-scale production when local (rural) labour and local raw material are used. Even the cost of Rs. 30/ - cannot be afforded by the class of people we are concerned with. However, incentives like subsidy and loans may be able to popularize the solar cookers if it becomes available at a cost say of Rs. 8-10. The data based on the following analysis will give an idea regarding the preferences of solar cookers over the traditional wood burning.

Total cost of a solar cooker (for a family of four)	Rs. 30.00
Subsidy by Government or allied agency (25%)	7.50
Loan by bank or allied agencies (50%)	15.00
The cost to be paid by the consumer for purchasing the cooker	8.50

The recurring expenses per month

Recurring expenditure (per month)	3.30	
Payment of loan plus interest (for 12 months only)	1.50	
Depreciation per month (10%)	0.25	
	5.05	per month for 1st year

After the first year, only Rs. 4.55 per month (since the bank loan has been repaid) for a family of four, i.e., average Rs. 1.14 per person per month.

This cost seems to be reasonable and may help to popularize these cookers in the light of the following data which indicates that its use will bring down the existing fuel bill to a considerable extent.

For a family of 4 persons, the average fuel bill @ 4 kg per day (i.e., 120 kg in terms of wood)	Rs. 24.00 (@Rs. 8 per 40 kg)
The savings in bill due to the use of solar cookers partially (for cooking rice and pulses - 1/4 of the total fuel/day)	Rs. 6.00

This will become more in the case of peoples who eat only rice especially in some parts of India. Thus the average saving may become (on a national scale) Rs. 7.50

Thus, a comparison indicates that per month, a saving of Rs. 7.50 against the expenditure of Rs. 4.00 for a family of four may be able to make this alternative acceptable although the cultural factor shall be important even in this case.

The increasing rate of desertification all over the world is one of the burning problems. In India, the Rajasthan area forms the major part of the arid and semi-arid zone. The desert here, like other places,is also moving fast. One of the reasons for this is now well-established: the human action e.g., overgrazing, use of brush and wood for fuel, etc. These practices, if continued will never allow any alternative effort to become effective. The different steps used to reclaim or save the land in these areas in India have been plantation of low transpiration plants and of plants which will check wind blow. In addition, efforts to reduce overgrazing have also been made by providing food/foder supply to inhabitants. However, on no account can they be asked to stop the use of wood as fuel. One alternative could be the use of solar cookers which may be able to solve this problem. In this reference, I would suggest that an effort to popularize the use of solar cookers in this area will be of great value since it will help in reducing ecological stress to a good extent. However, the popularization in this area may not be that easy as in other areas since the major population belongs to the category of the very poor, the category which we have excluded in this study. But due to the importance of this area from the ecological point of view, the subsidies to greater extent or even distribution of free solar cookers to the class of people under this category, i.e., poors, could be an alternative to popularize this practice. I may mention that the Central Institute of Semi-Arid Zone Research at Jodhpur is actively engaged in the development of appliances to harness the solar energy or arid and semi-arid regions.

CONCLUSION

In summary, I would like to mention that because of the scattered situations of the villages with relatively little population, it is very difficult to provide any centralized facilities, especially those involving sophisticated technology, to improve the living conditions of the inhabitants. A number of schemes have been tried, but failed mainly because of the economic factors and sometimes approach problems, i.e., lack of proper roads to reach the site. However, crude and self-contained technologies could be an alternative, especially since their energy demand is much less compared to the urban sector. The raw material and labour (both skilled and semi-skilled) required could be obtained locally which means large savings of energy required for transportation and labour charges. There is another very important benefit of trying a technology in rural sectors - that one can get the actual feedback in a short time for further analysis because of the involvement of a smaller population per village.

ACKNOWLEDGEMENT

The author gratefully acknowledges the suggestions of Professor Dhirendra Sharma of Centre for Studies of Science Policy, Jawaharlal Nehru University, New Delhi, India.

REFERENCES

1. Reddy, A.K.N. and Prasad, K.K. "Technological Alternatives and The Indian Energy Crisis", Economic and Political Weekly, XII, (33 & 34), 1465 (1971).

2. India: Reference Annual, 1976.

3. Makhijani, Arjun "Energy Policy For Rural India", Economic and Political Weekly, XII (33 & 34) 1451 (1977).

4. "Gobar Gas: Why and How" published by Directorate of Gobar Gas Scheme, Khadi & Village Industries Commission, Bombay, India, January 1977.

SOLAR ENERGY INTERNATIONAL PROGRESS

Pergamon Titles of Related Interest

Boer Sun II
Eggers-Lura Solar Energy for Domestic Heating and Cooling
Fazzolare and Smith Changing Energy Use Futures
Machovec Solar Energy Index
Messel Energy for Survival
deMontbrial Energy: The Countdown
Veziroglu Solar Energy and Conservation

Related Journals*

Energy
Energy Conversion
Solar Energy
Space Solar Power Review
Sun at Work in Britain
Sun World

*Specimen copies available upon request.

SOLAR ENERGY INTERNATIONAL PROGRESS

Proceedings of the International Symposium-Workshop on Solar Energy, 16-22 June 1978, Cairo, Egypt

EDITED BY
T. NEJAT VEZIROGLU

VOLUME 1

PERGAMON PRESS
New York • Oxford • Toronto • Paris • Frankfurt • Sydney

Pergamon Press Offices:

U.S.A.	Pergamon Press Inc., Maxwell House, Fairview Park, Elmsford, New York 10523, U.S.A.
U.K.	Pergamon Press Ltd., Headington Hill Hall, Oxford OX3 0BW, England
CANADA	Pergamon of Canada, Ltd., Suite 104, 150 Consumers Road, Willowdale, Ontario M2J 1P9, Canada
AUSTRALIA	Pergamon Press (Aust.) Pty. Ltd., P.O. Box 544, Potts Point, NSW 2011, Australia
FRANCE	Pergamon Press SARL, 24 rue des Ecoles, 75240 Paris, Cedex 05, France
FEDERAL REPUBLIC OF GERMANY	Pergamon Press GmbH, Hammerweg 6, Postfach 1305, 6242 Kronberg/Taunus, Federal Republic of Germany

Library of Congress Cataloging in Publication Data

International Symposium-Workshop on Solar Energy, Cairo, 1978.
Solar energy.

Presented by the Clean Energy Research Institute of the University of Miami, Coral Gables, Fla.
Bibliography: p.
Includes index.
1. Solar energy—Congresses. I. Veziroglu, T. Nejat. II. Miami, University of, Coral Gables, Fla. Clean Energy Research Institute. III. Title.
TJ810.I63 1978 621.47 79-13656
ISBN 0-08-025077-7

Printed in the United States of America

SOLAR ENERGY: INTERNATIONAL PROGRESS

Proceedings of the International Symposium-Workshop on Solar Energy, 16-22 June 1978, Cairo, Egypt.

Proceedings of the International Symposium-Workshop on Solar Energy, held in Cairo, Egypt on 16-22 June 1978, and presented by the Clean Energy Research Institute, School of Engineering and Architecture, University of Miami, Coral Gables, Florida, U.S.A.; hosted by the Solar Energy Laboratory, National Research Centre, Dokki, Cairo, Egypt; sponsored by the National Science Foundation, Washington, D.C., U.S.A.; in cooperation with the Industrial Development Centre for Arab States, the Egyptian Ministry for Electricity and Energy, Cairo, Egypt, the International College on Applied Physics, and the International Centre for Theoretical Physics, Trieste, Italy.

FOREWORD

According to reliable studies, it is clear that the world's fossil fuel resources - which meet ninety percent of the energy requirements today - are being consumed at a tremendous and ever-increasing rate, and that in a not-too-distant future of twenty to thirty years' time, their production will peak and then start decreasing. As we approach the dawn of the fossil fuel era, the amount of pollution, and for that matter the cost of pollution abatement, will increase since the low-sulphur high-quality coal and oil supplies will become depleted and will be replaced by lower quality fossil fuels. If we are going to improve the quality of life in the world, we must increase the total energy consumption per-capita and reduce the pollution of the environment. This is the greatest challenge to scientists, engineers, architects and energy planners in the last quarter of the twentieth century.

In deciding on the future energy sources, it is necessary that certain conditions be met if we are to avoid the problems associated with the fossil fuels. The new energy source or sources must be undepletable, and must not pollute or degrade the environment. There are several primary energy candidates to replace the fossil fuels, such as nuclear breeders, fusion reactors, geo-thermal and solar energy. When we compare them, solar energy - in both its direct and indirect forms - stands out as by far the best. It is undepletable and environmentally the most compatible.

On the other hand, there are also some problems with solar energy. It is a rather diffuse form of energy and is intermittently available. The intensity of solar energy changes with the time of the day, the seasons of the year and the location. These problems must be resolved if solar energy is going to make inroads into the vast energy market. It will take the best brains of the world and cooperation among them to resolve these problems quickly and effectively. This was the motivating force in organizing the International Symposium-Workshop on Solar Energy, 16-22 June 1978, Cairo, Egypt, which was attended by more than four hundred energy scientists and planners representing many countries of the world. The Symposium provided an international platform for the presentation, discussion and exchange of the latest information covering the various facets of solar energy - from the properties of solar radiation to storage, from energy conversion to energy planning and policies. The workshops were held to cover the various aspects of solar energy, including architectural and industrial applications, energy conversion, planning and commercialization.

The proceedings comprise of the full-length papers presented at the Symposium and recommended by the session chairpersons, as well as the workshop reports. The papers and workshop reports are grouped into eighteen chapters depending on the subject matter covered, and are presented in four volumes. In the case of a few papers, it was difficult to classify them by one subject area, as there were some overlaps. We have attempted to make the best choice possible.

Solar Energy: International Progress is expected to serve as a valuable reference covering the recent research and development studies in the world in the growing and important field of solar energy.

T. Nejat Veziroğlu
Editor

SYMPOSIUM COMMITTEE AND STAFF

SYMPOSIUM COMMITTEE

CHAIRMAN:

T. Nejat Veziroğlu
University of Miami
Coral Gables, Florida, U.S.A.

CO-CHAIRMEN:

Homer W. Hiser
University of Miami
Coral Gables, Florida, U.S.A.

Ibrahim A. Sakr
Egyptian National Research Centre
Cairo, Egypt

MEMBERS:

Mahmoud A.B. El-Koshairy
Egyptian Ministry of Electricity and Energy
Cairo, Egypt

E.M. El-Missiri
Industrial Development Centre for Arab States
Cairo, Egypt

A.A.M. Sayigh
University of Riyadh
Riyadh, Saudi Arabia

SYMPOSIUM STAFF

COORDINATORS:

Lucille Walter
Mary Winston
N.H. Helwa
A.I. Hegazi
A.M. El-Haggan
Deirdre Amari
Joanne Drost
Maria Eugenia Galis
Mary Connors

PUBLIC RELATIONS:

H.M. Abdel-Aal
A. Abdel Monem
S.A. Solman

MANUSCRIPT EDITOR:

Deirdre Amari

ACKNOWLEDGEMENTS

The Symposium Committee gratefully acknowledges the financial support of and extends a special thanks to the National Science Foundation, Washington, D.C., U.S.A. In addition, we gratefully acknowledge the cooperation provided by the Industrial Development Centre for Arab States, Cairo, Egypt; the Egyptian Ministry for Electricity and Energy, Cairo, Egypt; the International College on Applied Physics and the International Centre for Theoretical Physics, Trieste, Italy; and the support and service provided by the Solar Energy Laboratory, National Research Centre, Dokki, Cairo, and the Office of the Scientific Attache, Embassy of the United States, Cairo, Egypt.

Special thanks are also due to our authors whose participation and contributions provide the substance of the symposium as published herein.

And last, but not least, our debt of gratitude is owed to the session chairpersons and co-chairpersons and the workshop group leaders for the organization and execution of the technical sessions and the workshops. In acknowledgement, we list these symposium officials on the following pages.

The Symposium Committee

SESSION OFFICIALS

Session 3A - POLICY AND RESEARCH PROGRAMS

Session Chairman:	G. Beghi Joint Research Center of European Communities Ispra, Italy
Session Co-Chairman:	M. Saad UNESCO Middle East Section Cairo, Egypt

Session 3B - SOLAR RADIATION AND ENVIRONMENT

Session Chairman:	S. Kakac Orta Dogu Teknik Universitesi Ankara, Turkey
Session Co-Chairman:	M. Abdel Salam National Research Centre, Dokki Cairo, Egypt

Session 3C - COLLECTORS I

Session Chairman:	N. Mancini Istituto di Fisica Catania, Italy
Session Co-Chairman:	S. Awad Cairo University Cairo, Egypt

Session 6A - ECONOMICS AND RESOURCES

Session Chairman:	S. Warkov University of Connecticut Storrs, Connecticut, U.S.A.
Session Co-Chairman:	M. Saleh National Research Centre, Dokki Cairo, Egypt

Session 6B - THERMAL CONVERSION AND DESALINATION

Session Chairman: A.J. Parker, Jr.
Mueller Associates, Inc.
Baltimore, Maryland, U.S.A.

Session Co-Chairman: M.A. Sourour
Alexandria University
Alexandria, Egypt

Session 6C - COLLECTORS II

Session Chairman: K. Selcuk
Jet Propulsion Laboratory
Pasadena, California, U.S.A.

Session Co-Chairman: N. Helwa
National Research Centre, Dokki
Cairo, Egypt

Session 9A - BUILDINGS AND APPLICATIONS

Session Chairman: A. Bowen
University of Miami
Coral Gables, Florida, U.S.A.

Session Co-Chairman: A.A. Rizk
Ain Shams University
Abbassia, Cairo, Egypt

Session 9B - BIOLOGICAL AND CHEMICAL CONVERSION

Session Chairman: G. Atagunduz
Ege University
Izmir, Turkey

Session Co-Chairman: M. Hassib
National Research Centre, Dokki
Cairo, Egypt

Session 9C - THERMAL STORAGE AND WATER HEATING

Session Chairman: P.D. Chawla
Shahdara
New Delhi, India

Session Co-Chairman: M.M. Elkotb
Cairo University
Cairo, Egypt

Session 12A - RURAL AND AGRICULTURAL APPLICATIONS

Session Chairman: N. Lutkefedder
United States Department of Energy
Washington, D.C., U.S.A.

Session Co-Chairman: A.S. Elasfouri
Cairo University
Cairo, Egypt

Session 12B - ELECTRICITY PRODUCTION

Session Chairman: S. Kakac
Orta Dogu Teknik Universitesi
Ankara, Turkey

Session Co-Chairman: M.S. Abov Housien
Cairo University
Cairo, Egypt

Session 12C - HEATING, COOLING AND REFRIGERATION

Session Chairman: P. Smith
Berkeley, California, U.S.A.

Session Co-Chairman: M.M. Helal
Cairo University
Cairo, Egypt

WORKSHOPS

Workshop 1 - ARCHITECTURAL APPLICATIONS

Group Leader: A. Bowen
University of Miami
Coral Gables, Florida, U.S.A.

Workshop 2 - AGRICULTURAL AND INDUSTRIAL APPLICATONS

Group Leader: E. Bilgen
Ecole Polytechnic
Montreal, Quebec, Canada

Workshop 3 - ENERGY CONVERSION

Group Leader: Y. Marfaing
C.N.R.S.
Bellevue, France

Workshop 4 - SOLAR PLANNING

Group Leader: A.J. Parker, Jr.
Mueller Associates, Inc.
Baltimore, Maryland, U.S.A.

Workshop 5 - SOLAR ENERGY COMMERCIALIZATION

Group Leader: N.W. Lutkefedder
United States Department of Energy
Washington, D.C., U.S.A.

SOLAR ENERGY: INTERNATIONAL PROGRESS

TABLE OF CONTENTS

VOLUME I

Page No.

CHAPTER 1 SOLAR RADIATION

Plane Incident Solar Radiation, Effects of Slope and Orientations 1
J.I. Jimenez, C. Romeu-Nedwed, J.E. Llebot and J. Casas-Vezquez

Relationship Between Global and Net Radiation 14
A. Guerrini, A. Lavagnini and F. Vivona

Atmospheric Pollution and Solar Radiation 20
M.S. Abdel-Salam and N.A. Higazy

Solar Energy at Ground Surface 41
N.H. Helwa

Contribution to the Study of Nocturnal Radiation Very Near the Ground: Effect of a Circular Screen 56
J. Brodin, A. Flourens, V.V. Pham and C. Sinnasse

New Formulation for the Factor Ψ for the Middle East Region 81
E.M. Abd El-Salam

Computation of the Diffuse Solar Radiation for Egypt 90
E.M. Abd El-Salam

Solar Data Application to Egypt 114
M.S. Abdel Salam and N.A. Higazy

Global Radiation in Italy 127
A. Guerrini, A. Lavagnini and F. Vivona

On the Distribution of Solar Energy in Greece 140
G.J. Macris

CHAPTER 2 FLAT PLATE COLLECTORS

The Flat Plate Solar Collector: Its Steady-State and Transient State Behaviour 159
E. Hahne, N. Fisch and A. Arafa

Flat-Plate Solar Collectors - Influence of Design Material and Configuration 187
S-E. Ransmark

My New Types of Flat Plate Solar Collectors 215
R. Jankovic

Measurements on a Flat Thermal Solar Energy Collector with a Cellular Structure After Francia 239
A. Bahr, H. Piwecki and L. Rigolini

A Simulated Comparison of the Useful Energy Gain in a Fixed and a Fully Thermal Tracking Flat Plate Collector 258
P. Drago

Flat Plate Boosters for Elevating Temperatures 274
I.A. Sakr and A.I. Hegazi

A Black Liquid Flat Plate Collector for Economical Applications 298
R.S. Daryan and P.K.C. Pillai

Passive Solar Wall Collectors 308
A. Faist and J.B. Gay

Energy Analysis Method for a Solar Panel 324
B. Ciborro

CHAPTER 3 CONCENTRATING COLLECTORS

A Two-Dimensional Finite-Difference Solution for the Transient Thermal Behavior of a Tubular Solar Collector 328
F.L. Lansing

Optical and Thermal Performance of a Half-Circular, Cylindrical Concentrator 351
K. Bordoloi, T. Murray, F. Bynum and E. Chaves

Modeling a V-Trough Collector in Terms of Zero, Single or Multiple Thermal Capacitance Nodes 378
K. Bordoloi, C. Hoang and F. Bynum

High-Efficiency Solar Concentrator 392
F.L. Lansing and J. Dorman

The Scheme of Automatic Adjustment of Accuracy of the Inflatable Solar Collector 404
A. Gokhman and N. Ozboya

Thermo-Fluid Analysis of a Cylindrical Solar Heat Absorber 412
M.F. Khalil

An Experimental Investigation of Comparative Performance of a Header-Type Compound Parabolic Concentrator and a Flat-Plate Collector 431
K. Krishna Rao, S. Basheer Ahmed, R. Natarajan and M.C. Gupta

CHAPTER 4 COLLECTOR TECHNOLOGY AND RESEARCH

Interferometric Investigations of Convection in Slat, Flat Plate and Vee-Corrugated Solar Collectors 446
B.A. Meyer, K.R. Randall, M.M. El-Wakil and J.W. Mitchell

Study of a Multi-Pass Radiation Fluid-Heater Exposed to Non-Uniform Radiation 463
N.M. Rafat, M.M. Elkotb and M.F. El-Refaie

Simulations of Solar Captation Systems: Methodology and Applications to Distillation, Gas and Water Heating 490
R. Rocaries, G. Olalde, P. Granier, B. N'Doye and M. Daguenet

Low Cost Spiral Shaped Solar Collector 512
P.K.C. Pillai and R.C. Agarwal

Energy Requirements of an Aluminum Solar Collector 522
J.E. Robinson and W. Ashton

Optimum Inclination of Linear Reflectors 529
B.A. Khalefa and A.S. Hegazy

Selective Surface Coating with Black Chrome 536
T. Vurgun

Cadmium Stannate Selective Coatings Prepared by Spray Pyrolysis 540
O.P. Agnihotri and B.K. Gupta

Phase Change Collector Wall Versus Water Collector Wall 549
L. Bourdeau and A. Jaffrin

On The Evaluation of the Feasibility of Solar Collectors for Electrical Power Generation Applications 565
A. Gokhman and N. Ozboya

Effect of Solar Radiation on Outdoor Bus Bar Design 577
M.M. Abdel-Aziz

VOLUME II

CHAPTER 5 THERMAL STORAGE

Performance of a Storage Unit Coupled to a Flat Plate Solar Collector 593
S.B. Ahmed and V. Sriramulu

Transient Characteristics of Vertical Porous Heat Storage 604
N. Seki, S. Fukusako and H. Inaba

Seasonal Regeneration Through Underground Strata 623
T. Yokoyama, H. Umemiya, T. Teraoka, H. Watanabe, K. Katsuragi and K. Kasahara

Storage Tank Efficiency as Simulated in a Markovian Model of Meteorology 654
R. Lestienne

Some Aspects of Energy Storage Systems Using Glauber Salt Mixtures 666
D.Y.S. Lou and A.J. Harrison

A Direct Contact Thermal Energy Storage System 680
A.S. Elasfouri

CHAPTER 6 HEATING

Solar Heating Systems with Air Collectors and Rock Bed Storage 697
W.L. Dutre, L. Cypers, J. Berghmans and A. Debosscher

Influence of Heat Exchanger Effectiveness on the Performance of Thermosyphon Double Loop Water Heating 731
A.F. Orlando, L. Goldstein, Jr. and D. Magnoli

Hot Water Production by Means of a Solar Thermosyphon Loop 745
W.L. Dutre, L. Cypers, J. Berghmans and A. Debosscher

Effectiveness of Packed-Bed Solar Air Heaters 767
S.C. Gupta

Performance Comparison of Different Types of Solar Water Heaters Under Egyptian Conditions 778
S.W. Youssef and I.A. Sakr

Solar Water Heaters 793
D. Buxey

Latent Heat Diode Wall 802
L. Bordeau and A. Jaffrin

CHAPTER 7 COOLING

Comparison Between Two Absorption Cooling Systems of the "Open" Type Under Different Climate Conditions 827
M. Bolzan and R. Lazzarin

Optimisation of Cooling Systems in Tropical Zones 855
J. Flechon

Analysis of Ammonia-Water Intermittent Solar Refrigerator Operating with a Flat Plate Collector 870
A. Venkatesh and M.C. Gupta

A Thermodynamic Analysis of a Solar-Powered Jet Refrigeration System 886
F.L. Lansing and V.W. Chai

Development of a Solar Desiccant Dehumidifier 898
Z. Lavan, D. Gidaspow, M. Onischak, S. Chaturvedi, B. Mathipraksam and S. Perkari

Liquid Desiccant Solar Heat Pump for Developing Nations 932
H. Robison

CHAPTER 8 APPLICATIONS IN BUILDINGS

A Review of the Historical Developments of the Use of Passive Solar Systems 939
C.L. Robbins

Natural Lighting in Buildings: A Passive Solar System 960
C.L. Robbins

Radiative Cooling of Structures 983
V.K. Gupta and H.R. Tyagi

Solar Collector Related Research and Development in the United States for Heating and Cooling of Buildings 991
R.K. Collier

California State Energy Efficient State Office Building 1029
B.S. Setty

CHAPTER 9 BIOLOGICAL AND CHEMICAL CONVERSION

A Quinone-Assisted Photoformation of Energy-Rich Chemical Bonds 1056
S.W. Fox, T. Adachi and W. Stillwell

Ethanol, A Clean Burning Bio-Solar Fuel - Its Production and Utilization 1075
H.B. Mathur

Electrolytic Hydrogen Production and Its Potential Utilization for Ship Propulsion 1094
C.A. Schmidt

Hybrid Chemical Concepts for Storage of Solar Thermal Energy 1109
A.C. Vialaron

Use of Renewable Fuels in the Rural Development of India 1119
G.C. Pandey

VOLUME III

CHAPTER 10 DIRECT ELECTRICAL CONVERSION

Physics and Economics of Photovoltaic Solar Energy Conversion 1132
Y. Marfaing

Realization of P-N Junction Solar Cells by an Ion Implantation Doping Procedure 1154
J.C. Muller, M. Hage-Ali and P. Siffert

Solar Energy Conversion by Fluorescent Materials for Utilization in Space Communication Systems 1167
A.K. Sen and J.S. Sehra

New Possibility of Solar Energy Conversion by Means of Refractory Thermoelements 1184
S. Abdalla, J.M. Dusseau, F. Roche and J.M. Darolles

On Direct Conversion of Sunlight to Electricity 1192
M.A. Melehy

Performance of the Noval Solar Cell Substitute for the Future 1210
S.C. Rastogi

MOS Solar Cells 1216
H.M. Kizilyalli

Some Optical and Electrical Properties of Selenium-Cadmium Sulphide Photovoltaic Cells 1230
K. Abdel-Hady, F. El-Akkad and J. Al-Dabbagh

Optimization of Graded Doping CdTe Solar Cells 1242
A.S. Bouazzi

Saur-Vidyut Kosh - The Solar Cells 1257
M. Sharon, S.G. Sharan, B.M. Prasad, A.P. Bholagir and R.R. Pradhananga

CHAPTER 11 THERMAL POWER

Solar Thermal Power Plant: Thermodynamic Analysis 1279
A.A. Samuel, U.S.P. Shet and M.C. Gupta

Solar Power Units with Screw Expanders 1293
J.M. Merigoux and P. Pocard

New Developments for Future Solar Power Plants 1318
J. Lorenz, J. Feustel and M. Kraft

Power from the Sun by Thermodynamics - French Tower Type Projects 1329
M.J. Bignon

Steam and Gas Turbines for Small Outputs Using Solar Energy 1338
K. Bammert and H. Posentrup

Turbine Selection for Small Capacity Solar-Power Generation 1351
A. Mobarek, N. Rafat and M. Saad

A Helioelectric Farm 1368
O.J.M. Smith and P.S. Smith

A Central Receiver Solar Power System for Remote Locations 1408
R.J. Holl

Investigation of Pool Boiling Heat Transfer in the Energy Carrying Liquids of Solar Energy Power Plants 1476
M.S. Abdelhafiez and M.A. Abdelsalam

The Use of Solar Furnaces in Materials Research 1443
D. Suresh and P.K. Rohatgi

CHAPTER 12 DISTILLATION AND DESALINATION

Efficiency of a Solar Still as a Function of Its Main Parameters: Production Estimation for Valencia, Spain 1474
J. Riera, J.A. Martinez-Lozano, F. Tejerina and F. Tena-Sanguesa

Solar Water Supply and Purification System for Remote Areas and Villages 1497
U.R. Nejib and H.C. Hallez

Solar Water Recovery from the Air 1526
W.G. Bulang

Double Exposure Solar Still 1546
M.M. Mousa, F.M. Abdelfattah and I.A. Sakr

Direct Solar Distillation 1561
I.D. Abdelrazek and F.H. Hammad

Solar Energy for Desalination 1570
A.M. El-Nashar

Development Plans and Conceptual Designs for Multi-Purpose Solar Desalination Units (MSDU) 1587
H.K. Abdel-Aal

CHAPTER 13 AGRICULTURAL AND RURAL APPLICATIONS

Review on Application of Solar Energy for Drying Food Stuffs 1600
K.V. Mallan, T.V. Balasubramanian and R.K. Bhardwaj

Solar Energy Utilization for Grain Drying 1615
VR. Muthuveerappan

Studies on Utilisation of Solar Energy in Hay Making of Barseem 1634
R.D. Verma and M.K. Trehan

Solar Tobacco Drying Plant 1645
L. Andreotti, A. Ciminale, G. Marchetti, D. Ruggi and A. Serravezza

Crop and Timber Drying in Malaysia Proposed Plans for Studies 1654
K.S. Ong

Energy Consumptions and Possibilities for the Applications of Solar Energy in Italian Agriculture 1666
L. Andreotti, G. Marchetti, F. Passaro, D. Ruggi, A. Serravezza, A. Di Vecchia and A. Fanchiotti

Solar Energy Utilization for Agricultural Purposes in Egypt 1677
M. Hassib

Solar Powered Water Pump for the Rural Third World 1681
J. Boldt

A Solar Battery System and Its Use for Irrigation in Egypt 1692
M.G. Osman

Utilization of Solar Energy in Dairies 1703
R.D. Verma, S.C. Sarma, V.K. Nayar and R.K. Hohli

Global Radiation Within a Crop: Pear Trees 1713
A. Guerrini, A. Lavagnini and F. Vivona

VOLUME IV

CHAPTER 14 COMMERCIAL AND DOMESTIC UTILIZATION

Street Lighting by Sunlight 1723
J.G. Sakas

The First Solar Energy System in West Germany for Industrial Utilization 1742
I. Hoffman

A Simple Water Pump, Powered by Solar Energy 1748
K. Speidel

Solar Cookers 1762
D. Buxey

Solar Energy for Cooking in Egypt 1768
M.G. Osman

CHAPTER 15 WIND ENERGY

Sow Wind Turbines and Reap the Wind 1796
O.J.M. Smith

Alternative Energy for Domestic Hot Water: Wind of Solar? 1834
A.L. Weisbrich

Comfortable Heating and Cooling by Solar Reflectors and Wind 1850
S.A. Husain

Combination of Wind and Solar Energies at the Northeast Coast of Egypt 1862
I.A. Sakr, M.S. Abdel Hafiez and A.S. Hegazy

A Low Cost Tilting Blade Vertical Axis Windmill for Rural Irrigation 1891
V. Mukhopadhyay

CHAPTER 16 NATIONAL AND INTERNATIONAL PROGRAMS

Solar Energy: UNIDO Programme of Action for Developing Countries 1901
S. Rao

Contemporary Research and Development for the Use of Solar Energy in Thailand 1911
V. Bhanthumnavin

End Use Energy in Brazil and the Solar Technologies 1928
I.C. Macedo

French Solar Energy Program and Its Application for International Cooperation 1937
M.J. Pheline

Solar Collector Development in Rwanda 1950
F. Kalos, C. Ntakirutinka, P. Mpawenayo and G. Saunier

The Impact of Solar Energy Integration on the Power System of the Kingdom of Saudi Arabia 1962
F. Sultan and F. Taher

Solar Energy for the Common Man in the Arab World 1971
D.N. Singh

An Assessment of Need for Developing and Implementing Technical and Skilled Worker Training for the Solar Energy Industry 1976
C.G. Orsak. R. Barnstone, Y. Jani, H. Gibson and J. Morehouse

The Design of the Two-Year Associate Degree Curriculum for Training Solar Engineering Technicians 2003
M.Z. Lowenstein

CHAPTER 17 ECONOMICS AND POLICIES

Economic Considerations of Solar Energy 2021
J.D. Parker and J.A. Tipton

The Comparative Economics of Passive and Active Systems 2044
F. Roach, S. Noll and S. Ben-David

An Economic Analysis of Solar Energy Alternatives 2084
D.A. Huettner, B.J. Taylor, C.A. Ingene and J.F. Horrell

Impact of Solar Heating and Cooling of Buildings on Electric Utilities 2106
A.S. Debs

Economic Analyses of Solar Water Heater and Solar Stills in Thailand 2126
P. Wibulswas

Evaluation of Solar Thermal Power Plants Using Economic and Performance Simulations 2138
N. El-Gabawali

The American Experience with Solar Energy in Centralized Electricity Generation: Technology and Economics 2162
D.L. Hurwood

Commercialization of Solar Energy Devices in Developing Countries - Problems and Prospects 2189
R.O. Onyeonwu

The Economics of Solar Space Heating in Zambia 2202
G. Lewis

Solar Energy - Will It Conserve Our Non-Renewable Resources? 2210
S. Baron

The Role of Information Media in Public Understanding of Solar Energy 2230
S. Galal

Community Characteristics Predicting Public Interest in Household Solar 2239
S. Warkov

Accelerated Commercialization of Solar Energy - A New Priority for International Action 2251
L. Rishnell and N.D. Jorstad

Why Not Solar Energy? 2270
T. Yarman

CHAPTER 18 WORKSHOP REPORTS

Report 1: Architectural Applications 2286
Group Leader: A. Bowen

Report 2: Agricultural and Industrial Applications 2288
Group Leader: E. Bilgen

Report 3: Energy Conversion 2290
Group Leader: Y. Marfaing

Report 4: Solar Planning 2292
Group Leader: A.J. Parker, Jr.

Report 5: Solar Energy Commercialization 2304
Group Leader: N.W. Lutkefedder

PLANE INCIDENT SOLAR RADIATION. EFFECTS OF SLOPE AND ORIENTATION

J.I.Jiménez*, C.Romeu-Nedwed**, J.E.Llebot and J.Casas-Vázquez*
*Dpto. Termología, Fac. Ciencias, Univ. Autónoma de Barcelona, Bellaterra (Barcelona) Spain
**Instituto de Investigaciones Pesqueras, Paseo Nacional, s/n°
Barcelona-3 Spain

ABSTRACT

The incidence of solar radiation on a plane surface is function of latitude and longitude and of the surface's slope and orientation. The function relating the geographical coordinates with the incident energy is well known in the case of a surface with horizontal orientation, but efforts have yet to be done to determine the increase or decrease of captured energy when the surface is not horizontal. To this end and by means of a pyranometer ecuatorially mounted upon a modified theodolite system, measurements of the global and diffuse radiation flux at various tilt and azimuth angles have been made in clear days. The resulting data of several days of measurements have been compared with the ones obtained by TEMPS and COULSON [1] by a similar method. Furthemore the data have been used to evaluate the validity of ROBINSON'S [2] and DAVE'S [3] models. We consider the possibility of introducing some modifications in both models to achieve a better agreement with the experimental results. The mathematical formulation has been kept as simple and general as possible and distinction has been made between direct solar radiation, diffuse sky radiation and diffuse ground reflection by comparisib with the measurements obtained by pyranometers with fixed orientations and slopes, and with considerations of the surface's albedo.

INTRODUCTION

The problems concerning radiation balance, at present, has ra-- ther a poor and one-sided solution. The solar energy captured by a planes surface (solar collectors, roof of a house, etc.) and thus utilizable for energy conversion is a function of the position of the surface with respect to the earth surface and of its orientation. Because of the complexities associated to precise calculations of the flux of the solar radiation incident on slopes of differents orientations, as a rule, most authors have assumed that the radiation is isotropic in all cases where practical calculations of the skyligh and reflected radiation to the slopes are involved.

The nature of the flux of scattered and reflected radiation, and the effective long wave radiation of the slopes have been investigated with detail [4] and the precise theoretical formula for these calculations have been derived by various workers only for the mentioned hypothese that the radiation scattered and reflected is isotropically distributed. Nevertheless this assumption is very simple and the computation of scattered and reflected radiation fluxes of the slopes of different orientations, as well as the effective radiation of the slopes is rather complicated because the fluxes being considerably non-isotropic.

The non-isotropic character of the skyligh radiation and ground reflected radiation produces that the radiation flux captured for the plane if it is not oriented horizontally present maximum intensities near the Sun and minimum intensities in the direction normal to that of the Sun. Therefore, in the present paper, we investigate the dependance of the radiation fluxes upon the slopes and orientations. Similarly, in this study we consider the validity of models based on the isotropic assumption by comparison with measurements carried out in our laboratory. On the other hand, some considerations over the non-isotropic character of the diffuse radiation have been tested by utilizing the correction suggested by Temps and Coulson [1] in the model of Robinson [2] .

THEORETICAL REMARKS

The flux of radiation incident upon a tilted flat surface consist of the flux of direct solar radiation, flux of diffuse sky radiation and sun as well as sky radiation reflected by the ground underlying the atmosphere which is incident on the tilted surface. All these fluxes depend on the latitude of the place, inclination and altitude of the Sun and the receiving surface, cloud coverage and optical properties of the atmosphere and underlying surface.

Numerical simulation of the solar energy received by a tilted surface are being developed by a research group in the University Autonoma of Barcelona using several techniques developed for atmospheric radiation transfer studies, [5],[6],[7],[8].

Unfortunately, still now, no results are availables in order to obtain general information about the potential of the atmospheric modeling for solar energy studies, for what, as a simplification, an investigation with a roughly approximation have been carried out.

The portion of the extraterrestrial solar flux which is transmitted directly through the atmosphere can be estimated using the following expresion [1], [9]

$$F_D = (R_o/R)^2 \int_0^\infty F_{o\lambda} e^{-(\tau^R+\tau^A+\tau^W)\sec\theta} \qquad (1)$$

where F_D is the flux of directly transmitted radiation, R is the actual Sun-Earth distance and Ro is the mean value of R for one year, being the rate $(R_o/R)^2$ a correcting coefficient for the spectral solar constant $F_{o\lambda}$ (see [10] for values of the correcting coefficient). τ^R, τ^A and τ^W are the optical thickness for the molecular scattering, atmospheric aerosols and water vapor, respectively; θ is the zenith angle.

Following Kondratyev [9] $\tau^R = 0.00879\ \lambda^{-4.09}$, λ wavelength in micrometers. The aerosol optical thickness τ^A can be approximately computed from the assumption $\tau^A = \gamma\ \lambda^{-b}$ where γ is the turbidity coefficient which can be measured from the integral flux of solar radiation reaching the earth surface. The present accepted values for b shows that the magnitud of this parameter can change within the limits $2 \leqslant b \leqslant 4$. Finally, in order to determinate τ^W a vertical profile would be desirable. Not being for all the experiment days available we took the approximation developed by Hann [11] that is consistent with other author in the literature. For this author the amount of precipitable water in a vertical column through the atmosphere will be obtained by W = 0.25e where e is the water vapor pressure measured at the surface in millibars.

For computation of diffuse sky radiation flux the very simple method by Albrecht [12] is applied in our paper

$$F_d = K\ (F_{w,\gamma} - F_D)\ \sin\gamma \qquad (2)$$

where F_d is the flux of sky radiation, $F_{w,\gamma}$ is the flux of direct solar radiation at sun elevation γ which only has been attenuated by water vapor, K is a empirical coefficient given by

$$K = 0{,}5\ \sin^{1/3}\gamma \qquad (3)$$

and F_D is the direct flux radiation

As we want to evaluate both fluxes, the direct and the diffuse on a tilted surface we have for the direct

$$F_{D\varepsilon} = F_D \cos\psi \qquad (4)$$

where ε is the angle of the surface normal to the slopes makes with the zenith and ψ the angle of the surface normal with the Sun.

For the diffuse flux an assumption has been made about the sky radiation stating to be isotropically what is rarely valid. With a maximum near the horizon or near the sun and a minimum near the local zenith we can justify our isotropically distribution only for practical purposes with a relatively low intensity.

The exact expression for the diffuse flux of sky radiation is

$$F_{d\varepsilon} = \int_0^{2\pi} d\phi \int_{h(\phi)}^{\pi/2} I(h,\phi)\cos\alpha \cos h \, dh \qquad (5)$$

where ϕ is the azimuth, $I(h,\phi)$ the intensity, α the angle of incidence of radiation at the tilted plane and h the angular heigh, will have in the isotropically case the solution

$$F_{d\varepsilon} = F_d(1/2)(1 + \cos \varepsilon) \qquad (6)$$

Finally, the reflected radiation from the ground incident on a sloping surface can be isotropically distributed as well so

$$F_{g\varepsilon} = (F_D + F_d)\ A\ (1/2)\ (1-\cos \varepsilon) \qquad (7)$$

where A is the albedo of the surface.

The total flux will be given with adition of the three calculated components.

$$F_\varepsilon = F_{g\varepsilon} + F_{d\varepsilon} + F_{D\varepsilon} \qquad (8)$$

With the above mentioned method we can compute the flux of solar radiation on any tilted surface but only in the case of isotropically assumption for both, diffuse sky radiation and ground reflection. Following the technique of Temps and Coulson [1] to introduce some empirical modification to consider the evident non-isotropic character of the field of diffuse radiation we can consider that the anisotropy is manifest as a increase of intensity near the horizons, an increase of intensity in the circumsolar region of the sky and, finally, by an angular dependence of the albedo of the underlying surface. The empirical correction terms introduced in the expresions (6) and (7) and suggested by the mentioned authors have been verified with our data set and permit to write the diffuse sky radiation by the formula

$$F'_{d\varepsilon} = (1/2)\ F_d\ (1+\cos \varepsilon)[1+\sin^3(\varepsilon/2)][(1+\cos^2\alpha)(\sin^3 Z)\ ] \qquad (9)$$

where the factor $1+\sin^3(\varepsilon/2)$ take account the fact that the diffuse radiation is approximately 40% greater near the horizon that at the zenith and that the gradient is strongest at low elevation angles. The factor $[(1+\cos^2\alpha)(\sin^3 Z)]$ where Z is the solar zenith angle takes account the brightening of the sky in the vicinty of the Sun.

The ground reflection have been modified by introducing the factor $\{[1+\sin^2(Z/2)]\ [|\cos a|]\}$ to take account the variations in the albedo of surface with the angular position of the sensor. Here a is the azimuth angle with respect to that of the Sun. Then (7) becames

$$F'_{g\varepsilon} = (F_D+F_d)A(1/2)(1-\cos\varepsilon)\{[1+\sin^2(Z/2)][|\cos a|]\} \qquad (10)$$

and the modified total flux is

$$F'_{\varepsilon} = F_{D\varepsilon} + F'_{d\varepsilon} + F'_{g\varepsilon} \qquad (11)$$

The data set acquisited were used to assess the validity of the model described above and to prove the validity of the modifications by Temps and Coulson [1] for other geographic locations and similar atmospheric conditions.

MEASUREMENTS

The data necessary for the computations concerning the angular distribution of the energetic intensity of the scattered and reflected radiation were obtained by means of direct measurements carried out with Kipp-Zonen's pyranometer. The view angle is 2π stereoradian and measures the global irradiance (direct plus diffuse) over an horizontal surface. The general characteristic are the following

1.- Moll thermopile consisting of 14 constantan-manganina thermoelements
2.- Two glass dome of 30 and 50 mm in diameter filtering short--wave solar radiation
3.- Time constant 4 seconds

More details about these type of radiometers functioning and operation are given by most authors [9], [13]

Both solarimeters were convenienty calibrated before the experiments according to International Pyrheliometer scale 1956. During the experiment these instruments are recalibrated periodically by means of actinometric comparation. One of the pyranometer were used for comparation and it is mounted in fixed orientations and slopes for some time periodes. For the data reported here the pyranometer was mounted ecuatorially by use of an apropiately modified meteorological theodolite by which any desired orientation and slope of the sensor surface could be obtained. The system was installed over the roof of one of the University's building where the exposure of the instruments was good, without any appreciable obtacle for the incoming radiation. It was installed at a height of 2 meter over the roof level.

Outputs from the two sensors were recorded continuously on a BD-5 Micrograph four-channels automatic recorder on strip chart. Data was initially extracted manually from chart record but later two integrators was installed for completing the automatic data collection.

The complete system to carried out the measurements was installed, as already mentioned, at the roof of the University building and the set of measurements studied here was taken under conditions of clear skies on a serie that begun 20 January and finish 25 March 1978. Similar conditions in the atmosphere characteristic were considered in the selection of the days that compose our data set. Under these restriction four days are selected for the study that in all of them the precipitable water content vary between 0.7 and 0.9 cm and the turbidity coefficient between 0.04 and 0.05.

For each day we are analyzed a serie of data with the pyranometer oriented such that the surface normal to the sensor made angles of 0 to 90 with an increment of 10°except for 45° with the horizontal and 0 to 365 with an increment of 45° with the azimuth of the Sun. For each measure (two minutes) we carried out one with the sensor completely exposed to the Sun and one with the sensor shaded by a small (20 cm in diameter) opaque disk. The set of 64 points has been normalized for the changes in the Sun elevation during each serie. Simultaneously we measured the solar and diffuse radiation with selected slopes and orientations with other pyranometer, for comparation.

Data analysis

The data obtained with the above mentioned method for five days with similar meteorological conditions were used for construct maps for differents solar elevation. The fig 1 and 2 show the selected solar elevation $\gamma = 30°$. The fig. 1 shows the map for the global solar radiation incident on the sensor for the serie of slopes and orientations. The fig. 2 shows the diffuse radiation from the sky and ground when the sensor is shaded for the direct solar radiation.

We may see in the figure 1 that exist a little assymetry around the principal plane as a difference with the bibliography [1]. The change of flux from minimum to maximum is of an order of magnitude as a results of the great dominance of the flux of direct solar radiation over the skyligh and related with the proportionality of the solar radiation incident over the surface sensor with the $\cos\psi$, being ψ the angle between the surface normal and the direction of propagation of the radiation. In accordance with other authors [1], [2], the orientation of maximum flux appear when the zenith angle of the surface normal is

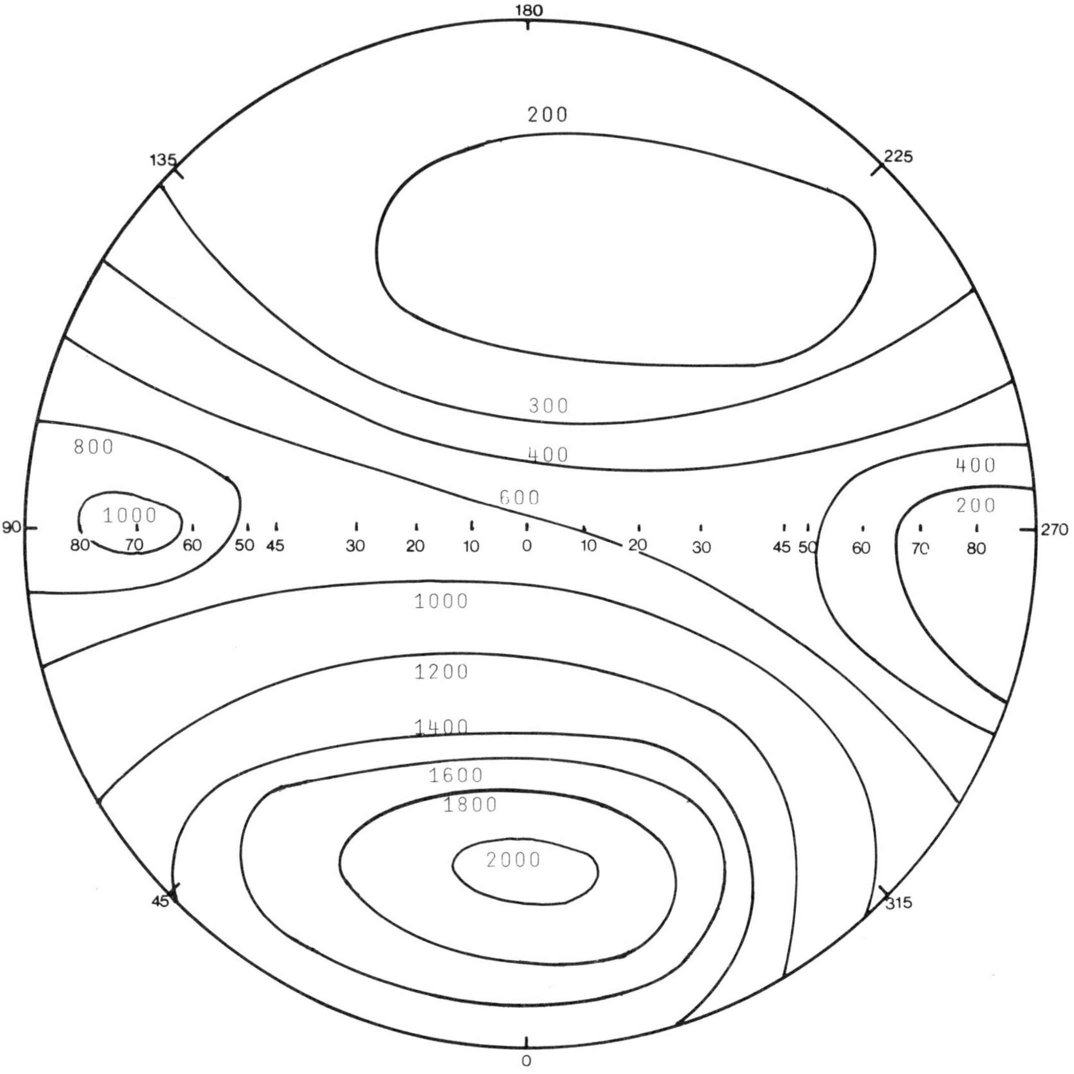

Fig. 1.- Total solar radiation on a pyranometer oriented at various combination of slopes and azimuth angles from clear skies. Slope is distance from center. Azimuth is with respect to the Sun. Units are W/m^2. Sun elevation 30°

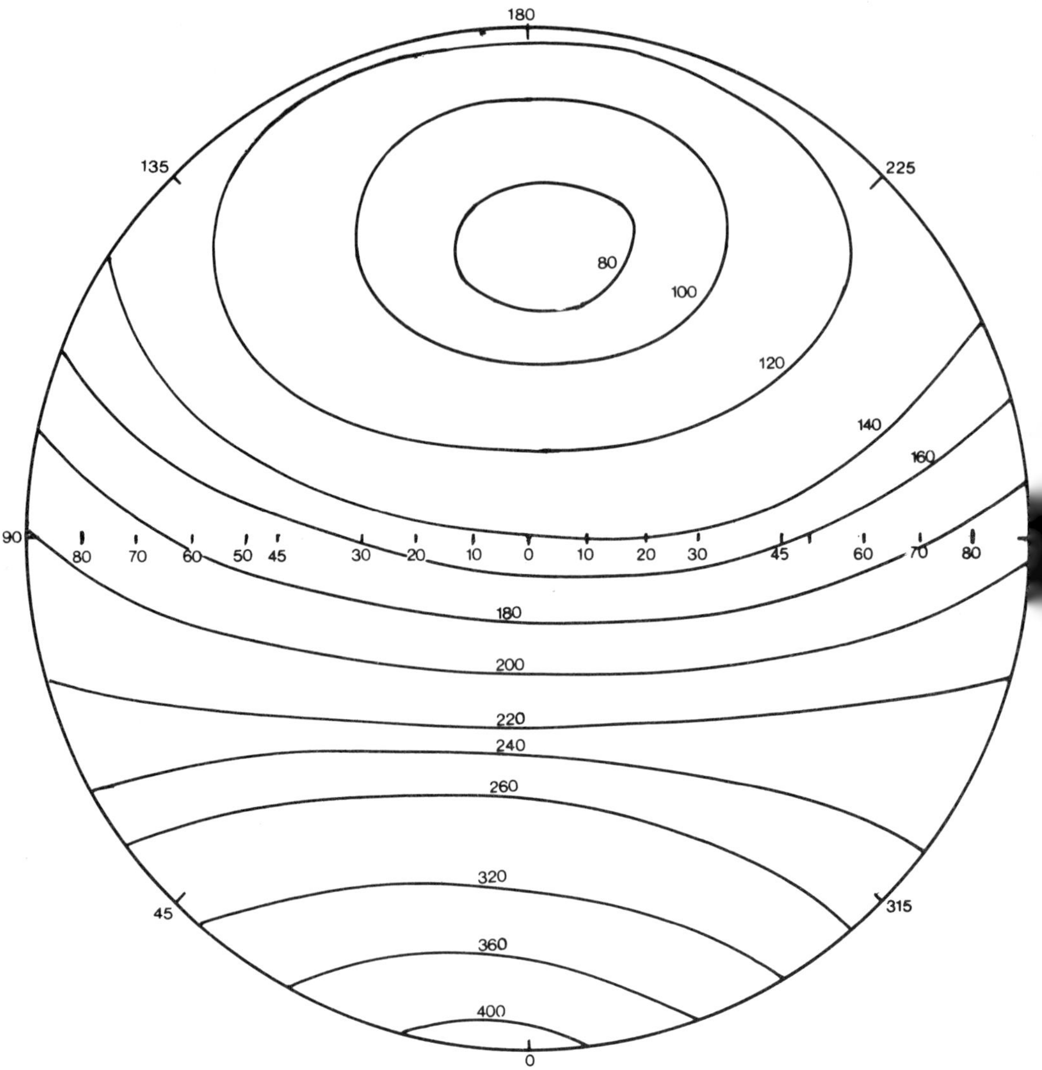

Fig. 2.- Diffuse radiation from sky and ground on a pyranometer as in Fig. 1

somewhat greater than that of the Sun, as a consequence of the increasing ground reflection and diffuse radiation near the horizon.

Figure 2 shows the map of diffuse radiation that shows the evident anisotropy of this type of radiation, comming from the sky and the ground. For a horizontal surface the diffuse radiation contributes to total flux approximately with a 14,5% of the total solar radiation incident on a horizontal surface, about 18% for a surface normal to the propagation of the solar beam and 100% for the surface oriented 180° with the orientation of the Sun. The anisotropy of this field is due principally to the well known horizon brightening of clear atmosphere and in our study, the anisotropy exhibited by the ground reflexion plays a very important part. All these results are consistent with the measurements founded in the literature.

Comparison with model calculations

A very important part of this work is the comparison of our data with model calculations based upon the Robinson model and in the modification that propose the excellent paper by Temps and Coulson [1]. These models have been before described and the data set obtained by measurements were used to evaluate the validity of these models. On the other hand, several modifications are studied for amelioration of the results of the models but any empirical modification produces substantially better results of the model proposed by Temps and Coulson with the assumption that exist three differents regions that generate the anisotropy in the diffuse radiation. The increasing of intensity near the horizon, the increasing of intensity in the circumsolar region and the angular dependence of the reflectivity of the ground surface can produce the observed anisotropy. As we mentioned above in the theoretical remarks the model modified introduce the factor $[1+\sin^3(\varepsilon/2)]$ for the variations in the intensity due to the slope of the surface and the factor $[1+\sin^2(Z/2)]\ [|\cos a|]$ for the anisotropy of the surface reflection, where a is the azimuth angle with respect to the Sun.

As is obvious, in the flux of direct radiation the two models gives the same results due to the fact that the modifications have been introducted in the diffuse radiation. As consequence we selected four azimuth angles to show the differences founded between the measured data and the computed data with the two models. The fig. 3 shows the diffuse radiation for 0°, 45°, 135° and 180°. Here we may see the good results obtained when the correction for the anisotropy were introduced. The modifications of the Robinson's model suggested by Temps and Coulson provides a better approximation to the experimental results

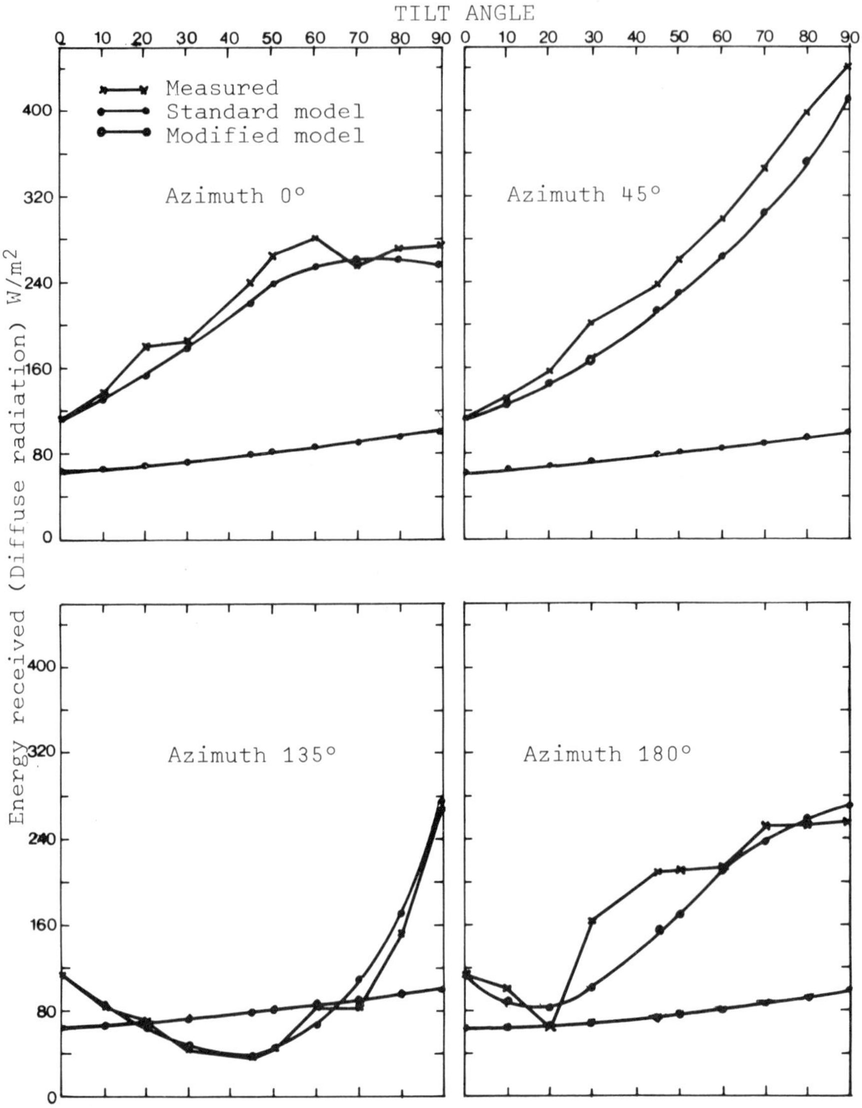

Fig. 3.- Diffuse radiation from sky and ground as a function of the tilt angle of the surface sensor for four azimuthal orientations. Comparison with models calculations.

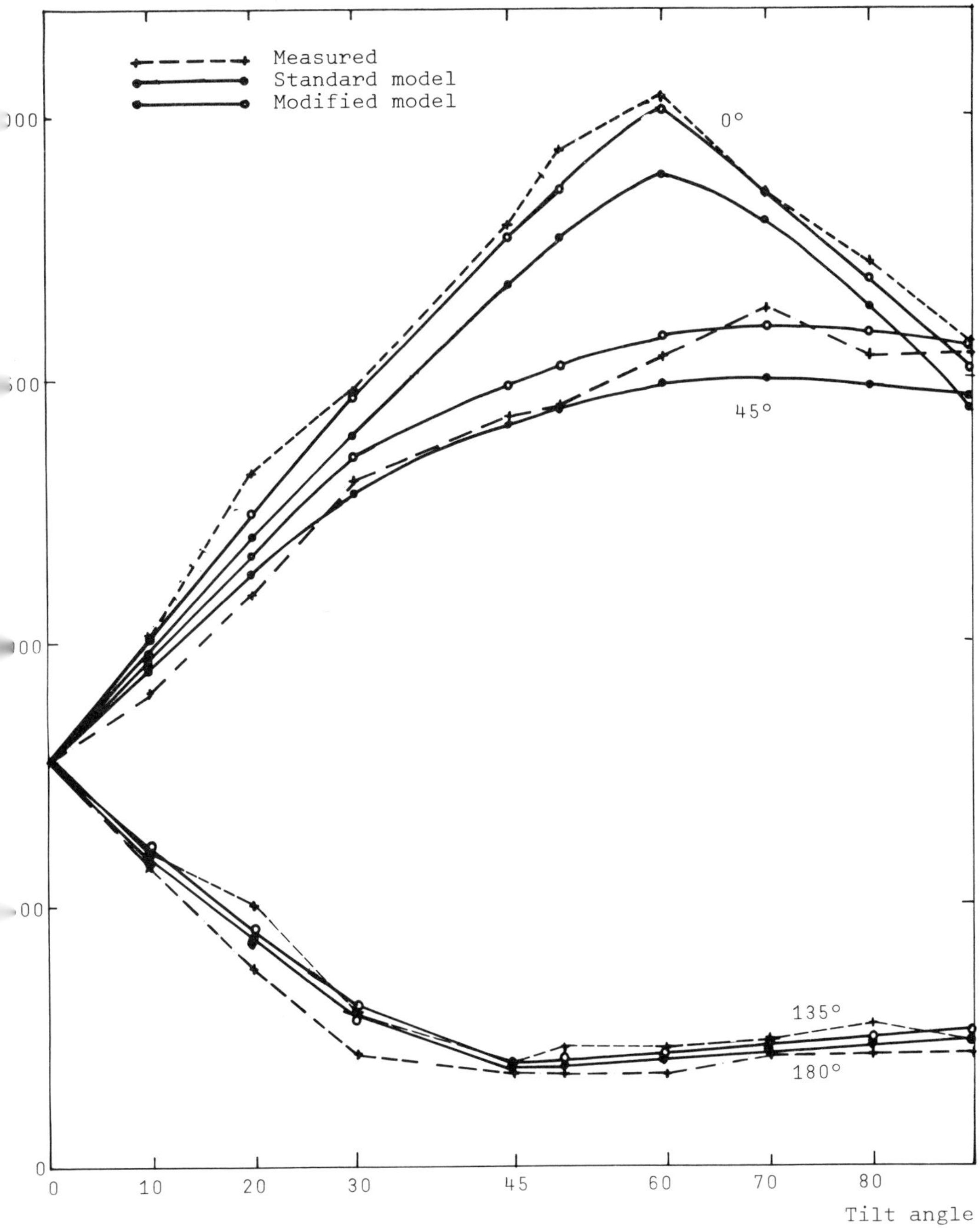

Fig. 4.- Total solar radiation as a function of the tilt angle of the surface sensor for four azimuthal orientations. Comparison with model calculations.

JIJ

that the original model.

The fig. 4 shows the results obtained for the total flux. We may see that the differences are very small and only in the azimuth of 0° and 45° the corrected model became a very important improvement of the Robinson's model.

In the comparison of figures 3 and 4 we can compute that the importance of the diffuse radiation is relatively small in the total flux of global radiation. For this reason, very much important in the 135 and 180°azimuths, the results of both models becomes practically the same, and in fig. 4 we represented only the results of 135° azimuth.

The very goods improvements founded in the 0° azimuth angle for the corrected model, permit a very important application for the flat-plate collectors for conversion of solar energy that it is possible to compute and predict the flux of solar radiation on a sloping surface for any orientation with good approximation.

CONCLUSIONS

To this work we may to extract inmediately a very important conclusion: The empirical corrections of the Robinson's model suggested by Temps and Coulson and calculated from data obtained in Davis (California) and for grass cover, may be applicated to a very different type of ground cover, as our asphalted floor in the roof of the University's building, and for a similar latitude, as them of Bellaterra (Barcelona). These results permit to exted the applicability of this method for many geographic locations, although, for this moment, only in similar atmospheric conditions.

A completely solution of the radiative transfer equation may be very much precise that these calculations and an adjustment of the constant involved will be necessary for the application of this method. At present, several studies are being carried out for the application of the Delta-Eddington approximation to achieve a general solution of the radiative transfer equation and on the other hand, several experiments are being carried out for all the meteorological conditions in order to improve the method used and for introducing some empirical modifications of the models to obtain better approximations in differents degrees of cloudiness.

TABLE 1

CORRELATION BETWEEN HOURLY DATA OF GLOBAL AND NET RADIATION

d/m/y	η	$\bar{R}_i$	$\bar{R}_n$	σ_{R_i}	σ_{R_n}	a	b	r
6/9/77	10	515	306	189	129	36	.52	.77
7	10	528	305	199	126	12	.56	.87
8	9	513	334	191	114	89	.48	.80
September	29	519	315	186	120	45	.52	.81
25/10/77	8	448	222	133	75	-11	.52	.92
26	8	361	175	142	90	-51	.63	.98
27	8	300	147	71	44	-25	.57	.93
28	8	373	204	119	81	-36	.64	.95
29	8	340	182	92	60	-34	.63	.97
October	40	365	186	119	73	-26	.58	.95
21/11/77	8	94	52	35	24	-10	.67	.97
22	8	218	107	139	73	- 3	.50	.96
23	7	247	123	62	48	-63	.75	.98
24	6	386	152	93	47	-28	.47	.93
November	29	225	105	134	61	9	.43	.94
19/12/77	6	435	116	177	37	42	.17	.82
20	6	346	110	82	37	-34	.41	.92
21	6	350	110	92	35	2	.31	.81
22	5	228	70	112	36	3	.29	.91
23	6	172	80	48	28	-16	.56	.96
December	29	309	98	141	37	31	.22	.82
24/1/78	5	201	102	149	80	- 6	.53	1.00
26	6	395	168	70	62	-124	.74	.83
27	7	356	147	107	64	-55	.57	.95
28	6	222	100	89	44	- 4	.47	.94
29	7	175	89	79	53	-22	.63	.95
January	31	272	122	130	65	- 4	.46	.92
22/2/78	8	521	246	148	92	-71	.61	.97
23	8	537	260	159	91	-49	.58	.96
24	7	376	186	151	89	-36	.59	1.00
25	8	150	83	56	30	7	.51	.95
February	31	397	194	203	104	- 5	.50	.98

20/3/78	9	259	126	100	49	2	.48	.97
21	9	539	277	157	108	-91	.68	.99
22	9	591	281	207	131	-94	.63	1.00
23	9	174	87	50	30	-14	.58	.98
24	9	462	239	272	170	-47	.62	1.00
March	45	405	202	234	133	-25	.56	.99
24/4/78	9	288	154	128	83	-29	.63	.98
25	10	380	207	192	122	-14	.58	.92
26	11	467	280	300	208	-40	.68	.99
27	10	413	242	245	169	-41	.69	.99
28	9	532	291	178	115	-26	.60	.92
29	9	555	337	226	160	-35	.67	.94
April	58	439	252	230	156	-36	.66	.97
September 77 April 78	292	376	192	200	126	-28	.59	.93

TABLE 2

CORRELATION BETWEEN MEAN HOURLY DATA OF GLOBAL AND NET RADIATION

	η	$\bar{R}_i$	$\bar{R}_n$	$\sigma_{\bar{R}_i}$	$\sigma_{\bar{R}_n}$	a	b	τ
September 77 April 78	37	363	181	135	83	-23	.56	.91

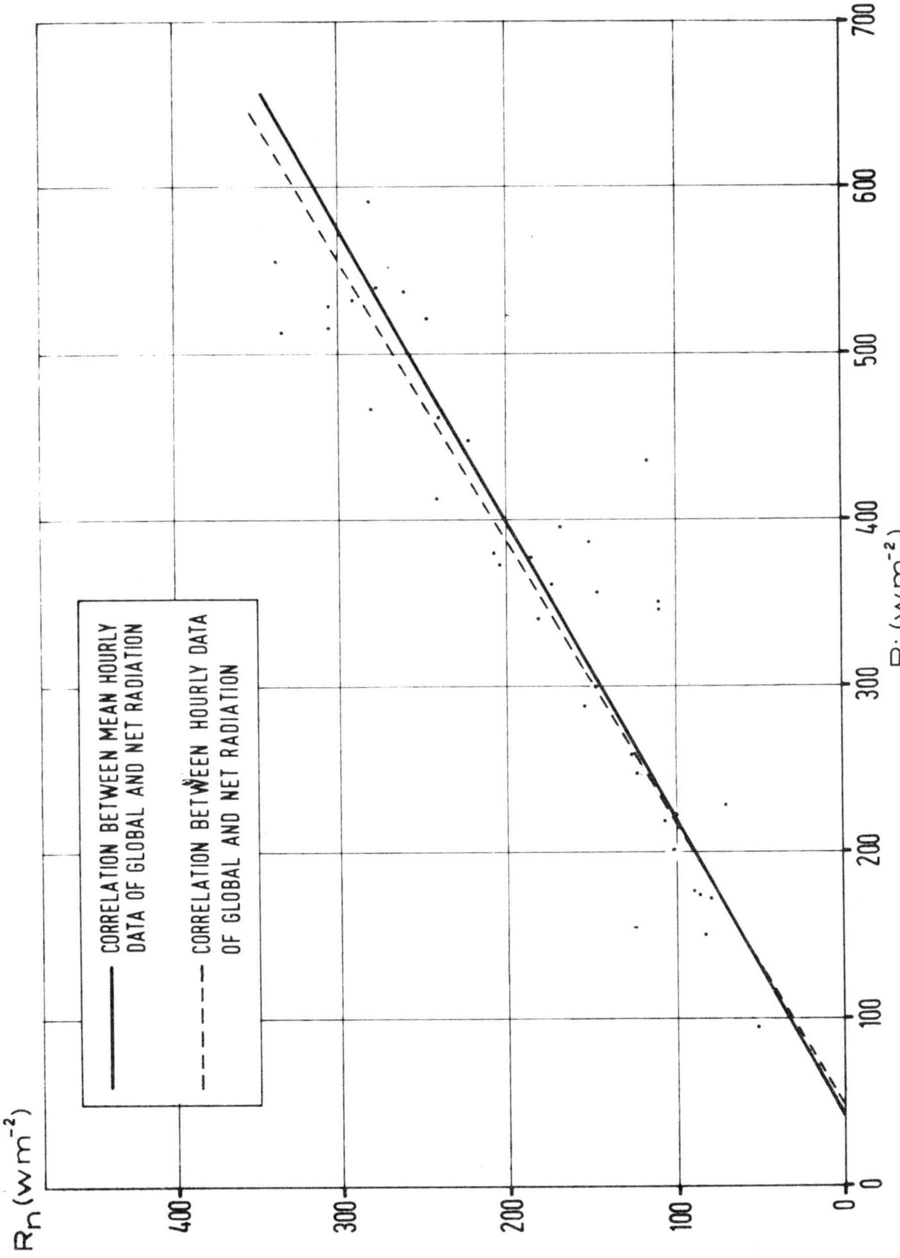

R_n (wm^{-2})
0
100
200
300
400
R_i (wm^{-2})
0
100
200
300
400
500
600
700
CORRELATION BETWEEN MEAN HOURLY DATA OF GLOBAL AND NET RADIATION
CORRELATION BETWEEN HOURLY DATA OF GLOBAL AND NET RADIATION

FIG. 1

The radiometer, covered by a double polyethylene cap, is composed by a 250 junction thermopile and by two plackned plates for detecting the thermal effect of the radiation. The radiometer is also equipped with a heater ring for avoiding the dew formation on the cap.
Either the pyranometer input or that relative to the radiometer were sent to an integrator and the data recorded hourly.
At the same time as the net and global radiation measurements, temperature and humidity were recorded by an Assmann psychrometer. The soil temperature at a 1,10,50 cm depth was obtained by a SIAP gas pressure thermometer and the heat flux into the soil was also measured by two soil heat flux plates built by the Middleton Instruments at a depth of 1 cm and 11 cm.

RESULTS

Measurements were effected for a total of 292 hours for 37 measurement days (on the average 8 hours a day); the analysis was carried out performing linear correlations among the hourly data of R_i and R_n referred to the same day, and those relative to the same month and those relative to the entire period of measurements.
The results are summarized in Table 1. A correlation was also made between the mean hourly values of R_i, R_n for the whole period of measurements. The results are reported in Table 2.
The R_i and R_n values are in W m-2. The symbols used have the following meaning: n : number of recordings; $\bar{R}_i, \bar{R}_n$: mean hourly values of R_i, R_n; σ_{R_i}, σ_{R_n} : standard deviations of R_i, R_n; a : intersection point of the correlation line ; b: regression coefficient; r : correlation coefficient.
We used only couples of data in which $R_n > 0$.
In fig.1 is reported the correlation line between the mean hourly values $\bar{R}_i, \bar{R}_n$ (with the data relative to it); for comparison the straight line relative to the single hourly values was also drawn.

COMMENTS

A first result emerging from this paper is that generally a good linear correlation exists between global and net radiation with minimum values during the months of September and December. This is probably due to the fact that in September the global radiation data used were those recorded by a Robitsch pyranometer at about 2 km away from the place where the net radiation measurement was carried out and this might have caused a certain error.
As for the low value of the correlation coefficient occurred in December, we are trying to find out whether this is due to any equipment factors or to any other one.
The value of b, representing the tangent of the angle that the correlation straight line forms with the x axis (R_i), seems to have a wave behaviour which may be however verified during an entire annual cycle.
The correlation straight lines between the hourly values of R_i, R_n and their mean values, relative to the entire measurement period are very similar. This allows to use the daily integrals instead of the hourly values where the hourly recording is not effected.

CONCLUSIONS

Linear correlations were carried out between hourly values and mean ones of the global and net radiation. Generally the correlation coefficients are satisfactory (> 0.90).
The monthly regression coefficients show a regular and presumably wave behaviour.

Acknowledgements

Translation : M. Mastrangelo; M. Roselli
Diagrams: P. Mollo
Setting up of measurement equipment: F. Belardinelli, I. Di Menno, C. Sciattella, M. Viterbini.
Measurements: F. Antonelli; I. Di Menno; M. Viterbini.

REFERENCES

1) Fitzpatrick E.A., and Stern W.R. - Net Radiation estimated from global radiation. UNESCO, 1973, Plant response to climatic factors, Proc. Uppsala Symp., 1970

2) Davies J.A. - A note on the relationship between net radiation and solar radiation. Quart. J. Roy. Met. Soc., 93, 1967, 109.

3) J.J. Jimemez, D. Segarra, J. Casas-Vazquez - The relationships between long-wave, net and solar radiation - COMPLES, Revue Inter: d'Heliotechnique, 1er semestre 1977, pp. 8-12.

ATMOSPHERIC POLLUTION AND SOLAR RADIATION

By

M.S. Abdel Salam and N.A. Higazy
National Research Centre, Cairo

ABSTRACT

Solar energy available at the earths surface chiefly depends on the received amount of solar radiation. The latter is, naturally, a very changeable climatic element.

Man, as ever with nature, deteriorately accelerated these changes through his industrialization and urbanization programmes.

Atmospheric turbidity was measured at three sites in the Helwan region during the months of February through December 1967. Results showed that the atmosphere over Helwan is very turbid. The turbidity factor varies between 3.86 in February (for air mass m = 1) to 6.62 in July (for air mass m = 2).

A comparison was made, for the air mass m = 1.7, between the monthly values of turbidity obtained in that study, and those averaged before industrialization, over the years 1922-1927. The turbidity factor were found to be almost doubled in 1967. This indicates that industrialization of the area has not only badly damaged, the Helwan atmospheric environment, but consequently reduced the received amounts of solar radiation.

Hourly measurements of total direct solar radiation received at Cairo earth's surface were also made during the years 1961-1962 and 1974-1976 at the roof of the National Research Center. Measurements made during the latter priod showed that the received amounts of solar radiation were sometimes decreased by up-to 21% of its corresponding values recorded during the 1961-1962 series of measurements. This decrease could only be attributed to the rapid increase in population and the consequent increase in their living facilities and activities which, in turn, increased the atmospheric pollution load. The maximum decrease was found to take place during autumn and winter times. These results were further supplemented through calculations of linke Turbidity Factor.

INTRODUCTION

Atmospheric turbidity is a good indicator for the degree of atmospheric contamination by aerosols. Its determination, therefore, is a simple toole for monitoring air pollution. Atmospheric turbidity is a rather complex factor as it depends on air mass, season of the year, latitude, and local atmospheric conditions. Its value must, therefore, show marked variation with time and place with a normal annual variation having its maximum value in summer and its minimum in winter. However, the greatest changes are those which arise from atmospheric pollution. For clean air above the mountains of Zugspitze (2962 m, the turbidity factor varies annually from 1.8 to 2.1, where as at Kew, near London, this value changes from 4.1 in winter to 5.1 in summer [1].

Linke and Boda [2] were the first to introduce this term of turbidity factor into atmospheric sciences, in an attempt to isolate the light extinction caused by suspended particles in the atmosphere from that resulting from the molecular scattering and gaseous absorption. Seven years later Angstrom [3] also reached the same objective through studying the atmospheric transmission of solar radiation and its relation to dust-content of the air. Voltz [4] constructed his sunphotometer which greatly simplified the measurements of atmospheric turbidity. Sunphotometers, then later in 1971, were introduced into the expanded United States network and by April 1973; it was in use in other countries of the world.

In Egypt, the first attempt to study the effects of dust-laden air on the incoming solar radiation, was made in 1961 [5] through a one year measurements of the received solar radiation and dust concentration at the roof of the National Research Center, Giza.

Suspended particulate matter were later found to appreciably decrease the amount of solar radiation received at the heavily industrial area of Helwan [6]. This finding led to the necessity for determining the turbidity factor in an attempt to estimate the changes in the Helwan environment caused by the recently established industries in the area.

Later measurements were also made of the direct solar radiation received at the site of the National Research Center during the period of May 1974 to April 1976. The objective was to estimate the effect of urbanization and increasing population density on the received amounts of the incoming solar radiation. Measurements were made at a height of about 30 meters above ground-level to avoid the effects of ground dust being picked up by wind or moving traffic.

All measurements made during these studie are tabulated as an annex to this paper.

ATMOSPHERIC TURBIDITY

Linke's turbidity factor τ is a measure of water vapour and dust content of atmosphere. It relates the extinction for total radiation over all wavelengths to the extinction of a dry (Rayleigh) atmosphere as follows

$$I = \frac{I}{F} I_o \exp. \left[- \tau \bar{a}_{R(m)} m \right] \qquad (1)$$

where I is the total solar radiation over all wavelengths measured with a normal incidence pyrheliometer.

F is the correction factor for the mean solar distance. It is already calculated for different days and latitudes in the IGY Instruction Manual [7].

I_o is the solar constant, taken as 1.98 cal/cm^2/min $\bar{a}_{R(m)}$ is the mean value over all wavelengths of the Rayleigh exinction coefficient.

m is the optical air mass.

In other words,

$$\tau = \frac{\log I_o - \log I - \log F}{m \, \bar{a}_{R(m)}} \qquad (2)$$

or $\tau = P(m) \, (\log_{10} I_o - \log_{10} I - \log_{10} F)$ (3)

where $P(m) = \left[\bar{a}_{R(m)} \, m \log e \right]^{-1}$

Values for P(m) for different air masses can be obtained from table 6 of the IGY Instruction Manual [7].

RESULTS AND DISCUSSION

The actinometer observations taken every hour on clear days, at three sites in Helwan during 1967, were plotted graphically for each day using Bauguer equation.

$$I = I_o \ e^{-am}$$

By interpolating and extrapolating the resulted straight lines, "I" was daily estimated for air masses m = 1, 1.7 and 2 and hence the linke turbidity factor was calculated. Obtained results from the three sites were then monthly averaged to represent conditions over Helwan area (Fig. 1) which shows that turbidity slightly increases with air mass. This is rather expected since the amount of incident solar radiation suffers more attenuation for higher air mass as a result of the relatively higher path-length of radiation. Results also show that, generally speaking, the values of turbidity throughout the 11 months of this study varied between 3.86 in February (for air mass "1" i.e. around noon time) to 6.62 in July, (for air mass "2" i.e. around 9 a.m. or 3 P.m). Compared with Ribinson's values, the atmosphere over Helwan is very turbid through-out the day hours in most months of the year. Robinson's values [7] considers the atmosphere as being clear when its turbidity factor equals 1.5 and very turbid when this value is 5.0. It is clearly noticed that for all air masses, the turbidity factor highly increases in warm months than in cold ones. This indicates that the atmosphere over this area is highly loaded with particulate matter in the warm months. It must therefore be of great interest to find out whether these huge amounts of suspended particulate matter are naturally dispersed into the atmosphere from the sandy and hilly lands surrounding the study area or as a result of the recent industrialization of the Helwan region.

I-A: Turbidity in the Helwan Atmosphere Before And After Industrialization

Continuous pyrheliometric measurements were made at Helwan observatory throughout the years 1922-1927 El-Nesr [8]. Using these date, the linke turbidity factor was calculated for every clear day, within these six years and for the air mass m = 1.7 (Table 2). It was therefore found that turbidity, within that period, varied between 2.1 and 3.5 with a slight maximum in summer than in winter. The turbidity factor varied from one year to another, during spring months, probably depending on the number and intensity of the passing Khamasin storms.

Figure [2] shows the monthly variation in the turbidity factor for the air mass m = 1.7 both for the mean values of the six years calculated by El-Nesr and for those calculated in the present study. It shows clearly that the 1967 monthly values of turbidity are almost double the values obtained from measurements made during the years 1922-1927. This clearly indicates the effects of Helwan industries on polluting the atmosphere with particulate matter and consequently reducing the amounts of solar radiation received at the earth's surface. Such effect is more evidenced by the marked variation found in the monthly values of turbidity during 1967. This marked variation could only be explained by the variation in the amounts of industrial pollutants discharged into the air and the monthly variation in the atmospheric capability for carrying and dispersing these dust particles.

II: The intensity of the total solar radiation received at normal incidence during various hours of clear days at the site of the NRC was also measured and then monthly and seasonally averaged for the years 1961-1962 and 1974-1976 Figs. 3,4 & 5). Assuming that weather elements were, in the average, almost the same at the corresponding months of the two study periods, the difference in the received amounts of the direct radiation could then only be attributed to the differences in the atmospheric contents of particulate matter.

Results show that, during the relatively cold months of autumn and winter, the received average amounts of direct radiation during the period of 1974-1976 were less than that found during 1961-1962. This decrease, therefore, must be largely due to the local atmospheric dust contents, resulting from the rapid urbanization taking place in this extensive residential district and the consequent increase of its population and their day-to-day chores of living activities. The highest decrease in the amounts of received direct radiation occurred during winter time while the summer values were almost identical for the two periods of study.

For a rather quantitative analysis, table 6 gives the intensity of total radiation during the period 1974-1976 expressed as a percentage of that received during the period 1961-1962. This shows that the decrease, during the 1974-1976 period of study, in the hourly averages of the total direct radiation received at Cairo, during winter time, varied between 79% to 91% of that received at the same site during 1961-1962. The corresponding figures for spring were 83% to 99% and for autumn were 79% to 94%.

These results were further reached through the calculations of linke Turbidity Factor using the hourly measurements made during the two periods of study. Table 7 shows the values of the calculated Linke Turbidity factor for the different hours of the day during the 1961-1962 period of study. The values varies between 2.9 in January and 5.2 in June. Table 8 shows that the corresponding values for the period 1974-1975 varies between 3.1 in February and 5.7 in April and August. Figure 6 graphically represents the mean monthly values of Linke Turbidity factor 1961-1962 and 1974-1975. It shows clearly that the 1974-1975 average values of turbidity are higher throughout all months of the year except for November. The maximum increase was found in September. Seasonally speaking, the increase in the value of the turbidity factor was 5 in winter, 3.7 in both spring and Summer and 5.3 in autumn.

CONCLUSION

The entire living systems of our universe rely for their continuous upon energy-releasing reactions. On the other hand, the natural prime and principal source of this energy is the sun, the radiation of which is the creative, mover and keeper of life on earth.

The present study is one of a series of studies which aim at a proper scientific assessment and control of environmental pollution hazards in the Arab Republic of Egypt. It shows that the amount of direct solar radiation reaching the ground is highly affected by the atmospheric content of particulate matter. The latter is continuously increasing through environmentally un-controlled development programmes of industrialization and urbanization [9].

Negligence of this factor of environmental impacts of development will, therefore, lead to the destruction of solar radiation; the principal ultimate source of light, heat and chemical energy needed for man's health and the production of his foods and fossil fuels. Besides, the power that could be obtained from the solar energy has been estimated to be around 100,000 times that of the world total existing electric power.

This, necessarily, calls for an urgent need and action to conserve and utilize such power as an environmently clean and inexhaustible source of energy.

In Egypt, efforts are now being spent in order to correct the environmental damages resulting from the ill-zoning of industrial activities and urban planning and also to avoid future wrong-doings in this respect. These efforts will no doubt be reflected, among other vital objectives, on the preservation of the amounts of incoming solar radiation.

REFERENCES

1 Sutton, O.G., Micrometeorgology Ed. Mc Graw-Hill Book Co., Now York-Toronto-London, P. 163. 1953.

2 Linke, F. and Boda K.: "Verschlage zur Berechnung des Trubungsgrades der Atmosphere aus den Messungen der Sonnenstrahlung" Met. Ztschr., 39, 161, (1922).

3 Angstrom, A., Solar and Terrestrial Radiation. Report to the International Commission for Solar Research on Actinometric Investigations of Solar and Atmospheric Radiation. Q.J.R. Met. Soc. 50, p. 121. (1924).

4 Voltz, F.: "Photometer mit Selen-Photoelement zur Spektralen Messung der Sonnestrahlung und zur Bestimmung der Wellenlangenabbhangigkeit der Dunsttrubung Archiv. Meteor. Geophys. Biokl., B10, pp. 100-131, (1959).

5 Abdel Salam, M.S.: "Air Pollution problems and Their Effects Upon Health, City Planning and Inhabitation programmes at Helwan Industrial Area" The Supreme Committee for Planning the Greater Cairo, Ministry of Housing, Cairo, 1968.

6 Higazy, N.A.: "Variation of Ultra-Violet Radiation with Dust Concentration" M.Sc. Thesis, Ein Shams University (1966).

7 IGY Instruction Manual: "Radiation Instruments and Measurements" Ann. Inter. Geophys. Year, 5, VI, pp. 367-466, Pergamon Press, New York (1958).

8 Robinson, N.: "Solar Radiation" Elsevier Publ. Co., New York, pp. 347, (1966).

9 El-Nesr, M.K.: "Sur L'etude Du Facteur. De Trouble De Linke et sa Variation Annelle En Egypte". J. de Mech. et de Phys. de L'Atmos., I, VII, Quai Branly, Paris (1963).

10 Abdel Salam, M.S.: "Assessment of Environmental Impacts of Industrialization in the Arab World" The fourth Conference on Industrial Development of the Arab Countries, Baghdad, Iraq, December (1976).

Table 1 : Linke Turbidity Factor for Different Air Mass at Helwan 1967 (Average for 3 Sites) (Clear days).

Air mass	Feb.	March	April	May	June	July	August	Sept.	Oct.	Novem.	Dec.
1	3.86	4.30	4.63	4.97	5.61	6.06	5.41	5.22	4.69	4.07	4.55
1.7	4.12	4.60	4.93	5.31	6.03	6.40	5.78	5.61	5.03	4.35	4.86
2	4.29	4.76	5.07	5.49	6.18	6.62	5.91	5.78	5.19	4.48	5.01

Table 2 : Linke Turbidity Factor at Helwan 1922-27 at m = 1·7

Year	Jan.	Feb.	March	April	May	June	July	Aug.	Sept.	Oct.	Nov.	Dec.
1922	2·19	2·34	2·25	2·63	2·75	3·06	2·93	2·95	2·93	2·62	2·40	2·34
1923	2·26	2·44	2·38	2·84	2·78	3·31	3·23	2·76	2·73	2·82	2·54	2·32
1924	2·15	2·48	2·83	2·75	2·94	3·29	3·05	2·95	2·81	2·87	2·47	2·20
1925	2·15	2·19	2·45	2·66	2·88	3·32	2·97	2·95	2·70	2·62	2·54	2·37
1926	2·43	2·59	2·91	3·28	3·15	2·95	3·44	2·93	2·79	2·50	2·35	2·35
1927	2·21	2·28	2·53	2·43	2·95	2·82	3·09	2·88	2·60	2·53	2·37	2·28
Mean	2·26	2·39	2·56	2·77	2·90	3·12	3·12	2·90	2·77	2·66	2·44	2·31

Table 3 : Hourly Rates of Total Solar Radiation Intensity (Ly/min) for Clear Sky Conditions at Giza (March 1961-Feb. 1962).

Local Time	March	April	May	June	July	Aug.	Sept.	Oct.	Nov.	Dec.	Jan.	Feb.
9	1.003	1.025	0.974	0.993	1.010	1.063	1.057	1.053	0.57	1.032	0.997	1.088
10	1.120	1.110	1.070	1.035	1.042	1.118	1.142	1.164	1.095	1.151	1.138	1.191
11	1.142	1.151	1.085	1.067	1.067	1.154	1.168	1.191	1.134	1.221	1.221	1.251
12	1.175	1.188	1.112	1.109	1.153	1.177	1.183	1.214	1.163	1.242	1.264	1.279
13	1.140	1.199	1.095	1.092	1.157	1.157	1.174	1.150	1.153	1.153	1.246	1.263
14	1.110	1.142	1.058	1.033	1.046	1.104	1.103	1.103	1.040	1.145	1.180	1.210
15	1.086	1.109	0.960	0.992	0.926	1.022	1.022	1.016	0.920	0.965	1.060	1.168
16	1.000	0.878	0.883	0.889	0.870	0.916	0.830	0.744	0.700	0.865	0.835	0.960

Table 4 : Hourly Rates of Total Solar Radiation Intensity (Ly /min) For Clear Sky Conditions at Giza (May 1974- April 1976.

Local Time	May	June	July	Aug.	Sept.	Oct.	Nov.	Dec.	Jan.	Feb.	March	April
8	0.8496	0.8157	0.9302	0.819	0.749	0.764	0.764	0.566	0.544	0.668	0.745	0.853
9	0.9816	0.9759	1.0625	0.940	0.953	0.939	0.947	0.838	0.807	0.964	0.978	0.946
10	1.0578	1.0614	1.102	1.061	1.027	1.043	1.022	0.921	0.908	1.064	1.090	1.001
11	1.0888	1.1043	1.153	1.127	1.090	1.092	1.085	1.072	1.035	1.097	1.148	1.092
12	1.1102	1.1248	1.164	1.130	1.101	1.092	1.117	1.098	1.136	1.192	1.155	1.105
13	1.0973	1.1199	1.134	1.100	1.082	1.079	1.070	1.044	1.141	1.157	1.131	1.102
14	1.0596	1.0715	1.088	1.097	1.037	1.016	0.977	0.984	1.067	1.040	1.052	1.017
15	0.9872	1.0121	1.040	1.008	0.951	0.899	0.806	0.772	1.000	1.020	0.999	0.956
16	-	0.8893	0.953	0.881	0.739	0.639	0.419	0.549	-	0.844	0.768	0.747

Table 5 : Hourly Rates of total Solar Radiation intensity (Ly /min) for different season at clear sky Conditions at Giza •

Local Time	Winter		Spring		Summer		Autumn	
	1961-62	1974-76	1961-62	1974-76	1961-62	1974-76	1961-62	1974-76
9	1•039	0•870	1•001	0•969	1•022	0•993	1•022	0•946
10	1•160	0•964	1•100	1•050	1•065	1•075	1•134	1•031
11	1•231	1•068	1•126	1•110	1•096	1•128	1•164	1•089
12	1•262	1•142	1•158	1•124	1•146	1•139	1•187	1•103
13	1•221	1•114	1•148	1•110	1•135	1•118	1•159	1•077
14	1•178	1•030	1•103	1•043	1•061	1•085	1•082	1•010
15	1•064	0•931	1•052	0•981	0•980	1•020	0•920	0•885
16	0•887	1•697	0•920	0•758	0•892	0•908	0•758	0•599

Table 6 : Percentage of Hourly Rates of Total Radiation at Normal Incidance During 1974-1976 to that During the year 1961-1962 at Cairo "N.R.C."

Local Time	Winter	Spring	Summer	Autumn
09	84	97	97	89
10	83	95	101	91
11	87	99	103	94
12	90	97	99	93
13	91	97	99	93
14	87	95	102	93
15	88	93	104	90
16	79	83	102	79

Table 7 : Linke Turbidity Factor for Different Hour of the Day at Giza 1961-1962 .

Local Time	Jan.	Feb.	March	April	May	June	July	August	Sept.	Oct.	Nov.	Dec.
8	-	-	-	-	-	4.1	4.2	4.5	4.0	-	-	-
9	3.5	-	4.8	4.7	4.7	4.6	4.3	4.5	4.1	4.0	4.4	3.2
10	3.8	3.4	4.0	4.6	4.6	4.9	4.4	4.6	4.2	4.2	4.0	3.5
11	3.5	3.6	4.3	4.7	4.8	5.2	4.6	4.8	4.2	4.4	4.1	3.6
12	3.4	3.7	4.4	4.6	4.9	5.0	4.6	4.9	4.3	4.3	4.1	3.6
13	3.4	3.6	4.7	4.7	5.2	5.0	4.6	4.9	4.3	4.3	4.0	3.0
14	3.1	3.6	4.3	4.4	5.1	5.1	4.5	4.9	4.2	4.2	4.0	3.9
15	3.1	3.5	4.2	4.3	5.3	4.9	4.5	4.8	4.2	3.9	4.0	3.2
16	2.9	3.4	4.5	4.4	5.2	4.5	4.5	4.6	3.9	4.2	4.8	-
17	-	-	-	4.6	4.8	4.7	4.1	4.5	3.6	-	-	-
18	-	-	-	-	4.7	4.2	4.8	-	-	-	-	-
Mean	3.3	3.5	4.4	4.6	4.9	4.8	4.5	4.7	4.1	4.2	4.2	3.4

Table 8 : Linke Turbidity Factor for Different Hour of the Day at Giza 1974-1975.

Month Hours	May 1974	June	July	Aug.	Sept.	Oct.	Nov.	Dec.	Jan. 1975	Feb.	March	April
8	4.9	5.6	4.3	5.3	4.9	4.2	3.2	3.6	3.7	4.4	4.3	4.6
9	4.9	5.3	4.6	5.5	5.4	4.5	4.0	3.7	3.7	4.4	4.3	5.0
10	5.1	5.4	4.7	5.2	5.6	4.5	4.3	4.0	3.9	4.8	4.3	5.5
11	4.9	5.5	4.7	5.7	5.4	4.8	4.4	3.6	3.9	5.1	4.7	5.4
12	5.1	5.5	4.7	5.1	5.5	4.9	4.3	3.7	3.8	4.1	4.6	5.6
13	5.1	5.3	4.8	5.4	5.3	4.9	4.4	3.9	3.7	4.1	4.9	5.4
14	5.2	5.3	4.7	5.1	5.4	4.8	4.5	3.8	4.0	4.2	4.8	5.7
15	5.1	5.3	4.5	5.0	5.2	4.8	4.3	3.6	3.6	3.4	4.5	5.5
16	-	4.9	4.3	4.9	5.3	4.9	3.9	3.3	3.7	3.1	5.3	5.7
Mean	5.0	5.3	4.6	5.2	5.3	4.7	4.1	3.7	3.8	4.2	4.6	5.4

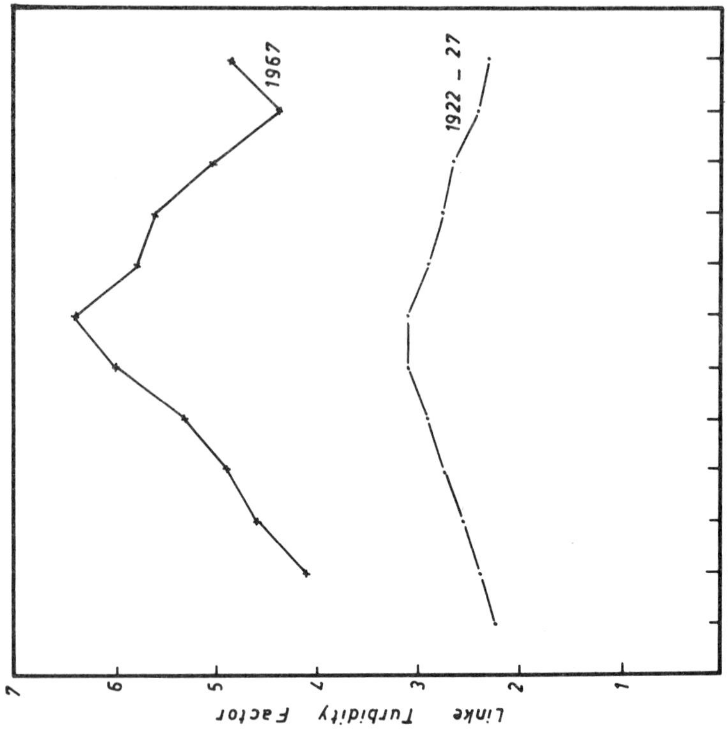

Fig.(2) Annual March of Linke Turbidity Factor for m = 1.7 at Helwan During 1922_27 and 1967.

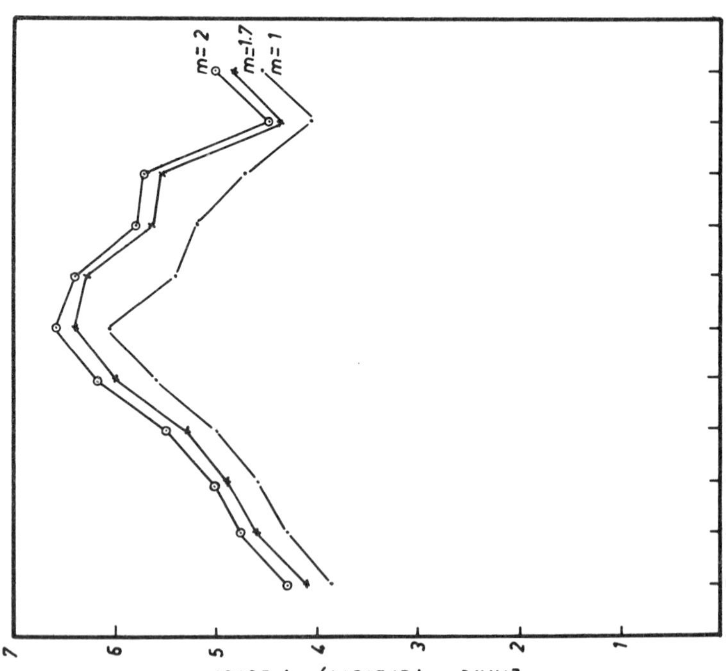

Fig(1) Annual March of Linke Turbidity Factor at Helwan During 1967 Different airmass (m)

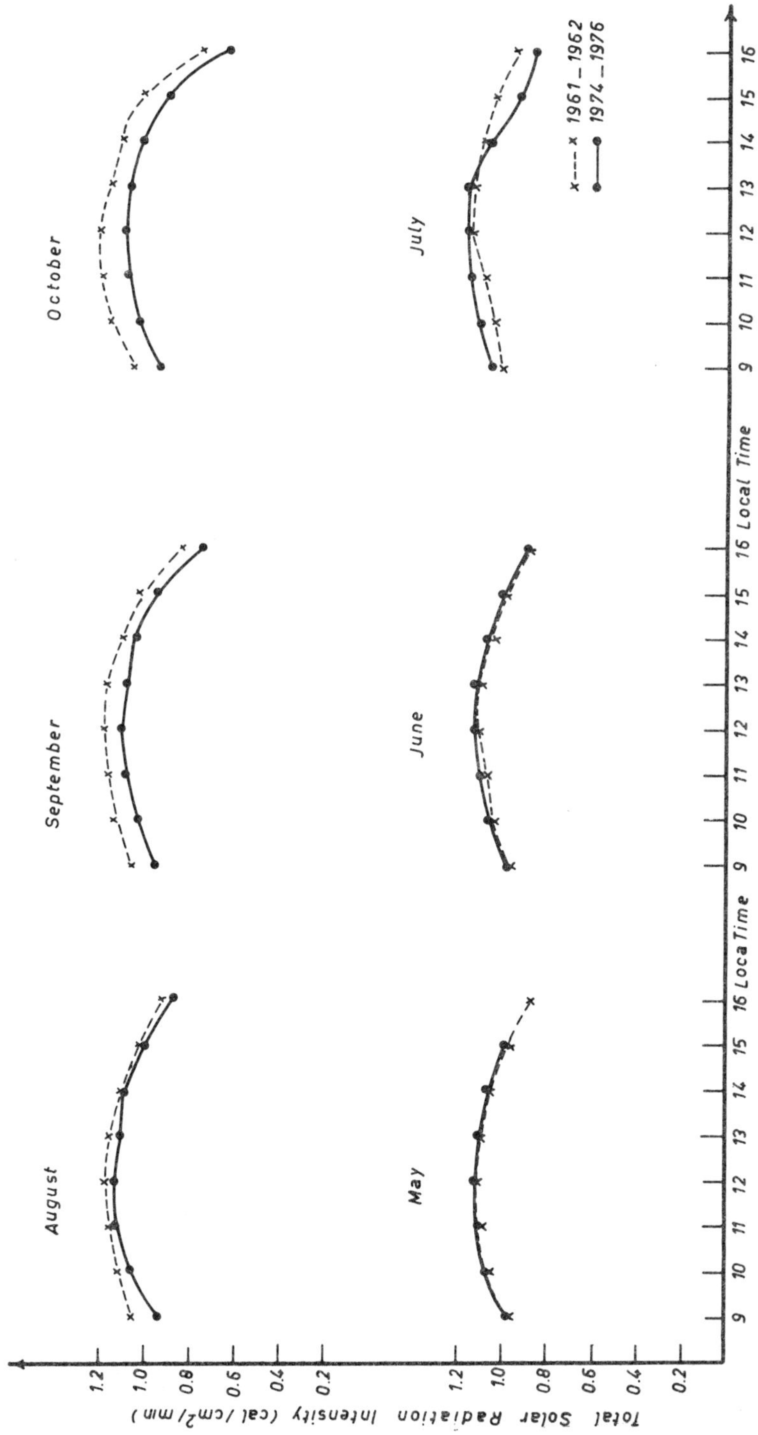

FIG(3) TOTAL RADIATION AT NORMAL INCIDENCE , CAIRO.

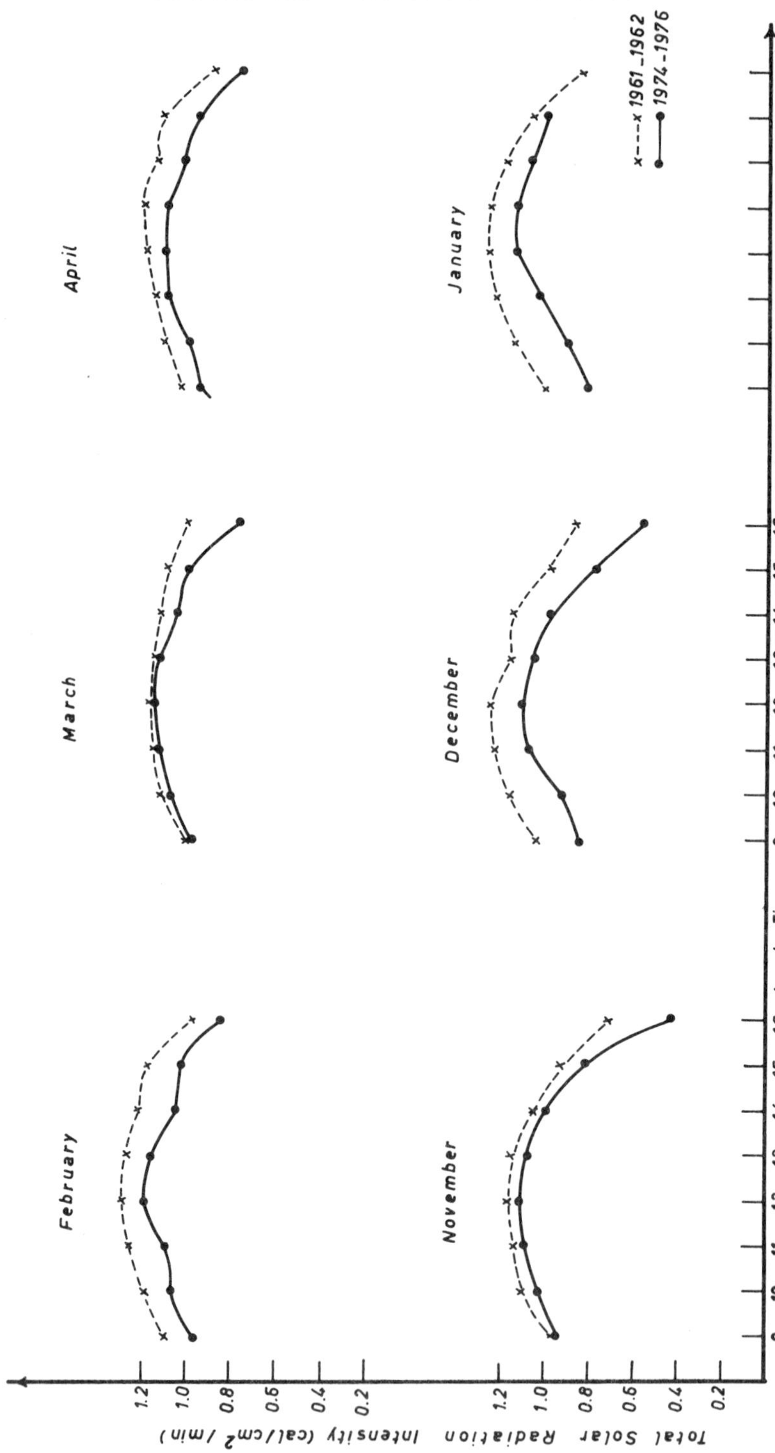

FIG. 4. TOTAL RADIATION AT NORMAL INCIDENCE, CAIRO.

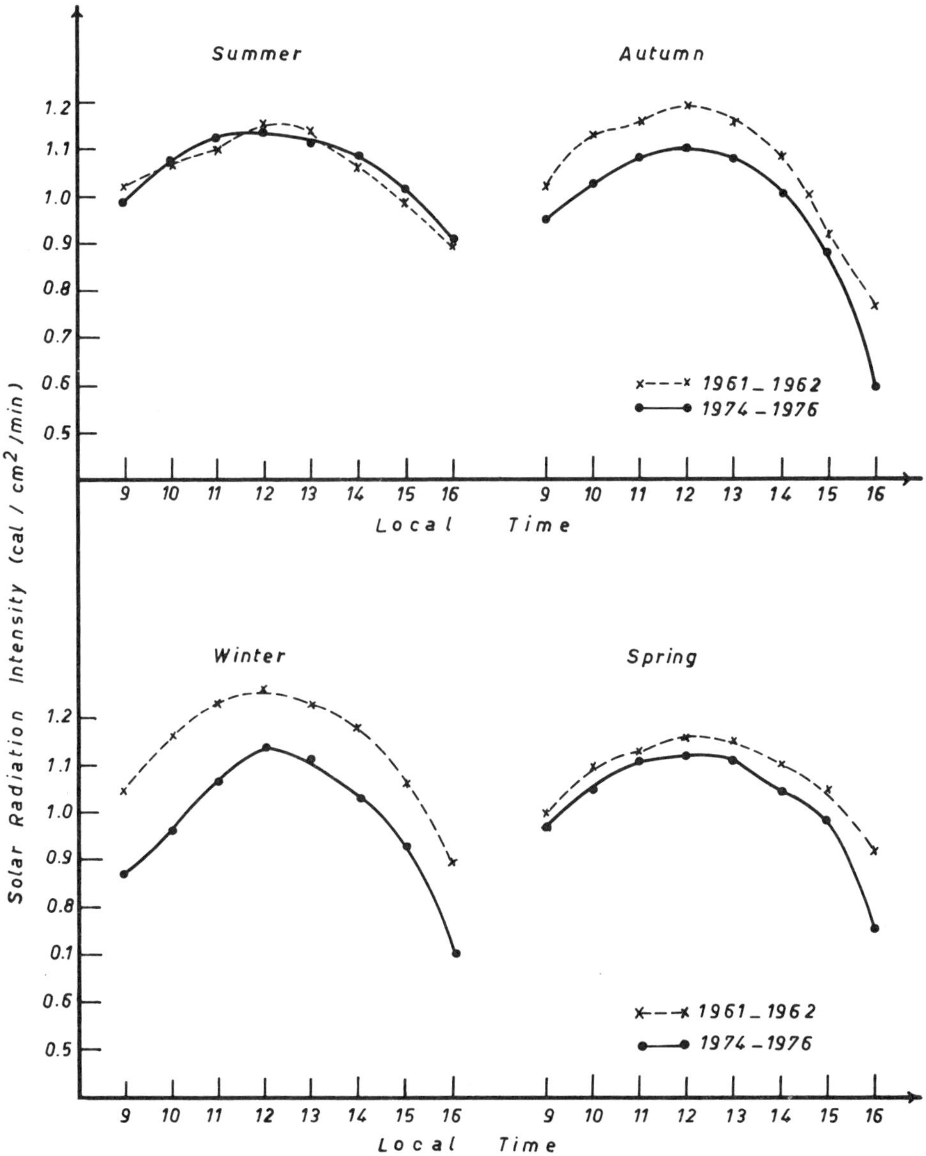

FIG. 5. TOTAL RADIATION AT NORMAL INCIDENCE CAIRO.

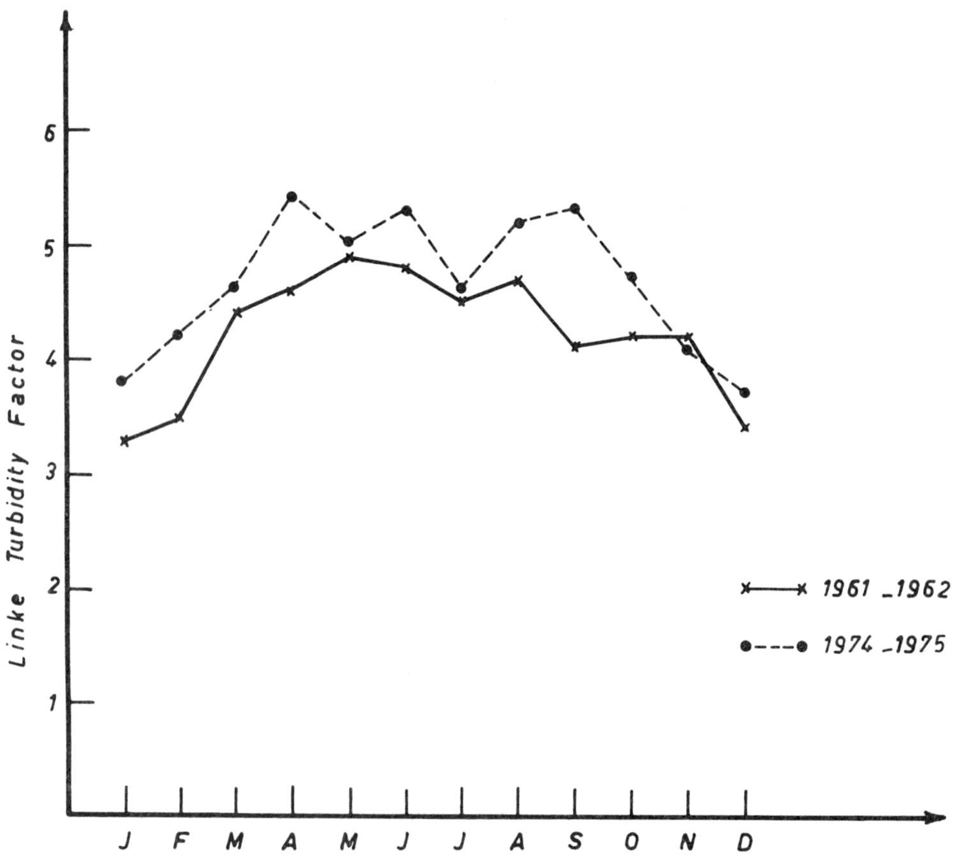

FIG. 6 MEAN MONTHLY VALUES OF LINKE TURBITIY FACTOR AT CAIRO.

SOLAR ENERGY AT GROUND SURFACE

N.H. Helwa
Solar Energy Laboratory
National Research Centre
Dokki, Cairo, Egypt

ABSTRACT

The utilization of solar energy is today an attractive problem. To make this utilization feasible for any purpose, it is necessary to measure and to calculate the amount of the solar radiation incident on different oriented surfaces. For example, to build a power generation system, we must know the amount of the direct solar radiation incident on a plane normal to the sun.

The planning of any solar project requires detailed information about the weather conditions and the hourly rates of incident solar radiation on different oriented surfaces for different months of the year. The work presented in this paper is mainly concerned with the study of these problems.

The total solar radiation incident on a horizontal surface, the ambient air temperature, the relative humidity, wind speed, and the direct solar radiation incident on a plane normal to the solar beam were recorded by meteorological stations in Egypt and in the Solar Energy Laboratory in Cairo. Some of these measurements and some other estimated data for different oriented surfaces were mapped in curves.

INTRODUCTION

A knowledge of the solar radiation reaching the earth's surface is a prerequisite for proper utilization of solar energy. The total energy from the sun falling on the earth is far more than that needed to do the world's work. Being an ever-continuing source of energy, the daily amount reaching the earth exceeds by many thousand-fold the total energy consumed daily by man. But, the use of solar energy is not practical in all locations, or, in all climates. The duration period of the sunlight, the frequency and predictability of cloudy weather, and the height of the sun in the sky, are all factors which must be considered. Any location between latitude 40°N and 40°S, except for regions where rains are too frequent, would be suitable. Egypt lies between latitude 20°N and 31°N, and has over 3500 sunshine hours per year.

The new program throughout the world is to utilize solar radiation energy for different purposes: for example, power generation, water desalination, water heating for domestic purposes, cooling, air conditioning, and so on.

ATTENUATION OF BEAM RADIATION

The solar radiation which reaches the outer limits of the atmosphere, and received in one minute by an area of one square cm, placed perpendicular to the ray, is known as the solar constant and is about 1353 w/m^2 [1].

The direct beam intensity is reduced as it passes through the atmosphere due to its absorption, by some gases as ozone, oxygen and carbon dioxide, by water vapour, and by dust. The direct solar energy I_D at normal incidence at the earth surface is defined by the equation

$$I_D = I_o T_t^m \tag{1}$$

where I_o is the solar radiation at normal incidence outside the atmosphere, T_t is the coefficient of transmission due to all the constituents of the atmosphere, and m, the thickness of the atmosphere, is known as the air mass and is equal to unity when the sun is in the zenith and the point of observation is at sea level. For other elevations, m is proportional to sec z where,

$$m = \sec z \tag{2}$$

where z is the zenithal angle of the sun. The total transmissivity could be written in the form:

$$T_t = T_a^{p/760} \cdot T_d^{d/800} \cdot T_w^{w/20} \tag{3}$$

where T_t, T_d, T_w are the transmission factors associated with atmospheric dust and water vapour scattering,

p = total pressure, mm
d = dust particles concentration at the ground, particles /cm^2, and
w = depth of precipitable water, cm.

Scattering By Water Vapour

The variation in the direct solar radiation due to absorption depends mainly on the actual amounts of water vapour, w, in the atmosphere, which is a function of relative humidity RH, and the ambient air temperature t_a, and is defined by the relation

$$\text{Relative humidity} = \frac{\text{actual vapour pressure } w}{\text{saturated vapour pressure at } t_a}$$

To account for the amount of water vapour in the atmosphere, the variations of the relative humidity and the ambient air temperature for the period from 1931 to 1960 and the the years 1976 and 1977 for Cairo was mapped in Figures 1 and 2. Figure 3 represents the hourly variation of the ambient air temperature t_a °C, the relative humidity RH, the actual amount of precipitable water in the atmosphere, w, divided by 20 in cm and the amount of the direct solar radiation for a day in March, 1976. This curve shows the dependence of the amount of the direct solar radiation upon weather conditions.

Scattering By Dust Particles

Solar energy is also depleted by dust particles and other solid particles mainly suspended in the atmosphere. This depletion results from a combination of scattering absorption and diffuse refuse reflection. True molecular scattering can take place only when the diameter of the solid particles is less than the wave length of the incident radiation.

Scattering By Air Molecules

The scattering associated with atmosphere mainly occurs by air molecules which are very small compared to the wavelengths of radiation significant in the solar energy spectrum. Scattering of the radiation by air molecules occurs in accordance with the theory of Rayleigh, which indicates that the scattering coefficient would vary approximately as λ, where λ is the wavelength of the radiation.

PATTERN OF SOLAR RADIATION

The measurement of solar radiation has been made for many years at a large number of stations throughout the world. The principle objective in these measurements was weather prediction. Solar data was also used for practical applications, in agriculture, and in the construction industry for the design of heating, air conditioning and lighting. The data available is that of the total amount of solar radiation incident at a horizontal surface, which is made up of two parts energy in direct beam, and the diffuse energy from the sky. The amount of direct energy depends on cloudiness and also on the solar altitude. The direct proportion is obviously the greatest on clear days. Fortunately, clear days prevail in Egypt. The sunlight falling on clouds is mainly diffused by scattering, some up and some down. Clouds do not absorb much energy. The diffused

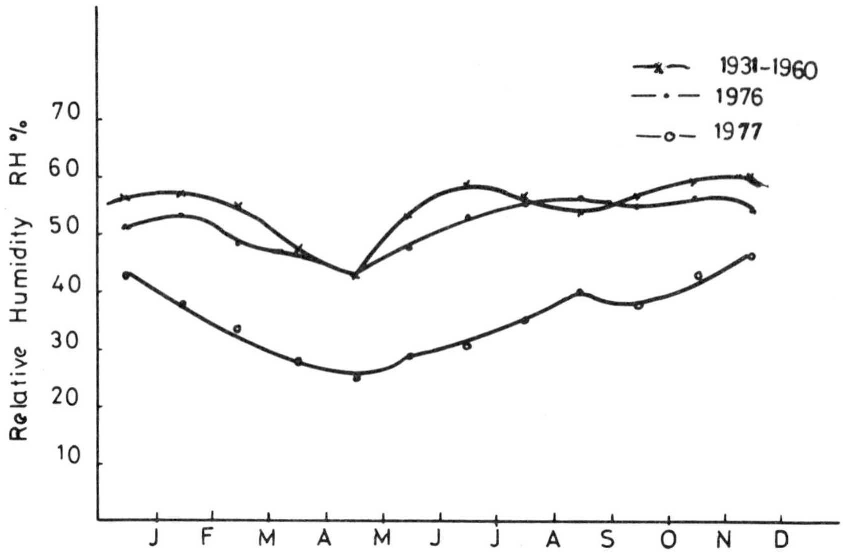

Figure 1 Distribution of Relative Humidity %

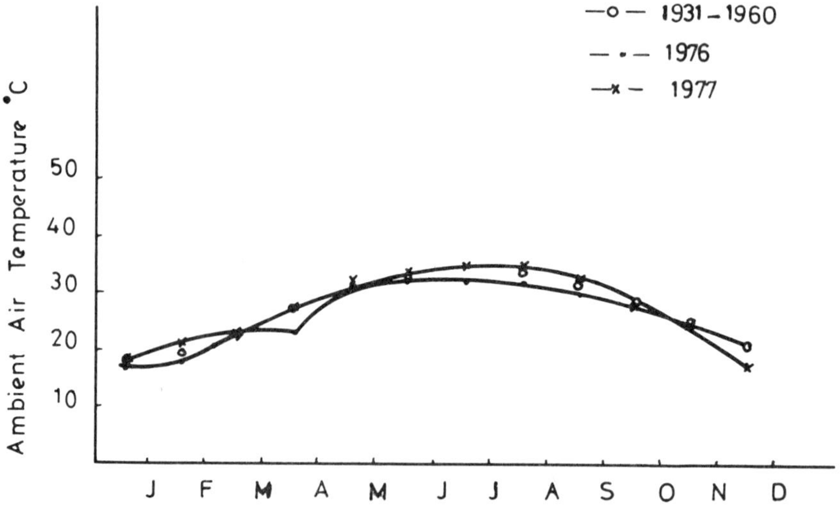

Figure 2 Distribution of Ambient Air Temperature °C

energy reaching the surface is the greatest on partially clouded days.

The total solar energy incident on a horizontal surface should be distinguished from direct solar radiation; which is the amount of solar radiation incident on a surface normal to the solar beam.

The average of the amount of solar radiation at horizontal surface across the main inhabited parts of the world is from 100 W/m^2 in high latitude climates to 300 W/m^2 in the Red Sea (averaged over 24 hours) [2].

Figure 4 shows the recorded mean monthly average of the solar radiation at a horizontal surface for Giza, 30°N, for the period from 1956 to 1960 and for the year 1976. The curves show that the maximum is in summer and that the maximum mean value over 8 hours is about 580 W/m^2. The instrument most widely used in solar radiation measurements is the pyranometer, which measures the global (sun + sky) radiation received on a horizontal surface. There are many models and types. All these instruments operate on the principle of the thermopile. Electrical voltage is generated which is recorded on a strip chart or in some other manner. For the direct solar radiation measurement, a pyroheliometer or actinometer is used. The sensor is a thermopile at the base of a cylindrical tube. The actinometer could easily measure the azmuthal and the zenithal angles of the sun which define its motion. The measurements of the amount of the direct solar radiation is for the construction and building of the solar concentrators, and, to define the part of the diffuse radiation received by a horizontal surface. Figure 5 shows the distribution of the average value of the direct solar radiation with time in January, June, September and December for the year 1976. The maximum value is at noon and is about 800 W/m^2.

Diffuse Radiation

As noted before, the components of the atmosphere scatter a portion of the solar radiation, and some of this scattered radiation reaches the ground, which is called diffuse radiation. To predict the amount of solar radiation incident on a certain plane, it is necessary, as a first step, to know the magnitude of this diffuse part. No separate measurements of diffuse radiation have been undertaken in Egypt. On the other hand, data in the literature of the amount of diffuse radiation is very limited [3,4]. The best information at the present time may be that contributed by Parmelee [5], for horizontal and vertical surfaces for different solar altitude and wall solar azimuth angles, such that it can be estimated for any vertical surface of any orientation. If the total solar radiation at a horizontal

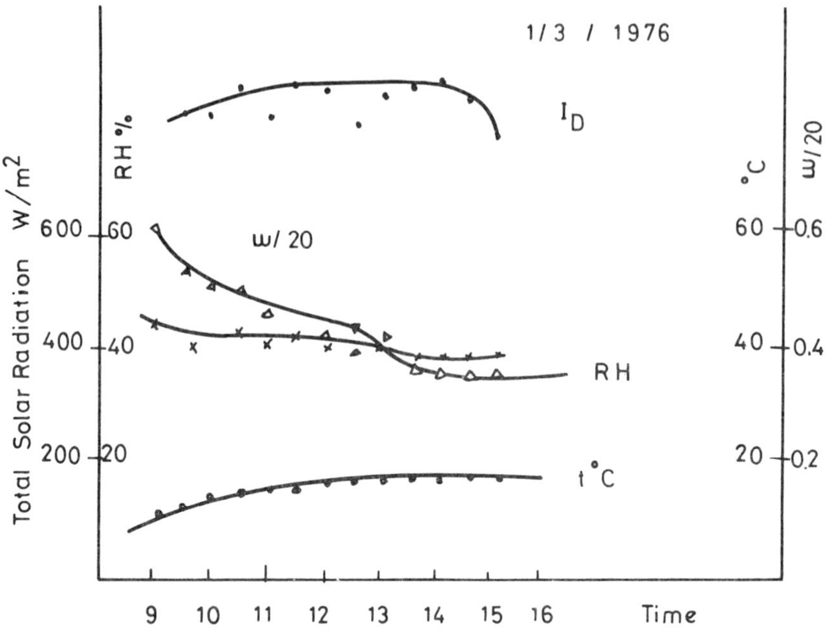

Figure 3 Hourly Variations of Direct Solar Radiation for a Day in March, 1976

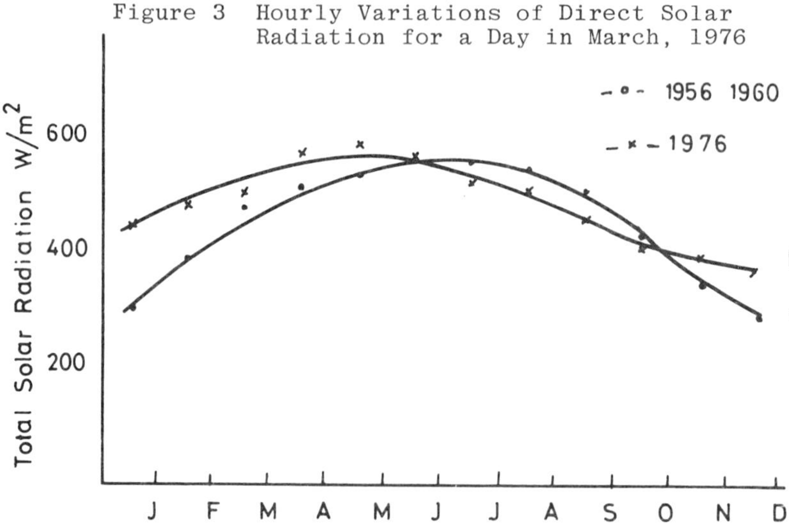

Figure 4 Recorded Monthly Average of Solar Radiation

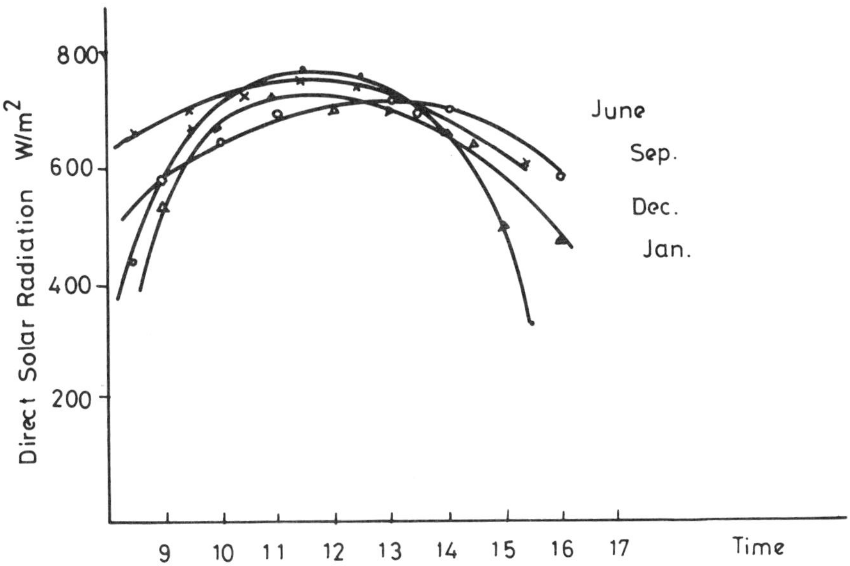

Figure 5 Distribution of the Average Value of the Direct Solar Radiation

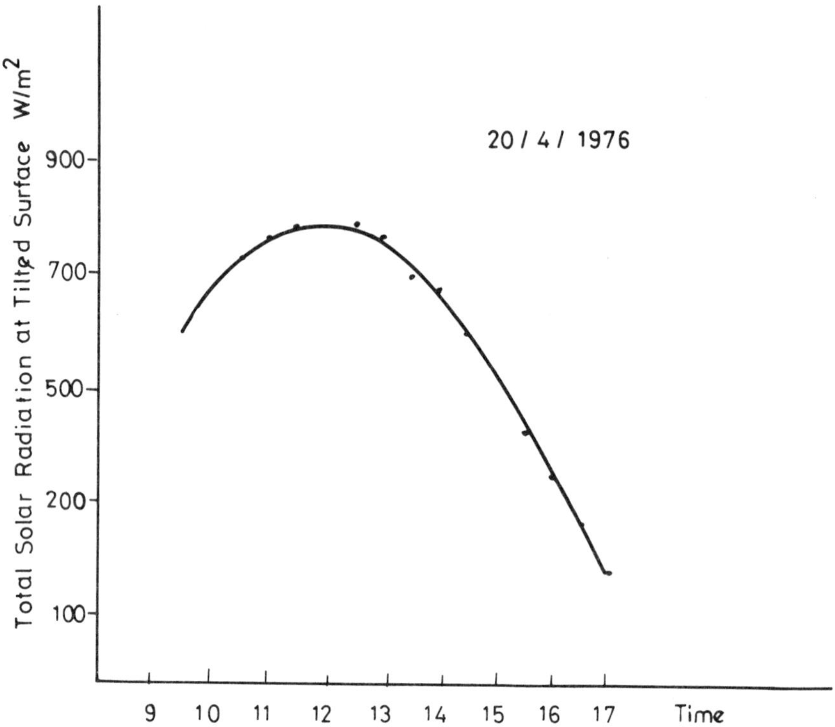

Figure 6 Calculated Amount of the Total Solar Radiation Incident for Giza, Egypt

surface and the direct radiation I_D are known, the amount of the diffuse radiation I_d at a horizontal surface could be estimated from the relation:

$$I_t = I_D \sin (Al) + I_d$$

where (Al) is the solar altitude which is equal to 90° zenith angle of the sun.

To account for the amount of diffuse radiation on an inclined surface, the following relation could be used [6]:

$$I_d(\text{tilted}) = H_d - (H_d - V_d) \frac{\alpha}{90}$$

where α is the tilted angle of the surface; H_d, V_d are the diffuse radiation at horizontal and vertical surfaces.

Solar Angles

The angles at which the sun's rays reach a surface on the earth change from day to day and from hour to hour, owing to the changing position of the earth relative to the sun and the rotation of the earth about its axis. The position of the sun is defined by the solar altitude and solar azimuth. The solar altitude is defined by the angle in a vertical plane between the sun ray and the horizontal. The solar azimuth is defined as the angle measured in a horizontal plane north the the horizontal projection of the sun's ray. These angles vary continuously from sunrise to sunset and are different for each day of the year. The declinations of the sun, which is the angular distance of the sun north or south of the celestial equator, is given from the equation by Cooper [7]:

$$S = 23.45 \sin (360 \frac{284 + n}{365})$$

where n is the number of the day from the first of January. The local hour angle is expressed in degrees, using 0° for solar noon, and each hour equaling 15° of longitude, with mornings positive and afternoon negative.

SOLAR RADIATION ON A SURFACE HAVING DIFFERENT ORIENTATIONS

To estimate the amount of solar radiation (direct and total) incident at a tilted surface with different orientations, three angles must be defined:

1. Altitude angle of the sun is defined by the equation $\sin Al = \sin(s) \sin (\Phi) + \cos (\delta) \cos (\Phi) \cos (h)$

2. The solar azimuth angle, AZ, and is defined by:

$$\sin AZ = \frac{\sin (AL \cos (h) - \cos (AL) \tan (\delta)}{\sin (h)}$$

3. The angle of incidence (θ) of solar radiation is defined by:

$$\cos\theta = \sin\delta \sin\Phi \cos\alpha - \sin\delta \cos\Phi \sin\alpha \cos p + \cos\delta \cos\Phi \cos\alpha \cos h + \cos\delta \sin\Phi \sin\alpha \cos p \cos h + \cos\delta \sin\alpha \sin p \sin h \quad (4)$$

where δ is the declination angle of the sun, latitude of the place; p the azimuth angle of the plane from the true south of the perpendicular; h is the hour angle and is the tilted angle of the surface.

For a surface sloped to the north or south, the angle can be derived from equation (4) by putting p, the azimuth angle of the surface equal to zero, and equation (4) reduces to

$$\cos\theta = \cos(\Phi - \alpha) \cos\delta \cos h + \sin(\Phi - \alpha) \sin\delta \quad (5)$$

Multiply the direct beam radiation I_D by cos θ and add the component of the diffuse radiation incident on the tilted surface.

In the design of solar equipment, as a solar water heater, the flat plate collector should be erected to receive the maximum possible radiation. It should be facing south and tilted with a certain angle. Figure 6 shows the calculated amount of the total solar radiation incident on a tilted surface facing south (angle of tilt is 30°) for Giza.

For a vertical surface, the angle of incidence of the solar radiation is given from equation (4) by putting the tilted angle of the surface equal to 90° so that the equation reduces to

$$\cos\theta = \cos\delta \sin\Phi \cos p \cos h + \cos\delta \sin p \sin h - \sin\delta \cos\Phi \cos p \quad (6)$$

The total solar radiation incident on a vertical surface facing different orientations was calculated from the measured data of the direct solar radiation and the diffuse radiation at a vertical surface produced by Parmelee. Figures 7-10 show these values for the months of January, June, September and December at 9, 12 and 13 o'clock. These curves show that the south surfaces received more energy in winter than in summer. Detailed monthly data for a horizontal south-facing surface for Giza are plotted in Figure 11. The large variation of mean monthly

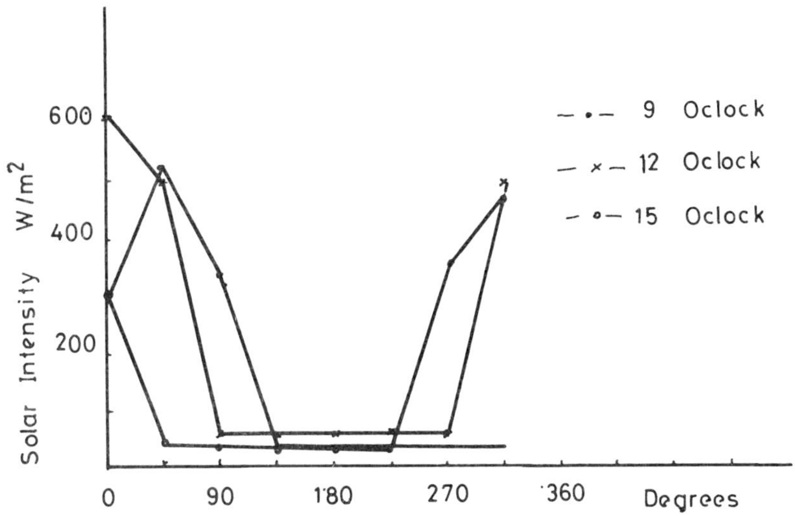

Figure 7 Total Radiation at Vertical Surface for Eight Different Orientations in January

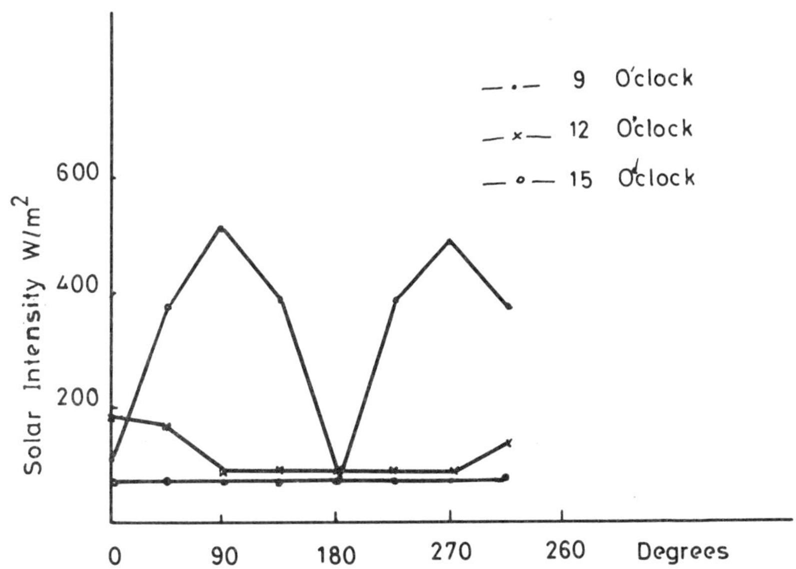

Figure 8 Total Radiation at Vertical Surface for Eight Different Orientations in June

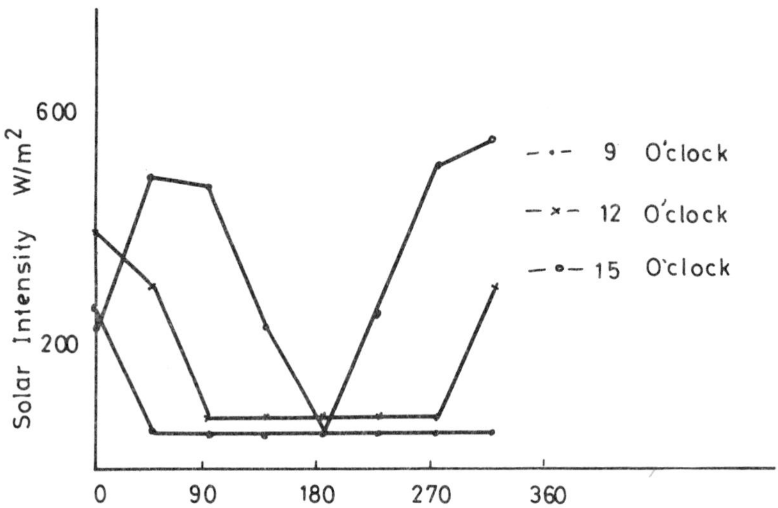

Figure 9 Total Solar Radiation at Vertical Surface for Eight Different Orientations in September

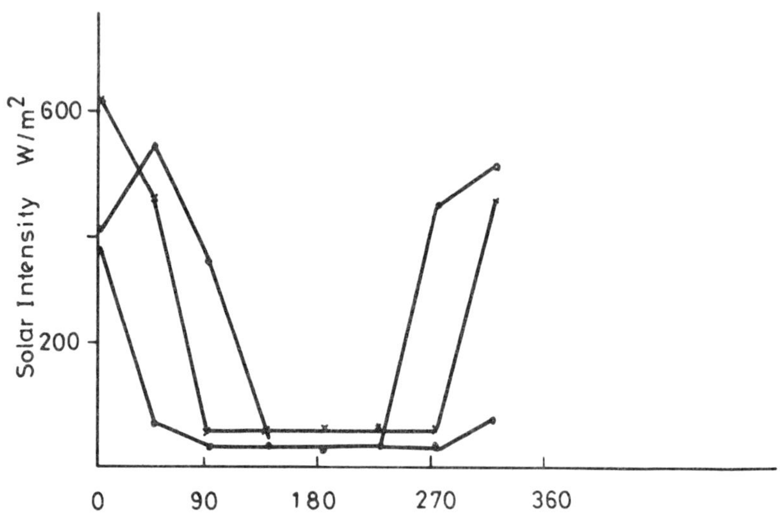

Figure 10 Total Solar Radiation at Vertical Surface for Eight Different Orientations in December

irradiance on a horizontal surface from summer to winter at high latitudes is apparent, as is the smaller variation in mean monthly irradiance on a vertical south-facing surface. There is, of course, always a large variability from day to day.

SUNSHINE DURATION

In addition to the knowledge of the solar radiation intensity received at any place, an important factor in solar energy utilization, which should be known, is the sunshine duration. It is received in Egypt with a Campbell stock sunline recorder. Egypt has the best location in that it receives a high rate of solar radiation and has a high duration, which comes to about 3500 hours per year. The rate of the total possible sunshine hours to the total actual hours (sunshine duration) is mapped for Cairo in the period from 1955 to 1960. The mapping in Figure 12 shows that the maximum sunshine duration is in September and comes to about 85%.

DIRECT USES OF SOLAR ENERGY

Direct uses of solar energy include many applications. There are many ways to utilize solar energy, by conversion to thermal or electrical energy. These include the heating of water and buildings, the cooling of buildings, the drying of agricultural and animal products, the desalination of seawater and electricity generation.

The applications listed are used in various parts of the world today, but this is not to say that the technology does not need further development. It means simply that some equipment is commercially available, and applications in developing countries during the next few years can be expected to involve currently available equipment. Within the next few years, only some of these applications can be reasonably expected to be adopted in rural areas in Egypt.

Heating By Solar Energy

One of the most important applications of solar energy is the production of hot water for domestic purposes. The flat plate solar heat collector is usually used. In Egypt, at the Solar Energy Laboratory, some types of solar water heaters have been constructed and examined and fruitful results have been obtained. Such systems are economically competitive with others; the economic advantage of this system, as compared to those using fossil fuels, will increase as the price of fuels rise.

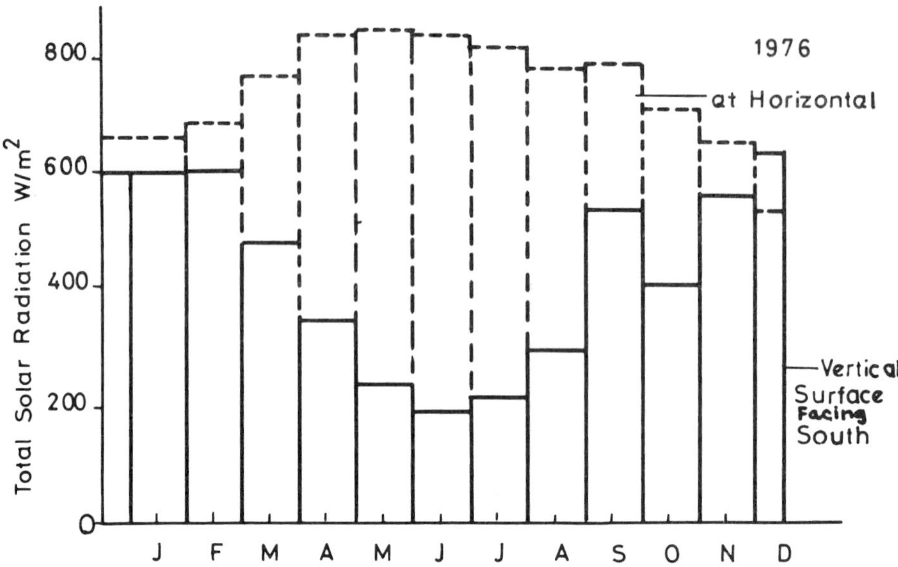

Figure 11 Distribution of Total Radiation at Horizontal Surface and The Estimated Values at Vertical Surface Facing South

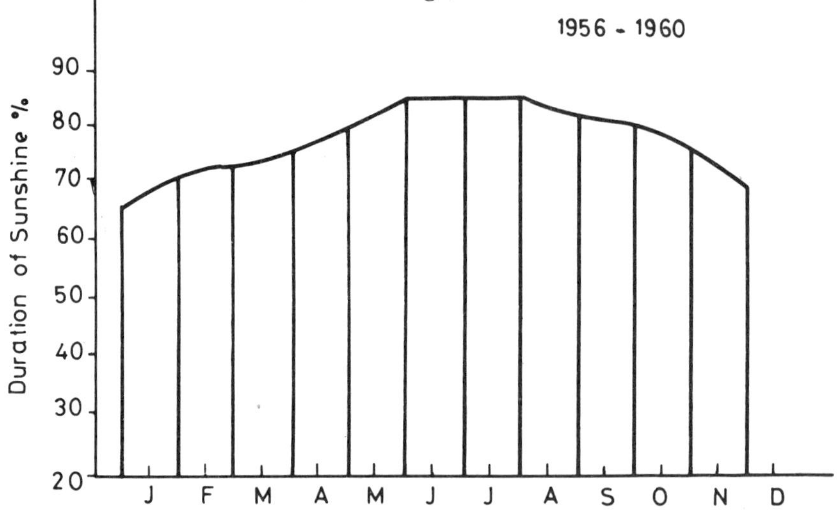

Figure 12 Duration of Sunshine

Desalination Of Sea Water

It is well known that the fresh water obtained from all main sources, such as rainfall, rivers and underground waters, represents only about 2% of the available water in the world [8].

The rapid increase of the fresh water consumption rate, due to the growth of population and the increase of the per capita consumption, makes the desalination of sea water very important for the extension in the arid region. During the past few years, some types of solar water stills have been designed and tested. Solar desalination eliminates most of the cost of electricity, chemicals and the like.

Drying Of Various Crops

Agricultural uses of solar energy include the drying of various crops, employing both direct exposure to sunlight and the use of heated air passed through or over the material to be dried. The cost of drying grains, fruits and other products by solar-heated air have not yet been reliably appraised. A prototype of an air heater for drying crops was designed and is now under testing and economic research in the Solar Energy Laboratory.

REFERENCES

1. Thekaeker, M.P. and Drummond, A.J., "Standard Value For The Solar Constant and Its Spectral Components", Nat. Phys. Sci. 229, 6, (1971).

2. Page, J.K., "Geographical Variations in the Climatic Factors Influencing Solar Building Design", Fourth Course on Solar Energy Conversion, 6-24 September 1977, Italy.

3. Hand, I.F., Heat & Vent. 47, 92-94, (1950).

4. Stagg, J.M., Geophys. Memoirs, 11, 86, (1950).

5. Parmelee, G.V., Heat, Pip. Air Cond., 26, 8 129-136, (1954).

6. Khalil, A.M., "Solar Energy In Egypt", Thesis, Cairo University, Cairo, Egypt, (1963).

7. Duffie & Beckman, Solar Energy Thermal Process, (1974).

8. Sakr, A.L., "Solar Energy Utilization", Lecture given at Cairo University, Cairo, Egypt.

CONTRIBUTION TO THE STUDY OF NOCTURNAL RADIATION VERY NEAR THE GROUND: EFFECT OF A CIRCULAR SCREEN

J. Brodin, A. Flourens, V.V. Pham and C. Sinnasse
Ecole Nationale Superieure de L'Aeronautique et de L'Espace
Toulouse, Cedex, France

ABSTRACT

Our contribution to the study of noctural radiation very near the ground is the determination of the practical parameters like the position of the screen with respect to the earth and to the top of plants, the degree of opacity to the nocturnal radiation of the ground* (single or doubled layered screens, thickness necessary to collect the nocturnal radiation of the ground during the night), in order to design the "gardener's palliasse", whose utilization is to protect certain vegetations against hardy weather.

The experimental results obtained coincide with the theory as elaborated by Brunt, Groen and Phillips. They point out amidst other properties, the existence of the optimum distance between the screen and the top of the plants. This distance is influenced by the ground.

I Introduction

The nocturnal radiation of a black body is the energy lost by its free radiation towards the interstellar space [I] .Two practical results can be deduced from this definition.If the black body (for example water) is allowed to radiate freely or if this radiation effect is intensified by the reflectors,the temperature of this body can be lowered,thus procuring a device to realize a cold source of the heat engine.On the other hand,if the black body is prevented from radiating (for example the green parts of the plants),then there is a possibility of protecting certain vegetations against hardy weather.

---

*ground: the nature of the ground is grass.

The second aspect of this problem is studied. The effect of a screen, with a single and double layer, on the radiated flux from the ground at a very low distance from it is estimated.

2 Physical model

It is known [2] that the heat transfer process at the ground surface is given by the equation

$$S - A - C_l + C_t = 0$$

S is the net flux statistically represented by the regression equation

$$S = C_s T^4 \ (I-a-b(p)^{I/2})$$

C_s: STEFAN-BOLTZMANN constant = 56.697 IO^{-I2} kW/m^{2}K^4

a=0.44 ; b=0.08 ; p: surface vapour pressure in mb

The transfer of heat by turbulence in the air

$$A = C_p \ d_a \ D \ (\partial T_p/\partial z)$$

where C_p: specific heat of air at constant pressure

d_a: air density

$D = D(z)$: the coefficient of eddy diffusivity

$\partial T_p/\partial z$: rate of change of potential temperature

C_l: the latent heat which is the rate of water condensation

C_t: heat conduction in the soil

$$C_t = -k_I(\partial T/\partial z)$$

where k_I is the thermal conductivity of the soil and $\partial T/\partial z$ is the rate of change of temperature of the soil

We have the equation (Annexe)

$$S - DC_p d_a(\partial T_p/\partial z) - C_l - k_I(\partial T/\partial z) = 0$$

Case:

$$A=0, \ C_l=0, \ z=0$$

According to BRUNT, $T_I - T = 2S_I(t)^{I/2}/C_I d_I(\pi K_I)^{I/2}$

where $T_I = T(z,0)$: temperature at sunset

S_I=S=constant

C_I: specific heat of soil

d_I: soil density

$K_I = k_I/C_I d_I$

Case:

$C_I=0$, $z=0$

According to PHILIPPS,

$$T_I-T = 2S_I(t)^{I/2}/ C_I d_I(\pi K_I)^{I/2} + C_p d_a(\pi D)^{I/2}$$

Remarks

Supposing $F(t,sol,air)=2(t)^{I/2}/C_I d_I(\pi K_I)^{I/2}+C_p d_a(\pi D)^{I/2}$

Then, $T_I^o - T = S_I F$

I) S_I changes with the sky state. For a clear sky, S_I is in the order of $II8W/m^2$ [3]. It changes with the base cloud height.

Cloud height	I.5 km	reduction factor	0.I4
"	3	"	0.25
"	7	"	0.80

2) S_I varies with temperature T_I^o (which can be considered as ambient temperature) and with surface vapour pressure p, according to the expression

$$S_I = C_s T_I^4 (I-0.44- 0.08 (p)^{I/2})$$

3) If the sky state is modified by placing a screen and if, at a given instant, the temperature difference between the two thermocouples (one placed below the screen, the other outside the screen) is measured, then

$$T_I^o - T' = F S_I' \qquad \text{(screen)}$$

$$T_I^o - T = F S_I$$

So, $T' - T = F (S_I - S_I')$

3 Experimental results and interpretations

The measurement scheme is given in fig I. The polyethylene circular screen is 346 mm in diameter and about 2 microns thick. The diameter is calculated in such a way that the thermocouple placed below the screen at a distance of I00mm subtends a semi solid angle of 60°. The choice of the screen width is done on economical basis.

The temperature differences T_I-T_2 (T_I and T_2 being the temperatures corresponding to the thermocouples Th_I and Th_2) and the ambiant temperature T_3(corresponding to the thermocouple Th_3) are recorded.

The parameters subjected to variations in the case of a single-layered screen are:

d_I: distance of the three thermocouples from the groŭnd

d : distance of the screen from the thermocouple Th_I

Figure 2 indicates the temperature T_3 which begins to decrease before sunset and which gets stabilized at about 3 or 4 A.M.

With the aid of the recordings of T_I-T_2 in function of time, the variation curves of T_I-T_2 in function of d at different moments when T_3 is probably stable(4,5,6,7 A.M.) are drawn. Figures 3,4,5,6 give a series of typical curves. T_3, corresponding to the measuring point, is also recorded to be used as reference temperature.

It is remarked that the thermocouple Th_I undergoes the radiation influence of the screen itself when it is in contact with the screen; that there is an optimum distance of the screen from the thermocouple Th_I corresponding to the maximum value of T_I-T_2 and that the temperature difference T_I-T_2 tends towards zero for higher values of d.

These remarks have guided us to study the temperature variation due to the screen in function of its distance (fig 7), the optimum distance variation in function of d_I (fig 8) and the variation of $(T_I-T_2)_{max}$ in function of d_I (fig 9).

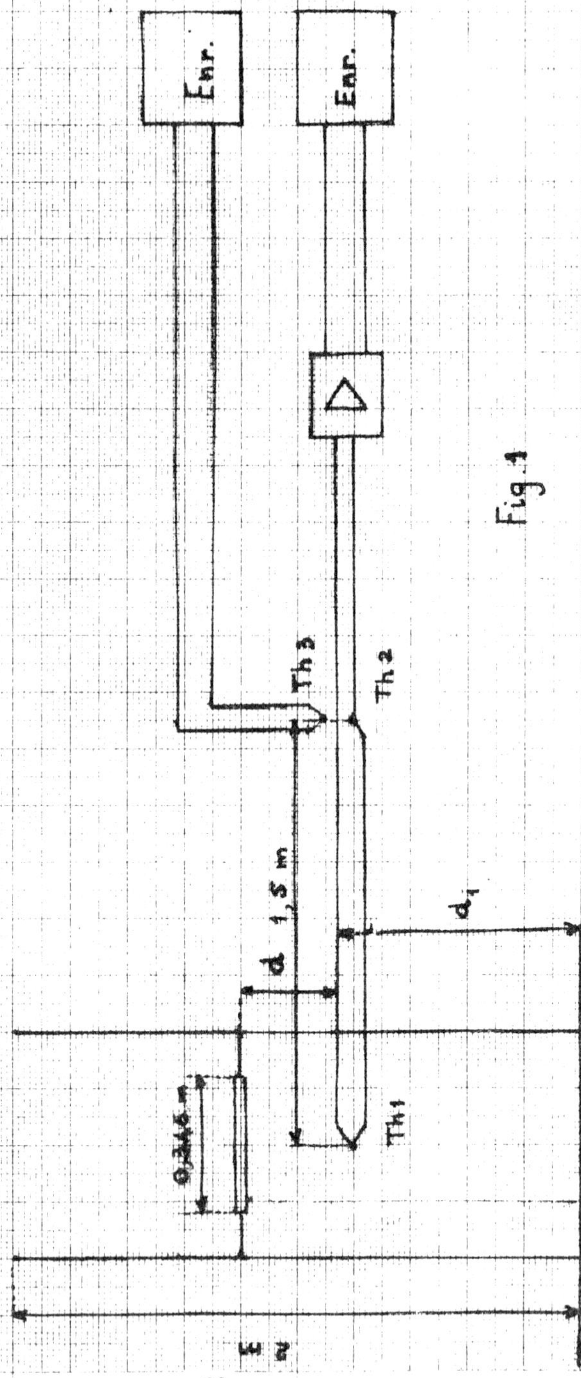

Fig. 1 Measurement Scheme

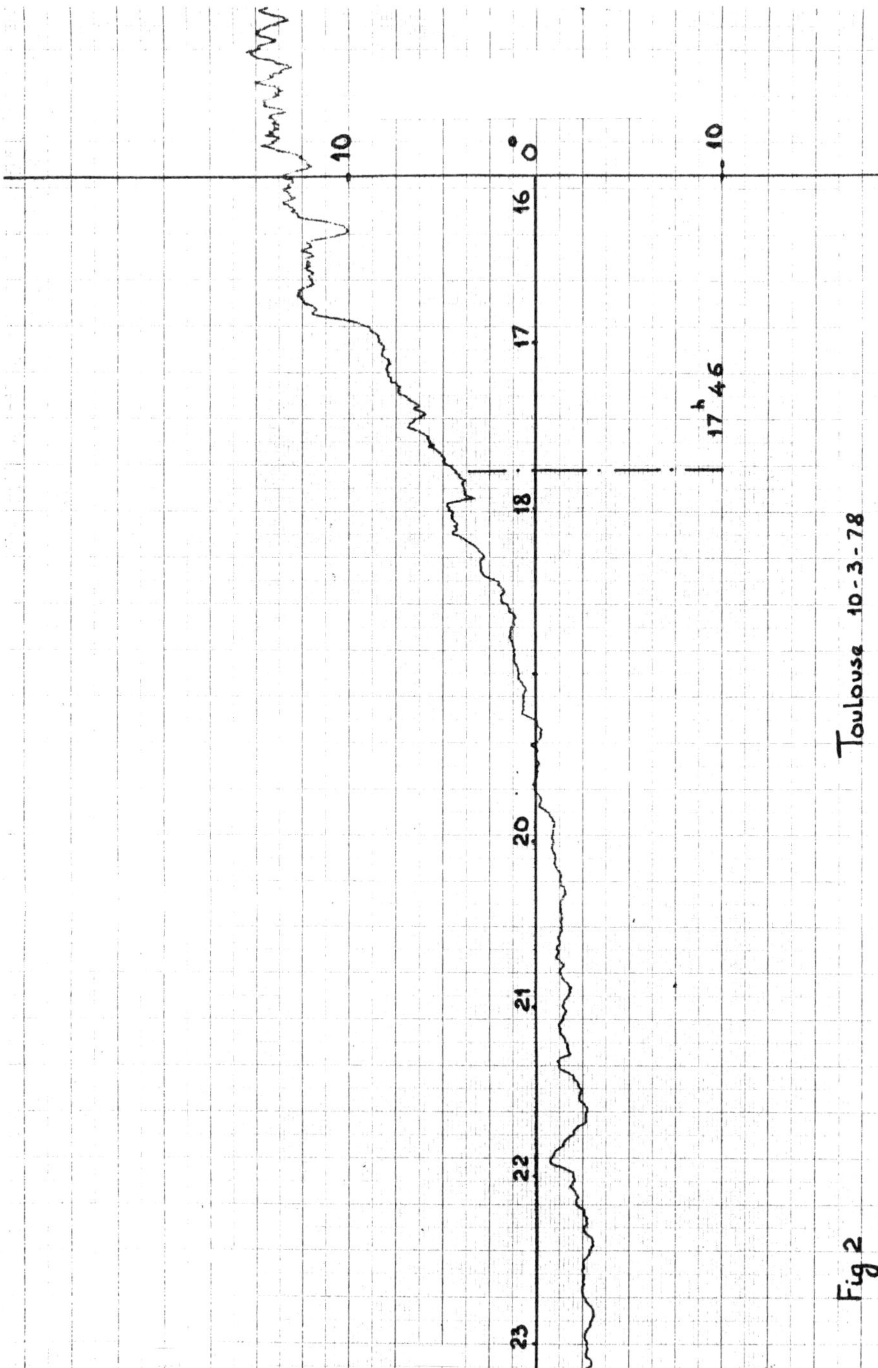

Fig. 2 Recording of the ambient temperature T_3

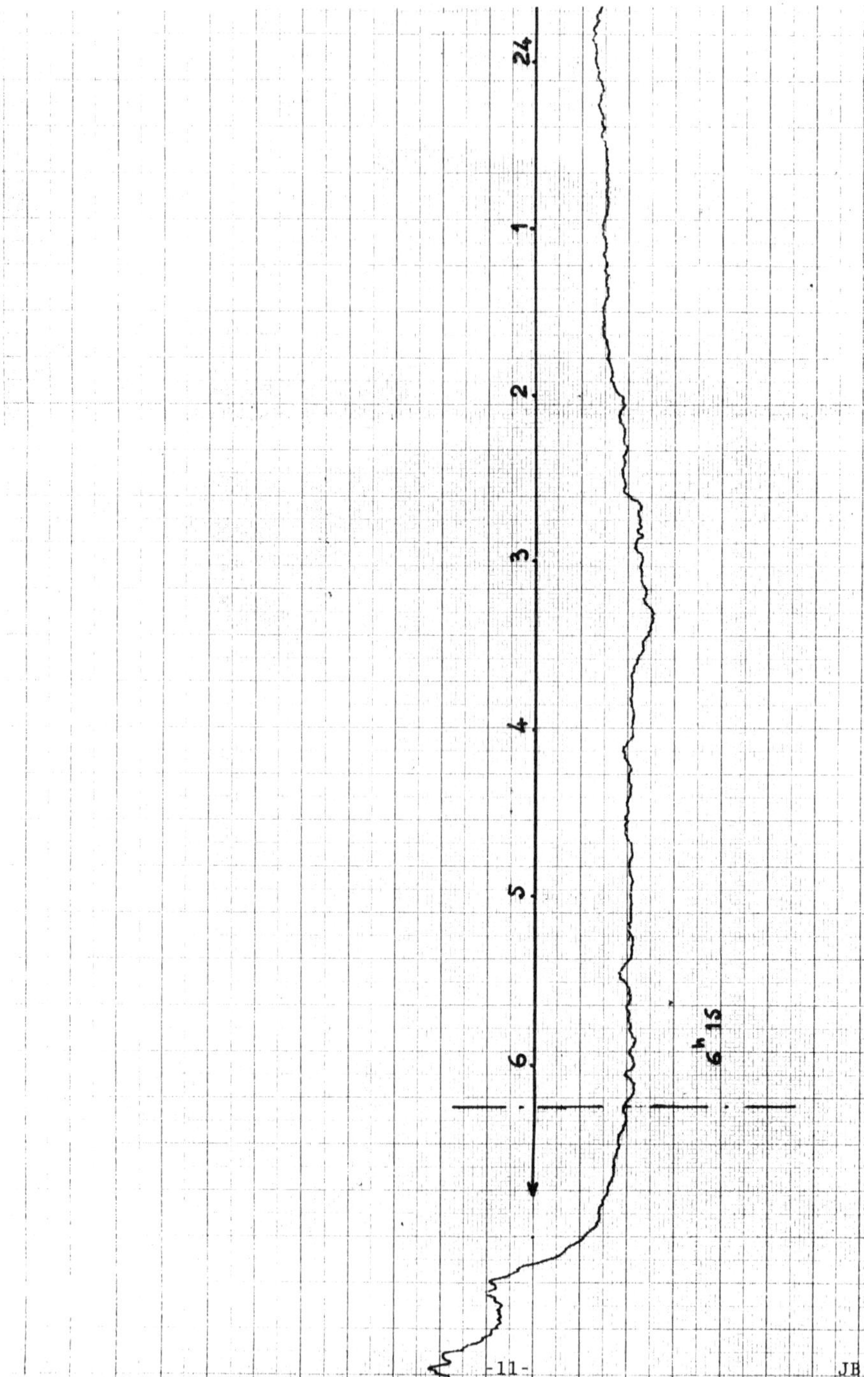

Fig. 3 $T_1 - T_2 = \Delta T$ as a function of d; parameter time t

-11-

JB

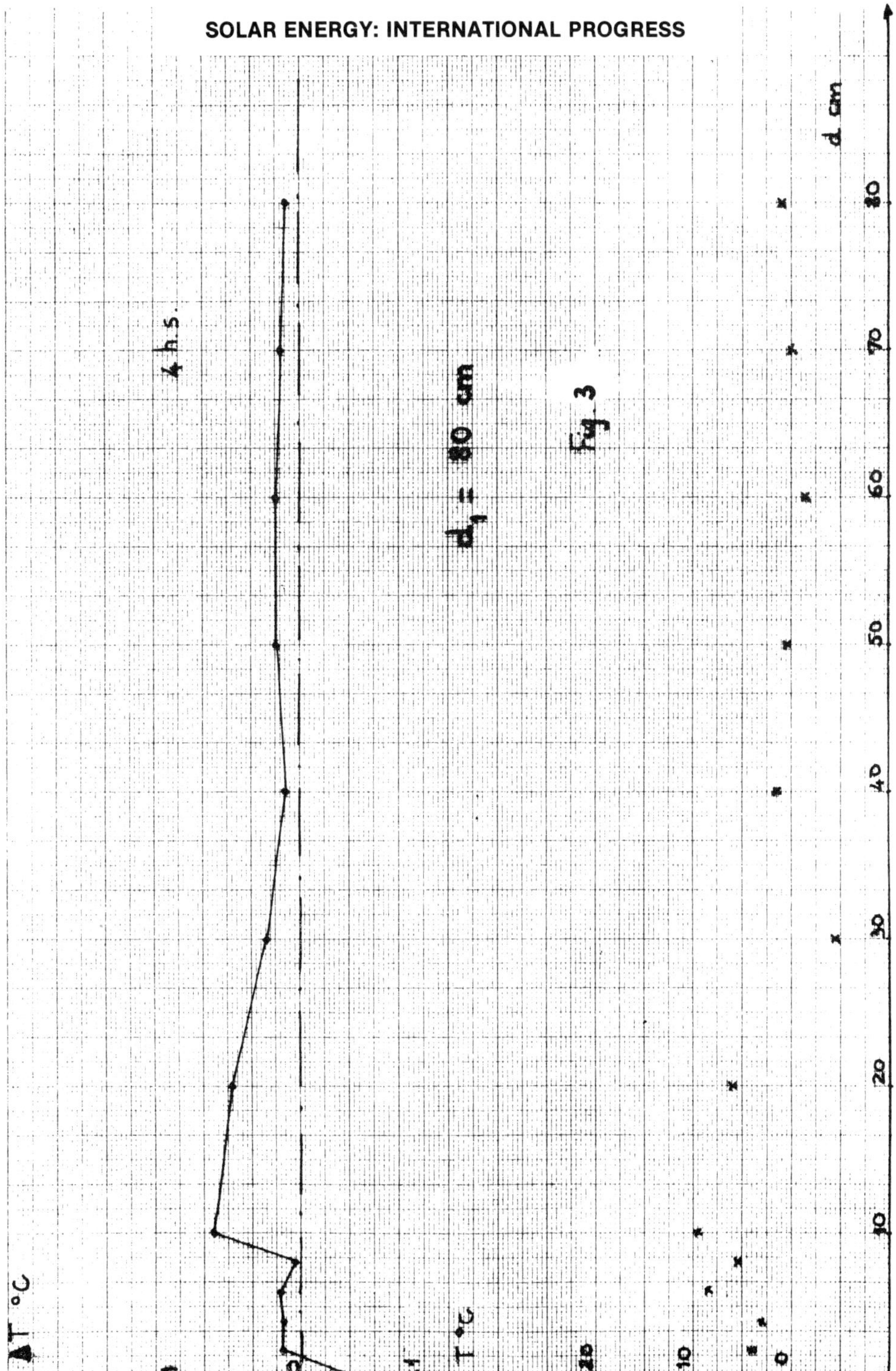

Fig. 3 $T_1 - T_2 = \Delta T$ as a function of d; parameter time t

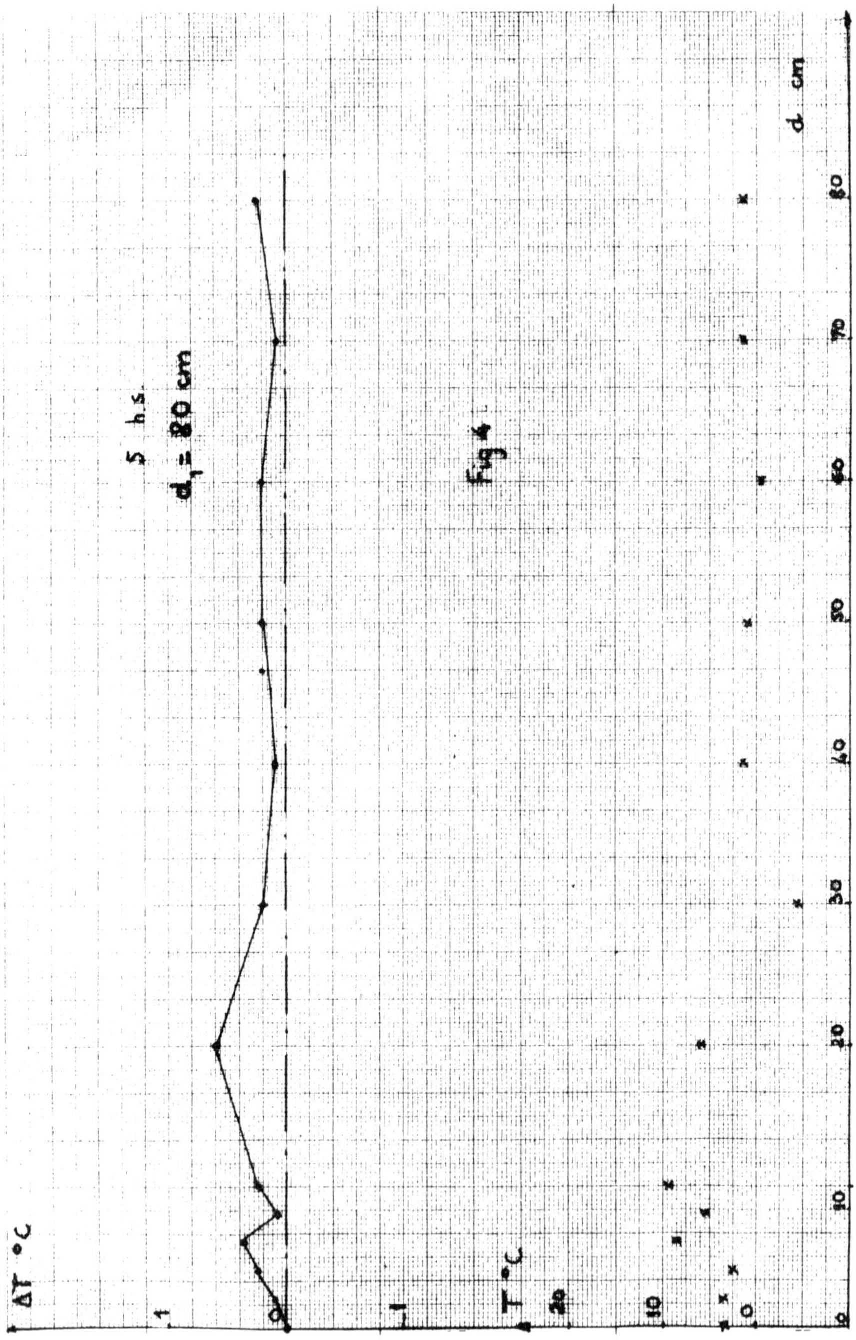

Fig. 4 $T_1 - T_2 = \Delta T$ as a function of d; parameter time t

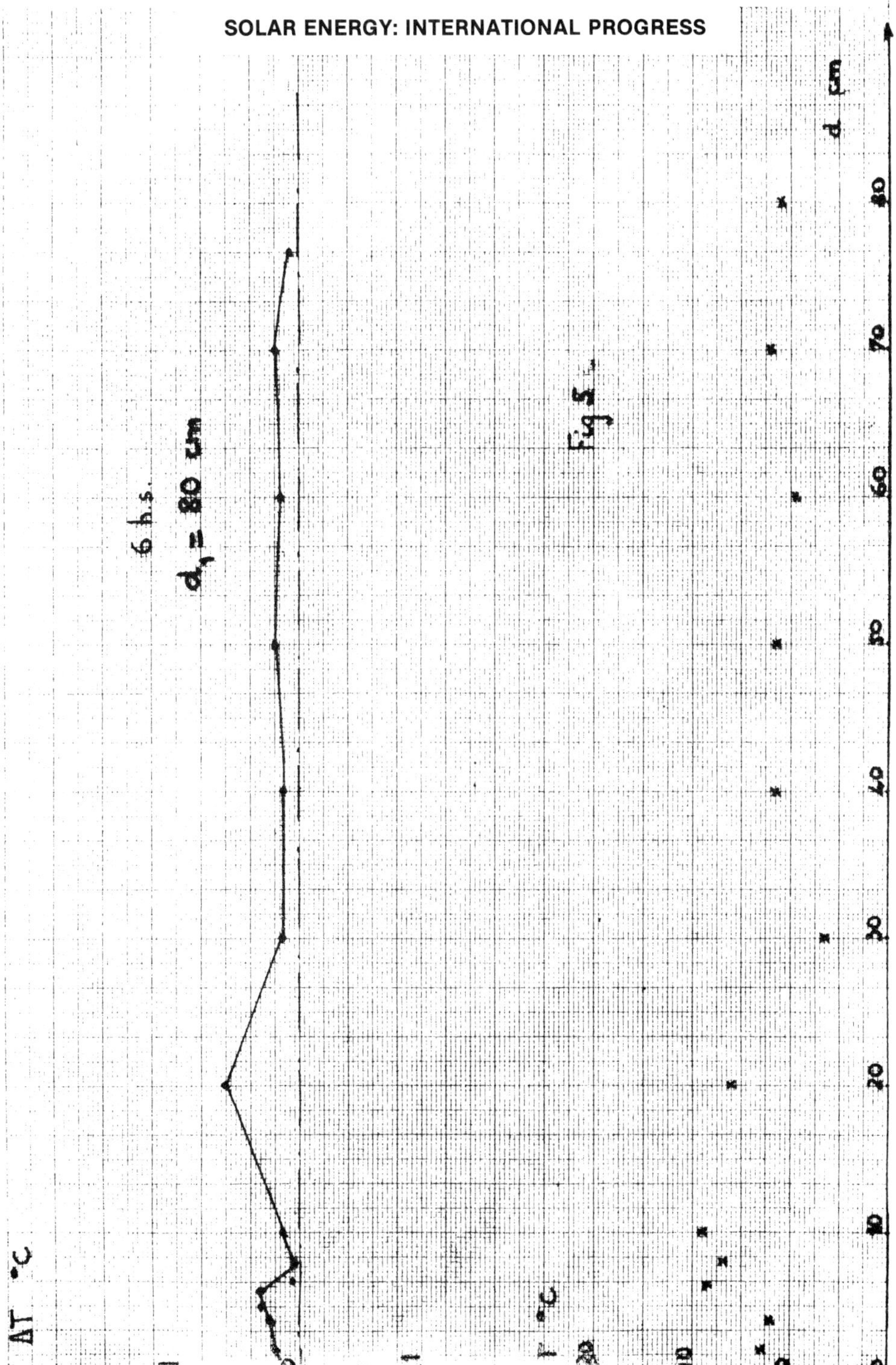

Fig. 5 $T_1 - T_2 = \Delta T$ as a function of d; parameter time t

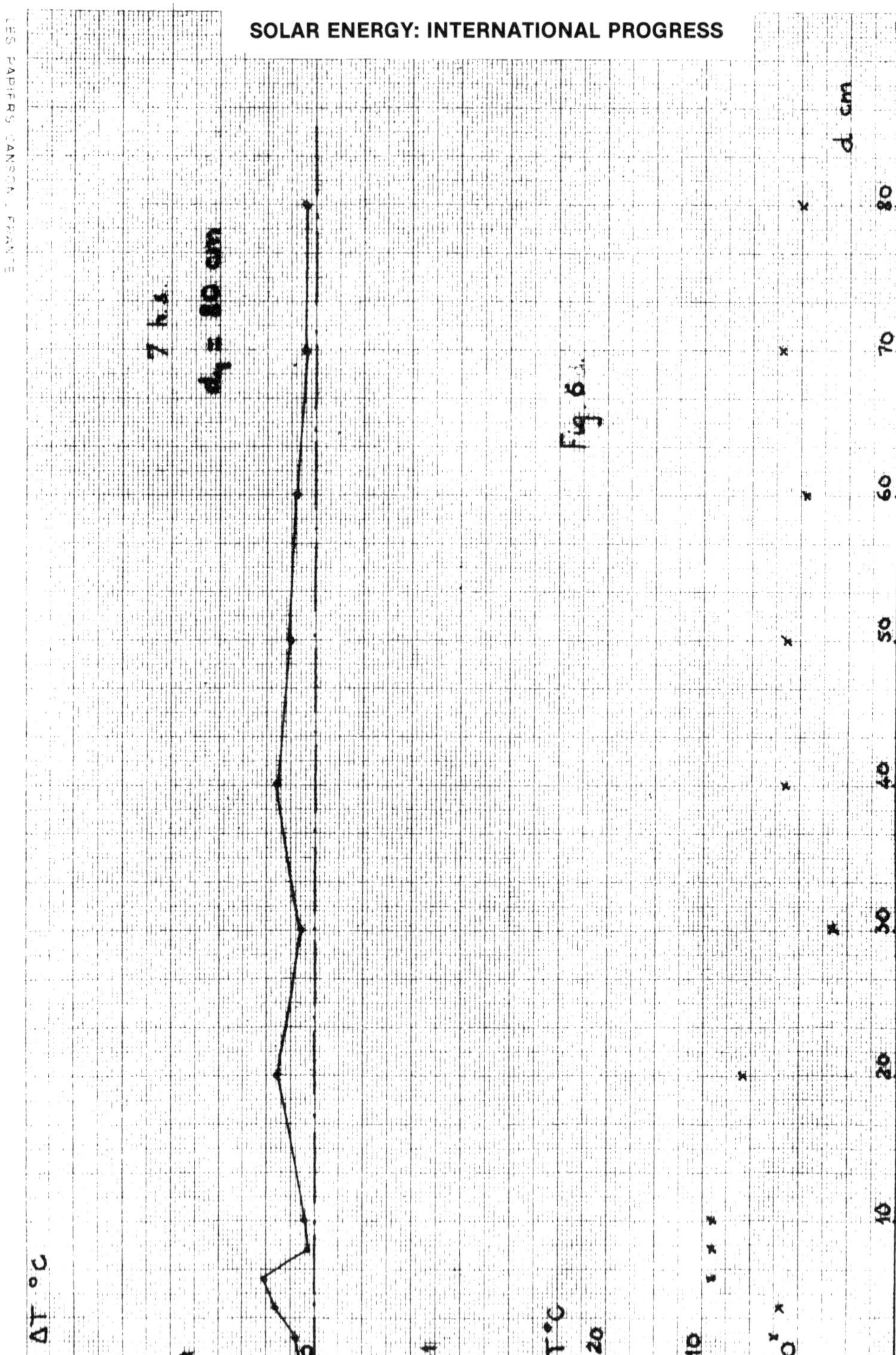

Fig. 6 $T_1 - T_2 = \Delta T$ as a function of d; parameter time t

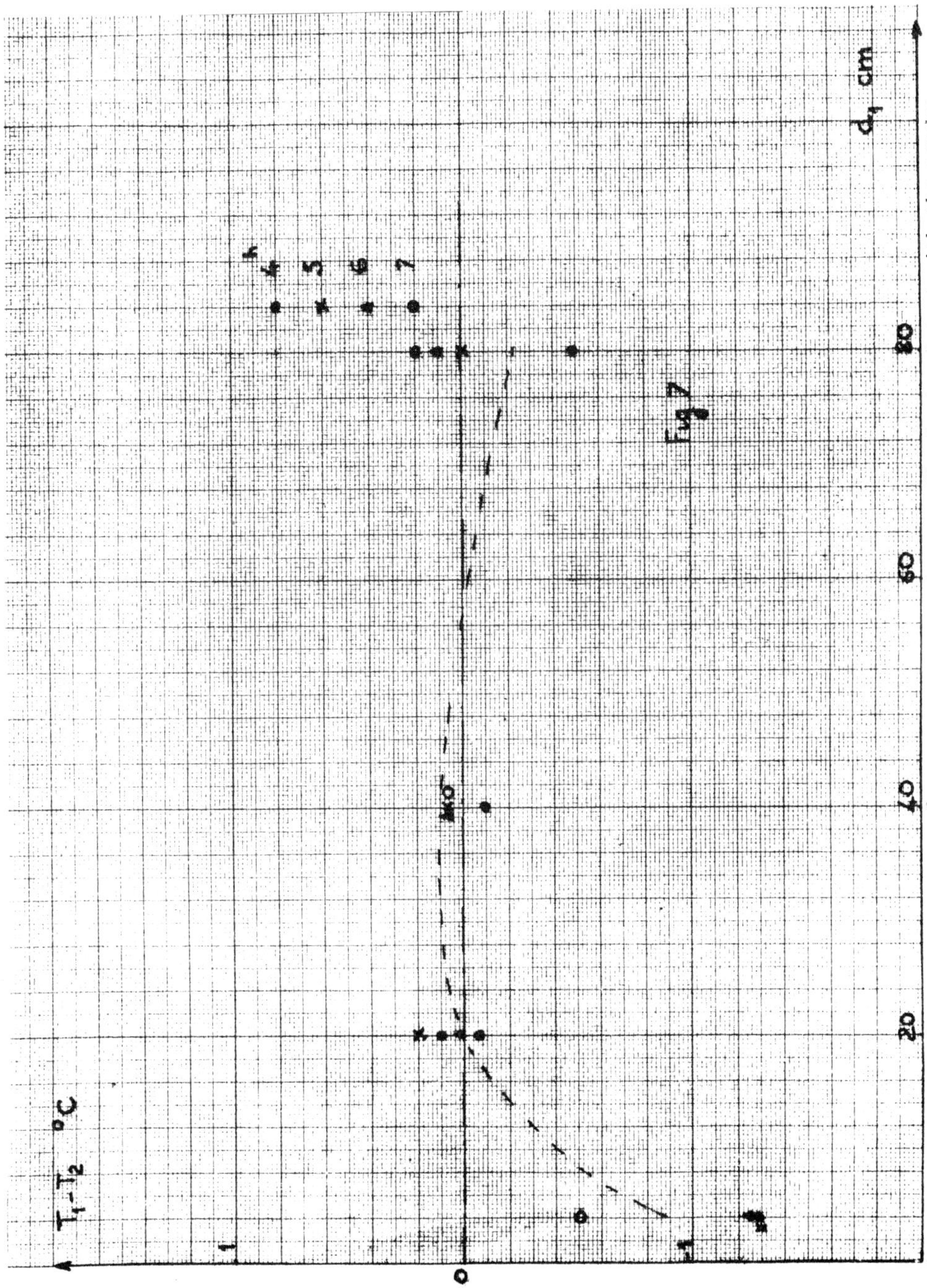

Fig. 7 $T_1 - T_2$ as a function of d_1; the thermocouple Th_1 is in contact with the screen

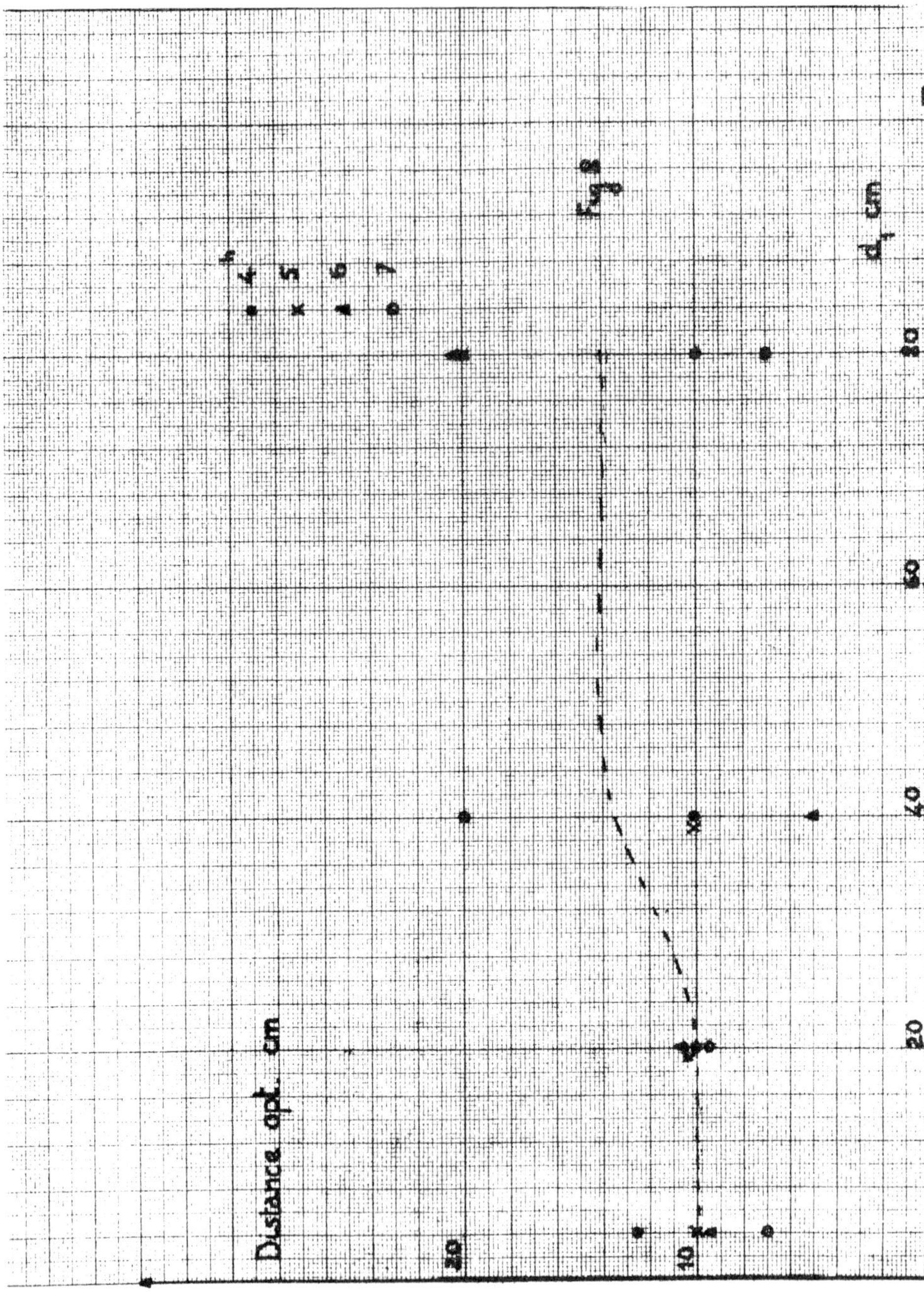

Fig. 8 d_{opt}, the optimum distance of the screen with respect to Th_1, as a function of d_1

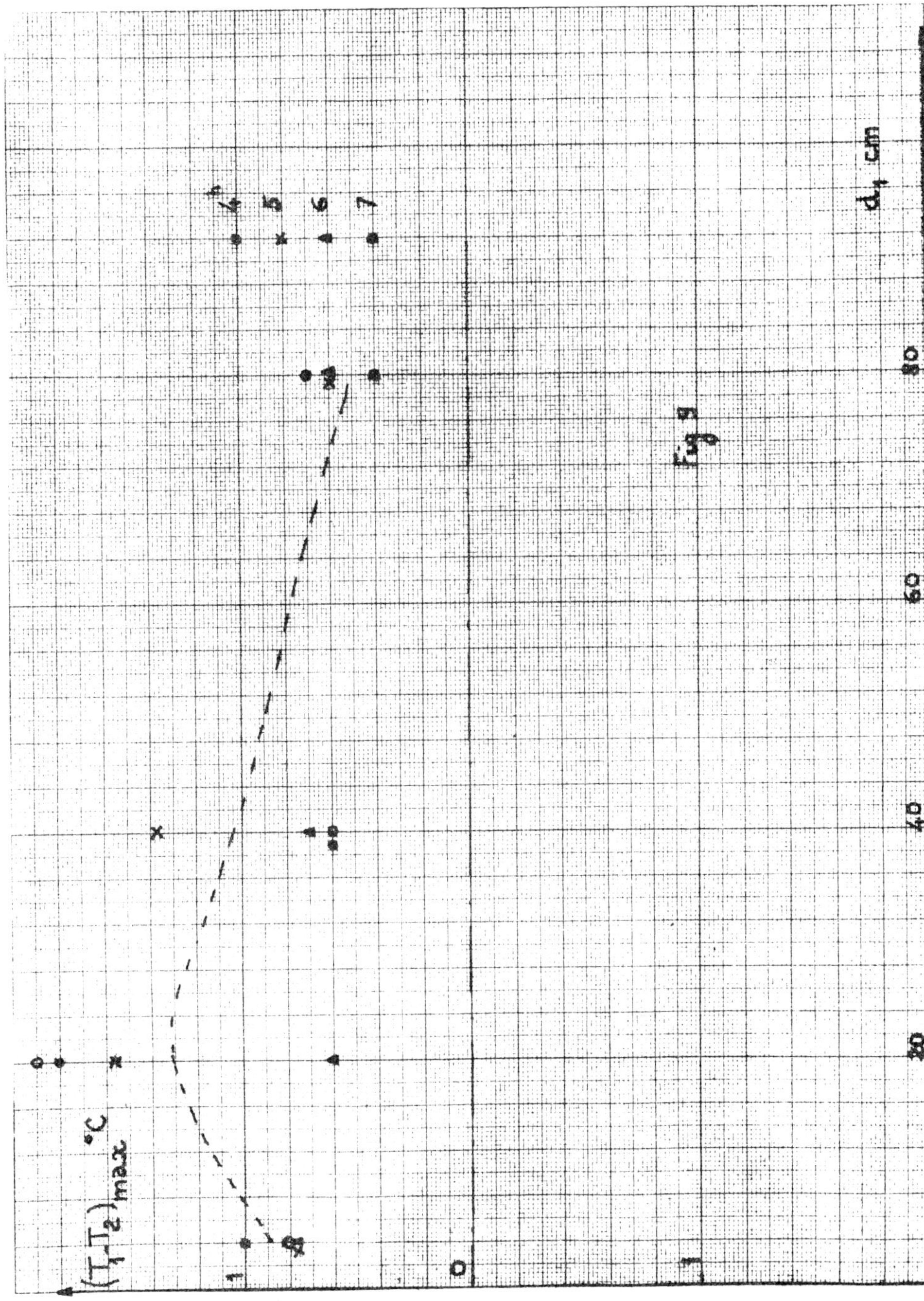

Fig. 9 $(T_1 - T_2)_{max}$ as a function of d_1

Though there is a dispersion in the results which are inherent to the measurements (measurement of very small temperature differences) and to the atmospheric disturbances (sky state, humidity, turbulence etc...) the nocturnal radiation of the ground and the screen effect are shown obviously (for example: a temperature gain).

From the above results, it is interesting to study the variation of $T_I - T_2$ in function of the ambiant temperature and the screen effect with double layer (for example: for $d_I = 80$ cm; $d = 20$ cm).

$T_I - T_2$ in function of the ambiant temperature seems to obey the law of proportionnality in T^4 (fig I0,II,I2,I3).

To study the screen effect with double layer, the temperature differences $T_I - T_2$ for a single and a double layered screen are compared. Figures I4,I5,I6,I7 show that there is a heat gain in the case of the double-layered screen compared to that with a single layer and this can perhaps be due to the opacity of the nocturnal radiation of the ground.

4 Conclusions

The importance of the nocturnal radiation of the ground and of the circular screen effect is being analyzed in this study. The practical parameters are: screen position with respect to the ground, screen position with respect to the thermocouple Th_I (the thermocouple represents the top of the plants), the degree of opacity of the nocturnal radiation of the ground.

The form and the nature of the screen, the thermal conductivity of the soil and that of the air layer are yet to be investigated.

Annexe

The heat transfer process at the ground surface is given by the equation

$$S - DC_p d_a \left(\partial T_p / \partial z\right) - C_1 - k_I \left(\partial T / \partial z\right) = 0 \qquad \text{A-I}$$

Case:

$$S = S_I = \text{constant}, \; DC_p d_a \left(\partial T_p / \partial z\right) = 0, \; C_1 = 0 \qquad \text{A-2}$$

The equation is reduced to

$$S - k_I(\partial T/\partial z) = 0 \quad \text{A-3}$$

For a layer very near the ground, the only heat transfer process is the heat conduction of the soil.

For a width dz,

$$C_t(z) = k_I\, \partial T/\partial z \quad \text{A-4}$$

$$C_t(z+dz) = k_I\, \partial T/\partial z + (\partial/\partial z)(k_I\, \partial T/\partial z)dz \quad \text{A-5}$$

So, the net heat gain of the layer is:

$$k_I(\partial^2 T/\partial z^2)dz = C_I d_I dz\ (\partial T/\partial t) \quad \text{A-6}$$

For a homogeneous soil (k_I, C_I, d_I are constants), we have

$$\partial T/\partial t = K_I(\partial^2 T/\partial z^2) \quad \text{A-7}$$

where $K_I = k_I/d_I C_I$

The boundary conditions are:

$$T(z,0) = T_I^o \text{ (temperature at sunset)} \quad \text{A-8}$$

$$k_I(\partial T/\partial z)_{z=0} = S_I$$

The solution is:

$$T = T_I^o - 2S_I(t)^{I/2} / C_I d_I (\pi K_I)^{I/2} \quad \text{A-9}$$

If S is considered as a function of temperature, the development in TAYLOR's series give:

$$S = S_I + (dS/dT)(T_I - T) \quad \text{A-I0}$$

The boundary conditions are:

$$T(z,0) = T_I^o \quad \text{A-II}$$

$$k_I(\partial T/\partial z)_{z=0} = S_I + (dS/dT)(T_I - T)$$

The solution is:

$$T = T_I^o - (S_I/(dS/dT))(I - \exp(t/t_I)(I - \phi(t/t_I)^{I/2})) \quad \text{A-I2}$$

where $t_I = C_I d_I K_I/(dS/dT)^2$,

$$\phi(x) = 2(\pi)^{-I/2}\int_0^x \exp(-s^2)ds$$

In order to compare A-I2 with A-9, the latter is put in the form: $T = T_I - 2S_I(t/\pi t_I)^{I/2} / (dS/dT)$ A-I3

Case:

$$S = S_I = \text{constant},\ C_1 = 0$$

For an atmospheric layer just above the ground surface, the temperature variation is given by:

$$\partial T/\partial t = 4.I(\partial S_I/\partial p) + (\partial/\partial z)(D(\partial T_p/\partial z)) \qquad \text{A-I4}$$

Supposing that D is independent of z, the above equation is found with the boundary conditions

$$T(z,0) = T_I^{o} \qquad \text{A-I5}$$

$$DC_p d_a(\partial T_p/\partial z)_{z=0} - k_I(\partial T/\partial z)_{z=0} = S_I$$

Supposing that the temperature variation at the boundary ground-air is continuous, then the solution is:

$$T = T_I - 2S_I(t)^{I/2} / (C_I d_I(\pi k_I)^{I/2} + C_p d_a(\pi D)^{I/2} \qquad \text{A-I6}$$

Bibliography

[I] H. Bouasse - Emission chaleur solaire Eclairage Théorie de l'émission. Librairie Delagrave I925

[2] G.J. Haltiner, F.L. Martin -Dynamical and physical meteorology Ed.Mc Graw-Hill I957

[3] A. Poggi - Introduction à la micrométéorologie, transferts d'énergie atmosphère-sol Ed. Masson I977

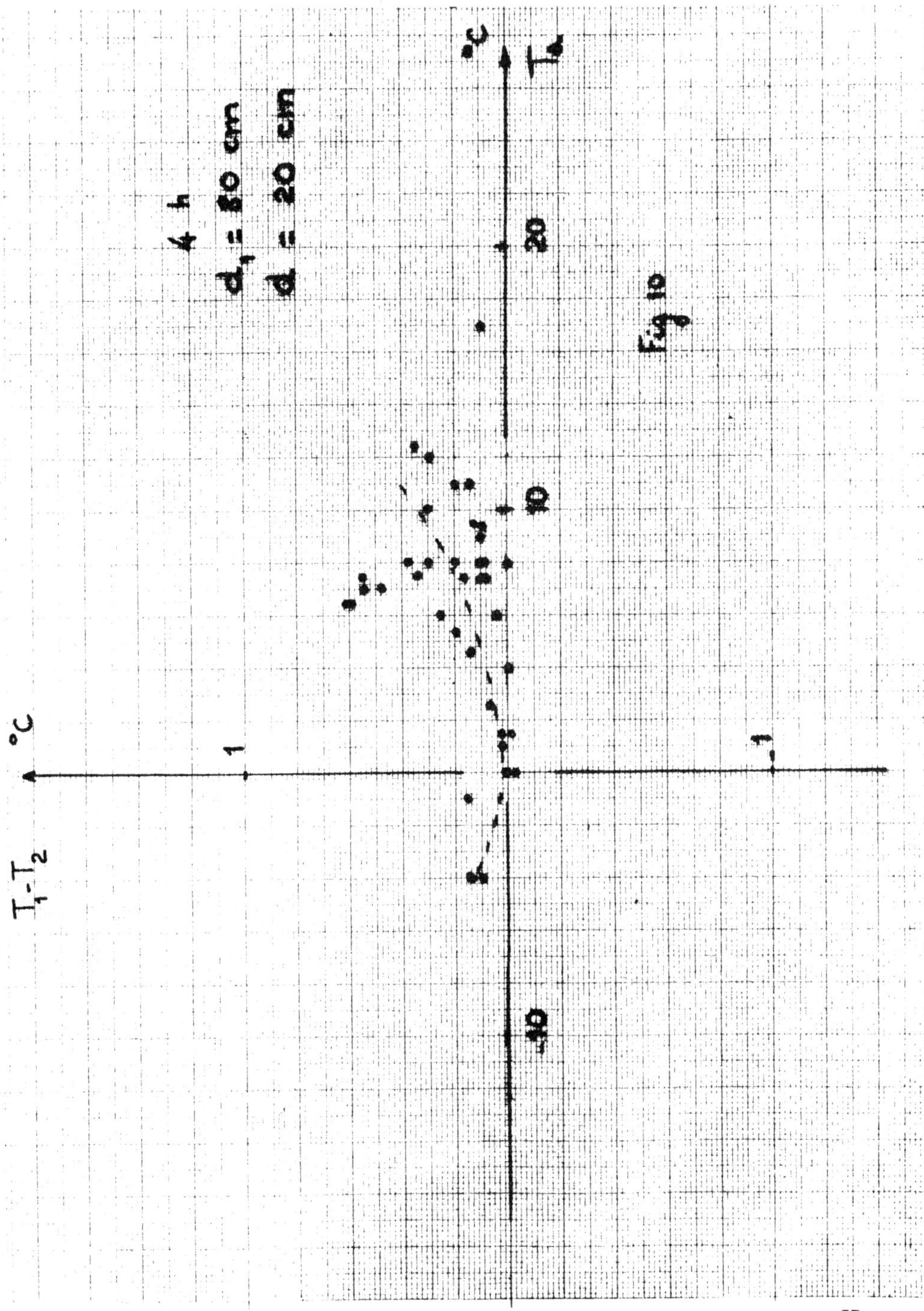

Fig. 10 $(T_1 - T_2)$ as a function of the ambient temperature

JB

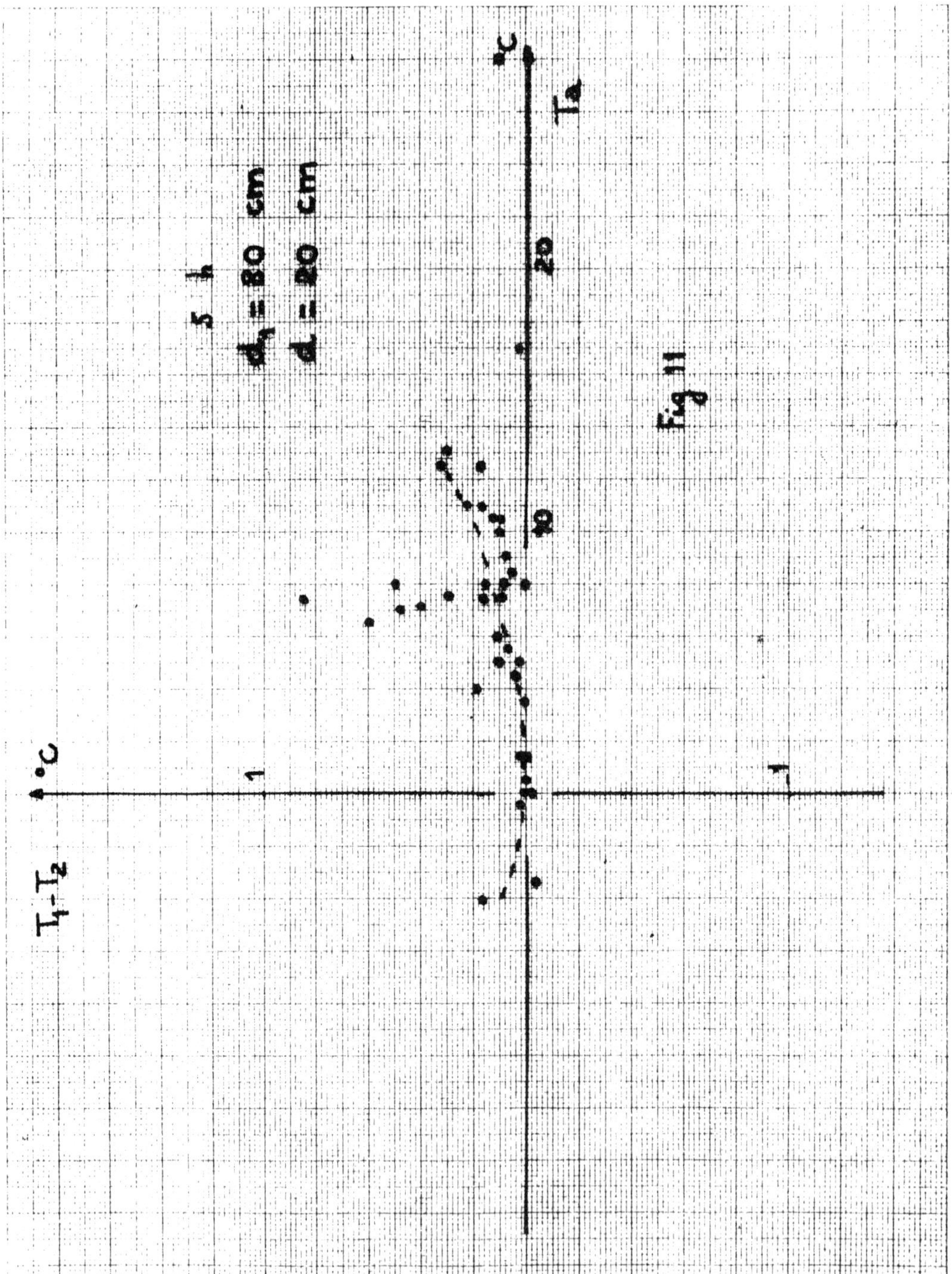

Fig. 11 $(T_1 - T_2)$ as a function of the ambient temperature

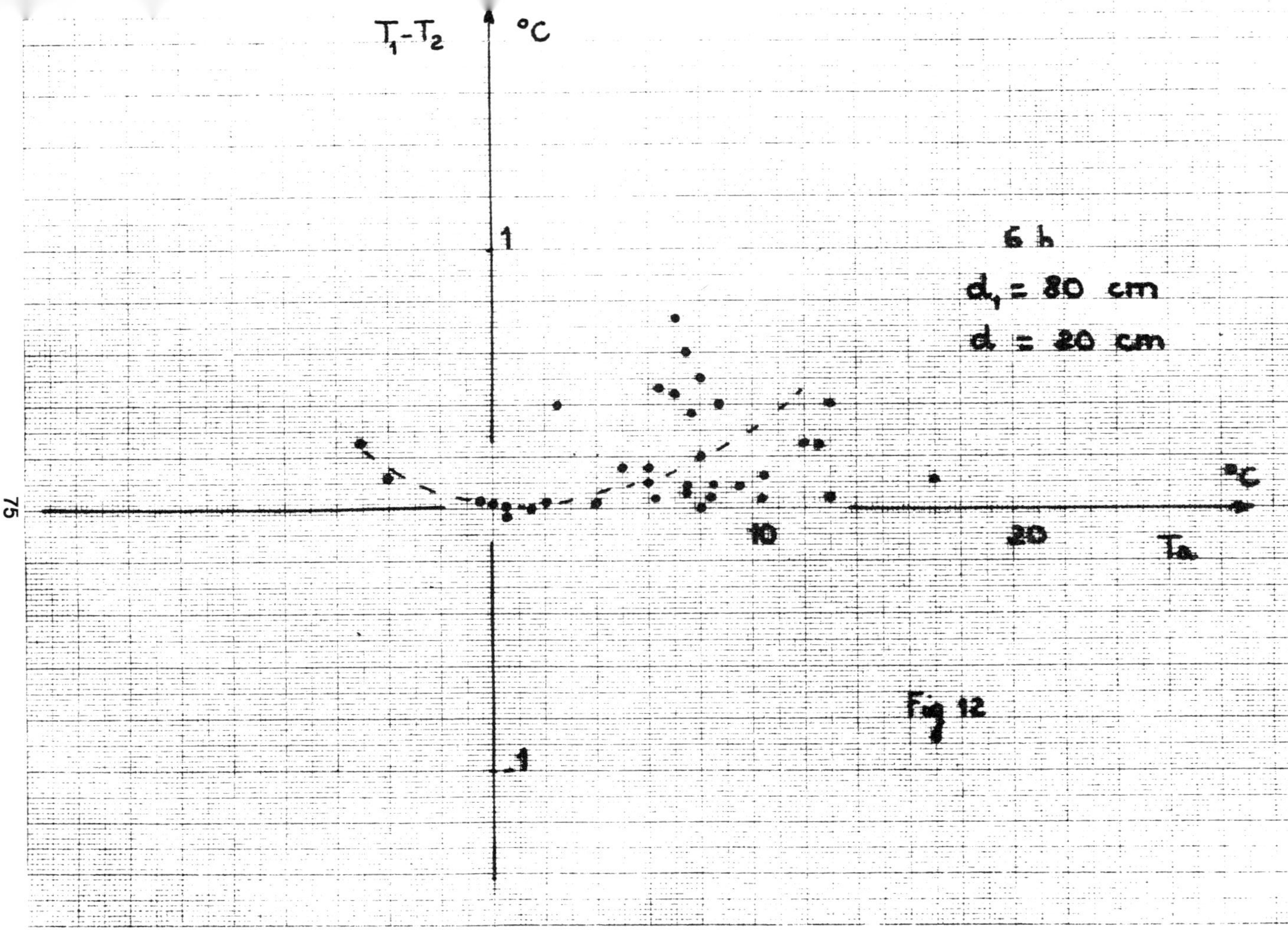

Fig. 12 (T_1 - T_2) as a function of the ambient temperature

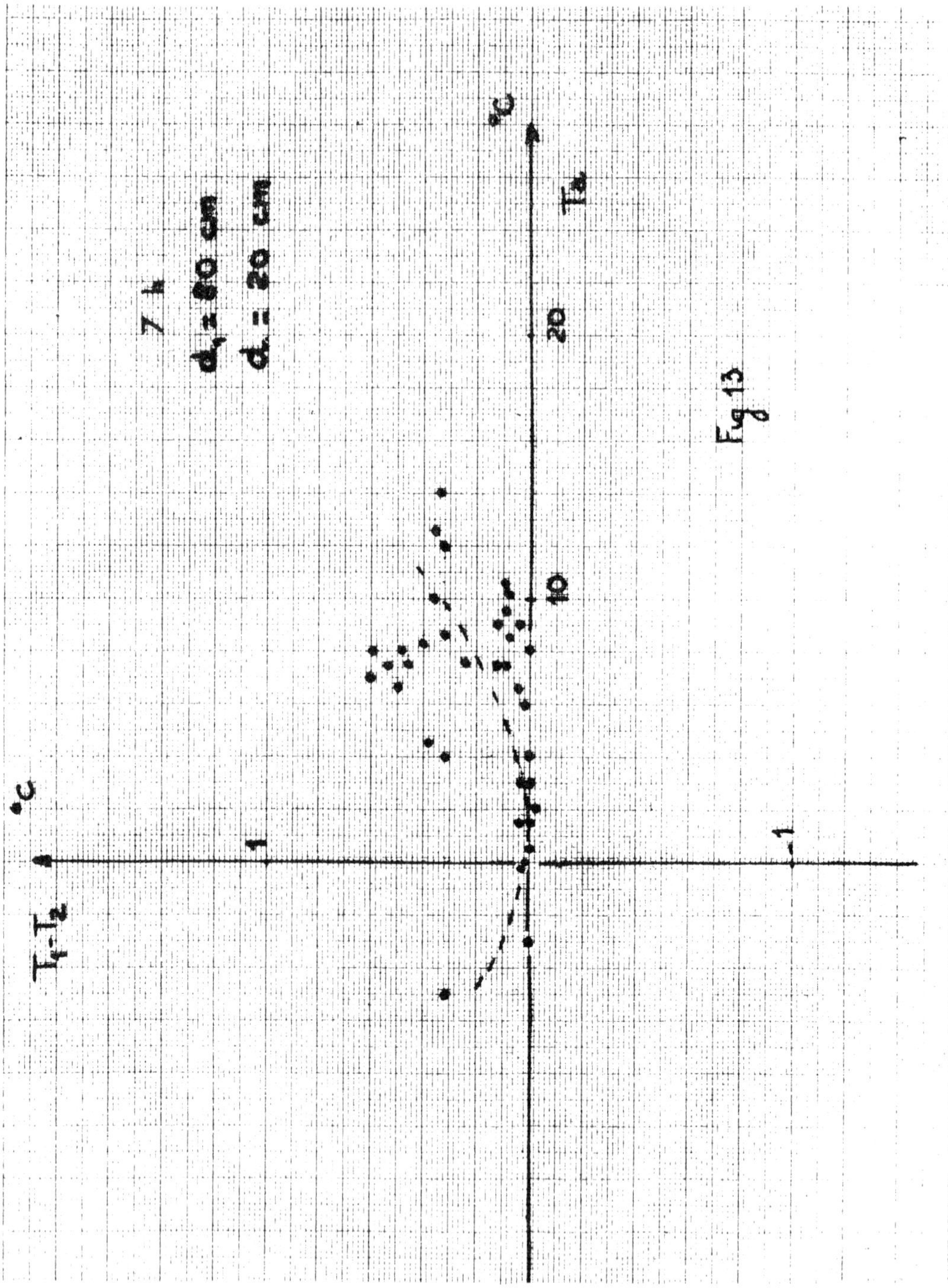

Fig. 13 $(T_1 - T_2)$ as a function of the ambient temperature

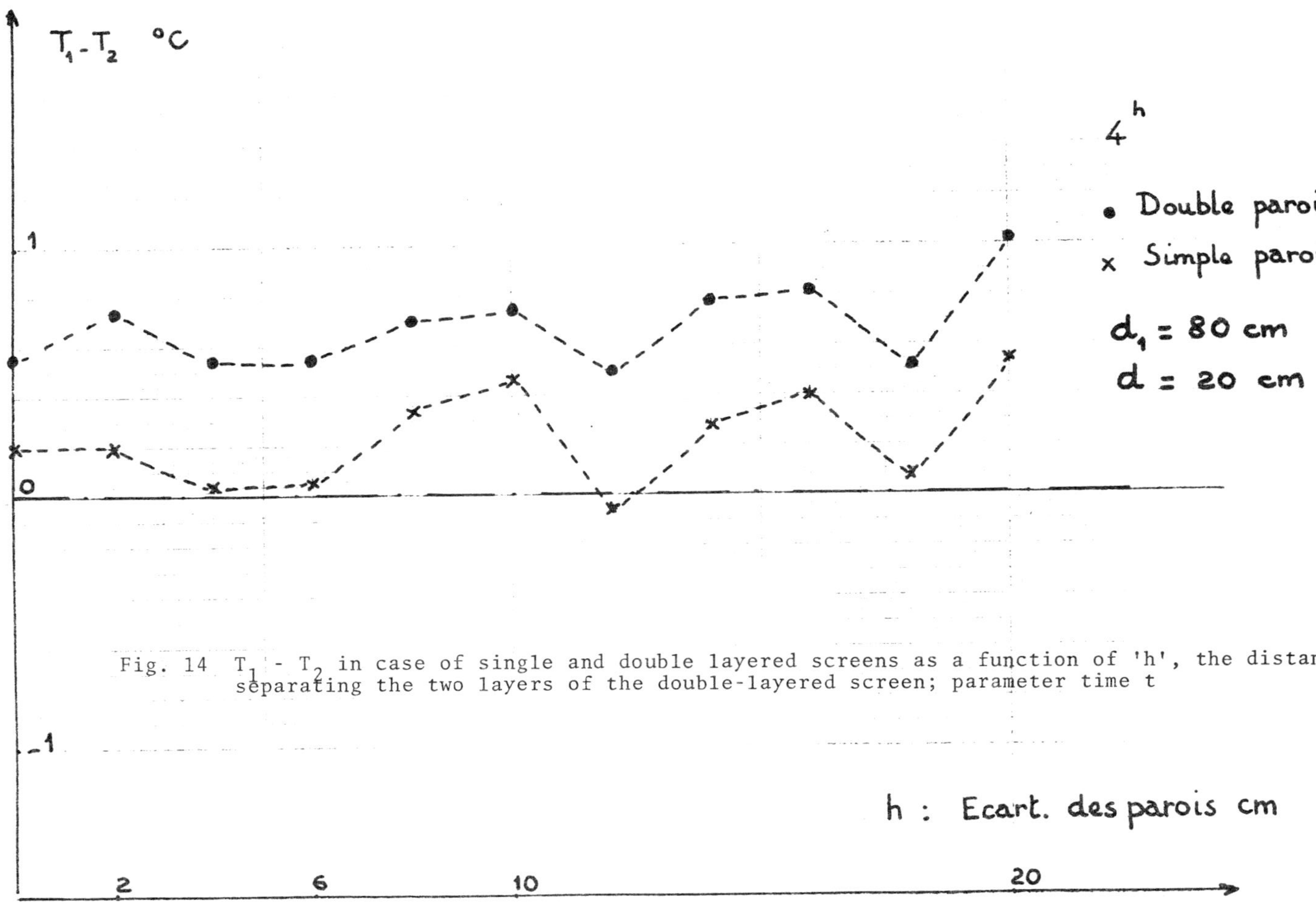

Fig. 14 $T_1 - T_2$ in case of single and double layered screens as a function of 'h', the distance separating the two layers of the double-layered screen; parameter time t

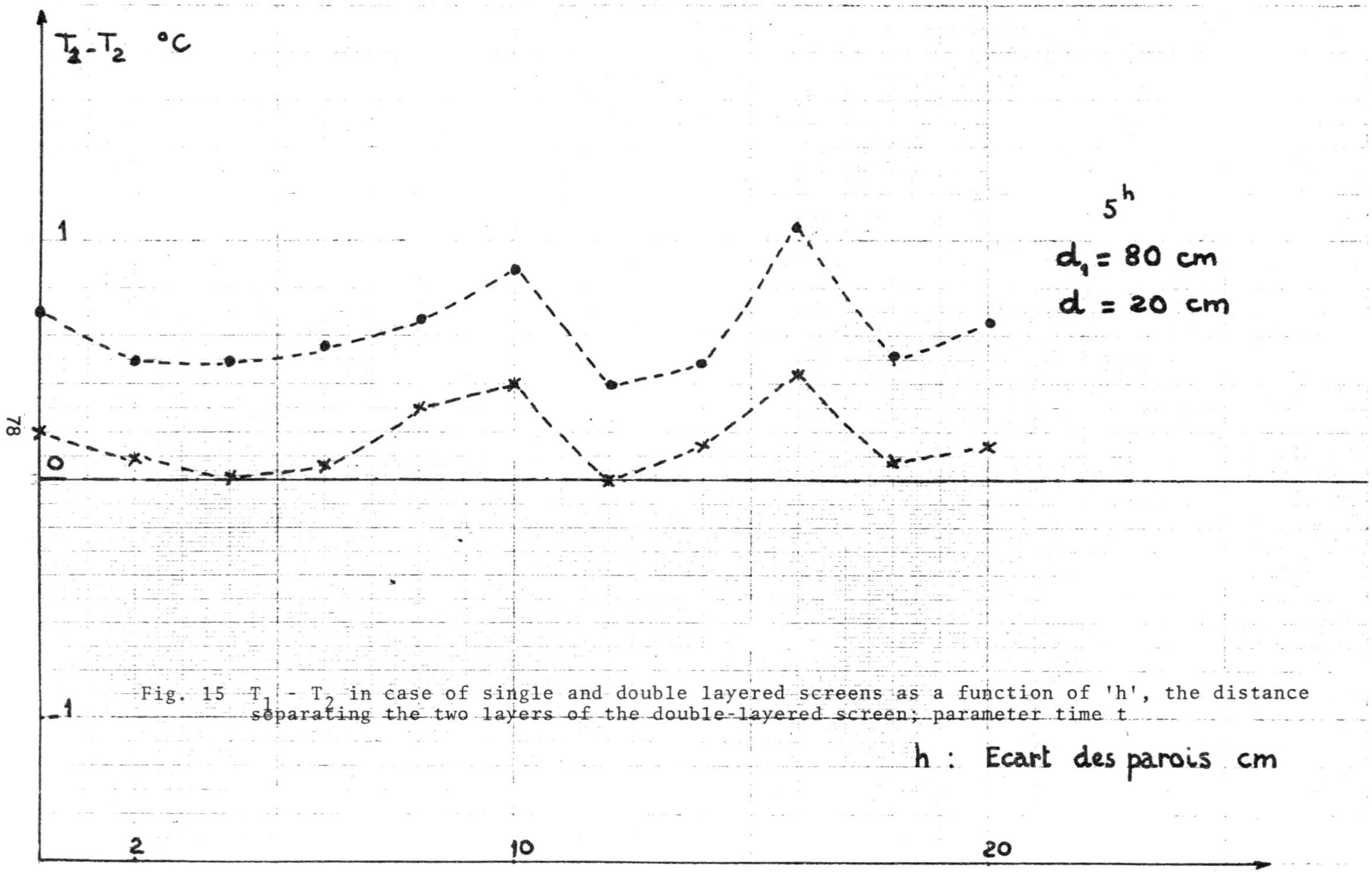

Fig. 15 $T_1 - T_2$ in case of single and double layered screens as a function of 'h', the distance separating the two layers of the double-layered screen; parameter time t

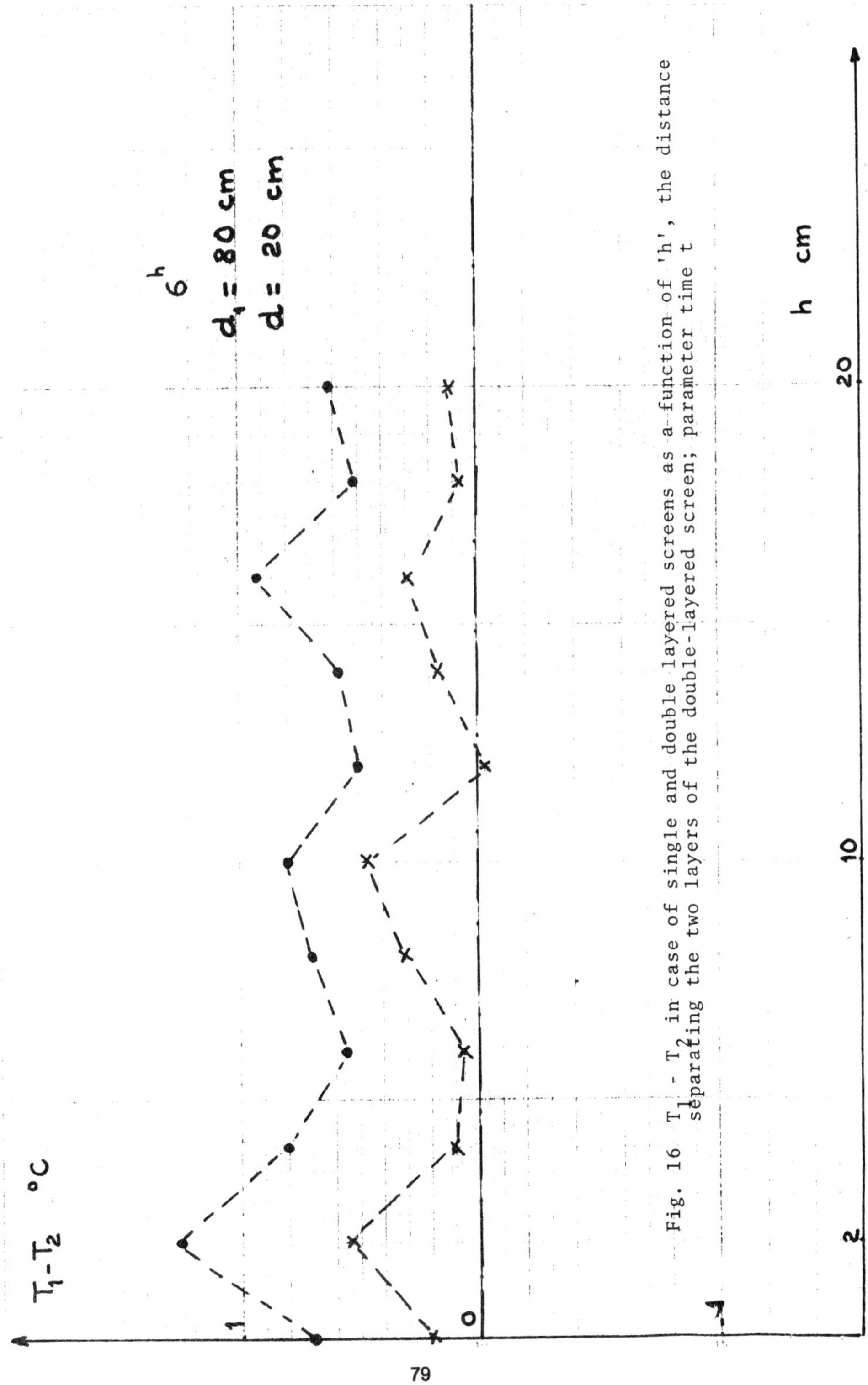

Fig. 16 $T_1 - T_2$ in case of single and double layered screens as a function of 'h', the distance separating the two layers of the double-layered screen; parameter time t

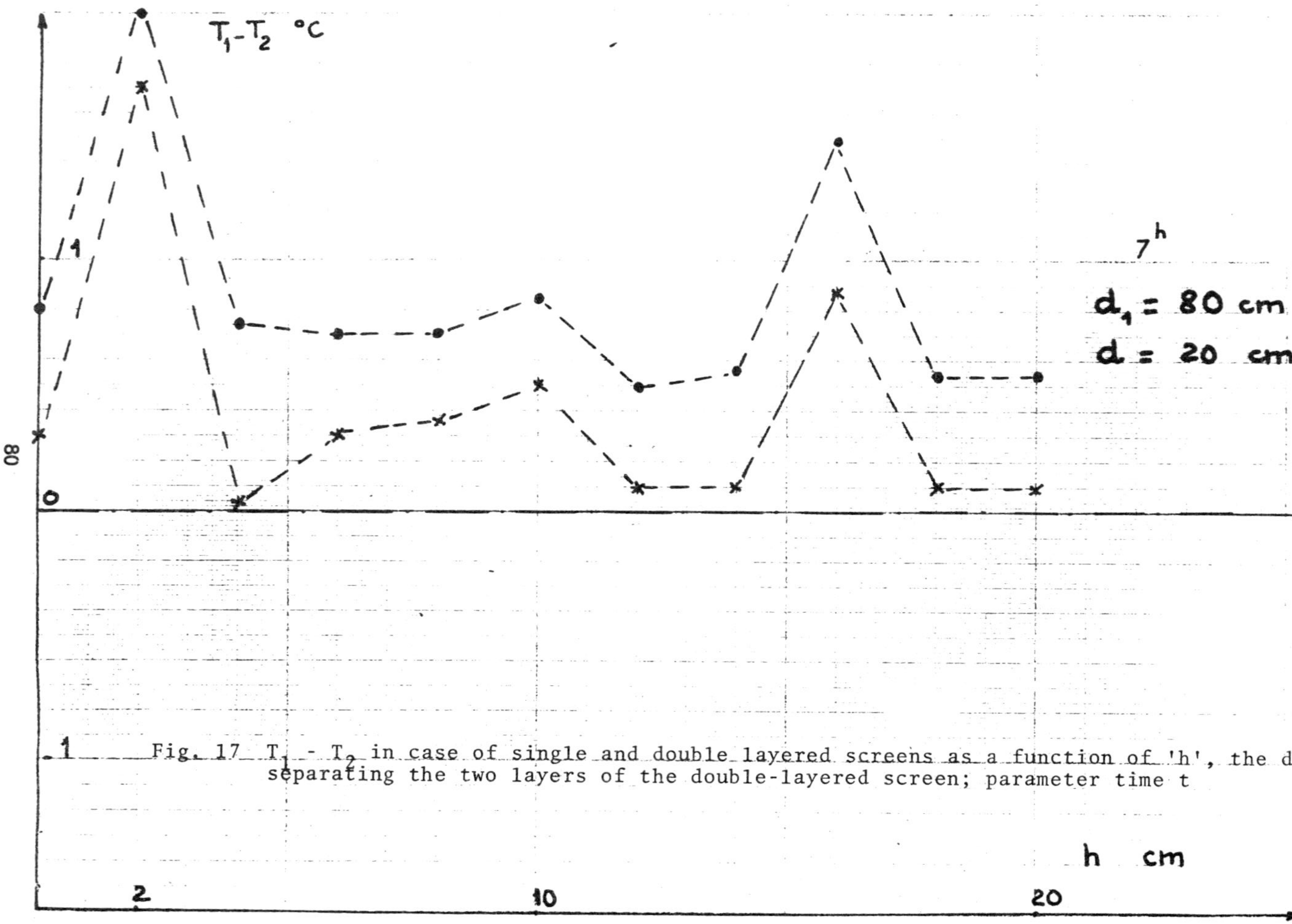

Fig. 17 $T_1 - T_2$ in case of single and double layered screens as a function of 'h', the distance separating the two layers of the double-layered screen; parameter time t

NEW FORMULATION FOR THE FACTOR Ψ FOR THE MIDDLE EAST REGION

E.M. Abd El-Salam
Cairo University
Giza, Egypt

ABSTRACT

Total solar radiation values were correlated with climatological data, ($S = D/D_O$, $t = r/M$ & R), and have been computed across the Middle East Region.

The Reddy's relation has been used first in some localities where the experimental values directly measured allowed a direct check of its validity in this region. It was found that by a modification of the seasonal factor Ψ, a much better fit for the results was achieved.

The new seasonal factor Ψ, for coastal, inland and hilly localities in the region were listed. The computed total solar radiation for all stations are used to map the solar intensity distribution over the entire Middle East region.

INTRODUCTION

The Middle East region is really the land of the sun. It is gifted by the ease of availability of solar radiation. It lies nearly with the longitudes 20°E and 60°E and latitudes between 5°N and 35°N.

In the generalized world map of the average solar energy received at the earth's surface, published by Landsberg [1] in 1961, the Middle East region appears crossed by the 160 K. $cal.cm^{-2}$ yr^{-1} and the 200 K. $cal.cm^{-2}$ yr^{-1} isolines.

In 1965, Lof et al. [2] made an attempt at mapping a world distribution of solar radiation. They prepared a series of 12 monthly world maps of the daily means of total solar radiation incident on a horizontal surface. The radiation levels corresponding to the region varies from 150 $cal.cm^{-2}$ day^{-1} at the north in December to 650 $cal.cm^{-2}$ day^{-1} at the south in June.

In 1973, Kettani [3] mapped the isolines of total incoming solar radiation received in the Arabian Peninsula, using the Reddy's complete formula [4]. In 1975, Sabbagh, Sayigh and Salam [5], also mapped the isolines for the Arabian Peninsula by using a newly developed formula making use of the climatological data collected over a widely distributed number of stations. They used only the Reddy's latitude and altitude factor K in the computations.

In this investigation, the Reddy's method for estimation of total solar radiation, with new values of the seasonal factor $\Psi_{i,j}$, suitable for this region, was used. The isolines with the monthly values were drawn, which could be of great use to the designing engineers working in the field of solar energy utilization.

COMPUTATION AND ANALYSIS OF THE RESULTS

In an attempt to draw the isolines for the Middle East region as a whole, the Reddy's formula was employed to compute the daily total solar radiation in many locations. The formula is

$$R_T = K \left[\frac{(1 + 0.8\ S)\ (1 - 0.2\ t)}{0.1\ (R)^{\frac{1}{2}}} \right] \text{cal.cm}^{-2}\ \text{day}^{-1}$$

where $K = (\lambda N + \Psi_{i,j} \cos \phi) \times 10^2$ cal.cm^{-2} day^{-1}, in which

ϕ = latitude of the place in degrees
λ = latitude factor = $0.2/(1 + 0.1\ \phi)$
N = mean length of the day during the month
$\Psi_{i,j}$ = seasonal factor, where i = 1, 2 and 3 refers to inland, coastal, and hilly stations, and j = 1, 2,....12, which stands for the months January to December, respectively.
S = the bright sunshine duration as a fraction of the possible, i.e., D/D_o.
t = r/M (r = number of rainy days during the month and M = the number of days in the month).
R = the mean humidity per day in the month.

For hilly stations, $K' = K + \varepsilon$, where $\varepsilon = 0.5\ H\ (12-N) \times 10^{-2}$, in which H is the height of the station above mean sea level in meters.

The Reddy's empirical relation was applied to 19 stations with direct measurements of total solar radiation. Verification of the results was obtained by comparing them with the values of the direct measurements. Table 1, as sample calculations, shows the general trend that the computed values were higher than the direct measured ones for initial and final months and lower for the middle months of the year.

TABLE 1

COMPARISON BETWEEN COMPUTED TOTAL SOLAR RADIATION AND ITS MEASURED VALUES (SAMPLE)

Months	Measured Q	Computed Q	"A" % dev.	Computed Q	"B" % dev.
J	251.9	304.1	+20.7	265.5	+5.4
F	329.3	359.1	+ 9.1	342.1	+3.7
M	406.4	380.1	- 6.5	443.3	+9.2
A	516.0	448.0	-13.2	481.9	-6.7
M	570.0	563.4	- 1.2	571.6	-2.9
J	588.5	600.9	+ 2.1	553.8	-6.3
J	591.1	591.4	+ 0.1	491.7	+0.1
A	543.1	535.5	- 1.4	539.7	-0.6
S	469.5	544.5	+16.0	475.4	+1.3
O	388.2	444.6	+14.5	385.1	-0.8
N	282.4	319.5	+13.2	272.2	-3.4
D	235.9	267.1	+13.2	223.6	-5.2

Computed "A" using Reddy's seasonal factors.

Computed "B" using the new seasonal factors.

TABLE 2

SEASONAL VARIATION FACTOR $\Psi_{i,j}$

	J	F	M	A	M	J	J	A	S	O	N	D
$\Psi_{1,j}$	1.04	1.35	1.64	1.96	2.12	2.15	2.27	2.16	1.97	1.41	1.09	0.88
$\Psi_{2,j}$	1.08	1.38	1.64	2.07	2.15	2.14	2.21	2.06	1.89	1.47	1.09	0.90
$\Psi_{3,j}$	1.37	1.67	2.09	2.27	2.30	2.45	2.37	2.29	2.23	1.89	1.50	1.23

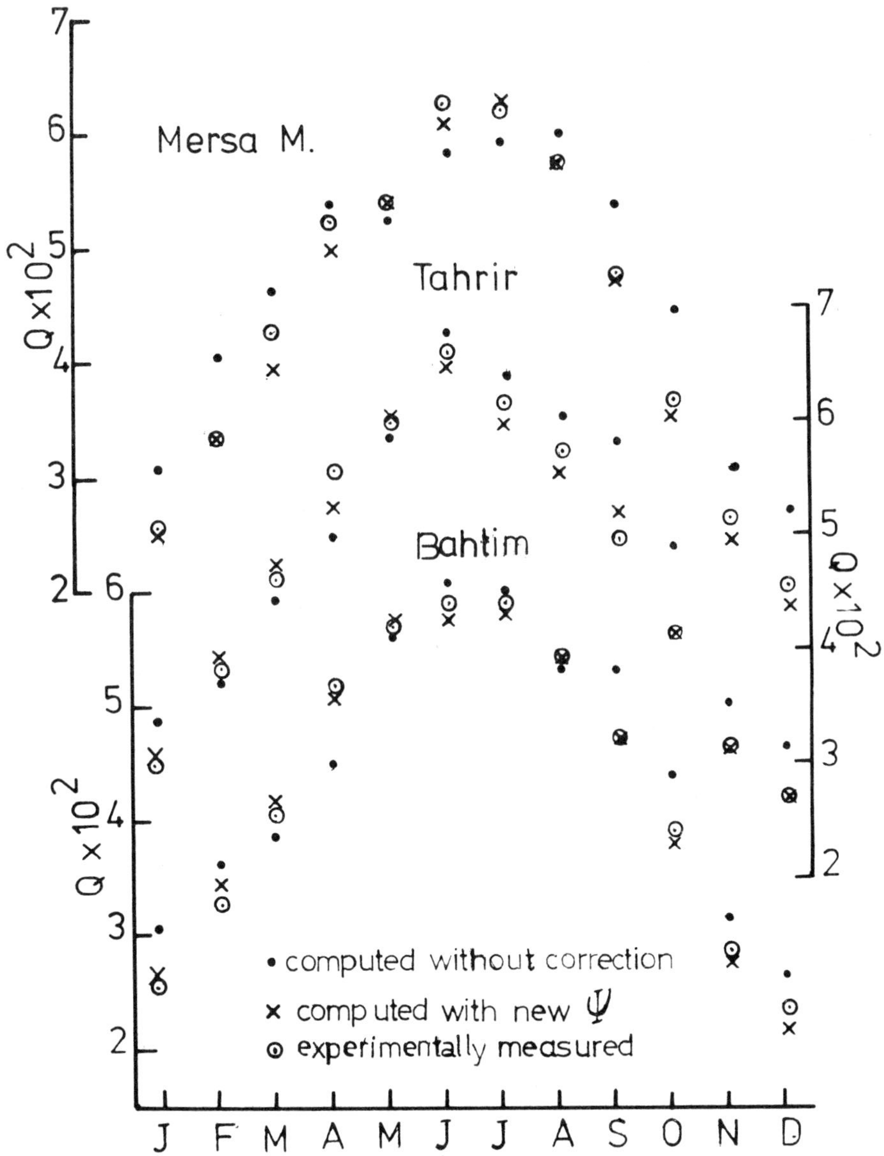

Figure 1 Comparison Between Measured And Computed Solar Radiation

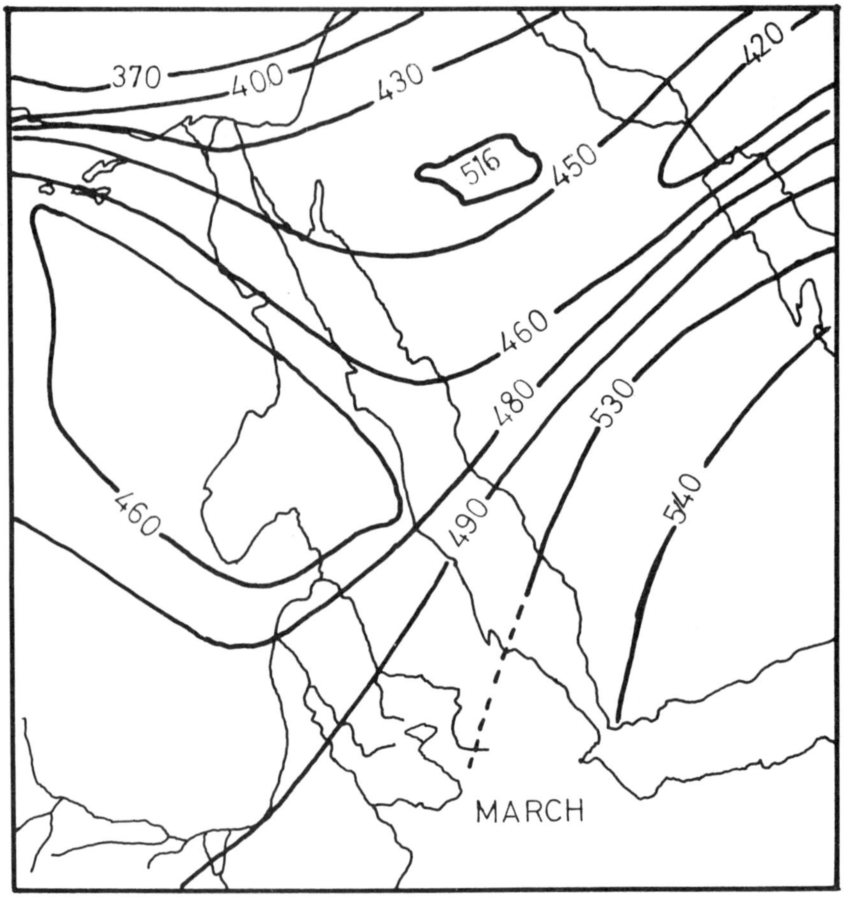

Figure 2 Map of Equinox and Solstices For the Month of March

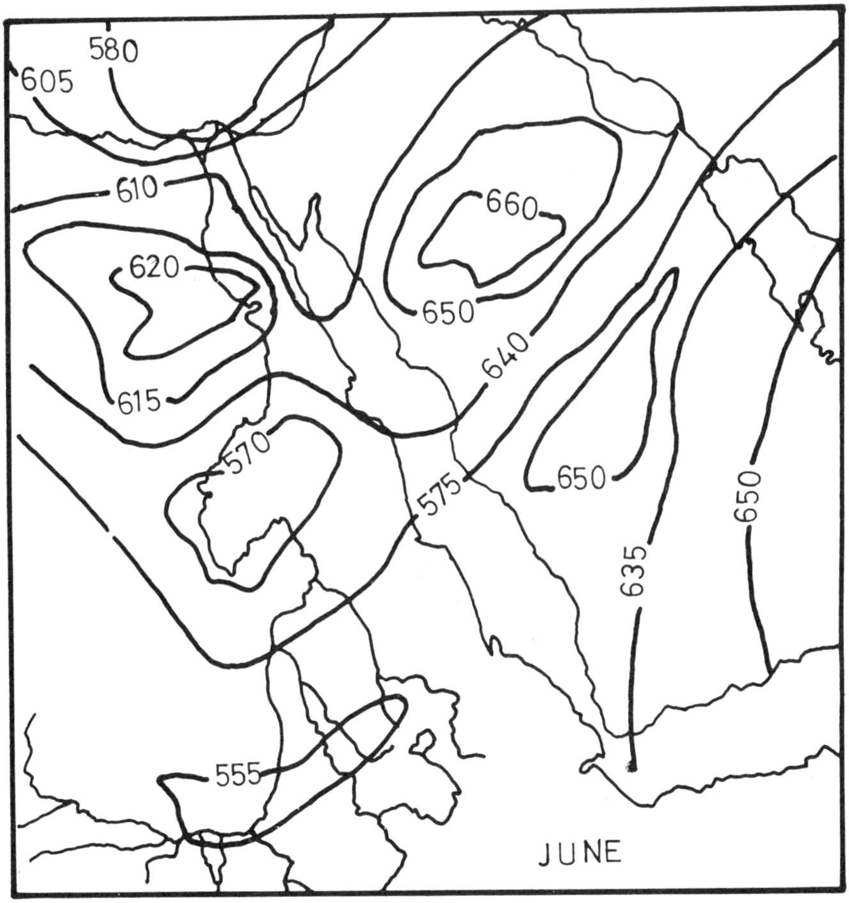

Fig. 3 Map of Equinox and Solstices For the Month of June

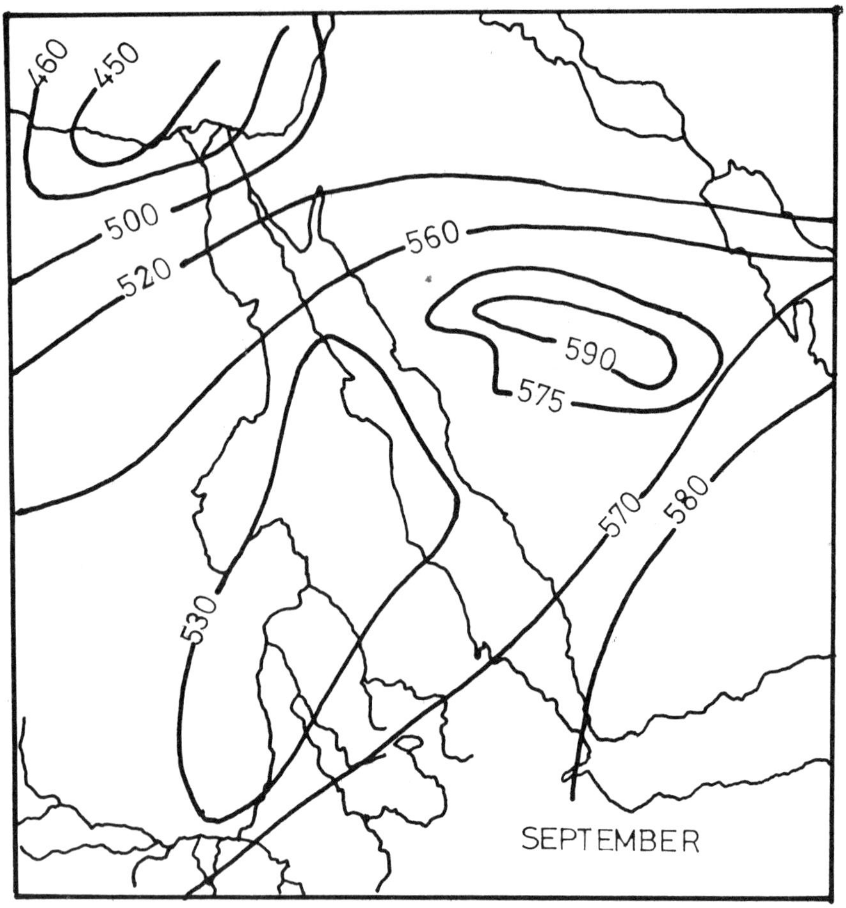

Fig. 4 Map of Equinox and Solstices For the Month of September

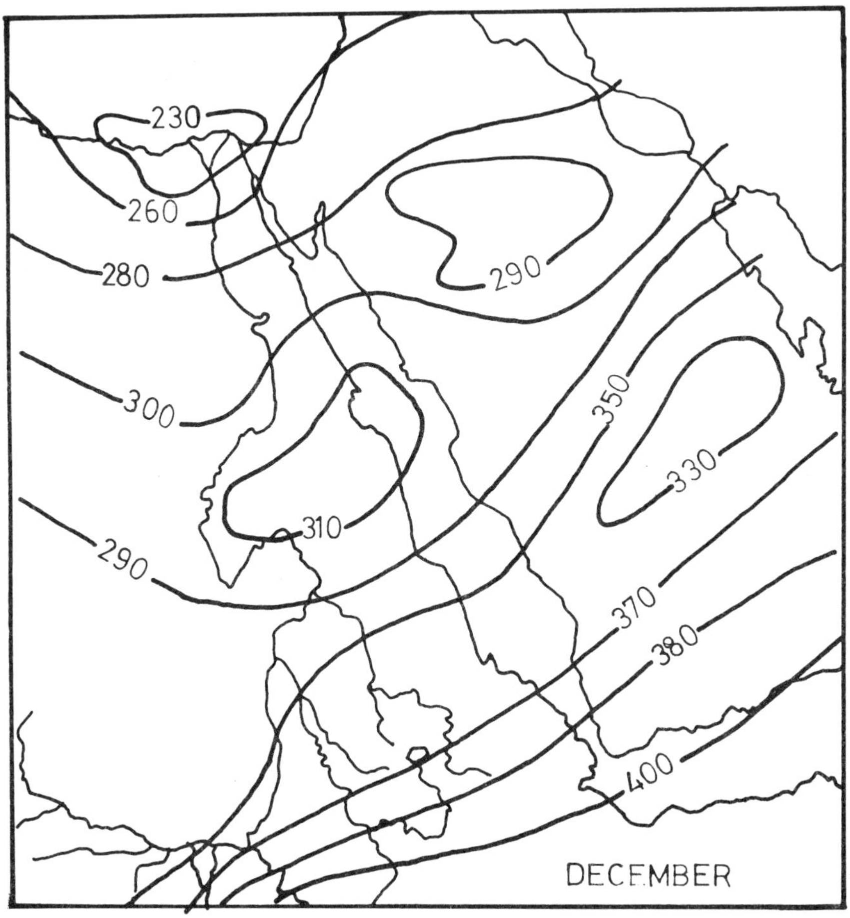

Fig. 5 Map of Equinox and Solstices For the Month of December

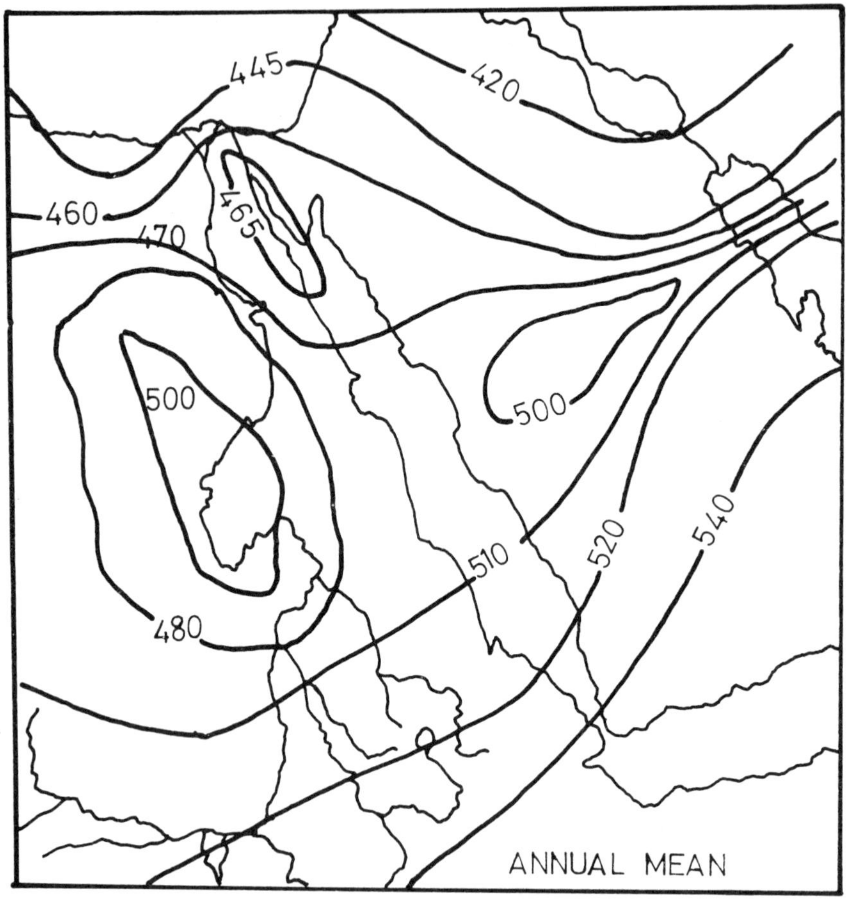

Fig. 6 Map of Yearly Average

Figure 1 represents the deviations of the values corresponding to three different stations as an example.

In this empirical relation, the seasonal factor $\Psi_{i,j}$ is a variable with monthly variations and has the greatest influence on the final values of the computed solar radiation. A method was adopted to modify that seasonal factor to minimize the sum of all deviations. Table 2 shows the values of the new seasonal factors for inland, coastal, and hilly localities, according to the meteorological and geographical characterisitcs of the Middle East region.

From the sample calculation given in Table 1, from Figure 1, and from similar ones for each station, it was observed that the computed values of the total solar radiation using the new seasonal factors are in better agreement with the direct measured ones. The mean percentage deviation was about 5% for the set of 288 data points.

Using Reddy's empirical relation, with the new seasonal factors, computations for the total solar radiation using the climatological data corresponding to about 40 different stations in the region was done. The isolines for the region with the monthly values were drawn. Four maps, corresponding to equinox and solstices, are shown in Figures 2 through 5, and a map of the yearly average is shown in Figure 6. It was observed that the yearly average of daily total solar radiation received in the region varies from 540 $cal.cm^{-2}$ day^{-1} to 420 $cal.cm^{-2}$ day^{-1}.

REFERENCES

1. Landsberg,H.E., "Solar Radiation at the Earth's Surface, Solar Energy, Vol. 5, No. 3, July-September 1961.

2. Lof, G.O.C. et al, "World Distribution of Solar Radiation", Solar Energy, Vol. 10, No. 1, January 1966.

3. Kettani, M.A., "Attempts at Mapping the Solar Intensity Distribution for the Arabian Peninsula",The Sun in the Service of Mankind", International Congress, Paris, France 1973.

4. Reddy, S.J., "An Empirical Method for Estimation of Solar Radiation", Solar Energy, Vol. 13, 1971.

5. Sabbagh, J.A., Sayigh, A.A.M. and Abd El-Salam, E.A., "Estimation of Total Solar Radiation from Meteorological Data", ISES Conference, Los Angeles, California, August 1975 - Solar Energy, Vol. 19, No. 3, 307, 1977.

COMPUTATION OF THE DIFFUSE SOLAR RADIATION FOR EGYPT

E.M. Abd El-Salam
Cairo University
Giza, Egypt

ABSTRACT

The recently measured diffuse solar radiation in stations at Cairo and Kharga is compared with the total solar radiation values for the same stations. Comparison of these results expressed as a fraction of total radiation with those from other stations in the region were established.

Relationships between the diffuse solar radiation as a fraction of the total Q_D/Q; the mean daily values of the clouidiness index Q/Q_o, the mean values of sunshine duration as a fraction of possible D/D_o and the mean amount of cloud cover X in oktas were established in the following forms:

$$Q_D/Q = A_1(Q/Q_o) + B_1$$

$$Q_D/Q = A_2(D/D_o) + B_2$$

$$Q_D/Q = A_3 X + B_3$$

where Q_D is the measured diffuse solar radiation,

Q is the total solar radiation,

Q_o is the extraterrestrial solar radiation,

D is the actual bright sunshine duration,

D_o is the total possible sunshine duration,

X is the mean amount of cloud cover, and

A_1, A_2, A_3, B_1, B_2 and B_3 are constants.

These relationships were compared with those obtained elsewhere, in order to justify their use in the regions where direct measurements are not available.

INTRODUCTION

In recent years, the number of stations recording the total solar radiation received on a horiztonal surface, Q, has increased rapidly. Many practical problems, however, require the knowledge of the magnitude of the diffuse component Q_D, as well as the total intensity. If both diffuse and total solar radiation are known, the equally important direct component can be obtained by subtraction.

Diffuse solar radiation is of great importance to architects, because of the estimation of the illumination and the amount of heat which is absorbed by various building surfaces which are not exposed to direct solar radiation. In agriculture, even illumination of the photosynthetic leaf surface of a crop is due to diffuse radiation, which in turn leads to a greater photosynthetic efficiency.

Additionally, diffuse solar radiation contributes significantly towards the collection of energy through such means as solar heating devices. This is more apparent in places where cloud cover persists for a long time during the day.

Despite diffuse solar radiation's obvious importance, it is measured at only a few locations. This has led to the development of various methods for estimating the diffuse beam when only the total is known. Even though various authors have suggested formulae for estimating total solar radiation [1-8], the estimation of the diffuse component has not been studied as extensively. Many of those, howver, consider only clear sky condition [9]. Liu and Jordan [7], developed a simple method that is applicable for all degrees of cloud cover. They equated the ratio of diffuse to total solar radiation Q_D/Q and the ratio of total to extraterrestrial radiation Q/Q_o.

They referred to the latter ratio as the "cloudiness index", since it is largely controlled by the amount of cloud cover. Stanhill [10], expressed the diffuse component as a fraction of the total, correlated between this fraction and the cloudiness index, the bright sunshine duration as a fraction of the possible and the amount of cloud cover. El-Salam and Sayigh [11], estimated the diffuse solar radiation for the Saudi Arabian Peninsula by a similar procedure.

The extraterrestrial solar radiation can be readily computed [12] or determined from published tables or monograms [13] for any station were the total radiation is measured. This allows the determination of the diffuse component from either graph or regression equations. Lui and Jordan suggested that their results would be applicable for practical use at stations with similar atmospheric dust content and surface albedo, as for which their relationships were developed.

In this study, an attempt was made to investigate, from the limited data available, the relationships between the diffuse and the total radiation and other more generally available climatological measurements; in order that they may be used for the computation of the diffuse solar radiation component for localities where only, any of, the total solar radiation; or the bright sunshine duration; or the amount of cloud cover is known.

Relationships between the diffuse solar radiation as a fraction of the total, and the bright sunshine duration as a fraction of the possible, for the stations at Cairo and Kharga, Egypt were established. These relationships were then used to estimate the diffuse component for all other localities in Egypt that do nothave such records.

The computed values of the diffuse solar radiation for all stations were correlated with the cloudiness index Q/Q_O, the bright sunshine duration as a fraction of the possible D/D_O, and the mean amount of cloud cover, X. These correlations will make it possible for the personnel working in the solar energy field to compute the diffuse solar radiation component whenever the other related climatological data are available.

EXPERIMENTAL DATA

The available measured diffuse solar radiation data for the stations of Cairo [1] and Kharga [14] was taken by the same procedure described by Drummond [15]. Analysis of these data were shown in Tables 1 and 2, from which it can be seen that the ratio of diffuse to total solar radiation in Cairo and Kharga are 34% and 25%, respectively. Also, the mean monthly diffuse solar radiation in each station is 152.3 and 128.8 gm. cal. $cm.^{-2}$. Over the year as a whole, 56.0 K cal. $cm.^{-2}yr^{-1}$ were received indirectly at Cairo and 47.0 K cal. $cm.^{-2}yr^{-1}$ at Kharga. Figures 1 and 2 show the mean monthly variation of Q, Q_D, Q_D/Q, D and X for both stations, Cairo and Kharga, respectively.

COMPUTATION OF THE DIFFUSE SOLAR RADIATION

Correlation between the diffuse solar radiation, as a fraction of the total for Cairo and Kharga; and, the bright sunshine duration expressed as a fraction of the total possible, were deduced, using the method of linear regressions. Figure 3a and 3b represent the plots of Q_D/Q vs. D/D_O for both stations; and the two linear relationships deduced were:

For Cairo $\quad Q_D/Q = -0.78\ (D/D_o) + 0.97$

$$(r) = -0.86 \qquad (1)$$

TABLE 1. MEAN VALUES OF TOTAL Q AND DIFFUSE Q_D SOLAR RADIATION ON HORIZONTAL SURFACE, AT CAIRO EGYPT. (gm cal. cm^{-2} day^{-1}).

	J	F	M	A	M	J	J	A	S	O	N	D	ANNUAL K cal $cm^{-2}yr^{-1}$
Mean Q	315	391	422	503	591	632	667	615	572	437	324	278	174.8
Mean Q_D	132	129	186	206	189	139	140	172	149	127	143	128	55.9
Q_D/Q	0.42	0.33	0.44	0.41	0.32	0.22	0.21	0.28	0.26	0.29	0.44	0.46	0.34
Mean D	7.24	8.67	8.02	9.16	11.3	13.3	13.3	11.5	11.1	9.82	7.10	6.57	9.74
Mean X	2.50	2.03	2.70	2.40	1.63	0.83	1.03	1.33	1.13	1.40	2.60	2.50	1.84

TABLE 2. MEAN VALUES OF TOTAL Q AND DIFFUSE Q_D SOLAR RADIATION ON HORIZONTAL SURFACE, AT KHARGA EGYPT. (gm cal. $cm^{-2}day^{-1}$)

	J	F	M	A	M	J	J	A	S	O	N	D	ANNUAL K cal $cm^{-2}Yr^{-1}$
Mean Q	350	408	483	571	573	603	633	599	548	463	386	334	181.0
Mean Q_D	98	94	146	177	166	159	158	145	137	111	100	90	48.1
Q_D/Q	0.28	0.23	0.30	0.31	0.29	0.26	0.25	0.24	0.25	0.24	0.26	0.27	0.27
Mean D	9.43	10.4	10.2	10.6	11.6	12.3	12.5	12.1	11.1	10.6	9.84	9.58	10.85
Mean X	1.4	0.6	1.1	0.4	0.5	0.0	0.1	0.1	0.2	0.1	1.0	1.1	0.55

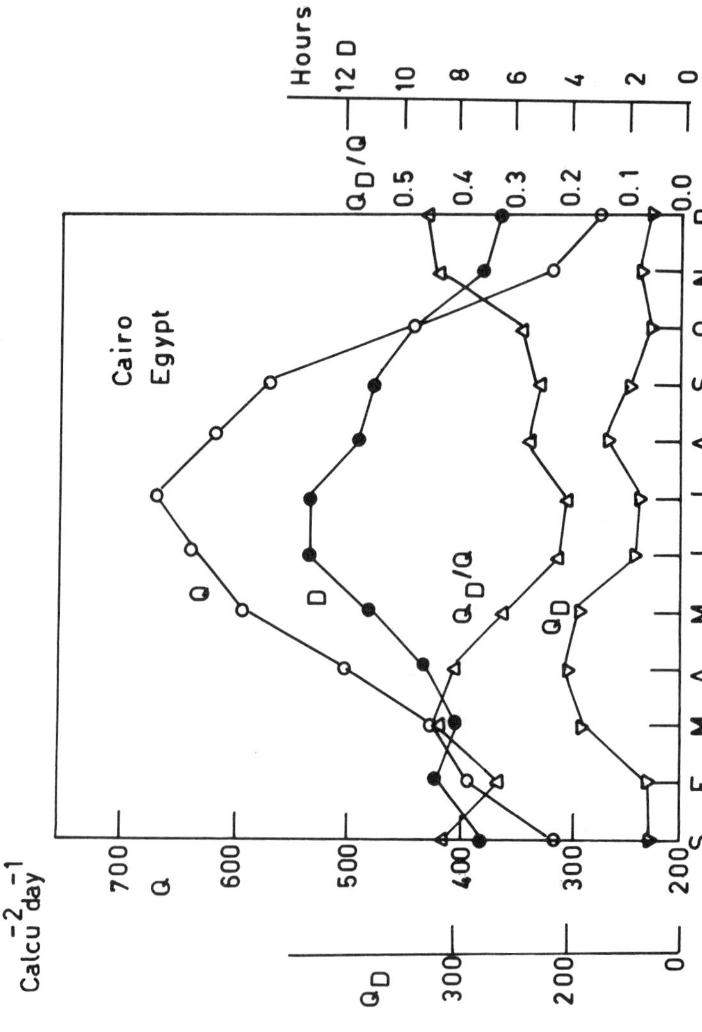

Fig.(1) Annual values of Q, Q_D, Q_D/Q, D for Cairo Egypt.

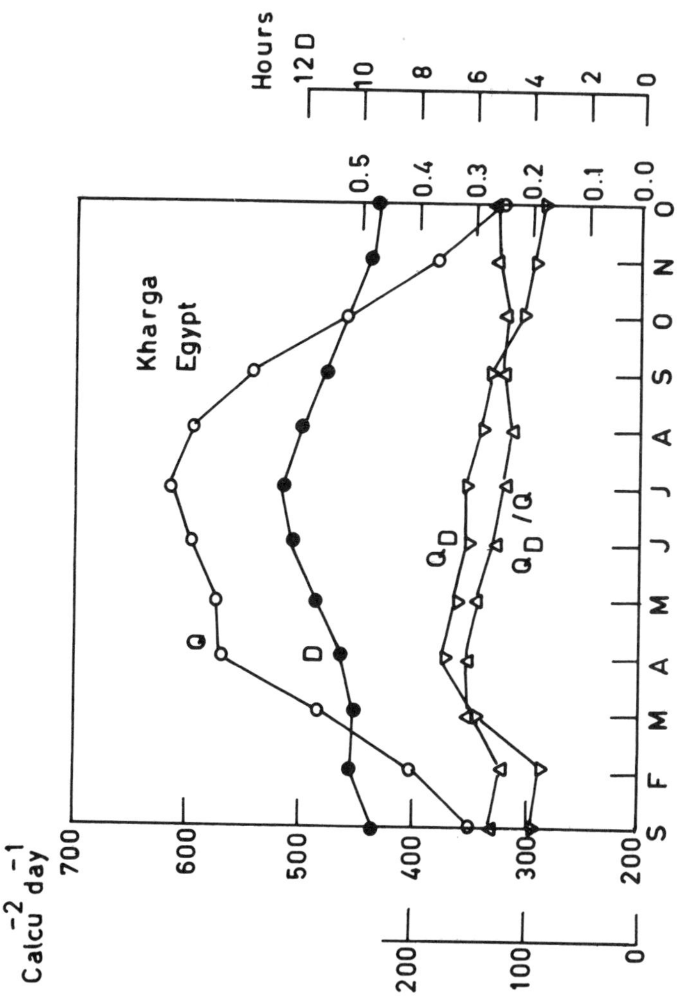

Fig. (2) Annual values of Q , Q_D , Q_D /Q , D for Kharga Egypt .

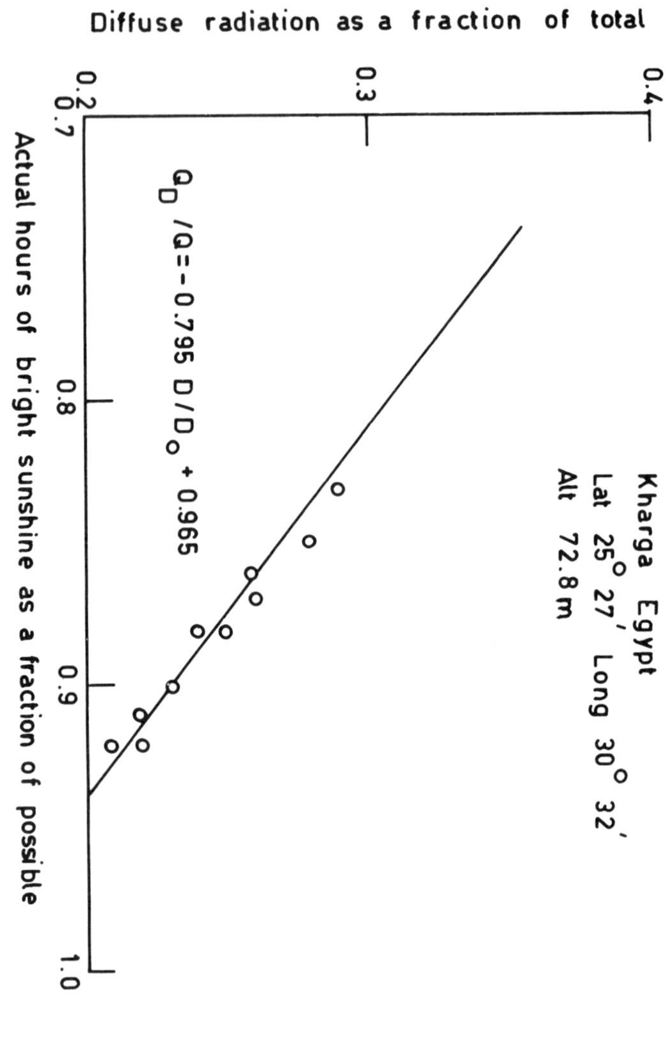

Fig. (3A) Relationship between Q_D/Q vs. D/D_o

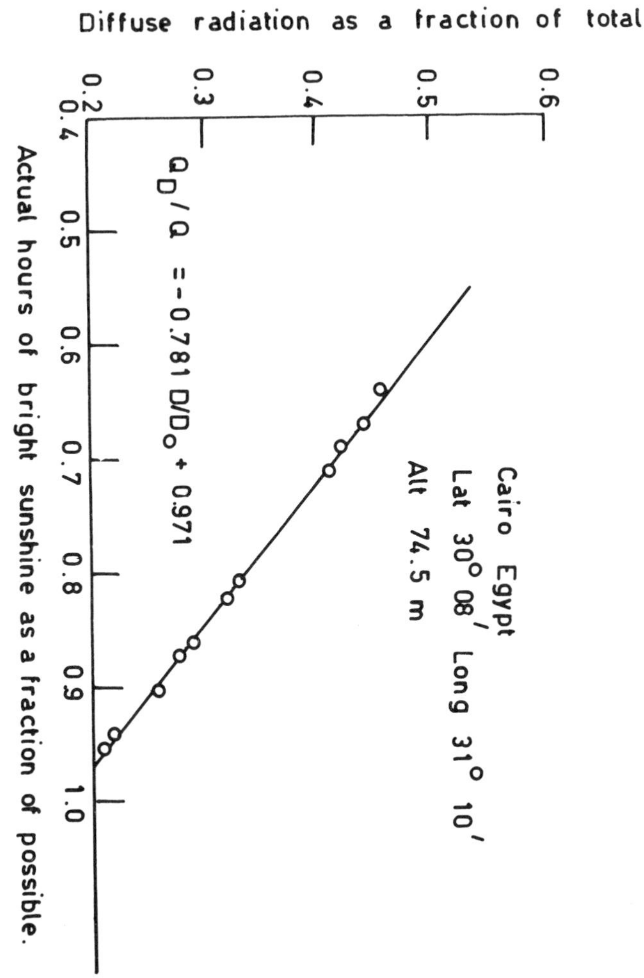

Fig. (3B) Relationship between Q_D/Q vs. D/D_o

For Kharga $Q_D/Q = -0.795\ (D/D_o) + 0.965$

$$(r) = -0.83 \qquad (2)$$

Using equations (1) and (2), the ratio of the diffuse to the total solar radiation for all stations recording the bright sunshine duration was computed. Equation (1) was used for stations of latitudes between 17°N and 27°N, while equation (2) was used for stations of latitudes between 27°N and 34°N. This was justified because the ratio Q_D/Q in various localities in the world does not vary a great deal from the mean value of 0.335 (as will be shown later).

The computed values of the diffuse component of solar radiation for the various stations in Egypt were obtained by estimating Q_D/Q from the above equations and then multiplying the values of Q, either measured or estimated by formula [2].

Table 3 shows the values of Q_D/Q and Q_o for all stations. The values of the extraterrestrial solar radiation on a horizontal surface Q_o were computed using the following equations:

$$Q_o = \frac{1}{m_2 - m_1} \sum_{n=m_1}^{m_2} (Q_o)_n \qquad (3)$$

where m_1 and m_2 are respectively, the days of the year at the start and the end of the month and $(Q_o)_n$ is the extraterrestrial radiation on the horizontal surface on day n of the year which is given by

$$(Q_o)_n = \frac{24}{\pi} I_{sc} \left[1 + 0.033 \cos\left(\frac{360\ n}{365}\right) \times \cos \phi \cos \delta \sin w_s + \left(w_s \frac{2\pi}{360}\right) \times \sin \phi \sin \delta\right] \qquad (4)$$

where I_{sc} is the solar constant, n is the 16th day of each month, ϕ is the latitude angle and δ is the solar dclination which can be approximately expressed as

$$\delta = 23.45° \sin 360\ (284 + n)/365 \qquad (5)$$

w_s is the sunset hour which is given by

$$\cos w_s = -\tan \phi \tan \delta \qquad (6)$$

Table 4 showns the station's coordinates, altitude, and the mean annual values of the diffuse and total solar radiation and their ratios.

TABLE 3. MEAN DAILY TOTAL, DIFFUSE AND EXTRATERRESTRIAL SOLAR RADIATION ($gm.cal.cm^{-2}day^{-1}$)

	J	F	M	A	M	J	J	A	S	O	N	D	Total Yearly K. cal. cm^{-2}
1-SALLUM													
Q	280	358	454	488	544	611	584	571	501	406	317	257	163.3
Q_D	129	168	191	234	228	183	158	177	175	174	168	162	65.3
Q_o	495	590	712	850	934	972	946	878	759	623	514	449	265.2
2-MERSA MATRUH													
Q	253	327	411	494	558	620	610	567	480	365	261	201	156.5
Q_D	96	114	160	178	246	180	171	147	139	117	102	101	53.3
Q_o	495	590	712	850	934	972	946	878	759	623	514	449	265.2
3-ALEXANDRIA													
Q	308	388	460	542	546	593	602	586	531	439	344	284	170.9
Q_D	139	122	179	173	186	190	169	152	143	136	114	116	55.3
Q_o	499	593	714	850	934	968	941	882	763	626	516	452	265.7
4-PORT SAID													
Q	380	442	504	535	557	574	585	593	575	515	410	328	182.4
Q_D	175	177	207	193	189	166	164	160	150	155	152	138	61.6
Q_o	495	590	712	850	934	972	946	878	759	623	514	449	265.2

TABLE 3 Cont.

	J	F	M	A	M	J	J	A	S	O	N	D	Total Yearly K.cal. cm^{-2}
5-TANTA													
Q	298	357	415	492	547	621	606	579	533	419	321	270	196.7
Q_D	113	121	158	172	186	168	176	168	144	122	116	108	47.8
Q_o	502	595	719	851	934	970	946	882	765	604	518	454	265.7
6-TAHRIR													
Q	246	332	413	503	546	611	559	524	444	363	264	220	152.8
Q_D	93	110	153	166	175	141	151	141	111	138	84	86	47.1
Q_o	502	595	719	851	934	970	946	882	765	604	518	454	265.7
7-BAHTIM													
Q	324	418	424	508	591	602	669	612	571	457	300	328	176.5
Q_D	146	163	178	153	189	144	167	202	154	133	138	141	58.0
Q_o	514	604	722	855	932	969	944	884	769	638	530	466	268.4
8-MINYA													
Q	369	407	428	481	569	639	676	653	628	489	393	329	184.3
Q_D	122	126	137	154	171	147	162	163	151	152	122	112	52.3
Q_o	545	628	733	865	932	969	944	889	781	659	559	492	273.6
9-ASSYOUT													
Q	324	409	441	519	569	619	658	634	591	476	419	325	181.9
Q_D	106	131	128	161	148	130	138	133	124	105	142	120	47.6
Q_o	554	698	741	867	932	956	941	891	788	667	571	502	270.8

TABLE 3 . Cont.

	J	F	M	A	M	J	J	A	S	O	N	D	Total Yearly K.cal. cm^{-2}
10-LUXOR													
Q	356	404	461	489	549	628	632	606	572	444	376	321	177.5
Q_D	117	133	148	156	115	119	133	127	114	89	113	122	45.2
Q_o	581	657	755	875	908	926	912	896	798	686	594	526	277.1
11-ASWAN													
Q	421	444	509	568	633	693	691	666	651	515	423	360	199.9
Q_D	76	160	143	142	120	132	152	160	117	149	131	126	48.9
Q_o	604	676	764	879	929	946	936	900	855	702	616	549	284.5
12-SIWA													
Q	361	405	393	449	522	637	678	652	618	467	386	328	179.3
Q_D	116	109	118	153	183	153	163	156	155	112	120	108	50.1
Q_o	530	616	614	774	886	931	965	943	860	728	544	477	269.6
13-HURGHADA													
Q	348	453	481	555	562	588	612	603	559	495	365	358	181.8
Q_D	97	122	164	167	163	141	147	145	149	119	110	104	44.9
Q_o	554	635	740	864	931	960	941	891	785	664	568	501	274.7

Q = Total Solar Radiation, Q_D= Diffuse Solar Radiation, Q_o=Extraterrestrial Rad.

TABLE 4. ANNUAL DIFFUSE, TOTAL SOLAR RADIATION & THEIR RATIOS ($K.cal.cm^{-2}\ yr^{-1}$)

Station	Station Coordinates			Annual Radiation		
	N	E	M S L m.	Diffuse	Total	Ratio
Sallum	31°32'	25°11'	6.0	65.3	163.3	0.40
Mersa M.	31 20	27 13	30.0	53.3	156.5	0.34
Alexandria	31 12	29 59	6.8	55.3	170.9	0.32
Port Said	31 17	32 14	6.1	61.6	182.4	0.34
Tanta	30 47	31 00	15.4	47.8	196.7	0.24
Tahrir	30 47	31 00	14.8	47.1	152.8	0.31
Bahtim	30 08	31 00	40.1	60.1	176.7	0.34
Minya	28 05	30 44	44.2	52.3	184.3	0.28
Assyout	27 11	31 06	69.6	47.6	181.9	0.26
Luxor	25 40	32 42	88.5	45.2	177.5	0.25
Aswan	23 58	32 47	193.5	48.9	199.9	0.24
Siwa	29 12	25 29	13.5	50.1	179.3	0.28
Hurghada	27 17	33 46	2.8	44.9	181.8	0.25

RESULTS AND DISCUSSION

The comparison between the annual values of diffuse and total solar radiation for a representative set of stations around the world are shown in Table 5. Except for Jerusalem [11], the mean ratio of annual diffuse to total radiation is about 0.328, which is in close agreement with the value of 0.30 given by Drummond [15] as the mean of the annual ratios of the diffuse to total radiation over most of South Africa. It is clear that there is a close agreement over large areas with varying conditions of cloud cover, atmospheric water vapour, and dust content. This agreement can be explained by the presence of one scattering agent to compensate for the lake of the others and vice versa. For example, in arid localities, small amounts of cloud cover are associated with large amounts of atmospheric dust content; while on other localities, none or large amounts of cloud cover are associated with a small content of atmospheric dust.

Comparing the ratio Q_D/Q for Cairo and Kharga with some other stations such as Gilat [10], Poona and Tananarive [16], a close agreement was found in spite of the varying conditions. On the other hand, comparing the computed values of Q_D/Q for several stations in Egypt, the same trend was also observed, as shown in Table 4.

Figures 1 and 2 show that the values of Q_D/Q for Cairo and Kharga exhibit a seasonal variation. In the period from November to April, the average ratios were 0.42 for Cairo and 0.24 for Kharga; while from May to October, the averages were 0.26 and 0.25, respectively. The mean daily diffuse radiation Q_D attains its maximum value at Cairo in April (206.2 cal.cm.$^{-2}$) and at Kharga in April (177.0 cal.cm.$^{-2}$). This is attributed mainly to the sand storms that prevail during this period.

The mean total radiation, on the other hand, reaches a maximum in July at both Cairo (667 cal.cm.$^{-2}$) and Kharga (633 cal.cm.$^{-2}$). However, the mean bright sunshine duration was not the least during April at Cairo (9.16 hours) and at Kharga (10.57 hours). Accordingly, the observed maximum values of Q_D are not only the direct result of cloud scatter, but also due to the increase in atmospheric turbidity. The mean daily diffuse radiation reaches its minimum value in Cairo and Kharga during both February and December, respectively.

Generally, diffuse radiation shows high values during the summer and the rainy seasons. This is attributed to the increased turbidity in the summer and cloud cover during the rainy season.

The computed values of the diffuse as a fraction of the total

TABLE 5. MEAN ANNUAL VALUES OF THE RATIO Q/Q_D AT REPRESENTATIVE SET OF STATIONS AROUND THE WORLD.

Station	Station Coordinates			Annual Radiation			Source of data
	N	E	Altit. M.	Diffuse	Total	Ratio	
Nice	43°54'	07°15'	20	39.5	133.2	0.30	(1)
Tashkent	41 20	69 18	478	47.8	142.2	0.34	(2)
Tananarive	18 54S	47 32	1310	70.0	194.4	0.36	(16)
Delhi	28 35	77 12	216	57.7	174.3	0.33	(16)
Poona	18 32	73 51	559	65.3	177.4	0.37	(16)
Gilat	31 20	34 40	150	64.1	178.1	0.36	(10)
Jerusalem	31 45	35 25	784	39.9	191.0	0.21	(3)
Qrendi	35 50	14 26	136	52.2	159.0	0.33	(2)
Khormask.	12 15	45 01	4	59.4	188.2	0.32	(2)
Riyadh	24 42	46 43	563	63.3	171.1	0.37	(11)
Hofuf	25 57	49 34	110	68.9	174.2	0.39	(11)
Cairo	30 08	31 00	75	58.0	176.5	0.33	
Kharga	25 27	30 32	73	42.1	177.9	0.25	

Source Of Data :-

(1) A.J.Drummond, Radiation and Thermal Balance,1958,Arid Zone Research,X,Climatology Rev. of Res.,56-74 ,U.N.E.S.C.O. Paris.

(2) I.G.Y. Radiation Observations Forms R-4,W.M.O.,Geneva, 1960.

(3) I.G.Y. Radiation Observations Forms R-2a,R-2b,W.M.O., Geneva, 1960.

solar radiation for all stations were compared with the more generally available climatological measurements, such as:

a) The total solar radiation Q as a fraction of the extraterrestrial radiation Q_o, i.e., the ratio Q/Q_o which is the cloudiness index.

b) The bright sunshine duration as a fraction of possible D/D_o.

c) The amount of cloud cover X in oktas.

Linear regressions have been fitted to the computed values of Q_D/Q by the method of least squares and are plotted in Figures 4, 5 and 6.

Figure 4 shows the relationship between the cloudiness index and Q_D/Q for the different latitudes. These relationships were:

1. For latitudes 17°N < ∅ < 27°N:

$$Q_D/Q = -0.983\ (Q/Q_o) + 1.02$$

$$(r) = -0.82$$

2. For latitudes 27°N < ∅ < 34°N:

$$Q_D/Q = -1.213\ (Q/Q_o) + 1.07$$

$$(r) = -0.86$$

These relationships can be compared with those found by Stanhill [10] from the measurements at Gilat; El-Salam and Sayigh for stations at Saudi Arabia [11]; by Chaudhary [17]; and by Liu and Jordan [7] from the Blue Hill Observatory measurements. As far as the daily averages, it was found to be somewhat lower than those found by Stanhill and Chaudhary and slightly higher than those of Liu and Jordan. This can be explained by the greater turbidity in the atmosphere.

Figure 5 shows the relationship between the bright sunshine duration as a fraction of the total possible and Q_D/Q. These relationships for the different latitudes were found to be

1. For latitudes 17°N < ∅ < 27°N:

$$Q_D/Q = -0.82\ (D/D_o) + 0.975$$

$$(r) = -0.84$$

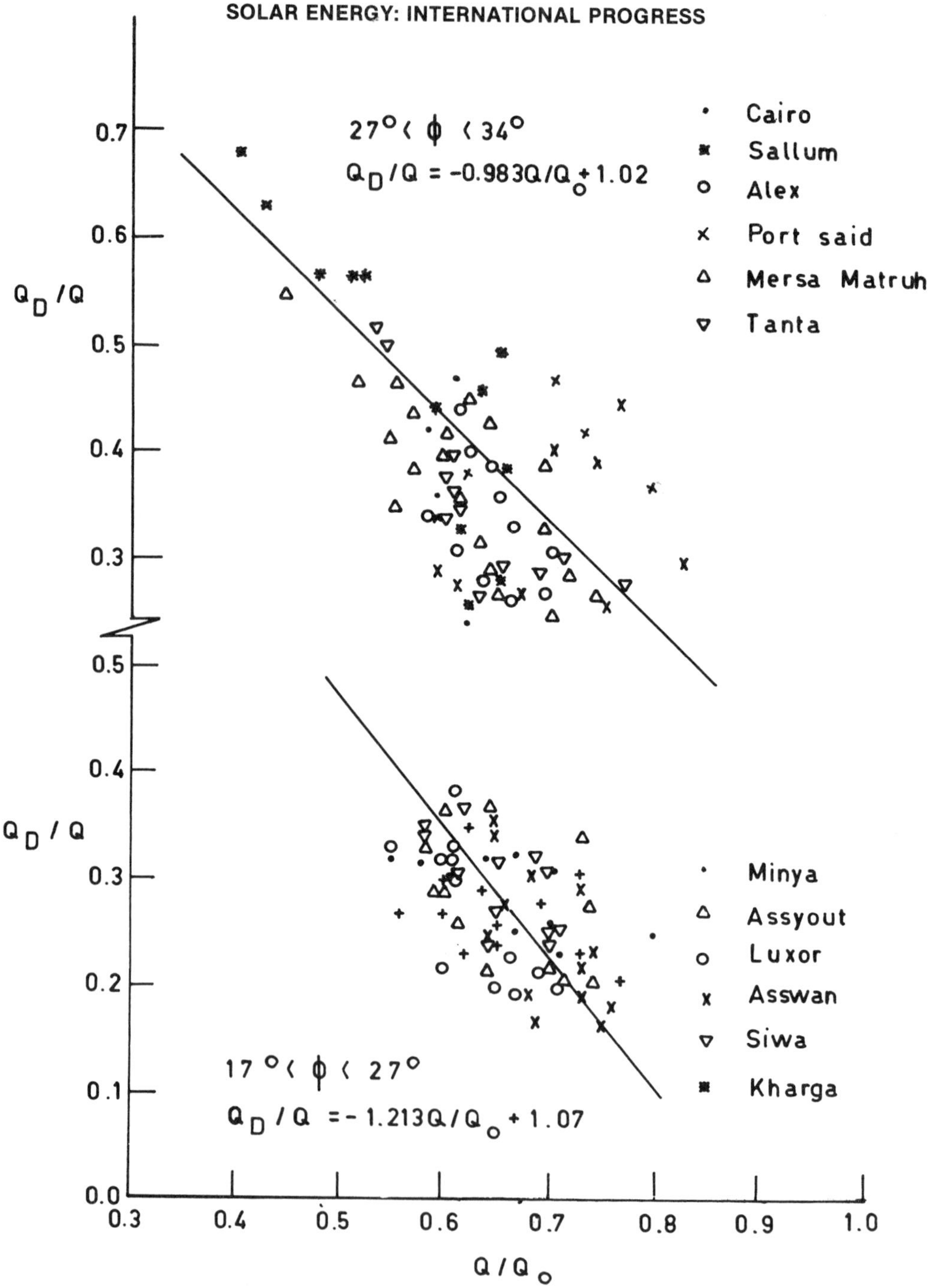

Fig.(4) Relationships between the cloudiness index and the ratio of diffuse to total radiation.

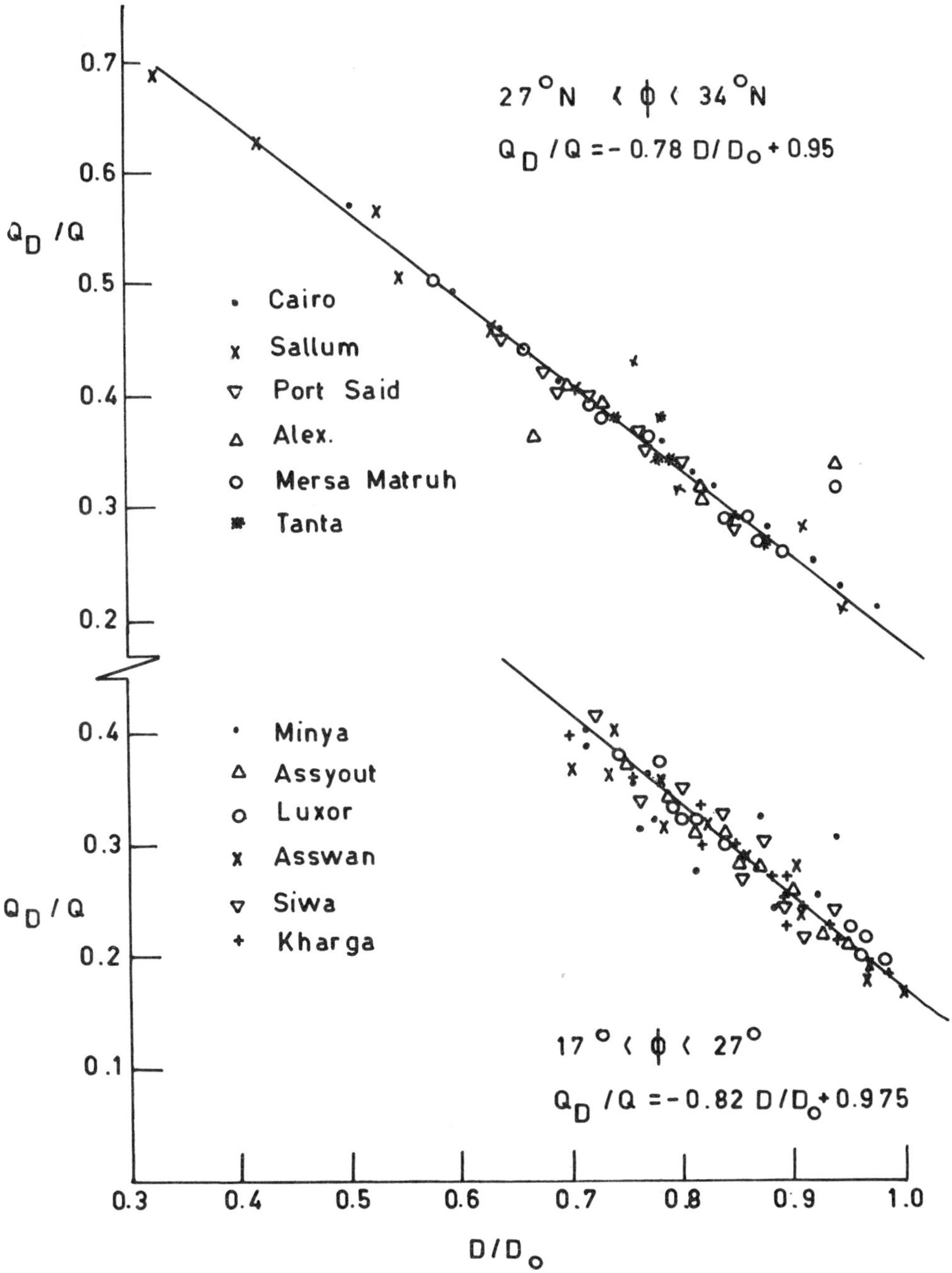

Fig. (5) Relationships between duration of sunshine hours as a fractions of possible and the ratio of diffuse total solar radiation.

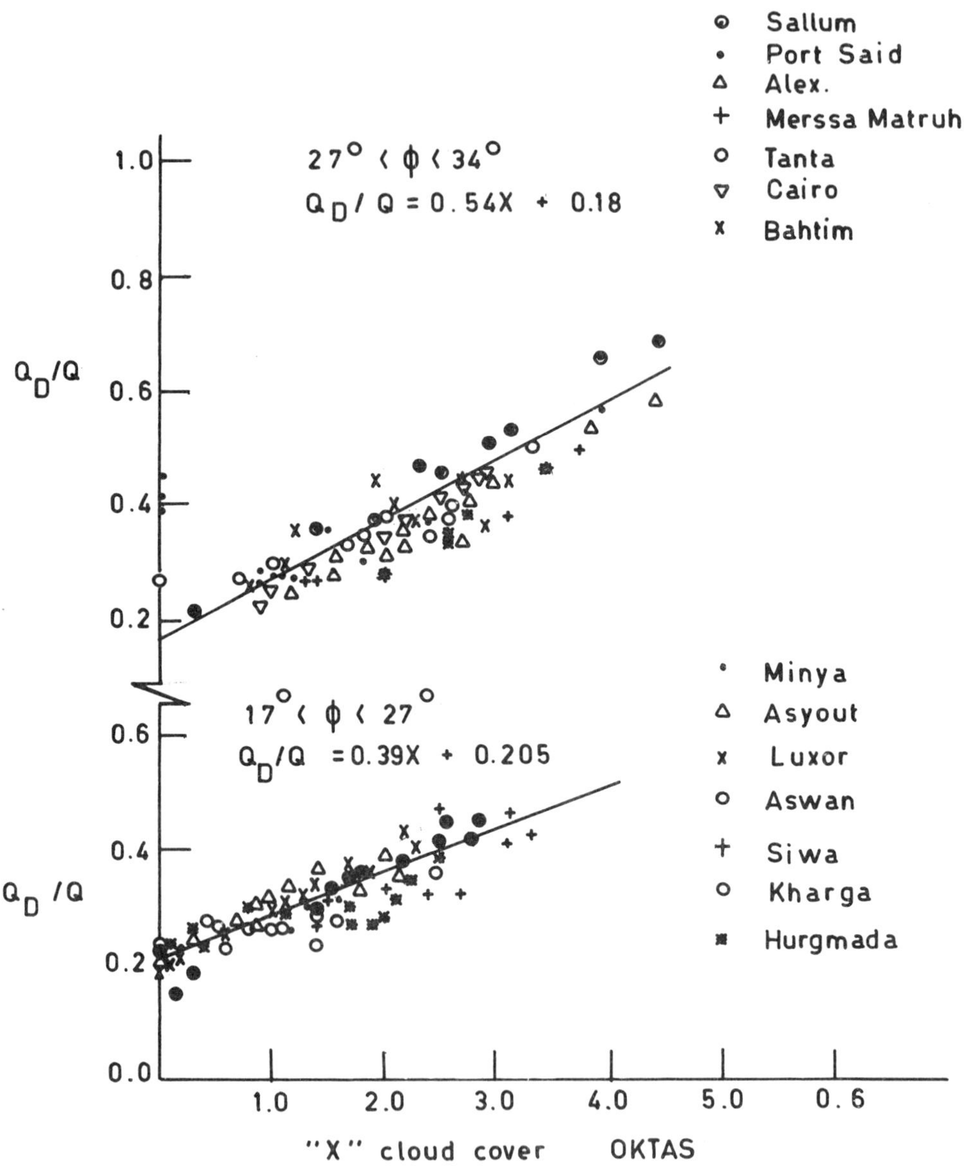

Fig. (6) Relationships between mean cloud cover and ratio of diffuse to total radiation.

2. For latitudes 27°N < ϕ < 34°N:

$$Q_D/Q = -0.78\ (D/D_o) + 0.95$$

$$(r) = -0.83$$

The above relationships found are in good agreement with those derived by Hand and Wollaston [18], by Wartena [19], and by El-Salam and Sayigh [11].

The correlation relationships between Q_D/Q and the mean cloud cover X, are shown in Figure 6. These relationships for different latitudes are:

1. For latitudes 17°N < ϕ < 27°N:

$$Q_D/Q = 0.39\ X + 0.205$$
$$(r) = 0.84$$

2. For latitudes 27°N < ϕ < 34°N:

$$Q_D/Q = 0.54\ X + 0.18$$

$$(r) = 0.82$$

From these relations, it was found that if the amount of cloud cover is zero, the diffuse radiation component averages to 0.193 of the total. This was found to be in good agreement with the four-year measurements at the Blue Hill Observatory [7], which showed that the diffuse component averaged to 0.18 of the total radiation for cloudless days. Also, similar results were confirmed by Choudhary's [17] analysis of the radiation measurements at New Delhi and those found by Mani and Chacko [16] at Poona.

Using the above relationships, an estimation of the diffuse solar radiation component for any location in Egypt can be done. This will be of use for personnel working on the design problems in the solar energy field.

REFERENCES

1. Moon, P., Proposed Standard Solar Radiation Curves for Engineering Use, H. Franklin Institute, 230, 583-617 (1940).

2. Sabbagh, J.A., Sayigh, A.A.M. and El-Salam, E.M. Abd, Estimation of Total Solar Radiation from Meterological Data, ISES Conference, Los Angeles, California, USA, August 1975 - Solar Energy 19, 3, 307 (1977).

3. Swartman, R.K. and Ogunladeo, Solar Radiation Estimates from Common Parameters, Solar Energy, Vol. II, 170-172 (1967).

4. Sabbagh, J.A., Sayigh, A.A.M. and El-Salam, E.M. Abd, Correlation of Solar Radiation and Sunshine Duration in Riyadh, Saudi Arabia, Pakistan J. Sci. Ind. Res., 16, (16) (1973).

5. Reddy, S.J., An Empirical Method for the Estimation of the Total Solar Radiation, Solar Energy, 13, 289 (1971).

6. El-Salam, E.M. Abd, The Relation Between Solar Radiation and Hours of Bright Sunshine for the Middle-East Region, (to be published).

7. Lui, B.Y.H. and Jordan, R.C., Interrelationships and Characteristics Distribution of Direct, Diffuse and Total Solar Radiation, Solar Energy, 4, 3, 1-19 (1960).

8. Ruth, D.W. and Chant, R.E., The Relationship of Diffuse Radiation to Total Radiation in Canada, Solar Energy, 18, 2, 153 (1976).

9. Parmelee, G.V., Irradiation of Vertical and Horizontal Surfaces by Diffuse Solar Radiation from Cloudless Skies, Heating, Piping and Air Conditioning, 26, 129 (1954).

10. Stanhill, G., Diffuse, Sky and Cloud Radiation in Israel, Solar Energy, 10, 2, 96-101 (1960).

11. El-Salam, E.M. Abd and Sayigh, A.A.M., Estimation of Diffuse Solar Radiation in the Arabian Peninsula, Proc. of the Symp. on Radiation in the Atmosphere, Garmisch-Partenkirchen, FRG, 19-28 August 1976, Ed. H.J. Bolle, Science Press Princeton, p. 607.

12. Klein, S.A., Calculation of Monthly Average Insolation on Tilted Surfaces, Solar Energy, 19, 4, 325-329 (1977).

13. List, R., Smithsonian Meteorological Tables, 6th Rev. Edn., 418-419, Smithsonian Institute, Washington, D.C., USA (1968).

14. The Egyptian Meteorological Authority, Cairo, Egypt.

15. Drummond, A.J., On The Measurements of Sky Radiation, Arch. Met., Wein B.7, 414-436 (1956).

16. Mani, A. and Chacko, O., Diffuse Solar Radiation at Delhi and Poona, India, J. Met. Geophys. 14, 4, 416-432 (1963).

17. Choudhary, N.K.D., Solar Radiation at New Delhi, Solar Energy, 7, 2, 44-51 (1963).

18. Hand, I.F. and Wollaston, F.A., Measurements of Diffuse Solar Radition at Blue Hill Observatory, U.S. Weather Bureau, Tech. Paper No. 18 (1952).

19. Wartena, L., The Climate and Evaporation from a Lake in Central Iraq, Meded Landbouwhogeschool, Wageningen, 59 (9), 1-90 (1959).

SOLAR DATA APPLICATION TO EGYPT

M.S. Abdel Salam and **N.A. Higazy**
Air Pollution Laboratory, National Research Center, Cairo

ABSTRACT

Economic utilization of the solar heat income requires, among other things, an accurate estimation of the amounts of incident solar radiation. These amounts are now being affected, sometimes appreciably, in different areas of the world by man-made atmospheric contaminants. The present investigation is one of a series of studies aiming at estimating the depletion, in the incoming direct solar radiation, caused by such contaminants.

Hourly measurements of the total incident direct radiation and its constituents in various spectral bands were taken during clear days of a twelve-month period at Giza, Egypt. It was found that the loss in the received amounts of total radiation at ground level, at noon time, reaches 42% of that incident at the outer boundary of the atmosphere. The highest loss (50%) occurs in the wavelengths shorter than 525 mμ. The corresponding percentages in the visible range 525 mμ - 710 mμ and in the infra-red range greater than 710 mμ in wavelength were 31% and 43% respectively. No appreciable changes were found in the monthly average values of these radiations.

More interesting features, however, were found through studying the monthly variations, for different air masses, in the received average amounts of total direct radiation and its constituents in the studied spectral bands. These variations were discussed in relation to atmospheric contaminants in the path traversed by the sun rays. The percentages of the received amounts of direct radiation to that theoretically incident at the outer boundary of the atmosphere were also calculated and discussed for different air masses.

A comparison was made between the findings of the present study and that of a similar one conducted by Moon in 1940 for the same latitude and under the same conditions. It was found that the received amounts of radiation at Giza are more depleted than that found by Moon. Such extra losses amounted to 6% for radiations of wavelengths less than 525 mμ, 12% in the wavelength band 525-710 mμ, and 11% in the wavelengths greater than 710 mμ.

It was, therefore, concluded that the amounts of received direct solar radiation, in the Egyptian highly industrialized and urbanized areas, would be appreciably depleted by atmospheric contaminants. A thorough widespread investigation of this problem, at the country level, is therefore urgently needed for maping the solar heat income at Egypt and, hence, the allocation of feasible areas for economically utilizing this environmentally clean and continuously renewable source of energy.

INTRODUCTION

The heat income from solar radiation at any given place above the earth's surface continuously changes with the time of the day and time of the year, depending mainly upon:

a- The solar output of radiation. This varies slightly with time depending on solar phenomena, particularly on the sunspot cycle. However, the thermally effective radiation that reaches the outer boundary of the earth's atmosphere is estimated [1] to constitute only 57% of the radiation emitted from the sun. The other 43% is reflected back to space.

-The intensity of solar radiation received normally at the outer boundary of the atmosphere, known as the solar constant, varies in the mean between 1.85 to 2.02 cal $cm^{-2}min.^{-1}$.

b- The distance from the sun, which depends on the time of the year and the latitude of the place under consideration.

c- The solar elevation defined as the inclination of sun rays to the plane of the horizon. Such elevation depends principally on the latitude of the place and time of the day;

d- The absorption and transmission of solar radiation through the earth's atmosphere. Such factor depends mainly on the length of the path traversed by sun-ray and on the atmospheric constituents within that path. The latter is very complicated as it is continuously changing depending on weather condition and the geography, topography and man's activity at the given place. However, the total loss in the energy received at the earth's surface due to absorption, scattering and diffuse reflection through a naturally clear atmosphere is only a small fraction of the energy falling on the outer limit of the atmosphere.

Objective of The Present Study:

One of the principal factors for economically utilizing the solar energy at a given place is the accurate estimation of the amounts of solar radiation received at that place. Such amounts are now being affected in some areas of the world by the continuously increasing amounts of pollutants dispersed into the atmosphere as a result of the various urban and industrial human activities. The present investigation is one of an integrated series of studies aiming at estimating the losses in the incoming solar radiation of Giza-Egypt (National Research Center site) caused by atmospheric contaminants. This particular study is concerned only with measurements made for days having cloudless skies and clear weather conditions.

It should, here, be noted that this observational site lies in a purely residential area, far from the direct effects of industrial discharges.

THE SOLAR SPECTRUM

The whole of the radiation from the sun [1] lies within limits of wavelengths 0.2 μ to 3 μ, and is subjected to losses due to absorption, scattering and reflection processes. This radiation is being rubbed off of all wavelengths less than 0.3 μ through absorption by the high concentrations of oxygen atoms at about 100 Km above ground level and the ozone layer at about 40 Km. Oxygen molecules also absorb at narrow bands centered around 0.69 μ and 0.76μ, but these have negligible effect on the incident radiation.

Carbon dioxide has two small absorption bands from 2.3 μ to 3 μ and 4.2 μ to 4.4 μ as well as another narrow band centered at about 15 μ. It, therefore, has a little effect on the solar beam as a whole.

The principal and most important absorption is due to water vapour content of the atmosphere. It has strong absorption bands at 1.91 μ and 2.03 μ.

Scattering and reflection of solar radiation at all wavelengths are being caused by atmospheric aerosols and even by the molecules of the different constituents of air.

METHODOLOGY

Measurements of the incident direct solar radiation were made for perfectly clear days during the twelve months of May 1974 through April 1975. The measuring instrument was placed at the roof of the National Research Center, at about 22 m above street level, to avoid the effects of ground dusts picked up by wind or moving traffic. The instrument used is the Linke and Feussner actinometer in combination with a millvoltmeter in which one scale division corresponds to 0.025 cal $cm^{-2}min^{-1}$. The actinometer was fitted with filters OG_1, RG_2 and RG_8 [2] that have transmissions in the range of wavelengths 525 mμ to 2800 mμ, 630 mμ to 2800 mμ and 710 mμ to 2800 mμ, respectively. Measurements obtained by these filters were reduced, within 0.5% accuracy, to 100% transmission through multiplying their values by the Davos Reduction Factor which has the values 1.109 for filter OG_1, 1.103 for RG_2 and 1.046 for the filter RG_8.

Daytime measurements were taken every hour for the direct total and filtered radiation. Calculations were then made to obtain the amounts of the hourly received radiation in the wavelength bands $\lambda \leq 525$ mμ, 525 mμ $< \lambda <$ 710 mμ, and $\lambda \geq 710$ mμ.

The graphical representation of these obtained results and that of the total direct radiation were used for calculating the received amounts of different radiation at various air masses by using the well known Bouguer equation which stated that

$$I = I_o \, e^{-am}$$

where I is the intensity of the measured radiation,
I_o the intensity of the corresponding radiation received at the outer limit of the atmosphere.
a the extinction coefficient integrated for all wavelengths of the spectrum.
and m is the air mass.

RESULTS AND DISCUSSION

Table (1) gives the monthly and yearly average intensities of measured total direct radiation and its component in the specified wavelength bands at noon-time. It shows that, at midday of clear days, the received direct radiation at Giza, in the average, does not vary much from one month to the other. Its value varies between 1.07 cal $cm^{-2}min^{-1}$ in September to 1.19 cal $cm^{-2}min^{-1}$ in July, with a yearly average of 1.14 cal. $cm^{-2}min^{-1}$. This yearly average of incoming energy, for clear

Table 1: Monthly average amounts of direct solar radiation received at various spectral bands at moon time May 1974 to April 1975 (cal/cm^2/min)

(Giza - Egypt)

Spectral region m. μ.	May	June	July	Aug.	Sep.	Oct.	Nov.	Dec.	Jan.	Feb.	Mar.	Apr.	Yearly average
$\lambda \leqslant 525$	0.26	0.26	0.28	0.25	0.25	0.26	0.25	0.25	0.30	0.32	0.32	0.32	0.28
$525 \leqslant \lambda \leqslant 710$	0.31	0.30	0.32	0.31	0.28	0.29	0.29	0.29	0.27	0.27	0.26	0.24	0.29
$\lambda \geqslant 710$	0.60	0.55	0.59	0.56	0.54	0.55	0.57	0.61	0.60	0.59	0.60	0.52	0.57
Measured Total Intensity	1.17	1.11	1.19	1.12	1.07	1.10	1.11	1.15	1.17	1.18	1.18	1.08	1.14

weather conditions, indicates that the total direct radiation incident at the outer boundary of the atmosphere (1.98 cal. $cm^{-2}min^{-1}$) is depleted by about 42% by the constituents of its atmospheric path till it reaches the ground at Giza. A percentage loss which seems rather very high at that time of the day when the incoming radiation is supposed to be at its maximum intensity and the least depleted one by atmospheric constituents. This loss is more than double, as compared with that found by Angstrom [3] in 1915 under the same conditions but before the world widespread urbanization and industrialization activities. The corresponding depletion figures in the amounts of direct radiation received in the studied spectral bands were 50% in the band less than 525 mμ, 31% in radiation having wavelengths between 525 mμ and 710 mμ and 43% in the wavelengths greater than 710 mμ. The highest loss occurs in the shorter wavelengths of the solar spectrum, thus indicating the loading of the atmosphere with ultra-fine particles. The atmospheric water vapour content shows its appreciable effect on radiation of wavelengths greater than 710 mμ. The 31% loss in the visible light range are principally due to relatively larger particles suspended in the surface layer of the atmosphere, most probably resulting from local and surrounding human activities.

The percentages of the amounts of radiation received in the different spectral bands (Table 2) to the total incoming direct radiation also, do not vary much in the average monthly values, being in the yearly average 24%, 25% and 51% for the above mentioned spectral bands, respectively.

Looking into the corresponding figures for different air masses (Table 3) the picture is rather different. Here, the variation in the received radiations, for the same air mass, varies distinctly from one month to another, reflecting, clearly, the effect of monthly variations in atmospheric constituents. The intensities of these radiations are, in general, higher in winter months of December and January and lower in summer months of August and September. The monthly and yearly average values of such intensities decreases rather appreciably with increasing air mass.

The percentages of radiation received in the different spectral bands (Table 4) shows also very interesting features. For radiation of wavelengths less than 525 mu, which is affected by the atmospheric content of very fine particles, the average monthly percentages are being maximum in March and minimum in August, for all air masses. This probably reflects the effect of natural ventilation and dispersion of atmospheric pollutants, in the atmosphere over Egypt [4]. Wind speeds are relatively higher in March and lower in August while atmospheric stability are more predominant from sunset to sunrise in August, leading to accumulation of pollutants during nightime. Strong convectional currents during daytime picks up huge amounts of ground

Table 2: Monthly distribution of average percentages of the direct solar radiation in the different wavelength bands at moontime to the total incident direct radiation (May 1974 to April 1975 at Giza).

Spectral region m.μ.	May	June	July	Aug.	Sep.	Oct.	Nov.	Dec.	Jan.	Feb.	Mar.	Apr.	Yearly average
$\lambda \leqslant 525$	22	23	23	22	23	24	23	22	26	27	27	30	24
$525 \leqslant \lambda \leqslant 710$	27	27	27	28	26	26	26	25	23	23	22	22	25
$\lambda \geqslant 710$	51	50	50	50	51	50	51	53	51	50	51	48	51

Table 3: Montly average distribution of the direct solar radiation in the various wavelength bands for different air masses (Giza May 1974 to April 1975) (in Cal cm^{-2} min^{-1})

Spectral region m. μ.	Air mass	May	June	July	Aug.	Sep.	Oct.	Nov.	Dec.	Jan.	Feb.	Mar.	Apr.	Yearly average
λ≤525	1.5	.190	.190	.209	.166	.180	.229	.240	.269	.313	.285	.277	.251	.233
	2	.131	.131	.151	.112	.124	.170	.182	.210	.258	.228	.218	.194	.176
	3	.064	.064	.079	.050	.059	.095	.105	.129	.176	.146	.138	.115	.102
	4	.030	.030	.041	.022	.027	.039	.045	.079	.120	.093	.086	.068	.057
525<λ<710	1.5	.260	.249	.261	.250	.240	.265	.274	.298	.282	.265	.239	.294	.265
	2	.218	.209	.222	.210	.199	.227	.239	.267	.247	.228	.198	.262	.227
	3	.157	.148	.161	.150	.136	.167	.178	.214	.189	.167	.136	.207	.168
	4	.114	.105	.117	.105	.093	.123	.133	.171	.145	.122	.094	.165	.124
λ≥710	1.5	.485	.433	.501	.472	.461	.513	.622	.682	.653	.583	.562	.466	.536
	2	.380	.327	.394	.369	.358	.409	.533	.599	.568	.486	.460	.357	.437
	3	.238	.186	.247	.223	.216	.265	.389	.463	.428	.339	.312	.214	.293
	4	.145	.108	.158	.139	.130	.172	.285	.359	.323	.237	.204	.129	.199
Measured Total Intensity	1.5	0.935	.872	0.971	.888	.881	1.007	1.136	1.249	1.248	1.133	1.078	1.011	.958
	2	0.729	.667	0.767	.691	.681	.806	.954	1.076	1.073	.942	.876	.813	.840
	3	0.459	.398	0.487	.423	.411	.527	.672	.806	.793	.652	.586	.536	.563
	4	0.289	.243	0.316	.266	.250	.334	.463	.609	.588	.452	.384	.362	.380

Table 4: Percentage distribution of the direct solar radiation in various wavelength bands for different air masses (Giza May 1974 to April 1975).

Spctral Region m.μ	Air Mass	May	June	July	Aug.	Sep.	Oct.	Nov.	Dec.	Jan.	Feb.	Mar.	Apr.	Yearly average
$\lambda \leqslant 525$	1.5	20	22	22	19	20	23	21	22	25	25	26	25	23
	2	18	20	20	16	18	21	19	20	24	24	25	24	21
	3	14	16	16	12	14	18	16	16	22	22	24	21	18
	4	10	12	13	9	11	12	10	13	20	20	22	19	14
$525 < \lambda < 710$	1.5	28	28	27	28	27	26	24	24	23	23	22	29	26
	2	30	31	29	31	29	28	25	25	23	24	23	32	28
	3	34	37	33	35	33	32	26	27	24	26	23	39	31
	4	39	43	37	39	37	37	29	28	25	27	25	46	34
$\lambda \geqslant 710$	1.5	52	50	52	53	52	51	55	54	52	52	52	46	52
	2	52	49	51	53	53	51	56	55	53	52	52	44	52
	3	52	47	51	53	53	50	58	57	54	52	53	40	52
	4	50	45	50	52	52	51	61	59	55	53	53	45	52

dust to the atmosphere. Turbulent mixing then diffuses these dust particles vertically and side ways from where it is abundant to where it is deficient. Dust haze then covers the region. This is usually a warm season phenomenon at daytime in Cairo. This picture is completely reversed for radiation of wavelengths between 525 mμ and 710 mμ. Here, the monthly and yearly average percentages of radiation increase with increasing air mass and are being higher in the warm months of April through October and lower in the other cold months of the year. This could be due to the prevalence of relatively more fresh winds in winter time and the consequent dispersion of large amounts of big particles in the atmosphere. This phenomenon; also, reflects the effect of the settling down of particles suspended in the atmosphere at higher air masses 4 when the convectional current die away, the wind speed is decreased and the thermal turbulence decreases noticeably. The more interesting feature is the steadiness of the monthly and yearly average values of percentages of the direct radiation constituent of wavelengths greater than 710 mμ, being 52% for yearly averages and for all air masses. This indicates that the total atmospheric content of water vapour changes within a very narrow range from month to month and from hour to hour, within the path of the incident direct solar beam.

CONCLUSION

The present study shows with no doubt that, even at residential areas far from the direct effects of industrial discharges and for perfectly clear days, the incoming direct solar radiation is highly depleted by atmospheric contaminants, at all wavelengths. It is, therefore, expected that in industrial cities, the amounts of this radiation would be more highly affected.

To study the order of magnitude of these losses in the present study area, the received amounts of radiation in the different spectral bands were calculated as percentages of the corresponding amounts theoretically received at the outer limit of the atmosphere. These (Table 5) show that, in the yearly average, such percentages amount to 42% for air mass 1.5 and goes down to 10% for air mass 4; for radiations in the wavelengths shorter than 525 mu. The corresponding figures for the spectral band 525 mμ - 710 mμ were 63% and 30%, and for wavelengths longer than 710 mμ were 53% and 20% , respectively.

The above percentages were further compared with the results of a similar study made by Moon in 1940 [5] for the same latitude and under the same conditions (Table 6). It shows that the radiation received at Giza is more depleted than that found by Moon. The extra losses were 6% for radiations of wavelengths less than 525 mμ, 12% in the wavelength band 525 - 710 and 11% in the wavelengths greater than 710 mμ.

Table 5: Percentage distribution of received solar radiation at the earth surface to that theoreticaly received at the outer atmosphere for different air masses (Giza May 1974 to April 1975).

Sepectral Region m. μ	Air mass	May	June	July	Aug.	Sep.	Oct.	Nov.	Dec.	Jan.	Feb.	Mar.	Apr.	Yearly avarege
$\lambda \leqslant 525$	1.5	34	34	37	30	32	41	43	48	56	51	50	45	42
I =.558	2	25	23	27	20	22	30	33	37	46	41	39	35	32
	3	11	11	14	9	11	17	19	23	32	26	25	21	18
Cal $cm^{-2} min^{-1}$	4	5	5	7	4	5	7	8	14	22	17	15	12	10
$525 > \lambda < 710$	1.5	62	59	62	60	57	63	65	71	67	63	57	70	63
I =.420	2	52	50	53	50	47	54	57	63	59	54	47	62	54
	3	37	35	38	36	32	40	42	51	45	40	32	49	40
Cal $cm^{-2} min^{-1}$	4	27	25	28	25	22	29	32	41	35	29	22	39	30
$\lambda \geqslant 710$	1.5	48	43	50	47	46	51	62	68	65	58	56	46	53
I =1.002	2	38	33	39	37	36	41	53	60	57	49	46	36	44
	3	23	19	25	22	22	26	39	46	43	34	31	21	29
Cal $cm^{-2} min^{-1}$	4	14	11	16	14	13	17	28	36	32	24	20	13	20

Table 6: Average percentages of incident direct solar radiation at vanous wavelength bands to that at the outside the atmosphere for air mass = 2.

Spectral region mμ	I_λ = Moon's recommended sea-level direct-beam sunshine altitude = 30° (m=2) 1940 $I_{o\lambda}$ =Johnson's solar constant outside atmosphere (1954) $I_\lambda/I_{o\lambda}$ %	I_λ = at Cairo-sea-level direct beam sunshine altitude=30° (m=2) $I_{o\lambda}$ =WMO-NO-299 Geneva (1971). $I_\lambda/I_{o\lambda}$ %	Different %
$\lambda \leq 510$	38 %		6 %
$\lambda \leq 525$		32 %	
$510 < \lambda < 700$	66 %		12 %
$525 < \lambda < 710$		54 %	
$\lambda \geq 700$	55 %		11 %
$\lambda \geq 710$		44 %	

It can, therefore, be concluded that a thorough widespread investigation of the problem of solar radiation depletion, at the country level, is urgently needed for maping the solar heat income at Egypt and, hence, for the allocation of feasible areas economically suitable for utilizing this environmentally clean and naturally renewable source of energy.

REFERENCES

[1] David Brunt: "Weather Study" Thomas Nelson and Sons Ltd, (1942).

[2] W. Schuepp: "Measurement of Atmospheric Turbidity" Met. Service, Leopoldville, Congo (1953).

[3] A. Angstrom: "A Study of the Radiation of the Atmosphere" Smithsonian Miscellaneous Collections, Vol. 65, No. 3, (1915).

[4] Abdel Salam, M.S.: "Natural Ventilation and Dispersion of Atmospheric Contaminants in Egypt" Under Publication.

[5] Parry Moon : "Proposed Standard Solar-Radiation Curves For Engineering Use" J. Franklin Inst., 230 (5), 583 - 617, (Nov. 1940).

§§§§§§§

§§

GLOBAL RADIATION IN ITALY

A. Guerrini , A. Lavagnini
and F. Vivona
Istituto di Fisica della
Atmosfera, del CNR
Roma, Italia

ABSTRACT

In this paper the results concerning the distribution of sunshine and global radiation in Italy are given. Data are relative to 130 stations of the Italian Hydrometeorological Services for the period 1900 - 1972. Measurements , equipments and methods , data processing and correlations between sunshine and global radiation intensity are also described.

INTRODUCTION

The systematic collection of the data existing in Italy refers to hours of sunshine and to global solar radiation, recorded at dif= ferent stations. These belong to Bodies such as : the Air Force Meteorological Service (AM) , the Agricultural Ecology Central Office (UCEA) , of the Ministry of Agriculture and Forestry , the Ministry of Public Works (LL.PP.) , the Agricultural Experimental Institute (ISA) , and also several Universities and Agencies for a total of 130 stations.

For the AM stations only,which are the more uniformily distribu= ted along the Italian peninsula, we have both sunshine and global radiation data from 1958 to 1972. For the other stations we have sunshine data recorded daily and continuously for a few decades.

AIM OF THE WORK

With the relevant number of data at our disposal we intend to :
a) make a sunshine map of Italy , using the mean values of daily sunshine hours , computed through all data recorded in each helio= phanographic station; and also with the same method make seasonal sunshine maps; b) study the sunshine behaviour with varying the

latitude, the longitude and the height above sea-level of the recording stations; c) establish a correlation between global radiation and sunshine recorded at the Air Force Meteorological Service stations in order to determine possibly, for the different areas, the equation which would better describe such a correlation.

With this work we intend also to obtain a global radiation map for the Italian peninsula , similarly to the sunshine one already planned. This could prove very interesting for exploiting solar energy from a mechanical , electrical , domestic , agricultural , building , tourist point of view.

RECORDING AND DATA PROCESSING

The daily diagrams were read and transferred on monthly tables where the following information were reported : a) daily absolute sunshine approximated to the tenth of an hour (F) ; b) absolute sunshine, relative to the four middle hours of each day from 10 AM to 2 PM, approximated to the tenth of an hour (F_1) ; c) a sunshine refractive index during the day (P) ; d) an index of the sunshine character (of continuity or intermittance) (C) ; e) daily absolute global radiation ($T_{eff.}$).

The sunshine and daily global radiation are here defined as "absolute" to distinguish them from their "relative" or "percentage" values. The latter are defined as the ratio between the recorded values and the maximum possible ones, which , as known, vary with the latitude and with the period of the year. During the data processing, the relative values were calculated in order to carry out comparisons among the data recorded at the stations located at different latitudes.

The cause to put into evidence,from the daily diagrams, the absolute partial sunshine F_1 , recorded between 10 AM and 2 PM of each day, lies in the fact that mainly within problems of agricultural , tourist , sanitary , etc. kind , it is of remarkable importance knowing for each location and for each season, the contribution to the total sunshine of the four middle hours of the day.

The data relative to point c) and d) were obtained trying to set the daily sunshine into one of the possible situation presented by the model (heliophanographic code) and giving it the corresponding number. In this work such data were not processed, because we intend to carry out a more detailed analysis of them.

From the montly tables, by summing up the daily data of each month we calcolated the montly absolute sunshine; such a computation was not done in the cases where the daily diagrams of the corresponding month were less than 2/3 or when the diagrams, even if equal to 2/3 of the days of the month, were distributed in such a way as not being representative of the sunshine monthly behaviour (for instance, when the data of the first or last 10 days of the month were missing). In the other cases of partially missing data, we substitued to each missing value the mean of the daily data available in the monthly tables.
The monthly values of the total absolute sunshine F and of the partial one F_1 were then transferred on punching cards, with the aid of a computer the seasonal absolute sunshine was calculated. Summing up the monthly data relative to each season and the annual absolute sunshine was also found adding the twelve monthly values. Their value expressed in full hours, is reported in the season tables or in the annual ones under the symbol F (keep in mind that winter covers December of the preceding year and January and February of the year to which is referred to).
In the case a monthly value of a season (or of a year) is not available, we substitute to it the mean of all sunshine values available relative to that month. If, instead more than one monthly value are missing the seasonal value is also considered missing. At the same manner we calculated the annual total sunshine as long as we had at our disposal 8 times 2/3 the monthly data.
To determine the relative or percentage sunshine F% we assumed that the heliophanograph would record the sunshine time only if the Sun height over the horizon would be over 5°. The maximum possible monthly sunshine was computed using data reported on the "Smithsonian Meteorological Tables", in the following manner: for the different latitudes the duration of the day ("daylight") was considered, relative to the days 1,9, 17, and 25 of each month and to first day of the following one. To each of these five values was subtracted the time taken by the sun to describe ten degrees of its virtual orbit (5° at sunrise and 5° at sunset) for each latitude and for the same days of the month. Of these five values thus obtained we made the aritmetic average and multiplying it by the number of days of each month (in hours and tenths of hours) we assumed the result as the "maximum possible monthly sunshine time" relative to each latitude. To determine the maximum possible monthly sunshine for latitudes not described in the above mentioned tables, linear interpolations were carried out. The data obtained, were digitalized to calculate the relative sunshine of each month. To get the seasonal and annual relative values we adopted the same method as for the corresponding absolute values. Such relative values are shown in the tables with F%.

It is perhaps convenient to note that to calculate the time the sun takes to make 10° along its virtual orbit, were used the data relative to the"civil twilight". This is defined as the time elapsing between sunrise (or sunset) and the instant in which the real position of the sun centre is 6° under the horizon.
To determine the relative global radiation for the Air Force Meteorological Service Stations, the values of maximum possible global radiation were used (that is, those we could possibly get without atmosphere) reported by Cuomo et al. An idea of the reliability of the heliophanographic data is given by a comparison carried out by the same authors on recordings relative to Bologna and Bolzano where the Air Force Met. Service and the Ministry of Public Works have their own recording station.

RESULTS AND COMMENTS

In table 1. is summarized the sunshine value and the monthly mean global radiation, seasonal and annual for each station. The regional daily mean sunshine values were found by carrying out the weighed average of the daily mean sunshine values concerning all stations existing in that area. We assumed as weight function the number of years of observation, relative to the station it self.
To find the values of daily mean sunshine for the four zones in which we subdivided Italy, we also made the weight average of the region values assuming as weight function the total number of years of observation in each region.
For an approximate analysis, Italy was subdivided into the three zones: continental Italy, Peninsular Italy, Insular Italy.
For a further analysis, Italy was in its turn subdivided in: central Italy and South Italy.
Table 2. shows the values of the regional daily mean absolute sunshine F, and the values of relative sunshine F%.
Lastly, the percentage values are also presented :

$$\Delta \%_0 = \frac{F_{summer} - F_{winter}}{F_{winter}}$$

Map n.1 shows a general view of the sunshine behaviour along the peninsula, with a general regional subdivision. The shaded areas (sunshine hours) were chosen in order to locate a difference of 0.5°C.
Looking at table 2 we can ,at first glance, make the following comment: the annual and regional sunshine show a maximum in Sicily and a minimum value in Lombardy (except in Summertime when the minimum occurrs in Piedmont). Such a phenomenon is probably due to the presence of clouds or fogs mostly in Lombardy rather than in the neighbouring regions.

Stazione ---→		67	68	69	70	7I	72	73	74	75	76	77	78	79	80	8I
Gennaio	F	97	I29	94	I09	95	I30	II3	II4	I26	86	I02	II2	89	I34	88
	F%	36	48	35	40	36	48	42	42	47	32	38	4I	33	49	32
Febbraio	F	II5	I40	94	II3	I2I	I3I	I22	I26	I36	II3	I29	I39	II9	I4I	I2I
	F%	42	5I	34	4I	44	48	44	47	49	4I	47	50	43	5I	43
Marzo	F	I45	I89	I37	I27	151	I73	I53	I68	I76	I46	I48	I68	I33	I90	I40
	F%	42	55	40	37	44	50	44	48	5I	42	43	48	38	55	40
Aprile	F	I75	226	I45	I37	I97	208	I63	I97	209	I79	I82	2II	I84	2I8	I84
	F%	46	60	38	36	52	55	43	52	55	48	48	56	49	58	49
Maggio	F	220	283	206	203	257	277	I89	258	27I	243	249	262	235	269	233
	F%	52	67	48	48	6I	65	45	6I	64	58	59	62	56	64	55
Giugno	F	244	302	230	2I5	266	287	245	287	287	252	268	28I	248	328	253
	F%	57	7I	54	50	62	67	58	67	68	59	63	66	59	77	60
Luglio	F	3I5	343	300	279	3I5	339	302	343	337	302	3I4	32I	293	369	30I
	F%	73	79	69	65	73	79	70	80	78	70	73	75	68	86	70
Agosto	F	294	309	278	263	294	3I3	27I	3I2	308	28I	299	307	282	340	288
	F%	73	77	69	66	73	78	68	78	77	70	75	77	70	85	72
Settembre	F	204	240	202	I93	2I6	240	202	230	24I	295	220	236	2I0	26I	2I3
	F%	58	69	56	62	62	69	58	69	69	59	63	68	75	75	6I
Ottobre	F	I48	206	I63	I77	I78	206	I62	I79		I56	I69	I94	I58	208	I5I
	F%	47	65	5I	56	56	65	5I	57	64	50	54	6I	50	66	48
Novembre	F	97	I30	92	93	II5	I2I	II6	I2I	I27	90	II3	I34	I02	I45	I00
	F%	36	49	35	35	43	45	44	45	48	34	42	50	38	54	37
Dicembre	F	I02	II0	75	92	85	II0	I06	98	I09	84	I04	I08	9I	I32	89
	F%	40	43	29	36	33	42	4I	38	42	32	40	42	35	5I	3I
Anno	F	5.9	7.I	5.5	5.5	6.3	7	5.9	6.7	6.9	5.8	6.3	6.8	5.9	7.5	6
	F%	50	6I	47	47	53	59	5I	57	59	50	54	58	50	64	50
Inverno	F	3I2	383	264	3I6	307	373	343	339	372	285	333	364	295	408	296
	σ (F)	53	47	46	66	53	39	85	57	35	40	42	42	33	56	40
	F%	39	48	33	39	38	46	43	42	46	35	4I	45	37	50	37
Primavera	F	540	698	489	467	605	658	506	623	655	568	580	64I	552	676	558
	σ (F)	6I	42	60	47	49	50	II4	68	44	52	35	37	73	6I	46
	F%	47	60	42	40	52	57	44	54	57	49	50	55	48	59	48
Estate	F	854	954	807	756	875	940	8I9	942	933	835	88I	909	822	I037	842
	σ (F)	66	43	42	57	43	38	8I	63	52	50	47	48	68	69	49
	F%	68	76	64	60	69	75	65	75	74	66	70	73	66	83	67
Autunno	F	448	577	457	463	509	567	480	530	570	45I	502	564	470	6I4	464
	σ (F)	57	46	43	78	67	48	76	53	5I	58	42	49	5I	58	56
	F%	47	6I	48	49	54	60	5I	56	60	47	53	60	50	65	49

TAB. 1

*INSOLAZIONE MEDIA GIORNALIERA REGIONALE ASSOLUTA **F** E RELATIVA **F%***

E VARIAZIONE $\frac{F_{estiva} - F_{invernale}}{F_{invernale}} = \Delta\%$

REGIONE	INV.		PRIM.		EST.		AUT.		ANNO		Δ%
	F	- F%	F	F%	F	F%	F	F%	F	F%	
Piemonte - - Val D'Aosta	3.7	43	5.9	47	7.7	55	4.6	44	5.5	47	108
Lombardia	2.6	30	5.7	46	8.0	57	4.1	39	5.2	43	208
Trentino	3.7	43	6.1	48	8.0	57	4.8	49	5.7	49	116
Veneto - Friuli Venezia Giulia	3.0	34	6.0	48	9.1	66	4.7	46	5.7	48	203
Emilia Romagna	3.0	34	6.3	51	9.3	67	4.8	46	5.9	49	210
Liguria	4.1	46	6.2	50	8.8	63	5.4	52	6.1	53	115
Toscana	3.6	41	6.3	50	9.4	68	5.4	52	6.0	53	161
Marche	2.8	32	6.1	49	9.6	70	4.8	46	5.8	49	243
Umbria	3.0	35	5.7	45	8.9	65	5.0	48	5.6	48	197
Lazio	4.0	45	6.6	52	10.0	73	5.9	59	6.6	57	150
Abruzzo e Molise	3.5	39	6.2	49	9.4	68	5.2	50	6.1	53	169
Campania	3.7	40	6.3	50	9.9	73	5.7	55	6.4	54	168
Puglie	3.9	42	6.5	52	9.9	73	5.7	54	6.5	55	154
Basilicata	3.5	38	5.7	50	9.4	69	5.3	50	6.0	51	169
Calabria	4.1	45	6.6	53	9.9	73	5.8	55	6.6	57	141
Sicilia	4.5	49	7.3	59	10.3	77	6.5	62	7.2	62	129
Sardegna	4.2	47	7.2	58	10.3	76	6.2	59	7.0	60	145

TAB. 2

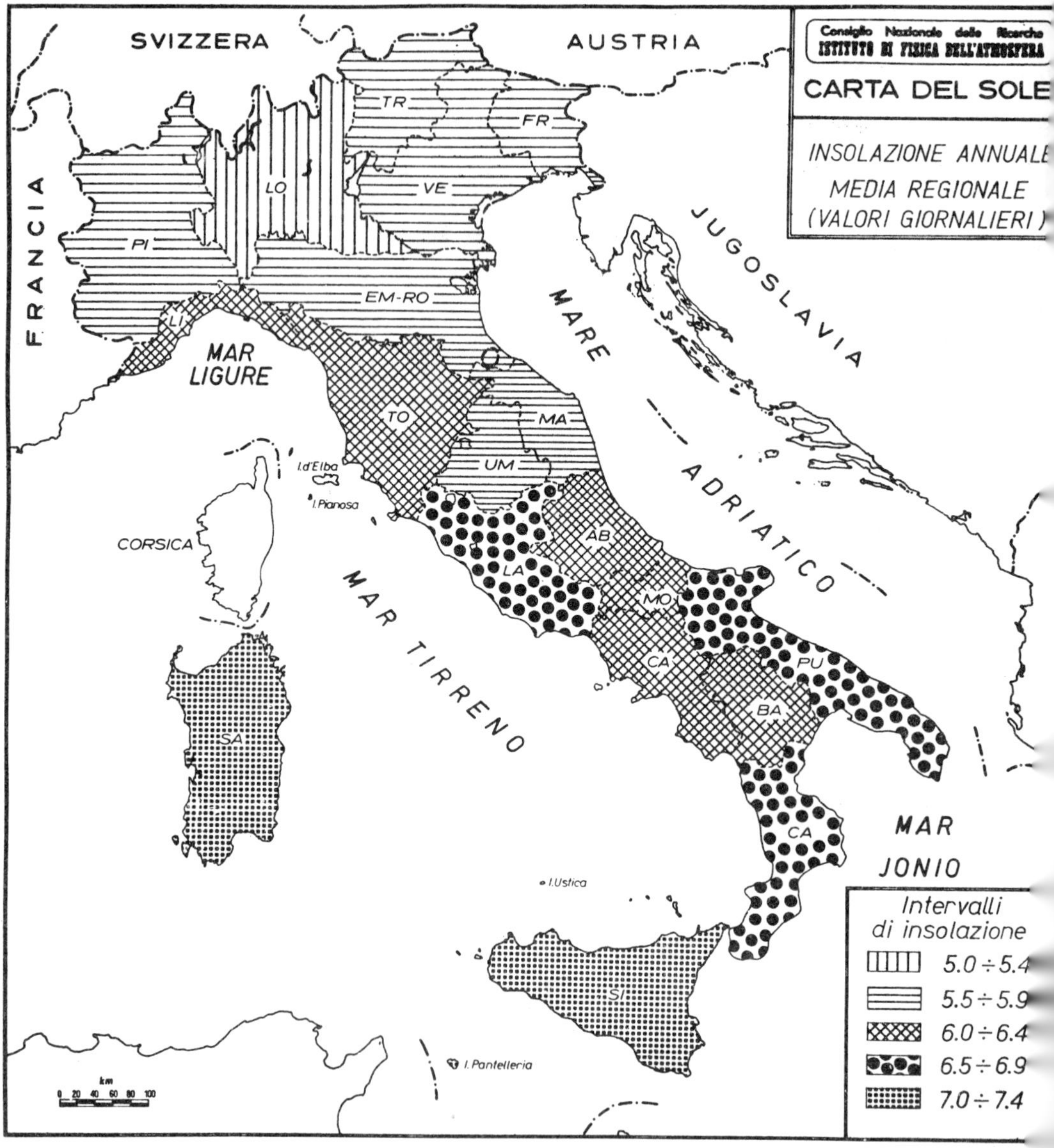
Consiglio Nazionale delle Ricerche
ISTITUTO DI FISICA DELL'ATMOSFERA
CARTA DEL SOLE
INSOLAZIONE ANNUALE
MEDIA REGIONALE
(VALORI GIORNALIERI)
SVIZZERA
AUSTRIA
FRANCIA
JUGOSLAVIA
MARE ADRIATICO
MAR LIGURE
MAR TIRRENO
MAR JONIO
CORSICA
I.d'Elba
I.Pianosa
I.Ustica
I.Pantelleria
TR
FR
LO
VE
PI
EM-RO
LI
TO
MA
UM
AB
LA
MO
CA
PU
BA
SA
CA
SI
km
Intervalli di insolazione
5.0÷5.4
5.5÷5.9
6.0÷6.4
6.5÷6.9
7.0÷7.4

Map n.2 shows the values of daily mean sunshine for the areas in which Italy was subdivided.
Analysing table 2 it derives that it has no sense in subdividing peninsular Italy into Central and South from a sunshine point of view. In fact, the difference from one zone to the other is within 0,1 ÷ 0,3 hours of sunshine, with a percentage variation of 2 ÷ 5 %.
Comparing the theoretical daily sunshine (i.e. the maximum possible for month, season and latitude) and also the percentage variation:

$$\frac{F_{summer} - F_{winter}}{F_{winter}}$$

the sunshine behaviour presents the following characteristics due to astronomical factors: a) during autumn and winter, sunshine increases with decreasing latitude; b) during spring and summer, sunshine decreases with decreasing latitude; c) the percentage variation:

$$\frac{F_{north} - F_{south}}{F_{south}}$$

will therefore be negative during autumn and winter, and will have values between -9% and -4%; d) the percentage variation:

$$\frac{F_{north} - F_{south}}{F_{south}}$$

is instead positive during spring and summer and will be +2% and +4%, respectively. From points c) and d) it can be inferred that from a theoretical point of view does not exist a large sunshine variation between two zones situated at extreme latitudes; e) the percentage variation:

$$(F_{summer} - F_{winter})/F_{winter}$$

shows an increasing with increasing latitude with values of 65% ÷ 44% . Practically, things are different. In fact, whatever the season, sunshine always increases with decreasing latitude. This result appears to be more evident if we take into account zones of latitudes and mean values with respect to the same latitude (similarly to the data shown on table 2).
The percentage variation: $(F_{north} - F_{south})/F_{south}$
is always negative, whatever the season and is -73% (winter), -28% (spring), -34% (Summer) and -59% (Autumn): these values are therefore much heigher than the theoretical ones.

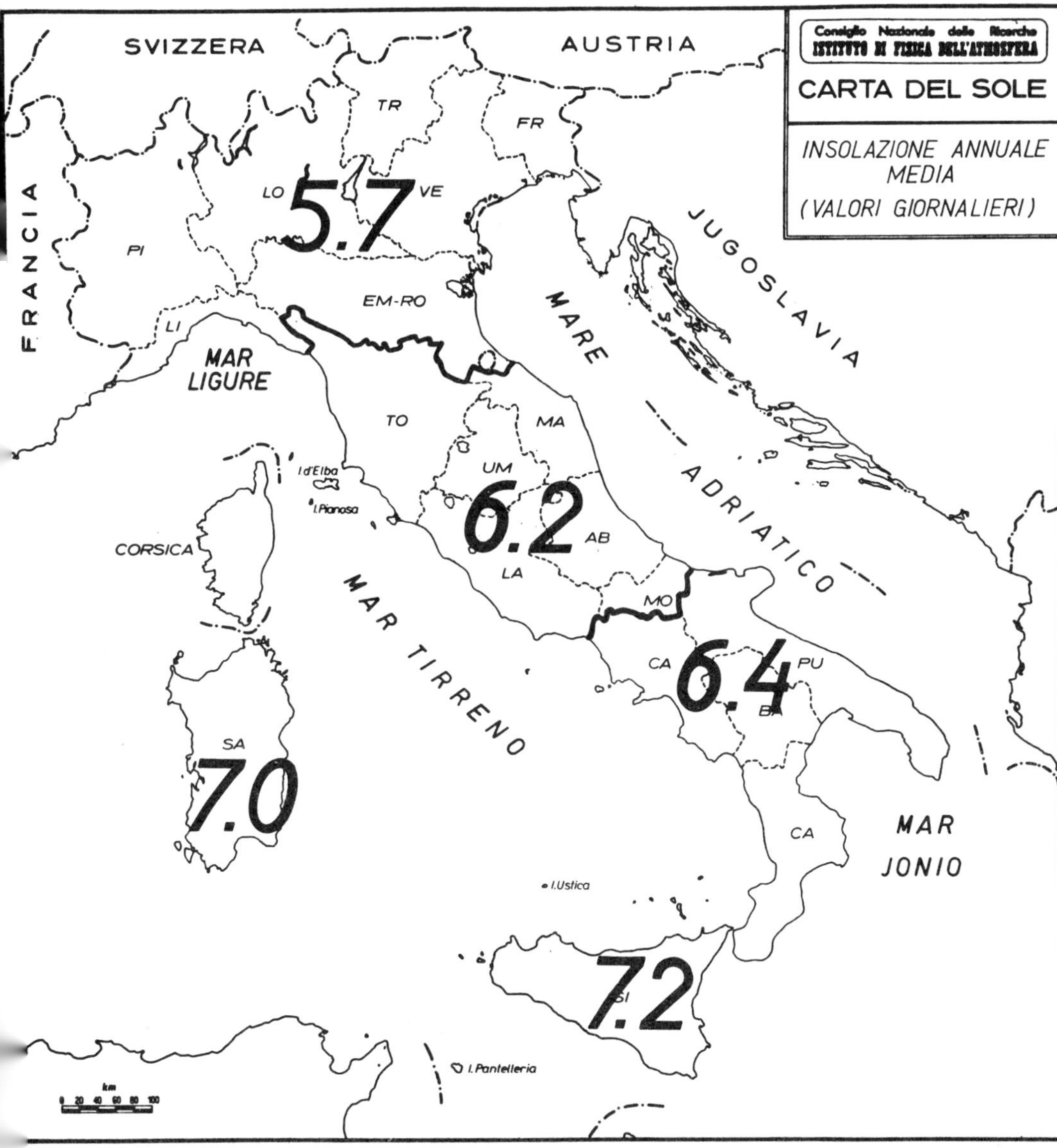
Consiglio Nazionale delle Ricerche
ISTITUTO DI FISICA DELL'ATMOSFERA
CARTA DEL SOLE
INSOLAZIONE ANNUALE MEDIA
(VALORI GIORNALIERI)
SVIZZERA
AUSTRIA
FRANCIA
JUGOSLAVIA
MARE ADRIATICO
MAR LIGURE
MAR TIRRENO
MAR JONIO
CORSICA
TR
FR
LO
VE
PI
EM-RO
LI
TO
MA
UM
AB
LA
MO
CA
PU
SA
CA
I.d'Elba
I.Pianosa
I.Ustica
I.Pantelleria
5.7
6.2
6.4
7.0
7.2
km

CORRELAZIONE TRA RADIAZIONE GLOBALE (G) ED INSOLAZIONE (F)

$$G = aF + b$$

N°staz.	a	σ(a)	b	r
2	60.5	2.7	-65.80	0.86
9	56.5	2.2	-37.28	0.89
11	76.8	5.2	-50.88	0.75
19	51.4	1.5	-25.96	0.93
22	52.5	1.5	-22.25	0.94
23	48.3	1.3	+23.87	0.94
32	58.0	2.3	-49.35	0.89
38	49.9	1.2	-44.23	0.95
42	56.7	1.3	-72.40	0.92
49	57.3	2.9	-31.46	0.84
55	61.0	2.2	-75.63	0.90
60	48.3	1.6	-27.86	0.92
61	51.3	1.3	-33.52	0.95
68	52.4	1.5	-65.30	0.93
70	47.9	1.3	+19.34	0.90
71	48.2	1.6	-16.74	0.91
72	62.3	1.9	-101.7	0.93
75	53.9	1.7	-49.94	0.92
78	51.0	1.6	-37.97	0.92
87	48.2	1.3	-48.09	0.94
90	47.4	1.7	-37.23	0.90
96	48.0	1.8	-44.83	0.90
98	56.3	1.4	-63.98	0.95
108	51.4	1.6	-30.34	0.92
113	53.4	1.7	-52.91	0.92
116	59.8	2.2	-75.14	0.90
118	51.3	1.6	-42.23	0.93
120	48.0	2.0	-33.92	0.88
124	50.3	1.8	-47.59	0.92
127	52.2	2.2	-67.77	0.88
128	50.9	1.7	-41.20	0.92

CORRELAZIONE TRA RADIAZIONE GLOBALE (G) ED INSOLAZIONE (F)

$$\frac{G}{G_{max}} = a' \frac{F}{F_{max}} + b'$$

N° Staz.	a'	σ(a')	b'	r
2	0.37	0.04	0.23	0.60
9	0.30	0.02	0.24	0.79
11	0.30	0.05	0.49	0.39
19	0.39	0.02	0.23	0.85
22	0.41	0.02	0.23	0.85
23	0.43	0.01	0.18	0.92
32	0.41	0.02	0.22	0.83
38	0.38	0.02	0.23	0.87
42	0.42	0.02	0.19	0.82
49	0.66	0.03	0.10	0.85
55	0.37	0.03	0.28	0.72
60	0.33	0.02	0.24	0.71
61	0.43	0.02	0.23	0.84
68	0.83	0.02	0.25	0.73
70	0.43	0.03	0.22	0.79
71	0.33	0.02	0.23	0.75
72	0.55	0.03	0.15	0.83
75	0.35	0.02	0.26	0.77
78	0.30	0.02	0.27	0.74
87	0.36	0.03	0.22	0.72
90	0.31	0.03	0.22	0.64
96	0.32	0.03	0.23	0.61
98	0.43	0.02	0.22	0.87
108	0.32	0.02	0.28	0.72
113	0.34	0.02	0.26	0.74
116	0.40	0.03	0.22	0.68
118	0.28	0.03	0.30	0.64
120	0.33	0.04	0.21	0.57
124	0.29	0.03	0.27	0.67
127	0.33	0.03	0.24	0.58
128	0.31	0.02	0.26	0.69

TAB. 4

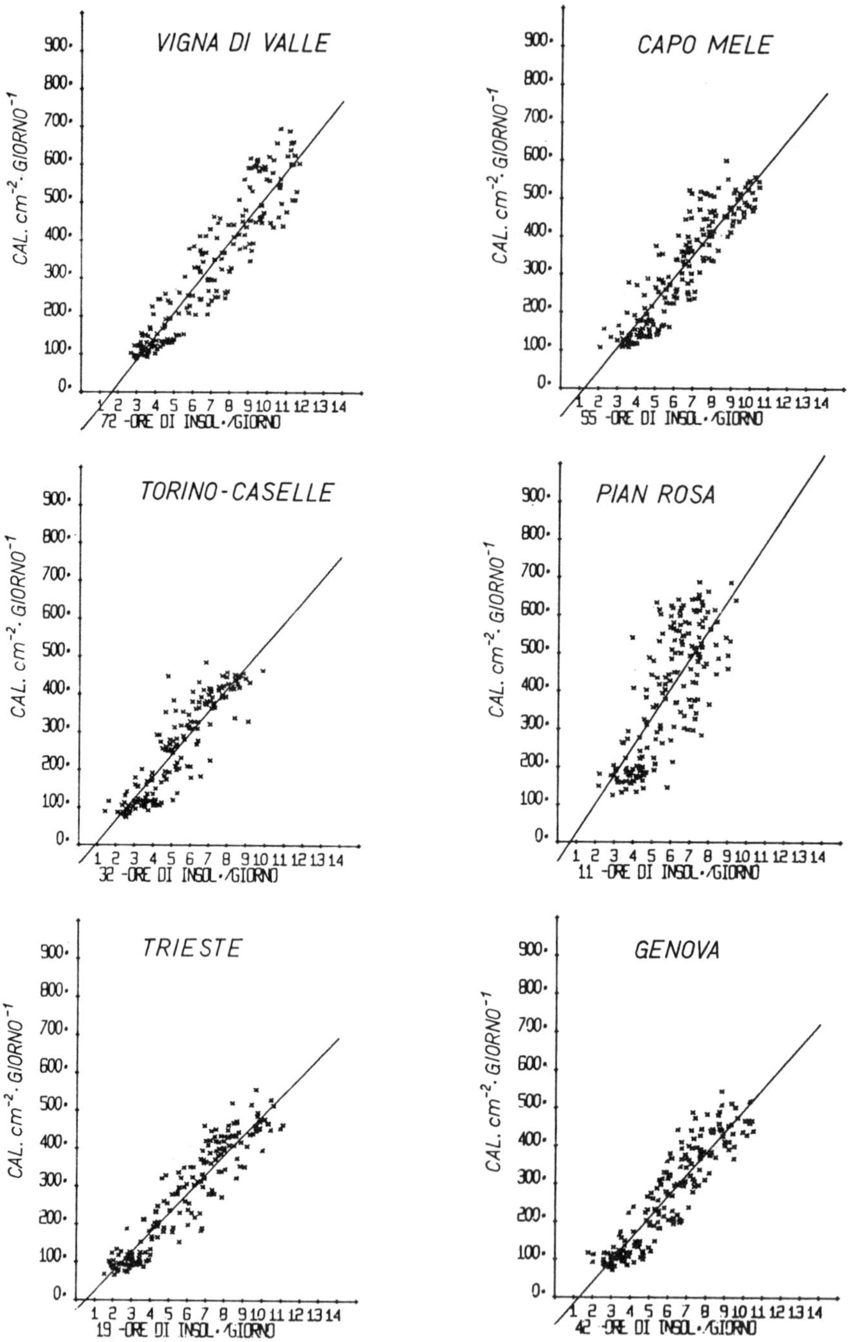
VIGNA DI VALLE
CAPO MELE
TORINO-CASELLE
PIAN ROSA
TRIESTE
GENOVA
CAL. cm^-2 · GIORNO^-1
900· 800· 700· 600· 500· 400· 300· 200· 100· 0·
1 2 3 4 5 6 7 8 9 10 11 12 13 14
72 -ORE DI INSOL·/GIORNO
55 -ORE DI INSOL·/GIORNO
32 -ORE DI INSOL·/GIORNO
11 -ORE DI INSOL·/GIORNO
19 -ORE DI INSOL·/GIORNO
42 -ORE DI INSOL·/GIORNO

FIG. 3

Such a behaviour is understandable if we take into account that the presence in Northern Italy of fogs and clouds, is larger than in Southern Italy. Such a difference in the cloudiness behaviour is more evident during winter and autumn, where we have a maximum percentage variation.

The percentage variation: $(F_{summer} - F_{winter}) / F_{winter}$

shows an uneven behaviour with the latitude and presents values of 108% ÷ 243 %. The highest values may be due to a very small summer cloudiness, together with a very large winter cloudiness. The lowest values are instead due to a summer cloudiness slightly lower than the winter one.

CORRELATION BETWEEN SUNSHINE AND GLOBAL RADIATION

For the Air Force Meteorological Service stations, of which we, have sunshine values F and global radiation G, we attempted an analysis of linear correlation between the daily mean number of sunshine hours and the daily mean value of global radiation (calorie x cm^{-2} x day^{-1}) relative to each month for each station and for 15 years, from 1958 to 1972.
The Albrecht formula was used:

$$G = b + a F$$

and made also a correlation between data of global radiation (G/G MAX) and of sunshine (F/F MAX) using the Angstrom relationship:

$$\frac{G}{G_{MAX}} = b + a \frac{F}{F_{MAX}}$$

In tables 3 and 4 are shown the regression coefficients a with the corresponding standard deviations $\sigma(a)$, and also the correlation coefficient r with the interception line b relative to the lines, as in the preceding formulae. It is evident that the Albrecht formula presents very good correlation coefficients whereas the Angstrom one does not give, except in a few cases, a sufficient reliability.
The scanty number of observation stations for ghe global radiation has not allowed us to interpret the correlation parameters as to characterize the single regions, in order to get a distribution of the global radiation along the peninsula, similarly to the sunshine one. In fact, attempts in this sense have given negative results.
Fig.3 represents the correlation lines relative to the Air Force Meteorological Service Stations.

ON THE DISTRIBUTION OF SOLAR ENERGY IN GREECE

G.J. Macris
Meteorological Institute National Observatory
Kallithea, Athens, Greece

ABSTRACT

In this study were calculated the average daily amounts of solar insolation on a horizontal surface for each month in ly per day.

These amounts are presented on maps by means of insolation lines and in Table III.

For these calculations, the observation of sunshine hours and solar radiation (for the period 1953-1973) were used, taken on the hill of the Meteorological Institute National Observatory of Athens.

A Sunshine Campbell Stokes recorder was used for the measurement of sunshine, while a Solarigraph Gorczinsky No. III and Epple pyrheliometers No. 3604 and 3034 were used for the measurement of solar radiation (for the periods 1953-1959 and 1960-1973 respectively).

INTRODUCTION

We used linear regression of the sunshine hours and total amount of solar energy in the form shown below:

$$Qm = Qo\ (a + b\ Sm/So) \qquad (1)$$

where

Qm is the total amount of solar radiation impinging on a horizontal surface per cm^2 at the earth's surface, expressed in ly per day.

Qo is the total amount of solar radiation impinging on a horizontal surface per cm^2 at the top of the atmosphere.

Sm is the average daily values of sunshine hours measured on a sunshine recorder, and

So is the maximum number of sunshine hours under perfectly clear sky conditions, with a and b as constants.

The values of Qm and Sm were obtained and calculated from the chart recorder paper of the actinometrical instruments of the Athens Observatory at an elevation of 107 meters.

In addition to the above actinometrical station, two other stations are in operation. One is located on the island of Thera at an elevation of 500 meters, and the other at the village of Magouliana in Arcadia at an elevation of 1200 meters. Both stations use the Robitzsh actinograph.

For the calculation of the average daily value of solar energy of the remaining areas of Greece by application of the above equalization [1], thirty stations were chosen from the National Meteorological Network of the country, which use the Campbell Stokes type Sunshine Recorder, operated during the period of 1960-1973.

Using sunshine hours and solar isolation values as recorded at the Athens Observatory, the constants a and b were determined for each month and the application of the least squares method.

TABLE III

STATIONS	Lat.	Long.	Jan.	Febr.	March.	April.	May.	June.	July.	Aug.	Sept.	Oct.	Nov.	Dec.	Year	M.O.	Rate
1. Timpakion	35°00′	24°45′	177	265	365	454	562	596	619	587	452	321	228	167	4793	399	452
2. Hierapetra	35°00′	25°44′	194	254	380	466	557	615	619	582	459	330	217	187	4860	405	432
3. Sitia	35°12′	26°06′	164	239	344	451	548	606	614	557	439	300	218	159	4639	387	455
4. Heracleon	35°29′	25°11′	159	237	340	447	557	601	632	568	442	314	211	160	4668	389	473
5. Hania	35°30′	24°07′	159	232	347	446	552	615	627	568	438	284	213	155	4636	386	472
6. Souda	35°43′	24°07′	166	237	351	446	562	610	627	563	434	282	207	157	4642	387	470
7. Cithira	36°09′	23°00′	178	234	340	436	534	600	604	554	417	295	221	165	4578	382	439
8. Rhodes	36°23′	28°07′	178	241	357	468	538	619	620	575	463	326	217	161	4763	397	459
9. Thera	36°25′	25°24′	169	236	335	468	524	587	612	552	434	318	209	145	4589	382	467
»			177*	232*	330*	451*	527*	565*	606*	524*	419*	301*	218*	152*	4490*	374*	442*
10. Milos	36°43′	24°27′	148	206	319	468	530	612	603	540	465	293	193	176	4553	379	464
11. Methoni	36°50′	21°42′	159	230	337	431	505	596	590	541	414	294	200	158	4455	371	432
12. Kalamai	37°04′	22°06′	159	230	330	425	506	592	588	538	412	297	207	159	4443	370	429
13. Naxos	37°06′	25°23′	148	225	330	421	505	562	603	516	416	289	206	144	4365	364	459
14. Magouliana	37°08′	22°08′	137*	198*	283*	350*	407*	441*	434*	415*	328*	239*	170*	121*	3523*	294*	313*
15. Syros	37°27′	24°67′	145	230	348	445	524	599	607	552	436	290	202	150	4528	377	462
16. Samos	37°44′	26°52′	159	229	340	441	529	599	603	552	438	297	200	140	4527	377	463
17. Corinthos	37°56′	22°57′	139	216	321	429	510	576	590	530	409	278	189	137	4324	360	453
18. Athens	37°58′	23°43′	161	241	331	448	538	599	612	563	416	284	198	151	4552	379	461
»			164*	240*	330*	456*	533*	588*	597*	536*	417*	291*	197*	148*	4497*	375*	449*
19. Araxos	38°10′	21°25′	152	224	334	427	515	547	567	522	401	275	185	140	4289	357	427
20. Chios	38°22′	26°09′	145	213	325	439	547	608	622	567	437	295	188	135	4521	377	487
21. Aliartos	38°23′	23°06′	130	204	300	427	501	561	576	526	387	251	176	124	4163	345	452
22. Lamia	38°54′	22°24′	148	220	319	433	514	585	593	541	395	254	179	133	4314	360	460
23. Skyros	38°54′	24°33′	121	206	287	441	558	626	630	549	402	264	171	115	4370	364	515
24. Mytilene	39°04′	26°35′	138	203	318	433	533	589	607	557	429	286	178	124	4395	366	483
25. Corfou	39°37′	19°55′	145	210	313	423	514	589	603	536	395	273	169	125	4295	358	478
26. Larissa	39°38′	22°25′	131	207	299	419	495	552	562	524	381	255	173	119	4117	343	443
27. Ioannina	39°42′	20°48′	139	200	295	415	468	542	562	507	371	252	149	111	4011	334	451
28. Lymnos	39°55′	25°15′	120	206	308	419	519	579	575	535	405	268	164	116	4214	351	459
29. Konitsa	40°03′	20°45′	146	203	293	383	472	524	543	498	359	248	162	122	3953	329	421
30. Sedes	40°33′	23°01′	136	213	291	417	505	521	566	497	374	256	167	119	4062	339	447
31. Seres	41°04′	23°34′	139	215	286	401	463	519	539	493	361	247	156	115	3934	328	424
32. Komotini	41°07′	25°24′	128	179	265	397	482	547	539	496	330	249	157	118	3887	324	421

TABLE III. It contains the geographical co-ordinates, the mean daily values of solar energy for each month in ly/day of the year and the rate of these values for all stations. It also ,contains yearly and mean yearly values of theme. In the stations Athens, Thira and Magouliana the values with (*) are measured values and the others calculated values from the equation (4).

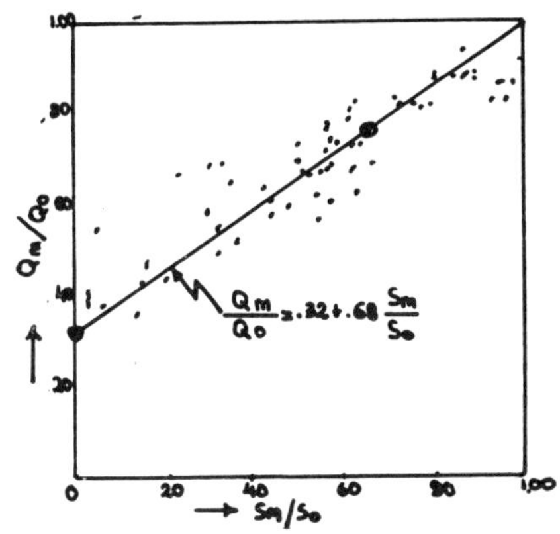

Fig. I Relation Between The Ratio of Sm/So of the Ratio Wm/Qo.

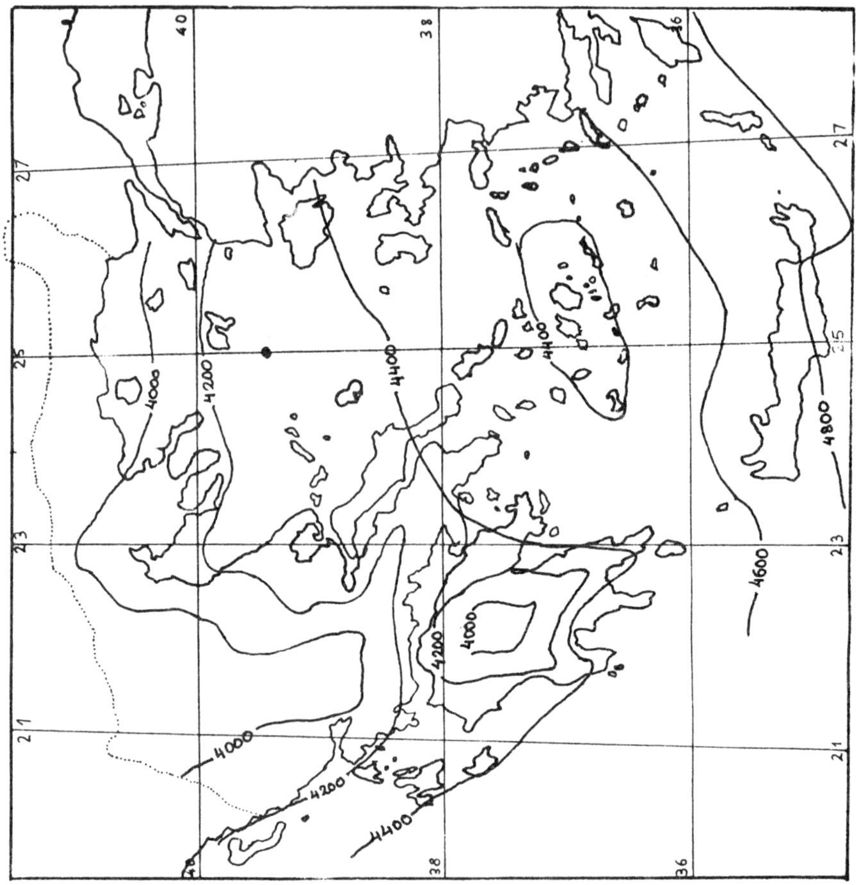

Fig. II Yearly Distribution of Solar Energy In "ly per year"

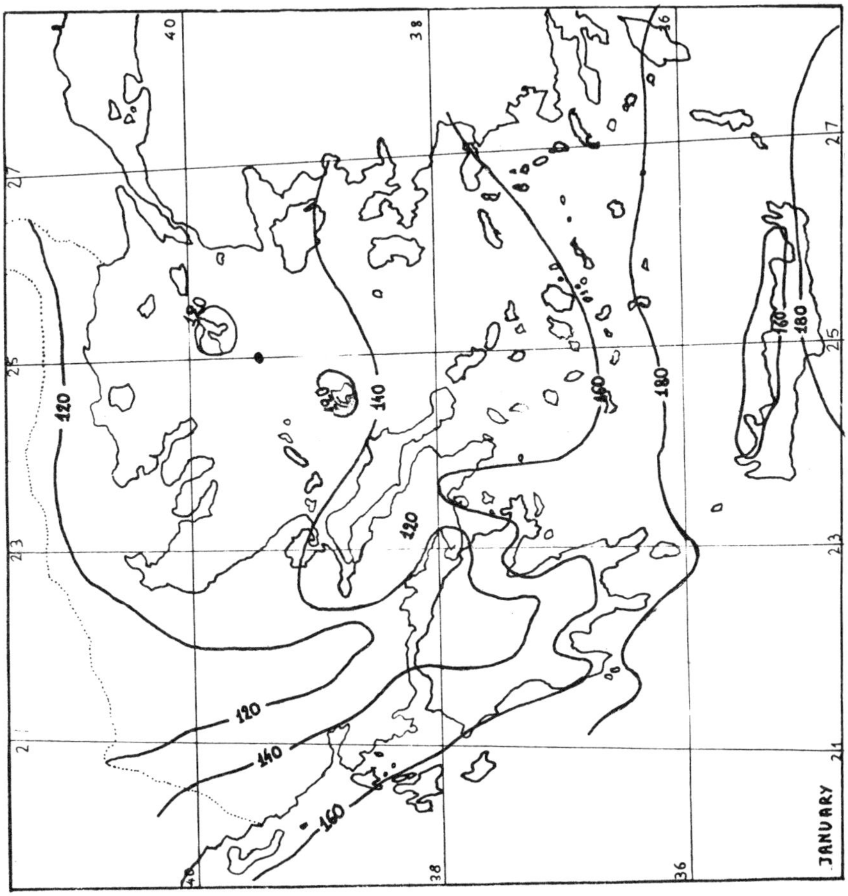

Fig. III Average Daily Solar Radiation Received On Horizontal Surface At The Ground in "ly per day" - January

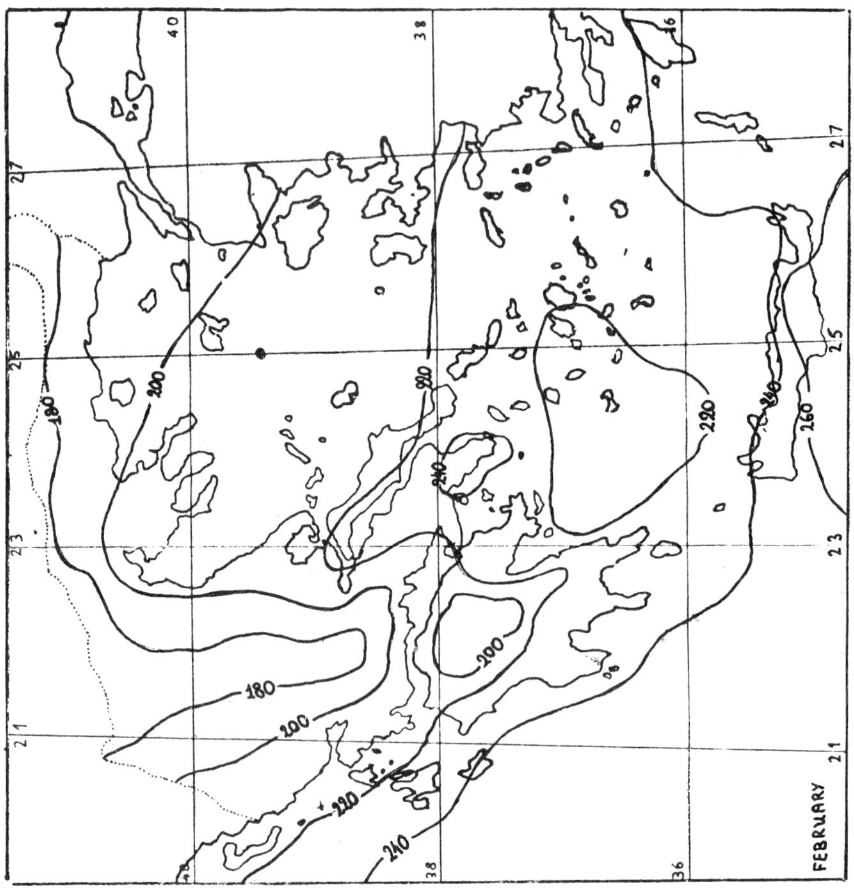

Fig. III Average Daily Solar Radiation Received On Horizontal Surface At The Ground in "ly per day" - February

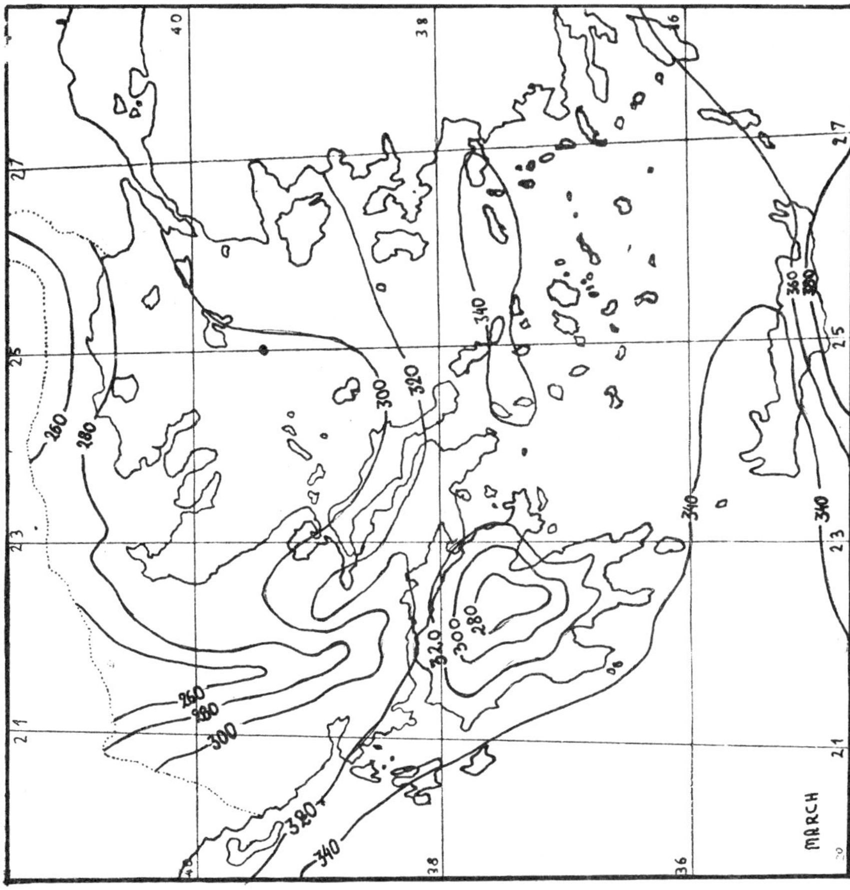

Fig. III Average Daily Solar Radiation Received On Horizontal Surface At The Ground in "1y per day" - March

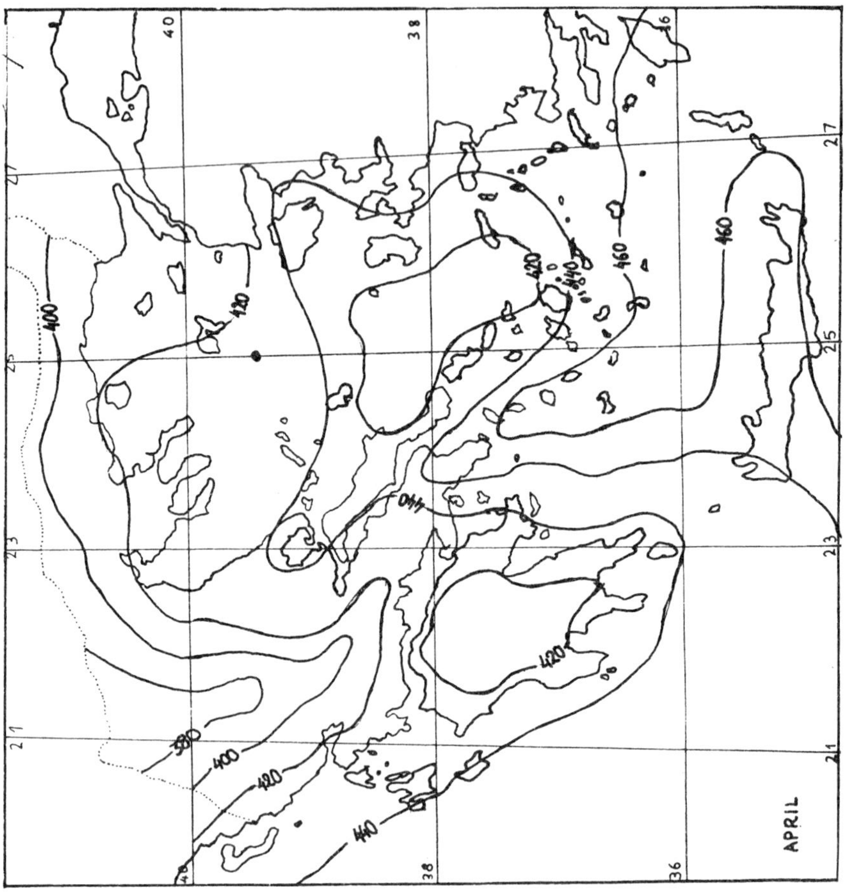

Fig. III Average Daily Solar Radiation Received On Horizontal Surface At The Ground in "ly per day" - April

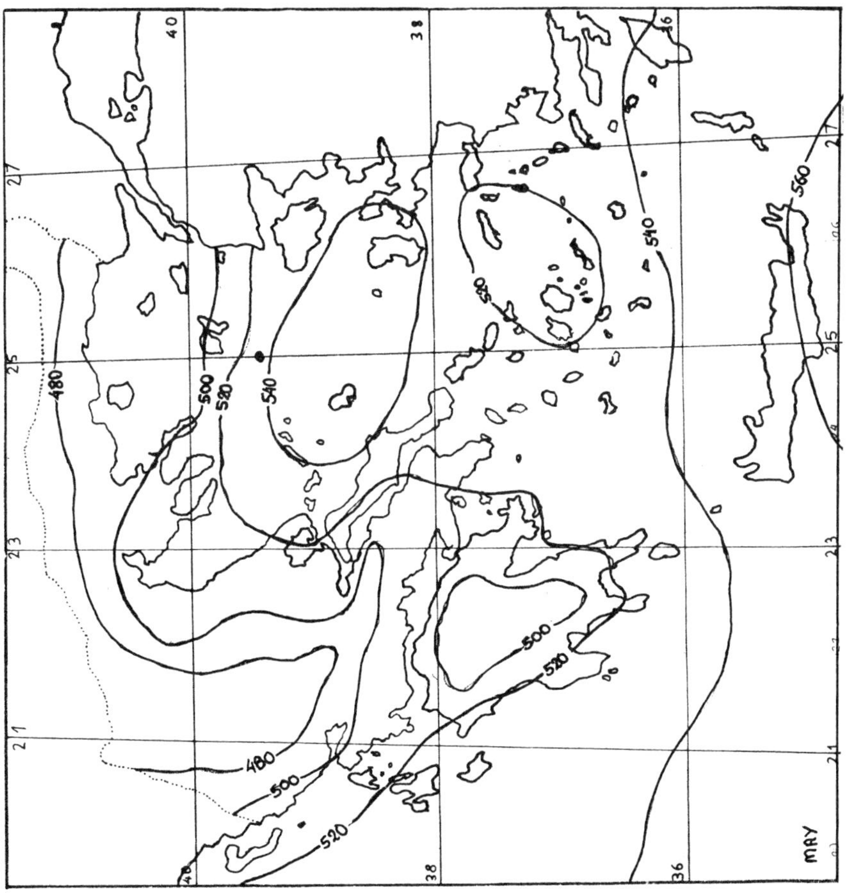

Fig. III Average Daily Solar Radiation Received On Horizontal Surface At The Ground in 'ly per day" - May

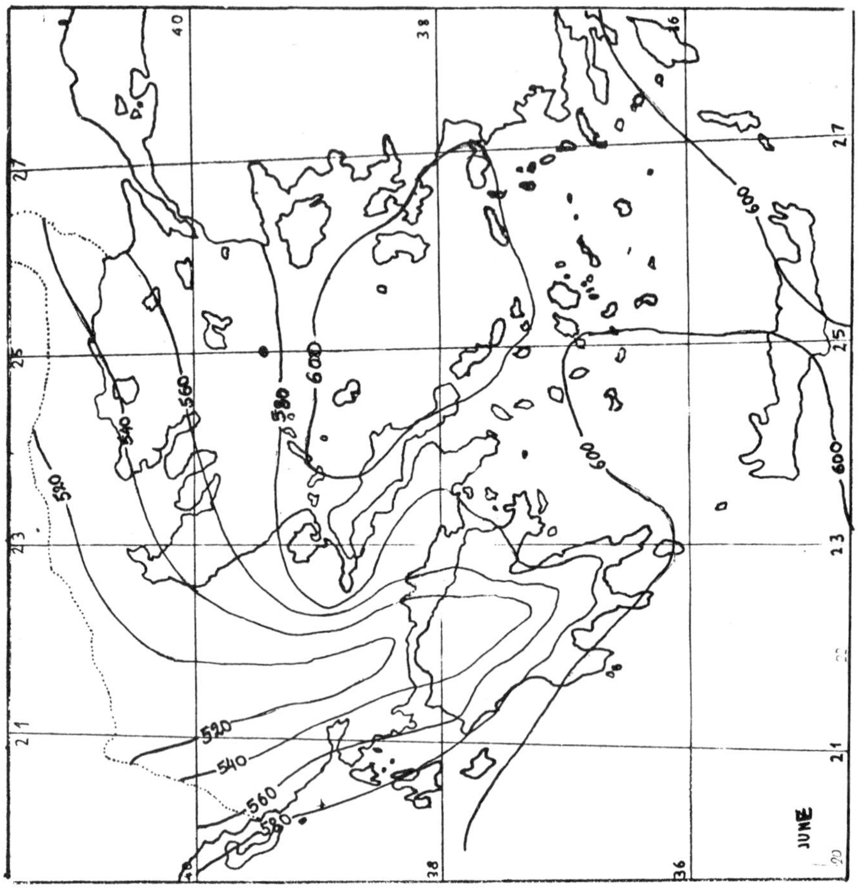

Fig. III Average Daily Solar Radiation Received On Horizontal Surface At The Ground in "ly per day" - June

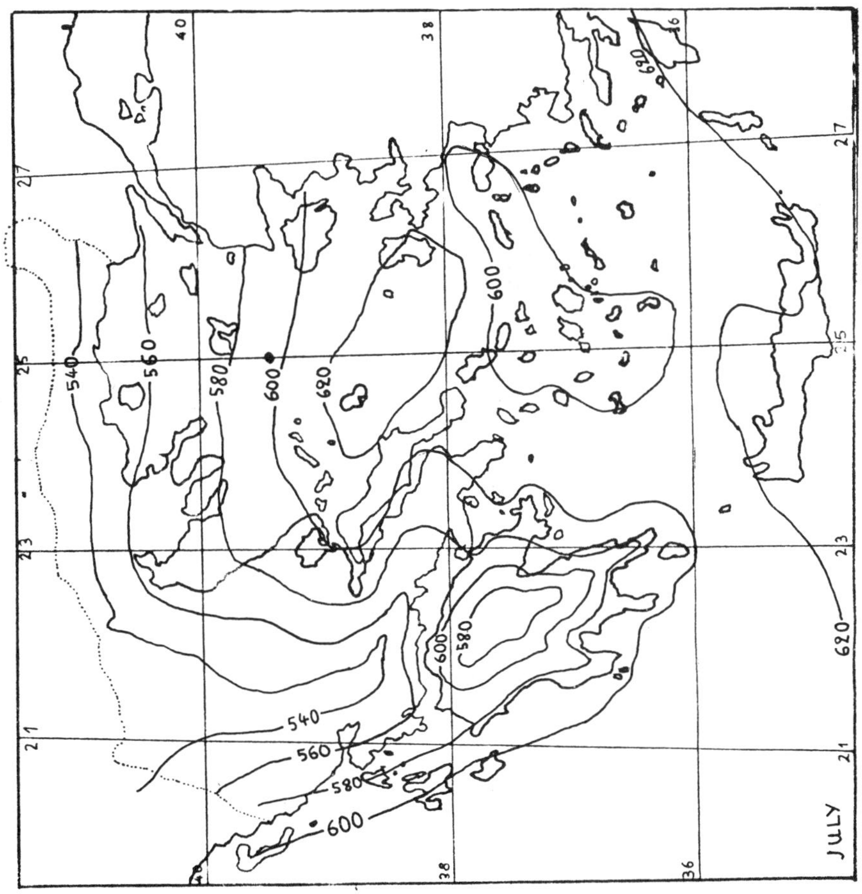

Fig. III Average Daily Solar Radiation Received On Horizontal Surface At The Ground in "ly per day" - July

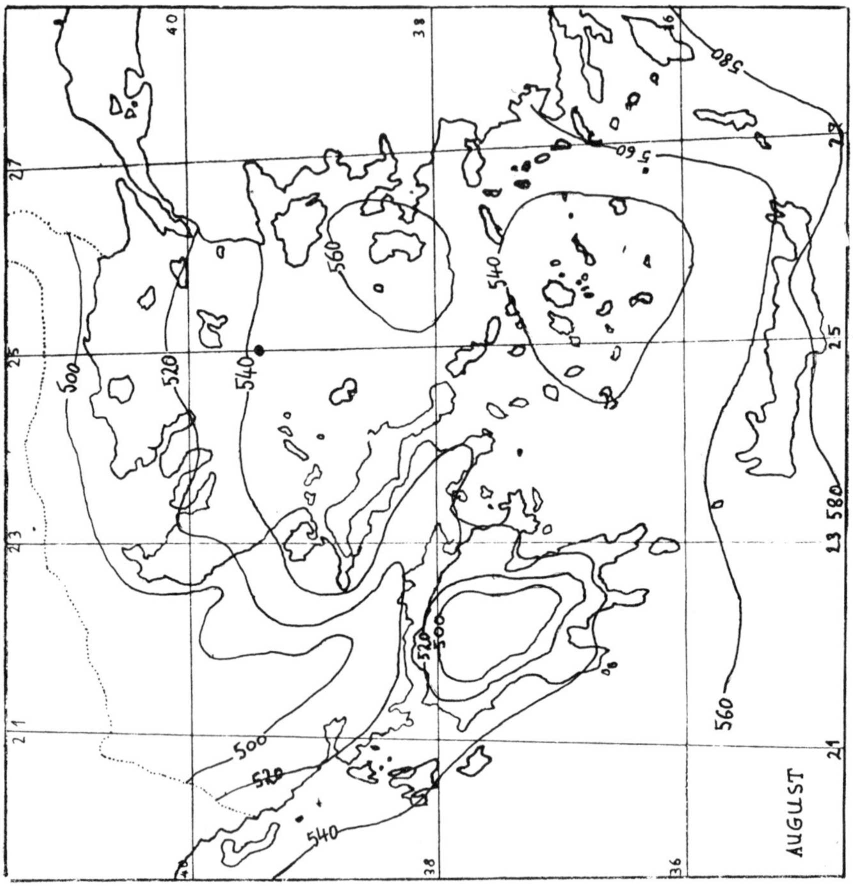

Fig. III ·Average Daily Solar Radiation Received On Horizontal Surface At The Ground in "ly per day" - August

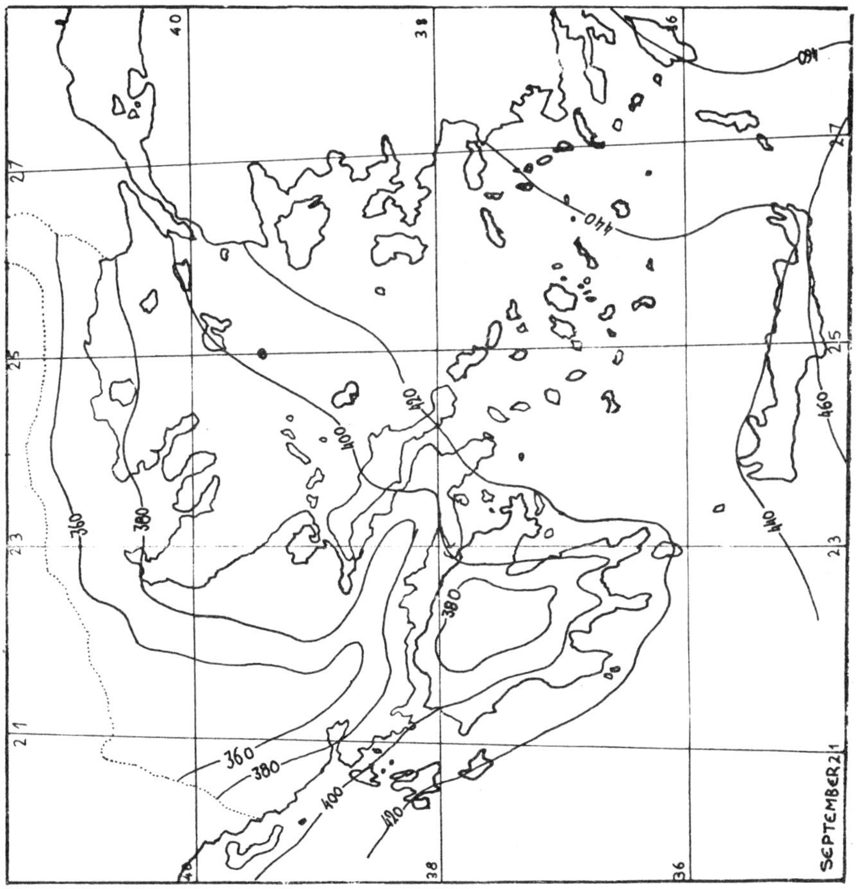

Fig. III Average Daily Solar Radiation Received On Horizontal Surface At The Ground in "ly per day" - September

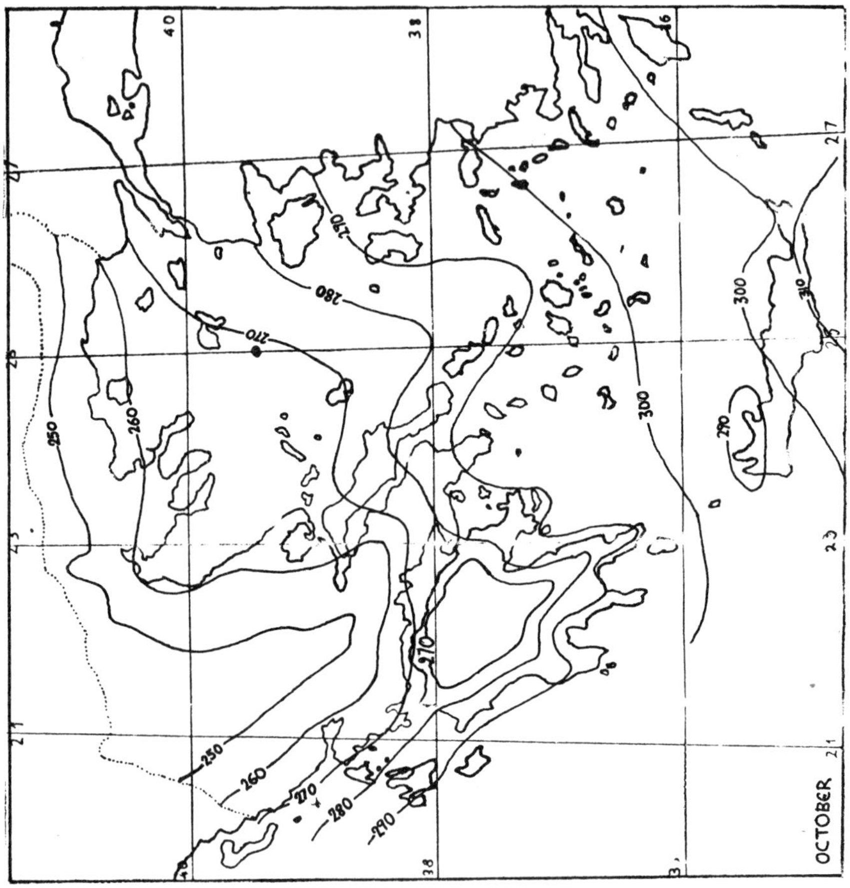

Fig. III Average Daily Solar Radiation Received On Horizontal Surface At The Ground in "ly per day" - October

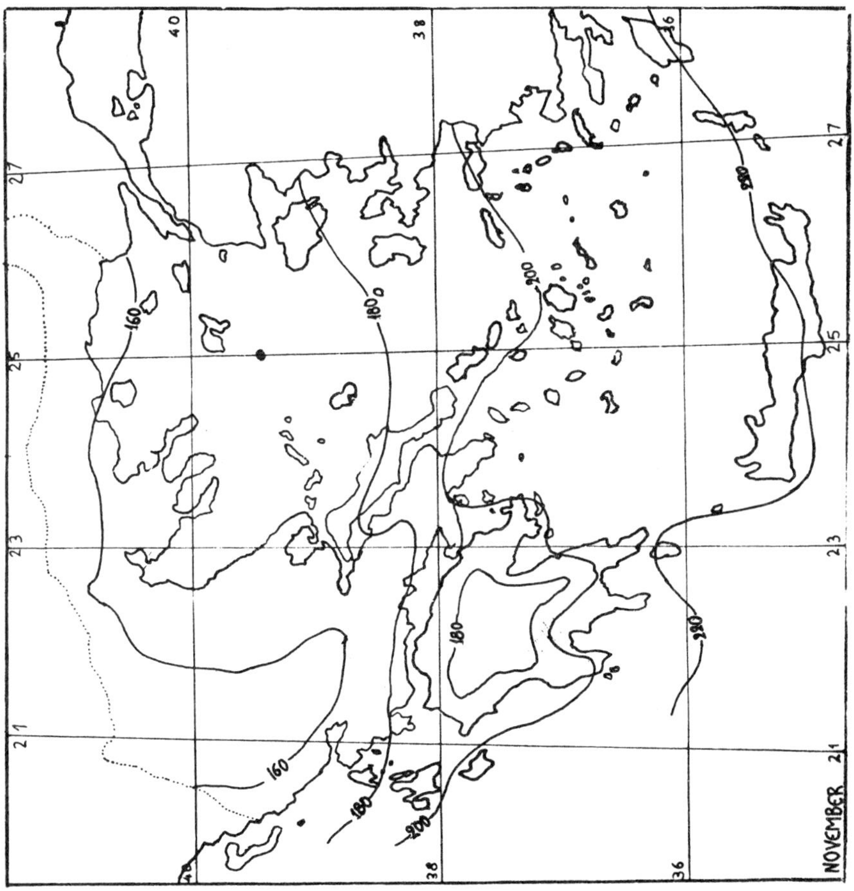

Fig. III Average Daily Solar Radiation Received On Horizontal Surface At The Ground in "ly per day" - November

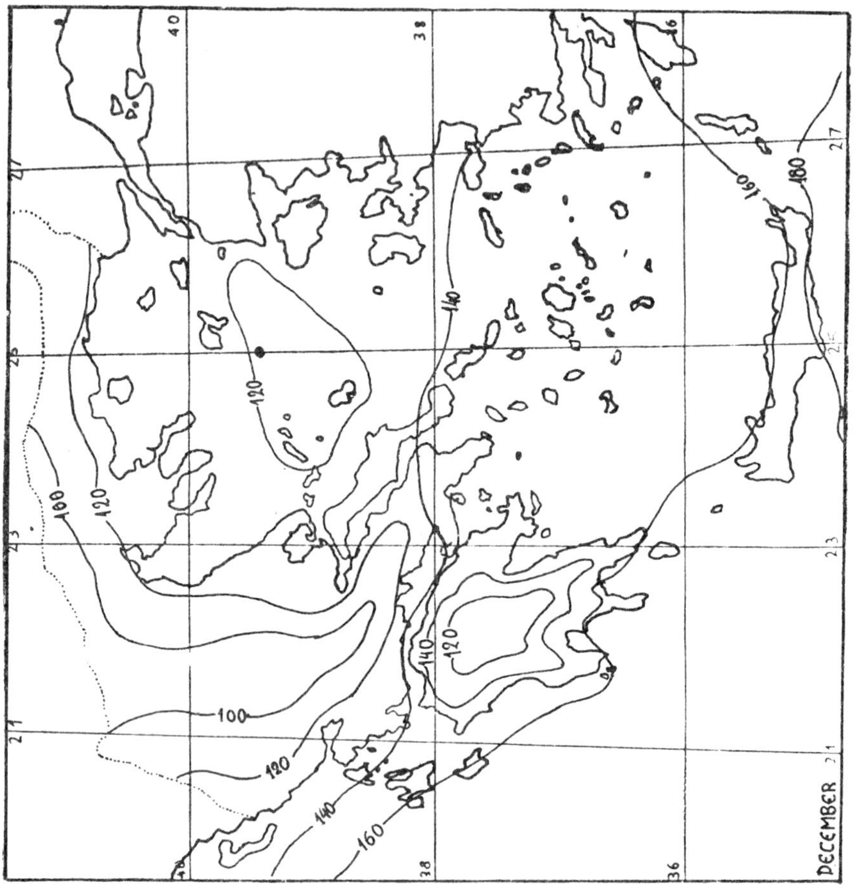

Fig. III Average Daily Solar Radiation Received On Horizontal Surface At The Ground in "ly per day" - December

REFERENCES

1. Angstrom, A. "Solar and Terrestrial Radiation" Quarterly Journal of the Royal Meteorological Society, V.50, No. 210, April 1924, pp. 121-125.

2. Angstrom, A. "On The Computation of Global Radiation from Records of Sunshine: Arkiv. for Geofusica, Band 2, Hefte 5, No. 22, 1956, pp. 471-479.

3. Bernhardt and Phillips, H., "Die Raumiliche unde Zeirleche Verteilung der Eintrahlung und der Strahlungsbilanz in Meersniveau, Teil 1, Die Einstrahlung Academie - Verlag, Berlin, 1958, p. 227.

4. Black, J. Conython, C. and Prescott, J., Quarterly Journal of the Royal Meteorological Society, V.80, 1954, p. 231.

5. Fritz, S. and MacDonald, T.H., "Average Solar Radiation in the United States", Heating and Ventilating, Vol. 46, No. 7, July 1949, pp. 61-64.

6. Fritz, S., "Solar Radiation on Cloudless Days" Heating and Ventilating, Vol. 46, No. 1, January 1949, pp. 69-74.

7. Haurwitz, B. "Daytime at Blue Hill Observatory in 1933, With Application to Turbidity in American Air Masses", Harvard University Press, Cambridge, Massachusetts, 1934, p. 31.

8. Hamon, R.W., Weiss, L.L. and Wilson, W.T., "Insolation as an Empirical Function of Daily Sunshine Duration", Monthly Weather Review, Vol. 82, No. 6, June 1954, pp. 141-146.

9. Kimball, H.H., "Measurements of Solar Radiation Intensity and the Depletion of the Atmosphere", Monthly Weather Review, Vol. 55, No. 4, April 1927, pp. 155-169.

10. Kimball, H.H., "Variations in the Total and Luminous Solar Radiation with Geographical Position in the U.S.", Monthly Weather Review, Vol. 47, No. 11, November 1919, pp. 769-795.

11. Karapiperis, L., Katsoulis, B., Papachristopoulos, K., "Contirbution to the Study of Sunshine in Greece", Hypomnemata of the National Observatory of Athens, S. II Meteorology, No. 38, Athens, 1974.

12. Karapiperis, L., "Recherches sur La Radiation Solarire a Athenes", Publication de L'Observatoire Astromique Royale D'Athenes, 1936, p. 131.

13. Macris, G.J., "Solar Energy and Sunshine Hours at Athens, Greece, Monthly Weather Review, Vol. 87, No. 1, 1959, pp. 29-32.

14. Mariolopoulos, E., "The Climate of Greece", Athens, Greece, 1938.

15. Mariolopoulos, E., and Livathinos, A., "Atlas Climatique de La Grece", Athens, Greece, 1938.

16. National Observatory Climatological Bulletine, Athens, Greece, 1953-1973.

17. Sham, P., "Total Solar and Sky Radiation in Hong Kong", Royal Observatory, Hong Kong Technical Note No. 22, 1964.

THE FLAT PLATE SOLAR COLLECTOR:
ITS STEADY-STATE AND TRANSIENT STATE BEHAVIOUR

E. Hahne, N. Fisch, A. Arafa
Institutsgemeinschaft Technische Nutzung
Solarer Energie
Institut für Thermodynamik und Wärmetechnik
Universität Stuttgart
Stuttgart, F.R.G.

ABSTRACT

For a satisfying performance of a collector, both its steady-state and transient state behaviour has to be taken into account. As criteria, the efficiency and effectivity of collectors were calculated here with respect to their great number of parameters. These were grouped into

1. collector parameters 2. operation parameters
3. meteorological parameters.

By their individual variation in computer programs, their particular effects are obtained as well as their relation to other parameters. The numerical results are compared to experimentally obtained data.
The agreement is considered satisfying with deviations of less than 10 % as an average for efficiency and less than 2 K for transient outlet temperatures of more than 30°C.

INTRODUCTION

Due to the periodic and intermittent supply of sun-shine the collection of solar energy consequently is periodic and intermittent with physical processes which are basically transient and only in special cases follow steady-state conditions.
In any case these processes depend upon a multitude of properties and conditions. Those which are not interrelated with each other are e.g. the geometric configuration of the absorber plate and the glass or plastic covers as well as their thermal and radiative properties. These as pertaining to the collector directly are named "collector parameters". Other parameters such as mass flow rate of the heat carrier and its temperature are considered as "operation parameters" and finally solar irradiation, ambient temperature, wind speed and sky temperature are grouped together as "meteorological parameters".
For the evaluation and rating of the combined action of all these parameters, the so-called thermal efficiency of a collector is considered. This is the ratio of actual useful energy collected to solar energy incident upon the collector.

In order to take account of time dependent irradiation a so-called effectivity is introduced. This is the ratio of actual useful energy collected within a certain time to the energy delivered during that time. Presently, it is one of the main tasks in national and international programs to investigate collectors in open air and laboratory test facilities in order to gain suitable test specifications. Due to the variety of parameters and the changing and uncontrollable nature of meteorological parameters such investigations are time-consuming and a wide scatter of results has to be tolerated.
So far it would appear premature to claim that we can do without these experiments, but we can support these by numerical treatment of the many parameters. Effects and trends upon the thermal efficiency are clearly shown and provide a basis for the construction and operation of a collector most suitable to given conditions.
This work is based upon a detailed numerical program for both steady-state and transient state performance. The results gained by this program theoretically are checked against experimental results gained in tests.

1. THERMAL EFFICIENCY

The definition of thermal efficiency is

$$\eta = \dot{Q}/JA \quad ^{+} \tag{1}$$

The collected useful heat flux $\dot{Q}$ can be expressed as the difference between the solar irradiation transmitted and absorbed $JA\tau_G^*\alpha_A^*$ and the heat lost to the surroundings $kA(T_A-T_\infty)$ ++ as

$$\dot{Q} = JA\,\tau_G{}^*\,\alpha_A{}^* - kA(T_A-T_\infty) \tag{2}$$

An average absorber plate temperature ϑ_A or T_A is difficult to measure and to determine. Measurements of a fluid inlet temperature ϑ_i or T_i and outlet temperature ϑ_o or T_o are easy and a mean temperature can be determined as $(T_i + T_o)/2$. Such a mean temperature does not necessarily coincide with the average absorber temperature and a correction factor F' is introduced for compensation.
With F' being the ratio of the collected useful heat flux Q to the useful heat flux if the entire collector were at a mean fluid temperature $(T_i+T_o)/2$ equation (2) becomes

$$\dot{Q} = JA\,\tau_G^*\,\alpha_A{}^*\,F' - kA\left[(T_i+T_o)/2 - T_\infty\right]\,F' \tag{3}$$

+ symbols are listed in the nomenclature
++ the temperature difference is expressed in thermodynamic temperatures instead of Centigrade temperatures, because radiative heat transfer is included.

Introducing equation (2) or (3) into (1) yields the efficiency of a collector as

$$\eta = \tau_G^* \alpha_A^* - k[T_A - T_\infty]/J \qquad (4a)$$

or

$$\eta = F'\tau_G^* \alpha_A^* - F'k[(T_i + T_o)/2 - T_\infty]/J \qquad (4b)$$

The efficiency may be plotted as shown in figure 1. In this figure the efficiency is calculated for a flat plate absorber of the pipe and sheet type, covered with one, two or three glass plates. Other parameters as listed in the figure being constant. A double or triple glazing which provides for smaller heat losses does not necessarily mean an increased efficiency. For small temperature differencies $\vartheta_A - \vartheta_\infty$ and a high solar radiation J, a single glass cover is superior. According to equation (4a) the intercept of the ordinate with the respective efficiency curve indicates the value of the transmission-absorptance product $\tau_G^* \alpha_A^*$. This product is larger when fewer glass plates have to be passed by irradiation.
The derivative $d\eta / d(T_A - T_\infty)/J$ gives the inclination of the curves in fig. 1 and according to equation (4a) equals the overall heat transfer coefficient k. It is obvious that a three glass cover provides a better insulation with a lower value of k and consequently a smaller inclination of respective curves. There is, however, a systematic scatter inherent in this type of plotting: e.g. for one glass cover in fig. 1 there are three different lines of efficiency according to which irradiation value J was taken for calculation. For the experimental determination of a collector efficiency curve this means that there will be a scatter in data points merely from the fact that the solar irradiation is not constant. These systematic deviations go on top of all the random deviations occurring in an experimental investigation.
This fanning of curves is due to the fact that the solar irradiation has a strong influence on the overall heat transfer coefficient which therefore is not a constant as equation (4a) might imply. The overall heat transfer coefficient k enhances both radiation and convection from the absorber plate, thus it depends on both T_A^4 and T_A. By numerical computation an abscissa value of

$$[(T_A/100)^{3.2} - (T_\infty/100)^{3.2}]/J$$

was obtained which unites respective irradiation curves into one curve as shown in figure 2.
Since broken exponents are unhandy and the comparative results presented in the further plots are not affected as long as

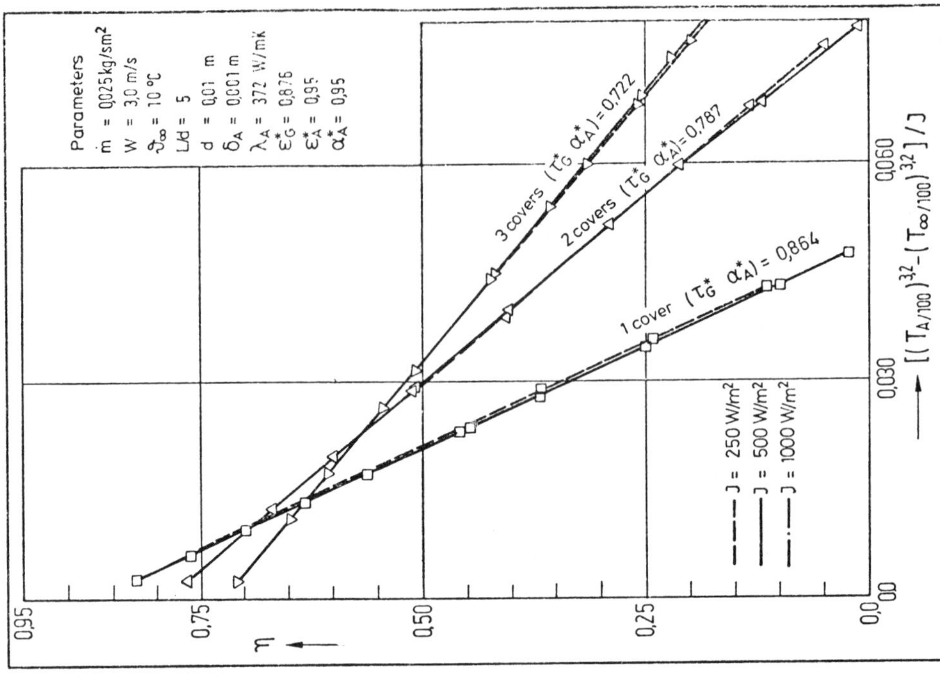

ITW Fig. 2 Effect of plotting

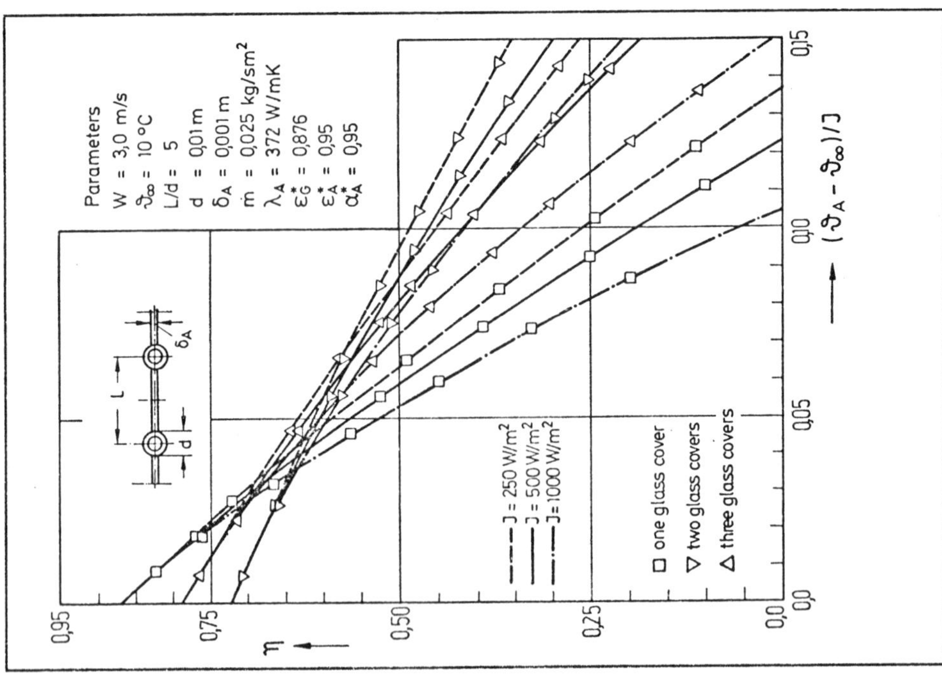

ITW Fig. 1

one irradiation value is used, abscissa values of the type

$$[(\vartheta_i + \vartheta_o)/2 - \vartheta_\infty)]/J$$

are further used.

2. NUMERICAL PROGRAM

2.1 Temperature Field

All calculations of heat losses (in steady-state) or temperature variations (in transient state) are based on the temperature field in the absorber plate. This field generally is three-dimensional. For the widely used pipe and sheet absorber geometry, a differential collector element is presented in figure 3. This element comprises a length L/2 as half the distance between two pipes; a depth of Δy, as an increment of the entire collector.
The two-dimensional temperature field in x- and y-direction on the absorber plate upper surface is shown in figure 4 for a section of half the sheet width (x-direction) and the entire sheet length (y-direction). The temperatures given in this figure were calculated for specific data; the most significant of these are indicated.
The two-dimensional temperature field in x- and z-direction within an absorber element is shown in figure 5. (Data used here are different from those of figure 4). The temperature variation within the absorber plate is found to be small enough for neglection. Thus, a two-dimensional temperature field for the absorber plate may be assumed usually (although in our calculations for transient processes the neglection was not made). There remains, of course, the temperature variation in z-direction from the absorber plate outside to the surroundings.

2.2 Mathematical Model

A number of elements shown in figure 3 are assumed to form a representative section of a collector. Each element is subdivided into a suitable number of nodes in x- and z-direction, as shown in figure 6. These nodes represent a lumped mass of a certain temperature.
The absorber sheet and pipe section is modelled by seven nodes while for each cover plate element and the back side insulation one node was found to suffice for each. The ambient temperature of either side of the collector is assumed to be the same.

All nodes are interconnected by various heat exchange mechanisms so that one heat balance equation is valid for each node and an overall heat balance holds for the entity of nodes.

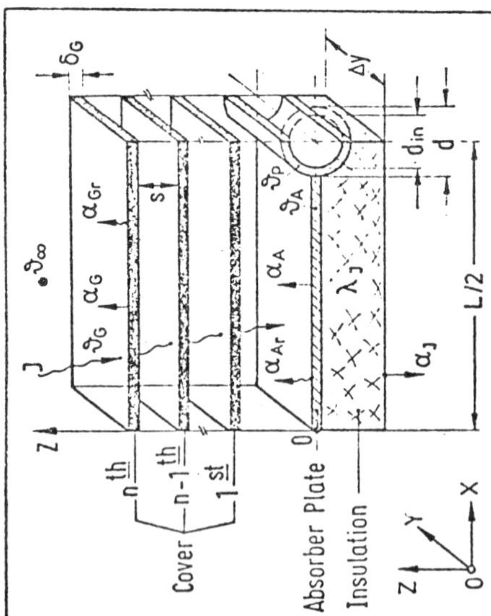

Fig. 3: Schematic Illustration of a Differential Element of the Various Components of a Flat-Plate Solar Collector

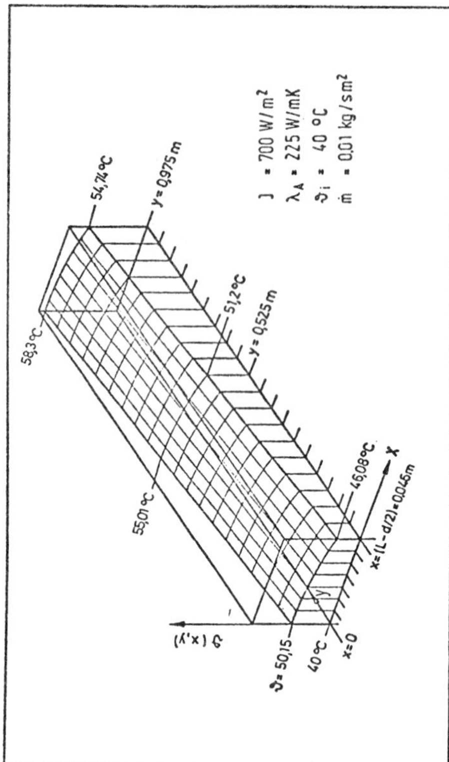

ITW UNI-STUTTGART	Temperature-field in x,y direction within an absorber fin	Fig. 4

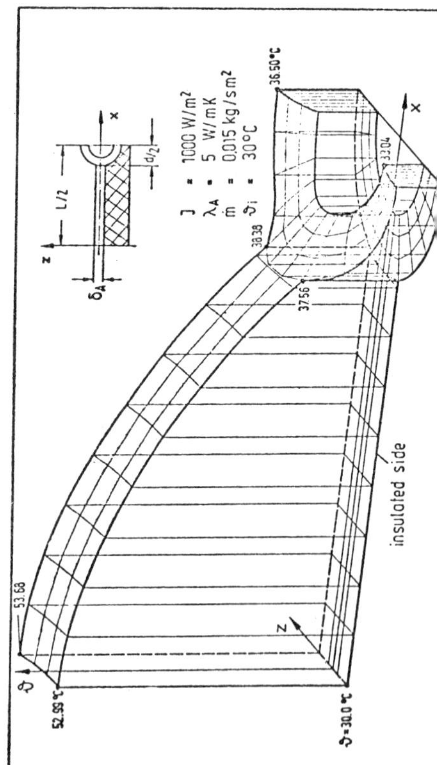

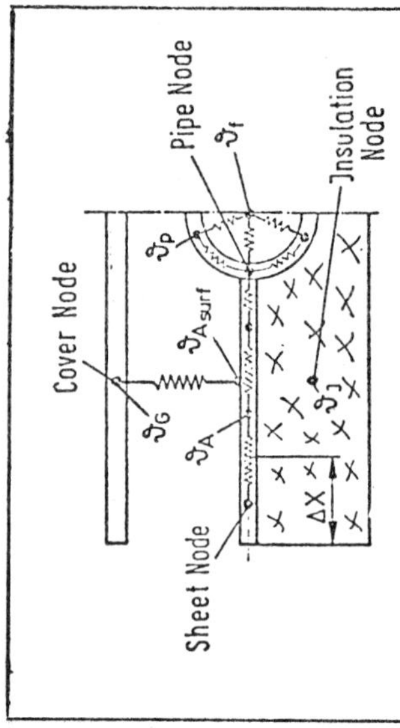

Fig. 6: Nodal Model

Starting from the initial and boundary conditions (given in chapter 2.3) the temperature field in the absorber plate, in the cover and insulation components is obtained by a stepwise and iterative solution of heat balance equations with $T = T(t,x,y)$ and $J = J(t)$.

2.3 Heat Balance Equations

A balance of thermal energy transmitted to and from each nodal mass, or stored in this, yields the following equations:

For a nodal point in the sheet of the absorber plate:

$$\delta_A \varrho_A c_A (\partial T_A/\partial t) = J\tau_G^* \alpha_A^* - k_A(T_A - T_{G1}) - k_J(T_A - T_J) + \\ + \lambda_A \delta_A (\partial^2 T_A/\partial x^2) + \partial^2 T_A/\partial y^2 \qquad (5)$$

For a nodal point in the pipe:

$$\frac{d^2 - d_{in}^2}{4} \cdot \frac{\pi}{d} \varrho_P c_P (\partial T_P/\partial t) = J\tau_G^* \alpha_A^* + \lambda_A \delta_A \frac{1}{d} (\partial T_A/\partial x)\Big|_{x=\frac{L-d}{2}} + \lambda_P \frac{d^2 - d_{in}^2}{4} \frac{\pi}{d} (\partial^2 T_P/\partial y^2) \\ - k_A(T_P - T_{G1}) - k_P \frac{1}{d}(T_P - T_f) - k_J(T_P - T_J) \qquad (6)$$

For the nodal point in the fluid:

$$\frac{d^2\pi}{4} \varrho_F c_F (\partial T_F/\partial t) = k_P(T_P - T_F) - \dot{m} c_F (\partial T_F/\partial y) \qquad (7)$$

For the nodal point in the insulation:

$$\delta_J \varrho_J c_J (\partial T_J/\partial t) = k_J (T_{A,surf} - T_J) - k_{J\infty}(T_J - T_\infty) \qquad (8)$$

with $T_{A,surf} = \frac{1}{L}\left(\int_0^{x=L-d} T_A dx + T_P d\right)$

For nodal points in the glass cover:
first cover, $j = 1$ (inside facing absorber);

$$\delta_{G1} \varrho_{G1} c_{G1} (\partial T_{G1}/\partial t) = k_A(T_{A,surf} - T_{G1}) - k_{G1}(T_{G1} - T_{G2}) \qquad (9a)$$

jth cover ($j=2$, - - - $n-1$)

$$\delta_{Gj}\,\varrho_{Gj}\,c_{Gj}(\partial T_{Gj}/\partial t) = k_{Gj}(T_{G,j-1} - T_{G,j}) - k_{G,j+1}(T_{Gj} - T_{G,j+1}) \tag{9b}$$

last cover j=n, outside (facing sky);

$$\delta_{Gn}\,\varrho_{Gn}\,c_{Gn}(\partial T_{Gn}/\partial t) = k_{Gn}(T_{G,n-1} - T_{G,n}) - \alpha_G(T_{Gn} - T_{\infty}) - \alpha_{Gr}(T_{Gn} - T_{sky}) \tag{9c}$$

For the nodal point "storage tank"

$$\varrho_f\,c_f(\partial T_S/\partial t) = \frac{\dot{M}c_f}{V_S}(T_o - T_S) - \frac{k_S A_S}{V_S}(T_S - T_{S\infty}) \tag{10}$$

The initial and boundary conditions are assumed as

$T(t=0,x,y) = T_{\infty}\ (t=0)$

$\partial T_A/\partial x\ (t,x=0,y) = 0$

$T_A(t,x=(L-d)/2,y) = T_P\ (t;\ x=(L-d/2),\ y)$

$T_A(t,x,y=0) = T_P(t,x,y=0) = T_F(t,x,y=0) = T_i(t)$

The overall heat transfer coefficients are defined as:

$k_A = 1/[\delta_A/2\lambda_A + \delta_{G1}/2\lambda_{G1} + 1/(\alpha_A + \alpha_{Ar})]$

$k_J = 1/[\delta_A/2\lambda_A + \delta_J/2\lambda_J]$

$k_P = 1/[1/d\pi\alpha_f + \ln(\frac{d+d_{in}}{2d_i})/2\pi\lambda_P]$

$k_{J\infty} = 1/[(\delta_J/2\lambda_J + 1/\alpha_J]$

$k_{Gj} = 1/[\delta_G/\lambda_G|_j + s/\lambda_{eff}|_j]\ ;\quad j = 1 \ldots n$

$k_s = 1/[1/\alpha_{s,inside} + (s/\lambda_s)_{wall} + 1/\alpha_{s,outside}]$

In case of a stagnant fluid in the pipe it is defined for k_P

$$k_P = 1/[1/2\pi\lambda_F + \ln(\frac{d+d_i}{2d})/2\pi\lambda_P]$$

The heat transfer coefficient by radiation is defined e.g. for the absorber as:

$$\alpha_{Ar} = \frac{\sigma}{\frac{1}{\varepsilon^*_A} + \frac{1}{\varepsilon^*_{G1}} - 1} \cdot \left[\frac{T_A^4 - T_{G1}^4}{T_A - T_{G1}} \right]$$

This definition holds correspondingly also for α_{Gr}. Convective heat transfer coefficients as used in the overall heat transfer coefficients are taken from literature /1,2,3,4/. These are indexed according to the heat exchanging area where they act.

3. STEADY-STATE PERFORMANCE

For steady-state conditions the left hand terms in equations (5) to (9c) become zero as $\partial T/\partial t = 0$.
The effects of various parameters are investigated individually varying one parameter and leaving the others constant.

3.1 Effects of Collector Parameters

For the construction of solar collectors those effects are of greatest interest which can be influenced by the choice of configuration, material or geometry.

3.1.1 Number of cover plates: The transparent covers are assumed to be thin glass plates with physical properties as indicated in the figures.
The effect of one, two or three covers upon the collector efficiency is shown in figure 7. For small values of excess temperatures of the absorber, i.e. $[(\vartheta_i+\vartheta_o)/2-\vartheta_\infty]/J < 0,025$ ("J" being constant as 500 W/m^2) one glass cover is superior to a multicover glassing. For one glass (n=1) the transmission-absorption product $\tau_G^* \alpha_A^* = 0,86$; for three covers this is 0,72 only. According to equation (4b) $F'\tau_G^x\alpha_A^x$ values can be read off from fig. 7 as intersection of the respective curve n with the ordinate. In this low temperature region, the decrease in transmission due to the increasing number of covers is dominant.
At high excess temperatures the insulating effect of the covers becomes dominant and a three plate cover is superior to one plate. Thus, the decision on how many glass covers are appropriate becomes a question of the anticipated absorber temperature.

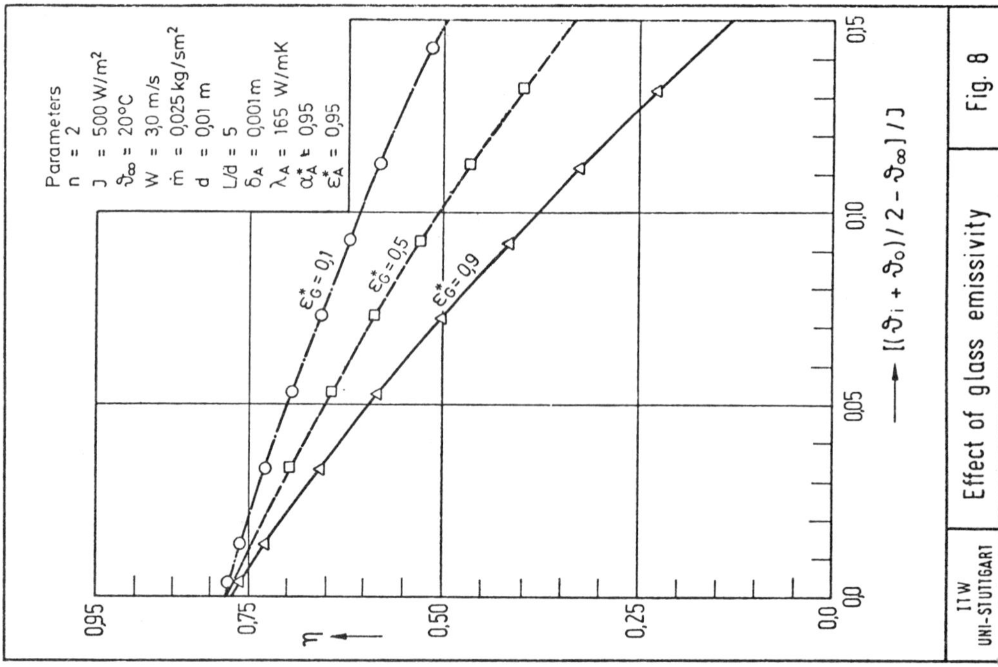
Parameters
n = 2
J = 500 W/m²
ϑ∞ = 20°C
W = 3,0 m/s
ṁ = 0,025 kg/sm²
d = 0,01 m
L/d = 5
δA = 0,001 m
λA = 165 W/mK
α*A = 0,95
ε*A = 0,95
ε*G = 0,1
ε*G = 0,5
ε*G = 0,9
η
0,95
0,75
0,50
0,25
0,0
0,0
0,05
0,10
0,15
[(ϑi + ϑo) / 2 - ϑ∞] / J
ITW
UNI-STUTTGART
Effect of glass emissivity
Fig. 8

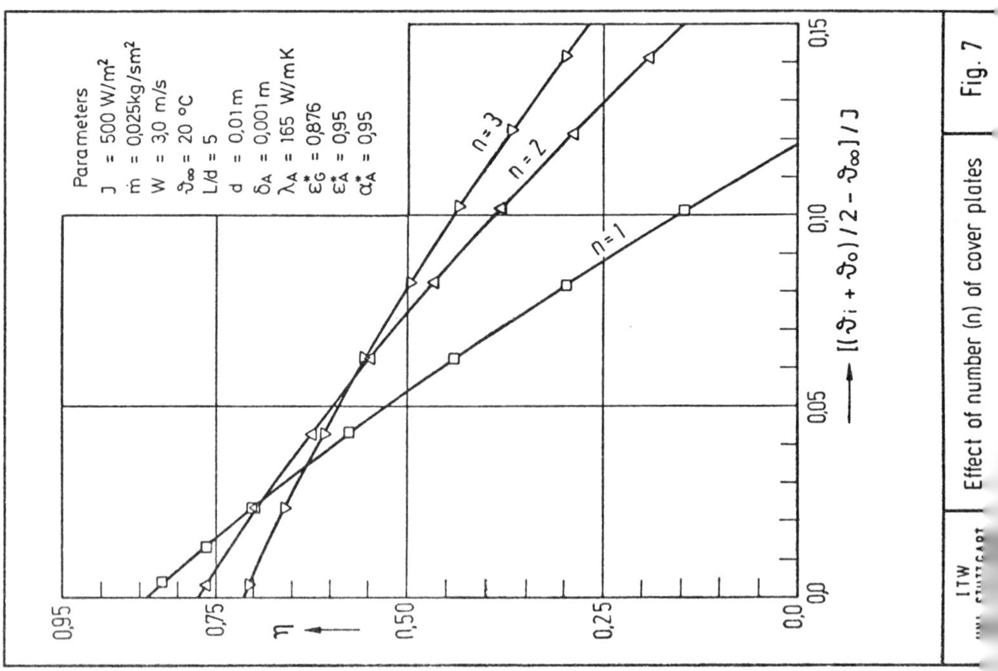
Parameters
J = 500 W/m²
ṁ = 0,025kg/sm²
W = 3,0 m/s
ϑ∞ = 20 °C
L/d = 5
d = 0,01 m
δA = 0,001 m
λA = 165 W/mK
ε*G = 0,876
ε*A = 0,95
α*A = 0,95
n = 3
n = 2
n = 1
η
0,95
0,75
0,50
0,25
0,0
0,0
0,05
0,10
0,15
[(ϑi + ϑo) / 2 - ϑ∞] / J
ITW
Effect of number (n) of cover plates
Fig. 7

3.1.2 Glass cover emissivity: The overall heat losses depend - among others - on the emissivity of the glass covers. Their influence upon collector efficiency is presented in figure 8. As a parameter, the infrared emissivity is varied between $\varepsilon_G^* = 0,1$ and $\varepsilon_G^* = 0,9$. The infrared transmission through the glass was assumed to be constant.
As can be expected, the heat losses from the glass covers decrease, and thus the efficiency increases, when the emissivity assumes small values.

3.1.3 Absorber plate material and geometry: Absorber plates may be manufactured of metallic or plastic materials. The significant physical property for the steady-state efficiency of a tube-sheet collector is the thermal conductivity of the absorber material. The effect of widely varying conductivities as for copper (λ= 372 W/Km) and polypropylene (λ= 0,22) is shown in figure 9, in combination with the effect of the number of covers.
From the point of thermal conductivity there is no difference in effects for copper and aluminium. Steel absorbers are slightly less effective, while plastic absorbers exhibit a drastic decline in efficiency for the assumed equal spacing of tubes L/d =5.
For a tube-sheet type absorber, the high thermal conductivities of metals allow for thin sheets; lower conductivities can be compensated by thicker fins. The combined influence of material λ_A and sheet thickness δ_A can be considered as $\lambda_A \cdot \delta_A$ to be representative for the heat conducted through the sheet. Its effect is shown in figure 10.
A variation of $\lambda_A \cdot \delta_A$ by two orders of magnitude, only results in an average change of efficiency of about 7 % as long as $\lambda_A \cdot \delta_A > 0,005$. Below this value, the decline in efficiency becomes relatively large with about additional 30 % when $\lambda_A \cdot \delta_A$ is lowered by an additional order of magnitude. It is obvious from figure 10 that a compensation of low conductivity by fin thickness in order to obtain equal efficiencies is impossible compensating copper (λ =372 W/K)with some plastic (e.g. λ=2,5 as a very well conducting plastic): The plastic sheet had to be 150times as thick as the copper, while for twice as thick a sheet in plastic a decline in efficiency of only 7 % might be tolerated.
Fig. 10 again holds for a tube spacing of L/d = 5. The effect of various tube spacings is presented in figure 11.
As can be expected the spacing is of great effect for plastic materials. A tube-tube arrangement (L/d=1.1) for a plastic absorber gives the same efficiency as an aluminium absorber; at L/d=10, however, the efficiency is low.

3.1.4 Absorber plate selectivity: Radiation heat losses from the absorber plate can be reduced when the emissivity of the absorber in the infrared is

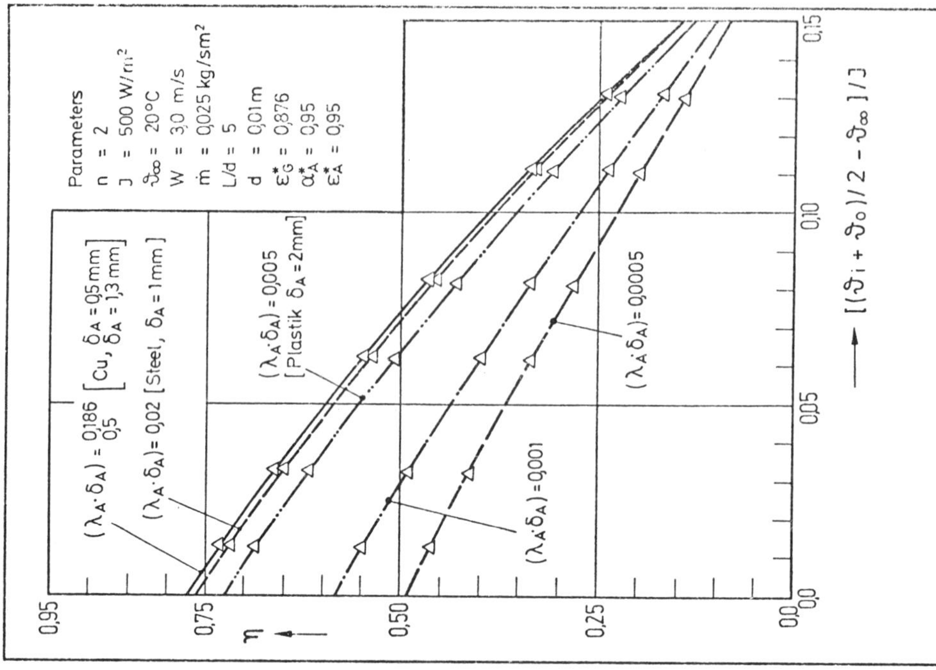

I T W UNI-STUTTGART | Effect of material and thickness | Fig. 10

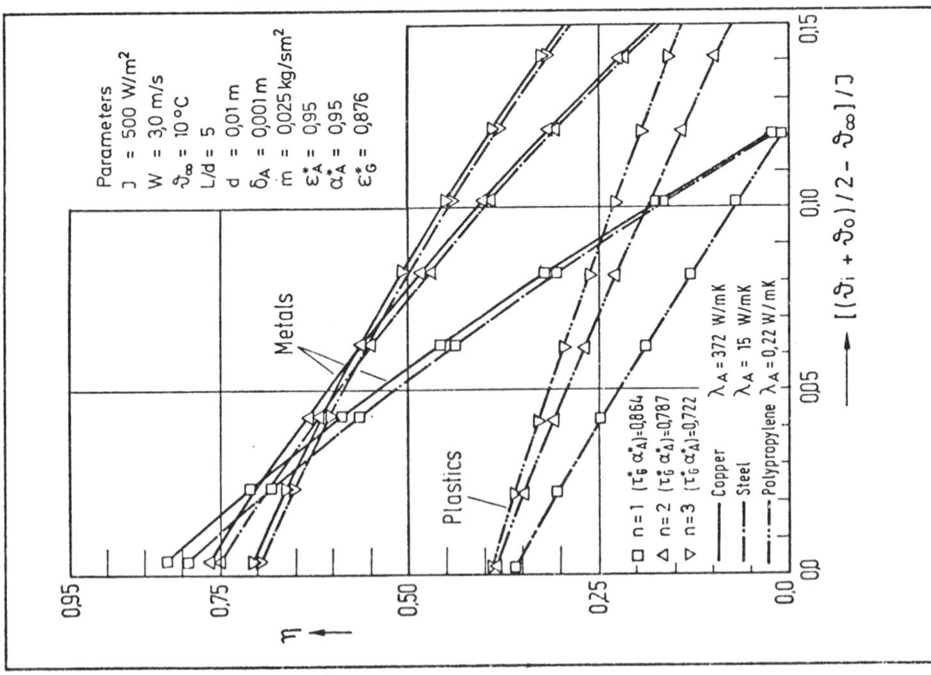

I T W | Effect of absorber plate material | Fig. 9

reduced. Its absorptivity in the short wave region, on the other hand, should be as high as possible. The ratio of absorptivity and emissivity is known as selectivity; its influence on collector efficiency is presented in figure 12.
Selective surfaces with $\alpha^*_A/\varepsilon^*_A > 1$ clearly yield higher efficiencies. This effect is especially advantageous for either high absorber temperatures or low irradiation rates.

3.2 Effects of Operation Parameters

This group of parameters enhances fluid inlet and outlet temperatures and mass flow rate. Like the collector parameters these operation parameters are subject to planned variation as the operation of a system can be adjusted to special requirements.
The effect of different mass flow rates upon collector efficiency is presented in figure 13. A specific mass flow rate normalized with the collector area is introduced, in order to provide comparability for different sized collectors.
The effect is represented by one single line and there is no difference for three different mass flow rates in this type of plot. For the three mass flow rates points are marked giving the same fluid inlet temperature of $\vartheta_i = 35^oC$. With a different mass flow rate the outlet temperatures differ and the mean absorber temperatures differ. Consequently the efficiency for a high mass flow exceeds the efficiency for low mass flow because heat losses are less.
The effect of the fluid inlet temperature can be observed from figure 14. Here the efficiency is plotted versus the inlet temperature with mass flow rate and number of covers being parameters. Again for high mass flow rates efficiency is higher than for small rates and three glass covers are more effective for high inlet temperatures than one cover.
Each family of curves converges into a common intersection with the abscissa. At this intersection temperature, the collector efficiency is zero which indicates zero heat gain i.e. heat losses compensate the irradiation. Such a "stagnation temperature" is also assumed by a collector with stagnant fluid. The stagnation temperature depends on irradiation and heat losses, so that values given in figure 12 are peculiar to the parameters assumed.

3.3 Effects of Meteorological Parameters

These effects are not controllable, yet their existence has to be obeyed for the proper selection of collectors for given weather conditions and for comparative outdoor tests.

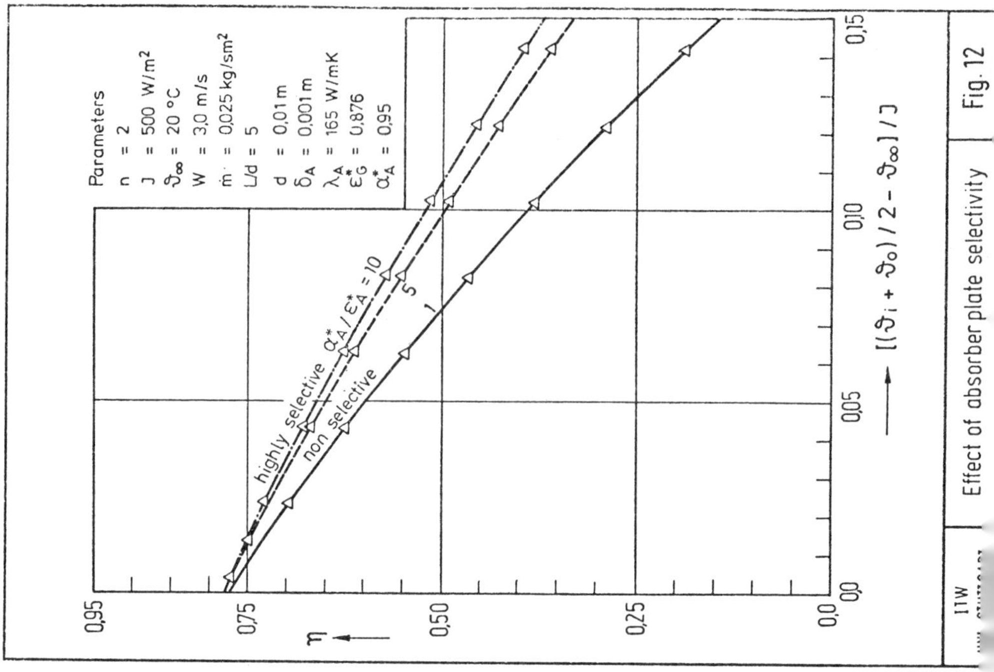

Fig. 12 Effect of absorber plate selectivity

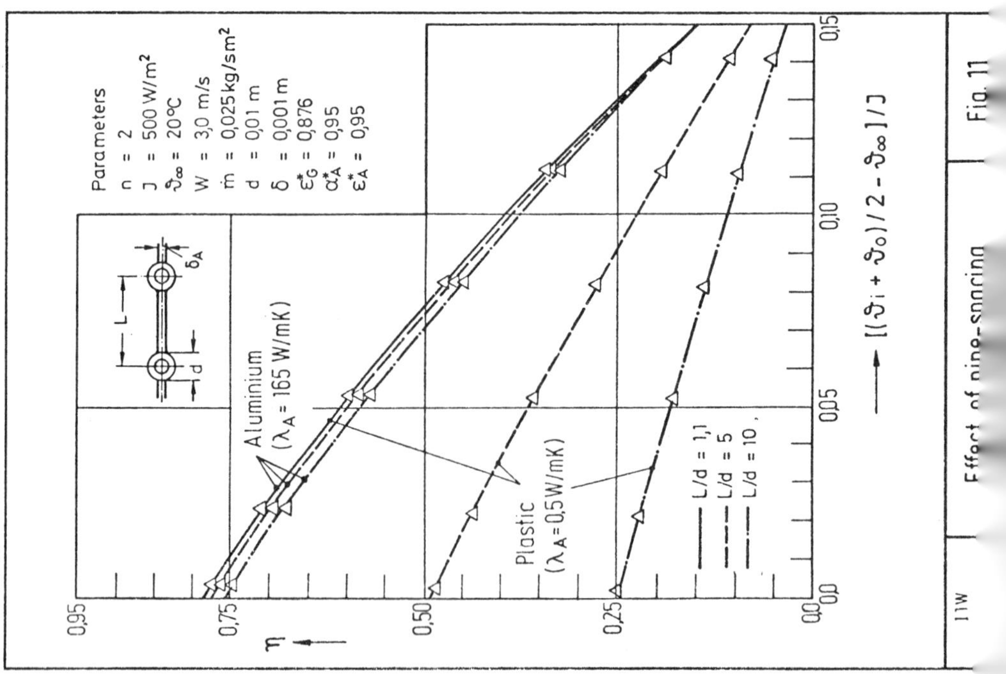

Fig. 11 Effect of pipe-spacing

3.3.1 Global irradiation: The effect of global irradiation upon efficiency is presented in figure 15. This plot with parameter curves of smaller irradiation giving higher efficiencies might be misleading. As an additional information the fluid inlet temperature is indicated. For equal inlet temperatures clearly the higher efficiencies are obtained for the higher irradiation.

3.3.2 Ambient temperature: The temperature of the surroundings affects the heat gain indirectly by way of heat losses. The effects are presented in figure 16. Again it might be misleading to find the zero ambient temperature resulting in better efficiencies than higher temperatures. For equal inlet temperatures, as they are marked in the figure, it can be observed that higher ambient temperatures lead to higher efficiencies since heat losses to the surroundings become smaller.

3.3.3 Wind speed: Wind influences the convective heat transfer from the outside cover plate. In figure 17 its effect for three different speed values are presented for a one and three cover configuration. The relative influence of wind upon a one plate cover is larger and thus wind effects are more distinct than for three cover plates. High wind speed and single glazing is tolerable, however, for low absorber temperatures and high irradiation.

A combination of effects of irradiation, ambient temperature and wind speed is shown in figure 18. The area enclosed between the curves indicates the possible region of scatter due to variations in meteorological parameters, for data obtained in outdoor tests.

For small abscissa values, it is observed that changes in irradiation cause little scatter. Thus, if for outdoor tests a large scatter is observed in this area, this is rather due to deficiencies in temperature measurements. The wide scatter for large abscissa values resulting from meteorological conditions calls for measurement specifications apt for those conditions.

4. TRANSIENT STATE PERFORMANCE

The effect of periodic and intermittent availability of solar irradiation becomes dominant when sun-shine is not abundant. In conjunction with thermal storage or for elaborate economic considerations, the dynamic behaviour of collectors due to time dependent changes of the energy supply has to be accounted for. This might result in changes of conclusions drawn only on steady-state performance. The final choice of a collector, at any rate, has to be based on the intended application under prevailing weather conditions.

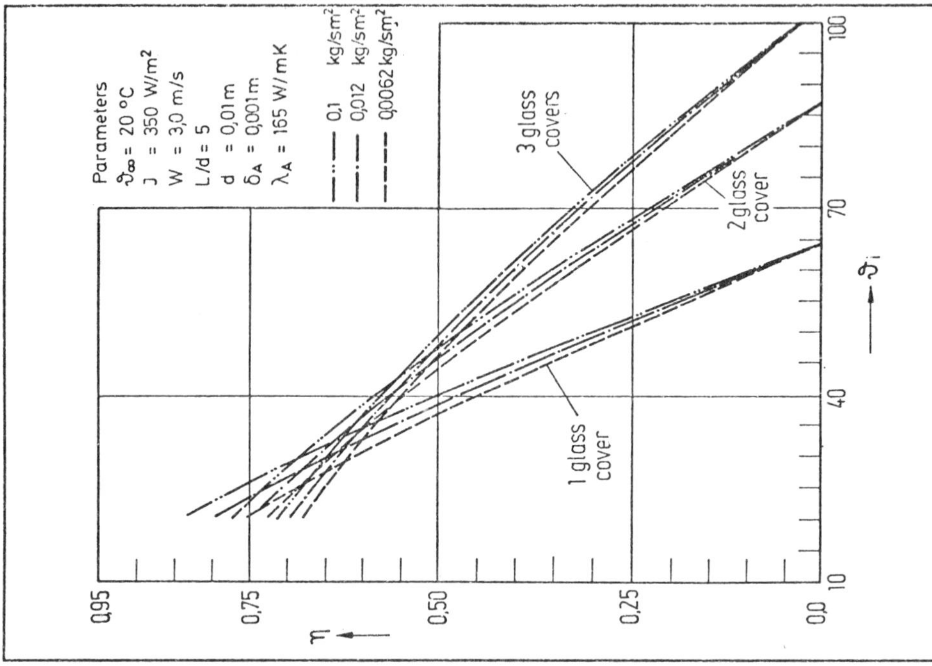

ITW UNI-STUTTGART | Effect of fluid inlet temperature and mass flow rate | Fig. 14

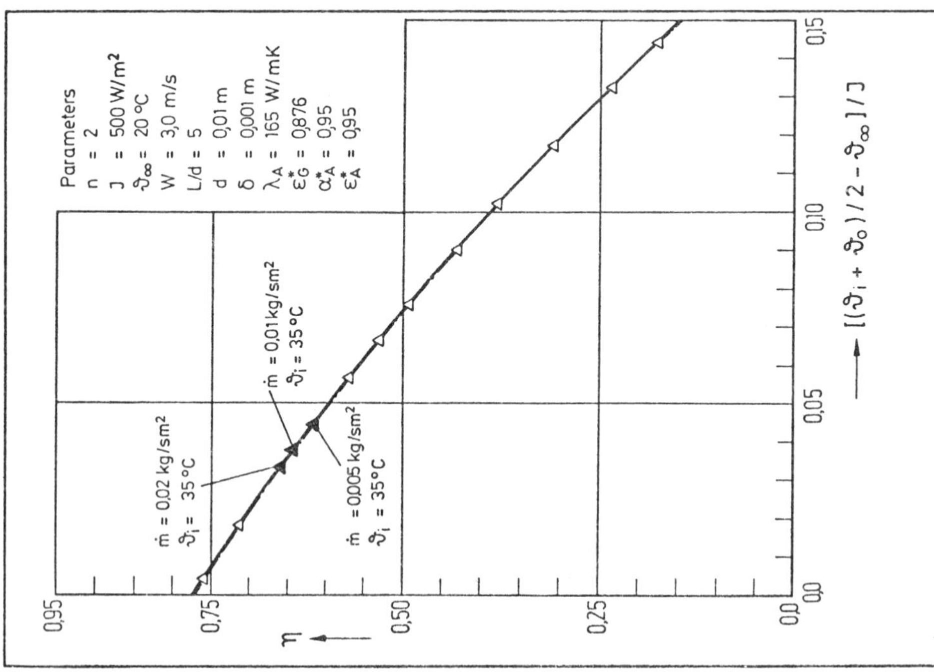

ITW UNI-STUTTGART | Effect of specific mass flow rate | Fig. 13

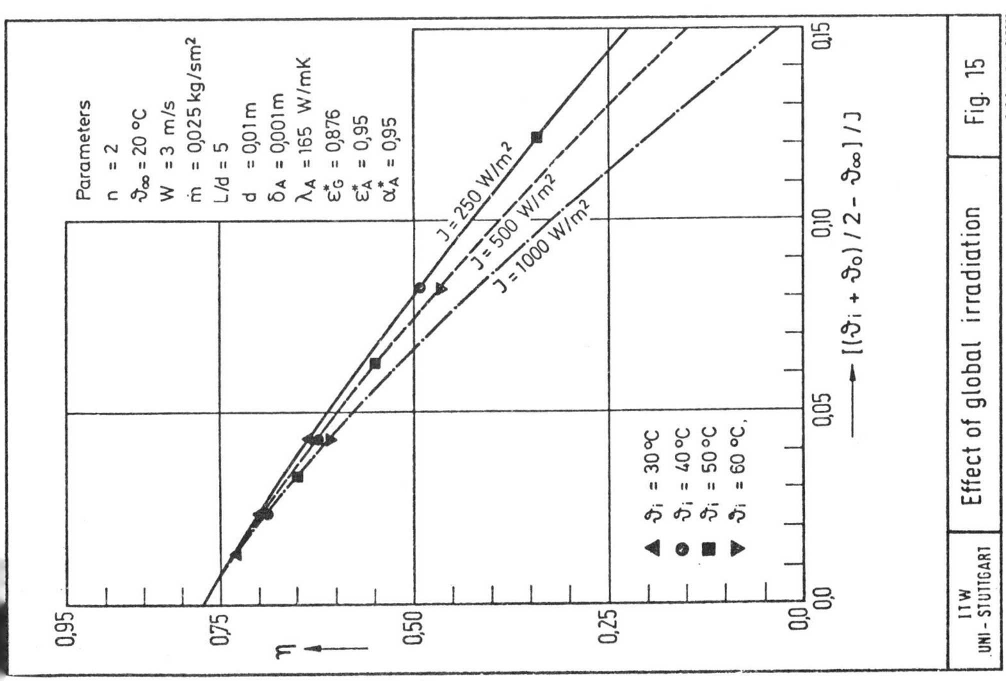

Effect of global irradiation Fig. 15

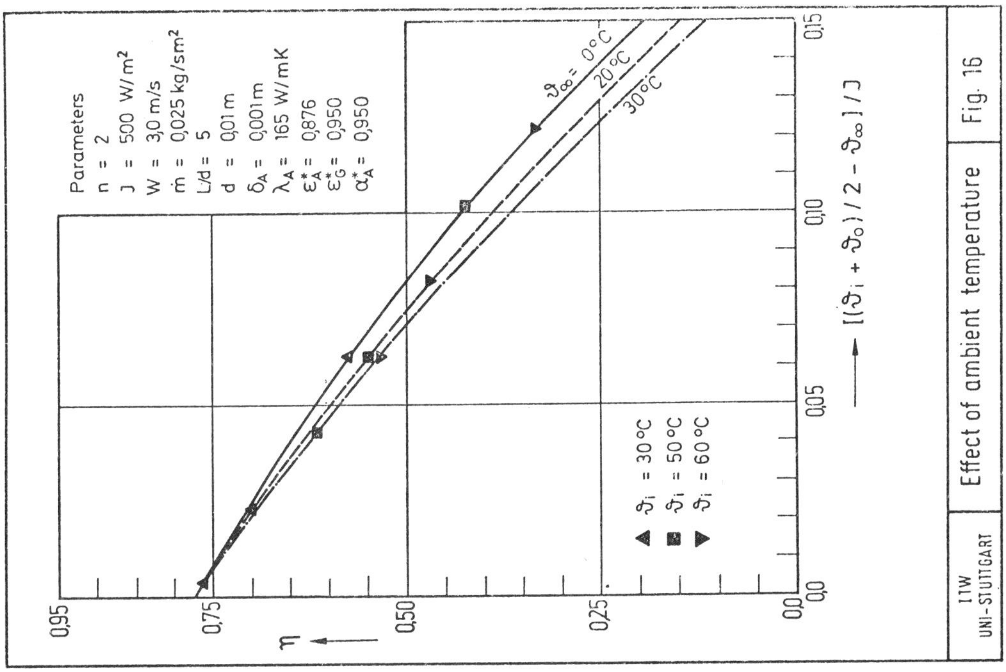

Effect of ambient temperature Fig. 16

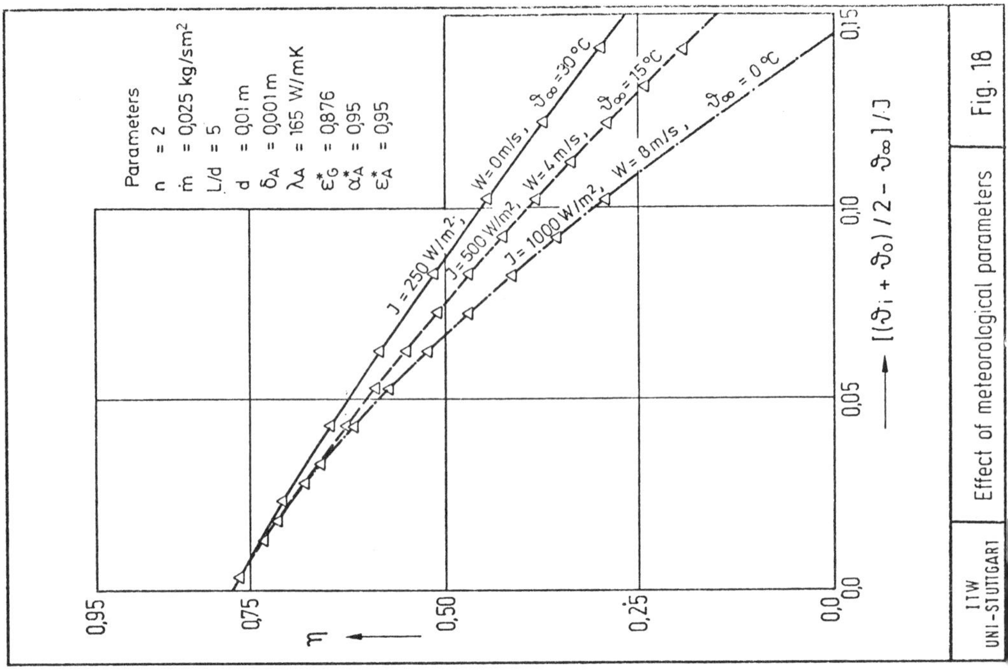

Fig. 18 Effect of meteorological parameters

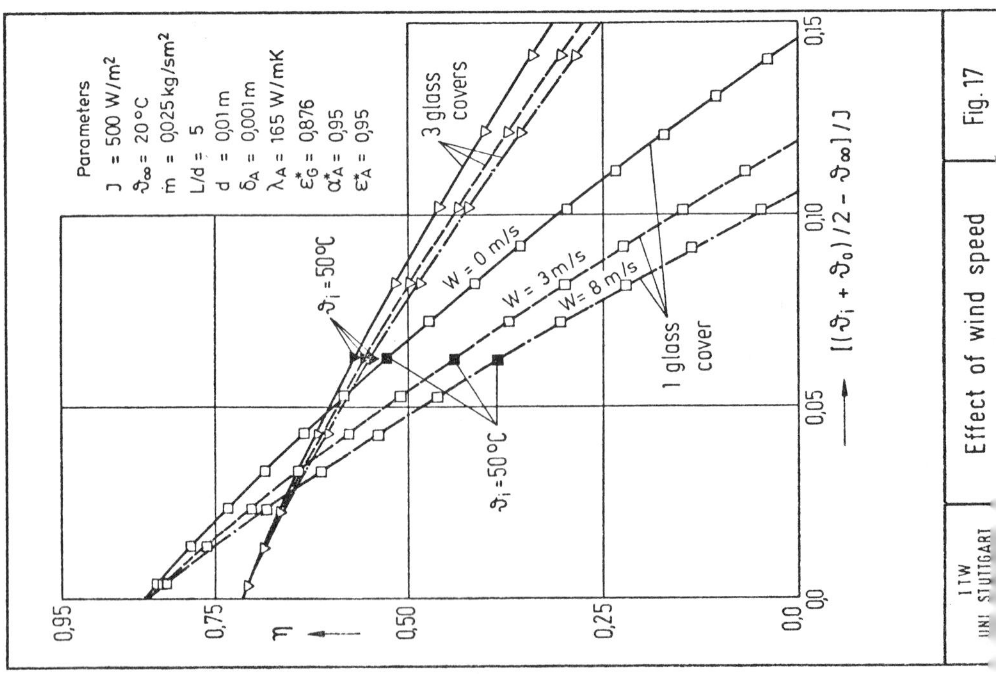

Fig. 17 Effect of wind speed

4.1 Effects of Collector Parameters

4.1.1 Number of cover plates: The time dependent rise of temperature for collectors with different glazing is presented in figure 19. Three different months are considered and the solar irradiation is sinusodial with intermittent shading periods of 10 minutes. These are shown in the curves as irregularities.

It is assumed that the collector has acquired ambient temperature during the night and - starting at the respective sun rise - will gradually warm up to 50 °C. This is supposed to be the temperature of a partially charged storage. During the warming period there is no water flow.

In December a single glazed collector starting with a temperature of 10 °C at 9 o'clock still has not reached the working temperature of 50 °C at 12 o'clock. For a two or three cover collector the warming time is about 2 hours in December, about 1 1/2 hours in June. In June a single glazed collector is faster than a three cover collector, due to increased reflections on two more covers.

4.1.2 Specific mass of collector: This effect is presented in figure 20. A time dependent change in temperature is shown for December; initial temperature is now 20 °C and warming should be to 50 °C.

A tube-sheet system with a spacing of L/d = 5 is approximately twice as fast as a tube-tube system having about 5times the mass of the former. The absorber material is of small influence for metals having about the same density and specific heat.

4.2 Effects of Meteorological Parameters

In order to investigate effects of periodic and intermittent energy supply, two different functions for solar irradiation were assumed. These are shown in figure 21. Function I represents a steady sinusodial supply distribution as considered normal for constant sun-shine. Function II is to simulate highly intermittent sun-shine which fluctuates between peak values of J_2= 800 W/m^2 and diffuse radiation values of J = 400 W/m^2.

Based on these functions the warming up interval was calculatedfor a collector of 20 °C initial temperature to reach 35 °C working temperature.

Further, an evaluation ratio for the time dependent utilization of energy was defined as an "effectivity"

$$\eta_{eff} = \frac{c_f \sum_{t=0}^{t} \dot{m}(t)\left[T_o(t) - T_i(t)\right]\Delta t}{A \sum_{t=0}^{t} J(t)\Delta t} \qquad (11)$$

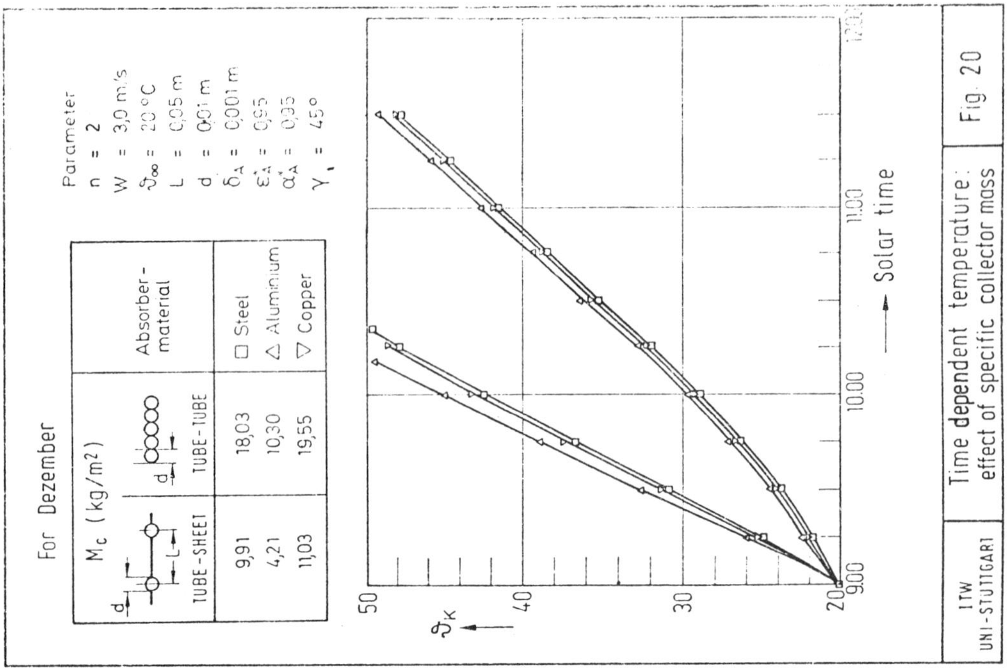

Time dependent temperature: effect of specific collector mass

Fig. 20

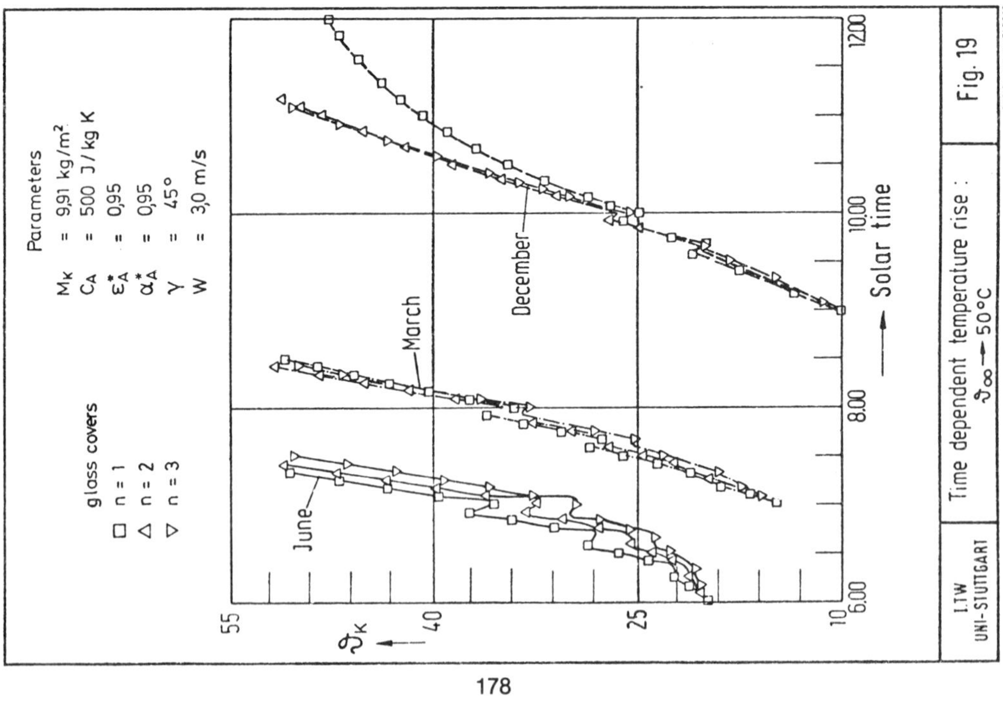

Time dependent temperature rise : $\vartheta_\infty \rightarrow 50°C$

Fig. 19

For steady-state performance equation (11) turns into equation (1). The effectivity was calculated for functions I and II, for a 6 hours' period and presented in figure 22. This figure comprises a comparative study of the effects of various parameters.

For a collector with design data given in the row "Const." parameters were varied one by one, leaving the others constant. These variations are listed under "Param.". The resulting width of variation is shown as a bar, unshaded for function I and shaded for II.
The resulting "Time of warming" is also presented in figure 22. As an example: for steady sinusodial irradiation, a tube-sheet steel collector with tube spacing L = 50 mm, non-selective absorber $\alpha^*_A / \varepsilon^*_A = 1$, tilted 45°, covered with one or two glass plates, needs 72 minutes to warm from 20°C to 35°C. With three plates it needs 78 minutes. The effectivity is 44 % with one cover, 49 % with two and 47 % with three covers.
A highly selective absorber $\alpha^*_A / \varepsilon^*_A = 12$ is best operated with only one glass cover. The absorber material has little influence when it is metal. Thermally low conducting plastics exhibit a drastic decrease in effectivity for a tube-sheet absorber. This immediately suggests for such materials to alter the design.

5. COMPARISON OF THEORETICAL AND EXPERIMENTAL RESULTS

The theoretical considerations as formulated in equations (5) to (10) and put in numerical programs were compared to experimental results obtained in our own and the DFVLR Stuttgart test stand. In these comparisons, both collector design data and measured meteorological data were entered into our computer programs. From these data the outlet temperature and efficiency was computed; η is plotted together with the corresponding measured values, in figure 23. The largest deviation in the respective linear average curves is 15 %.
A comparison of the computation method is given in figures 24 and 25. Here the outlet temperatures were computed for given inlet temperatures and meteorological and collector data in two different ways: either following the simplified steady-state program version with time dependent changes being neglected or according to the transient state version. Again these results are compared to measured data.

In figs. 24 and 25 a fluid temperature is plotted vs. time and measured and computed temperatures are presented as parameters. The measured global irradiation and the applied mass flow are shown in sub-diagrams.
In figure 24, the case of quite a regular, sinusodial irradiation is presented, with a prevailing constant mass flow

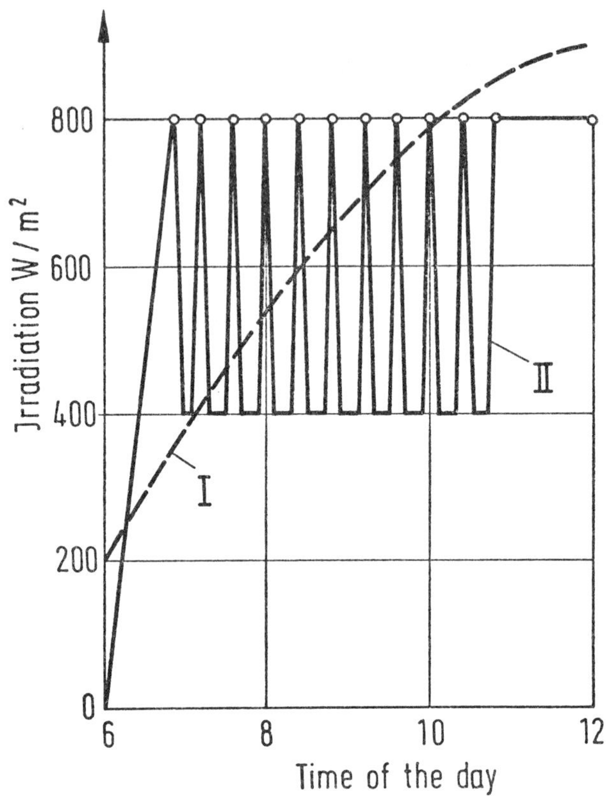

Gez. am , ersetzt durch:

ITW UNI-STUTTGART | Test functions for solar irradiation | Fig. 21

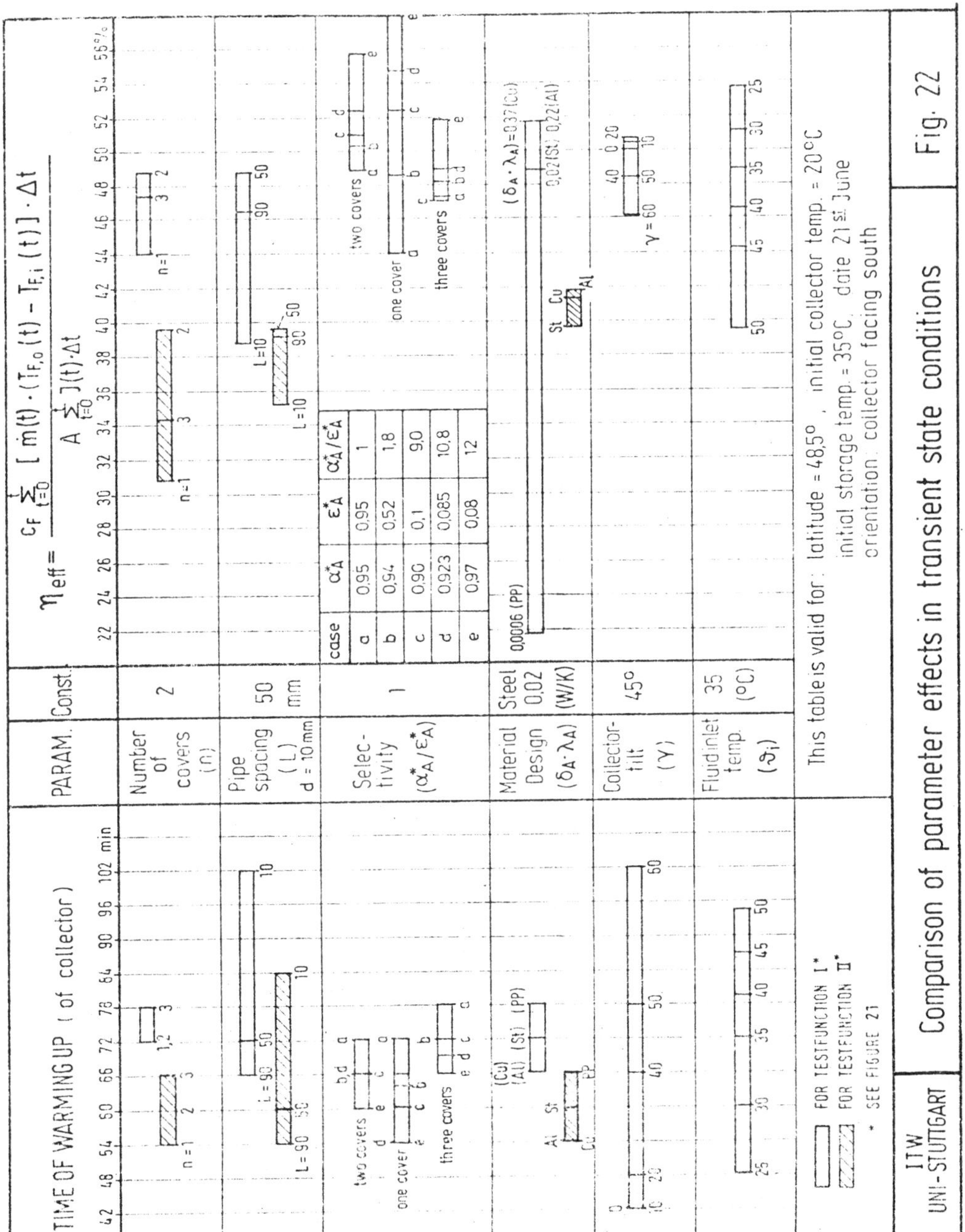

Comparison of parameter effects in transient state conditions

Fig. 22

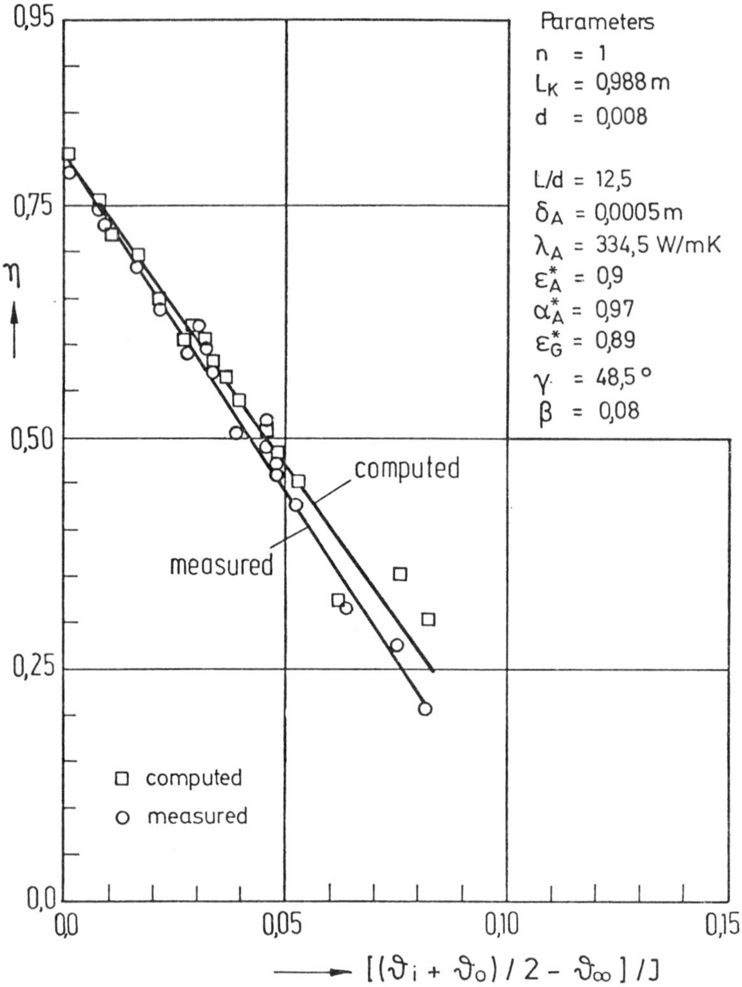

ITW UNI-STUTTGART	Comparison of measurements and theory	Fig. 23

GVO Bestell-Nr 515020

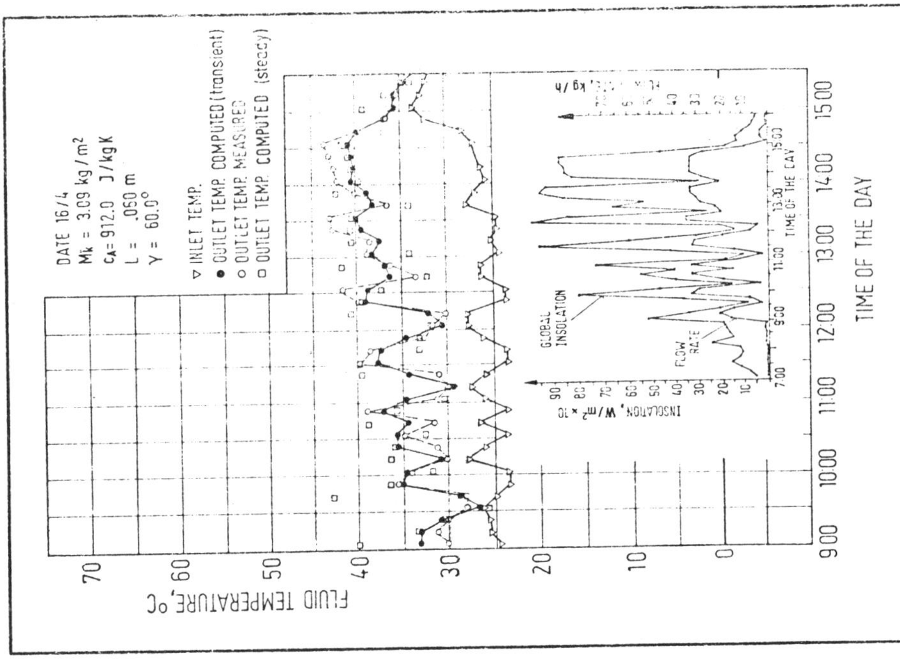

I.T.W. UNI-STUTTGART | COMPARISON BETWEEN COMPUTED AND EXPERIMENTAL RESULTS | FIGURE 25

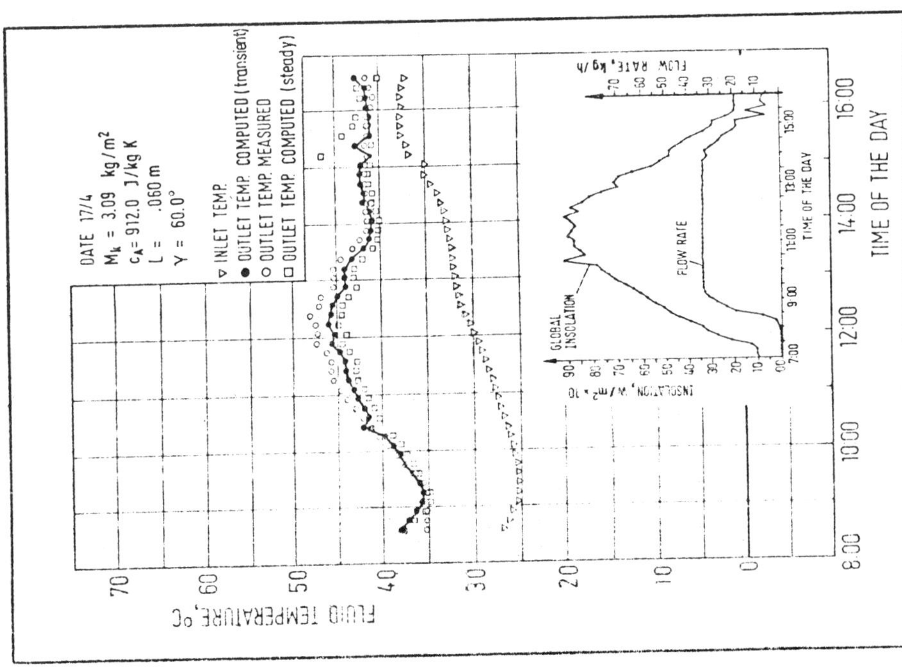

I.T.W. UNI-STUTTGART | COMPARISON BETWEEN COMPUTED AND EXPERIMENTAL RESULTS | FIGURE 24

and steadily increasing inlet temperature. Figure 25 shows a highly intermittent irradiation. The control of the heating system follows this intermittent supply and provides also highly fluctuating values of flow rate and inlet temperature.
In either case, the data from transient state computations give a better fit to the experimental results than those from steady-state calculations.

Taking into account the great many parameters involved, we consider the agreement between theoretical and experimental results as very satisfying.

6. CONCLUSIONS

Some of the following conclusions appear trivial, yet discussion in public and flat plate collectors in the market seem to justify and even necessitate their mention.

1. There is no collector which would fit all needs equally well.
2. Basis for the selection and production of an appropriate collector are the locally prevailing meteorological conditions and the intended application.
3. Cheap single-glazed collectors are not necessarily inferior to multi-glazed collectors.
4. Selective coatings of the absorber increase efficiency and effectivity and reduce warming up time.
5. The absorber plate material is of decisive influence. Proper design, however, can make low conductive material comparable to high conductive material.
6. Outdoor test results, plotted in the usual way,exhibit a systematic scatter simply due to the diurnal steady change in sunlight.
7. Numerical treatment seems well apt to produce predictions on collector performance in steady and transient state.

ACKNOWLEDGEMENT

This work is financed under the German "Bundesministerium für Forschung Technologie"solar research and development program ET 4045 A. The authors express their gratitude for this support.

NOMENCLATURE

A	heat exchanging area
c	specific heat
d	outer tube diameter
d_{in}	inner tube diameter
F'	correction factor
J	solar irradiation
k	overall heat transfer coefficient
L	pipe spacing
$\dot{m}$	specific mass flow through pipe
$\dot{M}$	total mass flow through storage tank
M_C	specific mass of collector
n	number of glass cover plates
$\dot{Q}$	heat flux
s	glass cover distance
t	time
T	thermodynamic temperature
T_{sky}	thermodynamic sky temperature
V_S	storage tank volume
w	wind speed
α	convective heat transfer coefficient
α_r	radiative heat transfer coefficient
α^*	absorptivity
γ	collector tilt
δ	thickness
η	efficiency
η_{eff}	effectivity
ε^*	emissivity
λ	thermal conductivity
λ_{eff}	effective thermal conductivity (including convection and radiation) of an air gap
ϑ	Centigrade temperature
τ^*	transmission coefficient
	density
σ	Stefan Boltzmann constant

Subscripts

A	absorber
f	fluid
G,1,2...	glass cover, one, two ...
i	inlet
J	insulation
o	outlet
P	pipe
S	storage tank
∞	ambient, surrounding

REFERENCES

1 J. A. Duffie and W. A. Beckmann:
Solar Energy Thermal Processes
Wiley, New York, 1974

2 ASHRAE Handbook of Fundamentals:
Air Conditioning Cooling Load,
pp. 385 - 394

3 H. Tabor:
Radiation, convection and conduction coefficients in solar collectors
Bull.Res.Counc. of Israel, Vol. 6C, 1958,
pp. 155 - 176

4 K. G. T. Hollands:
Free Convective Heat Transfer Across Inclined Air Layers
J. of Heat Transf., Vol. 98, 1976, pp. 189 - 193

5 H. Buchberg, T. Catton, D. K. Edwards:
Natural Convection in Enclosed Spaces - A Review of Application to Solar Energy Collectors,
J. of Heat Transf., Vol. 98, 1976, pp. 182 - 188

6 H. Hausen:
VDI-Wärmeatlas, 2nd edition, 1974, chapter Gb 1

FLAT-PLATE SOLAR COLLECTORS
Influence of Design Material and Configuration on Performance.

S-E. Ransmark
Lund Institute of Technology
Lund, Sweden

ABSTRACT

The rate of useful energy which can be collected from av flat-plate solar collector depends on many factors, such as: the absorbtance/emitance factor of the absorber, its heat conductivity, the heat capacitance of the collector, the mass flow rate, and the velocity of the fluid.

If, for instance, the number of risers in the absorber is increased a larger part of the absorber surface will be chilled directly by the fluid, and consequently the mean temperature of the surface, and thereby also the heat loss, will decrease. On the other hand, an increased number of parallel risers, for a given mass flow rate and a given pipe diameter, reduces the velocity of the fluid, which in turn causes a poorer heat transmission between plate and fluid. If the number of parallel risers is increased beyond a certain number, the useful gain per m^2 from the collector will no longer increase. Instead, the heat capacitance of the collector will increase, which is a disadvantage in areas where climate conditions are unfavourable due to variable cloudiness. The "optimum" number of risers will be different for different absorber materials.

For cases with variable values of insolation, mass flow rate, and temperature levels the efficiency of the solar collector is computed for absorbers of steel, aluminium, copper, and polyeten plastic. Also, riser diameter and number of risers are varied.

Attention is further payed to variations in circulation pump power as well as demands concerning the possibility of removing air from the absorber risers at various pipe diameters.

FLAT PLATE COLLECTOR

In a solar collector that is constructed with parallel risers between two headers the solar radiation will cause a temperature increase. This increase will be highest in the absorber plate between the risers and lowest within the risers. This is shown (schematically) in figure 1.

As the solar collectors' heat loss will increase when the temperatur rises, the temperature increase should be kept down.

It is obvious that the distance between the risers and the heat conductivity of the absorber material are important for the temperature increase. With decreasing distance the temperature in the absorber plate between the risers will decrease and approach the surface temperature of the risers. The same tendency will also be the result of an increased heat conductivity in the absorber material.

With a smaller distance more risers will be needed and more weldings have to be made, which means that the collector will get more expensive. If the collector plate, however, is made from Rollbond plate, (figure 2), the number of risers will nor have too much influence on the price.

Considering this the following comparison has been made only from a technical point of view.

The momentary useful gain from a solar collector can be calculated from the following equations [1].

$$P_a = (\tau \alpha) I_o \qquad (1)$$

P_a = useful gain for the absorber

I_o = total solar radiation reaching the first cover

$$(\tau \alpha) = \tau \alpha \left[1-(1-\alpha) \rho d\right]^{-1} \qquad (2)$$

α = absorber plate absorptance
τ = cover transmittance

$$\tau = \tau_a \cdot \tau_r$$

τ_a = cover transmittance due to absorption in the glass

$$\tau_a = e^{-KL}$$

Temperature distribution in absorber plate

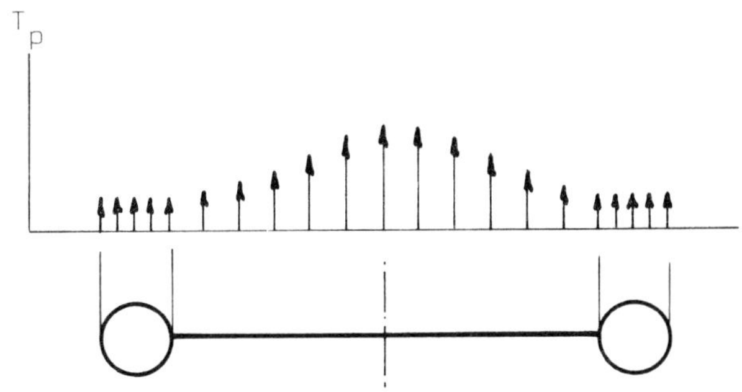

Figure 1

Rollbond plate

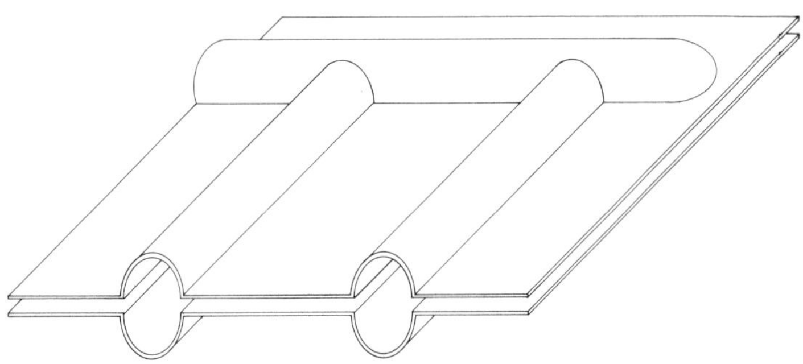

Figure 2

K = cover extinction coefficient

L = cover thickness

τ_r = cover transmittance due to reflection

$$\tau_r = \frac{1 - \rho}{1+(2N-1)\rho}$$

ρ = cover reflectance

$$\rho = \left(\frac{n-1}{n+1}\right)^2 \quad \text{(radiation at normal incidence)}$$

n = glassindex of refraction

N = number of covers

ρd = cover diffuse reflectance

$$P_n = U_L \; P_a \tag{3}$$

P_n = useful gain to the fluid

U_L = the overall loss coefficient

$$U_L = U_b + U_t$$

U_b = bottom loss coefficient

$$U_b = \frac{K}{L}$$

K = thermal conductivity of the insulation
L = Insulation thickness

$$U_t\ (45\ ^o) = \left[\frac{N}{(344/T_p)\left[(T_p-T_a)/(N+f)\right]^{0,31}} + \frac{1}{h_w}\right]^{-1} +$$

$$+ \frac{\sigma \, (T_p + T_a)(T_p^2 + T_a^2)}{\left[\varepsilon_p + 0,0425 \, N \, (1-\varepsilon_p)\right]^{-1} + \left[(2N+f-1)/\varepsilon_g\right] - N} \qquad (4)$$

N = number of covers

$$f = (1 - 0,04 \, h_w + 5,0 \times 10^{-4} \, h_w^2) \, (1 + 0,058 \, N)$$

ε_g = emittance of glass
ε_p = emittance of absorber plate
T_a = ambient temperature K
T_p = absorber plate temperature K
h_w = wind heat transfer coefficient
σ = Stefan-Boltymann constant

$U_{t\ (45\ ^o)}$ is valid only when the solar collector is tilled 45°

At other indications the following equation should be used:

$$U_{t\ (^o)} = U_{t\ (45\ ^o)} \times \left[1 - (s-45)(0,00259-0,00144 \ \varepsilon_p)\right]$$

s = tilt in degrees

The accuracy of U_t is $\pm$ 0,2 W_m^{-2} K^{-1} for absorber temperatures between 40 and 130 °C.

The average temperature of the absorber ($T_{p,\ m}$) can not be calculated directly. Neither is this possible for the average temperature of the fluid ($T_{f,m}$).

$T_{f,m}$ can be calculated as shown below if the initial temperature of the fluid ($T_{f,i}$) is known

$$T_{f,m} = T_{f,i} + \frac{P_n}{U_L \times F_R}\left[1 - \frac{F_R}{F'}\right]$$, where

$$F_R = \frac{\dot{m} \cdot C_p(T_{f,o} - T_{f,i})}{P_a - U_L(T_{f,i} - T_a)}$$

$\dot{m}$ = mass flow rate kg s^{-1} m^{-2}

c_p = fluid heat capacitivity

$T_{f,i}$ = fluid inlet temperature

P_a is given by equation (1)

T_a = ambient temperature

F' is obtained from

$$F' = \left[\frac{WU_L}{\pi \cdot d \cdot h_r} + \frac{W}{d + (W-d)F}\right]^{-1}$$ where

W = tube spacing (m)

d = tube diameter (m)

h_r = heat transfer coefficient between the fluid and the tube

$$F = \frac{\tanh\ m(W-d)/2}{m(W-d)/2}$$

$$m = \frac{U_L}{k \cdot \delta}^{1/2}$$

k = thermal conductivity of the plate

δ = plate thickness

F_R can also be written as:

$$F_R = \frac{\dot{m} \cdot c_p}{U_L} (1-e^{-[U_L \cdot F'/\dot{m}\, c_p]})$$

The average temperature of the absorber ($T_{p,m}$) is obtained from:

$$T_{p,m} = T_{f,m} + P_n R$$ where

p_n is given by equation (3)

$$R = (\pi \cdot d \cdot h_r \cdot N \cdot L)^{-1}$$

N = the number of risers

L = the length of the risers (m)

It is assumed that h_r varies with Reynolds' number as:

$$h_r = 2{,}524 \quad Re^{0{,}8}$$

By calculating above for different tube diameters (d = 5,10 and 20 mm), mass flow rates ($\dot{m}$ = 0,001, 0,01, 0,05 and 0,1 kg s^{-1} m^{-2}), thermal conductivities of the insulation (k = = 340, 220, 75 and 0,43 Wm^{-1} K^{-1}) corresponding to copper, aluminium, steel and polyeten) and for different d/W ratios for 2, 5, 10, 20, 50 and 100 tubes per meter and by keeping the other parameters constant it is possible to determine, what collector design that will give the highest momentary useful gain.

The values for the parameters that have been constant are:

I = 875 Wm^{-2}, which corresponds to the solar radiation in Lund (lat. 55.6 °N) June 21 at noon (sun altitude angel 57.7 °)

N = 2 (number of covers)

L = 0,004 (cover thickness)

$\alpha = \varepsilon g = 0{,}88$ (cover emittance)

$K = 16$ (cover extinction coefficient)

$n = 1{,}526$ (cover index of refraction)

$\rho d = 0{,}24$ (cover diffuse reflectance)

$\varepsilon p = 0{,}9$ (absorber emittance)

$A = 1$ (absorber area)

$\delta = 0{,}001$ (absorber thickness)

$L = 0{,}1$ (insulation thickness)

$K = 0{,}04$ (insulation thermal conductivity)

$s = 32{,}3$ (collector tilt)

$T_a = 19{,}2$ (ambient temperature)

$h_w = 24{,}7$ (heat transfer coefficient at 5 m/s)

The obtained values for the useful gains have been divided with the solar radiation to obtain a dimension free efficiency number. The calculated efficiency is of course only valid for one time of the year, but this is of no importance for this study as it is the form of the efficiency curve that is interesting and not the value.

The useful gain could consequently have been related to the radiation that reaches the absorber plate instead of the one that comes across the cover. This would give us another efficiency number. This number would have approached the vaule 1 with decreasing heat losses.

The pressure drops for the different solar collector designs have been calculated simultaneously with the efficiency from the following equations:

$\Delta p = LR + Zpd$

L = tube length, m

$$R = \frac{\lambda\ L}{d} \cdot pd$$

For Re < 2300:

$$\lambda_L = \frac{64}{Re}$$

For Re > 3500:

$$\lambda_T = \left[- 2 \log \left(\frac{1}{3,7} \cdot \frac{\varepsilon}{d} + \frac{2,51}{Re \sqrt{\lambda}} \right) \right]^{-2}$$

ε = tube roughness

For 2300 < Re < 3500:

$$\lambda = \frac{(3500-Re)\ \lambda_L + (Re - 2300)\ \lambda_T}{1200}$$

$$Z = \Sigma\ \zeta$$

ζ = resistance due to changes in direction and diamaters

$$p_d = \frac{\rho\ v^2}{2}$$

ρ = density, kg/m^3

v = mean velocity, m/s

For collector with parallel risers the value for $Z = \Sigma\ \zeta =$ = 3,0 and L = collector high + collector width and for collector with tube in serpentine ζ = 0,6 per 180° bend and L = number of tube meanders.

The results are shown in figures 3, 4 and 5.

If the solar collector is water cooled for example for 5 %, i.e d/W = 0,05, the best efficiency will be obtained with the smallest riser diameter and the largest diameter will give the poorest efficiency.

d/W = 0,05 means, however, that the diameters 0,005, 0,01 and 0,02 correspond to the distances 0,1, 0,2 and 0,4 m respectively.

If you on the other hand make a comparison using the same W value, i.e that the number of risers is the same per meter solar collector, you will get a reversed relation, but the difference will in this case be neglectable if the absorbers are made of metal.

FLAT-PLATE SOLAR COLLECTOR.

Efficiency and pressure drop as a function of d/W (diameter/riser spacing), mass flow rate and absorber material.

Data:

Insolation: 875 W/m^2

Absorber: 1 mm, $\alpha = \varepsilon = 0,9$

Cover: 2 x 4 mm

Insulation: 100 mm, k = 0,04 W/m K

T_a = 19,2 °C (ambient)

$T_{f,i}$ = 50 °C (fluid inlet)

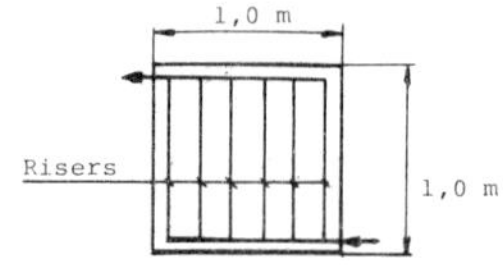

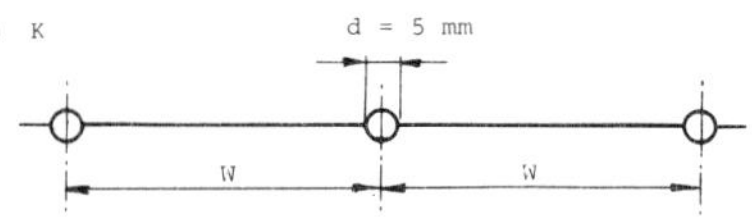

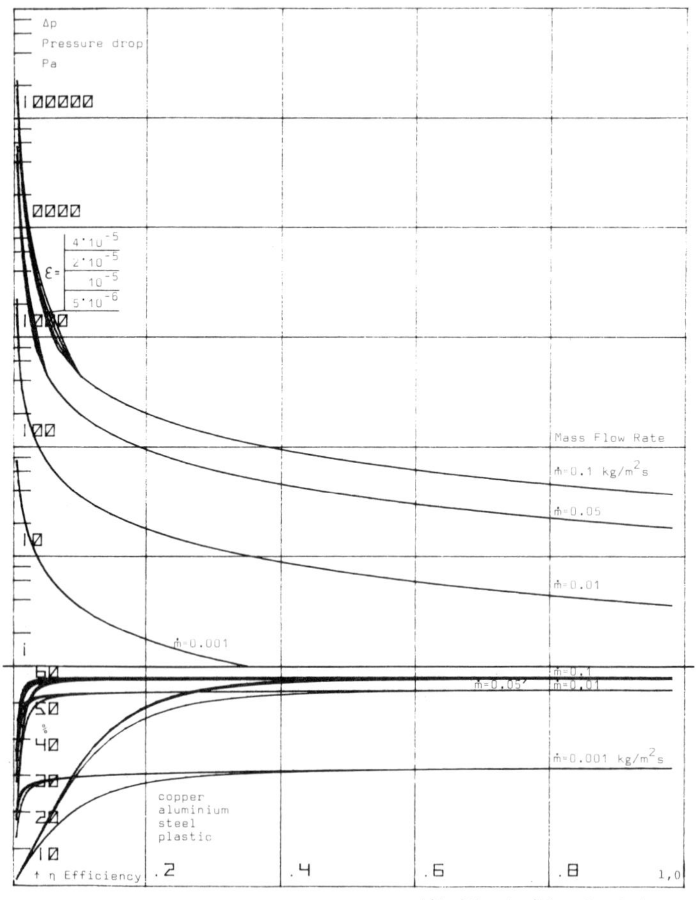

Figure 3
SER

FLAT-PLATE SOLAR COLLECTOR.

Efficiency and pressure drop as a function of d/W (diameter/riser spacing), mass flow rate and absorber material.

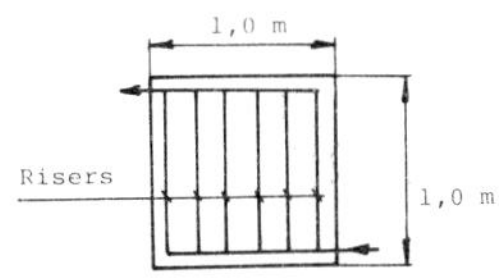

Data:

Insolation:	875 W/m^2
Absorber:	1 mm, α = ε = 0,9
Cover:	2 x 4 mm
Insulation:	100 mm, k = 0,04 W/k
T_a = 19,2 °C	(ambient)
$T_{f,i}$ = 50 °C	(fluid inlet)

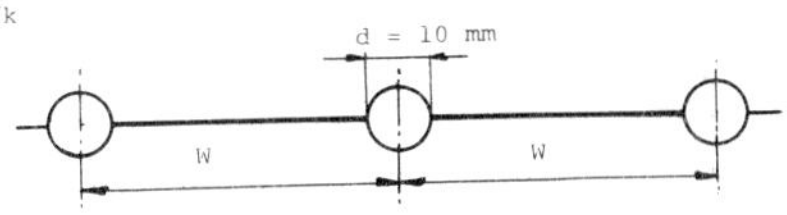

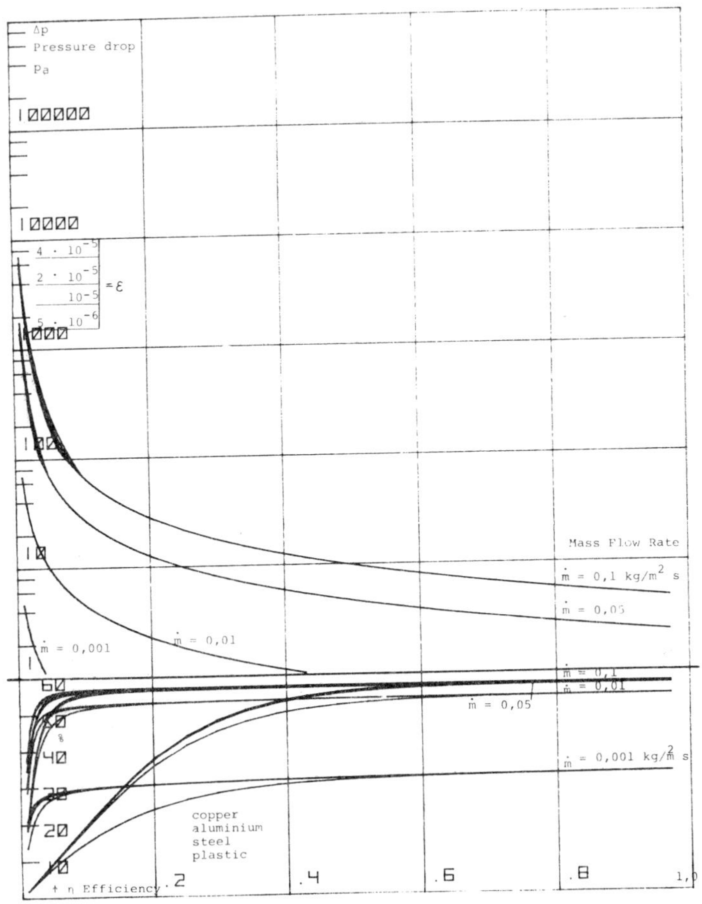

Figure 4

SER

FLAT-PLATE SOLAR COLLECTOR.

Efficiency and pressure drop as a function of d/W (diameter/riser spacing), mass flow rate and absorber material.

Data:

Insolation:	$875\ W/m^2$	
Absorber:	1 mm, $\alpha = \varepsilon = 0{,}9$	
Cover:	2 x 4 mm	
Insulation:	100 mm, k = 0,04 W/m K	
$T_A = 19{,}2\ ^oC$	(ambient)	
$T_{f,i} = 50\ ^oC$	(fluid inlet)	

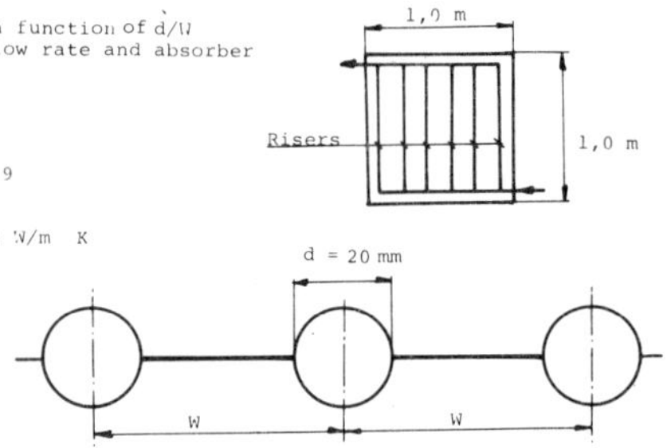

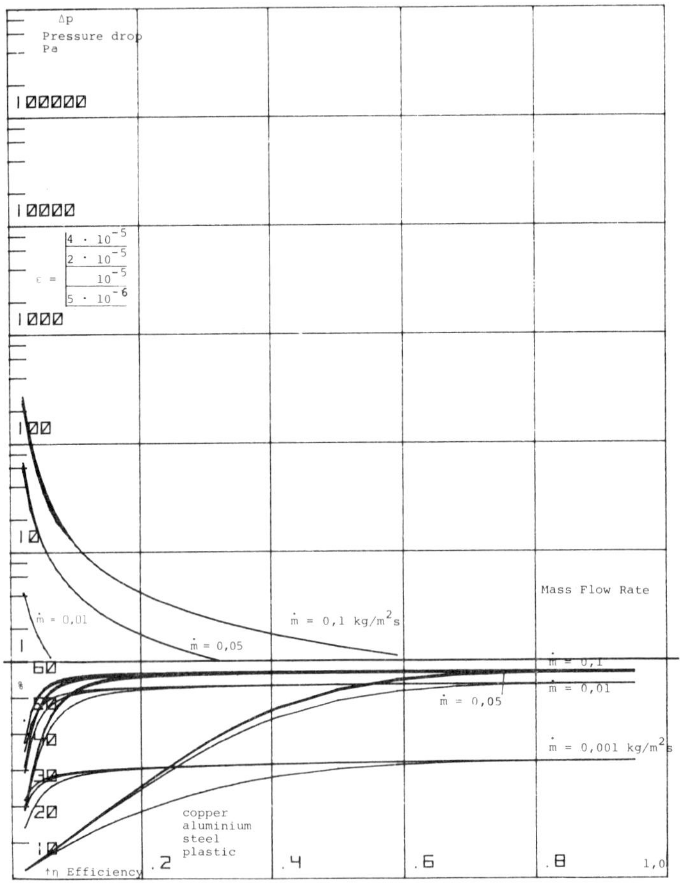

Figure 5 d/W (Diameter/Riser Spacing)

SER

With the same number of risers per meter the largest diameter will give the smallest uncooled surface, but the increase in efficiency caused by this is counteracted by the reduced heat transfer coefficient, which you get because of lower velocity in larger risers at constant mass flow rate.

Figures 3 through 5 show that absorbers of plastic must be water cooled up till at least 50 or 80 %, depending on the riser diameter in order to give maximum useful gain.

As the riser diameter has very little influence on the efficiency, if absorbers of metal are used, it could also be suitable to see how the efficiency varies with the number or risers. The same equations that were applied to give figures 3-5 will then give figures 6-8.

From the figure 8 you can see that, if the mass flow rate increases from 0,05 to 0,1 kg s^{-1} m^{-2}, the efficiency increase will be neglectable, while the pressure drop increases with a factor 4. This means that the pumping work needed to overcome the resistance in the solar collector will increase with a factor 8 that, if the number of risers is larger than 10 per meter, the efficiency will be unchanged.

From the figures 6 and 7 you can see that absorbers made of steel, aluminium or copper will give the same efficiency if the number of risers is larger than approximately 10 per meter.

If there were no either parameters that affected the efficiency of the solar collector other than the previously mentioned ones, it would be indifferent what diameter was chosen.

There are, however, two parameters that will direct the "optimal" diameter towards smaller dimensions. Those parameters are: 1) the temperature in the different risers has to be uniform, as the heat loss otherwise will increase, 2) the heat capacity has to be low to make the start-up losses small.

That an uneven temperature distribution will cause increased losses can be seen in figure 8. If the solar collector is designed for a mass flow rate of 0,05 kg s^{-1} m^{-2} and if

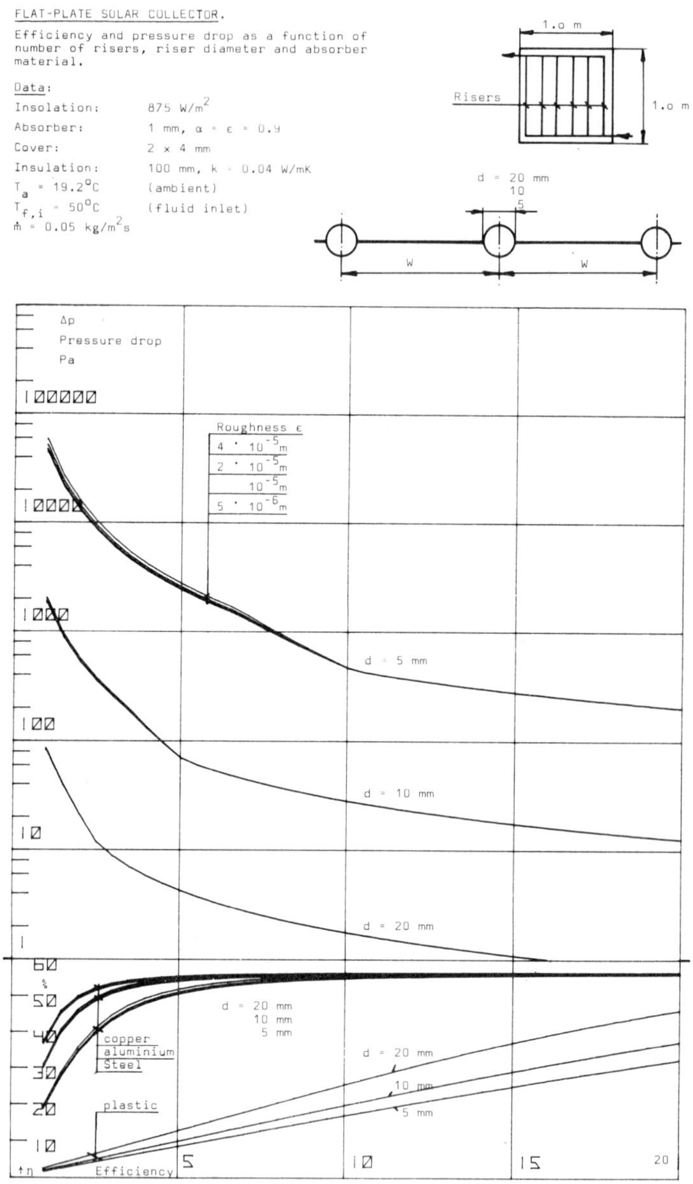
FLAT-PLATE SOLAR COLLECTOR.
Efficiency and pressure drop as a function of number of risers, riser diameter and absorber material.
Data:
Insolation: 875 W/m²
Absorber: 1 mm, α = ε = 0.9
Cover: 2 x 4 mm
Insulation: 100 mm, k = 0.04 W/mK
Ta = 19.2°C (ambient)
Tf,i = 50°C (fluid inlet)
ṁ = 0.05 kg/m²s
1.o m
Risers
1.o m
d = 20 mm
10
5
W
W
Δp
Pressure drop
Pa
100000
10000
1000
100
10
1
Roughness ε
4 · 10⁻⁵m
2 · 10⁻⁵m
10⁻⁵m
5 · 10⁻⁶m
d = 5 mm
d = 10 mm
d = 20 mm
60
%
50
40
30
20
10
η
d = 20 mm
10 mm
5 mm
copper
aluminium
Steel
plastic
d = 20 mm
10 mm
5 mm
Efficiency
5
10
15
20
Number of risers/m

Figure 6
SER

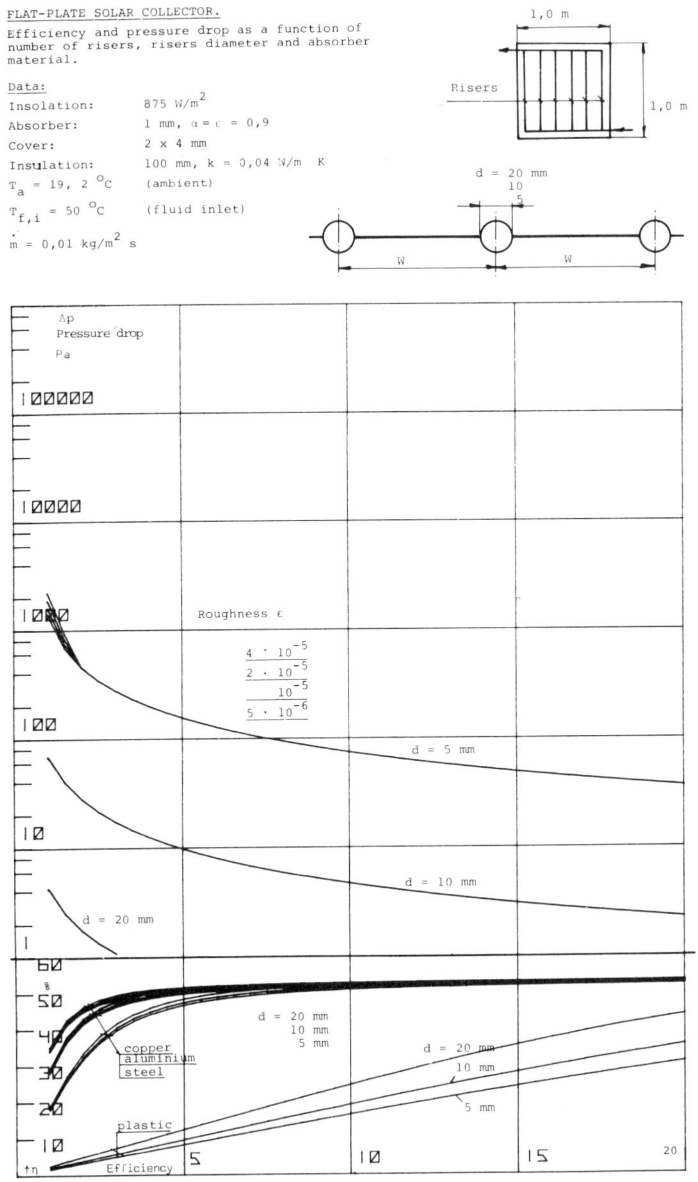

FLAT-PLATE SOLAR COLLECTOR.
Efficiency and pressure drop as a function of number of risers, risers diameter and absorber material.
Data:
Insolation: 875 W/m2
Absorber: 1 mm, α = ε = 0,9
Cover: 2 x 4 mm
Insulation: 100 mm, k = 0,04 W/m K
Ta = 19, 2 °C (ambient)
Tf,i = 50 °C (fluid inlet)
m = 0,01 kg/m2 s
1,0 m
Risers
1,0 m
d = 20 mm
10
5
W
W
Δp
Pressure drop
Pa
100000
10000
1000
100
10
1
Roughness ε
4 · 10-5
2 · 10-5
10-5
5 · 10-6
d = 5 mm
d = 10 mm
d = 20 mm
60
%
50
40
30
20
10
↑η
d = 20 mm
10 mm
5 mm
copper
aluminium
steel
plastic
Efficiency
5
10
15
20
Number or risers/m

Figure 7
SER

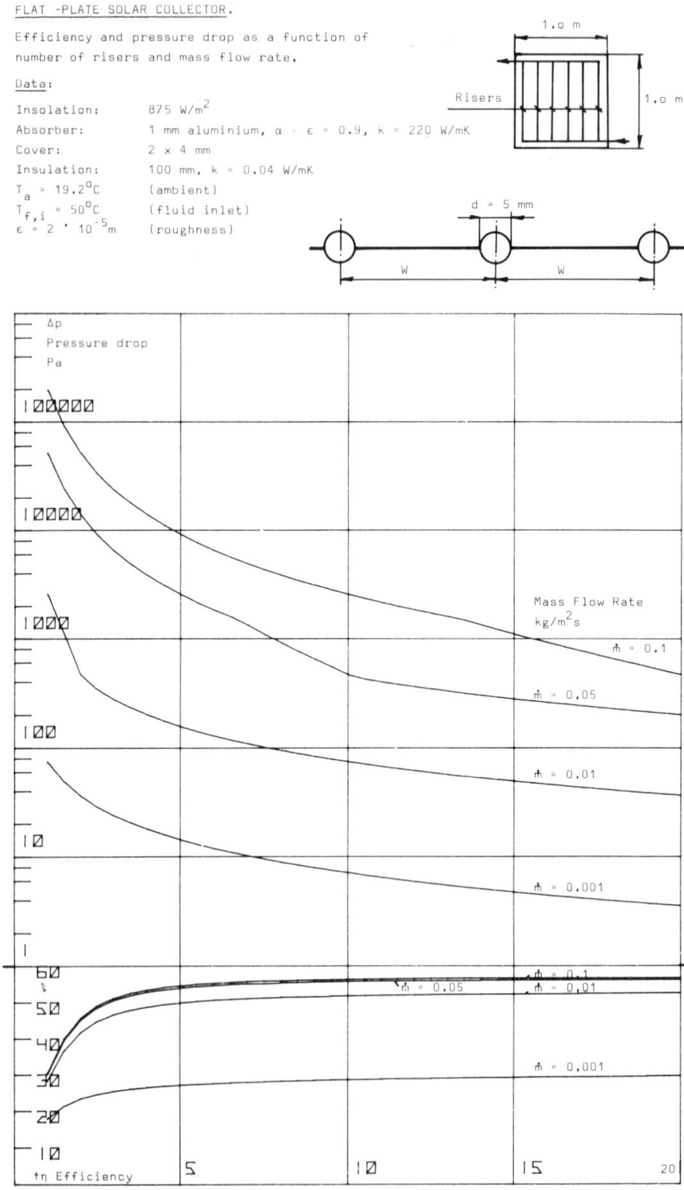
FLAT -PLATE SOLAR COLLECTOR.
Efficiency and pressure drop as a function of number of risers and mass flow rate.
Data:
Insolation: 875 W/m2
Absorber: 1 mm aluminium, α = ε = 0.9, k = 220 W/mK
Cover: 2 x 4 mm
Insulation: 100 mm, k = 0.04 W/mK
Ta = 19.2°C (ambient)
Tf,i = 50°C (fluid inlet)
ε = 2 · 10-5m (roughness)
1.o m
1.o m
Risers
d = 5 mm
W
W
Δp
Pressure drop
Pa
1ØØØØØ
1ØØØØ
1ØØØ
1ØØ
1Ø
1
Mass Flow Rate
kg/m2s
ṁ = 0.1
ṁ = 0.05
ṁ = 0.01
ṁ = 0.001
6Ø
%
5Ø
4Ø
3Ø
2Ø
1Ø
ṁ = 0.1
ṁ = 0.05
ṁ = 0.01
ṁ = 0.001
tη Efficiency
5
1Ø
15
20
Number of risers/m

Figure 8
SER

this varies within the collector with for example ± 0,04 kg s^{-1} m^{-2}, i.e between 0,01 and 0,09 kg s^{-1} m^{-2}, the efficiency decrease in the risers with the mass flow 0,01 would be mcuh larger than the corresonding increase in the risers that has the mass flow rate 0,09 kg s^{-1} m^{-2}. Hence the resulting efficiency with an uneven distribution would be less than with an even distribution.

To achieve uniform temperature, the mass flow rate in all the risers has to be the same, not just within the collector element but within the whole solar collector. To get a slefregulating mass flow distribution within the solar collector element, the resistance in the single riser has to be larger than in the headers. In order to obtain the same effect between the collector elements, the resistance in the elements has to be larger than in the tubes connecting the elements.

Uniform distribution within the collector element can be brought about through use of small enough risers, which will than cause a turbulent flow. When Reynolds' number, Re, is larger than 2300, turbulence starts, and when Re > 3500 it will be stable.

The Reynolds' numbers of the solar collectors according to figures 6-8 are shown in figure 9.

If you, for example, have decided to make a solar collector with ten risers per meter and a mass flow rate of 0,05 kg s^{-1} m-2, you can see from figure 9 that the risers may have a diameter of maxiumum 5 mm to obtain turbulence.

The manufacturer of soalr collectors can not always influence what dimensions will be chosen for the pipes that will connect the solar collector elements. Consequently the pressure drop over the element should be high enough, let us say 1000 Pa, to diminsh the influence from the different pipe dimensions that could be chosen.

If this condition, i.e pressure drop minimum 1000 Pa, is applied on above mentioned example, you will see from figure 6 that this is not fulfilled for risers with a diameter of 5 mm and, hence, even less for 10 mm or larger diameters, which normally are used for solar collectors.

From a practical point of view it is not feasible to make the risers too small as hard or polluted water can clog

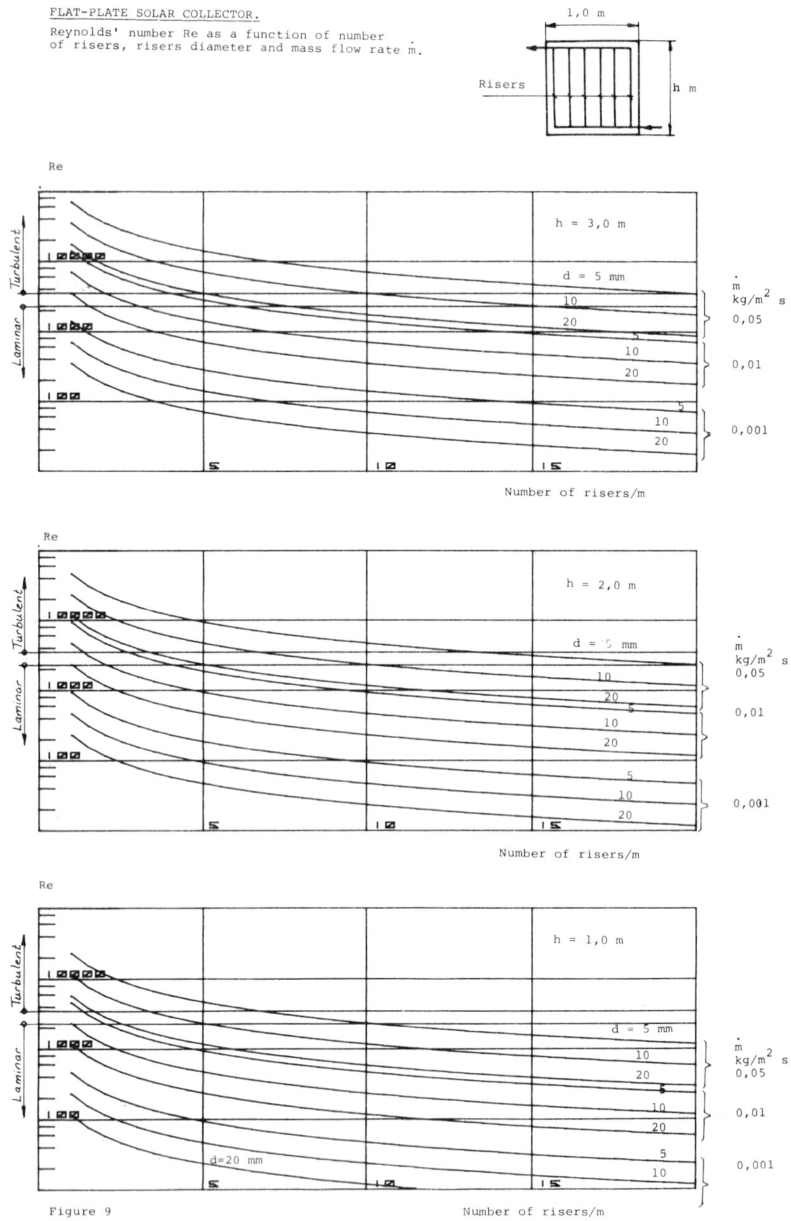
FLAT-PLATE SOLAR COLLECTOR.
Reynolds' number Re as a function of number of risers, risers diameter and mass flow rate ṁ.
1,0 m
Risers
h m
Re
Turbulent
Laminar
10000
1000
100
h = 3,0 m
d = 5 mm
10
20
5
10
20
5
10
20
ṁ kg/m² s
0,05
0,01
0,001
5
10
15
Number of risers/m
Re
h = 2,0 m
d = 5 mm
Number of risers/m
Re
h = 1,0 m
d = 5 mm
d=20 mm
Number of risers/m

Figure 9
SER

up the risers. Furthermore it might be difficult to evacuate the enclosed air from small tubes.

Nevertheless you have to be able to use diameters down to 5 mm as this diameter is commonly used for housing.

If the limit for the diameter is put at 5 mm and if a pressure drop of at least 1000 Pa is required with 10 risers per meter, this can be brought about either through a high enough mass flow rate or through making the solar collector elements higher than 1 m.

To increase the mass flow rate above 0,05 kg s^{-1} m^{-2} will only mean that more pumping work is required without giving a noticeable increase in efficiency. Therefore it might be more suitable to choose the second alternative with a higher collector element or to connect elements in a serie, which will give the same result.

Efficiency and pressure drop for 2 and 3 m high collector elements have been calculated and are shown in figures 10 and 11 respectively . With a height of 2 m compared to 1 m using the same mass flow rate, the velocity will be double as high and the pressure drop will be four times as high. For a height of 3 m the corresponding values will be 3 and 9 times as high.

It is clear from figure 10, that a 2 meter high collector element with 10 risers, diameter 5 mm, at a flow rate of 0,05 kg s^{-1} m^{-2} will give a pressure drop of approximately 3000 Pa. This element will consequently fulfill the requirements for a self-regulating flow distribution both within and between the elements.

Another way of getting the same velocity in all the risers within the collector element is to connect them in a serie instead of parallel to each other, Hence, the collector element will have a continous serpentine of horizontal tubes. In figure 12-14 the efficiency and the pressure drop for such a collector element are shown.

At a mass flow rate of 0,05 kg s^{-1} m^{-2}, the efficiency for a collector with parallel risers will be the same as for a tube serpentine collector. At lesser flow rates the latter collector will give a somewhat higher efficiency than the first one.

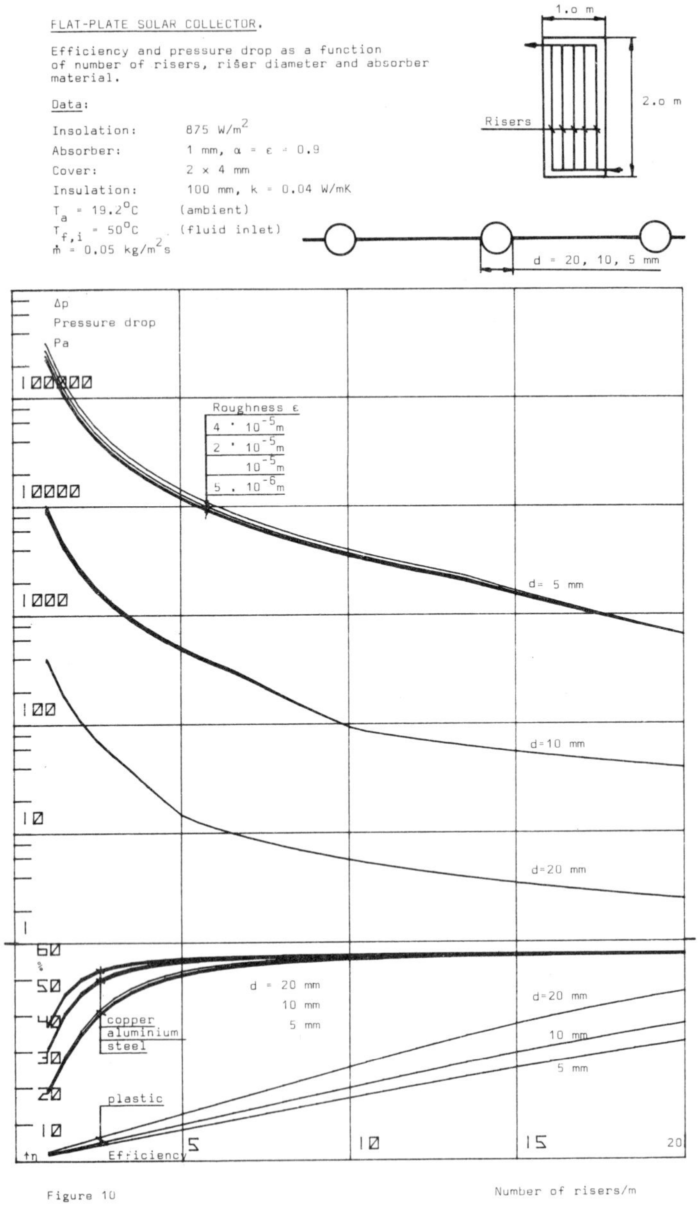

Figure 10
SER

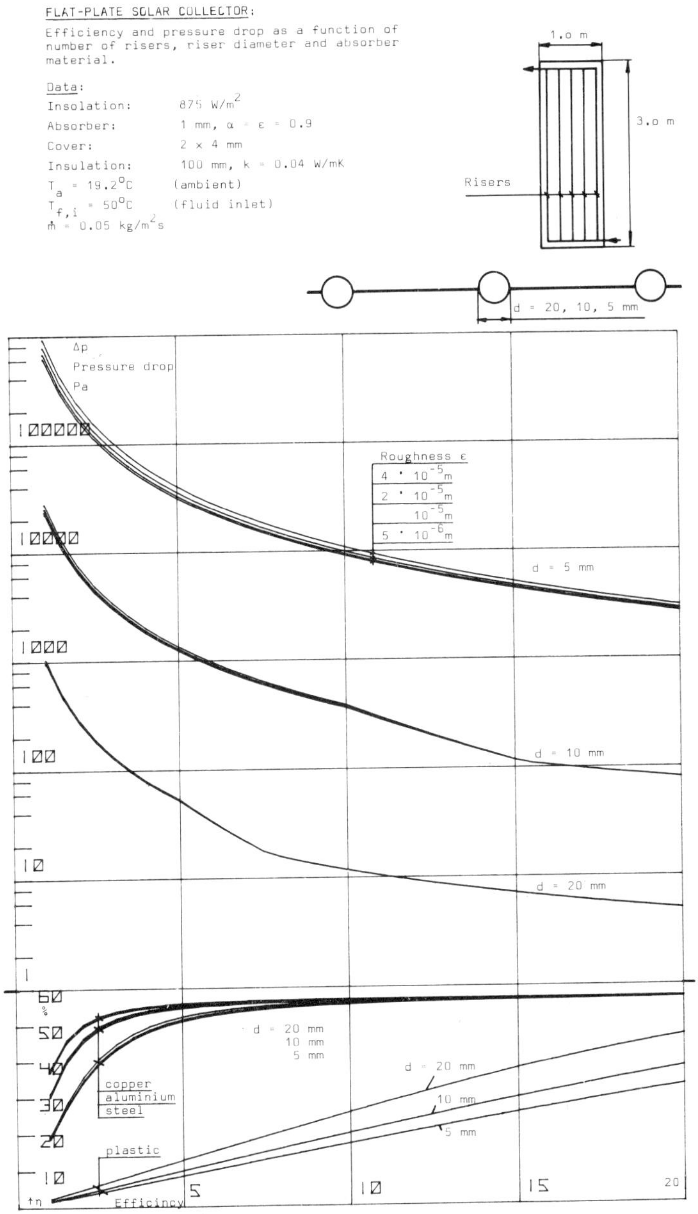

Figure 11
SER

FLAT-PLATE SOLAR COLLECTOR.

Efficiency and pressure drop as a function of number of tube meanders, tube diameter and absorber material.

Data:

Insolation: 875 W/m^2

Absorber: 1 mm, $\alpha = \epsilon = 0.9$

Cover: 2 x 4 mm

Insulation: 100 mm, k = 0.04 W/mK

T_a = 19.2°C (ambient)

$T_{f,i}$ = 50°C (fluid inlet)

$\dot{m}$ = 0.05 kg/m^2s

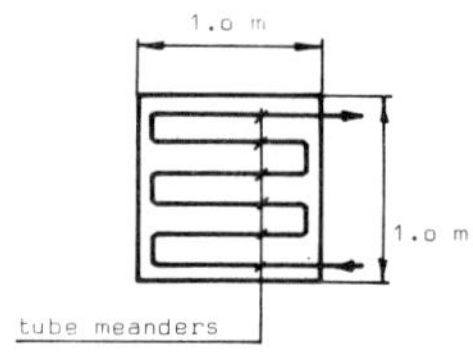

Tube Diameter
5, 10 and 20 mm

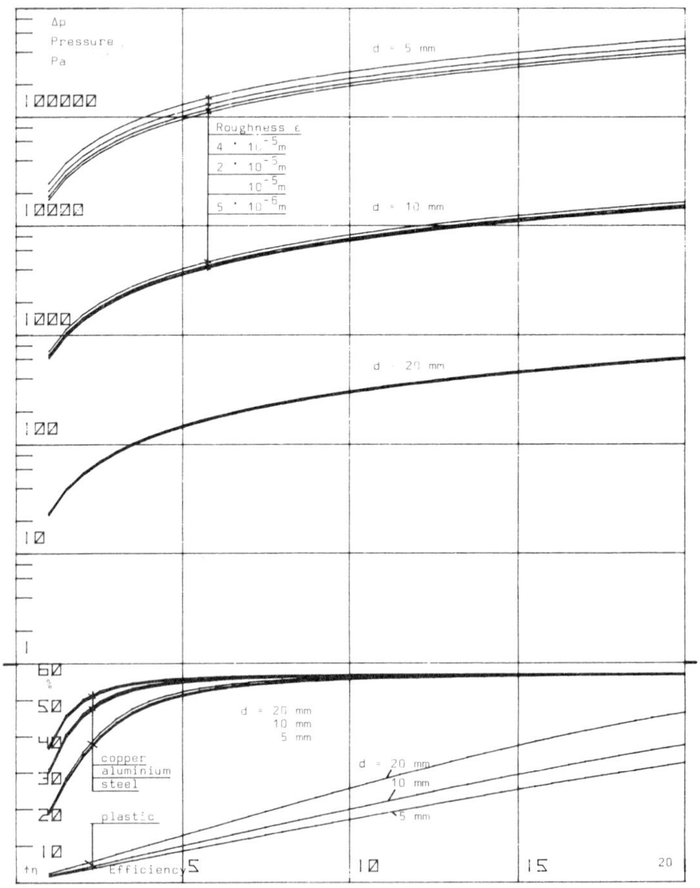

Figure 12
SER

Number of tube meanders

FLAT-PLATE SOLAR COLLECTOR.

Efficiency and pressure drop as a function of number of tube meanders, tube diameter and absorber material.

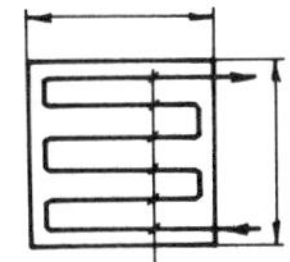

Data:

Insolation: 875 W/m^2

Absorber: 1 mm, $\alpha = \varepsilon = 0.9$

Cover: 2 x 4 mm

Insulation: 100 mm, k = 0.04 W/mK

$T_a = 19.2^oC$ (ambient)

$T_{f,i} = 50^oC$ (fluid inlet)

$\dot{m} = 0.01\ kg/m^2s$

Tube Diameter:
5, 10 and 20 mm

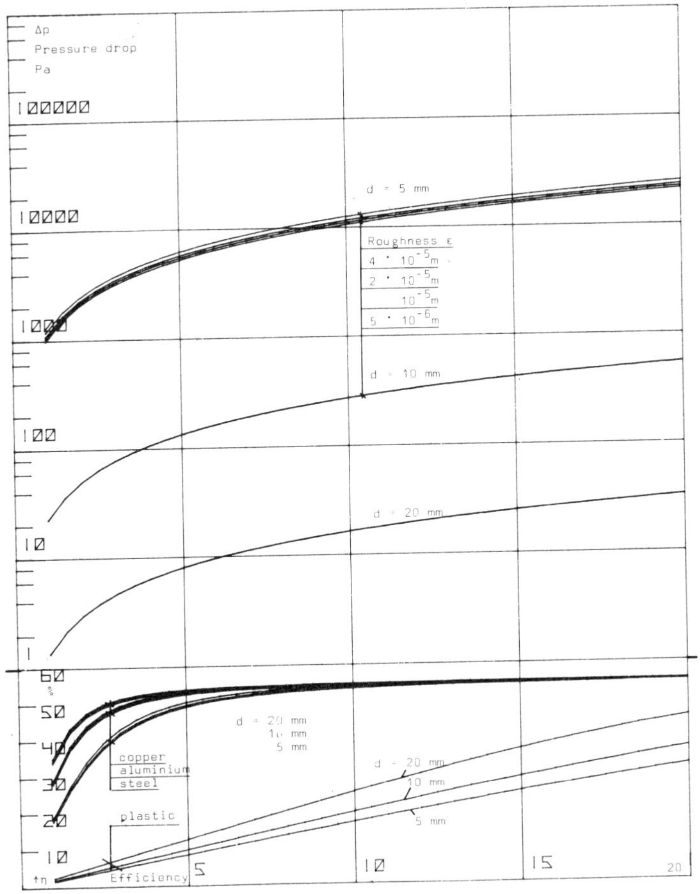

Figure 13
SER

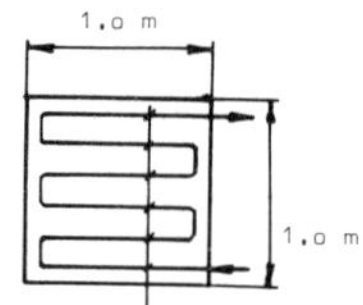

FLAT-PLATE SOLAR COLLECTOR.

Efficiency and pressure drop as a function of number of tube meanders, tube diameter and absorber material.

Data:

Insolation: 875 W/m^2

Absorber: 1 mm aluminium, $\alpha = \epsilon = 0.9$, $k = 220$ W/mK

Cover: 2 x 4 mm

Insulation: 100 mm, $k = 0.04$ W/mK

$T_a = 19.2^oC$ (ambient)

$T_{f,i} = 50^oC$ (fluid inlet)

$\epsilon = 2 \cdot 10^{-5}$m (roughness)

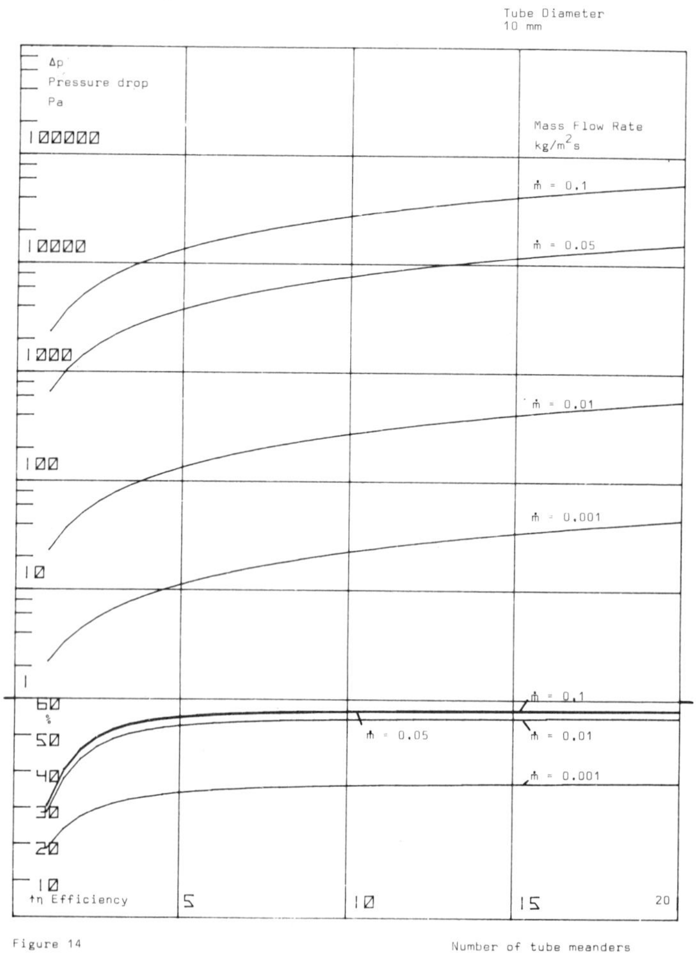

Figure 14
SER

The difference in pressure drop is on the contrary considerable. A solar collector with a 10 m long serpentine of 10 mm thick tubes will give a pressure drop of approximately 6000 Pa at a mass flow rate of 0,05 kg s^{-1} m^{-2}. If the tubes instead had been parallel to each other, the pressure drop would have been about 30 P_a.

The pressure drop is however of minor importance for the flow distribution within the collector element if serpentine connected, but in order to obtain a uniform distribution between the elements the pressure drop must be so high that the element itself regulates the distribution.

If, for the same reason as for solar collectors with parallel risers, a limit is put for the lowest acceptable pressure drop over the element at 1000 Pa you see from the figure 13 that with a flow rate of 0,01 kg s^{-1} m^{-2} and with 10 tube meanders per element only the tube diameter 5 mm will fulfill the requirement.

At a flow rate of 0,05 kg s^{-1} m^{-2} the requirement will be fulfilled both for 5 and 10 mm tubes. The pressure drop in a 5 mm tube serpentine is however unnecessarily high approximately 2 bars - why at this flow rate a diameter of 10 mm would have been better.

The diagrams showing the pressure drops have been calculated for 4 different surface roughnesses on the inside of the tubes ($5 \cdot 10^{-6}$, 10^{-5}, $2 \cdot 10^{-5}$ and $4 \cdot 10^{-5}$ m). This means that it is not necessary to consult the figure 9 to see if the design will cause turbulence, as this is shown directly in the pressure drop grahps. In the laminar area the pressure drop will not be influenced by the surface roughness, which is why the pressure drop grahps will join into one curve in this area. The theoretical pump work for the collectors in figure 6 - 14 are computed in figure 15.

The other factor that made it desirable to use smaller dimensions was that the heat capacity of the solar collector should be low. The smaller tube dimensions, the smaller amount of water the collector will contain which means that the heat capacity too will be low.

A number of scientists mention, however, that the heat capacity has little influence on the average efficiency of the solar collector and this might be true in a sunny climate.

FLAT-PLATE SOLAR COLLECTOR.

Theoretical pumping work for collectors according to figure 6-14.

Decreasing curves for collectors of parallel risers

Increasing curves for collectors of tube in serpentine.

$\varepsilon = 2 \cdot 10^{-5}$m (roughness)

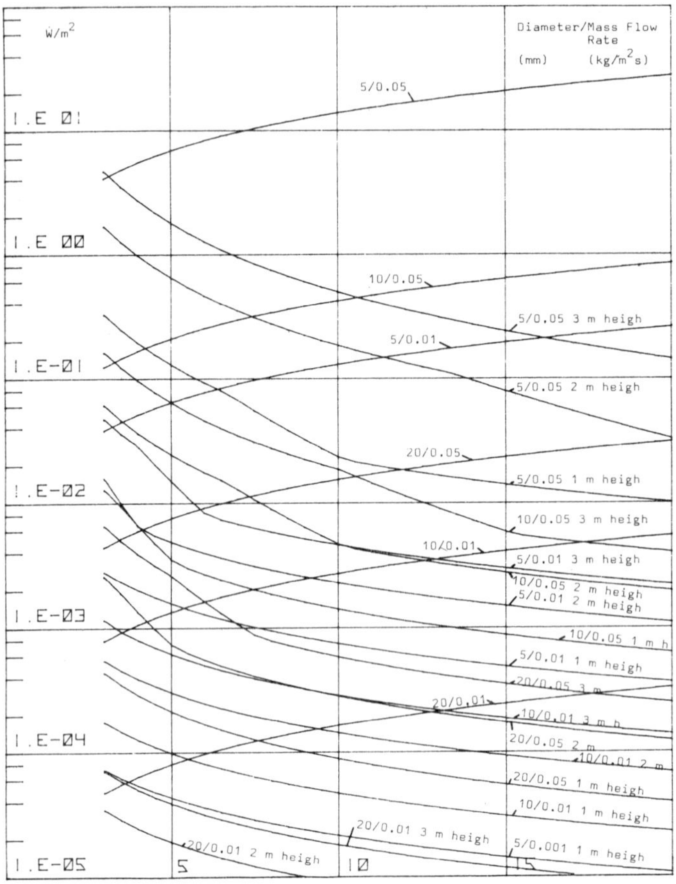

Figure 15
SER

Number of risers (tubes)/m width

Hill and Streed (2) show a graph by Schmidt (3) from which you can see that the start-up time for a typical tube-in sheet flat-plate collector with 2 covers is 10 to 11 minutes if the collector has an initial temperature of 21.1 ^{o}C and the solar radiation is 1104 W/m^2. In Sweden the temperature is rarely as hihg as 21.1 ^{o}C and the radiation is never as high as 1104 W/m^2, which is why the start-up time for a collector in Sweden has to be longer. If you assume that the average start-up time is 20 minutes, and furthermore that there will be 1000 start-ups per year (this can not be calculated from availabe meteorological data), you will have altogether 20 000 minutes = 333 hours of a total of 1500 sunny hours that will not give any useful gain. By this way of calculating, 20 % of the radiation should be lost due to the heat capacity of the solar collector. Any direct measurements of this have not been done, but we intend to do a separate study of this problems.

SUMMARY

A theoretical study has been made on what influence different materials, designs and mass flow rates will have on the efficiency of a flat-plate solar collector.

The study has shown

- that a mass flow rate higher than 0,05 kg s^{-1} m^{-2} will give minimal efficiency increase
- that with an absorber of metal the efficiency will be approximately the same for risers of 5, 10 and 20 mm diameter as long as the number of risers per meter is the same
- that with 10 or more risers per meter the efficiency will be approcimately the same for absorbers of copper, aluminium and steel
- that an absorber of plastic has to be water cooled till at least 50 % depending on the diameter of the risers. This is to ensure the same efficiency as if metal was used
- that parallel risers or tubes in serpentine give the same efficiency if the total tube length is the same and if the mass flow is 0,05 kg s^{-1} m^{-2} or larger

- that for parallel risers the riser diameter should be $\leq$ 5 mm, the riser height > 1,5 m and the mass flow $\geq$ 0,05 kg s^{-1} m^{-2} to ensure a uniform flow distribution
- that small dimensions will contribute by reducing the heat capacity, which will shorten the start-up time for the collector. This will increase the efficiency of the collector, at least in areas with varying cloudiness.

REFERENCES

J.A. Duffic and W.A. Beckman:
Solar Energy Thermal Processes
Wiley, New York (1974)

J.E. Hill and E. R. Streed:
A method of testing for rating solar collectors on thermal performance
Solar Energy 18, 421-429 (1976)

R. Schmidt:
Honeywell flat-plate collector code. Proc. of the Solar Heating and Cooling for Buildings Workshop
Washington, D.C., 21-23 March, 1973. Rep. No NSF/RANN/SE/ /GI-32488/73, University of Maryland (1973)

MY NEW TYPES OF FLAT PLATE SOLAR COLLECTORS

R. Jankovic
Beograd, Yugoslavia

ABSTRACT

This paper presents new types of flat plate solar collectors designed by the author, and collection efficiency factors for flat plate solar collectors with rectangular tubes between two plates.

The solar collectors described here are patented in the Yugoslav Patent Center at Belgrade.

We are now studying the first works on design, investigation and study procedure for practical use of these new types of flat plate solar collectors in the Univerzal Company.

DESCRIPTION

Figure 1 shows a cross-sectional view of the absorber of a semi-elliptic flat plate solar collector of the author's construction.

The length of L may be used in practical realization of this type of solar collector as:

$$L = 2.00 + 3.00\ L_1$$

where L_1 is the length of a circulating semi-elliptic channel, and W is the distance between the centers of the channel.

Figure 2 shows a cross-sectional view of the absorber of a trapezoidal flat plate solar collector, designed by the author as well. The relation for dimensions are the same as they are for the semi-elliptic absorber plate, from Figure 1.

Figure 3 shows a cross-sectional view of the flat plate solar collector with rectangular tube channels with a very small thickness of absorber plate. Equations are also given for the general dimensions which are important for main references in the actual construction of this type of flat plate solar collector.

The main principal views of channel configuration in the solar collector absorber plate are shown in Figure 4.

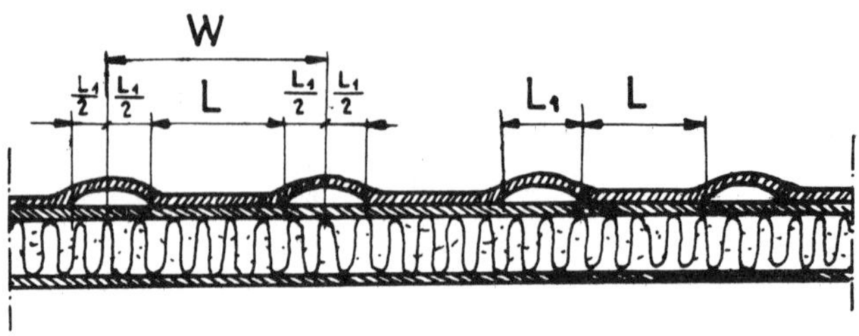

$$L = 2.00 \div 3.00\,L_1$$

$$W = L + 2\frac{L_1}{2} = L + L_1 = L_1 + (2.00 \div 3.00\,L_1)$$

(a)

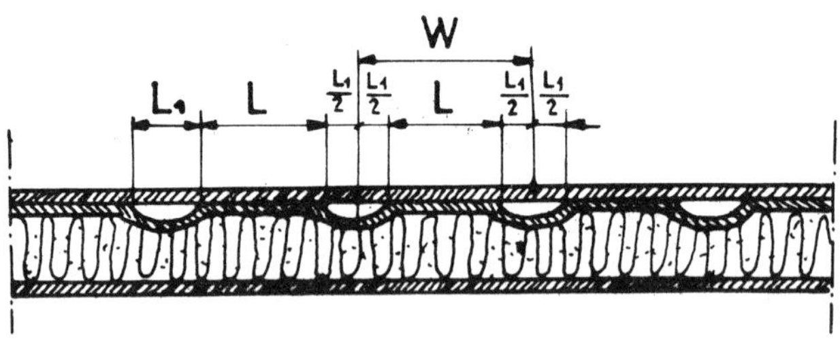

$$L = 2.00 \div 3.00\,L_1$$

$$W = L + \frac{L_1}{2} = L + L_1 = L_1 + (2.00 \div 3.00\,L_1)$$

may be $L = L_1$

(b)

Figure 1 Cross-Sectional View of the Absorber of a Semi-Elliptic Flat Plate Solar Collector

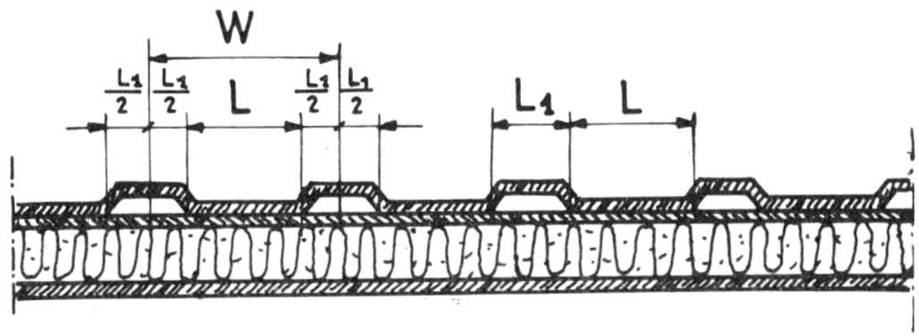

$$L = 2.00 \div 3.00\, L_1$$

$$W = L + 2\frac{L_1}{2} = L + L_1 = L_1 + (2.00 \div 3.00\, L_1)$$

(a)

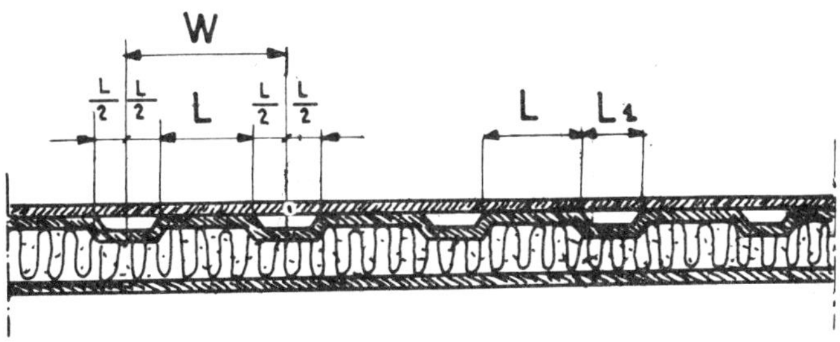

$$L = 2.00 \div 3.00\, L_1$$

$$W = L + \frac{L_1}{2} = L + L_1 = L_1 + (2.00 \div 3.00\, L_1)$$

may be $L = L_1$

(b)

Figure 2 Cross-Sectional View of the Absorber of a Trapezoidal Flat Plate Solar Collector

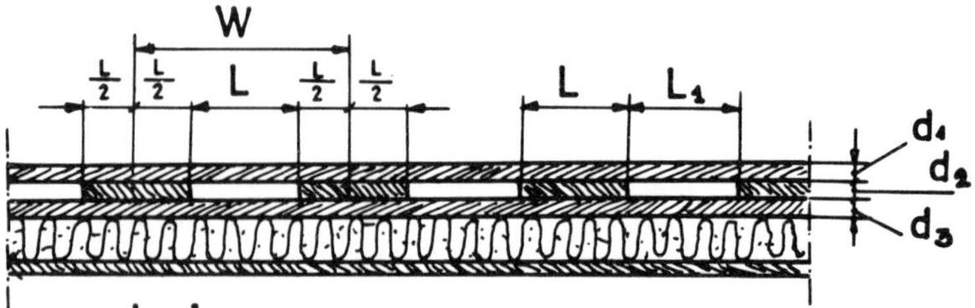

$$W = \frac{L}{2} + \frac{L}{2} + L_1 = L + L_1$$

1. $L = L_1$

$L_1 = W - L$

$W = L + L_1$

$L = W - L_1$

2. $L = \frac{L_1}{2}$

$L_1 = W - L = W - \frac{L_1}{2} =$

$= L + L_1 - \frac{L_1}{2} = 2L$

$W = L_1 + 2L$

3. $L = \frac{L_1}{4}$

$L_1 = W - L = W - \frac{L_1}{4} =$

$= L + L_1 - \frac{L_1}{4} = 4L$

$W = L_1 + 4L$

4. $L_1 = \frac{L}{2}$

$L = W - L_1 = W - \frac{L}{2} =$

$= L + L_1 - \frac{L}{2} = 2L_1$

$W = L + 2L_1$

5. $L_1 = \frac{L}{4}$

$L = W - L = W - \frac{L}{4} =$

$= L + L_1 - \frac{L}{4} = 4L_1$

$W = L + 4L_1$

Figure 3 Cross-Sectional View of the Flat Plate Solar Collector With Rectangular Tube Channels With a Very Small Thickness of Absorber Plate

The largest configuration of fluid is the configuration of channels in Figures 4a and 4b. A shorter configuration of fluid, with horizontal circulating channels in parallel configuration, is given in Figure 4d. And, the shortest configuration of fluid, with vertical circulating channels in parallel configuration, is shown in Figure 4c. The best configuration of circulatingchannels in the collector absorber plate are seen in Figures 4a and 4b.

Figures 5, 6 and 7 show a few details of practical construction of the flat plate solar collectors with and without a glass cover. Flat plate solar collectors without a glass cover have losses of heat, which are not very important for a great collector area, but are important and necessary for smaller areas.

In solar installations for pool heating with a large collector area where there is no need for a pool water temperature higher than 25-28°C, collectors without a glass cover may be used.

Figure 8 shows the main parameters of flat plate solar collectors with a rectangular channel. The collection efficiency factors for this type of solar collector is discussed in the next part of this paper.

In the future, the author was to give other calculations for other types of the flat plate solar collectors, such as:

(1) Design parameters and analytical determination for the author's type of flat plate solar collector.

(2) Practical studies and test procedure of flat plate solar collectors with rectangular circulating channels.

COLLECTION EFFICIENCY FACTORS FOR FLAT PLATE COLLECTORS WITH RECTANGULAR TUBES BETWEEN TWO PLATES

This case occurs frequently in liquid-heated solar heat collectors. The general arrangement is shown in Figure 8. A series of parallel rectangular tubes with dimensions L_1 and d_2, are spaced, W, on centers along the collector plate-absorber plate.

Rectangular tube dimensions may be calculated by an equivalent diameter of the tube, by the equation:

$$d_e = 0.637\ (d_2 + L_1) \tag{1}$$

Now wer have the useful heat gain from the sunshine striking the rectangular canals:

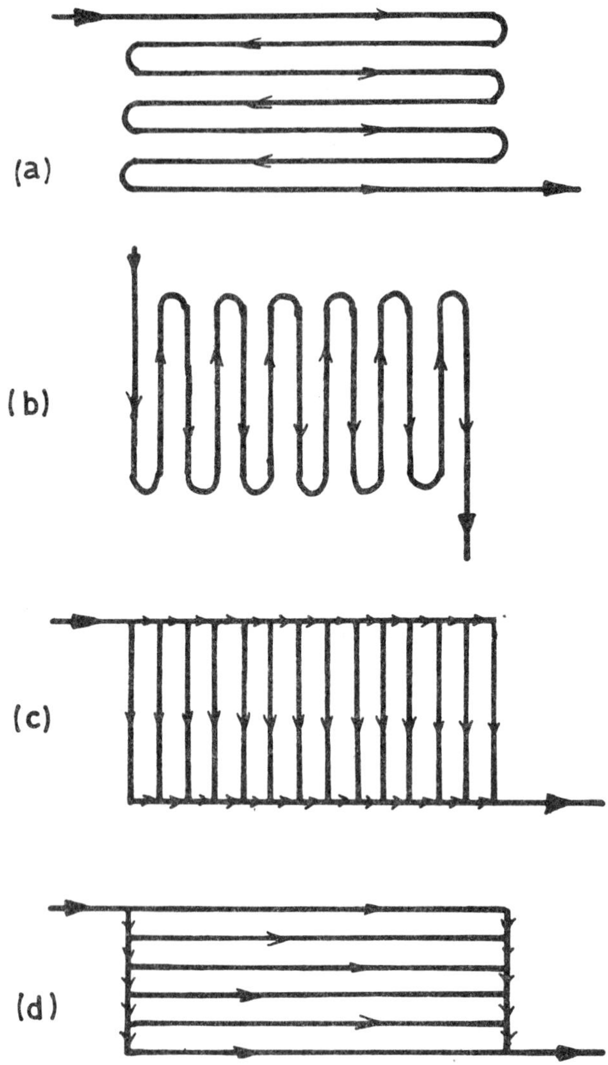

Figure 4 Principal Views of Channel Configuration in the Solar Collector Absorber Plate

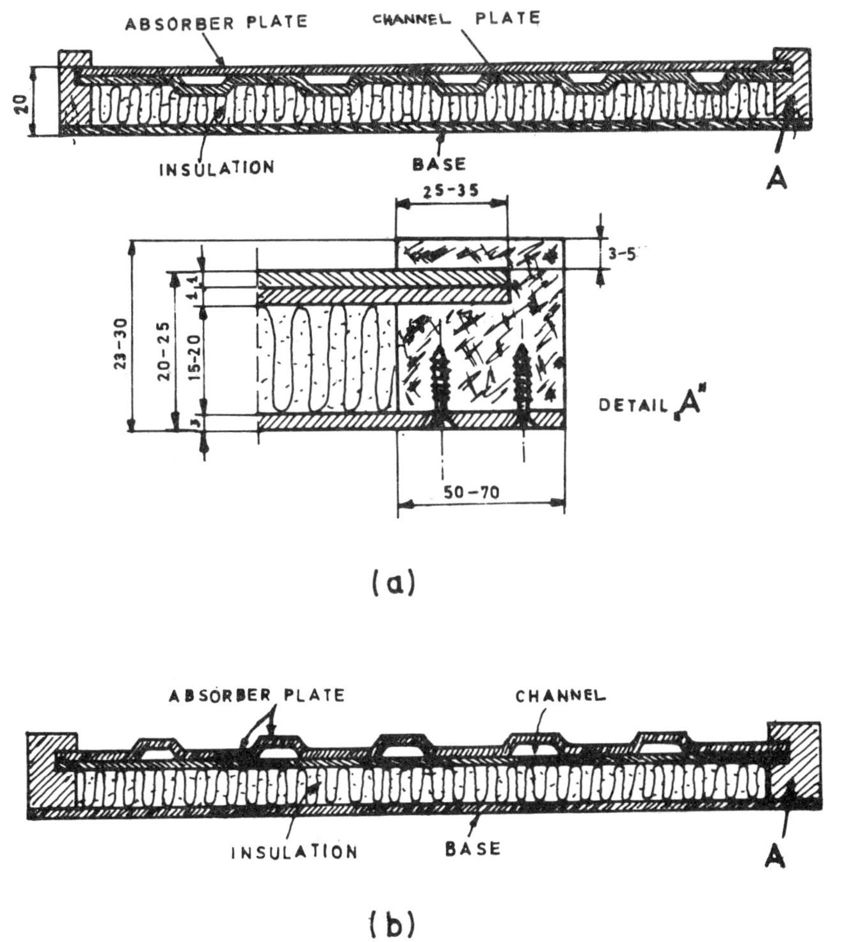

Figure 5 Details of Practical Construction of the Flat Plate Solar Collector With & Without a Glass Cover
ALL DIMENSIONS ARE IN mm

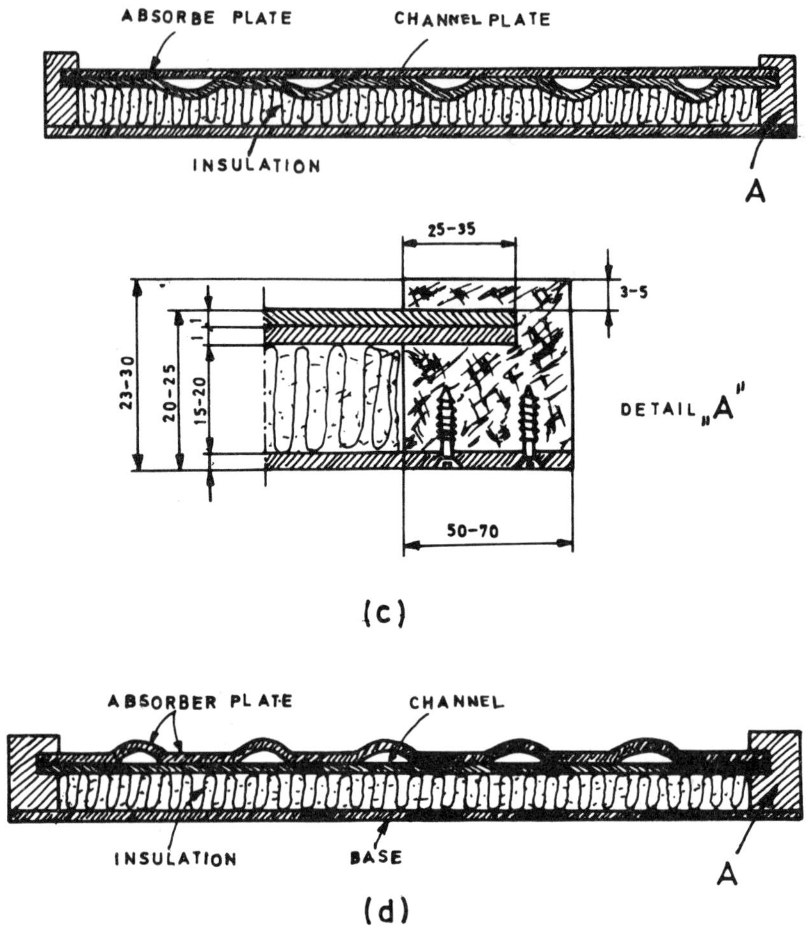

Figure 5 Details of Practical Construction of the Flat Plate Solar Collector With & Without a Glass Cover

ALL DIMENSIONS ARE IN mm

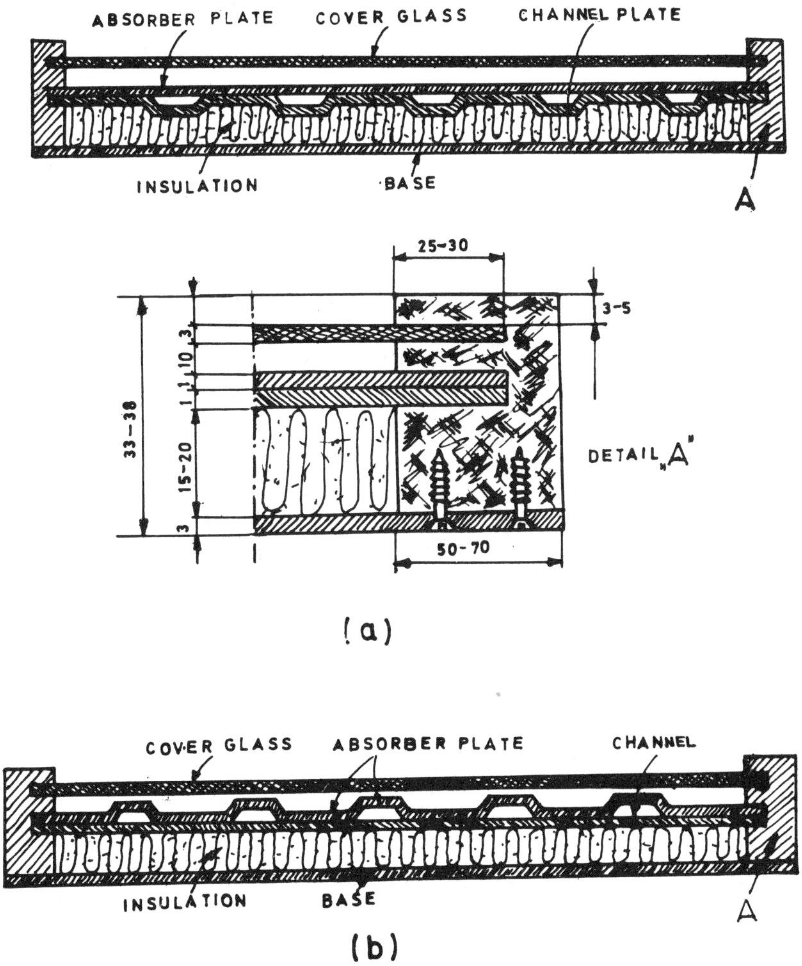

Figure 6 Details of Practical Construction of the Flat Plate Solar Collector With & Without a Glass Cover

ALL DIMENSIONS ARE IN mm

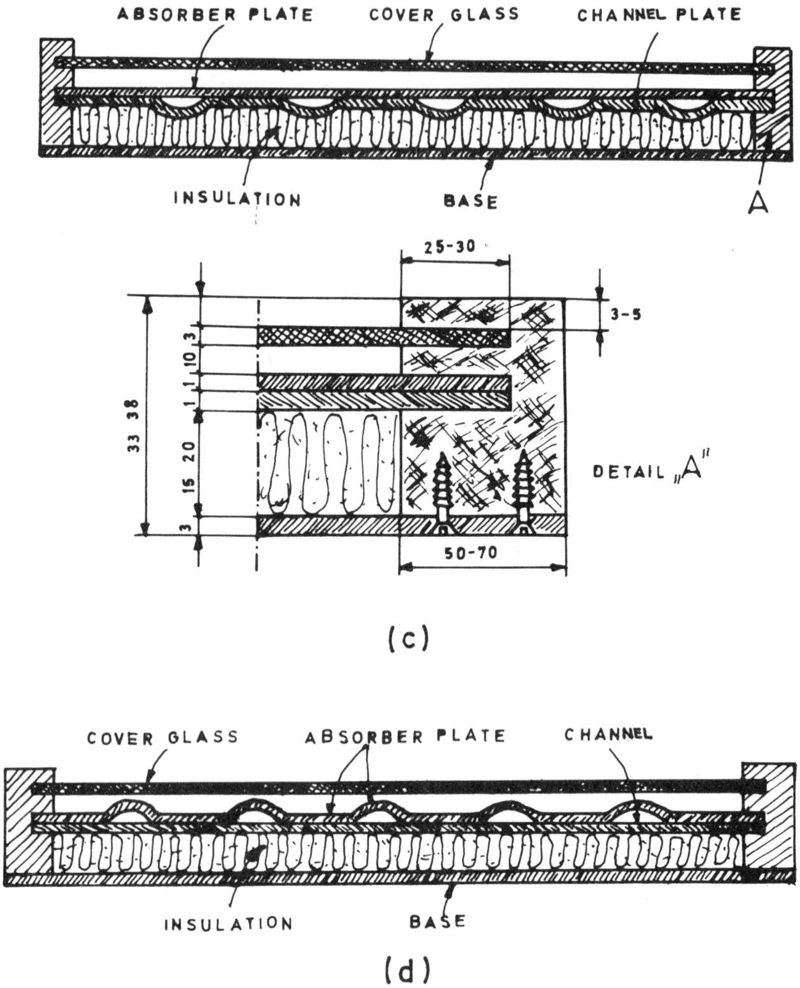

Figure 6 Details of Practical Construction of the Flat Plate Solar Collector With & Without a Glass Cover

ALL DIMENSIONS ARE IN mm

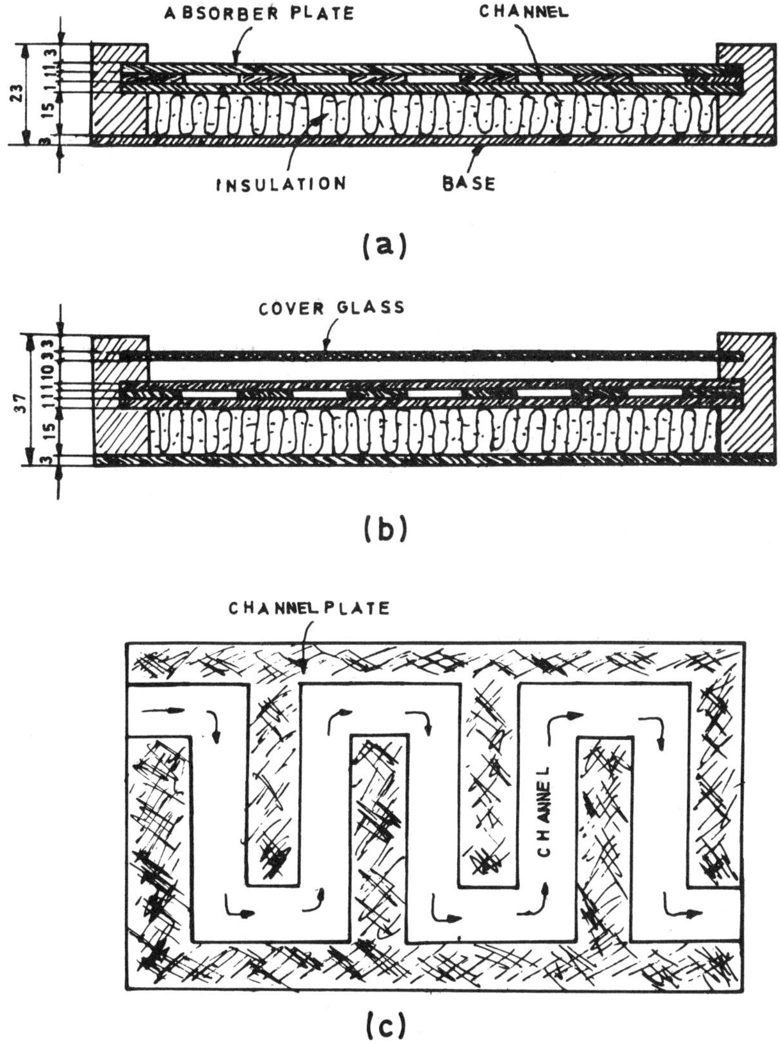

Figure 7 Details of Practical Construction of the Flat Plate Solar Collector With & Without a Glass Cover
ALL DIMENSIONS ARE IN mm

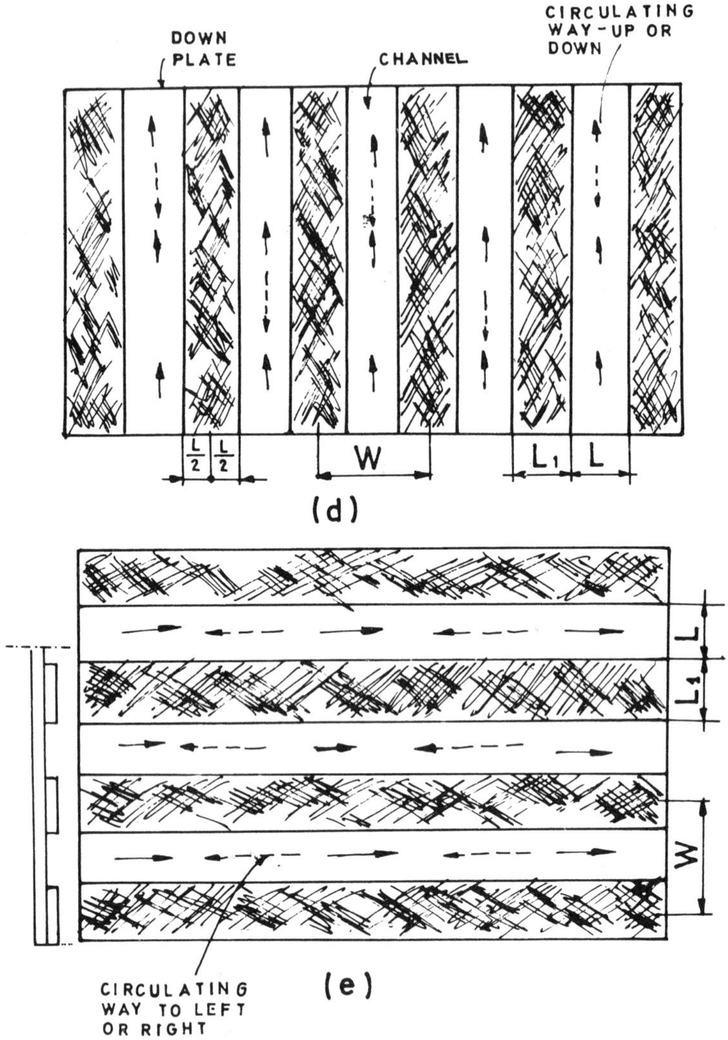

Figure 7 Details of Practical Construction of the Flat Plate Solar Collector With and Without a Glass Cover

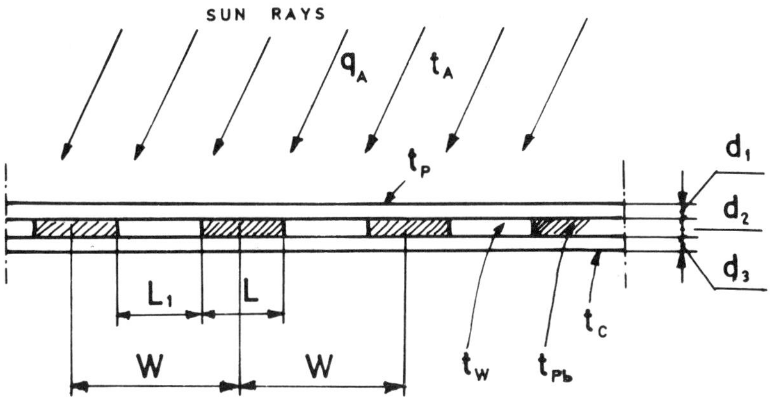

t_P - plate temperature °C

t_C - canel plate temperature °C

t_A - air temperature °C

t_{Pb} - chanel temperature °C

t_W - fluid temperature °C

Figure 8 The Main Parameters of the Flat Plate Solar Collector With a Rectangular Channel

$$q_{u\text{-}can} = \frac{0.637\ (d_2 + L_1)}{W}\ q_A - R - U_L(t_p - t_a) \qquad (2)$$

where

q_A = total sunshine absorption rate,

= average emissivity or absorptivity of surface for long wave radiation,

R = net radiation from a surface to the sky, the surface being a thermally exposed black horizontal plane at air temperature,

t_a = air temperature at °C

t_p = plate temperature at °C

For our use,

$$t_c = t_p \qquad (3)$$

Now, from Figures 3 and 8, we have:

1. For $L = L_1$

$$q_{U\text{-}CAN} = \frac{0.637(d_2+L_1)}{L_1 + L}\left[q_A - \varepsilon R - U_L(t_P - t_A)\right] \qquad (4)$$

2. For $L = L_1/2$

$$q_{U\text{-}CAN} = \frac{0.637(d_2+L_1)}{L_1 + 2L}\left[q_A - \varepsilon R - U_L(t_P - t_A)\right] \qquad (5)$$

3. For $L = L_1/4$

$$q_{U\text{-}CAN} = \frac{0.637(d_2+L_1)}{L_1 + 4L}\left[q_A - \varepsilon R - U_L(t_P - t_A)\right] \qquad (6)$$

4. For $L_1 = L/2$

$$q_{U\text{-}CAN} = \frac{0.637(d_2+L_1)}{L+2L_1}\left[q_A - \varepsilon R - U_L(t_A - t_P)\right] \tag{7}$$

5. For $L_1 = L/4$

$$q_{U\text{-}CAN} = \frac{0.637(d_2+L_1)}{L+4L_1}\left[q_A - \varepsilon R - U_L(t_P - t_A)\right] \tag{8}$$

The useful heat gain from the plate generally is:

$$q_{U\text{-}P} = \left[1 - \frac{0.637(d_2+L_1)}{W}\right] \cdot F\left[q_A - \varepsilon R - U_L(t_P - t_A)\right] \tag{9}$$

Where F = fin efficiency factor, dimensionless.

A consideration of the above equation, along with Figure 8, will show that the above F is the fin efficiency factor of a fin length, L, or from Figure 8:

$$\frac{W - L}{2} \tag{10}$$

or, calculating the diameter of channel equation (1):

$$\frac{W - d_e}{2} = \frac{W - 0.637\ (d_2 + L_1)}{2} \tag{11}$$

Heat transfer through the plate of absorber is:

$$q_{U\text{-}P} = \frac{0.637(d_2+L_1)}{W} \cdot U_C \cdot (t_P - t_{Pb}) \tag{12}$$

where: U_c = overall heat transfer coefficient for heat transfer through the plate absorber between plate and rectangular channel $W/hm^2 {}^\circ C$

t_{pb} = channel temperature °C

t_p = plate temperature °C

For our solar collector we have:

1. For $L = L_1$

$$q_{U-P} = \frac{0.637(d_2 + L_1)}{L_1 + L} \cdot U_c\,(t_P - t_{Pb}) \tag{13}$$

2. For $L = L_1/2$

$$q_{U-P} = \frac{0.637(d_2 + L_1)}{L_1 + 2L} \cdot U_c\,(t_P - t_{Pb}) \tag{14}$$

3. For $L = L_1/4$

$$q_{U-P} = \frac{0.637(d_2 + L_1)}{L_1 + 4L} \cdot U_c\,(t_P - t_{Pb}) \tag{15}$$

4. For $L_1 = L/2$

$$q_{U-P} = \frac{0.637(d_2 + L_1)}{L + 2L_1} \cdot U_c\,(t_P - t_{Pb}) \tag{16}$$

5. For $L_1 = L/4$

$$q_{U-P} = \frac{0.637(d_2 + L_1)}{L + 4L_1} \cdot U_c\,(t_P - t_{Pb}) \tag{17}$$

The heat transfer from absorber plate to fluid generally is:

$$q_{U\text{-}AP} = \frac{\pi \cdot 0.637\,(d_2 + L_1)}{W} \cdot C_{CONV} \cdot (t_{Pb} - t_W) \qquad (18)$$

or:

$$q_{U\text{-}AP} = \frac{3.142 \cdot 0.637\,(d_2 + L_1)}{W} \cdot U_{CONV} \cdot (t_{Pb} - t_W) =$$

$$= \frac{2.00 \cdot (d_2 + L_1)}{W} \cdot U_{CONV} \cdot (t_{Pb} - t_W) \qquad (18a)$$

where: $C_{conv.}$ = coefficient convection rectangular tube to circulating fluid

t_w = fluid temperature at a given point

And now:

1. For $L = L_1$

$$q_{U\text{-}AP} = \frac{2.00 \cdot (d_2 + L_1)}{L_1 + L} \cdot C_{CONV} \cdot (t_{Pb} - t_W) \qquad (19)$$

2. For $L = L_1/2$

$$q_{U-AP} = \frac{2\cdot(d_2 + L_1)}{L_1 + 2L}\cdot C_{CONV}\cdot(t_{Pb} - t_W) \tag{20}$$

3. For $L = L_1/4$

$$q_{U-AP} = \frac{2\cdot(d_2 + L_1)}{L_1 + 4L}\cdot C_{CONV}\cdot(t_{Pb} - t_W) \tag{21}$$

4. For $L_1 = L/2$

$$q_{U-AP} = \frac{2\cdot(d_2 + L_1)}{L + 2L_1}\cdot C_{CONV}\cdot(t_{Pb} - t_W) \tag{22}$$

5. For $L_1 = L/4$

$$q_{U-AP} = \frac{2\cdot(d_2 + L_1)}{L + 4L_1}\cdot C_{CONV}\cdot(t_{Pb} - t_W) \tag{23}$$

The total useful heat gain is:

$$q_u = q_{u-can} + q_{u-p} \tag{24}$$

With equation (4) and (9) we have:

1. For $L = L_1$

$$q_U = \frac{0.637(d_2+L_1)}{L_1+L}\left[q_A-\varepsilon R-U_L(t_P-t_A)\right]+$$

$$+\left[1-\frac{0.637(d_2+L_1)}{W}\right]\cdot F\cdot\left[q_A-\varepsilon R-U_L(t_P-t_A)\right] \tag{25}$$

2. For $L = L_1/2$

$$q_U = \frac{0.637(d_2+L_1)}{L_1+2L}\left[q_A-\varepsilon R-U_L(t_P-t_A)\right]+$$

$$+\left[1-\frac{0.637(d_2+L_1)}{W}\right]\cdot F\cdot\left[q_A-\varepsilon R-U_L(t_P-t_A)\right] \tag{26}$$

3. For $L = L_1/4$

$$q_U = \frac{0.637(d_2+L_1)}{L_1+4L}\left[q_A-\varepsilon R-U_L(t_P-t_A)\right]+$$

$$+\left[1-\frac{0.637(d_2+L_1)}{W}\right]\cdot F\cdot\left[q_A-\varepsilon R-U_L(t_P-t_A)\right] \tag{27}$$

4. For $L_1 = L/2$

$$q_U = \frac{0,637(d_2+L_1)}{L+2L_1}\left[q_A - \varepsilon R - U_L(t_P - t_A)\right] +$$

$$+\left[1-\frac{0,637(d_2+L_1)}{W}\right]\cdot F\cdot\left[q_A - \varepsilon R - U_L(t_P - t_A)\right] \tag{28}$$

5. For $L_1 = L/4$

$$q_U = \frac{0,637(d_2+L_1)}{L+4L_1}\left[q_A - \varepsilon R - U_L(t_P - t_A)\right] +$$

$$+\left[1-\frac{0,637(d_2+L_1)}{W}\right]\cdot F\cdot\left[q_A - \varepsilon R - U_L(t_P - t_A)\right] \tag{29}$$

The preceeding set of five equations (2), (9), (12), (18), and (24), contain five unknowns.

$$q_u,\ q_{u-can},\ q_{u-ap},\ t_{pb},\ t_w \tag{30}$$

We should like to eliminate all but the first of these and, furthermore, to determine if the solution for q_u of the forms is:

$$q_u = F'\left[q_A - \varepsilon R - U_L(t_w - t_A)\right] \tag{31}$$

Where F' is a function of the following variables only:

$$\frac{d}{W} = \frac{0{,}637(d_2+L_1)}{W}; \quad \left.\begin{array}{ll} L = L_1 & W = L_1 + L \\ L = \frac{L_1}{2} & W = L_1 + 2L \\ L = \frac{L_1}{4} & W = L_1 + 4L \\ L_1 = \frac{L}{2} & W = L + 2L_1 \\ L_1 = \frac{L}{4} & W = L + 4L_1 \end{array}\right\} \qquad (32)$$

$$\frac{\pi d C_{conv}}{W} = \frac{0.637(d_2+L_1)}{W} \cdot C_{conv}; \quad \left.\begin{array}{ll} L = L_1 & W = L_1 + L \\ L = \frac{L_1}{2} & W = L_1 + 2L \\ L = \frac{L_1}{4} & W = L_1 + 4L \\ L_1 = \frac{L}{2} & W = L + 2L_1 \\ L_1 = \frac{L}{4} & W = L + 4L_1 \end{array}\right\} \qquad (32b)$$

F, U_L, and U_c (32c)

The simultaneous of equations (2), (9), (12), (18), and (24) (with their derivations) involves fifth order determinants. The solution is straightforward, but the necessary calculations appeared too voluminous, even for insulation, in the notes. However, if the solution is carried through, it will be found that the equation for q_u is indeed of the form given by equation (31), and that the value for F' is:

$$F' = \frac{C_2(U_c+U_L)+C_1U_c}{\frac{U_L}{C_3}\left[C_2(U_c+U_L)+C_1U_c\right]+U_c+\frac{U_L \cdot C_2}{C_1}} \qquad (33)$$

see (Lit. 4), in which:

$$C_1 = \frac{d}{W}; \quad C_2 = (1 - \frac{d}{W} \cdot F; \quad C_3 = \frac{\pi d C}{W} \tag{34}$$

For our collectors we have:

$$C_1 = \frac{0,637(d_2 + L_1)}{W} \qquad \left.\begin{array}{l} W = L_1 + L \\ W = L_1 + 2L \\ W = L_1 + 4L \\ W = L + 2L_1 \\ W = L + 4L_1 \end{array}\right\} \tag{35}$$

$$C_2 = 1 - \frac{0,637(d_2 + L_1)}{W} \cdot F \qquad \left.\begin{array}{l} W = L_1 + L \\ W = L_1 + 2L \\ W = L_1 + 4L \\ W = L + 2L_1 \\ W = L + 4L_1 \end{array}\right\} \tag{36}$$

and for:

$$C_3 = \frac{2(d_2 + L_1)}{W} \cdot C_{conv.} \qquad \begin{array}{l} W = L + L \\ W = L + 2L \\ W = L + 4L \\ W = L + 2L \\ W = L + 4L \end{array} \tag{37}$$

Furthermore, it will be found that equation (33) can be manipulated into the following continuing fraction:

$$F' = \cfrac{1}{\cfrac{U_L}{C_3} + \cfrac{1}{C_1 + \cfrac{1}{\cfrac{U_L}{C_1\, U_c} + \cfrac{1}{C_2}}}} \tag{38}$$

or substituting values of (35), (36), and (37)

$$F' = \cfrac{1}{\cfrac{W\, U_L}{2(d_2+L_1)\, C_{CONV.}} + \cfrac{1}{\cfrac{0{,}637(d_2+L_1)}{W} + \cfrac{1}{\cfrac{W\, U_L}{0{,}637(d_2+L_1)U} + \cfrac{1}{\left[1 - \cfrac{0{,}637(d_2+L_1)}{W}\right]\cdot F}}}} \tag{39}$$

The equation (39) is the general solution for the new type of flat plate solar collector with rectangular-plate channels.

If $C_1 = 1$ from equation (35), equation (39) will be found to reduce to:

$$F' = \cfrac{1}{\cfrac{W\, U_L}{2\,(d_2 + L_1)\cdot C_{CONV}} + 1} \tag{40}$$

For the fin efficiency factor F = 1, equation (39) will be:

$$F' = \frac{1}{\dfrac{W\,U_L}{2(d_2+L_1)\cdot C_{CONV}} + \dfrac{\dfrac{W}{0.637(d_2+L_1)}\left[1-\dfrac{0.637(d_2+L_1)}{W}\right]+\dfrac{U_c}{U_L}}{\left[1-\dfrac{0.637(d_2+L_1)}{W}\right]+\dfrac{U_c}{U_L}}} \qquad (41)$$

Equation (41) may be used to estimate the efficiency factor of the flat plate solar collector with rectangular channels.

REFERENCES

1. Jankovic, R., New Solar Collector, Pat. Pend. No. 1238,9374.

2. Jankovic, R., Semi-Elliptic Flat Plate Solar Collector, Pat. Pend. No. 1239,9375.

3. Jankovic, R., New Flat Plate Solar Collector, Pat. Pend. No. 1240,9376.

4. Bliss, R.W., Jr., The Derivation of Several "Plate Efficiency Factors" Useful in the Design of Flat Plate Solar Heat Collectors, Solar Energy Laboratory, Institute of Atmospheric Physics, University of Arizona, Tuscon, Arizona, U.S.A.

5. Jankovic, R., The Coefficients of Heat Transfer by Convection at Solar Collectors, (to be published).

6. Jankovic, R., Conduction of Heat Through Cylindrical Solar Collectors, (to be published).

MEASUREMENTS ON A FLAT THERMAL SOLAR ENERGY COLLECTOR WITH A CELLULAR STRUCTURE AFTER FRANCIA

A. Bähr, H. Piwecki, L. Rigolini
Commission of the European Communities
Joint Research Center, Ispra Establishment, Italy

ABSTRACT

A thermal flat plate solar energy collector with a cellular structure after Francia has been tested in outdoor conditions. The results have been compared with measurements done at the same collector but with the cellular structure removed. Besides this measurements have been performed to determine separately the influence of the cellular structure on the internal heat transfer in the collector by radiation and by natural convection. The method applied - comparison of cover glass temperature in the selected working conditions - furthermore allows the determination of the influence of the absorption of solar radiation by the cover glass and the cellular structure on the effective power.

INTRODUCTION

There are several ways in which the efficiency of solar energy collectors may be improved. One of them consists of inserting a honeycomb-like cellular structure between the absorber plate and the cover of the entrance aperture, normally a glass plate / 1, 2, 3, 4 /. Such a cellular structure diminishes the heat transfer between the absorber plate and the cover plate and hence the heat losses of the collector in two ways: by reducing the heat transfer due to radiation and that due to natural convection. On the other side the intensity of solar radiation arriving at the absorber plate is reduced by reflection in the outside space and by absorption in the cellular structure.

The results are reported of an outdoor test of a collector with cellular structure[x)] and of supplementary measurements in the laboratory, which aimed to determine separately the listed effects of a cellular structure on the improvement of the total efficiency.

x) The collector tested has been developed and manufactured by Prof. G. Francia, Centro Studi di Matematica e Meccanica Applicata, Università di Genova.

OUTDOOR TESTS OF THE SOLAR COLLECTOR DEVELOPED BY FRANCIA

The collector with the dimensions listed in Fig. 1 has been mounted on a support, which enables it to be kept in any desired orientation with respect to the sun. It has been connected to a water loop with the necessary instrumentation for temperature and flow measurements.

The results of the measurements are listed in Table 1. Comparative measurements at the same collector but with the cellular structure removed are listed in Table 2. For a more simple comparison of the collector performance an efficiency related to a reference insolation intensity of 850 W/m^2 vertical to the collector plane has been calculated with two simplifying assumptions:

(1) the heat losses of the collector are proportional to the difference between absorber plate temperature and air temperature
(2) 82% (85% if the cellular structure is removed) of the incident power is accepted by the absorber plate.

One may calculate the heat losses of the collector at a defined average plate temperature T_a from measured values from

$$V = G \cdot 0,82 - Q \tag{1}$$

They are assumed to be independent of the intensity of incident solar radiation. By this the extractable power with an insolation intensity G_n is given as

$$\begin{aligned} Q_n &= G_n \cdot 0,82 - V \\ &= 0,82\,(G_n - G) + Q \end{aligned} \tag{2}$$

and the related efficiency (Fig. 3)

$$\eta_G = \frac{Q_n}{G_n} = \frac{0,82\,(G_n - G) + Q}{G_n} \tag{3}$$

The relatively large scatter of the measured values has several sources. Often it was not possible to approach sufficiently stable working conditions. Further meteorological effects such as variable air temperature, wind speed, sky temperature and part of diffuse radiation are not taken into consideration.

A		correction factor for the heat transfer by radiation
G	$[W/m^2]$	irradiation intensity sun
G_0, G_{25}	$[W/m^2]$	irradiation intensity at tilt angles 0° and 25° resp.
G_n	$[W/m^2]$	reference value irradiation intensity 850 W/m^2
f_z		ratio of cross section of the cellular structure to the entrance aperture $f_z = 0.1$
c		constant
l	$[m]$	distance absorber plate - cover glass
K	$[W/m^2\ C^o]$	calibration factor heat flux meter
Q	$[W/m^2]$	useful power per m^2 collector entrance aperture
Q_n	$[W/m^2]$	useful power if irradiation intensity is G_n
Q_0, Q_{25}	$[W/m^2]$	useful power at tilt angles 0° and 25° resp.
q_g	$[W/m^2]$	total heat flux density through the cover glass
$q_{\lambda a}, q_{\lambda z}, q_{cz}, q_r$		component heat flux densities : by conduction by air, by conduction by the cellular structure, by convection with and without cellular structure and by radiation
R_z		reduction factor for heat transfer by radiation
T_g	$[^oK]$	cover glass temperature
T_a	$[^oC]$	air temperature
T_m	$[^oC]$	average temperature of absorber plate
T_p	$[^oK]$	absorber plate temperature
ΔT_w	$[^oC]$	temperature drop on heat flux meter
V	$[W/m^2]$	heat losses per m^2 collector aperture
h	$[W/m^2\ ^oC]$	heat transfer coefficient
h_c, h_{cz}	$[W/m^2\ ^oC]$	convective heat transfer coefficient
$\varepsilon_g, \varepsilon_p$		emissivity of cover glass and absorber plate
η_G		collector efficiency related to $G_n = 850\ W/m^2$
k_a	$[W/m^2\ ^oC]$	thermal conductivity of air
k_z	$[W/m^2\ ^oC]$	thermal conductivity of cellular structure
σ	$[W/m^2\ T^4]$	radiation constant 5.7×10^{-8}

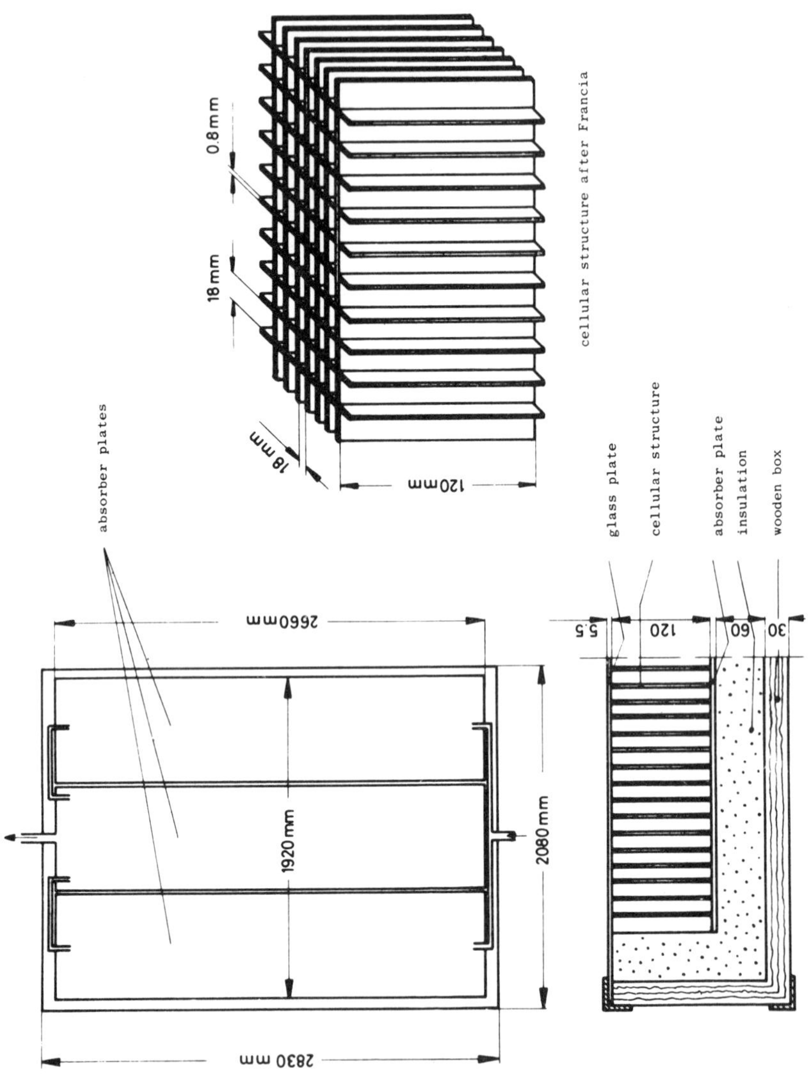

fig. 1 Collector with cellular structure after Francia

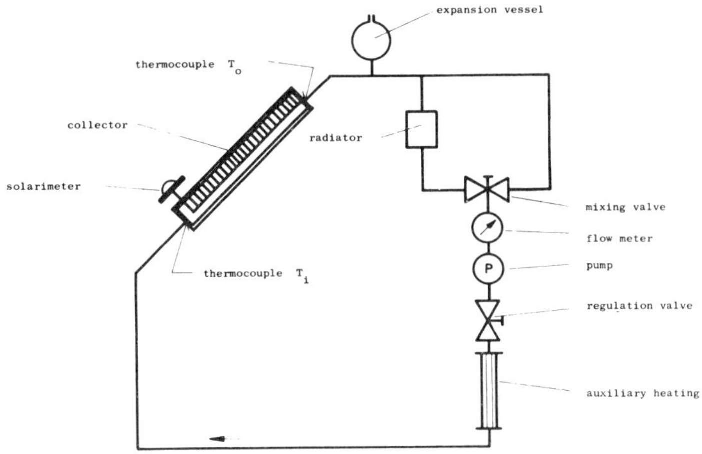

fig. 2 Test - installation

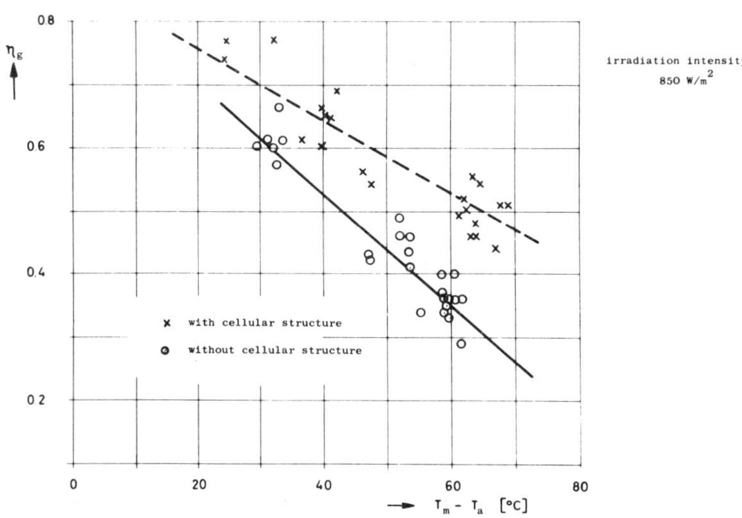

fig. 3 Effectivity in function of the working temperature

Table 1 Measurments on the collector with cellular structure

G [W/m^2]	T_a [^oC]	T_m [^oC]	Q [W/m^2]	$T_m - T_a$ [^oC]	η_G
820	28.0	96.8	400	69	0.51
830	27.5	95.3	410	68	0.51
790	27.5	92.0	400	65	0.54
980	29.0	70.9	680	42	0.69
940	29.0	68.9	630	40	0.66
900	28.0	67.8	590	40	0.66
710	27.0	89.9	340	63	0.53
840	29.0	92.4	380	63	0.43
950	28.0	91.9	470	64	0.46
930	30.0	91.1	470	61	0.49
820	30.0	96.7	340	67	0.44
970	29.5	61.0	610	32	0.61
970	30.5	66.9	610	37	0.61
820	31.0	92.4	410	61	0.52
870	31.0	93.1	430	62	0.50
860	29.3	76.8	460	48	0.54
600	29.5	53.4	410	24	0.74
810	29.0	92.6	360	64	0.48
590	28.5	68.3	280	40	0.60
860	31.0	76.9	480	46	0.56
940	29.5	53.8	710	24	0.77

Table 2 Measurements on the collector without cellular structure

G $[W/m^2]$	T_a $[^oC]$	T_m $[^oC]$	Q $[W/m^2]$	$T_m - T_a$ $[^oC]$	η_G
790	27.0	80.5	290	54	0.41
900	26,5	86.1	330	60	0.35
800	25.5	77.5	350	52	0.46
680	25.5	79.2	240	54	0.46
810	25.0	84.1	260	59	0.36
780	22.5	81.1	270	59	0.40
950	23.5	82.4	400	59	0.37
760	24.0	84.6	260	61	0.40
920	20.0	79.8	330	60	0.33
860	22.5	84.3	310	62	0.36
910	25.0	58.7	560	34	0.61
790	21.5	51.1	450	30	0.60
690	17.5	50.7	420	33	0.66
860	22.0	53.4	520	31	0.61
860	22.5	55.4	490	33	0.57
810	23.0	55.1	470	32	0.60
840	21.5	82.1	290	61	0.36
710	19.0	66.1	240	47	0.43
850	15.0	67.3	410	52	0.49
710	16.5	63.6	230	47	0.42
940	14.0	68.9	360	55	0.34

For collectors provided for fixed mounting the influence of tilt angle on the useful efficiency is of interest, specially in the present case if there is a medium with angular dependent absorption between cover glass and absorber plate. Some measurements are listed in Table 3.

In order to estimate the influence of the absorption in the cellular structure, the measured useful power Q_{25} at a tilt angle of 25° has been compared with a hypothetical power Q^x_{25} which has been calculated from the measured useful power at the tilt angle 0° and the radiation intensity at a tilt angle of 25°. In the calculation it is assumed that the absorption in the cellular structure would be the same as that at the tilt angle 0°. Following the argument for the determination of η_G one may write:

$$Q^x_{25} = Q_o - 0,82\ (G_o - G_{25}) \qquad (4)$$

The difference between Q_{25} and Q^x_{25} represents the decrease of collector power due to increased reflection at the cover glass and increased absorption in the cellular structure. Note that the relatively strong absorption of solar radiation by the thick wall cellular structure (Fig. 8) has only little effect on the useful power.

MEASUREMENTS IN THE LABORATORY IN ORDER TO DETERMINE THE EFFECT OF THE CELLULAR STRUCTURE ON THE INTERNAL HEAT TRANSFER IN THE COLLECTOR BY RADIATION AND NATURAL CONVECTION

Under stationary conditions the power output of a collector is given by the difference of absorbed solar radiation and the sum of all heat losses. In the present case, the valuation of a cellular structure, only the heat losses at the cover of the entrance aperture are of interest. All the other losses, such as those at the collector case and the connected installation, are not considered. They can be made small by a suitable thermal insulation.

The determining factor for the losses at the cover of the entrance aperture, normally a glass plate, is its temperature. It determines the heat transfer to the surroundings if the emissivity of the cover plate and the external heat transfer is defined (Table 4: heat balance of the cover glass).

The effect of the cellular structure on the heat transfer from the absorber plate to the cover glass can be determined in a simple way in a collector

Table 3 Influence of tilt angle on the useful power

G_O [W/m^2]	G_{25} [W/m^2]	$T_m - T_a$ [^oC]	Q_O [W/m^2]	Q_{25} [W/m^2]	$\frac{Q_O - Q_{25}}{Q_O}$	Q^*_{25} [W/m^2]	$Q^*_{25} - Q_{25}$ [W/m^2]	$\frac{Q^*_{25} - Q_{25}}{Q^*_{25}}$
990	890	54	500	370	0.26	420	50	0.12
980	880	53	480	350	0.26	400	50	0.12
940	860	39	550	440	0.21	490	50	0.10

Table 4

HEAT BALANCE OF THE COVER GLASS

HEAT FLOWING TO THE COVER GLASS	HEAT DELIVERED BY THE COVER GLASS
INTERNAL HEAT TRANSFER	
radiation, conduction, convection from the absorber plate	radiation of the cover glass to the absorber plate and the cellular structure
EXTERNAL HEAT TRANSFER	
heat, which is generated in the cover glass by absorption of solar radiation	radiation of the cover glass to the surroundings
heat radiation from the surroundings	conduction and convection to the surroundings

model with a heated absorber plate and a cooled cover glass. A comparison of the cover glass temperatures reached in four suitable experimental positions allows to determine separately the effect of the cellular structure on the heat transfer by radiation and natural convection (Fig. 4). In the experimental conditions, collector model horizontal, entrance aperture towards the bottom, natural convection is suppressed. The cover glass temperatures reached in the model with cellular structure and in a similar one but without cellular structure yield the effect on the heat transfer by radiation. With both models in the standard position - tilt angle 45° and entrance aperture above - the cover glass temperatures deliver the effect of the natural convection.

For the practical execution an installation has been chosen (Figs 5 and 6) in which the heat transfer to the surroundings is replaced by an equivalent thermal resistance acting as heat flux meter which is coupled to a heat sink. The material selected for the thermal resistance was acryl glass. Its thermal conductivity has a suitable value and it was available in homogeneous plates with the desired dimensions. The temperature drop across the heat flux meter yields the power flowing through the cover glass to the surroundings, the temperatures of the absorber plate and the cover glass determine the internal heat transfer. Neglecting the influence of the side walls one may write for the case, collector horizontal without cellular structure and with aperture towards the bottom

$$A\sigma (T_p^4 - T_g^4) + \frac{1}{l} \lambda_a (T_p - T_g) = K \, . \, \Delta T_w \qquad (5)$$

$$A = \frac{c}{\frac{1}{\varepsilon_p} + \frac{1}{\varepsilon_g} - 1} \qquad (6)$$

The factor A takes account of the emissivity of the absorber plate and the glass plate and c the heat exchange with the side walls.
For the collector in the same position but with the cellular structure installed if we assume a linear temperature profile between absorber plate and cover glass, the following holds:

$$R_z \, . \, A\sigma(T_p^4 - T_g^4) + \frac{1}{l} ((1 - f_z) \lambda_a - f_z \lambda_z)(T_p - T_g) = K \, . \, \Delta T_w \qquad (7)$$

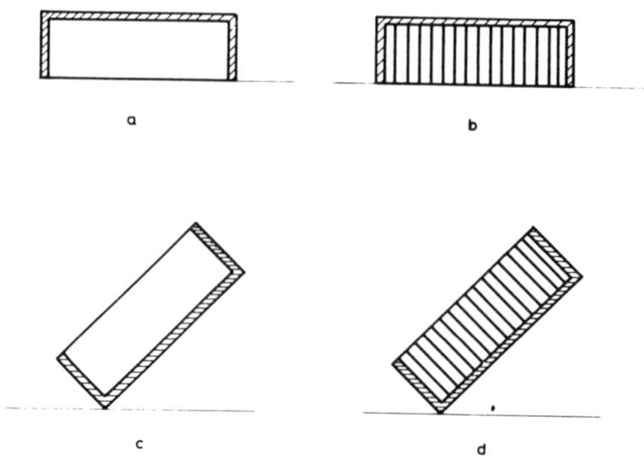

fig. 4 Positions of the collector model for the separate determination of the heat transfer by radiation and natural convection.

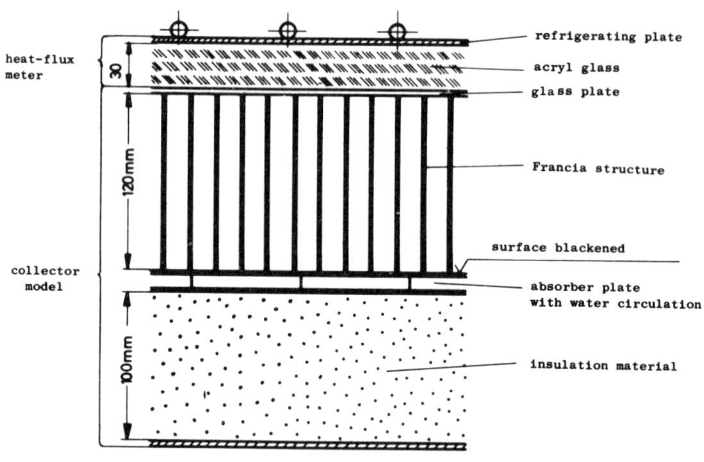

fig. 5 Collector model with heat-flux meter (detail)

The factor R_z indicates to what fraction the radiation heat transfer is reduced by the cellular structure. The relation (5) yields A, the relation (7) the value of R_z. The internal heat transfer by natural convection follows from the temperatures measured in the conditions of Figs 4c and 4d without cellular structure to

$$q_c = K \cdot \Delta T_w - A\sigma(T_p^4 - T_g^4) \tag{8}$$

and for the collector with cellular structure to

$$q_{cz} = K \Delta T_w - R_z A\sigma(T_p^4 - T_g^4) - \frac{1}{l} f_z \lambda_z (T_p - T_g) \tag{9}$$

The heat transfer via conduction in air appearing in the formulation of equations 4a and 4b is to be calculated. In the formulations of equations 4a and 4d the heat transfer by air conduction and by convection are not distinguished and are included in the values of q_c and q_{cz}. Hence we find heat transfer figures (valid for these collector dimensions) for the convective transfer.

$$h_c = \frac{q_c}{T_p - T_g} \qquad h_{cz} = \frac{q_{cz}}{T_p - T_g} \tag{10}$$

Table 5 shows the heat fluxes computed from the measured temperatures and the ratios h_c and R_z characterizing the efficiency of the cellular structure.

The factor R_z is determined solely by the properties of the cellular structure, on the contrary the heat fluxes depend also on the heat transfer to the surroundings. That means they hold for the conditions set by the thermal equivalent resistance. It has been dimensioned so that the heat flux in the laboratory measurements could be measured sufficiently precisely and so that it was comparable with those in real conditions.

As outlined in Table 5, both the heat transfer by radiation and that by natural convection are reduced by the cellular structure to approximately a quarter.

A comparison of the effect of the cellular structure on the cover glass temperature in the laboratory measurements and under real conditions shows that it is considerably smaller under real conditions. This is not only explained by the different properties of the equivalent resistance

Table 5 Influence of the cellular structure on the heat losses at the cover glass.

T_g : resulting cover glass temperature with an external thermal resistance of 0.18 $^oC\ m^2/W$ and a temperature of the heat sink of 15 ^oC.   A = 0.81; ε_p = 0.95; ε_g = 0.88

absorber plate temperature		50	70	90
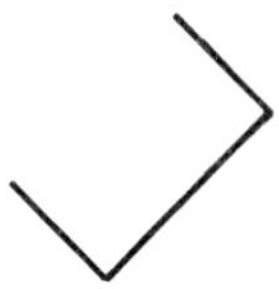	T_g ^oC	36.2	49.8	64.2
	q_g W/m^2	118	195	274
	q_r	81	138	195
	q_c	37	57	79
	T_g	33.3	46.2	60.5
	q_g	101	166	232
	q_r	96	160	220
	$q_{\lambda a}$	4	6	7
	T_g	24.0	30.8	38.5
	q_g	50	90	131
	q_r	33	57	84
	$q_{\lambda z}$	4	6	7
	q_{cz}	14	28	41
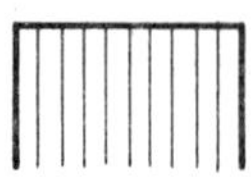	T_g	23.0	28.3	34.8
	q_g	46	74	111
	q_r	36	59	92
	$q_{\lambda z}$	4	6	7
	$q_{\lambda a}$	6	9	12
	R_z	0.23	0.22	0.23
	h_c $W/m^2\ ^oC$	2.7	2.8	3.1
	h_{cz} $W/m^2\ ^oC$	0.5	0.7	0.8

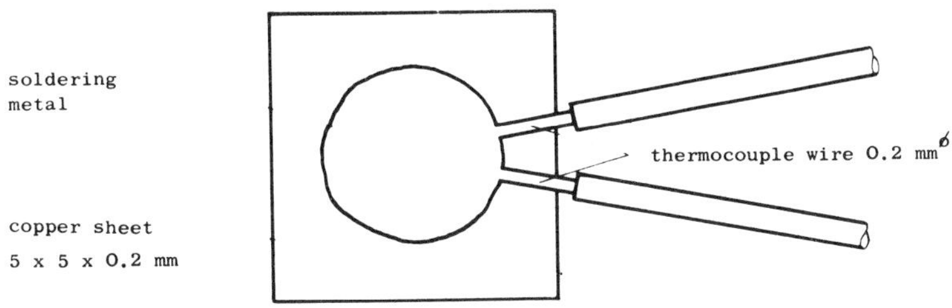

junction of a thermocouple

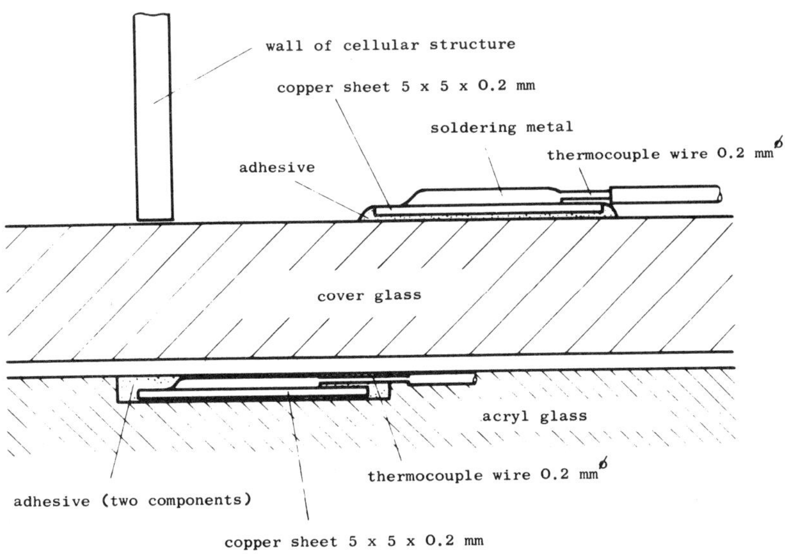

fig. 6 Positions of the thermocouples on the cover glass and on the heat-flux meter.

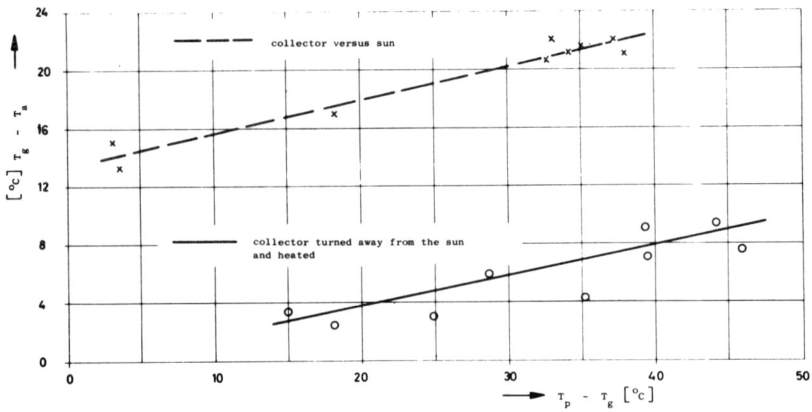

fig. 7 Influence of absorption of sun radiation in the cover glass and in the cellular structure on the temperature of the cover glass.

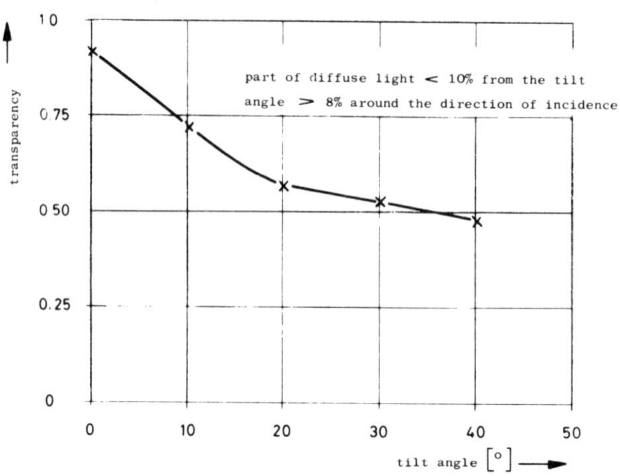

fig. 8 Transmission of the cellular structure for sun radiation in function of the tilt angle

Table 6

Transmission of some commercial glasses and the material for the cellular structure for solar radiation which is filtered by optical glass (10 mm thick).

test material	transmission	absorption
cover glass collector 5.5 mm	0.82	0.11
thick glass D 5 mm	0.88	0.041
thick glass A 5 mm	0.86	0.065
window glass B 3 mm	0.88	0.042
window glass I 3 mm	0.851	0.075
plastic material of the cellular structure 0.85 mm	0.856	0.07
acryl glass 2 mm	0.91	0.01

and the real heat transfer resistance to the surroundings but is also due to the absorption of solar radiation in the cover glass and in the cellular structure, if installed. In the present case the real output power achieved by the collector with cellular structure is smaller than had been anticipated after the laboratory measurements because of the relatively strong absorption of solar radiation in the cover glass and the cellular structure.

In order to estimate what fraction of the incident solar radiation is absorbed in the cover glass and in the cellular structure and what is the effect on the output power some supplementary measurements have been performed:

- comparison of the resulting cover glass temperatures at the Francia collector in real conditions and in modelled conditions, i. e. the absorber plate heated and the collector aperture turned away from the sun (Fig. 7);
- transparency of the cellular structure for solar radiation dependent on the tilt angle (Fig. 8);
- transparency of some commercially available glass plates and of the material of the cellular structure for solar radiation (Table 6).

From these measurements it can be seen that a considerable part of the incident solar radiation does not reach the absorber plate, nearly 30% at normal incidence and more than 50% at a tilt angle of 25°. Only a part of the power absorbed in the cover glass and in the cellular structure is lost because heating up of the cellular structure and of the cover glass reduces the heat losses of the absorber plate. A rough calculation shows that in the present case approximately 70% of the power absorbed in the cover glass and 35% of that absorbed in the cellular structure is directly transferred to the surroundings.

The distance of the two lines in Fig. 7 representing the cover glass temperatures of the irradiated or non irradiated collector respectively allows an estimation of what fraction of the solar radiation which is absorbed in the cover glass and the cellular structure is transferred to the surroundings. One calculates, assuming an effective thermal resistance between cover glass and surroundings of 0.1 m^2 °C/W (corresponding to an effective heat transfer coefficient of 10 W/m^2 °C), a power loss of approximately 140 W/m^2. This value contains the losses due to the absorption of UV radiation, in the cover glass.

From this results: The output power of the tested collector could be increased by installing a less absorbing cover glass. With a cover glass of the material D III in Table 6 - a commercially available "thick glass"- the output power of the tested collector should be 50 W/m^2 higher or about 10% at a working temperature of 55°C above the air temperature. And the relatively small effect of the absorption of solar radiation in the cellular structure explains the small decrease of the output power at greater tilt angles.

CONCLUSION

A cellular structure after Francia is a very effective structural element of solar energy collectors for reducing the heat transfer from the absorber plate to the cover glass. It effects on the collector tested an increase of output power by a factor of 1.1 or 1.5 at a temperature of the absorber plate of 25°C or 60°C respectively above ambient temperature and an irradiation intensity of 850 W/m^2.

From measurements performed with a collector model it can be seen that the internal heat transfer by radiation is reduced to a quarter. The heat transfer by natural convection is also reduced - always referring to the same difference of absorber plate and cover glass temperatures - to a quarter.

The method applied - comparison of the resulting cover glass temperatures under selected working conditions - further yields information on the effect of the absorption of solar radiation in the cover glass and in the cellular structure on the output power.

REFERENCES

1. Giovanni Francia
   Un nouveau collecteur de l'énergie rayonnante solaire - théorie et vérifications expérimentales
   Conférence des Nations Unies sur les sources nouvelles d'énergie
   E/conf. 35/S/71, 1961

2. Buchberg
   Performance characteristics of rectangular honeycomb solar thermal collectors
   Solar Energy Vol. 13 Nr. 2, May 1971, pp 193

3. P. Gallet
   Etudes des proprietés radiatives des structures cellulaires
   Thèse Université de Provence, Mars 1973

4. K.G.T. Hollands
   Studies on methods of reducing heat losses from flat plate solar collectors
   University of Waterloo, September 1975

A SIMULATED COMPARISON OF THE USEFUL ENERGY GAIN IN A FIXED AND A FULLY TRACKING FLAT PLATE COLLECTOR

P. Drago
UNITED STATES MERCHANT MARINE ACADEMY
KINGS POINT, NEW YORK
U.S.A.

ABSTRACT

A comparison is made in the useful energy gain of a flat plate collector facing due south and one that is fully tracking using actual hourly meterological data applied to a "theoretical" flat plate collector. The ratio of total hourly radiation on a tilted surface to that on a horizontal surface, R, is found for a tracking collector using R_b, the ratio of hourly beam radiation on a tilted surface to that on a horizontal surface, and a knowledge of the ratio of daily diffuse radiation to daily total radiation on a horizontal surface. A digital computer is used to calculate the daily useful energy gain using hourly values of solar radiation on a tilted surface, ambient temperature and wind velocity. The results show that the fully tracking flat plate collector outperforms the fixed flat plate collector by a wide enough margin to warrent further investigation.

INTRODUCTION

A study is presented here of the comparison in useful energy gain for a fixed flat plate collector facing due south and a flat plate collector that is allowed to rotate each hour in order that the maximum amount of beam radiation can be incident on its surface. The study was done at Kings Point, New York, over a 26 day period in January, 1977, using actual hourly meterological data applied to a "theoretical" flat plate collector. The period of time used for analysis is of special interest since it was the "coldest" recorded January in over fifty years at Kings Point, and it would be useful to compare and evaluate the performance of flat plate collectors under extreme cold conditions for this geographical area.

This paper is divided into two parts. In the first part, the ratio of hourly beam radiation on a tilted surface to that on a horizontal surface, R_b, is calculated for a fully tracking collector. The ratio of total hourly radiation on a tilted surface to that on a horizontal surface, R, is then found for a tracking collector using R_b and a knowledge of the ratio of daily diffuse radiation to daily total radiation on a horizontal surface.

In the second part of this paper a digital computer is used to calculate the daily useful energy gain using hourly values of solar radiation on a tilted surface, ambient temperature and wind velocity for four different flat plate collectors. Two of these, one with a single cover and one with a double cover, are fixed, and two similar collectors are fully tracking.

While a study has been made of the annual beam radiation on a fixed surface compared to a fully tracking surface by Eibling et al. [1], this has never been directly applied to a flat plate collector in a northern latitude (40.8°) in winter.

This report does not treat the technical or economic feasibility of allowing a flat plate collector, or a system of collectors, to be fully tracking. Obviously a system of this type would be limited to flat roof or ground installations. As a first step in determining the feasibility of a tracking system, however, a knowledge of the comparison of useful energy gain between a fixed and a tracking system would be an essential quantity.

1. Determination of R for a Tracking Collector

The angle of incidence of beam radiation on a tilted surface and the zenith angle of the sun can be found from the equations given by Duffie and Beckman [2], as

$$\begin{aligned}\cos\theta = {} & \sin\delta\sin\phi\cos S - \sin\delta\cos\phi\sin S\cos\gamma \\ & + \cos\delta\cos\phi\cos S\cos\omega \\ & + \cos\delta\sin\phi\sin S\cos\gamma\cos\omega \\ & + \cos\delta\sin S\sin\gamma\sin\omega\end{aligned} \tag{1}$$

$$\cos\theta_z = \sin\delta\sin\phi + \cos\delta\cos\phi\cos\omega \tag{2}$$

Using a method similar to that of Hottel and Woertz [3], a definition can be made for the ratio of beam radiation on a tilted surface to that on a horizontal surface, R_b. It can be seen from Figure 1 that the radiation on a tilted surface H_T and the radiation on a horizontal surface, H, are given in terms of the incoming radiation H_n, by

$$H_T = H_n \cos\theta \tag{3}$$

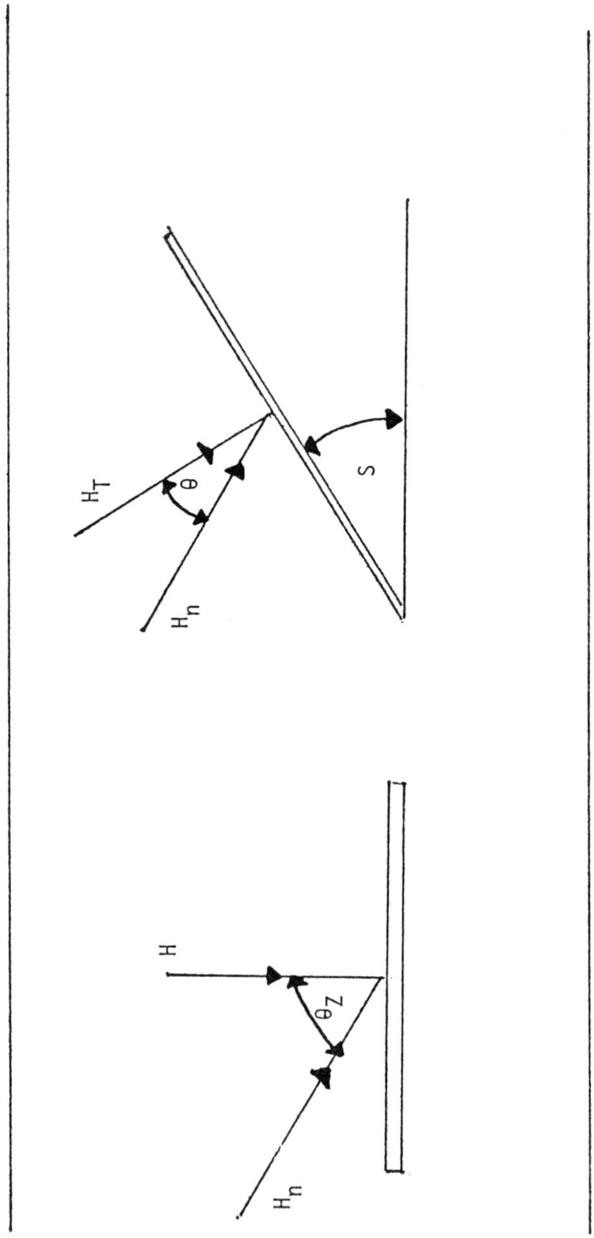

Figure 1. - Radiation on horizontal and tilted surfaces.

$$H = H_n \cos \theta_z \qquad (4)$$

The ratio, R_b, is then given by,

$$R_b = \frac{H_T}{H} = \frac{\cos \theta}{\cos \theta_z} \qquad (5)$$

From eqs. (1) and (5) it can be seen that R_b is dependent on the tilting angle, S, and the surface azimuth angle, γ. It is desirable to find the values S_m and γ_m which, for any declination, δ latitude, ϕ, and hour angle, ω, will give the maximum value of R_b. This will be the value of R_b for which θ is equal to zero (normal incidence).

For a fully tracking flat plate collector the normal to the flat plate must be normal to the solar beam. Therefore, the plates azimuth and altitude angles must coincide with the sun's aximuth angle and the compliment to the sun's altitude angle. These latter values are well known. Therefore the values S_m and γ_m can be found from the equations:

$$S_m = \cos^{-1}[\sin \delta \sin \phi + \cos \delta \cos \phi \cos \omega] \qquad (6)$$

$$\gamma_m = \sin^{-1}\left[\frac{\cos \delta \sin \omega}{\sin S_m}\right] \qquad (7)$$

These values of S_m and γ_m may then be used in eq. (5) to find the value of R_b for the tracking collector.

The values of R_b for a tracking collector were calculated using a digital computer for various values of hour angle and declination at a latitude of 40.8° (Kings Point, New York).

In order to find the ratio, R, of total radiation on a tilted surface to that on a horizontal surface a method is used similar to that of Liu and Jordan [4], and Klein, Duffie and Beckman [5].

The ratio of total daily radiation on a horizontal surface, H, to daily total extraterestrial radiation, H_0, is first found from the equation

$$K_T = \frac{H}{H_0} \qquad (8)$$

Fig. 7, in Liu and Jordan is then used to estimate the daily diffuse radiation, H_D, as a function of K_T. The hourly diffuse and total radiation on a horizontal surface can then be estimated from the daily diffuse and total radiation using Fig. 15 of Liu and Jordan.

The period of time used for study in this report contained eight hour days. It was found that for eight hour days the ratio of daily diffuse radiation to daily total radiation, H_0/H was very close to the ratio of hourly diffuse radiation to hourly total radiation H_{Dr}/H_r. For example for January 6, 1977 it was found that $H_D/H = 0.41$; while for the 2.5 hour from solar noon $H_{Dr}/H_r = 0.44$. The difference between the two values is well within the estimated error of the graph in Fig. 7 of Liu and Jordan. The value of H_D/H was therefore used for each hour of the day instead of H_{Dr}/H_r.

The total radiation for each hour can then be found from the equation

$$H_{(TOT)} = RH \tag{9}$$

where R is given as in Liu and Jordan by,

$$R = [1 - \frac{H_D}{H} R_b] + \frac{H_D}{H} [\frac{1+\cos S}{2}] + \frac{\rho}{2} [1-\cos S] \tag{10}$$

where ρ is the ground reflectance which is taken to be 0.2 in this report.

When eq. (10) is used to find R for a tracking collector, hourly values of R_b are found using γ_m and S_m for each hour. Also hourly values of S_m should be used in place of S in eq. (10). However it was found that R was accurate to within $\pm$ 0.03 if the "average" value of S during the period in question was used in the last two terms of eq. (10) instead of hourly values of S_m. For the 26 days in January used in this report the smallest value of S_m was 59° and the largest 80°, thus a value of 70° was used for S in the last two terms of eq. (10).

II. Determination of Useful Energy Gain

In order to measure the useful energy gain for a flat plate collector the Hottel and Whilier [6] collector equation is used. This expresses the rate of energy collection per square meter as

$$Q_u/A = F_R[HR(\tau\alpha)-U_L(T-T_a)] \quad (11)$$

where

$$F_R = \frac{(\dot{m}\ C_\rho)_c}{A\ U_L}\ [1-\exp(\frac{-F'U_LA}{(\dot{m}\ C_\rho)_c})] \quad (12)$$

The total loss coefficient, U_L, is given by

$$U_L = U_T + U_b + U_e \quad (13)$$

where U_T is the top loss coefficient, U_b the back loss coefficient and U_e the edge loss coefficient. U_T for a single cover collector is found from the method used in Duffie and Beckman [2] as,

$$U_T = [\frac{1}{h_{\rho-c}+h_{r,\rho-c}} + \frac{1}{h_W+h_{r,c-S}}]^{-1} \quad (14)$$

The quantity h_W, the wind loss coefficient, is given in terms of the wind velocity, V_W, in meters/sec., as,

$$h_W = 5.7 + 3.8\ V_W \quad (15)$$

Also the quantities $h_{\rho-c}$, $h_{r,\ \rho-c}$ and $h_{r,\ c-s}$ are given by the equations:

$$h_{r,\rho-c} = \frac{\sigma(T\rho^2+Tc^2)(T\rho+Tc)}{\frac{1}{\varepsilon_\rho}+\frac{1}{\varepsilon_c}-1} \quad (16)$$

$$h_{r,c-s} = \varepsilon_c\sigma\ (T_c^2 + T_s^2)(T_c + T_s) \quad (17)$$

$$h_{\rho-c} = [1-.0018((\bar{T}-10)][\frac{1.14\ \Delta T}{\ell^{0.07}}^{0.31}] \quad (18)$$

Using equations similar to eqs. (16, (17) and (18), U_T can be found for a two cover collector using the equation

$$U_T = \left[\frac{1}{h_{\rho-c_1} + h_{r,-c_1}} + \frac{1}{h_{c_1-c_2} + h_{r,c_1-c_2}} + \frac{1}{h_w + h_{r,c_{2-s}}}\right]^{-1} \quad (19)$$

It should be noted that eq. (18) assumes a plate tilt angle of S=45°.

However, using an equation developed by Klein, namely,

$$\frac{U_T(s)}{U_T(45)} = 1-(s-45)(0.00259-0.00144\varepsilon\rho) \quad (20)$$

it can be shown that for ε_ρ=.95, the top loss coefficient for any tilt angle, $U_T(s)$ is equal to the top loss coefficient, $U_T(45)$, to within an accuracy of ± 5.5%, when s varies between 0° and 90°. Therefore the value of $U_T(45)$ was used in this analysis.

The method used in computing the daily useful energy gain for the collector is similar to the method of Duffie and Beckman [2]. The hourly radiation on a horizontal surface at Kings Point, New York, was determined using an Eppley Pyronometer and a Chart recorder. The wind velocity and ambient temperature for each hour were found using appropriate sensing devices and chart recorders. It was shown in Duffie and Beckman [2], that when the plate emittance is high (ε=0.95), U_T is not a strong function of plate, cover and ambient temperature, but that it is a strong function of wind velocity, especially for a one cover collector. Therefore in finding U_T for each hour, the plate and cover temperature were considered fixed and only variations in h_w were taken into account. The fluid inlet temperature was also kept constant throughout the analysis.

Variations in the ($\tau\alpha$) product for a fixed collector are not considered. It has been shown in Duffie and Beckman [2] that for incident angles of 50° or less, the transmittance is almost constant. For measurements taken from 0800 to 1600 in January at Kings Point, the incident angle on a south facing collector tilted at 50.8° is never much greater than 50°. For a full tracking collector the incident angle will always be zero and thus ($\tau\alpha$) will remain constant.

The ratio of total radiation on a horizontal surface to that on a tilted surface, R, is found each hour using eq. (10) where R_b is found from equations (6), (7) and (5) for a tracking collector.

A computer program was set up to measure U_L as a function of wind velocity for each hour, R as a function of R_b and H_D/H for each hour, and finally use these, along with hourly data for H and T_a to find Q_u/A for each hour of the day. The daily useful energy gain was found by adding the hourly energy gains, where only positive hourly gains are considered. The daily collector efficiency can also be found from the equation used by Duffie and Beckman [2].

$$\eta = \frac{\Sigma\ Q_u/A}{\Sigma\ HR} \qquad (21)$$

Example

Using the data of Table 1 find $\frac{Q_u}{A}$ for a tracking collector for the 0.5 hour on Jan. 30, 1977 where:

$H = 486\ \frac{W}{m^2}$ $\qquad T_a = -9°c$

$R = 1.69$ $\qquad V_w = 6.8\ m/sec$

Using equations (14) thru (18) we obtain

$$h_w = 31.59\ \frac{w}{m^2}$$

$$h_{r,\rho c} = 6.24\ \frac{w}{m^2}$$

$$h_{\rho-c} = 3.19\frac{w}{m^2}$$

$$U_T = 7.5\ \frac{w}{m^2}$$

From eq. (13) we then obtain $U_L = 8.5\ \frac{w}{m^2}$

Then using eq. (11) we obtain

$$\frac{Q_u}{A} = .85\ [(486)(1.69)(.87)-8.5(60-(-9)] = 1.05\ \frac{w}{m^2}$$

Results

First, two identical single cover flat plate collectors are considered, one fixed, facing south, tilted at an angle of 50.8° and the other changing its azimuth angle γ_m, and tilting angle S_m, at hourly intervals in order to obtain new normal incidence of beam radiation. The parameters of these collectors are given in table (1).

Then two identical two cover flat plate collectors, one fixed as above, and one tracking, are considered. The parameters of these collectors are given in table (2).

Tables (3) and (4) show the parameters involved in calculating the daily useful energy gain for the collectors mentioned above for a "typical" day. During the testing period sunrise was at approximately 0730 and sunset at approximately 1630. Therefore only the hours between 0800 and 1600 were considered for useful energy gain. The hour angle, $\omega = 7.5°$ is used for the hourly interval of 1100 to 1200. The useful energy gains are shown for hourly intervals.

Table (5) shows the comparison in daily energy gain for a fixed and a fully tracking single cover collector for the entire 26 day testing period. The term Q_{uT}/Q_{uF} is the ratio of daily useful energy gain for a tracking collector to that of a fixed collector. The daily efficiency found from eq. (21) is also shown for each collector. The last column in the table shows the total energy gain for the testing period along with the average daily efficiency.

Table (6) shows the comparison in daily energy gain for a fixed and fully tracking double cover collector for the 26 day testing period.

Single cover collector parameters

F_R	0.85
T	60°C
$(\tau\alpha)$	0.87
ε_p	0.95
ε_c	0.9
U_b+U_e	$1\ \frac{W}{m^2}$
T_{plate}	65°C
T_{cover}	23°C
ℓ	2.5 cm

Table 1.

Table 2.

Double cover collector parameters.

F_R	0.85
T	60°C
$(\tau\alpha)$	0.75
ε_ρ	0.95
ε_{c_1}	0.88
ε_{c_2}	0.88
U_b+U_e	$1\ \frac{W}{m^2}$
T_{plate}	65°C
T_{cover_1}	45°C
T_{cover_2}	15°C

TABLE 3

	January 15, 1977		Double Cover			Fixed Collector			
H_D/H	ω degrees	γ degrees	S degrees	H $w \cdot m^{-2}$	R	T_a C°	V_n $m \cdot sec^{-1}$	U_L $w \cdot m^{-2}$	Q_u/A $w \cdot m^2$
0.43	52.5	0	50.8	111	2.19	-7	6.2	4.75	-
	37.5			202	1.75	-5	5.1	4.69	-
	22.5			397	1.61	-4	5.0	4.69	151
	7.5			446	1.55	-3	5.1	4.69	189
	-7.5			369	1.55	-3	3.5	4.58	119
	-22.5			348	1.61	-2	3.9	4.61	113
	-37.5			216	1.75	-2	3.5	4.58	-
	-52.5			62.8	2.19	-2	3.8	4.61	-
	JANUARY 15, 1977		Double Cover		Tracking Collector				
0.43	52.5	48.9	78.9	111	3.3	-7	6.2	4.75	-
	37.5	36.8	71.2	202	2.18	-5	5.1	4.69	21
	22.5	23.1	65.5	397	1.76	-4	5.0	4.69	190
	7.5	7.89	62.5	446	1.60	-3	5.1	4.69	204
	-7.5	-7.89	62.5	369	1.60	-3	3.5	4.58	131
	-22.5	-23.1	65.5	348	1.76	-2	3.9	4.61	148
	-37.5	-36.8	71.2	216	2.18	-2	3.5	4.58	58.1
	-52.5	-48.9	78.9	62.8	3.3	-2	3.8	4.61	-

January 30, 1977			Single Cover			Fixed Collector			
H_D/H	ω degrees	γ degrees	S degrees	H $W \cdot m^{-2}$	R	T_a C°	V_w $m \cdot sec^{-1}$	U_L $w \cdot m^{-1}$	Q_u/A $W \cdot m^{-2}$
0.26	52.5	0	50.8	97.5	2.19	-13	5.3	8.25	-
	37.5			341	1.84	-12	6.7	8.47	-
	22.5			446	1.71	-10	6.5	8.44	61.4
	7.5			487	1.66	-9	6.9	8.50	98.9
	-7.5			425	1.66	-8	6.5	8.44	33.1
	-22.5			362	1.71	-7	6.8	8.5	-
	-37.5			278	1.84	-8	7.0	8.53	-
	-52.5			181	2.19	-9	7.2	8.36	-
January 30, 1977			Single Cover		Tracking Collector				
0.26	52.5	50.9°	76.4°	97.5	3.4	-13	5.3	8.25	-
	37.5	38.5°	68.4°	34.1	2.32	-12	6.7	8.47	65
	22.5	24.2°	62.5°	446	1.84	-10	6.5	8.44	107
	7.5	8.31°	59.3°	486	1.69	-9	6.9	8.50	105
	-7.5	-8.31°	59.3°	425	1.69	-8	6.5	8.44	43
	-22.5	-24.2°	62.5°	362	1.84	-7	6.8	85	85
	-37.5	-38.5°	68.4°	278	2.32	-8	7.0	8.53	-
	-52.5	-50.9°	76.4°	181	3.4	-9	7.2	8.36	-

TABLE 5

SINGLE COVER

Day	H_D/H	HR Fixed $(K_J \cdot m^{-2} x 10^{-3})$	HR Tracking $(K_J \cdot m^{-2} x 10^{-3})$	Q_u/A Fixed $(K_J \cdot m^{-2} x 10^{-3})$	Q_u/A Tracking $(K_J \cdot m^{-2} x 10^{-3})$	EFF. Fixed %	EFF. TKG %	$\frac{Q_{uT}}{Q_{uF}}$
1/4	0.77	5.03	5.56	-	-	-	-	-
1/5	0.6	8.99	10.38	-	-	-	-	-
1/6	0.4	14.58	17.25	.835	2.2	5.7	12.6	2.66
1/7	0.92	1.83	1.92	-	-	-	-	-
1/8	0.53	10.84	12.84	-	-	-	-	-
1/9	0.37	15.17	17.64	1.379	2.3	9.1	13.7	1.72
1/10	0.92	1.80	1.88	-	-	-	-	-
1/11	0.5	11.10	13.11	-	-	-	-	-
1/12	0.67	6.96	7.89	-	-	-	-	-
1/13	0.33	16.13	18.86	.814	1.5	5.0	10.3	2.53
1/14	0.9	2.37	2.51	-	-	-	-	-
1/15	0.43	12.87	14.6	.378	.741	2.9	7.4	2.95
1/16	0.91	2.21	2.35	-	-	-	-	-
1/17	0.23	19.62	23.22	1.163	2.22	5.9	10.2	1.9
1/18	0.23	19.34	22.89	.950	1.81	4.9	9.2	2.2
1/19	0.23	19.14	22.86	2.153	3.53	11.3	17.5	1.9
1/20	0.23	18.95	23.15	2.592	4.82	13.7	21.3	1.9
1/21	0.51	10.59	12.53	-	-	-	-	-
1/24	0.63	7.29	8.33	-	-	-	-	-
1/25	0.53	9.77	10.87	.151	.24	1.5	4.2	3.07
1/26	0.24	17.93	21.16	2.135	3.39	11.9	16.6	1.58
1/27	0.24	17.96	21.4	1.278	2.69	7.1	14.1	2.4
1/28	0.44	11.93	14.0	.169	.655	1.4	4.9	4.08
1/29	0.19	20.21	24.01	1.433	2.78	7.1	12.5	2.2
1/30	0.26	16.73	19.48	.694	1.21	4.2	6.5	1.74
1/31	0.19	20.20	23.03	2.164	3.78	10.7	16.8	1.85
TOTAL		319.54	375.8	18.29	37.02	5.7	10.1	2.13

PD

DOUBLE COVER

Table 6

Day	H_D/H	HR Fixed $K_j \cdot m^{-2} x10^{-3}$	HR TRK $K_j \cdot m^{-2} x10^{-3}$	Q_u/A Fixed $K_j \cdot m^{-2} x10^{-3}$	Q_u/A Fixed $K_j \cdot m^{-2} x10^{-3}$	EFF. Fixed	EFF. TRK	$\frac{Q_{uT}}{Q_{uF}}$
1/4	0.77	5.03	5.56	-	-	-	-	-
1/5	0.6	8.99	10.38	.309	.605	3.4	5.8	1.96
1/6	0.4	14.58	17.25	2.833	4.21	19.4	24.4	1.47
1/7	0.92	1.83	1.92	-	-	-	-	-
1/8	0.53	10.84	12.71	.612	1.31	5.6	10.5	2.21
1/9	0.37	15.17	17.64	3.121	4.46	20.5	25.4	1.43
1/10	0.92	1.80	1.88	-	-	-	-	-
1/11	0.5	11.10	12.88	.425	.785	3.8	6.0	1.85
1/12	0.67	6.96	7.89	-	-	-	-	-
1/13	0.33	16.13	18.86	3.096	4.32	19.2	22.4	1.42
1/14	0.9	2.37	2.51	-	-	-	-	-
1/15	0.43	12.87	14.6	2.059	2.61	16.0	19.8	1.45
1/16	0.91	2.21	2.35	-	-	-	-	-
1/17	0.23	19.62	23.22	4.140	6.26	21.1	27.9	1.50
1/18	0.23	19.34	22.89	3.924	5.98	20.3	26.0	1.51
1/19	0.23	19.14	22.86	5.198	6.8	27	30	1.30
1/20	0.23	18.95	22.6	5.285	7.6	27.8	32.6	1.43
1/21	0.51	10.59	12.42	.436	0.94	4.1	8	2.31
1/24	0.63	7.29	8.21	.108	.126	1.5	1.9	1.5
1/25	0.53	9.77	10.87	1.264	1.56	12.9	13.4	1.25
1/26	0.24	17.93	21.16	4.450	6.34	24.8	30.0	1.43
1/27	0.24	17.96	21.4	4.140	6.34	23.2	30.0	1.54
1/28	0.44	11.93	14.0	1.490	2.15	12.5	15.2	1.41
1/29	0.19	20.21	24.01	4.532	6.73	22.4	28.7	1.55
1/30	0.26	16.73	19.48	3.197	4.57	19.1	22.8	1.35
1/31	0.19	20.20	23.83	5.083	7.3	25.2	30.6	1.42
TOTAL		319.54	380.36	55.70	81.9	17.4	21.8	1.47

Remarks

As might be expected the single cover fixed collector did not perform very well in a northern latitude in winter, its average daily efficiency being only 5.7%. The fully tracking single cover collector, although still not very adequate for this latitude, performed much better than the fixed collector with an average daily efficiency of 10.1% and a total useful energy gain which was 2.03 times greater than that for the fixed collector.

The double cover fixed collector, which is more practical for northern latitudes, had an average daily efficiency of 17.4%, while the tracking collector had an average daily efficiency of 21.8%. But what is more important, the tracking collector had a total useful energy gain which was 1.47 times greater than that for the fixed collector. It should be noted that this 47% increase in energy gain for the tracking collector resulted from only a 19% increase in total incident energy (HR) on the tracking collector surface over the fixed collector surface. This is partially due to the fact that a small increase in HR can result in positive energy gain in certain hourly intervals that would give no positive energy gain otherwise.

The results show clearly that a tracking collector system would have outperformed a fixed collector system by a wide enough margin to warrent further investigation into the feasibility of such a system. Tests should be made over time periods of at least a year to find the increase in yearly energy gain of a tracking system over a fixed system. Also a tracking collector system should be considered as part of an entire heating system to see what percent of load requirements could be satisfied by such a system as compared to a fixed collector system. A flat plate tracking system can also be compared to other types of collectors, like evacuated tube type collectors, for example, which have built into their geometry certain "tracking" characteristics. Finally it must be determined if the cost of a tracking system would be outweighed by the increased gain in useful energy of such a system.

References

1. J. A. Erbling, R. E. Thomas and B. A. Londry. Report from Battelle Memorial Institute to U. S. Department of the Interior, December, 1953. "An Investigation of Multiple-Effect Evaporation of Saline Waters by Steam from Solar Radiation."

2. J. A. Duffie and W. A. Beckman, Solar Energy Thermal Processes, Wiley, New York (1974)

3. H. C. Hottel and B. B. Woertz, Performance of Flat-Plate Solar Heat Collectors." Trans. ASME, 64, 91 (1942)

4. B. Y. H. Liu and R. C. Jordon, The long-term average performance of flat-plate solar energy collectors. Solar Energy 7(2), 53 (1963)

5. S. A. Klein, J. A. Duffie and W. A. Beckman, A design procedure for solar heating systems. Solar Energy 18(2) (1976)

6. H. C. Hottel and A. Whillier, Evaluation of flat plate solar collector performance. Trans. of the Conference on the Use of Solar Energy, II. Thermal Processes, 74-104, University of Arizona, Tempe, Arizona (1955)

FLAT PLATE BOOSTERS FOR ELEVATING TEMPERATURES

I.A. Sakr and A.I. Hegazi
Solar Energy Laboratory
National Research Centre
Dokki, Cairo, Egypt

ABSTRACT

A stationary concentrating solar flat-plate booster "SFPB" with a flat absorber suited for moderate temperature applications is presented in this paper. It has the features of a trapezoidal enclosure, with two reflecting sides, flat absorbing passage at the bottom and a glass cover at the top. Mathematical expressions are deduced for the geometrical parameter, "L, W, H" of the "SFPB" with apex angles $5.0 < \theta < \pi/2$. Concentration gain and effective transmission factors are deduced for different "SFPB" concerning the effect of beam deviation, apex angles and surface reflectivity. The year-round performance of different "SFPB" is computed for the Egyptian solar conditions. It is shown that concentration ratios up to 4 times that associated with temperatures up to 180 °C and effective transmission factors higher than 70% can be obtained with this class of solar boosters, depending on its apex angle and surface quality.

INTRODUCTION

Moderate temperature applications of solar energy, especially cooling and air conditioning, are essential to Egypt due to its tropical weather. Flat plate solar collectors cannot be used for such applications, as the attainable temperatures are not adequate to supply heat to an absorption refrigeration system, since these systems do not operate efficiently at temperatures below 100°C. It is believed that stationary moderate solar concentrators can play a good roll in the solution of such a problem. In the last few years, many papers [1-4] described different non-tracking solar concentrators such as "C.P.C." [2]; most of them need highly skilled manufacture and assembly, besides the reflecting wall lengths being very high compared to the collector aperture. The system proposed herein is similar to that proposed by Holland [1], but more attention is given to the effect of apex angles on reflection geometry; and, multiple reflections are considered. The analysis deals with channels having apex angles $0.0 < \theta < \pi/2$ which are capable of producing an effective concentration gain over the flat plate collectors [5]. According to the description given in [5], they can be classified to:

Double Reflection Channels $5° \leqslant \theta < 45°$

Single Reflection Channels $45° \leqslant \theta < \pi/2$

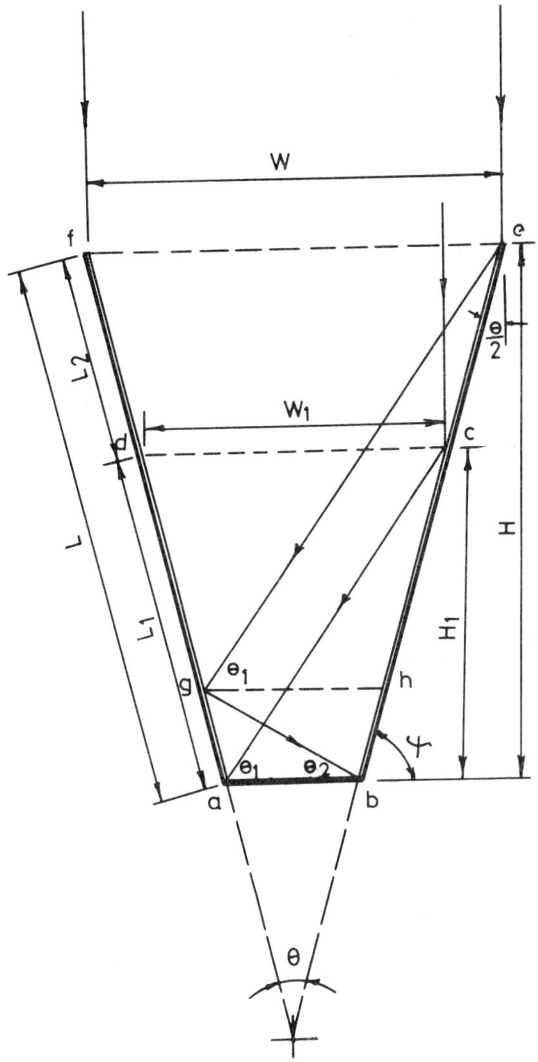

Figure 1 Geometry of "SFPB" Channel With Flat Absorber

Detailed analysis has been given to the geometry, effect of beam deviation, beam concentration and effective transmission of a similar system with a tubular absorber [5]. The work presented herein is an analytical evaluation of the similar parameter of "SFPB" with flat-absorber. In addition, the year-round performance of such solar boosters is computed and described.

GEOMETRY

A channel of the "SFPB" heater with apex angles (θ) and flat absorber is shown in Figure 1. The geometrical parameters (L, W, H) of the channel are deduced in dimensionless form, and referred to the width of the absorber, which is assumed to be unity. Two main assumptions are considered within this analysis: the first is that the absorber surface is uniformly radiated by the reflected radiation, while the second concerns the number of permissible reflections of the normal beam to reach the absorber surface. All "SFPB" channels with $\theta < 45°$ are classified as double reflection channels, and the normal beam received at the side tip of the aperture is considered to reach the same side corner of the absorber after two reflections. For apex angles $\theta > 45°$, the normal beam at the side tip is assumed to reach the opposite side top of the absorber after one reflection only, because the second reflection will be outside the channel [5].

Refer to Figure 1 and the sine law of reflection: the nth reflected beam is shown to be inclined to the absorber plane by an angle "$\pi/2 - n\theta$", therefore, the geometrical parameters of single reflection channels $\theta > 45°$ can be deduced from the trapezoidal channel abcd, where the solar beams at the top reach the absorber after one reflection inclined on an angle "$\theta_1 = \pi/2 - \theta$", thus:

$$L_1 = \cos\theta / \sin\frac{\theta}{2} \qquad (1)$$

and $$W_1 = 1 + 2\cos\theta \qquad (2)$$

$$H_1 = \cos\theta \cdot \cot\frac{\theta}{2} \qquad (3)$$

The geometrical parameters (L, W, H) of double reflection "SFPB" channels "$\theta < 45°$" can also be deduced from Figure 1. The solar beam received by the R.H.S. tip (e) is reflected to the other reflecting surface inclined "eg" inclined "$\pi/2 - 2\theta$", thus, the total length "L" of a double reflection "SFPB" can be given by:

$$L = \frac{\cos(2\theta)}{\sin\frac{(3\theta)}{2}} + \frac{\cos(\theta)\cdot\sin(\frac{5\theta}{2})}{\sin(\frac{3\theta}{2})\ \sin(\frac{\theta}{2})} \qquad (4)$$

and $$W = \sin(2.5\theta)/\sin\frac{\theta}{2} \qquad (5)$$

$$H = L\cdot\cos\frac{\theta}{2} \qquad (6)$$

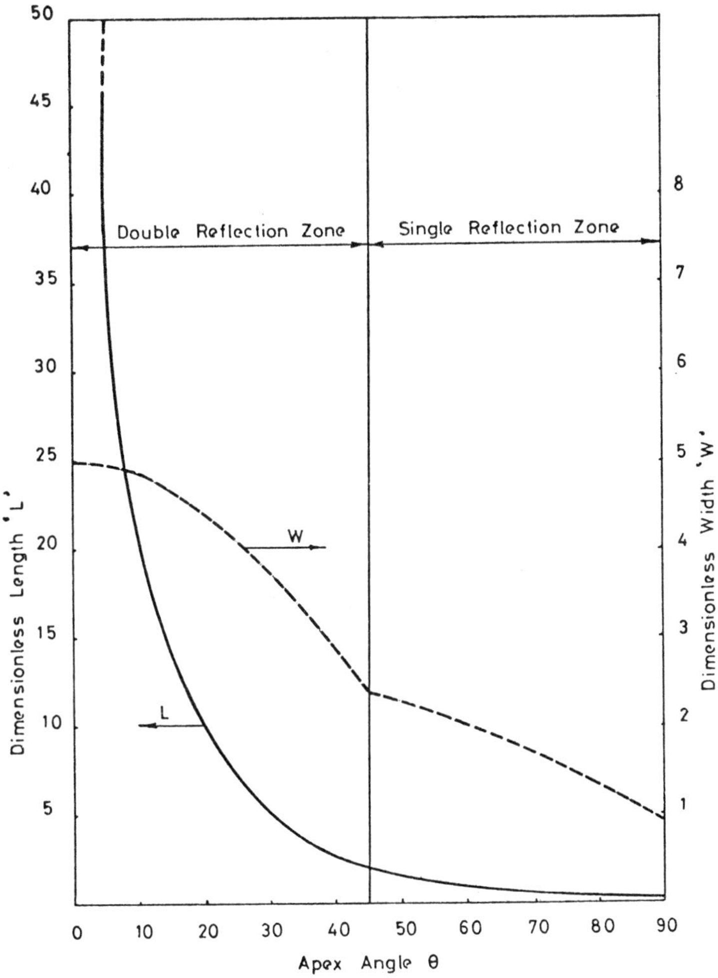

Figure 2 Geometrical Parameters L & W of "SFPB" With Flat Absorber

Figure 2 shows the value of the dimensionless geometrical parameters, referred to absorber width for both double reflection "SFPB" ($\theta < 45°$), and single reflection "SFPB" with $45° < \theta < \pi/2$. The deduced geometry of "SFPB" heaters shows a short length of reflecting surface "L" compared to other stationary concentrators [1], which reduce cost and weight. In addition, they add the advantage that, the deduced configurations help to reduce the convection loss from the absorber surface.*

EFFECT OF BEAM DEVIATION

The behaviour of any "SFPB" heater is deeply dependent on the direction of beam radiation impinging on its aperture. If the collector is subjected to normal beam radiation, the absorber will receive radiation directly, through one reflection from length "L_1" of both reflecting surfaces and via two reflections from a length "$L - L_1$" of each reflecting surface. In case the beam direction is deviated on angle "ϕ" from the normal, the behaviour of the "SFPB" to beam radiation is completely changed, depending on the value of beam deviation "ϕ" relative to apex angle "θ". The channels of "SFPB" can accept beam deviation up to "$\phi = \pi/2 - \theta$" at which all incident radiation is reflected back outside the collector enclosure. This acceptance range is divided into zones; each has its own characteristic behaviour, which are discussed in details in [5], and are given briefly here for both double and single reflection "SFPB" heaters to ensure the continuation of the subject.

Beam Deviation Zones Of Double Reflection "SFPB"

The behaviour of these channels differ according to four different beam deviation zones, which are as follows.

Zone $0.0 \leqslant \phi < \phi_1$: Figure 3 shows the change in the reflection behaviour of the reflecting walls A, B. If the solar beam is slightly deviated by an angle "ϕ" as shown in Figure 3a, the single reflection length of surface A "L_1" will be extended to

$$(L_1)^{A}_{\phi}$$

given by:

$$(L_1)^{A}_{\phi} = \cos(\theta - \phi)/\sin\left(\frac{\theta}{2} - \phi\right) \quad (7)$$

and hence

$$(L_2)^{A}_{\phi} = L - (L_1)^{A}_{\phi} \quad (8)$$

---

* The thermal analysis of such collectors will be presented in a separate paper.

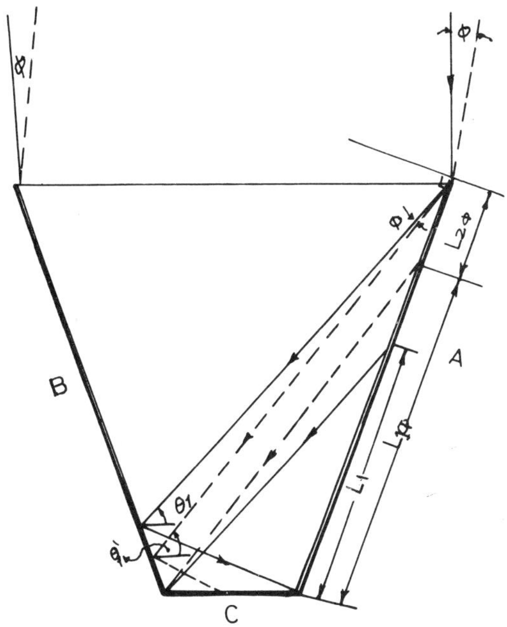

Figure 3a Behaviour of Surface "A"

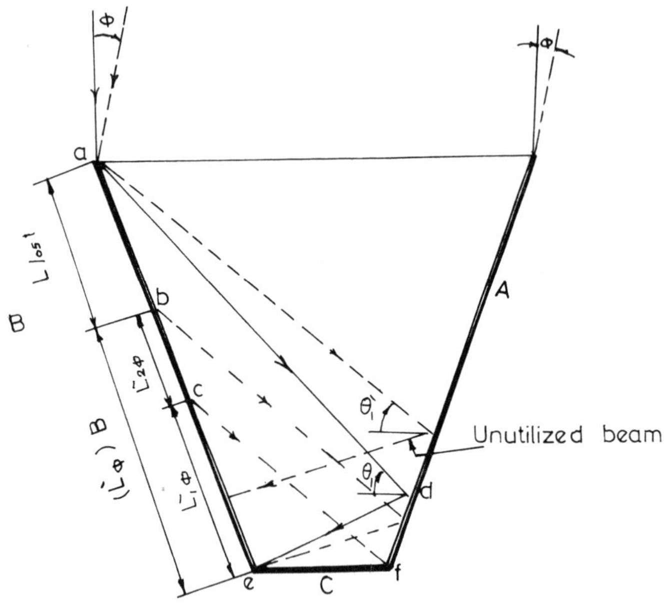

Figure 3b Behaviour of Surface "B"

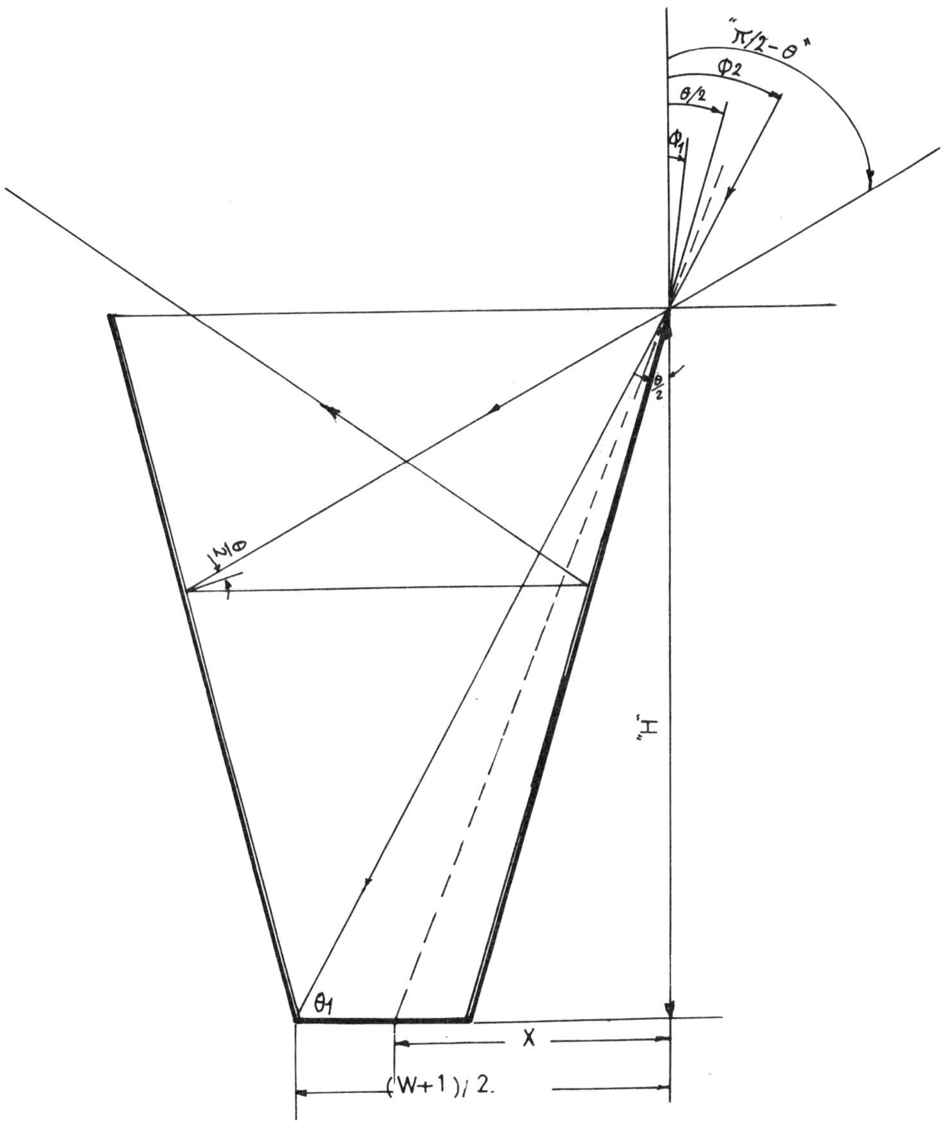

Figure 3c Beam Deviation Zones

As the solar beam deviates, as shown in Figure 3b, a part of the incident radiation at the upper part of the surface will not reach the absorber even after two reflections. Thus, both single and double reflection zones become shorter, so as to be:

$$(L)\,{}^{B}_{\phi} = 2\cos\left(\frac{3\theta}{2} + \phi\right) \cdot \cos\left(\frac{\theta}{2}\right) / \sin\left(\frac{\theta}{2} + \phi\right) \tag{9}$$

$$(L_1)\,{}^{B}_{\phi} = \cos(\theta + \phi)/\sin\left(\frac{\theta}{2} + \phi\right) \tag{10}$$

$$(L_2)\,{}^{B}_{\phi} = (L)\,{}^{B}_{\phi} - (L_1)\,{}^{B}_{\phi} \tag{11}$$

An angle $\phi - \phi_1$ will be reached, at which $(L_1\,{}^{A}_{\phi}$ will equate to the total length "L". It is denoted as the first critical angle and is given by:

$$\phi_1 = \tan^{-1}(2H/(W + 1)) + \left(\theta - \frac{\pi}{2}\right) \tag{12}$$

Zone $\phi_1 \leqslant \phi < \theta/2$: All surfaces, A, B, and the absorber, will have the same characteristics of the first zone, with the corresponding effective lengths, but surface A totally reflects its radiation through one reflection. Until ϕ reaches $\theta/2$, surface "A" will be completely shaded.

Zone $\theta/2 \leqslant \phi < \phi 2$: In this zone, surface A is completely shaded, the absorber is partially illuminated in width $(Li)_{\phi}$, and surface "B" has the same characteristics as much as $\phi < \pi/2 - 2\theta$,

where

$$\phi_2 = \tan^{-1}((W + 1)/2H) \tag{13}$$

$$(Li)_{\phi} = \frac{(W + 1)}{2} - L\cos\frac{\theta}{2} \cdot \tan\phi \tag{14}$$

Zone $\phi_2 \leqslant \phi < \pi/2 - \theta$: Only surface B is illuminated and reflects radiation to the absorber via one and two reflections as much as $\phi < \pi/2 - 2\theta$. If "ϕ" exceeds this value, it reflects radiation only through one reflection until ϕ reaches "$\pi/2 - \theta$" when it is no longer a collector.

Beam Deviation Zones Of Single Reflection SFPB

Only two deviation zones can be defined, according to Figure 4, as follows.

$0.0 \leqslant \phi < \theta/2$: The absorber receives radiation directly and via one reflection, both from the total length of surface A, and $(L_1)\,{}^{B}_{\phi}$, defined by equation (10).

$\theta/2 \leqslant \phi < \phi\ 2 = \theta$: Surface A is totally shaded and surface B is only effective if $\theta < \pi/3$. The absorber is partially illuminated on a length (Li_{ϕ}), given by equation (14).

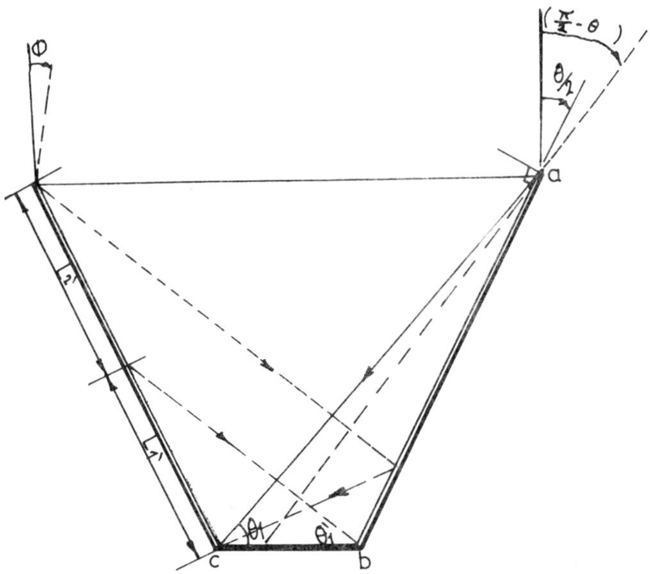

Figure 4 Deviation of Single Reflection Channel

The critical beam deviation angles ($\emptyset$) for different "SFPB" with a flat absorber are computed and shown in Table 2.

DIRECT BEAM CONCENTRATION GAIN FACTOR

The direct beam concentration gain factor is the ratio of concentrated heat intensity on the absorber surface to the intensity of beam radiation normal to "SFPB" surface, therefore:

$$C = q_{abs}/I_s \tag{15}$$

The absorber heat is the sum of the radiation received either directly or from surfaces A, B, concerning the solar beam intensities on each surface to be as follows:

$$I_s = I_N \cdot \cos\theta i \tag{16}$$

and $$I_A = I_B = I_s \sin\frac{\theta}{2} \tag{17}$$

Where terms are defined in the nomenclature and the effective lengths are referred to in equations (7-14). The concentration gain factor "C" is therefore deduced as a function of apex angle (θ), beam deviation "$\emptyset$", and surface reflectivity "ρ" as follows:

<u>Double Reflection "SFPB"</u>

$0.0 \leqslant \emptyset < \emptyset_1$:

$$C_{2_1} = 1. + \rho \sin\frac{\theta}{2}\left[\frac{\cos(\theta - \emptyset)}{\sin(\frac{\theta}{2} - \emptyset)} + \frac{\cos(\theta + \emptyset)}{\sin(\frac{\theta}{2} + \emptyset)}\right]$$

$$+ \rho^2 \sin\frac{\theta}{2}\left[L - \frac{\cos(\theta - \emptyset)}{\sin(\frac{\theta}{2} - \emptyset)} + VK \cdot \frac{\cos(2\theta + \emptyset)}{\sin(\frac{\theta}{2} + \emptyset)}\right] \tag{18}$$

$\emptyset_1 \leqslant \emptyset < \theta/2$:

$$C_{2_2} = 1. + \rho \sin\frac{\theta}{2}\left[L + \frac{\cos(\theta + \emptyset)}{\sin(\frac{\theta}{2} + \emptyset)}\right] + VK \cdot \rho^2 \sin\frac{\theta}{2}$$

$$\left[\frac{\cos(2\theta + \emptyset)}{\sin(\frac{\theta}{2} + \emptyset)}\right] \tag{19}$$

TABLE 1

CRITICAL BEAM DEVIATION ANGLES OF "SFPB" WITH FLAT ABSORBERS

Critical "Φ"°	Double Reflection "SFPB"				Single Reflection "SFPB"			
	θ = 10°	20°	30°	40°	50°	60°	70°	80°
ϕ_1	2.43	4.37	5.1	3.1	-	-	-	-
ϕ_2	7.58	15.6	24.9	36.93	-	-	-	-
θ/2	5.	10.	15.	20.	25	30	35	40
$(\frac{\pi}{2} - 2\theta)$	70	50	30	10	-	-	-	-
$(\frac{\pi}{2} - \theta)$	80	70	60	50	40	30	20	10
$(\phi_2) = 0$ single	-	-	-	-	50	60	70	80

TABLE 2

VARIATION OF BEAM CONCENTRATION GAIN OVER THE DAY = .9

	θ / τ	10	20	30	40	50	60	70	80
JUNE	8	0.38	2.61	2.07	2.16	1.83	1.62	1.37	1.1
	9	1.64	3.44	3.02	2.35	2.00	1.77	1.51	1.22
	10	4.15	3.86	3.32	2.5	2.14	1.88	1.6	1.3
	11	3.87	3.69	3.21	2.45	2.09	1.85	1.57	1.27
	12	3.79	3.62	3.16	2.43	2.07	1.83	1.56	1.26
MARCH	8	3.89	3.70	3.22	2.45	2.1	1.85	1.57	1.28
	9	4.18	3.87	3.33	2.50	2.14	1.89	1.61	1.3
	10	4.23	3.9	3.35	2.51	2.15	1.89	1.61	1.31
	11	4.68	3.82	3.3	2.49	2.13	1.88	1.6	1.3
	12	4.13	3.82	3.32	2.5	2.14	1.88	1.6	1.3
DECEMBER	8	0.46	1.06	2.76	2.2	1.88	1.66	1.41	1.13
	9	3.96	3.75	3.25	2.47	2.11	1.86	1.58	1.29
	10	3.68	3.56	3.12	2.4	2.05	1.81	1.54	1.25
	11	1.46	3.41	3.00	2.34	1.99	1.76	1.5	1.22
	12	1.02	3.35	2.96	2.31	1.97	1.75	1.49	1.20

$- \frac{\theta}{2} \leqslant \phi < \phi 2$

$$C_{2_3} = [\frac{(W + 1)}{2} - L \cdot \cos\frac{\theta}{2} \cdot \tan\phi] + \rho \sin\frac{\theta}{2} [\frac{\cos(\theta + \phi)}{\sin(\frac{\theta}{2} + \phi)}]$$

$$+ VK \cdot \rho^2 \cdot \sin\frac{\theta}{2} [\frac{\cos(2\theta + \phi)}{\sin(\frac{\theta}{2} + \phi)}] \quad (20)$$

$\phi_2 \leqslant \phi$ " $\frac{\P}{2} - \theta$ "

$$C_{2_4} = \rho \sin\frac{\theta}{2} [\frac{\cos(\theta + \phi)}{\sin(\frac{\theta}{2} + \phi)}] + VK \cdot \rho^2 \sin\frac{\theta}{2} [\frac{\cos(2\theta + \phi)}{\sin(\frac{\theta}{2} + \phi)}] \quad (21)$$

where VK = 1.0 unless $\phi >$ "$\frac{\P}{2} - 2\theta$"

Single Reflection "SFPB"

$0.0 \leqslant \phi < \theta/2$:

$$C_{11} = 1.0 + \rho \sin\frac{\theta}{2} \cdot [L + VK \frac{\cos(\theta + \phi)}{\sin(\frac{\theta}{2} + \phi)}] \quad (22)$$

$\frac{\theta}{2} \leqslant \phi < \phi_2$:

$$C_{12} = [\frac{W + 1}{2} - L \cos\frac{\theta}{2} \cdot \tan\phi] + VK \cdot \rho \sin\frac{\theta}{2} [\frac{\cos(\theta + \phi)}{\sin(\frac{\theta}{2} + \phi)}] \quad (23)$$

where Vk = 1.0, unless "$\phi \geqslant \frac{\P}{2} - \theta$".

Concentration gain factors are computed for a wide range of (θ, ϕ, ρ), and shown in Figures 5 and 6 for double and single reflection "SFPB" channels respectively, while Figure 7 shows the change of the effective transmission factors "E.T.F." with the beam deviation. It is calculated by "E.T.F. = C/W" as described in [5].

YEAR-ROUND PERFORMANCE OF "SFPB"

The performance of "SFPB" is mainly dependent on its geometry, quality of the reflecting surface, beam deviation angles, and the available solar radiation on its surface. The "SFPB" heaters are to be oriented with the channels elongated east-west and the heater tilted to an angle "S", which is adjusted monthly according to the following equation:

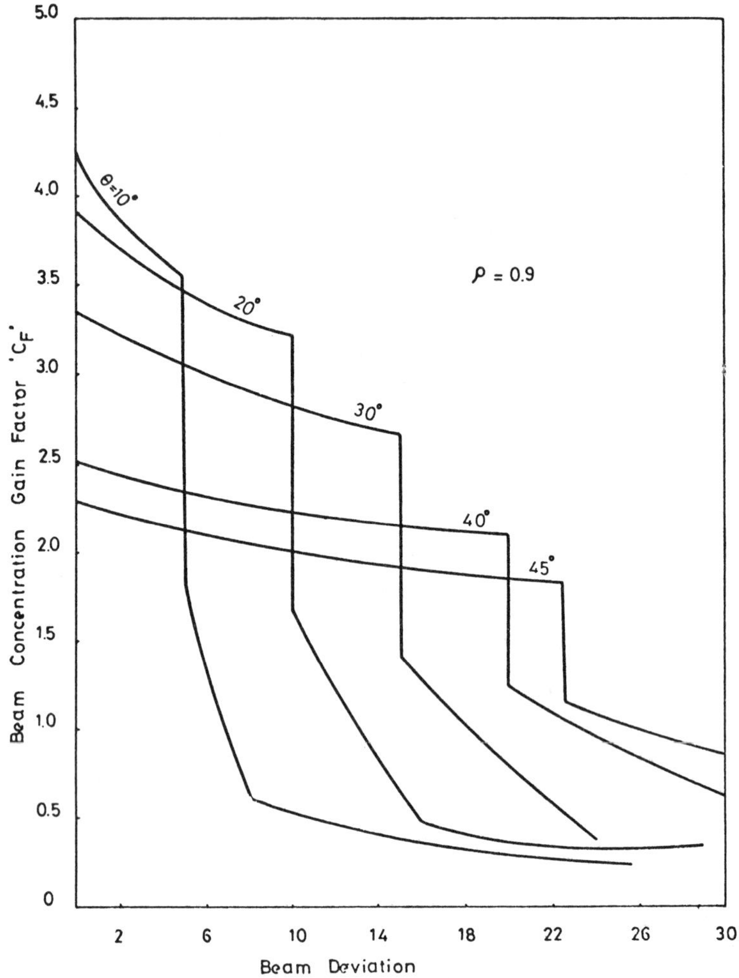

Figure 5 Concentration Gain Factor of Double Reflection Channels With Flat Absorber

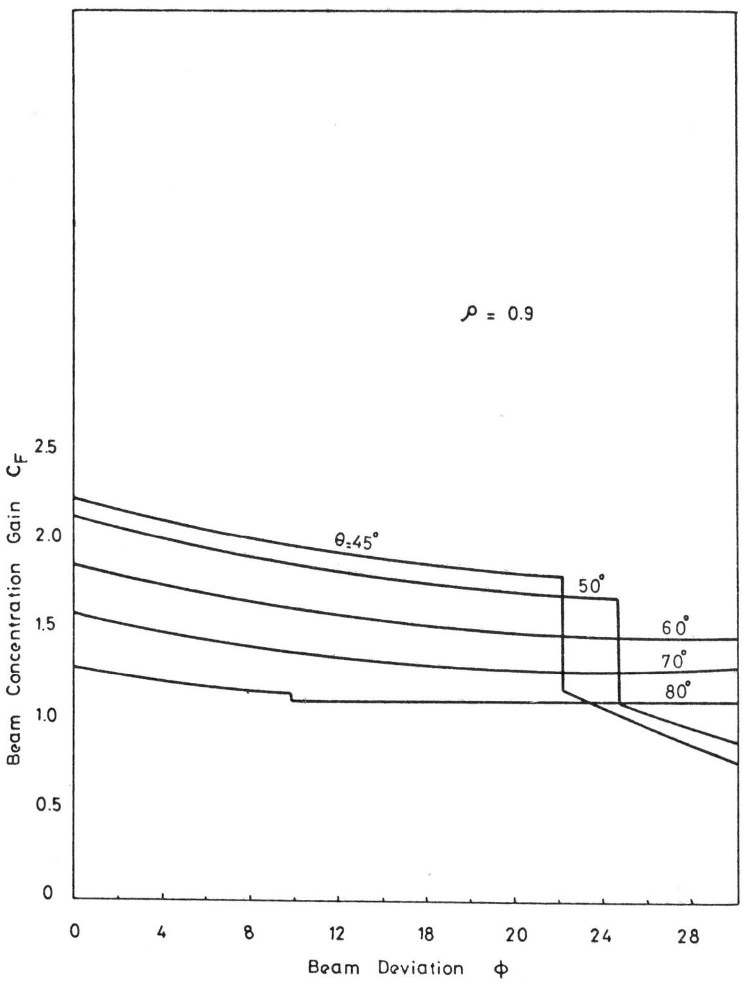

Figure 6 Beam Concentration Gain Factor of Single Reflection "SFPB" Channels With Flat Absorber

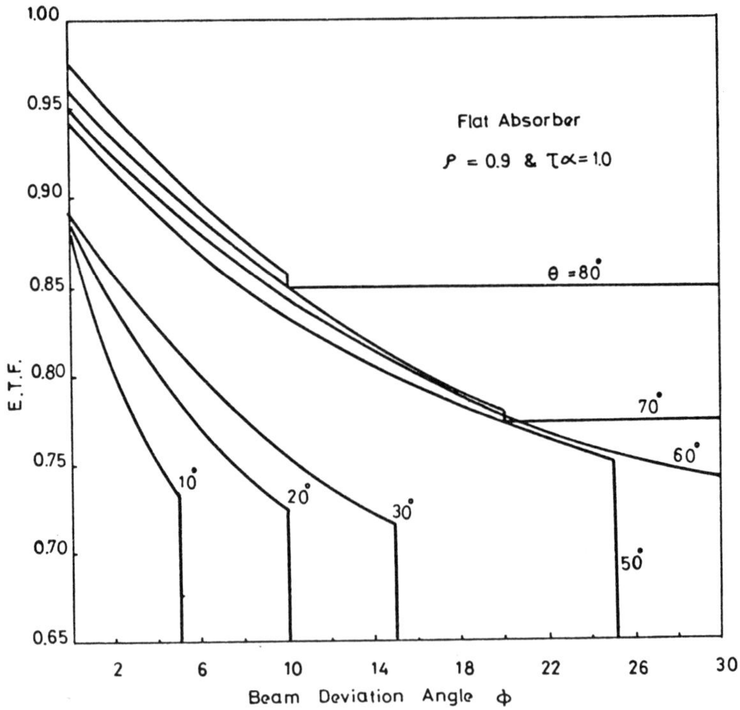

Figure 7 Effect of Beam Deviation Angle on Effective Transmission Factor (E.T.F.)

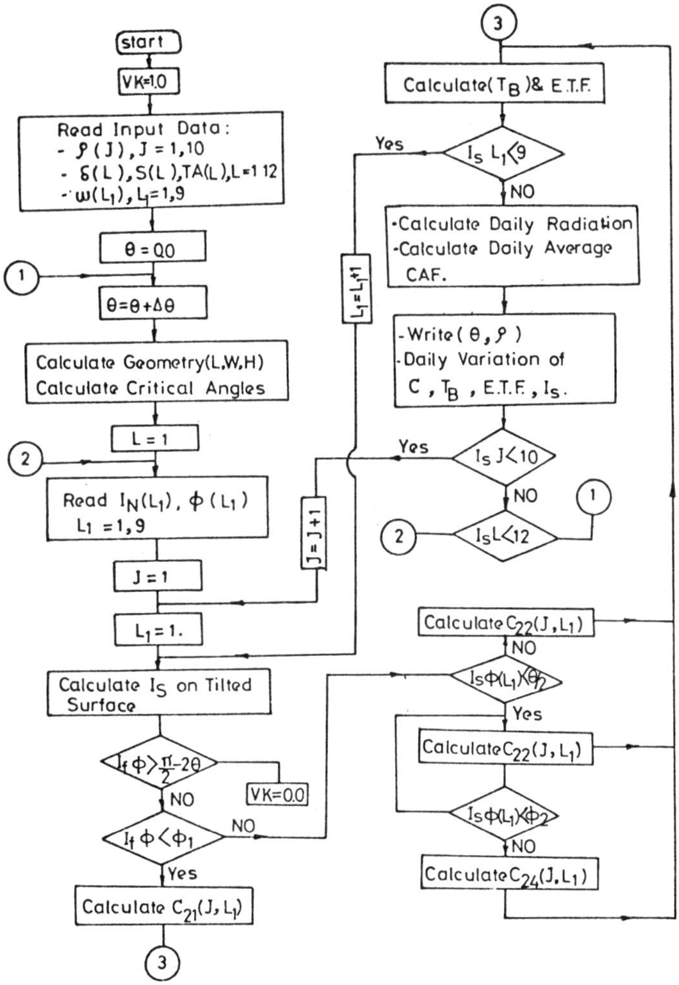

Figure 8 Flow Chart of Computer Program

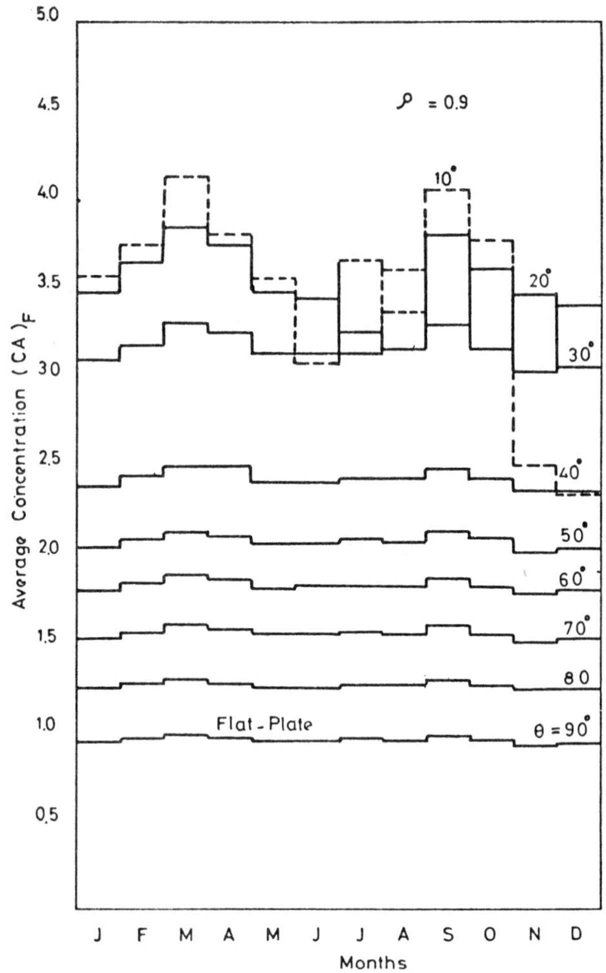

Figure 9 Daily Average Concentration Gain Factor for Different Channels With Flat Absorber

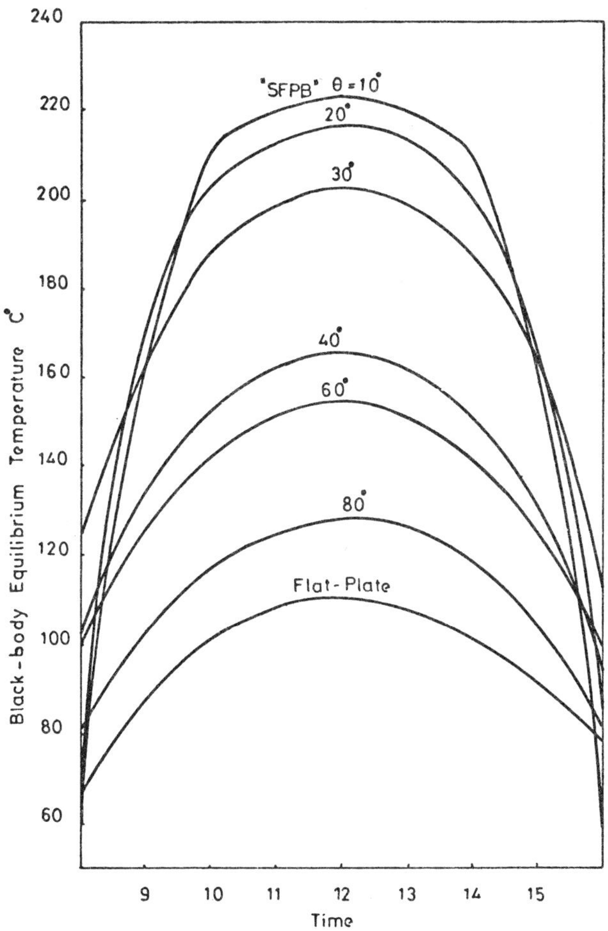

Figure 10 Attainable Temperature Variations of "SFPB" Channels With Flat Absorber

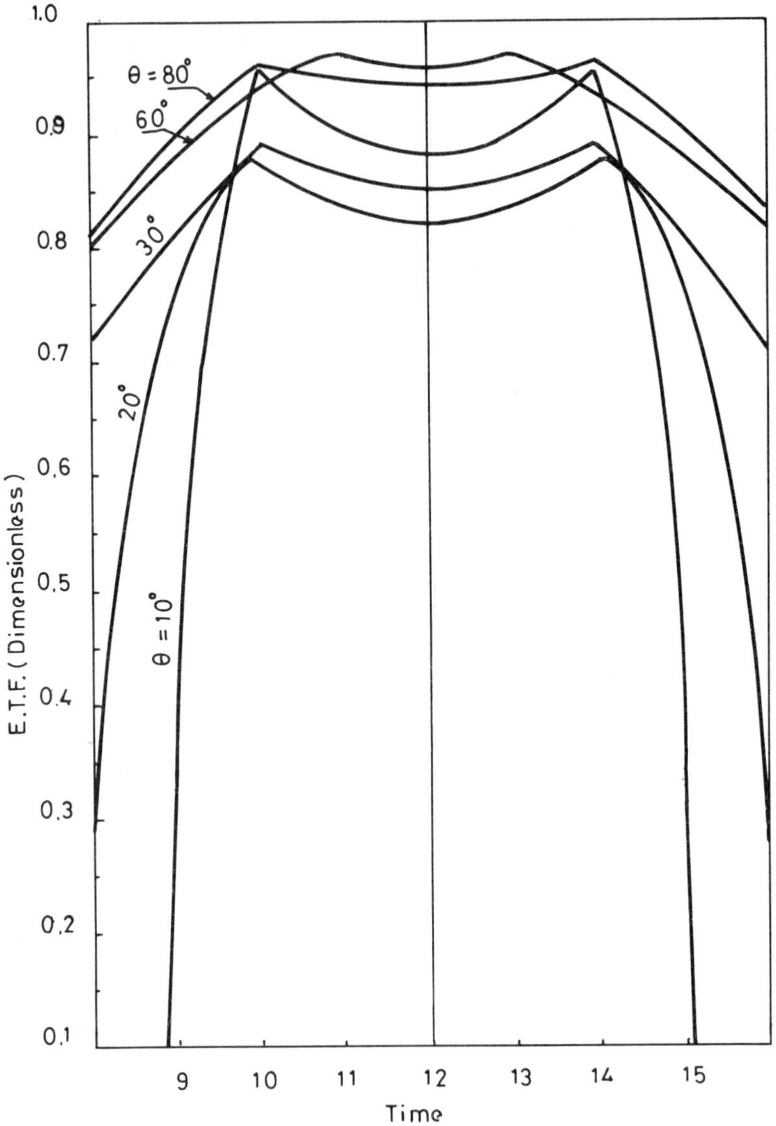

Figure 11 Attainable E.T.F. of Different "SFPB" Channels With Flat Absorber

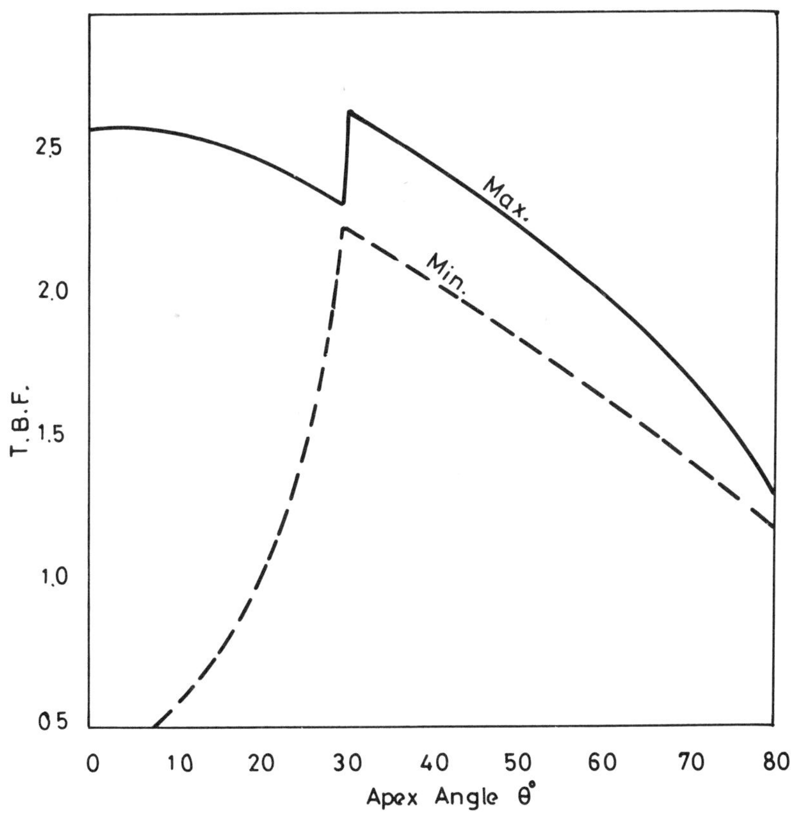

Figure 12 Variation of TBF of "SFPB" With Apex Angles of The Channel

$$S = (\gamma s + Z_n)/2 \qquad (24)$$

where the terms are defined in the nomenclature.

Using the solar radiation data at Cairo "at. 30°2'N", the proposed tilting routine, and the deduced equations for concentration gain and effective transmission, the year-round performance is computed. The maximum possible black body surface temperatures are also calculated within the computer program as follows:

$$T = \left(\frac{I_s \cdot C}{\sigma} + T_a^4\right)^{\frac{1}{4}} \qquad (25)$$

The flow chart of the computer program is shown in Figure 8. The concentration gain "C" of "SFPB" with 10° 0 90°, 0.9 is shown in Table 2, throughout the day; while Figure 9 shows the daily average concentrations over the year.

The attainable temperatures for different "SFPB" channels are found to be in the range of 160° to 200°C for double reflection channels, and higher than 100°C for single reflection channels. Figure 10 shows an example of the computed results during "June", while Figure 11 describes the variation of the E.T.F. over the day which is almost higher than 75%, as shown.

Temperature Boosting Factor "T.B.F."

The "T.B.F." is a factor defined as the ratio of the attainable temperature rise across a "SFPB" heater to that obtained by a flat plate collector, thus:

$$T.B.F. = \frac{T_{SFPB} - T_a}{T_{F.P} - T_{a.}} \qquad (26)$$

The values of T.B.F. are computed over the year for different "SFPB" channels. Figure 12 shows the limits between w, in which "T.B.F." are fluctuating for different channels during June. It can be seen that this variation is so severe that "$\theta < 20°$" exceeds the double for the double reflection channels and varies between 1.5 to 2.0 for single reflection channels.

REFERENCES

1. Winston, R., Principles of Solar Concentrators of Novel Design, Solar Energy 16, 1974.

2. Holland, D.G., A Concentrator For Thin Film Solar Cells, Solar Energy Vol. B, No. 2, 1971.

3. Winston, R., Principles of Cylindrical Concentrators For Solar Energy, Solar Energy 17, 1979.

4. Shapiro, M.M., Non-Focussing Solar Concentrators Of Easy Manufacture, Solar Energy 19, No. 2., 1977.

5. Sakr, I.A. and Hegazi, A.I., A Channeled Solar Flat Plate Booster, International Solar Energy Conference, New Delhi, India, 1978.

NOMENCLATURE

C	concentration gain factor, dimensionless
C_{2m}	concentration gain factor of double reflection channel at beam deviation zone No. m
I_s	intensity of solar radiation on the surface of "SFPB" heater, watt/m^2
H	dimensionless height of the "SFPB" channel
L	dimensionless length of the "SFPB" channel
$(L_1)^A_\emptyset$	the single reflection length on surface A at beam deviation = $\emptyset$, dimensionless
$(L)^B_\emptyset$	the total effective length of surface B at beam deviation $\emptyset$, dimensionless
$(L_i)_\emptyset$	the illuminated width of the absorber at beam deviation $\emptyset$, dimensionless
n	number of reflections, dimensionless
Ta	ambient air temperature, $^\circ K$
T	absorber surface temperature, $^\circ K$
W	dimensionless width of "SFPB" channel
Z_n	zenith angle of the solar beam at noon, degree
$\emptyset$	beam deviation angle, degree
σ	Stefan-Boltzmann constant
ρ	surface reflectivity

A BLACK LIQUID FLAT PLATE COLLECTOR
FOR ECONOMICAL APPLICATIONS

R.S. DARYAN AND P.K.C. PILLAI
Centre of Energy Studies
Department of Physics,
Indian Institute of Technology,
New Delhi-110029 (INDIA)

ABSTRACT:

A black liquid flat plate solar energy collector for economical applications has been designed and tested. A rectangular pattern of thin transparent glass tubing is used as the collector fluid container, in which a highly absorbent 'black' liquid flows and directly absorbs the solar energy. Here, the liquid is the hottest substance in the collector and does not require any metal in the design. The above structure is enclosed in a thermally insulated box with a two glass cover. The collector efficiency with different black liquids, in which black ingredient and working fluid are used in different proportion, were calculated and a comparison with other conventional type flat plate collector were made.

INTRODUCTION:

Solar energy collector is the heart of all solar energy applications. Flat-plate collectors of several design have been employed for applications like water heating, building air-conditioning (heating and cooling) crop drying, pumping of water, etc, where temperature requirement is below 100°C. Here the cost of the flat plate collectors is a major economic factor which prevents the wide range use of solar energy for photothermal applications. Economic studies suggest [1,2,3] that flat plate collector costing between $ 2 and $ 4 per feet are required to make solar energy competitive with other heating and cooling systems.

An attempt has been made here to develop a 'liquid' solar collector' which does not employ a blackened metal plate thereby reducing the cost of collector and simultaneously fulfilling the temperature requirements. In this collector, both direct and diffused solar radiation can be utilized thereby making such a system appropriate over a wide range of geographical areas and varying climating conditions. The collector is simple to design and can be used in rural areas as well.

Inspite of several advantages of this noval type of collector which meets both the cost and temperature requirements coupled with simplicity of fabrication, there is a very little li-

Fig.1: COLLECTOR DESIGN

terature [4,5,6] available on such collectors. Miller [4]was the first to propose a collector made of overlapped, partially blackened glass plate as air heater. Kusuda [5] in Japan used flat plate collectors made out of glass as solar water heaters. In 1975, Minardi et al [6] tested a collector made out of transparent plastic tubing in a spiral pattern and suggested to use glass tubing also. The main disadvantages with plastic are that it becomes opaque with exposure to solar radiation, it starts deforming at high temperatures, mainly under the no flow conditions, besides plastic may not be compatible with all black liquids. Keeping all these factors in mind we designed a flat-plate collector, rectangular pattern, of thin transparent glass tube (Figure 1). The collector was tested with two black liquids.

EXPERIMENTAL APPARATUS:

The experimental set up for the liquid collectors, which is a rectangular pattern of transparent glass tube of inner diameter 10 mm and of wall thickness 1 mm, is shown in Figure 1.In the glass structure, there is a 2 mm gap between the two consequent tubes along the length which are joined by a plastic tube using an epoxy resin, which can withstand a temperature upto 200°C. This glass structure is fixed onto a wooden frame which is in turn fixed into an insulated wooden box, 76 cm × 76 cm × 15 cm. The bottom and sides of the collector were insulated with 10 cm and 5 cm of glass wool respectively. The covers of 4 mm thickness and are separated by an air gap of 2.5 cm between the collector, the first cover and the second cover respectively. The light transmittance of the cover glass was approximately 94% in the visible region. The cover assembly as a whole has a light transmittance of about 88% in the solar region.

This whole collector assembly is fixed onto a angle iron stand with an arrangement to change the angle of tilt of the collector. A low wattage electric water pump was used to circulate the working fluid into the unit. Sensitive thermometers with least count of 0.2°C were employed, and a solarimeter was used for the solar radiation intensity on the collector surface. There was no arrangement to measure the plate temperature in the present set up.

To evaluate the overall performance of the collector, the black liquid at the inlet was supplied at a constant temperature. This constant temperature was maintained by an electric heater wrapped over a glass tube at the collector inlet.

BLACK LIQUIDS POTENTIAL CANDIDATE:

The success of the above liquid collector depends upon the black liquid itself. There are a number of black ingredients

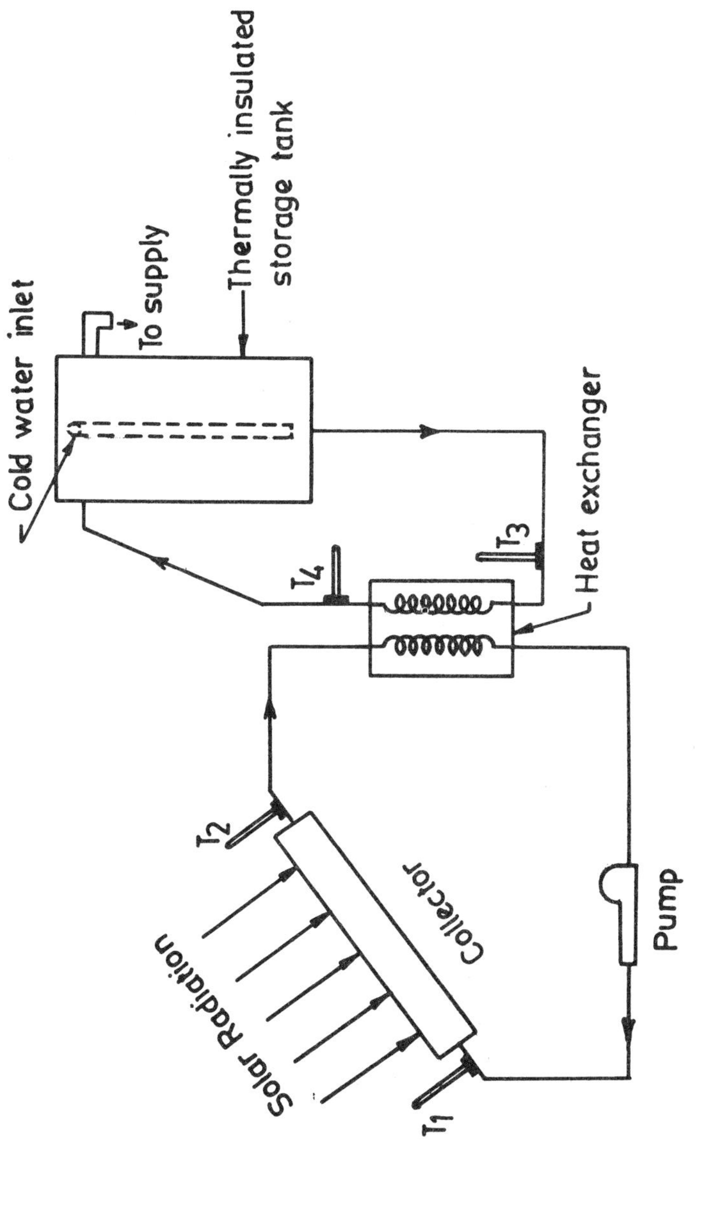

Fig. 2 SCHEMATIC DIAGRAM OF EXPERIMENTAL SETUP

which can be used with same working fluid to make a black liquid. Before we consider a liquid for testing it should meet some essential requirements such as spectral absorptivity and emissivity, specific heat, boiling and freezing point, effect on the collector structure, long term chemical stability and lastly the cost which is the main consideration in designing the collector. In the present study we have used two black ingredient with water and ethylene glycol as the working fluid respectively. In case of water we use a bblack colour, powdered, as the black ingredient, which is easily and cheaply available in the market. There were two disadvantagesiin using the black colour and water as the working fluid. The first was that the black colour (chemically stable) stained the wall of the collector glass tube and the second was that the water start boiling out around 98°C and so we could not measure the stagnant temperature of the collector. We represent this black liquid as (w + c) in the figures. In the second case we used black 'India-Ink' as the black ingredient with ethylene glycol as working fluid. We represent this black 'India Ink' in liquid as (I + E) in figures. This ink is chemically stable and inexpensive and it does not stain the wall of the collector used with ethylene glycol. The boiling temperature of ethyl glycol is 196°C and freezing temperature is -30°C. So this liquid is suitable to work at high temperatures.

There are some selective materials such as black antimony sulphide (Sb_2S_3) which can be used to increase the efficiency of the collector. The experimental work on the above selective liquids is in progress in our research laboratory. This selective liquids was first suggested by Minardi [6].

EXPERIMENTAL PROCEDURE:

All the experiments were conducted under clear sky conditions to eliminate variation in solar radiation intensity. In Figure 3 is shown the variation of solar radiation for clear sky, at Delhi, lattitude 23.5° North, on the surface of the collector inclined at an angle of 15° facing South, with time of the day. All experiments were conducted during the month of May, 1973. The flow rate was manually kept constant. We have calculated the overall efficiency of the collector with different black liquids and also the collector's efficiency at different inlet temperatures. This is shown in Figure 4.

RESULTS AND DISCUSSION:

The efficiency of a flat plate collector is given by the equation suggested by Whiller [7].

$$\eta = F' [f - \frac{U_L (T_f - T_a)}{Q_{ts}}] \tag{1}$$

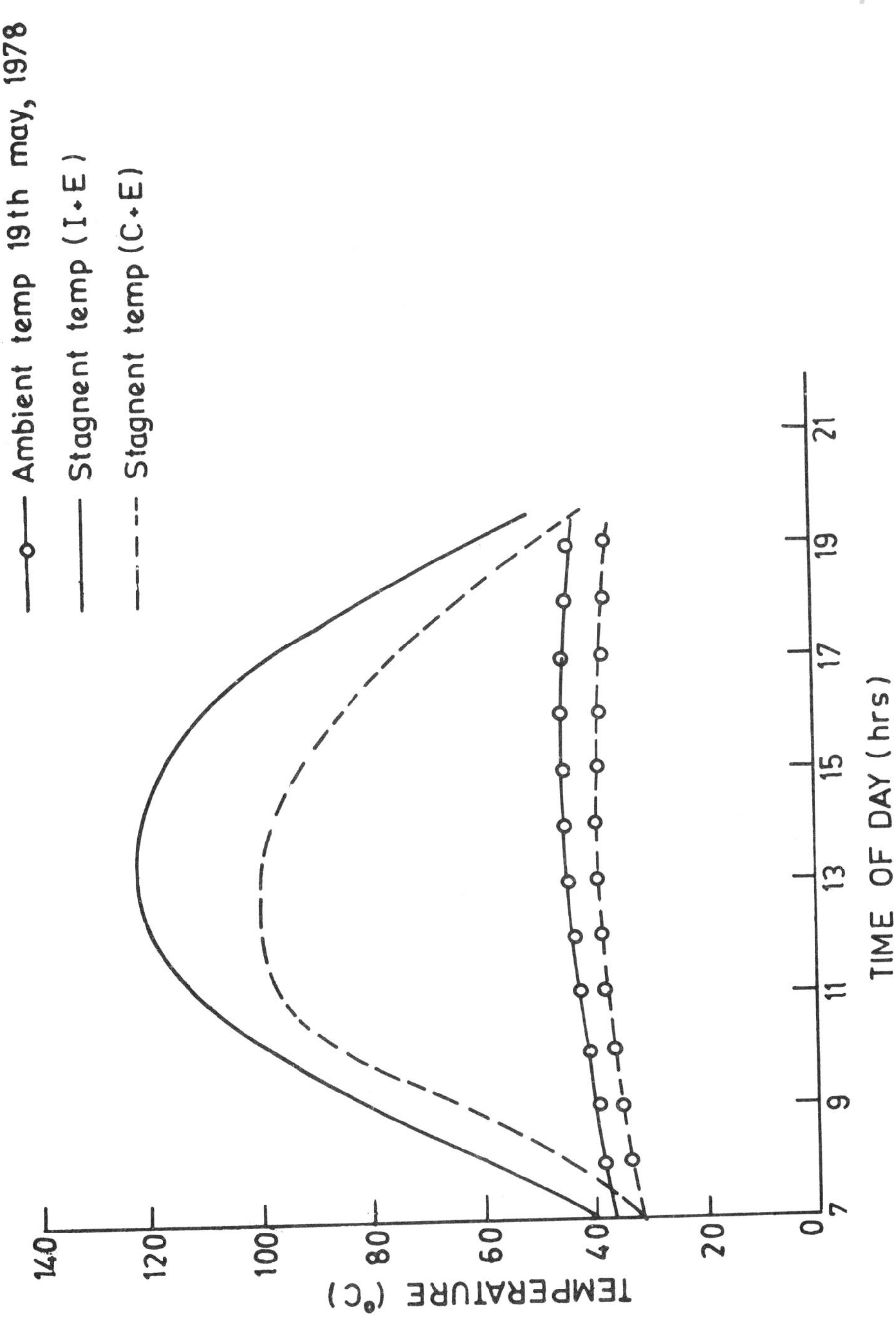

Fig. 3

TEMPERATURE VARIATION WITH DIFFERENT BLACK LIQUIDS IN THE COLLECTOR MAY, 1978

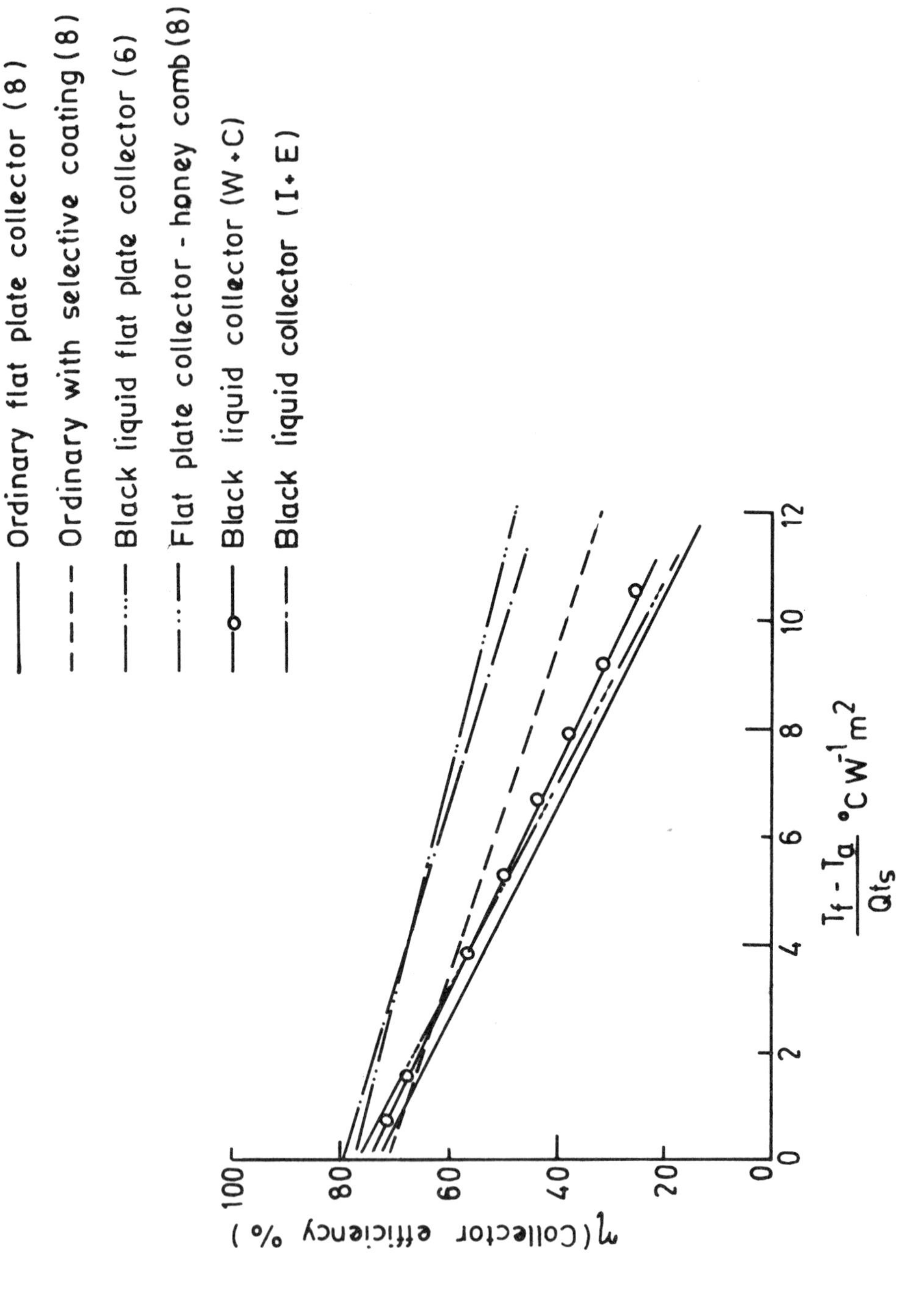

Fig. 4. PERFORMANCE OF THE BLACK LIQUID FLAT PLATE SOLAR

Here F', the plate efficiency factor, should be one [6] since the radiation is absorbed directly into the liquid. In terms of mass flow rate of the working fluid, the collector efficiency at steady state condition can be calculated from experimental measurements, i.e.,

$$\eta = \frac{m\ c_p(T - T_1)}{Q_{ts}} \tag{2}$$

Using the relation (2) we have calculated the instantanious efficiency of the collector and plotted it against $\frac{T_f - T_a}{Q_{ts}}$ (Fig. 4)

To calculate the average efficiency of the collector the inlet, outlet temperature and solar radiations were recorded on the collector surface ataan interval of 15 minutes. The maximum collector efficiency with two glass cover and the reflection by the glass tube of the collector structure was approximately 76% .

The performance of our liquid collector with water as working fluid (w + c) is comparable to the black liquid collector tested by Minardi [6].

It has been observed that the performance of the liquid (I+ E) collector with ethylene glycol as the working fluid was extremely good particularly at high temperature. A performance comparison of the above collector with different working fluids with other conventional type collector, with and without Honeycomb, is shown in figure 4.

From Figure 4 it is apparent that a black liquid (w + c) collector is equivalent to a conventional flat plate collector with selective surface and one glazing whereas black liquid (I + E) collector was comparable to a conventional flat plate collector with honey comb structure and one glazing.

CONCLUSION:

From Fig. 4 we observe that the black liquid collector, we have developed performed well as compared with conventional flat plate types and the one with black liquid collector. The performance of the collector in winter, may be improved by using an anti-reflection coating on the cover system, by using selective liquid and booster mirrors.

The black liquid flat plate collector differs from the conventional type flat plate collectors in the following ways.

1. In this type of collector solar radiation is observed directly by the liquid and does not requires any metal.

II The distinguish feature of this collector is the spiral

flow of the liquid which provides self insulating that is, cold liquid enters the collector from the outers edge and moves towards the centre of the collector along the flow channel. Therefore, as the liquid flows from the outer peripheral ring inwards its temperature gradually rises with the exposure to solar radiation. Under this condition the centre where the liquid leaves the collector will be the hottest point. Thus the radial heat losses from the centre is partially reobsorbed by the collector surface. Also due to the spiral flow arrangement the entire collector surface area is not subjected to high temperature, as in the case of conventional flat plate collector where the plate is at a uniform high temperature. Therefore the heat transfer losses through the edges and the heat transfer losses through the cover system by conduction and convection are reduced.

III A major problem in the conventional type flat plate collector is the carrosion of the liquid flow channel and the collector plate. This is eliminated in our collector by the use of glass tubes.

IV As this collector uses no metal in its construction, only glass tubes are used, in addition to insulation and frame, this is relatively cheaper than the metal flat plate collector.

ACKNOWLEDGEMENTS:

We acknowledge the help given by Mr. A.D. Sharma during the course of this study.

NOMENCLATURE:

F'	Plate efficiency factor.
f	Fraction of Q_{ts} absorbed.
Q_{ts}	Total solar radiation K.W/m^2.
T_a	Ambient temperature, ^oC.
T_f	Average collector fluid temperature, ^oC
T_i	Inlet collector fluid temperature, ^oC.
T_o	Outlet collector fluid temperature, ^oC.
U_L	Overall heat loss coefficient.
η	Collector efficiency, dimensionless.
m	mass flow rate through the collector K.gm/sec.
C_p	Specific heat of the fluid K.J./^oC.

REFERENCES:

1. R.S. TYBOUT and G.O.G. LOF, Solar House heating, Natural Resources Journal 10, (2) (April 1970).

2. L. Butz, S.A. Klein, W.A. Backman and J.A. Duffie, Digital simulation of solar heating and cooling system. Paper presented at the International Solar Energy Society U.S. Section Annual meeting, NASA-Lewis Research Centre, Cleveland, Ohio (Oct 1973).

3. Solar Energy as a national Energy Resource, NSF/NASA Solar Energy Panel, (December 1972).

4. K.W. Miller, Solar heat trap, Memorandum to office of production, Research and Development. War Production Board (12 July 1973).

5. T. Kuserda, Solar water heating in Japan, Proc. Solar heating and cooling for Buildings Workshop, Washington D.C. (Mar. 1973).

6. John E. Minardi and Henry N. Chuang, Performance of a 'Black' Liquid Flat Plate solar collector, solar energy 17, 1975.

7. A. Whiller, Solar Energy Collector and its utilization for house heating. Sc. D Thesis. Dept of Mech. Engg. M.I.T. (May 1953).

8. A.M. Sayrgh, A comparison among various flat plate collectors with Honey comb Structures, paper presented at the International Solar Energy Congress, New Delhi (Jan. 1978).

PASSIVE SOLAR WALL COLLECTORS

A. Faist and J. B. Gay
Ecole Polytechnique Fédérale de Lausanne
Laboratoire de Physique Théorique
Groupe Energie Solaire
Lausanne-Switzerland

ABSTRACT

The most important role of a traditional wall is to insulate the inside of a building from the outside. Thus the solar gains through the walls are ordinarily very poor.

To overcome this difficulty, we have studied a certain number of front covers to be used in passive solar wall systems (Trombe walls for example).

We studied 19 covers involving glass, translucid plastics and for some of them selective and non-selective metal plates.The measurements have been carried out in real climate on small concrete elements of about 0.5[m^2] (5 ft^2) front surface.A mathematical model has been developped and adapted to each of the 19 cases. By fitting the different measured curves, one gets accurately all the physical characteristics of the different covers. Knowing this, it is then possible to simulate the behaviour of these systems in any desired climate.

The result of our measurements are presented for the 19 differents covers. For four particular covers, we present also the result of a mathematical simulation based on meteorological data of two periods of the 1976-1977 winter in Ecublens near Lausanne (latitude 46.5 ^{o}N).

INTRODUCTION

The main purpose of a traditional wall is to insulate the inside of a building from the outside. For that reason, the main effort is made on the insulation. Thus, the solar gains through a conventional wall are ordinarely very poor.

To overcome this difficulty, we have studied different front covers and absorbing surfaces to be used in connection with thermal storage walls. The main purpose of this paper is to compare the thermal properties of the front covers and surfaces.

ABSORBING SURFACE

FRONT COVER	Rough concrete	Brown painted concrete	Black painted concrete	Black metal plate + concrete	Selective metal plate +
No cover	-	x	-	-	-
Single glass	x	-	-	-	x
Twofold glass	-	x	-	x	x
Single Kalwall	-	x	-	x	x
Twofold Kalwall	-	x	x	-	-
Kalwall + Tedlar	-	x	x	-	x
Twofold Tedlar	-	x	-	x	x
Glass + Tedlar	-	-	x	-	-
Single IR glass	-	x	-	-	-

Fig. 1. Combinations of front covers and absorbing surfaces on which measurements have been made (X)

EXPERIMENTAL SETUP

The measurements have been carried out in real climate, on small concrete elements of 0.5 [m^2] (about 5[ft^2]) front surface and 0.2 [m] (8") thickness, in front of which the different covers where placed. The back surfaces and the sides of the elements were well insulated. Thermocouples were mounted : near the front surface, in the middle and near the rear surface of the concrete elements. Temperatures have been measured automatically each half hour. Simultaneously the air temperature, the solar intensity, the wind velocity and the wind direction were also recorded.

We have studied 19 combinations of different front covers and absorbing surfaces. The Figure 1 summarizes these combinations. For the front cover, we compared glasses and translucid plastics (Kalwall [1] and Tedlar [2]). For the absorbing surface, we have looked at concrete of various colours and at metal plates (selective and non selective). The plates were mounted in front of the concrete elements.

FIRST MEASUREMENTS

A preliminary comparison between the different front covers can be obtained if we look only at the thermal losses of the elements during the night. As the thermal losses by the back and the sides of the elements can be evaluated, one can compute the coefficient of losses of the front surface. These values are summarized in the table 1. Figures 2a to 2e present typical fluctuations of the measured temperatures during clear days.

MATHEMATICAL DESCRIPTION

In order to be able to describe the thermal behaviour of the systems, and also to determine some parameters of the elements, we have to write and to solve the heat transfer equations. This can be achieved numerically using the method of the finite elements.

Type of front cover	Type of absorbing surface	k_f [W/m^2 ^oC]
No cover	concrete	15.04
Single glass	concrete	4.71
Single I.R. glass (3)	Brown painted concrete	2.65
Single Kalwall	" " "	5.02
Twofold glass	" " "	3.10
Twofold Kalwall	" " "	3.15
Twofold Kalwall	Black painted concrete	3.58
Kalwall + Tedlar	Brown painted concrete	3.25
Kalwall + Tedlar	Black painted concrete	3.35
Twofold Tedlar	Brown painted concrete	3.33
Glass + Tedlar	Black painted concrete	3.17
Single glass	Selective metal plate + concrete	2.65
Single Kalwall	Black metal plate + concrete	3.73
Single Kalwall	Selective metal plate + concrete	3.23
Twofold glass	Black metal plate + concrete	2.65
Twofold glass	Selective metal plate + concrete	1.88
Kalwall + Tedlar	Selective metal plate + concrete	1.92
Twofold Tedlar	Black metal plate + concrete	3.58
Twofold Tedlar	Selective metal plate + concrete	1.73

Table 1. Thermal losses of the front surface of the concrete elements (k_f) during the night.

Fig. 2. Typical fluctuations of the temperatures during a sunny day.

—□— Temperature near the front surface °C

—◇— Temperature in the middle of the concrete block

—✱— Temperature near the back surface

—— Outdoor temperature

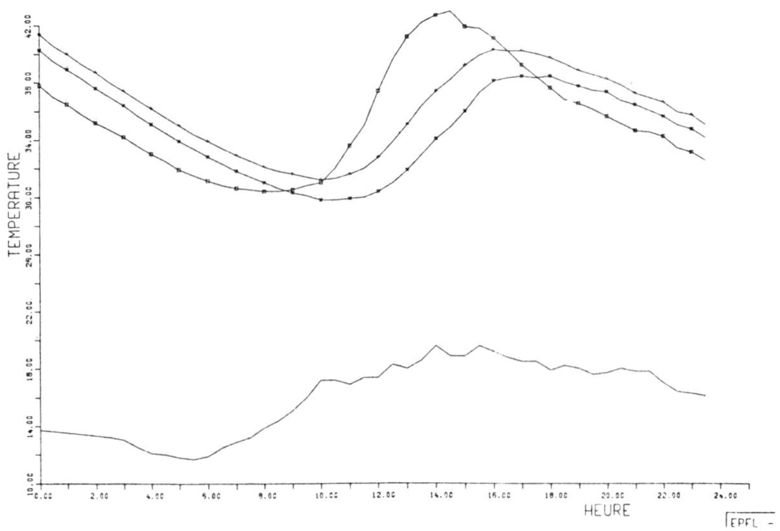

Fig. 2a. Single glass on rough concrete

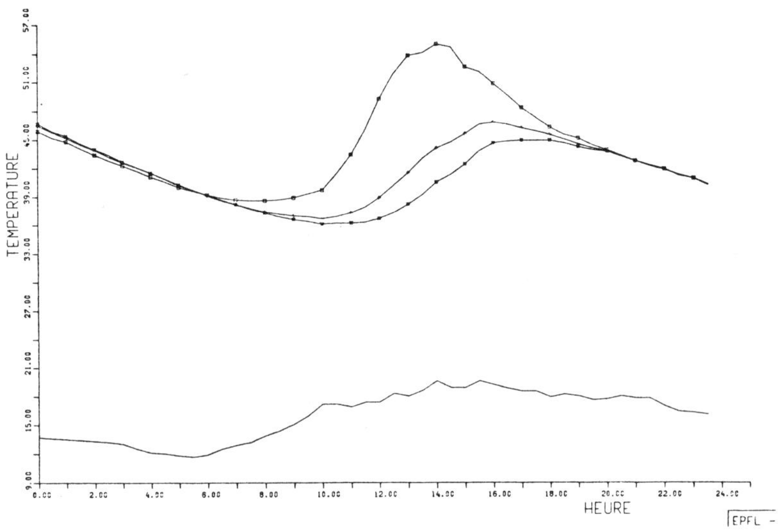

Fig. 2b. Twofold glass on brown painted concrete.

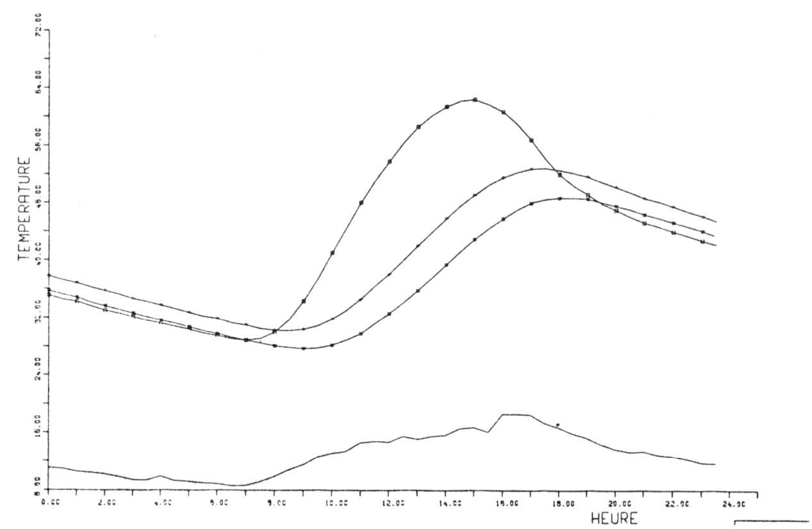

Fig. 2c. Single glass + Tedlar on black painted concrete

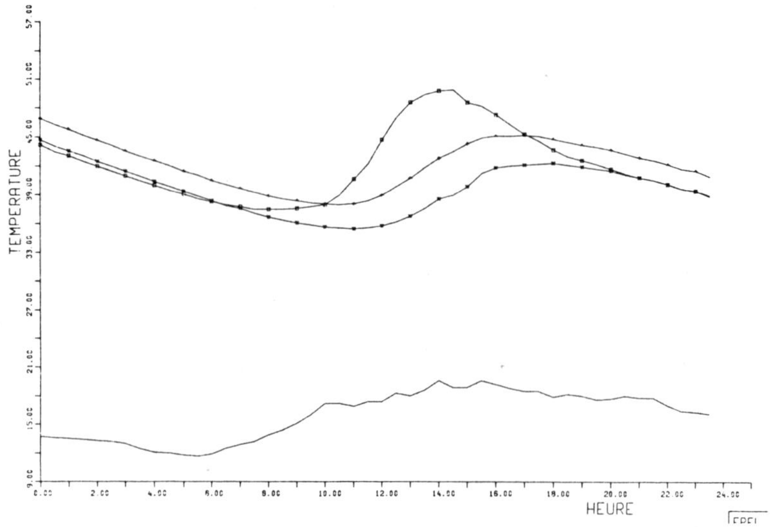

Fig. 2d. Twofold glass and non selective metal plate.

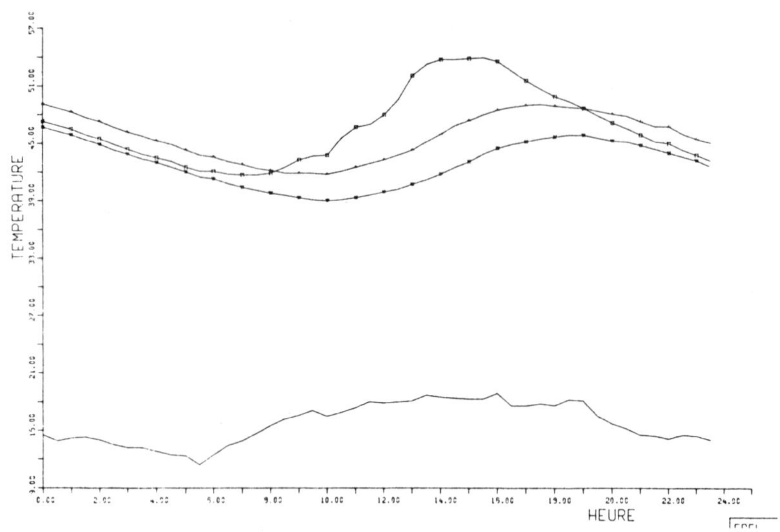

Fig. 2e. Twofold glass and selective metal plate.

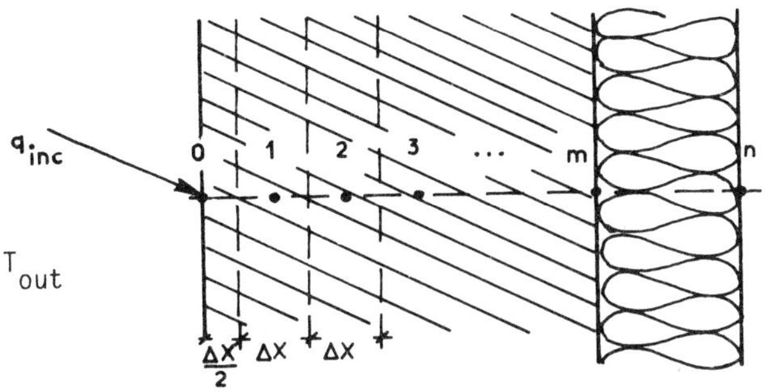

Fig. 3

We can divide the concrete block into m slices having a thickness Δx (see fig. 3 above), in the same way the back insulation can be cut into n-m slices.

The variation of the heat contained in the first slice is described by the following equation

$$\rho_1 c_1 \frac{\Delta x}{2} \frac{T_0^{t+1} - T_0^t}{\Delta t} = \frac{\lambda_1}{\Delta x}\left(T_1^t - T_0^t\right) + h_c\left(T_{out}^t - T_0^t\right) + q_0 \qquad (1)$$

where ρ_1 = densigy
c_1 = specific heat
λ_1 = thermal conduction } of the concrete
h_c = heat transfer coefficient at the surface
q_0 = $\alpha \cdot q_{inc}$ = radiative heat gain
Δt = considered time interval

The following slices are described by :

$$\rho_1 c_1 \Delta x \frac{T_i^{t+1} - T_i^t}{\Delta t} = \frac{\lambda_1}{\Delta x}\left(T_{i-1}^t - T_i^t\right) + \frac{\lambda_1}{\Delta x}\left(T_{i+1}^t - T_i^t\right) \qquad (2)$$

For the last slice, which is in contact with the back insulation, we can write :

$$\rho_1 c_1 \frac{\Delta x}{2} \frac{T_m^{t+1} - T_m^t}{\Delta t} = \frac{\lambda_1}{\Delta x}\left(T_{m-1}^t - T_m^t\right) + \frac{\lambda_2}{\Delta x}\left(T_{m+1} - T_m^t\right) \qquad (3)$$

(λ_2 being the thermal conduction coefficient of the insulating material).

These equations may be rewritten in a more convenient way using the following expressions :

$$M = \frac{\rho_1 c_1 \Delta x^2}{\lambda_1 \Delta t} = \frac{\Delta x^2}{a_1 \Delta t}$$

$$N = \frac{h_c \Delta x}{\lambda_1}$$

One gets finally the following set of equations :

$$T_0^{t+1} = \left(1 - \frac{2+2N}{M}\right) T_0^t + \frac{2N}{M} T_{out}^t + \frac{2}{M} T_1^t + \frac{2\Delta x}{\lambda_1 M} q_0 \qquad (1')$$

$$T_i^{t+1} = \left(1 - \frac{2}{M}\right) T_i^t + \frac{1}{M}\left(T_{i-1}^t + T_{i+1}^t\right) \qquad (2')$$

$$T_m^{t+1} = \left(1 - \frac{2}{M}\right) T_m^t + \frac{2}{M} T_{m-1}^t + \frac{2}{M'} T_{m+1}^t \qquad (3')$$

Knowing the computed evolution of the temperatures as a function of the time, knowing also the measured values, we can parametrize the considered systems. Using a fitting procedure, we can then determine the best values of the parameters.

For a reference block presenting a brown painted concrete surface without any cover, a good fit is found with the following values of the parameters :

λ_1	=	2.58 ± 0.09	[W/m °C]
c_1	=	0.27	[Wh/kg °C]
ρ_1	=	2122 ± 90	[kg/m^3]
α	=	0.705 ± 0.019	
h_c	=	15.0 ± 0.5	[W/m^2 °C]

Using the above values, one can evaluate the phase displacement of the heat peak as function of the depth in the concrete block. Performing this evaluation for a thickness of 0.1 [m] (4 inches), one gets :

$$\eta = \frac{T}{2\pi} \cdot \frac{x}{\delta} = \frac{24}{2\pi} \cdot \frac{0.1}{0.19} = 2.01 \ [h]$$

with : $T = 24$ [h] = periode
$X = 0.1$ [m] = considered position
$\delta = \sqrt{\frac{a \cdot T}{\pi}}$ = depth of penetration

The above value is in agreement with the measurements as it can be observed on figure 2.

The same fitting procedure can be applied to the elements with a front cover. One has just to use additional equations in order to describe the thermal and radiative effects of the covers.

As an example, we can consider a twofold glass cover mounted in front of a concrete element (see fig. 3).

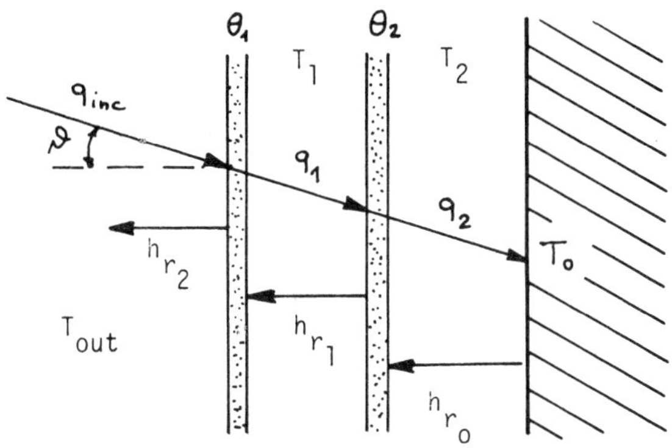

Fig. 3

Considering the figure 3, one can write the following equations for the successive glass (or plastic) panes and air spaces

$$\frac{\partial \theta_1}{\partial t} \rho_v c_v e_v = q_1^{abs} - h_e(\theta_1 - T_{out}) - h_i(\theta_1 - T_1) - h_{r_2}(\theta_1 - T_{out}) - h_{r_1}(\theta_1 - \theta_2) \quad (4)$$

$$\frac{\partial T_1}{\partial t} \rho_A c_A e_A = h_i(\theta_1 - T_1) + h_i(\theta_2 - T_1) \quad (5)$$

$$\frac{\partial \theta_2}{\partial t} \rho_v c_v e_v = q_2^{abs} - h_i(\theta_2 - T_1) - h_i(\theta_2 - T_2) - h_{r_1}(\theta_2 - \theta_1) - h_{r_0}(\theta_2 - T_0) \quad (6)$$

Fig. 5. Transmission factor of glass and plastics as function of the angle between the normal to the surface and the incident beam.

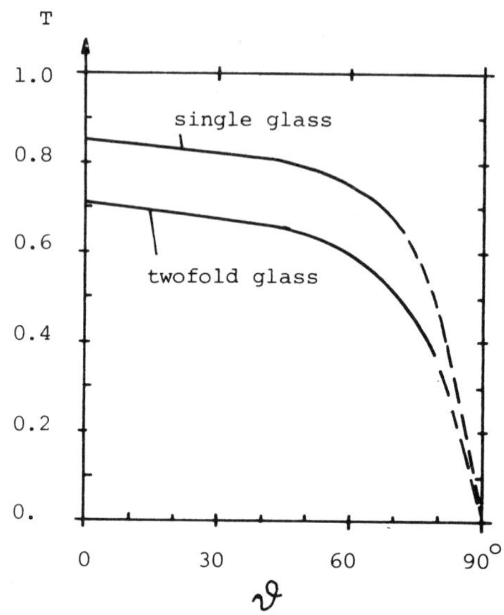

Fig. 5a. Measured transmission factor for single and twofold glass.

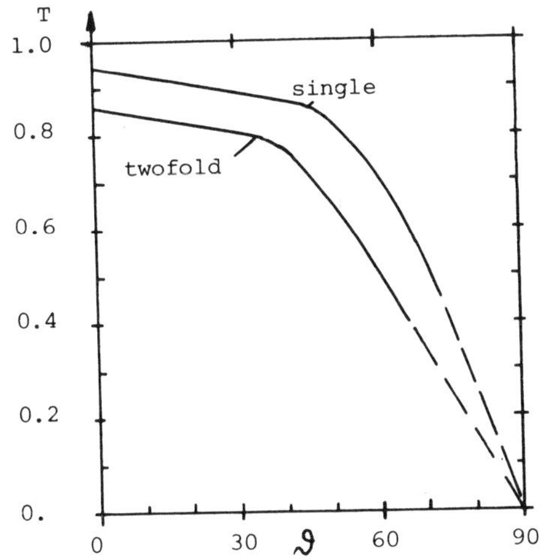

Fig. 5b. Measured transmission factor for Tedlar

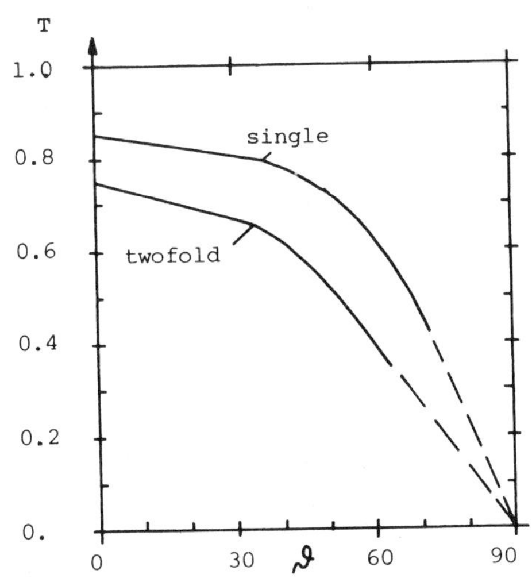

Fig. 5c. Measured transmission factor for Kalwall

$$\frac{\partial T_2}{\partial t}\rho_A c_A e_A = h_i(\theta_2 - T_2) + h_i(T_o - T_2) \qquad (7)$$

As temperatures of glasses vary slowly with time, one can omit the derivatives and rewrite the equations in a more convenient matrix form :

$$(h_e + h_i + h_{r_2} + h_{r_1})\theta_1 - h_i T_1 - h_{r_1}\theta_2 = q_1^{abs} + (h_c + h_{r_2})T_{out} \qquad (4')$$

$$h_1\theta_1 - 2h_i T_1 + h_i\theta_2 = 0 \qquad (5')$$

$$-h_{r_1}\theta_1 - h_i T_1 + (2h_i + h_{r_1} + h_{r_0})\theta_1 - h_i T_2 = q_2^{abs} + h_{r_0} T_o \qquad (6')$$

$$h_i\theta_2 - 2h_i T_2 = -h_i T_o \qquad (7')$$

where :

$$h_{r_2} = 4\varepsilon\sigma\bar{T}^3$$

$$h_{r_1} = 4\sigma\left(\frac{1}{1/\varepsilon_1 + 1/\varepsilon_2 - 1}\right)\bar{T}^3$$

The last parameter which has to be determined with accuracy is the amount of solar radiation which falls on the absorbing surface. This quantity is a function of the type of the cover, and of the angle (ϑ) between the normal to the cover and the incident beam.

We have measured experimentally the transmission of the covers as functions of ϑ. Some curves are given on figure 5. For the computations the curves have been parametrized and introduced into a small subroutine.

Using the described procedure, it was possible, on the basis of the experimental measurements to perform a fit for each combination of front cover and absorbing surface.

The following absorption factors were obtained :

Absorbing surface :	α
Rough concrete	0.51 ± 0.03
Brown concrete	0.705 ± 0.019
Black concrete	0.75 ± 0.03

Absorbing surface :	α
Black metal plate Black selective metal plate	0.93 ± 0.03 0.90 ± 0.03 (ε = 0.12 ± 0.04)

MATHEMATICAL SIMULATION

In order to compare the thermal characteristics of the different covers, we assume a south facing wall equiped with the considered cover and backed by a room in which a constant temperature is maintained.

Using real meteorological data, and assuming that the inner temperature in the room is 20 [°C] (68 °F), we have computed, half hour by half hour, the losses, respectively the gains, though the differently equiped walls.

Table 2 presents for two periods and for four typical front covers, the results of the simulation. The first period in February 1977 was characterized by cloudy days, the second one in March 1977, with sunny days. As a point of comparison the table gives also the thermal performance of a conventional, well insulated wall (K = 0.5 [W/m^2 °C] i.e. u = 0.09 [Btu/ft^2 h °F]).

We see that for the cloudy period, only a sophisticated front cover consisting of two glasses and a selective metal plate allows to maintain the heat losses below the values reached by a conventional wall. For the sunny period, any of the considered front covers lead to a heat gain.

CONCLUSION

The work presented here constitues only the first step of our study on passive solar wall collectors. In that step, we have compared different types of front covers and absorbing surfaces forgetting the rest of the whole system.

We have shown that for sunny days any of the considered covers give raise to heat gains. More care must be taken if we consider a longer period during the winter.In that case one has to take into account the thermal capacity of the wall and of the room in connection with the front cover and the absorbing surface.Next step will also be in a global study of the three functions:collection storage and distribution of the energy.

Days of 1977	Ext. temp. Text [°C]	Solar Energy [kWh/m²d]	Heat losses (-) or gains (+): Conventional wall K = 0.5 [w/m² °C]	Single glass + Concrete	Dobble Tedlar + Brown concrete	Dobble glass + Brown concrete	Dobble glass + Selective plate
No. 45	2.3	0.63	-213	-509	+ 14	+114	+307
46	3.8	0.40	-195	-841	- 542	+439	-311
47	3.5	1.61	-198	-713	- 428	-306	-175
48	3.3	0.24	-200	-826	- 512	-424	-298
49	6.7	2.29	-159	-571	- 210	-172	- 42
	3.9 [°C]	1.03 [kWh/m²d]	-965 [Wh/m²]	-3460 [Wh/m²]	- 1678 [Wh/m²]	-1227 [Wh/m²]	-519 [Wh/m²]
63	11.3	3.65	-104	+ 77	+ 394	+648	+831
64	8.7	5.23	-135	+358	+ 715	+1080	+1314
65	7.3	5.43	-152	+396	+ 843	+1254	+1531
66	7.9	5.28	-145	+346	+ 796	+1196	+1471
67	11.0	4.49	-108	+366	+ 769	+1147	+1397
	9.2 [°C]	4.82 [kWh/m²d]	-644 [Wh/m²]	+1543 [Wh/m²]	+3517 [Wh/m²]	+5325 [Wh/m²]	+6544 [Wh/m²]

Table 2 : Results of the mathematical simulation for two charachteristical periods of the Winter 1976-1977.

However, the results already obtained permitt a better approach of the first function :collection of solar energy.

REFERENCES

[1] Kalwall Sun-lite premium
Kalwall Corporation, Manchester (NH 03103)

[2] Tedlar PVF Film
Du Pont de Nemours International SA

[3] I.R. Reflecting Glass
Glaverbel SA - Bruxelles (B - 1170)

"ENERGY ANALYSIS METHOD FOR A SOLAR PANEL"

Bruno Ciborra - Politecnico di Milano
Istituto di Macchine
Milan, Italy

ABSTRACT

An energy balance is applied to a solar panel: it appears that pay-back energy time similar to that of nuclear energy.

Reflections on this fact are made for a right broadcast of solar energy.

INTRODUCTION

Coming from various papers on possibility of solar energy, we think useful to report shortly some results of a preliminary study made with energy analysis method on a solar panel for production on a little scale of hot water at medium temperature.

The choosen solar panel is made from the following material:
- 50 kg ferrous materials
- 6 kg of plastics (mock wool, and so on...)
- 30 kg of glass

and has a flat to solar rays of 1.6 m^2.

From our papers (see References) we got the energetic content of the various materials:

Ferrous materials	54418	Kjoule/kg
Plastic	83720	Kjoule/kg
Glass	62790	Kjoule/kg.

If we multiply these values for the pounds of the various materials we get the total amount given for the construction of the same panel.

SOLAR ENERGY: INTERNATIONAL PROGRESS

From medium values has seen an useful energy got from panel of about 6000-7000 Kjoule/day (zone of the research was in North Italy). In this art it is possible to calculate how much time must be on line this panel to give back the investment of energy.

The result of this calculation is very important also with a preliminary method and with medium values.

The found time of pay back is about two years very similar to that of nuclear power plants.

So the application of solar energy is requesting an energy investment not less of that of thermal plants.

Is not a case that U.S.A. with a very strong industrial structure is pushing into solar energy; in the same time developing countries are pushing into thermal (nuclear) power plants.

Very important in the side of solar energy, is the savings of fuel (fossil) during the use; this will be more sensitive for long times of use.

It will necessary bring a good planning of the consuming-structure in terms of time, geography, conserving the system, of level of heat requested.

The point about conserving the system is very important because a lost of energy properties of the warming panel (i.e. fouling, leakage) will break the global economy of the system.

Forgetting these points can bring a dangerous psycological trend (lost of popularity among just the first consumers) that can break the first promising steps on industrial scale of this new (but not too new) source of energy.

<u>ANNEXED Fig. 1</u>

In y axis is written the Megajoules of energy used to build the solar panel.

In x axis is written the number of days of use of solar panel, requested to give back the initial amount of energy of investment.

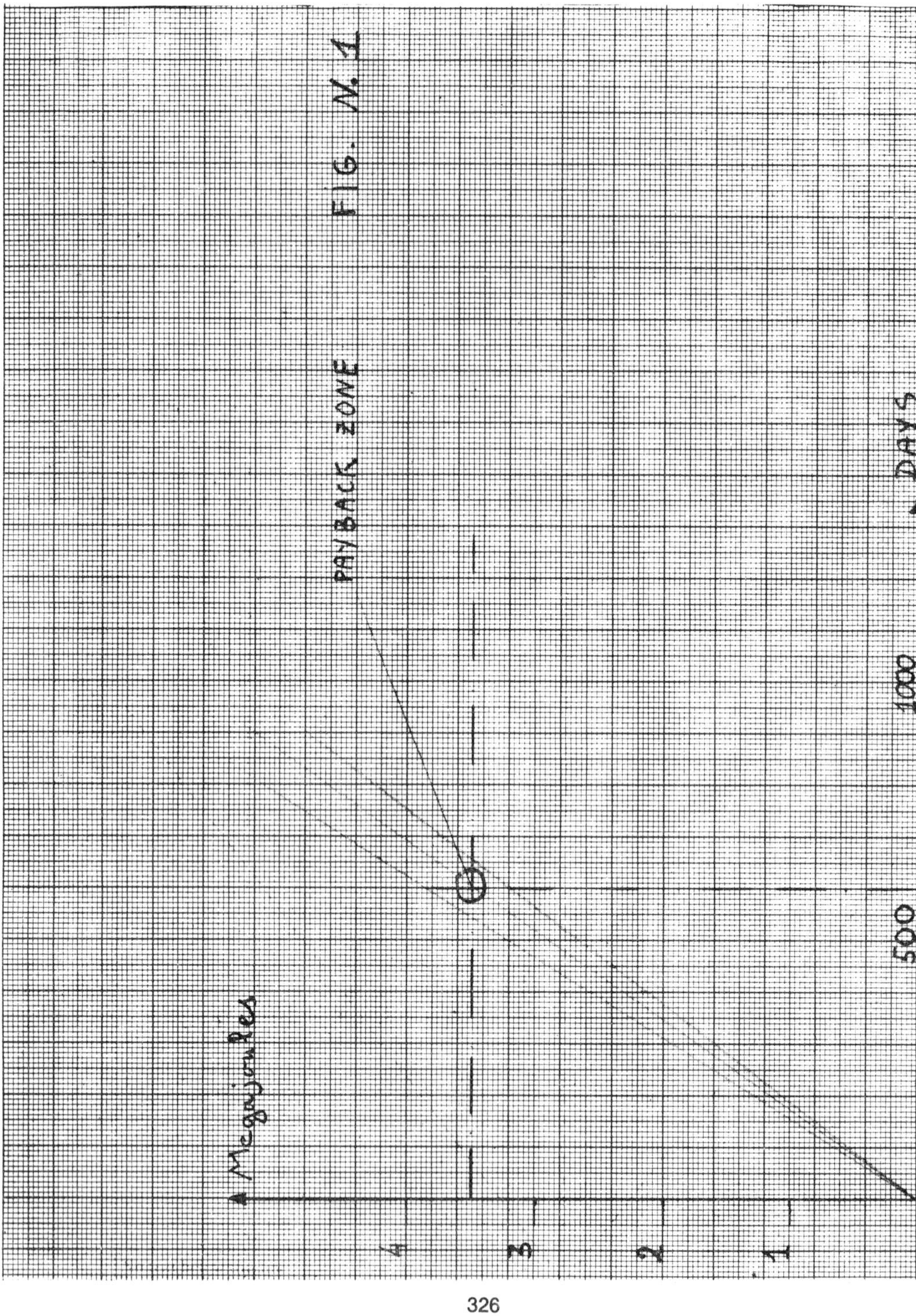
FIG. N. 1
PAYBACK ZONE
DAYS
1000
500
Megajoules
4
3
2
1

REFERENCES

Ciborra - Politecnico di Milano CLUP
"Introduzione al metodo dell'analisi energetica"
1978

Ciborra - Politecnico di Milano CLUP
"Nuove tecnologie e strategie per i problemi energetici"
1976

Chapman - Fuel's paradise
Peng in Book 1977 U.K.

Häfele - On Energy demand IIASA
Laxenburg Austria 1978

Thring - Energy in the 21st Century
Queen Mary College,U.K. 1977

A TWO-DIMENSIONAL FINITE-DIFFERENCE SOLUTION FOR THE TRANSIENT THERMAL BEHAVIOR OF A TUBULAR SOLAR COLLECTOR

F. L. Lansing
DSN Engineering Section
Jet Propulsion Laboratory
4800 Oak Grove Avenue
Pasadena, California U.S.A.

ABSTRACT

A numerical procedure was established using the finite-difference technique in the determination of the time-varying temperature distribution of a tubular solar collector under changing solar radiancy and ambient temperature. Three types of spatial discretization processes were considered and compared for their accuracy of computations and for selection of the shortest computer time and cost. The stability criteria of this technique were analyzed in detail to give the critical time increment to ensure stable computations. The results of the numerical analysis were in good agreement with the analytical solution previously reported. The numerical method proved to be a powerful tool in the investigation of the collector sensitivity to two different flow patterns and several flow control mechanisms.

I. INTRODUCTION

The need for an accurate prediction of the thermal behavior of solar collectors, to be used as a basic tool for their performance comparison, has become increasingly important with the rapid development of solar energy technology. For example, imposing changes on a specific solar collector such as introducing a new geometrical configuration, having a different control mechanism for its working fluid or developing new optical properties for the glazing and coating, require a sufficiently general analytical model which is both an adequate idealization of the physical system and capable of reasonably simple mathematical description.

The common procedure in analyzing the time-varying response of solar collectors has been based on a sequence of events of equal time intervals that range from 30 to 60 minutes, at the end of which the solar collector is assumed to have reached a quasi steady-state condition. This procedure cannot be generalized for all types of solar collectors and is restricted to those which possess very small thermal capacitance or very short response time (approximately 10 minutes) to the solar radiancy changes. The procedure would fit most flatplate collectors.

The solar collector under study has been selected as a candidate for Goldstone DSCC energy conservation projects. The time taken by this collector to reach quasi steady-state condition, measured experimentally by the solar simulator

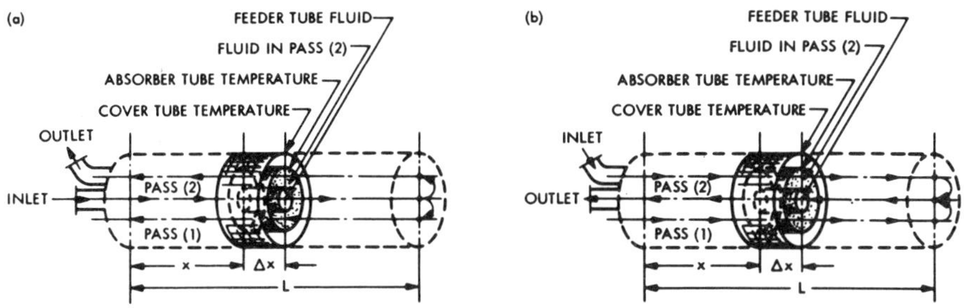

Fig. 1. Collector configuration (a) flow pattern ① (b) flow pattern ②

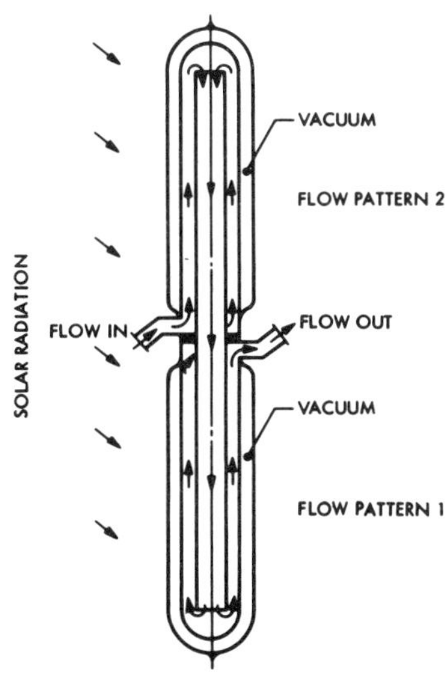

Fig. 2. Two collector units in series

team at NASA Lewis Research Center, was approximately one hour. Accordingly, a new investigation was required for an accurate procedure to study its transient thermal behavior.

The possibility of finding an exact analytical solution was not excluded from the investigation and was reported in a previous article (Ref. 1). The first results indicated that the problem can be solved by a closed algebraic form, but it is highly complicated. To facilitate the computations numerical solutions are preferred. The finite-difference method was then selected for the transient thermal study to supplement the analytical methods for additional coverage for this type of solar collector at all operating conditions.

II. NUMERICAL ANALYSIS

The numerical solution using the finite-difference method has become more popular in solving heat-transfer problems with the availability of high-speed digital computers and their associated technology. This method is used if a high degree of accuracy is required or if the problem is of considerable size and complexity.

In the finite-difference method, space and time discretization processes lead to a set of algebraic equations (nodal equations) instead of the partial differential equations that characterize the system. The nodal equations can be derived by purely mathematical methods or by energy consideration at the node of each spatial segment.

A. Collector Description and Flow Patterns

The tubular collector unit, as shown schematically in Figs. 1 and 2, is composed of three concentric tubes; an inner tube, an absorber tube, and a cover tube. The annulus space between the absorber and cover tubes is evacuated to minimize convection and conduction losses. The absorber tube surface is coated with a selective material to reduce the outward long-wave radiation losses. In flow pattern 1 , as shown in Fig. 1 (a), the circulating fluid starts from the open-end section of the inner tube. At the closed end of the collector, the fluid reverses its direction and passes in the annulus spacing between the inner and absorber tubes. In flow pattern 2 , as shown in Fig. 1 (b), the fluid path is reversed from the above. Each collector module consists of 24 collector units formed from flow patterns 1 and 2 in an alternate series as shown in Fig. 2. For more irradiancy augmentation, the set of collectors is mounted with lateral spaces separating them from each other with a highly reflective back reflector.

B. Program Procedure

Figure 3 illustrates one axisymmetric segment of the collector tube and the nodal and edge temperatures used throughout the analysis. The segment has a thickness ΔX and is bounded by sections (J) and (J + 1) perpendicular to the axis of symmetry. The nodal temperatures of the j th segment are located at the midsection of the segment and are denoted by TM_j. Four nodal temperatures are assigned to each segment, one for each of the following: the

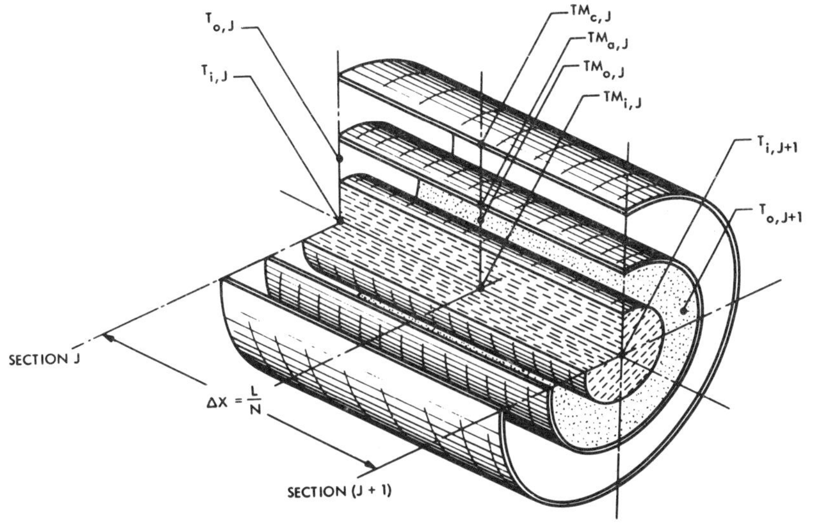

Fig. 3. Nodal and edge temperatures of the Jth segment

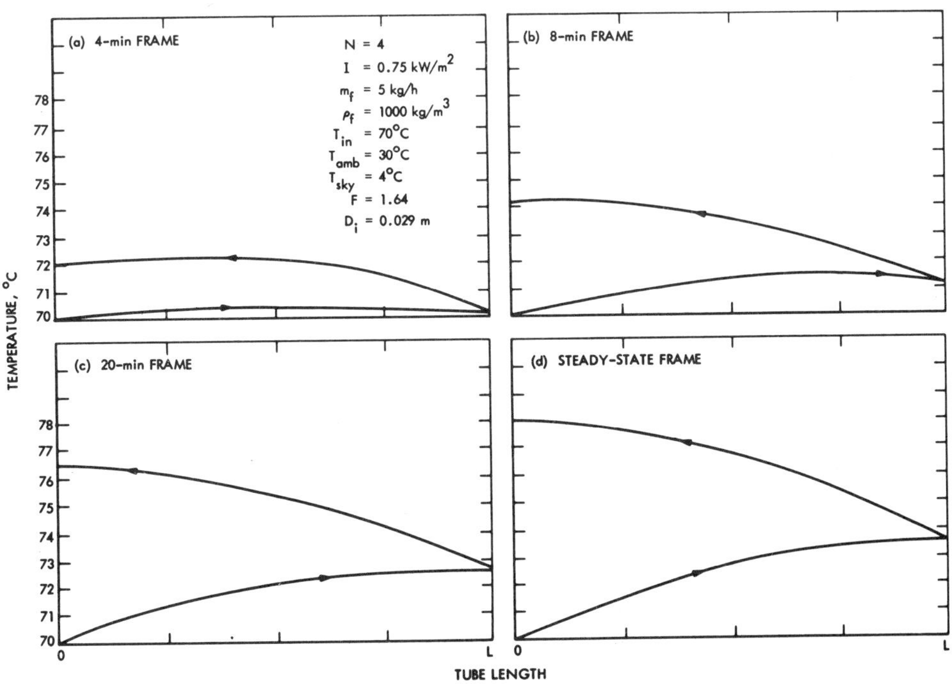

Fig. 4. Different time frames for one collector unit of flow pattern (1) showing the growth of the temperature profile

inner fluid pass, the outer fluid pass, the absorber tube surface, and the cover-tube surface temperatures. Because of the constrained boundary conditions at the open end and the closed end of each collector unit, as well as between the different flow patterns in series, the nodal temperatures of the fluid in the inner and outer passes only are converted back to the edge temperatures, after each computation process with an increment of time $\Delta\theta$. This results in a two-dimensional temperature distribution for the axisymmetric collector unit; in the axial direction and in the radial direction. No allowance is made for circumferential temperature differences between the upper tube's surface of the collector unit facing the Sun and the bottom tube's surface facing the back reflector. This means that the temperatures are uniform circumferentially.

The energy balance equations are listed in Appendix A for convenience. The present set of nodal and edge temperatures at time θ are then grouped together in a matrix format, and the corresponding future set of temperatures at time ($\theta + \Delta\theta$) is then computed and updated using matrix multiplication algebra. The computations proceed in the above sequence after preassigning an initial set of temperatures. The latter can be arbitrarily determined based on past experience or can be given as a result of measurements.

III. RESULTS AND CONCLUSIONS

One of the obstacles that is met during the execution of the computations is the numerical instability. In Appendix B, the necessary stability criteria are derived and presented for each flow pattern. This shows that the increment time $\Delta\theta$ and the spatial increment Δx cannot be arbitrarily chosen, but must be calculated from the above criteria to ensure stable computations. It can be shown from the critical stability condition Eq. (B-2), that the smaller the size of each segment, Δx , leads to a smaller allowable increment of time $\Delta\theta$ and a larger program execution time. This is due to the doubling of complexity; first due to the increased size of the matrices with larger number of segments, and second to the increased number of repetitions with smaller time intervals. Several program runs were made and the results are listed as follows:

(1) The time-varying temperature distribution of the working fluid in a unit collector at some given operating condition is plotted in Figs. 4 and 5 for flow patterns 1 and 2 , respectively. The collector fluid temperature is assumed to be uniform at 70°C before a sudden step of solar radiancy of 0.75 kW/m^2 is imposed and remains constant thereafter. In both flow patterns, the steady-state exit fluid temperature reaches the same value of 78°C as shown in frame (d) of Figs. 4 and 5. It is clear from the time frames (a), (b) and (c) of Figs. 4 and 5, the distinct difference in responding to changes of the solar radiation between the different flow patterns. The exit fluid temperature of flow pattern 1 responds by a much faster rate in the early stages since the fluid leaves the collector as soon as it absorbs the useful energy from the annulus area facing the Sun. On the other hand, the exit fluid temperature of flow pattern 2 is not significantly increased until the relatively colder fluid in the center tube is discharged. This results in a much higher fluid temperature at the closed end of the collector which may

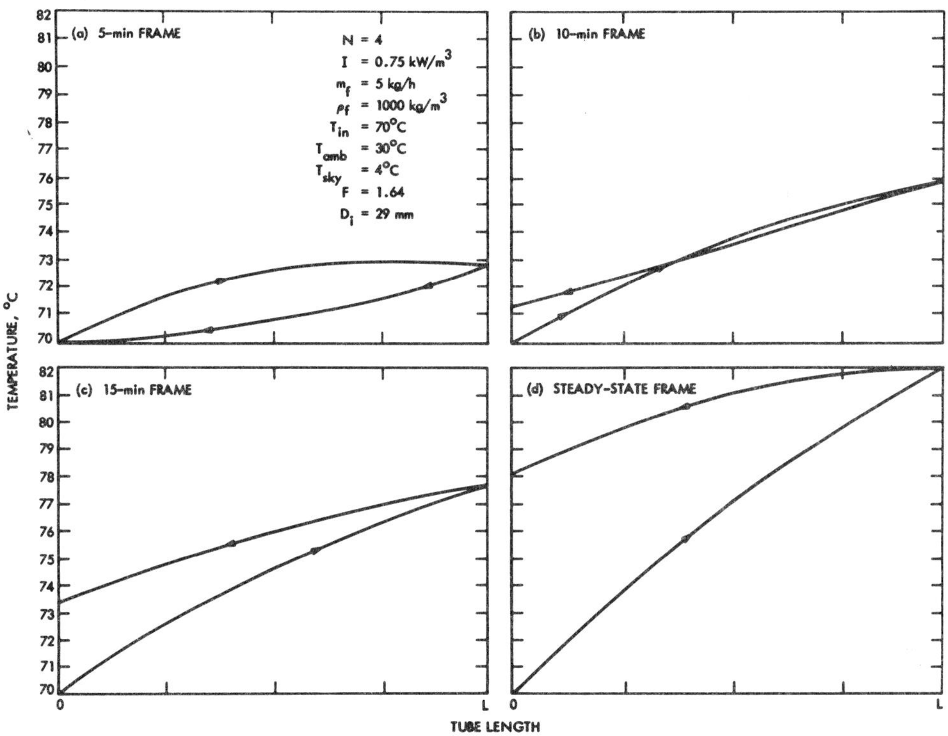

Fig. 5. Different time frames for one collector unit of flow pattern (2) showing the growth of the temperature profile

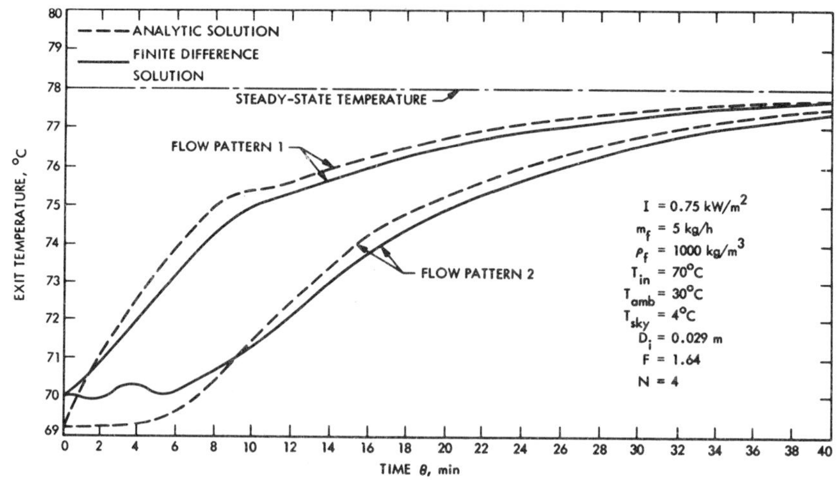

Fig. 6. Comparison between the finite difference solution and the analytical solution for a single tube and two different flow patterns

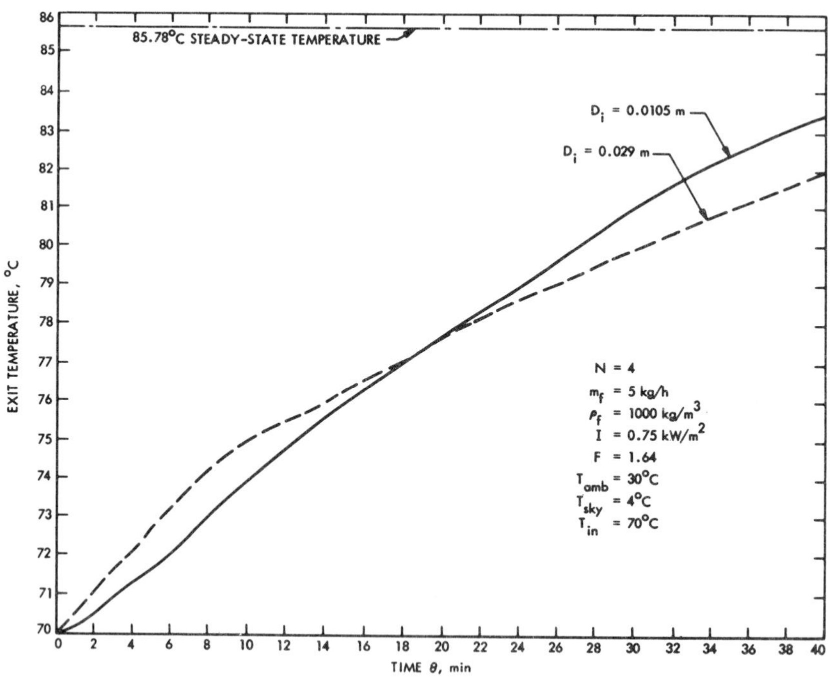

Fig. 7. Effect of increasing the inner tube diameter on thermal response for two collector units in series

cause some local degradations of the selective coating for this type of flow pattern.

(2) The exit fluid temperature obtained from the numerical results of the previous unit collector runs is plotted in Fig. 6. Superimposed on Fig. 6 is the analytical solution previously reported in Ref. (1) for the same operating conditions, optical and geometrical properties. Although the analytical solution was made with the assumption of zero tube thermal capacitance, it is in good agreement with the numerical solution which considers such a thermal capacitance with only a maximum deviation of $\pm$ 0.5°C. The deviation between the numerical and analytical solutions in the first 5 minutes in Fig. 6 is due to the different estimates of the initial temperature profile. The analytical solution considers an initial steady-state temperature profile with inlet fluid temperature of 70oC, which results in an exit temperature of 69.2°C, (Ref. 1), while the numerical solution arbitrarily assigns a uniform fluid temperature of 70°C in the two fluid passes.

(3) The finite-difference numerical solution is then applied to two collector units in series as shown in Fig. 2. In this case, the following two important criteria have been studied:
(a) The effect of changing the diameter of the inner tube on the rate of temperature growth is plotted as shown in Fig. 7. The manufacturer's specification of the inner diamerer, D_i, is 10.5 mm, which makes the fluid velocity in the inner fluid pass seventeen times larger than the fluid velocity in the annulus pass. The reason for that selection is to shorten the residence time inside the center tube for faster extraction of the hot fluid, though it causes an excessive impulse force at the closed-end of the collector.On the other hand, the minimization of these impulsive forces requires that the velocity in each fluid pass should be equal and the inner diameter in this case would be 29 mm. It is clear from Fig. 7 that reducing the diameter D_i will hasten the response of the collector after about 20 min, though it has no effect on the steady-state temperature that could be reached.
(b) The rate of cooling of the collector at zero solar radiation is illustrated in Fig. 8. The fluid in the two collector units is assumed uniform at 87°C when a sudden drop of the solar radiation to zero value is imposed and remains constant thereafter. The rate of cooling amounts to 2.9°C/h to an ambient temperature of 30°C. This shows the very low heat loss coefficient that this collector possesses over flat plate types. This is mainly due to the vacuum employed and the use of selective coatings.

(4) The application of the numerical solution has been extended further to study a complete module, consisting of 24 collector units connected in series with alternating flow patterns. Three different computer programs have been written corresponding to three different space discretization processes for N = 4, 2 and 1. This results in a substantial reduction of computer time and cost as the number of segments per tube is reduced. For example, the computer time is approximately in the ratio of 16:4:1 as the number of segments N goes from 4:2:1, respectively. This is due to the squared-reduction effects of matrices size and the number of time loops required. Regarding accuracy, it has been found that increasing the space

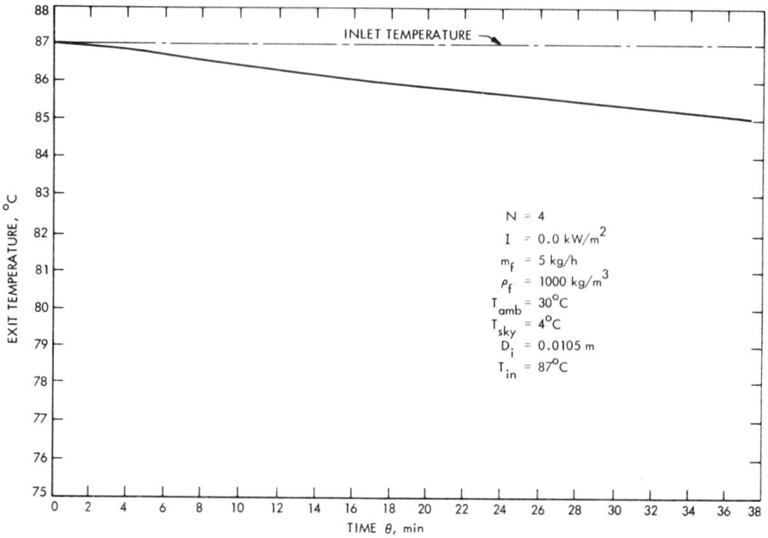

Fig. 8. Rate of cooling of two collector units in series at zero radiancy and constant inlet fluid temperature

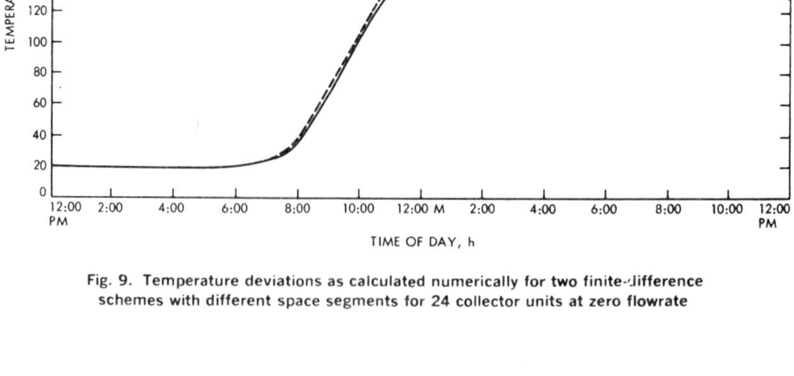

Fig. 9. Temperature deviations as calculated numerically for two finite-difference schemes with different space segments for 24 collector units at zero flowrate

segments has an insignificant effect as shown in Fig. 9. The temperature deviation is shown to be $\pm$ 1 percent as the number of segments N is reduced from 2 to 1 which was done to obtain a 75 percent reduction of the computer time and cost.

(5) The practical use of the tubular solar collector has indicated that the maximum end-to-end temperature rise across one module can reach as high as 50°C or approximately 2°C for each collector unit. This low temperature differential of the working fluid results in an insignificant temperature gradient across both the cover and absorber tube ends. Several module runs with three types of spatial divisions (four, two and one-segment per tube) indicate that the one-segment solution is more than adequate for the temperature profile prediction without sacrificing accuracy. The four-segment solution is not recommended as it consumes excessive computer time (at least 16 times that of a one-segment solution). The two-segment solution may or may not be used for further studies, depending on the allowable program funds.

(6) The collector circulating fluid may be stopped during operation at the following two time periods as a temperature vs. flow control technique: (a) from midnight to about 8:00 a.m. when the sun starts to rise, and (b) from the sunset at about 7:00 p.m. to midnight. The control system is designed to conserve what is collected during the sunshine hours. The actual hours of operation vary from day to day, depending on the type of application and the solar radiation time.

To allow for this, the numerical analysis program is provided with two extra computational controls. The first handles the periods of shut-down when all the 24 collector units behave as a single collector unit regardless of flow patterns, as occurs when the module end-to-end temperature difference is zero or 2°C at the maximum. The second computational control deals with sudden diurnal shut down after operation with a module end-to-end temperature difference exceeding 2°C.

(7) In Fig. 10 the results of six different flow control processes are plotted for a collector module (24 units in series) under variable solar radiancy and ambient temperatures. The total incident solar radiation falling on a plane surface at 20°C with the horizontal and facing south is registered and plotted for June 2, 1975 at the Goldstone DSCC, California. The accumulated solar irradiancy for that day was 10.465 kWh/m^2 and with a collector module projected area of 2.602 m^2, the total incident energy amounted to 27.228 kWh. The numerical analysis provided a valuable assistance in exposing the transient thermal behavior of the collector under several flow controls and the understanding of its important features. The flow control processes are as follows:

(a) Flow Control 1 : The fluid is left stagnant starting at midnight with a temperature of 20°C equal to the ambient temperature. During the sunshine period, the temperature rises and reaches a maximum of 230°C at around 4:00 a.m. under the given solar radiancy. The slow rate of cooling as discussed in item (3)b, will keep the fluid hot most of the time, and the temperature will reach about 130°C at the following midnight. Since no drainage of hot water is assumed in this control, the overall thermal collection efficiency will be zero.

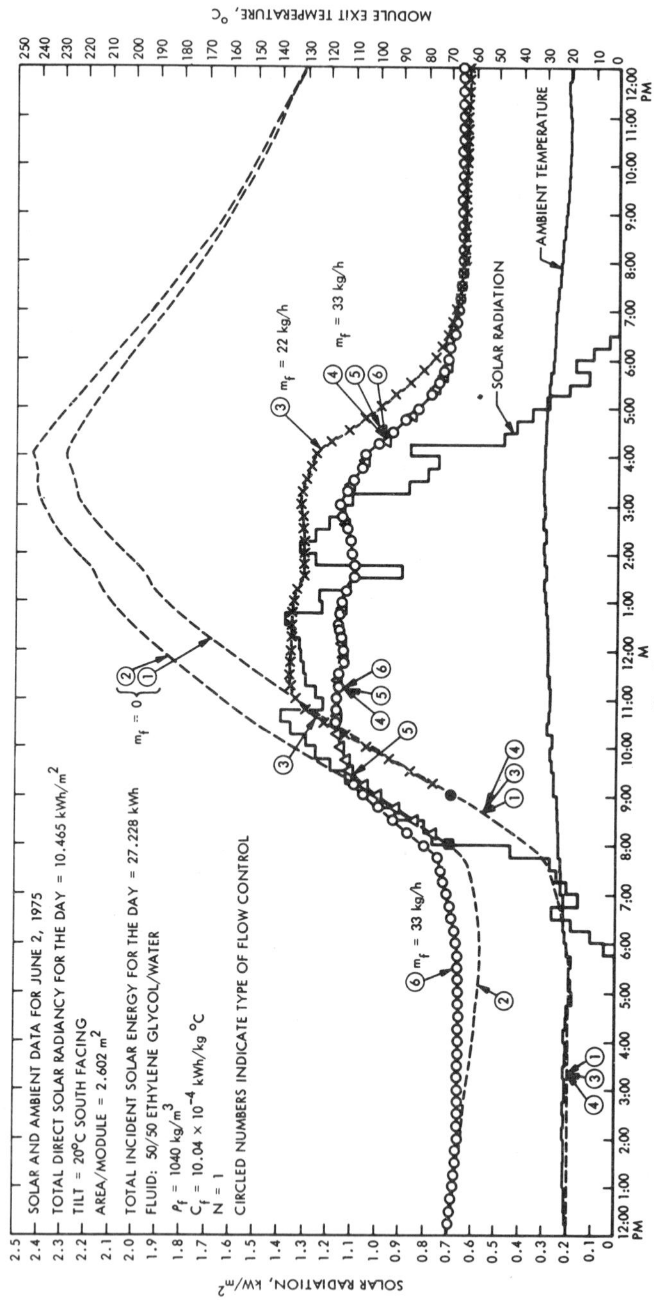

Fig. 10. Complete day performance of a collector module (24 units) under variable solar radiancy, ambient temperature and six flow control mechanisms

(b) Flow Control 2 : This is similar to flow control 1 except that the fluid is left stagnant starting midnight at a temperature of 70°C. At about 6:00 a.m. the fluid temperature drops by 13°C due to night cooling, and then starts to pick up some heat as the sun rises. The maximum temperature is found to be 244°C at 4:00 p.m. and it drops to 130°C at the following midnight as in flow control 1 . The overall thermal collection efficiency is also zero with this control. The results from flow controls 1 and 2 indicate that the no-flow maximum fluid temperature, though not significantly varying with the temperature of the residing fluid at midnight, can reach excessive values that cause severe thermal stresses and coating degradation problems.

(c) Flow Control 3 : This is a modification of flow control 1 whereby the fluid is left stagnant starting midnight with 20°C temperature and after reaching a temperature of 70°C at about 9 a.m., a fluid circulating pump is turned on and a uniform flow rate of 22 kg/h from the storage tank flows into the collector. The fluid temperature oscillates around 130°C for 5½ hours (10:30 a.m. to 4:00 p.m.) and then drops to 70°C at about 7:00 p.m. The peak of the accumulated thermal energy for the day occurs at 7:00 p.m. and reaches 36.23 percent of the total incident solar energy.

(d) Flow Control 4 : This is the same as flow control 3 except that the pump is rated at 33 kg/h. The fluid temperature thus oscillates about 110°C for 5½ h (10:30 a.m. to 4:00 p.m.); a temperature and a range most suitable for running an absorption type of air conditioner. The peak of the accumulated thermal energy for the day occurs at 6:30 p.m. and reaches 37.39 percent of the total incident solar energy.

(e) Flow Control 5 : This is a modification of flow control 2 whereby the fluid is left stagnant starting midnight with 70°C temperature and after reaching a temperature of 70°C at 8:00 a.m., a fluid circulating pump is turned on and a uniform flow rate of 33 kg/h from the storage tank flows into the collector. The fluid temperature oscillates around 110°C for about 7 h(9:15 a.m. to 4:00 p.m.) which is longer than the range given by flow control 4 . The peak of the accumulated thermal energy for the day occurs at 6:30 p.m. and reaches 42.34 percent of the total incident solar energy.

(f) Flow Control 6 : This is a continuous pumping operation for 24 h with a flow rate of 33 kg/h of the fluid at 70°C starting at midnight. The temperature drop for the period from midnight to 5:00 a.m. is only 5°C. As the Sun rises, the fluid temperature starts to increase and matches the temperatures reached by flow controls 4 and 5 . The peak of the accumulated thermal energy occurs at 6:30 p.m. and reaches 41.72 percent of the total incident solar energy.

It is clear from the above six trials of flow control that flow control 5 provides the highest accumulated thermal efficiency (42.34%) as well as the required hot temperature (110°C) for an airconditioning application over a period of about 7 h (9:15 a.m. to 4:00 p.m.).

Definition of Terms

C	Specific heat of tube material, kWh/kg·°C
C_f	Specific heat of flowing fluid, kWh/kg·°C
D	Tube diameter, m
F	Augumented radiation factor, ≥ 1
G_{mn}	Thermal conductance, kW/m^2
I	Direct solar radiancy, kW/m^2
L	Tube length, m
m_f	Fluid mass flow rate, kg/h
N	Number of space elements per tube,
Q_{in}	Nodal internal heat generation, kWh
r	Reflectivity,
T	Edge temperature, K
TM	Nodal temperature, K
T_{sky}	Sky temperature, K
T_{amb}	Ambient temperature, K
U_{ac}	Radiative heat transfer coefficient between the absorber and the cover tubes, kW/m^{2}·°C
U_{cv}	Convective heat transfer coefficient between the cover tube and the ambient air, kW/m^{2}·°C
U_{oi}	Overall heat transfer coefficient between the outer fluid and the inner fluid, kW/m^{2}·°C
U_{rd}	Radiative heat transfer coefficient between the cover tube and sky, kW/m^{2}·°C
V	Volume, m^3
X	Distance measured from the inlet tube section, m
θ	Time, h
α	Absorptivity
ϵ	Emissivity
τ	Transmissivity of cover tube
δ	Tube thickness, m
ρ	Density, kg/m^3
Δ	Increment

Definition of Terms

Subscripts

a absorber tube

c cover tube

f fluid

i inner fluid, in the center tube

o outer fluid, in the annulus area

ACKNOWLEDGEMENT

This paper presents the results of one phase of research carried out at the Jet Propulsion Laboratory, California Institute of Technology, under Contract No. NAS 7-100, sponsored by the National Aeronautics and Space Administration.

REFERENCES

1. Lansing, F. L., "The Transient Thermal Response of a Tubular Solar Collector," Technical Memorandum 33-781, Jet Propulsion Laboratory, Pasadena, Calif. July 15, 1976.

2. Razelos, P., "Methods of Obtaining Approximate Solutions," Handbook of Heat Transfer, edited by Rohsenow, W. M., and Hartnett, J. P., Sec. 4, McGraw Hill Book Co., N. Y., 1973.

APPENDIX A

NODAL ENERGY EQUATIONS

I. GENERAL REMARKS

The first step in obtaining the nodal equations is to subdivide the collector tubes into spatial segments by sections as shown in Fig. 3, locate a node in the center of each segment and then calculate the conductances between adjacent nodes. The energy balance for the mth node which is surrounded by n adjacent nodes, is then considered during a small time interval $\Delta\theta$. During this time interval, the temperatures of the n nodes are assumed to remain constant. Accordingly,

$$\rho_m C_m V_m \frac{\Delta T_m}{\Delta \theta} = \sum_n G_{mn} \left[T_n(\theta) - T_m(\theta) \right] + Q_{in} \qquad \text{(A-1)}$$

where V_m is the volume assigned to the segment whose node is m, G_{mn} is the therr con- ductance between nodes m, n, and Q_{in} is the rate of heat generation the node if any. The difference ΔT_m may be expressed in terms of either a forward, a backward or mid-difference (Ref. 2) which will yield different forr of Eq. (A-1). The simplest form is obtained using the "forward" difference ar is chosen for the present work, i.e.,

$$\Delta T_m = T_m(\theta + \Delta\theta) - T_m(\theta) \qquad \text{(A-2)}$$

In this case, the future temperature $T_m(\theta + \Delta\theta)$ is explicitly obtained from the knowledge of the present temperatures $T_m(\theta)$ and $T_n(\theta)$ at time θ. Combining Eqs. (A-1) and (A-2) gives the future temperature $T_m(\theta + \Delta\theta)$ as follov

$$T_m(\theta + \Delta\theta) =$$

$$T_m(\theta) + \left[\sum_n G_{mn} \left\{ T_n(\theta) - T_m(\theta) \right\} + Q_{in} \right] \frac{\Delta\theta}{\rho_m C_m V_m} \qquad \text{(A-3)}$$

It is important to note that the conductances G_{mn} include, in general, the energy conducted across the tube boundaries of the elements as well as the energy transported between neighboring nodes in case of an open system node with a fluid flowing across the segment (Lagrangian control volume approach).

II. BOUNDARY CONDITIONS

(1) For both flow patterns the closed-end temperature of the inner and the outer fluid must be equal.

(2) For flow pattern 1, the open-end temperature of the fluid in the inner tube is a known value.

(3) For flow pattern 2, the open-end temperature of the fluid in the annulu area is a known value.

(4) For a collector composed of two units of flow pattern 1 and 2 in series, the open end-to-end temperatures must be equal.

III. ENERGY EQUATIONS

The application of Eq. (A-3) of the inner fluid, outer fluid, absorber tube and the cover tube segments yield the following:

A. Inner Fluid

$$TM_{i,j}(\theta+\Delta\theta) = TM_{i,j}(\theta) + \left[G_{oi}\left\{TM_{o,j}(\theta) - TM_{i,j}(\theta)\right\} \pm G_{ff}\left\{T_{i,j+1}(\theta) - T_{i,j}(\theta)\right\}\right] \frac{\Delta\theta}{\rho_f C_f \frac{V_i}{N}} \qquad (A\text{-}4)$$

where the (+) sign is for flow pattern 2 and the (-) sign is for flow pattern 1 .

Expressing the inner fluid nodal temperatures in terms of the edge temperature as

$$TM_{i,j}(\theta) = \frac{T_{i,j}(\theta) + T_{i,j+1}(\theta)}{2}$$

$$TM_{o,j}(\theta) = \frac{T_{o,j}(\theta) + T_{o,j+1}(\theta)}{2} \qquad (A\text{-}5)$$

Then

$$\begin{aligned} TM_{i,j}(\theta+\Delta\theta) = {} & T_{i,j}(\theta)\left[0.5 - (0.5G_{oi}\ \angle\theta_i \pm G_{ff}\ \Delta\theta_i)\right] \\ & + T_{i,j+1}(\theta)\left[0.5 - G_{oi}\ \Delta\theta_i \pm G_{ff}\ \Delta\theta_i\right] \\ & + T_{o,j}(\theta)\left[0.5G_{oi}\cdot\Delta\theta_i\right] \\ & + T_{o,j+1}(\theta)\left[0.5G_{oi}\ \Delta\theta_i\right] \end{aligned} \qquad (A\text{-}6)$$

where

$$G_{oi} = U_{oi}\frac{\pi D_i L}{N}$$

$$G_{ff} = m_f C_f$$

$$V_i = \frac{\pi}{4} D_i^2 L \qquad (A\text{-}7)$$

$$\Delta_i = \frac{\Delta\theta\cdot N}{\rho_f C_f \cdot \frac{\pi}{4} D_i^2 L}$$

and the (+) sign is for flow pattern 2 and the (-) sign is for flow pattern 1

B. Outer Fluid

$$TM_{o,j}(\theta + \Delta\theta) = TM_{o,j}(\theta) + \Big\{ \Big[U_{ao}\frac{\pi D_a L}{N} \Big]\Big[TM_{a,j}(\theta) - TM_{o,j}(\theta)\Big] - \Big[U_{oi}\frac{\pi D_i L}{N}\Big]\Big[TM_{o,j}(\theta) - TM_{i,j}(\theta)\Big] \mp (m_f C_f)\Big[T_{o,j+1}(\theta) - T_{o,j}(\theta)\Big] \Big\} \cdot \frac{\Delta\theta}{\rho_f C_f \frac{V_o}{N}} \tag{A-8}$$

where the (+) sign is for flow pattern 1 and the (-) sign for flow pattern 2 .

Expressing the outer fluid nodal temperatures in terms of edge temperatures as

$$TM_{o,j}(\theta) = \frac{TM_{o,j}(\theta) + TM_{o,j+1}(\theta)}{2} \tag{A-9}$$

$$TM_{i,j}(\theta) = \frac{TM_{i,j}(\theta) + TM_{i,j-1}(\theta)}{2}$$

Then

$$\begin{aligned} TM_{o,j}(\theta + \Delta\theta) = {} & T_{o,j}(\theta)\Big\{ 0.5 - (0.5G_{ao}\cdot \Delta\theta_o + 0.5G_{oi}\Delta\theta_o \mp G_{ff}\Delta\theta_o) \Big\} \\ & + T_{o,j+1}(\theta)\Big\{ 0.5 - 0.5\, G_{ao}\,\Delta\theta_o - 0.5\, G_{oi}\Delta\theta_o \mp G_{ff}\,\Delta\theta_o \Big\} \\ & + TM_{a,j}(\theta)\Big\{ G_{ao}\cdot \Delta\theta_o \Big\} \\ & + T_{i,j}(\theta)\Big\{ 0.5G_{oi}\cdot\Delta\theta_o \Big\} \\ & + Ti_{,j+1}(\theta)\Big\{ 0.5G_{oi}\cdot\Delta\theta_o \Big\} \end{aligned} \tag{A-10}$$

where

$$G_{ao} = U_{ao}\cdot\frac{\pi D_a L}{N}$$

$$G_{oi} = U_{oi}\frac{\pi D_i L}{N}$$

$$G_{ff} = m_f C_f \tag{A-11}$$

$$V_o = \frac{\pi}{4}(D_a^2 - D_i^2)L$$

$$\Delta\theta_o = \frac{\Delta\theta\cdot N}{\rho_f C_f (D_a^2 - D_i^2)L}$$

and the (+) sign is for flow pattern 1 and the (-) sign is for flow pattern 2 .

C. Absorber Tube

$$TM_{a,j}(\theta + \Delta\theta) = TM_{a,j}(\theta) + \left\{ I \cdot F \cdot \frac{\alpha_a \tau_c}{1 - r_a r_c} \cdot \frac{D_c L}{N} - U_{ao} \frac{\pi D_a L}{N} \left[TM_{a,j}(\theta) - TM_{o,j}(\theta) \right] - U_{ac} \frac{\pi D_a L}{N} \left[TM_{a,j}(\theta) - TM_{c,j}(\theta) \right] \right\} \times \frac{\Delta\theta}{\rho_a C_a \frac{V_a}{N}} \qquad \text{(A-12)}$$

Expressing the outer fluid nodal temperatures in terms of edge temperatures as

$$TM_{o,j}(\theta) = \frac{T_{o,j}(\theta) + T_{o,j+1}(\theta)}{2} \qquad \text{(A-13)}$$

Hence

$$\begin{aligned} TM_{a,j}(\theta + \Delta\theta) = {} & TM_{a,j}(\theta) \left[1 - G_{ao} \Delta\theta_a - G_{ac} \Delta\theta_a \right] \\ & + TM_{c,j}(\theta) \cdot G_{ac} \cdot \Delta\theta_a \\ & + T_{o,j}(\theta)\ 0.5 G_{ao} \cdot \Delta\theta_a + T_{oj+1}(\theta) \cdot 0.5 G_{ao} \cdot \Delta\theta_a \\ & + \left[I \cdot F \cdot \frac{\alpha_a \tau_c}{1 - r_a r_c} \cdot \frac{D_c L}{N} \cdot \Delta\theta_a \right] \end{aligned} \qquad \text{(A-14)}$$

where

$$V_a = \pi D_a \delta_a L$$

$$\Delta\theta_a = \frac{\Delta\theta . N}{\rho_a C_a \cdot (\pi D_a \delta_a L)} \qquad \text{(A-15)}$$

D. Cover Tube

$$\begin{aligned} TM_{c,j}(\theta + \Delta\theta) = {} & TM_{c,j}(\theta) + \\ & + \left(\left\{ I.F \left[\alpha_c + \frac{\alpha_c r_a \tau_c}{1 - r_a r_c} \right] \cdot \frac{D_c L}{N} \right\} \right. \\ & + U_{ac} \frac{\pi D_a L}{N} \left[TM_{a,j}(\theta) - TM_{c,j}(\theta) \right] \\ & - U_{cv} \cdot \frac{\pi D_c L}{N} \left[TM_{c,j}(\theta) - T_{amb} \right] \\ & \left. - U_{rd} \cdot \frac{\pi D_c L}{N} \left[TM_{c,j}(\theta) - T_{sky} \right] \right) , \frac{\Delta\theta N}{\rho_c C_c V_c} \end{aligned} \qquad \text{(A-16)}$$

or

$$TM_{c,j}(\theta + \Delta\theta) = TM_{c,j}(\theta)\left\{1 - G_{ac}\cdot \Delta\theta_c - G_{cb}\cdot \Delta\theta_c - G_{cs}\cdot \Delta\theta_c\right\} + TM_{a,j}(\theta)\cdot G_{ac}\cdot \Delta\theta_c + \Delta\theta_c\left(I\cdot F\cdot\left\{ \alpha_c + \frac{\alpha_c r_a \tau_c}{1 - r_a r_c}\right\} \frac{D_c L}{N} + (G_{cb}\cdot T_{amb}) + (G_{cs}\cdot T_{sky})\right) \quad (A\text{-}17)$$

where

$$V_c = \pi D_c \delta_c L$$

$$\Delta\theta_c = \frac{\Delta\theta \cdot N}{\rho_c C_c(\pi D_c \delta_c L)}$$

$$G_{cb} = U_{cv}\frac{\pi D_c L}{N}$$

$$G_{cs} = \frac{U_{rd}\cdot \pi D_c L}{N} \quad (A\text{-}18)$$

APPENDIX B

CONVERGENCE AND STABILITY OF COMPUTATIONS

I. GENERAL REMARKS

In the numerical computations, the term error refers to the difference between the approximate solution obtained by the finite difference scheme and the exact solution of the original partial differential equations. Accordingly, there exists two types of errors.

(1) Truncation error due to the replacement of the derivatives by finite differences. This depends on the given initial and boundary conditions.

(2) Round-off error caused by carrying out the computations to a finite decimal place.

Apart from the above errors, the most serious problem associated with the finite difference scheme is the phenomena of instability. This means that the temperatures oscillate with an increasing magnitude as the computation progresses. Therefore, the space and time increments Δx , $\Delta\theta$, respectively, cannot be selected arbitrarily but rather constrained to satisfy certain stability criteria. For practical applications, the following elementary argument (Ref. 2) has been used for obtaining simple and adequate stability conditions.

The mth node temperature can be expressed in terms of the surrounding n nodes in the format:

$$T_m(\theta + \Delta\theta) = K_{mm}T_m(\theta) + \sum_n K_{mn}T_n(\theta) \qquad \text{(B-1)}$$

where the coefficients K_{mn} are set to be positive. If the coefficient K_{mm} was negative then the higher temperature $T_m(\theta)$ at the present time , the lower the temperature $T_m(\theta + \Delta\theta)$ at the future time $(\theta + \Delta\theta)$. This would be in violation of the thermodynamics principles. Therefore, the increment of time $\Delta\theta$ for a given space subdivision Δx should be chosen so that the coefficient K_{mm} is positive or at least zero. Applying the above simple rule to the finite difference equations developed earlier for each node, the following critical stability conditions are established.

II. CASE OF FLOW PATTERN 1

For the inner fluid temperature stability

$$0.5 - 0.5\,G_{oi}\,\Delta\theta_i - G_{ff}\,\Delta\theta_i \geq 0 \qquad \text{(B-2)}$$

For the outer fluid temperature stability

$$0.5 - 0.5G_{ao}\Delta\theta_o - 0.5G_{oi}\,\Delta\theta_o - G_{ff}\,\Delta\theta_o - 0.5G_{oi}\,\Delta\theta_i \geq 0 \qquad \text{(B-3)}$$

For the absorber tube temperature stability

$$1 - (G_{ao} + G_{ac})\,\Delta\theta_a \geq 0 \qquad \text{(B-4)}$$

For the cover tube temperature stability

$$1 - (G_{ac} + G_{cb} + G_{cs})\ \Delta\theta_c \geq 0 \qquad (B\text{-}5)$$

III. CASE OF FLOW PATTERN 2

For the inner fluid temperature stability

$$0.5 - G_{ff}\ \Delta\theta_i - 0.5G_{oi}\ \Delta\theta_i - 0.5G_{oi}\ \Delta\theta_o \geq 0 \qquad (B\text{-}6)$$

For the outer fluid temperature stability

$$0.5 - 0.5G_{ao}\ \Delta\theta_o - 0.5G_{oi}\ \Delta\theta_o - G_{ff}\ \Delta\theta_o \geq 0 \qquad (B\text{-}7)$$

For the absorber tube temperature stability

$$1 - (G_{ao} + G_{ac})\ \Delta\theta_a \geq 0 \qquad (B\text{-}8)$$

For the cover tube temperature stability

$$1 - (G_{ac} + G_{cb} + G_{cs})\ \Delta\theta_c \geq 0 \qquad (B\text{-}9)$$

IV. CASE OF A SERIES CONNECTION OF FLOW PATTERNS 1 and 2

The foregoing conditions are combined and listed as follows:

For the inner fluid temperatures stability, a comparison of conditions (B-2) and (B-3) shows that condition (B-4) predominates, thus leaving

$$0.5 - G_{ff}\ \Delta\theta_i - 0.5G_{oi}\ \Delta\theta_i - 0.5G_{oi}\ \Delta\theta_o \geq 0$$

or

$$\Delta\theta \leq \frac{\rho_f C_f}{(4U_{oi}/D_i) + (8NC_f m_f/\pi D_i^2 L) + \left\{ 4U_{oi}/(D_a^2/D_i)\left[1 - \left(\frac{D_i}{D_a}\right)^2\right]\right\}} \qquad (B\text{-}10)$$

For the outer fluid temperature stability, a comparison of conditions (B-3) and (B-7) shows that condition (B-7) predominates, thus

$$0.5 - 0.5G_{ao}\ \Delta\theta_o - 0.5G_{oi}\ \Delta\theta_o - G_{ff}\ \Delta\theta_o - 0.5G_{oi}\ \Delta\theta_i \geq 0$$

or

$$\Delta\theta < \frac{\rho_f C_f}{\frac{4U_{oi}}{D_i} + \frac{8Nm_f C_f}{\pi(D_a^2 - D_i^2)L} + \frac{4U_{oi}}{D_a \cdot \left(\frac{D_a}{D_i}\right) \cdot \left[1 - \left(\frac{D_i}{D_a}\right)^2\right]} + \frac{4U_{ao}}{D_a\left[1 - \left(\frac{D_i}{D_a}\right)^2\right]}} \qquad (B$$

For the absorber tube temperatures stability

$$1 - (G_{ao} + G_{ac})\ \Delta\theta_a \geq 0$$

or

$$\Delta\theta \leq \frac{\rho_a C_a \delta_a}{(U_{ao} + U_{ac})} \qquad \text{(B-12)}$$

For the cover tube temperature stability

$$1 - (G_{ac} + G_{cb} + G_{cs})\ \Delta\theta_c \geq 0$$

or

$$\Delta\theta \leq \frac{\rho_c C_c \delta_c}{U_{ac}\frac{D_a}{D_c} + U_{cv} + U_{rd}} \qquad \text{(B-13)}$$

V. EXAMPLE OF COMPUTATIONS FOR THE CRITICAL TIME INCREMENT

To determine which of the above four conditions, Eqs. (B-10), (B-11), (B-12) and (B-13) yield the most critical time increment $\Delta\theta$ to ensure stable temperature computations, the following numerical values are assigned from a typical running condition:

$D_i = 0.0105\text{m}$, $\delta_a = \delta_c = 0.004\text{m}$

$D_a = 0.041\text{m}$, $C_f = 11.634 \times 10^{-4}\text{kWh/Kg}\cdot{}^{o}\text{C}$ (water)

$D_c = 0.051\text{m}$, $C_c = C_a = 2.094 \times 10^{-4}\text{kWh/Kg}\cdot{}^{o}\text{C}$ (glass)

$L = 1.067\text{m}$, $\rho_f = 1000\ \text{kg/m}^3$(water)

$N = 4$ $\rho_c = \rho_a = 2468\ \text{kg/m}^3$(glass)

$m_f = 5\ \text{kg/h}$

$U_{oi} = 56.53 \times 10^{-3}\text{kW/m}^{2}\cdot{}^{o}\text{C}$

$U_{ao} = 113.06 \times 10^{-3}\text{kW/m}^{2}\cdot{}^{o}\text{C}$

$U_{ac} = 0.80 \times 10^{-3}\text{kW/m}^{2}\cdot{}^{o}\text{C}$

$U_{cv} = 18.62 \times 10^{-3}\text{kW/m}^{2}\cdot{}^{o}\text{C}$

$U_{rd} = 4.99 \times 10^{-3}\text{kW/m}^{2}\cdot{}^{o}\text{C}$

Condition (B-10) yields $\Delta\theta \leq 0.0022$ h (7.8 s)

Condition (B-11) yields $\Delta\theta \leq 0.0166$ h (59.7 s)

Condition (B-12) yields $\Delta\theta \leq 0.0182$ h (65.4 s)

Condition (B-13) yields $\Delta\theta \leq 0.0852$ h (306.7 s)

The above indicates that Condition (B-10) is the most critical condition that must be satisfied in order to ensure stable computations. It can be seen from Condition (B-10) that the resulting minimum increment of time is inversely proportional with the mass flowrate. In the limit, with zero fluid flowrate, Condition (B-10) yields $\Delta\theta \leq 0.0505$ h (181.8 s) and in this case Condition (B-12) will impose the minimum time required for stable computations.

With regard to accuracy, the round-off errors have little or no effect on the solution if the restrictions of stability are observed, while the truncation errors decrease as we select smaller time and space increments.

OPTICAL AND THERMAL PERFORMANCES OF A HALF-CIRCULAR, CYLINDRICAL CONCENTRATOR

K. Bordoloi, T. Murray, F. Bynum and E. Chaves
University of Louisville
Louisville, Kentucky 40208, U.S.A.

ABSTRACT

This paper deals with the investigation of the optical and thermal performances of a non-tracking, half-circular cylindrical solar concentrator. To date, the study of the optical and thermal properties has resulted in the selection of a class of receivers for the non-tracking mode in which the concentrator can be used. The receiver is designed to have placement and size to take into account the changing orientation of the incident solar radiation as the declination changes throughout the year. The design consideration of various types of receivers is under study, the chief among which are:(a) circular and (b) triangular.

Our study focuses primarily on:(a) the geometrical consideration of the concentrator and the receiver for the optimum absorption of sunlight, (b) the study of the optical parameters such as the concentration ratio and the average number of reflections as well as the case of multiple reflections and (c) the analysis of the collctor in terms of heat transfer coefficent, heat loss coefficient, useful energy gain and the efficiency of the collector. The results suggest that the concentrator performs satisfactorily even under non-tracking mode when compared with a flat-plate collector. The cost effectiveness of the use of such a concentrator favors comparably, if used for space heating and cooling.

INTRODUCTION

This paper describes the results of the investigation of a solar concentrator which is based on a cylindrical reflector having a half-circular cross section. The concentrator is analysed in its non-tracking mode. The receiver is designed to have a placement and size to take into account the changing orientation of the incident solar radiation as the declination changes throughout the year. The concentrator may be made as a part of an array of similar units. These units may either be connected in series or in parallel to form the whole system. The absorbers for this concentrator are mainly of two types:(a) circular and (b) triangular in cross section. Copper was used to construct the receiver. The selection of copper was primarily based on the information received about the corrosion problem encountered by various other groups when water was used as the heat transfer fluid.

The primary areas of present investigation are (a) an in-depth study of the geometry of the concentrator-receiver system with respect to the incoming solar radiaition, (b) detailed calculations of the optical parameters such as the average number of reflections from the reflector to the receiver and (c) modeling of the thermal performances of the collector in terms of heat transfer coefficient, heat loss coefficient, useful energy gain and the efficiency. These types of collectors are commonly referred to as "funneling collectors" or "mixed geometry reflection collectors"[1].

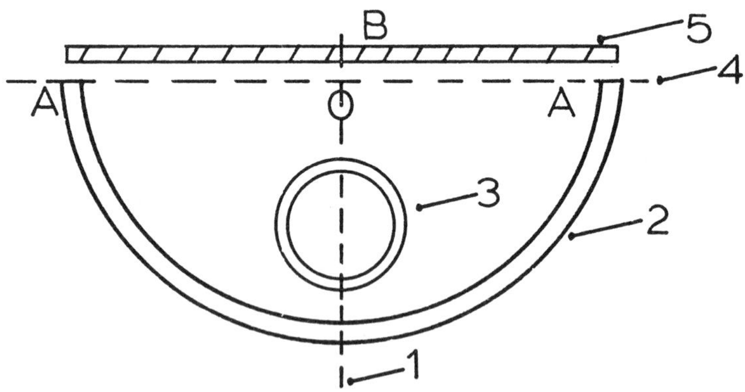

FIG. 1 THE VIEW OF THE COLLECTOR UNIT ALONG ITS LONGITUDINAL AXIS

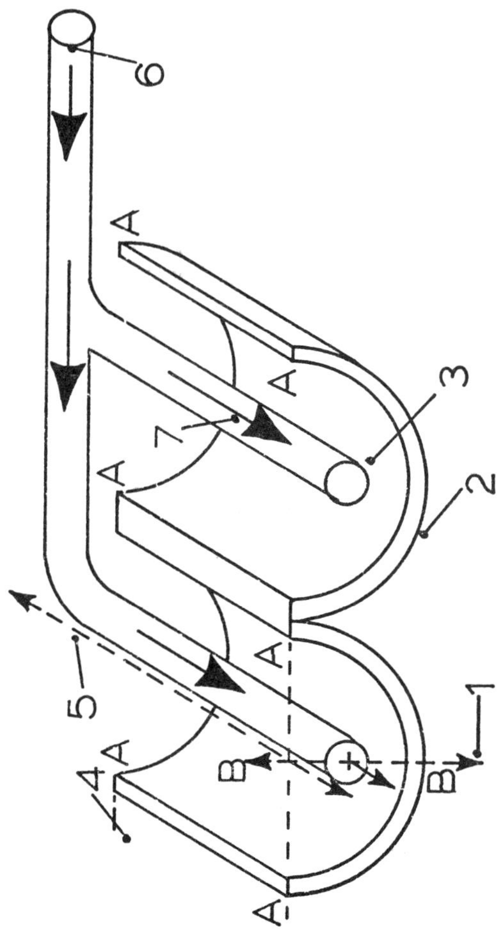

FIG. 2 COMPONENTS OF COLLECTOR IN ARRAY. (1) SYMMETRY AXIS BB, (2) REFLECTOR, (3) RECEIVER, (4) THE TOP PLANE, (5) LONGITUDINAL AXIS, (6) FLUID INLET AND (7) FLUID OUTLET

A prototype collector of approximately 10.88 ft^2 in size and a concentration ratio of 1.29 was built and tested against two commercial flat-plate collectors[2]. Water was pumped through the model at an average rate of 20.16 $lts/hr/m^2$. The temperature rise between the inlet and outlet water averaged 20.8°C for a net useful gain of 1695 $KJ/m^2/hr$, which compares favorably to the 1682 $KJ/m^2/hr$ registered for a "Revere" flat-plate collector and for a "Southeastern" flat-plate collector which registered 1469 $KJ/m^2/hr$. The tests were conducted in the month of August under varying weather conditions.

The most significant observation uptodate was that the experimental non-tracking concentrating collector performed as well as the flat-plate collectors under diffuse and direct radiation. The cost analysis performed on this concentrator suggests that the solar collectors based on similar principles to those employed on the model may be constructed and operated at lower costs and better thermal efficiencies than flat-plate collectors. This should be especially true if a triangular receiver were to be incorporated into the final design.

DESCRIPTION OF THE COLLECTOR'S GEOMETRY

The basic components of the collector are shown in Figures 1 and 2. The reflectors consist of several cylinders of half-circular cross section with concave surface facing the sun. The absorbing pipes are located inside the reflecting cylinders having the same length as the reflectors. The pipes are made of good conducting materials such as copper and have flat-black coating. There is a wide choice for construction materials for the reflectors. Among these are: polished aluminum sheets which have been formed to circular shape, or some non-metallic forms. These include highly reflecting mold blown plastic, reflecting mylar-a polyester made in extremely thin sheets of great tensile strength. Heavy aluminum foil, if properly bound for durability, will also be a good choice. The reflectors with their respective receivers are arranged in rows contained in a rectangular flat-box for the non-tracking setup. In some cases a sheet of transparent material such as sheet glass or sheet plastic or thin film plastic may be used to cover the box, admit radiation and reduce convective heat loss.

Figure 3 shows pertinent dimensions for considering the reflection of incoming vertical rays by an ideal reflector. In this figure one sees important angles defined, which are used to parameterize some of the equations relating reflection and collector geometry. AB= the aperature, O= the center of curvature, OD= the plane of symmetry and c, which for the moment we shall call the central angle, is th le between the plane of symmetry and OP, the normal drawn at the point of incidence The equations for the incident and the reflected rays are given in terms of OX and OY, for normal incidence on the collector.

$$OY = R/(2 \cos c), \text{ where } 0 \leq c < 60^{\circ} \quad (1)$$

$$OX = R \sin c \quad (2)$$

Figure 4 shows a pattern of reflection for a set of normal incident rays entering the collector. It also shows the envelope above which no reflected rays are found. The equations for the limiting line that separates the zone where the reflected

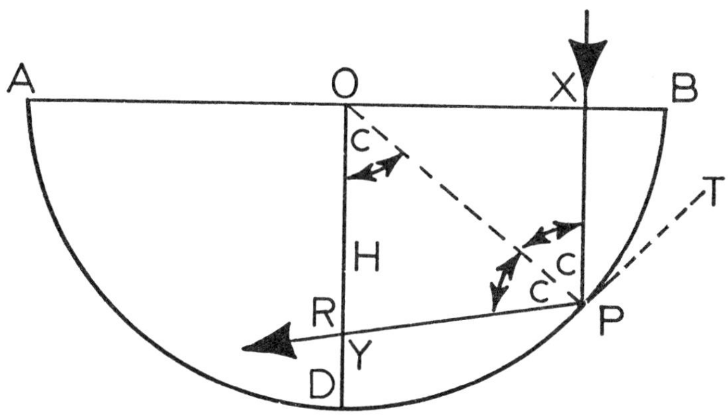

FIG. 3 REFLECTION OF VERTICAL RAYS FROM HALF-CIRCULAR REFLECTOR

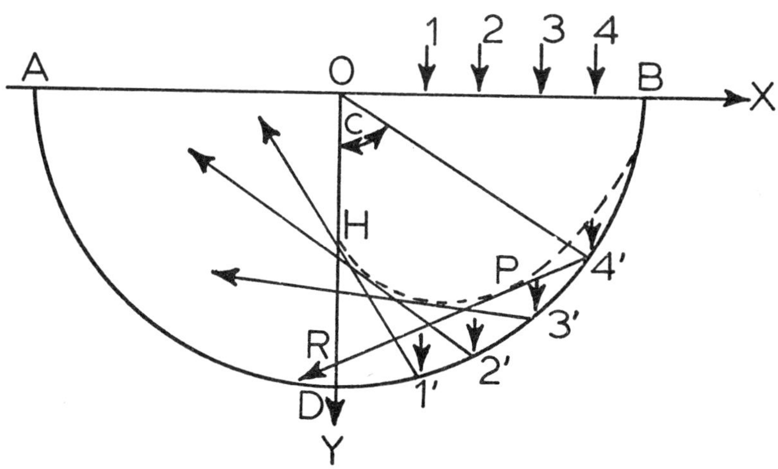

FIG. 4 LOCUS OF CURVES TANGENT TO RAYS REFLECTED BY THE CIRCULAR MIRROR

rays may be intercepted (cross-hatched area in Figure 4) is given as:

$$x = R \sin^3 c \quad (3)$$

$$y = R[-1/(2 \cos c) - \sin^3 c/\tan 2c)] \quad (4)$$

Due to symmetry what is true for one half of the collector is also true for the other half. One will note that the peak of this curve occurs on the plane of symmetry and appears to approach the limit OH, measured from 0, the center of curvature. OH = R/2. It should also be noted that the previous equations are derived for angle c less than 60°. Above this value, multiple reflection begins. 60°, for instance, is at the verge of creating a second reflection. That problem is treated separately.

In order to minimize the amount of tracking required for sunlight to continue entering the collector, the unit may be oriented such that the long axis of the cylinders and the pipes points in the east-west direction with the aperature plane, AB,as defined in Figures 3 and 4,slanted south. Under this arrangement the collector is illuminated throughout the day with intensity following a modified cosine curve with the maximum occurring roughly at noon. This accounts for the variation of the sunlight with the time of day. The sun angle also varies with the time of year at a particular hour during the day. Figure 5 indicates this variation of the sun's declination which is $\pm$ 23.45^o, the maximum at the solstices with its minimum being at the equinoxes.

Figure 6 shows how the pattern of reflection, as indicated by the previously mentioned envelope curve, is altered by incident rays having an angle d with respect to the reflector's plane of symmetry. The fact is that of a simple rotation through the angle'd'about 0 of the reflection envelope curve, if one neglects shading effects and multiple reflections. This is a fortunate circumstance which follows directly from our use of pure half-circular reflector. Figure 7 applies this finding to indicate the effect of extreme values in a solar declination angle which occurred during the year when the collector is oriented to receive normal sunlight incidence on noon of the equinoxes. Particula points of interest when one is designing receivers to fit this reflector and orientation strategy are the points D, H', H, and H". Between these points one will find the reflected rays. Thus they should be on or within the cross sectional shape of any receiver chosen.

DESIGN OF THE RECEIVER SYSTEM

The receiver design as presented here is based on east-west axis with a southward tilt and fully non-tracking for the collector. Once the choice of the collector's positioning has been selected the receiver's design must take into account these following conditions.

It must be able to direct solar radiaiton at all times of the year. This includes the summer and winter solstices at which time the extreme values of declination i.e., $\pm$ 23.45^o occur.

The concentration ratio as defined later, should be kept greater than unity. High concentration ratio is an index of good thermal performance and, hoped, of lower operating costs.

The receiver, obviously, must be of closed shape to allow the warmed fluid to flow through it.

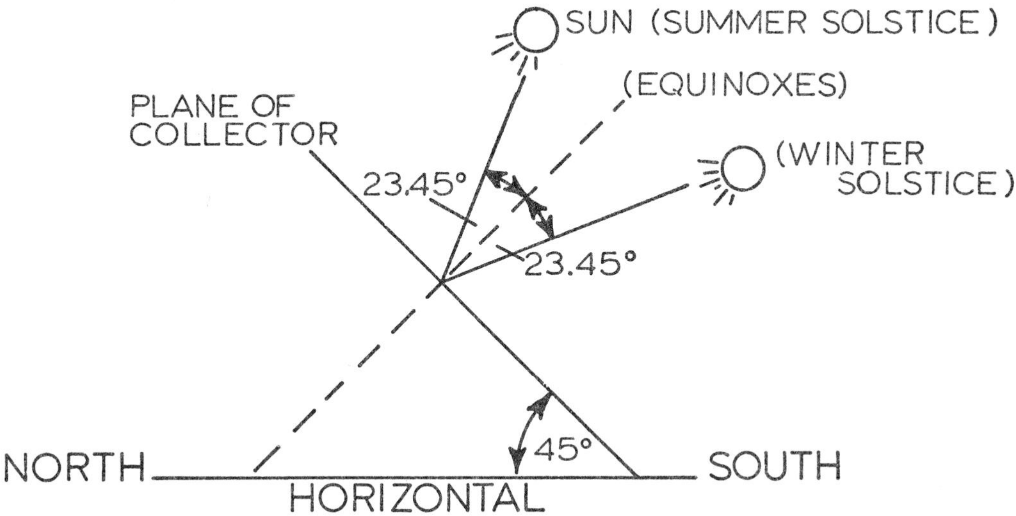

FIG. 5 EXAMPLE OF THE ORIENTATION OF THE COLLECTOR SHOWING MAXIMUM DEVIATION OF THE SUN'S DECLINATION ANGLE.

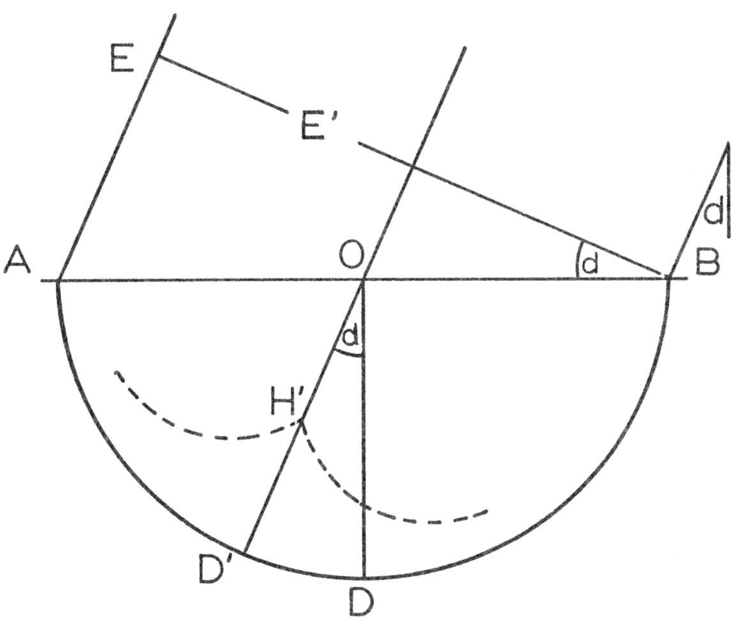

FIG. 6 THE EFFECT OF DECLINATION ANGLE ON THE PATTERN OF REFLECTION.

The size, shape, and materials of the collector should be selected so the installation, operation, and maintainance costs are at least comparable to or better than those for standard flat-plate collector.

THE GEOMETRY OF THE RECEIVER SYSTEM

In figure 8 where d, the maximum angle from which an entering solar ray will deviate from OD, the plane of symmetry, has been set equal to the extreme value of declination equal to 23.45^o, two candidate receiver designs have been drawn superimposed. One is a circular pipe and the other is a triangular pipe. They both pass through the three points D, H', nad H".

Triangular Receiver

The concentration ratio, using the simplest assumption for its calculation, may be defined as the ratio of the aperature to the reflector divided by the perimeter of the receiver pipe. For a triangular receiver the concentration ratio and the dimensions of the receiver are arrived at through the following steps [1].

The dimensions of a triangular receiver may be derived as function of the radius (R) and angle 'd' which was previously defined. R is the radius of the half-circular reflector. Considering Figure 8, and remembering that the line AB or, the plane of collection, is tilted toward the equator by an angle equal to the local latitude, and the angle d= 23.45^o. Then the following relationships follow directly from the geometry of Figure 8.

$$OA= (R/2)\cos d;\ AD= 1-(R/2) \cos d \quad (5$$

$$H'A= (R/2) \sin d \quad (6$$

$$H'H" \doteq R \sin d \quad (7$$

$$\tan \alpha= AH'/AD = \sin d/(2 - \cos d) \quad (8$$

$$H'D = H'A/ \sin \alpha = (R \sin d)/2 \sin \quad (9$$

If $d = 23.45^o$, then it is found:

$$OA =0.459R \quad (1$$

$$AD = 0.541R \quad (1$$

$$H'A = 0.199R \quad (1$$

$$H'H = 0.388R \quad (1$$

$$\alpha =20.183^o \quad (1$$

$$H'D = 0.577R \quad (1$$

The concentration ratio calculated by the ratio of AB to the perimeter of the triangle is:

$$C.R._T=2R/(H'H + 2 H'D) = 2R/ [0.398R + 2(0.577R)]= \underline{1.289} \quad (1$$

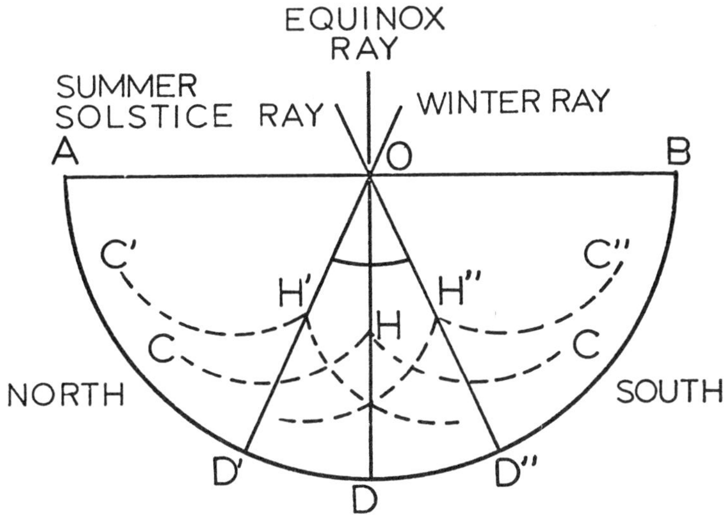

FIG. 7 THE PATTERN OF REFLECTION AT NOON TIME AT THE EQUINOXES AND SOLSTICES FOR THE CIRCULAR REFLECTOR.

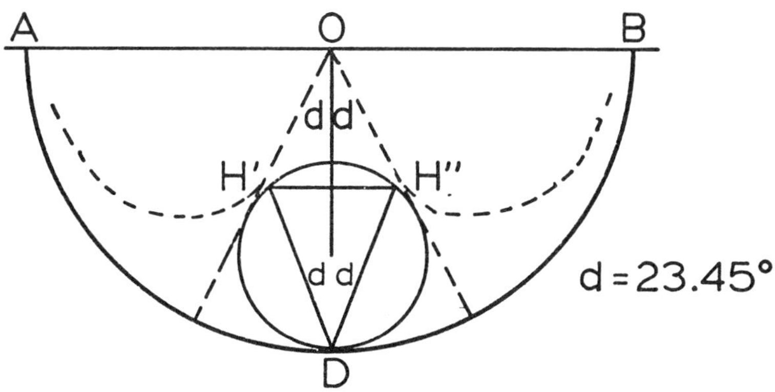

FIG. 8 CIRCULAR AND TRIANGULAR RECEIVERS SUPERIMPOSED FOR COMPARISON

Relationships (10) through (15) are useful as designing parameters if a triangular receiver is used. The C. R. =1.29 is the maximum possible value that can be obtained for a receiver inside a completely fixed half-circular reflector a and it compares favorably to a flat-plate collector. whose C.R. is equal to unity.

Figure 9 shows the main dimensions for the triangular receiver. Although a triangular receiver appears to be the optimum shape in terms of thermal performance and economy of materials, it may represent greater difficulties and higher costs to build triangular pipes and accesories for them unless they are manufactured in large quantities. The next step is to consider a different geometry for the ease of construction and therby trade off the thermal efficiency . Therefore, a circular receiver is considered next.

Circular Receiver

A circular shape that is circumscribed to the triangular geometry described above is expected to have a smaller C.R. because of its larger perimeter. Its expected advantages are based on the ease of construction and assembly offered by circular pipes.

Figure 10 and additionally from eqns 5 and 15 the following relationships hold:

CD = r = the radius of the circular receiver

$$SD = H'D/2 = (0.577R0/2 \qquad (16)$$

$$CD = SD/ \cos 20.18^{0} \qquad (17)$$

From equations 16 and 17 then we have

$$r = 0.3074R \qquad (18)$$

The C.R. is found by dividing 2R (the width of the collector) by the circle's perimeter.

$$C.R._{C} = 2R/[2\pi(0.3074R)] \qquad (19$$

The relationships developed above show that by using a circular pipe as a receiver it is possible to collect all radiation at a C.R. =1.036 which is slightly better than the C.R. of flat-plate collector (C.R. =1) and lower than and lower than the C.R. = 1.289 found for a triangular receiver.

Additional Improvement of C.R. by the Change of Reflector Shape

It is possible to increase the C.R. by approximately ten percent by introducing a minor modification of the circular reflector. In Figure 11 if the reflector section CB is changed into a tangent CB' at point C (where angle BOC equals approximately 23^{0}), the reflected beams striking between B' and C will be reflected to the receiver pipe even at an angle of maximum declination (23.45^{0}). This means that for the same size of receiver-pipe, the incoming radiation area is increased from AB to A'B', with a net gain close to 10%. If a circular

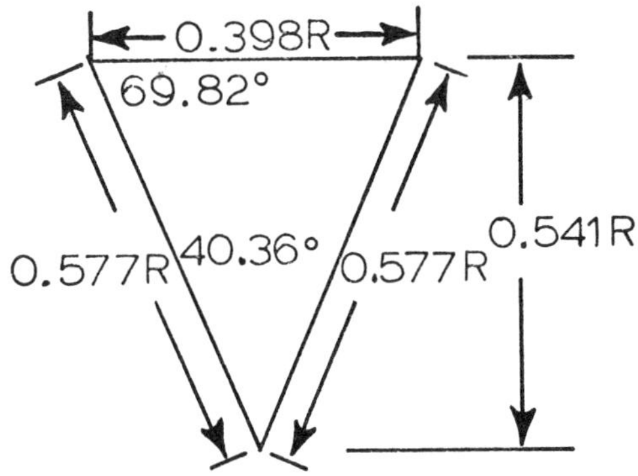

FIG. 9 GEOMETRY OF THE TRIANGULAR RECEIVER FOR THE NON-TRACKING RECEIVER

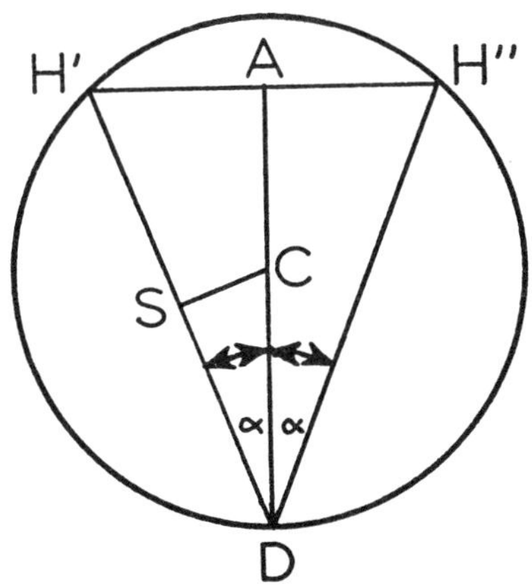

FIG. 10 THE GEOMETRY OF THE CIRCULAR RECEIVER

receiver is used then the C.R. is increased from 1.03 to about 1.13. This additional change is justified because it means less use of copper pipe for the receiver, less weight per unit area and a small but useful increase in the thermal efficiency of the collector. The geometry of this modification is discussed below.

Figure 12 shows the case when a sun beam strikes a plane forming an incident angle equal to 'd'; the plane is tilted by an angle equal to 'a' with respect to the vertical plane. The reflected beam forms an angle equal to 2a + d with the vertical line. Figure 13 shows a reflector that has been modified by making section CB' a tangent line instead of the previous arc CB. The coordinates of point B', X and Y are determined by the set of following equations:

$$Y = -\sin a + (x \quad \cos a)/\tan a \tag{20}$$

$$Y = -\cos(d/2) + (x + \sin (d/2))/\tan (2a + d) \tag{21}$$

Because of the fixed location of the collector, the maximum value being 23.45^{o}, and because the reflector beam is within an angle greater than zero, the angle 'a' must comply the following inequality

$$90 - 2a - d > 0, \text{ or } a < 33^{o} \tag{21}$$

If the above set of equations are solved for several angles'a', and keeping d= 23.45^{o}, several points (X,Y) are obtained, one for each value of 'a'. The general solution of the equation is given by:

$$X = [-1/2 \cos d + (1/2) \sin d / \tan(2a + d) + \sin a + \cos a?\tan a] X$$

$$[1/\tan a - 1/\tan(2a + d)]^{-1} \quad (22$$

Table I gives several values of X and Y for different angles 'a'. The bigger the value of X, the better the concentration ratio since more radiation is captured for a fixed size of the receiver. The value of X/R is a measurement of the factor that increases the previous C.R. if a circular reflector without modification were to be used. Therefore, the new C.R. is obtained by multiplying 1.0355 (C.R. for a circular receiver) times the value of X/R. Y gives the height of the reflector measured above the line AB(the top diameter of the circular reflector). The bigger the value of Y, the more reflecting material that is necessary.

Table I shows that the angle 'a' is about 13° for the concentration ratio to increa to about 1.24 provided we use a circular receiver. It would require a reflector unit with a height of 1.74R and 2.4R wide. These conditions determine the upper limits of concentration power for a non-tracking circular reflector with a circular pipe as an absorber. For the same reflector geometry, the C.R. of a triangular receiver would be about 1.60.

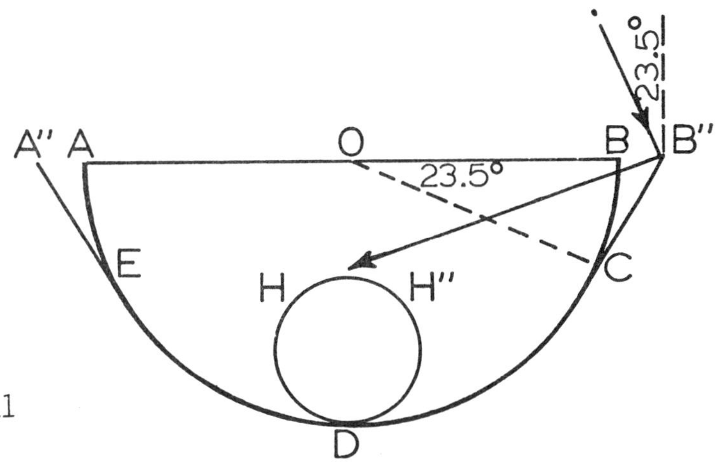

FIG. 11

THE CIRCULAR REFLECTOR MODIFIED TO INCREASE THE CONCENTRATION RATIO AND IN THIS CASE, THE ANGLE BOC = THE ANGLE OF MAXIMUM DECLINATION.

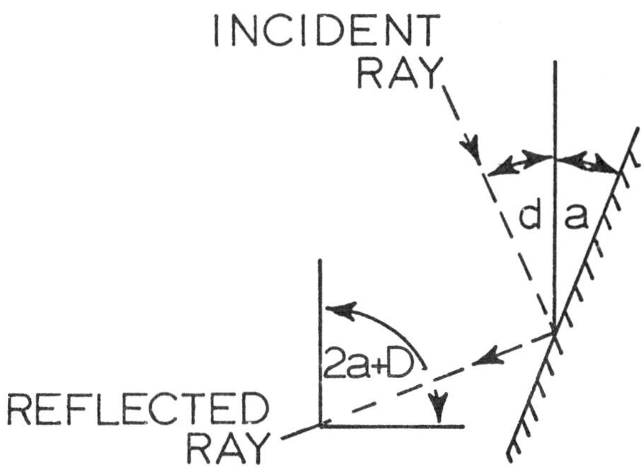

FIG. 12 REFLECTION OF RAYS OVER A TILTED SURFACE.

TABLE I
VARIATIONS ON C.R. WITH DIFFERENT REFLECTOR'S GEOMETRIES

a	X/R	Y/R	C.R.=1.0355X
5°	1.11079	1.22271	1.150
9°	1.18867	1.112514	1.231
10°	1.19379	1.0115276	1.236
11°	1.19660	0.9151144	1.239
11.725°=d/2	1.19729	0.84790	1.240
12°	1.19726	0.822957	1.240
13°	1.19594	0.734775	1.238
15°	1.1878	0.569355	1.230
20°	1.14094	0.210902	1.181
21°	1.12768	0.147287	1.168
22°	1.11337	0.086231	1.153
23°	1.09804	0.02752	1.137
23.45°=d	1.09083	0.00184	1.130
25°	1.06296	-0.086677	1.101
30°	0.967173	-0.324808	1.001

System Design Specifications

The collector that has been designed so far consists of a n ideal reflector and a pipe receiver that contains the heated fluid. Figure 14 shows the section of this geometry with the parameters needed to build it. The reflector is an ideal cylindrical surface whose section is the result of a circumference sector of radius 'R'(BB') combined with two tangent lines (AB, AB').

$$R = r/0.3074 = 3.253r = 1.627\ d_o \tag{23}$$

From table I a=12° produces the largest W, reflector width. Then the maximum concentration ratio for a circular receiver occured with a value 1.24.

$$W = (1.197)\ 2RX \tag{24}$$

When a=12°, then Y/R is .8229

$$H = (1 + Y/R \cdot R = 1.8229\ R = 2.965\ d_o \tag{25}$$

The length along the cross section of the reflector, L is given by;

$$L = 2\ \pi\ R\ [\ (180 - 2a)\ /360 + (Y/R + \text{Sin}\ a)\ /\ \cos a]$$

$$L = 4.83 = 15.71r = 7.86\ d_o \tag{26}$$

By selecting a convenient d_o and by using the above expressions for the cross section dimensions, given as numerical multiples of d_o when a=12° and d=23.45°, equals the maximum declination angle, then the useful collector cross section dimensions are fixed.

The receiver pipe for the model which was built had an inside diameter of one inch (2.847 cm) and therefore

R = 1.23 in. (4.631 cm)

W = 4.4 in. (11.18 cm)

H = 3.5 in. (8.51 cm)

L = 8.89 in. (22.58 cm)

L_c the collector length = 35.5 inches (2.96 ft) .902 meters)

A one eighth inch thick (.318 cm) Tedlar glass sheet was used as a cover.

The concentrator's nine cylindrical mirrors were placed beside each other and the pipes for their receivers were connected in series by joints and elbows. The pipes were painted flat black using non selective paint made from carbon black, copper oxides and manganese. Reputedly the paint is heat resistant to 1002° F (649° C).

The reflector material was paper backed aluminum foil supported at the peak between adjacent reflectors by a taut wire. The overall array of concentrators and cover glass was placed in a flat rectangular box made of one quarter inch plywood. With a width of 4.4 inches (11.18 cm), for each of the nine elements of the reflector array, there was 1.08 ft^2 (156 cm^2) of aperture for each reflector

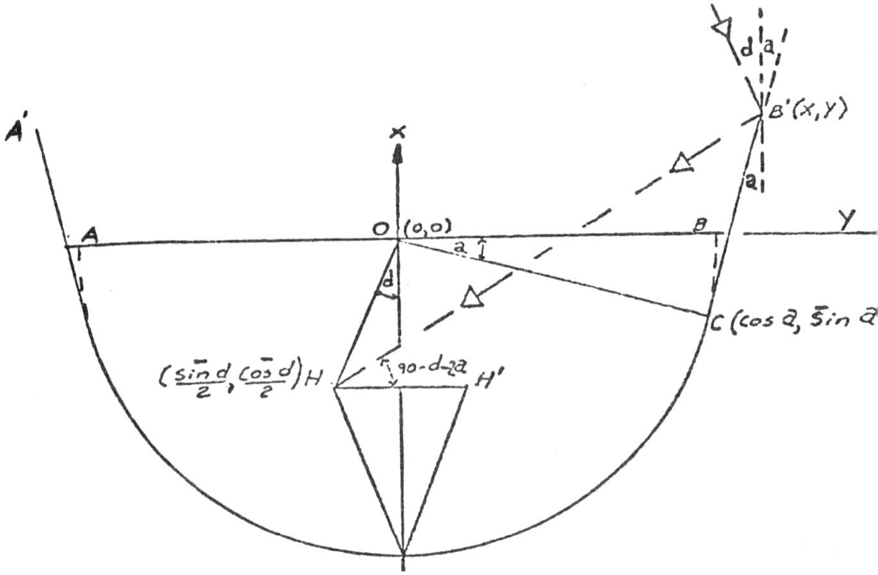

FIG. 13 REFLECTION OVER MODIFIED REFLECTOR WITH TRIANGULAR RECEIVER

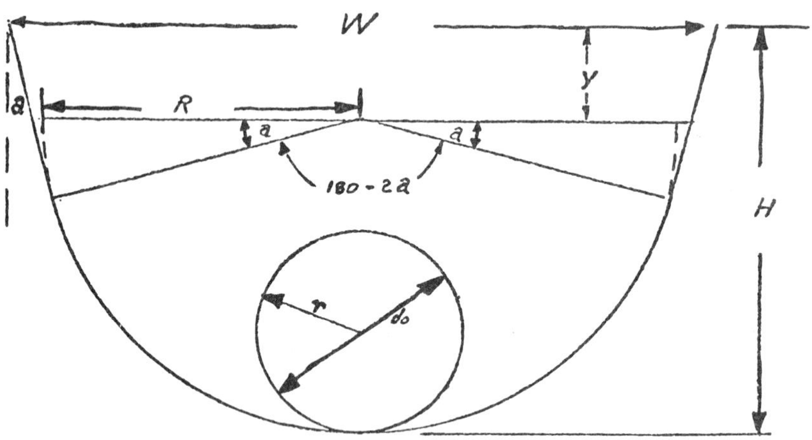

FIG. 14 GEOMETRY OF MODIFIED REFLECTOR& CIRCULAR RECIEVER WHICH WAS TESTED

and 9.76 to 10 ft^2 (.907 m^2) of collector area for the entire array. In addition to the 9 X 35.5 inches of pipe forming the receiver for each concentrator, there was 120.5 inches (306.1 cm) of pipe leading to, between and away from the concentrators. It was black and also exposed to sunlight. Multiplying it by d_o it's area could be found to read 9.38 ft2 (1.871 m^2) and added into the rest of the collector area which now becomes 10.9 ft^2 (.9071 m^2).

The measurements were taken of the performance of this concentrated collector during August of 1977 and are available to be correlated with the predictions of several versions of mathematical models which are constructed in order to theoretically calculate the behavior of Solar concentrators of this type.

THERMAL PERFORMANCE OF THE COLLECTOR

There are several published approaches, of different complexity, thoroughness or purpose in relation to the problem of calculating thermal performance of Solar concentrators. Singh and Cheema[3] provide a set of equations for the purpose of supporting their discussion of the selection of sets of optical parameters to be adjusted together when seeking to optimise several parabolic cylindrical reflector-receiver combinations. Thermal power gathered and fluid temperature were given as variables on a per unit concentrator lens basis and were not uniquely determined by the equation set given there.

Tanaka, et al [4-5] present an approach to thermal performance equation which is deterministic with respect to the above variables. That set of equations oriented more toward a strictly numerical solution than is the following one.

Finally the set of formulas summarized in Duffie and Beckman's text [6-7] were selected as the ones most closely fitting the requirements of our selected task. Those formulas, which will also be given below, are thorough and are closer to being (although not entirely) equations of closed form with respect to Q_u the useful thermal power yielded, receiver and circulating fluid temperatures.

The rate of Solar radiation reaching and being absorbed by the receiver is:

$$S = H_b \quad R_b \, \rho\gamma\alpha \tag{27}$$

H_b is the Solar radiation measured on a horizontal plane. R is a factor to adjust this measured value to the Solar radiation falling on the aperture plane of the reflector given the orientation scheme:

$$R_b = \cos\theta_i \, / \, \cos\theta_z \tag{28}$$

Subscript b denotes direct or beam Solar radiation. If one were to use the fact that the collector is tilted at a 45° degree angle toward the South rather than having a tilt equal to it's latitude and also were to use H, solar radiation measured on a horizontal plane then R_b would take on the form;

$$R_b = \cos\theta_t \, / \, \cos\theta_z \tag{28a}$$

In our case, however, the joint quantity $H_t R_t$, the rate of total Solar radiation received at a 45° tilt angle was the one measured and used for the calculations. The quantity $S/(H_b R_b)$ equal to $\rho\gamma\alpha$ or $\rho\gamma\tau\alpha$ is a measure of optical efficiency of the concentrator system. ρ is the specular reflectance of the reflector. α is the absorptance of the black coated receiver pipe. [6,8]

For aluminum base coatings ρ is given as follows: .86 for foil which is 99.5% pure; .87 for Al, SiO coating, clean: .89 for sputtered aluminum optical reflector. For Parsons Black paint α is .98. Emittance, ε, of the Parsons Black paint to infrared radiation, is .981 for temperatures in the range 240° to 460°K [8]. ε will appear in a later equation for radiative heat loss. γ is an optical intercept factor relating to the fraction of the distribution of direct and reflected radiation reaching the receiver. As Singh and Cheema note, concentration ratio C.R. and γ are competing factors so that as adjustments are made to raise one, the other usually falls [3]. Thus, for a later task, optimization of the energy collected, their suggestion is to maximize the product of these two variables. In the present work however, the receiver and reflector dimensions were selected so that γ is unity, i.e. all direct and reflected rays of sunlight are intercepted by the receiver. In fact the designs were chosen so that this would be true even for all the expected angles of declination encountered by the concentrator during the year (see Figs. 8 through 14 τ is a transmittance factor and is a ratio of the intensity of radiation reaching the receiver to that which reaches the collector concentrator aperture plane. It takes into account such intensity diminishing factors as losses during reflections from the reflector, noting that ρ is less than one for real reflector materials. It also accounts for losses by absorption and reflections at interfaces of transparent cover sheets or tube, if any. it is usually given as a product of each of these influences, where each is calculated separately while neglecting the others.

$$\tau = \tau_\rho \tau_r \tau_a \qquad (29)$$

τ_a, the transmittance due to optical absorption in the transparent cover, and τ_r, transmittance due to collection at transparent interfaces is given, for flat transparent covers.[9-10]. Rabl gives τ_ρ as approximately [1]:

$$\tau_\rho = \rho^{\langle n \rangle} \qquad (30)$$

where $\langle n \rangle$ is the average number of reflections. In the present paper initially equation 27 is used where in effect τ from equation 29 is set equal to 1, which means cover and reflection losses are neglected. The alternative is to set the equation 30 equal to 1, and multiply equation 27 by equation 29.

The useful energy gain for the fluid flowing through the receiver pipe is

$$Q_u = A_a F_R \left[S - \frac{A_r}{A_a} U_L (T_{f,i} - T_a) \right] = \dot{m} C_p (T_{f,o} - T_{f,i}) \qquad (3$$

where,

$$F_R = \frac{1}{C_1} [1 - \exp(-C_1 F')] \quad ; \quad C_1 = \frac{\dot{m} C_p}{A_r U_L} \qquad (3$$

A_a/A_c = C.R., the concentration ratio, as previously discussed.

$$A_r = l_c \pi d_o \text{ and } A_a = l_c W \qquad (33\ a,b)$$

where l_c, W are concentrator length and reflector width. In simulating performance of an array of concentrators one may wish to modify (33) by substituting $N_r \times l_c$ into (33) for l_c, where N_r is the number of side by side reflector receiver combinations, if they are linked in series, as is the case for the model discussed. The total receiver pipe length for the N_r receivers is a desirable basis for checking diameter/length ratios for flowing fluid, on which Reynolds number for turbulence or "fully developed flow" criterion depend.

If no transparent cover is present, U_L the overall heat loss coefficient is

$$U_L = (1 / h_w + 1 / h_r)^{-1} \qquad (34)$$

The collector efficiency factor, F', is

$$F' = U_0 / U_L \qquad (35)$$

where

$$U_0 = [1/U_L + \frac{d_o}{d_i h_i} + \frac{d_o}{2k_{pipe}} \ln(d_o/d_i)]^{-1} \qquad (36)$$

In Equations 33 through 36, the presence of d_o or πd_o or d_i indicates that the equations were originally derived for circular pipe receivers. For non-circular receiver pipes, modification of such equations may be made by using the product of length and actual outside perimeters for A_r, receiver area and by using the inside (and outside) hydraulic diameter to substitute for d_i and d_o, the inside and outside diameter, respectively. The hydraulic diameter is defined as the ratio of four times the cross-sectional area to the perimeter around the pipe.

The convective heat transfer coefficient between the inner surface of the receiver and the fluid flowing through the receiver is h_i. The value of h_i may be expressed as a function of the operating conditions in a manner similar to that used with other convective heat transfer coefficients, namely by applying the definiton of Nusselt number and by the use of one of the experimentally obtained correlations. In such correlations Nusselt number is given as a function of other non-dimensional numbers, such as Prandtl number or Reynold's number. The Prandtl number is the ratio of specific heat times viscosity to the thermal conductivity of the convecting fluid, whereas the Reynold's number is given by the following relation.

$$Re = \frac{v_f D_{H,i}}{\nu} \qquad (37)$$

v_f = fluid velocity, $D_{H,i}$ = hydraulic diameter inside of the tube (= d_i for circular pipe), and ν = the kinematic viscosity.

For our purpose we have modified the Reynold's number to include the mass flow rate of the fluid through the receiver pipe and the inside perimeter of the pipe, P_i. The modified Reynold's number therefore is:

$$Re = \frac{4\dot{m}}{\mu P_i} \quad \text{where } \mu = \text{the absolute viscosity.} \qquad (38)$$

$P_i = \pi d_i$ for a circular receiver.

Heat transfer coefficients for common geometries are given in many heat transfer textbooks [9-11]. For fully developed turbulant flow of liquids with the Prandtl number between 1.0 and 20 inside tubes, Kays recommends that we use the following relationship between the Prandtl number , the Reynold's number and the Nusselt number [6].

$$Nu= 0.0155\ Re^{0.83}\ Pr^{0.5} \tag{39}$$

For laminar flow inside short tubes, Kreith [9] gives the following equation which may be useful since the pipes in solar systems are often short, and fully developed conditions may not exist:

$$Nu = \frac{Re\ Pr\ D}{4L}\ \ln\left[1 - \frac{2.654}{Pr^{0.167}\ (Re\ Pr\ D/L)^{0.5}}\right]^{-1} \tag{40}$$

As one might expect, the D/L is the ratio of the flow length to the flow diameter. Equation 40 is valid for flow in tubes when L/D is less than 0.0048 Re, and in flat ducts when L/D is less than 0.0021 Re. In equation 34, U_L was shown to be dependent upon the heat transfer coefficient due to the wind speed past the collector. McAdams [11] expresses it as

$$h_{wind} = 5.7 + 3.8v \tag{41}$$

with the heat transfer coefficient expressed in Watts/m^2 °C and v in m/sec. The heat transfer coefficient h_i is now given as:

$$h_i= NuX\ k\ /L \tag{42}$$

$L = d_i/2$ where d_i is the inside diameter.

The above equations are used in our case where water is the circulating fluid and the receiver pipe is circular.

Multiple Reflection

A detailed study of the case of multiple reflection is presently underway. The multiple reflection indeed is a complex phenomena. However under certain simplifying assumptions we arrive at the following relation. Feferring to Figures 3 an[illegible] it was noted that for normal incidence and centarl angle c greater than 60°, multiple reflection occurs. The number of multiple reflections for a particular light ray increases without bound as the centarl angle c aproaches 90^0. The relat[illegible]ships between n, the number of reflections and the angle c are derived. When n is an integer and $c=c_v$= the nagle which is at the verge of increasing n by one additi[illegible] reflection, we have:

$$c_v/90^o = n/(n+ 1) \tag{43}$$

or, $$n = c_v/(90^o - c_v) \tag{44}$$

These equations hold for the the unmodified circular reflector, neglecting interception before the symmetry plane of the rays by the receiver.

TABLE II - Time, Inputs to simulation, Transmittance, and S

Time hours	H R W/m^2	S W/m^2	$T_{f,i}$ °K	T_a °K	τ
10.8	560.6	392.145	294.372	301.261	.713765
10.9667	595.137	416.561	298.594	302.039	.714225
11.1333	595.137	416.759	304.594	302.039	.714564
11.3	645.053	451.876	304.65	300.539	.714822
11.4667	656.397	459.944	304.261	301.761	.715011
11.7167	720.682	505.114	304.706	301.761	.715166
12	631.187	442.43	304.483	302.039	.715254
12.2167	760.072	532.743	305.594	302.817	.715215
12.3833	776.773	544.385	306.261	303.928	.715131
12.8	821.835	575.599	308.817	304.428	.714676
12.9667	838.568	587.068	310.65	304.317	.714371
13.2167	855.553	598.426	311.15	305.483	.713735
13.3833	869.419	607.635	312.372	304.706	.713161
13.6333	869.419	606.649	314.872	306.039	.712004
13.8	813.327	566.695	315.761	307.261	.71098
13.9667	869.419	604.687	317.706	307.817	.709701
14.1333	847.045	587.801	318.539	309.206	.708105
14.3833	849.881	587.145	319.706	309.15	.704955
14.55	931.182	640.808	320.483	308.539	.70221
14.7167	855.553	585.913	319.706	309.428	.698811
14.8	684.443	467.395	318.817	309.094	.69682
14.8833	420.687	286.367	315.261	308.706	.694606

TABLE III- Simulation Values of Temperatures for the Concentrator

Time hours	T_a °K	$T_{f,i}$ °K	$T_{f,m}$ °K	$T_{p,m}$ °K	$T_{f,o}$ °K
10.8	301.261	294.372	298.678	679.734	302.557
10.9667	302.039	298.594	302.801	675.24	306.594
11.1333	302.039	304.594	308.281	635.661	311.627
11.3	300.539	304.65	308.507	650.609	311.998
11.4667	301.761	304.261	308.319	667.783	311.982
11.7167	301.761	304.706	309.104	697.888	313.057
12	302.039	304.483	308.397	655.407	311.937
12.2167	302.817	305.594	310.237	720.03	314.397
12.3833	303.928	306.261	311.041	732.632	315.317
12.8	304.428	308.817	313.665	741.007	317.996
12.9667	304.317	310.65	315.413	735.428	319.672
13.2167	305.483	311.15	316.065	749.101	320.452
13.3833	304.706	312.372	317.179	740.911	321.474
13.6333	306.039	314.872	319.565	733.353	323.76
13.8	307.261	315.761	320.165	709.109	324.116
13.9667	307.817	317.706	322.286	726.379	326.385
14.1333	309.206	318.539	323.032	719.565	327.056
14.3833	309.15	319.706	324.088	711.09	328.018
14.55	308.539	320.483	325.163	737.695	329.344
14.7167	309.428	319.706	324.103	712.399	328.046
14.8	309.094	318.817	322.334	634.656	325.525
14.8833	308.706	315.261	317.551	522.816	319.67

RESULTS

We have made both theoretical and experimental observations of the thermal performanc of the concentrator. Very preliminary theoretical calculations of the optical performance were also done. Earlier in this paper, the optical properties were discussed at length. We then made a tentative assessment of how optical parameters will influence the concentrator performance. The effect that the glass cover will have on diminishing the amount of light reaching the receiver was considered in the calculations as was the influence of less than total reflection from non-ideal reflector surface. Rabl's expression (eqn.30) was used by assuming a nominal value for the average number of reflections. It was felt that the value 1.3 was reasonable, pending more investigation of that and related optical matters later. Inputing the parameters for the 1/8 th inch thick tampered glass cover, Deitz's procedure for determining the transmittance of flat glass covers as a function of the incidence angle of the sun light was used to complete the evaluation of equation 29. That result appears as the rightmost column of Table II. The result of Table II is composed of the basic time changing input which were used in the computer simulation of the concentrator performance.

In Table II the variables time, H.R, the solar radiation on a 45° southward tilted plane, S, the solar radiation reaching the receiver and being absorbed, $T_{f,i}$ the fluid inlet temperature, T_a, the ambient temperature, are listed. The values of these variables are typical for an August day in Louisville, Kentucky. The model of the collector was also tested during day in the month of August. The fixe inputs used to compile the data are listed below.

Latitude and collector tilt = 37.5° and 45°, respectively
Collector area = .907 sqm
Reflector and receiver length = .902 m
Width of reflector array containing nine reflectors =1.01 m and receivers connected in series

Single reflector width = .112 m
Flow rate/collector area = $\frac{401}{m^2\ hr}$ = .0111 kg/sec m^2

Reflectance of the reflector is .86 for 99.5% pure aluminum foil
Plate absorptance and emittance = .98 and .981
Inside and outside diameter of the receiver = .0254 m and .02397 m

In Table III further calculated results from simulations of the concentrator's performance are listed. Following the times column on Table II, T_a $T_{f,1}$ are again listed for comparison with the other temperatures of the Table, $T_{f,m}$ the mean fluid temperature $T_{p,m}$ the mean plate temperature and finally $T_{f,o}$ the fluid outlet temperature. We suspect that of the variables tabulated, mean plate temperature is being shown by the program at a higher value than is reasonable, due to a persistent program problem. Repeated program runs with slight logically chosen variations in program structure did however indicate the other variables remain stable and were affected to a much lesser extent than $T_{p,m}$. For this reason and due to favorable comparison to the experiment we feel the other values tabulated here according to our method of simulation are credible.

Table IV lists time, energy rates and instantaneous efficiency. After time, HR and S appear for comparison with useful energy gain Q_u, Q_u/A_c, H.R and the instantaneous efficiency. As stated earlier the glass cover was considered relative to optical losses but for simplicity U_L, The overall coefficient was found by equation 34. That equation is based on the absence of covers. F_R, The heat removal factor, F', the collector efficiency factor and U_L were calculated by a method in which the value fluctuated with time and changing conditions. The values of F_R and F' were close to .81 and .95. Despite the problem with plate temperature, U_L remained close to 16W / (m^2 hr ° k). Wind speed was assumed as the constant 5 m/sec.

Tables V and VI report experimental measurements which were performed on the model of this concentrator at the University of Louisville. Table V summarizes the full set of experiments to date while Table VI shows a typical example of one of these experiments. The model of this concentrator, having the cross section shown in fig. 14, was tested and compared with the performance of two other collectors. The other two collectors are commercially available flat plate collectors. f_3 / f_1 and g_3 / g_1 stand for the ratio of volume flow rate per aperture area of the model with respect to the Revere collector and ratio of useful energy gain per unit area or efficiency respectively. T_3 / T_1 is the ratio of the temperature rise of the circulating water between the model and the Revere collector. The subscript (1), (2), (3) are for the Revere, Southeastern and the Concentrator model collectors respectively.

The measurements taken show that the model, in spite of it s construction imperfections, worked at a comparable thermal efficiency under all conditions. The fact that the model worked as well as one of the best available commercial flat plate collectors is good evidence that it may be considered an alternative(to the others).

In general, at high flow rates per unit area, the model efficiency exceeded that for the flat plate collectors; because of the small temperature differences the thermal losses due to insulation deficiencies are not as significant as cases having low flow rates per unit area with high temperature differences. Better insulation will increase the efficiency at high temperatures.

In the experiments 4 through 7 the volume flow rate of water in the model concentrator was close to the 40 l / hr m^2 used in the simulation. The computer simulated results are similar to the above mentioned experiments, lying within the present range of uncertainty. The uncertainty arose due to the difference between the simulations and the experiments, such as solar radiation being different in each case and similar effects which have not been quantified at this time.

In conclusion, the half-circular cylindrical concentrator has performed exceptionally well under non-tracking mode. The cheaper construction makes it an attractive design for low temperature liquid heating applications. We are now studying other shapes of receivers for better heat collection. The results of this phase of work will be published at a future date.

TABLE IV Energy Rates and Instaneous Efficiency for the Simulation.

Time hrs	H.R watts/m^2	S w/m^2	Q_u watts	Q_u/A_c w/m^2	Efficiency
10.8	560.6	392.145	345.729	381.057	.67973
10.9667	595.137	416.561	337.91	372.439	.625803
11.1333	595.137	416.759	297.029	327.38	.550091
11.3	645.053	451.676	310.385	342.102	.530347
11.4667	656.397	459.944	326.138	359.463	.547631
11.7167	720.682	505.114	352.741	388.785	.539466
12	631.167	442.43	314.839	347.01	.549773
12.2167	760.072	532.743	371.801	409.792	.53915
12.3833	776.773	544.385	382.505	421.59	.542746
12.8	821.835	575.599	387.723	427.342	.519985
12.9667	838.568	587.068	381.075	420.014	.500871
13.2167	855.553	598.426	392.89	433.036	.506148
13.3833	869.419	607.635	384.447	423.731	.487373
13.6333	869.419	606.649	375.427	413.789	.475937
13.8	813.327	566.695	352.886	388.945	.478214
13.9667	869.419	604.687	366.63	404.093	.464785
14.1333	847.045	587.801	359.771	396.533	.468137
14.3833	849.881	587.145	351.123	387.002	.45536
14.55	931.182	640.808	374.287	412.532	.44302
14.7167	855.553	585.913	352.298	388.296	.453854
14.8	684.443	467.395	283.366	312.321	.456315
14.8833	420.687	286.367	186.235	205.265	.487929

TABLE V

SUMMARY OF EXPERIMENTS

	No.	DATE AUG/77	HOURS FROM: (pm)	HOURS TO: (pm)	TIME (MIN)	WATER FLOW f (LT/HR-M^2) (1)	(2)	(3)	TEMP. INCREASE T °C (1)	(2)	(3)
	1	4	12:43	3:18	150	18.07	15.68	12.60	19.3	19.5	26.2
	2	6	2:46	5:05	140	13.00	11.36	18.18	26.0	23.7	20.4
1)	3	7	12:50	2:15	95	19.48	10.84	12.30	16.8	19.7	18.5
2)	4	9	11:30am	2:30	180	94.8	129.6	91.7	3.97	2.61	6.07
3)	5	11	1:55	3:25	90	39.6	36.9	39.1	2.78	1.50	3.22
4)	6	13	3:05	5:40	145	38.0	79.4	39.9	8.25	3.71	9.30
	7	14	3:50	5:20	90	33.4	34.4	34.9	11.22	10.19	12.33
	8	16	1:00	3:30	150	17.26	14.70	14.89	15.10	15.87	16.39
	9	17	12:45	4:45	240	25.46	20.87	23.81	22.68	24.08	23.53
5)	Wt. Ave. of Exp. 1, 2, 7, 8, 9					21.09	18.51	20.16	19.81	20.10	20.78

	No.	DATE AUG/77	HOURS FROM: (pm)	HOURS TO: (pm)	TIME (MIN)	HEAT GAIN g KJoule/M^2-H (1)	(2)	(3)	RATIO f3/f1	T3/ T1	g3/g1
	1	4	12:43	3:18	150	1440	1329	1383	0.697	1.358	0.960
	2	6	2:45	5:05	140	1413	1125	1562	1.398	0.785	1.105
1)	3	7	12:50	2:15	95	1344	890	952	0.631	1.101	0.708
2)	4	9	11:30am	2:30	180	1633	1410	2322	0.967	1.529	1.422
3)	5	11	1:55	3:25	90	457	232	522	0.987	1.158	1.142
4)	6	13	3:05	5:40	145	1410	1242	1573	1.050	1.127	1.116
	7	14	3:50	5:20	90	1551	1435	1798	1.045	1.099	1.159
	8	16	1:00	3:30	150	1072	980	1012	0.863	1.085	0.944
	9	17	12:45	4:45	240	2419	2077	2356	0.935	1.037	0.974
5)	Wt. Ave. of Exp. 1, 2, 7, 8, 9					1682	1469	1695	0.972	1.070	1.01

TABLE VI

Experiment #7 August 14/77 3:50 to 5:20 pm

					°F	oz	sc	°F	oz	sc	°F	oz	sc
NO	HOUR	WEATHER	Tair	Tpool	T1	V1	τ1	T2	V2	τ2	T3	V3	τ3
1	3:50	PSNY	90	83	114	32	20	107	8	15	110	64	233
2	4:00	"	90	83	112	64	40	107	8	18	110	64	183
3	4:10	"		84	100	32	22	100	8	19	106	8	23
4	4:20	SNY		84	109	64	42	108	8	18	105	64	188
5	4:30	PSNY	88	83	102	64	34	106	8	19	105	64	190
6	4:40	SNY		83	105	64	35	102	32	54	105	64	190
7	4:50	"	90	84	104	64	35	102	32	55	107	64	193
8	5:00	"	89	83	98	64	36	99	8	14	104	64	195
9	5:10	PSNY		89	100	64	35	100	16	27	105	64	195
10	5:20	CDY		84	95	64	35	94	8	14	102	64	191

					f LT/M2-H			Δ T °C			g KJ/M^2-H		
NO	HOUR	WEATHER	Tair	Tpool	(1)	(2)	(3)	(1)	(2)	(3)	(1)	(2)	(3)
1	3:50	PSNY	90	83	31.14	35.48	28.95	17.22	13.33	15.00	2247	1982	1820
2	4:00	"	90	83	31.14	29.57	36.86	16.11	13.33	15.00	2102	1652	2317
3	4:10	"		84	28.31	28.01	36.67	8.89	8.89	12.22	1034	1043	1878
4	4:20	SNY		84	29.65	29.57	35.88	13.89	13.33	11.67	1726	1652	1759
5	4:30	PSNY	88	83	36.63	28.01	35.50	10.56	12.78	12.22	1621	1500	1818
6	4:40	SNY		83	35.59	39.43	35.50	12.22	10.56	12.22	1822	1743	1818
7	4:50	"	90	84	35.59	38.71	34.95	11.11	10.00	12.77	1657	1622	1890
8	5:00	"	89	83	34.60	38.02	34.59	7.22	5.00	10.55	1047	796	1529
9	5:10	PSNY		89	35.59	39.43	34.59	8.89	8.89	11.67	1326	1468	1691
10	5:20	CDY		84	35.59	38.02	35.32	6.11	5.56	10.00	911	886	1480
—	AVERAGE				33.38	34.43	34.88	11.22	10.19	12.33	1551	1435	1798

f3/f1=1.045 f2/f1=1.031 ΔT3/ T1=1.10 Δ T2/ T1=0.908 g3/g1=1.159 g2/g1=0.925

REFERENCES

1. Rabi, A., "Comparison of Solar Concentrators", Solar Energy, vol. 8, pp 93-111 (1976)

2. Unpublished M.S. THesis work of E. Chaves, University of Louisville, 1977.

3. Singh, P., and L. Cheema, "Performance and Optimization of a Cylindrical-Parabola Collector", Solar Energy, vol. 18, pp. 135-141 (1976).

4. Tanaka, et al., "An analysis of a Cylindrical Parbolic Focussing Collector for Distributed Collector Power System", Miami International Conference on Alternative Energy Sources, December 1977.

5. Tanaka, T., et al., Solar Energy, vol. 19, No. 4, 1977.

6. Duffie, J., and W. Beckman, Solar Energy Thermal Processes, John Wiley. New York, N.Y. 1974.

7. Duffie, J., and G. O. G. Lof, "Focusing Solar Collectors for Power Generation", Paper presented at World Power Conference, Melbourne (1962)

8. Honeywell Report 1973. No. 2. The National Science Foundation, NSF/RANN/SE/GI-34871/PR/73/2 (1973)

9. Dietz, A., Space Heating with Solar Energy, R. W. Hamilton, ed., MIT, 1954. "Diathermanous Materials and Properties of Surfaces".

10. Dietz, A., Introduction to the Utilization of Solar Energy, Zarem, A. M. and Erway, D. D. eds., New York, McGraw-Hill 1963. "Diathermanous Materials and Properties of Surfaces.

MODELING A V-TROUGH COLLECTOR IN TERMS OF ZERO, SINGLE OR MULTIPLE THERMAL CAPACITANCE NODES

K. Bordoloi, C. Hoang and F. Bynum
Department of Electrical Engineering
University of Louisville
Louisville, Kentucky 40208 U.S.A.

ABSTRACT

Previously a mathematical model for the thermal performance was developed for a V-trough collector. Essentially, it was an application to this configuration of an H-B-W zero capacitance node model. In this present paper this analysis is extended to take into account the effects of thermal capacitance. The collector involved is formed of a linear array of V-troughs, is non-concentrating in nature and it heats a stream of air going through the channels above the blackened heat-absorbing plate and below a transparent cover. Direct comparison is made of the results of calculations between the zero capacitance node model and the present version which includes the capacitance effects. In each case identical sets of time-changing inputs are used. The method of study is by the digital simulation of the collector's overall performance.

INTRODUCTION

We have previously investigated in great detail the thermal performance of a V-trough, forced circulation, air heating solar collector in terms of such design variables as details in shape of the cross section and the flow rate per collector area of the air flowing through the channels of the collector. The equations used are valid for an air flow above the blackened bottom and the sloping sides of the channels and below the transparent cover(s) (See fig. 1) The important parameters were expressed and evaluated as functions of the relative dimensions, other constructing details and operating conditions. The important dimensions of the V-trough collector under study are shown in Figure 2.

In our previous work we did not include the effects of thermal capacitance. The study consisted of the use of the zero capacitance or an assumed quasi steady state model of the solar collector originated by Hottel, Whillier and Bliss [1-3]. The solar collector is however exposed to continuously variable weather conditions as well as varying load and storage conditions. The weather conditions of interest include: (a) the rate of available solar radiation and its fluctuation, (b) ambient temperature and (c) wind speed. The University of Louisville Solar Energy Laboratory is equipped to make these measurements. The effect of load and storage conditions relative to the collector performance occurs through their influence upon the inlet temperature of the circulating fluid as it enters the collector. Later we will mention the procedures used to deal with the measured and assumed values of the above dynamic variables in the simulation technique used for the study of the overall performance of the collector.

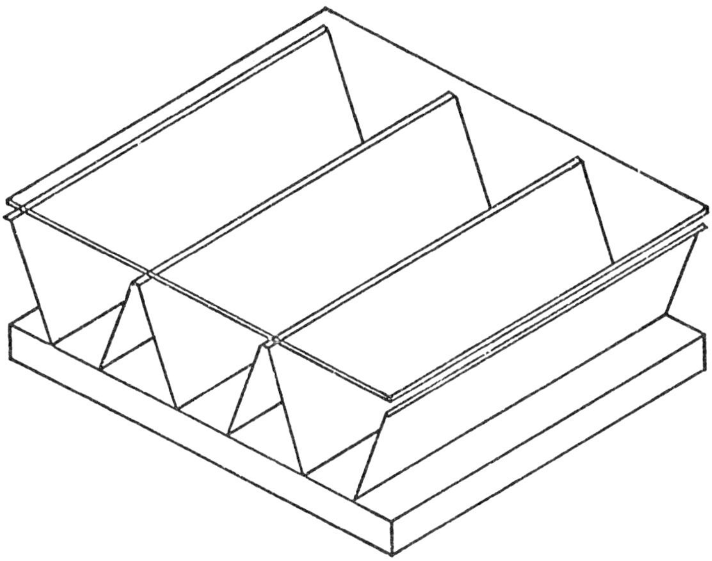

(a)

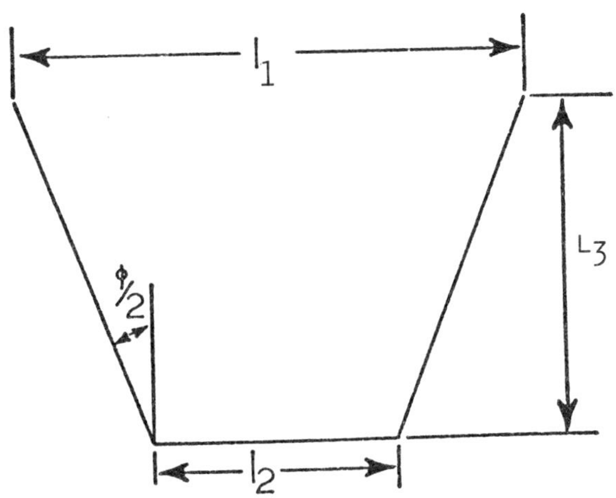

(b)

FIG.1 (A) SCHEMATICS OF A CLIPPED-V SOLAR ABSORBING TYPE COLLECTOR
(B) CROSS SECTION OF BLACKENED PLATE WITH DIMENSIONS & ANGLE

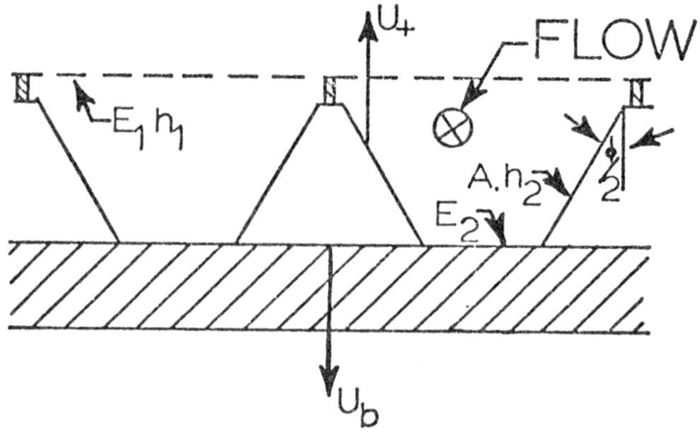

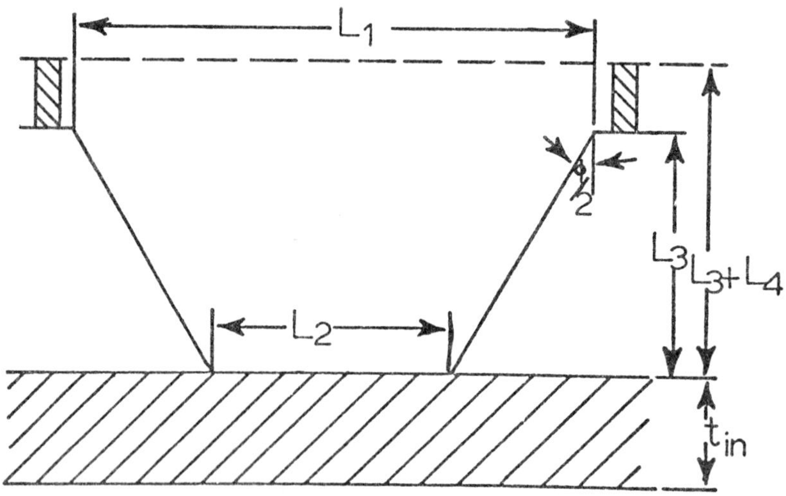

FIG. 2 DIMENSIONS OF CLIPPED-V CHANNEL. L_1=CHANNEL TOP WIDTH, L_4= COVER SUPPORT HEIGHT.

Under such variable operating conditions the thermal capacitance effects may be quite significant [4-6]. Several types of capacitance effects are (a) so called transients caused, for example, by abrupt shifts in the solar radiation rate due to passing clouds or (b) reduced performance due to time required for the collector to warm up each morning from ambient temperature [7]. To investigate these effects three mathematical models of this collector type are studied. First a quasi-steady-state model simulates the performance of the collector when it is considered to have zero capacitance. The second model accounts for the capacitance effects by using the assumption that a single value of thermal capacitance can be determined for the collector as a unit. Calculations are based upon a mean temperature for the collector such as mean plate temperature or mean temperature. A third model divides the collector into many nodes or isothermal segments. For each model the heat transfer coefficients, heat loss coefficients, collector efficiency factor etc, are calculated as functions of operating conditions.

We have insulation data at a much higher frequency than one per hour. The typical measurement period is at ten minute intervals, taken incidentally, at a 45° tilt typical of collectors in this region. Therefore, according to the discussion by Klein et al, [6] with this higher measurement frequency it becomes profitable to evaluate the effects of one or more capacitance nodes.

A number of modular computer program routines for the calculations involved for the zero capacitance node model are available from our previous studies [8-10] and they are extensively applied to this study together with additional required routines.

A ZERO CAPACITANCE NODE MODEL OF THE SOLAR COLLECTOR

The model that was proposed by Hottell, Whillier and Bliss [1-3] for reasons that were mentioned earlier is an especially convenient model for calculating thermal performance for flat plate and related types of solar collectors. Later this will be referred to simply as the zero capacitance node model. For the fluid circulating through the collector the useful energy gain or thermal power collected U_u, is expressed by this model as

$$Q_u = A_c \ F_r \ [S - U_L \ (T_{f,i} - T_a)] \quad (1)$$

where

$$F_r = \frac{1}{C_1} [1-\exp(- F' \ C_1)] \quad (2a)$$

$$C_1 = \frac{U_L}{GC_p} \quad (2b)$$

and $G = \dot{m} / A_c$ is the flow rate per collector area. For zero or more capacitance nodes, we specify that by definition, another of the common forms for Q_u contnues to hold.

$$Q_u = \dot{m} \ C_p \ (T_{f,o} - T_{f,i}) \quad (1a)$$

$$U_1 = \frac{U_t + U_b}{1 + \dfrac{U_t + U_b}{h_2 + h_r (1 + 1/A)}} \qquad (3)$$

and

$$F' = \frac{1}{1 + \dfrac{h_r\ U_t}{A h_2^2 + h_2 (h_r + A (U_t + h_r))}} \qquad (4)$$

in which $A = l_2 \ / \ l_1 + \ (1 - l_2 / l_1) / \sin \Phi / 2$ (5)

are the correct forms of the overall heat loss coefficient, U_L, and the collector efficiency factor, F', for our V-channel collector design. $U_b = k_{in}/t_{in}$ is the heat loss coefficient by conductivity through the insulated collector back, k_{in} & t_{in} are insulation thermal conductivity and thickness. h_r is the radiation heat transfer coefficient between heated plate and covers, while U_t is the combined radiative and convective coefficient for heat loss through the top covers. Both h_r and U_t are dependent upon plate temperature (the mean value), which in turn must be simply assumed, or better, iterativelly solved for by applying later equations relating that temperature to useful energy gain. h_2 is the convective heat transfer coefficient between heated plate and flowing air calculated on the basis of air flowing between a heated and unheated plate which are parallel with the same Reynolds Number as that for the flow along the channels of this collector. h_2, along with other convective heat transfer coefficients in general, is obtained by applying the usual definition of the Nusselt number ($N_u = h\ L/k$) and then using an empirical correlation which express N_u as a function of one or more other standard non dimensional groups. Such non dimensional groups include the Prandtl number ($P_r = C_p \mu / k$), Grashof number ($G_r = g\beta(\Delta T)L^3/\nu^2$) and the Reynolds number ($Re = v_f D_H / \nu$).

$$h_r \quad 4\sigma\bar{T}^3 \ / \ (1/\varepsilon_p + 1/\varepsilon_g - 1) \qquad (6)$$

when $\bar{T}$, a mean temperature for radiation between plate and cover, is in turn approximated by :

$$\bar{T} \simeq .5(T_p + T_c) \qquad \text{and} \quad T_c \simeq (T_p + T_a)/2 \qquad (6a)$$

with Ta, T_c, T_p ($=T_{p,m}$), being ambient, cover, and (mean) plate temperatures respectively, expressed in °K, to be compatible with Klein's emperical expression for U_t. [11]

$$U_t = \left(\frac{N}{(344/\bar{T}_p)[(T_p - T_a)/(N+f)]^{.31}} + \frac{1}{h_w} \right)^{-1} + \left(\frac{\sigma(T_p+T_a)(T_p^2 + T_a^2)}{[\varepsilon_p + .0425N(1 - \varepsilon_p)]^{-1} + [(2N + f - 1/\varepsilon_g] - N} \right) \qquad (7)$$

$$h_w = 5.7 + 3.8\ v_w \qquad (7a)$$

$$f = (\ 1 = .051N)\ \ (1 - .047\ h_w + .0005\ h_w^2\ ) \qquad (7b)$$

With v_w and h_w being wind speed (meters/sec), and wind heat transfer coefficient (watt/(m2 ^{o}K)).h_2 is determined from the correlation between Nu and Re for turbulent fully developed flow of air between parallel plates, one of them being heated. This correlation resulted from Kay's data [12-14]

$$Nu = 0.0158\ Re^{0.8} \qquad (8)$$

Also, as stated in our previous papers [8,9], the Reynolds number is redefined in terms of the geometry equations for the V-channel collector dimensions and apply the Nusselt number definition.

$$Re = \frac{4G\ lc}{\mu\ \ (P/l_1)} \qquad (8a)$$

$$h_2 = \frac{k_{air}\ Nu}{l_1\ (L/l_1)} \qquad (8b)$$

The equations for the mean plate and fluid temperatures, $T_{p,m}$, $T_{,f,m}$, are used in the iteration with Q_u, U_t, etc.

$$T_{f,m} = T_{f,i} + \frac{Q_u/A_c(\ 1 - F_R/F'\ )}{U_L F_R} \qquad (9)$$

$$T_{p,m} = T_{f,m} + R'_{p-f}\ Q_u\ /\ A_c \qquad (10\ a)$$

where for this air heating collector,

$$R'_{p-f} = 1/(A\ h_2) \qquad (10b)$$

Also, outlet fluid temperature, $T_{f,i}$, is expressed as:

$$T_{f,o} = \frac{Q_u/A_c}{G\ C_p} + T_{f,i} \qquad (11)$$

U_t is first evaluated with an assumed value for mean plate temperature, as is h_r. U_l, F', F_R, Q_u, then the mean fluid and plate temperatures are then evaluated. An absolute error comparison is made between the jth and (j+ 1)th value of $T_{,p,m}$ at a particular time step, with the iteration continued by plugging the new $T_{p,m}$ value back into U_t and repeating cycle, if the tolerance is not yet met. When that tolerance is met, the values of important changing variables are accepted and then the next time step is taken.

SINGLE AND CAPACITANCE NODE MODELS

When one is to compute the thermal performance for a solar collector using mathematical models based on one or more thermal capacitance nodes, then it is required that a set of N first order differential equations be solved. There will be one differential equation for each of the capacitance nodes. Those first order ordinary differential equations arise from an energy balance on each of the nodes. In them appears the derivative of the temperature of the node with respect to time.

A Single Capacitance Node Model

Close [15] and S. A. Klein et al. [6] have outlined a couple of single capacitance models. In their work zero capacitance models were applied to tube and sheet type liquid heating plastic collctors. This model attempts to treat the capacitance effects in a simple manner by using the assumption that a single capacitance may be usefully computed for the collector as an integral part of the whole unit. In this model a collector mean temperature and that single value capacitance are assigned to a single node which is representative of the entire unit.

There are several choices available when one is considering which is the temperature whose mean value will be the basis for the differential equation. As indicated by earlier equations there are such temperatures as the temperature of the fluid, the temperature of the plate and the cover temperature. Furthermore, they actually vary in the direction of the flow. For instance, the temperature of the fluid is predicted by the zero capacitance model to have a type of curve which may be called an exponentially rising toward an asymptotic value.

An energy balance for the one node model, when one uses the mean fluid temperature as the node temperature, yields the equation:

$$C_A \frac{dT_{f,m}}{dt} = F'[S - U_L(T_{f,m} - T_a)] - \frac{\dot{m}C_p}{A_c}[T_{f,o} - T_{f,i}] \tag{12}$$

A relationship between the inlet, outlet and the mean fluid temperatures is needed. Close had assumed a linear temperature distribution where $T_{f,m} = (T_{f,i} + T_{f,o})/2$. The approach which Klein uses, on the other hand, does not abandon the exponential saturation temperature profile which is predicted by the zero capacitance model. Under these assumptions and also taking into account the effect of the on and off pump control on the mean collector temperature, the energy balance equation assumes the following form.

$$C_A \frac{dT_{f,m}}{dt} = (1 - \gamma)[S - U_L(T_{f,m} - T_a)] + \gamma[F'[S - U_L(T_{f,m} - T_a)] - \frac{Q_u^z}{A_c} - 2\dot{m}C_p/A_c(T_{f,m} - T_{f,m}^z)]] \tag{13}$$

where γ=0 if the pump is off and 1 if the pump is on.

RESULTS

We have examined the behavior of the V-channel collector using computer programs which for the first case calculated zero-node model of the collector and in the second case made comparable calculations using a simplified single-node model. In both instances identical time changing data for a typical day were used. That data included the time of measurement, H.R, the solar radiation rate on a 45° tilted plane, T_a, the ambient temperature and $T_{f,i}$, the fluid inlet temperature. Our primary area of interest has been the collector itself. hence, $T_{f,i}$ was not produced by an elaboarate heating load model but instead was conveniently obtained as equivalent to the water inlet temperature to a bank of liquid heating flat plate collectors being tested at that time at the University of Louisville Solar Energy Laboratory. The data was gathered 11 August, 1976, which was the 242nd day of the year at which time the declination angle was 8.48^o. Table I shows a combination of times, the time changing input previously mentioned, and selected results from the zero capacitance model calculations, which may be directly compared with similar variables tabulated for the one capacitance model. The upper half of Table I , after the time column, lists the temperatures T_a, $T_{f,i}$, $T_{f,m}$, the mean fluid temperature, $T_{p,m}$, the mean plate temperature, and finally, $T_{f,o}$, the fluid outlet temparature for the zero capacitance model. The bottom half of the table, following the column of times, lists H.R, S, the solar radiation being absorbed by blackened plate, Q_u, the rate of useful energy gain by the flowing air, Q_u/A_c, the same quantity per collector area which will now have the same units as S and H.R. Finally, the last column shows the instantaneous efficiency of the collector.

The parameters used to compute the data in Tables I and II are;

Lattitude= 37.5^o
Collector tilt= 45°
Collector width= 0.612 m
Collector length= 2.41 m
No. of channels= 4
No. of glass covers= 2
Glass cover thickness= 0.23 cm
Refractive index= 1.526
Extinction coefficient of glass= 0.161/cm
Glass emittance= 0.88
Plate emittance= 0.17
Plate absorbtance= 0.89
Flow rate/collector area= 80 kg/m^2-hr
Channel half-angle= 45°
Channel bottom width/ channel top width= .8

Additionally, the parameters used in calulating the one thermal capcitance node model were:

Plate thickness= .5 mm
Cover plate, specific heat= 0.03822 KJ/Kg/^{o}K
Copper density= 8.795 gm/cm^3
Glass cover specific heat= .00748 KJ/Kg/^{o}K (value for SiO_2, pure)
Glass density= 2.515 gm/cm^3
Specific heat of air at const. pressure= 0.00101 KJ/Kg/^{o}K

Time hours	T_a °K	$T_{f,i}$ °K	$T^z_{f,m}$ °K	$T^z_{p,m}$ °K	$T^z_{f,o}$ °K
10.8	301.261	294.372	302.901	359.518	299.039
10.9667	302.039	298.594	307.452	366.81	303.44
11.1333	302.039	304.594	313.161	371.244	309.281
11.3	300.539	304.65	313.853	376.322	309.683
11.4667	301.761	304.261	313.701	377.745	309.423
11.7167	301.761	304.706	315.022	385.193	310.345
12	302.039	304.483	313.576	375.252	309.456
12.2167	302.817	305.594	316.455	390.539	311.531
12.3833	303.928	306.261	317.366	393.232	312.33
12.8	304.428	308.817	320.411	400.099	315.152
12.9667	304.317	310.65	322.357	403.129	317.046
13.2167	305.483	311.15	323.093	405.603	317.674
13.3833	304.706	312.372	324.384	407.578	318.934
13.6333	306.039	314.872	326.77	409.565	321.371
13.8	307.261	315.761	326.885	404.316	321.838
13.9667	307.817	317.706	329.454	411.642	324.122
14.1333	309.206	318.539	329.963	409.955	324.779
14.3833	309.15	319.706	331.014	410.2	325.883
14.55	308.539	320.483	332.688	418.594	327.148
14.7167	309.428	319.706	330.926	409.631	325.834
14.8	309.094	318.817	327.789	390.384	323.721
14.8833	308.706	315.261	320.837	359.239	318.312

Time hours	H R_2 W/m^2	S W/m^2	Q_u^z W	Q_u^z/A_{c_2} W/m^2	Efficiency
10.8	560.6	400.832	557.249	377.817	.67395
10.9667	595.137	425.963	578.632	392.314	.659199
11.1333	595.137	426.297	559.567	379.388	.637479
11.3	645.053	462.323	600.968	407.458	.631666
11.4667	656.397	470.655	616.34	417.88	.636628
11.7167	720.682	516.953	673.428	456.586	.633548
12	631.187	452.824	593.772	402.579	.637813
12.2167	760.072	545.241	708.839	480.595	.632302
12.3833	776.773	557.117	724.663	491.324	.632519
12.8	821.835	588.831	756.444	512.871	.624056
12.9667	838.568	600.399	763.78	517.845	.617535
13.2167	855.553	611.653	779.073	528.214	.617394
13.3833	869.419	620.723	783.514	531.224	.611011
13.6333	869.419	619.004	776.059	526.17	.605198
13.8	813.327	577.633	725.694	492.023	.604951
13.9667	869.419	615.538	766.234	519.509	.597536
14.1333	847.045	597.338	745.115	505.19	.596415
14.3833	849.881	594.645	737.586	500.086	.588418
14.55	931.182	647.045	795.802	539.556	.579431
14.7167	855.553	589.395	731.837	496.187	.579961
14.8	684.443	469.133	585.558	397.01	.580049
14.8833	420.687	286.724	364.35	247.03	.587208

Table I - Inputs and Zero Capacitance Node Model Results

Time hours	$T^z_{f,m}$ °K	$T_{f,m}$ °K	$T_{p,m}$ °K	$T_{f,o}$ °K
10.9667	307.452	304.371	361.461	303.44
11.1333	313.161	307.226	333.535	309.279
11.3	313.853	309.398	356.932	309.683
11.4667	313.701	311.242	381.096	309.423
11.7167	315.022	312.98	395.992	310.346
12	313.576	313.214	400.557	309.455
12.2167	316.455	314.302	401.904	311.532
12.3833	317.366	315.826	412.203	312.33
12.8	320.411	317.386	404.251	315.151
12.9667	322.357	319.576	410.41	317.046
13.2167	323.093	320.783	418.933	317.675
13.3833	324.384	322.376	424.575	318.934
13.6333	326.77	323.89	416.446	321.37
13.8	326.885	324.914	418.884	321.837
13.9667	329.454	326.484	417.06	324.123
14.1333	329.963	328.028	426.02	324.778
14.3833	331.014	329.058	425.649	325.88
14.55	332.688	330.326	432.502	327.149
14.7167	330.926	330.444	441.543	325.833
14.8	327.789	329.939	444.394	323.721
14.8833	320.837	329.939	479.042	318.311

Table II- Temperatures Based on One Capacitance Node Model and $T^z_{f,m}$

Klein has discussed the fact that the overall heat capacitance should actually, for a one capacitance model, ideally be calculated by a method more complex than a simple unweighted sum of the product of mass and specific heat for each collector component. Duffie and Beckman's book gives a procedure for taking a weighted sum for each of the covers plus a simple sum for the heated plate and the circulating fluid.

The method of solution of the differential equation between the time and the mean temperature was a first order method where Euler's method was used as a predictor and a trapezodal rule was used as the iterated corrector. The initial value for the mean fluid temperature used to start the differential equation was taken as equivalent to the value of the mean fluid temperature from the zero capacitance model at the first available unit of time. The differential equation was solved for subsequent time steps.

Table II starts with a column for time followed by mean fluid temperature from zero capacitance model followed by temperatures from the one capacitance model. These temperatures were $T_{f,m}$, T_{pm}, and $T_{f,o}$. Table III, after the time column, shows other outputs from the one capacitance model, relating to the useful energy gain. They are $Q_{u'}/A_c$, Q_u/A_c, Q_u and finally, the instantaneous collector efficiency. Useful energy gains for the one capacitance model were obtained in the following manner. After the mean fluid temerature in the one capacitance model had been solved then equation 9 involving mean fluid temperature is rearranged to solve for the value $Q_{u'}/A_c$. $Q_{u'}$ denotes that at this point F', U_L, and F_R still possess the zero capacitance values. Then the mean plate temperature(eqn. 10a) was found so that by the method described on the bottom of page 4 could be used. This will yield U_L, F', F_R, and finally, Q_u and efficiency of the one capacitance model.

Comarisons can be made now between the results on Tables II and III , assuming one capacitance node vs. those from Table I, assuming zero capacitance node. First, one sees that for the mean fluid temperature for the zero and one node model, respectively, that as expected the one node version lags in time behind that for the zero node version. Another substantial difference is seen in the variation of mean plate temperature from the two models. Again that for the one capacitance model lags in time behind the changes in its value for the zero capacitance model. Furthermore, its value for the one capacitance model reaches a much greater maximum than for the other model. For the zero capacitance model $T_{p,m}$ reaches a maximum at 418°K following very closely rises and falls in H.R. However, for the one capacitance model the temperature contains some dips but mostly continues to rise to a maximum of 479°K at the end of the run. This suggests that the one capacitance model is more realistic. As the temperature rises the quantities such as the rate of heat loss and also the heat loss coefficients themselves rise. This is to the detriment of the useful energy gain, i.e., the collector is less efficient. Such a pattern could be seen in comparing Table III with Table I.

Thus far we have culminated our study of the V-trough collector in terms of zero and one capacitance node model. More investigation is warranted in terms of extending the work to include multiple capaciatnce nodes. And also important, are needed experimental measurements which will be undertaken shortly.

Time hours	Q_u'/A_c W/m^2	Q_u/A_c W/m^2	Q_u W	Efficiency $=Q_u/A_c$ H R
10.9667	255.83	392.26	578.552	.659108
11.1333	116.515	379.286	559.417	.637308
11.3	210.213	407.488	601.012	.631712
11.4667	309.03	417.899	616.367	.636655
11.7167	366.218	456.629	673.492	.633607
12	386.542	402.529	593.698	.637733
12.2167	385.302	480.688	708.976	.632424
12.3833	423.179	491.326	724.666	.632522
12.8	379.046	512.846	756.407	.624026
12.9667	394.837	517.805	763.721	.617487
13.2167	426.017	528.225	779.089	.617407
13.3833	442.445	531.199	783.476	.610982
13.6333	398.804	526.084	775.932	.605098
13.8	404.846	491.913	725.532	.604816
13.9667	388.175	519.542	766.283	.597574
14.1333	419.596	505.127	745.022	.596341
14.3833	413.587	499.911	737.328	.588212
14.55	435.106	539.691	796.	.579576
14.7167	474.879	496.072	731.666	.579825
14.8	492.18	397.016	585.566	.580056
14.8833	650.295	246.945	364.224	.587005

Table III- Rates of Energy Gain and Instantaneous Efficiency

NOMENCLATURES

A	Correction factor used to modify U_L, F', heat transfer coeffients and the capacitances.
A_c	Collector area
C_p	Specific heat of a fluid at constant pressure
C_A	Effective thermal capacitance value per unit area of the total collec
F_R	Heat removal factor
F'	Collector efficiency factor.
f	A component in Klein's empirical equation.
G	Mass flow rate per collector area
h_2	Convective heat transfer coefficient between air and flat parallel plates.
h_r	Heat transfer coefficient by radiation between the heated plate and the transparent cover.
h_w	Heat transfer coefficient due to wind speed, v_w.
H.R	Rate of solar radiation falling on a tilted plane.
k_{air}	Thermal conductivity of the air
L	Reference length related to the plate spacing.
l_c	Collector length.
Nu	Nusselt number
N	No of glass covers.
p	Perimiter around cross section of a channel.
R'_{p-f}	Heat transfer resistance between the heated palte and air.
S	The rate of solar radiation reaching and being absorbed by the collector plate.
T	Temperature with appropriate subscripts: a= ambient, f,i= fluid inlet, f,m= fluid mean, f,o= fluid outlet and p,m= plate temp. The superscript z denotes zero capacitance model.
U_b	Heat loss coefficient by conduction through the insulated back cover.
α	Absorbtance of the coating on the plate.
ε	Emittances. subscript; p= plate, g=glass.

REFERENCES

1. Hottel, H., and A. Whillier, "Evaluation of Flat-plate Collector Performance", Trans. Conf. on the Use of Solar Energy, vol. 2, Part I, p 74. U. of Arizona Press(1958).

2. Whillier, A., "Solar Energy Collection and its Utilization for House Heating", Sc.D. Thesis, MIT (1953)

3. Bliss, R., "The derivations of Several 'PlateEfficiency Factors' Useful in the Design of Flat-plate Solar-heater Collectors". Solar Energy 3, 55 (1959).

4. Tabor, H., "Solar Energy Collector Design", Bull. Res. Council of Israel, vol. 5C, p 5 (1955)

5. Hottel, H., "Performance of Flat-Plate Solar Energy Collectors", Pro. Symposium on Space Heating with Solar Energy, M.I.T., 1950. pp 58-71.

6. Klein, S., et al., "Transient Considerations of Flat-Plate Solar Collectors", Trans. ASME, J. Eng. Power, April 1974. pp 109-113.

7. Duffie, J., and W. Beckman, Solar Energy Thermal Processes, John Wiley, New York, N.Y. 1974.

8. Bordoloi, K., and F. Bynum, "A Solar Collector Based on Linear Clipped-v Channel, to appear in the proc. of Soalr Heating and Cooling, Miami, 1976. Solar Energy Forum.

9. Bordoloi, K., Murray, T., and F. Bymum, "Further Modeling of a Solar Collector Based on Linear Clipped-V Channels" to appear in the Proc. of Alternative Energy Sources, Miami, 1977.

10. Bordoli, K., and F. Bynum, "Modeling of a Solar Collector Based on Linear V Troughs", submitted to the Journal of Solar Energy.

11. Klein, S., "The Effects of Thermal Capacitance Upon the Performance of Flat-Plate Solar Collector", M. S. Thesis, U. Wisconsin (1973).

12. Kays, W., Convective Heat and Mass Transfer McGraw-Hill, New York, N,Y. (1966).

13. Kreith, F., Principles of Heat Transfer, 3rd ed., Int. Textbook Co., Scranton Pa. (1973)

14. McAdams, W., Heat Transmission, 3rd ed., McGraw Hill, New York, N.Y., 1954.

15. Close, D., "A Design Approach for Solar Processes", Solar Energy Journal, Vol. 11, 1967. p 112.

HIGH-EFFICIENCY SOLAR CONCENTRATOR

F. L. Lansing and J. Dorman
DSN Engineering Section
Jet Propulsion Laboratory
Pasadena, California U.S.A.

ABSTRACT

A new type of solar concentrator is presented using liquid lenses and simple translational tracking mechanism. The concentrator achieves a 100:1 nominal concentration ratio and is compared in performance with a flat-plate collector having two sheets of glazing and non-selective coating. The results of the thermal analysis show that higher temperatures can be obtained with the concentrator than is possible with the non-concentrator flat-plate type. Furthermore, the thermal efficiency far exceeds that of the comparative flat-plate type for all operating conditions.

I. INTRODUCTION

One of the fundamental problems associated with the effort to replace conventional energy sources by solar energy is to achieve high-temperature collection at very low cost. High temperature is feasible if the solar energy is concentrated and delivered to a collecting medium such as a gas or liquid. The greater the energy concentration, the greater the temperature of the collecting medium and the more useful the energy collected for heating, airconditioning, and electric power generation. Unfortunately, low-cost flat-plate solar collectors cannot operate efficiently at temperatures above 100°C, and presently available higher temperature concentrators are very expensive (approximately 5 to 10 times the cost of flat-plate collectors).

This article describes a low-cost method of manufacturing a high-temperature solar concentrator. There are two main features of the design that are unique: (a) a liquid lens construction of optimum shape with 100:1 nominal concentration ratio, and (b) a translational rather than rotational tracking mechanism. There is a resemblance between the construction of this design to that of a simple, non-concentrating flat-plate solar collector having two sheets of glazing, and it is estimated that it will cost only approximately 25 percent more for the liquid lens construction.

II. DESCRIPTION

The main features of the solar concentrator are indicated in Fig. 1. The major elements of the device are the lens assembly, working medium tubes, and the tracking mechanism. The lens assembly forms a series of individual lenses arranged side by side and is fabricated from transparent plastic or glass, as illustrated in Fig. 2. The curvature of the upper and lower shells of the lens assembly is designed for minimum focal length and optimum concentration ratio. The space between shells is filled with liquid. The initial

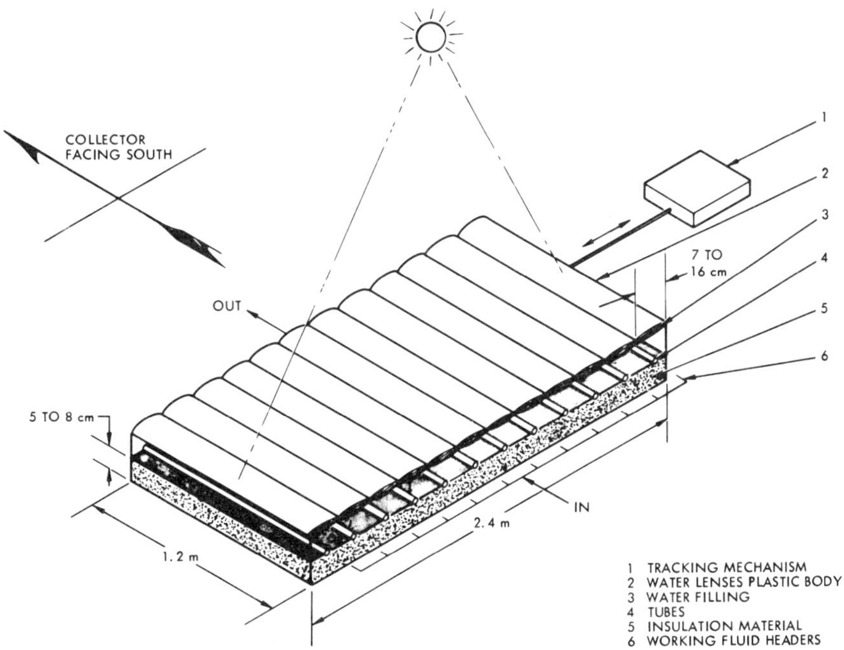

Fig. 1. Overall view of the solar concentrator

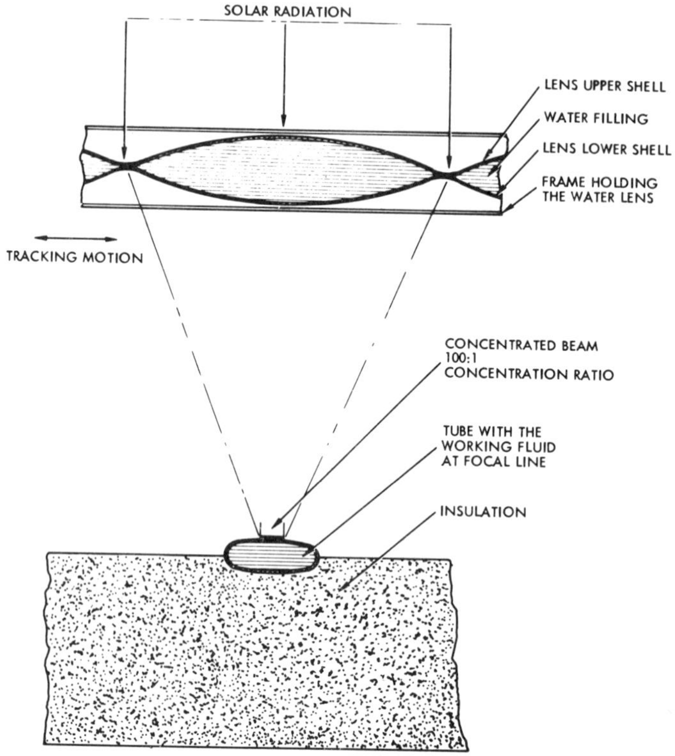

Fig. 2. Working principle of a "liquid" lens

design used water as the filling medium for its low cost. The refraction index of water (1.33) closely approximates that of glass (1.5) so that the lens optical transmissivity closely approximates that achieved by a solid lens manufactured from glass. It is foreseen that other transparent liquids or additives may enhance the performance of the concentrator for different applications. The lens assembly design also includes compensation for the expansion and the contraction of the liquid filler in addition to means for filling, draining, and cleaning (not shown in figures). The outer shells of the lens assembly are inexpensive to manufacture and light in weight for easy shipment and site installation.

The working medium tubes can be made of copper, steel, or aluminum and are placed on a thermal insulation layer at the focal lines of the individual lenses. There is no fin or plate connection of the tubes as is normal with flat-plate solar collectors. The tube surface is coated with a mat black paint for maximum absorption. A nominal concentration ratio of 100:1 was demonstrated experimentally on the single lens shown in Fig. 2. The resulting high temperatures can lead to a feasible application in electric power generation through solar-Rankine power cycles.

The function of the unique tracking mechanism adopted in this concentrator is to maintain the focal line at the working medium tubes independent of the Sun's incidence angle as indicated in Fig. 3. A simple drive system consisting of a fractional horsepower electric motor, reduction gear and an eccentric disk moves the lens assembly on its rollers in a translational motion with simple harmonic motion at a frequency of one cycle per day. Compensation for differences between summer and winter solstice and intermediate seasonal variations is achieved by eccentricity adjustments of the disk.

An important feature of the design is the absence of environmental impact. If water is chosen as the liquid filler for the lenses, it represents little or no hazard in case of leakage failure.

III. THERMAL ANALYSIS

This is performed by applying the first law of thermodynamics and the heat rate equations at steady-state condition to the unit concentrator shown in Fig. 2, which is composed of one 2-dimensional lens, a single tube at the lens focal line and the working fluid. After manipulating the energy balance equations for the lens, the tube, and the working fluid, the fluid temperature difference between the inlet and the outlet sections can be written as

$$(T_{out} - T_{in})_1 = \left\{ \alpha \ \tau_1 IR - U_{el}(T_{in} - T_a) \right\} \times \frac{1}{U_{el}} \left\{ 1 - \exp\left(\frac{-U_{el} U_{pf1}}{RG_f c_f (U_{el} + U_{pf1})} \right) \right\} \qquad (1)$$

where U_{pf1} is the tube surface-to-fluid conductance coefficient expressed by Whillier Ref.(1) for laminar water flow in tubes in the form

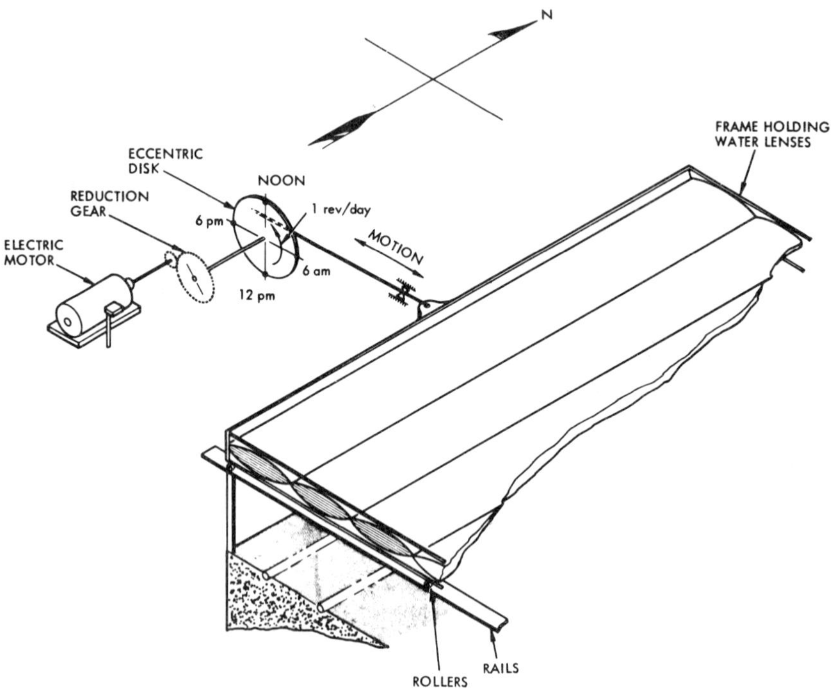

Fig. 3. Details of translational tracking mechanism

$$U_{pf1} = \frac{2.484}{d_i} + \left[\frac{0.0854 \left(\frac{m_f}{Ld_i}\right)}{1 + 0.0684\left(\frac{m_f}{L}\right)^{2/3}} \right] \tag{2}$$

The concentrator thermal efficiency is then given by

$$\eta_{th1} = \frac{G_f c_f (T_{out} - T_{in})_1}{I} \tag{3}$$

For comparison, the performance of a flat-plate type collector is given, based on analysis by Whillier Ref.(1). The comparative flat-plate design is shown in Fig. 4 and includes double glazing and a metallic plate integral with the tubes. The tube's dimensions, lateral spacing, and the optical properties of their coating material are taken the same as for the concentrator. The resulting fluid temperature difference can be written as

$$(T_{out} - T_{in})_2 = \left\{ \alpha \tau_2 I - U_{e2}(T_{in} - T_a) \right\} * \frac{1}{U_{e2}} \left\{ 1 - \exp \left(\frac{-U_{e2} U_{pf2}}{G_f c_f (U_{e2} + U_{pf2})} \right) \right\} \tag{4}$$

where U_{pf2} is the "equivalent" conductance coefficient between the fluid and the combined fin and tubes, given by the following relation

$$\frac{1}{U_{pf2}} = \frac{(d_o + 2\lambda)}{\pi d_i U_{pf1}} + \frac{2\lambda(1-F)}{U_{e2}(d_o + 2\lambda F)} \tag{5}$$

where F is the fin efficiency determined from

$$F = \frac{\tanh \lambda \sqrt{\frac{U_{e2}}{K\delta}}}{\lambda \sqrt{\frac{U_{e2}}{K\delta}}} \tag{6}$$

The corresponding thermal efficiency is then expressed as

$$\eta_{th2} = \frac{G_f c_f (T_{out} - T_{in})_2}{I} \tag{7}$$

Table 1 presents the results of a numerical comparison between the concentrator and the double-glazing collector, and is constructed from the following assumed operating conditions:

Solar radiation intensity $\qquad I = 700\ \text{kcal}/(\text{h.m}^2)$

Ambient air temperature $\qquad T_a = 20^{\circ}C$

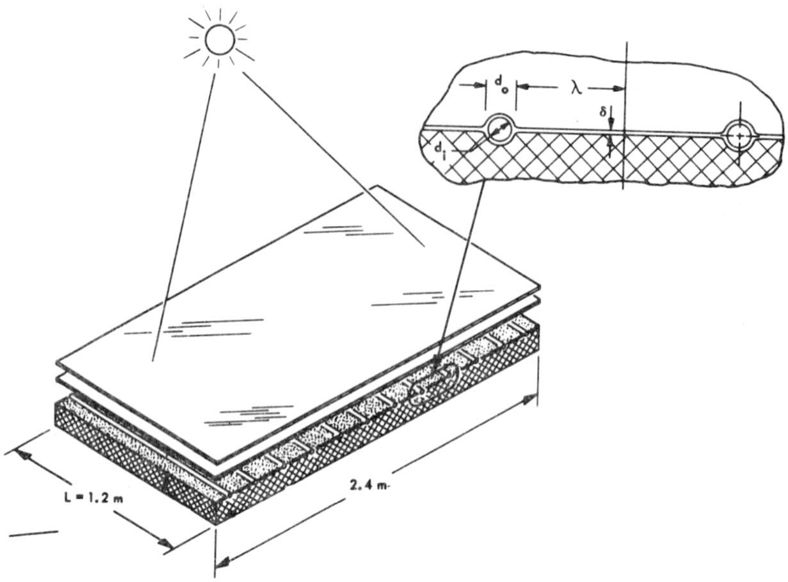

Fig. 4. Double-glazing flat-plate solar collector

Table 1. Numeric comparison between solar concentrator and analogous flat-plate double-glazing collector

G_f, kg/(h·m²)	Flow rate per tube, kg/h	Fluid temperature rise, $(T_{out} - T_{in})$ °C		Thermal efficiency η, %		% Improvement in efficiency
		Concentrator	Double-glazing	Concentrator	Double-glazing	
0	0	1062.6	142.9	00.0	00.0	0
2.5	0.48	174.7	99.1	62.4	35.4	76
5	0.96	91.3	63.8	65.2	45.6	43
10	1.92	46.7	36.6	66.7	52.4	27
20	3.84	23.6	19.7	67.5	56.4	20
30	5.76	15.8	13.5	67.8	57.9	17
50	9.6	9.5	8.3	68.1	59.2	15
100	19.2	4.8	4.2	68.4	60.5	13

Inlet fluid temperature	T_{in}	$= 30^{\circ}C$
Tube surface absorptivity	α	= 0.9
Tube surface emissivity	ϵ	= 0.9
Water-lens transmissivity[1]	τ_1	= 0.78
"Effective" transmissivity of double glazing[2]	τ_2	= 0.83
Water specific heat	c_f	$= 1$ kcal/kg$^{\circ}$C
Tube length	L	= 1.2m
Tube inner diameter	d_i	= 0.007m
Tube outer diameter	d_o	= 0.010m
Half width of fin	λ	= 0.075m
Plate thickness	δ	= 0.0015m
Tube material conductivity	K	= 45 kcal/(h.m$^{\circ}$C) (galvanized steel)
Lens/tube area ratio	R	= 16 (actual concentration ratio)
Number of lens/collector panels		15
Heat loss coefficient between tube (or plate) surface and ambient air[3]	U_{e1}	= 7.33 kcal/h.m$^{2\circ}$C
	U_{e2}	= 3.42 kcal/h.m$^{2\circ}$C

The above operating conditions were abstracted from the experimental results of flat-plate solar collectors to yield a practical evaluation of their comparison. The efficiency results are plotted as shown in Figs. 5 and 6.

IV. CONCLUSIONS

(1) The improved efficiency of the concentrator compared with the double-glazing collector at a given radiation intensity and a wide range of fluid mass

[1]Calculated based on a lens average thickness of 3.81 cm, a refraction index of 1.33 and an extinction coefficient of 0.08 cm^{-1}

[2]Based on 0.32 cm thickness for each panel, an extinction coefficient of 0.08 cm^{-1} and a refraction index of 1.526.

[3]U_{e1} is based on an average temperature of 120°C and U_{e2} is based on an average temperature of 70°C.

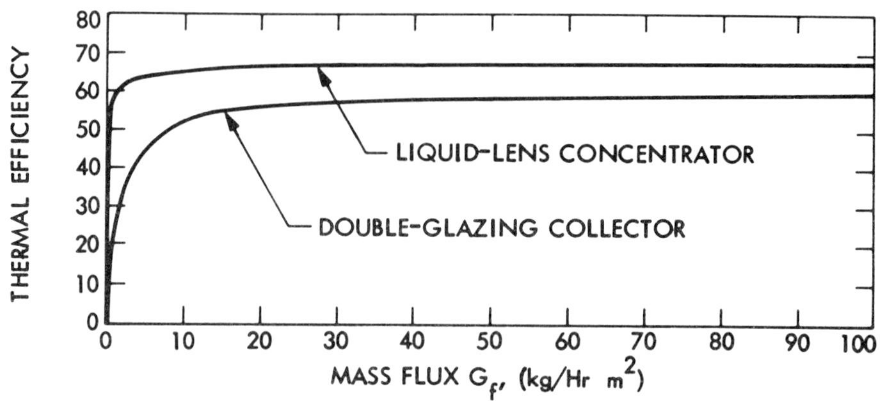

Fig. 5. Effect of varying the mass flux on the collector thermal efficiency

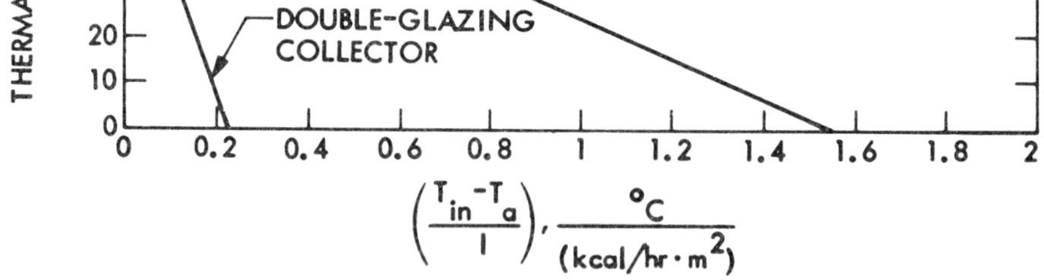

Fig. 6. Efficiency curves for a flow flux G_f = 30 kg/h • m²

flux is plotted in Fig. 5. The percentage improvement in efficiency is also included for comparison in Table 1. It appears that the thermal efficiency obtained from the concentrator exceeds that of the comparative double-glazing collector for all operating conditions. As an example, for a high temperature application where the fluid is heated by 175°C, the corresponding mass flux is 2.5 kg/(h.m^2) and an improvement in the thermal efficiency of 76 percent is obtained. On the other hand, for a high efficiency operation, Table 1 indicates that a large mass flux of 100 kg/(h.m^2) gives an efficiency improvement of 13 percent over the double-glazing collector, which is still considered a useful gain in collector performance.

(2) The performance curve expressing the thermal efficiency vs. the variable $(T_{in} - T_a)/I$ at a selected mass flux of 30 kg/(h.m^2) is shown in Fig. 6. The straight line relationship is evident from the fluid temperature rise expressions given by Eqs. (1) and (4). For the concentrator, the thermal efficiency can be written as

$$\eta_{th1} = 0.684 - 0.447 \left(\frac{T_{in} - T_a}{I}\right) \tag{8}$$

and that for the double-glazing collector as

$$\eta_{th2} = 0.619 - 2.835 \left(\frac{T_{in} - T_a}{I}\right) \tag{9}$$

The "zero-efficiency" condition, with a radiation intensity of 700 kcal/(h.m^2), corresponds to an inlet fluid temperature of 1072°C for the concentrator and 153°C for the double-glazing collector. This result shows that the maximum temperatures that can be reached by the concentrator are much higher than is possible with the comparative double-glazing type.

(3) The slope of the performance curves of Fig. 6, which is a measure of the heat loss, is shown to have a smaller value for the concentrator compared with the double-glazing collector. This is due to the fact that in the case of the concentrator the heat loss coefficient between the tube and the ambient air per unit area of the collector is given by U_{e1}/R, while that for the double-glazing type is U_{e2}. This means that the reduction in heat losses to the surroundings, brought by focusing the incident energy on a smaller area, outweighs its increase due to higher temperatures obtained.

(4) Practical examples of the utilization of the present concentrator in airconditioning and/or power generation applications, are taken from Table 1 as follows:

(a) For an electric-power generation application, a mass flux of 2.5 kg/(h.m^2) is chosen for each concentrator panel (1.2 X 2.4m) which results in an exit fluid temperature of 204.7°C. If a Rankine power cycle is operated between 204.7°C and an ambient temperature 20°C as its temperature limits at 50 percent relative efficiency to the Carnot cycle, it will give a thermal efficiency of 19 percent. This means that one concentrator panel can produce a mechanical work equivalent to 0.28 kW per panel for every hour of operation.

(b) It appears feasible for the airconditioning load of a normal residence (3 tons of refrigeration) to be carried out totally by solar energy. For example, if the airconditioner used is a mechanical vapor compression refrigeration unit that is driven by a solar-Rankine cycle it will consume about 3.6 kW. In this case, 12.8 concentrator panels are needed with a total area of 37 m^2 (400 ft^2).

On the other hand, if the airconditioner used is of the absorption type, a collector temperature in the order of 100°C is sufficient. A mass flux of 5 kg/(h.m^2) from Table 1 will result in a fluid temperature of 121.3°C, which is adequate for full-load operation (coefficient of performance is about 0.65). In this case, only 10.6 concentrator panels are needed for airconditioning the house with a total area of 30.6 m^2 (328 ft^2). Both cases indicate that the required collection area for airconditioning is adequate and does not exceed that of a south-facing roof in most houses.

DEFINITION OF TERMS

c_f fluid specific heat, kcal/kg.°C

d_i inner tube diameter, m

d_o outer tube diameter, m

F fin efficiency

G_f fluid mass flux (flow rate per unit collector area), kg/h.m^2

I solar radiation intensity, kcal/(h.m^2)

K conductivity, kcal/(h.m°C)

λ half width of fin, m

L tube length, m

m_f mass flow rate per tube, kg/hr

R lens area/concentrator tube projected area

T_{in} inlet fluid temperature, °C

T_{out} outlet fluid temperature, °C

T_a ambient temperature, °C

U_e "effective" heat loss coefficient between tube (or plate) surface and ambient, kcal/(hr·m^2·°C)

U_{pf} tube (or plate) surface to fluid conductance coefficient, kcal/h·m^2·°C)

α absorptivity

τ transmissivity

δ plate thickness, m

ϵ emissivity

η_{th} thermal efficiency

SUBSCRIPTS

1 concentrator

2 double-glazing collector

REFERENCE

1. Whillier, A., "Design Factors Influencing Solar Collector Performance," Low Temperature Engineering Application of Solar Energy, ASHRAE Publication, 1967, Chap. III, pp. 27-40.

ACKNOWLEDGEMENT

This paper presents the results of one phase of research carried out at the Jet Propulsion Laboratory, California Institute of Technology, under Contract No. NAS 7-100, sponsored by the National Aeronautics and Space Administration.

THE SCHEME OF AUTOMATIC ADJUSTMENT OF ACCURACY OF THE INFLATABLE SOLAR COLLECTOR

A. Gokhman and N. Ozboya
University of Miami
Coral Gables, Florida U.S.A.

ABSTRACT

The presented paper consists of the main principles of one possible scheme for automatic adjustment of the inflatable solar collector, in order to keep the shape of the main parabolic reflector and the position of the auxiliary parabolic mirror. The proposed scheme is based on measurements of the position of certain points and distances between certain points. For these measurements the optical scheme proposed in this paper is selected from various alternatives due to the high accuracy required.

INTRODUCTION

In order to get the highest efficiency from the inflatable collector, the accuracy of every part of the main parabolic reflector and the position of the parabolic auxiliary mirror should be at the highest level at all times. After the initial manual setting of the collector at the beginning of operation, the parts of the system should adjust themselves automatically for highest accuracy to compensate for the continuous changes due to sun-tracking, temperature variations, etc. We could keep the pressure constant inside the inflatable body but we cannot avoid the deformations of the plastic body because of temperature variations during the operation. This self-adjusting system should be reliable and should not add to the initial investment considerably. The proposed scheme is an optical adjusting system, which is superior to other available schemes considering the high accuracy required. We have still many alternatives in selecting the specific optical equipment. We can use ultraviolet laser equipment and detectors which are sensitive to radiation in 2000-2900 A^{o} range, which is outside the range covered by solar radiation. Other possibilities are to use regular or infrared laser equipment and detectors. In these cases we might have to use filters and pulses to differentiate the signal from the signals due to solar radiation. Further analysis will show the optimum scheme, but basically we will use one of these laser sources and detectors to adjust the auxiliary parabolic mirror and the main reflecting surface automatically and continuously.

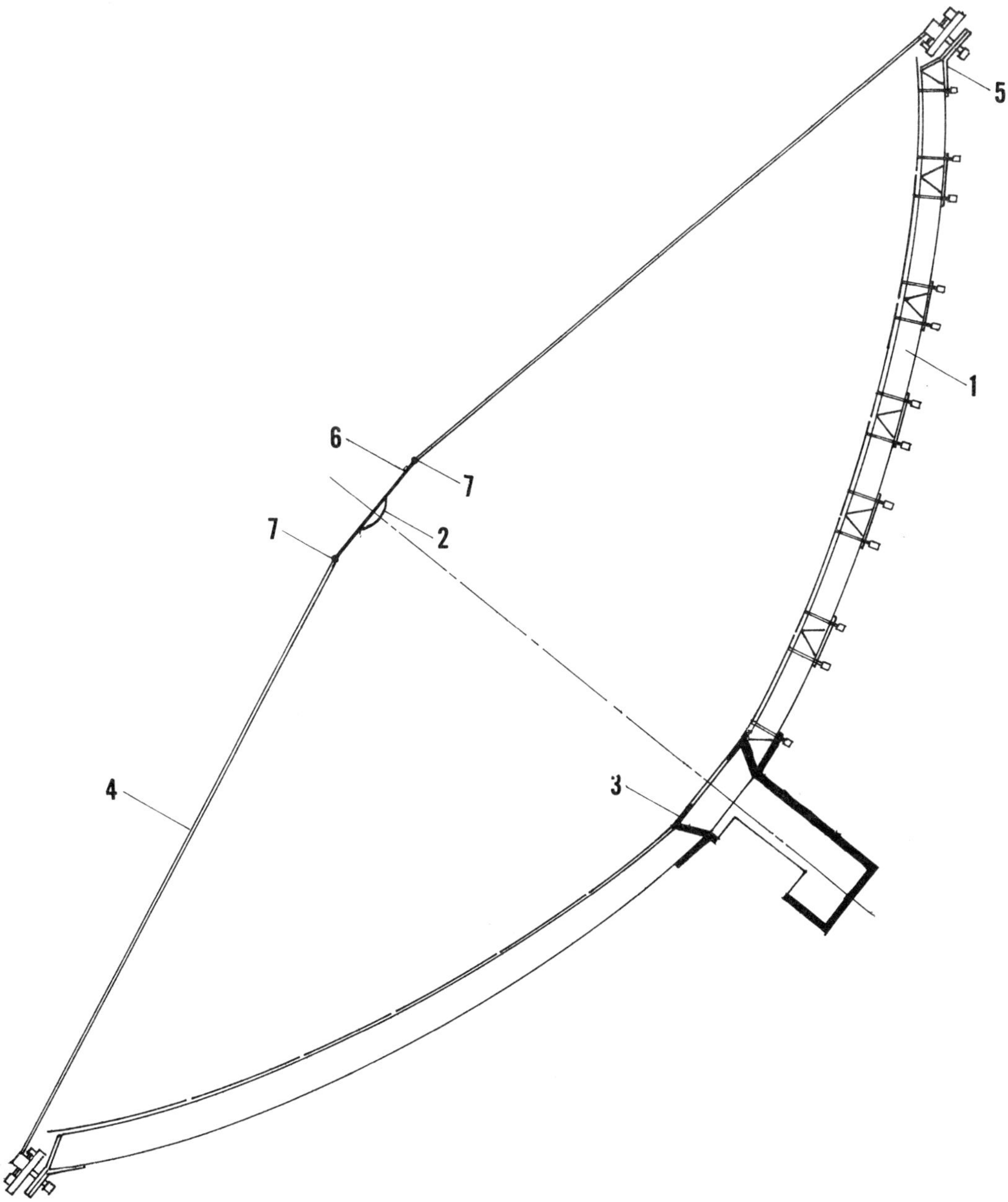

Fig.1 Schematic Drawing of Inflatable Solar Collector

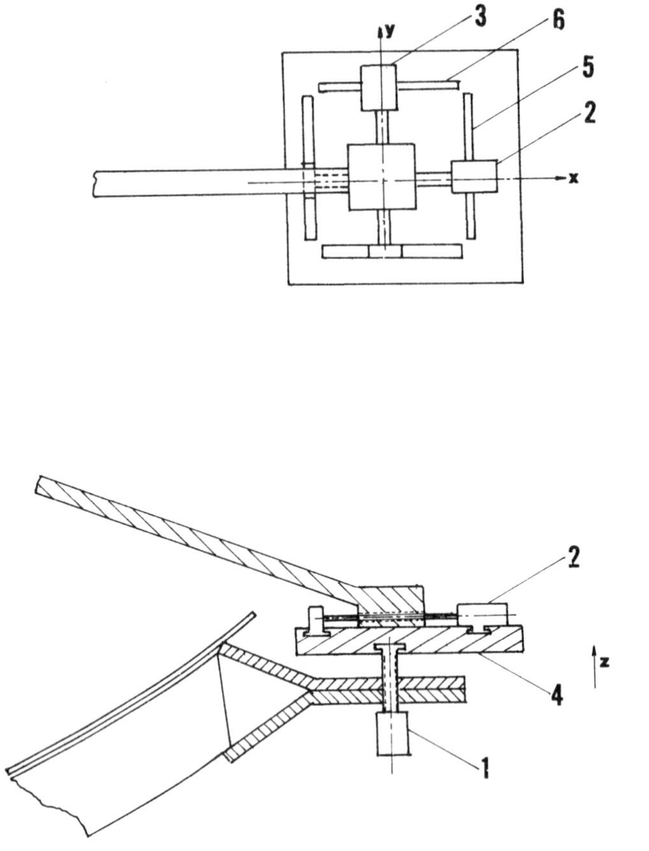

Fig.2 Connection of the Main Rod to the Partial Flange

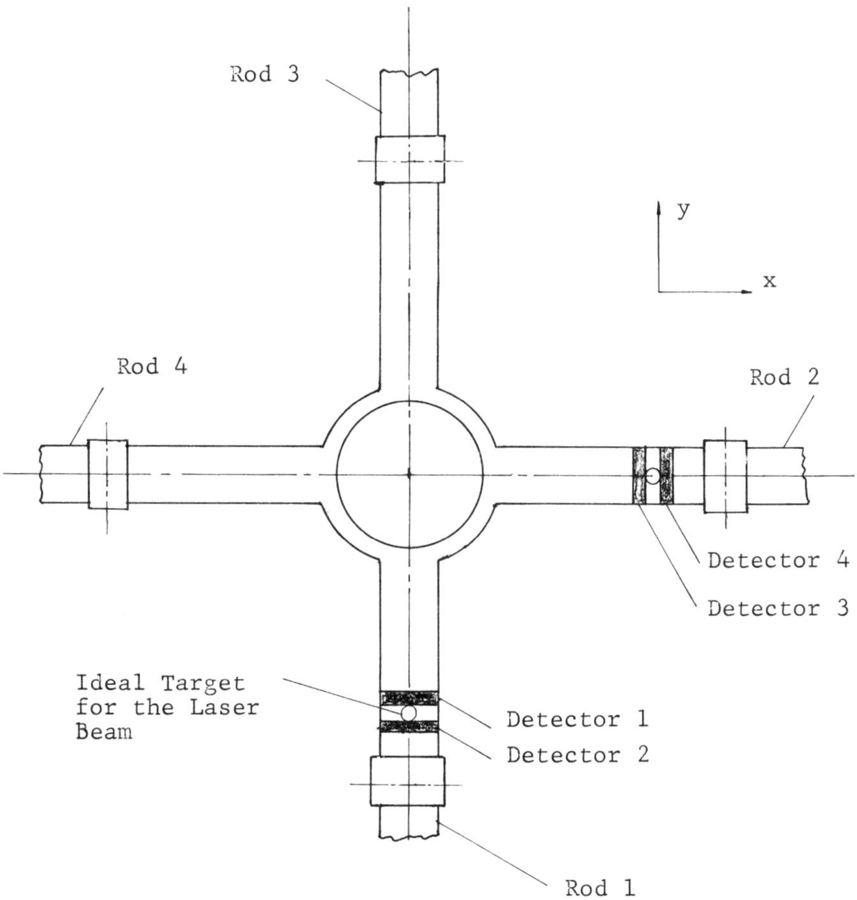

Fig.3 Auxiliary Parabolic Mirror
and Laser Detectors

SELF-ADJUSTING SYSTEM FOR THE AUXILIARY PARABOLIC MIRROR

Figure 1 shows schematically the way the auxiliary parabolic mirror is connected to the main body. In Figure 1, auxiliary parabolic mirror 2 is mounted in the frame 6. The frame 6 is connected by means of the spherical hinges 7 to connecting rods 4, which are connected to the partial flanges 5 of the body 1. The flange 3 at the vertex (base) of the collector, where the deflecting flat mirror is housed and the frame on which auxiliary parabolic mirror is placed, are rigid. These frames are relatively small and we can consider them as free from deformations. (These frames could be manufactured from special alloys which do not give significant deformations due to temperature variations during the operation hours.) Therefore we should make use of these frames to place our light sources and sensors as will be explained later. The frame on which the auxiliary parabolic mirror is placed should be parallel to the flange of the base at all times and also the parabolic mirror should be centered and have the same focal point with the main parabolic reflector. The ends of the rods at the partial flanges are connected to the body of the main parabolic reflector in a way that they can be moved in three orthogonal directions by 3 selsyns (servo-motors) at each connection. As seen in Figure 2, selsyn 1 gives vertical movements (z-direction) to the platform 4 on which selsyn 2 and selsyn 3 are placed. Selsyn 2 can move along the slot 5 and move the connecting rod in x-direction (normal direction). In the same way Selsyn 3 moves the connecting rod in tangential y - direction and is movable along the slot 6.

On the flange of the base, the laser source is mounted. The laser beam from the source is reflected in plane parallel to the flange of the base by a rotating mirror which is also mounted on the same flange. In order to satisfy the centralization of the auxiliary parabolic mirror and the base, we use two mirrors placed on the base flange at 90° interval. These mirrors reflect the laser beam coming from the rotating mirror in the vertical direction. On the bottom side of the auxiliary parabolic mirror frame, the detectors are placed corresponding to these two reflecting mirrors, as shown in figure 3.

In the ideal position, that is, when the auxiliary mirror is centered with respect to the base, the laser beams hit the gaps between the detectors. But if the laser beam hits any of the detectors, the current produced by the detector and amplifier will start the selsyns at the other ends of the connecting rods in x and y directions until the auxiliary parabolic mirror is centered again. For example, if the auxiliary parabolic mirror center is off in positive x direction, the laser beam will hit detector 4 and this detector will activate

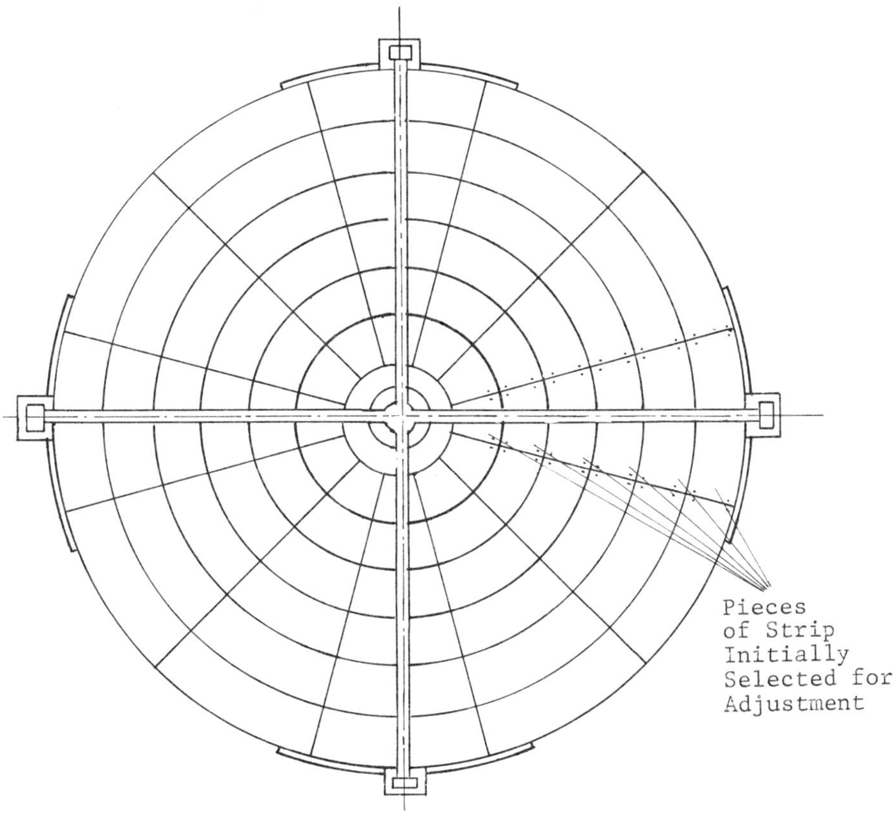

Fig.4 Top View of the Inflatable Collector

the selsyns to move the connecting rods 2 and 4 radially (in negative x direction) and rods 1 and 3 tangentially (again in negative x direction). If the center is off in both y and x directions, then two detectors will react and activate the selsyns in corresponding directions.

In order to satisfy the conditions that the auxiliary parabolic mirror frame and the base flange are parallel to each other and the auxiliary parabolic mirror has the same focal point as the main parabolic reflector, we use the laser system to determine the vertical distances between 4 points on the auxiliary parabolic mirror frame and the base flange. In this case, there are four additional mirrors on the base flange placed at 90° intervals, which reflect the laser beam from the rotating mirror vertically. On the bottom side of the auxiliary mirror frame, there are four three-dimensional corner mirrors exactly above the reflecting mirrors on the base flange. These corner mirrors reflect the laser beam downward by 180° whatever the angle of the auxiliary mirror frame with the base flange is. The returning beams are received by detectors on the base flange and the exact distances can be determined with continuous pulse method. Smaller or higher values of the distances will automatically activate the corresponding selsyns 1 connected to the ends of connecting rods, until all sides of the auxiliary mirror frame are at equal and predetermined distances from the base flange. Therefore, the auxiliary mirror frame becomes parallel to the base frame and has the same focal point as the main parabolic reflector.

SELF-ADJUSTMENT OF THE MAIN PARABOLIC REFLECTING SURFACE

The main reflecting surface consists of several pieces which form the surface of a paraboloid when they come together. These pieces are fastened to the body with bolts at four corners by means of spherical hinges. Each piece should be continuously adjusted, so that the reflected sun rays can hit the auxiliary parabolic mirror. In our case we start with the pieces composing one meridional strip and then adjusting the other pieces consequently. As the initial strip we select one which is under the shade of any of the four rods(see Figure 4). As shown in Figure 5, there is another three-dimensional corner mirror 1 mounted on a stand 2 perpendicular to the base flange 3 facing the strip we initially selected for adjustment. There are laser sources 4 and detectors at corners of each piece 5 of the strip. The laser beam reflects back from the corner mirror and hits the corresponding detector. Using the same principle, we can determine the distances from each corner to the corner mirror. If they are not equal to the predetermined value, the detector will activate the selsyns which are connected to the bolts fastening the

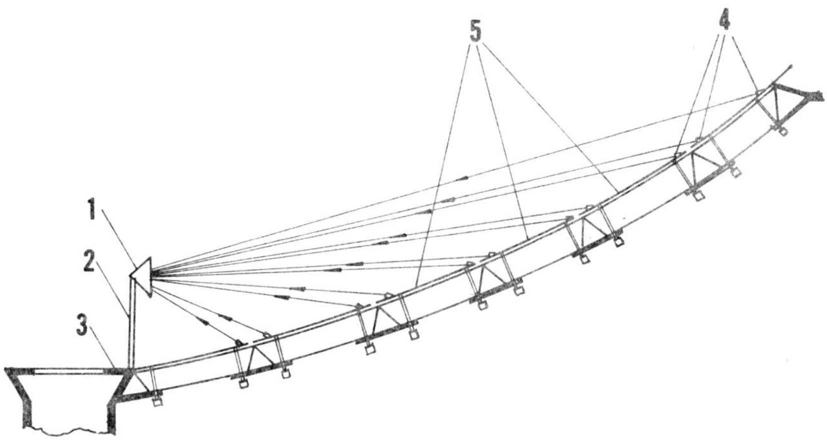

Fig.5 Laser Sources and Selsyns for Adjustment of the Parabolic Reflecting Surface

reflecting surface pieces so that they move the corners of these pieces up and down. The adjustment of the adjacent strips will be done by using light sources and photocells. The light sources will be placed under the pieces initially adjusted and the photocells placed under the adjacent pieces will activate the selsyns to move these pieces until they are in the correct position with respect to the initial ones. Since the light sources and the detectors will be in shade and close to each other, we can use ordinary light sources. So all the pieces will be adjusted using the adjacent one as reference.

The selsyns used in reflector-piece adjustment are very small and have very little capacities compared to the selsyns used in adjustment of the auxiliary mirror, because the loads on them are very little and they do not have to move the pieces instantly. It will be a continuous reaction as the collector moves and the temperature changes. Therefore those selsyns will not add much weight to the body and they will not be expensive. It is clear that problems of stability of our adjustment system is not a very serious one, since we do not have abrupt, discontinuous changes in the shape of the inflatable body. In this connection we can say that it is possible to develop another type of adjustment by regulating the pressure inside the tubes of the body. In that case, however, the tubes have to be separated into the pockets and each pocket has to have its own pressure.

THERMO-FLUID ANALYSIS OF A CYLINDRICAL SOLAR HEAT ABSORBER

M.F. Khalil
University of Alexandria
Alexandria, Egypt

ABSTRACT

In this paper, a theoretical analysis for the fluid flow through a cylindrical solar heat absorber is considered, taking into consideration the heat loss to the surroundings due to convection and radiation. The variation of fluid density and viscosity with temperature is also considered.

The continuity, momentum and energy equations in cylindrical co-ordinates, are reduced in such a way to describe the flow and temperature fields along the absorber. These equations are solved numerically using the Runge-Kutta method.

The effect of radiative and convective heat losses from the absorber is demonstrated. The effect of fluid thermal conductivity and absorber thermal capacity are considered on the fluid temperature rise, pressure drop for fluid flow along the absorber, and finally on the absorber efficiency.

INTRODUCTION

A cylindrical solar heat absorber, such as that schematically shown in Figure 1, is usually used in solar collectors and concentrators.

The analysis of solar heat absorbers of different kinds has received considerable attention recently. This is almost entirely due to the recent solar energy utilization as a solution for the fast depletion of other energy resources in the world.

Although a considerable amount of research work has been carried out on solar heat absorbers, relatively too little or no attention whatsoever has been paid to the effect of fluid flow and radiation losses on their performance.

Analysis for solar absorbers of different kinds are given by Cobble [1,2] by neglecting the effect of both radiation losses and fluid flow. The effect of radiation losses was introduced by Lumsadine [3,4] in a linearized form, which causes a large error, especially at high temperatures. Sakar [5] presented a similar analysis, but without linearization. Both neglected the variation of fluid properties.

It is the purpose of this paper to present a theoretical analysis for the fluid flow through a cylindrical solar heat absorber, taking into

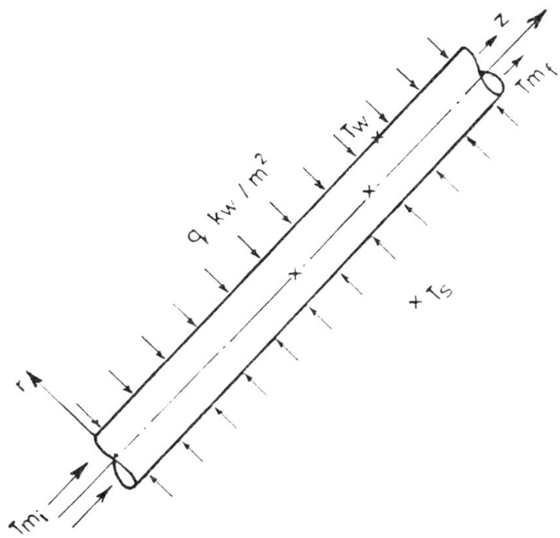

Figure 1 Cylindrical Solar Heat Absorber

consideration the heat loss to the surroundings due to convection and radiation. The variation of fluid density and viscosity with temperature is also considered.

ANALYSIS

Governing Equations

The equations governing the fluid flow through the absorber are obtained from the general equations of the flow of a viscous, incompressible, heat conducting fluid. These general equations as presented in [6], are the equations of continuity, motion and energy. The reduction of these equations to their final form will be carried out using the following assumptions:

(1) Fully developed axially symmetric laminar steady-state conditions prevail.

(2) The variation of pressure, density and viscosity across the flow are far less important than along it, so that p, ρ, and μ are functions of Z only.

(3) The velocity gradients across the flow are more important than along it; symbolically this means:

$$\frac{\partial w}{\partial r} >> \frac{\partial w}{\partial z}\ , \quad \frac{\partial^2 w}{\partial r^2} >> \frac{\partial^2 w}{\partial z \partial r} >> \frac{\partial^2 w}{\partial z^2}$$

(4) The second order temperature gradient across the flow are more important than along it.
Thus, $\frac{\partial^2 T}{\partial z^2}$ may be neglected.

(5) The body forces, other than those caused by gravity force, are neglected.

$$F_z = -g \sin \alpha$$

(6) The fluid enthalpy is a function of its temperature, and may be written as: $H = c_p T$.

(7) Solar radiation heat in, with convection and re-radiation out, at the surface.

(8) The heat resistance of the absorber surface is neglected.

(9) The temperature is a function of r and z only:

$T = T(r,z)$, with the boundary conditions:

$T(D/2,z) = T_w(z)$

$$\frac{\partial T}{\partial r}(D/2,\ z) = \frac{1}{K}[q - h\ (T_w - T_s) - C_r\ (T_w^4 - T_s^4)]$$

$$\frac{\partial T(o,z)}{\partial r} = 0.0$$

(10) The fluid bulk temperatures at inlet and absorber exit are constants.

(11) The dissipated energy due to fluid flow is neglected, the flow is subsonic [7].

After imposing the previous assumptions, the governing equations in cylindrical coordinates are reduced to the following forms:

Continuity equation:

$$\frac{\partial(\rho w)}{\partial z} = 0\ \ldots\ldots \tag{4}$$

Momentum equation:

$$0 = -\ \rho g \sin\alpha - \frac{dp}{dz} + \frac{\mu}{r}\frac{\partial}{\partial r}[r\ \frac{\partial w}{\partial r}] \tag{5}$$

Energy equation:

$$\rho\ C_p\ w\ \frac{\partial T}{\partial z} = \frac{1}{r}\frac{\partial}{\partial r}[K\ r\ \frac{\partial T}{\partial r}] \tag{6}$$

In this form, equation (5) may be integrated twice with respect to r to give the velocity distribution.

On applying the velocity boundary conditions,

$$w = 0 \quad \text{at} \quad r = D/2, \quad \frac{\partial w}{\partial r} = 0 \quad \text{at} \quad r = 0$$

the velocity expression is found to be:

$$w = \frac{D^2}{16\mu}\ (\rho g\ \sin\ \alpha + \frac{dp}{dr})\ (\frac{4r^2}{D^2} - 1) \tag{7}$$

and the mass flow rate, M, through the absorber, is evaluated from equation (7), through the integration

$$M = \int_o^{D/2} \rho\ w\ 2\pi\ r\ d\ r$$

$$M = - \frac{\pi D^4}{128} \frac{\rho}{\mu} [\rho g \sin \alpha + \frac{dp}{dz}] \quad (8)$$

Hence, velocity distribution and pressure gradient can be written respectively as,

$$w = \frac{8M}{\pi D^2 \rho} [1 - \frac{4r^2}{D^2}] \quad (9)$$

$$\frac{dp}{dz} = - \rho g \sin \alpha - \frac{128M}{\pi D^4} \frac{\mu}{\rho} \quad (10)$$

The temperature expression that satisfies condition (9) may be assumed in the following form:

$$T(r,z) = T_w + \frac{D}{4K} [q - h(T_w - T_s) - C_r (T_w^4 - T_s^4)] \cdot S$$

$$S = \frac{4r^2}{D^2} - 1 \quad (11)$$

and the mean temperature at any section is obtained using

$$T_m = \frac{1}{M} \int_0^{D/2} \rho\, w\, T\, 2\pi\, r dr$$

as:

$$T_m = T_w - \frac{D}{K} [q - h(T_w - T_s) - C_r (T_w^4 - T_s^4)] \quad (12)$$

This could be written after using equation (4), in the following form:

$$C_p \frac{\partial}{\partial z} \int_0^{D/2} 2\pi \rho\, W\, T\, r dr = 2\pi [K\, r \frac{\partial T}{\partial r}]^{\frac{D}{2}} \quad (13)$$

Substituting expressions for W and T, given by equations (9) and (11), into equation (13), the integration gives:

$$\frac{dT_m}{dz} = \frac{\pi D}{MC_p} [q - h(T_w - T_s) - C_r (T_w^4 - T_s^4)] \quad (14)$$

After introducing the new dimensionless variables in the form:

$$a = \frac{C_r T_s^4}{q}, \quad b = \frac{h T_s}{q}, \quad c = \frac{M C_p T_s}{\pi D L q}$$

$$d = \frac{6 K T_s}{qD}, \quad e = \frac{128 \mu_s M}{\pi D^4 g \sin \alpha \rho_s^2}, \quad z = z/L$$

$\theta_m = T_m/T_s$, $\theta_w = T_w/T_s$, $P = \frac{P}{\rho_s Lg \sin \alpha}$

$\rho' = \rho/\rho_s$, $\nu' = \nu/\nu_s$

equations (10), (11) and (14) become:

$$\frac{dp}{dz} = - \rho' - e \cdot \nu' \tag{15}$$

$$\theta_m = \theta_w - [1 - a(\theta_w^4 - 1) - b(\theta_w - 1)]/d \tag{16}$$

$$\frac{d\theta_m}{dz} = [1 - a(\theta_w^4 - 1) - b(\theta_w - 1)]/c \tag{17}$$

For the pressure and temperature ranges excepted in solar absorbers, the density and viscosity of flowing fluids may be considerable functions of temperatures only and, for water, can be accurately represented by:

$$\rho = 1.0000226 + 0.000817802\theta - 0.056175210\theta^2 + 0.013475540\theta 3$$
$$+ 0.85002778 \text{ X } 10^{-8}\theta^4 - 0.2550764 \text{ X } 10^{-10}\theta^5 - 0.64625692 \text{ X } 10^{-1}\theta^6 \tag{18}$$

$$\nu' = 1.7400385 - 4.61403610\theta + 5.6378240\theta^2 - 2.50782350\theta^3$$
$$+ 0.0002455829\theta^4 - 0.38252568 \text{ X } 10^{-10}\theta^5 + 0.23419579 \text{ X } 10^{-10}\theta^6$$

The equations obtained in the previous section are obviously too elaborate to permit an exact mathematical solution. Thus, any solution will have to be obtained by numerical means, the numerical.... treatment followed here is described:

(1) Assume the mean inlet temperature is θ_{mi}

(2) Use equation (16) through Newton-Raphson method to obtain wall temperature θ_w as

$$\theta_{w_{i+1}} = \theta_w - \frac{f}{f'} \qquad \text{or}$$

$$\theta_{w_{i+1}} = \frac{3a\ \theta^4_{w_i} + (1 + a + b - d \cdot \theta_m)}{4a\ \theta_{w_i} + b + d} \tag{19}$$

(3) After obtaining a better approximation for the value of θ_W at the inlet section, use equation (17) through the Runge-Kutta method to obtain the wall temperature distribution along the absorber.

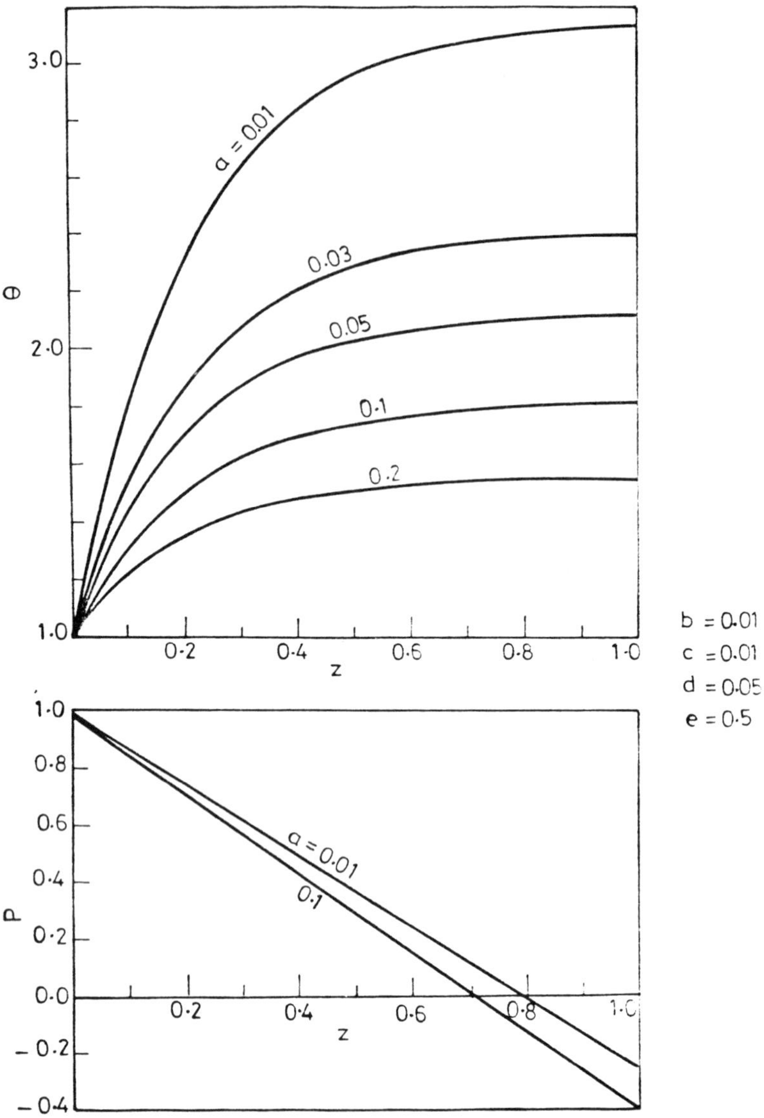

Figure 2 Effect of Parameter (a) On Fluid Temperature And Pressure Distribution

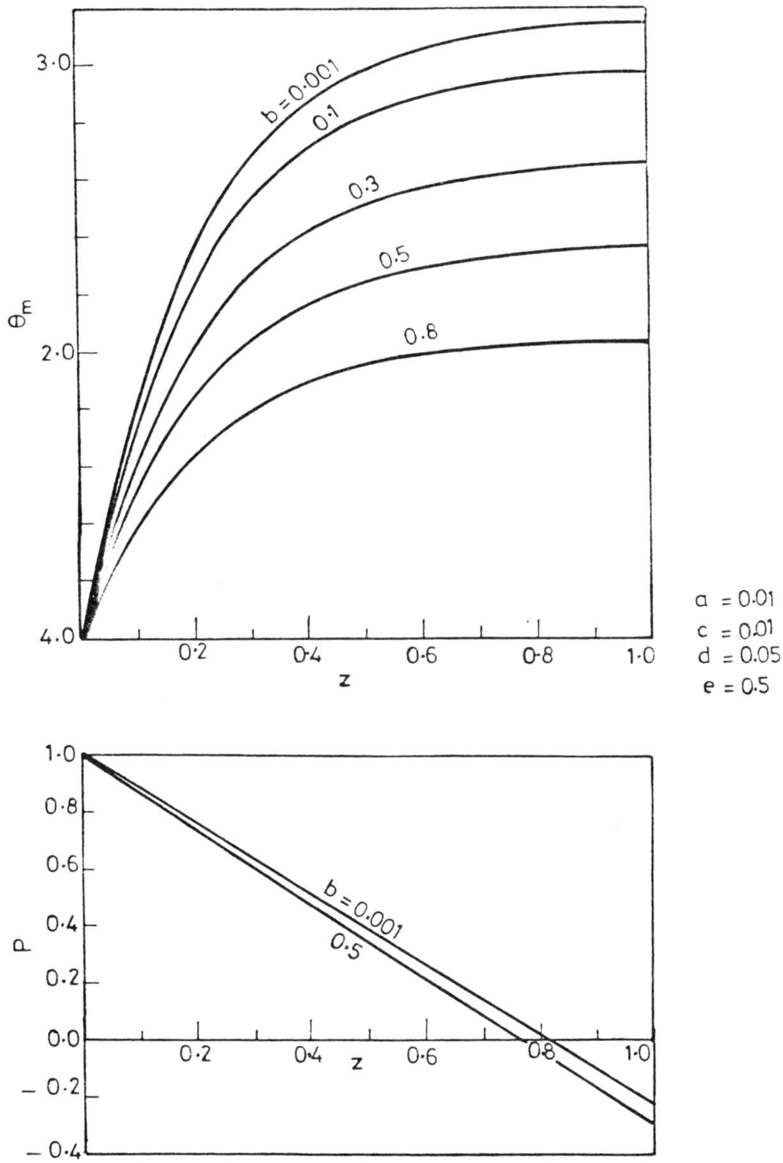

Figure 3 Effect of Parameter (b) On Fluid Temperature and Pressure Distribution

(4) Equation (16) is then used to obtain the mean temperature distribution.

(5) Equation (19) is then used to obtain the fluid density and viscosity at each station along the absorber.

(6) The pressure distribution along the absorber is then obtained throughout, using the Runge-Kutta method.

(7) After obtaining the pressure and temperature distribution, the efficiency of the absorber is determined from:

$$\eta = \frac{M\ C_p\ (T_{m_f} - T_{m_i})}{q\P\ D\ L}$$

or as

$$\eta = C(\theta_{m_f} - \theta_{m_i}) \qquad (20)$$

DISCUSSION

The dimensionless temperature and pressure distributions and absorber efficiency have been evaluated for a wide variety of cases. The computation has been carried out, using the Alexandria University PDP 11 Computer.

The numerical solution for the pressure and temperature distributions are presented in Figures 2 through 6. The effect of changing the radiation loss parameter "a" between 0.01 and 0.2 is shown in Figure 2. As "a" is decreased, the temperature at any section of the absorber increased, while the pressure decreased. This is due to the decrease in radiation loss and increase of frictional pressure drop.

Figure 3 provides the same results of pressure and temperature profiles, for convective loss parameter "b" ranging from 0.001 to 0.8.

The effect of absorber thermal capacity factor "c" on the temperature and pressure profiles is shown in Figure 4. It can be noticed that a decrease in the absorber thermal capacity factor "c", (by increasing L, D, q or decreasing M, C_p), rises up the temperature profile. On the contrary, a decrease in "c" lowers down the pressure profiles.

Figure 5 shows, that as the fluid conductivity factor "d" increases, the temperature and pressure profiles are elevated.

Figure 6 shows the effect of the fluid flow factor "e" on the pressure distribution. It is clear from the figure that the higher the value of "e", the lower the pressure distribution obtained. This is due to the increase in pressure loss due to fluid flow.

The effect of the parameters a, b, c, d and e on the temperature rise and pressure drop for the following fluid throughout the absorber is shown in

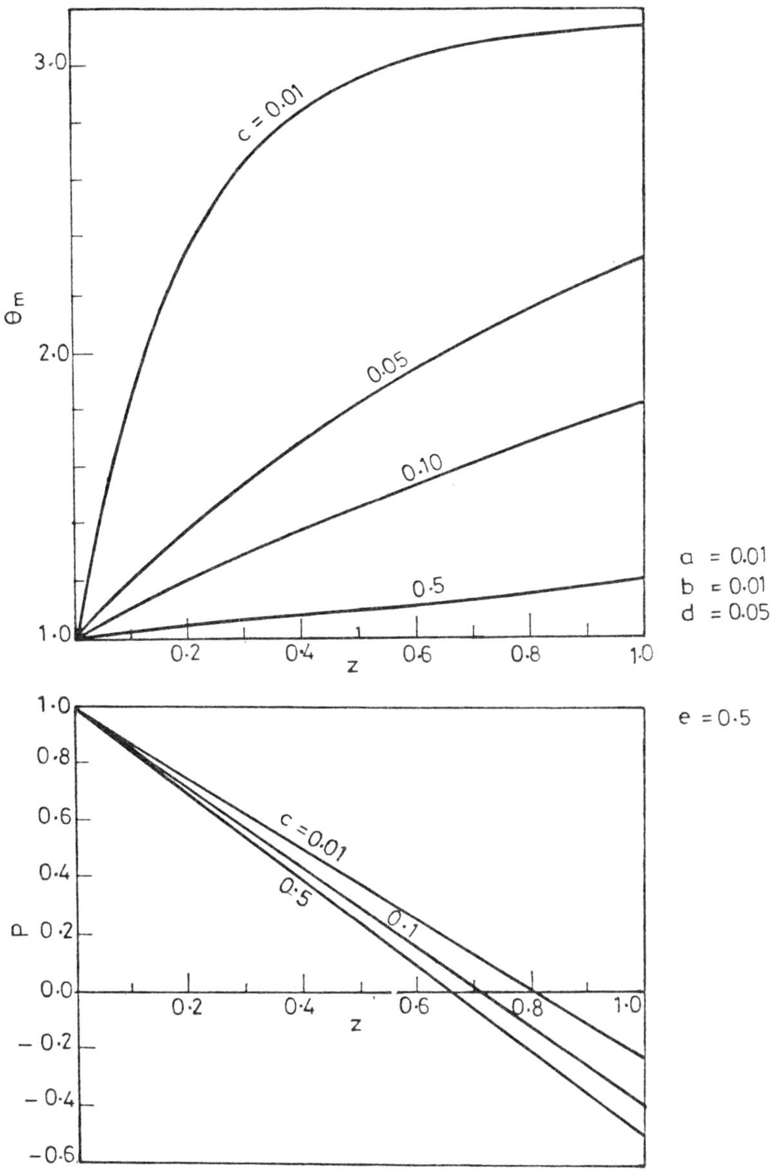

Figure 4 Effect of Parameter (c) On Fluid Temperature and Pressure Distribution

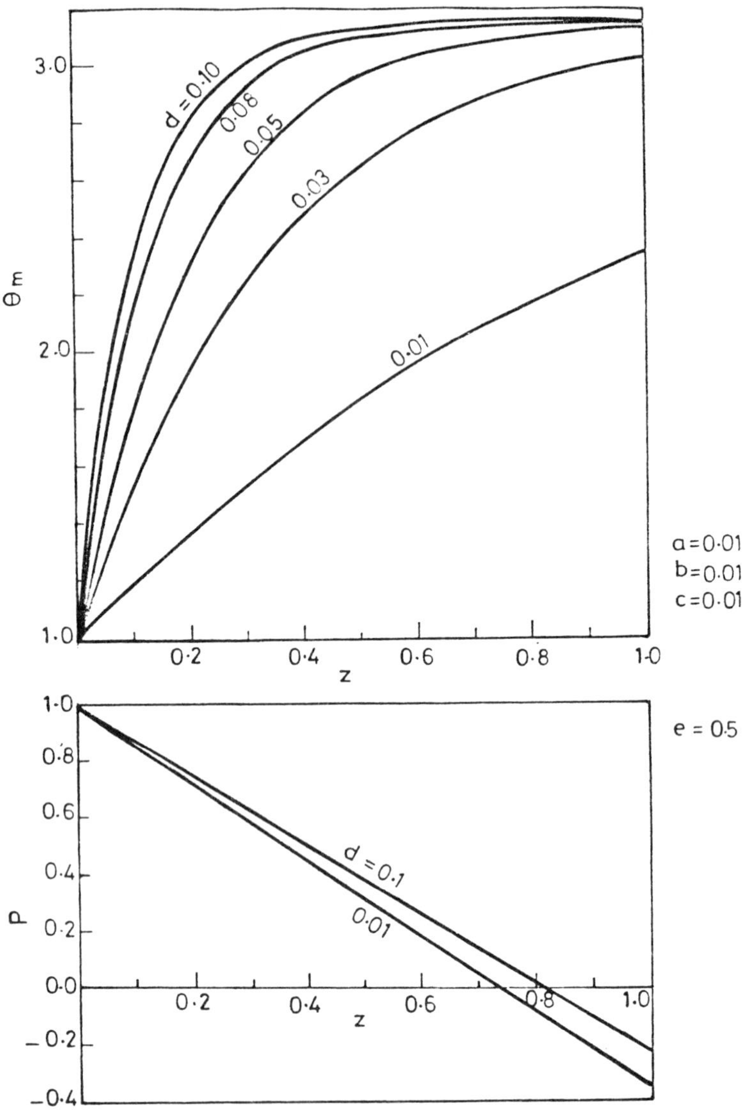

Figure 5 Effect of Parameter (d) On Fluid Temperature and Pressure Distribution

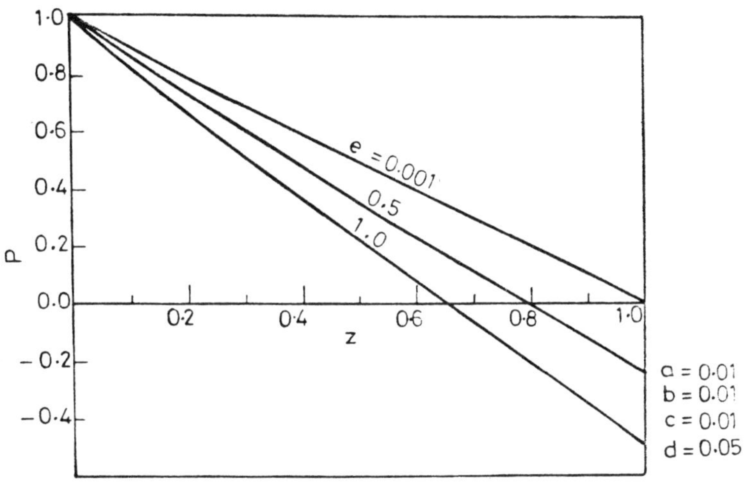

Figure 6 Effect of Parameter (e) On Pressure Distribution

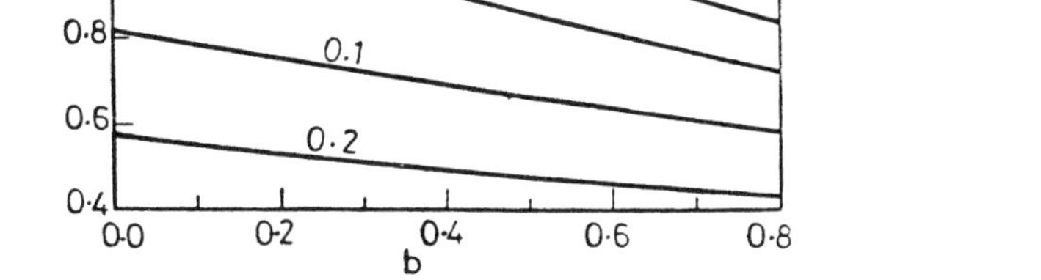

Figure 7 Effect of Paramter (a & b) On Temperature Rise

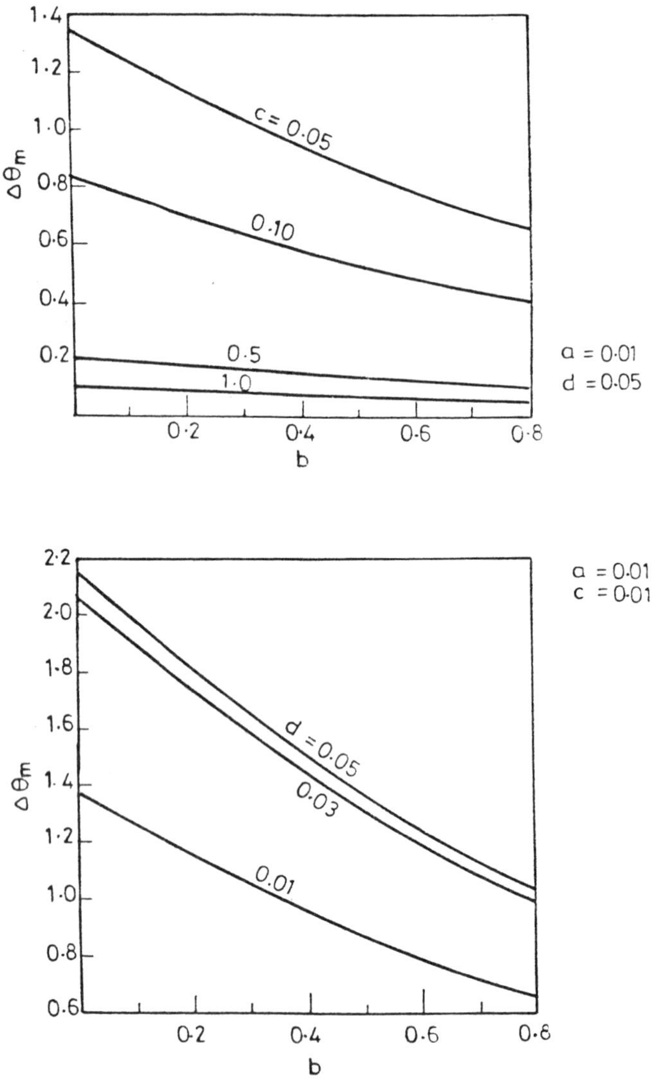

Figure 8 Effect of Parameter (c & d) On Temperature Rise

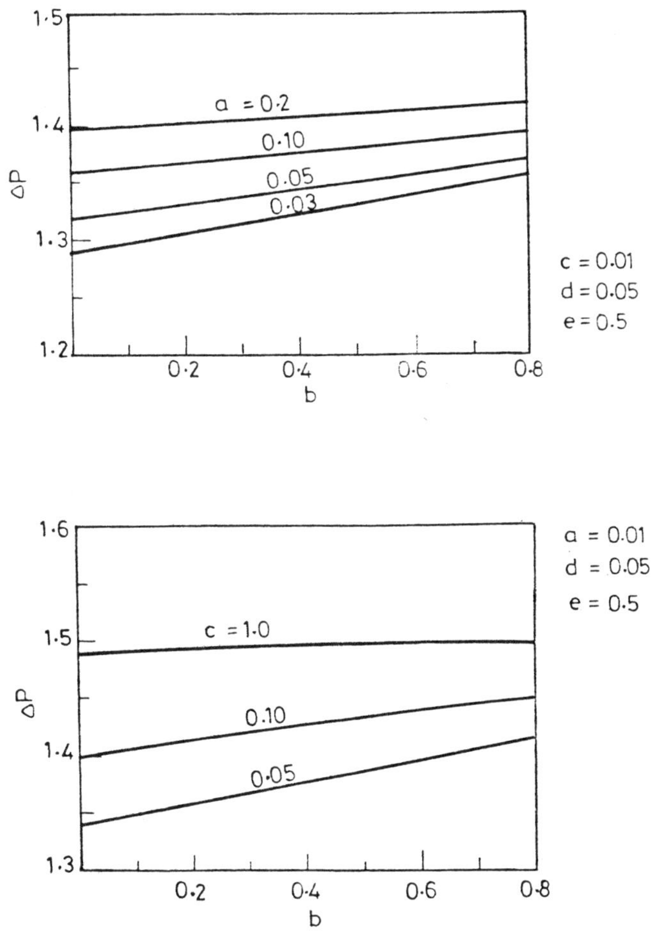

Figure 9 Effect of Parameter (a, b & c) On Pressure Drop

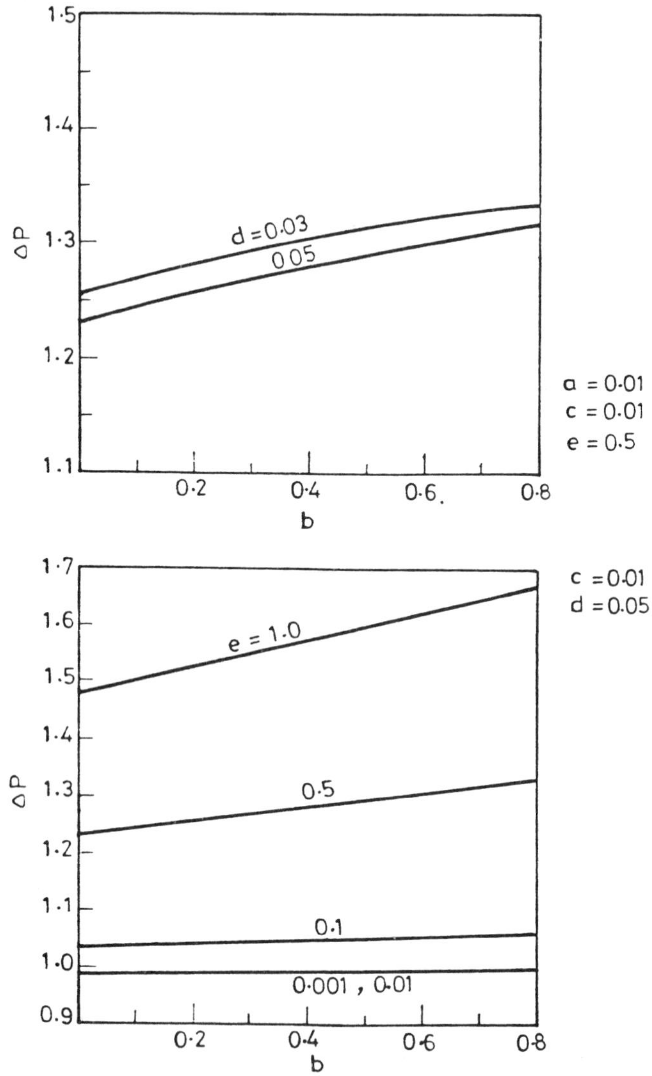

Figure 10 Effect of Parameter (d & e) On Pressure Drop

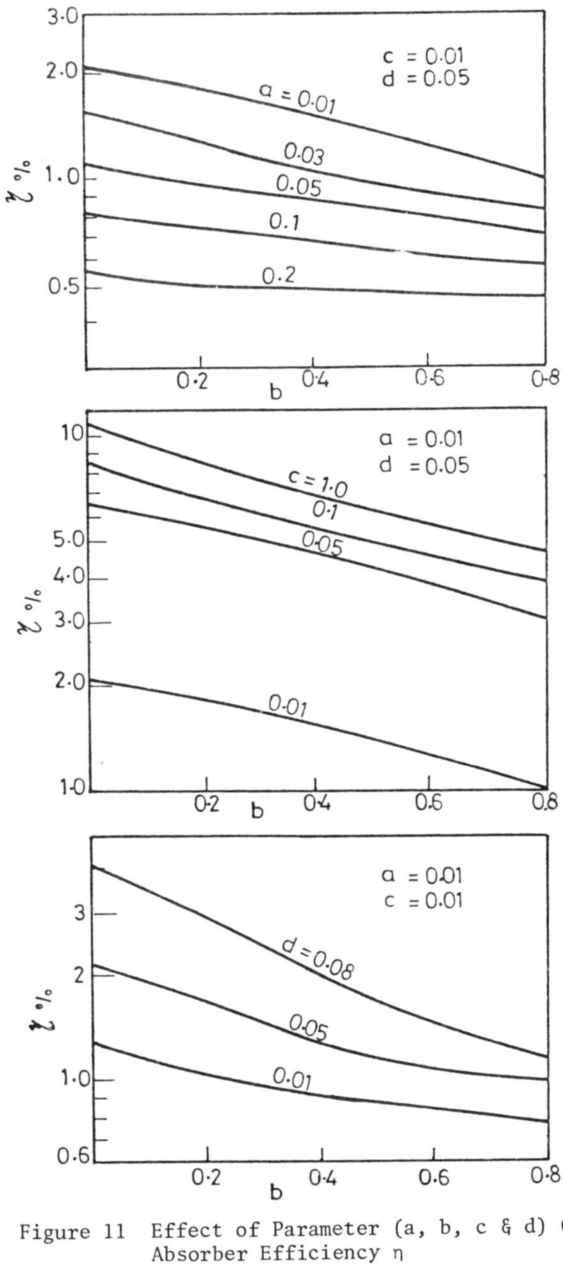

Figure 11 Effect of Parameter (a, b, c & d) On Absorber Efficiency η

Figures 7 through 10. From the figures, it can be concluded that the lower the values of radiation parameter loss "a", convection parameter loss "b" and lower absorber thermal capacity factor "c", the more the rise in fluid temperature. Also, a higher pressure drop is obtained at higher values of the previous parameters, plus the fluid flow parameter "e", and at lower values of fluid conductivity factor "d".

The effect of the previous factors on absorber efficiency is shown in Figure 11. It is clear from the figure that higher absorber efficiencies are obtained at lower values of: radiation loss parameter "a", convection loss parameter "b", and higher values of both thermal capacity factor "c" and fluid conductivity factor "d".

CONCLUSIONS

From the previous analysis and discussions, the following conclusions can be drawn:

(1) The highest fluid temperature rise is theoretically attainable with solar absorbers having lower values of: absorber thermal capacity, heat loss by both convection and radiation, but with higher values of fluid thermal conductivity factor.

(2) Higher absorber efficiency is obtained at lower values of heat loss by convection and radiation, with higher values of thermal capacity factor and fluid thermal conductivity factor.

(3) Higher pressure drop for fluid flow along the absorber is obtained at higher values of: heat loss by radiation and convection, absorber thermal capacity factor, fluid thermal conductivity factor, and higher fluid flow rate.

REFERENCES

1. Cobble, M.H., Heat Exchangers for Solar Concentrators. Solar Energy 3, (9), 18-21 (1963).

2. Cobble, M.H., Spherical Shell Heat Exchanger. New Mexico State University Engineering Experiment Station Bulletin No. 5 (1962).

3. Lumsdaine, E., Cylindrical-Annulus Heat Exchanger For Parabolic Cylindrical Solar Concentrator. South Dakota State University Engineering Experiment Station Bulletin No. 12 (1970).

4. Lumsdaine, E., Comparison of Solar Heat Exchangers. Solar Energy 17, 269-275 (1975).

5. Sakr, I.A. Abd-Elafiz, M.A. and Hegazy, A.S., Numerical Heat Analysis of Solar Heat Absorber Using Digital Computer. International Conference and Exhibition on Solar Building Technology, London, England,

July 25-29, 1977.

6. Yuan, S.W., Foundations of Fluid Mechanics. Prentice-Hall of India, Private, Limited, New Delhi, India (1969).

7. Kreith, F., Principles of Heat Transfer. 2nd Edition. International Textbook Company, Scranton, Pennsylvania, USA (1966).

NOMENCLATURE

a	radiation loss parameter
b	convection loss parameter
c	absorber thermal capacity factor
c_p	fluid specific heat
c_r	radiation loss coefficient = $-\varepsilon$
d	fluid thermal conductivity factor
D	absorber internal diameter
e	fluid flow factor
F_z	body force per unit volume in Z-direction
g	acceleration due to gravity
H	fluid enthalpy
h	effective overall convective coefficient of heat transfer between absorber and surroundings, based on absorber internal diameter
k	fluid thermal conductivity
L	absorber length
M	fluid mass flow rate
P	fluid pressure
p	dimensionless pressure = p = $\rho_s L \sin\alpha$
q	solar heat flus per unit area
r, Z	co-ordinates
Z	= Z/L
T	fluid temperatue
W	fluid flow velocity in the axial direction

Greek Symbols

α	absorber axis inclination angle
ε	emissivity
η	absorber efficiency
θ	ratio of fluid temperature to surroundings
σ	Stefan-Boltzmann constant
ν	fluid kinetic viscosity
ρ	fluid density
μ	fluid dynamic viscosity

Subscripts

m	mean
m_f	final mean
m_i	initial mean
s	surrounding
w	wall

AN EXPERIMENTAL INVESTIGATION OF COMPARATIVE PERFORMANCE OF A HEADER-TYPE COMPOUND PARABOLIC CONCENTRATOR AND A FLAT--PLATE COLLECTOR

K.Krishna Rao, S.Basheer Ahmed,
R.Natarajan and M.C.Gupta
Department of Mechanical Engineering
Indian Institute of Technology
Madras - 600 036 - INDIA

ABSTRACT

The comparative performance of a header-type compound parabolic concentrator and a flat-plate collector has been experimentally investigated. Two configurations of the collectors, one in which the absorber tubes were all oriented in the east-west direction, and another in which the absorber tubes of the FPC were oriented north-south, three different flow rates and three angles of tilt were considered. The hourly efficiency and useful thermal energy gain have been obtained as a function of the hour of day, and efficiency vs $\Delta T/I$ curves have been drawn. The results show that in configuration I, the CPC was more efficient than the FPC for about 2 to 3 hours around noon when kept at an angle equal to latitude or declination. The mean daily efficiency was about the same for both collectors. When kept horizontal the CPC was quite inefficient. In configuration II, the CPC was more efficient than the FPC for about two hours around noon when kept at the declination angle; when kept at an angle equal to the latitude the FPC had the better over-all performance.

INTRODUCTION

There is a multitude of designs of collectors for solar energy in current use, these have distinctly different performance characteristics. The novel design proposed by Winston [1] for concentrating solar energy using the principle of ideal cylindrical radiation collection, commonly referred to as Compound Parabolic Concentrators (CPC), has two major advantages over other concentrators: (i) There is no need for diurnal tracking; only seasonal adjustments are necessary. For low concentration factors of about 3 even the seasonal adjustments can be dispensed with. (ii) In addition to beam radiation it accepts diffuse radiation, unlike a focusing collector, thus allowing more energy to be collected during a year's operation. The CPCs have an advantage over the flat-plate collectors in that they can produce higher temperatures because of concentration from 3 to 10 without diurnal tracking. The potential applications of CPCs are in the areas of space heating and cooling, process heat, irrigation pumping, electric power generation, and photovoltaic conversion.

The information available in the literature on CPCs is very limi-

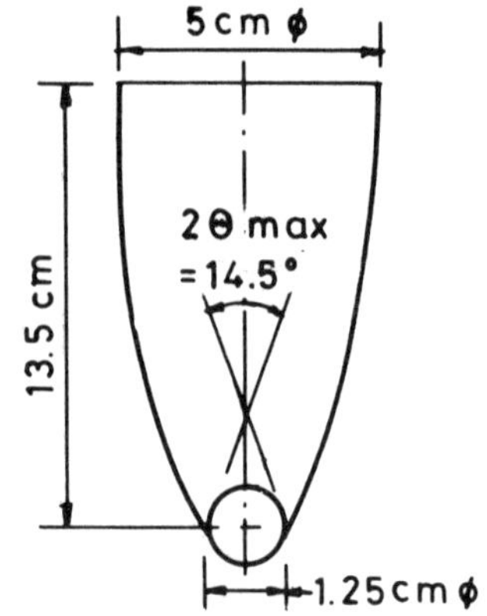

FIG.1(a). CONFIGURATION OF C.P.C

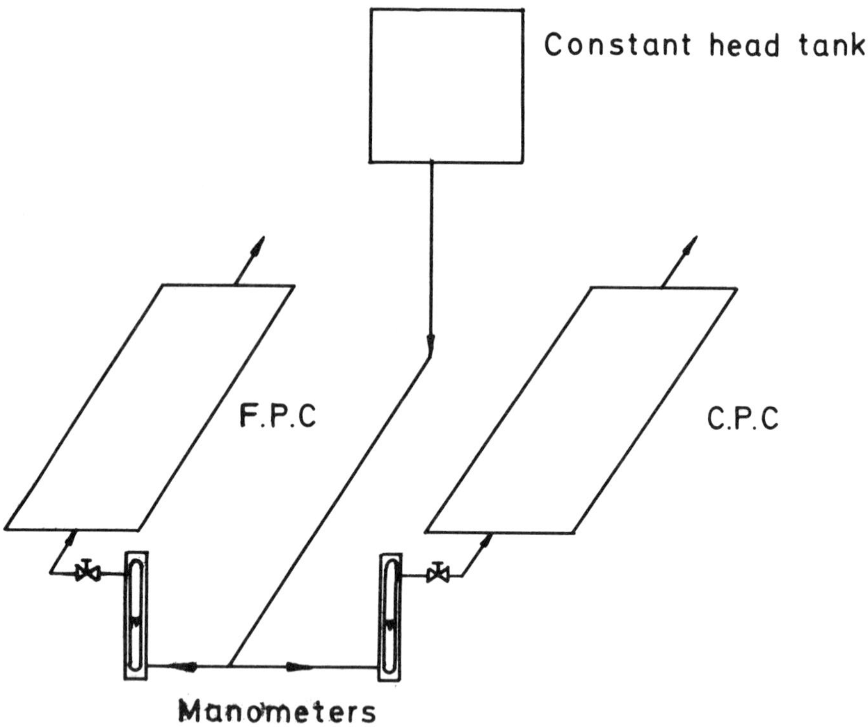

FIG.1(b). SCHEMATIC OF THE EXPERIMENTAL SET-UP

TABLE I SUMMARY OF TEST CONDITIONS

1. Configurations:	1. Both CPC and FPC absorber tubes along E-W.
	2. CPC absorber tubes along E-W; FPC absorber tubes along N-S
2. Flow rates:	5, 10, 15 l/h
3. Angles of tilt:	Latitude, horizontal, declination (-9.04° to -8.67°)
4. Days of year on which experiments were done	48 to 75
5. Latitude	13°

ted. The available information is mainly concerned with optical analysis of the concentrator. Rabl [2,3] has performed a comprehensive study of CPCs with particular reference to their optical performance characteristics. The experimental performance of the CPCs for thermal processes has not been investigated in detail.

PRESENT WORK

The present paper provides a comparison of the performance of a CPC with that of a conventional flat plate collector of the same exposed area.

EXPERIMENTAL SET-UP

The CPC has 10 parallel rows, each having an aperture area of 0.045^2 and an acceptance angle of 14.5° (Fig.1a). The concentrators are truncated to give a concentration of 3.7. The absorbers used are 12.5 mm dia. G.I. pipes which are welded to common headers. Parabolic aluminium plates are used as reflectors. The CPC is enclosed in a wooden box with double-glazing and glass wool insulation to minimize radiation and convection losses. The total exposed area of the CPC is $0.54\ m^2$. A conventional flat plate collector of the same overall size has also been fabricated using 12.5 mm G.I. pipes, 22 gage G.I. sheet, double glass covers and glass-wool insulation. Water was supplied at a constant head from an overhead tank to each collector at a fixed rate as measured by rotameters provided at inlet to the collectors.

Figure 1b shows a schematic of the experimental arrangement.

EXPERIMENTAL PROCEDURE

In the first set of experiments (Configuration I), the absorber tubes of both the collectors were maintained in the east-west direction, and the collectors were oriented facing south. They were tested for identical angles of tilt, viz., 0°, latitude (13°) and declination with respect to the horizontal. Three different flow rates of 5, 10 and 15 litres/h were employed. The water inlet temperature, outlet temperature and glazing temperature were measured with copper-constantan thermocouples. The solar insolation was measured with a Kipp and Zonen solarimeter. These measurements were all taken at intervals of ½ hour.

In the second set of experiments (Configuration II), the absorber tubes of the flat plate collector were oriented in the north-south direction. The two collectors were again tested for the same experimental conditions as before. Since the declination was getting close to zero in March when these experiments were being performed, for the second set of experiments, only two angles of tilt were considered, viz., horizontal and latitude.

RESULTS AND DISCUSSION

Figures 2, 3 and 4 show the comparative performance of the CPC and FPC for Configuration I, for three different angles of tilt and for three different flow rates. The period under investigation was characterized by stochastic insolation conditions, with frequent cloud cover, and this can be seen from the insolation curves.

Figure 2 shows that when the angle of tilt was equal to the latitude of the place, the CPC is more efficient than the FPC for only about two hours in the day around noon, and the temporal efficiency curves for both FPC and CPC follow closely the curve of hourly insolation. The efficiency curve for FPC is much flatter than that for CPC. Table II (a and b) shows in addition that the mean daily efficiency is approximately same for the two collectors for the three flow rates considered.

Figure 3 shows that when the collectors were kept horizontal, the FPC is more efficient than the CPC for all the three flow rates. But the mean daily efficiency for FPC is higher than that for CPC (by over 3), as shown in Table II (a and b). This shows that the CPC is quite inefficient when kept horizontal.

Figure 4 shows that when the angle of tilt was equal to the declination for the day (-9.04° to -8.67°), the CPC is more efficient than the FPC for about three hours around noon. The mean daily efficiency is about same for the two collectors.

Figures 2,3 and 4 and Table II (a and b) also show that maximum efficiency of the order of 50 and mean daily efficiency of about 45 for both collectors could be obtained when the angle of tilt was equal to declination, and for flow rates of 10 and 15 l/h.

Figures 5 and 6 show the comparative performance of the CPC and FPC for Configuration II, for three different flow rates. Since, as mentioned earlier, the declination was getting closer to zero, only two tilts, viz., horizontal and latitude were investigated.

Figure 5 shows that when the collectors were tilted at an angle equal to the latitude, the CPC was more efficient than FPC for less than an hour during the day. This is standard configuration for the FPC, and under these conditions it can be seen to be superior to the CPC for almost the entire day. Table II (a and b) shows that the maximum efficiency of the CPC is only marginally higher than that of the FPC.

Figure 6 shows that when the collectors were kept horizontal (which is now equal to the declination of the day), the CPC is more efficient than the FPC for nearly two hours around noon. Table II (a and b) shows that the mean daily efficiency is about the same for the two collectors.

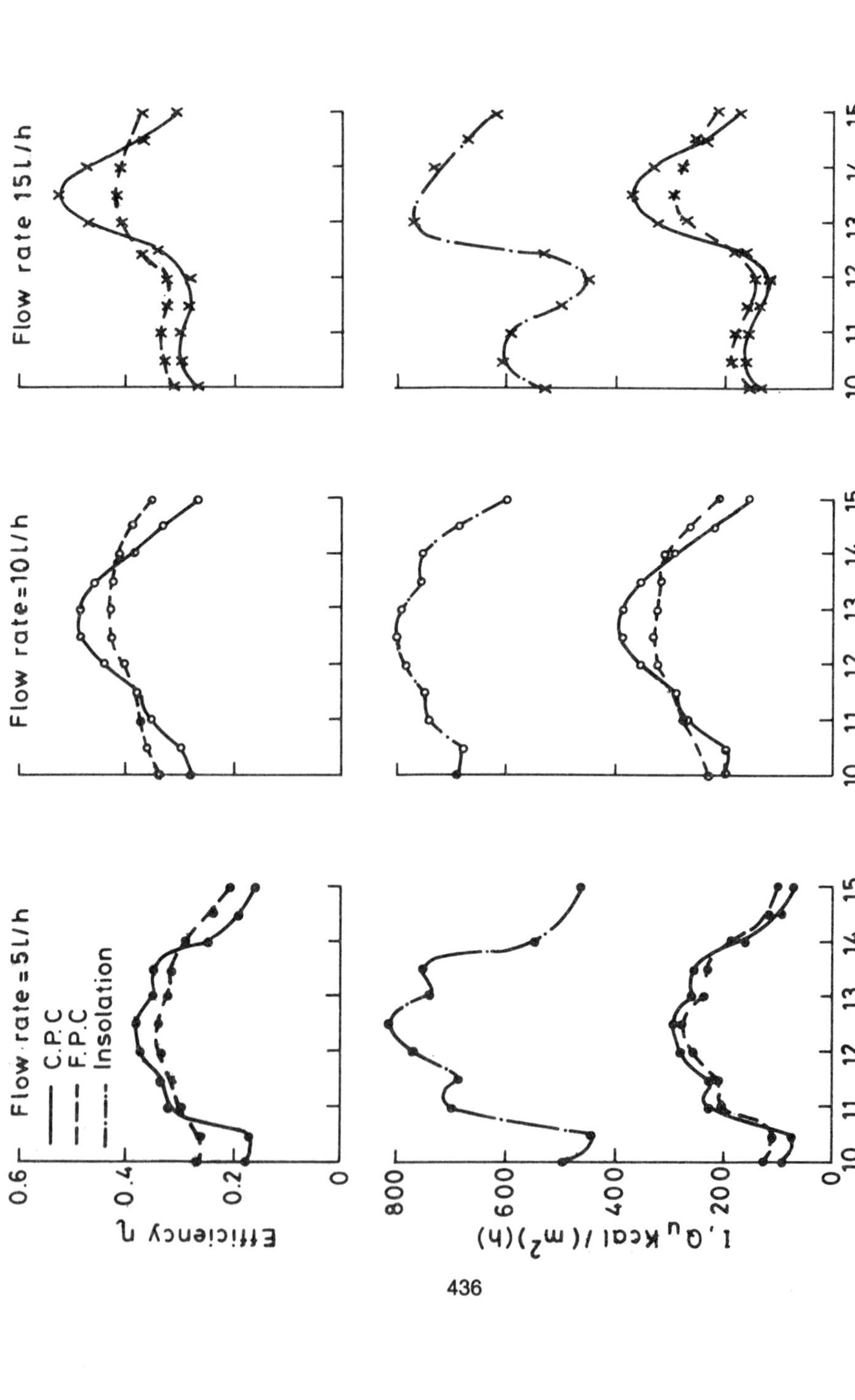

FIG. 2. COMPARATIVE PERFORMANCE OF C.P.C. AND F.P.C – ANGLE OF TILT = LATITUDE (13°) CONFIGURATION I

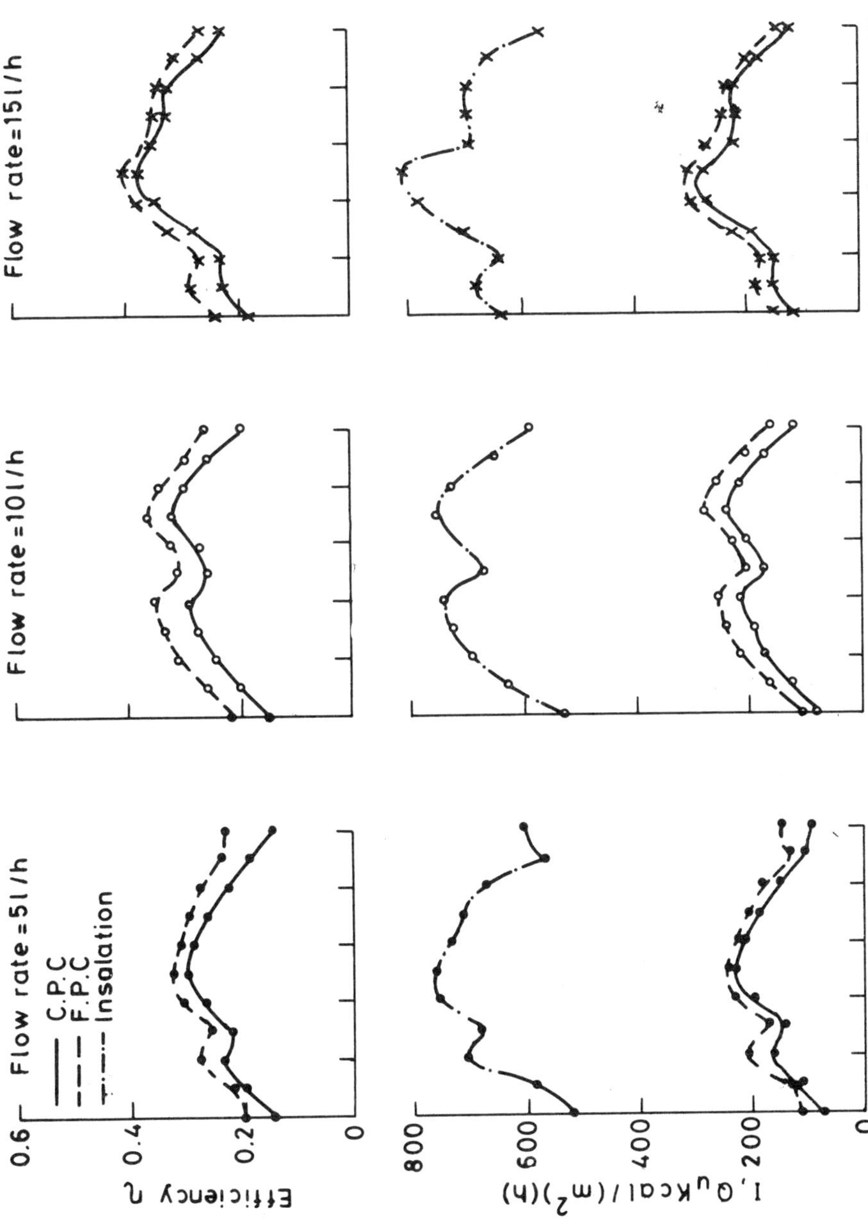

FIG. 3. COMPARATIVE PERFORMANCE OF C.P.C. AND F.P.C - ANGLE OF TILT = HORIZONTAL CONFIGURATION I

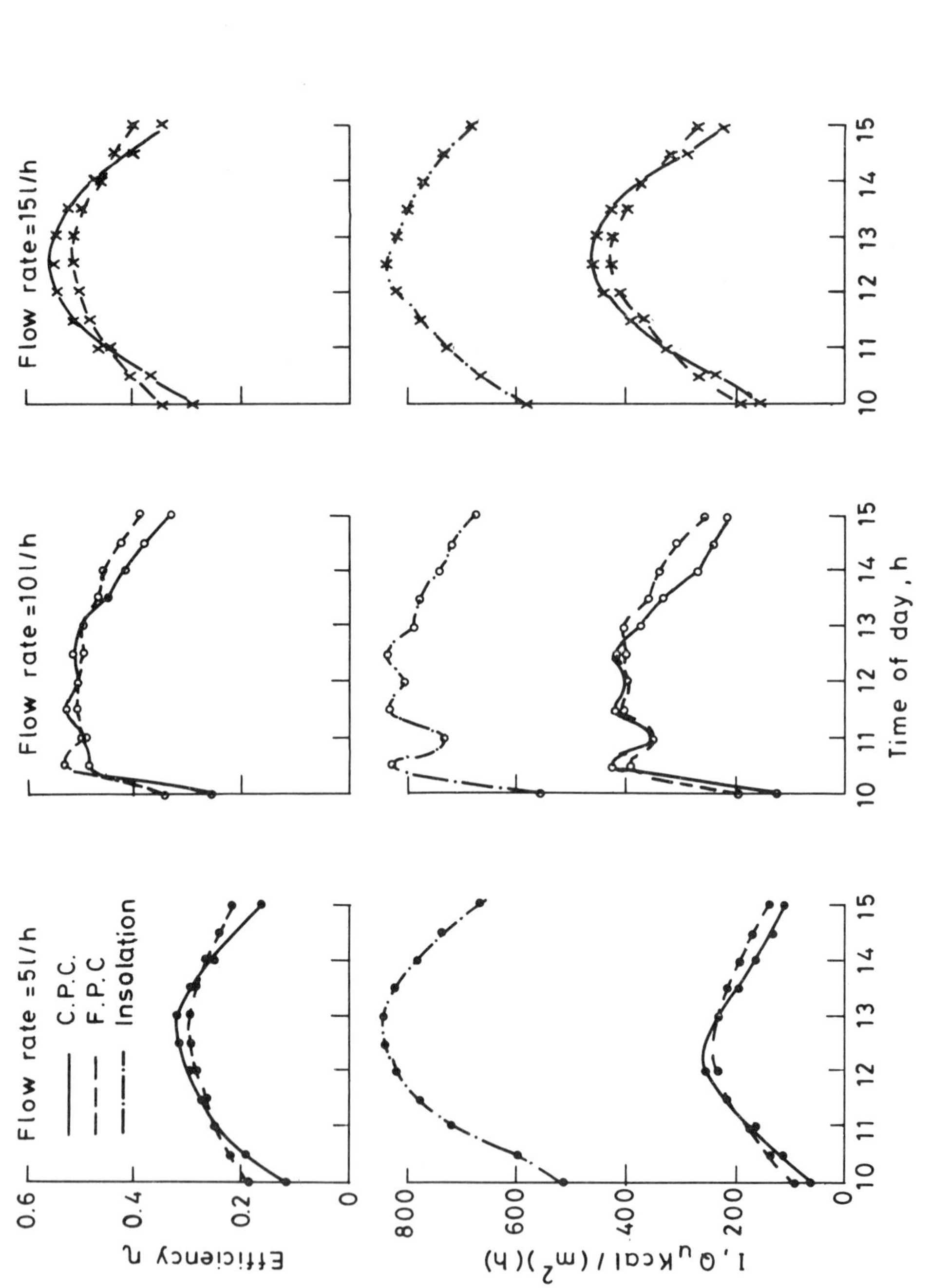

FIG. 4. COMPARATIVE PERFORMANCE OF C.P.C. AND F.P.C – ANGLE OF TILT = DECLINATION ($-9.04°$ to $-8.29°$)

TABLE IIa SUMMARY OF COMPARATIVE PERFORMANCE

CONFIGURATION I

(Both CPC and FPC absorber tubes along E-W)

	Latitude (13°)			Horizontal			Declination -8.69 to -9.04		
Day of the Year	48	52	51	54	55	56	60	58	59
($\delta-\beta$)	2°	2°	2°	10°	10°	10°	0	0	0
Flow rate (l/h)	5	10	15	5	10	15	5	10	15
Maximum efficiency									
CPC η	37%	48.5%	52.5%	30.5%	32.5%	35%	31%	52%	55%
FPC	34%	42.5%	41%	32%	37%	41%	29%	50%	51%
No. of hours of day during which CPC gives higher temperature than FPC	Nearly 3h	Nearly 2.2h	Nearly 2h	0	0	0	Nearly 3.2 mts.	Nearly 2.30 hr.	Nearly 3h

TABLE IIb SUMMARY OF COMPARATIVE PERFORMANCE

CONFIGURATION II

(CPC absorber tubes along E-W; FPC absorber tubes along N-S)

	Latitude (13°)			Horizontal		
Day of the Year	68	66	67	75	70	69
$(\delta-\beta)$	8°	8°	8°	0	0	0
Flow rate (l/h)	5	10	15	5	10	15
Maximum efficiency η						
CPC	36.5%	41.0%	44%	35%	51%	55%
FPC	35%	40%	43%	30.5%	48.5%	51.5%
No. of hours of day during which CPC gives higher temperature than FPC	0	20 mts.	30 mts.	Nearly 3 hrs.	Nearly 2.30 hrs.	Nearly 2 hrs.

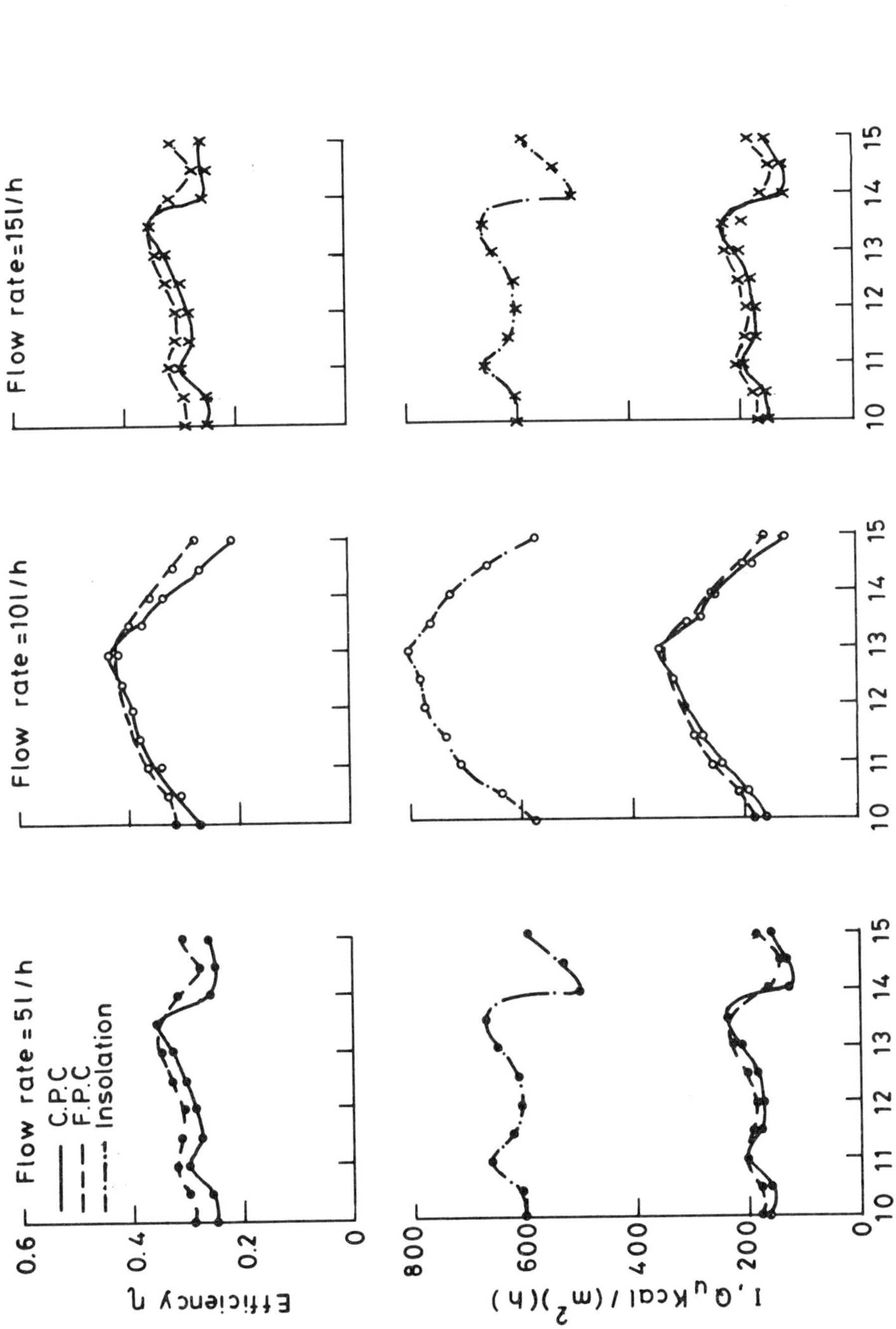

FIG.5. COMPARATIVE PERFORMANCE OF C.P.C AND F.P.C- ANGLE OF TILT = LATITUDE (13°) CONFIGURATION II

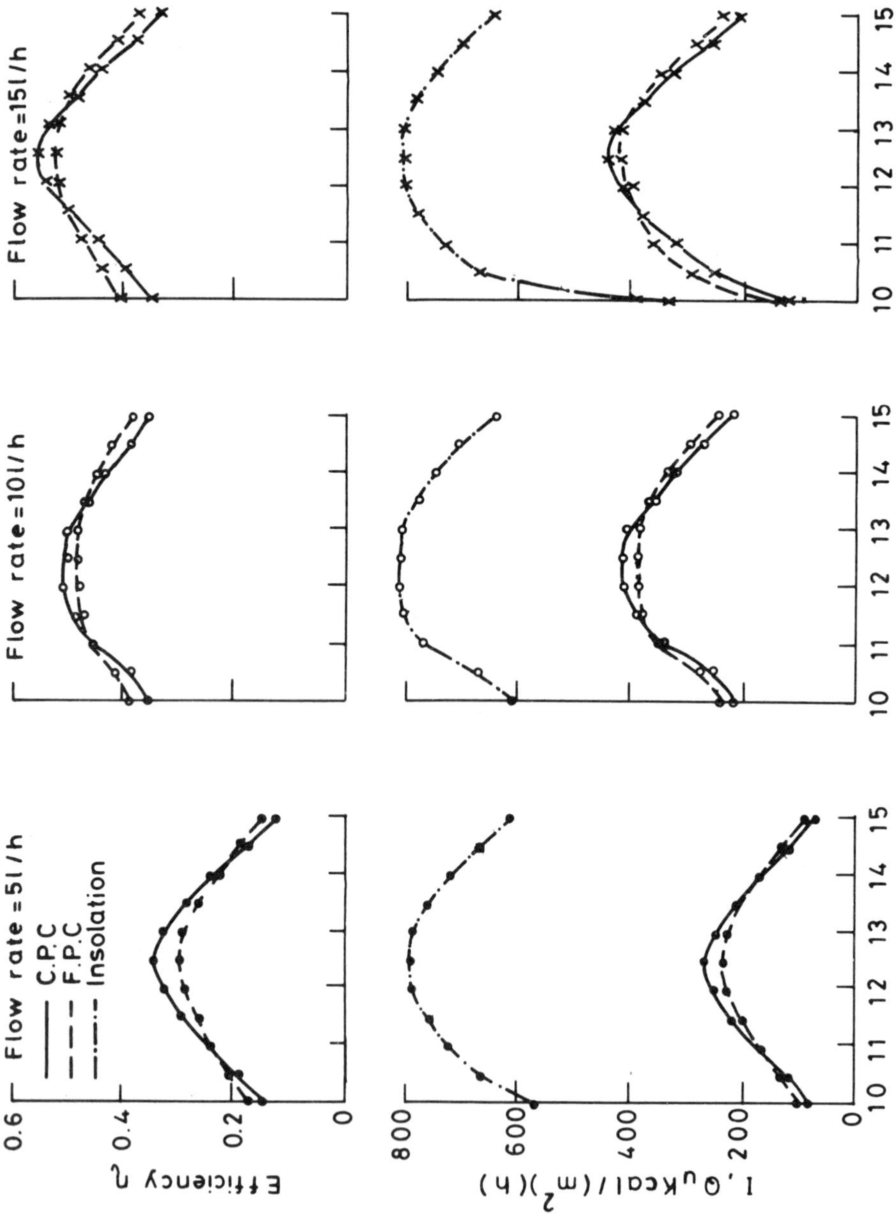

FIG. 6 COMPARATIVE PERFORMANCE OF C.P.C AND F.P.C- ANGLE OF TILT = HORIZONTAL

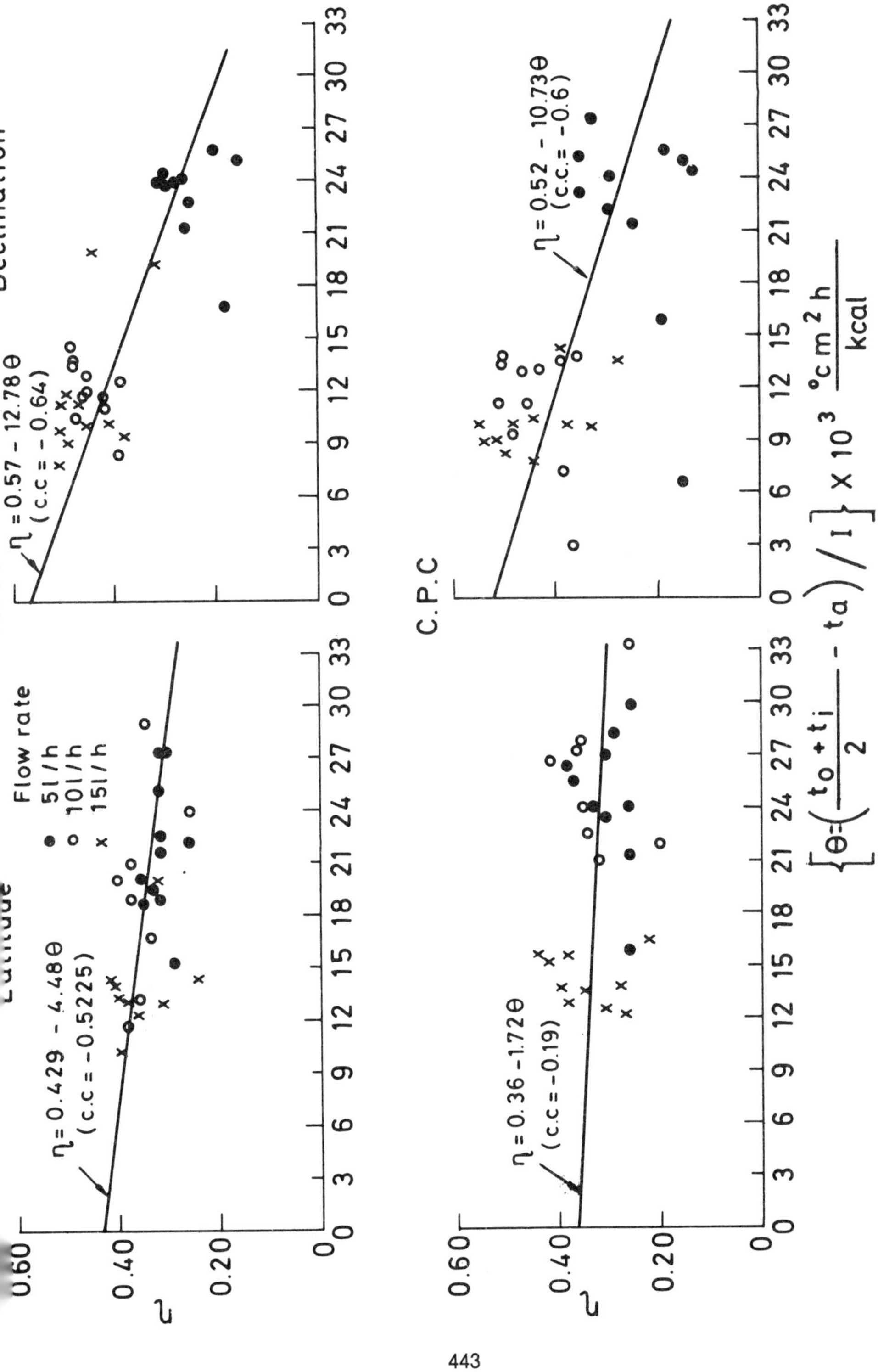

FIG.7. EFFICIENCY vs ΔT/I PLOTS FOR C.P.C & F.P.C (CONFIGURATION II)

In Configuration II, maximum efficiency of the order of 50 to 55% and mean daily efficiency of about 45% could be obtained for both CPC and FPC, when the collectors were tilted at an angle equal to the declination of the day.

The above results show that the performance of both CPC and FPC depend quite strongly on the declination of the day.

Figures 7a and 7b show the efficiency - $\Delta T/I$ curves for the FPC and CPC respectively. It can be seen that for both the collectors, the curves are steeper when the collectors were kept horizontal (and equal to declination). Even though the maximum efficiency is greater for this orientation, an angle of tilt equal to the latitude yields better performance at larger $\Delta T/I$. The equations representing least-square fits of the points, and the correlation coefficients are indicated in each graph.

CONCLUDING REMARKS

In Configuration I, the CPC was more efficient than the FPC for about two to three hours around noon when kept at angle equal to the latitude (13°) or to the declination of the day (-9.04° to -8.67°). The mean daily efficiency was about the same for both collectors. When kept horizontal the CPC was quite inefficient.

In Configuration II, the CPC was more efficient than the FPC for about two hours around noon when kept at an angle equal to the declination of the day (which was very close to the horizontal position during the tests). On the other hand, when tilted at an angle equal to latitude, the FPC had the better over-all performance.

The η vs $\Delta T/I$ curves for Configuration II show that the maximum η is slightly higher when the collectors were tilted at an angle equal to the declination, but the curves are much steeper now.

For the CPCs serpentine configuration may be expected to yield better performance than the present header-type configuration, and studies are under way to investigate their characteristics.

NOMENCLATURE

I	Total insolation on an inclined plane $Kcal/m^2/h$
Q_u	Useful heat collected $Kcal/m^2/h$
t_i	Inlet water temperature $^\circ C$
t_o	Outlet water temperature $^\circ C$
t_a	Ambient temperature $^\circ C$
η	Efficiency

θ $\left(\frac{t_o + t_i}{2} - t_a \right) / I \times 10^3$ - $\frac{{}^oC\ m^2}{Kcal}$

δ Declination - degrees

β Angle of tilt - degrees

REFERENCES

1. Winston,R., 'Solar concentrators of a Novel design', Solar Energy, 16, 89 (1974).
2. Rabl,A., 'Comparison of Solar Concentrators', Solar Energy, 18, 93-111 (1976).
3. Rabl,A., 'Optical and Thermal Properties of Compound Parabolic Concentrators', Solar Energy, 18, 497-511 (1976).

INTERFEROMETRIC INVESTIGATIONS OF CONVECTION IN SLAT, FLAT PLATE AND VEE-CORRUGATED SOLAR COLLECTORS

B.A. Meyer, K.R. Randall, M.M. El-Wakil and J.W. Mitchell
The University of Wisconsin
Madison, Wisconsin 53706 U.S.A.

ABSTRACT

Local and average heat-transfer coefficients for natural convection between parallel plates, parallel plates with slats forming small aspect ratio enclosures, and a vee-corrugated plate with a parallel flat plate have been experimentally studied using interferometric techniques. The experimental conditions are similar to those existing in solar collectors.

Measurements of the heat-transfer coefficient in the flat plate geometries were made over angles of inclination with respect to the horizontal ranging from 45° to vertical. The Grashof number range was 1×10^3 to 3×10^5. The aspect ratio (the ratio of plate length to spacing) ranged from .25 to 36.

For aspect ratios of 4 and less, the enclosure is partitioned by slats. Slat angle, measured with respect to the plates, of 45°, 60° and 90° were studied.

The local heat-transfer coefficients inside the vee cavity were determined and integrated to obtain average values. The vee opening angle examined was 60° and the enclosure angles of inclination from the horizontal were 0°, 45° and 60°. The vee aspect ratios (mean plate spacing to vee height) studied were .75, 1 and 2.

The results are useful in determining the top convective loss coefficient and therefore in minimizing losses and optimizing the design of solar collectors. The results indicate that the convective losses are strong functions of the aspect ratio for aspect ratios less than 4. Slat angles less than 90° help reduce convective losses. The convective losses for a vee-corrugated surface are higher than those for a flat plate.

INTRODUCTION

The free convection heat loss across inclined layers is of interest to designers of solar collectors. Reduction of heat loss from the absorber plate through the cover plates improves collector efficiency and allows smaller collector areas to be used. Though many relevant studies of free convection heat transfer in large aspect ratio flat collectors have been made, [1-8], adequate correlations do not yet exist for slat type collectors.

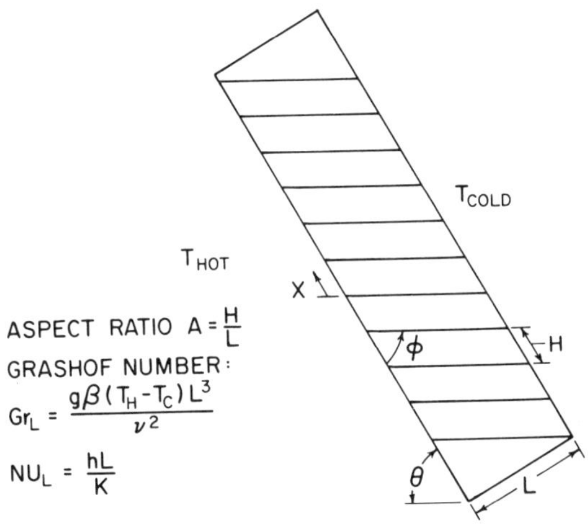

Fig. 1 Schematic of Slat Test Section.

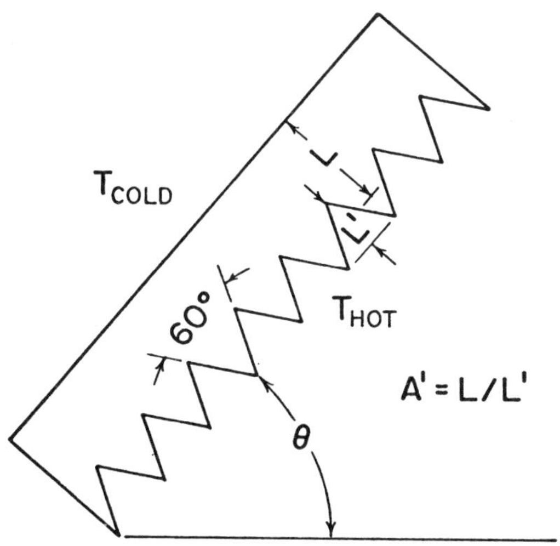

Fig. 2 Schematic of vee-corrugated enclosure.

The free convection heat-transfer in vee-corrugated enclosures is virtually unknown [13,14].

The "slat collector" described in Figure 1, is represented as an array of cells which are essentially individual and contiguous flat plate collectors, each having a small aspect ratio. Slat collectors are more complex in construction than conventional flat plate collectors, but experimental studies [9-11] have shown that, depending upon aspect ratio, a decrease in the convective heat transfer losses may be achieved.

The vee-corrugated collector shown schematically in Figure 2 has more desirable characteristics than the flat-plate or slat collector. These are an increased effective solar absorptance due to multiple reflections in the vee cavity and an increase in the heat transfer area between the absorber surface and the working fluid. Hollands [12] has shown that a vee opening angle of 60 degrees is most desirable in terms of the increase in effective absorptance of the surface.

In the present study, local values of the heat transfer coefficient along a heated surface corresponding to the absorber surface were obtained using interferometric techniques. The local heat transfer coefficients were integrated over the surface to determine average values.

The aspect ratio range studied in the flat-plate study (including the slat-type) was between .25 and 36. The angle of inclination of the enclosure ranged from 45 degrees with respect to the horizontal to the vertical. The Grashof number range studied was 1×10^3 to 3×10^5. Slat angles of 45°, 60°,and 90° were studied. These parameters are defined in Figure 1. The effect of the slat angle (ϕ) on the heat transfer coefficient has not previously been investigated. Slat angles less than 90° restrict upward convective currents and therefore lower the convective losses to the top cover.

The results for flat plate enclosures with aspect ratios in range of 9 to 36 have been previously determined and reported by Randall, et al. [8]. These results were obtained using the same test facility and data reduction methods as in the present study, and are included here to show the continuity over the entire aspect ratio range of 0.25 to 36.

The local and average heat-transfer coefficients over the vee surface were obtained for flow in the fully established regime. In this region the heat-transfer patterns in all adjacent vee cavities are identical to the one under study. This eliminates the necessity of examining the entire length and permits the study of one representative vee cavity. Figure 2 shows a 2-dimensional schematic of the enclosure. The vee aspect ratios, A', studied were .75, 1 and 2 at tilt angles, θ, of 0, 45 and 60 degrees.

TEST PROCEDURE

The test model for the flat plate collector consisted of an upper isothermal water cooled aluminum plate and a lower isothermal electrically heated aluminum plate. The plates are 45.7 cm long by 10.2 cm wide. The plate spacing was varied from .62 cm to 2.5 cm. The slat collector consisted of contiguous cells formed by slats between the same plates. The cell width was varied from .62 cm to 5.0 cm, resulting in the various aspect ratios. The cell walls are .18 mm thick and the ratio of the cell wall conductivity to the air conductivity is 2.5. Plate temperature differences (TH-TC) ranged between 20 C and 90 C. The maximum plate temperature variation is less than 0.5 C over the cold plate and 1.5 C over the hot plate.

Vee aspect ratios of .75, 1 and 2 were examined using the same apparatus as in the study of the flat plate geometry by replacing the flat heated surface by a vee-corrugated surface. Two vee surfaces of vee height 1.27 and 2.2 cm were used to obtain a wide range of Grashof numbers. Care was taken to obtain nearly isothermal surfaces. Temperature variations of .3 C and 2 C over the vee surface and plate length respectively were observed.

The temperature distribution in the enclosed air space was measured using a Mach-Zehnder interferometer. Figures 3 and 4 show interferograms typical of those obtained in this study. The dark and light lines represent iso-isotherms.

Analysis of interferograms allows determination of the local temperatures and therefore the temperature gradients at the wall. From these, the local heat transfer coefficients are obtained. The heat transfer coefficient can be determined from the interferograms within ±5%.

The local Nusselt number is evaluated at the mean enclosure temperature and based upon the mean plate spacing (L) and the total temperature difference (TH-TC). The corresponding value of the Grashof number is evaluated at the mean temperature.

LOCAL HEAT TRANSFER RESULTS

Flat-Plate and Slat-type Enclosures

Figures 3 through 5 are interferograms for different aspect ratios, slat angles and Grashof numbers. In all cases the hot plate is the left-hand surface. The isotherms visually show how the heat flux varies with position along the plate. Close spacing of isotherms at the walls indicates high rates of heat transfer.

Figure 3 demonstrates that there is more convective motion as aspect ratio increases to 2. At an aspect ratio of 0.5, there is little motion and the heat transfer is mainly by conduction. Figure 4 shows the convective

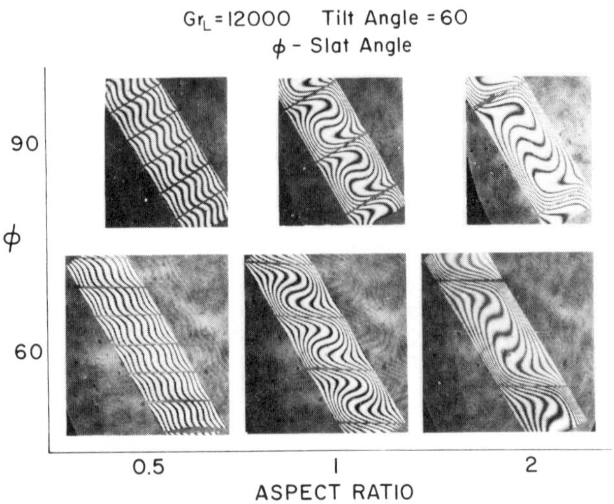

Fig. 3 Interferograms Showing the Effects of Aspect Ratio and Slat Angle.

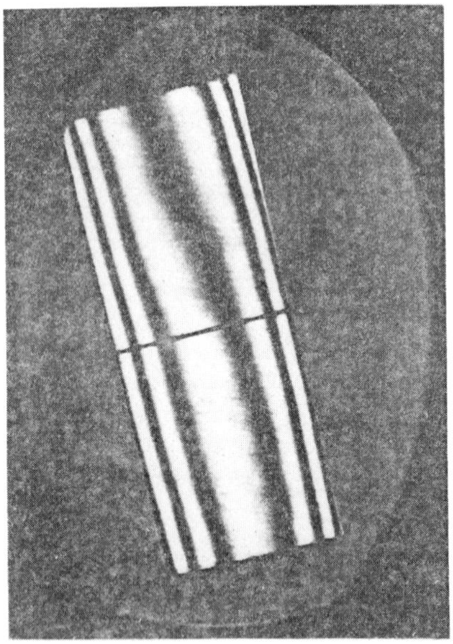

Fig. 4 Interferogram of Flat Plate at A = 24, Gr_L = 26000.

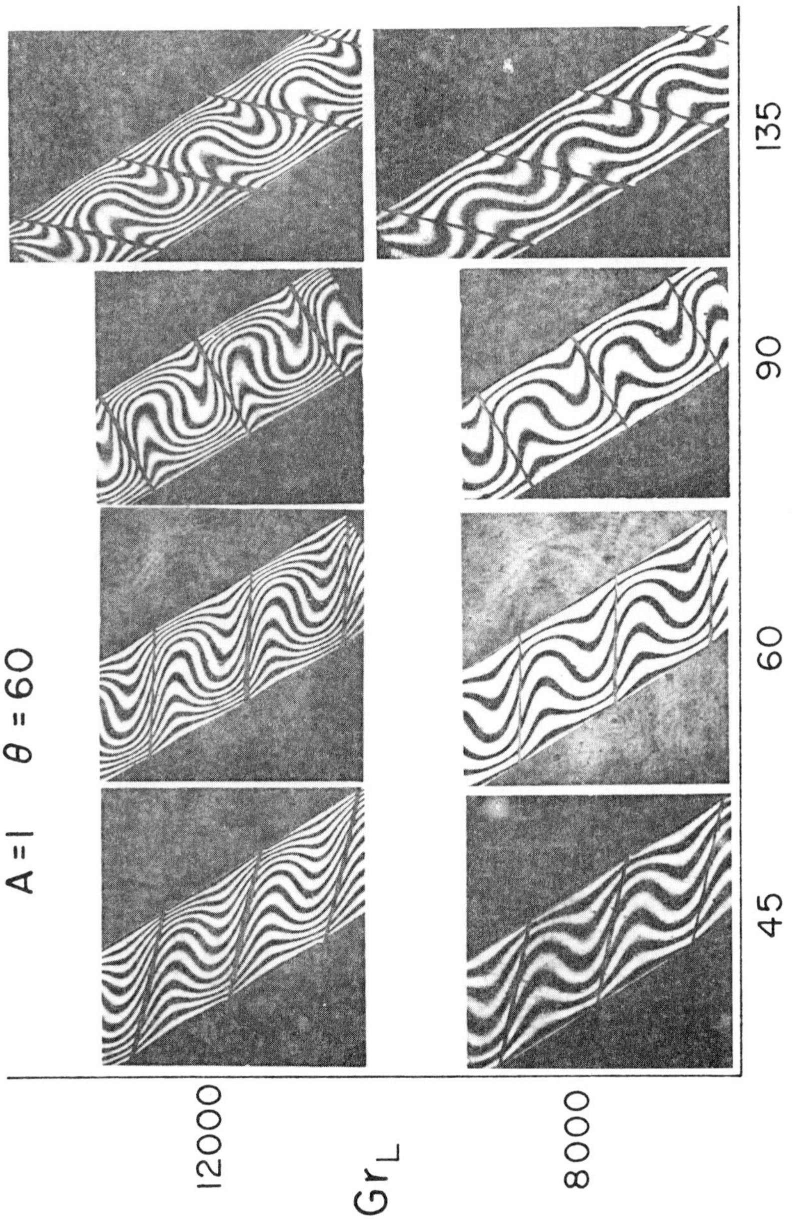

Fig. 5 Interferograms Showing the Effect of Slat Angle (φ).

motion for an aspect ratio of 24 is less than the convective motion for an aspect ratio of 2. The effect of increased Grashof number, as shown in Figure 5, is to increase convective motion. The effect of tilt angle is fairly small, as shown in Figure 8. As the slat angle ϕ decreases, there is a suppression of convection. This is shown in Figure 5.

Figure 6 shows the variation of the local Nusselt number with position for different aspect ratios. The local value is a strong function of aspect ratio and indicates enhanced convection at aspect ratios in the range of 1 to 2. As aspect ratio decreases from this range, convection is suppressed and the local value becomes more uniform and approaches unity. As aspect ratio increases beyond 2, the end effects occur over a smaller fraction of the area and Nusselt number again decreases.

Figure 7 shows the effect of the slat angle on the local Nusselt numbers. The local Nusselt numbers are generally lower as the slat angle is decreased. Decreasing slat angle tends to suppress convective currents and hence reduce the local convective heat transfer.

Figure 8 shows that changing the angle of inclination of the enclosure (θ) has little effect on the local and hence the average Nusselt number for an aspect ratio of 1. This result is typical of results at other Grashof numbers, aspect ratio, and slat angle. As shown by Randall et al., [8], the effect of the angle of inclination on larger aspect ratios is more pronounced.

Vee Corrugated Enclosure

Figure 11 shows a typical interferogram from the vee study at a vee aspect ratio of .75 and an enclosure tilt angle of 45 degrees. The fluid motion at aspect ratios of .75 and 1 and tilt angles of 45 and 60 degrees exhibited the single cell rotation evident on this figure with negligible interaction between adjacent cells and negligible base flow of the fluid up the heated surface and down the cold.

The local Nusselt numbers obtained from this interferometer picture are also shown along each face. The downslope face always exhibits a higher heat transfer rate indicating again a clockwise rotation within each cell. The local Nusselt numbers are based upon the beam plate spacing L and the projected area of the vee plate.

The flow patterns at 45 and 60 degrees are different than for the horizontal where the flow is cellular and two-dimensional with two counter rotating cells in each vee cavity. Here the heat transfer characteristics for each vee face are identical.

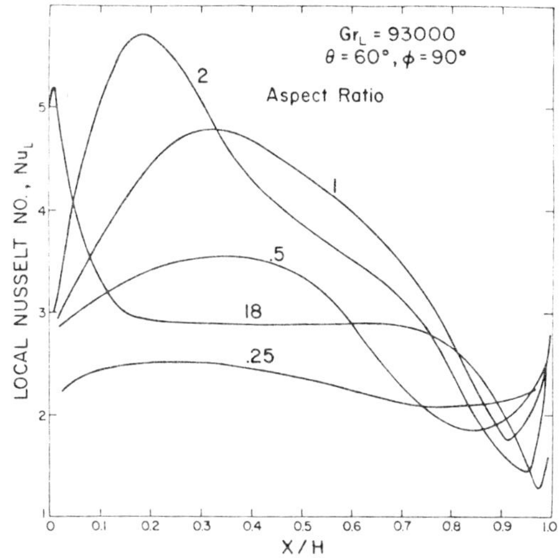

Fig. 6 Local Nusselt Numbers Along the Plate Surface for Different Aspect Ratios (A).

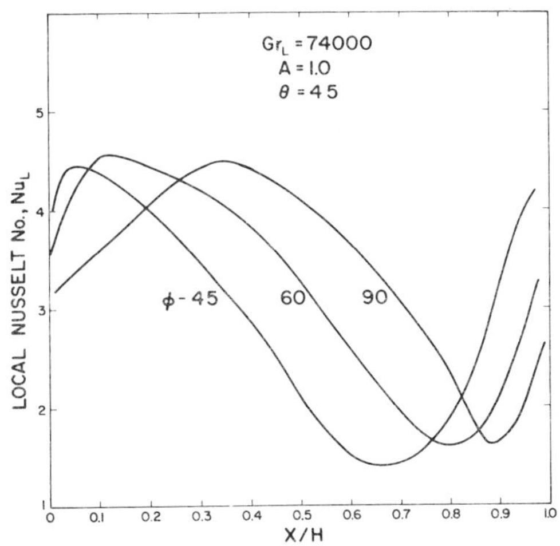

Fig. 7 Local Nusselt Number Along the Plate Surface for Different Slat Angles (φ).

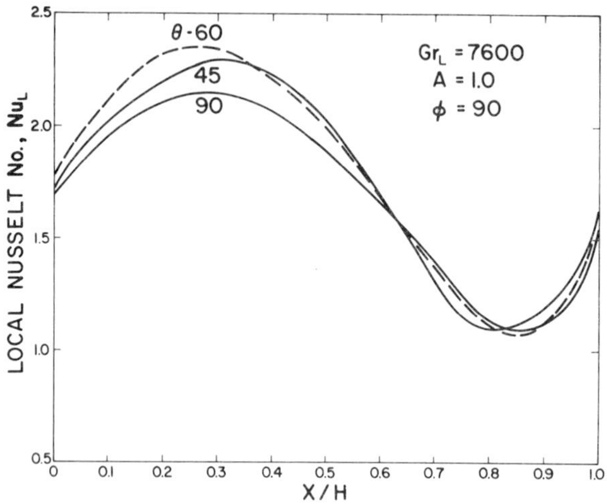

Fig. 8 Local Nusselt Numbers Along the Plate Surface for Different Tilt Angles (θ).

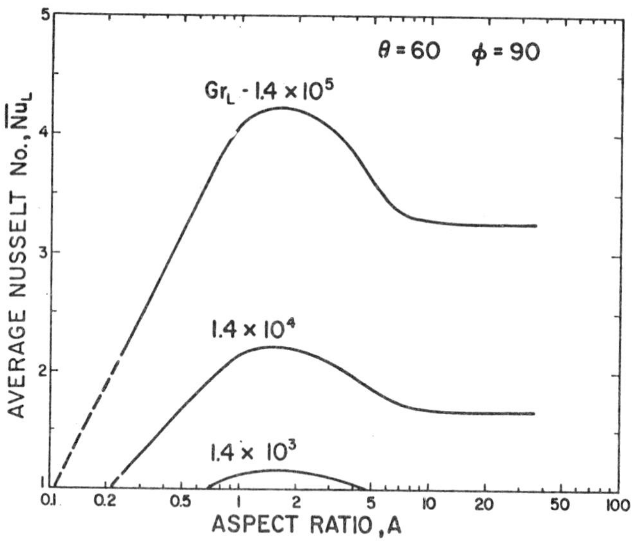

Fig. 9 Average Nusselt Number Versus Aspect Ratio for Different Rayleigh Numbers.

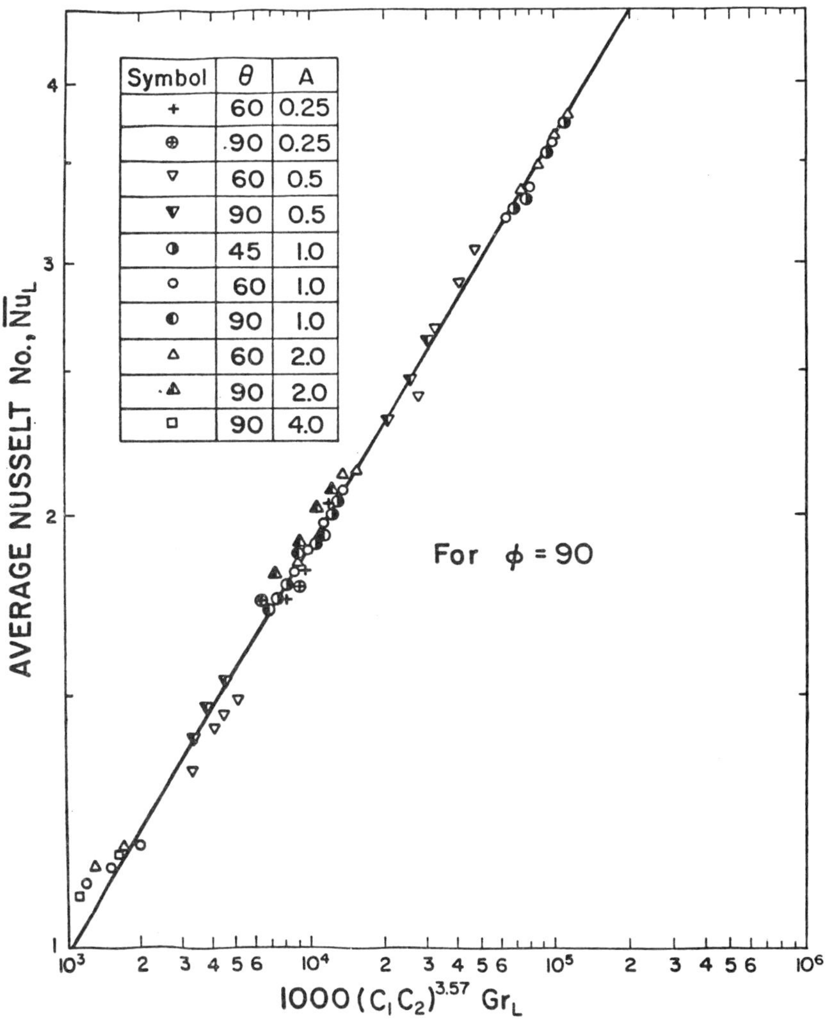

Fig. 10 Average Nusselt Number Correlation for Grashof Number, Aspect Ratio, and Tilt Angle.

AVERAGE HEAT TRANSFER RESULTS

Flat-Plate and Slat-type Enclosures

The average heat transfer coefficients are determined through integration of the local Nusselt numbers over the plate surface. The effect of the aspect ratio on the average heat transfer is governed by the penetration depth of the end disturbances. For very large aspect ratios these disturbances occupy a relatively small portion of the total and therefore have little effect on the average values. As the aspect ratio is decreased, the end disturbances occupy a larger proportion of the total area and have a larger effect on the average values.

The average Nusselt number as a function of the aspect ratio for various values of the Grashof number is shown in Figure 9. There is a maximum value of the average Nusselt number between aspect ratios of 1 and 2. This maximum can be attributed to the end disturbances which encompass the entire cell at these ratios. As the aspect ratio is decreased below 1, the viscous effects on the cell wall begin to dominate and hence retard the fluid motion which reduces the convective heat transfer. At large aspect ratios, the flow becomes fully established in the center with relatively lower average values.

The average Nusselt number was correlated with Grashof number, aspect ratio, and tilt angle for slat angles of 90 degrees using the form

$$\overline{Nu}_L = C_1 C_2 Gr_L \qquad (1)$$

This correlation is appropriate for $\overline{Nu}_L$ greater than unity; $\overline{Nu}_L = 1$ is the conduction limit for all situations. The coefficients C_1 and C_2 include the effects of aspect ratio and tilt angle, respectively. These are presented in Table 1. The correlation equation and empirically derived constants fit the data within ±5%. The results of the present study agree within ±10% of Kee's experimental results [11] for the aspect ratios in common with both studies. The adequacy of this correlation is shown in Figure 10, where all the data shown is for $\phi = 90$.

For slat angles (ϕ) other than 90° it was found that a correlation of the form

$$\overline{Nu}_L = C_3 Gr_L^{\,n} \qquad (2)$$

fit the experimental data to within ±5%. The constants C_3 and n are tabulated in Table 2 for the tested combinations of tilt angles, slat angles and aspect ratios.

These results demonstrate that reducing slat angles tends to decrease the average Nusselt number by suppressing convective currents. The magnitude of this effect can be appreciable.

Table 1. Constants for Equation (1).

C_1	Aspect Ratio Range
$0.0367 + 0.195\,A - 0.0846\,A^2$	$.25 \leq A \leq 1$
$0.1237 + 0.0318A - .0091\,A^2$	$1 \leq A \leq 2$
$0.1646 - 0.0068A$	$2 \leq A \leq 5$
.12	$9 \leq A \leq 36$
C_2	
$0.908+3.76x10^{-3}\theta-3.7x10^{-5}\theta^2$	$.25 \leq A \leq 4$
$[\cos(\theta-45)]^{.48*}$	$9 \leq A \leq 36$
*(θ in degrees)	

Table 2. Constants for Equation (2). (with slats)

θ,ϕ	A	C_3	n
60,60	0.5	0.103	0.29
60,60	1.0	0.069	0.343
60,60	2.0	0.124	0.295
45,45	1.0	0.063	0.34

Table 3. Constants for Equation (2) and vee-corrugated correlation.

θ	A'	C_3	n
0	.75	.060	.41
	1	.060	.41
	2	.043	.41
45	.75	.75	.36
	1	.082	.36
	2	.037	.41
60	.75	.162	.30
	1	.141	.30
	2	.027	.42

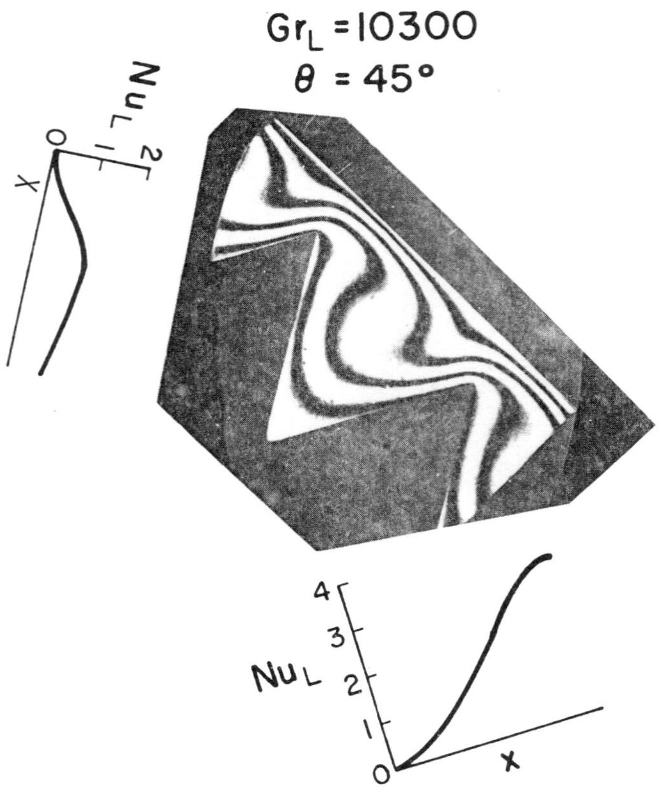

Fig. 11 Interferogram of Vee at $\theta = 45°$, $Gr_L = 10300$.

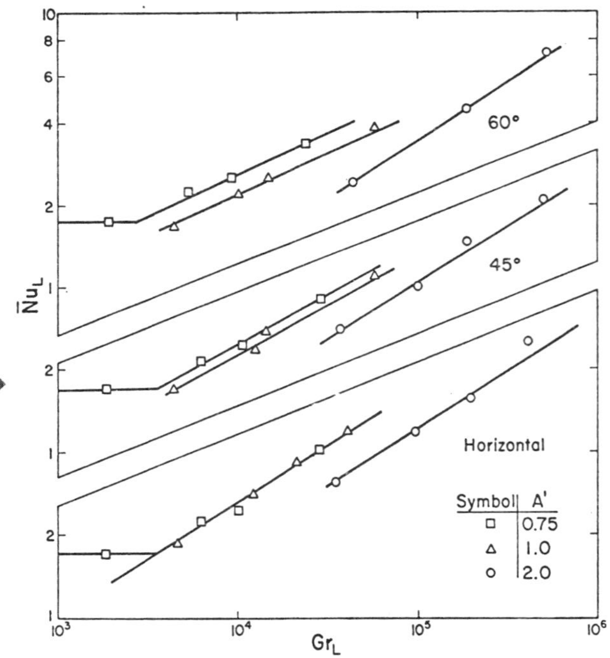

Fig. 12 Comparison of Vee Correlations with Data, $\theta = 0°, 45°, 60°$.

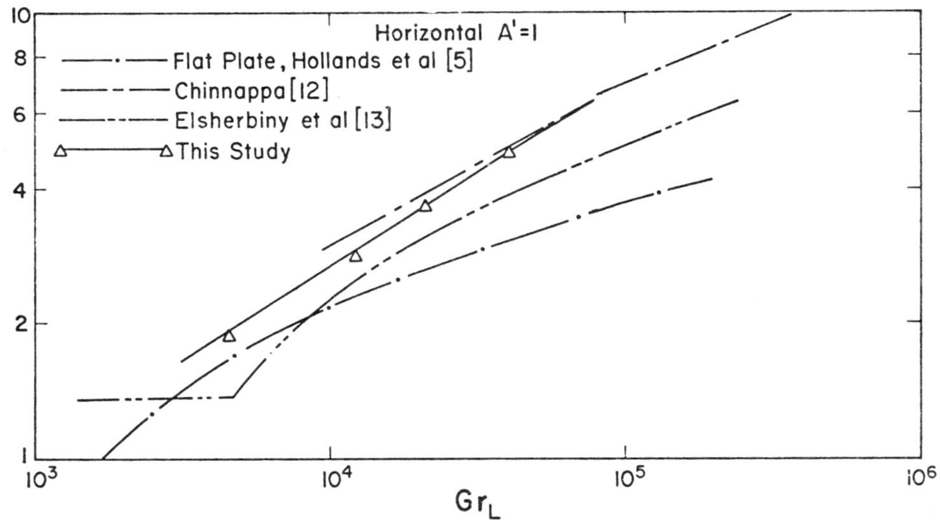

Fig. 13 Comparison of A' = 1 Results with Literature and Flat Plate.

Vee-Corrugated Enclosures

The local coefficients on the hot vee-corrugated surface are integrated to find the average. The result is an average Nusselt number based upon mean plate spacing and projected surface area which provides a valid comparison to the heat-transfer characteristics of the flat-plate goemetry.

As mentioned previously the heat transfer results were only examined for fully established flow, i.e., well removed from the end effects. A visual examination of the interferometer pictures of the ends of the enclosure shows that the entry length is considerably shorter than that of the flat-plate enclosure at the same operating conditions. The heat transfer for the range of parameters in this study was fully established within 2 vees of the end boundaries.

The average Nusselt number for 0, 45 and 60 degrees are shown in Figure 12. In all cases the average values for A' = 2 are considerably lower than those of .75 and 1. This is due to circulation of the base flow up the hot surface and down the cold in effect adding more resistance to the heat flow.

The increasing difference between .75 and one from horizontal to 60 degrees can also be attributed to the same effect. At larger tilt angles the large component of the buoyant force up the slope creates more interaction between adjacent cells and circulation of the flow.

A correlation of the form of equation (2) was found to fit the data to within ±5%. The constant C_3 and exponent n are tabulated in Table 3. These results show that small vee aspect ratios have considerably higher losses than flat-plate enclosure. Figure 13 shows a comparison of these results for A' = 1 and $\theta = 0$ degrees to those of Chinnappa [13] and Elsherbiny [14] and the flat plate correlation of Hollands et al. [5].

SUMMARY

An interferometric study was used to determine the local and average values of the Nusselt number in moderate and small aspect ratio flat-plate enclosures. The investigation was made over collector tilt angles from 45° to 90°, aspect ratios from .25 to 36 and slat angles from 45° to 90°. The results show that aspect ratios between .5 and 10 are undesirable to the collector designer from the point of view of convective loss. Collector geometries with slat angles less than 90° are effective in reducing convective losses.

Heat transfer rates in an enclosure bounded by a heated vee-corrugated surface on the bottom and a cooled flat surface on the top were also examined for vee aspect ratios of .75, 1 and 2 and tilt angles of 0, 45 and 60 degrees. The results show that the average heat transfer for this geometry is considerably higher than the flat-plate collector operating under similar conditions.

Complete optimization of solar collectors using slat cells on a vee corrugated absorber should also take into account the determination of the radiant absorption as a function of the above variables.

ACKNOWLEDGEMENTS

The work reported in this paper has been supported by the Department of Energy under contract E (11-1)-2941.

NOMENCLATURE

- A - aspect ratio, H/L
- C_1, C_2 - constants in Equation (1)
- C_3 - constant in Equation (2)
- g - acceleration due to gravity
- H - cell length
- k - thermal conductivity
- L - mean plate spacing
- n - constant in Equation (2)
- T_C - cold plate temperature
- T_H - hot plate temperature
- x - distance measured along the suraace
- h - local heat transfer coefficient
- $\bar{h}$ - average heat transfer coefficient
- α - thermal diffusivity
- β - coefficient of volumetric expansion
- ν - kinematic viscoisty
- θ - enclosure tilt andle from horizontal
- ϕ - slat angle
- Gr_L - $g\beta(T_H-T_C)L^3/\nu^2$ Grashof number based on L
- Nu_L - hL/k local Nusselt number based on L
- $\overline{Nu}_L$ - $\bar{h}L/k$ average Nusselt number based on L
- A' - vee aspect ratio L/L'
- L' - vee height

REFERENCES

1. Arnold, J.N., Catton, I. and Edwards, K.K.: "Experimental investigation of natural convection in inclined rectangular regions of differing aspect ratios", J. HEAT TRANSFER, Vol. 19, pp. 67-71, 1976.
2. Catton, I., Ayyaswamy, P. and Clever, R.M.: "Natural convection flow in a finite rectangular slot arbitrarily oriented with respect to the gravity vector", INT. J. HEAT MASS TRANSFER, Vol. 17, pp. 173-184, 1974.
3. Dropkin, D. and Somerscales, E.: "Heat transfer by natural convection in liquids confined by two parallel plates which are inclined at various angles with respect to the horizontal", J. HEAT TRANSFER, Trans. ASME, Vol. 87, pp. 77-84, 1965.

4. Hollands, K.G.T. and Konicek, L.: "Experimental study of the stability of differentially heated inclined air layers", INT. J. HEAT MASS TRANSFER, Vol. 16, pp. 1467-1476, 1973.
5. Hollands, K.G.T., Unny, T.E., Raithby, G. and Konicek, L.: "Free convection heat transfer across inclined air layers", J. HEAT TRANSFER, Trans. ASME, Vol. 98, pp. 189-193, 1976.
6. Ozoe, H. and Sayama, H.: "Natural convection in an inclined square channel", INT. J. HEAT MASS TRANSFER, Vol. 17, pp. 401-406, 1974.
7. Raithby, G.D., Hollands, K.G.T. and Unny, T.E.: "Free convection heat transfer across fluid layers of large aspect ratio", ASME Paper 76-HT-37.
8. Randall, K.R., El-Wakil, M.M., Mitchell, J.W.: "Natural convection characteristics of flat plate collectors", ISES Proceedings, Orlando, FL, 1977.
9. Arnold, J.N., Edwards, D.K., Catton, I.: "Effect of tilt and horizontal aspect ratio on natural convection in rectangular honeycomb solar collectors", ASME, Paper 76-HT-18, 1977.
10. Hollands, K.G.T.: "Studies of methods of reducing heat losses from flat-plate solar collectors", ERDA Annual Progress Report, C00-2597-3, University of Waterloo, Waterloo, Ontario, Canada, 1977.
11. Kee, Kay Hong, "An experimental study of natural convection in rectangular channels", MS thesis, The University of Melbourne, March 1977.
12. Hollands, K.G.T.: "Directional selectivity, emittance, and absorptance properties of vee-corrugated specular surfaces", SOLAR ENERGY, Vol. 7, No. 3, 1963.
13. Chinnappa, J.C.V.: "Free convection in air between a 60° vee-corrugated plate and a flat plate", INT. J. HEAT AND MASS TRANS., Vol. 13, 1970.
14. Elshevbiny, S.M., Hollands, K.G.T. and Raithby, G.D.: "Free convection across inclined air layers with one surface vee-corrugated", Presented at ASME Winter Annual Meeting, 1977.

STUDY OF A MULTI-PASS RADIATION FLUID-HEATER EXPOSED TO NON-UNIFORM RADIATION

N. M. Rafat , M. M. Elkotb and M. F. El-Refaie
Faculty of Engineering , Cairo University , Egypt .

ABSTRACT

The steady state performance of a multi-pass radiation fluid-heater is investigated when the distribution of the incident thermal radiation is non-uniform in the cross direction . Such a condition may be realized in a multi-pass solar collector combined with a parabolic reflection trough . The analysis and results presented herein are valid for any general form of radiant energy distribution .

Mathematical expressions are derived for the fluid and wall temperature distributions along the serpentine . These expressions are given in terms of a number of convenient dimensionless groups .

The performance of the heater is evaluated on two bases . A new effectiveness index is used , together with the conventional efficiency , to describe the performance .

When the heater passes are exposed to different radiation densities , the sequence of flow through the different passes will be of paramount importance . The effect of the flow arrangement on the performance is studied and the optimum flow pattern is determined .

INTRODUCTION

This work is an analytical study of the steady state performance of a multi-pass radiation fluid-heater when the distribution of the received radiant energy is non-uniform in the cross direction , i.e normal to the directions of flow . In the field of solar energy collection , this condition may be met with when a reflective radiation booster , such as a parabolic trough , is used to concentrate the radiation on the plane of the collector . The collector-reflector assembly is shown in Fig. 1 . The density of the received radiation is uniform along any longitudinal line .

The incident radiation intensity is assumed to be invariable with time . Therefore , when dealing with a solar heater , the solar radiation has to be averaged over a period of time [1 and 2] .

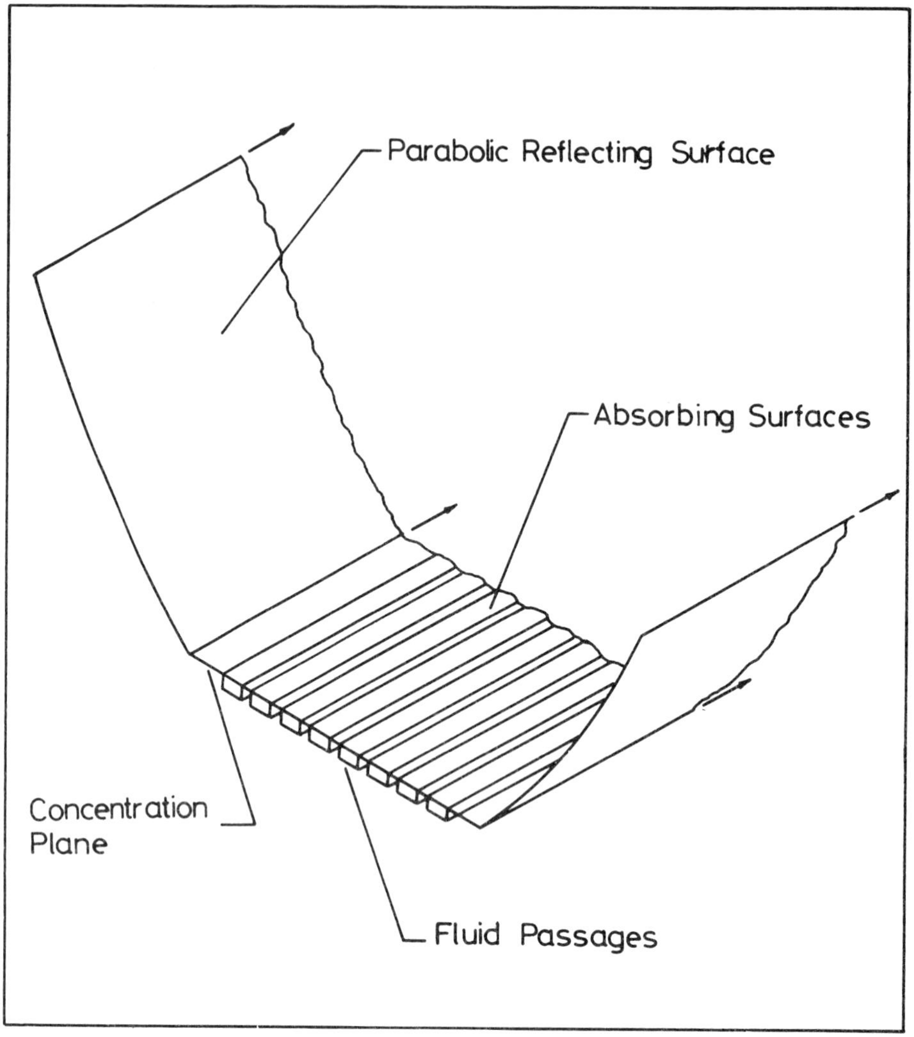

Fig. 1 A Multi-Pass Collector with a Reflective Parabolic Trough

The shape of the crosswise distribution of radiant flux depends on many factors imposed by the geometry , size , and quality of the reflecting surfaces , the part of the concentration plane occupied by the target heater , and the type and precision of sun-tracking system . In general , the distribution has a smooth continuous form [3] . The problem formulation and the derived expressions presented hereinafter are not limited to any particular distribution ; they hold for any general form of distribution that the received radiation may have .

The progressive change of the heated fluid temperature as well as the variation of the absorber material temperature along the different passes are investigated . The performance of the heater is considered from two points of view ; first the actual temperature rise of the fluid across the heater and second the fraction of the absorbed energy that is transferred to the fluid . Effectiveness and efficiency indices are used to describe the heater performance on these bases .

Mathematical expressions are derived for the temperature distributions and the performance indices in terms of the various design and working parameters and for any general form of radiant energy distribution . These expressions can be used for design purposes , for comparison between heaters , and for studying the effect of any particular parameter . Dimensionless groups are used to simplify the solution and to yield compact expressions .

In the type of heater under consideration the radiant flux varies from one pass to another . Therefore , the flow arrangement , i.e the sequence in which the fluid traverses the different passes , becomes an important factor which largely affects the performance of the heater . This effect is discussed and the optimum flow pattern is presented .

Throughout this work all calculations are based on the amount of radiant energy absorbed by the irradiated surfaces which represents the actual or net energy input . This is equal to the initially incident energy after being attenuated by passing through the glass covers and absorption by the absorber surface . The depletion of the incident radiation before being absorbed is beyond the scope of this work . However , if such effects are to be taken into consideration , they can be simply superimposed on the final derived expressions .

FORMULATION AND ANALYSIS OF THE PROBLEM

A diagramatic sketch of the heater under consideration is illustrated in Fig. 2 . The passes shown are rectangular ; however , any other form of cross section may be considered with only minor changes in the formulation procedure . Each pass

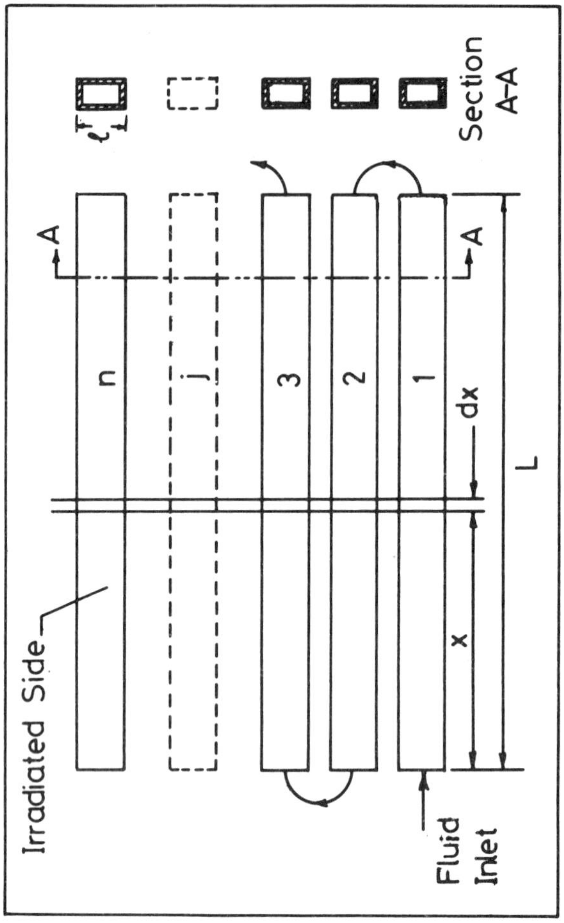

Fig. 2 Diagramatic Sketch of a Multi-Pass Heater

is assumed to be in perfect thermal isolation from other passes ; thus , heat can not be exchanged between adjacent passes. The radiant energy is received along the length L only ; accordingly , it may be assumed that no fluid temperature changes occur within the interconnection between any two successive passes .

The flow through the heater is considered to be a monophase flow and the specific heat of the fluid is supposed to have a mean value independent of temperature . In addition , the convection coefficient of heat transfer on the internal walls and the overall heat loss coefficient [4 and 5] on the outer irradiated surface are considered to have fixed values allover the exchange surfaces .

At any cross section along any pass , the fluid temperature is taken to be uniform and equal to the mixing-cup value . Similarly , the temperature of the passage wall is assumed to be constant all around the perimeter .

If the actual distribution of the radiant energy is considered, the problem will be too complicated because the radiation density will vary across the width of each pass . To avoid such complexity , the smooth continuous distribution is conceivably transformed to another one which has a step form such that the areas under the two distributions are the same ; this is explained in Fig. 3 for two different distributions . A necessary condition is that each step of the new distribution should span one pass or an integer number of passes . Thus , the radiation density is uniform allover each pass , or group of passes , and varies from one pass to another .

The radiant energy flux absorbed by any pass of order j will be denoted by I_j . The ratio of the absorbed fluxes of any two passes is defined as the Intensity Ratio of these two passes ; it is given the symbol R with the orders of the two passes as subscripts . For example

$$R_{j,k} = I_j / I_k \tag{1}$$

Throughout this work , the second subscript will be omitted whenever the divisor in eqn. (1) is the density of the first pass I_1 . Thus ,

$$R_j = I_j / I_1 \tag{2}$$

The equilibrium temperature $t_{e,j}$ of any pass , of order j , is defined as the maximum temperature which can be attained under no-flow conditions ; accordingly

$$t_{e,j} = t_a + I_j / U \tag{3}$$

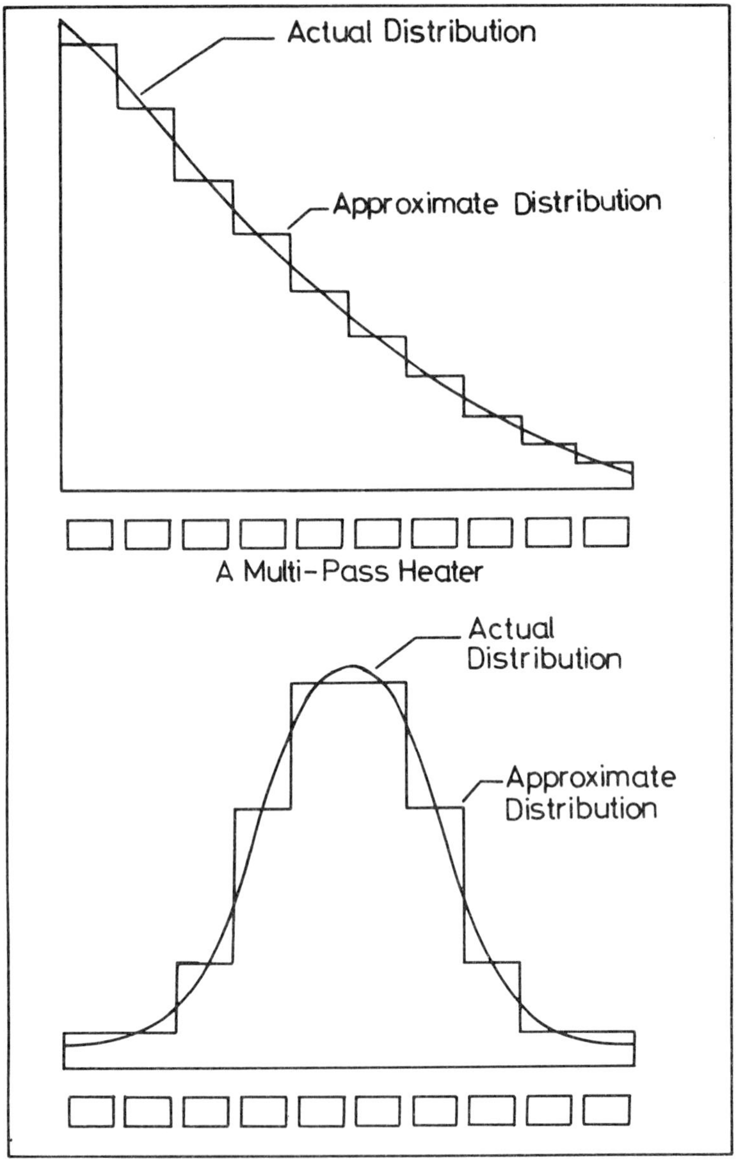

Fig. 3 Approximation of the Non-Uniform Radiation Distribution to a Step Form.

A dimensionless temperature T corresponding to any temperature t is defined, with reference to the absorbed radiant flux of any pass I_j, by the equation

$$T(j) = (t - t_a)/(t_{e,j} - t_a) \qquad (4)$$

$$= (t - t_a)\, U/I_j \qquad (5)$$

It can be seen that, for fixed values of t_a and I_j, the dimensionless temperature T(j) increases as the overall heat loss coefficient U increases. However, this does not imply any contradiction because the difference $(t_{e,j} - t_a)$ decreases in the same proportion.

It can be directly correlated from eqn. (5) that, if reference is changed to the absorbed flux of any other pass I_k, the dimensionless temperature can be converted by the simple relation

$$T(k) = T(j)\, I_j/I_k \qquad (6)$$

A dimensionless temperature considered with reference to I_1 will be simply denoted by T. Through the rest of this work the dimensionless temperature will be briefly referred to as temperature; whenever the natural or real temperature is meant, this will be stated explicitly.

Fluid and Wall Temperature Distributions

Considering a differential element of length along the heater, as shown in Fig. 2, the following two equations, expressing the conservation of energy and the convection of heat, can be written for any pass of order j

$$I_j \ell\, dx = U\,(t_{w,j} - t_a)\, \ell\, dx + (-1)^{j+1} \dot{m}\, c\, dt_{f,j} \qquad (7)$$

and

$$h\,(t_{w,j} - t_{f,j})\, p\, dx = (-1)^{j+1}\, \dot{m}\, c\, dt_{f,j} \qquad (8)$$

Three dimensionless quantities are defined by the following equations

$$\text{Heat Loss Number } N_L = \dot{m}\, c/U \ell L \qquad (9)$$

$$\text{Convection Number } N_c = \dot{m}\, c/h\, p\, L \qquad (10)$$

$$\text{Dimensionless Distance } X = x/L \qquad (11)$$

Transforming the real temperatures in eqns. (7) and (8) to dimensionless values, with reference to I_1, we get

$$1 = R_{1,j}\, T_{w,j} + (-1)^{j+1}\, N_L\, R_{1,j}\, (dT_{f,j}/dX) \qquad (12)$$

and

$$T_{w,j} - T_{f,j} = (-1)^{j+1} N_c \; (dT_{f,j}/dX) \tag{13}$$

Eliminating the wall temperature from eqns. (12) and (13) results in the following differential equation in the fluid temperature only

$$dT_{f,j}/dX + (-1)^{j+1} \; N \; T_{f,j} = (-1)^{j+1} \; R_j \; N \tag{14}$$

where N is the Characteristic Number defined by

$$N = (N_L + N_c)^{-1} \tag{15}$$

The boundary condition necessary for the completely defined solution of eqn. (14) is that the fluid temperatures at inlet to a certain pass and exit from the preceding one are equal ; this is mathematically expressed by

$$T_{f,j} \Big|_{X=\delta} = T_{f,(j-1)} \Big|_{X=\delta}$$

where δ can only have one of the two values zero or one for odd and even passes respectively

$$\delta = (1 + (-1)^j)/2 \tag{16}$$

Integrating eqn. (14) , with the boundary condition given above, results in

$$T_{f,j} = R_j + \left\{ \sum_{i=1}^{j-1} (R_{j-i} - R_{j-i+1}) \text{Exp.} \left[-N(i - \Psi + (-1)^{j+1} . X) \right] \right\} - (1-T_i) \text{ Exp.} \left[-N(j - \Psi + (-1)^{j+1} X) \right] \tag{17}$$

where Ψ is defined by

$$\Psi = 1 - \delta \tag{18}$$

Equation (17) can be used to calculate the fluid temperature at any position , defined by X , along any pass , defined by j. This equation can be written in the more compact form

$$T_{f,j} = R_j + \left\{ \sum_{i-1}^{j-1} (R_{j-i} - R_{j-i+1}) \text{ Exp.} \left[-N(i - \Psi + (-1)^{j+1} X) \right] \right\} - (1 - T_i) \; e^{-NP} \tag{19}$$

where P is the Position Function defined by

$$P = (-1)^{j+1} X + j - (1-(-1)^j)/2$$

Thus , the value of P , at any point , is the ratio of the actual distance travelled by the fluid through the heater , up to this point , to the pass length L . It is to be noted that the position function increases in the same direction of X along odd passes and in the opposite direction along even passes .

The fluid temperature at exit from any pass is given by

$$T_{ex,j} = T_{f,j}\Big|_{X=\Psi}$$

Thus , it can be found , from eqn. (19) , that

$$T_{ex,j}= R_j + \left\{\sum_{i=1}^{n-1}(R_{j-i}-R_{j-i+1})\ e^{-iN}\right\}-(1-T_i)\ e^{-jN} \quad (20)$$

and the output temperature of an n-pass heater is given by

$$T_{ex}= R_n + \left\{\sum_{i=1}^{n-1}(R_{n-i} - R_{n-i+1}).\ e^{-iN}\right\} -(1-T_i)e^{-nN} \quad (21)$$

A relation between the fluid and wall temperatures , at any section along the serpentine , can be obtained by substituting for $dT_{f,j}/dX$ from eqn. (14) into eqn. (13) ; this gives

$$T_{w,j}= N\ (N_c\ R_j + N_L\ T_{f,j}) \quad (22)$$

Now , the wall temperature can be expressed as a function of position by substituting for $T_{f,j}$ from eqn. (19) into eqn.(22)

$$T_{w,j}= R_j+ N\ N_L \left\{\left\{\sum_{i=1}^{j-1}(R_{j-i}-R_{j-i+1})\mathrm{Exp.}\left[-N(i-\Psi +(-1)^{j+1}\ X\ )\right]\right\} -(1-T_i)\ e^{-N\ P}\right\} \quad (23)$$

The real temperatures in eqns. (7) and (8) could have been transformed to dimensionless values with reference to the absorbed energy flux of any other pass , of arbitrary order k , I_k instead of I_1 . Following the same steps of solution , the fluid and wall temperature distributions would have taken the forms

$$T(k)_{f,j}= R_{j,k}+ \left\{\sum_{i=1}^{j-1}(R_{(j-i),k} -R_{(j-i+1),k})\ \mathrm{Exp.}\left[-N.(i-\Psi +(-1)^{j+1}X)\right]\right\} -(R_{1,k}-T(k)_i)e^{-N\ P} \quad (24)$$

and

$$T(k)_{w,j}= R_{j,k}+ N\ N_L\left\{\left\{\sum_{i=1}^{j-1}(R_{(j-i),k}-R_{(j-i+1),k})\ \mathrm{Exp.}\left[-N(i-\Psi +(-1)^{j+1}X)\right]\right\} -(R_{1,k}-T(k)_i)e^{-N\ P}\right\} \quad (25)$$

Equations (24) and (25) could be directly deduced from applying the conversion relation of eqn. (6) to eqns. (19) and (23). It is important to point out that , for the same heater and under the same condition different dimensionless temperature distributions may be obtained depending on the reference equilibrium temperature or absorbed radiation flux . However , when transforming back to real or natural values all these dimensionless distributions will revert to the same actual temperature distributions.

The fluid and wall temperature distributions , as calculated from eqns. (19) and (23) , are drawn in Fig. 4 along a ten-pass heater . The distributions are shown for three different radiation distributions having the same maximum and minimum densities . The convection and heat loss numbers are fixed, for the three cases and the real initial fluid temperature is equal to that of the ambient air . The wall temperature distributions are shown in Figs. 4.b and 4.c up to the fifth pass only in order to have clear figures . It must be noted that the absorbed radiation flux on the first pass I_1, which was taken as a reference for dimensionless temperatures in eqns. (19) and (23) , is not the same for the three distributions of Fig. 4 . Therefore , a fluid or wall temperature obtained from any one of the three figures can not be directly compared with the corresponding value for any of the other two radiation distributions . These dimensionless temperatures have to be transformed to a common reference or to real values before any comparison can be made .

A very important result can be deduced from eqn. (19) or (24). The fluid temperature , at any position along the path , is a function of the characteristic number N , i.e it depends on the sum (N_L+N_c) . Thus , for a fixed value of N , the relative magnitudes of N_c and N_L would have no effect on the fluid temperature . On the other hand , the wall temperature at any point , as expressed by eqn. (23) or (25) , is dependant on the individual values of N_L and N_c . Accordingly , for any particular radiation distribution , a fixed value of the characteristic number , and a specified fluid inlet temperature, there can be only one fluid temperature distribution along the serpentine . At the same time , there will be any one of an infinite number of possible wall temperature distributions depending on the relative values of N_L and N_c . The property common to all these wall temperature distributions is the constant value of the sum (N_L+N_c) , i.e constant N .

The temperature distributions along the different passes are completely dependant on the form of the radiant energy distribution over the heater . Therefore , no specific features can be generalized for all possible radiation , and consequently temperature , distributions . However , the radiation distributions may be broadly classified into three main forms; the names , or terms , used with ordinary two-fluid recuperators

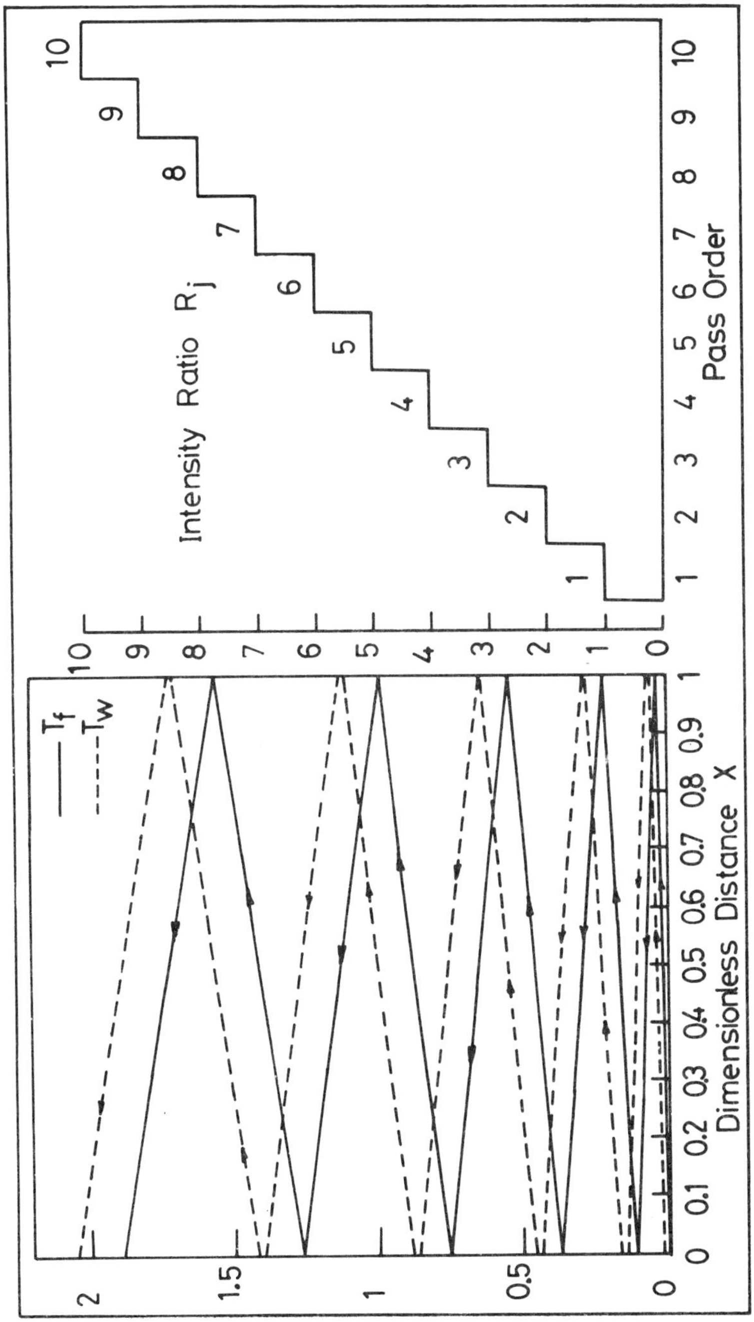

Fig. 4.a Example of the Fluid and Wall Temperature Distributions

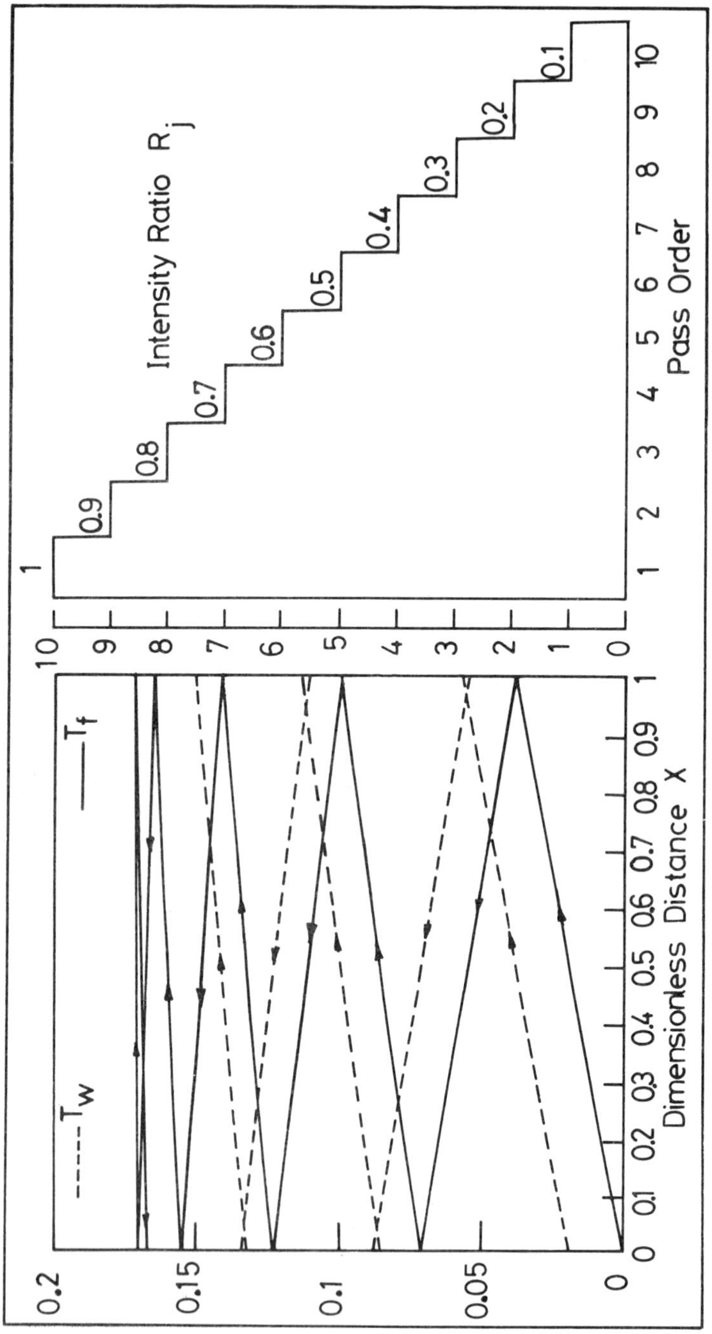

Fig. 4.b Example of the Fluid and Wall Temperature Distributions

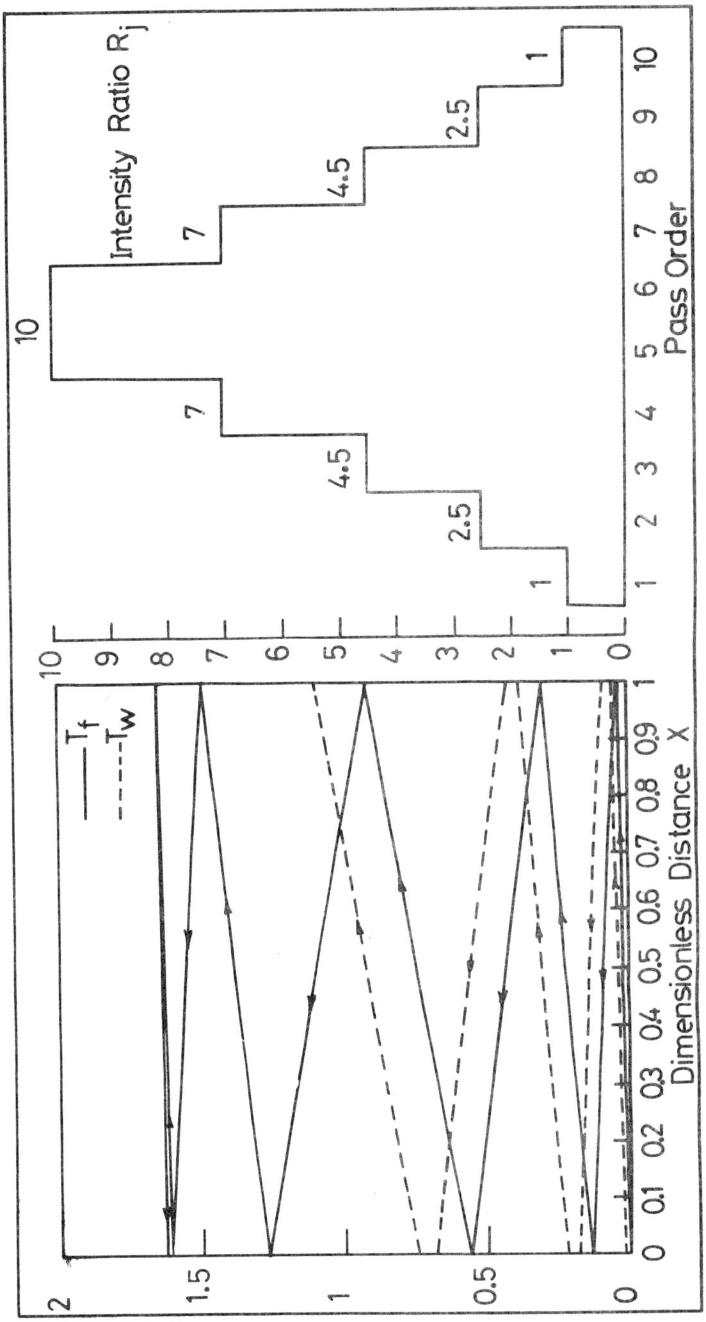

Fig. 4.c Examle of the Fluid and Wall Temperture Distributions

will be employed to describe the direction of fluid flow relative to the radiant energy distribution .

1-The flux density increases continuously as the pass order increases ; such as the case in Fig. 4.a . As the heated fluid advances from one pass to the proceeding one it will be exposed to a higher radiation density ; therefore, this arrangement will be referred to as Counter Flow .

2- The flux density decreases continuously with the increase of the pass order ; this is shown in Fig. 4.b . The fluid is progressively exposed to lower radiation densities ; accordingly , the flow under such a condition will be known as Parallel Flow .

3- Any combination of counter and parallel flow arrangements ; such as the case in Fig. 4.c .

In a counter flow heater , see Fig. 4.a , the fluid temperature increases continuously from inlet to exit of the heater . The wall temperature is also increasing along all passes . At any position along any pass , the wall temperature is higher than the corresponding fluid temperature .

When parallel flow is realized in some or all passes of the heater , as shown in Figs. 4.c and 4.b respectively , it is possible to have the fluid temperature increasing up to a maximum value after which it decreases again as the distance travelled through the heater , i.e the position function P , increases . This phenomenon can be seen along the tenth pass in Figs. 4b and 4.c . When the fluid temperature decreases in a particular pass , the rate of change $dT_{f,j}/dX$ will be negative or positive depending on whether the order of this pass is odd or even respectively . Thus , it can be concluded , from eqn. (14) , that the fluid temperature will begin to drop in any pass , of order j , whenever

$$T_{f,j} > R_j$$

It should be noted that the intensity ratio R_j is numerically equal to the dimensionless temperature , with reference to I_1, corresponding to the real equilibrium temperature of the pass number j , $t_{e,j}$.

The behaviour of the fluid and wall temperatures along a certain pass depends on the value of the fluid temperature at entrance to this pass relative to the corresponding intensity ratio . The fluid temperature at inlet to the jth pass will be denoted by $T_{i,j}$. There are three possible cases

1- $T_{i,j} < R_j$ It can be found , from eqn. (22) , that the wall temperature at inlet to the pass is higher than $T_{i,j}$. Both the fluid and wall temperatures increase continuously along the pass . The wall-to-fluid temperature difference decreases as the fluid progresses through the pass , i.e with the increase of the position function ; this results in a continuously decreasing value of $|dT_{f,j}/dx|$. Theoretically , the fluid and wall temperatures will both acquire the value R_j if the heater is infinitely long or the characteristic number is infinitely large .

2- $T_{i,j} = R_j$. The fluid and wall temperatures are equal and constant along the pass . They are equal to R_j , i.e the equilibrium temperature of the pass under consideration .

3- $T_{i,j} > R_j$. This means that the fluid temperature at inlet to the pass is higher than its equilibrium temperature . Therefore , the fluid temperature decreases along the pass . The wall temperature is also decreasing and its value at any section is lower than the corresponding fluid temperature . If the fluid temperature decreases , within the length of the pass , to the value R_j , i.e $T_{e,j}$, then both the fluid and wall temperatures will remain constant and equal to R_j along the remaining part of the pass.

The analysis given above for the temperature behaviours along any pass shows that , the point at which the fluid temperature drop starts , if such a point does exist , is necessarily at the beginning of a pass . This point may appear earlier for lower values of N_L and/or N_c , higher initial fluid temperature T_i , and steeper distribution of radiant energy .

It can be found from eqn. (21) that , in a counter flow heater, the fluid output temperature is always and under all conditions increasing with the increase of the characteristic number N , i.e the decrease of N_L and/or N_c . In any other type of flow , the variation of the output temperature with the characteristic number depends on the form of the energy distribution, the number of passes , and the initial fluid temperature T_i .

Effectiveness and Efficiency of the Heater

It is useful to define an index which describes the performance of a radiation heater from the point of view of the fluid temperature rise . The heater Effectiveness E is defined to provide a measure for the actual fluid temperature rise across the heater as a ratio to an arbitrarily specified theoretical temperature rise . Thus ,

$$E = (t_{ex} - t_i)/(t_{th} - t_i) \tag{26}$$

where t_{th} is the reference theoretical maximum temperature . A suitable temperature value which may be assigned to t_{th} in eqn. (26) is the equilibrium temperature $t_{e,max}$ of the particular pass subjected to the maximum radiant flux ; e.g $t_{e,10}$ in Fig. 4.a or $t_{e,1}$ in Fig. 4.b . Accordingly ,

$$E = (t_{ex} - t_i)/(t_{e,max} - t_i) = (T_{ex} - T_i)/(R_{max} - T_i) \tag{27}$$

where R_{max} is the maximum value of intensity ratio I_{max}/I_1. Substituting for T_{ex} from eqn. (21) results in

$$E = \left\{R_n + \left[\sum_{i=1}^{n-1} (R_{n-i} - R_{n-i+1})e^{-iN}\right] - (1-T_i)\, e^{-nN} - T_i\right\} / (R_{max} - T_i) \qquad (28)$$

It can be seen that the effectiveness , as defined by eqn.(27) and expressed by eqn. (28) is dependant on the form of radiation distribution and the initial temperature of the incoming fluid . Thus , the appraisal of the performance of a certain heater or the comparison between different heaters becomes more complicated . Moreover , if it is required to construct effectiveness charts , similar to those of recuperators and regenerators , they will have to be repeated for different values of initial temperature and different forms of radiant energy distribution .

In order to have the effectiveness independent of neither the initial fluid temperature nor the radiation distribution , a new temperature , other than $t_{e,max}$, had to be sought and substituted for t_{th} in eqn. (26) . Consider that the non-uniform radiation distribution , to which the heater is exposed , is imaginarily replaced by a uniformly distributed radiation such that the real fluid temperature at heater exit , t_{ex} , remain unaltered . This fictitious uniform radiation will be known as the Equivalent Radiation and its density will be denoted by I_{eq} . The real equilibrium temperature corresponding to I_{eq} will be referred to as the Equivalent Equilibrium Temperature $t_{e,eq}$. Considering the imaginary case of uniform irradiation , the exit temperature is found , from eqns. (21) and (6) , to be

$$T_{ex} = R_{eq} - (R_{eq} - T_i)\, e^{-n\,N} \qquad (29)$$

where R_{eq} is equal to I_{eq}/I_1 . In the actual case of exposure to the non-uniform radiation , the exit fluid temperature is given by eqn. (21) . Equating the right hand sides of eqns. (21) and (29) we get

$$I_{eq} = \left\{\sum_{i=0}^{n-1} I_{n-i}\; e^{-i\,N}\right\} (1 - e^{-N})/(1 - e^{-n\,N}) \qquad (30)$$

Consequently

$$T_{e,eq} = \left\{\sum_{i=0}^{n-1} R_{n-i}\; e^{-i\,N}\right\} (1 - e^{-N})/(1 - e^{-n\,N}) \qquad (31)$$

Now , the equivalent equilibrium temperature can be cosidered as the maximum temperature which the fluid may theoretically attain . Thus , it would be more logical to substitute $t_{e,eq}$, rather than $t_{e,max}$, for t_{th} in eqn. (26) . Accordingly , the effectiveness is defined by

$$E = (t_{ex} - t_i)/(t_{e,eq} - t_i) = (T_{ex} - T_i)/(T_{e,eq} - T_i) \quad (32)$$

Substituting for T_{ex} and $T_{e,eq}$ from eqns. (21) and (31) respectively , we get

$$E = 1 - e^{-n N} \quad (33)$$

Thus , the definition of the effectiveness by eqn. (32) satisfies the condition of being independent of the initial fluid temperature and the form of radiation distribution . The effectiveness , as given by eqn. (33) , depends only on the number of passes and the characteristic number . An effectiveness chart is shown in Fig. 5 .

It must be kept in mind that , the dependence on the initial fluid temperature and the form of radiation distribution was shifted to the theoretical , or reference , temperature rise, i.e the denomenator of eqn. (32) . It will be taken into account when calculating the actual temperature rise corresponding to an effectiveness obtained from Fig. 5 or eqn. (33).

The importance of the concept of the equivalent equilibrium temperature and the definition of E by eqn. (32) lies in the possibility of gathering all possible effectiveness values in one graph ; Fig. 5 .

It can be seen from eqn. (33) or Fig. 5 that , the effectiveness , as defined by eqn. (32) , increases with the increase of the characteristic number , i.e the decrease of N_L and/or N_c . It also increases upon increasing the number of passes . However , and as can be found from eqn. (21) , the variation of the exit fluid temperature with the characteristic number or the number of passes can not be , as readily , described by a general trend which would be common to all heaters under different irradiation conditions . This is because the effect of these two parameters on the equivalent equilibrium temperature is completely dependant on the form of radiation distribution , eqn. (31) .

The efficiency of the heater is defined as the fraction of the absorbed radiant energy that is imparted to the flowing fluid. Thus ,

$$\eta = \dot{m}\, c\, (t_{ex} - t_i)/\ell L\, (I_1 + I_2 + \ldots + I_n) \quad (34)$$

Transforming to dimensionless temperatures

$$\eta = N_L\, (T_{ex} - T_i)/\sum_{i=1}^{n} R_i \quad (35)$$

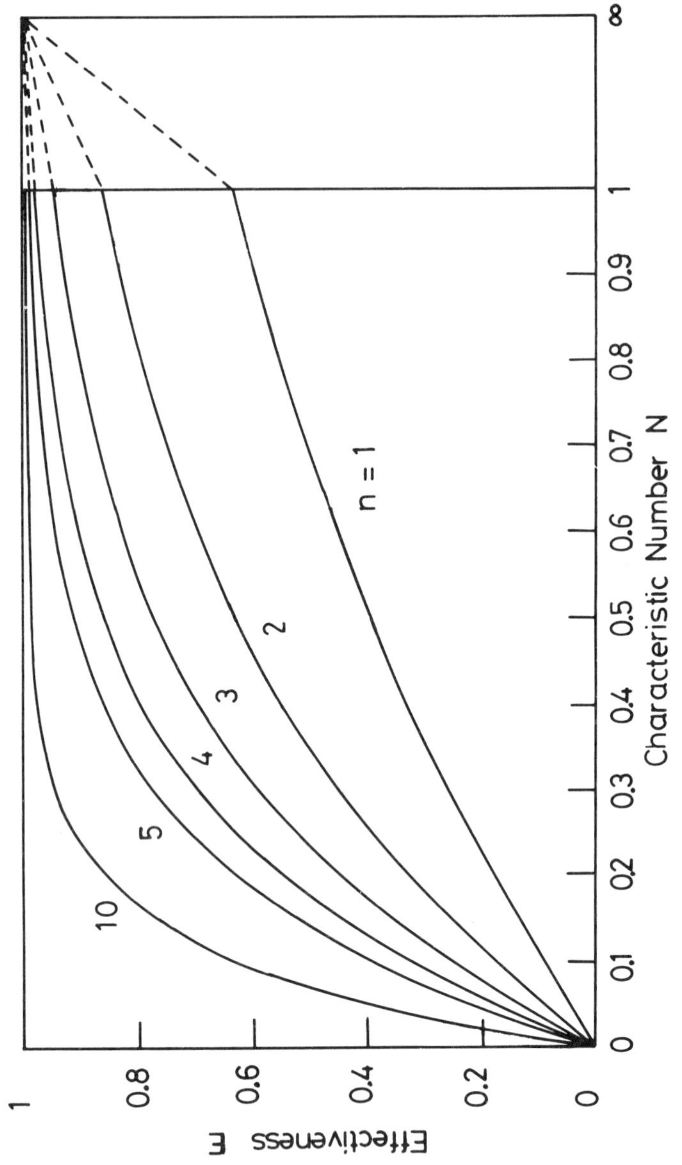

Fig. 5 Effectiveness Chart

The efficiency can also be written in terms of the effectiveness by substituting from eqn. (32) into eqn. (35)

$$\eta = N_L \, (T_{e,eq} - T_i) \, E / \sum_{i=1}^{n} R_i \tag{36}$$

Equation (35) or (36) implies that the efficiency is dependant not only on the characteristic number , as was the case for the effectiveness , but also on the individual values of the heat loss and convection numbers . Accordingly , a number of heaters having the same number of passes , exposed to the same radiation distribution , and fed by fluid at the same initial temperature can have the same effectiveness while their efficiencies are different ; this can be explained from eqn. (15) .

The trend of variation of the efficiency with either the heat loss number or the convection number can be investigated by differentiating eqn. (36) and going through a rather long mathematical procedure . It can be shown that the efficiency is always increasing upon increasing the heat loss number ; this is true for all possible radiation distributions and under all working conditions . On the other hand , the change of the efficiency with the convection number does not have a fixed direction ; it depends on the shape of the radiation distribution and the value of the characteristic number . The efficiency increases with the increase of N_c if the value of the characteristic number is large and the fluid is flowing in a parallel or combined flow relative to a sharply drooping energy distribution . Considering the practical range of characteristic number values which may be encountered in solar collectors , the efficiency will be usually changing in the opposite direction of N_c for almost all radiation distributions .

The initial temperature of the fluid is a very important parameter regarding the heater efficiency . It is obvious from eqn. (36) that , a lower efficiency will be attained for a higher value of T_i .

The variation of the efficiency with the number of passes does not have a fixed monotonic trend . It depends on the initial fluid temperature , the characteristic number , the form of radiation distribution , the number of added or removed passes, and the initial and final numbers of passes . It can be shown, using eqn. (35) , that increasing the number of passes from n_1 to n_2 would result in an increase of efficiency if the following condition was satisfied

$$(T_{ex,n2} - T_i)/(T_{ex,n1} - T_i) > 1 + \left(\sum_{k=n1+1}^{n2} R_k\right)/\left(\sum_{k=1}^{n1} R_k\right)$$

where $T_{ex,n1}$ and $T_{ex,n2}$ are the output fluid temperatures for n_1 and n_2 passes respectively . On the other hand , if the left hand side of the inequality is smaller the efficiency wil decrease upon increasing the number of passes from n_1 to n_2. If the two sides are equal, the efficiency will not change. A relation between the two temperatures $T_{ex,n1}$ and $T_{ex,n2}$ can be deduced from eqn. (21); it has the form

$$T_{ex,n2} = (1 - e^{-N}) \left\{ \sum_{k=0}^{n2-n1-1} R_{n2-k}\, e^{-kN} \right\} + T_{ex,n1} e^{-(n2-n1)N}$$

Arrangement of Fluid Flow through the Heater

Since the radiant flux is different over the various passes, it is very important to decide upon a flow sequence for the fluid through the different passes, i.e an arrangement for pass ordering. Consider the simple case of a non-uniform radiation distribution with monotonic change over an n-pass heater as shown in Fig. 6. The figure shows two possible methods or directions for ordering the passes. The first is a counter flow arrangement in which the cold fluid enters the heater through the pass exposed to the lowest radiant flux. The second is a parallel flow pattern where the fluid is first admitted to the pass subjected to the highest flux. It is required to find out which of these two arrangements would result in better performance.

The effectiveness is given for both arrangements by eqn. (33), i.e it has a fixed value independent of the direction of flow. Accordingly, and from eqn. (32), the output fluid temperature will be higher for the type of flow which possesses a higher value of the equivalent equilibrium temperature, i.e higher equivalent radiation. And the problem reduces to that of finding which of the two types of flow will result in a higher equivalent radiation.

The absorbed radiant flux for any pass, of order j, will be denoted by I_j or $\bar{I}_j$ for counter or parallel flow respectively. The equivalent radiation for the case of counter flow is given, from eqn. (30), by

$$I_{eq} = \left\{ I_n + I_{n-1}e^{-N} + I_{n-2}e^{-2N} + \ldots + I_1 e^{-(n-1)N} \right\} \cdot (1 - e^{-N})/(1 - e^{-nN}) \qquad (37)$$

Similarly, the equivalent radiation for the parallel flow pattern can be written as

$$\bar{I}_{eq} = \left\{ \bar{I}_n + \bar{I}_{n-1}e^{-N} + \bar{I}_{n-2}e^{-2N} + \ldots + \bar{I}_1 e^{-(n-1)N} \right\} \cdot (1 - e^{-N})/(1 - e^{-nN}) \qquad (38)$$

It can be seen from Fig. 6 that

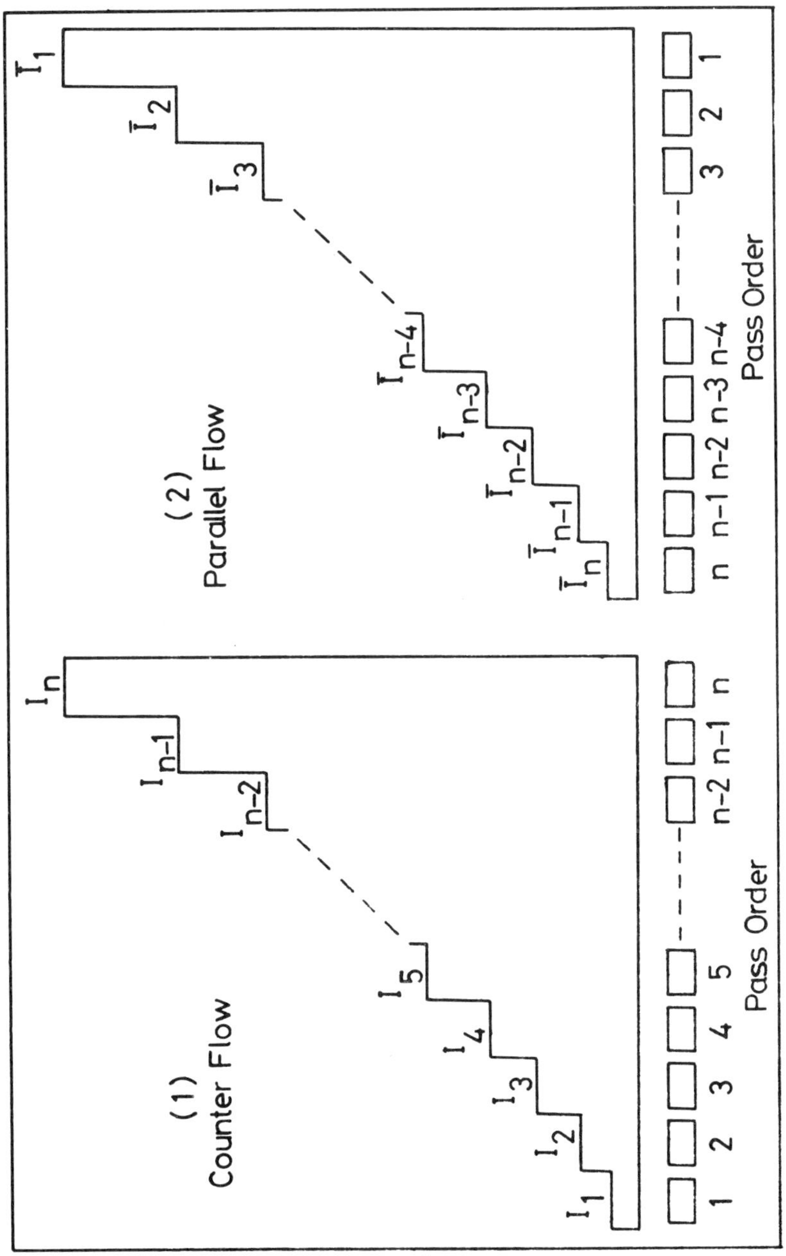

Fig. 6 Counter and Parallel Flow Arrangements for the Same Radiation Distribution

$$\bar{I}_j = I_{n-j+1}$$

Thus , eqn. (38) can changed to

$$\bar{I}_{eq} = \left\{ I_1 + I_2\, e^{-N} + I_3\, e^{-2N} + \ldots + I_n\, e^{-(n-1)\,N} \right\}. \quad (1 - e^{-N})/(1 - e^{-n\,N}) \qquad (39)$$

Subtracting eqn. (39) from eqn.(37) , we get

$$I_{eq} - \bar{I}_{eq} = \left\{ (I_n - I_1)\,\{1 - e^{-(n-1)N}\} + (I_{n-1} - I_2).\,\{e^{-N} - e^{-(n-2)N}\} + (I_{n-2} - I_3)\,\{e^{-2N} - e^{-(n-3)N}\} + \ldots\ldots \right\}\,(1 - e^{-N})/(1 - e^{-nN}) \qquad (40)$$

The right hand side of eqn. (40) is always a positive quantity which indicates that

$$I_{eq} > \bar{I}_{eq}$$

i.e under the same conditions the counter flow arrangement will always have a higher equivalent radiation than that attained with parallel flow .

The heater efficiency , as expressed by eqn. (36) , increases with the increase of the equivalent equilibrium temperature . This means that , counter flow results not only in a higher output temperature but also in a better efficiency .

The ratio of the two equivalent radiation values , corresponding to the counter and parallel types of flow , is plotted in Fig. 7 . In this figure , one particular form of radiation distribution is considered ; that is when the absorbed radiation densities on the successive passes are represented by an arithmetic progression of the form I , I(1+b) , I(1+2b) , ..., I(1+(n-1)b) .

From the analysis given above , together with Fig. 7 , we can conclude that , it is always recommended to order the passes such that the fluid flows through the heater in a counter flow regime . This recommendation becomes more important in the following cases

1- For large number of passes .

2- When the radiant energy distribution has a step form , i.e when the change of radiant flux from pass to pass is large .

3- At high values of the characteristic number .

When the distribution of the radiant energy is not monotonic , such as the case in Fig. 4.C , it is impossible to order all passes in one direction and still maintain the favourable counter flow pattern . Two suggested flow arrangements , which

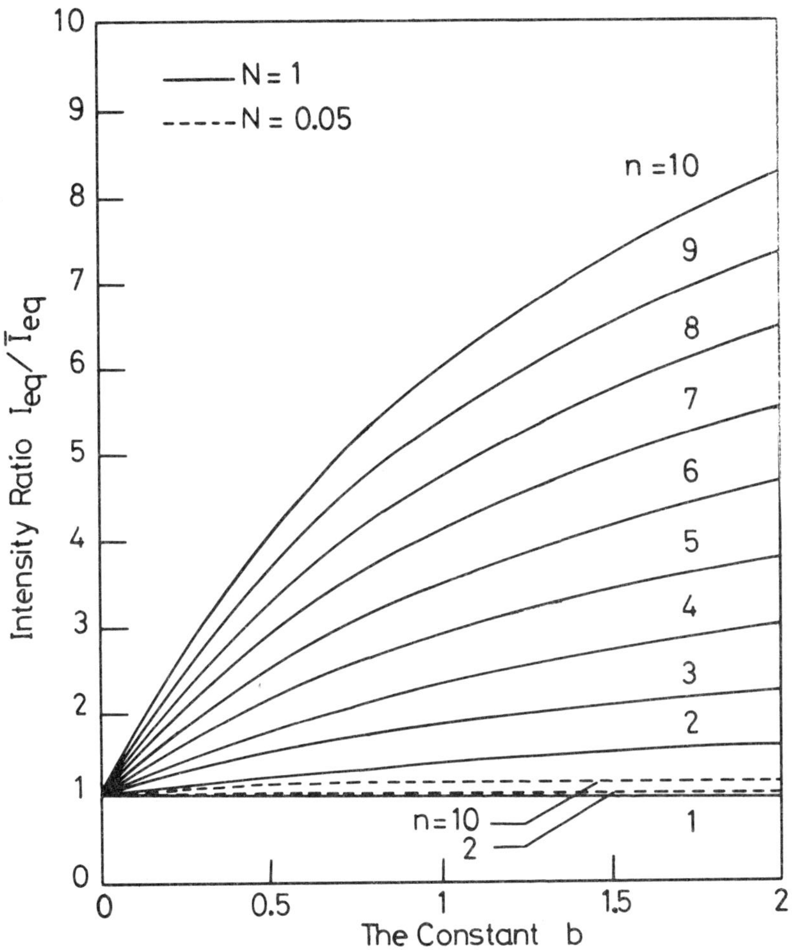

Fig. 7 Ratio of the Equivalent Radiations of Counter and Parallel Flow Systems for a Radiation Distribution Represented by I, I(1+b), I(1+2b), ----------, I(1+(n-1)b)

satisfy the required counter flow condition, are shown in Fig.8 .

CONCLUSIONS

The crosswise radiation distribution has a very strong influence on the performance of a multi-pass heater . The mathematical expressions derived herein are valid for any form of distribution .

Similar to ordinary recuperators , the flow of fluid through the heater can be classified into counter flow , parallel flow, or a combination of both . In counter flow heaters , the fluid will be progressively exposed to higher radiation intensities as it flows through succeeding passes . The reverse is true for parallel flow heaters , i.e the radiant flux density decreases continuously as the pass order increases .

The effectiveness is a very important and significant performance index . Together with the efficiency , they give a complete picture of the heater performance . It must be kept in mind that an efficient heater is not necessarily an effective one .

Introducing the new concept of equivalent radiation , or equivalent equilibrium temperature , it became possible to have the effectiveness independent of neither the initial fluid temperature nor the form of radiant energy distribution . Accordingly, one effectiveness chart , Fig. 5 , can be used for all heaters under different irradiation conditions . This chart can be very helpful for the comparison of different heaters or for the appraisal of a certain heater on a temperature rise basis .

When the fluid flow through all or some of the passes is a parallel flow , the fluid temperature may , sometimes , decrease again after acquiring a maximum value . Whenever such a phenomenon occurs , the drop will start at inlet to a pass , i.e after traversing an integer number of passes . In such a case the output fluid temperature will decrease upon further increasing the number of passes although the effectiveness , as given by eqn. (33) or Fig. 5 , will increase . This is explained by the decrease of the reference or theoretical maximum temperature, viz. the equivalent equilibrium temperature .

For any particular heater exposed to a certain radiation distribution and having a specified value of the characteristic number , there can be only one , intrinsic , fluid temperature distribution along the serpentine . At the same time there may exist any one of an infinite number of possible wall temperature distributions . This explains why a number of heaters may all have the same effectiveness while their efficiencies are different .

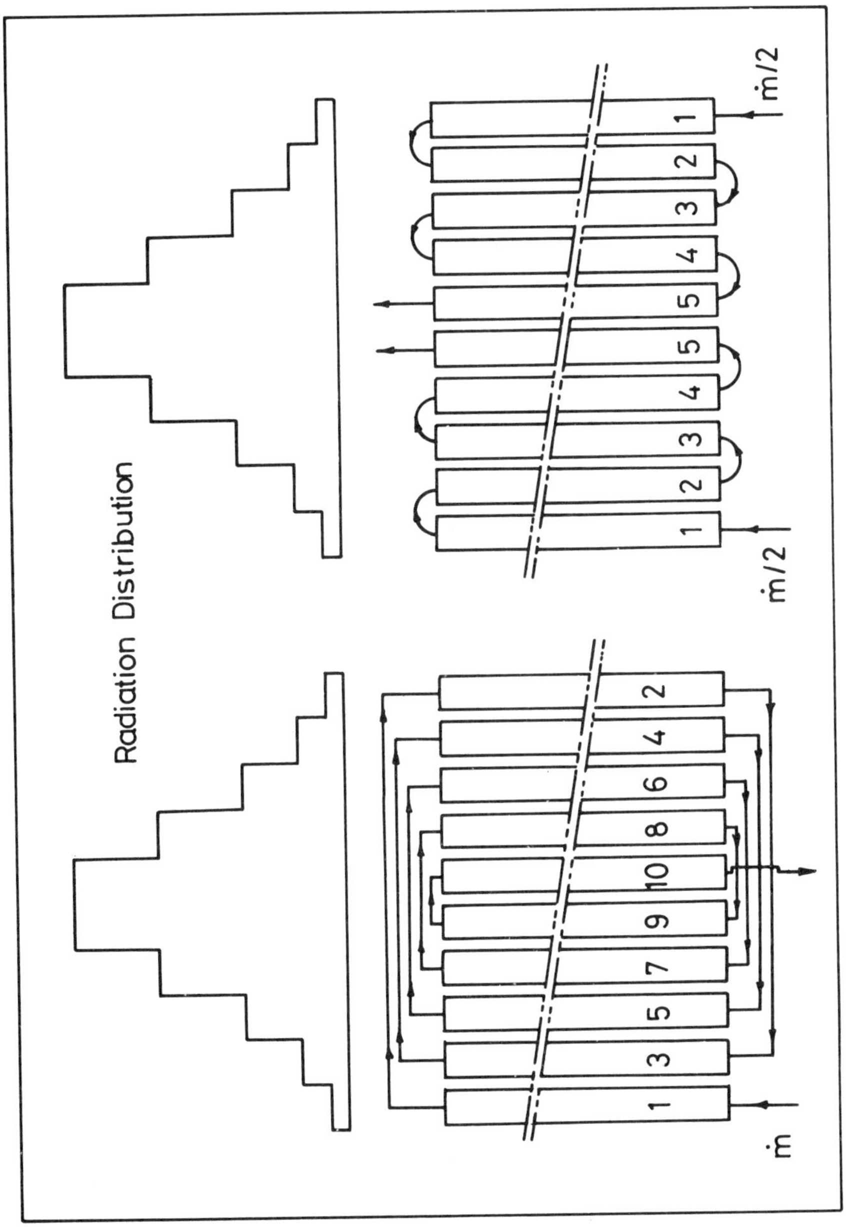

Fig. 8 Two Suggested Flow Arrangements to be Used with Double Sloping Distributions

As the convection number decreases both the fluid output temperature and the heater efficiency will increase ; thus , it is recommended to keep its value as low as possible . The values of the stream heat capasity ($\dot{m}$ c) and the heater length may often be dictated by the application requirements and space limitations . Therefore , it is of utmost necessity to have a high value of the product (h.p) . Accordingly , rectangular fluid passages may be used , instead of circular tubes , since they possess longer perimeter for the same flow area and they also show higher values of the internal film coefficient . It may also be useful to use corrugated walls , internal fins, or turbulence promoters .

The efficiency of the heater is always and under all conditions increasing upon increasing the value of N_L . On the contrary, the variation of the effectiveness is in opposite direction of N_L .

It is highly recommended to order the passes such as to have the fluid always flowing in a counter flow pattern relative to the radiant energy distribution . This results in an output temperature and an efficiency higher than those attained with a parallel flow arrangement . The superiority of counter flow is more obvious for

1- Large number of passes
2- Steep distribution of radiant energy
3- High value of the characteristic number .

In a properly designed heater , i.e one which satisfies the counter flow condition , the output temperature increases continuously upon increasing the number of passes . However , this may result in a lower efficiency .

NOMENCLATURE

c	Specific heat of the heated fluid , joule/kg^{o}c .
E	Effectiveness of the heater , dimensionless .
h	Convection film coefficient of heat transfer inside the passage , W/m^2 ^{o}c .
I	Radiant energy absorbed per unit time and surface area, W/m^2.
j	Order of an arbitrary pass .
k	Order of an arbitrary pass .
ℓ	Irradiated width of the pass , m .
L	Length of the pass , m .
$\dot{m}$	Mass rate of flow , kg/hr .
n	Number of passes .
N	Characteristic number , dimensionless .
N_c	Convection number , dimensionless .
N_L	Heat loss number , dimensionless .
p	Internal perimeter of the fluid passage , m .
P	Position function , dimensionless .

R_j	Intensity ratio , defined by eqn. (2) , dimensionless .
$R_{j,k}^j$	Intensity ratio , defined by eqn. (1) , dimensionless .
t	Real or natural temperature , °c .
T	Dimensionless temperature , defined with reference to I_1; dimensionless .
T(j)	Dimensionless temperature , defined with reference to I_j, dimensionless .
U	Overall heat loss coefficient , $W/m^2\ ^oc$.
x	Distance along the heater , m .
X	Dimensionless distance , dimensionless .

Greek Letters

δ	A variable defined by eqn. (16) , dimensionless .
η	Efficiency of the heater , dimensionless .
ψ	A variable defined by eqn. (18) , dimensionless .

Subscripts

a	Ambient air .
e	Equilibrium .
eq	Equivalent .
ex	At exit from the heater .
ex, j	At exit from the jth pass .
f	Fluid .
i	At inlet to heater .
i, j	At inlet to the jth pass .
j	Pertaining to the jth pass .
max	Maximum .
w	Wall .

REFERENCES

1- H.C. Hottel and B.B. Woertz , The performance of Flat-Plate Solar-Heat Collectors . Transactions of the ASME , pp 64-91,1942.

2- H. Birnbreier , Flat Plate Solar Collectors . Workshop proceedings , Engineering Aspects of Solar Energy Utilization , Apr. 1977 , Stuttgart University , West Germany .

3- K.E. Hassan and M.F. El-Refaie , Theoretical Performance of Cylindrical Parabolic Solar Concentrators . Solar Energy , vol. 15 , pp 219-244 , 1973 .

4- H.C. Hottel and D.D. Erway , Collection of Solar Energy , in Introduction to the Utilization of Solar Energy , edited by A.M. Zarem and D.D. Erway . McGraw-Hill Book Company , 1963 .

5- A. Whillier and G. Saluja , Effect of Materials and Construction Details on the Thermal Performance of Solar Water Heaters . Solar Energy , vol. 9 , No 1 , 1965 .

SIMULATIONS OF SOLAR CAPTATION SYSTEMS : METHODOLODY AND APPLICATIONS TO DISTILLATION, GAS AND WATER HEATING

F. ROCARIES, G. OLALDE, P. GRANIER
B. N'DOYE and M. DAGUENET
Laboratoire de thermodynamique et Energetique - Centre Universitaire
Cedex, France

ABSTRACT

The authors study heat transfer phenomena with numerical methods. Because The experimental difficulties, it is usefull of course to simulate the behaviour of solar systems. That, is available for high temperatures systems (gas reheaters in focusing devices) ; for solar stills ; flat plate collectors, etc. Simulation enables an accurate previsions of the behaviour, and a systematical study of all the parameters effects. This last study point out the major parameters, to simplify the model. The present study regards : solar stills, various flat collectors, high temperatures gas re-heaters. The authors give appropriate organigrams and graphical results.

PRELIMINARIES :

A team of research workers of the Thermodynamic and Energetic Laboratory of PERPIGNAN have been working for early two years on informatic processing model making in the solar captation systems. The researches had to be oriented in this way because the transfer problems inherent in these phenomena were so complicated and because the cost of the manufactury of the demonstration model was to high (the only acceptable scale for this kind of problem is scale 1)
The work already done cane be classified under two chapters :
- Solar captation by glasshouses effect
- Solar captation by concentration and heating at high temperatures.
This work takes shape in a series of computer programs. These programs will have a double function :
1) a study function which allows us to study the influence of each parametre in systems where a large number interfere and when possible to verify the basic assumptions of these models experimentaly and eventualy to bring them together.
2) a simulation function which studies on the temporal evolution of a complex system. The captation or the heating represent in this case only one part of the system. The runing of this system is then represented (as) quickly in an environment made with a card file of meteorological data. We use the FORTRAN IV language in these programs. We are going to study the two apects of our research in this paper.

1. CAPTATION BY GLASSHOUSES EFFECT

Here we use the low temperature (less than 100°C). In this particular

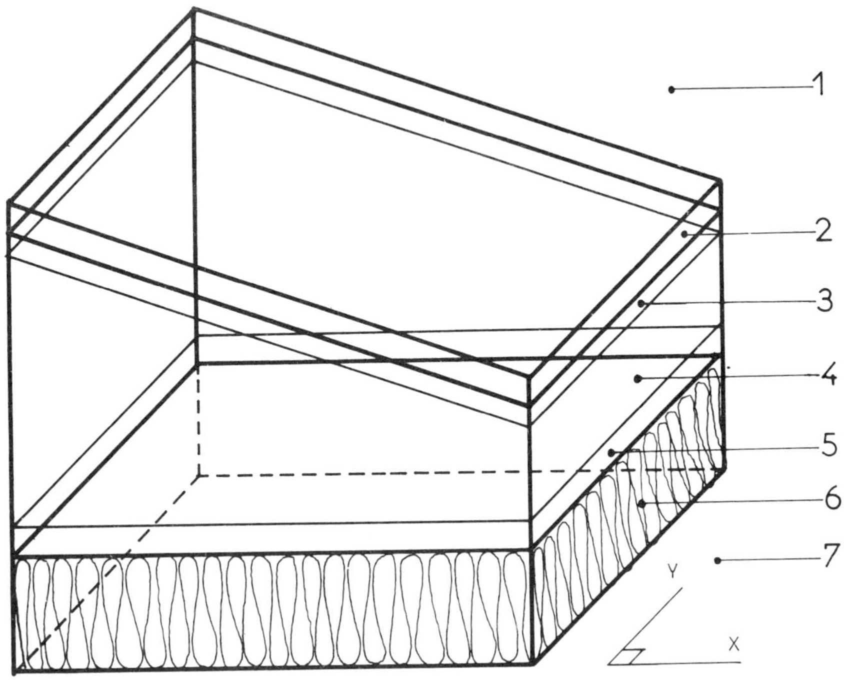

1. SUPERIOR AMBIANT AIR
2. COVER
3. CONDENSATION FILM
4. SPACE
5. FILM OF STREAMING WATER
6. BACK INSULATION
7. INFERIOR AMBIANT AIR

FIG. 1 a

Fig. 1a Scheme of a Water Collector By Running Onto the Black Plane

$$\cdot\ P_a = P_u + P_{iner} + P_{ph} + P_{pb}$$

$$\cdot\ P_{iner} = M.C_p \frac{\partial T_{wat}}{\partial t}$$

$$\cdot\ P_{ph} = (T_{wat} - T_{ext}) \cdot \frac{1}{R_{ch}}$$

$$\cdot\ P_{pb} = (T_{pn} - T_{ext}) \cdot \frac{1}{R_{cb}}$$

P_a is the solar power absorbed by the black plate
P_u is the useful power recuperated by the water
P_{iner} is the power of inerty (transitory working)
P_{ph} is the power lost on the front face
P_{pb} is the power lost on the back face
M is the equivalent mass of water from the collector
C_p is the specific heat of the water

R_{ch} is the thermal resistance to the top
R_{cb} is the thermal resistance to the bottom
T means temperatures
In this series of unknown process we are above all interested, in P_u which is the useful power retrieved by the water, we are then going to modify the system in order to obtain P_u. That is to say

$$\cdot\ P_{iner} = M\ C_p \frac{\partial T_{wat}}{\partial t}$$

$$\cdot\ P_{ph} = (T_{wat} - T_{ext}) \cdot \frac{1}{R_{ch}}$$

$$\cdot\ P_{pb} = (T_{pn} - T_{ext}) \cdot \frac{1}{R_{cb}}$$

$$\cdot\ P_u = P_a - P_{iner} - P_{ph} - P_{pb}$$

This system of equations admits an equation with variable coefficients. As a matter of fact if we can say that R_{cb} (which is a resistance of conduction) is independant from the temperatures applied to its limits, it is not the case of R_{ch}.
Let"s look at fig. 1 b which represents the equivalent circuit we notice that R_{ch} is the sum of resistances which all depend on the temperatures.

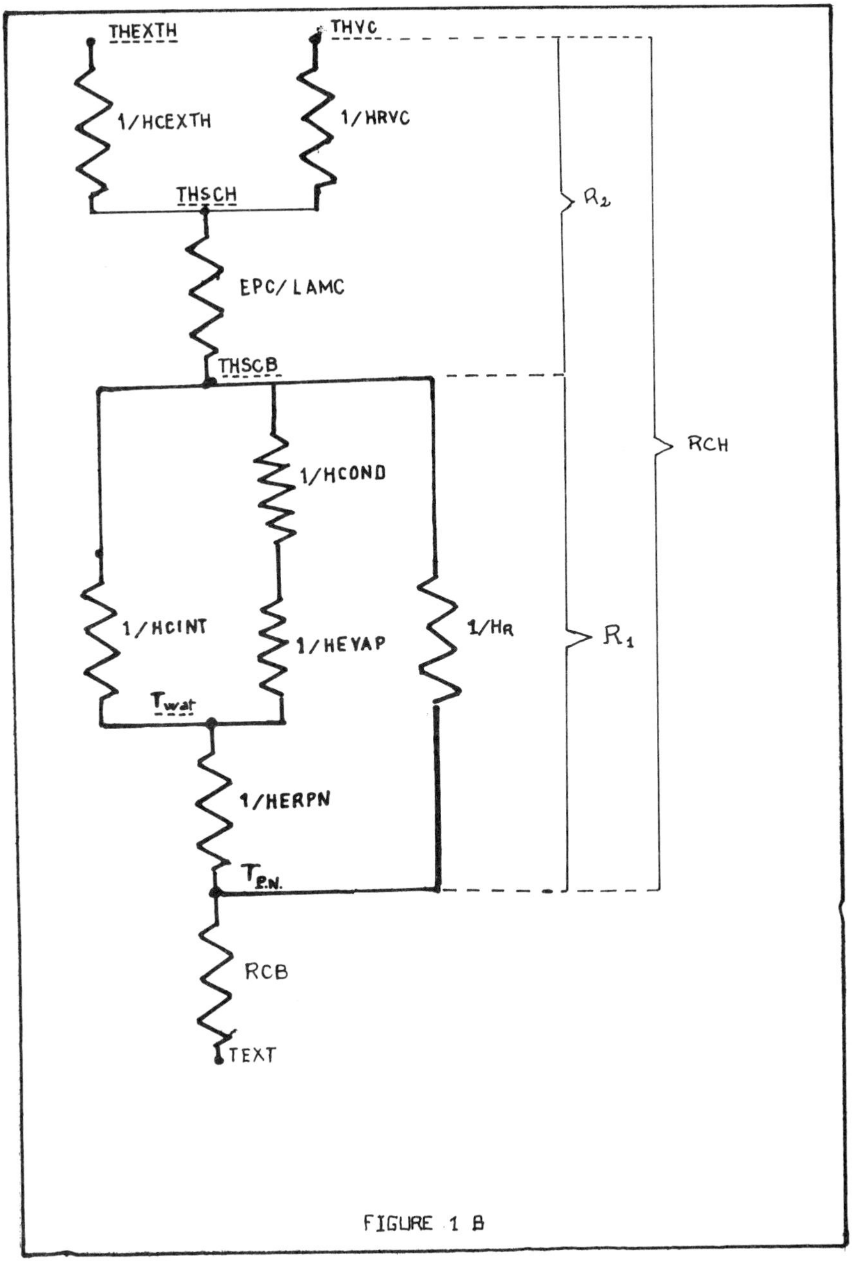

Fig. 1b Network Representing a Collector Section

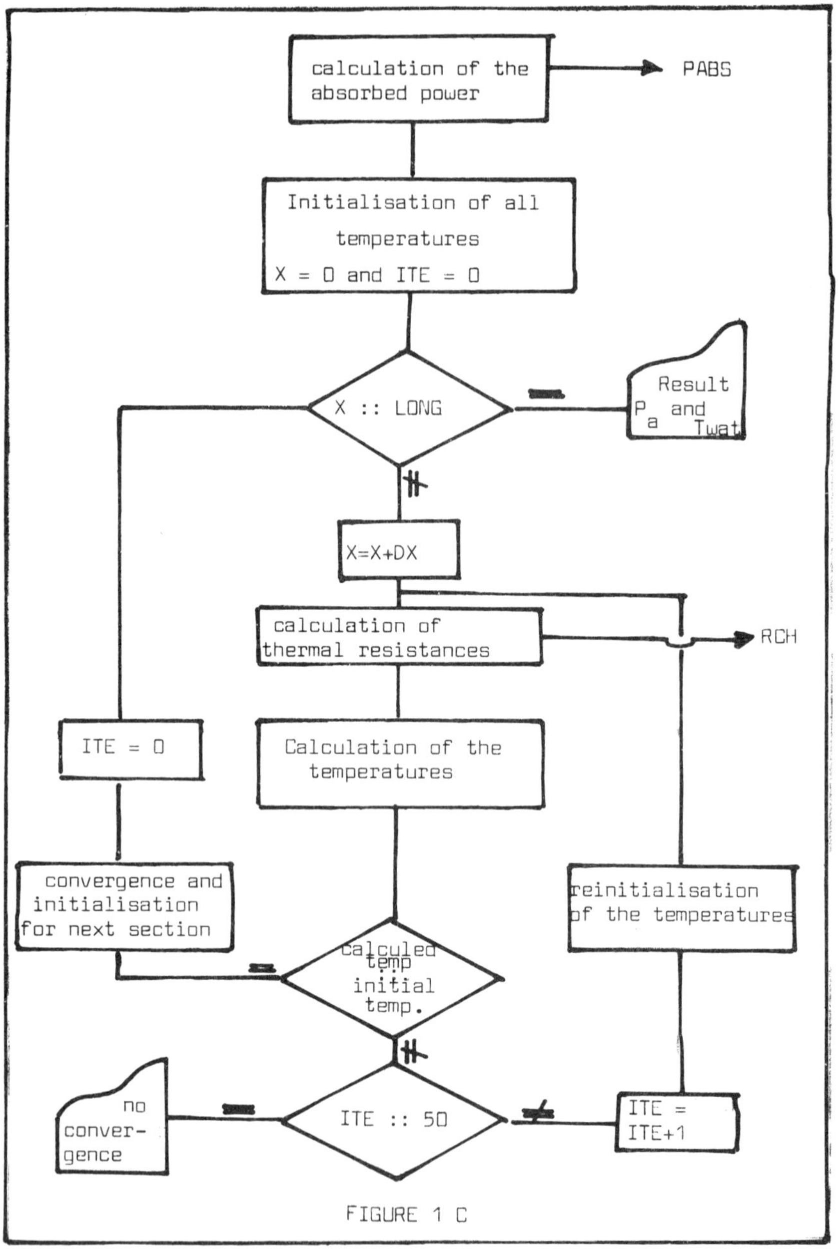

Fig. 1c Brief Organigram of the Program

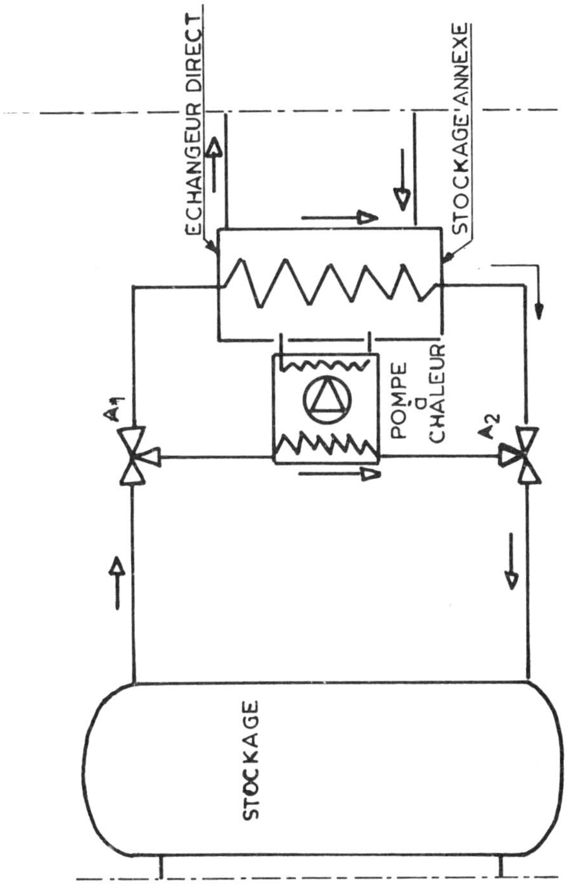

Fig n°1c Chauffage par Pompe à Chaleur

Fig. 1c Heating With Heat Pump

category of captors we can group the water collectors, air collectors, solar stills, and glasshouses. We took a particular interest in the water and air collectors and solar stills too.

1.1. Methodology of calculation

The exemple we have chosen to illustrate our enterprise with is the water collector by runing of the water on to the blackbody. Let's look at a diagram of that type of water collector, we can see that from a pure optical point of view the system constitued by this kind of collector is far from being simple. The presence of phenomena of evaporation followed by a recondensation on the inferior face of the transparent cover presents us first of all with a series of problems :

- we have three media with different refractive indexes (air, cover and water)
- it's hard for us to know the nature of the condensated water on the glass.

In order to work out this latter problem, we have adopted the following hypothesis :

- we have considered that the condensation is a thin film of a given (or calculated) thickness, parallele to the surface of the cover. (This supposes of course a perfect wetability from the cover and a regular flow of condensation water).
- moreover since it is almost impossible for us to find (should it exist) a generic term for the series formed by the various transmissions - reflexions we have gathered a series of terms such as :

$$R = (1 - \rho_1)\ (1 - \rho_2)\ (1 - \rho_3)^2$$

in order to obtain a similar transmission. This summation replaces the analytical formula usual to conditions semi-transparent the formulation of which is :

$$= \frac{1 - \rho_1}{1 + (2n - 1)\rho_1}$$

n represents the number of covers

ρ_1 represents the coefficient of reflexion air-cover

ρ_2 represents the coefficient of reflexion cover-water

ρ_3 represents the coefficient of reflexion water-air

This first problem being worked out we shall have to face other problems concerning the evaluation of the losses from the front surface.

As a matter of fact if we make the electrical analogy of a cell DX long from our collector (cf. fig. 1 b) we can see the neat difference between that and a normal collector.

It we write the balance of flows in the set, we shall have.

$$R_1 = \frac{h_{cond} + h_{evap}}{h_{cond} \times h_{evap} + (h_{cond} + h_{evap})\,(h_r + h_{conv})}$$

and all the h are factors of the temperatures in the limits as we can see it in table I. Then R_{ch} depends on the temperatures of the water, on the covers and exterior factors. We are there confronted with an analytically insoluble problem we are going to work out by the numerical way by iterations. There follows our plan :

- sorting out the values of all the intermediatry temperatures to obtain a balance in the system as acurate as possible.
- the convergence of the process is made by a process of a linear surrelaxation applied to the temperatures and is considered as finished when the difference between two . successive calculations will let some variations inferior to $\frac{1}{100}$ Kelvin (which gives us a relative precision of less than $5 \cdot 10^{-5}$) appear.

The working out of the captation problem in that case is the slow resolving of a number N of sets DX wide, N so that N . DX = L which is the length of the collector.
Then it is worked out by a program written in FORTRAN IV language. A brief organigram of this program accompanies this paper (fig. 1 c).
As we can see on this organigram that most of the annex calculations have been made by different units.
So unit PABS which calculates the factors of transmissions for the scattered radiation as well as for the total radiation, it is R_{ch} which calculates all the thermal resistances on the front face of the collector. This modular aspect with one basic program allows us to make all the calculations necessary to the various systems already studied or being studied.

1.2. Applications

This system of modelisation and the resulting programs have allowed us to work out a certain amount of problems as far as the solar system using glasshouse effect is concerned.
Here we define three main axes :

- the water collector and its utilisation
- the air collector and its utilisation
- the stills with glasshouse effect static or dynamic.

1.2.1 Water collectors

This way of captation is frequently used now at present as well for the production of sanutary hot water as for central heating. Lately we have developped the models of this type of collectors. First the streaming collector which is the exemple of our methoddogical study, secondly the more classical collector (tube solar collector etc.) with these models we have ben able to develop a whole series of simulations for the central heating

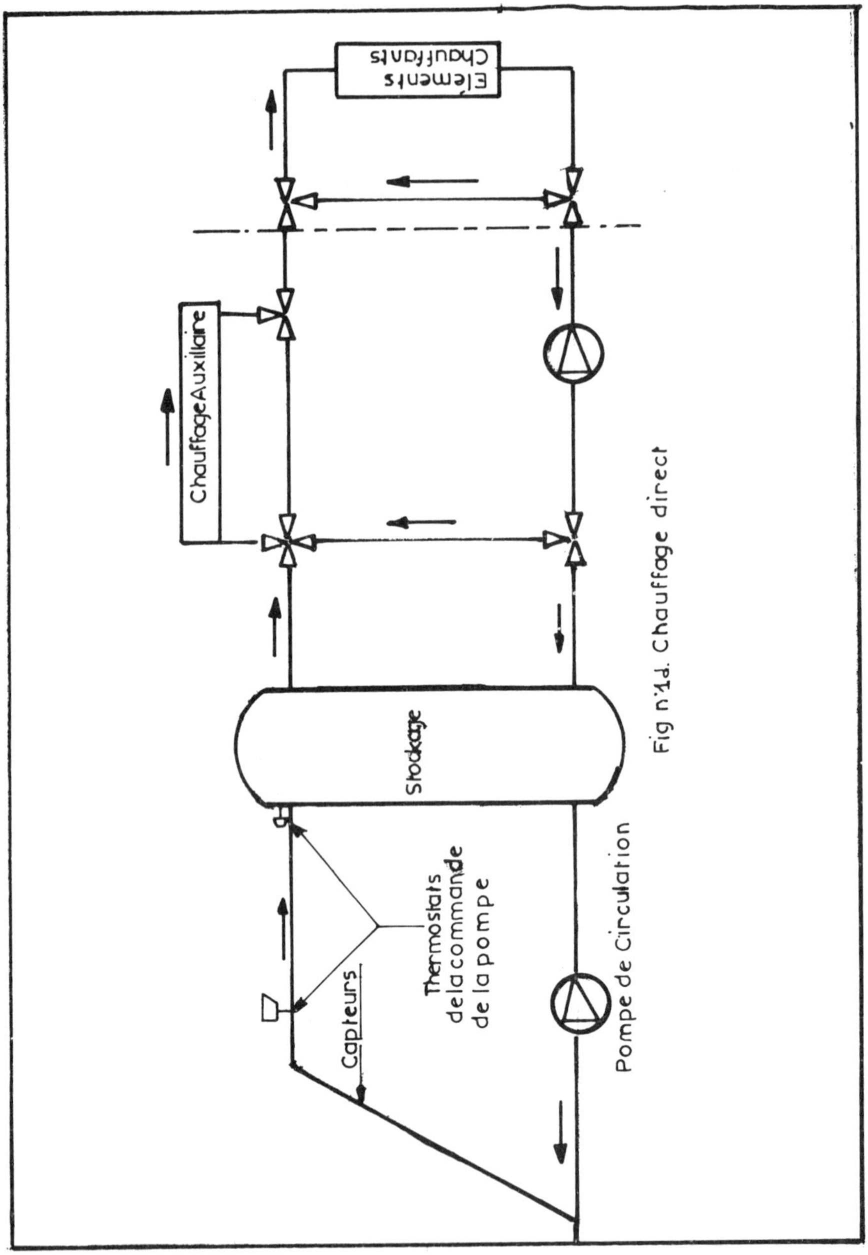

Fig n°1d. Chauffage direct

Fig. 1d Direct Heating

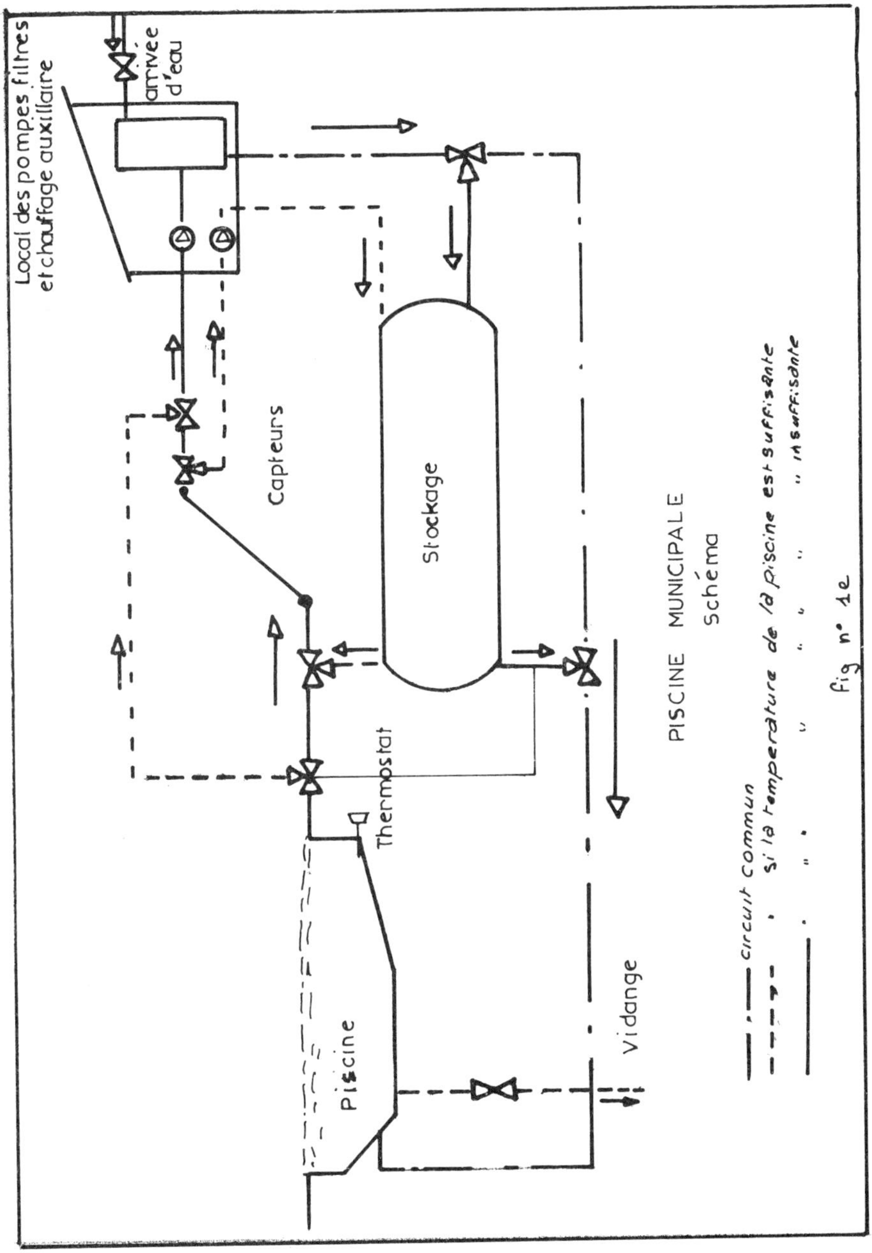

Fig. 1e Water Heating For a Swimming Pool

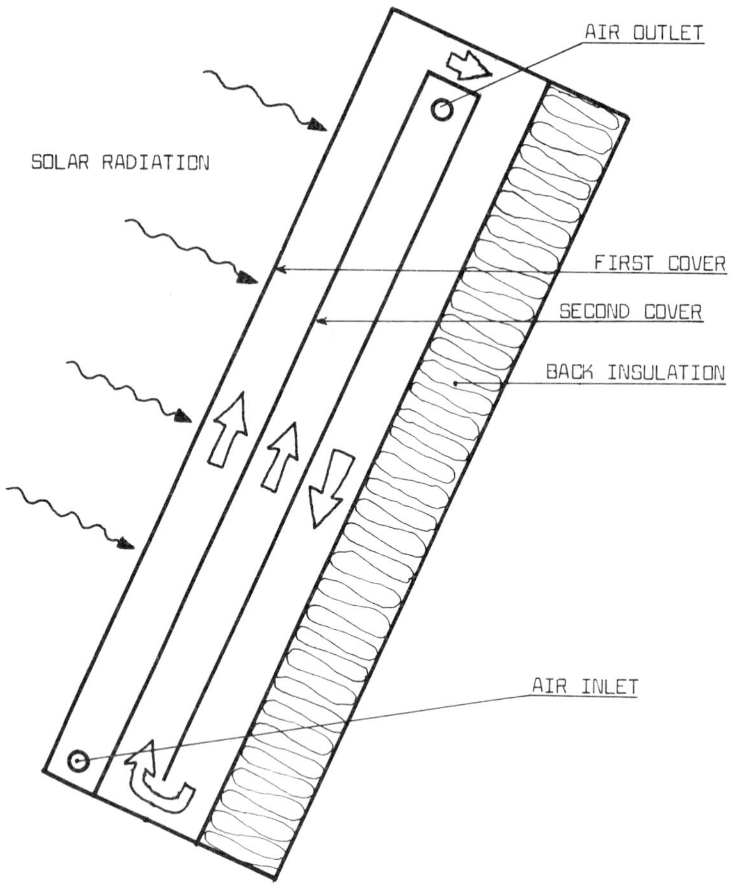

FIG. 1 g : An exemple of an air collector model with multiple circulations.

and for production of hot water for swimming poob [I] .
This kind of study allows us to sumulate dusing one more years the working of rather complex systems, as we can see in fig. Id, Ic, If. These simulations use real meteorological data as basic data. The card file is made out from datas available in FRANCE given by the national meteorology department of course in such calculations the part constitued by the collectors represents only a part of the whole but it is one of the most delicate parts and it is surely the most important.

1.2.2. The air collectors

In concurrence with t the water collectors, the air collectors have an important role to play, especially in domestic heating and in the drying problems. The present study on this kind of collectors is still at the stage of the model.
As a matter of fact this model must represent insolators which use a non sophisticated technology and there fore simple material (different profiles for the absorber and the covers).
In the model which has been used the number of covers is limited to 2 and the air can run into all the available canals (fig. 1 g) in any way. Besides these various configurations of flowing the way of circulation may be natural (chimney effect) or forced (mecanical ventilation).
In the case of the solar dryer the air is of course wet and the study on the content of moisture constitutes one important factor of this simulation. We notice that the methodology for resdution is (above all in the case of the forced circulation) very close to the one used for water collectors.

1.2.3. The distillers

Here we shall speak of the distillers with glasshouse effect. Among those distillers we are going to distinguish the static and the dynamic distillers.
- Static distillers :
In this type of distillers we distinguish the "hot box" and the distillers in "cascade".
They are all made of an enclosure with a black side for the "bot box" (a series of black-steps for the "cascade") on which a certain amount of brine remains. This brine evaporates and the water condenses it self on the top dosing glass of the enclosure. The black face rests on an isulater which minimises the losses from the back side. As far as the hot boxes are concerned it seems difficult to increase their productiveness by enbarging the difference of temperature existing between the brine and the water from condensation. This is possible for example in making the water flow between the two sheets of glass of the cover. It is different with the cascade distiller. We can try to optimise the relation, the width of the steps on the hight of the steps, and tohe average hight of the enclosure above the steps. Our programs of modelisation answer this type of question.

- Dynamic Distiller

There we find the same elements as the ones described above but the brine in this case flows in thin films, on the black face or on a porous ground like a bent wick.

We distinguih.

The co-current wher the brine and the condensation circulate in the same way. The counter-current where the brine and the condensation circulate in opposite ways. The crossed-current where the brine and the condensation circulate in perpendicular ways. The stills with a wick and with double condensation where the brine flows in a black wick and where there is a condensation on the glasses situated above and under the wick [2] [3]. In all cases the aim is to calculate the inward efficiency of the distillers.

$$\eta = \frac{Q_e}{G_r}$$

Q_e = is the heat used for the evaporation

G_r = is the solar energy absorbed really

Il we use the same methodology as the example already mentioned to work out these problem (we in fact notice the great similarity existing between the collector with flowing water and the co-current distiller). The results obtained are radically differen. In the case of the collectors the interest was to extract the useful power carried by the water, here we must extract the power of evaporation. These models allow us to compare the action of the different types of distillers in the given meteorological surrounding, thus we can select.

The most active, ofr, for a giving type we can Know the optimal valve of parameters on which the user can act.

2. GAS HEATING IN HIGH TEMPERATURES IN THE HEARTH OF A CONCENTRATOR :

2.1 Nomenclature (cf. table I)

2.2 Introduction

Until now the studies in this domain deal with the specific problem of the heating of a non absorbant gas going through a porous or an alveolar ground exposed to a concentrated radiative flow furnished by the hearth of a solar concentrator.

We thought of different cases concerning the relative sense of the radiative flow and of the gas current and the geometric configuration of the porous ground (methodical and antimethodical circulations, cylindrical absorbers, in cylindrical crouwn, spherical).

The distribution of temperatures in the fluid and the solid is analytically and numerically calculated [4] . We also study the influence of parametre such as : porousness, diametre of grains, nature of the material,... This type of captation destinated for the heating of gas at a high tempe-

rature by the solar way is very interesting in chemistry (reactors) as well as in thermodynamic (production of steam, electricity, hydrogene, etc.).
The different studies made allow us to foresee the experimental conditions necessary to bring a gas to a given temperature, to determine the conditions for which the heat transfer is optimal and to know the quantity of radiant energy necessary to obtain the chosen profile of the temperatures.

2.3 Study of one case

Here we give some grafical results (fig. 2a, 2b, 2c, 2d, 2e) about the case of the cylindrical porous ground, the radiative flow and the flowing of gas in opposed ways. The main assumptions are :
We suppose that the gas is transparent to the radiation and goes through an homogeneous bed of spheres contained in a receptacle with sufficient cuts in order to lesten the effects of the side.

2.3.1 Equation of the problem

We propose to resolve the following differential system :

$$\rho C_p u \frac{d\theta}{dX} = \frac{1-\xi}{\xi} \frac{h}{\zeta}(T-\theta) + \lambda_F \frac{d^2\theta}{dX^2} \qquad (1)$$

$$\lambda_S^* \frac{d^2T}{dX^2} = \frac{h}{\zeta}(T-\theta)$$

$$X = 0, \qquad T = T_0 = \text{CSTE} \quad (3) \text{ and } \frac{d\theta}{dX} = 0 \quad (4)$$

$$X = L, \qquad \theta = \theta_L = \text{CSTE} \quad (5) \text{ and } \frac{dT}{dX} = 0 \quad (6)$$

$$\alpha\Phi = \varepsilon\sigma(T_0^4 - T_a^4) + \varepsilon\sigma(T_L^4 - T_a^4) - \lambda_S \left.\frac{dT}{dX}\right|_{X=0} \quad (7)$$

For the small Reynold's numbers, we have the relation [5] :

$$Nu = 0.262 \times 10^{-3} \left(\frac{1-\xi}{\xi^3}\right)^{0.75} Re^{0.75}$$

For the large Reynold's numbers, we have :

$$Nu = 0.066 \left(\frac{1-\xi}{\xi^3}\right)^{0.75} Re^{0.75}$$

The passing from one domain to another seems to be done for one critical value verifying the relation approximatly :

$$\frac{1-\xi}{\xi^3} Re = 600.$$

The solution is given by :

$$T^+ = A_0 + A_1 \exp(p_1 X^+) + A_2 \exp(p_2 X^+) + A_3 \exp(p_3 X^+) \qquad (23)$$

$$\theta^+ = A_0 + a_1 A_1 \exp(p_1 X^+) + a_2 A_2 \exp(p_2 X^+) + a_3 A_3 \exp(p_3 X^+) \qquad (24)$$

TABLE I : NOMENCLATURE

= density

Cp = specific heat

U = main velocity of the fluid

$^*_s$ = apparent thermal conductivity of the solid

$_F$ = thermal conductivity of the fluid

θ = temperature of the fluid

T = temperature of the solid

T_a = ambiant temperature

= absolute viscosity

h = heat transfer coefficient by convection between fluid and solid

= porousness

= total energetic radiation

= Stefan-Boltzmann constant

= absorptance of the face exposed to the radiation

= emittance of the face exposed to the radiation

= characteristic length depending on the form of the grins, ratio of volume to area ; equal to D/6 for a ground constitued by spheres.

D = diametre of a grain

L = study distance

h_{cond} = heat transfer coefficient by condensation

h_{evap} = heat transfer coefficient by evaporation

h_r = heat transfer coefficient by radiation

h_{conv} = heat transfer coefficient by convection

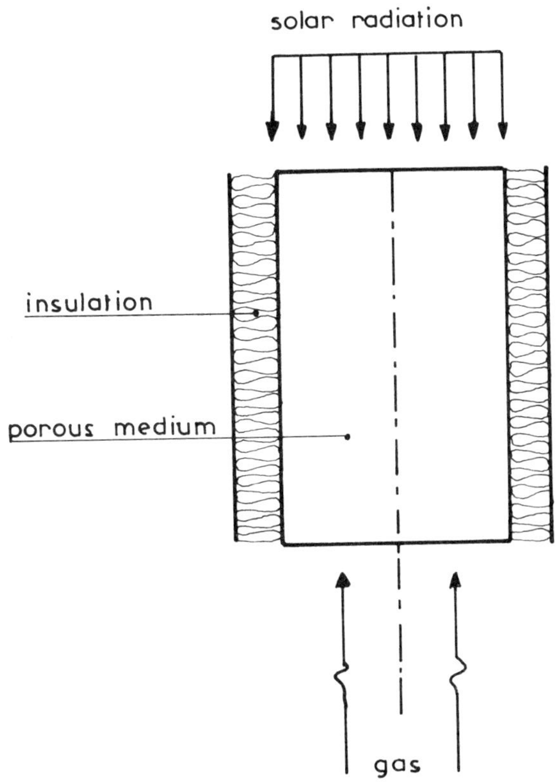

FIG. 2a

Fig. 2a Scheme of a Gas Re-Heater

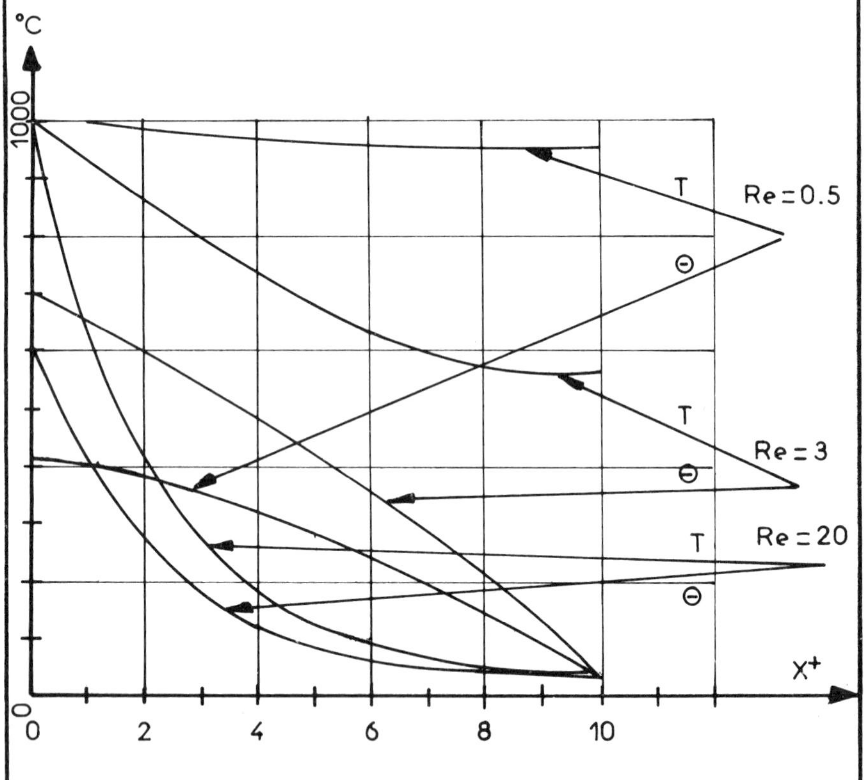

FIG. 2b

Fig. 2b Temperature Variations and θ As A Function Of X^+ At Various Values of Re

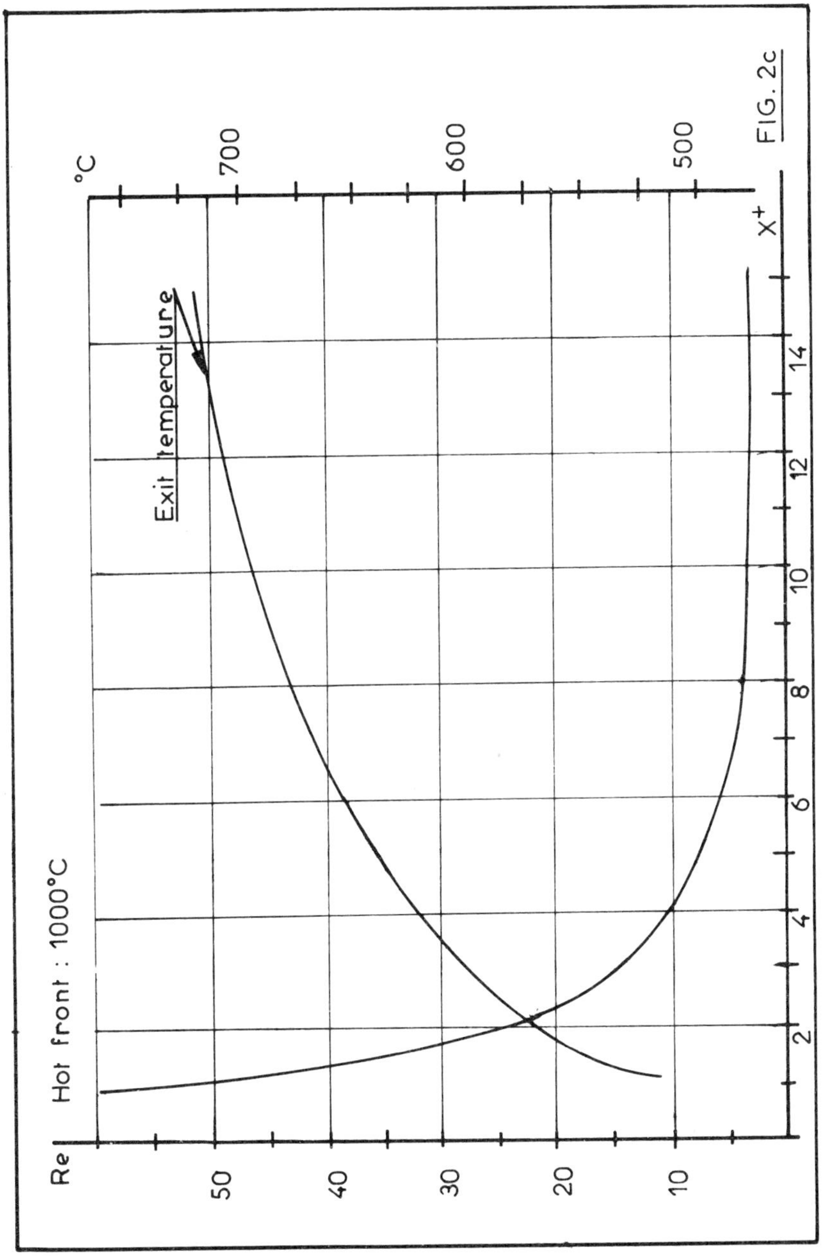

Fig. 2c Obtaining of the Best Reynold's Number

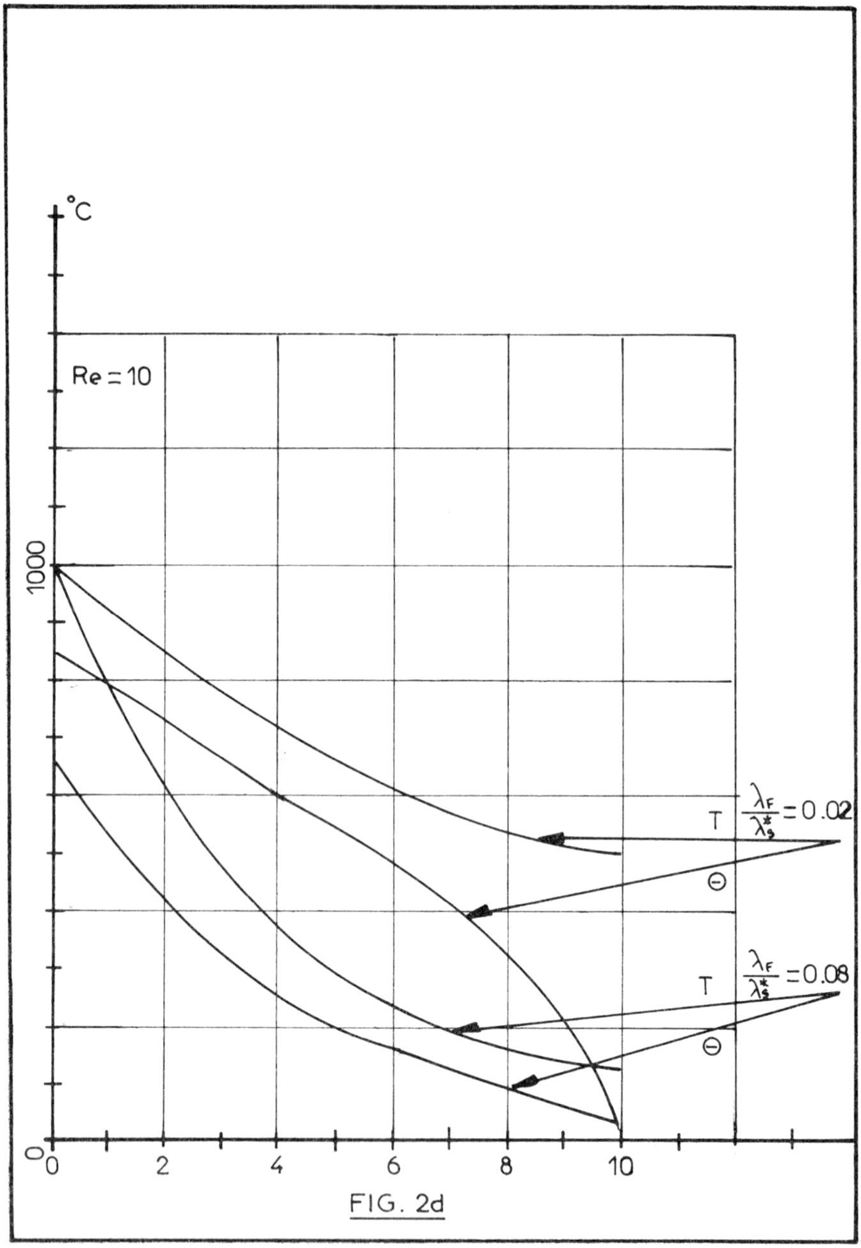

Fig. 2d Temperature Variations T and θ as a Function of X^+ for Two Values of Conductivities Ratio

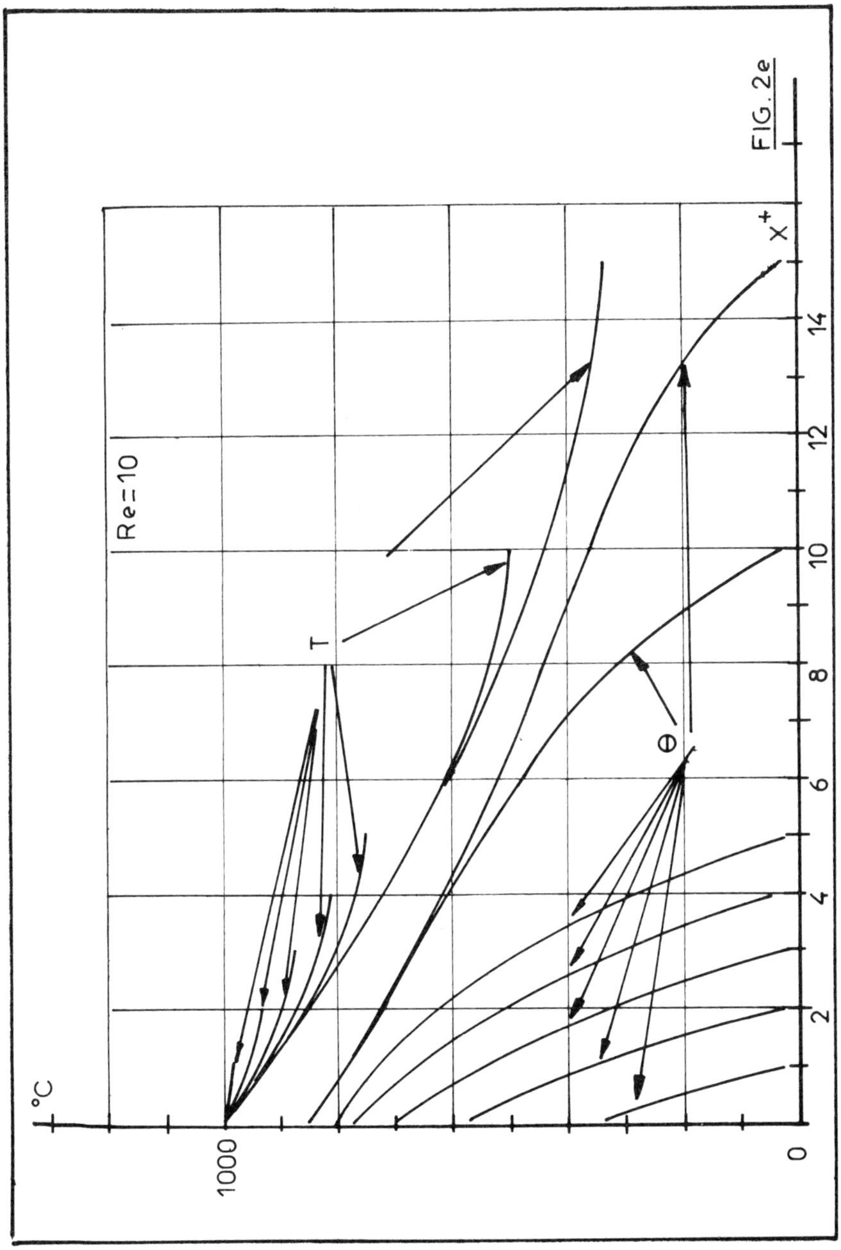

Fig. 2e Temperature Variations T and θ as a Function of X^+ as Various Values of L

with : $T^+ = \frac{T - T_0}{\theta_L - T_0}$ (8) , $\theta^+ = \frac{\theta - T_0}{\theta_L - T_0}$ (9) , $X^+ = \frac{X}{D}$ (10) , $Pr = \frac{\mu C_p}{\lambda_F}$ (11)

$Nu = \frac{hD}{\lambda_F}$ (12) , $Re = \frac{u D \rho}{\mu}$ (13), $a_1 = 1 - \frac{p_1^2 \lambda_s^*}{6 Nu \lambda_F}$ (25)

$a_2 = 1 - \frac{p_2^2 \lambda_s^*}{6 Nu \lambda_F}$ (26), $a_3 = 1 - \frac{p_3^2 \lambda_s^*}{6 Nu \lambda_F}$ (27), $A_0 = -A_1 - A_2 - A_3$ (28)

$A_1 = \frac{1 - A_2 \alpha_2 - A_3 \alpha_3}{\alpha_1}$ (29), $A_2 = \frac{a_1 P_1 - A_3 \beta_2}{\beta_1}$ (30), $A_3 = \frac{a_1 P_1 \beta_3 - P_1 P_1^+ \beta_1}{\beta_2 P_3 - \beta_1 \beta_4}$ (31)

and : $P_1^+ = \exp(p_1 L^+)$ (32) , $P_2^+ = \exp(p_2 L^+)$ (33) , $P_3^+ = \exp(p_3 L^+)$ (34)

$\alpha_1 = a_1 P_1^+ - 1$ (35) , $\alpha_2 = a_2 P_2^+ - 1$ (36) , $\alpha_3 = a_3 P_3^+ - 1$ (37)

$\beta_1 = a_1 p_1 \alpha_2 - a_2 p_2 \alpha_1$ (38) , $\beta_2 = a_1 p_1 \alpha_3 - a_3 p_3 \alpha_1$ (39)

$\beta_3 = \alpha_2 p_1 P_1^+ - \alpha_1 p_2 P_2^+$ (40) , $\beta_4 = \alpha_3 p_1 P_1^+ - \alpha_1 p_3 P_3^+$ (41)

2.3.2 Discussions about the results

In all the following cases the initial temperature of the solid is 1000°C, the initial temperature of the fluid is 20°C and the ambiant temperature is 20°C. We have fixed :

$Pr = 0.7$ $\xi = 0.4$ $\varepsilon = 0.6$ $\alpha = 0.6$

The fig. 2 b represents the variation of the profile of the temperatures T and θ of the solid and the fluid respectively according to X^+ for $Re = 0.5$, $Re = 3.0$ and $Re = 20$. The solid material is limited to $X^+ = 10$. The thermal conductivity λ_s^* and λ_F are fixed respectively to 2.0 and 0.04 kcal/h.m.k

On the fig. 2c we show that, everything otherwise equal, an optimal Reynold's number exists for each value of X^+ for which the temperature θ of the gas on the exit face is maximum

The fig. 2 d represents the profiles of the temperatures T and θ according to X^+ for $Re = 10$, $X^+ = 10$ and for two values of the ratio of the thermal conductivities $\frac{\lambda_F}{\lambda_s^*} = 0.02$ and $\frac{\lambda_F}{\lambda_s^*} = 0.08$. We notice that the temperature of the gas on the exit face increases with λ_s^*.

On the fig. 2e we represent for $Re = 10$, $\lambda_F = 0.04$ kcal/h.m.k and $\lambda_s^* = 2.0$ kcal/h.m.k the variation of the profile of the temperatures T and θ according to X^+ for seven values of X^+. We notice that the more X^+ increases, the more the temperature θ of the gas increases to a maximum value. In this case this value is reached from $X^+ = 10$.

- 8 -

REFERENCES

1 - F. ROCARIES -
Thèse de docteur-ingénieur, 3ème cycle (PERPIGNAN - France).
Modélisation informatique d'un système de chauffage solaire pour l'habitat.

2 - P. GRANIER, B. N'DOYE, F. ROCARIES, M. DAGUENET -
Modélisation sur ordinateur de systèmes de distillation : méthodologie et applications.
Communication à la XVIe Rencontre internationale de la C.O.M.P.L.E.S., TUNIS, octobre 1977.

3 - B. N'DOYE -
La distillation solaire (1ère partie).
Cours post-maîtrise U.N.E.S.C.O. Laboratoire de Thermodynamique et Energétique de l'Université de PERPIGNAN (France).

4 - G. CHARET -
Cours d'analyse numérique.
Société d'Enseignement Supérieur 75007 PARIS

5 - P. CALVET -
Echanges thermiques dans les lits poreux.
I.F.C.E. - Journées de la Transmission de la chaleur - Juin 1961.

LOW COST SPIRAL SHAPED SOLAR COLLECTOR

P.K.C. Pillai and R.C. Agarwal
Centre of Energy Studies
Department of Physics
Indian Institute of Technology
New Delhi-110029.(India).

ABSTRACT

The details of the design and development of an inexpensive solar collector system are given in this paper. A spiral-shaped plastic tube coated with a black paint constitutes the solar absorber in the above collector. The above structure is enclosed in a wooden box with 4" thick glass wool insulation at the bottom and also provided with a glass cover at the top. Water is used as a working fluid. For high temperature applications with higher efficiencies, plastic tubing is replaced by a metallic tubing. Studies on the characteristics of this solar collector with and without selective absorptive coatings are also being made.

INTRODUCTION

The flat-plate solar energy collector essentially consists of a blackened absorber plate to which are attached tubes or fins for energy removal. The blackened side exposed to the sun is covered by one or more parallel air spaced glass plates to reduce heat losses to the surroundings. As the operating temperature of the collector goes up, convection and radiation losses from the collector to the surroundings increase correspondingly and the amount of energy available for utilization becomes quite small.

Kusuda [1] and Minardi [2] have developed some collectors completely made of glass or plastic for solar water heaters. They have used black liquid as the working fluid, which directly absorbs the solar radiation. The main drawback with plastic is that it becomes opaque with the exposure to sun light, thus the efficiency of the solar collector is considerably reduced with time. Besides this a motor is required for recycling of the working fluid. Various configurations with glass tubings are possible: Circular spiral, rectangular, zig-zag etc. In the case of circular spiral and rectangular patterns, a significant amount of collector area is wasted due to technical difficulties in making such patterns of glass.

Details of a spiral shaped solar collector system which overcomes the above shortcomings are presented in this paper. A spiral shaped plastic tube coated with a black paint constitutes the solar absorber in the above collector. The tube size is chosen to provide uniform flow for effective transfer of heat to the working fluid. As the

Fig.1. Experimental Installation

PKC

liquid moves progressively towards the centre, its temperature gradually rises and in the process it absorbs a part of the heat transferred along a radial direction from the centre of the collector, thus reducing the heat losses.

ECONOMICS OF A SOLAR COLLECTOR

For solar energy to be competitive with other conventional energy sources, collector costs must be quite low. In the conventional flat_plate collectors, tubes or fins are attached to the blackened absorber plate for heat removal. The usual method for bonding the tubes to the absorber plate is either soft soldering or gas welding, both of them are very expensive. Besides this the soldered or gas welded bond may develop cracks due to thermal stresses associated with heating and cooling of the structure. Thus the conventional flat-plate solar collectors require technically skilled workers for usual repairing of the collector. It turns out that even under ideal conditions, it takes long time for a square foot of collector to absorb a dollar's worth of heat. Obviously, for solar energy to be competitive with other conventional energy sources, the investment per square foot divided by the life in years must be only a fraction of a dollar. Naturally, the solution can only result when efficient collector design is combined with cheap long lasting easily available materials. Economic studies suggest that the metal flat-plate solar collector costs 50 dollars or more per sq. meter [3]. In this paper details of a collector system are given which is economically competitive.

DETAILS OF CONSTRUCTION

The experimental installation is shown in Fig.1 and the details of the construction of the spiral shaped solar collector are given in Figs. 2 and 3. It is designed to be flexible so that the tilt and the orientation can be changed as required. The absorber is a spiral shaped plastic tube coated with a black paint. For the experimental purpose, a plastic tubing of 3/8" diameter was used for preparing a 3.1 sq.foot area of structure for the collector. The absorber system is enclosed in a wooden box with 4" thick glass wool insulation at the bottom. 1" thick insulation was also provided at the sides to reduce the side losses. One glass sheet of 1/8" thickness is provided at the top at a distance of 1" from the absorber plate to reduce the heat loss from the absorber plate due to re-radiation and convection.

EXPERIMENTAL SET-UP

The experimental set-up is shown in Fig.1. For the evaluation of the day-long performance of the collector, water was taken from a constant head supply tank to provide water at a constant flow rate. The flow was controlled by means of a manually adjusted control valve. Plastic tube was used to connect the inlet end of the collector to the storage tank, which was kept in the sun. As a result of this, the water inlet temperature was slightly above that of the ambient temperature. Temperatures of the ambient, inlet and outlet water were measured by calibrated copper-constantan thermocouples connected to a D.C. Microvoltmeter.

To evaluate the overall performance of the collector, water at the inlet end was supplied from a constant temperature bath. Temperature of the bath was controlled by a water heater immersion rod connected to a variac. To reduce the effect of wind over the collector performance during testing, the collector was kept at such a site where wind had no appreciable effect over the collector performance.

EXPERIMENTAL OBSERVATIONS

Experiments were carried out to test the performance of the spiral shaped plastic tube solar collector with and without selective coating. The selective coating had $\alpha(S) = 0.96$ and $\epsilon(212^{o}F) = 0.8$. In both the cases rise in water temperature with the variation in solar flux during the day was studied. For this temperature and flow measurements were taken after every half an hour from 0830 to 1700 hours during the month of April 1978 in New Delhi, India. Temperature profiles for the black painted and selectively coated spiral shaped plastic tube solar collector are given in Figs. 4 and 5 respectively. From the data of Fig.4, it is found that the average collector efficiency of the spiral shaped solar collector with black paint is 68.6% with water heated from 98.6°F to 136.2°F. From the data of Fig.5, it is found that the average collection efficiency of the spiral shaped solar collector with selective coating ($\alpha(S) = 0.96$, $\epsilon(212^{o}F) = 0.8$ is 73.6%, with water heated from 97.9°F to 139.3°F. Whereas for metal-plate types of flat plate collectors the average collector efficiency is 64.8%, with water heated from 101.3°F to 136.4°F [4] . Fig. 6 shows the variation in rise in water temperature with the variation of radiation intensity during the day for spiral shaped solar collector with and without selective coating. Two curves are obtained for each type of collector as shown in Fig. 6. The lower curves are obtained when the radiation intensity goes up. The upper curves are obtained when the radiation intensity after reaching a certain maximum goes down. Different slopes in the forenoon and the afternoon for the same collector and same radiation intensity, are due to the thermal inertia of the solar collector. Thermal inertia implies that a solar collector has a certain heat capacity, set by the mass and types of materials used to fabricate a unit area of collector. In general, a collector cools down to the environmental temperature during the night. One must each day put in enough thermal energy to bring this mass upto the operating temperature of the collector.

The stagnation temperatures were also measured in the case of the above collectors with and without selective coating. In the case of

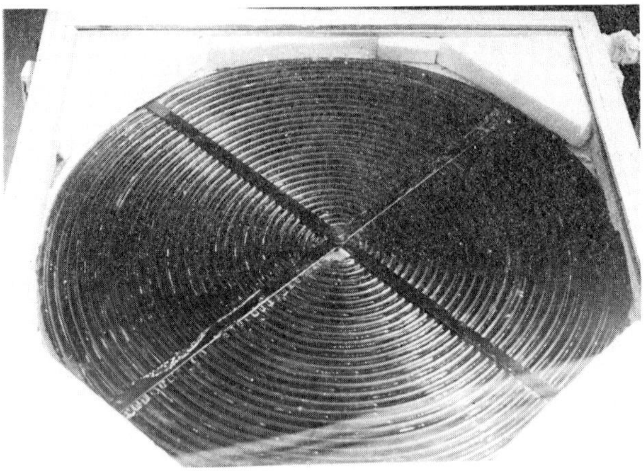

Fig.2 Top view of the spiral shaped solar collector absorber plate.

Fig. 3. Cross - Sectional View of The Spiral - Shaped Solar Collector

Details of construction of the spiral shaped solar collector

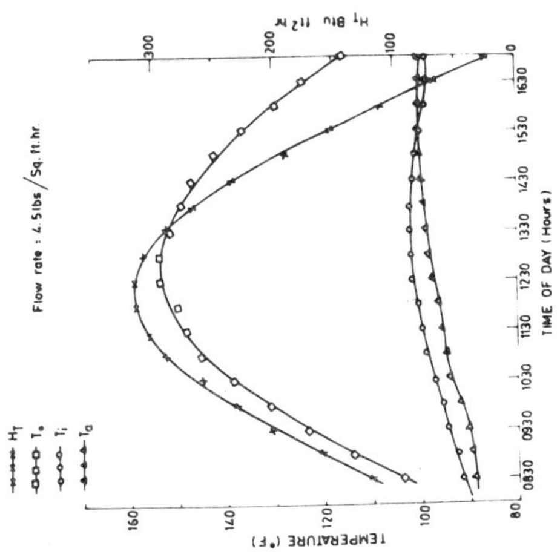

Fig 4 Temperature Profile For A Black Painted Spiral - Shaped Plastic Tube Solar Collector

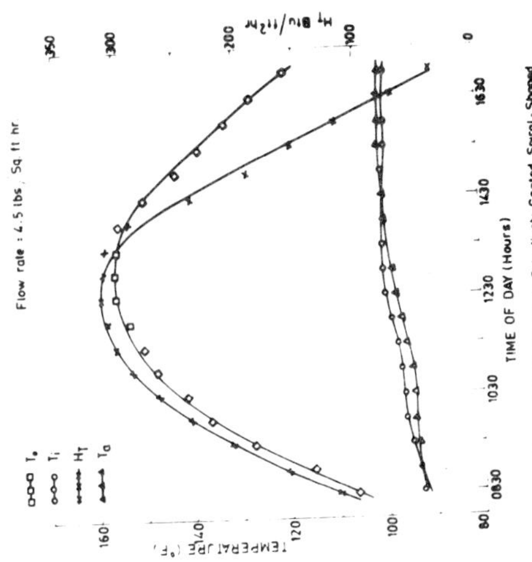

Fig 5. Temperature Profile For A Selectively Coated Spiral - Shaped Plastic Tube Solar Collector

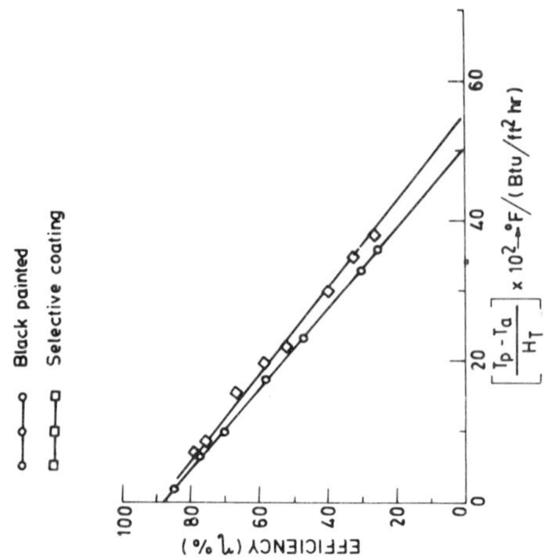

Fig.7. Collector Performance

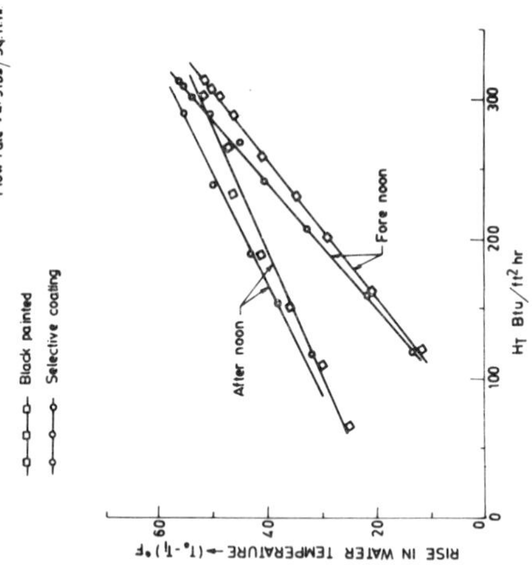

Fig. 6. Temperature Rise Vs Radiation Intensity For Spiral - Shaped Solar Collector

the black painted spiral shaped plastic tube solar collector, the stagnation temperature was 186.8°F, while in the case of selective coating the stagnation temperature was 196.3°F. Each collector was tested for several days and efficiencies were computed by using the equation (1)

$$\eta(\%) = \frac{\dot{m}\, C_p(T_o - T_i)}{H_T} \times 10^2 \qquad (1)$$

For the spiral shaped plastic tube solar collector, collector efficiencies with and without selective coating are shown in Fig.7.

CONCLUSION

The spiral shaped solar collector differs from the convential flat-plate solar collectors in the following ways:

(1) Possibility of corrosion of the collector has been completely eliminated by using a plastic tubing absorber system.

(2) The fabrication process is very simple and economical and requires no skilled workers or welding units as in the case of other flat-plate solar collectors. In the metal flat-plate solar collectors the usual method for bonding the tubes to the absorber plate is either soft soldering or gas welding, both are expensive. Besides this the soldered or gas welded bond may develop cracks due to thermal stresses associated with heating and cooling of the structure. However, in the case of the plastic structure no such problem arises.

(3) Because of the spiral-flow pattern of the liquid progressively towards the centre, the hottest part of the collector is in the centre. As the liquid moves progressively towards the centre, its temperature gradually rises and in the process it absorbs a part of the heat transferred along a radial direction from the centre of the collector, thus reducing the heat losses from the absorber plate.

(4) The above collector system is very light and hence its transportation is easier. Besides this it can be installed on light structures. But for metallic flat-plate solar collectors, strong foundations are required. The spiral-shaped plastic tube solar collector cost may be substantially reduced if lighter, cheap and easily available materials are used for providing insulation at the bottom of the absorber plate. For example, in place of glass wool, air inflated plastic pads may be used.

Evaluation of the collector performance shows that the spiral shaped plastic tube solar collector is more efficient and economically competitive with other conventional flat-plate solar collectors. By using metallic tubing, still higher temperatures and higher efficiencies may be attained. This type of solar collector may be used to produce domestic hot water, provide heat for an industrial process, heating buildings, regenerate an absorption cooling unit or drive a turbine with suitable modifications.

ACKNOWLEDGEMENT

One of the authors (RCA) expresses his thanks to Dr.S.D. Loiwal, Principal, M.M. College, Modinagar for granting leave for taking up an F.I.P. fellowship under the U.G.C. which helped him in pursuing his studies.

NOMENCLATURE

C_p	Specific heat of the working fluid.
$\dot{m}$	Mass flow rate of the working fluid.
T_a	Ambient air temperature.
T_i	Inlet collector fluid temperature.
T_o	Outlet collector fluid temperature.
T_p	Average of inlet and outlet collector fluid temperatures.
H_T	Total solar radiation on the tilted collector.
η	Percentage collector efficiency.
$\epsilon(T)$	Total hemispherical emittance of the absorber surface.
$\alpha(S)$	Solar absorptivity of the coating.

REFERENCES

1. T.Kusuda, Solar Water Heating in Japan, Proc. Solar Heating and Cooling for Buildings Workshop, Washington, D.C. (March 1973).

2. John E. Minardi and Henry N. Chuang, Performance of a Black Liquid Flat-Plate Solar Collector, Solar Energy, Vol.17, 1975.

3. Theodore B. Taylor, Prospects For World-Wide Use of Solar Energy, Jan. 1978.

4. R.S. Chauhan and V. Kadambi, Performance of a Collector-Cum-Storage Type of Solar Water Heaters, Solar Energy, Vol. 18, No.4, 1976.

ENERGY REQUIREMENTS OF AN ALUMINUM SOLAR COLLECTOR

J.E. Robinson and W. Ashton
Department of Man-Environment Studies,
University of Waterloo,
Waterloo, Ontario, Canada N2L 3G1

ABSTRACT

It has been assumed that active solar collectors of many types will become much more extensively used in the future as a way of enhancing the gain of solar energy for space heating. However, while the economic costs of active collector installation have been examined by many, the energy implications of collector construction and installation, especially when some collectors are being manufactured from such energy intensive materials as aluminum, have not been investigated.

From this energy analysis of an aluminum solar collector installed in New Brunswick Power's new office building in Shediac, Canada, it appears that the collector system uses almost as much energy in its manufacture and installation as it will produce in four years of its expected lifetime if the steel storage tank and particularly the aluminum are not recycled. On the contrary, if recycling of materials takes place, the collector will access substantially more energy than it consumes over a period of the first two years.

Thus, the analysis of collector systems for their total energy implications should be undertaken, as a program to accelerate mass production of collectors in a period of substantial energy supply difficulties conceivably could exacerbate the problem for a number of years. As well, the energy viability of one technology e.g. aluminum solar collectors, for moving towards a "conserver society" can be enhanced significantly by parallel developments in others, e.g. recycling.

SOLAR ENERGY - A "FREE" ENERGY SOURCE

Active solar collectors of many types are anticipated to become more extensively used in the future as a way of providing hot water and space heating at an economical cost. This change in technologies is heralded as part of a shift towards a "conserver society".[1] In this shift, solar collectors are seen as one of the principal means of substituting renewable energy forms in place of the oil and natural gas, and perhaps some electricity, currently used.

Manufacturing Process		Energy Required
BAUXITE		
BAYER PROCESS	- crushed - filtration - liquor - calcination	heat 1200°C gas, oil, coal electricity
SMELTING	- reduction cell - molten skim - degassing	heat 1000°C gas, oil, coal electricity
RECYCLING	- recovery - separation - shredded - kiln dried - smelting	heat 800°C gas, oil electricity
SHEET INGOT ROLLING	- gauging - rolling - embossing - cutting	heat water
PRODUCT		
PACKAGING		
SHIPPING		

Fig. 1 Aluminum Manufacturing Process

Operation of a solar collector requires very little non-renewable energy, hence solar collectors are often considered a "free" energy source.

However, some energy is bound up in manufacturing the solar collector. When solar collectors began to be manufactured from aluminum, one of the most energy-intensive materials to manufacture, it seemed prudent to analyze the energy consumption of an installation of an aluminum solar collector to ascertain the length of time which would be required to recover the energy invested in manufacture.

For this case study, the recently constructed New Brunswick Electric Power Commission building in Shediac, Canada was selected. The building incorporates a major installation of 1260 square feet of aluminum water-heating solar collectors. Also, this project was part of a Commission monitoring program in conjunction with two other buildings without active solar collectors--one conventional in Buctouche, and one using an air-to-air heat pump in Sackville. The Shediac building, which contains 7000 square feet of office space and 5000 square feet of warehouse space, is heated using two water-to-air heat pumps working from a 4000 gallon water heat storage and supplemented by electrical resistance heating.

Each of the 105 solar panels measures two feet by six feet, and for purposes of the analysis, are composed of three major components: anodized aluminum, single plate glass, and fibreglass insulation. There are numerous other minor components on a panel which consume energy, but the amounts were considered negligible: for example, nails and fittings and weatherstripping. In addition, the heat pumps and ducting are not included because some type of heating system would be required in the building with or without the solar collector system.

ENERGY ANALYSIS OF SOLAR PANEL COMPONENTS

Aluminum Absorber Plate

Summarized in Fig. 1 is the complex process of manufacturing aluminum from bauxite and an indication of the types of energy required which make aluminum one of the most energy-intensive manufacturing industries. Estimates of the energy requirements to manufacture one pound of aluminum from bauxite range from 102,000 BTU to 126,000 BTU with an average of 113,000 BTU.[2]

Figure One also illustrates that recycled aluminum uses substantially less energy. In fact, recycling requires only four to five percent of the total energy of virgin aluminum; for example, about 5100 BTU per pound.[3] In spite of this difference in energy consumption, it has been estimated that only about 8 per cent of aluminum is presently being recycled.[4]

As shown in Table 1 and Table 2, the absorber plates made of 0.06" 1100 alloy produced by Reynold's Aluminum, Virginia, U.S.A., would consume 96,700 BTU/ft^2 per panel or 122,000,000 BTU for the entire collector surface.[5] If this collector were manufactured using recycled aluminum, the totals would be only 4,330 BTU/ft^2 or 5,490,000 BTU total for all the aluminum content.

Anodizing each square foot of the absorber plates requires 1.25 kwh. or 4270 BTU. This means that 5,370,000 BTU are required in total for the panels.

Glass and Fibreglass

An estimate of the energy requirement for glass production, including raw material mining and transportation, is 8250 BTU per pound.[6] Transporting a pound of finished glass uses about 252 BTU, making a total of 8500 BTU per pound of glass.[7] As each panel consists of one-eighth inch glass, each panel uses about 16.5 lb. of glass.[8] Thus, the total energy required per square foot of panel is 11,700 BTU and the total for the system is 14,800,000 BTU.

Information on energy costs of fibreglass was not available, consequently it was assumed the energy requirements would be no greater than that of glass. Each solar panel contains 2.5 inch fibreglass batt insulation. The total amount of energy used to produce the fibreglass would be approximately 1,910 BTU/ft^2 or 2,400,000 BTU for the entire project.[9]

Other System Components

P.V.C. piping, fibreglass pipe wrap and the steel storage tank are required for the operation of this solar heating system. Approximately 200 feet of three inch and 540 feet of two inch P.V.C. piping (schedule 80) require about 1,700,000 BTU.

One inch fibreglass pipe wrap surrounding all this pipe has an energy cost of about 302,000 BTU.

The 4000 imperial gallon plate steel tank weighing about 2.5 tons requires 116,500,000 BTU if the steel were produced from iron ore, and 49,650,000 BTU if the steel were produced from shredded scrap in an electric furnace.[10]

Total Energy Requirements

The total energy requirements for the solar collector system revealed in Table 2, are approximately 2.63×10^8 BTU without metal recycling, and only 7.97×10^7 BTU with metal recycling.

When related to the collector area, the energy required without and with metal recycling are 2.09×10^5 BTU/ft^2 and 6.3×10^4 BTU/ft^2 respectively.

TABLE 1

Manufacturing Energy Requirements
of an Aluminum Solar Panel

Component	Total BTU/ft^2 (all virgin material)	Total BTU/ft^2 (recycled metal)
Aluminum Absorber Plate*	96,700 (77%)	4,330 (20%)
Anodizing	4,270 (3%)	4,270 (19%)
Glass	11,700 (19%)	11,700 (53%)
Fibreglass	1,910 (1%)	1,910 (9%)
TOTAL	115,000 (100%)	22,200 (100%)

*Using aluminum from bauxite

TABLE 2

Manufacturing Energy Requirements
for the Shediac Aluminum Solar Collector System

Component	Total BTU (all virgin material)	Total BTU (recycled metal)
Aluminum Absorber Plate*	122,000,000 BTU (46%)	5,490,000 BTU (7%)
Anodizing	5,370,000 BTU (2%)	5,370,000 BTU (7%)
Glass	14,800,000 BTU (6%)	14,800,000 BTU (19%)
Fibreglass	2,400,000 BTU (1%)	2,400,000 BTU (3%)
Piping	1,700,000 BTU (0.7%)	1,700,000 BTU (2%)
Fibreglass Pipe Wrap	302,000 BTU (0.1%)	302,000 BTU (0.4%)
Steel Tank*	116,500,000 BTU (44%)	49,650,000 BTU (62%)
TOTAL SYSTEM	2.63×10^8 BTU (100%)	7.97×10^7 BTU (100%)
TOTAL SYSTEM/COLLECTOR AREA	2.09×10^5 BTU/ft^2	6.3×10^4 BTU/ft^2

*Using metal from ore

SOLAR ENERGY GAIN

As a result of technical problems with the monitoring instruments at the Shediac building, exact measurement of the heat gain from the collector system was not available. However, using data from a similar building in Sackville and computer projections for the Shediac building, an estimate of the reduced heating load attributable to the solar collectors is 73 x 10^6 BTU annually.[11]

It can be seen that approximately four years of operation will be required to regain the energy consumed in manufacture of the solar collector system from raw materials. An accelerated mass production of collectors in a period of energy supply difficulties could conceivably exacerbate the problem in the short run. To check this, the quantities of the various forms of energy consumed by this system should be estimated for various production programs. Comparisons then could be made with alternative collector systems using different materials and present production technology.

Also, apparent from Table 2, is that the use of recycled metal substantially reduces the energy pay-back period to just over one year. In this case, then, the energy viability of one new technology, a solar collector system, is enhanced significantly by parallel developments in others such as metal recycling.

REFERENCES

1. Science Council of Canada. Canada as a Conserver Society: Resource Uncertainties and the Need for New Technologies.

2. The following references were used to establish energy required for processing aluminum:

   A. Hurlich, "An Overview of Materials Shortages", p. 7.

   G. Bourcier, Reynolds Aluminum. Private communication.

   H. Mokino. Consumer Goods: A Thermodynamic Analysis of Packaging. Transport and Shortage.

   R. Stephen Berry and Margaret F. Fels. The Production and Consumption of Automobiles, p. 86.

3. R. Stephen Berry and Margaret F. Fels, p. 19.

4. Reynolds Aluminum. Recovery of Non-Ferrous Metals from Solid Wastes.

5. Estimated alloy weight is 0.86 pounds per square foot; G. Bourcier,

Manager, Resource Recovery Products, Reynolds Aluminum, Virginia. Private communication.

6. G. Boettner, Corning Works, New York. Private communication.

7. Both of these references estimate the energy required to produce glass:

   B. Hannon. System Energy and Recycling.

   R. Stephen Berry and Margaret F. Fels. The Production and Consumption of Automobiles.

8. Chemical Engineering Handbook. Weight of glass is 130 to 140 pounds per square foot.

9. The density of insulation was estimated at 0.224 pounds per square foot by the authors.

10. Using energy cost values from R. Stephen Berry and Margaret F. Fels, pp. 18, 30.

11. Personal communication with officials of New Brunswick Power Commission, July 6, 1978. Monitoring data is expected to be available in 1979.

BIBLIOGRAPHY

Berry, R. Stephen and Margaret F. Fels. The Production and Consumption of Automobiles. Illinois: University of Chicago, 1972. A report to the Illinois Institute for Environmental Quality.

Hannon, B. System Energy and Recycling. Illinois: Centre for Advanced Computation, University of Illinois, 1972.

Hurlich, A. "An Overview of Material Shortage", A.S.M. News. September 1973, pp. 7-8.

Mokino, H. Consumer Goods: A Thermodynamic Analysis of Packaging, Transport and Shortage. Illinois: Illinois Institute for Environmental Quality.

Reynolds Aluminum. Recovery of Non-Ferrous Metals from Solid Wastes.

Science Council of Canada. Canada as a Conserver Society: Resource Uncertainties and the Need for New Technologies. Ottawa: Supply and Services Canada, 1977.

OPTIMUM INCLINATION OF LINEAR REFLECTORS

B.A. Khalefa and A.S. Hegazy
Menoufia University
Shebin El Koum, Egypt

ABSTRACT

This paper presents a computer model for determining the optimum angle of inclination of a linear reflector for a specified site and a certain time period. This optimum angle of inclination is selected on the basis of maximum solar radiation energy reflected from the concentrator and received by its target during the testing time period.

This proposed model was applied to some sites in Egypt for certain time periods of the month and year.

INTRODUCTION

One of the main approaches to solar thermal power generation is the solar farm, with large numbers of linear reflectors focusing solar radiation on long pipes which collect the heat. Accordingly, the linear reflectors play a very important role in energy production using solar radiation.

The linear reflector is usually fixed at an angle of certain inclination to the horizontal, (from 0 to 90 degrees), and facing the equator. It rotates about one axis with the azimuth angle of the sun. This angle of inclination has a big effect on heat reflected by the concentrator and received by its target.

This paper presents a methodology for finding the optimum angle of inclination of a linear refelctor for a specified site and a certain time period (year, month, etc.). The optimization is based on maximum solar radiation energy reflected by the concentrator and incident on its target.

THERMAL ANALYSIS OF LINEAR CONCENTRATORS

For the purpose of obtaining an expression for the intensity of solar radiation reflected by a linear reflector and incident on its target, the following analyses are made. To simplify the analysis, we assume that:

1. The optical and thermal losses are neglected.
2. The target heat absorber is cylindrical.
3. The absorber is long and wide enough to receive all solar radiation energy reflected by the concentrator.

Thus, for a site on the northern hemisphere, suppose that, beam solar radiation of intensity I_N on normal surface having zenith angle Z is allowed to fall on a linear reflector surface. Also, suppose that, this reflector has an angle of inclination to the horizontal θ, and faces north-south. (see Figure 1)

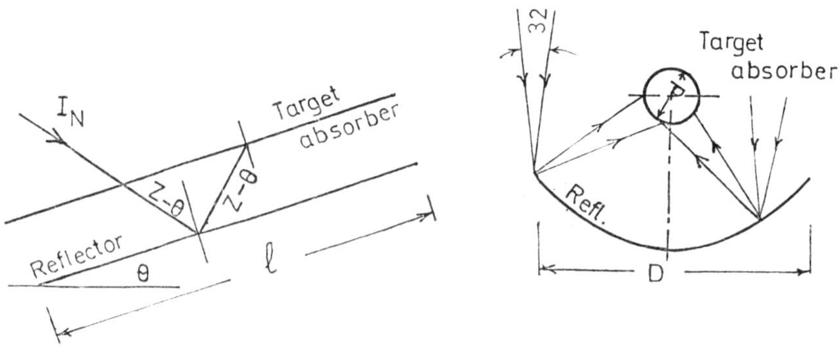

Figure 1 Reflection Properties of Solar Radiation from a Linear Concentration

The solar radiation power reflected by the concentrator surface Q_r could be found as:

$$Q_r = I_N D\ell \tag{1}$$

The solar radiation power received by the concentrator target Q_a may be given by:

$$Q_a = I_N D\ell \, \mathrm{Cos}(Z - \theta) \tag{2}$$

The intensity of solar radiation received by the concentrator target I_O could be found from:

$$I_O = I_N \frac{D\ell}{d\ell} \mathrm{Cos}(Z - \theta)$$

$$= I \, C_a \frac{\mathrm{Cos}\ (Z - \theta)}{\mathrm{Cos}\ Z} \tag{3}$$

The previous equation shows clearly that the intensity of solar radiation incident on the concentrator target is mainly affected by the concentrator angle of inclination.

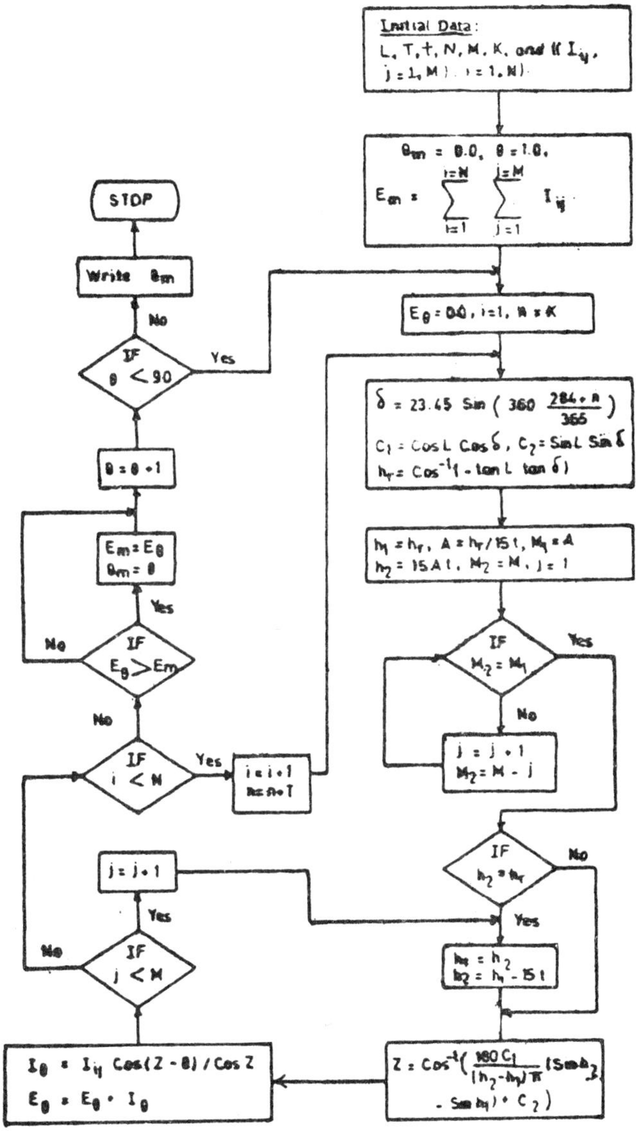

Figure 2 Flow Chart For Obtaining the Optimum Angle of Inclination of a Linear Reflector

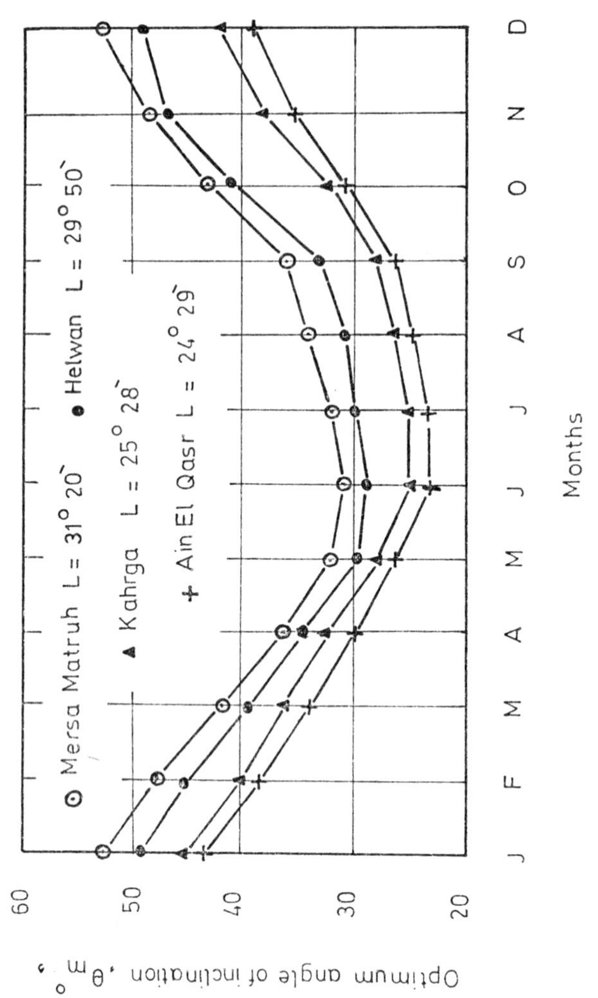

Figure 3 Monthly Optimum Angles of Inclination of a Linear Reflector for Some Sites in Egypt, 1969.

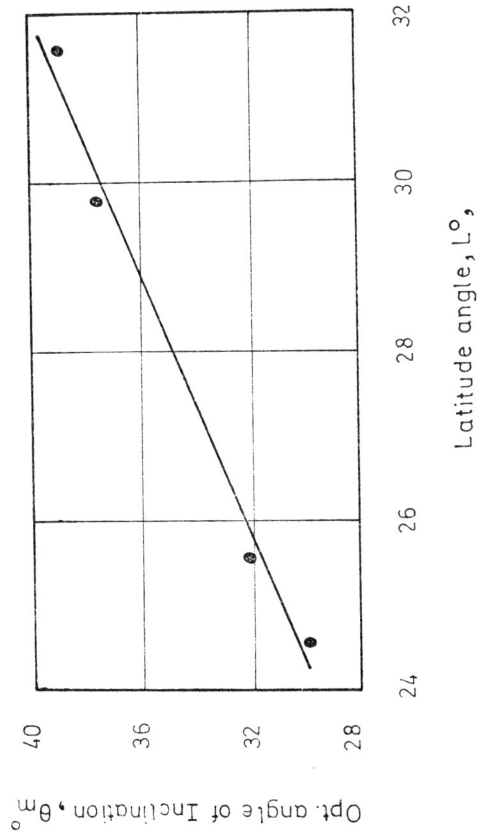

Figure 4 Annual Optimum Angle of Inclination of a Linear Reflector for Some Sites in Egypt, 1969.

MODEL CONSTRUCTION

For obtaining the optimum angle of inclination of a linear reflector, based on maximum solar radiation energy reflected by the concentrator and incident on its target for a certain time period and at a specified site, the following computer model has been constructed, see Figure 2. This model computes the necessary average solar angles for each time interval of the solar radiation records at the site of question [2], see Figure 2. From this it computes the intensity of solar radiation energy incident on the reflector target during the testing time period for each angle of inclination from 0 to 90 degrees using equation (3), and considering $C_a = 1.0$. Then, a comparison is made to select the optimum angle of inclination at which maximum solar radiation energy is incident on the concentrator target.

ILLUSTRATIVE EXAMPLES

The previous proposed model was run for four sites in Egypt, using the data of monthly averages of hourly solar radiation intensities on a horizontal surface which were recorded by their meteorological stations in the year 1969 [3]. The optimum angles of inclination of these sites for each month were determined as given in Figure 3. This figure shows that the optimum angle of inclination of a linear reflector is high for winter months where the zenith angles are also high. On the contrary, this optimum angle is low for summer months where the zenith angles are low.

Also, the optimum angles of inclination of these sites based on a year time period were evaluated as shown in Figure 4. This figure reveals that this optimum angle varies for each site. But, in general, we can fairly conclude that this angle may be found for a site in Egypt by adding 6 degrees to its latitude angle.

CONCLUSION

A computer model has been proposed for determining the optimum angle of inclination of a linear reflector for a specified location and a certain time period. The optimization is based on maximum solar radiation energy reflected by the concentrator and incident on its target.

Applying this proposed model to some sites in Egypt shows that the optimum angle of inclination is high for winter months where the zenith angles are high. But, it is low for summer months where the zenith angles are low. Also, it shows that the optimum angle of inclination of a linear reflector for a site in Egypt based on a year time period is higher than the latitude angle of this site by 6 degrees.

REFERENCES

1. Russell, J.L., "Investigation of a Central Station Solar Power Plant", Proceedings of the Solar Thermal Conversion Workshop, Washington, D.C., USA, January 1973.

2. Beckman, W.A. and Duffie, J.A., "Solar Energy Thermal Processes", A Wiley Interscience Publication, John Wiley and Sons, 1974.

3. Meteorological Department at Cairo, Egypt, "Monthly Weather Report", 1969.

NOMENCLATURE

C_a	reflector concentration ratio
d	target absorber diameter
D	reflector diamter
h_r	sunrise hour angle
I	beam solar radiation intensity on a horizontal surface
I_N	beam solar radiation intensity on a normal surface
I_θ	beam solar radiation intensity received by the reflector target
K	number of the day of beginning the test from 1 January
ℓ	reflector length
L	latitude angle
M	maximum number of solar radiation intensity records per day from sunrise to solar noon
N	number of days of measuring solar radiation intensity
t	time interval of measuring each solar radiation intensity
T	number of days between two consequent days of measuring solar radiation intensities
Z	zenith angle
δ	declination angle
θ	reflector angle of inclination to the horizontal
θ_m	reflector optimum angle of inclination to the horizontal

SELECTIVE SURFACE COATING WITH BLACK CHROME

T. Vurgun
M.T.A. Solar Energy Research Center
Marmaris, Turkey

ABSTRACT

The M.T.A. Marmaris Solar Energy Research Center has determined that electroplated black chrome has desirable solar selective properties. Black chrome electroplated coating has high absorbtance in the solar spectrum and low emissivity in the thermal spectrum.

Additionally, black chrome coating has been used for decorative finishing purposes. Black chrome is an important coating because the current extensive use of black chrome in the electroplating industry is a durable, and lower cost material used as a solar selective coating.

The spectral reflectance of the black chrome and black painted on steel were measured. The results of the spectral measurements of reflectance showed that the black chrome gave better results than the black paints.

INTRODUCTION

Black chrome plating was studied for the steel-dull nickel-black chrome system. Results of the varying current densities and plating times are given in Table 1. The optimal optical properties were (α = 0.94 and ε = 0.13) which were achieved using a current density of 80 Amp/dm^2 for six minutes.

The low emittance obtained showed that the black chrome coatings were thin. Absorbtance may be increased while keeping the emittance small by increasing the black chrome coating thickness.

DESCRIPTION OF BLACK CHROME PREPARATION

This research used black chrome plated steel panels as 25 x 30 cm. test panels, which were 0.50 mm. The panels were plated in the following sequence.

1. Ten minutes 80°C clean by hot alkaline cleaner.

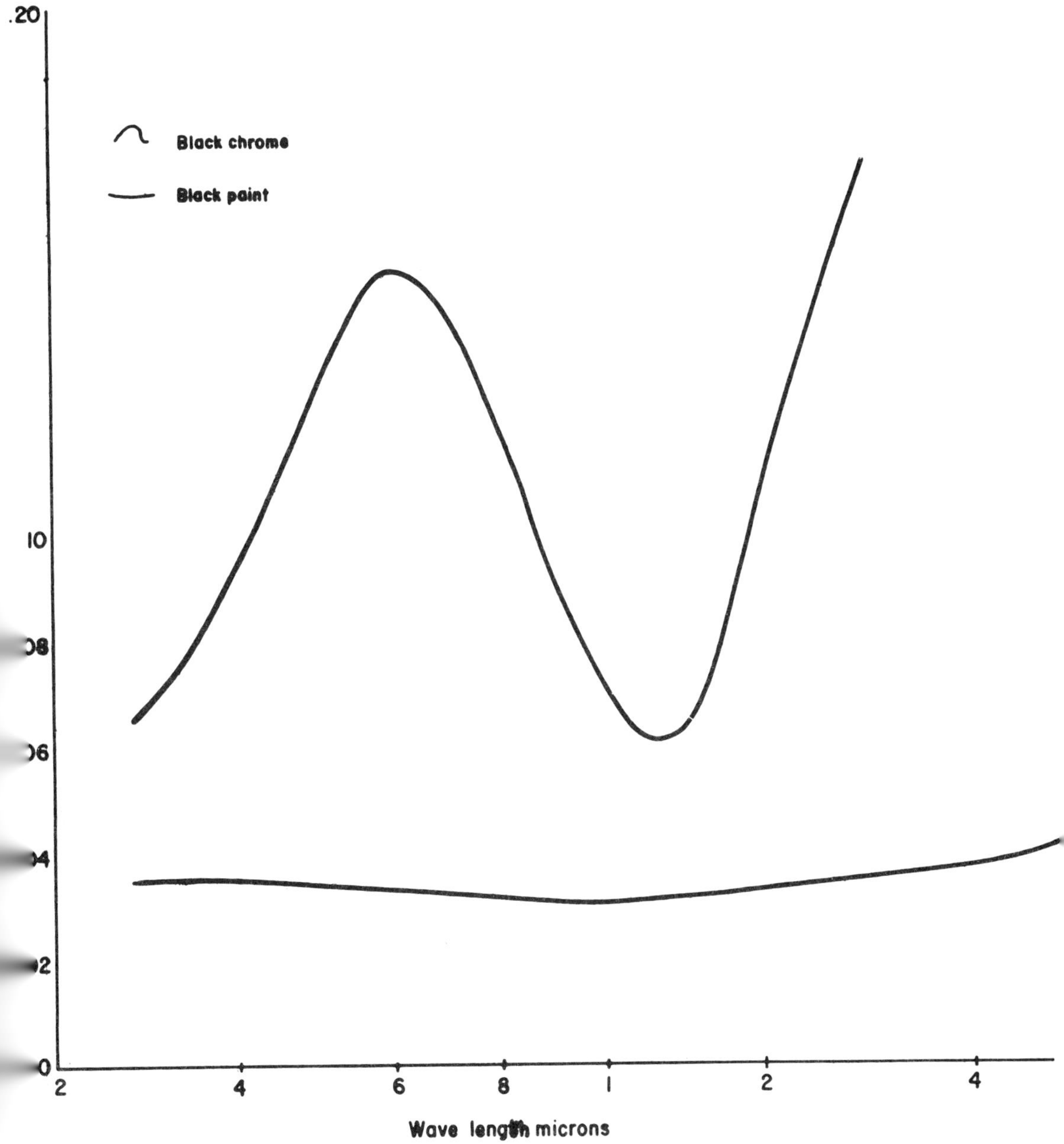

Figure 1 The Results of the Measurements

TABLE 1

EFFECT OF CURRENT DENSITY AND PLATING TIME ON THE OPTICAL PROPERTIES OF DULL NICKEL/BLACK CHROME COATINGS

	Current Density (Amp/dm^2)					
Plating Time	80		120		160	
	α	ϵ	α	ϵ	α	ϵ
2	-	-	-	-	-	-
3	-	-	-	-	0.94	0.14
4	-	-	0.93	0.13	0.93	0.13
5	-	-	0.93	0.14	0.89	0.16
6	0.94	0.13	0.92	0.15	-	-
7	0.93	0.14	-	-	-	-
8	0.93	0.14	-	-	-	-
9	0.91	0.15	-	-	-	-

2. Clean by electrolytic alkaline cleaner 66 Amp./7.5 dm^2, two cycle with acid
3. Dull nickel plated for fifteen minutes at 45 Amp./7.5 dm^2, to deposit approximately twelve of nickel
4. Water rinsed
5. Alcohol rinsed
6. Air dried

The dull nickel and black chrome solution was prepared in accordance with Reference [2].

The spectral reflectance 0.2 to 4 μm of the black chrome and black painted panels were measured with a N.G.1. 217 spectroreflectometer. The results of the measurements are shown in Figure 1.

REFERENCES

1. Edwards, D.K. Gier, J.T., Nelson, K.E. and Roddick, R.D., "Spectral and Directional Thermal Radiation Characteristics of Selective Surfaces for Solar Collectors".
2. Metal Finishing, Metals and Plastics Publications, Inc. (1977).
3. Tabor, H., "Selective Radiation Wave Length Discrimination", Bull. Research Counsel of Israel, (1956).
4. Electroplating Handbook.
5. McDonald, D.E., "Spectral Reflectance Properties of Black Chrome for Use as a Solar Selective Coating", (1974).

CADMIUM STANNATE SELECTIVE COATINGS PREPARED BY SPRAY PYROLYSIS

O.P.Agnihotri and B.K.Gupta,
Center of Energy Studies and
Materials Research Laboratory,
Physics Department,
Indian Institute of Technology, Delhi
New Delhi-110029, India.

ABSTRACT

The infrared emission from solar collectors can be minimised by use of transparent conducting coatings which will allow the visible radiation to go through but reflect the infrared radiation back to the collector. Thermal radiation is the dominating loss mechanism at high operating temperatures and has to be kept at low levels if these collectors are to become economically competitive.

Cd_2SnO_4 is an n-type defect semiconductor in which oxygen vacancies provide the donor states. The free electron concentration exceeds 10^{21} cm^{-3} so that infrared reflectivities are high and the thin coatings will act as transparent heat mirrors. We have prepared cadmium stannate selective coatings by a pyrolytic method in which two parts of 0.2M $CdCl_2.5H_2O$ solution in ethyl alcohol and one part 0.2M solution of $SnCl_4.3H_2O$ in acetic acid has been sprayed onto a heated substrate of glass and quartz. The films were transparent in the visible portion of the spectrum and exhibited interference colours. The absorption data is interpreted due to allowed direct transition of energy gap of 2.8 eV and an indirect interband transition of energy gap of 2.2 eV. The films have sheet resistance of about 100 ohms per square and conductivity is about 1000 Ω^{-1} cm^{-1}.

INTRODUCTION

There has been tremendous interest during the last few years in transparent conductor materials because of their application in solar energy conversion, display technology and electronics [1,2]. For applications requiring a sheet resistivity of 100 ohms per square tin oxide films can be used. Resistance below 10 ohm per square became accessible with the recent development of indium tin oxide coatings[3]. Cadmium stannate has been found to be an outstanding transparent conductor material for high and low sheet resistance including the 1 ohm per square range[1]. Because of its large Burstein shift and relatively high mobility, Cd_2SnO_4 becomes more transparent to visible light as its conductivity increases. Because of their ratio of conductivity to absorption coefficient and because they are strongly adherent, hard and stable they have potential applications as transparent electrodes. These films can also find applications in semiconductor device because of their large band gap and small charge carrier effective mass.

Cd_2SnO_4 powder was prepared by mixing a 2:1 mole ratio of CdO and SnO_2 and heating the oxides at 1050°C for 3-6 hours. Alternatively, a stoichiometric mixed Cd-Sn hydrous oxide can be precipitated with base (NH_4OH) from an aqueous solution of the metal chlorides[2]. The potential utility of cadmium stannate for transparent electrodes was first demonstrated by Nozik[2] who measured electrical and optical properties of amorphous cadmium stannate films prepared by rf sputtering from a polycrystalline Cd_2SnO_4 target onto quartz or pyrex substrates. Haacke[1] prepared polycrystalline cadmium stannate films on various transparent substrates by rf sputtering and achieved high electrical conductivities and high optical transmissions. Inspite of these advancements the optical properties of this material have not been investigated in detail. Nozik has determined the bandgap of this material in the amorphous state. In this investigation we have prepared polycrystalline Cd_2SnO_4 by a chemical spray deposition process and investigated its intrinsic absorption edge.

MATERIAL PREPARATION:

The starting solution was prepared from two parts 0.2M $CdCl_2.5H_2O$ solution in ethyl alcohol and one part 0.2M $SnCl_4.3H_2O$ solution in acetic acid. The mixed solution was sparyed by a double nozzle sprayer onto heated substrate. The volumetric flow of the solution was 30 ml per minute and on each substrate the spraying time was about 20 sec. i.e. 10 ml solution was sprayed for each substrate. During the film preparation, which is done under normal atmospheric conditions, an uncontrollable amount of oxygen is introduced into the films and acts as electron trap. To eliminate this contribution all samples were annealed at about 300°C under high vacuum (10^{-5} torr). The details of the technique of producing thin layers by pyrolytic method have been published elsewhere[4,5]. The films were transparent in the visible portion of the spectrum. They exhibited interference colours and very little scattering. The thickness of the films varied from a few tengths to nearly one micron. The transmission of films was measured by a ratio recording technique which consisted in measuring directly the ratio of transmissions of films of different thicknesses. A thin sample was placed in the reference beam and a thicker sample was placed in sample beam. The net result is equivalent to measuring the transmission of a hypothetical thin film without reflectance losses of a thickness equal to the difference in thickness of those actually used. Percentage transmission of cadmium stannate is shown in Fig. 1.

EXPERIMENTAL RESULTS AND DISCUSSIONS

The energy dependence of the absorption coefficient is of the form

$$\alpha = \alpha_0 \, (h\nu - \Delta E)^x \quad \ldots \quad \ldots \quad (1)$$

where $h\nu$ is the photon energy and ΔE is the energy band gap. The value of x = 1/2 for allowed transitions and 3/2 for forbidden transitions. The transmission of a thin film on a substrate is closely approximated by

$$T = (1-R_1)(1-R_2)(1-R_3)(1+k^2/n^2)\, e^{-\alpha \cdot d} \quad \ldots \quad (2)$$

where R_1, R_2, R_3 are the reflection coefficients at the air thin film, thin films-substrate and substrate - air interfaces respectively. The parameters k,n and d are the extinction coefficient, refractive index and thickness of thin film respectively. For $T < 0.10$, the error introduced in calculation using equation (2) is less than 0.1%. The ratio of transmission of two films of different thickness, which is obtained directly from the ratio recording technique is given by

$$T_{1-2} = e^{-\alpha \cdot \Delta d} \quad \ldots \quad \ldots \quad (3)$$

where T_{1-2} is the ratio of transmissions and d is thickness difference of the two films. One can then accurately calculate $\alpha \cdot \Delta d$ without determining Δd, Eq.(1) can be written to the form

$$\alpha \cdot \Delta d = \alpha_0' \, (h\nu - \Delta E)^x \quad \ldots \quad (4)$$

where α_0' is equal to $\alpha_0 \cdot \Delta d$. One can then proceed to determine x by plotting $(\alpha \cdot \Delta d)^{1/x}$ vs., $h\nu$ to obtain a straight line with a slope of α_0' the abcissa intercept giving the energy band gap ΔE. A satisfactory fit to equation (4) is obtained with cadmium stannate for x = 1/2 corresponding to allowed transition. This is shown in the Fig. 2 for two

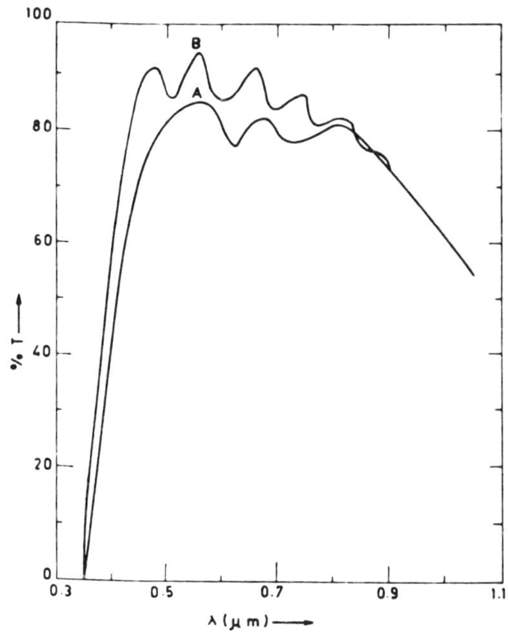

FIG. 1. Optical transmission versus wavelength of Cd_2SnO_4 film. Curve B is taken from Ref. 1. Curve A is obtained in the present investigation.

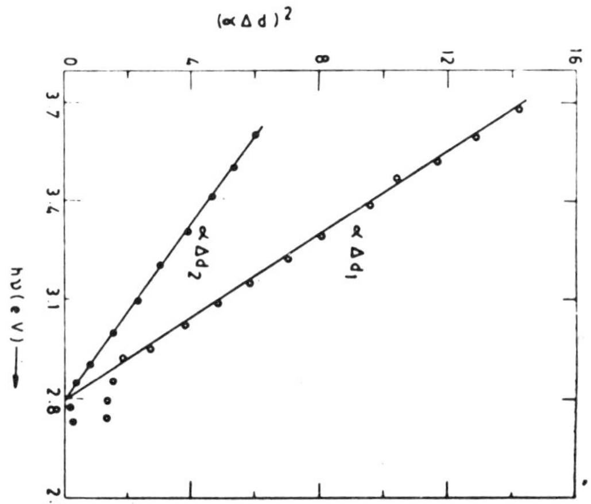

FIG. 2. Square of the absorption coefficient of cadmium stannate thin films as a function of photon energy.

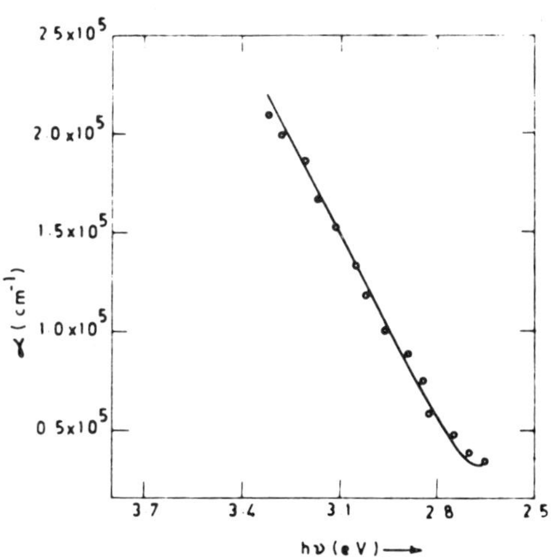

FIG. 3. Absorption coefficient as a function of photon energy in the region of direct band-to-band transition.

combinations of three films of different thicknesses. It is seen that two straight lines are obtained with different slopes corresponding to the different Δd's but having a common intercept at 2.8 eV. The absorption coefficient shown in Fig. 3 was calculated using approximate values of Δ d from interference fringes in the visible. The absorption coefficient reaches values exceeding 10^5 cm^{-1} at the higher photon energies, which is expected for direct allowed transitions.

The energy gap is approximately 2.8 eV deduced above is further suggested by a maximum in the reflectivity curve (Fig.4). The maximum in the reflectivity at about 2.8 eV must be due solely to a maximum in the refractive index because the extination coefficient at that photon energy is quite small. In turn the inter-relations of the optical constants predict a maximum in the refractive index where $d\alpha/d\nu$ has a maximum. This maximum should, of course, occur at $h\nu = \Delta E$ if the absorption is described by eq. (I). We conclude that allowed transitions are observed in Cadmium stannate at room temperature with an energy gap of 2.8 eV.

INDIRECT INTERBAND TRANSITION:

The energy dependence of the absorption coefficient for an indirect inter band transition is crystalline semiconductors, wherein crystal momentum ($\vec{k}$) is conserved by emissions or absorption of phonons is given by

$$\alpha = \alpha_0 (h\nu - \Delta E)^2 \qquad \ldots \quad \ldots \quad (5)$$

Hence, a plot of $(\alpha \cdot h\nu)^{1/2}$ = vs $h\nu$ should be linear, and the intercept of the line on the abcissa at $(\alpha h\nu)^{1/2} = 0$ yields the optical gap ΔE. The result of this plot is shown in Fig. 5. Where $\Delta E = 2.2$eV Eq. 5 also describes the energy dependence of amorphous semiconductors. In the work of Nazik[2] a plot of $(\alpha \hbar\omega)^{1/2}$ vs. $\hbar\omega$ yielded a value of 2.06 eV for amorphous Cd_2SnO_4. In view of the occurence of a direct band to band transition at 2.8 eV., the transition at 2.2 eV in the present investigation has been identified as an indirect interband transition. The direct transition occurs at a higher energy as compared to the 2.06 eV transition in amorphous Cd_2SnO_4. The gap obtained from $(\alpha \hbar\omega)^{1/2}$ vs. $\hbar\omega$ plot is an accurate measure of the lowest energy needed to excite

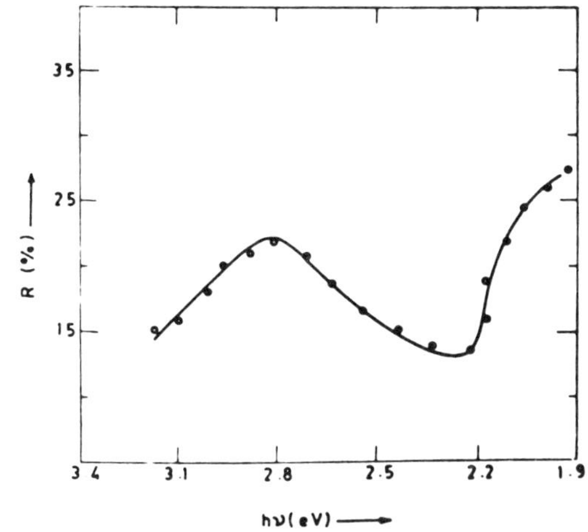

FIG. 4. Reflectivity as a function of photon energy in the 1.9–3.0-eV region.

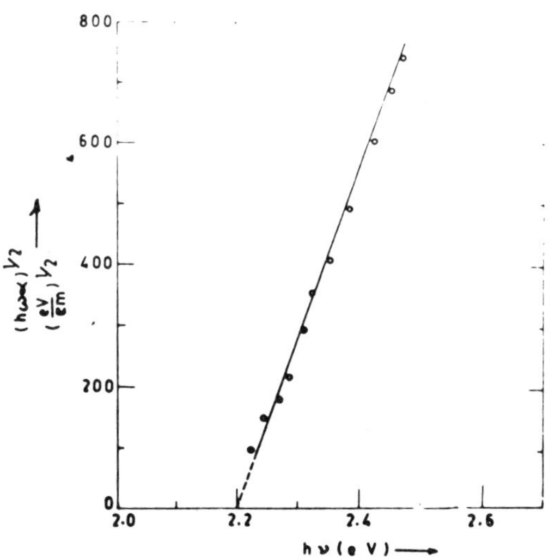

FIG. 5. Indirect interband transition at 2.2 eV.

electrons across the gap. In the amorphous material, this means a transition of electrons from the localized states at the valence band edge to extended states in the conduction band, or a transition of electrons from extended states in the valence band to localized states at the conduction band edge. This type of transition generally occurs at lower energies than the band to band transition in the polycrystalline material. This leads to the amorphous material starting to absorb at lower energies and yielding a shallower absorption edge.

The cadmium stannate films can be used as transparent heat reflectors which are transparent to sunlight (0.3-2.5 μm wavelength) but highly reflective beyond 2.5 μm. Heat reflectors are positioned in front of the absorbing surface and reflects the emitted energy back to the absorber. The parameters determining the performance of a heat reflective coating in a solar collector are total solar transmission T_s and infrared reflectivity R_I. The equivalent parameters α_s (solar absorptivity) and ε_T (total infrared emissinity) are also used to facilitate convenient comparison with selective absorber surfaces. In the present investigation α_s, the solar absorptivity is 0.82 and $\varepsilon_T = 0.09$ at 5 μm. The material is stable in air and at operating temperatures of 300 to 400°C and does not degrade at operating temperatures.

ELECTRICAL PROPERTIES;

The conduction mechanism in thin Cd_2SnO_4 films involves oxygen or anion vacancies. These vacancies are incorporated either during film growth or after deposition by annealing the films in an inert atmosphere or oxygen free environment such as vacuum. Annealing in a reducing atmosphere of forming gas or a partial pressure of hydrogen in an inert gas can also be used to produce oxygen vacancies. In the present investigation the films were annealed under vacuum at 300°C. The annealed films have sheet resistance of about 100 ohms per square and conductivity is about 1000 Ω^{-1} cm^{-1}. In sputter coated Cd_2SnO_4 films resistivity values approaching 3×10^{-4} ohm cm have been obtained. The conductivities observed in spray coated Cd_2SnO_4 do not reach these high values. Here the best values are in the range 500-1000 Ω^{-1} cm^{-1}. Hall measurements with spray coated Cd_2SnO_4 yielded mobilities of about 10-20 $cm^2/V.Sec.$ which are lower than the mobilities measured in sputtered coatings and are presumably due to greater density of lattice imperfections in sprayed films. In annealed films the mobility is found to increase and is about 30 $cm^2/V.Sec.$

CONCLUSIONS

Cadmium stannate selective coatings have been prepared by a low cost pyrolytic process. The absorption data is interpreted due to an allowed direct transition of energy gap of 2.8 eV and an indirect interband transition of energy gap of 2.2 eV. The absorption coefficient reaches values exceeding 10^5 cm^{-1} at the higher photon energies which is expected for direct allowed transitions. The direct energy gap of 2.8 eV is further suggested by maximum in the reflectivity curve. The coatings have optical properties which are suitable for their application in solar collectors. Annealing of the films at 300°C under vacuum improves the electrical characteristics.

ACKNOWLEDGEMENTS

We are grateful to Professor M.S.Sodha, Head of the Centre of Energy Studies for the provision of special facilities which made this work possible.

REFERENCES

1. G.Haacke, Appl. Phys. Lett. 28, 622 (1976).

2. A.J.Nazik, Phys. Rev. B.6, 453 (1972).

3. D.B.Fraser and H.D.Cook, J.Elect. Chem. Soc., 119, 1368 (1972).

4. A.Raza, O.P.Agnihotri and B.K.Gupta, J.Phys. D.10, 1871 (1977).

5. B.K.Gupta and O.P.Agnihotri, Sol.Stat.Comm., 23, 295 (1977).

6. P.Lloyd, Thin Solid Films, 41, 113 (1977).

FIGURE CAPTIONS

1. Optical transmission vs. wavelength of Cd_2SnO_4 film. Curve B is taken from Reference 1. Curve A is obtained in the present investigation.

2. Square of the absorption coefficient of cadmium stannate thin films as a function of photon energy.

3. Absorption coefficient as a function of photon energy in the region of direct band to band transition.

4. Reflectivity as a function of photon energy in the 1.9 to 3.0 eV region.

5. Indirect interband transition at 2.2 eV.

PHASE CHANGE COLLECTOR WALL VERSUS WATER COLLECTOR WALL

L. Bourdeau and A. Jaffrin
Laboratoire CNRS d'Ecothermique Solaire
Observatoire de Nice - 06 FRANCE -

ABSTRACT

Two experimental cells have been constructed to experiment under true weather conditions the thermal behavior of collector - storage walls containing a phase changing material or plain water. Recorded tempera--tures show that the latent heat wall collects solar energy with a much larger efficiency (up to 100%) and stores twice more heat than the water wall of same volume, while acting as a very good temperature filter.

INTRODUCTION

A reliable latent heat material based on calcium chloride hexahydrate was recently developed by the "Laboratoire CNRS d'Ecothermique Solaire" in Nice. Manufactured under the name "chliarolithe" (1), this material offers the following physical properties : melting point 28° C, latent heat of fusion 32 Kcal/Kg, density 1500 Kg/m3 in the liquide phase, apparent change in specific volume after solidification 10%, thermal conductivity 0.12cal/sec- m-° C, specific heat 0,6 Kcal/Kg-°C.
Its mechanical and thermal stabilities were tested over 6000 phase change cycles.
Some applications of this latent heat material to the field of space heating and temperature control have already been studied in laboratory experiments (2,3). The present communication reports on the first kind of experiment performed on a larger scale and in real weather to get a practical estimate of the real benefits that can be derived from the use of latent heat materials in buildings, in the presence of large temperature gradients and fluctuating thermal conditions usually obser--ved in outdoor conditions. It also gave the possibility to check the feasability of producing large quantities of chliarolithe on a construc--tion site with ordinary building equipments.
The experiment consits in testing two distinct collector - storage walls which receive direct solar radiation behind a single glazing and which store calories either in the form of sensible heat or in the form of latent heat. It was considered that the most severe test for the latent heat material would be to compare its thermal performan--ces with those of water (the highest specific heat of usual materials combined with a high thermal conductivity due to free internal convec--tion). Each collector wall acts as the solar collector, heat storage and heating system of a small house module located behind. The front glazing, the collector wall and the house unit constitute an experi--mental cell. Two identical experimental cells were constructed.

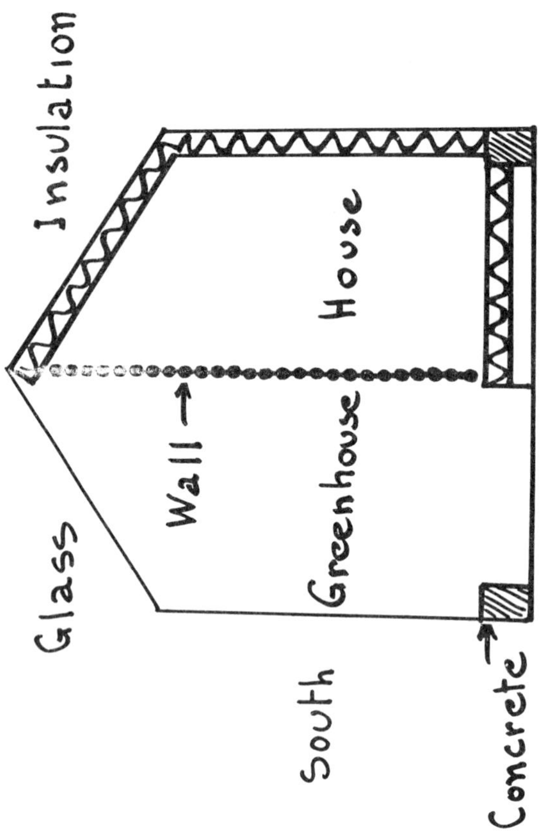

Fig. I : Experimental cell elevation.

DESCRIPTION OF THE EXPERIMENTAL CELLS

For commodity reasons, the use of concrete as building material was avoided, and ready-made materials like glass panels and aluminium profiles were the basic elements of construction of the cells. The glass panels were sealed with silicon ciment to prevent any uncontrol--led air leak. Fig. I shows the geometry of a cell and the position of the collector wall inside a cell. The front part of the cell, facing the south, is treated as a green-house : the east south and west sides are made of 3 mm single glass panels supported by aluminium profiles and the heat exchanges between the greenhouse and the outside are purely radiative and conductive. The north side of the greenhouse is the collector wall. Behind the collector wall, a volume identical to that of the greenhouse is treated as a house module : the exterior surface is still made of glass and aluminium profiles, but a strong insulation is placed on the inside. This insulation covers all east north and west sides of the house module, as well as the ceiling and the floor; it consists in IO cm of expansed polystyrene to fight conduc--tion losses (or gains). An optional portion of I square meter of insulation can be removed to provide controlled losses. Between the house module and the greenhouse module stands the collector wall. The collector wall consists of an array of 26 horizontal gray PVC plastic tubes of diameter 5 cm and thickness I.2 mm, closed at both ends. The I.6 cm gap between two successive tubes is sealed with a stripe of 2 cm thick black neoprene foam. This foam prevents direct heat conduc--tion between the greenhouse and the house module; the heat conduction has to take place through to tubes. The gap close to the floor and the gap next to the ceiling are optionally left open to create a natural thermo-convection between the greenhouse volume and the house volume. The collector wall was initially designed to operate in the peak of winter and to be protected by a second glazing leaving a IO cm air space in front of its surface. The thermo-circulation would then have been entirely between the collector wall and the house module. But as complete measurements could not be obtained before the end of the winter, the double glazing was discarded to reduce over heating. Therefore the thermocirculation, when it is allowed, only partly affects the house module in the present configuration. For the same reason, the surface of the collector wall was not painted in black. Still, as the two experimental cells operate in parallel, the comparison of their thermal performances is not affected by these modifications. Fig. 2 shows a view of the completed experimental cells located on the site of Nice Observatory, on a south west facing slope which receive solar beams from 9 a.m. solar time to sunset. Table I summarizes the relevant areas and volumes of the greenhouse module and the house module in a cell.

In terms of thermal inertia, the greenhouse module and the house module behave quite differently. In the greenhouse, the glass and aluminium structure, as well as the concrete blocs of the base and the ground contribute to give some thermal inertia. In the house module, the pre--sence of insulation on all sides except on the collector-wall side make the thermal inertia quite negligeable. 44 Kg of concrete blocs were pi--led up in this module to add some inertia; but the main contribution comes from the collector wall.

TABLE I : Volume and surfaces involved in an experimental cell.

	Greenhouse	Collector wall	House module
Volume	2,80 m^3	0,07 m^3	2,80 m^3
Floor area	I,70 m^2	0,07 m^2	I,70 m^2
Glass surface			
north	—	—	0 or I m^2
south	4,20 m^2	—	—
east	I,70 m^2	—	—
west	I,70 m^2	—	—
collector area			
horizontal	I,70 m^2	—	—
vertical	—	2,80 m^2	—

Fig. 2 : Global view showing the experimental cells in operation.

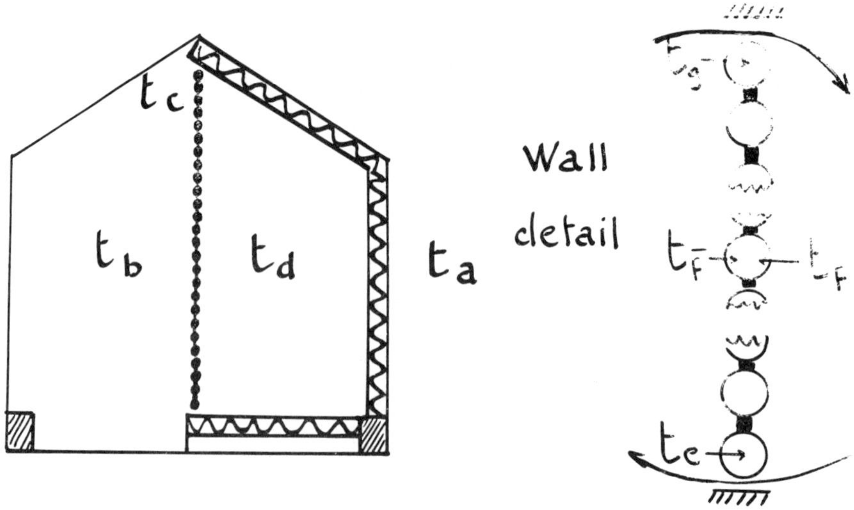

Fig. 3 : Elevation showing the position of temperature probes;
3 a : air temperature probes;
3 b : storage material probes.

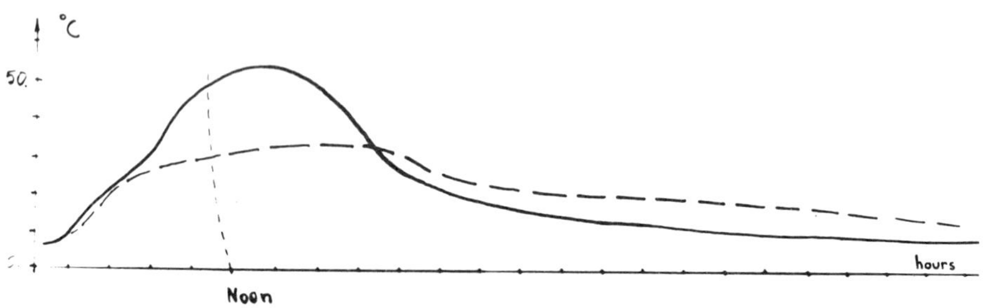

Fig. 4 : Thermal behavior of a cell with empty storage tanks (full line) compared with that of a cell equiped with IOO Kg of chliarolithe (dashed line). Data recorded on sunny days in january.

DATA ACQUISITION

An early run was made in january 78 in the first completed cell by means of mechanical temperature recorders (dilatometer). In that run only the external temperature and the temperature inside the house module were recorded.
During the complete runs performed in april 78 on the two cells, I4 different data were simultaneously recorded : the global solar radiation incident on the vertical collector wall and I3 different temperatures. The location of the temperature probes is indicated in Fig. 3. The temperatures are measured by means of CTN thermistances (I30Ω at 25° C) linearized by parallel resistances. The precision is of the order of ±0.5° C on a 50° C range. The data are collected on a MECI Speedomax recorder every 30 minutes. The global solar radiation measurement makes use of a Kipp and Zonen pyranometer.

DESCRIPTION OF THE COLLECTED DATA

Preliminary run .

Some simple minded temperature measurements were performed in the course of the construction of the cells. First an empty collector-wall was placed in a cell to check the temperature variations in the presence of minimum thermal inertial. Fig. 4 shows how a bright sun in january can bring the internal temperature up to 55° C during a short period, but this temperature goes down to 8° C in the night. Then the tubes were filled with I00 Kg of chliarolithe and the temperatures were again reccorded on a sunny day (dashed line) This time the bump of temperature is flattened out and appears under the form of a wide maximum at 34°. The cooling phase shows a long period where the temperature stays between 22° and I7° C as the result of the solidification of the chliarolithe. In this preliminary run , the gaps between the tubes were not get filled and a strong thermal convection existed between the greenhouse module and the house module, which explains the rapid temperature drop at night.

Complete runs with two cells in parallel.

These complete runs took place during the month of april 78. Various weather conditions were encountered during that period. The result will be presented in three different situations a) partly clear day, b) very clear day, c) rainy day. The thermal configurations of the cells were any one of the followings : α) open air vents and controlled heat loss through the north window, β) closed air vents and controlled heat loss through the north window, γ) closed air vents and insulated north window.

Partly clear day

This weather condition was the most favorable condition for the present size of collector walls : the amount of heat received by the walls did not exceed too much the thermal capacity of the latent heat material, so that the latent heat wall worked at its best. Fig.5a shows the temperatures recorded on april 3 rd in the latent heat cell and Fig. 5 b the ones in the water cell, with configuration "α" (see above).

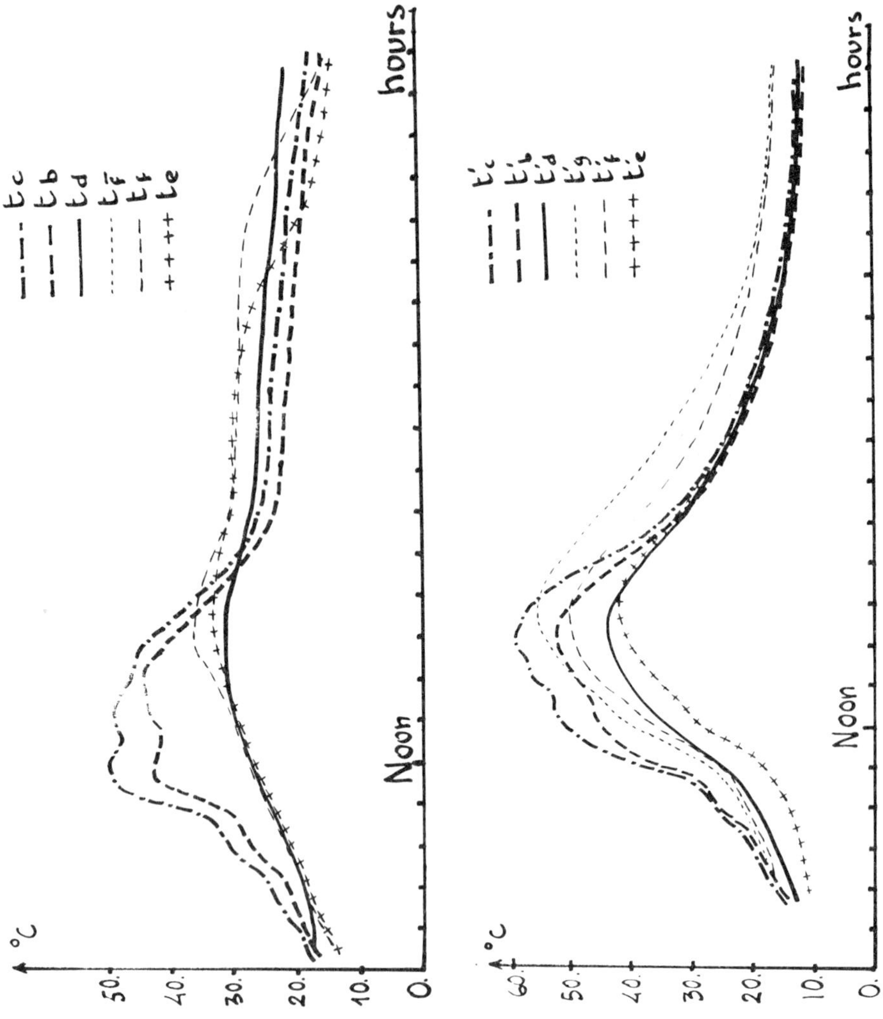

Fig. 5 : Temperature records on a partly sunny day (april 3rd) fol--lowing a similar day;
5 a : latent heat cell
5 b : water cell
Thermal configuration : open air vents and north window.

It can be checked from curves t_f or t_d that the chliarolithe went through complcte fusion with little over-heating. The behavior of temperatures in the water cell is quite simple : temperatures first rise rapidly up , then the rise slows down because heat losses increase sharply with the temperatures. In the cooling process, the behavior is a hyperbo--lic decay typical of high conductivity medium.In that day the maximum air temperature approches 60° C in the water cell; the house module temperature reached 43° C. Both temperatures dropped to I0° C at the end of the night. The probes placed inside the water tubes show that there exists a strong temperature gradient between the bottom and the top part of the collector wall (I5° C over I,80 meters). This is due to the na--tural convection of hot air along the wall which carries additional heat to the top.
The behavior of temperatures in the phase change cell is quite diffe--rent (Fig. 5a). The house module temperature rises more slowly because of the limited heat conduction through the chliarolithe. It then stops around 30° C (I3° C below that of the water cell); it stays between 30° C and 25° C for the major part of the night. Similarly the front air temperatures t_b and t_e stop increasing around II a.m. solar time, because the chliarolithe starts melting. These temperatures stay at their maximum for a large period until the incident solar energy vani--shes. During that period a certain temperature gradient is establi--shed across the chliarolithe tubes,such that the rate of storing calories in the form of latent heat balances the rate of incident energy (minus losses). This temperature gradient could have been smaller had the thermal conductivity of the chliarolithe been larger; but it is already sufficiently low to maintain all temperatures I0 to I5° C below those of the water cell.
During the cooling phase the difference between the water cell and the phase change cell is even more drastic : as the chliarolithe inside the tubes starts solidifying at 28° C, the various air temperatures stop their decay and stay relatively constant. In this way, a new temperature gra--dient (opposite to the previous one) is created across the tubes such that the constant heat flux emitted by the latent heat tubes exactly balances the heat loss of the cell. In the house module this temperatu--re gradient is small because losses are small; in the greenhouse modu--le it is of course larger and the front air temperature stabilizes around 20° C. At the end of the night, the chliarolithe is completely solidified and all temperatures start droping like in the sensible heat cell. Fig. 6 a and 6 b represent the temperature variations under simi--lar weather conditions (april I9 th) and a different configuration of the cells : closed air vents and reduced losses in the house module. The pattern is similar to that of fig. 5 except for a certain shift of the house module temperatures which now stay much higher than the greenhouse temperatures during the cooling phase.

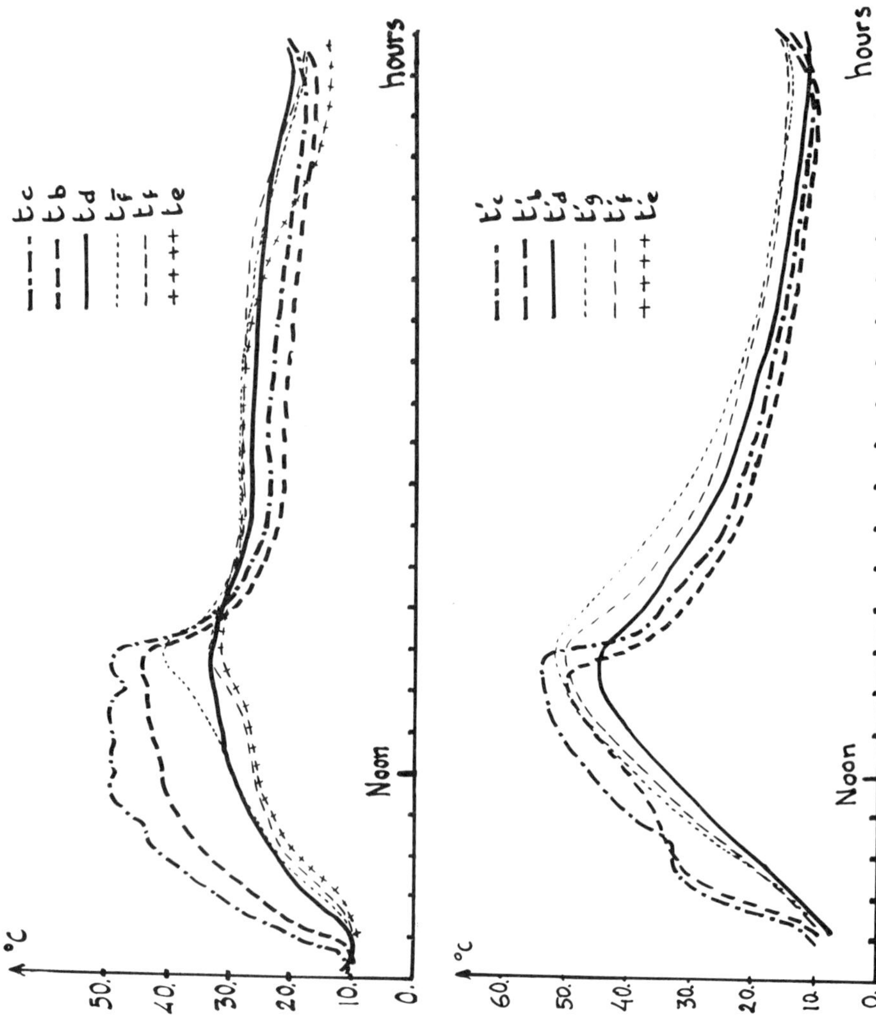

Fig. 6 : Temperature records on a partly sunny day (april 19th) fol-
-lowing a cool day.
6 a : latent heat cell
6 b : water cell
Thermal configuration : closed air vents and insulated north window.

Very clear day

This is a situation where the amount of incident solar energy in april clearly exceeds the heat capacity of the latent heat wall which was designed to operate in january. More precisely, the vertical wall does not receive so much more energy itself, but the ground of the green-house, which also acts as a moderately efficient collector, receives in april three times more energy than in january.
The result of a very sunny day (april 4th) is shown in Fig. 7. Temperatures observed in the water cell are qualitatively very close to those of a partly sunny day; the only difference is a higher amplitude for the variations and a larger temperature gradient along the wall. But a new feature appears in the curves of the latent heat cell. One can describe this feature by saying that a bump corresponding to sensible heat storage is added on top of the curves shown in Fig. 5 a or 6 a : in fact, after the complete fusion of the chliarolithe had occured, more energy was received which was used to over-heat the material.
During the cooling phase, the liquid chliarolithe first cools down, like the water, and then the process of solidification takes place and all temperatures behave like in Fig. 5 a or 6 a with a long plateau during the night. Fig. 8 summarizes the comparison between the two cells.

Rainy day

This is of course not a good condition for experimenting a solar system. But the results still deserve to be shown because they prove that a passive heating system is not unefficient in the presence of some dif-fuse radiation. Fig. 9 shows the house module temperatures in the two cells during april Irst. First comment : both walls operate as sensible heat storage and they store about the same amount of heat. Second comment : as the chliarolithe is not a good heat conductor , there is a time delay for the temperature response in the latent heat house module. Third comment : because of this time delay, the temperature stays above I5° C for most of the night. This is after all not a bad performance.

QUANTITATIVE COMPARISONS

The present discussion is based on the data of april I9 th shown in Fig. 6.
The analysis of the incident solar energy received on that day reads as follows :

Incident energy on a vertical south facing surface	:	4 000 Kcal/m^2
Incident energy on a horizontal surface	:	5 800 Kcal/m^2
Energy absorbed by a 2.80 m^2 collector wall of absorbtivity 0.7	:	7 840 Kcal
Energy absorbed by I.70 m^2 of ground of absorbtivity 0.6	:	5 900 Kcal
Total energy received by the cell	:	2I 000 Kcal
Total energy absorbed by the cell	:	I3 740 Kcal

(The difference is due to the low absorbtivity of the surfaces).
Part of this energy is stored in the form of sensible heat or latent heat, the other part being lost by radiation or convection.

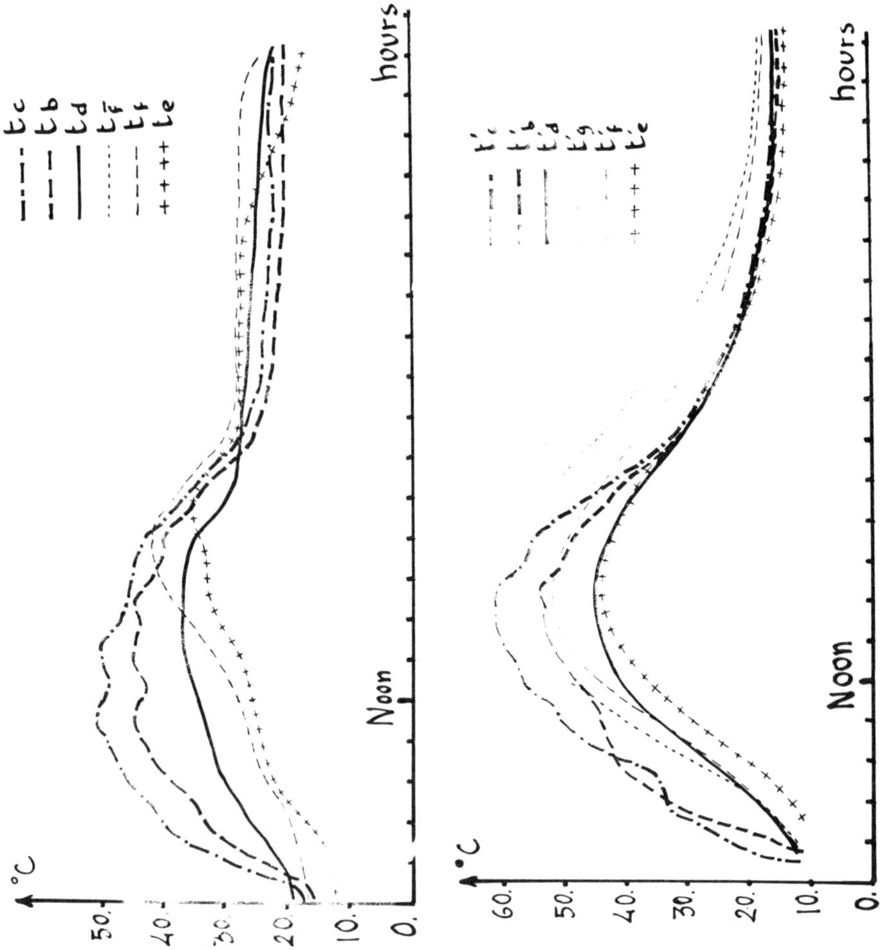

Fig. 7 : Temperature records on a very sunny day (april 4th).
7 a : latent heat cell;
7 b : water cell.

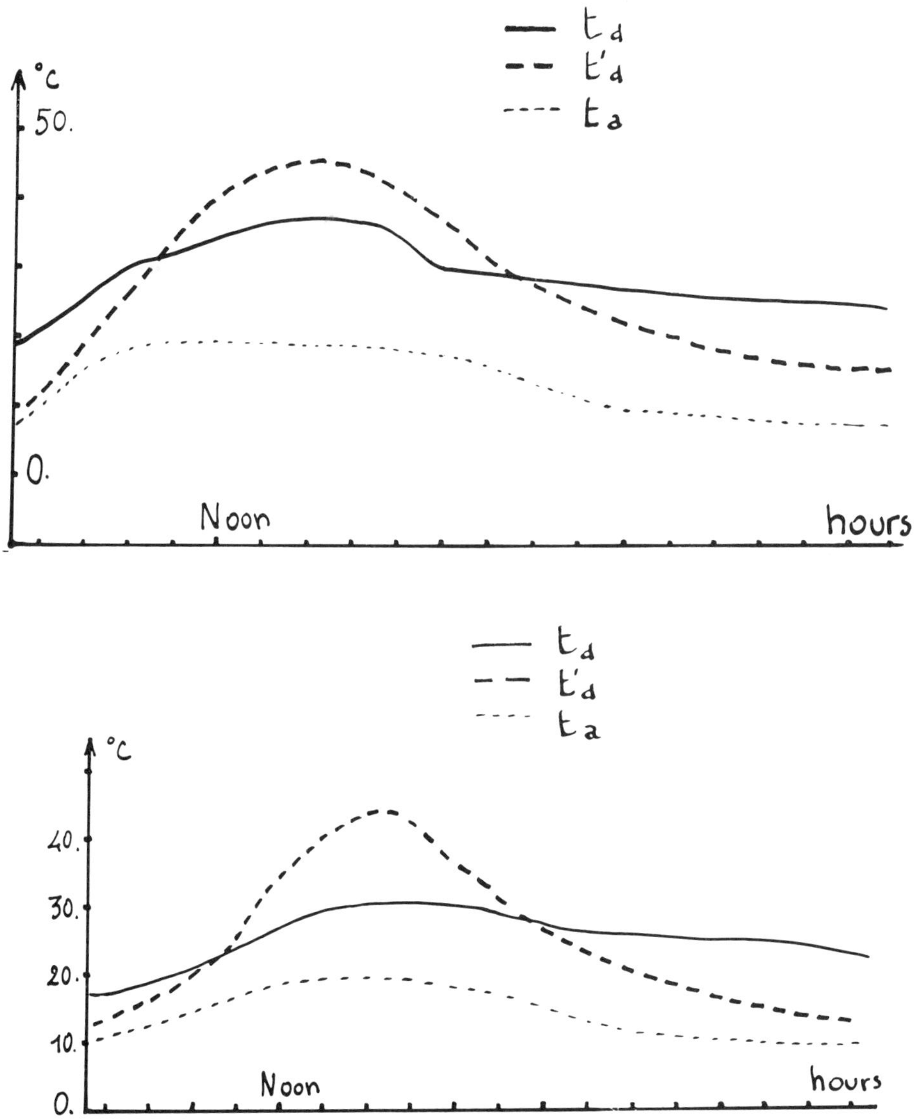

Fig. 8 : Compared behavior of house modules in sunny days.
8 a : april 3rd;
8 b : april 4th

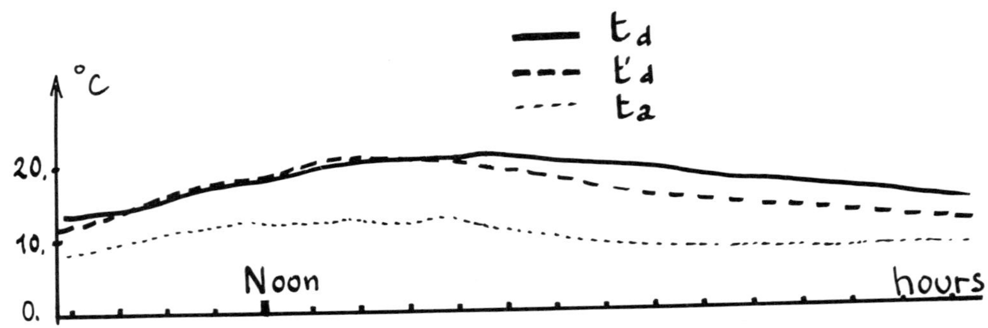

Fig. 9 : Temperature records on a rainy day (april Ist) full line : latent heat house temperature; dashed line : water house temperature.

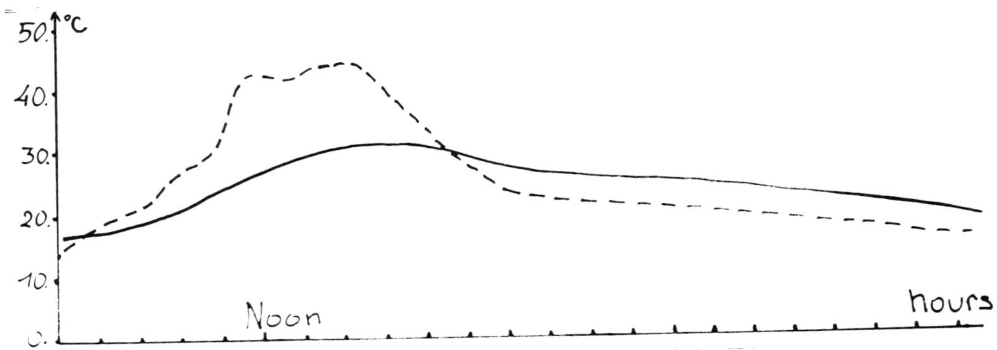

Fig. IO: Variations of latent heat house temperature compared with corresponding greenhouse temperature in a quasi-permanent regime (april 3rd). The latent heat wall operates as an efficient filter.

Estimation of the sensible heat stored in the water cell :

67 Kg of water heated from I0° C to 45° C	2 345 Kcal
25 Kg of massive structure of wall from I0° to 45°	200 Kcal
22 Kg of glass and aluminium in greenhouse from I0° to 30°	I80 Kcal
44 Kg of concrete blocs heated from I0° to 40°	350 Kcal
I76 Kg of concrete blocs in the base of the greenhouse + ground, partly heated from I0° to 35°	400 Kcal
Total sensible heat	3 475 Kcal
Stored Fraction of incident energy	I7%
Stored Fraction of absorbed energy	25%

Estimation of the sensible heat and latent heat stored in the phase change cell :

I00 Kg of chliarolithe with 32 Kcal/Kg	3 200 Kcal
I00 Kg of chliarolithe from I0° to 32° C	I 400 Kcal
25 Kg of massive structure of wall from I0° to 40°	I70 Kcal
22 Kg of glass and aluminium from I0° to 20°	90 Kcal
44 Kg of concrete blocs from I0° to 30°	230 Kcal
I76 Kg of concrete blocs of greenhouse + ground partly heated from I0° to 30°	320 Kcal
Total stored heat	5 4I0 Kcal
Stored fraction of incident energy	26%
Stored fraction of absorbed energy	39%

This quantitative estimate shows that the use of a latent heat material melting at a relatively low temperature is profitable in terms of collec--tor efficiency : as the over all thermal conductivity of the phase change material in its envelope is not too low, the incoming heat flux does not require a prohibitively high temperature gradient to penetrate the material. In case one would really like to operate with a smaller gradient, it would be necessary to choose a different geometry for the storage elements so that a larger surface of exchange could be obliquely offered to the same amount of solar radiation .

A second remark is that the actual amount of heat stored in the phase change material is much larger than the one stored in the water : in term of pure latent heat it is already I.37 time larger; including both latent and sensible heat, the total amount of stored heat is twice that of the water wall.

In other words, a water wall able to store the same amount of heat as the present phase change wall should be twice thicker. But a water wall able to collect the same amount of heat should even be bigger since this condition requires a larger collector efficiency and consequently a lower gradient in the greenhouse.

In fact one can estimate that the water wall must be 2.5 times thicker than the phase change wall in order to have the same performance during the solar collection.

Now let us consider the phase of heat restitution. The temperatures in the phase change cell are characterized by a long horizontal pla--teau whereas the temperature in the water cell exibit a typical hyper--bolic decay which corresponds to the very high apparent conductivity of water due to free internal convection.
The reason for the long plateau in the latent heat cell is twofold : a larger heat storage and a constant and moderate (28° C) temperature during the solidification of the storage material .
There are therefore moderate heat losses from the phase change cell in the beginning of the cooling process, a period during which the water cell loses a large fraction of its heat.
A quantitative estimation of the thermal losses can be made in the water cell. 3500 Kcal are lost during a I2 hour period; the temperature difference between the inside and the outside of the cell decreases exponentially from 30° C to 5° C. This is consistent with a coefficient K=4 Kcal/h/m²° C which describes the heat transfer through the green--house glass area in the absence of external wind.
Similarly the plateau of solidification lasts for 8 hours in the phase change cell and corresponds to a constant heat flux of 400 Kcal/h origi--nated from the latent heat storage. This is again consistent with the coefficient K= 4 Kcal/h m²° C and a constant air temperature gradient of I3° C between inside and outside.
The net result is that the temperature in the house module (and also in the greenhouse to a smaller extent) is relatively stable over a 24 hour period : the fluctuations do not exceed I5° C in a permanent regime between consecutive sunny days as fig. I0 shows.
On the other hand the corresponding temperature fluctuations observed in the water cell regularly exceed 30° C; and even a much thicker water wall, as quoted above, would not filtrate the temperature as the phase change wall does it, since the heat restitution imposes cooling down the water.
As a conclusion, we believe that the present experimental gives strong arguments in favor of the phase change material called chliarolithe : it has shown its ability (over a four month period of cycles) to impro--ve the capture efficiency, increase the filter effect of a collector storage wall which is a simple prototype of passive solar heating element.

REFERENCES :

I - ANVAR -CNRS Report n° 77.36.216
2 - " Latent heat diode wall "; L. Bourdeau and A. Jaffrin ; Section 9 A of present Conference .

ON THE EVALUATION OF THE FEASIBILITY OF SOLAR COLLECTORS FOR ELECTRICAL POWER GENERATION APPLICATIONS

A. Gokhman and N. Ozboya
University of Miami
Coral Gables, Florida U.S.A.

ABSTRACT

In this paper certain technical parameters have been proposed for economical evaluation of solar collectors for electric power generation applications. In the analysis only high temperature concentrating collectors have been considered. These parameters could be useful for government agencies and power companies in order to compare various solar collector subsystems quickly and correctly.

INTRODUCTION

Because of the recent extensive research in the area of developing feasible solar electrical plants (SEP) in the U.S. and the other countries, it is vitally important to have a proper system of evaluation and comparison of different types of solar collectors from an economical point of view. This system should also be employed, if the comparison of particular SEP with electrical plants utilizing nuclear or fossil fuels, is desired. According to some researchers, the efficiency of the collector is the most important criterion in the evaluation of the collector. Though some others deny completely any importance of the efficiency (Ref 1)

So now it is the proper time to establish a scientific system of evaluation. When the solar collectors are considered as a subsystem of an electric power plant, the efficiency of the solar collector itself is not an independent parameter, (of course not in the sense of dimensional analysis) since the sun does not charge for the energy it gives, unlike OPEC countries. In the comparison it is not necessary to include the initial capital investment for fuel burning equipment and the daily fuel cost for SEP in contrast to electrical plants utilizing nuclear or fossil fuels.

THE MAIN OUTPUT PARAMETERS OF SOLAR COLLECTORS

In this paper, only high temperature concentrating solar collectors, which do not produce any power in cloudy weather, are taken into consideration. The solar collectors are considered to be a subsystem of a solar electric power plant which also has energy storage, and energy conversion subsystems.

The authors think that the following five parameters are necessary to evaluate the feasibility of certain collector types or system of collectors at a particular geographical location for certain weather conditions and the requirements of the electrical system (utility grid).

1. The cost of 1 kilowatt of maximum available thermal power output of the collector, $K_{\$}$

$$K_{\$} = \frac{C_c}{P_{max}} \quad (\$/kW)$$

Where:

C_c is the cost of collector which does not include the cost of land or the cost of any component of the steam turbine unit.

P_{max} is the largest available thermal power output of the collector when it is placed at this altitude at noon of the day with highest position of the sun.
It is given by:

$$P_{max} = I_{max} \cdot A \cdot \eta_{max}$$

where:

I_{max} is the largest possible insolation (kW/m^2) at that time.

η_{max} is the maximum efficiency

A is the active area of the collector

2. The ratio K_e of the available thermal energy of the collector during a particular day of the year when the collector is placed at a certain altitude to P_{max}

$$K_e = \frac{E_d}{P_{max}} \text{ (hr)}$$

Where:

$$E_d = \int_0^{24} P_c dt$$

P_c is the instantaneous thermal power output of the collector:

$$P_c = I \cdot A \cdot \eta$$

where:

I is the instantaneous insolation (kW/m^2)

A is the active area of the collector

η is the instantaneous efficiency of the collector.

3. The bulk temperature, T_b, of the working fluid.

4. The life cycle of the collector, t_ℓ

5. The required area of the land and the quality of the required land; solid dry land, desert, swamps, etc.

These five parameters describe sufficiently every single type of collector. Now let us explain these parameters further:

1. $K_\$$ depends on the power capacity of the collector as well as the latitude (since P_{max} depens on the latitude). For a certain type of collector this coefficient may decrease with increase in capacity, for another type it may increase. For example, for a concentrating parabolic dish type collector made from metal, this coefficient will increase with an increase in capacity, because more material per kW thermal has to be used in order to have enough resistance for deflections and stresses. This parameter has to be introduced as a topogram, as shown in Figure 1, in which one axis is the maximum thermal power output of the collector, the other axis is the latitude of the selected location and the curves are the constant $K_\$$ lines. The same topogram can also be introduced as

in Figure 2, where one axis gives K_S, the other one gives the latitude and the curves give constant maximum power output lines.

2. K_e depends only on the latitude and the day of the year for the solar collectors of the same type. In Figure 3 this parameter is introduced as a topogram - the axes are the latitude and the time of the year; and the curves are the constant K_e lines. Or by analogy to K_S, K_e can be represented as shown in Figure 4.

3. T_b will define the available steam turbine equipment and furthermore the efficiency and the cost of this equipment.

4. The life cycle of the collector, t_ℓ is necessary for comparison of various solar collectors.

5. The cost of required land is also a very important factor and can affect the feasibility of certain types of collector. The same collector could be feasible in relatively cheap land, but fail to be feasible in some other areas. Some collectors such as the inflatable solar collector do not require solid land. This collector can be placed in swamps, deserts and even on the ocean surface. This feature of the inflatable solar collector makes it more attractive from an economical point of view.

These five parameters enable one to conduct economical evaluation of the collector as shown in the following example. A solar electric power plant with pumped storage (solar complex electrical plant)(SCEP) will be considered (Reference 2). It is clear that utilization of solar electrical plant without any storage directly in a utility grid is not possible since the sun is not a controllable source of energy. Consequently, one can comprehend a solar electrical plant only in combination with some kind of storage, which has the ability of controlled output. So the example which will be discussed is typical for solar energy applications. Let us compare two different types of collectors which form the collector subsystem of a solar electric power plant, when the plant participates in the same load curve of utility grid during the worst month. The worst month is defined as the month, among the months during which SCEP is in operation, that has the highest power demand from the utility grid. Of course, the worst months of the system in Northern U.S. are December and January, but it is absolutely not economical to assign certain requirements to SCEP during that time since the solar radiation is very low, the number of cloudy days is very high and the equipment will be very costly. This worst month has to be found by an economical analysis which

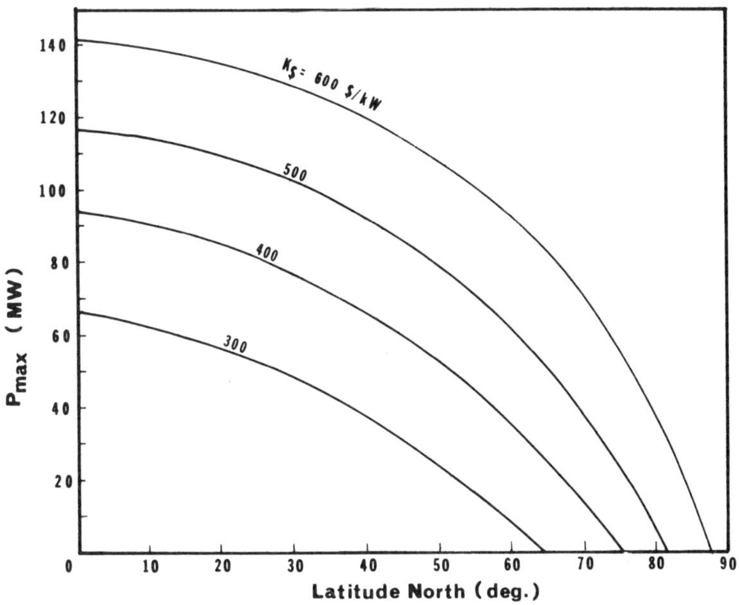

FIGURE 1

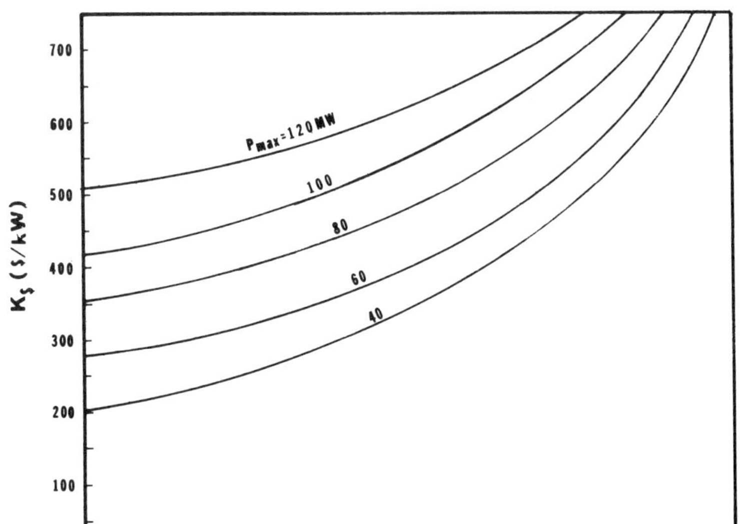

FIGURE 2

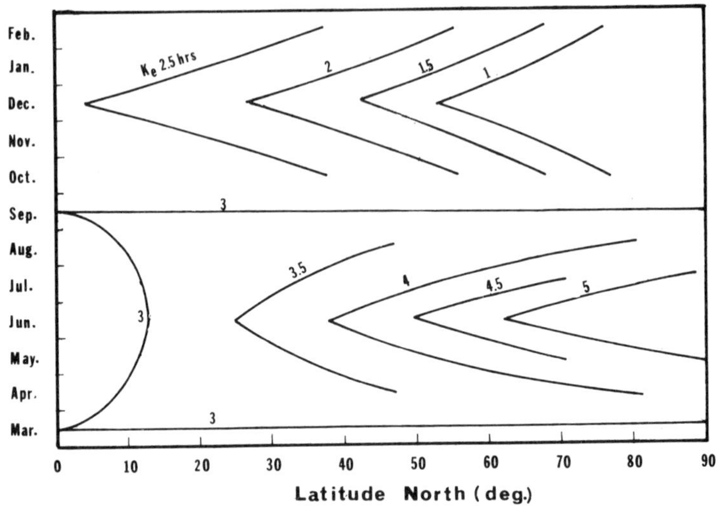

FIGURE 3

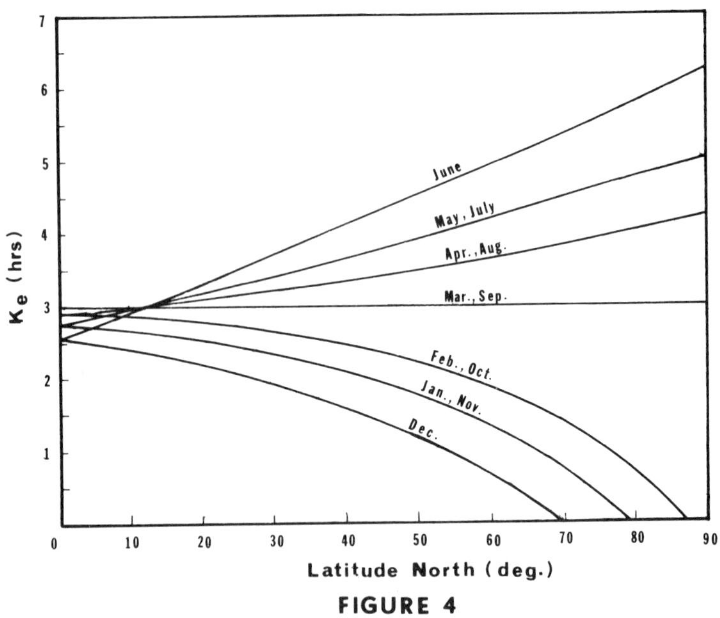

FIGURE 4

takes the features of the collector into consideration. For example, the power participation of SCEP to the utility grid is not feasible in January, February, and March; the first month it becomes feasible is April and the last month is September. After April until September the power demand from SCEP is lower and the sun's radiation is higher. Therefore, the worst months can be taken as April and September. This does not mean that the SCEP will not work between September and April, but there is no responsibility of SCEP to provide the utility grid with certain power.

Let us take April as the worst month. Now it is necessary to determine the energy requirement of the pumped storage plant (PSP) of the Solar Electric Plant (SEP) in order to participate in the desirable zone of the load curve of the electrical system for this month. As an example, the zone of participation has been selected, as shown in Figure 5.

So, the energy output of PSP is

$$E''_p = \int_0^{24} P_p dt$$

where:

P_p is the instantaneous power of the PSP (see Figure 6).

The energy which PSP has to spend to accomplish the requirements of the system is:

$$E'_p = \int_0^{24} \frac{P_p}{\eta_H} dt$$

where:

η_H is the instantaneous efficiency of the hydraulic turbines and the generators of PSP.

Therefore SEP has to supply PSP with the amount of energy equal to E'_p

But it should be considered that not everyday in April is sunny. Only sunny days can be used for the accepted period of regulation of PSP. The period of regulation cannot be one day since if it is not sunny, SCEP will not accomplish the participation requirements of the system. Therefore, the period of regulation of pumped storage plant has to be at least one week. Of course, it is better to have a month of regulation, but it will increase the volume of the storage reservoir. So, the period of regulation is also an economical parameter which has to be found by economical analysis.

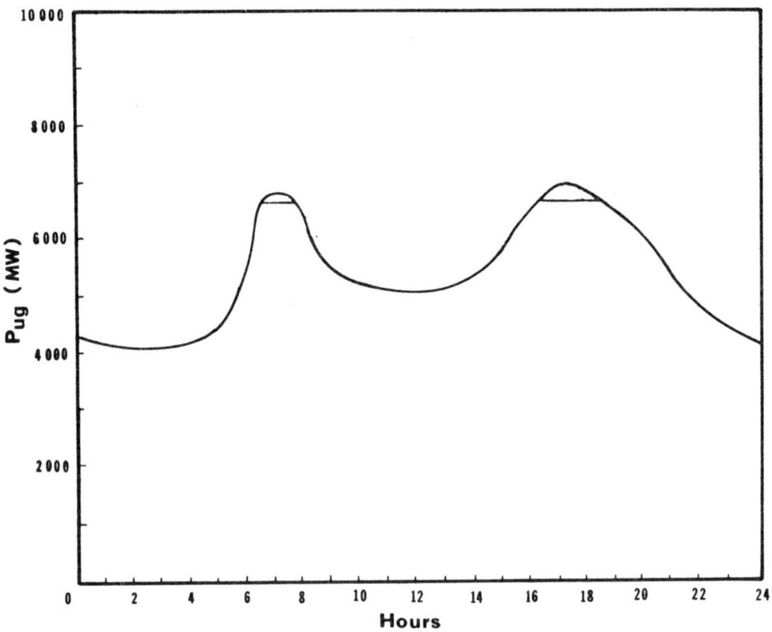

FIGURE 5: Typical Daily Load Curve of Utility Grid for Worst Month

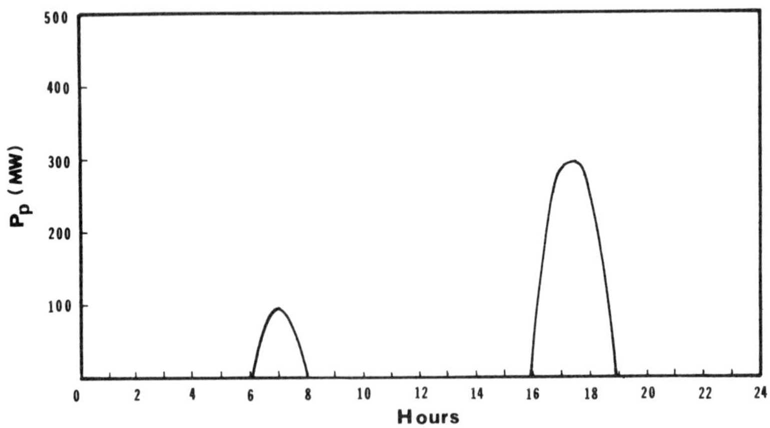

FIGURE 6: Power Curve of SCEP

It is necessary to develop a curve of probability of sunny days for every month, and for the certain location of SCEP, in analogy to the curves of probability of river flux of each year in hydrology. The curve shown in Figure 7 gives the sunny day ratio, S

$$S = \frac{N_s}{N}$$

where:

N is the number of days in that month

and

N_s is the number of sunny days.

This graph has to be prepared according to the meteorological observations during certain number of years.

In hydroelectrical applications, the analysis is usually based on the value of river flux ratio corresponding to 90 percent probability. The same procedure is proposed in the case of solar electrical economical analysis, but this value of probability has to be accepted by U.S. agencies and become the conventional value (and it may be 50 percent, 60 percent, etc).

For this case, for 90 percent probability, the corresponding value of sunny day ratio is 0.4 (S = 0.4). So the amount of energy which must be supplied to PSP by SEP in one day during the period of regulation is

$$E_p = \frac{E'_p}{S} = \frac{E'_p}{0.4}$$

and also

$$E_p = \int_0^{24} P_c \cdot \eta_s \cdot \eta_p \cdot dt$$

where:

P_c is the instantaneous thermal power output of the collector subsystem,

η_s is the instantaneous efficiency of steam-turbine and generator unit, and

η_p is the efficiency of the pumps of PSP.

The knowledge of working fluid temperature, T_b makes it possible to select the type of steam turbine and generator.

The necessary output of the collector subsystem is

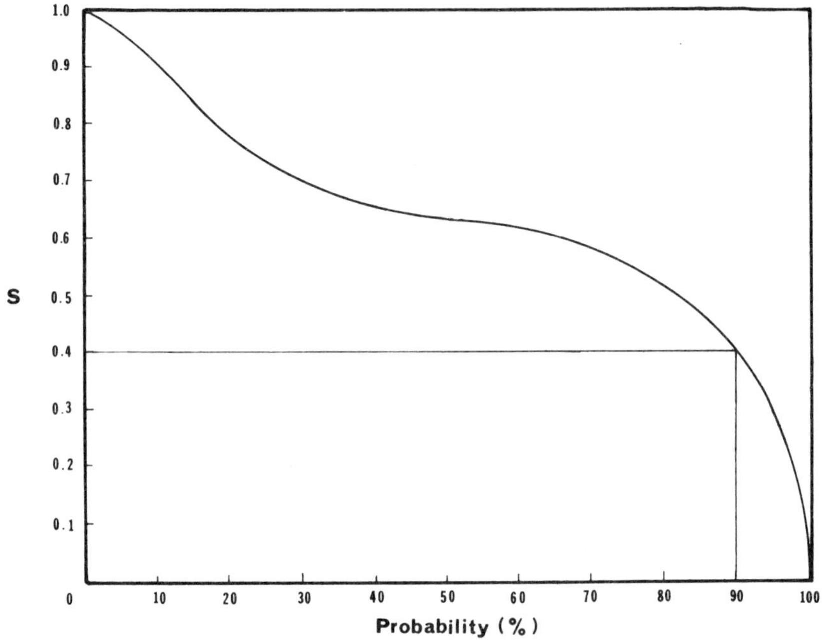

FIGURE 7: Probability Curve for Sunny Days

$$E_c = \int_0^{24} P_c dt$$

In order to calculate E_c, mean values of η_s and η_p can be used:

$$E_c = \frac{E_p}{\bar{\eta}_s \cdot \bar{\eta}_p}$$

where:

$\bar{\eta}_s$ and $\bar{\eta}_p$ are the average efficiencies of the steam turbine and generator unit and the pumps respectively.

The value of P_{max} can be found by taking the value of K_e from Figure 3 or 4.

$$P_{max} = \frac{E_c}{K_e}$$

The cost of the collector subsystem can be calculated by:

$$C_c = P_{max} \cdot K_{\$}$$

where:

$K_{\$}$ can be determined from the proper topogram (Figure 1 or 2)

It is also possible to calculate the area of necessary land which is approximately proportional to P_{max}. Once P_{max} is determined, the cost of steam turbine and generator, as well as the savings in fuel per year, S_f, because of the participation of our collector in the electrical system, can be easily calculated.

So, for each type of collector working with the same pumped storage plant and with the same degree of participation in the utility grid during the worst month of operation, we can find the capital investment, K, to the solar plant with this type of collector and the savings of fuel per year, S_f, in the system. We also know the life of these collectors. Now the economists can easily compare the feasibility of these collectors, using the proper methodology utilized in U.S., including inflation, interest rate, fuel cost changes, maintenance costs, etc.

The same type of economical evaluation makes it possible to compare various solar electrical plants with conventional power plants utilizing fossil or nuclear fuels. Of course, this type of analysis is necessary in the final stage, when the power company evaluates the possibility of construction of the solar plant in the system.

For the preliminary comparison of the different systems of solar collectors at this stage of research, it is necessary to keep in mind that the most important parameter is $K_\$$ and the quality of occupied land. So, generally the collector with the smallest value of $K_\$$ and the lowest requirements of land has to be more feasible if this collector does not have a drastically small parameter K_e in comparison with other collectors. For example, a collector might have very small $K_\$$ and very low requirement of land, but it can produce power only for one hour during the day, because it cannot track the sun. Of course, this collector, despite its small $K_\$$ and low requirement of land, could not be feasible in comparison with others which can work six hours.

As it is clear from the presented material, the efficiency of the collector does not participate directly in the evaluation of the economical feasibility of the collector. Therefore, the tendency to receive the biggest possible efficiency of the collector, no matter how, is false. However, the efficiency defines P_c and P_{max}. So, if it is possible to increase the efficiency without substantial increase in the cost of the collector, the collector becomes more attractive because of the smaller $K_\$$ and bigger K_e. But the greater efficiency is the problem of designer and researcher and does not participate directly in any economical evaluation.

We hope that the suggested parameters for feasibility analysis will give the capability to DOE to compare quickly and correctly different schemes of solar collectors and find the best direction for future development.

References:

1. W. A. Shurcliff, Open Letter to Marvin Shapiro, Solar Age, p.47, February 1978.

2. A. Gokhman, "Complex Solar Electrical Plant of Large Capacity", to be published in the Proceedings of Miami International Conference on Alternative Energy Sources, Miami Beach, Florida, December 1977.

EFFECT OF SOLAR RADIATION ON OUTDOOR BUS BAR DESIGN

M. M. Abdel-Aziz
Cairo University
Cairo, Egypt

ABSTRACT

The analysis of heat transfer for outdoor bus bar, in case of natural and forced convection are presented, for different practical bus bar shapes. The effect of solar radiation on bus bar ampacity are presented for clear and industrial atmosphere. Graphs are given for indecation of bus bar ampacity derating due to solar heat gain, specially for circular, square and rectangular shapes. Optimization analysis is suggested for future bus bar design.

INTRODUCTION

There are several factors which must be taken into consideration for suitable design of bus bars. The first important factor is the mechanical stresses acting on bus bars due to temperature rise and short circuit currents. The bus bar must be designed to overcome short circuit forces between different phases at the werest conditions. In other hand, the resultant stresses acting on bus bar cross section, must be below the design stress of bus bar material. The second important factor is temperature rise of bus bar i.e. at the werest atmospheric conditions, the temperature rise of bus bar must be less or equal to the specified maximum allowable value.

This paper deals mainly with the effect of solar radiation on the current carrying capacity of circular, square, and rectangular bus bars. Specified conditions are assumed for the obtained relations between bus bar ampacity and dimensions. At any other conditions, the corrosponding relations could be easily obtained. Graphs are given to clearify solar radiation effect on the proper choice of the required bus bar.

ANALYSIS OF HEAT TRANSFER

As the current pass through bus bar, losses will be generated inside the conductor. These losses depend on wheather that current is direct or alternating one. The a.c.losses depend on frequancy and type of conductor material, and is higher than the d.c. one by the skin effect factor. The generated heat is transfered to the surronding air mainly by convection and radiation. In other hand the bus bar will be subjected to solar radiation, hence it will absorp considerable quantity of heat . So to keep the maximum permissible temperature rise at the same level, the heat generated inside the bus bar must be reduced, i.e. the bus bar current must be reduced otherwise overloading will occur.

Solar heat gain depends on, bus bar dimensions, atmospheric conditions, geographic location and hieght above sea level . Here we shall apply calculations of solar heat gain at sea level and at 30° north latitude.

Heat transfer by convection may be classified into forced and natural convection.Each type of convective heat transfer depends on geometrical bus bar surface,shape.Forced convection analysis may be applied if the bus bar surface is subjected to moving air has a velocity of 2 ft/sec.[1] parallel to or in any other direction to the surface. Natural convection analysis may be applied if some surfaces of bus bar may be sheilded from direct exposure to wind or when the spacing between conductors is greater than 20% of the major dimension of conductor or 1 inch whichever is smaller. This assumption is based on the fact that the boundary layer for mass transfer is roughly 10% of the length of the flow path.

Heat transfer by radiation depends on the absolute conductor temperature, absolute ambient temperature and conductor material.

In the following items the equations govern heat transfer from bus bar surface will be presented specially for circular square,and rectangular copper bus bars.

1-Circular Bus Bars

a-Forced Convection Loss

Assuming air velocity of 2ft/sec. and 1 atmosphere pressure, the convective heat loss (watt/ft) for a cylindrical shape at least 1 inch diameter is given by [2] :

$$q_c = 0.01\, d^{-0.4}\, A \Delta t \qquad (1)$$

where:

A=surface area inch^2/ft

Δt= difference in temp. in c° between conductor surface and ambient air.
(the same relation will be used for diameters less than linch approximetly)

For all the presented analysis (Δt)may be taken equal to 50C° (t_c=95C° and t_a=45C°)

$$q_c=18.85\ d^{0.6} \quad (2)$$

b-Natural Convection Loss

The natural convection formula for cylinders is

$$q_c=0.0022(\Delta t)^{1.25}\ d^{-0.25}\ A \quad \text{watt/ft} \quad (3)$$

For Δt=50C°

$$q_c=11.027\ d^{0.75} \quad \text{watt/ft} \quad (4)$$

c-Radiation Loss

The basic equation of radiation is given by:

$$q_r=36.9\text{x}10^{-12}\epsilon\ A(\ T_c^{\ 4} - T_a^{\ 4}\ )\ \text{watt/ft} \quad (5)$$

where :
= emissivity corrosponding to temperature of interest (it is assumed approximatly that emissivity at $T_c$equal to absorptivity at T_a). For copper bus bars ϵ=0.8 : 0.85 and 0.5 for aluminum bus bars.
T_c=Temp. of conductor K° (95+273)
T_a=Temp. of air K° (45+273)
A =Surface area inch^2/ft

For ϵ =0.8

$$q_r=9.03\ d \quad \text{watt/ft} \quad (6)$$

2-Square bus bars

a-Forced Convection Loss

The forced convection loss is given by [1] :

$$q_c = 0.0085 \sqrt{\frac{V}{\ell}} \; A \, \Delta t \qquad \text{watt/ft} \tag{7}$$

where:

V= air velocity ft/sec.
ℓ= length of surface over which air flows inches

For V=2 ft/sec.
and Δt=50 C°

$$q_c = 28.8 \, \ell^{1/2} \qquad \text{watt/ft} \tag{8}$$

b- Natural Convection Loss

For vertical or upward facing surfaces the convective heat loss is given by :

$$q_c = 0.0022(\Delta t)^{1.25} \ell^{-0.25} A \tag{9}$$

and for surfaces facing down :

$$q_c = 0.0011(\Delta t)^{1.25} \ell^{-0.25} A \tag{10}$$

where:

ℓ=length of conductor surface (width or thickness in inches)

The total natural convection loss for Δ t=50 C° and squre bus bar is :

$$q_c = 12.29 \, \ell^{0.75} \qquad \text{watt/ft} \tag{11}$$

c-Radiation Loss

Applying equation (5) the radiation loss for square bus bar is given by:

$$q_r = 11.49 \, \ell \qquad \text{watt/ft} \tag{12}$$

3-Rectangular Bus Bars

a-Forced Convection Loss

Equation (4) will be applied here but it will be assumed that thickness of bus bar is equal to 0.1 width(ℓ)

$$q_c = 19\,\ell^{1/2} \quad \text{watt/ft} \tag{13}$$

b-Natural Convection Loss

With the same previous assumption and by applying equations (9) and (10), the total convection loss for rectangular bus bar, is given by:

$$q_c = 7.96\,\ell^{0.75} \quad \text{watt/ft} \tag{14}$$

c-Radiation Loss

Applying equation (5), the radiation loss of rectangular bus bar is:

$$q_r = 6.32\,\ell \quad \text{watt/ft} \tag{15}$$

SOLAR HEAT GAIN

The heat gain from incident solar radiation [1] is estimated as follows:

$$q_s = 0.00695\,\epsilon' \, Q_s A' \, k(\sin\theta) \tag{16}$$

where:

ϵ' =coefficient of solar absorption, usually somewhat higher than emittance, but generally taken as equal to that used for radiation loss .

θ =effective angle of incidence of sun ,and equal to :

$\cos^{-1} [\cos H_c \cos(Z_c - Z_1)]$

q_s = solar heat gain in watt/ft

where:

H_c=altitude of sun , degrees

Z_c=azimuth of sun ,degrees

Z_1=azimuth of conductor line , degrees
0 or 180 for N-S
90 or 270 for E-W

A` = projected area of conductor ,square inches /ft (area casting shadow)

Q_s=total solar and sky radiated heat on a surface normal to suns rays, watt/sq.ft.

k =multiplying factor of heat for high altitudes

In cases where solar heat input is high, it is important to consider whether solar heating will peak during the time the maximum current load is on the circuit. If not , the estimate of the solar load should be reduced accordingly in order to arrive at the most cost-effective conductor size.

The projected area of a flat surface is the area of its shadow on a plane normal to the direction of the sun's rays,e.g., per foot of conductor.

A` = 12 sin γ x conductor size

where:

γ = angle between plane of the conductor surface and sun's altitude.

For a vertical surface

$\gamma = 90- H_c$

For a horizontal surface

$\gamma = H_c$

The values of projected area are as follow:

A`=12d for round tube or bar (17)

A`=12ℓ [sin(90 − H_c) + sin H_c] for square bar (18)

and A`=12 [ℓ sin(90 −H_c) + w sin H_c] (19)

for rectangular bus bar (ℓ x w)

The calculations of solar heat gain for clear and industrial atmosphere have been applied for 30° north latitude(H_c=30°) at 2 P.M., assuming the bus bar lies on a E-W line at sea level. (table I) The following equations are obtained :

TABLE I

DATA FOR CALCULATING SOLAR HEAT GAIN

Altitude and Azimuth in Degrees of the Sun at Various Latitudes Declination 23.0° Northern Hemisphere. June 10 and July 3

Degrees North Latitude	10.00 A.M		12.00 N.		2.00 P.M.	
	Hc	Zc	Hc	Zc	Hc	Zc
20	62	78	87	0	62	282
25	62	88	88	180	62	272
30	62	98	83	180	62	262
35	61	107	78	180	61	253
40	60	115	73	180	60	245
45	57	122	68	180	57	238
50	54	128	63	180	54	232
60	47	137	53	180	47	223
70	40	143	43	180	40	217

Total Heat Received by a Surface at Sea Level Normal to the Sun's Rays

Solar Altitude Degrees HC	Q_s watt/sq.ft Clear Atmosphere	Industrial Atmosphere
5	21.7	12.6
10	40.2	22.3
15	54.2	30.5
20	64.4	39.2
25	71.5	46.6
30	77.0	53.0
35	81.5	57.5
40	84.8	61.5
45	87.4	64.5
50	90.0	67.5
60	92.9	71.6
70	95.0	75.2
80	95.8	77.4
90	96.4	78.9

Solar Heat Multiplying Factor(K) for High Altitudes

Elevation above Sea Level, feet	K
0	1.00
5,000	1.15
10,000	1.25
15,000	1.30

ii MMAA

Circular Bus Bar

$q_s = 6.41\ d$ watt/ft (clear atm.) (20)

$q_s = 5.2\ d$ watt/ft (industrial atm.) (21)

Square Bus Bar

$q_s = 8.65\ \ell$ watt/ft (clear atm.) (22)

$q_s = 7.02\ \ell$ watt/ft (industrial atm.) (23)

Rectangular Bus Bar

$q_s = 3.58\ \ell$ watt/ft (clear atm.) (24)

$q_s = 2.90\ \ell$ watt/ft (industrial atm.), $w=0.1\ell$ (25)

BUS BAR AMPACITY

The ampacity has been calculated according to the following equation :

Heat generated + Solar heat gain = Heat dissipated

$$I^2R_{a.c} = q_c + q_r - q_s \tag{26}$$

The copper bus bar resistance is calculated by :

$$R = \frac{8.145x10^{-4}}{C' A_1} \left(1 + \frac{0.0039C'}{100}(t_c - 20)\right) \quad \text{Ohm/ft} \tag{27}$$

where :

C' = conductivity as % IACS

A_1 = cross section area inch^2

t_c = cond. temp. C°

For t_c = 95C° and C'=92%

$$R = \frac{1.08x10^{-5}}{A_1} \quad \text{Ohm/ft}$$

the skin effect factor at 50 HZ is assumed to be 1.2 approx.

$$R_{a.c} = \frac{12.97 \times 10^{-6}}{A_1} \quad \text{Ohms/ft} \tag{28}$$

Applying the previous analysis the following equations of bus bar ampacity calculations have been obtained.(in amps.)

Circular Bus Bar

Forced Convection

$$I_{without} = 250(18.85d^{2.6} + 9.03d^3)^{1/2}$$

$$I_{clean} = 250(18.85d^{2.6} + 2.62\, d^3)^{1/2}$$

$$I_{ind.} = 250(18.85d^{2.6} + 3.83d^3)^{1/2}$$

Natural Convection

$$I_{without} = 250(11.03d^{2.75} + 9.03d^3)^{1/2}$$

$$I_{clean} = 250(11.03d^{2.75} + 2.62d^3)^{1/2}$$

$$I_{ind.} = 250(11.03d^{2.75} + 3.83d^3)^{1/2}$$

Square Bus Bar

Forced Convection

$$I_{without} = 280(28.8\ell^{2.5} + 11.49\ell^3)^{1/2}$$

$$I_{clean} = 280\,(28.8\ell^{2.5} + 2.84\,\ell^3)^{1/2}$$

$$I_{ind.} = 280\,(28.8\ell^{2.5} + 4.47\,\ell^3)^{1/2}$$

Natural Convection

$$I_{without} = 280\,(12.29\ell^{2.75} + 11.49\ell^3)^{1/2}$$

$$I_{clean} = 280\,(12.29\ell^{2.75} + 2.84\ell^3)^{1/2}$$

$$I_{ind.} = 280\,(12.29\ell^{2.75} + 4.47\ell^3)^{1/2}$$

Rectangular Bus bar

Forced Convection

$$I_{without} = 90(19\ell^{2.5} + 6.32\ell^{3})^{1/2}$$

$$I_{clean} = 90(19\ell^{2.5} + 2.74\ell^{3})^{1/2}$$

$$I_{ind.} = 90(19\ell^{2.5} + 3.42\ell^{3})^{1/2}$$

Natiral Convection

$$I_{without} = 90(7.96\ell^{2.75} + 6.32\ell^{3})^{1/2}$$

$$I_{clean} = 90(7.96\ell^{2.75} + 2.47\ell^{3})^{1/2}$$

$$I_{ind.} = 90(7.96\ell^{2.75} + 3.42\ell^{3})^{1/2}$$

The results of ampacity calculations for circular, square, and rectangular bus bars (copper) are shown in figures(1to 5)

REQUIREMENTS OF OPTIMUM BUS BAR DESIGN

The following points represent the main requirments for optimum bus bar design :

1- The bus bar cross section must be designed to overcome mechanical stresses due to short circuit currents.

2- The bus bar surface area must be high enough to dissipate large generated heat, in other hand that area must be small to avoid solar heat gain.

3- The bus bar resistance must be very small to avoid large generated heat inside the conductors.

4- Overall cost of bus bar material must be of minimum value for the required economical considerations.

Details of such optimization analysis , taking into consideration all the previous mentioned points, will be given in future paper.

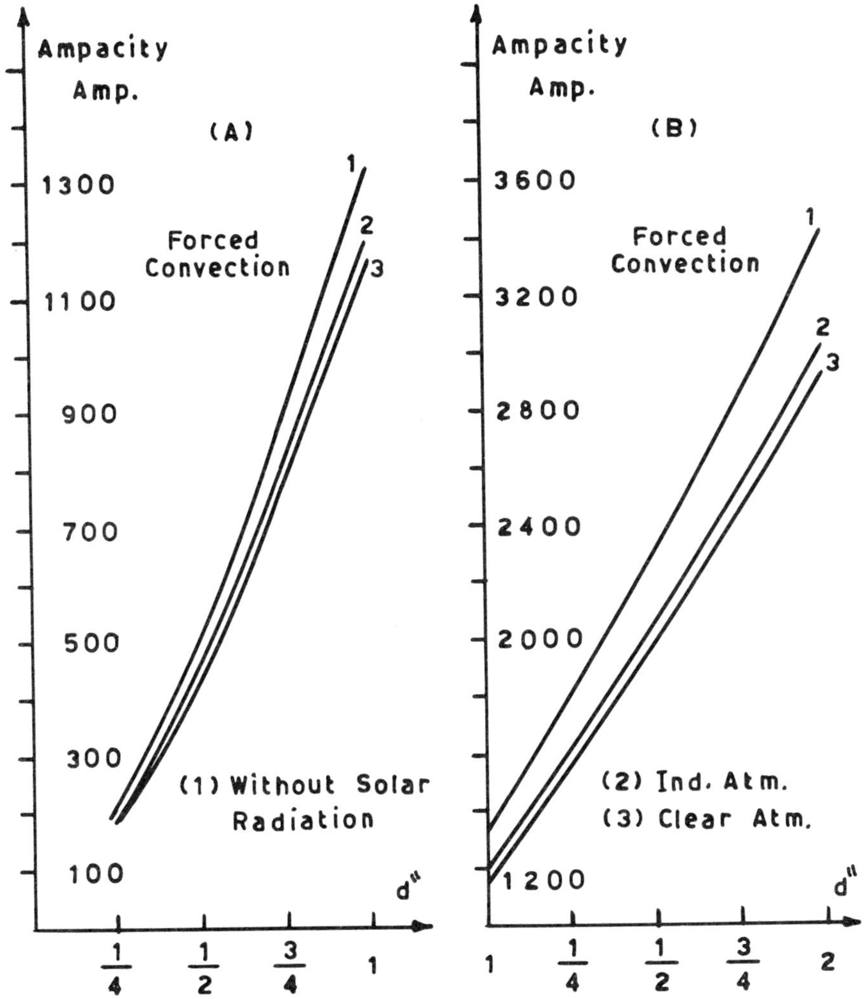

FIG. (1)- AMPACITY OF CIRCULAR BUS BAR

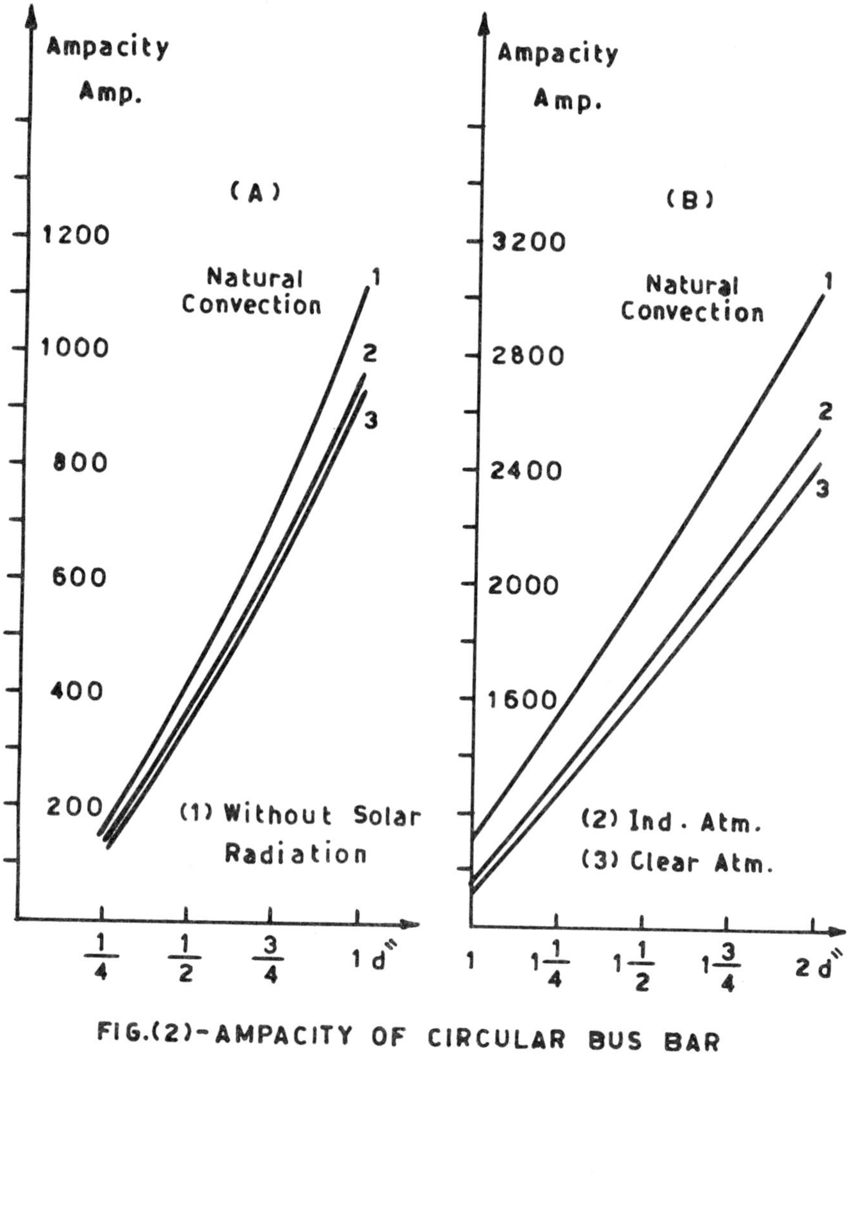

FIG.(2)-AMPACITY OF CIRCULAR BUS BAR

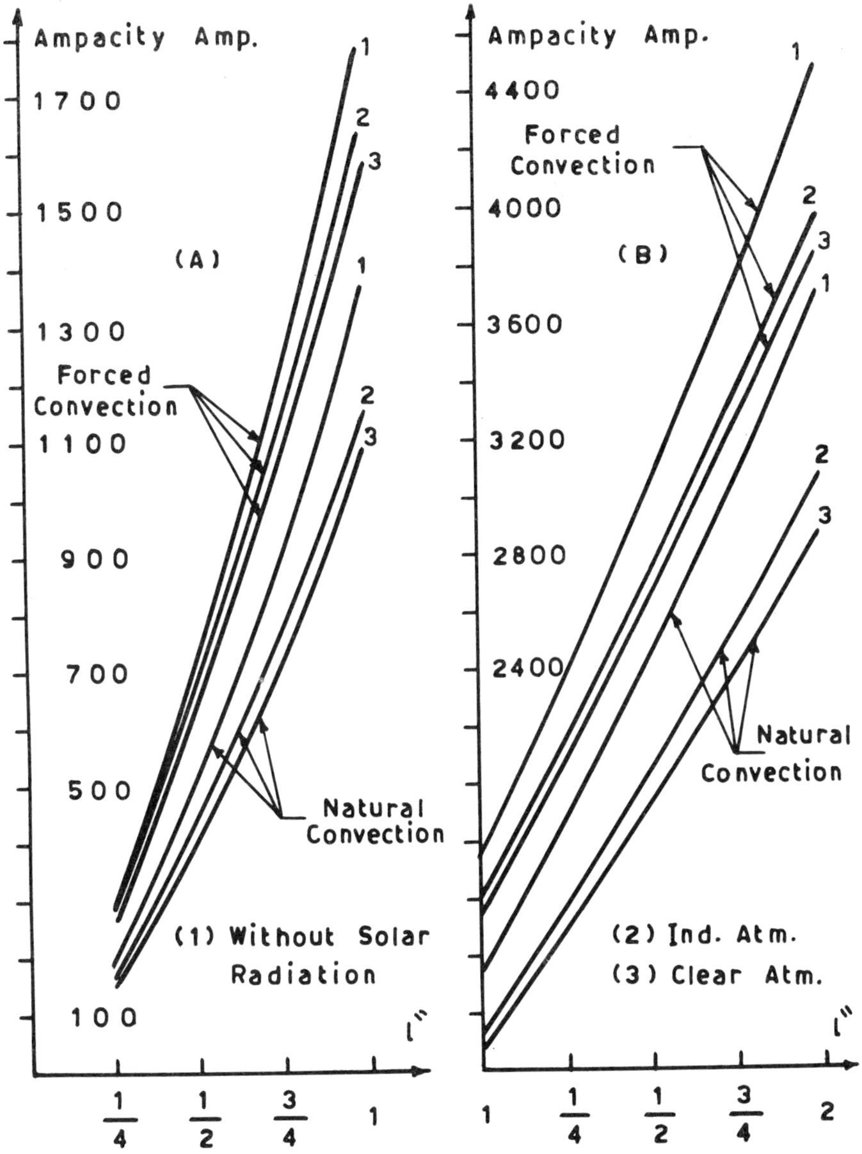

FIG.(3) AMPACITY OF SQUARE BUS BAR

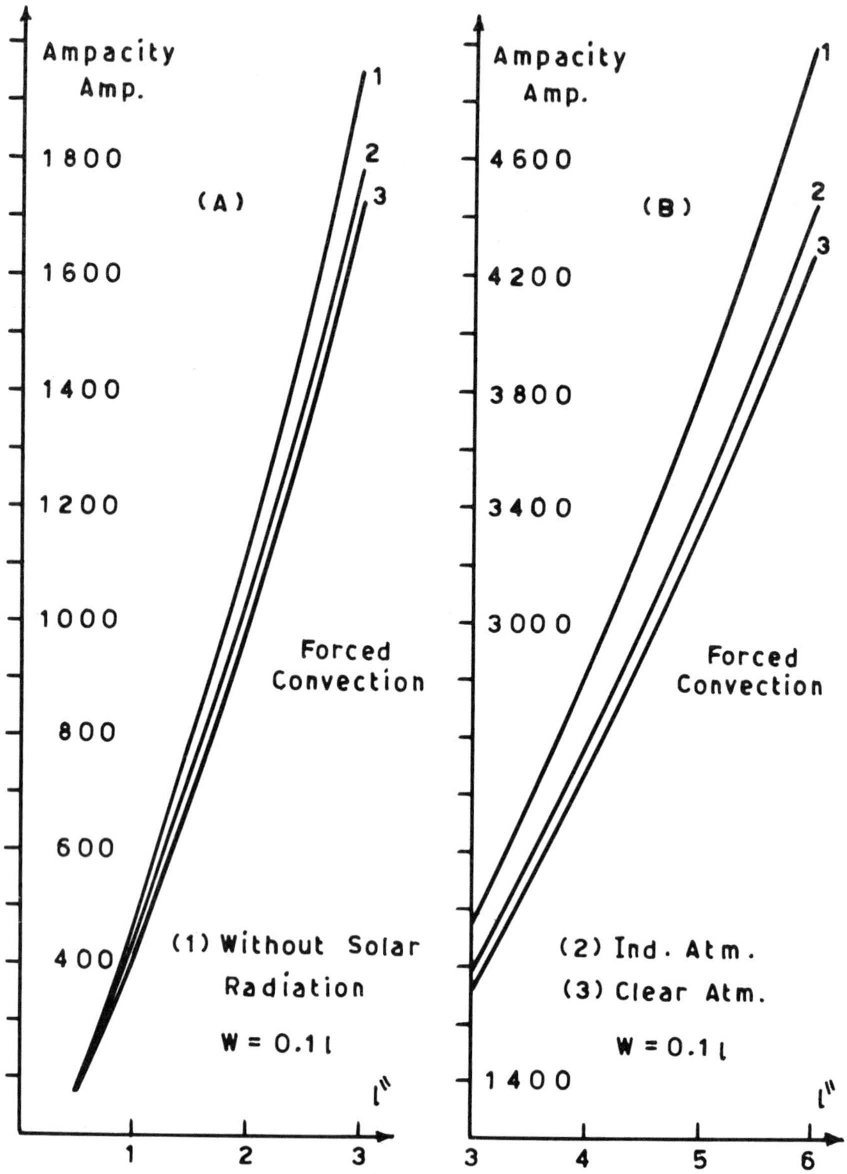

FIG. (4) AMPACITY OF RECTANGULAR BUS BAR

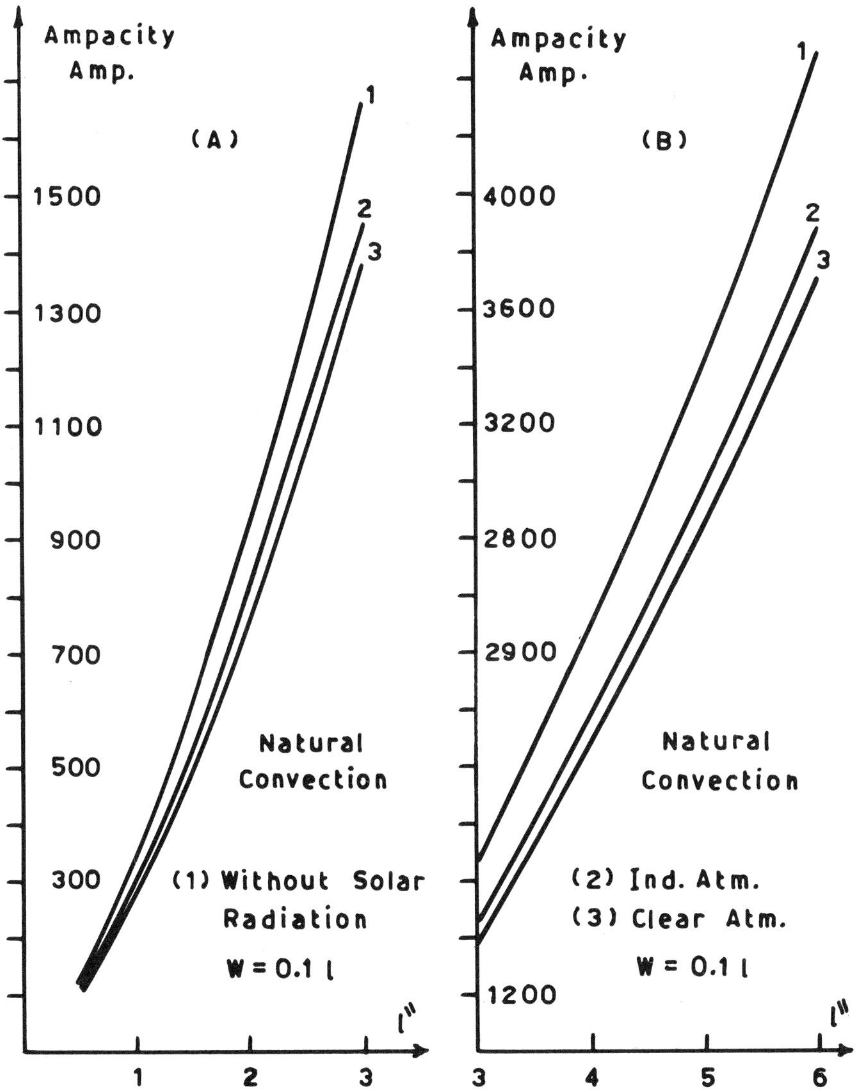

FIG. (5) AMPACITY OF RECTANGULAR BUS BAR

REFERENCE

1- M.Prager, D.L.Pemberton; A.G.Crraig Jr.,and N.A.Bleshman. "Thermal Considerations For Outdoor Bus Conductor Design" Paper No. F 76 205-5 presented at IEEE§PES Winter Meeting Tesla Symposium New York,N.Y.,Jan. 25-30 1976.

2- McAdams. W.H.Heat Transmission McGraw-Hill BOOK CO.N.Y. 1954.

CONCLUSION

Bus bar ampacity has been calculated taking into consideration different types of heat transfer . Simplified relations for calculating copper bus bars are obtained for natural and forced convection, for clear and industrial atmosphere. The obtained results show that,a great effect of solar heat gain on bus bars ampacity, occurs at natural convection, and increases as bus bar dimensions increase. Also that effect is large for squre bus bar of large cross section.As an example ; 3"x3" squre bus bar ampacity will decrease by 24% at natural convection mode of heat transfer,due to solar heat gain.So , an accurate analysis of solar heat gain, for any type of bus bars, is required to avoid the increase in temperature rise beyound the maximum allowable value.